Interest Boxes

Organic Chemistry

Fourth Edition

Paula Yurkanis Bruice

University of California, Santa Barbara

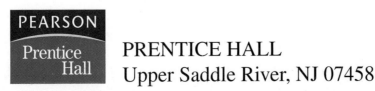

PRENTICE HALL
Upper Saddle River, NJ 07458

Library of Congress Cataloging-in-Publication Data

Bruice, Paula Yurkanis
Organic chemistry/Paula Yurkanis Bruice.--4th ed.
p. cm.
Includes index.
ISBN 0-13-140748-1
1. Chemistry, Organic. I. Title.

QD251.3 .B78 2004
547--dc21 2002192688

Senior Editor: Nicole Folchetti
Editor in Chief, Development: Ray Mullaney
Editor in Chief, Physical Sciences: John Challice
Development Editor: Carol Pritchard-Martinez
Media Editor: Paul Draper
Project Manager: Kristen Kaiser
Art Director: Maureen Eide
Assistant Art Director: John Christiana
Executive Managing Editor: Kathleen Schiaparelli
Assistant Managing Editor, Science Media: Nicole Bush
Assistant Managing Editor, Science Supplements: Becca Richter
Executive Marketing Manager: Steve Sartori
National Sales Director for Key Markets: David Theisen
Media Production Editor: Elizabeth Wright
Editorial Assistant: Nancy Bauer
Creative Director: Carole Anson
Director, Creative Services: Paul Belfanti
Vice President ESM Production and Manufacturing: David W. Riccardi
Manufacturing Manager: Trudy Pisciotti
Manufacturing Buyer: Alan Fischer
Managing Editor, AV Production & Management: Patty Burns
AV Art Editor: Connie Long
Art Studio: Artworks: Production Manager: Ronda Whitson; Manager, Production Technologies: Matthew Haas; Illustration Supervisor: Kathryn Anderson; Illustrators: Royce Copenheaver, Jay McElroy, Daniel Knopsnyder, Mark Landis, Stacy Smith; Quality Assurance Supervisor: Pamela Taylor; Quality Assurance: Kenneth Mooney, Timothy Nguyen, Cathy Shelley
Contributing Art Studios: Imagineering and Wavefunction / Richard Johnson
Spectra: Reproduced by permission of Aldrich Chemical Co.
Interior Designer: Joseph Sengotta
Photo Editor: Cynthia Vincenti
Photo Researcher: Truitt & Marshall
Cover Designer: Maureen Eide
Cover Researcher: Karen Sanatar
Cover Photo: Vincent Van Gogh (1853–1890), *Garden in Autumn*, also known as *The Public Park*.
© Copyright Giraudon/Art Resource, NY
Production Services/Composition: Preparé, Inc.

Pearson Education, Inc.
Upper Saddle River, NJ 07458

Printed in the United States of America
10 9 8 7 6 5 4 3 2

ISBN 0-13-140748-1

Pearson Education Ltd., *London*
Pearson Education Australia Pty. Ltd., *Sydney*
Pearson Education Singapore, Pte. Ltd.
Pearson Education North Asia Ltd., *Hong Kong*
Pearson Education Canada, Inc., *Toronto*
Pearson Educacíon de Mexico, S.A. de C.V.
Pearson Education—Japan, *Tokyo*
Pearson Education Malaysia, Pte. Ltd.

Organic Chemistry

To Meghan, Kenton, and Alec
with love and immense respect,
and to Tom, my best friend

Brief Contents

Contents

2 An Introduction to Organic Compounds: Nomenclature, Physical Properties, and Representation of Structure 60

Part 2 Hydrocarbons, Stereochemistry, and Resonance 109

3 Alkenes: Structure, Nomenclature, and an Introduction to Reactivity • Thermodynamics and Kinetics 111

4 Reactions of Alkenes 141

7 Electron Delocalization and Resonance • More About Molecular Orbital Theory 263

8 Reactions of Dienes • Ultraviolet and Visible Spectroscopy 298

9 Reactions of Alkanes • Radicals 336

Part 3 Substitution and Elimination Reactions 359

Part 5 Aromatic Compounds 593

15 Aromaticity • Reactions of Benzene 594

16 Reactions of Substituted Benzenes 622

Part 6 Carbonyl Compounds and Amines 669

17 Carbonyl Compounds I: Nucleophilic Acyl Substitution 670

18 Carbonyl Compounds II: Nucleophilic Acyl Addition, Nucleophilic Acyl Substitution, and Nucleophilic Addition–Elimination • Reactions of α,β-Unsaturated Carbonyl Compounds 731

23 Amino Acids, Peptides, and Proteins 959

24 Catalysis 999

25 The Organic Mechanisms of the Coenzymes • Metabolism 1033

26 Lipids 1075

27 Nucleosides, Nucleotides, and Nucleic Acids 1106

Part 8 Special Topics in Organic Chemistry 1145

28 Synthetic Polymers 1146

Preface

To the Instructor

My guiding principle in writing this book was to create a text that focuses on the student, one that presents the material in a way that encourages students to think about what they have already learned and then apply this knowledge in a new setting. Some students look upon organic chemistry as a course they have to endure simply because it is part of their course of study. Others may get the impression that learning organic chemistry is akin to learning a foreign "language"—a diverse collection of molecules and reactions—the language of a country that they will never visit. I hope, however, that as students proceed through their study of organic chemistry, they see that it is a subject that unfolds and grows and allows them to use what they learn at the beginning of the course to predict what follows. Countering the impression that the study of organic chemistry necessarily involves simply memorizing molecules and reactions, this book revolves around shared features and unifying concepts, and it emphasizes principles that can be applied again and again. I want students to learn how to apply what they have learned to a new setting, reasoning their way to a solution, rather than memorizing a multitude of facts. I also want to encourage students to see that organic chemistry is integral to biology as well as to their daily lives.

From the comments I have received from colleagues and students using previous editions, the book is working in the way I had hoped. As much as I enjoy having faculty tell me that their students are scoring higher than ever before on tests, nothing is more rewarding than hearing from the students themselves. Many students have generously attributed their success in organic chemistry to this book—not giving themselves nearly enough credit for how hard they studied to achieve that success. And they always seem surprised that they have come to love "orgo" (on the East Coast) or "o-chem" (on the West Coast). I also hear from many premedical students who say that the book gave them the permanent understanding of organic chemistry that allowed them to find the organic chemistry section of the MCAT to be the easiest part.

In striving to make this fourth edition even more useful to students, I have relied on constructive comments from many of you. For these, I am extremely grateful. I also have kept a journal of questions that students asked when they came to my office. These questions let me know what sections in the book needed clarifying and what answers in the *Study Guide and Solutions Manual* needed more in-depth explanations. Most important, this analysis showed me where new problems could reduce the chance that students using the new edition would ask these same questions. Because I teach large classes, I have a vested interest in foreseeing potential confusion before it arises. In this edition, many sections have been rewritten to optimize readability and comprehension. A variety of new in-chapter and end-of-chapter problems will help students master organic chemistry through problem solving. There are also new interest boxes to show students the relevance of organic chemistry and there are additional margin notes to remind students of important concepts and principles.

I hope you find the fourth edition even more appealing to your students. As always, I am eager to hear your comments—positive comments are the most fun, but critical comments are the most useful.

A Functional Group Approach with a Mechanistic Organization That Ties Together Synthesis and Reactivity

The organization of this book is designed to discourage rote memorization. The presentation of functional groups is organized around mechanistic similarities—electrophilic additions, radical substitutions, nucleophilic substitutions, eliminations, electrophilic aromatic substitutions, nucleophilic acyl substitutions, and nucleophilic additions. This organization allows a lot of material to be understood based on unifying principles of reactivity.

Instead of discussing the synthesis of a functional group when its reactivity is discussed—reactions that generally have little to do with one an other—I discuss the synthesis of the compounds that are formed as a result of the functional group's reactivity. In Chapter 4, for example, students learn about the reactions of alkenes, but they do *not* learn at this point about the synthesis of alkenes. Instead, they learn about the synthesis of alkyl halides, alcohols, ethers, and alkanes—the compounds formed when alkenes react. Because alkenes are synthesized from the reactions of alkyl halides and alcohols, the synthesis of alkenes is covered when the reactions of alkyl halides and alcohols are discussed. Tying together the reactivity of a functional group and the synthesis of compounds resulting from the functional group's reactivity prevents the student from having to memorize lists of unrelated reactions. It also results in a certain economy of presentation, allowing more material to be covered in less time.

Although memorizing different ways a particular functional group can be prepared is hard on students and can be somewhat counterproductive to their mastering organic chemistry, it is useful to have a compilation of reactions that yield a particular functional group when designing multistep syntheses. For this purpose, the different reactions that yield a particular functional group are compiled in Appendix IV. As students learn how to design syntheses, they appreciate the importance of reactions that change the carbon skeleton of a molecule; these reactions are compiled in Appendix V.

A Modular Format

Different people teach differently, so I have tried to make the book as modular as possible. The spectroscopy chapters (Chapters 13 and 14) have been composed so that they may be covered at any time. For those who prefer to teach spectroscopy at the beginning of the course—or in a separate laboratory course—I placed a table of the functional groups at the beginning of Chapter 13. For those who prefer to cover carbonyl chemistry earlier in the course, Part 6 (Carbonyl Compounds and Amines) can be covered before Part 5 (Aromatic Compounds). I anticipate that most instructors will cover the first 23 chapters during a year-long course and will then choose among the remaining chapters, depending on personal preference and the interests of the students enrolled in the course. Those teaching students whose interests are primarily in the biological sciences might be more inclined to cover Chapter 24 (Catalysis), Chapter 25 (The Organic Mechanisms of the Coenzymes • Metabolism), Chapter 26 (Lipids), and Chapter 27 (Nucleosides, Nucleotides, and Nucleic Acids). Those teaching courses designed for chemistry or engineering majors may decide to include Chapter 28 (Synthetic Polymers) and Chapter 29 (Pericyclic Reactions). The book ends with a chapter on drug discovery and design—a topic that in my experience interests students sufficiently that they will choose to read it on their own, even if it is not covered in the course.

A Bioorganic Emphasis

Today, many students taking organic chemistry are interested in the biological sciences. Therefore, bioorganic material is introduced throughout this text to allow students to see that organic chemistry and biochemistry are not separate entities, but two parts of a continuum of knowledge. Once students learn how such things as electron delocalization, leaving-group tendency, electrophilicity, and nucleophilicity affect the reactions of simple organic compounds, they can appreciate how these same factors are involved in the reactions of more complicated organic molecules such as enzymes, nucleic acids, and vitamins. I have found that the economy of presentation afforded by the first 21 chapters of this text (explained previously) makes it possible to devote more time to bioorganic topics.

In the first two-thirds of the book, the bioorganic material is limited primarily to the later sections of the chapters; as a result the material is available to the curious student, but the instructor is not compelled to introduce bioorganic topics into the course. For example, after the stereochemistry of organic reactions is presented, the stereoselectivity of enzymatic reactions is discussed; after alkyl halides are discussed, biological compounds use to methylate substrates are examined; after the methods chemists use to activate carboxylic acids are presented, the methods that cells use to activate these acids are explained; after condensation reactions are discussed, examples of biological condensation reactions are shown.

The six chapters of Part 7 (Chapters 22–27) specifically address bioorganic chemistry. They contain more chemistry than one would expect to find in a biochemistry text. Chapter 24 on catalysis, for example, explains the various modes of catalysis that occur in organic reactions and then shows that these are identical to the modes of catalysis that occur in enzymatic reactions. All of this is presented in a way that allows students to understand the lightning-fast rates of enzymatic reactions. Chapter 25 on coenzymes emphasizes the role of vitamin B_1 as an electron delocalizer, vitamin K as a strong base, vitamin B_{12} as a radical initiator, biotin as a compound that can transfer a carboxyl group; and how the many different reactions of vitamin B_6 are controlled by the overlap of *p* orbitals. Chapter 26 on lipids, covers the mechanisms of prostaglandin formation (allowing students to understand how aspirin works), fat breakdown, and terpene biosynthesis. Chapter 27 on nucleic acids explains mechanistically such topics as the chemical function of ATP. (Its role is not to provide a magic shot of energy that allows an endothermic reaction to take place. Rather, its role is to provide a reaction pathway involving a good leaving group for a reaction that cannot occur because of a poor leaving group.) Students learn that DNA contains thymine instead of uracil because of imine hydrolysis, and they see how DNA strands are synthesized in the laboratory. Thus, these chapters do not replicate what will be covered in a biochemistry course, but rather they bridge the two disciplines, allowing students to see that a knowledge of organic chemistry is central to the understanding of biological processes.

With the conviction that learning should be fun, biologically oriented material found in interest boxes covers intriguing asides: for example, why Dalmatians are the only mammals that excrete uric acid; why life is based on carbon instead of silicon; how a microorganism has learned to use industrial waste as a source of carbon; and the chemistry associated with SAMe, a product prominently displayed in health food stores.

An Early and Consistent Emphasis on Organic Synthesis

Students are introduced to synthetic chemistry and retrosynthetic analysis early in the book (Chapters 4 and Chapter 6, respectively), so they can use this technique throughout the course as they design multistep syntheses. Spread throughout the book are eight special sections on synthesis design, each with a different focus. For example, one section emphasizes the proper choice of reagents and reaction conditions to maximize the yield of the target molecule (Chapter 11), one discusses making new carbon–carbon bonds (Chapter 19), and one focuses on controlling stereochemistry (Chapter 20). The use of combinatorial methods in organic synthesis is described in Chapter 30, which focuses on drug discovery and design.

Pedagogical Features

Margin Notes and Boxed Material to Engage the Student

Margin notes and biographical sketches appear throughout the text. The margin notes remind students of important principles, and the biographical sketches give students some appreciation of the history of chemistry and the people who contributed to that history. Box Features have been included to connect chemistry to real life (for example, Semisynthetic Drugs, Measuring Toxicity, Chimney Sweeps and Cancer, Ultraviolet Light and Sunscreens, and Penicillin and Drug Resistance) or to give the student extra help (for example, Calculating Kinetic Parameters, A Few Words About Curved Arrows, and Incipient Primary Carbocations).

Summaries and Voice Boxes to Help the Student

Each chapter concludes with a Summary to help students synthesize the key points of the chapter and a list of cross-referenced Key Terms. Chapters that cover reactions conclude with a Summary of Reactions. Illustration annotations (voice boxes) are found throughout the book to help students focus on points being discussed.

Problems, Solved Problems, and Problem-Solving Strategies

The book contains more than 1600 problems. The answers to all the problems (along with explanations, when needed) are in the accompanying *Study Guide and Solutions Manual*, which I authored for consistency in language with that of the text. The problems within each chapter are primarily drill problems to allow students to test themselves on material covered just before moving on to the next section. Solutions to selected problems provide insight into how to solve problems. Short answers are provided at the end of the book for problems marked with a diamond so students can quickly test their understanding. Most chapters also contain at least one Problem-Solving Strategy that teaches students how to approach certain kinds of problems. For example, the Problem-Solving Strategy in Chapter 10 teaches students how to determine whether a reaction will be more apt to take place by an S_N1 or an S_N2 pathway. Each Problem-Solving Strategy is followed by an exercise giving the student an opportunity to use the problem-solving skill just learned.

The end-of-chapter problems vary in difficulty. The initial problems are drill problems that integrate material from the entire chapter. These provide a greater challenge by requiring the student to think in terms of all the material in the chapter, rather than individual sections. The problems become more challenging as the student proceeds, often reinforcing concepts covered in prior chapters. The net effect is to progressively build both problem-solving ability and confidence.

Companion Website

WWW icons in the margins identify 3-D molecules, movies, and interactive animations on the Companion Website (*http://chem.prenhall.com/bruice*) that are pertinent to the material being discussed. Each chapter on the Website has four practice exercises and two quizzes—a total of about 80 multiple-choice questions. Although I had never been a fan of this type of question, the quality of these questions has changed my mind.

Art Program: Rich in Three-Dimensional, Computer-Generated Structures

This edition continues to present energy-minimized, three-dimensional structures throughout the text that give students an appreciation of the three-dimensional shapes of organic molecules. Color is used to highlight and organize the information, not simply for show. This edition makes greater use of highlights to focus on points of interest. I have attempted to make the color consistent (for example, mechanism arrows are always red), but there is no need for a student to memorize a color palette.

A Closer Look at Coverage and Organization

This book is divided into eight parts. Each part starts with a brief overview so students can understand where they are "going." The first chapter in Part 1 provides a summary of the material that students need to recall from general chemistry. The sections on acids and bases emphasize the relationship between acidity and the stability of the conjugate base, a theme recurrently raised throughout the text. (Acids and bases are covered even more extensively in the *Study Guide and Solutions Manual.*) In Chapter 2, students learn how to name five classes of organic compounds—those that will be the products of the reactions in the chapters that immediately follow. Chapter 2 also covers topics that are necessary before the study of reactions can begin—structures, conformations, and physical properties of organic compounds.

The seven chapters of Part 2 deal with hydrocarbons, stereochemistry, and resonance. Chapter 3 sets the stage for the study of organic reactions by giving the students the knowledge of thermodynamics and kinetics they will need as they progress through the course. Rate equations are derived in an appendix for those who wish to give their students a more mathematical treatment, and the *Study Guide and Solutions Manual* contains a section on calculating kinetic parameters. Chapter 3 also introduces students to the concept of curved arrows. (The *Study Guide and Solutions Manual* contains an extensive exercise on "electron pushing." I have found this exercise to be very successful in making my students comfortable with a topic that should be easy, but somehow perplexes even the best of them unless they have sufficient practice.)

I lead off with the reactions of alkenes, because of their simplicity. Chapter 4 covers a wide variety of reactions, but they all have similar mechanisms—an electrophile adds to the least substituted sp^2 carbon and a nucleophile adds to the other sp^2 carbon. The many reactions differ only in the nature of the electrophile and the nucleophile. Because organic chemistry is all about the interactions of electrophiles and nucleophiles, it makes sense to begin the study of organic reactions by introducing students to a variety of electrophiles and nucleophiles. The reactions in Chapter 4 are discussed without regard to stereochemistry because I have found that students do well as long as only one new concept is introduced at a time. Chapter 5 reviews the isomers introduced in Chapters 2 and 3 (constitutional isomers and cis–trans isomers) and then discusses isomers that result from having a chirality center—one kind of stereogenic center. The *Study Guide and Solutions Manual* has a model-building exercise that encourages students to open their box of molecular models and see how model building can help them learn. In addition, the book's Companion Web site gives students the opportunity to manipulate many molecules in three dimensions. Now that the students are comfortable with both isomers and electrophilic addition reactions, the two topics are considered together at the end of Chapter 5, where the stereochemistry of the addition reactions covered in Chapter 4 is presented. Chapter 6 covers alkynes. This chapter builds the students' confidence because of the similarity of the material to that in Chapter 4.

Understanding electron delocalization is vitally important in organic chemistry, so this subject is covered in its own chapter (Chapter 7). This chapter offers a continuation of the introduction to this topic found in Chapter 1. Chapter 8 moves on to the reactions of dienes, allowing students to apply the concept of electron delocalization to the electrophilic addition reactions that were mastered in Chapters 4 and 6. Chapter 8 also introduces UV/Vis spectroscopy, describing how chemists use this tool to gain information about the structures, physical properties, and reactivity of certain organic compounds. Chapter 9 covers the reactions of alkanes. Students learn that alkanes are largely unreactive because of the absence of a functional group, underscoring the importance of a functional group to chemical reactivity.

Part 3 covers substitution and elimination reactions at an sp^3 hybridized carbon. First, the substitution reactions of alkyl halides are discussed in Chapter 10. Chapter 11 discusses elimination reactions of alkyl halides and then goes on to consider competition between substitution and elimination. Chapter 12 covers substitution and elimination reactions at an sp^3 hybridized carbon when a group other than a halogen is the leaving group—reactions of alcohols, ethers, epoxides, arene oxides, thiols, and sulfides. Organometallic compounds and transition metal-catalyzed coupling reactions are also introduced in Chapter 12.

The two chapters in Part 4 discuss mass spectrometry and IR spectroscopy (Chapter 13) and NMR spectroscopy (Chapter 14). Each spectral technique is written as a stand-alone topic so it can be covered independently at any time during the course. The first of these chapters opens with a table of functional groups for those who want to cover spectroscopy before students have been introduced to all the functional groups.

Part 5 is all about aromatic compounds. Chapter 15 covers aromaticity and the reactions of benzene. Chapter 16 discusses the reactions of substituted benzenes. Because not all organic courses cover the same amount of material in a semester, Chapters 13–16 have been strategically placed to come near the end of the first semester. Ending a semester before Chapter 13, 14, or 15, or after Chapter 16 will not interfere with the flow of information.

Part 6 discusses primarily the chemistry of carbonyl compounds. Some instructors were initially skeptical about discussing carboxylic acids and their derivatives before aldehydes and ketones. However, when the opinions of those who had used previous editions of the book were sought, consensus emerged that this is a preferred order of treatment. Chapter 17 starts the treatment of carbonyl chemistry by discussing the reactions of carboxylic acids and their derivatives with oxygen and nitrogen nucleophiles. In this way, students are introduced to carbonyl chemistry by learning how tetrahedral intermediates partition. The first part of Chapter 18 discusses the reactions of carboxylic acid derivatives, aldehydes, and ketones with strong (carbon and hydrogen) nucleophiles. By studying all these carbonyl compounds together, students see how the reactions of aldehydes and ketones differ from those of carboxylic acid derivatives. Then, when they move on to study the formation and hydrolysis of imines, enamines, and acetals in the second part of Chapter 18, they can easily understand these mechanisms because they are well versed in how tetrahedral intermediates partition. Over the years I have experimented with my classes, and I believe this is the most effective and easiest way to teach carbonyl chemistry. That being said, Chapters 17 and 18 can easily be switched as long as Section 18.4 and Section 18.5 are skipped and then covered with Chapter 17. Chapter 19 deals with reactions at the α-carbon of carbonyl compounds. Chapter 20 revisits reduction reactions and discusses oxidation reactions. I have found that students are much better able to understand oxidation reactions when they are presented together as a unit, rather than being introduced one at a time as each functional group is introduced. Because reduction reactions appear in earlier chapters, students have an opportunity to extend their knowledge within the framework of what they already know. Chapter 21 reviews material on amines that was covered in previous chapters—structure and physical properties, acid–base properties, nomenclature, reactivity, and synthesis—and then goes on to cover these topics in greater detail. The chapter concludes with a discussion of heterocyclic compounds.

Part 7 covers bioorganic topics. Part 8 covers synthetic polymers, pericyclic reactions, and drug discovery and design.

Changes to This Edition

Content and Organization

In response to user feedback and comments from reviewers, the coverage of molecular orbital theory has been increased. For example, molecular orbital theory is used to explain why alkyl halides undergo back-side attack, how carbonyl compounds react, why amides are unreactive, and so on. In response to popular demand, there is a new chapter on amines and new sections on such topics as transition metal coupling reactions and chiral catalysts. Much of the book has been rewritten to aid student understanding. In particular, the sections on acid–base chemistry, α,β-carbonyl compounds, and the Wittig and Diels–Alder reactions have been expanded and clarified. The introduction to alkenes is now covered in two chapters. Chapter 3 introduces students to chemical reactions, curved arrows, and thermodynamics and kinetics. Chapter 4 now focuses on electrophilic addition reactions, allowing students to see that the many reactions in this chapter all follow similar mechanisms: The electrophile adds to the sp^2 carbon that is bonded to the greater number of hydrogens, and the nucleophile adds to the other sp^2 carbon.

Synthesis Sections

This edition has eight "Designing a Synthesis" sections, including a new one on controlling stereochemistry and one that discusses disconnections, synthons, and synthetic equivalents. There are more synthesis problems, including some that encourage students to approach the problem using retrosynthetic analysis.

Pedagogical Elements

In response to student feedback, each chapter now concludes with a summary. These have been strongly endorsed by my students who have tried them out this past year. This edition includes more voice boxes to aid student learning, especially to point out important steps in mechanisms. In response to student comments, there are many more margin notes that succinctly recap key points and facilitate review. More than a dozen new interest boxes are in this edition, including one on environmental adaptation, one on omega fatty acids, and even one on how a banana slug knows what to eat.

Problem Sets

Much of the focus of this revision was on the problems, both in-chapter and end-of-chapter. There are new solved problems, new problem-solving strategies, new problems designed to show the common reactivity of the functional groups, and new problems to stress that electrophiles react with nucleophiles. Some of the many topics for which there are new problems include molecular orbital theory, acid–base equilibria, and stereochemistry. Many of the new problems are spectroscopy problems to enhance the integration of spectroscopy throughout the course.

For the Instructor

Instructor's Resource CD-ROM (0-13-141012-1) by John Hogan, Louisiana State University, and Irene Lee, Case Western Reserve University. This lecture resource built specifically for *Organic Chemistry, Fourth Edition*, features nearly all of the art from the text in JPEG and PDF formats, two pre-built PowerPoint presentations, hundreds of 3-D images of molecules, animations of selected organic reactions, and interactive tutorials on key concepts—all in one, convenient resource.

Transparency Pack (0-13-141008-3) by Paula Yurkanis Bruice, University of California, Santa Barbara. This set features 275 full-color images from the text.

Test Item File (0-13-141000-8) by Gary Hollis, Roanoke College. Includes a selection of 2,300 multiple choice, short answer, and essay test questions.

TestGen (0-13-141006-7) This computerized version of the Test Item File allows professors to create and tailor exams to their needs, and includes tools for course management and administering tests over a local area network.

Course Management Options Prentice Hall offers content cartridges for online, text-specific course management systems depending on your preferred platform (WebCT, BlackBoard, and Pearson Educations' own CourseCompass). Hundreds of text-specific problems, media activities, chemistry-related Web destinations, and much more are provided. Visit *www.prenhall.com/demo* for details on how to communicate with your students, customize content to meet your course needs, create quizzes and tests, track grades, and many more online options.

For the Student

Study Guide and Solutions Manual (0-13-141010-5) by Paula Yurkanis Bruice. This Study Guide and Solutions Manual contains complete and detailed explanations of the solutions to the problems in the text. In addition, you'll find a section on advanced acid/base chemistry with an additional set of problems, an 18-page tutorial on "pushing electrons," exercises on building molecular models and calculating kinetic parameters, as well as 23 practice tests.

Molecular Modeling Workbook (0-13-141040-7) Features SpartanView™ and SpartanBuild™ software. This workbook includes a software tutorial and numerous challenging exercises students can tackle to solve problems involving structure building and analysis using the tools included in the two pieces of Spartan software. Available free when packaged with the text; just contact your Prentice Hall representative for details.

Companion Website (*http://chem.prenhall.com/bruice*) Built to complement *Organic Chemistry, Fourth Edition,* as part of an integrated course package, the easy-to-use Companion Website features the following modules for each chapter:

- The **Tutorial Gallery**, containing highly interactive tutorials with feedback, helps students learn about reaction mechanisms and other important organic chemistry topics.
- An **Animation Gallery** of important mechanisms and reactions is included.
- A **Molecule Gallery**, where students can access 3-D (Chime) renderings of most of the important molecules in the text. Students can rotate the molecule in three dimensions, change its representation, and explore its structure in detail.
- **Practice Exercises and Quizzes** give students access to 2,000 new exercises to test their understanding of the content. Each question includes a hint, a cross-reference back to the text, and detailed feedback.
- Applications, in the **Current Topics** section, that provide students with current news articles that enrich their understanding of organic chemistry.
- Dynamic **Web Destinations**, which are a valuable source of supplemental information. These can also be used for online research projects.

Organic ChemIST (0-13-033832-X) Written by Robert M. Hanson, this premium Website offers a full short course in organic chemistry, built with highly interactive tutorials, animations, 3D molecular models, and self-assessment questions with instant feedback. Linked from the Companion Website for *Organic Chemistry*, it requires a Premium Access Code to view the entire site.

ChemOffice Student CD This software includes ChemDraw LTD and Chem3D LTD. A free workbook is available on the Companion Website that includes a software tutorial and numerous exercises for each chapter of the text. The software is sold at the instructor's discretion and is available at a substantial discount if purchased with the textbook.

Prentice Hall Molecular Model Kit (0-205-508136-3) This best-selling model kit allows students to build space-filling and ball-and-stick models of common organic molecules. It allows accurate depiction of double and triple bonds, including heteroatomic molecules (which some model kits cannot handle well).

Prentice Hall Framework Molecular Model Kit (0-13-330076-5) This model kit allows students to build scale models that show the mutual relations of atoms in organic molecules, including precise interatomic distances and bond angles. This is the most accurate model kit available.

Acknowledgements

It gives me great pleasure to acknowledge the dedicated efforts of many good friends who made this book a reality. I am especially grateful for the comments and suggestions of Ron Magid of the University of Tennessee, whose kind prodding has made an enormous difference to this edition. The many contributions of Ed Skibo of Arizona State University; Paul Papadopoulos of the University of New Mexico; Ron Starkey of the University of Wisconsin, Green Bay; and Francis Klein of Creighton University persist in this edition. Particular thanks go to David Yerzley, M.D., for his assistance with the section on MRI; Warren Hehre of Wavefunction, Inc., and Alan Shusterman of Reed College for their advice on the electrostatic potential maps that appear in the book; John Perona of the University of California, Santa Barbara, who provided some of the molecular images used in this edition; and Jeremy Davis who created the art that appears on pages 992 and 1016. Others whose contributions have been important to the development of this edition are Heinz Roth of Rutgers University and Ron Kluger of the University of Toronto. I also want to acknowledge the efforts of the many users of previous editions who generously gave their time to help me make this a better book. My particular thanks go to Tom Pettus and Curt Anderson, University of California, Santa Barbara; Barry Coddens and Terry Sheppard, Northwestern University; Barbara Schowen, University of Kansas; Jack Kirsch, University of California, Berkeley; Jim Long, University of Oregon; Tom Tidwell, University of Toronto; Peter Wagner, Michigan State University; Paige Phillips, Virginia Polytechnic Institute and State University; Barbara Mayer, California State University, Fresno; George Clemans, Bowling Green State University; Tom Nalli, Winona State University; Vincent Spaziano, St. Louis University; Paul Kropp, University of North Carolina, Chapel Hill; and Stanley Kudzin, State University of New York, New Paltz. I am also very grateful to my students, who pointed out sections that needed clarification, worked the problems, and searched for errors.

The following reviewers have played an enormously important role in the development of this textbook.

Fourth Edition Manuscript Reviewers

Merritt Andrus, *Brigham Young University*
Daniel Appella, *Northwestern University*
George Bandik, *University of Pittsburgh*
Daniel Blanchard, *Kutztown University*
Ron Blankespoor, *Calvin College*
Paul Buonora, *California State University, Long Beach*
Robert Chesnut, *Eastern Illinois University*
Michael Chong, *University of Waterloo*
Robert Coleman, *Ohio State University*
David Collard, *Georgia Institute of Technology*
Debbie Crans, *Colorado State University*
Malcolm Forbes, *University of North Carolina, Chapel Hill*
Deepa Godambe, *Harper College*
Fathi Halaweish, *South Dakota State University*
Steve Hardinger, *University of California, Los Angeles*
Alvan Hengge, *Utah State University*
Steve Holmgren, *University of Montana*
Nichole Jackson, *Odessa College*
Carl Kemnitz, *California State University, Bakersfield*
Keith Krumpe, *University of North Carolina, Asheville*
Michael Kurz, *Illinois State University*
Li, Yuzhuo, *Clarkson University*
Janis Louie, *University of Utah*
Charles Lovelette, *Columbus State University*
Ray Lutgring, *University of Evansville*
Janet Maxwell, *Angelo State University*
Mark McMills, *Ohio University*
Andrew Morehead, *University of Maryland*
John Olson, *Augustana University*
Brian Pagenkopf, *University of Texas, Austin*
Joanna Petridou, *Spokane Falls Community College*
Michael Rathke, *Michigan State University*

Christopher Roy, *Duke University*
Tomikazu Sasaki, *University of Washington*
David Soriano, *University of Pittsburgh*
Jon Stewart, *University of Florida*
John Taylor, *Rutgers University*
Carl Wamser, *Portland State University*
Marshall Werner, *Lake Superior State University*
Catherine Woytowicz, *George Washington University*
Zhaohui Sunny Zhou, *Washington State University*

Critique Reviewers

Neil Allison, *University of Arkansas*
Joseph W. Bausch, *Villanova University*
Dana Chatellier, *University of Delaware*
Steven Fleming, *Brigham Young University*
Malcolm Forbes, *University of North Carolina, Chapel Hill*
Charlie Garner, *Baylor University*
Andrew Knight, *Loyola University*
Joe LeFevre, *State University of New York, Oswego*
Charles Liotta, *Georgia Institute of Technology*
Andrew Morehead, *University of Maryland*
Richard Pagni, *University of Tennessee*
Jimmy Rogers, *University of Texas, Arlington*
Richard Theis, *Oregon State University*
Peter J. Wagner, *Michigan State University*
John Williams, *Temple University*
Catherine Woytowicz, *George Washington University*

Accuracy Reviewers

Bruce Banks, *University of North Carolina, Greensboro*
Debra Bautista, *Eastern Kentucky University*
Vladimir Benin, *University of Dayton*
Linda Betz, *Widener University*
Anthony Bishop, *Amherst College*
Phil Brown, *Brigham Young University*
Sushama Dandekar, *University of North Texas*
S. Todd Deal, *Georgia Southern University*
Michael Detty, *University of Buffalo*
Matthew Dintzner, *DePaul University*
Nicholas Drapela, *Oregon State University*
Jeffrey Elbert, *University of Northern Iowa*
Mark Forman, *Saint Joseph's University*
Joe Fox, *University of Delaware*
Anne Gaquere, *State University of West Georgia*
Charles Garner, *Baylor University*
Scott Goodman, *Buffalo State College*
Steven Graham, *St. John's University*
Christian Hamann, *Albright College*
Cliff Harris, *Albion College*
Alfred Hortmann, *Washington University*
Floyd Klavetter, *Indiana University of Pennsylvania*
Thomas Lectka, *Johns Hopkins University*
Len MacGillivray, *University of Iowa*
Jerry Manion, *University of Central Arkansas*
Przemyslaw Maslak, *Pennsylvania State University*
Michael McKinney, *Marquette University*
Alex Nickon, *Johns Hopkins University*
Patrick O'Connor, *Rutgers University*
Kenneth Overly, *Providence College*
Cass Parker, *Clark Atlanta University*
Marchland Philip, *University of North Texas*
Christopher Roy, *Duke University*
Susan Schelble, *University of Colorado, Denver*
Chris Spilling, *University of Missouri*
Janet Stepanek, *Colorado College*

Previous Edition Manuscript Reviewers

Mahamed Asgar Ali, *Howard University*
Shelby R. Anderson, *Trinity College*
John Barbas, *Valdosta State University*
Rick Bolesta, *Mt. Hood Community College*
Joyce C. Brockwell, *Northwestern University*
Thomas A. Bryson, *University of South Carolina*
Paul Buonora, *University of Scranton*
George B. Clemans, *Bowling Green State University*
Barry A. Coddens, *Northwestern University*
John Cullen, *Monroe Community College*
Mark R. DeCamp, *University of Michigan, Dearborn*
Michael R. Detty, *State University of New York, Buffalo*
John DiCesare, *University of Tulsa*
Veljko Dragojlovic, *Nova Southeastern University*
Jeffrey Elbert, *South Dakota State University*
Jan M. Fleischer, *Earlham College*
Warren P. Giering, *Boston University*
Michael M. Haley, *University of Oregon*
James E. Hanson, *Seton Hall University*
David Harpp, *McGill University*
John Isidor, *Montclair State University*
Richard Johnson, *University of New Hampshire*
Dennis Lehman, *Harold Washington College*
William Loffredo, *East Stroudsberg University*
James W. Long, *University of Oregon*
Jerry Manion, *University of Central Arkansas*
Amanda Martin-Esker, *Northwestern University*
John N. Marx, *Texas Tech University*
Anthony Masulaitis, *Jersey City State College*
Barbara J. Mayer, *California State University, Fresno*
Robert McClelland, *University of Toronto*
Gary W. Morrow, *University of Dayton*
Thomas W. Nalli, *Winona State University*
Abby Parrill, *University of Memphis*
Lawrence Principe, *Johns Hopkins University*
Michael Rathke, *Michigan State University*
J. Ty Redd, *Southern Utah University*
Todd Richmond, *The Claremont Colleges*
Anthony Sky, *Lawrence Technical University*
Homer A. Smith, Jr., *University of Richmond*
Andrew B. Turner, *Saint Vincent College*
T. K. Vinod, *Western Illinois University*
Maria Vogt, *Bloomfield College*

Special thanks to those who provided valuable feedback to the author and/or publisher:

Joseph Bausch, *Villanova University*
Josef Krause, *Niagara University*
Thomas Johnson, *University of Georgia*

I am deeply grateful to my editor, Nicole Folchetti, whose creative talents guided this book. She gave me the encouragement I needed to make deadlines with more patience than I deserved and found the reviewers who contributed to making this book as good as it could be. I also want to thank the other talented and dedicated people at Prentice Hall whose contributions made this book a reality. Much thanks goes to Ray Mullaney, editor in chief of book development, who kept me on track with a great deal of patience. I am deeply grateful to Carol Pritchard-Martinez, the development editor, who checked every word and every drawing that went into this edition—no matter how garbled my sentences, she was somehow able to fix them. I am also grateful to Fran Daniele, who worked with a very difficult schedule—and with good humor—to bring this new edition through the production process. A huge thank-you goes to Paul Draper, media editor, whose talents have guided the creation of the Companion Website. Finally, I'd like to thank David Theisen, National Sales Director for Key Markets, whose keen understanding of the book, and indefatigable efforts, have brought it to the attention of a vast portion of the organic chemistry community.

I particularly want to thank the many wonderful and talented students I have had over the years that taught me how to be a teacher. And I want to thank my children, from whom I may have learned the most.

To make this textbook as user-friendly as possible, I would appreciate any comments that will help me achieve this goal in future editions. If you find sections that could be clarified or expanded, or examples that could be added, please let me know. Finally, this edition (and the accompanying *Solutions Manual and Study Guide*) has been painstakingly combed for typographical errors. Any that remain are my responsibility; if you find any, please send me a quick e-mail so that errors can be corrected in future printings.

Paula Yurkanis Bruice
University of California, Santa Barbara
pybruice@chem.ucsb.edu

For the Student

Welcome to organic chemistry! You are about to embark on an exciting journey. This book has been written for you—one who is encountering the subject for the first time—to make your journey both stimulating and enjoyable. You should start by familiarizing yourself with the book. The material on the inside of the front and back covers is information that you may want to refer to many times during the course. The Chapter Summaries, Key Terms, and Summaries of Reactions at the end of the chapters, and the Glossary at the end of the book are useful study aids—take advantage of them! Also look at the Appendices to see what kind of information is provided there. The electrostatic potential maps and the molecular models throughout the book are there to give you an appreciation of what molecules look like in three dimensions and how electronic charge is distributed within the molecule. Look at the margin notes as you read a chapter—they emphasize important points.

Work all the problems within each chapter. These are drill problems that allow you to check whether you have mastered the material. Some of them are solved for you in the text. Others—those marked with a diamond—have short answers provided at the end of the book. Don't overlook the Problem-Solving Strategies sprinkled throughout the text. These provide practical suggestions on the best way to approach important problem types. Read these Problem-Solving Strategies carefully, and refer back to them when you are working the end-of-chapter problems.

Work as many end-of-chapter problems as you can. The more problems you work, the more comfortable you will be with the subject and the more prepared you will be for the material in subsequent chapters. Do not let any problem frustrate you. If you cannot figure out the answer in a reasonable amount of time, turn to the *Study Guide and Solutions Manual* to learn how you should have approached the problem. Later on, go back and try to work the problem again, this time on your own. Be sure to visit the Companion Website (*www.prenhall.com/bruice*) to try out the tutorials, study the 3-D molecules, and take the tests.

The most important thing for you to remember in organic chemistry is DO NOT GET BEHIND. Organic chemistry consists of a lot of simple steps—each one is very easy to master. But the subject can become overwhelming if you don't keep up.

Before many of the theories and mechanisms were worked out, organic chemistry was a discipline that could be mastered only through memorization. Fortunately, that is no longer true. You will find many common threads that allow you to use what you have learned in one situation to predict what will happen in other situations. So, as you read the book and study your notes, always try to understand why each thing happens. If the reasons behind the reactivity are understood, most reactions can be predicted. Approaching the class with the misconception that you must memorize hundreds of unrelated reactions, could be your downfall. There is simply too much material to memorize! Reasoning, not memorization, provides the necessary foundation on which to lay subsequent material. From time to time, some memorization will be required. Some fundamental rules have to be memorized, and you will have to memorize the common names of some organic compounds. But the latter should not be a problem. After all, your friends have common names and you've been able to learn them.

Students who take organic chemistry to gain entrance into medical school sometimes wonder why medical schools pay so much attention to how they do in organic chemistry. The importance of organic chemistry is not in the subject matter alone. Mastering organic chemistry requires a thorough understanding of fundamentals and the ability to use these fundamentals to analyze, classify, and predict. This parallels the study of medicine: a physician uses an understanding of fundamentals to analyze, classify, and diagnose.

Good luck in your study. I hope you enjoy your course in organic chemistry and learn to appreciate the logic of the discipline. If you have any comments about the book or any suggestions about how it can be improved for the students who will follow you, I would love to hear from you. Positive comments are the most fun, but critical comments are the most useful.

Paula Yurkanis Bruice
pybruice@chem.ucsb.edu

Highlights of *Organic Chemistry, Fourth Edition*

In the Fourth Edition of *Organic Chemistry*, Paula Bruice adds new material and enhances many hallmark features while staying true to the text's main goal—encouraging students to understand the "why" of organic chemistry. Highlights of the Fourth Edition are listed below.

Organization and Approach

- **The functional groups have been organized around mechanistic similarities.** When a functional group is presented, its reactivity is discussed, but not its synthesis. Instead, synthesis is discussed in terms of the products that are formed as a result of the functional group's reactivity. This organization fosters an understanding of unifying principles of reactivity, discourages rote memorization, and allows more material to be presented in a shorter period of time. For a complete list of the methods that can be used to synthesize a particular functional group, see Appendix IV.

- **A strong bioorganic flavor throughout the text** encourages students to recognize that organic chemistry and biochemistry are not separate entities, but two parts of a continuum of knowledge. This material is found in special interest boxes, specific chapter sections, and chapters that focus on bioorganic topics. A complete list of the special interest boxes is located opposite the inside front cover.

- **A new chapter on amines** pulls together important information on the subject in a single location. See Chapter 21.

- **Alkenes,** previously covered in Chapter 3, are now covered in two chapters for a slower-paced presentation of this important functional group. See Chapters 3 and 4.

Emphasis on Problem Solving

- **Solved Problems** throughout the text carefully walk students through the steps involved in solving a particular type of problem.

- **Problem-Solving Strategies** in many chapters teach students how to approach a variety of problems, organize their thoughts, and improve their problem-solving abilities. Every strategy is followed by an exercise that allows students to practice the strategy just discussed.

- **New and revised end-of-chapter problems**, many which focus on overarching principles and concepts that are especially difficult for students.

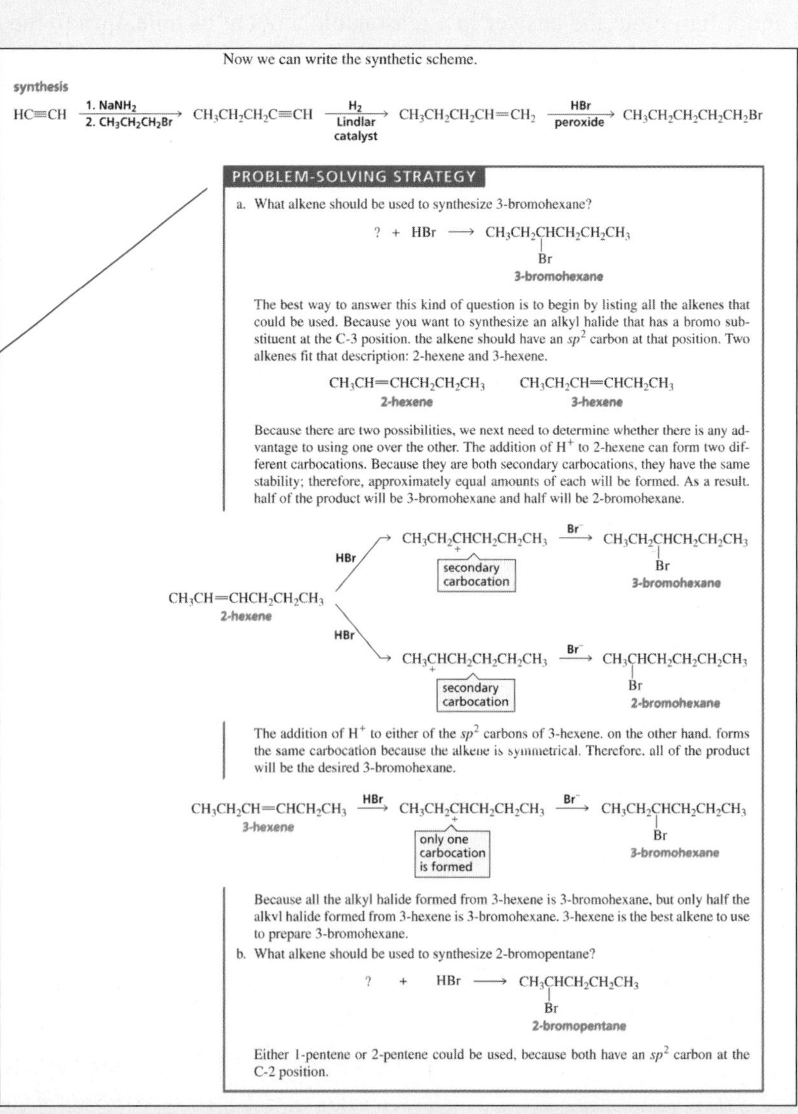

Now we can write the synthetic scheme.

synthesis

$$HC{\equiv}CH \xrightarrow[2.\ CH_3CH_2CH_2Br]{1.\ NaNH_2} CH_3CH_2CH_2C{\equiv}CH \xrightarrow[\text{Lindlar catalyst}]{H_2} CH_3CH_2CH_2CH{=}CH_2 \xrightarrow[\text{peroxide}]{HBr} CH_3CH_2CH_2CH_2CH_2Br$$

PROBLEM-SOLVING STRATEGY

a. What alkene should be used to synthesize 3-bromohexane?

$$? + HBr \longrightarrow CH_3CH_2CH(Br)CH_2CH_2CH_3$$

3-bromohexane

The best way to answer this kind of question is to begin by listing all the alkenes that could be used. Because you want to synthesize an alkyl halide that has a bromo substituent at the C-3 position, the alkene should have an sp^2 carbon at that position. Two alkenes fit that description: 2-hexene and 3-hexene.

$CH_3CH{=}CHCH_2CH_2CH_3$ (2-hexene) $\quad CH_3CH_2CH{=}CHCH_2CH_3$ (3-hexene)

Because there are two possibilities, we next need to determine whether there is any advantage to using one over the other. The addition of H^+ to 2-hexene can form two different carbocations. Because they are both secondary carbocations, they have the same stability; therefore, approximately equal amounts of each will be formed. As a result, half of the product will be 3-bromohexane and half will be 2-bromohexane.

$$CH_3CH{=}CHCH_2CH_2CH_3 \text{ (2-hexene)} \xrightarrow{HBr} CH_3CH_2\overset{+}{C}HCH_2CH_2CH_3 \text{ (secondary carbocation)} \xrightarrow{Br^-} CH_3CH_2CH(Br)CH_2CH_2CH_3 \text{ (3-bromohexane)}$$

$$CH_3CH{=}CHCH_2CH_2CH_3 \text{ (2-hexene)} \xrightarrow{HBr} CH_3\overset{+}{C}HCH_2CH_2CH_2CH_3 \text{ (secondary carbocation)} \xrightarrow{Br^-} CH_3CH(Br)CH_2CH_2CH_2CH_3 \text{ (2-bromohexane)}$$

The addition of H^+ to either of the sp^2 carbons of 3-hexene, on the other hand, forms the same carbocation because the alkene is symmetrical. Therefore, all of the product will be the desired 3-bromohexane.

$$CH_3CH_2CH{=}CHCH_2CH_3 \text{ (3-hexene)} \xrightarrow{HBr} CH_3CH_2\overset{+}{C}HCH_2CH_2CH_3 \text{ (only one carbocation is formed)} \xrightarrow{Br^-} CH_3CH_2CH(Br)CH_2CH_2CH_3 \text{ (3-bromohexane)}$$

Because all the alkyl halide formed from 3-hexene is 3-bromohexane, but only half the alkyl halide formed from 3-hexene is 3-bromohexane, 3-hexene is the best alkene to use to prepare 3-bromohexane.

b. What alkene should be used to synthesize 2-bromopentane?

$$? + HBr \longrightarrow CH_3CH(Br)CH_2CH_2CH_3$$

2-bromopentane

Either 1-pentene or 2-pentene could be used, because both have an sp^2 carbon at the C-2 position.

HIGHLIGHTS

Enhanced Pedagogy

- Nearly 120 **Special Interest Boxes** throughout the text cover bioorganic and other applied topics to enliven the subject being discussed.

BLOOD: A BUFFERED SOLUTION

Blood is the fluid that transports oxygen to all the cells of the human body. The normal pH of human blood is 7.35 to 7.45. Death will result if this pH decreases to a value less than ~6.8 or increases to a value greater than ~8.0 for even a few seconds. Oxygen is carried to cells by a protein in the blood called hemoglobin. When hemoglobin binds O_2, hemoglobin loses a proton, which would make the blood more acidic if it did not contain a buffer to maintain its pH.

$$HbH^+ + O_2 \rightleftharpoons HbO_2 + H^+$$

A carbonic acid/bicarbonate (H_2CO_3/HCO_3^-) buffer is used to control the pH of blood. An important feature of this buffer is that carbonic acid decomposes to CO_2 and H_2O:

$$CO_2 + H_2O \rightleftharpoons \underset{\text{carbonic acid}}{H_2CO_3} \rightleftharpoons \underset{\text{bicarbonate}}{HCO_3^-} + H^+$$

Cells need a constant supply of O_2, with very high levels required during periods of strenuous exercise. When O_2 is consumed by cells, the hemoglobin equilibrium shifts to the left to release more O_2, so the concentration of H^+ decreases. At the same time, increased metabolism during exercise produces large amounts of CO_2. This shifts the carbonic acid/bicarbonate equilibrium to the right, which increases the concentration of H^+. Significant amounts of lactic acid are also produced during exercise, and this further increases the concentration of H^+.

Receptors in the brain respond to the increased concentration of H^+ and trigger a reflex that increases the rate of breathing. This increases the releases of oxygen to the cells and the elimination CO_2 by exhalation. Both processes decrease the concentration of H^+ in the blood.

- **Biographical Sketches** give students an appreciation of the history of chemistry and the people who contributed to that history.

August Wilhelm von Hofmann (1818–1892) *was born in Germany. He first studied law and then changed to chemistry. He founded the German Chemical Society. Hofmann taught at the Royal College of Chemistry in London for 20 years and then returned to Germany to teach at the University of Berlin. He was one of the founders of the German dye industry. Married four times—he was left a widower three times—he had 11 children.*

- **Designing a Synthesis** sections appear throughout the text and help students learn to efficiently design multistep syntheses. Many multistep problems include the synthesis of compounds that students recognize such as Novocaine®, Valium®, and ketoprofen.

6.11 Designing a Synthesis I: An Introduction to Multistep Synthesis

For each reaction we have studied so far, we have seen *why* the reaction occurs, *how* occurs, and the products that are formed. A good way to review these reactions is design syntheses, because in doing so, you have to be able to recall many of the rea tions you have learned.

Synthetic chemists consider time, cost, and yield in designing syntheses. In the i terest of time, a well-designed synthesis should require as few steps (sequential rea tions) as possible, and those steps should each involve a reaction that is easy to car out. If two chemists in a pharmaceutical company were each asked to prepare a ne drug, and one synthesized the drug in three simple steps while the other used 20 dif cult steps, which chemist would not get a raise? In addition, each step in the synthes should provide the greatest possible yield of the desired product, and the cost of t starting materials must be considered. The more reactant needed to synthesize o gram of product, the more expensive it is to produce. Sometimes it is preferable to d

- **Margin Notes**, which have been increased by 10%, emphasize core ideas and remind students of important principles to help them grasp concepts in the text.

Designing a synthesis by working backward from product to reactant is not simply a technique taught to organic chemistry students. It is used so frequently by experienced synthetic chemists that it has been given a name—**retrosynthetic analysis**. Chemists use open arrows to indicate they are working backward. Typically the reagents needed to carry out each step are not included until the reaction is written in the forward direction. For example, the route to the synthesis of the previous ketone can be arrived at by the following retrosynthetic analysis.

retrosynthetic analysis

$$CH_3CH_2\overset{O}{\overset{\|}{C}}CH_2CH_2CH_3 \Longrightarrow CH_3CH_2C{\equiv}CCH_2CH_3 \Longrightarrow CH_3CH_2C{\equiv}CH$$

If atoms have the same atomic number, but different mass numbers, the one with the greater mass number has the higher priority.

- **Voice Balloons**, greatly enhanced in this edition, clarify for students important aspects of a reaction or graphic.

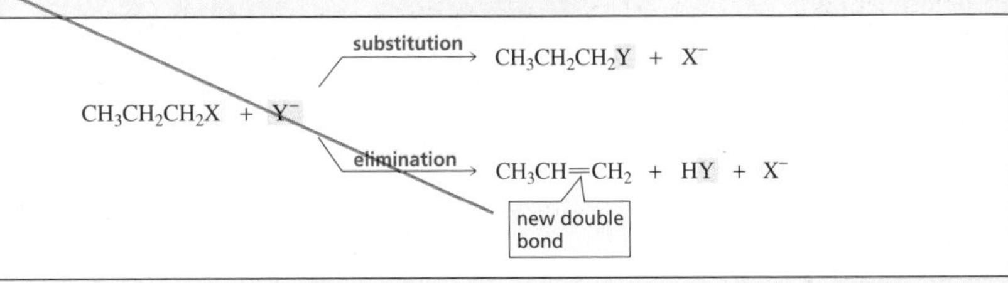

- **Summary of Reactions** sections list reactions covered in the chapter for student review. Cross-references make it easy to locate the sections covering specific reaction types.

- **NEW! End-of-Chapter Summaries** review the major concepts of the chapter in a concise, narrative format.

Summary

Localized electrons belong to a single atom or are confined to a bond between two atoms. **Delocalized electrons** are shared by more than two atoms; they result when a *p* orbital overlaps the *p* orbitals of more than one adjacent atom. Electron delocalization occurs only if all the atoms sharing the delocalized electrons lie in or close to the same plane.

Benzene is a planar molecule. Each of its six carbon atoms is sp^2 hybridized, with bond angles of 120°. A *p* orbital of each carbon overlaps the *p* orbitals of both adjacent carbons. nance energy of the compound. Allylic and benzylic cations (and radicals) have delocalized electrons, so they are more stable than similarly substituted carbocations (and radicals) with localized electrons. Donation of a lone pair is called **resonance electron donation**.

Electron delocalization can affect the nature of the product formed in a reaction and the pK_a of a compound. A carboxylic acid and a phenol are more acidic than an alcohol such as ethanol, and a protonated aniline is more acidic than

- **Key Terms**. A list of important terms appears at the end of every chapter to serve as a convenient reference.

Key Terms

allylic carbon (p. 278)
allylic cation (p. 278)
antibonding molecular orbital (p. 286)
aromatic compounds (p. 292)
asymmetric molecular orbital (p. 288)
benzylic carbon (p. 278)
delocalized electrons (p. 263)
electron delocalization (p. 275)
highest occupied molecular orbital (HOMO) (p. 289)
linear combination of atomic orbitals (LCAO) (p. 287)
resonance (p. 275)
resonance contributor (p. 267)
resonance electron donation (p. 282)
resonance energy (p. 275)
resonance hybrid (p. 268)
resonance structure (p. 267)

Visualization

One of the greatest challenges facing organic chemistry students lies in the often abstract nature of the subject. To help you better visualize important concepts, we have developed striking artwork on paper and in the Molecule Galleries at the Companion Website.

Mechanisms

- **Accurate and Complete Mechanisms.** This text includes hundreds of complete mechanisms integrated into the text prose to promote true understanding, not just memorization.

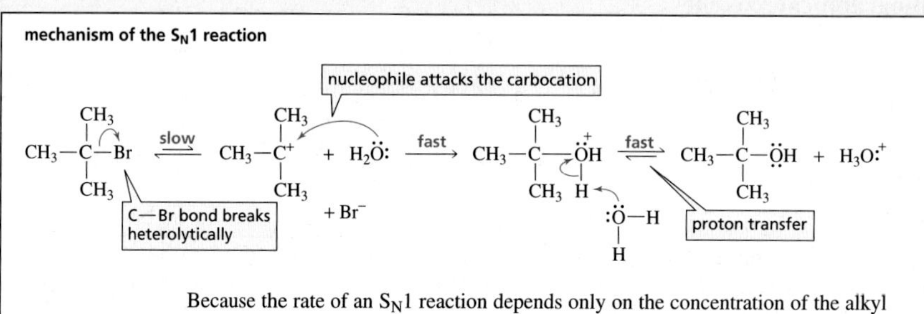

Because the rate of an S_N1 reaction depends only on the concentration of the alkyl halide, the first step must be the slow and rate-determining step. The nucleophile, therefore, is not involved in the rate-determining step, so its concentration has no

Electrostatic Potential Maps

- **Electrostatic potential maps**—found throughout the textbook—help students visualize the electronic structure of molecules and atoms to give them a greater understanding of why and how reactions occur, and help them to better understand why certain molecules and ions behave the way they do.

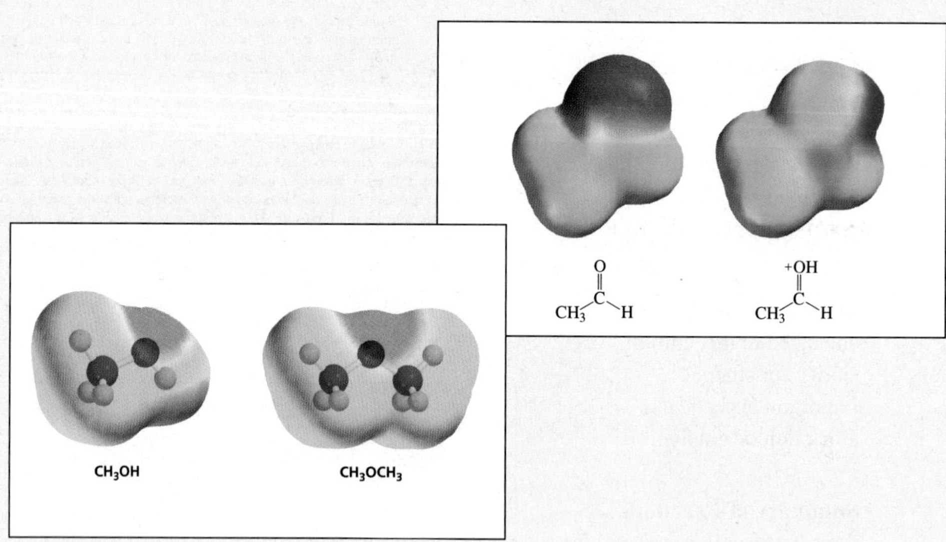

Molecular Art

- Energy-minimized, three-dimensional structures appear throughout the text to give students a more accurate appreciation for the "shapes" of organic molecules.

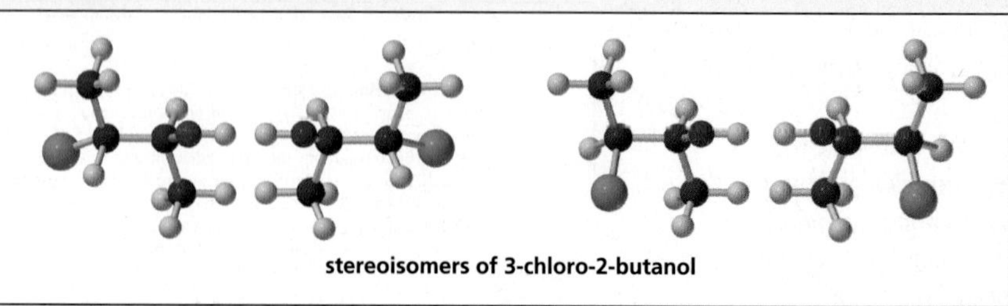

stereoisomers of 3-chloro-2-butanol

Media Resources

Companion Website

http://chem.prenhall.com/bruice

- **Interactive Tutorial and Animation Galleries** highlight central concepts in each chapter and illustrate key mechanisms. The Tutorials often allow students to make incorrect choices and then explain why there is a better answer.

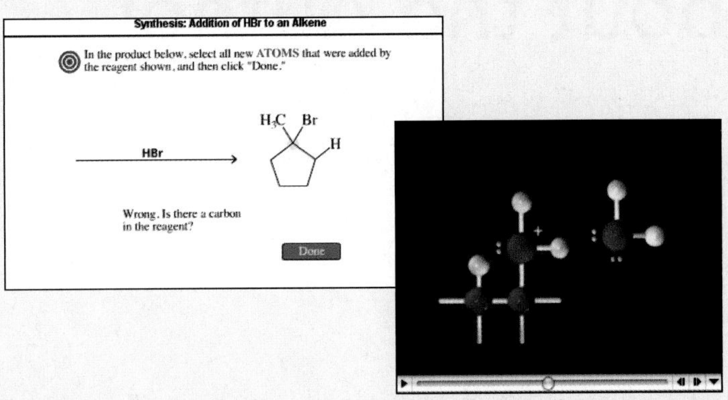

- **Molecule Galleries** feature hundreds of 3-D molecular models of compounds noted in the chapter. Students can rotate and compare models, change their representation, and examine electrostatic potential map surfaces—a unique feature of learning organic chemistry on the Web.

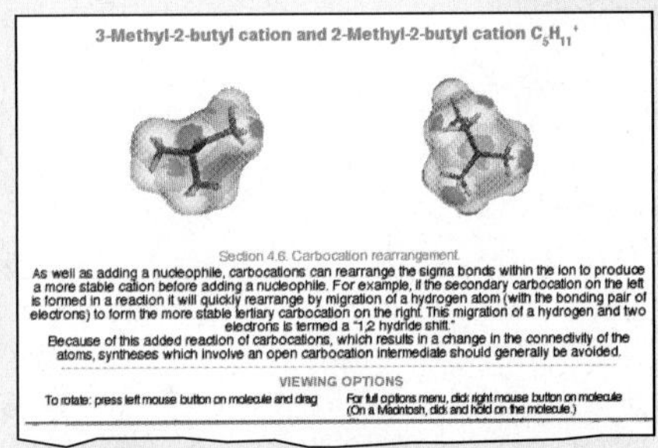

- **Practice Exercises and Quizzes** offer 1800 new exercises to test the student's understanding of the material. Each question includes a hint with a cross-reference to a text reading, and detailed feedback.

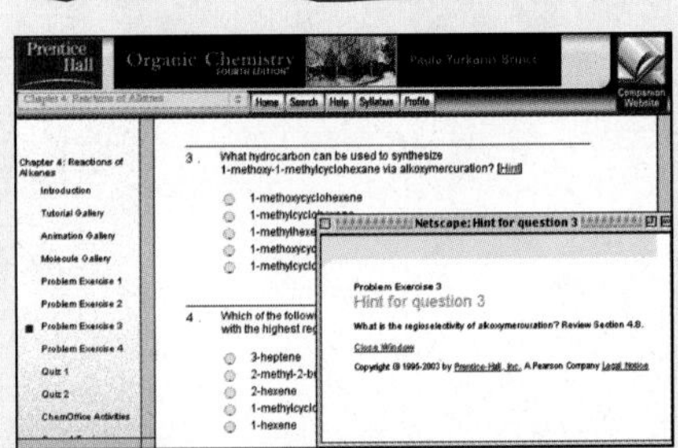

- **Organic ChemIST,** a companion premium Website, offers students an online, media-rich short course in organic chemistry. Written by Robert M. Hanson (St. Olaf College), it uses interactive tutorials, animations, molecular models, graded self-assessment questions, and more, all in an unparalleled study guide which students can use for preview or review of the material. A Premium Access Code is required for access to this site.

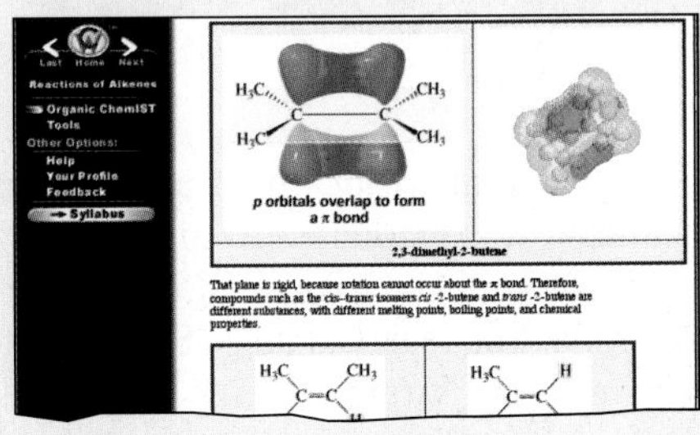

Instructor Resource CD-ROM

- Hundreds of assets from the textbook and from the online resources are made available to instructors in one, simple CD-ROM, including images and tables, interactive tutorials, animations, and 3-D molecular models. Static art is provided in JPEG files for easy importing, ready-made in PowerPoint presentations (with and without lecture notes), or Adobe PDF format for high-resolution printing. Web-ready objects will run directly from the CD-ROM in your Web browser, even if you aren't connected to the Internet. The included MediaPortfolio browser allows you to browse through thumbnails or search for items by keyword, title, or description.

About the Author

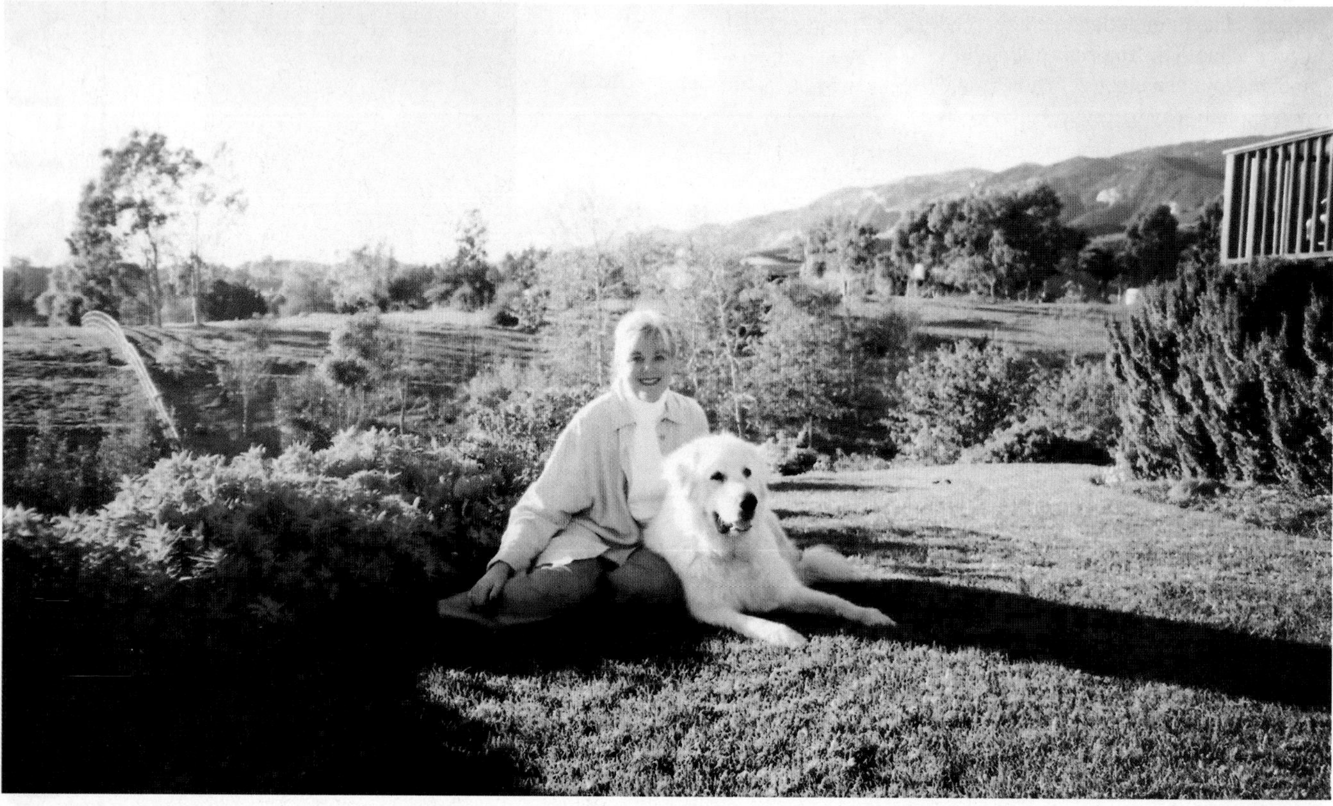

Paula Bruice and Zeus

Paula Yurkanis Bruice was raised primarily in Massachusetts, Germany, and Switzerland and was graduated from the Girls' Latin School in Boston. She received an A.B. from Mount Holyoke College and a Ph.D. in chemistry from the University of Virginia. She received an NIH postdoctoral fellowship for study in biochemistry at the University of Virginia Medical School, and she held a postdoctoral appointment in the Department of Pharmacology at Yale Medical School.

She is a member of the faculty at the University of California, Santa Barbara, where she has received the Associated Students Teacher of the Year Award, the Academic Senate Distinguished Teaching Award, and two Mortar Board Professor of the Year Awards. Her research interests concern the mechanism and catalysis of organic reactions, particularly those of biological significance. Paula has a daughter and a son who are physicians and a son who is a lawyer. Her main hobbies are reading mystery/suspense novels and her pets (three dogs, two cats, and a parrot).

PART ONE

An Introduction to the Study of Organic Chemistry

The first two chapters of the text cover a variety of topics that you need to get started with your study of organic chemistry.

Chapter 1 reviews the topics from general chemistry that will be important to your study of organic chemistry. The chapter starts with a description of the structure of atoms and then proceeds to a description of the structure of molecules. Molecular orbital theory is introduced. Acid–base chemistry, which is central to understanding many organic reactions, is reviewed. You will see how the structure of a molecule affects its acidity and how the acidity of a solution affects molecular structure.

To discuss organic compounds, you must be able to name them and visualize their structures when you read or hear their names. In **Chapter 2**, you will learn how to name five different classes of organic compounds. This will give you a good understanding of the basic rules followed in naming compounds. Because the compounds examined in the chapter are either the reactants or the products of many of the reactions presented in the next 10 chapters, you will have the opportunity to review the nomenclature of these compounds as you proceed through those chapters. The structures and physical properties of these compounds will be compared and contrasted, which makes learning about them a little easier than if each compound were presented separately. Because organic chemistry is a study of compounds that contain carbon, the last part of Chapter 2 discusses the spatial arrangement of the atoms in both chains and rings of carbon atoms.

1 Electronic Structure and Bonding • Acids and Bases

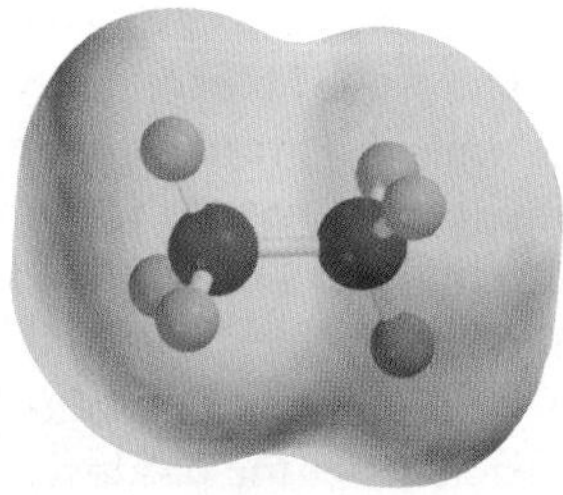

Ethane

Ethene

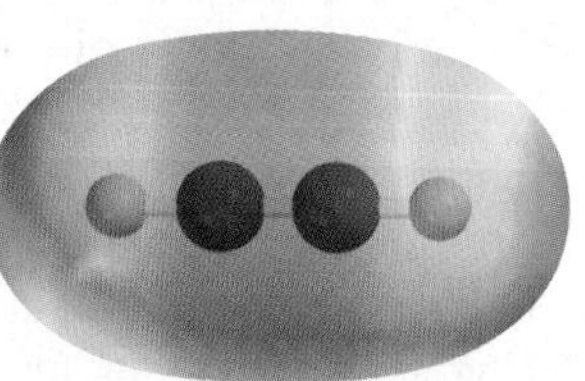

Ethyne

To stay alive, early humans must have been able to tell the difference between two kinds of materials in their world. "You can live on roots and berries," they might have said, "but you can't live on dirt. You can stay warm by burning tree branches, but you can't burn rocks."

Jöns Jakob Berzelius (1779–1848) *not only coined the terms "organic" and "inorganic," but also invented the system of chemical symbols still used today. He published the first list of accurate atomic weights and proposed the idea that atoms carry an electric charge. He purified or discovered the elements cerium, selenium, silicon, thorium, titanium, and zirconium.*

By the eighteenth century, scientists thought they had grasped the nature of that difference, and in 1807, Jöns Jakob Berzelius gave names to the two kinds of materials. Compounds derived from living organisms were believed to contain an unmeasurable vital force—the essence of life. These he called "organic." Compounds derived from minerals—those lacking that vital force—were "inorganic."

Because chemists could not create life in the laboratory, they assumed they could not create compounds with a vital force. With this mind-set, you can imagine how surprised chemists were in 1828 when Friedrich Wöhler produced urea—a compound known to be excreted by mammals—by heating ammonium cyanate, an inorganic mineral.

German chemist **Friedrich Wöhler (1800–1882)** *began his professional life as a physician and later became a professor of chemistry at the University of Göttingen. Wöhler codiscovered the fact that two different chemicals could have the same molecular formula. He also developed methods of purifying aluminum—at the time, the most expensive metal on Earth—and beryllium.*

$$\overset{+}{N}H_4\ \overset{-}{O}CN \xrightarrow{\text{heat}} H_2N{-}\overset{\overset{\displaystyle O}{\|}}{C}{-}NH_2$$

ammonium cyanate → urea

For the first time, an "organic" compound had been obtained from something other than a living organism and certainly without the aid of any kind of vital force. Clearly, chemists needed a new definition for "organic compounds." **Organic compounds** are now defined as *compounds that contain carbon*.

Why is an entire branch of chemistry devoted to the study of carbon-containing compounds? We study organic chemistry because just about all of the molecules that make life possible—proteins, enzymes, vitamins, lipids, carbohydrates, and nucleic acids—contain carbon, so the chemical reactions that take place in living systems, including our own bodies, are organic reactions. Most of the compounds found in nature—those we rely on for food, medicine, clothing (cotton, wool, silk), and energy (natural gas, petroleum)—are organic as well. Important organic compounds are not, however, limited to the ones we find in nature. Chemists have learned to synthesize millions of organic compounds never found in nature, including synthetic fabrics, plastics, synthetic rubber, medicines, and even things like photographic film and Super glue. Many of these synthetic compounds prevent shortages of naturally occurring products. For example, it has been estimated that if synthetic materials were not available for clothing, all of the arable land in the United States would have to be used for the production of cotton and wool just to provide enough material to clothe us. Currently, there are about 16 million known organic compounds, and many more are possible.

What makes carbon so special? Why are there so many carbon-containing compounds? The answer lies in carbon's position in the periodic table. Carbon is in the center of the second row of elements. The atoms to the left of carbon have a tendency to give up electrons, whereas the atoms to the right have a tendency to accept electrons (Section 1.3).

the second row of the periodic table

Because carbon is in the middle, it neither readily gives up nor readily accepts electrons. Instead, it shares electrons. Carbon can share electrons with several different kinds of atoms, and it can also share electrons with other carbon atoms. Consequently, carbon is able to form millions of stable compounds with a wide range of chemical properties simply by sharing electrons.

When we study organic chemistry, we study how organic compounds react. When an organic compound reacts, some old bonds break and some new bonds form. Bonds form when two atoms share electrons, and bonds break when two atoms no longer share electrons. How readily a bond forms and how easily it breaks depend on the particular electrons that are shared, which, in turn, depend on the atoms to which the electrons belong. So if we are going to start our study of organic chemistry at the beginning, we must start with an understanding of the structure of an atom—what electrons an atom has and where they are located.

1.1 The Structure of an Atom

An atom consists of a tiny dense nucleus surrounded by electrons that are spread throughout a relatively large volume of space around the nucleus. The nucleus contains positively charged protons and neutral neutrons, so it is positively charged. The electrons are negatively charged. Because the amount of positive charge on a proton equals the amount of negative charge on an electron, a neutral atom has an equal number of protons and electrons. Atoms can gain electrons and thereby become negatively charged, or they can lose electrons and become positively charged. However, the number of protons in an atom does not change.

Protons and neutrons have approximately the same mass and are about 1800 times more massive than an electron. This means that most of the mass of an atom is in its nucleus. However, most of the *volume* of an atom is occupied by its electrons, and that is where our focus will be because it is the electrons that form chemical bonds.

Louis Victor Pierre Raymond duc de Broglie (1892–1987) *was born in France and studied history at the Sorbonne. During World War I, he was stationed in the Eiffel Tower as a radio engineer. Intrigued by his exposure to radio communications, he returned to school after the war, earned a Ph.D. in physics, and became a professor of theoretical physics at the Faculté des Sciences at the Sorbonne. He received the Nobel Prize in physics in 1929, five years after obtaining his degree, for his work that showed electrons to have properties of both particles and waves. In 1945, he became an adviser to the French Atomic Energy Commissariat.*

The **atomic number** of an atom equals the number of protons in its nucleus. The atomic number is also the number of electrons that surround the nucleus of a neutral atom. For example, the atomic number of carbon is 6, which means that a neutral carbon atom has six protons and six electrons. Because the number of protons in an atom does not change, the atomic number of a particular element is always the same—all carbon atoms have an atomic number of 6.

The **mass number** of an atom is the *sum* of its protons and neutrons. Not all carbon atoms have the same mass number, because, even though they all have the same number of protons, they do not all have the same number of neutrons. For example, 98.89% of naturally occurring carbon atoms have six neutrons—giving them a mass number of 12—and 1.11% have seven neutrons—giving them a mass number of 13. These two different kinds of carbon atoms (^{12}C and ^{13}C) are called isotopes. **Isotopes** have the same atomic number (i.e., the same number of protons), but different mass numbers because they have different numbers of neutrons. The chemical properties of isotopes of a given element are nearly identical.

Naturally occurring carbon also contains a trace amount of ^{14}C, which has six protons and eight neutrons. This isotope of carbon is radioactive, decaying with a half-life of 5730 years. (The half-life is the time it takes for one-half of the nuclei to decay.) As long as a plant or animal is alive, it takes in as much ^{14}C as it excretes or exhales. When it dies, it no longer takes in ^{14}C, so the ^{14}C in the organism slowly decreases. Therefore, the age of an organic substance can be determined by its ^{14}C content.

The **atomic weight** of a naturally occurring element is the average weighted mass of its atoms. Because an *atomic mass unit (amu)* is defined as exactly 1/12 of the mass of ^{12}C, the atomic mass of ^{12}C is 12.0000 amu; the atomic mass of ^{13}C is 13.0034 amu. Therefore, the atomic weight of carbon is 12.011 amu $(0.9889 \times 12.0000 + 0.0111 \times 13.0034 = 12.011)$. The **molecular weight** is the sum of the atomic weights of all the atoms in the molecule.

PROBLEM 1◆

Oxygen has three isotopes with mass numbers of 16, 17, and 18. The atomic number of oxygen is eight. How many protons and neutrons does each of the isotopes have?

1.2 The Distribution of Electrons in an Atom

Erwin Schrödinger (1887–1961) *was teaching physics at the University of Berlin when Hitler rose to power. Although not Jewish, Schrödinger left Germany to return to his native Austria, only to see it taken over later by the Nazis. He moved to the School for Advanced Studies in Dublin and then to Oxford University. In 1933, he shared the Nobel Prize in physics with Paul Dirac, a professor of physics at Cambridge University, for mathematical work on quantum mechanics.*

Electrons are moving continuously. Like anything that moves, electrons have kinetic energy, and this energy is what counters the attractive force of the positively charged protons that would otherwise pull the negatively charged electrons into the nucleus. For a long time, electrons were perceived to be particles—infinitesimal "planets" orbiting the nucleus of an atom. In 1924, however, a French physicist named Louis de Broglie showed that electrons also have wavelike properties. He did this by combining a formula developed by Einstein that relates mass and energy with a formula developed by Planck relating frequency and energy. The realization that electrons have wavelike properties spurred physicists to propose a mathematical concept known as quantum mechanics.

Quantum mechanics uses the same mathematical equations that describe the wave motion of a guitar string to characterize the motion of an electron around a nucleus. The version of quantum mechanics most useful to chemists was proposed by Erwin Schrödinger in 1926. According to Schrödinger, the behavior of each electron in an atom or a molecule can be described by a **wave equation**. The solutions to the Schrödinger equation are called **wave functions** or **orbitals**. They tell us the *energy* of the electron and the *volume of space* around the nucleus where an electron is most likely to be found.

An orbital tells us the energy of the electron and the volume of space around the nucleus where an electron is most likely to be found.

According to quantum mechanics, the electrons in an atom can be thought of as occupying a set of concentric shells that surround the nucleus. The first shell is the one

ALBERT EINSTEIN

Albert Einstein (1879–1955) was born in Germany. When he was in high school, his father's business failed and his family moved to Milan, Italy. Einstein had to stay behind because German law required compulsory military service after finishing high school. Einstein wanted to join his family in Italy. His high school mathematics teacher wrote a letter saying that Einstein could have a nervous breakdown without his family and also that there was nothing left to teach him. Eventually, Einstein was asked to leave the school because of his disruptive behavior. Popular folklore says he left because of poor grades in Latin and Greek, but his grades in those subjects were fine.

Einstein was visiting the United States when Hitler came to power, so he accepted a position at the Institute for Advanced Study in Princeton, becoming a U.S. citizen in 1940. Although a lifelong pacifist, he wrote a letter to President Roosevelt warning of ominous advances in German nuclear research. This led to the creation of the Manhattan Project, which developed the atomic bomb and tested it in New Mexico in 1945.

closest to the nucleus. The second shell lies farther from the nucleus, and even farther out lie the third and higher numbered shells. Each shell contains subshells known as **atomic orbitals**. Each atomic orbital has a characteristic shape and energy and occupies a characteristic volume of space, which is predicted by the Schrödinger equation. An important point to remember is that *the closer the atomic orbital is to the nucleus, the lower is its energy.*

The closer the orbital is to the nucleus, the lower is its energy.

The first shell consists of only an *s* atomic orbital; the second shell consists of *s* and *p* atomic orbitals; the third shell consists of *s*, *p*, and *d* atomic orbitals; and the fourth and higher shells consist of *s*, *p*, *d*, and *f* atomic orbitals (Table 1.1).

Each shell contains one *s* atomic orbital. The second and higher shells—in addition to their *s* orbital—each contain three *degenerate p* atomic orbitals. **Degenerate orbitals** are orbitals that have the same energy. The third and higher shells—in

Table 1.1 Distribution of Electrons in the First Four Shells That Surround the Nucleus

	First shell	Second shell	Third shell	Fourth shell
Atomic orbitals	*s*	*s*, *p*	*s*, *p*, *d*	*s*, *p*, *d*, *f*
Number of atomic orbitals	1	1, 3	1, 3, 5	1, 3, 5, 7
Maximum number of electrons	2	8	18	32

MAX KARL ERNST LUDWIG PLANCK

Max Planck (1858–1947) was born in Germany, the son of a professor of civil law. He was a professor at the Universities of Munich (1880–1889) and Berlin (1889–1926). Two of his daughters died in childbirth, and one of his sons was killed in action in World War I. In 1918, Planck received the Nobel Prize in physics for his development of quantum theory. He became president of the Kaiser Wilhelm Society of Berlin—later renamed the Max Planck Society—in 1930. Planck felt that it was his duty to remain in Germany during the Nazi era, but he never supported the Nazi regime. He unsuccessfully interceded with Hitler on behalf of his Jewish colleagues and, as a consequence, was forced to resign from the presidency of the Kaiser Wilhelm Society in 1937. A second son was accused of taking part in the plot to kill Hitler and was executed. Planck lost his home to Allied bombings. He was rescued by Allied forces during the final days of the war.

addition to their *s* and *p* orbitals—also contain five degenerate *d* atomic orbitals, and the fourth and higher shells also contain seven degenerate *f* atomic orbitals. Because a maximum of two electrons can coexist in an atomic orbital (see the Pauli exclusion principle, below), the first shell, with only one atomic orbital, can contain no more than two electrons. The second shell, with four atomic orbitals—one *s* and three *p*—can have a total of eight electrons. Eighteen electrons can occupy the nine atomic orbitals—one *s*, three *p*, and five *d*—of the third shell, and 32 electrons can occupy the 16 atomic orbitals of the fourth shell. In studying organic chemistry, we will be concerned primarily with atoms that have electrons only in the first and second shells.

The **ground-state electronic configuration** of an atom describes the orbitals occupied by the atom's electrons when they are all in the available orbitals with the lowest energy. If energy is applied to an atom in the ground state, one or more electrons can jump into a higher energy orbital. The atom then would be in an **excited-state electronic configuration**. The ground-state electronic configurations of the 11 smallest atoms are shown in Table 1.2. (Each arrow—whether pointing up or down—represents one electron.) The following principles are used to determine which orbitals electrons occupy:

1. The **aufbau principle** (*aufbau* is German for "building up") tells us the first thing we need to know to be able to assign electrons to the various atomic orbitals. According to this principle, an electron always goes into the available orbital with the lowest energy. The relative energies of the atomic orbitals are as follows:

   $1s < 2s < 2p < 3s < 3p < 4s < 3d < 4p < 5s < 4d < 5p <$
   $6s < 4f < 5d < 6p < 7s < 5f$

   Because a 1*s* atomic orbital is closer to the nucleus, it is lower in energy than a 2*s* atomic orbital, which is lower in energy—and is closer to the nucleus—than a 3*s* atomic orbital. Comparing atomic orbitals in the same shell, we see that an *s* atomic orbital is lower in energy than a *p* atomic orbital, and a *p* atomic orbital is lower in energy than a *d* atomic orbital.

2. The **Pauli exclusion principle** states that (a) no more than two electrons can occupy each atomic orbital, and (b) the two electrons must be of opposite spin. It is called an exclusion principle because it states that only so many electrons can occupy any particular shell. Notice in Table 1.2 that spin in one direction is designated by an upward-pointing arrow, and spin in the opposite direction by a downward-pointing arrow.

As a teenager, Austrian **Wolfgang Pauli (1900–1958)** *wrote articles on relativity that caught the attention of Albert Einstein. Pauli went on to teach physics at the University of Hamburg and at the Zurich Institute of Technology. When World War II broke out, he immigrated to the United States, where he joined the Institute for Advanced Study at Princeton.*

TABLE 1.2 The Ground-State Electronic Configurations of the Smallest Atoms

Atom	Name of element	Atomic number	1s	2s	$2p_x$	$2p_y$	$2p_z$	3s
H	Hydrogen	1	↑					
He	Helium	2	↑↓					
Li	Lithium	3	↑↓	↑				
Be	Beryllium	4	↑↓	↑↓				
B	Boron	5	↑↓	↑↓	↑			
C	Carbon	6	↑↓	↑↓	↑	↑		
N	Nitrogen	7	↑↓	↑↓	↑	↑	↑	
O	Oxygen	8	↑↓	↑↓	↑↓	↑	↑	
F	Fluorine	9	↑↓	↑↓	↑↓	↑↓	↑	
Ne	Neon	10	↑↓	↑↓	↑↓	↑↓	↑↓	
Na	Sodium	11	↑↓	↑↓	↑↓	↑↓	↑↓	↑

From these first two rules, we can assign electrons to atomic orbitals for atoms that contain one, two, three, four, or five electrons. The single electron of a hydrogen atom occupies a $1s$ atomic orbital, the second electron of a helium atom fills the $1s$ atomic orbital, the third electron of a lithium atom occupies a $2s$ atomic orbital, the fourth electron of a beryllium atom fills the $2s$ atomic orbital, and the fifth electron of a boron atom occupies one of the $2p$ atomic orbitals. (The subscripts x, y, and z distinguish the three $2p$ atomic orbitals.) Because the three p orbitals are degenerate, the electron can be put into any one of them. Before we can continue to larger atoms—those containing six or more electrons—we need Hund's rule:

Friedrich Hermann Hund (1896–1997) *was born in Germany. He was a professor of physics at several German universities, the last being the University of Göttingen. He spent a year as a visiting professor at Harvard University. In February 1996, the University of Göttingen held a symposium to honor Hund on his 100th birthday.*

3. **Hund's rule** states that when there are degenerate orbitals—two or more orbitals with the same energy—an electron will occupy an empty orbital before it will pair up with another electron. In this way, electron repulsion is minimized. The sixth electron of a carbon atom, therefore, goes into an empty $2p$ atomic orbital, rather than pairing up with the electron already occupying a $2p$ atomic orbital. (See Table 1.2.) The seventh electron of a nitrogen atom goes into an empty $2p$ atomic orbital, and the eighth electron of an oxygen atom pairs up with an electron occupying a $2p$ atomic orbital rather than going into a higher energy $3s$ orbital.

Using these three rules, the locations of the electrons in the remaining elements can be assigned.

PROBLEM 2◆

Potassium has an atomic number of 19 and one unpaired electron. What orbital does the unpaired electron occupy?

PROBLEM 3◆

Write electronic configurations for chlorine (atomic number 17), bromine (atomic number 35), and iodine (atomic number 53).

1.3 Ionic, Covalent, and Polar Bonds

In trying to explain why atoms form bonds, G. N. Lewis proposed that *an atom is most stable if its outer shell is either filled or contains eight electrons and it has no electrons of higher energy*. According to Lewis's theory, an atom will give up, accept, or share electrons in order to achieve a filled outer shell or an outer shell that contains eight electrons. This theory has come to be called the **octet rule**.

Lithium (Li) has a single electron in its $2s$ atomic orbital. If it loses this electron, the lithium atom ends up with a filled outer shell—a stable configuration. Removing an electron from an atom takes energy—called the **ionization energy**. Lithium has a relatively low ionization energy—the drive to achieve a filled outer shell with no electrons of higher energy causes it to lose an electron relatively easily. Sodium (Na) has a single electron in its $3s$ atomic orbital. Consequently, sodium also has a relatively low ionization energy because, when it loses an electron, it is left with an outer shell of eight electrons. Elements (such as lithium and sodium) that have low ionization energies are said to be **electropositive**—they readily lose an electron and thereby become positively charged. The elements in the first column of the periodic table are all electropositive—each readily loses an electron because each has a single electron in its outermost shell.

Electrons in inner shells (those below the outermost shell) are called **core electrons**. Core electrons do not participate in chemical bonding. Electrons in the outermost shell are called **valence electrons**, and the outermost shell is called the valence shell. Carbon, for example, has two core electrons and four valence electrons (Table 1.2).

Lithium and sodium each have one valence electron. Elements in the same column of the periodic table have the same number of valence electrons, and because the number of valence electrons is the major factor determining an element's chemical properties, elements in the same column of the periodic table have similar chemical properties. Thus, the chemical behavior of an element depends on its electronic configuration.

PROBLEM 4

Compare the ground-state electronic configurations of the following atoms, and check the relative positions of the atoms in Table 1.3 on p. 10.

a. carbon and silicon
b. oxygen and sulfur
c. fluorine and bromine
d. magnesium and calcium

When we draw the electrons around an atom, as in the following equations, core electrons are not shown; only valence electrons are shown. Each valence electron is shown as a dot. Notice that when the single valence electron of lithium or sodium is removed, the resulting atom—now called an ion—carries a positive charge.

$$\text{Li}\cdot \longrightarrow \text{Li}^+ + e^-$$

$$\text{Na}\cdot \longrightarrow \text{Na}^+ + e^-$$

Fluorine has seven valence electrons (Table 1.2). Consequently, it readily acquires an electron in order to have an outer shell of eight electrons. When an atom acquires an electron, energy is released. Elements in the same column as fluorine (e.g., chlorine, bromine, and iodine) also need only one electron to have an outer shell of eight, so they, too, readily acquire an electron. Elements that readily acquire an electron are said to be **electronegative**—they acquire an electron easily and thereby become negatively charged.

$$:\ddot{\underset{\cdot\cdot}{\text{F}}}\cdot + e^- \longrightarrow :\ddot{\underset{\cdot\cdot}{\text{F}}}:^-$$

$$:\ddot{\underset{\cdot\cdot}{\text{Cl}}}\cdot + e^- \longrightarrow :\ddot{\underset{\cdot\cdot}{\text{Cl}}}:^-$$

Ionic Bonds

Because sodium gives up an electron easily and chlorine acquires an electron readily, when sodium metal and chlorine gas are mixed, each sodium atom transfers an electron to a chlorine atom, and crystalline sodium chloride (table salt) is formed as a result. The positively charged sodium ions and negatively charged chloride ions are independent species held together by the attraction of opposite charges (Figure 1.1). A **bond** is an attractive force between two atoms. Attractive forces between opposite charges are called **electrostatic attractions**. A **bond** that is the result of only electrostatic attractions is called an ionic bond. Thus, an **ionic bond** is formed when there is a *transfer of electrons*, causing one atom to become a positively charged ion and the other to become a negatively charged ion.

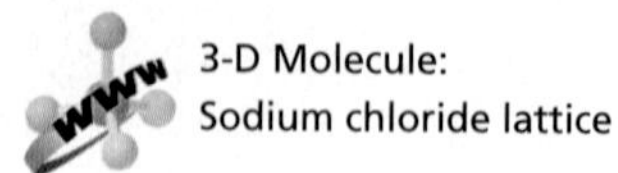

3-D Molecule: Sodium chloride lattice

a.

b.

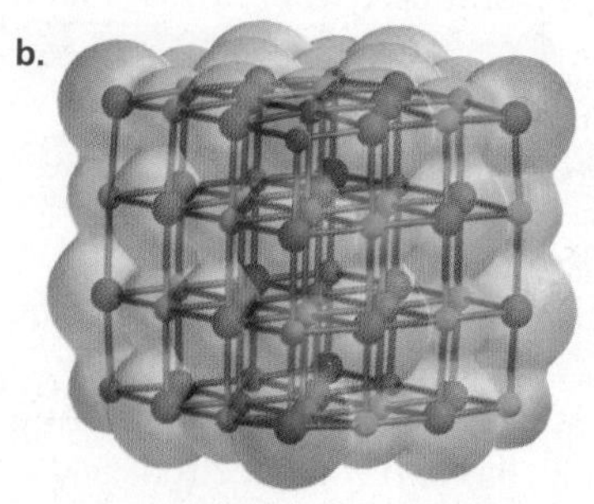

Figure 1.1 ▶ (a) Crystalline sodium chloride. (b) The electron-rich chloride ions are red and the electron-poor sodium ions are blue. Each chloride ion is surrounded by six sodium ions, and each sodium ion is surrounded by six chloride ions. Ingore the "bonds" holding the balls together; they are there only to keep the model from falling apart.

ionic bond

$$\begin{array}{ccc} :\ddot{\underset{..}{Cl}}:^- & Na^+ & :\ddot{\underset{..}{Cl}}:^- \\ Na^+ & :\ddot{\underset{..}{Cl}}:^- & Na^+ \\ :\ddot{\underset{..}{Cl}}:^- & Na^+ & :\ddot{\underset{..}{Cl}}:^- \end{array}$$

sodium chloride

Sodium chloride is an example of an ionic compound. **Ionic compounds** are formed when an element on the left side of the periodic table (an electropositive element) transfers one or more electrons to an element on the right side of the periodic table (an electronegative element).

Covalent Bonds

Instead of giving up or acquiring electrons, an atom can achieve a filled outer shell by sharing electrons. For example, two fluorine atoms can each attain a filled shell of eight electrons by sharing their unpaired valence electrons. A bond formed as a result of *sharing electrons* is called a **covalent bond**.

a covalent bond

$$:\ddot{\underset{..}{F}}\cdot + \cdot\ddot{\underset{..}{F}}: \longrightarrow :\ddot{\underset{..}{F}}:\ddot{\underset{..}{F}}:$$

Two hydrogen atoms can form a covalent bond by sharing electrons. As a result of covalent bonding, each hydrogen acquires a stable, filled outer shell (with two electrons).

$$H\cdot + \cdot H \longrightarrow H:H$$

Similarly, hydrogen and chlorine can form a covalent bond by sharing electrons. In doing so, hydrogen fills its only shell and chlorine achieves an outer shell of eight electrons.

$$H\cdot + \cdot\ddot{\underset{..}{Cl}}: \longrightarrow H:\ddot{\underset{..}{Cl}}:$$

A hydrogen atom can achieve a completely empty shell by losing an electron. Loss of its sole electron results in a positively charged **hydrogen ion**. A positively charged hydrogen ion is called a **proton** because when a hydrogen atom loses its valence electron, only the hydrogen nucleus—which consists of a single proton—remains. A hydrogen atom can achieve a filled outer shell by gaining an electron, thereby forming a negatively charged hydrogen ion, called a **hydride ion**.

$$\underset{\text{a hydrogen atom}}{H\cdot} \longrightarrow \underset{\text{a proton}}{H^+} + e^-$$

$$\underset{\text{a hydrogen atom}}{H\cdot} + e^- \longrightarrow \underset{\text{a hydride ion}}{H:^-}$$

Shown is a bronze sculpture of **Albert Einstein** *on the grounds of the National Academy of Sciences in Washington, DC. The statue measures 21 feet from the top of the head to the tip of the feet and weighs 7000 pounds. In his left hand, Einstein holds the mathematical equations that represent his three most important contributions to science: the photoelectric effect, the equivalency of energy and matter, and the theory of relativity. At his feet is a map of the sky.*

Because oxygen has six valence electrons, it needs to form two covalent bonds to achieve an outer shell of eight electrons. Nitrogen, with five valence electrons, must form three covalent bonds, and carbon, with four valence electrons, must form four covalent bonds to achieve a filled outer shell. Notice that all the atoms in water, ammonia, and methane have filled outer shells.

$$2\,H\cdot + \cdot\ddot{\underset{.}{O}}: \longrightarrow \underset{\text{water}}{\begin{array}{c} H:\ddot{\underset{..}{O}}: \\ H \end{array}}$$

$$3\,H\cdot + \cdot\ddot{\underset{.}{N}}\cdot \longrightarrow \underset{\text{ammonia}}{\begin{array}{c} H:\ddot{\underset{..}{N}}:H \\ H \end{array}}$$

$$4\,H\cdot + \cdot\dot{\underset{.}{C}}\cdot \longrightarrow \underset{\text{methane}}{\begin{array}{c} H \\ H:\ddot{\underset{..}{C}}:H \\ H \end{array}}$$

Polar Covalent Bonds

In the F—F and H—H covalent bonds shown previously, the atoms that share the bonding electrons are identical. Therefore, they share the electrons equally; that is, each electron spends as much time in the vicinity of one atom as in the other. An even (nonpolar) distribution of charge results. Such a bond is called a **nonpolar covalent bond**.

In contrast, the bonding electrons in hydrogen chloride, water, and ammonia are more attracted to one atom than another because the atoms that share the electrons in these molecules are different and have different electronegativities. **Electronegativity** is the tendency of an atom to pull bonding electrons toward itself. The bonding electrons in hydrogen chloride, water, and ammonia molecules are more attracted to the atom with the greater electronegativity. This results in a polar distribution of charge. A **polar covalent bond** is a covalent bond between atoms of different electronegativities. The electronegativities of some of the elements are shown in Table 1.3. Notice that electronegativity increases as you go from left to right across a row of the periodic table or up any of the columns.

A polar covalent bond has a slight positive charge on one end and a slight negative charge on the other. Polarity in a covalent bond is indicated by the symbols $\delta+$ and $\delta-$, which denote partial positive and partial negative charges, respectively. The negative end of the bond is the end that has the more electronegative atom. The greater the difference in electronegativity between the bonded atoms, the more polar the bond will be.

δ+ δ−
H—C̈l:

δ+ δ−
H—Ö:
|
H
δ+

δ+ δ− δ+
H—N̈—H
|
H
δ+

The direction of bond polarity can be indicated with an arrow. By convention, the arrow points in the direction in which the electrons are pulled, so the head of the arrow is at the negative end of the bond; a short perpendicular line near the tail of the arrow marks the positive end of the bond.

+⟶
H—C̈l:

TABLE 1.3 The Electronegativities of Selected Elements[a]

IA	IIA	IB	IIB	IIIA	IVA	VA	VIA	VIIA
H 2.1								
Li 1.0	Be 1.5			B 2.0	C 2.5	N 3.0	O 3.5	F 4.0
Na 0.9	Mg 1.2			Al 1.5	Si 1.8	P 2.1	S 2.5	Cl 3.0
K 0.8	Ca 1.0							Br 2.8
								I 2.5

increasing electronegativity (→ across a row)

increasing electronegativity (↑ up a column)

[a]Electronegativity values are relative, not absolute. As a result, there are several scales of electronegativities. The electronegativities listed here are from the scale devised by Linus Pauling.

You can think of ionic bonds and nonpolar covalent bonds as being at the opposite ends of a continuum of bond types. An ionic bond involves no sharing of electrons. A nonpolar covalent bond involves equal sharing. Polar covalent bonds fall somewhere in between, and the greater the difference in electronegativity between the atoms forming the bond, the closer the bond is to the ionic end of the continuum. C—H bonds are relatively nonpolar, because carbon and hydrogen have similar electronegativities (electronegativity difference = 0.4; see Table 1.3). N—H bonds are relatively polar (electronegativity difference = 0.9), but not as polar as O—H bonds (electronegativity difference = 1.4). The bond between sodium and chloride ions is closer to the ionic end of the continuum (electronegativity difference = 2.1), but sodium chloride is not as ionic as potassium fluoride (electronegativity difference = 3.2).

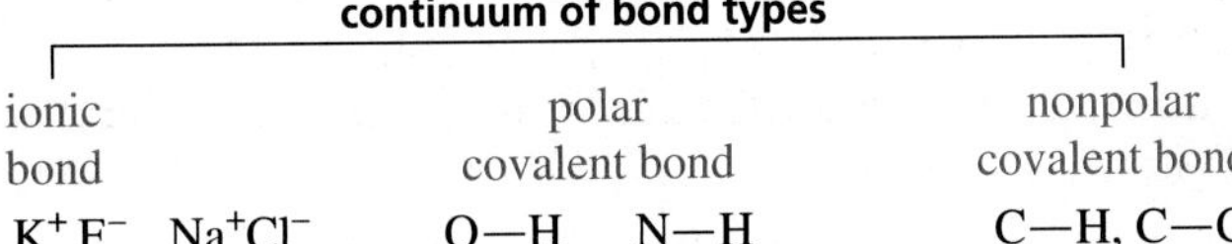

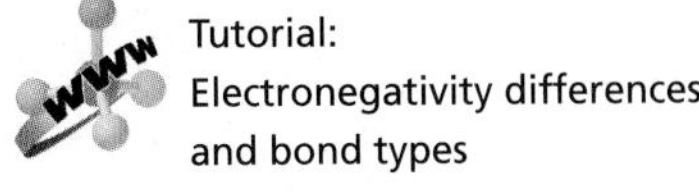

PROBLEM 5◆

Which of the following has

a. the most polar bond?

b. the least polar bond?

NaI LiBr Cl_2 KCl

Understanding bond polarity is critical to understanding how organic reactions occur, because a central rule that governs the reactivity of organic compounds is that *electron-rich atoms or molecules are attracted to electron-deficient atoms or molecules.* **Electrostatic potential maps** (often simply called potential maps) are models that show how charge is distributed in the molecule under the map. Therefore, these maps show the kind of electrostatic attraction an atom or molecule has for another atom or molecule, so you can use them to predict chemical reactions. The potential maps for LiH, H_2, and HF are shown below.

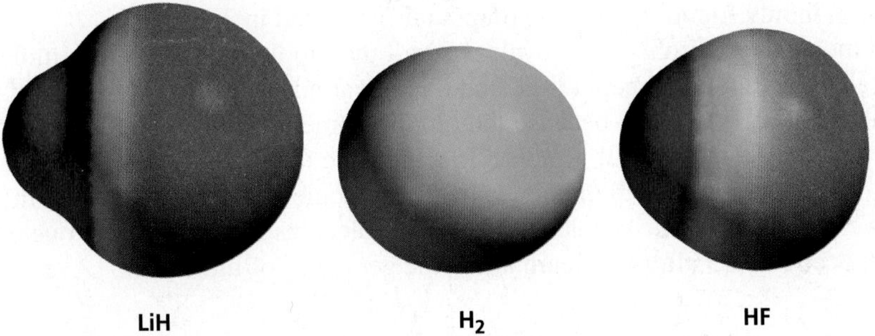

The colors on a potential map indicate the degree to which a molecule or an atom in a molecule attracts charged particles. Red—signifying the most negative electrostatic potential—is used for regions that attract positively charged molecules most strongly, and blue is used for areas with the most positive electrostatic potential—that is, regions that attract negatively charged molecules most strongly. Other colors indicate intermediate levels of attraction.

red < orange < yellow < green < blue

most negative electrostatic potential **most positive electrostatic potential**

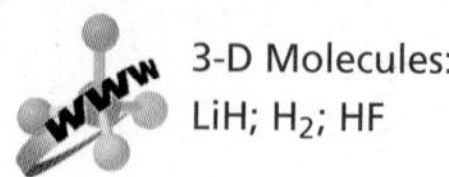
3-D Molecules: LiH; H_2; HF

The colors on a potential map can also be used to estimate charge distribution. For example, the potential map for LiH indicates that the hydrogen atom is more negatively charged than the lithium atom. By comparing the three maps, we can tell that the hydrogen in LiH is more negatively charged than a hydrogen in H_2, and the hydrogen in HF is more positively charged than a hydrogen in H_2.

A molecule's size and shape are determined by the number of electrons in the molecule and by the way they move. Because a potential map roughly marks the "edge" of the molecule's electron cloud, the map tells us something about the relative size and shape of the molecule. Notice that a given kind of atom can have different sizes in different molecules. The negatively charged hydrogen in LiH is bigger than a neutral hydrogen in H_2, which, in turn, is bigger than the positively charged hydrogen in HF.

PROBLEM 6◆

After examining the potential maps for LiH, HF, and H_2, answer the following questions:

a. Which compounds are polar?
b. Why does LiH have the largest hydrogen?
c. Which compound has the most positively charged hydrogen?

A polar bond has a **dipole**—it has a negative end and a positive end. The size of the dipole is indicated by the dipole moment, which is given the Greek letter μ. The **dipole moment** of a bond is equal to the magnitude of the charge (e) on the atom (either the partial positive charge or the partial negative charge, because they have the same magnitude) times the distance between the two charges (d):

$$\text{dipole moment} = \mu = e \times d$$

A dipole moment is reported in a unit called a **debye (D)** (pronounced de-bye). Because the charge on an electron is 4.80×10^{-10} electrostatic units (esu) and the distance between charges in a polar bond is on the order of 10^{-8} cm, the product of charge and distance is on the order of 10^{-18} esu cm. A dipole moment of 1.5×10^{-18} esu cm can be more simply stated as 1.5 D. The dipole moments of some bonds commonly found in organic compounds are listed in Table 1.4.

In a molecule with only one covalent bond, the dipole moment of the molecule is identical to the dipole moment of the bond. For example, the dipole moment of hydrogen chloride (HCl) is 1.1 D because the dipole moment of the single H—Cl bond is 1.1 D. The dipole moment of a molecule with more than one covalent bond depends on the dipole moments of all the bonds in the molecule and the geometry of the molecule. We will examine the dipole moments of molecules with more than one covalent bond in Section 1.15 after you learn about the geometry of molecules.

Peter Debye (1884–1966) *was born in the Netherlands. He taught at the universities of Zürich (succeeding Einstein), Leipzig, and Berlin, but returned to his homeland in 1939 when the Nazis ordered him to become a German citizen. Upon visiting Cornell to give a lecture, he decided to stay in the country, and he became a U.S. citizen in 1946. He received the Nobel Prize in chemistry in 1936 for his work on dipole moments and the properties of solutions.*

Table 1.4 The Dipole Moments of Some Commonly Encountered Bonds

Bond	Dipole moment (D)	Bond	Dipole moment (D)
H—C	0.4	C—C	0
H—N	1.3	C—N	0.2
H—O	1.5	C—O	0.7
H—F	1.7	C—F	1.6
H—Cl	1.1	C—Cl	1.5
H—Br	0.8	C—Br	1.4
H—I	0.4	C—I	1.2

PROBLEM 7 SOLVED

Determine the partial negative charge on the oxygen atom in a C=O bond. The bond length is 1.22 Å* and the bond dipole moment is 2.30 D.

SOLUTION If there were a full negative charge on the oxygen atom, the dipole moment would be

$$(4.80 \times 10^{-10}\,\text{esu})(1.22 \times 10^{-8}\,\text{cm}) = 5.86 \times 10^{-18}\,\text{esu cm} = 5.86\,\text{D}$$

Knowing that the dipole moment is 2.30 D, we calculate that the partial negative charge on the oxygen atom is about 0.4:

$$\frac{2.30}{5.86} = 0.39$$

PROBLEM 8

Use the symbols $\delta+$ and $\delta-$ to show the direction of polarity of the indicated bond in each of the following compounds (for example, $\overset{\delta+}{H_3C}-\overset{\delta-}{OH}$).

a. HO—H
b. F—Br
c. H_3C-NH_2
d. H_3C-Cl
e. HO—Br
f. $H_3C-MgBr$
g. I—Cl
h. H_2N-OH

1.4 Representation of Structure

Lewis Structures

The chemical symbols we have been using, in which the valence electrons are represented as dots, are called **Lewis structures**. Lewis structures are useful because they show us which atoms are bonded together and tell us whether any atoms possess *lone-pair electrons* or have a *formal charge*.

The Lewis structures for H_2O, H_3O^+, HO^-, and H_2O_2 are shown below.

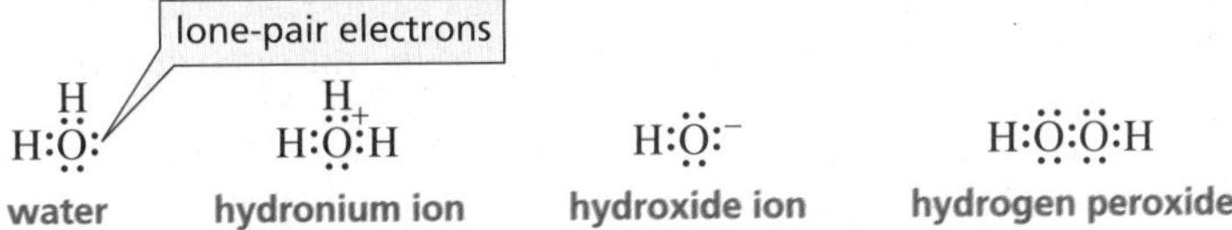

When you draw a Lewis structure, make sure that hydrogen atoms are surrounded by no more than two electrons and that C, O, N, and halogen (F, Cl, Br, I) atoms are surrounded by no more than eight electrons—they must obey the octet rule. Valence electrons not used in bonding are called **nonbonding electrons** or **lone-pair electrons**.

Once the atoms and the electrons are in place, each atom must be examined to see whether a charge should be assigned to it. A positive or a negative charge assigned to an atom is called a *formal charge*; the oxygen atom in the hydronium ion has a formal charge of +1, and the oxygen atom in the hydroxide ion has a formal charge of −1. A **formal charge** is the *difference* between the number of valence electrons an atom has when it is not bonded to any other atoms and the number of electrons it "owns" when it is bonded. An atom "owns" all of its lone-pair electrons and half of its bonding (shared) electrons.

formal charge = number of valence electrons − (number of lone-pair electrons + 1/2 number of bonding electrons)

American chemist **Gilbert Newton Lewis (1875–1946)** *was born in Weymouth, Massachusetts, and received a Ph.D. from Harvard in 1899. He was the first person to prepare "heavy water," which has deuterium atoms in place of the usual hydrogen atoms (D_2O versus H_2O). Because heavy water can be used as a moderator of neutrons, it became important in the development of the atomic bomb. Lewis started his career as a professor at the Massachusetts Institute of Technology and joined the faculty at the University of California, Berkeley, in 1912.*

* The angstrom (Å) is not a Système International unit. Those who opt to adhere strictly to SI units can convert it into picometers: 1 picometer (pm) = 10^{-12} m; 1 Å = 10^{-10} m = 100 pm. Because the angstrom continues to be used by many organic chemists, we will use angstroms in this book.

Movie: Formal charge

For example, an oxygen atom has six valence electrons (Table 1.2). In water (H_2O), oxygen "owns" six electrons (four lone-pair electrons and half of the four bonding electrons). Because the number of electrons it "owns" is equal to the number of its valence electrons ($6 - 6 = 0$), the oxygen atom in water has no formal charge. The oxygen atom in the hydronium ion (H_3O^+) "owns" five electrons: two lone-pair electrons plus three (half of six) bonding electrons. Because the number of electrons it "owns" is one less than the number of its valence electrons ($6 - 5 = 1$), its formal charge is $+1$. The oxygen atom in hydroxide ion (HO^-) "owns" seven electrons: six lone-pair electrons plus one (half of two) bonding electron. Because it "owns" one more electron than the number of its valence electrons ($6 - 7 = -1$), its formal charge is -1.

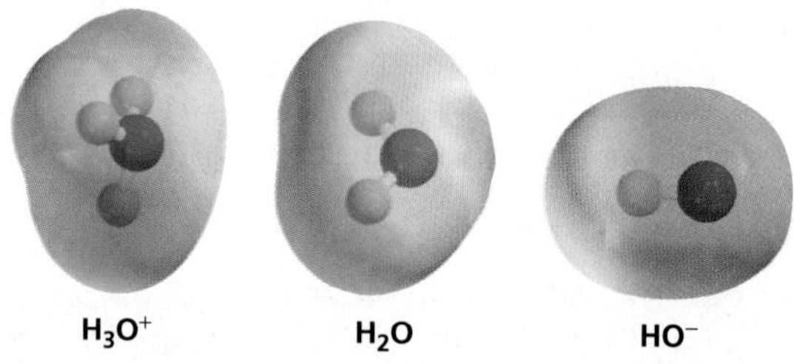

H_3O^+ H_2O HO^-

PROBLEM 9◆

A formal charge is a bookkeeping device. It does not necessarily indicate that the atom has greater or less electron density than other atoms in the molecule without formal charges. You can see this by examining the potential maps for H_2O, H_3O^+, and HO^-.

a. Which atom bears the formal negative charge in the hydroxide ion?
b. Which atom is the more negative in the hydroxide ion?
c. Which atom bears the formal positive charge in the hydronium ion?
d. Which atom is the most positive in the hydronium ion?

Knowing that nitrogen has five valence electrons (Table 1.2), convince yourself that the appropriate formal charges have been assigned to the nitrogen atoms in the following Lewis structures:

H:N̈:H H	H H:N:H (+) H	H:N̈:⁻ H	H:N̈:N̈:H H H
ammonia	ammonium ion	amide anion	hydrazine

Carbon has four valence electrons. Take a moment to confirm why the carbon atoms in the following Lewis structures have the indicated formal charges:

H H:C:H H	H H:C⁺ H	H H:C:⁻ H	H H:C· H	H H H:C:C:H H H
methane	methyl cation a carbocation	methyl anion a carbanion	methyl radical	ethane

A species containing a positively charged carbon atom is called a **carbocation**, and a species containing a negatively charged carbon atom is called a **carbanion**. (Recall that a *cation* is a positively charged ion and an *anion* is a negatively charged ion.) Carbocations were formerly called carbonium ions, so you will see this term in older chemical literature. A species containing an atom with a single unpaired electron is called a **radical** (often called a **free radical**). Hydrogen has one valence electron, and each halogen (F, Cl, Br, I) has seven valence electrons, so the following species have the indicated formal charges:

H^+	$H:^-$	H·	:Br:⁻	:Br·	:Br:Br:	:Cl:Cl:
hydrogen ion	**hydride ion**	**hydrogen radical**	**bromide ion**	**bromine radical**	**bromine**	**chlorine**

In studying the molecules in this section, notice that when the atoms don't bear a formal charge or an unpaired electron, hydrogen and the halogens each have *one* covalent bond, oxygen always has *two* covalent bonds, nitrogen always has *three* covalent bonds, and carbon has *four* covalent bonds. Notice that (except for hydrogen) the sum of the number of bonds and lone pairs is four: The halogens, with one bond, have three lone pairs; oxygen, with two bonds, has two lone pairs; and nitrogen, with three bonds, has one lone pair. Atoms that have more bonds or fewer bonds than the number required for a neutral atom will have either a formal charge or an unpaired electron. These numbers are very important to remember when you are first drawing structures of organic compounds because they provide a quick way to recognize when you have made a mistake.

H—	:F— :Cl— :I— :Br—	:O—	—N—	—C—
one bond	**one bond**	**two bonds**	**three bonds**	**four bonds**

In the Lewis structures for CH_2O_2, HNO_3, CH_2O, CO_3^{2-}, and N_2, notice that each atom has a complete octet (except hydrogen, which has a filled outer shell) and that each atom has the appropriate formal charge. (In drawing the Lewis structure for a compound that has two or more oxygen atoms, avoid oxygen–oxygen single bonds. These are weak bonds, and few compounds have them.)

:O: over H:C:O:H H:O:N:O:⁻ (with :O above N, + on N) H:C:H (with :O above C) ⁻:O:C:O:⁻ (with :O above C) :N⋮⋮N:

A pair of shared electrons can also be shown as a line between two atoms. Compare the preceding structures with the following ones:

H—C(=O)—O—H H—O—N⁺(=O)—O:⁻ H—C(=O)—H ⁻:O—C(=O)—O:⁻ :N≡N:

Suppose you are asked to draw a Lewis structure. In this example, we will use HNO_2.

1. Determine the total number of valence electrons (1 for H, 5 for N, and 6 for each O = 1 + 5 + 12 = 18).
2. Use the number of valence electrons to form bonds and fill octets with lone-pair electrons.
3. If after all the electrons have been assigned, any atom (other than hydrogen) does not have a complete octet, use a lone pair to form a double bond.
4. Assign a formal charge to any atom whose number of valence electrons is not equal to the number of its lone-pair electrons plus one-half its bonding electrons. (None of the atoms in HNO_2 has a formal charge.)

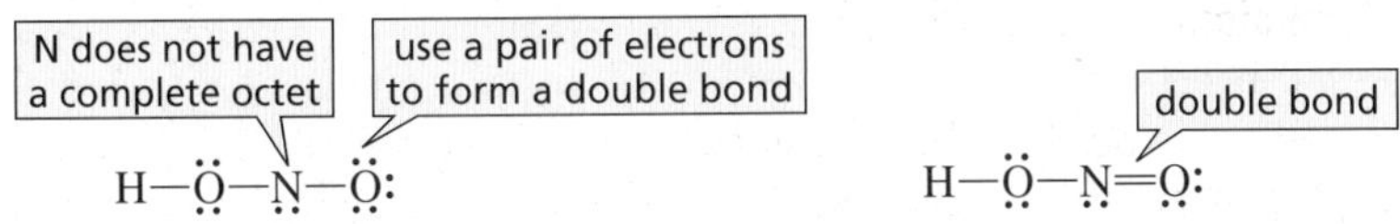

18 electrons have been assigned

double bond

H—O—N=O:

by using one of oxygen's lone pairs to form a double bond, N gets a complete octet

Kekulé Structures

In **Kekulé structures**, the bonding electrons are drawn as lines and the lone-pair electrons are usually left out entirely, unless they are needed to draw attention to some chemical property of the molecule. (Although lone-pair electrons may not be shown, you should remember that neutral nitrogen, oxygen, and halogen atoms always have them: one pair in the case of nitrogen, two pairs in the case of oxygen, and three pairs in the case of a halogen.)

```
   O                                               H             H
   ‖                                               |             |
H—C—O—H        H—C≡N        H—O—N=O           H—C—H         H—C—N—H
                                                   |             |  |
                                                   H             H  H
```

Condensed Structures

Frequently, structures are simplified by omitting some (or all) of the covalent bonds and listing atoms bonded to a particular carbon (or nitrogen or oxygen) next to it with a subscript to indicate the number of such atoms. These kinds of structures are called **condensed structures**. Compare the preceding structures with the following ones:

HCO_2H $\quad$ HCN $\quad$ HNO_2 $\quad$ CH_4 $\quad$ CH_3NH_2

You can find more examples of condensed structures and the conventions commonly used to create them in Table 1.5. Notice that since none of the molecules in Table 1.5 have a formal charge or an unpaired electron, each C has four bonds, each N has three bonds, each O has two bonds, and each H or halogen has one bond.

Table 1.5 Kekulé and Condensed Structures

Kekulé structure	Condensed structures
Atoms bonded to a carbon are shown to the right of the carbon. Atoms other than H can be shown hanging from the carbon.	
H H H H H H H—C—C—C—C—C—C—H H Br H H Cl H	$CH_3CHBrCH_2CH_2CHClCH_3$ or $CH_3CHCH_2CH_2CHCH_3$ (with Br hanging from C-2 and Cl hanging from C-5)
Repeating CH_2 groups can be shown in parentheses.	
H H H H H H H—C—C—C—C—C—C—H H H H H H H	$CH_3CH_2CH_2CH_2CH_2CH_3$ or $CH_3(CH_2)_4CH_3$
Groups bonded to a carbon can be shown (in parentheses) to the right of the carbon, or hanging from the carbon.	
H H H H H H H—C—C—C—C—C—C—H H H CH_3 H OH H	$CH_3CH_2CH(CH_3)CH_2CH(OH)CH_3$ or $CH_3CH_2CHCH_2CHCH_3$ (with CH_3 and OH hanging)
Groups bonded to the far-right carbon are not put in parentheses.	
H H CH_3 H H H—C—C—C—C—C—H H H CH_3 H OH	$CH_3CH_2C(CH_3)_2CH_2CH_2OH$ or $CH_3CH_2CCH_2CH_2OH$ (with CH_3 above and CH_3 below the central C)

Table 1.5 (continued)

Kekulé structure	Condensed structures
Two or more identical groups considered bonded to the "first" atom on the left can be shown (in parentheses) to the left of that atom, or hanging from the atom.	
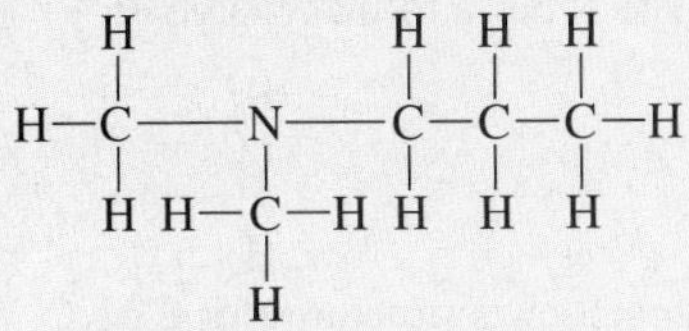 H–C(H)(H)–N(–C(H)(H)–H)–C(H)(H)–C(H)(H)–C(H)(H)–H	$(CH_3)_2NCH_2CH_2CH_3$ or 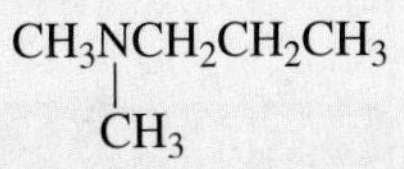 $CH_3NCH_2CH_2CH_3$ with CH_3 hanging from N
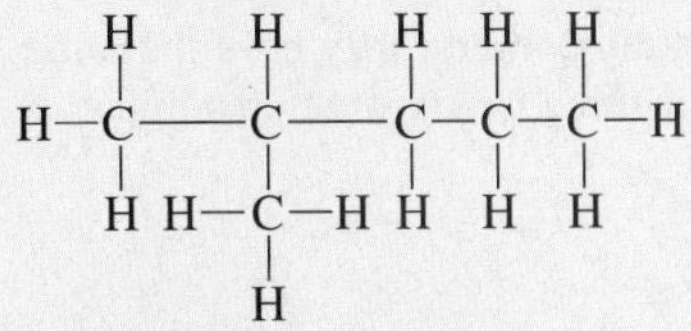 H–C(H)(H)–C(H)(–C(H)(H)–H)–C(H)(H)–C(H)(H)–C(H)(H)–H	$(CH_3)_2CHCH_2CH_2CH_3$ or 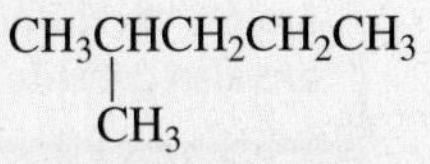 $CH_3CHCH_2CH_2CH_3$ with CH_3 hanging from CH
An oxygen doubly bonded to a carbon can be shown hanging off the carbon or to the right of the carbon.	
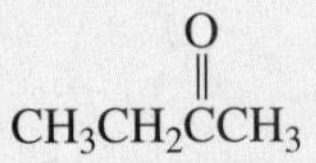 $CH_3CH_2\overset{O}{\overset{\Vert}{C}}CH_3$ or $CH_3CH_2COCH_3$ or $CH_3CH_2C(=O)CH_3$	
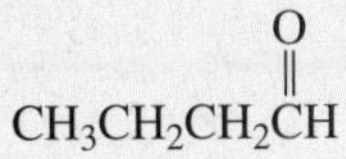 $CH_3CH_2CH_2\overset{O}{\overset{\Vert}{C}}H$ or $CH_3CH_2CH_2CHO$ or $CH_3CH_2CH_2CH=O$	
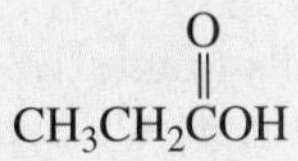 $CH_3CH_2\overset{O}{\overset{\Vert}{C}}OH$ or $CH_3CH_2CO_2H$ or CH_3CH_2COOH	
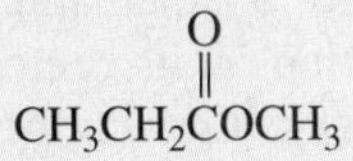 $CH_3CH_2\overset{O}{\overset{\Vert}{C}}OCH_3$ or $CH_3CH_2CO_2CH_3$ or $CH_3CH_2COOCH_3$	

PROBLEM 10 SOLVED

Draw the Lewis structure for each of the following:

a. NO_3^-
b. NO_2^+
c. NO_2^-
d. CO_2
e. HCO_3^-
f. N_2
g. $CH_3NH_3^+$
h. $^+C_2H_5$
i. $^-CH_3$
j. NaOH
k. NH_4Cl
l. Na_2CO_3

SOLUTION TO 10a The only way we can arrange one N and three O's and avoid O—O single bonds is to place the three O's around the N. The total number of valence electrons is 23 (5 for N, and 6 for each of the three O's). Because the species has one negative charge, we must add 1 to the number of valence electrons, for a total of 24. We then use the 24 electrons to form bonds and fill octets with lone-pair electrons.

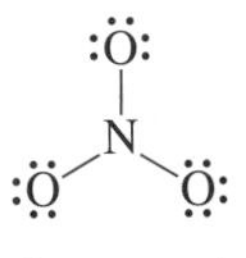

:Ö:
|
N
:Ö̤ Ö̤:

When all 24 electrons have been assigned, we see that N does not have a complete octet. We complete N's octet by using one of oxygen's lone pairs to from a double bond. (It doesn't make any difference which oxygen atom we choose.) When we check each atom to see whether it has a formal charge, we find that two of the O's are negatively charged and the N is positively charged, for an overall charge of −1.

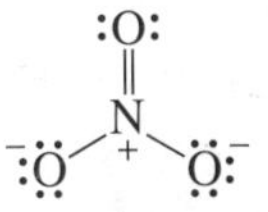

:O:
‖
N⁺
:Ö̤⁻ Ö̤:⁻

SOLUTION TO 10b The total number of valence electrons is 17 (5 for N and 6 for each of the two O's). Because the species has one positive charge, we must subtract 1 from the number of valence electrons, for a total of 16. The 16 electrons are used to form bonds and fill octets with lone-pair electrons.

$$:\ddot{\underset{\cdot\cdot}{O}}—N—\ddot{\underset{\cdot\cdot}{O}}:$$

Two double bonds are necessary to complete N's octet. The N has a formal charge of +1.

$$:\underset{\cdot\cdot}{O}=\overset{+}{N}=\underset{\cdot\cdot}{O}:$$

PROBLEM 11

a. Draw two Lewis structures for C_2H_6O. b. Draw three Lewis structures for C_3H_8O.

(*Hint:* The two Lewis structures in part a are **constitutional isomers**; they have the same atoms, but differ in the way the atoms are connected. The three Lewis structures in part b are also constitutional isomers.)

PROBLEM 12

Expand the following condensed structures to show the covalent bonds and lone-pair electrons:

a. $CH_3NHCH_2CH_3$
b. $(CH_3)_2CHCl$
c. $(CH_3)_2CHCHO$
d. $(CH_3)_3C(CH_2)_3CH(CH_3)_2$

1.5 Atomic Orbitals

We have seen that electrons are distributed into different atomic orbitals (Table 1.2). An **orbital** is a three-dimensional region around the nucleus where there is a high probability of finding an electron. But what does an orbital look like? Mathematical calculations indicate that the *s* atomic orbital is a sphere with the nucleus at its center, and experimental evidence supports this theory. The **Heisenberg uncertainty principle** states that both the precise location and the momentum of an atomic particle cannot be simultaneously determined. This means that we can never say precisely where an electron is—we can only describe its probable location. Thus, when we say that an electron occupies a 1*s* atomic orbital, we mean that there is a greater than 90% probability that the electron is in the space defined by the sphere.

Because the average distance from the nucleus is greater for an electron in a 2*s* atomic orbital than for an electron in a 1*s* atomic orbital, a 2*s* atomic orbital is represented by a larger sphere. Consequently, the average electron density in a 2*s* atomic orbital is less than the average electron density in a 1*s* atomic orbital.

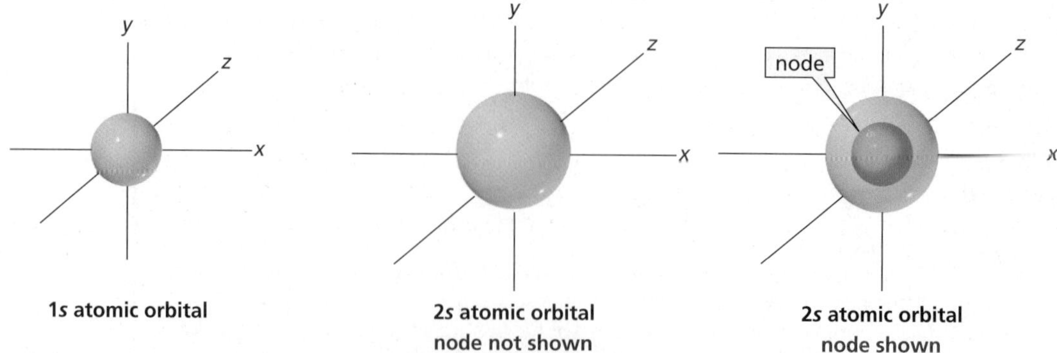

1*s* atomic orbital **2*s* atomic orbital node not shown** **2*s* atomic orbital node shown**

An electron in a 1*s* atomic orbital can be anywhere within the 1*s* sphere, but a 2*s* atomic orbital has a region where the probability of finding an electron falls to zero. This is called a **node**, or, more precisely—since the absence of electron density is at one set distance from nucleus—a **radial node**. So a 2*s* electron can be found anywhere within the 2*s* sphere—including the region of space defined by the 1*s* sphere—except in the node.

To understand why nodes occur, you need to remember that electrons have both particlelike and wavelike properties. A node is a consequence of the wavelike properties of an electron. Consider the following two types of waves: traveling waves and standing waves. Traveling waves move through space; light is an example of a traveling wave. A standing wave, in contrast, is confined to a limited space. A vibrating string of a guitar is an example of a standing wave—the string moves up and down, but does not travel through space. If you were to write a wave equation for the guitar string, the wave function would be $(+)$ in the region above where the guitar string is at rest and $(-)$ in the region below where the guitar string is at rest—the regions are of opposite phase. The region where the guitar string has no transverse displacement is called a *node*. A **node** is the region where a standing wave has an amplitude of zero.

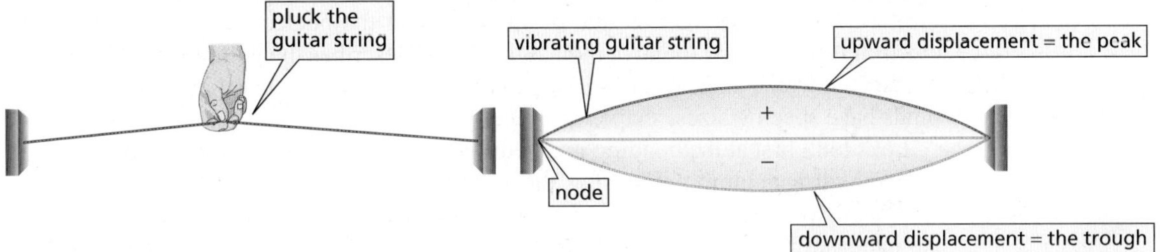

An electron behaves like a standing wave, but—unlike the wave created by a vibrating guitar string—it is three dimensional. This means that the node of a 2*s* atomic orbital is actually a surface—a spherical surface within the 2*s* atomic orbital. Because the electron wave has zero amplitude at the node, there is zero probability of finding an electron at the node.

Unlike *s* atomic orbitals that resemble spheres, *p* atomic orbitals have two lobes. Generally, the lobes are depicted as teardrop-shaped, but computer-generated representations reveal that they are shaped more like doorknobs. Like the vibrating guitar string, the lobes are of opposite phase, which can be designated by plus $(+)$ and minus $(-)$ signs or by two different colors. (In this context, $+$ and $-$ do not indicate charge, just the phase of the orbital.) The node of the *p* atomic orbital is a plane that passes through the center of the nucleus, bisecting its two lobes. This is called a **nodal plane**. There is zero probability of finding an electron in the nodal plane of the *p* orbital.

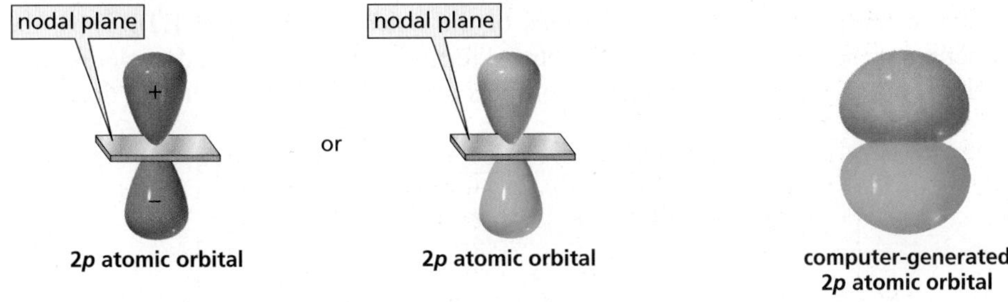

2*p* atomic orbital **2*p* atomic orbital** **computer-generated 2*p* atomic orbital**

In Section 1.2, you saw that there are three degenerate *p* atomic orbitals. The p_x orbital is symmetrical about the *x*-axis, the p_y orbital is symmetrical about the *y*-axis, and the p_z orbital is symmetrical about the *z*-axis. This means that each *p* orbital is perpendicular to the other two *p* orbitals. The energy of a 2*p* atomic orbital is slightly greater than that of a 2*s* atomic orbital because the average location of an electron in a 2*p* atomic orbital is farther away from the nucleus.

Degenerate orbitals are orbitals that have the same energy.

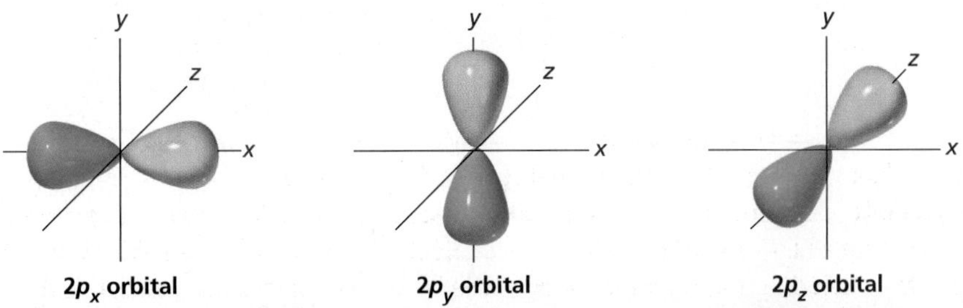

$2p_x$ orbital **$2p_y$ orbital** **$2p_z$ orbital**

1.6 An Introduction to Molecular Orbital Theory

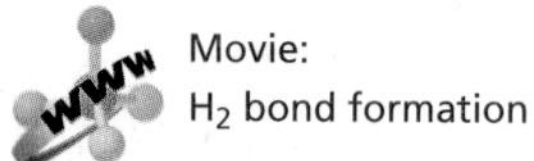

Movie:
H_2 bond formation

How do atoms form covalent bonds in order to form molecules? The Lewis model, which describes how atoms attain a complete octet by sharing electrons, tells us only part of the story. A drawback of the model is that it treats electrons like particles and does not take into account their wavelike properties.

Molecular orbital (MO) theory combines the tendency of atoms to fill their octets by sharing electrons (the Lewis model) with their wavelike properties—assigning electrons to a volume of space called an orbital. According to MO theory, covalent bonds result from the combination of atomic orbitals to form **molecular orbitals**—orbitals that belong to the whole molecule rather than to a single atom. Like an atomic orbital that describes the volume of space around the nucleus of an atom where an electron is likely to be found, a molecular orbital describes the volume of space around a molecule where an electron is likely to be found. Like atomic orbitals, molecular orbitals have specific sizes, shapes, and energies.

Let's look first at the bonding in a hydrogen molecule (H_2). As the 1*s* atomic orbital of one hydrogen atom approaches the 1*s* atomic orbital of a second hydrogen atom, they begin to overlap. As the atomic orbitals move closer together, the amount of overlap increases until the orbitals combine to form a molecular orbital. The covalent bond that is formed when the two *s* atomic orbitals overlap is called a **sigma (σ) bond.** A σ bond is cylindrically symmetrical—the electrons in the bond are symmetrically distributed about an imaginary line connecting the centers of the two atoms joined by the bond. (The term σ comes from the fact that cylindrically symmetrical molecular orbitals possess σ symmetry.)

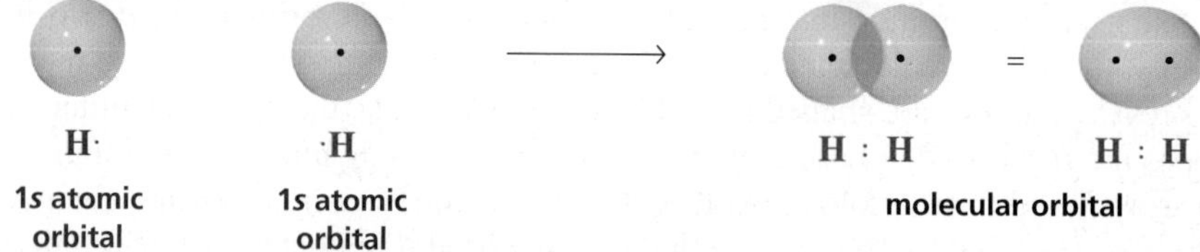

During bond formation, energy is released as the two orbitals start to overlap, because the electron in each atom not only is attracted to its own nucleus but also is attracted to the positively charged nucleus of the other atom (Figure 1.2). Thus, the attraction of the negatively charged electrons for the positively charged nuclei is what holds the atoms together. The more the orbitals overlap, the more the energy decreases

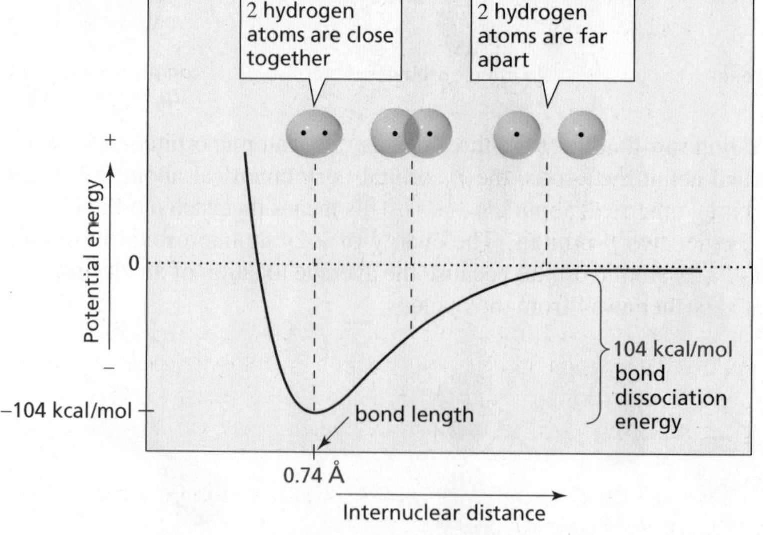

Figure 1.2 ▶ The change in energy that occurs as two 1s atomic orbitals approach each other. The internuclear distance at minimum energy is the length of the H—H covalent bond.

until the atoms approach each other so closely that their positively charged nuclei start to repel each other. This repulsion causes a large increase in energy. We see that maximum stability (i.e., minimum energy) is achieved when the nuclei are a certain distance apart. This distance is the **bond length** of the new covalent bond. The length of the H—H bond is 0.74 Å.

Maximum stability corresponds to minimum energy.

As Figure 1.2 shows, energy is released when a covalent bond forms. When the H—H bond forms, 104 kcal/mol (or 435 kJ/mol)* of energy is released. Breaking the bond requires precisely the same amount of energy. Thus, the **bond strength**—also called the **bond dissociation energy**—is the energy required to break a bond, or the energy released when a bond is formed. Every covalent bond has a characteristic bond length and bond strength.

Orbitals are conserved—the number of molecular orbitals formed must equal the number of atomic orbitals combined. In describing the formation of an H—H bond, however, we combined two atomic orbitals, but discussed only one molecular orbital. Where is the other molecular orbital? It is there, but it contains no electrons.

Atomic orbitals can combine in two different ways: constructively and destructively. They can combine in a constructive, additive manner, just as two light waves or sound waves may reinforce each other (Figure 1.3). This is called a **σ (sigma) bonding molecular orbital**. Atomic orbitals can also combine in a destructive way, canceling each other. The cancellation is similar to the darkness that occurs when two light waves cancel each other or to the silence that occurs when two sound waves cancel each other (Figure 1.3). This destructive type of interaction is called a **σ^* antibonding molecular orbital**. An antibonding orbital is indicated by an asterisk (*).

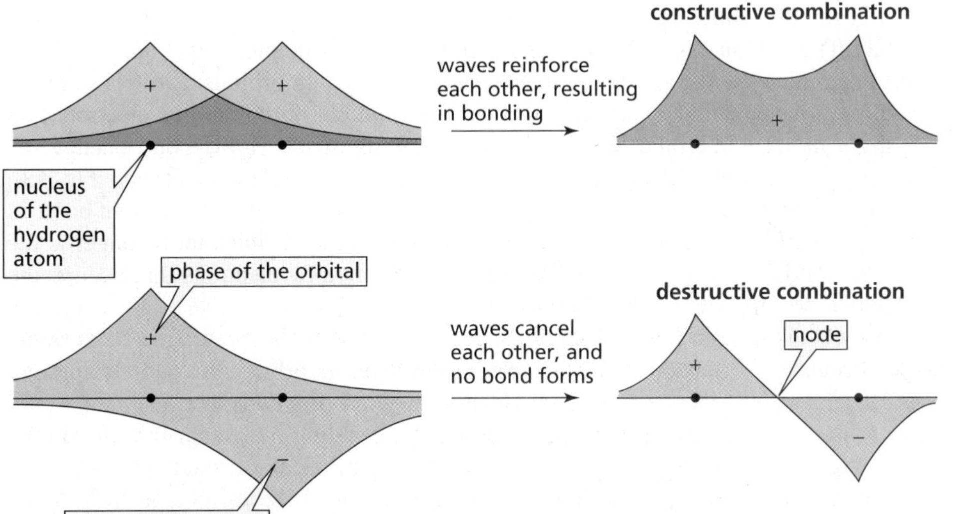

Figure 1.3
The wave functions of two hydrogen atoms can interact to reinforce, or enhance, each other (top) or can interact to cancel each other (bottom). Note that waves that interact constructively are in-phase, whereas waves that interact destructively are out-of-phase.

The σ bonding molecular orbital and σ^* antibonding molecular orbital are shown in the molecular orbital diagram in Figure 1.4. In an MO diagram, the energies are represented as horizontal lines; the bottom line is the lowest energy level, the top line the highest energy level. We see that any electrons in the bonding orbital will most likely be found between the nuclei. This increased electron density between the nuclei is what binds the atoms together. Because there is a node between the nuclei in the antibonding molecular orbital, any electrons that are in that orbital are more likely to be found anywhere except between the nuclei, so the nuclei are more exposed to one another and will be forced apart by electrostatic repulsion. Thus, electrons that occupy this orbital detract from, rather than aid, the formation of a bond between the atoms.

*1 kcal = 4.184 kJ. Joules are the Système International (SI) units for energy, although many chemists use calories. We will use both in this book.

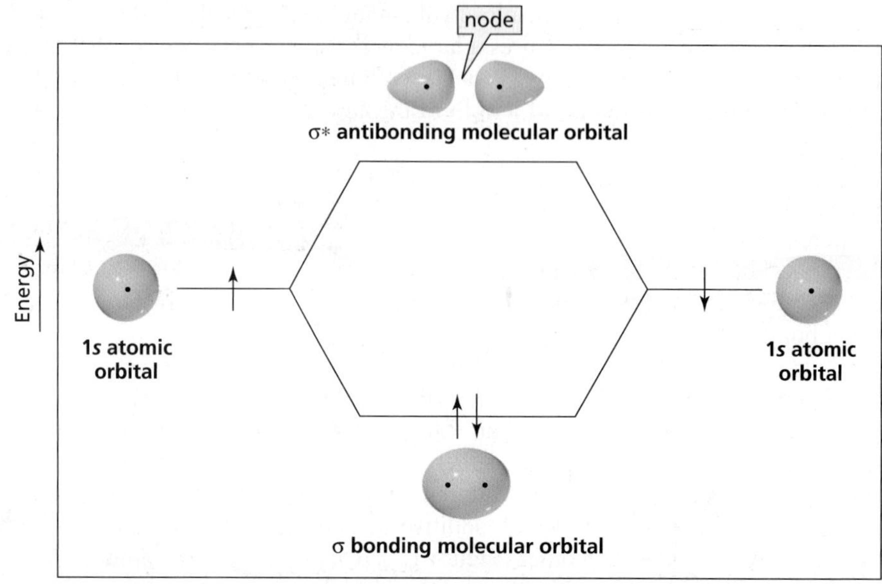

Figure 1.4 ▶
Atomic orbitals of H· and molecular orbitals of H_2. Before covalent bond formation, each electron is in an atomic orbital. After covalent bond formation, both electrons are in the bonding molecular orbital. The antibonding molecular orbital is empty.

When two atomic orbitals overlap, two molecular orbitals are formed—one lower in energy and one higher in energy than the atomic orbitals.

The MO diagram shows that the bonding molecular orbital is more stable—is lower in energy—than the individual atomic orbitals. This is because the more nuclei an electron "feels," the more stable it is. The antibonding molecular orbital, with less electron density between the nuclei, is less stable—is of higher energy—than the atomic orbitals.

After the MO diagram is constructed, the electrons are assigned to the molecular orbitals. The aufbau principle and the Pauli exclusion principle, which apply to electrons in atomic orbitals, also apply to electrons in molecular orbitals: Electrons always occupy available orbitals with the lowest energy, and no more than two electrons can occupy a molecular orbital. Thus, the two electrons of the H—H bond occupy the lower energy bonding molecular orbital (Figure 1.4), where they are attracted to both positively charged nuclei. It is this electrostatic attraction that gives a covalent bond its strength. Therefore, the greater the overlap of the atomic orbitals, the stronger is the covalent bond. The strongest covalent bonds are formed by electrons that occupy the molecular orbitals with the lowest energy.

The MO diagram in Figure 1.4 allows us to predict that H_2^+ would not be as stable as H_2 because H_2^+ has only one electron in the bonding orbital. We can also predict that He_2 does not exist: Because each He atom would bring two electrons, He_2 would have four electrons—two filling the lower energy bonding molecular orbital and the remaining two filling the higher energy antibonding molecular orbital. The two electrons in the antibonding molecular orbital would cancel the advantage to bonding gained by the two electrons in the bonding molecular orbital.

In-phase overlap forms a bonding MO; out-of-phase overlap forms an antibonding MO.

PROBLEM 13♦

Predict whether or not He_2^+ exists.

Two *p* atomic orbitals can overlap either end-on or side-to-side. Let's first look at end-on overlap. End-on overlap forms a σ bond. If the overlapping lobes of the *p* orbitals are in-phase (a blue lobe of one *p* orbital overlaps a blue lobe of the other *p* orbital), a σ bonding molecular orbital is formed (Figure 1.5). The electron density of the σ bonding molecular orbital is concentrated between the nuclei, which causes the back lobes (the nonoverlapping lobes) of the molecular orbital to be quite small. The σ bonding molecular orbital has two nodes—a nodal plane passing through each of the nuclei.

If the overlapping lobes of the *p* orbitals are out-of-phase (a blue lobe of one *p* orbital overlaps a green lobe of the other *p* orbital), a σ^* antibonding molecular orbital is

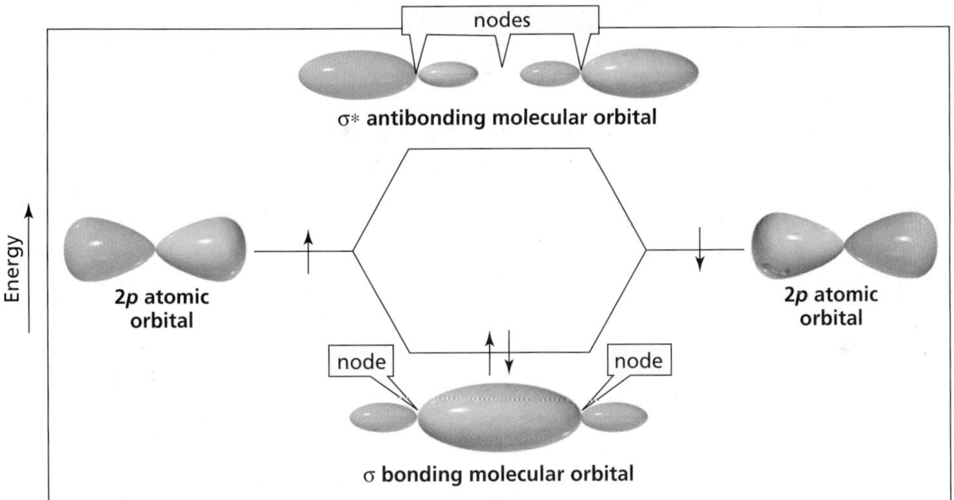

Figure 1.5
End-on overlap of two p orbitals to form a σ bonding molecular orbital and a σ^* antibonding molecular orbital.

formed. The σ^* antibonding molecular orbital has *three* nodes. (Notice that after each node, the phase of the molecular orbital changes.)

Unlike the σ bond formed as a result of end-on overlap, side-to-side overlap of two p atomic orbitals forms a **pi (π) bond** (Figure 1.6). Side-to-side overlap of two in-phase p atomic orbitals forms a π bonding molecular orbital, whereas side-to-side overlap of two out-of-phase p orbitals forms a π^* antibonding molecular orbital. The π bonding molecular orbital has one node—a nodal plane that passes through both nuclei. The π^* antibonding molecular orbital has two nodal planes. Notice that σ bonds are cylindrically symmetrical, but π bonds are not.

Side-to-side overlap of two *p* atomic orbitals forms a π bond. All other covalent bonds in organic molecules are σ bonds.

The extent of overlap is greater when p orbitals overlap end-on than when they overlap side-to-side. This means that a σ bond formed by the end-on overlap of p orbitals is stronger than a π bond formed by the side-to-side overlap of p orbitals. It also means that a σ bonding molecular orbital is more stable than a π bonding molecular orbital because the stronger the bond, the more stable it is. Figure 1.7 shows a molecular orbital diagram of two identical atoms using their three degenerate atomic orbitals to form three bonds—one σ bond and two π bonds.

A σ bond is stronger than a π bond.

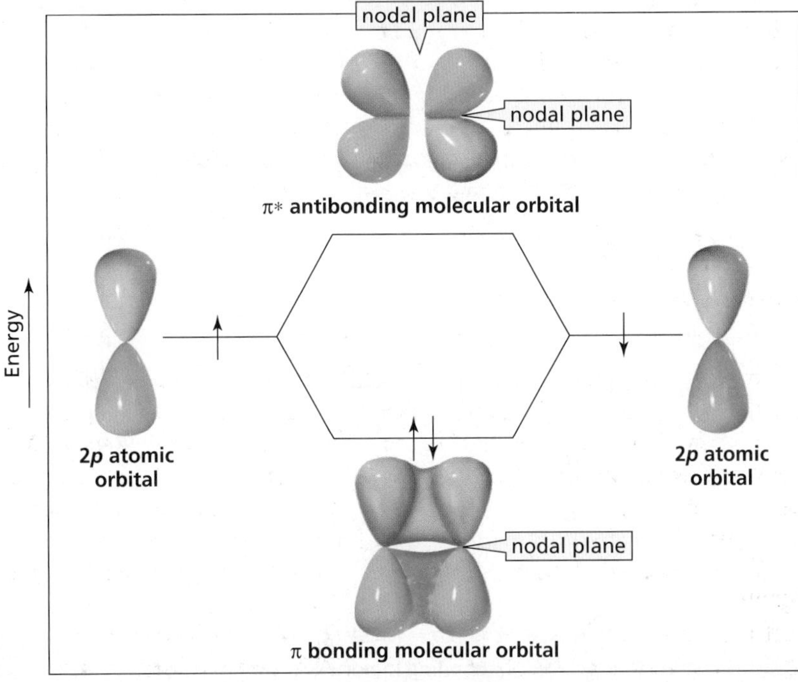

Figure 1.6
Side-to-side overlap of two parallel p orbitals to form a π bonding molecular orbital and a π^* antibonding molecular orbital.

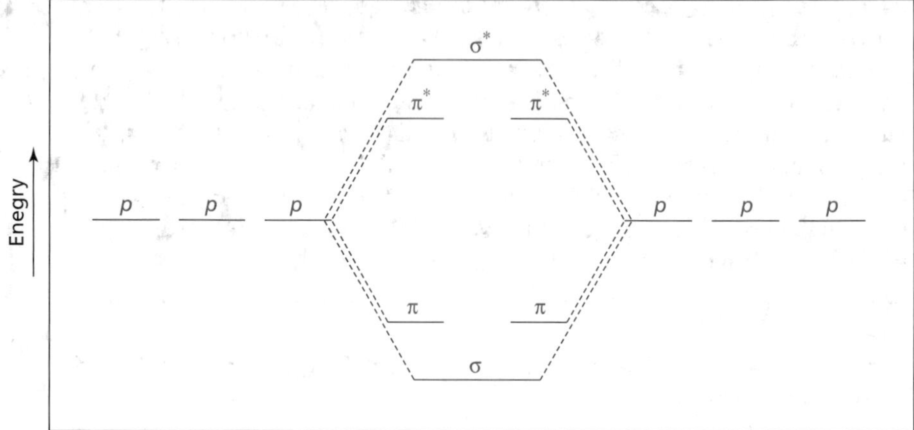

Figure 1.7 ▶
p Orbitals can overlap end-on to form σ bonding and σ^* antibonding molecular orbitals, or can overlap side-to-side to form π bonding and π^* antibonding molecular orbitals. The relative energies of the molecular orbitals are $\sigma < \pi < \pi^* < \sigma^*$.

Now let's look at the molecular orbital diagram for side-to-side overlap of a *p* orbital of carbon with a *p* orbital of oxygen—the orbitals are the same, but they belong to different atoms (Figure 1.8). When the two *p* atomic orbitals combine to form molecular orbitals, they do so unsymmetrically. The atomic orbital of the more electronegative atom contributes more to the bonding molecular orbital, and the atomic orbital of the less electronegative atom contributes more to the antibonding molecular orbital. This means that if we were to put electrons in the bonding MO, they would be more apt to be around the oxygen atom than around the carbon atom. Thus, both the Lewis theory and molecular orbital theory tell us that the electrons shared by carbon and oxygen are not shared equally—the oxygen atom of a carbon–oxygen bond has a partial negative charge and the carbon atom has a partial positive charge.

Organic chemists find that the information obtained from MO theory, where valence electrons occupy bonding and antibonding molecular orbitals, does not always yield the needed information about the bonds in a molecule. The **valence-shell electron-pair repulsion (VSEPR) model** combines the Lewis concept of shared electron pairs and lone-pair electrons with the concept of atomic orbitals and adds a third principle: *the minimization of electron repulsion*. In this model, atoms share electrons by overlapping

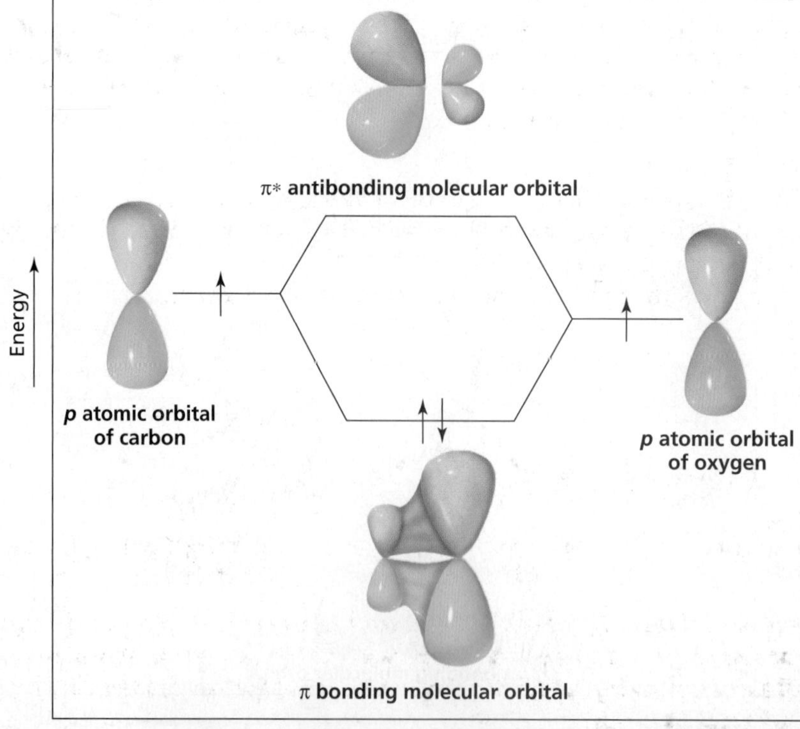

Figure 1.8 ▶
Side-to-side overlap of a *p* orbital of carbon with a *p* orbital of oxygen to form a π bonding molecular orbital and a π^* antibonding molecular orbital.

their atomic orbitals, and because electron pairs repel each other, the bonding electrons and lone-pair electrons around an atom are positioned as far apart as possible.

Because organic chemists generally think of chemical reactions in terms of the changes that occur in the bonds of the reacting molecules, the VSEPR model often provides the easiest way to visualize chemical change. However, the model is inadequate for some molecules because it does not allow for antibonding orbitals. We will use both the MO and the VSEPR models in this book. Our choice will depend on which model provides the best description of the molecule under discussion. We will use the VSEPR model in Sections 1.7–1.13.

Indicate the kind of molecular orbital (σ, σ^*, π, or π^*) that results when the orbitals are combined as indicated:

a. 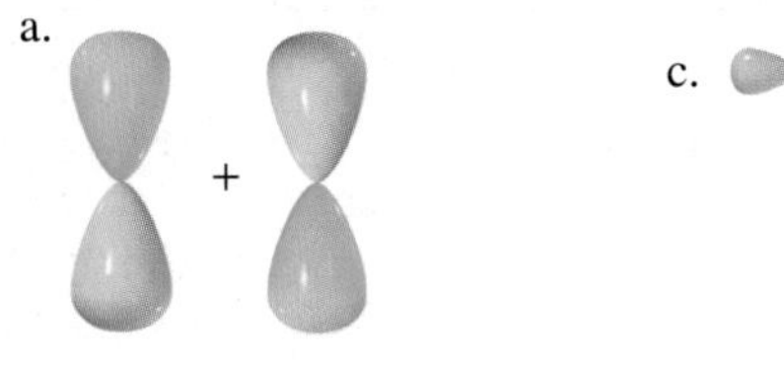

c. 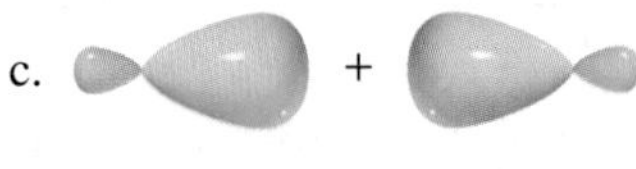

b. 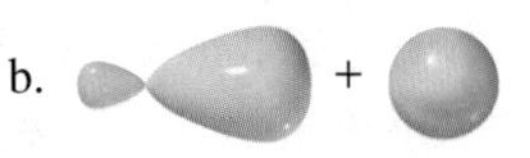

d. 

1.7 Bonding in Methane and Ethane: Single Bonds

We will begin the discussion of bonding in organic compounds by looking at the bonding in methane, a compound with only one carbon atom. Then we will examine the bonding in ethane (a compound with two carbons and a carbon–carbon single bond), in ethene (a compound with two carbons and a carbon–carbon double bond), and in ethyne (a compound with two carbons and a carbon–carbon triple bond).

Next, we will look at bonds formed by atoms other than carbon that are commonly found in organic compounds—bonds formed by oxygen, nitrogen, and the halogens. Because *the orbitals used in bond formation determine the bond angles in a molecule*, you will see that if we know the bond angles in a molecule, we can figure out which orbitals are involved in bond formation.

Bonding in Methane

Methane (CH_4) has four covalent C—H bonds. Because all four bonds have the same length and all the bond angles are the same (109.5°), we can conclude that the four C—H bonds in methane are identical.

Four different ways to represent a methane molecule are shown here.

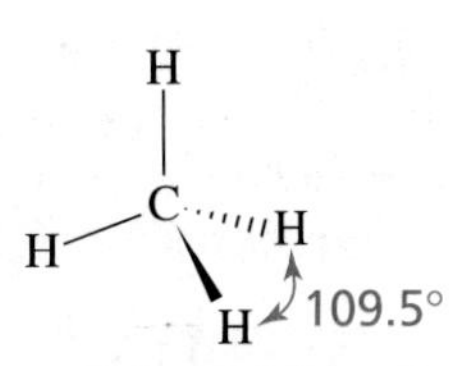

perspective formula of methane

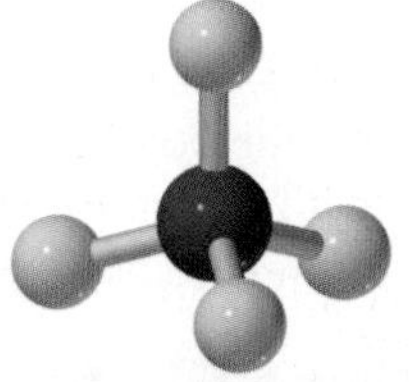

ball-and-stick model of methane

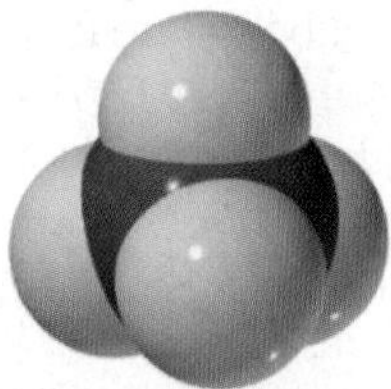

space-filling model of methane

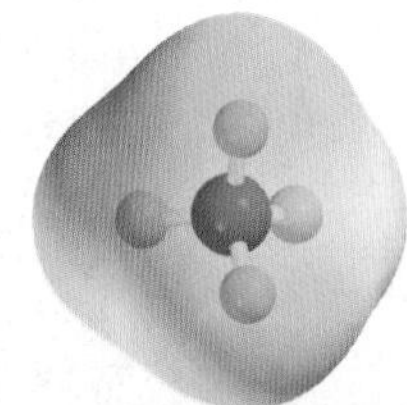

electrostatic potential map for methane

In a perspective formula, bonds in the plane of the paper are drawn as solid lines, bonds protruding out of the plane of the paper toward the viewer are drawn as solid wedges, and those protruding back from the plane of the paper away from the viewer are drawn as hatched wedges.

The potential map of methane shows that neither carbon nor hydrogen carries much of a charge: There are neither red areas, representing partially negatively charged atoms, nor blue areas, representing partially positively charged atoms. (Compare this map with the potential map for water on p. 14). The absence of partially charged atoms can be explained by the similar electronegativities of carbon and hydrogen, which cause carbon and hydrogen to share their bonding electrons relatively equally. Methane is a **nonpolar molecule**.

You may be surprised to learn that carbon forms four covalent bonds since you know that carbon has only two unpaired electrons in its ground-state electronic configuration (Table 1.2). But if carbon were to form only two covalent bonds, it would not complete its octet. Now we need to come up with an explanation that accounts for carbon's forming four covalent bonds.

If one of the electrons in the 2*s* orbital were promoted into the empty 2*p* atomic orbital, the new electronic configuration would have four unpaired electrons; thus, four covalent bonds could be formed. Let's now see whether this is feasible energetically.

Because a *p* orbital is higher in energy than an *s* orbital, promotion of an electron from an *s* orbital to a *p* orbital requires energy. The amount of energy required is 96 kcal/mol. The formation of four C—H bonds releases 420 kcal/mol of energy because the bond dissociation energy of a single C—H bond is 105 kcal/mol. If the electron were not promoted, carbon could form only two covalent bonds, which would release only 210 kcal/mol. So, by spending 96 kcal/mol (or 402 kJ/mol) to promote an electron, an extra 210 kcal/mol (or 879 kJ/mol) is released. In other words, promotion is energetically advantageous (Figure 1.9).

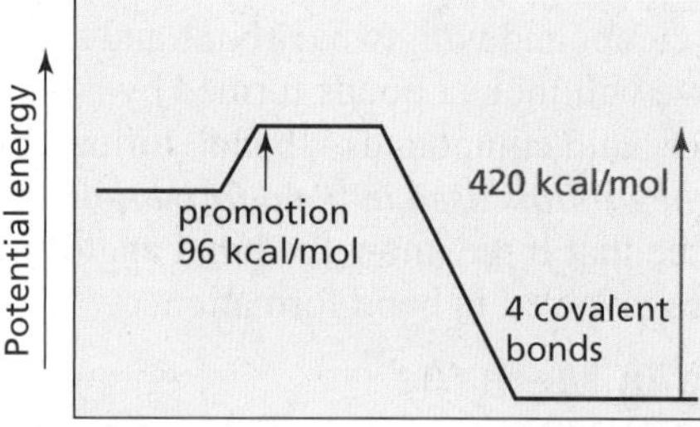

Figure 1.9 ▶
As a result of electron promotion, carbon forms four covalent bonds and releases 420 kcal/mol of energy. Without promotion, carbon would form two covalent bonds and release 210 kcal/mol of energy. Because it requires 96 kcal/mol to promote an electron, the overall energy advantage of promotion is 114 kcal/mol.

Linus Carl Pauling (1901–1994) *was born in Portland, Oregon. A friend's home chemistry laboratory sparked Pauling's early interest in science. He received a Ph.D. from the California Institute of Technology and remained there for most of his academic career. He received the Nobel Prize in chemistry in 1954 for his work on molecular structure. Like Einstein, Pauling was a pacifist, winning the 1964 Nobel Peace Prize for his work on behalf of nuclear disarmament.*

We have managed to account for the observation that carbon forms four covalent bonds, but what accounts for the fact that the four C—H bonds in methane are identical? Each has a bond length of 1.10 Å, and breaking any one of the bonds requires the same amount of energy (105 kcal/mol, or 439 kJ/mol). If carbon used an *s* orbital and three *p* orbitals to form these four bonds, the bond formed with the *s* orbital would be different from the three bonds formed with *p* orbitals. How can carbon form four identical bonds, using one *s* and three *p* orbitals? The answer is that carbon uses *hybrid orbitals*.

Hybrid orbitals are mixed orbitals—they result from combining orbitals. The concept of combining orbitals, called **orbital hybridization**, was first proposed by Linus Pauling in 1931. If the one *s* and three *p* orbitals of the second shell are combined and then apportioned into four equal orbitals, each of the four resulting orbitals will be one part *s* and three parts *p*. This type of mixed orbital is called an sp^3 (stated "*s*-*p*-three" not "*s*-*p*-cubed") orbital. (The superscript 3 means that three *p* orbitals were mixed with one *s* orbital to form the hybrid orbitals.) Each sp^3 orbital has 25% *s* character and 75% *p* character. The four sp^3 orbitals are degenerate—they have the same energy.

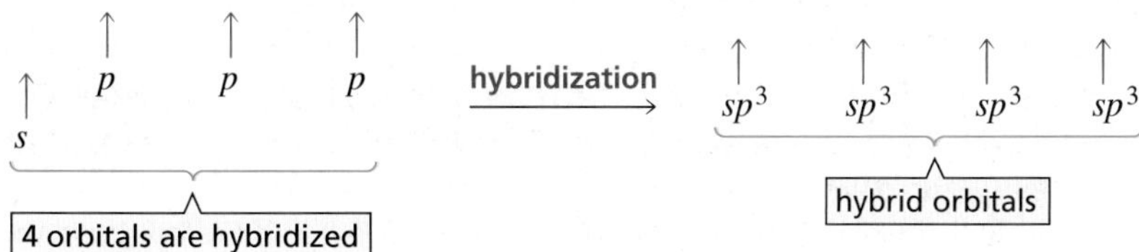

Like a *p* orbital, an sp^3 orbital has two lobes. The lobes differ in size, however, because the *s* orbital adds to one lobe of the *p* orbital and subtracts from the other lobe of the *p* orbital (Figure 1.10). The stability of an sp^3 orbital reflects its composition; it is more stable than a *p* orbital, but not as stable as an *s* orbital (Figure 1.11). The larger lobe of the sp^3 orbital is used in covalent bond formation.

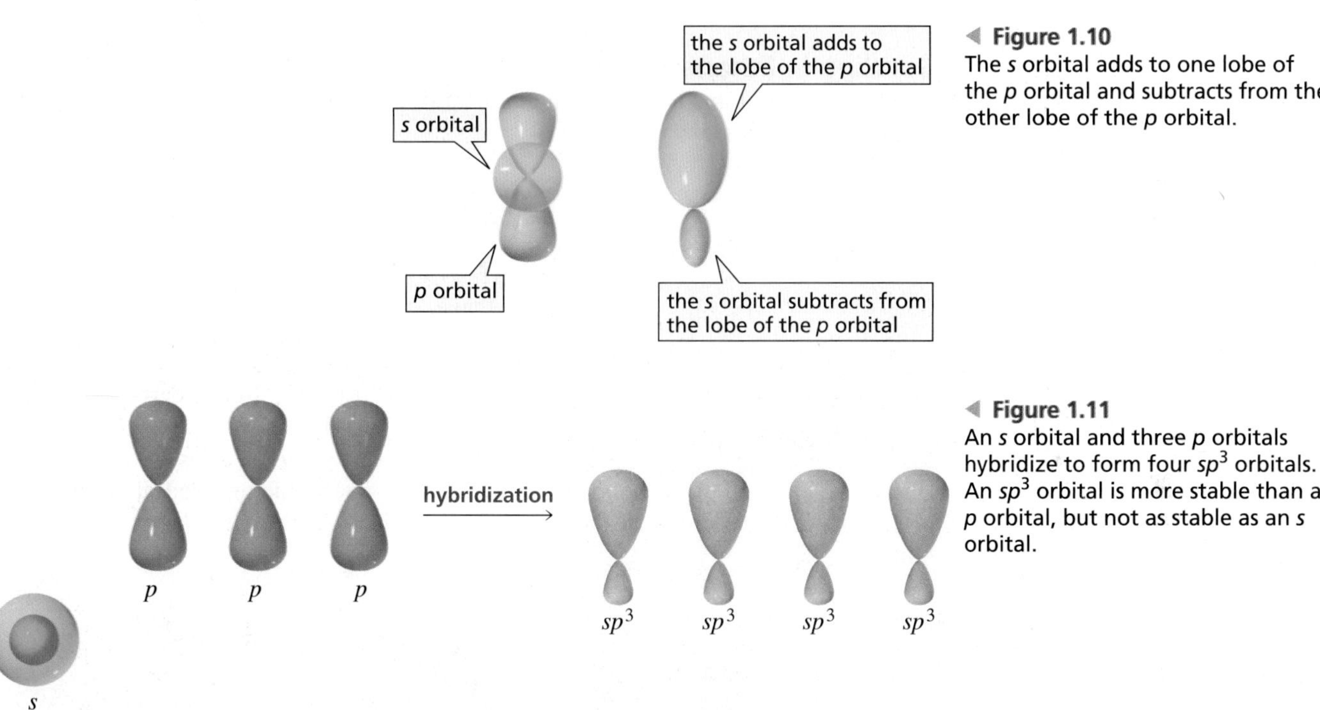

Figure 1.10
The *s* orbital adds to one lobe of the *p* orbital and subtracts from the other lobe of the *p* orbital.

Figure 1.11
An *s* orbital and three *p* orbitals hybridize to form four sp^3 orbitals. An sp^3 orbital is more stable than a *p* orbital, but not as stable as an *s* orbital.

The four sp^3 orbitals arrange themselves in space in a way that allows them to get as far away from each other as possible (Figure 1.12a). This occurs because electrons repel each other and getting as far from each other as possible minimizes the repulsion (Section 1.6). When four orbitals spread themselves into space as far from each other as possible, they point toward the corners of a regular tetrahedron (a pyramid with four

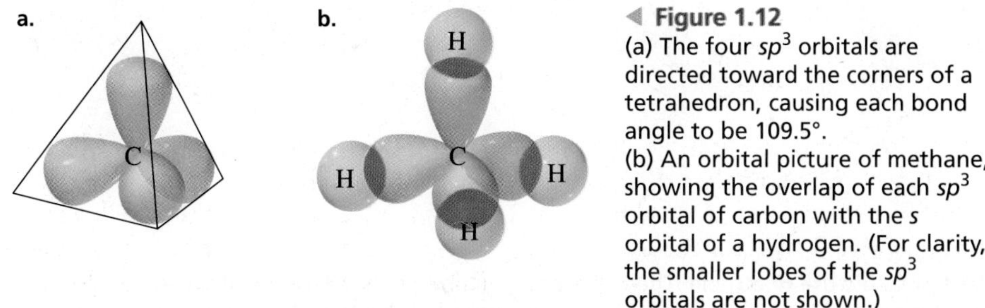

Figure 1.12
(a) The four sp^3 orbitals are directed toward the corners of a tetrahedron, causing each bond angle to be 109.5°.
(b) An orbital picture of methane, showing the overlap of each sp^3 orbital of carbon with the *s* orbital of a hydrogen. (For clarity, the smaller lobes of the sp^3 orbitals are not shown.)

Electron pairs spread themselves into space as far from each other as possible.

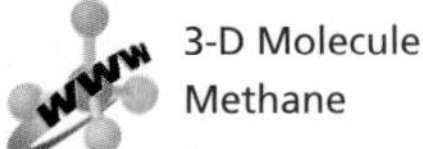

3-D Molecule: Methane

faces, each an equilateral triangle). Each of the four C—H bonds in methane is formed from overlap of an sp^3 orbital of carbon with the s orbital of a hydrogen (Figure 1.12b). This explains why the four C—H bonds are identical.

The angle formed between any two bonds of methane is 109.5°. This bond angle is called the **tetrahedral bond angle**. A carbon, such as the one in methane, that forms covalent bonds using four equivalent sp^3 orbitals is called a **tetrahedral carbon**.

The postulation of hybrid orbitals may appear to be a theory contrived just to make things fit—and that is exactly what it is. Nevertheless, it is a theory that gives us a very good picture of the bonding in organic compounds.

Note to the student

It is important to understand what molecules look like in three dimensions. As you study each chapter, make sure to visit the Web site www.prenhall.com/bruice and look at the three-dimensional representations of molecules that can be found in the molecule gallery that accompanies the chapter.

Bonding in Ethane

The two carbon atoms in ethane are tetrahedral. Each carbon uses four sp^3 orbitals to form four covalent bonds:

```
    H  H
    |  |
H—C—C—H
    |  |
    H  H
```

ethane

One sp^3 orbital of one carbon overlaps an sp^3 orbital of the other carbon to form the C—C bond. Each of the remaining three sp^3 orbitals of each carbon overlaps the s orbital of a hydrogen to form a C—H bond. Thus, the C—C bond is formed by sp^3–sp^3 overlap, and each C—H bond is formed by sp^3–s overlap (Figure 1.13). Each of the bond angles in ethane is nearly the tetrahedral bond angle of 109.5°, and the length of the C—C bond is 1.54 Å. Ethane, like methane, is a nonpolar molecule.

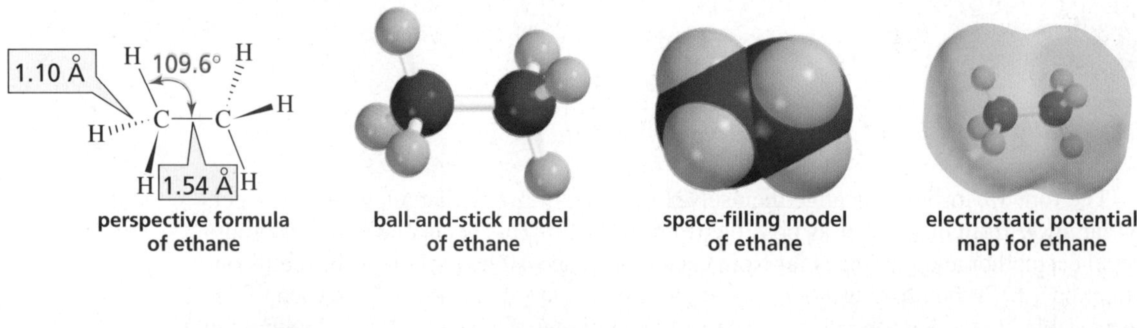

perspective formula of ethane **ball-and-stick model of ethane** **space-filling model of ethane** **electrostatic potential map for ethane**

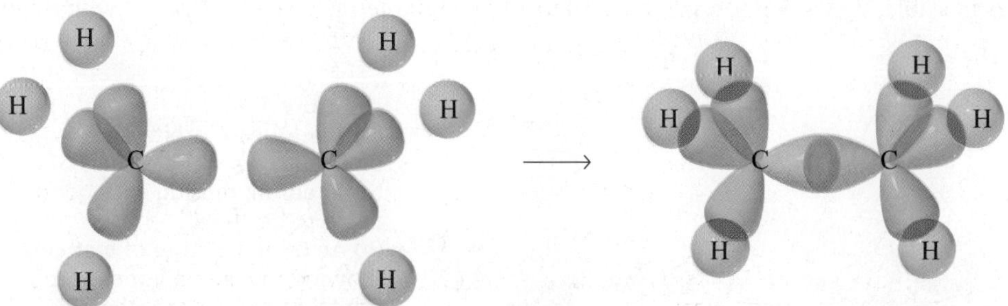

▲ **Figure 1.13**
An orbital picture of ethane. The C—C bond is formed by sp^3–sp^3 overlap, and each C—H bond is formed by sp^3–s overlap. (The smaller lobes of the sp^3 orbitals are not shown.)

All the bonds in methane and ethane are sigma (σ) bonds because they are all formed by the end-on overlap of atomic orbitals. All **single bonds** found in organic compounds are sigma bonds.

All single bonds found in organic compounds are sigma bonds.

PROBLEM 15◆

What orbitals are used to form the 10 covalent bonds in propane ($CH_3CH_2CH_3$)?

3-D Molecule: Ethane

The MO diagram illustrating the overlap of an sp^3 orbital of one carbon with an sp^3 orbital of another carbon (Figure 1.14) is similar to the MO diagram for the end-on overlap of two p orbitals, which should not be surprising since sp^3 orbitals have 75% p character.

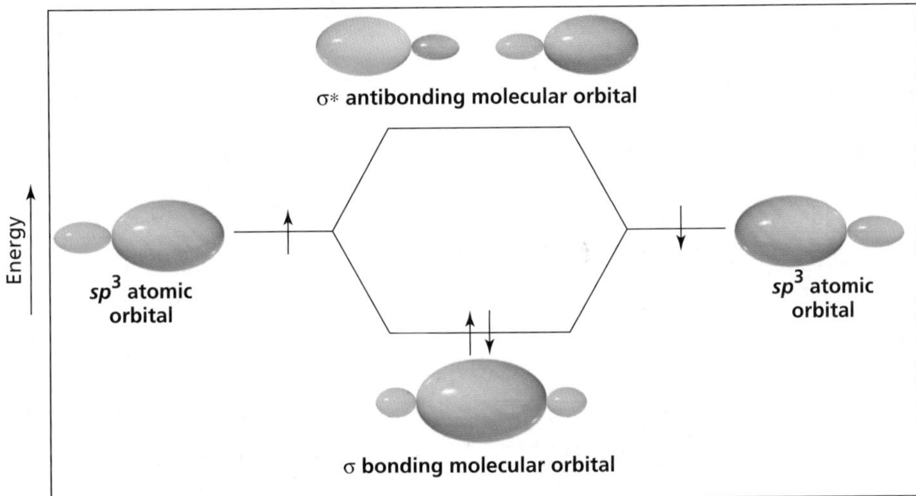

Figure 1.14
End-on overlap of two sp^3 orbitals to form a σ bonding molecular orbital and a σ^* antibonding molecular orbital.

1.8 Bonding in Ethene: A Double Bond

Each of the carbon atoms in ethene (also called ethylene) forms four bonds, but each is bonded to only three atoms:

$$\mathrm{H_2C{=}CH_2}$$

ethene
(ethylene)

To bond to three atoms, each carbon hybridizes three atomic orbitals. Because three orbitals (an s orbital and two of the p orbitals) are hybridized, three hybrid orbitals are obtained. These are called sp^2 orbitals. After hybridization, each carbon atom has three degenerate sp^2 orbitals and one p orbital:

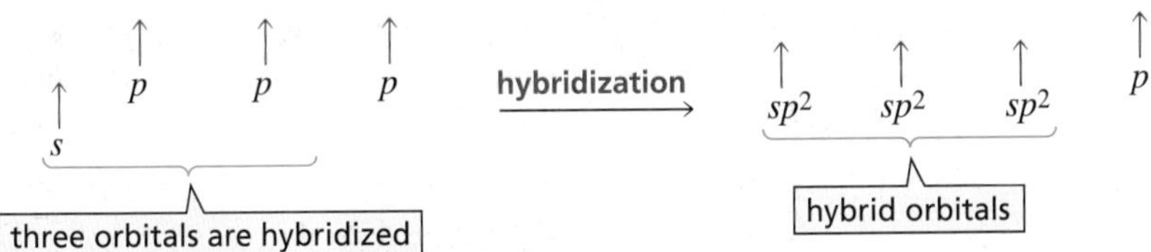

To minimize electron repulsion, the three sp^2 orbitals need to get as far from each other as possible. Therefore, the axes of the three orbitals lie in a plane, directed toward the corners of an equilateral triangle with the carbon nucleus at the center. This means that the bond angles are all close to 120°. Because the sp^2 hybridized carbon

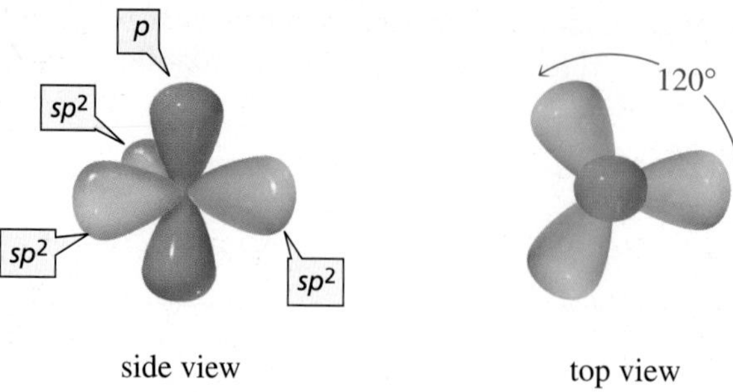

Figure 1.15 ▶
An sp^2 hybridized carbon. The three degenerate sp^2 orbitals lie in a plane. The unhybridized *p* orbital is perpendicular to the plane. (The smaller lobes of the sp^2 orbitals are not shown.)

atom is bonded to three atoms that define a plane, it is called a **trigonal planar carbon**. The unhybridized *p* orbital is perpendicular to the plane defined by the axes of the sp^2 orbitals (Figure 1.15).

The carbons in ethene form two bonds with each other. This is called a **double bond**. The two carbon–carbon bonds in the double bond are not identical. One of the bonds results from the overlap of an sp^2 orbital of one carbon with an sp^2 orbital of the other carbon; this is a sigma (σ) bond because it is formed by end-on overlap (Figure 1.16a). Each carbon uses its other two sp^2 orbitals to overlap the *s* orbital of a hydrogen to form the C—H bonds. The second carbon–carbon bond results from side-to-side overlap of the two unhybridized *p* orbitals. Side-to-side overlap of *p* orbitals forms a pi (π) bond (Figure 1.16b). Thus, one of the bonds in a double bond is a σ bond and the other is a π bond. All the C—H bonds are σ bonds.

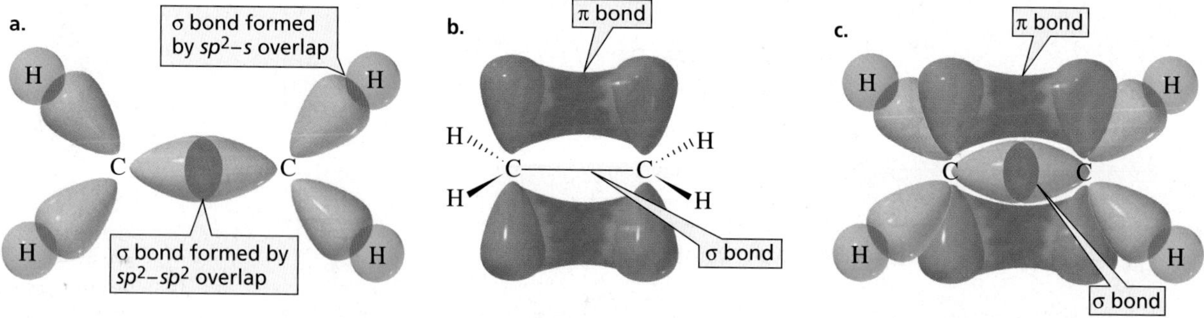

▲ **Figure 1.16**
(a) One C—C bond in ethene is a σ bond formed by sp^2–sp^2 overlap, and the C—H bonds are formed by sp^2–*s* overlap. (b) The second C—C bond is a π bond formed by side-to-side overlap of a *p* orbital of one carbon with a *p* orbital of the other carbon. (c) There is an accumulation of electron density above and below the plane containing the two carbons and four hydrogens.

The two *p* orbitals that overlap to form the π bond must be parallel to each other for maximum overlap to occur. This forces the triangle formed by one carbon and two hydrogens to lie in the same plane as the triangle formed by the other carbon and two hydrogens. This means that all six atoms of ethene lie in the same plane, and the electrons in the *p* orbitals occupy a volume of space above and below the plane (Figure 1.16c). The electrostatic potential map for ethene shows that it is a nonpolar molecule with an accumulation of negative charge (the orange area) above the two carbons. (If you could turn the potential map over, a similar accumulation of negative charge would be found on the other side.)

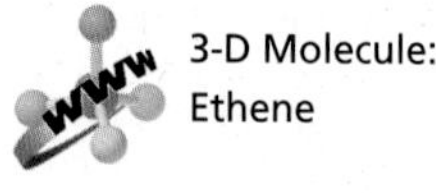

3-D Molecule: Ethene

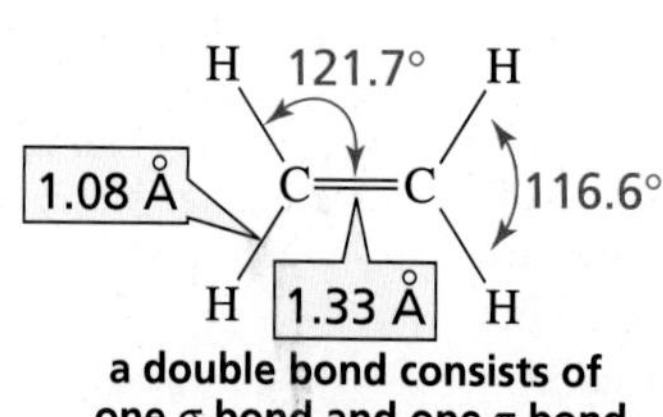

a double bond consists of one σ bond and one π bond

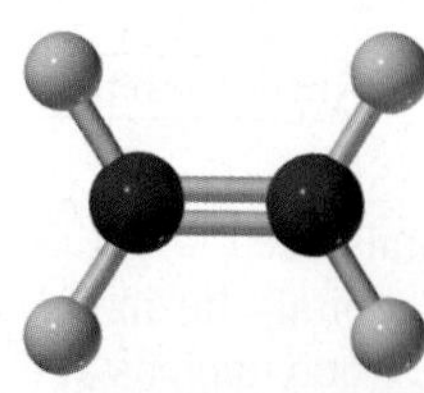
ball-and-stick model of ethene

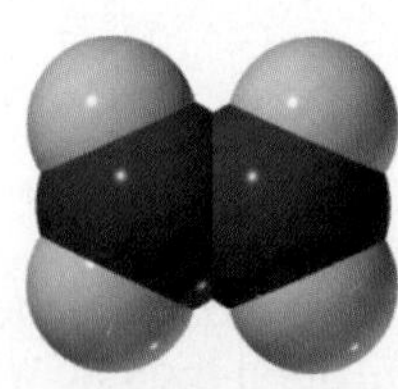
space-filling model of ethene

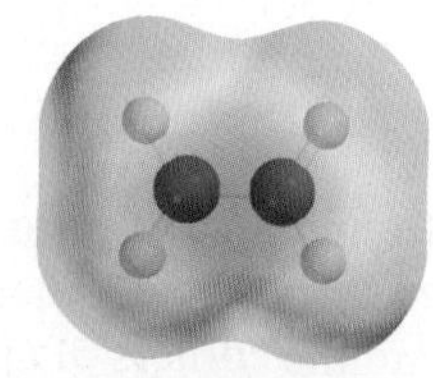
electrostatic potential map for ethene

Four electrons hold the carbons together in a carbon–carbon double bond; only two electrons bind the atoms in a carbon–carbon single bond. This means that a carbon–carbon double bond is stronger (174 kcal/mol or 728 kJ/mol) and shorter (1.33 Å) than a carbon–carbon single bond (90 kcal/mol or 377 kJ/mol, and 1.54 Å).

DIAMOND, GRAPHITE, AND BUCKMINSTERFULLERENE: SUBSTANCES CONTAINING ONLY CARBON ATOMS

Diamond is the hardest of all substances. Graphite, in contrast, is a slippery, soft solid most familiar to us as the "lead" in pencils. Both materials, in spite of their very different physical properties, contain only carbon atoms. The two substances differ solely in the nature of the carbon–carbon bonds holding them together. Diamond consists of a rigid three-dimensional network of atoms, with each carbon bonded to four other carbons via sp^3 orbitals. The carbon atoms in graphite, on the other hand, are sp^2 hybridized, so each bonds to only three other carbon atoms. This trigonal planar arrangement causes the atoms in graphite to lie in flat, layered sheets that can shear off of neighboring sheets. You experience this when you write with a pencil: Sheets of carbon atoms shear off, leaving a thin trail of graphite. There is a third substance found in nature that contains only carbon atoms: buckminsterfullerene. Like graphite, buckminsterfullerene contains only sp^2 hybridized carbons, but instead of forming planar sheets, the sp^2 carbons in buckminsterfullerene form spherical structures. (Buckminsterfullerene is discussed in more detail in Section 15.2.)

1.9 Bonding in Ethyne: A Triple Bond

The carbon atoms in ethyne (also called acetylene) are each bonded to only two atoms—a hydrogen and another carbon:

$$\mathrm{H{-}C{\equiv}C{-}H}$$

ethyne (acetylene)

Because each carbon forms covalent bonds with two atoms, only two orbitals (an *s* and a *p*) are hybridized. Two degenerate *sp* orbitals result. Each carbon atom in ethyne, therefore, has two *sp* orbitals and two unhybridized *p* orbitals (Figure 1.17).

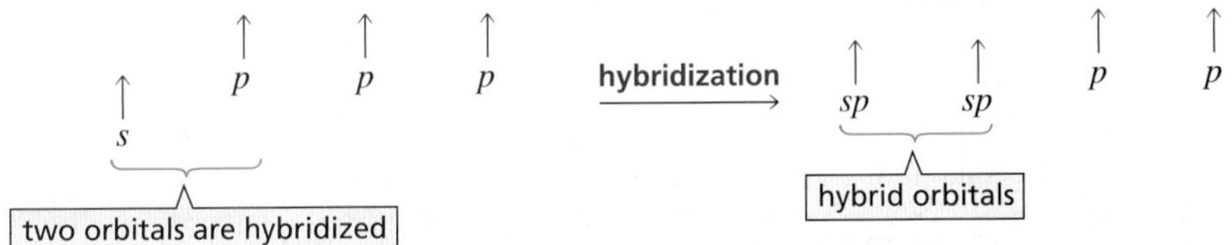

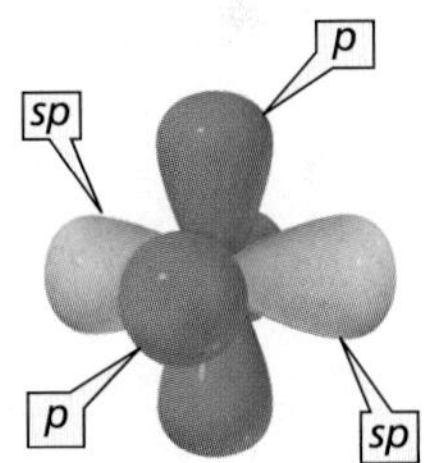

▲ Figure 1.17
An *sp* hybridized carbon. The two *sp* orbitals are oriented 180° away from each other, perpendicular to the two unhybridized *p* orbitals. (The smaller lobes of the *sp* orbitals are not shown.)

One of the *sp* orbitals of one carbon in ethyne overlaps an *sp* orbital of the other carbon to form a carbon–carbon σ bond. The other *sp* orbital of each carbon overlaps the *s* orbital of a hydrogen to form a C—H σ bond (Figure 1.18a). To minimize electron

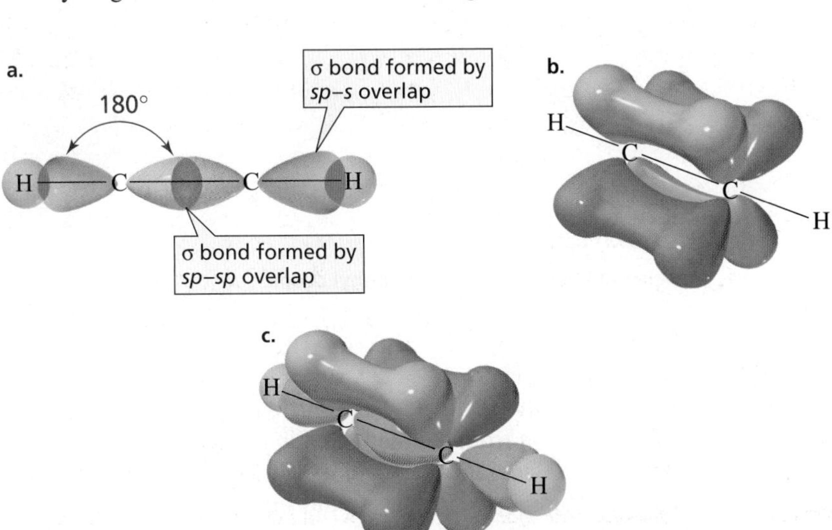

◀ Figure 1.18
(a) The C—C σ bond in ethyne is formed by *sp–sp* overlap, and the C—H bonds are formed by *sp–s* overlap. The carbon atoms and the atoms bonded to them are in a straight line. (b) The two carbon–carbon π bonds are formed by side-to-side overlap of the *p* orbitals of one carbon with the *p* orbitals of the other carbon. (c) The triple bond has an electron-dense region above and below and in front of and in back of the internuclear axis of the molecule.

repulsion, the two *sp* orbitals point in opposite directions. Consequently, the bond angles are 180°.

The two unhybridized *p* orbitals are perpendicular to each other, and both are perpendicular to the *sp* orbitals. Each of the unhybridized *p* orbitals engages in side-to-side overlap with a parallel *p* orbital on the other carbon, with the result that two π bonds are formed (Figure 1.18b). The overall result is a triple bond. **A triple bond consists of one σ bond and two π bonds.** Because the two unhybridized *p* orbitals on each carbon are perpendicular to each other, there is a region of high electron density above and below, *and* in front of and in back of, the internuclear axis of the molecule (Figure 1.18c). The potential map for ethyne shows that negative charge accumulates in a cylinder that wraps around the egg-shaped molecule.

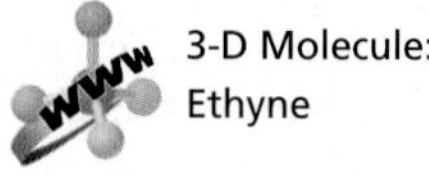
3-D Molecule: Ethyne

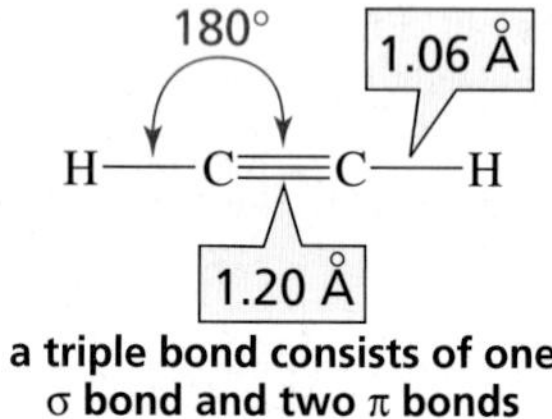

a triple bond consists of one σ bond and two π bonds

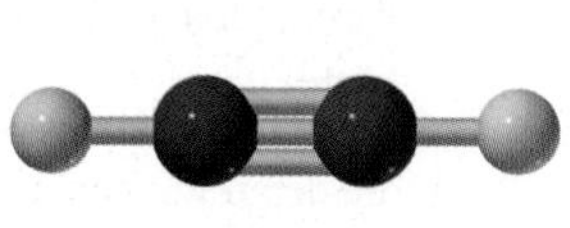
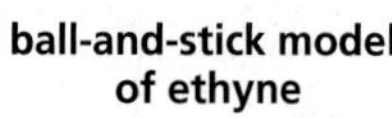
ball-and-stick model of ethyne

space-filling model of ethyne

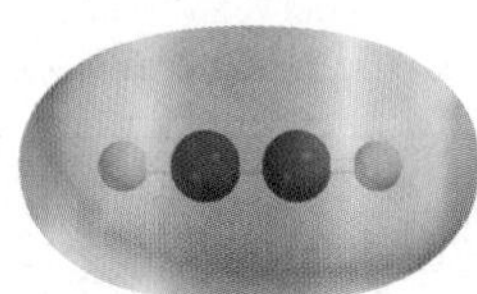
electrostatic potential map for ethyne

Because the two carbon atoms in a triple bond are held together by six electrons, a triple bond is stronger (231 kcal/mol or 967 kJ/mol) and shorter (1.20 Å) than a double bond.

1.10 Bonding in the Methyl Cation, the Methyl Radical, and the Methyl Anion

Not all carbon atoms form four bonds. A carbon with a positive charge, a negative charge, or an unpaired electron forms only three bonds. Now we will see what orbitals carbon uses when it forms three bonds.

The Methyl Cation ($^+CH_3$)

The positively charged carbon in the methyl cation is bonded to three atoms, so it hybridizes three orbitals—an *s* orbital and two *p* orbitals. Therefore, it forms its three covalent bonds using sp^2 orbitals. Its unhybridized *p* orbital remains empty. The positively charged carbon and the three atoms bonded to it lie in a plane. The *p* orbital stands perpendicular to the plane

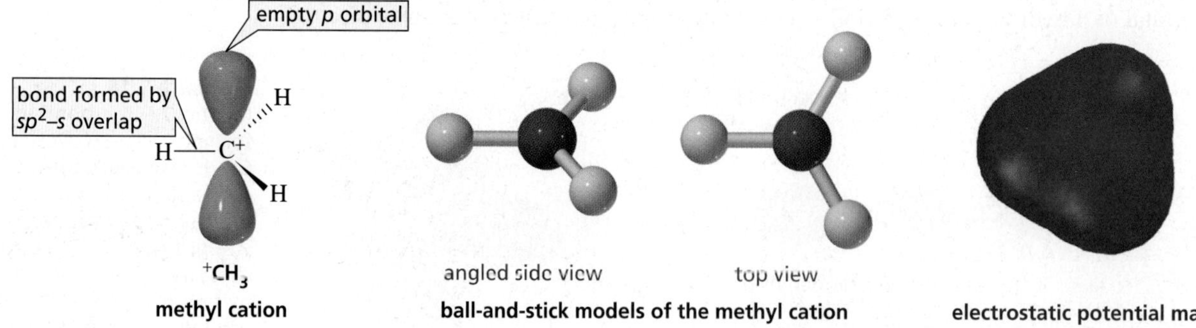

methyl cation

ball-and-stick models of the methyl cation

electrostatic potential map for the methyl cation

The Methyl Radical ($\cdot CH_3$)

The carbon atom in the methyl radical is also sp^2 hybridized. The methyl radical differs by one unpaired electron from the methyl cation. That electron is in the *p* orbital. Notice the similarity in the ball-and-stick models of the methyl cation and the methyl radical. The potential maps, however, are quite different because of the additional electron in the methyl radical.

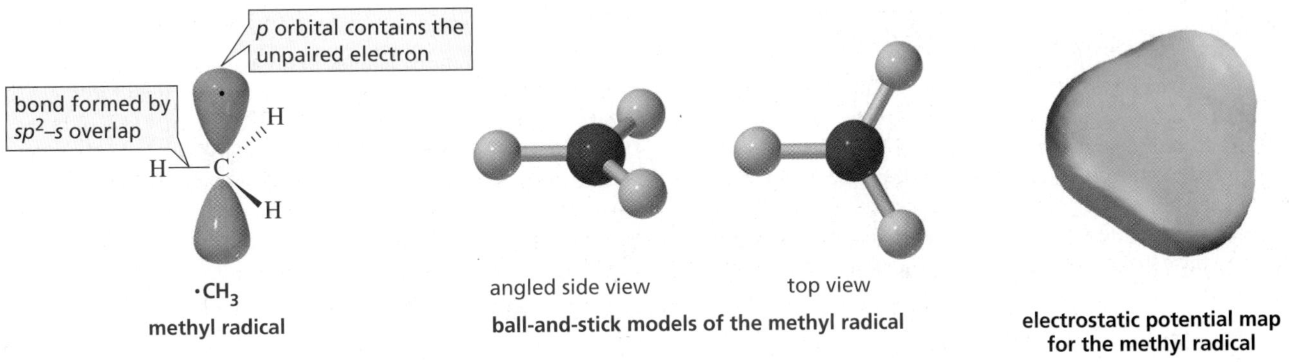

$\cdot CH_3$
methyl radical

ball-and-stick models of the methyl radical

electrostatic potential map for the methyl radical

The Methyl Anion ($\ddot{:}CH_3$)

The negatively charged carbon in the methyl anion has three pairs of bonding electrons and one lone pair. The four pairs of electrons are farthest apart when the four orbitals containing the bonding and lone-pair electrons point toward the corners of a tetrahedron. In other words, a negatively charged carbon is sp^3 hybridized. In the methyl anion, three of carbon's sp^3 orbitals each overlap the s orbital of a hydrogen, and the fourth sp^3 orbital holds the lone pair.

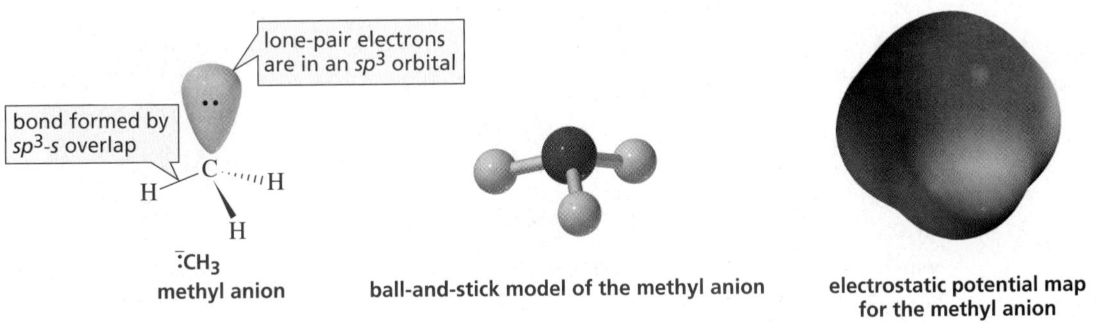

$\ddot{:}CH_3$
methyl anion

ball-and-stick model of the methyl anion

electrostatic potential map for the methyl anion

Take a moment to compare the potential maps for the methyl cation, the methyl radical, and the methyl anion.

1.11 Bonding in Water

The oxygen atom in water (H_2O) forms two covalent bonds. Because oxygen has two unpaired electrons in its ground-state electronic configuration (Table 1.2), it does not need to promote an electron to form the number (two) of covalent bonds required to achieve an outer shell of eight electrons (i.e., to complete its octet). If we assume that oxygen uses p orbitals to form the two O—H bonds, as predicted by oxygen's ground-state electronic configuration, we would expect a bond angle of about 90° because the two p orbitals are at right angles to each other. However, the experimentally observed bond angle is 104.5°. How can we explain the observed bond angle? Oxygen must use hybrid orbitals to form covalent bonds—just as carbon does. The s orbital and the three p orbitals must hybridize to produce four sp^3 orbitals.

The bond angles in a molecule indicate which orbitals are used in bond formation.

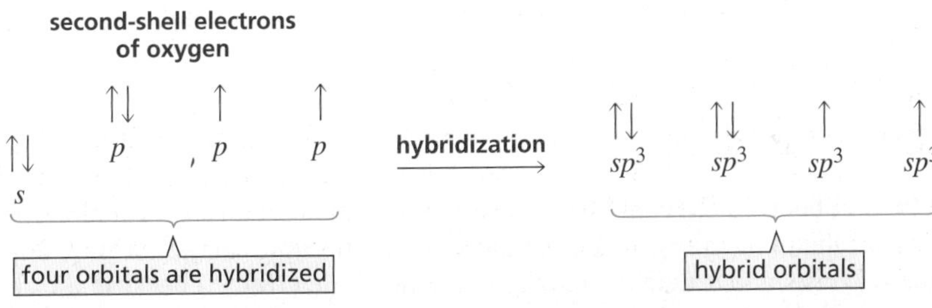

WATER—A UNIQUE COMPOUND

Water is the most abundant compound found in living organisms. Its unique properties have allowed life to originate and evolve. Its high heat of fusion (the heat required to convert a solid to a liquid) protects organisms from freezing at low temperatures because a lot of heat must be removed from water to freeze it. Its high heat capacity (the heat required to raise the temperature of a substance a given amount) minimizes temperature changes in organisms, and its high heat of vaporization (the heat required to convert a liquid to a gas) allows animals to cool themselves with a minimal loss of body fluid. Because liquid water is denser than ice, ice formed on the surface of water floats and insulates the water below. That is why oceans and lakes don't freeze from the bottom up. It is also why plants and aquatic animals can survive when the ocean or lake they live in freezes.

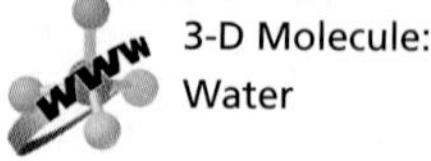

Each of the two O—H bonds is formed by the overlap of an sp^3 orbital of oxygen with the s orbital of a hydrogen. A lone pair occupies each of the two remaining sp^3 orbitals.

The bond angle in water is a little smaller (104.5°) than the tetrahedral bond angle (109.5°) in methane, presumably because each lone pair "feels" only one nucleus, which makes the lone pair more diffuse than the bonding pair that "feels" two nuclei and is therefore relatively confined between them. Consequently, there is more electron repulsion between lone-pair electrons, causing the O—H bonds to squeeze closer together, thereby decreasing the bond angle.

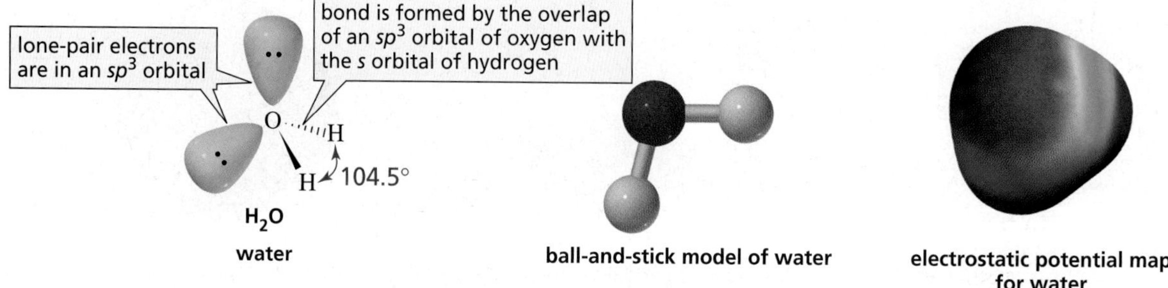

Compare the potential map for water with that for methane. Water is a polar molecule; methane is nonpolar.

PROBLEM 16◆

The bond angles in H_3O^+ are greater than ________ and less than ________.

1.12 Bonding in Ammonia and in the Ammonium Ion

The experimentally observed bond angles in NH_3 are 107.3°. The bond angles indicate that nitrogen also uses hybrid orbitals when it forms covalent bonds. Like carbon and oxygen, the one s and three p orbitals of the second shell of nitrogen hybridize to form four degenerate sp^3 orbitals:

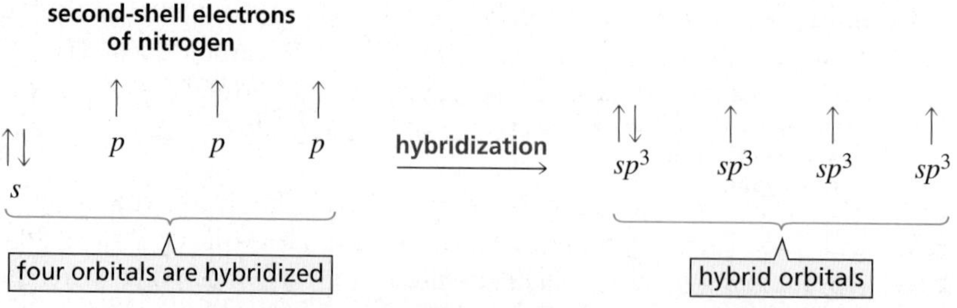

The N—H bonds in NH_3 are formed from the overlap of an sp^3 orbital of nitrogen with the s orbital of a hydrogen. The single lone pair occupies an sp^3 orbital. The bond angle (107.3°) is smaller than the tetrahedral bond angle (109.5°) because the electron repulsion between the relatively diffuse lone pair and the bonding pairs is greater than the electron repulsion between two bonding pairs. Notice that the bond angles in NH_3 (107.3°) are larger than the bond angles in H_2O (104.5°) because nitrogen has only one lone pair, whereas oxygen has two lone pairs.

3-D Molecule: Ammonia

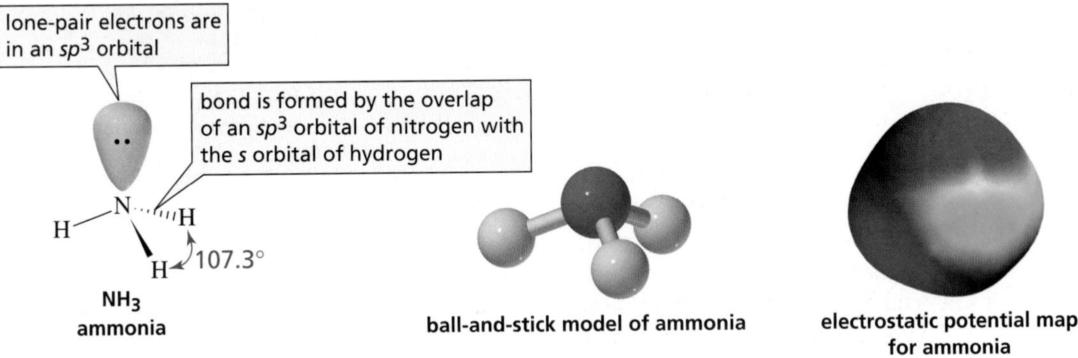

NH_3
ammonia

ball-and-stick model of ammonia

electrostatic potential map for ammonia

Because the ammonium ion ($^+NH_4$) has four identical N—H bonds and no lone pairs, all the bond angles are 109.5°—just like the bond angels in methane.

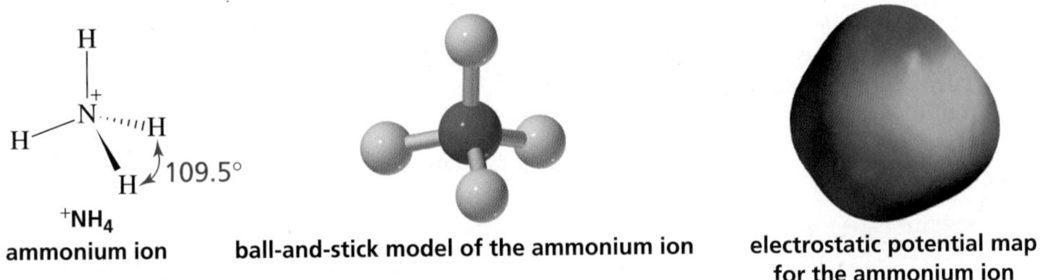

$^+NH_4$
ammonium ion

ball-and-stick model of the ammonium ion

electrostatic potential map for the ammonium ion

PROBLEM 17◆

According to the potential map for the ammonium ion, which atom(s) is (are) most positively charged?

PROBLEM 18◆

Compare the potential maps for methane, ammonia, and water. Which is the most polar molecule? Which is the least polar?

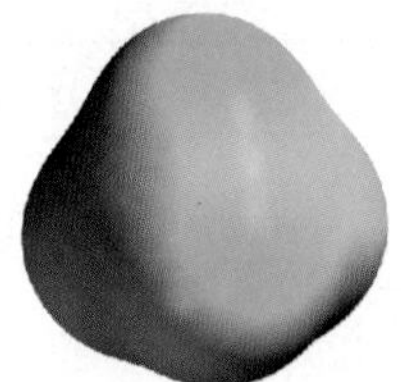

electrostatic potential map for methane

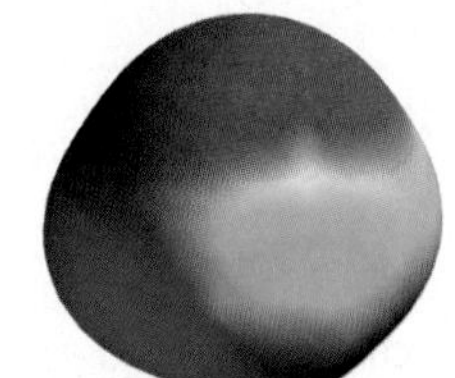

electrostatic potential map for ammonia

electrostatic potential map for water

1.13 Bonding in the Hydrogen Halides

Fluorine, chlorine, bromine, and iodine are collectively known as the halogens. HF, HCl, HBr, and HI are called hydrogen halides. Bond angles will not help us determine the orbitals involved in a hydrogen halide bond, as they did with other molecules, because hydrogen halides have only one bond. We do know, however, that bonding electrons and lone-pair electrons position themselves to minimize electron repulsion (Section 1.6). If the three lone pairs were in sp^3 orbitals, they would be farther apart than they would be if one pair resided in an *s* orbital and the other two pairs resided in

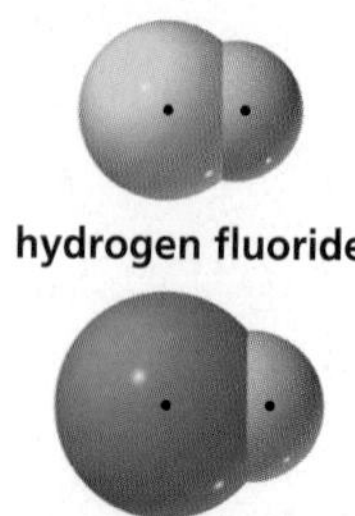

hydrogen fluoride

hydrogen chloride

hydrogen bromide

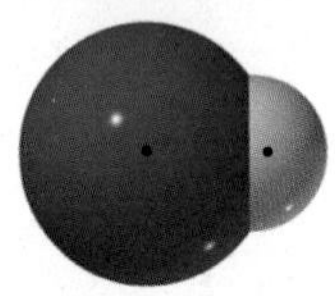

hydrogen iodide

p orbitals. Therefore, we will assume that the hydrogen–halogen bond is formed by the overlap of an sp^3 orbital of the halogen with the *s* orbital of hydrogen.

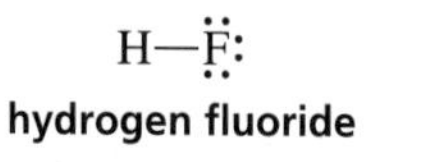

hydrogen fluoride

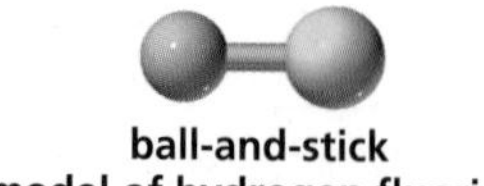

ball-and-stick model of hydrogen fluoride

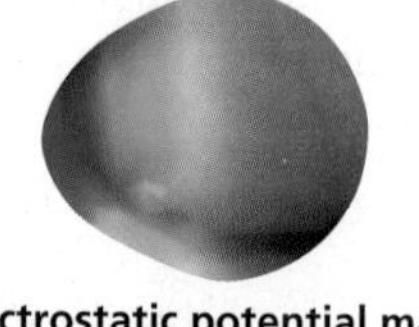

electrostatic potential map for hydrogen fluoride

In the case of fluorine, the sp^3 orbital used in bond formation belongs to the second shell of electrons. In chlorine, the sp^3 orbital belongs to the third shell of electrons. Because the average distance from the nucleus is greater for an electron in the third shell than for an electron in the second shell, the average electron density is less in a $3sp^3$ orbital than in a $2sp^3$ orbital. This means that the electron density in the region where the *s* orbital of hydrogen overlaps the sp^3 orbital of the halogen decreases as the size of the halogen increases (Figure 1.19). Therefore, the hydrogen–halogen bond becomes longer and weaker as the size (atomic weight) of the halogen increases (Table 1.6).

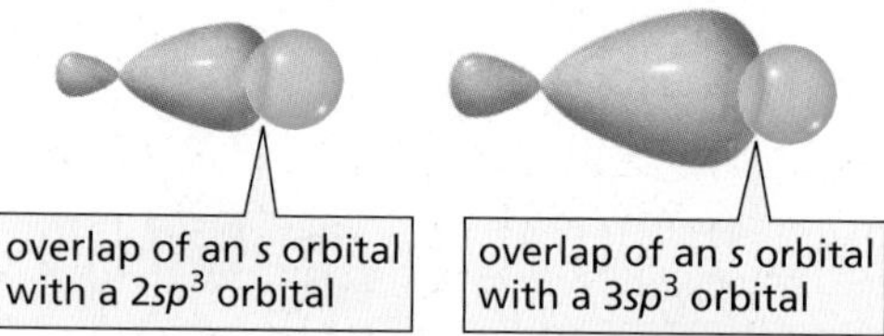

Figure 1.19 ▶
There is greater electron density in the region of overlap of an *s* orbital with a $2sp^3$ orbital than in the region of overlap of an *s* orbital with a $3sp^3$ orbital.

The shorter the bond, the stronger it is.

Table 1.6 Hydrogen–Halogen Bond Lengths and Bond Strengths

Hydrogen halide	Bond length (Å)	Bond strength kcal/mol	Bond strength kJ/mol
H—F	0.917	136	571
H—Cl	1.2746	103	432
H—Br	1.4145	87	366
H—I	1.6090	71	298

PROBLEM 19◆

a. Predict the relative lengths and strengths of the bonds in Cl_2 and Br_2.
b. Predict the relative lengths and strengths of the bonds in HF, HCl, and HBr.

1.14 Summary: Orbital Hybridization, Bond Lengths, Bond Strengths, and Bond Angles

All single bonds are σ bonds. All double bonds are composed of one σ bond and one π bond. All triple bonds are composed of one σ bond and two π bonds. The easiest way to determine the hybridization of a carbon, oxygen, or nitrogen atom is to look at the number of π bonds it forms: If it forms no π bonds, it is sp^3 hybridized; if it forms one π bond, it is sp^2 hybridized; if it forms two π bonds, it is *sp* hybridized. The exceptions are carbocations and carbon radicals, which are sp^2 hybridized—not because they form a π bond, but because they have an empty or half-filled *p* orbital (Section 1.10).

The hybridization of a C, O, or N is $sp^{(3 - \text{the number of } \pi \text{ bonds})}$.

Compound	Hybridization (left to right)
CH_3-NH_2	sp^3, sp^3
$(CH_3)_2C=N-NH_2$	sp^3, sp^2, sp^2, sp^3
$CH_3-C\equiv N:$	sp^3, sp, sp
CH_3-OH	sp^3, sp^3
$CH_3-C(=O)-OH$	sp^3, sp^2, sp^3 (carbonyl O ← sp^2)
$:O=C=O:$	sp^2, sp, sp^2

In comparing the lengths and strengths of carbon–carbon single, double, and triple bonds, we see that the more bonds holding two carbon atoms together, the shorter and stronger is the carbon–carbon bond (Table 1.7). Triple bonds are shorter and stronger than double bonds, which are shorter and stronger than single bonds.

A double bond (a σ bond plus a π bond) is stronger than a single bond (a σ bond), but it is not twice as strong. We can conclude, therefore, that a π bond is weaker than a σ bond. This is what we would expect, because the end-on overlap that forms σ bonds is better than the side-to-side overlap that forms π bonds (Section 1.6).

A π bond is weaker than a σ bond.

The data in Table 1.7 indicate that a C—H σ bond is shorter and stronger than a C—C σ bond. This is because the *s* orbital of hydrogen is closer to the nucleus than is the sp^3 orbital of carbon. Consequently, the nuclei are closer together in a bond formed by sp^3–s overlap than they are in a bond formed by sp^3–sp^3 overlap. In addition to being shorter, a C—H bond is stronger than a C—C bond because there is greater electron density in the region of overlap of an sp^3 orbital with the *s* orbital than in the region of overlap of an sp^3 orbital with an sp^3 orbital.

The greater the electron density in the region of orbital overlap, the stronger is the bond.

The length and strength of a C—H bond depend on the hybridization of the carbon atom to which the hydrogen is attached. The more *s* character in the orbital used by carbon to form the bond, the shorter and stronger is the bond—again, because an *s* orbital is closer to the nucleus than is a *p* orbital. So a C—H bond formed by an *sp* hybridized carbon (50% *s*) is shorter and stronger than a C—H bond formed by an sp^2 hybridized carbon (33.3% *s*), which in turn is shorter and stronger than a C—H bond formed by an sp^3 hybridized carbon (25% *s*).

The more *s* character, the shorter and stronger is the bond.

The bond angle also depends on the orbital used by carbon to form the bond. The greater the amount of *s* character in the orbital, the larger is the bond angle. For example, *sp* hybridized carbons have bond angles of 180°, sp^2 hybridized carbons have bond angles of 120°, and sp^3 hybridized carbons have bond angles of 109.5°.

The more *s* character, the larger is the bond angle.

Table 1.7 Comparison of the Bond Angles and the Lengths and Strengths of the Carbon–Carbon and Carbon–Hydrogen Bonds in Ethane, Ethene, and Ethyne

Molecule	Hybridization of carbon	Bond angles	Length of C—C bond (Å)	Strength of C—C bond (kcal/mol)	Strength of C—C bond (kJ/mol)	Length of C—H bond (Å)	Strength of C—H bond (kcal/mol)	Strength of C—H bond (kJ/mol)
ethane	sp^3	109.5°	1.54	90	377	1.10	101	423
ethene	sp^2	120°	1.33	174	720	1.08	111	466
H—C≡C—H ethyne	sp	180°	1.20	231	967	1.06	131	548

You may wonder how an electron "knows" what orbital it should go into. In fact, electrons know nothing about orbitals. They simply arrange themselves around atoms in the most stable manner possible. It is chemists who use the concept of orbitals to explain this arrangement.

PROBLEM 20◆

Which of the bonds of a carbon–carbon double bond has more effective orbital–orbital overlap, the σ bond or the π bond?

PROBLEM 21◆

Why would you expect a C—C σ bond formed by sp^2–sp^2 overlap to be stronger than a σ bond formed by sp^3–sp^3 overlap?

PROBLEM 22

a. What is the hybridization of each of the carbon atoms in the following compound?

$$CH_3\underset{\displaystyle CH_3}{\underset{|}{C}}HCH{=}CHCH_2C{\equiv}CCH_3$$

b. What is the hybridization of each of the carbon, oxygen, and nitrogen atoms in the following compounds?

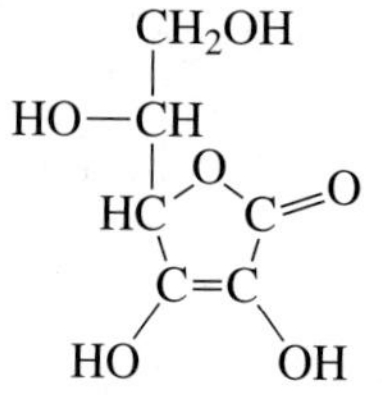

vitamin C

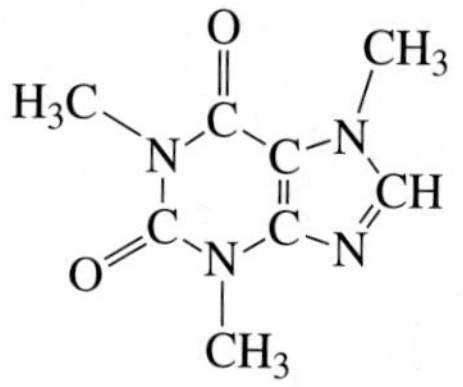

caffeine

PROBLEM 23

Describe the orbitals used in bonding and the bond angles in the following compounds. (*Hint:* see Table 1.7).

a. BeH_2 b. BH_3 c. CCl_4 d. CO_2 e. HCOOH f. N_2

1.15 Dipole Moments of Molecules

In Section 1.3, we saw that for molecules with one covalent bond, the dipole moment of the bond is identical to the dipole moment of the molecule. For molecules that have more than one covalent bond, the geometry of the molecule must be taken into account because both the *magnitude* and the *direction* of the individual bond dipole moments (the vector sum) determine the overall dipole moment of the molecule. Symmetrical molecules, therefore, have no dipole moment. For example, let's look at the dipole moment of carbon dioxide (CO_2). Because the carbon atom is bonded to two atoms, it uses *sp* orbitals to form the C—O σ bonds. The remaining two *p* orbitals on carbon form the two C—O π bonds. The individual carbon–oxygen bond dipole moments cancel each other—because *sp* orbitals form a bond angle of 180°—giving carbon dioxide a dipole moment of zero D. Another symmetrical molecule is carbon tetrachloride (CCl_4). The four atoms bonded to the sp^3 hybridized carbon atom are identical and project symmetrically out from the carbon atom. Thus, as with CO_2, the symmetry of the molecule causes the bond dipole moments to cancel. Methane also has no dipole moment.

O=C=O
carbon dioxide
μ = 0 D

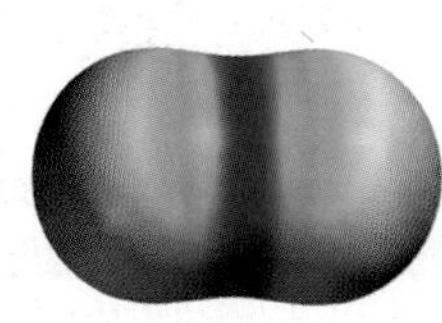

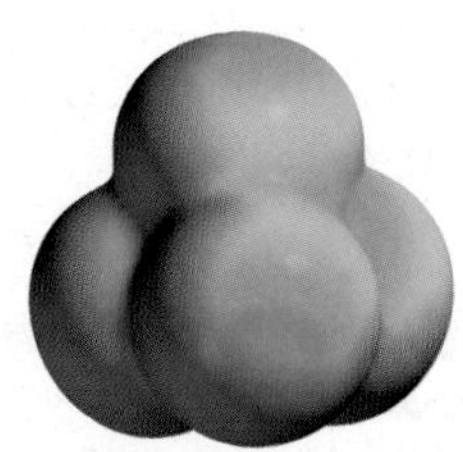

carbon tetrachloride
μ = 0 D

The dipole moment of chloromethane (CH_3Cl) is greater (1.87 D) than the dipole moment of the C—Cl bond (1.5 D) because the C—H dipoles are oriented so that they reinforce the dipole of the C—Cl bond—they are all in the same relative direction. The dipole moment of water (1.85 D) is greater than the dipole moment of a single O—H bond (1.5 D) because the dipoles of the two O—H bonds reinforce each other. The lone-pair electrons also contribute to the dipole moment. Similarly, the dipole moment of ammonia (1.47 D) is greater than the dipole moment of a single N—H bond (1.3 D).

chloromethane
μ = 1.87 D

water
μ = 1.85 D

ammonia
μ = 1.47 D

PROBLEM 24

Account for the difference in the shape and color of the potential maps for ammonia and the ammonium ion in Section 1.12.

PROBLEM 25◆

Which of the following molecules would you expect to have a dipole moment of zero? To answer parts g and h, you may need to consult your answers to Problem 23 a and b.

a. CH_3CH_3
b. $H_2C{=}O$
c. CH_2Cl_2
d. NH_3
e. $H_2C{=}CH_2$
f. $H_2C{=}CHBr$
g. $BeCl_2$
h. BF_3

1.16 An Introduction to Acids and Bases

Early chemists called any compound that tasted sour an acid (from *acidus*, Latin for "sour"). Some familiar acids are citric acid (found in lemons and other citrus fruits), acetic acid (found in vinegar), and hydrochloric acid (found in stomach acid—the sour taste associated with vomiting). Compounds that neutralize acids, such as wood ashes and other plant ashes, were called bases, or alkaline compounds ("ash" in Arabic is *al kalai*). Glass cleaners and solutions designed to unclog drains are alkaline solutions.

The definitions of "acid" and "base" that we use now were provided by Brønsted and Lowry in 1923. In the Brønsted–Lowry definitions, an **acid** is a species that donates a proton, and a **base** is a species that accepts a proton. (Remember that positively charged hydrogen ions are also called protons.) In the following reaction, hydrogen chloride (HCl) meets the Brønsted–Lowry definition of an acid because it donates a proton to water. Water meets the definition of a base because it accepts a proton from HCl. Water can accept a proton because it has two lone pairs. Either lone pair can form a covalent bond with a proton. In the reverse reaction, H_3O^+ is an acid because it donates a proton to Cl^-, and Cl^- is a base because it accepts a proton from H_3O^+.

Born in Denmark, **Johannes Nicolaus Brønsted (1879–1947)** *studied engineering before he switched to chemistry. He was a professor of chemistry at the University of Copenhagen. During World War II, he became known for his anti-Nazi position, and in 1947 he was elected to the Danish parliament. He died before he could take his seat.*

Thomas M. Lowry (1874–1936) *was born in England, the son of an army chaplain. He earned a Ph.D. at Central Technical College, London (now Imperial College). He was head of chemistry at Westminster Training College and, later, at Guy's Hospital in London. In 1920, he became a professor of chemistry at Cambridge University.*

$$\underset{\text{an acid}}{H\ddot{\underset{..}{Cl}}:} + \underset{\text{a base}}{H_2\ddot{O}:} \rightleftharpoons \underset{\text{a base}}{:\ddot{\underset{..}{Cl}}:^-} + \underset{\text{an acid}}{H_3\ddot{O}^+}$$

According to the Brønsted–Lowry definitions, any species that has a hydrogen can potentially act as an acid, and any compound possessing a lone pair can potentially act as a base. Both an acid and a base must be present in a *proton-transfer reaction*, because an acid cannot donate a proton unless a base is present to accept it. **Acid–base reactions** are often called **proton-transfer reactions**.

When a compound loses a proton, the resulting species is called its **conjugate base**. Thus, Cl^- is the conjugate base of HCl, and H_2O is the conjugate base of H_3O^+. When a compound accepts a proton, the resulting species is called its **conjugate acid**. Thus, HCl is the conjugate acid of Cl^-, and H_3O^+ is the conjugate acid of H_2O.

In a reaction involving ammonia and water, ammonia (NH_3) is a base because it accepts a proton, and water is an acid because it donates a proton. Thus, HO^- is the conjugate base of H_2O, and ($^+NH_4$) is the conjugate acid of (NH_3). In the reverse reaction, ammonium ion ($^+NH_4$) is an acid because it donates a proton, and hydroxide ion (HO^-) is a base because it accepts a proton.

$$\underset{\text{a base}}{\ddot{N}H_3} + \underset{\text{an acid}}{H_2\ddot{O}:} \rightleftharpoons \underset{\text{an acid}}{^+NH_4} + \underset{\text{a base}}{H\ddot{\underset{..}{O}}:^-}$$

Notice that water can behave as either an acid or a base. It can behave as an acid because it has a proton that it can donate, but it can also behave as a base because it has a lone pair that can accept a proton. In Section 1.17, we will see how we know that water acts as a base in the first reaction in this section and acts as an acid in the second reaction.

Acidity is a measure of the tendency of a compound to give up a proton. **Basicity** is a measure of a compound's affinity for a proton. A strong acid is one that has a strong tendency to give up its proton. This means that its conjugate base must be weak because it has little affinity for the proton. A weak acid has little tendency to give up its proton, indicating that its conjugate base is strong because it has a high affinity for the proton. Thus, the following important relationship exists between an acid and its conjugate base: *The stronger the acid, the weaker is its conjugate base.* For example, since HBr is a stronger acid than HCl, we know that Br^- is a weaker base than Cl^-.

The stronger the acid, the weaker is its conjugate base.

PROBLEM 26◆

a. Draw the conjugate acid of each of the following:
1. NH_3 2. Cl^- 3. HO^- 4. H_2O

b. Draw the conjugate base of each of the following:
1. NH_3 2. HBr 3. HNO_3 4. H_2O

PROBLEM 27

a. Write an equation showing CH_3OH reacting as an acid with NH_3 and an equation showing it reacting as a base with HCl.

b. Write an equation showing NH_3 reacting as an acid with HO^- and an equation showing it reacting as a base with HBr.

1.17 Organic Acids and Bases; pK_a and pH

When a strong acid such as hydrogen chloride is dissolved in water, almost all the molecules dissociate (break into ions), which means that products are favored at equilibrium. When a much weaker acid, such as acetic acid, is dissolved in water, very few molecules dissociate, so reactants are favored at equilibrium. Two half-headed arrows are used to designate equilibrium reactions. A longer arrow is drawn toward the species favored at equilibrium.

$$HCl + H_2O \rightleftharpoons H_3O^+ + Cl^-$$

hydrogen chloride

$$CH_3COOH + H_2O \rightleftharpoons H_3O^+ + CH_3COO^-$$

acetic acid

Whether a reversible reaction favors reactants or products at equilibrium is indicated by the **equilibrium constant** of the reaction, K_{eq}. Remember that brackets are used to indicate concentration in moles/liter [i.e., molarity (M)].

$$HA + H_2O \rightleftharpoons H_3O^+ + A^-$$

$$K_{eq} = \frac{[H_3O^+][A^-]}{[H_2O][HA]}$$

The degree to which an acid (HA) dissociates is normally determined in a dilute solution, so the concentration of water remains nearly constant. The equilibrium expression, therefore, can be rewritten using a new constant called the **acid dissociation constant, K_a**.

$$K_a = \frac{[H_3O^+][A^-]}{[HA]} = K_{eq}[H_2O]$$

The acid dissociation constant is the equilibrium constant multiplied by the molar concentration of water (55.5 M).

The larger the acid dissociation constant, the stronger is the acid—that is, the greater is its tendency to give up a proton. Hydrogen chloride, with an acid dissociation constant of 10^7, is a stronger acid than acetic acid, with an acid dissociation constant of only 1.74×10^{-5}. For convenience, the strength of an acid is generally indicated by its **pK_a** value rather than its K_a value, where

The stronger the acid, the smaller is its pK_a.

$$pK_a = -\log K_a$$

The pK_a of hydrogen chloride is −7 and the pK_a of acetic acid, a much weaker acid, is 4.76. Notice that the smaller the pK_a, the stronger is the acid.

very strong acids	$pK_a < 1$
moderately strong acids	$pK_a = 1–5$
weak acids	$pK_a = 5–15$
extremely weak acids	$pK_a > 15$

Unless otherwise stated, the pK_a values given in this text indicate the strength of the acid *in water*. Later (in Section 10.10), you will see how the pK_a of an acid is affected when the solvent is changed.

The **pH** of a solution indicates the concentration of positively charged hydrogen ions in the solution. The concentration can be indicated as $[H^+]$ or, because a hydrogen ion in water is solvated, as $[H_3O^+]$. The lower the pH, the more acidic is the solution.

$$pH = -\log [H_3O^+]$$

Acidic solutions have pH values less than 7; basic solutions have pH values greater than 7. The pH values of some commonly encountered solutions are shown in the margin. The pH of a solution can be changed simply by adding acid or base to the solution.

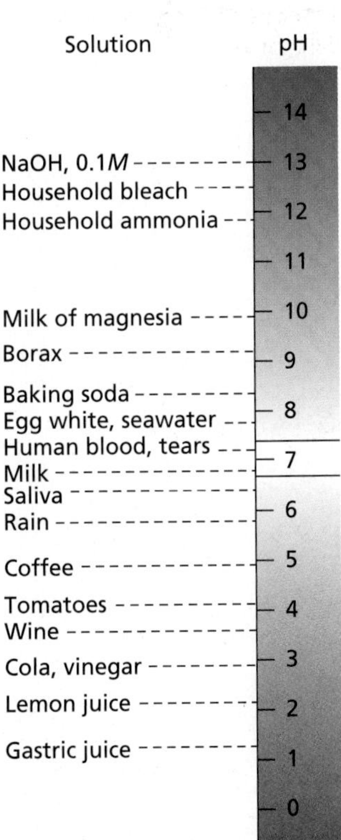

Do not confuse pH and pK_a: The pH scale is used to describe the acidity of a *solution*; the pK_a is characteristic of a particular *compound*, much like a melting point or a boiling point—it indicates the tendency of the compound to give up its proton.

PROBLEM 28◆

a. Which is a stronger acid, one with a pK_a of 5.2 or one with a pK_a of 5.8?

b. Which is a stronger acid, one with an acid dissociation constant of 3.4×10^{-3} or one with an acid dissociation constant of 2.1×10^{-4}?

PROBLEM 29◆

An acid has a K_a of 4.53×10^{-6} in water. What is its K_{eq}? ($[H_2O] = 55.5$ M)

The importance of organic acids and bases will become clear when we discuss how and why organic compounds react. The most common organic acids are carboxylic acids—compounds that have a COOH group. Acetic acid and formic acid are examples of carboxylic acids. Carboxylic acids have pK_a values ranging from about 3 to 5. (They are moderately strong acids.) The pK_a values of a wide variety of organic compounds are given in Appendix II.

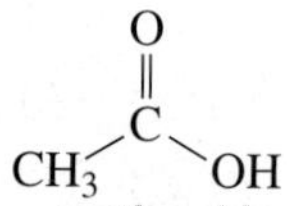

CH_3COOH (O=C(OH)–CH_3)
acetic acid
$pK_a = 4.76$

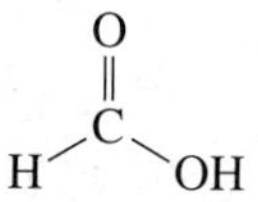

$HCOOH$ (O=C(OH)–H)
formic acid
$pK_a = 3.75$

Alcohols—compounds that have an OH group—are much weaker organic acids, with pK_a values close to 16. Methanol and ethanol are examples of alcohols.

CH_3OH
methanol
$pK_a = 15.5$

CH_3CH_2OH
ethanol
$pK_a = 15.9$

We have seen that water can behave both as an acid and as a base. An alcohol behaves similarly: It can behave as an acid and donate a proton, or as a base and accept a proton.

$$\underset{\text{an acid}}{CH_3OH} + HO^- \rightleftharpoons CH_3O^- + H_2O$$

$$\underset{\text{a base}}{CH_3OH} + H_3O^+ \rightleftharpoons CH_3\overset{+}{O}H_2 + H_2O$$

A carboxylic acid can behave as an acid and donate a proton, or as a base and accept a proton.

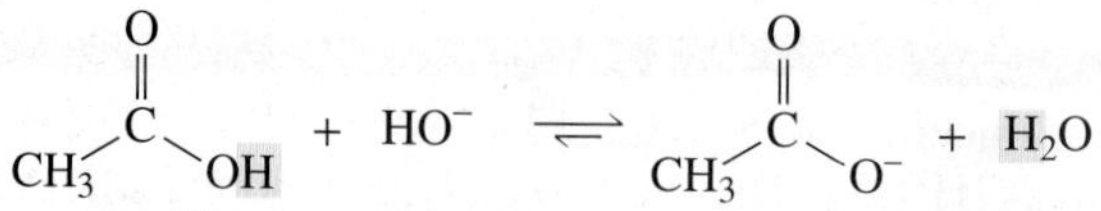

$$\underset{\text{an acid}}{CH_3C(=O)OH} + HO^- \rightleftharpoons CH_3C(=O)O^- + H_2O$$

$$\underset{\text{a base}}{CH_3C(=O)OH} + H_3O^+ \rightleftharpoons CH_3C(={}^+OH)OH + H_2O$$

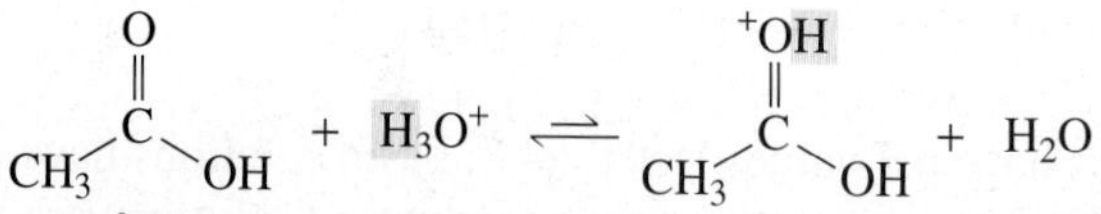

A *protonated* compound is a compound that has gained an additional proton. A protonated alcohol or a protonated carboxylic acid is a very strong acid. For example, protonated methanol has a pK_a of -2.5, protonated ethanol has a pK_a of -2.4, and protonated acetic acid has a pK_a of -6.1.

$CH_3\overset{+}{O}H$ (with H on O)
protonated methanol
p$K_a = -2.5$

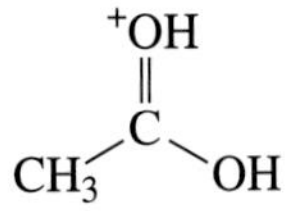

$CH_3CH_2\overset{+}{O}H$ (with H on O)
protonated ethanol
p$K_a = -2.4$

$CH_3C(={}^{+}OH)OH$
protonated acetic acid
p$K_a = -6.1$

An amine can behave as an acid and donate a proton, or as a base and accept a proton. Compounds with NH_2 groups are amines.

$$\underset{\text{an acid}}{CH_3NH_2} + HO^- \rightleftharpoons CH_3\overset{-}{N}H + H_2O$$

$$\underset{\text{a base}}{CH_3NH_2} + H_3O^+ \rightleftharpoons CH_3\overset{+}{N}H_3 + H_2O$$

Amines, however, have such high pK_a values that they rarely behave as acids. Ammonia also has a high pK_a.

CH_3NH_2
methylamine
p$K_a = 40$

NH_3
ammonia
p$K_a = 36$

Amines are much more likely to act as bases. In fact, amines are the most common organic bases. Instead of talking about the strength of a base in terms of its pK_b value, it is easier to talk about the strength of its conjugate acid as indicated by its pK_a value, remembering that the stronger the acid, the weaker is its conjugate base. For example, protonated methylamine is a stronger acid than protonated ethylamine, which means that methylamine is a weaker base than ethylamine. Notice that the pK_a values of protonated amines are about 10 to 11.

$CH_3\overset{+}{N}H_3$
protonated methylamine
p$K_a = 10.7$

$CH_3CH_2\overset{+}{N}H_3$
protonated ethylamine
p$K_a = 11.0$

It is important to know the approximate pK_a values of the various classes of compounds we have discussed. An easy way to remember them is in units of five, as shown in Table 1.8. (R is used when the particular carboxylic acid or amine is not specified.)

Table 1.8 Approximate pK_a Values

p$K_a < 0$	p$K_a \sim 5$	p$K_a \sim 10$	p$K_a \sim 15$
$R\overset{+}{O}H_2$ a protonated alcohol	$RC(=O)OH$ a carboxylic acid	$R\overset{+}{N}H_3$ a protonated amine	ROH an alcohol
$RC(={}^{+}OH)OH$ a protonated carboxylic acid			H_2O water
H_3O^+ protonated water			

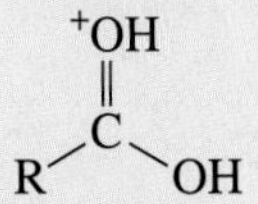

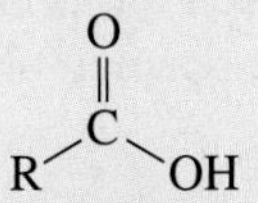

Be sure to learn the approximate pK_a values given in Table 1.8.

Protonated alcohols, protonated carboxylic acids, and protonated water have pK_a values less than 0, carboxylic acids have pK_a values of about 5, protonated amines have pK_a values of about 10, and alcohols and water have pK_a values of about 15. These values are also listed inside the back cover of this book for easy reference.

Now let's see how we knew that water acts as a base in the first reaction in Section 1.16 and as an acid in the second reaction. To determine which of the reactants will be the acid, we need to compare their pK_a values: The pK_a of hydrogen chloride is −7 and the pK_a of water is 15.7. Because hydrogen chloride is the stronger acid, it will donate a proton to water. Water, therefore, is a base in this reaction. When we compare the pK_a values of the two reactants of the second reaction, we see that the pK_a of ammonia is 36 and the pK_a of water is 15.7. In this case, water is the stronger acid, so it donates a proton to ammonia. Water, therefore, is an acid in this reaction.

Strong reacts to give weak.

In determining the position of equilibrium for an acid–base reaction (i.e., whether reactants or products are favored at equilibrium), remember that the equilibrium favors *reaction* of the strong acid and strong base and *formation* of the weak acid and weak base. In other words, *strong reacts to give weak*. Thus, the equilibrium lies away from the stronger acid and toward the weaker acid.

$$\underset{\substack{\textbf{stronger acid}\\ pK_a = 4.8}}{CH_3C(=O)OH} + \underset{\textbf{stronger base}}{NH_3} \rightleftharpoons \underset{\textbf{weaker base}}{CH_3C(=O)O^-} + \underset{\substack{\textbf{weaker acid}\\ pK_a = 9.4}}{\overset{+}{N}H_4}$$

$$\underset{\substack{\textbf{weaker acid}\\ pK_a = 15.9}}{CH_3CH_2OH} + \underset{\textbf{weaker base}}{CH_3NH_2} \rightleftharpoons \underset{\textbf{stronger base}}{CH_3CH_2O^-} + \underset{\substack{\textbf{stronger acid}\\ pK_a = 10.7}}{CH_3\overset{+}{N}H_3}$$

PROBLEM 30

a. For each of the acid–base reactions in Section 1.17, compare the pK_a values of the acids on either side of the equilibrium arrows and convince yourself that the position of equilibrium is in the direction indicated. (The pK_a values you need can be found in Section 1.17 or in Problem 31.)

b. Do the same thing for the equilibria in Section 1.16. (The pK_a of $^+NH_4$ is 9.4.)

The precise value of the equilibrium constant can be calculated by dividing the K_a of the reactant acid by the K_a of the product acid.

$$K_{eq} = \frac{K_a \text{ reactant acid}}{K_a \text{ product acid}}$$

Thus, the equilibrium constant for the reaction of acetic acid with ammonia is 4.0×10^4, and the equilibrium constant for the reaction of ethanol with methylamine is 6.3×10^{-6}. The calculations are as follows:

reaction of acetic acid with ammonia:

$$K_{eq} = \frac{10^{-4.8}}{10^{-9.4}} = 10^{4.6} = 4.0 \times 10^4$$

reaction of ethanol with methylamine:

$$K_{eq} = \frac{10^{-15.9}}{10^{-10.7}} = 10^{-5.2} = 6.3 \times 10^{-6}$$

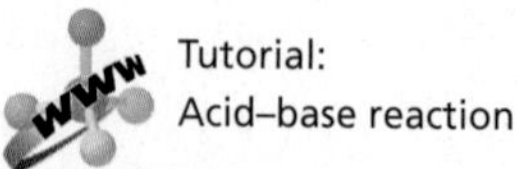

Tutorial:
Acid–base reaction

PROBLEM 31◆

a. Which is a stronger base, CH_3COO^- or $HCOO^-$? (The pK_a of CH_3COOH is 4.8; the pK_a of HCOOH is 3.8.)

b. Which is a stronger base, HO^- or $^-NH_2$? (The pK_a of H_2O is 15.7; the pK_a of NH_3 is 36.)

c. Which is a stronger base, H_2O or CH_3OH? (The pK_a of H_3O^+ is −1.7; the pK_a of $CH_3OH_2^+$ is −2.5.)

PROBLEM 32◆

Using the pK_a values in Section 1.17, rank the following species in order of decreasing base strength:

$$CH_3NH_2 \quad CH_3NH^- \quad CH_3OH \quad CH_3O^- \quad CH_3\overset{\overset{\displaystyle O}{\|}}{C}O^-$$

PROBLEM 33◆

Calculate the equilibrium constant for the acid–base reactions between the following pairs of reactants.

a. $HCl + H_2O$

b. $CH_3COOH + H_2O$

c. $CH_3NH_2 + H_2O$

d. $CH_3\overset{+}{N}H_3 + H_2O$

1.18 The Effect of Structure on pK_a

The strength of an acid is determined by the stability of the conjugate base that is formed when the acid gives up its proton. The more stable the base, the stronger is its conjugate acid. A stable base is a base that readily bears the electrons it formerly shared with a proton. In other words, stable bases are weak bases—they don't share their electrons well. So we can say, *the weaker the base, the stronger is its conjugate acid*, or *the more stable the base, the stronger is its conjugate acid.*

The weaker the base, the stronger is its conjugate acid.

Stable bases are weak bases.

The more stable the base, the stronger is its conjugate acid.

The elements in the second row of the periodic table are all about the same size, but they have very different electronegativities. The electronegativities increase across the row from left to right. Of the atoms shown, carbon is the least and fluorine is the most electronegative.

relative electronegativities: $C < N < O < F$ (F: most electronegative)

If we look at the bases formed when hydrogens are attached to these elements, we see that the stabilities of the bases also increase from left to right because the more electronegative atom is better able to bear its negative charge.

relative stabilities: $^-CH_3 < {}^-NH_2 < HO^- < F^-$ (F^-: most stable)

The stronger acid is the acid that forms the more stable conjugate base, so HF is the strongest acid and methane is the weakest acid (Table 1.9).

relative acidities: $CH_4 < NH_4 < H_2O < HF$ (HF: strongest acid)

Table 1.9 The pK_a Values of Some Simple Acids

CH_4	NH_3	H_2O	HF
pK_a = 50	pK_a = 36	pK_a = 15.7	pK_a = 3.2
		H_2S	HCl
		pK_a = 7.0	pK_a = −7
			HBr
			pK_a = −9
			HI
			pK_a = −10

When atoms are similar in size, the stronger acid will have its proton attached to the more electronegative atom.

We therefore can conclude that when the atoms are similar in size, the more acidic compound will have its hydrogen attached to the more electronegative atom.

The effect that the electronegativity of the atom bonded to a hydrogen has on the acidity of that hydrogen can be appreciated when the pK_a values of alcohols and amines are compared. Because oxygen is more electronegative than nitrogen, an alcohol is more acidic than an amine.

CH_3OH	CH_3NH_2
methanol	**methylamine**
pK_a = 15.5	**pK_a = 40**

Similarly, a protonated alcohol is more acidic than a protonated amine.

$CH_3\overset{+}{O}H_2$	$CH_3\overset{+}{N}H_3$
protonated methanol	**protonated methylamine**
pK_a = −2.5	**pK_a = 10.7**

In comparing atoms that are very different in size, the *size* of the atom is more important than its *electronegativity* in determining how well it bears its negative charge. For example, as we proceed down a column in the periodic table, the elements get larger and their electronegativity *decreases*, but the stability of the base increases, so the strength of the conjugate acid *increases*. Thus, HI is the strongest acid of the hydrogen halides, even though iodine is the least electronegative of the halogens.

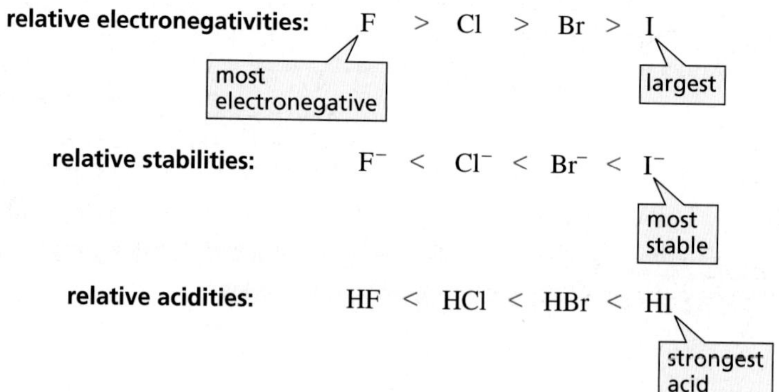

Why does the size of an atom have such a significant effect on the stability of the base and, therefore, on the acidity of a hydrogen attached to it? The valence electrons of F^- are in a $2sp^3$ orbital, the valence electrons of Cl^- are in a $3sp^3$ orbital, those of Br^- are in a $4sp^3$ orbital, and those of I^- are in a $5sp^3$ orbital. The volume of space occupied by a $3sp^3$ orbital is significantly greater than the volume of space occupied by a $2sp^3$ orbital because a $3sp^3$ orbital extends out farther from the nucleus. Because its negative charge is spread over a larger volume of space, Cl^- is more stable than F^-.

When atoms are very different in size, the stronger acid will have its proton attached to the largest atom.

Thus, as the halide ion increases in size, its stability increases because its negative charge is spread over a larger volume of space—its electron density decreases. Therefore, HI is the strongest acid of the hydrogen halides because I^- is the most stable halide ion, even though iodine is the least electronegative of the halogens (Table 1.9). The potential maps illustrate the large difference in size of the halide ions:

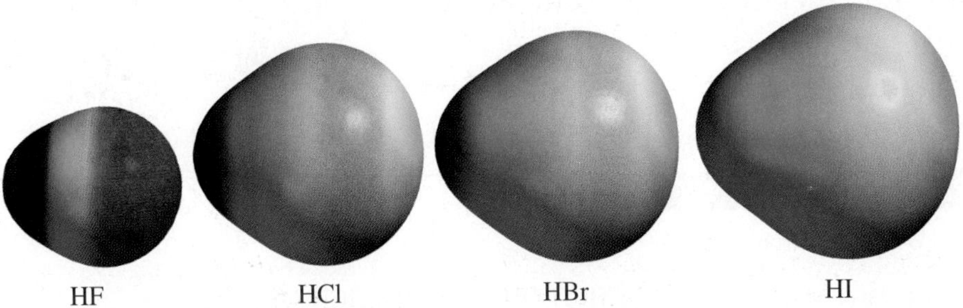

In summary, as we move across a row of the periodic table, the orbitals have approximately the same volume, so it is the electronegativity of the element that determines the stability of the base and, therefore, the acidity of a proton bonded to that base. As we move down a column of the periodic table, the volume of the orbitals increases. The increase in volume causes the electron density of the orbital to decrease. The electron density of the orbital is more important than electronegativity in determining the stability of the base and, therefore, the acidity of its conjugate acid. That is, *the lower the electron density, the more stable is the conjugate base and the stronger is its conjugate acid.*

Although the acidic proton of each of the following five carboxylic acids is attached to an oxygen atom, the five compounds have different acidities:

CH_3COOH	ICH_2COOH	$BrCH_2COOH$	$ClCH_2COOH$	FCH_2COOH
pK_a = 4.76	pK_a = 3.15	pK_a = 2.86	pK_a = 2.81	pK_a = 2.66

This difference indicates that there must be a factor—other than the nature of the atom to which the hydrogen is bonded—that affects acidity.

From the pK_a values of the five carboxylic acids, we see that replacing one of the hydrogen atoms of the CH_3 group with a halogen atom affects the acidity of the compound. (Chemists call this *substitution*, and the new atom is called a *substituent*.) All the halogens are more electronegative than hydrogen (Table 1.3). An electronegative halogen atom pulls the bonding electrons towards itself. Pulling electrons through sigma (σ) bonds is called **inductive electron withdrawal.** If we look at the conjugate base of a carboxylic acid, we see that inductive electron withdrawal will stabilize it by *decreasing the electron density* about the oxygen atom. Stabilizing a base increases the acidity of its conjugate acid.

$$Br \leftarrow CH_2 \leftarrow C(=O) \leftarrow O^-$$

inductive electron withdrawal

As the pK_a values of the five carboxylic acids show, inductive electron withdrawal increases the acidity of a compound. The greater the electron-withdrawing ability (electronegativity) of the halogen substituent, the more the acidity is increased because the more its conjugate base is stabilized.

The effect of a substituent on the acidity of a compound decreases as the distance between the substituent and the oxygen atom increases.

$CH_3CH_2CH_2CH(Br)COOH$ $pK_a = 2.97$

$CH_3CH_2CH(Br)CH_2COOH$ $pK_a = 4.01$

$CH_3CH(Br)CH_2CH_2COOH$ $pK_a = 4.59$

$CH_2(Br)CH_2CH_2CH_2COOH$ $pK_a = 4.71$

PROBLEM-SOLVING STRATEGY

a. Which is a stronger acid?

$CH_3CH(F)CH_2OH$ or $CH_3CH(Br)CH_2OH$

When you are asked to compare two items, pay attention to how they differ; ignore where they are the same. These two compounds differ only in the halogen atom that is attached to the middle carbon of the molecule. Because fluorine is more electronegative than bromine, there is greater electron withdrawal from the oxygen atom in the fluorinated compound. The fluorinated compound, therefore, will have the more stable conjugate base, so it will be the stronger acid.

b. Which is a stronger acid?

$CH_3CCl_2CH_2OH$ or $CH_2(Cl)CH(Cl)CH_2OH$

These two compounds differ in the location of one of the chlorine atoms. Because the chlorine in the compound on the left is closer to the O—H bond than is the chlorine in the compound on the right, the chlorine is more effective at withdrawing electrons from the oxygen atom. Thus, the compound on the left will have the more stable conjugate base, so it will be the stronger acid.

Now continue on to Problem 34.

PROBLEM 34◆

For each of the following compounds, indicate which is the stronger acid:

a. $CH_3OCH_2CH_2OH$ or $CH_3CH_2CH_2CH_2OH$

b. $CH_3CH_2CH_2\overset{+}{N}H_3$ or $CH_3CH_2CH_2\overset{+}{O}H_2$

c. $CH_3OCH_2CH_2CH_2OH$ or $CH_3CH_2OCH_2CH_2OH$

d. CH_3COCH_2OH or CH_3CH_2COOH

PROBLEM 35◆

List the following compounds in order of decreasing acidity:

$CH_3CH(F)CH_2OH$ $CH_3CH_2CH_2OH$ $CH_2(Cl)CH_2CH_2OH$ $CH_3CH(Cl)CH_2OH$

PROBLEM 36◆

For each of the following compounds, indicate which is the stronger base:

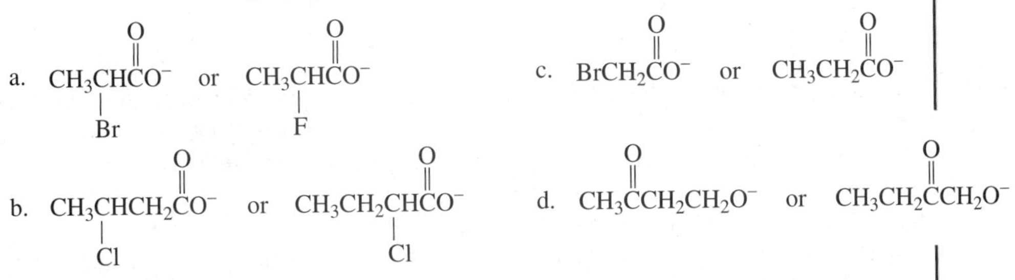

a. $CH_3\underset{\underset{Br}{|}}{C}H\overset{\overset{O}{\|}}{C}O^-$ or $CH_3\underset{\underset{F}{|}}{C}H\overset{\overset{O}{\|}}{C}O^-$

b. $CH_3\underset{\underset{Cl}{|}}{C}HCH_2\overset{\overset{O}{\|}}{C}O^-$ or $CH_3CH_2\underset{\underset{Cl}{|}}{C}H\overset{\overset{O}{\|}}{C}O^-$

c. $BrCH_2\overset{\overset{O}{\|}}{C}O^-$ or $CH_3CH_2\overset{\overset{O}{\|}}{C}O^-$

d. $CH_3\overset{\overset{O}{\|}}{C}CH_2CH_2O^-$ or $CH_3CH_2\overset{\overset{O}{\|}}{C}CH_2O^-$

PROBLEM 37 SOLVED

HCl is a weaker acid than HBr. Why, then, is $ClCH_2COOH$ a stronger acid than $BrCH_2COOH$?

SOLUTION To compare the acidities of HCl and HBr, we need to compare the stabilities of Cl^- and Br^-. Because we know that size is more important than electronegativity in determining stability, we know that Br^- is more stable than Cl^-. Therefore, HBr is a stronger acid than HCl. In comparing the acidities of the two carboxylic acids, we need to compare the stabilities of $RCOO^-$ and $R'COO^-$. (An O—H bond is broken in both compounds.) Therefore, the only factor to be considered is the electronegativities of the atoms that are pulling electrons away from the oxygen atom in the conjugate bases. Because Cl is more electronegative than Br, Cl is better at inductive electron withdrawal. Thus, it is better at stabilizing the base that is formed when the proton leaves.

PROBLEM 38◆

a. Which of the halide ions (F^-, Cl^-, Br^-, I^-) is the strongest base?
b. Which is the weakest base?

PROBLEM 39◆

a. Which is more electronegative, oxygen or sulfur?
b. Which is a stronger acid, H_2O or H_2S?
c. Which is a stronger acid, CH_3OH or CH_3SH?

PROBLEM 40◆

Using the table of pK_a values given in Appendix II, answer the following:

a. Which is the most acidic organic compound in the table?
b. Which is the least acidic organic compound in the table?
c. Which is the most acidic carboxylic acid in the table?
d. Which is more electronegative, an sp^3 hybridized oxygen or an sp^2 hybridized oxygen? (*Hint:* Pick a compound in Appendix II with a hydrogen attached to an sp^2 oxygen and one with a hydrogen attached to an sp^3 oxygen, and compare their pK_a values.)
e. What are the relative electronegativities of sp^3, sp^2, and sp hybridized nitrogen atoms?
f. What are the relative electronegativities of sp^3, sp^2, and sp hybridized carbon atoms?
g. Which is more acidic, HNO_3 or HNO_2? Why?

1.19 An Introduction to Delocalized Electrons and Resonance

We have seen that a carboxylic acid has a pK_a of about 5, whereas the pK_a of an alcohol is about 15. Because a carboxylic acid is a much stronger acid than an alcohol, we know that a carboxylic acid has a considerably more stable conjugate base.

CH_3COOH (CH_3–C(=O)–O–H)
$pK_a = 4.76$

CH_3CH_2O–H
$pK_a = 15.9$

There are two factors that cause the conjugate base of a carboxylic acid to be more stable than the conjugate base of an alcohol. First, a carboxylate ion has a doubly bonded oxygen in place of two hydrogens of the alkoxide ion. Inductive electron withdrawal by this electronegative oxygen decreases the electron density of the ion. Second, the electron density is further decreased by *electron delocalization.*

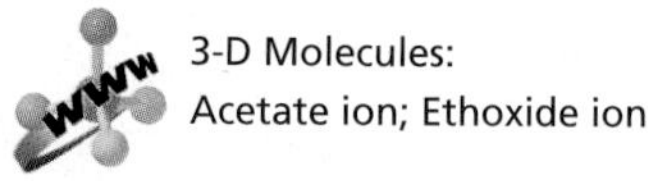

3-D Molecules:
Acetate ion; Ethoxide ion

When an alcohol loses a proton, the negative charge resides on its single oxygen atom—the electrons are *localized.* In contrast, when a carboxylic acid loses a proton, the negative charge is shared by both oxygen atoms because the electrons are *delocalized.* **Delocalized electrons** do not belong to a single atom, nor are they confined to a bond between two atoms. Delocalized electrons are shared by more than two atoms. The two structures shown for the conjugate base of acetic acid are called **resonance contributors.** Neither resonance contributor represents the actual structure of the conjugate base. The actual structure—called a **resonance hybrid**—is a composite of the two resonance contributors. The double-headed arrow between the two resonance contributors is used to indicate that the actual structure is a hybrid. Notice that the two resonance contributors differ only in the location of their π electrons and lone-pair electrons—all the atoms stay in the same place. In the resonance hybrid, the negative charge is shared equally by the two oxygen atoms, and both carbon–oxygen bonds are the same length—they are not as long as a single bond, but they are longer than a double bond. A resonance hybrid can be drawn by using dotted lines to show the delocalized electrons.

Delocalized electrons are shared by more than two atoms.

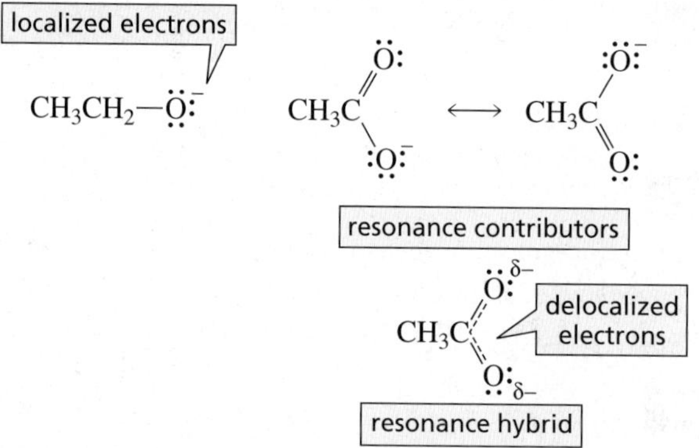

3-D Molecule:
Acetic acid

The following potential maps show that there is less electron density on the oxygen atoms in the carboxylate ion (orange region) than on the oxygen atom of the alkoxide ion (red region):

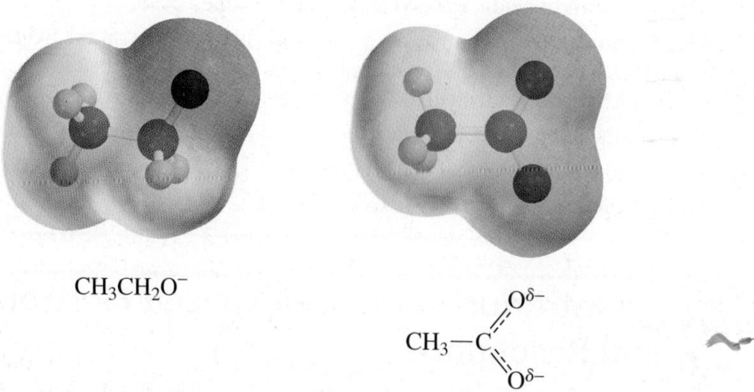

Thus, the combination of inductive electron withdrawal and the ability of two atoms to share the negative charge decrease the electron density, making the conjugate base of the carboxylic acid more stable than the conjugate base of the alcohol.

We will discuss delocalized electrons in greater detail in Chapter 7. By that time, you will be thoroughly comfortable with compounds that have only localized electrons, and you can then further explore how delocalized electrons affect the stability and reactivity of organic compounds.

PROBLEM 41

Which compound would you expect to be a stronger acid? Why?

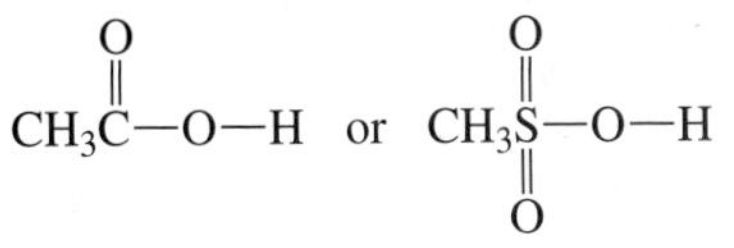

$$CH_3\overset{\overset{\large O}{\|}}{C}-O-H \quad \text{or} \quad CH_3\underset{\underset{\large O}{\|}}{\overset{\overset{\large O}{\|}}{S}}-O-H$$

PROBLEM 42◆

Draw resonance contributors for the following compounds:

a. 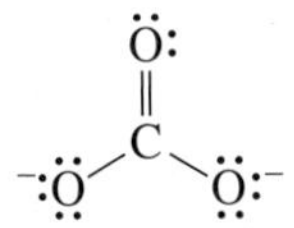 CO_3^{2-}: $\ddot{O}{:}$ double-bonded to C, with C bonded to $^-{:}\ddot{O}{:}$ and $\ddot{O}{:}^-$

b. 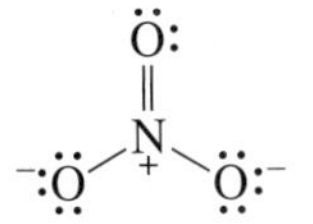 NO_3^-: $\ddot{O}{:}$ double-bonded to N^+, with N^+ bonded to $^-{:}\ddot{O}{:}$ and $\ddot{O}{:}^-$

1.20 The Effect of pH on the Structure of an Organic Compound

Whether a given acid will lose a proton in an aqueous solution depends on the pK_a of the acid and on the pH of the solution. The relationship between the two is given by the **Henderson–Hasselbalch equation**. This is an extremely useful equation because it tells us whether a compound will exist in its acidic form (with its proton retained) or in its basic form (with its proton removed) at a particular pH.

the Henderson–Hasselbalch equation

$$pK_a = pH + \log\frac{[HA]}{[A^-]}$$

The Henderson–Hasselbalch equation tells us that when the pH of a solution equals the pK_a of the compound that undergoes dissociation, the concentration of the compound in its acidic form [HA] will equal the concentration of the compound in its basic form $[A^-]$ (because log 1 = 0). If the pH of the solution is less than the pK_a of the compound, the compound will exist primarily in its acidic form. If the pH of the solution is greater than the pK_a of the compound, the compound will exist primarily in its basic form. In other words, *compounds exist primarily in their acidic forms in solutions that are more acidic than their* pK_a *values and primarily in their basic forms in solutions that are more basic than their* pK_a *values.*

A compound will exist primarily in its acidic form if the pH of the solution is less than its pK_a.

A compound will exist primarily in its basic form if the pH of the solution is greater than its pK_a.

If we know the pH of the solution and the pK_a of the compound, the Henderson–Hasselbalch equation allows us to calculate precisely how much of the compound will be in its acidic form, and how much will be in its basic form. For example, when a compound with a pK_a of 5.2 is in a solution of pH 5.2, half the compound will be in the acidic form and the other half will be in the basic form (Figure 1.20). If the pH is one unit less than the pK_a of the compound (pH = 4.2), there will be 10 times more compound present in the acidic form than in the basic form (because log 10 = 1). If the pH is two units less than the pK_a of the compound (pH = 3.2), there will be 100 times more compound present in the acidic form than in the basic form (because log 100 = 2). If the pH is 6.2, there will be 10 times more compound in the basic form than in the acidic form, and at pH = 7.2 there will be 100 times more compound present in the basic form than in the acidic form.

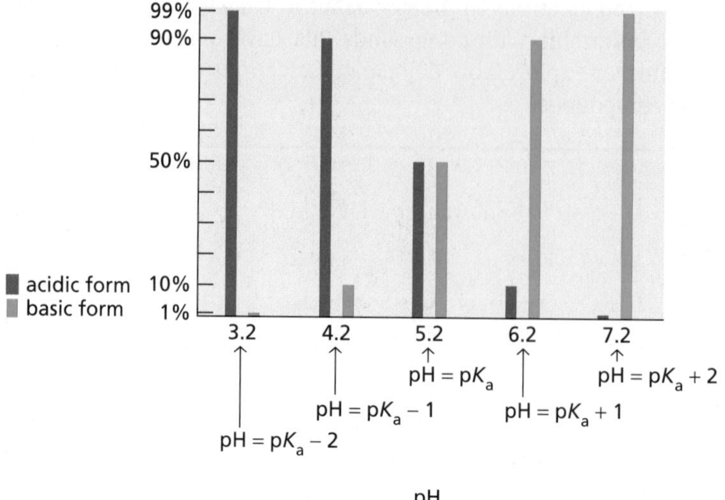

Figure 1.20 ▶ The relative amounts of a compound with a $\mathrm{p}K_\mathrm{a}$ of 5.2 in the acidic and basic forms at different pH values.

The Henderson–Hasselbalch equation can be very useful in the laboratory when compounds need to be separated from each other. Water and diethyl ether are not miscible liquids and, therefore, will form two layers when combined. The ether layer will lie above the more dense water layer. Charged compounds are more soluble in water, whereas neutral compounds are more soluble in diethyl ether. Two compounds, such as a carboxylic acid (RCOOH) with a $\mathrm{p}K_\mathrm{a}$ of 5.0 and a protonated amine (RNH_3^+) with a $\mathrm{p}K_\mathrm{a}$ of 10.0, dissolved in a mixture of water and diethyl ether, can be separated by adjusting the pH of the water layer. For example, if the pH of the water layer is 2, the carboxylic acid and the amine will both be in their acidic forms because the pH of the water is less than the $\mathrm{p}K_\mathrm{a}$'s of both compounds. The acidic form of a carboxylic acid is neutral, whereas the acidic form of an amine is charged. Therefore, the carboxylic acid will be more soluble in the ether layer, whereas the protonated amine will be more soluble in the water layer.

ether
water

acidic form		basic form
RCOOH	⇌	$RCOO^- + H^+$
$R\overset{+}{N}H_3$	⇌	$RNH_2 + H^+$

For the most effective separation, it is best if the pH of the water layer is at least two units away from the $\mathrm{p}K_\mathrm{a}$ values of the compounds being separated. Then the relative amounts of the compounds in their acidic and basic forms will be at least 100:1 (Figure 1.20).

DERIVATION OF THE HENDERSON–HASSELBALCH EQUATION

The Henderson–Hasselbalch equation can be derived from the expression that defines the acid dissociation constant:

$$K_\mathrm{a} = \frac{[H_3O^+][A^-]}{[HA]}$$

Taking the logarithms of both sides of the equation and then, in the next step, multiplying both sides of the equation by −1, we obtain

$$\log K_\mathrm{a} = \log[H_3O^+] + \log\frac{[A^-]}{[HA]}$$

and

$$-\log K_\mathrm{a} = -\log[H_3O^+] - \log\frac{[A^-]}{[HA]}$$

Substituting and remembering that when a fraction is inverted, the sign of its log changes, we get

$$\mathrm{p}K_\mathrm{a} = \mathrm{pH} + \log\frac{[HA]}{[A^-]}$$

BLOOD: A BUFFERED SOLUTION

Blood is the fluid that transports oxygen to all the cells of the human body. The normal pH of human blood is 7.35 to 7.45. Death will result if this pH decreases to a value less than ~6.8 or increases to a value greater than ~8.0 for even a few seconds. Oxygen is carried to cells by a protein in the blood called hemoglobin. When hemoglobin binds O_2, hemoglobin loses a proton, which would make the blood more acidic if it did not contain a buffer to maintain its pH.

$$\text{HbH}^+ + O_2 \rightleftharpoons \text{HbO}_2 + \text{H}^+$$

A carbonic acid/bicarbonate (H_2CO_3/HCO_3^-) buffer is used to control the pH of blood. An important feature of this buffer is that carbonic acid decomposes to CO_2 and H_2O:

$$CO_2 + H_2O \rightleftharpoons \underset{\text{carbonic acid}}{H_2CO_3} \rightleftharpoons \underset{\text{bicarbonate}}{HCO_3^-} + H^+$$

Cells need a constant supply of O_2, with very high levels required during periods of strenuous exercise. When O_2 is consumed by cells, the hemoglobin equilibrium shifts to the left to release more O_2, so the concentration of H^+ decreases. At the same time, increased metabolism during exercise produces large amounts of CO_2. This shifts the carbonic acid/bicarbonate equilibrium to the right, which increases the concentration of H^+. Significant amounts of lactic acid are also produced during exercise, and this further increases the concentration of H^+.

Receptors in the brain respond to the increased concentration of H^+ and trigger a reflex that increases the rate of breathing. This increases the release of oxygen to the cells and the elimination of CO_2 by exhalation. Both processes decrease the concentration of H^+ in the blood.

The Henderson–Hasselbalch equation is also useful when one is working with buffer solutions. A **buffer solution** is a solution that maintains a nearly constant pH when small amounts of acid or base are added to it. Buffer solutions are discussed in detail in the Study Guide and Solutions Manual (see Special Topic I).

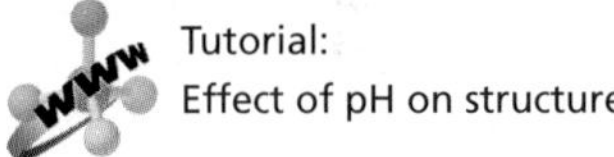

Tutorial:
Effect of pH on structure

PROBLEM 43◆

As long as the pH is greater than __________, more than 50% of a protonated amine with a pK_a of 10.4 will be in its neutral, nonprotonated form.

PROBLEM 44◆ SOLVED

a. At what pH will 99% of a compound with a pK_a of 8.4 be in its basic form?
b. At what pH will 91% of a compound with a pK_a of 3.7 be in its acidic form?
c. At what pH will 9% of a compound with a pK_a of 5.9 be in its basic form?
d. At what pH will 50% of a compound with a pK_a of 7.3 be in its basic form?
e. At what pH will 1% of a compound with a pK_a of 7.3 be in its acidic form?

SOLUTION TO 44a If 99% is in the basic form and 1% is in the acidic form, the Henderson–Hasselbalch equation becomes

$$pK_a = pH + \log\frac{1}{99}$$
$$8.4 = pH + \log .01$$
$$8.4 = pH - 2.0$$
$$pH = 10.4$$

There is a faster way to get the answer: If about 100 times more compound is present in the basic form than in the acidic form, the pH will be two units more basic than the pK_a. Thus, pH = 8.4 + 2.0 = 10.4.

SOLUTION TO 44b If 91% is in the acidic form and 9% is in the basic form, there is about 10 times more compound present in the acidic form. Therefore, the pH is one unit more acidic than the pK_a. Thus, pH = 3.7 − 1.0 = 2.7.

PROBLEM 45◆

a. Indicate whether a carboxylic acid (RCOOH) with a pK_a of 5 will be mostly charged or mostly neutral in solutions with the following pH values:

1. pH = 1
2. pH = 3
3. pH = 5
4. pH = 7
5. pH = 9
6. pH = 11
7. pH = 13

b. Answer the same question for a protonated amine ($R\overset{+}{N}H_3$) with a pK_a of 9.

c. Answer the same question for an alcohol (ROH) with a pK_a of 15.

PROBLEM 46◆

For each compound in 1 and 2, indicate the pH at which

a. 50% of the compound will be in a form that possesses a charge.

b. more than 99% of the compound will be in a form that possesses a charge.

1. CH_3CH_2COOH (pK_a = 4.9)
2. $CH_3\overset{+}{N}H_3$ (pK_a = 10.7)

PROBLEM 47

For each of the following compounds, shown in their acidic forms, draw the form in which it will predominate in a solution of pH = 7:

a. CH_3COOH (pK_a = 4.76)

b. $CH_3CH_2\overset{+}{N}H_3$ (pK_a = 11.0)

c. H_3O^+ (pK_a = −1.7)

d. CH_3CH_2OH (pK_a = 15.9)

e. $CH_3CH_2\overset{+}{O}H_2$ (pK_a = −2.5)

f. $\overset{+}{N}H_4$ (pK_a = 9.4)

g. HC≡N (pK_a = 9.1)

h. HNO_2 (pK_a = 3.4)

i. HNO_3 (pK_a = −1.3)

j. HBr (pK_a = −9)

1.21 Lewis Acids and Bases

Lewis base: Have pair, will share.
Lewis acid: Need two from you.

In 1923, G. N. Lewis offered new definitions for the terms "acid" and "base." He defined an acid as a species that accepts a share in an electron pair and a base as a species that donates a share in an electron pair. All proton-donating acids fit the Lewis definition because all proton-donating acids lose a proton and the proton accepts a share in an electron pair.

$$H^+ + :NH_3 \rightleftharpoons H{-}\overset{+}{N}H_3$$

acid — accepts a share in a pair of electrons

base — donates a share in a pair of electrons

The Lewis definition of an acid is much broader than the Brønsted–Lowry definition because it is not limited to compounds that donate protons. According to the Lewis definition, compounds such as aluminum chloride ($AlCl_3$), boron trifluoride (BF_3), and borane (BH_3) are acids because they have unfilled valence orbitals and thus can accept a share in an electron pair. These compounds react with a compound that has a lone pair just as a proton reacts with ammonia, but they are not proton-donating acids. Thus, the Lewis definition of an acid includes all proton-donating acids and some additional acids that do not have protons. Throughout this text, the term "acid" is used to mean a proton-donating acid, and the term **"Lewis acid"** is used to refer to non-proton-donating acids such as $AlCl_3$ or BF_3. All bases are **Lewis bases** because they have a pair of electrons that they can share, either with an atom such as aluminum or boron or with a proton.

the curved arrow indicates where the pair of electrons starts from and where it ends up

$$\mathrm{AlCl_3} + \mathrm{CH_3\ddot{O}CH_3} \rightleftharpoons \mathrm{Cl_3\overset{-}{Al}{-}\overset{+}{O}(CH_3){-}CH_3}$$

aluminum trichloride — a Lewis acid; dimethyl ether — a Lewis base

$$\mathrm{BH_3} + \mathrm{:NH_3} \rightleftharpoons \mathrm{H_3\overset{-}{B}{-}\overset{+}{N}H_3}$$

borane — a Lewis acid; ammonia — a Lewis base

PROBLEM 48

What is the product of each of the following reactions?

a. $ZnCl_2 + CH_3OH \rightleftharpoons$

b. $FeBr_3 + Br^- \rightleftharpoons$

c. $AlCl_3 + Cl^- \rightleftharpoons$

d. $BF_3 + HCH(=O) \rightleftharpoons$

PROBLEM 49

Show how each of the following compounds reacts with HO^-:

a. CH_3OH b. $^+NH_4$ c. $CH_3\overset{+}{N}H_3$ d. BF_3 e. $^+CH_3$ f. $FeBr_3$ g. $AlCl_3$ h. CH_3COOH

Summary

Organic compounds are compounds that contain carbon. The **atomic number** of an atom equals the number of protons in its nucleus. The **mass number** of an atom is the sum of its protons and neutrons. **Isotopes** have the same atomic number, but different mass numbers.

An **atomic orbital** indicates where there is a high probability of finding an electron. The closer the atomic orbital is to the nucleus, the lower is its energy. **Degenerate orbitals** have the same energy. Electrons are assigned to orbitals following the **aufbau principle**, the **Pauli exclusion principle**, and **Hund's rule**.

The **octet rule** states that an atom will give up, accept, or share electrons in order to fill its outer shell or attain an outer shell with eight electrons. **Electropositive** elements readily lose electrons; **electronegative** elements readily acquire electrons. The **electronic configuration** of an atom describes the orbitals occupied by the atom's electrons. Electrons in inner shells are called **core electrons**; electrons in the outermost shell are called **valence electrons**. **Lone-pair electrons** are valence electrons that are not used in bonding. Attractive forces between opposite charges are called **electrostatic attractions**. An **ionic bond** is formed by a transfer of electrons; a **covalent bond** is formed by sharing electrons. A polar covalent bond has a **dipole**, measured by a **dipole moment**. The **dipole moment** of a molecule depends on the magnitudes and directions of the bond dipole moments.

Lewis structures indicate which atoms are bonded together and show **lone pairs** and **formal charges**. A **carbocation** has a positively charged carbon, a **carbanion** has a negatively charged carbon, and a **radical** has an unpaired electron.

According to **molecular orbital (MO) theory**, covalent bonds result when atomic orbitals combine to form **molecular orbitals**. Atomic orbitals combine to give a **bonding MO** and a higher energy **antibonding MO**. Cylindrically symmetrical bonds are called **sigma (σ) bonds**; **pi (π) bonds** form when p orbitals overlap side-to-side. Bond strength is measured by the **bond dissociation energy**. A σ bond is stronger than a π bond. All **single**

bonds in organic compounds are σ bonds, a **double bond** consists of one σ bond and one π bond, and a **triple bond** consists of one σ bond and two π bonds. Triple bonds are shorter and stronger than double bonds, which are shorter and stronger than single bonds. To form four bonds, carbon promotes an electron from a 2*s* to a 2*p* orbital. C, N, and O form bonds using **hybrid orbitals**. The **hybridization** of C, N, or O depends on the number of π bonds the atom forms: No π bonds means that the atom is sp^3 **hybridized**, one π bond indicates that it is sp^2 **hybridized**, and two π bonds signifies that it is ***sp* hybridized**. Exceptions are carbocations and carbon radicals, which are sp^2 hybridized. The more *s* character in the orbital used to form a bond, the shorter and stronger the bond is and the larger the bond angle is. Bonding and lone-pair electrons around an atom are positioned as far apart as possible.

An **acid** is a species that donates a proton, and a **base** is a species that accepts a proton. A **Lewis acid** is a species that accepts a share in an electron pair; a **Lewis base** is a species that donates a share in an electron pair.

Acidity is a measure of the tendency of a compound to give up a proton. **Basicity** is a measure of a compound's affinity for a proton. The stronger the acid, the weaker is its conjugate base. The strength of an acid is given by the **acid dissociation constant (K_a)**. Approximate pK_a values are as follows: protonated alcohols, protonated carboxylic acids, protonated water < 0; carboxylic acids ~ 5; protonated amines ~ 10; alcohols and water ~ 15. The **pH** of a solution indicates the concentration of positively charged hydrogen ions in the solution. In **acid–base reactions**, the equilibrium favors reaction of the strong and formation of the weak.

The strength of an acid is determined by the stability of its conjugate base: The more stable the base, the stronger is its conjugate acid. When atoms are similar in size, the more acidic compound has its hydrogen attached to the more electronegative atom. When atoms are very different in size, the more acidic compound has its hydrogen attached to the larger atom. **Inductive electron withdrawal** increases acidity; acidity decreases with increasing distance between the electron-withdrawing substituent and the ionizing group.

Delocalized electrons are electrons shared by more than two atoms. A compound with delocalized electrons has **resonance**. The **resonance hybrid** is a composite of the **resonance contributors**, which differ only in the location of their lone-pair and π electrons.

The **Henderson–Hasselbalch equation** gives the relationship between pK_a and pH: A compound exists primarily in its acidic form in solutions more acidic than its pK_a value and primarily in its basic form in solutions more basic than its pK_a value.

Key Terms

acid (p. 39)
acid–base reaction (p. 40)
acid dissociation constant (K_a) (p. 41)
acidity (p. 40)
antibonding molecular orbital (p. 21)
atomic number (p. 4)
atomic orbital (p. 5)
atomic weight (p. 4)
aufbau principle (p. 6)
base (p. 39)
basicity (p. 40)
bond (p. 8)
bond dissociation energy (p. 21)
bond length (p. 21)
bonding molecular orbital (p. 21)
bond strength (p. 21)
buffer solution (p. 53)
carbanion (p. 14)
carbocation (p. 14)
condensed structure (p. 16)
conjugate acid (p. 40)
conjugate base (p. 40)
constitutional isomer (p. 18)
core electrons (p. 7)
covalent bond (p. 9)
degenerate orbitals (p. 5)
delocalized electrons (p. 50)
debye (D) (p. 12)
dipole (p. 12)
dipole moment (μ) (p. 12)
double bond (p. 30)
electronegative (p. 8)
electronegativity (p. 10)
electropositive (p. 7)
electrostatic attraction (p. 8)
electrostatic potential map (p. 11)
equilibrium constant (p. 41)
excited-state electronic configuration (p. 6)
formal charge (p. 13)
free radical (p. 14)
ground-state electronic configuration (p. 6)
Heisenberg uncertainty principle (p. 18)
Henderson–Hasselbalch equation (p. 51)
Hund's rule (p. 7)
hybrid orbital (p. 26)
hydride ion (p. 9)
hydrogen ion (p. 9)
inductive electron withdrawal (p. 47)
ionic bond (p. 8)
ionic compound (p. 9)
ionization energy (p. 7)
isotopes (p. 4)
Kekulé structure (p. 16)
Lewis acid (p. 54)
Lewis base (p. 54)
Lewis structure (p. 13)
lone-pair electrons (p. 13)
mass number (p. 4)
molecular weight (p. 4)
molecular orbital (p. 20)
molecular orbital (MO) theory (p. 20)
node (p. 18)
nodal plane (p. 19)
nonbonding electrons (p. 13)
nonpolar covalent bond (p. 10)
nonpolar molecule (p. 26)
octet rule (p. 7)
orbital (p. 4)
orbital hybridization (p. 26)
organic compound (p. 2)
Pauli exclusion principle (p. 6)
pH (p. 41)
pi (π) bond (p. 23)
pK_a (p. 41)
polar covalent bond (p. 10)
proton (p. 9)
proton-transfer reaction (p. 40)
quantum mechanics (p. 4)
radial node (p. 18)
radical (p. 18)
resonance (p. 51)
resonance contributors (p. 50)
resonance hybrid (p. 50)
sigma (σ) bond (p. 20)

sigma (σ) bonding molecular orbital (p. 21)
single bond (p. 29)
tetrahedral bond angle (p. 28)
tetrahedral carbon (p. 28)
trigonal planar carbon (p. 30)
triple bond (p. 32)
valence electrons (p. 7)
valence-shell electron-pair repulsion (VSEPR) model (p. 24)
wave equation (p. 4)
wave functions (p. 4)

Problems

50. Draw a Lewis structure for each of the following species:
    a. H_2CO_3
    b. CO_3^{2-}
    c. H_2CO
    d. N_2H_4
    e. CH_3NH_2
    f. $CH_3N_2^+$
    g. CO_2
    h. NO^+
    i. H_2NO^-

51. Give the hybridization of the central atom of each of the following species, and tell whether the bond arrangement around it is linear, trigonal planar, or tetrahedral:
    a. NH_3
    b. BH_3
    c. $^-CH_3$
    d. $\cdot CH_3$
    e. $^+NH_4$
    f. $^+CH_3$
    g. HCN
    h. $C(CH_3)_4$
    i. H_3O^+

52. Draw the condensed structure of a compound that contains only carbon and hydrogen atoms and that has
    a. three sp^3 hybridized carbons.
    b. one sp^3 hybridized carbon and two sp^2 hybridized carbons.
    c. two sp^3 hybridized carbons and two sp hybridized carbons.

53. Predict the indicated bond angles:
    a. the C—N—H bond angle in $(CH_3)_2NH$
    b. the C—N—C bond angle in $(CH_3)_2NH$
    c. the C—N—C bond angle in $(CH_3)_2\overset{+}{N}H_2$
    d. the C—O—C bond angle in CH_3OCH_3
    e. the C—O—H bond angle in CH_3OH
    f. the H—C—H bond angle in $H_2C{=}O$
    g. the F—B—F bond angle in $^-BF_4$
    h. the C—C—N bond angle in $CH_3C{\equiv}N$
    i. the C—C—N bond angle in $CH_3CH_2NH_2$

54. Give each atom the appropriate formal charge:
    a. $H{:}\ddot{\underset{..}{O}}{:}$
    b. $H{:}\ddot{\underset{..}{O}}\cdot$
    c. $CH_3—N(CH_3)_2—CH_3$ (N bonded to four CH_3 groups)
    d. $H—\ddot{N}—H$
    e. $H—\ddot{C}—H$
    f. $H—N(H)_2—B(H)_2—H$ (N bonded to three H and B; B bonded to three H and N)
    g. $H—\ddot{C}(H)—H$
    h. $CH_3—\ddot{O}(H)—CH_3$

55. Draw the ground-state electronic configuration for:
    a. Ca
    b. Ca^{2+}
    c. Ar
    d. Mg^{2+}

56. Write the Kekulé structure for each of the following compounds:
    a. CH_3CHO
    b. CH_3OCH_3
    c. CH_3COOH
    d. $(CH_3)_3COH$
    e. $CH_3CH(OH)CH_2CN$
    f. $(CH_3)_2CHCH(CH_3)CH_2C(CH_3)_3$

57. Show the direction of the dipole moment in each of the following bonds (use the electronegativities given in Table 1.3):
    a. $CH_3—Br$
    b. $CH_3—Li$
    c. $HO—NH_2$
    d. I—Br
    e. $CH_3—OH$
    f. $(CH_3)_2N—H$

58. What is the hybridization of the indicated atom in each of the following molecules?
    a. $CH_3CH{=}CH_2$
    b. $CH_3\overset{O}{\overset{\|}{C}}CH_3$
    c. CH_3CH_2OH
    d. $CH_3C{\equiv}N$
    e. $CH_3CH{=}NCH_3$
    f. $CH_3OCH_2CH_3$

59. a. Which of the indicated bonds in each molecule is shorter?
 b. Indicate the hybridization of the C, O, and N atoms in each of the molecules.

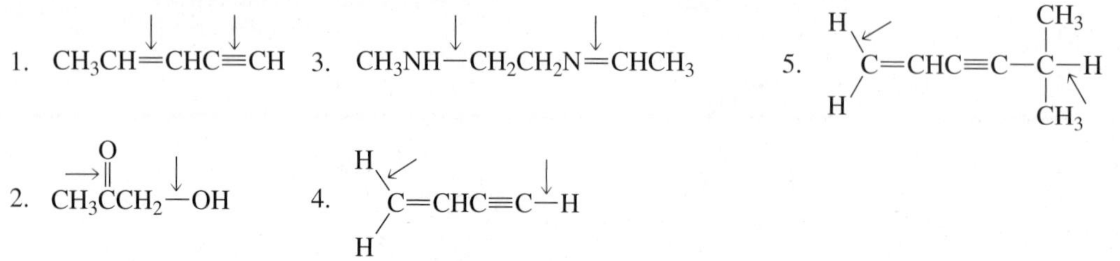

1. $CH_3CH{=}CHC{\equiv}CH$
2. $CH_3\overset{O}{\overset{\|}{C}}CH_2{-}OH$
3. $CH_3NH{-}CH_2CH_2N{=}CHCH_3$
4. $H_2C{=}CHC{\equiv}C{-}H$
5. $H_2C{=}CHC{\equiv}C{-}C(CH_3)_2{-}H$

60. For each of the following compounds, draw the form in which it will predominate at pH = 3, pH = 6, pH = 10, and pH = 14:

a. CH_3COOH pK_a = 4.8
b. $CH_3CH_2\overset{+}{N}H_3$ pK_a = 11.0
c. CF_3CH_2OH pK_a = 12.4

61. Which of the following molecules have tetrahedral bond angles?

H_2O H_3O^+ $^+CH_3$ BF_3 NH_3 $^+NH_4$ $^-CH_3$

62. Do the sp^2 hybridized carbons and the indicated atoms lie in the same plane?

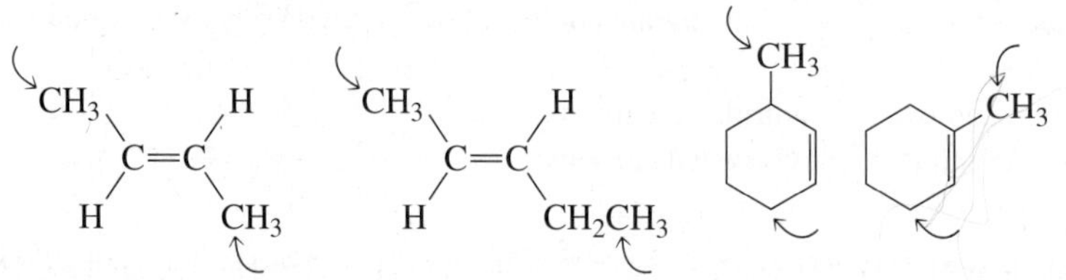

$CH_3(H)C{=}C(H)CH_3$ $CH_3(H)C{=}C(H)CH_2CH_3$ (methylcyclohexene structures with CH_3)

63. Give the products of the following acid–base reactions, and indicate whether reactants or products are favored at equilibrium (use the pK_a values that are given in Section 1.17):

a. $CH_3\overset{O}{\overset{\|}{C}}OH + CH_3O^- \rightleftharpoons$
b. $CH_3CH_2OH + {}^-NH_2 \rightleftharpoons$
c. $CH_3\overset{O}{\overset{\|}{C}}OH + CH_3NH_2 \rightleftharpoons$
d. $CH_3CH_2OH + HCl \rightleftharpoons$

64. For each of the following molecules, indicate the hybridization of each carbon atom and give the approximate values of all the bond angles:

a. $CH_3C{\equiv}CH$ b. $CH_3CH{=}CH_2$ c. $CH_3CH_2CH_3$

65. a. Estimate the pK_a value of each of the following acids without using a calculator (i.e., between 3 and 4, between 9 and 10, etc.):
 1. nitrous acid (HNO_2), $K_a = 4.0 \times 10^{-4}$
 2. nitric acid (HNO_3), $K_a = 22$
 3. bicarbonate (HCO_3^-), $K_a = 6.3 \times 10^{-11}$
 4. hydrogen cyanide (HCN), $K_a = 7.9 \times 10^{-10}$
 5. formic acid (HCOOH), $K_a = 2.0 \times 10^{-4}$
 b. Determine the pK_a values, using a calculator.
 c. Which is the strongest acid?

66. a. List the following carboxylic acids in order of decreasing acidity:
 1. $CH_3CH_2CH_2COOH$ $K_a = 1.52 \times 10^{-5}$
 2. $CH_3CH_2\underset{Cl}{CH}COOH$ $K_a = 1.39 \times 10^{-3}$
 3. $ClCH_2CH_2CH_2COOH$ $K_a = 2.96 \times 10^{-5}$
 4. $CH_3\underset{Cl}{CH}CH_2COOH$ $K_a = 8.9 \times 10^{-5}$
 b. How does the presence of an electronegative substituent such as Cl affect the acidity of a carboxylic acid?
 c. How does the location of the substituent affect the acidity of a carboxylic acid?

67. Draw a Lewis structure for each of the following species:

a. $CH_3N_2^+$ b. CH_2N_2 c. N_3^- d. N_2O (arranged NNO)

68. a. For each of the following pairs of reactions, indicate which one has the more favorable equilibrium constant (that is, which one most favors products):

1. $CH_3CH_2OH + NH_3 \rightleftharpoons CH_3CH_2O^- + \overset{+}{N}H_4$
   or
   $CH_3OH + NH_3 \rightleftharpoons CH_3O^- + \overset{+}{N}H_4$
2. $CH_3CH_2OH + NH_3 \rightleftharpoons CH_3CH_2O^- + \overset{+}{N}H_4$
   or
   $CH_3CH_2OH + CH_3NH_2 \rightleftharpoons CH_3CH_2O^- + CH_3\overset{+}{N}H_3$

b. Which of the four reactions has the most favorable equilibrium constant?

69. The following compound has two isomers:

$$ClCH{=}CHCl$$

One isomer has a dipole moment of 0 D, and the other has a dipole moment of 2.95 D. Propose structures for the two isomers that are consistent with these data.

70. Knowing that pH + pOH = 14 and that the concentration of water in a solution of water is 55.5 M, show that the pK_a of water is 15.7. (*Hint:* pOH = $-\log[HO^-]$.)

71. Water and diethyl ether are immiscible liquids. Charged compounds dissolve in water, and uncharged compounds dissolve in ether. $C_6H_{11}COOH$ has a pK_a of 4.8 and $C_6H_{11}\overset{+}{N}H_3$ has a pK_a of 10.7.
   a. What pH would you make the water layer in order to cause both compounds to dissolve in it?
   b. What pH would you make the water layer in order to cause the acid to dissolve in the water layer and the amine to dissolve in the ether layer?
   c. What pH would you make the water layer in order to cause the acid to dissolve in the ether layer and the amine to dissolve in the water layer?

72. How could you separate a mixture of the following compounds? The reagents available to you are water, ether, 1.0 M HCl, and 1.0 M NaOH. (*Hint:* See problem 71.)

COOH (benzene ring)	$\overset{+}{N}H_3Cl^-$ (benzene ring)	OH (benzene ring)	Cl (benzene ring)	$\overset{+}{N}H_3Cl^-$ (cyclohexane ring)
$pK_a = 4.17$	$pK_a = 4.60$	$pK_a = 9.95$		$pK_a = 10.66$

73. Using molecular orbital theory, explain why shining light on Br_2 causes it to break apart into atoms, but shining light on H_2 does not break the molecule apart.

74. Show that $K_{eq} = \dfrac{K_a \text{ reactant acid}}{K_a \text{ product acid}} = \dfrac{[\text{products}]}{[\text{reactants}]}$

75. Carbonic acid has a pK_a of 6.1 at physiological temperature. Is the carbonic acid/bicarbonate buffer system that maintains the pH of the blood at 7.3 better at neutralizing excess acid or excess base?

76. a. If an acid with a pK_a of 5.3 is in an aqueous solution of pH 5.7, what percentage of the acid is present in the acidic form?
   b. At what pH will 80% of the acid exist in the acidic form?

77. Calculate the pH values of the following solutions:
   a. a 1.0 M solution of acetic acid (pK_a = 4.76)
   b. a 0.1 M solution of protonated methylamine (pK_a = 10.7)
   c. a solution containing 0.3 M HCOOH and 0.1 M $HCOO^-$ (pK_a of HCOOH = 3.76)

For help in answering Problems 75–77, see Special Topic I in the Study Guide and Solutions Manual.

2 An Introduction to Organic Compounds

Nomenclature, Physical Properties, and Representation of Structure

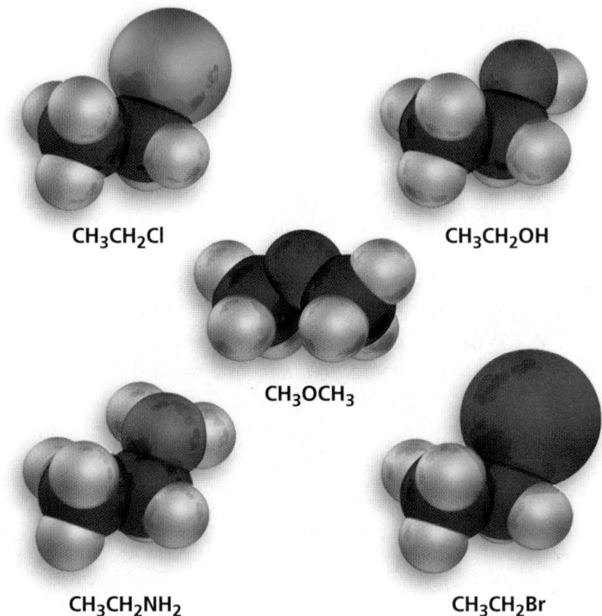

This book organizes organic chemistry according to how organic compounds react. In studying how compounds react, we must not forget that whenever a compound undergoes a reaction, a new compound is synthesized. In other words, while we are learning how organic compounds react, we will simultaneously be learning how to synthesize new organic compounds.

$$\underset{\text{Y is reacting}}{\text{Y}} \longrightarrow \underset{\text{Z is being synthesized}}{\text{Z}}$$

The main classes of compounds that are synthesized by the reactions you will study in Chapters 3–11 are alkanes, alkyl halides, ethers, alcohols, and amines. As you learn how to synthesize compounds, you will need to be able to refer to them by name, so you will begin your study of organic chemistry by learning how to name these five classes of compounds.

First you will learn how to name alkanes because they form the basis for the names of almost all organic compounds. **Alkanes** are composed of only carbon atoms and hydrogen atoms and contain only single bonds. Compounds that contain only carbon and hydrogen are called **hydrocarbons**, so an alkane is a hydrocarbon that has only single bonds. Alkanes in which the carbons form a continuous chain with no branches are called **straight-chain alkanes**. The names of several straight-chain alkanes are given in Table 2.1. It is important that you learn the names of at least the first 10.

The family of alkanes shown in the table is an example of a homologous series. A **homologous series** (*homos* is Greek for "the same as") is a family of compounds in which each member differs from the next by one **methylene (CH_2) group**. The members of a homologous series are called **homologs**. Propane ($CH_3CH_2CH_3$) and butane ($CH_3CH_2CH_2CH_3$) are homologs.

If you look at the relative numbers of carbon and hydrogen atoms in the alkanes listed in Table 2.1, you will see that the general molecular formula for an alkane is

C_nH_{2n+2}, where n is any integer. So, if an alkane has one carbon atom, it must have four hydrogen atoms; if it has two carbon atoms, it must have six hydrogens.

We have seen that carbon forms four covalent bonds and hydrogen forms only one covalent bond (Section 1.4). This means that there is only one possible structure for an alkane with molecular formula CH_4 (methane) and only one structure for an alkane with molecular formula C_2H_6 (ethane). We examined the structures of these compounds in Section 1.7. There is also only one possible structure for an alkane with molecular formula C_3H_8 (propane).

name	Kekulé structure	condensed structure	ball-and-stick model
methane	$\begin{array}{c} \text{H} \\ \text{H}-\text{C}-\text{H} \\ \text{H} \end{array}$	CH_4	
ethane	$\begin{array}{c} \text{H}\ \ \text{H} \\ \text{H}-\text{C}-\text{C}-\text{H} \\ \text{H}\ \ \text{H} \end{array}$	CH_3CH_3	
propane	$\begin{array}{c} \text{H}\ \ \text{H}\ \ \text{H} \\ \text{H}-\text{C}-\text{C}-\text{C}-\text{H} \\ \text{H}\ \ \text{H}\ \ \text{H} \end{array}$	$CH_3CH_2CH_3$	
butane	$\begin{array}{c} \text{H}\ \ \text{H}\ \ \text{H}\ \ \text{H} \\ \text{H}-\text{C}-\text{C}-\text{C}-\text{C}-\text{H} \\ \text{H}\ \ \text{H}\ \ \text{H}\ \ \text{H} \end{array}$	$CH_3CH_2CH_2CH_3$	

As the number of carbons in an alkane increases beyond three, the number of possible structures increases. There are two possible structures for an alkane with molecular formula C_4H_{10}. In addition to butane—a straight-chain alkane—there is a branched butane called isobutane. Both of these structures fulfill the requirement that each carbon forms four bonds and each hydrogen forms only one bond.

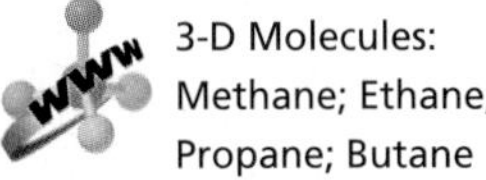

3-D Molecules: Methane; Ethane; Propane; Butane

Compounds such as butane and isobutane that have the same molecular formula but differ in the order in which the atoms are connected are called **constitutional isomers**—their molecules have different constitutions. In fact, isobutane got its name because it is an "iso"mer of butane. The structural unit—a carbon bonded to a hydrogen and two CH_3 groups—that occurs in isobutane has come to be called "iso." Thus, the name isobutane tells you that the compound is a four-carbon alkane with an iso structural unit.

$$CH_3CH_2CH_2CH_3$$
butane

$$\begin{array}{l} CH_3CHCH_3 \\ \quad\ \ \vert \\ \quad\ CH_3 \end{array}$$
isobutane

$$\begin{array}{l} CH_3CH- \\ \quad\ \ \vert \\ \quad\ CH_3 \end{array}$$
an "iso" structural unit

Table 2.1 Nomenclature and Physical Properties of Straight-Chain Alkanes

Number of carbons	Molecular formula	Name	Condensed structure	Boiling point (°C)	Melting point (°C)	Density[a] (g/mL)
1	CH_4	methane	CH_4	−167.7	−182.5	
2	C_2H_6	ethane	CH_3CH_3	−88.6	−183.3	
3	C_3H_8	propane	$CH_3CH_2CH_3$	−42.1	−187.7	
4	C_4H_{10}	butane	$CH_3CH_2CH_2CH_3$	−0.5	−138.3	
5	C_5H_{12}	pentane	$CH_3(CH_2)_3CH_3$	36.1	−129.8	0.5572
6	C_6H_{14}	hexane	$CH_3(CH_2)_4CH_3$	68.7	−95.3	0.6603
7	C_7H_{16}	heptane	$CH_3(CH_2)_5CH_3$	98.4	−90.6	0.6837
8	C_8H_{18}	octane	$CH_3(CH_2)_6CH_3$	127.7	−56.8	0.7026
9	C_9H_{20}	nonane	$CH_3(CH_2)_7CH_3$	150.8	−53.5	0.7177
10	$C_{10}H_{22}$	decane	$CH_3(CH_2)_8CH_3$	174.0	−29.7	0.7299
11	$C_{11}H_{24}$	undecane	$CH_3(CH_2)_9CH_3$	195.8	−25.6	0.7402
12	$C_{12}H_{26}$	dodecane	$CH_3(CH_2)_{10}CH_3$	216.3	−9.6	0.7487
13	$C_{13}H_{28}$	tridecane	$CH_3(CH_2)_{11}CH_3$	235.4	−5.5	0.7546
⋮	⋮	⋮	⋮	⋮	⋮	⋮
20	$C_{20}H_{42}$	eicosane	$CH_3(CH_2)_{18}CH_3$	343.0	36.8	0.7886
21	$C_{21}H_{44}$	heneicosane	$CH_3(CH_2)_{19}CH_3$	356.5	40.5	0.7917
⋮	⋮	⋮	⋮	⋮	⋮	⋮
30	$C_{30}H_{62}$	triacontane	$CH_3(CH_2)_{28}CH_3$	449.7	65.8	0.8097

[a] Density is temperature dependent. The densities given are those determined at 20 °C ($d^{20°}$).

There are three alkanes with molecular formula C_5H_{12}. Pentane is the straight-chain alkane. Isopentane, as its name indicates, has an iso structural unit and five carbon atoms. The third isomer is called neopentane. The structural unit with a carbon surrounded by four other carbons is called "neo."

$CH_3CH_2CH_2CH_2CH_3$
pentane

$CH_3CHCH_2CH_3$
|
CH_3
isopentane

CH_3
|
CH_3CCH_3
|
CH_3
neopentane

CH_3
|
$CH_3CCH_2—$
|
CH_3
a "neo" structural unit

There are five constitutional isomers with molecular formula C_6H_{14}. We are now able to name three of them (hexane, isohexane, and neohexane), but we cannot name the other two without defining names for new structural units. (For now, ignore the names written in blue.)

$CH_3CH_2CH_2CH_2CH_2CH_3$
common name: hexane
systematic name: hexane

$CH_3CHCH_2CH_2CH_3$
|
CH_3
isohexane
2-methylpentane

CH_3
|
$CH_3CCH_2CH_3$
|
CH_3
neohexane
2,2-dimethylbutane

$CH_3CH_2CHCH_2CH_3$
|
CH_3
3-methylpentane

$CH_3CH—CHCH_3$
| |
CH_3 CH_3
2,3-dimethylbutane

There are nine alkanes with molecular formula C_7H_{16}. We can name only two of them (heptane and isoheptane) without defining new structural units. Notice that neoheptane cannot be used as a name because three different heptanes have a carbon that is bonded to four other carbons and *a name must specify only one compound.*

A compound can have more than one name, but a name must specify only one compound.

$CH_3CH_2CH_2CH_2CH_2CH_2CH_3$
common name: **heptane**
systematic name: **heptane**

$CH_3CHCH_2CH_2CH_2CH_3$ (with CH_3 on C-2)
isoheptane
2-methylhexane

$CH_3CH_2CHCH_2CH_2CH_3$ (with CH_3 on C-3)
3-methylhexane

$CH_3CH{-}CHCH_2CH_3$ (with CH_3 on C-2 and C-3)
2,3-dimethylpentane

$CH_3CHCH_2CHCH_3$ (with CH_3 on C-2 and C-4)
2,4-dimethylpentane

$CH_3CCH_2CH_2CH_3$ (with two CH_3 on C-2)
2,2-dimethylpentane

$CH_3CH_2CCH_2CH_3$ (with two CH_3 on C-3)
3,3-dimethylpentane

$CH_3CH_2CHCH_2CH_3$ (with CH_2CH_3 on C-3)
3-ethylpentane

$CH_3C{-}CHCH_3$ (with two CH_3 on C-2 and one CH_3 on C-3)
2,2,3-trimethylbutane

The number of constitutional isomers increases rapidly as the number of carbons in an alkane increases. For example, there are 75 alkanes with molecular formula $C_{10}H_{22}$ and 4347 alkanes with molecular formula $C_{15}H_{32}$. To avoid having to memorize the names of thousands of structural units, chemists have devised rules that name compounds on the basis of their structures. That way, only the rules have to be learned. Because the name is based on the structure, these rules make it possible to deduce the structure of a compound from its name.

This method of nomenclature is called **systematic nomenclature**. It is also called **IUPAC nomenclature** because it was designed by a commission of the International Union of Pure and Applied Chemistry (abbreviated IUPAC and pronounced "eye-you-pack") at a meeting in Geneva, Switzerland, in 1892. The IUPAC rules have been continually revised by the commission since then. Names such as isobutane and neopentane—nonsystematic names—are called **common names** and are shown in red in this text. The systematic or IUPAC names are shown in blue. Before we can understand how a systematic name for an alkane is constructed, we must learn how to name alkyl substituents.

2.1 Nomenclature of Alkyl Substituents

Removing a hydrogen from an alkane results in an **alkyl substituent** (or an alkyl group). Alkyl substituents are named by replacing the "ane" ending of the alkane with "yl." The letter "R" is used to indicate any alkyl group.

CH_3-	CH_3CH_2-	$CH_3CH_2CH_2-$	$CH_3CH_2CH_2CH_2-$
a methyl group	**an ethyl group**	**a propyl group**	**a butyl group**

$R-$
any alkyl group

If a hydrogen of an alkane is replaced by an OH, the compound becomes an **alcohol**; if it is replaced by an NH_2, the compound becomes an **amine**; and if it is replaced by a halogen, the compound becomes an **alkyl halide**.

R—OH	R—NH_2	R—X	X = F, Cl, Br, or I
an alcohol	**an amine**	**an alkyl halide**	

An alkyl group name followed by the name of the class of the compound (alcohol, amine, etc.) yields the common name of the compound. The following examples show how alkyl group names are used to build common names:

CH_3OH	$CH_3CH_2NH_2$	$CH_3CH_2CH_2Br$	$CH_3CH_2CH_2CH_2Cl$
methyl alcohol	**ethylamine**	**propyl bromide**	**butyl chloride**
CH_3I	CH_3CH_2OH	$CH_3CH_2CH_2NH_2$	$CH_3CH_2CH_2CH_2OH$
methyl iodide	**ethyl alcohol**	**propylamine**	**butyl alcohol**

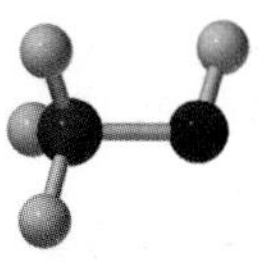

methyl alcohol

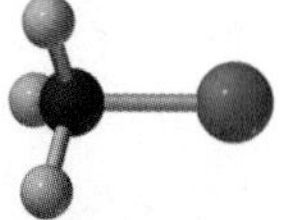

methyl chloride

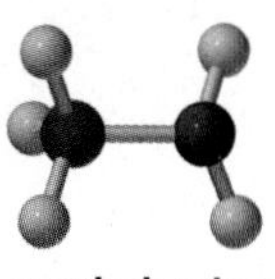

methylamine

Notice that there is a space between the name of the alkyl group and the name of the class of compound, except in the case of amines.

Two alkyl groups—a propyl group and an isopropyl group—contain three carbon atoms. A propyl group is obtained when a hydrogen is removed from a *primary carbon* of propane. A **primary carbon** is a carbon that is bonded to only one other carbon. An isopropyl group is obtained when a hydrogen is removed from the *secondary carbon* of propane. A **secondary carbon** is a carbon that is bonded to two other carbons. Notice that an isopropyl group, as its name indicates, has its three carbon atoms arranged as an iso structural unit.

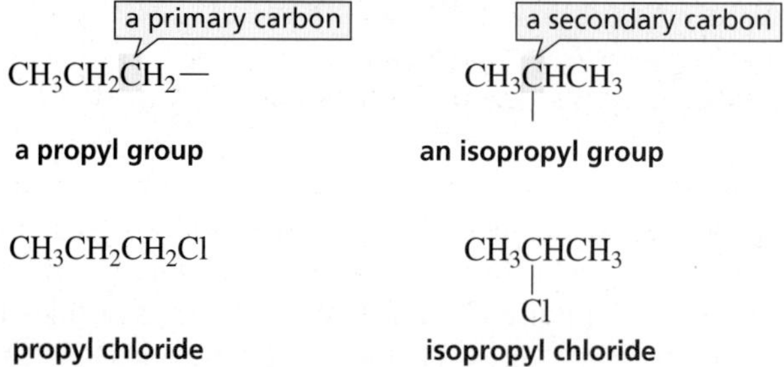

Build models of the two representations of isopropyl chloride, and convince yourself that they represent the same compound.

Molecular structures can be drawn in different ways. Isopropyl chloride, for example, is drawn here in two ways. Both represent the same compound. At first glance, the two-dimensional representations appear to be different: The methyl groups are across from one another in one structure and at right angles in the other. The structures are identical, however, because carbon is tetrahedral. The four groups bonded to the central carbon—a hydrogen, a chlorine, and two methyl groups—point to the corners of a tetrahedron. If you rotate the three-dimensional model on the right 90° in a clockwise direction, you should be able to see that the two models are the same. (You can simulate this rotation on the Web site www.prenhall.com/bruice by visiting the Molecule Gallery in Chapter 2.)

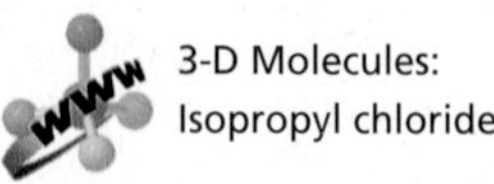

3-D Molecules:
Isopropyl chloride

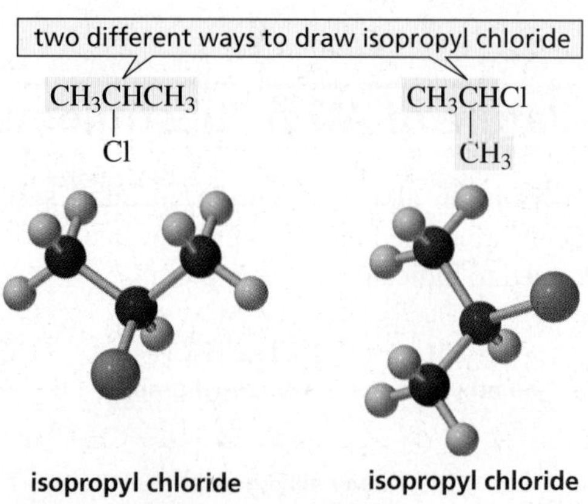

There are four alkyl groups that contain four carbon atoms. The butyl and isobutyl groups have a hydrogen removed from a primary carbon. A *sec*-butyl group has a hydrogen removed from a secondary carbon (*sec*-, often abbreviated *s*-, stands for secondary), and a *tert*-butyl group has a hydrogen removed from a tertiary carbon (*tert*-, sometimes abbreviated *t*-, stands for tertiary). A **tertiary carbon** is a carbon that is bonded to three other carbons. Notice that the isobutyl group is the only group with an iso structural unit.

A primary carbon is bonded to one carbon, a secondary carbon is bonded to two carbons, and a tertiary carbon is bonded to three carbons.

a primary carbon	a primary carbon	a secondary carbon	a tertiary carbon
$CH_3CH_2CH_2CH_2-$	$CH_3CH(CH_3)CH_2-$	$CH_3CH_2CH(CH_3)-$	$CH_3C(CH_3)_2-$
a butyl group	**an isobutyl group**	**a *sec*-butyl group**	**a *tert*-butyl group**

A name of a straight-chain alkyl group often has the prefix "*n*" (for "normal"), to emphasize that its carbon atoms are in an unbranched chain. If the name does not have a prefix such as "*n*" or "iso," it is assumed that the carbons are in an unbranched chain.

Tutorial:
Alkyl group nomenclature

$CH_3CH_2CH_2CH_2Br$
butyl bromide
or
***n*-butyl bromide**

$CH_3CH_2CH_2CH_2F$
pentyl fluoride
or
***n*-pentyl fluoride**

Like the carbons, the hydrogens in a molecule are also referred to as primary, secondary, and tertiary. **Primary hydrogens** are attached to primary carbons, **secondary hydrogens** to secondary carbons, and **tertiary hydrogens** to tertiary carbons.

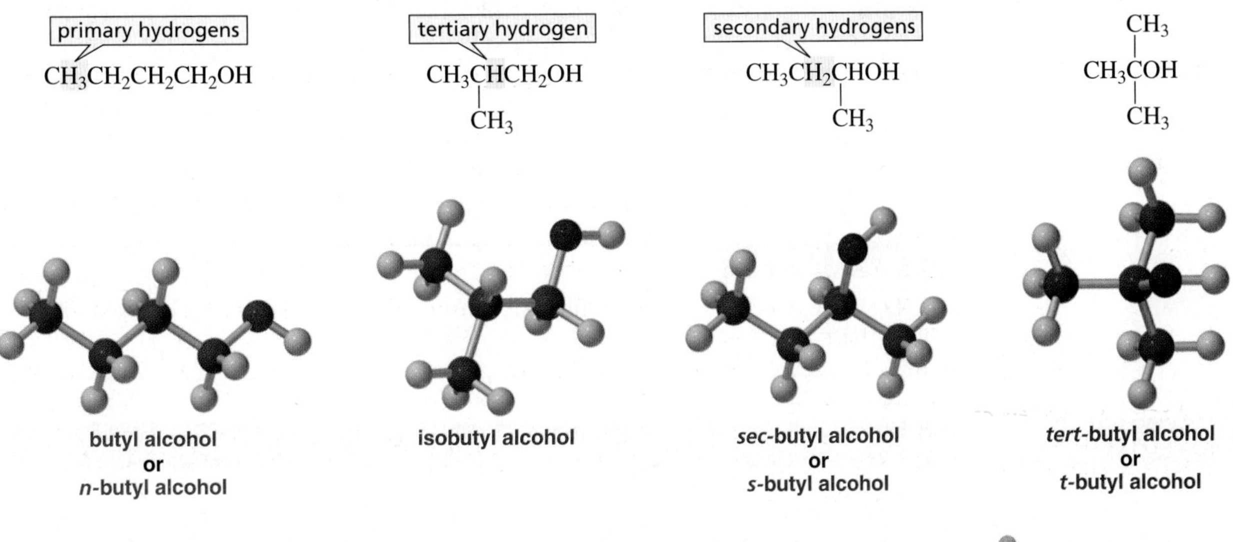

butyl alcohol
or
***n*-butyl alcohol**

isobutyl alcohol

***sec*-butyl alcohol**
or
***s*-butyl alcohol**

***tert*-butyl alcohol**
or
***t*-butyl alcohol**

Because a chemical name must specify only one compound, the only time you will see the prefix "*sec*" is in *sec*-butyl. The name "*sec*-pentyl" cannot be used because pentane has two different secondary carbon atoms. Therefore, there are two different alkyl groups that result from removing a hydrogen from a secondary carbon of pentane. Because the name would specify two different compounds, it is not a correct name.

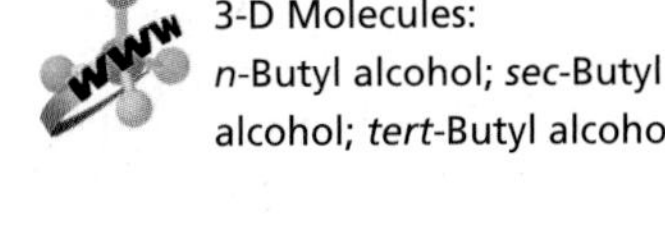

3-D Molecules:
n-Butyl alcohol; *sec*-Butyl alcohol; *tert*-Butyl alcohol

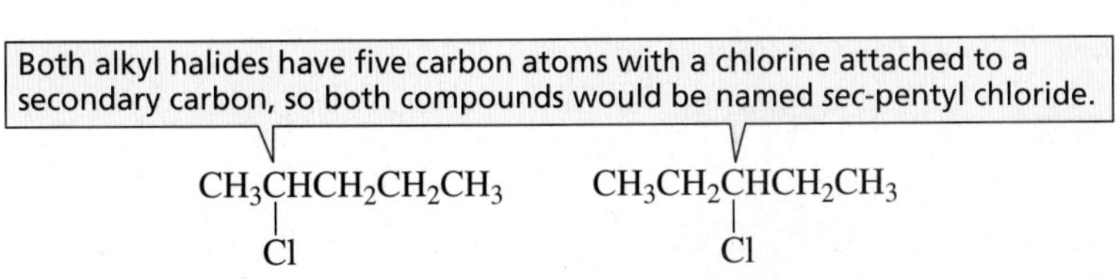

Both alkyl halides have five carbon atoms with a chlorine attached to a secondary carbon, so both compounds would be named sec-pentyl chloride.
CH3CHCH2CH2CH3
Cl
CH3CH2CHCH2CH3
Cl

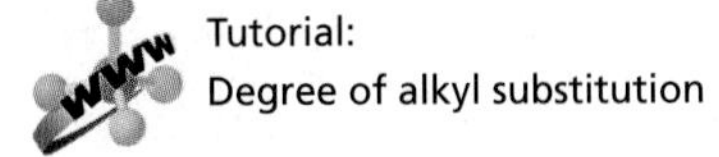

Tutorial:
Degree of alkyl substitution

The prefix "*tert*" is found in *tert*-butyl and *tert*-pentyl because each of these substituent names describes only one alkyl group. The name "*tert*-hexyl" cannot be used because it describes two different alkyl groups. (In older literature, you might find "amyl" used instead of "pentyl" to designate a five-carbon alkyl group.)

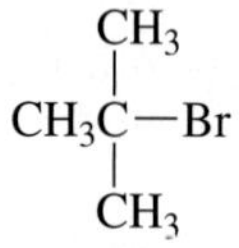

$CH_3C(CH_3)_2{-}Br$
tert-butyl bromide

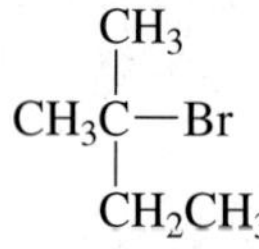

$CH_3C(CH_3)(CH_2CH_3){-}Br$
tert-pentyl bromide

$CH_3CH_2C(CH_2CH_3)(CH_3){-}Br$

$CH_3CH_2CH_2C(CH_3)_2{-}Br$

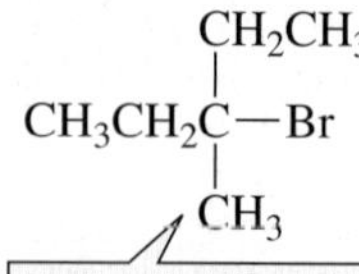

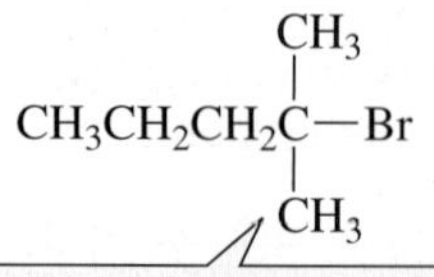

Both alkyl bromides have six carbon atoms with a bromine attached to a tertiary carbon, so both compounds would be named *tert*-hexyl bromide.

If you examine the following structures, you will see that whenever the prefix "iso" is used, the iso structural unit will be at one end of the molecule and any group replacing a hydrogen will be at the other end:

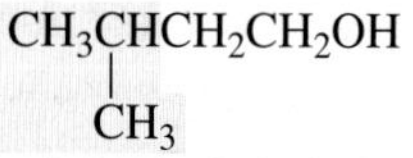

$CH_3CH(CH_3)CH_2CH_2OH$
isopentyl alcohol
or
isoamyl alcohol

$CH_3CH(CH_3)CH_2CH_2CH_2Cl$
isohexyl chloride

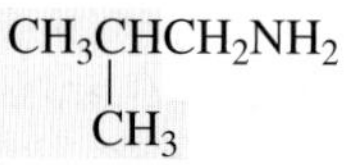

$CH_3CH(CH_3)CH_2NH_2$
isobutylamine

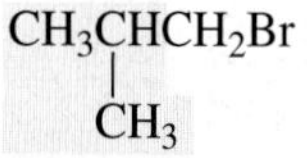

$CH_3CH(CH_3)CH_2Br$
isobutyl bromide

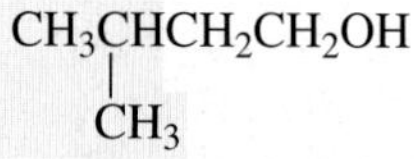

$CH_3CH(CH_3)CH_2CH_2OH$
isopentyl alcohol

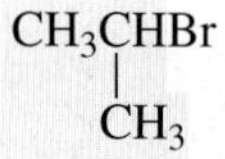

$CH_3CH(CH_3)Br$
isopropyl bromide

Notice that an iso group has a methyl group on the next-to-the-last carbon in the chain. Notice also that *all* isoalkyl compounds have the substituent (OH, Cl, NH_2, etc.) on a primary carbon, except for isopropyl, which has the substituent on a secondary carbon. The isopropyl group could have been called a *sec*-propyl group. Either name would have been appropriate because the group has an iso structural unit and a hydrogen has been removed from a secondary carbon. Chemists decided to call it isopropyl, however, which means that "*sec*" is used only for *sec*-butyl.

Alkyl group names are used so frequently that you should learn them. Some of the most common alkyl group names are compiled in Table 2.2 for your convenience.

PROBLEM 1◆

Draw the structures and name the four constitutional isomers with molecular formula C_4H_9Br.

Table 2.2 Names of Some Alkyl Groups

Name	Structure	Name	Structure	Name	Structure
methyl	$CH_3{-}$	*sec*-butyl	$CH_3CH_2CH(CH_3){-}$	neopentyl	$CH_3C(CH_3)_2CH_2{-}$
ethyl	$CH_3CH_2{-}$				
propyl	$CH_3CH_2CH_2{-}$			hexyl	$CH_3CH_2CH_2CH_2CH_2CH_2{-}$
isopropyl	$CH_3CH(CH_3){-}$	*tert*-butyl	$CH_3C(CH_3)_2{-}$	isohexyl	$CH_3CH(CH_3)CH_2CH_2CH_2{-}$
butyl	$CH_3CH_2CH_2CH_2{-}$	pentyl	$CH_3CH_2CH_2CH_2CH_2{-}$		
isobutyl	$CH_3CH(CH_3)CH_2{-}$	isopentyl	$CH_3CH(CH_3)CH_2CH_2{-}$		

PROBLEM 2◆

Which of the following can be used to verify that carbon is tetrahedral?

a. Methyl bromide does not have constitutional isomers.
b. Tetrachloromethane does not have a dipole moment.
c. Dibromomethane does not have constitutional isomers.

PROBLEM 3◆

Write a structure for each of the following compounds:

a. isopropyl alcohol
b. isopentyl fluoride
c. *sec*-butyl iodide
d. neopentyl chloride
e. *tert*-butylamine
f. *n*-octyl bromide

2.2 Nomenclature of Alkanes

The systematic name of an alkane is obtained using the following rules:

1. Determine the number of carbons in the longest continuous carbon chain. This chain is called the **parent hydrocarbon**. The name that indicates the number of carbons in the parent hydrocarbon becomes the alkane's "last name." For example, a parent hydrocarbon with eight carbons would be called *octane*. The longest continuous chain is not always a straight chain; sometimes you have to "turn a corner" to obtain the longest continuous chain.

First, determine the number of carbons in the longest continuous chain.

$\overset{8}{C}H_3\overset{7}{C}H_2\overset{6}{C}H_2\overset{5}{C}H_2\overset{4}{C}H\overset{3}{C}H_2\overset{2}{C}H_2\overset{1}{C}H_3$ with CH_3 on C4

4-methyloctane

$\overset{8}{C}H_3\overset{7}{C}H_2\overset{6}{C}H_2\overset{5}{C}H_2\overset{4}{C}HCH_2CH_3$ with $\underset{3}{C}H_2\underset{2}{C}H_2\underset{1}{C}H_3$ on C4

4-ethyloctane

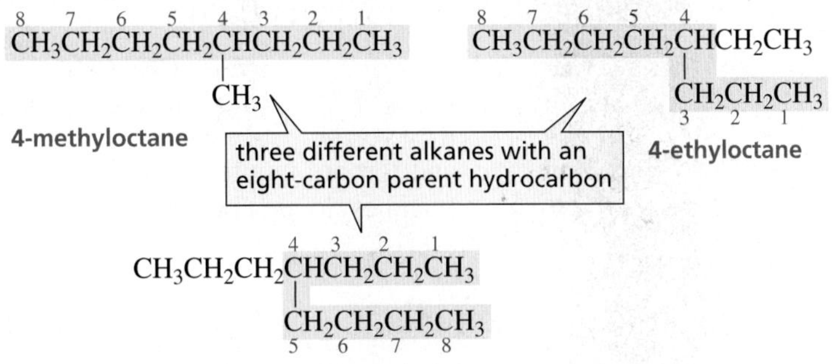

three different alkanes with an eight-carbon parent hydrocarbon

$CH_3CH_2CH_2\overset{4}{C}H\overset{3}{C}H_2\overset{2}{C}H_2\overset{1}{C}H_3$ with $\underset{5}{C}H_2\underset{6}{C}H_2\underset{7}{C}H_2\underset{8}{C}H_3$ on C4

4-propyloctane

2. The name of any alkyl substituent that hangs off the parent hydrocarbon is cited before the name of the parent hydrocarbon, together with a number to designate the carbon to which the alkyl substituent is attached. The chain is numbered in the direction that gives the substituent as low a number as possible. The substituent's name and the name of the parent hydrocarbon are joined in one word, and there is a hyphen between the number and the substituent's name.

Number the chain so that the substituent gets the lowest possible number.

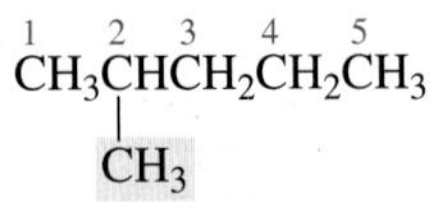

$\overset{1}{C}H_3\overset{2}{C}H\overset{3}{C}H_2\overset{4}{C}H_2\overset{5}{C}H_3$ with CH_3 on C2

2-methylpentane

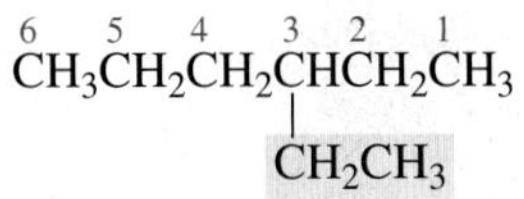

$\overset{6}{C}H_3\overset{5}{C}H_2\overset{4}{C}H_2\overset{3}{C}H\overset{2}{C}H_2\overset{1}{C}H_3$ with CH_2CH_3 on C3

3-ethylhexane

$\overset{1}{C}H_3\overset{2}{C}H_2\overset{3}{C}H_2\overset{4}{C}H\overset{5}{C}H_2\overset{6}{C}H_2\overset{7}{C}H_2\overset{8}{C}H_3$ with $CHCH_3$ (bearing CH_3) on C4

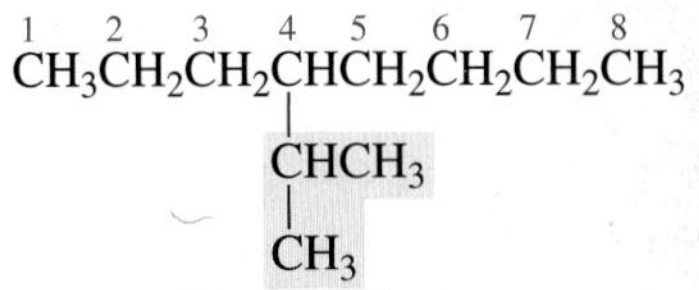

4-isopropyloctane

Numbers are used only for systematic names, never for common names.

Notice that only systematic names have numbers; common names never contain numbers.

$$\begin{array}{c} CH_3 \\ | \\ CH_3CHCH_2CH_2CH_3 \end{array}$$

common name: isohexane
systematic name: 2-methylpentane

Substituents are listed in alphabetical order.

3. If more than one substituent is attached to the parent hydrocarbon, the chain is numbered in the direction that will result in the lowest possible number in the name of the compound. The substituents are listed in alphabetical (not numerical) order, with each substituent getting the appropriate number. In the following example, the correct name (5-ethyl-3-methyloctane) contains a 3 as its lowest number, while the incorrect name (4-ethyl-6-methyloctane) contains a 4 as its lowest number:

$$CH_3CH_2\underset{\displaystyle CH_3}{\underset{|}{C}}HCH_2\underset{\displaystyle CH_2CH_3}{\underset{|}{C}}HCH_2CH_2CH_3$$

5-ethyl-3-methyloctane
not
4-ethyl-6-methyloctane
because 3 < 4

A number and a word are separated by a hyphen; numbers are separated by a comma.

If two or more substituents are the same, the prefixes "di," "tri," and "tetra" are used to indicate how many identical substituents the compound has. The numbers indicating the locations of the identical substituents are listed together, separated by commas. Notice that there must be as many numbers in a name as there are substituents.

$$CH_3CH_2\underset{\displaystyle CH_3}{\underset{|}{C}}HCH_2\underset{\displaystyle CH_3}{\underset{|}{C}}HCH_3$$

2,4-dimethylhexane

$$CH_3CH_2CH_2\overset{\displaystyle CH_3}{\overset{|}{\underset{\displaystyle CH_3}{\underset{|}{C}}}}-\overset{\displaystyle CH_3}{\overset{|}{\underset{\displaystyle CH_3}{\underset{|}{C}}}}CH_2CH_3$$

3,3,4,4-tetramethylheptane

di, tri, tetra, *sec*, and *tert* are ignored in alphabetizing.

iso, neo, and cyclo are not ignored in alphabetizing.

The prefixes di, tri, tetra, *sec*, and *tert* are ignored in alphabetizing substituent groups, but the prefixes iso, neo, and cyclo are not ignored.

$$CH_3CH_2\overset{\displaystyle CH_2CH_3}{\overset{|}{\underset{\displaystyle CH_2CH_3}{\underset{|}{C}}}}CH_2CH_2\underset{\displaystyle CH_2CH_3}{\underset{|}{C}}H\overset{\displaystyle CH_3}{\overset{|}{C}}HCH_2CH_2CH_3$$

3,3,6-triethyl-7-methyldecane

$$CH_3CH_2CH_2\underset{\displaystyle \underset{\displaystyle CH_3}{\underset{|}{CHCH_3}}}{\underset{|}{C}}HCH_2CH_2\overset{\displaystyle CH_3}{\overset{|}{C}}HCH_3$$

5-isopropyl-2-methyloctane

4. When both directions lead to the same lowest number for one of the substituents, the direction is chosen that gives the lowest possible number to one of the remaining substituents.

$$CH_3\overset{\displaystyle CH_3}{\overset{|}{\underset{\displaystyle CH_3}{\underset{|}{C}}}}CH_2\underset{\displaystyle CH_3}{\underset{|}{C}}HCH_3$$

2,2,4-trimethylpentane
not
2,4,4-trimethylpentane
because 2 < 4

$$CH_3CH_2\underset{\displaystyle CH_3}{\underset{|}{C}}H\overset{\displaystyle CH_3}{\overset{|}{C}}HCH_2\overset{\displaystyle CH_2CH_3}{\overset{|}{C}}HCH_2CH_3$$

6-ethyl-3,4-dimethyloctane
not
3-ethyl-5,6-dimethyloctane
because 4 < 5

5. If the same substituent numbers are obtained in both directions, the first group cited receives the lower number.

$$\underset{\text{Br}}{\overset{\text{Cl}}{CH_3CHCHCH_3}}$$

2-bromo-3-chlorobutane
not
3-bromo-2-chlorobutane

$$CH_3CH_2\overset{CH_2CH_3}{CH}CH_2\underset{CH_3}{CH}CH_2CH_3$$

3-ethyl-5-methylheptane
not
5-ethyl-3-methylheptane

Only if the same set of numbers is obtained in both directions does the first group cited get the lower number.

6. If a compound has two or more chains of the same length, the parent hydrocarbon is the chain with the greatest number of substituents.

3-ethyl-2-methylhexane (two substituents)

not
3-isopropylhexane (one substituent)

In the case of two hydrocarbon chains with the same number of carbons, choose the one with the most substituents.

7. Names such as "isopropyl," "*sec*-butyl," and "*tert*-butyl" are acceptable substituent names in the IUPAC system of nomenclature, but systematic substituent names are preferable. Systematic substituent names are obtained by numbering the alkyl substituent starting at the carbon that is attached to the parent hydrocarbon. This means that the carbon that is attached to the parent hydrocarbon is always the number 1 carbon of the substituent. In a compound such as 4-(1-methylethyl)octane, the substituent name is in parentheses; the number inside the parentheses indicates a position on the substituent, whereas the number outside the parentheses indicates a position on the parent hydrocarbon.

$$CH_3CH_2CH_2CH_2\underset{\underset{CH_3}{{}^1CH^2CH_3}}{CH}CH_2CH_2CH_3$$

4-isopropyloctane
or
4-(1-methylethyl)octane

$$CH_3CH_2CH_2CH_2\underset{\underset{CH_3}{{}^1CH_2{}^2CH^3CH_3}}{CH}CH_2CH_2CH_2CH_2CH_3$$

5-isobutyldecane
or
5-(2-methylpropyl)decane

Some substituents have only a systematic name.

$$CH_3CH_2CH_2CH_2\underset{\underset{CH_3}{CH_3{}^1CH^2CH^3CH_3}}{CH}CH_2\overset{CH_2CH_2CH_3}{CH}CH_2CH_2CH_3$$

6-(1,2-dimethylpropyl)-4-propyldecane

$$CH_3\overset{CH_3}{CH}\underset{CH_3}{CH}CH_2\underset{CH_2CH_2CH_2CH_2CH_3}{CH}{}^1CH_2{}^2\overset{CH_3}{CH}{}^3CH_2{}^4CH_3$$

2,3-dimethyl-5-(2-methylbutyl)decane

These rules will allow you to name thousands of alkanes, and eventually you will learn the additional rules necessary to name many other kinds of compounds. The rules are important if you want to look up a compound in the scientific literature, because it usually will be listed by its systematic name. Nevertheless, you must still learn common names because they have been in existence for so long and are so entrenched in chemists' vocabulary that they are widely used in scientific conversation and are often found in the literature.

Look at the systematic names (the ones written in blue) for the isomeric hexanes and isomeric heptanes at the beginning of this chapter to make sure you understand how they are constructed.

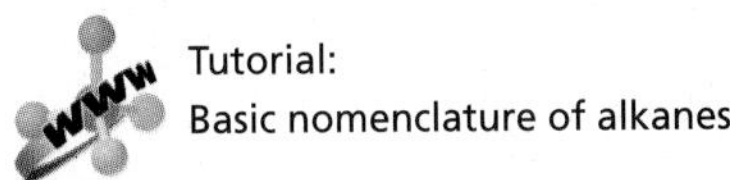

Tutorial:
Basic nomenclature of alkanes

PROBLEM 4◆

Draw the structure of each of the following compounds:

a. 2,3-dimethylhexane
b. 4-isopropyl-2,4,5-trimethylheptane
c. 4,4-diethyldecane
d. 2,2-dimethyl-4-propyloctane
e. 4-isobutyl-2,5-dimethyloctane
f. 4-(1,1-dimethylethyl)octane

PROBLEM 5 SOLVED

a. Draw the 18 isomeric octanes.
b. Give each isomer its systematic name.
c. How many isomers have common names?
d. Which isomers contain an isopropyl group?
e. Which isomers contain a *sec*-butyl group?
f. Which isomers contain a *tert*-butyl group?

SOLUTION TO 5a Start with the isomer with an eight-carbon continuous chain. Then draw isomers with a seven-carbon continuous chain plus one methyl group. Next, draw isomers with a six-carbon continuous chain plus two methyl groups or one ethyl group. Then draw isomers with a five-carbon continuous chain plus three methyl groups or one methyl group and one ethyl group. Finally, draw a four-carbon continuous chain with four methyl groups. (You will be able to tell whether you have drawn duplicate structures by your answers to 3b because if two structures have the same systematic name, they are the same compound.)

PROBLEM 6◆

Give the systematic name for each of the following compounds:

a.
$$\begin{array}{l} \quad\quad\quad CH_3 \quad\quad CH_3 \\ CH_3CH_2\overset{|}{C}HCH_2\overset{|}{C}CH_3 \\ \quad\quad\quad\quad\quad\quad\quad | \\ \quad\quad\quad\quad\quad\quad CH_3 \end{array}$$

b. $CH_3CH_2C(CH_3)_3$

c. $CH_3CH_2C(CH_2CH_3)_2CH_2CH_2CH_3$

d.
$$\begin{array}{l} \quad\quad\quad\quad\quad\quad\quad CH_3 \\ \quad\quad\quad\quad\quad\quad\quad\quad | \\ CH_3CHCH_2CH_2CHCH_3 \\ \quad\quad\quad | \\ \quad\quad CH_2CH_3 \end{array}$$

e. $CH_3CH_2C(CH_2CH_3)_2CH(CH_3)CH(CH_2CH_2CH_3)_2$

f.
$$\begin{array}{l} CH_3CH_2CH_2CHCH_2CH_2CH_3 \\ \quad\quad\quad\quad\quad | \\ \quad\quad\quad CH_3CHCH_2CH_3 \end{array}$$

g.
$$\begin{array}{l} \quad\quad\quad CH_3 \; CH_2CH_2CH_3 \\ \quad\quad\quad\; | \quad\quad | \\ CH_3C\!-\!CHCH_2CH_3 \\ \quad\quad\quad\; | \\ \quad\quad CH_2CH_2CH_3 \end{array}$$

h.
$$\begin{array}{l} CH_3CH_2CH_2CH_2CHCH_2CH_2CH_3 \\ \quad\quad\quad\quad\quad\quad\quad | \\ \quad\quad\quad\quad\quad\quad CH(CH_3)_2 \end{array}$$

PROBLEM 7

Draw the structure and give the systematic name of a compound with a molecular formula C_5H_{12} that has

a. only primary and secondary hydrogens
b. only primary hydrogens
c. one tertiary hydrogen
d. two secondary hydrogens

2.3 Nomenclature of Cycloalkanes

Cycloalkanes are alkanes with their carbon atoms arranged in a ring. Because of the ring, a cycloalkane has two fewer hydrogens than an acyclic (noncyclic) alkane with the same number of carbons. This means that the general molecular formula for a cycloalkane is C_nH_{2n}. Cycloalkanes are named by adding the prefix "cyclo" to the alkane name that signifies the number of carbon atoms in the ring.

cyclopropane cyclobutane cyclopentane cyclohexane

Cycloalkanes are almost always written as **skeletal structures**. Skeletal structures show the carbon–carbon bonds as lines, but do not show the carbons or the hydrogens bonded to carbons. Atoms other than carbon and hydrogens bonded to atoms other than carbon are shown. Each vertex in a skeletal structure represents a carbon. It is understood that each carbon is bonded to the appropriate number of hydrogens to give the carbon four bonds.

cyclopropane cyclobutane cyclopentane cyclohexane

Acyclic molecules can also be represented by skeletal structures. In a skeletal structure of an acyclic molecule, the carbon chains are represented by zigzag lines. Again, each vertex represents a carbon, and carbons are assumed to be present where a line begins or ends.

butane 2-methylhexane 3-methyl-4-propylheptane 6-ethyl-2,3-dimethylnonane

The rules for naming cycloalkanes resemble the rules for naming acyclic alkanes:

1. In the case of a cycloalkane with an attached alkyl substituent, the ring is the parent hydrocarbon unless the substituent has more carbon atoms than the ring. In that case, the substituent is the parent hydrocarbon and the ring is named as a substituent. There is no need to number the position of a single substituent on a ring.

If there is only one substituent on a ring, do not give that substituent a number.

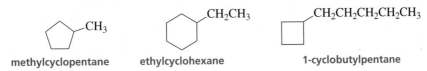

methylcyclopentane ethylcyclohexane 1-cyclobutylpentane

2. If the ring has two different substituents, they are cited in *alphabetical order* and the number 1 position is given to the substituent cited first.

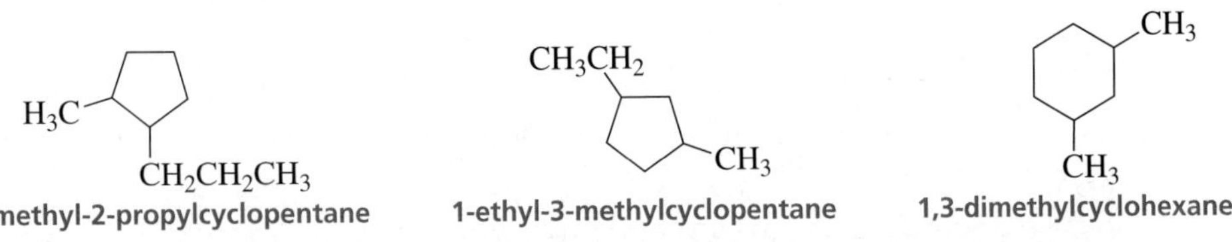

1-methyl-2-propylcyclopentane 1-ethyl-3-methylcyclopentane 1,3-dimethylcyclohexane

3. If there are more than two substituents on the ring, they are cited in alphabetical order. The substituent given the number 1 position is the one that results in a second

substituent getting as low a number as possible. If two substituents have the same low number, the ring is numbered—either clockwise or counterclockwise—in the direction that gives the third substituent the lowest possible number. For example, the correct name of the following compound is 4-ethyl-2-methyl-1-propylcyclohexane, not 5-ethyl-1-methyl-2-propylcyclohexane:

Tutorial: Advanced alkane nomenclature

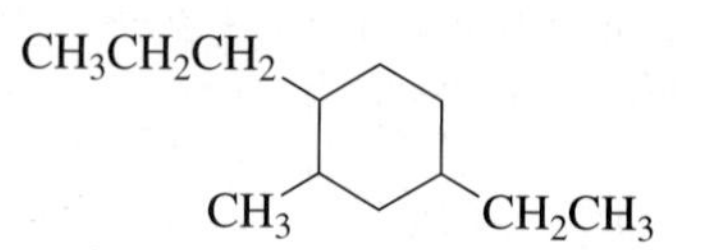

4-ethyl-2-methyl-1-propylcyclohexane
not
1-ethyl-3-methyl-4-propylcyclohexane
because 2 < 3
not
5-ethyl-1-methyl-2-propylcyclohexane
because 4 < 5

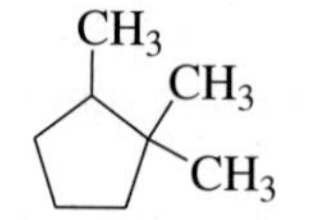

1,1,2-trimethylcyclopentane
not
1,2,2-trimethylcyclopentane
because 1 < 2
not
1,1,5-trimethylcyclopentane
because 2 < 5

PROBLEM 8◆

Convert the following condensed structures into skeletal structures (remember that condensed structures show atoms but few, if any, bonds, whereas skeletal structures show bonds but few, if any, atoms):

a. $CH_3CH_2CH_2CH_2CH_2CH_2OH$

b. $CH_3CH_2\underset{}{\overset{CH_3}{C}}HCH_2\overset{CH_3}{C}HCH_2CH_3$

c. $CH_3\overset{CH_3}{C}HCH_2CH_2\underset{Br}{C}HCH_3$

d. $CH_3CH_2CH_2CH_2OCH_3$

PROBLEM 9◆

Give the systematic name for each of the following compounds:

a. 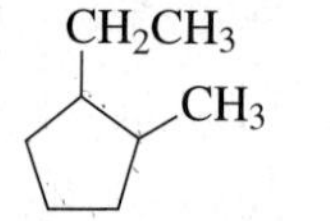

b. 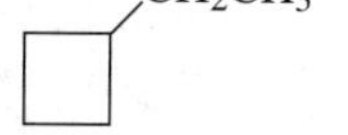

c. 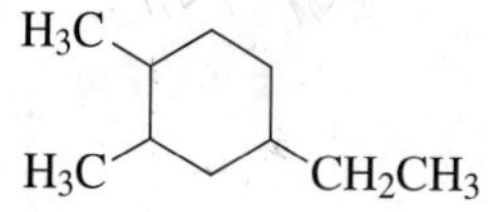

d. 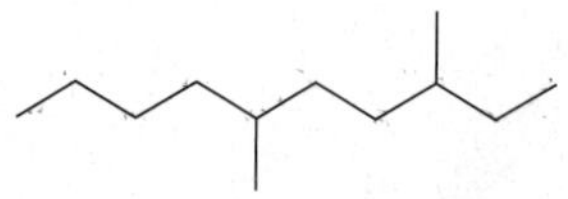

e. 

f. 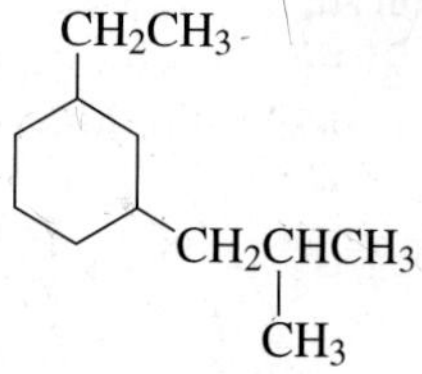

g. 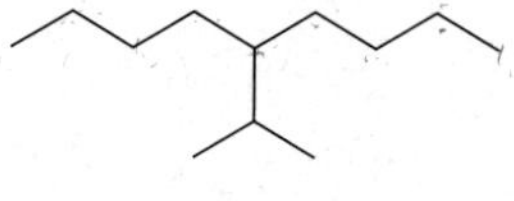

h. 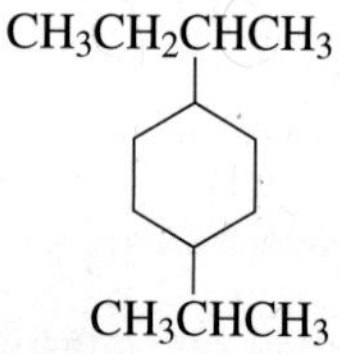

2.4 Nomenclature of Alkyl Halides

Alkyl halides are compounds in which a hydrogen of an alkane has been replaced by a halogen. Alkyl halides are classified as primary, secondary, or tertiary, depending on the carbon to which the halogen is attached. **Primary alkyl halides** have a halogen bonded to a primary carbon, **secondary alkyl halides** have a halogen bonded to a secondary carbon, and **tertiary alkyl halides** have a halogen bonded to a tertiary carbon (Section 2.1). The lone-pair electrons on the halogens are generally not shown unless they are needed to draw your attention to some chemical property of the atom.

a primary carbon	a secondary carbon	a tertiary carbon
$R{-}CH_2{-}Br$	$R{-}CH{-}R$ (Br on CH)	$R{-}C{-}R$ (R above, Br below C)
a primary alkyl halide	**a secondary alkyl halide**	**a tertiary alkyl halide**

The number of alkyl groups attached to the carbon to which the halogen is bonded determines whether an alkyl halide is primary, secondary, or tertiary.

The common names of alkyl halides consist of the name of the alkyl group, followed by the name of the halogen—with the "ine" ending of the halogen name replaced by "ide" (i.e., fluoride, chloride, bromide, iodide).

	CH_3Cl	CH_3CH_2F	CH_3CHI (with CH_3 on CH)	CH_3CH_2CHBr (with CH_3 on CH)
common name:	**methyl chloride**	**ethyl fluoride**	**isopropyl iodide**	***sec*-butyl bromide**
systematic name:	**chloromethane**	**fluoroethane**	**2-iodopropane**	**2-bromobutane**

In the IUPAC system, alkyl halides are named as substituted alkanes. The substituent prefix names for the halogens end with "o" (i.e., "fluoro," "chloro," "bromo," "iodo"). Therefore, alkyl halides are often called haloalkanes.

$CH_3CH_2CHCH_2CH_2CHCH_3$ (with CH_3 on C-3 and Br on C-6)

2-bromo-5-methylheptane

$CH_3CCH_2CH_2CH_2CH_2Cl$ (with two CH_3 on C-2)

1-chloro-5,5-dimethylhexane

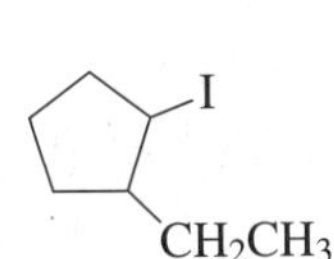

1-ethyl-2-iodocyclopentane

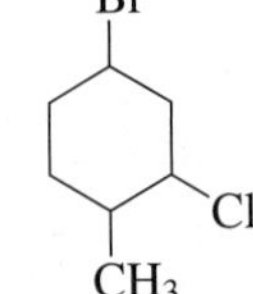

4-bromo-2-chloro-1-methylcyclohexane

CH_3F
methyl fluoride

CH_3Cl
methyl chloride

CH_3Br
methyl bromide

CH_3I
methyl iodide

PROBLEM 10◆

Give two names for each of the following compounds, and tell whether each alkyl halide is primary, secondary, or tertiary:

a. $CH_3CH_2CHCH_3$ (with Cl on CH)

b. $CH_3CHCH_2CH_2CH_2Cl$ (with CH_3 on CH)

c. cyclohexane ring bearing Br

d. CH_3CHCH_3 (with F on CH)

PROBLEM 11

Draw the structures and provide systematic names for a–c by substituting a chlorine for a hydrogen of methylcyclohexane:

a. a primary alkyl halide b. a tertiary alkyl halide c. three secondary alkyl halides

2.5 Nomenclature of Ethers

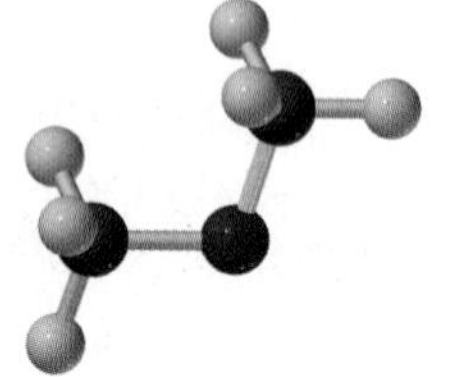

dimethyl ether

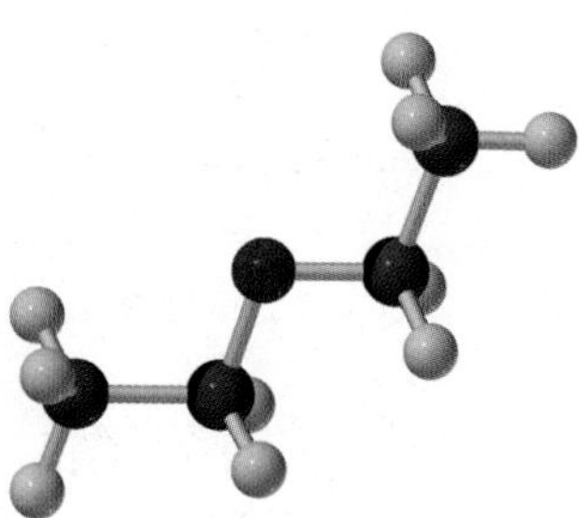

diethyl ether

Chemists sometimes neglect the prefix "di" when they name symmetrical ethers. Try not to make this oversight a habit.

Ethers are compounds in which an oxygen is bonded to two alkyl substituents. If the alkyl substituents are identical, the ether is a **symmetrical ether**. If the substituents are different, the ether is an **unsymmetrical ether**.

R—O—R
a symmetrical ether

R—O—R′
an unsymmetrical ether

The common name of an ether consists of the names of the two alkyl substituents (in alphabetical order), followed by the word "ether." The smallest ethers are almost always named by their common names.

$CH_3OCH_2CH_3$
ethyl methyl ether

$CH_3CH_2OCH_2CH_3$
diethyl ether
often called ethyl ether

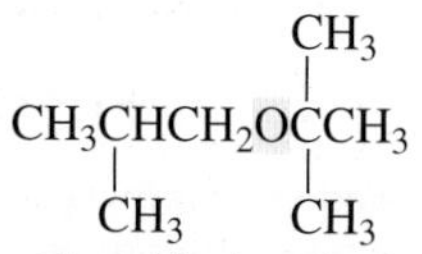

$CH_3CH(CH_3)CH_2OC(CH_3)_2CH_3$
tert-butyl isobutyl ether

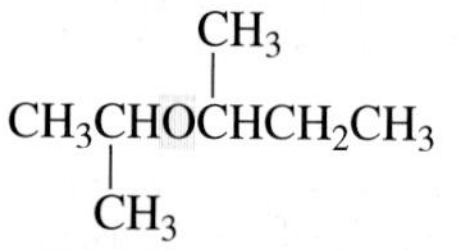

$CH_3CH(CH_3)OCH(CH_3)CH_2CH_3$
sec-butyl isopropyl ether

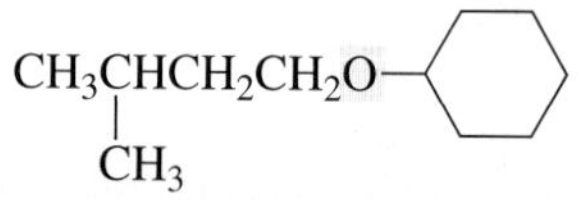

$CH_3CH(CH_3)CH_2CH_2O$–cyclohexyl
cyclohexyl isopentyl ether

The IUPAC system names an ether as an alkane with an RO substituent. The substituents are named by replacing the "yl" ending in the name of the alkyl substituent with "oxy."

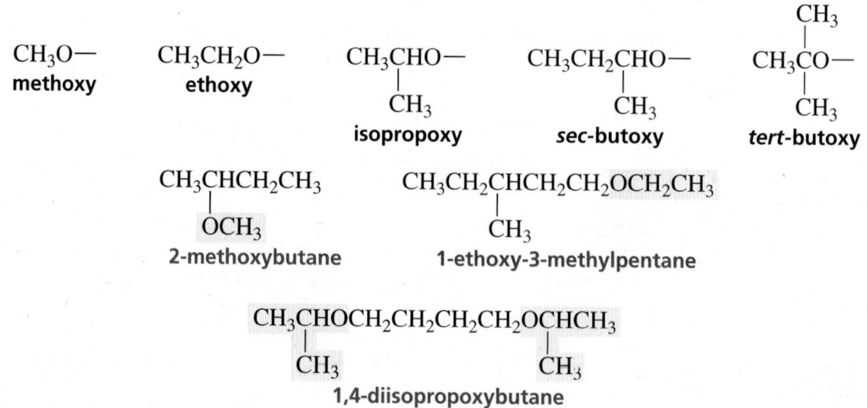

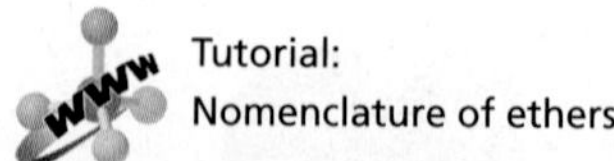

Tutorial:
Nomenclature of ethers

PROBLEM 12◆

a. Give the systematic (IUPAC) name for each of the following ethers:

1. $CH_3OCH_2CH_3$
2. $CH_3CH_2OCH_2CH_3$
3. 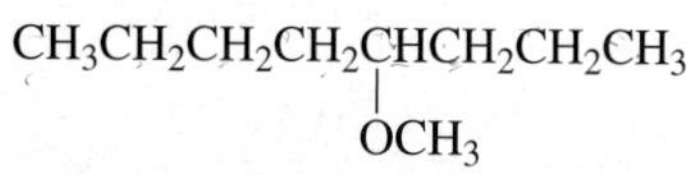 $CH_3CH_2CH_2CH_2CH(OCH_3)CH_2CH_2CH_3$
4. $CH_3CH_2CH_2OCH_2CH_2CH_2CH_3$

5. $CH_3CHOCHCH_2CH_2CH_3$ (with CH_3 on the first CH and CH_3 on the second CH)

6. $CH_3CHOCH_2CH_2CHCH_3$ (with CH_3 on the first CH and CH_3 on the last CH)

b. Do all of these ethers have common names?

c. What are their common names?

2.6 Nomenclature of Alcohols

Alcohols are compounds in which a hydrogen of an alkane has been replaced by an OH group. **Alcohols** are classified as **primary**, **secondary**, or **tertiary**, depending on whether the OH group is bonded to a primary, secondary, or tertiary carbon—the same way alkyl halides are classified.

The number of alkyl groups attached to the carbon to which the OH group is attached determines whether an alcohol is primary, secondary, or tertiary.

$R{-}CH_2{-}OH$ — a primary alcohol

$R{-}CH(R){-}OH$ — a secondary alcohol

$R{-}C(R)(R){-}OH$ — a tertiary alcohol

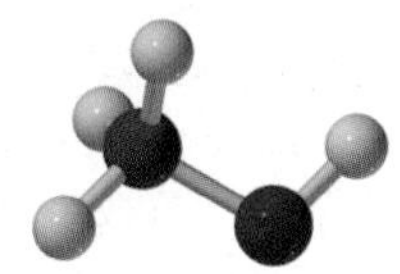

methyl alcohol

The common name of an alcohol consists of the name of the alkyl group to which the OH group is attached, followed by the word "alcohol."

CH_3CH_2OH	$CH_3CH_2CH_2OH$	CH_3CHOH (with CH_3 on CH)	CH_3CCH_2OH (with CH_3 above and below C)
ethyl alcohol	propyl alcohol	isopropyl alcohol	neopentyl alcohol

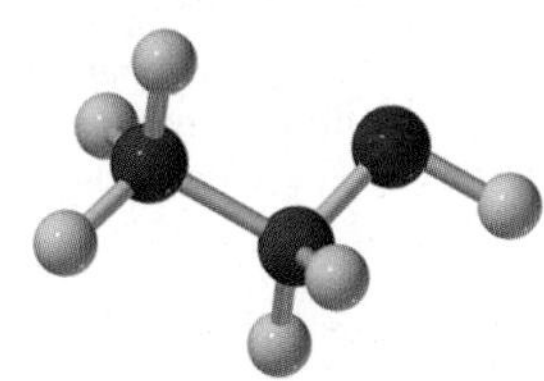

ethyl alcohol

The **functional group** is the center of reactivity in a molecule. In an alcohol, the OH is the functional group. The IUPAC system uses a suffix to denote certain functional groups. The systematic name of an alcohol, for example, is obtained by replacing the "e" at the end of the name of the parent hydrocarbon with the suffix "ol."

CH_3OH	CH_3CH_2OH
methanol	ethanol

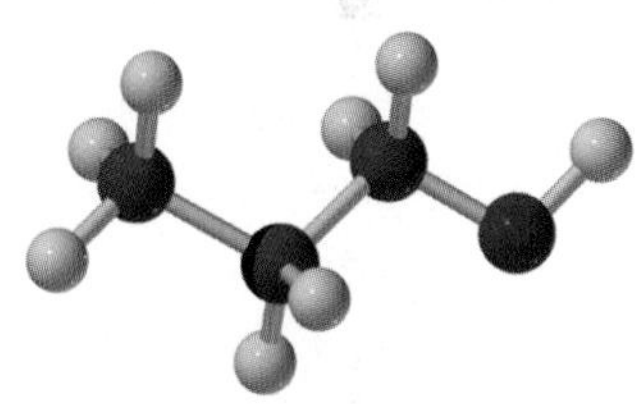

propyl alcohol

When necessary, the position of the functional group is indicated by a number immediately preceding the name of the alcohol or immediately preceding the suffix. The most recently approved IUPAC names are those with the number immediately preceding the suffix. However, names with the number preceding the name of the alcohol have been in use for a long time, so those are the ones most likely to appear in the literature, on reagent bottles, and on standardized tests. They will also be the ones that appear most often in this book.

$CH_3CH_2CHCH_2CH_3$ (with OH on CH)

3-pentanol
or
pentan-3-ol

The following rules are used to name a compound that has a functional group suffix:

1. The parent hydrocarbon is the longest continuous chain *containing the functional group.*

2. The parent hydrocarbon is numbered in the direction that gives the *functional group suffix the lowest possible number.*

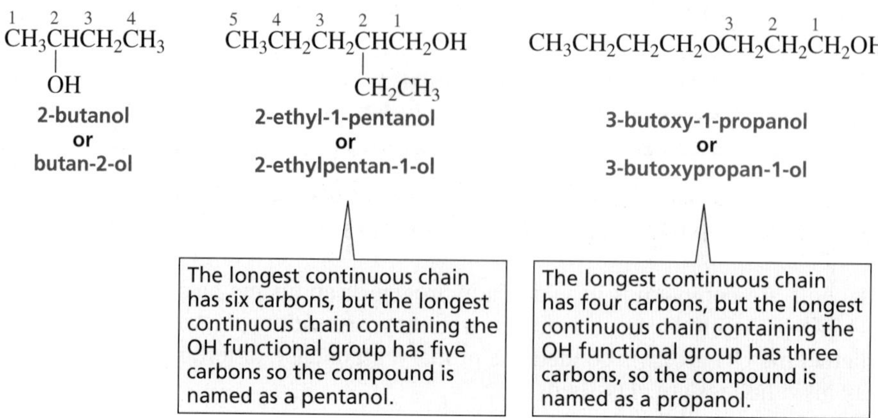

When there is only a substituent, the substituent gets the lowest possible number.

When there is only a functional group suffix, the functional group suffix gets the lowest possible number.

When there is both a functional group suffix and a substituent, the functional group suffix gets the lowest possible number.

3. If there is a functional group suffix and a substituent, the functional group suffix gets the lowest possible number.

$\overset{1}{H}O\overset{}{C}H_2\overset{2}{C}H_2\overset{3}{C}H_2Br$ — **3-bromo-1-propanol**

$Cl\overset{4}{C}H_2\overset{3}{C}H_2\overset{2}{C}H(OH)\overset{1}{C}H_3$ — **4-chloro-2-butanol**

$\overset{5}{C}H_3\overset{4}{C}(CH_3)_2\overset{3}{C}H_2\overset{2}{C}H(OH)\overset{1}{C}H_3$ — **4,4-dimethyl-2-pentanol**

4. If the same number for the functional group suffix is obtained in both directions, the chain is numbered in the direction that gives a substituent the lowest possible number. Notice that a number is not needed to designate the position of a functional group suffix in a cyclic compound, because it is assumed to be at the 1-position.

$CH_3CH(Cl)CH(OH)CH_2CH_3$ — **2-chloro-3-pentanol** not **4-chloro-3-pentanol**

$CH_3CH_2CH_2CH(OH)CH_2CH(CH_3)CH_3$ — **2-methyl-4-heptanol** not **6-methyl-4-heptanol**

(cyclohexane ring with CH_3 and OH) — **3-methylcyclohexanol** not **5-methylcyclohexanol**

5. If there is more than one substituent, the substituents are cited in alphabetical order.

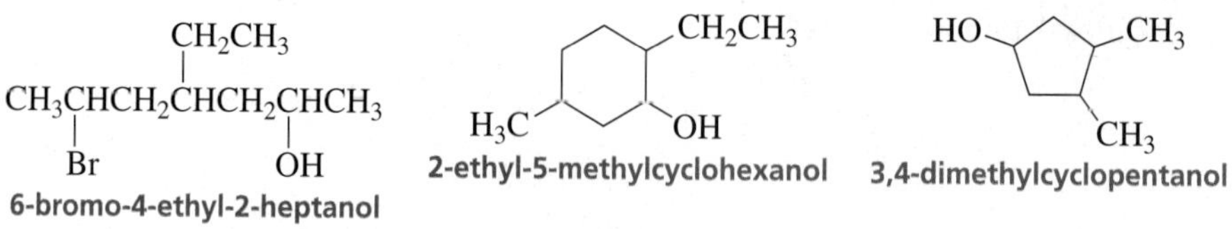

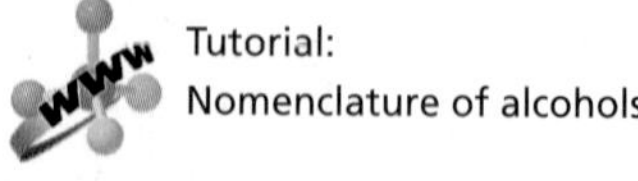

Tutorial:
Nomenclature of alcohols

Remember that the name of a substituent is stated *before* the name of the parent hydrocarbon, and the functional group suffix is stated *after* the name of the parent hydrocarbon.

[substituent] [parent hydrocarbon] [functional group suffix]

PROBLEM 13

Draw the structures of a homologous series of alcohols that have from one to six carbons, and then give each of them a common name and a systematic name.

PROBLEM 14◆

Give each of the following compounds a systematic name, and indicate whether each is a primary, secondary, or tertiary alcohol:

a. $CH_3CH_2CH_2CH_2CH_2OH$

b. 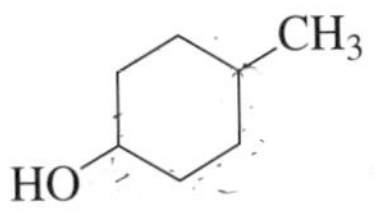 (cyclohexane ring with HO and CH_3 substituents)

c. $CH_3C(CH_3)(OH)CH_2CH_2CH_2Cl$

d. $CH_3CH(CH_3)CH_2CH(OH)CH_2CH_3$

e. $CH_3CH(CH_3)CH_2CH(OH)CH_2CH(CH_3)CH_2CH_3$

f. 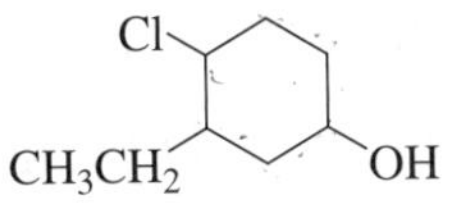 (cyclohexane ring with Cl, CH_3CH_2, and OH substituents)

PROBLEM 15◆

Write the structures of all the tertiary alcohols with molecular formula $C_6H_{14}O$, and give each a systematic name.

2.7 Nomenclature of Amines

Amines are compounds in which one or more of the hydrogens of ammonia have been replaced by alkyl groups. Smaller amines are characterized by their fishy odors. Fermented shark, for example, a traditional dish in Iceland, smells exactly like triethylamine. There are **primary amines, secondary amines**, and **tertiary amines**. The classification depends on how many alkyl groups are bonded to the nitrogen. Primary amines have one alkyl group bonded to the nitrogen, secondary amines have two, and tertiary amines have three.

NH_3	$R{-}NH_2$	$R{-}NH{-}R$	$R{-}N(R){-}R$
ammonia	a primary amine	a secondary amine	a tertiary amine

The number of alkyl groups attached to the nitrogen determines whether an amine is primary, secondary, or tertiary.

Notice that the number of alkyl groups *attached to the nitrogen* determines whether an amine is primary, secondary, or tertiary. For an alkyl halide or an alcohol, on the other hand, the number of alkyl groups *attached to the carbon* to which the halogen or the OH is bonded determines the classification (Sections 2.4 and 2.6).

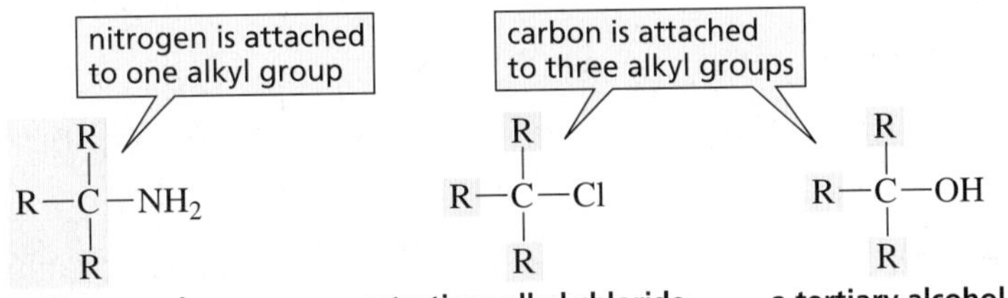

nitrogen is attached to one alkyl group

carbon is attached to three alkyl groups

$R_3C{-}NH_2$	$R_3C{-}Cl$	$R_3C{-}OH$
a primary amine	a tertiary alkyl chloride	a tertiary alcohol

The common name of an amine consists of the names of the alkyl groups bonded to the nitrogen, in alphabetical order, followed by "amine." The entire name is written as one word (unlike the common names of alcohols, ethers, and alkyl halides, in which "alcohol," "ether," and "halide" are separate words).

CH_3NH_2 **methylamine**	$CH_3NHCH_2CH_2CH_3$ **methylpropylamine**	$CH_3CH_2NHCH_2CH_3$ **diethylamine**
CH_3 (on N) CH_3NCH_3 **trimethylamine**	CH_3 (on N) $CH_3NCH_2CH_2CH_2CH_3$ **butyldimethylamine**	CH_3 (on N) $CH_3CH_2NCH_2CH_2CH_3$ **ethylmethylpropylamine**

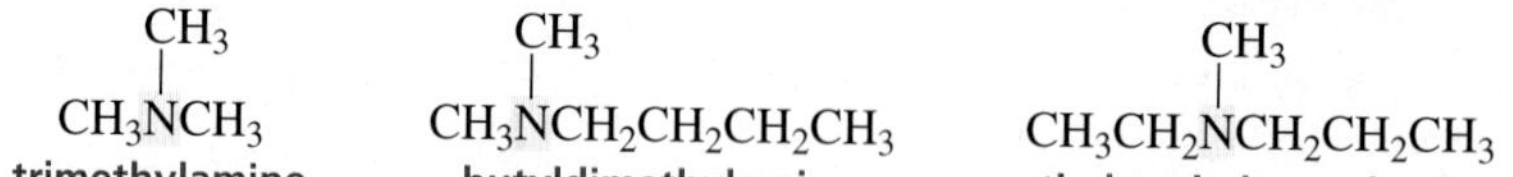

The IUPAC system uses a suffix to denote the amine functional group. The "e" at the end of the name of the parent hydrocarbon is replaced by "amine"—similar to the way in which alcohols are named. A number identifies the carbon to which the nitrogen is attached. The number can appear before the name of the parent hydrocarbon or before "amine." The name of any alkyl group bonded to nitrogen is preceded by an "*N*" (in italics) to indicate that the group is bonded to a nitrogen rather than to a carbon.

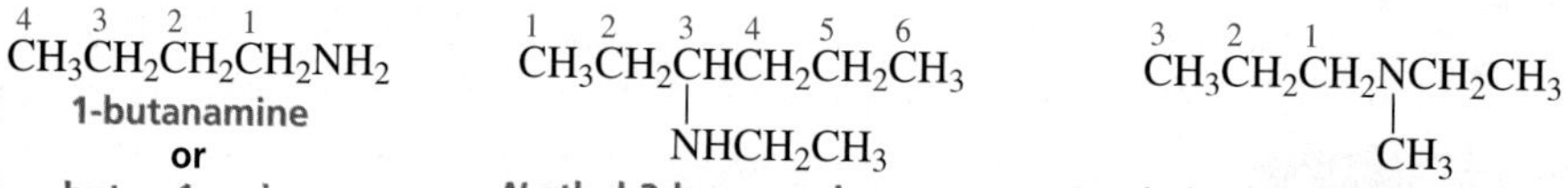

4 3 2 1 $CH_3CH_2CH_2CH_2NH_2$ **1-butanamine** **or** **butan-1-amine**	1 2 3 4 5 6 $CH_3CH_2CHCH_2CH_2CH_3$ $NHCH_2CH_3$ (on C-3) ***N*-ethyl-3-hexanamine** **or** ***N*-ethylhexan-3-amine**	3 2 1 $CH_3CH_2CH_2NCH_2CH_3$ CH_3 (on N) ***N*-ethyl-*N*-methyl-1-propanamine** **or** ***N*-ethyl-*N*-methylpropan-1-amine**

The substituents—regardless of whether they are attached to the nitrogen or to the parent hydrocarbon—are listed in alphabetical order, and then a number or an "*N*" is assigned to each one. The chain is numbered in the direction that gives the functional group suffix the lowest possible number.

4 3 2 1 $CH_3CHCH_2CH_2NHCH_3$ Cl (on C-3) **3-chloro-*N*-methyl-1-butanamine**	CH_3 (on C-5) 1 2 3 4 5 6 $CH_3CH_2CHCH_2CHCH_3$ $NHCH_2CH_3$ (on C-3) ***N*-ethyl-5-methyl-3-hexanamine**
Br (on C-4) 5 4 3 2 1 $CH_3CHCH_2CHCH_3$ CH_3NCH_3 (on C-2) **4-bromo-*N*,*N*-dimethyl-2-pentanamine**	cyclohexane ring with CH_2CH_3 and $NHCH_2CH_2CH_3$ on adjacent carbons **2-ethyl-*N*-propylcyclohexanamine**

Nitrogen compounds with four alkyl groups bonded to the nitrogen—thereby giving the nitrogen a positive formal charge—are called **quaternary ammonium salts**. Their names consist of the names of the alkyl groups in alphabetical order, followed by "ammonium" (all in one word), and then the name of the counterion as a separate word.

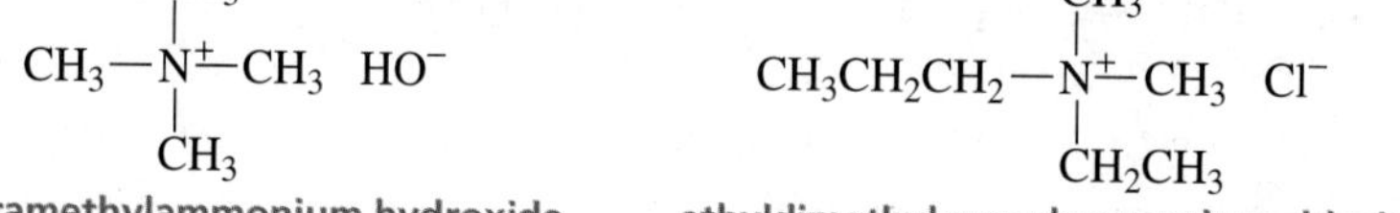

CH_3 $CH_3—\overset{+}{N}—CH_3$ HO^- CH_3 **tetramethylammonium hydroxide**	CH_3 $CH_3CH_2CH_2—\overset{+}{N}—CH_3$ Cl^- CH_2CH_3 **ethyldimethylpropylammonium chloride**

Table 2.3 summarizes the ways in which alkyl halides, ethers, alcohols, and amines are named.

Table 2.3 Summary of Nomenclature

	Systematic name	Common name
Alkyl halide	substituted alkane CH_3Br bromomethane CH_3CH_2Cl chloroethane	alkyl group to which halogen is attached, plus *halide* CH_3Br methyl bromide CH_3CH_2Cl ethyl chloride
Ether	substituted alkane CH_3OCH_3 methoxymethane $CH_3CH_2OCH_3$ methoxyethane	alkyl groups attached to oxygen, plus *ether* CH_3OCH_3 dimethyl ether $CH_3CH_2OCH_3$ ethyl methyl ether
Alcohol	functional group suffix is *ol* CH_3OH methanol CH_3CH_2OH ethanol	alkyl group to which OH is attached, plus *alcohol* CH_3OH methyl alcohol CH_3CH_2OH ethyl alcohol
Amine	functional group suffix is *amine* $CH_3CH_2NH_2$ ethanamine $CH_3CH_2CH_2NHCH_3$ *N*-methyl-1-propanamine	alkyl groups attached to N, plus *amine* $CH_3CH_2NH_2$ ethylamine $CH_3CH_2CH_2NHCH_3$ methylpropylamine

PROBLEM 16◆

Give common and systematic names for each of the following compounds:

a. $CH_3CH_2CH_2CH_2CH_2CH_2NH_2$

b. $CH_3CH_2CH_2NHCH_2CH_2CH_2CH_3$

c. $CH_3CH(CH_3)CH_2NHCH(CH_3)CH_2CH_3$

d. $CH_3CH_2CH_2N(CH_2CH_3)CH_2CH_3$

e. 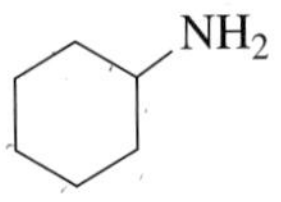 (cyclohexane ring bearing NH_2)

PROBLEM 17◆

Draw the structure of each of the following compounds:

a. 2-methyl-*N*-propyl-1-propanamine
b. *N*-ethylethanamine
c. 5-methyl-1-hexanamine
d. methyldipropylamine
e. *N,N*-dimethyl-3-pentanamine
f. cyclohexylethylmethylamine

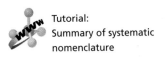

Tutorial:
Summary of systematic nomenclature

PROBLEM 18◆

For each of the following compounds, give the systematic name and the common name (for those that have common names), and indicate whether the amines are primary, secondary, or tertiary:

a. $CH_3CH(CH_3)CH_2CH_2CH_2CH_2CH_2NH_2$

b. $CH_3CH_2CH_2NHCH_2CH_2CH(CH_3)CH_3$

c. $(CH_3CH_2)_2NCH_3$

d. 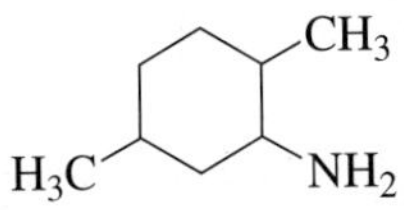 (cyclohexane ring bearing CH_3, NH_2, and H_3C substituents)

2.8 Structures of Alkyl Halides, Alcohols, Ethers, and Amines

The C—X bond (where X denotes a halogen) of an alkyl halide is formed from the overlap of an sp^3 orbital of carbon with an sp^3 orbital of the halogen (Section 1.13). Fluorine uses a $2sp^3$ orbital, chlorine a $3sp^3$ orbital, bromine a $4sp^3$ orbital, and iodine a $5sp^3$ orbital. Because the electron density of the orbital decreases with increasing volume, the C—X bond becomes longer and weaker as the size of the halogen increases (Table 2.4). Notice that this is the same trend shown by the H—X bond (Table 1.6, page 36).

The oxygen of an alcohol has the same geometry it has in water (Section 1.11). In fact, an alcohol molecule can be thought of as a water molecule with an alkyl group in place of one of the hydrogens. The oxygen atom in an alcohol is sp^3 hybridized, as it is in water. One of the sp^3 orbitals of oxygen overlaps an sp^3 orbital of a carbon, one sp^3 orbital overlaps the s orbital of a hydrogen, and the other two sp^3 orbitals each contain a lone pair.

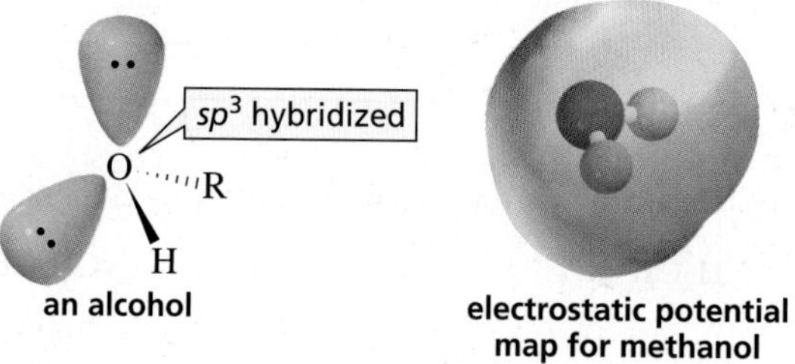

an alcohol

electrostatic potential map for methanol

The oxygen of an ether also has the same geometry it has in water. An ether molecule can be thought of as a water molecule with alkyl groups in place of both hydrogens.

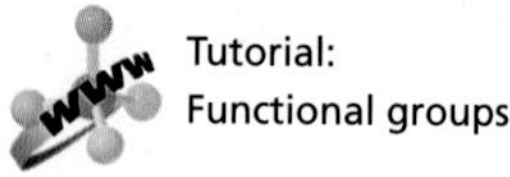
Tutorial:
Functional groups

Table 2.4 Carbon–Halogen Bond Lengths and Bond Strengths

	Orbital interactions	Bond lengths	Bond strength kcal/mol	Bond strength kJ/mol
$H_3C—F$	C, F	1.39 Å	108	451
$H_3C—Cl$	C, Cl	1.78 Å	84	350
$H_3C—Br$	C, Br	1.93 Å	70	294
$H_3C—I$	C, I	2.14 Å	57	239

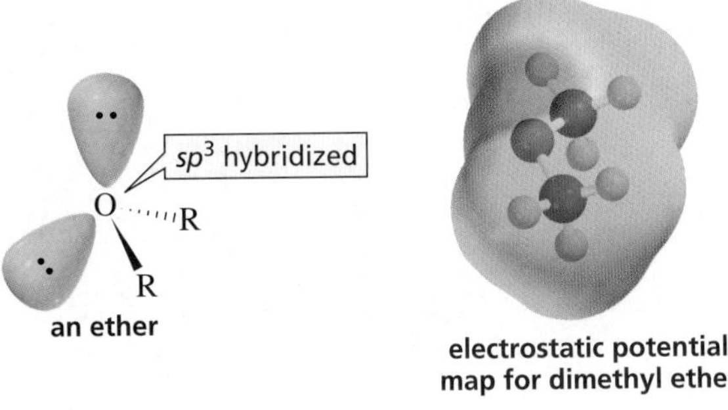

an ether

electrostatic potential map for dimethyl ether

The nitrogen of an amine has the same geometry it has in ammonia (Section 1.12). One, two, or three hydrogens may be replaced by alkyl groups. The number of hydrogens replaced by alkyl groups determines whether the amine is primary, secondary, or tertiary (Section 2.7).

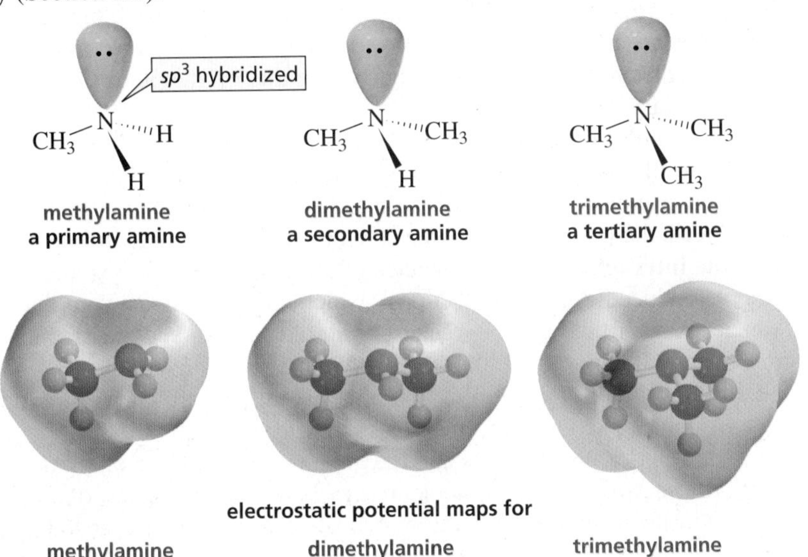

methylamine
a primary amine

dimethylamine
a secondary amine

trimethylamine
a tertiary amine

electrostatic potential maps for
methylamine dimethylamine trimethylamine

3-D Molecules: Methylamine; Dimethylamine; Trimethylamine

PROBLEM 19◆

Predict the approximate size of the following angles. (*Hint*: See Sections 1.11 and 1.12.)

a. the C—O—C bond angle in an ether
b. the C—N—C bond angle in a secondary amine
c. the C—O—H bond angle in an alcohol
d. the C—N—C bond angle in a quaternary ammonium salt

2.9 Physical Properties of Alkanes, Alkyl Halides, Alcohols, Ethers, and Amines

Boiling Points

The **boiling point (bp)** of a compound is the temperature at which the liquid form of the compound becomes a gas (vaporizes). In order for a compound to vaporize, the forces that hold the individual molecules close to each other in the liquid must be overcome. This means that the boiling point of a compound depends on the strength of the attractive forces between the individual molecules. If the molecules are held together by strong forces, it will take a lot of energy to pull the molecules away from each other and the compound will have a high boiling point. In contrast, if the molecules are held together by weak forces, only a small amount of energy will be needed to pull the molecules away from each other and the compound will have a low boiling point.

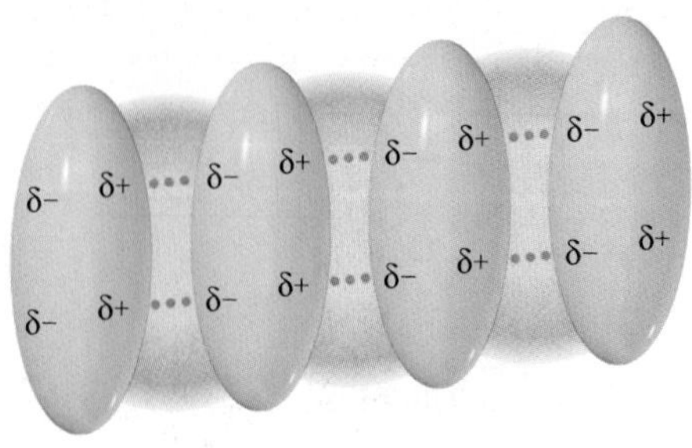

Figure 2.1 ▶ Van der Waals forces are induced-dipole–induced-dipole interactions.

Johannes Diderik van der Waals (1837–1923) *was a Dutch physicist. He was born in Leiden, the son of a carpenter, and was largely self-taught when he entered the University of Leiden, where he earned a Ph.D. He was a professor of physics at the University of Amsterdam from 1877 to 1903. He won the 1910 Nobel Prize for his research on the gaseous and liquid states of matter.*

Relatively weak forces hold alkane molecules together. Alkanes contain only carbon and hydrogen atoms. Because the electronegativities of carbon and hydrogen are similar, the bonds in alkanes are nonpolar. Consequently, there are no significant partial charges on any of the atoms in an alkane.

It is, however, only the average charge distribution over the alkane molecule that is neutral. Electrons are moving continuously, so at any instant the electron density on one side of the molecule can be slightly higher than that on the other side, giving the molecule a temporary dipole.

A temporary dipole in one molecule can induce a temporary dipole in a nearby molecule. As a result, the negative side of one molecule ends up adjacent to the positive side of another molecule, as shown in Figure 2.1. Because the dipoles in the molecules are induced, the interactions between the molecules are called **induced-dipole–induced-dipole interactions**. The molecules of an alkane are held together by these induced-dipole–induced-dipole interactions, which are known as **van der Waals forces**. Van der Waals forces are the weakest of all the intermolecular attractions.

In order for an alkane to boil, the van der Waals forces must be overcome. The magnitude of the van der Waals forces that hold alkane molecules together depends on the area of contact between the molecules. The greater the area of contact, the stronger are the van der Waals forces and the greater is the amount of energy needed to overcome those forces. If you look at the homologous series of alkanes in Table 2.1, you will see that the boiling points of alkanes increase as their size increases. This relationship holds because each additional methylene group increases the area of contact between the molecules. The four smallest alkanes have boiling points below room temperature (room temperature is about 25 °C), so they exist as gases at room temperature. Pentane (bp = 36.1 °C) is the smallest alkane that is a liquid at room temperature.

Because the strength of the van der Waals forces depends on the area of contact between the molecules, branching in a compound lowers its boiling point because it reduces the area of contact. A branched compound has a more compact, nearly spherical shape. If you think of the unbranched alkane pentane as a cigar and branched neopentane as a tennis ball, you can see that branching decreases the area of contact between molecules: Two cigars make contact over a greater area than do two tennis balls. Thus, if two alkanes have the same molecular weight, the more highly branched alkane will have a lower boiling point.

$CH_3CH_2CH_2CH_2CH_3$
pentane
bp = 36.1 °C

$CH_3CH(CH_3)CH_2CH_3$
isopentane
bp = 27.9 °C

$CH_3C(CH_3)_2CH_3$
neopentane
bp = 9.5 °C

The boiling points of the compounds in any homologous series increase as their molecular weights increase because of the increase in van der Waals forces. So the boiling points of the compounds in a homologous series of ethers, alkyl halides, alcohols, and amines increase with increasing molecular weight. (See Appendix I.)

The boiling points of these compounds, however, are also affected by the polar character of the C—Z bond (where Z denotes N, O, F, Cl, or Br) because nitrogen, oxygen, and the halogens are more electronegative than the carbon to which they are attached.

$$\mathrm{R{-}\overset{|}{\underset{|}{C}}^{\delta+}{-}Z^{\delta-}} \qquad Z = N, O, F, Cl, \text{ or } Br$$

The magnitude of the charge differential between the two bonded atoms is indicated by the bond dipole moment (Section 1.3).

$H_3C{-}NH_2$ 0.2 D $H_3C{-}O{-}CH_3$ 0.7 D $H_3C{-}I$ 1.2 D $H_3C{-}OH$ 0.7 D

$H_3C{-}F$ 1.6 D $H_3C{-}Br$ 1.4 D $H_3C{-}Cl$ 1.5 D

The dipole moment of a bond is equal to the magnitude of the charge on one of the bonded atoms times the distance between the bonded atoms.

Molecules with dipole moments are attracted to one another because they can align themselves in such a way that the positive end of one dipole is adjacent to the negative end of another dipole. These electrostatic attractive forces, called **dipole–dipole interactions**, are stronger than van der Waals forces, but not as strong as ionic or covalent bonds.

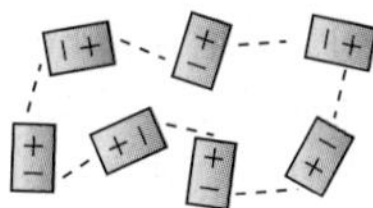

Ethers generally have higher boiling points than alkanes of comparable molecular weight because both van der Waals forces and dipole–dipole interactions must be overcome for an ether to boil (Table 2.5).

More extensive tables of physical properties can be found in Appendix I.

cyclopentane
bp = 49.3 °C

tetrahydrofuran
bp = 65 °C

As the table shows, alcohols have much higher boiling points than alkanes or ethers of comparable molecular weight because, in addition to van der Waals forces and the dipole–dipole interactions of the C—O bond, alcohols can form **hydrogen bonds**. A hydrogen bond is a special kind of dipole–dipole interaction that occurs between a hydrogen that is bonded to an oxygen, a nitrogen, or a fluorine and the lone-pair electrons of an oxygen, nitrogen, or fluorine in another molecule.

The length of the covalent bond between oxygen and hydrogen is 0.96 Å. The hydrogen bond between an oxygen of one molecule and a hydrogen of another molecule is almost twice as long (1.69–1.79 Å), which means that a hydrogen bond is not as strong as an O—H covalent bond. A hydrogen bond, however, is stronger than other

Table 2.5 Comparative Boiling Points (°C)

Alkanes	Ethers	Alcohols	Amines
$CH_3CH_2CH_3$ −42.1	CH_3OCH_3 −23.7	CH_3CH_2OH 78	$CH_3CH_2NH_2$ 16.6
$CH_3CH_2CH_2CH_3$ −0.5	$CH_3OCH_2CH_3$ 10.8	$CH_3CH_2CH_2OH$ 97.4	$CH_3CH_2CH_2NH_2$ 47.8
$CH_3CH_2CH_2CH_2CH_3$ 36.1	$CH_3CH_2OCH_2CH_3$ 34.5	$CH_3CH_2CH_2CH_2OH$ 117.3	$CH_3CH_2CH_2CH_2NH_2$ 77.8

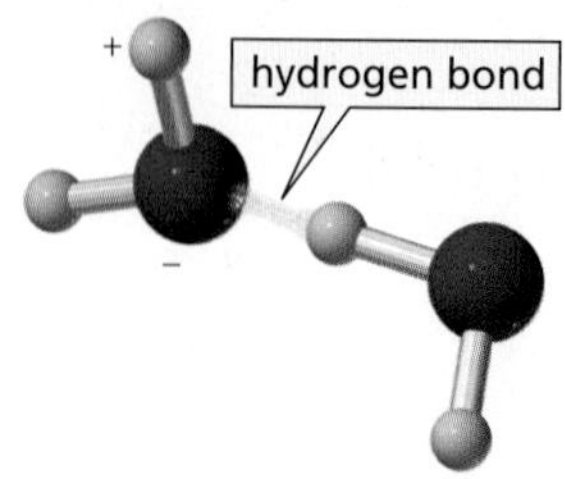

hydrogen bonding in water

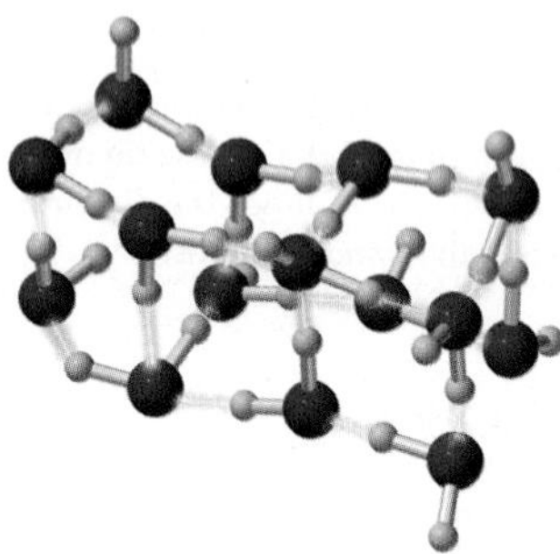

dipole–dipole interactions. The strongest hydrogen bonds are linear—the two electronegative atoms and the hydrogen between them lie on a straight line.

H—Ö:·····H—Ö:·····H—Ö:·····H—Ö:
hydrogen bond (H···Ö between the rows) 1.69 – 1.79 Å
H—Ö:·····H—Ö:·····H—Ö:·····H—Ö:
0.96 Å (O—H)

hydrogen bond
H—N̈—H·····:N—H·····:N—H·····:N—H (each N also bonded to two further H)

hydrogen bonds
H—F̈:·····H—F̈:·····H—F̈:

Although each individual hydrogen bond is weak—requiring about 5 kcal/mol (or 21 kJ/mol) to break—there are many such bonds holding alcohol molecules together. The extra energy required to break these hydrogen bonds is the reason alcohols have much higher boiling points than either alkanes or ethers with similar molecular weights.

The boiling point of water illustrates the dramatic effect hydrogen bonding has on boiling points. Water has a molecular weight of 18 and a boiling point of 100 °C. The alkane nearest in size is methane, with a molecular weight of 16. Methane boils at −167.7 °C.

Primary and secondary amines also form hydrogen bonds, so these amines have higher boiling points than alkanes with similar molecular weights. Nitrogen is not as electronegative as oxygen, however, which means that the hydrogen bonds between amine molecules are weaker than the hydrogen bonds between alcohol molecules. An amine, therefore, has a lower boiling point than an alcohol with a similar molecular weight (Table 2.5).

Because primary amines have two N—H bonds, hydrogen bonding is more significant in primary amines than in secondary amines. Tertiary amines cannot form hydrogen bonds between their own molecules because they do not have a hydrogen attached to the nitrogen. Consequently, if you compare amines with the same molecular weight and similar structures, you will find that primary amines have higher boiling points than secondary amines and secondary amines have higher boiling points than tertiary amines.

$CH_3CH_2CH(CH_3)CH_2NH_2$	$CH_3CH_2CH(CH_3)NHCH_3$	$CH_3CH_2N(CH_3)CH_2CH_3$
a primary amine bp = 97 °C	**a secondary amine** bp = 84 °C	**a tertiary amine** bp = 65 °C

PROBLEM-SOLVING STRATEGY

a. Which of the following compounds will form hydrogen bonds between its molecules?

1. $CH_3CH_2CH_2OH$ 2. $CH_3CH_2CH_2SH$ 3. $CH_3OCH_2CH_3$

b. Which of these compounds form hydrogen bonds with a solvent such as ethanol?

In solving this type of question, start by defining the kind of compound that will do what is being asked.

a. A hydrogen bond forms when a hydrogen that is attached to an O, N, or F of one molecule interacts with a lone pair on an O, N, or F of another molecule. Therefore, a compound that will form hydrogen bonds with itself must have a hydrogen bonded to an O, N, or F. Only compound 1 will be able to form hydrogen bonds with itself.

b. Ethanol has an H bonded to an O, so it will be able to form hydrogen bonds with a compound that has a lone pair on an O, N, or F. Compounds 1 and 3 will be able to form hydrogen bonds with ethanol.

Now continue on to Problem 20.

PROBLEM 20◆

a. Which of the following compounds will form hydrogen bonds between its molecules?
1. $CH_3CH_2CH_2COOH$
2. $CH_3CH_2N(CH_3)_2$
3. $CH_3CH_2CH_2CH_2Br$
4. $CH_3CH_2CH_2NHCH_3$
5. $CH_3CH_2OCH_2CH_2OH$
6. $CH_3CH_2CH_2CH_2F$

b. Which of the preceding compounds form hydrogen bonds with a solvent such as ethanol?

PROBLEM 21

Explain why
a. H_2O has a higher boiling point than CH_3OH (65 °C).
b. H_2O has a higher boiling point than NH_3 (−33 °C).
c. H_2O has a higher boiling point than HF (20 °C).

PROBLEM 22◆

List the following compounds in order of decreasing boiling point:

OH   HO OH OH   OH OH NH_2

PROBLEM 23

List the following compounds in order of decreasing boiling point:

a. $CH_3CH_2CH_2CH_2CH_2CH_2Br$ $CH_3CH_2CH_2CH_2Br$ $CH_3CH_2CH_2CH_2CH_2Br$

b. $CH_3CH(CH_3)CH_2CH_2CH_2CH_2CH_3$ $CH_3C(CH_3)_2{-}C(CH_3)_2CH_3$

$CH_3CH_2CH_2CH_2CH_2CH_2CH_2CH_3$ $CH_3CH_2CH_2CH_2CH_2CH_2CH_2CH_2CH_3$

c. $CH_3CH_2CH_2CH_2CH_3$ $CH_3CH_2CH_2CH_2OH$ $CH_3CH_2CH_2CH_2Cl$
$CH_3CH_2CH_2CH_2CH_2OH$

Both van der Waals forces and dipole–dipole interactions must be overcome in order for an alkyl halide to boil. As the halogen atom increases in size, the size of its electron cloud increases. As a result, both the van der Waals contact area and the *polarizability* of the electron cloud increase.

Polarizability indicates how readily an electron cloud can be distorted. The larger the atom, the more loosely it holds the electrons in its outermost shell, and the more they can be distorted. The more polarizable the atom, the stronger are the van der Waals interactions. Therefore, an alkyl fluoride has a lower boiling point than an alkyl chloride with the same alkyl group. Similarly, alkyl chlorides have lower boiling points than alkyl bromides, which have lower boiling points than alkyl iodides (Table 2.6).

Table 2.6 Comparative Boiling Points of Alkanes and Alkyl Halides (°C)

	Y				
	H	F	Cl	Br	I
$CH_3{-}Y$	−161.7	−78.4	−24.2	3.6	42.4
$CH_3CH_2{-}Y$	−88.6	−37.7	12.3	38.4	72.3
$CH_3CH_2CH_2{-}Y$	−42.1	−2.5	46.6	71.0	102.5
$CH_3CH_2CH_2CH_2{-}Y$	−0.5	32.5	78.4	101.6	130.5
$CH_3CH_2CH_2CH_2CH_2{-}Y$	36.1	62.8	107.8	129.6	157.0

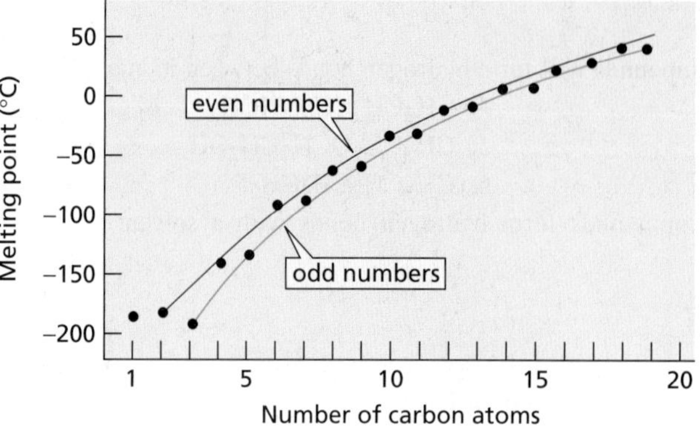

Figure 2.2 ▶
Melting points of straight-chain alkanes. Alkanes with even numbers of carbon atoms fall on a melting-point curve that is higher than the melting-point curve for alkanes with odd numbers of carbon atoms.

Melting Points

The **melting point (mp)** is the temperature at which a solid is converted into a liquid. If you examine the melting points of the alkanes in Table 2.1, you will see that the melting points increase (with a few exceptions) in a homologous series as the molecular weight increases. The increase in melting point is less regular than the increase in boiling point because *packing* influences the melting point of a compound. **Packing** is a property that determines how well the individual molecules in a solid fit together in a crystal lattice. The tighter the fit, the more energy is required to break the lattice and melt the compound.

In Figure 2.2, you can see that the melting points of alkanes with even numbers of carbon atoms fall on a smooth curve (the red line). The melting points of alkanes with odd numbers of carbon atoms also fall on a smooth curve (the green line). The two curves do not overlap, however, because alkanes with an odd number of carbon atoms pack less tightly than alkanes with an even number of carbon atoms. Alkanes with an odd number of carbon atoms pack less tightly because the methyl groups at the ends of their chains can avoid those of another chain only by increasing the distance between their chains. Consequently, alkane molecules with odd numbers of carbon atoms have lower intermolecular attractions and correspondingly lower melting points.

Solubility

The general rule that explains **solubility** on the basis of the polarity of molecules is that "like dissolves like." In other words, polar compounds dissolve in polar solvents, and nonpolar compounds dissolve in nonpolar solvents. This is because a polar solvent such as water has partial charges that can interact with the partial charges on a polar compound. The negative poles of the solvent molecules surround the positive pole of the polar solute, and the positive poles of the solvent molecules surround the negative pole of the polar solute. Clustering of the solvent molecules around the solute molecules separates solute molecules from each other, which is what makes them dissolve. The interaction between a solvent and a molecule or an ion dissolved in that solvent is called **solvation**.

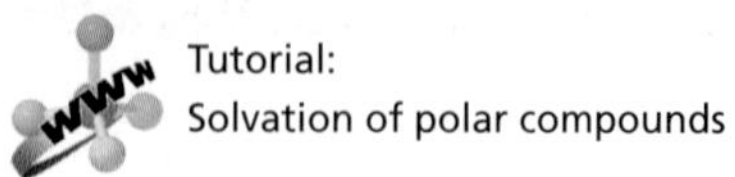
Tutorial:
Solvation of polar compounds

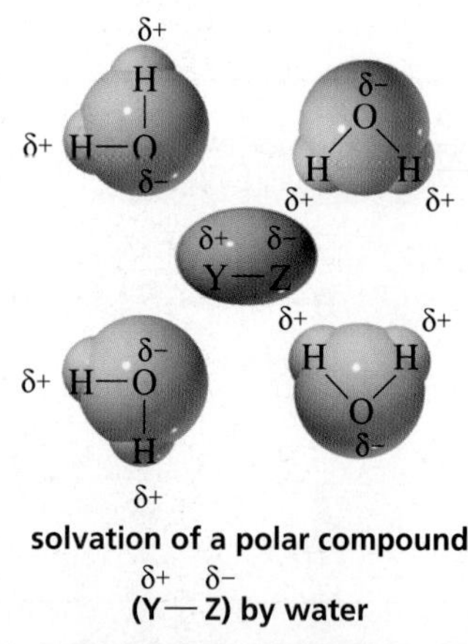

solvation of a polar compound
$\overset{\delta+}{\text{Y}}$—$\overset{\delta-}{\text{Z}}$ (Y—Z) by water

Because nonpolar compounds have no net charge, polar solvents are not attracted to them. In order for a nonpolar molecule to dissolve in a polar solvent such as water, the nonpolar molecule would have to push the water molecules apart, disrupting their hydrogen bonding, which is strong enough to exclude the nonpolar compound. In contrast, nonpolar solutes dissolve in nonpolar solvents because the van der Waals interactions between solvent and solute molecules are about the same as between solvent–solvent and solute–solute molecules.

Alkanes are nonpolar, which causes them to be soluble in nonpolar solvents and insoluble in polar solvents such as water. The densities of alkanes (Table 2.1) increase with increasing molecular weight, but even a 30-carbon alkane such as triacontane (density at 20 °C = 0.8097 g/mL) is less dense than water (density at 20 °C = 0.9982 g/mL). This means that a mixture of an alkane and water will separate into two distinct layers, with the less dense alkane floating on top. The Alaskan oil spill of 1989, the Persian Gulf spill of 1991, and the even larger spill off the northwest coast of Spain in 2002 are large-scale examples of this phenomenon. (Crude oil is primarily a mixture of alkanes.)

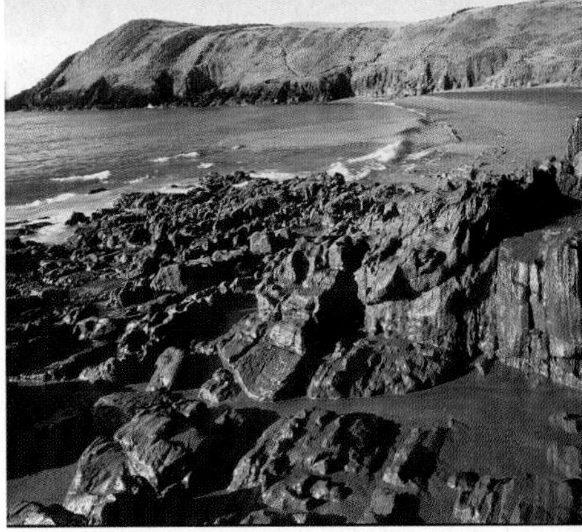
▲ Oil from a 70,000-ton oil spill in 1996 off the coast of Wales.

An alcohol has both a nonpolar alkyl group and a polar OH group. So is an alcohol molecule nonpolar or polar? Is it soluble in a nonpolar solvent, or is it soluble in water? The answer depends on the size of the alkyl group. As the alkyl group increases in size, it becomes a more significant fraction of the alcohol molecule and the compound becomes less and less soluble in water. In other words, the molecule becomes more and more like an alkane. Four carbons tend to be the dividing line at room temperature. Alcohols with fewer than four carbons are soluble in water, but alcohols with more than four carbons are insoluble in water. In other words, an OH group can drag about three or four carbons into solution in water.

The four-carbon dividing line is only an approximate guide because the solubility of an alcohol also depends on the structure of the alkyl group. Alcohols with branched alkyl groups are more soluble in water than alcohols with nonbranched alkyl groups with the same number of carbons, because branching minimizes the contact surface of the nonpolar portion of the molecule. So *tert*-butyl alcohol is more soluble than *n*-butyl alcohol in water.

Similarly, the oxygen atom of an ether can drag only about three carbons into solution in water (Table 2.7). We have already seen (photo on page 52) that diethyl ether—an ether with four carbons—is not soluble in water.

Low-molecular-weight amines are soluble in water because amines can form hydrogen bonds with water. Comparing amines with the same number of carbons, we find that primary amines are more soluble than secondary amines because primary amines have two hydrogens that can engage in hydrogen bonding. Tertiary amines, like primary and secondary amines, have lone-pair electrons that can accept hydrogen bonds, but unlike primary and secondary amines, tertiary amines do not have hydrogens to donate for hydrogen bonds. Tertiary amines, therefore, are less soluble in water than are secondary amines with the same number of carbons.

Alkyl halides have some polar character, but only the alkyl fluorides have an atom that can form a hydrogen bond with water. This means that alkyl fluorides are the most water soluble of the alkyl halides. The other alkyl halides are less soluble in water than ethers or alcohols with the same number of carbons (Table 2.8).

Table 2.7 Solubilities of Ethers in Water

2 C's	CH_3OCH_3	soluble
3 C's	$CH_3OCH_2CH_3$	soluble
4 C's	$CH_3CH_2OCH_2CH_3$	slightly soluble (10 g/100 g H_2O)
5 C's	$CH_3CH_2OCH_2CH_2CH_3$	minimally soluble (1.0 g/100 g H_2O)
6 C's	$CH_3CH_2CH_2OCH_2CH_2CH_3$	insoluble (0.25 g/100 g H_2O)

Table 2.8 Solubilities of Alkyl Halides in Water

CH_3F very soluble	CH_3Cl soluble	CH_3Br slightly soluble	CH_3I slightly soluble
CH_3CH_2F soluble	CH_3CH_2Cl slightly soluble	CH_3CH_2Br slightly soluble	CH_3CH_2I slightly soluble
$CH_3CH_2CH_2F$ slightly soluble	$CH_3CH_2CH_2Cl$ slightly soluble	$CH_3CH_2CH_2Br$ slightly soluble	$CH_3CH_2CH_2I$ slightly soluble
$CH_3CH_2CH_2CH_2F$ insoluble	$CH_3CH_2CH_2CH_2Cl$ insoluble	$CH_3CH_2CH_2CH_2Br$ insoluble	$CH_3CH_2CH_2CH_2I$ insoluble

PROBLEM 24◆

Rank the following groups of compounds in order of decreasing solubility in water:

a. $CH_3CH_2CH_2OH$ $CH_3CH_2CH_2CH_2Cl$ $CH_3CH_2CH_2CH_2OH$ $HOCH_2CH_2CH_2OH$

b. CH_3 (methylcyclopentane) NH_2 (cyclopentylamine) OH (cyclopentanol)

PROBLEM 25◆

In which of the following solvents would cyclohexane have the lowest solubility: 1-pentanol, diethyl ether, ethanol, or hexane?

2.10 Conformations of Alkanes: Rotation About Carbon–Carbon Bonds

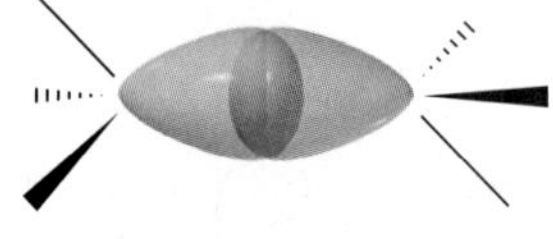

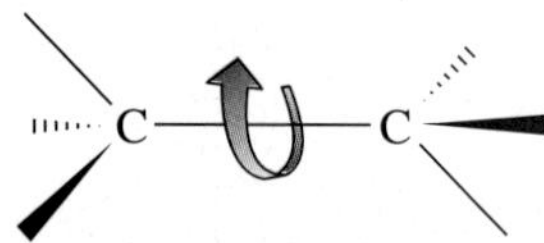

▲ **Figure 2.3**
A carbon–carbon bond is formed by the overlap of cylindrically symmetrical sp^3 orbitals. Therefore, rotation about the bond can occur without changing the amount of orbital overlap.

We have seen that a carbon–carbon single bond (a σ bond) is formed when an sp^3 orbital of one carbon overlaps an sp^3 orbital of a second carbon (Section 1.7). Because σ bonds are cylindrically symmetrical (i.e., symmetrical about an imaginary line connecting the centers of the two atoms joined by the σ bond), rotation about a carbon–carbon single bond can occur without any change in the amount of orbital overlap (Figure 2.3). The different spatial arrangements of the atoms that result from rotation about a single bond are called **conformations**. A specific conformation is called a **conformer**.

When rotation occurs about the carbon–carbon bond of ethane, two extreme conformations can result—a *staggered conformation* and an *eclipsed conformation*. An infinite number of conformations between these two extremes are also possible.

Compounds are three dimensional, but we are limited to a two-dimensional sheet of paper when we show their structures. Perspective formulas, sawhorse projections, and Newman projections are methods chemists commonly use to represent on paper the three-dimensional spatial arrangements of the atoms that result from rotation about a σ bond. In a **perspective formula**, solid lines are used for bonds that lie in the plane of the paper, solid wedges for bonds protruding out from the plane of the paper, and hatched wedges for bonds extending behind the paper. In a **sawhorse projection**, you are looking at the carbon–carbon bond from an oblique angle. In a **Newman projection**, you are looking down the length of a particular carbon–carbon bond. The carbon in front is represented by the point at which three bonds intersect, and the carbon in back is represented by a circle. The three lines emanating from each of the carbons represent its other three bonds. In discussing the conformations of alkanes, we will use Newman projections because they are easy to draw and they do a good job of representing the spatial relationships of the substituents on the two carbon atoms.

Melvin S. Newman (1908–1993) *was born in New York. He received a Ph.D. from Yale University in 1932 and was a professor of chemistry at Ohio State University from 1936 to 1973.*

	staggered conformation for rotation about the carbon–carbon bond in ethane		eclipsed conformation for rotation about the carbon–carbon bond in ethane
perspective formulas		60° ⇌	
sawhorse projections		60° ⇌	
Newman projections		60° ⇌	

The electrons in a C—H bond will repel the electrons in another C—H bond if the bonds get too close to each other. The **staggered conformation**, therefore, is the most stable conformation of ethane because the C—H bonds are as far away from each other as possible. The **eclipsed conformation** is the least stable conformation because in no other conformation are the C—H bonds as close to one another. The extra energy of the eclipsed conformation is called *torsional strain*. **Torsional strain** is the name given to the repulsion felt by the bonding electrons of one substituent as they pass close to the bonding electrons of another substituent. The investigation of the various conformations of a compound and their relative stabilities is called **conformational analysis.**

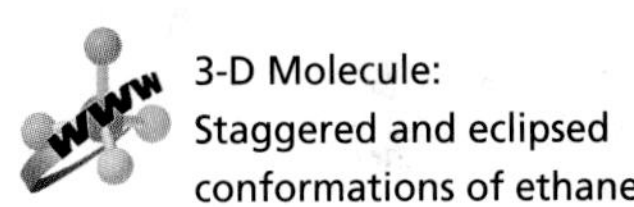

Rotation about a carbon–carbon single bond is not completely free because of the energy difference between the staggered and eclipsed conformers. The eclipsed conformer is higher in energy, so an energy barrier must be overcome when rotation about the carbon–carbon bond occurs (Figure 2.4). However, the barrier in ethane is small

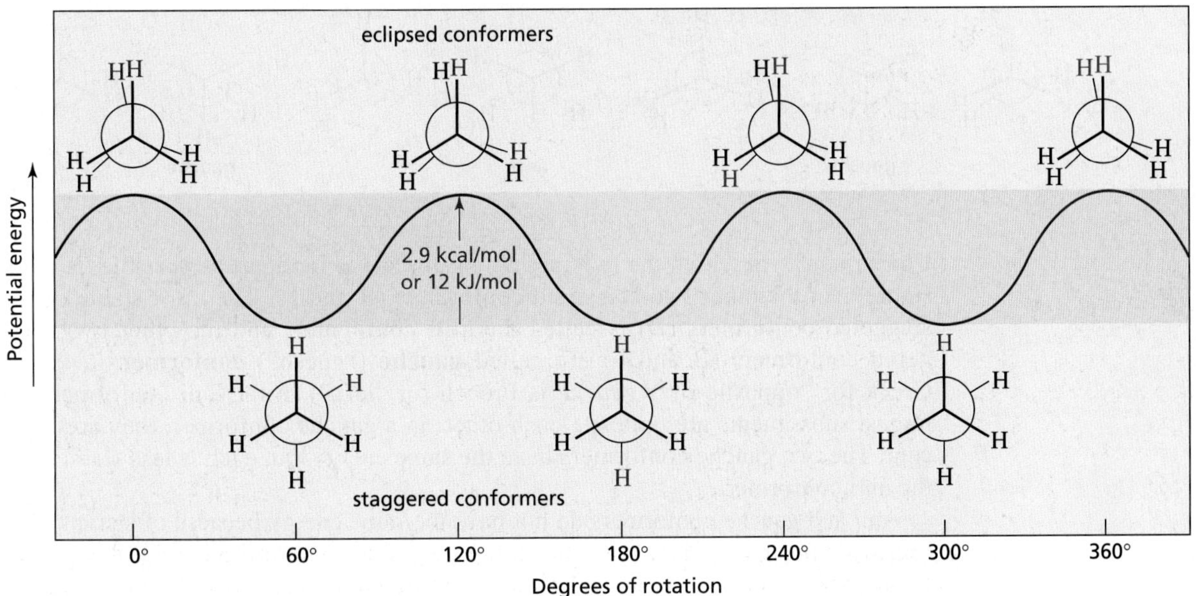

▲ **Figure 2.4**
Potential energy of ethane as a function of the angle of rotation about the carbon–carbon bond.

enough (2.9 kcal/mol or 12 kJ/mol) to allow the conformers to interconvert millions of times per second at room temperature. Because the conformers interconvert, they cannot be separated.

Figure 2.4 shows the potential energies of all the conformers of ethane obtained during one complete 360° rotation. Notice that the **staggered conformers** are at energy minima, whereas the **eclipsed conformers** are at energy maxima.

Butane has three carbon–carbon single bonds, and the molecule can rotate about each of them. In the following figure, staggered and eclipsed conformers are drawn for rotation about the C-1—C-2 bond:

the C-2—C-3 bond

$\overset{1}{C}H_3-\overset{2}{C}H_2-\overset{3}{C}H_2-\overset{4}{C}H_3$

butane

the C-1—C-2 bond

the C-3—C-4 bond

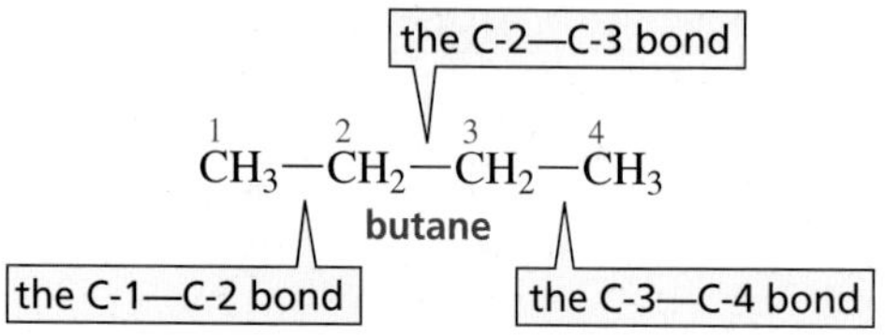

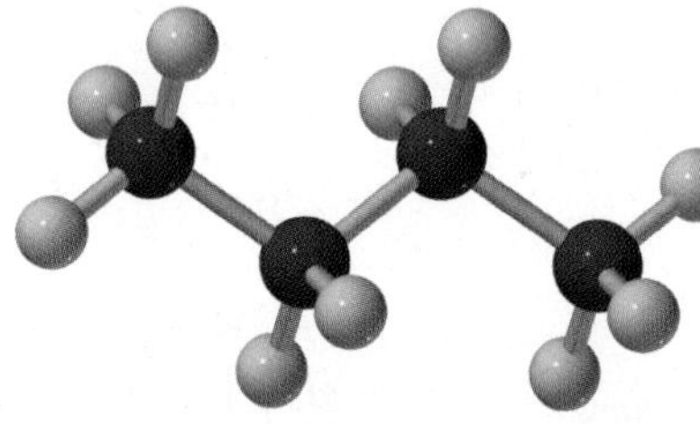

ball-and-stick model of butane

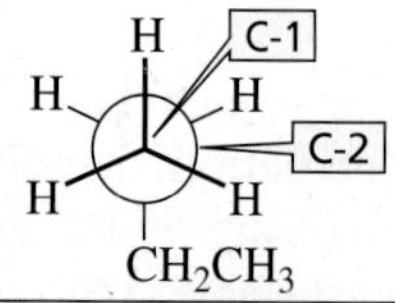

staggered conformation for rotation about the C-1—C-2 bond in butane

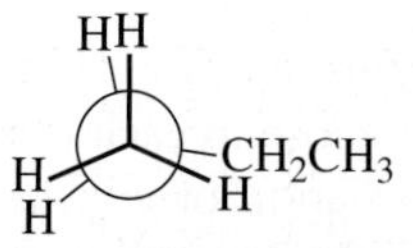

eclipsed conformation for rotation about the C-1—C-2 bond in butane

Note that the carbon in the foreground in a Newman projection has the lower number. Although the staggered conformers resulting from rotation about the C-1—C-2 bond in butane all have the same energy, the staggered conformers resulting from rotation about the C-2—C-3 bond do not have the same energy. The staggered conformers for rotation about the C-2—C-3 bond in butane are shown below.

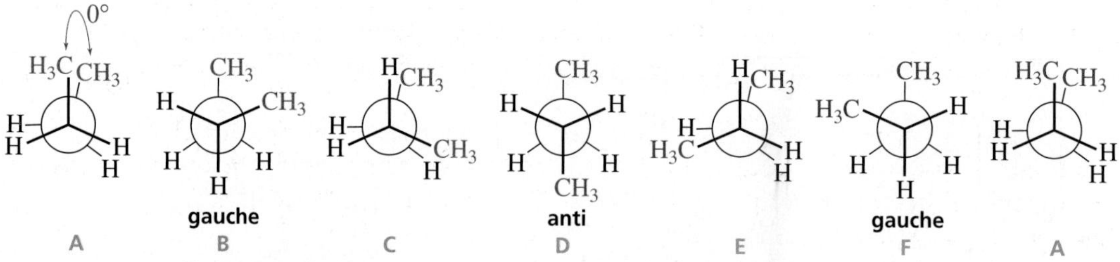

0°

A, gauche B, C, anti D, E, gauche F, A

Conformer D, in which the two methyl groups are as far apart as possible, is more stable than the other two staggered conformers (B and F). The most stable of the staggered conformers (D) is called the **anti conformer**, and the other two staggered conformers (B and F) are called **gauche** ("goesh") **conformers**. (*Anti* is Greek for "opposite of"; *gauche* is French for "left.") In the anti conformer, the largest substituents are opposite each other; in a gauche conformer, they are adjacent. The two gauche conformers have the same energy, but each is less stable than the anti conformer.

Anti and gauche conformers do not have the same energy because of steric strain. **Steric strain** is the strain (i.e., the extra energy) put on a molecule when atoms or groups are too close to one another, which results in repulsion between the electron

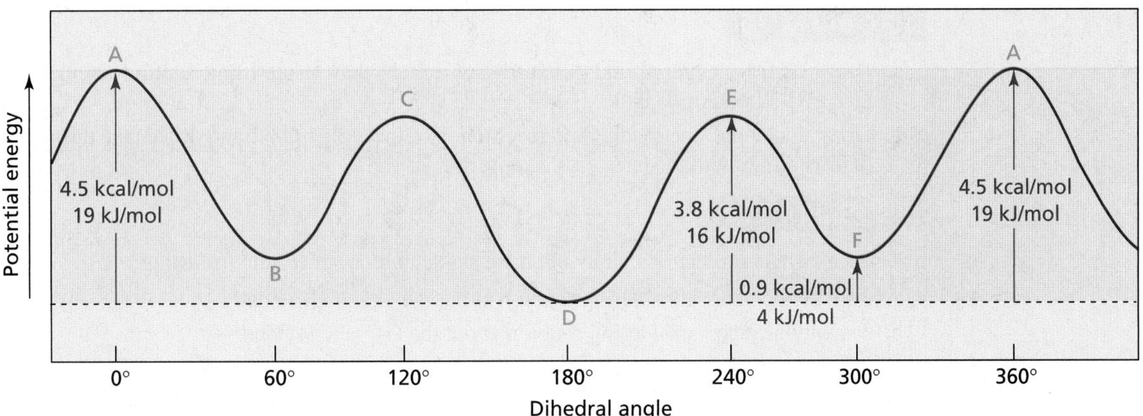

▲ **Figure 2.5**
Potential energy of butane as a function of the degree of rotation about the C-2—C-3 bond. Green letters refer to the conformers (A–F) shown on page 90.

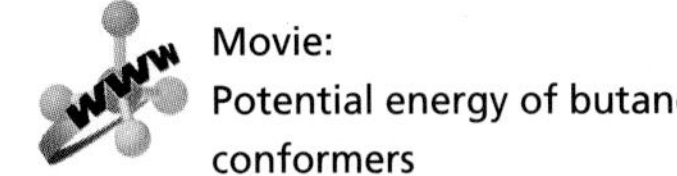
Movie: Potential energy of butane conformers

clouds of these atoms or groups. For example, there is more steric strain in a gauche conformer than in the anti conformer because the two methyl groups are closer together in a gauche conformer. This type of steric strain is called a **gauche interaction**.

The eclipsed conformers resulting from rotation about the C-2—C-3 bond in butane also have different energies. The eclipsed conformer in which the two methyl groups are closest to each other (A) is less stable than the eclipsed conformers in which they are farther apart (C and E). The energies of the conformers obtained from rotation about the C-2—C-3 bond of butane are shown in Figure 2.5. (The dihedral angle is the angle between the CH_3—C—C and C—C—CH_3 planes. Therefore, the conformer in which one methyl group stands directly in front of the other—the least stable conformer—has a dihedral angle of 0°.) All the eclipsed conformers have both torsional and steric strain—torsional strain due to bond–bond repulsion and steric strain due to the closeness of the groups. In general, steric strain in molecules increases as the size of the group increases.

Because there is continuous rotation about all the carbon–carbon single bonds in a molecule, organic molecules with carbon–carbon single bonds are not static balls and sticks—they have many interconvertible conformers. The conformers cannot be separated, however, because their small energy difference allows them to interconvert rapidly.

The relative number of molecules in a particular conformation at any one time depends on the stability of the conformation: The more stable the conformation, the greater is the fraction of molecules that will be in that conformation. Most molecules, therefore, are in staggered conformations, and more molecules are in an anti conformation than in a gauche conformation. The tendency to assume a staggered conformation causes carbon chains to orient themselves in a zigzag fashion, as shown by the ball-and-stick model of decane.

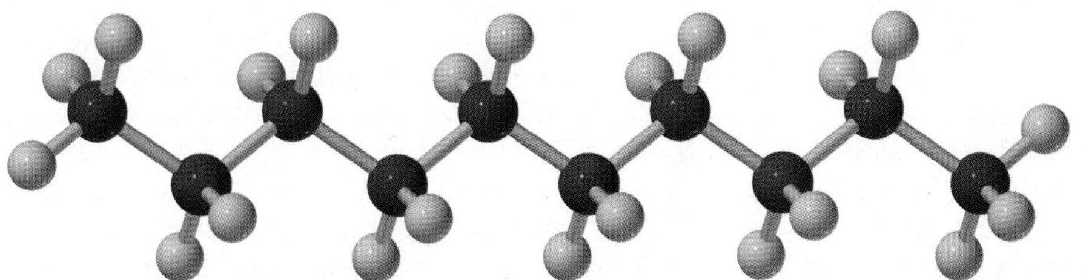

ball-and-stick model of decane

3-D Molecule: Decane

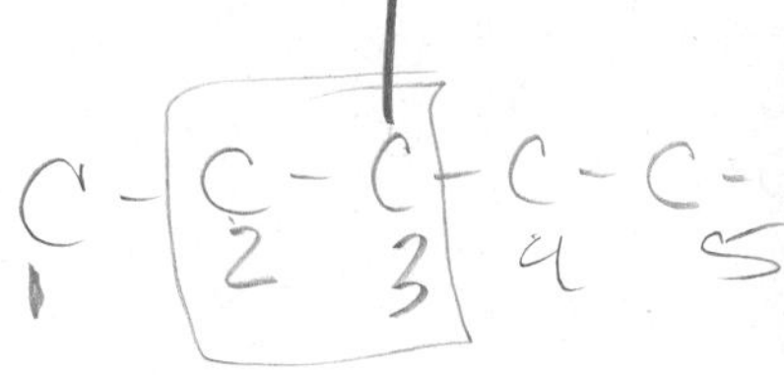

PROBLEM 26

a. Draw all the staggered and eclipsed conformers that result from rotation about the C-2—C-3 bond of pentane.
b. Draw a potential-energy diagram for rotation of the C-2—C-3 bond of pentane through 360°, starting with the least stable conformer.

PROBLEM 27◆

Using Newman projections, draw the most stable conformer for the following:
a. 3-methylpentane, considering rotation about the C-2—C-3 bond
b. 3-methylhexane, considering rotation about the C-3—C-4 bond
c. 3,3-dimethylhexane, considering rotation about the C-3—C-4 bond

2.11 Cycloalkanes: Ring Strain

Early chemists observed that cyclic compounds found in nature generally had five- or six-membered rings. Compounds with three- and four-membered rings were found much less frequently. This observation suggested that compounds with five- and six-membered rings were more stable than compounds with three- or four-membered rings.

In 1885, the German chemist Adolf von Baeyer proposed that the instability of three- and four-membered rings was due to angle strain. We know that, ideally, an sp^3 hybridized carbon has bond angles of 109.5° (Section 1.7). Baeyer suggested that the stability of a cycloalkane could be predicted by determining how close the bond angle of a planar cycloalkane is to the ideal tetrahedral bond angle of 109.5°. The angles in an equilateral triangle are 60°. The bond angles in cyclopropane, therefore, are compressed from the ideal bond angle of 109.5° to 60°, a 49.5° deviation. This deviation of the bond angle from the ideal bond angle causes strain called **angle strain**.

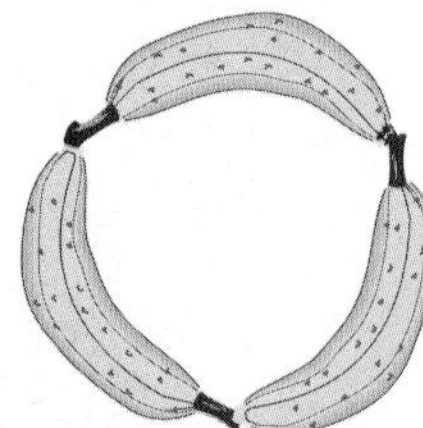

banana bonds

The angle strain in a three-membered ring can be appreciated by looking at the orbitals that overlap to form the σ bonds in cyclopropane (Figure 2.6). Normal σ bonds are formed by the overlap of two sp^3 orbitals that point directly at each other. In cyclopropane, overlapping orbitals cannot point directly at each other. Therefore, the orbital overlap is less effective than in a normal C—C bond. The less effective orbital overlap is what causes angle strain, which in turn causes the C—C bond to be weaker than a normal C—C bond. Because the C—C bonding orbitals in cyclopropane can't point directly at each other, they have shapes that resemble bananas and, consequently, are often called **banana bonds**. In addition to possessing angle strain, three-membered rings have torsional strain because all the adjacent C—H bonds are eclipsed.

The bond angles in planar cyclobutane would have to be compressed from 109.5° to 90°, the bond angle associated with a planar four-membered ring. Planar cyclobutane would then be expected to have less angle strain than cyclopropane because the bond angles in cyclobutane are only 19.5° away from the ideal bond angle.

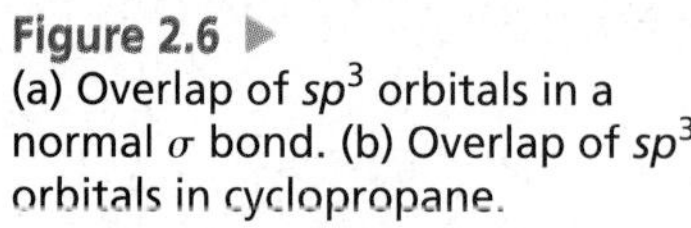

Figure 2.6 ▶ (a) Overlap of sp^3 orbitals in a normal σ bond. (b) Overlap of sp^3 orbitals in cyclopropane.

PROBLEM 28◆

The bond angles in a regular polygon with n sides are equal to

$$180° - \frac{360°}{n}$$

a. What are the bond angles in a regular octagon?
b. In a regular nonagon?

HIGHLY STRAINED HYDROCARBONS

Organic chemists have been able to synthesize some highly strained cyclic hydrocarbons, such as bicyclo[1.1.0]butane, cubane, and prismane.[1] Philip Eaton, the first to synthesize cubane, recently also synthesized octanitrocubane—cubane with an NO_2 group bonded to each of the eight corners. This compound is expected to be the most powerful explosive known.[2]

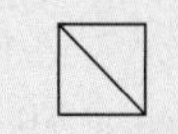
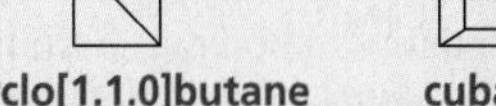
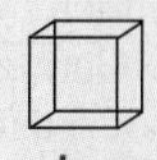

bicyclo[1.1.0]butane   cubane   prismane

[1]Bicyclo[1.1.0]butane was synthesized by David Lemal, Fredric Menger, and George Clark at the University of Wisconsin (*Journal of the American Chemical Society*, 1963, *85*, 2529). Cubane was synthesized by Philip Eaton and Thomas Cole, Jr., at the University of Chicago (*Journal of the American Chemical Society*, 1964, *86*, 3157). Prismane was synthesized by Thomas Katz and Nancy Acton at Columbia University (*Journal of the American Chemical Society*, 1973, *95*, 2738).

[2]Mao-Xi Zhang, Philip Eaton, and Richard Gilardi, *Angew. Chem. Int. Ed.*, 2000, *39 (2)*, 401.

Baeyer predicted that cyclopentane would be the most stable of the cycloalkanes because its bond angles (108°) are closest to the ideal tetrahedral bond angle. He predicted that cyclohexane, with bond angles of 120°, would be less stable and that as the number of sides in the cycloalkanes increases, their stability would decrease.

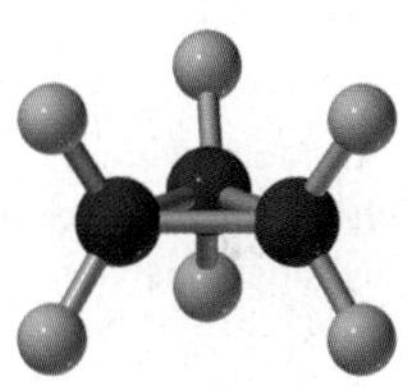
cyclopropane

"planar" cyclopentane
bond angles = 108°

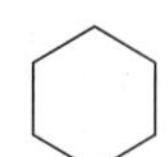
"planar" cyclohexane
bond angles = 120°

"planar" cycloheptane
bond angles = 128.6°

Contrary to what Baeyer predicted, cyclohexane is more stable than cyclopentane. Furthermore, cyclic compounds do not become less and less stable as the number of sides increases. The mistake Baeyer made was to assume that all cyclic molecules are planar. Because three points define a plane, the carbons of cyclopropane must lie in a plane. The other cycloalkanes, however, are not planar. Cyclic compounds twist and bend in order to attain a structure that minimizes the three different kinds of strain that can destabilize a cyclic compound:

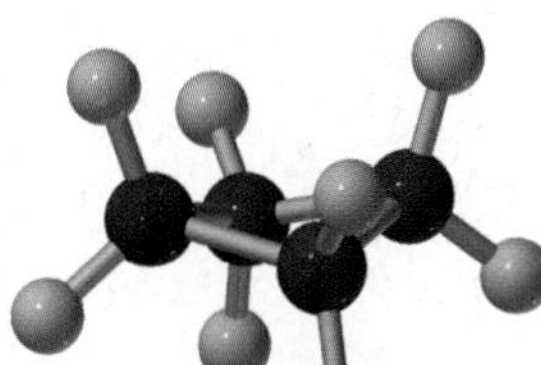
cyclobutane

1. *Angle strain* is the strain induced in a molecule when the bond angles are different from the ideal tetrahedral bond angle of 109.5°.
2. *Torsional strain* is caused by repulsion between the bonding electrons of one substituent and the bonding electrons of a nearby substituent.
3. *Steric strain* is caused by atoms or groups of atoms approaching each other too closely.

Although planar cyclobutane would have less angle strain than cyclopropane, it could have more torsional strain because it has eight pairs of eclipsed hydrogens, compared with the six pairs of cyclopropane. So cyclobutane is not a planar molecule—it is a bent molecule. One of its methylene groups is bent at an angle of about 25° from the plane defined by the other three carbon atoms. This increases the angle strain, but the increase is more than compensated for by the decreased torsional strain as a result of the adjacent hydrogens not being as eclipsed, as they would be in a planar ring.

If cyclopentane were planar, as Baeyer had predicted, it would have essentially no angle strain, but its 10 pairs of eclipsed hydrogens would be subject to considerable torsional strain. So cyclopentane puckers, allowing the hydrogens to become nearly staggered. In the process, however, it acquires some angle strain. The puckered form of cyclopentane is called the *envelope conformation* because the shape resembles a squarish envelope with the flap up.

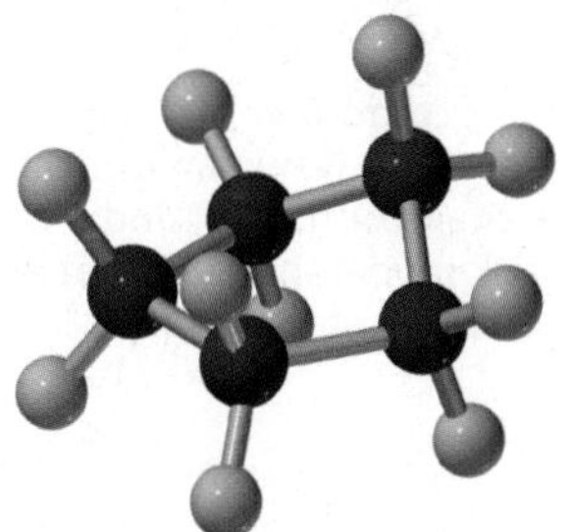
cyclopentane

VON BAEYER AND BARBITURIC ACID

Johann Friedrich Wilhelm Adolf von Baeyer (1835–1917) was born in Germany. He discovered barbituric acid—the first of a group of sedatives known as barbiturates—in 1864 and named it after a woman named Barbara. Who Barbara was is not certain. Some say she was his girlfriend, but because Baeyer discovered barbituric acid in the same year that Prussia defeated Denmark, some believe he named the acid after Saint Barbara, the patron saint of artillerymen. Baeyer was the first to synthesize indigo, the dye used in the manufacture of blue jeans. He was a professor of chemistry at the University of Strasbourg and later at the University of Munich. He received the Nobel Prize in chemistry in 1905 for his work in synthetic organic chemistry.

barbituric acid

indigo

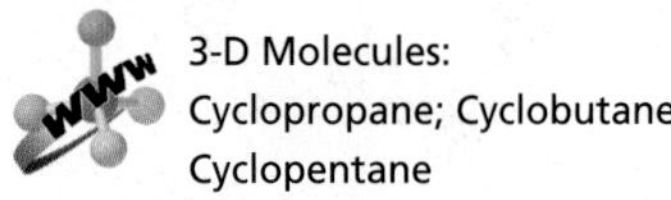

3-D Molecules: Cyclopropane; Cyclobutane; Cyclopentane

PROBLEM 29◆

The effectiveness of a barbiturate as a sedative is related to its ability to penetrate the nonpolar membrane of a cell. Which of the following barbiturates would you expect to be the more effective sedative?

CH_3CH_2, $CH_3(CH_2)_4CH_2$ — hexethal

CH_3CH_2, CH_3CH_2 — barbital

2.12 Conformations of Cyclohexane

The cyclic compounds most commonly found in nature contain six-membered rings because such rings can exist in a conformation that is almost completely free of strain. This conformation is called the **chair conformation** (Figure 2.7). In the chair conformer of cyclohexane, all the bond angles are 111°, which is very close to the ideal tetrahedral bond angle of 109.5°, and all the adjacent bonds are staggered. The chair conformer is such an important conformer that you should learn how to draw it:

1. Draw two parallel lines of the same length, slanted upward. Both lines should start at the same height.

Figure 2.7 ▶ The chair conformer of cyclohexane, a Newman projection of the chair conformer, and a ball-and-stick model showing that all the bonds are staggered.

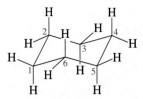

chair conformer of cyclohexane

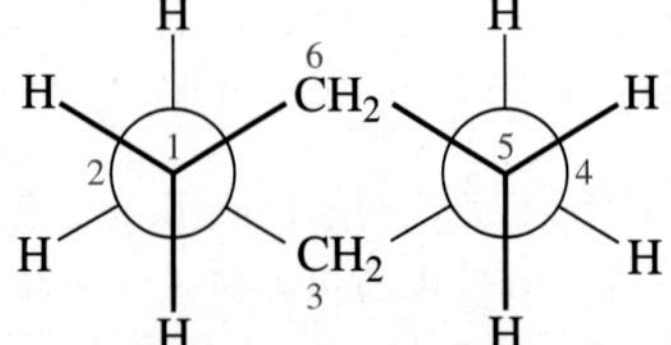

Newman projection of the chair conformer

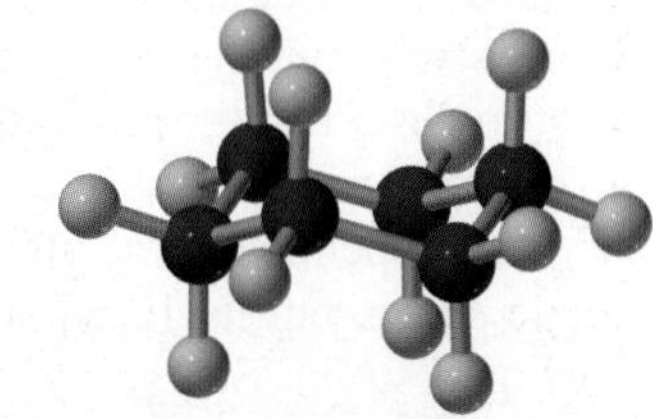

ball-and-stick model of the chair conformer of cyclohexane

2. Connect the tops of the lines with a V; the left-hand side of the V should be slightly longer than the right-hand side. Connect the bottoms of the lines with an inverted V; the lines of the V and the inverted V should be parallel. This completes the framework of the six-membered ring.

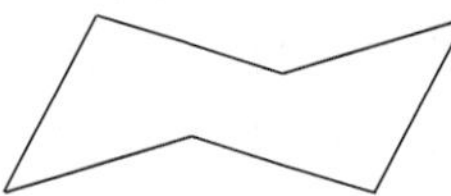

3. Each carbon has an axial bond and an equatorial bond. The **axial bonds** (red lines) are vertical and alternate above and below the ring. The axial bond on one of the uppermost carbons is up, the next is down, the next is up, and so on.

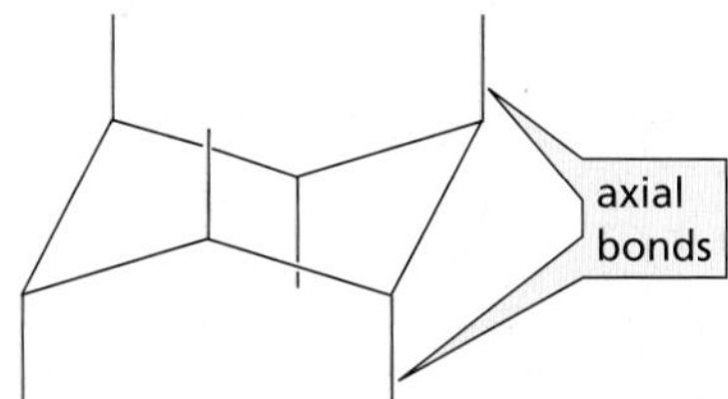

4. The **equatorial bonds** (red lines with blue balls) point outward from the ring. Because the bond angles are greater than 90°, the equatorial bonds are on a slant. If the axial bond points up, the equatorial bond on the same carbon is on a downward slant. If the axial bond points down, the equatorial bond on the same carbon is on an upward slant.

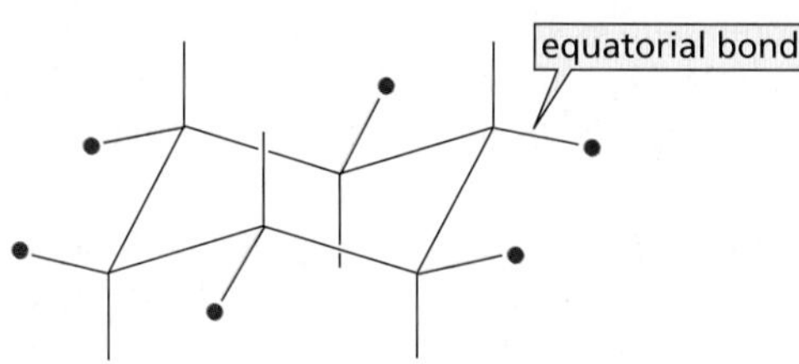

Notice that each equatorial bond is parallel to two ring bonds (two carbons over) and parallel to the opposite equatorial bond.

Remember that cyclohexane is viewed on edge. The lower bonds of the ring are in front and the upper bonds of the ring are in back.

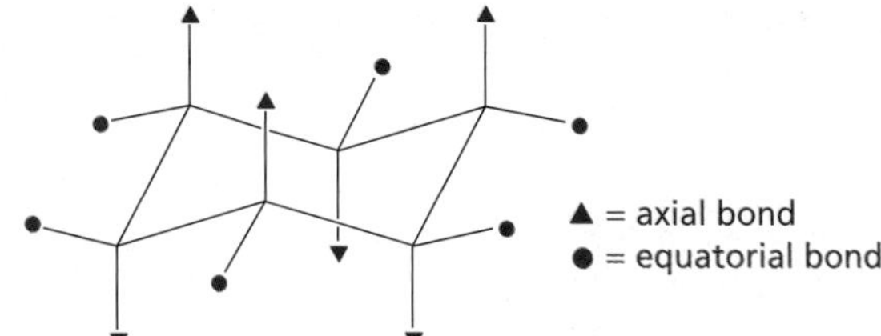

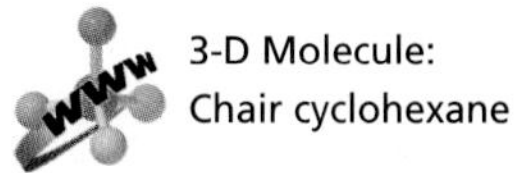

3-D Molecule: Chair cyclohexane

PROBLEM 30

Draw 1,2,3,4,5,6-hexamethylcyclohexane with

a. all the methyl groups in axial positions
b. all the methyl groups in equatorial positions

Table 2.9 Heats of Formation and Total Strain Energies of Cycloalkanes

	Heat of formation		"Strainless" heat of formation		Total strain energy	
	(kcal/mol)	(kJ/mol)	(kcal/mol)	(kJ/mol)	(kcal/mol)	(kJ/mol)
Cylopropane	+12.7	53.1	−14.6	−61.1	27.3	114.2
Cyclobutane	+6.8	28.5	−19.7	−82.4	26.5	110.9
Cyclopentane	−18.4	−77.0	−24.6	−102.9	6.2	25.9
Cyclohexane	−29.5	−123.4	−29.5	−123.4	0	0
Cycloheptane	−28.2	−118.0	−34.4	−143.9	6.2	25.9
Cyclooctane	−29.7	−124.3	−39.4	−164.8	9.7	40.6
Cyclononane	−31.7	−132.6	−44.3	−185.4	12.6	52.7
Cyclodecane	−36.9	−154.4	−49.2	−205.9	12.3	51.5
Cycloundecane	−42.9	−179.5	−54.1	−226.4	11.2	46.9

If we assume that cyclohexane is completely free of strain, we can calculate the total strain energy (angle strain + torsional strain + steric strain) of the other cycloalkanes. Taking the *heat of formation* of cyclohexane (Table 2.9) and dividing by 6 for its six CH_2 groups gives us a value of −4.92 kcal/mol (or −20.6 kJ/mol) for a "strainless" CH_2 group $(-29.5/6 = -4.92)$. (The **heat of formation** is the heat given off when a compound is formed from its elements under standard conditions.) We can now calculate the heat of formation of a "strainless" cycloalkane by multiplying the number of CH_2 groups in its ring by −4.92 kcal/mol. The total strain in the compound is the difference between its "strainless" heat of formation and its actual heat of formation (Table 2.9). For example, cyclopentane has a "strainless" heat of formation of $(5)(-4.92) = -24.6$ kcal/mol. Because its actual heat of formation is −18.4 kcal/mol, cyclopentane has a total strain energy of 6.2 kcal/mol $[-18.4 - (-24.6) = 6.2]$. (Multiplying by 4.184 converts kcal into kJ.)

PROBLEM 31◆

Calculate the total strain energy of cycloheptane.

Bonds that are equatorial in one chair conformer are axial in the other chair conformer.

Cyclohexane rapidly interconverts between two stable chair conformations because of the ease of rotation about its carbon–carbon bonds. This interconversion is known as **ring flip** (Figure 2.8). When the two chair conformers interconvert, bonds that are equatorial in one chair conformer become axial in the other chair conformer and vice versa.

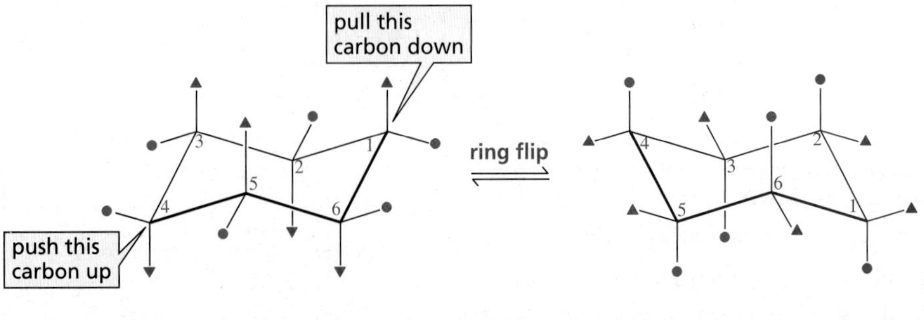

Figure 2.8 ▶ The bonds that are axial in one chair conformer are equatorial in the other chair conformer. The bonds that are equatorial in one chair conformer are axial in the other chair conformer.

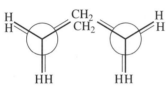

flagpole hydrogens

boat conformer of cyclohexane

Newman projection of the boat conformer

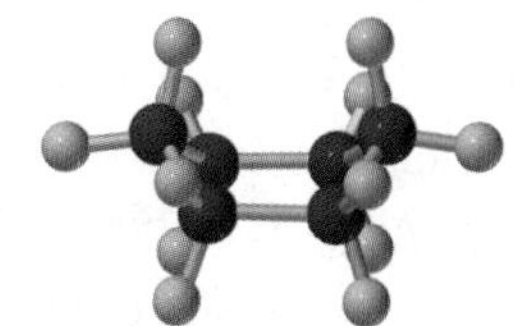

ball-and-stick model of the boat conformer of cyclohexane

▲ **Figure 2.9**
The boat conformer of cyclohexane, a Newman projection of the boat conformer, and a ball-and-stick model showing that some of the bonds are eclipsed.

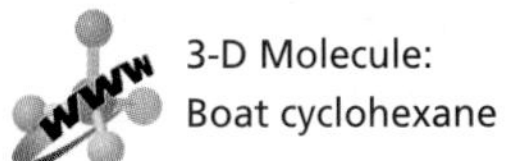

3-D Molecule: Boat cyclohexane

Cyclohexane can also exist in a **boat conformation**, shown in Figure 2.9. Like the chair conformer, the boat conformer is free of angle strain. However, the boat conformer is not as stable as the chair conformer because some of the bonds in the boat conformer are eclipsed, giving it torsional strain. The boat conformer is further destabilized by the close proximity of the **flagpole hydrogens** (the hydrogens at the "bow" and "stern" of the boat), which causes steric strain.

The conformations that cyclohexane can assume when interconverting from one chair conformer to the other are shown in Figure 2.10. To convert from the boat conformer to one of the chair conformers, one of the topmost carbons of the boat conformer must be pulled down so that it becomes the bottommost carbon. When the carbon is pulled down just a little, the **twist-boat** (or **skew-boat**) **conformer** is obtained. The twist-boat conformer is more stable than the boat conformer because there is less eclipsing and, consequently, less torsional strain and the flagpole hydrogens have moved away from each other, thus relieving some of the steric strain. When the carbon is pulled down to the point where it is in the same plane as the sides of the boat, the very unstable **half-chair conformer** is obtained. Pulling the carbon down farther produces the *chair conformer.* The graph in Figure 2.10 shows the energy of a cyclohexane molecule as it interconverts from one chair conformer to the other; the energy barrier for interconversion is 12.1 kcal/mol (50.6 kJ/mol). From this value, it can be calculated that cyclohexane undergoes 10^5 ring flips per second at room temperature. In other words, the two chair conformers are in rapid equilibrium.

Build a model of cyclohexane, and convert it from one chair conformer to the other. To do this, pull the topmost carbon down and push the bottommost carbon up.

Because the chair conformers are the most stable of the conformers, at any instant more molecules of cyclohexane are in chair conformations than in any other conformation. It has been calculated that, for every thousand molecules of cyclohexane in

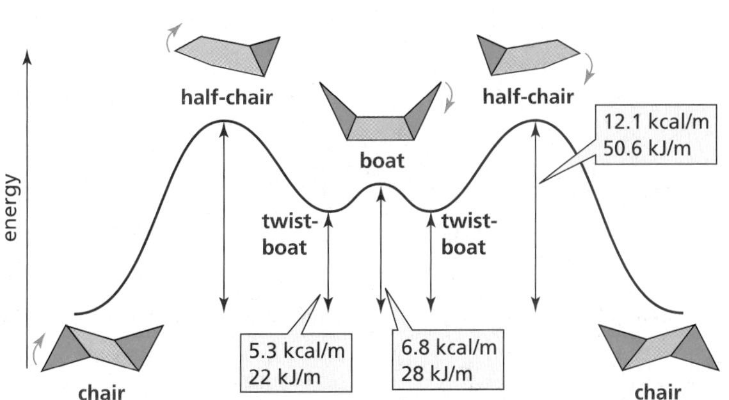

◀ **Figure 2.10**
The conformers of cyclohexane—and their relative energies—as one chair conformer interconverts to the other chair conformer.

Go to the Web site for three-dimensional representations of the conformers of cyclohexane.

a chair conformation, no more than two molecules are in the next most stable conformation—the twist-boat.

2.13 Conformations of Monosubstituted Cyclohexanes

Unlike cyclohexane, which has two equivalent chair conformers, the two chair conformers of a monosubstituted cyclohexane such as methylcyclohexane are not equivalent. The methyl substituent is in an equatorial position in one conformer and in an axial position in the other (Figure 2.11), because substituents that are equatorial in one chair conformer are axial in the other (Figure 2.8).

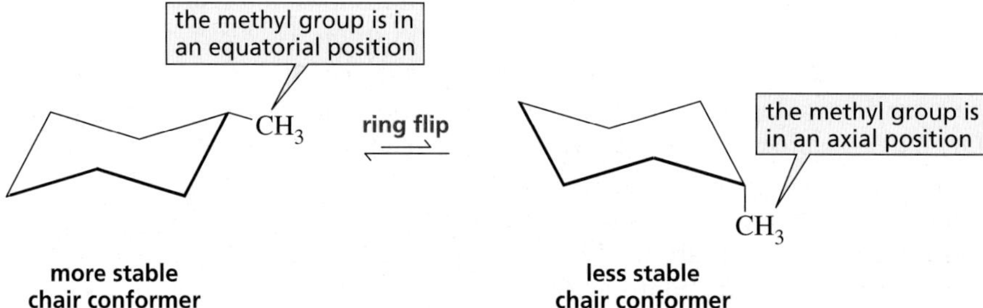
the methyl group is in an equatorial position
CH_3
ring flip
the methyl group is in an axial position
CH_3
more stable chair conformer
less stable chair conformer

Figure 2.11 ▶ A substituent is in an equatorial position in one chair conformer and in an axial position in the other. The conformer with the substituent in the equatorial position is more stable.

The chair conformer with the methyl substituent in an equatorial position is the more stable conformer because a substituent has more room and, therefore, fewer steric interactions when it is in an equatorial position. This can be best understood by examining Figure 2.12, which shows that when the methyl group is in an equatorial position, it is anti to the C-3 and C-5 carbons. Therefore, the substituent extends into space, away from the rest of the molecule.

In contrast, when the methyl group is in an axial position, it is gauche to the C-3 and C-5 carbons (Figure 2.13). As a result, there are unfavorable steric interactions between the axial methyl group and both the axial substituent on C-3 and the axial substituent on C-5 (in this case, hydrogens). In other words, the three axial bonds on the same side of the ring are parallel to each other, so any axial substituent will be relatively close to the axial substituents on the other two carbons. Because the interacting substituents are on 1,3-positions relative to each other, these unfavorable steric interactions are called **1,3-diaxial interactions**. If you take a few minutes to build models, you will see that a substituent has more room if it is in an equatorial position than if it is in an axial position.

Build a model of methylcyclohexane, and convert it from one chair conformer to the other.

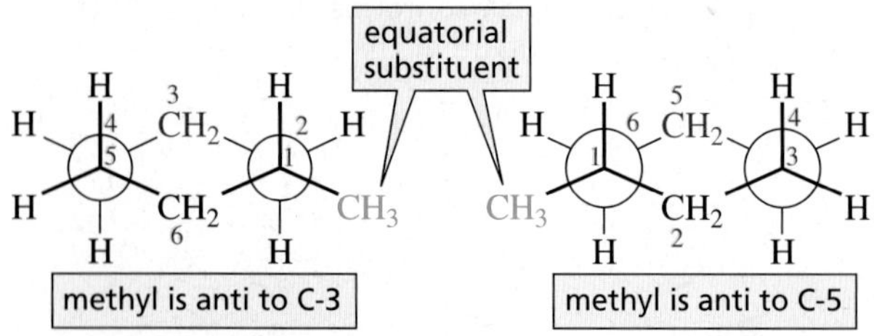

equatorial substituent

methyl is anti to C-3

methyl is anti to C-5

▲ **Figure 2.12**
An equatorial substituent on the C-1 carbon is anti to the C-3 and C-5 carbons.

axial substituent

methyl is gauche to C-3

methyl is gauche to C-5

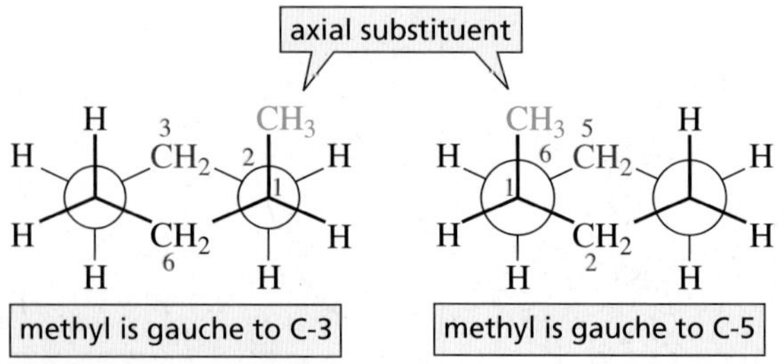

▲ **Figure 2.13**
An axial substituent on the C-1 carbon is gauche to the C-3 and C-5 carbons.

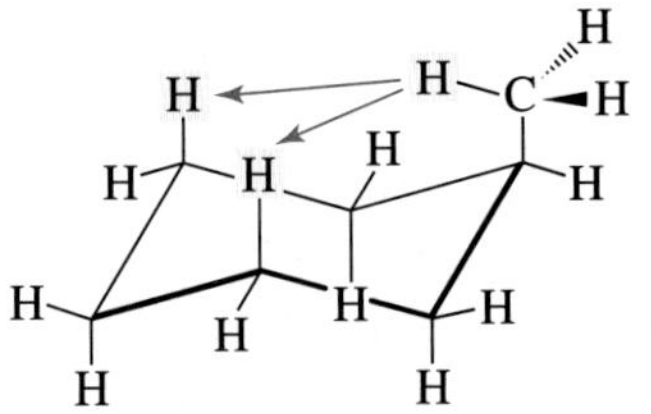

1,3-diaxial interactions

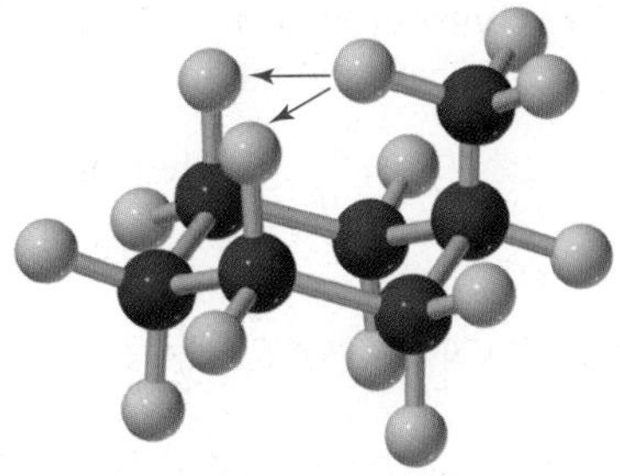

ball-and-stick model

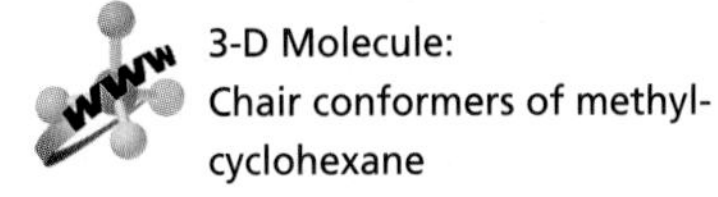

The gauche conformer of butane and the axial-substituted conformer of methylcyclohexane are compared in Figure 2.14. Notice that the gauche interaction is the same in both—an interaction between a methyl group and a hydrogen bonded to a carbon gauche to the methyl group. Butane has one such gauche interaction and methylcyclohexane has two.

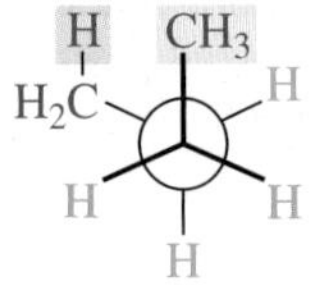

gauche butane

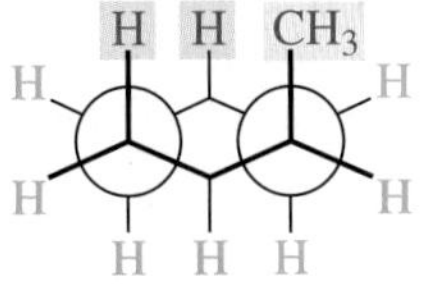

axial methylcyclohexane

◀ Figure 2.14
The steric strain of gauche butane is the same as the steric strain between an axial methyl group and one of its axial hydrogens. Butane has one gauche interaction between a methyl group and a hydrogen; methylcyclohexane has two.

In Section 2.10, we saw that the gauche interaction between the methyl groups of butane caused the gauche conformer to be 0.9 kcal/mol (3.8 kJ/mol) less stable than the anti conformer. Because there are two such gauche interactions in the chair conformer of methylcyclohexane when the methyl group is in an axial position, this chair conformer is 1.8 kcal/mol (7.5 kJ/mol) less stable than the chair conformer with the methyl group in the equatorial position.

Because of the difference in stability of the two chair conformers, at any one time more monosubstituted cyclohexane molecules will be in the chair conformer with the substituent in the equatorial position than in the chair conformer with the substituent in the axial position. The relative amounts of the two chair conformers depend on the substituent (Table 2.10). The substituent with the greater bulk in the area of the 1,3-diaxial hydrogens will have a greater preference for the equatorial position because it will have stronger 1,3-diaxial interactions. For example, the equilibrium constant

Table 2.10 Equilibrium Constants for Several Monosubstituted Cyclohexanes at 25 °C

Substituent	Axial $\overset{K_{eq}}{\rightleftharpoons}$ Equatorial	Substituent	Axial $\overset{K_{eq}}{\rightleftharpoons}$ Equatorial
H	1	CN	1.4
CH_3	18	F	1.5
CH_3CH_2	21	Cl	2.4
$(CH_3)_2CH$	35	Br	2.2
$(CH_3)_3C$	4800	I	2.2
		HO	5.4

The larger the substituent on a cyclohexane ring, the more the equatorial-substituted conformer will be favored.

(K_{eq}) for the conformers of methylcyclohexane indicates that 95% of methylcyclohexane molecules have the methyl group in the equatorial position at 25 °C:

$$K_{eq} = \frac{[\text{equatorial conformer}]}{[\text{axial conformer}]} = \frac{18}{1}$$

$$\%\text{ of equatorial conformer} = \frac{[\text{equatorial conformer}]}{[\text{equatorial conformer}] + [\text{axial conformer}]} \times 100$$

$$\%\text{ of equatorial conformer} = \frac{18}{18 + 1} \times 100 = 95\%$$

In the case of *tert*-butylcyclohexane, where the 1,3-diaxial interactions are even more destabilizing because a *tert*-butyl group is larger than a methyl group, more than 99.9% of the molecules have the *tert*-butyl group in the equatorial position.

PROBLEM 32◆

The chair conformer of fluorocyclohexane is 0.25 kcal/mol (1.0 kJ/mol) more stable when the fluoro substituent is in the equatorial position than when it is in the axial position. How much more stable is the anti conformer of 1-fluoropropane compared with a gauche conformer?

PROBLEM 33◆

From the data in Table 2.10, calculate the percentage of molecules of cyclohexanol that have the OH group in the equatorial position.

PROBLEM 34

Bromine is a larger atom than chlorine, but the equilibrium constants in Table 2.10 indicate that a chloro substituent has a greater preference for the equatorial position. Suggest an explanation for this fact.

2.14 Conformations of Disubstituted Cyclohexanes

If there are two substituents on a cyclohexane ring, both substituents have to be taken into account when determining which of the two chair conformers is the more stable. Let's start by looking at 1,4-dimethylcyclohexane. First of all, note that there are two different dimethylcyclohexanes. One has both methyl substituents on the *same side* of the cyclohexane ring; it is called the **cis isomer** (*cis* is Latin for "on this side"). The other has the two methyl substituents on *opposite sides* of the ring; it is called the **trans isomer** (*trans* is Latin for "across"). *cis*-1,4-Dimethylcyclohexane and *trans*-1,4-dimethylcyclohexane are called **geometric isomers** or **cis–trans isomers**: They have the same atoms, and the atoms are linked in the same order, but they differ in the spatial arrangement of the atoms.

The cis isomer has its substituents on the same side of the ring.

The trans isomer has its substituents on opposite sides of the ring.

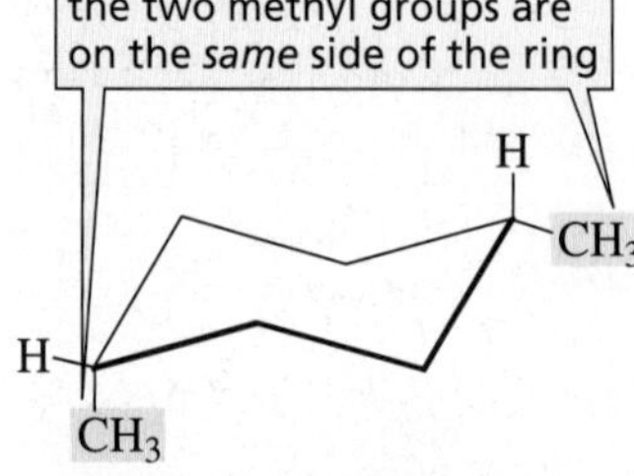
the two methyl groups are on the *same* side of the ring
H
CH_3
H
CH_3
cis-1,4-dimethylcyclohexane

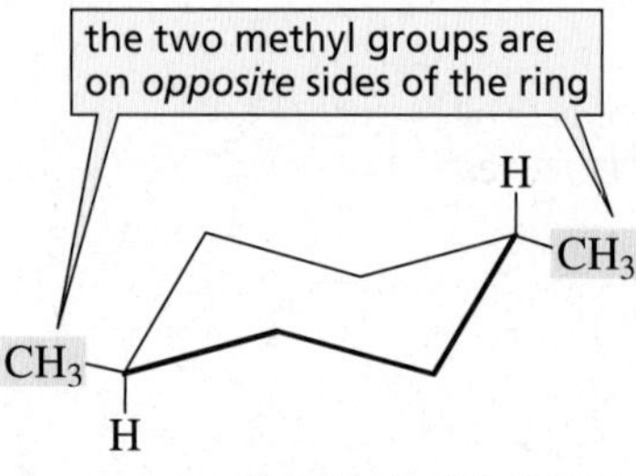
the two methyl groups are on *opposite* sides of the ring
H
CH_3
CH_3
H
trans-1,4-dimethylcyclohexane

First we will determine which of the two chair conformers of *cis*-1,4-dimethylcyclohexane is more stable. One chair conformer has one methyl group in an equatorial position and one methyl group in an axial position. The other chair conformer also has one methyl group in an equatorial position and one methyl group in an axial position. Therefore, both chair conformers are equally stable.

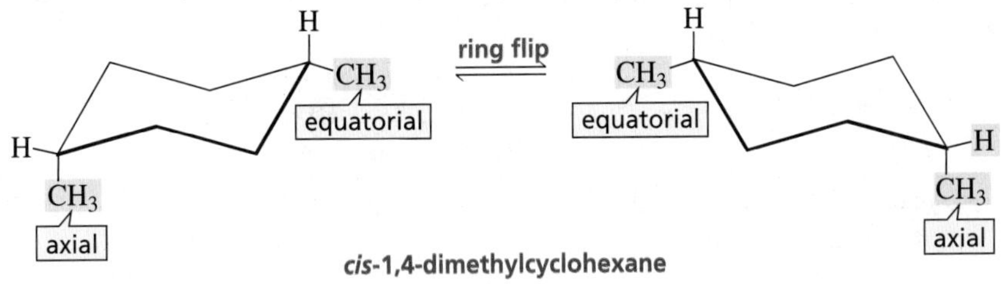

cis-1,4-dimethylcyclohexane

In contrast, the two chair conformers of *trans*-1,4-dimethylcyclohexane have different stabilities because one has both methyl substituents in equatorial positions and the other has both methyl groups in axial positions.

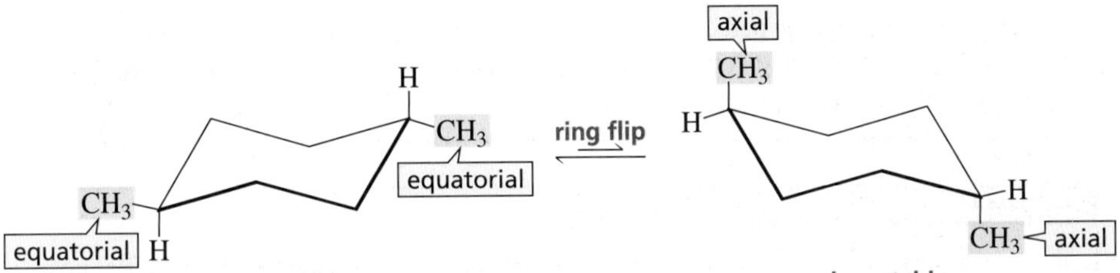

more stable

less stable

trans-1,4-dimethylcyclohexane

The chair conformer with both substituents in axial positions has four 1,3-diaxial interactions, causing it to be about 4 × 0.9 kcal/mol = 3.6 kcal mol (15.1 kJ/mol) less stable than the chair conformer with both methyl groups in equatorial positions. We can, therefore, predict that *trans*-1,4-dimethylcyclohexane will exist almost entirely in the more stable diequatorial conformation.

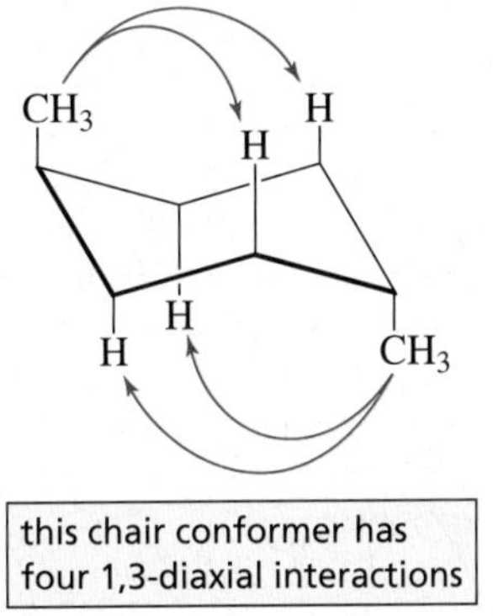

Now let's look at the geometric isomers of 1-*tert*-butyl-3-methylcyclohexane. Both substituents of the cis isomer are in equatorial positions in one conformer and in axial positions in the other conformer. The conformer with both substituents in equatorial positions is more stable.

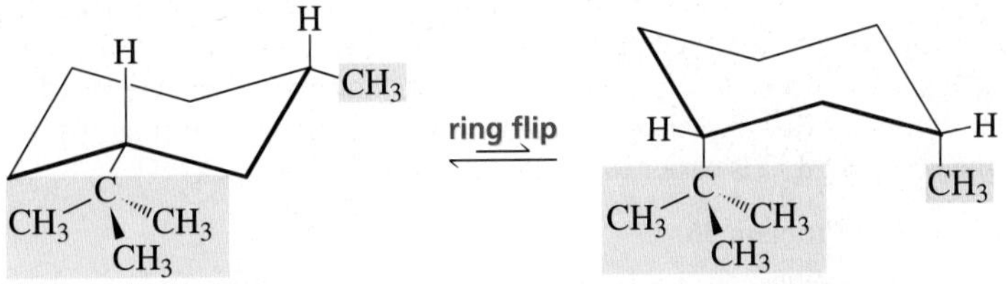

more stable

less stable

cis-1-*tert*-butyl-3-methylcyclohexane

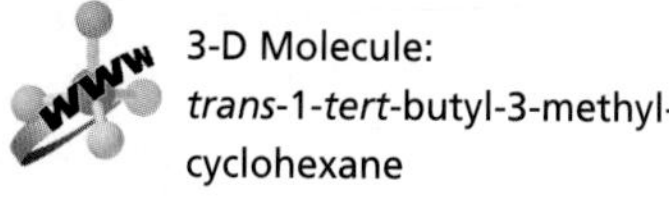

Both conformers of the trans isomer have one substituent in an equatorial position and the other in an axial position. Because the *tert*-butyl group is larger than the methyl group, the 1,3-diaxial interactions will be stronger when the *tert*-butyl group is in the axial position. Therefore, the conformer with the *tert*-butyl group in the equatorial position is more stable.

ring flip

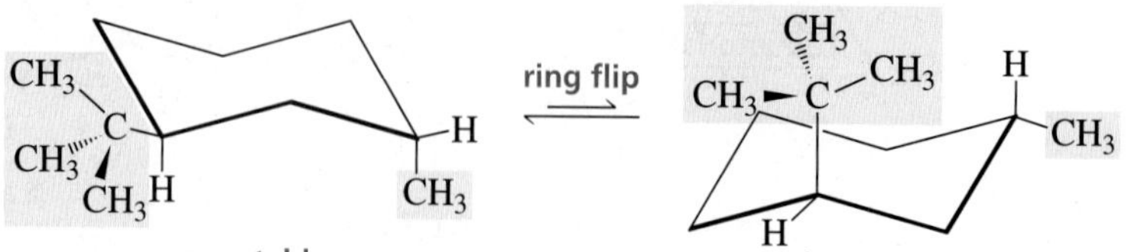

more stable

less stable

***trans*-1-*tert*-butyl-3-methylcyclohexane**

PROBLEM 35◆

Which will have a higher percentage of the diequatorial-substituted conformer, compared with the diaxial-substituted conformer: *trans*-1,4-dimethylcyclohexane or *cis*-1-*tert*-butyl-3-methylcyclohexane?

PROBLEM-SOLVING STRATEGY

Is the conformer of 1,2-dimethylcyclohexane with one methyl group in an equatorial position and the other in an axial position the cis isomer or the trans isomer?

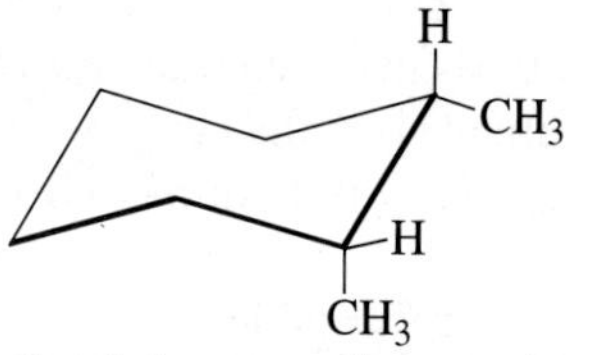

Is this the cis isomer or the trans isomer?

To solve this kind of problem you need to determine whether the two substituents are on the same side of the ring (cis) or on opposite sides of the ring (trans). If the bonds bearing the substituents are both pointing upward or both pointing downward, the compound is the cis isomer; if one bond is pointing upward and the other downward, the compound is the trans isomer. Because the conformer in question has both methyl groups attached to downward-pointing bonds, it is the cis isomer.

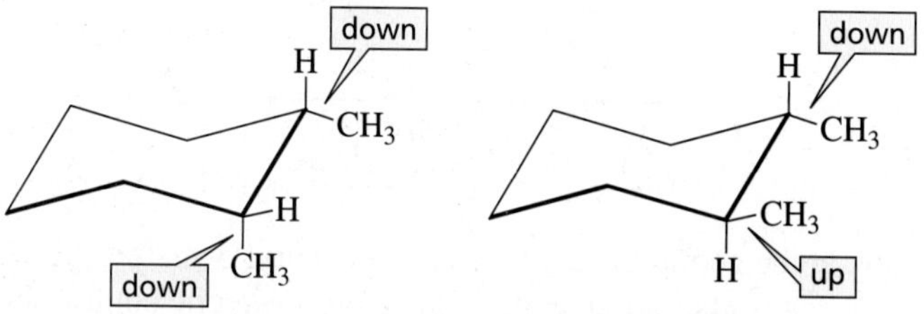

down

down

down

up

the cis isomer

the trans isomer

The isomer that is the most misleading when it is drawn in two dimensions is a *trans*-1,2-disubstituted isomer. At first glance, the methyl groups of *trans*-1,2-dimethylcyclohexane appear to be oriented in the same direction, so you might think that the compound is the cis isomer. Closer inspection shows, however, that one bond is pointed upward and the other downward, so we know that it is the trans isomer. (If you build a model of the compound, it is easier to see that it is the trans isomer.)

Now continue on to Problem 36.

PROBLEM 36◆

Determine whether each of the following compounds is a cis isomer or a trans isomer:

a.
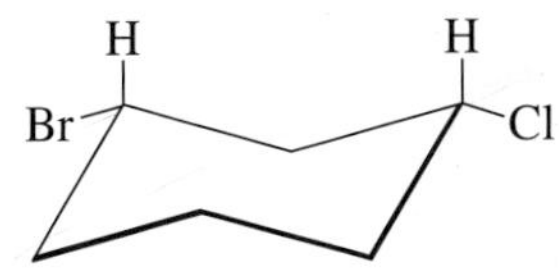

d.
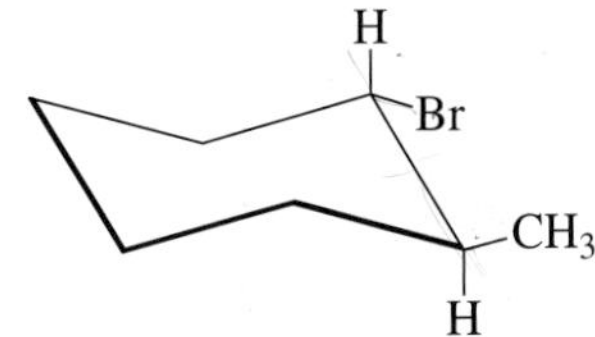

b.
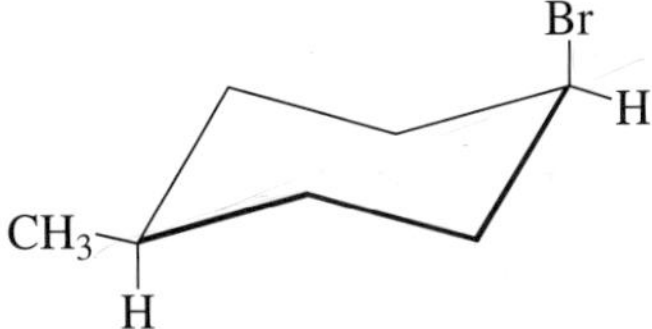

e.
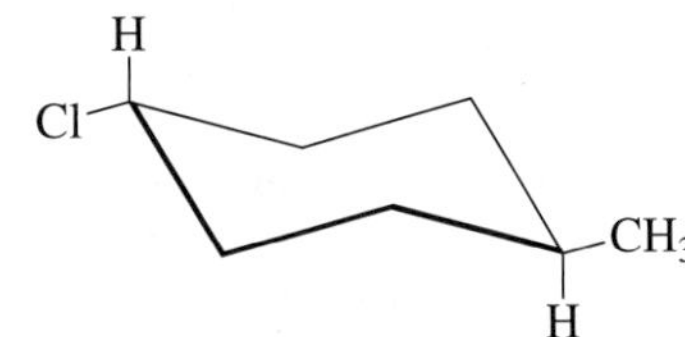

c.
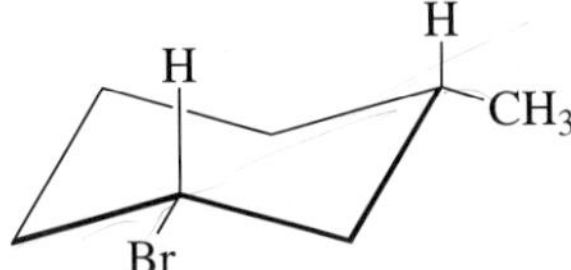

f.
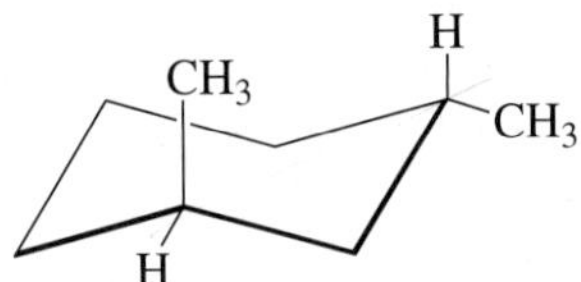

PROBLEM 37 SOLVED

a. Draw the more stable chair conformer of *cis*-1-ethyl-2-methylcyclohexane.

b. Draw the more stable conformer of *trans*-1-ethyl-2-methylcyclohexane.

c. Which is more stable, *cis*-1-ethyl-2-methylcyclohexane or *trans*-1-ethyl-2-methylcyclohexane?

SOLUTION TO 37a If the two substituents of a 1,2-disubstituted cyclohexane are to be on the same side of the ring, one must be in an equatorial position and the other must be in an axial position. The more stable chair conformer is the one in which the larger of the two substituents (the ethyl group) is in the equatorial position.

PROBLEM 38◆

For each of the following disubstituted cyclohexanes, indicate whether the substituents in the two chair conformers would be both equatorial in one chair conformer and both axial in the other *or* one equatorial and one axial in each of the chair conformers:

a. *cis*-1,2-
b. *trans*-1,2-
c. *cis*-1,3-
d. *trans*-1,3-
e. *cis*-1,4-
f. *trans*-1,4-

PROBLEM 39◆

a. Calculate the energy difference between the two chair conformers of *trans*-1,4-dimethylcyclohexane.

b. What is the energy difference between the two chair conformers of *cis*-1,4-dimethylcyclohexane?

2.15 Conformations of Fused Rings

When two cyclohexane rings are fused together, the second ring can be considered to be a pair of substituents bonded to the first ring. As with any disubstituted cyclohexane, the two substituents can be either cis or trans. If the cyclohexane rings are drawn in their chair conformations, the trans isomer (with one substituent bond pointing upward and the other downward) will have both substituents in the equatorial position. The cis isomer will have one substituent in the equatorial position and one substituent in the axial position. **Trans-fused** cyclohexane rings, therefore, are more stable than **cis-fused** cyclohexane rings.

trans-decalin
trans-fused rings
more stable

cis-decalin
cis-fused rings
less stable

Summary

Alkanes are **hydrocarbons** that contain only single bonds. Their general molecular formula is C_nH_{2n+2}. **Constitutional isomers** have the same molecular formula, but their atoms are linked differently. Alkanes are named by determining the number of carbons in their **parent hydrocarbon**—the longest continuous chain. **Substituents** are listed in alphabetical order, with a number to designate their position on the chain. When there is only a **substituent**, the substituent gets the lowest possible number; when there is only a **functional group suffix**, the functional group suffix gets the lowest possible number; when there is both a functional group suffix and a substituent, the functional group suffix gets the lowest possible number. The **functional group** is the center of reactivity in a molecule.

Alkyl halides and **ethers** are named as substituted alkanes. **Alcohols** and **amines** are named using a functional group suffix. **Systematic names** can contain numbers; **common names** never do. A compound can have more than one name, but a name must specify only one compound. Whether alkyl halides or alcohols are **primary**, **secondary**, or **tertiary** depends on whether the X (halogen) or OH group is bonded to a primary, secondary, or tertiary carbon. A **primary carbon** is bonded to one carbon, a **secondary carbon** is bonded to two carbons, and a **tertiary carbon** is bonded to three carbons. Whether amines are **primary**, **secondary**, or **tertiary** depends on the number of alkyl groups bonded to the nitrogen. Compounds with four alkyl groups bonded to nitrogen are called **quaternary ammonium salts**.

The oxygen of an alcohol has the same geometry it has in water; the nitrogen of an amine has the same geometry it has in ammonia. The greater the attractive forces between molecules—**van der Waals forces, dipole–dipole interactions, hydrogen bonds**—the higher is the **boiling point** of the compound. A **hydrogen bond** is an interaction between a hydrogen bonded to an O, N, or F and the lone pair of an O, N, or F in another molecule. The boiling point increases with increasing molecular weight of the **homolog**. Branching lowers the boiling point. **Polarizability** indicates the ease with which an electron cloud can be distorted: Larger atoms are more polarizable.

Polar compounds dissolve in **polar solvents**, and **nonpolar compounds** dissolve in **nonpolar solvents**. The interaction between a solvent and a molecule or an ion dissolved in that solvent is called **solvation**. The oxygen of an alcohol or an ether can drag about three or four carbons into solution in water.

Rotation about a C—C bond results in two extreme **conformations** that rapidly interconvert: **staggered** and **eclipsed**. A **staggered conformer** is more stable than an **eclipsed conformer** because of **torsional strain**—repulsion between pairs of bonding electrons. There can be two different staggered conformers: The **anti conformer** is more stable than the **gauche conformer** because of **steric strain**—repulsion between the electron clouds of atoms or groups. The steric strain in a gauche conformer is called a **gauche interaction**.

Five- and six-membered rings are more stable than three- and four-membered rings because of **angle strain** that results when bond angles deviate from the ideal bond angle of 109.5°. In a process called **ring flip**, cyclohexane rapidly interconverts between two stable chair conformations. **Bonds** that are **axial** in one chair conformer are **equatorial** in the other and vice versa. The chair conformer with a substituent in the equatorial position is more stable, because there is more room in an equatorial position. A substituent in an axial position experiences unfavorable **1,3-diaxial interactions**. In the case of disubstituted cyclohexanes, the more stable conformer will have its larger substituent in the equatorial position. A **cis isomer** has its two substituents on the same side of the ring; **a trans isomer** has its substituents on opposite sides of the ring. Cis and trans isomers are called **geometric isomers** or **cis–trans isomers**. Cyclohexane rings are more stable if they are **trans fused** than **cis fused**.

Key Terms

alcohol (p. 64)
alkane (p. 60)
alkyl halide (p. 64)
alkyl substituent (p. 63)
amine (p. 64)
angle strain (p. 92)
anti conformer (p. 90)
axial bond (p. 95)
banana bond (p. 92)
boat conformation (p. 97)
boiling point (bp) (p. 81)
chair conformation (p. 94)
cis fused (p. 104)
cis isomer (p. 100)
cis–trans isomers (p. 100)
common name (p. 63)
conformation (p. 88)
conformational analysis (p. 89)
conformer (p. 88)
constitutional isomers (p. 61)
cycloalkane (p. 71)
1,3-diaxial interaction (p. 98)
dipole–dipole interaction (p. 83)
eclipsed conformation (p. 89)
eclipsed conformer (p. 90)
equatorial bond (p. 95)
ether (p. 74)
flagpole hydrogen (p. 97)
functional group (p. 75)
gauche conformer (p. 90)
gauche interaction (p. 91)
geometric isomers (p. 100)
half-chair conformer (p. 97)
heat of formation (p. 96)
homolog (p. 60)
homologous series (p. 60)
hydrocarbon (p. 60)
hydrogen bond (p. 83)
induced-dipole–induced-dipole interaction (p. 82)
IUPAC nomenclature (p. 63)
melting point (mp) (p. 86)
methylene (CH_2) group (p. 60)
Newman projection (p. 88)
packing (p. 86)
parent hydrocarbon (p. 67)
perspective formula (p. 88)
polarizability (p. 84)
primary alcohol (p. 75)
primary alkyl halide (p. 73)
primary amine (p. 77)
primary carbon (p. 64)
primary hydrogen (p. 65)
quaternary ammonium salt (p. 78)
ring flip (p. 96)
sawhorse projection (p. 88)
secondary alcohol (p. 75)
secondary alkyl halide (p. 73)
secondary amine (p. 77)
secondary carbon (p. 64)
secondary hydrogen (p. 65)
skeletal structure (p. 71)
skew-boat conformer (p. 97)
solubility (p. 86)
solvation (p. 86)
staggered conformation (p. 89)
staggered conformers (p. 90)
steric strain (p. 90)
straight-chain alkane (p. 60)
symmetrical ether (p. 74)
systematic nomenclature (p. 63)
tertiary alcohol (p. 75)
tertiary alkyl halide (p. 73)
tertiary amine (p. 77)
tertiary carbon (p. 65)
tertiary hydrogen (p. 65)
torsional strain (p. 89)
trans fused (p. 104)
trans isomer (p. 100)
twist-boat conformer (p. 97)
unsymmetrical ether (p. 74)
van der Waals forces (p. 82)

Problems

40. Write a structural formula for each of the following compounds:
   a. *sec*-butyl *tert*-butyl ether
   b. isoheptyl alcohol
   c. *sec*-butylamine
   d. neopentyl bromide
   e. 1,1-dimethylcyclohexane
   f. 4,5-diisopropylnonane
   g. triethylamine
   h. cyclopentylcyclohexane
   i. 4-*tert*-butylheptane
   j. 5,5-dibromo-2-methyloctane
   k. 1-methylcyclopentanol
   l. 3-ethoxy-2-methylhexane
   m. 5-(1,2-dimethylpropyl)nonane
   n. 3,4-dimethyloctane

41. Give the systematic name for each of the following compounds:

   a. Br
   $CH_3CHCH_2CH_2CHCH_2CH_2CH_3$
   CH_3

   b. $(CH_3)_3CCH_2CH_2CH_2CH(CH_3)_2$

   c. CH_3
   $CH_3CHCH_2CHCHCH_3$
   CH_3 CH_3

   d. $(CH_3CH_2)_4C$

   e. $BrCH_2CH_2CH_2CH_2CH_2NHCH_2CH_3$

   f. $CH_3CHCH_2CHCH_2CH_3$
   CH_3 OH

   g. $CH_3CH_2CHOCH_2CH_3$
   $CH_2CH_2CH_2CH_3$

   h. CH_3
   Br

   i. CH_3
   NCH_3

   j. CH_2CH_3
   OH

   k. $CH_3OCH_2CH_2CH_2OCH_3$

42. a. How many primary carbons does the following structure have?

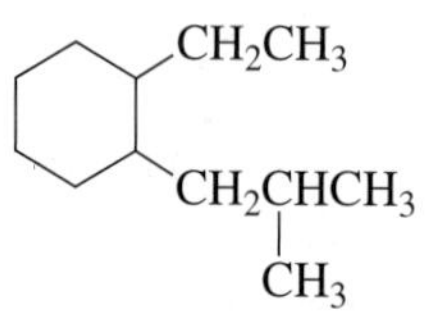

(cyclohexane ring bearing CH_2CH_3 and CH_2CHCH_3 with a CH_3 branch)

b. How many secondary carbons does the structure have?
c. How many tertiary carbons does it have?

43. Which of the following conformers of isobutyl chloride is the most stable?

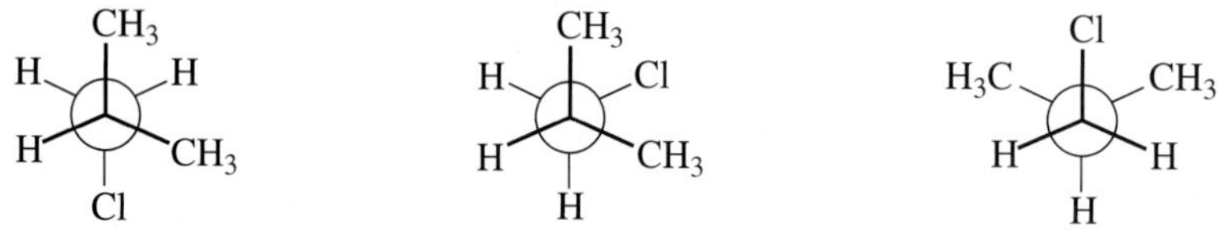

(Newman projections: front CH_3, H, H / back H, CH_3, Cl; front CH_3, H, H / back Cl, CH_3, H; front Cl, H_3C, CH_3 / back H, H, H)

44. Draw the structural formula of an alkane that has
a. six carbons, all secondary
b. eight carbons and only primary hydrogens
c. seven carbons with two isopropyl groups

45. Give two names for each of the following compounds:

a. $CH_3CH_2CH_2OCH_2CH_3$

b. $CH_3CHCH_2CH_2CH_2OH$ with CH_3 on C2

c. $CH_3CH_2CHCH_3$ with NH_2

d. $CH_3CH_2CHCH_3$ with Cl

e. $CH_3CHCH_2CH_2CH_3$ with CH_3

f. CH_3CBr with CH_3 and CH_2CH_3

g. 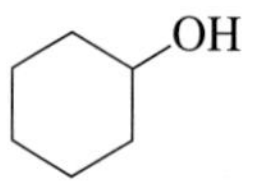 cyclohexane–OH

h. 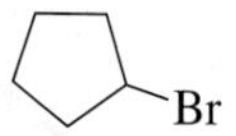 cyclopentane–Br

i. CH_3CHNH_2 with CH_3

j. $CH_3CH_2CH(CH_3)NHCH_2CH_3$

46. Which of the following pairs of compounds has
a. the higher boiling point: 1-bromopentane or 1-bromohexane?
b. the higher boiling point: pentyl chloride or isopentyl chloride?
c. the greater solubility in water: 1-butanol or 1-pentanol?
d. the higher boiling point: 1-hexanol or 1-methoxypentane?
e. the higher melting point: hexane or isohexane?
f. the higher boiling point: 1-chloropentane or 1-pentanol?
g. the higher boiling point: 1-bromopentane or 1-chloropentane?
h. the higher boiling point: diethyl ether or butyl alcohol?
i. the greater density: heptane or octane?
j. the higher boiling point: isopentyl alcohol or isopentylamine?
k. the higher boiling point: hexylamine or dipropylamine?

47. Ansaid® and Motrin® belong to the group of drugs known as nonsteroidal anti-inflammatory drugs (NSAIDs). Both are only slightly soluble in water, but one is a little more soluble than the other. Which of the drugs has the greater solubility in water?

(Ansaid®: biphenyl with F, bearing $CH(CH_3)COOH$; Motrin®: $CH_3CH(CH_3)CH_2$–benzene–$CH(CH_3)COOH$)

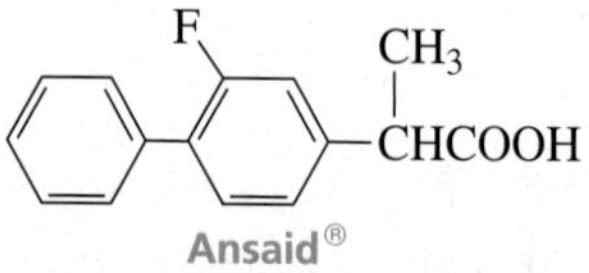

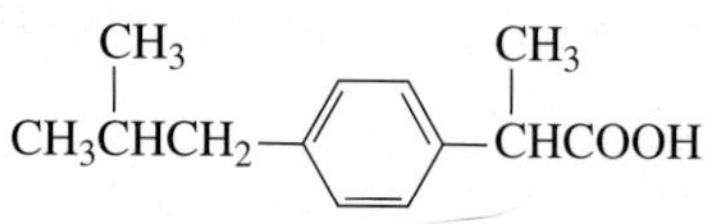

Ansaid®　　Motrin®

48. Al Kane was given the structural formulas of several compounds and was asked to give them systematic names. How many did Al name correctly? Correct those that are misnamed.

a. 4-bromo-3-pentanol
b. 2,2-dimethyl-4-ethylheptane
c. 5-methylcyclohexanol
d. 1,1-dimethyl-2-cyclohexanol
e. 5-(2,2-dimethylethyl)nonane
f. isopentyl bromide
g. 3,3-dichlorooctane
h. 5-ethyl-2-methylhexane
i. 1-bromo-4-pentanol
j. 3-isopropyloctane
k. 2-methyl-2-isopropylheptane
l. 2-methyl-*N*,*N*-dimethyl-4-hexanamine

49. Which of the following conformers has the highest energy?

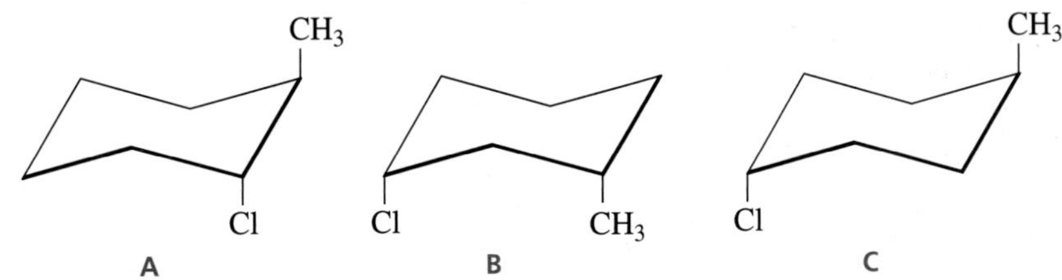

50. Give systematic names for all the alkanes with molecular formula C_7H_{16} that do not have any secondary hydrogens.

51. Draw skeletal structures of the following compounds:
a. 5-ethyl-2-methyloctane
b. 1,3-dimethylcyclohexane
c. 2,3,3,4-tetramethylheptane
d. propylcyclopentane
e. 2-methyl-4-(1-methylethyl)octane
f. 2,6-dimethyl-4-(2-methylpropyl)decane

52. For rotation about the C-3—C-4 bond of 2-methylhexane:
a. Draw the Newman projection of the most stable conformer.
b. Draw the Newman projection of the least stable conformer.
c. About which other carbon–carbon bonds may rotation occur?
d. How many of the carbon–carbon bonds in the compound have staggered conformers that are all equally stable?

53. Which of the following structures represents a cis isomer?

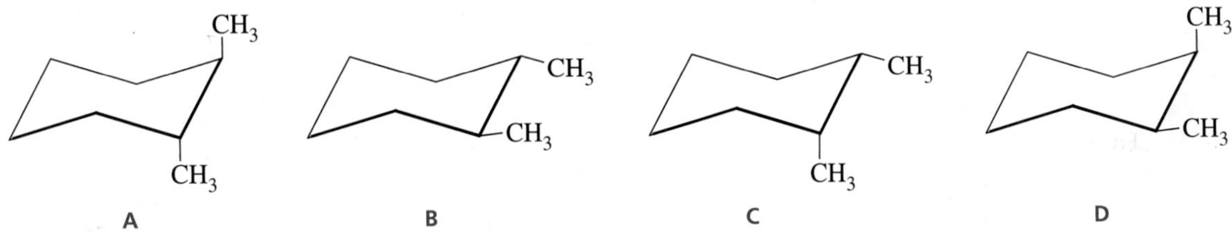

54. Draw all the isomers that have the molecular formula $C_5H_{11}Br$. (*Hint:* There are eight such isomers.)
a. Give the systematic name for each of the isomers.
b. Give a common name for each isomer that has one.
c. How many isomers do not have common names?
d. How many of the isomers are primary alkyl halides?
e. How many of the isomers are secondary alkyl halides?
f. How many of the isomers are tertiary alkyl halides?

55. Give the systematic name for each of the following compounds:

a.

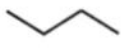

b. 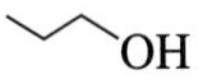 OH

c. 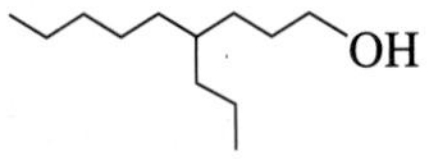 OH

d.

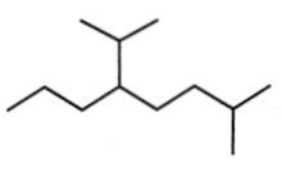

e. 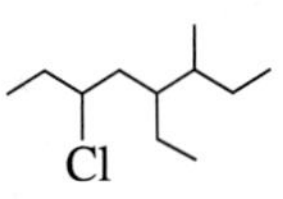 Cl

f. 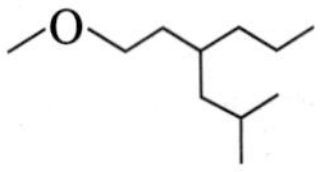 O

g. 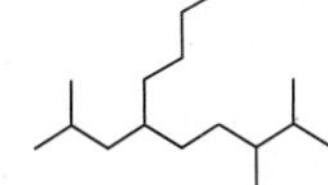

h. 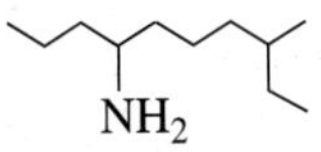 NH$_2$

56. Draw the two chair conformers of each compound, and indicate which conformer is more stable:
a. *cis*-1-ethyl-3-methylcyclohexane
b. *trans*-1-ethyl-2-isopropylcyclohexane
c. *trans*-1-ethyl-2-methylcyclohexane
d. *trans*-1-ethyl-3-methylcyclohexane
e. *cis*-1-ethyl-3-isopropylcyclohexane
f. *cis*-1-ethyl-4-isopropylcyclohexane

57. Why are alcohols of lower molecular weight more soluble in water than those of higher molecular weight?

58. The most stable conformer of *N*-methylpiperidine is shown on p. 108.
a. Draw the other chair conformer.

b. Which takes up more space, the lone-pair electrons or the methyl group?

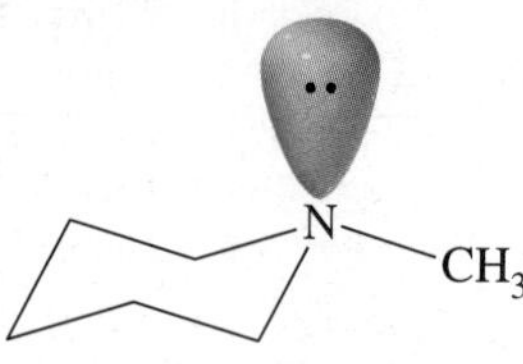

N-methylpiperidine

59. How many ethers have the molecular formula $C_5H_{12}O$? Give the structural formula and systematic name for each. What are their common names?

60. Draw the most stable conformer of the following molecule:

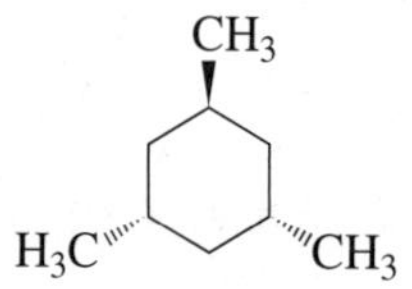

CH_3
H_3C CH_3

61. Give the systematic name for each of the following compounds:

a. $CH_3CH_2CHCH_2CH_2CHCH_3$ (with $NHCH_3$ on C-3 and CH_3 on C-6)

$NHCH_3$ CH_3

b. $CH_3CH_2CHCH_2CHCH_2CH_3$ (with $CHCH_3$–CH_3 on C-3 and CH_3 on C-5)

CH_3
$CHCH_3$
CH_3

c. $CH_3CHCHCH_2CH_2CH_2Cl$ (with CH_2CH_3 on C-2 and Cl on C-3)

CH_2CH_3
Cl

d. $CH_3CH_2CH_2CH_2CHCH_2CH_2CH_2CH_3$ (with $CH_3CCH_2CH_3$, bearing CH_3, on C-5)

$CH_3CCH_2CH_3$
CH_3

e. $CH_3CH_2CH_2CH_2CH_2CHCH_2CHCH_2CH_3$ (with CH_2CH_3 on C-8 and CH_2–CH_3CCH_3–CH_2CH_3 on C-6)

CH_2CH_3
CH_2
CH_3CCH_3
CH_2CH_3

62. Calculate the energy difference between the two chair conformers of *trans*-1,2-dimethylcyclohexane.

63. The most stable form of glucose (blood sugar) is a six-membered ring in a chair conformation with its five substituents all in equatorial positions. Draw the most stable form of glucose by putting the OH groups on the appropriate bonds in the chair conformer.

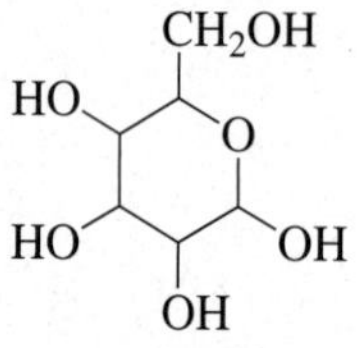

CH_2OH
HO O
HO OH
OH
glucose

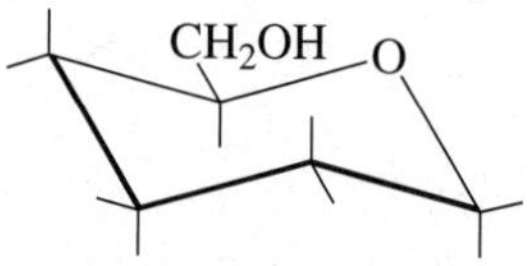

CH_2OH O

64. Explain the following facts:
a. 1-Hexanol has a higher boiling point than 3-hexanol.
b. Diethyl ether has very limited solubility in water, but tetrahydrofuran is essentially completely soluble.

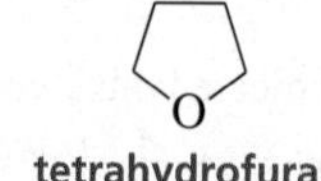

tetrahydrofuran

65. One of the chair conformers of *cis*-1,3-dimethylcyclohexane has been found to be 5.4 kcal/mol (or 23 kJ/mol) less stable than the other. How much steric strain does a 1,3-diaxial interaction between two methyl groups introduce into the conformer?

66. Calculate the amount of steric strain in each of the chair conformers of 1,1,3-trimethylcyclohexane. Which conformer would predominate at equilibrium?

PART TWO

Hydrocarbons, Stereochemistry, and Resonance

Five of the next seven chapters cover the reactions of hydrocarbons—compounds that contain only carbon and hydrogen. The other two chapters treat topics that are so important to the study of organic reactions that each deserves its own chapter. The first of these is stereochemistry and the second is electron delocalization and resonance.

Chapter 3 begins by looking at the structure and nomenclature of alkenes—*hydrocarbons that contain carbon–carbon double bonds*. Then some fundamental principles that govern the reactions of organic compounds are introduced. You will learn how to draw curved arrows to show how electrons move during the course of a reaction as new covalent bonds are formed and existing covalent bonds are broken. This chapter also discusses the principles of thermodynamics and kinetics—principles that are central to an understanding of how and why organic reactions take place.

Organic compounds can be divided into families, and fortunately, all members of a family react in the same way. In **Chapter 4**, you will learn how the family of compounds known as alkenes reacts and what kinds of products are formed from the reactions. Although many different reactions are covered, you will see that they all take place by a similar pathway.

Chapter 5 is all about stereochemistry. Here, you will learn about the different kinds of isomers that are possible for organic compounds. Then you will revisit the reactions that you learned in Chapter 4, to determine whether the products of those reactions can exist as isomers and, if so, which isomers are formed.

Chapter 6 covers the reactions of alkynes—*hydrocarbons that contain carbon–carbon triple bonds*. Because alkenes and alkynes both have reactive carbon–carbon π bonds, you will discover that their reactions have many similarities. This chapter will also introduce you to some of the techniques chemists use to design syntheses of organic compounds, and you will then have your first opportunity to design a multistep synthesis.

In **Chapter 7**, you will learn more about delocalized electrons and the concept known as resonance—topics you were introduced to in Chapter 1. Then you will see how electron delocalization affects some of the things with which you are already familiar—acidity, the stability of carbocations and radicals, and the reactions of alkenes.

In Chapter 8, you will learn about the reactions of dienes—*hydrocarbons that have two carbon–carbon double bonds*. You will see that if the two double bonds in a diene are sufficiently separated, most reactions that occur are identical to those of alkenes (Chapter 4). If, however, the double bonds are separated by only one single bond, electron delocalization (Chapter 7) plays an important role in the reactions of the compound. Notice how this chapter combines many of the concepts and theories introduced in previous chapters.

Chapter 9 covers the reactions of alkanes—*hydrocarbons that contain only single bonds*. In the previous chapters, you saw that when an organic compound reacts, the weakest bond in the molecule is the one that breaks first. Alkanes, however, have only strong bonds. Knowing this, you can correctly predict that alkanes undergo reactions only under extreme conditions.

3 Alkenes

Structure, Nomenclature, and an Introduction to Reactivity

Thermodynamics and Kinetics

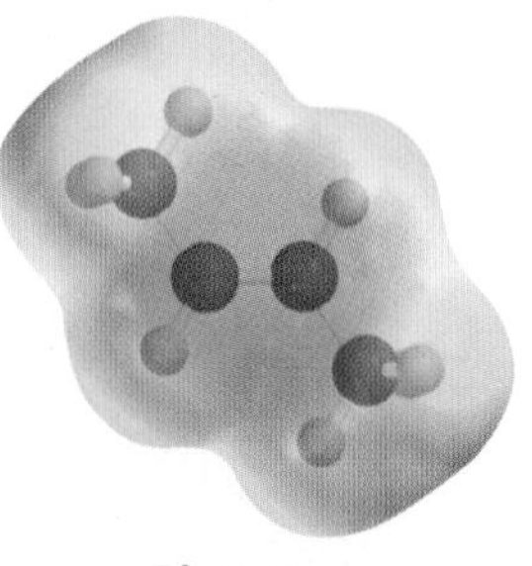

E isomer of 2-butene

Z isomer of 2-butene

In Chapter 2, you learned that alkanes are hydrocarbons that contain only carbon–carbon *single* bonds. Hydrocarbons that contain a carbon–carbon *double* bond are called **alkenes**. Early chemists noted that an oily substance was formed when ethene ($H_2C{=}CH_2$), the smallest alkene, reacted with chlorine. On the basis of this observation, alkenes were originally called *olefins* (oil forming).

Alkenes play many important roles in biology. Ethene, for example, is a plant hormone—a compound that controls the plant's growth and other changes in its tissues. Ethene affects seed germination, flower maturation, and fruit ripening.

Insects communicate by releasing pheromones—chemical substances that other insects of the same species detect with their antennae. There are sex, alarm, and trail pheromones, and many of these are alkenes. Interfering with an insect's ability to send or receive chemical signals is an environmentally safe way to control insect populations. For example, traps with synthetic sex attractants have been used to capture such crop-destroying insects as the gypsy moth and the boll weevil. Many of the flavors and fragrances produced by certain plants also belong to the alkene family.

▲ Ethene is the hormone that causes tomatoes to ripen.

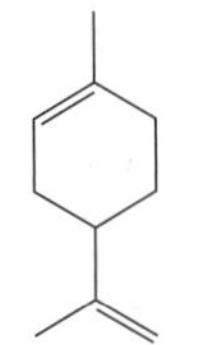

limonene
from lemon and orange oils

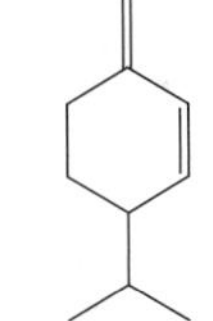

β-phellandrene
oil of eucalyptus

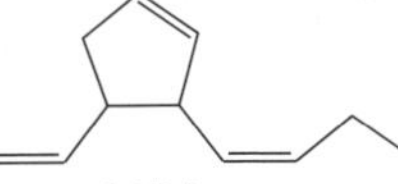

multifidene
sex attractant of brown algae

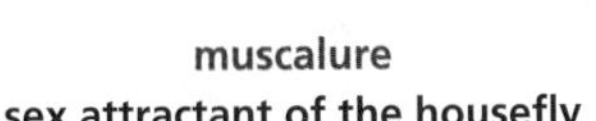

muscalure
sex attractant of the housefly

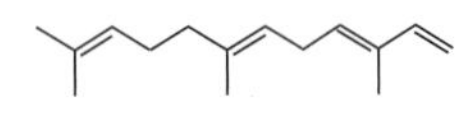

α-farnesene
found in the waxy coating on apple skins

3-D Molecules: Limonene; β-Phellandrene; Multifidene

3.1 Molecular Formula and the Degree of Unsaturation

The general molecular formula for a hydrocarbon is C_nH_{2n+2}, minus two hydrogens for every π bond or ring present in the molecule.

In Chapter 2, you learned that the general molecular formula for a noncyclic alkane is C_nH_{2n+2}. You also learned that the general molecular formula for a cyclic alkane is C_nH_{2n} because the cyclic structure reduces the number of hydrogens by two. Noncyclic compounds are also called **acyclic** compounds ("*a*" is Greek for "non" or "not").

The general molecular formula for an *acyclic alkene* is also C_nH_{2n} because, as a result of the carbon–carbon double bond, an alkene has two fewer hydrogens than an alkane with the same number of carbon atoms. Thus, the general molecular formula for a *cyclic alkene* must be C_nH_{2n-2}. We can, therefore, make the following statement: *The general molecular formula for a hydrocarbon is* C_nH_{2n+2}, *minus two hydrogens for every π bond and/or ring in the molecule.*

$CH_3CH_2CH_2CH_2CH_3$	$CH_3CH_2CH_2CH{=}CH_2$	(cyclopentane)	(cyclopentene)
an alkane	an alkene	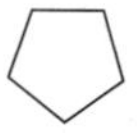 a cyclic alkane	 a cyclic alkene
C_5H_{12}	C_5H_{10}	C_5H_{10}	C_5H_8
C_nH_{2n+2}	C_nH_{2n}	C_nH_{2n}	C_nH_{2n-2}

Therefore, if we know the molecular formula of a hydrocarbon, we can determine how many rings and/or π bonds it has because, for every *two* hydrogens that are missing from the general molecular formula C_nH_{2n+2}, a hydrocarbon has either a π bond or a ring. For example, a compound with a molecular formula of C_8H_{14} needs four more hydrogens to become C_8H_{18} ($C_8H_{2\times8+2}$). Consequently, the compound has either (1) two double bonds, (2) a ring and a double bond, (3) two rings, or (4) a triple bond. Remember that a triple bond consists of two π bonds and a σ bond (Section 1.9).

Several compounds with molecular formula C_8H_{14}

$CH_3CH{=}CH(CH_2)_3CH{=}CH_2$ $\quad$ $CH_3(CH_2)_5C{\equiv}CH$ $\quad$ (1-ethylcyclohexene, CH_2CH_3) $\quad$ (bicyclic compound)

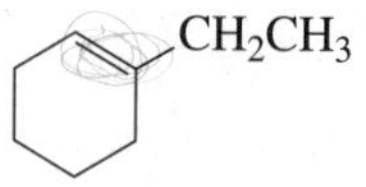

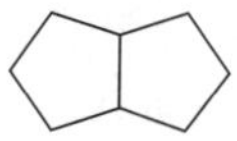

Because alkanes contain the maximum number of carbon–hydrogen bonds possible—that is, they are saturated with hydrogen—they are called **saturated hydrocarbons**. In contrast, alkenes are called **unsaturated hydrocarbons**, because they have fewer than the maximum number of hydrogens. The total number of π bonds and rings in an alkene is called its **degree of unsaturation.**

$CH_3CH_2CH_2CH_3$	$CH_3CH{=}CHCH_3$
a saturated hydrocarbon	an unsaturated hydrocarbon

PROBLEM 1◆ SOLVED

Determine the molecular formula for each of the following:

a. a 5-carbon hydrocarbon with one π bond and one ring
b. a 4-carbon hydrocarbon with two π bonds and no rings
c. a 10-carbon hydrocarbon with one π bond and two rings
d. an 8-carbon hydrocarbon with three π bonds and one ring

SOLUTION TO 1a For a 5-carbon hydrocarbon with no π bonds and no rings, $C_nH_{2n+2} = C_5H_{12}$. A 5-carbon hydrocarbon with a degree of unsaturation of 2 has four fewer hydrogens, because two hydrogens are subtracted for every π bond or ring present in the molecule. Its molecular formula, therefore, is C_5H_8.

PROBLEM 2◆ SOLVED

Determine the degree of unsaturation for the hydrocarbons with the following molecular formulas:

a. $C_{10}H_{16}$ b. $C_{20}H_{34}$ c. C_8H_{16} d. $C_{12}H_{20}$ e. $C_{40}H_{56}$

SOLUTION TO 2a For a 10-carbon hydrocarbon with no π bonds and no rings, $C_nH_{2n+2} = C_{10}H_{22}$. Thus, a 10-carbon compound with molecular formula $C_{10}H_{16}$ has six fewer hydrogens. Its degree of unsaturation, therefore, is 3.

PROBLEM 3

Determine the degree of unsaturation, and then draw possible structures, for compounds with the following molecular formulas:

a. C_3H_6 b. C_3H_4 c. C_4H_6

3.2 Nomenclature of Alkenes

The systematic (IUPAC) name of an alkene is obtained by replacing the "ane" ending of the corresponding alkane with "ene." For example, a two-carbon alkene is called ethene and a three-carbon alkene is called propene. Ethene also is frequently called by its common name: ethylene.

	$H_2C{=}CH_2$	$CH_3CH{=}CH_2$		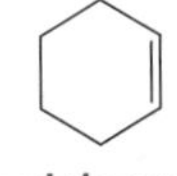
systematic name:	**ethene**	**propene**	**cyclopentene**	**cyclohexene**
common name:	**ethylene**	**propylene**		

Most alkene names need a number to indicate the position of the double bond. (The four names above do not, because there is no ambiguity.) The IUPAC rules you learned in Chapter 2 apply to alkenes as well:

Number the longest continuous chain containing the functional group in the direction that gives the functional group suffix the lowest possible number.

1. The longest continuous chain containing the functional group (in this case, the carbon–carbon double bond) is numbered in a direction that gives the functional group suffix the lowest possible number. For example, 1-butene signifies that the double bond is between the first and second carbons of butene; 2-hexene signifies that the double bond is between the second and third carbons of hexene.

$$\overset{4}{C}H_3\overset{3}{C}H_2\overset{2}{C}H{=}\overset{1}{C}H_2 \qquad \overset{1}{C}H_3\overset{2}{C}H{=}\overset{3}{C}H\overset{4}{C}H_3 \qquad \overset{1}{C}H_3\overset{2}{C}H{=}\overset{3}{C}H\overset{4}{C}H_2\overset{5}{C}H_2\overset{6}{C}H_3$$

1-butene **2-butene** **2-hexene**

$$\overset{6}{C}H_3\overset{5}{C}H_2\overset{4}{C}H_2\overset{3}{C}H_2\overset{2}{C}(={}^{1}CH_2)CH_2CH_2CH_3$$

2-propyl-1-hexene

the longest continuous chain has eight carbons but the longest continuous chain containing the functional group has six carbons, so the parent name of the compound is hexene

Notice that 1-butene does not have a common name. You might be tempted to call it "butylene," which is analogous to "propylene" for propene, but butylene is not an appropriate name. A name must be unambiguous, and "butylene" could signify either 1-butene or 2-butene.

2. The name of a substituent is cited before the name of the longest continuous chain containing the functional group, together with a number to designate the carbon to which the substituent is attached. Notice that the chain is still numbered in the direction that gives the *functional group suffix* the lowest possible number.

$\overset{1}{C}H_3\overset{2}{C}H{=}\overset{3}{C}H\overset{4}{C}(CH_3)H\overset{5}{C}H_3$
4-methyl-2-pentene

$CH_3\overset{3}{C}(\overset{2}{C}H_2\overset{1}{C}H_3){=}\overset{4}{C}H\overset{5}{C}H_2\overset{6}{C}H_2\overset{7}{C}H_3$
3-methyl-3-heptene

$CH_3CH_2CH_2CH_2CH_2O\overset{4}{C}H_2\overset{3}{C}H_2\overset{2}{C}H{=}\overset{1}{C}H_2$
4-pentoxy-1-butene

Substituents are cited in alphabetical order.

3. If a chain has more than one substituent, the substituents are cited in alphabetical order, using the same rules for alphabetizing that you learned in Section 2.2. (The prefixes *di*, *tri*, *sec*, and *tert* are ignored in alphabetizing, but *iso*, *neo*, and *cyclo* are not ignored.) The appropriate number is then assigned to each substituent.

3,6-dimethyl-3-octene

$\overset{7}{C}H_3\overset{6}{C}H_2\overset{5}{C}H(Br)\overset{4}{C}H(Cl)\overset{3}{C}H_2\overset{2}{C}H{=}\overset{1}{C}H_2$
5-bromo-4-chloro-1-heptene

4. If the same number for the alkene functional group suffix is obtained in both directions, the correct name is the name that contains the lowest substituent number. For example, 2,5-dimethyl-4-octene is a 4-octene whether the longest continuous chain is numbered from left to right or from right to left. If you number from left to right, the substituents are at positions 4 and 7, but if you number from right to left, they are at positions 2 and 5. Of those four substituent numbers, 2 is the lowest, so the compound is named 2,5-dimethyl-4-octene and *not* 4,7-dimethyl-4-octene.

$CH_3CH_2CH_2C(CH_3){=}CHCH_2CH(CH_3)CH_3$
2,5-dimethyl-4-octene
not
4,7-dimethyl-4-octene
because 2 < 4

$CH_3CH(Br)CH{=}C(CH_3)CH_2CH_3$
2-bromo-4-methyl-3-hexene
not
5-bromo-3-methyl-3-hexene
because 2 < 3

5. In cyclic alkenes, a number is not needed to denote the position of the functional group, because the ring is always numbered so that the double bond is between carbons 1 and 2.

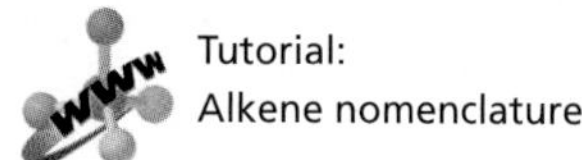
Tutorial:
Alkene nomenclature

3-ethylcyclopentene **4,5-dimethylcyclohexene** **4-ethyl-3-methylcyclohexene**

In cyclohexenes, the double bond is between C-1 and C-2, regardless of whether you move around the ring clockwise or counterclockwise. Therefore, you move around the ring in the direction that puts the lowest substituent number into the name, *not* in the direction that gives the lowest *sum* of the substituent numbers. For example, 1,6-dichlorocyclohexene is *not* called 2,3-dichlorocyclohexene because 1,6-dichlorocyclohexene has the lowest substituent number (1), even though it does not have the lowest sum of the substituent numbers $(1 + 6 = 7$ versus $2 + 3 = 5)$.

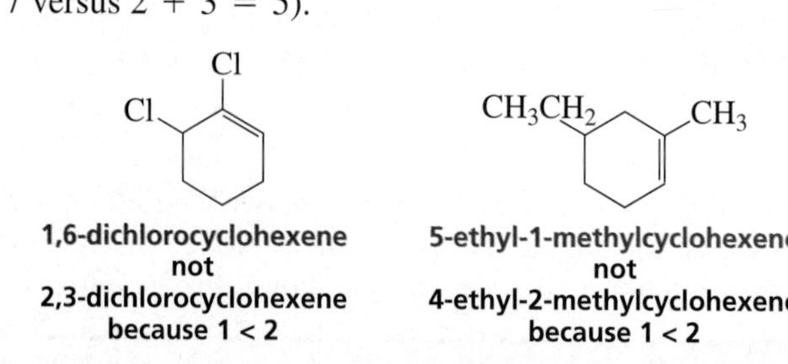

1,6-dichlorocyclohexene
not
2,3-dichlorocyclohexene
because 1 < 2

5-ethyl-1-methylcyclohexene
not
4-ethyl-2-methylcyclohexene
because 1 < 2

6. If both directions lead to the same number for the alkene functional group suffix and the same low number(s) for one or more of the substituents, then those substituents are ignored and the direction is chosen that gives the lowest number to one of the remaining substituents.

$$\underset{\displaystyle CH_3}{CH_3\overset{}{C}H}CH_2CH{=}\underset{\displaystyle CH_2CH_3}{C}CH_2\overset{\displaystyle Br}{C}HCH_3$$

2-bromo-4-ethyl-7-methyl-4-octene
not
7-bromo-5-ethyl-2-methyl-4-octene
because 4 < 5

(cyclohexene ring bearing Cl, CH_3 and Br)

6-bromo-3-chloro-4-methylcyclohexene
not
3-bromo-6-chloro-5-methylcyclohexene
because 4 < 5

The sp^2 carbons of an alkene are called **vinylic carbons**. An sp^3 carbon that is adjacent to a vinylic carbon is called an **allylic carbon**.

vinylic carbons

$$RCH_2{-}CH{=}CH{-}CH_2R$$

allylic carbons

Two groups containing a carbon–carbon double bond are used in common names—the **vinyl group** and the **allyl group**. The vinyl group is the smallest possible group that contains a vinylic carbon; the allyl group is the smallest possible group that contains an allylic carbon. When "allyl" is used in nomenclature, the substituent must be attached to the allylic carbon.

	$H_2C{=}CH{-}$ **the vinyl group**	$H_2C{=}CHCH_2{-}$ **the allyl group**
	$H_2C{=}CHCl$	$H_2C{=}CHCH_2Br$
systematic name:	**chloroethene**	**3-bromopropene**
common name:	**vinyl chloride**	**allyl bromide**

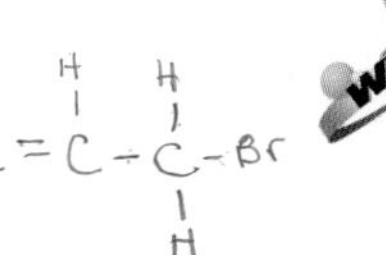

Tutorial: Common names of alkyl groups

PROBLEM 4◆

Draw the structure for each of the following compounds:

a. 3,3-dimethylcyclopentene
b. 6-bromo-2,3-dimethyl-2-hexene
c. ethyl vinyl ether
d. allyl alcohol

PROBLEM 5◆

Give the systematic name for each of the following compounds:

a. $CH_3\underset{\displaystyle CH_3}{C}HCH{=}CHCH_3$

b. $CH_3CH_2\underset{\displaystyle CH_3}{C}{=}\overset{\displaystyle CH_3}{\underset{\displaystyle Cl}{C}}CHCH_3$

c. Br (bromomethyl-substituted cyclopentene ring)

d. $BrCH_2CH_2CH{=}\underset{\displaystyle CH_2CH_3}{C}CH_3$

e. (cyclohexene ring with two CH_3 groups) CH_3 CH_3

f. $CH_3CH{=}CHOCH_2CH_2CH_2CH_3$

3.3 The Structure of Alkenes

The structure of the smallest alkene (ethene) was described in Section 1.8. Other alkenes have similar structures. Each double-bonded carbon of an alkene has three sp^2 orbitals that lie in a plane with angles of 120°. Each of these sp^2 orbitals overlaps an orbital of another atom to form a σ bond. Thus, one of the carbon–carbon bonds in a double bond is a σ bond, formed by the overlap of an sp^2 orbital of one carbon with an sp^2 orbital of the other carbon. The second carbon–carbon bond in the double bond (the π bond) is formed from side-to-side overlap of the remaining p orbitals of the sp^2 carbons. Because three points determine a plane, each sp^2 carbon and the two atoms singly bonded to it lie in a plane. In order to achieve maximum orbital–orbital overlap, the two p orbitals must be parallel to each other. Therefore, all six atoms of the double-bond system are in the same plane.

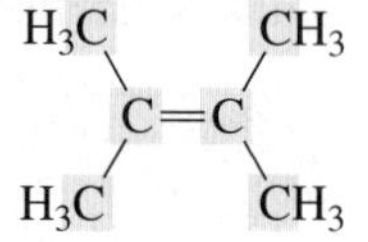

$(H_3C)_2C{=}C(CH_3)_2$

the six carbon atoms are in the same plane

It is important to remember that the π bond represents the cloud of electrons that is above and below the plane defined by the two sp^2 carbons and the four atoms bonded to them.

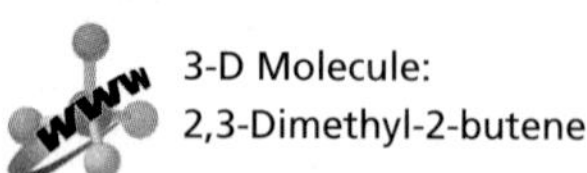

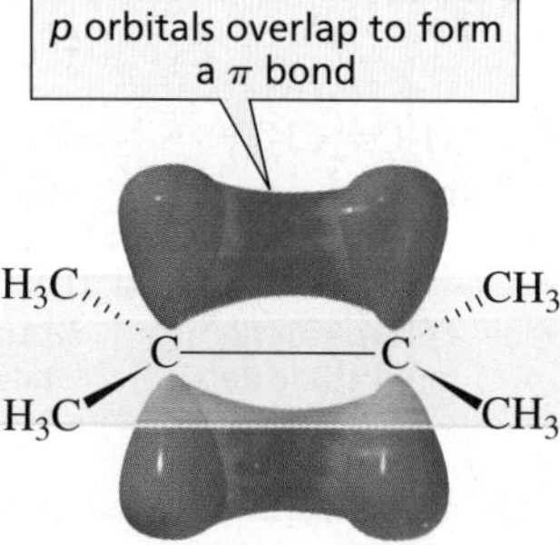

PROBLEM 6♦

For each of the following compounds, tell how many of its carbon atoms lie in the same plane:

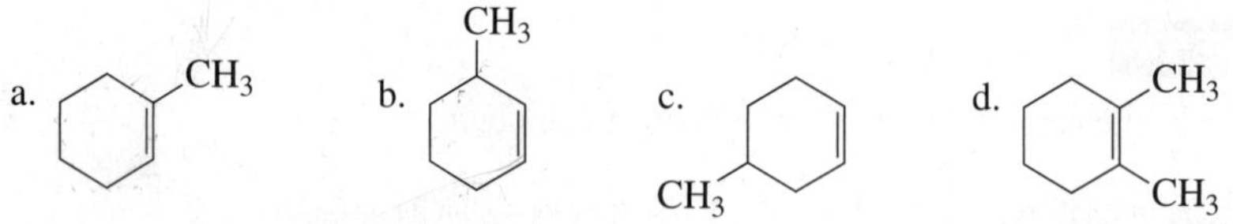

a. 1-methylcyclohexene (CH_3 on the double-bonded carbon)   b. cyclohexene with CH_3 on the allylic carbon   c. cyclohexene with CH_3 on a ring carbon away from the double bond   d. 1,2-dimethylcyclohexene (CH_3, CH_3)

3.4 Cis–Trans Isomerism

Because the two p orbitals that form the π bond must be parallel to achieve maximum overlap, rotation about a double bond does not readily occur. If rotation were to occur, the two p orbitals would no longer overlap and the π bond would break (Figure 3.1). The barrier to rotation about a double bond is 63 kcal/mol. Compare this to the barrier to rotation (2.9 kcal/mol) about a carbon–carbon single bond (Section 2.10).

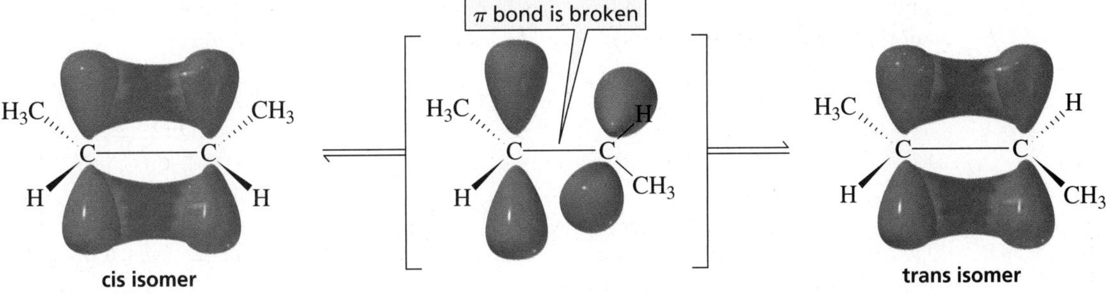

▲ **Figure 3.1**
Rotation about the carbon–carbon double bond would break the π bond.

Because there is an energy barrier to rotation about a carbon–carbon double bond, an alkene such as 2-butene can exist in two distinct forms: The hydrogens bonded to the sp^2 carbons can be on the same side of the double bond or on opposite sides of the double bond. The isomer with the hydrogens on the same side of the double bond is called the **cis isomer**, and the isomer with the hydrogens on opposite sides of the double bond is called the **trans isomer**. A pair of isomers such as *cis*-2-butene and *trans*-2-butene is called **cis–trans isomers** or **geometric isomers**. This should remind you of the cis–trans isomers of 1,2-disubstituted cyclohexanes you encountered in Section 2.13—the cis isomer had its substituents on the same side of the ring, and the trans isomer had its substituents on opposite sides of the ring. Cis–trans isomers have the same molecular formula, but differ in the way their atoms are arranged in space (Section 2.14).

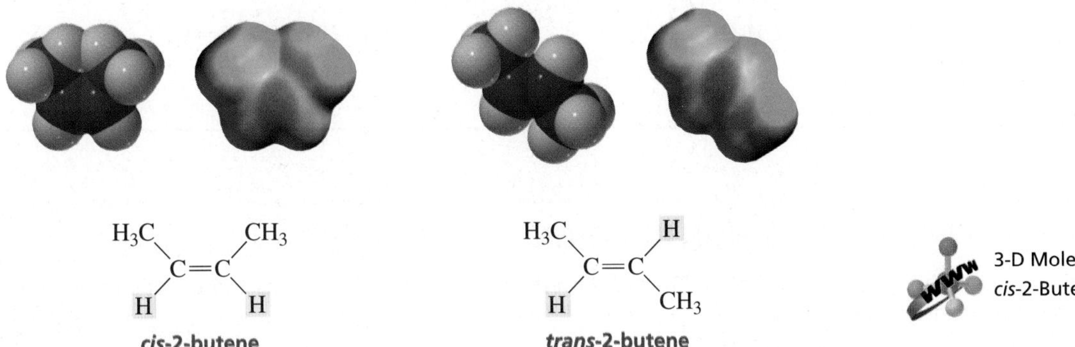

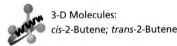

3-D Molecules:
cis-2-Butene; *trans*-2-Butene

If one of the sp^2 carbons of the double bond is attached to two identical substituents, there is only one possible structure for the alkene. In other words, cis and trans isomers are not possible for an alkene that has identical substituents attached to one of the double-bonded carbons.

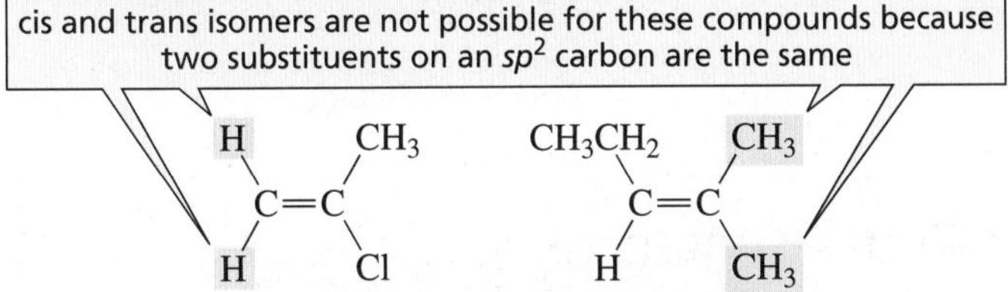

3-D Molecule:
2-Methyl-2-pentene

Because of the energy barrier to rotation about a double bond, cis and trans isomers cannot interconvert (except under conditions extreme enough to overcome the barrier and break the π bond). This means that they can be separated from each other. In other words, the two isomers are different compounds with different physical properties, such as different boiling points and different dipole moments. Notice that *trans*-2-butene and *trans*-1,2-dichloroethene have dipole moments (μ) of zero because the bond dipole moments cancel (Section 1.15).

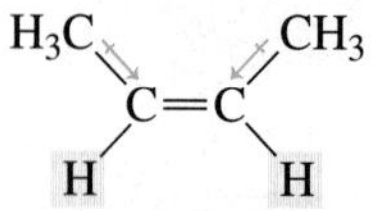

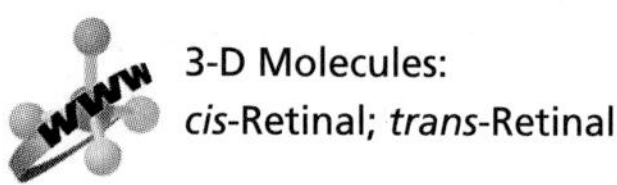

cis-2-butene bp = 3.7 °C μ = 0.33 D

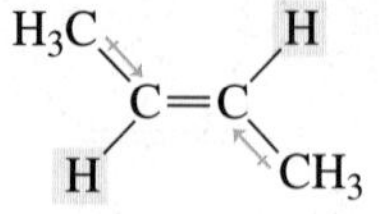

trans-2-butene bp = 0.9 °C μ = 0 D

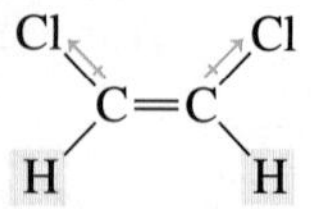

cis-1,2-dichloroethene bp = 60.3 °C μ = 2.95 D

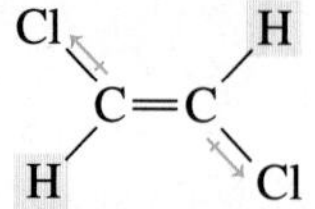

trans-1,2-dichloroethene bp = 47.5 °C μ = 0 D

Cis and trans isomers can be interconverted (in the absence of any added reagents) only when the molecule absorbs sufficient heat or light energy to cause the π bond to break, because once the π bond is broken, rotation can occur about the remaining σ bond (Section 2.10). Cis–trans interconversion, therefore, is not a practical laboratory process.

3-D Molecules: *cis*-Retinal; *trans*-Retinal

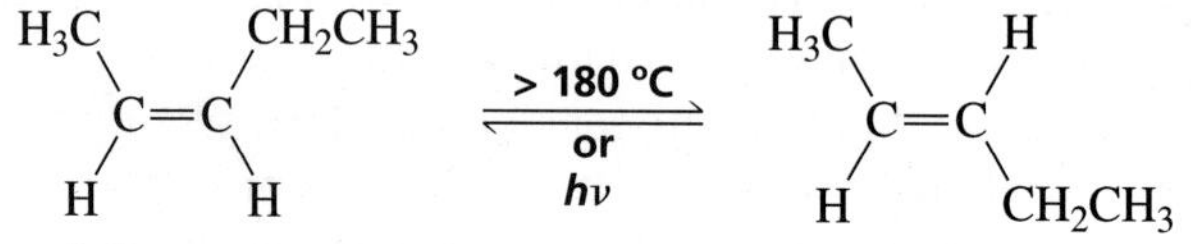

cis-2-pentene $\rightleftharpoons$ (> 180 °C or $h\nu$) trans-2-pentene

PROBLEM 7◆

a. Which of the following compounds can exist as cis–trans isomers?

b. For those compounds, draw and label the cis and trans isomers.

1. $CH_3CH{=}CHCH_2CH_3$
2. $CH_3C(CH_3){=}CHCH_3$
3. $CH_3CH{=}CHCH_3$
4. $CH_3CH_2CH{=}CH_2$

CIS–TRANS INTERCONVERSION IN VISION

When rhodopsin absorbs light, a double bond interconverts between the cis and trans forms. This process plays an important role in vision (Section 26.7).

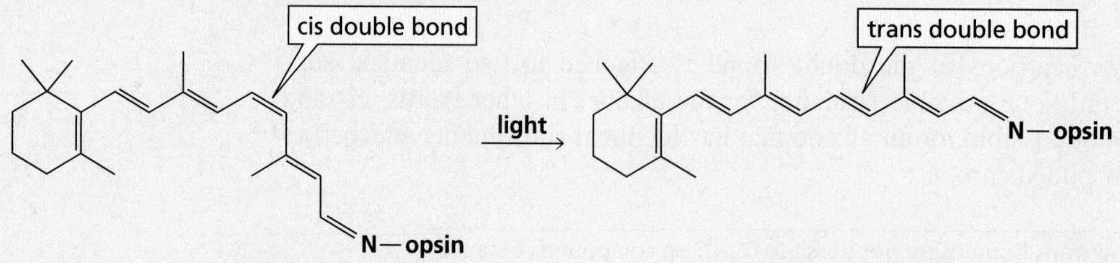

cis double bond — rhodopsin (N—opsin) —light→ metarhodopsin II (*trans*-rhodopsin) (N—opsin) — trans double bond

—OPSIN —OPSIN

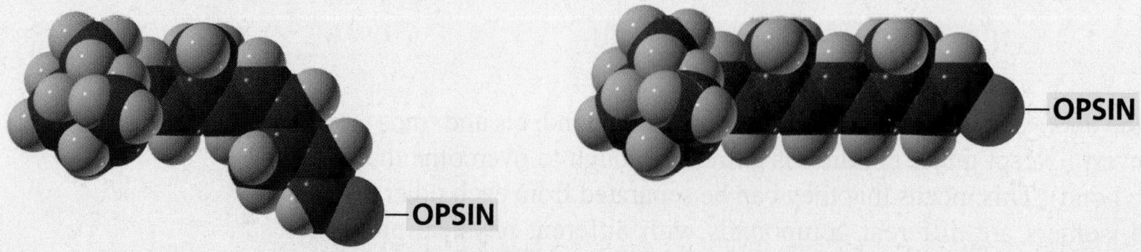

PROBLEM 8◆

Which of the following compounds have a dipole moment of zero?

A: H(H)C=C(Cl)Cl B: H(Cl)C=C(H)H C: H(Cl)C=C(Cl)H D: H(Cl)C=C(H)Cl

3.5 The *E,Z* System of Nomenclature

As long as each of the sp^2 carbons of an alkene is bonded to only one substituent, we can use the terms cis and trans to designate the structure of the alkene. *If the hydrogens are on the same side of the double bond, it is the cis isomer; if they are on opposite sides of the double bond, it is the trans isomer.* But how would you designate the isomers of a compound such as 1-bromo-2-chloropropene?

Br(H)C=C(Cl)CH₃ Br(H)C=C(CH₃)Cl

Which isomer is cis and which is trans?

For a compound such as 1-bromo-2-chloropropene, the cis–trans system of nomenclature cannot be used because there are four different substituents on the two vinylic carbons. The *E,Z* system of nomenclature was devised for these kinds of situations.*

To name an isomer by the *E,Z* system, we first determine the relative priorities of the two groups bonded to one of the sp^2 carbons and then the relative priorities of the two groups bonded to the other sp^2 carbon. (The rules for assigning relative priorities are explained next.) If the high-priority groups are on the same side of the double bond, the isomer has the *Z* configuration (*Z* is for *zusammen*, German for "together"). If the high-priority groups are on opposite sides of the double bond, the isomer has the *E* configuration (*E* is for *entgegen*, German for "opposite").

The *Z* isomer has the high-priority groups on the same side.

low priority / low priority / C=C / high priority / high priority — the *Z* isomer

low priority / high priority / C=C / high priority / low priority — the *E* isomer

The relative priorities of the two groups bonded to an sp^2 carbon are determined using the following rules:

- **Rule 1.** The relative priorities of the two groups depend on the atomic numbers of the atoms that are bonded directly to the sp^2 carbon. The greater the atomic number, the higher is the priority.

  For example, in the following compounds, one of the sp^2 carbons is bonded to a Br and to an H:

high priority: Br, Cl — Br(H)C=C(Cl)CH₃ — the *Z* isomer

high priority: Br, Cl — Br(H)C=C(CH₃)Cl — the *E* isomer

*IUPAC prefers the *E* and *Z* designations because they can be used for all alkene isomers. Many chemists, however, continue to use the cis and trans designations for simple molecules.

The greater the atomic number of the atom bonded to an sp^2 carbon, the higher is the priority of the substituent.

Br has a greater atomic number than H, so **Br** has a higher priority than **H**. The other sp^2 carbon is bonded to a Cl and to a C. Cl has the greater atomic number, so **Cl** has a higher priority than **C**. (Notice that you use the atomic number of C, not the mass of the CH_3 group, because the priorities are based on the atomic numbers of atoms, *not* on the masses of groups.) The isomer on the left has the high-priority groups (Br and Cl) on the same side of the double bond, so it is the ***Z* isomer**. (Zee groups are on Zee Zame Zide.) The isomer on the right has the high-priority groups on opposite sides of the double bond, so it is the ***E* isomer**.

- **Rule 2.** If the two substituents bonded to an sp^2 carbon start with the same atom (there is a tie), you must move outward from the point of attachment and consider the atomic numbers of the atoms that are attached to the "tied" atoms.

  In the following compounds, one of the sp^2 carbons is bonded both to a Cl and to the C of a CH_2Cl group:

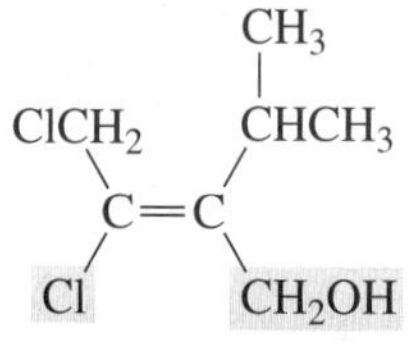

```
                 CH3                     CH3
                 |                       |
     ClCH2       CHCH3        Cl         CHCH3
          \     /               \       /
           C=C                   C=C
          /     \               /     \
        Cl       CH2OH      ClCH2      CH2OH
```

the *Z* isomer 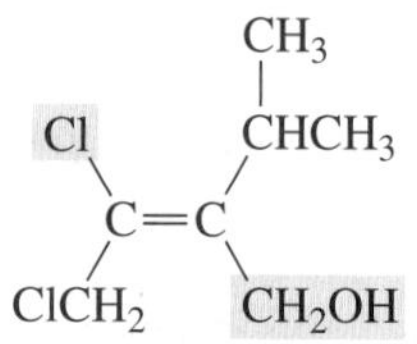   the *E* isomer

If the atoms attached to sp^2 carbons are the same, the atoms attached to the "tied" atoms are compared; the one with the greatest atomic number belongs to the group with the higher priority.

Cl has a greater atomic number than C, so the Cl group has the higher priority. Both atoms bonded to the other sp^2 carbon are C's (from a CH_2OH group and an $CH(CH_3)_2$ group), so there is a tie at this point. The C of the CH_2OH group is bonded to **O**, **H**, and **H**, and the C of the $CH(CH_3)_2$ group is bonded to **C**, **C**, and **H**. Of these six atoms, O has the greatest atomic number, so $\mathbf{CH_2OH}$ has a higher priority than $\mathbf{CH(CH_3)_2}$. (Note that you do not add the atomic numbers; you take the single atom with the greatest atomic number.) The *E* and *Z* isomers are as shown.

- **Rule 3.** If an atom is doubly bonded to another atom, the priority system treats it as if it were singly bonded to two of those atoms. If an atom is triply bonded to another atom, the priority system treats it as if it were singly bonded to three of those atoms.

  For example, one of the sp^2 carbons in the following pair of isomers is bonded to a CH_2CH_3 group and to a $CH{=}CH_2$ group:

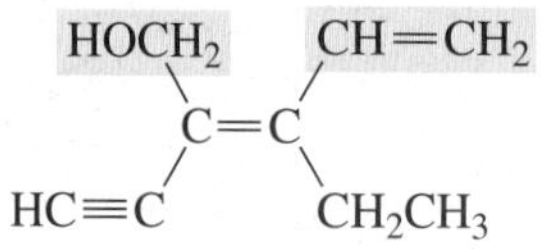

```
     HOCH2      CH=CH2          HOCH2      CH2CH3
          \    /                     \    /
           C=C                        C=C
          /    \                     /    \
     HC≡C       CH2CH3          HC≡C       CH=CH2
```

the *Z* isomer 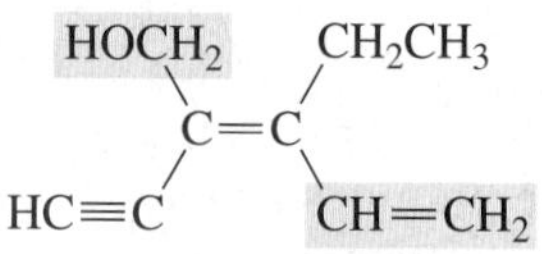   the *E* isomer

If an atom is doubly bonded to another atom, treat it as if it were singly bonded to two of those atoms.

If an atom is triply bonded to another atom, treat it as if it were singly bonded to three of those atoms.

Cancel atoms that are identical in the two groups; use the remaining atoms to determine the group with the higher priority.

Because the atoms immediately bonded to the sp^2 carbon are both C's, there is a tie. The first carbon of the CH_2CH_3 group is bonded to **C**, **H**, and **H**. The first carbon of the $CH{=}CH_2$ group is bonded to an H and doubly bonded to a C. Therefore, it is considered to be bonded to **C**, **C**, and **H**. One C cancels in each of the two groups, leaving H and H in the CH_2CH_3 group and C and H in the $CH{=}CH_2$ group. C has a greater atomic number than H, so $\mathbf{CH{-}CH_2}$ has a higher priority than $\mathbf{CH_2CH_3}$. Both atoms that are bonded to the other sp^2 carbon are C's, so there is a tie there as well. The triple-bonded C is considered to be bonded to **C**, **C**, and **C**; the other C is bonded to **O**, **H**, and **H**. Of the six atoms, O has the greatest atomic number, so $\mathbf{CH_2OH}$ has a higher priority than $\mathbf{C{\equiv}CH}$.

- **Rule 4.** In the case of isotopes (atoms with the same atomic number, but different mass numbers), the mass number is used to determine the relative priorities.

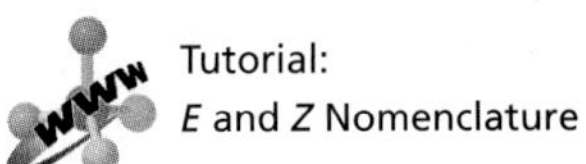

In the following structures, for example, one of the sp^2 carbons is bonded to a deuterium (D) and a hydrogen (H):

$$\underset{\text{the } Z \text{ isomer}}{\begin{array}{c} \text{H} \quad\quad \text{CHCH}_3(\text{CH}_3) \\ \text{C}=\text{C} \\ \text{D} \quad\quad \text{CH}=\text{CH}_2 \end{array}} \qquad \underset{\text{the } E \text{ isomer}}{\begin{array}{c} \text{H} \quad\quad \text{CH}=\text{CH}_2 \\ \text{C}=\text{C} \\ \text{D} \quad\quad \text{CHCH}_3(\text{CH}_3) \end{array}}$$

the *Z* isomer

the *E* isomer

D and H have the same atomic number, but D has a greater mass number, so **D** has a higher priority than **H**. The C's that are bonded to the other sp^2 carbon are *both* bonded to **C**, **C**, and **H**, so you must go out to the next set of atoms to break the tie. The second carbon of the $CH(CH_3)_2$ group is bonded to **H**, **H**, and **H**, whereas the second carbon of the $CH{=}CH_2$ group is bonded to **H**, **H**, and **C**. Therefore, $\mathbf{CH{=}CH_2}$ has a higher priority than $\mathbf{CH(CH_3)_2}$.

If atoms have the same atomic number, but different mass numbers, the one with the greater mass number has the higher priority.

Notice in all these examples that you never count the atom attached to the σ bond from which you originate. In differentiating between the $CH(CH_3)_2$ and $CH{=}CH_2$ groups in the last example, you do, however, count the atom attached to a π bond from which you originate. In the case of a triple bond, you count the atoms bonded to both π bonds from which you originate.

PROBLEM 9

Draw and label the *E* and *Z* isomers for each of the following compounds:

1. $CH_3CH_2CH{=}CHCH_3$

2. $CH_3CH_2\underset{|}{\overset{}{C}}{=}CHCH_2CH_3$ (with Cl on the C)

3. $CH_3CH_2CH_2CH_2$ on C of $CH_3CH_2C{=}CCH_2Cl$, with $CHCH_3$ / CH_3 on the second C

4. $HOCH_2CH_2C{=}CC{\equiv}CH$, with $O{=}CH$ on the first C and $C(CH_3)_3$ on the second C

PROBLEM 10

Draw the structure of (*Z*)-3-isopropyl-2-heptene.

3.6 How Alkenes React • Curved Arrows

There are many millions of organic compounds. If you had to memorize how each of them reacts, studying organic chemistry would be a horrendous experience. Fortunately, organic compounds can be divided into families, and all the members of a family react in similar ways. What determines the family an organic compound belongs to is its functional group. The **functional group** is a structural unit that acts as the center of reactivity of a molecule. You will find a table of common functional groups inside the back cover of this book. You are already familiar with the functional group of an alkene: the carbon–carbon double bond. All compounds with a carbon–carbon double bond react in similar ways, whether the compound is a small molecule like ethene or a large molecule like cholesterol.

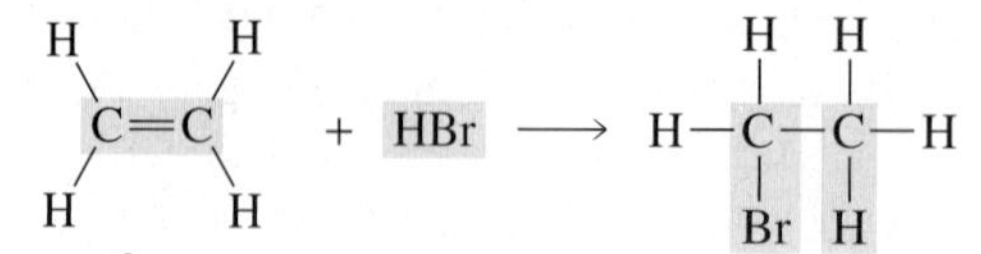

H H H H
C=C + HBr ⟶ H—C—C—H
H H Br H
ethene

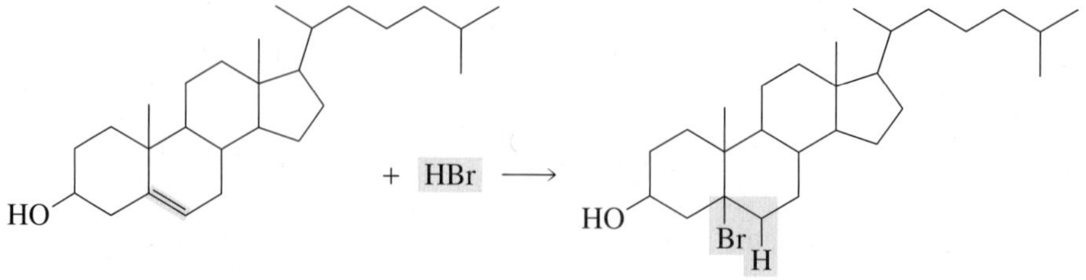

HO + HBr ⟶ HO Br H
cholesterol

To further reduce the need to memorize, one needs to understand *why* a functional group reacts the way it does. It is not sufficient to know that a compound with a carbon–carbon double bond reacts with HBr to form a product in which the H and Br atoms have taken the place of the π bond; we need to understand *why* the compound reacts with HBr. In each chapter that discusses the reactivity of a particular functional group, we will see how the nature of the functional group allows us to predict the kind of reactions it will undergo. Then, when you are confronted with a reaction you have never seen before, knowledge of how the structure of the molecule affects its reactivity will help you predict the products of the reaction.

In essence, organic chemistry is about the interaction between electron-rich atoms or molecules and electron-deficient atoms or molecules. It is these forces of attraction that make chemical reactions happen. From this follows a very important rule that determines the reactivity of organic compounds: *Electron-rich atoms or molecules are attracted to electron-deficient atoms or molecules.* Each time you study a new functional group, remember that the reactions it undergoes can be explained by this very simple rule.

Electron-rich atoms or molecules are attracted to electron-deficient atoms or molecules.

Therefore, to understand how a functional group reacts, you must first learn to recognize electron-deficient and electron-rich atoms and molecules. An electron-deficient atom or molecule is called an **electrophile**. An electrophile can have an atom that can accept a pair of electrons, or it can have an atom with an unpaired electron and, therefore, is in need of an electron to complete its octet. Thus, an electrophile looks for electrons. Literally, "electrophile" means "electron loving" (*phile* is the Greek suffix for "loving").

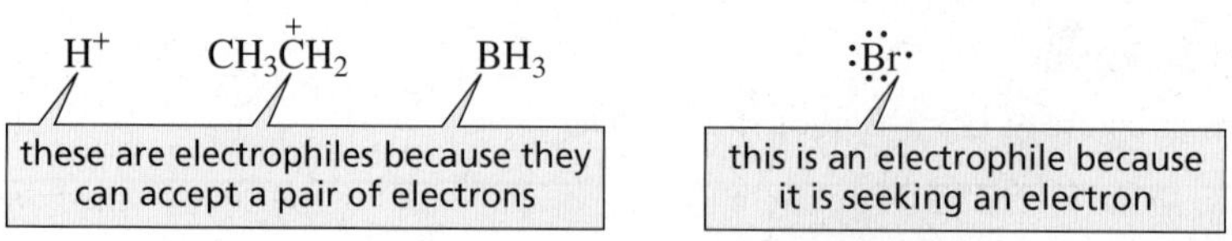

H^+ $CH_3\overset{+}{C}H_2$ BH_3 — these are electrophiles because they can accept a pair of electrons

:Br· — this is an electrophile because it is seeking an electron

A nucleophile reacts with an electrophile.

An electron-rich atom or molecule is called a **nucleophile**. A nucleophile has a pair of electrons it can share. Some nucleophiles are neutral and some are negatively charged. Because a nucleophile has electrons to share and an electrophile is seeking electrons, it should not be surprising that they attract each other. Thus, the preceding rule can be restated as *a nucleophile reacts with an electrophile.*

$HO:^-$ $:Cl:^-$ CH_3NH_2 $H_2O:$ — these are nucleophiles because they have a pair of electrons to share

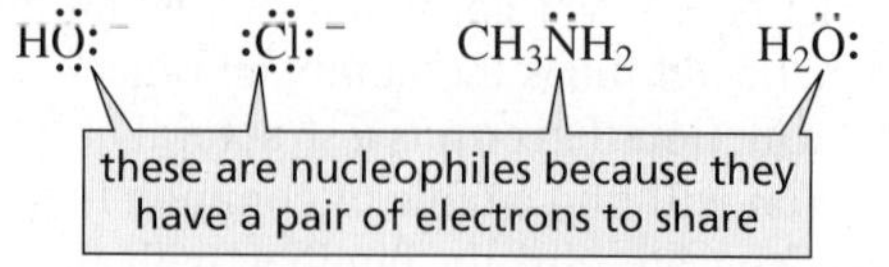

Because an electrophile accepts a pair of electrons, it is sometimes called a *Lewis acid*. Because a nucleophile has a pair of electrons to share, it is sometimes called a *Lewis base* (Section 1.21).

We have seen that a π bond is weaker than a σ bond (Section 1.14). The π bond, therefore, is the bond that is most easily broken when an alkene undergoes a reaction. We also have seen that the π bond of an alkene consists of a cloud of electrons above and below the σ bond. As a result of this cloud of electrons, an alkene is an electron-rich molecule—it is a nucleophile. (Notice the relatively electron-rich orange area in the electrostatic potential maps for *cis*- and *trans*-2-butene in Section 3.4.) We can, therefore, predict that an alkene will react with an electrophile and, in the process, the π bond will break. So if a reagent such as hydrogen bromide is added to an alkene, the alkene will react with the partially positively charged hydrogen of hydrogen bromide and a carbocation will be formed. In the second step of the reaction, the positively charged carbocation (an electrophile) will react with the negatively charged bromide ion (a nucleophile) to form an alkyl halide.

$$CH_3CH{=}CHCH_3 + \overset{\delta+}{H}-\overset{\delta-}{Br} \longrightarrow CH_3\overset{+}{C}H-\underset{\displaystyle H}{\underset{|}{C}}HCH_3 + Br^- \longrightarrow CH_3\underset{\displaystyle Br}{\underset{|}{C}}H-\underset{\displaystyle H}{\underset{|}{C}}HCH_3$$

a carbocation

2-bromobutane
an alkyl halide

Curved arrows show the flow of electrons; they are drawn from an electron-rich center to an electron-deficient center.

An arrowhead with two barbs signifies the movement of two electrons.

An arrowhead with one barb signifies the movement of one electron.

The description of the step-by-step process by which reactants (e.g., alkene + HBr) are changed into products (e.g., alkyl halide) is called the **mechanism of the reaction**. To help us understand a mechanism, curved arrows are drawn to show how the electrons move as new covalent bonds are formed and existing covalent bonds are broken. In other words, the curved arrows show which bonds are formed and which are broken. Because the curved arrows show how the electrons flow, *they are drawn from an electron-rich center* (at the tail of the arrow) *to an electron-deficient center* (at the point of the arrow). An arrowhead with two barbs represents the simultaneous movement of two electrons (an electron pair). An arrowhead with one barb represents the movement of one electron. These are called "curved" arrows to distinguish them from the "straight" arrows used to link reactants with products in chemical reactions.

$$CH_3CH{=}CHCH_3 + \overset{\delta+}{H}-\overset{\delta-}{\ddot{B}r}: \longrightarrow CH_3\overset{+}{C}H-\underset{\displaystyle H}{\underset{|}{C}}HCH_3 + :\ddot{B}r:^-$$

For the reaction of 2-butene with HBr, an arrow is drawn to show that the two electrons of the π bond of the alkene are attracted to the partially positively charged hydrogen of HBr. The hydrogen, however, is not free to accept this pair of electrons because it is already bonded to a bromine, and hydrogen can be bonded to only one atom at a time (Section 1.4). Therefore, as the π electrons of the alkene move toward the hydrogen, the H—Br bond breaks, with bromine keeping the bonding electrons. Notice that the π electrons are pulled away from one carbon, but remain attached to the other. Thus, the two electrons that formerly formed the π bond now form a σ bond between carbon and the hydrogen from HBr. The product of this first step in the reaction is a carbocation because the sp^2 carbon that did not form the new bond with hydrogen no longer shares the pair of π electrons. It is, therefore, positively charged.

In the second step of the reaction, a lone pair on the negatively charged bromide ion forms a bond with the positively charged carbon of the carbocation. Notice that both steps of the reaction involve *the reaction of an electrophile with a nucleophile.*

$$CH_3\overset{+}{C}H-\underset{\displaystyle H}{\underset{|}{C}}HCH_3 + :\ddot{B}r:^- \longrightarrow CH_3\underset{\displaystyle :\ddot{B}r:}{\underset{|}{C}}H-\underset{\displaystyle H}{\underset{|}{C}}HCH_3$$

Solely from the knowledge that an electrophile reacts with a nucleophile and a π bond is the weakest bond in an alkene, we have been able to predict that the product of the reaction of 2-butene and HBr is 2-bromobutane. Overall, the reaction involves the *addition* of 1 mole of HBr to 1 mole of the alkene. The reaction, therefore, is called an **addition reaction**. Because the first step of the reaction involves the addition of an electrophile (H^+) to the alkene, the reaction is more precisely called an **electrophilic addition reaction**. Electrophilic addition reactions are the characteristic reactions of alkenes.

At this point, you may think that it would be easier just to memorize the fact that 2-bromobutane is the product of the reaction, without trying to understand the mechanism that explains why 2-bromobutane is the product. Keep in mind, however, that the number of reactions you encounter is going to increase substantially, and it will be impossible to memorize them all. If you strive to understand the mechanism of each reaction, the unifying principles of organic chemistry will become apparent, making mastery of the material much easier and a lot more fun.

It will be helpful to do the exercise on drawing curved arrows in the Study Guide/Solution Manual (Special Topic III).

PROBLEM 11◆

Which of the following are electrophiles, and which are nucleophiles?

H^- $AlCl_3$ CH_3O^- $CH_3C{\equiv}CH$ $CH_3\overset{+}{C}HCH_3$ NH_3

A FEW WORDS ABOUT CURVED ARROWS

1. Make certain that the arrows are drawn in the direction of the electron flow and never against the flow. This means that an arrow will always be drawn away from a negative charge and/or toward a positive charge.

correct: $CH_3-C(-\ddot{O}:^-)(CH_3)-\ddot{Br}: \longrightarrow CH_3-C(=\ddot{O})CH_3 + :\ddot{Br}:^-$

incorrect: $CH_3-C(-\ddot{O}:^-)(CH_3)-\ddot{Br}: \longrightarrow CH_3-C(=\ddot{O})CH_3 + :\ddot{Br}:^-$

correct: $CH_3-\overset{+}{\ddot{O}}(H)-H \longrightarrow CH_3-\ddot{O}-H + H^+$

incorrect: $CH_3-\overset{+}{\ddot{O}}(H)-H \longrightarrow CH_3-\ddot{O}-H + H^+$

2. Curved arrows are drawn to indicate the movement of electrons. Never use a curved arrow to indicate the movement of an atom. For example, you can't use an arrow as a lasso to remove the proton, as shown here:

correct: $CH_3C(=\overset{+}{O}-H)CH_3 \longrightarrow CH_3C(=\ddot{O})CH_3 + H^+$

incorrect: $CH_3C(=\overset{+}{O}-H)CH_3 \longrightarrow CH_3C(=\ddot{O})CH_3 + H^+$

3. The arrow starts at the electron source. It does not start at an atom. In the following example, the arrow starts at the electrons of the π bond, not at a carbon atom:

correct: $CH_3CH{=}CHCH_3 + H-\ddot{Br}: \longrightarrow CH_3\overset{+}{C}H-CH(H)CH_3 + :\ddot{Br}:^-$

incorrect: $CH_3CH{=}CHCH_3 + H-\ddot{Br}: \longrightarrow CH_3\overset{+}{C}H-CH(H)CH_3 + :\ddot{Br}:^-$

PROBLEM 12

Use curved arrows to show the movement of electrons in each of the following reaction steps:

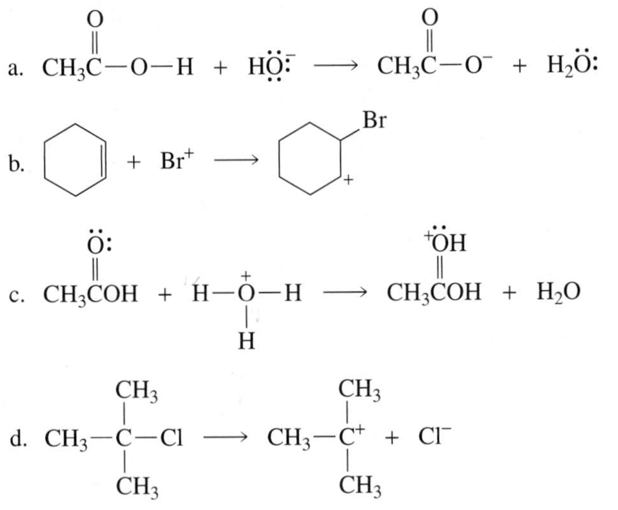

a. $CH_3C(=O)—O—H + HO{:}^- \longrightarrow CH_3C(=O)—O^- + H_2O{:}$

b. cyclohexene $+ Br^+ \longrightarrow$ bromocyclohexyl cation (Br, +)

c. $CH_3C(=\ddot{O}{:})OH + H—\overset{+}{O}(H)—H \longrightarrow CH_3C(=\overset{+}{O}H)OH + H_2O$

d. $CH_3—C(CH_3)_2—Cl \longrightarrow CH_3—C^+(CH_3)_2 + Cl^-$

3.7 Thermodynamics and Kinetics

Before we can understand the energy changes that take place in a reaction such as the addition of HBr to an alkene, we must have an understanding of *thermodynamics*, which describes a reaction at equilibrium, and an appreciation of *kinetics*, which deals with the rates of chemical reactions.

If we consider a reaction in which Y is converted to Z, the *thermodynamics* of the reaction tells us the relative amounts of Y and Z that are present when the reaction has reached equilibrium, whereas the *kinetics* of the reaction tells us how fast Y is converted into Z.

$$Y \rightleftharpoons Z$$

Reaction Coordinate Diagrams

The mechanism of a reaction describes the various steps that are believed to occur as reactants are converted into products. A **reaction coordinate diagram** shows the energy changes that take place in each of the steps of the mechanism. In a reaction coordinate diagram, the total energy of all species is plotted against the progress of the reaction. A reaction progresses from left to right as written in the chemical equation, so the energy of the reactants is plotted on the left-hand side of the x-axis and the energy of the products is plotted on the right-hand side. A typical reaction coordinate diagram is shown in Figure 3.2. The diagram describes the reaction of A—B with C to form A and B—C. Remember that *the more stable the species, the lower is its energy.*

The more stable the species, the lower is its energy.

$$\underbrace{A—B + C}_{\text{reactants}} \rightleftharpoons \underbrace{A + B—C}_{\text{products}}$$

As the reactants are converted into products, the reaction passes through a *maximum* energy state called a **transition state**. The structure of the transition state lies somewhere between the structure of the reactants and the structure of the products. Bonds that break and bonds that form, as reactants are converted to products, are partially broken and partially formed in the transition state. Dashed lines are used to show partially broken or partially formed bonds.

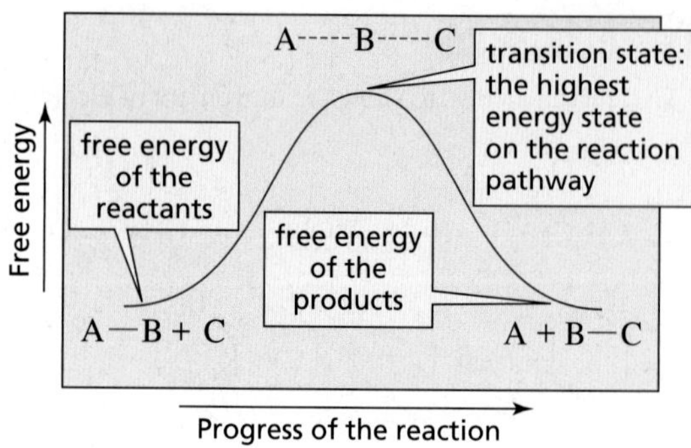

Figure 3.2 ▶ A reaction coordinate diagram. The dashed lines in the transition state indicate bonds that are partially formed or partially broken.

Thermodynamics

The field of chemistry that describes the properties of a system at equilibrium is called **thermodynamics**. The relative concentrations of reactants and products at equilibrium can be expressed numerically as an equilibrium constant, K_{eq} (Section 1.17). For example, in a reaction in which m moles of A react with n moles of B to form s moles of C and t moles of D, K_{eq} is equal to the relative concentrations of products and reactants at equilibrium.

$$m\,\mathbf{A} + n\,\mathbf{B} \rightleftharpoons s\,\mathbf{C} + t\,\mathbf{D}$$

$$K_{eq} = \frac{[\text{products}]}{[\text{reactants}]} = \frac{[\mathrm{C}]^s[\mathrm{D}]^t}{[\mathrm{A}]^m[\mathrm{B}]^n}$$

The more stable the compound, the greater is its concentration at equilibrium.

The relative concentrations of products and reactants at equilibrium depend on their relative stabilities: *The more stable the compound, the greater is its concentration at equilibrium.* Thus, if the products are more stable (have a lower free energy) than the reactants (Figure 3.3a), there will be a higher concentration of products than reactants at equilibrium, and K_{eq} will be greater than 1. On the other hand, if the reactants are more stable than the products (Figure 3.3b), there will be a higher concentration of reactants than products at equilibrium, and K_{eq} will be less than 1.

Several thermodynamic parameters are used to describe a reaction. The difference between the free energy of the products and the free energy of the reactants under standard conditions is called the **Gibbs free-energy change** ($\Delta G°$). The symbol ° indicates standard conditions—all species at a concentration of 1 M, a temperature of 25 C°, and a pressure of 1 atm.

$$\Delta G° = (\text{free energy of the products}) - (\text{free energy of the reactants})$$

From this equation, we can see that $\Delta G°$ will be negative if the products have a lower free energy—are more stable—than the reactants. In other words, the reaction will

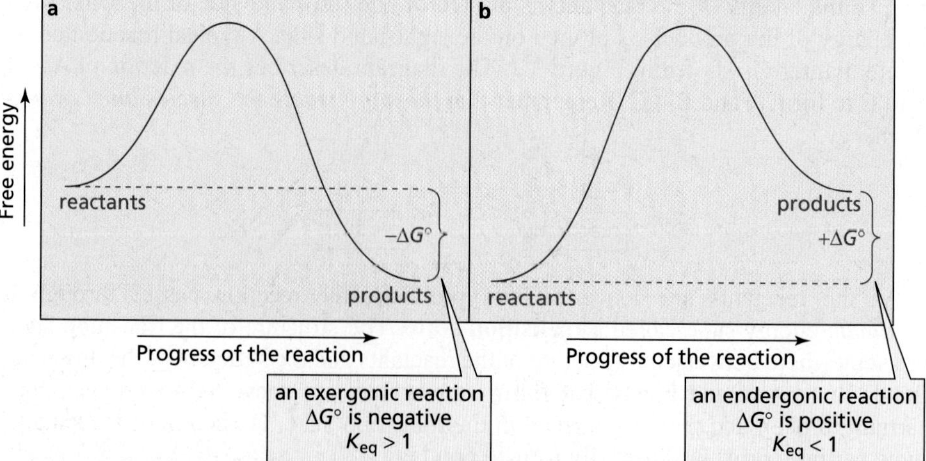

Figure 3.3 ▶ Reaction coordinate diagrams for (a) a reaction in which the products are more stable than the reactants (an exergonic reaction) and (b) a reaction in which the products are less stable than the reactants (an endergonic reaction).

release more energy than it will consume. It will be an **exergonic reaction** (Figure 3.3a). If the products have a higher free energy—are less stable—than the reactants, $\Delta G°$ will be positive, and the reaction will consume more energy than it will release; it will be an **endergonic reaction** (Figure 3.3b). (Notice that the terms *exergonic* and *endergonic* refer to whether the reaction has a negative $\Delta G°$ or a positive $\Delta G°$, respectively. Do not confuse these terms with *exothermic* and *endothermic*, which are defined later.)

Therefore, whether reactants or products are favored at equilibrium can be indicated either by the equilibrium constant (K_{eq}) or by the change in free energy ($\Delta G°$). These two quantities are related by the equation

When products are favored at equilibrium, $\Delta G°$ is negative and K_{eq} is greater than 1.

When reactants are favored at equilibrium, $\Delta G°$ is positive and K_{eq} is less than 1.

$$\Delta G° = -RT \ln K_{eq}$$

where R is the gas constant (1.986×10^{-3} kcal mol^{-1} K^{-1}, or 8.314×10^{-3} kJ mol^{-1} K^{-1}, because 1 kcal = 4.184 kJ) and T is the temperature in degrees Kelvin (K = °C + 273; therefore, 25° C = 298 K). (By solving Problem 13, you will see that even a small difference in $\Delta G°$ gives rise to a large difference in the relative concentrations of products and reactants.)

PROBLEM 13♦

a. Which of the monosubstituted cyclohexanes in Table 2.10 has a negative $\Delta G°$ for the conversion of an axial-substituted chair conformer to an equatorial-substituted chair conformer?
b. Which monosubstituted cyclohexane has the most negative value of $\Delta G°$?
c. Which monosubstituted cyclohexane has the greatest preference for the equatorial position?
d. Calculate $\Delta G°$ for conversion of "axial" methylcyclohexane to "equatorial" methylcyclohexane at 25 °C.

PROBLEM 14 SOLVED

a. The $\Delta G°$ for conversion of "axial" fluorocyclohexane to "equatorial" fluorocyclohexane at 25 °C is −0.25 kcal/mol (or −1.05 kJ/mol). Calculate the percentage of fluorocyclohexane molecules that have the fluoro substituent in the equatorial position.
b. Do the same calculation for isopropylcyclohexane (whose $\Delta G°$ at 25 °C is −2.1 kcal/mol, or −8.8 kJ/mol).
c. Why does isopropylcyclohexane have a greater percentage of the conformer with the substituent in the equatorial position?

SOLUTION TO 14a

$$\underset{\textbf{axial}}{\text{fluorocyclohexane}} \rightleftharpoons \underset{\textbf{equatorial}}{\text{fluorocyclohexane}}$$

$$\Delta G° = -0.25 \text{ kcal/mol at } 25 \text{ °C}$$

$$\Delta G° = -RT \ln K_{eq}$$

$$-0.25 \frac{\text{kcal}}{\text{mol}} = -1.986 \times 10^{-3} \frac{\text{kcal}}{\text{mol K}} \times 298 \text{ K} \times \ln K_{eq}$$

$$\ln K_{eq} = 0.422$$

$$K_{eq} = 1.53 = \frac{[\text{fluorocyclohexane}]\ \textbf{equatorial}}{[\text{fluorocyclohexane}]\ \textbf{axial}} = \frac{1.53}{1}$$

Now we must determine the percentage of the total that is equatorial:

$$\frac{[\text{fluorocyclohexane}]\ \textbf{equatorial}}{[\text{fluorocyclohexane}]\ \textbf{equatorial} + [\text{fluorocyclohexane}]\ \textbf{axial}} = \frac{1.53}{1.53 + 1} = \frac{1.53}{2.53} = .60 \text{ or } 60\%$$

Josiah Willard Gibbs (1839–1903) *was born in New Haven, Connecticut, the son of a Yale professor. In 1863, he obtained the first Ph.D. awarded by Yale in engineering. After studying in France and Germany, he returned to Yale to become a professor of mathematical physics. His work on free energy received little attention for more than 20 years because few chemists could understand his mathematical treatment and because Gibbs published it in* Transactions of the Connecticut Academy of Sciences, *a relatively obscure journal. In 1950, he was elected to the Hall of Fame for Great Americans.*

The gas constant R *is thought to be named after* **Henri Victor Regnault (1810–1878)**, *who was commissioned by the French Minister of Public Works in 1842 to redetermine all the physical constants involved in the design and operation of the steam engine. Regnault was known for his work on the thermal properties of gases. Later, while studying the thermodynamics of dilute solutions, van't Hoff (p. 194) found that* R *could be used for all chemical equilibria.*

The Gibbs standard free-energy change (ΔG°) has an enthalpy (ΔH°) component and an entropy (ΔS°) component:

$$\Delta G^\circ = \Delta H^\circ - T\Delta S^\circ$$

The **enthalpy** term (ΔH°) is the heat given off or the heat consumed during the course of a reaction. Atoms are held together by bonds. Heat is given off when bonds are formed, and heat is consumed when bonds are broken. Thus, ΔH° is a measure of the bond-making and bond-breaking processes that occur as reactants are converted into products.

$$\Delta H^\circ = (\text{energy of the bonds being broken}) - (\text{energy of the bonds being formed})$$

If the bonds that are formed in a reaction are stronger than the bonds that are broken, more energy will be released as a result of bond formation than will be consumed in the bond-breaking process, and ΔH° will be negative. A reaction with a negative ΔH° is called an **exothermic reaction**. If the bonds that are formed are weaker than those that are broken, ΔH° will be positive. A reaction with a positive ΔH° is called an **endothermic reaction**.

Entropy (ΔS°) is defined as the degree of disorder. It is a measure of the freedom of motion of the system. Restricting the freedom of motion of a molecule decreases its entropy. For example, in a reaction in which two molecules come together to form a single molecule, the entropy in the product will be less than the entropy in the reactants because two individual molecules can move in ways that are not possible when the two are bound together in a single molecule. In such a reaction, ΔS° will be negative. In a reaction in which a single molecule is cleaved into two separate molecules, the products will have greater freedom of motion than the reactant, and ΔS° will be positive.

$$\Delta S^\circ = (\text{freedom of motion of the products}) - (\text{freedom of motion of the reactants})$$

PROBLEM 15◆

a. For which reaction will ΔS° be more significant?
   1. $A \rightleftharpoons B$ or $A + B \rightleftharpoons C$
   2. $A + B \rightleftharpoons C$ or $A + B \rightleftharpoons C + D$
b. For which reaction will ΔS° be positive?

Entropy is the degree of disorder of a system.

The formation of products with stronger bonds and with greater freedom of motion causes ΔG° to be negative.

A reaction with a negative ΔG° has a favorable ($K_{eq} > 1$) equilibrium constant; that is, the reaction is favored as written from left to right because the products are more stable than the reactants. If you examine the expression for the Gibbs standard free-energy change, you will find that negative values of ΔH° and positive values of ΔS° contribute to make ΔG° negative. In other words, *the formation of products with stronger bonds and with greater freedom of motion causes ΔG° to be negative.*

PROBLEM 16◆

a. For a reaction with $\Delta H^\circ = -12$ kcal mol^{-1} and $\Delta S^\circ = 0.01$ kcal mol^{-1}, calculate the ΔG° and the equilibrium constant **(1)** at 30 °C and **(2)** at 150 °C.
b. How does ΔG° change as T increases?
c. How does K_{eq} change as T increases?

Values of ΔH° are relatively easy to calculate, so organic chemists frequently evaluate reactions only in terms of that quantity. However, you can ignore the entropy term only if the reaction involves only a small change in entropy, because then the $T\Delta S^\circ$ term will be small and the value of ΔH° will be very close to the value of ΔG°. Ignoring the entropy term can be a dangerous practice, however, because many organic reactions occur with a significant change in entropy or occur at high temperatures and so have significant $T\Delta S°$ terms. It is permissible to use $\Delta H°$ values to *approximate* whether a reaction occurs with a favorable equilibrium constant, but if a precise answer is needed, $\Delta G°$ values must be used. When $\Delta G°$ values are used to construct reaction coordinate diagrams, the *y*-axis is free energy; when $\Delta H°$ values are used, the *y*-axis is potential energy.

Values of $\Delta H°$ can be calculated from bond dissociation energies (Table 3.1). For example, the $\Delta H°$ for the addition of HBr to ethene is calculated as shown here:

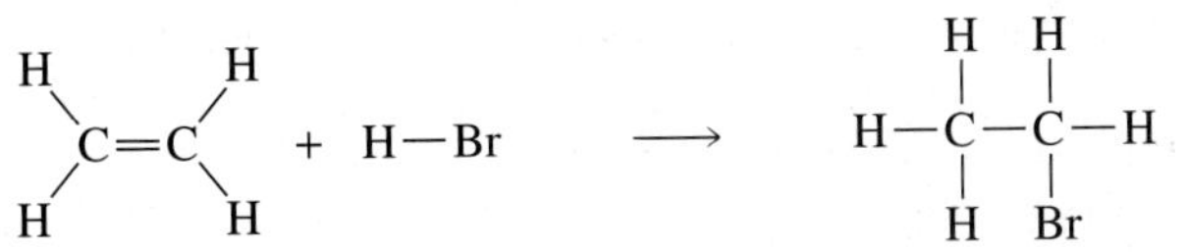

$$H_2C{=}CH_2 + H{-}Br \longrightarrow H{-}\underset{H}{\overset{H}{C}}{-}\underset{Br}{\overset{H}{C}}{-}H$$

bonds being broken		bonds being formed	
π bond of ethene	$DH° = 63$ kcal/mol	C—H	$DH° = 101$ kcal/mol
H—Br	$DH° = 87$ kcal/mol	C—Br	$DH° = 72$ kcal/mol
	$DH°_{total} = 150$ kcal/mol		$DH°_{total} = 173$ kcal/mol

$$\begin{aligned}\Delta H° \text{ for the reaction} &= DH° \text{ for bonds being broken} - DH° \text{ for bonds being formed}\\ &= 150 \text{ kcal/mol} - 173 \text{ kcal/mol}\\ &= -23 \text{ kcal/mol}\end{aligned}$$

Table 3.1 Homolytic Bond Dissociation Energies Y—Z → Y· + ·Z

	$DH°$			$DH°$	
Bond	**kcal/mol**	**kJ/mol**	**Bond**	**kcal/mol**	**kJ/mol**
$CH_3{-}H$	105	439	H—H	104	435
$CH_3CH_2{-}H$	101	423	F—F	38	159
$CH_3CH_2CH_2{-}H$	101	423	Cl—Cl	58	242
$(CH_3)_2CH{-}H$	99	414	Br—Br	46	192
$(CH_3)_3C{-}H$	97	406	I—I	36	150
			H—F	136	571
$CH_3{-}CH_3$	90.1	377	H—Cl	103	432
$CH_3CH_2{-}CH_3$	89.0	372	H—Br	87	366
$(CH_3)_2CH{-}CH_3$	88.6	371	H—I	71	298
$(CH_3)_3C{-}CH_3$	87.5	366			
			$CH_3{-}F$	115	481
$H_2C{=}CH_2$	174	728	$CH_3{-}Cl$	84	350
$HC{\equiv}CH$	231	966	$CH_3CH_2{-}Cl$	85	356
			$(CH_3)_2CH{-}Cl$	85	356
HO—H	119	497	$(CH_3)_3C{-}Cl$	85	356
$CH_3O{-}H$	105	439	$CH_3{-}Br$	72	301
$CH_3{-}OH$	92	387	$CH_3CH_2{-}Br$	72	301
			$(CH_3)_2CH{-}Br$	74	310
			$(CH_3)_3C{-}Br$	73	305
			$CH_3{-}I$	58	243
			$CH_3CH_2{-}I$	57	238

S. J. Blanksby and G. B. Ellison *Acc. Chem. Res.*, **2003**, *36*, 255.

The bond dissociation energy is indicated by the special term $DH°$. Recall from Section 2.3 that the barrier to rotation about the π bond of ethene is 63 kcal/mol. In other words, it takes 63 kcal/mol to break the *π bond.*

The value of −23 kcal/mol for $\Delta H°$—calculated by subtracting the $\Delta H°$ for the bonds being formed from the $\Delta H°$ for the bonds being broken—indicates that the addition of HBr to ethene is an exothermic reaction. But does this mean that the $\Delta G°$ for the reaction is also negative? In other words, is the reaction exergonic as well as exothermic? Because $\Delta H°$ has a significant negative value (−23 kcal/mol), you can assume that $\Delta G°$ is also negative. If the value of $\Delta H°$ were close to zero, you could no longer assume that $\Delta H°$ has the same sign as $\Delta G°$.

Keep in mind that two assumptions are being made when using $\Delta H°$ values to predict values of $\Delta G°$. The first is that the entropy change in the reaction is small, causing $T\Delta S°$ to be close to zero and, therefore, the value of $\Delta H°$ to be very close to the value of $\Delta G°$; the second is that the reaction is taking place in the gas phase.

When reactions are carried out in solution, which is the case for the vast majority of organic reactions, the solvent molecules can interact with the reagents and with the products. Polar solvent molecules cluster around a charge (either a full charge or a partial charge) on a reactant or product, so that the negative poles of the solvent molecules surround the positive charge and the positive poles of the solvent molecules surround the negative charge. The interaction between a solvent and a species (a molecule or ion) in solution is called **solvation**.

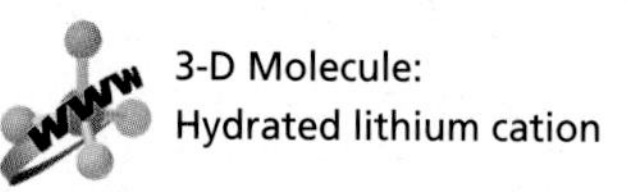

3-D Molecule: Hydrated lithium cation

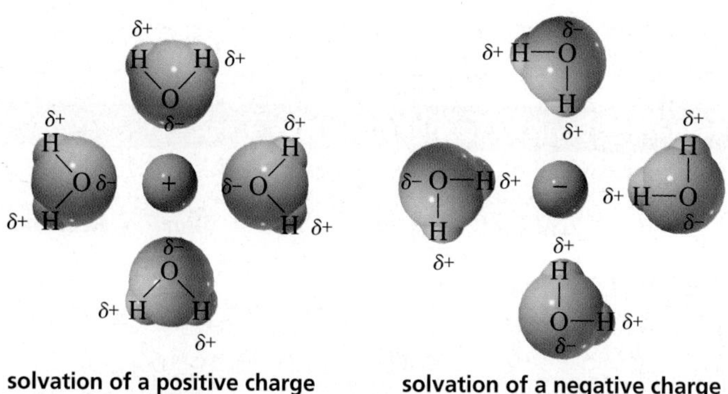

solvation of a positive charge by water

solvation of a negative charge by water

Solvation can have a large effect on both the $\Delta H°$ and the $\Delta S°$ of a reaction. For example, in a reaction in which a polar reagent is solvated, the $\Delta H°$ for breaking the dipole–dipole interactions between the solvent and the reagent has to be taken into account, and in a reaction in which a polar product is solvated, the $\Delta H°$ for forming the dipole–dipole interactions between the solvent and the product has to be taken into account. In addition, solvation of a polar reagent or a polar product by a polar solvent can greatly reduce the freedom of motion of the solvent molecules, and this will affect the value of $\Delta S°$.

PROBLEM 17◆

a. Using the bond dissociation energies in Table 3.1, calculate the $\Delta H°$ for the addition of HCl to ethene.
b. Calculate the $\Delta H°$ for the addition of H_2 to ethene.
c. Are the reactions exothermic or endothermic?
d. Do you expect the reactions to be exergonic or endergonic?

Kinetics

Knowing whether a given reaction is exergonic or endergonic will not tell you how fast the reaction occurs, because the $\Delta G°$ of a reaction tells you only the difference between the stability of the reactants and the stability of the products; it does not tell you anything about the energy barrier of the reaction, which is the energy "hill" that must be climbed for the reactants to be converted into products. The higher the energy barrier, the slower is the reaction. **Kinetics** is the field of chemistry that studies the rates of chemical reactions and the factors that affect those rates.

The energy barrier of a reaction, indicated in Figure 3.4 by $\Delta G^\ddagger$, is called the **free energy of activation**. It is the difference between the free energy of the transition state and the free energy of the reactants:

$$\Delta G^\ddagger = \text{(free energy of the transition state)} - \text{(free energy of the reactants)}$$

The smaller the $\Delta G^\ddagger$, the faster is the reaction. Thus, *anything that destabilizes the reactant or stabilizes the transition state will make the reaction go faster.*

Like ΔG°, $\Delta G^\ddagger$ has both an enthalpy component and an entropy component. Notice that any quantity that refers to the transition state is represented by the double-dagger superscript (‡):

$$\Delta G^\ddagger = \Delta H^\ddagger - T\Delta S^\ddagger$$

$$\Delta H^\ddagger = \text{(enthalpy of the transition state)} - \text{(enthalpy of the reactants)}$$

$$\Delta S^\ddagger = \text{(entropy of the transition state)} - \text{(entropy of the reactants)}$$

Some exergonic reactions have small free energies of activation and therefore can take place at room temperature (Figure 3.4a). In contrast, some exergonic reactions have free energies of activation that are so large that the reaction cannot take place without adding energy above that provided by the existing thermal conditions (Figure 3.4b). Endergonic reactions can also have either small free energies of activation, as in Figure 3.4c, or large free energies of activation, as in Figure 3.4d.

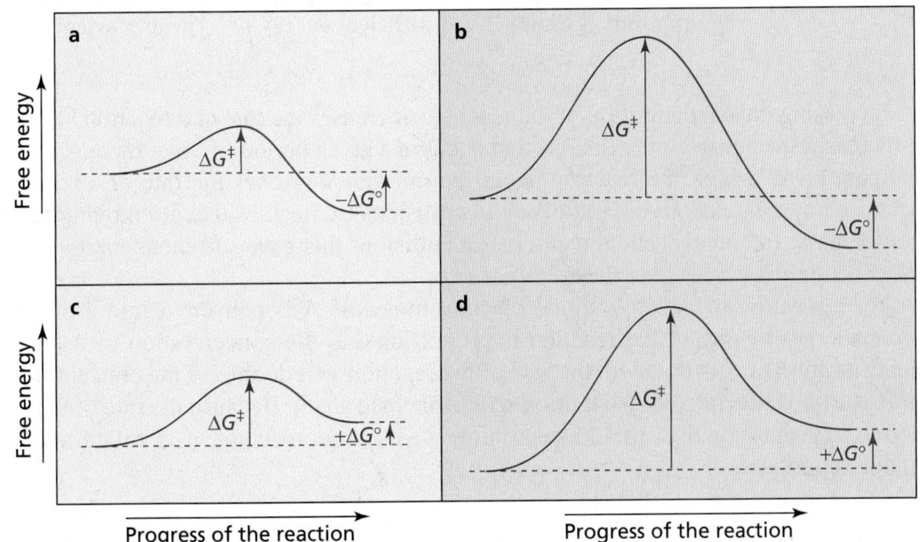

Figure 3.4
Reaction coordinate diagrams for (a) a fast exergonic reaction, (b) a slow exergonic reaction, (c) a fast endergonic reaction, and (d) a slow endergonic reaction. (The four reaction coordinates are drawn on the same scale.)

Notice that ΔG° relates to the *equilibrium constant* of the reaction, whereas $\Delta G^\ddagger$ relates to the *rate* of the reaction. The **thermodynamic stability** of a compound is indicated by ΔG°. If ΔG° is negative, for example, the product is *thermodynamically stable* compared with the reactant, and if ΔG° is positive, the product is *thermodynamically unstable* compared with the reactant. The **kinetic stability** of a compound is indicated by $\Delta G^\ddagger$. If $\Delta G^\ddagger$ for a reaction is large, the compound is *kinetically stable* because it does not undergo that reaction rapidly. If $\Delta G^\ddagger$ is small, the compound is *kinetically unstable*—it undergoes the reaction rapidly. Generally, when chemists use the term "stability," they are referring to thermodynamic stability.

PROBLEM 18◆

a. Which of the reactions in Figure 3.4 has a thermodynamically stable product?
b. Which of the reactions in Figure 3.4 has the most kinetically stable product?
c. Which of the reactions in Figure 3.4 has the least kinetically stable product?

PROBLEM 19

Draw a reaction coordinate diagram for a reaction in which

a. the product is thermodynamically unstable and kinetically unstable.
b. the product is thermodynamically unstable and kinetically stable.

The rate of a chemical reaction is the speed at which the reacting substances are used up or the speed at which the products are formed. The rate of a reaction depends on the following factors:

1. *The number of collisions that take place between the reacting molecules in a given period of time.* The greater the number of collisions, the faster is the reaction.
2. *The fraction of the collisions that occur with sufficient energy to get the reacting molecules over the energy barrier.* If the free energy of activation is small, more collisions will lead to reaction than if the free energy of activation is large.
3. *The fraction of the collisions that occur with the proper orientation.* For example, 2-butene and HBr will react only if the molecules collide with the hydrogen of HBr approaching the π bond of 2-butene. If collision occurs with the hydrogen approaching a methyl group of 2-butene, no reaction will take place, regardless of the energy of the collision.

$$\textbf{rate of a reaction} = \left(\begin{array}{c}\textbf{number of collisions}\\ \textbf{per unit of time}\end{array}\right) \times \left(\begin{array}{c}\textbf{fraction with}\\ \textbf{sufficient energy}\end{array}\right) \times \left(\begin{array}{c}\textbf{fraction with}\\ \textbf{proper orientation}\end{array}\right)$$

Increasing the concentration of the reactants increases the rate of a reaction because it increases the number of collisions that occur in a given period of time. Increasing the temperature at which the reaction is carried out also increases the rate of a reaction because it increases both the frequency of collisions (molecules that are moving faster collide more frequently) and the number of collisions that have sufficient energy to get the reacting molecules over the energy barrier.

For a reaction in which a single reactant molecule A is converted into a product molecule B, the rate of the reaction is proportional to the concentration of A. If the concentration of A is doubled, the rate of the reaction will double; if the concentration of A is tripled, the rate of the reaction will triple; and so on. Because the rate of this reaction is proportional to the concentration of only *one* reactant, it is called a **first-order reaction**.

$$\mathbf{A} \longrightarrow \mathbf{B}$$

$$\text{rate} \propto [\text{A}]$$

We can replace the proportionality symbol ($\propto$) with an equals sign if we use a proportionality constant k, which is called a **rate constant**. The rate constant of a first-order reaction is called a **first-order rate constant**.

$$\text{rate} = k[\text{A}]$$

A reaction whose rate depends on the concentrations of *two* reactants is called a **second-order reaction**. If the concentration of either A or B is doubled, the rate of the reaction will double; if the concentrations of both A and B are doubled, the rate of the reaction will quadruple; and so on. In this case, the rate constant k is a **second-order rate constant**.

$$\mathbf{A} + \mathbf{B} \longrightarrow \mathbf{C} + \mathbf{D}$$

$$\text{rate} = k[\text{A}][\text{B}]$$

A reaction in which two molecules of A combine to form a molecule of B is also a second-order reaction: If the concentration of A is doubled, the rate of the reaction will quadruple.

$$\mathbf{A} + \mathbf{A} \longrightarrow \mathbf{B}$$

$$\text{rate} = k[\text{A}]^2$$

Do not confuse the *rate constant* of a reaction (k) with the *rate* of a reaction. The *rate constant* tells us how easy it is to reach the transition state (how easy it is to get over the energy barrier). Low energy barriers are associated with large rate constants (Figures 3.4a and 3.4c), whereas high energy barriers have small rate constants (Figures 3.4b and 3.4d). The reaction *rate* is a measure of the amount of product that is formed per unit of time. The preceding equations show that the *rate* is the product of the *rate constant and the concentration(s)* of the reactants. *Thus, reaction rates depend on concentration, whereas rate constants are independent of concentration.* Therefore, when we compare two reactions to see which one occurs more readily, we must compare their rate constants and not their concentration-dependent rates of reaction. (Appendix III explains how rate constants are determined.)

The smaller the rate constant, the slower is the reaction.

Although rate constants are independent of concentration, they depend on temperature. The **Arrhenius equation** relates the rate constant of a reaction to the experimental energy of activation and to the temperature at which the reaction is carried out. A good rule of thumb is that an increase of 10 °C in temperature will double the rate constant for a reaction and, therefore, double the rate of the reaction.

Swedish chemist **Svante August Arrhenius (1859–1927)** *received a Ph.D. from the University of Uppsala. Threatened with a low passing grade on his dissertation because his examiners did not understand his thesis on ionic dissociation, he sent the work to several scientists, who subsequently defended it. His dissertation earned Arrhenius the 1903 Nobel Prize in chemistry. He was the first to describe the "greenhouse" effect, predicting that as concentrations of atmospheric carbon dioxide (CO_2) increase, so will Earth's surface temperature (Section 9.0).*

The Arrhenius equation:

$$k = Ae^{-E_a/RT}$$

where k is the rate constant, E_a is the experimental energy of activation, R is the gas constant (1.986×10^{-3} kcal mol^{-1} K^{-1}, or 8.314×10^{-3} kJ mol^{-1} K^{-1}), T is the absolute temperature (K), and A is the frequency factor. The frequency factor accounts for the fraction of collisions that occur with the proper orientation for reaction. The term $e^{-E_a/RT}$ corresponds to the fraction of the collisions that have the minimum energy (E_a) needed to react. Taking the logarithm of both sides of the Arrhenius equation, we obtain

$$\ln k = \ln A - \frac{E_a}{RT}$$

Problem 43 on page 140 shows how this equation is used to calculate kinetic parameters.

THE DIFFERENCE BETWEEN $\Delta G^\ddagger$ AND E_a

Do not confuse the *free energy of activation*, $\Delta G^\ddagger$, with the **experimental energy of activation**, E_a, in the Arrhenius equation. The free energy of activation ($\Delta G^\ddagger = \Delta H^\ddagger - T\Delta S^\ddagger$) has both an enthalpy component and an entropy component, whereas the experimental energy of activation ($E_a = \Delta H^\ddagger + RT$) has only an enthalpy component since the entropy component is implicit in the A term in the Arrhenius equation. Therefore, the experimental energy of activation is an approximate energy barrier to a reaction. The true energy barrier to a reaction is given by $\Delta G^\ddagger$ because some reactions are driven by a change in enthalpy and some by a change in entropy, but most by a change in both enthalpy and entropy.

PROBLEM 20 SOLVED

At 30 °C, the second-order rate constant for the reaction of methyl chloride and HO^- is $1.0 \times 10^{-5}\ M^{-1}s^{-1}$.

a. What is the rate of the reaction when $[CH_3Cl] = 0.10$ M and $[HO^-] = 0.10$ M?

b. If the concentration of methyl chloride is decreased to 0.01 M, what effect will this have on the *rate* of the reaction?

c. If the concentration of methyl chloride is decreased to 0.01 M, what effect will this have on the *rate constant* of the reaction?

SOLUTION TO 20a The rate of the reaction is given by

$$\text{rate} = k[\text{methyl chloride}]\,[HO^-]$$

Substituting the given rate constant and reactant concentrations yields

$$\begin{aligned}\text{rate} &= 1.0 \times 10^{-5}\ M^{-1}s^{-1}\ [0.10\ M][0.10\ M]\\ &= 1.0 \times 10^{-7}\ Ms^{-1}\end{aligned}$$

PROBLEM 21◆

The rate constant for a reaction can be increased by _____ the stability of the reactant or by _____ the stability of the transition state.

PROBLEM 22◆

From the Arrhenius equation, predict how

a. increasing the experimental activation energy will affect the rate constant of a reaction.

b. increasing the temperature will affect the rate constant of a reaction.

How are the rate constants for a reaction related to the equilibrium constant? At equilibrium, the rate of the forward reaction must be equal to the rate of the reverse reaction because the amounts of reactants and products are not changing:

$$\mathbf{A} \underset{k_{-1}}{\overset{k_1}{\rightleftharpoons}} \mathbf{B}$$

forward rate = reverse rate

$$k_1[\mathbf{A}] = k_{-1}[\mathbf{B}]$$

Therefore,

$$K_{eq} = \frac{k_1}{k_{-1}} = \frac{[\mathbf{B}]}{[\mathbf{A}]}$$

From this equation, we can see that the equilibrium constant for a reaction can be determined from the relative concentrations of the products and reactants at equilibrium or from the relative rate constants for the forward and reverse reactions. The reaction shown in Figure 3.3a has a large equilibrium constant because the products are much more stable than the reactants. We could also say that it has a large equilibrium constant because the rate constant of the forward reaction is much greater than the rate constant of the reverse reaction.

PROBLEM 23◆

a. Which reaction has the greater equilibrium constant, one with a rate constant of 1×10^{-3} for the forward reaction and a rate constant of 1×10^{-5} for the reverse reaction or one with a rate constant of 1×10^{-2} for the forward reaction and a rate constant of 1×10^{-3} for the reverse reaction?

b. If both reactions start with a reactant concentration of 1 M, which reaction will form the most product?

Reaction Coordinate Diagram for the Addition of HBr to 2-Butene

We have seen that the addition of HBr to 2-butene is a two-step reaction (Section 3.6). The structure of the transition state for each of the steps is shown below in brackets. Notice that the bonds that break and the bonds that form during the course of the reaction are partially broken and partially formed in the transition state—indicated by dashed lines. Similarly, atoms that either become charged or lose their charge during the course of the reaction are partially charged in the transition state. Transition states are shown in brackets with a double-dagger superscript.

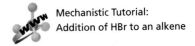

Mechanistic Tutorial: Addition of HBr to an alkene

$$CH_3CH{=}CHCH_3 + HBr \longrightarrow \left[\begin{array}{c} CH_3\overset{\delta+}{C}H{\cdots}CHCH_3 \\ \vdots \\ H \\ \vdots \\ \overset{\delta-}{Br} \end{array} \right]^{\ddagger} \longrightarrow CH_3\underset{+}{C}HCH_2CH_3 + Br^-$$

transition state

$$CH_3\underset{+}{C}HCH_2CH_3 + Br^- \longrightarrow \left[\begin{array}{c} CH_3\overset{\delta+}{C}HCH_2CH_3 \\ \vdots \\ \overset{\delta-}{Br} \end{array} \right]^{\ddagger} \longrightarrow \begin{array}{c} CH_3CHCH_2CH_3 \\ | \\ Br \end{array}$$

transition state

A reaction coordinate diagram can be drawn for each of the steps in the reaction (Figure 3.5). In the first step of the reaction, the alkene is converted into a carbocation that is less stable than the reactants. The first step, therefore, is endergonic ($\Delta G°$ is positive). In the second step of the reaction, the carbocation reacts with a nucleophile to form a product that is more stable than the carbocation reactant. This step, therefore, is exergonic ($\Delta G°$ is negative).

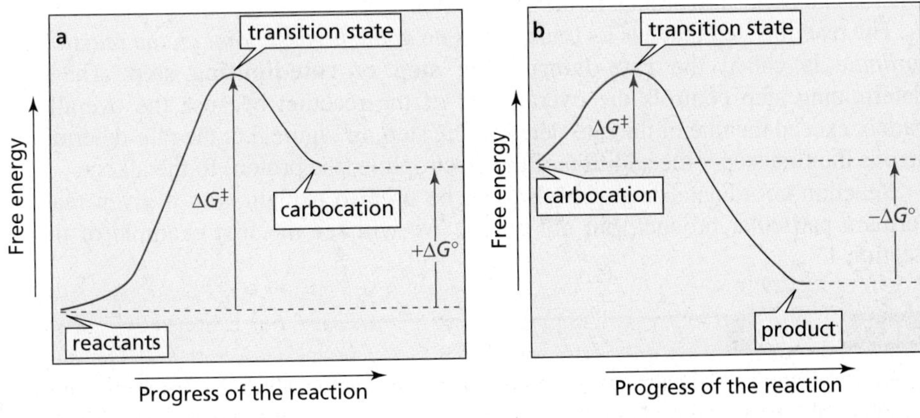

◀ **Figure 3.5**
Reaction coordinate diagrams for the two steps in the addition of HBr to 2-butene: (a) the first step; (b) the second step.

Because the product of the first step is the reactant in the second step, we can hook the two reaction coordinate diagrams together to obtain the reaction coordinate diagram for the overall reaction (Figure 3.6). The $\Delta G°$ for the overall reaction is the difference between the free energy of the final products and the free energy of the initial reactants. The figure shows that $\Delta G°$ for the overall reaction is negative. Therefore, the overall reaction is exergonic.

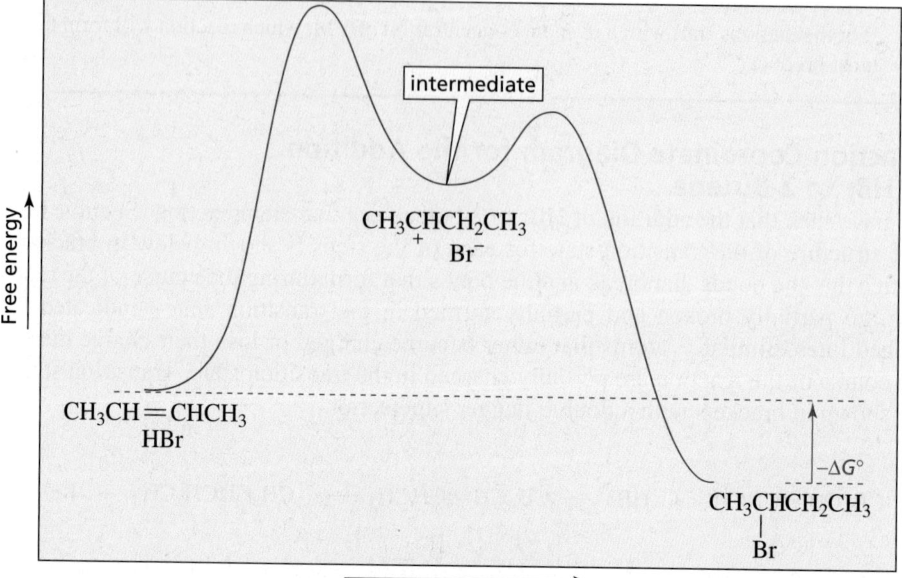

Figure 3.6 ▶ Reaction coordinate diagram for the addition of HBr to 2-butene.

Transition states have partially formed bonds. Intermediates have fully formed bonds.

A chemical species that is the product of one step of a reaction and is the reactant for the next step is called an **intermediate**. The carbocation intermediate in this reaction is too unstable to be isolated, but some reactions have more stable intermediates that can be isolated. **Transition states**, in contrast, represent the highest-energy structures that are involved in the reaction. They exist only fleetingly and can never be isolated. Do not confuse transition states with intermediates: *Transition states have partially formed bonds, whereas intermediates have fully formed bonds.*

We can see from the reaction coordinate diagram that the free energy of activation for the first step of the reaction is greater than the free energy of activation for the second step. In other words, the rate constant for the first step is smaller than the rate constant for the second step. This is what you would expect because the molecules in the first step of this reaction must collide with sufficient energy to break covalent bonds, whereas no bonds are broken in the second step.

The reaction step that has its transition state *at the highest point on the reaction coordinate* is called the **rate-determining step** or **rate-limiting step**. The rate-determining step controls the overall rate of the reaction because the overall rate cannot exceed the rate of the rate-determining step. In Figure 3.6, the rate-determining step is the first step—the addition of the electrophile (the proton) to the alkene.

Reaction coordinate diagrams also can be used to explain why a given reaction forms a particular product, but not others. We will see the first example of this in Section 4.3.

PROBLEM 24

Draw a reaction coordinate diagram for a two-step reaction in which the first step is endergonic, the second step is exergonic, and the overall reaction is endergonic. Label the reactants, products, intermediates, and transition states.

PROBLEM 25◆

a. Which step in the following reaction has the greatest free energy of activation?

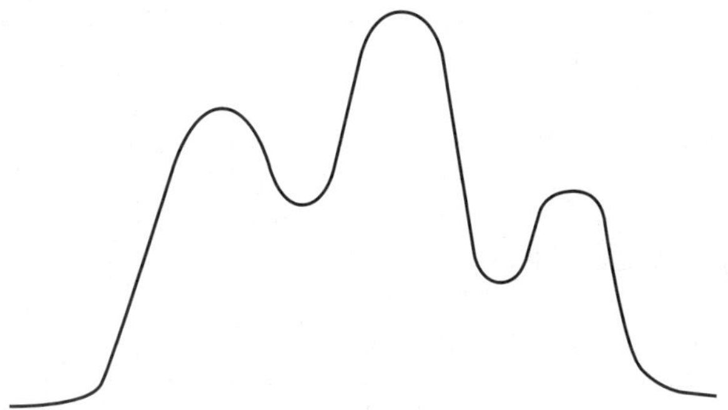

b. Is the first-formed intermediate more apt to revert to reactants or go on to form products?
c. Which step in the reaction sequence is rate determining?

PROBLEM 26◆

Draw a reaction coordinate diagram for the following reaction in which C is the most stable and B the least stable of the three species and the transition state going from A to B is more stable than the transition state going from B to C:

$$A \underset{k_{-1}}{\overset{k_1}{\rightleftharpoons}} B \underset{k_{-2}}{\overset{k_2}{\rightleftharpoons}} C$$

a. How many intermediates are there?
b. How many transition states are there?
c. Which step has the greater rate constant in the forward direction?
d. Which step has the greater rate constant in the reverse direction?
e. Of the four steps, which has the greatest rate constant?
f. Which is the rate-determining step in the forward direction?
g. Which is the rate-determining step in the reverse direction?

Summary

Alkenes are hydrocarbons that contain a double bond. The double bond is the **functional group** or center of reactivity of the alkene. The **functional group suffix** of an alkene is "ene." The general molecular formula for a hydrocarbon is C_nH_{2n+2}, minus two hydrogens for every π bond or ring in the molecule. The number of π bonds and rings is called the **degree of unsaturation**. Because alkenes contain fewer than the maximum number of hydrogens, they are called **unsaturated hydrocarbons**.

Because of restricted rotation about the double bond, an alkene can exist as **cis–trans isomers**. The **cis isomer** has its hydrogens on the same side of the double bond; the **trans isomer** has its hydrogens on opposite sides of the double bond. The **Z isomer** has the high-priority groups on the same side of the double bond; the ***E* isomer** has the high-priority groups on opposite sides of the double bond. The relative priorities depend on the atomic numbers of the atoms bonded directly to the sp^2 carbons.

All compounds with a particular **functional group** react similarly. Due to the cloud of electrons above and below its π bond, an alkene is an electron-rich molecule, or **nucleophile**. Nucleophiles are attracted to electron-deficient atoms or molecules, called **electrophiles**. Alkenes undergo **electrophilic addition reactions**. The description of the step-by-step process by which reactants are changed into products is called the **mechanism of the reaction**. **Curved arrows** show which bonds are formed and which are broken and the direction of the electron flow that accompany these changes.

Thermodynamics describes a reaction at equilibrium; **kinetics** describes how fast the reaction occurs. A **reaction coordinate diagram** shows the energy changes that take place in a reaction. The more stable the species, the lower is its energy. As reactants are converted into products, a reaction passes through a maximum energy **transition state**. An **intermediate** is the product of one step of a reaction and the reactant for the next step. Transition states have partially formed bonds; intermediates have fully formed bonds. The **rate-determining step** has its transition state at the highest point on the reaction coordinate.

The relative concentrations of reactants and products at equilibrium are given by the equilibrium constant K_{eq}. The more stable the compound, the greater is its concentration at equilibrium. If products are more stable than reactants, K_{eq} is >1, $\Delta G°$ is negative, and the reaction is **exergonic**; if reactants are more stable than products, $K_{eq} < 1$, $\Delta G°$ is positive, and the reaction is **endergonic**. $\Delta G° = \Delta H° - T\Delta S°$; $\Delta G°$ is the Gibbs free-energy change. $\Delta H°$ is the change in **enthalpy**—the heat given off or consumed as a result of bond making and bond breaking. An **exothermic reaction** has a negative $\Delta H°$; an **endothermic reaction** has a positive $\Delta H°$. $\Delta S°$ is the change in **entropy**—the change in the degree of disorder of the system. A reaction with a negative $\Delta G°$ has a **favorable equilibrium constant**: The formation of products with stronger bonds and greater freedom of motion causes $\Delta G°$ to be negative. $\Delta G°$ and K_{eq} are related by the formula $\Delta G° = -RT \ln K_{eq}$. The interaction between a solvent and a species in solution is called **solvation**.

The **free energy of activation**, $\Delta G^{\ddagger}$, is the energy barrier of a reaction. It is the difference between the free energy of the reactants and the free energy of the transition state. The smaller the $\Delta G^{\ddagger}$, the faster is the reaction. Anything that destabilizes the reactant or stabilizes the transition state makes the reaction go faster. **Kinetic stability** is given by $\Delta G^{\ddagger}$, **thermodynamic stability** by $\Delta G°$. The **rate** of a reaction depends on the concentration of the reactants, the temperature, and the rate constant. The **rate constant**, which is independent of concentration, indicates how easy it is to reach the transition state. A **first-order reaction** depends on the concentration of one reactant, a **second-order reaction** on the concentration of two reactants.

Key Terms

acyclic (p. 112)
addition reaction (p. 124)
alkene (p. 111)
allyl group (p. 115)
allylic carbon (p. 115)
Arrhenius equation (p. 133)
cis isomer (p. 117)
cis–trans isomers (p. 117)
degree of unsaturation (p. 112)
E isomer (p. 120)
electrophile (p. 122)
electrophilic addition reaction (p. 124)
endergonic reaction (p. 127)
endothermic reaction (p. 128)
enthalpy (p. 128)
entropy (p. 128)
exergonic reaction (p. 127)
exothermic reaction (p. 128)
experimental energy of activation (p. 133)
first-order rate constant (p. 132)
first-order reaction (p. 132)
free energy of activation (p. 131)
functional group (p. 121)
geometric isomers (p. 117)
Gibbs free-energy change (p. 126)
intermediate (p. 136)
kinetics (p. 131)
kinetic stability (p. 131)
mechanism of the reaction (p. 123)
nucleophile (p. 122)
rate constant (p. 132)
rate-determining step (p. 136)
rate-limiting step (p. 136)
reaction coordinate diagram (p. 125)
saturated hydrocarbon (p. 112)
second-order rate constant (p. 133)
second-order reaction (p. 132)
solvation (p. 130)
thermodynamics (p. 126)
thermodynamic stability (p. 131)
trans isomer (p. 117)
transition state (p. 125)
unsaturated hydrocarbon (p. 112)
vinyl group (p. 115)
vinylic carbon (p. 115)
Z isomer (p. 120)

Problems

27. Give the systematic name for each of the following compounds:

a. 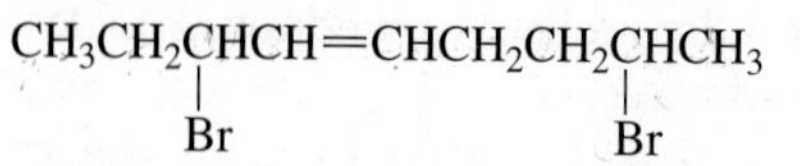

b. 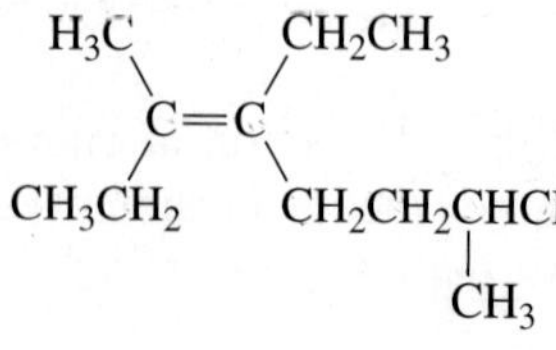

c. 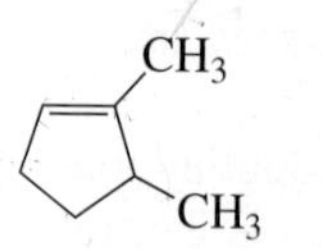

d. 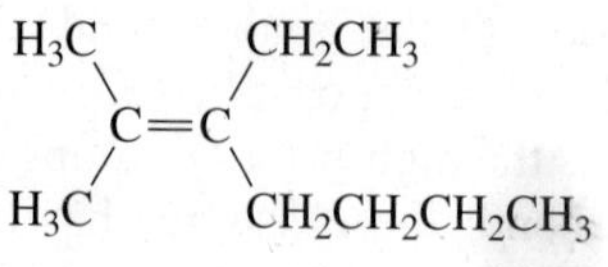

28. Give the structure of a hydrocarbon that has six carbon atoms and
    a. three vinylic hydrogens and two allylic hydrogens.
    b. three vinylic hydrogens and one allylic hydrogen.
    c. three vinylic hydrogens and no allylic hydrogens.

29. Draw the structure for each of the following:
    a. (Z)-1,3,5-tribromo-2-pentene
    b. (Z)-3-methyl-2-heptene
    c. (E)-1,2-dibromo-3-isopropyl-2-hexene
    d. vinyl bromide
    e. 1,2-dimethylcyclopentene
    f. diallylamine

30. a. Give the structures and the systematic names for all alkenes with molecular formula C_6H_{12}, ignoring cis–trans isomers. (*Hint:* There are 13.)
    b. Which of the compounds have *E* and *Z* isomers?

31. Name the following compounds:

a. 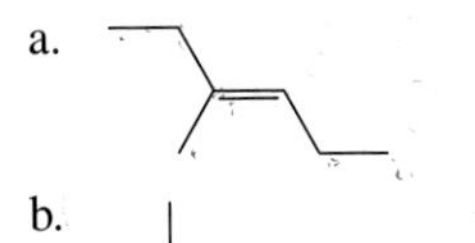

b.

c. Br

d.

e. 

f.

32. Draw curved arrows to show the flow of electrons responsible for the conversion of the reactants into the products:

$$\mathrm{H{-}\ddot{\underset{..}{O}}{:}^{-}} + \mathrm{H{-}C(H)(H){-}C(H)(Br){-}H} \longrightarrow \mathrm{H_2O} + \mathrm{(H)(H)C{=}C(H)(H)} + \mathrm{Br^-}$$

33. In a reaction in which reactant A is in equilibrium with product B at 25 °C, what are the relative amounts of A and B present at equilibrium if $\Delta G°$ at 25 °C is
    a. 2.72 kcal/mol?
    b. 0.65 kcal/mol?
    c. −2.72 kcal/mol?
    d. −0.65 kcal/mol?

34. Several studies have shown that β-carotene, a precursor of vitamin A, may play a role in preventing cancer. β-Carotene has a molecular formula of $C_{40}H_{56}$ and contains two rings and no triple bonds. How many double bonds does it have?

35. Tell whether each of the following compounds has the *E* or the *Z* configuration:

a. H_3C, CH_2CH_3 (top); C=C; CH_3CH_2, CH_2CH_2Cl (bottom)

b. H_3C, $CH(CH_3)_2$ (top); C=C; $HC{\equiv}C$, $CH_2CH{=}CH_2$ (bottom)

c. H_3C, CH_2Br (top); C=C; Br, $CH_2CH_2CH_2CH_3$ (bottom)

d. $CH_3C(=O)$, CH_2Br (top); C=C; $HOCH_2$, CH_2CH_2Cl (bottom)

36. Squalene, a hydrocarbon with molecular formula $C_{30}H_{50}$, is obtained from shark liver. (*Squalus* is Latin for "shark.") If squalene is an acyclic compound, how many π bonds does it have?

37. Assign relative priorities to each set of substituents:
    a. $-Br$, $-I$, $-OH$, $-CH_3$
    b. $-CH_2CH_2OH$, $-OH$, $-CH_2Cl$, $-CH{=}CH_2$
    c. $-CH_2CH_2CH_3$, $-CH(CH_3)_2$, $-CH{=}CH_2$, $-CH_3$
    d. $-CH_2NH_2$, $-NH_2$, $-OH$, $-CH_2OH$
    e. $-COCH_3$, $-CH{=}CH_2$, $-Cl$, $-C{\equiv}N$

38. Molly Kule was a lab technician who was asked by her supervisor to add names to the labels on a collection of alkenes that showed only structures on the labels. How many did Molly get right? Correct the incorrect names.
    a. 3-pentene
    b. 2-octene
    c. 2-vinylpentane
    d. 1-ethyl-1-pentene
    e. 5-ethylcyclohexene
    f. 5-chloro-3-hexene
    g. 5-bromo-2-pentene
    h. (*E*)-2-methyl-1-hexene
    i. 2-methylcyclopentene
    j. 2-ethyl-2-butene

39. Given the following reaction coordinate diagram for the reaction of A to give D, answer the following questions:

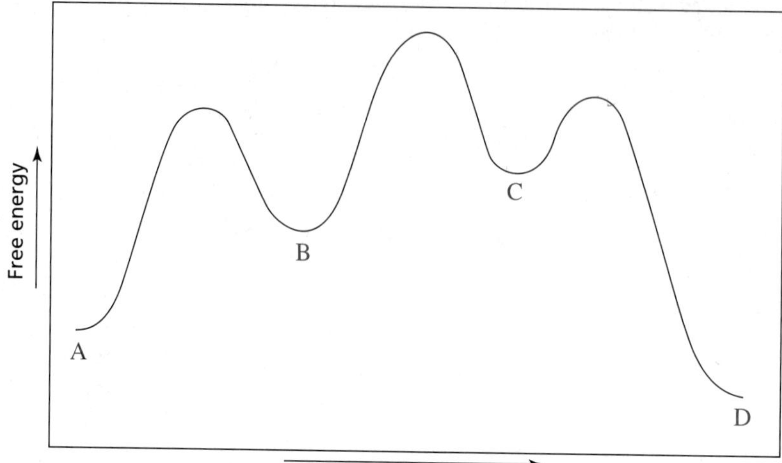

a. How many intermediates are there in the reaction?
b. How many transition states are there?
c. What is the fastest step in the reaction?
d. Which is more stable, A or D?
e. What is the reactant of the rate-determining step?
f. Is the first step of the reaction exergonic or endergonic?
g. Is the overall reaction exergonic or endergonic?

40. a. What is the equilibrium constant of a reaction that is carried out at 25 °C (298 K) with $\Delta H^\circ = 20$ kcal/mol and $\Delta S^\circ = 25$ cal K^{-1} mol^{-1}?
b. What is the equilibrium constant of the same reaction carried out at 125 °C?

41. a. For a reaction that is carried out at 25 °C, how much must ΔG° change in order to increase the equilibrium constant by a factor of 10?
b. How much must ΔH° change if $\Delta S^\circ = 0$ cal K^{-1} mol^{-1}?
c. How much must ΔS° change if $\Delta H^\circ = 0$ kcal mol^{-1}?

42. Given that the twist-boat conformer of cyclohexane is 3.8 kcal/mole (or 15.9 kJ/mol) higher in free energy than the chair conformer, calculate the percentage of twist-boat conformers present in a sample of cyclohexane at 25 °C. Does your answer agree with the statement made in Section 2.12 about the relative number of molecules in these two conformations?

CALCULATING KINETIC PARAMETERS

In order to calculate E_a, $\Delta H^\ddagger$, and $\Delta S^\ddagger$ for a reaction, rate constants for the reaction must be obtained at several temperatures:

- E_a can be obtained from the Arrhenius equation from the slope of a plot of ln k versus $1/T$, because

$$\ln k_2 - \ln k_1 = -E_a/R\left(\frac{1}{T_2} - \frac{1}{T_1}\right)$$

- At a given temperature, $\Delta H^\ddagger$ can be determined from E_a because $\Delta H^\ddagger = E_a - RT$.
- $\Delta G^\ddagger$, in kJ/mol, can be determined from the following equation, which relates $\Delta G^\ddagger$ to the rate constant at a given temperature:

$$-\Delta G^\ddagger = RT \ln \frac{kh}{Tk_B}$$

In this equation, h is Planck's constant (6.62608×10^{-34} Js) and k_B is Boltzmann's constant (1.38066×10^{-23} JK^{-1}).

- The entropy of activation can be determined from the other two kinetic parameters via the formula $\Delta S^\ddagger = (\Delta H^\ddagger - \Delta G^\ddagger)/T$.

Use this information to answer Problem 43.

43. Rate constants for a reaction were determined at five temperatures. From the following data, calculate the experimental energy of activation and then calculate $\Delta G^\ddagger$, $\Delta H^\ddagger$, and $\Delta S^\ddagger$ for the reaction at 30 °C:

Temperature	Observed rate constant
31.0 °C	$2.11 \times 10^{-5}\ s^{-1}$
40.0 °C	$4.44 \times 10^{-5}\ s^{-1}$
51.5 °C	$1.16 \times 10^{-4}\ s^{-1}$
59.8 °C	$2.10 \times 10^{-4}\ s^{-1}$
69.2 °C	$4.34 \times 10^{-4}\ s^{-1}$

4 Reactions of Alkenes

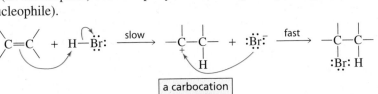

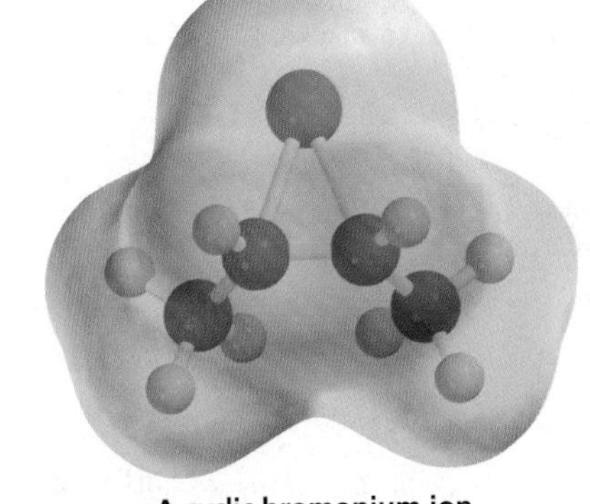

A cyclic bromonium ion

We have seen that an alkene such as 2-butene undergoes an **electrophilic addition reaction** with HBr (Section 3.6). The first step of the reaction is a relatively slow addition of the electrophilic proton to the nucleophilic alkene to form a carbocation intermediate. In the second step, the positively charged carbocation intermediate (an electrophile) reacts rapidly with the negatively charged bromide ion (a nucleophile).

$$\text{C=C} + \text{H—Br:} \xrightarrow{\text{slow}} \text{—}\overset{+}{\text{C}}\text{—}\underset{\text{H}}{\text{C}}\text{—} + \text{:Br:}^{-} \xrightarrow{\text{fast}} \text{—}\underset{\text{:Br:}}{\text{C}}\text{—}\underset{\text{H}}{\text{C}}\text{—}$$

a carbocation intermediate

In this chapter, we will look at a wide variety of alkene reactions. You will see that some of the reactions form carbocation intermediates like the one formed when HBr reacts with an alkene, some form other kinds of intermediates, and some don't form an intermediate at all. At first, the reactions covered in this chapter might appear to be quite different, but you will see that they all occur by similar mechanisms. So as you study each reaction, notice the feature that all alkene reactions have in common: *The relatively loosely held π electrons of the carbon–carbon double bond are attracted to an electrophile. Thus, each reaction starts with the addition of an electrophile to one of the* sp^2 *carbons of the alkene and concludes with the addition of a nucleophile to the other* sp^2 *carbon*. The end result is that the π bond breaks and the *sp*^{2} carbons form new σ bonds with the electrophile and the nucleophile.

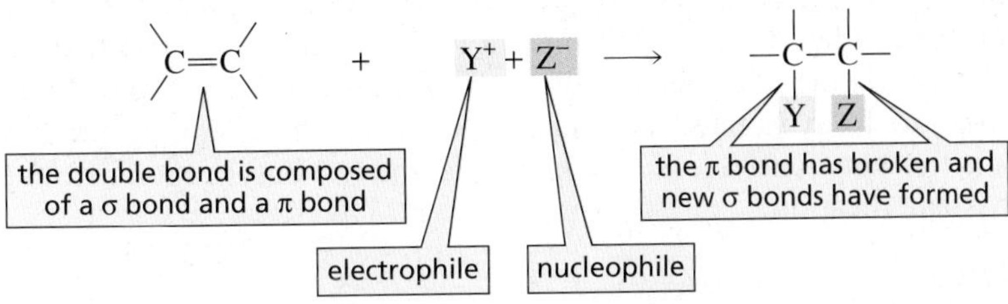
C=C + Y⁺ + Z⁻ → —C—C— (Y, Z)
the double bond is composed of a σ bond and a π bond
the π bond has broken and new σ bonds have formed
electrophile
nucleophile

This reactivity makes alkenes an important class of organic compounds because they can be used to synthesize a wide variety of other compounds. For example, alkyl halides, alcohols, ethers, and alkanes all can be synthesized from alkenes by electrophilic addition reactions. The particular product obtained depends only on the *electrophile* and the *nucleophile* used in the addition reaction.

4.1 Addition of Hydrogen Halides

If the electrophilic reagent that adds to an alkene is a hydrogen halide (HF, HCl, HBr, or HI), the product of the reaction will be an alkyl halide:

$$\underset{\text{ethene}}{CH_2{=}CH_2} + HCl \longrightarrow \underset{\text{ethyl chloride}}{CH_3CH_2Cl}$$

Synthetic Tutorial: Addition of HBr to an alkene.

$$\underset{\text{2,3-dimethyl-2-butene}}{(H_3C)_2C{=}C(CH_3)_2} + HBr \longrightarrow \underset{\text{2-bromo-2,3-dimethylbutane}}{CH_3CH(CH_3){-}CBr(CH_3)CH_3}$$

$$\text{cyclohexene} + HI \longrightarrow \text{iodocyclohexane}$$

Because the alkenes in the preceding reactions have the same substituents on both of the sp^2 carbons, it is easy to determine the product of the reaction: The electrophile (H^+) adds to one of the sp^2 carbons, and the nucleophile (X^-) adds to the other sp^2 carbon. It doesn't make any difference which sp^2 carbon the electrophile attaches to, because the same product will be obtained in either case.

But what happens if the alkene does not have the same substituents on both of the sp^2 carbons? Which sp^2 carbon gets the hydrogen? For example, does the addition of HCl to 2-methylpropene produce *tert*-butyl chloride or isobutyl chloride?

$$\underset{\text{2-methylpropene}}{CH_3C(CH_3){=}CH_2} + HCl \longrightarrow \underset{\textit{tert}\text{-butyl chloride}}{CH_3C(CH_3)(Cl)CH_3} \quad \text{or} \quad \underset{\text{isobutyl chloride}}{CH_3CH(CH_3)CH_2Cl}$$

To answer this question, we need to look at the **mechanism of the reaction.** Recall that the first step of the reaction—the addition of H^+ to an sp^2 carbon to form either the *tert*-butyl cation or the isobutyl cation—is the rate-determining step (Section 3.7). If there is any difference in the rate of formation of these two carbocations, the one that is formed faster will be the preferred product of the first step. Moreover, because carbocation formation is rate determining, the particular carbocation that is formed in the first step determines the final product of the reaction. That is, if the *tert*-butyl cation is formed, it will react rapidly with Cl^- to form *tert*-butyl chloride. On the other hand, if the isobutyl cation is formed, it will react rapidly with Cl^- to form isobutyl chloride. It turns out that the only product of the reaction is *tert*-butyl chloride, so we know that the *tert*-butyl cation is formed faster than the isobutyl cation.

$$CH_3C(CH_3){=}CH_2 + HCl \longrightarrow CH_3\overset{+}{C}(CH_3)CH_3 \xrightarrow{Cl^-} CH_3C(CH_3)(Cl)CH_3$$

tert-butyl cation → *tert*-butyl chloride, only product formed

$$CH_3C(CH_3){=}CH_2 + HCl \not\longrightarrow CH_3CH(CH_3)\overset{+}{C}H_2 \xrightarrow{Cl^-} CH_3CH(CH_3)CH_2Cl$$

isobutyl cation → isobutyl chloride, not formed

The question now is, Why is the *tert*-butyl cation formed faster than the isobutyl cation? To answer this, we need to take a look at the factors that affect the stability of carbocations and, therefore, the ease with which they are formed.

George Olah *was born in Hungary in 1927 and received a doctorate from the Technical University of Budapest in 1949. The Hungarian revolution caused him to emigrate to Canada in 1956, where he worked as a scientist at the Dow Chemical Company until he joined the faculty at Case Western Reserve University in 1965. In 1977, he became a professor of chemistry at the University of Southern California. In 1994, he received the Nobel Prize for his work on carbocations.*

4.2 Carbocation Stability

Carbocations are classified according to the number of alkyl substituents that are bonded to the positively charged carbon: A **primary carbocation** has one such substituent, a **secondary carbocation** has two, and a **tertiary carbocation** has three. The stability of a carbocation increases as the number of alkyl substituents bonded to the positively charged carbon increases. Thus, tertiary carbocations are more stable than secondary carbocations, and secondary carbocations are more stable than primary carbocations. Notice that when we talk about the stabilities of carbocations, we mean their *relative* stabilities: Carbocations are not stable species; even the relatively stable tertiary carbocation is not stable enough to isolate.

relative stabilities of carbocations

most stable → $R_3C^+ > R_2HC^+ > RH_2C^+ > H_3C^+$ ← least stable

a tertiary carbocation > a secondary carbocation > a primary carbocation > methyl cation

Why does the stability of a carbocation increase as the number of alkyl substituents bonded to the positively charged carbon increases? Alkyl groups decrease the concentration of positive charge on the carbon—and decreasing the concentration of positive charge increases the stability of the carbocation. Notice that the blue—recall that blue represents electron-deficient atoms—is most intense for the least stable methyl cation and is least intense for the most stable *tert*-butyl cation.

The greater the number of alkyl substituents bonded to the positively charged carbon, the more stable the carbocation is.

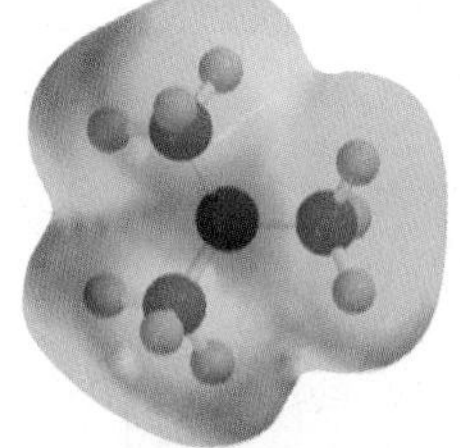

electrostatic potential map for the *tert*-butyl cation

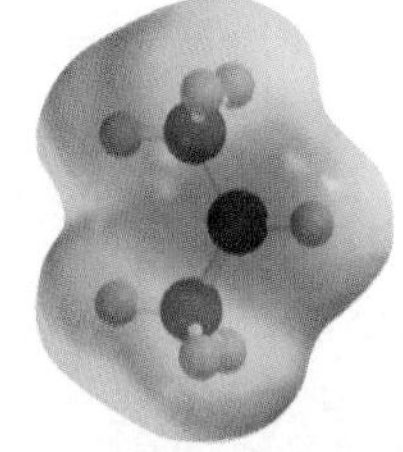

electrostatic potential map for the isopropyl cation

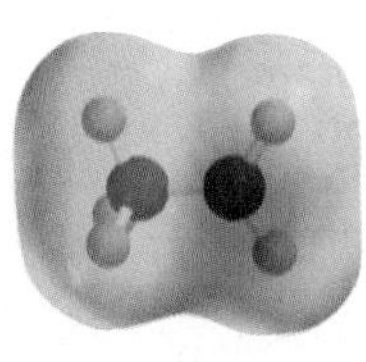

electrostatic potential map for the ethyl cation

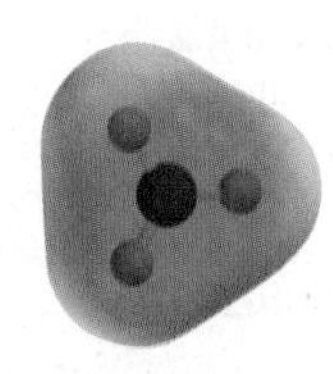

electrostatic potential map for the methyl cation

Carbocation stability: 3° > 2° > 1°

How do alkyl groups decrease the concentration of positive charge on the carbon? Recall that the positive charge on a carbon signifies an empty p orbital (Section 1.10). Figure 4.1 shows that in the ethyl cation, the orbital of an adjacent C—H σ bond can overlap with the empty p orbital. No such overlap is possible in the methyl cation. Movement of electrons from the σ bond orbital toward the vacant p orbital of the ethyl cation decreases the charge on the sp^2 carbon and causes a partial positive charge to develop on the carbon bonded by the σ bond. Therefore, the positive charge is no longer localized solely on one atom, but is spread out over a greater volume of space. This dispersion of the positive charge stabilizes the carbocation because a charged species is more stable if its charge is spread out (delocalized) over more than one atom (Section 1.19). Delocalization of electrons by the overlap of a σ bond orbital with an empty p orbital is called **hyperconjugation**. The simple molecular orbital diagram in Figure 4.2 is another way to explain the stabilization achieved by the overlap of a filled C—H σ bond orbital with an empty p orbital.

Figure 4.1 ▶
Stabilization of a carbocation by hyperconjugation: The electrons of an adjacent C—H bond in the ethyl cation spread into the empty p orbital. Hyperconjugation cannot occur in a methyl cation.

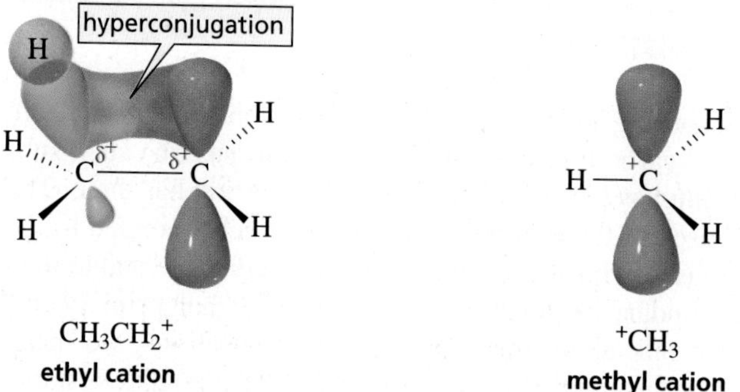

Hyperconjugation occurs only if the σ bond orbital and the empty p orbital have the proper orientation. The proper orientation is easily achieved because there is free rotation about a carbon–carbon σ bond (Section 2.10). In the case of the *tert*-butyl cation, nine C—H σ bond orbitals can potentially overlap with the empty p orbital of the positively charged carbon. The isopropyl cation has six such orbitals, and the ethyl cation has three. Therefore, there is greater stabilization through hyperconjugation in the tertiary *tert*-butyl cation than in the secondary isopropyl cation and greater stabilization in the secondary isopropyl cation than in the primary ethyl cation.

most stable ▶ $CH_3-\overset{CH_3}{\underset{CH_3}{C^+}}$ > $CH_3-\overset{CH_3}{\underset{H}{C^+}}$ > $CH_3-\overset{H}{\underset{H}{C^+}}$ ◀ least stable

tert-butyl cation — isopropyl cation — ethyl cation

Figure 4.2 ▶
A molecular orbital diagram showing the stabilization achieved by overlapping the electrons of a filled C—H bond with an empty p orbital.

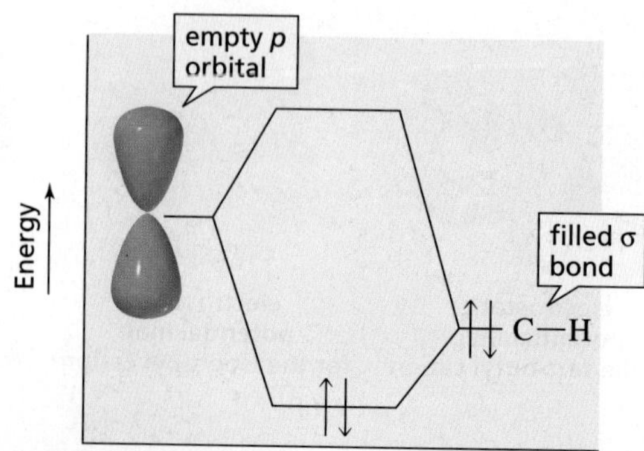

PROBLEM 1◆

List the carbocations in order of decreasing stability.

a. $CH_3CH_2\overset{+}{C}(CH_3)CH_3$ $CH_3CH_2\overset{+}{C}HCH_3$ $CH_3CH_2CH_2\overset{+}{C}H_2$

b. $CH_3CH(Cl)CH_2\overset{+}{C}H_2$ $CH_3CH(CH_3)CH_2\overset{+}{C}H_2$ $CH_3CH(F)CH_2\overset{+}{C}H_2$

PROBLEM 2◆

a. How many C—H bond orbitals are available for overlap with the vacant *p* orbital in the methyl cation?
b. Which is more stable, a methyl cation or an ethyl cation?

4.3 The Structure of the Transition State

Knowing something about the structure of a transition state is important when you are trying to predict the products of a reaction. In Section 3.7, you saw that the structure of the transition state lies between the structure of the reactants and the structure of the products. But what do we mean by "between"? Does the structure of the transition state lie *exactly* halfway between the structures of the reactants and products (as in II in the following diagram), or does it resemble the reactants more closely than it resembles the products (as in I), or is it more like the products than the reactants (as in III)?

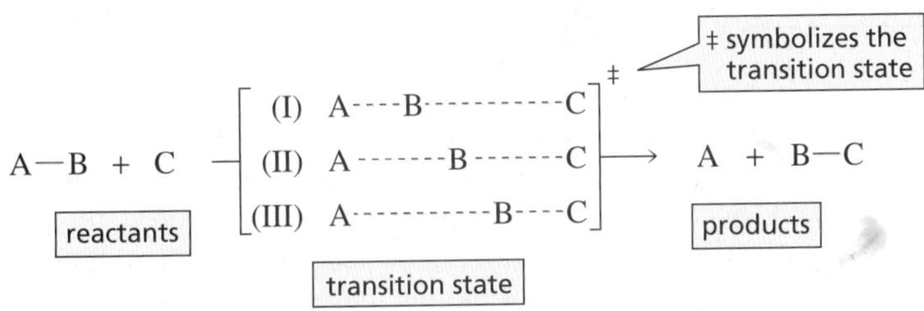

George Simms Hammond *was born in Maine in 1921. He received a B.S. from Bates College in 1943 and a Ph.D. from Harvard University in 1947. He was a professor of chemistry at Iowa State University and at the California Institute of Technology and was a scientist at Allied Chemical Co.*

According to the **Hammond postulate**, *the transition state is more similar in structure to the species to which it is more similar in energy.* In the case of an exergonic reaction, the transition state (I) is more similar in energy to the reactant than to the product (Figure 4.3, curve I). Therefore, the structure of the transition state will more

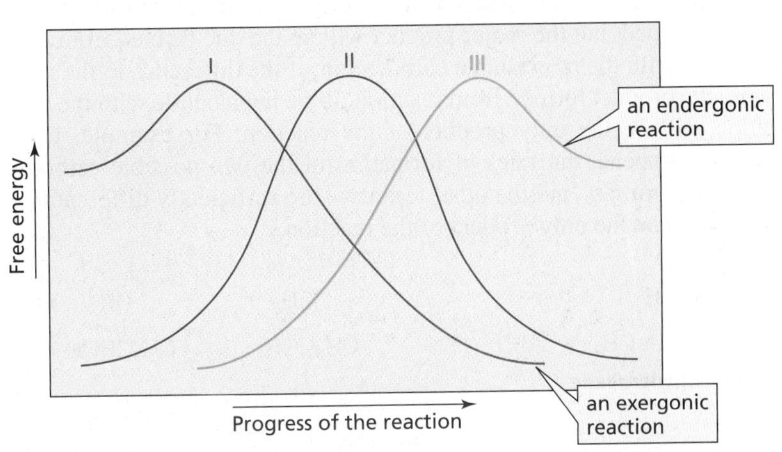

◀ **Figure 4.3**
Reaction coordinate diagrams for reactions with (I) an early transition state, (II) a midway transition state, and (III) a late transition state.

closely resemble the structure of the reactant than that of the product. In an endergonic reaction (Figure 4.3, curve III), the transition state (III) is more similar in energy to the product, so the structure of the transition state will more closely resemble the structure of the product. Only when the reactant and the product have identical energies (Figure 4.3, curve II) would we expect the structure of the transition state (II) to be exactly halfway between the structures of the reactant and the product.

Now we can understand why the *tert*-butyl cation is formed faster than the isobutyl cation when 2-methylpropene reacts with HCl. Because the formation of a carbocation is an endergonic reaction (Figure 4.4), the structure of the transition state will resemble the structure of the carbocation product. This means that the transition state will have a significant amount of positive charge on a carbon. We know that the *tert*-butyl cation (a tertiary carbocation) is more stable than the isobutyl cation (a primary carbocation). The same factors that stabilize the positively charged carbocation product stabilize the partially positively charged transition state. Therefore, the transition state leading to the *tert*-butyl cation is more stable than the transition state leading to the isobutyl cation. Because the amount of positive charge in the transition state is not as great as the amount of positive charge in the carbocation product, the difference in the stabilities of the two transition states is not as great as the difference in the stabilities of the two carbocation products (Figure 4.4).

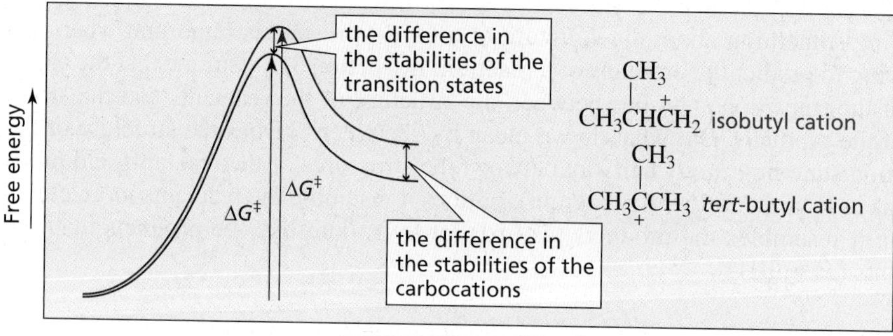

Figure 4.4 ▶ Reaction coordinate diagram for the addition of H^+ to 2-methylpropene to form the primary isobutyl cation and the tertiary *tert*-butyl cation.

We have seen that the rate of a reaction is determined by the free energy of activation, which is the difference between the free energy of the transition state and the free energy of the reactant (Section 3.7). The more stable the transition state, the smaller is the free energy of activation, and therefore, the faster is the reaction. Because the free energy of activation for the formation of the *tert*-butyl cation is less than that for the formation of the isobutyl cation, the *tert*-butyl cation will be formed faster. Thus, in an electrophilic addition reaction, the more stable carbocation will be the one that is formed more rapidly.

Because the formation of the carbocation is the rate-limiting step of the reaction, the relative rates of formation of the two carbocations determine the relative amounts of the products that are formed. If the difference in the rates is small, both products will be formed, but the major product will be the one formed from reaction of the nucleophile with the more stable carbocation. If the difference in the rates is sufficiently large, the product formed from reaction of the nucleophile with the more stable carbocation will be the only product of the reaction. For example, when HCl adds to 2-methylpropene, the rates of formation of the two possible carbocation intermediates—one primary and the other tertiary—are sufficiently different to cause *tert*-butyl chloride to be the only product of the reaction.

$$\underset{\textbf{2-methylpropene}}{CH_3\overset{\overset{\large CH_3}{|}}{C}{=}CH_2} + HCl \longrightarrow \underset{\text{only product formed}}{CH_3\overset{\overset{\large CH_3}{|}}{\underset{\underset{\large Cl}{|}}{C}}CH_3} \qquad \underset{\text{not formed}}{CH_3\overset{\overset{\large CH_3}{|}}{C}HCH_2Cl}$$

PROBLEM 3◆

For each of the following reaction coordinate diagrams, tell whether the structure of the transition state will more closely resemble the structure of the reactants or the structure of the products:

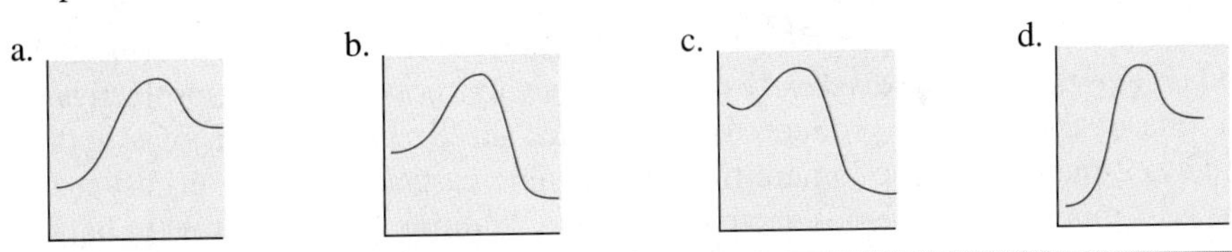

4.4 Regioselectivity of Electrophilic Addition Reactions

When an alkene that does not have the same substituents on its sp^2 carbons undergoes an electrophilic addition reaction, the electrophile can add to two different sp^2 carbons. We have just seen that the major product of the reaction is the one obtained by adding the electrophile to the sp^2 carbon that results in the formation of the more stable carbocation (Section 4.3). For example, when propene reacts with HCl, the proton can add to the number-1 carbon (C-1) to form a secondary carbocation, or it can add to the number-2 carbon (C-2) to form a primary carbocation. The secondary carbocation is formed more rapidly because it is more stable than the primary carbocation. (Primary carbocations are so unstable that they form only with great difficulty.) The product of the reaction, therefore, is 2-chloropropane.

$$CH_3\overset{2}{C}H{=}\overset{1}{C}H_2 \xrightarrow{HCl} CH_3\overset{+}{C}HCH_3 \xrightarrow{Cl^-} CH_3CH(Cl)CH_3$$

a secondary carbocation; 2-chloropropane

$$CH_3CH{=}CH_2 \xrightarrow{HCl} \not\rightarrow CH_3CH_2\overset{+}{C}H_2$$

a primary carbocation

The sp^2 carbon that does *not* become attached to the proton is the carbon that is positively charged in the carbocation.

The major product obtained from the addition of HI to 2-methyl-2-butene is 2-iodo-2-methylbutane; only a small amount of 2-iodo-3-methylbutane is obtained. The major product obtained from the addition of HBr to 1-methylcyclohexene is 1-bromo-1-methylcyclohexane. In both cases, the more stable tertiary carbocation is formed more rapidly than the less stable secondary carbocation, so the major product of each reaction is the one that results from forming the tertiary carbocation.

$$CH_3CH{=}C(CH_3)CH_3 + HI \longrightarrow CH_3CH_2C(CH_3)(I)CH_3 + CH_3CH(CH_3)CH(I)CH_3$$

2-methyl-2-butene; 2-iodo-2-methylbutane, major product; 2-iodo-3-methylbutane, minor product

1-methylcyclohexene + HBr ⟶ 1-bromo-1-methylcyclohexane (major product) + 1-bromo-2-methylcyclohexane (minor product)

The two different products in each of these reactions are called *constitutional isomers*. **Constitutional isomers** have the same molecular formula, but differ in how their atoms are connected. A reaction (such as either of those just shown) in which two or more constitutional isomers could be obtained as products, but one of them predominates, is called a **regioselective reaction**.

There are degrees of regioselectivity: A reaction can be *moderately regioselective, highly regioselective,* or *completely regioselective*. In a completely regioselective reaction, one of the possible products is not formed at all. The addition of a hydrogen halide to 2-methylpropene (where the two possible carbocations are tertiary and primary) is more highly regioselective than the addition of a hydrogen halide to 2-methyl-2-butene (where the two possible carbocations are tertiary and secondary) because the two carbocations formed from 2-methyl-2-butene are closer in stability.

Regioselectivity is the preferential formation of one constitutional isomer over another.

The addition of HBr to 2-pentene is not regioselective. Because the addition of a proton to either of the sp^2 carbons produces a secondary carbocation, both carbocation intermediates have the same stability, so both will be formed equally easily. Thus, approximately equal amounts of the two alkyl halides will be formed.

$$\underset{\textbf{2-pentene}}{CH_3CH{=}CHCH_2CH_3} + HBr \longrightarrow \underset{\textbf{2-bromopentane}}{CH_3\overset{\overset{\large Br}{|}}{C}HCH_2CH_2CH_3} + \underset{\textbf{3-bromopentane}}{CH_3CH_2\overset{\overset{\large Br}{|}}{C}HCH_2CH_3}$$

We now have seen that if we want to predict the major product of an electrophilic addition reaction, we must first determine the relative stabilities of the two possible carbocation intermediates. In 1865, when carbocations and their relative stabilities were not yet known, Vladimir Markovnikov published a paper in which he described a way to predict the major product obtained from the addition of a hydrogen halide to an unsymmetrical alkene. His shortcut is known as **Markovnikov's rule**: "When a hydrogen halide adds to an unsymmetrical alkene, the addition occurs in such a manner that the halogen attaches itself to the double-bonded carbon atom of the alkene bearing the lesser number of hydrogen atoms." Because H^+ is the first species that adds to the alkene, most chemists rephrase Markovnikov's rule as follows: *"The* **hydrogen** *adds to the* sp^2 *carbon that is bonded to the greater number of hydrogens."* Although Markovnikov devised his rule only for the addition of hydrogen halides, chemists now use it for any addition reaction that involves adding a hydrogen to one of the sp^2 carbons. As you study the alkene reactions in this chapter, you will see that not all of them follow Markovnikov's rule. Those that do—that is, the hydrogen *does add* to the sp^2 carbon that is bonded to the greater number of hydrogens—are called **Markovnikov addition reactions**. Those that do not follow Markovnikov's rule—that is, the hydrogen *does not add* to the sp^2 carbon that is bonded to the greater number of hydrogens—are called **anti-Markovnikov addition reactions**.

Now that we understand the mechanisms of alkenes reactions, we can devise a rule that applies to *all* alkene electrophilic addition reactions: *The* **electrophile** *adds to the* sp^2 *carbon that is bonded to the greater number of hydrogens*. This is the rule you should remember because all electrophilic addition reactions follow this rule, so it will keep you from having to memorize which reactions follow Markovnikov's rule and which ones don't.

The electrophile adds to the sp^2 carbon that is bonded to the greater number of hydrogens.

Using the rule that the electrophile adds to the sp^2 carbon bonded to the greater number of hydrogens is simply a quick way to determine the relative stabilities of the intermediates that could be formed in the rate-determining step. You will get the same answer, whether you identify the major product of an electrophilic addition reaction by using the rule or whether you identify it by determining relative carbocation stabilities. In the following reaction for example, H^+ is the electrophile:

$$CH_3CH_2\overset{2}{C}H{=}\overset{1}{C}H_2 + HCl \longrightarrow CH_3CH_2\overset{\overset{\large Cl}{|}}{C}HCH_3$$

We can say that H^+ adds preferentially to C-1 because C-1 is bonded to two hydrogens, whereas C-2 is bonded to only one hydrogen. Or we can say that H^+ adds to C-1 because that results in the formation of a secondary carbocation, which is more stable than the primary carbocation that would be formed if H^+ added to C-2.

Vladimir Vasilevich Markovnikov (1837–1904) *was born in Russia, the son of an army officer. He was a professor of chemistry at Kazan, Odessa, and Moscow Universities. By synthesizing rings containing four carbons and seven carbons, he disproved the notion that carbon could form only five- and six-membered rings.*

PROBLEM 4◆

What would be the major product obtained from the addition of HBr to each of the following compounds?

a. $CH_3CH_2CH{=}CH_2$

b. $CH_3CH{=}C(CH_3)CH_3$ (drawn as $CH_3CH{=}\overset{CH_3}{\overset{|}{C}}CH_3$)

c. 1-methylcyclopentene (cyclopentene ring bearing CH_3 on the double-bond carbon)

d. $CH_2{=}\overset{CH_3}{\overset{|}{C}}CH_2CH_2CH_3$

e. methylenecyclohexane (cyclohexane ring with exocyclic $=CH_2$)

f. $CH_3CH{=}CHCH_3$

PROBLEM-SOLVING STRATEGY

a. What alkene should be used to synthesize 3-bromohexane?

$$? \ + \ HBr \longrightarrow CH_3CH_2\underset{Br}{\underset{|}{C}}HCH_2CH_2CH_3$$

3-bromohexane

The best way to answer this kind of question is to begin by listing all the alkenes that could be used. Because you want to synthesize an alkyl halide that has a bromo substituent at the C-3 position, the alkene should have an sp^2 carbon at that position. Two alkenes fit the description: 2-hexene and 3-hexene.

$CH_3CH{=}CHCH_2CH_2CH_3$ **2-hexene** $\qquad CH_3CH_2CH{=}CHCH_2CH_3$ **3-hexene**

Because there are two possibilities, we next need to determine whether there is any advantage to using one over the other. The addition of H^+ to 2-hexene can form two different carbocations. Because they are both secondary carbocations, they have the same stability; therefore, approximately equal amounts of each will be formed. As a result, half of the product will be 3-bromohexane and half will be 2-bromohexane.

$$CH_3CH{=}CHCH_2CH_2CH_3 \text{ (2-hexene)} \xrightarrow{HBr} CH_3CH_2\overset{+}{C}HCH_2CH_2CH_3 \text{ (secondary carbocation)} \xrightarrow{Br^-} CH_3CH_2\underset{Br}{\underset{|}{C}}HCH_2CH_2CH_3 \text{ (3-bromohexane)}$$

$$CH_3CH{=}CHCH_2CH_2CH_3 \text{ (2-hexene)} \xrightarrow{HBr} CH_3\overset{+}{C}HCH_2CH_2CH_2CH_3 \text{ (secondary carbocation)} \xrightarrow{Br^-} CH_3\underset{Br}{\underset{|}{C}}HCH_2CH_2CH_2CH_3 \text{ (2-bromohexane)}$$

The addition of H^+ to either of the sp^2 carbons of 3-hexene, on the other hand, forms the same carbocation because the alkene is symmetrical. Therefore, all of the product will be the desired 3-bromohexane.

$$CH_3CH_2CH{=}CHCH_2CH_3 \text{ (3-hexene)} \xrightarrow{HBr} CH_3CH_2\overset{+}{C}HCH_2CH_2CH_3 \text{ (only one carbocation is formed)} \xrightarrow{Br^-} CH_3CH_2\underset{Br}{\underset{|}{C}}HCH_2CH_2CH_3 \text{ (3-bromohexane)}$$

Because all the alkyl halide formed from 3-hexene is 3-bromohexane, but only half the alkyl halide formed from 2-hexene is 3-bromohexane, 3-hexene is the best alkene to use to prepare 3-bromohexane.

b. What alkene should be used to synthesize 2-bromopentane?

$$? \quad + \quad HBr \longrightarrow \underset{\underset{\textbf{2-bromopentane}}{Br}}{CH_3\underset{|}{C}HCH_2CH_2CH_3}$$

Either 1-pentene or 2-pentene could be used because both have an sp^2 carbon at the C-2 position.

$$\underset{\textbf{1-pentene}}{CH_2{=}CHCH_2CH_2CH_3} \qquad \underset{\textbf{2-pentene}}{CH_3CH{=}CHCH_2CH_3}$$

When H^+ adds to 1-pentene, one of the carbocations that could be formed is secondary and the other is primary. A secondary carbocation is more stable than a primary carbocation, which is so unstable that little, if any, will be formed. Thus, 2-bromopentane will be the only product of the reaction.

$$\underset{\textbf{1-pentene}}{CH_2{=}CHCH_2CH_2CH_3} \xrightarrow{HBr} CH_3\overset{+}{C}HCH_2CH_2CH_3 \xrightarrow{Br^-} \underset{\underset{\textbf{2-bromopentane}}{Br}}{CH_3\underset{|}{C}HCH_2CH_2CH_3}$$

$$CH_2{=}CHCH_2CH_2CH_3 \xrightarrow{HBr} \not\rightarrow \overset{+}{C}H_2CH_2CH_2CH_2CH_3$$

When H^+ adds to 2-pentene, on the other hand, each of the two carbocations that can be formed is secondary. Both are equally stable, so they will be formed in approximately equal amounts. Thus, only about half of the product of the reaction will be 2-bromopentane. The other half will be 3-bromopentane.

$$\underset{\textbf{2-pentene}}{CH_3CH{=}CHCH_2CH_3} \xrightarrow{HBr} CH_3\overset{+}{C}HCH_2CH_2CH_3 \xrightarrow{Br^-} \underset{\underset{\textbf{2-bromopentane}}{Br}}{CH_3\underset{|}{C}HCH_2CH_2CH_3}$$

$$CH_3CH{=}CHCH_2CH_3 \xrightarrow{HBr} CH_3CH_2\overset{+}{C}HCH_2CH_3 \xrightarrow{Br^-} \underset{\underset{\textbf{3-bromopentane}}{Br}}{CH_3CH_2\underset{|}{C}HCH_2CH_3}$$

Because all the alkyl halide formed from 1-pentene is 2-bromopentane, but only half the alkyl halide formed from 2-pentene is 2-bromopentane, 1-pentene is the best alkene to use to prepare 2-bromopentane.

Now continue on to answer the questions in Problem 5.

PROBLEM 5◆

What alkene should be used to synthesize each of the following alkyl bromides?

a. $CH_3C(CH_3)(Br)CH_3$

b. $C_6H_{11}{-}CH_2CH(Br)CH_3$ (cyclohexyl)

c. $C_6H_{11}{-}C(CH_3)(Br)CH_3$ (cyclohexyl)

d. 1-bromo-1-ethylcyclohexane (CH_2CH_3 and Br on the same ring carbon)

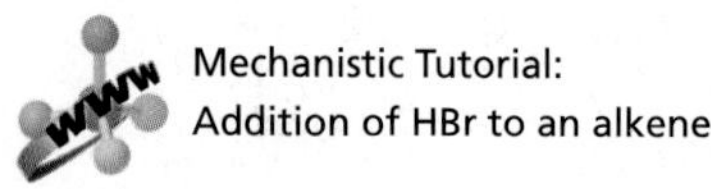

Mechanistic Tutorial: Addition of HBr to an alkene

PROBLEM 6◆

The addition of HBr to which of the following alkenes is more highly regioselective?

a. $CH_3CH_2\overset{CH_3}{\overset{|}{C}}=CH_2$ or $CH_3\overset{CH_3}{\overset{|}{C}}=CHCH_3$

b. methylenecyclohexane ($=CH_2$ on a cyclohexane ring) or 1-methylcyclohexene (CH_3 on a cyclohexene ring)

4.5 Addition of Water and Addition of Alcohols

Addition of Water

When water is added to an alkene, no reaction takes place, because there is no electrophile present to start a reaction by adding to the nucleophilic alkene. The O—H bonds of water are too strong—water is too weakly acidic—to allow the hydrogen to act as an electrophile for this reaction.

$$CH_3CH{=}CH_2 + H_2O \longrightarrow \text{no reaction}$$

If, however, an acid (e.g., H_2SO_4 or HCl) is added to the solution, a reaction will occur because the acid provides an electrophile. The product of the reaction is an alcohol. The addition of water to a molecule is called **hydration**, so we can say that an alkene will be *hydrated* in the presence of water and acid.

$$CH_3CH{=}CH_2 + H_2O \overset{H^+}{\rightleftharpoons} \underset{\text{OH}}{CH_3CH}{-}\underset{\text{H}}{CH_2}$$

2-propanol

H_2SO_4 ($pK_a = -5$) and HCl ($pK_a = -7$) are strong acids, so they dissociate almost completely in an aqueous solution (Section 1.17). The acid that participates in the reaction, therefore, is most apt to be a hydronium ion (H_3O^+).

$$H_2SO_4 + H_2O \rightleftharpoons \underset{\text{hydronium ion}}{H_3O^+} + HSO_4^-$$

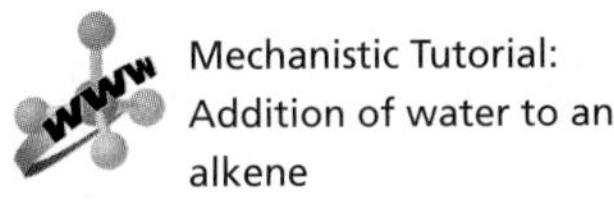

The first two steps of the mechanism for the acid-catalyzed addition of water to an alkene are essentially the same as the two steps of the mechanism for the addition of a hydrogen halide to an alkene: The electrophile (H^+) adds to the sp^2 carbon that is bonded to the greater number of hydrogens, and the nucleophile (H_2O) adds to the other sp^2 carbon.

mechanism for the acid-catalyzed addition of water

$$CH_3CH{=}CH_2 + H{-}\overset{+}{\ddot{O}}H(H) \underset{\text{slow}}{\rightleftharpoons} CH_3\overset{+}{C}HCH_3 + H_2\ddot{O}: \underset{\text{fast}}{\rightleftharpoons} CH_3\underset{:\overset{+}{O}H-H}{CH}CH_3$$

addition of the electrophile

addition of the nucleophile

a protonated alcohol

H_2O removes a proton, regenerating the acid catalyst

$H_2\ddot{O}$: fast

$$CH_3\underset{:\ddot{O}H}{CH}CH_3 + H_3\overset{+}{O}:$$

an alcohol

As we saw in Section 3.7, the addition of the electrophile to the alkene is relatively slow, and the subsequent addition of the nucleophile to the carbocation occurs rapidly. The reaction of the carbocation with a nucleophile is so fast that the carbocation combines with whatever nucleophile it collides with first. In the previous hydration reaction, there were two nucleophiles in solution: water and the counterion of the acid (e.g., Cl^-) that was used to start the reaction. (Notice that HO^- is not a nucleophile in this reaction because there is no appreciable concentration of HO^- in an acidic solution.)* Because the concentration of water is much greater than the concentration of the counterion, the carbocation is much more likely to collide with water. The product of the collision is a protonated alcohol. Because the pH of the solution is greater than the pK_a of the protonated alcohol (remember that protonated alcohols are very strong acids; see Sections 1.17 and 1.19), the protonated alcohol loses a proton, and the final product of the addition reaction is an alcohol. A reaction coordinate diagram for the reaction is shown in Figure 4.5.

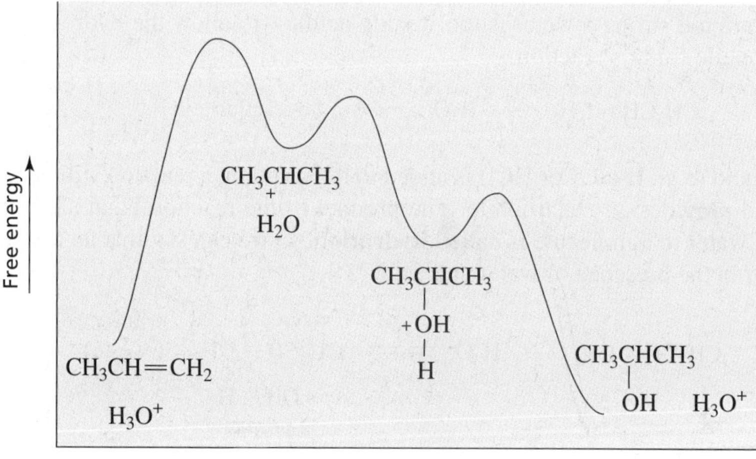

Figure 4.5 ▶ A reaction coordinate diagram for the acid-catalyzed addition of water to an alkene.

A proton adds to the alkene in the first step, but a proton is returned to the reaction mixture in the final step. Overall, a proton is not consumed. A species that increases the rate of a reaction and is not consumed during the course of the reaction is called a **catalyst**. Catalysts increase the reaction rate by decreasing the activation energy of the reaction (Section 3.7). Catalysts do *not* affect the equilibrium constant of the reaction. In other words, a catalyst increases the *rate* at which a product is formed, but does not affect the *amount* of product formed. The catalyst in the hydration of an alkene is an acid, so the reaction is said to be an **acid-catalyzed reaction**.

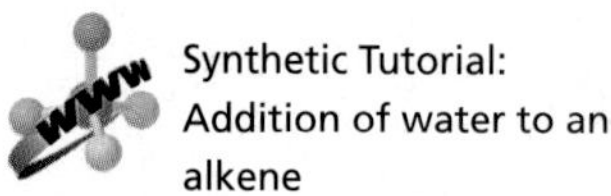

Synthetic Tutorial: Addition of water to an alkene

PROBLEM 7◆

The pK_a of a protonated alcohol is about -2.5 and the pK_a of an alcohol is about 15. Therefore, as long as the pH of the solution is greater than _____ and less than _____, more than 50% of 2-propanol (the product of the previous reaction) will be in its neutral, nonprotonated form.

PROBLEM 8◆

Use Figure 4.5 to answer the following questions about the acid-catalyzed hydration of an alkene:

*At a pH of 4, for example, the concentration of HO^- is 1×10^{-10} M, whereas the concentration of water in a dilute aqueous solution is 55.5 M.

a. How many transition states are there?

b. How many intermediates are there?

c. Which is more stable, the protonated alcohol or the neutral alcohol?

d. Of the six steps in the forward and reverse directions, which are the two fastest?

PROBLEM 9

Give the major product obtained from the acid-catalyzed hydration of each of the following alkenes:

a. $CH_3CH_2CH_2CH{=}CH_2$

b. 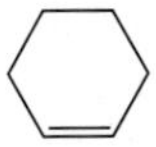 (cyclohexene)

c. $CH_3CH_2CH_2CH{=}CHCH_3$

d. 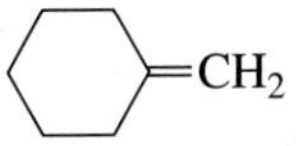 (methylenecyclohexane) $=CH_2$

Addition of Alcohols

Alcohols react with alkenes in the same way that water does. Like the addition of water, the addition of an alcohol requires an acid catalyst. The product of the reaction is an ether.

$$CH_3CH{=}CH_2 + CH_3OH \overset{H^+}{\rightleftharpoons} \underset{\substack{|\\ OCH_3}}{CH_3CH}{-}\underset{\substack{|\\ H}}{CH_2}$$

2-methoxypropane

The mechanism for the acid-catalyzed addition of an alcohol is essentially the same as the mechanism for the acid-catalyzed addition of water.

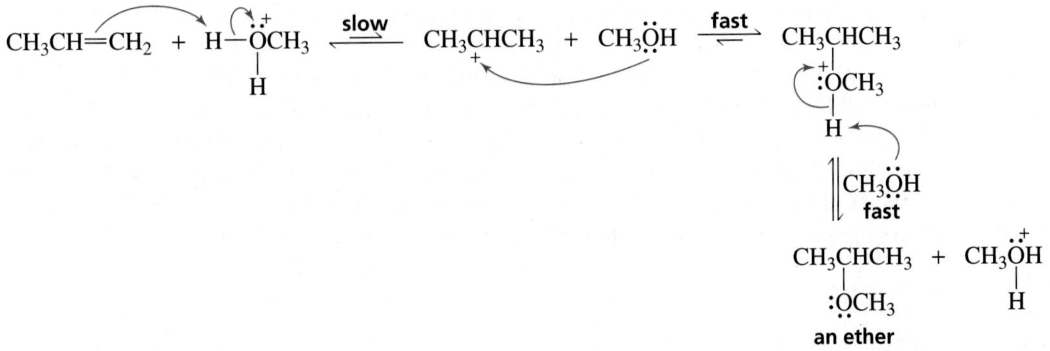

$$CH_3CH{=}CH_2 + H{-}\overset{+}{\underset{\substack{|\\ H}}{\ddot{O}}}CH_3 \underset{}{\overset{\text{slow}}{\rightleftharpoons}} CH_3\overset{+}{C}HCH_3 + CH_3\ddot{O}H \overset{\text{fast}}{\rightleftharpoons} \underset{\substack{|\\ :\overset{+}{O}CH_3\\ |\\ H}}{CH_3CHCH_3}$$

$$\Big\Downarrow \; CH_3\ddot{O}H \;\; \text{fast}$$

$$\underset{\substack{|\\ :\ddot{O}CH_3}}{CH_3CHCH_3} + CH_3\overset{+}{\underset{\substack{|\\ H}}{\ddot{O}H}}$$

an ether

PROBLEM 10

a. Give the major product of each of the following reactions:

1. $CH_3\overset{\substack{CH_3\\ |}}{C}{=}CH_2 + HCl \longrightarrow$

2. $CH_3\overset{\substack{CH_3\\ |}}{C}{=}CH_2 + HBr \longrightarrow$

3. $CH_3\overset{\substack{CH_3\\ |}}{C}{=}CH_2 + H_2O \xrightarrow{H^+}$

4. $CH_3\overset{\substack{CH_3\\ |}}{C}{=}CH_2 + CH_3OH \xrightarrow{H^+}$

b. What do all the reactions have in common?

c. How do all the reactions differ?

PROBLEM 11

How could the following compounds be prepared, using an alkene as one of the starting materials?

a. 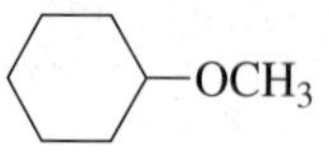 cyclohexyl–OCH_3

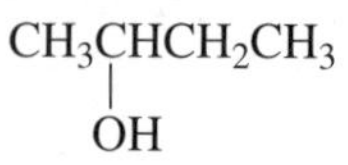

b. 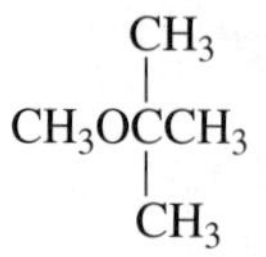 $CH_3OC(CH_3)_2CH_3$

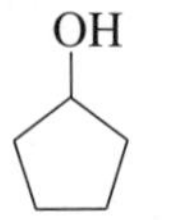

c. 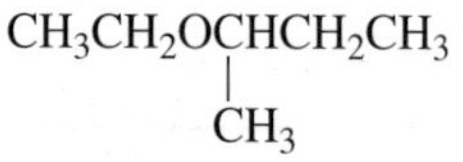 $CH_3CH_2OCH(CH_3)CH_2CH_3$

d. $CH_3CH(OH)CH_2CH_3$

e. cyclopentanol (cyclopentane–OH)

f. 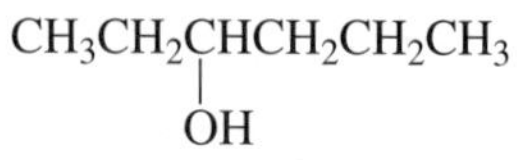 $CH_3CH_2CH(OH)CH_2CH_2CH_3$

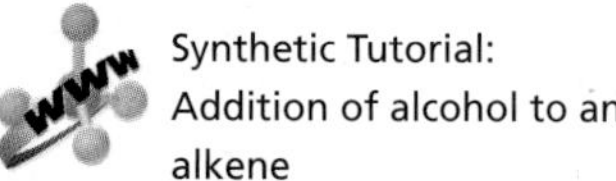

Synthetic Tutorial: Addition of alcohol to an alkene

PROBLEM 12

Propose a mechanism for the following reaction (remember to use curved arrows when showing a mechanism):

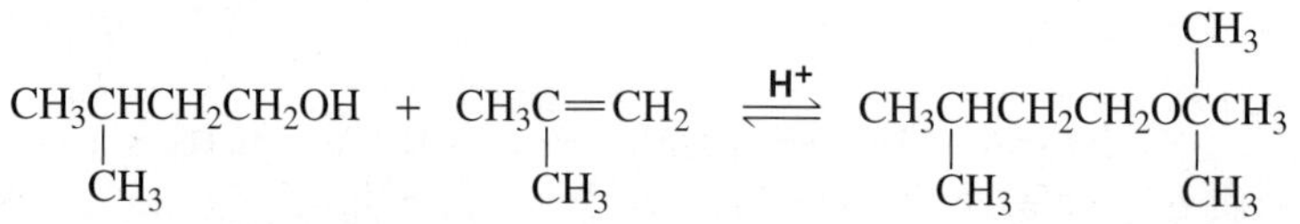

$$CH_3CH(CH_3)CH_2CH_2OH + CH_3C(CH_3){=}CH_2 \overset{H^+}{\rightleftharpoons} CH_3CH(CH_3)CH_2CH_2OC(CH_3)_2CH_3$$

4.6 Rearrangement of Carbocations

Some electrophilic addition reactions give products that are clearly not the result of the addition of an electrophile to the sp^2 carbon bonded to the greater number of hydrogens and the addition of a nucleophile to the other sp^2 carbon. For example, the addition of HBr to 3-methyl-1-butene forms 2-bromo-3-methylbutane (minor product) and 2-bromo-2-methylbutane (major product). 2-Bromo-3-methylbutane is the product you would expect from the addition of H^+ to the sp^2 carbon bonded to the greater number of hydrogens and Br^- to the other sp^2 carbon. 2-Bromo-2-methylbutane is an "unexpected" product, even though it is the major product of the reaction.

Frank (Rocky) Clifford Whitmore (1887–1947) *was born in Massachusetts. He received a Ph.D. from Harvard University and was a professor of chemistry at Minnesota, Northwestern, and Pennsylvania State Universities. Whitmore never slept a full night; when he got tired, he took a one-hour nap. Consequently, he had the reputation of being an indefatigable worker; 20-hour workdays were common. He generally had 30 graduate students working in his lab at a time, and he wrote an advanced textbook that was considered a milestone in the field of organic chemistry.*

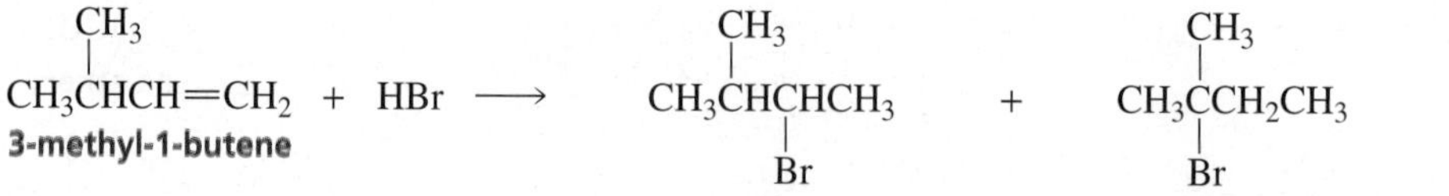

$$CH_3CH(CH_3)CH{=}CH_2 + HBr \longrightarrow CH_3CH(CH_3)CH(Br)CH_3 + CH_3C(CH_3)(Br)CH_2CH_3$$

3-methyl-1-butene → **2-bromo-3-methylbutane** (**minor product**) + **2-bromo-2-methylbutane** (**major product**)

In another example, the addition of HCl to 3,3-dimethyl-1-butene forms both 3-chloro-2,2-dimethylbutane (an "expected" product) and 2-chloro-2,3-dimethylbutane (an "unexpected product"). Again, the unexpected product is obtained in greater yield.

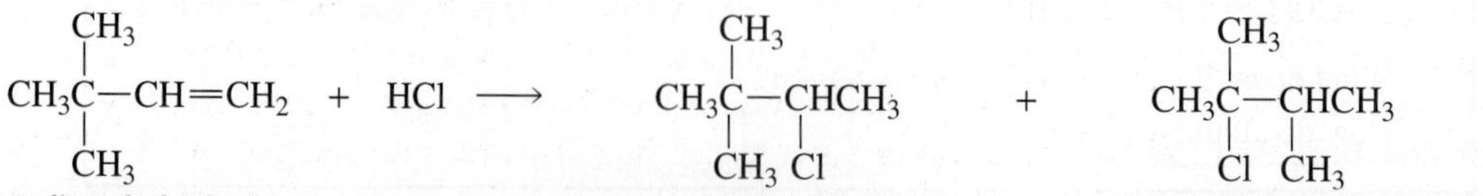

$$CH_3C(CH_3)_2{-}CH{=}CH_2 + HCl \longrightarrow CH_3C(CH_3)_2{-}CH(Cl)CH_3 + CH_3C(Cl)(CH_3){-}CH(CH_3)CH_3$$

3,3-dimethyl-1-butene → **3-chloro-2,2-dimethylbutane** (**minor product**) + **2-chloro-2,3-dimethylbutane** (**major product**)

F. C. Whitmore was the first to suggest that the unexpected product results from a *rearrangement* of the carbocation intermediate. Not all carbocations rearrange. In fact, none of the carbocations that we have seen up to this point rearranges. Carbocations rearrange only if they become more stable as a result of the rearrangement. For example, when an electrophile adds to 3-methyl-1-butene, a *secondary* carbocation is formed initially. However, the secondary carbocation has a hydrogen that can shift with its pair of electrons to the adjacent positively charged carbon, creating a more stable *tertiary* carbocation.

$CH_3CH(CH_3)-CH{=}CH_2$ + H—Br ⟶ $CH_3C(H)(CH_3)-\overset{+}{C}HCH_3$ ⟶ $CH_3\overset{+}{C}(CH_3)-CH_2CH_3$

3-methyl-1-butene

a 1,2-hydride shift

a secondary carbocation

a tertiary carbocation

addition to the unrearranged carbocation — Br^- ⟶ $CH_3CH(CH_3)-CH(Br)CH_3$ minor product

addition to the rearranged carbocation — Br^- ⟶ $CH_3C(CH_3)(Br)-CH_2CH_3$ major product

As a result of **carbocation rearrangement**, two alkyl halides are formed—one from the addition of the nucleophile to the unrearranged carbocation and one from the addition to the rearranged carbocation. The major product is the rearranged one. Because a shift of a hydrogen with its pair of electrons is involved in the rearrangement, it is called a hydride shift. (Recall that $H{:}^-$ is a hydride ion.) More specifically it is called a **1,2-hydride shift** because the hydride ion moves from one carbon to an *adjacent* carbon. (Notice that this does not mean that it moves from C-1 to C-2.)

3,3-Dimethyl-1-butene adds an electrophile to form a *secondary* carbocation. In this case, a methyl group can shift with its pair of electrons to the adjacent positively charged carbon to form a more stable *tertiary* carbocation. This kind of shift is called a **1,2-methyl shift**. (It should have been called a 1,2-methide shift to make it analogous to a 1,2-hydride shift, but, for some reason, it was not so named.)

$CH_3C(CH_3)_2-CH{=}CH_2$ + H—Cl ⟶ $CH_3C(CH_3)(CH_3)-\overset{+}{C}HCH_3$ ⟶ $CH_3\overset{+}{C}(CH_3)-CH(CH_3)CH_3$

3,3-dimethyl-1-butene

a 1,2-methyl shift

a secondary carbocation

a tertiary carbocation

addition to the unrearranged carbocation — Cl^- ⟶ $CH_3C(CH_3)(H_3C)-CH(Cl)CH_3$ minor product

addition to the rearranged carbocation — Cl^- ⟶ $CH_3C(CH_3)(Cl)-CH(CH_3)CH_3$ major product

A shift involves only the movement of a species from one carbon to an adjacent electron-deficient carbon; 1,3-shifts normally do not occur. Furthermore, if the rearrangement does not lead to a more stable carbocation, then a carbocation rearrangement does not occur. For example, when a proton adds to 4-methyl-1-pentene, a secondary carbocation is formed. A 1,2-hydride shift would form a different secondary carbocation. Because both carbocations are equally stable, there is no energetic advantage to the shift. Consequently, rearrangement does not occur, and only one alkyl halide is formed.

Rearrangement involves a change in the way the atoms are connected.

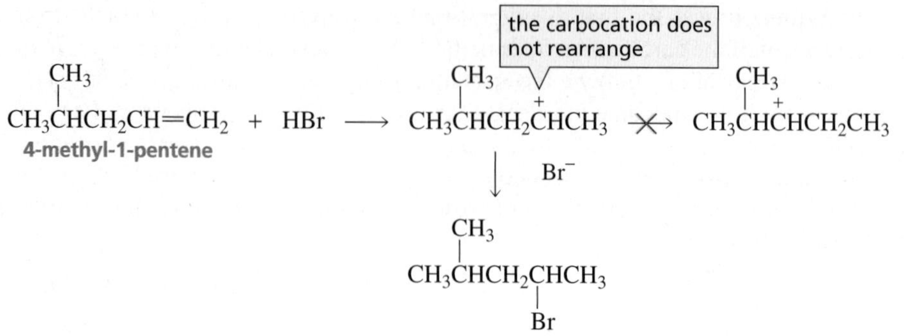

the carbocation does not rearrange

$$CH_3CH(CH_3)CH_2CH{=}CH_2 + HBr \longrightarrow CH_3CH(CH_3)CH_2\overset{+}{C}HCH_3 \not\longleftrightarrow CH_3\overset{+}{C}(CH_3)CH_2CH_2CH_3$$

4-methyl-1-pentene

$$\xrightarrow{Br^-} CH_3CH(CH_3)CH_2CH(Br)CH_3$$

Carbocation rearrangements also can occur by *ring expansion*, another type of 1,2-shift. In the following example, a secondary carbocation is formed initially:

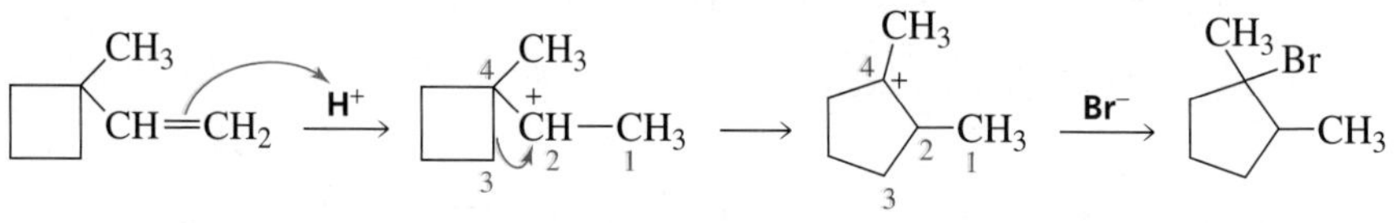

$$\text{(1-methyl-1-vinylcyclobutane)} \xrightarrow{H^+} \text{(cyclobutyl)}\overset{+}{C}H{-}CH_3 \longrightarrow \text{(1,2-dimethylcyclopentyl cation)} \xrightarrow{Br^-} \text{1-bromo-1,2-dimethylcyclopentane}$$

Ring expansion leads to a more stable carbocation—it is tertiary rather than secondary, and a five-membered ring has less angle strain than a four-membered ring (Section 2.11).

In subsequent chapters, you will study other reactions that involve the formation of carbocation intermediates. Keep in mind that *whenever a reaction leads to the formation of a carbocation, you must check its structure for the possibility of rearrangement.*

PROBLEM 13 SOLVED

Which of the following carbocations would you expect to rearrange?

a. 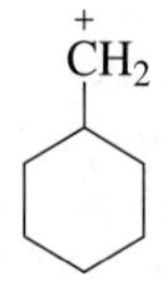 cyclohexyl-$\overset{+}{C}H_2$

b. 1-methylcyclohexyl cation (CH_3, + on ring carbon)

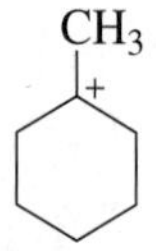

c. 2-methylcyclohexyl cation (CH_3, + on adjacent ring carbon)

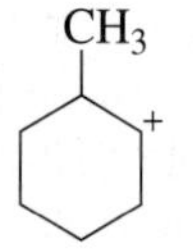

d. $CH_3CH_2\overset{+}{C}HCH_3$

e. 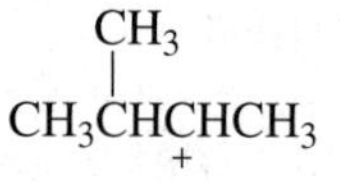 $CH_3CH(CH_3)\overset{+}{C}HCH_3$

SOLUTION

a. This carbocation will rearrange because a 1,2-hydride shift will convert a primary carbocation into a tertiary carbocation.

$$\text{(H)cyclohexyl-}\overset{+}{C}H_2 \longrightarrow \text{1-methylcyclohexyl cation}$$

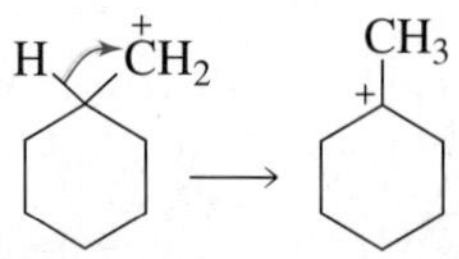

b. This carbocation will not rearrange because it is tertiary and its stability cannot be improved by a carbocation rearrangement.

c. This carbocation will rearrange because a 1,2-hydride shift will convert a secondary carbocation into a tertiary carbocation.

$$\text{2-methylcyclohexyl cation} \longrightarrow \text{1-methylcyclohexyl cation}$$

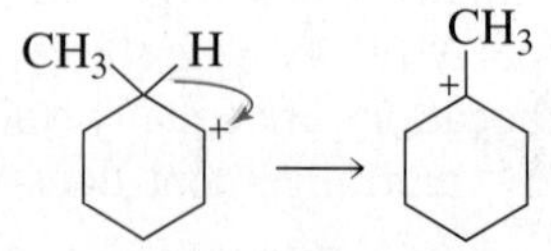

d. This carbocation will not rearrange because it is a secondary carbocation and a carbocation rearrangement would yield another secondary carbocation.

e. This carbocation will rearrange because a 1,2-hydride shift will convert a secondary carbocation into a tertiary carbocation.

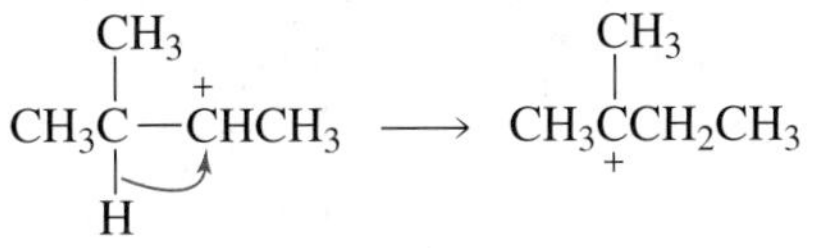

$$CH_3\overset{CH_3}{\underset{H}{C}}-\overset{+}{C}HCH_3 \longrightarrow CH_3\overset{CH_3}{\underset{+}{C}}CH_2CH_3$$

PROBLEM 14

Give the major product(s) obtained from the reaction of each of the following with HBr:

a. 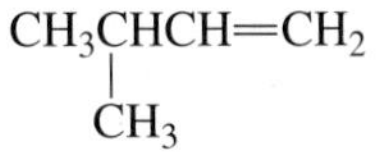 $CH_3\underset{CH_3}{C}HCH{=}CH_2$

b. (methylenecyclohexane: cyclohexane ring with exocyclic $=CH_2$)

c. $CH_3\underset{CH_3}{C}HCH_2CH{=}CH_2$

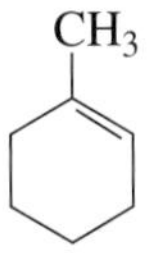

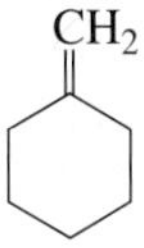

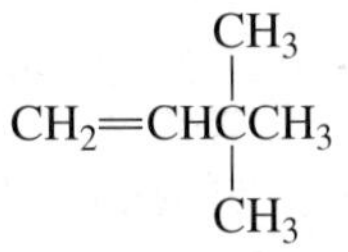

d. (1-methylcyclohexene: cyclohexene ring with CH_3 on the double-bond carbon)

e. $CH_2{=}CH\overset{CH_3}{\underset{CH_3}{C}}CH_3$

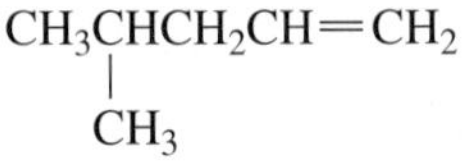

f. 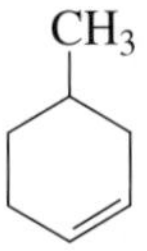 (cyclohexene ring with CH_3 on a ring carbon not part of the double bond)

In any reaction that forms a carbocation intermediate, always check to see if the carbocation will rearrange.

4.7 Addition of Halogens

The halogens Br_2 and Cl_2 add to alkenes. This may be surprising because it is not immediately apparent that an electrophile—which is necessary to start an electrophilic addition reaction—is present.

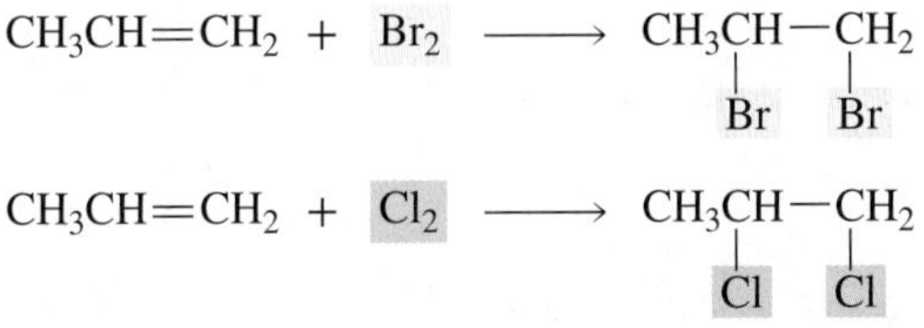

$$CH_3CH{=}CH_2 + Br_2 \longrightarrow CH_3\underset{Br}{C}H-\underset{Br}{C}H_2$$

$$CH_3CH{=}CH_2 + Cl_2 \longrightarrow CH_3\underset{Cl}{C}H-\underset{Cl}{C}H_2$$

However, the bond joining the two halogen atoms is relatively weak (see the bond dissociation energies listed in Table 3.1) and, therefore, easily broken. When the π electrons of the alkene approach a molecule of Br_2 or Cl_2, one of the halogen atoms accepts the electrons and releases the shared electrons to the other halogen atom. Therefore, in an electrophilic addition reaction, Br_2 behaves as if it were Br^+ and Br^-, and Cl_2 behaves as if it were Cl^+ and Cl^-.

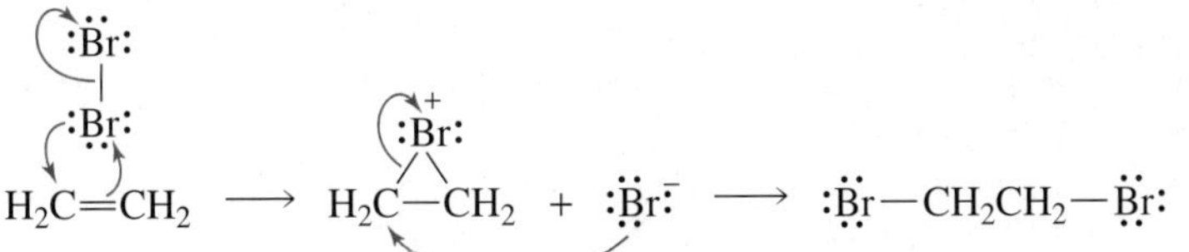

$$H_2C{=}CH_2 + :\ddot{Br}-\ddot{Br}: \longrightarrow H_2C\overset{:\overset{+}{Br}:}{\frown}CH_2 + :\ddot{Br}:^- \longrightarrow :\ddot{Br}-CH_2CH_2-\ddot{Br}:$$

a bromonium ion

1,2-dibromoethane
a vicinal dibromide

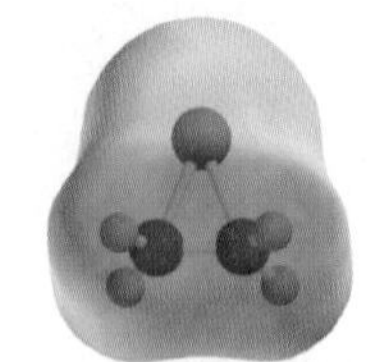

cyclic bromonium ion of ethene

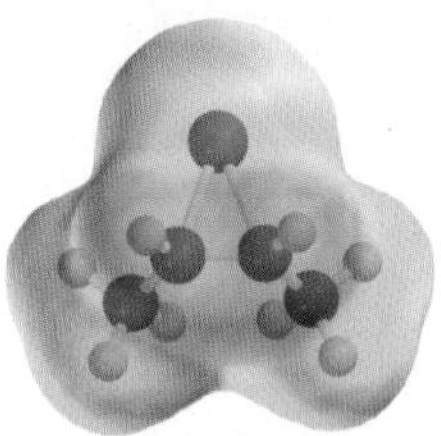

cyclic bromonium ion of *cis*-2-butene

The product of the first step is not a carbocation; rather, it is a cyclic bromonium ion because bromine's electron cloud is close enough to the other sp^2 carbon to engage in bond formation. The cyclic bromonium ion is more stable than the carbocation would have been, since all the atoms (except hydrogen) in the bromonium ion have complete octets, whereas the positively charged carbon of the carbocation does not have a complete octet. (To review the octet rule, see Section 1.3.)

$$H_2C{=}CH_2 + :\ddot{Br}{-}\ddot{Br}: \;\not\longrightarrow\; H_2\overset{+}{C}{-}CH_2(\ddot{Br}:) + :\ddot{Br}:^-$$

less stable

$$H_2C{=}CH_2 + :\ddot{Br}{-}\ddot{Br}: \longrightarrow \text{cyclic } H_2C{-}CH_2 \text{ bridged by } :\overset{+}{Br}: \; + :\ddot{Br}:^-$$

more stable

In the second step of the reaction, Br^- attacks a carbon atom of the bromonium ion. This releases the strain in the three-membered ring and forms a *vicinal dibromide*. **Vicinal** indicates that the two bromine atoms are on adjacent carbons (*vicinus* is the Latin word for "near"). The electrostatic potential maps for the cyclic bromonium ions show that the electron-deficient region (the blue area) encompasses the carbons, even though the formal positive charge is on the bromine atom.

When Cl_2 adds to an alkene, a cyclic chloronium ion intermediate is formed. The final product of the reaction is a vicinal dichloride.

$$CH_3\underset{}{\overset{CH_3}{\overset{|}{C}}}{=}CH_2 + Cl_2 \xrightarrow{CH_2Cl_2} CH_3\underset{\underset{Cl}{|}}{\overset{CH_3}{\overset{|}{C}}}CH_2Cl$$

2-methylpropene → 1,2-dichloro-2-methylpropane, a vicinal dichloride

Because a carbocation is not formed when Br_2 or Cl_2 adds to an alkene, carbocation rearrangements do not occur in these reactions.

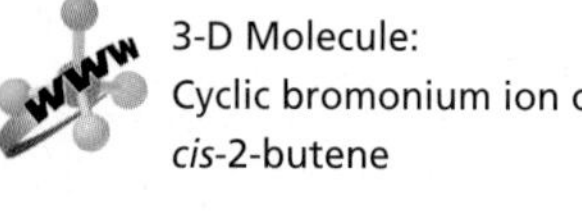

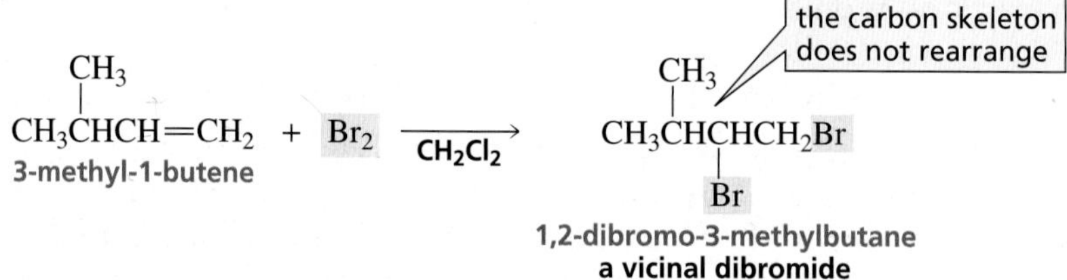

PROBLEM 15◆

What would have been the product of the preceding reaction if HBr had been used as a reagent instead of Br_2?

PROBLEM 16

a. How does the first step in the reaction of ethene with Br_2 differ from the first step in the reaction of ethene with HBr?

b. To understand why Br^- attacks a carbon atom of the bromonium ion rather than the positively charged bromine atom, draw the product that would be obtained if Br^- *did* attack the bromine atom.

Reactions of alkenes with Br_2 or Cl_2 are generally carried out by mixing the alkene and the halogen in an inert solvent, such as dichloromethane (CH_2Cl_2), that readily dissolves both reactants, but does not participate in the reaction. The foregoing reactions illustrate the way in which organic reactions are typically written. The reactants are placed to the left of the reaction arrow, and the products are placed to the right of the arrow. The reaction conditions, such as the solvent, the temperature, or any required catalyst, are written above or below the arrow. Sometimes reactions are written by placing only the organic (carbon-containing) reagent on the left-hand side of the arrow and writing the other reagent(s) above or below the arrow.

$$CH_3CH{=}CHCH_3 \xrightarrow[CH_2Cl_2]{Cl_2} CH_3\underset{Cl}{C}H\underset{Cl}{C}HCH_3$$

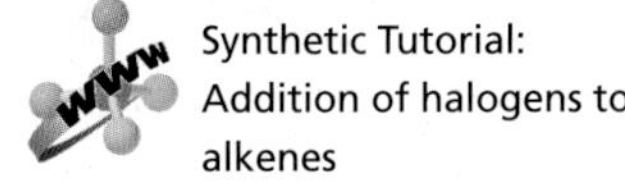
Synthetic Tutorial: Addition of halogens to alkenes

F_2 and I_2 are halogens, but they are not used as reagents in electrophilic addition reactions. Fluorine reacts explosively with alkenes, so the electrophilic addition of F_2 is not a synthetically useful reaction. The addition of I_2 to an alkene is a thermodynamically unfavorable reaction: The vicinal diiodides are unstable at room temperature, decomposing back to the alkene and I_2.

$$CH_3CH{=}CHCH_3 + I_2 \underset{CH_2Cl_2}{\rightleftharpoons} CH_3\underset{I}{C}H\underset{I}{C}HCH_3$$

If H_2O rather than CH_2Cl_2 is used as the solvent, the major product of the reaction will be a vicinal halohydrin. A **halohydrin** (or more specifically, a bromohydrin or a chlorohydrin) is an organic molecule that contains both a halogen and an OH group. In a vicinal halohydrin, the halogen and the OH group are bonded to adjacent carbons.

$$\underset{\text{propene}}{CH_3CH{=}CH_2} + Br_2 \xrightarrow{H_2O} \underset{\substack{\text{a bromohydrin}\\\text{major product}}}{CH_3\underset{OH}{C}HCH_2Br} + \underset{\text{minor product}}{CH_3\underset{Br}{C}HCH_2Br} + HBr$$

$$\underset{\text{2-methyl-2-butene}}{CH_3CH{=}\overset{CH_3}{C}CH_3} + Cl_2 \xrightarrow{H_2O} \underset{\substack{\text{a chlorohydrin}\\\text{major product}}}{CH_3\underset{Cl}{C}H\underset{OH}{\overset{CH_3}{C}}CH_3} + \underset{\text{minor product}}{CH_3\underset{Cl}{C}H\underset{Cl}{\overset{CH_3}{C}}CH_3} + HCl$$

The mechanism for halohydrin formation involves the formation of a cyclic bromonium ion (or chloronium ion) in the first step of the reaction, because Br^+ (or Cl^+) is the only electrophile in the reaction mixture. In the second step, the bromonium ion rapidly reacts with whatever nucleophile it bumps into. In other words, the electrophile and nucleophile do not have to come from the same molecule. There are two nucleophiles present in solution: H_2O and Br^-. Because H_2O is the solvent, its concentration far exceeds that of Br^-. Consequently, the bromonium ion is more likely to collide with a molecule of water than with Br^-. The protonated halohydrin that is formed is a strong acid (Section 1.19), so it loses a proton.

mechanism for halohydrin formation

$$CH_3CH{=}CH_2 + :Br{-}Br: \xrightarrow{\text{slow}} CH_3CH{-}CH_2 \text{ (bridged by } :Br^+:) + :Br:^- \xrightarrow[\text{fast}]{H_2\ddot{O}:} CH_3\underset{:\overset{+}{O}H-H}{C}HCH_2{-}Br: \xrightarrow[\text{fast}]{H_2\ddot{O}:} CH_3\underset{:\ddot{O}H}{C}HCH_2{-}Br: + H_3\overset{+}{O}:$$

How can we explain the regioselectivity of the preceding addition reaction? The electrophile (Br^+) ends up on the sp^2 carbon bonded to the greater number of hydrogens because, in the transition state for the second step of the reaction, the breaking of the C—Br bond has occurred to a greater extent than has the formation of the C—O bond. As a result, there is a partial positive charge on the carbon that is attacked by the nucleophile.

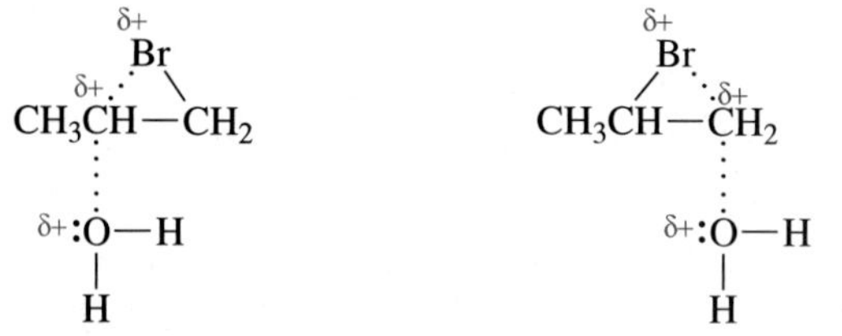

$$\begin{array}{cc} \overset{\delta+}{Br} \text{ bridging } CH_3\overset{\delta+}{C}H-CH_2 \text{ with } \overset{\delta+}{:\ddot{O}}(H)-H \text{ at CH} & \overset{\delta+}{Br} \text{ bridging } CH_3CH-\overset{\delta+}{C}H_2 \text{ with } \overset{\delta+}{:\ddot{O}}(H)-H \text{ at } CH_2 \end{array}$$

more stable transition state | less stable transition state

Therefore, the more stable transition state is achieved by adding the nucleophile to the most substituted sp^2 carbon—the one bonded to the *lesser number* of hydrogens—because the partial positive charge will be on a secondary carbon rather than on a primary carbon. Thus, this reaction also follows the general rule for electrophilic addition reactions: The electrophile adds to the sp^2 carbon that is bonded to the greater number of hydrogens. In this case, the electrophile is Br^+.

Do not memorize the products of alkene addition reactions. Instead, for each reaction, ask yourself, "What is the electrophile?" and "What nucleophile is present in the greatest concentration?"

When nucleophiles other than H_2O are added to the reaction mixture, they, too, change the product of the reaction, just as water changed the product of Br_2 addition from a vicinal dibromide to a vicinal bromohydrin. Because the concentration of the added nucleophile will be greater than the concentration of the halide ion generated from Br_2 or Cl_2, the added nucleophile will be the nucleophile most likely to participate in the second step of the reaction. (Ions such as Na^+ and K^+ cannot form covalent bonds, so they do not react with organic compounds. They serve only as counterions to negatively charged species, so their presence generally is ignored in writing chemical equations.)

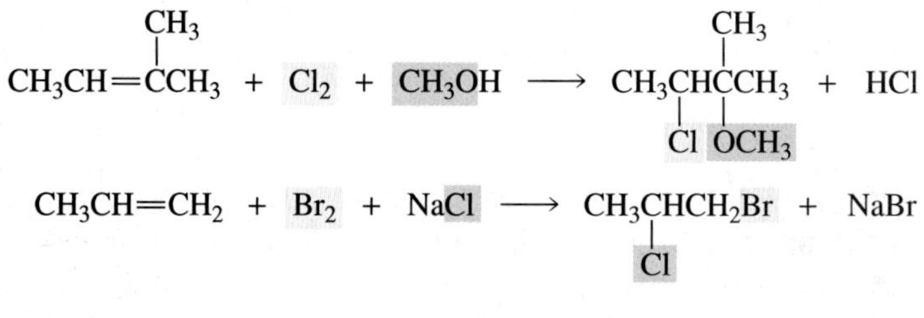

$$CH_3CH=\overset{CH_3}{\overset{|}{C}}CH_3 + Cl_2 + CH_3OH \longrightarrow CH_3\underset{Cl}{\underset{|}{C}H}\overset{CH_3}{\overset{|}{\underset{OCH_3}{\underset{|}{C}}}}CH_3 + HCl$$

$$CH_3CH=CH_2 + Br_2 + NaCl \longrightarrow CH_3\underset{Cl}{\underset{|}{C}H}CH_2Br + NaBr$$

PROBLEM 17

There are two nucleophiles in each of the following reactions:

a. 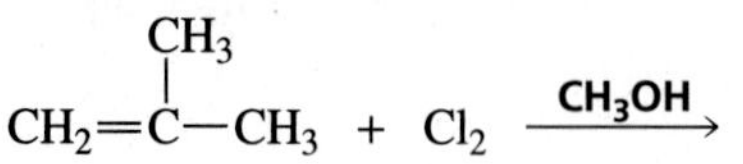 $CH_2=\overset{CH_3}{\overset{|}{C}}-CH_3 + Cl_2 \xrightarrow{CH_3OH}$

b. $CH_2=CHCH_3 + 2\,NaI + HBr \longrightarrow$

c. $CH_3CH=CHCH_3 + HCl \xrightarrow{H_2O}$

d. $CH_3CH=CHCH_3 + HBr \xrightarrow{CH_3OH}$

For each reaction, explain why there is a greater concentration of one nucleophile than the other. What will be the major product of each reaction?

PROBLEM 18

Why are Na^+ and K^+ unable to form covalent bonds?

PROBLEM 19◆

What will be the product of the addition of I–Cl to 1-butene? [*Hint:* Chlorine is more electronegative than iodine (Table 1.3).]

Synthetic Tutorial: Halohydrin reaction

PROBLEM 20◆

What would be the major product obtained from the reaction of Br_2 with 1-butene if the reaction were carried out in

a. dichloromethane?
b. water?
c. ethyl alcohol?
d. methyl alcohol?

4.8 Oxymercuration–Reduction and Alkoxymercuration–Reduction

In Section 4.5, you learned that water adds to an alkene if an acid catalyst is present. This is the way alkenes are converted into alcohols industrially. However, under normal laboratory conditions, water is added to an alkene by a procedure known as **oxymercuration–reduction**. The addition of water by oxymercuration–reduction has two advantages over acid-catalyzed addition: It does not require acidic conditions that are harmful to many organic molecules, and because carbocation intermediates are not formed, carbocation rearrangements do not occur.

In oxymercuration, the alkene is treated with mercuric acetate in aqueous tetrahydrofuran (THF). When reaction with that reagent is complete, sodium borohydride is added to the reaction mixture. (The numbers 1 and 2 in front of the reagents above and below the arrow in the chemical equation indicate two sequential reactions; the second reagent is not added until reaction with the first reagent is completely over.)

$$\mathrm{R{-}CH{=}CH_2} \xrightarrow[\textbf{2. NaBH}_4]{\textbf{1. Hg(OAc)}_2\textbf{, H}_2\textbf{O/THF}} \mathrm{R{-}\underset{\displaystyle OH}{\underset{|}{CH}}{-}CH_3}$$

In the first step of the oxymercuration mechanism, the electrophilic mercury of mercuric acetate adds to the double bond. (Two of mercury's 5*d* electrons are shown.) Because carbocation rearrangements do not occur, we can conclude that the product of the addition reaction is a cyclic mercurinium ion rather than a carbocation. The reaction is analogous to the addition of Br_2 to an alkene to form a cyclic bromonium ion.

mechanism for oxymercuration

$$\mathrm{CH_3CH{=}CH_2} + \mathrm{:Hg(OAc){-}OAc} \longrightarrow \mathrm{CH_3\overset{\displaystyle Hg^+(OAc)}{CH{-}CH_2}} + \mathrm{AcO^-} \xrightarrow{\mathbf{H_2\ddot{O}:}} \mathrm{CH_3\underset{\displaystyle H{-}\overset{+}{:OH}}{CH}CH_2{-}Hg{-}OAc} \xrightarrow{:\ddot{O}Ac} \mathrm{CH_3\underset{\displaystyle :\ddot{O}H}{CH}CH_2{-}Hg{-}OAc} + \mathrm{AcOH}$$

$$\mathrm{AcO^-} = \mathrm{CH_3\overset{\displaystyle O}{\overset{\|}{C}}O^-}$$

In the second step of the reaction, water attacks the more substituted carbon of the mercurinium ion—the one bonded to the lesser number of hydrogens—for the same reason that it attacks the more substituted carbon of the bromonium ion in the halohydrin reaction (Section 4.7). That is, attacking at the more substituted carbon leads to the more stable transition state.

$$\underset{\text{more stable transition state}}{\begin{array}{c} OAc \\ \delta+| \\ Hg \\ \delta+ \; \cdot\cdot \; \backslash \\ CH_3CH-CH_2 \\ \vdots \\ \delta+\,O-H \\ | \\ H \end{array}} \qquad \underset{\text{less stable transition state}}{\begin{array}{c} OAc \\ \delta+| \\ Hg \\ / \; \cdot\cdot \; +\delta \\ CH_3CH-CH_2 \\ \vdots \\ \delta+\,O-H \\ | \\ H \end{array}}$$

Sodium borohydride ($NaBH_4$) converts the C—Hg bond into a C—H bond. A reaction that increases the number of C—H bonds or decreases the number of C—O, C—N, or C—X bonds in a compound (where X denotes a halogen), is called a **reduction reaction**. Consequently, the reaction with sodium borohydride is a reduction reaction. The mechanism of the reduction reaction is not fully understood, although it is known that the intermediate is a radical.

Reduction increases the number of C—H bonds or decreases the number of C—O, C—N, or C—X bonds.

$$CH_3\underset{OH}{\underset{|}{C}}HCH_2-Hg-OAc \xrightarrow{NaBH_4} CH_3\underset{OH}{\underset{|}{C}}HCH_3 + Hg + AcO^-$$

The overall reaction (oxymercuration–reduction) forms the same product that would be formed from the acid-catalyzed addition of water: The hydrogen adds to the sp^2 carbon bonded to the greater number of hydrogens, and OH adds to the other sp^2 carbon.

We have seen that alkenes react with alcohols in the presence of an acid catalyst to form ethers (Section 4.5). Just as the addition of water works better in the presence of mercuric acetate than in the presence of a strong acid, the addition of an alcohol works better in the presence of mercuric acetate. [Mercuric trifluoroacetate, $Hg(O_2CCF_3)_2$, works even better.] This reaction is called **alkoxymercuration–reduction.**

$$\underset{\text{1-methylcyclohexene}}{\text{(cyclohexene ring with } CH_3)} \xrightarrow[\text{2. } NaBH_4]{\text{1. } Hg(O_2CCF_3)_2,\ CH_3OH} \underset{\substack{\text{1-methoxy-1-methylcyclohexane} \\ \text{an ether}}}{\text{(cyclohexane ring with } H_3C \text{ and } OCH_3)}$$

The mechanisms for oxymercuration and alkoxymercuration are essentially identical; the only difference is that water is the nucleophile in oxymercuration and an alcohol is the nucleophile in alkoxymercuration. Therefore, the product of oxymercuration–reduction is an alcohol, whereas the product of alkoxymercuration–reduction is an ether.

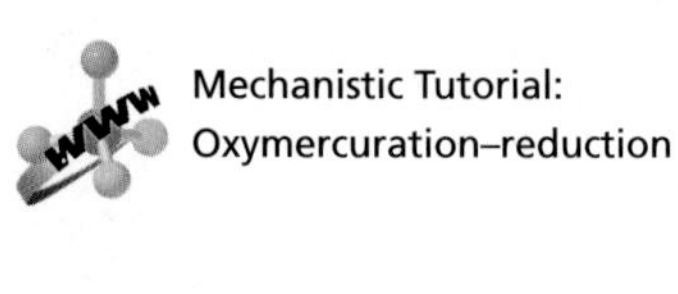
Mechanistic Tutorial: Oxymercuration–reduction

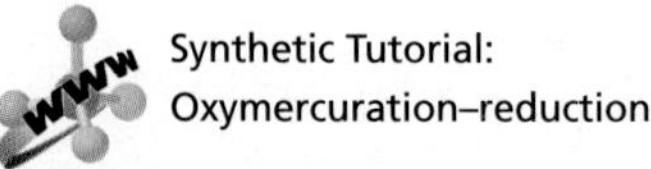
Synthetic Tutorial: Oxymercuration–reduction

PROBLEM 21

How could the following compounds be synthesized from an alkene?

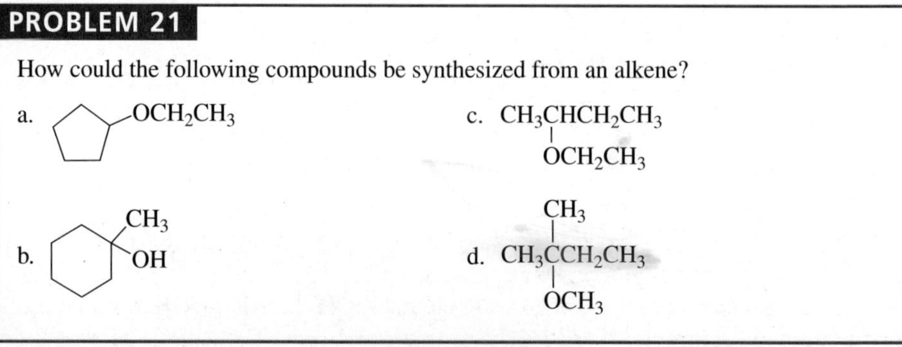

PROBLEM 22

How could the following compounds be synthesized from 3-methyl-1-butene?

a. $CH_3C(CH_3)(OH)CH_2CH_3$

b. $CH_3CH(CH_3)CH(OH)CH_3$

4.9 Addition of Borane: Hydroboration–Oxidation

An atom or a molecule does not have to be positively charged to be an electrophile. Borane (BH_3), a neutral molecule, is an electrophile because boron has only six shared electrons in its valence shell. Boron, therefore, readily accepts a pair of electrons in order to complete its octet. Thus, alkenes undergo electrophilic addition reactions with borane serving as the electrophile. When the addition reaction is over, an aqueous solution of sodium hydroxide and hydrogen peroxide is added to the reaction mixture, and the resulting product is an alcohol. The addition of borane to an alkene, followed by reaction with hydroxide ion and hydrogen peroxide, is called **hydroboration–oxidation**. The overall reaction was first reported by H. C. Brown in 1959.

Herbert Charles Brown *was born in London in 1921 and was brought to the United States by his parents at age two. He received a Ph.D. from the University of Chicago and has been a professor of chemistry at Purdue University since 1947. For his studies on boron-containing organic compounds, he shared the 1979 Nobel Prize in chemistry with G. Wittig.*

$$CH_2{=}CH_2 \xrightarrow[\text{2. } HO^-,\ H_2O_2,\ H_2O]{\text{1. } BH_3/THF} \underset{H}{CH_2}{-}\underset{OH}{CH_2}$$

an alcohol

The alcohol that is formed from the hydroboration–oxidation of an alkene has the H and OH groups on opposite carbons, compared with the alcohol that is formed from the acid-catalyzed addition of water (Section 4.5). In other words, the reaction violates Markovnikov's rule. Thus, hydroboration–oxidation is an anti-Markovnikov addition reaction. However, you will see that the general rule for electrophilic addition reactions is not violated: *The electrophile adds to the* sp^2 *carbon that is bonded to the greater number of hydrogens*. The reaction violates Markovnikov's rule because his rule states where the hydrogen adds, so the rule applies only if the electrophile is a hydrogen. The reaction does not violate the general rule because, as you will see, H^+ is not the electrophile in hydroboration–oxidation; BH_3 is the electrophile and H^- is the nucleophile. This shows why it is better to understand the mechanism of a reaction than to memorize rules. *The first step in the mechanism of all alkene reactions is the same*: the addition of an electrophile to the *sp*^{2} carbon that is bonded to the greater number of hydrogens.

$$\underset{\text{propene}}{CH_3CH{=}CH_2} \xrightarrow[\text{2. } HO^-,\ H_2O_2,\ H_2O]{\text{1. } BH_3/THF} \underset{\text{1-propanol}}{CH_3CH_2CH_2OH}$$

$$\underset{\text{propene}}{CH_3CH{=}CH_2} \xrightarrow[H_2O]{H_2SO_4} \underset{\text{2-propanol}}{CH_3\underset{OH}{C}HCH_3}$$

Because diborane (B_2H_6), the source of borane, is a flammable, toxic, explosive gas, a solution of borane—prepared by dissolving diborane in an ether such as THF—is a more convenient and less dangerous reagent. One of the ether oxygen's lone pairs satisfies boron's requirement for an additional two electrons: The ether is a Lewis base

BORANE AND DIBORANE

Borane exists primarily as a colorless gas called diborane. Diborane is a **dimer**—a molecule formed by joining two identical molecules. Because boron is surrounded by only six electrons, it has a strong tendency to acquire an additional electron pair. Two boron atoms, therefore, share the two electrons in a hydrogen–boron bond in unusual half-bonds. The hydrogen–boron bonds in diborane are shown as dotted lines to indicate that the bond is made up of fewer than the normal two electrons.

$$2\ BH_3 \rightleftharpoons B_2H_6$$

borane **diborane**

3-D Molecule: Diborane

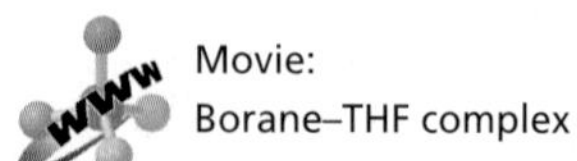
Movie: Borane–THF complex

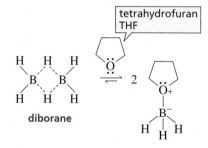

and borane is a Lewis acid. So the reagent actually used as the source of BH_3 for the first step of hydroboration–oxidation is a borane–THF complex.

tetrahydrofuran THF

$$B_2H_6 + 2\ \text{THF} \rightleftharpoons 2\ \text{THF}^{+}\text{—}\overset{-}{B}H_3$$

diborane

To understand why the hydroboration–oxidation of propene forms 1-propanol, we must look at the mechanism of the reaction. The boron atom of borane is electron deficient, so borane is the electrophile that reacts with the nucleophilic alkene. As boron accepts the π electrons and forms a bond with one sp^2 carbon, it donates a hydride ion to the other sp^2 carbon. In all the addition reactions that we have seen up to this point, the electrophile adds to the alkene in the first step and the nucleophile adds to the positively charged intermediate in the second step. In contrast, the addition of the electrophilic boron and the nucleophilic hydride ion to the alkene take place in one step. Therefore, an intermediate is not formed.

$$CH_3CH{=}CH_2 + H{-}BH_2 \longrightarrow CH_3CH(H){-}CH_2(BH_2)$$

an alkylborane

The addition of borane to an alkene is an example of a *concerted* reaction. A **concerted reaction** is a reaction in which all the bond-making and bond-breaking processes occur in a single step. The addition of borane to an alkene is also an example of a *pericyclic* reaction. (*Pericyclic* means "around the circle.") A **pericyclic reaction** is a concerted reaction that takes place as the result of a cyclic rearrangement of electrons.

The electrophilic boron adds to the sp^2 carbon that is bonded to the greater number of hydrogens. The electrophiles that we have looked at previously (e.g., H^+) also added to the sp^2 carbon bonded to the greater number of hydrogens, in order to form the most stable carbocation intermediate. Given that an intermediate is not formed in this concerted reaction, how can we explain the regioselectivity of the reaction. Why does boron add preferentially to the sp^2 carbon bonded to the greater number of hydrogens?

If we examine the two possible transition states for the addition of borane, we see that the C—B bond has formed to a greater extent than has the C—H bond. Consequently, the sp^2 carbon that does not become attached to boron has a partial positive charge. The partial positive charge is on a secondary carbon if boron adds to the sp^2 carbon bonded to the greater number of hydrogens. The partial positive charge is on a primary carbon if boron adds to the other sp^2 carbon. So, even though a carbocation intermediate is not formed, a carbocation-like transition state is formed. Thus, the addition of borane and the addition of an electrophile such as H^+ take place at the same

sp^2 carbon for the same reason: to form the more stable carbocation or carbocation-like transition state.

addition of BH_3

$CH_3-\overset{\delta+}{C}(H)\cdots C(H)-H$ with $H\cdots BH_2^{\delta-}$ — more stable transition state

$CH_3-C(H)\cdots\overset{\delta+}{C}(H)-H$ with $H_2B^{\delta-}\cdots H$ — less stable transition state

addition of HBr

$CH_3-\overset{\delta+}{C}(H)\cdots C(H)-H$ with $Br^{\delta-}\cdots H$ — more stable transition state

$CH_3-C(H)\cdots\overset{\delta+}{C}(H)-H$ with $H\cdots Br^{\delta-}$ — less stable transition state

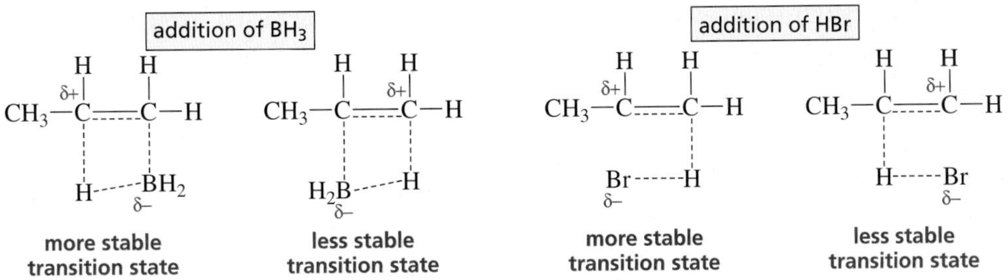

The alkylborane formed in the first step of the reaction reacts with another molecule of alkene to form a dialkylborane, which then reacts with yet another molecule of alkene to form a trialkylborane. In each of these reactions, boron adds to the sp^2 carbon bonded to the greater number of hydrogens and the nucleophilic hydride ion adds to the other sp^2 carbon.

$$CH_3CH{=}CH_2 + \underset{\text{an alkylborane}}{R-BH_2} \longrightarrow \underset{\text{a dialkylborane}}{CH_3\underset{|}{\overset{}{C}}H(H)-CH_2-BH-R}$$

$$CH_3CH{=}CH_2 + \underset{\text{a dialkylborane}}{R-BH(R)} \longrightarrow \underset{\text{a trialkylborane}}{CH_3CH(H)-CH_2-B(R)-R}$$

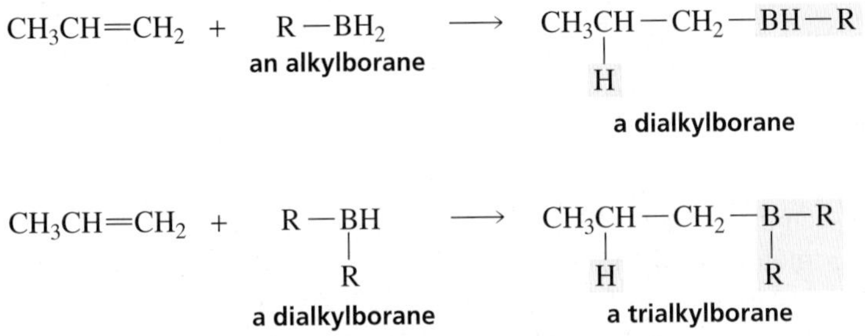

The alkylborane (RBH_2) is a bulkier molecule than BH_3 because R is a larger substituent than H. The dialkylborane with two R groups (R_2BH) is even bulkier than the alkylborane. Thus, there are now two reasons for the alkylborane and the dialkylborane to add to the sp^2 carbon that is bonded to the greater number of hydrogens: first, to achieve the *most stable carbocation-like transition state*, and second, because there is *more room* at this carbon for the bulky group to attach itself. **Steric effects** are space-filling effects. **Steric hindrance** refers to bulky groups at the site of the reaction that make it difficult for the reactants to approach each other. Steric hindrance associated with the alkylborane—and particularly with the dialkylborane—causes the addition to occur at the sp^2 carbon that is bonded to the greater number of hydrogens because that is the least sterically hindered of the two sp^2 carbons. Therefore, in each of the three successive additions to the alkene, boron adds to the sp^2 carbon that is bonded to the greater number of hydrogens and H^- adds to the other sp^2 carbon.

When the hydroboration reaction is over, aqueous sodium hydroxide and hydrogen peroxide are added to the reaction mixture. Notice that both hydroxide ion and hydroperoxide ion are reagents in the reaction:

$$HOOH + HO^- \rightleftharpoons HOO^- + H_2O$$

The end result is replacement of boron by an OH group. Because replacing boron by an OH group is an *oxidation reaction*, the overall reaction is called hydroboration–oxidation. An **oxidation reaction** increases the number of C—O, C—N, or C—X bonds in a compound (where X denotes a halogen), or it decreases the number of C—H bonds.

$$R-B(R)(R) \xrightarrow{HO^-,\ H_2O_2,\ H_2O} 3\ R-OH + BO_3^{3-}$$

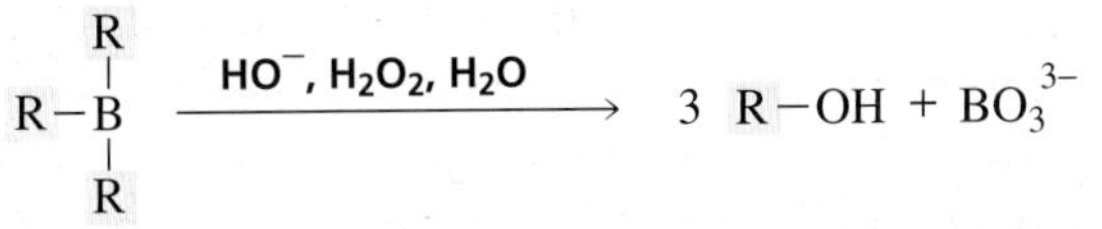

Oxidation decreases the number of C—H bonds or increases the number of C—O, C—N, or C—X bonds.

The mechanism of the oxidation reaction shows that a hydroperoxide ion (a Lewis base) reacts with R_3B (a Lewis acid). Then, a 1,2-alkyl shift displaces a hydroxide ion. These two steps are repeated two more times. Then, hydroxide ion (a Lewis base) reacts with $(RO)_3B$ (a Lewis acid), and an alkoxide ion is eliminated. Protonation of the alkoxide ion forms the alcohol. These three steps are repeated two more times.

$$R_3B + {}^-\ddot{O}OH \longrightarrow R_3\overset{-}{B}{-}O{-}OH \longrightarrow R_2B{-}OR + HO^- \xrightarrow{\text{repeat the two preceding steps two times}} (RO)_3B$$

$$(RO)_3B + {}^-\ddot{O}H \longrightarrow (RO)_3\overset{-}{B}{-}OH \longrightarrow R\ddot{O}^- + (RO)_2B{-}OH \longrightarrow ROH + (RO)_2B{-}O^- \xrightarrow{\text{repeat the three preceding steps two times}} 3\,ROH + BO_3^{3-}$$

We have seen that, in the overall hydroboration–oxidation reaction, 1 mole of BH_3 reacts with 3 moles of alkene to form 3 moles of alcohol. The OH ends up on the sp^2 carbon that was bonded to the greater number of hydrogens because it replaces boron, which was the original electrophile in the reaction.

$$3\,CH_3CH{=}CH_2 + BH_3 \xrightarrow{\text{THF}} (CH_3CH_2CH_2)_3B \xrightarrow[H_2O]{HO^-,\ H_2O_2} 3\,CH_3CH_2CH_2OH + BO_3^{3-}$$

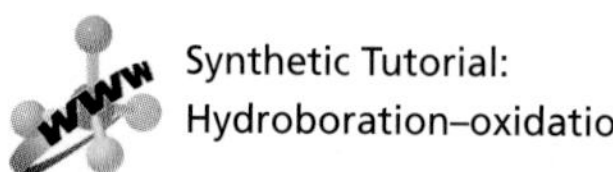
Synthetic Tutorial: Hydroboration–oxidation

Because carbocation intermediates are not formed in the hydroboration reaction, carbocation rearrangements do not occur.

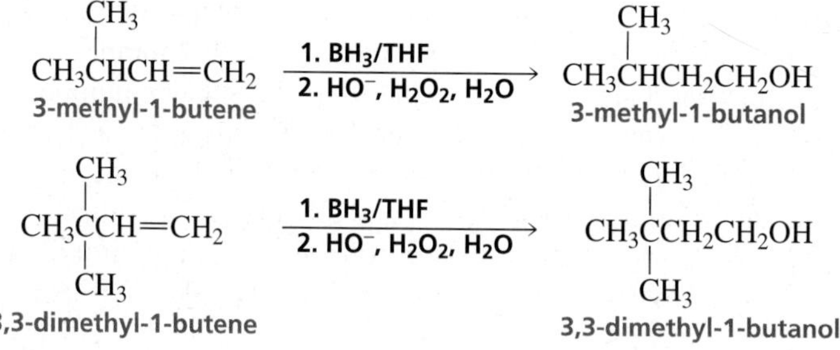

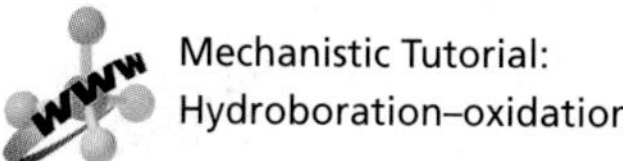
Mechanistic Tutorial: Hydroboration–oxidation

PROBLEM 23◆

How many moles of BH_3 are needed to react with 2 moles of 1-pentene?

PROBLEM 24◆

What product would be obtained from hydroboration–oxidation of the following alkenes?

a. 2-methyl-2-butene

b. 1-methylcyclohexene

PROBLEM-SOLVING STRATEGY

A **carbene** is an unusual carbon-containing species. It has a carbon with a lone pair of electrons and an empty orbital. The empty orbital makes the carbene highly reactive. The simplest carbene, methylene (:CH_2), is generated by heating diazomethane. Propose a mechanism for the following reaction:

$$:\overset{-}{C}H_2-\overset{+}{N}\equiv N \xrightarrow[H_2C=CH_2]{\Delta} \text{cyclopropane} + N_2$$

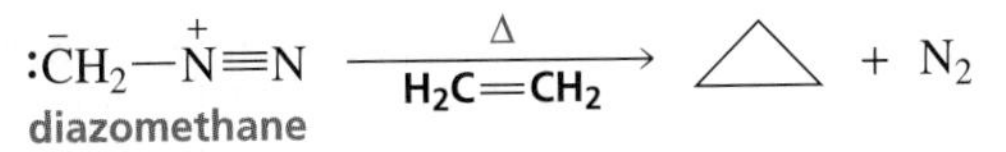

diazomethane

The information provided is all you need to write a mechanism. First, because you know the structure of methylene, you can see that it can be generated by breaking the C—N bond of diazomethane. Second, because methylene has an empty orbital, it is an electrophile and, therefore, will react with ethene (a nucleophile). Now, the question is, What nucleophile reacts with the other sp^2 carbon of the alkene? Because you know that cyclopropane is the product of the reaction, you also know that the nucleophile must be the lone-pair electrons of methylene.

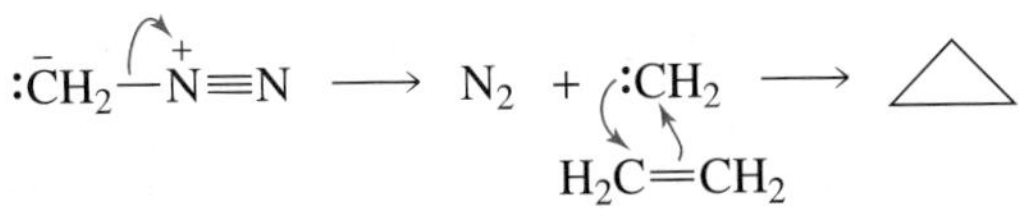

$$:\overset{-}{C}H_2-\overset{+}{N}\equiv N \longrightarrow N_2 + :CH_2 \;\; (H_2C=CH_2) \longrightarrow \text{cyclopropane}$$

(*Note:* Diazomethane is a gas that must be handled with great care because it is both explosive and toxic.)

Now continue on to Problem 25.

PROBLEM 25

Propose a mechanism for the following reaction:

$$CH_2=\underset{CH_3}{C}CH_2\underset{CH_3}{C}HCH_2OH \xrightarrow{H_2SO_4} \text{2,2,4-trimethyltetrahydrofuran}$$

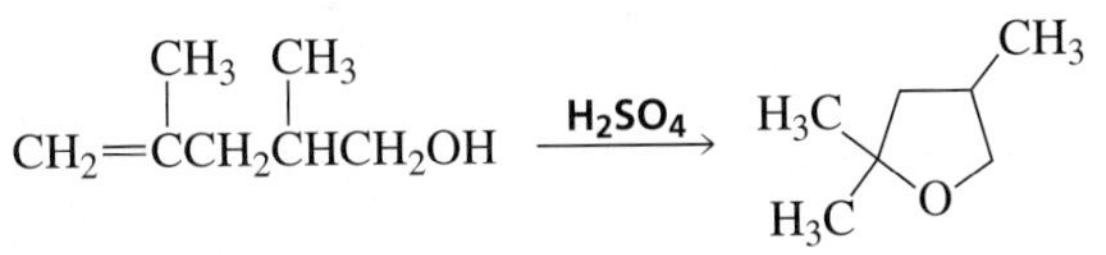

4.10 Addition of Radicals • The Relative Stabilities of Radicals

The addition of HBr to 1-butene forms 2-bromobutane. But what if you wanted to synthesize 1-bromobutane? The formation of 1-bromobutane requires the anti-Markovnikov addition of HBr. If an alkyl peroxide (ROOR) is added to the reaction mixture, the product of the addition reaction will be the desired 1-bromobutane. Thus, the presence of a peroxide causes the *anti-Markovnikov* addition of HBr.

$$\underset{\text{1-butene}}{CH_3CH_2CH=CH_2} + HBr \longrightarrow \underset{\text{2-bromobutane}}{CH_3CH_2\underset{|}{\overset{Br}{C}}HCH_3}$$

$$\underset{\text{1-butene}}{CH_3CH_2CH=CH_2} + HBr \xrightarrow{\text{peroxide}} \underset{\text{1-bromobutane}}{CH_3CH_2CH_2CH_2Br}$$

A peroxide reverses the order of addition because it changes the mechanism of the reaction in a way that causes $Br^{\cdot}$ to be the electrophile. Markovnikov's rule is not followed because it applies only when the electrophile is a hydrogen. The general rule—that the electrophile adds to the sp^2 carbon bonded to the greater number of hydrogens—*is* followed, however, because $Br^{\cdot}$ is the electrophile when HBr adds to an alkene in the presence of a peroxide.

When a bond breaks such that both of its electrons stay with one of the atoms, the process is called **heterolytic bond cleavage** or **heterolysis**. When a bond breaks such that each of the atoms retains one of the bonding electrons, the process is called **homolytic bond cleavage** or **homolysis**. Remember that an arrowhead with two barbs signifies the movement of two electrons, whereas an arrowhead with one barb—sometimes called a fishhook—signifies the movement of a single electron.

Radical stability: tertiary > secondary > primary

An arrowhead with two barbs signifies the movement of two electrons.

heterolytic bond cleavage

$$H-\ddot{B}r: \longrightarrow H^+ + :\ddot{B}r:^-$$

homolytic bond cleavage

$$H-\ddot{B}r: \longrightarrow H\cdot + \cdot\ddot{B}r:$$

An alkyl peroxide can be used to reverse the order of addition of H and Br to an alkene. The akyl peroxide contains a weak oxygen–oxygen single bond that is readily broken homolytically in the presence of light or heat to form *radicals*. A **radical** (also called a **free radical**) is a species with an unpaired electron.

$$R\ddot{O}-\ddot{O}R \xrightarrow[\Delta]{\text{light or}} 2\,R\ddot{O}\cdot$$

an alkyl peroxide — alkoxyl radicals

A radical is highly reactive because it seeks an electron to complete its octet. The alkoxyl radical completes its octet by removing a hydrogen atom from a molecule of HBr, forming a bromine radical.

$$R-\ddot{O}\cdot + H-\ddot{B}r: \longrightarrow R-\ddot{O}-H + \cdot\ddot{B}r:$$

a bromine radical

The bromine radical now seeks an electron to complete its octet. Because the double bond of an alkene is electron rich, the bromine radical completes its octet by combining with one of the electrons of the π bond of the alkene to form a C—Br bond. The second electron of the π bond is the unpaired electron in the resulting alkyl radical. If the bromine radical adds to the sp^2 carbon of 1-butene that is bonded to the greater number of hydrogens, a secondary alkyl radical is formed. If the bromine radical adds to the other sp^2 carbon, a primary alkyl radical is formed. Like carbocations, radicals are stabilized by electron-donating alkyl groups, so a **tertiary alkyl radical** is more stable than a **secondary alkyl radical**, which in turn is more stable than a **primary alkyl radical**. The bromine radical, therefore, adds to the sp^2 carbon that is bonded to the greater number of hydrogens, thereby forming the more stable (in this case) secondary radical. The alkyl radical that is formed removes a hydrogen atom from another molecule of HBr to produce a molecule of the alkyl halide product and another bromine radical. Because the first species that adds to the alkene is a radical ($Br^{\cdot}$), the addition of HBr in the presence of a peroxide is called a **radical addition reaction**.

$$:\ddot{B}r\cdot + CH_2{=}CHCH_2CH_3 \longrightarrow \underset{:\ddot{B}r:}{\underset{|}{C}H_2}\dot{C}HCH_2CH_3$$

an alkyl radical

An arrowhead with one barb signifies the movement of one electron.

$$\underset{Br}{\underset{|}{C}H_2}\dot{C}HCH_2CH_3 + H-\ddot{B}r: \longrightarrow \underset{Br}{\underset{|}{C}H_2}-\underset{H}{\underset{|}{C}H}CH_2CH_3 + \cdot\ddot{B}r:$$

When HBr reacts with an alkene in the absence of a peroxide, the electrophile—the first species to add to the alkene—is H^+. In the presence of a peroxide, the electrophile is $Br^{\cdot}$. In both cases, the electrophile adds to the sp^2 carbon that is bonded to the greater number of hydrogens, so both reactions follow the general rule for electrophilic addition reactions: *The electrophile adds to the* sp^2 *carbon that is bonded to the greater number of hydrogens.*

Because the addition of HBr in the presence of a peroxide forms a radical intermediate rather than a carbocation intermediate, the intermediate does not rearrange. Radicals do not rearrange as readily as carbocations.

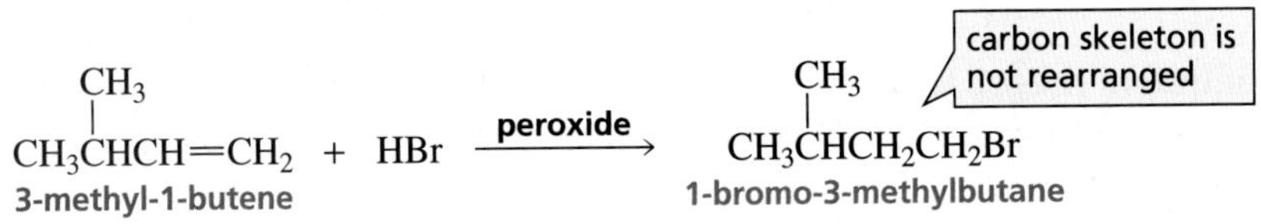

Radical intermediates do not rearrange.

$$CH_3CH(CH_3)CH{=}CH_2 + HBr \xrightarrow{\text{peroxide}} CH_3CH(CH_3)CH_2CH_2Br$$

3-methyl-1-butene → 1-bromo-3-methylbutane (carbon skeleton is not rearranged)

As just mentioned, the relative stabilities of primary, secondary, and tertiary alkyl radicals are in the same order as the relative stabilities of primary, secondary, and tertiary carbocations. However, energy differences between the radicals are quite a bit smaller than between the carbocations.

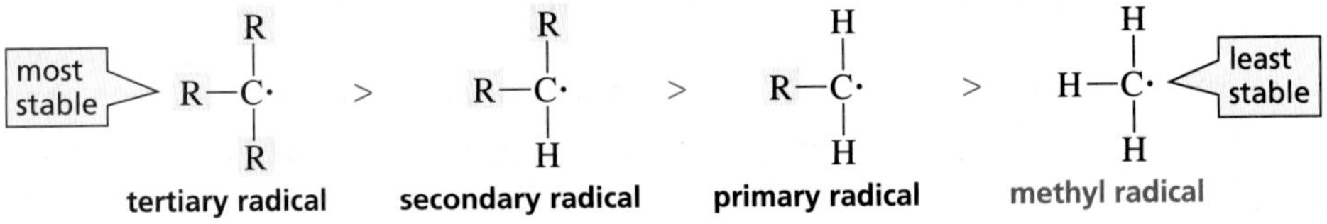

most stable → $R_3C\cdot$ (tertiary radical) > $R_2HC\cdot$ (secondary radical) > $RH_2C\cdot$ (primary radical) > $H_3C\cdot$ (methyl radical) ← least stable

The relative stabilities of primary, secondary, and tertiary alkyl radicals are reflected in the transition states leading to their formation (Section 4.3). Consequently, the more stable the radical, the less energy is required to make it. This explains why the bromine radical adds to the sp^2 carbon of 1-butene that is bonded to the greater number of hydrogens to form a secondary alkyl radical, rather than adding to the other sp^2 carbon to form a primary alkyl radical. The secondary radical is more stable than the primary radical; therefore, the energy barrier to its formation is lower.

The following mechanism for the addition of HBr to an alkene in the presence of a peroxide involves seven steps. The steps can be divided into initiation steps, propagation steps, and termination steps:

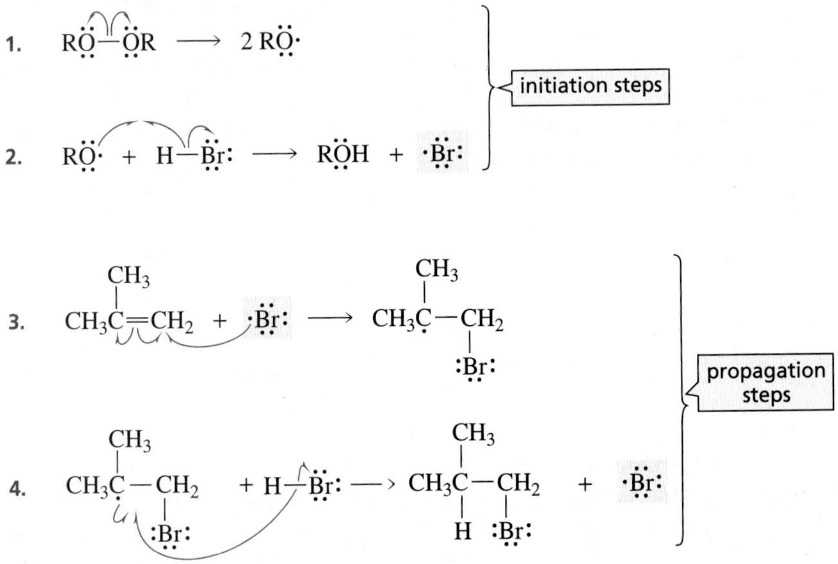

initiation steps

1. $R\ddot{O}{-}\ddot{O}R \longrightarrow 2\,R\ddot{O}\cdot$
2. $R\ddot{O}\cdot + H{-}\ddot{Br}: \longrightarrow R\ddot{O}H + \cdot\ddot{Br}:$

propagation steps

3. $CH_3C(CH_3){=}CH_2 + \cdot\ddot{Br}: \longrightarrow CH_3\dot{C}(CH_3){-}CH_2\ddot{Br}:$
4. $CH_3\dot{C}(CH_3){-}CH_2\ddot{Br}: + H{-}\ddot{Br}: \longrightarrow CH_3CH(CH_3){-}CH_2\ddot{Br}: + \cdot\ddot{Br}:$

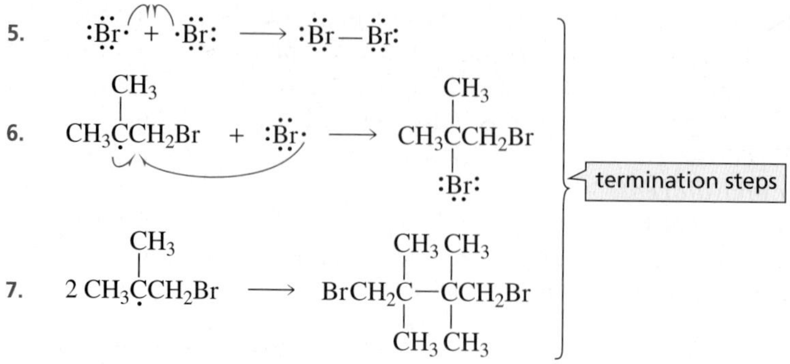

5. $:\ddot{Br}\cdot + \cdot\ddot{Br}: \longrightarrow :\ddot{Br}-\ddot{Br}:$

6. $CH_3C(CH_3)CH_2Br$ (radical) $+ :\ddot{Br}\cdot \longrightarrow CH_3C(CH_3)(Br)CH_2Br$

7. $2\ CH_3C(CH_3)CH_2Br$ (radical) $\longrightarrow BrCH_2C(CH_3)_2-C(CH_3)_2CH_2Br$

termination steps

- **Initiation Steps.** The first step is an **initiation step** because it creates radicals. The second step is also an initiation step because it forms the chain-propagating radical ($Br^{\cdot}$).
- **Propagation Steps.** Steps 3 and 4 are **propagation steps**. In step 3, a radical ($Br^{\cdot}$) reacts to produce another radical. In step 4, the radical produced in the first propagation step reacts to form the radical ($Br^{\cdot}$) that was the reactant in the first propagation step. The two propagation steps are repeated over and over. Hence, the reaction is called a **radical chain reaction**. A propagation step is a step that propagates the chain.
- **Termination Steps.** Steps 5, 6, and 7 are **termination steps**. In a termination step, two radicals combine to produce a molecule in which all the electrons are paired, thus ending the role of those radicals in the radical chain reaction. Any two radicals present in the reaction mixture can combine in a termination step, so radical reactions produce a mixture of products.

PROBLEM 26

Write out the propagation steps that occur when HBr adds to 1-methylcyclohexene in the presence of a peroxide.

An alkyl peroxide is a **radical initiator** because it creates radicals. Without a peroxide, the preceding radical reaction would not occur. Any reaction that occurs in the presence of a radical initiator, but does not occur in its absence, must take place by a mechanism that involves radicals as intermediates. Any compound that can readily undergo homolysis—dissociate to form radicals—can act as a radical initiator. Examples of radical initiators are shown in Table 28.3.

Whereas radical initiators cause radical reactions to occur, **radical inhibitors** have the opposite effect: They trap radicals as they are formed, preventing reactions that take place by mechanisms involving radicals. How radical inhibitors trap radicals is discussed in Section 9.8.

A peroxide has no effect on the addition of HCl or HI to an alkene. In the presence of a peroxide, addition occurs just as it does in the absence of a peroxide.

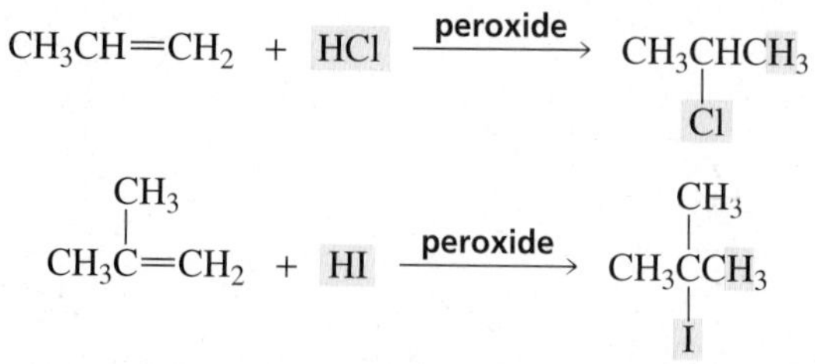

$$CH_3CH{=}CH_2 + HCl \xrightarrow{\text{peroxide}} CH_3CH(Cl)CH_3$$

$$CH_3C(CH_3){=}CH_2 + HI \xrightarrow{\text{peroxide}} CH_3C(CH_3)(I)CH_3$$

Why is the **peroxide effect** observed for the addition of HBr, but not for the addition of HCl or HI? This question can be answered by calculating the $\Delta H°$ for the two propagation steps in the radical chain reaction (using the bond dissociation energies in Table 3.1).

$Cl\cdot + CH_2{=}CH_2 \longrightarrow ClCH_2\dot{C}H_2$ $\Delta H° = 63 - 85 = -22$ kcal/mol (or -91 kJ/mol) exothermic

$ClCH_2\dot{C}H_2 + HCl \longrightarrow ClCH_2CH_3 + Cl\cdot$ $\Delta H° = 103 - 101 = +2$ kcal/mol (or $+8$ kJ/mol)

$Br\cdot + CH_2{=}CH_2 \longrightarrow BrCH_2\dot{C}H_2$ $\Delta H° = 63 - 72 = -9$ kcal/mol (or -38 kJ/mol)

$BrCH_2\dot{C}H_2 + HBr \longrightarrow BrCH_2CH_3 + Br\cdot$ $\Delta H° = 87 - 101 = -14$ kcal/mol (or -59 kJ/mol)

exothermic

$I\cdot + CH_2{=}CH_2 \longrightarrow ICH_2\dot{C}H_2$ $\Delta H° = 63 - 57 = +6$ kcal/mol (or $+25$ kJ/mol)

$ICH_2\dot{C}H_2 + HI \longrightarrow ICH_2CH_3 + I\cdot$ $\Delta H° = 71 - 101 = -30$ kcal/mol (or -126 kJ/mol) exothermic

For the radical addition of HCl, the first propagation step is exothermic and the second is endothermic. For the radical addition of HI, the first propagation step is endothermic and the second is exothermic. Only for the radical addition of HBr are both propagation steps exothermic. In a radical reaction, the steps that propagate the chain reaction compete with the steps that terminate it. Termination steps are always exothermic, because only bond making (and no bond breaking) occurs. Therefore, only when both propagation steps are exothermic can propagation compete with termination. When HCl or HI adds to an alkene in the presence of a peroxide, any chain reaction that is initiated is terminated rather than propagated because propagation cannot compete with termination. Consequently, the radical chain reaction does not occur, and all we have is ionic addition (H^+ followed by Cl^- or I^-).

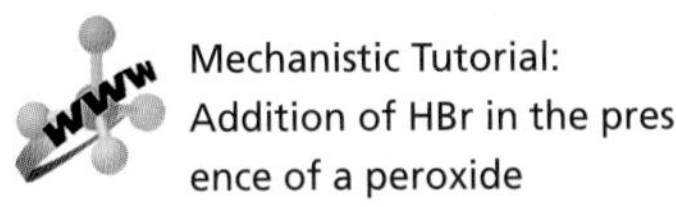

Mechanistic Tutorial: Addition of HBr in the presence of a peroxide

4.11 Addition of Hydrogen • The Relative Stabilities of Alkenes

In the presence of a metal catalyst such as platinum, palladium, or nickel, hydrogen (H_2) adds to the double bond of an alkene to form an alkane. Without the catalyst, the energy barrier to the reaction would be enormous because the H—H bond is so strong (Table 3.1). The catalyst decreases the energy of activation by breaking the H—H bond. Platinum and palladium are used in a finely divided state adsorbed on charcoal (Pt/C, Pd/C). The platinum catalyst is frequently used in the form of PtO_2, which is known as Adams catalyst.

$$CH_3CH{=}CHCH_3 + H_2 \xrightarrow{Pt/C} CH_3CH_2CH_2CH_3$$

2-butene → butane

$$CH_3\overset{\overset{\large CH_3}{|}}{C}{=}CH_2 + H_2 \xrightarrow{Pd/C} CH_3\overset{\overset{\large CH_3}{|}}{C}HCH_3$$

2-methylpropene → 2-methylpropane

cyclohexene + H_2 $\xrightarrow{Ni}$ cyclohexane

The addition of hydrogen is called **hydrogenation**. Because the preceding reactions require a catalyst, they are examples of **catalytic hydrogenation**. The metal catalysts are insoluble in the reaction mixture and therefore are classified as **heterogeneous catalysts**. A heterogeneous catalyst can easily be separated from the reaction mixture by filtration. It can then be recycled, which is an important property, since metal catalysts tend to be expensive.

Roger Adams (1889–1971) *was born in Boston. He received a Ph.D. from Harvard University and was a professor of chemistry at the University of Illinois. He and Sir Alexander Todd (Section 27.1) clarified the structure of tetrahydrocannabinol (THC), the active ingredient of the marijuana plant. Adams's research showed that the test commonly used at that time by the Federal Bureau of Narcotics to detect marijuana was actually detecting an innocuous companion compound.*

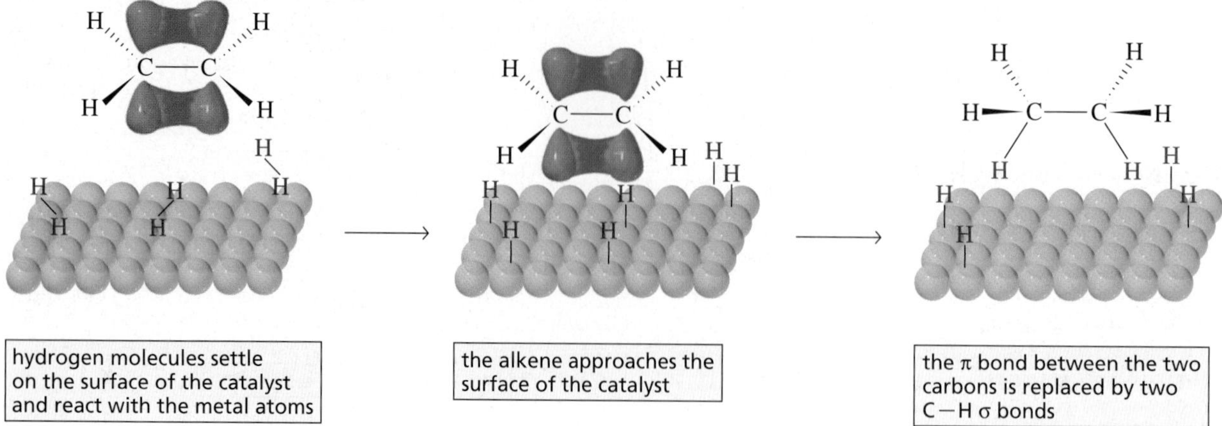
hydrogen molecules settle on the surface of the catalyst and react with the metal atoms
the alkene approaches the surface of the catalyst
the π bond between the two carbons is replaced by two C—H σ bonds

▲ **Figure 4.6**
Catalytic hydrogenation of an alkene.

The details of the mechanism of catalytic hydrogenation are not completely understood. We know that hydrogen is adsorbed on the surface of the metal and that the alkene complexes with the metal by overlapping its own *p* orbitals with vacant orbitals of the metal. Breaking the π bond of the alkene and the σ bond of H_2 and forming the C—H σ bonds all occur on the surface of the metal. The alkane product diffuses away from the metal surface as it is formed (Figure 4.6).

The heat released in a hydrogenation reaction is called the **heat of hydrogenation**. It is customary to give it a positive value. Hydrogenation reactions, however, are exothermic (they have negative $\Delta H°$ values). So the heat of hydrogenation is the positive value of the $\Delta H°$ of the reaction.

Reaction	heat of hydrogenation	$\Delta H°$ kcal/mol	$\Delta H°$ kJ/mol
$CH_3C(CH_3){=}CHCH_3 + H_2 \xrightarrow{\text{Pt/C}} CH_3CH(CH_3)CH_2CH_3$ (2-methyl-2-butene)	26.9 kcal/mol	−26.9	−113
$CH_2{=}C(CH_3)CH_2CH_3 + H_2 \xrightarrow{\text{Pt/C}} CH_3CH(CH_3)CH_2CH_3$ (2-methyl-1-butene)	28.5 kcal/mol	−28.5	−119
$CH_3CH(CH_3)CH{=}CH_2 + H_2 \xrightarrow{\text{Pt/C}} CH_3CH(CH_3)CH_2CH_3$ (3-methyl-1-butene)	30.3 kcal/mol	−30.3	−127

Because we do not know the precise mechanism of a hydrogenation reaction, we cannot draw a reaction coordinate diagram for it. We can, however, draw a diagram showing the relative energies of the reactants and products (Figure 4.7). The preceding three catalytic hydrogenation reactions all form the same alkane product, so the energy of the *product* is the same for each reaction. The three reactions, however, have different heats of hydrogenation, so the three *reactants* must have different energies. For example, 3-methyl-1-butene releases the most heat, so it must be the *least* stable (have the greatest energy) of the three alkenes. In contrast, 2-methyl-2-butene releases the least heat, so it must be the *most* stable of the three alkenes. Notice that the greater the stability of a compound, the lower is its energy and the smaller is its heat of hydrogenation.

The most stable alkene has the smallest heat of hydrogenation.

If you look at the structures of the three alkene reactants in Figure 4.7, you will see that the most stable alkene has two alkyl substituents bonded to one of the sp^2 carbons and one alkyl substituent bonded to the other sp^2 carbon, for a total of three alkyl substituents (three methyl groups) bonded to its two sp^2 carbons. The alkene of intermediate stability has a total of two alkyl substituents (a methyl group and an ethyl group)

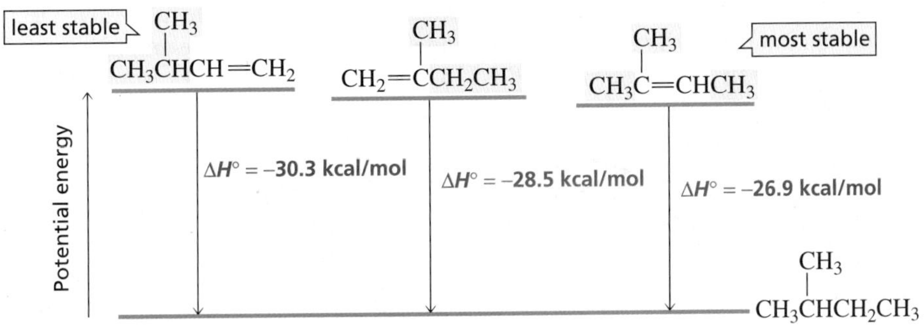

◀ **Figure 4.7**
The relative energy levels (stabilities) of three alkenes that can be catalytically hydrogenated to 2-methylbutane.

bonded to its sp^2 carbons, and the least stable of the three alkenes has only one alkyl substituent (an isopropyl group) bonded to its sp^2 carbons. Thus, it is apparent that alkyl substituents bonded to the sp^2 carbons of an alkene have a stabilizing effect on the alkene. We can, therefore, make the following statement: *The more alkyl substituents bonded to the* sp^2 *carbons of an alkene, the greater is its stability.* (Some students find it easier to look at the number of hydrogens bonded to the sp^2 carbons. In terms of hydrogens, the statement is, *the fewer hydrogens bonded to the* sp^2 *carbons of an alkene, the greater is its stability.*)

The fewer hydrogens bonded to the sp^2 carbons of an alkene, the more stable it is.

Alkyl substituents stabilize both alkenes *and* carbocations.

relative stabilities of alkyl-substituted alkenes

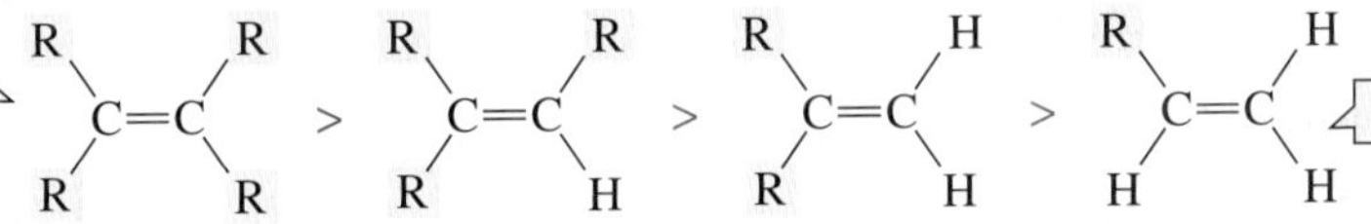

most stable $R_2C{=}CR_2$ > $R_2C{=}CHR$ > $R_2C{=}CH_2$ > $RHC{=}CH_2$ least stable

PROBLEM 27◆

The same alkane is obtained from the catalytic hydrogenation of both alkene A and alkene B. The heat of hydrogenation of alkene A is 29.8 kcal/mol (125 kJ/mol), and the heat of hydrogenation of alkene B is 31.4 kcal/mol (131 kJ/mol). Which alkene is more stable?

PROBLEM 28◆

a. Which of the following compounds is the most stable?

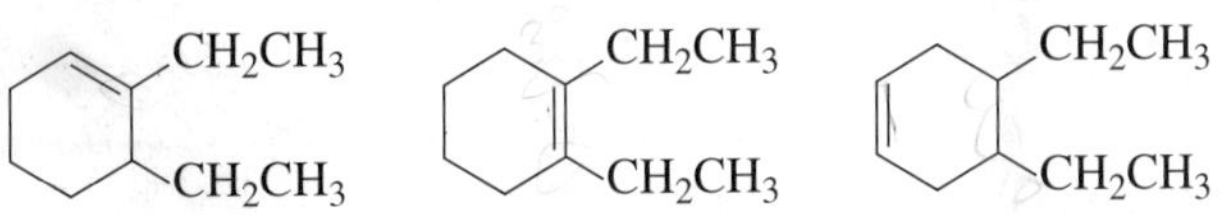

[three isomeric diethylcyclohexenes, each bearing CH_2CH_3 and CH_2CH_3 groups]

b. Which is the least stable?

c. Which has the smallest heat of hydrogenation?

Platinum and palladium are expensive metals, so the accidental finding by **Paul Sabatier (1854–1941)** *that nickel, a much cheaper metal, can catalyze hydrogenation reactions made hydrogenation a feasible large-scale industrial process. The conversion of plant oils to margarine is one such hydrogenation reaction. Sabatier was born in France and was a professor at the University of Toulouse. He shared the 1912 Nobel Prize in chemistry with Victor Grignard (p. 468).*

Both *trans*-2-butene and *cis*-2-butene have two alkyl groups bonded to their sp^2 carbons, but *trans*-2-butene has a smaller heat of hydrogenation. This means that the trans isomer, in which the large substituents are farther apart, is more stable than the cis isomer, in which the large substituents are closer together.

Reaction	heat of hydrogenation	$\Delta H°$ kcal/mol	$\Delta H°$ kJ/mol
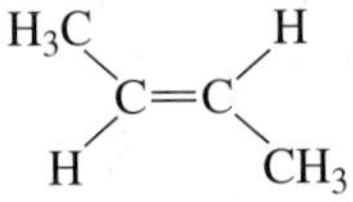 *trans*-2-butene ($H_3C(H)C{=}C(H)CH_3$) + H_2 $\xrightarrow{Pd/C}$ $CH_3CH_2CH_2CH_3$	27.6	−27.6	−115
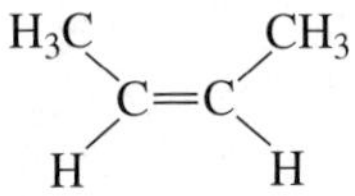 *cis*-2-butene ($H_3C(H)C{=}C(H)CH_3$) + H_2 $\xrightarrow{Pd/C}$ $CH_3CH_2CH_2CH_3$	28.6	−28.6	−120

When the large substituents are on the same side of the molecule, their electron clouds can interfere with each other, causing strain in the molecule and making it less stable. You saw in Section 2.11 that this kind of strain is called *steric strain.* When the large substituents are on opposite sides of the molecule, their electron clouds cannot interact, and the molecule has less steric strain.

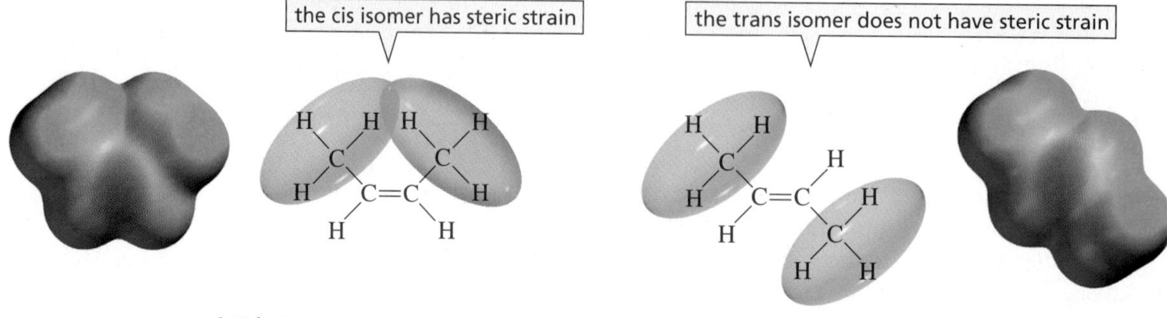
the cis isomer has steric strain
the trans isomer does not have steric strain

cis-2-butene

trans-2-butene

The heat of hydrogenation of *cis*-2-butene, in which the two alkyl substituents are on the *same side* of the double bond, is similar to that of 2-methylpropene, in which the two alkyl substituents are on the *same carbon*. The three dialkyl-substituted alkenes are all *less* stable than a trialkyl-substituted alkene and are all *more* stable than a monoalkyl-substituted alkene.

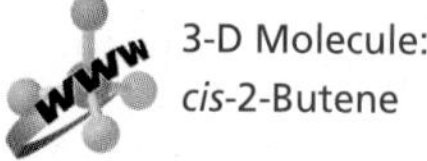

3-D Molecule: *cis*-2-Butene

3-D Molecule: *trans*-2-Butene

relative stabilities of dialkyl-substituted alkenes

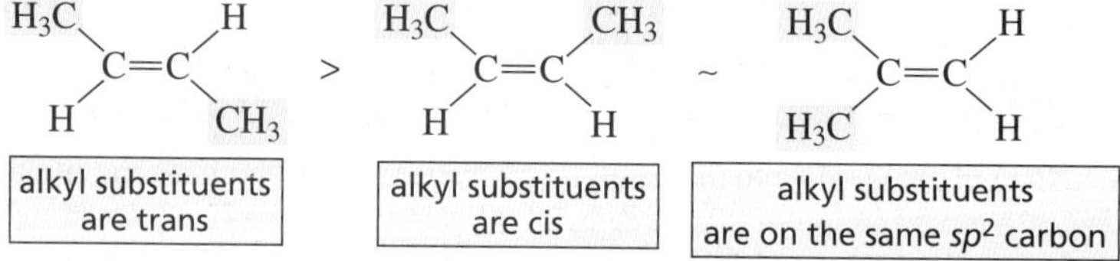
alkyl substituents are trans
>
alkyl substituents are cis
~
alkyl substituents are on the same sp^2 carbon

PROBLEM 29◆

Rank the following compounds in order of decreasing stability:
trans-3-hexene, *cis*-3-hexene, 1-hexene, *cis*-2,5-dimethyl-3-hexene

4.12 Reactions and Synthesis

This chapter has been concerned with the reactions of alkenes. You have seen why alkenes react, the kinds of reagents with which they react, the mechanisms by which the reactions occur, and the products that are formed. It is important to remember that when you are studying reactions, you are simultaneously studying synthesis. When you learn that compound A reacts with a certain reagent to form compound B, you are learning not only about the reactivity of A, but also about one way that compound B can be synthesized.

$$A \longrightarrow B$$

For example, you have seen that alkenes can add many different reagents and that, as a result of adding these reagents, compounds such as alkyl halides, vicinal dihalides, halohydrins, alcohols, ethers, and alkanes are synthesized.

Although you have seen how alkenes react and have learned about the kinds of compounds that are synthesized when alkenes undergo reactions, you have not yet seen how alkenes are synthesized. Reactions of alkenes involve the *addition* of atoms (or groups of atoms) to the two sp^2 carbons of the double bond. Reactions that lead to the synthesis of alkenes are exactly the opposite—they involve the *elimination* of atoms (or groups of atoms) from two adjacent sp^3 carbons.

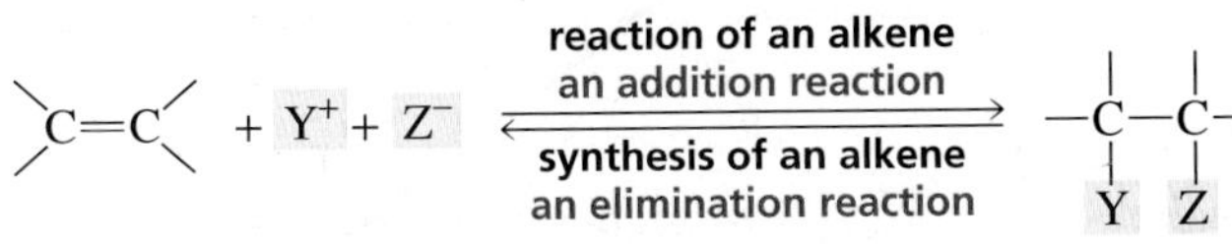

$$>C=C< + Y^+ + Z^- \underset{\text{synthesis of an alkene / an elimination reaction}}{\overset{\text{reaction of an alkene / an addition reaction}}{\rightleftharpoons}} -\overset{|}{\underset{Y}{C}}-\overset{|}{\underset{Z}{C}}-$$

You will learn how alkenes are synthesized when you study compounds that undergo elimination reactions. The various reactions that result in the synthesis of alkenes are listed in Appendix IV.

PROBLEM 30 SOLVED

Starting with an alkene, indicate how each of the following compounds can be synthesized:

a. 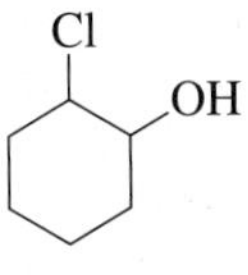 Cl, OH (on cyclohexane) b. 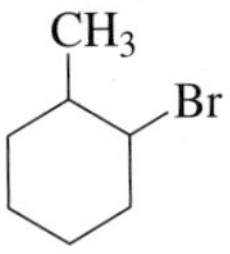 CH_3, Br (on cyclohexane) c. 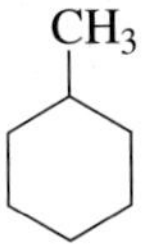 CH_3 (on cyclohexane)

SOLUTION

a. The only alkene that can be used for this synthesis is cyclohexene. To get the desired substituents on the ring, cyclohexene must react with Cl_2 in an aqueous solution so that water will be the nucleophile.

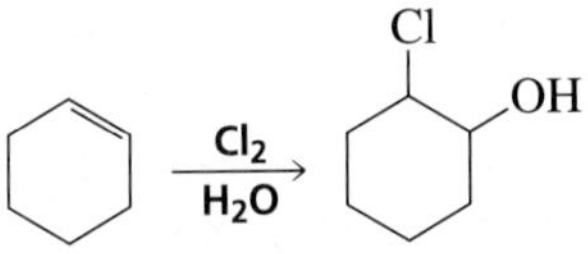

$$\text{cyclohexene} \xrightarrow[H_2O]{Cl_2} \text{Cl, OH-cyclohexane}$$

b. The alkene that should be used here is 1-methylcyclohexene. To get the substituents in the desired locations, the electrophile in the reaction must be a bromine radical. Therefore, the reagents required to react with 1-methylcyclohexene are HBr and a peroxide.

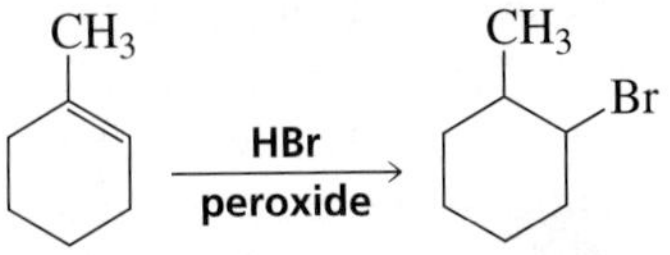

$$CH_3\text{-cyclohexene} \xrightarrow[\text{peroxide}]{HBr} CH_3, Br\text{-cyclohexane}$$

c. In order to synthesize an alkane from an alkene, the alkene must undergo catalytic hydrogenation. Several alkenes could be used for this synthesis.

$$CH_2 \text{ or } CH_3 \text{ or } CH_3 \text{ or } CH_3 \xrightarrow[Pd/C]{H_2} CH_3$$

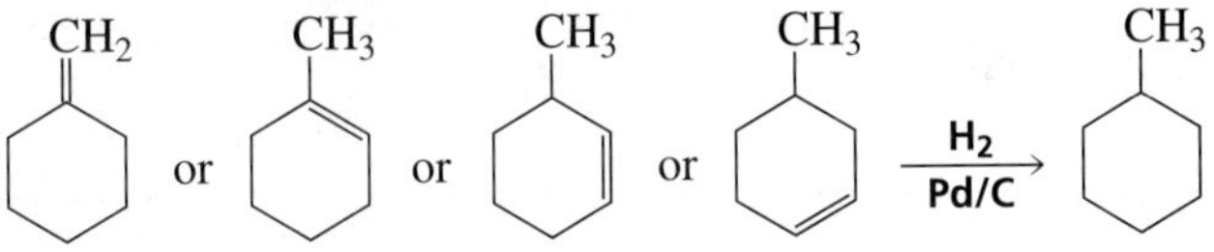

PROBLEM 31

Why should 3-methylcyclohexene not be used as the starting material in Problem 30b?

PROBLEM 32

Starting with an alkene, indicate how each of the following compounds can be synthesized:

a. CH_3CHOCH_3 (with CH_3 on the CH)

b. $CH_3CH_2CHCHCH_3$ (with Br on each CH)

c. 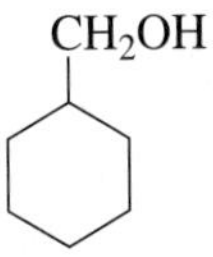 CH_2OH (on cyclohexane)

d. 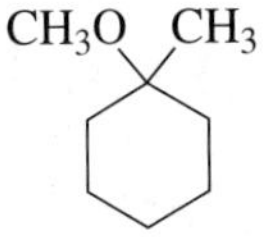 CH_3O CH_3 (on cyclohexane)

e. 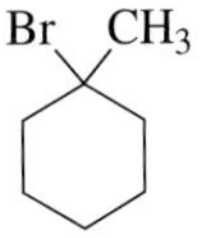 Br CH_3 (on cyclohexane)

f. 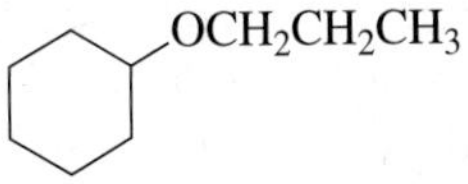 $OCH_2CH_2CH_3$ (on cyclohexane)

Summary

Alkenes undergo **electrophilic addition reactions**. Each reaction starts with the addition of an electrophile to one of the sp^2 carbons and concludes with the addition of a nucleophile to the other sp^2 carbon. In all electrophilic addition reactions, the *electrophile* adds to the sp^2 carbon bonded to the greater number of hydrogens. **Markovnikov's rule** states that the hydrogen adds to the sp^2 carbon of the alkene bonded to the greater number of hydrogens. While all addition reactions add the electrophile to the sp^2 carbon bonded to the greater number of hydrogens, they all do not follow Markovnikov's rule, because hydrogen is not always the electrophile. **Hydroboration–oxidation** and the **addition of HBr** in the presence of a **peroxide** are **anti-Markovnikov additions.**

The addition of hydrogen halides and the acid-catalyzed addition of water and alcohols form **carbocation intermediates. Hyperconjugation** causes **tertiary carbocations** to be more stable than **secondary carbocations**, which are more stable than **primary carbocations**. A carbocation will rearrange if it becomes more stable as a result of the rearrangement. **Carbocation rearrangements** occur by **1,2-hydride shifts, 1,2-methyl shifts**, and **ring expansion.** HBr in the presence of a peroxide forms a **radical intermediate**. Radical intermediates do not rearrange. **Hydroboration–oxidation** is a **concerted reaction** and does not form an **intermediate**.

Oxymercuration, alkoxymercuration, and the addition of Br_2 and Cl_2 form **cyclic intermediates**. Oxymercuration and alkoxymercuration are followed by a reduction reaction. **Reduction** increases the number of C—H bonds or decreases the number of C—O, C—N, or C—X bonds (where X denotes a halogen). **Hydroboration** is followed by an oxidation reaction. **Oxidation** decreases the number of C—H bonds or increases the number of C—O, C—N, or C—X bonds (where, again, X denotes a halogen).

The **Hammond postulate** states that a **transition state** is more similar in structure to the species to which it is more similar in energy. Thus, the more stable product will have the more stable transition state and will lead to the major product of the reaction. **Regioselectivity** is the preferential formation of one **constitutional isomer** over another.

In **heterolytic bond cleavage,** a bond breaks such that both electrons in the bond stay with one of the atoms; in **homolytic bond cleavage,** a bond breaks such that each of the atoms retains one of the bonding electrons. An alkyl peroxide is a **radical initiator** because it creates radicals. **Radical addition reactions** are **chain reactions** with **initiation, propagation,** and **termination steps.** Radicals are stabilized by electron-donating alkyl groups. Thus, a **tertiary alkyl radical** is more stable than a **secondary alkyl radical**, which is more stable than a **primary alkyl radical.** A **peroxide** reverses the order of addition of H and Br because it causes $Br^{\cdot}$, instead of H^+, to be the electrophile. The peroxide **effect** is observed only for the addition of HBr.

The addition of H_2 to a reaction is called **hydrogenation.** The **heat of hydrogenation** is the heat released in a hydrogenation reaction. The *greater* the *stability* of a compound, the *lower* is its *energy* and the *smaller* is its *heat of hydrogenation*. The more alkyl substituents bonded to the sp^2 carbons of an alkene, the greater is its stability. Hence, carbocations, alkyl radicals, and alkenes are all stabilized by **alkyl substituents. Trans alkenes** are more stable than **cis alkenes** because of steric strain.

Electrophilic addition reactions of alkenes lead to the **synthesis** of **alkyl halides, vicinal dihalides, halohydrins, alcohols, ethers,** and **alkanes.**

Summary of Reactions

As you review the reactions of alkenes, keep in mind the feature that is common to all of them: The first step of each reaction is the addition of an electrophile to the sp^2 carbon of the alkene that is bonded to the greater number of hydrogens.

1. Electrophilic addition reactions

a. Addition of hydrogen halides (H^+ is the electrophile; Section 4.1)

$$RCH{=}CH_2 + HX \longrightarrow \underset{\displaystyle X}{RCHCH_3}$$

HX = HF, HCl, HBr, HI

b. Addition of hydrogen bromide in the presence of a peroxide ($Br^{\cdot}$ is the electrophile; Section 4.10)

$$RCH{=}CH_2 + HBr \xrightarrow{\text{peroxide}} RCH_2CH_2Br$$

c. Addition of halogen (Br^+ or Cl^+ is the electrophile; Section 4.7)

$$RCH{=}CH_2 + Cl_2 \xrightarrow[CH_2Cl_2]{} \underset{\displaystyle Cl}{RCHCH_2Cl}$$

$$RCH{=}CH_2 + Br_2 \xrightarrow[CH_2Cl_2]{} \underset{\displaystyle Br}{RCHCH_2Br}$$

$$RCH{=}CH_2 + Br_2 \xrightarrow[H_2O]{} \underset{\displaystyle OH}{RCHCH_2Br}$$

d. Acid-catalyzed addition of water and alcohols (H^+ is the electrophile; Section 4.5)

$$RCH{=}CH_2 + H_2O \overset{H^+}{\rightleftharpoons} \underset{\displaystyle OH}{RCHCH_3}$$

$$RCH{=}CH_2 + CH_3OH \overset{H^+}{\rightleftharpoons} \underset{\displaystyle OCH_3}{RCHCH_3}$$

e. Addition of water and alcohols: oxymercuration–reduction and alkoxymercuration–reduction (Hg is the electrophile and is replaced by H in the second step; Section 4.8)

$$RCH{=}CH_2 \xrightarrow[\text{2. NaBH}_4]{\text{1. Hg(OAc)}_2\text{, H}_2\text{O,THF}} \underset{\displaystyle OH}{RCHCH_3}$$

$$RCH{=}CH_2 \xrightarrow[\text{2. NaBH}_4]{\text{1. Hg(O}_2\text{CCF}_3)_2\text{, CH}_3\text{OH}} \underset{\displaystyle OCH_3}{RCHCH_3}$$

f. Hydroboration–oxidation (B is the electrophile and is replaced in the second step by OH; Section 4.9)

$$RCH{=}CH_2 \xrightarrow[\text{2. HO}^-\text{, H}_2\text{O}_2\text{, H}_2\text{O}]{\text{1. BH}_3\text{/THF}} RCH_2CH_2OH$$

2. Addition of hydrogen ($H^{\cdot}$ is the electrophile; Section 4.11)

$$RCH{=}CH_2 + H_2 \xrightarrow{\text{Pd/C, Pt/C, or Ni}} RCH_2CH_3$$

Key Terms

acid-catalyzed reaction (p. 152)
alkoxymercuration–reduction (p. 162)
carbene (p. 167)
carbocation rearrangement (p. 155)
catalyst (p. 152)
catalytic hydrogenation (p. 171)
concerted reaction (p. 164)
constitutional isomers (p. 148)
dimer (p. 164)
electrophilic addition reaction (p. 141)
free radical (p. 168)
halohydrin (p. 159)
Hammond postulate (p. 145)
heat of hydrogenation (p. 172)
heterogeneous catalyst (p. 171)
heterolysis (p. 168)
heterolytic bond cleavage (p. 168)
homolysis (p. 168)
homolytic bond cleavage (p. 168)
hydration (p. 151)
hydroboration–oxidation (p. 163)
hydrogenation (p. 171)
hyperconjugation (p. 144)
initiation step (p. 170)
Markovnikov's rule (p. 148)
mechanism of the reaction (p. 142)
1,2-hydride shift (p. 155)
1,2-methyl shift (p. 155)
oxymercuration–reduction (p. 161)
pericyclic reaction (p. 164)
peroxide effect (p. 170)
primary alkyl radical (p. 168)
primary carbocation (p. 143)
propagation step (p. 170)
radical (p. 168)
radical addition reaction (p. 168)
radical chain reaction (p. 170)
radical inhibitor (p. 170)
radical initiator (p. 170)
reduction reaction (p. 162)
regioselective reaction (p. 148)
secondary alkyl radical (p. 168)
secondary carbocation (p. 143)
steric effects (p. 165)
steric hindrance (p. 165)
termination step (p. 170)
tertiary alkyl radical (p. 168)
tertiary carbocation (p. 143)

Problems

33. Give the major product of each of the following reactions:

a. 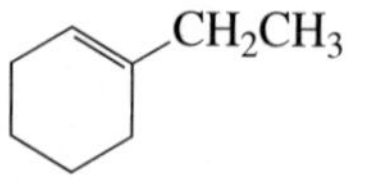 CH_2CH_3 + HBr $\longrightarrow$

b.
$$\begin{array}{l} \quad\quad CH_3 \\ \quad\quad | \\ CH_2{=}CCH_2CH_3 \end{array} \quad + \quad HBr \longrightarrow$$

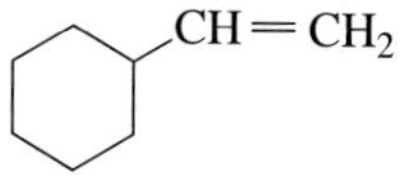

c. $CH{=}CH_2$ + HBr $\longrightarrow$

d.
$$\begin{array}{l} \quad\quad\quad\quad CH_3 \\ \quad\quad\quad\quad | \\ CH_3CH_2CCH{=}CH_2 \\ \quad\quad\quad\quad | \\ \quad\quad\quad\quad CH_3 \end{array} \quad + \quad HBr \longrightarrow$$

34. Identify the electrophile and the nucleophile in each of the following reaction steps. Then draw curved arrows to illustrate the bond-making and bond-breaking processes.

a.
$$CH_3\overset{}{\underset{+}{C}}HCH_3 \quad + \quad :\ddot{\underset{..}{Cl}}:^- \longrightarrow \begin{array}{l} CH_3CHCH_3 \\ \quad\quad | \\ \quad\ :\ddot{\underset{..}{Cl}}: \end{array}$$

b.
$$CH_3CH{=}CH_2 \quad + \quad :\ddot{\underset{..}{Br}}\cdot \longrightarrow CH_3\dot{C}HCH_2\ddot{\underset{..}{Br}}:$$

c.
$$\begin{array}{l} \quad\quad CH_3 \\ \quad\quad | \\ CH_3C^+ \\ \quad\quad | \\ \quad\quad CH_3 \end{array} \quad + \quad CH_3\ddot{\underset{..}{O}}H \longrightarrow \begin{array}{l} \quad\quad CH_3 \\ \quad\quad | \\ CH_3C{-}\overset{+}{\ddot{O}}CH_3 \\ \quad\quad |\quad\ | \\ \quad\quad CH_3\ H \end{array}$$

35. What will be the major product of the reaction of 2-methyl-2-butene with each of the following reagents?

a. HBr
b. HBr + peroxide
c. HI
d. HI + peroxide
e. ICl
f. H_2/Pd
g. Br_2 + excess NaCl
h. $Hg(OAc)_2$, H_2O followed by $NaBH_4$
i. H_2O + trace HCl
j. Br_2/CH_2Cl_2
k. Br_2/H_2O
l. Br_2/CH_3OH
m. BH_3/THF, followed by H_2O_2/HO^-
n. $Hg(O_2CCF_3)_2$ + CH_3OH, followed by $NaBH_4$

36. Which of the following compounds is the most stable?

3,4-dimethyl-2-hexene; 2,3-dimethyl-2-hexene; 4,5-dimethyl-2-hexene

Which would you expect to have the largest heat of hydrogenation? Which would you expect to have the lowest heat of hydrogenation?

37. When 3-methyl-1-butene reacts with HBr, two alkyl halides are formed: 2-bromo-3-methylbutane and 2-bromo-2-methylbutane. Propose a mechanism that explains the formation of these products.

38. Problem 30 in Chapter 3 asked you to give the structures of all alkenes with molecular formula C_6H_{12}. Use those structures to answer the following questions:

a. Which of the compounds is the most stable?
b. Which of the compounds is the least stable?

39. Draw curved arrows to show the flow of electrons responsible for the conversion of reactants into products.

a. $CH_3-C(O^-)(CH_3)-OCH_3 \longrightarrow CH_3-C(=O)-CH_3 + CH_3O^-$

b. $CH_3C{\equiv}C-H + {}^-NH_2 \longrightarrow CH_3C{\equiv}C^- + NH_3$

c. $CH_3CH_2-Br + CH_3O^- \longrightarrow CH_3CH_2-OCH_3 + Br^-$

40. Give the reagents that would be required to carry out the following syntheses:

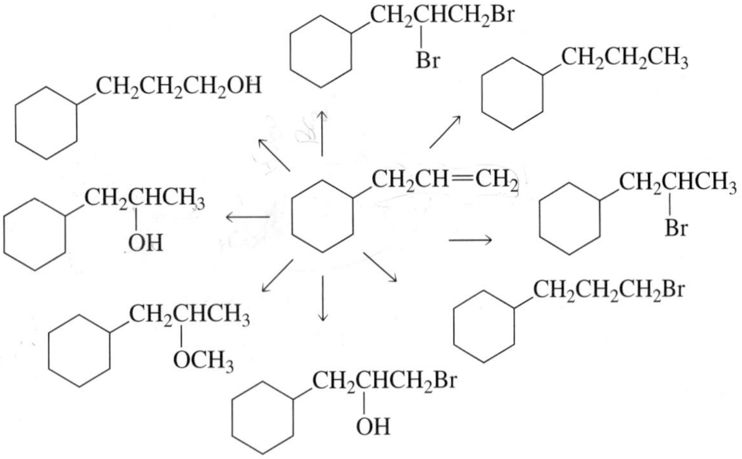

Starting material: cyclohexyl-$CH_2CH{=}CH_2$

Products: cyclohexyl-$CH_2CHBrCH_2Br$; cyclohexyl-$CH_2CH_2CH_3$; cyclohexyl-$CH_2CH_2CH_2OH$; cyclohexyl-$CH_2CH(OH)CH_3$; cyclohexyl-$CH_2CHBrCH_3$; cyclohexyl-$CH_2CH_2CH_2Br$; cyclohexyl-$CH_2CH(OCH_3)CH_3$; cyclohexyl-$CH_2CH(OH)CH_2Br$

41. For each pair of bonds, which has the greater strength? Briefly explain why.

a. CH_3-Cl or CH_3-Br

b. $CH_3CH_2CH_2-H$ or $CH_3CH(H)CH_3$

c. CH_3-CH_3 or $CH_3-CH_2CH_3$

d. $I-Br$ or $Br-Br$

42. Give the major product of each of the following reactions (substrate: 1-methylcyclohexene):

a. $\xrightarrow{HCl}$

b. $\xrightarrow[CH_3OH]{Br_2}$

c. $\xrightarrow{H_2O}$

d. $\xrightarrow[H_2O]{H^+}$

e. $\xrightarrow[\text{peroxide}]{HBr}$

f. $\xrightarrow{HBr}$

g. $\xrightarrow[H_2O]{Cl_2}$

h. $\xrightarrow[CH_2Cl_2]{Cl_2}$

i. $\xrightarrow[CH_3OH]{H^+}$

j. $\xrightarrow[\text{peroxide}]{HCl}$

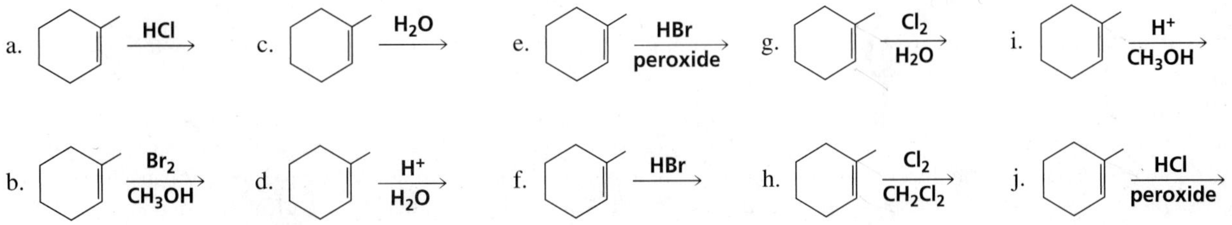

43. Using any alkene and any other reagents, how would you prepare the following compounds?

a. cyclohexane

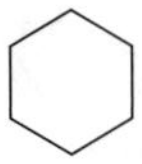

b. $CH_3CH_2CH_2CHClCH_3$

c. cyclohexyl-CH_2Br

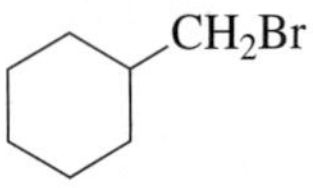

d. cyclohexyl-$CH_2CH(OH)CH_3$

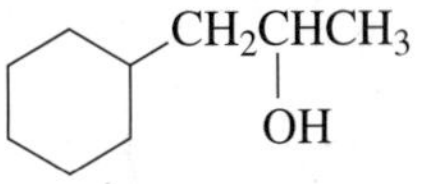

e. $CH_3CH_2CHBrCH(OH)CH_2CH_3$

f. $CH_3CH_2CHBrCHClCH_2CH_3$

44. There are two alkenes that react with HBr to give 1-bromo-1-methylcyclohexane.
a. Identify the alkenes.
b. Will both alkenes give the same product when they react with HBr/peroxide?
c. With HCl?
d. With HCl/peroxide?

45. For each of the following pairs, indicate which member is the more stable:

a. $CH_3\overset{CH_3}{\underset{+}{C}}CH_3$ or $CH_3\overset{+}{C}HCH_2CH_3$

b. $CH_3\overset{CH_3}{\underset{CH_3}{C}}\dot{C}H_2$ or $CH_3\dot{C}HCH_2CH_3$

c. $CH_3\overset{+}{C}HCH_3$ or $CH_3\overset{+}{C}HCH_2Cl$

d. $CH_3CH_2CH_2\dot{C}HCH_3$ or $CH_3CH_2CH_2CH_2\dot{C}H_2$

e. $CH_3\overset{CH_3}{C}{=}CHCH_2CH_3$ or $CH_3CH{=}CH\overset{CH_3}{C}HCH_3$

f. CH_3 or CH_3

g. $CH_3\overset{+}{C}H\underset{Cl}{C}HCH_3$ or $CH_3\overset{+}{C}HCH_2\underset{Cl}{C}H_2$

46. The second-order rate constant (in units of $M^{-1}s^{-1}$) for acid-catalyzed hydration at 25 °C is given for each of the following alkenes:

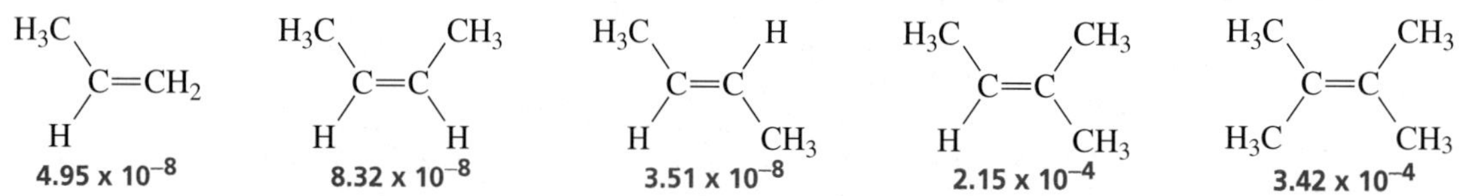

a. Calculate the relative rates of hydration of the alkenes.
b. Why does (*Z*)-2-butene react faster than (*E*)-2-butene?
c. Why does 2-methyl-2-butene react faster than (*Z*)-2-butene?
d. Why does 2,3-dimethyl-2-butene react faster than 2-methyl-2-butene?

47. Which compound has the greater dipole moment?

a. $\overset{Cl}{\underset{H}{}}C{=}C\overset{H}{\underset{Cl}{}}$ or $\overset{H}{\underset{H}{}}C{=}C\overset{H}{\underset{Cl}{}}$

b. $\overset{Cl}{\underset{H}{}}C{=}C\overset{H}{\underset{CH_3}{}}$ or $\overset{Cl}{\underset{H}{}}C{=}C\overset{H}{\underset{H}{}}$

c. $\overset{Cl}{\underset{H}{}}C{=}C\overset{H}{\underset{CH_3}{}}$ or $\overset{Cl}{\underset{H}{}}C{=}C\overset{CH_3}{\underset{H}{}}$

48. a. What five-carbon alkene will form the same product whether it reacts with HBr in the *presence* of a peroxide or with HBr in the *absence* of a peroxide?
b. Give three alkenes containing six carbon atoms that form the same product, whether they react with HBr in the *presence* of a peroxide or with HBr in the *absence* of a peroxide.

49. Mark Onikoff was about to turn in the products he had obtained from the reaction of HI with 3,3,3-trifluoropropene when he realized that the labels had fallen off his flasks. He didn't know which label belonged to which flask. Another student in the next lab told him that because the product obtained by following Markovnikov's rule was 1,1,1-trifluoro-2-iodopropane, he should put that label on the flask containing the most product and label the flask with the least product 1,1,1-trifluoro-3-iodopropane, the anti-Markovnikov product. Should Mark follow the student's advice?

50. a. Propose a mechanism for the following reaction (show all curved arrows):

$$CH_3CH_2CH{=}CH_2 + CH_3OH \xrightarrow{H^+} CH_3CH_2\underset{OCH_3}{C}HCH_3$$

b. Which step is the rate-determining step?
c. What is the electrophile in the first step?
d. What is the nucleophile in the first step?
e. What is the electrophile in the second step?
f. What is the nucleophile in the second step?

51. a. What product is obtained from the reaction of HCl with 1-butene? With 2-butene?
    b. Which of the two reactions has the greater free energy of activation?
    c. Which of the two alkenes reacts more rapidly with HCl?
    d. Which compound reacts more rapidly with HCl, (*Z*)-2-butene or (*E*)-2-butene?

52. a. How many alkenes could you treat with H_2/Pt in order to prepare methylcyclopentane?
    b. Which of the alkenes is the most stable?
    c. Which of the alkenes has the smallest heat of hydrogenation?

53. a. Propose a mechanism for the following reaction:

CH3 C=CH2 (on cyclobutane) + HBr ⟶ Br CH3 CH3 (cyclopentane) + CH3 CH3 Br (cyclopentane)

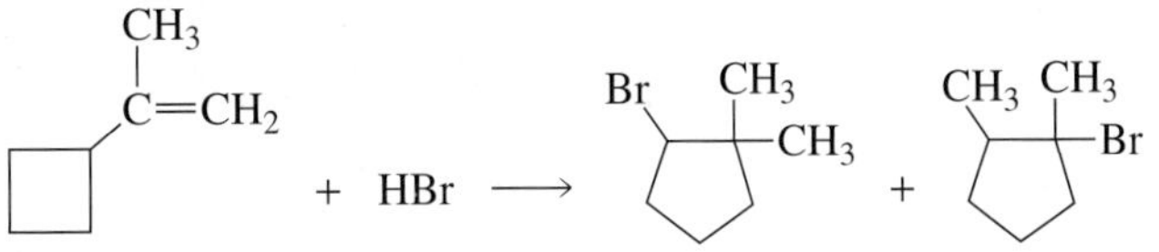

    b. Is the initially formed carbocation primary, secondary, or tertiary?
    c. Is the rearranged carbocation primary, secondary, or tertiary?
    d. Why does the rearrangement occur?

54. When the following compound is hydrated in the presence of acid, the unreacted alkene is found to have retained the deuterium atoms:

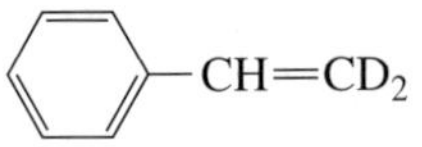

C_6H_5–CH=CD_2

What does the preceding statement tell you about the mechanism of hydration?

55. Propose a mechanism for each of the following reactions:

a. 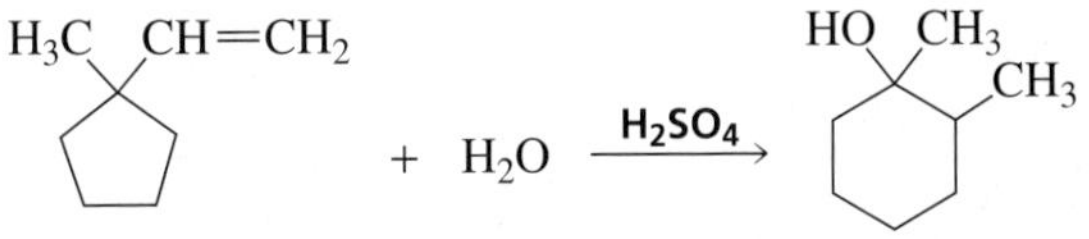 H_3C CH=CH_2 (cyclopentane) + H_2O $\xrightarrow{H_2SO_4}$ HO CH_3 CH_3 (cyclohexane)

b. 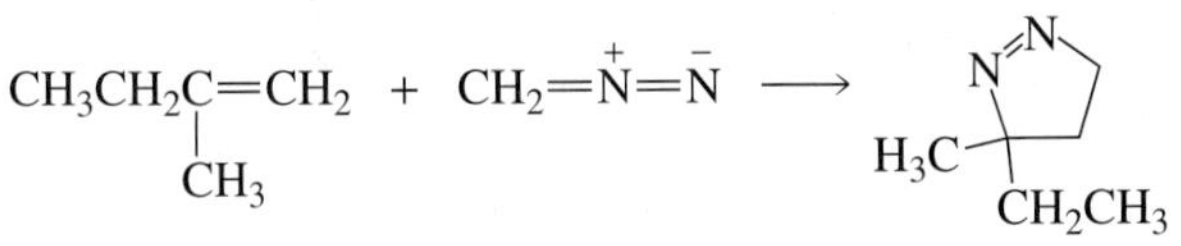 $CH_3CH_2\underset{\displaystyle CH_3}{C}{=}CH_2$ + $CH_2{=}\overset{+}{N}{=}\overset{-}{N}$ ⟶ N=N ring, H_3C, CH_2CH_3

c. cyclooctadiene + $HCCl_3$ $\xrightarrow{\text{peroxide}}$ bicyclic product with CCl_3

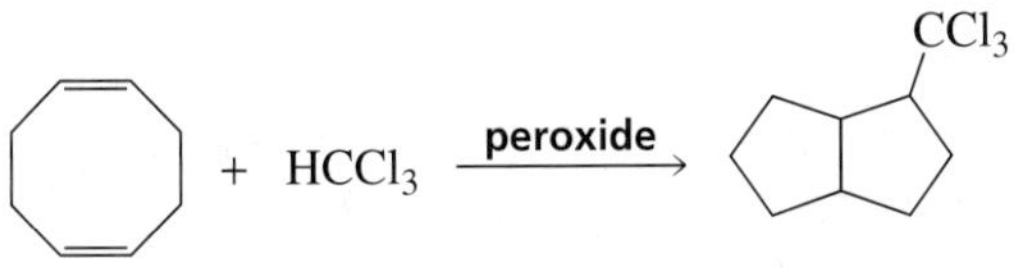

56. a. Dichlorocarbene can be generated by heating chloroform with HO^-. Propose a mechanism for the reaction.

$$\underset{\text{chloroform}}{CHCl_3} + HO^- \xrightarrow{\Delta} \underset{\text{dichlorocarbene}}{Cl_2C:} + H_2O + Cl^-$$

    b. Dichlorocarbene can also be generated by heating sodium trichloroacetate. Propose a mechanism for the reaction.

$$\underset{\text{sodium trichloroacetate}}{Cl_3C\overset{O}{\overset{\|}{C}}O^-\ Na^+} \xrightarrow{\Delta} Cl_2C: + CO_2 + Na^+\ Cl^-$$

5 Stereochemistry

The Arrangement of Atoms in Space; The Stereochemistry of Addition Reactions

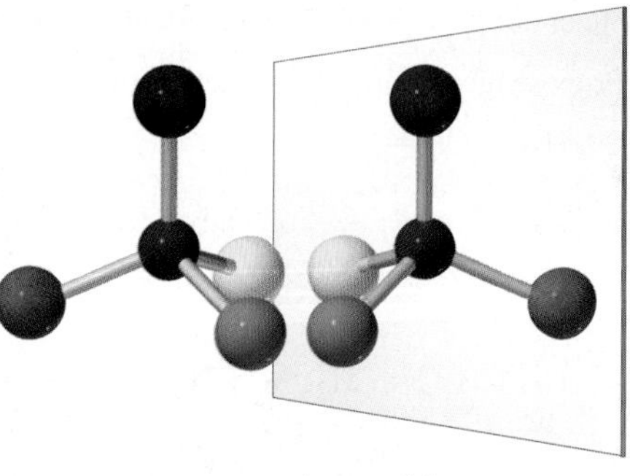

nonsuperimposable mirror images

Compounds that have the same molecular formula but are not identical are called **isomers**. Isomers fall into two main classes: *constitutional isomers* and *stereoisomers*. **Constitutional isomers** differ in the way their atoms are connected (Section 2.0). For example, ethanol and dimethyl ether are constitutional isomers because they have the same molecular formula, C_2H_6O, but the atoms in each compound are connected differently. The oxygen in ethanol is bonded to a carbon and to a hydrogen, whereas the oxygen in dimethyl ether is bonded to two carbons.

constitutional isomers

CH_3CH_2OH ethanol	and	CH_3OCH_3 dimethyl ether		$CH_3CH_2CH_2CH_2Cl$ 1-chlorobutane	and	$CH_3CH_2CH(Cl)CH_3$ 2-chlorobutane
$CH_3CH_2CH_2CH_2CH_3$ pentane	and	$CH_3CH(CH_3)CH_2CH_3$ isopentane		$CH_3C(=O)CH_3$ acetone	and	$CH_3CH_2CH(=O)$ propionaldehyde

Unlike the atoms in constitutional isomers, the atoms in stereoisomers are connected in the same way. **Stereoisomers** (also called **configurational isomers**) differ in the way their atoms are arranged in space. Stereoisomers are different compounds that do not readily interconvert. Therefore, they can be separated. There are two kinds of stereoisomers: **cis–trans isomers** and isomers that contain chirality (ky-RAL-i-tee) centers.

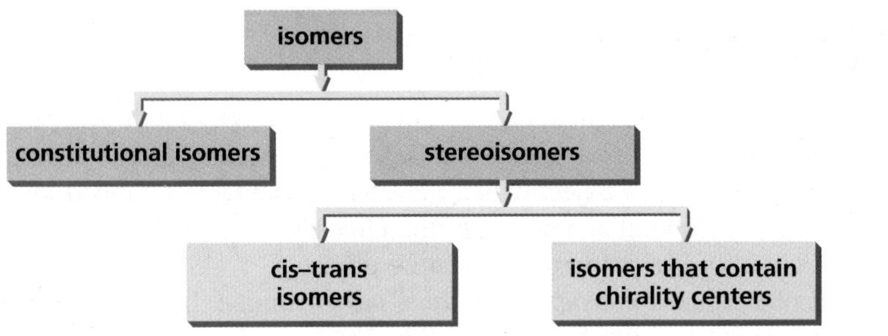

Movie: Isomerism

PROBLEM 1◆

a. Draw three constitutional isomers with molecular formula C_3H_8O.
b. How many constitutional isomers can you draw for $C_4H_{10}O$?

5.1 Cis–Trans Isomers

Cis–trans isomers (also called **geometric isomers**) result from restricted rotation (Section 3.4). Restricted rotation can be caused either by a double bond or by a cyclic structure. As a result of the restricted rotation about a carbon–carbon double bond, an alkene such as 2-pentene can exist as cis and trans isomers. The **cis isomer** has the hydrogens on the *same side* of the double bond, whereas the **trans isomer** has the hydrogens on *opposite sides* of the double bond.

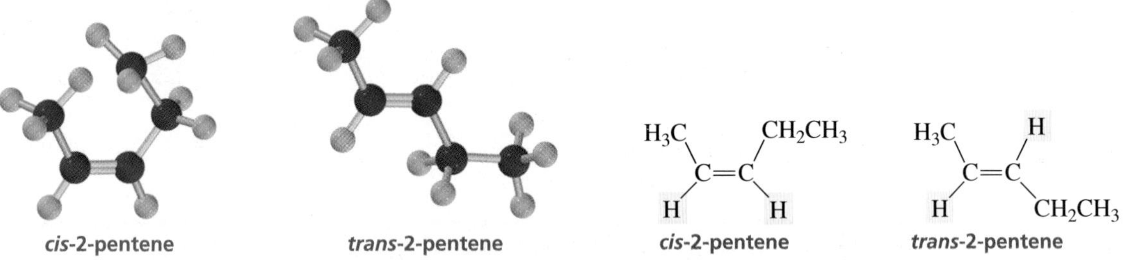

cis-2-pentene *trans*-2-pentene *cis*-2-pentene *trans*-2-pentene

Cyclic compounds can also have cis and trans isomers (Section 2.14). The cis isomer has the hydrogens on the same side of the ring, whereas the trans isomer has the hydrogens on opposite sides of the ring.

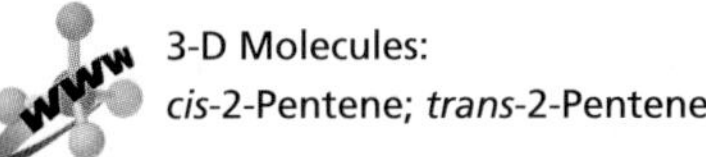

3-D Molecules: *cis*-2-Pentene; *trans*-2-Pentene

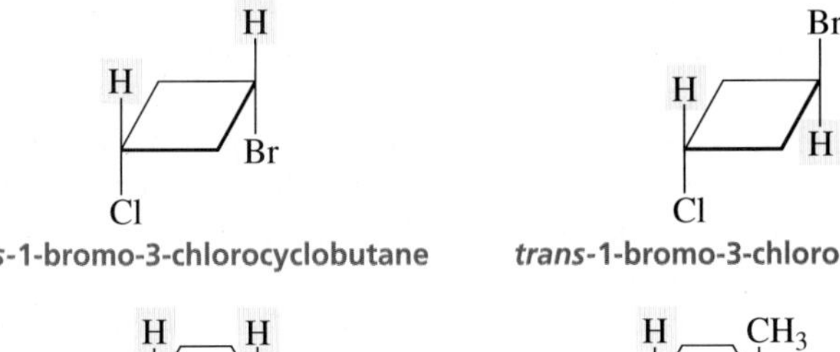

cis-1-bromo-3-chlorocyclobutane *trans*-1-bromo-3-chlorocyclobutane

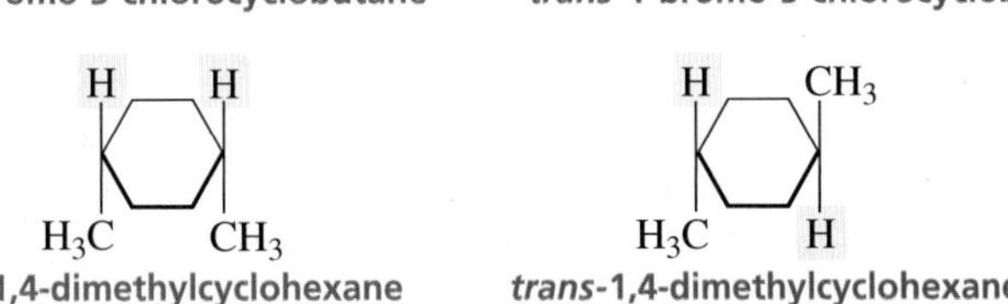

cis-1,4-dimethylcyclohexane *trans*-1,4-dimethylcyclohexane

PROBLEM 2

Draw the cis and trans isomers for the following compounds:

a. 1-ethyl-3-methylcyclobutane
b. 2-methyl-3-heptene
c. 1-bromo-4-chlorocyclohexane
d. 1,3-dibromocyclobutane

5.2 Chirality

Why can't you put your right shoe on your left foot? Why can't you put your right glove on your left hand? It is because hands, feet, gloves, and shoes have right-handed and left-handed forms. An object with a right-handed and a left-handed form is said to be **chiral** (ky-ral). "Chiral" comes from the Greek word *cheir*, which means "hand." Notice that chirality is a property of an entire object.

A chiral object has a *nonsuperimposable mirror image*. In other words, its mirror image is not the same as itself. A hand is chiral because if you look at your left hand in a mirror, you do not see your left hand; you see your right hand (Figure 5.1). In contrast, a chair is not chiral—it looks the same in the mirror. Objects that are not chiral are said to be **achiral**. An achiral object has a *superimposable mirror image*. Some other achiral objects would be a table, a fork, and a glass.

PROBLEM 3♦

a. Name five capital letters that are chiral. b. Name five capital letters that are achiral.

Figure 5.1 ▶ Using a mirror to test for chirality. A chiral object is not the same as its mirror image—they are nonsuperimposable. An achiral object is the same as its mirror image—they are superimposable.

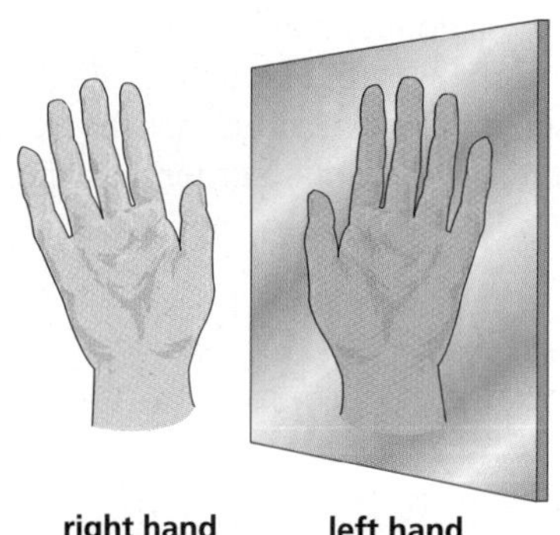

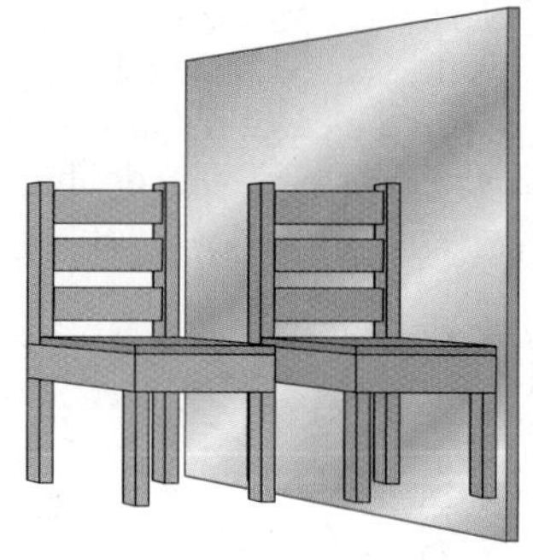

5.3 Asymmetric Carbons, Chirality Centers, and Stereocenters

A molecule with one asymmetric carbon is chiral.

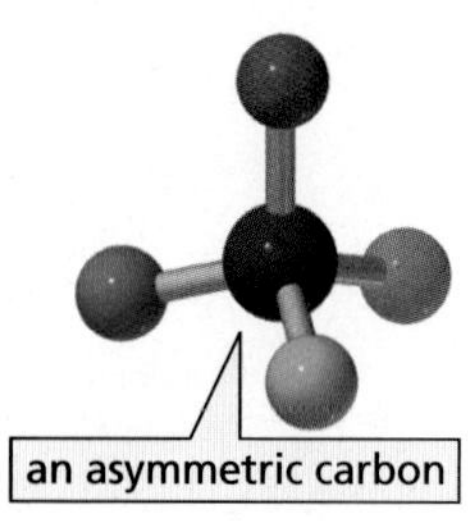

Not only can objects be chiral, molecules can be chiral, too. The feature that most often is the cause of chirality in a molecule is an *asymmetric carbon*. (Other features that cause chirality are relatively uncommon and are beyond the scope of this book. You can, however, see one of these in Problem 88.)

An **asymmetric carbon** is a carbon atom that is bonded to four different groups. The asymmetric carbon in each of the following compounds is indicated by an asterisk. For example, the starred carbon in 4-octanol is an asymmetric carbon because it is bonded to four different groups (H, OH, $CH_2CH_2CH_3$, and $CH_2CH_2CH_2CH_3$). Notice that the difference in the groups bonded to the asymmetric carbon is not necessarily right next to the asymmetric carbon. For example, the propyl and butyl groups are different even though the point at which they differ is somewhat removed from the asymmetric carbon. The starred carbon in 2,4-dimethylhexane is an asymmetric carbon because it is bonded to four different groups—methyl, ethyl, isobutyl, and hydrogen.

$CH_3CH_2CH_2\overset{*}{C}HCH_2CH_2CH_2CH_3$ with OH on the starred carbon

4-octanol

$CH_3\overset{*}{C}HCH_2CH_3$ with Br on the starred carbon

2-bromobutane

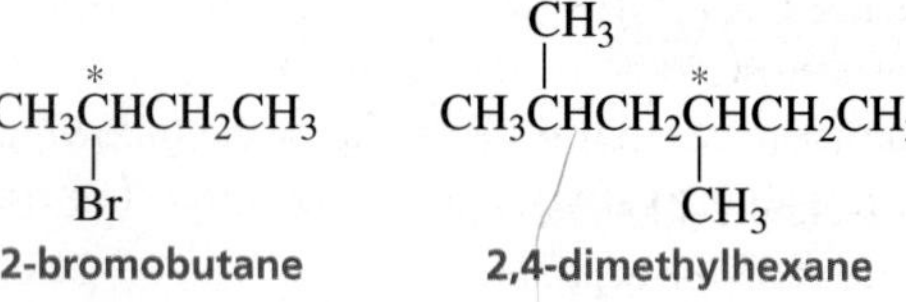

2,4-dimethylhexane

Notice that the only carbons that can be asymmetric carbons are sp^3 hybridized carbons; sp^2 and sp hybridized carbons cannot be asymmetric carbons because they cannot have four groups attached to them.

An asymmetric carbon is also known as a **chirality center**. We will see that atoms other than carbon, such as nitrogen and phosphorus, can be chirality centers—when they are bonded to four different atoms or groups (Section 5.17). In other words, an asymmetric carbon is just one kind of chirality center. A chirality center also belongs to a broader group known as *stereocenters*. Stereocenters will be defined in Section 5.5.

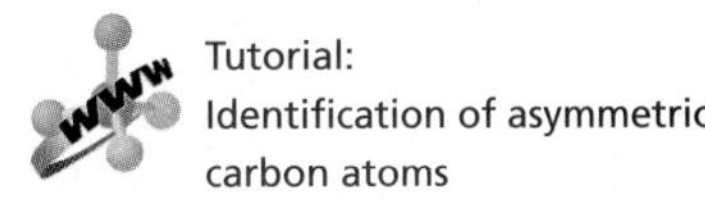

Tutorial:
Identification of asymmetric carbon atoms

PROBLEM 4◆

Which of the following compounds have asymmetric carbons?

a. $CH_3CH_2CHCH_3$ (with Cl on the CH)

b. $CH_3CH_2CHCH_3$ (with CH_3 on the CH)

c. $CH_3CH_2CCH_2CH_2CH_3$ (with CH_3 and Br on the central C)

d. CH_3CH_2OH

e. $CH_3CH_2CHCH_2CH_3$ (with Br on the CH)

f. $CH_2{=}CHCHCH_3$ (with NH_2 on the CH)

PROBLEM 5 SOLVED

Tetracycline is called a broad-spectrum antibiotic because it is active against a wide variety of bacteria. How many asymmetric carbons does tetracycline have?

SOLUTION First, locate all the sp^3 hybridized carbons in tetracycline. (They are numbered in red.) Only sp^3 hybridized carbons can be asymmetric carbons, because an asymmetric carbon must have four different groups attached to it. Tetracycline has nine sp^3 hybridized carbons. Four of them (#1, #2, #5, and #8) are not asymmetric carbons because they are not bonded to four different groups. Tetracycline, therefore, has five asymmetric carbons.

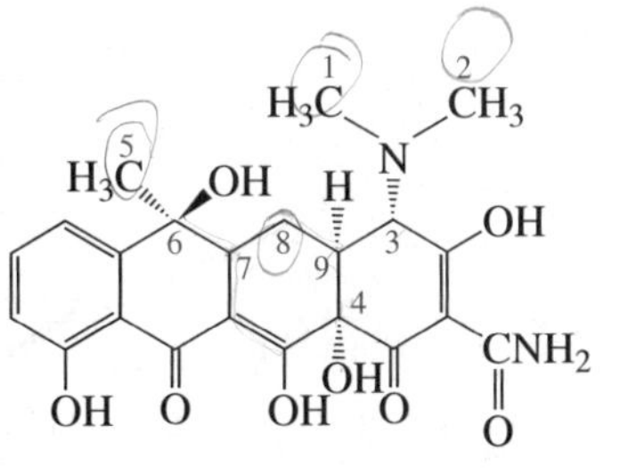

tetracycline

5.4 Isomers with One Asymmetric Carbon

A compound with one asymmetric carbon, such as 2-bromobutane, can exist as two different stereoisomers. The two isomers are analogous to a left and a right hand. Imagine a mirror between the two isomers; notice how they are mirror images of each other. The two stereoisomers are nonsuperimposable mirror images—they are different molecules.

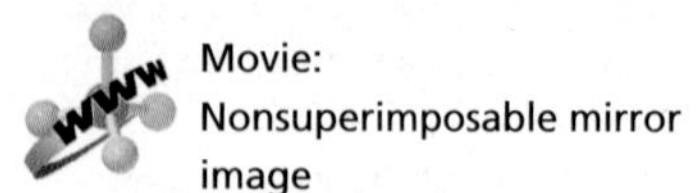
Movie: Nonsuperimposable mirror image

$$CH_3\overset{*}{C}HCH_2CH_3$$
Br

2-bromobutane

Br, CH_3CH_2–C–H, CH_3 | mirror | Br, H–C–CH_2CH_3, CH_3

the two isomers of 2-bromobutane
enantiomers

Take a break and convince yourself that the two 2-bromobutane isomers are not identical, by building ball-and-stick models using four different-colored balls to represent the four different groups bonded to the asymmetric carbon. Try to superimpose them.

Nonsuperimposable mirror-image molecules are called **enantiomers** (from the Greek *enantion*, which means "opposite"). The two stereoisomers of 2-bromobutane are enantiomers. A molecule that has a nonsuperimposable mirror image, like an object that has a nonsuperimposable mirror image, is chiral. Each of the enantiomers is chiral. A molecule that has a superimposable mirror image, like an object that has a superimposable mirror image, is achiral. To see that the achiral moleule is superimposable on its mirror image (i.e., they are identical molecules), mentally rotate the achiral molecule clockwise. Notice that chirality is a property of the entire molecule.

A chiral molecule has a nonsuperimposable mirror image.

An achiral molecule has a superimposable mirror image.

Br, CH_3CH_2–C–H, CH_3 (a chiral molecule) | Br, H–C–CH_2CH_3, H_3C (nonsuperimposable mirror image)
enantiomers

Br, CH_3CH_2–C–CH_3, CH_3 (an achiral molecule) | Br, H_3C–C–CH_2CH_3, H_3C (superimposable mirror image)
identical molecules

PROBLEM 6◆

Which of the compounds in Problem 4 can exist as enantiomers?

5.5 Drawing Enantiomers

Chemists draw enantiomers using either *perspective formulas* or *Fischer projections*.

This book has been written in a way that allows you to use either perspective formulas or Fischer projections. Most chemists use perspective formulas. If you choose to use perspective formulas, you can ignore all the Fischer projections in the book.

Perspective formulas show two of the bonds to the asymmetric carbon in the plane of the paper, one bond as a solid wedge protruding out of the paper, and the fourth bond as a hatched wedge extending behind the paper. You can draw the first enantiomer by putting the four groups bonded to the asymmetric carbon in any order. Draw the second enantiomer by drawing the mirror image of the first enantiomer.

perspective formulas of the enantiomers of 2-bromobutane

The solid wedges represent bonds that point out of the plane of the paper toward the viewer.

The hatched wedges represent bonds that point back from the plane of the paper away from the viewer.

Make certain when you draw a perspective formula that the two bonds in the plane of the paper are adjacent to one another; neither the solid wedge nor the hatched wedge should be drawn between them.

A shortcut—called a **Fischer projection**—for showing the three-dimensional arrangement of groups bonded to an asymmetric carbon was devised in the late 1800s by Emil Fischer. A Fischer projecton represents an asymmetric carbon as the point of intersection of two perpendicular lines; horizontal lines represent the bonds that project out of the plane of the paper toward the viewer, and vertical lines represent the bonds that extend back from the plane of the paper away from the viewer. The carbon chain always is drawn vertically with C-1 at the top of the chain.

asymmetric carbon

Fischer projections of the enantiomers of 2-bromobutane

In a Fischer projection horizontal lines project out of the plane of the paper toward the viewer and vertical lines extend back from the plane of the paper away from the viewer.

To draw enantiomers using a Fischer projection, draw the first enantiomer by arranging the four atoms or groups bonded to the asymmetric carbon in any order. Draw the second enantiomer by interchanging two of the atoms or groups. It does not matter which two you interchange. (Make models to convince yourself that this is true.) It is best to interchange the groups on the two horizontal bonds because the enantiomers then look like mirror images on paper.

Note that interchanging two atoms or groups gives you the enantiomer—whether you are drawing perspective formulas or Fischer projections. Interchanging two atoms or groups a second time, brings you back to the original molecule.

A **stereocenter** (or stereogenic center) is an atom at which the interchange of two groups produces a stereoisomer. Therefore, both *asymmetric carbons*—where the interchange of two groups produces an enantiomer and the carbons where the interchange of two groups converts a cis isomer to a trans isomer (or a *Z* isomer to an *E* isomer)—are stereocenters.

a stereocenter

a stereocenter

a stereocenter

Emil Fischer (1852–1919) *was born in a village near Cologne, Germany. He became a chemist against the wishes of his father, a successful merchant, who wanted him to enter the family business. He was a professor of chemistry at the Universities of Erlangen, Würzburg, and Berlin. In 1902 he received the Nobel Prize in chemistry for his work on sugars. During World War I, he organized German chemical production. Two of his three sons died in that war.*

PROBLEM 7

Draw enantiomers for each of the following compounds using:

a. perspective formulas
b. Fischer projections

1. $CH_3CH(Br)CH_2OH$
2. $ClCH_2CH_2CH(CH_3)CH_2CH_3$
3. $CH_3CH(CH_3)CH(OH)CH_3$

5.6 Naming Enantiomers: The *R,S* System of Nomenclature

Robert Sidney Cahn (1899–1981), *was born in England and received an M.A. from Cambridge University and a doctorate in natural philosophy in France. He edited the* Journal of the Chemical Society *(London).*

Sir Christopher Ingold (1893–1970) *was born in Ilford, England, and was knighted by Queen Elizabeth II. He was a professor of chemistry at Leeds University (1924–1930) and at University College, London (1930–1970).*

Vladimir Prelog (1906–1998) *was born in Sarajevo, Bosnia. In 1929 he received a Dr. Ing. degree from the Institute of Technology in Prague, Czechoslovakia. He taught at the University of Zagreb from 1935 until 1941, when he fled to Switzerland just ahead of the invading German army. He was a professor at the Swiss Federal Institute of Technology (ETH). For his work that contributed to an understanding of how living organisms carry out chemical reactions, he shared the 1975 Nobel Prize in chemistry with John Cornforth (page 231).*

We need a way to name the individual stereoisomers of a compound such as 2-bromobutane so that we know which stereoisomer we are talking about. In other words, we need a system of nomenclature that indicates the **configuration** (arrangement) of the atoms or groups about the asymmetric carbon. Chemists use the letters *R* and *S* to indicate the configuration about an asymmetric carbon. For any pair of enantiomers with one asymmetric carbon, one will have the ***R* configuration** and the other will have the ***S* configuration**. The *R,S* system was devised by Cahn, Ingold, and Prelog.

Let's first look at how we can determine the configuration of a compound if we have a three-dimensional model of the compound.

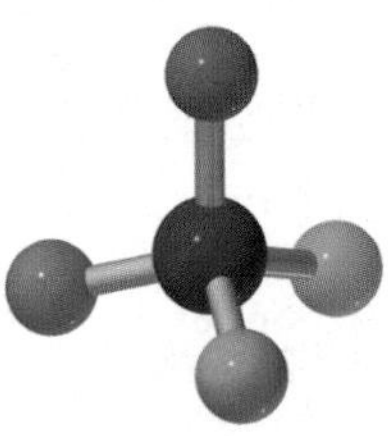

1. *Rank the groups (or atoms) bonded to the asymmetric carbon in order of priority.* The atomic numbers of the atoms directly attached to the asymmetric carbon determine the relative priorities. The higher the atomic number, the higher the priority. (This should remind you of the way relative priorities are determined for the *E,Z* system of nomenclature because the system of priorities was originally devised for the *R,S* system of nomenclature and was later borrowed for the *E,Z* system. You may want to revisit Section 3.5 to review how relative priorities are determined before you proceed with the *R,S* system.

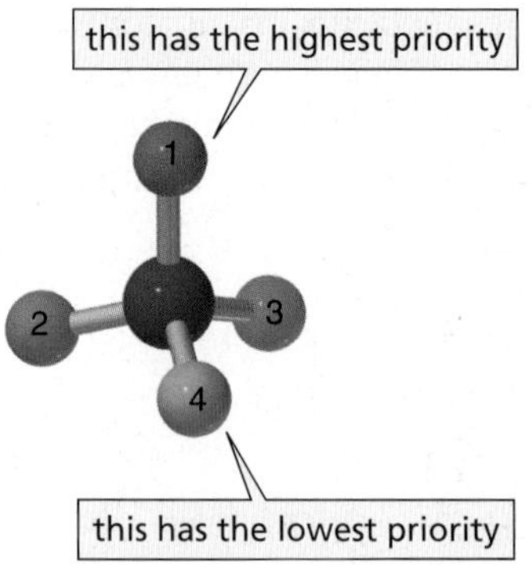

The molecule is oriented so the group with the lowest priority points away from the viewer. If an arrow drawn from the highest priority group to the next highest priority group points clockwise, the molecule has the *R* configuration.

2. *Orient the molecule so that the group (or atom) with the lowest priority (4) is directed away from you. Then draw an imaginary arrow from the group (or atom) with the highest priority (1) to the group (or atom) with the next highest priority (2).* If the arrow points clockwise, the asymmetric carbon has the *R* configuration (*R* is for *rectus*, which is Latin for "right"). If the arrow points counterclockwise, the asymmetric carbon has the *S* configuration (*S* is for *sinister*, which is Latin for "left").

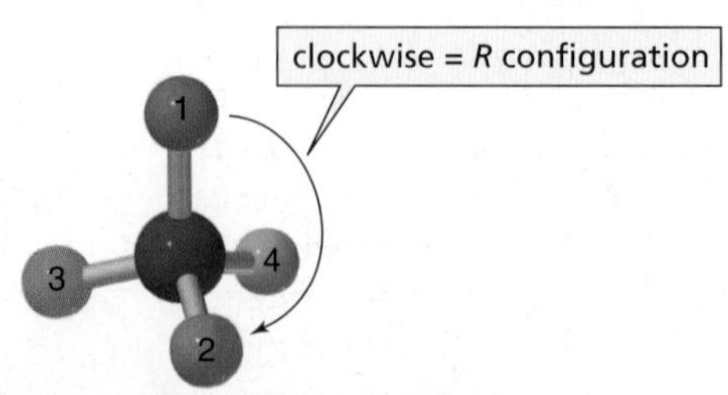

If you forget which is which, imagine driving a car and turning the steering wheel clockwise to make a right turn or counterclockwise to make a left turn.

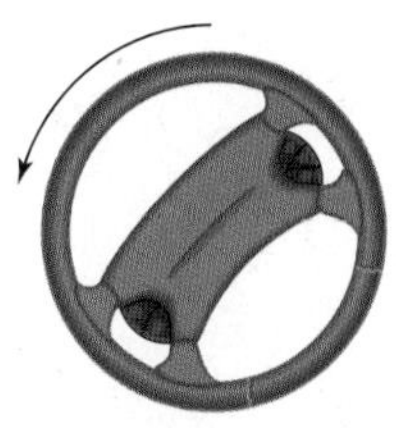
left turn

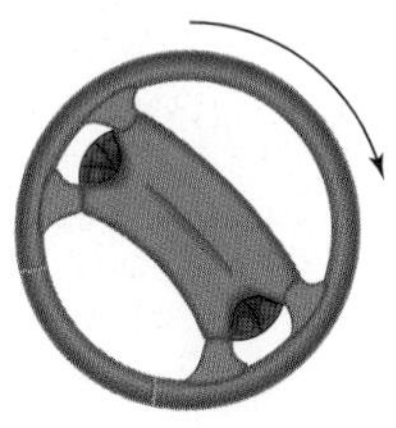
right turn

If you are able to easily visualize spatial relationships, the above two rules are all you need to determine whether the asymmetric carbon of a molecule written on a two-dimensional piece of paper has the *R* or the *S* configuration. Just mentally rotate the molecule so that the group (or atom) with the lowest priority (4) is directed away from you, then draw an imaginary arrow from the group (or atom) with the highest priority to the group (or atom) with the next highest priority.

If you have trouble visualizing spatial relationships and you don't have access to a model, the following will allow you to determine the configuration about an asymmetric carbon without having to mentally rotate the molecule.

First, let's look at how you can determine the configuration of a compound drawn as a perspective formula. As an example, we will determine which of the enantiomers of 2-bromobutane has the *R* configuration and which has the *S* configuration.

Br, CH_3CH_2, C, H, CH_3 Br, H, C, CH_2CH_3, CH_3

the enantiomers of 2-bromobutane

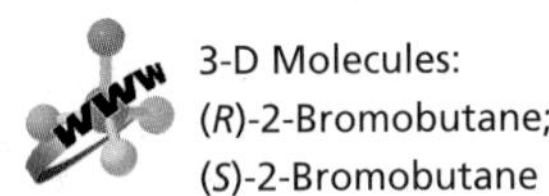
3-D Molecules: (*R*)-2-Bromobutane; (*S*)-2-Bromobutane

1. Rank the groups (or atoms) that are bonded to the asymmetric carbon in order of priority. In the following pair of enantiomers, bromine has the highest priority (1), the ethyl group has the second highest priority (2), the methyl group is next (3), and hydrogen has the lowest priority (4). (Revisit Section 3.5 if you don't understand how these priorities are assigned.)

Br (1), CH_3CH_2 (2), C, H (4), CH_3 (3)   Br (1), H (4), C, CH_2CH_3 (2), CH_3 (3)

2. If the group (or atom) with the lowest priority is bonded by a hatched wedge, draw an arrow from the group (or atom) with the highest priority (1) to the group (or atom) with the second highest priority (2). If the arrow points clockwise, the compound has the *R* configuration, and if it points counterclockwise, the compound has the *S* configuration.

Br (1), CH_3CH_2 (2), C, H (4), CH_3 (3)   Br (1), H (4), C, CH_2CH_3 (2), CH_3 (3)

(*S*)-2-bromobutane   **(*R*)-2-bromobutane**

Clockwise specifies *R* if the lowest priority substituent is on a hatched wedge.

3. If the group with the lowest priority (4) is NOT bonded by a hatched wedge, then switch two groups so group 4 is bonded by a hatched wedge. Then proceed as in step #2 (above): Draw an arrow from the group (or atom) with the highest priority (1) to the group (or atom) with the second highest priority (2). Because you have switched two groups, you are now determining the configuration of the enantiomer of the original molecule. So if the arrow points clockwise, the enantiomer (with the switched groups) has the *R* configuration, which means the original molecule has the *S* configuration. In contrast, if the arrow points counterclockwise, the enantiomer (with the switched groups) has the *S* configuration, which means the original molecule has the *R* configuration.

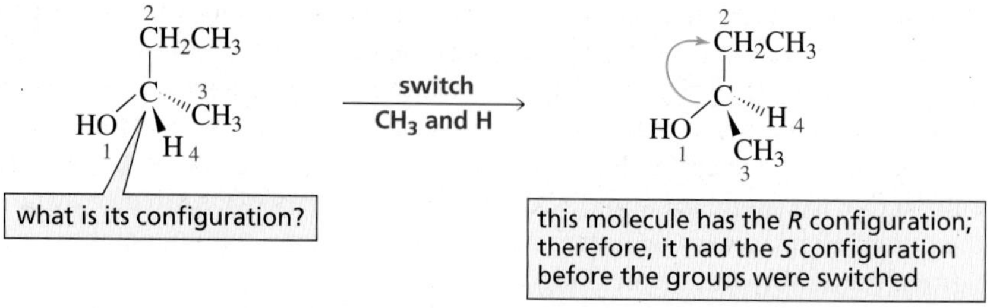

what is its configuration?

switch CH_3 and H

this molecule has the *R* configuration; therefore, it had the *S* configuration before the groups were switched

4. In drawing the arrow from group 1 to group 2, you can draw past the group with the lowest priority (4), but never draw past the group with the next lowest priority (3).

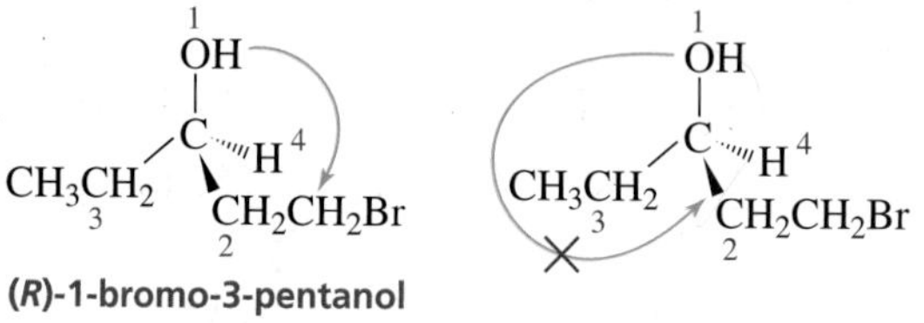

(*R*)-1-bromo-3-pentanol

Now let's see how to determine the configuration of a compound drawn as a Fischer projection.

1. Rank the groups (or atoms) that are bonded to the asymmetric carbon in order of priority.

2. Draw an arrow from the group (or atom) with the highest priority (1) to the group (or atom) with the next highest priority (2). If the arrow points clockwise, the enantiomer has the *R* configuration; if it points counterclockwise, the enantiomer has the *S* configuration, *provided that the group with the lowest priority (4) is on a vertical bond.*

Clockwise specifies *R* if the lowest priority substituent is on a vertical bond.

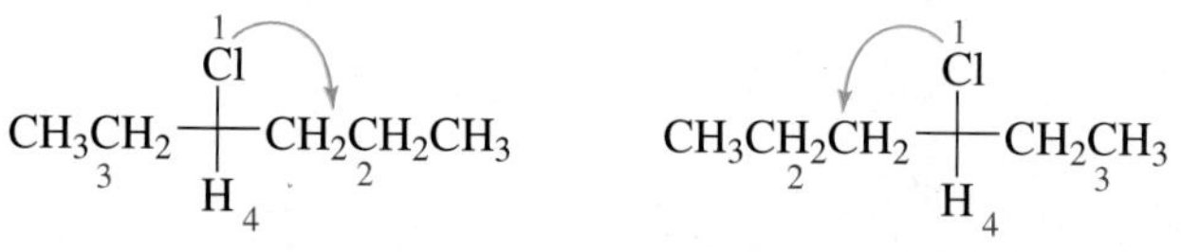

(*R*)-3-chlorohexane **(*S*)-3-chlorohexane**

3. If the group (or atom) with the lowest priority is on a *horizontal* bond, the answer you get from the direction of the arrow will be the opposite of the correct answer. For example, if the arrow points clockwise, suggesting that the asymmetric carbon has the *R* configuration, it actually has the *S* configuration; if the arrow points counterclockwise, suggesting that the asymmetric carbon has the *S* configuration, it actually has the *R* configuration. In the following example, the group with the lowest priority is on a horizontal bond, so clockwise signifies the *S* configuration, not the *R* configuration.

Clockwise specifies *S* if the lowest priority substituent is on a horizontal bond.

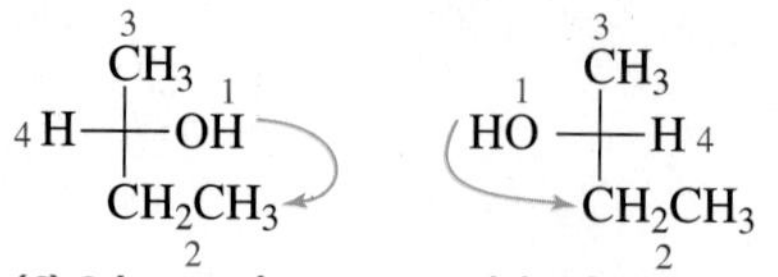

(*S*)-2-butanol **(*R*)-2-butanol**

4. In drawing the arrow from group 1 to group 2, you can draw past the group (or atom) with the lowest priority (4), but never draw past the group (or atom) with the next lowest priority (3).

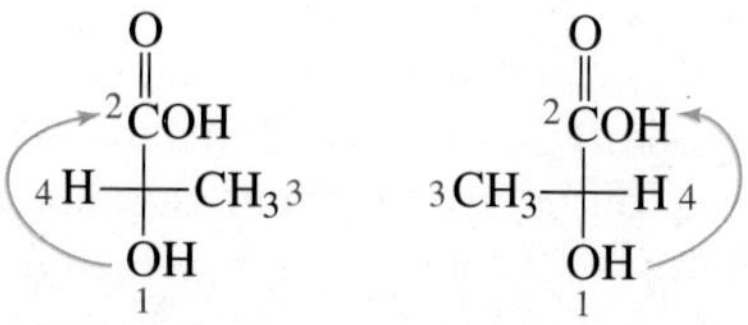

(*S*)-lactic acid **(*R*)-lactic acid**

It is easy to tell whether two molecules are enantiomers (nonsuperimposable) or identical molecules (superimposable) if you have molecular models of the molecules—just see whether the models superimpose. If, however, you are working with structures on a two-dimensional piece of paper, the easiest way to determine whether two molecules are enantiomers or identical molecules is by determining their configurations. If one has the *R* configuration and the other has the *S* configuration, they are enantiomers. If they both have the *R* configuration or both have the *S* configuration, they are identical molecules.

When comparing two Fischer projections to see if they are the same or different, never rotate one 90° or turn one over, because this is a quick way to get a wrong answer. A Fischer projection can be rotated 180° in the plane of the paper, but this is the only way to move it without risking an incorrect answer.

PROBLEM 8◆

Indicate whether each of the following structures has the *R* or the *S* configuration:

a. 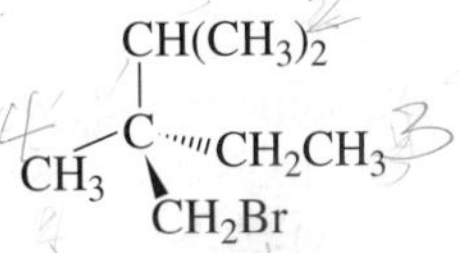 C 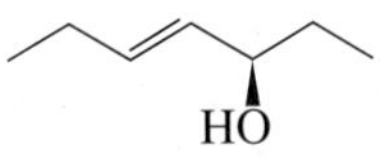 bearing $CH(CH_3)_2$ (up), CH_3, CH_2CH_3 (hashed), CH_2Br (wedge)

b. 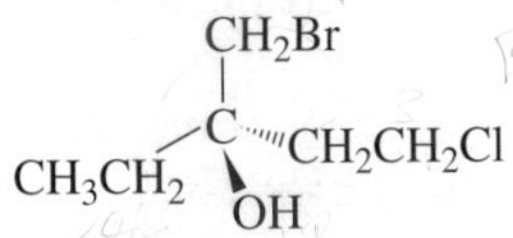 C bearing CH_2Br (up), CH_3CH_2, CH_2CH_2Cl (hashed), OH (wedge)

c. (skeletal structure with HO on wedge)

d. 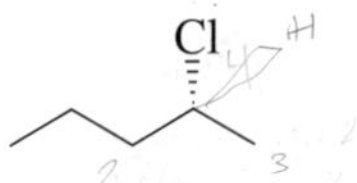 (skeletal structure with Cl on hash)

PROBLEM 9◆ SOLVED

Do the following structures represent identical molecules or a pair of enantiomers?

a. 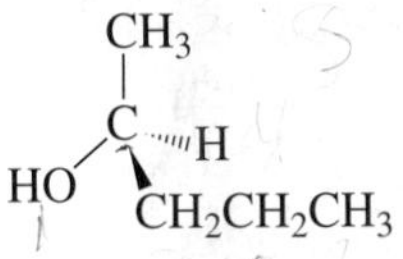 C bearing CH_3 (up), HO, H (hashed), $CH_2CH_2CH_3$ (wedge) and C bearing OH (up), $CH_3CH_2CH_2$, CH_3 (hashed), H (wedge)

b. C bearing CH_2Br (up), CH_3, Cl (hashed), CH_2CH_3 (wedge) and C bearing Cl (up), CH_3CH_2, CH_3 (hashed), CH_2Br (wedge)

c. C bearing CH_2Br (up), H, OH (hashed), CH_3 (wedge) and C bearing H (up), HO, CH_3 (hashed), CH_2Br (wedge)

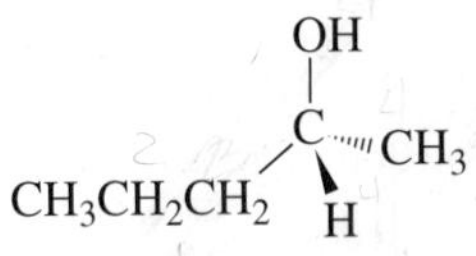

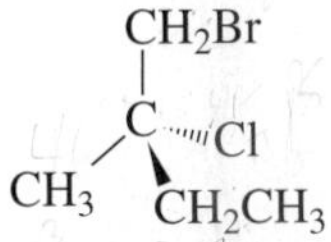

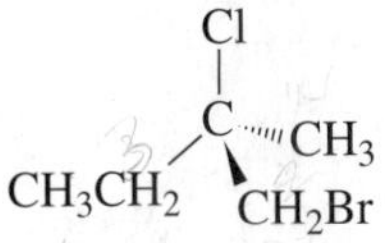

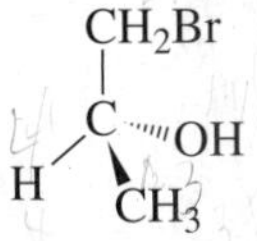

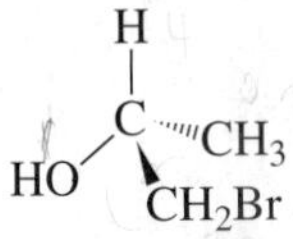

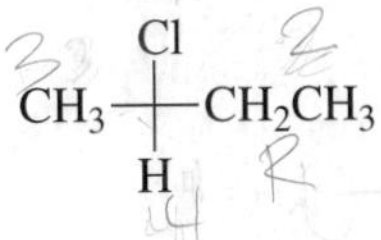

d. Fischer projection: Cl (top), CH_3 (left), CH_2CH_3 (right), H (bottom) and Fischer projection: CH_3 (top), H (left), Cl (right), CH_2CH_3 (bottom)

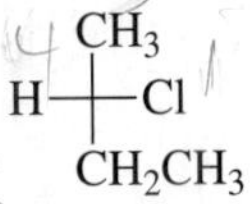

SOLUTION TO 9a The first structure shown in part (a) has the *S* configuration, and the second structure has the *R* configuration. Because they have opposite configurations, the structures represent a pair of enantiomers.

PROBLEM 10◆

Assign relative priorities to the following groups:

a. $-CH_2OH$ $-CH_3$ $-CH_2CH_2OH$ $-H$

b. $-CH{=}O$ $-OH$ $-CH_3$ $-CH_2OH$

c. $-CH(CH_3)_2$ $-CH_2CH_2Br$ $-Cl$ $-CH_2CH_2CH_2Br$

d. $-CH{=}CH_2$ $-CH_2CH_3$ —(phenyl) $-CH_3$

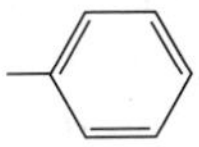

PROBLEM 11◆

Indicate whether each of the following structures has the *R* or the *S* configuration:

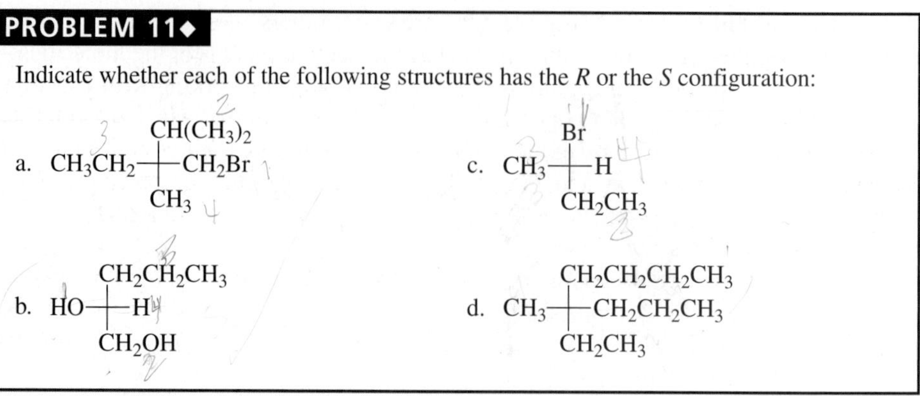

5.7 Optical Activity

Enantiomers share many of the same properties—they have the same boiling points, the same melting points, and the same solubilities. In fact, all the physical properties of enantiomers are the same except those that stem from how groups bonded to the asymmetric carbon are arranged in space. One of the properties that enantiomers do not share is the way they interact with polarized light.

Born in Scotland, **William Nicol (1768–1851)** *was a professor at the University of Edinburgh. He developed the first prism that produced plane-polarized light. He also developed methods to produce thin slices of materials for use in microscopic studies.*

What is polarized light? Normal light consists of electromagnetic waves that oscillate in all directions. **Plane-polarized light** (or simply polarized light), in contrast, oscillates only in a single plane passing through the path of propagation. Polarized light is produced by passing normal light through a polarizer such as a polarized lens or a Nicol prism.

Joseph Achille Le Bel (1847–1930), *a French chemist, inherited his family's fortune, which enabled him to establish his own laboratory. He and van't Hoff independently arrived at the reason for the optical activity of certain molecules. Although van't Hoff's explanation was more precise, both chemists are given credit for the work.*

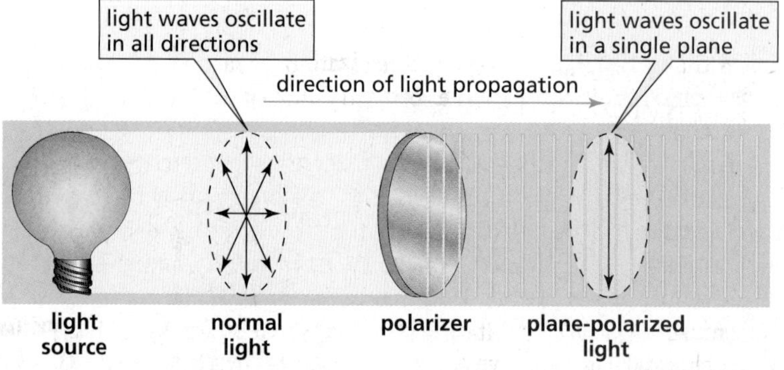

You experience the effect of a polarized lens with polarized sunglasses. Polarized sunglasses allow only light oscillating in a single plane to pass through them, so they block reflections (glare) more effectively than nonpolarized sunglasses.

In 1815, the physicist Jean-Baptiste Biot discovered that certain naturally occurring organic substances such as camphor and oil of turpentine are able to rotate the plane of polarization. He noted that some compounds rotated the plane of polarization clockwise and others counterclockwise, while some did not rotate the plane of polarization at all. He predicted that the ability to rotate the plane of polarization was attributable to some asymmetry in the molecules. Van't Hoff and Le Bel later determined that the molecular asymmetry was associated with compounds having one or more asymmetric carbons.

▲ When light is filtered through two polarized lenses at a 90° angle to one other, no light is transmitted through them.

When polarized light passes through a solution of achiral molecules, the light emerges from the solution with its plane of polarization unchanged. *An achiral compound does not rotate the plane of polarization. It is optically inactive.*

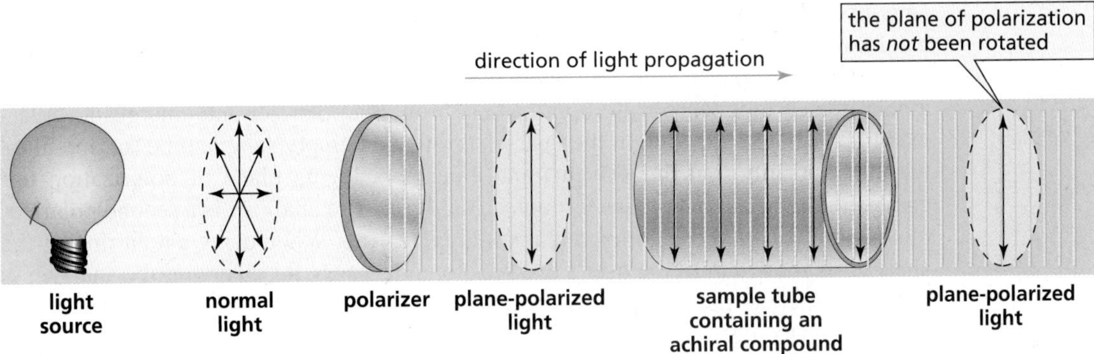

However, when polarized light passes through a solution of a chiral compound, the light emerges with its plane of polarization changed. Thus, *a chiral compound rotates the plane of polarization*. A chiral compound will rotate the plane of polarization clockwise or counterclockwise. If one enantiomer rotates the plane of polarization clockwise, its mirror image will rotate the plane of polarization exactly the same amount counterclockwise.

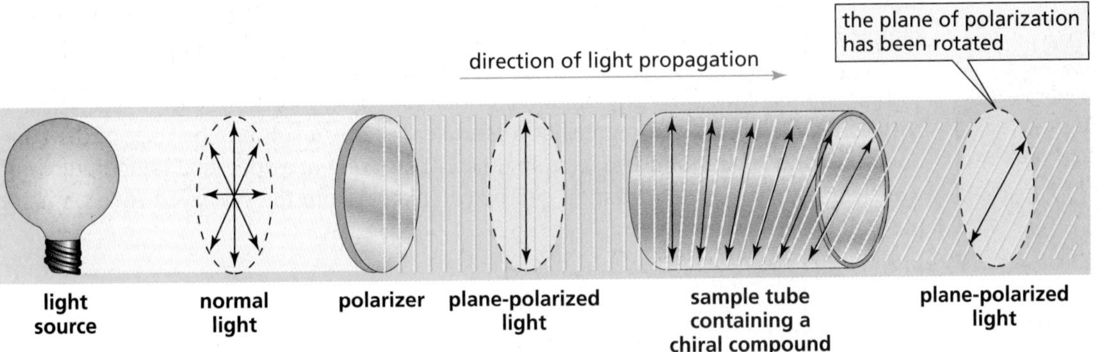

A compound that rotates the plane of polarization is said to be **optically active**. In other words, chiral compounds are optically active and achiral compounds are **optically inactive**.

If an optically active compound rotates the plane of polarization clockwise, it is called **dextrorotatory**, indicated by (+). If an optically active compound rotates the plane of polarization counterclockwise, it is called **levorotatory**, indicated by (−). *Dextro* and *levo* are Latin prefixes for "to the right" and "to the left," respectively. Sometimes lowercase *d* and *l* are used instead of (+) and (−).

Do not confuse (+) and (−) with *R* and *S*. The (+) and (−) symbols indicate the direction in which an optically active compound rotates the plane of polarization, whereas *R* and *S* indicate the arrangement of the groups about an asymmetric carbon. Some compounds with the *R* configuration are (+) and some are (−).

Some molecules with the *R* configuration are (+), and some molecules with the *R* configuration are (−).

The degree to which an optically active compound rotates the plane of polarization can be measured with an instrument called a **polarimeter** (Figure 5.2). Because the

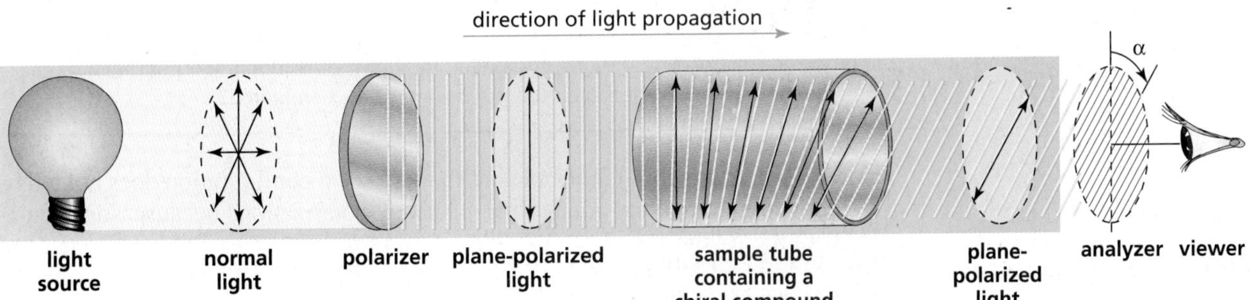

▼ Figure 5.2
Schematic of a polarimeter.

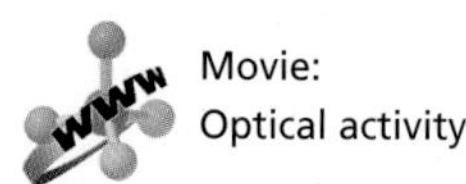

Movie:
Optical activity

amount of rotation will vary with the wavelength of the light used, the light source for a polarimeter must produce monochromatic (single wavelength) light. Most polarimeters use light from a sodium arc (called the sodium D-line; wavelength = 589 nm). In a polarimeter, monochromatic light passes through a polarizer and emerges as polarized light. The polarized light then passes through an empty sample tube (or one filled with an optically inactive solvent) and emerges with its plane of polarization unchanged. The light then passes through an analyzer. The analyzer is a second polarizer mounted on an eyepiece with a dial marked in degrees. When using a polarimeter, the analyzer is rotated until the user's eye sees total darkness. At this point the analyzer is at a right angle to the first polarizer, so no light passes through. This analyzer setting corresponds to zero rotation.

The sample to be measured is then placed in the sample tube. If the sample is optically active, it will rotate the plane of polarization. The analyzer will no longer block all the light, so light reaches the user's eye. The user then rotates the analyzer again until no light passes through. The degree to which the analyzer is rotated can be read from the dial and represents the difference between an optically inactive sample and the optically active sample. This is called the **observed rotation** (α); it is measured in degrees. The observed rotation depends on the number of optically active molecules the light encounters in the sample. This, in turn, depends on the concentration of the sample and the length of the sample tube. The observed rotation also depends on the temperature and the wavelength of the light source.

Jacobus Hendricus van't Hoff (1852–1911), *a Dutch chemist, was a professor of chemistry at the University of Amsterdam and later at the University of Berlin. He received the first Nobel Prize in chemistry (1901) for his work on solutions.*

Born in France, **Jean-Baptiste Biot (1774–1862)** *was imprisoned for taking part in a street riot during the French Revolution. He became a professor of mathematics at the University of Beauvais and later a professor of physics at the Collège de France. He was awarded the Legion of Honor by Louis XVIII. (Also see p. 212.)*

Each optically active compound has a characteristic specific rotation. The **specific rotation** is the number of degrees of rotation caused by a solution of 1.0 g of the compound per mL of solution in a sample tube 1.0 dm long at a specified temperature and wavelength. The specific rotation can be calculated from the observed rotation using the following formula:

$$[\alpha]_{\lambda}^{T} = \frac{\alpha}{l \times c}$$

where $[\alpha]$ is the specific rotation; T is temperature in °C; λ is the wavelength of the incident light (when the sodium D-line is used, λ is indicated as D); α is the observed rotation; l is the length of the sample tube in decimeters; and c is the concentration of the sample in grams per milliliter of solution.

For example, one enantiomer of 2-methyl-1-butanol has been found to have a specific rotation of +5.75°. Because its mirror image rotates the plane of polarization the same amount but in the opposite direction, the specific rotation of the other enantiomer must be −5.75°.

CH_2OH, CH_3, H, CH_2CH_3 — **(*R*)-2-methyl-1-butanol**

$[\alpha]_{D}^{20\,°C} = +5.75°$

CH_2OH, H, CH_3, CH_2CH_3 — **(*S*)-2-methyl-1-butanol**

$[\alpha]_{D}^{20\,°C} = -5.75°$

PROBLEM 12◆

The observed rotation of 2.0 g of a compound in 50 mL of solution in a polarimeter tube 50-cm long is +13.4°. What is the specific rotation of the compound?

Knowing whether a chiral molecule has the *R* or the *S* configuration does not tell us the direction the compound rotates the plane of polarization, because some compounds with the *R* configuration rotate the plane to the right (+) and some rotate the plane to the left (−). We can tell by looking at the structure of a compound whether it has the *R* or the *S* configuration, but the only way we can tell whether a compound is dextrorotatory (+) or levorotatory (−) is to put the compound in a polarimeter. For example, (*S*)-lactic acid and (*S*)-sodium lactate have the same configuration, but (*S*)-lactic acid is dextrorotatory whereas (*S*)-sodium lactate is levorotatory. When we know the direction an optically active compound rotates the plane of polarization, we can incorporate (+) or (−) into its name.

CH_3, C, H, HO, COOH — (*S*)-(+)-lactic acid

CH_3, C, H, HO, COO^-Na^+ — (*S*)-(−)-sodium lactate

PROBLEM 13◆

a. Is (*R*)-lactic acid dextrorotatory or levorotatory?
b. Is (*R*)-sodium lactate dextrorotatory or levorotatory?

A mixture of equal amounts of two enantiomers—such as (*R*)-(−)-lactic acid and (*S*)-(+)-lactic acid—is called a **racemic mixture** or a **racemate**. Racemic mixtures do not rotate the plane of polarized light. They are optically inactive because for every molecule in a racemic mixture that rotates the plane of polarization in one direction, there is a mirror-image molecule that rotates the plane in the opposite direction. As a result, the light emerges from a racemic mixture with its plane of polarization unchanged. The symbol (±) is used to specify a racemic mixture. Thus, (±)-2-bromobutane indicates a mixture of (+)-2-bromobutane and an equal amount of (−)-2-bromobutane.

PROBLEM 14◆

(*S*)-(+)-Monosodium glutamate (MSG) is a flavor enhancer used in many foods. Some people have an allergic reaction to MSG (headache, chest pains, and an overall feeling of weakness). "Fast food" often contains substantial amounts of MSG, and it is widely used in Chinese food as well. MSG has a specific rotation of +24°.

$COO^-\ Na^+$, C, H, $^-OOCCH_2CH_2$, $^+NH_3$

(*S*)-(+)-monosodium glutamate

a. What is the specific rotation of (*R*)-(−)-monosodium glutamate?
b. What is the specific rotation of a racemic mixture of MSG?

5.8 Optical Purity and Enantiomeric Excess

Whether a particular sample consists of a single enantiomer or a mixture of enantiomers can be determined by its *observed specific rotation*. For example, an **enantiomerically pure** sample—meaning only one enantiomer is present—of (*S*)-(+)-2-bromobutane will have an *observed specific rotation* of +23.1° because the *specific rotation* of (*S*)-(+)-2-bromobutane is +23.1°. If, however, the sample of 2-bromobutane has an observed specific rotation of 0°, we will know that the compound is a racemic mixture. If the observed specific rotation is positive but less than +23.1°, we will know that we have a mixture of enantiomers and the mixture contains more of the enantiomer with the *S* configuration than the enantiomer with the *R* configuration. From the observed specific rotation, we can calculate the **optical purity (op)** of the mixture.

$$\text{optical purity} = \frac{\text{observed specific rotation}}{\text{specific rotation of the pure enantiomer}}$$

For example, if a sample of 2-bromobutane has an observed specific rotation of $+9.2°$, its optical purity is 0.40. In other words, it is 40% optically pure—40% of the mixture consists of an excess of a single enantiomer.

$$\text{optical purity} = \frac{+9.2°}{+23.1°} = 0.40 \text{ or } 40\%$$

Because the observed specific rotation is positive, we know that the solution contains excess (*S*)-(+)-2-bromobutane. The **enantiomeric excess (ee)** tells us how much excess (*S*)-(+)-2-bromobutane is in the mixture. As long as the compound is chemically pure, enantiomeric excess and optical purity will be the same.

$$\textbf{enantiomeric excess} = \frac{\textbf{excess of a single enantiomer}}{\textbf{entire mixture}} \times \textbf{100\%}$$

$$= \frac{\textbf{40}}{\textbf{100}} = \textbf{40\%}$$

If the mixture has a 40% enantiomeric excess, 40% of the mixture is excess *S* enantiomer and 60% is a racemic mixture. Half of the racemic mixture plus the amount of excess *S* enantiomer equals the amount of the *S* enantiomer present in the mixture. Thus, 70% of the mixture is the *S* enantiomer $(1/2 \times 60 + 40)$ and 30% is the *R* enantiomer.

PROBLEM 15◆

(+)-Mandelic acid has a specific rotation of $+158°$. What would be the observed specific rotation of each of the following mixtures?

a. 25% (−)-mandelic acid and 75% (+)-mandelic acid
b. 50% (−)-mandelic acid and 50% (+)-mandelic acid
c. 75% (−)-mandelic acid and 25% (+)-mandelic acid

PROBLEM 16◆

Naproxen, a nonsteroidal anti-inflammatory drug, is the active ingredient in Aleve. Naproxen has a specific rotation of $+66°$ in chloroform. One commercial preparation results in a mixture that is 97% optically pure.

a. Does naproxen have the *R* or the *S* configuration?
b. What percent of each enantiomer is obtained from the commercial preparation?

PROBLEM 17 SOLVED

A solution prepared by mixing 10 mL of a 0.10 M solution of the *R* enantiomer and 30 mL of a 0.10 M solution of the *S* enantiomer was found to have an observed specific rotation of $+4.8°$. What is the specific rotation of each of the enantiomers?

SOLUTION One (10.0 mL × 0.10 M) millimole (mmol) of the *R* enantiomer is mixed with 3 (30.0 mL × 0.10 M) mmol of the *S* enantiomer; 1 mmol of the *R* enantiomer plus 1 mmol of the *S* enantiomer will form 2 mmol of a racemic mixture. There will be 2 mmol of *S* enantiomer left over. Therefore, 2 mmol out of 4 mmol is excess *S* enantiomer $(2/4 = 0.50)$. The solution is 50% optically pure.

$$\text{optical purity} = 0.50 = \frac{\text{observed specific rotation}}{\text{specific rotation of the pure enantiomer}}$$

$$0.50 = \frac{+4.8°}{x}$$

$$x = +9.6°$$

The *S* enantiomer has a specific rotation of $+9.6°$; the *R* enantiomer has a specific rotation of $-9.6°$.

5.9 Isomers with More than One Asymmetric Carbon

Many organic compounds have more than one asymmetric carbon. The more asymmetric carbons a compound has, the more stereoisomers are possible for the compound. If we know how many asymmetric carbons a compound has, we can calculate the maximum number of stereoisomers for that compound: *a compound can have a maximum of* 2^n *stereoisomers* (provided it doesn't have any other stereocenters), *where* n *equals the number of asymmetric carbons*. For example, 3-chloro-2-butanol has two asymmetric carbons. Therefore, it can have as many as four ($2^2 = 4$) stereoisomers. The four stereoisomers are shown both as perspective formulas and as Fischer projections.

$CH_3\overset{*}{C}H\overset{*}{C}HCH_3$ (with Cl on C-3 and OH on C-2)

3-chloro-2-butanol

erythro enantiomers (1, 2) threo enantiomers (3, 4)

perspective formulas of the stereoisomers of 3-chloro-2-butanol (staggered)

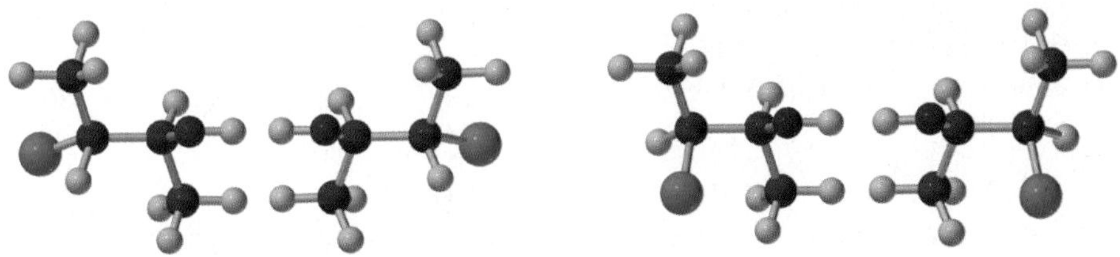

stereoisomers of 3-chloro-2-butanol

3-D Molecules: (2*S*,3*S*)-3-Chloro-2-butanol; (2*S*,3*S*)-3-Chloro-2-butanol; (2*S*,3*S*)-3-Chloro-2-butanol; (2*R*,3*R*)-3-Chloro-2-butanol

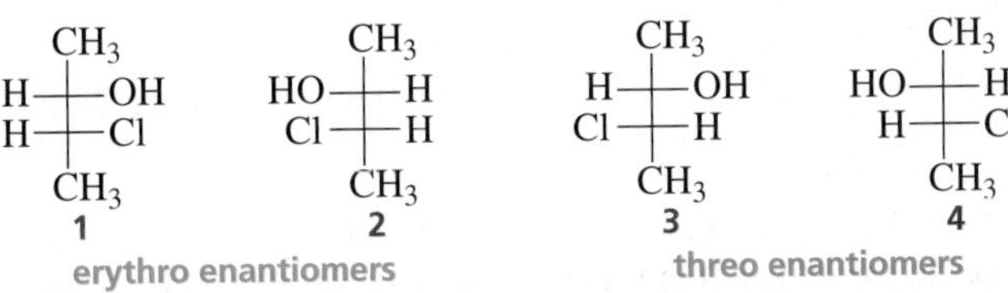

erythro enantiomers threo enantiomers

Fisher projections of the stereoisomers of 3-chloro-2-butanol

The four stereoisomers of 3-chloro-2-butanol consist of two pairs of enantiomers. Stereoisomers **1** and **2** are nonsuperimposable mirror images. They, therefore, are enantiomers. Stereoisomers **3** and **4** are also enantiomers. Stereoisomers **1** and **3** are not identical, and they are not mirror images. Such stereoisomers are called diastereomers. **Diastereomers** are stereoisomers that are not enantiomers. Numbers **1** and **4**, **2** and **3**, and **2** and **4** are also diastereomers. (Cis–trans isomers are also considered to be diastereomers because they are stereoisomers that are not enantiomers.)

Diastereomers are stereoisomers that are not enantiomers.

Enantiomers have identical physical properties (except for the way they interact with polarized light) and identical chemical properties—they react at the same rate with a given achiral reagent. Diastereomers have different physical properties (different melting points, different boiling points, different solubilities, different specific rotations, and so on) and different chemical properties—they react with the same achiral reagent at different rates.

When Fischer projections are drawn for stereoisomers with two adjacent asymmetric carbons (such as those for 3-chloro-2-butanol), the enantiomers with similar groups on the same side of the carbon chain are called the **erythro enantiomers** (Section 22.3). Those with similar groups on opposite sides are called the **threo enantiomers**. Therefore, **1** and **2** are the erythro enantiomers of 3-chloro-2-butanol (the hydrogens are on the same side), whereas **3** and **4** are the threo enantiomers. In each of the Fischer projections shown here, the horizontal bonds project out of the paper toward the viewer and the vertical bonds extend behind the paper away from the viewer. Groups can rotate freely about the carbon–carbon single bonds, but Fischer projections show the stereoisomers in their eclipsed conformations.

A Fischer projection does not show the three-dimensional structure of the molecule, and it represents the molecule in a relatively unstable eclipsed conformation. Most chemists, therefore, prefer to use perspective formulas because they show the molecule's three-dimensional structure in a stable, staggered conformation, so they provide a more accurate representation of structure. When perspective formulas are drawn to show the stereoisomers in their less stable eclipsed conformations, it can easily be seen—as the eclipsed Fischer projections show—that the erythro isomers have similar groups on the same side. We will use both prespective formulas and Fischer projections to depict the arrangement of groups bonded to an asymmetric carbon.

1   2 — erythro enantiomers   3   4 — threo enantiomers

perspective formulas of the stereoisomers of 3-chloro-2-butanol (eclipsed)

PROBLEM 18

The following compound has only one asymmetric carbon. Why then does it have four stereoisomers?

$$CH_3CH_2\overset{*}{C}HCH_2CH{=}CHCH_3 \quad (\text{Br on the starred carbon})$$

PROBLEM 19◆

a. Stereoisomers with two asymmetric carbons are called ______ if the configuration of both asymmetric carbons in one isomer is the opposite of the configuration of the asymmetric carbons in the other isomer.

b. Stereoisomers with two asymmetric carbons are called ______ if the configuration of both asymmetric carbons in one isomer is the same as the configuration of the asymmetric carbons in the other isomer.

c. Stereoisomers with two asymmetric carbons are called ______ if one of the asymmetric carbons has the same configuration in both isomers and the other asymmetric carbon has the opposite configuration in the two isomers.

PROBLEM 20◆

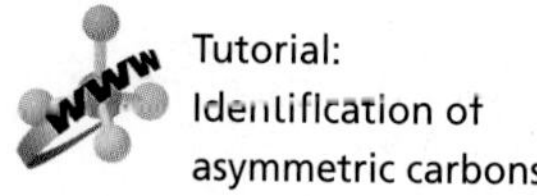

Tutorial: Identification of asymmetric carbons

a. How many asymmetric carbons does cholesterol have?

b. What is the maximum number of stereoisomers that cholesterol can have?

c. How many of these stereoisomers are found in nature?

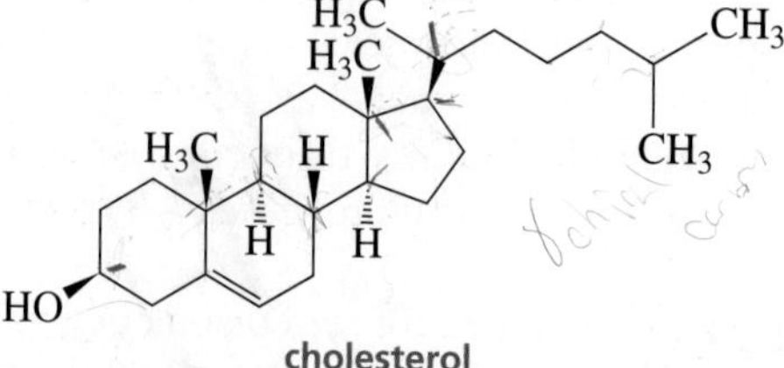

cholesterol

PROBLEM 21

Draw the stereoisomers of 2,4-dichlorohexane. Indicate pairs of enantiomers and pairs of diastereomers.

3-D Molecules: (1*R*,2*S*)-1-Bromo-2-methylcyclopentane; (1*S*,2*R*)-1-Bromo-2-methylcyclopentane; (1*R*,2*R*)-1-Bromo-2-methylcyclopentane; (1*S*,2*S*)-1-Bromo-2-methylcyclopentane

1-Bromo-2-methylcyclopentane also has two asymmetric carbons and four stereoisomers. Because the compound is cyclic, the substituents can be in either the cis or the trans configuration. The cis isomer exists as a pair of enantiomers, and the trans isomer exists as a pair of enantiomers.

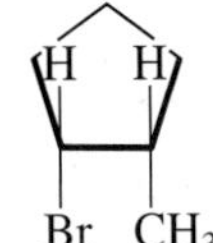

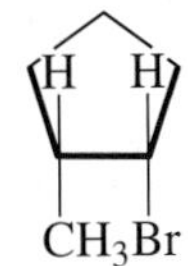

cis-1-bromo-2-methylcyclopentane

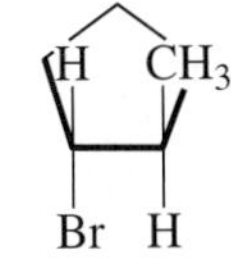

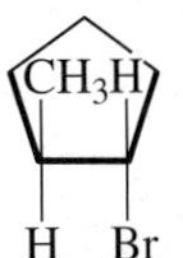

trans-1-bromo-2-methylcyclopentane

1-Bromo-3-methylcyclobutane does not have any asymmetric carbons. C-1 has a bromine and a hydrogen attached to it, but its other two groups ($—CH_2CH(CH_3)CH_2—$) are identical; C-3 has a methyl group and a hydrogen attached to it, but its other two groups ($—CH_2CH(Br)CH_2—$) are identical. Because the compound does not have a carbon with four different groups attached to it, it has only two stereoisomers, the cis isomer and the trans isomer. The cis and trans isomers do not have enantiomers.

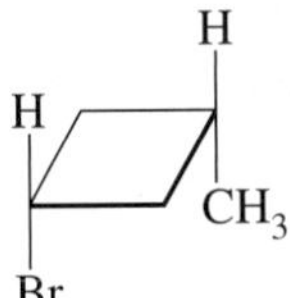

cis-1-bromo-3-methylcyclobutane

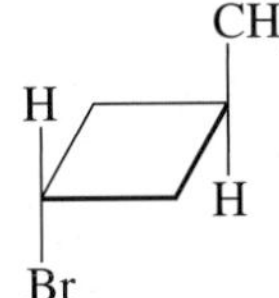

trans-1-bromo-3-methylcyclobutane

1-Bromo-3-methylcyclohexane has two asymmetric carbons. The carbon that is bonded to a hydrogen and a bromine is also bonded to two different carbon-containing groups ($—CH_2CH(CH_3)CH_2CH_2CH_2—$ and $—CH_2CH_2CH_2CH(CH_3)CH_2—$), so it is an asymmetric carbon. The carbon that is bonded to a hydrogen and a methyl group is also bonded to two different carbon-containing groups, so it is also an asymmetric carbon.

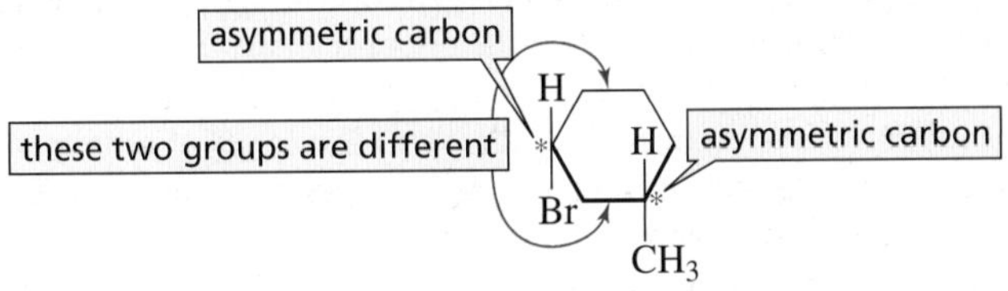

Because the compound has two asymmetric carbons, it has four stereoisomers. Enantiomers can be drawn for the cis isomer, and enantiomers can be drawn for the trans isomer.

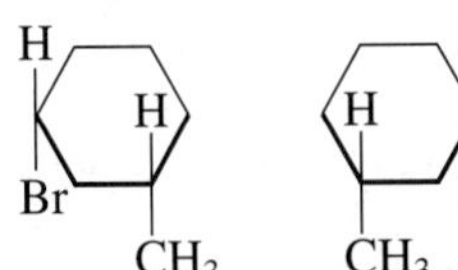

cis-1-bromo-3-methylcyclohexane

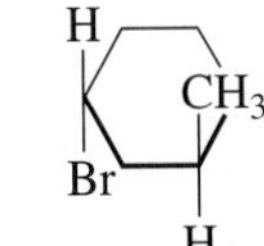

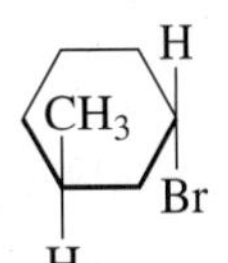

trans-1-bromo-3-methylcyclohexane

1-Bromo-4-methylcyclohexane has no asymmetric carbons. Therefore, the compound has only one cis isomer and one trans isomer.

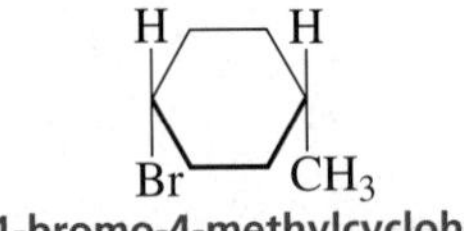

cis-1-bromo-4-methylcyclohexane

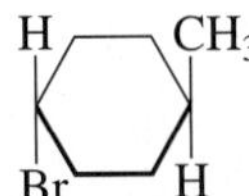

trans-1-bromo-4-methylcyclohexane

PROBLEM 22

Draw all possible stereoisomers for each of the following compounds:

a. 2-chloro-3-hexanol
b. 2-bromo-4-chlorohexane
c. 2,3-dichloropentane
d. 1,3-dibromopentane

PROBLEM 23

Draw the stereoisomers of 1-bromo-3-chlorocyclohexane.

PROBLEM 24◆

Of all the possible cyclooctanes that have one chloro substituent and one methyl substituent, which ones do not have any asymmetric carbons?

PROBLEM 25

Draw a diastereomer for each of the following.

a. 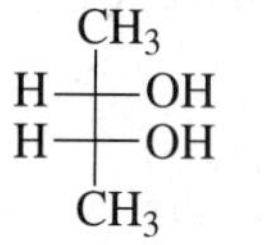 Fischer projection: CH_3 top; H—C—OH; H—C—OH; CH_3 bottom

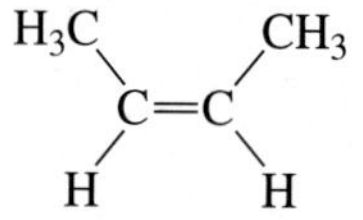

b. 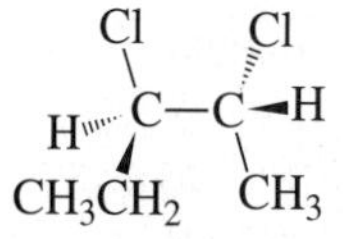 Cl, H, CH_3CH_2 on C—C bearing Cl, H, CH_3

c. H_3C and CH_3; C=C; H and H

d. 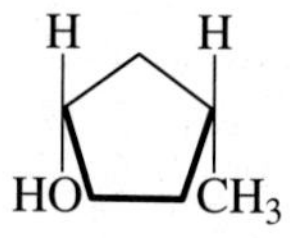 cyclopentane with H, HO and H, CH_3

5.10 Meso Compounds

In the examples we have just seen, each compound with two asymmetric carbons has four stereoisomers. However, some compounds with two asymmetric carbons have only three stereoisomers. This is why we emphasized in Section 5.9 that the *maximum* number of stereoisomers a compound with *n* asymmetric carbons can have (provided it doesn't have any other stereocenters) is 2^n, instead of stating that a compound with *n* asymmetric carbons has 2^n stereoisomers.

An example of a compound with two asymmetric carbons that has only three stereoisomers is 2,3-dibromobutane.

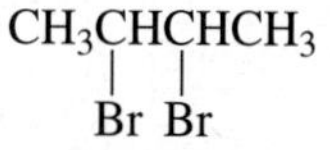

$CH_3CHCHCH_3$ (Br on C2 and C3)

2,3-dibromobutane

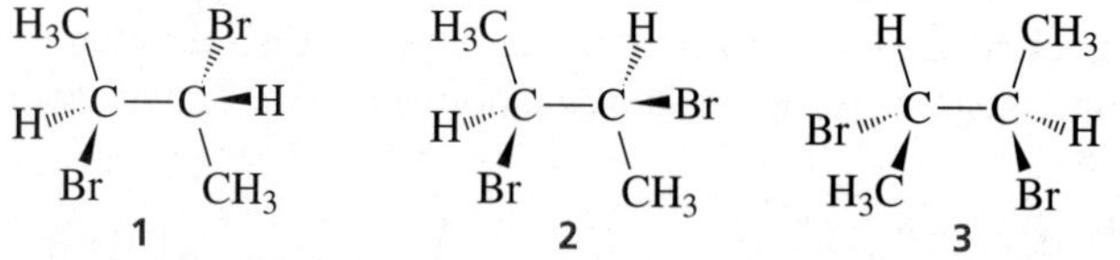

1: H_3C, H, Br — C—C — Br, H, CH_3
2: H_3C, H, Br — C—C — H, Br, CH_3
3: H, Br, H_3C — C—C — CH_3, H, Br

perspective formulas of the stereoisomers of 2,3-dibromobutane (staggered)

The "missing" isomer is the mirror image of **1** because **1** and its mirror image are the same molecule. This can be seen more clearly if you look either at the perspective formulas drawn in their eclisped conformations or at the Fischer projections.

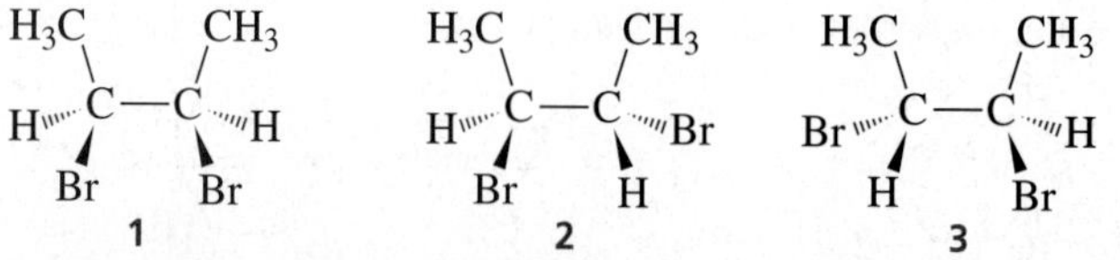

1: H_3C, H, Br — C—C — CH_3, H, Br
2: H_3C, H, Br — C—C — CH_3, Br, H
3: H_3C, Br, H — C—C — CH_3, H, Br

perspective formulas of the stereoisomers of 2,3-dibromobutane (eclipsed)

1 **2** **3**

Fischer projections of the stereoisomers of 2,3-dibromobutane

It is obvious that **1** and its mirror image are identical when looking at the perspective formula in the eclipsed conformation. To convince yourself that the Fischer projection of **1** and its mirror image are identical, rotate the mirror image by 180°. *(Remember, you can move Fischer projections only by rotating them 180° in the plane of the paper.)*

superimposable mirror image **superimposable mirror image**

Stereoisomer **1** is called a *meso compound.* Even though a **meso** (mee-zo) **compound** has asymmetric carbons, it is an achiral molecule because it is superimposable on its mirror image. *Mesos* is the Greek word for "middle." A meso compound is achiral—when polarized light is passed through a solution of a meso compound, the plane of polarization is not rotated. A meso compound can be recognized by the fact that it has two or more asymmetric carbons and a plane of symmetry. *If a compound has a plane of symmetry, it will not be optically active even though it has asymmetric carbons.* A plane of symmetry cuts the molecule in half, and one-half is the mirror image of the other half. Stereoisomer **1** has a **plane of symmetry**, which means that it does *not* have a nonsuperimposable mirror image—it does not have an enantiomer.

A meso compound has two or more asymmetric carbons and a plane of symmetry.

A meso compound is achiral.

A chiral compound cannot have a plane of symmetry.

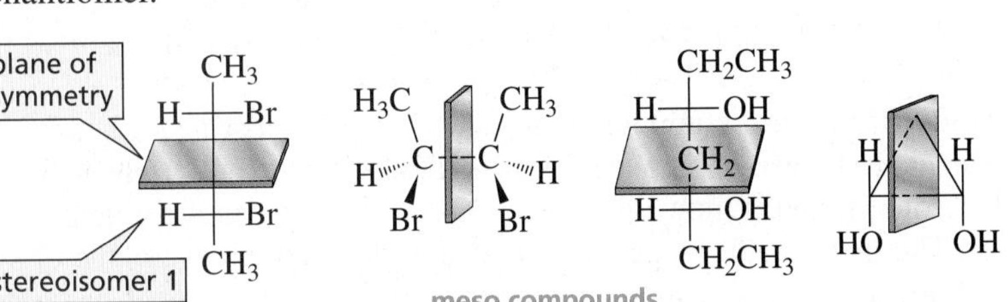

meso compounds

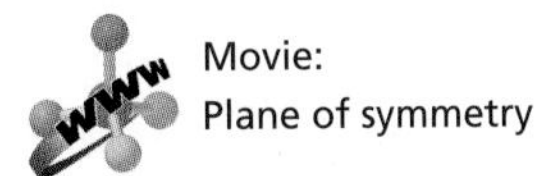

Movie: Plane of symmetry

It is easy to recognize when a compound with two asymmetric carbons has a stereoisomer that is a meso compound—the four atoms or groups bonded to one asymmetric carbon are identical to the four atoms or groups bonded to the other asymmetric carbon. A compound with the same four atoms or groups bonded to two different asymmetric carbons will have three stereoisomers: One will be a meso compound, and the other two will be enantiomers.

If a compound with two asymmetric carbons has the same four groups bonded to each of the asymmetric carbons, one of its stereoisomers will be a meso compound.

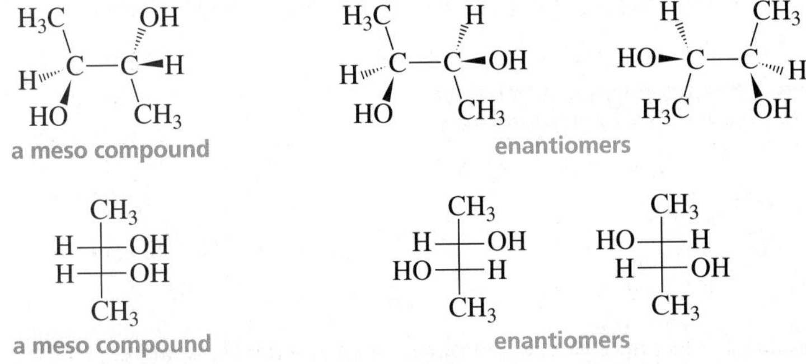

a meso compound   enantiomers

a meso compound   enantiomers

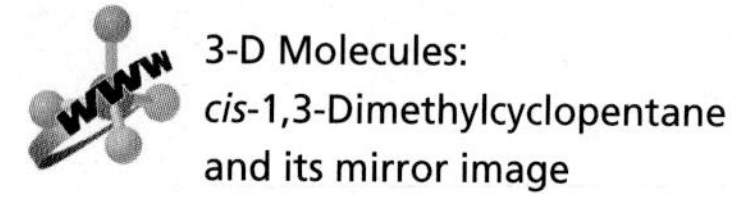

3-D Molecules: *cis*-1,3-Dimethylcyclopentane and its mirror image

In the case of cyclic compounds, the cis isomer will be the meso compound and the trans isomer will exist as enantiomers.

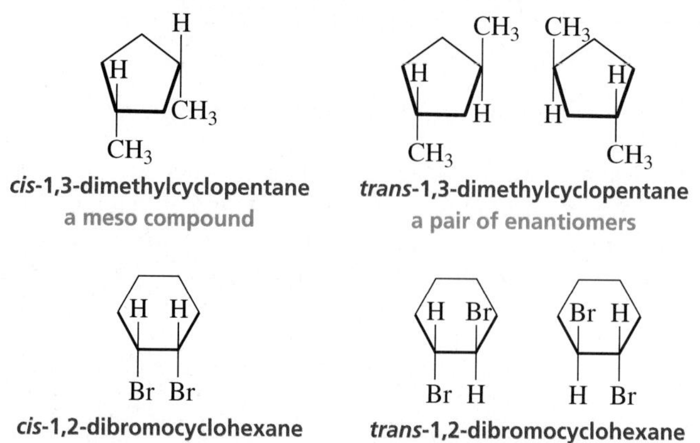

The preceding structure for *cis*-1,2-dibromocyclohexane suggests that the compound has a plane of symmetry. Cyclohexane, however, is not a planar hexagon—it exists preferentially in the chair conformation, and the chair conformer of *cis*-1,2-dibromocyclohexane does not have a plane of symmetry. Only the much less stable boat conformer of *cis*-1,2-dibromocyclohexane has a plane of symmetry. Then, is *cis*-1,2-dibromocyclohexane a meso compound? The answer is yes. As long as any one conformer of a compound has a plane of symmetry, the compound will be achiral, and an achiral compound with two asymmetric carbons is a meso compound.

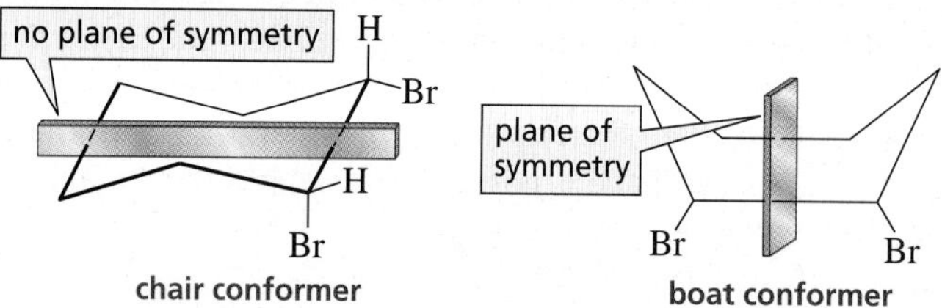

This holds for acyclic compounds as well. We have just seen that 2,3-dibromobutane is an achiral meso compound because it has a plane of symmetry. To see that it had a plane of symmetry, however, we had to look at a relatively unstable eclipsed conformer. The more stable staggered conformer does not have a plane of symmetry. 2,3-Dibromobutane is still a meso compound, however, because it has a conformer that has a plane of symmetry.

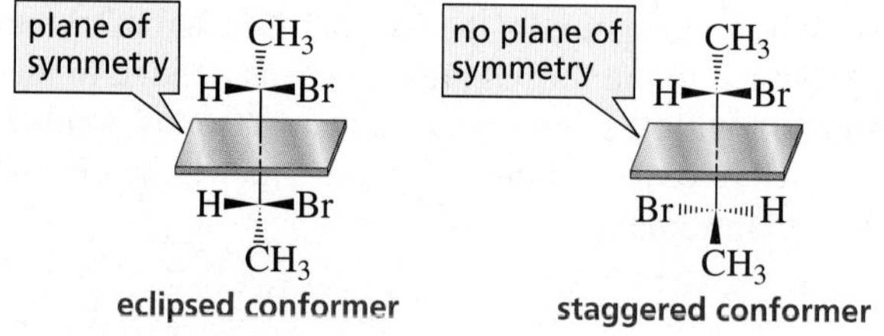

PROBLEM-SOLVING STRATEGY

Which of the following compounds has a stereoisomer that is a meso compound?

a. 2,3-dimethylbutane
b. 3,4-dimethylhexane
c. 2-bromo-3-methylpentane
d. 1,3-dimethylcyclohexane
e. 1,4-dimethylcyclohexane
f. 1,2-dimethylcyclohexane
g. 3,4-diethylhexane
h. 1-bromo-2-methylcyclohexane

Check each compound to see if it has the necessary requirements to have a stereoisomer that is a meso compound. That is, does it have two asymmetric carbons with the same four substituents attached to each of the asymmetric carbons?

Compounds A, E, and G do *not* have a stereoisomer that is a meso compound because they don't have any asymmetric carbons.

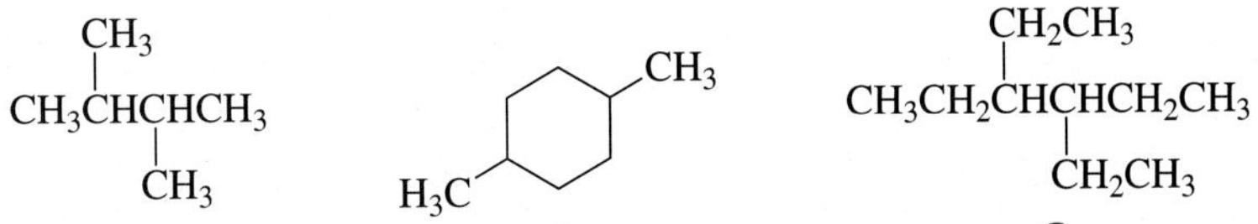

$CH_3CH(CH_3)CH(CH_3)CH_3$ (A) — 1,4-dimethylcyclohexane: H_3C, CH_3 (E) — $CH_3CH_2CH(CH_2CH_3)CH(CH_2CH_3)CH_2CH_3$ (G)

Compounds C and H each have two asymmetric carbons. They do *not* have a stereoisomer that is a meso compound because each of the asymmetric carbons is *not* bonded to the same four substituents.

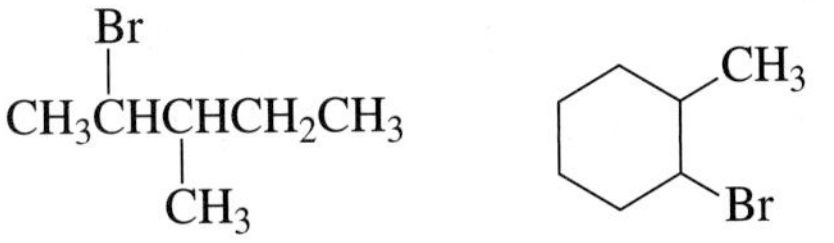

$CH_3CH(Br)CH(CH_3)CH_2CH_3$ (C) — cyclohexane with CH_3, Br (H)

Compounds B, D, and F have a stereoisomer that is a meso compound—they have two asymmetric carbons and each asymmetric carbon is bonded to the same four atoms or groups.

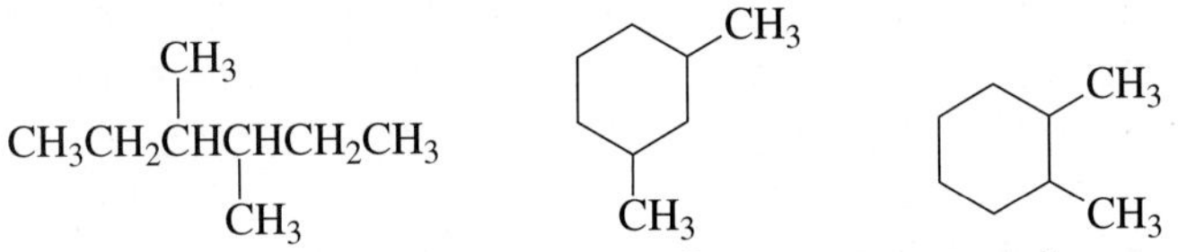

$CH_3CH_2CH(CH_3)CH(CH_3)CH_2CH_3$ (B) — cyclohexane with CH_3, CH_3 (D) — cyclohexane with CH_3, CH_3 (F)

The isomer that is the meso compound is the one with a plane of symmetry when an acyclic compound is drawn in its eclipsed conformation (B), or when a cyclic compound is drawn with a planar ring (D and F).

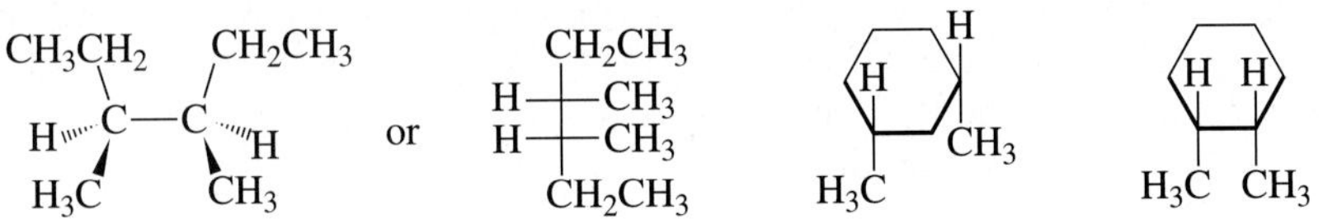

B: CH_3CH_2, CH_2CH_3, H, H, H_3C, CH_3 (eclipsed) or Fischer projection: CH_2CH_3 / H–CH_3 / H–CH_3 / CH_2CH_3

D: H, H, H_3C, CH_3 — F: H H, H_3C CH_3

Now continue on to Problem 26.

PROBLEM 26◆

Which of the following compounds has a stereoisomer that is a meso compound?

a. 2,4-dibromohexane
b. 2,4-dibromopentane
c. 2,4-dimethylpentane
d. 1,3-dichlorocyclohexane
e. 1,4-dichlorocyclohexane
f. 1,2-dichlorocyclobutane

PROBLEM 27 SOLVED

Which of the following are chiral?

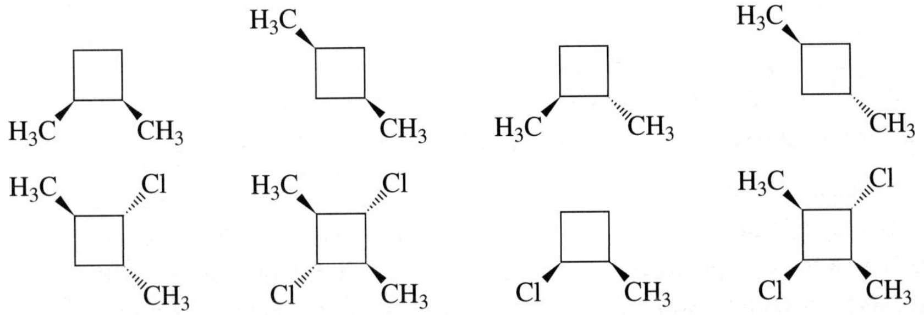

(cyclobutane structures with H_3C / CH_3 and H_3C / Cl / CH_3 / Cl substituents)

SOLUTION To be chiral, a molecule must not have plane of symmetry. Therefore, only the following compounds are chiral.

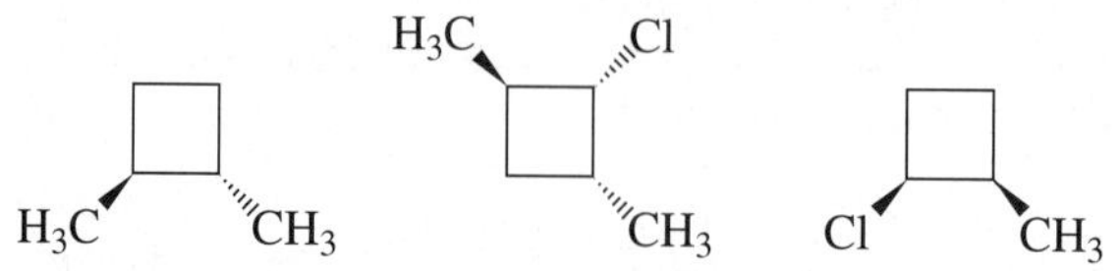

In the top row of compounds, only the third compound is chiral. The first, second, and fourth compounds each have a plane of symmetry. In the bottom row of compounds, the first and third compounds are chiral. The second and fourth compounds each have a plane of symmetry.

PROBLEM 28

Draw all the stereoisomers for each of the following compounds:

a. 1-bromo-2-methylbutane
b. 1-chloro-3-methylpentane
c. 2-methyl-1-propanol
d. 2-bromo-1-butanol
e. 3-chloro-3-methylpentane
f. 3-bromo-2-butanol
g. 3,4-dichlorohexane
h. 2,4-dichloropentane
i. 2,4-dichloroheptane
j. 1,2-dichlorocyclobutane
k. 1,3-dichlorocyclohexane
l. 1,4-dichlorocyclohexane
m. 1-bromo-2-chlorocyclobutane
n. 1-bromo-3-chlorocyclobutane

5.11 The *R,S* System of Nomenclature for Isomers with More than One Asymmetric Carbon

If a compound has more than one asymmetric carbon, the steps used to determine whether an asymmetric carbon has the *R* or the *S* configuration must be applied to each of the asymmetric carbons individually. As an example, let's name one of the stereoisomers of 3-bromo-2-butanol.

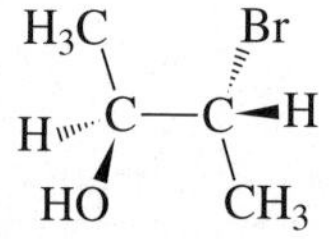

a stereoisomer of 3-bromo-2-butanol

First, we will determine the configuration at C-2. The OH group has the highest priority, the C-3 carbon (the C attached to Br, C, H) has the next highest priority, CH_3 is next, and H has the lowest priority. Because the group with the lowest priority is bonded by a hatched wedge, we can immediately draw an arrow from the group with the highest priority to the group with the next highest priority. Because that arrow points counterclockwise, the configuration at C-2 is *S*.

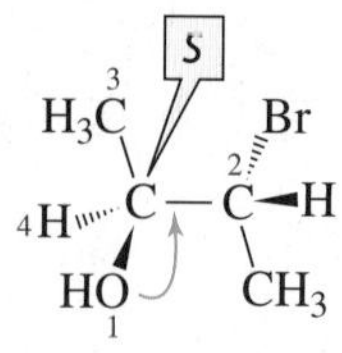

Now we need to determine the configuration at C-3. Because the group with the lowest priority (H) is not bonded by a hatched wedge, we must put it there by temporarily switching two groups.

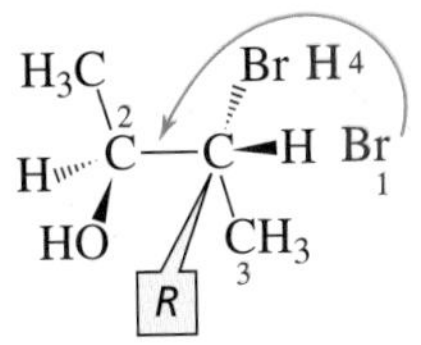

The arrow going from the highest priority group (Br) to the next highest priority group (the C attached to O, C, H) points counterclockwise, suggesting it has the *S* configuration. However, because we switched two groups before we drew the arrow, C-3 has the opposite configuration—it has the *R* configuration. Thus, the isomer is named (2*S*,3*R*)-3-bromo-2-butanol.

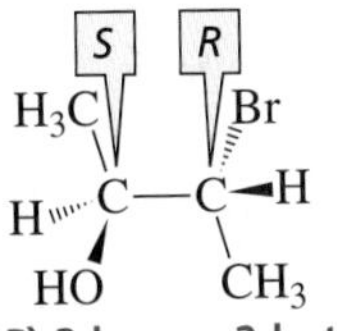

(2*S*,3*R*)-3-bromo-2-butanol

Fischer projections with two asymmetric carbons can be named in a similar manner. Just apply the steps to each asymmetric carbon that you learned for a Fischer projection with one asymmetric carbon. For C-2, the arrow from the group with the highest priority to the group with the next highest priority points clockwise, suggesting it has the *R* configuration. But because the group with the lowest priority is on a horizontal bond, we can conclude that C-2 has the *S* configuration (Section 5.6).

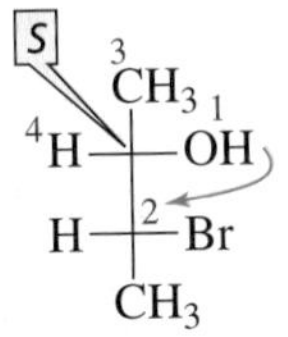

By repeating these steps for C-3, you will find that it has the *R* configuration. Thus, the isomer is named (2*S*,3*R*)-3-bromo-2-butanol.

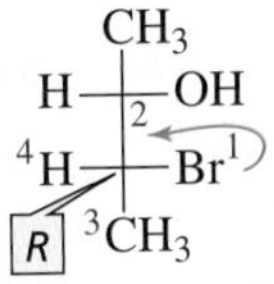

(2*S*,3*R*)-3-bromo-2-butanol

The four stereoisomers of 3-bromo-2-butanol are named as shown here. Take a few minutes to verify their names.

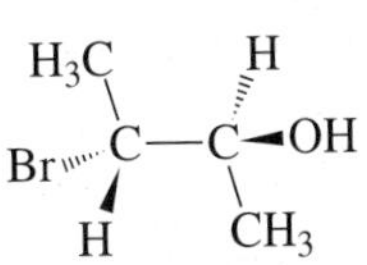

(2*S*,3*R*)-3-bromo-2-butanol 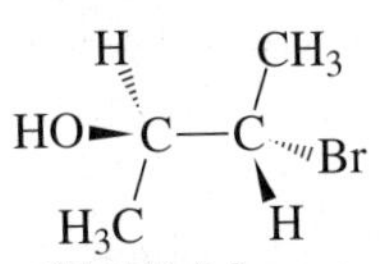 **(2*R*,3*S*)-3-bromo-2-butanol** 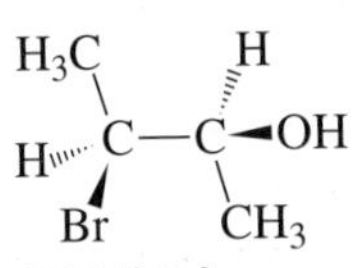 **(2*S*,3*S*)-3-bromo-2-butanol** 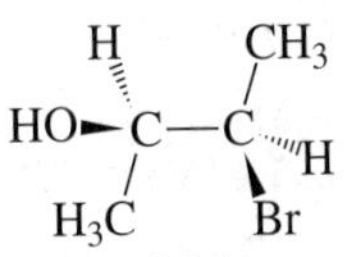 **(2*R*,3*R*)-3-bromo-2-butanol**

perspective formulas of the stereoisomers of 3-bromo-2-butanol

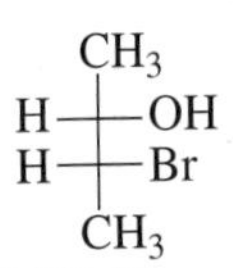

(2*S*,3*R*)-3-bromo-2-butanol 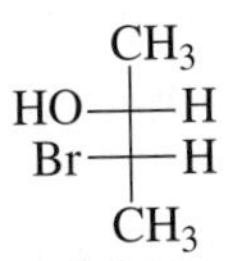 **(2*R*,3*S*)-3-bromo-2-butanol** 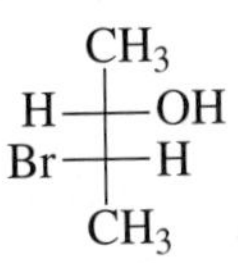 **(2*S*,3*S*)-3-bromo-2-butanol** 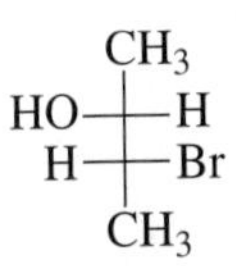 **(2*R*,3*R*)-3-bromo-2-butanol**

Fischer projections of the stereoisomers of 3-bromo-2-butanol

Notice that enantiomers have the opposite configuration at both asymmetric carbons, whereas diastereomers have the same configuration at one asymmetric carbon and the opposite configuration at the other.

PROBLEM 29

Draw and name the four stereoisomers of 1,3-dichloro-2-butanol using:

a. perspective formulas

b. Fischer projections

Tartaric acid has three stereoisomers because each of its two asymmetric carbons has the same set of four substituents. The meso compound and the pair of enantiomers are named as shown.

(2*R*,3*S*)-tartaric acid
a meso compound

(2*R*,3*R*)-tartaric acid

(2*S*,3*S*)-tartaric acid

a pair of enantiomers

perspective formulas of the stereoisomers of tartaric acid

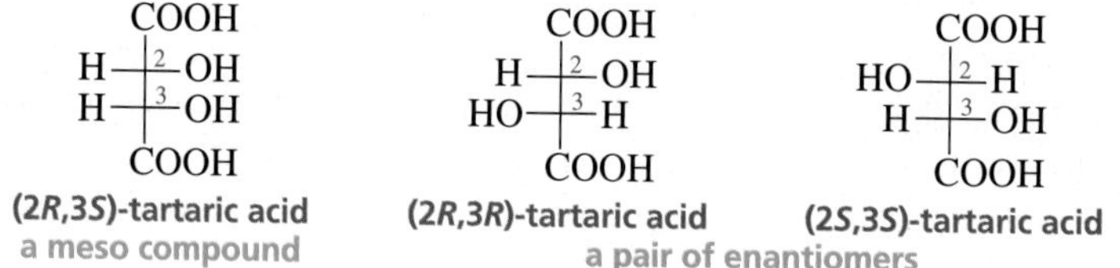

(2*R*,3*S*)-tartaric acid
a meso compound

(2*R*,3*R*)-tartaric acid

(2*S*,3*S*)-tartaric acid

a pair of enantiomers

Fischer projections of the stereoisomers of tartaric acid

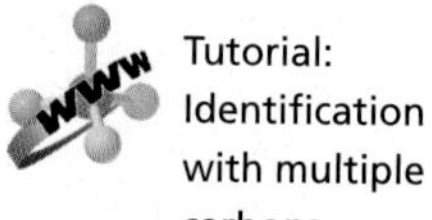

Tutorial:
Identification of stereoisomers with multiple asymmetric carbons

The physical properties of the three stereoisomers of tartaric acid are listed in Table 5.1. The meso compound and either one of the enantiomers are diastereomers. Notice that the physical properties of the enantiomers are the same, whereas the physical properties of the diastereomers are different. Also notice that the physical properties of the racemic mixture differ from the physical properties of the enantiomers.

Table 5.1 Physical Properties of the Stereoisomers of Tartaric Acid

	Melting point, °C	$[\alpha]_D^{25\,°C}$	Solubility, g/100 g H_2O at 15 °C
(2*R*,3*R*)-(+)-Tartaric acid	170	+11.98°	139
(2*S*,3*S*)-(−)-Tartaric acid	170	−11.98°	139
(2*R*,3*S*)-Tartaric acid	140	0°	125
(±)-Tartaric acid	206	0°	139

PROBLEM 30◆

Chloramphenicol is a broad-spectrum antibiotic that is particularly useful against typhoid fever. What is the configuration of each asymmetric carbon in chloramphenicol?

chloramphenicol

PROBLEM-SOLVING STRATEGY

Draw perspective formulas for the following compounds:

a. (*R*)-2-butanol

b. (2*S*,3*R*)-3-chloro-2-pentanol

a. First draw the compound—ignoring the configuration at the asymmetric carbon—so you know what groups are bonded to the asymmetric carbon.

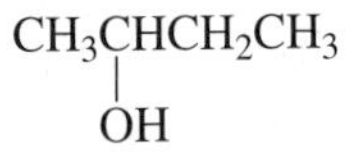

$CH_3CH(OH)CH_2CH_3$

2-butanol

Draw the bonds about the asymmetric carbon.

Put the group with the lowest priority on the hatched wedge. Put the group with the highest priority on any remaining bond.

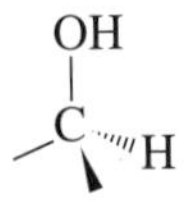

Because you have been asked to draw the *R* enantiomer, draw an arrow clockwise from the group with the highest priority to the next available bond and put the group with the next highest priority on that bond.

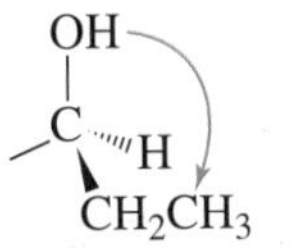

Put the remaining substituent on the last available bond.

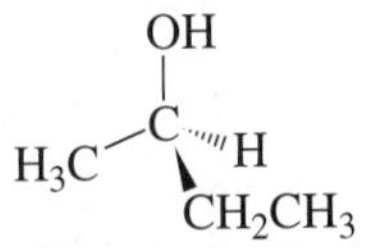

(*R*)-2-butanol

b. First draw the compound, ignoring the configuration at the asymmetric carbon.

$CH_3CH(OH)CH(Cl)CH_2CH_3$

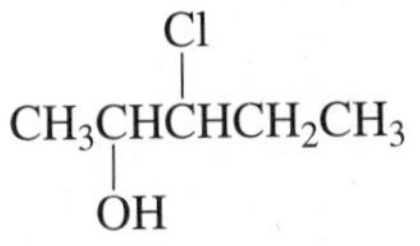

3-chloro-2-pentanol

Draw the bonds about the asymmetric carbons.

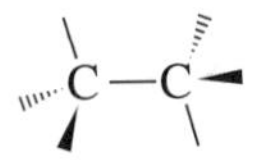

For each asymmetric carbon, put the group with the lowest priority on the hatched wedge.

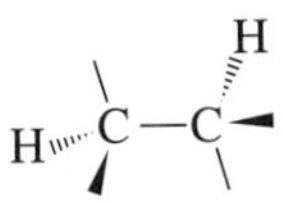

For each asymmetric carbon, put the group with the highest priority on a bond such that an arrow points clockwise (if you want the *R* configuration) or counterclockwise (if you want the *S* configuration) to the group with the next highest priority.

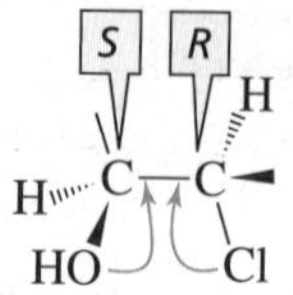

Put the remaining substituents on the last available bonds.

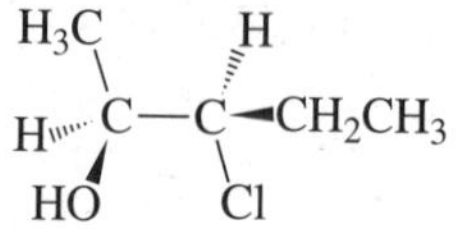

(2S,3R)-3-chloro-2-pentanol

Now continue on to Problem 31.

PROBLEM 31

Draw perspective formulas for the following compounds:

a. (*S*)-3-chloro-1-pentanol

b. (2*R*,3*R*)-2,3-dibromopentane

c. (2*S*,3*R*)-3-methyl-2-pentanol

d. (*R*)-1,2-dibromobutane

PROBLEM 32◆

For many centuries, the Chinese have used extracts from a group of herbs known as ephedra to treat asthma. Chemists have been able to isolate a compound from these herbs, which they named ephedrine, a potent dilator of air passages in the lungs.

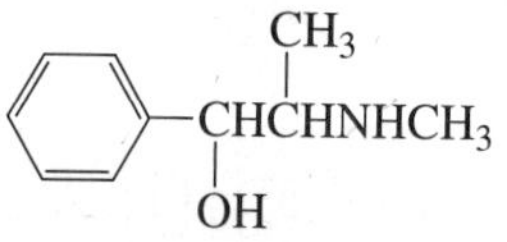

ephedrine

a. How many stereoisomers are possible for ephedrine?

b. The stereoisomer shown here is the one that is pharmacologically active. What is the configuration of each of the asymmetric carbons?

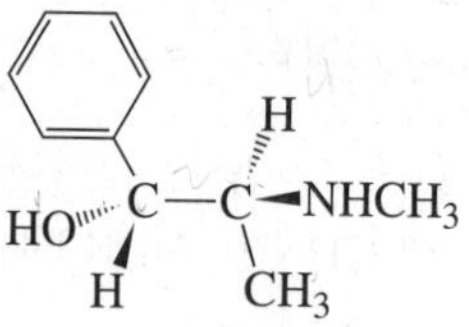

PROBLEM 33◆

Name the following compounds:

a.

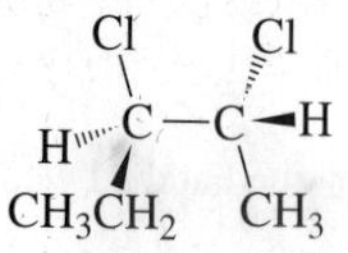

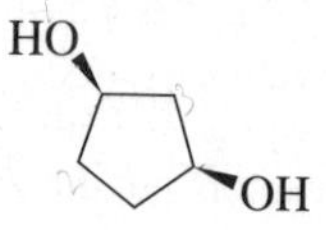

b.

c.

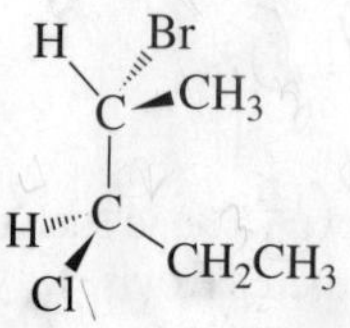

d.

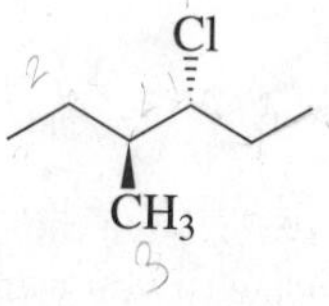

5.12 Reactions of Compounds that Contain an Asymmetric Carbon

When a compound that contains an asymmetric carbon undergoes a reaction, what happens to the configuration of the asymmetric carbon depends on the reaction. If the reaction does not break any of the four bonds to the asymmetric carbon, then the relative positions of the groups bonded to the asymmetric carbon will not change. For example, when (*S*)-1-chloro-3-methylhexane reacts with hydroxide ion, OH substitutes for Cl. The reactant and the product have the same **relative configuration** because the reaction does not break any of the bonds to the asymmetric carbon.

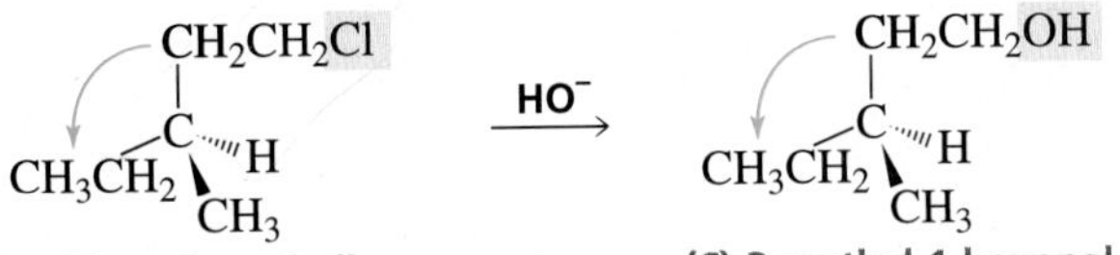

If a reaction does not break a bond to the asymmetric carbon, the reactant and the product will have the same relative configurations.

CH_2CH_2Cl, CH_3CH_2, H, CH_3 on C $\xrightarrow{HO^-}$ CH_2CH_2OH, CH_3CH_2, H, CH_3 on C

(*S*)-1-chloro-3-methylhexane (*S*)-3-methyl-1-hexanol

A word of warning: If the four groups bonded to the asymmetric carbon maintain their relative positions, it does not necessarily mean that an *S* reactant will always yield an *S* product as occurred in the preceding reaction. In the following example, the groups maintained their relative positions during the reaction. Therefore, the reactant and the product have the same *relative configurations*. However, the reactant has the *S* configuration, whereas the product has the *R* configuration. Although, the groups maintained their relative positions, their relative priorities—as defined by the Cahn–Ingold–Prelog rules—changed (Section 5.5). The change in priorities—not the change in positions of the groups—is what caused the *S* reactant to become an *R* product.

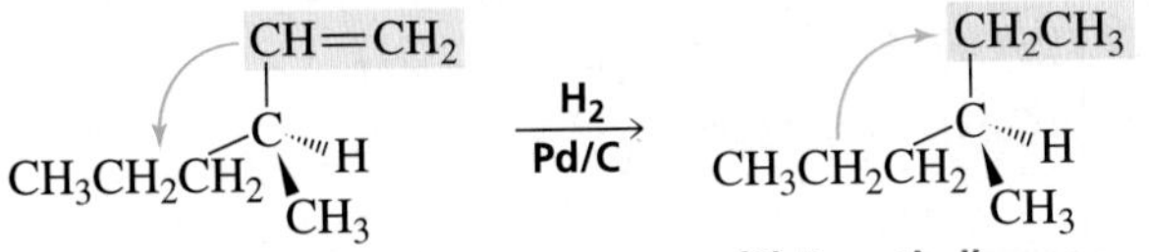

$CH{=}CH_2$, $CH_3CH_2CH_2$, H, CH_3 on C $\xrightarrow[Pd/C]{H_2}$ CH_2CH_3, $CH_3CH_2CH_2$, H, CH_3 on C

(*S*)-3-methylhexene (*R*)-3-methylhexane

The reactant and product in this example have the same relative configuration, but they have different **absolute configurations**—the reactant has the *S* configuration, whereas the product has the *R* configuration. The actual configuration is called the absolute configuration to indicate that the configuration is known in an absolute sense rather than in a relative sense. Knowing the *absolute configuration* of a compound means that you know whether it has the *R* or the *S* configuration. Knowing that two compounds have the same *relative configuration* means that they have the same relative positions of their substituents.

We have just seen that if the reaction does not break any of the bonds to the asymmetric carbon, the reactant and product will have the same relative configuration. In contrast, if the reaction *does break* a bond to the asymmetric carbon, the product can have the same relative configuration as the reactant or it can have the opposite relative configuration. Which of the products is actually formed depends on the mechanism of the reaction. Therefore, we cannot predict what the configuration of the product will be unless we know the mechanism of the reaction.

CH_2CH_3, CH_3, H, Y on C $\xrightarrow{Z^-}$ CH_2CH_3, CH_3, H, Z on C + CH_2CH_3, H, CH_3, Z on C + Y^-

has the same relative configuration as the reactant

has a relative configuration opposite to that of the reactant

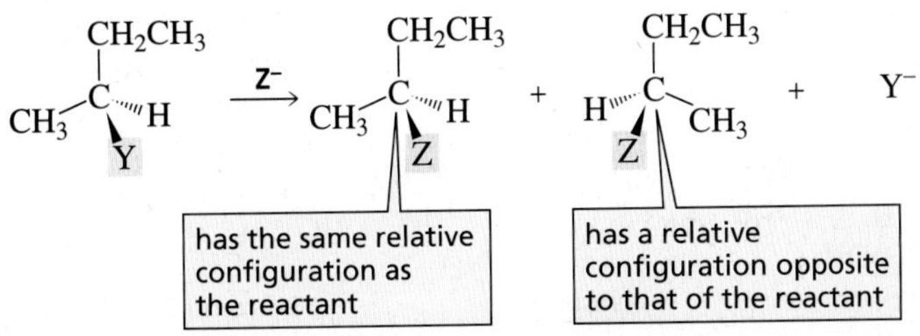

If a reaction does break a bond to the asymmetric carbon, you cannot predict the configuration of the product unless you know the mechanism of the reaction.

PROBLEM 34 SOLVED

(S)-(−)-2-Methyl-1-butanol can be converted to (+)-2-methylbutanoic acid without breaking any of the bonds to the asymmetric carbon. What is the configuration of (−)-2-methylbutanoic acid?

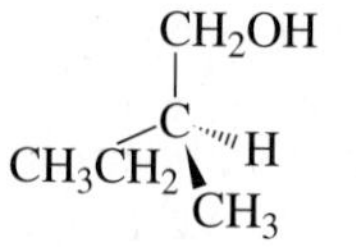

CH_2OH / C / CH_3CH_2, H, CH_3
(S)-(−)-2-methyl-1-butanol

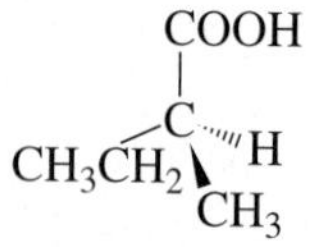

COOH / C / CH_3CH_2, H, CH_3
(+)-2-methylbutanoic acid

SOLUTION We know that (+)-2-methylbutanoic acid has the relative configuration shown because it was formed from (S)-(−)-2-methyl-1-butanol without breaking any bonds to the asymmetric carbon. Therefore, we know that (+)-2-methylbutanoic acid has the S configuration. We can conclude then that (−)-2-methylbutanoic acid has the R configuration.

PROBLEM 35◆

The stereoisomer of 1-iodo-2-methylbutane with the S configuration rotates the plane of polarized light counterclockwise. The following reaction results in an alcohol that rotates the plane of polarized light clockwise. What is the configuration of (−)-2-methyl-1-butanol?

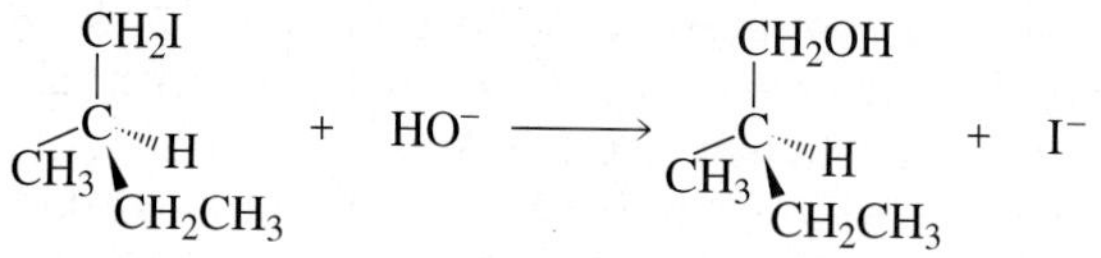

CH_2I / C / CH_3, H, CH_2CH_3 + HO^- ⟶ CH_2OH / C / CH_3, H, CH_2CH_3 + I^-

5.13 The Absolute Configuration of (+)-Glyceraldehyde

Glyceraldehyde has one chirality center and, therefore, has two stereoisomers. The absolute configuration of glyceraldehde was not known until 1951. Until then, chemists did not know whether (+)-glyceraldehyde had the R or the S configuration, although they had arbitrarily decided that it had the R configuration. They had a 50–50 chance of being correct.

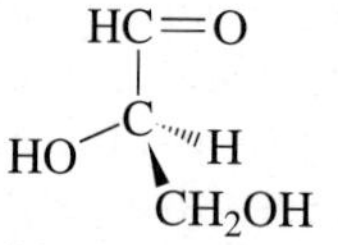

HC=O / C / HO, H, CH_2OH
(R)-(+)-glyceraldehyde

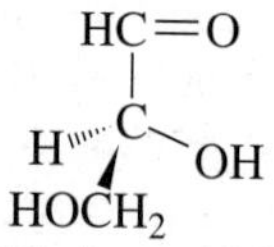

HC=O / C / H, OH, $HOCH_2$
(S)-(−)-glyceraldehyde

The configurations of many organic compounds were "determined" by synthesizing them from (+)- or (−)-glyceraldehyde or by converting them to (+)- or (−)-glyceraldehyde, always using reactions that did not break any of the bonds to the asymmetric carbon. For example, (−)-lactic could be related to (+)-glyceraldehyde through the following reactions. Thus the configuration of (−)-lactic was assumed to be that shown below. Because it was assumed that (+)-glyceraldehyde was the R enantiomer, the configurations assigned to these molecules were relative configurations, not absolute configurations. They were relative to (+)-glyceraldehyde, and were based on the *assumption* that (+)-glyceraldehyde had the R configuration.

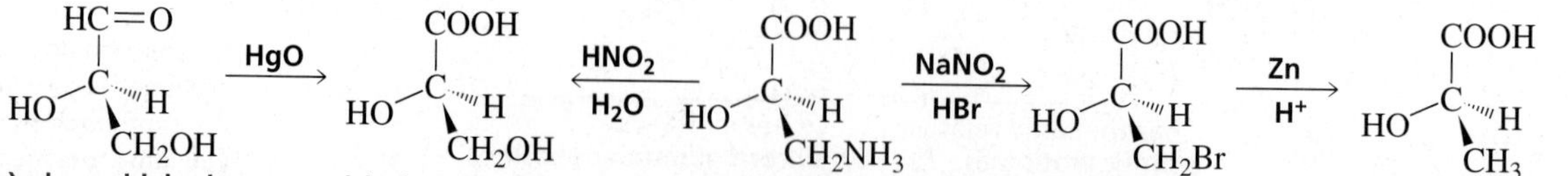

HC=O / C / HO, H, CH_2OH (+)-glyceraldehyde —HgO→ COOH / C / HO, H, CH_2OH (−)-glyceric acid ←HNO_2, H_2O— COOH / C / HO, H, CH_2NH_3 (+)-isoserine —$NaNO_2$, HBr→ COOH / C / HO, H, CH_2Br (−)-3-bromo-2-hydroxypropanoic acid —Zn, H^+→ COOH / C / HO, H, CH_3 (−)-lactic acid

In 1951, the Dutch chemists J. M. Bijvoet, A. F. Peerdeman, and A. J. van Bommel, using X-ray crystallography and a new technique known as anomalous dispersion, determined that the sodium rubidium salt of (+)-tartaric acid had the *R,R* configuration. Because (+)-tartaric acid could be synthesized from (−)-glyceraldehyde, (−)-glyceraldehyde had to be the *S* enantiomer. The assumption, therefore, that (+)-glyceraldehyde had the *R* configuration was correct!

HC=O / HO—H / CH$_2$OH (*S*) — several steps → — COOH / H—OH (*R*) / HO—H (*R*) / COOH

(−)-glyceraldehyde (+)-tartaric acid

The work of these chemists immediately provided absolute configurations for all those compounds whose relative configurations had been determined by relating them to (+)-glyceraldehyde. Thus, (−)-lactic acid has the configuration shown above. If (+)-glyceraldehyde had been the *S* enantiomer, (−)-lactic acid would have had the opposite configuration.

PROBLEM 36◆

What is the absolute configuration of the following?

a. (−)-glyceric acid
b. (+)-isoserine
c. (−)-glyceraldehyde
d. (+)-lactic acid

PROBLEM 37◆

Which of the following is true?

a. If two compounds have the same relative configuration, they will have the same absolute configuration.
b. If two compounds have the same relative configuration and you know the absolute configuration of either one of them, you can determine the absolute configuration of the other.
c. If two compounds have the same relative configuration, you can determine the absolute configuration of only one of them.
d. An *R* reactant always forms an *S* product.

5.14 Separating Enantiomers

Enantiomers cannot be separated by the usual separation techniques such as fractional distillation or crystallization because their identical boiling points and solubilities cause them to distill or crystallize simultaneously. Louis Pasteur was the first to separate a pair of enantiomers successfully. While working with crystals of sodium ammonium tartrate, he noted that the crystals were not identical—some of the crystals were "right-handed" and some were "left-handed." He painstakingly separated the two kinds of crystals with a pair of tweezers. He found that a solution of the right-handed crystals rotated the plane of polarized light clockwise, whereas a solution of the left-handed crystals rotated the plane of polarized light counterclockwise.

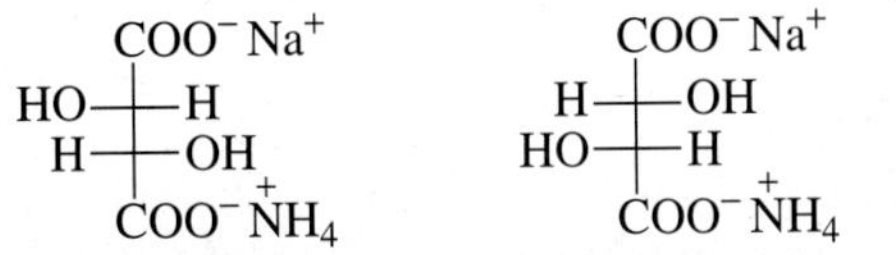

COO^-Na^+ / HO—H / H—OH / $COO^-\overset{+}{N}H_4$

sodium ammonium tartrate
left-handed crystals

COO^-Na^+ / H—OH / HO—H / $COO^-\overset{+}{N}H_4$

sodium ammonium tartrate
right-handed crystals

The French chemist and microbiologist **Louis Pasteur (1822–1895)** *was the first to demonstrate that microbes cause specific diseases. Asked by the French wine industry to find out why wine often went sour while aging, he showed that microorganisms cause grape juice to ferment, producing wine, and cause wine to slowly become sour. Gently heating the wine after fermentation, a process called pasteurization, kills the organisms so they cannot sour the wine.*

Pasteur was only 26 years old at the time and was unknown in scientific circles. He was concerned about the accuracy of his observations because a few years earlier, the well-known German organic chemist Eilhardt Mitscherlich had reported that crystals of the same salt were all identical. Pasteur immediately reported his findings to Jean-Baptiste Biot and repeated the experiment with Biot present. Biot was convinced that Pasteur had successfully separated the enantiomers of sodium ammonium tartrate. Pasteur's experiment also created a new chemical term. Tartaric acid is obtained from grapes, so it was also called racemic acid (*racemus* is Latin for "a bunch of grapes"). When Pasteur found that tartaric acid was actually a mixture of enantiomers, he called it a "racemic mixture." Separation of enantiomers is called the **resolution of a racemic mixture**.

Later, chemists recognized how lucky Pasteur had been. Sodium ammonium tartrate forms asymmetric crystals only under certain conditions—precisely the conditions that Pasteur had employed. Under other conditions, the symmetrical crystals that had fooled Mitscherlich are formed. But to quote Pasteur, "Chance favors the prepared mind."

Separating enantiomers by hand, as Pasteur did, is not a universally useful method to resolve a racemic mixture because few compounds form asymmetric crystals. A more commonly used method is to convert the enantiomers into diastereomers. Diastereomers can be separated because they have different physical properties. After separation, the individual diastereomers are converted back into the original enantiomers.

For example, because an acid reacts with a base to form a salt, a racemic mixture of a carboxylic acid reacts with a naturally occurring optically pure (a single enantiomer) base to form two diastereomeric salts. Morphine, strychnine, and brucine are examples of naturally occurring chiral bases commonly used for this purpose. The chiral base exists as a single enantiomer because when a chiral compound is synthesized in a living system, generally only one enantiomer is formed (Section 5.20). When an *R*-acid reacts with an *S*-base, an *R*,*S*-salt will be formed; when an *S*-acid reacts with an *S*-base, an *S*,*S*-salt will be formed.

COOH (HO, H, CH_3) *R*-acid + COOH (H, OH, CH_3) *S*-acid — enantiomers —(*S*-base)→ COO^- *S*-baseH^+ (HO, H, CH_3) *R*,*S*-salt + COO^- *S*-baseH^+ (H, OH, CH_3) *S*,*S*-salt — diastereomers

separate

COO^- *S*-baseH^+ (HO, H, CH_3) *R*,*S*-salt —HCl→ *S*-baseH^+ + COOH (HO, H, CH_3) *R*-acid

COO^- *S*-baseH^+ (H, OH, CH_3) *S*,*S*-salt —HCl→ COOH (H, OH, CH_3) *S*-acid + *S*-baseH^+

separated enantiomers

Eilhardt Mitscherlich (1794–1863), *a German chemist, first studied medicine so he could travel to Asia—a way to satisfy his interest in Oriental languages. He later became fascinated by chemistry. He was a professor of chemistry at the University of Berlin and wrote a successful chemistry textbook that was published in 1829.*

One of the asymmetric carbons in the *R*,*S*-salt is identical to an asymmetric carbon in the *S*,*S*-salt, and the other asymmetric carbon in the *R*,*S*-salt is the mirror image of an asymmetric carbon in the *S*,*S*-salt. Therefore, the salts are diastereomers and have different physical properties, so they can be separated. After separation, they can be converted back to the carboxylic acids by adding a strong acid such as HCl. The chiral base can be separated from the carboxylic acid and used again.

Enantiomers can also be separated by a technique called **chromatography**. In this method, the mixture to be separated is dissolved in a solvent and the solution is passed through a column packed with material that tends to adsorb organic compounds. If the chromatographic column is packed with *chiral* material, the two enantiomers can be expected to move through the column at different rates because they will have different affinities for the chiral material—just as a right hand prefers a right-hand glove—so one enantiomer will emerge from the column before the other. The chiral material is an example of a **chiral probe**—it can distinguish between enantiomers. A polarimeter is another example of a chiral probe (Section 5.7). In the next section you will see two kinds of biological molecules that are chiral probes—enzymes and receptors, both of which are proteins.

▲ Crystals of potassium hydrogen tartrate. Grapes are unusual in that they produce large quantities of tartaric acid, whereas most fruits produce citric acid. © Dr. Jeremy Burgess/Science Photo Library/Photo Researchers, Inc.

5.15 Discrimination of Enantiomers by Biological Molecules

Enzymes

Enantiomers can be separated easily if they are subjected to reaction conditions that cause only one of them to react. Enantiomers have the same chemical properties, so they react with *achiral* reagents at the same rate. Thus, hydroxide ion (an achiral reagent) reacts with (*R*)-2-bromobutane at the same rate that it reacts with (*S*)-2-bromobutane. However, *chiral* molecules recognize only one enantiomer, so if a synthesis is carried out using a chiral reagent or a chiral catalyst, only one enantiomer will undergo the reaction. One example of a **chiral catalyst** is an *enzyme*. An **enzyme** is a protein that catalyzes a chemical reaction. The enzyme D-amino acid oxidase, for example, catalyzes only the reaction of the *R* enantiomer and leaves the *S* enantiomer unchanged. The product of the enzyme-catalyzed reaction can be easily separated from the unreacted enantiomer. If you imagine an enzyme to be a right-hand glove and the enantiomers to be a pair of hands, the enzyme typically binds only one enantiomer because only the right hand fits into the right-hand glove.

An achiral reagent reacts identically with both enantiomers. A sock, which is achiral, fits on either foot.

A chiral reagent reacts differently with each enantiomer. A shoe, which is chiral, fits on only one foot.

$$\underset{R\text{ enantiomer}}{R\text{–}C(COO^-)(H)NH_2} + \underset{S\text{ enantiomer}}{H_2N\text{–}C(COO^-)(H)\text{–}R} \xrightarrow{\text{D-amino acid oxidase}} \underset{\text{oxidized } R\text{ enantiomer}}{{}^-OOC(R)C{=}NH} + \underset{\text{unreacted } S\text{ enantiomer}}{H_2N\text{–}C(COO^-)(H)\text{–}R}$$

The problem of having to separate enantiomers can be avoided if a synthesis is carried out that forms one of the enantiomers preferentially. Non-enzymatic chiral catalysts are being developed that will synthesize one enantiomer in great excess over the other. If a reaction is carried out with a reagent that does not have an asymmetric carbon and forms a product with an asymmetric carbon, a racemic mixture of the product will be formed. For example, the catalytic hydrogenation of 2-ethyl-1-pentene forms equal amounts of the two enantiomers because H_2 can be delivered equally easily to both faces of the double bond (Section 5.18).

$$(CH_3CH_2)(CH_3CH_2CH_2)C{=}CH_2 \xrightarrow[\text{Pd/C}]{H_2} \underset{\substack{(R)\text{-3-methylhexane}\\50\%}}{CH_3CH_2\text{–}C(CH_3CH_2CH_2)(H)(CH_3)} + \underset{\substack{(S)\text{-3-methylhexane}\\50\%}}{CH_3CH_2\text{–}C(CH_3CH_2CH_2)(H)(CH_3)}$$

If, however, the metal is complexed to a chiral organic molecule, H_2 will be delived to only one face of the double bond. One such chiral catalyst—using Ru(II) as the metal and BINAP (2,2′-bis(diphenylphosphino)-1,1′-binaphthyl) as the chiral molecule—has been used to synthesize (*S*)-naproxen, the active ingredient in Aleve and several other over-the-counter nonsteroidal anti-inflammatory drugs, in greater than 98% enantiomeric excess.

$$\xrightarrow[\textbf{(R)-BINAP–Ru(II)}]{H_2}$$

(*S*) naproxen
>98% ee

PROBLEM 38◆

What percent of naproxen is obtained as the *S* enantiomer in the above synthesis?

Receptors

A receptor is a protein that binds a particular molecule. Because a receptor is chiral, it will bind one enantiomer better than the other. In Figure 5.3 the receptor binds the *R* enantiomer but it does not bind the *S* enantiomer.

Because a receptor typically recognizes only one enantiomer, different physiological properties may be associated with each enantiomer. Receptors located on the exterior of nerve cells in the nose, for example, are able to perceive and differentiate the estimated 10,000 smells to which they are exposed. (*R*)-(−)-carvone is found in spearmint oil, and (*S*)-(+)-carvone is the main constituent of caraway seed oil. The reason these two enantiomers have such different odors is that each fits into a different receptor.

(*R*)-(−)-carvone
spearmint oil
$[\alpha]_D^{20\,°C} = -62.5°$

(*S*)-(+)-carvone
caraway seed oil
$[\alpha]_D^{20\,°C} = +62.5°$

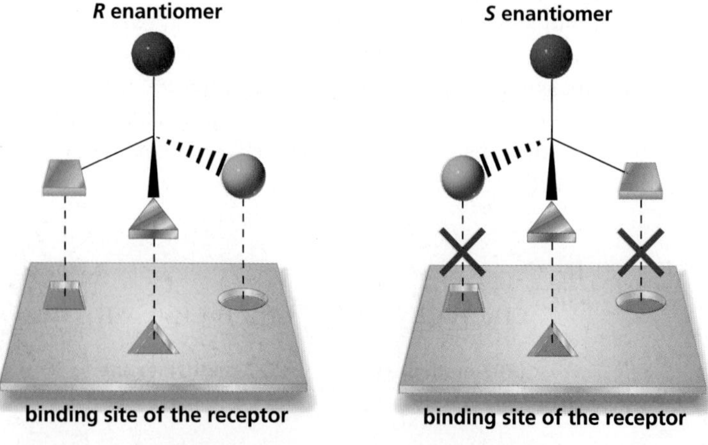

Figure 5.3 ▶
Schematic diagram showing why only one enantiomer is bound by a receptor. One enantiomer fits into the binding site and one does not.

Many drugs exert their physiological activity by binding to cellular receptors. If the drug has an asymmetric carbon, the receptor can preferentially bind one of the enantiomers. Thus, enantiomers can have the same physiological activities, different degrees of the same activity, or very different activities.

THE ENANTIOMERS OF THALIDOMIDE

Thalidomide was approved as a sedative for use in Europe and Canada in 1956. It was not approved for use in the United States because some neurological side effects had been noted. The dextrorotatory isomer has stronger sedative properties, but the commercial drug was a racemic mixture. However, it wasn't recognized that the levorotatory isomer was highly teratogenic—it causes horrible birth defects—until it was noticed that women who were given the drug during the first three months of pregnancy gave birth to babies with a wide variety of defects, such as deformed limbs. It was eventually determined that the dextrorotatory isomer also has mild teratogenic activity and that both enantiomers racemize in vivo. Thus, it is not clear whether giving those women only the dextrorotatory isomer would have decreased the severity of the birth defects. Thalidomide recently has been approved—with restrictions—to treat leprosy as well as melanomas.

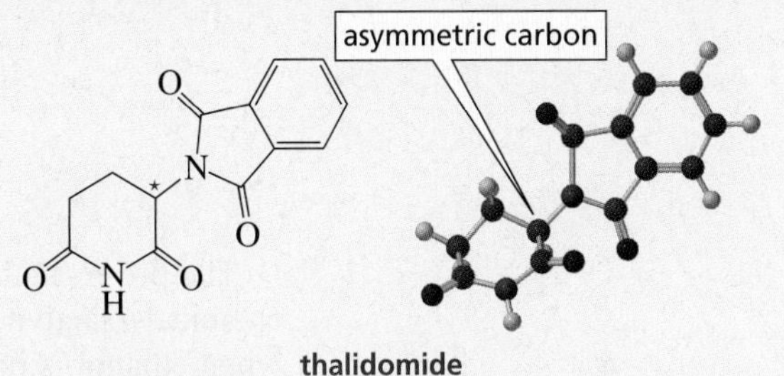

thalidomide

CHIRAL DRUGS

Until relatively recently, most drugs have been marketed as racemic mixtures because of the high cost of separating the enantiomers. In 1992, the Food and Drug Administration (FDA) issued a policy statement encouraging drug companies to use recent advances in synthetic and separation techniques to develop single enantiomer drugs. Now one-third of all drugs sold are single enantiomers. A new treatment for asthma is a single-enantiomer drug called Singulair. The antidepresanets Zoloft® and Paxil® (single enantiomers) are cutting into Prozac®'s (a racemate) market. Testing of single-enantiomer drugs is simpler because if a drug is sold as a racemate, the FDA requres that both enantiomers be tested. Testing has shown that (*S*)-(+)-ketamine is four times more potent an anesthetic than (*R*)-(−)-ketamine and, even more important, the disturbing side effects appear to be associated only with the (*R*)-(−)-enantiomer. The activity of ibuprofen, the popular analgesic marketed as Advil®, Nuprin®, and Motrin®, resides primarily in the (*S*)-(+)-enantiomer. The FDA has some concern about approving the drug as a single enantiomer because of potential drug overdoses. Can people who are used to taking two pills be convinced to take only one? Heroin addicts can be maintained with (−)-α-acetylmethadol for a 72-hour period compared to 24 hours with racemic methadone. This means less frequent visits to the clinic, and a single dose can get an addict through an entire weekend. Another reason for the increase in single-enantiomeric drugs is that drug companies may be able to extend their patents by developing a drug as a single enantiomer that was marketed previously as a racemate.

5.16 Enantiotopic Hydrogens, Diastereotopic Hydrogens, and Prochiral Carbons

If a carbon is bonded to two hydrogens and to two different groups, the two hydrogens are called **enantiotopic hydrogens**. For example, the two hydrogens (H_a and H_b) in the CH_2 group of ethanol are enantiotopic hydrogens because the other two groups bonded to the carbon (CH_3 and OH) are not identical. Replacing an enantiotopic hydrogen by a deuterium (or any other atom or group other than CH_3 or OH) forms a chiral molecule.

enantiotopic hydrogen → H_a

$$CH_3-\underset{\displaystyle H_b}{\overset{\displaystyle H_a}{\overset{|}{\underset{|}{C}}}}-OH$$

enantiotopic hydrogen → H_b

The carbon to which the enantiotopic hydrogens are attached is called a **prochiral carbon** because it will become a chirality center (an asymmetric carbon) if one of the hydrogens is replaced by a deuterium (or any group other than CH_3 or OH). If the H_a hydrogen is replaced by a deuterium, the asymmetric carbon will have the *R* configuration. Thus, the H_a hydrogen is called the **pro-*R*-hydrogen**. The H_b hydrogen is called the **pro-*S*-hydrogen** because if it is replaced by a deuterium, the asymmetric carbon will have the *S* configuration. The molecule containing the prochiral carbon is called a prochiral molecule because it would become a chiral molecule if one of the enantiotopic hydrogens is replaced.

prochiral carbon

H_a — pro-*R*-hydrogen

CH_3—C—OH

H_b — pro-*S*-hydrogen

The pro-*R*- and pro-*S*-hydrogens are chemically equivalent, so they have the same chemical reactivity and cannot be distinguished by achiral reagents. For example, when ethanol is oxidized by pyridinium chlorochromate (PCC) to acetaldehyde, one of the enantiotopic hydrogens is removed. (PCC is discussed in Section 20.2.) Because the two hydrogens are chemically equivalent, half the product results from removing the H_a hydrogen and the other half results from removing the H_b hydrogen.

$CH_3C(H_a)(H_b)OH$ —(PCC, pyridine)→ $CH_3C(=O)H_b$ (50%) + $CH_3C(=O)H_a$ (50%)

Enantiotopic hydrogens, however, are not chemically equivalent toward chiral reagents. An enzyme can distinguish between them because an enzyme is chiral (Section 5.15). For example, when the oxidation of ethanol to acetaldehyde is catalyzed by the enzyme alcohol dehydrogenase, only one of the enantiotopic hydrogens (H_a) is removed.

$CH_3C(H_a)(H_b)OH$ —(alcohol dehydrogenase)→ $CH_3C(=O)H_b$ (100%)

If a carbon is bonded to two hydrogens and replacing each of them in turn with deuterium (or another group) creates a pair of diastereomers, the hydrogens are called **diastereotopic hydrogens**.

diastereotopic hydrogens

Fischer projection: CH_3 (top); H_a—C—H_b; H—C—Br; CH_3 (bottom)

H_a and H_b are diastereotopic hydrogens

replace H_a with a D → CH_3; D—C—H_b; H—C—Br; CH_3

replace H_b with a D → CH_3; H_a—C—D; H—C—Br; CH_3

diastereomers

Unlike enantiotopic hydrogens, diastereotopic hydrogens do not have the same reactivity with achiral reagents. For example, in Chapter 11 we will see that because *trans*-2-butene is more stable than *cis*-2-butene (Section 4.11), removal of H_b and Br to form *trans*-2-butene occurs more rapidly than removal of H_a and Br to form *cis*-2-butene.

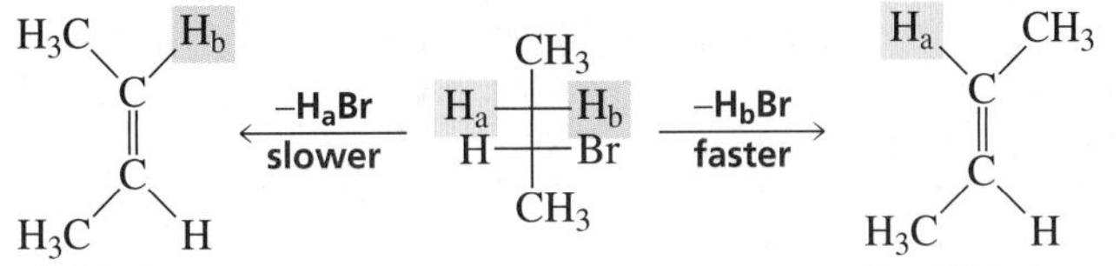

cis-2-butene ← $-H_aBr$ slower — [CH_3; H_a–C–H_b; H–C–Br; CH_3] — $-H_bBr$ faster → *trans*-2-butene

PROBLEM 39◆

Tell whether the H_a and H_b hydrogens in each of the following compounds are enantiotopic, diastereotopic, or neither.

a. 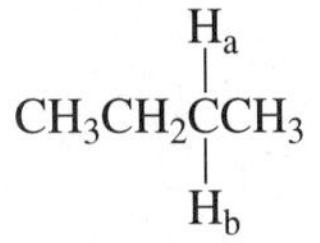 $CH_3CH_2CCH_3$ (C bearing H_a and H_b)

b. (cyclobutane with H, H_a, Br, H_b)

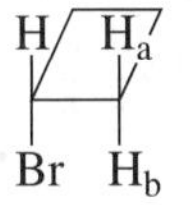

c. 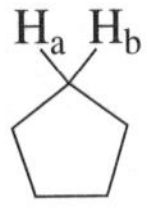 (cyclopentane with H_a H_b)

d. 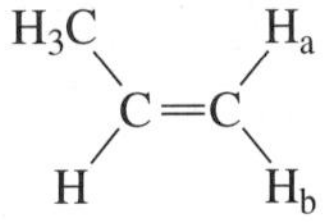 H_3C(H)C=C(H_a)(H_b)

5.17 Nitrogen and Phosphorus Chirality Centers

Atoms other than asymmetric carbons can be chirality centers. When an atom such as nitrogen or phosphorus has four different groups or atoms attached to it and it has a tetrahedral geometry, it is a chirality center. A compound with a chirality center can exist as enantiomers, and the enantiomers can be separated.

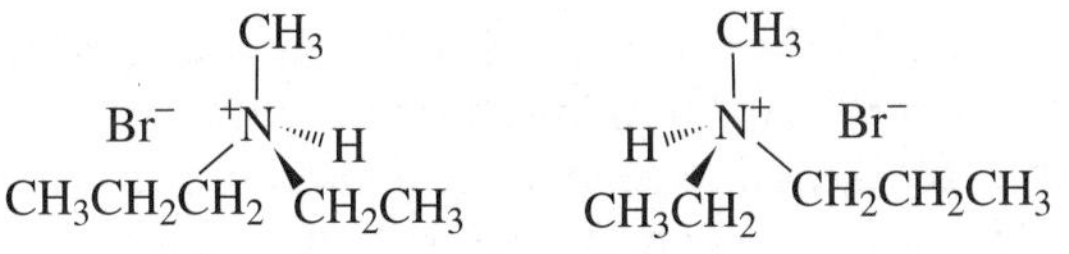

enantiomers

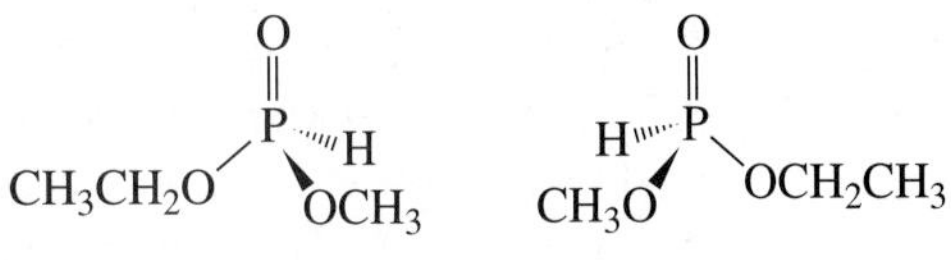

enantiomers

If one of the four "groups" attached to nitrogen is a lone pair, the enantiomers cannot be separated because they interconvert rapidly at room temperature. This is called **amine inversion** (Section 21.2). One way to picture amine inversion is to compare it to an umbrella that turns inside out in a windstorm.

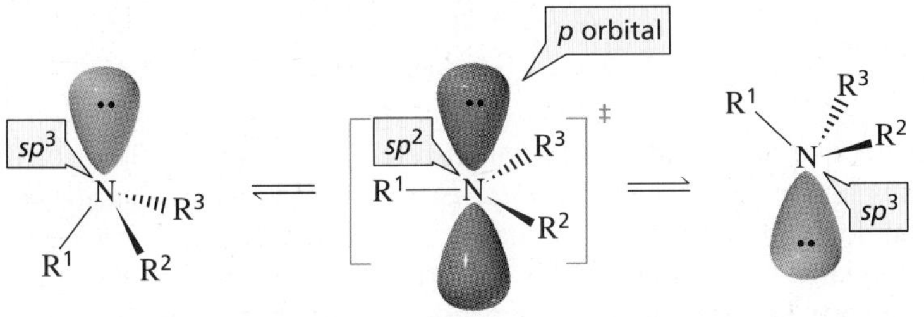

p orbital

sp^3

sp^2

sp^3

transition state

PROBLEM 40

Compound A has two stereoisomers, but compounds B and C exist as single compounds. Explain.

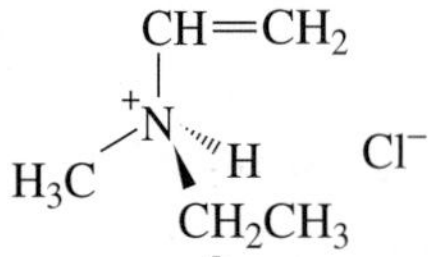

A: $H_3C-\overset{+}{N}(CH{=}CH_2)(H)(CH_2CH_3)\ Cl^-$

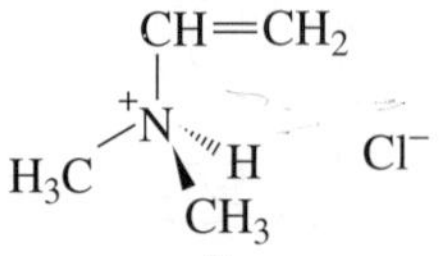

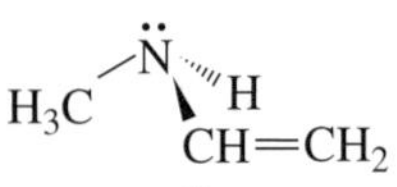

B: $H_3C-\overset{+}{N}(CH{=}CH_2)(H)(CH_3)\ Cl^-$

C: $H_3C-\ddot{N}(H)(CH{=}CH_2)$

5.18 Stereochemistry of Reactions: Regioselective, Stereoselective, and Stereospecific Reactions

In Chapter 4 we saw that alkenes undergo electrophilic addition reactions, and we looked at the different kinds of reagents that add to alkenes. We also examined the step-by-step process by which each reaction occurs (the mechanism of the reaction), and we determined what products are formed. However, we did not consider the stereochemistry of the reactions.

Stereochemistry is the field of chemistry that deals with the structures of molecules in three dimensions. When we study the stereochemistry of a reaction, we are concerned with the following questions:

1. If a reaction product can exist as two or more stereoisomers, does the reaction produce a single stereoisomer, a set of particular stereoisomers, or all possible stereoisomers?
2. If stereoisomers are possible for the reactant, do all stereoisomers react to form the same stereoisomeric product, or does each reactant form a different stereoisomer or a different set of stereoisomers?

Before we examine the stereochemistry of electrophilic addition reactions, we need to become familiar with some terms used in describing the stereochemistry of a reaction.

A regioselective reaction forms more of one constitutional isomer than of another.

In Section 4.4 we saw that a **regioselective** reaction is one in which two *constitutional isomers* can be obtained as products but more of one is obtained than of the other. In other words, a regioselective reaction selects for a particular constitutional isomer. Recall that a reaction can be *moderately regioselective*, *highly regioselective*, or *completely regioselective* depending on the relative amounts of the constitutional isomers formed in the reaction.

a regioselective reaction

constitutional isomers

$$A \longrightarrow B + C$$

more B is formed than C

Stereoselective is a similar term, but it refers to the preferential formation of a *stereoisomer* rather than a *constitutional isomer*. If a reaction that generates a carbon–carbon double bond or an asymmetric carbon in a product forms one stereoisomer preferentially over another, it is a stereoselective reaction. In other words, it selects for a particular stereoisomer. Depending on the degree of preference for a particular stereoisomer, a reaction can be described as being *moderately stereoselective*, *highly stereoselective*, or *completely stereoselective*.

a stereoselective reaction

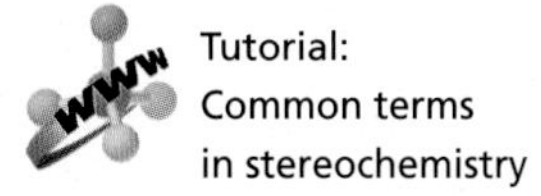

$$A \longrightarrow \underbrace{B + C}_{\text{stereoisomers}}$$

more B is formed than C

A stereoselective reaction forms more of one stereoisomer than of another.

A reaction is **stereospecific** if the reactant can exist as stereoisomers and each stereoisomeric reactant leads to a different stereoisomeric product or a different set of stereoisomeric products.

stereospecific reactions

stereoisomers { A ⟶ B ; C ⟶ D } stereoisomers

In a stereospecific reaction each stereoisomer forms a different stereoisomeric product or a different set of stereoisomeric products.

In the preceding reaction, stereoisomer A forms stereoisomer B but does not form D, so the reaction is stereoselective in addition to being stereospecific. *All stereospecific reactions, therefore, are also stereoselective. All stereoselective reactions are not stereospecific*, however, because there are stereoselective reactions in which the reactant does not have a carbon–carbon double bond or an asymmetric carbon, so it cannot exist as stereoisomers.

A stereospecific reaction is also stereoselective. A stereoselective reaction is not necessarily stereospecific.

5.19 Stereochemistry of Electrophilic Addition Reactions of Alkenes

Now that you are familiar with electrophilic addition reactions and with stereoisomers, we can combine the two topics and look at the stereochemistry of electrophilic addition reactions. In other words, we will look at the stereoisomers that are formed in the electrophilic addition reactions that were discussed in Chapter 4.

In Chapter 4 we saw that when an alkene reacts with an electrophilic reagent such as HBr, the major product of the addition reaction is the one obtained by adding the electrophile (H^+) to the sp^2 carbon bonded to the greater number of hydrogens and adding the nucleophile (Br^-) to the other sp^2 carbon. For example, the major product obtained from the reaction of propene with HBr is 2-bromopropane. This particular product does not have stereoisomers because it does not have an asymmetric carbon. Therefore, we do not have to be concerned with the stereochemistry of this reaction.

$$CH_3CH{=}CH_2 \xrightarrow{\text{HBr}} CH_3\overset{+}{C}HCH_3 \;\; Br^- \longrightarrow CH_3CH(Br)CH_3$$

propene → 2-bromopropane, major product

If, however, the reaction creates a product with an asymmetric carbon, we need to know which stereoisomers are formed. For example, the reaction of HBr with 1-butene forms 2-bromobutane, a compound with an asymmetric carbon. What is the configuration of the product? Do we get the *R* enantiomer, the *S* enantiomer, or both?

$$CH_3CH_2CH{=}CH_2 \xrightarrow{\text{HBr}} CH_3CH_2\overset{+}{C}HCH_3 \;\; Br^- \longrightarrow CH_3CH_2CH(Br)CH_3$$

1-butene → 2-bromobutane (asymmetric carbon)

In discussing the stereochemistry of electrophilic addition reactions, we will look first at reactions that form a product with one asymmetric carbon. Then we will look at reactions that form a product with two asymmetric carbons.

Addition Reactions that Form One Asymmetric Carbon

When a reactant that does not have an asymmetric carbon undergoes a reaction that forms a product with *one* asymmetric carbon, the product will be a racemic mixture. For example, the reaction of 1-butene with HBr forms identical amounts of (*R*)-2-bromobutane and (*S*)-2-bromobutane. Thus, an electrophilic addition reaction that forms a compound with one asymmetric carbon from a reactant without any asymmetric carbons is not stereoselective because it does not select for a particular stereoisomer. Why is this so?

You will see why a racemic mixture is obtained if you examine the structure of the carbocation formed in the first step of the reaction. The positively charged carbon is sp^2 hybridized, so the three atoms to which it is bonded lie in a plane (Section 1.10). When the bromide ion approaches the positively charged carbon from above the plane, one enantiomer is formed, but when it approaches from below the plane, the other enantiomer is formed. Because the bromide ion has equal access to both sides of the plane, identical amounts of the *R* and *S* enantiomers are obtained from the reaction.

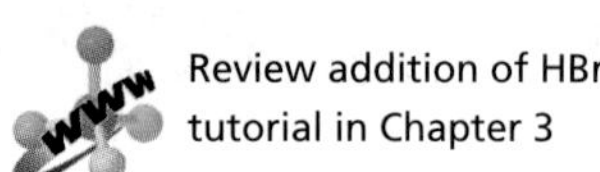
Review addition of HBr tutorial in Chapter 3

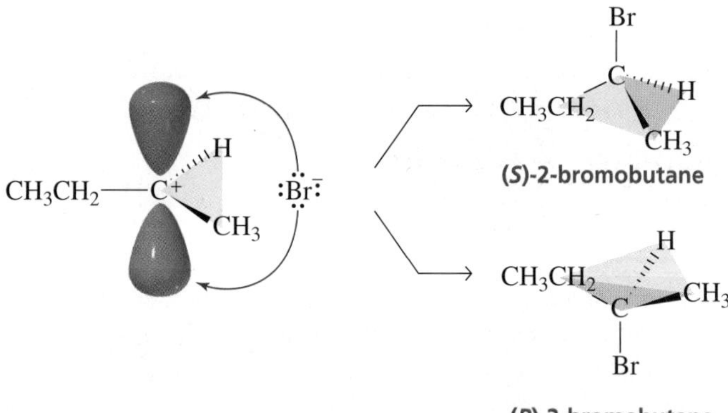

PROBLEM 41◆

a. Is the reaction of 2-butene with HBr regioselective?
b. Is it stereoselective?
c. Is it stereospecific?
d. Is the reaction of 1-butene with HBr regioselective?
e. Is it stereoselective?
f. Is it stereospecific?

When HBr adds to 2-methyl-1-butene in the presence of a peroxide, the product has one asymmetric carbon. Therefore, we can predict that identical amounts of the *R* and *S* enantiomers are formed.

$$\underset{\textbf{2-methyl-1-butene}}{CH_3CH_2\overset{\overset{CH_3}{|}}{C}{=}CH_2} + HBr \xrightarrow{\textbf{peroxide}} \underset{\textbf{1-bromo-2-methylbutane}}{CH_3CH_2\overset{\overset{CH_3}{|}}{\underset{*}{C}}HCH_2Br}$$

asymmetric carbon

The product is a racemic mixture because the carbon in the radical intermediate that bears the unpaired electron is sp^2 hybridized. This means that the three atoms bonded to it are all in the same plane (Section 1.10). Consequently, identical amounts of the *R* enantiomer and the *S* enantiomer will be obtained because HBr has equal access to both sides of the radical. (Although the unpaired electron is shown to be in the top lobe of the *p* orbital, it actually is spread out equally through both lobes.)

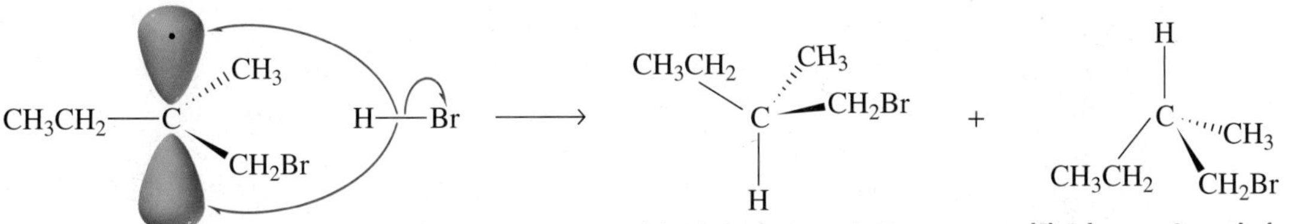

CH_3CH_2—C(CH_3)(CH_2Br)· + H—Br ⟶ **(*R*)-1-bromo-2-methylbutane** + **(*S*)-1-bromo-2-methylbutane**

PROBLEM 42

What stereoisomers are obtained from each of the following reactions?

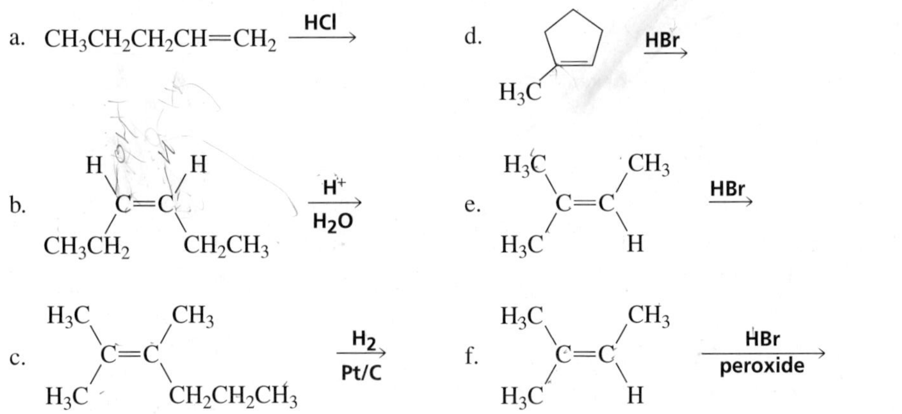

a. $CH_3CH_2CH_2CH{=}CH_2 \xrightarrow{\text{HCl}}$

b. (*cis*) $CH_3CH_2CH{=}CHCH_2CH_3$ (H, H on same side) $\xrightarrow[\text{H}_2\text{O}]{\text{H}^+}$

c. $(H_3C)_2C{=}C(CH_3)CH_2CH_2CH_3 \xrightarrow[\text{Pt/C}]{\text{H}_2}$

d. 1-methylcyclopentene $\xrightarrow{\text{HBr}}$

e. $(H_3C)_2C{=}CHCH_3 \xrightarrow{\text{HBr}}$

f. $(H_3C)_2C{=}CHCH_3 \xrightarrow[\text{peroxide}]{\text{HBr}}$

If a reaction creates an asymmetric carbon in a compound that already has an asymmetric carbon (and is a single enantiomer of that compound), a pair of diastereomers will be formed. For example, let's look at the following reaction. Because none of the bonds to the asymmetric carbon in the reactant is broken during the addition of HBr, the configuration of this asymmetric carbon does not change. The bromide ion can approach the planar carbocation intermediate from either the top or the bottom in the process of creating the new asymmetric carbon, so two stereoisomers result. The stereoisomers are diastereomers because one of the asymmetric carbons has the same configuration in both isomers and the other has opposite configurations in the two isomers.

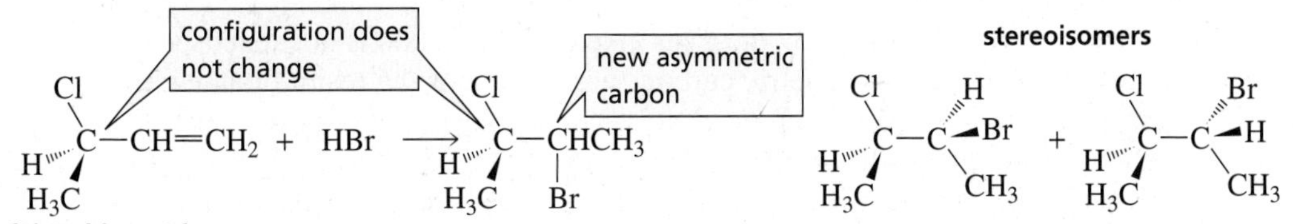

$CH_3CHClCH{=}CH_2$ **(*R*)-3-chloro-1-butene** + HBr ⟶ $CH_3CHCl{-}CHBrCH_3$ ⟶ stereoisomers (diastereomers)

configuration does not change

new asymmetric carbon

stereoisomers

diastereomers

Because the products of the preceding reaction are diastereomers, the transition states that lead to them are also diastereomeric. The two transition states, therefore, will not have the same stability, so different amounts of the two diastereomers will be formed. The reaction is stereoselective—more of one stereoisomer is formed than of the other.

Addition Reactions that Form Products with Two Asymmetric Carbons

When a reactant that does not have an asymmetric carbon undergoes a reaction that forms a product with *two* asymmetric carbons, the stereoisomers that are formed depend on the mechanism of the reaction.

1. Addition Reactions that Form a Carbocation Intermediate or a Radical Intermediate

If two asymmetric carbons are created as the result of an addition reaction that forms a carbocation intermediate, four stereoisomers can be obtained as products.

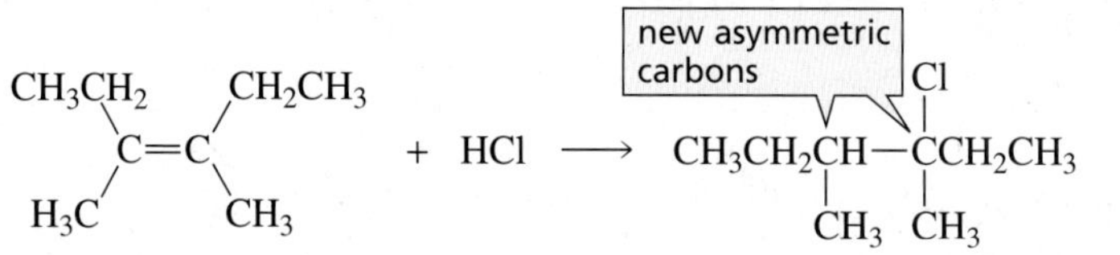

$$(CH_3CH_2)(H_3C)C{=}C(CH_2CH_3)(CH_3) + HCl \longrightarrow CH_3CH_2CH(CH_3){-}CCl(CH_3)CH_2CH_3$$

new asymmetric carbons

cis-3,4-dimethyl-3-hexene → **3-chloro-3,4-dimethylhexane**

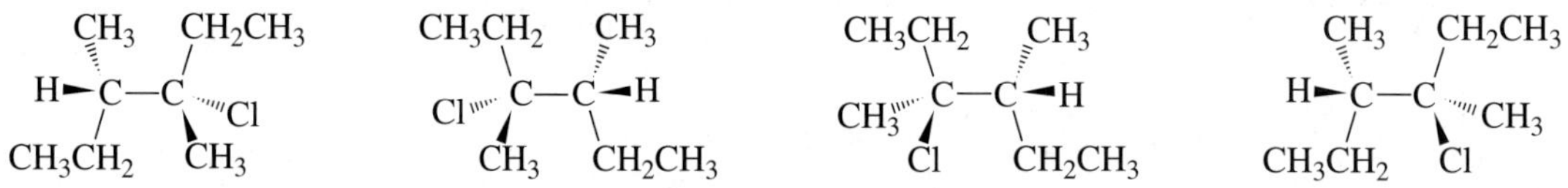

perspective formulas of the stereoisomers of the product

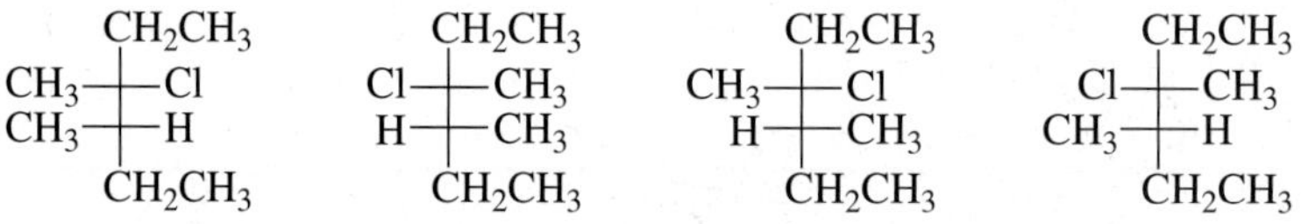

Fischer projections of the stereoisomers of the product

CH_2CH_3	CH_2CH_3	CH_2CH_3	CH_2CH_3
CH_3—C—Cl	Cl—C—CH_3	CH_3—C—Cl	Cl—C—CH_3
CH_3—C—H	H—C—CH_3	H—C—CH_3	CH_3—C—H
CH_2CH_3	CH_2CH_3	CH_2CH_3	CH_2CH_3

In the first step of the reaction, the proton can approach the plane containing the double-bonded carbons of the alkene from above or below to form the carbocation. Once the carbocation is formed, the chloride ion can approach the positively charged carbon from above or below. As a result, four stereoisomers are obtained as products: The proton and the chloride ion can add from above-above, above-below, below-above, or below-below. When the two substituents add to the same side of the double bond, the addition is called **syn addition**. When the two substituents add to opposite sides of the double bond, the addition is called **anti addition**. Both syn and anti addition occur in alkene addition reactions that take place by way of a carbocation intermediate. Because the four stereoisomers formed by the cis alkene are identical to the four stereoisomers formed by the trans alkene, the reaction is not stereospecific.

Similarly, if two asymmetric carbons are created as the result of an addition reaction that forms a radical intermediate, four stereoisomers can be formed because both syn and anti addition are possible. And because the stereoisomers formed by the cis isomer are identical to those formed by the trans isomer, the reaction is not stereospecific—in a stereospecific reaction, the cis and trans isomers form different stereoisomers.

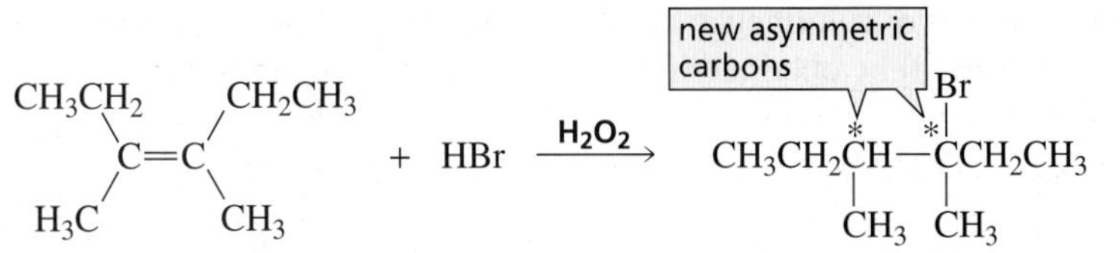

new asymmetric carbons

$CH_3CH_2(CH_3)C{=}C(CH_3)CH_2CH_3 + HBr \xrightarrow{H_2O_2} CH_3CH_2\overset{*}{C}H(CH_3)-\overset{*}{C}Br(CH_3)CH_2CH_3$

cis-3,4-dimethyl-3-hexene — 3-bromo-3,4-dimethylhexane

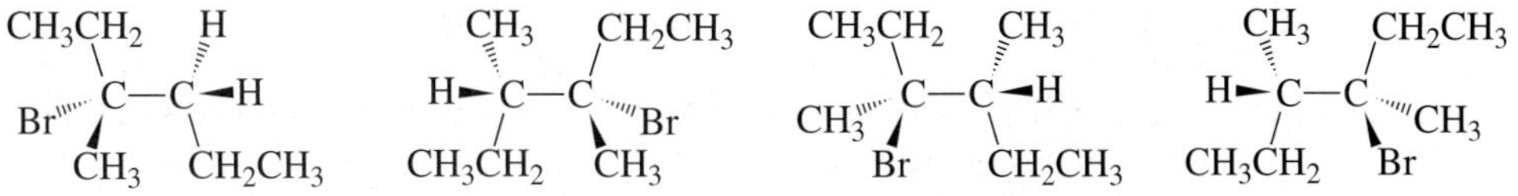

perspective formulas of the stereoisomers of the product

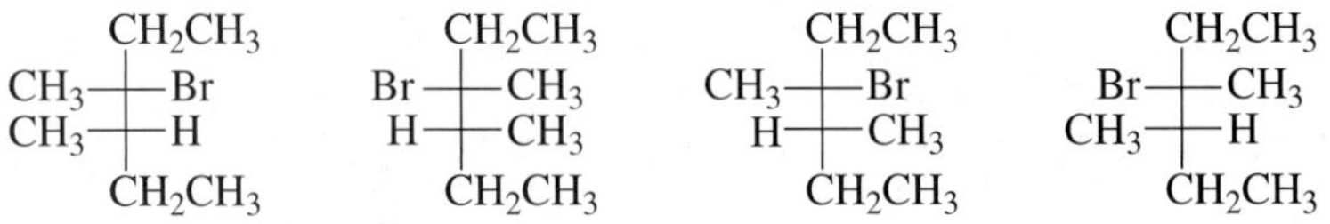

Fischer projections of the stereoisomers of the product

2. Stereochemistry of Hydrogen Addition

In a catalytic hydrogenation, the alkene sits on the surface of the metal catalyst and both hydrogen atoms add to the same side of the double bond (Section 4.11). Therefore, the addition of H_2 to an alkene is a syn addition reaction.

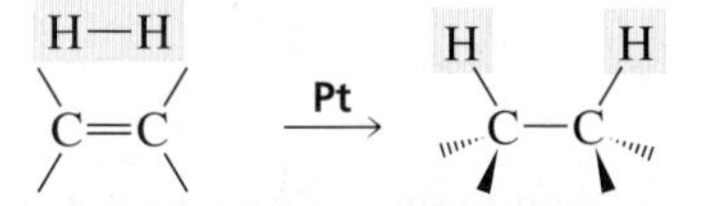

addition of H_2 is a syn addition

If addition of hydrogen to an alkene forms a product with two asymmetric carbons, only two of the four possible stereoisomers are obtained because only syn addition can occur. (The other two stereoisomers would have to come from anti addition.) One stereoisomer results from addition of both hydrogens from above the plane of the double bond, and the other stereoisomer results from addition of both hydrogens from below the plane. The particular pair of stereoisomers that is formed depends on whether the reactant is a *cis*-alkene or a *trans*-alkene. Syn addition of H_2 to a *cis*-alkene forms only the erythro enantiomers. (In Section 5.8, we saw that the erythro enantiomers are the ones with identical groups on the same side of the carbon chain in the eclisped conformers.)

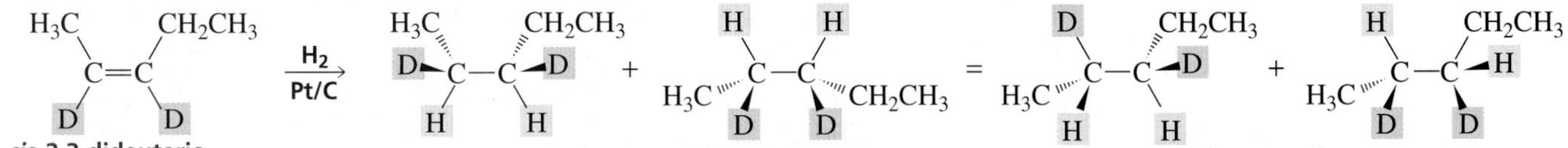

cis-2,3-dideuterio-2-pentene $\xrightarrow[Pt/C]{H_2}$

erythro enantiomers
perspective formulas
(eclipsed conformers)

erythro enantiomers
perspective formulas
(staggered conformers)

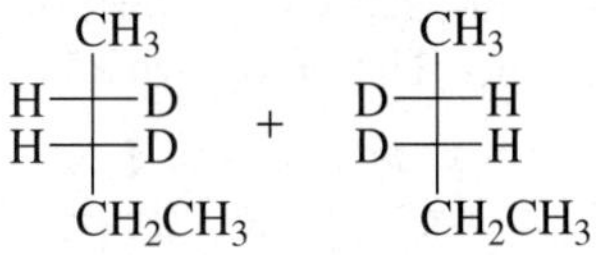

erythro enantiomers
Fischer projections

If you have trouble determining the configuration of a product, make a model.

If each of the two asymmetric carbons is bonded to the same four substituents, a meso compound will be obtained instead of the erythro enantiomers.

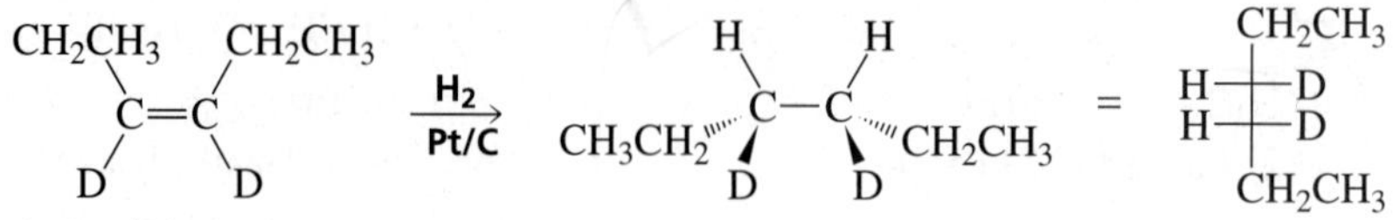

cis-2,3-dideuterio-2-pentene $\xrightarrow[\text{Pt/C}]{H_2}$ meso compound

In contrast, syn addition of H_2 to a *trans*-alkene forms only the threo enantiomers. Thus the addition of hydrogen is a stereospecific reaction—the product obtained from addition to the cis isomer is different from the product obtained from addition to the trans isomer. It is also a stereoselective reaction because all possible isomers are not formed; for example, only the threo enantiomers are formed in the following reaction.

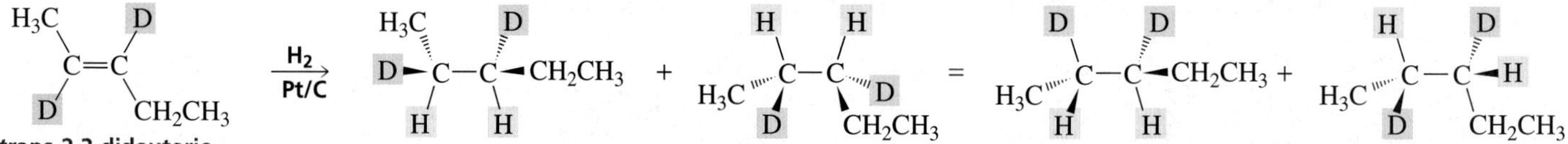

trans-2,3-dideuterio-2-pentene $\xrightarrow[\text{Pt/C}]{H_2}$

threo enantiomers perspective formulas (eclipsed conformers)

threo enantiomers perspective formulas (staggered conformers)

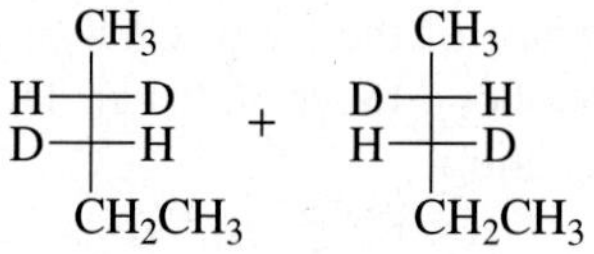

threo enantiomers Fischer projections

Cyclic alkenes with fewer than eight carbons in the ring, such as cyclopentene and cyclohexene, can exist only in the cis configuration because they do not have enough carbons to incorporate a trans double bond. Therefore, it is not necessary to use the cis designation with their names. Both cis and trans isomers are possible for rings containing eight or more carbons, however, so the configuration of the compound must be specified in its name.

cis-cyclooctene *trans*-cyclooctene

Therefore, the addition of H_2 to a cyclic alkene with fewer than eight ring atoms forms only the cis enantiomers.

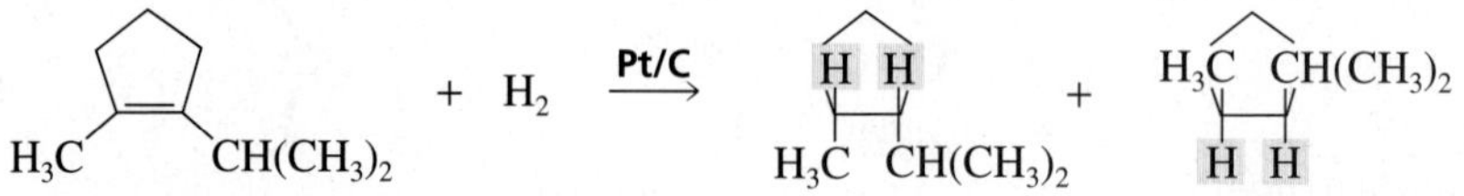

1-isopropyl-2-methylcyclopentene + H_2 $\xrightarrow{\text{Pt/C}}$

Each of the two asymmetric carbons in the product of the following reaction is bonded to the same four substituents. Therefore, syn addition forms a meso compound.

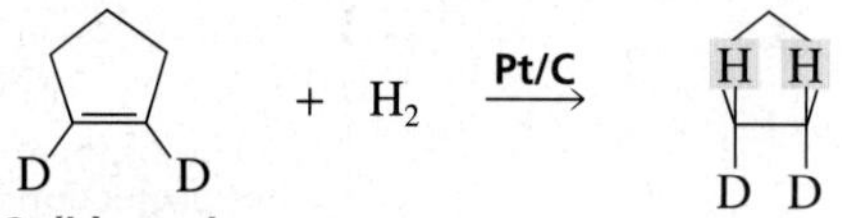

1,2-dideuteriocyclopentene + H_2 $\xrightarrow{\text{Pt/C}}$

PROBLEM 43

a. What stereoisomers are formed in the following reaction?

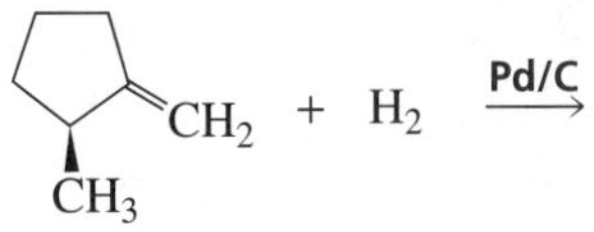

CH_2 + H_2 $\xrightarrow{Pd/C}$

CH_3

b. What stereoisomer is formed in greater yield?

3. Stereochemistry of Hydroboration–Oxidation

The addition of borane to an alkene is a concerted reaction. The boron and the hydride ion add to the two sp^2 carbons of the double bond at the same time (Section 4.9). Because the two species add simultaneously, they must add to the same side of the double bond. So the addition of borane to an alkene, like the addition of hydrogen, is a syn addition.

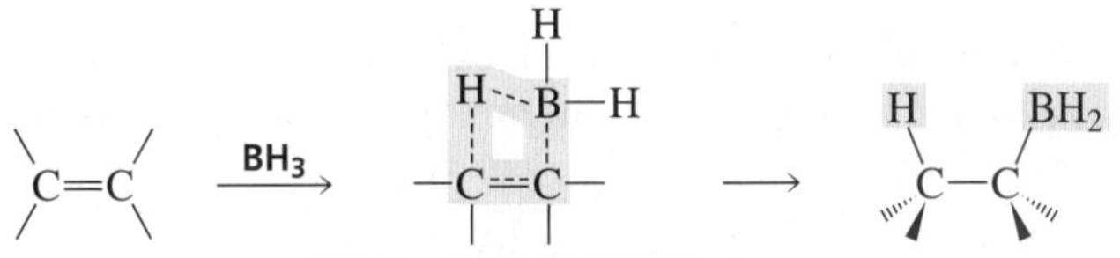

C=C $\xrightarrow{BH_3}$ H---B—H (H) / —C=C— $\longrightarrow$ H, BH_2 C—C

syn addition of borane

When the alkylborane is oxidized by reaction with hydrogen peroxide and hydroxide ion, the OH group ends up in the same position as the boron group it replaces. Consequently, the overall hydroboration–oxidation reaction amounts to a syn addition of water to a carbon–carbon double bond.

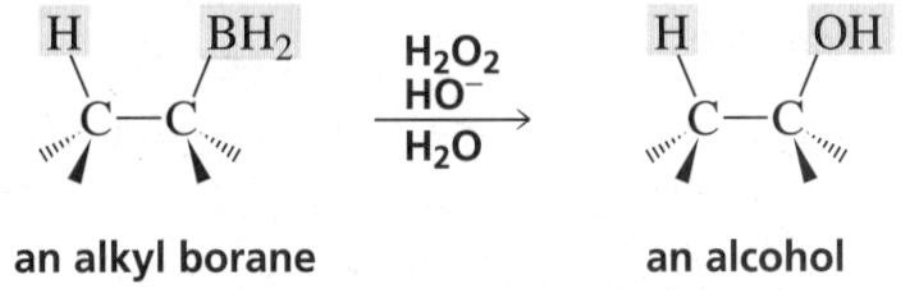

H, BH_2 C—C $\xrightarrow[H_2O]{H_2O_2,\ HO^-}$ H, OH C—C

an alkyl borane — an alcohol

hydroboration–oxidation is a syn addition of water

Because only syn addition occurs, hydroboration–oxidation is stereoselective—only two of the four possible stereoisomers are formed. As we saw when we looked at the addition of H_2 to a cycloalkene, syn addition results in the formation of only the pair of enantiomers that has the added groups on the same side of the ring.

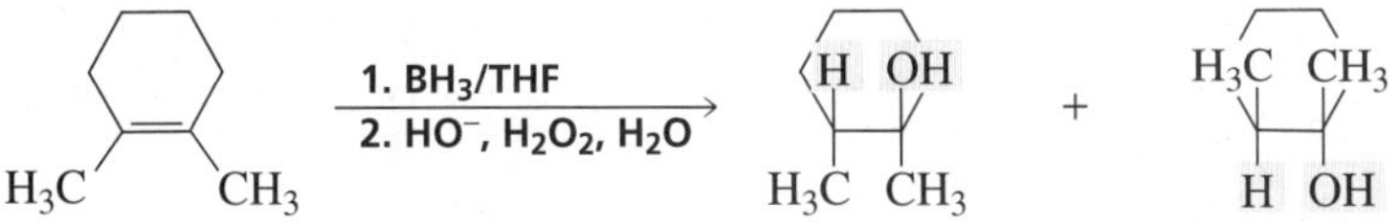

H_3C CH_3 $\xrightarrow[2.\ HO^-,\ H_2O_2,\ H_2O]{1.\ BH_3/THF}$ H OH / H_3C CH_3 + H_3C CH_3 / H OH

PROBLEM 44◆

What stereoisomers would be obtained from hydroboration–oxidation of the following compounds?

a. cyclohexene
b. 1-ethylcyclohexene
c. 1,2-dimethylcyclopentene
d. *cis*-2-butene

4. Addition Reactions that Form a Bromonium Ion Intermediate

If two asymmetric carbons are created as the result of an addition reaction that forms a bromonium ion intermediate, only one pair of enantiomers will be formed. In other words, the addition of Br_2 is a stereoselective reaction. The addition of Br_2 to the cis alkene forms only the threo enantiomers.

cis-2-pentene + Br_2 $\xrightarrow{CH_2Cl_2}$

threo enantiomers
perspective formulas

threo enantiomers
Fischer projections

Similarly, the addition of Br_2 to the trans alkene forms only the erythro enantiomers. Because the cis and trans isomers form different products, the reaction is stereospecific as well as stereoselective.

trans-2-pentene + Br_2 $\xrightarrow{CH_2Cl_2}$

erythro enantiomers
perspective formulas

erythro enantiomers
Fischer projections

Because the addition of Br_2 to the cis alkene forms the threo enantiomers, we know that addition of Br_2 is an example of anti addition because if syn addition had occurred to the cis alkene, the erythro enantiomers would have been formed. The addition of Br_2 is anti because the reaction intermediate is a cyclic bromonium ion (Section 4.7). Once the bromonium ion is formed, the bridged bromine atom blocks that side of the ion. As a result, the negatively charged bromide ion must approach from the opposite side (following either the green arrows *or* the red arrows). Thus, the two bromine atoms add to opposite sides of the double bond. Because only anti addition of Br_2 can occur, only two of the four possible stereoisomers are obtained.

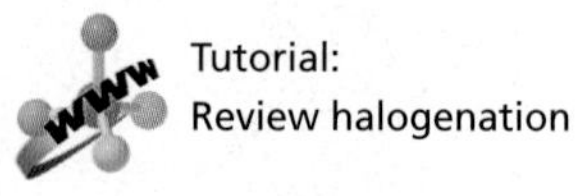
Tutorial:
Review halogenation

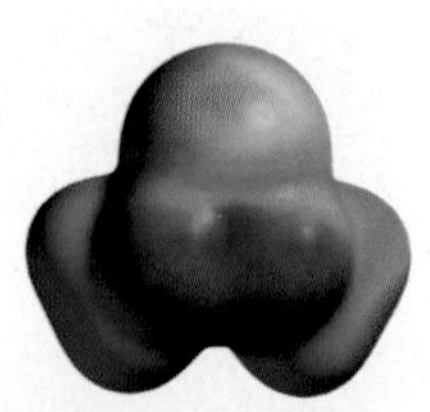

the bromonium ion formed from the reaction of Br_2 with *cis*-2-butene

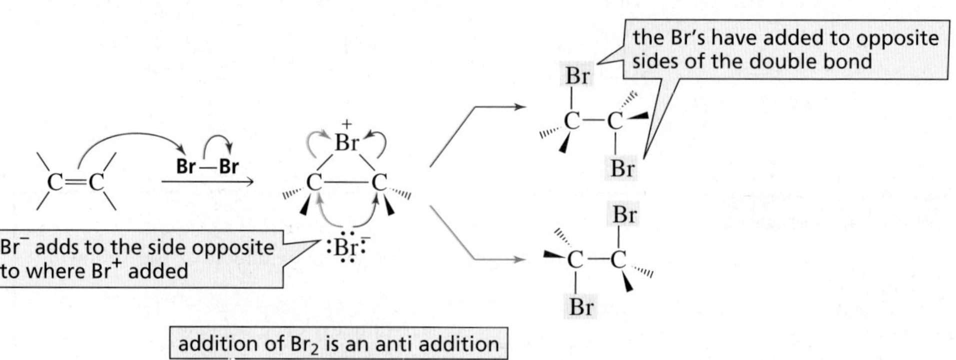
Br—Br
Br
Br
Br
Br
Br
:Br:
Br⁻ adds to the side opposite to where Br⁺ added
the Br's have added to opposite sides of the double bond
addition of Br₂ is an anti addition

PROBLEM 45

Reaction of 2-ethyl-1-pentene with Br_2, with H_2/Pt, or with BH_3 followed by $HO^- + H_2O_2$ leads to a racemic mixture. Explain why a racemic mixture is obtained in each case.

PROBLEM 46

How could you prove, using a sample of *trans*-2-butene, that addition of Br_2 forms a cyclic bromonium ion intermediate rather than a carbocation intermediate?

If the two asymmetric carbons in the product each have the same four substituents, the erythro isomers are identical and constitute a meso compound. Therefore, addition of Br_2 to *trans*-2-butene forms a meso compound.

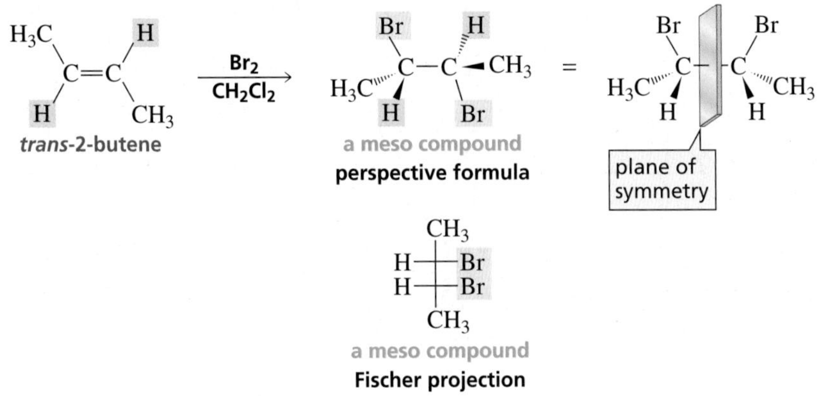

trans-2-butene

$\xrightarrow[CH_2Cl_2]{Br_2}$

a meso compound
perspective formula

plane of symmetry

a meso compound
Fischer projection

Because only anti addition occurs, addition of Br_2 to a cyclohexene forms only the enantiomers that have the added bromine atoms on opposite sides of the ring.

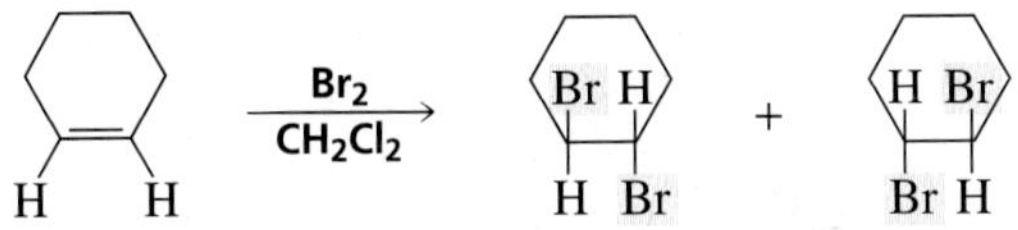

$\xrightarrow[CH_2Cl_2]{Br_2}$

One way to determine what stereoisomers are obtained from many reactions that create a product with two asymmetric carbons is the mnemonic **CIS-SYN-ERYTHRO**, which is easy to remember because all three terms mean "on the same side." You can change any two of the terms but you can't change just one. (For example, **TRANS-ANTI-ERYTHRO**, **TRANS-SYN-ERYTHREO**, and **CIS-ANTI-THREO** are allowed, but **TRANS-SYN-ERYTHRO** is not allowed.) So if you have a trans reactant that undergoes addition of Br_2 (which is anti), the erythro products are obtained. This mnemonic will work for all reactions that have a product with a structure that can be described by erythro or threo.

A summary of the stereochemistry of the products obtained from addition reactions to alkenes is given in Table 5.2.

PROBLEM-SOLVING STRATEGY

Give the configuration of the products obtained from the following reactions:

a. 1-butene + HCl
b. 1-butene + HBr + peroxide
c. 3-methyl-3-hexene + HBr + peroxide
d. *cis*-3-heptene + Br_2
e. *trans*-3-heptene + Br_2
f. *trans*-3-hexene + Br_2

Table 5.2 Stereochemistry of Alkene Addition Reactions

Reaction	Type of addition	Stereoisomers formed
Addition reactions that create one asymmetric carbon in the product		1. If the reactant does not have an asymmetric carbon, a pair of enantiomers will be obtained (equal amounts of the *R* and *S* isomers). 2. If the reactant has an asymmetric carbon, unequal amounts of a pair of diastereomers will be obtained.
Addition reactions that create two asymmetric carbons in the product		
Addition of reagents that form a carbocation or radical intermediate	syn and anti	Four stereoisomers can be obtained[a] (the cis and trans isomers form the same products)
Addition of H_2 Addition of borane	syn	cis ⟶ erythro enantiomers[a] trans ⟶ threo enantiomers
Addition of Br_2	anti	cis ⟶ threo enantiomers trans ⟶ erythro enantiomers[a]

[a] If the two asymmetric carbons have the same substituents, a meso compound will be obtained instead of the pair of erythro enantiomers.

Start by drawing the product without regard to its configuration, to check whether the reaction has created any asymmetric carbons. Then determine the configuration of the products, paying attention to the configuration (if any) of the reactant, how many asymmetric carbons are formed, and the mechanism of the reaction. Let's start with a.

a. $CH_3CH_2CHCH_3$ (with Cl on the third carbon)

The product has one asymmetric carbon, so equal amounts of the *R* and *S* enantiomers will be formed.

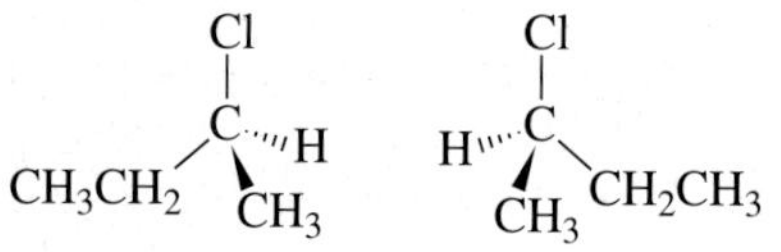

[Structures: the two enantiomers of $CH_3CH_2CHClCH_3$]

b. $CH_3CH_2CH_2CH_2Br$

The product does not have an asymmetric carbon, so it has no stereoisomers.

c. $CH_3CH_2CHCHCH_2CH_3$ (with CH_3 on the third carbon and Br on the fourth carbon)

Two asymmetric carbons have been created in the product. Because the reaction forms a radical intermediate, two pairs of enantiomers are formed.

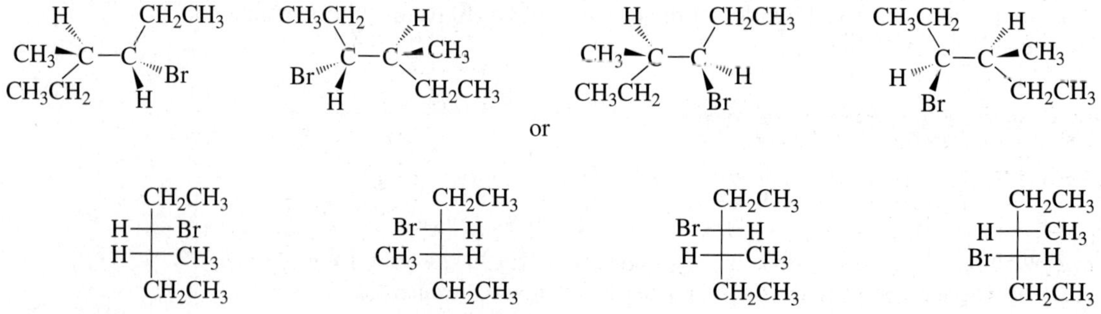

[Structures: four stereoisomers shown as perspective formulas]

or

[Structures: the same four stereoisomers shown as Fischer projections]

d. $CH_3CH_2CHCHCH_2CH_2CH_3$ (with Br on C3 and C4)

Two asymmetric carbons have been created in the product. Because the reactant is cis and addition of Br_2 is anti, the threo enantiomers are formed.

CH₃CH₂ / C—C / CH₂CH₂CH₃ (H, Br on C3; Br, H on C4) and CH₃CH₂CH₂ / C—C / CH₂CH₃ (Br, H; H, Br)

or

Fischer projections: CH_2CH_3; H—Br; Br—H; $CH_2CH_2CH_3$ and CH_2CH_3; Br—H; H—Br; $CH_2CH_2CH_3$

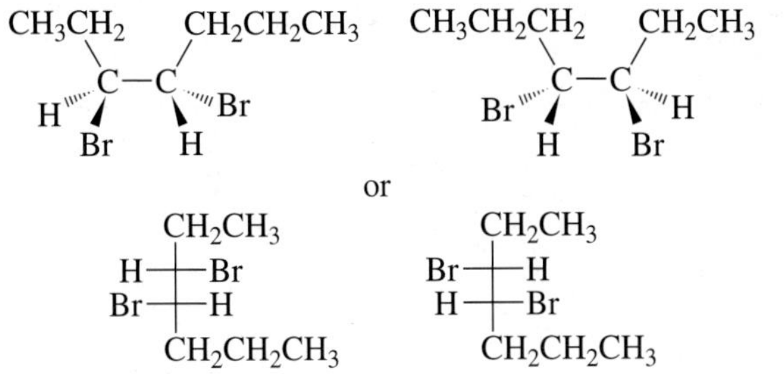

e. $CH_3CH_2CHCHCH_2CH_2CH_3$ (with Br on C3 and C4)

Two asymmetric carbons have been created in the product. Because the reactant is trans and addition of Br_2 is anti, the erythro enantiomers are formed.

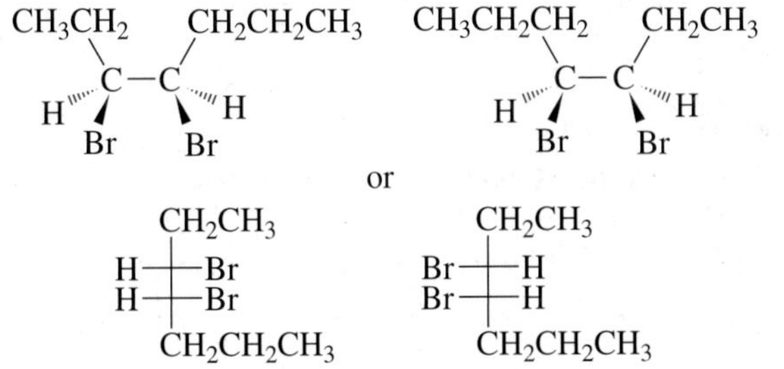

CH₃CH₂ / C—C / CH₂CH₂CH₃ (H, Br; H, Br) and CH₃CH₂CH₂ / C—C / CH₂CH₃ (H, Br; H, Br)

or

Fischer projections: CH_2CH_3; H—Br; H—Br; $CH_2CH_2CH_3$ and CH_2CH_3; Br—H; Br—H; $CH_2CH_2CH_3$

f. $CH_3CH_2CHCHCH_2CH_3$ (with Br on C3 and C4)

Two asymmetric carbons have been created in the product. Because the reactant is trans and addition of Br_2 is anti, one would expect the erythro enantiomers. However, the two asymmetric carbons are bonded to the same four groups, so the erythro product is a meso compound. Thus, only one stereoisomer is formed.

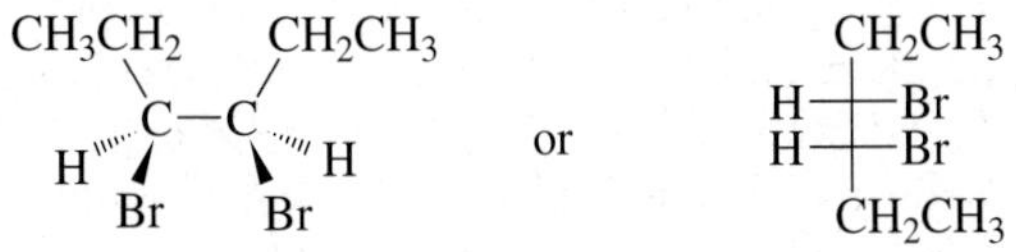

CH₃CH₂ / C—C / CH₂CH₃ (H, Br; H, Br) or Fischer projection: CH_2CH_3; H—Br; H—Br; CH_2CH_3

Now continue on to Problem 47.

PROBLEM 47

Give the configuration of the products obtained from the following reactions:

a. *trans*-2-butene + HBr + peroxide
b. (Z)-3-methyl-2-pentene + HBr
c. (Z)-3-methyl-2-pentene + HBr + peroxide
d. *cis*-3-hexene + HBr
e. *cis*-2-pentene + Br_2
f. 1-hexene + Br_2

PROBLEM 48

When Br_2 adds to an alkene with different substituents on each of the two sp^2 carbons, such as *cis*-2-heptene, identical amounts of the two threo enantiomers are obtained even though Br^- is more likely to attack the less sterically hindered carbon atom of the bromonium ion. Explain why identical amounts of the stereoisomers are obtained.

PROBLEM 49

a. What products would be obtained from the addition of Br_2 to cyclohexene if the solvent were H_2O instead of CH_2Cl_2?
b. Propose a mechanism for the reaction.

PROBLEM 50

What stereoisomers would you expect to obtain from each of the following reactions?

a. 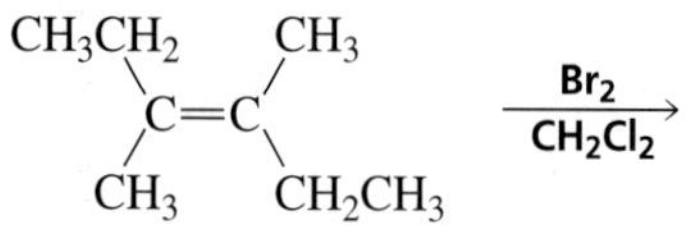 $(CH_3CH_2)(CH_3)C{=}C(CH_3)(CH_2CH_3)$ $\xrightarrow[CH_2Cl_2]{Br_2}$

b. $(CH_3CH_2)(H_3C)C{=}C(CH_3)(CH_2CH_3)$ $\xrightarrow[Pt/C]{H_2}$

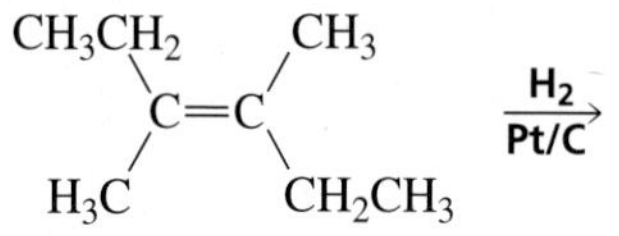

c. 1,2-dimethylcyclopentene (ring substituents CH_3, CH_3) $\xrightarrow[CH_2Cl_2]{Br_2}$

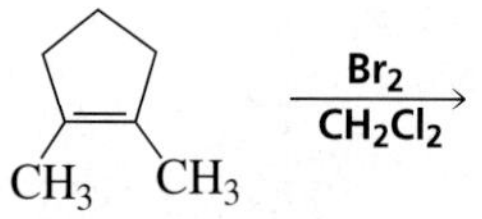

d. 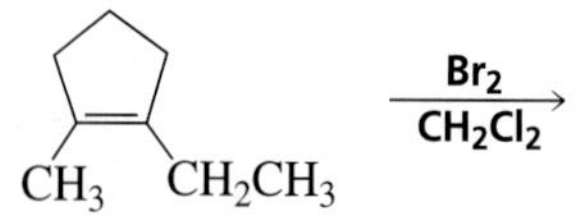 1-ethyl-2-methylcyclopentene (ring substituents CH_3, CH_2CH_3) $\xrightarrow[CH_2Cl_2]{Br_2}$

e. 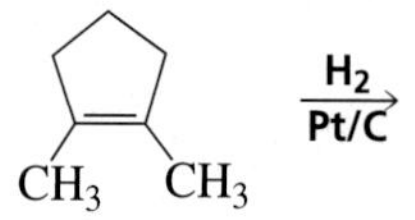 1,2-dimethylcyclopentene (ring substituents CH_3, CH_3) $\xrightarrow[Pt/C]{H_2}$

f. 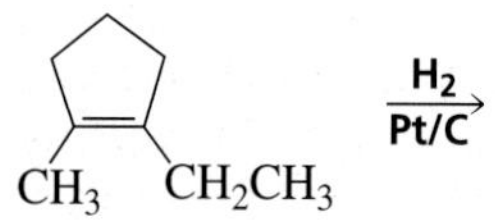 1-ethyl-2-methylcyclopentene (ring substituents CH_3, CH_2CH_3) $\xrightarrow[Pt/C]{H_2}$

PROBLEM 51◆

a. What is the major product obtained from the reaction of propene and Br_2 plus excess Cl^-?
b. Indicate the relative amounts of the stereoisomers obtained.

5.20 Stereochemistry of Enzyme-Catalyzed Reactions

The chemistry associated with living organisms is called **biochemistry**. When you study biochemistry, you study the structures and functions of the molecules found in the biological world and the reactions involved in the synthesis and degradation of these molecules. Because the compounds in living organisms are organic compounds, it is not surprising that many of the reactions encountered in organic chemistry are also seen when you study the chemistry of biological systems. Living cells do not contain molecules such as Cl_2, HBr, or BH_3, so you would not expect to find the addition of such reagents to alkenes in biological systems. However, living cells do contain water and acid catalysts, so some alkenes found in biological systems undergo the acid-catalyzed addition of water (Section 4.5).

The organic reactions that occur in biological systems are catalyzed by enzymes. Enzyme-catalyzed reactions are almost always completely stereoselective. In other words, enzymes catalyze reactions that form only a single stereoisomer. For example, the enzyme fumarase, which catalyzes the addition of water to fumarate, forms only (*S*)-malate—the *R* enantiomer is not formed.

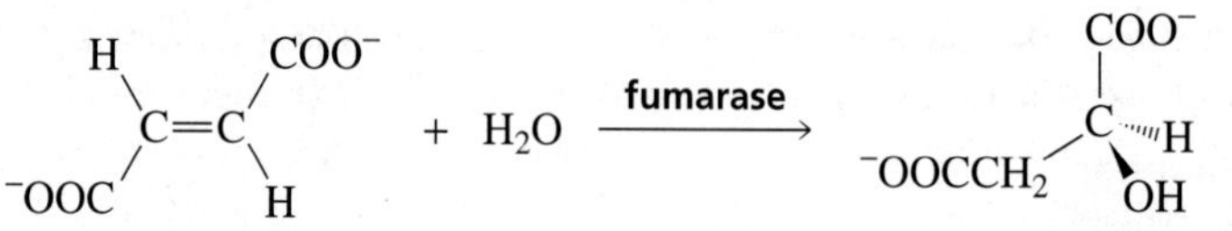

$$(H)(^-OOC)C{=}C(COO^-)(H) + H_2O \xrightarrow{\text{fumarase}} {}^-OOCCH_2\text{–}C(COO^-)(H)(OH)$$

fumarate (*S*)-malate

An enzyme-catalyzed reaction forms only one stereoisomer because the binding site of an enzyme is chiral. The chiral binding site restricts delivery of reagents to only one side of the functional group of the reactant. Consequently, only one stereoisomer is formed.

Enzyme-catalyzed reactions are also stereospecific—an enzyme typically catalyzes the reaction of only one stereoisomer. For example, fumarase catalyzes the addition of water to fumarate (the trans isomer) but not to maleate (the cis isomer).

$$\underset{\text{maleate}}{{}^{-}OOC(H)C{=}C(H)COO^{-}} + H_2O \xrightarrow{\text{fumarase}} \text{no reaction}$$

An enzyme is able to differentiate between stereoisomeric reactants because of its chiral binding site. The enzyme will bind only the stereoisomer whose substituents are in the correct positions to interact with substituents in the chiral binding site (Figure 5.3). Other stereoisomers do not have substituents in the proper positions, so they cannot bind efficiently to the enzyme. An enzyme's stereospecificity can be likened to a right-handed glove which fits only the right hand.

For studies on the stereochemistry of enzyme-catalyzed reactions, **Sir John Cornforth** *received the Nobel Prize in chemistry in 1975 (sharing it with Vladimir Prelog, page 188). Born in Australia in 1917, he studied at the University of Sydney and received a Ph.D. from Oxford. His major research was carried out in laboratories at Britain's Medical Research Council and in laboratories at Shell Research Ltd. He was knighted in 1977.*

Fundamental work on the stereochemistry of enzyme-catalyzed reactions was done by **Frank H. Westheimer**. *Westheimer was born in Baltimore in 1912 and received his graduate training at Harvard University. He served on the faculty of the University of Chicago and subsequently returned to Harvard as a professor of chemistry.*

PROBLEM 52◆

a. What would be the product of the reaction of fumarate and H_2O if H^+ were used as a catalyst instead of fumarase?
b. What would be the product of the reaction of maleate and H_2O if H^+ were used as a catalyst instead of fumarase?

Summary

Stereochemistry is the field of chemistry that deals with the structures of molecules in three dimensions. Compounds that have the same molecular formula but are not identical are called **isomers**; they fall into two classes: constitutional isomers and stereoisomers. **Constitutional isomers** differ in the way their atoms are connected. **Stereoisomers** differ in the way their atoms are arranged in space. There are two kinds of stereoisomers: **cis–trans isomers** and isomers that contain chirality centers.

A **chiral** molecule has a nonsuperimposable mirror image. An **achiral** molecule has a superimposable mirror image. The feature that is most often the cause of chirality is an asymmetric carbon. An **asymmetric carbon** is a carbon bonded to four different atoms or groups. An asymmetric carbon is also known as a **chirality center**. Nitrogen and phosphorus atoms can also be chirality centers. Nonsuperimposable mirror-image molecules are called **enantiomers**. **Diastereomers** are stereoisomers that are not enantiomers. Enantiomers have identical physical and chemical properties; diastereomers have different physical and chemical properties. An achiral reagent reacts identically with both enantiomers; a chiral reagent reacts differently with each enantiomer. A mixture of equal amounts of two enantiomers is called a **racemic mixture**.

A **stereocenter** is an atom at which the interchange of two groups produces a stereoisomer: asymmetric carbons—where the interchange of two groups produces an enantiomer—and the carbons where the interchange of two groups converts a cis isomer to a trans isomer (or an *E* isomer to a *Z* isomer)—are stereocenters. The letters ***R*** and ***S*** indicate the **configuration** about an asymmetric carbon. If one molecule has the *R* and the other has the *S* configuration, they are enantiomers; if they both have the *R* or both have the *S* configuration, they are identical.

Chiral compounds are **optically active**—they rotate the plane of polarized light; achiral compounds are **optically inactive**. If one enantiomer rotates the plane of polarization clockwise (+), its mirror image will rotate the plane of polarization the same amount counterclockwise (−). Each optically active compound has a characteristic **specific rotation**. A **racemic mixture** is optically inactive. A **meso compound** has two or more asymmetric carbons and a plane of symmetry; it is an achiral molecule. A compound with the same four groups bonded to two different asymmetric carbons will have three stereoisomers, a meso compound and a pair of enantiomers. If a reaction does not break any bonds to the asymmetric carbon, the reactant and product will have the same **relative configuration**—their substituents will have the same relative positions. The **absolute configuration** is the actual configuration. If a reaction does break a bond to the asymmetric carbon, the configuration of the product will depend on the mechanism of the reaction.

A **regioselective** reaction selects for a particular constitutional isomer; a **stereoselective** reaction selects for a particular stereoisomer. A reaction is **stereospecific** if the reactant can exist as stereoisomers and each stereoisomeric reactant leads to a different stereoisomeric product or a different set of stereoisomeric products. When a reactant that does not have an asymmetric carbon forms a product with one asymmetric carbon, the product will be a racemic mixture.

In **syn addition** the two substituents add to the same side of the double bond; in **anti addition** they add to opposite sides of the double bond. Both syn and anti addition occur in electrophilic addition reactions that take place by way of a carbocation or a radical intermediate. Addition of H_2 to an alkene is a syn addition reaction. Hydroboration–oxidation is overall a syn addition of water. Addition of Br_2 is an anti addition reaction. An enzyme-catalyzed reaction forms only one stereoisomer; an enzyme typically catalyzes the reaction of only one stereoisomer.

Key Terms

absolute configuration (p. 209)
achiral (p. 184)
amine inversion (p. 217)
anti addition (p. 222)
asymmetric carbon (p. 184)
biochemistry (p. 230)
chiral (p. 184)
chiral catalyst (p. 213)
chiral probe (p. 213)
chirality center (p. 185)
cis isomer (p. 183)
cis–trans isomers (p. 182)
chromatography (p. 213)
configuration (p. 188)
configurational isomers (p. 182)
constitutional isomers (p. 182)
dextrorotatory (p. 193)
diastereomer (p. 197)
diastereotopic hydrogens (p. 216)
enantiomer (p. 186)
enantiomerically pure (p. 195)
enantiomeric excess (ee) (p. 196)
enantiotopic hydrogens (p. 215)
enzyme (p. 213)
erythro enantiomers (p. 198)
Fischer projection (p. 187)
geometric isomers (p. 183)
isomers (p. 182)
levorotatory (p. 193)
meso compound (p. 201)
observed rotation (p. 194)
optical purity (op) (p. 195)
optically active (p. 193)
optically inactive (p. 193)
perspective formula (p. 186)
plane of symmetry (p. 201)
plane-polarized light (p. 192)
polarimeter (p. 193)
prochiral carbon (p. 216)
pro-*R* hydrogen (p. 216)
pro-*S* hydrogen (p. 216)
racemate (p. 195)
racemic mixture (p. 195)
R configuration (p. 188)
regioselective (p. 218)
relative configuration (p. 209)
resolution of a racemic mixture (p. 212)
S configuration (p. 188)
specific rotation (p. 194)
stereocenter (p. 187)
stereochemistry (p. 218)
stereoisomers (p. 182)
stereoselective (p. 218)
stereospecific (p. 219)
syn addition (p. 222)
threo enantiomers (p. 198)
trans isomer (p. 183)

Problems

53. Neglecting stereoisomers, give the structures of all compounds with molecular formula C_5H_{10}. Which ones can exist as stereoisomers?

54. Draw all possible stereoisomers for each of the following compounds. State if no stereoisomers are possible.
   a. 1-bromo-2-chlorocyclohexane
   b. 2-bromo-4-methylpentane
   c. 1,2-dichlorocyclohexane
   d. 2-bromo-4-chloropentane
   e. 3-heptene
   f. 1-bromo-4-chlorocyclohexane
   g. 1,2-dimethylcyclopropane
   h. 4-bromo-2-pentene
   i. 3,3-dimethylpentane
   j. 3-chloro-1-butene
   k. 1-bromo-2-chlorocyclobutane
   l. 1-bromo-3-chlorocyclobutane

55. Name each of the following compounds using *R,S* and *E,Z* (Section 3.5) designations where necessary:

a. CH_2CH_3 | $C{=}CH_2$ | CH_2 | H—C—CH_3 | CH_2CH_3

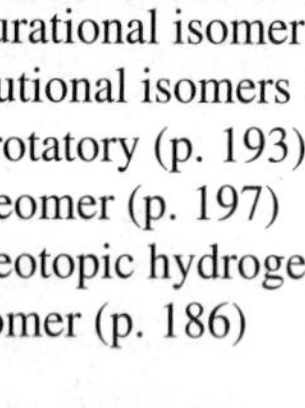

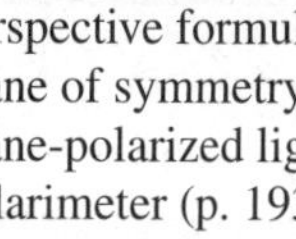

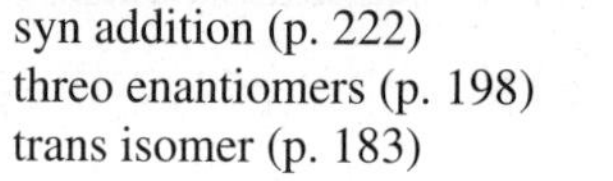

b. (F)(Cl)C=C(I)(Br)

c. (F)(Cl)C=C(H)(Br)

d. ($CH_3CH(CH_3)$)(H_3C)C=C($CH_2CH_2CH_2CH_3$)(CH_2CH_2Cl)

e. [skeletal structure]—Br

f. ($BrCH_2CH_2$)(Br)C=C($CH_2CH_2CH(CH_3)CH_3$)(CH_3)

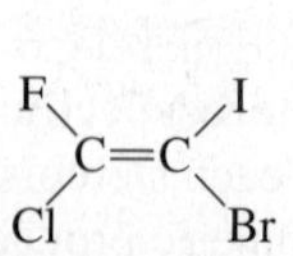

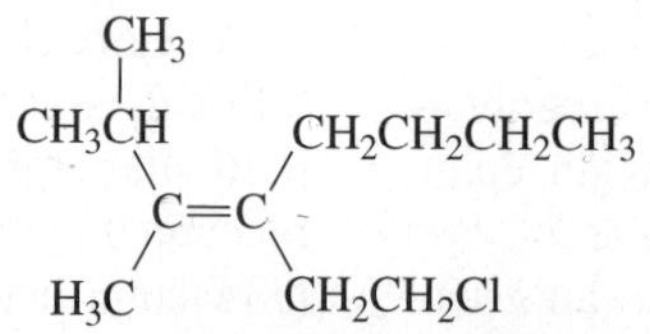

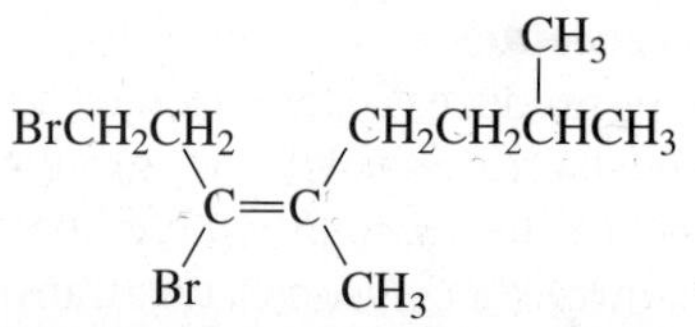

g.
CH$_2$OH
HO—C—CH$_3$
CH$_2$CH$_2$CH$_2$OH

h.
CH$_3$
HO—C—H
H—C—Cl
CH$_2$CH$_3$

56. Mevacor^{TM} is used clinically to lower serum cholesterol levels. How many asymmetric carbons does Mevacor^{TM} have?

HO, O, O, O, CH$_3$, CH$_3$, H$_3$C

Mevacor^{TM}

57. Indicate whether each of the following pairs of compounds are identical or are enantiomers, diastereomers, or constitutional isomers:

a. (H Cl / H Cl) and (H H / Cl Cl)

b. (Cl H / H Cl) and (H Cl / Cl H)

c. H Br / Br H and H H / Br Br

d.
CH$_3$
HO—C—H
H—C—Cl
CH$_2$CH$_3$
and
CH$_2$CH$_3$
HO—C—H
H—C—Cl
CH$_3$

e. H, CH$_3$ C=C H$_3$C, Br and H$_3$C, CH$_3$ C=C H, Br

f. CH$_3$ (cyclohexane) and CH$_2$CH$_3$ (cyclopentane)

g.
CH$_2$OH
H—C—CH$_3$
CH$_2$CH$_3$
and
CH$_2$CH$_3$
CH$_3$—C—H
CH$_2$OH

h.
CH$_2$Cl
CH$_3$CH$_2$—C—CH$_3$
H
and
CH$_2$CH$_3$
CH$_3$—C—CH$_2$Cl
H

i. (H H / Cl Cl) and (H Cl / Cl H)

j. H Br / Br H and Br H / H Br

k. H CH$_3$ / H CH$_3$ and H$_3$C H / H CH$_3$

l.
CH$_3$
HO—C—H
H—C—Cl
CH$_3$
and
CH$_3$
H—C—OH
Cl—C—H
CH$_3$

m. H$_3$C, H C=C H$_3$C, Br and H, CH$_3$ C=C Br, CH$_3$

n. CH$_3$, CH$_3$, H$_3$C and CH$_3$, CH$_3$, H$_3$C

o.
CH$_3$
Cl—C—H
CH$_3$—C—H
CH$_2$CH$_3$
and
CH$_2$CH$_3$
H—C—CH$_3$
H—C—Cl
CH$_3$

p.
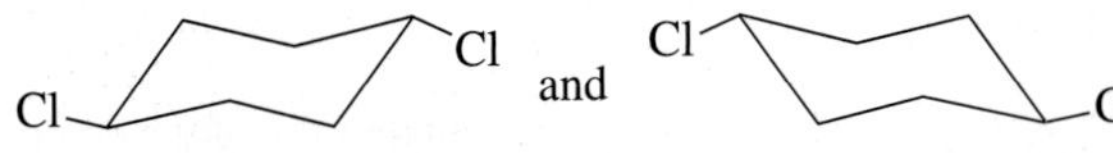

58. a. Give the product(s) that would be obtained from the reaction of *cis*-2-butene and *trans*-2-butene with each of the following reagents. If the products can exist as stereoisomers, show which stereoisomers are obtained.
   1. HCl
   2. BH_3/THF followed by HO^-, H_2O_2
   3. HBr + peroxide
   4. Br_2 in CH_2Cl_2
   5. $Br_2 + H_2O$
   6. H_2/Pt/C
   7. $HCl + H_2O$
   8. $HCl + CH_3OH$

   b. With which reagents do the two alkenes react to give different products?

59. Which of the following compounds have an achiral stereoisomer?
   a. 2,3-dichlorobutane
   b. 2,3-dichloropentane
   c. 2,3-dichloro-2,3-dimethylbutane
   d. 1,3-dichlorocyclopentane
   e. 1,3-dibromocyclobutane
   f. 2,4-dibromopentane
   g. 2,3-dibromopentane
   h. 1,4-dimethylcyclohexane
   i. 1,2-dimethylcyclopentane
   j. 1,2-dimethylcyclobutane

60. Give the products and their configurations obtained from the reaction of 1-ethylcyclohexene with the following reagents:
   a. HBr
   b. HBr + peroxide
   c. H_2, Pt/C
   d. BH_3/THF followed by HO^-, H_2O_2
   e. Br_2/CH_2Cl_2

61. Citrate synthase, one of the enzymes in the series of enzyme-catalyzed reactions known as the Krebs cycle, catalyzes the synthesis of citric acid from oxaloacetic acid and acetyl-CoA. If the synthesis is carried out with acetyl-CoA that has radioactive carbon (^{14}C) in the indicated position, the isomer shown here is obtained.

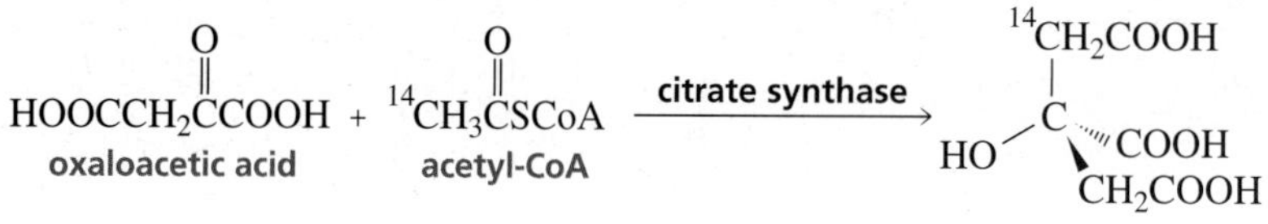

$$\underset{\text{oxaloacetic acid}}{HOOCCH_2\overset{O}{\overset{\|}{C}}COOH} + \underset{\text{acetyl-CoA}}{{}^{14}CH_3\overset{O}{\overset{\|}{C}}SCoA} \xrightarrow{\text{citrate synthase}} \underset{\text{citric acid}}{HO\text{–}C({}^{14}CH_2COOH)(COOH)(CH_2COOH)}$$

   a. Which stereoisomer of citric acid is synthesized, *R* or *S*?
   b. Why is the other stereoisomer not obtained?
   c. If the acetyl-CoA used in the synthesis does not contain ^{14}C, will the product of the reaction be chiral or achiral?

62. Give the products of the following reactions. If the products can exist as stereoisomers, show which stereoisomers are obtained.
   a. *cis*-2-pentene + HCl
   b. *trans*-2-pentene + HCl
   c. 1-ethylcyclohexene + H_3O^+
   d. 1,2-diethylcyclohexene + H_3O^+
   e. 1,2-dimethylcyclohexene + HCl
   f. 1,2-dideuteriocyclohexene + H_2, Pt/C
   g. 3,3-dimethyl-1-pentene + Br_2/CH_2Cl_2
   h. (*E*)-3,4-dimethyl-3-heptene + H_2, Pt/C
   i. (*Z*)-3,4-dimethyl-3-heptene + H_2, Pt/C
   j. 1-chloro-2-ethylcyclohexene + H_2, Pt/C

63. The specific rotation of (*R*)-(+)-glyceraldehyde is +8.7°. If the observed specific rotation of a mixture of (*R*)-glyceraldehyde and (*S*)-glyceraldehyde is +1.4°, what percent of glyceraldehyde is present as the *R* enantiomer?

64. Indicate whether each of the following structures is (*R*)-2-chlorobutane or (*S*)-2-chlorobutane. (Use models, if necessary.)

   a. 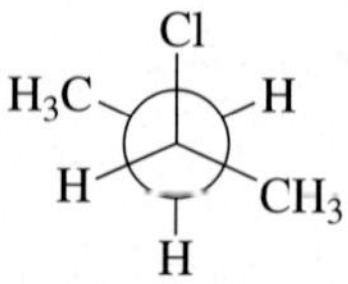 Newman projection: front Cl, H_3C, H; back H, CH_3, H
   b. 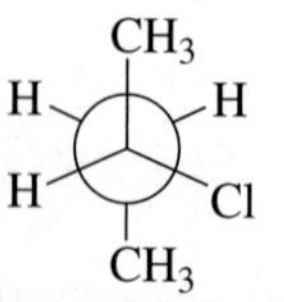 Newman projection: front CH_3, H, H; back H, Cl, CH_3
   c. Fischer projection: CH_3 (top), H (left), Cl (right), CH_2CH_3 (bottom)
   d. 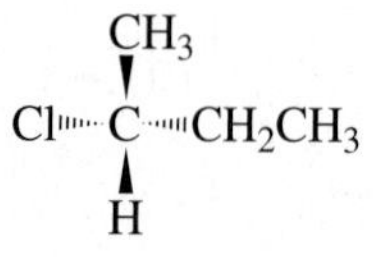 CH_3 (top), H (bottom), Cl–C–CH_2CH_3
   e. (skeletal structure with Cl)
   f. 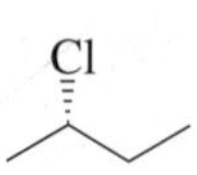 (skeletal structure with Cl)

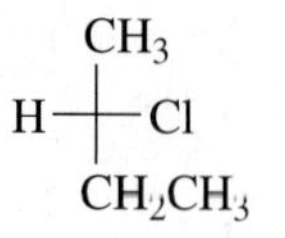

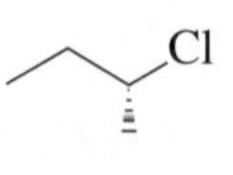

65. A solution of an unknown compound (3.0 g of the compound in 20 mL of solution), when placed in a polarimeter tube 2.0 dm long, was found to rotate the plane of polarized light 1.8° in a counterclockwise direction. What is the specific rotation of the compound?

66. Butaclamol is a potent antipsychotic that has been used clinically in the treatment of schizophrenia. How many asymmetric carbons does Butaclamol™ have?

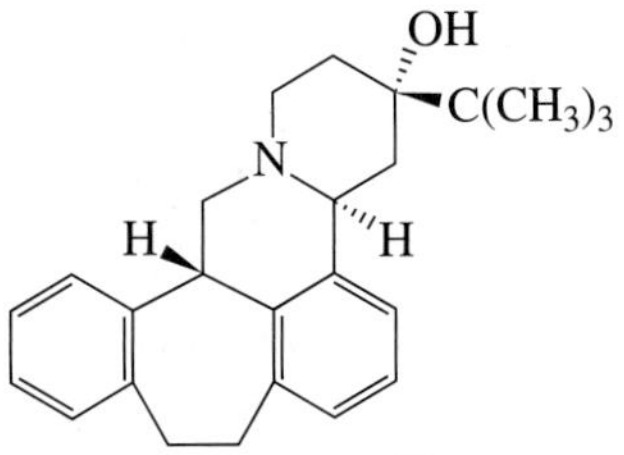

Butaclamol™

67. Which of the following objects are chiral?
   a. a mug with DAD written on one side
   b. a mug with MOM written on one side
   c. a mug with DAD written opposite the handle
   d. a mug with MOM written opposite the handle
   e. an automobile
   f. a wineglass
   g. a nail
   h. a screw

68. Explain how *R* and *S* are related to (+) and (−).

69. Give the products of the following reactions. If the products can exist as stereoisomers, show which stereoisomers are obtained.
   a. *cis*-2-pentene + Br_2/CH_2Cl_2
   b. *trans*-2-pentene + Br_2/CH_2Cl_2
   c. 1-butene + HCl
   d. 1-butene + HBr + peroxide
   e. *trans*-3-hexene + Br_2/CH_2Cl_2
   f. *cis*-3-hexene + Br_2/CH_2Cl_2
   g. 3,3-dimethyl-1-pentene + HBr
   h. *cis*-2-butene + HBr + peroxide
   i. (*Z*)-2,3-dichloro-2-butene + H_2, Pt/C
   j. (*E*)-2,3-dichloro-2-butene + H_2, Pt/C
   k. (*Z*)-3,4-dimethyl-3-hexene + H_2, Pt/C
   l. (*E*)-3,4-dimethyl-3-hexene + H_2, Pt/C

70. a. Draw all possible stereoisomers for the following compound.

$$\mathrm{HOCH_2\underset{|}{\overset{}{C}}H-\underset{|}{C}H-\underset{|}{C}HCH_2OH}$$
$$\mathrm{OH\quad OH\quad OH}$$

   b. Which isomers are optically inactive (will not rotate plane-polarized light)?

71. Indicate the configuration of the asymmetric carbons in the following molecules:

   a. 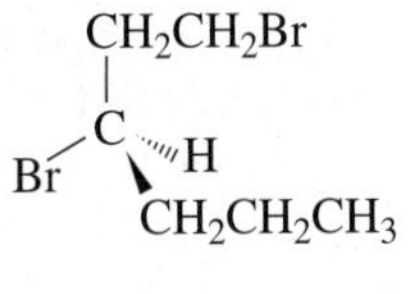 CH_2CH_2Br, Br, H, $CH_2CH_2CH_3$
   b. 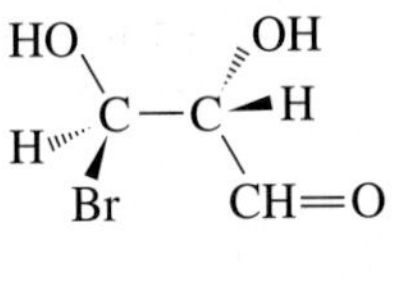 HO, OH, H, H, Br, CH=O
   c. 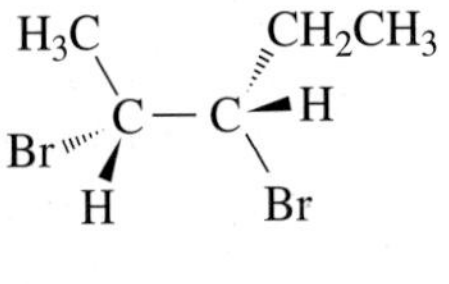 H_3C, CH_2CH_3, Br, H, H, Br

72. a. Draw all the isomers with molecular formula C_6H_{12} that contain a cyclobutane ring. (*Hint:* There are seven.)
   b. Name the compounds without specifying the configuration of any asymmetric carbons.
   c. Identify:
      1. constitutional isomers
      2. stereoisomers
      3. cis–trans isomers
      4. chiral compounds
      5. achiral compounds
      6. meso compounds
      7. enantiomers
      8. diastereomers

73. A compound has a specific rotation of −39.0°. A solution of the compound (0.187 g/mL) has an observed rotation of −6.52° when placed in a polarimeter tube 10 cm long. What is the percent of each enantiomer in the solution?

74. Draw structures for each of the following molecules:
   a. (*S*)-1-bromo-1-chlorobutane
   b. (2*R*,3*R*)-2,3-dichloropentane
   c. an achiral isomer of 1,2-dimethylcyclohexane
   d. a chiral isomer of 1,2-dibromocyclobutane
   e. two achiral isomers of 3,4,5-trimethylheptane

75. The enantiomers of 1,2-dimethylaziridine can be separated even though one of the "groups" attached to nitrogen is a lone pair. Explain.

enantiomers of 1,2-dimethylaziridine

76. Of the possible monobromination products shown for the following reaction, circle any that would not be formed.

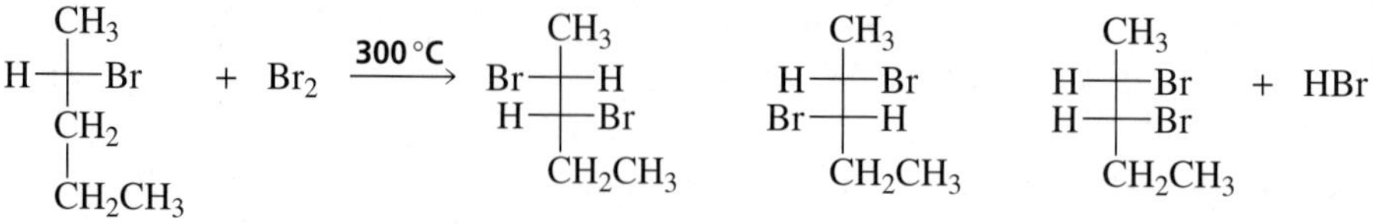

$$\begin{array}{c} CH_3 \\ H\!-\!\!\!+\!\!\!-\!Br \\ CH_2 \\ | \\ CH_2CH_3 \end{array} + Br_2 \xrightarrow{300\,°C} \begin{array}{c} CH_3 \\ Br\!-\!\!\!+\!\!\!-\!H \\ H\!-\!\!\!+\!\!\!-\!Br \\ CH_2CH_3 \end{array} \quad \begin{array}{c} CH_3 \\ H\!-\!\!\!+\!\!\!-\!Br \\ Br\!-\!\!\!+\!\!\!-\!H \\ CH_2CH_3 \end{array} \quad \begin{array}{c} CH_3 \\ H\!-\!\!\!+\!\!\!-\!Br \\ H\!-\!\!\!+\!\!\!-\!Br \\ CH_2CH_3 \end{array} + HBr$$

77. A sample of (*S*)-(+)-lactic acid was found to have an optical purity of 72%. How much *R* isomer is present in the sample?

78. Is the following compound optically active?

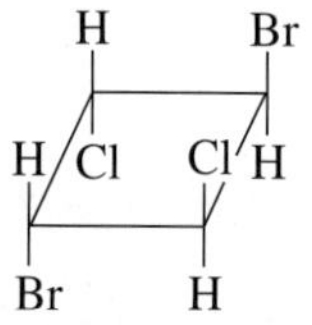

79. Give the products of the following reactions and their configurations:

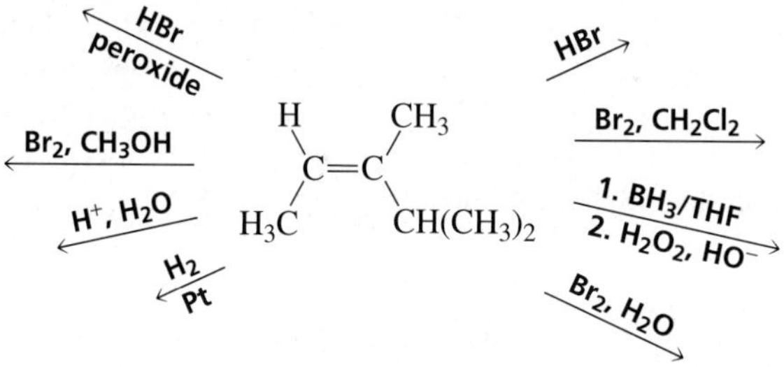

Reagents shown around (*E*)-$CH_3(H)C{=}C(CH_3)CH(CH_3)_2$: HBr, peroxide; Br_2, CH_3OH; H^+, H_2O; H_2, Pt; HBr; Br_2, CH_2Cl_2; 1. BH_3/THF 2. H_2O_2, HO^-; Br_2, H_2O

80. a. Using the wedge-and-dash notation, draw the nine stereoisomers of 1,2,3,4,5,6-hexachlorocyclohexane.
b. From the nine stereoisomers, identify one pair of enantiomers.
c. Draw the most stable conformation of the most stable stereoisomer.

81. Sherry O. Eismer decided that the configuration of the asymmetric carbons in a sugar such as D-glucose could be determined rapidly by simply assigning the *R* configuration to an asymmetric carbon with an OH group on the right and the *S* configuration to an asymmetric carbon with an OH group on the left. Is she correct? (We will see in Chapter 22 that the "D" in D-glucose means the OH group on the bottommost asymmetric carbon is on the right.)

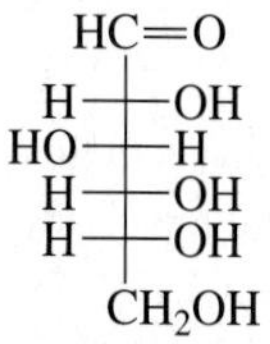

$$\begin{array}{c} HC{=}O \\ H\!-\!\!\!+\!\!\!-\!OH \\ HO\!-\!\!\!+\!\!\!-\!H \\ H\!-\!\!\!+\!\!\!-\!OH \\ H\!-\!\!\!+\!\!\!-\!OH \\ CH_2OH \end{array}$$

D-glucose

82. Cyclohexene exists only in the cis form, while cyclodecene exists in both the cis and trans forms. Explain. (*Hint:* Molecular models are helpful for this problem.)

83. When fumarate reacts with D_2O in the presence of the enzyme fumarase, only one isomer of the product is formed. Its structure is shown. Is the enzyme catalyzing a syn or an anti addition of D_2O?

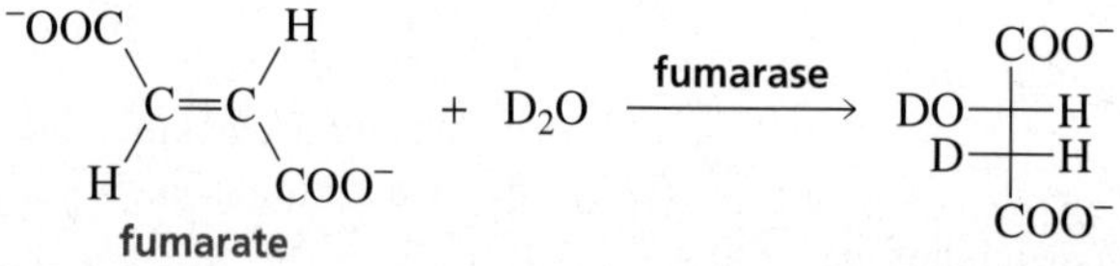

$$\text{fumarate} + D_2O \xrightarrow{\text{fumarase}} \begin{array}{c} COO^- \\ DO\!-\!\!\!+\!\!\!-\!H \\ D\!-\!\!\!+\!\!\!-\!H \\ COO^- \end{array}$$

fumarate

84. Two stereoisomers are obtained from the reaction of HBr with (*S*)-4-bromo-1-pentene. One of the stereoisomers is optically active, and the other is not. Give the structures of the stereoisomers, indicating their absolute configurations, and explain the difference in optical properties.

85. When (*S*)-(+)-1-chloro-2-methylbutane reacts with chlorine, one of the products formed is (−)-1,4-dichloro-2-methylbutane. Does this product have the *R* or the *S* configuration?

86. Indicate the configuration of the asymmetric carbons in the following molecules:

a. 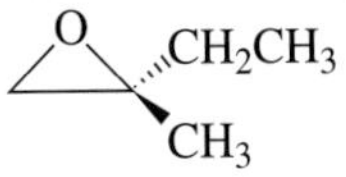 O CH_2CH_3 CH_3

b. 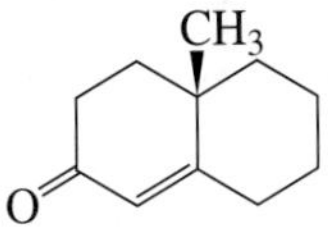 CH_3 O

c. 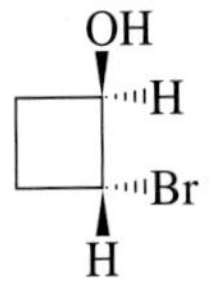 OH H Br H

87. a. Draw the two chair conformers for each of the stereoisomers of *trans*-1-*tert*-butyl-3-methylcyclohexane.
b. For each pair, indicate which conformer is more stable.

88. a. Do the following compounds have any asymmetric carbons?
1. $CH_2{=}C{=}CH_2$ 2. $CH_3CH{=}C{=}CHCH_3$
b. Are they chiral? (*Hint:* Make models.)

6 Reactions of Alkynes • Introduction to Multistep Synthesis

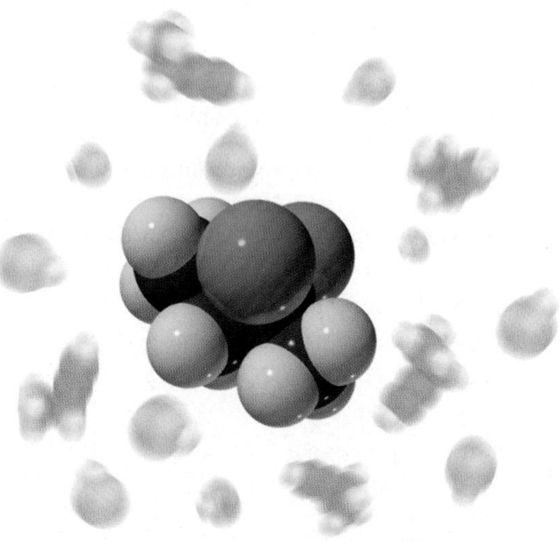

1-butyne + 2HCl ⟶ 2,2-dichlorobutane

Alkynes are hydrocarbons that contain a carbon–carbon triple bond. Because of its triple bond, an alkyne has four fewer hydrogens than the corresponding alkane. Therefore, the general molecular formula for an acyclic (noncyclic) alkyne is C_nH_{2n-2}, and that for a cyclic alkyne is C_nH_{2n-4}.

There are only a few naturally occurring alkynes. Examples include capillin, which has fungicidal activity, and ichthyothereol, a convulsant used by the Amazon Indians for poisoned arrowheads. A class of naturally occurring compounds called enediynes has been found to have powerful antibiotic and anticancer properties. These compounds all have a nine- or ten-membered ring that contains two triple bonds separated by a double bond. Some enediynes are currently in clinical trials.

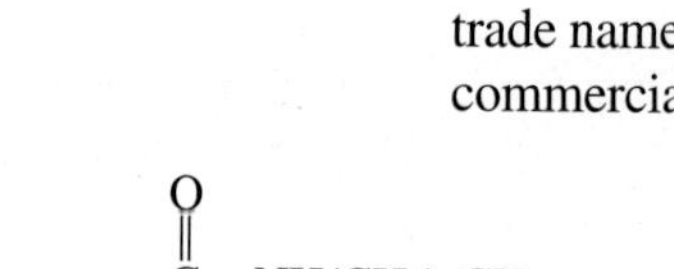

$CH_3C{\equiv}C{-}C{\equiv}C{-}C(=O){-}C_6H_5$
capillin

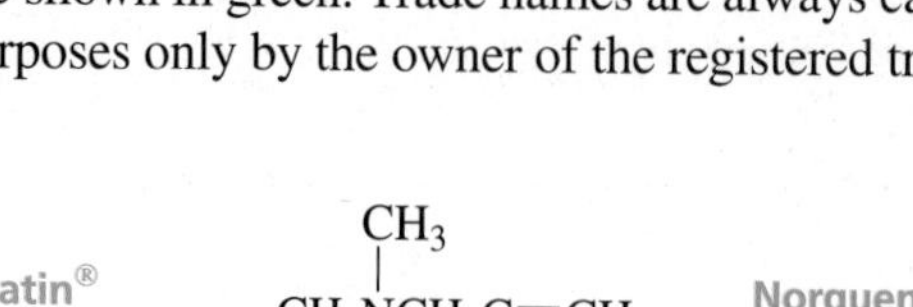

$CH_3C{\equiv}C{-}C{\equiv}C{-}C{\equiv}C{-}CH{=}CH{-}$(HO-substituted tetrahydropyran ring)
ichthyothereol

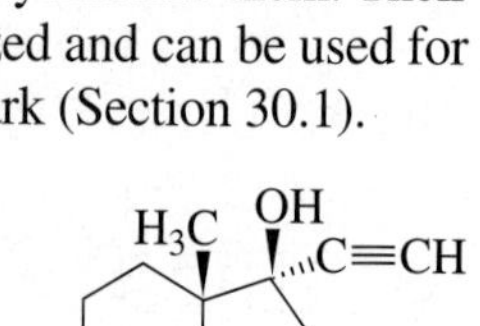

R^1, R^2, R^3, R^4, R^5
an enediyne

A few drugs contain alkyne functional groups, but they are not naturally occurring compounds. They exist only because chemists have been able to synthesize them. Their trade names are shown in green. Trade names are always capitalized and can be used for commercial purposes only by the owner of the registered trademark (Section 30.1).

Parsal®
Sinovial® 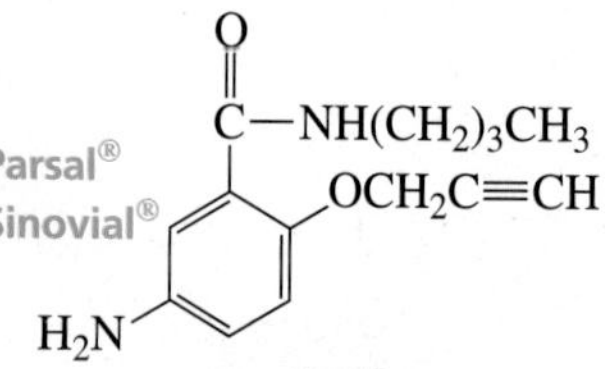
$C(=O){-}NH(CH_2)_3CH_3$, $OCH_2C{\equiv}CH$, H_2N
parsalmide
an analgesic

Eudatin®
Supirdyl® 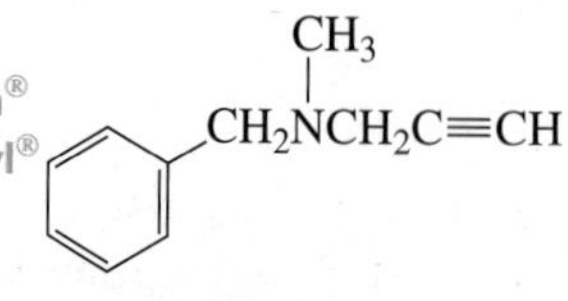
$CH_2N(CH_3)CH_2C{\equiv}CH$
pargyline
an antihypertensive

Norquen®
Ovastol® 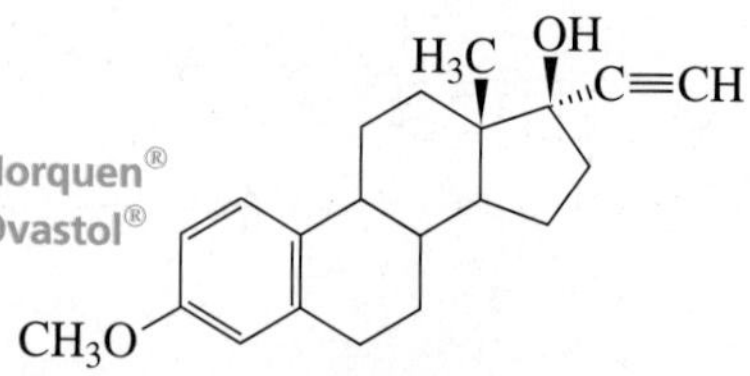
H_3C, OH, $C{\equiv}CH$, CH_3O
mestranol
a component in oral contraceptives

Acetylene ($HC{\equiv}CH$), the common name for the smallest alkyne, may be a familiar word because of the oxyacetylene torch used in welding. Acetylene is supplied to the torch from one high-pressure gas tank, and oxygen is supplied from another. Burning acetylene produces a high-temperature flame capable of melting and vaporizing iron and steel.

PROBLEM 1◆

What is the molecular formula for a cyclic alkyne with 14 carbons and two triple bonds?

6.1 Nomenclature of Alkynes

The systematic name of an alkyne is obtained by replacing the "ane" ending of the alkane name with "yne." Analogous to the way compounds with other functional groups are named, the longest continuous chain containing the carbon–carbon triple bond is numbered in the direction that gives the alkyne functional group suffix as low a number as possible. If the triple bond is at the end of the chain, the alkyne is classified as a **terminal alkyne**. Alkynes with triple bonds located elsewhere along the chain are called **internal alkynes**. For example, 1-butyne is a terminal alkyne, whereas 2-pentyne is an internal alkyne.

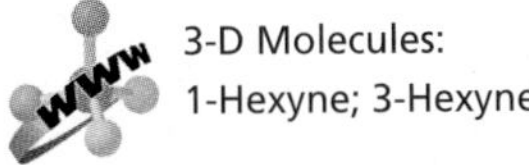

3-D Molecules: 1-Hexyne; 3-Hexyne

1-hexyne
a terminal alkyne

	$HC{\equiv}CH$	$\overset{4}{C}H_3\overset{3}{C}H_2\overset{2}{C}{\equiv}\overset{1}{C}H$	$\overset{1}{C}H_3\overset{2}{C}{\equiv}\overset{3}{C}\overset{4}{C}H_2\overset{5}{C}H_3$	$CH_3\overset{4}{C}H(\overset{5}{C}H_2\overset{6}{C}H_3)\overset{3}{C}{\equiv}\overset{2}{C}\overset{1}{C}H_3$
Systematic:	**ethyne**	**1-butyne**	**2-pentyne**	**4-methyl-2-hexyne**
Common:	**acetylene**	**ethylacetylene**	**ethylmethylacetylene**	***sec*-butylmethylacetylene**
		a terminal alkyne	**an internal alkyne**	

In common nomenclature, alkynes are named as *substituted acetylenes*. The common name is obtained by citing the names of the alkyl groups, in alphabetical order, that have replaced the hydrogens of acetylene. Acetylene is an unfortunate common name for the smallest alkyne because its "ene" ending is characteristic of a double bond rather than a triple bond.

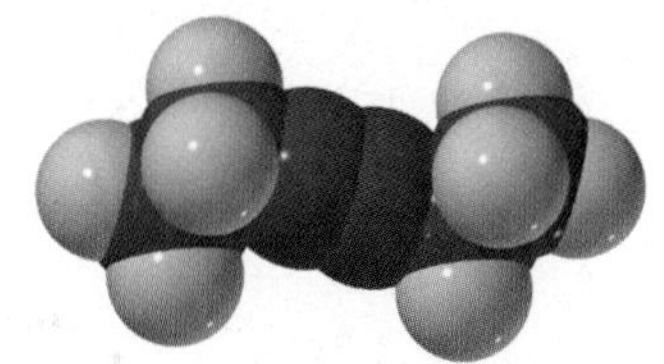

3-hexyne
an internal alkyne

If the same number for the alkyne functional group suffix is obtained counting from either direction along the carbon chain, the correct systematic name is the one that contains the lowest substituent number. If the compound contains more than one substituent, the substituents are listed in alphabetical order.

A substituent receives the lowest possible number only if there is no functional group suffix or if the same number for the functional group suffix is obtained in both directions.

$$\overset{1}{C}H_3\overset{2}{C}H(Cl)\overset{3}{C}H(Br)\overset{4}{C}{\equiv}\overset{5}{C}\overset{6}{C}H_2\overset{7}{C}H_2\overset{8}{C}H_3$$

3-bromo-2-chloro-4-octyne
***not* 6-bromo-7-chloro-4-octyne**
because 2 < 6

$$\overset{6}{C}H_3\overset{5}{C}H(CH_3)\overset{4}{C}{\equiv}\overset{3}{C}\overset{2}{C}H_2\overset{1}{C}H_2Br$$

1-bromo-5-methyl-3-hexyne
***not* 6-bromo-2-methyl-3-hexyne**
because 1 < 2

The triple-bond-containing propargyl group is used in common nomenclature. It is analogous to the double-bond-containing allyl group that you saw in Section 3.2.

$HC{\equiv}CCH_2{-}$	$H_2C{=}CHCH_2{-}$
propargyl group	**allyl group**
$HC{\equiv}CCH_2Br$	$H_2C{=}CHCH_2OH$
propargyl bromide	**allyl alcohol**

PROBLEM 2◆

Draw the structure for each of the following compounds.

a. 1-chloro-3-hexyne
b. cyclooctyne
c. isopropylacetylene
d. propargyl chloride
e. 4,4-dimethyl-1-pentyne
f. dimethylacetylene

PROBLEM 3◆

Draw the structures and give the common and systematic names for the seven alkynes with molecular formula C_6H_{10}.

PROBLEM 4◆

Give the systematic name for each of the following compounds:

a. $BrCH_2CH_2C{\equiv}CCH_3$

b. $CH_3CH_2\underset{\displaystyle Br}{CH}C{\equiv}CCH_2\underset{\displaystyle Cl}{CH}CH_3$

c. $CH_3OCH_2C{\equiv}CCH_2CH_3$

d. $CH_3CH_2\underset{\displaystyle CH_2CH_2CH_3}{CH}C{\equiv}CH$

PROBLEM 5

Which would you expect to be more stable, an internal alkyne or a terminal alkyne? Why?

6.2 Physical Properties of Unsaturated Hydrocarbons

All hydrocarbons have similar physical properties. In other words, alkenes and alkynes have physical properties similar to those of alkanes (Section 2.9). All are insoluble in water and all are soluble in solvents with low polarity such as benzene and ether. They are less dense than water and, like other homologous series, have boiling points that increase with increasing molecular weight (Table 6.1). Alkynes are more linear than alkenes, and a triple bond is more polarizable than a double bond (Section 2.9). These two features cause alkynes to have stronger van der Waals interactions. As a result, an alkyne has a higher boiling point than an alkene containing the same number of carbon atoms.

Internal alkenes have higher boiling points than terminal alkenes. Similarly, internal alkynes have higher boiling points than terminal alkynes. Notice that the boiling point of *cis*-2-butene is slightly higher than that of *trans*-2-butene because the cis isomer has a small dipole moment, whereas the dipole moment of the trans isomer is zero (Section 3.4).

Table 6.1 Boiling Points of the Smallest Hydrocarbons

	bp (°C)		bp (°C)		bp (°C)
CH_3CH_3 ethane	−88.6	$H_2C{=}CH_2$ ethene	−104	$HC{\equiv}CH$ ethyne	−84
$CH_3CH_2CH_3$ propane	−42.1	$CH_3CH{=}CH_2$ propene	−47	$CH_3C{\equiv}CH$ propyne	−23
$CH_3CH_2CH_2CH_3$ butane	−0.5	$CH_3CH_2CH{=}CH_2$ 1-butene	−6.5	$CH_3CH_2C{\equiv}CH$ 1-butyne	8
$CH_3(CH_2)_3CH_3$ pentane	36.1	$CH_3CH_2CH_2CH{=}CH_2$ 1-pentene	30	$CH_3CH_2CH_2C{\equiv}CH$ 1-pentyne	39
$CH_3(CH_2)_4CH_3$ hexane	68.7	$CH_3CH_2CH_2CH_2CH{=}CH_2$ 1-hexene	63.5	$CH_3CH_2CH_2CH_2C{\equiv}CH$ 1-hexyne	71
		$CH_3CH{=}CHCH_3$ cis-2-butene	3.7	$CH_3C{\equiv}CCH_3$ 2-butyne	27
		$CH_3CH{=}CHCH_3$ trans-2-butene	0.9	$CH_3CH_2C{\equiv}CCH_3$ 2-pentyne	55

6.3 The Structure of Alkynes

The structure of ethyne was discussed in Section 1.9. Each carbon is *sp* hybridized, so each has two *sp* orbitals and two *p* orbitals. One *sp* orbital overlaps the *s* orbital of a hydrogen, and the other overlaps an *sp* orbital of the other carbon. Because the *sp* orbitals are oriented as far from each other as possible to minimize electron repulsion, ethyne is a linear molecule with bond angles of 180° (Section 1.9).

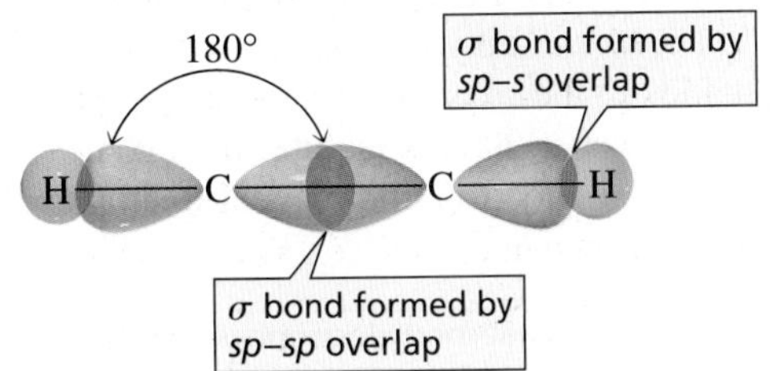

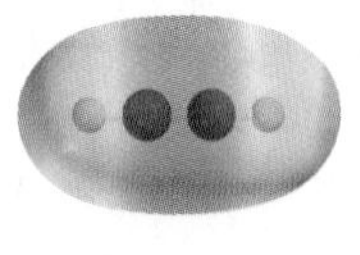

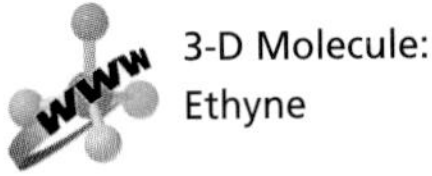

3-D Molecule: Ethyne

The two remaining *p* orbitals on each carbon are oriented at right angles to one another and to the *sp* orbitals (Figure 6.1). Each of the two *p* orbitals on one carbon overlaps the parallel *p* orbital on the other carbon to form two π bonds. One pair of overlapping *p* orbitals results in a cloud of electrons above and below the σ bond, and the other pair results in a cloud of electrons in front of and behind the σ bond. The electrostatic potential map of ethyne shows that the end result can be thought of as a cylinder of electrons wrapped around the σ bond.

A triple bond is composed of a σ bond and two π bonds.

a.

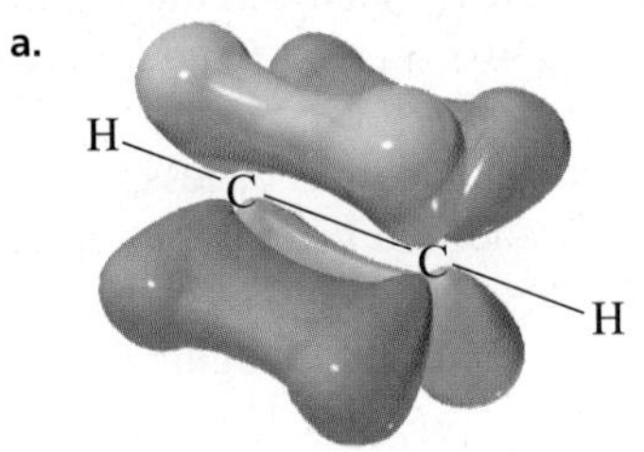

b.

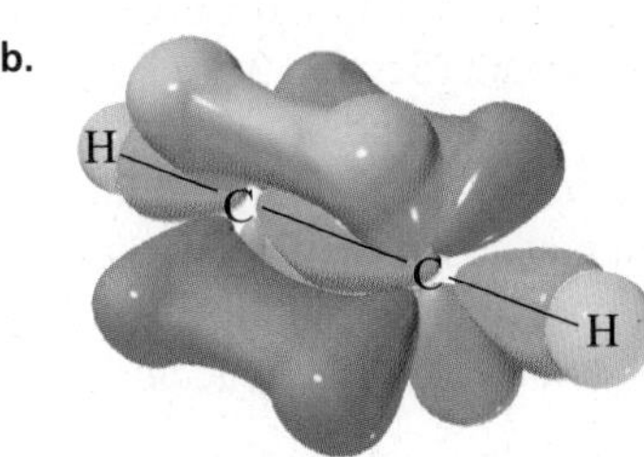

Figure 6.1
(a) Each of the two π bonds of a triple bond is formed by side-to-side overlap of a *p* orbital of one carbon with a parallel *p* orbital of the adjacent carbon.
(b) A triple bond consists of a σ bond formed by *sp–sp* overlap (yellow) and two π bonds formed by *p–p* overlap (blue and purple).

We have seen that a carbon–carbon triple bond is shorter and stronger than a carbon–carbon double bond, which in turn, is shorter and stronger than a carbon–carbon single bond. We have also seen that a carbon–carbon π bond is weaker than a carbon–carbon σ bond (Section 1.14). The relatively weak π bonds allow alkynes to react easily. Like alkenes, alkynes are stabilized by electron-donating alkyl groups. Internal alkynes, therefore, are more stable than terminal alkynes. We now have seen that *alkyl groups stabilize alkenes, alkynes, carbocations,* and *alkyl radicals.*

PROBLEM 6◆

What orbitals are used to form the carbon–carbon σ bond between the highlighted carbons?

a. $CH_3CH{=}CHCH_3$
b. $CH_3CH{=}CHCH_3$
c. $CH_3CH{=}C{=}CH_2$
d. $CH_3C{\equiv}CCH_3$
e. $CH_3C{\equiv}CCH_3$
f. $CH_2{=}CHCH{=}CH_2$
g. $CH_3CH{=}CHCH_2CH_3$
h. $CH_3C{\equiv}CCH_2CH_3$
i. $CH_2{=}CHC{\equiv}CH$

6.4 How Alkynes React

With a cloud of electrons completely surrounding the σ bond, an alkyne is an electron-rich molecule. In other words, it is a nucleophile and consequently it will react with electrophiles. For example, if a reagent such as HCl is added to an alkyne, the relatively weak π bond will break because the π electrons are attracted to the electrophilic proton. In the second step of the reaction, the positively charged carbocation intermediate reacts rapidly with the negatively charged chloride ion.

$$CH_3C{\equiv}CCH_3 + H{-}\ddot{\underset{..}{Cl}}: \longrightarrow CH_3\overset{+}{C}{=}CHCH_3 + :\ddot{\underset{..}{Cl}}:^- \longrightarrow CH_3\overset{Cl}{\overset{|}{C}}{=}CHCH_3$$

Thus alkynes, like alkenes, undergo electrophilic addition reactions. We will see that the same electrophilic reagents that add to alkenes also add to alkynes and that—again like alkenes—electrophilic addition to a *terminal* alkyne is regioselective: When an electrophile adds to a terminal alkyne, it adds to the *sp* carbon that is bonded to the hydrogen. The addition reactions of alkynes, however, have a feature that alkenes do not have: Because the product of the addition of an electrophilic reagent to an alkyne is an alkene, a second electrophilic addition reaction can occur.

a second electrophilic addition reaction occurs

$$CH_3C{\equiv}CCH_3 \xrightarrow{HCl} CH_3\overset{Cl}{\overset{|}{C}}{=}CHCH_3 \xrightarrow{HCl} CH_3\overset{Cl}{\underset{Cl}{\overset{|}{\underset{|}{C}}}}CH_2CH_3$$

Alkynes are less reactive than alkenes in electrophilic addition reactions.

An alkyne is *less* reactive than an alkene. This might at first seem surprising because an alkyne is less stable than an alkene (Figure 6.2). However, reactivity depends on $\Delta G^\ddagger$, which in turn depends on the stability of the reactant *and* the stability of the transition state (Section 3.7). For an alkyne to be both less stable and less reactive than an alkene, two conditions must hold: The transition state for the first step (the rate-limiting step) of an electrophilic addition reaction for an alkyne must be less stable than the transition state for the first step of an electrophilic addition reaction for an alkene, *and* the difference in the stabilities of the transition states must be greater than the difference in the stabilities of the reactants so that $\Delta G^\ddagger_{alkyne} > \Delta G^\ddagger_{alkene}$ (Figure 6.2).

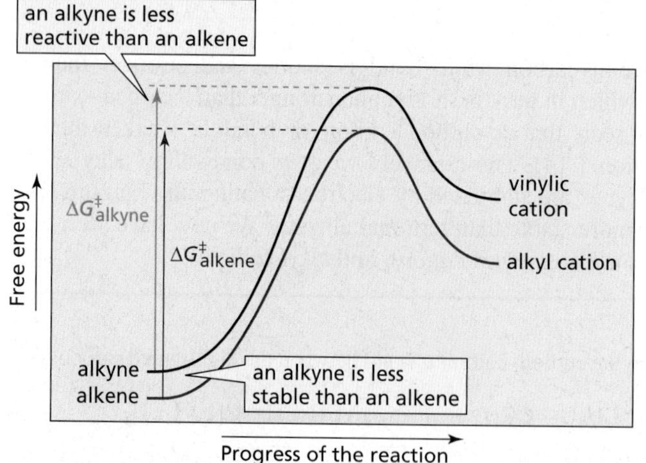

Figure 6.2 ▶ Comparison of the free energies of activation for the addition of an electrophile to an alkyne and to an alkene. Because an alkyne is less reactive than an alkene toward electrophilic addition, we know that $\Delta G^\ddagger$ for the reaction of an alkyne is greater than the $\Delta G^\ddagger$ for the reaction of an alkene.

PROBLEM 7◆

Under what circumstances can you assume that the less stable reactant will be the more reactive reactant?

Why is the transition state for the first step of an electrophilic addition reaction for an alkyne less stable than that for an alkene? The Hammond postulate predicts that the structure of the transition state will resemble the structure of the intermediate (Section 4.3). The intermediate formed when a proton adds to an alkyne is a vinylic cation, whereas the intermediate formed when a proton adds to an alkene is an alkyl cation. A **vinylic cation** has a positive charge on a vinylic carbon. A vinylic cation is less stable than a similarly substituted alkyl cation. In other words, a primary vinylic cation is less stable than a primary alkyl cation, and a secondary vinylic cation is less stable than a secondary alkyl cation.

relative stabilities of carbocations

most stable R_3C^+ > R_2CH^+ > $RCH{=}\overset{+}{C}{-}R$ ≈ RCH_2^+ > $RCH{=}\overset{+}{C}{-}H$ ≈ CH_3^+ least stable

a tertiary carbocation; a secondary carbocation; a secondary vinylic cation; a primary carbocation; a primary vinylic cation; the methyl cation

A vinylic cation is less stable because the positive charge is on an *sp* carbon, which we will see is more electronegative than the sp^2 carbon of an alkyl cation (Section 6.9). Therefore, a vinylic cation is less able to bear a positive charge. In addition, hyperconjugation is less effective in stabilizing the charge on a vinylic cation than on an alkyl cation (Section 4.2).

6.5 Addition of Hydrogen Halides and Addition of Halogens

We have just seen that an alkyne is a nucleophile and that in the first step of the reaction of an alkyne with a hydrogen halide, the electrophilic H^+ adds to the alkyne. If the alkyne is a *terminal* alkyne, the H^+ will add to the *sp* carbon bonded to the hydrogen, because the secondary vinylic cation that results is more stable than the primary vinylic cation that would be formed if the H^+ added to the other *sp* carbon. (Recall that alkyl groups stabilize positively charged carbon atoms; Section 4.2).

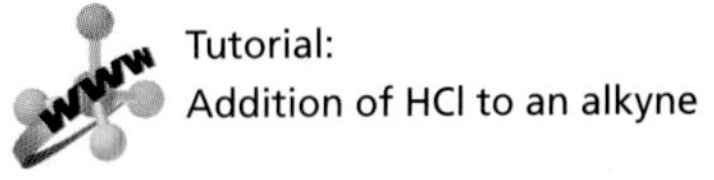

$$CH_3CH_2C{\equiv}CH \xrightarrow{\text{HBr}} CH_3CH_2\overset{+}{C}{=}CH(H) \quad Br^- \longrightarrow CH_3CH_2C(Br){=}CH(H)$$

1-butyne; 2-bromo-1-butene, a halo-substituted alkene

The electrophile adds to the *sp* carbon of a terminal alkyne that is bonded to the hydrogen.

$CH_3CH_2\overset{+}{C}{=}CH_2$ — a secondary vinylic cation

$CH_3CH_2CH{=}\overset{+}{C}H$ — a primary vinylic cation

Although the addition of a hydrogen halide to an alkyne can generally be stopped after the addition of one equivalent* of hydrogen halide, a second addition reaction will take place if excess hydrogen halide is present. The product of the second addition reaction is a **geminal dihalide**, a molecule with two halogens on the same carbon. "Geminal" comes from *geminus*, which is Latin for "twin."

$$CH_3CH_2C(Br){=}CH_2 \xrightarrow{\text{HBr}} CH_3CH_2CBr_2CH_3$$

electrophile adds here (at the CH_2 carbon)

2-bromo-1-butene; 2,2-dibromobutane, a geminal dihalide

*One equivalent means the same number of moles as the other reactant.

When the second equivalent of hydrogen halide adds to the double bond, the electrophile (H^+) adds to the sp^2 carbon bonded to the greater number of hydrogens—as predicted by the rule that governs electrophilic addition reactions (Section 4.4). The carbocation that results is more stable than the carbocation that would have been formed if H^+ had added to the other sp^2 carbon because bromine can share the positive charge with carbon by sharing one of its lone pairs (Section 1.19).

bromine shares a lone pair with carbon

$$CH_3CH_2\overset{:\ddot{Br}:}{\underset{+}{C}}-CH_3 \longleftrightarrow CH_3CH_2\overset{:\ddot{Br}^+}{\overset{\|}{C}}-CH_3$$

carbocation formed by adding the electrophile to the sp^2 carbon bonded to the greater number of hydrogens

The addition of a hydrogen halide to an alkyne can be stopped after the addition of one equivalent of HBr or HCl because an alkyne is more reactive than the halo-substituted alkene that is the product of the first addition reaction. The halogen substituent withdraws electrons inductively (through the σ bond), thereby decreasing the nucleophilic character of the double bond.

In describing the mechanism for addition of a hydrogen halide, we have shown that the intermediate is a vinylic cation. This mechanism may not be completely correct. A secondary vinylic cation is about as stable as a primary carbocation, and generally primary carbocations are too unstable to be formed. Some chemists, therefore, think that a *pi-complex* rather than a vinylic cation is formed as an intermediate.

3-D Molecule: Pi-complex

$$\begin{array}{c}\delta-\ Cl \\ \vdots \\ \delta+\ H \\ / \ \backslash \\ HC{\equiv}CH\end{array}$$

a pi-complex

Support for the intermediate's being a pi-complex comes from the observation that many (but not all) alkyne addition reactions are stereoselective. For example, the addition of HCl to 2-butyne forms only (*Z*)-2-chloro-2-butene, which means that only anti addition of H and Cl occurs. Clearly, the nature of the intermediate in alkyne addition reactions is not completely understood.

$$\underset{\textbf{2-butyne}}{CH_3C{\equiv}CCH_3} \xrightarrow{HCl} \underset{\textbf{(Z)-2-chloro-2-butene}}{\overset{H\qquad\qquad CH_3}{\underset{H_3C\qquad\qquad Cl}{C{=}C}}}$$

Addition of a hydrogen halide to an *internal* alkyne forms two geminal dihalides because the initial addition of the proton can occur with equal ease to either of the *sp* carbons.

$$\underset{\textbf{2-pentyne}}{CH_3CH_2C{\equiv}CCH_3} + \underset{\textbf{excess}}{HCl} \longrightarrow \underset{\textbf{2,2-dichloropentane}}{CH_3CH_2CH_2\overset{Cl}{\underset{Cl}{C}}CH_3} + \underset{\textbf{3,3-dichloropentane}}{CH_3CH_2\overset{Cl}{\underset{Cl}{C}}CH_2CH_3}$$

If, however, the same group is attached to each of the *sp* carbons of the internal alkyne, only one geminal dihalide is obtained.

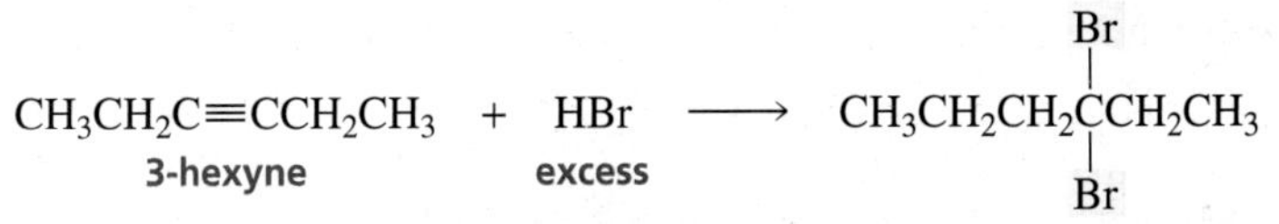

$$\underset{\text{3-hexyne}}{CH_3CH_2C{\equiv}CCH_2CH_3} + \underset{\text{excess}}{HBr} \longrightarrow \underset{\text{3,3-dibromohexane}}{CH_3CH_2CH_2CBr_2CH_2CH_3}$$

An alkyl peroxide has the same effect on the addition of HBr to an alkyne that it has on the addition of HBr to an alkene (Section 4.10)—it reverses the order of addition because the peroxide causes Br$^{\cdot}$ to become the electrophile.

$$\underset{\text{1-butyne}}{CH_3CH_2C{\equiv}CH} + HBr \xrightarrow{\text{peroxide}} \underset{\text{1-bromo-1-butene}}{CH_3CH_2CH{=}CHBr}$$

The mechanism of the reaction is the same as that for the addition of HBr to an alkene in the presence of a peroxide. That is, the peroxide is a radical initiator and creates a bromine radical (an electrophile). If the alkyne is a terminal alkyne, the bromine radical adds to the *sp* carbon bonded to the hydrogen; if it is an internal alkyne, the bromide radical can add with equal ease to either of the *sp* carbons. The resulting vinylic radical abstracts a hydrogen atom from HBr and regenerates the bromine radical. Any two radicals can combine in a termination step.

mechanism for the addition of HBr in the presence of a peroxide

$$RO{-}OR \longrightarrow 2\,RO\cdot$$

$$RO\cdot + H{-}Br \longrightarrow RO{-}H + \cdot Br$$

$$CH_3CH_2C{\equiv}CH + \cdot Br \longrightarrow CH_3CH_2\dot{C}{=}CH(Br)$$

$$CH_3CH_2\dot{C}{=}CH(Br) + H{-}Br \longrightarrow CH_3CH_2C(H){=}CHBr + \cdot Br$$

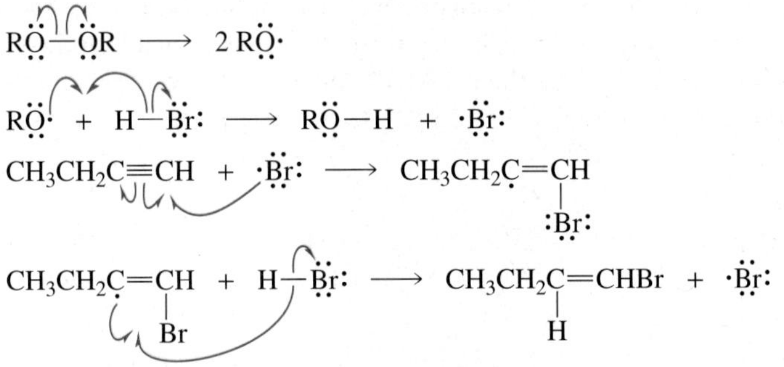

The halogens Cl_2 and Br_2 also add to alkynes. In the presence of excess halogen, a second addition reaction occurs. Typically the solvent is CH_2Cl_2.

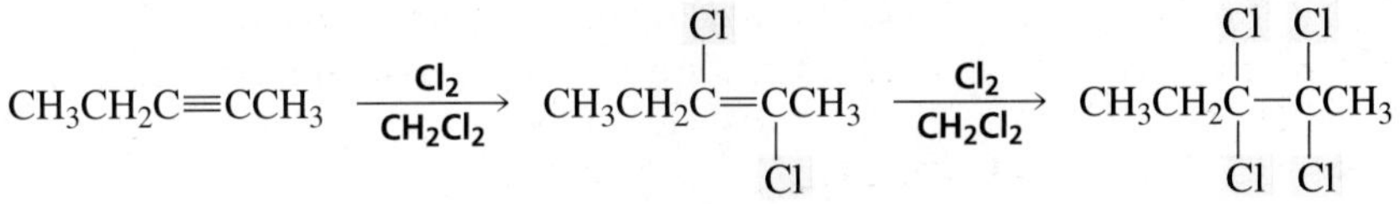

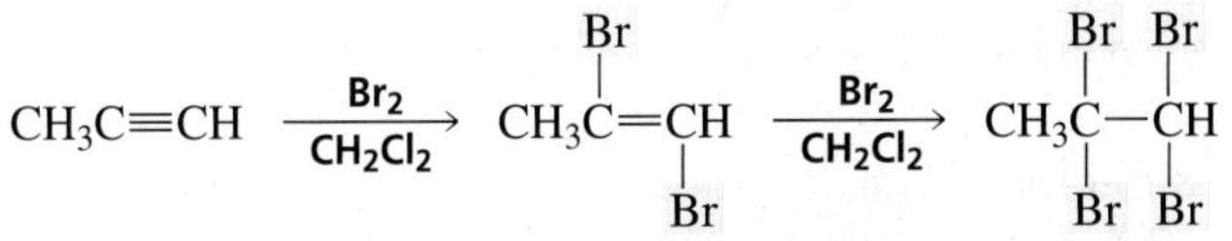

$$CH_3CH_2C{\equiv}CCH_3 \xrightarrow[CH_2Cl_2]{Cl_2} CH_3CH_2C(Cl){=}C(Cl)CH_3 \xrightarrow[CH_2Cl_2]{Cl_2} CH_3CH_2CCl_2{-}CCl_2CH_3$$

$$CH_3C{\equiv}CH \xrightarrow[CH_2Cl_2]{Br_2} CH_3C(Br){=}CH(Br) \xrightarrow[CH_2Cl_2]{Br_2} CH_3CBr_2{-}CHBr_2$$

PROBLEM 8◆

Give the major product of each of the following reactions:

a. $HC{\equiv}CCH_3 \xrightarrow[\text{peroxide}]{HBr}$

b. $HC{\equiv}CCH_3 \xrightarrow{HBr}$

c. $HC{\equiv}CCH_3 \xrightarrow{\text{excess HBr}}$

d. $CH_3C{\equiv}CCH_3 \xrightarrow[CH_2Cl_2]{Br_2}$

e. $CH_3C{\equiv}CCH_3 \xrightarrow{\text{excess HBr}}$

f. $CH_3C{\equiv}CCH_2CH_3 \xrightarrow{\text{excess HBr}}$

PROBLEM 9♦

From what you know about the stereochemistry of alkene addition reactions, predict the configurations of the products that would be obtained from the reaction of 2-butyne with the following:

a. one equivalent of Br_2 in CH_2Cl_2

b. one equivalent of HBr + peroxide

6.6 Addition of Water

In Section 4.5, we saw that alkenes undergo the acid-catalyzed addition of water. The product of the reaction is an alcohol.

$$\underset{\text{1-butene}}{CH_3CH_2CH{=}CH_2} + H_2O \xrightarrow{H_2SO_4} \underset{\text{2-butanol}}{CH_3CH_2\underset{OH}{\underset{|}{C}H}{-}\underset{H}{\underset{|}{C}H_2}}$$

Alkynes also undergo the acid-catalyzed addition of water. The product of the reaction is an *enol.* An **enol** has a carbon–carbon double bond and an OH group bonded to one of the sp^2 carbons. (The ending "ene" signifies the double bond, and "ol" the OH. When the two endings are joined, the final e of "ene" is dropped to avoid two consecutive vowels, but it is pronounced as if the *e* were there, "ene-ol.")

$$CH_3C{\equiv}CCH_3 + H_2O \xrightarrow{H_2SO_4} \underset{\text{an enol}}{CH_3\overset{OH}{\overset{|}{C}}{=}CHCH_3} \rightleftharpoons \underset{\text{a ketone}}{CH_3\overset{O}{\overset{\|}{C}}{-}CH_2CH_3}$$

The enol immediately rearranges to a *ketone*. A carbon doubly bonded to an oxygen is called a **carbonyl** ("car-bo-nil") **group**. A **ketone** is a compound that has two alkyl groups bonded to a carbonyl group. An **aldehyde** is a compound that has at least one hydrogen bonded to a carbonyl group.

$$\underset{\text{a carbonyl group}}{\overset{O}{\overset{\|}{C}}} \qquad \underset{\text{a ketone}}{R{-}\overset{O}{\overset{\|}{C}}{-}R} \qquad \underset{\text{an aldehyde}}{R{-}\overset{O}{\overset{\|}{C}}{-}H}$$

A ketone and an enol differ only in the location of a double bond and a hydrogen. The ketone and enol are called **keto–enol tautomers**. **Tautomers** ("taw-toe-mers") are isomers that are in rapid equilibrium. Interconversion of the tautomers is called **tautomerization**. We will examine the mechanism of this reaction in Chapter 19. For now, the important thing to remember is that the keto and enol tautomers come to equilibrium in solution, and the keto tautomer, because it is usually much more stable than the enol tautomer, predominates at equilibrium.

$$\underset{\text{keto tautomer}}{RCH_2{-}\overset{O}{\overset{\|}{C}}{-}R} \rightleftharpoons \underset{\text{enol tautomer}}{RCH{=}\overset{OH}{\overset{|}{C}}{-}R}$$

tautomerization

Addition of water to an internal alkyne that has the same group attached to each of the *sp* carbons forms a single ketone as a product. But if the two groups are not identical, two ketones are formed because the initial addition of the proton can occur to either of the *sp* carbons.

$$CH_3CH_2C{\equiv}CCH_2CH_3 + H_2O \xrightarrow{H_2SO_4} CH_3CH_2\overset{\overset{O}{\|}}{C}CH_2CH_2CH_3$$

$$CH_3C{\equiv}CCH_2CH_3 + H_2O \xrightarrow{H_2SO_4} CH_3\overset{\overset{O}{\|}}{C}CH_2CH_2CH_3 + CH_3CH_2\overset{\overset{O}{\|}}{C}CH_2CH_3$$

Terminal alkynes are less reactive than internal alkynes toward the addition of water. Terminal alkynes will add water if mercuric ion (Hg^{2+}) is added to the acidic mixture. The mercuric ion acts as a catalyst to increase the rate of the addition reaction.

$$CH_3CH_2C{\equiv}CH + H_2O \xrightarrow[HgSO_4]{H_2SO_4} \underset{\text{an enol}}{CH_3CH_2\overset{\overset{OH}{|}}{C}{=}CH_2} \rightleftharpoons \underset{\text{a ketone}}{CH_3CH_2\overset{\overset{O}{\|}}{C}{-}CH_3}$$

The first step in the mercuric-ion-catalyzed hydration of an alkyne is formation of a cyclic mercurinium ion. (Two of the electrons in mercury's filled 5*d* atomic orbital are shown.) This should remind you of the cyclic bromonium and mercurinium ions formed as intermediates in electrophilic addition reactions of alkenes (Sections 4.7 and 4.8). In the second step of the reaction, water attacks the most substituted carbon of the cyclic intermediate (Section 4.8). Oxygen loses a proton to form a mercuric enol, which immediately rearranges to a mercuric ketone. Loss of the mercuric ion forms an enol, which rearranges to a ketone. Notice that the overall addition of water follows both the general rule for electrophilic addition reactions and Markovnikov's rule: The electrophile (H^+ in the case of Markovnikov's rule) adds to the *sp* carbon bonded to the greater number of hydrogens.

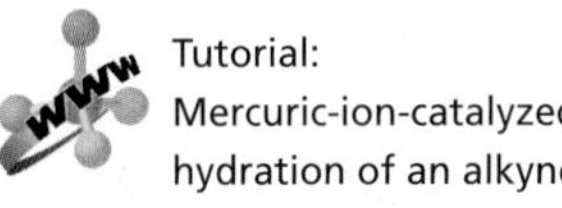
Tutorial: Mercuric-ion-catalyzed hydration of an alkyne

mechanism for the mercuric-ion-catalyzed hydration of an alkyne

$$CH_3C{\equiv}CH \xrightarrow{:Hg^{2+}} CH_3\overset{Hg^{2+}}{C{=}CH} \xrightarrow{H_2\ddot{O}:} CH_3\underset{\overset{+|}{:OH}\,|\,H}{C}{=}\overset{Hg^+}{CH} \xrightarrow{H_2\ddot{O}:} \underset{\text{a mercuric enol}}{CH_3\underset{:\ddot{O}H}{C}{=}\overset{Hg^+}{CH}} + H_3\overset{+}{O}:$$

$$\rightleftharpoons \underset{\text{a mercuric ketone}}{CH_3\overset{}{\underset{\|O}{C}}{-}CH_2{-}Hg^+} \xrightarrow{H{-}\ddot{O}^+{-}H\,(H)} \underset{OH}{CH_3C{=}CH_2} + H_2O + Hg^{2+} \rightleftharpoons CH_3\underset{\|O}{C}CH_3$$

PROBLEM 10♦

What ketones would be formed from the acid-catalyzed hydration of 3-heptyne?

PROBLEM 11♦

Which alkyne would be the best reagent to use for the synthesis of each of the following ketones?

a. $CH_3\overset{\overset{O}{\|}}{C}CH_3$ b. $CH_3CH_2\overset{\overset{O}{\|}}{C}CH_2CH_2CH_3$ c. $CH_3\overset{\overset{O}{\|}}{C}$–cyclohexyl

PROBLEM 12◆

Draw all the enol tautomers for each of the ketones in Problem 11.

6.7 Addition of Borane: Hydroboration–Oxidation

Borane adds to alkynes in the same way it adds to alkenes. That is, one mole of BH_3 reacts with three moles of alkyne to form one mole of boron-substituted alkene (Section 4.9). When the addition reaction is over, aqueous sodium hydroxide and hydrogen peroxide are added to the reaction mixture. The end result, as in the case of alkenes, is replacement of the boron by an OH group. The enol product immediately rearranges to a ketone.

3-D Molecule: Disiamylborane

$$3\ CH_3C{\equiv}CCH_3 + BH_3 \xrightarrow{THF} (H_3C)(H)C{=}C(CH_3)(BR_2) \xrightarrow[H_2O]{HO^-,\ H_2O_2} 3\ (H_3C)(H)C{=}C(CH_3)(OH) \rightleftharpoons 3\ CH_3CH_2\overset{O}{\overset{\|}{C}}CH_3$$

boron-substituted alkene; an enol

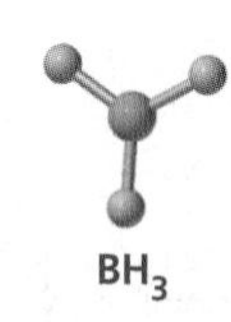
BH₃

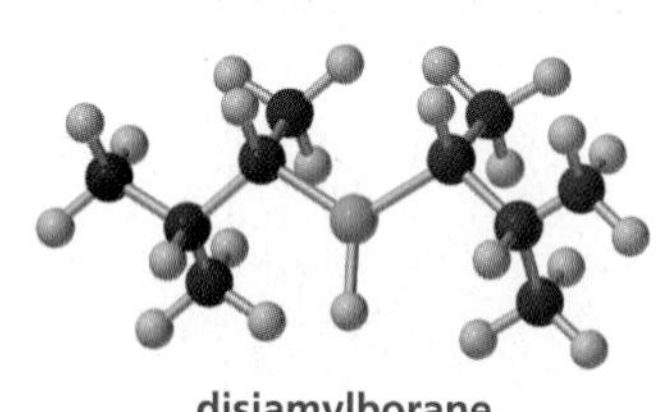
disiamylborane

In order to obtain the enol as the product of the addition reaction, only one equivalent of BH_3 can be allowed to add to the alkyne. In other words, the reaction must stop at the alkene stage. In the case of internal alkynes, the substituents on the boron-substituted alkene prevent the second addition from occurring. However, there is less steric hindrance in a terminal alkyne, so it is harder to stop the addition reaction at the alkene stage. A special reagent called disiamylborane has been developed for use with terminal alkynes ("siamyl" stands for secondary iso **amyl**; amyl is a common name for a five-carbon fragment). The bulky alkyl groups of disiamylborane prevent a second addition to the boron-substituted alkene. So borane can be used to hydrate internal alkynes, but disiamylborane is preferred for the hydration of terminal alkynes.

$$CH_3CH_2C{\equiv}CH + \left(\begin{matrix} CH_3\ \ CH_3 \\ |\quad\ \ | \\ CH_3CH{-}CH{-} \end{matrix}\right)_2 BH \longrightarrow (CH_3CH_2)(H)C{=}C(H)\,B\left(\begin{matrix} CH_3\ \ CH_3 \\ |\quad\ \ | \\ {-}CH{-}CHCH_3 \end{matrix}\right)_2 \xrightarrow[H_2O]{HO^-,\ H_2O_2} (CH_3CH_2)(H)C{=}C(H)(OH) \rightleftharpoons CH_3CH_2CH_2\overset{O}{\overset{\|}{C}}H$$

bis(1,2-dimethylpropyl)borane
disiamylborane

an enol

The addition of borane (or disiamylborane) to a terminal alkyne exhibits the same regioselectivity seen in borane addition to an alkene. Boron, with its electron-seeking empty orbital, adds preferentially to the *sp* carbon bonded to the hydrogen. In the second step of the hydroboration–oxidation reaction, boron is replaced by an OH group. The overall reaction is *anti-Markovnikov* addition because the hydrogen *does not* add to the *sp* carbon bonded to the greater number of hydrogens. In hydroboration–oxidation, H^+ is not the electrophile, $H{:}^-$ is the nucleophile. The reaction, however, *does* follow the general rule for electrophilic addition reactions: The electrophile (BH_3) adds to the *sp* carbon bonded to the greater number of hydrogens. Consequently, mercuric-ion-catalyzed addition of water to a terminal alkyne produces a *ketone* (the carbonyl group

Addition of water to a terminal alkyne forms a ketone.

is *not* on the terminal carbon), whereas hydroboration–oxidation of a terminal alkyne produces an *aldehyde* (the carbonyl group *is* on the terminal carbon).

Hydroboration–oxidation of a terminal alkyne forms an aldehyde.

$$CH_3C{\equiv}CH \xrightarrow[HgSO_4]{H_2O,\ H_2SO_4} CH_3\overset{OH}{\overset{|}{C}}{=}CH_2 \rightleftharpoons CH_3\overset{O}{\overset{\|}{C}}CH_3$$

a ketone

$$CH_3C{\equiv}CH \xrightarrow[2.\ HO^-,\ H_2O_2,\ H_2O]{1.\ \text{disiamylborane}} CH_3CH{=}\overset{OH}{\overset{|}{C}H} \rightleftharpoons CH_3CH_2\overset{O}{\overset{\|}{C}}H$$

an aldehyde

PROBLEM 13◆

Give the products of (1) mercuric-ion-catalyzed addition of water and (2) hydroboration–oxidation for the following:

a. 1-butyne b. 2-butyne c. 2-pentyne

PROBLEM 14◆

There is only one alkyne that forms an aldehyde when it undergoes either acid- or mercuric-ion-catalyzed addition of water. Identify the alkyne.

6.8 Addition of Hydrogen

Hydrogen adds to an alkyne in the presence of a metal catalyst such as palladium, platinum, or nickel in the same manner that it adds to an alkene (Section 4.11). It is difficult to stop the reaction at the alkene stage because hydrogen readily adds to alkenes in the presence of these efficient metal catalysts. The product of the hydrogenation reaction, therefore, is an alkane.

$$\underset{\text{alkyne}}{CH_3CH_2C{\equiv}CH} \xrightarrow[Pt/C]{H_2} \underset{\text{alkene}}{CH_3CH_2CH{=}CH_2} \xrightarrow[Pt/C]{H_2} \underset{\text{alkane}}{CH_3CH_2CH_2CH_3}$$

$(CH_3COO^-)_2Pb^{2+}$

lead(II) acetate

quinoline

The reaction can be stopped at the alkene stage if a "poisoned" (partially deactivated) metal catalyst is used. The most commonly used partially deactivated metal catalyst is Lindlar catalyst, which is prepared by precipitating palladium on calcium carbonate and treating it with lead(II) acetate and quinoline. This treatment modifies the surface of the palladium, making it much more effective at catalyzing the addition of hydrogen to a triple bond than to a double bond.

Because the alkyne sits on the surface of the metal catalyst and the hydrogens are delivered to the triple bond from the surface of the catalyst, only syn addition of hydrogen occurs (Section 5.19). Syn addition of hydrogen to an internal alkyne forms a *cis alkene*.

Herbert H. M. Lindlar *was born in Switzerland in 1909 and received a Ph.D. from the University of Bern. He worked at Hoffmann-La Roche and Co. in Basel, Switzerland, and he authored many patents. His last patent was a procedure for isolating the carbohydrate xylose from the waste produced in paper mills.*

$$\underset{\text{2-pentyne}}{CH_3CH_2C{\equiv}CCH_3} + H_2 \xrightarrow[\text{catalyst}]{\text{Lindlar}} \underset{\textit{cis}\text{-2-pentene}}{\overset{H\qquad H}{CH_3CH_2\text{–}C{=}C\text{–}CH_3}}$$

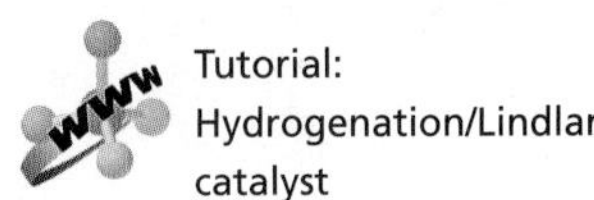
Tutorial: Hydrogenation/Lindlar catalyst

Internal alkynes can be converted into *trans alkenes* using sodium (or lithium) in liquid ammonia. The reaction stops at the alkene stage because sodium (or lithium) reacts more rapidly with triple bonds than with double bonds. Ammonia (bp = −33 °C) is a gas at room temperature, so it is kept in the liquid state by using a dry ice/acetone mixture (bp = −78 °C).

$$CH_3C{\equiv}CCH_3 \xrightarrow[NH_3\text{ (liq)},\ -78\ °C]{\text{Na or Li}} \text{trans-2-butene}$$

2-butyne → *trans*-2-butene

The first step in the mechanism of this reaction is transfer of the *s* orbital electron from sodium (or lithium) to an *sp* carbon to form a **radical anion**—a species with a negative charge and an unpaired electron. (Recall that sodium and lithium have a strong tendency to lose the single electron in their outer-shell *s* orbital; Section 1.3.) The radical anion is such a strong base that it can remove a proton from ammonia. This results in the formation of a **vinylic radical**—the unpaired electron is on a vinylic carbon. Another single-electron transfer from sodium (or lithium) to the vinylic radical forms a vinylic anion. The vinylic anion abstracts a proton from another molecule of ammonia. The product is the trans alkene.

$CH_3-C{\equiv}C-CH_3$ + Na· ⟶ $CH_3-\dot{C}{=}\ddot{C}^{-}-CH_3$ (a radical anion + Na^+) $\xrightarrow{H-NH_2}$ a vinylic radical + $^-NH_2$ $\xrightarrow{Na\cdot}$ a vinylic anion + Na^+ $\xrightarrow{H-NH_2}$ a trans alkene + $^-NH_2$

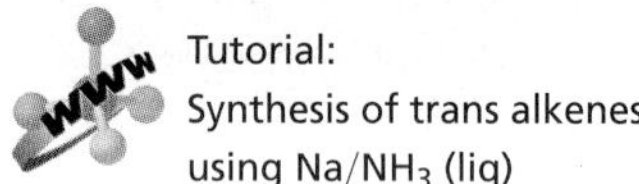

Tutorial:
Synthesis of trans alkenes using Na/NH3 (liq)

The vinylic anion can have either the cis or the trans configuration. The cis and trans configurations are in equilibrium, but the equilibrium favors the more stable trans configuration because in this configuration the bulky alkyl groups are as far from each other as possible.

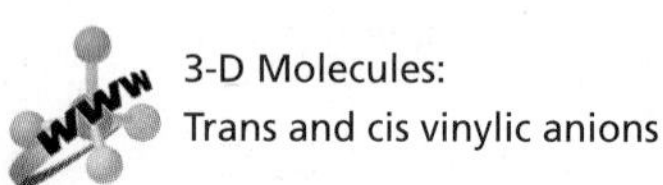

trans vinylic anion more stable ⇌ cis vinylic anion less stable

PROBLEM 15◆

What alkyne would you start with and what reagents would you use if you wanted to synthesize:

a. pentane? b. *cis*-2-butene? c. *trans*-2-pentene? d. 1-hexene?

6.9 Acidity of a Hydrogen Bonded to an *sp* Hybridized Carbon

Carbon forms nonpolar covalent bonds with hydrogen because carbon and hydrogen, having similar electronegativities, share their bonding electrons almost equally. However, not all carbon atoms have the same electronegativity. An *sp* hybridized carbon is more electronegative than an sp^2 hybridized carbon, which is more electronegative than an sp^3 hybridized carbon, which is just slightly more electronegative than a hydrogen. (Chapter 1, Problem 40d, e, and f.)

sp Hybridized carbons are more electronegative than sp^2 hybridized carbons, which are more electronegative than sp^3 hybridized carbons.

relative electronegativities of carbon atoms

most electronegative > $sp > sp^2 > sp^3$ < least electronegative

Why does the type of hybridization affect the electronegativity of the carbon atom? Electronegativity is a measure of the ability of an atom to pull the bonding electrons toward itself. Thus, the most electronegative carbon atom will be the one with its bonding electrons closest to the nucleus. The average distance of a 2*s* electron from the nucleus is less than the average distance of a 2*p* electron from the nucleus. Therefore, the electrons in an *sp* hybrid orbital (50% *s* character) are closer, on average, to the nucleus than those in an sp^2 hybrid orbital (33.3% *s* character). In turn, sp^2 electrons are closer to the nucleus than sp^3 electrons (25% *s* character). The *sp* hybridized carbon, therefore, is the most electronegative.

In Section 1.18 we saw that the acidity of a hydrogen attached to some second-row elements depends on the electronegativity of the atom to which the hydrogen is attached. The greater the electronegativity of the atom, the greater the acidity of the hydrogen—the stronger the acid. (Don't forget, the stronger the acid, the lower its pK_a.)

relative electronegativities N < O < F < most electronegative

relative acid strengths NH_3 < H_2O < HF < strongest acid

$pK_a = 36$ $pK_a = 15.7$ $pK_a = 3.2$

Because the electronegativity of carbon atoms follows the order $sp > sp^2 > sp^3$, ethyne is a stronger acid than ethene, and ethene is a stronger acid than ethane.

HC≡CH	$H_2C{=}CH_2$	CH_3CH_3
ethyne	ethene	ethane
$pK_a = 25$	$pK_a = 44$	$pK_a = 50$

We can compare the acidities of these compounds with the acidities of hydrogens attached to other second-row elements.

relative acid strengths

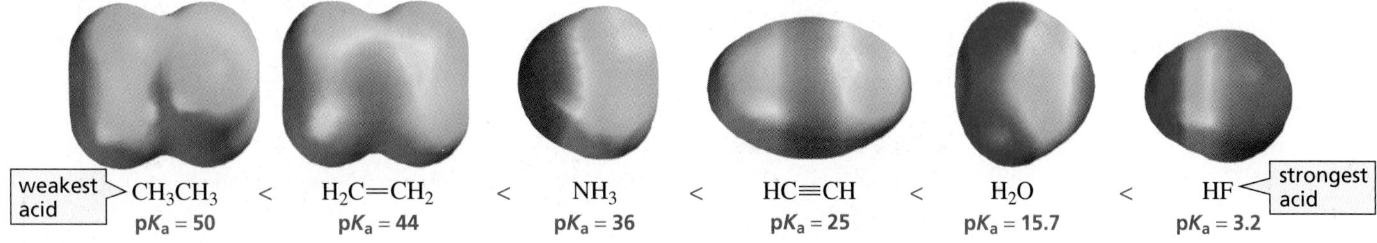

weakest acid > CH_3CH_3 < $H_2C{=}CH_2$ < NH_3 < HC≡CH < H_2O < HF < strongest acid

$pK_a = 50$ $pK_a = 44$ $pK_a = 36$ $pK_a = 25$ $pK_a = 15.7$ $pK_a = 3.2$

The corresponding conjugate bases of these compounds have the following relative base strengths because *the stronger the acid, the weaker is its conjugate base.*

relative base strengths

strongest base > $CH_3CH_2^-$ > $H_2C{=}CH^-$ > H_2N^- > $HC{\equiv}C^-$ > HO^- > F^- < weakest base

In order to remove a proton from an acid (in a reaction that strongly favors products), the base that removes the proton must be stronger than the base that is generated as a result of proton removal (Section 1.17). In other words, you must start with a stronger base than the base that will be formed. The amide ion ($^-NH_2$) can remove a

The stronger the acid, the weaker its conjugate base.

hydrogen bonded to an *sp* carbon of a terminal alkyne to form a carbanion called an **acetylide ion**, because the amide ion is a stronger base than the acetylide ion.

To remove a proton from an acid in a reaction that favors products, the base that removes the proton must be stronger than the base that is formed.

$$\mathrm{RC{\equiv}CH} + {}^{-}\mathrm{NH_2} \rightleftharpoons \mathrm{RC{\equiv}C^-} + \mathrm{NH_3}$$

RC≡CH	$^{-}NH_2$ amide ion	RC≡C⁻ acetylide ion	NH_3
stronger acid	stronger base	weaker base	weaker acid

If hydroxide ion were used to remove a hydrogen bonded to an *sp* carbon, the reaction would strongly favor the reactants because hydroxide ion is a much weaker base than the acetylide ion that would be formed.

$$\mathrm{RC{\equiv}CH} + \mathrm{HO^-} \rightleftharpoons \mathrm{RC{\equiv}C^-} + \mathrm{H_2O}$$

RC≡CH	HO^- hydroxide anion	RC≡C⁻ acetylide anion	H_2O
weaker acid	weaker base	stronger base	stronger acid

The amide ion cannot remove a hydrogen bonded to an sp^2 or an sp^3 carbon. Only a hydrogen bonded to an *sp* carbon is sufficiently acidic to be removed by the amide ion. Consequently, a hydrogen bonded to an *sp* carbon sometimes is referred to as an "acidic" hydrogen. The "acidic" property of terminal alkynes is one way their reactivity differs from that of alkenes. Be careful not to misinterpret what is meant when we say that a hydrogen bonded to an *sp* carbon is "acidic." It is more acidic than most other carbon-bound hydrogens but it is much less acidic than a hydrogen of a water molecule, and water is only a very weakly acidic compound ($pK_a = 15.7$).

SODIUM AMIDE AND SODIUM

Take care not to confuse sodium amide ($Na^+\,{}^-NH_2$) with sodium (Na) in liquid ammonia. Sodium amide is the strong base used to remove a proton from a terminal alkyne. Sodium is used as a source of electrons in the reduction of an internal alkyne to a trans alkene (Section 6.8).

PROBLEM 16◆

Any base whose conjugate acid has a pK_a greater than ________ can remove a proton from a terminal alkyne to form an acetylide ion (in a reaction that favors products).

PROBLEM 17◆

Which carbocation in each of the following pairs is more stable?

a. $CH_3\overset{+}{C}H_2$ or $H_2C{=}\overset{+}{C}H$ b. $H_2C{=}\overset{+}{C}H$ or $HC{\equiv}\overset{+}{C}$

PROBLEM 18◆

Explain why sodium amide cannot be used to form a carbanion from an alkane in a reaction that favors products?

PROBLEM-SOLVING STRATEGY

a. List the following compounds in order of decreasing acidity:

$$CH_3CH_2\overset{+}{N}H_3 \qquad CH_3CH{=}\overset{+}{N}H_2 \qquad CH_3C{\equiv}\overset{+}{N}H$$

To compare the acidities of a group of compounds, first look at how they differ. These three compounds differ in the hybridization of the nitrogen to which the acidic hydrogen is attached. Now, recall what you know about hybridization and acidity. You know that hybridization of an atom affects its electronegativity (*sp* is more electronegative than sp^2, and sp^2 is more electronegative than sp^3); and you know that the more electronegative the atom to which a hydrogen is attached, the more acidic the hydrogen. Now you can answer the question.

relative acidities $CH_3C{\equiv}\overset{+}{N}H > CH_3CH{=}\overset{+}{N}H_2 > CH_3CH_2\overset{+}{N}H_3$

b. Draw the conjugate bases and list them in order of decreasing basicity.
First remove a proton from each acid to get the structures of the conjugate bases, and then recall that the stronger the acid, the weaker its conjugate base.

relative basicities $CH_3CH_2NH_2 > CH_3CH{=}NH > CH_3C{\equiv}N$

Now continue on to Problem 19.

PROBLEM 19◆

List the following species in order of decreasing basicity:

a. $CH_3CH_2CH{=}\bar{C}H$ $\quad CH_3CH_2C{\equiv}C^-$ $\quad CH_3CH_2CH_2\bar{C}H_2$

b. $CH_3CH_2O^-$ $\quad F^-$ $\quad CH_3C{\equiv}C^-$ $\quad {}^-NH_2$

6.10 Synthesis Using Acetylide Ions

Reactions that form carbon–carbon bonds are important in the synthesis of organic compounds because without such reactions, we could not convert molecules with small carbon skeletons into molecules with larger carbon skeletons. Instead, the product of a reaction would always have the same number of carbons as the starting material.

One reaction that forms a carbon–carbon bond is the reaction of an acetylide ion with an alkyl halide. Only primary alkyl halides or methyl halides should be used in this reaction.

$$CH_3CH_2C{\equiv}C^- + CH_3CH_2CH_2Br \longrightarrow \underset{\textbf{3-heptyne}}{CH_3CH_2C{\equiv}CCH_2CH_2CH_3} + Br^-$$

The mechanism of this reaction is well understood. Bromine is more electronegative than carbon, and as a result, the electrons in the C—Br bond are not shared equally by the two atoms. This means that the C—Br bond is polar, with a partial positive charge on carbon and a partial negative charge on bromine. The negatively charged acetylide ion (a nucleophile) is attracted to the partially positively charged carbon (an electrophile) of the alkyl halide. As the electrons of the acetylide ion approach the carbon to form the new carbon–carbon bond, they push out the bromine and its bonding electrons because carbon can bond to no more than four atoms at a time. This is an example of an *alkylation reaction*. An **alkylation reaction** attaches an alkyl group to the starting material.

$$CH_3CH_2C{\equiv}\ddot{C}^- + CH_3CH_2\overset{\delta+}{C}H_2{-}\overset{\delta-}{Br} \longrightarrow CH_3CH_2C{\equiv}CCH_2CH_2CH_3 + Br^-$$

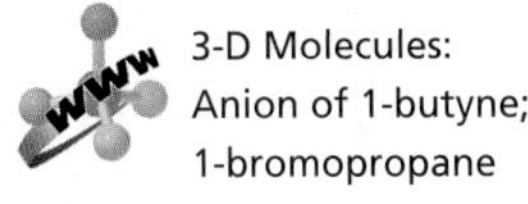

3-D Molecules: Anion of 1-butyne; 1-bromopropane

The mechanism of this and similar reactions will be discussed in greater detail in Chapter 10. At that time we will also see why the reaction works best with primary alkyl halides and methyl halides.

Simply by choosing an alkyl halide of the appropriate structure, terminal alkynes can be converted into internal alkynes of any desired chain length.

$$\underset{\textbf{1-pentyne}}{CH_3CH_2CH_2C{\equiv}CH} \xrightarrow[\textbf{2. } CH_3CH_2CH_2CH_2CH_2Cl]{\textbf{1. } NaNH_2} \underset{\textbf{4-decyne}}{CH_3CH_2CH_2C{\equiv}CCH_2CH_2CH_2CH_2CH_3}$$

PROBLEM 20 SOLVED

A chemist wants to synthesize 4-decyne but cannot find any 1-pentyne, the starting material used in the synthesis just described. How else can 4-decyne be synthesized?

SOLUTION The *sp* carbons of 4-decyne are bonded to a propyl group and to a pentyl group. Therefore to obtain 4-decyne, the acetylide ion of 1-pentyne can react with a pentyl halide or the acetylide ion of 1-heptyne can react with a propyl halide. Since 1-pentyne is not available, the chemist should use 1-heptyne and a propyl halide.

$$\underset{\text{1-heptyne}}{CH_3CH_2CH_2CH_2CH_2C{\equiv}CH} \xrightarrow[\text{2. } CH_3CH_2CH_2Cl]{\text{1. } NaNH_2} \underset{\text{4-decyne}}{CH_3CH_2CH_2C{\equiv}CCH_2CH_2CH_2CH_2CH_3}$$

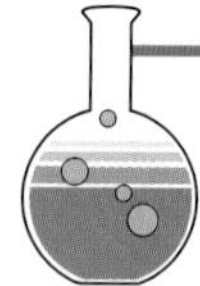

6.11 Designing a Synthesis I: An Introduction to Multistep Synthesis

For each reaction we have studied so far, we have seen *why* the reaction occurs, *how* it occurs, and the products that are formed. A good way to review these reactions is to design syntheses, because in doing so, you have to be able to recall many of the reactions you have learned.

Synthetic chemists consider time, cost, and yield in designing syntheses. In the interest of time, a well-designed synthesis should require as few steps (sequential reactions) as possible, and those steps should each involve a reaction that is easy to carry out. If two chemists in a pharmaceutical company were each asked to prepare a new drug, and one synthesized the drug in three simple steps while the other used 20 difficult steps, which chemist would not get a raise? In addition, each step in the synthesis should provide the greatest possible yield of the desired product, and the cost of the starting materials must be considered. The more reactant needed to synthesize one gram of product, the more expensive it is to produce. Sometimes it is preferable to design a synthesis involving several steps if the starting materials are inexpensive, the reactions are easy to carry out, and the yield of each step is high. This would be better than designing a synthesis with fewer steps that require expensive starting materials and reactions that are more difficult or give lower yields. At this point you don't know how much chemicals cost or how difficult it is to carry out certain reactions. So, for the time being, when you design a synthesis, just try to find the route with the fewest steps.

The following examples will give you an idea of the type of thinking required for the design of a successful synthesis. This kind of problem will appear repeatedly throughout this book because working such problems is a good way to learn organic chemistry.

Example 1. Starting with 1-butyne, how could you make the following ketone? You can use any organic and inorganic reagents.

$$\underset{\text{1-butyne}}{CH_3CH_2C{\equiv}CH} \xrightarrow{?} CH_3CH_2\overset{\overset{\displaystyle O}{\|}}{C}CH_2CH_2CH_3$$

Many chemists find that the easiest way to design a synthesis is to work backward. Instead of looking at the starting material and deciding how to do the first step of the synthesis, look at the product and decide how to do the last step. The product is a ketone. At this point the only reaction you know that forms a ketone is the addition of water (in the presence of a catalyst) to an alkyne. If the alkyne used in the reaction has identical substituents on each of the *sp* carbons, only one ketone will be obtained. Thus, 3-hexyne is the best alkyne to use for the synthesis of the desired ketone.

$$\underset{\text{3-hexyne}}{CH_3CH_2C{\equiv}CCH_2CH_3} \xrightarrow[H_2SO_4]{H_2O} CH_3CH_2\overset{OH}{\overset{|}{C}}{=}CHCH_2CH_3 \rightleftharpoons CH_3CH_2\overset{O}{\overset{\|}{C}}CH_2CH_2CH_3$$

3-Hexyne can be obtained from the starting material by removing a proton from the *sp* carbon, followed by alkylation. To obtain the desired product, a two-carbon alkyl halide must be used in the alkylation reaction.

$$\underset{\text{1-butyne}}{CH_3CH_2C{\equiv}CH} \xrightarrow[\text{2. } CH_3CH_2Br]{\text{1. } NaNH_2} \underset{\text{3-hexyne}}{CH_3CH_2C{\equiv}CCH_2CH_3}$$

Designing a synthesis by working backward from product to reactant is not simply a technique taught to organic chemistry students. It is used so frequently by experienced synthetic chemists that it has been given a name—**retrosynthetic analysis**. Chemists use open arrows to indicate they are working backward. Typically, the reagents needed to carry out each step are not included until the reaction is written in the forward direction. For example, the route to the synthesis of the previous ketone can be arrived at by the following retrosynthetic analysis.

Elias James Corey *coined the term "retrosynthetic analysis." He was born in Massachusetts in 1928 and is a professor of chemistry at Harvard University. He received the Nobel Prize in chemistry in 1990 for his contribution to synthetic organic chemistry.*

retrosynthetic analysis

$$CH_3CH_2\overset{O}{\overset{\|}{C}}CH_2CH_2CH_3 \Longrightarrow CH_3CH_2C{\equiv}CCH_2CH_3 \Longrightarrow CH_3CH_2C{\equiv}CH$$

Once the complete sequence of reactions has been worked out by retrosynthetic analysis, the synthetic scheme can be shown by reversing the steps and including the reagents required for each step.

synthesis

$$CH_3CH_2C{\equiv}CH \xrightarrow[\text{2. } CH_3CH_2Br]{\text{1. } NaNH_2} CH_3CH_2C{\equiv}CCH_2CH_3 \xrightarrow[H_2SO_4]{H_2O} CH_3CH_2\overset{O}{\overset{\|}{C}}CH_2CH_2CH_3$$

Example 2. Starting with ethyne, how could you make 1-bromopentane?

$$\underset{\text{ethyne}}{HC{\equiv}CH} \xrightarrow{?} \underset{\text{1-bromopentane}}{CH_3CH_2CH_2CH_2CH_2Br}$$

A primary alkyl halide can be prepared from a terminal alkene (using HBr in the presence of a peroxide). A terminal alkene can be prepared from a terminal alkyne, and a terminal alkyne can be prepared from ethyne and an alkyl halide with the appropriate number of carbons.

retrosynthetic analysis

$$CH_3CH_2CH_2CH_2CH_2Br \Longrightarrow CH_3CH_2CH_2CH{=}CH_2 \Longrightarrow CH_3CH_2CH_2C{\equiv}CH \Longrightarrow HC{\equiv}CH$$

Now we can write the synthetic scheme.

synthesis

$$HC{\equiv}CH \xrightarrow[\text{2. } CH_3CH_2CH_2Br]{\text{1. } NaNH_2} CH_3CH_2CH_2C{\equiv}CH \xrightarrow[\text{Lindlar catalyst}]{H_2} CH_3CH_2CH_2CH{=}CH_2 \xrightarrow[\text{peroxide}]{HBr} CH_3CH_2CH_2CH_2CH_2Br$$

Example 3. How could 2,6-dimethylheptane be prepared from an alkyne and an alkyl halide? (The prime on R′ signifies that R and R′ are different alkyl groups.)

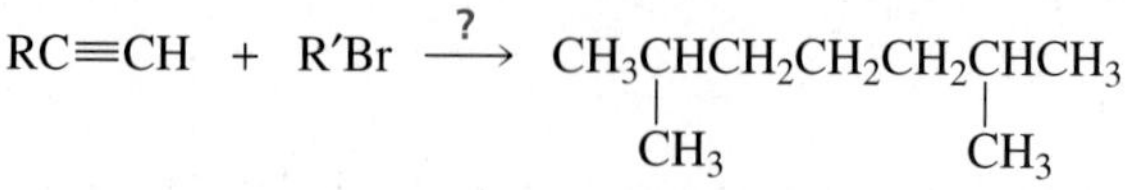

$$RC{\equiv}CH + R'Br \xrightarrow{?} CH_3\underset{\displaystyle CH_3}{\underset{|}{C}}HCH_2CH_2CH_2\underset{\displaystyle CH_3}{\underset{|}{C}}HCH_3$$

2,6-dimethylheptane

2,6-Dimethyl-3-heptyne is the only alkyne that will form the desired alkane upon hydrogenation. This alkyne can be dissected in two different ways. In one case the alkyne could be prepared from the reaction of an acetylide ion with a primary alkyl halide (isobutyl bromide); in the other case the acetylide ion would have to react with a secondary alkyl halide (isopropyl bromide).

retrosynthetic analysis

$$CH_3\underset{\displaystyle CH_3}{\underset{|}{C}}HCH_2CH_2CH_2\underset{\displaystyle CH_3}{\underset{|}{C}}HCH_3 \Longrightarrow CH_3\underset{\displaystyle CH_3}{\underset{|}{C}}HCH_2C{\equiv}C\underset{\displaystyle CH_3}{\underset{|}{C}}HCH_3$$

$$CH_3\underset{\displaystyle CH_3}{\underset{|}{C}}HCH_2Br + HC{\equiv}C\underset{\displaystyle CH_3}{\underset{|}{C}}HCH_3 \quad \text{or} \quad CH_3\underset{\displaystyle CH_3}{\underset{|}{C}}HBr + HC{\equiv}CCH_2\underset{\displaystyle CH_3}{\underset{|}{C}}HCH_3$$

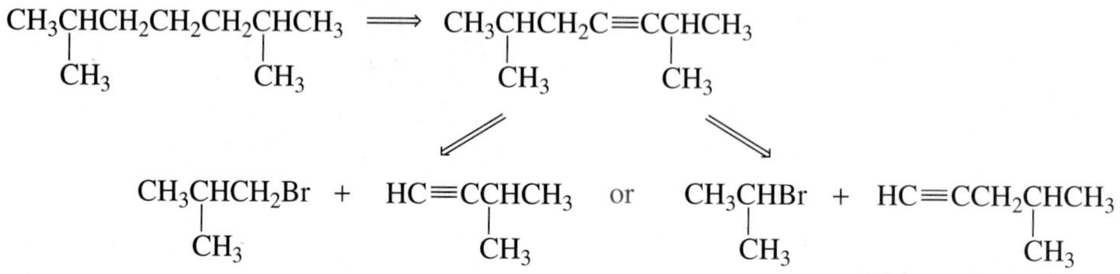

Because we know that the reaction of an acetylide ion with an alkyl halide works best with primary alkyl halides and methyl halides, we know how to proceed:

synthesis

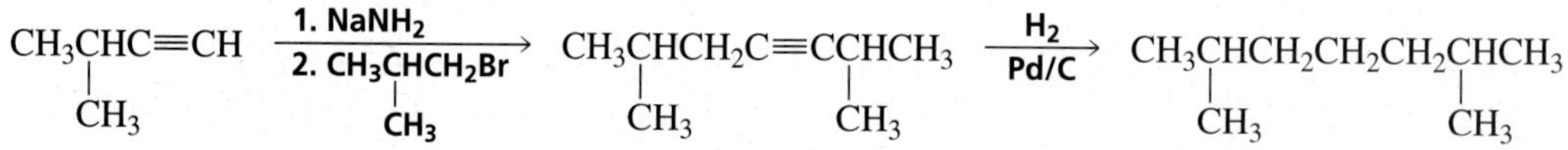

$$CH_3\underset{\displaystyle CH_3}{\underset{|}{C}}HC{\equiv}CH \xrightarrow[\text{2. } CH_3\underset{CH_3}{C}HCH_2Br]{\text{1. } NaNH_2} CH_3\underset{\displaystyle CH_3}{\underset{|}{C}}HCH_2C{\equiv}C\underset{\displaystyle CH_3}{\underset{|}{C}}HCH_3 \xrightarrow[Pd/C]{H_2} CH_3\underset{\displaystyle CH_3}{\underset{|}{C}}HCH_2CH_2CH_2\underset{\displaystyle CH_3}{\underset{|}{C}}HCH_3$$

Example 4. How could you carry out the following synthesis using the given starting material?

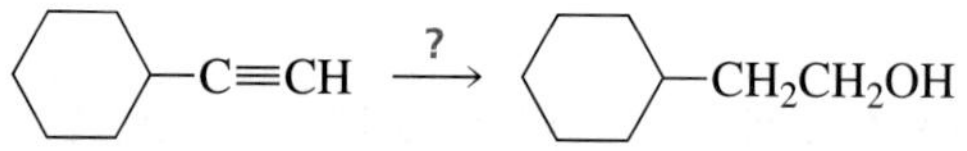

$$C_6H_{11}{-}C{\equiv}CH \xrightarrow{?} C_6H_{11}{-}CH_2CH_2OH$$

An alcohol can be prepared from an alkene, and an alkene can be prepared from an alkyne.

retrosynthetic analysis

$$C_6H_{11}{-}CH_2CH_2OH \Longrightarrow C_6H_{11}{-}CH{=}CH_2 \Longrightarrow C_6H_{11}{-}C{\equiv}CH$$

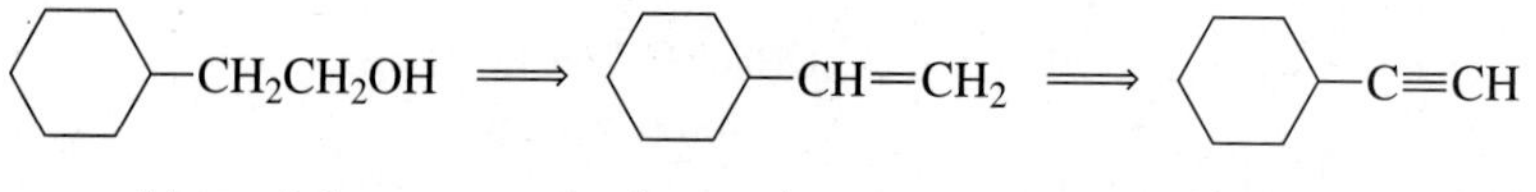

You can use either of the two methods you know to convert an alkyne into an alkene, because this alkene does not have cis–trans isomers. Hydroboration–oxidation must be used to convert the alkene into the desired alcohol, because acid-catalyzed addition of water would not form the desired alcohol.

synthesis

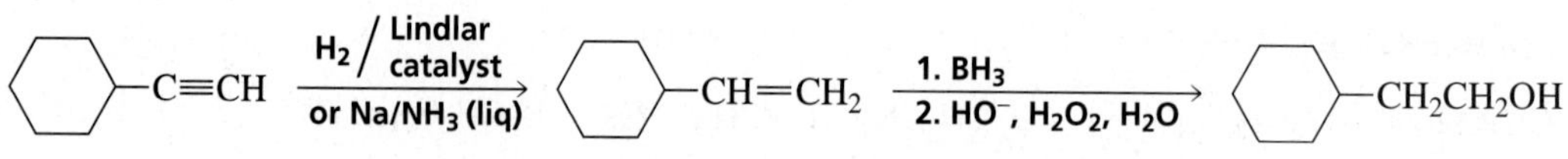

$$C_6H_{11}{-}C{\equiv}CH \xrightarrow[\text{or } Na/NH_3\text{ (liq)}]{H_2/\text{Lindlar catalyst}} C_6H_{11}{-}CH{=}CH_2 \xrightarrow[\text{2. } HO^-,\ H_2O_2,\ H_2O]{\text{1. } BH_3} C_6H_{11}{-}CH_2CH_2OH$$

Example 5. How could you prepare (*E*)-2-pentene from ethyne?

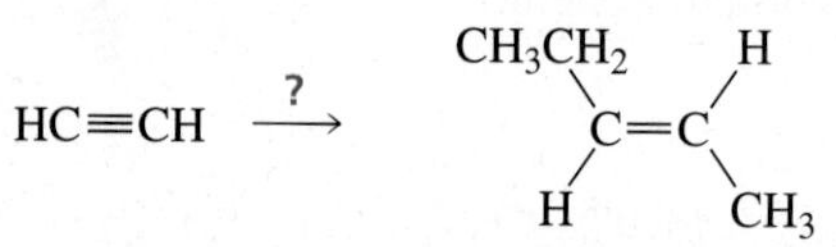

$$HC{\equiv}CH \xrightarrow{?} \text{(}CH_3CH_2\text{)(H)}C{=}C\text{(H)(}CH_3\text{)}$$

(*E*)-2-pentene

A trans alkene can be prepared from an internal alkyne. The alkyne needed to synthesize the desired alkene can be prepared from 1-butyne and a methyl halide. 1-Butyne can be prepared from ethyne and an ethyl halide.

retrosynthetic analysis

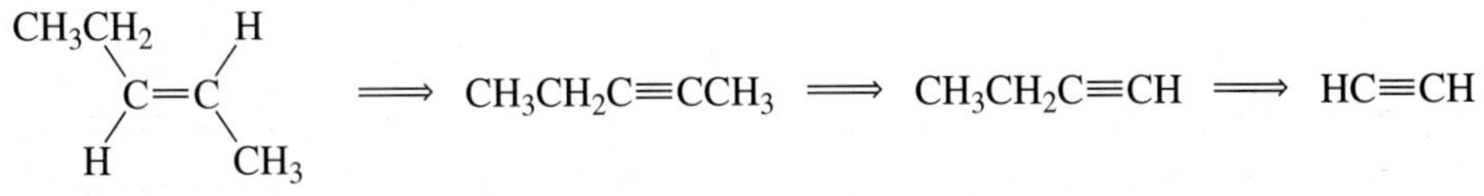

$$\text{(trans) } CH_3CH_2(H)C{=}C(H)CH_3 \Longrightarrow CH_3CH_2C{\equiv}CCH_3 \Longrightarrow CH_3CH_2C{\equiv}CH \Longrightarrow HC{\equiv}CH$$

synthesis

$$HC{\equiv}CH \xrightarrow[\text{2. } CH_3CH_2Br]{\text{1. } NaNH_2} CH_3CH_2C{\equiv}CH \xrightarrow[\text{2. } CH_3Br]{\text{1. } NaNH_2} CH_3CH_2C{\equiv}CCH_3 \xrightarrow[Na]{NH_3\text{ (liq)}} \text{(trans) } CH_3CH_2(H)C{=}C(H)CH_3$$

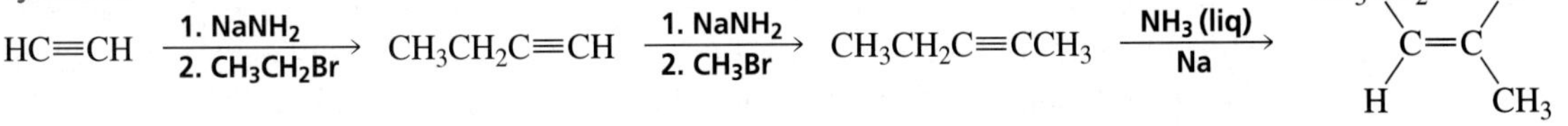

Example 6. How could you prepare 3,3-dibromohexane from reagents that contain no more than two carbon atoms?

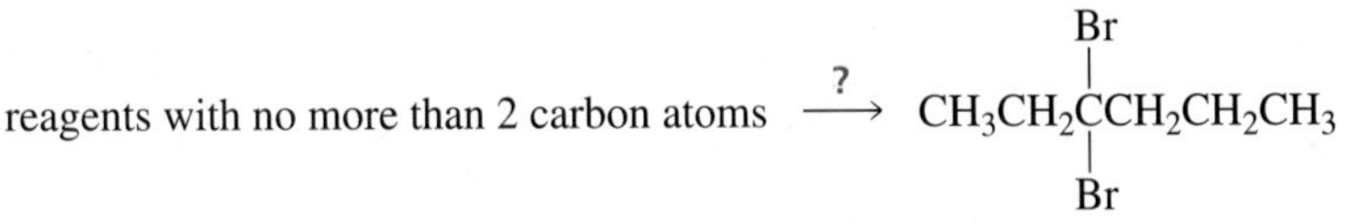

$$\text{reagents with no more than 2 carbon atoms} \xrightarrow{?} CH_3CH_2CBr_2CH_2CH_2CH_3$$

3,3-dibromohexane

A geminal dibromide can be prepared from an alkyne. 3-Hexyne is the desired alkyne because it will form one geminal bromide, whereas 2-hexyne would form two different geminal dibromides. 3-Hexyne can be prepared from 1-butyne and ethyl bromide, and 1-butyne can be prepared from ethyne and ethyl bromide.

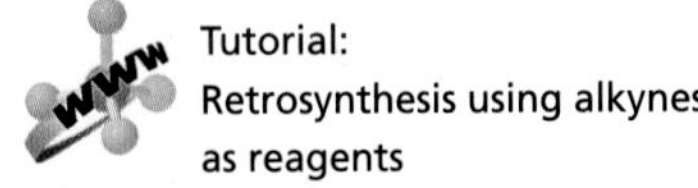

Tutorial: Retrosynthesis using alkynes as reagents

retrosynthetic analysis

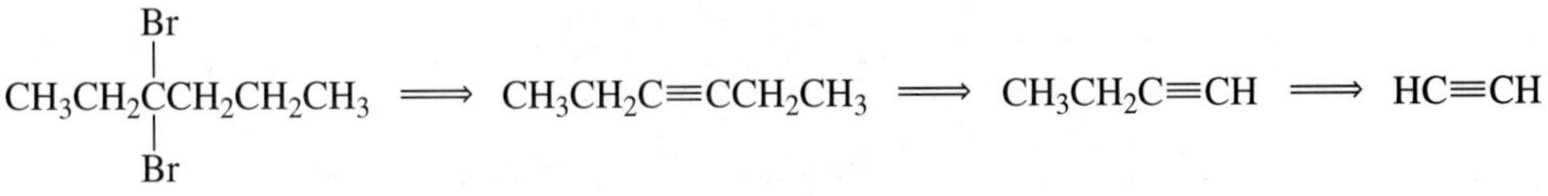

$$CH_3CH_2CBr_2CH_2CH_2CH_3 \Longrightarrow CH_3CH_2C{\equiv}CCH_2CH_3 \Longrightarrow CH_3CH_2C{\equiv}CH \Longrightarrow HC{\equiv}CH$$

synthesis

$$HC{\equiv}CH \xrightarrow[\text{2. } CH_3CH_2Br]{\text{1. } NaNH_2} CH_3CH_2C{\equiv}CH \xrightarrow[\text{2. } CH_3CH_2Br]{\text{1. } NaNH_2} CH_3CH_2C{\equiv}CCH_2CH_3 \xrightarrow{\text{excess HBr}} CH_3CH_2CBr_2CH_2CH_2CH_3$$

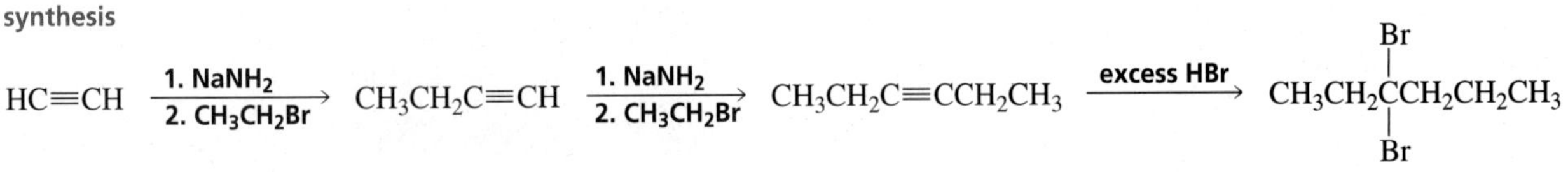

PROBLEM 21

Starting with acetylene, how could the following compounds be synthesized?

a. $CH_3CH_2CH_2C{\equiv}CH$

b. $CH_3CH{=}CH_2$

c. 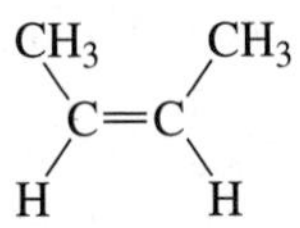 (cis) $CH_3(H)C{=}C(H)CH_3$

d. 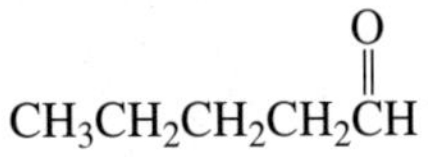 $CH_3CH_2CH_2CH_2CHO$ ($CH_3CH_2CH_2CH_2\overset{O}{\overset{\|}{C}}H$)

e. 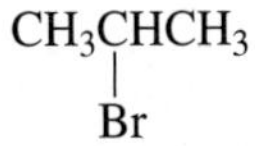 $CH_3CHBrCH_3$

f. 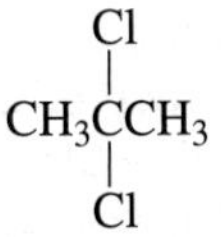 $CH_3CCl_2CH_3$

6.12 Commercial Use of Ethyne

Most of the ethyne produced commercially is used as a starting material for **polymers** that we encounter daily, such as vinyl flooring, plastic piping, Teflon, and acrylics. Polymers are large molecules that are made by linking together many small molecules. The small molecules used to make polymers are called *monomers*. The mechanisms by which monomers are converted into polymers are discussed in Chapter 28.

Poly(vinyl chloride), the polymer produced from the polymerization of vinyl chloride, is known as PVC. Straight-chain poly(vinyl chloride) is hard and rather brittle; branched poly(vinyl chloride) is the soft elastic vinyl commonly used both as a substitute for leather and in the manufacture of such things as garbage bags and shower curtains (Section 28.2). Poly(acrylonitrile) looks like wool when it is made into fibers. It is marketed under the trade names Orlon® (DuPont), Creslan® (Sterling Fibers), and Acrilan® (Monsanto).

$$HC{\equiv}CH + HCl \longrightarrow \underset{\substack{\text{vinyl chloride}\\ \text{a monomer}}}{H_2C{=}CHCl} \xrightarrow{\text{polymerization}} \underset{\substack{\text{poly(vinyl chloride)}\\ \text{PVC}\\ \text{a polymer}}}{\left[CH_2-\underset{\substack{|\\ Cl}}{CH} \right]_n}$$

$$HC{\equiv}CH + HCN \longrightarrow \underset{\substack{\text{acrylonitrile}\\ \text{a monomer}}}{H_2C{=}CHCN} \xrightarrow{\text{polymerization}} \underset{\substack{\text{poly(acrylonitrile)}\\ \text{Orlon®}\\ \text{a polymer}}}{\left[CH_2-\underset{\substack{|\\ CN}}{CH} \right]_n}$$

ETHYNE CHEMISTRY OR THE FORWARD PASS?

Father Julius Arthur Nieuwland (1878–1936) did much of the early work on the polymerization of ethyne. He was born in Belgium and settled with his parents in South Bend, Indiana, two years later. He became a priest and a professor of botany and chemistry at the University of Notre Dame, where Knute Rockne—the inventor of the forward pass—worked for him as a research assistant. Rockne also taught chemistry at Notre Dame, but when he received an offer to coach the football team, he switched fields, in spite of Father Nieuwland's attempts to convince him to continue his work as a scientist.

Knute Rockne in his uniform during the year he was captain of the Notre Dame football team.

Summary

Alkynes are hydrocarbons that contain a carbon–carbon triple bond. A triple bond can be thought of as a cylinder of electrons wrapped around the σ bond. The functional group suffix of an alkyne is "yne." A **terminal alkyne** has the triple bond at the end of the chain; an **internal alkyne** has the triple bond located elsewhere along the chain. Internal alkynes, with two alkyl substituents bonded to the *sp* carbons, are more stable than terminal alkynes. We now have seen that *alkyl groups stabilize alkenes, alkynes, carbocations,* and *alkyl radicals.*

An alkyne is *less* reactive than an alkene because a **vinylic cation** is less stable than a similarly substituted alkyl cation. Like alkenes, alkynes undergo electrophilic addition reactions. The same reagents that add to alkenes add to alkynes. Electrophilic addition to a *terminal* alkyne is regioselective; in all electrophilic addition reactions to terminal alkynes, the *electrophile* adds to the *sp* carbon that is bonded to the hydrogen because the intermediate formed—a secondary vinylic cation—is more stable than a primary vinylic cation. If excess reagent is available, alkynes can undergo a second addition reaction with hydrogen halides and halogens because the product of the first reaction is an alkene. An alkyl peroxide has the same effect on the addition of HBr to an alkyne that it has on the addition of HBr to an alkene—it reverses the order of addition because the peroxide causes $Br^{\cdot}$ to become the electrophile.

When an alkyne undergoes the acid-catalyzed addition of water, the product of the reaction is an enol. The enol immediately rearranges to a ketone. A **ketone** is a compound that has two alkyl groups bonded to a **carbonyl** (C=O) **group**. An **aldehyde** is a compound that has at least one hydrogen bonded to a carbonyl group. The ketone and enol are called **keto–enol tautomers**; they differ in the location of a double bond and a hydrogen. Interconversion of the tautomers is called **tautomerization**. The keto tautomer predominates at equilibrium. Terminal alkynes add water if mercuric ion is added to the acidic mixture. In hydroboration–oxidation, H^+ is not the electrophile, $H:^-$ is the nucleophile. Consequently, mercuric-ion-catalyzed addition of water to a terminal alkyne produces a *ketone*, whereas hydroboration–oxidation of a terminal alkyne produces an *aldehyde*.

Hydrogen adds to an alkyne in the presence of a metal catalyst (Pd, Pt, or Ni) to form an alkane. Addition of hydrogen to an internal alkyne in the presence of Lindlar catalyst forms a *cis alkene*. Sodium in liquid ammonia converts an internal alkyne to a *trans alkene*.

Electronegativity decreases with decreasing percentage of *s* character in the orbital. Thus the electronegativities of carbon atoms decrease in the order: $sp > sp^2 > sp^3$. Ethyne is, therefore, a stronger acid than ethene, and ethene is a stronger acid than ethane. An amide ion can remove a hydrogen bonded to an *sp* carbon of a terminal alkyne because it is a stronger base than the **acetylide ion** that is formed. The acetylide ion can undergo an alkylation reaction with a methyl halide or a primary alkyl halide to form an internal alkyne. An **alkylation reaction** attaches an alkyl group to the starting material.

Designing a synthesis by working backward is called **retrosynthetic analysis**. Chemists use open arrows to indicate they are working backward. The reagents needed to carry out each step are not included until the reaction is written in the forward direction.

Most ethyne produced commercially is for the synthesis of monomers used in the synthesis of polymers. **Polymers** are large molecules that are made by linking together many small molecules called monomers.

Summary of Reactions

1. Electrophilic addition reactions
   a. Addition of hydrogen halides (H^+ is the electrophile; Section 6.5)

$$RC{\equiv}CH \xrightarrow{HX} RC(X){=}CH_2 \xrightarrow{\text{excess HX}} RCX_2{-}CH_3$$

$$HX = HF, HCl, HBr, HI$$

   b. Addition of hydrogen bromide in the presence of a peroxide ($Br^{\cdot}$ is the electrophile; Section 6.5)

$$RC{\equiv}CH + HBr \xrightarrow{\text{peroxide}} RCH{=}CHBr$$

   c. Addition of halogens (Section 6.5)

$$RC{\equiv}CH \xrightarrow[CH_2Cl_2]{Cl_2} RC(Cl){=}CH(Cl) \xrightarrow[CH_2Cl_2]{Cl_2} RCCl_2{-}CHCl_2$$

$$RC{\equiv}CCH_3 \xrightarrow[CH_2Cl_2]{Br_2} RC(Br){=}C(Br)CH_3 \xrightarrow[CH_2Cl_2]{Br_2} RCBr_2{-}CBr_2CH_3$$

Tutorial: Common terms in the reactions of alkynes

   d. Addition of water/hydroboration–oxidation (Sections 6.6 and 6.7)

$$\underset{\text{an internal alkyne}}{RC{\equiv}CR'} \xrightarrow[\text{1. } BH_3/THF \quad \text{2. } HO^-, H_2O_2, H_2O]{H_2O, H_2SO_4 \text{ or}} RC({=}O)CH_2R' + RCH_2C({=}O)R'$$

$$\underset{\text{a terminal alkyne}}{RC{\equiv}CH} \xrightarrow[HgSO_4]{H_2O, H_2SO_4} RC(OH){=}CH_2 \rightleftharpoons RC({=}O)CH_3$$

$$\underset{\text{a terminal alkyne}}{RC{\equiv}CH} \xrightarrow[\text{2. } HO^-, H_2O_2, H_2O]{\text{1. disiamylborane}} RCH{=}CH(OH) \rightleftharpoons RCH_2C({=}O)H$$

2. Addition of hydrogen (Section 6.8)

$$RC{\equiv}CR' + 2\,H_2 \xrightarrow[\text{or Ni}]{\text{Pd/C, Pt/C,}} RCH_2CH_2R'$$

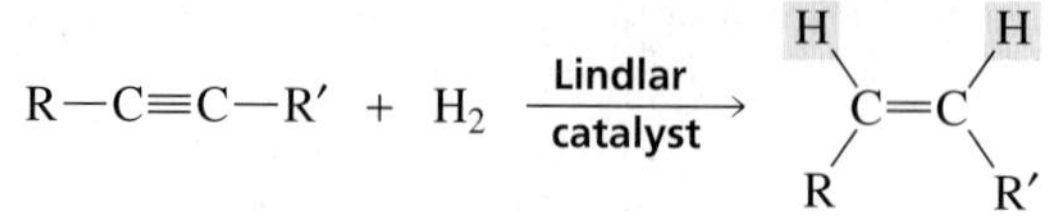

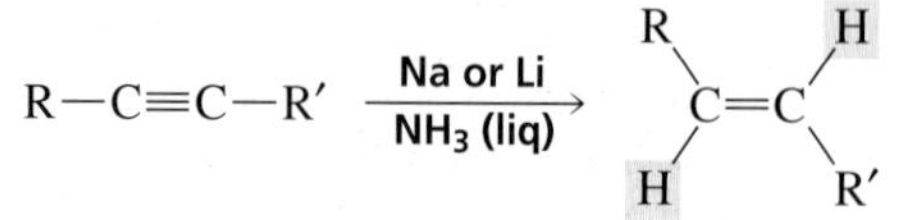

$$R{-}C{\equiv}C{-}R' + H_2 \xrightarrow[\text{catalyst}]{\text{Lindlar}} \text{(cis) } (R)(H)C{=}C(H)(R')$$

$$R{-}C{\equiv}C{-}R' \xrightarrow[\text{NH}_3\text{ (liq)}]{\text{Na or Li}} \text{(trans) } (R)(H)C{=}C(H)(R')$$

3. Removal of a proton, followed by alkylation (Sections 6.9 and 6.10)

$$RC{\equiv}CH \xrightarrow{\text{NaNH}_2} RC{\equiv}C^- \xrightarrow{\text{R'CH}_2\text{Br}} RC{\equiv}CCH_2R'$$

Key Terms

acetylide ion (p. 252)
aldehyde (p. 246)
alkylation reaction (p. 253)
alkynes (p. 238)
carbonyl group (p. 246)
enol (p. 246)
geminal dihalide (p. 243)
internal alkyne (p. 239)
keto–enol tautomers (p. 246)
ketone (p. 246)
polymers (p. 258)
radical anion (p. 250)
retrosynthetic analysis (p. 255)
tautomerization (p. 246)
tautomers (p. 246)
terminal alkyne (p. 239)
vinylic cation (p. 243)
vinylic radical (p. 250)

Problems

22. Give the major product obtained from reaction of each of the following with excess HCl:
 a. $CH_3CH_2C{\equiv}CH$ b. $CH_3CH_2C{\equiv}CCH_2CH_3$ c. $CH_3CH_2C{\equiv}CCH_2CH_2CH_3$

23. Draw a structure for each of the following:
 a. 2-hexyne
 b. 5-ethyl-3-octyne
 c. methylacetylene
 d. vinylacetylene
 e. methoxyethyne
 f. *sec*-butyl-*tert*-butylacetylene
 g. 1-bromo-1-pentyne
 h. propargyl bromide
 i. diethylacetylene
 j. di-*tert*-butylacetylene
 k. cyclopentylacetylene
 l. 5,6-dimethyl-2-heptyne

24. Identify the electrophile and the nucleophile in each of the following reaction steps. Then draw curved arrows to illustrate the bond-making and -breaking processes.

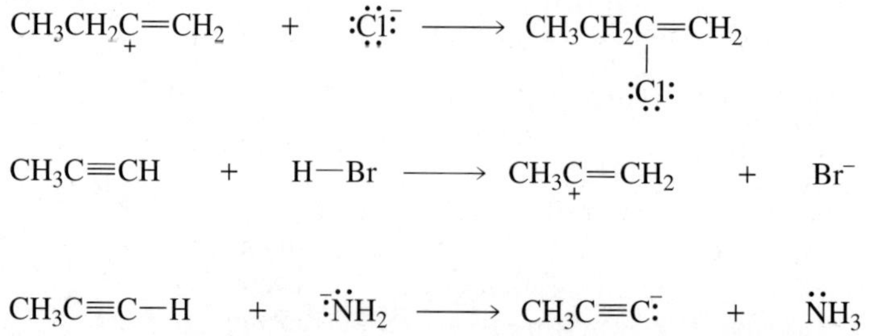

$$CH_3CH_2\overset{+}{C}{=}CH_2 + :\ddot{Cl}:^- \longrightarrow CH_3CH_2C(:\ddot{Cl}:){=}CH_2$$

$$CH_3C{\equiv}CH + H{-}Br \longrightarrow CH_3\overset{+}{C}{=}CH_2 + Br^-$$

$$CH_3C{\equiv}C{-}H + :\ddot{N}H_2^- \longrightarrow CH_3C{\equiv}C:^- + \ddot{N}H_3$$

25. Give the systematic name for each of the following structures:

a. $CH_3C{\equiv}CCH_2\underset{\underset{\displaystyle Br}{|}}{C}HCH_3$

b. $CH_3C{\equiv}CCH_2\underset{\underset{\displaystyle CH_2CH_2CH_3}{|}}{C}HCH_3$

c. $CH_3C{\equiv}CCH_2\overset{\displaystyle CH_3}{\overset{|}{\underset{\underset{\displaystyle CH_3}{|}}{C}}}CH_3$

d. $CH_3\underset{\underset{\displaystyle Cl}{|}}{C}HCH_2C{\equiv}C\underset{\underset{\displaystyle CH_3}{|}}{C}HCH_3$

26. What reagents could be used to carry out the following syntheses?

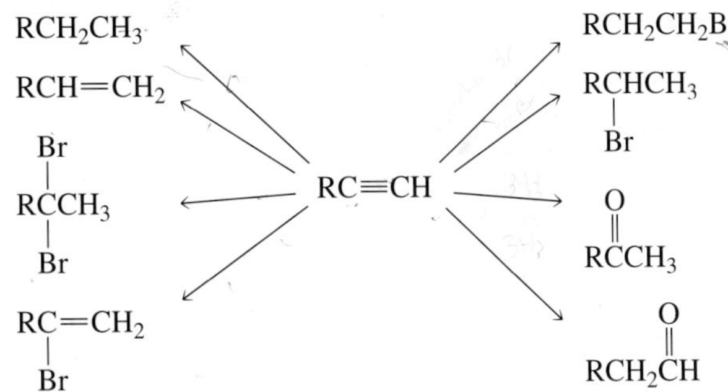

27. Al Kyne was given the structural formulas of several compounds and was asked to give them systematic names. How many did Al name correctly? Correct those that are misnamed.

a. 4-ethyl-2-pentyne
b. 1-bromo-4-heptyne
c. 2-methyl-3-hexyne
d. 3-pentyne

28. Draw the structures and give the common and systematic names for alkynes with molecular formula C_7H_{12}.

29. How could the following compounds be synthesized, starting with a hydrocarbon that has the same number of carbon atoms as the desired product?

a. $CH_3CH_2CH_2CH_2\overset{\displaystyle O}{\overset{\|}{C}}H$

b. $CH_3CH_2CH_2CH_2OH$

c. $CH_3CH_2CH_2\overset{\displaystyle O}{\overset{\|}{C}}CH_2CH_2CH_2CH_3$

d. $CH_3CH_2CH_2CH_2CH_2Br$

30. What reagents would you use for the following syntheses?

a. (*Z*)-3-hexene from 3-hexyne
b. (*E*)-3-hexene from 3-hexyne
c. hexane from 3-hexyne

31. What is the molecular formula of a hydrocarbon that has 1 triple bond, 2 double bonds, 1 ring, and 32 carbons?

32. What will be the major product of the reaction of 1 mol of propyne with each of the following reagents?

a. HBr (1 mol)
b. HBr (2 mol)
c. Br_2(1 mol)/CH_2Cl_2
d. Br_2(2 mol)/CH_2Cl_2
e. aqueous H_2SO_4, $HgSO_4$
f. disiamylborane followed by H_2O_2/HO^-
g. HBr + H_2O_2
h. excess H_2/Pt
i. H_2/Lindlar catalyst
j. sodium in liquid ammonia
k. sodium amide
l. product of Problem k followed by 1-chloropentane

33. Answer Problem 32, using 2-butyne as the starting material instead of propyne.

34. a. Starting with isopropylacetylene, how could you prepare the following alcohols?
   1. 2-methyl-2-pentanol
   2. 4-methyl-2-pentanol

b. In each case a second alcohol would also be obtained. What alcohol would it be?

35. How many of the following names are correct? Correct the incorrect names.

a. 4-heptyne
b. 2-ethyl-3-hexyne
c. 4-chloro-2-pentyne
d. 2,3-dimethyl-5-octyne
e. 4,4-dimethyl-2-pentyne
f. 2,5-dimethyl-3-hexyne

36. Which of the following pairs are keto–enol tautomers?

a. $CH_3CH_2CH{=}CHCH_2OH$ and $CH_3CH_2CH_2CH_2\overset{O}{\overset{\|}{C}}H$

b. $CH_3\overset{OH}{\overset{|}{C}}HCH_3$ and $CH_3\overset{O}{\overset{\|}{C}}CH_3$

c. $CH_3CH_2CH{=}CHOH$ and $CH_3CH_2CH_2\overset{O}{\overset{\|}{C}}H$

d. $CH_3CH_2CH_2CH{=}CHOH$ and $CH_3CH_2CH_2\overset{O}{\overset{\|}{C}}CH_3$

e. $CH_3CH_2CH_2\overset{OH}{\overset{|}{C}}{=}CH_2$ and $CH_3CH_2CH_2\overset{O}{\overset{\|}{C}}CH_3$

37. Using ethyne as the starting material, how can the following compounds be prepared?

a. $CH_3\overset{O}{\overset{\|}{C}}H$

b. $CH_3CH_2\underset{Br}{\underset{|}{C}}HCH_2Br$

c. $CH_3\overset{O}{\overset{\|}{C}}CH_3$

d.

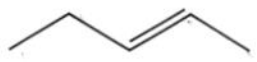

e.

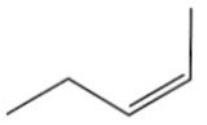

f.

38. Give the stereoisomers obtained from the reaction of 2-butyne with the following reagents:

a. 1. H_2/Lindlar catalyst 2. Br_2/CH_2Cl_2
b. 1. Na/NH_3(liq) 2. Br_2/CH_2Cl_2
c. 1. Cl_2/CH_2Cl_2 2. Br_2/CH_2Cl_2

39. Draw the keto tautomer for each of the following:

a. $CH_3CH{=}\overset{OH}{\overset{|}{C}}CH_3$

b. $CH_3CH_2CH_2\overset{OH}{\overset{|}{C}}{=}CH_2$

c. 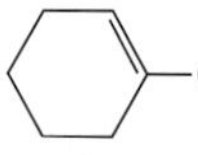 —OH

d. 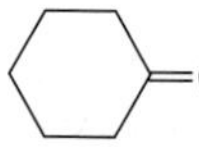 =CHOH

40. Show how each of the following compounds could be prepared using the given starting material, any necessary inorganic reagents, and any necessary organic compound that has no more than four carbon atoms:

a. $HC{\equiv}CH \longrightarrow CH_3CH_2CH_2CH_2\overset{O}{\overset{\|}{C}}CH_3$

b. $HC{\equiv}CH \longrightarrow CH_3CH_2\underset{Br}{\underset{|}{C}}HCH_3$

c. $HC{\equiv}CH \longrightarrow CH_3CH_2CH_2\underset{OH}{\underset{|}{C}}HCH_3$

d. 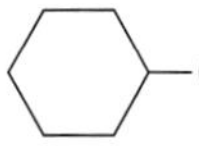 $-C{\equiv}CH \longrightarrow$ $-CH_2\overset{O}{\overset{\|}{C}}H$

e. 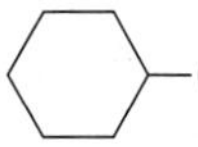 $-C{\equiv}CH \longrightarrow$ $-\overset{O}{\overset{\|}{C}}CH_3$

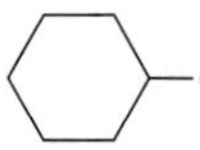

f. 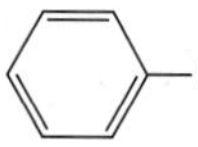 $-C{\equiv}CCH_3 \longrightarrow$ H, H, C=C, CH_3

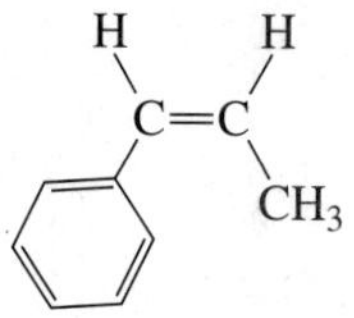

41. Dr. Polly Meher was planning to synthesize 3-octyne by adding 1-bromobutane to the product obtained from the reaction of 1-butyne with sodium amide. Unfortunately, however, she had forgotten to order 1-butyne. How else can she prepare 3-octyne?

42. a. Explain why a single pure product is obtained from hydroboration–oxidation of 2-butyne, whereas two products are obtained from hydroboration–oxidation of 2-pentyne.
b. Name two other internal alkynes that will yield only one product on hydroboration–oxidation.

43. Give the configurations of the products obtained from the following reactions:

a. $CH_3CH_2C{\equiv}CCH_2CH_3 \xrightarrow[\text{2. } D_2\text{, Pd/C}]{\text{1. Na, } NH_3\text{(liq)}}$

b. $CH_3CH_2C{\equiv}CCH_2CH_3 \xrightarrow[\text{2. } D_2\text{, Pd/C}]{\text{1. } H_2\text{/Lindlar catalyst}}$

44. In Section 6.4 it was stated that hyperconjugation is less effective in stabilizing the charge on a vinylic cation than the charge on an alkyl cation. Why do you think this is so?

7 Electron Delocalization and Resonance • More About Molecular Orbital Theory

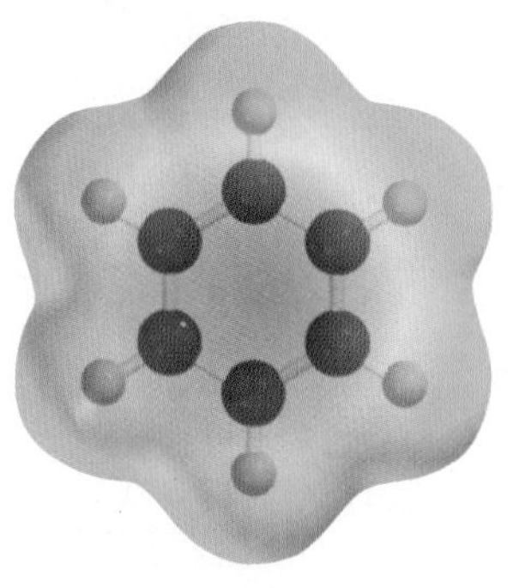

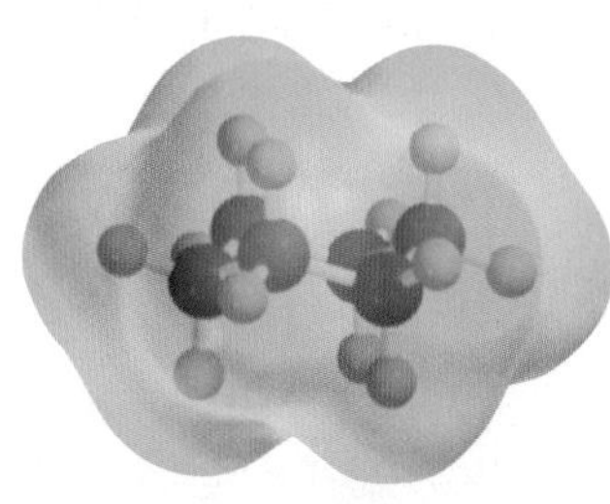

benzene

cyclohexane

As you continue your study of organic chemistry, you will notice that the concept of having delocalized electrons is invoked frequently to explain the behavior of organic compounds. For example, in Chapter 8 you will see that having delocalized electrons causes certain dienes to form products that would not be expected on the basis of what you have learned about electrophilic addition reactions in Chapters 3–6. Electron delocalization is such an important concept that this entire chapter is devoted to it.

Electrons that are restricted to a particular region are called **localized electrons**. Localized electrons either belong to a single atom or are confined to a bond between two atoms.

$CH_3-\ddot{N}H_2$ (localized electrons) $\qquad CH_3-CH{=}CH_2$ (localized electrons)

Not all electrons are confined to a single atom or bond. Many organic compounds contain *delocalized* electrons. **Delocalized electrons** neither belong to a single atom nor are confined to a bond between two atoms, but are shared by three or more atoms. You were first introduced to delocalized electrons in Section 1.19, where you saw that the two electrons represented by the π bond of the COO^- group are shared by three atoms—the carbon and both oxygen atoms. The dashed lines indicate that the two electrons are delocalized over three atoms.

$CH_3C(\ddot{O}:^{\delta-})(\ddot{O}:^{\delta-})$ — delocalized electrons

In this chapter, you will learn to recognize compounds that contain delocalized electrons and to draw structures that represent the electron distribution in molecules with delocalized electrons. You will also be introduced to some of the special characteristics of compounds that have delocalized electrons. You will then be able to understand the wide-ranging effects that delocalized electrons have on the reactivity of organic compounds. We begin by taking a look at benzene, a compound whose properties chemists could not explain until they recognized that electrons in organic molecules could be delocalized.

7.1 Delocalized Electrons: The Structure of Benzene

The structure of benzene puzzled early organic chemists. They knew that benzene had a molecular formula of C_6H_6, that it was an unusually stable compound, and that it did not undergo the addition reactions characteristic of alkenes (Section 3.6). They also knew the following facts:

1. When a different atom is substituted for one of the hydrogen atoms of benzene, only one product is obtained.
2. When the substituted product undergoes a second substitution, three products are obtained.

$$C_6H_6 \xrightarrow[\text{with an X}]{\text{replace a hydrogen}} C_6H_5X \xrightarrow[\text{with an X}]{\text{replace a hydrogen}} C_6H_4X_2 + C_6H_4X_2 + C_6H_4X_2$$

one monosubstituted compound

three disubstituted compounds

For every *two* hydrogens that are missing from the general molecular formula C_nH_{2n+2}, a hydrocarbon has either a π bond or a ring.

What kind of structure might we predict for benzene if we knew only what the early chemists knew? The molecular formula (C_6H_6) tells us that benzene has eight fewer hydrogens than an acyclic (noncyclic) alkane with six carbons ($C_nH_{2n+2} = C_6H_{14}$). Benzene, therefore, has a degree of unsaturation of four (Section 3.1). This means that benzene is either an acyclic compound with four π bonds, a cyclic compound with three π bonds, a bicyclic compound with two π bonds, a tricyclic compound with one π bond, or a tetracyclic compound.

Because only one product is obtained regardless of which of the six hydrogens is replaced with another atom, we know that all the hydrogens must be identical. Two structures that fit these requirements are shown here:

$$CH_3C{\equiv}C{-}C{\equiv}CCH_3$$

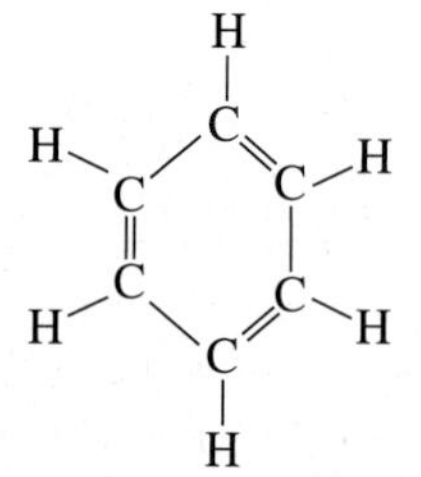

Neither of these structures is consistent with the observation that three compounds are obtained if a second hydrogen is replaced with another atom. The acyclic structure yields two disubstituted products.

$$CH_3C{\equiv}C{-}C{\equiv}CCH_3 \xrightarrow[\text{with Br's}]{\text{replace 2 H's}} CH_3C{\equiv}C{-}C{\equiv}CCHBr_{(|Br)} \quad \text{and} \quad BrCH_2C{\equiv}C{-}C{\equiv}CCH_2Br$$

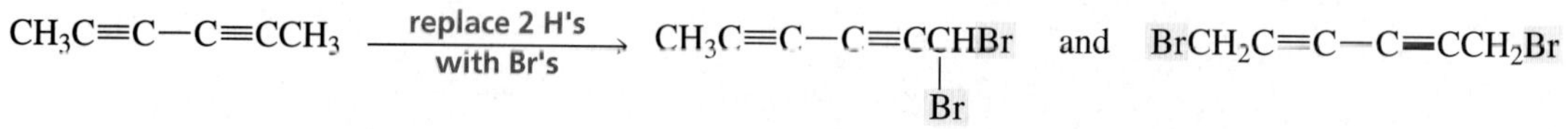

The cyclic structure, with alternating single and slightly shorter double bonds, yields four disubstituted products—a 1,3-disubstituted product, a 1,4-disubstituted product, and two 1,2-disubstituted products—because the two substituents can be placed either on two adjacent carbons joined by a single bond or on two adjacent carbons joined by a double bond.

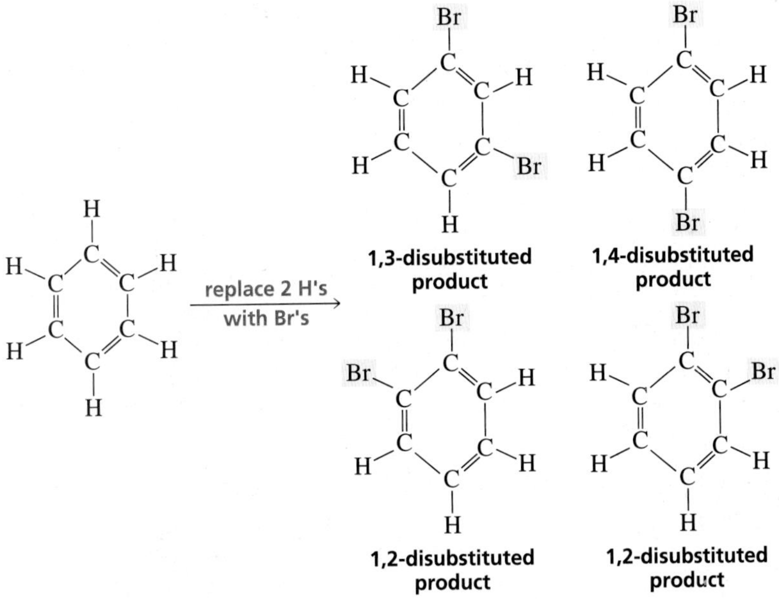

replace 2 H's with Br's

1,3-disubstituted product

1,4-disubstituted product

1,2-disubstituted product

1,2-disubstituted product

3-D Molecules: 1,2-Difluorobenzene; 1,3-Difluorobenzene; 1,4-Difluorobenzene

In 1865, the German chemist Friedrich Kekulé suggested a way of resolving this dilemma. He proposed that benzene was not a single compound, but a mixture of two compounds in rapid equilibrium.

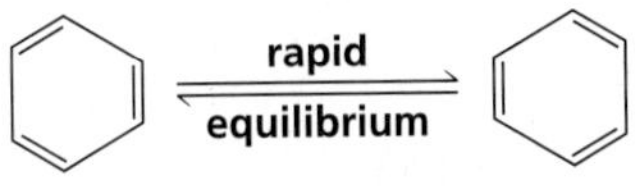

rapid equilibrium

Kekulé structures of benzene

Kekule's proposal explained why only three disubstituted products are obtained when a monosubstituted benzene undergoes a second substitution. According to Kekulé, there actually *are* four disubstituted products, but the two 1,2-disubstituted products interconvert too rapidly to be distinguished and separated from each other.

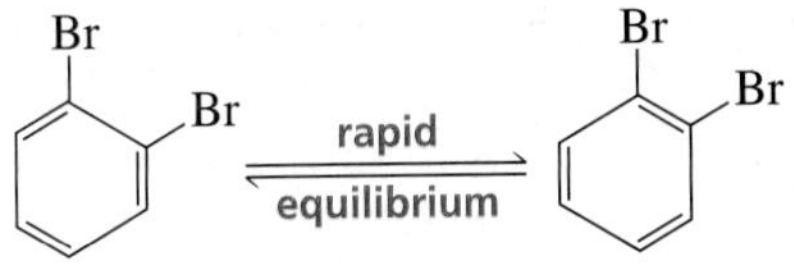

rapid equilibrium

The Kekulé structures of benzene account for the molecular formula of benzene and for the number of isomers obtained as a result of substitution. However, they fail to account for the unusual stability of benzene and for the observation that the double bonds of benzene do not undergo the addition reactions characteristic of alkenes. That benzene had a six-membered ring was confirmed in 1901, when Paul Sabatier (Section 4.11) found that the hydrogenation of benzene produced cyclohexane. This, however, still did not solve the puzzle of benzene's structure.

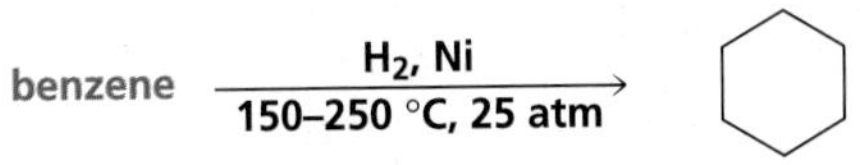

benzene $\xrightarrow[150\text{–}250\ ^\circ\text{C, 25 atm}]{H_2,\ \text{Ni}}$ cyclohexane

Controversy over the structure of benzene continued until the 1930s, when the new techniques of X-ray and electron diffraction produced a surprising result: They showed that *benzene is a planar molecule and the six carbon–carbon bonds have the same length*. The length of each carbon–carbon bond is 1.39 Å, which is shorter than a carbon–carbon single bond (1.54 Å) but longer than a carbon–carbon double bond

KEKULÉ'S DREAM

Friedrich August Kekulé von Stradonitz (1829–1896) was born in Germany. He entered the University of Giessen to study architecture, but switched to chemistry after taking a course in the subject. He was a professor of chemistry at the University of Heidelberg, at the University of Ghent in Belgium, and then at the University of Bonn. In 1890, he gave an extemporaneous speech at the twenty-fifth-anniversary celebration of his first paper on the cyclic structure of benzene. In this speech, he claimed that he had arrived at the Kekulé structures as a result of dozing off in front of a fire while working on a textbook. He dreamed of chains of carbon atoms twisting and turning in a snakelike motion, when suddenly the head of one snake seized hold of its own tail and formed a spinning ring. Recently, the veracity of his snake story has been questioned by those who point out that there is no written record of the dream from the time he experienced it in 1861 until the time he related it in 1890. Others counter that dreams are not the kind of evidence one publishes in scientific papers, although it is not uncommon for scientists to report moments of creativity through the subconscious, when they were not thinking about science. Also, Kekulé warned against publishing dreams when he said, "Let us learn to dream, and perhaps then we shall learn the truth. But let us also beware not to publish our dreams until they have been examined by the wakened mind." In 1895, he was made a nobleman by Emperor William II. This allowed him to add "von Stradonitz" to his name. Kekulé's students received three of the first five Nobel Prizes in chemistry: van't Hoff in 1901 (page 194), Fischer in 1902 (page 187), and Baeyer in 1905 (page 94).

Friedrich August Kekulé von Stradonitz

(1.33 Å; Section 1.14). In other words, benzene does not have alternating single and double bonds.

If the carbon–carbon bonds all have the same length, they must also have the same number of electrons between the carbon atoms. This can be so, however, only if the π electrons of benzene are delocalized around the ring, rather than each pair of π electrons being localized between two carbon atoms. To better understand the concept of delocalized electrons, we'll now take a close look at the bonding in benzene.

PROBLEM 1◆

a. How many monosubstituted products would each of the following compounds have? (Notice that each compound has the same molecular formula as benzene.)
   1. $HC{\equiv}CC{\equiv}CCH_2CH_3$ 2. $CH_2{=}CHC{\equiv}CCH{=}CH_2$
b. How many disubstituted products would each of the preceding compounds have? (Do not include stereoisomers.)
c. How many disubstituted products would each of the compounds have if stereoisomers are included?

Sir James Dewar (1842–1923) *was born in Scotland, the son of an innkeeper. After studying under Kekulé, he became a professor at Cambridge University and then at the Royal Institution in London. Dewar's most important work was in the field of low-temperature chemistry. He used double-walled flasks with evacuated space between the walls in order to reduce heat transmission. These flasks are now called Dewar flasks—better known to nonchemists as thermos bottles.*

Albert Ladenburg (1842–1911) *was born in Germany. He was a professor of chemistry at the University of Kiel.*

PROBLEM 2

Between 1865 and 1890, other possible structures were proposed for benzene, two of which are shown here:

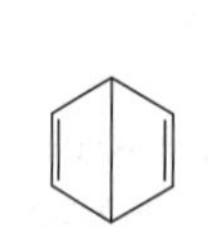

Dewar benzene

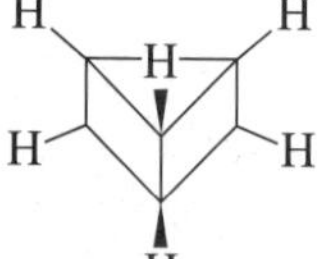

Ladenburg benzene

Considering what nineteenth-century chemists knew about benzene, which is a better proposal for the structure of benzene, Dewar benzene or Ladenburg benzene? Why?

7.2 The Bonding in Benzene

Benzene is a planar molecule. Each of its six carbon atoms is sp^2 hybridized. An sp^2 hybridized carbon has bond angles of 120°—identical to the size of the angles of a planar hexagon. Each of the carbons in benzene uses two sp^2 orbitals to bond to two other carbons; the third sp^2 orbital overlaps the *s* orbital of a hydrogen (Figure 7.1a). Each carbon also has a *p* orbital at right angles to the sp^2 orbitals. Because benzene is planar, the six *p* orbitals are parallel (Figure 7.1b). The *p* orbitals are close enough for side-to-side overlap, so each *p* orbital overlaps the *p* orbitals on both adjacent carbons. As a result, the overlapping *p* orbitals form a continuous doughnut-shaped cloud of electrons above, and another doughnut-shaped cloud of electrons below, the plane of the benzene ring (Figure 7.1c). The electrostatic potential map (Figure 7.1d) shows that all the carbon–carbon bonds have the same electron density.

Each of the six π electrons, therefore, is localized neither on a single carbon nor in a bond between two carbons (as in an alkene). Instead, each π electron is shared by all six carbons. The six π electrons are delocalized—they roam freely within the doughnut-shaped clouds that lie over and under the ring of carbon atoms. Consequently, benzene can be represented by a hexagon containing either dashed lines or a circle, to symbolize the six delocalized π electrons.

3-D Molecule: Benzene

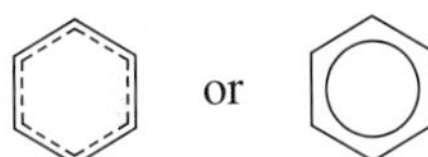

This type of representation makes it clear that there are no double bonds in benzene. We see now that Kekulé's structure for benzene was pretty close to the correct structure. The actual structure of benzene is a Kekulé structure with delocalized electrons.

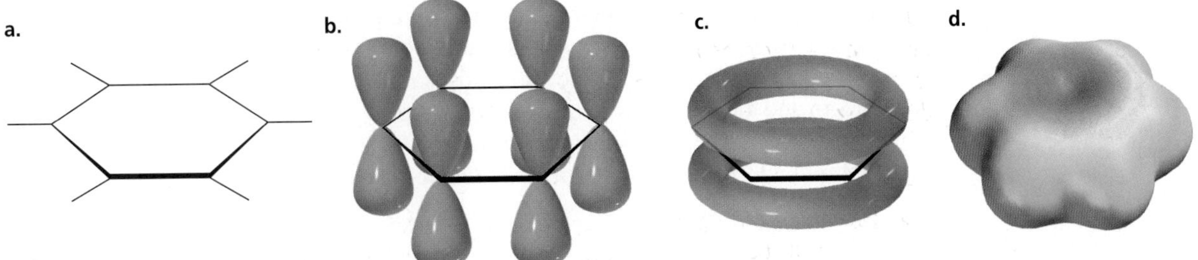

▲ **Figure 7.1**
(a) The carbon–carbon and carbon–hydrogen σ bonds in benzene.
(b) The *p* orbital on each carbon of benzene can overlap with two adjacent *p* orbitals.
(c) The clouds of electrons above and below the plane of the benzene ring.
(d) The electrostatic potential map for benzene.

7.3 Resonance Contributors and the Resonance Hybrid

A disadvantage to using dashed lines to represent delocalized electrons is that they do not tell us how many π electrons are present in the molecule. For example, the dashed lines inside the hexagon in the representation of benzene indicate that the π electrons are shared equally by all six carbons and that all the carbon–carbon bonds have the same length, but they do not show how many π electrons are in the ring. Consequently, chemists prefer to use structures with localized electrons to approximate the actual structure that has delocalized electrons. The approximate structure with localized electrons is called a **resonance contributor**, a **resonance structure**, or a **contributing**

resonance structure. The actual structure with delocalized electrons is called a **resonance hybrid**. Notice that it is easy to see that there are six π electrons in the ring of each resonance contributor.

resonance contributor $\longleftrightarrow$ resonance contributor

resonance hybrid

Resonance contributors are shown with a double-headed arrow between them. The double-headed arrow does *not* mean that the structures are in equilibrium with one another. Rather, it indicates that the actual structure lies somewhere between the structures of the resonance contributors. Resonance contributors are merely a convenient way to show the π electrons; they do not depict any real electron distribution. For example, the bond between C-1 and C-2 in benzene is not a double bond, although the resonance contributor on the left implies that it is. Nor is it a single bond, as represented by the resonance contributor on the right. Neither of the contributing resonance structures accurately represents the structure of benzene. The actual structure of benzene—the resonance hybrid—is given by the average of the two resonance contributors.

Electron delocalization is shown by double-headed arrows ($\leftrightarrow$). Equilibrium is shown by two arrows pointing in opposite directions ($\rightleftharpoons$).

The following analogy illustrates the difference between resonance contributors and the resonance hybrid. Imagine that you are trying to describe to a friend what a rhinoceros looks like. You might tell your friend that a rhinoceros looks like a cross between a unicorn and a dragon. The unicorn and the dragon don't really exist, so they are like the resonance contributors. They are not in equilibrium: A rhinoceros does not jump back and forth between the two resonance contributors, looking like a unicorn one minute and a dragon the next. The rhinoceros is real, so it is like the resonance hybrid. The unicorn and the dragon are simply ways to represent what the actual structure—the rhinoceros—looks like. *Resonance contributors, like unicorns and dragons, are imaginary, not real. Only the resonance hybrid, like the rhinoceros, is real.*

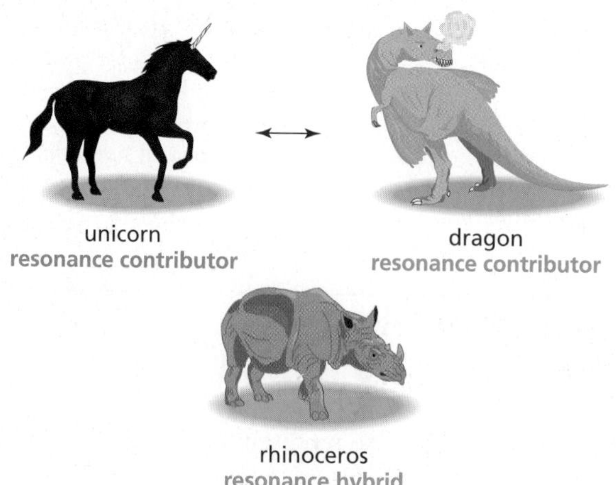

unicorn
resonance contributor

dragon
resonance contributor

rhinoceros
resonance hybrid

Electron delocalization occurs only if all the atoms sharing the delocalized electrons lie in or close to the same plane, so that their p orbitals can effectively overlap. For example, cyclooctatetraene is not planar, but tub shaped. Because the p orbitals cannot overlap, each pair of π electrons is *localized* between two carbons instead of being *delocalized* over the entire ring of eight carbons.

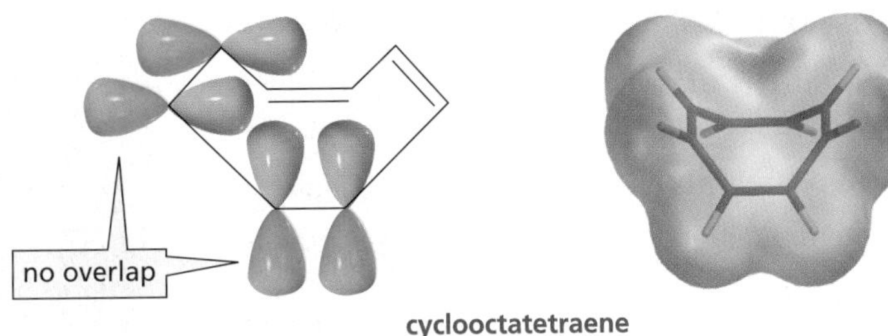

cyclooctatetraene

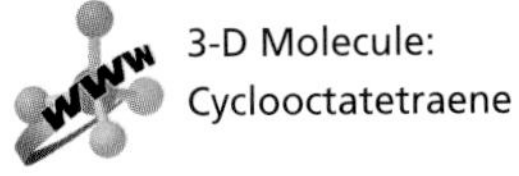

7.4 Drawing Resonance Contributors

We have seen that an organic compound with delocalized electrons is generally represented as a structure with localized electrons, so that we will know how many π electrons are present in the molecule. For example, nitroethane is represented as having a nitrogen–oxygen double bond and a nitrogen–oxygen single bond.

$$CH_3CH_2-\overset{+}{N}(=O)-O^-$$

nitroethane

However, the two nitrogen–oxygen bonds in nitroethane are identical; they each have the same bond length. A more accurate description of the molecule's structure is obtained by drawing the two resonance contributors. Both resonance contributors show the compound with a nitrogen–oxygen double bond and a nitrogen–oxygen single bond, but to show that the electrons are delocalized, the double bond in one contributor is the single bond in the other.

$$CH_3CH_2-\overset{+}{N}(=O)-O^- \longleftrightarrow CH_3CH_2-\overset{+}{N}(-O^-)=O$$

resonance contributor **resonance contributor**

The resonance hybrid shows that the p orbital of nitrogen overlaps the p orbital of each oxygen. In other words, the two electrons are shared by three atoms. The resonance hybrid also shows that the two nitrogen–oxygen bonds are identical and that the negative charge is shared by both oxygen atoms. Although the resonance contributors tell us where the formal charges reside in a molecule and give us the approximate bond orders, we need to visualize and mentally average both resonance contributors to appreciate what the actual molecule—the resonance hybrid—looks like.

Delocalized electrons result from a *p* orbital overlapping the *p* orbitals of more than one adjacent atom.

$$CH_3CH_2-\overset{+}{N}(\cdots\overset{\delta-}{O})\cdots\overset{\delta-}{O}$$

resonance hybrid

Rules for Drawing Resonance Contributors

In drawing resonance contributors, the electrons in one resonance contributor are moved to generate the next resonance contributor. As you draw resonance contributors, keep in mind the following constraints:

1. Only electrons move. The nuclei of the atoms never move.
2. The only electrons that can move are π electrons (electrons in π bonds) and lone-pair electrons.
3. The total number of electrons in the molecule does not change, and neither do the numbers of paired and unpaired electrons.

The electrons can be moved in one of the following ways:

1. Move π electrons toward a positive charge or toward a π bond (Figures 7.2 and 7.3).
2. Move lone-pair electrons toward a π bond (Figure 7.4).
3. Move a single nonbonding electron toward a π bond (Figure 7.5).

Figure 7.2
Resonance contributors are obtained by moving π electrons toward a positive charge.

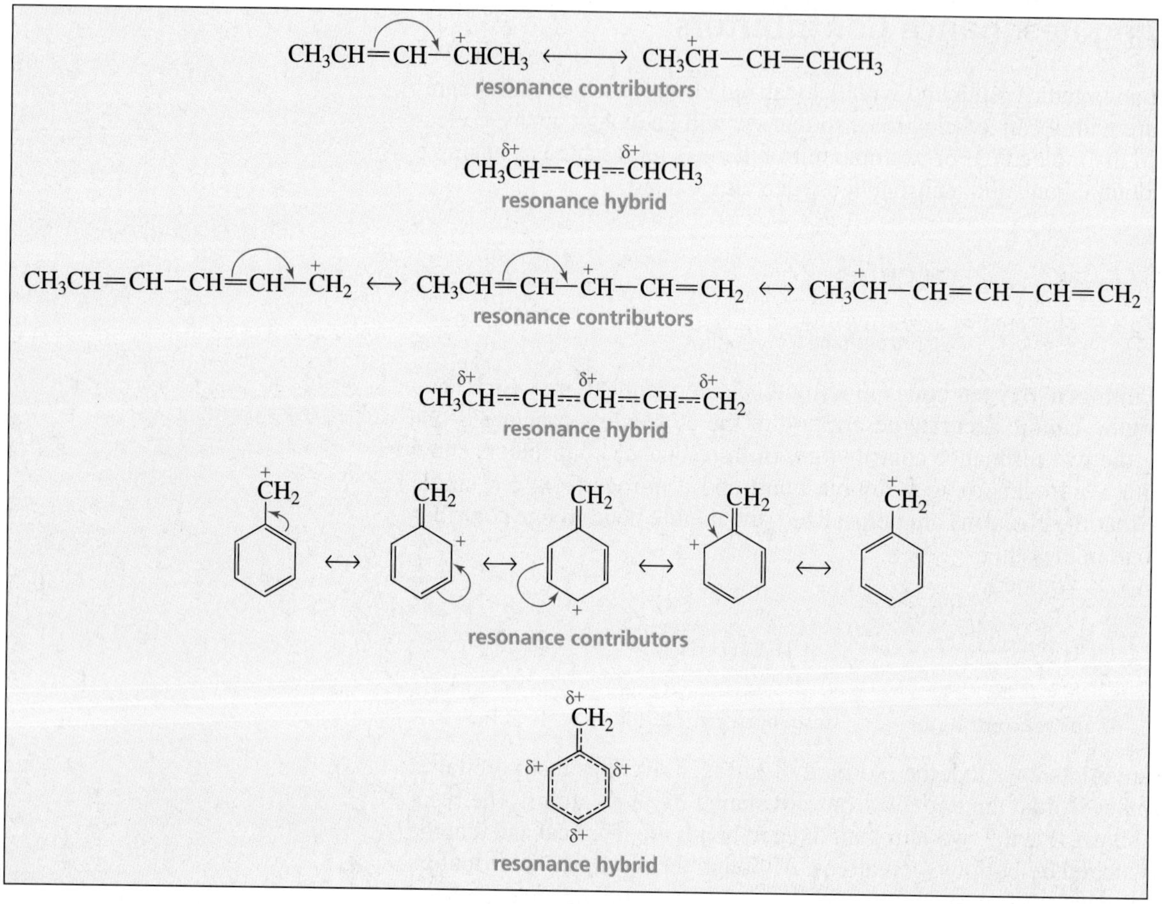

Figure 7.3
Resonance contributors are obtained by moving π electrons toward a π bond. (In the second example, the red arrows lead to the resonance contributor on the right, and the blue arrows lead to the resonance contributor on the left.)

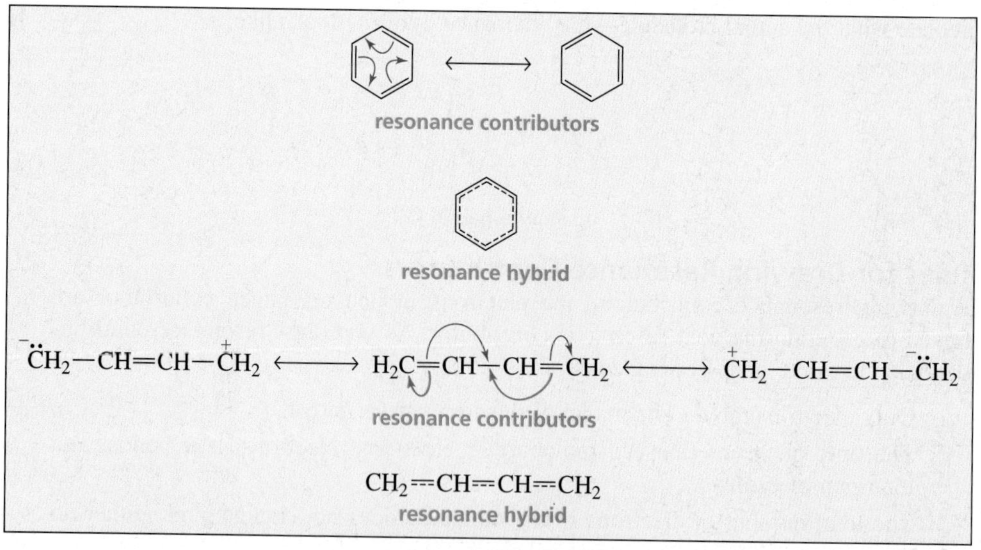

Notice that in all cases, the electrons are moved toward an sp^2 hybridized atom. Remember that an sp^2 hybridized carbon is either a double-bonded carbon (it can accommodate the new electrons by breaking a π bond) or a carbon that has a positive charge or an unpaired electron (Sections 1.8 and 1.10.) Electrons cannot be moved toward an sp^3 hybridized carbon because it cannot accommodate any more electrons.

Because electrons are neither added to nor removed from the molecule when resonance contributors are drawn, each of the resonance contributors for a particular compound must have the same net charge. If one resonance structure has a net charge of -1, all the others must also have net charges of -1; if one has a net charge of 0, all the others must also have net charges of 0. (A net charge of 0 does not necessarily mean that there is no charge on any of the atoms: A molecule with a positive charge on one atom and a negative charge on another atom has a net charge of 0.)

Radicals can also have delocalized electrons if the unpaired electron is on a carbon that is adjacent to an sp^2 hybridized atom. The arrows in Figure 7.5 are single barbed because they denote the movement of only one electron (Section 3.6).

One way to recognize compounds with delocalized electrons is to compare them with similar compounds in which all the electrons are localized. In the following example, the compound on the left has delocalized electrons because the lone-pair

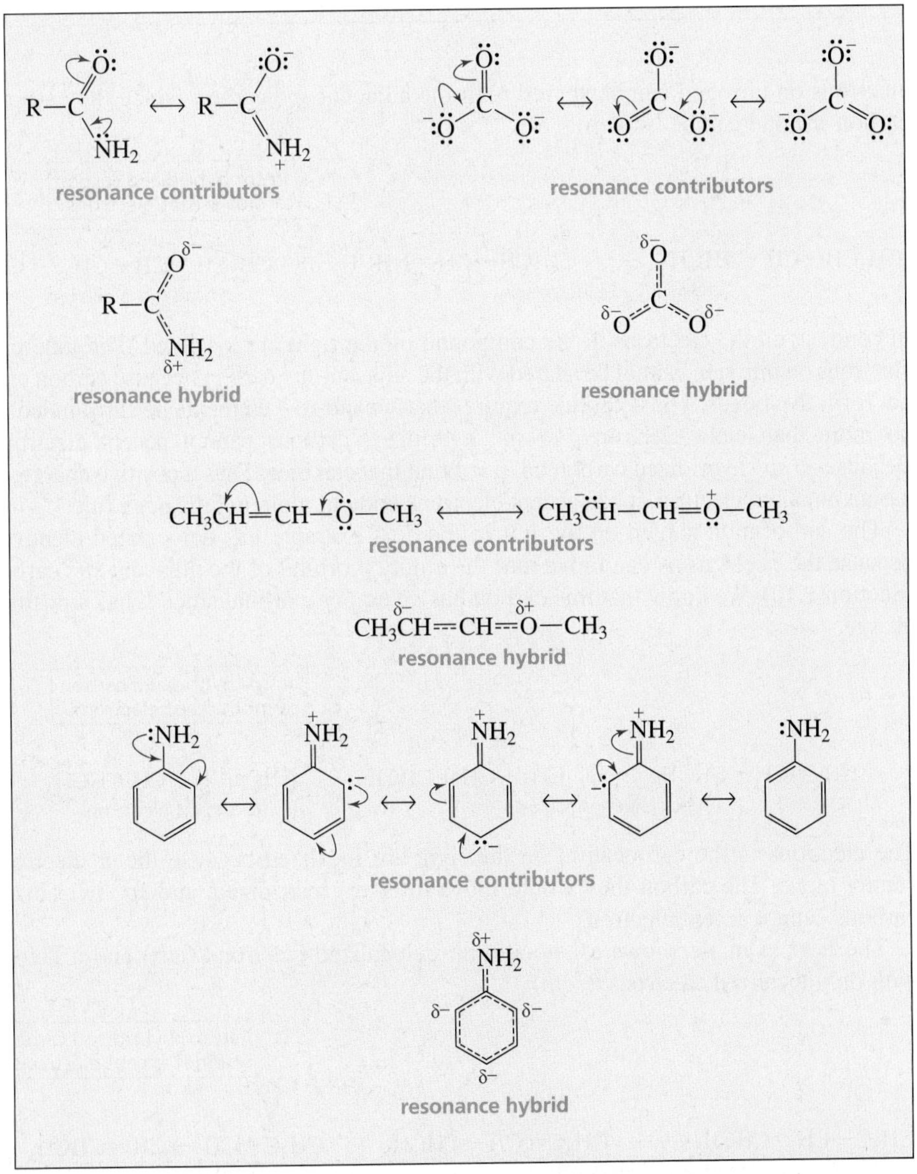

Figure 7.4
Resonance contributors are obtained by moving a lone pair toward a π bond.

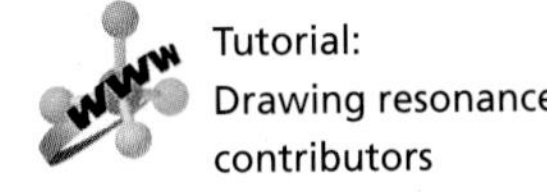
Tutorial:
Drawing resonance contributors

Figure 7.5 ▶
Resonance structures for an allylic radical and for the benzyl radical.

$CH_3-\dot{C}H=CH-\dot{C}H_2 \longleftrightarrow CH_3-\dot{C}H-CH=CH_2$
resonance contributors

$CH_3-\overset{\delta\cdot}{C}H\text{---}CH\text{---}\overset{\delta\cdot}{C}H_2$
resonance hybrid

resonance contributors

resonance hybrid

electrons on nitrogen can be shared with the adjacent sp^2 carbon (since the carbon–carbon π bond can be broken):

$CH_3CH=CH-\ddot{N}HCH_3 \longleftrightarrow CH_3\bar{\ddot{C}}H-CH=\overset{+}{N}HCH_3$
delocalized electrons

an sp^3 hybridized carbon cannot accept electrons
$CH_3CH=CH-CH_2-\ddot{N}H_2$
localized electrons

In contrast, all the electrons in the compound on the right are localized. The lone-pair electrons on nitrogen cannot be shared with the adjacent sp^3 carbon because carbon cannot form five bonds. The octet rule requires that second-row elements be surrounded by no more than eight electrons, so sp^3 hybridized carbons cannot accept electrons. Because an sp^2 hybridized carbon has a π bond that can break, has a positive charge, or has an unpaired electron, it can accept electrons without violating the octet rule.

The carbocation shown on the left in the next example has delocalized electrons because the π electrons can move into the empty *p* orbital of the adjacent sp^2 carbon (Section 1.10). We know that this carbon has an empty *p* orbital since it has a positive charge.

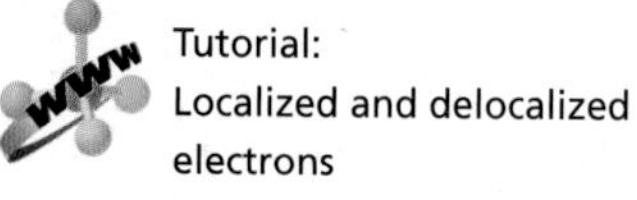
Tutorial:
Localized and delocalized electrons

$CH_2=CH-\overset{+}{C}HCH_3 \longleftrightarrow \overset{+}{C}H_2-CH=CHCH_3$
delocalized electrons

an sp^3 hybridized carbon cannot accept electrons
$CH_2=CH-CH_2\overset{+}{C}HCH_3$
localized electrons

The electrons in the carbocation on the right are localized because the π electrons cannot move. The carbon they would move to is sp^3 hybridized, and sp^3 hybridized carbons cannot accept electrons.

The next example shows a ketone with delocalized electrons (left) and a ketone with only localized electrons (right):

$CH_3\overset{\overset{\ddot{O}:}{\|}}{C}-CH=CHCH_3 \longleftrightarrow CH_3\overset{\overset{:\ddot{O}:^-}{|}}{C}=CH-\overset{+}{C}HCH_3$
delocalized electrons

an sp^3 hybridized carbon cannot accept electrons
$CH_3\overset{\overset{\ddot{O}:}{\|}}{C}-CH_2-CH=CHCH_3$
localized electrons

PROBLEM 3◆

a. Predict the relative bond lengths of the three carbon–oxygen bonds in the carbonate ion (CO_3^{2-}).
b. What would you expect the charge to be on each oxygen atom?

PROBLEM 4

a. Which of the following compounds have delocalized electrons?

1. 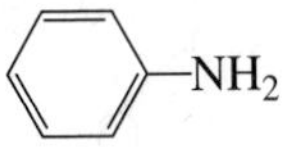 (benzene ring)–NH_2
2. (benzene ring)–CH_2NH_2
3. (pyrylium ring, $\overset{+}{O}$)
4. $CH_2{=}CHCH_2CH{=}CH_2$
5. (pyrrole ring, N–H)
6. $CH_3CH{=}CHCH{=}CH\overset{+}{C}H_2$
7. $CH_3CH_2NHCH_2CH{=}CH_2$

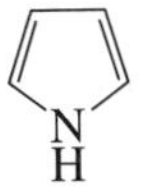

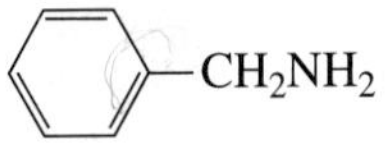

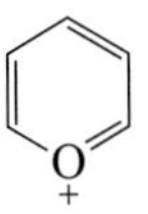

b. Draw the contributing resonance structures for these compounds.

7.5 Predicted Stabilities of Resonance Contributors

All resonance contributors do not necessarily contribute equally to the resonance hybrid. The degree to which each resonance contributor contributes depends on its predicted stability. Because resonance contributors are not real, their stabilities cannot be measured. Therefore, the stabilities of resonance contributors have to be predicted based on molecular features that are found in real molecules. *The greater the predicted stability of the resonance contributor, the more it contributes to the resonance hybrid; and the more it contributes to the resonance hybrid, the more similar the contributor is to the real molecule*. The examples that follow illustrate these points.

The greater the predicted stability of the resonance contributor, the more it contributes to the structure of the resonance hybrid.

The two resonance contributors for a carboxylic acid are labeled **A** and **B**. Structure **B** has separated charges. A molecule with **separated charges** is a molecule with a positive charge and a negative charge that can be neutralized by the movement of electrons. We can predict that resonance contributors with separated charges are relatively unstable because it takes energy to keep opposite charges separated. Since structure **A** does not have separated charges, it is predicted to have a considerably greater stability. Since structure **A** is predicted to be more stable than structure **B**, structure **A** makes a greater contribution to the resonance hybrid; that is, the resonance hybrid looks more like **A** than like **B**.

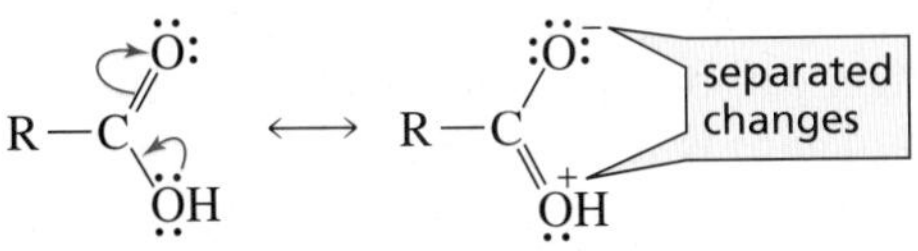

$$\mathrm{R{-}C({=}\ddot{O}{:}){-}\ddot{O}H} \longleftrightarrow \mathrm{R{-}C({-}\ddot{O}{:}^{-}){=}\overset{+}{O}H}$$

separated changes

A B

a carboxylic acid

The two resonance contributors for a carboxylate ion are shown next.

$$\mathrm{R{-}C({=}\ddot{O}{:}){-}\ddot{O}{:}^{-}} \longleftrightarrow \mathrm{R{-}C({-}\ddot{O}{:}^{-}){=}\ddot{O}{:}}$$

C D

a carboxylate ion

Structures **C** and **D** are predicted to be equally stable and therefore are expected to contribute equally to the resonance hybrid.

When electrons can be moved in more than one direction, they are always moved toward the more electronegative atom. For example, structure **G** in the next example results from moving the π electrons toward oxygen—the most electronegative atom in the molecule. Structure **E** results from moving the π electrons away from oxygen.

$$\left[CH_3\overset{\overset{+\ddot{O}:}{|}}{C}{=}CH{-}\overset{\bar{\cdot\cdot}}{C}H_2 \right] \longleftrightarrow CH_3\overset{\overset{\ddot{O}:}{\|}}{C}{-}CH{=}CH_2 \longleftrightarrow CH_3\overset{\overset{:\ddot{O}:^-}{|}}{C}{=}CH{-}\overset{+}{C}H_2$$

E F G

resonance contributor obtained by moving π electrons away from the more electronegative atom

an insignificant resonance contributor

resonance contributor obtained by moving π electrons toward the more electronegative atom

We can predict that structure **G** will make only a small contribution to the resonance hybrid because it has separated charges as well as an atom with an incomplete octet. Structure **E** also has separated charges and an atom with an incomplete octet, but its predicted stability is even less than that of structure **G** because it has a positive charge on the electronegative oxygen. Its contribution to the resonance hybrid is so insignificant that we do not need to include it as one of the resonance contributors. The resonance hybrid, therefore, looks very much like structure **F**.

The only time resonance contributors obtained by moving electrons away from the more electronegative atom should be shown is when that is the only way the electrons can be moved. In other words, movement of the electrons away from the more electronegative atom is better than no movement at all, because electron delocalization makes a molecule more stable (Section 7.6). For example, the only resonance contributor that can be drawn for the following molecule requires movement of the electrons away from oxygen:

$$CH_2{=}CH{-}\ddot{\underset{\cdot\cdot}{O}}CH_3 \longleftrightarrow \overset{\bar{\cdot\cdot}}{C}H_2{-}CH{=}\overset{+}{\underset{\cdot\cdot}{O}}CH_3$$

H I

Structure **I** is predicted to be relatively unstable because it has separated charges and its most electronegative atom is the atom with the positive charge. Therefore, the structure of the resonance hybrid is similar to structure **H**, with only a small contribution from structure **I**.

Of the two contributing resonance structures for the enolate ion, structure **J** has a negative charge on carbon and structure **K** has a negative charge on oxygen. Oxygen is more electronegative than carbon, so oxygen can accommodate the negative charge better. Consequently, structure **K** is predicted to be more stable than structure **J**. The resonance hybrid, therefore, more closely resembles structure **K**; that is, the resonance hybrid has a greater concentration of negative charge on the oxygen atom than on the carbon atom.

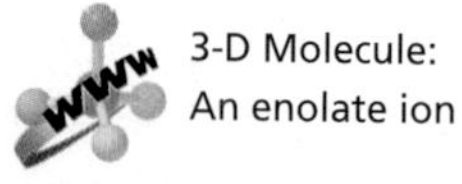

3-D Molecule: An enolate ion

$$R{-}\overset{\overset{:\ddot{O}}{\|}}{C}{-}\overset{\cdot\cdot}{C}^-HCH_3 \longleftrightarrow R{-}\overset{\overset{:\ddot{O}:^-}{|}}{C}{=}CHCH_3$$

J K

an enolate ion

We can summarize the features that decrease the predicted stability of a contributing resonance structure as follows:

1. an atom with an incomplete octet
2. a negative charge that is not on the most electronegative atom or a positive charge that is not on the least electronegative (most electropositive) atom
3. charge separation

When we compare the relative stabilities of contributing resonance structures, each of which has only one of these features, an atom with an incomplete octet (feature 1) generally makes a structure more unstable than does either feature 2 or feature 3.

PROBLEM 5 SOLVED

Draw contributing resonance structures for each of the following species, and rank the structures in order of decreasing contribution to the hybrid:

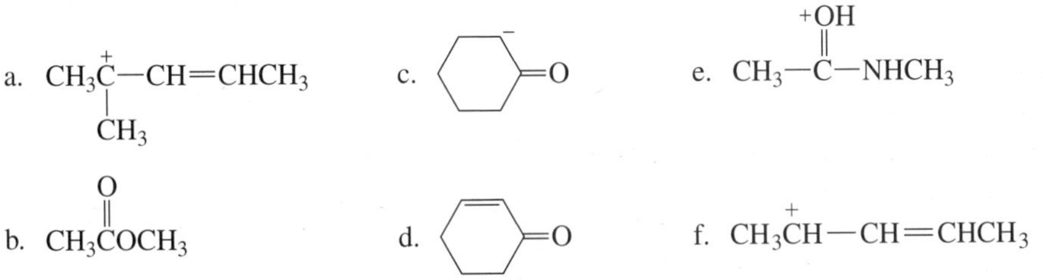

a. $CH_3\overset{+}{C}(CH_3)-CH{=}CHCH_3$

b. $CH_3C(=O)OCH_3$

c. (cyclohexane ring with a negative charge on the carbon adjacent to C=O)=O

d. (cyclohexene ring)=O

e. $CH_3-C(={}^{+}OH)-NHCH_3$

f. $CH_3\overset{+}{C}H-CH{=}CHCH_3$

SOLUTION TO 5a Structure **A** is more stable than structure **B** because the positive charge is on a tertiary carbon in **A** and on a secondary carbon in **B**.

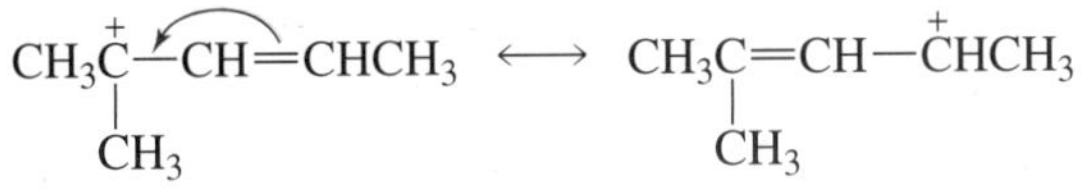

$$\underset{\textbf{A}}{CH_3\overset{+}{C}(CH_3)-CH{=}CHCH_3} \longleftrightarrow \underset{\textbf{B}}{CH_3C(CH_3){=}CH-\overset{+}{C}HCH_3}$$

PROBLEM 6

Draw the resonance hybrid for each of the species in Problem 5.

7.6 Resonance Energy

A compound with delocalized electrons is more stable than it would be if all of its electrons were localized. The extra stability a compound gains from having delocalized electrons is called **delocalization energy** or **resonance energy**. **Electron delocalization** gives a compound **resonance**, so saying that a compound is *stabilized by electron delocalization* is the same as saying that it is *stabilized by resonance*. Since the resonance energy tells us how much more stable a compound is as a result of having delocalized electrons, it is frequently called *resonance stabilization*.

The resonance energy is a measure of how much more stable a compound with delocalized electrons is than it would be if its electrons were localized.

To understand the concept of resonance energy better, let's take a look at the resonance energy of benzene. In other words, let's see how much more stable benzene (with three pairs of delocalized π electrons) is than the unknown, unreal, hypothetical compound "cyclohexatriene" (with three pairs of localized π electrons).

The $\Delta H°$ for the hydrogenation of cyclohexene, a compound with one localized double bond, has been determined experimentally to be −28.6 kcal/mol. We would then expect the $\Delta H°$ for the hydrogenation of "cyclohexatriene," a hypothetical compound with three localized double bonds, to be three times that of cyclohexene; that is, $3 \times (-28.6) = -85.8$ kcal/mol (Section 4.11).

cyclohexene $+ \ H_2 \longrightarrow$ $\Delta H° = -28.6$ kcal/mol (-120 kJ/mol) **experimental**

cyclohexene

"cyclohexatriene" $+ \ 3\,H_2 \longrightarrow$ $\Delta H° = -85.8$ kcal/mol (-359 kJ/mol) **calculated**

"cyclohexatriene"
hypothetical

When the $\Delta H°$ for the hydrogenation of benzene was determined experimenally, it was found to be -49.8 kcal/mol, much less than that calculated for hypothetical "cyclohexatriene."

benzene $+ \ 3\,H_2 \longrightarrow$ $\Delta H° = -49.8$ kcal/mol (-208 kJ/mol) **experimental**

benzene

Because the hydrogenation of "cyclohexatriene" and the hydrogenation of benzene both form cyclohexane, the difference in the $\Delta H°$ values can be accounted for only by a difference in the energies of "cyclohexatriene" and benzene. Figure 7.6 shows that benzene must be 36 kcal/mol (or 151 kJ/mole) more stable than "cyclohexatriene" because the experimental $\Delta H°$ for the hydrogenation of benzene is 36 kcal/mol less than that calculated for "cyclohexatriene."

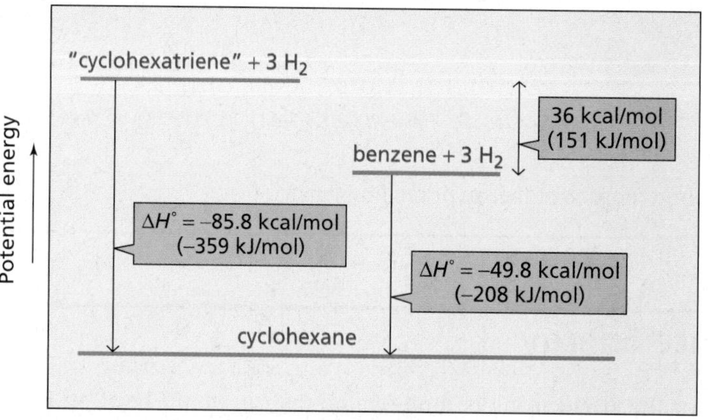

Figure 7.6 ▶
The difference in the energy levels of "cyclohexatriene" + hydrogen versus cyclohexane and the difference in the energy levels of benzene + hydrogen versus cyclohexane.

Because benzene and "cyclohexatriene" have different energies, they must be different compounds. Benzene has six delocalized π electrons, whereas hypothetical "cyclohexatriene" has six localized π electrons. The difference in their energies is the resonance energy of benzene. The resonance energy tells us *how much more stable a compound with delocalized electrons is than it would be if its electrons were localized.* Benzene, with six delocalized π electrons, is 36 kcal/mol more stable than hypothetical "cyclohexatriene," with six localized π electrons. Now we can understand why nineteenth-century chemists, who didn't know about delocalized electrons, were puzzled by benzene's unusual stability (Section 7.1).

Since the ability to delocalize electrons increases the stability of a molecule, we can conclude that *a resonance hybrid is more stable than the predicted stability of any of its resonance contributors.* The resonance energy associated with a compound that has delocalized electrons depends on the number *and* predicted stability of the resonance contributors: *The greater the number of relatively stable resonance*

A resonance hybrid is more stable than any of its resonance contributors is predicted to be.

contributors, the greater is the resonance energy. For example, the resonance energy of a carboxylate ion with two relatively stable resonance contributors is significantly greater than the resonance energy of a carboxylic acid with only one relatively stable resonance contributor.

The greater the number of relatively stable resonance contributors, the greater is the resonance energy.

R—C(=O:)—ÖH (relatively stable) ⇌ R—C(=O:)—Ö:⁻ (relatively stable) + H^+

↕ ↕

R—C(—:Ö:⁻)=ÖH⁺ (relatively unstable) R—C(—:Ö:⁻)=O: (relatively stable)

resonance contributors of a carboxylic acid

resonance contributors of a carboxylate ion

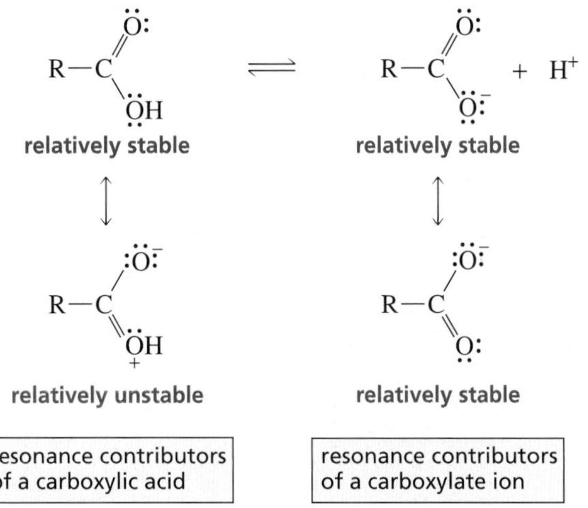

Notice that it is the number of *relatively stable* resonance contributors—not the total number of resonance contributors—that is important in determining the resonance energy. For example, the resonance energy of a carboxylate ion with two relatively stable resonance contributors is greater than the resonance energy of the compound in the following example because even though this compound has three resonance contributors, only one of them is relatively stable:

$\overset{-}{C}H_2-CH=CH-\overset{+}{C}H_2 \longleftrightarrow CH_2=CH-CH=CH_2 \longleftrightarrow \overset{+}{C}H_2-CH=CH-\overset{-}{C}H_2$

relatively unstable relatively stable relatively unstable

The more nearly equivalent the resonance contributors are in structure, the greater is the resonance energy. The carbonate dianion is particularly stable because it has three equivalent resonance contributors.

The more nearly equivalent the resonance contributors are in structure, the greater is the resonance energy.

[carbonate dianion: three equivalent resonance contributors, each with one C=O and two C—O⁻, connected by ⟷]

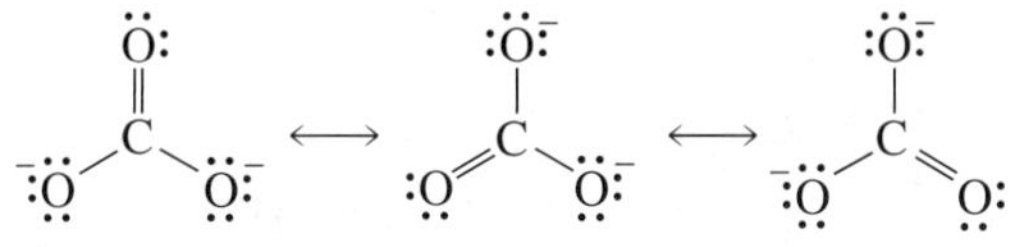

We can now summarize what we know about contributing resonance structures:

1. The greater the predicted stability of a resonance contributor, the more it contributes to the resonance hybrid.
2. The greater the number of relatively stable resonance contributors, the greater is the resonance energy.
3. The more nearly equivalent the resonance contributors, the greater is the resonance energy.

PROBLEM-SOLVING STRATEGY

Which carbocation is more stable?

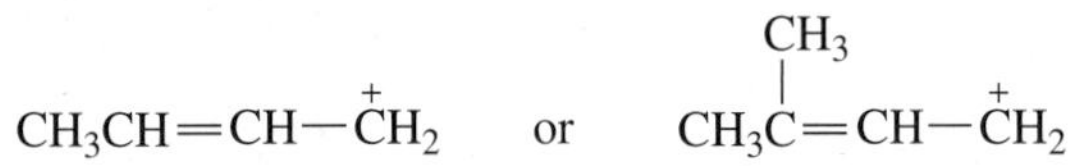

$CH_3CH=CH-\overset{+}{C}H_2$ or $CH_3C(CH_3)=CH-\overset{+}{C}H_2$

Start by drawing the resonance contributors for each carbocation.

$$CH_3CH{=}CH{-}\overset{+}{C}H_2 \longleftrightarrow CH_3\overset{+}{C}H{-}CH{=}CH_2 \qquad CH_3\overset{CH_3}{\overset{|}{C}}{=}CH{-}\overset{+}{C}H_2 \longleftrightarrow CH_3\underset{+}{\overset{CH_3}{\overset{|}{C}}}{-}CH{=}CH_2$$

Then compare the predicted stabilities of the set of resonance contributors for each carbocation.

Each carbocation has two resonance contributors. The positive charge of the carbocation on the left is shared by a primary carbon and a secondary carbon. The positive charge of the carbocation on the right is shared by a primary carbon and a tertiary carbon. Because a tertiary carbon is more stable than a secondary carbon, the carbocation on the right is more stable—it has greater resonance energy.

Now continue on to Problem 7.

PROBLEM 7◆

Which species is more stable?

a. $CH_3CH_2\overset{CH_2}{\overset{\|}{C}}\dot{C}H_2$ or $CH_3CH_2CH{=}CH\dot{C}H_2$

b. $CH_3\overset{O}{\overset{\|}{C}}CH{=}CH_2$ or $CH_3\overset{O}{\overset{\|}{C}}CH{=}CHCH_3$

c. $CH_3\overset{O^-}{\overset{|}{C}}HCH{=}CH_2$ or $CH_3\overset{O^-}{\overset{|}{C}}{=}CHCH_3$

d. $CH_3{-}\overset{^+NH_2}{\overset{\|}{C}}{-}NH_2$ or $CH_3{-}\overset{^+OH}{\overset{\|}{C}}{-}NH_2$

7.7 Stability of Allylic and Benzylic Cations

Allylic and benzylic cations have delocalized electrons, so they are more stable than similarly substituted carbocations with localized electrons. An **allylic cation** is a carbocation with the positive charge on an allylic carbon; an **allylic carbon** is a carbon adjacent to an sp^2 carbon of an alkene. A **benzylic cation** is a carbocation with the positive charge on a benzylic carbon; a **benzylic carbon** is a carbon adjacent to an sp^2 carbon of a benzene ring.

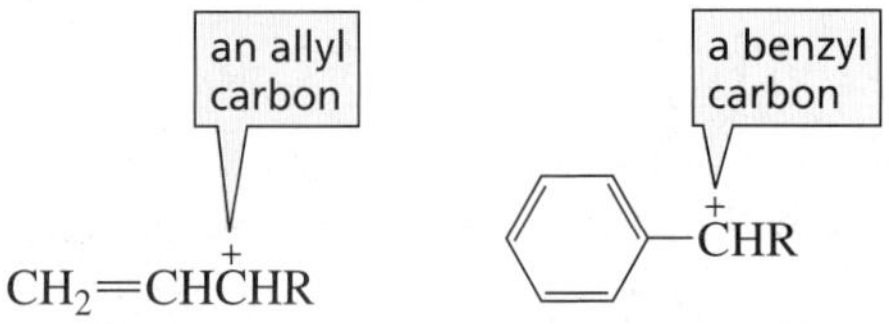

an allyl carbon

$CH_2{=}CH\overset{+}{C}HR$

an allylic cation

a benzyl carbon

$C_6H_5{-}\overset{+}{C}HR$

a benzylic cation

The *allyl cation* is an unsubstituted allylic cation, and the *benzyl cation* is an unsubstituted benzylic cation.

$CH_2{=}CH\overset{+}{C}H_2$

the allyl cation

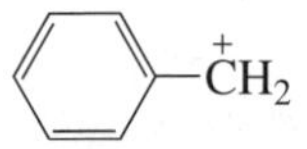

$C_6H_5{-}\overset{+}{C}H_2$

the benzyl cation

An allylic cation has two resonance contributors. The positive charge is not localized on a single carbon, but is shared by two carbons.

$$RCH{=}CH{-}\overset{+}{C}H_2 \longleftrightarrow R\overset{+}{C}H{-}CH{=}CH_2$$

an allylic cation

A benzylic cation has five resonance contributors. Notice that the positive charge is shared by four carbons.

a benzylic cation

Not all allylic and benzylic cations have the same stability. Just as a tertiary alkyl carbocation is more stable than a secondary alkyl carbocation, a tertiary allylic cation is more stable than a secondary allylic cation, which in turn is more stable than the (primary) allyl cation. Similarly, a tertiary benzylic cation is more stable than a secondary benzylic cation, which is more stable than the (primary) benzyl cation.

relative stabilities

most stable $CH_2{=}CH{-}\overset{+}{C}(R){-}R$ > $CH_2{=}CH\overset{+}{C}H{-}R$ > $CH_2{=}CH\overset{+}{C}H_2$

tertiary allylic cation **secondary allylic cation** **allyl cation**

most stable $C_6H_5{-}\overset{+}{C}(R){-}R$ > $C_6H_5{-}\overset{+}{C}H{-}R$ > $C_6H_5{-}\overset{+}{C}H_2$

tertiary benzylic cation **secondary benzylic cation** **benzyl cation**

Because the allyl and benzyl cations have delocalized electrons, they are more stable than other primary carbocations. (Indeed, they have about the same stability as secondary alkyl carbocations.) We can add the benzyl and allyl cations to the group of carbocations whose relative stabilities were shown in Sections 4.2 and 6.4.

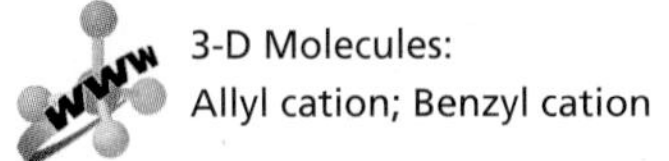

3-D Molecules:
Allyl cation; Benzyl cation

relative stabilities of carbocations

most stable R_3C^+ > $C_6H_5\overset{+}{C}H_2$ ≈ $CH_2{=}CH\overset{+}{C}H_2$ ≈ R_2HC^+ > RH_2C^+ > H_3C^+ > $CH_2{=}\overset{+}{C}H$ least stable

a tertiary carbocation · **benzyl cation** · **allyl cation** · **a secondary carbocation** · **a primary carbocation** · **methyl cation** · **vinyl cation**

Notice that it is the *primary* benzyl and the *primary* allyl cations that have about the same stability as *secondary* alkyl carbocations. Secondary benzylic and allylic cations, as well as tertiary benzylic and allylic cations, are even more stable than primary benzyl and allyl cations.

PROBLEM 8◆

Which carbocation in each of the following pairs is more stable?

a. $CH_3O\overset{+}{C}H_2$ or $CH_3NH\overset{+}{C}H_2$

b. (phenyl-substituted tertiary cation) or (2,6-di-tert-butylphenyl-substituted tertiary cation)

c. $CH_3OCH_2\overset{+}{C}H_2$ or $CH_3O\overset{+}{C}H_2$

d. (cyclohex-1-enyl)$\overset{+}{C}HCH_3$ or (cyclohex-2-enyl)$\overset{+}{C}HCH_3$

e. $CH_2{=}\overset{+}{C}(OCH_3)CH_2$ or $CH_3OCH{=}CH\overset{+}{C}H_2$

7.8 Stability of Allylic and Benzylic Radicals

An allylic radical has an unpaired electron on an allylic carbon and, like an allylic cation, has two contributing resonance structures.

$$R\dot{C}H{-}CH{=}CH_2 \longleftrightarrow RCH{=}CH{-}\dot{C}H_2$$

an allylic radical

A benzylic radical has an unpaired electron on a benzylic carbon and, like a benzylic cation, has five contributing resonance structures.

$$C_6H_5\dot{C}HR \longleftrightarrow \cdot C_6H_5{=}CHR \longleftrightarrow \cdot C_6H_5{=}CHR \longleftrightarrow \cdot C_6H_5{=}CHR \longleftrightarrow C_6H_5\dot{C}HR$$

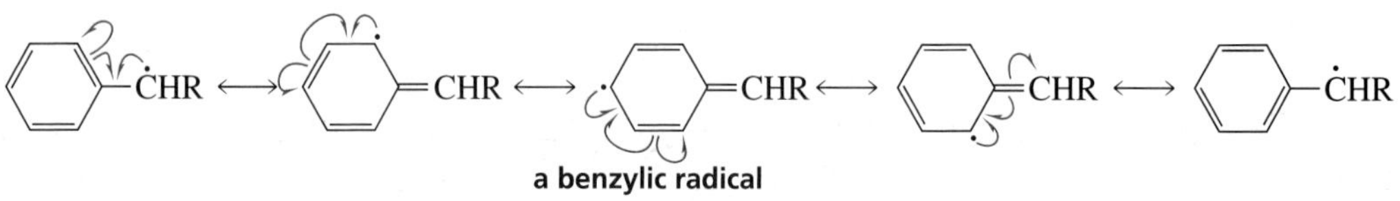

a benzylic radical

Because of their delocalized electrons, allyl and benzyl radicals are both more stable than other primary radicals. They are even more stable than tertiary radicals.

relative stabilities of radicals

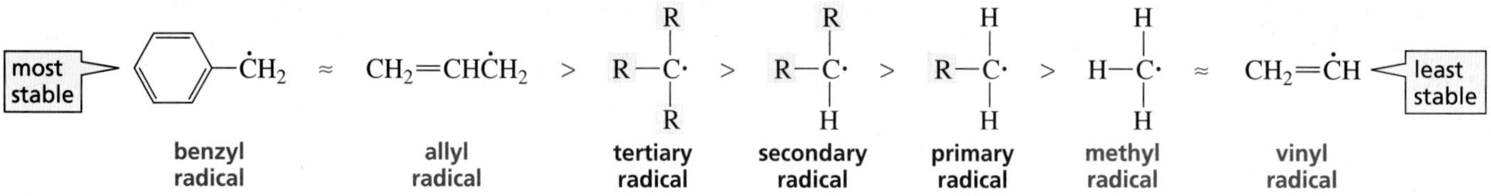

most stable $C_6H_5\dot{C}H_2 \approx CH_2{=}CH\dot{C}H_2 > R_3C\cdot > R_2HC\cdot > RH_2C\cdot > H_3C\cdot \approx CH_2{=}\dot{C}H$ least stable

benzyl radical, allyl radical, tertiary radical, secondary radical, primary radical, methyl radical, vinyl radical

7.9 Some Chemical Consequences of Electron Delocalization

Our ability to predict the correct product of an organic reaction often depends upon recognizing when organic molecules have delocalized electrons. For example, in the following reaction, both sp^2 carbons of the alkene are bonded to the same number of hydrogens:

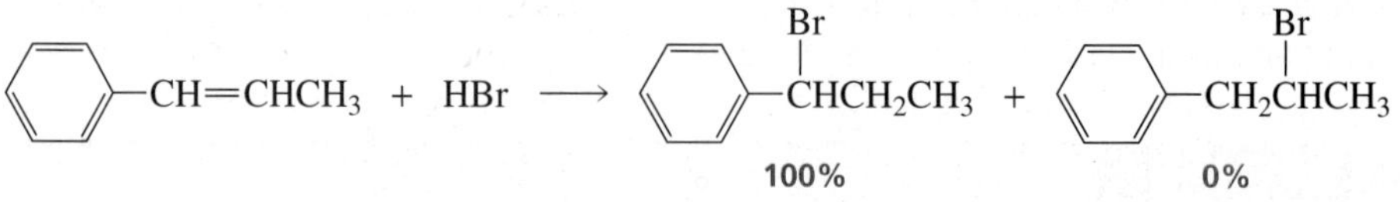

$$C_6H_5CH{=}CHCH_3 + HBr \longrightarrow C_6H_5CH(Br)CH_2CH_3 + C_6H_5CH_2CH(Br)CH_3$$

100%   0%

Therefore, the rule that tells us to add the electrophile to the sp^2 carbon bonded to the greater number of hydrogens (or Markovnikov's rule that tells us where to add the proton) predicts that approximately equal amounts of the two addition products will be formed. When the reaction is carried out, however, only one of the products is obtained.

The rules lead us to an incorrect prediction of the reaction product because they do not take electron delocalization into consideration. They presume that both carbocation intermediates are equally stable since they are both secondary carbocations. The rules do not take into account the fact that one intermediate is a secondary alkyl carbocation and the other is a secondary benzylic cation. Because the secondary benzylic cation is stabilized by electron delocalization, it is formed more readily, so only one product is obtained.

$C_6H_5-\overset{+}{C}HCH_2CH_3$ — **a secondary benzylic cation**

$C_6H_5-CH_2\overset{+}{C}HCH_3$ — **a secondary carbocation**

This example serves as a warning. Neither the rule indicating which sp^2 carbon the electrophile becomes attached to nor Markovnikov's rule can be used for reactions in which the carbocations can be stabilized by electron delocalization. In such cases, you must look at the relative stabilities of the individual carbocations to predict the product of the reaction.

Here is another example of how electron delocalization can affect the outcome of a reaction:

$$C_6H_5-CH_2CH{=}CH_2 \xrightarrow{\text{HBr}} C_6H_5-CH\overset{+}{C}HCH_3 \xrightarrow{\text{carbocation rearrangement}} C_6H_5-\overset{+}{C}HCH_2CH_3 \xrightarrow{\text{Br}^-} C_6H_5-\underset{\text{Br}}{CH}CH_2CH_3$$

a secondary carbocation **a secondary benzylic carbocation**

The addition of a proton to the alkene forms a secondary alkyl carbocation. A carbocation rearrangement occurs because a 1,2-hydride shift leads to a more stable secondary benzylic cation (Section 4.6). It is electron delocalization that causes the benzylic secondary cation to be more stable than the initially formed secondary carbocation. Had we neglected electron delocalization, we would not have anticipated the carbocation rearrangement, and we would not have correctly predicted the product of the reaction.

The relative rates at which alkenes **A**, **B**, and **C** undergo an electrophilic addition reaction with a reagent such as HBr illustrate the effect that delocalized electrons can have on the reactivity of a compound.

relative reactivities toward addition of HBr

$$\underset{\textbf{A}}{CH_2{=}C(CH_3)\ddot{O}CH_3} > \underset{\textbf{B}}{CH_2{=}C(CH_3)CH_3} > \underset{\textbf{C}}{CH_2{=}C(CH_3)CH_2\ddot{O}CH_3}$$

A is the most reactive of the three alkenes. The addition of a proton to the sp^2 carbon bonded to the greater number of hydrogens—recall that this is the rate-limiting step of an electrophilic addition reaction—forms a carbocation intermediate with a positive charge that is shared by carbon and oxygen. Being able to share the positive charge with another atom increases the stability of the carbocation—and, therefore, makes it easier to form. In contrast, the positive charge on the carbocation intermediates formed by **B** and **C** is localized on a single atom.

$$CH_2{=}C(CH_3)\ddot{O}CH_3 \longleftrightarrow {}^-CH_2-C(CH_3){=}\overset{+}{O}CH_3$$

$$\xrightarrow{\text{HBr}} CH_3-\overset{+}{C}(CH_3)\ddot{O}CH_3 \longleftrightarrow CH_3-C(CH_3){=}\overset{+}{O}CH_3 \quad + \quad Br^-$$

resonance electron donation

resonance electron donation

B reacts with HBr more rapidly than **C** does, because the carbocation formed by **C** is destabilized by the OCH_3 group that withdraws electrons inductively (through the σ bonds) from the positively charged carbon of the carbocation intermediate.

$$CH_2{=}C(CH_3)CH_3 \xrightarrow{HBr} CH_3{-}\overset{+}{C}(CH_3)CH_3 + Br^-$$

$$CH_2{=}C(CH_3)CH_2\ddot{O}CH_3 \xrightarrow{HBr} CH_3{-}\overset{+}{C}(CH_3)CH_2\ddot{O}CH_3 + Br^-$$

inductive electron withdrawal

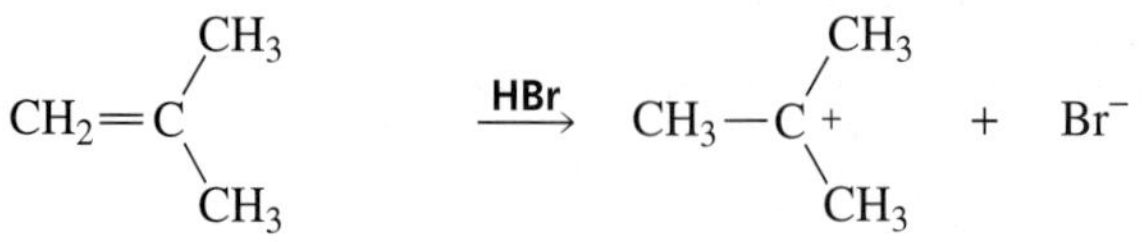

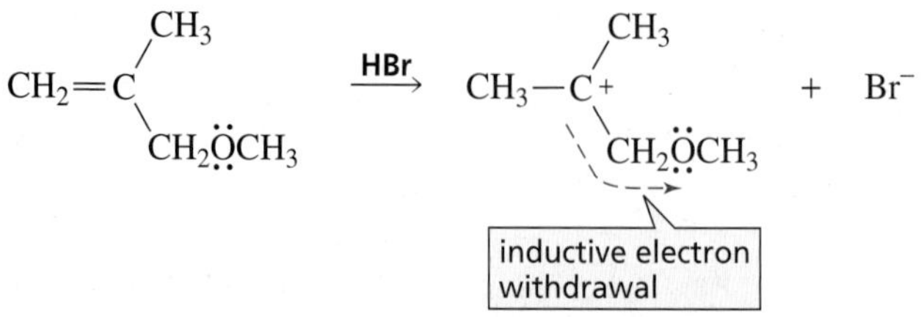

Notice that the OCH_3 group in **C** can only withdraw electrons inductively, whereas the OCH_3 group in **A** is positioned so that in addition to withdrawing electrons inductively, it can donate a lone pair to stabilize the carbocation. This is called **resonance electron donation**. Because stabilization by resonance electron donation outweighs destabilization by inductive electron withdrawal, the overall effect of the OCH_3 group in **A** is stabilization of the carbocation intermediate.

PROBLEM 9 SOLVED

Predict the sites on each of the following compounds where the reaction can occur:

a. $CH_3CH{=}CHOCH_3 + H^+$

b. 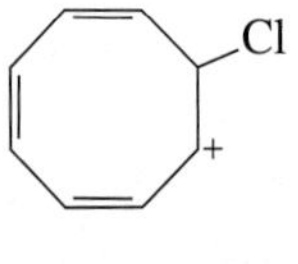 (cyclooctatrienyl cation bearing Cl) $+ HO^-$

c. 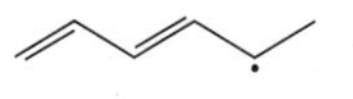 (conjugated triene radical) $+ Br^{\bullet}$

d. 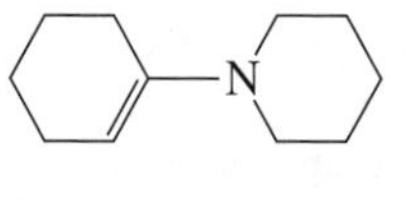 (cyclohexenyl–N-piperidine) $+ H^+$

SOLUTION TO 9a The contributing resonance structures show that there are two sites that can be protonated: the lone pair on oxygen and the lone pair on carbon.

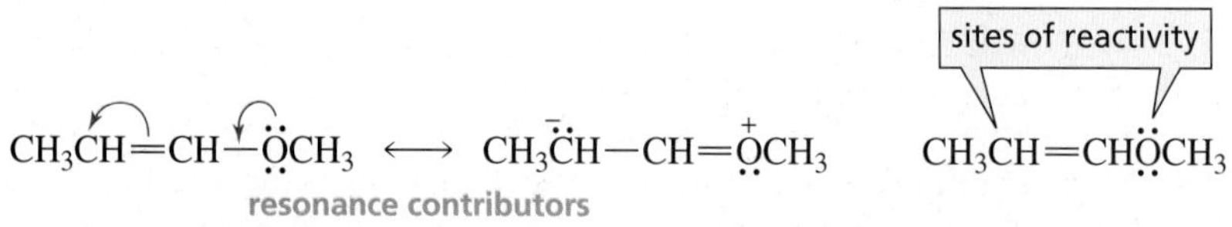

$$CH_3CH{=}CH{-}\ddot{O}CH_3 \longleftrightarrow CH_3\overset{-}{\ddot{C}}H{-}CH{=}\overset{+}{\ddot{O}}CH_3$$

resonance contributors

$CH_3CH{=}CH\ddot{O}CH_3$ — sites of reactivity

7.10 The Effect of Electron Delocalization on pK_a

We have seen that a carboxylic acid is a much stronger acid than an alcohol because the conjugate base of a carboxylic acid is considerably more stable than the conjugate base of an alcohol (Section 1.19). (Recall that the stronger the acid, the more stable is its conjugate base.) For example, the pK_a of acetic acid is 4.76, whereas the pK_a of ethanol is 15.9.

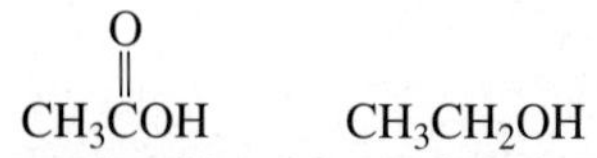

$CH_3\overset{O}{\overset{\|}{C}}OH$	CH_3CH_2OH
acetic acid	ethanol
pK_a = 4.76	pK_a = 15.9

In Section 1.19, you saw that the difference in stability of the two conjugate bases is attributable to two factors. First, the carboxylate ion has a double-bonded oxygen atom in place of two hydrogens of the alkoxide ion. Electron withdrawal by the electronegative oxygen atom stabilizes the ion by decreasing the electron density of the negatively charged oxygen.

Electron withdrawal increases the stability of an anion.

$$\underset{\text{a carboxylate ion}}{CH_3\overset{O}{\overset{\|}{C}}O^-} \qquad \underset{\text{an alkoxide ion}}{CH_3CH_2O^-}$$

The other factor responsible for the increased stability of the carboxylate ion is its *greater resonance energy* relative to that of its conjugate acid. The carboxylate ion has greater resonance energy than a carboxylic acid does, because the ion has two equivalent resonance contributors that are predicted to be relatively stable, whereas the carboxylic acid has only one (Section 7.6). Therefore, loss of a proton from a carboxylic acid is accompanied by an increase in resonance energy—in other words, an increase in stability (Figure 7.7).

$$\underset{\text{relatively stable}}{CH_3C(=O)OH} \longleftrightarrow \underset{\text{relatively unstable}}{CH_3C(O^-)=\overset{+}{O}H} \qquad \underset{\text{relatively stable}}{CH_3C(=O)O^-} \longleftrightarrow \underset{\text{relatively stable}}{CH_3C(O^-)=O}$$

resonance contributors of a carboxylic acid

resonance contributors of a carboxylate ion

In contrast, all the electrons in an alcohol—such as ethanol—and its conjugate base are localized, so loss of a proton from an alcohol is not accompanied by an increase in resonance energy.

$$\underset{\text{ethanol}}{CH_3CH_2OH} \rightleftharpoons CH_3CH_2O^- + H^+$$

Phenol, a compound in which an OH group is bonded to an sp^2 carbon of a benzene ring, is a stronger acid than an alcohol such as ethanol or cyclohexanol, compounds

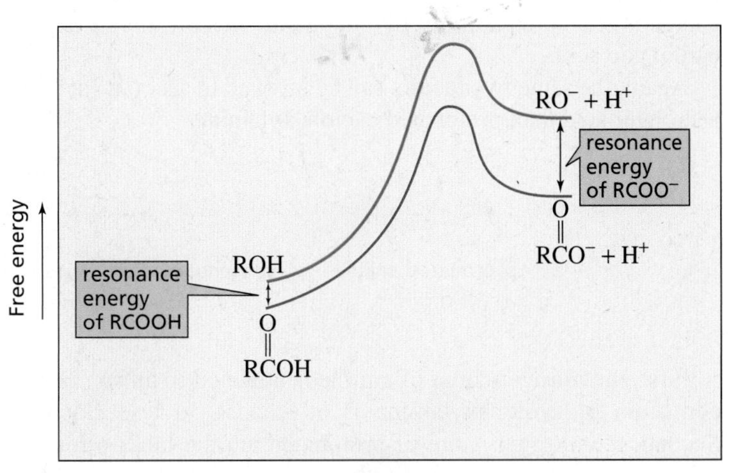

Figure 7.7
One factor that makes a carboxylic acid more acidic than an alcohol is the greater resonance energy of the carboxylate ion, compared with that of the carboxylic acid, which increases the K_a (and therefore decreases the pK_a).

in which an OH group is bonded to an sp^3 carbon. The same factors responsible for the greater acidity of a carboxylic acid compared with an alcohol cause phenol to be more acidic than an alcohol such as cyclohexanol—stablization of phenol's conjugate base by *electron withdrawal* and by *increased resonance energy.*

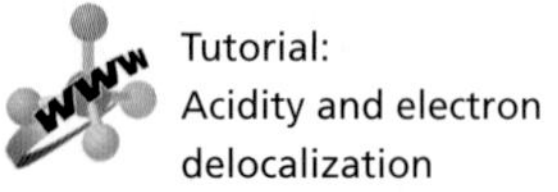

phenol $pK_a = 10$ cyclohexanol $pK_a = 16$ CH_3CH_2OH ethanol $pK_a = 16$

The OH group of phenol is attached to an sp^2 carbon that is more electonegative than the sp^3 carbon to which the OH group of cyclohexanol is attached (Section 6.9). Greater *inductive electron withdrawal* by the sp^2 carbon stabilizes the conjugate base by decreasing the electron density of its negatively charged oxygen. While both phenol and the phenolate ion have delocalized electrons, the resonance energy of the phenolate ion is greater than that of phenol because three of phenol's resonance contributors have separated charges. The loss of a proton from phenol, therefore, is accompanied by an increase in resonance energy. In contrast, neither cyclohexanol nor its conjugate base has delocalized electrons, so loss of a proton is not accompanied by an increase in resonance energy.

phenol

phenolate ion

$+ \ H^+$

Electron withdrawal from the oxygen in the phenolate ion is not as great as in the carboxylate ion. In addition, the increased resonance energy resulting from loss of a proton is not as great in a phenolate ion as in a carboxylate ion, where the negative charge is shared equally by two oxygens. Phenol, therefore, is a weaker acid than a carboxylic acid.

Again, the same two factors can be invoked to account for why protonated aniline is a stronger acid than protonated cyclohexylamine.

protonated aniline $pK_a = 4.60$ protonated cyclohexylamine $pK_a = 11.2$

First, the nitrogen atom of aniline is attached to an sp^2 carbon, whereas the nitrogen atom of cyclohexylamine is attached to a less electronegative sp^3 carbon. Second, the nitrogen atom of protonated aniline lacks a lone pair that can be delocalized. When it loses a proton, however, the lone pair that formerly held the proton can be delocalized. Loss of a proton, therefore, is accompanied by an increase in resonance energy.

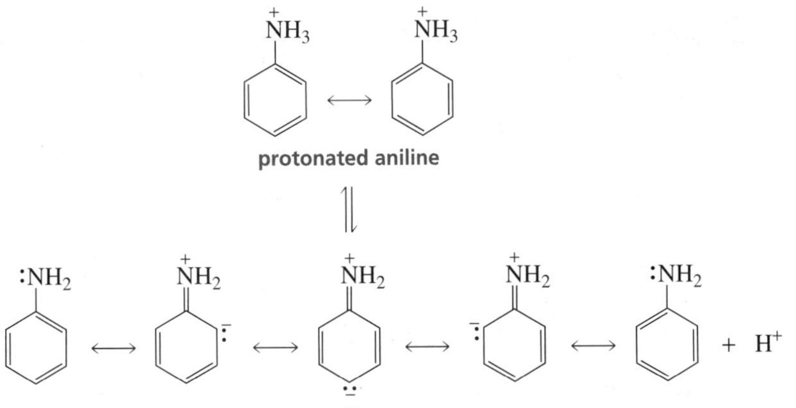

protonated aniline

aniline

An amine such as cyclohexylamine has no delocalized electrons either in the protonated form or in the unprotonated form, so proton loss is not associated with a change in the amine's resonance energy.

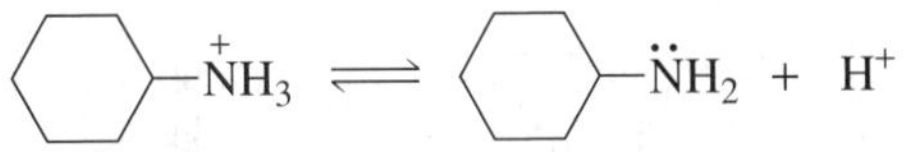

We can now add phenol and protonated aniline to the classes of organic compounds whose approximate pK_a values you should know (Table 7.1). They are also listed inside the back cover for easy reference.

Table 7.1 Approximate pK_a Values

p$K_a < 0$	p$K_a \approx 5$	p$K_a \approx 10$	p$K_a \approx 15$
$R\overset{+}{O}H$ (with H)	RCOH (C=O)	$R\overset{+}{N}H_3$	ROH
$^+$OH ‖ RCOH	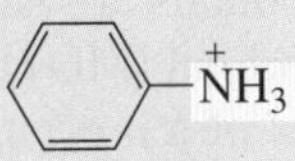 $C_6H_5\overset{+}{N}H_3$	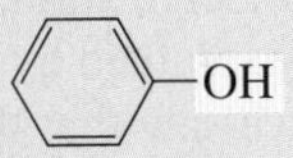 C_6H_5OH	H_2O
H_3O^+			

PROBLEM 10 SOLVED

Which of the following would you predict to be the stronger acid?

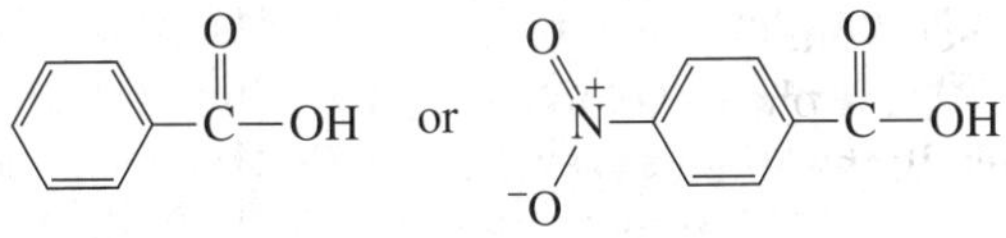

SOLUTION The nitro-substituted compound is the stronger acid because the nitro substituent can withdraw electrons inductively (through the σ bonds) and it can withdraw electrons by resonance (through the π bonds). We have seen that electron-withdrawing substituents increase the acidity of a compound by stabilizing its conjugate base.

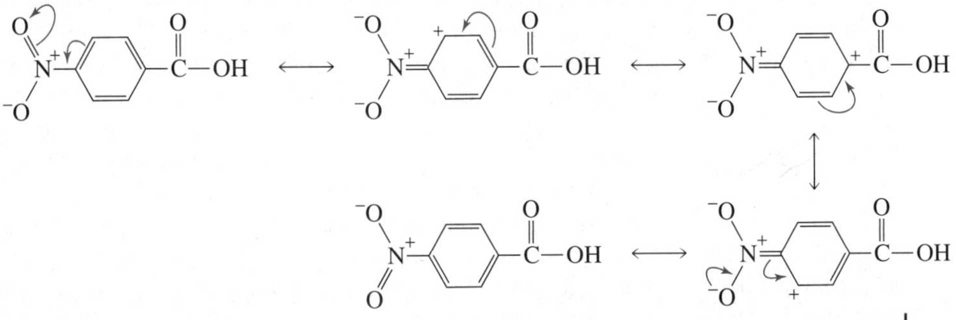

PROBLEM 11◆

Which is a stronger acid?

a. $CH_3CH_2CH_2OH$ or $CH_3CH{=}CHOH$

b. $HC(=O)CH_2OH$ or $CH_3C(=O)OH$

c. $CH_3CH{=}CHCH_2OH$ or $CH_3CH{=}CHOH$

d. $CH_3CH_2CH_2\overset{+}{N}H_3$ or $CH_3CH{=}CH\overset{+}{N}H_3$

PROBLEM 12◆

Which is a stronger base?

a. ethylamine or aniline

b. ethylamine or ethoxide ion ($CH_3CH_2O^-$)

c. phenolate ion or ethoxide ion

PROBLEM 13◆

Rank the following compounds in order of decreasing acid strength:

C_6H_5–OH C_6H_5–CH_2OH C_6H_5–COOH

7.11 A Molecular Orbital Description of Stability

We have used contributing resonance structures to show why compounds are stabilized by electron delocalization. Why compounds are stabilized by electron delocalization can also be explained by molecular orbital (MO) theory.

Take a few minutes to review Section 1.6.

In Section 1.5, we saw that the two lobes of a *p* orbital have opposite phases. We also saw that when two in-phase *p* orbitals overlap, a covalent bond is formed, and when two out-of-phase *p* orbitals overlap, they cancel each other and produce a node between the two nuclei (Section 1.6). A *node* is a region where there is zero probability of finding an electron.

Let's review how the π molecular orbitals of ethene are constructed. An MO description of ethene is shown in Figure 7.8. The two *p* orbitals can be either in-phase or out-of-phase. (The different phases are indicated by different colors.) Notice that the number of orbitals is conserved—the number of molecular orbitals equals the number of atomic orbitals that produced them. Thus, the two atomic *p* orbitals of ethene overlap to produce two molecular orbitals. Side-to-side overlap of in-phase *p* orbitals (lobes of the same color) produces a **bonding molecular orbital** designated ψ_1 (the Greek letter psi). The bonding molecular orbital is lower in energy than the *p* atomic orbitals, and it encompasses both carbons. In other words, each electron in the bonding molecular orbital spreads over both carbon atoms.

Side-to-side overlap of out-of-phase *p* orbitals produces an **antibonding molecular orbital**, ψ_2, which is higher in energy than the *p* atomic orbitals. The antibonding molecular orbital has a node between the lobes of opposite phases. The bonding MO arises from *constructive overlap* of the atomic orbitals, whereas the antibonding MO arises from *destructive overlap*. In other words, the overlap of in-phase orbitals holds atoms together—it is a bonding interaction—whereas the overlap of out-of-phase orbitals pulls atoms apart—it is an antibonding interaction.

The π electrons are placed in molecular orbitals according to the same rules that govern the placement of electrons in atomic orbitals (Section 1.2): the aufbau principle (orbitals are filled in order of increasing energy), the Pauli exclusion principle (each

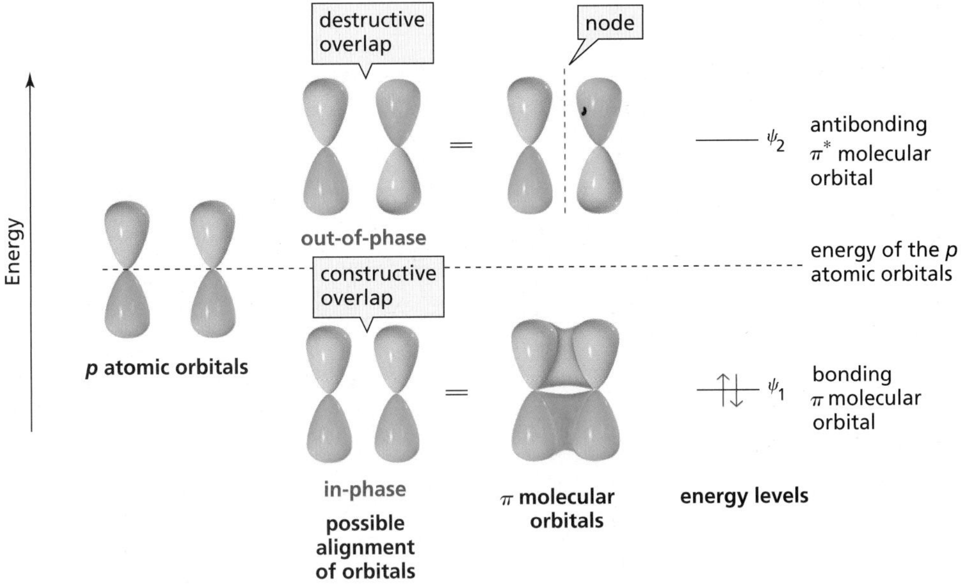

▲ **Figure 7.8**
The distribution of electrons in ethene. Overlapping of in-phase *p* orbitals produces a bonding molecular orbital that is lower in energy than the *p* atomic orbitals. Overlapping of out-of-phase *p* orbitals produces an antibonding molecular orbital that is higher in energy than the *p* atomic orbitals.

orbital can hold two electrons of opposite spin), and Hund's rule (an electron will occupy an empty degenerate orbital before it will pair up with an electron that is already present in an orbital).

1,3-Butadiene and 1,4-Pentadiene

The π electrons in 1,3-butadiene are delocalized over four sp^2 carbons. In other words, there are four carbons in the π system. A molecular orbital description of 1,3-butadiene is shown in Figure 7.9.

$$\overset{-}{C}H_2-CH=CH-\overset{+}{C}H_2 \longleftrightarrow CH_2=CH-CH=CH_2 \longleftrightarrow \overset{+}{C}H_2-CH=CH-\overset{-}{C}H_2$$

resonance contributors

$$CH_2 \overset{\text{---}}{=} CH \overset{\text{---}}{=} CH \overset{\text{---}}{=} CH_2$$

resonance hybrid

Each of the four carbons contributes one *p* atomic orbital, and the four *p* atomic orbitals combine to produce four π molecular orbitals: ψ_1, ψ_2, ψ_3, and ψ_4. Thus, a molecular orbital results from the **linear combination of atomic orbitals (LCAO)**. Half of the MOs are bonding (π) MOs (ψ_1 and ψ_2), and the other half are antibonding (π^*) MOs (ψ_3 and ψ_4), and they are given the designations ψ_1, ψ_2, ψ_3, and ψ_4, in order of increasing energy. The energies of the bonding and antibonding MOs are symmetrically distributed above and below the energy of the *p* atomic orbitals.

Notice that as the MOs increase in energy, the number of nodes increases and the number of bonding interactions decreases. The lowest-energy MO (ψ_1) has only the node that bisects the *p* orbitals—it has no nodes between the nuclei because all the blue lobes overlap on one face of the molecule and all the green lobes overlap on the other face; ψ_1 has three bonding interactions; ψ_2 has one node between the nuclei and two bonding interactions (for a net of one bonding interaction); ψ_3 has two nodes between the nuclei and one bonding interaction (for a net of one antibonding interaction); and ψ_4 has three nodes between the nuclei—three antibonding interactions—and no bonding interactions. The four π electrons of 1,3-butadiene reside in ψ_1 and ψ_2.

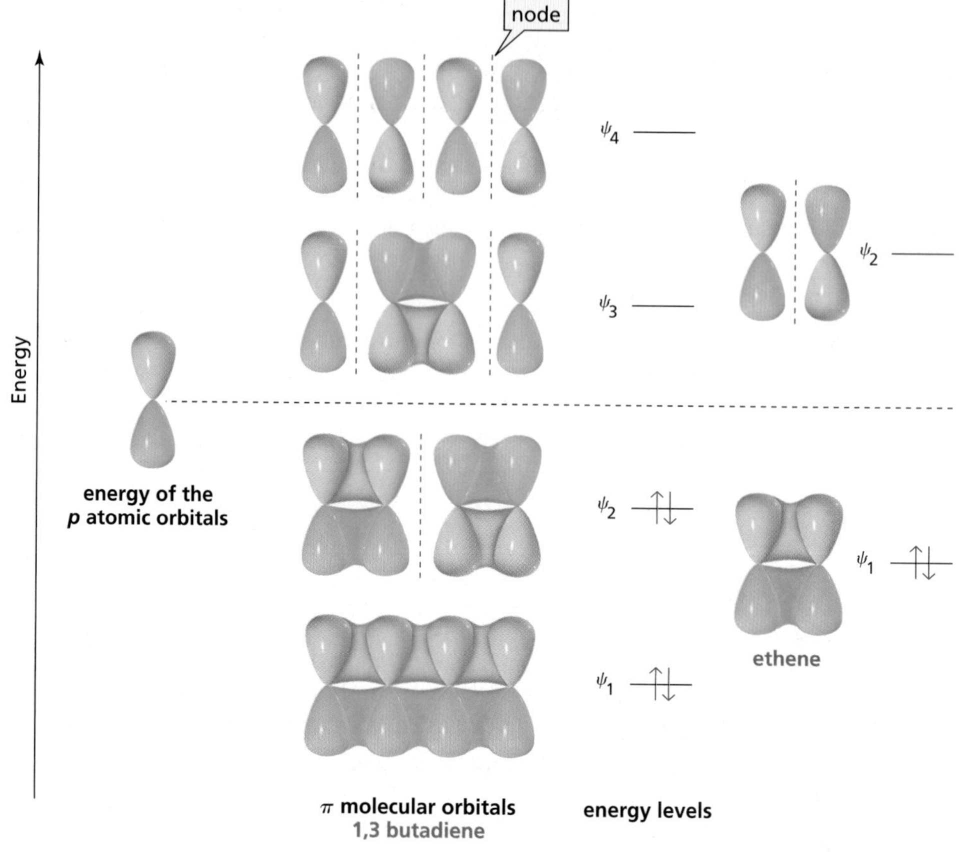

Figure 7.9 ▶ Four *p* atomic orbitals overlap to produce four molecular orbitals in 1,3-butadiene, and two *p* atomic orbitals overlap to produce two molecular orbitals in ethene. In both compounds, the bonding MOs are filled and the antibonding MOs are empty.

1,3-Butadiene's lowest-energy MO (ψ_1) is particularly stable because it has three bonding interactions and its two electrons are delocalized over all four nuclei—they encompass all the carbons in the π system. The MO next in energy (ψ_2) is also a bonding MO because it has one more bonding interaction than antibonding interaction; it is not as strongly bonding or as low in energy as ψ_1. These two bonding MOs show that the greatest π electron density in a compound with two double bonds joined by one single bond is between C-1 and C-2 and between C-3 and C-4, but there is some π electron density between C-2 and C-3—just as the resonance contributors show. They also show why 1,3-butadiene is most stable in a planar confomation: If 1,3-butadiene weren't planar, there would be little or no overlap between C-2 and C-3. Overall, ψ_3 is an antibonding MO: It has one more antibonding interaction than bonding interaction, but it is not as strongly antibonding as ψ_4, which has no bonding interactions and three antibonding interactions.

Both ψ_1 and ψ_3 are **symmetric molecular orbitals**; they have a plane of symmetry, so one half is the mirror image of the other half. Both ψ_2 and ψ_4 are *fully asymmetric*; they do not have a plane of symmetry, but would have one if one half of the MO were turned upside down. Notice that as the MOs increase in energy, they alternate from being symmetric to asymmetric.

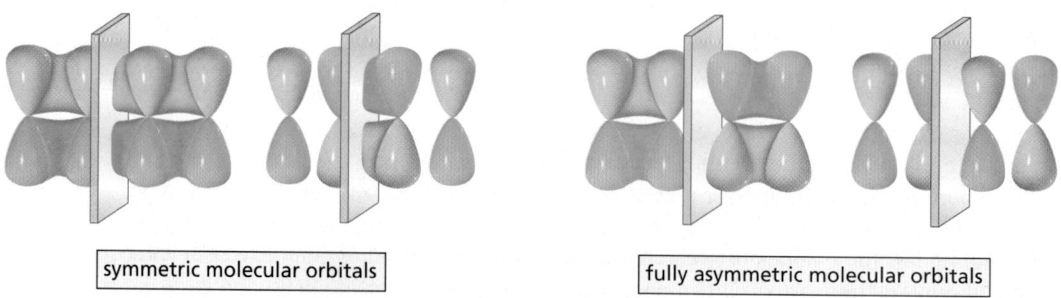

The energies of the MOs of 1,3-butadiene and ethene are compared in Figure 7.9. Notice that the average energy of the electrons in 1,3-butadiene is lower than the electrons in ethene. This lower energy is the resonance energy. In other words, 1,3-butadiene is stabilized by electron delocalization (resonance).

PROBLEM 14◆

What is the total number of nodes in the ψ_3 and ψ_4 molecular orbitals of 1,3-butadiene?

The highest-energy molecular orbital of 1,3-butadiene that contains electrons is ψ_2. Therefore, ψ_2 is called the **highest occupied molecular orbital (HOMO)**. The lowest-energy molecular orbital of 1,3-butadiene that does not contain electrons is ψ_3; ψ_3 is called the **lowest unoccupied molecular orbital (LUMO)**.

HOMO = the highest occupied molecular orbital.

LUMO = the lowest unoccupied molecular orbital.

The molecular orbital description of 1,3-butadiene shown in Figure 7.9 represents the electronic configuration of the molecule in its ground state. If the molecule absorbs light of an appropriate wavelength, the light will promote an electron from its HOMO to its LUMO (from ψ_2 to ψ_3). The molecule then is in an excited state (Section 1.2). The excitation of an electron from the HOMO to the LUMO is the basis of ultraviolet and visible spectroscopy (Section 8.9).

PROBLEM 15◆

Answer the following questions for the π molecular orbitals of 1,3-butadiene:

a. Which are the bonding MOs and which are the antibonding MOs?
b. Which MOs are symmetric and which are asymmetric?
c. Which MO is the HOMO and which is the LUMO in the ground state?
d. Which MO is the HOMO and which is the LUMO in the excited state?
e. What is the relationship between the HOMO and the LUMO and symmetric and asymmetric orbitals?

Now let's look at the π molecular orbitals of 1,4-pentadiene.

$CH_2{=}CHCH_2CH{=}CH_2$
1,4-pentadiene

$-CH_2-$

1,4-Pentadiene, like 1,3-butadiene, has four π electrons. However, unlike the delocalized π electrons in 1,3-butadiene, the π electrons in 1,4-pentadiene are completely separate from one another. In other words, the electrons are localized. The molecular orbitals of 1,4-pentadiene have the same energy as those of ethene—a compound with one pair of localized π electrons. Thus, molecular orbital theory and contributing resonance structures are two different ways to show that the π electrons in 1,3-butadiene are delocalized and that electron delocalization stabilizes a molecule.

The Allyl Cation, the Allyl Radical, and the Allyl Anion

Let's now look at the molecular orbitals of the allyl cation, the allyl radical, and the allyl anion.

$CH_2{=}CH{-}\overset{+}{C}H_2$ the allyl cation $CH_2{=}CH{-}\dot{C}H_2$ the allyl radical $CH_2{=}CH{-}\overset{\ddot{-}}{C}H_2$ the allyl anion

The three p atomic orbitals of the allyl group combine to produce three π molecular orbitals: ψ_1, ψ_2, and ψ_3 (Figure 7.10). The bonding MO (ψ_1) encompasses all the

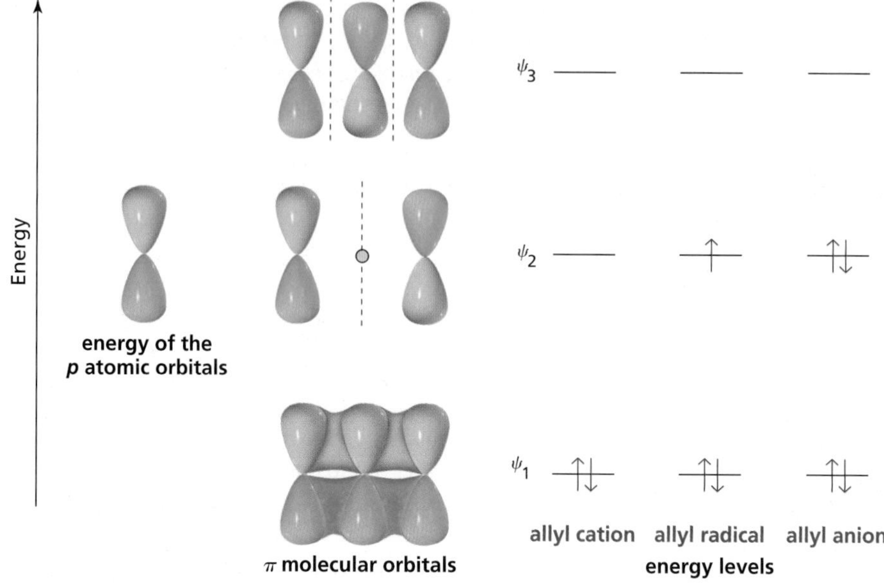

Figure 7.10 ▶
The distribution of electrons in the molecular orbitals of the allyl cation, the allyl radical, and the allyl anion. Three *p* atomic orbitals overlap to produce three π molecular orbitals.

carbons in the π system. In an acyclic π system, the number of bonding MOs always equals the number of antibonding MOs. Therefore, when there is an odd number of MOs, one of them must be a **nonbonding molecular orbital**. In an allyl system, ψ_2 is a nonbonding MO. We have seen that as the energy of the MO increases, the number of nodes increases. Consequently, the ψ_2 MO must have a node—in addition to the one that ψ_1 has that bisects the *p* orbitals. The only symmetrical position for a node in ψ_2 is for it to pass through the middle carbon. (You also know that it needs to pass through the middle carbon because that is the only way ψ_2 can be fully asymmetric, which it must be, since ψ_1 and ψ_3 are symmetric.)

You can see from Figure 7.10 why ψ_2 is called a nonbonding molecular orbital: There is no overlap between the *p* orbital on the middle carbon and the *p* orbital on either of the end carbons. Notice that a nonbonding MO has the same energy as the isolated *p* atomic orbitals. The third molecular orbital (ψ_3) is an antibonding MO.

The two π electrons of the allyl cation are in the bonding MO, which means that they are spread over all three carbons. Consequently, the two carbon–carbon bonds in the allyl cation are identical, with each having some double-bond character. The positive charge is shared equally by the end carbon atoms, which is another way of showing that the stability of the allyl cation is due to electron delocalization.

$$CH_2{=}CH{-}\overset{+}{C}H_2 \longleftrightarrow \overset{+}{C}H_2{-}CH{=}CH_2 \qquad \overset{\delta+}{C}H_2{\overset{\cdots}{-}}CH{\overset{\cdots}{-}}\overset{\delta+}{C}H_2$$

resonance contributors of the allyl cation — resonance hybrid

The allyl radical has two electrons in the bonding π molecular orbital, so these electrons are spread over all three carbon atoms. The third electron is in the nonbonding MO. The molecular orbital diagram shows that the third electron is shared equally by the end carbons, with none of the electron density on the middle carbon. This is in agreement with what the resonance contributors show: Only the end carbons have radical character.

$$CH_2{=}CH{-}\dot{C}H_2 \longleftrightarrow \dot{C}H_2{-}CH{=}CH_2 \qquad \overset{\delta\cdot}{C}H_2{\overset{\cdots}{-}}CH{\overset{\cdots}{-}}\overset{\delta\cdot}{C}H_2$$

resonance contributors of the allyl radical — resonance hybrid

Finally, the allyl anion has two electrons in the nonbonding MO. These two electrons are shared equally by the end carbon atoms. This again agrees with what the resonance contributors show.

$CH_2{=}CH{-}\bar{C}H_2 \longleftrightarrow \bar{C}H_2{-}CH{=}CH_2$

resonance contributors of the allyl anion

$\overset{\delta-}{C}H_2 \overset{---}{=} CH \overset{---}{=} \overset{\delta-}{C}H_2$

resonance hybrid

1,3,5-Hexatriene and Benzene

1,3,5-Hexatriene, with six carbon atoms, has six *p* atomic orbitals.

$CH_2{=}CH{-}CH{=}CH{-}CH{=}CH_2$

1,3,5-hexatriene

$CH_2 \overset{---}{=} CH \overset{---}{=} CH \overset{---}{=} CH \overset{---}{=} CH \overset{---}{=} CH_2$

resonance hybrid of 1,3,5-hexatriene

The six *p* atomic orbitals combine to produce six π molecular orbitals: ψ_1, ψ_2, ψ_3, ψ_4, ψ_5, and ψ_6 (Figure 7.11). Half of the MOs (ψ_1, ψ_2, and ψ_3) are bonding MOs, and the other half (ψ_4, ψ_5, and ψ_6) are antibonding. 1,3,5-Hexatriene's six π electrons occupy the three bonding MOs (ψ_1, ψ_2, and ψ_3), and two of the electrons (those in ψ_1) are delocalized over all six carbons. Thus, molecular orbital theory and resonance contributors are two different ways of showing that the π electrons in 1,3,5-hexatriene are delocalized. Notice in the figure that as the MOs increase in energy, the number of nodes increases, the number of bonding interactions decreases, and the MOs alternate from being symmetric to being asymmetric.

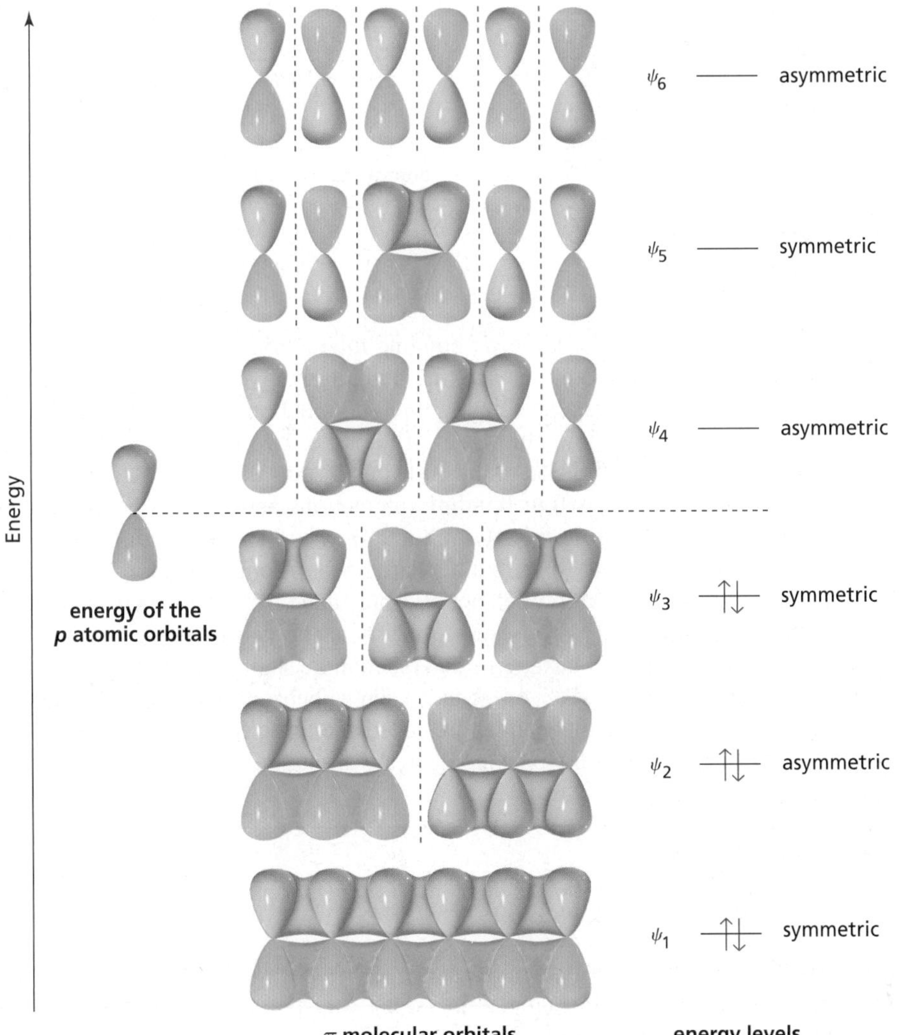

◀ **Figure 7.11**
Six *p* atomic orbitals overlap to produce the six π molecular orbitals of 1,3,5-hexatriene. The six electrons occupy the three bonding molecular orbitals ψ_1, ψ_2, and ψ_3.

PROBLEM 16◆

Answer the following questions for the π molecular orbitals of 1,3,5-hexatriene:

a. Which are the bonding MOs and which are the antibonding MOs?
b. Which MOs are symmetric and which are asymmetric?
c. Which MO is the HOMO and which is the LUMO in the ground state?
d. Which MO is the HOMO and which is the LUMO in the excited state?
e. What is the relationship between the HOMO and the LUMO and symmetric and asymmetric orbitals.

Like 1,3,5-hexatriene, benzene has a six-carbon π system. The six-carbon π system in benzene, however, is cyclic. The six *p* atomic orbitals combine to produce six π molecular orbitals (Figure 7.12). Three of the MOs are bonding (ψ_1, ψ_2, and ψ_3) and three are antibonding (ψ_4, ψ_5, and ψ_6). Benzene's six π electrons occupy the three lowest-energy MOs (the bonding MOs). The two electrons in ψ_1 are delocalized over the six carbon atoms. The method used to determine the relative energies of the MOs of compounds with cyclic π systems is described in Section 15.6.

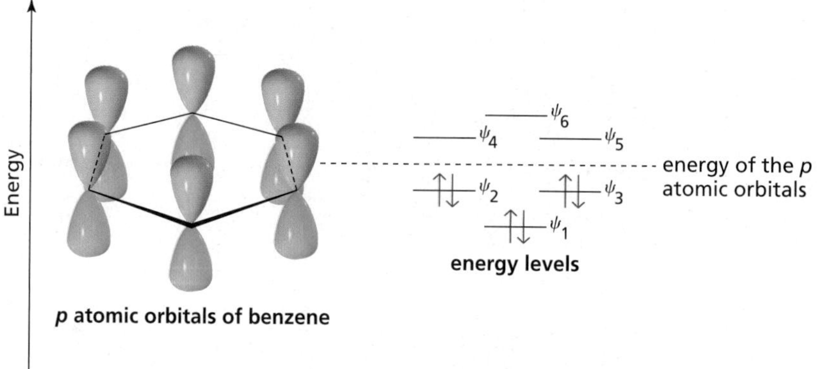

Figure 7.12 ▶ Benzene has six π molecular orbitals, three bonding (ψ_1, ψ_2, ψ_3) and three antibonding (ψ_4, ψ_5, ψ_6). The six π electrons occupy the three bonding molecular orbitals.

Figure 7.13 shows that there are *six* bonding interactions in the lowest-energy MO (ψ_1) of benzene—one more than in the lowest-energy MO of 1,3,5-hexatriene (Figure 7.12). In other words, putting the three double bonds into a ring is accompanied by an increase in stabilization. Each of the other two bonding MOs of benzene (ψ_2 and ψ_3) has a node in addition to the node that bisects the *p* orbitals. These two orbitals are degenerate: ψ_2 has four bonding interactions and two antibonding interactions, for a net of two bonding interactions, and ψ_3 also has two bonding interactions. Thus, ψ_2 and ψ_3 are bonding MOs, but they are not as strongly bonding as ψ_1.

The energy levels of the MOs of ethene, 1,3-butadiene, 1,3,5-hexatriene, and benzene are compared in Figure 7.14. You can see that benzene is a particularly stable molecule—more stable than 1,3,5-hexatriene and much more stable than a molecule with one or more isolated double bonds. Compounds such as benzene that are unusually stable because of large delocalization energies are called **aromatic compounds**. The structural features that cause a compound to be aromatic are discussed in Section 15.1.

PROBLEM 17◆

How many bonding interactions are there in the ψ_1 and ψ_2 molecular orbitals of the following compounds?

a. 1,3-butadiene
b. 1,3,5,7-octatetraene

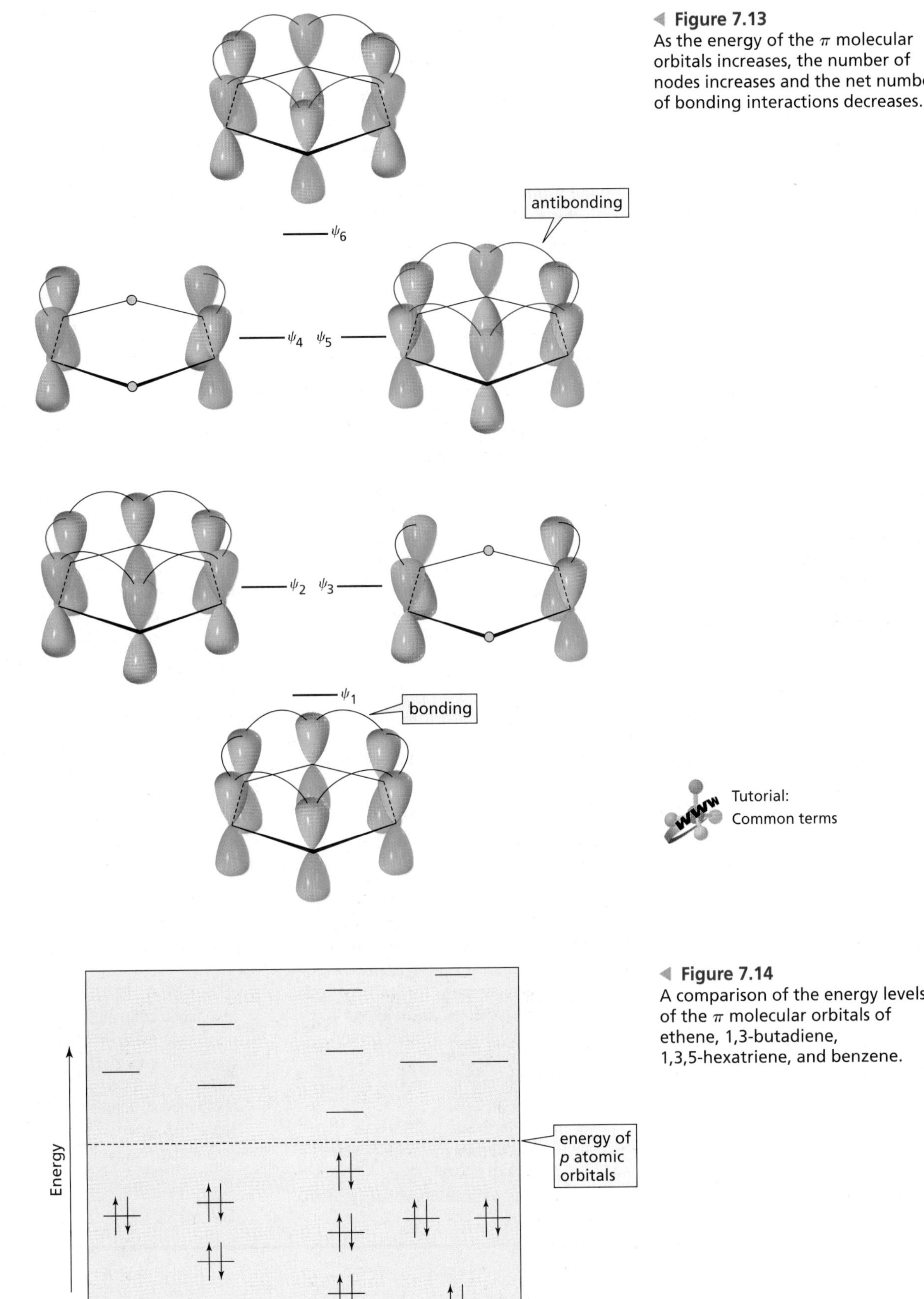

◀ **Figure 7.13**
As the energy of the π molecular orbitals increases, the number of nodes increases and the net number of bonding interactions decreases.

Tutorial:
Common terms

◀ **Figure 7.14**
A comparison of the energy levels of the π molecular orbitals of ethene, 1,3-butadiene, 1,3,5-hexatriene, and benzene.

Summary

Localized electrons belong to a single atom or are confined to a bond between two atoms. **Delocalized electrons** are shared by more than two atoms; they result when a *p* orbital overlaps the *p* orbitals of more than one adjacent atom. Electron delocalization occurs only if all the atoms sharing the delocalized electrons lie in or close to the same plane.

Benzene is a planar molecule. Each of its six carbon atoms is sp^2 hybridized, with bond angles of 120°. A *p* orbital of each carbon overlaps the *p* orbitals of both adjacent carbons. The six π electrons are shared by all six carbons. Compounds such as benzene that are unusually stable because of large **delocalization energies** are called **aromatic compounds**.

Chemists use **resonance contributors**—structures with localized electrons—to approximate the actual structure of a compound that has delocalized electrons: **the resonance hybrid**. To draw resonance contributors, move only π electrons, lone pairs, or unpaired electrons toward an sp^2 hybridized atom. The total number of electrons and the numbers of paired and unpaired electrons do not change.

The greater the **predicted stability** of the resonance contributor, the more it contributes to the hybrid and the more similar it is to the real molecule. The predicted stability is decreased by (1) an atom with an incomplete octet, (2) a negative (positive) charge not on the most electronegative (electropositive) atom, or (3) charge separation. A resonance hybrid is more stable than the predicted stability of any of its resonance contributors.

The extra stability a compound gains from having delocalized electrons is called **resonance energy**. It tells us how much more stable a compound with delocalized electrons is than it would be if its electrons were localized. The greater the number of relatively stable resonance contributors and the more nearly equivalent they are, the greater is the resonance energy of the compound. Allylic and benzylic cations (and radicals) have delocalized electrons, so they are more stable than similarly substituted carbocations (and radicals) with localized electrons. Donation of a lone pair is called **resonance electron donation**.

Electron delocalization can affect the nature of the product formed in a reaction and the pK_a of a compound. A carboxylic acid and a phenol are more acidic than an alcohol such as ethanol, and a protonated aniline is more acidic than a protonated amine because electron withdrawal stabilizes their conjugate bases and the loss of a proton is accompanied by an increase in resonance energy.

A **molecular orbital** results from the **linear combination of atomic orbitals**. The number of orbitals is conserved: The number of molecular orbitals equals the number of atomic orbitals that produced them. Side-to-side overlap of in-phase *p* orbitals produces a **bonding molecular orbital**, which is more stable than the atomic orbitals. Side-to-side overlap of out-of-phase *p* orbitals produces an **antibonding molecular orbital**, which is less stable than the atomic orbitals. The **highest occupied molecular orbital (HOMO)** is the highest-energy MO that contains electrons. The **lowest unoccupied molecular orbital (LUMO)** is the lowest-energy MO that does not contain electrons.

As the MOs increase in energy, the number of **nodes** increases and the number of bonding interactions decreases, and they alternate from being **symmetric** to **asymmetric**. When there is an odd number of molecular orbitals, one must be a **nonbonding molecular orbital**. **Molecular orbital theory** and contributing resonance structures both show that electrons are delocalized and that electron delocalization makes a molecule more stable.

Key Terms

allylic carbon (p. 278)
allylic cation (p. 278)
antibonding molecular orbital (p. 286)
aromatic compounds (p. 292)
asymmetric molecular orbital (p. 288)
benzylic carbon (p. 278)
benzylic cation (p. 278)
bonding molecular orbital (p. 286)
contributing resonance structure (p. 267)
delocalization energy (p. 275)
delocalized electrons (p. 263)
electron delocalization (p. 275)
highest occupied molecular orbital (HOMO) (p. 289)
linear combination of atomic orbitals (LCAO) (p. 287)
localized electrons (p. 263)
lowest unoccupied molecular orbital (LUMO) (p. 289)
nonbonding molecular orbital (p. 290)
resonance (p. 275)
resonance contributor (p. 267)
resonance electron donation (p. 282)
resonance energy (p. 275)
resonance hybrid (p. 268)
resonance structure (p. 267)
separated charges (p. 273)
symmetric molecular orbital (p. 288)

Problems

18. Which of the following compounds have delocalized electrons?

a. $CH_2{=}CHC(=O)CH_3$

b. 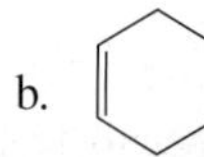

c. 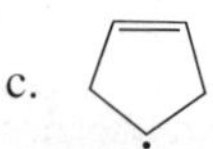

d.

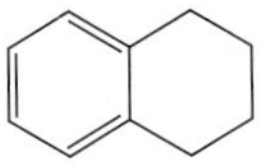

e. $CH_2{=}CHCH_2CH{=}CH_2$

f. 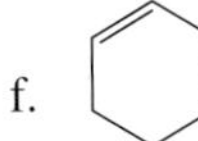

g.

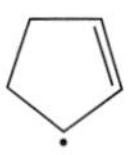

h. $CH_3CH_2NHCH_2CH{=}CHCH_3$

i. $CH_3CH_2NHCH{=}CHCH_3$

j.

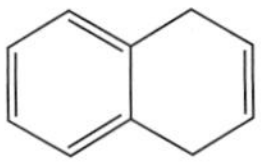

k. $CH_3\overset{+}{C}(CH_3)CH_2CH{=}CH_2$

l. $CH_3CH_2\overset{+}{C}HCH{=}CH_2$

m. $CH_3CH{=}CHOCH_2CH_3$

19. a. Draw resonance contributors for the following species, showing all the lone pairs:
    1. CH_2N_2 2. N_2O 3. NO_2^-

    b. For each species, indicate the most stable resonance contributor.

20. Draw resonance contributors for the following ions:

a. 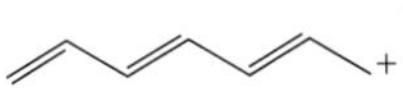 b. 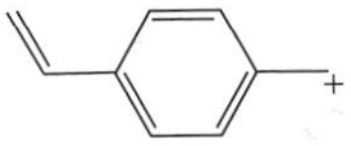 c. 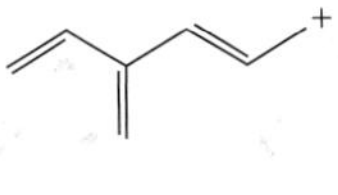 d.

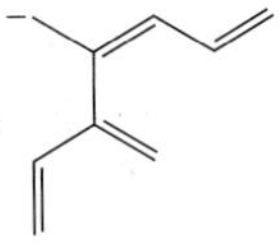

21. Are the following pairs of structures resonance contributors or different compounds?

a. 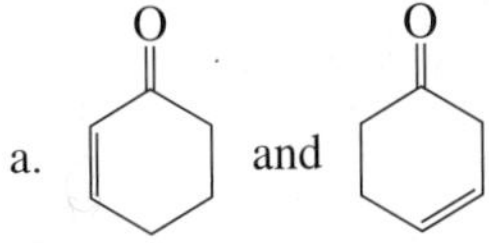

b. $CH_3CH{=}CH\dot{C}HCH{=}CH_2$ and $CH_3\dot{C}HCH{=}CHCH{=}CH_2$

c. $CH_3\overset{O}{\overset{\|}{C}}CH_2CH_3$ and $CH_3\overset{OH}{\overset{|}{C}}{=}CHCH_3$

d. 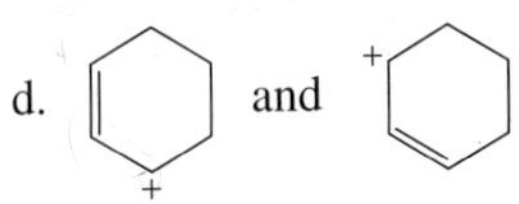

e. $CH_3\overset{+}{C}HCH{=}CHCH_3$ and $CH_3CH{=}CHCH_2\overset{+}{C}H_2$

22. a. Draw resonance contributors for the following species. Do not include structures that are so unstable that their contributions to the resonance hybrid would be negligible. Indicate which species are major contributors and which are minor contributors to the resonance hybrid.

1. $CH_3CH{=}CHOCH_3$
2. 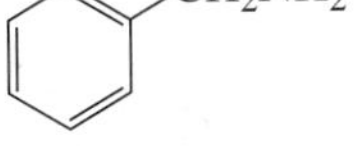 (benzene ring with CH_2NH_2)
3. $CH_3\bar{C}HC{\equiv}N$
4. 
5. 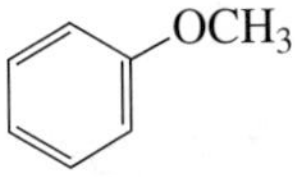 (benzene ring with OCH_3)
6. 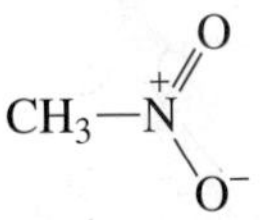 $CH_3{-}\overset{+}{N}(=O)O^-$
7. $CH_3CH_2\overset{O}{\overset{\|}{C}}OCH_2CH_3$
8. $CH_3CH{=}CHCH{=}CH\dot{C}H_2$
9. $H\overset{O}{\overset{\|}{C}}NHCH_3$
10. $CH_3CH{=}CH\overset{+}{C}H_2$
11. CO_3^{2-}
12. $H\overset{O}{\overset{\|}{C}}CH{=}CH\bar{C}H_2$
13. 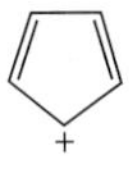
14. 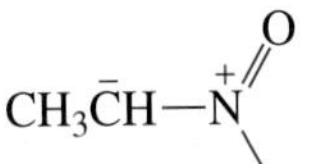 $CH_3\bar{C}H{-}\overset{+}{N}(=O)O^-$
15. 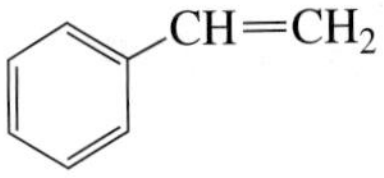 (benzene ring with $CH{=}CH_2$)
16. 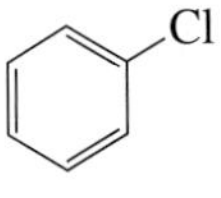 (benzene ring with Cl)
17. $CH_3\overset{O}{\overset{\|}{C}}\bar{C}H\overset{O}{\overset{\|}{C}}CH_3$
18. $\bar{C}H_2\overset{O}{\overset{\|}{C}}OCH_2CH_3$

b. Do any of the species have resonance contributors that all contribute equally to the resonance hybrid?

23. Which resonance contributor makes the greater contribution to the resonance hybrid?

a. $CH_3\overset{+}{C}HCH{=}CH_2$ or $CH_3CH{=}CH\overset{+}{C}H_2$

b.  [methylcyclopentenyl cation structures] or

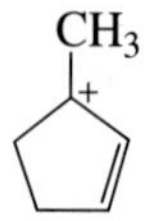

c. 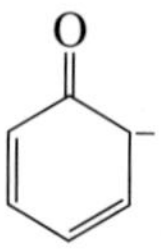 [cyclohexadienone anion] or [phenoxide]

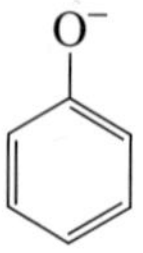

24. a. Which oxygen atom has the greater electron density?

$CH_3\overset{O}{\overset{\|}{C}}OCH_3$

b. Which compound has the greater electron density on its nitrogen atom?

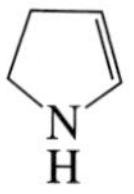

[N–H pyrroline structures] or

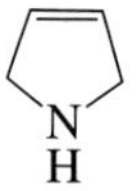

c. Which compound has the greater electron density on its oxygen atom?

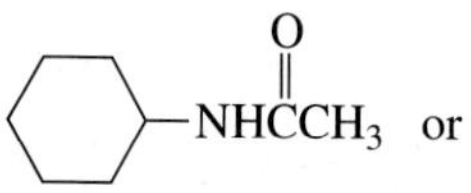

$C_6H_{11}-NH\overset{O}{\overset{\|}{C}}CH_3$ or $C_6H_5-NH\overset{O}{\overset{\|}{C}}CH_3$

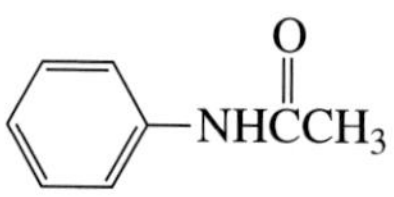

25. Which can lose a proton more readily, a methyl group bonded to cyclohexane or a methyl group bonded to benzene?

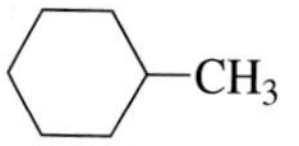

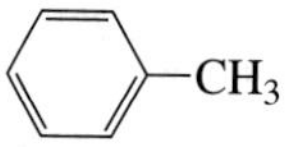

$C_6H_{11}-CH_3$ $C_6H_5-CH_3$

26. The triphenylmethyl cation is so stable that a salt such as triphenylmethyl chloride can be isolated and stored. Why is this carbocation so stable?

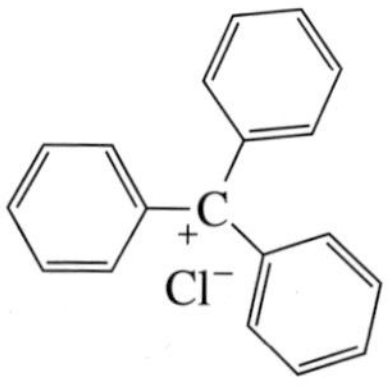

$(C_6H_5)_3\overset{+}{C}\ Cl^-$

triphenylmethyl chloride

27. Draw the contributing resonance structures for the following anion, and rank them in order of decreasing stability:

$CH_3CH_2\ddot{O}-\overset{\ddot{O}:}{\overset{\|}{C}}-\bar{C}H-C{\equiv}N:$

28. Rank the following compounds in order of decreasing acidity:

[three skeletal structures]

29. Which species is more stable?

a. $CH_3CH_2O^-$ or $CH_3\overset{O}{\overset{\|}{C}}O^-$

b. $CH_3\overset{O}{\overset{\|}{C}}\bar{C}HCH_2\overset{O}{\overset{\|}{C}}H$ or $CH_3\overset{O}{\overset{\|}{C}}\bar{C}H\overset{O}{\overset{\|}{C}}CH_3$

c. $CH_3\bar{C}HCH_2\overset{O}{\overset{\|}{C}}CH_3$ or $CH_3CH_2\bar{C}H\overset{O}{\overset{\|}{C}}CH_3$

d. $CH_3\overset{NH_2}{\overset{|}{C}}HCH_3$ or $CH_3\overset{NH}{\overset{\|}{C}}NH_2$

e. $CH_3\bar{C}(CH_3)-\overset{O}{\overset{\|}{C}}H$ or $CH_3\bar{C}(CH_3)-\overset{CH_2}{\overset{\|}{C}}H$

f. 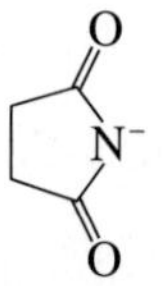 [succinimide anion] or [pyrrolidinone anion]

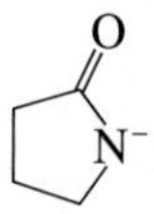

30. Which species in each of the pairs in Problem 29 is the stronger base?

31. We saw in Chapter 6 that ethyne reacts with an equivalent amount of HCl to form vinyl chloride. In the presence of excess HCl, the final reaction product is 1,1-dichloroethane. Why is 1,1-dichloroethane formed rather than 1,2-dichloroethane?

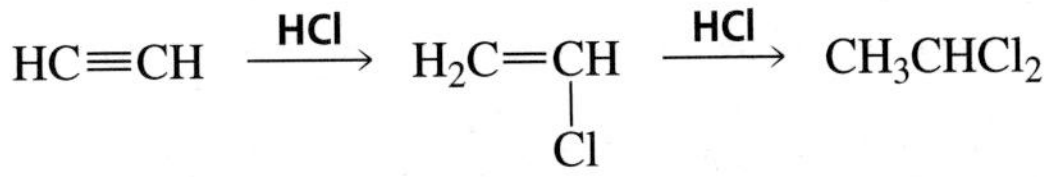

$$HC{\equiv}CH \xrightarrow{HCl} H_2C{=}CHCl \xrightarrow{HCl} CH_3CHCl_2$$

32. Why is the resonance energy of pyrrole (21 kcal/mol) greater than the resonance energy of furan (16 kcal/mol)?

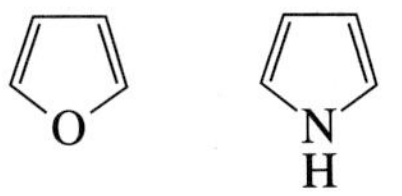

furan pyrrole

33. Rank the following compounds in order of decreasing acidity of the indicated hydrogen:

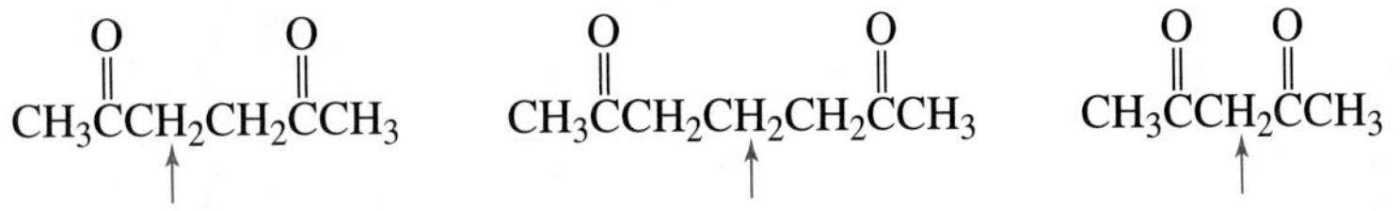

$CH_3COCH_2CH_2COCH_3$ $\qquad CH_3COCH_2CH_2CH_2COCH_3$ $\qquad CH_3COCH_2COCH_3$

34. Explain why the electrophile adds to the sp^2 carbon bonded to the greater number of hydrogens in reaction **a**, but not in reaction **b**.

a. $CH_2{=}CHF + HF \longrightarrow CH_3CHF_2$ b. $CH_2{=}CHCF_3 + HF \longrightarrow FCH_2CH_2CF_3$

35. The acid dissociation constant (K_a) for loss of a proton from cyclohexanol is 1×10^{-16}.
   a. Draw an energy diagram for loss of a proton from cyclohexanol.

$$C_6H_{11}OH \underset{}{\overset{K_a = 1 \times 10^{-16}}{\rightleftharpoons}} C_6H_{11}O^- + H^+$$

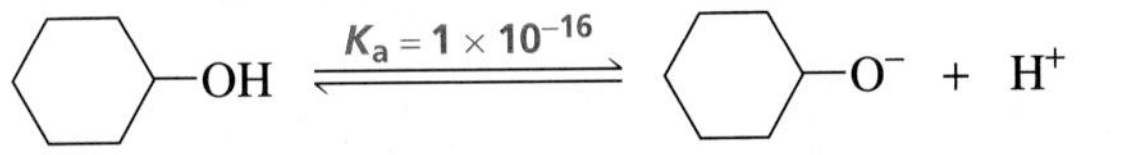

   b. Draw the contributing resonance structures for phenol.
   c. Draw the contributing resonance structures for the phenolate ion.
   d. Draw an energy diagram for loss of a proton from phenol on the same plot with the energy diagram for loss of a proton from cyclohexanol.

$$C_6H_5OH \rightleftharpoons C_6H_5O^- + H^+$$

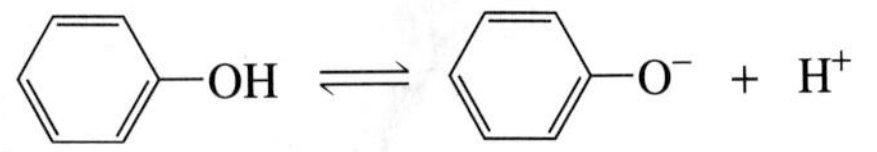

   e. Which has a greater K_a, cyclohexanol or phenol?
   f. Which is a stronger acid, cyclohexanol or phenol?

36. Protonated cyclohexylamine has a $K_a = 1 \times 10^{-11}$. Using the same sequence of steps as in Problem 35, determine which is a stronger base, cyclohexylamine or aniline.

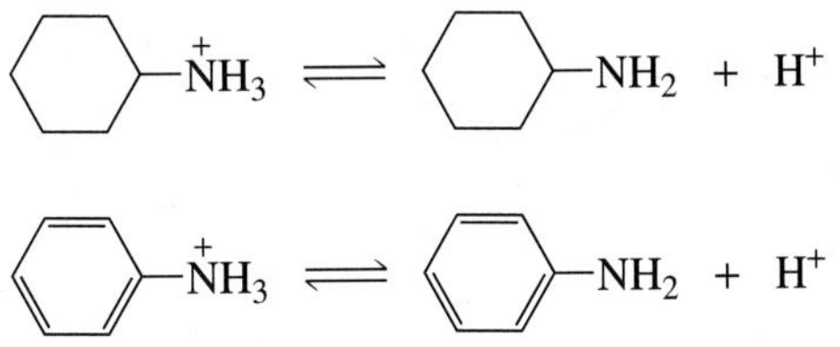

$$C_6H_{11}\overset{+}{N}H_3 \rightleftharpoons C_6H_{11}NH_2 + H^+$$

$$C_6H_5\overset{+}{N}H_3 \rightleftharpoons C_6H_5NH_2 + H^+$$

37. Answer the following questions for the π molecular orbitals of 1,3,5,7-octatetraene:
   a. How many π MOs does the compound have?
   b. Which are the bonding MOs and which are the antibonding MOs?
   c. Which MOs are symmetric and which are asymmetric?
   d. Which MO is the HOMO and which is the LUMO in the ground state?
   e. Which MO is the HOMO and which is the LUMO in the excited state?
   f. What is the relationship between HOMO and LUMO and symmetric and asymmetric orbitals?
   g. How many nodes does the highest-energy π molecular orbital of 1,3,5,7-octatetraene have between the nuclei?

8 Reactions of Dienes • Ultraviolet and Visible Spectroscopy

retinol
vitamin A

In this chapter, we will look at the reactions of compounds that have two double bonds. Hydrocarbons with two double bonds are called **dienes**, and those with three double bonds are called **trienes**. **Tetraenes** have four double bonds, and **polyenes** have many double bonds. Although we will be concerned mainly with the reactions of dienes, the same considerations apply to hydrocarbons that contain more than two double bonds.

α-cadinene
oil of citronella
a diene

β-selinene
oil of celery
a diene

zingiberene
oil of ginger
a triene

β-carotene
a polyene

Double bonds can be *conjugated*, *isolated*, or *cumulated*. **Conjugated double bonds** are separated by one single bond. **Isolated double bonds** are separated by more than one single bond. In other words, the double bonds are isolated from each other. **Cumulated double bonds** are adjacent to each other. Compounds with cumulated double bonds are called **allenes**.

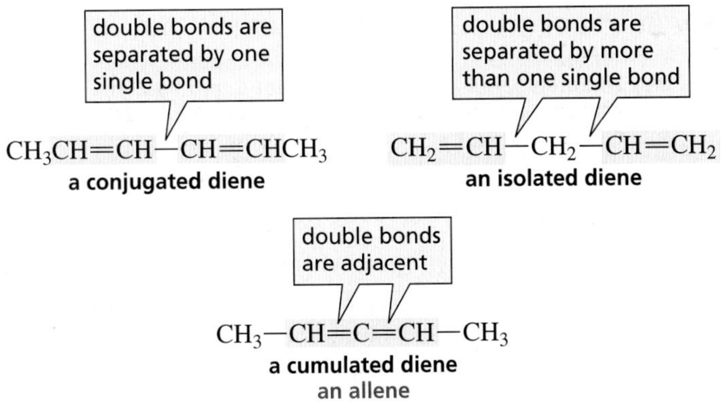

double bonds are separated by one single bond

$CH_3CH{=}CH{-}CH{=}CHCH_3$
a conjugated diene

double bonds are separated by more than one single bond

$CH_2{=}CH{-}CH_2{-}CH{=}CH_2$
an isolated diene

double bonds are adjacent

$CH_3{-}CH{=}C{=}CH{-}CH_3$
a cumulated diene
an allene

8.1 Nomenclature of Alkenes with More than One Functional Group

To arrive at the systematic name of a diene, we first identify the longest continuous chain that contains both double bonds by its alkane name and then replace the "ne" ending with "diene." The chain is numbered in the direction that gives the double bonds the lowest possible numbers. The numbers indicating the locations of the double bonds are cited either before the name of the parent compound or before the suffix. Substituents are cited in alphabetical order. Propadiene, the smallest member of the class of compounds known as allenes, is frequently called allene.

$CH_2{=}C{=}CH_2$
systematic: propadiene
common: allene

$\overset{1}{C}H_2{=}\overset{2}{C}(CH_3){-}\overset{3}{C}H{=}\overset{4}{C}H_2$
2-methyl-1,3-butadiene
or
2-methylbuta-1,3-diene
isoprene

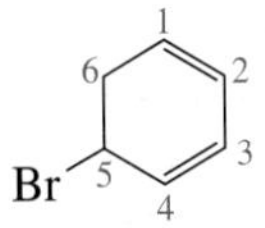

5-bromo-1,3-cyclohexadiene
or
5-bromocyclohexa-1,3-diene

$\overset{6}{C}H_3\overset{5}{C}H{=}\overset{4}{C}H\overset{3}{C}H_2\overset{2}{C}(CH_3){=}\overset{1}{C}H_2$
2-methyl-1,4-hexadiene
or
2-methylhexa-1,4-diene

$\overset{1}{C}H_3\overset{2}{C}(CH_3){=}\overset{3}{C}H\overset{4}{C}H{=}\overset{5}{C}(CH_2CH_3)\overset{6}{C}H_2\overset{7}{C}H_3$
5-ethyl-2-methyl-2,4-heptadiene
or
5-ethyl-2-methylhepta-2,4-diene

To name an alkene in which the second functional group is not another double bond—and is named with a functional group suffix—choose the longest continuous chain containing both functional groups, and cite both designations at the end of the name. The "ene" ending is cited first, with the terminal "e" omitted in order to avoid two adjacent vowels. The location of the first cited functional group is usually given before the name of the parent chain. The location of the second functional group is cited immediately before its suffix.

If the functional groups are a *double bond* and a *triple bond*, the chain is numbered in the direction that yields the lowest number in the name of the compound. Thus, the lower number is given to the alkene suffix in the compound on the left and to the alkyne suffix in the compound on the right.

When the functional groups are a double bond and a triple bond, the chain is numbered to get the lowest possible number in the name of the compound, regardless of which functional group gets the lower number.

$\overset{7}{C}H_3\overset{6}{C}H{=}\overset{5}{C}H\overset{4}{C}H_2\overset{3}{C}H_2\overset{2}{C}{\equiv}\overset{1}{C}H$
5-hepten-1-yne
***not* 2-hepten-6-yne**
because 1 < 2

$\overset{1}{C}H_2{=}\overset{2}{C}H\overset{3}{C}H_2\overset{4}{C}H_2\overset{5}{C}{\equiv}\overset{6}{C}\overset{7}{C}H_3$
1-hepten-5-yne
***not* 6-hepten-2-yne**
because 1 < 2

$CH_2{=}CHCH(CH_2CH_2CH_2CH_3)C{\equiv}CCH_3$ (numbered 1–6 along the main chain)
3-butyl-1-hexen-4-yne

the longest continuous chain has eight carbons, but the 8-carbon chain does not contain both functional groups; therefore, the compound is named as a hexenyne because the longest continuous chain containing both functional groups has six carbons

If the same low number is obtained in both directions, the chain is numbered in the direction that gives the double bond the lower number.

If there is a tie between a double bond and a triple bond, the double bond gets the lower number.

$\overset{1}{C}H_3\overset{2}{C}H{=}\overset{3}{C}H\overset{4}{C}{\equiv}\overset{5}{C}\overset{6}{C}H_3$
2-hexen-4-yne
***not* 4-hexen-2-yne**

$\overset{6}{H}C{\equiv}\overset{5}{C}\overset{4}{C}H_2\overset{3}{C}H_2\overset{2}{C}H{=}\overset{1}{C}H_2$
1-hexen-5-yne
***not* 5-hexen-1-yne**

The relative priorities of the functional group suffixes are shown in Table 8.1. If the second functional group suffix has a higher priority than the alkene, the chain is numbered in the direction that assigns the lowest possible number to the functional group with the higher priority.

A chain is numbered to give the lowest possible number to the functional group with the highest priority.

$CH_2{=}CHCH_2OH$
2-propen-1-ol
***not* 1-propen-3-ol**

$CH_3C(CH_3){=}CHCH_2CH_2OH$
4-methyl-3-penten-1-ol

$CH_2{=}CHCH_2CH_2CH_2CH(NH_2)CH_3$
6-hepten-2-amine

$CH_3CH_2CH_2CH_2CH_2CH(OH)CH{=}CH_2$
1-octen-3-ol

6-methyl-2-cyclohexenol

3-cyclohexenamine

Table 8.1 Priorities of Functional Group Suffixes

C=O > OH > NH_2 > C=C = C≡C

the double bond is given priority over a triple bond only when there is a tie

← increasing priority

HOW A BANANA SLUG KNOWS WHAT TO EAT

Many species of mushrooms synthesize 1-octen-3-ol, which serves as a repellent that drives off predatory slugs. Such mushrooms can be recognized by small bite marks on their caps, where the slug started to nibble before the volatile compound was released. Humans are not put off by the smell because, to them, 1-octen-3-ol smells like a mushroom. 1-Octen-3-ol also has antibacterial properties that may protect the mushroom from organisms that attempt to invade the wound made by the slug. Not surprisingly, the species of mushroom that banana slugs commonly eat cannot synthesize 1-octen-3-ol.

PROBLEM 1◆

Give the systematic name for each of the following compounds:

a. 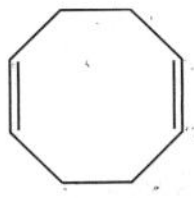

e. 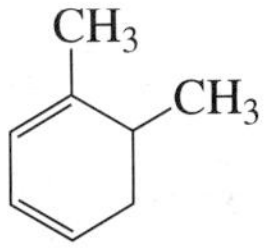

b. $CH_2{=}CHCH_2C{\equiv}CCH_2CH_3$

f. $HOCH_2CH_2C{\equiv}CH$

c. 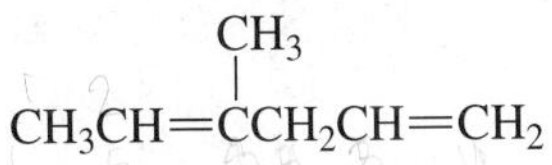

g. $CH_3CH{=}CHCH{=}CHCH{=}CH_2$

d. 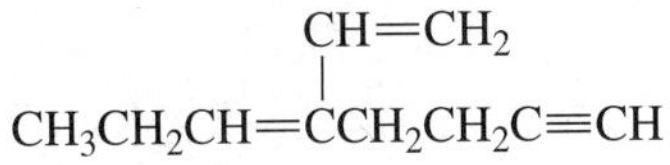

h. 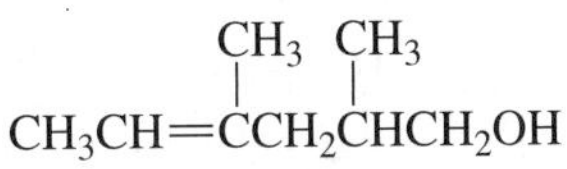

8.2 Configurational Isomers of Dienes

A diene such as 1-chloro-2,4-heptadiene has four configurational isomers because each of the double bonds can have either the *E* or the *Z* configuration. Thus, there are *E-E*, *Z-Z*, *E-Z*, and *Z-E* isomers. The rules for determining the *E* and *Z* configurations were given in Section 3.5. Recall that the *Z* isomer has the high-priority groups on the same side.

The *Z* isomer has the high-priority groups on the same side.

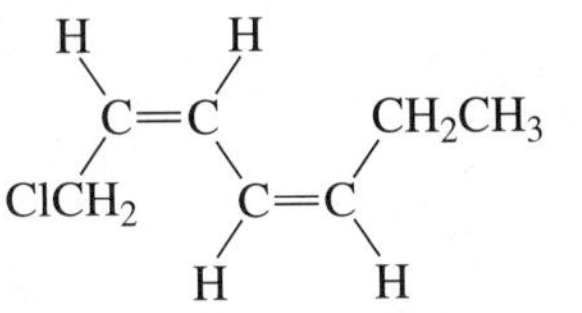

(2*Z*,4*Z*)-1-chloro-2,4-heptadiene

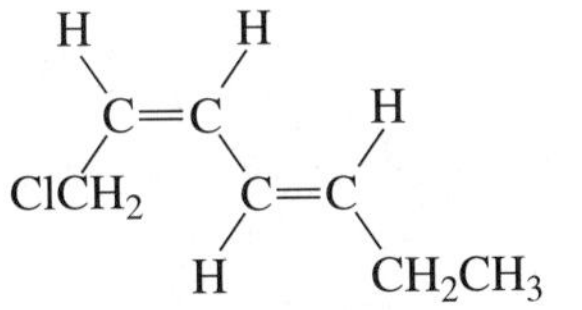

(2*Z*,4*E*)-1-chloro-2,4-heptadiene

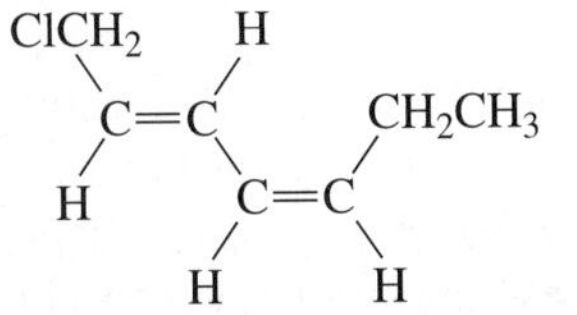

(2*E*,4*Z*)-1-chloro-2,4-heptadiene

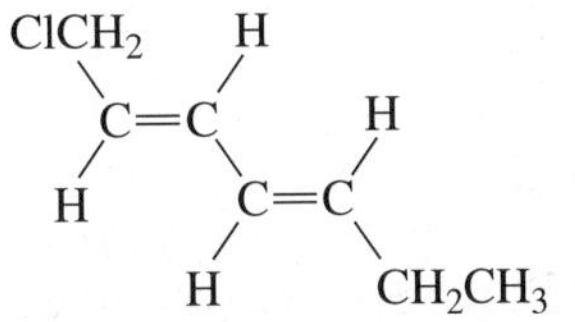

(2*E*,4*E*)-1-chloro-2,4-heptadiene

Molecules:
(1*Z*,3*Z*)-1,4-Difluoro-1,3-butadiene;
(1*E*,3*E*)-1,4-Difluoro-1,3-butadiene;
(1*Z*,3*E*)-1,4-Difluoro-1,3-butadiene

PROBLEM 2

Draw the configurational isomers for the following compounds, and name each one:

a. 2-methyl-2,4-hexadiene b. 2,4-heptadiene c. 1,3-pentadiene

8.3 Relative Stabilities of Dienes

In Section 4.11, we saw that the relative stabilities of substituted alkenes can be determined by their relative values of $-\Delta H°$ for catalytic hydrogenation. Remember that the least stable alkene has the greatest $-\Delta H°$ value; the least stable alkene gives off the most heat when it is hydrogenated, because it has more energy to begin with. The $-\Delta H°$ for hydrogenation of 2,3-pentadiene (a cumulated diene) is greater than that of

The least stable alkene has the greatest $-\Delta H°$ value.

1,4-pentadiene (an isolated diene) which, in turn, is greater than than that of 1,3-pentadiene (a conjugated diene).

3-D Molecules: 2,3-Pentadiene; 1,4-Pentadiene; 1,3-Pentadiene

$$\underset{\substack{\textbf{2,3-pentadiene} \\ \textbf{a cumulated diene}}}{CH_3CH{=}C{=}CHCH_3} + 2\,H_2 \xrightarrow{\text{Pt}} CH_3CH_2CH_2CH_2CH_3 \qquad \Delta H^\circ = -70.5 \text{ kcal/mol } (-295 \text{ kJ/mol})$$

$$\underset{\substack{\textbf{1,4-pentadiene} \\ \textbf{an isolated diene}}}{CH_2{=}CHCH_2CH{=}CH_2} + 2\,H_2 \xrightarrow{\text{Pt}} CH_3CH_2CH_2CH_2CH_3 \qquad \Delta H^\circ = -60.2 \text{ kcal/mol } (-252 \text{ kJ/mol})$$

$$\underset{\substack{\textbf{1,3-pentadiene} \\ \textbf{a conjugated diene}}}{CH_2{=}CHCH{=}CHCH_3} + 2\,H_2 \xrightarrow{\text{Pt}} CH_3CH_2CH_2CH_2CH_3 \qquad \Delta H^\circ = -54.1 \text{ kcal/mol } (-226 \text{ kJ/mol})$$

From the relative $-\Delta H^\circ$ values for the three pentadienes, we can conclude that conjugated dienes are more stable than isolated dienes, which are more stable than cumulated dienes.

relative stabilities of dienes

most stable → conjugated diene > isolated diene > cumulated diene ← least stable

Why is a conjugated diene such as 1,3-pentadiene more stable than an isolated diene? Two factors contribute to the stability of a conjugated diene. One is the *hybridization of the orbitals* forming the carbon–carbon single bonds. The carbon–carbon single bond in 1,3-butadiene is formed from the overlap of an sp^2 orbital with another sp^2 orbital, whereas the carbon–carbon single bonds in 1,4-pentadiene are formed from the overlap of an sp^3 orbital with an sp^2 orbital.

single bond formed by sp^2–sp^2 overlap

$$\underset{\textbf{1,3-butadiene}}{CH_2{=}CH{-}CH{=}CH_2}$$

single bonds formed by sp^3–sp^2 overlap

$$\underset{\textbf{1,4-pentadiene}}{CH_2{=}CH{-}CH_2{-}CH{=}CH_2}$$

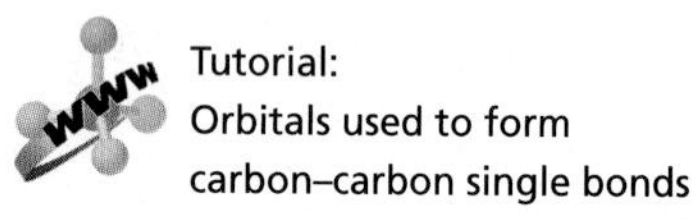

In Section 1.14, you saw that the length and strength of a bond depend on how close the electrons in the bonding orbital are to the nucleus: *The closer the electrons are to the nucleus, the shorter and stronger is the bond.* Because a 2*s* electron is closer, on average, to the nucleus than is a 2*p* electron, a bond formed by sp^2–sp^2 overlap is shorter and stronger than one formed by sp^3–sp^2 overlap (Table 8.2). (An sp^2 orbital has 33.3% *s* character, whereas an sp^3 orbital has 25% *s* character.) Thus, a conjugated diene has one stronger single bond than an isolated diene, and stronger bonds cause a compound to be more stable.

Electron delocalization also causes a conjugated diene to be more stable than an isolated diene. The π electrons in each of the double bonds of an isolated diene are *localized* between two carbons. In contrast, the π electrons in a conjugated diene are *delocalized.* As you discovered in Section 7.6, electron delocalization stabilizes a molecule. Both the resonance hybrid and the molecular orbital diagram of 1,3-butadiene in Figure 7.9 show that the single bond in 1,3-butadiene is not a pure single bond, but has partial double-bond character as a result of electron delocalization.

$$\overset{-}{C}H_2{-}CH{=}CH{-}\overset{+}{C}H_2 \longleftrightarrow CH_2{=}CH{-}CH{=}CH_2 \longleftrightarrow \overset{+}{C}H_2{-}CH{=}CH{-}\overset{-}{C}H_2$$

resonance contributors

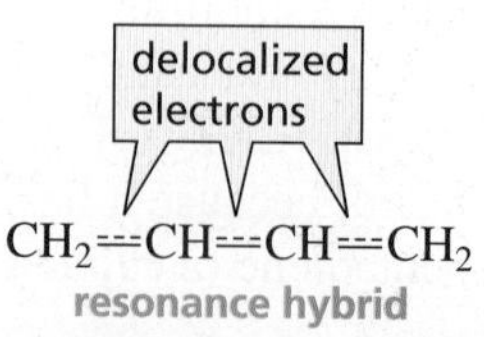

Table 8.2 Dependence of the Length of a Carbon–Carbon Single Bond on the Hybridization of the Orbitals Used in Its Formation

Compound	Hybridization	Bond length (Å)
H_3C-CH_3	sp^3–sp^3	1.54
$H_3C-C(H)=CH_2$	sp^3–sp^2	1.50
$H_2C=C(H)-C(H)=CH_2$	sp^2–sp^2	1.47
$H_3C-C\equiv CH$	sp^3–sp	1.46
$H_2C=C(H)-C\equiv CH$	sp^2–sp	1.43
$HC\equiv C-C\equiv CH$	sp–sp	1.37

PROBLEM 3◆

Name the following dienes and rank them in order of increasing stability. (Alkyl groups stabilize dienes in the same way that they stabilize alkenes; see Section 4.11.)

$CH_3CH=CHCH=CHCH_3$ $\quad CH_2=CHCH_2CH=CH_2$ $\quad CH_3C(CH_3)=CHCH=C(CH_3)CH_3$ $\quad CH_3CH=CHCH=CH_2$

We have just seen that a conjugated diene is *more* stable than an isolated diene. Now we need to see why a cumulated diene is *less* stable than an isolated diene. Cumulated dienes are unlike other dienes in that the central carbon is *sp* hybridized since it has two π bonds. In contrast, all the double-bonded carbons of isolated dienes and conjugated dienes are sp^2 hybridized. The *sp* hybridization gives the cumulated dienes unique properties. For example, the $-\Delta H°$ for hydrogenation of allene is similar to the $-\Delta H°$ for the hydrogenation of propyne, a compound with two *sp* hybridized carbons.

$$\underset{\text{allene}}{CH_2=C=CH_2} + 2\,H_2 \xrightarrow{Pt} CH_3CH_2CH_3 \qquad \Delta H° = -70.5 \text{ kcal/mol } (-295 \text{ kJ/mol})$$

$$\underset{\text{propyne}}{CH_3C\equiv CH} + 2\,H_2 \xrightarrow{Pt} CH_3CH_2CH_3 \qquad \Delta H° = -69.9 \text{ kcal/mol } (-292 \text{ kJ/mol})$$

Allenes have an unusual geometry. One of the *p* orbitals of the central carbon of allene overlaps a *p* orbital of an adjacent sp^2 carbon. The second *p* orbital of the central carbon overlaps a *p* orbital of the other sp^2 carbon (Figure 8.1a). The two *p* orbitals of the central carbon are perpendicular. Therefore, the plane containing one H—C—H group is perpendicular to the plane containing the other H—C—H group. Thus, a substituted allene such as 2,3-pentadiene has a nonsuperimposable mirror image (Figure 8.1b), so it is a chiral molecule, even though it does not have any asymmetric carbons (Section 5.4). We will not consider the reactions of cumulated dienes, because they are rather specialized and are more appropriately covered in an advanced course in organic chemistry.

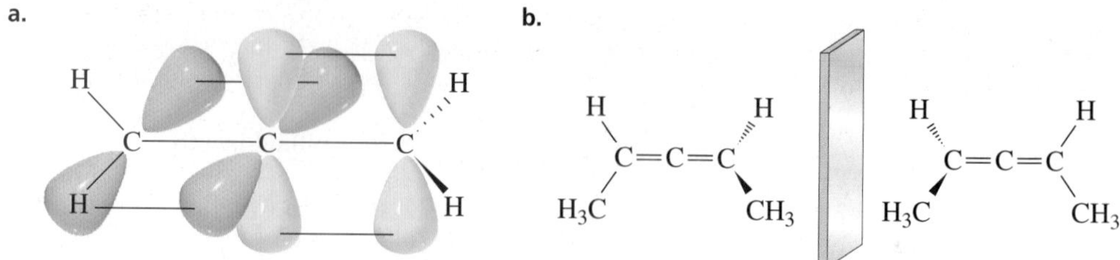

▲ **Figure 8.1**
(a) Double bonds are formed by *p* orbital–*p* orbital overlap. The two *p* orbitals on the central carbon are perpendicular, causing allene to be a nonplanar molecule.
(b) 2,3-Pentadiene has a nonsuperimposable mirror image. It is, therefore, a chiral molecule, even though it does not have an asymmetric carbon.

8.4 How Dienes React

Dienes, like alkenes and alkynes, are nucleophiles because of the electron density of their π bonds. Therefore, they react with electrophilic reagents. Like alkenes and alkynes, dienes undergo electrophilic addition reactions.

Until now, we have been concerned with the reactions of compounds that have only one functional group. Compounds with two or more functional groups exhibit reactions characteristic of the individual functional groups if the groups are sufficiently separated from each other. If they are close enough to allow electron delocalization, however, one functional group can affect the reactivity of the other. Therefore, we will see that the reactions of isolated dienes are the same as the reactions of alkenes, but the reactions of conjugated dienes are a little different because of electron delocalization.

8.5 Electrophilic Addition Reactions of Isolated Dienes

The reactions of isolated dienes are just like those of alkenes. If an excess of the electrophilic reagent is present, two independent addition reactions will occur, each following the rule that applies to all electrophilic addition reactions: The electrophile adds to the sp^2 carbon that is bonded to the greater number of hydrogens.

$$\underset{\text{1,5-hexadiene}}{CH_2{=}CHCH_2CH_2CH{=}CH_2} + \underset{\text{excess}}{HBr} \longrightarrow CH_3\underset{\underset{Br}{|}}{C}HCH_2CH_2\underset{\underset{Br}{|}}{C}HCH_3$$

The reaction proceeds exactly as we would predict from our knowledge of the mechanism for the reaction of alkenes with electrophilic reagents. The electrophile (H^+) adds to the electron-rich double bond in a manner that produces the more stable carbocation (Section 4.4). The bromide ion then adds to the carbocation. Because there is an excess of the electrophilic reagent, both double bonds undergo addition.

mechanism for the reaction of 1,5-hexadiene with excess HBr

$$CH_2{=}CHCH_2CH_2CH{=}CH_2 + H{-}\ddot{\underset{..}{Br}}: \longrightarrow CH_3\overset{+}{C}HCH_2CH_2CH{=}CH_2 \;\; + :\ddot{\underset{..}{Br}}:^- \longrightarrow CH_3\underset{\underset{Br}{|}}{C}HCH_2CH_2CH{=}CH_2 \;\; H{-}\ddot{\underset{..}{Br}}:$$

$$\downarrow$$

$$CH_3\underset{\underset{Br}{|}}{C}HCH_2CH_2\underset{\underset{Br}{|}}{C}HCH_3 \longleftarrow CH_3\underset{\underset{Br}{|}}{C}HCH_2CH_2\overset{+}{C}HCH_3 \;\; + :\ddot{\underset{..}{Br}}:^-$$

If there is only enough electrophilic reagent to add to one of the double bonds, it will add preferentially to the more reactive double bond. For example, in the reaction of 2-methyl-1,5-hexadiene with HCl, addition of HCl to the double bond on the left forms a secondary carbocation, whereas addition of HCl to the double bond on the right forms a tertiary carbocation. Because the transition state leading to formation of a tertiary carbocation is more stable than that leading to a secondary carbocation, the tertiary carbocation is formed faster (Section 4.4). So in the presence of a limited amount of HCl, the major product of the reaction will be 5-chloro-5-methyl-1-hexene.

$$CH_2{=}CHCH_2CH_2\overset{CH_3}{\overset{|}{C}}{=}CH_2 + HCl \longrightarrow CH_2{=}CHCH_2CH_2\overset{CH_3}{\underset{Cl}{C}}CH_3$$

2-methyl-1,5-hexadiene 1 mol; HCl 1 mol; 5-chloro-5-methyl-1-hexene, major product

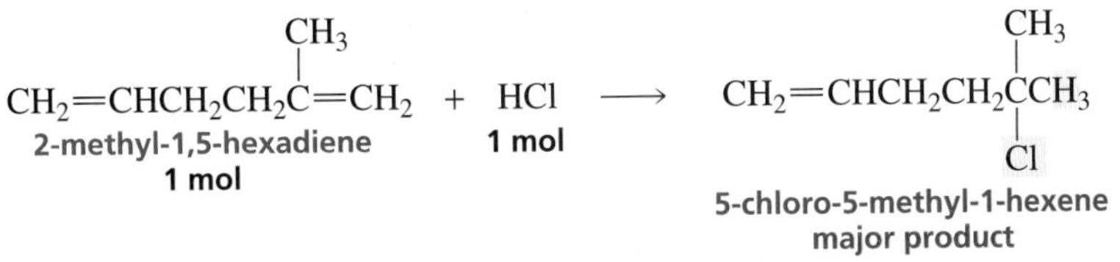

PROBLEM 4◆

Which of the double bonds in zingiberene (whose structure is given on the first page of this chapter) is the most reactive in an electrophilic addition reaction?

PROBLEM 5

What stereoisomers are obtained from the two reactions in Section 8.5? (*Hint:* Review Section 5.19.)

PROBLEM 6

Give the major product of each of the following reactions, and show the stereoisomers that would be obtained (equivalent amounts of reagents are used in each case):

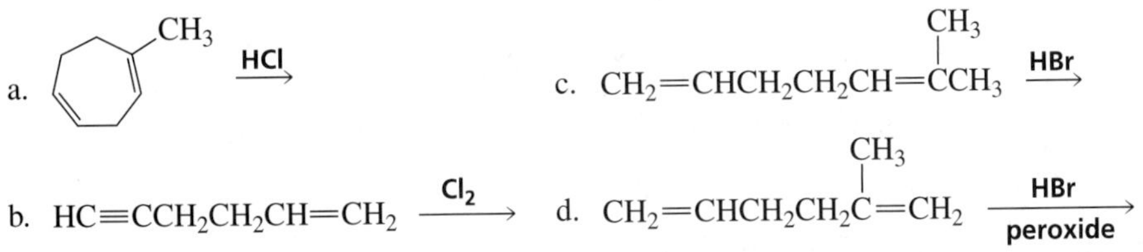

a. 1-methyl-1,3-cycloheptadiene (ring with CH_3) $\xrightarrow{HCl}$

b. $HC{\equiv}CCH_2CH_2CH{=}CH_2 \xrightarrow{Cl_2}$

c. $CH_2{=}CHCH_2CH_2CH{=}\overset{CH_3}{\overset{|}{C}}CH_3 \xrightarrow{HBr}$

d. $CH_2{=}CHCH_2CH_2\overset{CH_3}{\overset{|}{C}}{=}CH_2 \xrightarrow[\text{peroxide}]{HBr}$

8.6 Electrophilic Addition Reactions of Conjugated Dienes

If a conjugated diene, such as 1,3-butadiene, reacts with a limited amount of electrophilic reagent so that addition can occur at only one of the double bonds, two addition products are formed. One is a 1,2-addition product, which is a result of addition at the 1- and 2-positions. The other is a 1,4-addition product, the result of addition at the 1- and 4-positions.

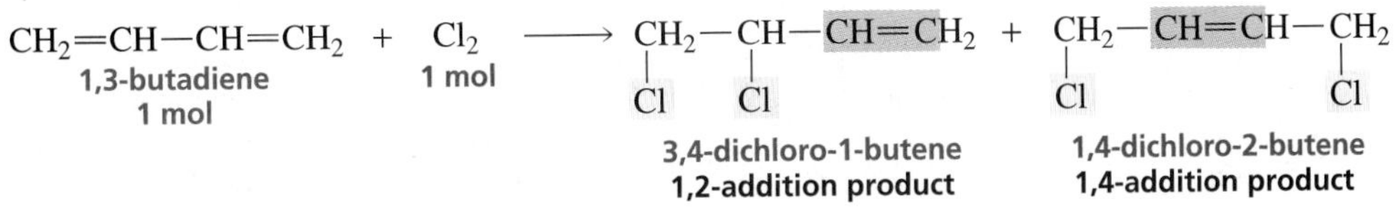

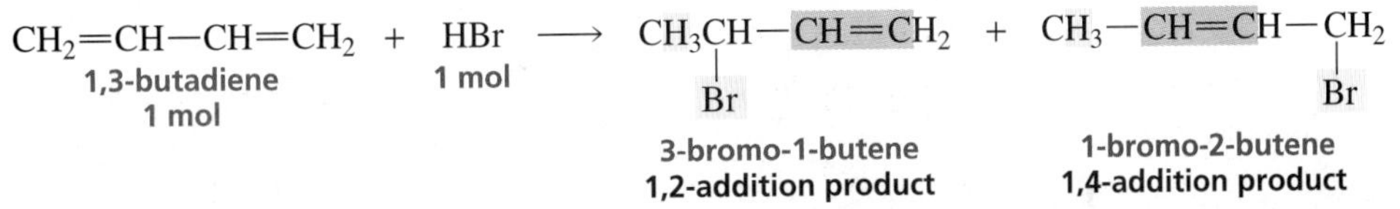

$$CH_2{=}CH{-}CH{=}CH_2 + Cl_2 \longrightarrow \underset{Cl}{CH_2}{-}\underset{Cl}{CH}{-}CH{=}CH_2 + \underset{Cl}{CH_2}{-}CH{=}CH{-}\underset{Cl}{CH_2}$$

1,3-butadiene 1 mol; Cl_2 1 mol; 3,4-dichloro-1-butene, 1,2-addition product; 1,4-dichloro-2-butene, 1,4-addition product

$$CH_2{=}CH{-}CH{=}CH_2 + HBr \longrightarrow CH_3\underset{Br}{CH}{-}CH{=}CH_2 + CH_3{-}CH{=}CH{-}\underset{Br}{CH_2}$$

1,3-butadiene 1 mol; HBr 1 mol; 3-bromo-1-butene, 1,2-addition product; 1-bromo-2-butene, 1,4-addition product

Addition at the 1- and 2-positions is called **1,2-addition** or **direct addition**. Addition at the 1- and 4-positions is called **1,4-addition** or **conjugate addition**. On the basis of our knowledge of how electrophilic reagents add to double bonds, we expect the 1,2-addition product to form. That the 1,4-addition product also forms may be surprising because not only did the reagent not add to adjacent carbons, but a double bond has changed its position. The double bond in the 1,4-product is between the 2- and 3-positions, while the reactant had a single bond in this position.

When we talk about addition at the 1- and 2-positions or at the 1- and 4-positions, the numbers refer to the four carbons of the conjugated system. Thus, the carbon in the 1-position is one of the sp^2 carbons at the end of the conjugated system—it is not necessarily the first carbon in the molecule.

$$\mathrm{R{-}\overset{1}{C}H{=}\overset{2}{C}H{-}\overset{3}{C}H{=}\overset{4}{C}H{-}R}$$

the conjugated system

$$\mathrm{CH_3CH{=}CH{-}CH{=}CHCH_3 \xrightarrow{Br_2} CH_3\underset{Br}{C}H{-}\underset{Br}{C}H{-}CH{=}CHCH_3 + CH_3\underset{Br}{C}H{-}CH{=}CH{-}\underset{Br}{C}HCH_3}$$

2,4-hexadiene → **4,5-dibromo-2-hexene** (**1,2-addition product**) + **2,5-dibromo-3-hexene** (**1,4-addition product**)

To understand why both 1,2-addition and 1,4-addition products are obtained from the reaction of a conjugated diene with a limited amount of electrophilic reagent, we must look at the mechanism of the reaction. In the first step of the addition of HBr to 1,3-butadiene, the electrophilic proton adds to C-1, forming an allylic cation. (Recall that an allylic cation has a positive charge on a carbon that is next to a double-bonded carbon.) The π electrons of the allylic cation are delocalized—the positive charge is shared by two carbons. Notice that because 1,3-butadiene is symmetrical, adding to C-1 is the same as adding to C-4. The proton does not add to C-2 or C-3 because doing so would form a primary carbocation. The π electrons of a primary carbocation are localized; thus, it is not as stable as the delocalized allylic cation.

mechanism for the reaction of 1,3-butadiene with HBr

$$\mathrm{CH_2{=}CH{-}CH{=}CH_2 + H{-}\ddot{B}r{:} \longrightarrow CH_3{-}\overset{+}{C}H{-}CH{=}CH_2 \longleftrightarrow CH_3{-}CH{=}CH{-}\overset{+}{C}H_2}$$

1,3-butadiene; $+\ :\ddot{Br}:^-$ an allylic cation $+\ :\ddot{Br}:^-$

$$\mathrm{\not\to\ \overset{+}{C}H_2{-}CH_2{-}CH{=}CH_2}$$ a primary carbocation

$$\mathrm{\downarrow\ CH_3{-}\underset{Br}{C}H{-}CH{=}CH_2 + CH_3{-}CH{=}CH{-}\underset{Br}{C}H_2}$$

3-bromo-1-butene (**1,2-addition product**) + **1-bromo-2-butene** (**1,4-addition product**)

The contributing resonance structures of the allylic cation show that the positive charge on the carbocation is not localized on C-2, but is shared by C-2 and C-4. Consequently, in the second step of the reaction, the bromide ion can attack either C-2 (direct addition) or C-4 (conjugate addition) to form the 1,2-addition product or the 1,4-addition product, respectively.

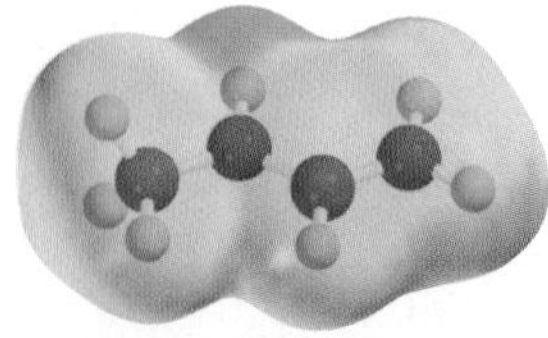

$$\mathrm{CH_3{-}\overset{\delta+}{C}H{\cdots}CH{\cdots}\overset{\delta+}{C}H_2}$$

As we look at more examples, notice that the first step in all electrophilic additions to conjugated dienes is addition of the electrophile to one of the sp^2 carbons at the end of the conjugated system. This is the only way to obtain a carbocation that is stabilized by resonance (i.e., by electron delocalization). If the electrophile were to add to one of the internal sp^2 carbons, the resulting carbocation would not be stabilized by resonance.

Let's review by comparing addition to an isolated diene with addition to a conjugated diene. The carbocation formed by addition of an electrophile to an isolated diene is not stabilized by resonance. The positive charge is localized on a single carbon, so only direct (1,2) addition occurs.

An isolated diene undergoes only 1,2-addition.

addition to an isolated diene

$$CH_2{=}CHCH_2CH_2CH{=}CH_2 \xrightarrow{HBr} CH_3\overset{+}{C}HCH_2CH_2CH{=}CH_2 + Br^- \longrightarrow CH_3\underset{Br}{C}HCH_2CH_2CH{=}CH_2$$

1,5-hexadiene — addition of the electrophile — addition of the nucleophile — **5-bromo-1-hexene**

The carbocation formed by addition of an electrophile to a conjugated diene, in contrast, is stabilized by resonance. The positive charge is shared by two carbons, and as a result, both direct (1,2-) and conjugate (1,4-) addition occur.

A conjugated diene undergoes both 1,2- and 1,4-addition.

addition to a conjugated diene

$$CH_3CH{=}CH{-}CH{=}CHCH_3 \xrightarrow{HBr} CH_3CH_2{-}\overset{+}{C}H{-}CH{=}CHCH_3 \longleftrightarrow CH_3CH_2{-}CH{=}CH{-}\overset{+}{C}HCH_3 \quad + Br^-$$

2,4-hexadiene — addition of the electrophile — the carbocation is stabilized by electron delocalization — addition of the nucleophile

$$CH_3CH_2{-}\underset{Br}{C}H{-}CH{=}CHCH_3 \qquad CH_3CH_2{-}CH{=}CH{-}\underset{Br}{C}HCH_3$$

4-bromo-2-hexene
1,2-addition product

2-bromo-3-hexene
1,4-addition product

If the conjugated diene is not symmetrical, the major products of the reaction are those obtained by adding the electrophile to whichever terminal sp^2 carbon results in formation of the more stable carbocation. For example, in the reaction of 2-methyl-1,3-butadiene with HBr, the proton adds preferentially to C-1 because the positive charge on the resulting carbocation is shared by a tertiary allylic and a primary allylic carbon. Adding the proton to C-4 would form a carbocation with the positive charge shared by a secondary allylic and a primary allylic carbon. Because addition to C-1 forms the more stable carbocation, 3-bromo-3-methyl-1-butene and 1-bromo-3-methyl-2-butene are the major products of the reaction.

$$\overset{1}{C}H_2{=}\overset{CH_3}{C}{-}CH{=}\overset{4}{C}H_2 + HBr \longrightarrow CH_3{-}\underset{Br}{\overset{CH_3}{C}}{-}CH{=}CH_2 + CH_3{-}\overset{CH_3}{C}{=}CH{-}\underset{Br}{C}H_2$$

2-methyl-1,3-butadiene — **3-bromo-3-methyl-1-butene** — **1-bromo-3-methyl-2-butene**

$$CH_3\overset{CH_3}{\underset{+}{C}}{-}CH{=}CH_2 \longleftrightarrow CH_3\overset{CH_3}{C}{=}CH{-}\underset{+}{C}H_2$$

carbocation formed by adding H⁺ to C-1

$$CH_2{=}\overset{CH_3}{C}{-}\underset{+}{C}HCH_3 \longleftrightarrow \underset{+}{C}H_2{-}\overset{CH_3}{C}{=}CHCH_3$$

carbocation formed by adding H⁺ to C-4

PROBLEM 7

What products would be obtained from the reaction of 1,3,5-hexatriene with one equivalent of HBr? Ignore stereoisomers.

PROBLEM 8◆

Give the products of the following reactions, ignoring stereoisomers (equivalent amounts of reagents are used in each case):

a. $CH_3CH{=}CH{-}CH{=}CHCH_3 \xrightarrow{Cl_2}$

b. $CH_3CH{=}CH{-}\underset{\displaystyle CH_3}{\underset{|}{C}}{=}CHCH_3 \xrightarrow{HBr}$

c. $CH_3CH{=}\underset{\displaystyle CH_3}{\underset{|}{C}}{-}\overset{\displaystyle CH_3}{\overset{|}{C}}{=}CHCH_3 \xrightarrow{HBr}$

d. (cyclopentadiene) $\xrightarrow{Br_2}$

8.7 Thermodynamic Versus Kinetic Control of Reactions

When a conjugated diene undergoes an electrophilic addition reaction, two factors—the temperature at which the reaction is carried out and the structure of the reactant—determine whether the 1,2-addition product or the 1,4-addition product will be the major product of the reaction.

The thermodynamic product is the most stable product.

The kinetic product is the product that is formed most rapidly.

When a reaction produces more than one product, the product that is formed most rapidly is called the **kinetic product**, and the most stable product is called the **thermodynamic product**. Reactions that produce the kinetic product as the major product are said to be *kinetically controlled*. Reactions that produce the thermodynamic product as the major product are said to be *thermodynamically controlled*.

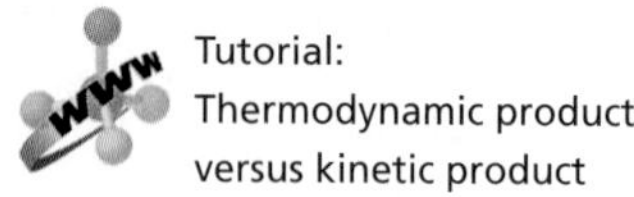
Tutorial: Thermodynamic product versus kinetic product

For many organic reactions, the most stable product is the one that is formed most rapidly. In other words, the kinetic product and the thermodynamic product are one and the same. Electrophilic addition to 1,3-butadiene is an example of a reaction in which the kinetic product and the thermodynamic product are *not* the same: The 1,2-addition product is the kinetic product, and the 1,4-addition product is the thermodynamic product.

$$\underset{\textbf{1,3-butadiene}}{CH_2{=}CHCH{=}CH_2} + HBr \longrightarrow \underset{\substack{\textbf{1,2-addition product}\\ \textbf{kinetic product}}}{\underset{\displaystyle Br}{\underset{|}{CH_3CHCH{=}CH_2}}} + \underset{\substack{\textbf{1,4-addition product}\\ \textbf{thermodynamic product}}}{CH_3CH{=}CH\underset{\displaystyle Br}{\underset{|}{CH_2}}}$$

The kinetic product predominates when the reaction is irreversible.

For a reaction in which the kinetic and thermodynamic products are not the same, the product that predominates depends on the conditions under which the reaction is carried out. If the reaction is carried out under sufficiently mild (low-temperature) conditions to cause the reaction to be *irreversible*, the major product will be the *kinetic product*. For example, when addition of HBr to 1,3-butadiene is carried out at −80 °C, the major product is the 1,2-addition product.

$$CH_2{=}CHCH{=}CH_2 + HBr \xrightarrow{-80\ °C} \underset{\substack{\textbf{1,2-addition product}\\ \textbf{80\%}}}{\underset{\displaystyle Br}{\underset{|}{CH_3CHCH{=}CH_2}}} + \underset{\substack{\textbf{1,4-addition product}\\ \textbf{20\%}}}{CH_3CH{=}CH\underset{\displaystyle Br}{\underset{|}{CH_2}}}$$

If, on the other hand, the reaction is carried out under sufficiently vigorous (high-temperature) conditions to cause the reaction to be *reversible*, the major product will be the *thermodynamic product*. When the same reaction is carried out at 45 °C, the major product is the 1,4-addition product. Thus, the 1,2-addition product is the kinetic product (it is formed more rapidly), and the 1,4-addition product is the thermodynamic product (it is the more stable product).

The thermodynamic product predominates when the reaction is reversible.

$$CH_2{=}CHCH{=}CH_2 + HBr \xrightarrow{45\,°C} \underset{\substack{\textbf{1,2-addition product}\\ 15\%}}{\underset{\text{Br}}{CH_3\underset{|}{C}HCH{=}CH_2}} + \underset{\substack{\textbf{1,4-addition product}\\ 85\%}}{CH_3CH{=}CH\underset{\text{Br}}{\underset{|}{C}H_2}}$$

A reaction coordinate diagram helps explain why different products predominate under different reaction conditions (Figure 8.2). The first step of the addition reaction—addition of a proton to C-1—is the same whether the 1,2-addition product or the 1,4-addition product is being formed. It is the second step of the reaction that determines whether the nucleophile (Br^-) attacks C-2 or C-4. Because the 1,2-addition product is formed more rapidly, we know that the transition state for its formation is more stable than the transition state for formation of the 1,4-addition product. This is the first time we have seen a reaction in which the less stable product has the more stable transition state!

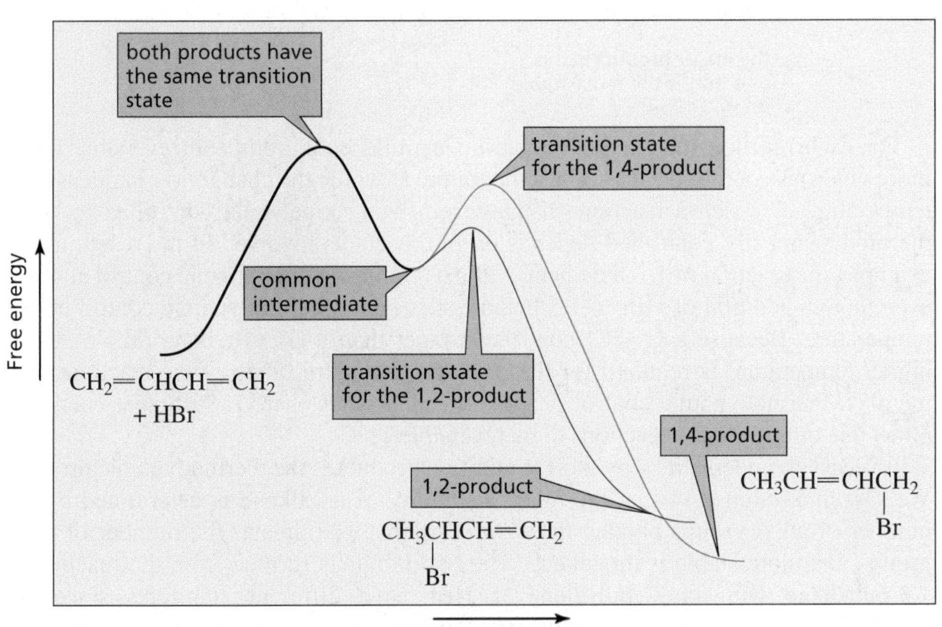

Figure 8.2
Reaction coordinate diagram for the addition of HBr to 1,3-butadiene.

At low temperatures (−80 °C), there is enough energy for the reactants to overcome the energy barrier for the first step of the reaction and therefore form the intermediate, and there is enough energy for the intermediate to form the two addition products. However, there is not enough energy for the reverse reaction to occur: The products cannot overcome the large energy barriers separating them from the intermediate. Consequently, at −80 °C, the relative amounts of the two products obtained reflect the relative energy barriers to the second step of the reaction. The energy barrier to formation of the 1,2-addition product is lower than the energy barrier to formation of the 1,4-addition product, so the major product is the 1,2-addition product.

In contrast, at 45 °C, there is enough energy for one or more of the products to go back to the intermediate. The intermediate is called a **common intermediate** because it is an intermediate that both products have in common. The ability to return to a common intermediate allows the products to interconvert. Because the products can interconvert, the relative amounts of the two products at equilibrium depend on their relative stabilities. The thermodynamic product reverses less readily because it has a higher energy barrier to the common intermediate, so it gradually comes to predominate in the product mixture.

Thus, when a reaction is *irreversible* under the conditions employed in the experiment, it is said to be under *kinetic control*. When a reaction is under **kinetic control**, the relative amounts of the products *depend on the rates* at which they are formed.

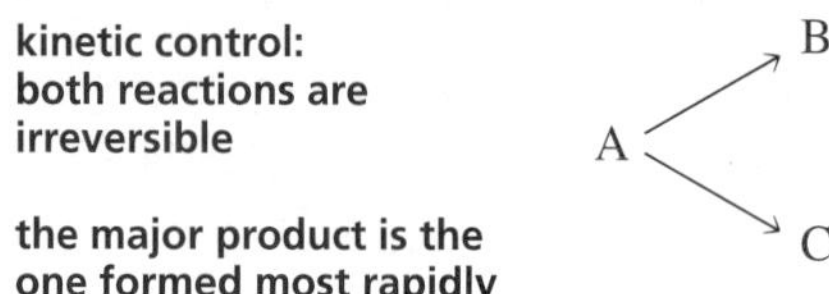
kinetic control:
both reactions are
irreversible
the major product is the
one formed most rapidly
A
B
C

A reaction is said to be under *thermodynamic control* when there is sufficient energy to allow it to be *reversible*. When a reaction is under **thermodynamic control**, the relative amounts of the products *depend on their stabilities*. Because a reaction must be reversible to be under thermodynamic control, thermodynamic control is also called **equilibrium control**.

thermodynamic control:
one or both reactions
are reversible

the major product is the
one that is the most stable

A ⇌ B, A ⇌ C or A ⇌ B, A → C

For each reaction that is irreversible under mild conditions and reversible under more vigorous conditions, there is a temperature at which the changeover happens. The temperature at which a reaction changes from being kinetically controlled to being thermodynamically controlled depends on the reactants involved in the reaction. For example, the reaction of 1,3-butadiene with HCl remains under kinetic control at 45 °C even though addition of HBr to 1,3-butadiene is under thermodynamic control at that temperature. Because a C—Cl bond is stronger than a C—Br bond (Table 3.1), a higher temperature is required for the products to undergo the reverse reaction. (Remember, thermodynamic control is achieved only when there is sufficient energy to allow one or both of the reactions to be reversible.)

It is easy to understand why the 1,4-addition product is the thermodynamic product. We saw in Section 4.11 that the relative stability of an alkene is determined by the number of alkyl groups bonded to its sp^2 carbons: The greater the number of alkyl groups, the more stable is the alkene. The two products formed from the reaction of 1,3-butadiene with one equivalent of HBr have different stabilities since the 1,2-addition product has one alkyl group bonded to its sp^2 carbons, whereas the 1,4-product has two alkyl groups bonded to its sp^2 carbons. The 1,4-addition product, therefore, is more stable than the 1,2-addition product. Thus, the 1,4-addition product is the thermodynamic product.

$CH_3CHCH{=}CH_2$ (Br on C-2)
1,2-addition product
kinetic product

$CH_3CH{=}CHCH_2$ (Br on C-1)
1,4-addition product
thermodynamic product

Now we need to see why the 1,2-addition product is formed faster. In other words, why is the transition state for formation of the 1,2-addition product more stable than the transition state for formation of the 1,4-addition product? For many years, chemists thought it was because the transition state for formation of the 1,2-addition product resembles the contributing resonance structure in which the positive charge is on a secondary allylic carbon. In contrast, the transition state for formation of the 1,4-addition product resembles the contributing resonance structure in which the positive charge is on a less stable primary allylic carbon.

secondary allylic carbon — primary allylic carbon

$$CH_2{=}CHCH{=}CH_2 \xrightarrow{HBr} CH_3\overset{+}{C}HCH{=}CH_2 + Br^- \longleftrightarrow CH_3CH{=}CH\overset{+}{C}H_2 + Br^-$$

$$\left[\begin{array}{c} CH_3\overset{\delta+}{C}HCH{=}CH_2 \\ \overset{\delta-}{:Br:} \end{array}\right]^{\ddagger} \qquad \left[\begin{array}{c} CH_3CH{=}CH\overset{\delta+}{C}H_2 \\ \overset{\delta-}{:Br:} \end{array}\right]^{\ddagger}$$

transition state for formation of the 1,2-addition product — **transition state for formation of the 1,4-addition product**

However, when the reaction of 1,3-pentadiene + DCl is carried out under kinetic control, essentially the same relative amounts of 1,2- and 1,4-addition products are obtained as are obtained from the kinetically controlled reaction of 1,3-butadiene + HBr. The transition states for formation of the 1,2- and 1,4-addition products from 1,3-pentadiene should both have the same stability because both resemble a contributing resonance structure in which the positive charge is on a secondary allylic carbon. Why, then, is the 1,2-addition product still formed faster?

$$\underset{\textbf{1,3 pentadiene}}{CH_2{=}CHCH{=}CHCH_3} + DCl \xrightarrow{-78\ °C} \underset{\substack{\textbf{1,2-addition product} \\ \textbf{78\%}}}{\underset{D\quad Cl}{CH_2CHCH{=}CHCH_3}} + \underset{\substack{\textbf{1,4-addition product} \\ \textbf{22\%}}}{\underset{D\qquad\qquad Cl}{CH_2CH{=}CHCHCH_3}}$$

When the π electrons of the diene abstract D^+ from a molecule of undissociated DCl, the chloride ion can better stabilize the positive charge at C-2 than at C-4 simply because when the chloride ion is first produced, it is closer to C-2 than to C-4. So it is a *proximity effect* that causes the 1,2-addition product to be formed faster. A **proximity effect** is an effect caused by one species being close to another.

secondary allylic cation

$$\underset{D}{CH_2}{-}\overset{}{\underset{+}{C}H}{-}CH{=}CHCH_3 \;\; Cl^- \longleftrightarrow \underset{D}{CH_2}{-}CH{=}CH{-}\underset{+}{C}HCH_3 \;\; Cl^-$$

Cl^- is closer to C-2 than to C-4

PROBLEM 9◆

a. Why does deuterium add to C-1 rather than to C-4 in the preceding reaction?
b. Why was DCl rather than HCl used in the reaction?

PROBLEM 10◆

a. When HBr adds to a conjugated diene, what is the rate-determining step?
b. When HBr adds to a conjugated diene, what is the product-determining step?

Because the greater proximity of the nucleophile to C-2 contributes to the faster rate of formation of the 1,2-addition product, the 1,2-addition product is the kinetic product for essentially all conjugated dienes. Do *not* assume, however, that the 1,4-addition product is *always* the thermodynamic product. The structure of the conjugated diene is what ultimately determines the thermodynamic product. For example, the 1,2-addition product is both the kinetic product and the thermodynamic product in the reaction of 4-methyl-1,3-pentadiene with HBr, because not only is the 1,2-product formed faster, it is more stable than the 1,4-product.

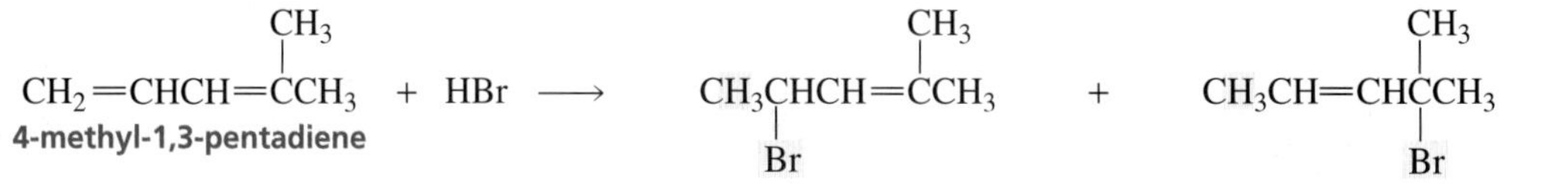

$$CH_2{=}CHCH{=}C(CH_3)CH_3 + HBr \longrightarrow CH_3CH(Br)CH{=}C(CH_3)CH_3 + CH_3CH{=}CHC(CH_3)(Br)CH_3$$

4-methyl-1,3-pentadiene

4-bromo-2-methyl-2-pentene — **1,2-addition product**, **kinetic product**, **thermodynamic product**

4-bromo-4-methyl-2-pentene — **1,4-addition product**

The 1,2- and 1,4-addition products obtained from the reaction of 2,4-hexadiene with HCl have the same stability—both have the same number of alkyl groups bonded to their sp^2 carbons. Thus, neither product is themodynamically controlled.

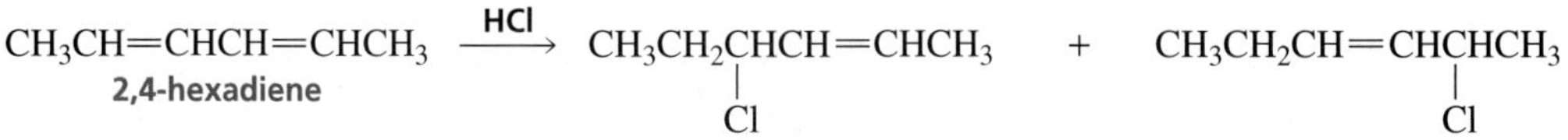

$$CH_3CH{=}CHCH{=}CHCH_3 \xrightarrow{HCl} CH_3CH_2CH(Cl)CH{=}CHCH_3 + CH_3CH_2CH{=}CHCH(Cl)CH_3$$

2,4-hexadiene

4-chloro-2-hexene — **1,2-addition product**

2-chloro-3-hexene — **1,4-addition product**

the products have the same stability

PROBLEM 11 SOLVED

For each of the following reactions, (1) give the major 1,2- and 1,4-addition products and (2) indicate which is the kinetic product and which is the thermodynamic product:

a. (3-methylenecyclohexene, CH_2) + HCl ⟶

b. $CH_3CH{=}CHC(CH_3){=}CH_2$ + HCl ⟶

c. (1-methyl-1,3-cyclohexadiene, CH_3) + HCl ⟶

d. (1-propenylcyclohexene, $CH{=}CHCH_3$) + HCl ⟶

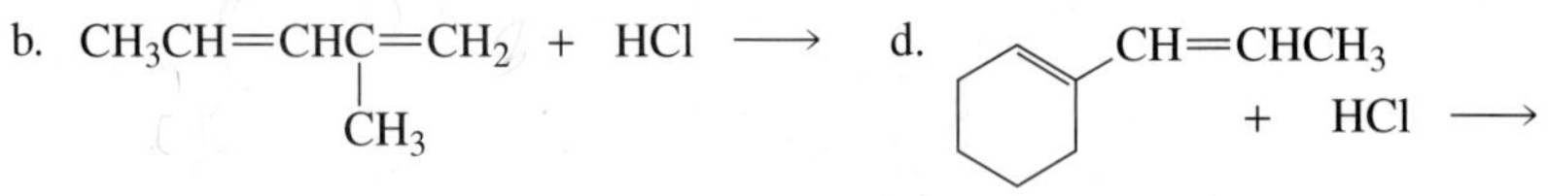

SOLUTION TO 11a First we need to determine which of the terminal sp^2 carbons of the conjugated system is going to be the C-1 carbon. The proton will be more apt to add to the indicated sp^2 carbon because the carbocation that is formed shares its positive charge with a tertiary allylic and a secondary allylic carbon. If the proton were to add to the sp^2 carbon at the other end of the conjugated system, the carbocation that would be formed would be less stable because its positive charge would be shared by a primary allylic and a secondary allylic carbon. Therefore, 3-chloro-3-methylcyclohexene is the 1,2-addition product and 3-chloro-1-methylcyclohexene is the 1,4-addition product. 3-Chloro-3-methylcyclohexene is the kinetic product because of the chloride ion's proximity to C-2, and 3-chloro-1-methylcyclohexene is the thermodynamic product because its more highly substituted double bond makes it more stable.

H⁺ adds here

CH_2 $\xrightarrow{HCl}$ CH_3 (+) $\longleftrightarrow$ CH_3 (+) $\longrightarrow$ Cl CH_3 + CH_3 Cl

+ Cl^-

3-chloro-3-methylcyclohexene — kinetic product

3-chloro-1-methylcyclohexene — thermodynamic product

8.8 The Diels–Alder Reaction: A 1,4-Addition Reaction

Otto Paul Hermann Diels (1876–1954) *was born in Germany, the son of a professor of classical philology at the University of Berlin. He received a Ph.D. from that university, working with Emil Fischer. Diels was a professor of chemistry at the University of Berlin and later at the University of Kiel. Two of his sons died in World War II. He retired in 1945 after his home and laboratory were destroyed in bombing raids. He received the 1950 Nobel Prize in chemistry, sharing it with his former student Kurt Alder.*

Reactions that create new carbon–carbon bonds are very important to synthetic organic chemists because it is only through such reactions that small carbon skeletons can be converted into larger ones (Section 6.10). The Diels–Alder reaction is particularly important because it creates *two* new carbon–carbon bonds in a way that forms a cyclic molecule. In recognition of the importance of this reaction to synthetic organic chemistry, Otto Diels and Kurt Alder received the Nobel Prize in chemistry in 1950.

In a **Diels–Alder reaction**, a conjugated diene reacts with a compound containing a carbon–carbon double bond. The latter compound is called a **dienophile** because it "loves a diene." (Δ signifies heat.)

$$CH_2{=}CH{-}CH{=}CH_2 + CH_2{=}CH{-}R \xrightarrow{\Delta}$$ (cyclohexene with R)

conjugated diene | dienophile

This reaction may not look like any reaction that you have seen before, but it is simply a 1,4-addition of an electrophile and a nucleophile to a conjugated diene. However, unlike the other 1,4-addition reactions you have seen—where the electrophile adds to the diene in the first step and the nucleophile adds to the carbocation in the second step—the Diels–Alder reaction is a **concerted reaction**: The addition of the electrophile and the nucleophile occurs in a single step. The reaction initially looks odd because the electrophile and the nucleophile that add to the conjugated diene are the adjacent sp^2 carbons of a double bond. As with other 1,4-addition reactions, the double bond in the product is between C-2 and C-3 of what was the conjugated diene.

nucleophile

electrophile

diene — four π electrons

dienophile — two π electrons

transition state — six π electrons ($\ddagger$)

new σ bond

new double bond

new σ bond

The Diels–Alder reaction is a 1,4-addition of a dienophile to a conjugated diene.

The Diels–Alder reaction is another example of a **pericyclic reaction**—a reaction that takes place in one step by a cyclic shift of electrons (Section 4.9). It is also a **cycloaddition reaction**—a reaction in which two reactants form a cyclic product. More precisely, the Diels–Alder reaction is a **[4 + 2] cycloaddition reaction** because, of the six π electrons involved in the cyclic transition state, *four* come from the conjugated diene and *two* come from the dienophile. The reaction, in essence, converts two π bonds into two σ bonds.

Kurt Alder (1902–1958) *was born in a part of Germany that is now Poland. After World War I, he and his family moved to Germany. They were expelled from their home region when it was ceded to Poland. After receiving his Ph.D. under Diels in 1926, Alder continued working with him, and in 1928 they discovered the Diels–Alder reaction. Alder was a professor of chemistry at the University of Kiel and at the University of Cologne. He received the 1950 Nobel Prize in chemistry, sharing it with his mentor, Otto Diels.*

The reactivity of the dienophile is increased if one or more electron-withdrawing groups are attached to its sp^2 carbons.

electron-withdrawing group

$$\underset{\text{1,3-butadiene}}{H_2C{=}CH{-}CH{=}CH_2} + CH_2{=}CH{-}\overset{O}{\overset{\|}{C}}CH_3 \xrightarrow{30\ °C} \text{4-acetylcyclohexene}$$

a 1,4-addition reaction to 1,3-butadiene

An electron-withdrawing group—such as a carbonyl group (C$=$O) or a cyano (C$\equiv$N) group—withdraws electrons from the double bond. This puts a partial positive charge on one of its sp^2 carbons (Figure 8.3), making the Diels–Alder reaction easier to initiate.

$$CH_2{=}CH{-}\overset{O}{\overset{\|}{C}}CH_3 \longleftrightarrow \overset{+}{C}H_2{-}CH{=}\overset{O^-}{\overset{|}{C}}CH_3$$

resonance contributors of the dienophile

$$\overset{\delta+}{C}H_2{\overset{---}{=}}CH{\overset{---}{=}}\overset{\delta-O}{\overset{\|}{C}}CH_3$$

resonance hybrid

The partially positively charged sp^2 carbon of the dienophile can be likened to the electrophile that is attacked by π electrons from C-1 of the conjugated diene. The other sp^2 carbon of the dienophile is the nucleophile that adds to C-4 of the diene.

A Molecular Orbital Description of the Diels–Alder Reaction

The two new σ bonds that are formed in a Diels–Alder reaction result from a transfer of electron density between the reactants. Molecular orbtial theory provides an insight into this process. When electrons are transferred between molecules, we must use the **HOMO (highest occupied molecular orbital)** of one reactant and the **LUMO (lowest unoccupied molecular orbital)** of the other because only an empty orbital can accept electrons. It doesn't matter whether we use the HOMO of the dienophile and the LUMO of the diene or the HOMO of the diene and the LUMO of the dienophile. We just need to use the HOMO of one and the LUMO of the other.

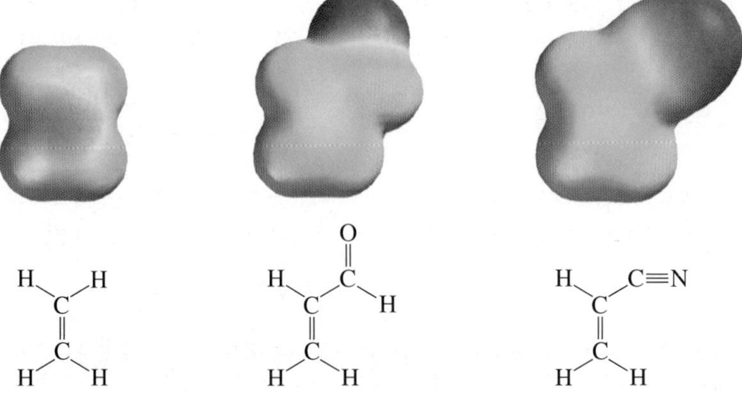

Figure 8.3 ▶
By comparing the electrostatic potential maps you can see that an electron-withdrawing substituent decreases the electron density of the carbon–carbon double bond.

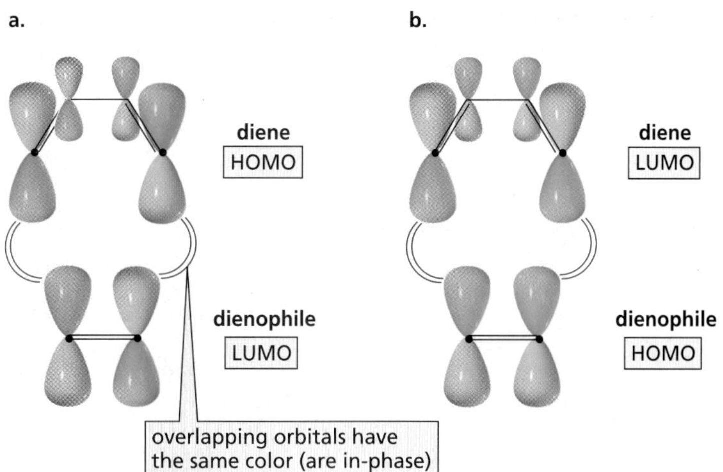

◀ **Figure 8.4**
The new σ bonds formed in a Diels–Alder reaction result from overlap of in-phase orbitals. (a) Overlap of the HOMO of the diene and the LUMO of the dienophile. (b) Overlap of the HOMO of the dienophile and the LUMO of the diene.

To construct the HOMO and LUMO needed to illustrate the transfer of electrons in a Diels–Alder reaction, we need to look at Figures 7.8 and 7.9 on pages 287 and 288. We see that the HOMO of the diene and the LUMO of the dienophile are asymmetric (Figure 8.4a) and that the LUMO of the diene and the HOMO of the dienophile are symmetric (Figure 8.4b).

A pericyclic reaction such as the Diels–Alder reaction can be described by a theory called the *conservation of orbital symmetry*. This simple theory says that pericyclic reactions occur as a result of the overlap of in-phase orbitals. The phase of the orbitals in Figure 8.4 is indicated by their color. Thus, each new σ bond formed in a Diels–Alder reaction must be created by the overlap of orbitals of the same color. Because two new σ bonds are formed, we need to have four orbitals in the correct place and with the correct color (symmetry). The figure shows that, regardless of which pair of HOMO and LUMO we choose, the overlapping orbitals have the same color. In other words, a Diels–Alder reaction occurs with relative ease. The Diels–Alder reaction and other cycloaddition reactions are discussed in greater detail in Section 29.4.

PROBLEM 12

Explain why a [2 + 2] cycloaddition reaction will not occur at room temperature. Recall that at room temperature all molecules have a ground-state electronic configuration; that is, the electrons are in the available orbitals with the lowest energy.

Stereochemistry of the Diels–Alder Reaction

If a Diels–Alder reaction creates an aymmetric carbon in the product, identical amounts of the *R* and *S* enantiomers will be formed. In other words, the product will be a racemic mixture (Section 5.19).

$$CH_2{=}CH{-}CH{=}CH_2 \; + \; CH_2{=}CH{-}C{\equiv}N \xrightarrow{\Delta}$$

asymmetric carbon

(cyclohexene with H and C≡N on the asymmetric carbon) + (enantiomer with C≡N and H)

The Diels–Alder reaction is a syn addition reaction with respect to both the diene and the dienophile: One face of the diene adds to one face of the dienophile. Therefore, if the substituents in the *dienophile* are cis, they will be cis in the product; if the substituents in the *dienophile* are trans, they will be trans in the product.

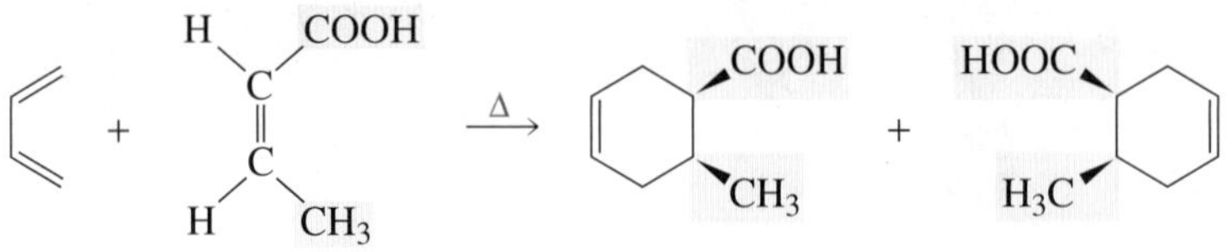

cis dienophile → cis products

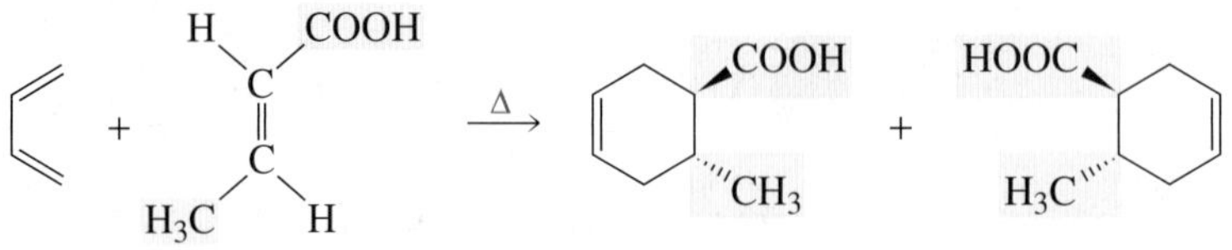

trans dienophile → trans products

The substituents in the *diene* will also maintain their relative configurations in the products. Notice that compounds containing carbon–carbon triple bonds can also be used as dienophiles in Diels–Alder reactions to prepare compounds with two isolated double bonds.

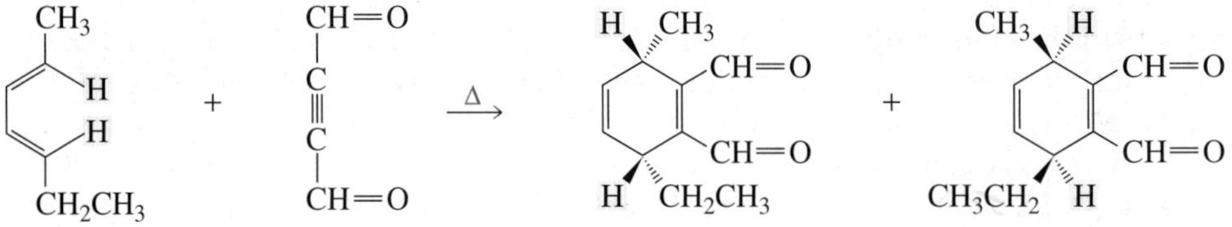

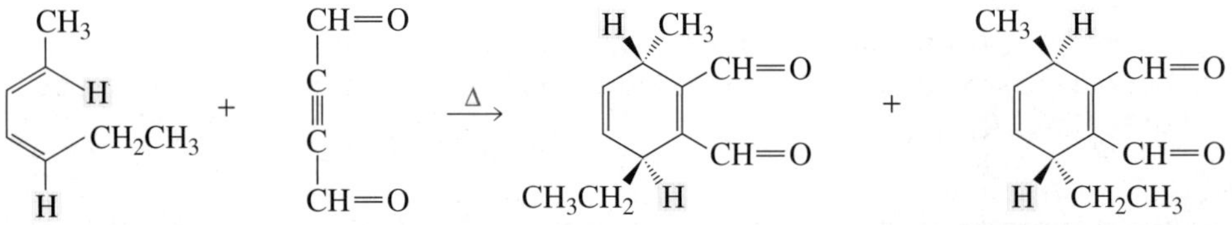

Each of the four previous reactions forms a product with two new asymmetric carbons. Thus, each product has four stereoisomers. Because only syn addition occurs, each reaction forms only two of the stereoisomers (Section 5.19). The Diels–Alder reaction is stereospecific—the configuration of the reactants is maintained during the course of the reaction—because the reaction is concerted. The stereochemistry of the reaction will be discussed in more detail in Section 29.4.

A wide variety of cyclic compounds can be obtained by varying the structures of the conjugated diene and the dienophile.

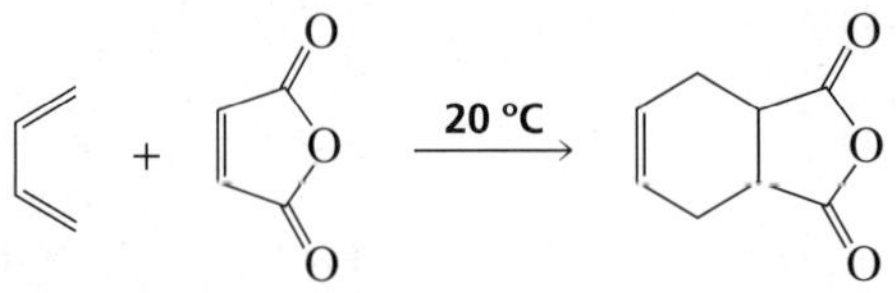

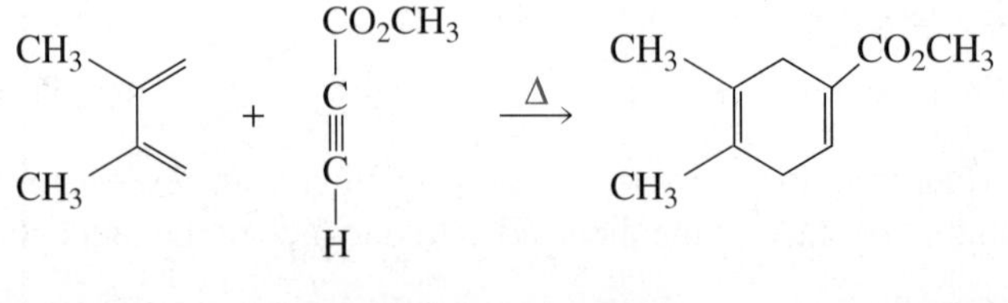

20 °C

If the dienophile has two carbon–carbon double bonds, two successive Diels–Alder reactions can occur if excess diene is available.

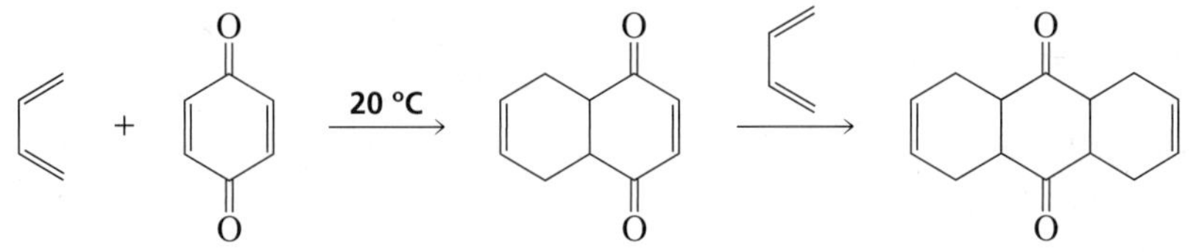

20 °C

PROBLEM 13◆

Give the products of each of the following reactions:

a. $CH_2{=}CH{-}CH{=}CH_2 + CH_3\overset{O}{\overset{\|}{C}}{-}C{\equiv}C{-}\overset{O}{\overset{\|}{C}}CH_3 \xrightarrow{\Delta}$

b. $CH_2{=}CH{-}CH{=}CH_2 + HC{\equiv}C{-}C{\equiv}N \xrightarrow{\Delta}$

c. $CH_2{=}\underset{}{C}(CH_3){-}C(CH_3){=}CH_2$ + (maleic anhydride) $\longrightarrow$

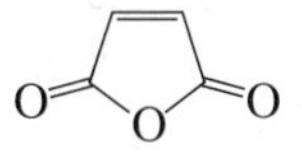

d. $CH_3\overset{CH_3}{\overset{|}{C}}{=}CH{-}CH{=}\overset{CH_3}{\overset{|}{C}}CH_3$ + (benzoquinone) $\longrightarrow$

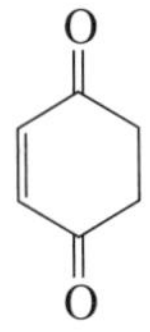

PROBLEM 14◆

Explain why the following are not optically active:

a. the product obtained from the reaction of 1,3-butadiene with *cis*-1,2-dichloroethene
b. the product obtained from the reaction of 1,3-butadiene with *trans*-1,2-dichloroethene

Predicting the Product when Both Reagents Are Unsymmetrically Substituted

In each of the preceding Diels–Alder reactions, only one product is formed (ignoring stereoisomers) because at least one of the reacting molecules is symmetrically substituted. If both the diene and the dienophile are unsymmetrically substituted, however, two products are possible:

$$CH_2{=}CHCH{=}CHOCH_3 + CH_2{=}CH\overset{O}{\overset{\|}{C}}H \longrightarrow$$

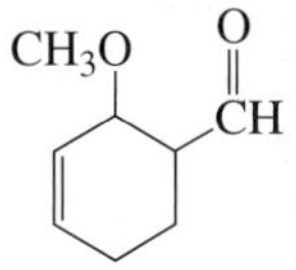

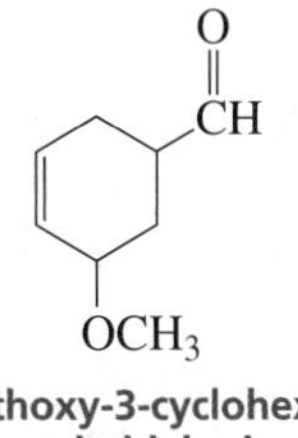

neither compound is symmetric

2-methoxy-3-cyclohexenecarbaldehyde + 5-methoxy-3-cyclohexenecarbaldehyde

Two products are possible because the reactants can be aligned in two different ways.

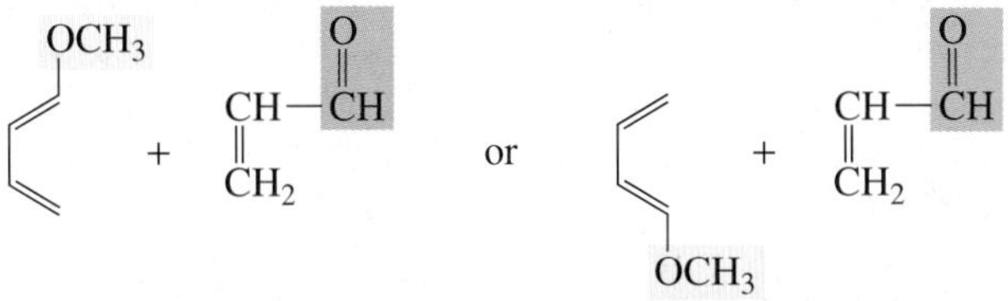

OCH_3 (diene) + $CH_2{=}CH{-}\overset{O}{\overset{\|}{C}}H$ or (diene) OCH_3 + $CH_2{=}CH{-}\overset{O}{\overset{\|}{C}}H$

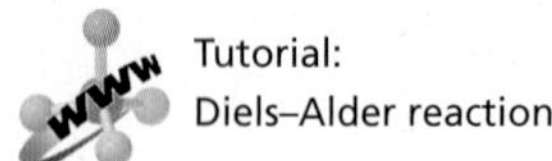
Tutorial: Diels–Alder reaction

Which of the two products will be formed (or will be formed in greater yield) depends on the charge distribution in each of the reactants. To determine the charge distribution, we need to draw contributing resonance structures. In the preceding example, the methoxy group of the diene is capable of donating electrons by resonance. As a result, the terminal carbon atom bears a partial negative charge. The aldehyde group of the dienophile, on the other hand, withdraws electrons by resonance, so its terminal carbon has a partial positive charge.

$$CH_2{=}CH{-}CH{=}CH{-}\ddot{O}CH_3 \longleftrightarrow \overset{-}{\ddot{C}}H_2{-}CH{=}CH{-}CH{=}\overset{+}{O}CH_3$$

resonance contributors of the diene

$$CH_2{=}CH{-}\overset{\ddot{O}:}{\overset{\|}{C}}H \longleftrightarrow \overset{+}{C}H_2{-}CH{=}\overset{:\ddot{O}:^-}{\overset{|}{C}}H$$

resonance contributors of the dienophile

The partially positively charged carbon atom of the dienophile will bond preferentially to the partially negatively charged carbon of the diene. Therefore, 2-methoxy-3-cyclohexenecarbaldehyde will be the major product.

$$\overset{\delta+}{CH_3O}\text{-diene}^{\delta-} + \overset{\delta-}{O}{=}CH{-}CH{=}\overset{\delta+}{CH_2} \longrightarrow \text{2-methoxy-3-cyclohexenecarbaldehyde } (CH_3O,\ \overset{O}{\overset{\|}{C}}H)$$

PROBLEM 15◆

What would be the major product if the methoxy substituent in the preceding reaction were bonded to C-2 of the diene rather than to C-1?

PROBLEM 16◆

Give the products of each of the following reactions (ignore stereoisomers):

a. $CH_2{=}CH{-}CH{=}CH{-}CH_3 + HC{\equiv}C{-}C{\equiv}N \xrightarrow{\Delta}$

b. $CH_2{=}CH{-}\underset{\displaystyle CH_3}{\underset{|}{C}}{=}CH_2 + HC{\equiv}C{-}C{\equiv}N \xrightarrow{\Delta}$

Conformations of the Diene

We saw in Section 7.11 that a conjugated diene such as 1,3-butadiene is most stable in a planar confomation. A conjugated diene can exist in two different planar conformations: an ***s*-cis conformation** and an ***s*-trans conformation**. By "*s*-cis," we mean that the double bonds are cis about the single bond (s = single). The *s*-trans conformation is little more stable (2.3 kcal or 9.6 kJ) than the *s*-cis conformation because the close proximity of the hydrogens causes some steric strain (Section 2.10). The rotational barrier between the *s*-cis and *s*-trans conformations (4.9 kcal/mol or 20.5 kJ/mole) is low enough to allow them to interconvert rapidly at room temperature.

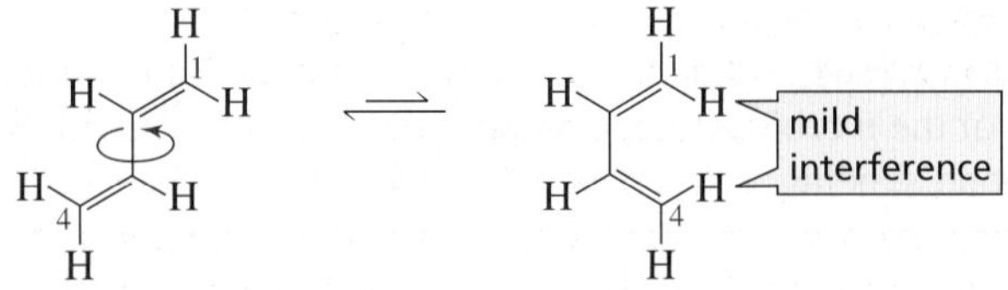

s-trans conformation **s-cis conformation**

mild interference

3-D Molecules: *s*-trans Conformation of Butadiene; *s*-cis Conformation of Butadiene

In order to participate in a Diels–Alder reaction, the conjugated diene must be in an *s*-cis conformation because in an *s*-trans conformation, the number 1 and number 4 carbons are too far apart to react with the dienophile. A conjugated diene that is locked in an *s*-trans conformation cannot undergo a Diels–Alder reaction.

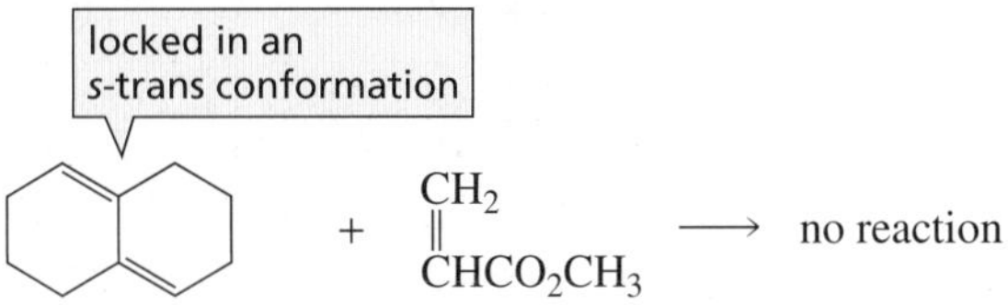

locked in an *s*-trans conformation

$$+ \; CH_2{=}CHCO_2CH_3 \longrightarrow \text{no reaction}$$

A conjugated diene that is locked in an *s*-cis conformation, such as 1,3-cyclopentadiene, is highly reactive in a Diels–Alder reaction. When the diene is a cyclic compound, the product of a Diels–Alder reaction is a **bridged bicyclic compound**—a compound that contains two rings that share two nonadjacent carbons.

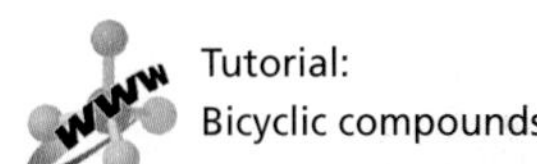

locked in an *s*-cis conformation

both rings share these carbons

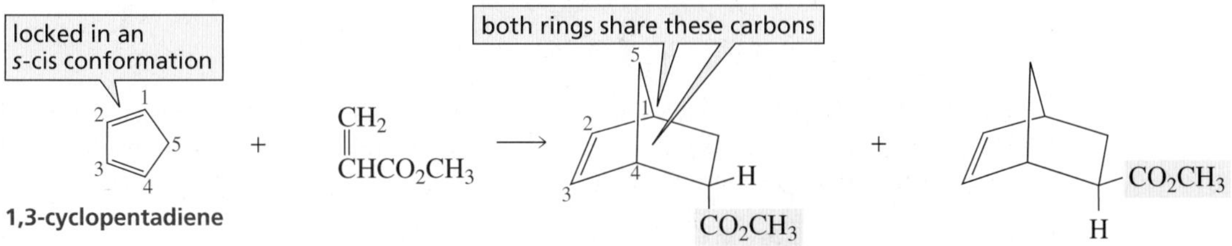

1,3-cyclopentadiene + $CH_2{=}CHCO_2CH_3$ ⟶ 81% + 19%

bridged bicyclic compounds

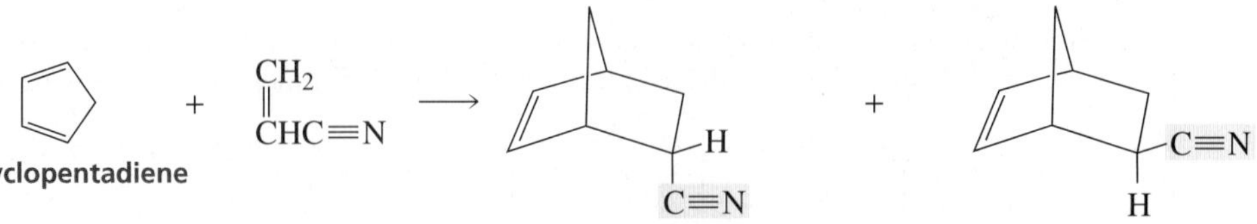

1,3-cyclopentadiene + $CH_2{=}CHC{\equiv}N$ ⟶ 60% + 40%

bridged bicyclic compounds

There are two possible configurations for bridged bicyclic compounds, because the substituent (R) can point either toward the double bond (the **endo** configuration) or away from the double bond (the **exo** configuration).

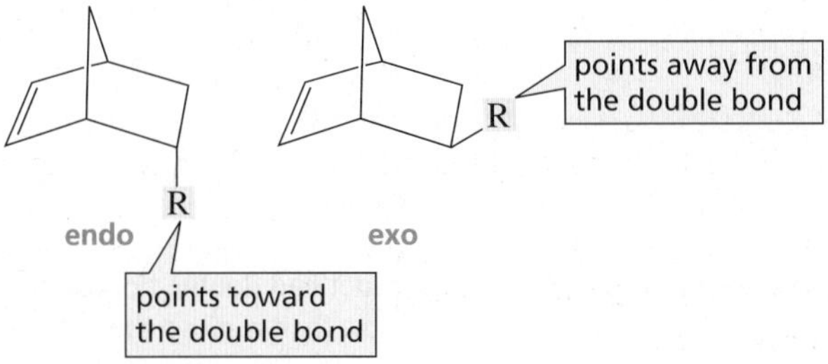

endo — points toward the double bond

exo — points away from the double bond

When the dienophile has π electrons (other than the π electrons of its carbon–carbon double bond), more of the endo product is formed.

The endo product is formed faster when the dienophile has a substituent with π electrons. It has been suggested that this is due to stabilization of the transition state by the interaction of the *p* orbitals of the dienophile's substituent with the *p* orbitals of the new double bond's being formed in what was the diene. This interaction, called *secondary orbital overlap*, can occur only if the substituent with the *p* orbitals lies underneath (endo) rather than away from (exo) the six-membered ring.

PROBLEM 17◆

Which of the following conjugated dienes would not react with a dienophile in a Diels–Alder reaction?

a. 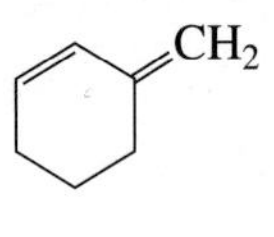 CH_2

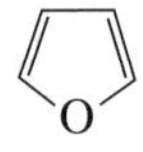

b.

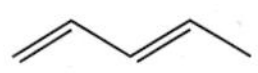

c. 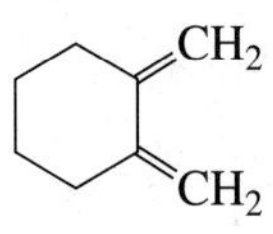 CH_2 CH_2

d.

e. O

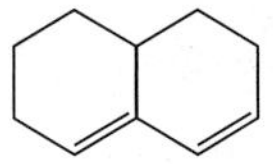

f. 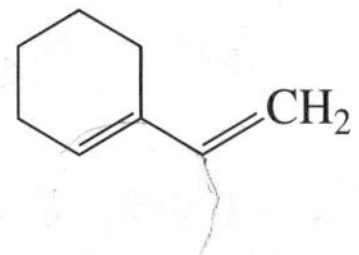 CH_2

PROBLEM 18 SOLVED

List the following dienes in order of decreasing reactivity in a Diels–Alder reaction:

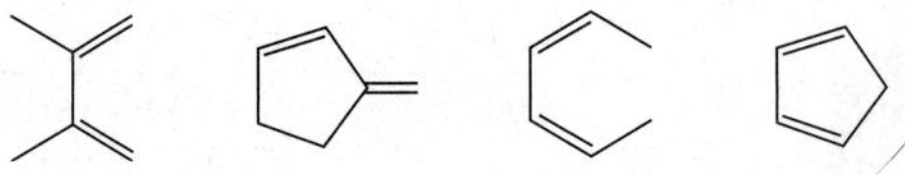

SOLUTION The most reactive diene has the double bonds locked in an *s*-cis conformation, whereas the least reactive diene cannot achieve the required *s*-cis conformation because it is locked in an *s*-trans conformation.

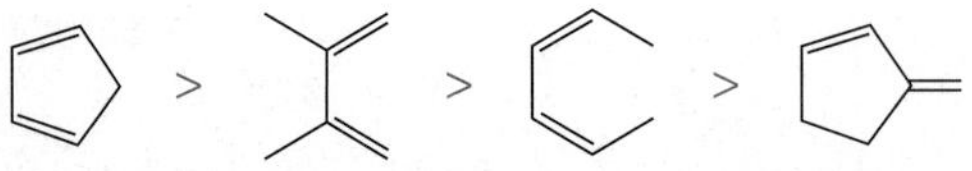

\> \> \>

2,3-Dimethyl-1,3-butadiene and 2,4-hexadiene are of intermediate reactivity because they can exist in both *s*-cis and *s*-trans conformations. 2,4-Hexadiene is less apt to be in the required *s*-cis conformation because of steric interference between the terminal methyl groups. Consequently, 2,4-hexadiene is less reactive than 2,3-dimethyl-1,3-butadiene.

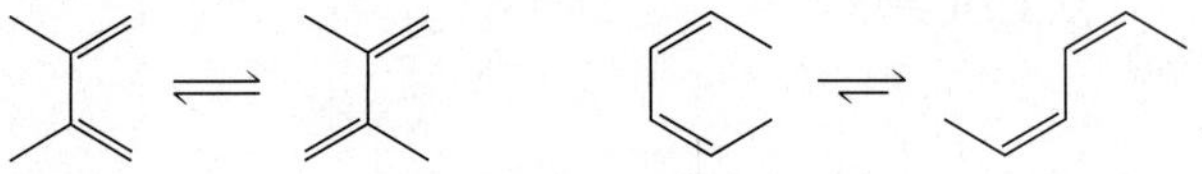

s-cis *s*-trans *s*-cis *s*-trans

PROBLEM-SOLVING STRATEGY

What diene and what dienophile were used to synthesize the following compound?

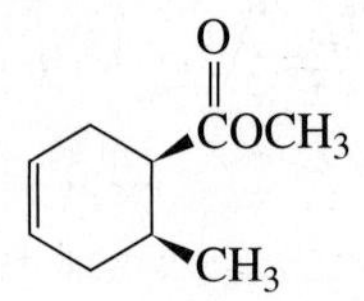

O ‖ $COCH_3$ CH_3

The diene that was used to form the cyclic product had double bonds on either side of the double bond in the product, so put in the π bonds, then remove the π bond between them.

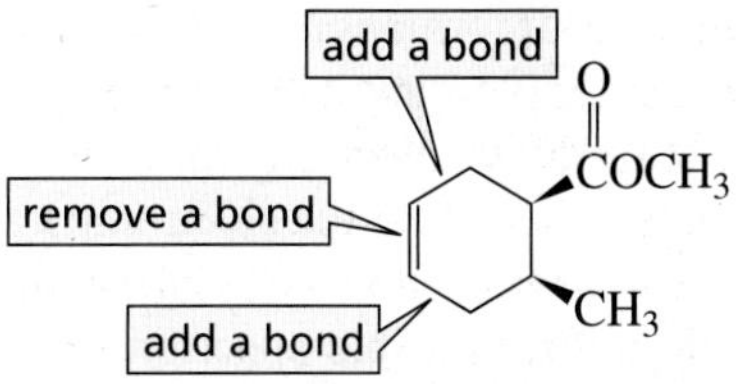

The new σ bonds are now on either side of the double bonds.

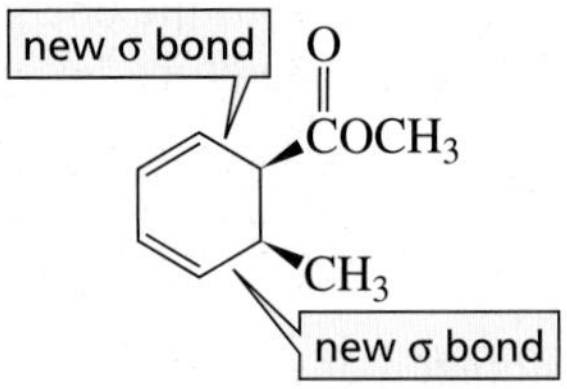

Erasing these σ bonds and putting a π bond between the two carbons that were attached by the σ bonds gives the diene and the dienophile.

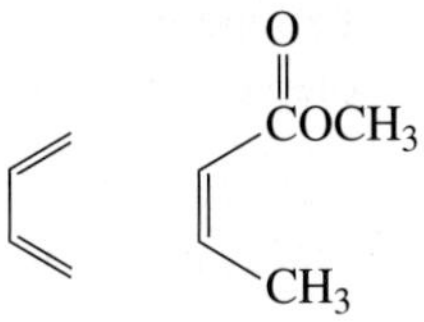

Now continue on to Problem 19.

PROBLEM 19◆

What diene and what dienophile should be used to synthesize the following compounds?

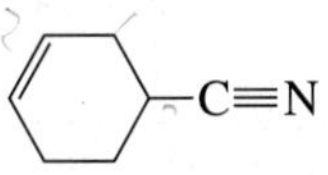

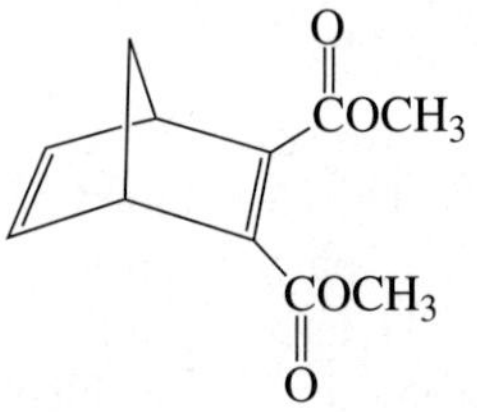

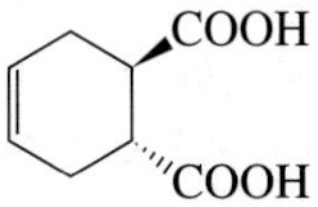

8.9 Ultraviolet and Visible Spectroscopy

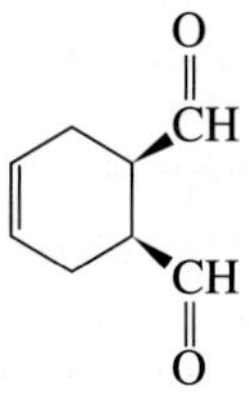

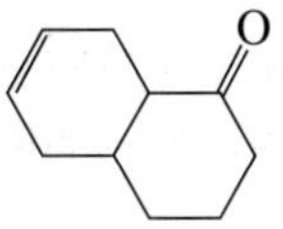

Spectroscopy is the study of the interaction between matter and **electromagnetic radiation**—radiant energy that displays the properties of both particles and waves. Several different spectrophotometric techniques are used to identify compounds. Each employs a different type of electromagnetic radiation. We will start here by looking at ultraviolet and visible (UV/Vis) spectroscopy. We will look at infrared (IR) spectroscopy in Chapter 13 and nuclear magnetic resonance (NMR) spectroscopy in Chapter 14.

UV/Vis spectroscopy provides information about compounds with conjugated double bonds. Ultraviolet light and visible light have just the right energy to cause an electronic transition—the promotion of an electron from one orbital to another of higher energy. Depending on the energy needed for the electronic transition, a molecule will absorb either ultraviolet or visible light. If it absorbs **ultraviolet light**, a UV spectrum is obtained; if it absorbs **visible light**, a visible spectrum is obtained. Ultraviolet light is electromagnetic radiation with wavelengths ranging from 180 to 400 nm (nanometers); visible light has wavelengths ranging from 400 to 780 nm. (One nanometer is 10^{-9} m, or 10 Å.) **Wavelength** (λ) is inversely related to the energy: The shorter the wavelength, the greater is the energy. Ultraviolet light, therefore, has greater energy than visible light.

The shorter the wavelength, the greater is the energy of the radiation.

$$E = \frac{hc}{\lambda}$$

h = Planck's constant
c = velocity of light
λ = wavelength

The normal electronic configuration of a molecule is known as its **ground state**—all the electrons are in the lowest-energy molecular orbitals. When a molecule absorbs light of an appropriate wavelength and an electron is promoted to a higher energy molecular orbital, the molecule is then in an **excited state**. Thus, an **electronic transition** is the promotion of an electron to a higher energy MO. The relative energies of the bonding, nonbonding, and antibonding molecular orbitals are shown in Figure 8.5.

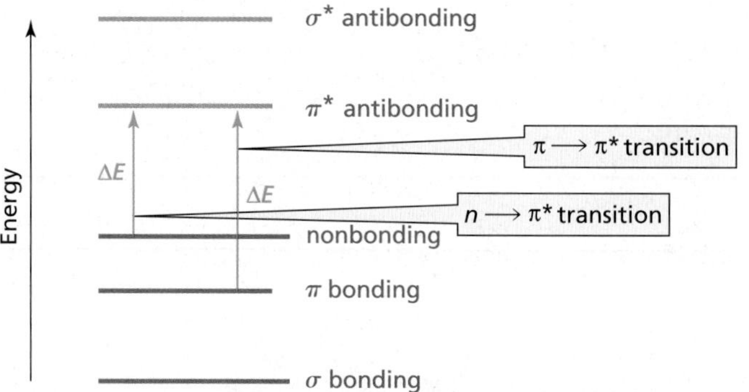

Figure 8.5 ▶ Relative energies of the bonding, nonbonding, and antibonding orbitals.

Ultraviolet and visible light have sufficient energy to cause only the two electronic transitions shown in Figure 8.5. The electronic transition with the lowest energy is the promotion of a nonbonding (lone-pair) electron (n) into a π^* antibonding molecular orbital. This is called an $n \longrightarrow \pi^*$ (stated as "n to π star") transition. The higher energy electronic transition is the promotion of an electron from a π bonding molecular orbital into a π^* antibonding molecular orbital, known as a $\pi \longrightarrow \pi^*$ (stated as "π to π star") transition. This means that *only organic compounds with π electrons can produce UV/Vis spectra.*

The UV spectrum of acetone is shown in Figure 8.6. Acetone has both π electrons and lone-pair electrons. Thus, there are two **absorption bands**: one for the $\pi \longrightarrow \pi^*$

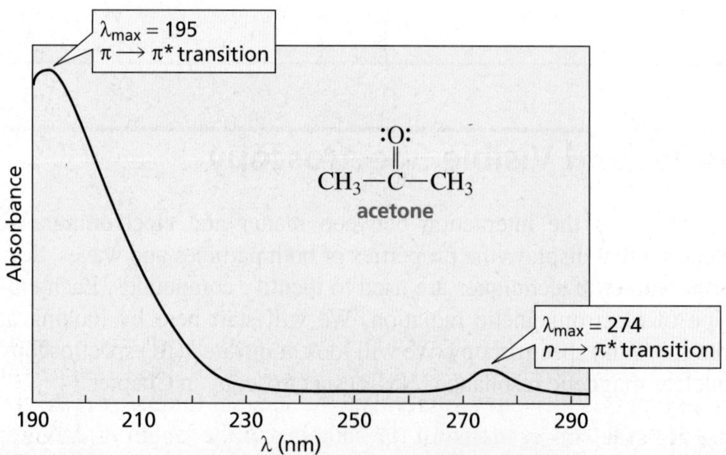

Figure 8.6 ▶ The UV spectrum of acetone.

transition and one for the $n \rightarrow \pi^*$ transition. The λ_{max} (stated as "lambda max") is the wavelength corresponding to the highest point (maximum absorbance) of the absorption band. For the $\pi \rightarrow \pi^*$ transition, $\lambda_{max} = 195$ nm; for the $n \rightarrow \pi^*$ transition, $\lambda_{max} = 274$ nm. We know that the $\pi \rightarrow \pi^*$ transition in Figure 8.6 corresponds to the λ_{max} at the shorter wavelength because that transition requires more energy than the $n \rightarrow \pi^*$ transition.

A **chromophore** is that part of a molecule that absorbs UV or visible light. The carbonyl group is the chromophore of acetone. The following four compounds all have the same chromophore, so they all have approximately the same λ_{max}.

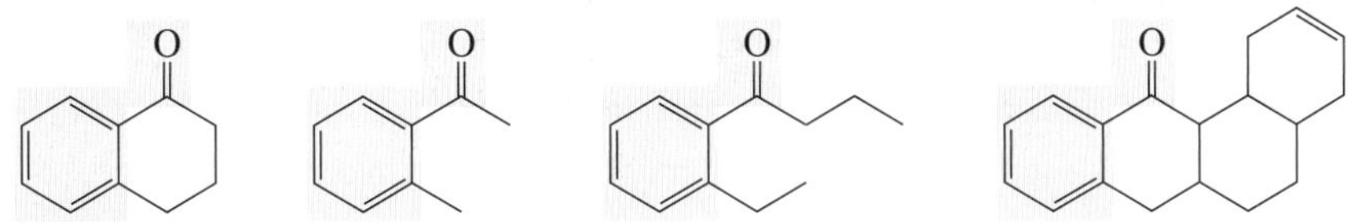

O O O O

PROBLEM 20

Explain why diethyl ether does not have a UV spectrum, even though it has lone-pair electrons.

ULTRAVIOLET LIGHT AND SUNSCREENS

Exposure to ultraviolet light stimulates specialized cells in the skin to produce a black pigment known as melanin, which causes the skin to look tan. Melanin absorbs UV light, so it protects our bodies from the harmful effects of the sun. If more UV light reaches the skin than the melanin can absorb, the light will burn the skin and can cause photochemical reactions that can result in skin cancer (Section 29.6). UV-A is the lowest-energy UV light (315 to 400 nm) and does the least biological damage. Fortunately, most of the more dangerous, higher-energy UV light, UV-B (290 to 315 nm) and UV-C (180 to 290 nm), is filtered out by the ozone layer in the stratosphere. That is why there is such great concern about the apparent thinning of the ozone layer (Section 9.9).

Applying a sunscreen can protect skin against UV light. Some sunscreens contain an inorganic component, such as zinc oxide, that reflects the light as it reaches the skin. Others contain a compound that absorbs UV light. PABA was the first commercially available UV-absorbing sunscreen. PABA absorbs UV-B light, but is not very soluble in oily skin lotions. Less polar compounds, such as Padimate O, are now commonly used. Recent research has shown that sunscreens that absorb only UV-B light do not give adequate protection against skin cancer; both UV-A and UV-B protection are needed. Giv Tan F absorbs both UV-B and UV-A light, so it gives better protection.

The amount of protection provided by a particular sunscreen is indicated by its SPF (sun protection factor). The higher the SPF, the greater is the protection.

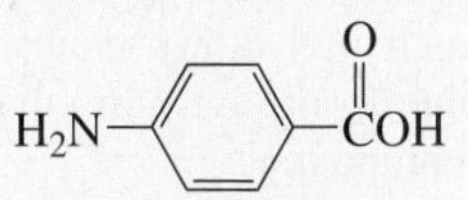

H_2N–C_6H_4–$\overset{O}{\overset{\|}{C}}OH$

para-aminobenzoic acid
PABA

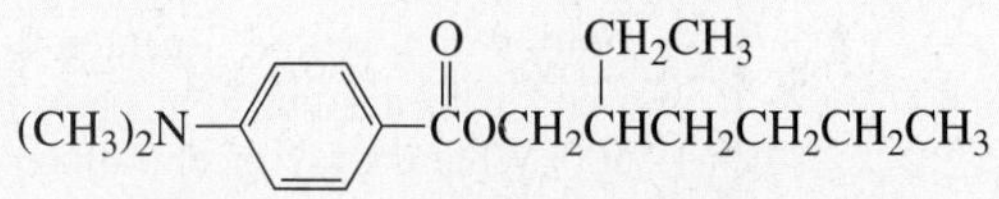

$(CH_3)_2N$–C_6H_4–$\overset{O}{\overset{\|}{C}}OCH_2\underset{}{C}HCH_2CH_2CH_2CH_3$ (with CH_2CH_3 on CH)

2-ethylhexyl 4-(dimethylamino)benzoate
Padimate O

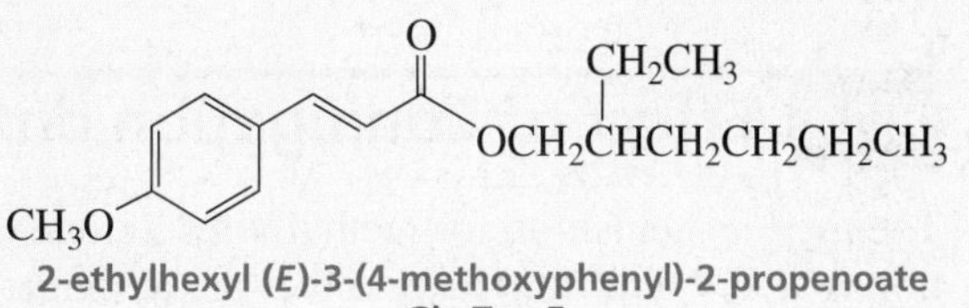

CH_3O–C_6H_4–CH=CH–$\overset{O}{\overset{\|}{C}}OCH_2CHCH_2CH_2CH_2CH_3$ (with CH_2CH_3 on CH)

2-ethylhexyl (*E*)-3-(4-methoxyphenyl)-2-propenoate
Giv Tan F

8.10 The Beer–Lambert Law

Wilhelm Beer and Johann Lambert independently proposed that at a given wavelength, the absorbance of a sample depends on the amount of absorbing species that the light encounters as it passes through a solution of the sample. In other words, absorbance depends on both the concentration of the sample and the length of the light

Wilhelm Beer (1797–1850) *was born in Germany. He was a banker whose hobby was astronomy. He was the first to make a map of the darker and lighter areas of Mars.*

Johann Heinrich Lambert (1728–1777), *a German-born mathematician, was the first to make accurate measurements of light intensities and to introduce hyperbolic functions into trigonometry.*

Although the equation that relates absorbance, concentration, and light path bears the names of Beer and Lambert, it is believed that **Pierre Bouguer (1698–1758)**, *a French mathematician, first formulated the relationship in 1729.*

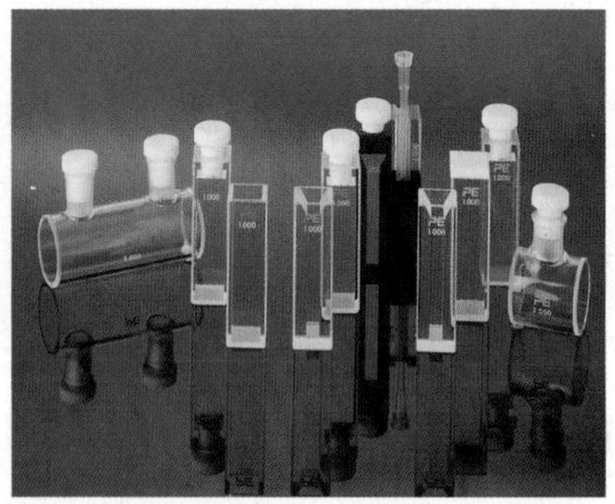

▲ Cells used in UV/Vis spectroscopy.

path through the sample. The relationship among absorbance, concentration, and length of the light path is known as the **Beer–Lambert law** and is given by

$$A = cl\varepsilon$$

where

$$A = \text{absorbance of the sample} = \log\frac{I_0}{I}$$

I_0 = intensity of the radiation entering the sample

I = intensity of the radiation emerging from the sample

c = concentration of the sample, in moles/liter

l = length of the light path through the sample, in centimeters

ε = molar absorptivity (liter mol^{-1} cm^{-1})

The **molar absorptivity** (formerly called the extinction coefficient) of a compound is a constant that is characteristic of the compound at a particular wavelength. It is the absorbance that would be observed for a 1.00 M solution in a cell with a 1.00-cm path length. The molar absorptivity of acetone, for example, is 9000 at 195 nm and 13.6 at 274 nm. The solvent in which the sample is dissolved when the spectrum is taken is reported because molar absorptivity is not exactly the same in all solvents. So the UV spectrum of acetone in hexane would be reported as λ_{max} 195 nm (ε_{max} = 9000, hexane); λ_{max} 274 nm (ε_{max} = 13.6, hexane). Because absorbance is proportional to concentration, the concentration of a solution can be determined if the absorbance and molar absorptivity at a particular wavelength are known.

The two absorption bands of acetone in Figure 8.6 are very different in size because of the difference in molar absorptivity at the two wavelengths. Small molar absorptivities are characteristic of $n \rightarrow \pi^*$ transitions, so these transitions can be difficult to detect. Consequently, $\pi \rightarrow \pi^*$ transitions are usually more useful in UV/Vis spectroscopy.

In order to obtain a UV or visible spectrum, the solution is placed in a cell. Most cells have 1-cm path lengths. Either glass or quartz cells can be used for visible spectra, but quartz cells must be used for UV spectra because glass absorbs UV light.

PROBLEM 21◆

A solution of 4-methyl-3-penten-2-one in ethanol shows an absorbance of 0.52 at 236 nm in a cell with a 1-cm light path. Its molar absorptivity in ethanol at that wavelength is 12,600. What is the concentration of the compound?

8.11 Effect of Conjugation on λ_{max}

The $n \rightarrow \pi^*$ transition for methyl vinyl ketone is at 324 nm, and the $\pi \rightarrow \pi^*$ transition is at 219 nm. Both λ_{max} values are at longer wavelengths than the corresponding λ_{max} values of acetone because methyl vinyl ketone has two conjugated double bonds.

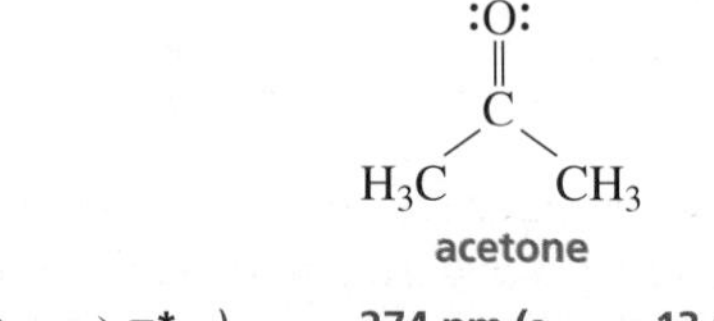

acetone

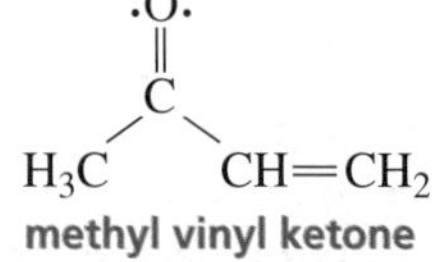

methyl vinyl ketone

	acetone	methyl vinyl ketone
$n \rightarrow \pi^*$	λ_{max} = 274 nm (ε_{max} = 13.6)	λ_{max} = 331 nm (ε_{max} = 25)
$\pi \rightarrow \pi^*$	λ_{max} = 195 nm (ε_{max} = 9000)	λ_{max} = 203 nm (ε_{max} = 9600)

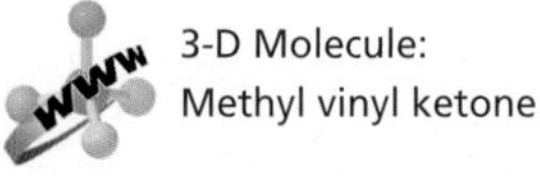

3-D Molecule: Methyl vinyl ketone

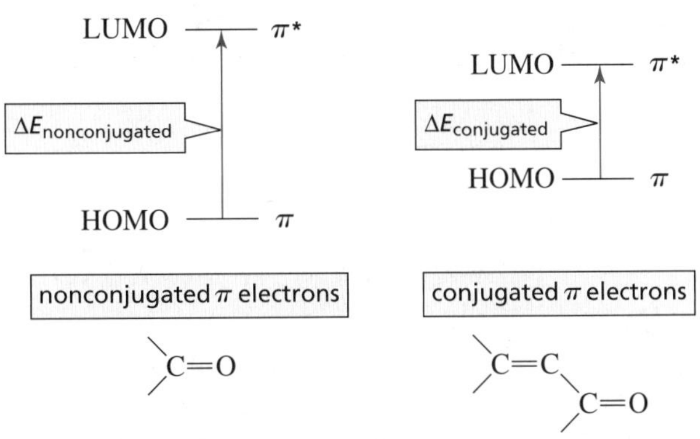

Figure 8.7
Conjugation raises the energy of the HOMO and lowers the energy of the LUMO.

Conjugation raises the energy of the HOMO and lowers the energy of the LUMO, so less energy is required for an electronic transition in a conjugated system than in a nonconjugated system (Figure 8.7). The more conjugated double bonds there are in a compound, the less energy is required for the electronic transition, and therefore the longer is the wavelength at which the electronic transition occurs.

The λ_{max} values of the $\pi \rightarrow \pi^*$ transition for several conjugated dienes are shown in Table 8.3. Notice that both the λ_{max} and the molar absorptivity increase as the number of conjugated double bonds increases. Thus, the λ_{max} of a compound can be used to predict the number of conjugated double bonds in the compound.

The λ_{max} increases as the number of conjugated double bonds increases.

Table 8.3 Values of λ_{max} and ε for Ethylene and Conjugated Dienes

Compound	λ_{max} (nm)	ε (M^{-1} cm^{-1})
$H_2C{=}CH_2$	165	15,000
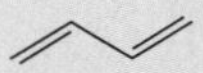	217	21,000
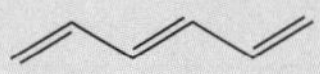	256	50,000
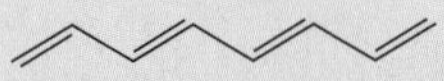	290	85,000
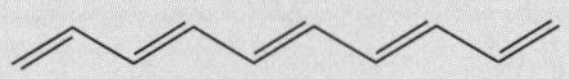	334	125,000
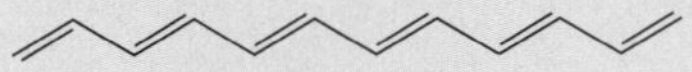	364	138,000

If a compound has enough conjugated double bonds, it will absorb visible light ($\lambda_{max} > 400$ nm) and the compound will be colored. β-Carotene, a precursor of vitamin A, is an orange substance found in carrots, apricots, and sweet potatoes. Lycopene is red and is found in tomatoes, watermelon, and pink grapefruit.

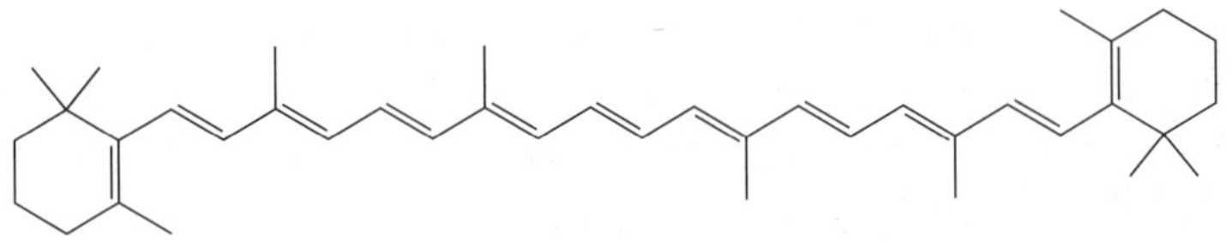

β-carotene
λ_{max} = 455 nm

3-D Molecule: 1,3,5,7-Octatetraene

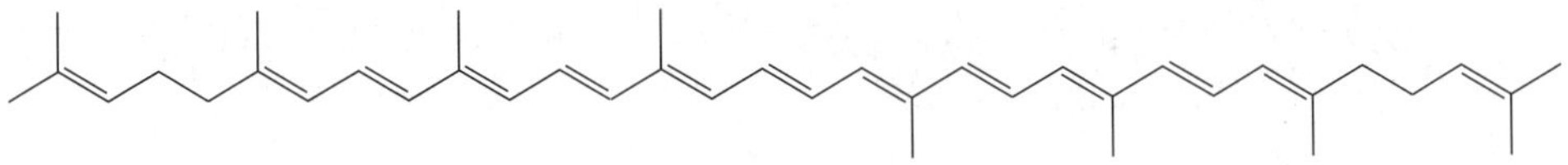

lycopene
λ_{max} = 474 nm

An **auxochrome** is a substituent that when attached to a chromophore, alters the λ_{max} and the intensity of the absorption, usually increasing both; OH and NH_2 groups are auxochromes. The lone-pair electrons on oxygen and nitrogen are available for interaction with the π electron cloud of the benzene ring, and such an interaction increases λ_{max}. Because the anilinium ion does not have an auxochrome, its λ_{max} is similar to that of benzene.

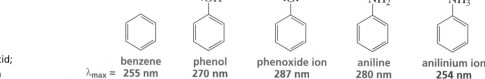

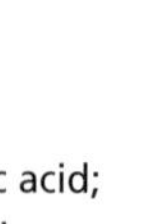

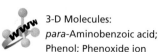

3-D Molecules: *para*-Aminobenzoic acid; Phenol; Phenoxide ion

	benzene	phenol	phenoxide ion	aniline	anilinium ion
		$\ddot{O}H$	$\ddot{O}^-$	$\ddot{N}H_2$	$\overset{+}{N}H_3$
λ_{max} =	255 nm	270 nm	287 nm	280 nm	254 nm

Removing a proton from phenol and thereby forming phenoxide ion (also called phenolate ion) increases the λ_{max} because the resulting ion has an additional lone pair. Protonating aniline (and thereby forming the anilinium ion) decreases the λ_{max}, because the lone pair is no longer available to interact with the π cloud of the benzene ring. Because wavelengths of red light are longer than those of blue light (Figure 13.11 on page 497), a shift to a longer wavelength is called a **red shift**, and a shift to a shorter wavelength is called a **blue shift**. Deprotonation of phenol results in a red shift, whereas protonation of aniline produces a blue shift.

red shift →

← blue shift

200 nm — 400 nm

PROBLEM 22◆

Rank the compounds in order of decreasing λ_{max}:

benzene; biphenyl; C_6H_5—CH=CH—C_6H_5; C_6H_5—CH=CH_2

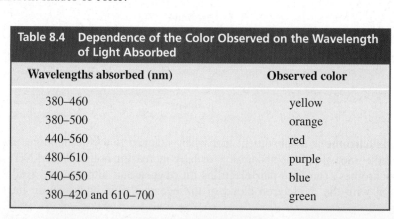

8.12 The Visible Spectrum and Color

White light is a mixture of all wavelengths of visible light. If any color is removed from white light, the remaining light appears colored. So if a compound absorbs visible light, the compound will appear colored. Its color depends on the color of the light transmitted to the eye. In other words, it depends on the color produced from the wavelengths of light that are *not* absorbed.

The relationship between the wavelengths of the light absorbed and the color observed is shown in Table 8.4. Notice that two absorption bands are necessary to produce green. Most colored compounds have fairly broad absorption bands; vivid colors have narrow absorption bands. The human eye is able to distinguish more than a million different shades of color!

R = CH_3 in chlorophyll *a*

R = $\overset{O}{\overset{\|}{C}}H$ in chlorophyll *b*

Chlorophyll *a* and *b* are the pigments that make plants look green. These highly conjugated compounds absorb nongreen light. Therefore, plants reflect green light.

Table 8.4 Dependence of the Color Observed on the Wavelength of Light Absorbed

Wavelengths absorbed (nm)	Observed color
380–460	yellow
380–500	orange
440–560	red
480–610	purple
540–650	blue
380–420 and 610–700	green

Azobenzenes (benzene rings connected by an N=N bond) have an extended conjugated system that causes them to absorb light from the visible region of the spectrum. Some substituted azobenzenes are used commercially as dyes. Varying the extent of conjugation and the substituents attached to the conjugated system provides a large number of different colors. Notice that the only difference between butter yellow and methyl orange is an $SO_3^-\ Na^+$ group. Methyl orange is a commonly used acid–base indicator. When margarine was first produced, it was colored with butter yellow to make it look more like butter. (White margarine would not have been very appetizing.) This dye was abandoned after it was found to be carcinogenic. β-Carotene is currently used to color margarine (page 325).

methyl orange
an azobenzene

butter yellow
an azobenzene

▲ Lycopene, β-carotene, and anthocyanins are found in the leaves of trees, but their characteristic colors are usually obscured by the green color of chlorophyll. In the fall when chlorophyll degrades, the colors become apparent.

PROBLEM 23

a. At pH = 7, one of the following ions is purple and the other is blue. Which is which?

b. What would be the difference in the colors of the compounds at pH = 3?

ANTHOCYANINS: A COLORFUL CLASS OF COMPOUNDS

A class of highly conjugated compounds called anthocyanins is responsible for the red, purple, and blue colors of many flowers (poppies, peonies, cornflowers), fruits (cranberries, rhubarb, strawberries, blueberries), and vegetables (beets, radishes, red cabbage).

In a neutral or basic solution, the monocyclic fragment of the anthocyanin is not conjugated with the rest of the molecule. The anthocyanin is a colorless compound because it does not absorb visible light. In an acidic environment, the OH group becomes protonated and water is eliminated. (Recall that water, being a weak base, is a good leaving group.) Loss of water results in the third ring's becoming conjugated with the rest of the molecule. As a result of the extended conjugation, the anthocyanin absorbs visible light with wavelengths between 480 and 550 nm. The exact wavelength of light absorbed depends on the substituents (R) on the anthocyanin. Thus, the flower, fruit, or vegetable appears red, purple, or blue, depending on what the R groups are. You can see this if you change the pH of cranberry juice so that it is no longer acidic.

anthocyanin
(three rings are conjugated)
red, blue, or purple

(conjugation is disrupted)
colorless

(conjugation is disrupted)
colorless

R = H, OH, or OCH_3
R' = H, OH, or OCH_3

8.13 Uses of UV/VIS Spectroscopy

Reaction rates are commonly measured using UV/Vis spectroscopy. The rate of any reaction can be measured, as long as one of the reactants or one of the products absorbs UV or visible light at a wavelength at which the other reactants and products have little or no absorbance. For example, the anion of nitroethane has a λ_{max} at 240 nm, but neither H_2O (the other product) nor the reactants show any significant absorbance at that wavelength. In order to measure the rate at which hydroxide ion removes a proton from nitroethane (i.e., the rate at which the nitroethane anion is formed), the UV spectrophotometer is adjusted to measure absorbance at 240 nm as a function of time instead of absorbance as a function of wavelength. Nitroethane is added to a quartz cell containing a basic solution, and the rate of the reaction is determined by monitoring the increase in absorbance at 240 nm (Figure 8.8).

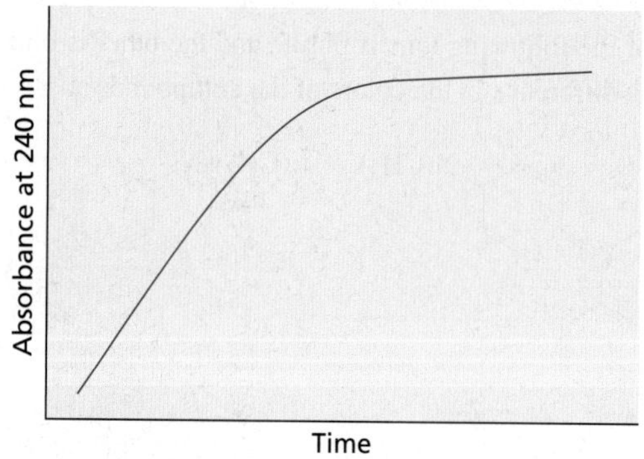

Figure 8.8 ▶
The rate at which a proton is removed from nitroethane is determined by monitoring the increase in absorbance at 240 nm.

$$\underset{\text{nitroethane}}{CH_3CH_2NO_2} + HO^- \rightleftharpoons \underset{\substack{\text{nitroethane anion} \\ \lambda_{max} = 240 \text{ nm}}}{CH_3\bar{C}HNO_2} + H_2O$$

Absorbance at 240 nm

Time

The enzyme lactate dehydrogenase catalyzes the reduction of pyruvate by NADH to form lactate. NADH is the only species in the reaction mixture that absorbs light at 340 nm, so the rate of the reaction can be determined by monitoring the decrease in absorbance at 340 nm (Figure 8.9).

Figure 8.9 ▶
The rate of reduction of pyruvate by NADH is measured by monitoring the decrease in absorbance at 340 nm.

$$\underset{\text{pyruvate}}{CH_3\overset{O}{\overset{\|}{C}}COO^-} + \underset{\lambda_{max} = 340 \text{ nm}}{NADH} + H^+ \xrightarrow{\text{lactate dehydrogenase}} \underset{\text{lactate}}{CH_3\overset{OH}{\overset{|}{C}}HCOO^-} + NAD^+$$

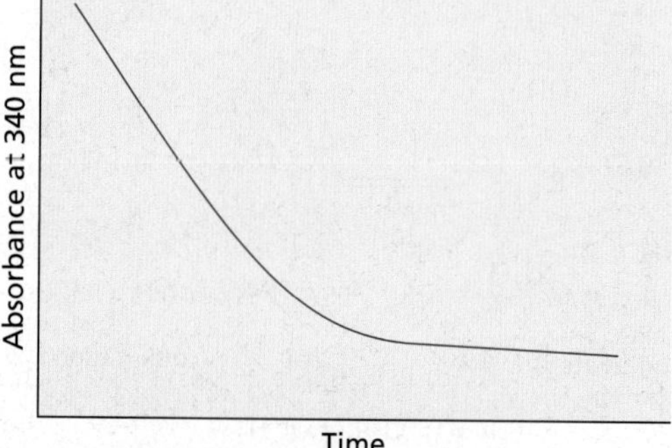

PROBLEM 24

Describe how one could determine the rate of the alcohol dehydrogenase catalyzed oxidation of ethanol by NAD^+ (Section 20.11).

The pK_a of a compound can be determined by UV/Vis spectroscopy if either the acidic form or the basic form of the compound absorbs UV or visible light. For example, the phenoxide ion has a λ_{max} at 287 nm. If the absorbance at 287 nm is determined as a function of pH, the pK_a of phenol can be ascertained by determining the pH at which exactly one-half the increase in absorbance has occurred (Figure 8.10). At this pH, half of the phenol has been converted into phenoxide. Recall from Section 1.20 that the Henderson–Hasselbalch equation states that the pK_a of the compound is the pH at which half the compound exists in its acidic form and half exists in its basic form ([HA] = [A^-]).

UV spectroscopy can also be used to estimate the nucleotide composition of DNA. The two strands of DNA are held together by both A–T base pairs and G–C base pairs (Section 27.4). When DNA is heated, the strands break apart. Single-stranded DNA has a greater molar absorptivity at 260 nm than does double-stranded DNA. The melting temperature (T_m) of DNA is the midpoint of an absorbance-versus-temperature curve (Figure 8.11). For double-stranded DNA, T_m increases with increasing numbers of G–C base pairs because they are held together by three hydrogen bonds, whereas A–T base pairs are held together by only two hydrogen bonds. Therefore, T_m can be used to estimate the number of G–C base pairs. These are just a few examples of the many uses of UV/Vis spectroscopy.

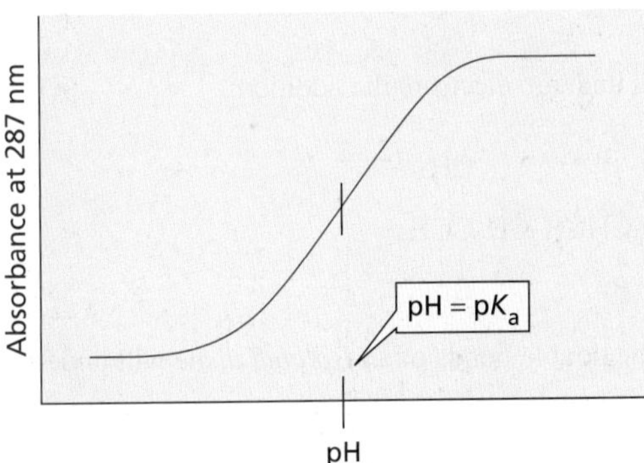

▲ **Figure 8.10**
The absorbance of an aqueous solution of phenol as a function of pH.

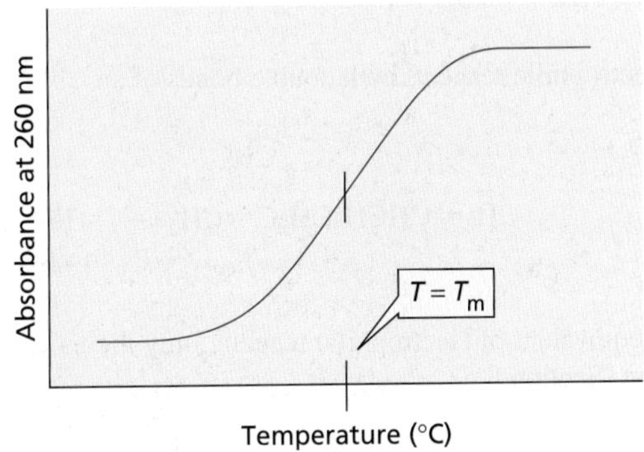

▲ **Figure 8.11**
The absorbance of a solution of DNA as a function of temperature.

Summary

Dienes are hydrocarbons with two double bonds. **Conjugated double bonds** are separated by one single bond. **Isolated double bonds** are separated by more than one single bond. **Cumulated double bonds** are adjacent to each other. A conjugated diene is more stable than an **isolated diene**, which is more stable than a cumulated diene. The least stable alkene has the greatest $-\Delta H°$ value. A diene can have as many as four configurational isomers: *E-E*, *Z-Z*, *E-Z*, and *Z-E*.

An isolated diene, like an alkene, undergoes only 1,2-addition. If there is only enough electrophilic reagent to add to one of the double bonds, it will add preferentially to the more reactive bond. A conjugated diene reacts with a limited amount of electrophilic reagent to form a **1,2-addition**

product and a **1,4-addition product**. The first step is addition of the electrophile to one of the sp^2 carbons at the end of the conjugated system.

When a reaction produces more than one product, the product formed most rapidly is the **kinetic product**; the most stable product is the **thermodynamic product**. If the reaction is carried out under mild conditions so that it is irreversible, the major product will be the kinetic product; if the reaction is carried out under vigorous conditions so that it is reversible, the major product will be the thermodynamic product. When a reaction is under **kinetic control**, the relative amounts of the products depend on the rates at which they are formed; when a reaction is under **thermodynamic control**, the relative amounts of the products depend on their stabilities. A **common intermediate** is an intermediate that both products have in common.

In a Diels–Alder reaction, a conjugated diene reacts with a dienophile to form a cyclic compound; in this concerted [4 + 2] cycloaddition reaction, two new σ bonds are formed at the expense of two π bonds. The conjugated diene must be in an ***s*-cis conformation**. The reactivity of the dienophile is increased by electron-withdrawing groups attached to the sp^2 carbons. The **HOMO** of one reactant and the **LUMO** of the other are used to show the transfer of electrons between the molecules. According to the conservation of orbital symmetry, **pericyclic reactions** occur as a result of the overlap of in-phase orbitals. The Diels–Alder reaction is stereospecific; it is a syn addition reaction with respect to both the diene and the dienophile. If both the diene and the dienophile are unsymmetrically substituted, two products are possible because the reactants can be aligned in two different ways. In **bridged bicyclic compounds**, a substituent can be **endo** or **exo**; endo is favored if the dienophile's substituent has π electrons.

Ultraviolet and **visible (UV/Vis) spectroscopy** provide information about compounds with conjugated double bonds. UV light is higher in energy than visible light; the shorter the wavelength, the greater is the energy. UV and visible light cause $n \rightarrow \pi^*$ and $\pi \rightarrow \pi^*$ **electronic transitions**; $\pi \rightarrow \pi^*$ transistions have larger molar absorptivities. A **chromophore** is that part of a molecule that absorbs UV or visible light. The **Beer–Lambert law** is the relationship among absorbance, concentration, and length of the light path: A = *cle*. The more conjugated double bonds there are in a compound, the less energy is required for the electronic transition and the longer is the λ_{max} at which it occurs. Reaction rates and pK_a values are commonly measured using UV/Vis spectroscopy.

Summary of Reactions

1. If there is excess electrophilic reagent, both double bonds of an *isolated diene* will undergo electrophilic addition.

$$CH_2{=}CHCH_2CH_2\overset{CH_3}{\overset{|}{C}}{=}CH_2 + \underset{\text{excess}}{HBr} \longrightarrow CH_3\underset{Br}{\underset{|}{C}H}CH_2CH_2\underset{Br}{\underset{|}{\overset{CH_3}{\overset{|}{C}}}}CH_3$$

If there is only one equivalent of electrophilic reagent, only the most reactive of the double bonds of an *isolated diene* will undergo electrophilic addition (Section 8.5).

$$CH_2{=}CHCH_2CH_2\overset{CH_3}{\overset{|}{C}}{=}CH_2 + HBr \longrightarrow CH_2{=}CHCH_2CH_2\underset{Br}{\underset{|}{\overset{CH_3}{\overset{|}{C}}}}CH_3$$

2. *Conjugated dienes* undergo 1,2- and 1,4-addition with one equivalent of an electrophilic reagent (Section 8.6).

$$RCH{=}CHCH{=}CHR + HBr \longrightarrow \underset{\textbf{1,2-addition product}}{RCH_2\underset{Br}{\underset{|}{C}H}CH{=}CHR} + \underset{\textbf{1,4-addition product}}{RCH_2CH{=}CH\underset{Br}{\underset{|}{C}H}R}$$

3. *Conjugated dienes* undergo 1,4-cycloaddition with a dienophile (a Diels–Alder reaction) (Section 8.8).

$$CH_2{=}CH{-}CH{=}CH_2 + CH_2{=}CH{-}\overset{O}{\overset{\|}{C}}{-}R \xrightarrow{\Delta} \text{(cyclohexene)}{-}\overset{O}{\overset{\|}{C}}{-}R$$

Key Terms

Problems

25. Give the systematic name for each of the following compounds:

a. $CH_3C{\equiv}CCH_2CH_2CH_2CH{=}CH_2$

b. 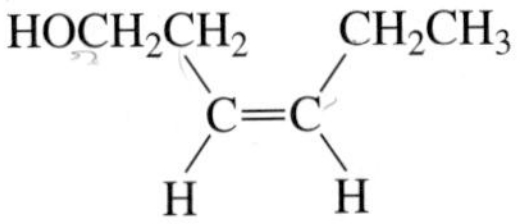 $HOCH_2CH_2$ and H on one carbon, CH_2CH_3 and H on the other: $C{=}C$

c. $CH_3CH_2C{\equiv}CCH_2CH_2C{\equiv}CH$

d. 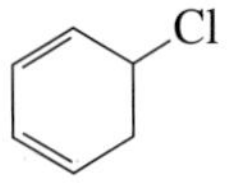 Cl

e. 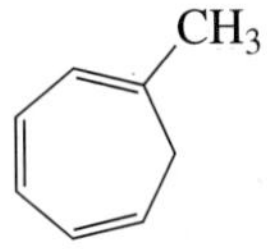 CH_3

26. Draw structures for the following compounds:

a. (2*E*,4*E*)-1-chloro-3-methyl-2,4-hexadiene
b. (3*Z*,5*E*)-4-methyl-3,5-nonadiene
c. (3*Z*,5*Z*)-4,5-dimethyl-3,5-nonadiene
d. (3*E*,5*E*)-2,5-dibromo-3,5-octadiene

27. a. How many linear dienes are there with molecular formula C_6H_{10}? (Ignore cis–trans isomers.)
b. How many of the linear dienes that you found in part a are conjugated dienes?
c. How many are isolated dienes?

28. Which compound would you expect to have the greater $-\Delta H^\circ$ value, 1,2-pentadiene or 1,4-pentadiene?

29. In Chapter 3, you learned that α-farnesene is found in the waxy coating of apple skins. What is its systematic name?

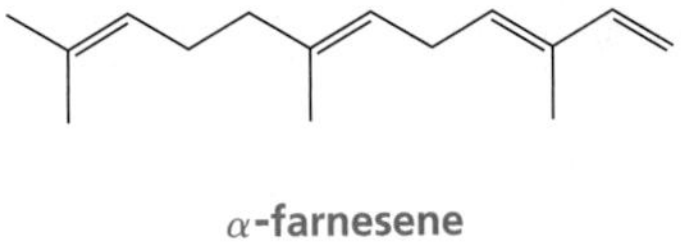

α-farnesene

30. Diane O'File treated 1,3-cyclohexadiene with Br_2 and obtained two products (ignoring stereoisomers). Her lab partner treated 1,3-cyclohexadiene with HBr and was surprised to find that he obtained only one product (ignoring stereoisomers). Account for these results.

31. Give the major product of each of the following reactions. One equivalent of each reagent is used:

a. 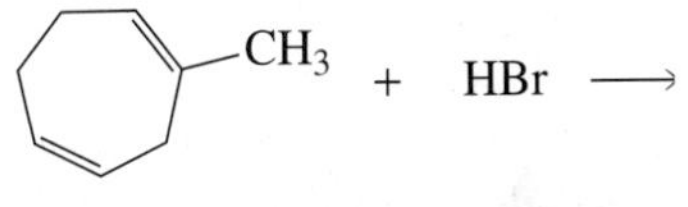 CH_3 + HBr ⟶

b. 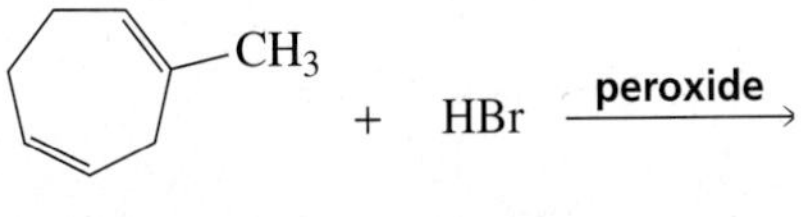 CH_3 + HBr $\xrightarrow{\text{peroxide}}$

c. 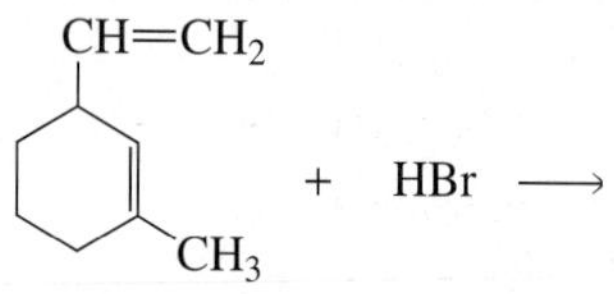 CH=CH$_2$ (on ring, with CH$_3$) + HBr ⟶

d. 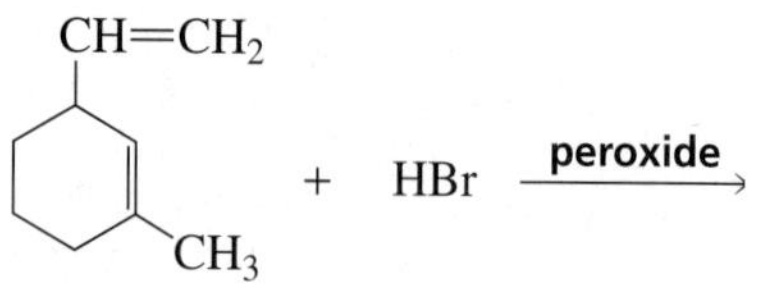 CH=CH$_2$ (on ring, with CH$_3$) + HBr $\xrightarrow{\text{peroxide}}$

32. a. Give the products obtained from the reaction of 1 mole HBr with 1 mole 1,3,5-hexatriene.
    b. Which product(s) will predominate if the reaction is under kinetic control?
    c. Which product(s) will predominate if the reaction is under thermodynamic control?

33. 4-Methyl-3-penten-2-one has two absorption bands in its UV spectrum, one at 236 nm and one at 314 nm.
    a. Why are there two absorption bands?
    b. Which band shows the greater absorbance?

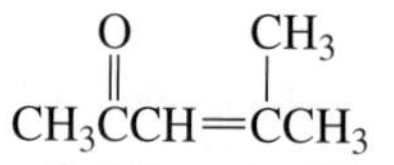

$CH_3\overset{O}{\overset{\|}{C}}CH{=}\overset{CH_3}{\overset{|}{C}}CH_3$

4-methyl-3-penten-2-one

34. How could the following compounds be synthesized using a Diels–Alder reaction?

a. 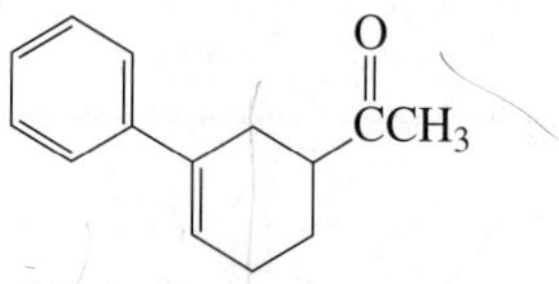 (structure with $\overset{O}{\overset{\|}{C}}CH_3$)

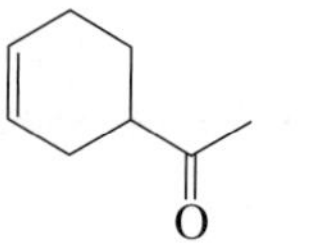

b. 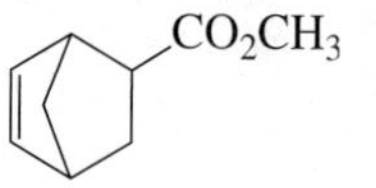 (structure with CO_2CH_3)

c. (structure)

d. 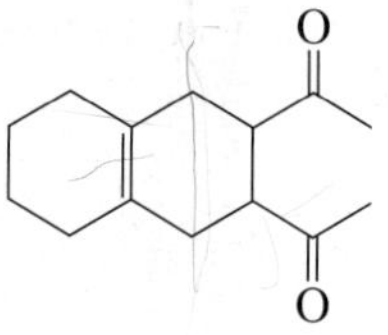 (structure)

35. a. How could each of the following compounds be prepared from a hydrocarbon in a single step?

1. 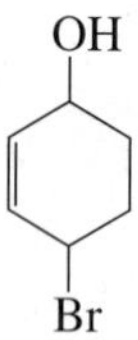 (structure with OH, Br)

2. 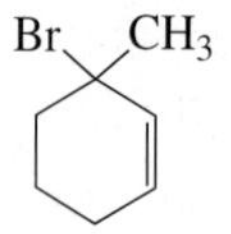 (structure with Br, CH_3)

3. (structure with CH_3, HO, CH_3)

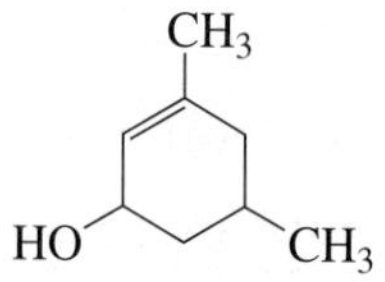

    b. What other organic compound would be obtained from each synthesis?

36. How would the following substituents affect the rate of a Diels–Alder reaction?
    a. an electron-donating substituent in the diene
    b. an electron-donating substituent in the dienophile
    c. an electron-withdrawing substituent in the diene

37. Give the major products obtained from the reaction of one equivalent of HCl with the following:
    a. 2,3-dimethyl-1,3-pentadiene
    b. 2,4-dimethyl-1,3-pentadiene
    For each reaction, indicate the kinetic and thermodynamic products.

38. Give the product or products that would be obtained from each of the following reactions:

a. 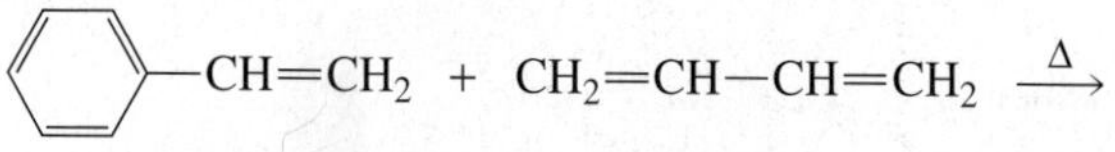 (benzene ring)–$CH{=}CH_2$ + $CH_2{=}CH{-}CH{=}CH_2$ $\xrightarrow{\Delta}$

b. 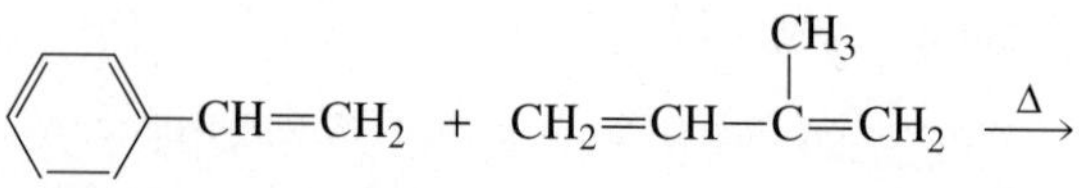 (cyclohexene ring)–$CH{=}CH_2$ + $CH_2{=}CH{-}\overset{CH_3}{\overset{|}{C}}{=}CH_2$ $\xrightarrow{\Delta}$

c. 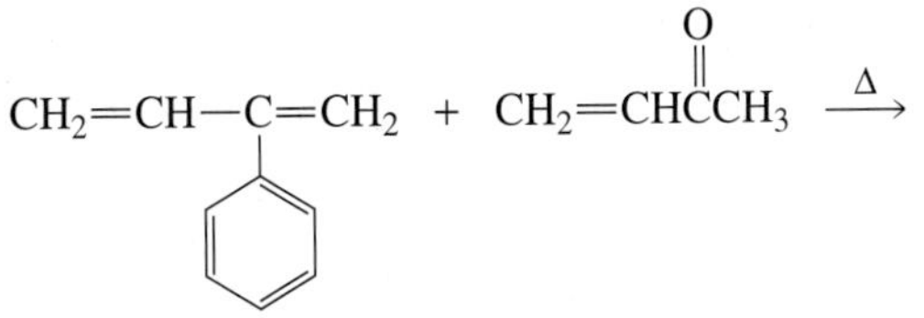 $CH_2{=}CH{-}\underset{\displaystyle C_6H_5}{C}{=}CH_2 + CH_2{=}CH\overset{\displaystyle O}{\overset{\|}{C}}CH_3 \xrightarrow{\Delta}$

d. 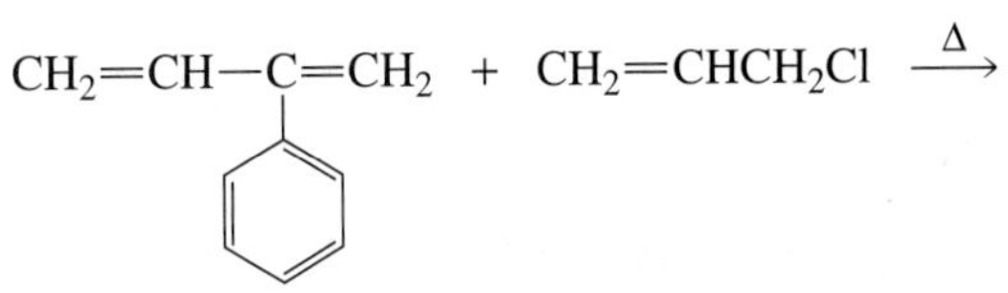 $CH_2{=}CH{-}\underset{\displaystyle C_6H_5}{C}{=}CH_2 + CH_2{=}CHCH_2Cl \xrightarrow{\Delta}$

39. How could one use UV spectroscopy to distinguish between the compounds in each of the following pairs?

a. 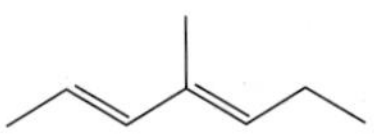 [skeletal structure] and [skeletal structure]

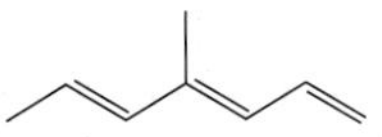

b. $CH_2{=}CHCH{=}CHCH{=}CH_2$ and $CH_2{=}CHCH{=}CH\overset{\displaystyle O}{\overset{\|}{C}}CH_3$

c. 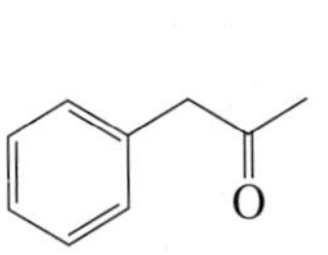 [skeletal structure] and [skeletal structure]

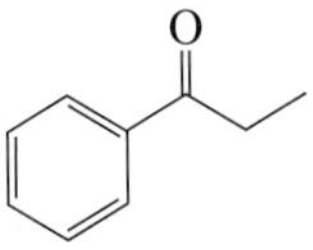

d. 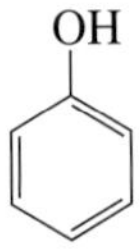 [benzene ring with OH] and [benzene ring with OCH_3]

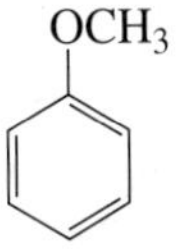

40. What two sets of a conjugated diene and a dienophile could be used to prepare the following compound?

$\overset{\displaystyle O}{\overset{\|}{C}}CH_3$

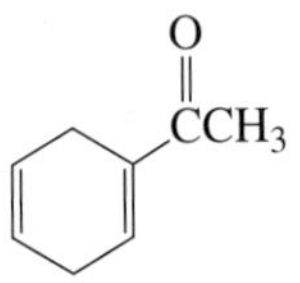

41. a. Which dienophile is more reactive in a Diels–Alder reaction?

1. $CH_2{=}CH\overset{\displaystyle O}{\overset{\|}{C}}H$ or $CH_2{=}CHCH_2\overset{\displaystyle O}{\overset{\|}{C}}H$ 2. $CH_2{=}CH\overset{\displaystyle O}{\overset{\|}{C}}H$ or $CH_2{=}CHCH_3$

b. Which diene is more reactive in a Diels–Alder reaction?

$CH_2{=}CHCH{=}CHOCH_3$ or $CH_2{=}CHCH{=}CHCH_2OCH_3$

42. Cyclopentadiene can react with itself in a Diels–Alder reaction. Draw the endo and exo products.

43. Which diene and which dienophile could be used to prepare each of the following compounds?

a. 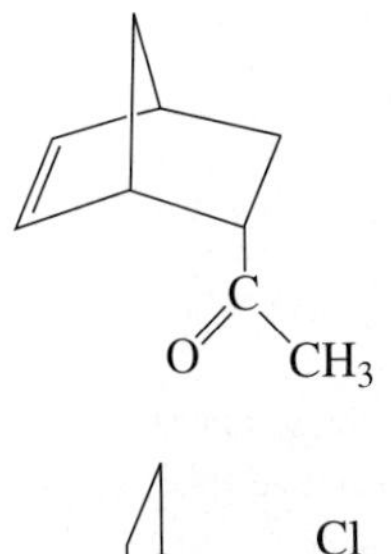 [bicyclic structure with C(=O)CH_3]

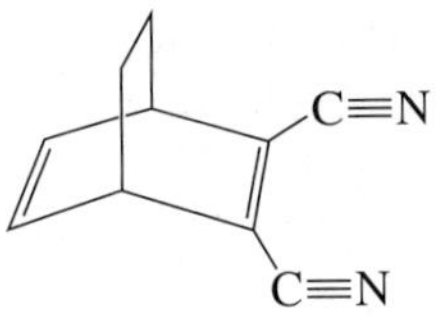

b. 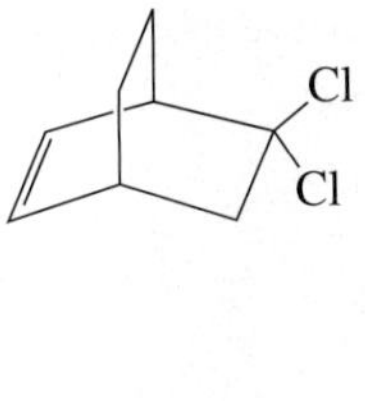 [bicyclic structure with Cl, Cl]

c. [bicyclic structure with C≡N, C≡N]

d. 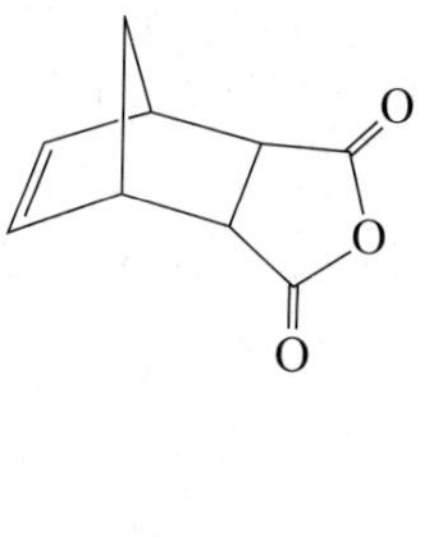 [bicyclic structure with fused anhydride ring: O, O, O]

44. a. Give the products of the following reaction:

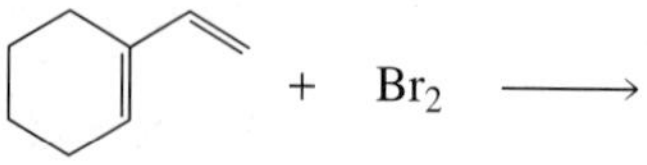

+ Br_2 ⟶

b. How many stereoisomers of each product could be obtained?

45. As many as 18 different Diels–Alder products could be obtained by heating a mixture of 1,3-butadiene and 2-methyl-1,3-butadiene. Identify the products.

46. Many credit-card slips do not have carbon paper. Nevertheless, when you sign the slip, an imprint of your signature is made on the bottom copy. The carbonless paper contains tiny capsules that are filled with the colorless compound whose structure is shown here:

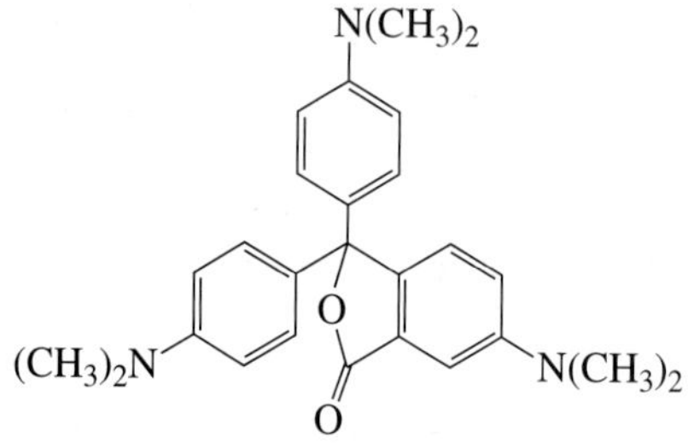

$N(CH_3)_2$

$(CH_3)_2N$ O $N(CH_3)_2$

O

When you press on the paper, the capsules burst and the colorless compound comes into contact with the acid-treated paper, forming a highly colored compound. What is the structure of the colored compound?

47. On the same graph, draw the reaction coordinate for the addition of one equivalent of HBr to 2-methyl-1,3-pentadiene and one equivalent of HBr to 2-methyl-1,4-pentadiene. Which reaction is faster?

48. While attempting to recrystallize maleic anhydride, Professor Nots O. Kareful dissolved it in freshly distilled cyclopentadiene rather than in freshly distilled cyclopentane. Was his recrystallization successful?

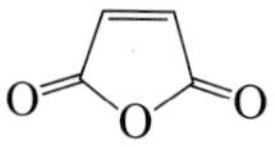

O O O

maleic anhydride

49. A solution of ethanol has been contaminated with benzene—a technique employed to make ethanol unfit to drink. Benzene has a molar absorptivity of 230 at 260 nm in ethanol, and ethanol shows no absorbance at 260 nm. How could the concentration of benzene in the solution be determined?

50. Reverse Diels–Alder reactions can occur at high temperatures. Why are high temperatures required?

51. The following equilibrium is driven to the right if the reaction is carried out in the presence of maleic anhydride:

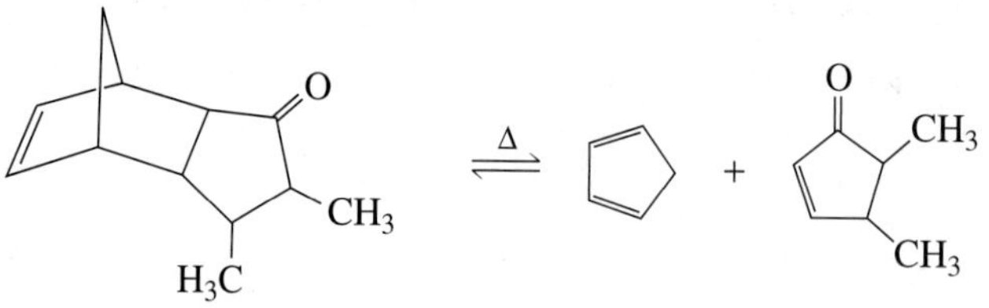

O CH_3 H_3C $\overset{\Delta}{\rightleftharpoons}$ + O CH_3 CH_3

What is the function of maleic anhydride?

52. In 1935, J. Bredt, a German chemist, proposed that a bicycloalkene cannot have a double bond at a bridgehead carbon unless one of the rings contains at least eight carbon atoms. This is known as Bredt's rule. Explain why there cannot be a double bond at this position.

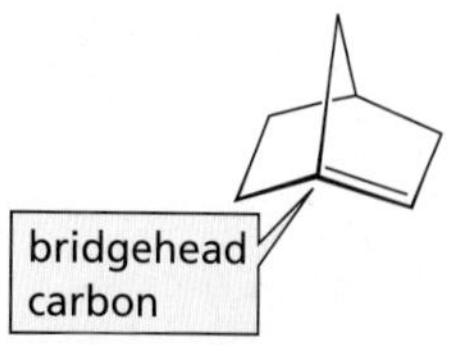

bridgehead carbon

53. The experiment shown below and discussed in Section 8.8, showed that the proximity of the chloride ion to C-2 in the transition state caused the 1,2-addition product to be formed faster than the 1,4-addition product:

$$CH_2{=}CHCH{=}CHCH_3 \quad + \quad DCl \xrightarrow{-78\,°C} \underset{\substack{| \quad\; | \\ D \quad Cl}}{CH_2CHCH{=}CHCH_3} \quad + \quad \underset{\substack{| \qquad\qquad\quad | \\ D \qquad\qquad\quad Cl}}{CH_2CH{=}CHCHCH_3}$$

a. Why was it important for the investigators to know that the reaction was being carried out under kinetic control?

b. How could the investigators determine that the reaction was being carried out under kinetic control?

54. A student wanted to know whether the greater proximity of the nucleophile to the C-2 carbon in the transition state was what caused the 1,2-addition product to be formed faster when 1,3-butadiene reacts with HCl. Therefore, he decided to investigate the reaction of 2-methyl-1,3-cyclohexadiene with HCl. His friend, Noel Noall, told him that he should use 1-methyl-1,3-cyclohexadiene instead. Should he follow his friend's advice?

55. a. Methyl orange (whose structure is given in Section 8.12) is an acid–base indicator. In solutions of pH < 4, it is red, and in solutions of pH > 4, it is yellow. Account for the change in color.

b. Phenolphthalein is also an indicator, but it exhibits a much more dramatic color change. In solutions of pH < 8.5, it is colorless, and in solutions of pH > 8.5, it is deep red-purple. Account for the change in color.

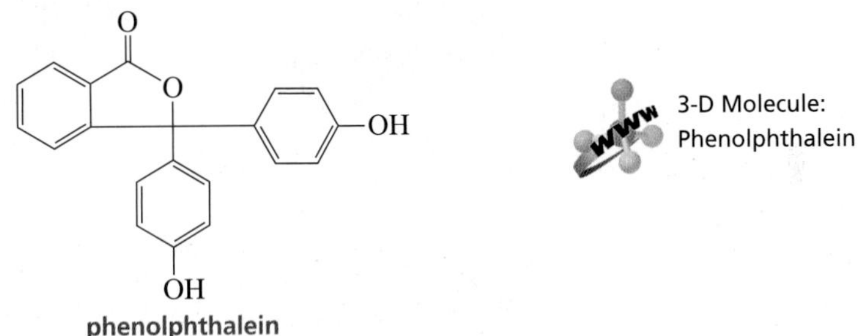

phenolphthalein

9 Reactions of Alkanes • Radicals

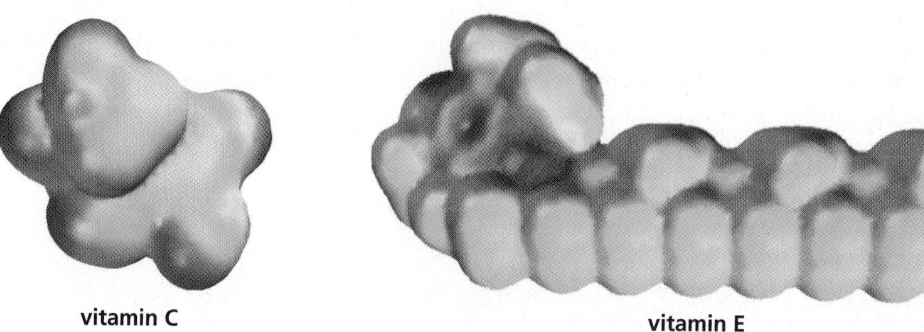

vitamin C vitamin E

We have seen that there are three classes of hydrocarbons: *alkanes*, which contain only single bonds; *alkenes*, which contain double bonds; and *alkynes*, which contain triple bonds. We examined the chemistry of alkenes in Chapter 4 and the chemistry of alkynes in Chapter 6. Now we will take a look at the chemistry of alkanes.

Alkanes are called **saturated hydrocarbons** because they are saturated with hydrogen. In other words, they do not contain any double or triple bonds. A few examples of alkanes are shown here. Their nomenclature is discussed in Section 2.2.

$CH_3CH_2CH_2CH_3$
butane

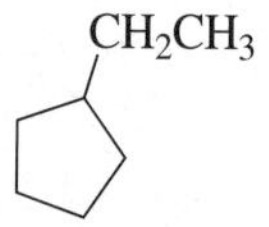

CH_2CH_3
ethylcyclopentane

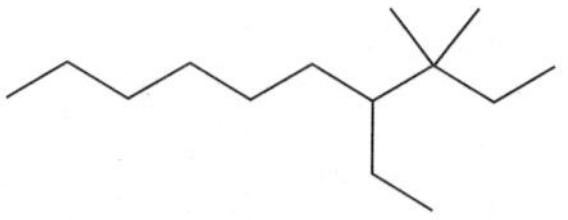

4-ethyl-3,3-dimethyldecane

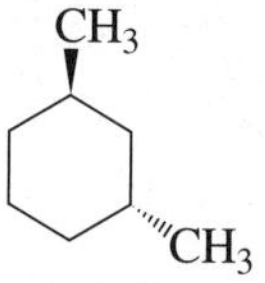

CH_3
CH_3
trans-1,3-dimethylcyclohexane

Alkanes are widespread both on Earth and on other planets. The atmospheres of Jupiter, Saturn, Uranus, and Neptune contain large quantities of methane (CH_4), the smallest alkane, which is an odorless and flammable gas. In fact, the blue colors of Uranus and Neptune are due to methane in their atmospheres. Alkanes on Earth are found in natural gas and petroleum, which are formed by the decomposition of plant and animal material that has been buried for long periods in the Earth's crust, an environment with little oxygen. Natural gas and petroleum, therefore, are known as *fossil fuels.*

Natural gas consists of about 75% methane. The remaining 25% is composed of small alkanes such as ethane, propane, and butane. In the 1950s, natural gas replaced coal as the main energy source for domestic and industrial heating in many parts of the United States.

Petroleum is a complex mixture of alkanes and cycloalkanes that can be separated into fractions by distillation. The fraction that boils at the lowest temperature (hydrocarbons containing three and four carbons) is a gas that can be liquefied under pressure. This gas is used as a fuel for cigarette lighters, camp stoves, and barbecues. The fraction that boils at somewhat higher temperatures (hydrocarbons containing 5 to 11 carbons) is gasoline; the next fraction (9 to 16 carbons) includes kerosene and jet fuel. The fraction with 15 to 25 carbons is used for heating oil and diesel oil, and the highest-boiling fraction is used for lubricants and greases. The nonpolar nature of these compounds is what gives them their oily feel. After distillation, a nonvolatile residue called asphalt or tar is left behind.

The 5- to 11-carbon fraction that is used for gasoline is, in fact, a poor fuel for internal combustion engines and requires a process known as catalytic cracking to become a high-performance gasoline. Catalytic cracking converts straight-chain hydrocarbons that are poor fuels into branched-chain compounds that are high-performance fuels. Originally, cracking (also called *pyrolysis*) involved heating gasoline to very high temperatures in order to obtain hydrocarbons with three to five carbons. Modern cracking methods use catalysts to accomplish the same thing at much lower temperatures. The small hydrocarbons are then catalytically recombined to form highly branched hydrocarbons.

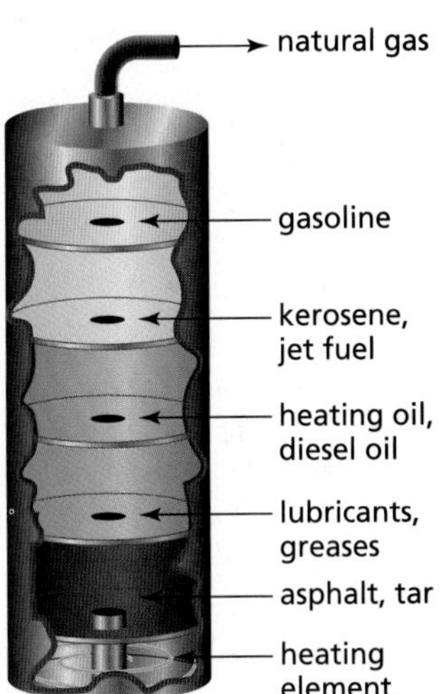

OCTANE NUMBER

When poor fuels are used in an engine, combustion can be initiated before the spark plug fires. A pinging or knocking may then be heard in the running engine. As the quality of the fuel improves, the engine is less likely to knock. The quality of a fuel is indicated by its octane number. Straight-chain hydrocarbons have low octane numbers and make poor fuels. Heptane, for example, with an arbitrarily assigned octane number of 0, causes engines to knock badly. Branched-chain fuels burn more slowly—thereby reducing knocking—because they have more primary hydrogens. Consequently, branched-chain alkanes have high octane numbers. 2,2,4-Trimethylpentane, for example, does not cause knocking and has arbitrarily been assigned an octane number of 100.

$CH_3CH_2CH_2CH_2CH_2CH_2CH_3$
heptane
octane number = 0

$CH_3C(CH_3)_2CH_2CH(CH_3)CH_3$
2,2,4-trimethylpentane
octane number = 100

The octane rating of a gasoline is determined by comparing its knocking with the knocking of mixtures of heptane and 2,2,4-trimethylpentane. The octane number given to the gasoline corresponds to the percent of 2,2,4-trimethylpentane in the matching mixture. The term "octane number" originated from the fact that 2,2,4-trimethylpentane contains eight carbons.

FOSSIL FUELS: A PROBLEMATIC ENERGY SOURCE

We face three major problems as a consequence of our dependence on fossil fuels for energy. First, fossil fuels are a nonrenewable resource and the world's supply is continually decreasing. Second, a group of Middle Eastern and South American countries controls a large portion of the world's supply of petroleum. These countries have formed a cartel known as the *Organization of Petroleum Exporting Countries* (*OPEC*), which controls both the supply and the price of crude oil. Political instability in any OPEC country can seriously affect the world oil supply. Third, burning fossil fuels increases the concentrations of CO_2 and SO_2 in the atmosphere. Scientists have established experimentally that atmospheric SO_2 causes "acid rain," a threat to the Earth's plants and, therefore, to our food and oxygen supplies. The atmospheric CO_2 concentration has increased 20% in the last 10 years, causing scientists to predict an increase in the Earth's temperature as a result of the absorption of infrared radiation by CO_2 (the so-called *greenhouse effect*). A steady increase in the temperature of the Earth would have devastating consequences, including the formation of new deserts, massive crop failure, and the melting of polar ice caps with a concomitant rise in sea level. Clearly, what we need is a renewable, nonpolitical, nonpolluting, and economically affordable source of energy.

9.1 The Low Reactivity of Alkanes

The double and triple bonds of alkenes and alkynes are composed of strong σ bonds and relatively weak π bonds. We have seen that the reactivity of *alkenes* and *alkynes* is the result of an electrophile being attracted to the cloud of electrons that constitutes the π bond.

Alkanes have only strong σ bonds. Because the carbon and hydrogen atoms of an alkane have approximately the same electronegativity, the electrons in the C—H and C—C σ bonds are shared equally by the bonding atoms. Consequently, none of the atoms in an alkane have any significant charge. This means that neither nucleophiles nor electrophiles are attracted to them. Because they have only strong σ bonds and atoms with no partial charges, alkanes are very unreactive compounds. Their failure to undergo reactions prompted early organic chemists to call them **paraffins**, from the Latin *parum affinis,* which means "little affinity" (for other compounds).

9.2 Chlorination and Bromination of Alkanes

Alkanes do react with chlorine (Cl_2) or bromine (Br_2) to form alkyl chlorides or alkyl bromides. These **halogenation reactions** take place only at high temperatures or in the presence of light (symbolized by $h\nu$). They are the only reactions that alkanes undergo—with the exception of **combustion**, a reaction with oxygen that takes place at high temperatures and converts alkanes to carbon dioxide and water.

$$CH_4 + Cl_2 \xrightarrow[\text{or } h\nu]{\Delta} \underset{\text{methyl chloride}}{CH_3Cl} + HCl$$

$$CH_3CH_3 + Br_2 \xrightarrow[\text{or } h\nu]{\Delta} \underset{\text{ethyl bromide}}{CH_3CH_2Br} + HBr$$

The mechanism for the halogenation of an alkane is well understood. The high temperature (or light) supplies the energy required to break the Cl—Cl or Br—Br bond *homolytically*. **Homolytic bond cleavage** is the **initiation step** of the reaction because it creates the radical that is used in the first propagation step (Section 4.10). Recall that an arrowhead with one barb signifies the movement of one electron (Section 3.6).

$$:\ddot{\underset{..}{Cl}}-\ddot{\underset{..}{Cl}}: \xrightarrow[\text{or } h\nu]{\Delta} 2\ :\ddot{\underset{..}{Cl}}\cdot$$

homolytic cleavage

$$:\ddot{\underset{..}{Br}}-\ddot{\underset{..}{Br}}: \xrightarrow[\text{or } h\nu]{\Delta} 2\ :\ddot{\underset{..}{Br}}\cdot$$

A **radical** (often called a **free radical**) is a species containing an atom with an unpaired electron. A radical is highly reactive because it wants to acquire an electron to complete its octet. In the mechanism for the monochlorination of methane, the chlorine radical formed in the initiation step abstracts a hydrogen atom from methane, forming HCl and a methyl radical. The methyl radical abstracts a chlorine atom from Cl_2, forming methyl chloride and another chlorine radical, which can abstract a hydrogen atom from another molecule of methane. These two steps are called **propagation steps** because the radical created in the first propagation step reacts in the second propagation step to produce a radical that can repeat the first propagation step. Thus, the two propagation steps are repeated over and over. The first propagation step is the rate-determining step of the overall reaction. Because the reaction has radical intermediates and repeating propagation steps, it is called a **radical chain reaction**.

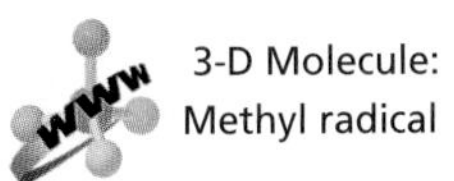
3-D Molecule: Methyl radical

mechanism for the monochlorination of methane

$$:\ddot{\underset{..}{Cl}}-\ddot{\underset{..}{Cl}}: \xrightarrow[\text{or } h\nu]{\Delta} 2\ :\ddot{\underset{..}{Cl}}\cdot \quad \boxed{\text{initiation step}}$$

$$:\ddot{\underset{..}{Cl}}\cdot + H-CH_3 \longrightarrow H\ddot{\underset{..}{Cl}}: + \underset{\text{a methyl radical}}{\cdot CH_3}$$

$$\cdot CH_3 + :\ddot{\underset{..}{Cl}}-\ddot{\underset{..}{Cl}}: \longrightarrow CH_3Cl + :\ddot{\underset{..}{Cl}}\cdot$$

propagation steps

$$:\ddot{\underset{..}{Cl}}\cdot + :\ddot{\underset{..}{Cl}}\cdot \longrightarrow Cl_2$$

$$\cdot CH_3 + \cdot CH_3 \longrightarrow CH_3CH_3$$

$$:\ddot{\underset{..}{Cl}}\cdot + \cdot CH_3 \longrightarrow CH_3Cl$$

termination steps

Any two radicals in the reaction mixture can combine to form a molecule in which all the electrons are paired. The combination of two radicals is called a **termination step** because it helps bring the reaction to an end by decreasing the number of radicals available to propagate the reaction. The radical chlorination of alkanes other than methane follows the same mechanism. A radical chain reaction, with its characteristic initiation, propagation, and termination steps, was first described in Section 4.10.

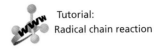

Tutorial: Radical chain reaction

The reaction of an alkane with chlorine or bromine to form an alkyl halide is called a **radical substitution reaction** because radicals are involved as intermediates and the end result is the substitution of a halogen atom for one of the hydrogen atoms of the alkane.

In order to maximize the amount of monohalogenated product obtained, a radical substitution reaction should be carried out in the presence of excess alkane. Excess alkane in the reaction mixture increases the probability that the halogen radical will collide with a molecule of alkane rather than with a molecule of alkyl halide—even toward the end of the reaction, by which time a considerable amount of alkyl halide will have been formed. If the halogen radical abstracts a hydrogen from a molecule of alkyl halide rather than from a molecule of alkane, a dihalogenated product will be obtained.

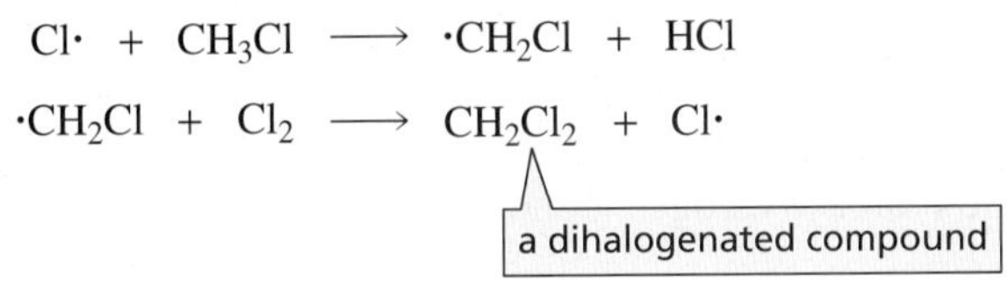

$$Cl\cdot + CH_3Cl \longrightarrow \cdot CH_2Cl + HCl$$

$$\cdot CH_2Cl + Cl_2 \longrightarrow CH_2Cl_2 + Cl\cdot$$

Bromination of alkanes follows the same mechanism as chlorination.

mechanism for the monobromination of ethane

$$Br-Br \xrightarrow[\text{or } h\nu]{\Delta} 2\ Br\cdot \quad \boxed{\text{initiation step}}$$

$$Br\cdot + H-CH_2CH_3 \longrightarrow CH_3\dot{C}H_2 + HBr$$

$$CH_3\dot{C}H_2 + Br-Br \longrightarrow CH_3CH_2Br + Br\cdot$$

propagation steps

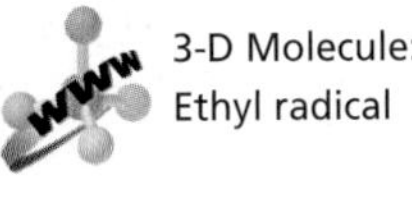
3-D Molecule: Ethyl radical

$$Br\cdot + Br\cdot \longrightarrow Br_2$$

$$CH_3\dot{C}H_2 + CH_3\dot{C}H_2 \longrightarrow CH_3CH_2CH_2CH_3$$

$$CH_3\dot{C}H_2 + Br\cdot \longrightarrow CH_3CH_2Br$$

termination steps

PROBLEM 1

Show the initiation, propagation, and termination steps for the monochlorination of cyclohexane.

PROBLEM 2

Write the mechanism for the formation of carbon tetrachloride, CCl_4, from the reaction of methane with $Cl_2 + h\nu$.

PROBLEM 3 SOLVED

If cyclopentane reacts with more than one equivalent of Cl_2 at a high temperature, how many dichlorocyclopentanes would you expect to obtain as products?

SOLUTION Seven dichlorocyclopentanes would be obtained as products. Only one isomer is possible for the 1,1-dichloro compound. The 1,2- and 1,3-dichloro compounds have two asymmetric carbons. Each has three stereoisomers because the cis isomer is a meso compound and the trans isomer is a pair of enantiomers.

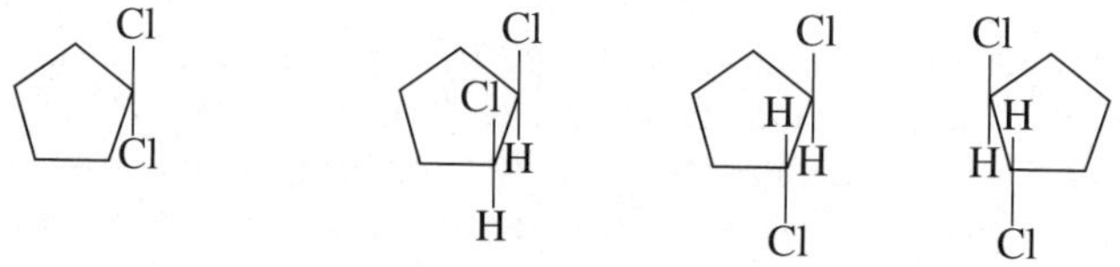

meso compound enantiomers

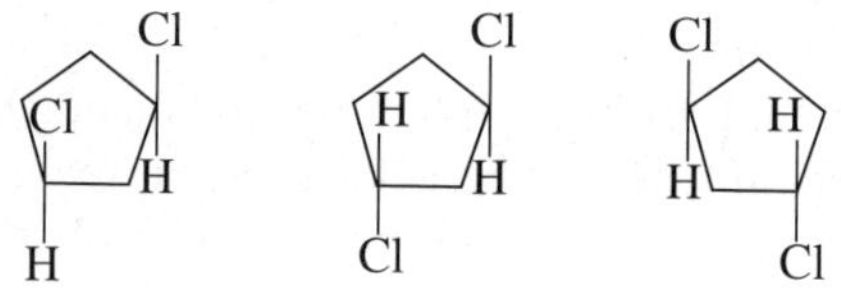

meso compound enantiomers

9.3 Factors that Determine Product Distribution

Two different alkyl halides are obtained from the monochlorination of butane. Substitution of a hydrogen bonded to one of the terminal carbons produces 1-chlorobutane, whereas substitution of a hydrogen bonded to one of the internal carbons forms 2-chlorobutane.

$$\underset{\text{butane}}{CH_3CH_2CH_2CH_3} + Cl_2 \xrightarrow{h\nu} \underset{\substack{\text{1-chlorobutane}\\ \text{expected} = 60\%\\ \text{experimental} = 29\%}}{CH_3CH_2CH_2CH_2Cl} + \underset{\substack{\text{2-chlorobutane}\\ \text{expected} = 40\%\\ \text{experimental} = 71\%}}{CH_3CH_2\overset{\overset{Cl}{|}}{C}HCH_3} + HCl$$

The expected (statistical) distribution of products is 60% 1-chlorobutane and 40% 2-chlorobutane because six of butane's 10 hydrogens can be substituted to form 1-chlorobutane, whereas only four can be substituted to form 2-chlorobutane. This assumes, however, that all of the C—H bonds in butane are equally easy to break. Then, the relative amounts of the two products would depend only on the probability of a chlorine radical colliding with a primary hydrogen, compared with its colliding with a secondary hydrogen. When we carry out the reaction in the laboratory and analyze the product, however, we find that it is 29% 1-chlorobutane and 71% 2-chlorobutane. Therefore, probability alone does not explain the regioselectivity of the reaction. Because more 2-chlorobutane is obtained than expected and *the rate-determining step of the overall reaction is hydrogen atom abstraction*, we conclude that it must be easier to abstract a hydrogen atom from a secondary carbon than from a primary carbon.

Alkyl radicals have different stabilities (Section 4.10), and the more stable the radical, the more easily it is formed because the stability of the radical is reflected in the stability of the transition state leading to its formation. Consequently, it is easier to remove a hydrogen atom from a secondary carbon to form a secondary radical than it is to remove a hydrogen atom from a primary carbon to form a primary radical.

relative stabilities of alkyl radicals

most stable → $R_3C\cdot$ (tertiary radical) > $R_2HC\cdot$ (secondary radical) > $RH_2C\cdot$ (primary radical) > $H_3C\cdot$ (methyl radical) ← least stable

When a chlorine radical reacts with butane, it can abstract a hydrogen atom from an internal carbon, thereby forming a secondary alkyl radical, or it can abstract a hydrogen atom from a terminal carbon, thereby forming a primary alkyl radical. Because it is easier to form the more stable secondary alkyl radical, 2-chlorobutane is formed faster than 1-chlorobutane.

$$CH_3CH_2CH_2CH_3 \xrightarrow{Cl\cdot} CH_3CH_2\dot{C}HCH_3 \text{ (a secondary alkyl radical)} + HCl \xrightarrow{Cl_2} CH_3CH_2CH(Cl)CH_3 \text{ (2-chlorobutane)} + Cl\cdot$$

$$CH_3CH_2CH_2CH_3 \xrightarrow{Cl\cdot} CH_3CH_2CH_2\dot{C}H_2 \text{ (a primary alkyl radical)} + HCl \xrightarrow{Cl_2} CH_3CH_2CH_2CH_2Cl \text{ (1-chlorobutane)} + Cl\cdot$$

After experimentally determining the amount of each chlorination product obtained from various hydrocarbons, chemists were able to conclude that *at room temperature* it is 5.0 times easier for a chlorine radical to abstract a hydrogen atom from a tertiary carbon than from a primary carbon, and it is 3.8 times easier to abstract a hydrogen atom from a secondary carbon than from a primary carbon. (See Problem 17.) The precise ratios differ at different temperatures.

relative rates of alkyl radical formation by a chlorine radical at room temperature

tertiary > secondary > primary
5.0 3.8 1.0

← increasing rate of formation

To determine the relative amounts of products obtained from radical chlorination of an alkane, both *probability* (the number of hydrogens that can be abstracted that will lead to the formation of the particular product) and *reactivity* (the relative rate at which a particular hydrogen is abstracted) must be taken into account. When both factors are considered, the calculated amounts of 1-chlorobutane and 2-chlorobutane agree with the amounts obtained experimentally.

relative amount of 1-chlorobutane

number of hydrogens × reactivity

$6 \times 1.0 = 6.0$

$$\text{percent yield} = \frac{6.0}{21} = 29\%$$

relative amount of 2-chlorobutane

number of hydrogens × reactivity

$4 \times 3.8 = 15$

$$\text{percent yield} = \frac{15}{21} = 71\%$$

The percent yield of each alkyl halide is calculated by dividing the relative amount of the particular product by the sum of the relative amounts of all the alkyl halide products ($6 + 15 = 21$).

Radical monochlorination of 2,2,5-trimethylhexane results in the formation of five monochlorination products. Because the relative amounts of the five alkyl halides total 35 ($9.0 + 7.6 + 7.6 + 5.0 + 6.0 = 35$), the percent yield of each product can be calculated as follows:

CH_3 CH_3
$CH_3CCH_2CH_2CHCH_3$ + Cl_2 $\xrightarrow{\Delta}$
CH_3
2,2,5-trimethylhexane

CH_2Cl CH_3
$CH_3CCH_2CH_2CHCH_3$ +
CH_3
$9 \times 1.0 = 9.0$
$\frac{9.0}{35} = 26\%$

CH_3 CH_3
$CH_3C{-}CHCH_2CHCH_3$ +
CH_3 Cl
$2 \times 3.8 = 7.6$
$\frac{7.6}{35} = 22\%$

CH_3 CH_3
$CH_3CCH_2CHCHCH_3$ +
CH_3 Cl
$2 \times 3.8 = 7.6$
$\frac{7.6}{35} = 22\%$

CH_3 CH_3
$CH_3CCH_2CH_2CCH_3$ +
CH_3 Cl
$1 \times 5.0 = 5.0$
$\frac{5.0}{35} = 14\%$

CH_3 CH_3
$CH_3CCH_2CH_2CHCH_2Cl$ + HCl
CH_3
$6 \times 1.0 = 6.0$
$\frac{6.0}{35} = 17\%$

Because radical chlorination of an alkane can yield several different monosubstitution products as well as products that contain more than one chlorine atom, it is not the best method for synthesizing an alkyl halide. Addition of a hydrogen halide to an alkene (Section 4.1) or conversion of an alcohol to an alkyl halide (a reaction we will study in Chapter 12) is a much better way to make an alkyl halide. Radical halogenation of an alkane is nevertheless still a useful reaction because it is the only way to convert an inert alkane into a reactive compound. In Chapter 10, we will see that once the halogen is introduced into the alkane, it can be replaced by a variety of other substituents.

PROBLEM 4

When 2-methylpropane is monochlorinated in the presence of light at room temperature, 36% of the product is 2-chloro-2-methylpropane and 64% is 1-chloro-2-methylpropane. From these data, calculate how much easier it is to abstract a hydrogen atom from a tertiary carbon than from a primary carbon under these conditions.

PROBLEM 5◆

How many alkyl halides can be obtained from monochlorination of the following alkanes? Neglect stereoisomers.

a. $CH_3CH_2CH_2CH_2CH_3$

b.
CH_3 CH_3
$CH_3CHCH_2CH_2CHCH_3$

c.
CH_3
$CH_3CHCH_2CH_2CH_3$

d. (cyclohexane)

e. (ring with CH_3)

f. (ring with CH_3 and CH_3)

g.
CH_3 CH_3
$CH_3CCH_2CCH_3$
CH_3 CH_3

h.
CH_3 CH_3
$CH_3C{-}CCH_3$
CH_3 CH_3

i.
CH_3 CH_3
$CH_3CCH_2CHCH_3$
CH_3

PROBLEM 6◆

Calculate the percent yield of each product obtained in Problems 5a, b, and c if chlorination is carried out in the presence of light at room temperature.

9.4 The Reactivity–Selectivity Principle

The relative rates of radical formation when a bromine radical abstracts a hydrogen atom are different from the relative rates of radical formation when a chlorine radical abstracts a hydrogen atom. At 125 °C, a bromine radical abstracts a hydrogen atom from a tertiary carbon 1600 times faster than from a primary carbon and abstracts a hydrogen atom from a secondary carbon 82 times faster than from a primary carbon.

relative rates of radical formation by a bromine radical at 125 °C

tertiary	>	secondary	>	primary
1600		82		1

← increasing rate of formation

When a bromine radical is the hydrogen-abstracting agent, the differences in reactivity are so great that the reactivity factor is vastly more important than the probability factor. For example, radical bromination of butane gives a 98% yield of 2-bromobutane, compared with the 71% yield of 2-chlorobutane obtained when butane is chlorinated (Section 9.3). In other words, bromination is more highly regioselective than chlorination.

A bromine radical is less reactive and more selective than a chlorine radical.

$$CH_3CH_2CH_2CH_3 + Br_2 \xrightarrow{h\nu} \underset{\substack{\textbf{1-bromobutane} \\ \textbf{2\%}}}{CH_3CH_2CH_2CH_2Br} + \underset{\substack{\textbf{2-bromobutane} \\ \textbf{98\%}}}{CH_3CH_2\overset{\overset{\displaystyle Br}{|}}{C}HCH_3} + HBr$$

Similarly, bromination of 2,2,5-trimethylhexane gives an 82% yield of the product in which bromine replaces the tertiary hydrogen. Chlorination of the same alkane results in a 14% yield of the tertiary alkyl chloride (Section 9.3).

$$\underset{\textbf{2,2,5-trimethylhexane}}{CH_3\overset{\overset{\displaystyle CH_3}{|}}{\underset{\underset{\displaystyle CH_3}{|}}{C}}CH_2CH_2\overset{\overset{\displaystyle CH_3}{|}}{C}HCH_3} + Br_2 \xrightarrow{h\nu} \underset{\substack{\textbf{2-bromo-2,5,5-trimethylhexane} \\ \textbf{82\%}}}{CH_3\overset{\overset{\displaystyle CH_3}{|}}{\underset{\underset{\displaystyle CH_3}{|}}{C}}CH_2CH_2\overset{\overset{\displaystyle CH_3}{|}}{\underset{\underset{\displaystyle Br}{|}}{C}}CH_3} + HBr$$

PROBLEM 7◆

Carry out the calculations that predict that

a. 2-bromobutane will be obtained in 98% yield.

b. 2-bromo-2,5,5,-trimethylhexane will be obtained in 82% yield.

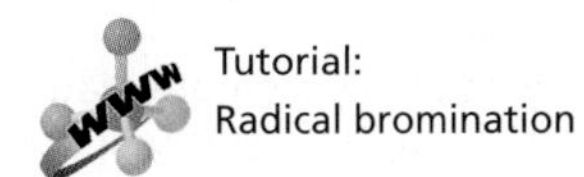

Tutorial: Radical bromination

Why are the relative rates of radical formation so different when a bromine radical rather than a chlorine radical is used as the hydrogen-abstracting reagent? To answer this question, we must compare the $\Delta H°$ values for the formation of primary, secondary, and tertiary radicals when a chlorine radical is used, as opposed to when a bromine radical is used. These $\Delta H°$ values can be calculated using the bond dissociation energies in Table 3.1 on p. 129. (Remember that $\Delta H°$ is equal to the energy of the bond being broken minus the energy of the bond being formed.)

Reaction	$\Delta H°$ (kcal/mol)	$\Delta H°$ (kJ/mol)
$Cl\cdot + CH_3CH_2CH_3 \longrightarrow CH_3CH_2\dot{C}H_2 + HCl$	101 − 103 = −2	−8
$Cl\cdot + CH_3CH_2CH_3 \longrightarrow CH_3\dot{C}HCH_3 + HCl$	99 − 103 = −4	−17
$Cl\cdot + CH_3CH(CH_3)CH_3 \longrightarrow CH_3\dot{C}(CH_3)CH_3 + HCl$	97 − 103 = −6	−25

Reaction	$\Delta H°$ (kcal/mol)	$\Delta H°$ (kJ/mol)
$Br\cdot + CH_3CH_2CH_3 \longrightarrow CH_3CH_2\dot{C}H_2 + HBr$	101 − 87 = 14	59
$Br\cdot + CH_3CH_2CH_3 \longrightarrow CH_3\dot{C}HCH_3 + HBr$	99 − 87 = 12	50
$Br\cdot + CH_3CH(CH_3)CH_3 \longrightarrow CH_3\dot{C}(CH_3)CH_3 + HBr$	97 − 87 = 10	42

We must also be aware that bromination is a much slower reaction than chlorination. The activation energy for abstraction of a hydrogen atom by a bromine radical has been found experimentally to be about 4.5 times greater than that for abstraction of a hydrogen atom by a chlorine radical. Using the calculated $\Delta H°$ values and the experimental activation energies, we can draw reaction coordinate diagrams for the formation of primary, secondary, and tertiary radicals by chlorine radical abstraction (Figure 9.1a) and by bromine radical abstraction (Figure 9.1b).

Because the reaction of a chlorine radical with an alkane to form a primary, secondary, or tertiary radical is exothermic, the transition states resemble the reactants more than they resemble the products (see the Hammond postulate, Section 4.3). The

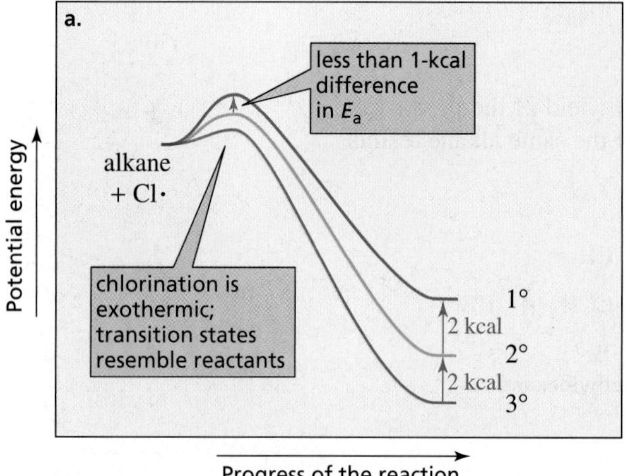

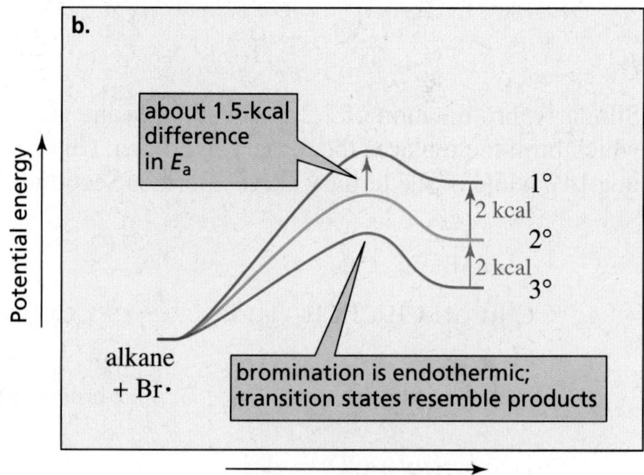

▲ **Figure 9.1**
(a) Reaction coordinate diagrams for the formation of primary, secondary, and tertiary alkyl radicals as a result of abstraction of a hydrogen atom by a chlorine radical. The transition states have relatively little radical character because they resemble the reactants.
(b) Reaction coordinate diagrams for the formation of primary, secondary, and tertiary alkyl radicals as a result of abstraction of a hydrogen atom by a bromine radical. The transition states have a relatively high degree of radical character because they resemble the products.

reactants all have approximately the same energy, so there will be only a small difference in the activation energies for removal of a hydrogen atom from a primary, secondary, or tertiary carbon. In contrast, the reaction of a bromine radical with an alkane is endothermic, so the transition states resemble the products more than they resemble the reactants. Because there is a significant difference in the energies of the product radicals—depending on whether they are primary, secondary, or tertiary—there is a significant difference in the activation energies. Therefore, a chlorine radical makes primary, secondary, and tertiary radicals with almost equal ease, whereas a bromine radical has a clear preference for formation of the easiest-to-form tertiary radical (Figure 9.1). In other words, because a bromine radical is relatively unreactive, it is highly selective about which hydrogen atom it abstracts. In contrast, the much more reactive chlorine radical is considerably less selective. These observations illustrate the **reactivity–selectivity principle**, which states that *the greater the reactivity of a species, the less selective it will be.*

The more reactive a species is, the less selective it will be.

Because chlorination is relatively nonselective, it is a useful reaction only when there is just one kind of hydrogen in the molecule.

$$\text{cyclohexane} + Cl_2 \xrightarrow{h\nu} \text{chlorocyclohexane} + HCl$$

PROBLEM 8◆

If 2-methylpropane is brominated at 125 °C in the presence of light, what percent of the product will be 2-bromo-2-methylpropane? Compare your answer with the percent given in Problem 4 for chlorination.

PROBLEM 9◆

Take the same alkanes whose percentages of monochlorination products you calculated in Problem 6, and calculate what the percentages of monobromination products would be if bromination were carried out at 125 °C.

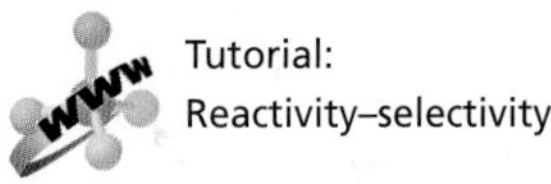

Tutorial:
Reactivity–selectivity

By comparing the $\Delta H°$ values for the sum of the two propagating steps for the monohalogenation of methane, we can understand why alkanes undergo chlorination and bromination but not iodination and why fluorination is too violent a reaction to be useful.

$$F\cdot + CH_4 \longrightarrow \cdot CH_3 + HF \qquad 105 - 136 = -31$$
$$\cdot CH_3 + F_2 \longrightarrow CH_3F + F\cdot \qquad 38 - 108 = -70$$
$$\Delta H° = -101 \text{ kcal/mol} \quad (\text{or } -423 \text{ kJ/mol})$$

$$Cl\cdot + CH_4 \longrightarrow \cdot CH_3 + HCl \qquad 105 - 103 = 2$$
$$\cdot CH_3 + Cl_2 \longrightarrow CH_3Cl + Cl\cdot \qquad 58 - 84 = -26$$
$$\Delta H° = -24 \text{ kcal/mol} \quad (\text{or } -100 \text{ kJ/mol})$$

$$Br\cdot + CH_4 \longrightarrow \cdot CH_3 + HBr \qquad 105 - 87 = 18$$
$$\cdot CH_3 + Br_2 \longrightarrow CH_3Br + Br\cdot \qquad 46 - 70 = -24$$
$$\Delta H° = -6 \text{ kcal/mol} \quad (\text{or } -25 \text{ kJ/mol})$$

$$I\cdot + CH_4 \longrightarrow \cdot CH_3 + HI \qquad 105 - 71 = 34$$
$$\cdot CH_3 + I_2 \longrightarrow CH_3I + I\cdot \qquad 36 - 57 = -21$$
$$\Delta H° = 13 \text{ kcal/mol} \quad (\text{or } 54 \text{ kJ/mol})$$

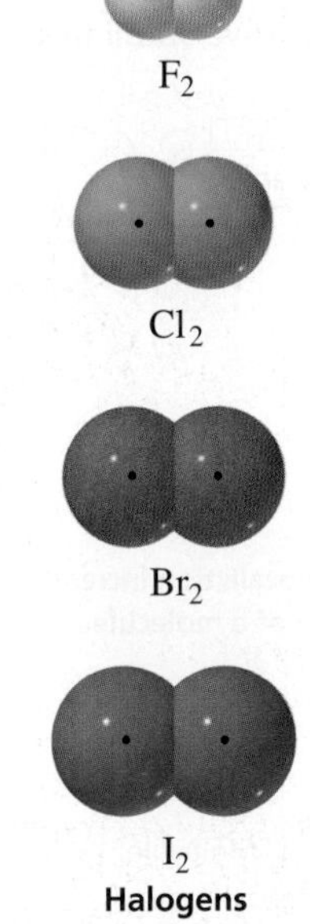

Halogens

The fluorine radical is the most reactive of the halogen radicals, and it reacts violently with alkanes ($\Delta H° = -31$ kcal/mol). In contrast, the iodine radical is the least reactive of the halogen radicals. In fact, it is so unreactive ($\Delta H° = 34$ kcal/mol) that it is unable to abstract a hydrogen atom from an alkane. Consequently, it reacts with another iodine radical and reforms I_2.

PROBLEM-SOLVING STRATEGY

Would chlorination or bromination of methylcyclohexane produce a greater yield of 1-halo-1-methylcyclohexane?

To solve this kind of problem, first draw the structures of the compounds being discussed.

$$\text{methylcyclohexane} \xrightarrow[h\nu]{X_2} \text{1-halo-1-methylcyclohexane}$$

1-Halo-1-methylcyclohexane is a tertiary alkyl halide, so the question becomes, "Will bromination or chlorination produce a greater yield of a tertiary alkyl halide?" Because bromination is more selective, it will produce a greater yield of the desired compound. Chlorination will form some of the tertiary alkyl halide, but it will also form significant amounts of primary and secondary alkyl halides.

Now continue on to Problem 10.

PROBLEM 10◆

a. Would chlorination or bromination produce a greater yield of 1-halo-2,3-dimethylbutane?
b. Would chlorination or bromination produce a greater yield of 2-halo-2,3-dimethylbutane?
c. Would chlorination or bromination be a better way to make 1-halo-2,2-dimethylpropane?

9.5 Radical Substitution of Benzylic and Allylic Hydrogens

Benzyl and allyl radicals are more stable than alkyl radicals because their unpaired electrons are delocalized. We have seen that electron delocalization increases the stability of a molecule (Sections 7.6 and 7.8).

relative stabilities of radicals

most stable: $C_6H_5\dot{C}H_2 \approx CH_2{=}CH\dot{C}H_2 > R_3C\cdot > R_2HC\cdot > RH_2C\cdot > CH_2{=}\dot{C}H \approx H_3C\cdot$:least stable

benzyl radical, allyl radical, tertiary radical, secondary radical, primary radical, vinyl radical, methyl radical

Electron delocaliztion increases the stability of a molecule.

We know that the more stable the radical, the faster it can be formed. This means that a hydrogen bonded to either a benzylic carbon or an allylic carbon will be preferentially substituted in a halogenation reaction. As we saw in Section 9.4, bromination is more highly regioselective than chlorination, so the percent of substitution at the benzylic or allylic carbon is greater for bromination.

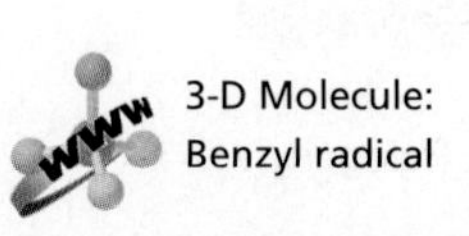

$$C_6H_5CH_2CH_3 + X_2 \xrightarrow{\Delta} C_6H_5CHXCH_3 + HX \quad (X = Cl \text{ or } Br)$$

benzylic substituted product

$$CH_3CH{=}CH_2 + X_2 \xrightarrow{\Delta} \underset{|}{X}CH_2CH{=}CH_2 + HX$$

allylic substituted product

N-Bromosuccinimide (NBS) is frequently used to brominate allylic positions because it allows a radical substitution reaction to be carried out without subjecting the reactant to a relatively high concentration of Br_2 that could add to its double bond.

cyclohexene + *N*-bromosuccinimide (NBS) $\xrightarrow[\text{peroxide}]{h\nu \text{ or } \Delta}$ 3-bromocyclohexene + succinimide

N-bromosuccinimide
NBS

succinimide

The reaction involves initial homolytic cleavage of the N—Br bond in NBS. This generates the bromine radical needed to initiate the radical reaction. Light or heat and a radical initiator such as a peroxide are used to promote the reaction. The bromine radical abstracts an allylic hydrogen to form HBr and an allylic radical. The allylic radical reacts with Br_2, forming the allyl bromide and a chain-propagating bromine radical.

$$Br\cdot + \text{cyclohexene} \longrightarrow HBr + \text{cyclohexenyl radical} \xrightarrow{Br_2} \text{3-bromocyclohexene} + Br\cdot$$

The Br_2 used in the second step of the preceding reaction sequence is produced in low concentration from the reaction of NBS with HBr.

$$\text{NBS (N—Br)} + HBr \longrightarrow \text{succinimide (N—H)} + Br_2$$

The advantage to using NBS to brominate allylic positions is that neither Br_2 or HBr is formed at a high enough concentration to add to the double bond.

When a radical abstracts a hydrogen atom from an allylic carbon, the unpaired electron of the allylic radical is shared by two carbons. In other words, the allylic radical has two resonance contributors. In the following reaction, only one substitution product is formed, because the groups attached to the sp^2 carbons are the same in both resonance contributors:

Tutorial: Radicals: common terms

$$Br\cdot + CH_3CH{=}CH_2 \longrightarrow \dot{C}H_2CH{=}CH_2 \longleftrightarrow CH_2{=}CH\dot{C}H_2 + HBr$$

$$\downarrow Br_2$$

$$BrCH_2CH{=}CH_2 + Br\cdot$$

3-bromopropene

If, however, the groups attached to the two sp^2 carbons of the allylic radical are *not* the same in both resonance contributors, two substitution products are formed:

3-D Molecule: Allyl radical

$$Br\cdot + CH_3CH_2CH{=}CH_2 \longrightarrow CH_3\dot{C}HCH{=}CH_2 \longleftrightarrow CH_3CH{=}CH\dot{C}H_2 + HBr$$

$$\downarrow Br_2$$

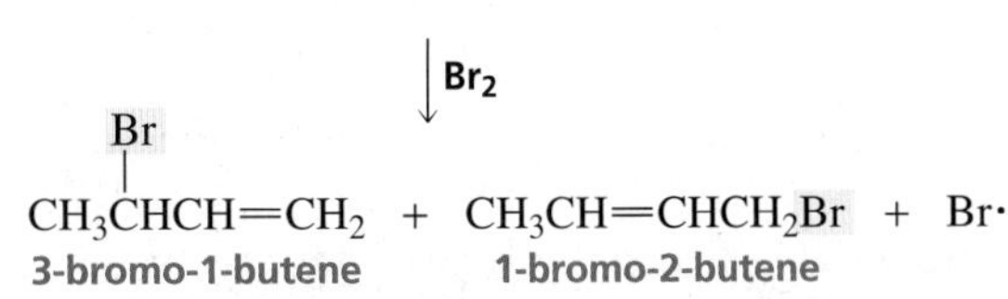

$$CH_3\underset{}{\overset{Br}{C}}HCH{=}CH_2 + CH_3CH{=}CHCH_2Br + Br\cdot$$

3-bromo-1-butene 1-bromo-2-butene

PROBLEM 11

Two products are formed when methylenecyclohexane reacts with NBS. Explain how each is formed.

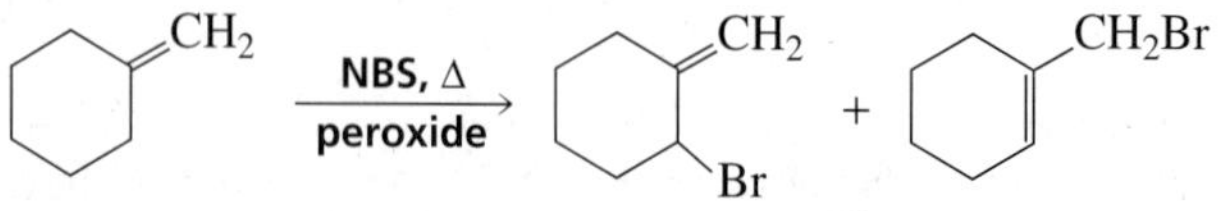

(methylenecyclohexane, CH_2) $\xrightarrow[\text{peroxide}]{\text{NBS, }\Delta}$ (2-bromomethylenecyclohexane, CH_2, Br) + (1-cyclohexenyl CH_2Br)

PROBLEM 12 SOLVED

How many allylic substituted bromoalkenes are formed from the reaction of 2-pentene with NBS? Ignore stereoisomers.

SOLUTION The bromine radical will abstract a secondary allylic hydrogen from C-4 of 2-pentene in preference to a primary allylic hydrogen from C-1. The resonance contributors of the resulting radical intermediate have the same groups attached to the sp^2 carbons, so only one bromoalkene is formed. Because of the high selectivity of the bromine radical, an insignificant amount of radical will be formed by abstraction of a hydrogen from the less reactive primary allylic position.

$$CH_3CH{=}CHCH_2CH_3 \xrightarrow[\text{peroxide}]{\text{NBS, }\Delta} CH_3CH{=}CH\dot{C}HCH_3 \longleftrightarrow CH_3\dot{C}HCH{=}CHCH_3 + HBr$$

$$\downarrow$$

$$CH_3CH{=}CHCHCH_3 \text{ (Br on C-4)}$$

9.6 Stereochemistry of Radical Substitution Reactions

If a reactant does not have an asymmetric carbon and a radical substitution reaction forms a product with an asymmetric carbon, a racemic mixture will be obtained.

$$CH_3CH_2CH_2CH_3 + Br_2 \xrightarrow{h\nu} CH_3CH_2\underset{Br}{\underset{|}{C}}HCH_3 + HBr$$

an asymmetric carbon

configuration of the products

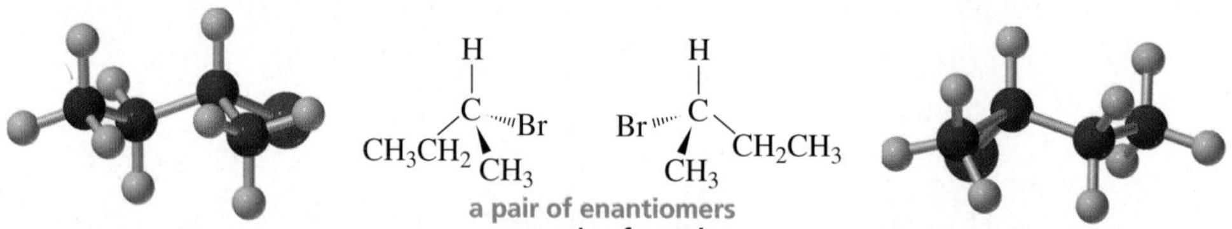

a pair of enantiomers
perspective formulas

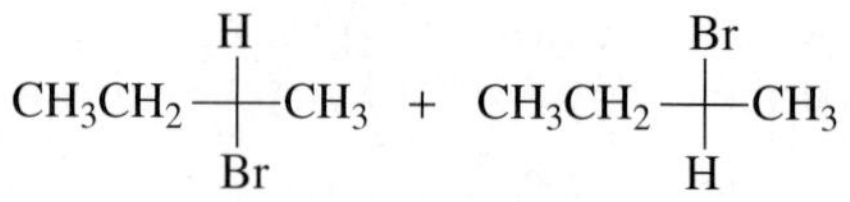

$$CH_3CH_2\text{—}\underset{Br}{\overset{H}{+}}\text{—}CH_3 \quad + \quad CH_3CH_2\text{—}\underset{H}{\overset{Br}{+}}\text{—}CH_3$$

a pair of enantiomers
Fischer projections

To understand why both enantiomers are formed, we must look at the propagation steps of the radical substitution reaction. In the first propagation step, the bromine radical removes a hydrogen atom from the alkane, creating a radical intermediate. The carbon bearing the unpaired electron is sp^2 hybridized; therefore, the three atoms to which it is bonded lie in a plane. In the second propagation step, the incoming halogen has equal access to both sides of the plane. As a result, both the *R* and *S* enantiomers are formed. Identical amounts of the *R* enantiomer and the *S* enantiomer are obtained, so the reaction is not stereoselective.

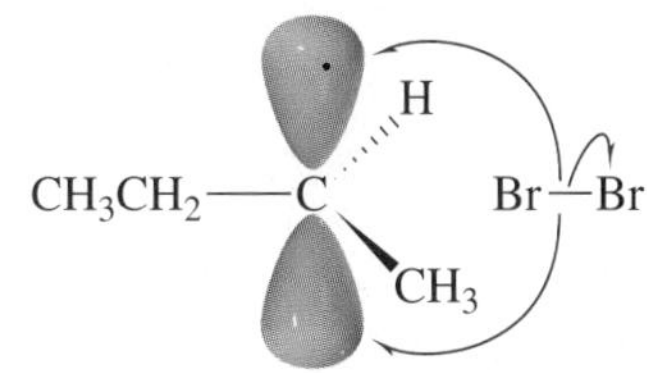

a radical intermediate

Notice that the stereochemical outcome of a *radical substitution reaction* is identical to the stereochemical outcome of a *radical addition reaction* (Section 5.19). This is because both reactions form a radical intermediate, and it is the reaction of the intermediate that determines the configuration of the products.

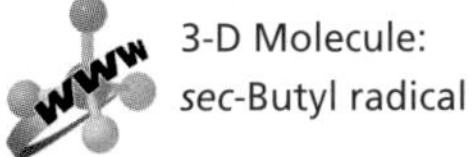

3-D Molecule: *sec*-Butyl radical

Identical amounts of the *R* and *S* enantiomers are also obtained if a hydrogen bonded to an asymmetric carbon is substituted by a halogen. Breaking the bond to the asymmetric carbon destroys the configuration at the asymmetric carbon and forms a planar radical intermediate. The incoming halogen has equal access to both sides of the plane, so identical amounts of the two enantiomers are obtained.

CH_2CH_3 / C(H)(CH_3)(phenyl) $\xrightarrow{Br\cdot}$ CH_2CH_3 / Ċ(CH_3)(phenyl) + HBr $\xrightarrow{Br_2}$ CH_2CH_3 / C(Br)(CH_3)(phenyl) + CH_2CH_3 / C(Br)(H_3C)(phenyl) + Br·

What happens if the reactant already has an asymmetric carbon and the radical substitution reaction creates a second asymmetric carbon? In this case, a pair of diastereomers will be formed in unequal amounts.

configuration of the products

configuration does not change

H_3C(H)(Cl)C—C(Br)(H)(CH_2CH_3) + H_3C(H)(Cl)C—C(Br)(H)(CH_2CH_3)

a pair of diastereomers
perspective formulas

configuration does not change

CH_3 / H—C—Cl / H—C—Br / CH_2CH_3 + CH_3 / H—C—Cl / Br—C—H / CH_2CH_3

a pair of diastereomers
Fischer projections

Diastereomers are formed because the new asymmetric carbon created in the product can have either the *R* or the *S* configuration, but the configuration of the asymmetric carbon in the reactant will be unchanged in the product because none of the bonds to that carbon are broken during the course of the reaction.

original asymmetric carbon

CH_3 / H—C—Cl / CH_2 / CH_2CH_3 + Br_2 $\xrightarrow{h\nu}$ CH_3 / H—C—Cl / CH—Br / CH_2CH_3 + HBr

new asymmetric carbon

More of one diastereomer will be formed than the other because the transition states leading to their formation are diastereomeric and, therefore, do not have the same energy.

PROBLEM 13◆

a. What hydrocarbon with molecular formula C_4H_{10} forms only two monochlorinated products? Both products are achiral.

b. What hydrocarbon with the same molecular formula as in part a forms only three monochlorinated products? One is achiral and two are chiral.

9.7 Reactions of Cyclic Compounds

Cyclic compounds undergo the same reactions as acyclic compounds. For example, cyclic alkanes, like acyclic alkanes, undergo radical substitution reactions with chlorine or bromine.

$$\text{cyclopentane} + Cl_2 \xrightarrow{h\nu} \text{chlorocyclopentane} + HCl$$

chlorocyclopentane

$$\text{cyclohexane} + Br_2 \xrightarrow{h\nu} \text{bromocyclohexane} + HBr$$

bromocyclohexane

Cyclic alkenes undergo the same reactions as acyclic alkenes.

$$\text{cyclohexene} + HBr \longrightarrow \text{bromocyclohexane}$$

bromocyclohexane

$$\text{cyclopentene} + NBS \xrightarrow[\text{peroxide}]{\Delta} \text{3-bromocyclopentene}$$

3-bromocyclopentene

In other words, *the reactivity of a compound usually depends solely on its functional group, not on whether it is cyclic or acyclic.*

PROBLEM 14◆

a. Give the major product(s) of the reaction of 1-methylcyclohexene with the following reagents, ignoring stereoisomers:
   1. NBS/Δ/peroxide 2. Br_2/CH_2Cl_2 3. HBr 4. HBr/peroxide
b. Give the configuration of the products.

CYCLOPROPANE

Cyclopropane is one notable exception to the generalization that cyclic and acyclic compounds undergo the same reactions. Although it is an alkane, cyclopropane undergoes electrophilic addition reactions as if it were an alkene.

$$\text{cyclopropane} + HBr \longrightarrow CH_3CH_2CH_2Br$$

$$\text{cyclopropane} + Cl_2 \xrightarrow{FeCl_3} ClCH_2CH_2CH_2Cl$$

$$\text{cyclopropane} + H_2 \xrightarrow[80\ °C]{Ni} CH_3CH_2CH_3$$

Cyclopropane is more reactive than propene toward addition of acids such as HBr and HCl, but is less reactive toward addition of Cl_2 and Br_2, so a Lewis acid ($FeCl_3$ or $FeBr_3$) is needed to catalyze halogen addition (Section 1.21).

It is the strain in the small ring that makes it possible for cyclopropane to undergo electrophilic addition reactions (Section 2.11). Because of the 60° bond angles in the three-membered ring, the sp^3 orbitals cannot overlap head-on; this decreases the effectiveness of the orbital overlap. Thus, the C—C "banana bonds" in cyclopropane are considerably weaker than normal C—C σ bonds (Figure 2.6 on p. 92). Consequently, three-membered rings undergo ring-opening reactions with electrophilic reagents.

$$\text{cyclopropane} + XY \longrightarrow H_2C(X)\text{—}CH_2\text{—}CH_2(Y)$$

9.8 Radical Reactions in Biological Systems

For a long time, scientists assumed that radical reactions were not important in biological systems because a large amount of energy—heat or light—is required to initiate a radical reaction and it is difficult to control the propagation steps of the chain reaction once initiation occurs. However, it is now widely recognized that there are biological reactions that involve radicals. Instead of being generated by heat or light, the radicals are formed by the interaction of organic molecules with metal ions. These radical reactions take place in the active site of an enzyme. Containing the reaction in a specific site allows the reaction to be controlled.

One biological reaction that involves radicals is the one responsible for the conversion of toxic hydrocarbons to less toxic alcohols. Carried out in the liver, the hydroxylation of the hydrocarbon is catalyzed by an iron-porphyrin-containing enzyme called cytochrome P_{450} (Section 12.8). An alkyl radical intermediate is created when $Fe^{V}O$ abstracts a hydrogen atom from an alkane. In the next step, $Fe^{IV}OH$ dissociates homolytically into Fe^{III} and $HO^{\cdot}$, and the $HO^{\cdot}$ immediately combines with the radical intermediate to form the alcohol.

$$Fe^{V}{=}O + H{-}\overset{|}{\underset{|}{C}}{-} \longrightarrow Fe^{IV}{-}OH + \cdot\overset{|}{\underset{|}{C}}{-} \longrightarrow Fe^{III} + HO{-}\overset{|}{\underset{|}{C}}{-}$$

an alkane — a radical intermediate — an alcohol

This reaction can also have the opposite toxicological effect. That is, instead of converting a toxic hydrocarbon into a less toxic alcohol, substituting an OH for an H in some compounds causes a nontoxic compound to become toxic. Therefore, compounds that are nontoxic *in vitro* are not necessarily nontoxic *in vivo*. For example, studies done on animals showed that substituting an OH for an H caused methylene chloride (CH_2Cl_2) to become a carcinogen when it is inhaled.

Another important biological reaction shown to involve a radical intermediate is the conversion of a ribonucleotide into a deoxyribonucleotide. The biosynthesis of ribonucleic acid (RNA) requires ribonucleotides, whereas the biosynthesis of deoxyribonucleic acid (DNA) requires deoxyribonucleotides (Section 27.1). The first step in the conversion of a ribonucleotide to the deoxyribonucleotide needed for DNA biosynthesis involves abstraction of a hydrogen atom from the ribonucleotide to form

DECAFFEINATED COFFEE AND THE CANCER SCARE

Animal studies showing that methylene chloride becomes a carcinogen when inhaled caused some concern because methylene chloride was the solvent used to extract caffeine from coffee beans in the manufacture of decaffeinated coffee. However, when methylene chloride was added to drinking water fed to laboratory rats and mice, researchers found no toxic effects. They observed no toxicological responses of any kind either in rats that had consumed an amount of methylene chloride equivalent to the amount that would be ingested by drinking 120,000 cups of decaffeinated coffee per day or in mice that had consumed an amount equivalent to drinking 4.4 million cups of decaffeinated coffee per day. In addition, no increased risk of cancer was found in a study of thousands of workers exposed daily to inhaled methylene chloride. (Studies done on humans do not always agree with those done on animals.) Because of the initial concern, however, researchers sought alternative methods for extracting caffeine from coffee beans. Extraction by liquid CO_2 at supercritical temperatures and pressures was found to be a better method since it extracts caffeine without simultaneously extracting some of the flavor compounds that are removed when methylene chloride is used. There is essentially no difference in flavor between regular coffee and coffee decaffeinated with CO_2.

a radical intermediate. The radical that abstracts the hydrogen is formed as a result of the interaction between Fe(III) and an amino acid at the active site of the enzyme.

several steps

a ribonucleotide → a radical intermediate → a deoxyribonucleotide

Unwanted radicals in biological systems must be destroyed before they have an opportunity to cause damage to cells. Cell membranes, for example, are susceptible to the same kind of radical reactions that cause butter to become rancid (Section 26.3). Imagine the state of your cell membranes if radical reactions could occur readily. Radical reactions in biological systems also have been implicated in the aging process. Unwanted radical reactions are prevented by **radical inhibitors**—compounds that destroy reactive radicals by creating unreactive radicals or compounds with only paired electrons. Hydroquinone is an example of a radical inhibitor. When hydroquinone traps a radical, it forms semiquinone, which is stabilized by electron delocalization and is, therefore, less reactive than other radicals. Furthermore, semiquinone can trap another radical and form quinone, a compound whose electrons are all paired.

hydroquinone + R· (reactive radical) → semiquinone + RH → quinone + RH

Two examples of radical inhibitors that are present in biological systems are vitamin C and vitamin E. Like hydroquinone, they form relatively stable radicals. Vitamin C (also called ascorbic acid) is a water-soluble compound that traps radicals formed in the aqueous environment of the cell and in blood plasma. Vitamin E (also called α-tocopherol) is a water-insoluble (hence fat-soluble) compound that traps radicals formed in nonpolar membranes. Why one vitamin functions in aqueous environments and the other in nonaqueous environments should be apparent from their structures and electrostatic potential maps, which show that vitamin C is a relatively polar compound, whereas vitamin E is nonpolar.

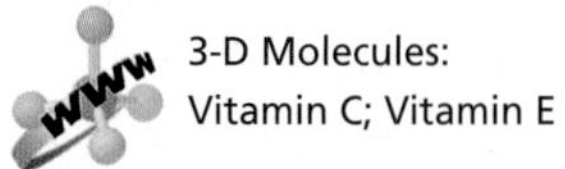

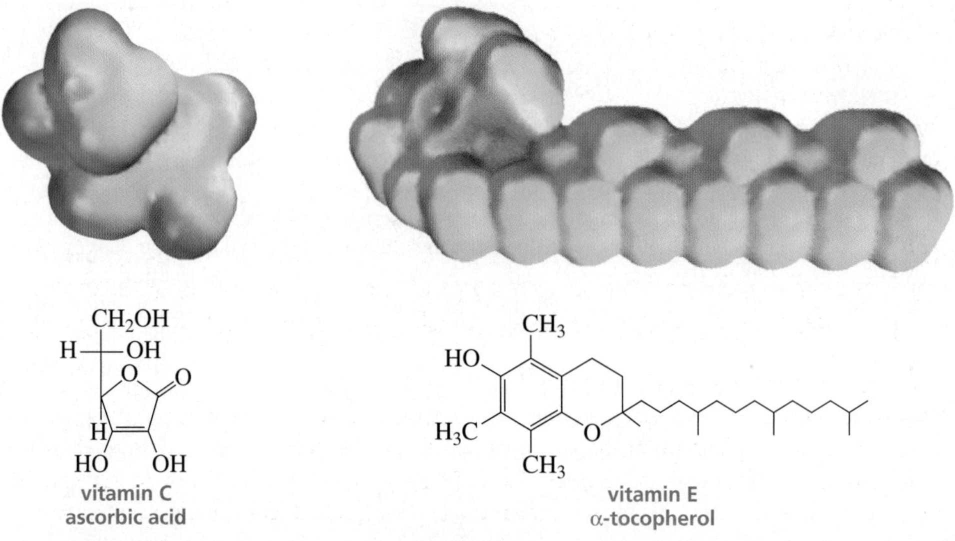

vitamin C
ascorbic acid

vitamin E
α-tocopherol

FOOD PRESERVATIVES

Radical inhibitors that are present in food are known as *preservatives* or *antioxidants*. They preserve food by preventing unwanted radical reactions. Vitamin E is a naturally occurring preservative found in vegetable oil. BHA and BHT are synthetic preservatives that are added to many packaged foods.

food preservatives

9.9 Radicals and Stratospheric Ozone

In 1995, the Nobel Prize in chemistry was awarded to **Sherwood Rowland, Mario Molina,** *and* **Paul Crutzen** *for their pioneering work in explaining the chemical processes responsible for the depletion of the ozone layer in the stratosphere. Their work demonstrated that human activities could interfere with global processes that support life. This was the first time that a Nobel Prize was presented for work in the environmental sciences.*

Ozone (O_3), a major constituent of smog, is a health hazard at ground level. In the stratosphere, however, a layer of ozone shields the Earth from harmful solar radiation. The greatest concentration of ozone occurs between 12 and 15 miles above the Earth's surface. The ozone layer is thinnest at the equator and densest towards the poles. Ozone is formed in the atmosphere from the interaction of molecular oxygen with very short wavelength ultraviolet light.

$$O_2 \xrightarrow{h\nu} O + O$$

$$O + O_2 \longrightarrow \underset{\text{ozone}}{O_3}$$

The stratospheric ozone layer acts as a filter for biologically harmful ultraviolet radiation that otherwise would reach the surface of the Earth. Among other effects, high-energy short-wavelength ultraviolet light can damage DNA in skin cells, causing mutations that trigger skin cancer (Section 29.6). We owe our very existence to this protective ozone layer. According to current theories of evolution, life could not have developed on land in the absence of this ozone layer. Instead, life would have had to remain in the ocean, where water screens out the harmful ultraviolet radiation.

Since about 1985, scientists have noted a precipitous drop in stratospheric ozone over Antarctica. This area of ozone depletion, known as the "ozone hole," is unprecedented in the history of ozone observations. Scientists subsequently noted a similar decrease in ozone over Arctic regions, and in 1988 they detected a depletion of ozone over the United States for the first time. Three years later, scientists determined that the rate of ozone depletion was two to three times faster than originally anticipated. Many in the scientific community blame recently observed increases in cataracts and skin cancer as well as diminished plant growth on the ultraviolet radiation that has penetrated the reduced ozone layer. It has been predicted that erosion of the protective ozone layer will cause an additional 200,000 deaths from skin cancer over the next 50 years.

Strong circumstantial evidence implicates synthetic chlorofluorocarbons (CFCs)—alkanes in which all the hydrogens have been replaced by fluorine and chlorine, such as $CFCl_3$ and CF_2Cl_2—as a major cause of ozone depletion. These gases, known commercially as Freons®, have been used extensively as cooling fluids in refrigerators and air conditioners. They were also once widely used as propellants in aerosol

F. Sherwood Rowland *was born in Ohio in 1927. He received a Ph.D. from the University of Chicago and is a professor of chemistry at the University of California, Irvine.*

Mario Molina *was born in Mexico in 1943 and subsequently became a U.S. citizen. He received a Ph.D. from the University of California, Berkeley, and then became a postdoctoral fellow in Rowland's laboratory. He is currently a professor of earth, atmospheric, and planetary sciences at the Massachusetts Institute of Technology.*

Paul Crutzen *was born in Amsterdam in 1933. He was trained as a meteorologist and became interested in stratospheric chemistry and statospheric ozone in particular. He is a professor at the Max Planck Institute for Chemistry in Mainz, Germany.*

THE CONCORDE AND OZONE DEPLETION

Supersonic aircraft cruise in the lower stratosphere, and their jet engines convert molecular oxygen and nitrogen into nitrogen oxides such as NO and NO_2. Like CFCs, nitrogen oxides react with stratospheric ozone. Fortunately, the supersonic Concorde, built jointly by England and France, makes only a limited number of flights each week.

Polar stratospheric clouds increase the rate of ozone destruction. These clouds form over Antarctica during the cold winter months. Ozone depletion in the Arctic is less severe because it generally does not get cold enough for the polar stratospheric clouds to form.

spray cans (deodorant, hair spray, etc.) because of their odorless, nontoxic, and nonflammable properties and because they are chemically inert and thus do not react with the contents of the can. Such use now, however, has been banned.

Chlorofluorocarbons remain very stable in the atmosphere until they reach the stratosphere. There they encounter wavelengths of ultraviolet light that cause the homolytic cleavage that generates chlorine radicals.

$$\text{F—C(Cl)(F)—Cl} \xrightarrow{h\nu} \text{F—}\dot{\text{C}}\text{(Cl)(F)} + \text{Cl}\cdot$$

Movie: Chlorofluorocarbons and ozone

The chlorine radicals are the ozone-removing agents. They react with ozone to form chlorine monoxide radicals and molecular oxygen. The chlorine monoxide radicals then react with ozone to regenerate chlorine radicals. These two propagating steps are repeated over and over, destroying a molecule of ozone in each step. It has been calculated that each chlorine atom destroys 100,000 ozone molecules!

$$\text{Cl}\cdot + O_3 \longrightarrow \text{ClO}\cdot + O_2$$
$$\text{ClO}\cdot + O_3 \longrightarrow \text{Cl}\cdot + 2\,O_2$$

▶ Growth of the Antarctic ozone hole, located mostly over the continent of Antarctica, since 1979. The images were made from data supplied by total ozone-mapping spectrometers (TOMS). The color scale depicts the total ozone values in Dobson units. The lowest ozone densities are represented by dark blue.

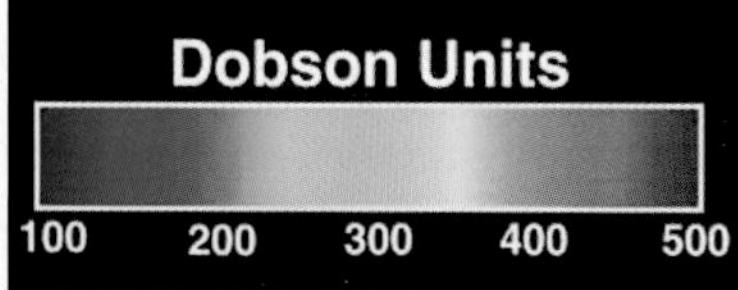

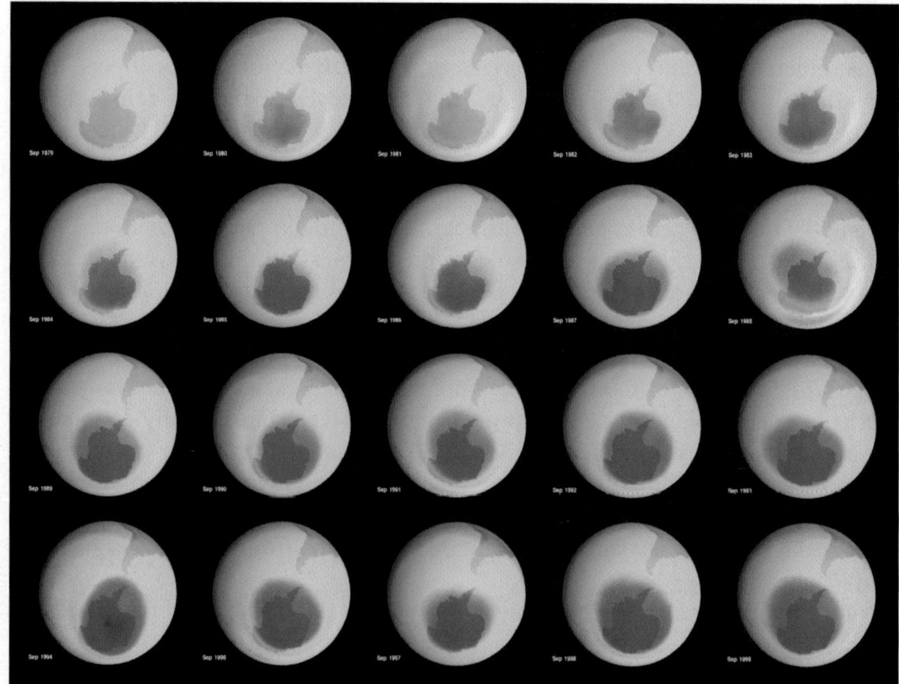

Summary

Alkanes are called **saturated hydrocarbons** because they do not contain any double or triple bonds. Since they also have only strong σ bonds and atoms with no partial charges, alkanes are very unreactive. Alkanes do undergo **radical substitution reactions** with chlorine (Cl_2) or bromine (Br_2) at high temperatures or in the presence of light, to form alkyl chlorides or alkyl bromides. The substitution reaction is a **radical chain reaction** with **initiation**, **propagation**, and **termination steps**. Unwanted radical reactions are prevented by **radical inhibitors**—compounds that destroy reactive radicals by creating unreactive radicals or compounds with only paired electrons.

The rate-determining step of the radical substitution reaction is hydrogen atom abstraction to form a **radical**. The relative rates of radical formation are benzylic ~ allyl > 3° > 2° > 1° > vinyl ~ methyl. To determine the relative amounts of products obtained from the radical halogenation of an alkane, both probability and the relative rate at which a particular hydrogen is abstracted must be taken into account. The **reactivity–selectivity principle** states that the more reactive a species is, the less selective it will be. A bromine radical is *less reactive* than a chlorine radical, so a bromine radical is *more selective* about which hydrogen atom it abstracts. *N*-Bromosuccinimide (NBS) is used to brominate allylic positions. Cyclic compounds undergo the same reactions as acyclic compounds.

If a reactant does not have an asymmetric carbon, and a radical substitution reaction forms a product with an asymmetric carbon, a racemic mixture will be obtained. A racemic mixture is also obtained if a hydrogen bonded to an asymmetric carbon is substituted by a halogen. If a radical substitution reaction creates an asymmetric carbon in a reactant that already has an asymmetric carbon, a pair of diastereomers will be obtained in unequal amounts.

Some biological reactions involve radicals formed by the interaction of organic molecules with metal ions. The reactions take place in the active site of an enzyme.

Strong circumstantial evidence implicates synthetic chlorofluorocarbons as being responsible for the diminishing ozone layer. The interaction of these compounds with UV light generates chlorine radicals, which are the ozone-removing agents.

Summary of Reactions

1. *Alkanes* undergo radical substitution reactions with Cl_2 or Br_2 in the presence of heat or light (Sections 9.2, 9.3, and 9.4).

$$\underset{\text{excess}}{CH_3CH_3} + Cl_2 \xrightarrow{\Delta \text{ or } h\nu} CH_3CH_2Cl + HCl$$

$$\underset{\text{excess}}{CH_3CH_3} + Br_2 \xrightarrow{\Delta \text{ or } h\nu} CH_3CH_2Br + HBr$$

bromination is more selective than chlorination

2. *Alkyl-substituted benzenes* undergo radical halogenation at the benzylic position (Section 9.5).

$$C_6H_5\text{-}CH_2R + Br_2 \xrightarrow{h\nu} C_6H_5\text{-}\underset{\displaystyle Br}{CHR} + HBr$$

3. *Alkenes* undergo radical halogenation at the allylic positions. NBS is used for radical bromination at the allylic position (Section 9.5).

$$\text{cyclopentene} + NBS \xrightarrow[\text{peroxide}]{\Delta \text{ or } h\nu} \text{3-bromocyclopentene (Br)}$$

$$RCH_2CH{=}CH_2 + NBS \xrightarrow[\text{peroxide}]{\Delta \text{ or } h\nu} \underset{\displaystyle Br}{RCH}CH{=}CH_2 + RCH{=}CH\underset{\displaystyle Br}{CH_2} + HBr$$

Key Terms

alkane (p. 336)
combustion (p. 338)
free radical (p. 338)
halogenation reaction (p. 338)
homolytic bond cleavage (p. 338)
initiation step (p. 338)
paraffin (p. 338)
propagation step (p. 338)
radical (p. 338)
radical chain reaction (p. 338)
radical inhibitor (p. 352)
radical substitution reaction (p. 339)
reactivity–selectivity principle (p. 345)
saturated hydrocarbon (p. 336)
termination step (p. 339)

Problems

15. Give the product(s) of each of the following reactions, ignoring stereoisomers:

a. $CH_2{=}CHCH_2CH_2CH_3 + Br_2 \xrightarrow{h\nu}$

b. $CH_3C(CH_3){=}CHCH_3 + NBS \xrightarrow[\text{peroxide}]{\Delta}$

c. $CH_3CH_2CH(CH_3)CH_2CH_2CH_3 + Br_2 \xrightarrow{h\nu}$

d. 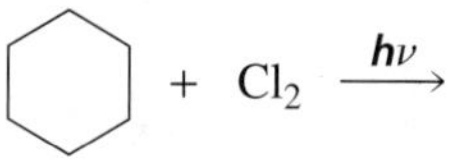 cyclohexane $+ Cl_2 \xrightarrow{h\nu}$

e. 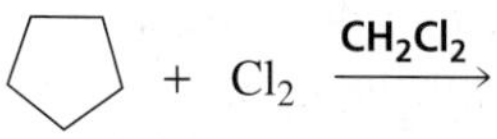 cyclopentane $+ Cl_2 \xrightarrow{CH_2Cl_2}$

f. 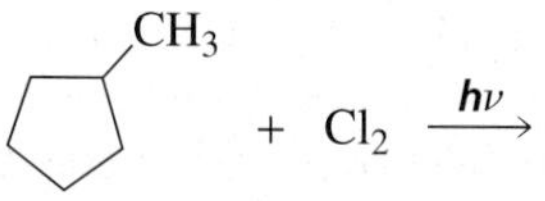 methylcyclopentane (CH_3) $+ Cl_2 \xrightarrow{h\nu}$

16. a. An alkane with molecular formula C_5H_{12} forms only one monochlorinated product when heated with Cl_2. Give the systematic name of this alkane.
b. An alkane with molecular formula C_7H_{16} forms seven monochlorinated products (ignore stereoisomers) when heated with Cl_2. Give the systematic name of this alkane.

17. Dr. Al Cahall wanted to determine experimentally the relative ease of removal of a hydrogen atom from a tertiary, a secondary, and a primary carbon by a chlorine radical. He allowed 2-methylbutane to undergo chlorination at 300 °C and obtained as products 36% 1-chloro-2-methylbutane, 18% 2-chloro-2-methylbutane, 28% 2-chloro-3-methylbutane, and 18% 1-chloro-3-methylbutane. What values did he obtain for the relative ease of removal of tertiary, secondary, and primary hydrogen atoms by a chlorine radical under the conditions of his experiment?

18. At 600 °C, the ratio of the relative rates of formation of a tertiary, a secondary, and a primary radical by a chlorine radical is 2.6 : 2.1 : 1. Explain the change in the degree of regioselectivity compared with what Dr. Al Cahall found in Problem 17.

19. Iodine (I_2) does not react with ethane even though I_2 is more easily cleaved homolytically than the other halogens. Explain.

20. Give the major product of each of the following reactions, ignoring stereoisomers:

a.  cyclopentene $+ NBS \xrightarrow[\text{peroxide}]{\Delta}$

b. $CH_2{=}CHCH_2CH_2CH_3 + NBS \xrightarrow[\text{peroxide}]{\Delta}$

c.  3-methylcyclohexene (CH_3) $+ NBS \xrightarrow[\text{peroxide}]{\Delta}$

d. $CH_3CH(CH_3)CH_3 + Cl_2 \xrightarrow{h\nu}$

e. $CH_3CH(CH_3)CH_3 + Br_2 \xrightarrow{h\nu}$

f. 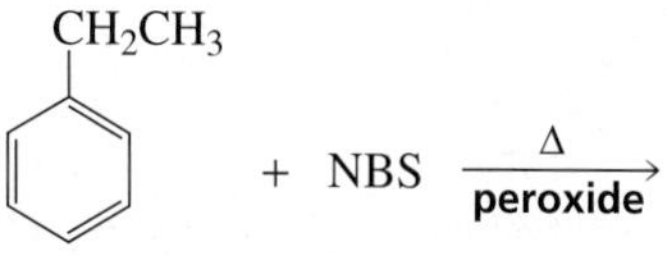 ethylbenzene (CH_2CH_3) $+ NBS \xrightarrow[\text{peroxide}]{\Delta}$

g.  cyclopentene $+ NBS \xrightarrow[\text{peroxide}]{\Delta}$

h. 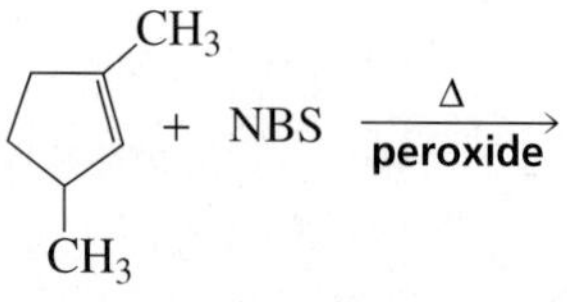 1,3-dimethylcyclopentane (CH_3, CH_3) $+ NBS \xrightarrow[\text{peroxide}]{\Delta}$

21. The deuterium kinetic isotope effect for chlorination of an alkane is defined in the following equation:

$$\textbf{deuterium kinetic isotope effect} = \frac{\textbf{rate of homolytic cleavage of a C—H bond by Cl}\cdot}{\textbf{rate of homolytic cleavage of a C—D bond by Cl}\cdot}$$

Predict whether chlorination or bromination would have a greater deuterium kinetic isotope effect.

22. a. How many monobromination products would be obtained from the radical bromination of methylcyclohexane? Ignore stereoisomers.
b. Which product would be obtained in greatest yield? Explain.
c. How many monobromination products would be obtained if all stereoisomers are included?

23. a. Propose a mechanism for the following reaction:

$$CH_3CH_3 + CH_3-\underset{CH_3}{\overset{CH_3}{C}}-OCl \xrightarrow{\Delta} CH_3CH_2Cl + CH_3-\underset{CH_3}{\overset{CH_3}{C}}-OH$$

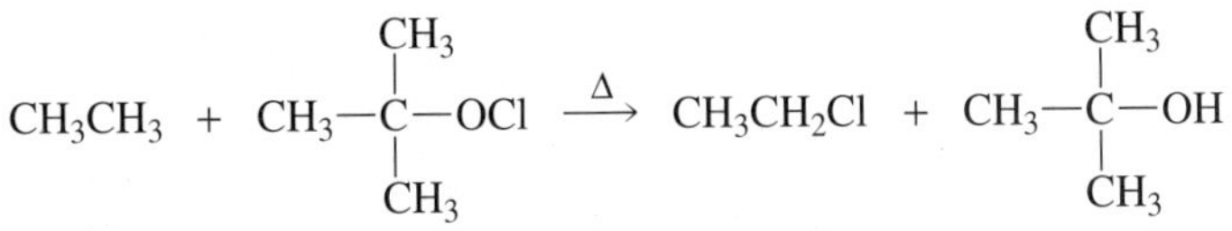

b. Given that the $\Delta H°$ value for the reaction is −42 kcal/mol and the bond dissociation energies for the C—H, C—Cl, and O—H bonds are 101, 82, and 102 kcal/mol, respectively, calculate the bond dissociation energy of the O—Cl bond.
c. Which set of propagation steps is more likely to occur?

24. a. Calculate the $\Delta H°$ value for the following reaction:

$$CH_4 + Cl_2 \xrightarrow{h\nu} CH_3Cl + HCl$$

b. Calculate the sum of the $\Delta H°$ values for the following two propagation steps:

$$CH_3-H + \cdot Cl \longrightarrow \cdot CH_3 + H-Cl$$
$$\cdot CH_3 + Cl-Cl \longrightarrow CH_3-Cl + \cdot Cl$$

c. Why do both calculations give you the same value of $\Delta H°$?

25. A possible alternative mechanism to that shown in Problem 24 for the monochlorination of methane would involve the following propagation steps:

$$CH_3-H + \cdot Cl \longrightarrow CH_3-Cl + \cdot H$$
$$\cdot H + Cl-Cl \longrightarrow H-Cl + \cdot Cl$$

How do you know that the reaction does not take place by this mechanism?

26. Explain why the rate of bromination of methane is decreased if HBr is added to the reaction mixture.

PART THREE

Substitution and Elimination Reactions

The three chapters in Part Three discuss the reactions of compounds that have an electron-withdrawing atom or group—a potential leaving group—bonded to an sp^3 hybridized carbon. These compounds can undergo substitution and/or elimination reactions.

Chapter 10 discusses the substitution reactions of alkyl halides. Of the different compounds that undergo substitution and elimination reactions, alkyl halides are examined first because they have relatively good leaving groups. You will also see the kinds of compounds biological organisms use in place of alkyl halides, since alkyl halides are not readily available in nature.

Chapter 11 covers the elimination reactions of alkyl halides. Because alkyl halides can undergo both substitution and elimination reactions, this chapter also discusses the factors that determine whether a given alkyl halide will undergo a substitution reaction, an elimination reaction, or both substitution and elimination reactions.

Chapter 12 discusses compounds other than alkyl halides that undergo substitution and elimination reactions. You will see that because alcohols and ethers have relatively poor leaving groups compared with the leaving groups of alkyl halides, alcohols and ethers must be activated before the groups can be substituted or eliminated. Several methods commonly used to activate leaving groups will be examined. The reactions of thiols and sulfides will be compared with those of alcohols and ethers. By looking at the reactions of epoxides, you will see how ring strain can affect leaving ability. You will also see how the carcinogenicity of arene oxides is related to carbocation stability. Finally, this chapter will introduce you to organometallic compounds, a class of compounds that is very important to synthetic organic chemists.

10 Substitution Reactions of Alkyl Halides

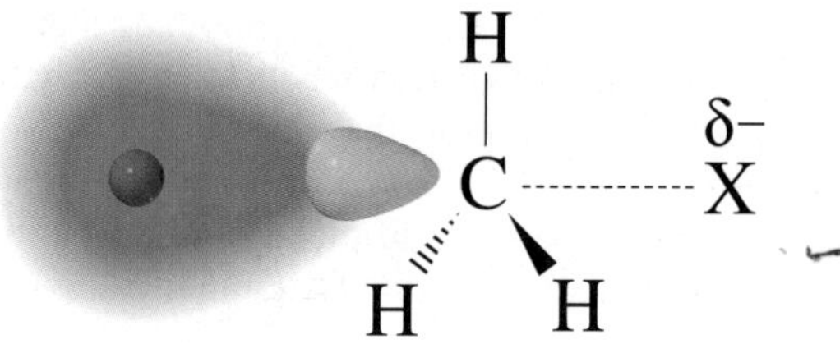

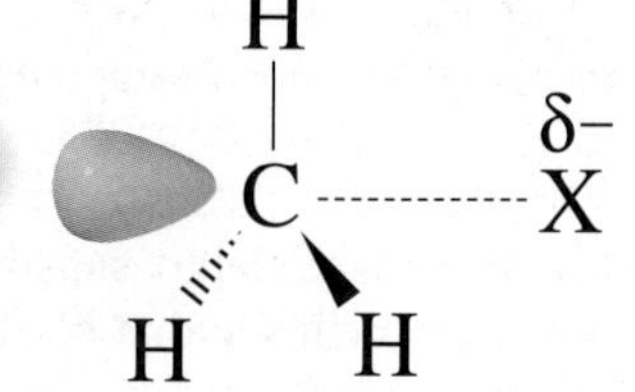

Organic compounds that have an electronegative atom or group bonded to an sp^3 hybridized carbon undergo substitution reactions and/or elimination reactions. In a **substitution reaction**, the electronegative atom or group is replaced by another atom or group. In an **elimination reaction**, the electronegative atom or group is eliminated, along with a hydrogen from an adjacent carbon. The atom or group that is *substituted* or *eliminated* in these reactions is called a **leaving group**.

$$RCH_2CH_2X + Y^- \xrightarrow{\text{a substitution reaction}} RCH_2CH_2Y + X^-$$

$$RCH_2CH_2X + Y^- \xrightarrow{\text{an elimination reaction}} RCH{=}CH_2 + HY + X^-$$

(X = the leaving group)

This chapter focuses on the substitution reactions of alkyl halides—compounds in which the leaving group is a halide ion (F^-, Cl^-, Br^-, or I^-). The nomenclature of alkyl halides was discussed in Section 2.4.

alkyl halides

R—F	R—Cl	R—Br	R—I
an alkyl fluoride	**an alkyl chloride**	**an alkyl bromide**	**an alkyl iodide**

In Chapter 11, we will discuss the elimination reactions of alkyl halides and the factors that determine whether substitution or elimination will prevail when an alkyl halide undergoes a reaction.

Alkyl halides are a good family of compounds with which to start our study of substitution and elimination reactions because they have relatively good leaving groups; that is, the halide ions are easily displaced. We will then be prepared to discuss, in

SURVIVAL COMPOUNDS

Several marine organisms, including sponges, corals, and algae, synthesize organohalides (halogen-containing organic compounds) that they use to deter predators. For example, red algae synthesize a toxic, foul-tasting organohalide that keeps predators from eating them. One predator, however, that is not deterred is a mollusk called a sea hare. After consuming red algae, a sea hare converts the original organohalide into a structurally similar compound it uses for its own defense. Unlike other mollusks, a sea hare does not have a shell. Its method of defense is to surround itself with a slimy material that contains the organohalide, thereby protecting itself from carnivorous fish.

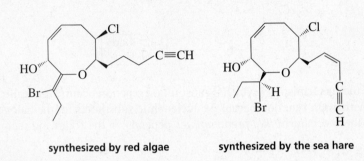

synthesized by red algae

synthesized by the sea hare

a sea hare

Chapter 12, the substitution and elimination reactions of compounds with leaving groups that are more difficult to displace.

Substitution reactions are important in organic chemistry because they make it possible to convert readily available alkyl halides into a wide variety of other compounds. Substitution reactions are also important in the cells of plants and animals. Because cells exist in predominantly aqueous environments and alkyl halides are insoluble in water, biological systems use compounds in which the group that is replaced is more polar than a halogen and therefore more soluble in water. The reactions of some of these biological compounds are discussed in this chapter.

10.1 How Alkyl Halides React

A halogen is more electronegative than carbon. Consequently, the two atoms do not share their bonding electrons equally. Because the more electronegative halogen has a larger share of the electrons, it has a partial negative charge and the carbon to which it is bonded has a partial positive charge.

$$\overset{\delta+}{RCH_2}-\overset{\delta-}{X} \qquad X = F, Cl, Br, I$$

It is the polar carbon–halogen bond that causes alkyl halides to undergo substitution and elimination reactions. There are two important mechanisms for the substitution reaction:

1. A nucleophile is attracted to the partially positively charged carbon (an electrophile). As the nucleophile approaches the carbon and forms a new bond, the carbon–halogen bond breaks heterolytically (the halogen takes both of the bonding electrons).

$$\ddot{N}u^- + -\overset{\delta+}{C}-\overset{\delta-}{X} \longrightarrow -C-Nu + X^-$$

substitution product

2. The carbon–halogen bond breaks heterolytically without any assistance from the nucleophile, forming a carbocation. The carbocation—an electrophile—then reacts with the nucleophile to form the substitution product.

$$-\overset{\delta+|}{\underset{|}{C}}-\overset{\delta-}{X} \longrightarrow -\overset{|}{\underset{|}{C^+}} + X^-$$

$$-\overset{|}{\underset{|}{C^+}} + \ddot{N}u^- \longrightarrow -\overset{|}{\underset{|}{C}}-Nu$$

substitution product

Regardless of the mechanism by which a substitution reaction occurs, it is called a **nucleophilic substitution reaction** because a nucleophile substitutes for the halogen. We will see that the mechanism *that predominates* depends on the following factors:

- the structure of the alkyl halide
- the reactivity of the nucleophile
- the concentration of the nucleophile
- the solvent in which the reaction is carried out

10.2 The Mechanism of an S_N2 Reaction

How is the mechanism of a reaction determined? We can learn a great deal about the mechanism of a reaction by studying its **kinetics**—the factors that affect the rate of the reaction.

The rate of a nucleophilic substitution reaction such as the reaction of methyl bromide with hydroxide ion depends on the concentrations of both reagents. If the concentration of methyl bromide in the reaction mixture is doubled, the rate of the nucleophilic substitution reaction doubles. If the concentration of hydroxide ion is doubled, the rate of the reaction also doubles. If the concentrations of both reactants are doubled, the rate of the reaction quadruples.

$$\underset{\text{methyl bromide}}{CH_3Br} + HO^- \longrightarrow \underset{\text{methyl alcohol}}{CH_3OH} + Br^-$$

When you know the relationship between the rate of a reaction and the concentration of the reactants, you can write a **rate law** for the reaction. Because the rate of the reaction of methyl bromide with hydroxide ion is dependent on the concentration of both reactants, the rate law for the reaction is

$$\textbf{rate} \propto \textbf{[alkyl halide][nucleophile]}$$

As we saw in Section 3.7, a proportionality sign ($\propto$) can be replaced by an equals sign and a proportionality constant. The proportionality constant, in this case k, is called the **rate constant**. The rate constant describes how difficult it is to overcome the energy barrier of the reaction—how hard it is to reach the transition state. The larger the rate constant, the easier it is to reach the transition state (see Figure 10.3 on p. 365).

$$\textbf{rate} = k\textbf{[alkyl halide][nucleophile]}$$

Because the rate of this reaction depends on the concentration of two reactants, the reaction is a **second-order reaction** (Section 3.7).

The rate law tells us which molecules are involved in the transition state of the rate-determining step of the reaction. From the rate law for the reaction of methyl bromide with hydroxide ion, we know that both methyl bromide and hydroxide ion are involved in the rate-determining transition state.

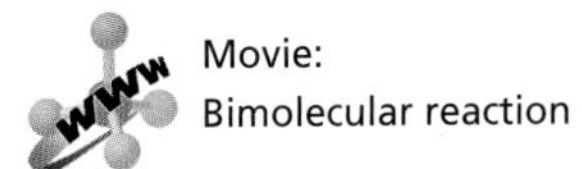

Movie: Bimolecular reaction

The reaction of methyl bromide with hydroxide ion is an example of an **S_N2 reaction**, where "S" stands for substitution, "N" for nucleophilic, and "2" for bimolecular. **Bimolecular** means that two molecules are involved in the rate-determining step. In 1937, Edward Hughes and Christopher Ingold proposed a mechanism for an S_N2 reaction. Remember that a mechanism describes the step-by-step process by which reactants are converted into products. It is a theory that fits the experimental evidence that has been accumulated concerning the reaction. Hughes and Ingold based their mechanism for an S_N2 reaction on the following three pieces of experimental evidence:

Edward Davies Hughes (1906–1963) *was born in North Wales. He earned two doctoral degrees: a Ph.D. from the University of Wales and a D.Sc. from the University of London, working with Sir Christopher Ingold. He was a professor of chemistry at University College, London.*

Sir Christopher Ingold (1893–1970) *was born in Ilford, England. In addition to determining the mechanism of the S_N2 reaction, he was a member of the group that developed nomenclature for enantiomers.* *(See p. 188.)* *He also participated in developing the theory of resonance.*

1. The rate of the reaction depends on the concentration of the alkyl halide *and* on the concentration of the nucleophile. This means that both reactants are involved in the transition state of the rate-determining step.
2. When the hydrogens of methyl bromide are successively replaced with methyl groups, the rate of the reaction with a given nucleophile becomes progressively slower (Table 10.1).
3. The reaction of a chiral alkyl halide in which the halogen is bonded to an asymmetric carbon leads to the formation of only one stereoisomer, and the configuration of the asymmetric carbon in the product is inverted relative to its configuration in the reacting alkyl halide.

Hughes and Ingold proposed that an S_N2 reaction is a *concerted* reaction—it takes place in a single step, so no intermediates are formed. The nucleophile attacks the carbon bearing the leaving group and displaces the leaving group.

mechanism of the S_N2 reaction

$$H\ddot{O}:^- + CH_3-\ddot{Br}: \longrightarrow CH_3-OH + :\ddot{Br}:^-$$

Because a productive collision is a collision in which the nucleophile hits the carbon on the side opposite the side bonded to the leaving group, the carbon is said to undergo **back-side attack**. Why does the nucleophile attack from the back side? The simplest explanation is that the leaving group blocks the approach of the nucleophile to the front side of the molecule.

Table 10.1 Relative Rates of S_N2 Reactions for Several Alkyl Halides

$$R-Br + Cl^- \xrightarrow{S_N2} R-Cl + Br^-$$

Alkyl halide	Class of alkyl halide	Relative rate
CH_3-Br	methyl	1200
CH_3CH_2-Br	primary	40
$CH_3CH_2CH_2-Br$	primary	16
$CH_3CH(CH_3)-Br$	secondary	1
$(CH_3)_3C-Br$	tertiary	too slow to measure

Molecular orbital theory also explains back-side attack. Recall from Section 8.9 that to form a bond, the LUMO (lowest unoccupied molecular orbital) of one species must interact with the HOMO (highest occupied molecular orbital) of the other. When the nucleophile approaches the alkyl halide, the filled nonbonding molecular orbital (the HOMO) of the nucleophile must interact with the empty σ^* antibonding molecular orbital (the LUMO) associated with the C—Br bond. Figure 10.1a shows that back-side attack involves a bonding interaction between the nucleophile and the larger lobe of σ^*. Compare this with what happens when the nucleophile approaches the front side of the carbon (Figure 10.1b): Both a bonding and an antibonding interaction occur, and the two cancel each other. Consequently, the best overlap of the interacting orbitals is achieved through back-side attack. In fact, a nucleophile always approaches an sp^3 hybridized carbon from the back side. [We saw back-side attack previously in the reaction of a bromide ion with a cyclic bromonium ion (Section 5.19).]

A nucleophile always approaches an sp^3 hybridized carbon on its back side.

a. Back-side attack

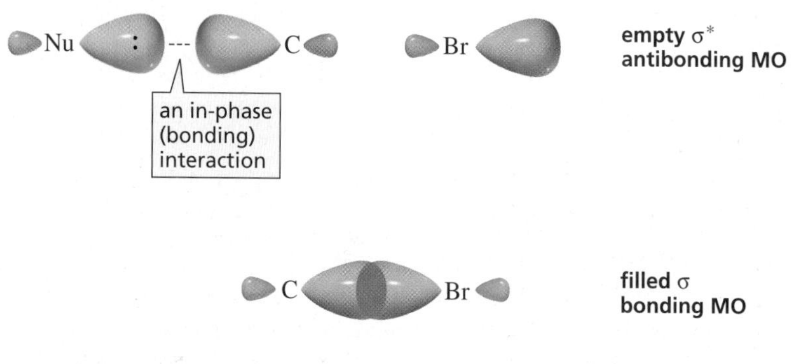

b. Front-side attack

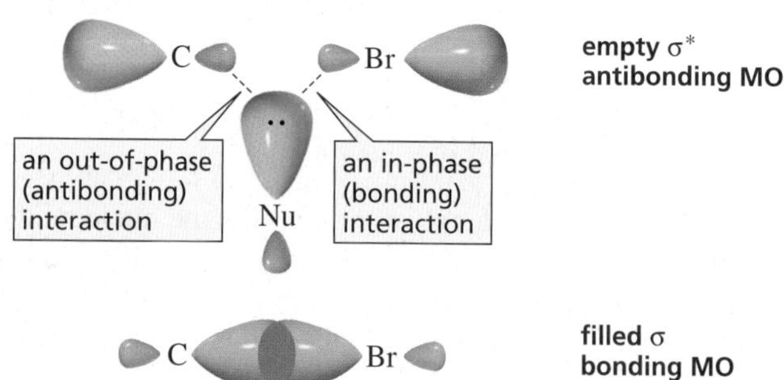

Figure 10.1 ▶
(a) Back-side attack results in a bonding interaction between the HOMO (the filled nonbonding orbital) of the nucleophile and the LUMO (the empty σ^* antibonding orbital) of C—Br. (b) Front-side attack would result in both a bonding and an antibonding interaction that would cancel out.

How does Hughes and Ingold's mechanism account for the three observed pieces of experimental evidence? The mechanism shows the alkyl halide and the nucleophile coming together in the transition state of the one-step reaction. Therefore, increasing the concentration of either of them makes their collision more probable. Thus, the reaction will follow second-order kinetics, exactly as observed.

$$\mathrm{HO^-} + \mathrm{C{-}Br} \longrightarrow \left[\overset{\delta-}{\mathrm{HO}}\text{---}\mathrm{C}\text{---}\overset{\delta-}{\mathrm{Br}}\right]^{\ddagger} \longrightarrow \mathrm{HO{-}C} + \mathrm{Br^-}$$

transition state

Because the nucleophile attacks the back side of the carbon that is bonded to the halogen, bulky substituents attached to this carbon will make it harder for the nucleophile to get to that side and will therefore decrease the rate of the reaction (Figure 10.2). This explains why substituting methyl groups for the hydrogens in methyl bromide progressively slows the rate of the substitution reaction (Table 10.1). It is the bulk of the alkyl groups that is responsible for the difference in reactivity.

Steric effects are caused by groups occupying a certain volume of space. A steric effect that decreases reactivity is called **steric hindrance**. Steric hindrance results when groups are in the way at the reaction site. Steric hindrance causes alkyl halides to

Viktor Meyer (1848–1897) *was born in Germany. To prevent him from becoming an actor, his parents persuaded him to enter the University of Heidelberg, where he earned a Ph.D. in 1867 at the age of 18. He was a professor of chemistry at the Universities of Stuttgart and Heidelberg. He coined the term "stereochemistry" for the study of molecular shapes and was the first to describe the effect of steric hindrance on a reaction.*

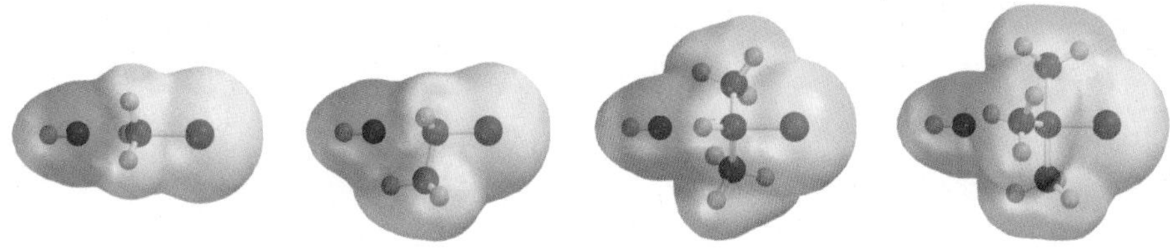

◀ **Figure 10.2**
The approach of HO^- to a methyl halide, a primary alkyl halide, a secondary alkyl halide, and a tertiary alkyl halide. Increasing the bulk of the substituents bonded to the carbon that is undergoing nucleophilic attack decreases access to the back side of the carbon, thereby decreasing the rate of an S_N2 reaction.

have the following relative reactivities in an S_N2 reaction because, *generally*, primary alkyl halides are less hindered than secondary alkyl halides, which, in turn, are less hindered than tertiary alkyl halides:

relative reactivities of alkyl halides in an S_N2 reaction

most reactive > methyl halide > 1° alkyl halide > 2° alkyl halide > 3° alkyl halide < least reactive

Tertiary alkyl halides cannot undergo S_N2 reactions.

The three alkyl groups of a tertiary alkyl halide make it impossible for the nucleophile to come within bonding distance of the tertiary carbon, so tertiary alkyl halides are unable to undergo S_N2 reactions. The reaction coordinate diagrams for the S_N2 reactions of *unhindered* methyl bromide and a *sterically hindered* secondary alkyl bromide show that steric hindrance raises the energy of the transition state, slowing the reaction (Figure 10.3).

The rate of an S_N2 reaction depends not only on the *number* of alkyl groups attached to the carbon that is undergoing nucleophilic attack, but also on their size. For example, while ethyl bromide and propyl bromide are both primary alkyl halides, ethyl bromide is more than twice as reactive in an S_N2 reaction because the bulkier group on the carbon undergoing nucleophilic attack in propyl bromide provides more steric hindrance to back-side attack. Also, although neopentyl bromide is a primary alkyl halide, it undergoes S_N2 reactions very slowly because its single alkyl group is unusually bulky.

$$CH_3\overset{CH_3}{\underset{CH_3}{C}}CH_2Br$$
neopentyl bromide

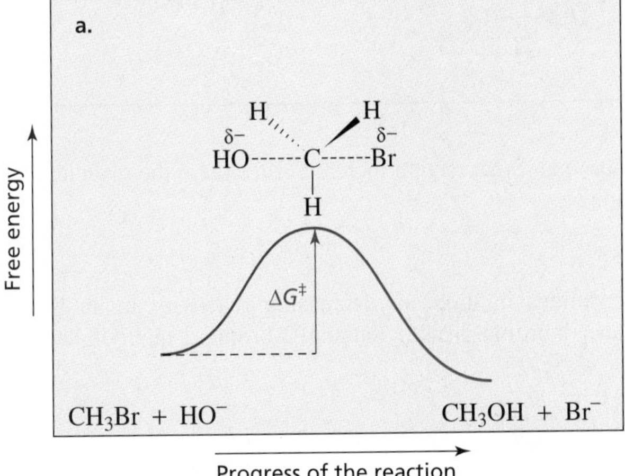

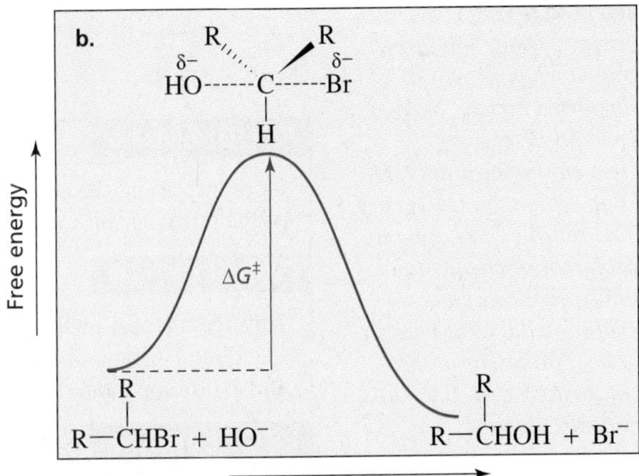

▲ **Figure 10.3**
Reaction coordinate diagrams for (a) the S_N2 reaction of methyl bromide with hydroxide ion; (b) an S_N2 reaction of a sterically hindered secondary alkyl bromide with hydroxide ion.

Figure 10.4 shows that as the nucleophile approaches the back side of the carbon of methyl bromide, the C—H bonds begin to move away from the nucleophile and its attacking electrons. By the time the transition state is reached, the C—H bonds are all in the same plane and the carbon is pentacoordinate (fully bonded to three atoms and partially bonded to two) rather than tetrahedral. As the nucleophile gets closer to the carbon and the bromine moves farther away from it, the C—H bonds continue to move in the same direction. Eventually, the bond between the carbon and the nucleophile is fully formed, and the bond between the carbon and bromine is completely broken, so that the carbon is once again tetrahedral.

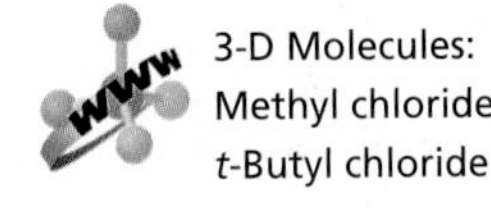

3-D Molecules: Methyl chloride; *t*-Butyl chloride

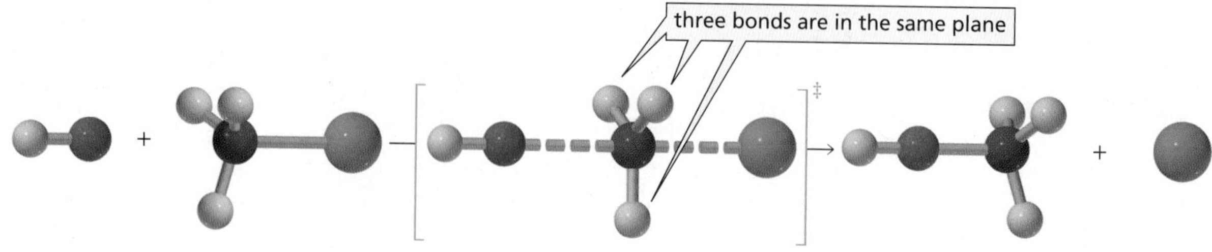
three bonds are in the same plane
+
→
+

▲ **Figure 10.4**
An S_N2 reaction between hydroxide ion and methyl bromide.

Tutorial: S_N2

The carbon at which substitution occurs has inverted its configuration during the course of the reaction, just as an umbrella has a tendency to invert in a windstorm. This **inversion of configuration** is called a *Walden inversion*, after Paul Walden, who first discovered that the configuration of a compound was inverted in an S_N2 reaction.

Because an S_N2 reaction takes place with inversion of configuration, only one substitution product is formed when a chiral alkyl halide—whose halogen atom is bonded to an asymmetric carbon—undergoes an S_N2 reaction. The configuration of that product is inverted relative to the configuration of the alkyl halide. For example, the substitution product of the reaction of hydroxide ion with (*R*)-2-bromopentane is (*S*)-2-pentanol. Therefore, the proposed mechanism accounts for the observed configuration of the product.

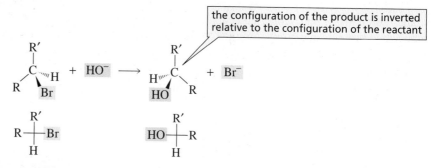

Paul Walden (1863–1957) *was born in Cesis, Latvia, the son of a farmer. His parents died when he was a child. He supported himself at Riga University and St. Petersburg University by working as a tutor. He received a Ph.D. from the University of Leipzig and returned to Latvia to become a professor of chemistry at Riga University. Following the Russian Revolution, he went back to Germany to be a professor at the University of Rostock and, later, at the University of Tübingen.*

the configuration of the product is inverted relative to the configuration of the reactant

R′, R, H, Br on C + HO^- ⟶ R′, R, H, HO on C + Br^-

R′ / R—+—Br / H   HO—+—R / R′ / H

PROBLEM 1◆

Does increasing the energy barrier to an S_N2 reaction increase or decrease the magnitude of the rate constant for the reaction?

PROBLEM 2◆

Arrange the following alkyl bromides in order of decreasing reactivity in an S_N2 reaction: 1-bromo-2-methylbutane, 1-bromo-3-methylbutane, 2-bromo-2-methylbutane, and 1-bromopentane.

PROBLEM 3◆ SOLVED

What product would be formed from the S_N2 reaction of

a. 2-bromobutane and hydroxide ion?
b. (*R*)-2-bromobutane and hydroxide ion?
c. (*S*)-3-chlorohexane and hydroxide ion?
d. 3-iodopentane and hydroxide ion?

SOLUTION TO 3a The product is 2-butanol. Because the reaction is an S_N2 reaction, we know that the configuration of the product is inverted relative to the configuration of the reactant. The configuration of the reactant is not specified, however, so we cannot specify the configuration of the product.

the configuration is not specified

$$CH_3\underset{\displaystyle Br}{\underset{|}{C}}HCH_2CH_3 + HO^- \longrightarrow CH_3\underset{\displaystyle OH}{\underset{|}{C}}HCH_2CH_3 + Br^-$$

10.3 Factors Affecting S_N2 Reactions

The Leaving Group

If an alkyl iodide, an alkyl bromide, an alkyl chloride, and an alkyl fluoride (all with the same alkyl group) were allowed to react with the same nucleophile under the same conditions, we would find that the alkyl iodide is the most reactive and the alkyl fluoride is the least reactive.

	relative rates of reaction
$HO^- + RCH_2I \longrightarrow RCH_2OH + I^-$	30,000
$HO^- + RCH_2Br \longrightarrow RCH_2OH + Br^-$	10,000
$HO^- + RCH_2Cl \longrightarrow RCH_2OH + Cl^-$	200
$HO^- + RCH_2F \longrightarrow RCH_2OH + F^-$	1

The only difference among these four reactions is the nature of the leaving group. From the relative reaction rates, we can see that the iodide ion is the best leaving group and the fluoride ion is the worst. This brings us to an important rule in organic chemistry—one that you will see frequently: *The weaker the basicity of a group, the better is its leaving ability.* The reason leaving ability depends on basicity is because *weak bases are stable bases*—they readily bear the electrons they formerly shared with a proton (Section 1.18). Because weak bases don't share their electrons well, a weak base is not bonded as strongly to the carbon as a strong base would be and a weaker bond is more easily broken (Section 1.13).

The weaker the base, the better it is as a leaving group.

Stable bases are weak bases.

We have seen that the halide ions have the following relative basicities (or relative stabilities) because larger atoms are better able to stabilize their negative charge (Section 1.18):

relative basicities of the halide ions

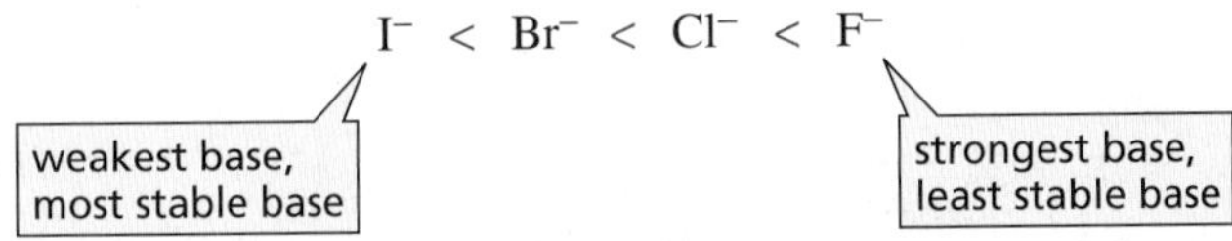

Because stable (weak) bases are better leaving groups, the halide ions have the following relative leaving abilities:

relative leaving abilities of the halide ions

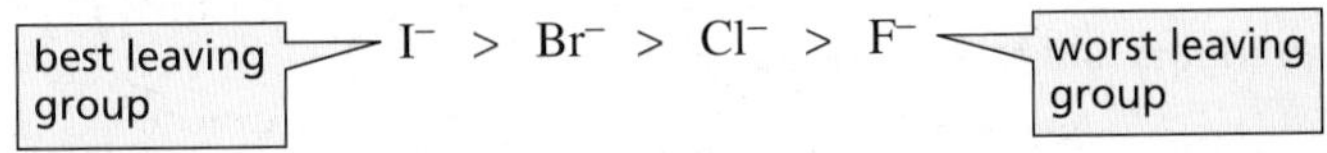

Therefore, alkyl iodides are the most reactive of the alkyl halides, and alkyl fluorides are the least reactive. In fact, the fluoride ion is such a strong base that alkyl fluorides essentially do not undergo S_N2 reactions.

relative reactivities of alkyl halides in an S_N2 reaction

most reactive → RI > RBr > RCl > RF ← least reactive

In Section 10.1, we saw that it is the polar carbon–halogen bond that causes alkyl halides to undergo substitution reactions. Carbon and iodine, however, have the same electronegativity. (See Table 1.3 on p. 10.) Why, then, does an alkyl halide undergo a substitution reaction? We know that larger atoms are more polarizable than smaller atoms. (Recall from Section 2.9 that polarizability is a measure of how easily an atom's electron cloud can be distorted.) The high polarizability of the large iodine atom causes it to react as if it were polar even though, on the basis of the electronegativity of the atoms, the bond is nonpolar.

The Nucleophile

When we talk about atoms or molecules that have lone-pair electrons, sometimes we call them bases and sometimes we call them nucleophiles. What is the difference between a base and a nucleophile?

Basicity is a measure of how well a compound (a **base**) shares its lone pair with a proton. The stronger the base, the better it shares its electrons. Basicity is measured by an *equilibrium constant* (the acid dissociation constant, K_a) that indicates the tendency of the conjugate acid of the base to lose a proton (Section 1.17).

Nucleophilicity is a measure of how readily a compound (a **nucleophile**) is able to attack an electron-deficient atom. Nucleophilicity is measured by a *rate constant* (k). In the case of an S_N2 reaction, nucleophilicity is a measure of how readily the nucleophile attacks an sp^3 hybridized carbon bonded to a leaving group.

When comparing molecules *with the same attacking atom,* there is a direct relationship between basicity and nucleophilicity: Stronger bases are better nucleophiles. For example, a species with a negative charge is a stronger base *and* a better nucleophile than a species with the same attacking atom that is neutral. Thus HO^- is a stronger base and a better nucleophile than H_2O.

stronger base, better nucleophile		weaker base, poorer nucleophile
HO^-	>	H_2O
CH_3O^-	>	CH_3OH
$^-NH_2$	>	NH_3
$CH_3CH_2NH^-$	>	$CH_3CH_2NH_2$

When comparing molecules *with attacking atoms of approximately the same size,* the stronger bases are again the better nucleophiles. The atoms across the second row of the periodic table have approximately the same size. If hydrogens are attached to the second-row elements, the resulting compounds have the following relative acidities (Section 1.18):

relative acid strengths

weakest acid → $NH_3 < H_2O < HF$

Consequently, the conjugate bases have the following relative base strengths and nucleophilicities:

relative base strengths and relative nucleophilicities

strongest base → $^-NH_2 > HO^- > F^-$ ← best nucleophile

Note that the amide anion is the strongest base, as well as the best nucleophile.

When comparing molecules *with attacking atoms that are very different in size,* another factor comes into play—the polarizability of the atom. Because the electrons are farther away in the larger atom, they are not held as tightly and can, therefore, move more freely toward a positive charge. As a result, the electrons are able to overlap from farther away with the orbital of carbon, as shown in Figure 10.5. This results in a greater degree of bonding in the transition state, making it more stable. Whether the greater polarizability of the larger atoms makes up for their decreased basicity depends on the conditions under which the reaction is carried out.

more bonding

transition state

little bonding

transition state

◀ **Figure 10.5**
An iodide ion is larger and more polarizable than a fluoride ion. Therefore, the relatively loosely held electrons of the iodide ion can overlap from farther away with the orbital of carbon undergoing nucleophilic attack. The tightly bound electrons of the fluoride ion cannot start to overlap until the atoms are closer together.

If the reaction is carried out in the gas phase, the direct relationship between basicity and nucleophilicity is maintained—the stronger bases are still the best nucleophiles. If, however, the reaction is carried out in a protic solvent—meaning the solvent molecules have a hydrogen bonded to an oxygen or to a nitrogen—the relationship between basicity and nucleophilicity becomes inverted. The largest atom is the best nucleophile even though it is the weakest base. Therefore, iodide ion is the best nucleophile of the halide ions in a protic solvent.

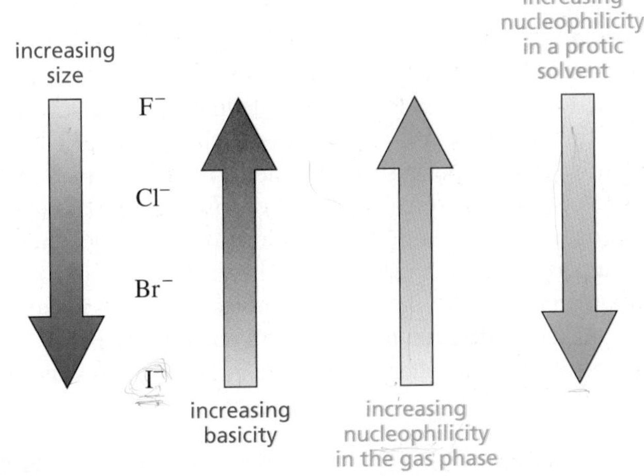

A protic solvent contains a hydrogen bonded to an oxygen or a nitrogen; it is a hydrogen bond donor.

PROBLEM 4◆

a. Which is a stronger base, RO^- or RS^-?
b. Which is a better nucleophile in an aqueous solution?

The Effect of the Solvent on Nucleophilicity

Why, in a protic solvent, is the smallest atom the poorest nucleophile even though it is the strongest base? *How does a protic solvent make strong bases less nucleophilic?* When a negatively charged species is placed in a protic solvent, the ion becomes solvated (Section 2.9). Protic solvents are hydrogen bond donors. The solvent molecules arrange themselves so that their partially positively charged hydrogens point toward the negatively charged species. The interaction between the ion and the dipole of the protic solvent is called an **ion–dipole interaction**. Because the solvent shields the nucleophile, at least one of the ion–dipole interactions must be broken before the nucleophile can participate in an S_N2 reaction. Weak bases interact weakly with protic solvents, whereas strong bases interact more strongly because they are better at sharing their electrons. It is easier, therefore, to break the ion–dipole interactions between

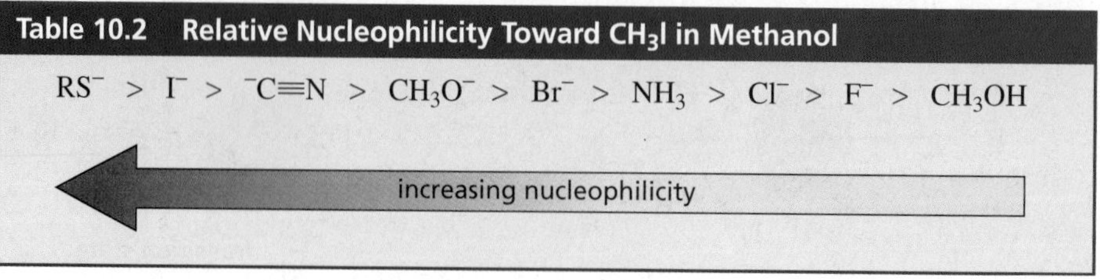
Table 10.2 Relative Nucleophilicity Toward CH_3I in Methanol

$RS^- > I^- > {}^-C{\equiv}N > CH_3O^- > Br^- > NH_3 > Cl^- > F^- > CH_3OH$

increasing nucleophilicity

an iodide ion (a weak base) and the solvent than between a fluoride ion (a stronger base) and the solvent. As a result, iodide ion is a better nucleophile than fluoride ion in a protic solvent (Table 10.2).

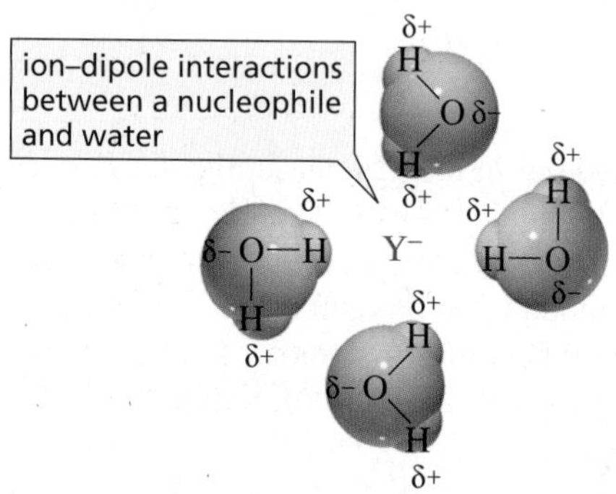
ion–dipole interactions between a nucleophile and water
Y⁻

An aprotic solvent does not contain a hydrogen bonded to either an oxygen or a nitrogen; it is not a hydrogen bond donor.

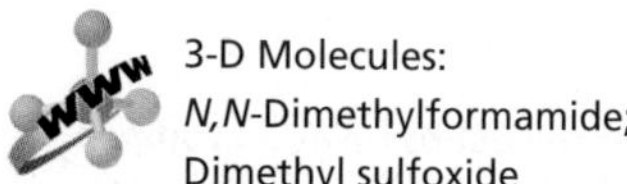
3-D Molecules: *N,N*-Dimethylformamide; Dimethyl sulfoxide

Fluoride ion would be a better nucleophile in a *nonpolar solvent* than in a polar solvent because there would be no ion–dipole interactions between the ion and the nonpolar solvent. However, ionic compounds are insoluble in most nonpolar solvents, but they can dissolve in aprotic polar solvents, such as dimethylformamide (DMF) or dimethylsulfoxide (DMSO). An aprotic polar solvent is not a hydrogen bond donor because it does not have a hydrogen attached to an oxygen or to a nitrogen, so there are no positively charged hydrogens to form ion–dipole interactions. The molecules of an aprotic polar solvent have a partial negative charge on their surface that can solvate cations, but the partial positive charge is on the *inside* of the molecule, which makes it less accessible. The relatively "naked" anion can be a powerful nucleophile in an aprotic polar solvent. Fluoride ion, therefore, is a better nucleophile in DMSO than it is in water.

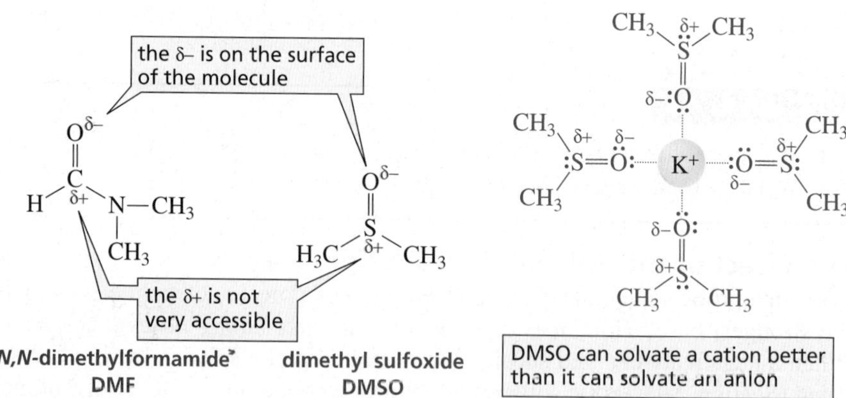
the δ– is on the surface of the molecule
the δ+ is not very accessible
K⁺
DMSO can solvate a cation better than it can solvate an anion

***N,N*-dimethylformamide DMF** **dimethyl sulfoxide DMSO**

PROBLEM 5◆

Indicate whether each of the following solvents is protic or aprotic:

a. chloroform ($CHCl_3$)
b. diethyl ether ($CH_3CH_2OCH_2CH_3$)
c. acetic acid (CH_3COOH)
d. hexane [$CH_3(CH_2)_4CH_3$]

Nucleophilicity Is Affected by Steric Effects

Base strength is relatively unaffected by steric effects because a base removes a relatively unhindered proton. The strength of a base depends only on how well the base shares its electrons with a proton. Thus, *tert*-butoxide ion is a stronger base than ethoxide ion since *tert*-butanol ($pK_a = 18$) is a weaker acid than ethanol ($pK_a = 15.9$).

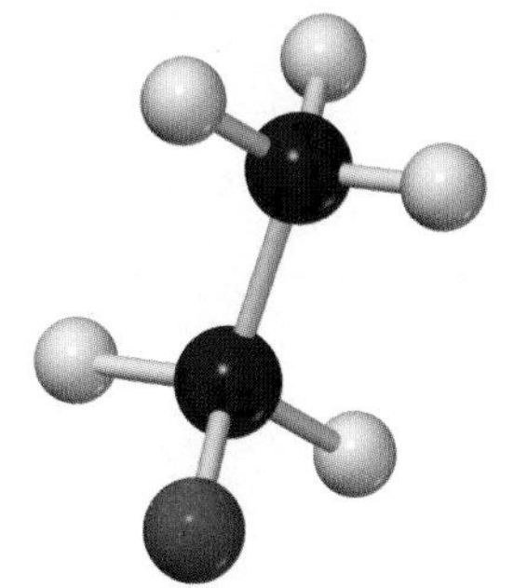

ethoxide ion

$CH_3CH_2O^-$	$CH_3C(CH_3)_2O^-$
ethoxide ion better nucleophile	*tert*-butoxide ion stronger base

Steric effects, on the other hand, do affect nucleophilicity. A bulky nucleophile cannot approach the back side of a carbon as easily as a less sterically hindered nucleophile can. Thus, the bulky *tert*-butoxide ion, with its three methyl groups, is a poorer nucleophile than ethoxide ion even though *tert*-butoxide ion is a stronger base.

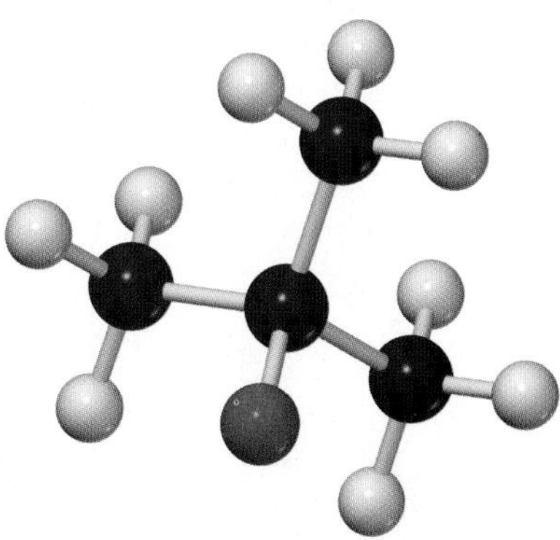

tert-butoxide ion

PROBLEM 6 SOLVED

List the following species in order of *decreasing* nucleophilicity in an aqueous solution:

$$C_6H_5O^- \quad CH_3OH \quad HO^- \quad CH_3\overset{\overset{O}{\|}}{C}O^- \quad CH_3S^-$$

SOLUTION Let's first divide the nucleophiles into groups. We have one nucleophile with a negatively charged sulfur, three with negatively charged oxygens, and one with a neutral oxygen. We know that in the polar aqueous solvent, the one with the negatively charged sulfur is the most nucleophilic because sulfur is larger than oxygen. We also know that the poorest nucleophile is the one with the neutral oxygen atom. So now, to complete the problem, we need to rank the three nucleophiles with negatively charged oxygens in order of the pK_a's of their conjugate acids. A carboxylic acid is a stronger acid than phenol, which is a stronger acid than water (Section 7.10). Because water is the weakest acid, its conjugate base is the strongest base and the best nucleophile. Thus, the relative nucleophilicities are

$$CH_3S^- > HO^- > C_6H_5O^- > CH_3\overset{\overset{O}{\|}}{C}O^- > CH_3OH$$

PROBLEM 7◆

For each of the following pairs of S_N2 reactions, indicate which reaction occurs faster:

a. $CH_3CH_2Br + H_2O$ or $CH_3CH_2Br + HO^-$

b. $CH_3CH(CH_3)CH_2Br + HO^-$ or $CH_3CH_2CH(CH_3)Br + HO^-$

c. $CH_3CH_2Cl + CH_3O^-$ or $CH_3CH_2Cl + CH_3S^-$ **(in ethanol)**

d. $CH_3CH_2Cl + I^-$ or $CH_3CH_2Br + I^-$

10.4 The Reversibility of an S_N2 Reaction

Many different kinds of nucleophiles can react with alkyl halides. Therefore, a wide variety of organic compounds can be synthesized by means of S_N2 reactions.

$$CH_3CH_2Cl + HO^- \longrightarrow CH_3CH_2OH + Cl^-$$
an alcohol

$$CH_3CH_2Br + HS^- \longrightarrow CH_3CH_2SH + Br^-$$
a thiol

$$CH_3CH_2I + RO^- \longrightarrow CH_3CH_2OR + I^-$$
an ether

$$CH_3CH_2Br + RS^- \longrightarrow CH_3CH_2SR + Br^-$$
a thioether

$$CH_3CH_2Cl + {}^-NH_2 \longrightarrow CH_3CH_2NH_2 + Cl^-$$
a primary amine

$$CH_3CH_2Br + {}^-C{\equiv}CR \longrightarrow CH_3CH_2C{\equiv}CR + Br^-$$
an alkyne

$$CH_3CH_2I + {}^-C{\equiv}N \longrightarrow CH_3CH_2C{\equiv}N + I^-$$
a nitrile

It may seem that the reverse of each of these reactions can also occur via nucleophilic substitution. In the first reaction, for example, ethyl chloride reacts with hydroxide ion to form ethyl alcohol and a chloride ion. The reverse reaction would appear to satisfy the requirements for a nucleophilic substitution reaction, given that chloride ion is a nucleophile and ethyl alcohol has an HO^- leaving group. But ethyl alcohol and chloride ion do *not* react.

Why does a nucleophilic substitution reaction take place in one direction, but not in the other? We can answer this question by comparing the leaving tendency of Cl^- in the forward direction and the leaving tendency of HO^- in the reverse direction. Comparing leaving tendencies involves comparing basicities. Most people find it easier to compare the acid strengths of the conjugate acids (Table 10.3), so we will compare the acid strengths of HCl and H_2O. HCl is a much stronger acid than H_2O, which means that Cl^- is a much weaker base than HO^-. (Remember, the stronger the acid, the weaker is its conjugate base.) Because it is a weaker base, Cl^- is a better leaving group. Consequently, HO^- can displace Cl^- in the forward reaction, but Cl^- cannot displace HO^- in the reverse reaction. The reaction proceeds in the direction that allows the stronger base to displace the weaker base (the better leaving group).

An S_N2 reaction proceeds in the direction that allows the stronger base to displace the weaker base.

If the difference between the basicities of the nucleophile and the leaving group is not very large, the reaction will be reversible. For example, in the reaction of ethyl bromide with iodide ion, Br^- is the leaving group in one direction and I^- is the leaving group in the other direction. The reaction is reversible because the pK_a values of the conjugate acids of the two leaving groups are similar (pK_a of HBr = −9; pK_a of HI = −10; see Table 10.3).

$$CH_3CH_2Br + I^- \rightleftharpoons CH_3CH_2I + Br^-$$

an S_N2 reaction is reversible when the basicities of the leaving groups are similar

You can drive a reversible reaction toward the desired products by removing one of the products as it is formed. Recall that the concentrations of the reactants and products at equilibrium are governed by the equilibrium constant of the reaction

Table 10.3 The Acidities of the Conjugate Acids of Some Leaving Groups

Acid	pK_a	Conjugate base (leaving group)
HI	−10.0	I^-
HBr	−9.0	Br^-
HCl	−7.0	Cl^-
H_2SO_4	−5.0	$^-OSO_3H$
$CH_3\overset{+}{O}H_2$	−2.5	CH_3OH
H_3O^+	−1.7	H_2O
$C_6H_5SO_3H$ (benzenesulfonic acid)	−0.6	$C_6H_5SO_3^-$
HF	3.2	F^-
$CH_3\overset{O}{\overset{\parallel}{C}}OH$	4.8	$CH_3\overset{O}{\overset{\parallel}{C}}O^-$
H_2S	7.0	HS^-
$HC{\equiv}N$	9.1	$^-C{\equiv}N$
$\overset{+}{N}H_4$	9.4	NH_3
CH_3CH_2SH	10.5	$CH_3CH_2S^-$
$(CH_3)_3\overset{+}{N}H$	10.8	$(CH_3)_3N$
CH_3OH	15.5	CH_3O^-
H_2O	15.7	HO^-
$HC{\equiv}CH$	25	$HC{\equiv}C^-$
NH_3	36	$^-NH_2$
H_2	~40	H^-

(Section 3.7). **Le Châtelier's principle** states that *if an equilibrium is disturbed, the system will adjust to offset the disturbance*. In other words, if the concentration of product C is decreased, A and B will react to form more C and D to maintain the value of the equilibrium constant.

If an equilibrium is disturbed, the system will adjust to offset the disturbance.

$$A + B \rightleftharpoons C + D$$

$$K_{eq} = \frac{[C][D]}{[A][B]}$$

For example, the reaction of ethyl chloride with methanol is reversible because the difference between the basicities of the nucleophile and the leaving group is not very large. If the reaction is carried out in a neutral solution, the protonated product will lose a proton (Section 1.20), disturbing the equilibrium and driving the reaction toward the products.

$$CH_3CH_2Cl + CH_3OH \rightleftharpoons CH_3CH_2\overset{+}{O}(H)CH_3 + Cl^- \xrightarrow{\text{fast}} CH_3CH_2OCH_3 + H^+$$

Henri Louis Le Châtelier (1850–1936) *was born in France. He studied mining engineering and was particularly interested in learning how to prevent explosions. His interest in mine safety is understandable, considering that his father was France's inspector-general of mines. Le Châtelier's research into preventing explosions led him to study heat and its measurement, which in turn focused his interest on thermodynamics.*

PROBLEM 8 SOLVED

What product is obtained when ethylamine reacts with excess methyl iodide in a basic solution of potassium carbonate?

$$CH_3CH_2\ddot{N}H_2 + \underset{\text{excess}}{CH_3{-}I} \xrightarrow{K_2CO_3} ?$$

SOLUTION Methyl iodide and ethylamine undergo an S_N2 reaction. The product of the reaction is a secondary amine, which is predominantly in its basic (neutral) form since the reaction is carried out in a basic solution. The secondary amine can undergo an S_N2 reaction with another equivalent of methyl iodide, forming a tertiary amine. The tertiary amine can react with methyl iodide in yet another S_N2 reaction. The final product of the reaction is a quaternary ammonium iodide.

$$CH_3CH_2\ddot{N}H_2 + CH_3{-}I \longrightarrow CH_3CH_2\overset{+}{N}H_2CH_3\ I^- \xrightleftharpoons{K_2CO_3} CH_3CH_2\ddot{N}HCH_3$$

$$CH_3CH_2\ddot{N}HCH_3 \xrightarrow{CH_3-I} CH_3CH_2\overset{+}{N}H(CH_3)CH_3\ I^- \xrightleftharpoons{K_2CO_3} CH_3CH_2\ddot{N}(CH_3)CH_3 \xrightarrow{CH_3-I} CH_3CH_2\overset{+}{N}(CH_3)_2CH_3\ I^-$$

PROBLEM 9

a. Explain why the reaction of an alkyl halide with ammonia gives a low yield of primary amine.

b. Explain why a much better yield of primary amine is obtained from the reaction of an alkyl halide with azide ion ($^-N_3$), followed by catalytic hydrogenation. (*Hint:* An alkyl azide is not nucleophilic.)

$$CH_3CH_2CH_2Br \xrightarrow{^-N_3} \underset{\text{an alkyl azide}}{CH_3CH_2CH_2N{=}\overset{+}{N}{=}N^-} \xrightarrow[\text{Pt}]{H_2} CH_3CH_2CH_2NH_2 + N_2$$

PROBLEM 10

Using the pK_a values listed in Table 10.3, convince yourself that each of the reactions on page 372 proceeds in the direction shown.

PROBLEM 11◆

What is the product of the reaction of ethyl bromide with each of the following nucleophiles?

a. CH_3OH b. $^-N_3$ c. $(CH_3)_3N$ d. $CH_3CH_2S^-$

PROBLEM 12

The reaction of an alkyl chloride with potassium iodide is generally carried out in acetone to maximize the amount of alkyl iodide that is formed. Why does the solvent increase the yield of alkyl iodide? (*Hint:* Potassium iodide is soluble in acetone, but potassium chloride is not.)

WHY CARBON INSTEAD OF SILICON?

There are two reasons living organisms are composed primarily of carbon, oxygen, hydrogen, and nitrogen: the *fitness* of these elements for specific roles in life processes and their *availability* in the environment. Of the two reasons, fitness was probably more important than availability because carbon rather than silicon became the fundamental building block of living organisms, even though silicon is just below carbon in the periodic table and, as the following table shows, is more than 140 times more abundant than carbon in the Earth's crust.

	Abundance (atoms/100 atoms)	
Element	**In living organisms**	**In Earth's crust**
H	49	0.22
C	25	0.19
O	25	47
N	0.3	0.1
Si	0.03	28

Why are hydrogen, carbon, oxygen, and nitrogen so fit for the roles they play in living organisms? First and foremost, they are among the smallest atoms that form covalent bonds, and carbon, oxygen, and nitrogen can also form multiple bonds. Because the atoms are small and can form multiple bonds, they form strong bonds that give rise to stable molecules. The compounds that make up living organisms must be stable and, therefore, slow to react if the organisms are to survive.

Silicon has almost twice the diameter of carbon, so silicon forms longer and weaker bonds. Consequently, an S_N2 reaction at silicon would occur much more rapidly than an S_N2 reaction at carbon. Moreover, silicon has another problem. The end product of carbon metabolism is CO_2. The analogous product of silicon metabolism would be SiO_2. Because silicon is only singly bonded to oxygen in SiO_2, silicon dioxide molecules polymerize to form quartz (sea sand). It is hard to imagine that life could exist, much less proliferate, were animals to exhale sea sand instead of CO_2!

10.5 The Mechanism of an S_N1 Reaction

Given our understanding of S_N2 reactions, we would expect the rate of reaction of *tert*-butyl bromide with water to be relatively slow because water is a poor nucleophile and *tert*-butyl bromide is sterically hindered to attack by a nucleophile. It turns out, however, that the reaction is surprisingly fast. In fact, it is over one million times faster than the reaction of methyl bromide—a compound with no steric hindrance—with water (Table 10.4). Clearly, the reaction must be taking place by a mechanism different from that of an S_N2 reaction.

$$(CH_3)_3C{-}Br + H_2O \longrightarrow (CH_3)_3C{-}OH + HBr$$

tert-butyl bromide → *tert*-butyl alcohol

As we have seen, a study of the kinetics of a reaction is one of the first steps undertaken when one is investigating the mechanism of a reaction. If we were to investigate the kinetics of the reaction of *tert*-butyl bromide with water, we would find that doubling the concentration of the alkyl halide doubles the rate of the reaction. We would also find that changing the concentration of the nucleophile has no effect on the rate of the reaction. Knowing that the rate of this nucleophilic substitution reaction depends only on the concentration of the alkyl halide, we can write the following rate law for the reaction:

$$\textbf{rate} = k[\textbf{alkyl halide}]$$

Because the rate of the reaction depends on the concentration of only one reactant, the reaction is a **first-order reaction**.

The rate law for the reaction of *tert*-butyl bromide with water differs from the rate law for the reaction of methyl bromide with hydroxide ion (Section 10.2), so the two reactions must have different mechanisms. We have seen that the reaction between methyl bromide and hydroxide ion is an S_N2 reaction. The reaction between *tert*-butyl bromide and water is an S_N**1 reaction**, where "S" stands for substitution, "N" stands for nucleophilic, and "1" stands for unimolecular. **Unimolecular** means that only one molecule is involved in the rate-determining step. The mechanism of an S_N1 reaction is based on the following experimental evidence:

1. The rate law shows that the rate of the reaction depends only on the concentration of the alkyl halide. This means that we must be observing a reaction whose rate-determining transition state involves only the alkyl halide.
2. When the methyl groups of *tert*-butyl bromide are successively replaced by hydrogens, the rate of the S_N1 reaction decreases progressively (Table 10.4). This is opposite to the order of reactivity exhibited by alkyl halides in S_N2 reactions (Table 10.1).
3. The reaction of an alkyl halide in which the halogen is bonded to an asymmetric carbon forms two stereoisomers: one with the same relative configuration at the asymmetric carbon as the reacting alkyl halide, the other with the inverted configuration.

Table 10.4 Relative Rates of S_N1 Reactions for Several Alkyl Bromides (solvent is H_2O, nucleophile is H_2O)

Alkyl bromide	Class of alkyl bromide	Relative rate
$(CH_3)_2C(CH_3){-}Br$ ($CH_3C{-}Br$ with two CH_3)	tertiary	1,200,000
$CH_3CH(CH_3){-}Br$	secondary	11.6
$CH_3CH_2{-}Br$	primary	1.00*
$CH_3{-}Br$	methyl	1.05*

*Although the rate of the S_N1 reaction of this compound with water is 0, a small rate is observed as a result of an S_N2 reaction.

3-D Molecules: *t*-Butyl bromide; *t*-Butyl cation; protonated *t*-Butyl alcohol; *t*-Butyl alcohol

Unlike an S_N2 reaction, where the leaving group departs and the nucleophile approaches *at the same time*, the leaving group in an S_N1 reaction departs *before* the nucleophile approaches. In the first step of an S_N1 reaction of an alkyl halide, the carbon–halogen bond breaks heterolytically, with the halogen retaining the previously shared pair of electrons, and a carbocation intermediate is formed. In the second step, the nucleophile reacts rapidly with the carbocation to form a protonated alcohol. Whether the alcohol product will exist in its protonated (acidic) form or neutral (basic) form depends on the pH of the solution. At pH = 7, the alcohol will exist predominantly in its neutral form (Section 1.20).

mechanism of the S_N1 reaction

$$(CH_3)_3C{-}Br \xrightleftharpoons{\text{slow}} (CH_3)_3C^+ + Br^- ;\quad (CH_3)_3C^+ + H_2\ddot{O}: \xrightarrow{\text{fast}} (CH_3)_3C{-}\overset{+}{\ddot{O}}H_2 \xrightleftharpoons{\text{fast}} (CH_3)_3C{-}\ddot{O}H + H_3O^+$$

C—Br bond breaks heterolytically

nucleophile attacks the carbocation

proton transfer

Because the rate of an S_N1 reaction depends only on the concentration of the alkyl halide, the first step must be the slow and rate-determining step. The nucleophile, therefore, is not involved in the rate-determining step, so its concentration has no

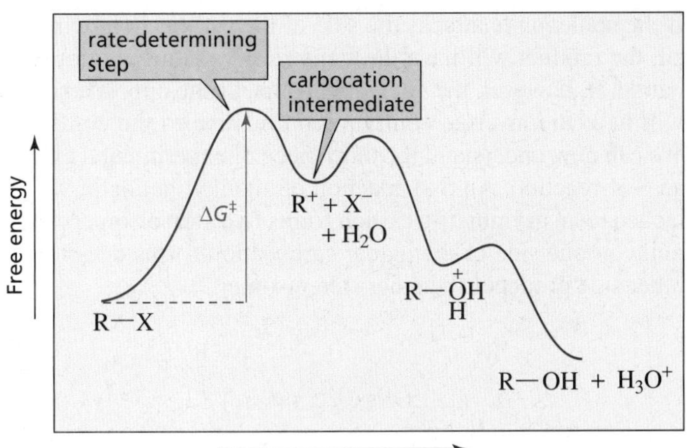

◀ **Figure 10.6**
Reaction coordinate diagram for an S_N1 reaction.

effect on the rate of the reaction. If you look at the reaction coordinate diagram in Figure 10.6, you will see why increasing the rate of the second step will not make an S_N1 reaction go any faster.

How does the mechanism for an S_N1 reaction account for the three pieces of experimental evidence? First, because the alkyl halide is the only species that participates in the rate-determining step, the mechanism agrees with the observation that the rate of the reaction depends on the concentration of the alkyl halide and does not depend on the concentration of the nucleophile.

Second, the mechanism shows that a carbocation is formed in the rate-determining step of an S_N1 reaction. We know that a tertiary carbocation is more stable, and is therefore easier to form, than a secondary carbocation, which in turn is more stable and easier to form than a primary carbocation (Section 4.2). Tertiary alkyl halides, therefore, are more reactive than secondary alkyl halides, which are more reactive than primary alkyl halides in an S_N1 reaction. This relative order of reactivity agrees with the observation that the rate of an S_N1 reaction decreases as the methyl groups of *tert*-butyl bromide are successively replaced by hydrogens (Table 10.4).

relative reactivities of alkyl halides in an S_N1 reaction

most reactive ▷ 3° alkyl halide > 2° alkyl halide > 1° alkyl halide ◁ least reactive

Actually, primary carbocations and methyl cations are so unstable that primary alkyl halides and methyl halides do not undergo S_N1 reactions. (The very slow reactions reported for ethyl bromide and methyl bromide in Table 10.4 are actually S_N2 reactions.)

Primary alkyl halides and methyl halides cannot undergo S_N1 reactions.

The positively charged carbon of the carbocation intermediate is sp^2 hybridized, and the three bonds connected to an sp^2 hybridized carbon are in the same plane. In the second step of the S_N1 reaction, the nucleophile can approach the carbocation from either side of the plane.

C—Br

Br^- ⇌ $-Br^-$

a $\overset{+}{C}$ b

$H_2\ddot{O}$: :$\ddot{O}H_2$

inverted configuration relative to the configuration of the alkyl halide

same configuration as the alkyl halide

$H^+ +$ HO—C ⇌ $H\overset{+}{O}$—C (H) a b C—$\overset{+}{O}H$ (H) ⇌ C—OH $+ H^+$

Tutorial: S_N1

If the nucleophile attacks the side of the carbon from which the leaving group departed, the product will have the same relative configuration as that of the reacting alkyl halide. If, however, the nucleophile attacks the opposite side of the carbon, the product will have the inverted configuration relative to the configuration of the alkyl halide. We can now understand the third piece of experimental evidence for the mechanism of an S_N1 reaction: An S_N1 reaction of an alkyl halide in which the leaving group is attached to an asymmetric carbon forms two stereoisomers because attack of the nucleophile on one side of the planar carbocation forms one stereoisomer and attack on the other side produces the other stereoisomer.

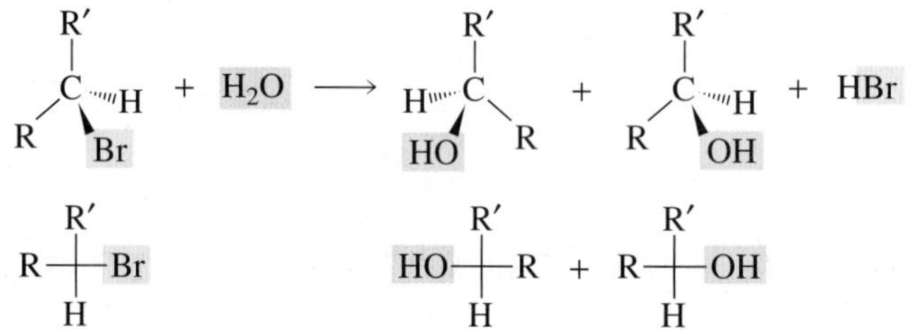

R′ C H R Br + H_2O ⟶ R′ H C HO R + R′ C H R OH + HBr

R′ R—|—Br H   HO—|—R (R′, H) + R—|—OH (R′, H)

if the leaving group in an S_N1 reaction is attached to an asymmetric carbon, a pair of enantiomers will be formed as products

PROBLEM 13◆

Arrange the following alkyl bromides in order of decreasing reactivity in an S_N1 reaction: isopropyl bromide, propyl bromide, *tert*-butyl bromide, methyl bromide.

PROBLEM 14

Why are the rates of the S_N2 reactions of ethyl bromide and methyl bromide given in Table 10.4 so slow?

10.6 Factors Affecting S_N1 Reactions

The Leaving Group

Because the rate-determining step of an S_N1 reaction is the dissociation of the alkyl halide to form a carbocation, two factors affect the rate of an S_N1 reaction: the ease with which the leaving group dissociates from the carbon and the stability of the carbocation that is formed. In the preceding section, we saw that tertiary alkyl halides are more reactive than secondary alkyl halides, which are more reactive than primary alkyl halides. This is because the more substituted the carbocation is, the more stable it is and therefore the easier it is to form. But how do we rank the relative reactivity of a series of alkyl halides with different leaving groups that dissociate to form the same carbocation? As in the case of the S_N2 reaction, there is a direct relationship between basicity and leaving ability in the S_N1 reaction: The weaker the base, the less tightly it is bonded to the carbon and the easier it is to break the carbon–halogen bond. As a result, an alkyl iodide is the most reactive and an alkyl fluoride is the least reactive of the alkyl halides in both S_N1 and S_N2 reactions.

relative reactivities of alkyl halides in an S_N1 reaction

most reactive > RI > RBr > RCl > RF < least reactive

The Nucleophile

The nucleophile reacts with the carbocation that is formed in the rate-determining step of an S_N1 reaction. Because the nucleophile comes into play *after* the rate-determining step, the reactivity of the nucleophile has no effect on the rate of an S_N1 reaction (Figure 10.6).

In most S_N1 reactions, the solvent is the nucleophile. For example, the relative rates given in Table 10.4 are for the reactions of alkyl halides with water in water. Water serves as both the nucleophile and the solvent. Reaction with a solvent is called **solvolysis**. Thus, each rate in Table 10.4 is for the solvolysis of the indicated alkyl bromide in water.

Carbocation Rearrangements

A carbocation intermediate is formed in an S_N1 reaction. In Section 4.6, we saw that a carbocation will rearrange if it becomes more stable in the process. If the carbocation formed in an S_N1 reaction can rearrange, S_N1 and S_N2 reactions of the same alkyl halide can produce different constitutional isomers as products, since a carbocation is not formed in an S_N2 reaction and therefore the carbon skeleton cannot rearrange. For example, the product obtained when OH is substituted for the Br of 2-bromo-3-methylbutane by an S_N1 reaction is different from the product obtained by an S_N2 reaction. When the reaction is carried out under conditions that favor an S_N1 reaction, the initially formed secondary carbocation undergoes a 1,2-hydride shift to form a more stable tertiary carbocation.

When a reaction forms a carbocation intermediate, always check for the possibility of a carbocation rearrangement.

$CH_3CH(CH_3)CHBrCH_3$ (2-bromo-3-methylbutane)

S_N1, H_2O: $\rightarrow CH_3CH(CH_3)\overset{+}{C}HCH_3 + Br^-$ (secondary carbocation) $\xrightarrow{\text{1,2-hydride shift}} CH_3\overset{+}{C}(CH_3)CH_2CH_3$ (tertiary carbocation) $\xrightarrow{H_2O} CH_3C(CH_3)(\overset{+}{O}H_2)CH_2CH_3 \overset{H_2O}{\rightleftharpoons} CH_3C(CH_3)(OH)CH_2CH_3 + H_3O^+$ (2-methyl-2-butanol)

S_N2, HO^-: $\rightarrow CH_3CH(CH_3)CH(OH)CH_3 + Br^-$ (3-methyl-2-butanol)

The product obtained from the reaction of 3-bromo-2,2-dimethylbutane with a nucleophile also depends on the conditions under which the reaction is carried out. The carbocation formed under conditions that favor an S_N1 reaction will undergo a 1,2-methyl shift. Because a carbocation is not formed under conditions that favor an S_N2 reaction, the carbon skeleton does not rearrange. In Sections 10.9 and 10.10, we will see that we can exercise some control over whether an S_N1 or an S_N2 reaction takes place, by selecting appropriate reaction conditions.

$CH_3C(CH_3)_2CHBrCH_3$ (3-bromo-2,2-dimethylbutane)

S_N1, H_2O: $\rightarrow CH_3C(CH_3)_2\overset{+}{C}HCH_3 + Br^-$ (secondary carbocation) $\xrightarrow{\text{1,2-methyl shift}} CH_3\overset{+}{C}(CH_3)CH(CH_3)CH_3$ (tertiary carbocation) $\xrightarrow{H_2O} CH_3C(CH_3)(\overset{+}{O}H_2)CH(CH_3)CH_3 \overset{H_2O}{\rightleftharpoons} CH_3C(CH_3)(OH)CH(CH_3)CH_3 + H_3O^+$ (2,3-dimethyl-2-butanol)

S_N2, HO^-: $\rightarrow CH_3C(CH_3)_2CH(OH)CH_3 + Br^-$ (3,3-dimethyl-2-butanol)

PROBLEM 15◆

Arrange the following alkyl halides in order of decreasing reactivity in an S_N1 reaction: 2-bromopentane, 2-chloropentane, 1-chloropentane, 3-bromo-3-methylpentane.

PROBLEM 16◆

Which of the following alkyl halides form a substitution product in an S_N1 reaction that is different from the substitution product formed in an S_N2 reaction?

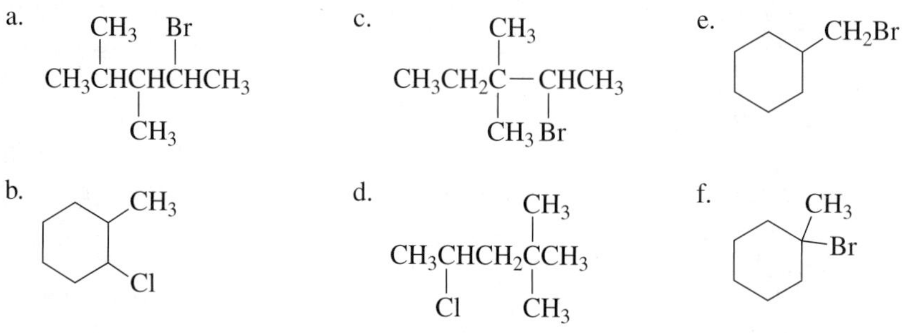

PROBLEM 17

Two substitution products result from the reaction of 3-chloro-3-methyl-1-butene with sodium acetate (CH_3COO^- Na^+) in acetic acid under S_N1 conditions. Identify the products.

10.7 More About the Stereochemistry of S_N2 and S_N1 Reactions

The Stereochemistry of S_N2 Reactions

The reaction of 2-bromopropane with hydroxide ion forms a substitution product without any asymmetric carbons. The product, therefore, has no stereoisomers.

2-bromopropane 2-propanol

The reaction of 2-bromobutane with hydroxide ion forms a substitution product with an asymmetric carbon. The product, therefore, can exist as enantiomers.

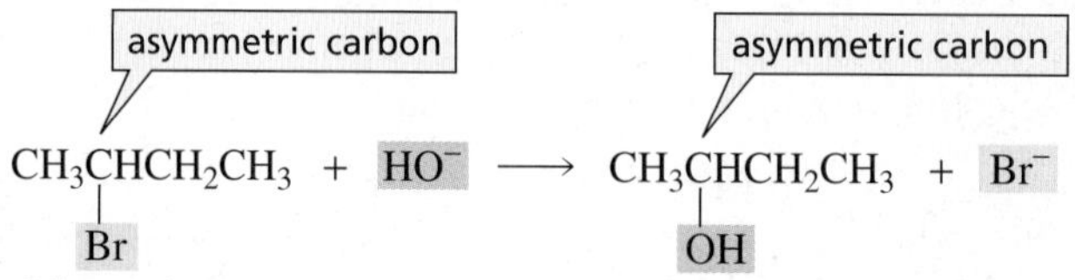

2-bromobutane 2-butanol

We cannot specify the configuration of the product formed from the reaction of 2-bromobutane with hydroxide ion unless we know the configuration of the alkyl halide and whether the reaction is an S_N2 or an S_N1 reaction. For example, if we know that the reaction is an S_N2 reaction and that the reactant has the *S* configuration, we know that the product will be (*R*)-2-butanol because in an S_N2 reaction, the incoming nucleophile attacks the back side of the carbon that is attached to the halogen (Section 10.2). Therefore, the product will have a configuration that is inverted relative to that of the reactant. (Recall that an S_N2 reaction takes place with *inversion of configuration*).

the configuration is inverted relative to that of the reactant

$$CH_2CH_3\text{-}C(H)(CH_3)(Br) + HO^- \xrightarrow{S_N2\ \text{conditions}} H\text{-}C(CH_2CH_3)(CH_3)(HO) + Br^-$$

(*S*)-2-bromobutane → (*R*)-2-butanol

An S_N2 reaction takes place with inversion of configuration.

The Stereochemistry of S_N1 Reactions

In contrast to the S_N2 reaction, the S_N1 reaction of (*S*)-2-bromobutane forms two substitution products—one with the same relative configuration as the reactant and the other with the inverted configuration. In an S_N1 reaction, the leaving group leaves before the nucleophile attacks. This means that the nucleophile is free to attack either side of the planar carbocation. If it attacks the side from which the bromide ion left, the product will have the same relative configuration as the reactant. If it attacks the opposite side, the product will have the inverted configuration.

product with inverted configuration — product with retained configuration

$$CH_2CH_3\text{-}C(H)(CH_3)(Br) + H_2O \xrightarrow{S_N1\ \text{conditions}} H\text{-}C(CH_2CH_3)(CH_3)(HO) + CH_2CH_3\text{-}C(H)(CH_3)(OH) + HBr$$

(*S*)-2-bromobutane → (*R*)-2-butanol + (*S*)-2-butanol

An S_N1 reaction takes place with racemization.

Although you might expect that equal amounts of both products should be formed in an S_N1 reaction, a greater amount of the product with the inverted configuration is obtained in most cases. Typically, 50–70% of the product of an S_N1 reaction is the inverted product. If the reaction leads to equal amounts of the two stereoisomers, the reaction is said to take place with **complete racemization**. When more of one of the products is formed, the reaction is said to take place with **partial racemization**.

Saul Winstein was the first to explain why extra inverted product generally is formed in an S_N1 reaction. He postulated that dissociation of the alkyl halide initially results in the formation of an **intimate ion pair**. In an intimate ion pair, the bond between the carbon and the leaving group has broken, but the cation and anion remain next to each other. This species then forms a *solvent-separated ion pair*—an ion pair in which one or more solvent molecules have come between the cation and the anion. Further separation between the two results in dissociated ions.

Saul Winstein (1912–1969) *was born in Montreal, Canada. He received a Ph.D. from the California Institute of Technology and was a professor of chemistry at the University of California, Los Angeles, from 1942 until his death.*

$$R{-}X \longrightarrow R^+\ X^- \longrightarrow R^+\ \text{(solvent)}\ X^- \longrightarrow R^+ \qquad X^-$$

undissociated molecule — intimate ion pair — solvent-separated ion pair — dissociated ions

The nucleophile can attack any of these four species. If the nucleophile attacks only the completely dissociated carbocation, the product will be completely racemized. If the nucleophile attacks the carbocation of either the intimate ion pair or the solvent-separated ion pair, the leaving group will be in position to partially block the approach of the nucleophile to that side of the carbocation and more of the product with the inverted configuration will be obtained. (Notice that if the nucleophile attacks the undissociated molecule, the reaction will be an S_N2 reaction and all of the product will have the inverted configuration.)

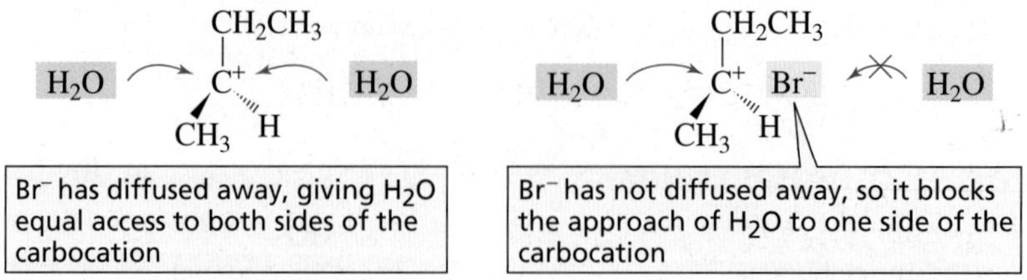

Br⁻ has diffused away, giving H_2O equal access to both sides of the carbocation

Br⁻ has not diffused away, so it blocks the approach of H_2O to one side of the carbocation

Movies: S_N1 inversion; S_N1 retention

The difference between the products obtained from an S_N1 reaction and from an S_N2 reaction is a little easier to visualize in the case of cyclic compounds. For example, when *cis*-1-bromo-4-methylcyclohexane undergoes an S_N2 reaction, only the trans product is obtained because the carbon bonded to the leaving group is attacked by the nucleophile only on its back side.

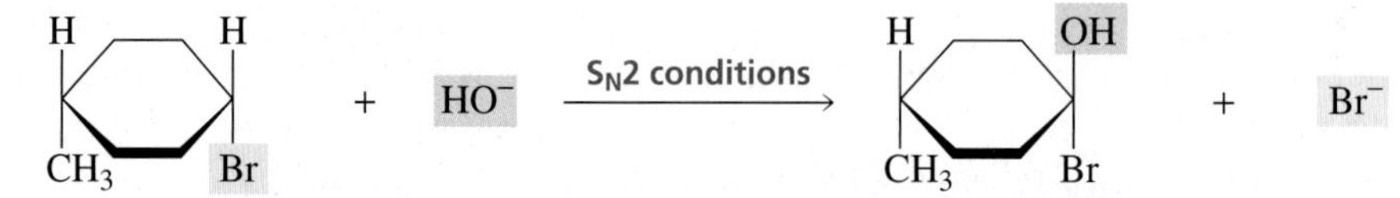

cis-1-bromo-4-methylcyclohexane + HO^- —(S_N2 conditions)→ *trans*-4-methylcyclohexanol + Br^-

However, when *cis*-1-bromo-4-methylcyclohexane undergoes an S_N1 reaction, both the cis and the trans products are formed because the nucleophile can approach the carbocation intermediate from either side.

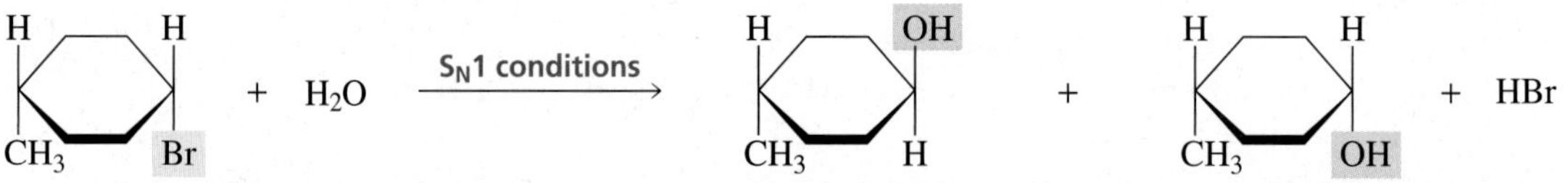

cis-1-bromo-4-methylcyclohexane + H_2O —(S_N1 conditions)→ *trans*-4-methylcyclohexanol + *cis*-4-methylcyclohexanol + HBr

PROBLEM 18◆

If the products of the preceding reaction are not obtained in equal amounts, which stereoisomer will be present in excess?

PROBLEM 19

Give the products that will be obtained from the following reactions if

a. the reaction is carried out under conditions that favor an S_N2 reaction
b. the reaction is carried out under conditions that favor an S_N1 reaction
   1. *trans*-1-bromo-4-methylcyclohexane + sodium methoxide/methanol
   2. *cis*-1-chloro-3-methylcyclobutane + sodium hydroxide/water

PROBLEM 20◆

Which of the following reactions will go faster if the concentration of the nucleophile is increased?

a. 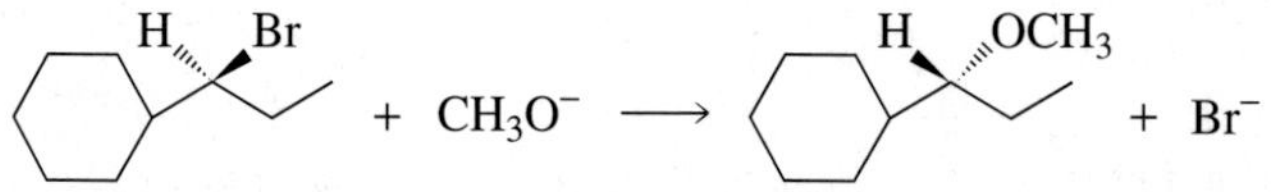 (cyclohexyl)CH(Br)CH₂CH₃ + CH_3O^- ⟶ (cyclohexyl)CH(OCH_3)CH₂CH₃ + Br^-

b. 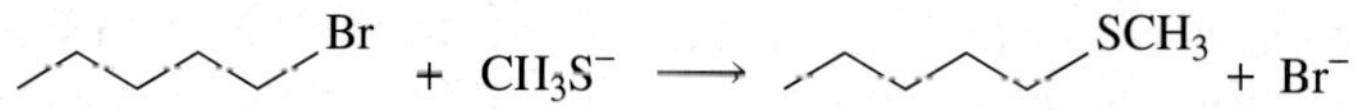 $CH_3(CH_2)_6CH_2Br$ + CH_3S^- ⟶ $CH_3(CH_2)_6CH_2SCH_3$ + Br^-

c. 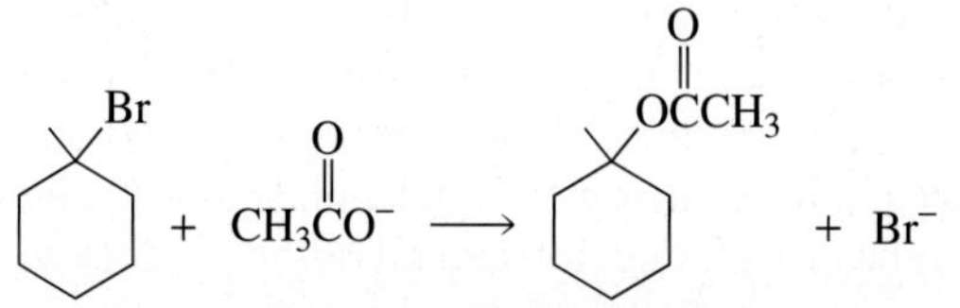 1-bromo-1-methylcyclohexane + $CH_3CO_2^-$ ⟶ 1-methylcyclohexyl acetate ($OCCH_3$) + Br^-

10.8 Benzylic Halides, Allylic Halides, Vinylic Halides, and Aryl Halides

Our discussion of substitution reactions, to this point, has been limited to methyl halides and primary, secondary, and tertiary alkyl halides. But what about benzylic, allylic, vinylic, and aryl halides? Let's first consider benzylic and allylic halides. Benzylic and allylic halides readily undergo S_N2 reactions unless they are tertiary. Tertiary benzylic and tertiary allylic halides, like other tertiary halides, are unreactive in S_N2 reactions because of steric hindrance.

$$C_6H_5CH_2Cl + CH_3O^- \xrightarrow{S_N2 \text{ conditions}} C_6H_5CH_2OCH_3 + Cl^-$$

benzyl chloride → benzyl methyl ether

$$CH_3CH{=}CHCH_2Br + HO^- \xrightarrow{S_N2 \text{ conditions}} CH_3CH{=}CHCH_2OH + Br^-$$

1-bromo-2-butene, an allylic halide → 2-buten-1-ol

Benzylic and allylic halides readily undergo S_N1 reactions because they form relatively stable carbocations. While primary alkyl halides (such as CH_3CH_2Br and $CH_3CH_2CH_2Br$) cannot undergo S_N1 reactions because their carbocations are too unstable, primary benzylic and primary allylic halides readily undergo S_N1 reactions because their carbocations are stabilized by electron delocalization (Section 7.7).

Benzylic and allylic halides undergo S_N1 and S_N2 reactions.

$$C_6H_5CH_2Cl \overset{S_N1}{\rightleftharpoons} C_6H_5\overset{+}{C}H_2 + Cl^- \xrightarrow{CH_3OH} C_6H_5CH_2OCH_3 + H^+$$

$$CH_2{=}CHCH_2Br \overset{S_N1}{\rightleftharpoons} CH_2{=}CH\overset{+}{C}H_2 + Br^- \longleftrightarrow \overset{+}{C}H_2CH{=}CH_2 \xrightarrow{H_2O} CH_2{=}CHCH_2OH + H^+$$

If the resonance contributors of the allylic carbocation intermediate have different groups bonded to their sp^2 carbons, two substitution products will be obtained.

3-D Molecule: Benzyl cation

$$CH_3CH{=}CHCH_2Br \overset{S_N1}{\rightleftharpoons} CH_3CH{=}CH\overset{+}{C}H_2 \longleftrightarrow CH_3\overset{+}{C}HCH{=}CH_2 + Br^-$$

$$CH_3CH{=}CH\overset{+}{C}H_2 \xrightarrow{H_2O} CH_3CH{=}CHCH_2OH + H^+$$

$$CH_3\overset{+}{C}HCH{=}CH_2 \xrightarrow{H_2O} CH_3\underset{OH}{C}HCH{=}CH_2 + H^+$$

3-D Molecule: Allyl cation

Vinylic halides and aryl halides do not undergo S_N2 or S_N1 reactions. They do not undergo S_N2 reactions because, as the nucleophile approaches the back side of the sp^2 carbon, it is repelled by the π electron cloud of the double bond or the aromatic ring.

Vinylic and aryl halides undergo neither S_N1 nor S_N2 reactions.

a nucleophile is repelled by the π electron cloud

$$\underset{H}{\overset{R}{>}}C{=}C\underset{H}{\overset{Cl}{<}} \quad \text{✗ nucleophile}$$

a vinylic halide

$$C_6H_5Br \quad \text{✗ nucleophile}$$

an aryl halide

There are two reasons that vinylic halides and aryl halides do not undergo S_N1 reactions. First, vinylic and aryl cations are even more unstable than primary carbocations (Section 10.5) because the positive charge is on an *sp* carbon. Since *sp* carbons are more electronegative than the sp^2 carbons that carry the positive charge of alkyl carbocations, *sp* carbons are more resistant to becoming positively charged. Second, we have seen that sp^2 carbons form stronger bonds than do sp^3 carbons (Section 1.14). As a result, it is harder to break the carbon–halogen bond when the halogen is bonded to an sp^2 carbon.

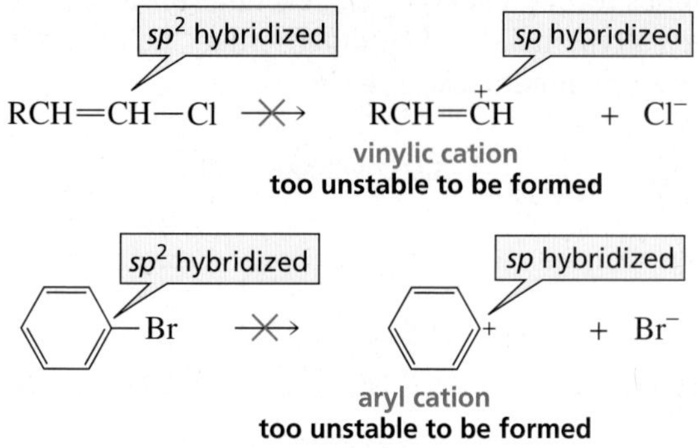

sp^2 hybridized — *sp* hybridized

$$RCH{=}CH{-}Cl \not\rightarrow RCH{=}\overset{+}{C}H + Cl^-$$

vinylic cation
too unstable to be formed

sp^2 hybridized — *sp* hybridized

$$C_6H_5{-}Br \not\rightarrow C_6H_5^+ + Br^-$$

aryl cation
too unstable to be formed

PROBLEM-SOLVING STRATEGY

Which alkyl halide would you expect to be more reactive in an S_N1 solvolysis reaction?

$$CH_3\ddot{O}{-}CH{=}CH{-}CH_2Br \quad \text{or} \quad CH_2{=}C(\ddot{O}CH_3){-}CH_2Br$$

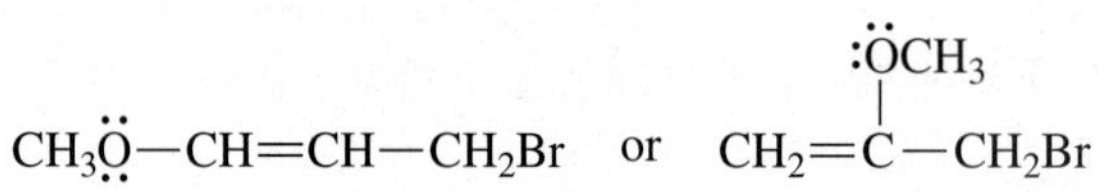

When asked to determine the relative reactivities of two compounds, we need to compare the $\Delta G^{\ddagger}$ values of their rate-determining steps. The faster reacting compound will be the one with the smallest difference between its free energy and the free energy of its rate-determining transition state; that is, the faster reacting compound will have the *smaller* $\Delta G^{\ddagger}$ value. Both alkyl halides have approximately the same stability, so the difference in their reaction rates will be due to the difference in stability of the transition states of their rate-determining steps. The rate-determining step is carbocation formation, so the compound that forms the more stable carbocation will be the one that has the faster rate of solvolysis. The compound on the left forms the more stable carbocation; it has three resonance contributors, whereas the other carbocation has only two resonance contributors.

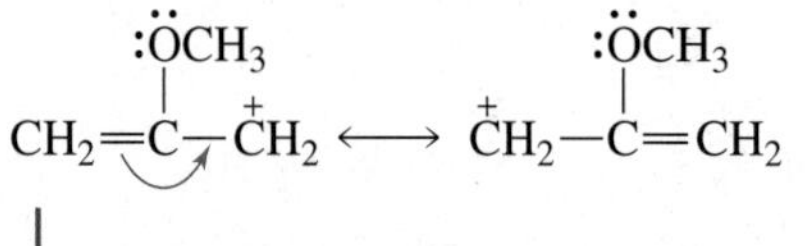

$$CH_3\ddot{O}{-}CH{=}CH{-}\overset{+}{C}H_2 \longleftrightarrow CH_3\ddot{O}{-}\overset{+}{C}H{-}CH{=}CH_2 \longleftrightarrow CH_3\overset{+}{O}{=}CH{-}CH{=}CH_2$$

$$CH_2{=}C(\ddot{O}CH_3){-}\overset{+}{C}H_2 \longleftrightarrow \overset{+}{C}H_2{-}C(\ddot{O}CH_3){=}CH_2$$

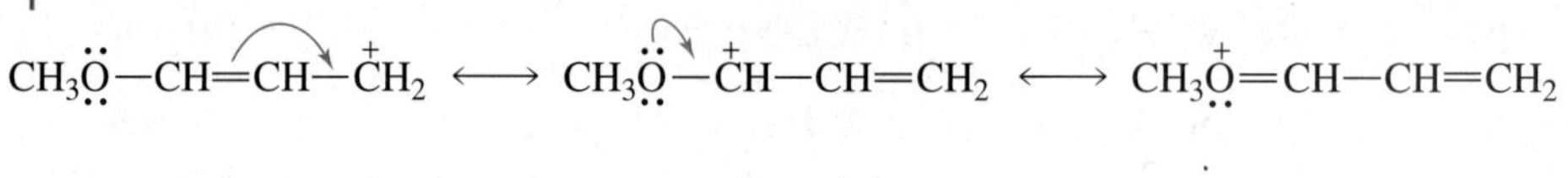

Now continue on to Problems 21–23.

PROBLEM 21◆

Which alkyl halide would you expect to be more reactive in an S_N1 solvolysis reaction?

$$CH_3CH(Br)CH_2{-}CH{=}CH{-}CH_2CH_3 \quad \text{or} \quad CH_3CH_2CH(Br){-}CH{=}CH{-}CH_3$$

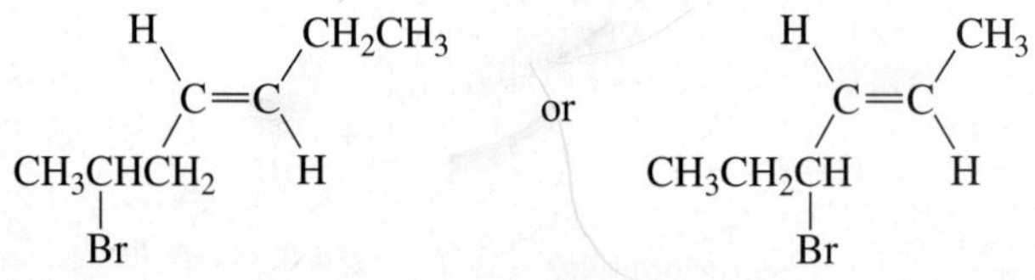

PROBLEM 22◆

Which alkyl halide would you expect to be more reactive in an S_N2 reaction with a given nucleophile? In each case, you can assume that both alkyl halides have the same stability.

a. $CH_3CH_2CH_2Br$ or $CH_3CH_2CH_2I$

b. $CH_3CH_2CH_2Cl$ or CH_3OCH_2Cl

c. $CH_3CH_2\overset{CH_3}{\overset{|}{C}}HBr$ or $CH_3CH_2\overset{CH_2CH_3}{\overset{|}{C}}HBr$

d. $CH_3CH_2CH_2\overset{CH_3}{\overset{|}{C}}HBr$ or $CH_3CH_2\overset{CH_3}{\overset{|}{C}}HCH_2Br$

e. 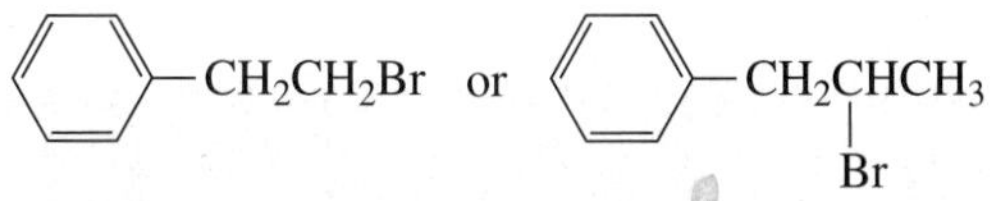 $C_6H_5{-}CH_2CH_2Br$ or $C_6H_5{-}CH_2\underset{Br}{\underset{|}{C}}HCH_3$

f. 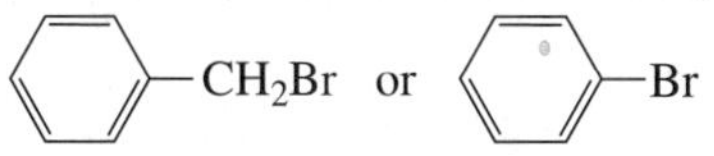 $C_6H_5{-}CH_2Br$ or $C_6H_5{-}Br$

g. $CH_3CH{=}\underset{Br}{\underset{|}{C}}CH_3$ or $CH_3CH{=}CH\underset{Br}{\underset{|}{C}}HCH_3$

PROBLEM 23◆

For each of the pairs in Problem 22, which compound would be more reactive in an S_N1 reaction?

PROBLEM 24◆

Give the products obtained from the following reactions and show their configurations:

a. under conditions that favor an S_N2 reaction

b. under conditions that favor an S_N1 reaction

$$CH_3CH{=}CHCH_2Br \;+\; CH_3O^- \xrightarrow{CH_3OH}$$

10.9 Competition Between S_N2 and S_N1 Reactions

The characteristics of S_N2 and S_N1 reactions are summarized in Table 10.5. Remember that the "2" in "S_N2" and the "1" in "S_N1" refer to the molecularity—how many molecules are involved in the rate-determining step. Thus, the rate-determining step of an S_N2 reaction is bimolecular, whereas the rate-determining step of an S_N1 reaction is

Table 10.5 Comparison of S_N2 and S_N1 Reactions

S_N2	S_N1
A one-step mechanism	A stepwise mechanism that forms a carbocation intermediate
A bimolecular rate-determining step	A unimolecular rate-determining step
No carbocation rearrangements	Carbocation rearrangements
Product has inverted configuration relative to the reactant	Products have both retained and inverted configurations relative to the reactant
Reactivity order: methyl > 1° > 2° > 3°	Reactivity order: 3° > 2° > 1° > methyl

unimolecular. These numbers do not refer to the number of steps in the mechanism. In fact, just the opposite is true: An S_N2 reaction proceeds by a *one*-step concerted mechanism, and an S_N1 reaction proceeds by a *two*-step mechanism with a carbocation intermediate.

We have seen that methyl halides and primary alkyl halides undergo only S_N2 reactions because methyl cations and primary carbocations, which would be formed in an S_N1 reaction, are too unstable to be formed. Tertiary alkyl halides undergo only S_N1 reactions because steric hindrance makes them unreactive in an S_N2 reaction. Secondary alkyl halides as well as benzylic and allylic halides (unless they are tertiary) can undergo both S_N1 and S_N2 reactions because they form relatively stable carbocations and the steric hindrance associated with these alkyl halides is generally not very great. Vinylic and aryl halides do not undergo either S_N1 or S_N2 reactions. These results are summarized in Table 10.6.

Table 10.6 Summary of the Reactivity of Alkyl Halides in Nucleophilic Substitution Reactions

Methyl and 1° alkyl halides	S_N2 only
Vinylic and aryl halides	Neither S_N1 nor S_N2
2° alkyl halides	S_N1 and S_N2
1° and 2° benzylic and 1° and 2° allylic halides	S_N1 and S_N2
3° alkyl halides	S_N1 only
3° benzylic and 3° allylic halides	S_N1 only

When an alkyl halide can undergo both an S_N1 reaction *and* an S_N2 reaction, both reactions take place simultaneously. The conditions under which the reaction is carried out determine which of the reactions predominates. Therefore, we have some experimental control over which reaction takes place.

When an alkyl halide can undergo substitution by both mechanisms, what conditions favor an S_N1 reaction? What conditions favor an S_N2 reaction? These are important questions to synthetic chemists because an S_N2 reaction forms a single substitution product, whereas an S_N1 reaction can form two substitution products if the leaving group is bonded to an asymmetric carbon. An S_N1 reaction is further complicated by carbocation rearrangements. In other words, an S_N2 reaction is a synthetic chemist's friend, but an S_N1 reaction can be a synthetic chemist's nightmare.

When the *structure* of the alkyl halide allows it to undergo both S_N2 and S_N1 reactions, three conditions determine which reaction will predominate: (1) the *concentration* of the nucleophile, (2) the *reactivity* of the nucleophile, and (3) the *solvent* in which the reaction is carried out. To understand how the concentration and the reactivity of the nucleophile affect whether an S_N2 or an S_N1 reaction predominates, we must examine the rate laws for the two reactions. The rate constants have been given subscripts that indicate the reaction order.

Rate law for an S_N2 reaction = k_2 [alkyl halide] [nucleophile]

Rate law for S_N1 reaction = k_1 [alkyl halide]

The rate law for the reaction of an alkyl halide that can undergo both S_N2 and S_N1 reactions simultaneously is the sum of the individual rate laws.

$$\textbf{rate} = k_2\textbf{[alkyl halide][nucleophile]} + k_1\textbf{[alkyl halide]}$$

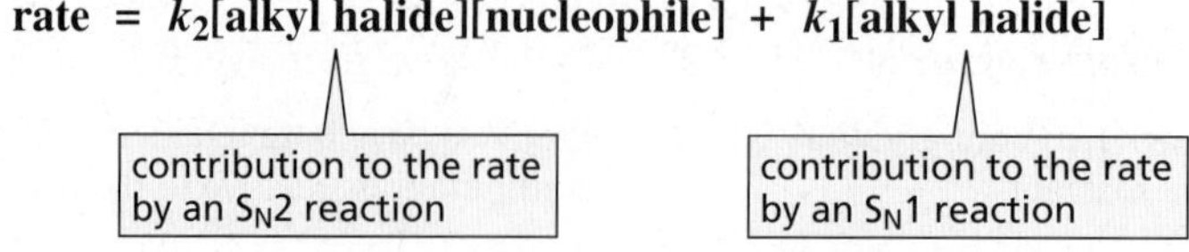

From the rate law, you can see that increasing the *concentration* of the nucleophile increases the rate of an S_N2 reaction but has no effect on the rate of an S_N1 reaction.

Therefore, when both reactions occur simultaneously, increasing the concentration of the nucleophile increases the fraction of the reaction that takes place by an S_N2 pathway. In contrast, decreasing the concentration of the nucleophile decreases the fraction of the reaction that takes place by an S_N2 pathway.

The slow (and only) step of an S_N2 reaction is attack of the nucleophile on the alkyl halide. Increasing the *reactivity* of the nucleophile increases the rate of an S_N2 reaction by increasing the value of the rate constant (k_2), because more reactive nucleophiles are better able to displace the leaving group. The slow step of an S_N1 reaction is the dissociation of the alkyl halide. The carbocation formed in the slow step reacts rapidly in a second step with any nucleophile present in the reaction mixture. Increasing the rate of the fast step does not affect the rate of the prior slow, carbocation-forming step. This means that increasing the reactivity of the nucleophile has no effect on the rate of an S_N1 reaction. A good nucleophile, therefore, favors an S_N2 reaction over an S_N1 reaction. A poor nucleophile favors an S_N1 reaction, not by increasing the rate of the S_N1 reaction itself, but by decreasing the rate of the competing S_N2 reaction. In summary:

- An S_N2 reaction is favored by a high concentration of a good nucleophile.
- An S_N1 reaction is favored by a low concentration of a nucleophile or by a poor nucleophile.

Look back at the S_N1 reactions in previous sections and notice that they all use poor nucleophiles (H_2O, CH_3OH), whereas the S_N2 reactions use good nucleophiles (HO^-, CH_3O^-). In other words, a poor nucleophile is used to encourage an S_N1 reaction, and a good nucleophile is used to encourage an S_N2 reaction. In Section 10.10, we will look at the third factor that influences whether an S_N2 or an S_N1 reaction will predominate—the solvent in which the reaction is carried out.

PROBLEM-SOLVING STRATEGY

This problem will give you practice determining whether a substitution reaction will take place by an S_N1 or an S_N2 pathway. (Keep in mind that good nucleophiles encourage S_N2 reactions, whereas poor nucleophiles encourage S_N1 reactions.)

Give the configuration(s) of the substitution product(s) that will be obtained from the reactions of the following secondary alkyl halides with the indicated nucleophile:

a. CH_3–C(CH_2CH_3)(H)(Br) + CH_3O^- (**high concentration**) ⟶ H–C(CH_2CH_3)(CH_3)(CH_3O)

Because a high concentration of a good nucleophile is used, we can predict that the reaction is an S_N2 reaction. Therefore, the product will have the inverted configuration relative to the configuration of the reactant. (An easy way to draw the inverted product is to draw the mirror image of the reacting alkyl halide and then put the nucleophile in the same location as the leaving group.)

b. CH_3–C(CH_2CH_3)(H)(Br) + CH_3OH ⟶ CH_3–C(CH_2CH_3)(H)(OCH_3) + H–C(CH_2CH_3)(CH_3)(CH_3O)

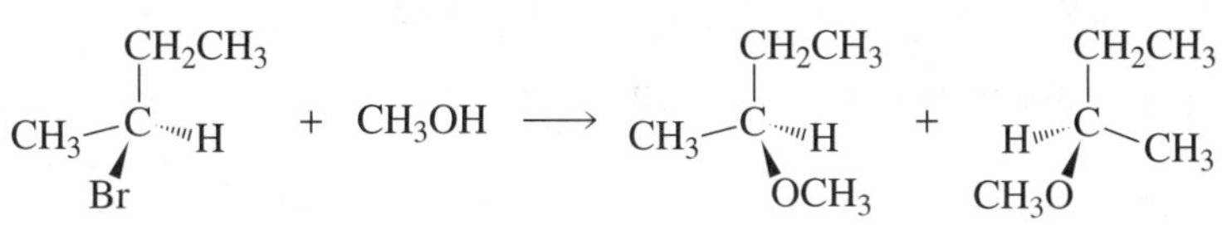

Because a poor nucleophile is used, we can predict that the reaction is an S_N1 reaction. Therefore, we will obtain two substitution products, one with the retained configuration and one with the inverted configuration, relative to the configuration of the reactant.

c. $CH_3CH_2CHCH_2CH_3$ (with I on C-3) + CH_3OH ⟶ $CH_3CH_2CHCH_2CH_3$ (with OCH_3 on C-3)

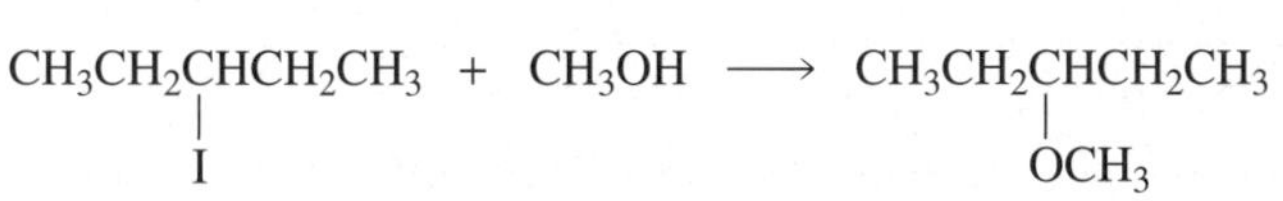

The poor nucleophile allows us to predict that the reaction is an S_N1 reaction. However, the product does not have an asymmetric carbon, so it does not have stereoisomers.

(The same substitution product would have been obtained if the reaction had been an S_N2 reaction.)

d.  $CH_3CH_2CHCH_3$ (with Cl on C-2) + HO^- (**high concentration**) $\longrightarrow$ $CH_3CH_2CHCH_3$ (with OH on C-2)

Because a high concentration of a good nucleophile is employed, we can predict that the reaction is an S_N2 reaction. Therefore, the product will have the inverted configuration relative to the configuration of the reactant. But since the configuration of the reactant is not indicated, we do not know the configuration of the product.

e. 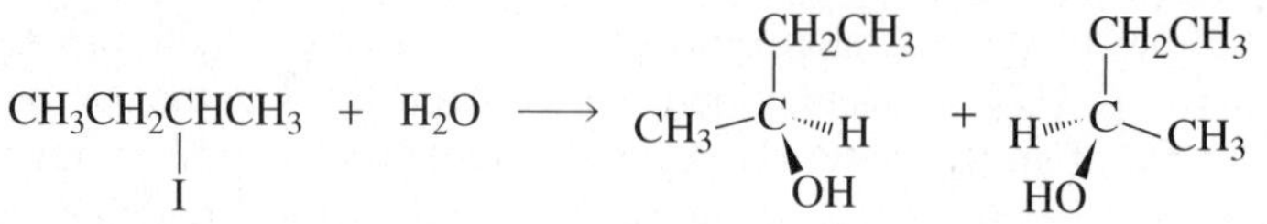 $CH_3CH_2CHCH_3$ (with I on C-2) + H_2O $\longrightarrow$ CH_3—C(CH_2CH_3)(H)(OH) + H—C(CH_2CH_3)(CH_3)(HO)

Because a poor nucleophile is employed, we can predict that the reaction is an S_N1 reaction. Therefore, both stereoisomers will be formed, regardless of the configuration of the reactant.

Now continue on to Problem 25.

PROBLEM 25

Give the configuration(s) of the substitution product(s) that will be obtained from the reactions of the following secondary alkyl halides with the indicated nucleophile:

a. 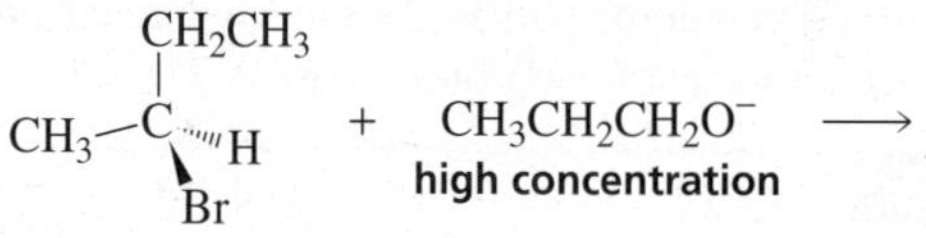 CH_3—C(CH_2CH_3)(H)(Br) + $CH_3CH_2CH_2O^-$ (**high concentration**) $\longrightarrow$

b. CH_3—C($CH_2CH_2CH_3$)(H)(Cl) + NH_3 $\longrightarrow$

c. cyclohexane ring with H, CH_3 and Cl, H substituents + CH_3O^- (**high concentration**) $\longrightarrow$

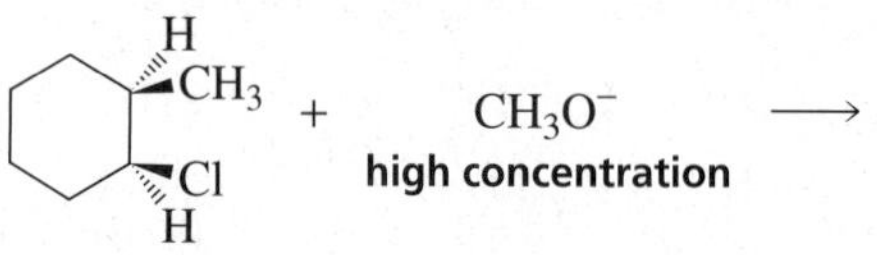

d. 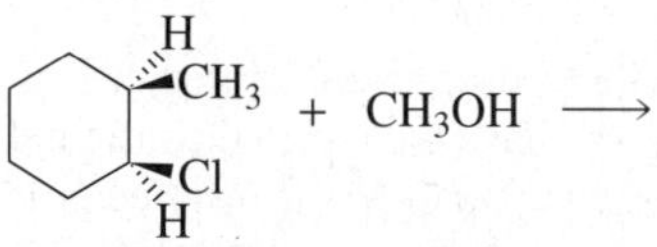 cyclohexane ring with H, CH_3 and Cl, H substituents + CH_3OH $\longrightarrow$

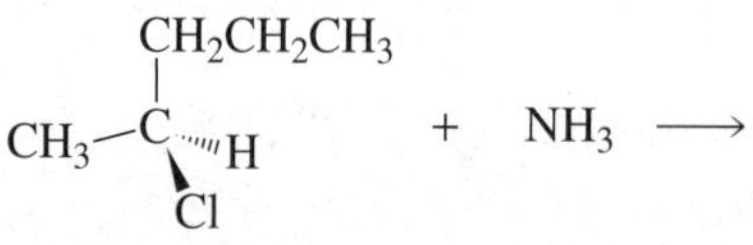

e. 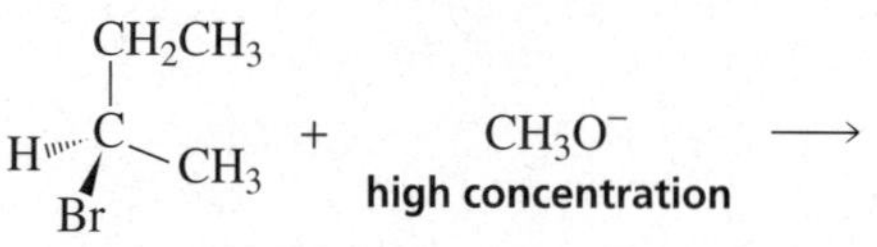 H—C(CH_2CH_3)(CH_3)(Br) + CH_3O^- (**high concentration**) $\longrightarrow$

f. 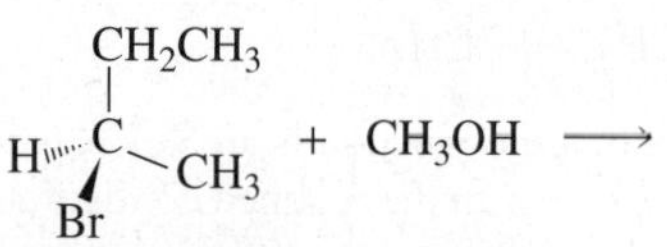 H—C(CH_2CH_3)(CH_3)(Br) + CH_3OH $\longrightarrow$

PROBLEM 26 SOLVED

The rate law for the substitution reaction of 2-bromobutane and HO^- in 75% ethanol and 25% water at 30 °C is

$$\text{rate} = 3.20 \times 10^{-5}[\text{2-bromobutane}][\text{HO}^-] + 1.5 \times 10^{-6}[\text{2-bromobutane}]$$

What percentage of the reaction takes place by an S_N2 pathway when these conditions are met?

a. $[HO^-] = 1.00$ M b. $[HO^-] = 0.001$ M

SOLUTION TO 26a

$$\text{percentage by } S_N2 = \frac{S_N2}{S_N2 + S_N1} \times 100$$

$$= \frac{3.20 \times 10^{-5}[\text{2-bromobutane}](1.00) \times 100}{3.20 \times 10^{-5}[\text{2-bromobutane}](1.00) + 1.5 \times 10^{-6}[\text{2-bromobutane}]}$$

$$= \frac{3.20 \times 10^{-5}}{3.20 \times 10^{-5} + 0.15 \times 10^{-5}} \times 100 = \frac{3.20 \times 10^{-5}}{3.35 \times 10^{-5}} \times 100$$

$$= 96\%$$

10.10 The Role of the Solvent in S_N2 and S_N1 Reactions

The solvent in which a nucleophilic substitution reaction is carried out also influences whether an S_N2 or an S_N1 reaction will predominate. Before we can understand how a particular solvent favors one reaction over another, however, we must understand how solvents stabilize organic molecules.

The **dielectric constant** of a solvent is a measure of how well the solvent can insulate opposite charges from one another. Solvent molecules insulate charges by clustering around a charge, so that the positive poles of the solvent molecules surround negative charges while the negative poles of the solvent molecules surround positive charges. Recall that the interaction between a solvent and an ion or a molecule dissolved in that solvent is called *solvation* (Section 2.9). When an ion interacts with a polar solvent, the charge is no longer localized solely on the ion, but is spread out to the surrounding solvent molecules. Spreading out the charge stabilizes the charged species.

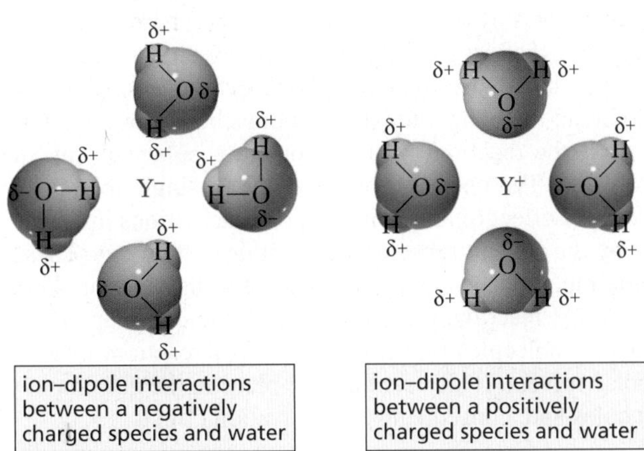

ion–dipole interactions between a negatively charged species and water

ion–dipole interactions between a positively charged species and water

Polar solvents have high dielectric constants and thus are very good at insulating (solvating) charges. Nonpolar solvents have low dielectric constants and are poor insulators. The dielectric constants of some common solvents are listed in Table 10.7. In this table, solvents are divided into two groups: protic solvents and aprotic solvents. Recall that **protic solvents** contain a hydrogen bonded to an oxygen or a nitrogen, so protic solvents are hydrogen bond donors. **Aprotic solvents**, on the other hand, do not have a hydrogen bonded to an oxygen or a nitrogen, so they are *not* hydrogen bond donors.

Table 10.7 The Dielectric Constants of Some Common Solvents

Solvent	Structure	Abbreviation	Dielectric constant (ε, at 25 °C)	Boiling point (°C)
Protic solvents				
Water	H_2O	—	79	100
Formic acid	HCOOH	—	59	100.6
Methanol	CH_3OH	MeOH	33	64.7
Ethanol	CH_3CH_2OH	EtOH	25	78.3
tert-Butyl alcohol	$(CH_3)_3COH$	*tert*-BuOH	11	82.3
Acetic acid	CH_3COOH	HOAc	6	117.9
Aprotic solvents				
Dimethyl sulfoxide	$(CH_3)_2SO$	DMSO	47	189
Acetonitrile	CH_3CN	MeCN	38	81.6
Dimethylformamide	$(CH_3)_2NCHO$	DMF	37	153
Hexamethylphosphoric acid triamide	$[(CH_3)_2N]_3PO$	HMPA	30	233
Acetone	$(CH_3)_2CO$	Me_2CO	21	56.3
Dichloromethane	CH_2Cl_2	—	9.1	40
Tetrahydrofuran	(five-membered ring with O)	THF	7.6	66
Ethyl acetate	$CH_3COOCH_2CH_3$	EtOAc	6	77.1
Diethyl ether	$CH_3CH_2OCH_2CH_3$	Et_2O	4.3	34.6
Benzene	(benzene ring)	—	2.3	80.1
Hexane	$CH_3(CH_2)_4CH_3$	—	1.9	68.7

Stabilization of charges by solvent interaction plays an important role in organic reactions. For example, when an alkyl halide undergoes an S_N1 reaction, the first step is dissociation of the carbon–halogen bond to form a carbocation and a halide ion. Energy is required to break the bond, but with no bonds being formed, where does the energy come from? If the reaction is carried out in a polar solvent, the ions that are produced are solvated. The energy associated with a single ion–dipole interaction is small, but the additive effect of all the ion–dipole interactions involved in stabilizing a charged species by the solvent represents a great deal of energy. These ion–dipole interactions provide much of the energy necessary for dissociation of the carbon–halogen bond. So in an S_N1 reaction, the alkyl halide does not fall apart spontaneously, but rather, polar solvent molecules pull it apart. An S_N1 reaction, therefore, cannot take place in a nonpolar solvent. It also cannot take place in the gas phase, where there are no solvent molecules and, consequently, no solvation effects.

SOLVATION EFFECTS

The tremendous amount of energy that is provided by solvation can be appreciated by considering the energy required to break the crystal lattice of sodium chloride (table salt) (Figure 1.1). In the absence of a solvent, sodium chloride must be heated to more than 800 °C to overcome the forces that hold the oppositely charged ions together. However, sodium chloride readily dissolves in water at room temperature because solvation of the Na^+ and Cl^- ions by water provides the energy necessary to separate the ions.

The Effect of the Solvent on the Rate of a Reaction

Increasing the polarity of the solvent will decrease the rate of the reaction if one or more reactants in the rate-determining step are charged.

Increasing the polarity of the solvent will increase the rate of the reaction if none of the reactants in the rate-determining step is charged.

One simple rule describes how a change in solvent will affect the rate of most chemical reactions: *Increasing the polarity of the solvent will decrease the rate of the reaction if one or more reactants in the rate-determining step are charged and will increase the rate of the reaction if none of the reactants in the rate-determining step is charged.*

Now let's see why this rule is true. The rate of a reaction depends on the difference between the free energy of the reactants and the free energy of the transition state in the rate-determining step of the reaction. We can, therefore, predict how changing the polarity of the solvent will affect the rate of a reaction simply by looking at the charge on the reactant(s) of the rate-determining step and the charge on the transition state of the rate-determining step, to determine which of these species will be more stabilized by a polar solvent. *The greater the charge on the solvated molecule, the stronger the interaction with a polar solvent and the more the charge will be stabilized.*

Therefore, if the charge on the reactants is greater than the charge on the transition state, a polar solvent will stabilize the reactants more than it will stabilize the transition state, increasing the difference in energy ($\Delta G^\ddagger$) between them. Consequently, *increasing the polarity of the solvent will decrease the rate of the reaction*, as shown in Figure 10.7.

On the other hand, if the charge on the transition state is greater than the charge on the reactants, a polar solvent will stabilize the transition state more than it will stabilize the reactants. Therefore, *increasing the polarity of the solvent* will decrease the difference in energy ($\Delta G^\ddagger$) between them, which *will increase the rate of the reaction*, as shown in Figure 10.8.

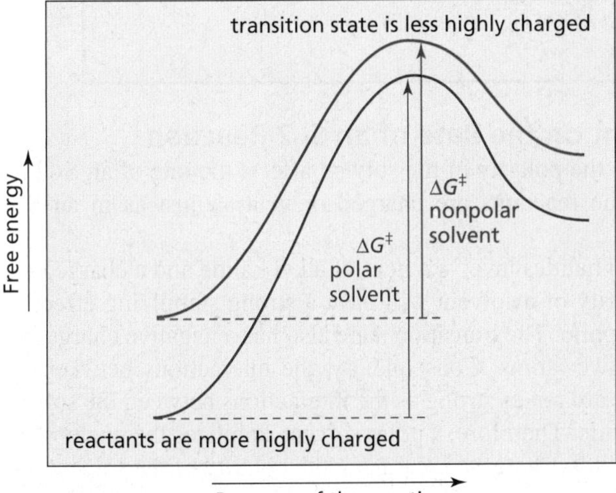

▲ **Figure 10.7**
Reaction coordinate diagram for a reaction in which the charge on the reactants is greater than the charge on the transition state.

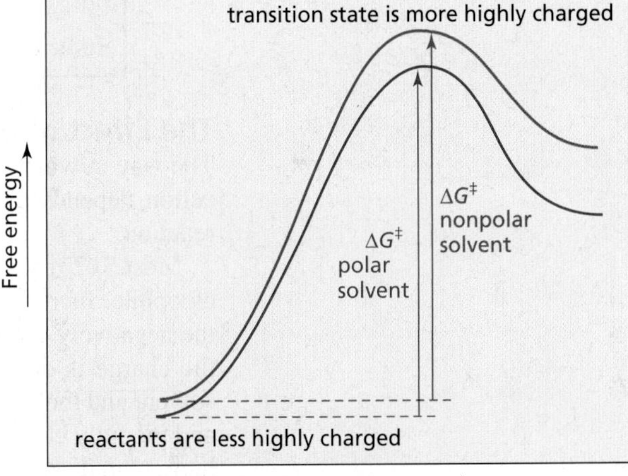

▲ **Figure 10.8**
Reaction coordinate diagram for a reaction in which the charge on the transition state is greater than the charge on the reactants.

The Effect of the Solvent on the Rate of an S_N1 Reaction

Now let's see how increasing the polarity of the solvent affects the rate of an S_N1 reaction of an alkyl halide. The alkyl halide is the only reactant in the rate-determining step of an S_N1 reaction. It is a neutral molecule with a small dipole moment. The rate-determining transition state has a greater charge because as the carbon–halogen bond breaks, the carbon becomes more positive and the halogen becomes more negative. Since the charge on the transition state is greater than the charge on the reactant, increasing the polarity of the solvent will stabilize the transition state more than the reactant and will increase the rate of the S_N1 reaction (Figure 10.8 and Table 10.8).

rate-determining step of an S_N1 reaction

$$\overset{\delta+}{C}-\overset{\delta-}{X} \longrightarrow \left[\overset{\delta+}{C}\text{-----}\overset{\delta-}{X}\right]^{\ddagger} \longrightarrow -C^{+} \quad X^{-}$$

reactant transition state products

In Chapter 12, we will see that compounds other than alkyl halides undergo S_N1 reactions. As long as the compound undergoing an S_N1 reaction is neutral, increasing the polarity of the solvent will *increase* the rate of the S_N1 reaction because the polar solvent will stabilize the dispersed charges on the transition state more than it will stabilize the relatively neutral reactant (Figure 10.8). If, however, the compound undergoing an S_N1 reaction is charged, increasing the polarity of the solvent will *decrease* the rate of the reaction because the more polar solvent will stabilize the full charge on the reactant to a greater extent than it will stabilize the dispersed charge on the transition state (Figure 10.7).

Table 10.8 The Effect of the Polarity of the Solvent on the Rate of Reaction of *tert*-Butyl Bromide in an S_N1 Reaction

Solvent	Relative rate
100% water	1200
80% water / 20% ethanol	400
50% water / 50% ethanol	60
20% water / 80% ethanol	10
100% ethanol	1

The Effect of the Solvent on the Rate of an S_N2 Reaction

The way in which a change in the polarity of the solvent affects the rate of an S_N2 reaction depends on whether the reactants are charged or neutral, just as in an S_N1 reaction.

Most S_N2 reactions of alkyl halides involve a neutral alkyl halide and a charged nucleophile. Increasing the polarity of a solvent will have a strong stabilizing effect on the negatively charged nucleophile. The transition state also has a negative charge, but the charge is dispersed over two atoms. Consequently, the interactions between the solvent and the transition state are not as strong as the interactions between the solvent and the fully charged nucleophile. Therefore, a polar solvent stabilizes the nucleophile more than it stabilizes the transition state, so increasing the polarity of the solvent will decrease the rate of the reaction (Figure 10.7).

If, however, the S_N2 reaction involves an alkyl halide and a neutral nucleophile, the charge on the transition state will be larger than the charge on the neutral reactants, so increasing the polarity of the solvent will increase the rate of the substitution reaction (Figure 10.8).

In summary, the way in which a change in solvent affects the rate of a substitution reaction does not depend on the mechanism of the reaction. It depends *only* on whether a reactant in the rate-determining step is charged: *If a reactant in the rate-determining step is charged, increasing the polarity of the solvent will decrease the rate of the reaction. If none of the reactants in the rate-determining step is charged, increasing the polarity of the solvent will increase the rate of the reaction.*

In considering the solvation of charged species by a polar solvent, the polar solvents we considered were hydrogen bond donors (protic polar solvents) such as water and alcohols. Some polar solvents—for example, *N,N*-dimethylformamide (DMF), dimethyl sulfoxide (DMSO), and hexamethylphosphoric acid triamide (HMPA)—are not hydrogen bond donors (they are aprotic polar solvents) (Table 10.7).

Ideally, one would like to carry out an S_N2 reaction with a negatively charged nucleophile in a nonpolar solvent because the localized negative charge on the reactant is greater than the dispersed negative charge on the transition state. However, a negatively charged nucleophile generally will not dissolve in a nonpolar solvent, so an aprotic polar solvent is used. Because aprotic polar solvents are not hydrogen bond donors, they are less effective than polar protic solvents in solvating negative charges; indeed, we have seen that they solvate negative charges particularly poorly (Section 10.3). Thus, the rate of an S_N2 reaction involving a negatively charged nucleophile will be greater in an aprotic polar solvent than in a protic polar solvent. Consequently, an aprotic polar solvent is the solvent of choice for an S_N2 reaction in which the nucleophile is negatively charged, whereas a protic polar solvent is used if the nucleophile is a neutral molecule.

An S_N2 reaction of an alkyl halide is favored by a high concentration of a good nucleophile in an aprotic polar solvent.

We have now seen that when an alkyl halide can undergo both S_N2 and S_N1 reactions, the S_N2 reaction will be favored by a high concentration of a good (negatively charged) nucleophile in an aprotic polar solvent, whereas the S_N1 reaction will be favored by a poor (neutral) nucleophile in a protic polar solvent.

An S_N1 reaction of an alkyl halide is favored by a low concentration of a poor nucleophile in a protic polar solvent.

ENVIRONMENTAL ADAPTATION

The microorganism *Xanthobacter* has learned to use the alkyl halides that reach the ground as industrial pollutants as a source of carbon. The microorganism synthesizes an enzyme that uses the alkyl halide as a starting material to produce other carbon-containing compounds that it needs. This enzyme has a several nonpolar groups at its active site—a pocket in the enzyme where the reaction it catalyzes takes place. The first step of the enzyme-catalyzed reaction is an S_N2 reaction. Because the nucleophile is charged, the reaction will be faster in a nonpolar solvent. The nonpolar groups on the surface of the enzyme provide this nonpolar environment.

PROBLEM 27◆

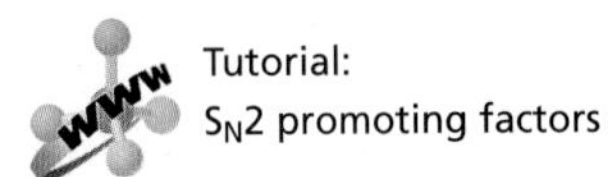

How will the rate of each of the following S_N2 reactions change if the polarity of a protic polar solvent is increased?

a. $CH_3CH_2CH_2CH_2Br + HO^- \longrightarrow CH_3CH_2CH_2CH_2OH + Br^-$

b. $CH_3\overset{+}{S}(CH_3)CH_3 + CH_3O^- \longrightarrow CH_3OCH_3 + CH_3SCH_3$

c. $CH_3CH_2I + NH_3 \longrightarrow CH_3CH_2\overset{+}{N}H_3I^-$

PROBLEM 28◆

Which reaction in each of the following pairs will take place more rapidly?

a. $CH_3Br + HO^- \longrightarrow CH_3OH + Br^-$
$CH_3Br + H_2O \longrightarrow CH_3OH + HBr$

b. $CH_3I + HO^- \longrightarrow CH_3OH + I^-$
$CH_3Cl + HO^- \longrightarrow CH_3OH + Cl^-$

c. $CH_3Br + NH_3 \longrightarrow CH_3\overset{+}{N}H_3 + Br^-$
$CH_3Br + H_2O \longrightarrow CH_3OH + Br^-$

d. $CH_3Br + HO^- \xrightarrow{\text{DMSO}} CH_3OH + Br^-$
$CH_3Br + HO^- \xrightarrow{\text{EtOH}} CH_3OH + Br^-$

e. $CH_3Br + NH_3 \xrightarrow{Et_2O} CH_3\overset{+}{N}H_3 + Br^-$

$CH_3Br + NH_3 \xrightarrow{EtOH} CH_3\overset{+}{N}H_3 + Br^-$

PROBLEM 29 SOLVED

Most of the pK_a values given throughout this text are values determined in water. How would the pK_a values of the following classes of compounds change if they were determined in a solvent less polar than water: carboxylic acids, alcohols, phenols, ammonium ions (RNH_3^+), and anilinium ions ($C_6H_5NH_3^+$)?

SOLUTION A pK_a is the negative logarithm of an equilibrium constant, K_a (Section 1.17). Because we are determining how changing the polarity of a solvent affects an equilibrium constant, we must look at how changing the polarity of the solvent affects the stability of the reactants and products.

$$K_a = \frac{[B^-][H^+]}{[HB]} \qquad K_a = \frac{[B][H^+]}{[HB^+]}$$

a neutral acid   a positively charged acid

Carboxylic acids, alcohols, and phenols are neutral in their acidic forms (HB) and charged in their basic forms (B^-). A polar protic solvent will stabilize B^- and H^+ more than it will stabilize HB, thereby increasing K_a. Therefore, K_a will be larger in water than in a less polar solvent, which means that the pK_a value will be lower in water. So the pK_a values of carboxylic acids, alcohols, and phenols determined in a less polar solvent than water will be higher than those determined in water.

Ammonium ions and anilinium ions are charged in their acidic forms (HB^+) and neutral in their basic forms (B). A polar solvent will stabilize HB^+ and H^+ more than it will stabilize B. Because HB^+ is stabilized slightly more than H^+, K_a will decrease, which means that the pK_a value will be higher in water than in a less polar solvent. So the pK_a values of ammonium ions and anilinium ions determined in a less polar solvent than water will be slightly lower than those determined in water.

PROBLEM 30◆

Would you expect acetate ion ($CH_3CO_2^-$) to be a more reactive nucleophile in an S_N2 reaction carried out in methanol or in dimethyl sulfoxide?

PROBLEM 31◆

Under which of the following reaction conditions would (*R*)-2-chlorobutane form the most (*R*)-2-butanol: HO^- in 50% water and 50% ethanol or HO^- in 100% ethanol?

10.11 Biological Methylating Reagents

If an organic chemist wanted to put a methyl group on a nucleophile, methyl iodide would most likely be the methylating agent used. Of the methyl halides, methyl iodide has the most easily displaced leaving group because I^- is the weakest base of the halide ions. In addition, methyl iodide is a liquid, so it is easier to handle than methyl bromide or methyl chloride. The reaction would be a simple S_N2 reaction.

$$\ddot{N}u + CH_3\text{—}I \longrightarrow CH_3\text{—}Nu + I^-$$

In a living cell, however, methyl iodide is not available. It is only slightly soluble in water, so is not found in the predominantly aqueous environments of biological systems. Instead, biological systems use *S*-adenosylmethionine (SAM) and N^5-methyltetrahydrofolate as methylating agents, both of which are soluble in water. Although they look much more complicated than methyl iodide, they perform the same function—the transfer of a methyl group to a nucleophile. Notice that the methyl group in each of these methylating agents is attached to a positively charged atom. This means that the methyl groups are attached to very good leaving groups, so biological methylation can take place at a reasonable rate.

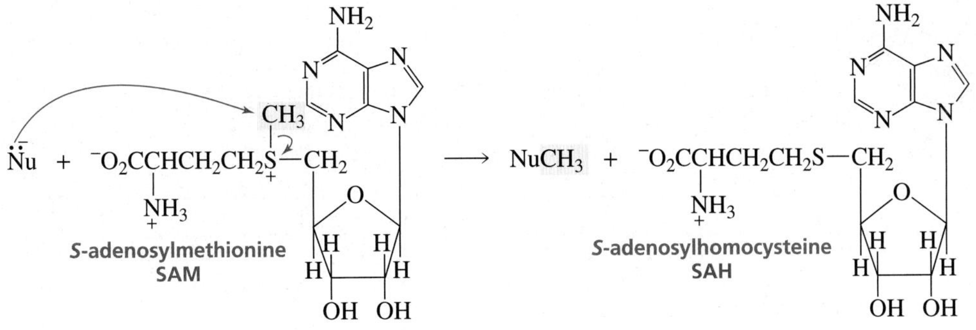

$\ddot{Nu}^- + {}^-O_2CCHCH_2CH_2\overset{+}{S}(CH_3)–CH_2$–(adenosine) ⟶ $NuCH_3 + {}^-O_2CCHCH_2CH_2S–CH_2$–(adenosine)

S-adenosylmethionine
SAM

S-adenosylhomocysteine
SAH

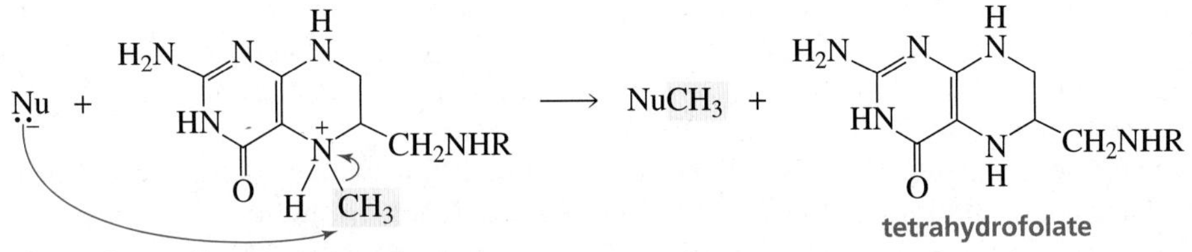

Nu^- + *N*[5]-methyltetrahydrofolate ⟶ $NuCH_3$ + tetrahydrofolate

N[5]-methyltetrahydrofolate

tetrahydrofolate

Biological systems use SAM to convert norepinephrine (noradrenaline) into epinephrine (adrenaline). The reaction is a simple methylation reaction. Norepinephrine and epinephrine are hormones that are released into the bloodstream in response to stress. Epinephrine is the more potent hormone of the two.

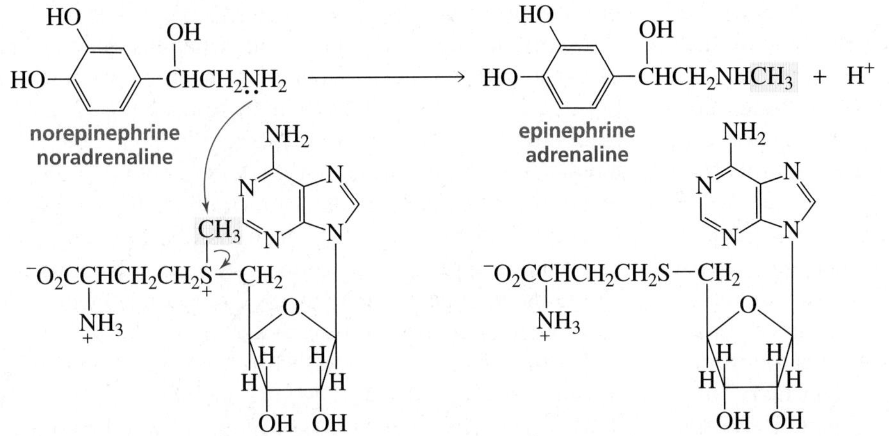

norepinephrine
noradrenaline

epinephrine
adrenaline

$+ H^+$

The conversion of phosphatidylethanolamine, a component of cell membranes, into phosphatidylcholine requires three methylations by three equivalents of SAM. (Biological cell membranes will be discussed in Section 26.4; the use of *N*[5]-methyltetrahydrofolate as a biological methylating agent is discussed in more detail in Section 25.8.)

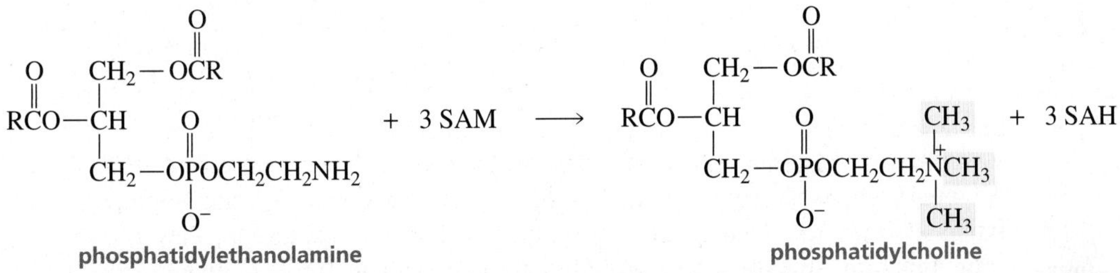

phosphatidylethanolamine + 3 SAM ⟶ phosphatidylcholine + 3 SAH

phosphatidylethanolamine

phosphatidylcholine

S-ADENOSYLMETHIONINE: A NATURAL ANTIDEPRESSANT

S-Adenosylmethionine is sold in many health food and drug stores as a treatment for depression and arthritis. It is marketed under the name SAMe (pronounced Sammy). Although SAMe has been used clinically in Europe for more than two decades, it has not been rigorously evaluated in the United States and is not approved by the FDA. It can be sold, however, because the FDA does not prohibit the sale of most naturally occurring substances, as long as the marketer does not make therapeutic claims. SAMe has also been found to be effective in the treatment of liver diseases—diseases caused by alcohol and the hepatitis C virus. The attenuation of liver injuries is accompanied by increased levels of glutathione in the liver. Glutathione is an important antioxidant (Section 23.8). SAM is required for the synthesis of cysteine (an amino acid), which, in turn, is required for the synthesis of glutathione.

Summary

Alkyl halides undergo two kinds of **nucleophilic substitution reactions**: S_N2 and S_N1. In both reactions, a nucleophile substitutes for a halogen, which is called a **leaving group**. An S_N2 reaction is bimolecular—two molecules are involved in the rate-limiting step; an S_N1 reaction is unimolecular—one molecule is involved in the rate-limiting step.

The rate of an **S_N2 reaction** depends on the concentration of both the alkyl halide and the nucleophile. An S_N2 reaction is a one-step reaction: The nucleophile attacks the back side of the carbon that is attached to the halogen. The reaction proceeds in the direction that allows the stronger base to displace the weaker base; it is reversible only if the difference between the basicities of the nucleophile and the leaving group is small. The rate of an S_N2 reaction depends on steric hindrance: The bulkier the groups at the back side of the carbon undergoing attack, the slower is the reaction. Tertiary carbocations, therefore, cannot undergo S_N2 reactions. An S_N2 reaction takes place with **inversion of configuration**.

The rate of an **S_N1 reaction** depends only on the concentration of the alkyl halide. The halogen departs in the first step, forming a carbocation that is attacked by a nucleophile in the second step. Therefore, carbocation rearrangements can occur. The rate of an S_N1 reaction depends on the ease of carbocation formation. Tertiary alkyl halides, therefore, are more reactive than secondary alkyl halides since tertiary carbocations are more stable than secondary carbocations. Primary carbocations are so unstable that primary alkyl halides cannot undergo S_N1 reactions. An S_N1 reaction takes place with racemization. Most S_N1 reactions are **solvolysis** reactions: The solvent is the nucleophile.

The rates of both S_N2 and S_N1 reactions are influenced by the nature of the leaving group. Weak bases are the best leaving groups because weak bases are best able to accommodate the negative charge. Thus, the weaker the basicity of the leaving group, the faster the reaction will occur. Therefore, the relative reactivities of alkyl halides that differ only in the halogen atom are RI > RBr > RCl > RF in both S_N2 and S_N1 reactions.

Basicity is a measure of how well a compound shares its lone pair with a proton. **Nucleophilicity** is a measure of how readily a compound is able to attack an electron-deficient atom. In comparing molecules with the same attacking atom or with attacking atoms of the same size, the stronger base is a better nucleophile. If the attacking atoms are very different in size, the relationship between basicity and nucleophilicity depends on the solvent. In protic solvents, stronger bases are poorer nucleophiles because of **ion–dipole interactions** between the ion and the solvent.

Methyl halides and primary alkyl halides undergo only S_N2 reactions, tertiary alkyl halides undergo only S_N1 reactions, vinylic and aryl halides undergo neither S_N2 nor S_N1 reactions, secondary alkyl halides and benzylic and allylic halides (unless they are tertiary) undergo both S_N1 and S_N2 reactions. When the structure of the alkyl halide allows it to undergo both S_N2 and S_N1 reactions, the S_N2 reaction is favored by a high concentration of a good nucleophile in an aprotic polar solvent, while the S_N1 reaction is favored by a poor nucleophile in a protic polar solvent.

Protic solvents (H_2O, ROH) are hydrogen bond donors; **aprotic solvents** (DMF, DMSO) are not hydrogen bond donors. The **dielectric constant** of a solvent tells how well the solvent insulates opposite charges from one another. Increasing the polarity of the solvent will decrease the rate of the reaction if one or more reactants in the rate-determining step are charged and will increase the rate of the reaction if none of the reactants in the rate-determining step is charged.

Summary of Reactions

1. S_N2 reaction: a one-step mechanism

$$\ddot{\mathrm{Nu}}^- + -\overset{|}{\underset{|}{\mathrm{C}}}-\mathrm{X} \longrightarrow -\overset{|}{\underset{|}{\mathrm{C}}}-\mathrm{Nu} + \mathrm{X}^-$$

Relative reactivities of alkyl halides: $CH_3X > 1° > 2° > 3°$.

Only the inverted product is formed.

2. S_N1 reaction: a two-step mechanism with a carbocation intermediate

$$-\overset{|}{\underset{|}{C}}-X \longrightarrow -\overset{|}{\underset{|}{C^+}} + X^- \xrightarrow{\ddot{Nu}^-} -\overset{|}{\underset{|}{C}}-Nu$$

Relative reactivities of alkyl halides: 3° > 2° > 1° > CH_3X.

Both the inverted and noninverted products are formed.

Key Terms

aprotic solvent (p. 389)
back-side attack (p. 363)
base (p. 368)
basicity (p. 368)
bimolecular (p. 363)
complete racemization (p. 381)
dielectric constant (p. 389)
elimination reaction (p. 360)
first-order reaction (p. 375)
intimate ion pair (p. 381)
inversion of configuration (p. 366)
ion–dipole interaction (p. 369)
kinetics (p. 362)
leaving group (p. 360)
Le Châtelier's principle (p. 373)
nucleophile (p. 368)
nucleophilicity (p. 368)
nucleophilic substitution reaction (p. 362)
partial racemization (p. 381)
protic solvent (p. 389)
rate constant (p. 362)
rate law (p. 362)
second-order reaction (p. 362)
S_N1 reaction (p. 376)
S_N2 reaction (p. 363)
solvolysis (p. 379)
steric effects (p. 364)
steric hindrance (p. 364)
substitution reaction (p. 360)
unimolecular (p. 376)

Problems

32. Give the product of the reaction of methyl bromide with each of the following nucleophiles:
    a. HO^- b. $^-NH_2$ c. H_2S d. HS^- e. CH_3O^- f. CH_3NH_2

33. a. Indicate how each of the following factors affects an S_N1 reaction:
    b. Indicate how each of the following factors affects an S_N2 reaction:
    1. the structure of the alkyl halide
    2. the reactivity of the nucleophile
    3. the concentration of the nucleophile
    4. the solvent

34. Which is a better nucleophile in methanol?
    a. H_2O or HO^-
    b. NH_3 or $^-NH_2$
    c. H_2O or H_2S
    d. HO^- or HS^-
    e. I^- or Br^-
    f. Cl^- or Br^-

35. For each of the pairs in Problem 34, indicate which is a better leaving group.

36. What nucleophiles could be used to react with butyl bromide to prepare the following compounds?
    a. $CH_3CH_2CH_2CH_2OH$
    b. $CH_3CH_2CH_2CH_2OCH_3$
    c. $CH_3CH_2CH_2CH_2SH$
    d. $CH_3CH_2CH_2CH_2SCH_2CH_3$
    e. $CH_3CH_2CH_2CH_2NHCH_3$
    f. $CH_3CH_2CH_2CH_2C{\equiv}N$
    g. $CH_3CH_2CH_2CH_2O\overset{O}{\overset{\|}{C}}CH_3$
    h. $CH_3CH_2CH_2CH_2C{\equiv}CCH_3$

37. Rank the following compounds in order of *decreasing* nucleophilicity:
    a. $CH_3\overset{O}{\overset{\|}{C}}O^-$, $CH_3CH_2S^-$, $CH_3CH_2O^-$ in methanol
    b. (benzene ring)$-O^-$ and (cyclohexane ring)$-O^-$ in DMSO
    c. H_2O and NH_3 in methanol
    d. Br^-, Cl^-, I^- in methanol

38. The pK_a of acetic acid in water is 4.76 (Section 1.17). What effect would a decrease in the polarity of the solvent have on the pK_a? Why?

39. For each of the following reactions, give the substitution products; if the products can exist as stereoisomers, show what stereoisomers are obtained:
   a. (*R*)-2-bromopentane + high concentration of CH_3O^-
   b. (*R*)-2-bromopentane + CH_3OH
   c. *trans*-1-chloro-2-methylcyclohexane + high concentration of CH_3O^-
   d. *trans*-1-chloro-2-methylcyclohexane + CH_3OH
   e. 3-bromo-2-methylpentane + CH_3OH
   f. 3-bromo-3-methylpentane + CH_3OH

40. Would you expect methoxide ion to be a better nucleophile if it were dissolved in CH_3OH or if it were dissolved in dimethyl sulfoxide (DMSO)? Why?

41. Which reaction in each of the following pairs will take place more rapidly?

a. Cl $\xrightarrow{CH_3S^-}$ S + Cl⁻

Cl $\xrightarrow{(CH_3)_2CHS^-}$ S + Cl⁻

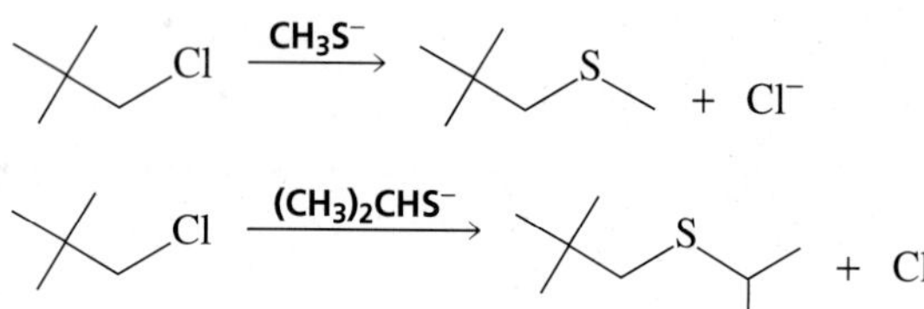

b. Cl $\xrightarrow{HO^-}$ OH + Cl⁻

O Cl $\xrightarrow{HO^-}$ O OH + Cl⁻

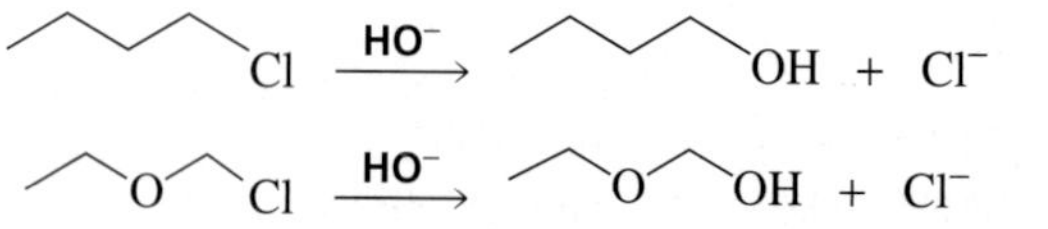

c. Cl $\xrightarrow{H_2O}$ OH + HCl

Cl $\xrightarrow{H_2O}$ OH + HCl

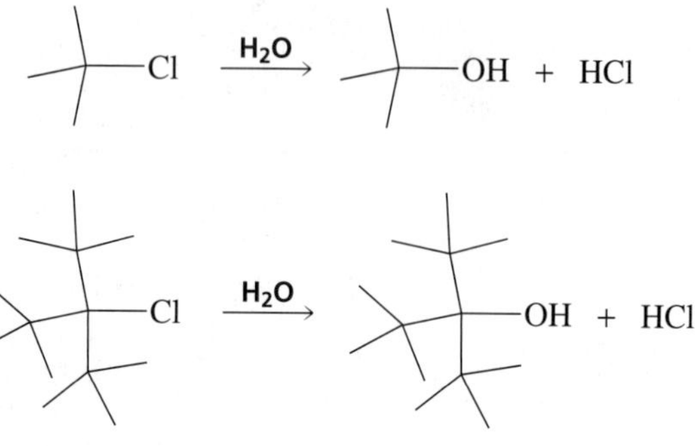

d. $(CH_3)_3CBr \xrightarrow{H_2O} (CH_3)_3COH + HBr$

$(CH_3)_3CBr \xrightarrow{CH_3CH_2OH} (CH_3)_3COCH_2CH_3 + HBr$

42. Which of the following compounds would you expect to be more reactive in an S_N2 reaction?

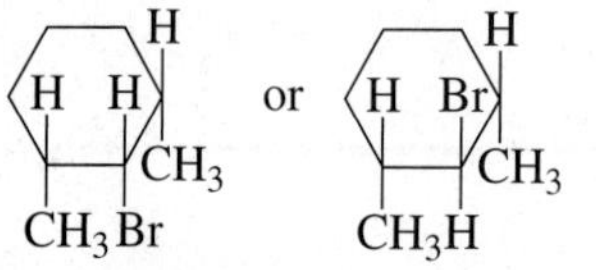

H, H H, CH₃, CH₃ Br or H, H Br, CH₃, CH₃ H

43. In Section 10.11, we saw that *S*-adenosylmethionine (SAM) methylates the nitrogen atom of noradrenaline to form adrenaline, a more potent hormone. If SAM methylates an OH group on the benzene ring instead, it completely destroys noradrenaline's activity. Give the mechanism for the methylation of the OH group by SAM.

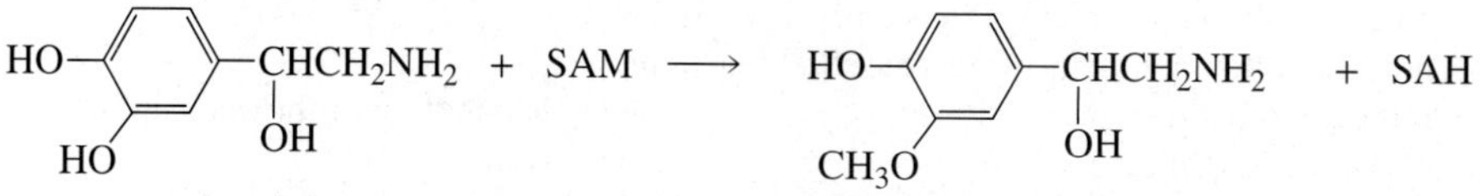

HO, HO —$CHCH_2NH_2$ (OH) + SAM ⟶ HO, CH_3O —$CHCH_2NH_2$ (OH) + SAH

norepinephrine **a biologically inactive compound**

44. For each of the following reactions, give the substitution products; if the products can exist as stereoisomers, show what stereoisomers are obtained:
   a. (2*S*,3*S*)-2-chloro-3-methylpentane + high concentration of CH_3O^-
   b. (2*S*,3*R*)-2-chloro-3-methylpentane + high concentration of CH_3O^-
   c. (2*R*,3*S*)-2-chloro-3-methylpentane + high concentration of CH_3O^-
   d. (2*R*,3*R*)-2-chloro-3-methylpentane + high concentration of CH_3O^-
   e. 3-chloro-2,2-dimethylpentane + CH_3CH_2OH
   f. benzyl bromide + CH_3CH_2OH

45. Give the substitution products obtained when each of the following compounds is added to a solution of sodium acetate in acetic acid.
   a. 2-chloro-2-methyl-3-hexene
   b. 3-bromo-1-methylcyclohexene

46. The rate of reaction of methyl iodide with quinuclidine was measured in nitrobenzene, and then the rate of reaction of methyl iodide with triethylamine was measured in the same solvent.
   a. Which reaction had the larger rate constant?
   b. The same experiment was done using isopropyl iodide instead of methyl iodide. Which reaction had the larger rate constant?
   c. Which alkyl halide has the larger $k_{triethylamine}/k_{quinuclidine}$ ratio?

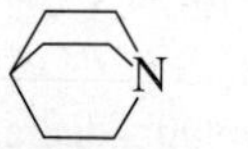

N **quinuclidine**

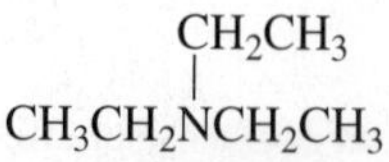

$CH_3CH_2NCH_2CH_3$ (with CH_2CH_3 on N) **triethylamine**

47. Only one bromoether (ignoring stereoisomers) is obtained from the reaction of the following alkyl dihalide with methanol:

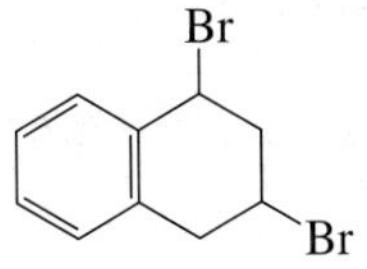

Give the structure of the ether.

48. Starting with cyclohexane, how could the following compounds be prepared?
   a. cyclohexyl bromide
   b. methoxycyclohexane
   c. cyclohexanol

49. For each of the following reactions, give the substitution products, assuming that all the reactions are carried out under S_N2 conditions; if the products can exist as stereoisomers, show what stereoisomers are formed:
   a. (3*S*,4*S*)-3-bromo-4-methylhexane + CH_3O^-
   b. (3*S*,4*R*)-3-bromo-4-methylhexane + CH_3O^-
   c. (3*R*,4*R*)-3-bromo-4-methylhexane + CH_3O^-
   d. (3*R*,4*S*)-3-bromo-4-methylhexane + CH_3O^-

50. Explain why tetrahydrofuran can solvate a positively charged species better than diethyl ether can.

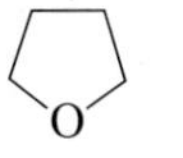

tetrahydrofuran

$CH_3CH_2OCH_2CH_3$
diethyl ether

51. Propose a mechanism for each of the following reactions:

a. 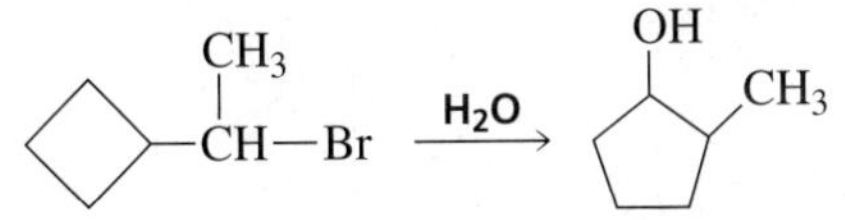 (cyclobutyl)$CH(CH_3)$—Br $\xrightarrow{H_2O}$ 1-methylcyclopentanol

b. 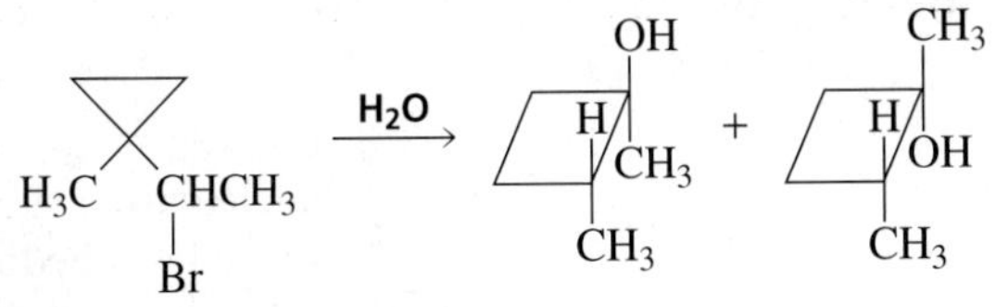 (cyclopropyl with H_3C and $CHCH_3$—Br) $\xrightarrow{H_2O}$ products (cyclobutanes with H, OH, CH_3, CH_3) + (H, CH_3, OH, CH_3)

52. Which of the following will react faster in an S_N1 reaction?

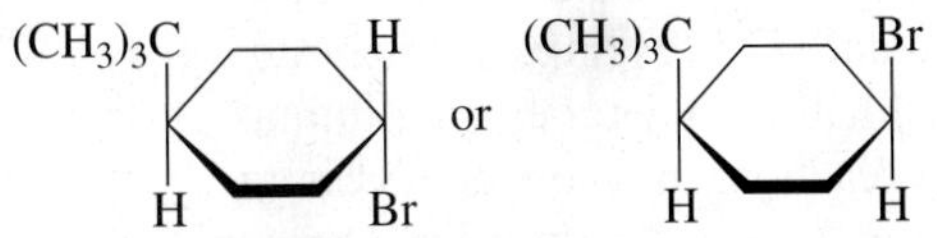

$(CH_3)_3C$, H, H, Br (cyclohexane) or $(CH_3)_3C$, Br, H, H (cyclohexane)

53. Alkyl halides have been used as insecticides since the discovery of DDT in 1939. DDT was the first compound to be found that had a high toxicity to insects and a relatively low toxicity to mammals. In 1972, DDT was banned in the United States because it is a long-lasting compound and its widespread use caused accumulation in substantial concentrations in wildlife. Chlordane is an alkyl halide insecticide that is used to protect wooden buildings from termites. Chlordane can be synthesized from two reactants in a one-step reaction. One of the reactants is hexachlorocyclopentadiene. What is the other reactant? (*Hint:* See Section 8.8.)

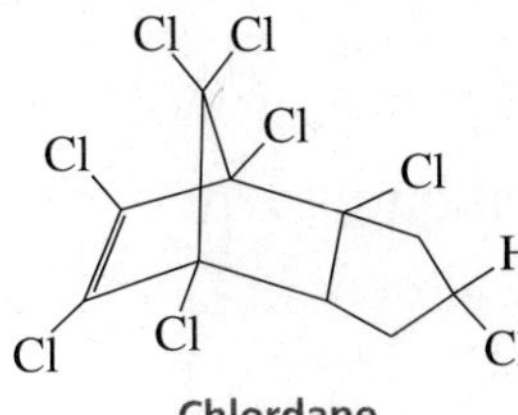

Chlordane

54. Explain why the following alkyl halide does not undergo a substitution reaction, regardless of the conditions under which the reaction is run:

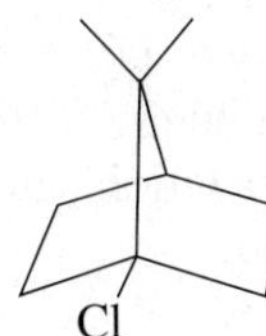

11 Elimination Reactions of Alkyl Halides • Competition Between Substitution and Elimination

$CH_3CH_2CHCH_2CH_3$ (with Cl on C3) + CH_3O^- → (E)-H_3C–CH=CH–CH_2CH_3 + (Z)-H_3C–CH=CH–CH_2CH_3

In addition to undergoing the nucleophilic substitution reactions described in Chapter 10, alkyl halides undergo *elimination* reactions. In an **elimination reaction**, groups are eliminated from a reactant. For example, when an alkyl halide undergoes an elimination reaction, the halogen (X) is removed from one carbon and a proton is removed from an adjacent carbon. A double bond is formed between the two carbons from which the atoms are eliminated. Therefore, *the product of an elimination reaction is an alkene.*

$$CH_3CH_2CH_2X + Y^- \xrightarrow{\text{substitution}} CH_3CH_2CH_2Y + X^-$$

$$CH_3CH_2CH_2X + Y^- \xrightarrow{\text{elimination}} CH_3CH{=}CH_2 + HY + X^-$$

new double bond

In this chapter, we will first look at the elimination reactions of alkyl halides. Then we will examine the factors that determine whether an alkyl halide will undergo a substitution reaction, an elimination reaction, or both a substitution and an elimination reaction.

INVESTIGATING NATURALLY OCCURRING ORGANOHALIDES

Organohalides isolated from marine organisms have been found to have interesting and potent biological activity. Compounds produced in nature are called *natural products*. *Bengamide A* is a natural product that comes from an orange encrusting sponge. This compound, as well as a host of analogs, has unique antitumor properties that are currently being exploited in the development of new anticancer drugs. *Jaspamide*, also found in a sponge, modulates the formation and depolymerization of actin microtubules. Microtubules are found in all cells and are used for motile events, such as transportation of vesicles, migration, and cell division. Jaspamide is being used to further our understanding of these processes. Notice that each of these naturally occurring compounds has six asymmetric carbons.

cyclocinamide A

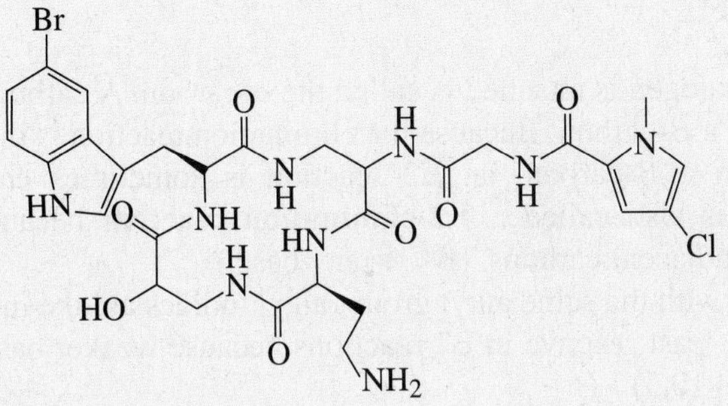

jasplankinolide

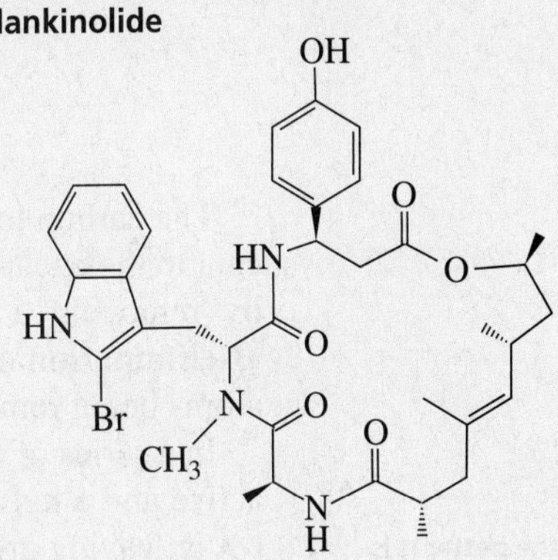

11.1 The E2 Reaction

Just as there are two important nucleophilic substitution reactions—S_N1 and S_N2—there are two important elimination reactions: E1 and E2. The reaction of *tert*-butyl bromide with hydroxide ion is an example of an **E2 reaction**; "E" stands for *elimination* and "2" stands for *bimolecular*. The product of an elimination reaction is an alkene.

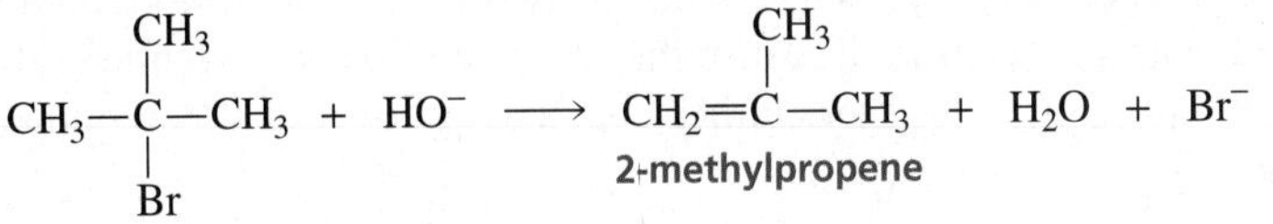

$$CH_3-\underset{Br}{\overset{CH_3}{C}}-CH_3 + HO^- \longrightarrow CH_2=\overset{CH_3}{C}-CH_3 + H_2O + Br^-$$

tert-butyl bromide → **2-methylpropene**

The rate of an E2 reaction depends on the concentrations of both *tert*-butyl bromide and hydroxide ion. It is, therefore, a second-order reaction (Section 10.2).

rate = *k*[alkyl halide][base]

The rate law tells us that both *tert*-butyl bromide and hydroxide ion are involved in the rate-determining step of the reaction. The following mechanism agrees with the observed second-order kinetics:

mechanism of the E2 reaction

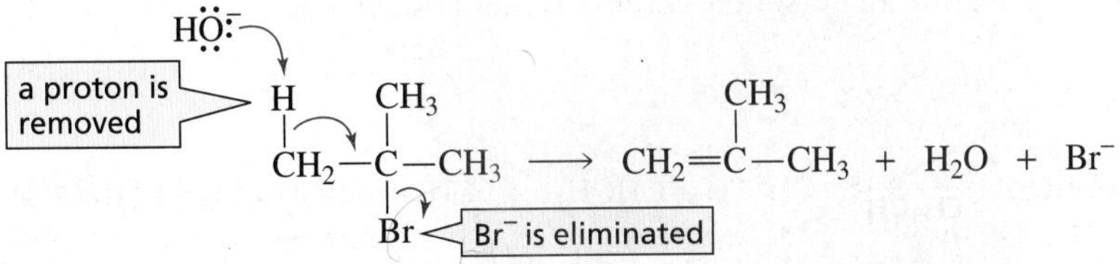

$$H\ddot{O}:^- \quad \underset{\text{Br}}{\overset{\text{H}\quad CH_3}{CH_2-C-CH_3}} \longrightarrow CH_2=\overset{CH_3}{C}-CH_3 + H_2O + Br^-$$

a proton is removed

Br^- is eliminated

We see that an E2 reaction is a concerted, one-step reaction: The proton and the bromide ion are removed in the same step, so no intermediate is formed.

In an E2 reaction of an alkyl halide, a base removes a proton from a carbon that is adjacent to the carbon bonded to the halogen. As the proton is removed, the electrons that the hydrogen shared with carbon move toward the carbon bonded to the halogen. As these electrons move toward the carbon, the halogen leaves, taking its bonding electrons with it. The electrons that were bonded to the hydrogen in the reactant have formed a π bond in the product. Removal of a proton and a halide ion is called **dehydrohalogenation**.

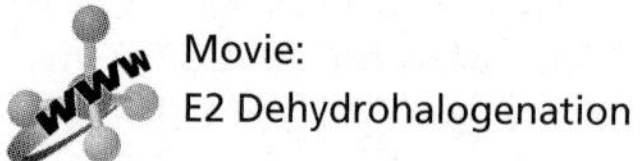

a base → $B:^-$ H; α-carbon; β-carbon

$$\text{RCH}-\text{CHR (Br)} \longrightarrow \text{RCH}{=}\text{CHR} + \text{BH} + \text{Br}^-$$

The carbon to which the halogen is attached is called the α-carbon. A carbon adjacent to the α-carbon is called a β-carbon. Because the elimination reaction is initiated by removing a proton from a β-carbon, an E2 reaction is sometimes called a **β-elimination reaction**. It is also called a **1,2-elimination reaction** because the atoms being removed are on adjacent carbons. ($B:^-$ is any base.)

In a series of alkyl halides with the same alkyl group, alkyl iodides are the most reactive and alkyl fluorides the least reactive in E2 reactions because weaker bases are better leaving groups (Section 10.3).

The weaker the base, the better it is as a leaving group.

relative reactivities of alkyl halides in an E2 reaction

most reactive → RI > RBr > RCl > RF ← least reactive

11.2 The Regioselectivity of the E2 Reaction

An alkyl halide such as 2-bromopropane has two β-carbons from which a proton can be removed in an E2 reaction. Because the two β-carbons are identical, the proton can be removed with equal ease from either one. The product of this elimination reaction is propene.

β-carbons

$$CH_3CH(Br)CH_3 + CH_3O^- \longrightarrow CH_3CH{=}CH_2 + CH_3OH + Br^-$$

2-bromopropane → **propene**

In contrast, 2-bromobutane has two structurally different β-carbons from which a proton can be removed. So when 2-bromobutane reacts with a base, two elimination products are formed: 2-butene and 1-butene. This E2 reaction is *regioselective* because more of one constitutional isomer is formed than the other.

β-carbons

$$CH_3CH(Br)CH_2CH_3 + CH_3O^- \xrightarrow{CH_3OH} CH_3CH{=}CHCH_3 + CH_2{=}CHCH_2CH_3 + CH_3OH + Br^-$$

2-bromobutane → **2-butene 80% (mixture of *E* and *Z*)** + **1-butene 20%**

What is the regioselectivity of an E2 reaction? In other words, what are the factors that dictate which of the two elimination products will be formed in greater yield?

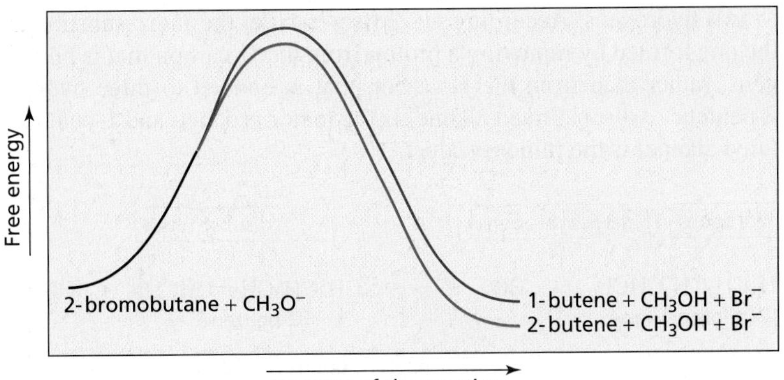

Figure 11.1
Reaction coordinate diagram for the E2 reaction of 2-bromobutane and methoxide ion.

To answer this question, we must determine which of the alkenes is formed more easily—that is, which is formed faster. The reaction coordinate diagram for the E2 reaction of 2-bromobutane is shown in Figure 11.1.

In the transition state leading to an alkene, the C—H and C—Br bonds are partially broken and the double bond is partially formed (partially broken and partially formed bonds are indicated by dashed lines), giving the transition state an alkene-like structure. Because the transition state has an alkene-like structure, any factors that stabilize the alkene will also stabilize the transition state leading to its formation, allowing the alkene to be formed faster. The difference in the rate of formation of the two alkenes is not very great. Consequently, both products are formed, but the *more stable* of the two alkenes will be the major product of the reaction.

The major product of an E2 reaction is the most stable alkene.

$$\overset{\delta-}{OCH_3}\cdots H\cdots CH_3CH\text{---}CHCH_3\cdots \overset{\delta-}{Br}$$

transition state leading to 2-butene

more stable

$$\overset{\delta-}{OCH_3}\cdots H\cdots CH_2\text{---}CHCH_2CH_3\cdots \overset{\delta-}{Br}$$

transition state leading to 1-butene

less stable

We know that the stability of an alkene depends on the number of alkyl substituents bonded to its sp^2 carbons: The greater the number of substituents, the more stable is the alkene (Section 4.11). Therefore, 2-butene, with a total of two methyl substituents bonded to its sp^2 carbons, is more stable than 1-butene, with one ethyl substituent.

The reaction of 2-bromo-2-methylbutane with hydroxide ion produces both 2-methyl-2-butene and 2-methyl-1-butene. Because 2-methyl-2-butene is the more substituted alkene (it has a greater number of alkyl substituents bonded to its sp^2 carbons), it is the more stable of the two alkenes and, therefore, is the major product of the elimination reaction.

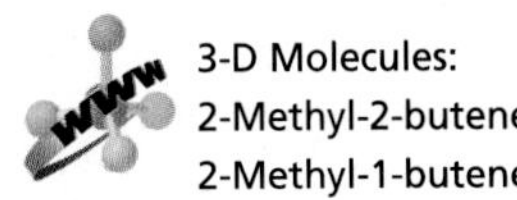

3-D Molecules: 2-Methyl-2-butene; 2-Methyl-1-butene

$$CH_3C(CH_3)(Br)CH_2CH_3 + HO^- \xrightarrow{H_2O} CH_3C(CH_3){=}CHCH_3 + CH_2{=}C(CH_3)CH_2CH_3 + H_2O + Br^-$$

2-bromo-2-methylbutane → 2-methyl-2-butene 70% + 2-methyl-1-butene 30%

Alexander M. Zaitsev, a nineteenth-century Russian chemist, devised a shortcut to predict the more substituted alkene product. He pointed out *that the more substituted alkene product is obtained when a proton is removed from the β-carbon that is bonded to the fewest hydrogens*. This is called Zaitsev's rule. In 2-chloropentane, for example, one β-carbon is bonded to three hydrogens and the other β-carbon is bonded

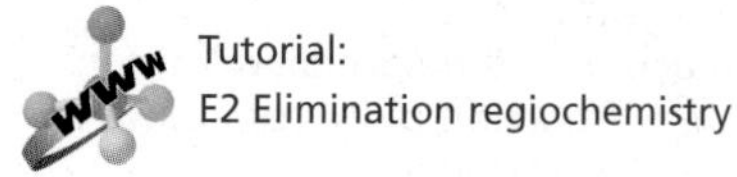

to two hydrogens. According to **Zaitsev's rule**, the more substituted alkene will be the one formed by removing a proton from the β-carbon that is bonded to two hydrogens, rather than from the β-carbon that is bonded to three hydrogens. Therefore, 2-pentene (a disubstituted alkene) is the major product, and 1-pentene (a monosubstituted alkene) is the minor product.

$$CH_3CH_2CH_2\underset{}{\overset{Cl}{C}}HCH_3 + HO^- \longrightarrow CH_3CH_2CH{=}CHCH_3 + CH_3CH_2CH_2CH{=}CH_2$$

2 β-hydrogens; 3 β-hydrogens; disubstituted; monosubstituted

2-chloropentane → **2-pentene** 67% (mixture of *E* and *Z*) + **1-pentene** 33%

Alexander M. Zaitsev (1841–1910) *was born in Kazan, Russia. The German transliteration, Saytzeff, is sometimes used for his family name. He received a Ph.D. from the University of Leipzig in 1866. He was a professor of chemistry first at the University of Kazan and later at the University of Kiev.*

Because elimination from a tertiary alkyl halide typically leads to a more substituted alkene than does elimination from a secondary alkyl halide, and elimination from a secondary alkyl halide generally leads to a more substituted alkene than does elimination from a primary alkyl halide, the relative reactivities of alkyl halides in an E2 reaction are as follows:

relative reactivities of alkyl halides in an E2 reaction

tertiary alkyl halide > secondary alkyl halide > primary alkyl halide

$$RCH_2CR(R)Br \longrightarrow RCH{=}CR(R) \qquad RCH_2CHRBr \longrightarrow RCH{=}CHR \qquad RCH_2CH_2Br \longrightarrow RCH{=}CH_2$$

three alkyl substituents; two alkyl substituents; one alkyl substituent

The most stable alkene is generally (but not always) the most substituted alkene.

Zaitsev's rule leads to the most substituted alkene.

Keep in mind that the major product of an E2 reaction is the *more stable alkene,* and Zaitsev's rule is just a shortcut to determine which of the possible alkene products is the *more substituted alkene*. The more substituted alkene is not, however, always the more stable alkene. In the following reactions, the conjugated alkene is the more stable alkene even though it is not the most substituted alkene. The major product of each reaction is, therefore, the conjugated alkene because, being more stable, it is more easily formed.

$$CH_2{=}CHCH_2CH(Cl)CH(CH_3)CH_3 \xrightarrow{HO^-} CH_2{=}CHCH{=}CHCH(CH_3)CH_3 + CH_2{=}CHCH_2CH{=}C(CH_3)CH_3 + H_2O + Cl^-$$

4-chloro-5-methyl-1-hexene → **5-methyl-1,3-hexadiene** (a conjugated diene, major product) + **5-methyl-1,4-hexadiene** (an isolated diene, minor product)

$$C_6H_5CH_2CH(Br)CH(CH_3)CH_3 \xrightarrow{HO^-} C_6H_5CH{=}CHCH(CH_3)CH_3 + C_6H_5CH_2CH{=}C(CH_3)CH_3 + H_2O + Br^-$$

2-bromo-3-methyl-1-phenylbutane → **3-methyl-1-phenyl-1-butene** (the double bond is conjugated with the benzene ring, major product) + **3-methyl-1-phenyl-2-butene** (the double bond is not conjugated with the benzene ring, minor product)

Zaitsev's rule cannot be used to predict the major products of the foregoing reactions because it does not take into account the fact that conjugated double bonds are more stable than isolated double bonds (Section 8.3). Therefore, if there is a double bond or a benzene ring in the alkyl halide, do not use Zaitsev's rule to predict the major (most stable) product of an elimination reaction.

In some elimination reactions, the *less stable alkene* is the major product. For example, if the base in an E2 reaction is sterically bulky and the approach to the alkyl halide is sterically hindered, the base will preferentially remove the most accessible hydrogen. In the following reaction, it is easier for the bulky *tert*-butoxide ion to remove one of the more exposed terminal hydrogens, which leads to formation of the less substituted alkene. Because the less substituted alkene is more easily formed, it is the major product of the reaction.

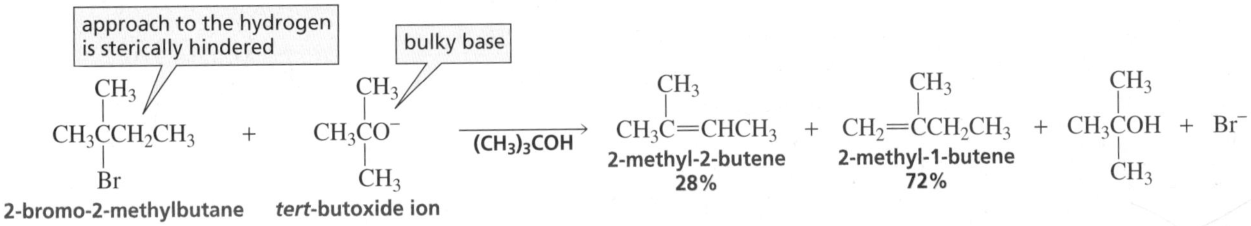

approach to the hydrogen is sterically hindered

bulky base

$$CH_3C(CH_3)(Br)CH_2CH_3 + CH_3C(CH_3)_2O^- \xrightarrow{(CH_3)_3COH} CH_3C(CH_3){=}CHCH_3 + CH_2{=}C(CH_3)CH_2CH_3 + CH_3C(CH_3)_2OH + Br^-$$

2-bromo-2-methylbutane; *tert*-butoxide ion; 2-methyl-2-butene 28%; 2-methyl-1-butene 72%

The data in Table 11.1 show that when a sterically hindered alkyl halide undergoes an E2 reaction with a variety of alkoxide ions, the percentage of the less substituted alkene increases as the size of the base increases.

If the alkyl halide is not sterically hindered and the base is only moderately hindered, the major product will still be the more stable product. For example, the major

Table 11.1 Effect of the Steric Properties of the Base on the Distribution of Products in an E2 Reaction

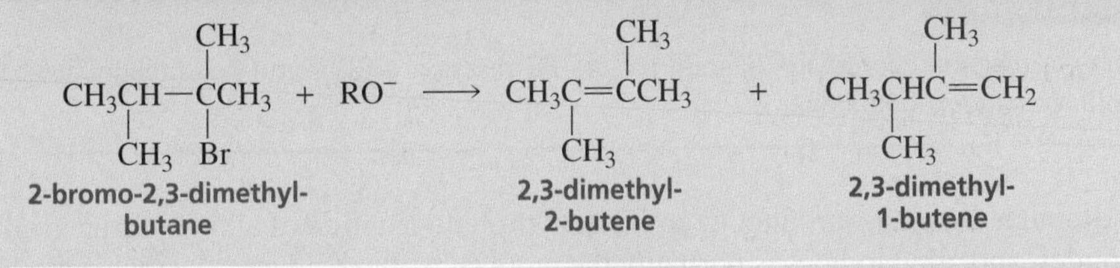

$$CH_3CH(CH_3){-}C(CH_3)(Br)CH_3 + RO^- \longrightarrow CH_3C(CH_3){=}C(CH_3)CH_3 + CH_3CH(CH_3)C(CH_3){=}CH_2$$

2-bromo-2,3-dimethylbutane; 2,3-dimethyl-2-butene; 2,3-dimethyl-1-butene

Base	More substituted product	Less substituted product
$CH_3CH_2O^-$	79%	21%
$CH_3C(CH_3)_2O^-$	27%	73%
$CH_3C(CH_3)(CH_2CH_3)O^-$	19%	81%
$CH_3CH_2C(CH_2CH_3)_2O^-$	8%	92%

product obtained from the reaction of 2-iodobutane and *tert*-butoxide ion is 2-butene. In other words, it takes a lot of hindrance for the less stable product to be the major product.

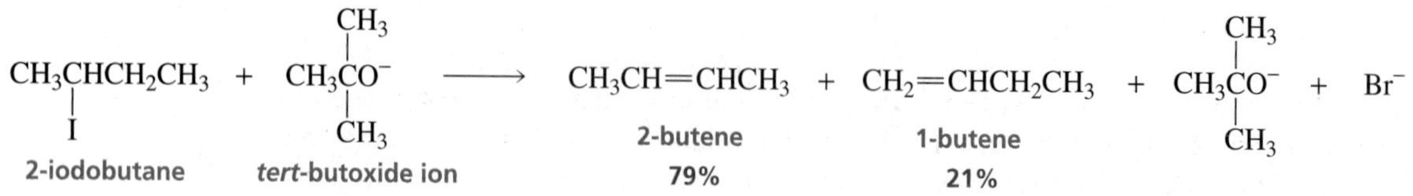

$$\underset{\text{2-iodobutane}}{\underset{\text{I}}{CH_3CHCH_2CH_3}} + \underset{\textit{tert}\text{-butoxide ion}}{(CH_3)_3CO^-} \longrightarrow \underset{\substack{\text{2-butene}\\ 79\%}}{CH_3CH{=}CHCH_3} + \underset{\substack{\text{1-butene}\\ 21\%}}{CH_2{=}CHCH_2CH_3} + (CH_3)_3COH + Br^-$$

PROBLEM 1◆

Which of the alkyl halides is more reactive in an E2 reaction?

a. 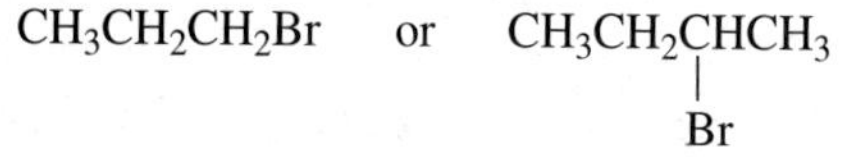 $CH_3CH_2CH_2Br$ or $CH_3CH_2CHBrCH_3$

b. 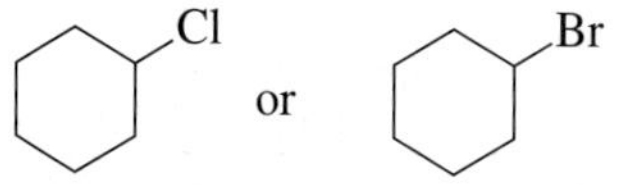 chlorocyclohexane or bromocyclohexane

c. 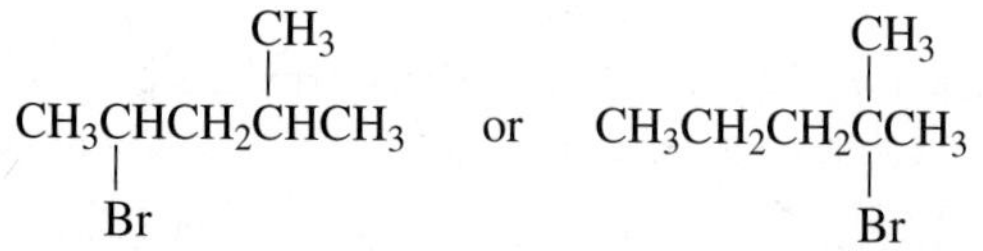 $CH_3CHBrCH_2CH(CH_3)CH_3$ or $CH_3CH_2CH_2CBr(CH_3)CH_3$

d. 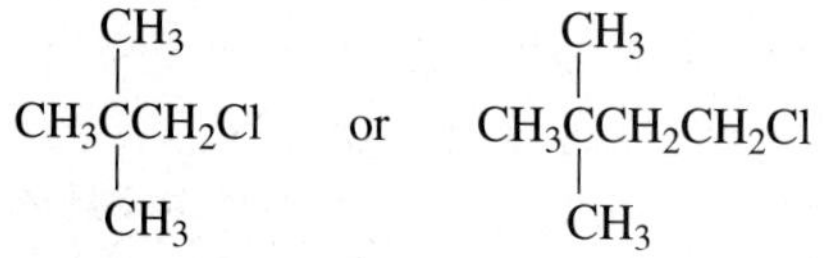 $CH_3C(CH_3)_2CH_2Cl$ or $CH_3C(CH_3)_2CH_2CH_2Cl$

PROBLEM 2

Draw a reaction coordinate diagram for the E2 reaction of 2-bromo-2,3-dimethylbutane with sodium *tert*-butoxide.

Although the major product of an E2 dehydrohalogenation of alkyl chlorides, alkyl bromides, and alkyl iodides is normally the more substituted alkene, the major product of the E2 dehydrohalogenation of alkyl fluorides is the less substituted alkene (Table 11.2).

Table 11.2 Products Obtained from the E2 Reaction of CH_3O^- and 2-Halohexanes

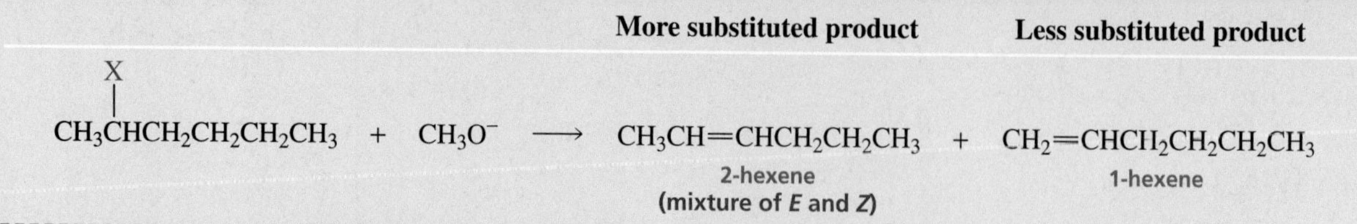

$$CH_3CHXCH_2CH_2CH_2CH_3 + CH_3O^- \longrightarrow \underset{\substack{\text{2-hexene}\\ \text{(mixture of } E \text{ and } Z)}}{CH_3CH{=}CHCH_2CH_2CH_3} + \underset{\text{1-hexene}}{CH_2{=}CHCH_2CH_2CH_2CH_3}$$

Leaving group	Conjugate acid	pK_a	More substituted product	Less substituted product
X = I	HI	−10	81%	19%
X = Br	HBr	−9	72%	28%
X = Cl	HCl	−7	67%	33%
X = F	HF	3.2	30%	70%

$$\underset{\text{2-fluoropentane}}{CH_3\overset{\overset{F}{|}}{C}HCH_2CH_2CH_3} + \underset{\text{methoxide ion}}{CH_3O^-} \xrightarrow{CH_3OH} \underset{\substack{\text{2-pentene}\\ 30\%\\ \text{(mixture of } E \text{ and } Z)}}{CH_3CH{=}CHCH_2CH_3} + \underset{\substack{\text{1-pentene}\\ 70\%}}{CH_2{=}CHCH_2CH_2CH_3} + CH_3OH + F^-$$

When a hydrogen and a chlorine, bromine, or iodine are eliminated from an alkyl halide, the halogen starts to leave as soon as the base begins to remove the proton. Consquently, a negative charge does not build up on the carbon that is losing the proton. Thus, the transition state resembles an alkene rather than a carbanion (Section 11.1). The fluoride ion, however, is the strongest base of the halide ions and, therefore, the poorest leaving group. So when a base begins to remove a proton from an alkyl fluoride, there is less tendency for the fluoride ion to leave than for the other halide ions to do so. As a result, negative charge develops on the carbon that is losing the proton, causing the transition state to resemble a carbanion rather than an alkene. To determine which of the carbanion-like transition states is more stable, we must determine which carbanion would be more stable.

$\delta^-OCH_3 \cdots H \cdots \overset{\delta-}{C}H_2CH(F)CH_2CH_2CH_3$ (carbanion-like transition state)

transition state leading to 1-pentene
more stable

$\delta^-OCH_3 \cdots H \cdots CH_3CH(F)\overset{\delta-}{C}HCH_2CH_3$

transition state leading to 2-pentene
less stable

We have seen that because they are positively charged, carbocations are stabilized by electron-donating groups. Recall that alkyl groups stabilize carbocations because alkyl groups are more electron donating than a hydrogen. Thus, tertiary carbocations are the most stable and methyl cations are the least stable (Section 4.2).

relative stabilities of carbocations

most stable R_3C^+ (tertiary carbocation) > R_2HC^+ (secondary carbocation) > RH_2C^+ (primary carbocation) > H_3C^+ (methyl cation) least stable

Carbanions, on the other hand, are negatively charged, so they are destabilized by alkyl groups. Therefore, methyl anions are the most stable and tertiary carbanions are the least stable.

Carbocation stability: 3° is more stable than 1°. Carbanion stability: 1° is more stable than 3°.

relative stabilities of carbanions

least stable $R_3C{:}^-$ (tertiary carbanion) < $R_2HC{:}^-$ (secondary carbanion) < $RH_2C{:}^-$ (primary carbanion) < $H_3C{:}^-$ (methyl anion) most stable

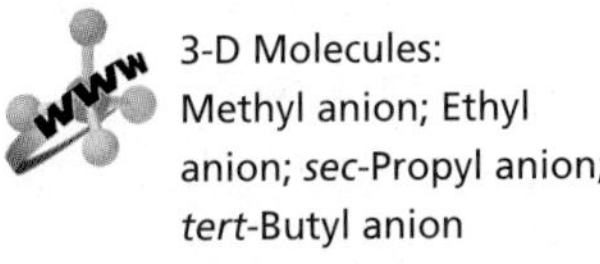

3-D Molecules: Methyl anion; Ethyl anion; *sec*-Propyl anion; *tert*-Butyl anion

The developing negative charge in the transition state leading to 1-pentene is on a primary carbon, which is more stable than the transition state leading to 2-pentene, in which the developing negative charge is on a secondary carbon. Because the transition state leading to 1-pentene is more stable, 1-pentene is formed more rapidly and is the major product of the E2 reaction of 2-fluoropentane.

The data in Table 11.2 show that as the halide ion increases in basicity (decreases in leaving ability), the yield of the more substituted alkene product decreases. However, the more substituted alkene remains the major elimination product in all cases, except when the halogen is fluorine.

We can summarize by saying that *the major product of an E2 elimination reaction is the more stable alkene* except *if the reactants are sterically hindered or the leaving group is poor* (e.g., a fluoride ion), in which case the major product will be the less stable alkene. In Section 11.6, you will see that the more stable product is not always the major product in the case of certain cyclic compounds.

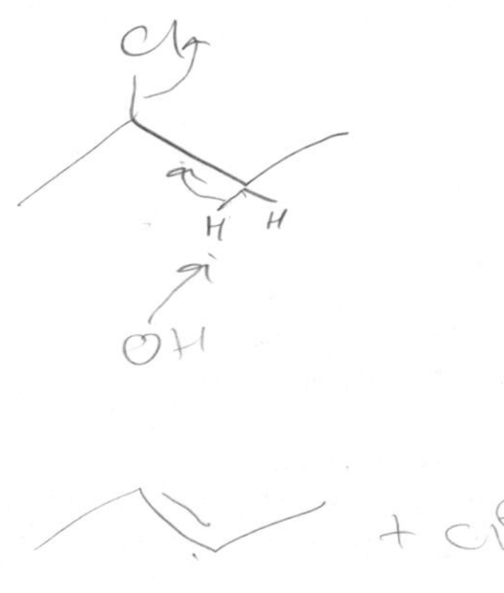

PROBLEM 3◆

Give the major elimination product obtained from an E2 reaction of each of the following alkyl halides with hydroxide ion:

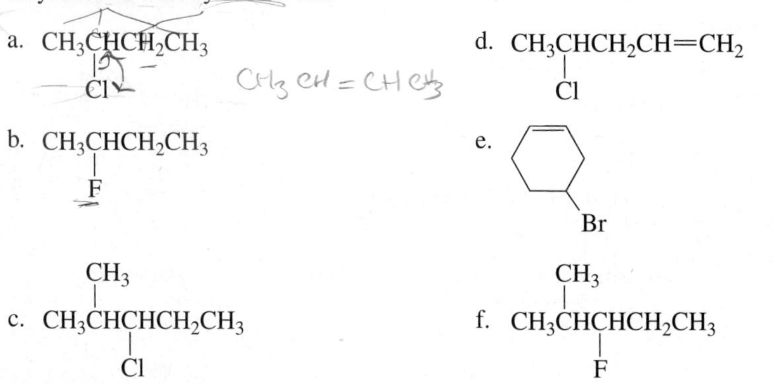

a. $CH_3CHCH_2CH_3$ (Cl on C-2)

b. $CH_3CHCH_2CH_3$ (F on C-2)

c. $CH_3CHCHCH_2CH_3$ (CH_3 on C-2, Cl on C-3)

d. $CH_3CHCH_2CH{=}CH_2$ (Cl on C-2)

e. cyclohexene ring with Br

Br

f. $CH_3CHCHCH_2CH_3$ (CH_3 on C-2, F on C-3)

PROBLEM 4◆

Which alkyl halide would you expect to be more reactive in an E2 reaction?

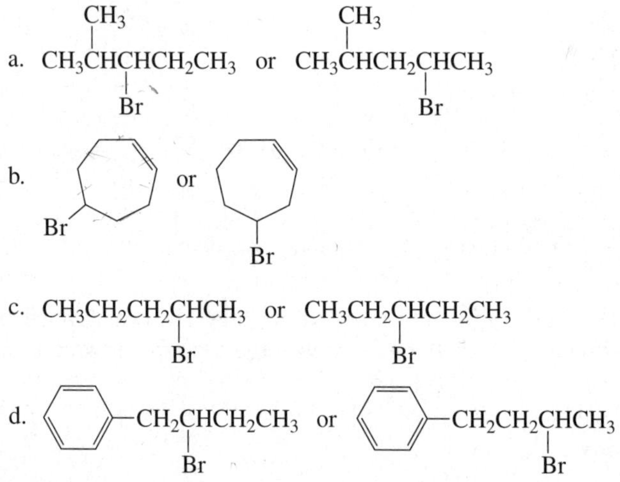

a. $CH_3CHCHCH_2CH_3$ (CH_3 on C-2, Br on C-3) or $CH_3CHCH_2CHCH_3$ (CH_3 on C-2, Br on C-4)

b. cycloheptene ring with Br or cycloheptene ring with Br

c. $CH_3CH_2CH_2CHCH_3$ (Br on C-2) or $CH_3CH_2CHCH_2CH_3$ (Br on C-3)

d. C_6H_5–$CH_2CHCH_2CH_3$ (Br on C-2) or C_6H_5–$CH_2CH_2CHCH_3$ (Br on C-3)

11.3 The E1 Reaction

The second kind of elimination reaction that alkyl halides can undergo is an E1 elimination. The reaction of *tert*-butyl bromide with water to form 2-methylpropene is an example of an **E1 reaction**; "E" stands for elimination and "1" stands for unimolecular.

$$\underset{\textit{tert-}\text{butyl bromide}}{CH_3-\overset{\displaystyle CH_3}{\overset{|}{\underset{\underset{\displaystyle Br}{|}}{C}}}-CH_3} + H_2O \longrightarrow \underset{\text{2-methylpropene}}{CH_2=\overset{\displaystyle CH_3}{\overset{|}{C}}-CH_3} + H_3O^+ + Br^-$$

An E1 reaction is a first-order elimination reaction because the rate of the reaction depends only on the concentration of the alkyl halide.

rate = *k*[alkyl halide]

We know, then, that only the alkyl halide is involved in the rate-determining step of the reaction. Therefore, there must be at least two steps in the reaction.

Movie:
E1 Elimination

mechanism of the E1 reaction

$$CH_3-\overset{\displaystyle CH_3}{\overset{|}{\underset{\underset{\displaystyle Br}{|}}{C}}}-CH_3 \underset{}{\overset{\text{slow}}{\rightleftharpoons}} \underset{\underset{\displaystyle H}{|}}{CH_2}-\overset{\displaystyle CH_3}{\overset{|}{\underset{+}{C}}}-CH_3 + Br^- \xrightarrow{\text{fast}} CH_2=\overset{\displaystyle CH_3}{\overset{|}{C}}-CH_3 + H_3O^+$$

$H_2\ddot{O}:$

the alkyl halide dissociates, forming a carbocation

the base removes a proton from a β-carbon

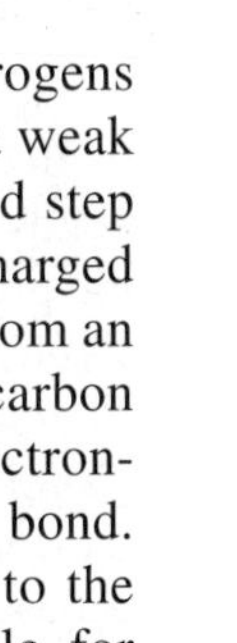

3-D Molecules:
t-Butyl chloride;
t-Butyl cation

The foregoing mechanism shows that an E1 reaction has two steps. In the first step, the alkyl halide dissociates heterolytically, producing a carbocation. In the second step, the base forms the elimination product by removing a proton from a carbon that is adjacent to the positively charged carbon (i.e., from the β-carbon). This mechanism agrees with the observed first-order kinetics. The first step of the reaction is the rate-determining step. Therefore, increasing the concentration of the base—which comes into play only in the second step of the reaction—has no effect on the rate of the reaction.

We have seen that the pK_a of a compound such as ethane that has hydrogens attached only to sp^3 hybridized carbons is 50 (Section 6.9). How, then, can a weak base like water remove a proton from an sp^3 hybridized carbon in the second step of the reaction? First of all, the pK_a is greatly reduced by the postively charged carbon that can accept the electrons left behind when the proton is removed from an adjacent carbon. Second, the carbon adjacent to the positively charged carbon shares the positive charge as a result of hyperconjugation, and this electron-withdrawing positive charge also increases the acidity of the C—H bond. Recall that hyperconjugation—where the σ electrons in the bond adjacent to the positively charged carbon spread into the empty p orbital—is responsible for the greater stability of a tertiary carbocation, compared with a secondary carbocation (Section 4.2).

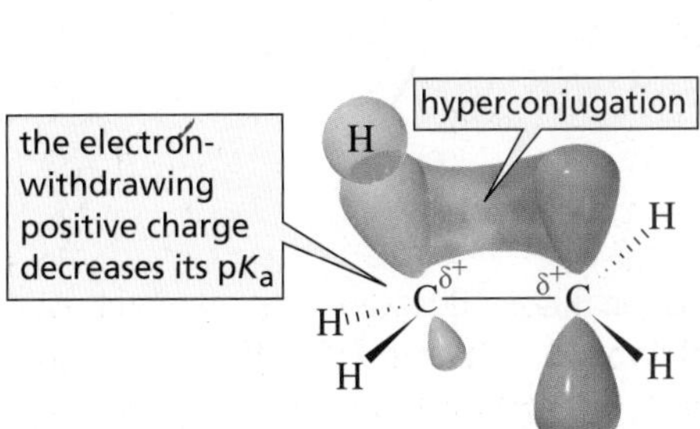

When two elimination products can be formed in an E1 reaction, the major product is generally the more substituted alkene.

$$\underset{\text{2-chloro-2-methylbutane}}{CH_3CH_2\overset{\overset{CH_3}{|}}{\underset{\underset{Cl}{|}}{C}}CH_3} + H_2O \longrightarrow \underset{\substack{\text{2-methyl-2-butene}\\ \text{major product}}}{CH_3CH{=}\overset{\overset{CH_3}{|}}{C}CH_3} + \underset{\substack{\text{2-methyl-1-butene}\\ \text{minor product}}}{CH_3CH_2\overset{\overset{CH_3}{|}}{C}{=}CH_2} + H_3O^+ + Cl^-$$

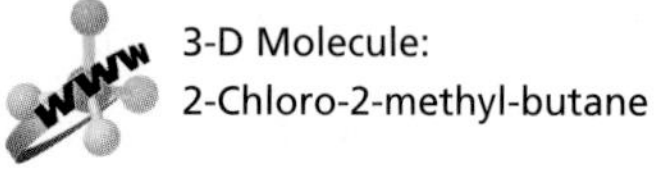

The more substituted alkene is the more stable of the two alkenes formed in the preceding reaction, and therefore, it has the more stable transition state leading to its formation (Figure 11.2). As a result, the more substituted alkene is formed more rapidly, so it is the major product. To obtain the more substituted alkene, the hydrogen is removed from the β-carbon bonded to the fewest hydrogens, in accordance with Zaitsev's rule.

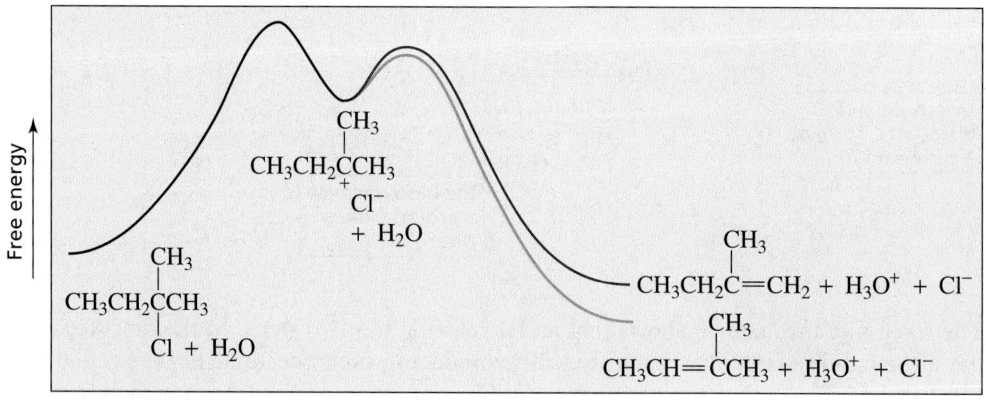

▲ **Figure 11.2**
Reaction coordinate diagram for the E1 reaction of 2-chloro-2-methylbutane. The major product is the more substituted alkene because its greater stability causes the transition state leading to its formation to be more stable.

Because the first step is the rate-determining step, the rate of an E1 reaction depends on both the ease with which the carbocation is formed *and* how readily the leaving group leaves. The more stable the carbocation, the easier it is formed because more stable carbocations have more stable transition states leading to their formation. Therefore, the relative reactivities of a series of alkyl halides with the same leaving group parallel the relative stabilities of the carbocations. A tertiary benzylic halide is the most reactive alkyl halide because a tertiary benzylic cation—the most stable carbocation—is the easiest to form (Sections 7.7 and 10.8).

relative reactivities of alkyl halides in an E1 reaction = relative stabilities of carbocations

3° benzylic ≈ 3° allylic > 2° benzylic ≈ 2° allylic ≈ 3° > 1° benzylic ≈ 1° allylic ≈ 2° > 1° > vinyl

most stable (3° benzylic) … least stable (vinyl)

Notice that a tertiary alkyl halide and a weak base were chosen to illustrate the E1 reaction in this section, whereas a tertiary alkyl halide and a strong base were used to illustrate the E2 reaction in Section 11.1. Tertiary alkyl halides are more reactive than secondary alkyl halides, which, in turn, are more reactive than primary alkyl halides in both E1 and E2 reactions (Section 11.2).

We have seen that the weakest bases are the best leaving groups (Section 10.3). Therefore, for series of alkyl halides with the same alkyl group, alkyl iodides are the most reactive and alkyl fluorides the least reactive in E1 reactions.

relative reactivities of alkyl halides in an E1 reaction

most reactive → RI > RBr > RCl > RF ← least reactive

← increasing reactivity

Because the E1 reaction forms a carbocation intermediate, the carbon skeleton can rearrange before the proton is lost, if rearrangement leads to a more stable carbocation. For example, the secondary carbocation that is formed when a chloride ion dissociates from 3-chloro-2-methyl-2-phenylbutane undergoes a 1,2-methyl shift to form a more stable tertiary benzylic cation, which then undergoes deprotonation to form the alkene.

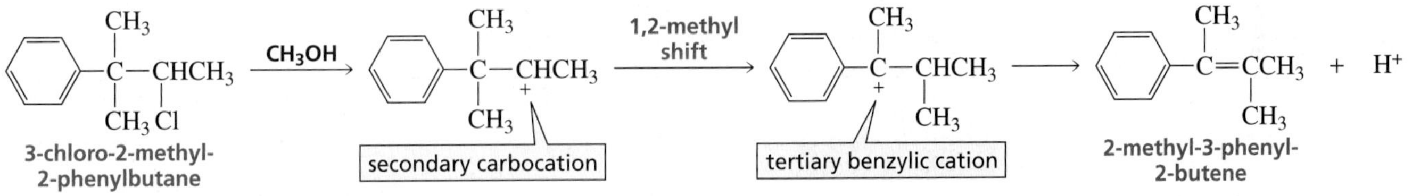

$$C_6H_5C(CH_3)_2CHClCH_3 \xrightarrow{CH_3OH} C_6H_5C(CH_3)_2\overset{+}{C}HCH_3 \xrightarrow{\text{1,2-methyl shift}} C_6H_5\overset{+}{C}(CH_3)CH(CH_3)CH_3 \longrightarrow C_6H_5C(CH_3){=}C(CH_3)CH_3 + H^+$$

3-chloro-2-methyl-2-phenylbutane; secondary carbocation; tertiary benzylic cation; 2-methyl-3-phenyl-2-butene

In the following reaction, the initially formed secondary carbocation undergoes a 1,2-hydride shift to form a more stable secondary allylic cation:

$$CH_3CH{=}CHCH_2CHBrCH_2CH_3 \underset{}{\overset{CH_3OH}{\rightleftharpoons}} CH_3CH{=}CHCH_2\overset{+}{C}HCH_2CH_3 \xrightarrow{\text{1,2-hydride shift}} CH_3CH{=}CH\overset{+}{C}HCH_2CH_2CH_3$$

5-bromo-2-heptene; secondary carbocation; secondary allylic cation

$$\downarrow$$

$$H^+ + CH_3CH{=}CHCH{=}CHCH_2CH_3$$

2,4-heptadiene

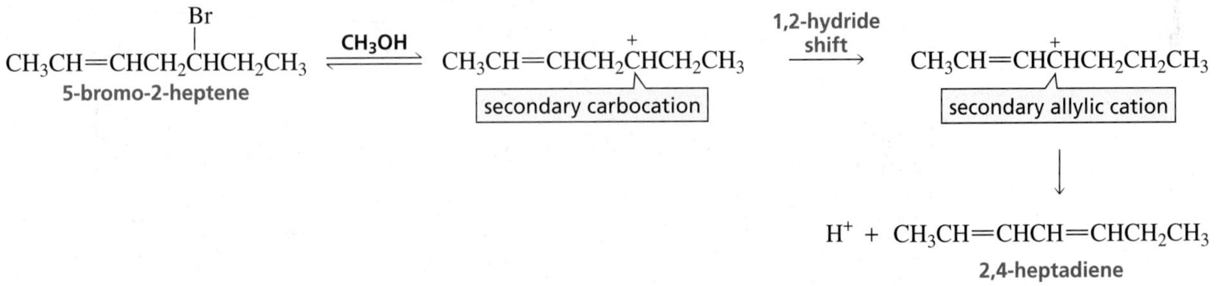

PROBLEM 5◆

Three alkenes are formed from the E1 reaction of 3-bromo-2,3-dimethylpentane. Give the structures of the alkenes, and rank them according to the amount that would be formed. (Ignore stereoisomers.)

PROBLEM 6

If 2-fluoropentane were to undergo an E1 reaction, would you expect the major product to be the one predicted by Zaitsev's rule? Explain.

PROBLEM-SOLVING STRATEGY

Propose a mechanism for the following reaction:

$$\text{4,4-dimethyl-1-methylenecyclohexa-2,5-diene} \xrightarrow{H_3O^+} \text{1,2,4-trimethylbenzene}$$

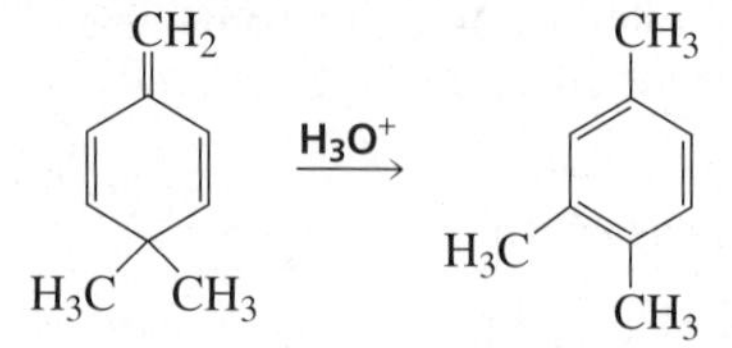

Because the given reagent is an acid, start by protonating the molecule at the position that allows the most stable carbocation to be formed. By protonating the CH_2 group, a tertiary allylic carbocation is formed in which the positive charge is delocalized over two other carbons. Then move the π electrons so that the 1,2-methyl shift required to obtain the product can take place. Loss of a proton gives the final product.

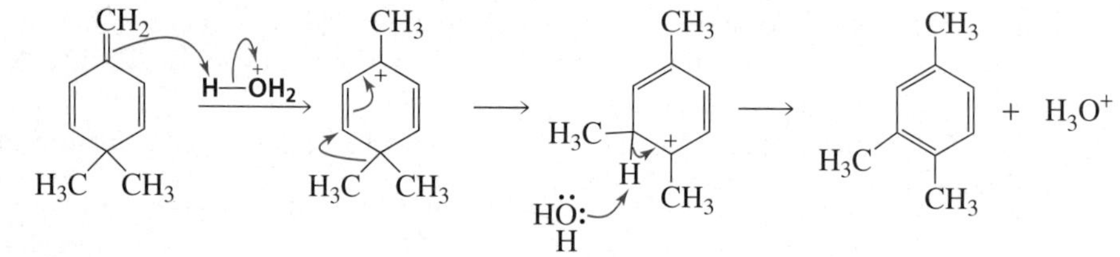

$$\xrightarrow{H-\overset{+}{O}H_2} \longrightarrow \longrightarrow + H_3O^+$$

Now continue on to Problem 7.

PROBLEM 7

Propose a mechanism for the following reaction:

$$\text{cyclobutyl–}C(CH_3)_2Cl \xrightarrow{CH_3OH} \text{1,2-dimethylcyclopentene}$$

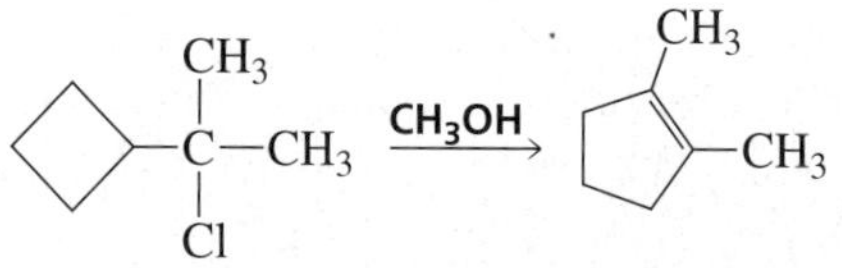

11.4 Competition Between E2 and E1 Reactions

Primary alkyl halides undergo only E2 elimination reactions. They cannot undergo E1 reactions because of the difficulty encountered in forming primary carbocations. *Secondary* and *tertiary* alkyl halides undergo both E2 and E1 reactions (Table 11.3).

For those alkyl halides that can undergo both E2 and E1 reactions, the E2 reaction is favored by the same factors that favor an S_N2 reaction and the E1 reaction is favored by the same factors that favor an S_N1 reaction. Thus, *an E2 reaction is favored by a high concentration of a strong base and an aprotic polar solvent (e.g., DMSO or DMF), whereas an E1 reaction is favored by a weak base and a protic polar solvent (e.g., H_2O or ROH).* How the solvent affects the mechanism of the reaction was discussed in Section 10.10.

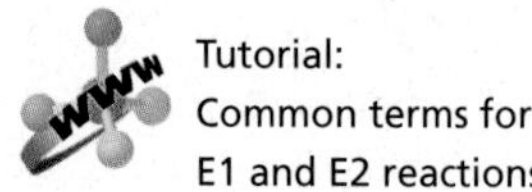

Tutorial: Common terms for E1 and E2 reactions

Table 11.3 Summary of the Reactivity of Alkyl Halides in Elimination Reactions

Primary alkyl halide	E2 only
Secondary alkyl halide	E1 and E2
Tertiary alkyl halide	E1 and E2

PROBLEM 8◆

For each of the following reactions, (1) indicate whether elimination will occur via an E2 or an E1 reaction, and (2) give the major elimination product of each reaction, ignoring stereoisomers:

a. $CH_3CH_2CHCH_3$ (with Br on C-2) $\xrightarrow[\text{DMSO}]{CH_3O^-}$

b. $CH_3CH_2CHCH_3$ (with Br on C-2) $\xrightarrow{CH_3OH}$

c. CH_3CCH_3 (with CH_3 and Cl on the central C) $\xrightarrow{H_2O}$

d. CH_3CCH_3 (with CH_3 and Cl on the central C) $\xrightarrow[\text{DMF}]{HO^-}$

e. $CH_3C{-}CHCH_3$ (with two CH_3 on C-2 and Br on C-3) $\xrightarrow{CH_3CH_2OH}$

f. $CH_3C{-}CHCH_3$ (with two CH_3 on C-2 and Br on C-3) $\xrightarrow[\text{DMSO}]{CH_3CH_2O^-}$

PROBLEM 9◆

The rate law for the reaction of HO^- with *tert*-butyl bromide to form an elimination product in 75% ethanol/25% water at 30 °C is the sum of the rate laws for the E2 and E1 reactions:

$$\textbf{rate} = 7.1 \times 10^{-5}[\textit{tert}\text{-butyl bromide}][HO^-] + 1.5 \times 10^{-5}[\textit{tert}\text{-butyl bromide}]$$

What percentage of the reaction takes place by the E2 pathway when these conditions exist?

a. $[HO^-] = 5.0$ M

b. $[HO^-] = 0.0025$ M

11.5 Stereochemistry of E2 and E1 Reactions

Stereochemistry of the E2 Reaction

An E2 reaction involves the removal of two groups from adjacent carbons. It is a concerted reaction because the two groups are eliminated in the same step. The bonds to the groups to be eliminated (H and X) must be in the same plane because the sp^3 orbital of the carbon bonded to H and the sp^3 orbital of the carbon bonded to X become overlapping *p* orbitals in the alkene product. Therefore, the orbitals must overlap in the transition state. This overlap is optimal if the orbitals are parallel (i.e., in the same plane).

There are two ways in which the C—H and C—X bonds can be in the same plane: They can be parallel to one another either on the same side of the molecule (**syn-periplanar**) or on opposite sides of the molecule **(anti-periplanar)**.

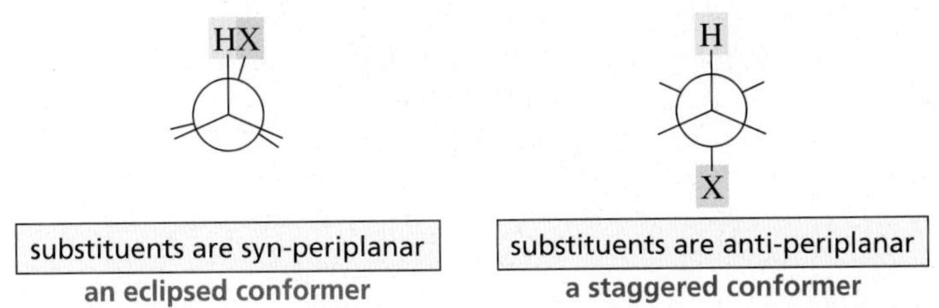

If an elimination reaction removes two substituents from the same side of the C—C bond, the reaction is called a **syn elimination**. If the substituents are removed from opposite sides of the C—C bond, the reaction is called an **anti elimination**. Both types of elimination can occur, but syn elimination is a much slower reaction, so anti elimination is highly favored in an E2 reaction. One reason anti elimination is favored is that syn elimination requires the molecule to be in an eclipsed conformation, whereas anti elimination requires it to be in a more stable, staggered conformation.

An E2 reaction involves anti elimination.

Sawhorse projections reveal another reason that anti elimination is favored. In syn elimination, the electrons of the departing hydrogen move to the *front* side of the carbon bonded to X, whereas in anti elimination, the electrons move to the *back* side of the carbon bonded to X. We have seen that displacement reactions involve back-side attack because the best overlap of the interacting orbitals is achieved through back-side attack (Section 10.2). Finally, anti elimination avoids the repulsion the electron-rich base experiences when it is on the same side of the molecule as the electron-rich departing halide ion.

base H X

syn elimination
front-side attack

base H X

anti elimination
back-side attack

In Section 11.1, we saw that an E2 reaction is *regioselective,* which means that more of one constitutional isomer is formed than the other. For example, the major product formed from the E2 elimination of 2-bromopentane is 2-pentene.

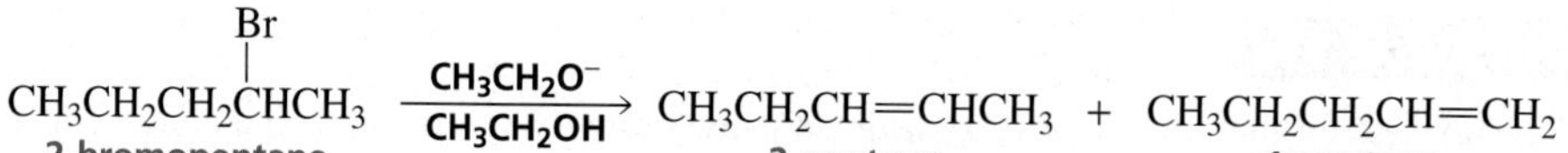

$$\underset{\text{2-bromopentane}}{CH_3CH_2CH_2CH(Br)CH_3} \xrightarrow[CH_3CH_2OH]{CH_3CH_2O^-} \underset{\substack{\text{2-pentene}\\ \text{major product}\\ \text{(mixture of } E \text{ and } Z)}}{CH_3CH_2CH{=}CHCH_3} + \underset{\substack{\text{1-pentene}\\ \text{minor product}}}{CH_3CH_2CH_2CH{=}CH_2}$$

The E2 reaction is also *stereoselective,* which means that more of one stereoisomer is formed than the other. For example, the 2-pentene obtained as the major product from the elimination reaction of 2-bromopentane can exist as a pair of stereoisomers, and more (*E*)-2-pentene is formed than (*Z*)-2-pentene.

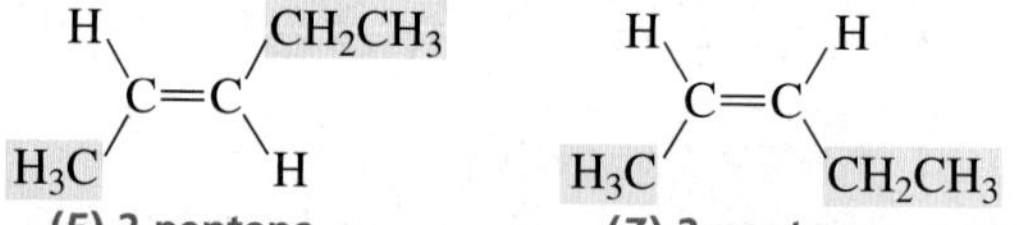

H, CH_2CH_3 / H_3C, H — C=C
(*E*)-2-pentene
41%

H, H / H_3C, CH_2CH_3 — C=C
(*Z*)-2-pentene
14%

When two hydrogens are bonded to the β-carbon, the major product of an E2 reaction is the alkene with the bulkiest substituents on opposite sides of the double bond.

We can make the following general statement about the stereoselectivity of E2 reactions: If the reactant has two hydrogens bonded to the carbon from which a hydrogen is to be removed, both the *E* and *Z* products will be formed because there are two conformers in which the groups to be eliminated are anti. The alkene with the *bulkiest groups on opposite sides of the double bond* will be formed in greater yield because it is the more stable alkene.

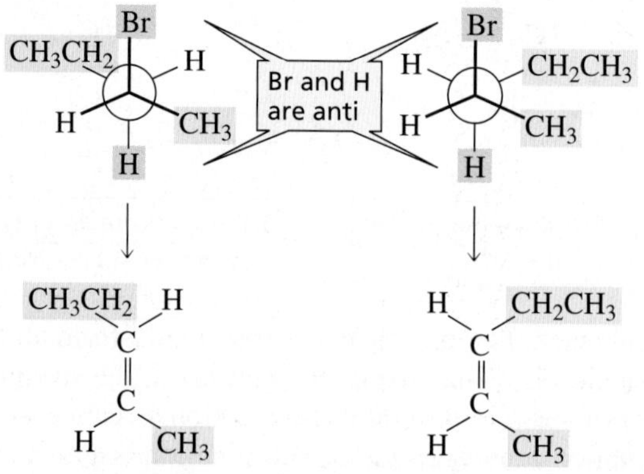

Br, CH_3CH_2, H, H, CH_3, H

Br and H are anti

Br, H, CH_2CH_3, H, CH_3, H

↓

CH_3CH_2, H / C=C / H, CH_3

(*E*)-2-pentene
more stable

↓

H, CH_2CH_3 / C=C / H, CH_3

(*Z*)-2-pentene
less stable

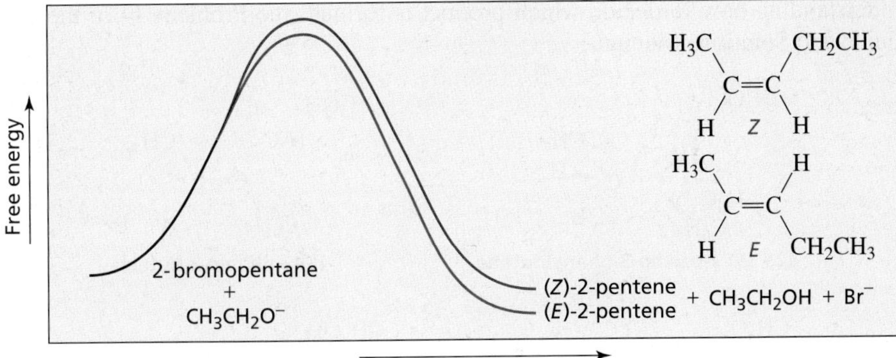

Figure 11.3
Reaction coordinate diagram for the E2 reaction of 2-bromopentane and ethoxide ion.

The more stable alkene has the more stable transition state and therefore is formed more rapidly (Figure 11.3). In Section 4.11, we saw that the alkene with the bulkiest groups on the *same* side of the double bond is less stable because the electron clouds of the large substituents can interfere with each other, causing steric strain.

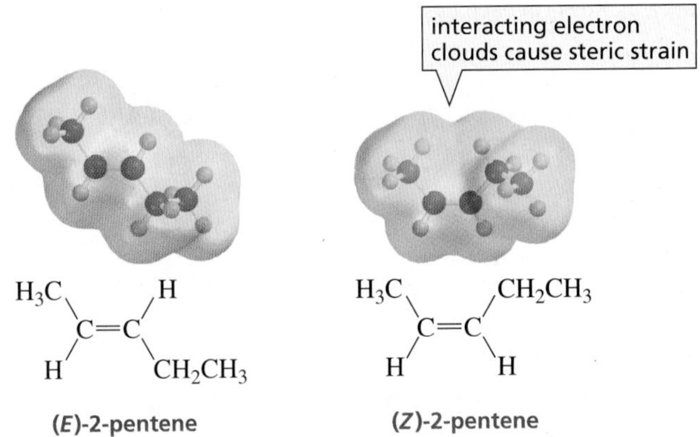

Elimination of HBr from 3-bromo-2,2,3-trimethylpentane leads predominantly to the *E* isomer because this stereoisomer has the methyl group, the bulkiest group on one sp^2 carbon, opposite the *tert*-butyl group, the bulkiest group on the other sp^2 carbon.

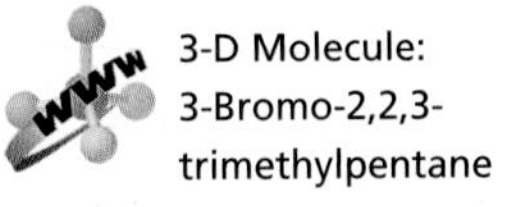
3-D Molecule: 3-Bromo-2,2,3-trimethylpentane

$$\mathrm{CH_3CH_2{-}C(Br)(CH_3){-}C(CH_3)_2{-}CH_3} \xrightarrow[\mathrm{CH_3CH_2OH}]{\mathrm{CH_3CH_2O^-}} \mathrm{(H)(H_3C)C{=}C(C(CH_3)_3)(CH_3)} + \mathrm{(H_3C)(H)C{=}C(C(CH_3)_3)(CH_3)}$$

3-bromo-2,2,3-trimethylpentane

(*E*)-3,4,4-trimethyl-2-pentene major product

(*Z*)-3,4,4-trimethyl-2-pentene minor product

If the β-carbon from which a hydrogen is to be removed is bonded to only one hydrogen, there is only one conformer in which the groups to be eliminated are anti. Therefore, only one alkene product can be formed. The particular isomer that is formed depends on the configuration of the reactant. For example, anti elimination of HBr from (2*S*,3*S*)-2-bromo-3-phenylbutane forms the *E* isomer, whereas anti elimination of HBr from (2*S*,3*R*)-2-bromo-3-phenylbutane forms the *Z* isomer. Notice that the groups that are not eliminated retain their relative positions. (If you have trouble

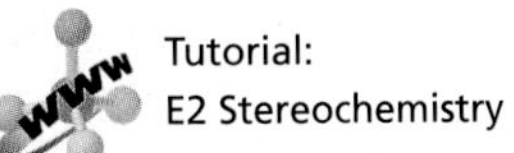
Tutorial: E2 Stereochemistry

understanding how to decide which product is formed, see Problem 44 in the Study Guide and Solutions Manual.)

When only one hydrogen is bonded to the β-carbon, the major product of an E2 reaction depends on the structure of the alkene.

Molecular models can be helpful whenever complex stereochemistry is involved.

$CH_3\ddot{O}:^-$ + (2*S*,3*S*)-2-bromo-3-phenylbutane ⟶ $CH_3\ddot{O}H$ + (*E*)-2-phenyl-2-butene + Br^-

$CH_3\ddot{O}:^-$ + (2*S*,3*R*)-2-bromo-3-phenylbutane ⟶ $CH_3\ddot{O}H$ + (*Z*)-2-phenyl-2-butene + Br^-

PROBLEM 10◆

a. Determine the major product that would be obtained from an E2 reaction of each of the following alkyl halides (in each case, indicate the configuration of the product):

1. $CH_3CH_2CH(Br)CH(CH_3)CH_3$

2. $CH_3CH_2CH(Cl)CH_2-C_6H_5$

3. $CH_3CH_2CH(Cl)CH_2CH{=}CH_2$

b. Does the product obtained depend on whether you started with the *R* or *S* enantiomer of the reactant?

Stereochemistry of the E1 Reaction

We have seen that an E1 elimination reaction takes place in two steps. The leaving group leaves in the first step, and a proton is lost from an adjacent carbon in the second step, following Zaitsev's rule in order to form the more stable alkene. The carbocation formed in the first step is planar, so the electrons from a departing proton can move toward the positively charged carbon from *either side*. Therefore, both syn and anti elimination can occur.

The major product of an E1 reaction is the alkene with the bulkiest substituents on opposite sides of the double bond.

Because both syn and anti elimination can occur in an E1 reaction, both the *E* and *Z* products are formed, regardless of whether the β-carbon from which the proton is removed is bonded to one or two hydrogens. The major product is the one with the bulkiest groups on opposite sides of the double bond, because it is the more stable alkene.

$CH_3CH_2CH(Br)CH_3$ ⟶ $CH_3CH_2\overset{+}{C}HCH_3$ + Br^- ⟶ (*E*)-2-butene (major product) + (*Z*)-2-butene (minor product) + H^+

β-carbon has two hydrogens

$CH_3CH_2CH(CH_3)C(Cl)(CH_3)CH_2CH_3$ ⟶ $CH_3CH_2CH(CH_3)\overset{+}{C}(CH_3)CH_2CH_3$ + Cl^- ⟶ (*E*)-3,4-dimethyl-3-hexene (major product) + (*Z*)-3,4-dimethyl-3-hexene (minor product) + H^+

β-carbon has one hydrogen

In contrast, we just saw that an E2 reaction forms both the *E* and *Z* products only if the β-carbon from which the proton is removed is bonded to two hydrogens. If it is bonded to only one hydrogen, only one product is obtained because anti elimination is favored.

PROBLEM 11◆ SOLVED

For each of the following alkyl halides, determine the major product that is formed when that alkyl halide undergoes an E1 reaction:

a. $CH_3CH_2CH_2CH_2CH(Br)CH_3$

b. $CH_3CH_2CH_2C(CH_3)(Cl)CH_3$

c. $CH_3CH_2CH_2CH(I)CH(CH_3)CH_2CH_3$

d. 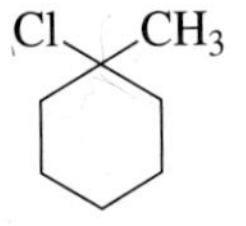 1-chloro-1-methylcyclohexane (Cl, CH_3 on the same ring carbon)

SOLUTION TO 11a First, we must consider the regiochemistry of the reaction: More 2-hexene will be formed than 1-hexene because 2-hexene is more stable. Next, we must consider the stereochemistry of the reaction: Of the 2-hexene that is formed, more (*E*)-2-hexene will be formed than (*Z*)-2-hexene because (*E*)-2-hexene is more stable. Thus, (*E*)-2-hexene is the major product of the reaction.

$$CH_3CH_2CH_2CH_2CH(Br)CH_3 \xrightarrow{E1} \underset{\text{2-hexene}}{CH_3CH_2CH_2CH{=}CHCH_3} + \underset{\text{1-hexene}}{CH_3CH_2CH_2CH_2CH{=}CH_2}$$

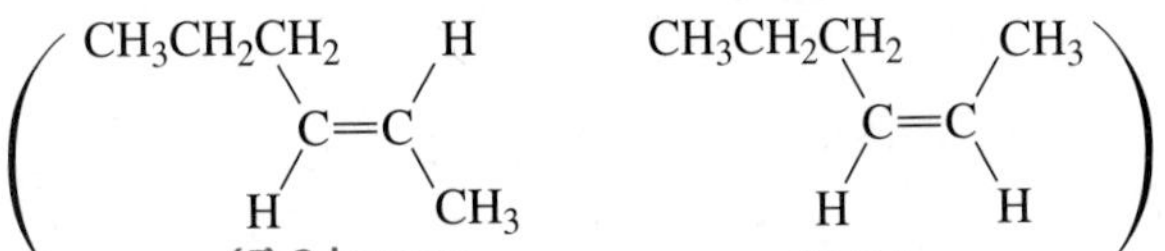

(*E*)-2-hexene: $CH_3CH_2CH_2(H)C{=}C(H)CH_3$ (trans)

(*Z*)-2-hexene: $CH_3CH_2CH_2(H)C{=}C(H)CH_3$ (cis)

11.6 Elimination from Cyclic Compounds

E2 Elimination from Cyclic Compounds

Elimination from cyclic compounds follows the same stereochemical rules as elimination from open-chain compounds. To achieve the anti-periplanar geometry that is preferred for an E2 reaction, the two groups that are being eliminated from a cyclic compound must be trans to one another. In the case of six-membered rings, the groups being eliminated will be anti-periplanar only if *both are in axial positions*.

In an E2 reaction of a substituted cyclohexane, the groups being eliminated must both be in axial positions.

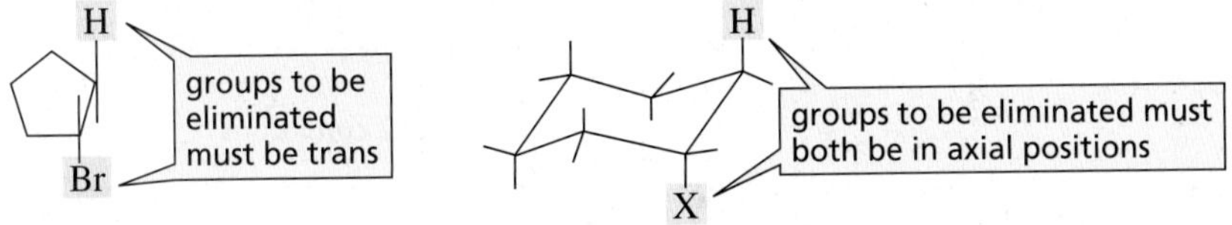

H, Br — groups to be eliminated must be trans

H, X — groups to be eliminated must both be in axial positions

The more stable conformer of chlorocyclohexane does not undergo an E2 reaction, because the chloro substituent is in an equatorial position. (Recall from Section 2.13 that the more stable conformer of a monosubstituted cyclohexane is the one in which the substituent is in an equatorial position because there is more room for a substituent in that position.) The less stable conformer, with the chloro substituent in the axial position, readily undergoes an E2 reaction.

3-D Molecules:
Axial chlorocyclohexane;
Equatorial chlorocyclohexane

Tutorial:
E2 Eliminations from cyclic compounds

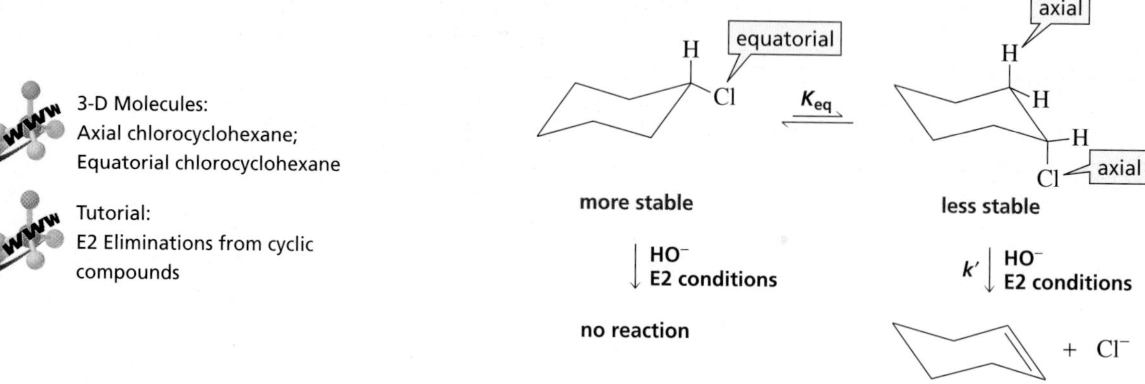

Because one of the two conformers does not undergo an E2 reaction, the rate of an elimination reaction is affected by the stability of the conformer that does undergo the reaction. The rate constant of the reaction is given by $k'K_{eq}$. Therefore, the reaction is faster if K_{eq} is large (i.e., if elimination takes place by way of the more stable conformer). If elimination has to take place by way of the less stable conformer, K_{eq} will be small. For example, neomenthyl chloride undergoes an E2 reaction with ethoxide ion about 200 times faster than menthyl chloride does. The conformer of neomenthyl chloride that undergoes elimination is the *more* stable conformer because when the Cl and H are in the required axial positions, the methyl and isopropyl groups are in the equatorial positions.

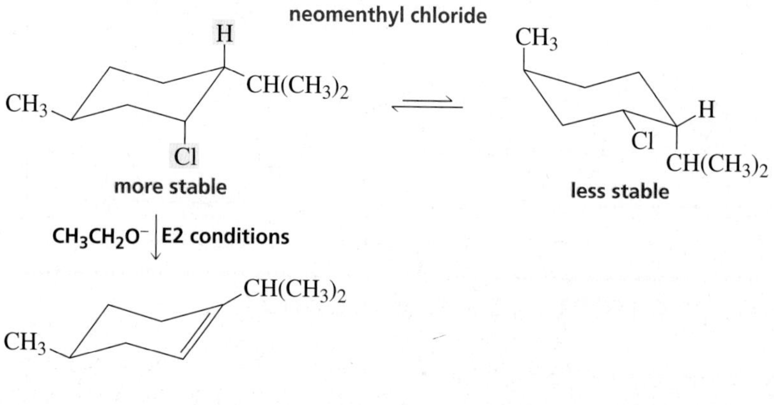

neomenthyl chloride

CH_3 H $CH(CH_3)_2$ Cl — more stable ⇌ CH_3 H Cl $CH(CH_3)_2$ — less stable

$CH_3CH_2O^-$ | E2 conditions

CH_3 $CH(CH_3)_2$

Sir Derek H. R. Barton (1918–1998) *was the first to point out that the chemical reactivity of substituted cyclohexanes was controlled by their conformation. Barton was born in Gravesend, Kent, England. He received a Ph.D. and a D.Sc. from Imperial College, London, in 1942 and became a faculty member there three years later. Subsequently, he was a professor at the University of London, the University of Glasgow, the Institut de Chimie des Substances Naturelles, and Texas A & M University. He received the 1969 Nobel Prize in chemistry for his work on the relationship of the three-dimensional structures of organic compounds to their chemical reactivity. Barton was knighted by Queen Elizabeth II in 1972.*

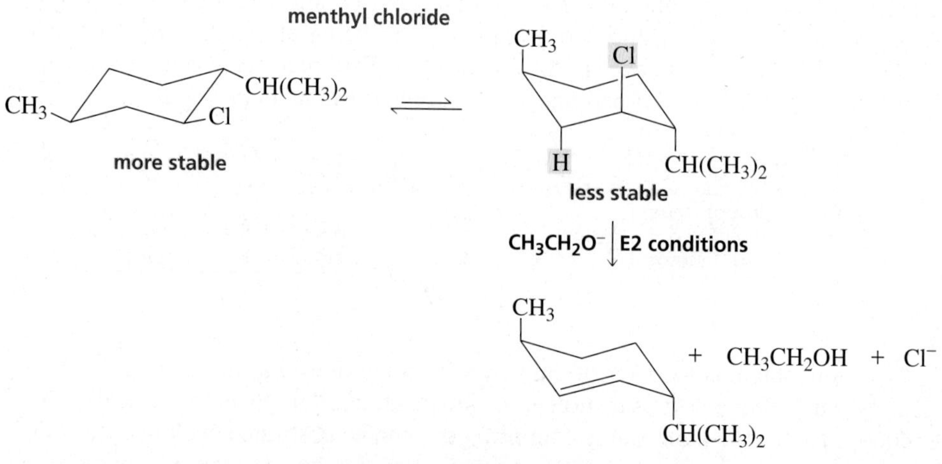

menthyl chloride

CH_3 $CH(CH_3)_2$ Cl — more stable ⇌ CH_3 Cl H $CH(CH_3)_2$ — less stable

$CH_3CH_2O^-$ | E2 conditions

CH_3 $CH(CH_3)_2$ + CH_3CH_2OH + Cl^-

In contrast, the conformer of menthyl chloride that undergoes elimination is the *less* stable conformer because when the Cl and H are in the required axial positions, the methyl and isopropyl groups are also in axial positions.

Notice that when menthyl chloride or *trans*-1-chloro-2-methylcyclohexane undergoes an E2 reaction, the hydrogen that is eliminated is not removed from the β-carbon bonded to the fewest hydrogens. This may seem like a violation of Zaitsev's rule, but the rule states that when there is *more than one* β-carbon from which a hydrogen can be removed, the hydrogen is removed from the β-carbon bonded to the fewest hydrogens. In the preceding two reactions, the hydrogen that is removed has to be in an axial position and only one β-carbon has a hydrogen in an axial position. Therefore, that hydrogen is the one that is removed even though it is not bonded to the β-carbon with the fewest hydrogens.

a.

b.

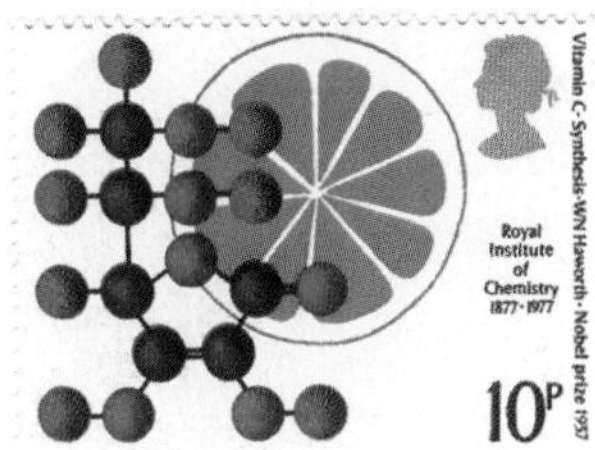

c.

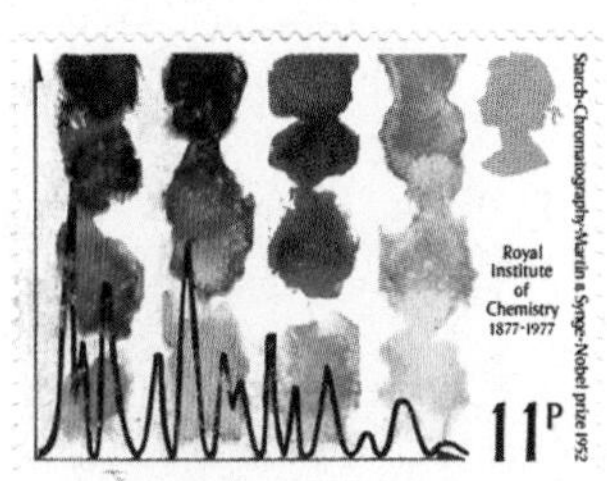

d.

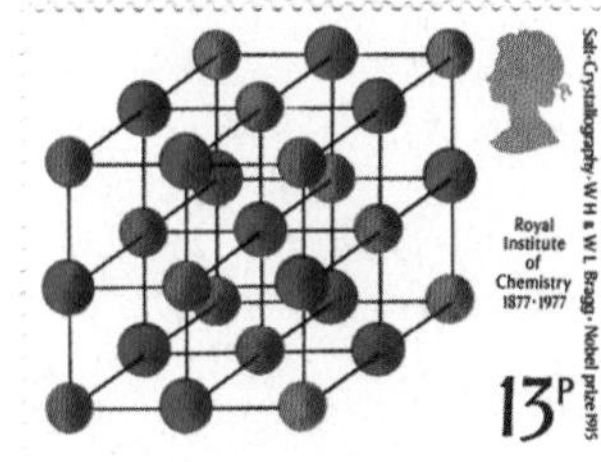

▲ Stamps issued in honor of English Nobel laureates: (a) Sir Derek Barton for conformational analysis, 1969; (b) Sir Walter Haworth for the synthesis of vitamin C, 1937; (c) A. J. P. Martin and Richard L. M. Synge for chromatography, 1952; (d) William H. Bragg and William L. Bragg for crystallography, 1915 (the only father and son to receive a Nobel Prize).

axial
equatorial
trans-**1-chloro-2-methylcyclohexane**
more stable
axial
less stable
HO^- **E2 conditions**
$+ H_2O + Cl^-$

> **PROBLEM 12**
>
> Why do *cis*-1-bromo-2-ethylcyclohexane and *trans*-1-bromo-2-ethylcyclohexane form different major products when they undergo an E2 reaction?
>
> **PROBLEM 13**
>
> Which isomer reacts more rapidly in an E2 reaction, *cis*-1-bromo-4-*tert*-butylcyclohexane or *trans*-1-bromo-4-*tert*-butylcyclohexane? Explain your choice.

E1 Elimination from Cyclic Compounds

When a substituted cyclohexane undergoes an E1 reaction, the two groups that are eliminated do not have to both be in axial positions, because the elimination reaction is not concerted. In the following reaction, a carbocation is formed in the first step. It then loses a proton from the adjacent carbon that is bonded to the fewest hydrogens—in other words, Zaitsev's rule is followed.

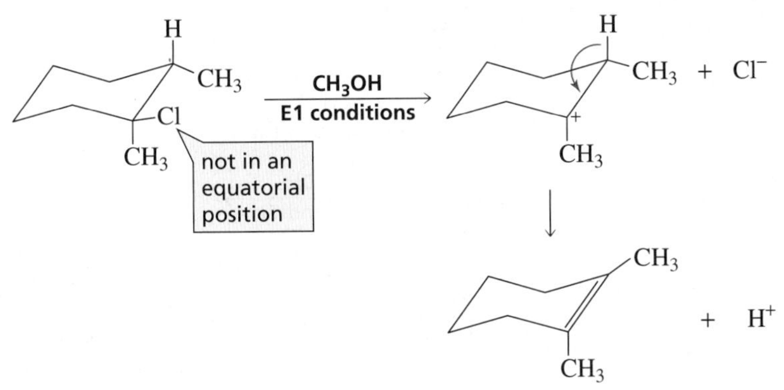

An E1 reaction involves both syn and anti elimination.

Because a carbocation is formed in an E1 reaction, you must check for the possibility of a carbocation rearrangement before you use Zaitsev's rule to determine the elimination product. In the following reaction, the secondary carbocation undergoes a 1,2-hydride shift, in order to form a more stable tertiary carbocation.

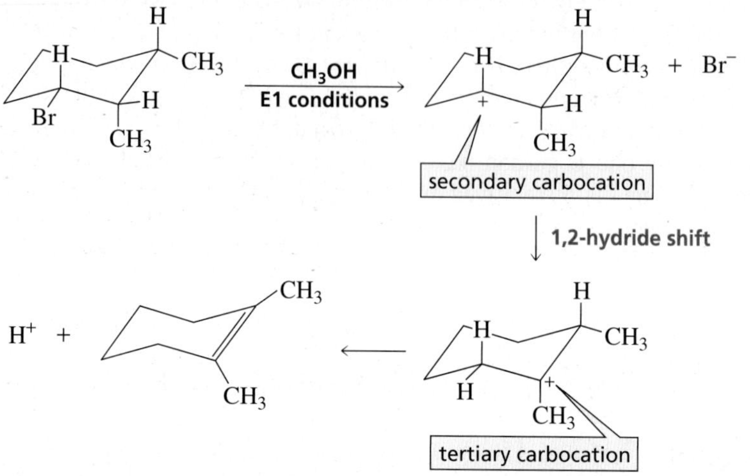

$\xrightarrow[\text{E1 conditions}]{CH_3OH}$ $+ \ Br^-$

secondary carbocation

1,2-hydride shift

$H^+ \ +$

tertiary carbocation

Table 11.4 summarizes the stereochemical outcome of substitution and elimination reactions.

Table 11.4 Stereochemistry of Substitution and Elimination Reactions

Mechanism	Products
S_N1	Both stereosiomers (*R* and *S*) are formed (more inverted than retained).
E1	Both *E* and *Z* stereoisomers are formed (more of the stereoisomer with the bulkiest groups on opposite sides of the double bond).
S_N2	Only the inverted product is formed.
E2	Both *E* and *Z* stereoisomers are formed (more of the stereoisomer with the bulkiest groups on opposite sides of the double bond is formed) unless the β-carbon of the reactant is bonded to only one hydrogen, in which case only one stereoisomer is formed, with a configuration that depends on the configuration of the reactant.

PROBLEM 14

Give the substitution and elimination products for the following reactions, showing the configuration of each product:

a. (*S*)-2-chlorohexane $\xrightarrow[S_N2/E2 \text{ conditions}]{CH_3O^-}$

b. (*S*)-2-chlorohexane $\xrightarrow[S_N1/E1 \text{ conditions}]{CH_3OH}$

c. *trans*-1-chloro-2-methylcyclohexane $\xrightarrow[S_N2/E2 \text{ conditions}]{CH_3O^-}$

d. *trans*-1-chloro-2-methylcyclohexane $\xrightarrow[S_N1/E1 \text{ conditions}]{CH_3OH}$

e. $CH_3CH_2C(CH_3)(H)Br$ $\xrightarrow[S_N2/E2 \text{ conditions}]{CH_3O^-}$

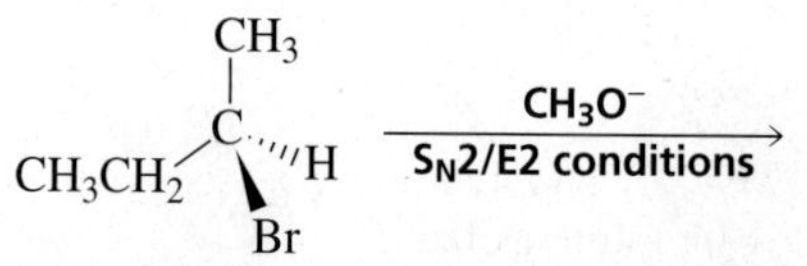

f. $CH_3CH_2C(CH_3)(H)Br$ $\xrightarrow[S_N1/E1 \text{ conditions}]{CH_3OH}$

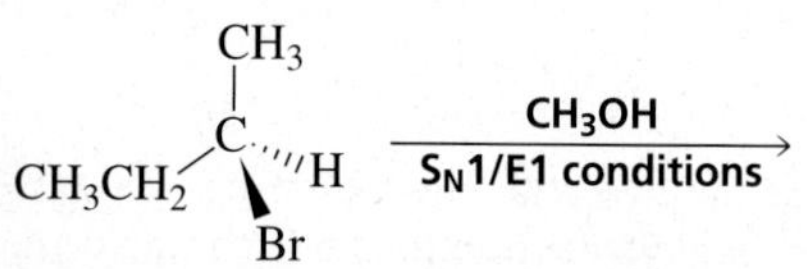

11.7 A Kinetic Isotope Effect

A *mechanism* is a model that accounts for all the experimental evidence that is known about a reaction. For example, the mechanisms of the S_N1, S_N2, E1, and E2 reactions are based on a knowledge of the rate law of the reaction, the relative reactivities of the reactants, and the structures of the products.

Another piece of experimental evidence that is helpful in determining the mechanism of a reaction is the **deuterium kinetic isotope effect**—the ratio of the rate constant observed for a compound containing hydrogen to the rate constant observed for an identical compound in which one or more of the hydrogens has been replaced by deuterium, an isotope of hydrogen. Recall that the nucleus of a deuterium has one proton and one neutron, whereas the nucleus of a hydrogen contains only a proton (Section 1.1).

$$\textbf{deuterium kinetic isotope effect} = \frac{k_H}{k_D} = \frac{\textbf{rate constant for H-containing reactant}}{\textbf{rate constant for D-containing reactant}}$$

The chemical properties of deuterium and hydrogen are similar; however, a C—D bond is about 1.2 kcal/mol (5 kJ/mol) stronger than a C—H bond. Therefore, it is more difficult to break a C—D bond than a corresponding C—H bond.

When the rate constant (k_H) for elimination of HBr from 1-bromo-2-phenylethane is compared with the rate constant (k_D) for elimination of DBr from 2-bromo-1,1-dideuterio-1-phenylethane (determined under identical conditions), k_H is found to be 7.1 times greater than k_D. The deuterium kinetic isotope effect, therefore, is 7.1. The difference in the reaction rates is due to the difference in energy required to break a C—H bond compared with a C—D bond.

$$C_6H_5CH_2CH_2Br + CH_3CH_2O^- \xrightarrow[CH_3CH_2OH]{k_H} C_6H_5CH{=}CH_2 + Br^- + CH_3CH_2OH$$

1-bromo-2-phenylethane

$$C_6H_5CD_2CH_2Br + CH_3CH_2O^- \xrightarrow[CH_3CH_2OH]{k_D} C_6H_5CD{=}CH_2 + Br^- + CH_3CH_2OD$$

2-bromo-1,1-dideuterio-1-phenylethane

Because the deuterium kinetic isotope effect is greater than unity for this reaction, we know that the C—H (or C—D) bond must be broken in the rate-determining step—a fact that is consistent with the mechanism proposed for an E2 reaction.

PROBLEM 15◆

List the following compounds in order of decreasing reactivity in an E2 reaction:

$(CH_3)_3C{-}Cl$ $(CD_3)_3C{-}F$ $(CH_3)_3C{-}F$ $(CH_3)_3C{-}Br$ $CH_3C(CD_3)_2{-}F$

PROBLEM 16◆

If the two reactions described in this section were E1 elimination reactions, what value would you expect to obtain for the deuterium kinetic isotope effect?

11.8 Competition Between Substitution and Elimination

You have seen that alkyl halides can undergo four types of reactions: S_N2, S_N1, E2, and E1. At this point, it may seem a bit overwhelming to be given an alkyl halide and a nucleophile/base and asked to predict the products of the reaction. So we need to organize what we know about the reactions of alkyl halides to make it a little easier to predict the products of any given reaction.

First we must decide whether the reaction conditions favor S_N2/E2 or S_N1/E1 reactions. (Recall that the conditions that favor an S_N2 reaction also favor an E2 reaction and the conditions that favor an S_N1 reaction also favor an E1 reaction.) The decision is easy if the reactant is a *primary* alkyl halide—it undergoes only S_N2/E2 reactions. Primary carbocations are too unstable to be formed, so primary alkyl halides cannot undergo S_N1/E1 reactions.

If the reactant is a *secondary* or a *tertiary* alkyl halide, it may undergo either S_N2/E2 or S_N1/E1 reactions, depending on the reaction conditions. S_N2/E2 reactions are favored by a high concentration of a good nucleophile/strong base, whereas S_N1/E1 reactions are favored by a poor nucleophile/weak base (Sections 10.9 and 11.4). In addition, the solvent in which the reaction is carried out can influence the mechanism (Section 10.10).

Having decided whether the conditions favor S_N2/E2 reactions or S_N1/E1 reactions, we must next decide how much of the product will be the substitution product and how much will be the elimination product. The relative amounts of substitution and elimination products will depend on whether the alkyl halide is primary, secondary, or tertiary, and on the nature of the nucleophile/base. This is discussed in the next section and is summarized later in Table 11.6.

S_N2/E2 Conditions

Let's first consider conditions that lead to S_N2/E2 reactions (a high concentration of a good nucleophile/strong base). The negatively charged species can act as a nucleophile and attack the back side of the α-carbon to form the substitution product, or it can act as a base and remove a β-hydrogen, leading to the elimination product. Thus, the two reactions compete with each other. Notice that both reactions occur for the same reason—the electron-withdrawing halogen gives the carbon to which it is bonded a partial positive charge.

$$CH_3-CH_2-Br \;\; (HO^-) \longrightarrow CH_3CH_2OH + Br^-$$

substitution product

$$CH_2(H)-CH_2-Br \;\; (HO^-) \longrightarrow CH_2{=}CH_2 + H_2O + Br^-$$

elimination product

The relative reactivities of alkyl halides in S_N2 and E2 reactions are shown in Table 11.5. Because a *primary* alkyl halide is the most reactive in an S_N2 reaction and the least reactive in an E2 reaction, a primary alkyl halide forms principally the substitution product in a reaction carried out under conditions that favor S_N2/E2 reactions. In other words, substitution wins the competition.

Table 11.5 Relative Reactivities of Alkyl Halides

In an S_N2 reaction: $1° > 2° > 3°$	In an S_N1 reaction: $3° > 2° > 1°$
In an E2 reaction: $3° > 2° > 1°$	In an E1 reaction: $3° > 2° > 1°$

a primary alkyl halide

$CH_3CH_2CH_2Br$ + CH_3O^- $\xrightarrow{CH_3OH}$ $CH_3CH_2CH_2OCH_3$ + $CH_3CH{=}CH_2$ + CH_3OH + Br^-

propyl bromide — methyl propyl ether 90% — propene 10%

However, if either the primary alkyl halide or the nucleophile/base is sterically hindered, the nucleophile will have difficulty getting to the back side of the α-carbon. As a result, elimination will win the competition, so the elimination product will predominate.

A bulky base encourages elimination over substitution.

the primary alkyl halide is sterically hindered

$CH_3CH(CH_3)CH_2Br$ + CH_3O^- $\xrightarrow{CH_3OH}$ $CH_3CH(CH_3)CH_2OCH_3$ + $CH_3C(CH_3){=}CH_2$ + CH_3OH + Br^-

1-bromo-2-methylpropane — isobutyl methyl ether 40% — 2-methylpropene 60%

the nucleophile is sterically hindered

$CH_3CH_2CH_2CH_2CH_2Br$ + $(CH_3)_3CO^-$ $\xrightarrow{(CH_3)_3COH}$ $CH_3CH_2CH_2CH_2CH_2OC(CH_3)_3$ + $CH_3CH_2CH_2CH{=}CH_2$ + $(CH_3)_3COH$ + Br^-

1-bromopentane — *tert*-butyl pentyl ether 15% — 1-pentene 85%

A *secondary* alkyl halide can form both substitution and elimination products under $S_N2/E2$ conditions. The relative amounts of the two products depend on the base strength and the bulk of the nucleophile/base. *The stronger and bulkier the base, the greater the percentage of the elimination product.* For example, acetic acid is a stronger acid ($pK_a = 4.76$) than ethanol ($pK_a = 15.9$), which means that acetate ion is a weaker base than ethoxide ion. The elimination product is the main product formed from the reaction of 2-chloropropane with the strongly basic ethoxide ion, whereas no elimination product is formed with the weakly basic acetate ion. The percentage of elimination product produced would be increased further if the bulky *tert*-butoxide ion were used instead of ethoxide ion (Section 10.3).

A weak base encourages substitution over elimination.

a secondary alkyl halide

a strong base favors the elimination product

$CH_3CHClCH_3$ + $CH_3CH_2O^-$ $\xrightarrow{CH_3CH_2OH}$ $CH_3CH(OCH_2CH_3)CH_3$ + $CH_3CH{=}CH_2$ + CH_3CH_2OH + Cl^-

2-chloropropane — ethoxide ion — 2-ethoxypropane 25% — propene 75%

$CH_3CHClCH_3$ + CH_3COO^- $\xrightarrow{\text{acetic acid}}$ $CH_3CH(OCOCH_3)CH_3$ + Cl^-

2-chloropropane — acetate ion — isopropyl acetate 100%

a weak base favors the substitution product

A *tertiary* alkyl halide is the least reactive of the alkyl halides in an S_N2 reaction and the most reactive in an E2 reaction (Table 11.5). Consequently, only the elimination product is formed when a tertiary alkyl halide reacts with a nucleophile/base under $S_N2/E2$ conditions.

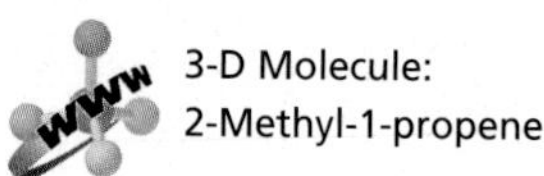

a tertiary alkyl halide

$$\underset{\text{2-bromo-2-methylpropane}}{CH_3\overset{CH_3}{\underset{CH_3}{C}}Br} + CH_3CH_2O^- \xrightarrow{CH_3CH_2OH} \underset{\text{2-methylpropene}}{CH_3\overset{CH_3}{C}{=}CH_2} + CH_3CH_2OH + Br^-$$

PROBLEM 17◆

How would you expect the ratio of substitution product to elimination product formed from the reaction of propyl bromide with CH_3O^- in methanol to change when the nucleophile is changed to CH_3S^-?

PROBLEM 18

Only a substitution product is obtained when the following compound is treated with sodium methoxide:

CH_3 Br CH_3

Explain why an elimination product is not obtained.

S_N1/E1 Conditions

Now let's look at what happens when conditions favor S_N1/E1 reactions (a poor nucleophile/weak base). In S_N1/E1 reactions, the alkyl halide dissociates to form a carbocation, which can then either combine with the nucleophile to form the substitution product or lose a proton to form the elimination product.

$$-\underset{H}{\overset{|}{C}}-\overset{|}{C}-Br \longrightarrow -\underset{H}{\overset{|}{C}}-\overset{|}{C^+} + Br^-$$

$$\xrightarrow[\text{substitution}]{H_2O} -\overset{|}{C}-\underset{H}{\overset{|}{C}}-OH + H_3O^+$$

$$\xrightarrow[\text{elimination}]{H_2O} \; C{=}C \; + H_3O^+$$

Alkyl halides have the same order of reactivity in S_N1 reactions as they do in E1 reactions because both reactions have the same rate-determining step—dissociation of the alkyl halide (Table 11.5). This means that all alkyl halides that react under S_N1/E1 conditions will give both substitution and elimination products. Remember that primary alkyl halides do not undergo S_N1/E1 reactions because primary carbocations are too unstable to be formed.

Table 11.6 summarizes the products obtained when alkyl halides react with nucleophiles/bases under S_N2/E2 and S_N1/E1 conditions.

Table 11.6 Summary of the Products Expected in Substitution and Elimination Reactions

Class of alkyl halide	S_N2 versus E2	S_N1 versus E1
Primary alkyl halide	Primarily substitution, unless there is steric hindrance in the alkyl halide or nucleophile, in which case elimination is favored	Cannot undergo S_N1/E1 reactions
Secondary alkyl halide	Both substitution and elimination; the stronger and bulkier the base, the greater is the percentage of elimination	Both substitution and elimination
Tertiary alkyl halide	Only elimination	Both substitution and elimination

PROBLEM 19◆

Indicate whether the alkyl halides listed will give mostly substitution products, mostly elimination products, or about equal amounts of substitution and elimination products when they react with the following:

a. methanol under $S_N1/E1$ conditions
b. sodium methoxide under $S_N2/E2$ conditions

1. 1-bromobutane
2. 1-bromo-2-methylpropane
3. 2-bromobutane
4. 2-bromo-2-methylpropane

PROBLEM 20◆

1-Bromo-2,2-dimethylpropane has difficulty undergoing either S_N2 or S_N1 reactions.

a. Explain why.
b. Can it undergo E2 and E1 reactions?

11.9 Substitution and Elimination Reactions in Synthesis

When substitution or elimination reactions are used in synthesis, care must be taken to choose reactants and reaction conditions that will maximize the yield of the desired product. In Section 10.4, you saw that nucleophilic substitution reactions of alkyl halides can lead to a wide variety of organic compounds. For example, ethers are synthesized by the reaction of an alkyl halide with an alkoxide ion. This reaction, discovered by Alexander Williamson in 1850, is still considered one of the best ways to synthesize an ether.

Alexander William Williamson (1824–1904) *was born in London to Scottish parents. As a child, he lost an arm and the use of an eye. He was midway through his medical education when he changed his mind and decided to study chemistry. He received a Ph.D. from the University of Geissen in 1846. In 1849, he became a professor of chemistry at University College, London.*

Williamson ether synthesis

$$\underset{\text{alkyl halide}}{R{-}Br} + \underset{\text{alkoxide ion}}{R{-}O^-} \longrightarrow \underset{\text{ether}}{R{-}O{-}R} + Br^-$$

The alkoxide ion (RO^-) for the **Williamson ether synthesis** is prepared by using sodium metal or sodium hydride (NaH) to remove a proton from an alcohol.

$$ROH + Na \longrightarrow RO^- + Na^+ + \tfrac{1}{2}H_2$$

$$ROH + NaH \longrightarrow RO^- + Na^+ + H_2$$

The Williamson ether synthesis is a nucleophilic substitution reaction. It is an S_N2 reaction because it requires a high concentration of a good nucleophile. If you want to synthesize an ether such as butyl propyl ether, you have a choice of starting materials: You can use either a propyl halide and butoxide ion or a butyl halide and propoxide ion.

$$\underset{\text{propyl bromide}}{CH_3CH_2CH_2Br} + \underset{\text{butoxide ion}}{CH_3CH_2CH_2CH_2O^-} \longrightarrow \underset{\text{butyl propyl ether}}{CH_3CH_2CH_2OCH_2CH_2CH_2CH_3} + Br^-$$

$$\underset{\text{butyl bromide}}{CH_3CH_2CH_2CH_2Br} + \underset{\text{propoxide ion}}{CH_3CH_2CH_2O^-} \longrightarrow \underset{\text{butyl propyl ether}}{CH_3CH_2CH_2OCH_2CH_2CH_2CH_3} + Br^-$$

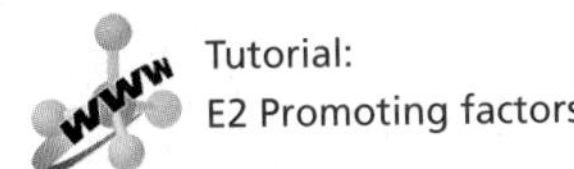

If, however, you want to synthesize *tert*-butyl ethyl ether, the starting materials must be an ethyl halide and *tert*-butoxide ion. If you tried to use a *tert*-butyl halide and ethoxide ion as reactants, you would obtain the elimination product and little or no ether because the reaction of a tertiary alkyl halide under $S_N2/E2$ conditions forms primarily the elimination product. So, in carrying out a Williamson ether synthesis, the less hindered alkyl group should be provided by the alkyl halide and the more hindered alkyl group should come from the alkoxide.

$$\underset{\text{ethyl bromide}}{CH_3CH_2Br} + \underset{\textit{tert}\text{-butoxide ion}}{(CH_3)_3CO^-} \longrightarrow \underset{\textit{tert}\text{-butyl ethyl ether}}{CH_3CH_2OC(CH_3)_3} + \underset{\text{ethene}}{CH_2{=}CH_2} + (CH_3)_3COH + Br^-$$

$$\underset{\text{ethoxide ion}}{CH_3CH_2O^-} + \underset{\textit{tert}\text{-butyl bromide}}{(CH_3)_3CBr} \longrightarrow \underset{\text{2-methylpropene}}{CH_2{=}C(CH_3)CH_3} + CH_3CH_2OH + Br^-$$

In synthesizing an ether, the less hindered group should be provided by the alkyl halide.

PROBLEM 21◆

What other organic product will be formed when the alkyl halide used in the synthesis of butyl propyl ether is

a. propyl bromide?
b. butyl bromide?

PROBLEM 22

How could the following ethers be prepared, using an alkyl halide and an alcohol?

a. $CH_3CH_2CH(CH_3)OCH_2CH_2CH_3$

b. cyclohexyl–$OC(CH_3)_3$

c. phenyl–CH_2O–phenyl

d. $CH_3CH_2OCH_2CH(CH_3)CH_2CH_2CH_3$

In Section 6.10, you saw that alkynes can be synthesized by the reaction of an acetylide anion with an alkyl halide.

$$CH_3CH_2C{\equiv}C^- + CH_3CH_2CH_2Br \longrightarrow CH_3CH_2C{\equiv}CCH_2CH_2CH_3 + Br^-$$

Now that you know that this is an S_N2 reaction (the alkyl halide reacts with a high concentration of a good nucleophile), you can understand why it is best to use primary alkyl halides and methyl halides in the reaction. These alkyl halides are the only ones that form primarily the desired substitution product. A tertiary alkyl halide would form only the elimination product, and a secondary alkyl halide would form mainly the elimination product because the acetylide ion is a very strong base.

If you want to synthesize an alkene, you should choose the most hindered alkyl halide in order to maximize the elimination product and minimize the substitution product. For example, 2-bromopropane would be a better starting material than 1-bromopropane for the synthesis of propene because the secondary alkyl halide would give a higher yield of the desired elimination product and a lower yield of the competing substitution product. The percentage of alkene could be further increased by using a sterically hindered base such as *tert*-butoxide ion instead of hydroxide ion.

$$\underset{\text{2-bromopropane}}{CH_3CH(Br)CH_3} + HO^- \longrightarrow \underset{\text{major product}}{CH_3CH{=}CH_2} + \underset{\text{minor product}}{CH_3CH(OH)CH_3} + H_2O + Br^-$$

$$\underset{\text{1-bromopropane}}{CH_3CH_2CH_2Br} + HO^- \longrightarrow \underset{\text{minor product}}{CH_3CH{=}CH_2} + \underset{\text{major product}}{CH_3CH_2CH_2OH} + H_2O + Br^-$$

To synthesize 2-methyl-2-butene from 2-bromo-2-methylbutane, you would use $S_N2/E2$ conditions (a high concentration of HO^- in an aprotic polar solvent) because a tertiary alkyl halide gives *only* the elimination product under those conditions. If $S_N1/E1$ conditions were used (a low concentration of HO^- in water), both elimination and substitution products would be obtained.

$$CH_3C(CH_3)(Br)CH_2CH_3 + HO^- \xrightarrow{S_N2/E2 \text{ conditions}} CH_3C(CH_3){=}CHCH_3 + H_2O + Br^-$$

$$CH_3C(CH_3)(Br)CH_2CH_3 + HO^- \xrightarrow{S_N1/E1 \text{ conditions}} CH_3C(CH_3){=}CHCH_3 + CH_3C(CH_3)(OH)CH_2CH_3 + H_2O + Br^-$$

2-bromo-2-methylbutane; 2-methyl-2-butene; 2-methyl-2-butanol

PROBLEM 23◆

Identify the three products that are formed when 2-bromo-2-methylpropane is dissolved in a mixture of 80% ethanol and 20% water.

PROBLEM 24

a. What products (including stereoisomers if applicable) would be formed from the reaction of 3-bromo-2-methylpentane with HO^- under $S_N2/E2$ conditions and under $S_N1/E1$ conditions?
b. Answer the same question for 3-bromo-3-methylpentane.

11.10 Consecutive E2 Elimination Reactions

Alkyl dihalides can undergo two consecutive dehydrohalogenations, giving products that contain two double bonds. In the following example, Zaitsev's rule predicts the most stable product of the first dehydrohalogenation, but not the most stable product of the second. The reason Zaitsev's rule fails in the second reaction is that a conjugated diene is more stable than an isolated diene.

$$CH_3CH(CH_3)CH(Cl)CH_2CH(Cl)CH(CH_3)CH_3 \xrightarrow{HO^-} CH_3C(CH_3){=}CHCH_2CH(Cl)CH(CH_3)CH_3 \xrightarrow{HO^-} CH_3C(CH_3){=}CHCH{=}CHCH(CH_3)CH_3 + 2\,H_2O + 2\,Cl^-$$

3,5-dichloro-2,6-dimethylheptane; 5-chloro-2,6-dimethyl-2-heptene; 2,6-dimethyl-2,4-heptadiene

If the two halogens are on the same carbon (geminal dihalides) or on adjacent carbons (vicinal dihalides), the two consecutive E2 dehydrohalogenations can result in the formation of a triple bond. This is how alkynes are commonly synthesized.

$$RCH_2CR(Br)_2 \xrightarrow{^-NH_2} RCH{=}CR(Br) \xrightarrow{^-NH_2} RC{\equiv}CR + 2\,NH_3 + 2\,Br^-$$

a geminal dibromide; a vinylic bromide; an alkyne

$$RCH(Cl)CH(Cl)R \xrightarrow{^-NH_2} RCH{=}CR(Cl) \xrightarrow{^-NH_2} RC{\equiv}CR + 2\,NH_3 + 2\,Br^-$$

a vicinal dichloride; a vinylic chloride; an alkyne

The vinylic halide intermediates in the preceding reactions are relatively unreactive. Consequently, a very strong base, such as $^-NH_2$, is needed for the second elimination. If a weaker base, such as HO^-, is used at room temperature, the reaction will stop at the vinylic halide and no alkyne will be formed.

Because a vicinal dihalide is formed from the reaction of an alkene with Br_2 or Cl_2, you have just learned how to convert a double bond into a triple bond.

$$\underset{\text{2-butene}}{CH_3CH{=}CHCH_3} \xrightarrow[CH_2Cl_2]{Br_2} CH_3\underset{Br}{CH}\underset{Br}{CH}CH_3 \xrightarrow{^-NH_2} CH_3CH{=}\underset{Br}{C}CH_3 \xrightarrow{^-NH_2} \underset{\text{2-butyne}}{CH_3C{\equiv}CCH_3} + 2\,NH_3 + 2\,Br^-$$

PROBLEM 25

Why isn't a cumulated diene formed in the preceding reaction?

11.11 Intermolecular Versus Intramolecular Reactions

A molecule with two functional groups is called a **bifunctional molecule**. If the two functional groups are able to react with each other, two kinds of reactions can occur. If we take as an example a molecule whose two functional groups are a nucleophile and a leaving group, the nucleophile of one molecule can displace the leaving group of a second molecule of the compound. Such a reaction is called an intermolecular reaction. *Inter* is Latin for "between," so an **intermolecular reaction** takes place between two molecules. If the product of each intermolecular reaction subsequently reacts with another bifunctional molecule, a polymer will be formed (Chapter 28).

an intermolecular reaction

$$BrCH_2(CH_2)_nCH_2\ddot{\underset{\cdot\cdot}{O}}{:}^- \quad Br{-}CH_2(CH_2)_nCH_2\ddot{\underset{\cdot\cdot}{O}}{:}^- \longrightarrow BrCH_2(CH_2)_nCH_2\ddot{\underset{\cdot\cdot}{O}}CH_2(CH_2)_nCH_2\ddot{\underset{\cdot\cdot}{O}}{:}^- + Br^-$$

Alternatively, the nucleophile of a molecule can displace the leaving group of the same molecule, thereby forming a cyclic compound. Such a reaction is called an intramolecular reaction. *Intra* is Latin for "within," so an **intramolecular reaction** takes place within a single molecule.

an intramolecular reaction

$$Br{-}CH_2(CH_2)_nCH_2\ddot{\underset{\cdot\cdot}{O}}{:}^- \longrightarrow H_2C\underset{\ddot{\underset{\cdot\cdot}{O}}}{\overset{(CH_2)_n}{\frown}}CH_2 + Br^-$$

Which reaction is more likely to occur—an intermolecular reaction or an intramolecular reaction? The answer depends on the *concentration* of the bifunctional molecule and the *size of the ring* that will be formed in the intramolecular reaction.

The intramolecular reaction has an advantage: The reacting groups are tethered together, so they don't have to diffuse through the solvent to find a group with which to react. Therefore, a low concentration of reactant favors an intramolecular reaction because the two functional groups have a better chance of finding one another if they are in the same molecule. A high concentration of reactant helps compensate for the advantage gained by tethering.

How much of an advantage an intramolecular reaction has over an intermolecular reaction depends on the size of the ring that is formed—that is, on the length of the tether. If the intramolecular reaction forms a five- or six-membered ring, it will be highly favored over the intermolecular reaction because five- and six-membered rings are stable and, therefore, easily formed.

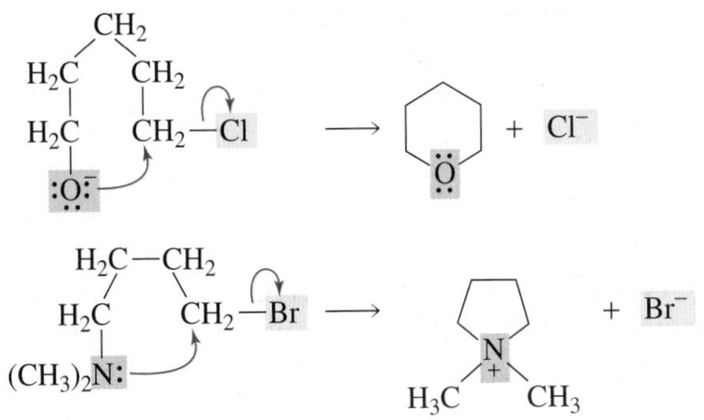

H_2C—CH_2—CH_2—CH_2—CH_2—Cl with :Ö:⁻ ⟶ (cyclic ether, Ö) + Cl^-

$(CH_3)_2N$:—H_2C—H_2C—CH_2—CH_2—Br ⟶ (N^+ ring with H_3C, CH_3) + Br^-

Three- and four-membered rings are strained, so they are less stable than five- and six-membered rings and, therefore, are less easily formed. The higher activation energy for three- and four-membered ring formation cancels some of the advantage gained by tethering.

Three-membered ring compounds are formed more easily than four-membered ring compounds. To form a cyclic ether, the nucleophilic oxygen atom must be oriented so that it can attack the back side of the carbon bonded to the halogen. Rotation about a C—C bond in the molecule forms conformations with the groups pointed away from one another. The molecule leading to formation of a three-membered ring ether has one C—C bond that can rotate, whereas the molecule leading to formation of a four-membered ring has two C—C bonds that can rotate. Therefore, molecules that form three-membered rings are more apt to have their reacting groups in the conformation required for reaction.

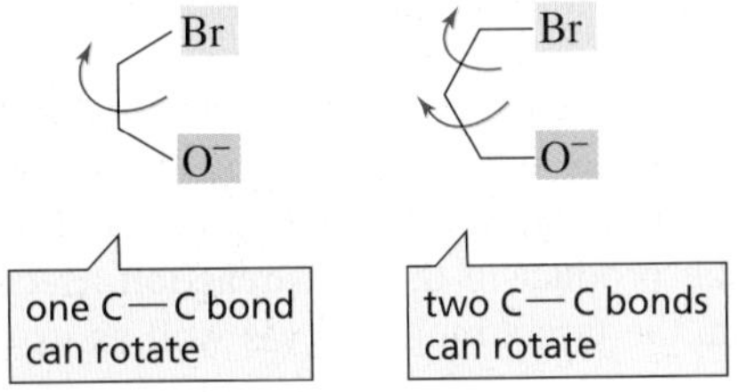

Br, O^- — one C—C bond can rotate

Br, O^- — two C—C bonds can rotate

The likelihood of the reacting groups finding each other decreases sharply when the groups are in compounds that would form seven-membered and larger rings. Therefore, the intramolecular reaction becomes less favored as the ring size increases beyond six members.

PROBLEM 26◆

Which compound, upon treatment with sodium hydride, would form a cyclic ether more rapidly?

a. 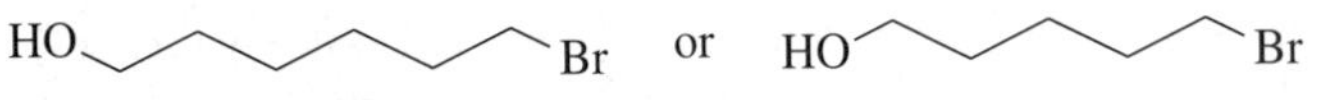 HO⁀⁀⁀Br or HO⁀⁀⁀Br

b. HO⁀⁀Br or HO⁀⁀Br

c. 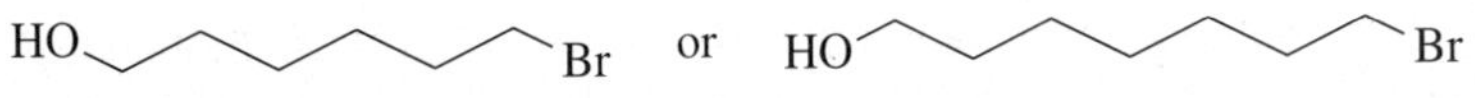 HO⁀⁀⁀Br or HO⁀⁀⁀Br

11.12 Designing a Synthesis II: Approaching the Problem

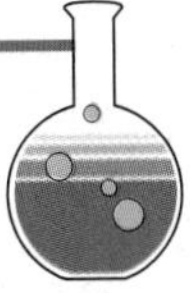

When you are asked to design a synthesis, one way to approach the problem is to look at the given starting material to see whether there is an obvious series of reactions that can get you started on the pathway to the **target molecule** (the desired product). Sometimes this is the best way to approach a *simple* synthesis. The examples that follow will give you practice in designing a successful synthesis.

Example 1. How could you prepare 1,3-cyclohexadiene from cyclohexane?

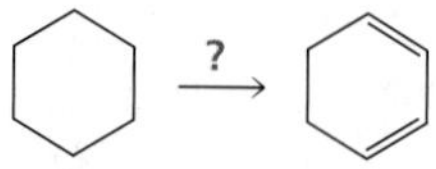

$\xrightarrow{?}$

Since the only reaction an alkane can undergo is halogenation, deciding what the first reaction should be is easy. An E2 reaction using a high concentration of a strong and bulky base to encourage elimination over substitution will form cyclohexene. Therefore, *tert*-butoxide ion is used as the base and *tert*-butyl alcohol is used as the solvent. Bromination of cyclohexene will give an allylic bromide, which will form the desired target molecule by undergoing another E2 reaction.

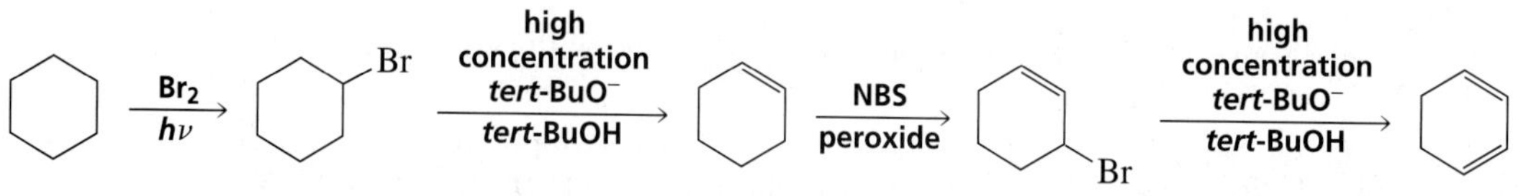

$\xrightarrow[h\nu]{Br_2}$ Br $\xrightarrow[\textit{tert}\text{-BuOH}]{\text{high concentration } \textit{tert}\text{-BuO}^-}$ $\xrightarrow[\text{peroxide}]{\text{NBS}}$ Br $\xrightarrow[\textit{tert}\text{-BuOH}]{\text{high concentration } \textit{tert}\text{-BuO}^-}$

Example 2. Starting with methylcyclohexane, how could you prepare the following vicinal *trans*-dihalide?

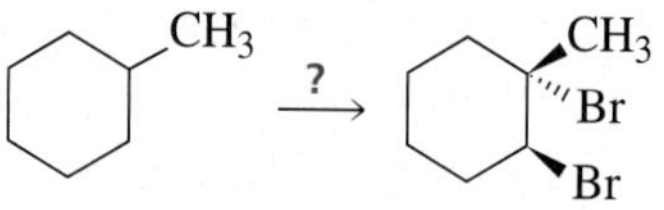

CH_3 $\xrightarrow{?}$ CH_3 Br Br

Again, since the starting material is an alkane, the first reaction must be a radical substitution. Bromination leads to selective substitution of the tertiary hydrogen. Under E2 conditions, tertiary alkyl halides undergo only elimination, so there will be no competing substitution product in the next reaction. Because addition of Br_2 involves only anti addition, the target molecule (as well as its enantiomer) is obtained.

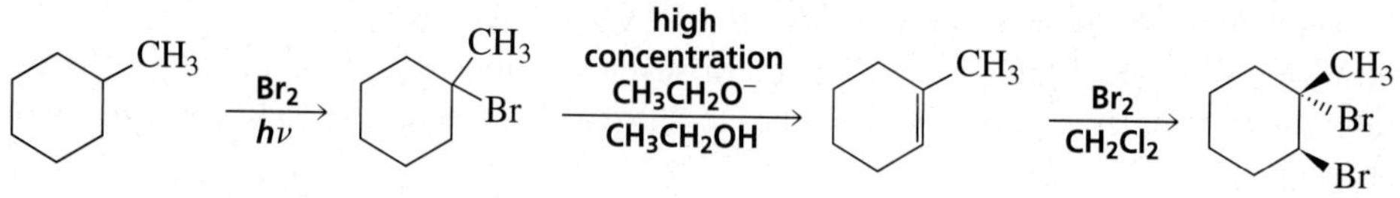

CH_3 $\xrightarrow[h\nu]{Br_2}$ CH_3 Br $\xrightarrow[CH_3CH_2OH]{\text{high concentration } CH_3CH_2O^-}$ CH_3 $\xrightarrow[CH_2Cl_2]{Br_2}$ CH_3 Br Br

As you saw in Section 6.11, working backward can be a useful way to design a synthesis—particularly when the starting material does not clearly indicate how to proceed. Look at the target molecule and ask yourself how it could be prepared. Once you have an answer, work backward to the next compound, asking yourself how *it* could be prepared. Keep working backward one step at a time, until you get to a readily available starting material. This technique is called *retrosynthetic analysis*.

Example 3. How could you prepare ethyl methyl ketone from 1-bromobutane?

$$CH_3CH_2CH_2CH_2Br \xrightarrow{?} CH_3CH_2\overset{\overset{\large O}{\|}}{C}CH_3$$

At this point, you know only one way to synthesize a ketone—the addition of water to an alkyne (Section 6.6). The alkyne can be prepared from two successive E2 reactions of a vicinal dihalide, which in turn can be synthesized from an alkene. The desired alkene can be prepared from the given starting material by an elimination reaction.

retrosynthetic analysis

$$\underset{\text{target molecule}}{CH_3CH_2\overset{O}{\overset{\|}{C}}CH_3} \Longrightarrow CH_3CH_2C{\equiv}CH \Longrightarrow CH_3CH_2\underset{Br}{\underset{|}{C}}HCH_2Br \Longrightarrow CH_3CH_2CH{=}CH_2 \Longrightarrow CH_3CH_2CH_2CH_2Br$$

Now the reaction sequence can be written in the forward direction, indicating the reagents needed to carry out each reaction. A bulky base is used in the elimination reaction in order to maximize the amount of elimination product.

synthesis

$$CH_3CH_2CH_2CH_2Br \xrightarrow[\textit{tert}\text{-BuOH}]{\text{high concentration } \textit{tert}\text{-BuO}^-} CH_3CH_2CH{=}CH_2 \xrightarrow[CH_2Cl_2]{Br_2} CH_3CH_2\underset{Br}{\underset{|}{C}}HCH_2Br \xrightarrow{^-NH_2} CH_3CH_2C{\equiv}CH \xrightarrow[H_2SO_4,\ HgSO_4]{H_2O} \underset{\text{target molecule}}{CH_3CH_2\overset{O}{\overset{\|}{C}}CH_3}$$

Example 4. How could the following cyclic ether be prepared from the given starting material?

$$BrCH_2CH_2CH_2CH{=}CH_2 \xrightarrow{?} \text{2-methyltetrahydrofuran (O ring, } CH_3\text{)}$$

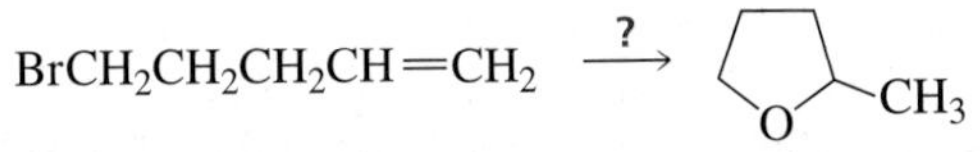

In order to obtain a cyclic ether, the two groups required for ether synthesis (the alkyl halide and the alcohol) must be in the same molecule. To obtain a five-membered ring, the carbons bearing the two groups must be separated by two additional carbons. Addition of water to the given starting material will give the required bifunctional compound.

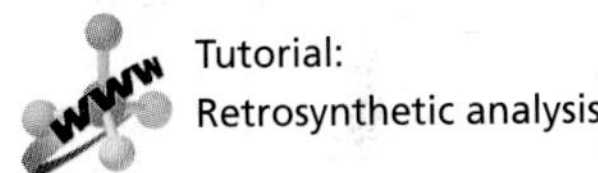

retrosynthetic analysis

$$\underset{\text{target molecule}}{\text{(cyclic ether, O, } CH_3)} \Longrightarrow BrCH_2CH_2CH_2\underset{OH}{\underset{|}{C}}HCH_3 \Longrightarrow BrCH_2CH_2CH_2CH{=}CH_2$$

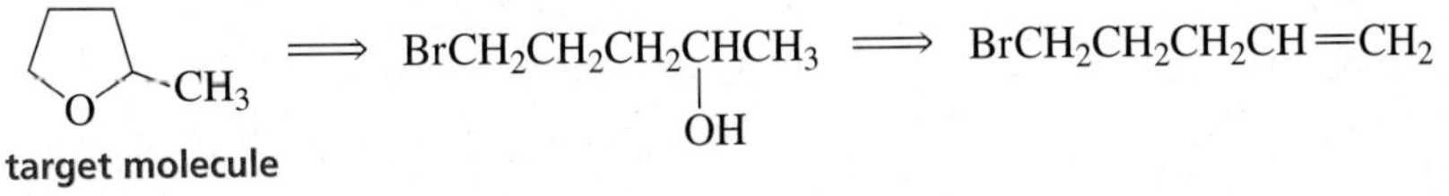

synthesis

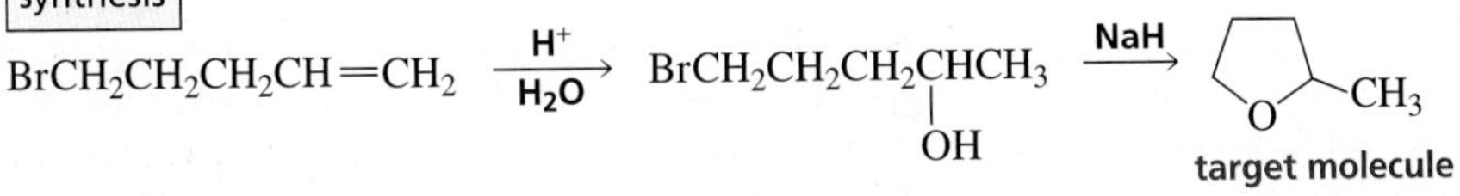

$$BrCH_2CH_2CH_2CH{=}CH_2 \xrightarrow[H_2O]{H^+} BrCH_2CH_2CH_2\underset{OH}{\underset{|}{C}}HCH_3 \xrightarrow{NaH} \underset{\text{target molecule}}{\text{(cyclic ether, O, } CH_3)}$$

PROBLEM 27

How could you have prepared the target molecule in the preceding synthesis using 4-penten-1-ol as the starting material? Which synthesis would give you a higher yield of the target molecule?

PROBLEM 28

Design a multistep synthesis to show how each of the following compounds could be prepared from the given starting material:

a. 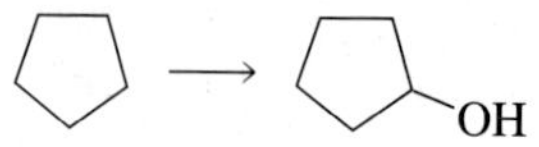 cyclopentane ⟶ cyclopentanol (OH)

b.  cyclohexane ⟶ cyclohexene oxide (O)

c. 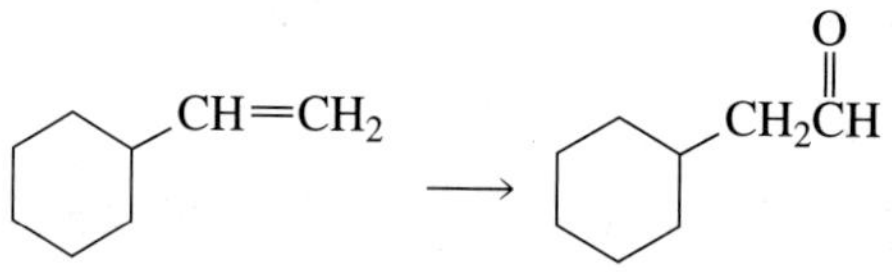 cyclohexyl–$CH{=}CH_2$ ⟶ cyclohexyl–$CH_2\overset{O}{\overset{\|}{C}}H$

d. $CH_3CH_2CH_2CH_2Br$ ⟶ $CH_3CH_2\overset{O}{\overset{\|}{C}}CH_2CH_2CH_3$

e. $BrCH_2CH_2CH_2CH_2Br$ ⟶ cyclohexyl–CH_2CH_3

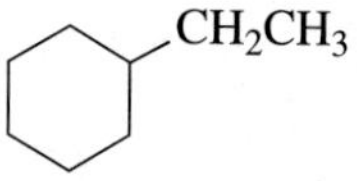

Summary

In addition to undergoing nucleophilic substitution reactions, alkyl halides undergo **β-elimination reactions**: The halogen is removed from one carbon and a proton is removed from an adjacent carbon. A double bond is formed between the two carbons from which the atoms are eliminated. Therefore, the product of an elimination reaction is an alkene. Removal of a proton and a halide ion is called **dehydrohalogenation**. There are two important β-elimination reactions, E1 and E2.

An **E2 reaction** is a concerted, one-step reaction; the proton and the halide ion are removed in the same step, so no intermediate is formed. In an **E1 reaction**, the alkyl halide dissociates, forming a carbocation. In a second step, a base removes a proton from a carbon that is adjacent to the positively charged carbon. Because the E1 reaction forms a carbocation intermediate, the carbon skeleton can rearrange before the proton is lost.

Primary alkyl halides undergo only E2 elimination reactions. Secondary and tertiary alkyl halides undergo both E2 and E1 reactions. For alkyl halides that can undergo both E2 and E1 reactions, the E2 reaction is favored by the same factors that favor an S_N2 reaction—a high concentration of a strong base and an aprotic polar solvent—and the E1 reaction is favored by the same factors that favor an S_N1 reaction—a weak base and a protic polar solvent.

An E2 reaction is regioselective. The major product is the more stable alkene, unless the reactants are sterically hindered or the leaving group is poor. The more stable alkene is generally (but not always) the more substituted alkene. The more substituted alkene is predicted by Zaitsev's rule: It is the alkene formed when a proton is removed from the β-carbon that is bonded to the fewest hydrogens.

An E2 reaction is stereoselective: If the β-carbon has two hydrogens, both E and Z products will be formed, but the one with the bulkiest groups on opposite sides of the double bond is more stable and will be formed in greater yield. **Anti elimination** is favored in an E2 reaction. If the β-carbon has only one hydrogen, only one alkene is formed, since there is only one conformer in which the groups to be eliminated are anti. If the reactant is a cyclic compound, the two groups to be eliminated must be trans to one another; in the case of six-membered rings, both groups must be in axial positions. Elimination is more rapid when H and X are diaxial in the more stable conformer.

An E1 reaction is regioselective. The major product is the most stable alkene, which is generally the most substituted alkene. An E1 reaction is stereoselective. The major product is the alkene with the bulkiest groups on opposite sides of the double bond. The carbocation formed in the first step can undergo both syn and anti elimination; therefore, the two groups to be eliminated in a cyclic compound do not have to be trans or both in axial positions. Alkyl substitution increases the stability of a carbocation and decreases the stability of a carbanion.

Predicting which products are formed when an alkyl halide undegoes a reaction begins with determining whether the conditions favor S_N2/E2 or S_N1/E1 reactions. When S_N2/E2 reactions are favored, primary alkyl halides form primarily substitution products unless the nucleophile/base is sterically hindered, in which case elimination products predominate. Secondary alkyl halides form both substitution and elimination products; the stronger and bulkier the base, the greater is the percentage of the elimination product. Tertiary alkyl halides form only elimination products. When S_N1/E1 conditions are favored, secondary and tertiary alkyl halides form both substitution and elimination products; primary alkyl halides do not undergo S_N1/E1 reactions.

If the two halogens are on the same or adjacent carbons, two consecutive E2 dehydrohalogenations can result in the formation of a triple bond. The **Williamson ether synthesis** involves the reaction of an alkyl halide with an alkoxide ion. If the two functional groups of a **bifunctional molecule** can react with each other, both **intermolecular** and **intramolecular reactions** can occur. The reaction that is more likely to occur depends on the concentration of the bifunctional molecule and the size of the ring that will be formed in the intramolecular reaction.

Summary of Reactions

1. E2 reaction: a one-step mechanism

$$B^- + -\overset{H}{\underset{|}{\overset{|}{C}}}-\underset{|}{\overset{|}{C}}-X \longrightarrow \;\rangle C{=}C\langle\; + BH + X^-$$

Relative reactivities of alkyl halides: 3° > 2° > 1°

Anti elimination; both *E* and *Z* stereoisomers are formed. The isomer with the bulkiest groups on opposite sides of the double bond will be formed in greater yield. If the β-carbon from which the hydrogen is removed is bonded to only one hydrogen, only one elimination product is formed. Its configuration depends on the configuration of the reactant.

2. E1 reaction: a two-step mechanism with a carbocation intermediate

$$-\underset{H}{\overset{|}{C}}-\underset{|}{\overset{|}{C}}-X \longrightarrow -\underset{H}{\overset{|}{C}}-\underset{|}{\overset{|}{C}}^{+} \xrightarrow{B} \;\rangle C{=}C\langle\; + \overset{+}{B}H$$
$$+ X^-$$

Relative reactivities of alkyl halides: 3° > 2° > 1°

Anti and syn elimination; both *E* and *Z* stereoisomers are formed. The isomer with the bulkiest groups on opposite sides of the double bond will be formed in greater yield.

Competing S_N2 and E2 Reactions
- Primary alkyl halides: primarily substitution
- Secondary alkyl halides: substitution and elimination
- Tertiary alkyl halides: only elimination

Competing S_N1 and E1 Reactions
- Primary alkyl halides: cannot undergo S_N1 or E1 reactions
- Secondary alkyl halides: substitution and elimination
- Tertiary alkyl halides: substitution and elimination

Key Terms

anti elimination (p. 413)
anti-periplanar (p. 413)
bifunctional molecule (p. 428)
dehydrohalogenation (p. 402)
deuterium kinetic isotope effect (p. 421)
elimination reaction (p. 400)
β-elimination reaction (p. 402)
E1 reaction (p. 408)
E2 reaction (p. 401)
1,2-elimination reaction (p. 402)
intermolecular reaction (p. 428)
intramolecular reaction (p. 428)
syn elimination (p. 413)
syn-periplanar (p. 413)
target molecule (p. 429)
Williamson ether synthesis (p. 425)
Zaitsev's rule (p. 404)

Problems

29. Give the major product obtained when each of the following alkyl halides undergoes an E2 reaction:

a. $CH_3CH(Br)CH_2CH_3$

b. $CH_3CH(Cl)CH_2CH_3$

c. $CH_3CH(Cl)CH_2CH_2CH_3$

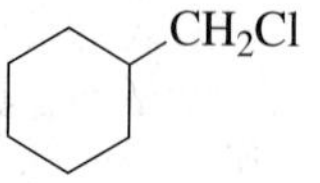

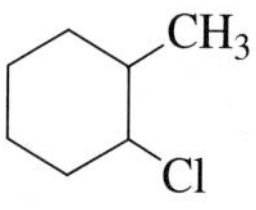

d. 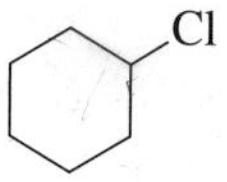 chlorocyclohexane (Cl)

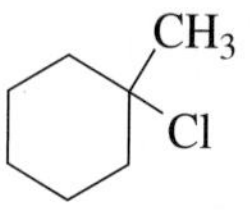

e. cyclohexyl–CH_2Cl

f. 1-chloro-1-methylcyclohexane (CH_3, Cl)

g. 1-chloro-2-methylcyclohexane (CH_3, Cl)

30. Give the major product obtained when the alkyl halides in Problem 29 undergo an E1 reaction.

31. a. Indicate how each of the following factors affects an E1 reaction:
 1. the structure of the alkyl halide
 2. the strength of the base
 3. the concentration of the base
 4. the solvent

 b. Indicate how each of the same factors affects an E2 reaction.

32. Dr. Don T. Doit wanted to synthesize the anesthetic 2-ethoxy-2-methylpropane. He used ethoxide ion and 2-chloro-2-methylpropane for his synthesis and ended up with very little ether. What was the predominant product of his synthesis? What reagents should he have used?

33. Which reactant in each of the following pairs will undergo an elimination reaction more rapidly? Explain your choice.

a. $(CH_3)_3CCl \xrightarrow[H_2O]{HO^-}$

 $(CH_3)_3CI \xrightarrow[H_2O]{HO^-}$

b. 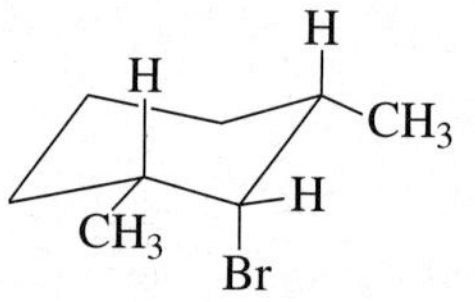 chair cyclohexane with H, H, CH_3, H, CH_3, Br (Br axial) $\xrightarrow[CH_3OH]{CH_3O^-}$

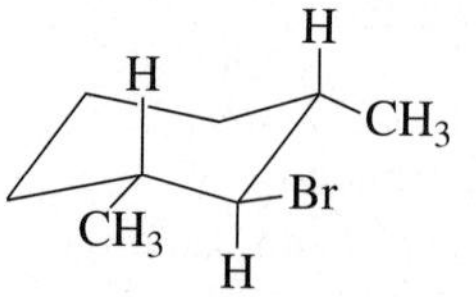

 chair cyclohexane with H, H, CH_3, Br, CH_3, H (Br equatorial) $\xrightarrow[CH_3OH]{CH_3O^-}$

34. For each of the following reactions, give the elimination products; if the products can exist as stereoisomers, indicate which isomers are obtained.

a. (*R*)-2-bromohexane + high concentration of HO^-
b. (*R*)-2-bromohexane + H_2O
c. *trans*-1-chloro-2-methylcyclohexane + high concentration of CH_3O^-
d. *trans*-1-chloro-2-methylcyclohexane + CH_3OH
e. 3-bromo-3-methylpentane + high concentration of HO^-
f. 3-bromo-3-methylpentane + H_2O

35. Indicate which of the compounds in each pair will give a higher substitution/elimination ratio when it reacts with isopropyl bromide:

a. ethoxide ion or *tert*-butoxide ion
b. ^-OCN or ^-SCN
c. Cl^- or Br^-
d. CH_3S^- or CH_3O^-

36. Rank the following compounds in order of decreasing reactivity in an E2 reaction:

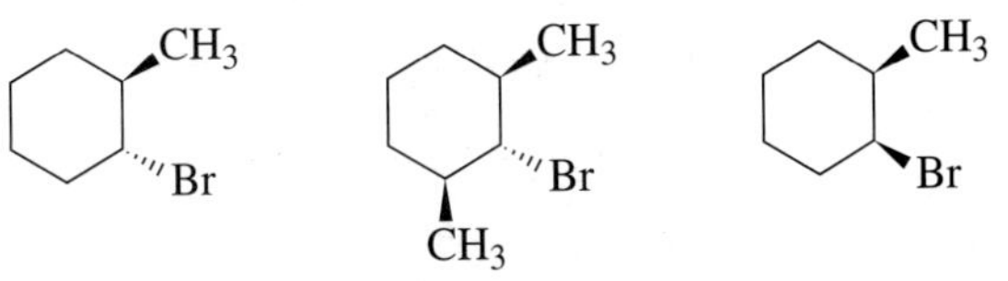

37. Using the given starting material and any necessary organic or inorganic reagents, indicate how the desired compounds could be synthesized:

a. 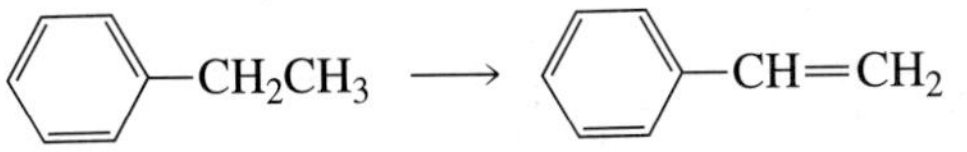 $C_6H_5CH_2CH_3 \longrightarrow C_6H_5CH{=}CH_2$

d. 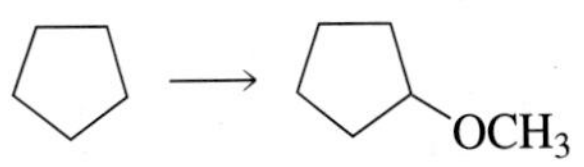 cyclopentane $\longrightarrow$ methoxycyclopentane (OCH_3)

b. $CH_3CH_2CH{=}CH_2 \longrightarrow CH_2CH_2CH_2CH_2NH_2$

e. $CH_3CH_2CH{=}CH_2 \longrightarrow CH_2{=}CHCH{=}CH_2$

c. $HOCH_2CH_2CH{=}CH_2 \longrightarrow$ tetrahydrofuran

38. When 2-bromo-2,3-dimethylbutane reacts with a base under E2 conditions, two alkenes (2,3-dimethyl-1-butene and 2,3-dimethyl-2-butene) are formed.
a. Which of the bases shown would give the highest percentage of the 1-alkene?
b. Which would give the highest percentage of the 2-alkene?

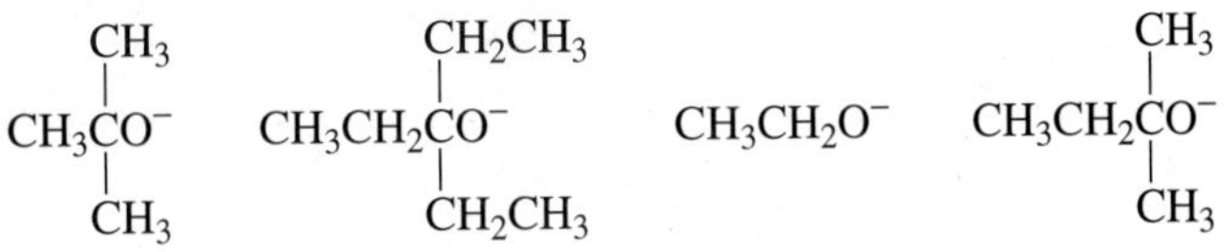

$(CH_3)_3CO^-$ $\quad CH_3CH_2C(CH_2CH_3)_2O^-$ $\quad CH_3CH_2O^-$ $\quad CH_3CH_2C(CH_3)_2O^-$

39. a. Give the structure of the products obtained from the reaction of each enantiomer of *cis*-1-chloro-2-isopropylcyclopentane with a high concentration of sodium methoxide in methanol.
b. Are all the products optically active?
c. How would the products differ if the starting material were the trans isomer? Are all these products optically active?
d. Will the cis enantiomers or the trans enantiomers form substitution products more rapidly?
e. Will the cis enantiomers or the trans enantiomers form elimination products more rapidly?

40. Starting with cyclohexane, how could the following compounds be prepared?
a. *trans*-1,2-dichlorocyclohexane
b. 2-cyclohexenol

41. The rate constant of an intramolecular reaction depends on the size of the ring (n) that is formed. Explain the relative rates of formation of the cyclic secondary ammonium ions.

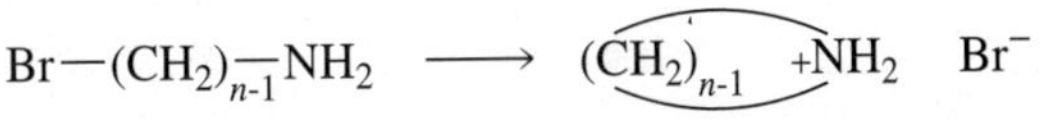

$$Br{-}(CH_2)_{n-1}{-}NH_2 \longrightarrow (CH_2)_{n-1}\,\overset{+}{N}H_2 \quad Br^-$$

n =	3	4	5	6	7	10
relative rate:	1×10^{-1}	2×10^{-3}	100	1.7	3×10^{-3}	1×10^{-8}

42. *cis*-1-Bromo-4-*tert*-butylcyclohexane and *trans*-1-bromo-4-*tert*-butylcyclohexane both react with sodium ethoxide in ethanol to give 4-*tert*-butylcyclohexene. Explain why the cis isomer reacts much more rapidly than the trans isomer.

43. Cardura®, a drug used to treat hypertension, is synthesized as follows:

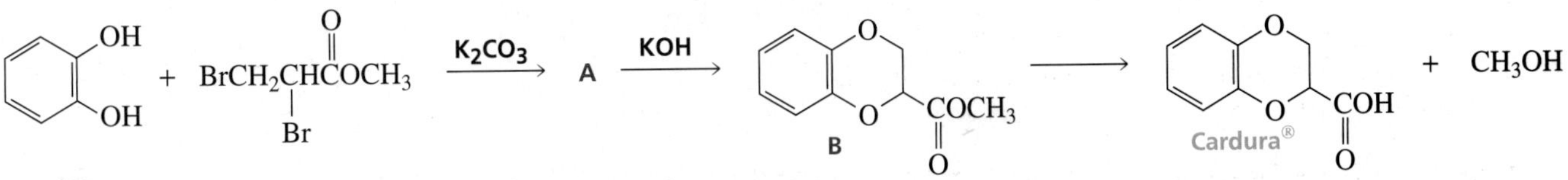

catechol + $BrCH_2CH(Br)COCH_3$ (with C=O) $\xrightarrow{K_2CO_3}$ A $\xrightarrow{KOH}$ B (with $COCH_3$, C=O) $\longrightarrow$ Cardura® (with COH, C=O) + CH_3OH

Identify the intermediate (A), and show the mechanism for its formation. Also, show the mechanism for the conversion of A to B. Which step do you think will occur most rapidly? Why? (The mechanism for the conversion of B to the final product is explained in Chapter 17.)

44. For each of the following reactions, give the elimination products; if the products can exist as stereoisomers, indicate which stereoisomers are obtained.
    a. (2*S*,3*S*)-2-chloro-3-methylpentane + high concentration of CH_3O^-
    b. (2*S*,3*R*)-2-chloro-3-methylpentane + high concentration of CH_3O^-
    c. (2*R*,3*S*)-2-chloro-3-methylpentane + high concentration of CH_3O^-
    d. (2*R*,3*R*)-2-chloro-3-methylpentane + high concentration of CH_3O^-
    e. 3-chloro-3-ethyl-2,2-dimethylpentane + high concentration of $CH_3CH_2O^-$

45. Which of the following hexachlorocyclohexanes is the least reactive in an E2 reaction?

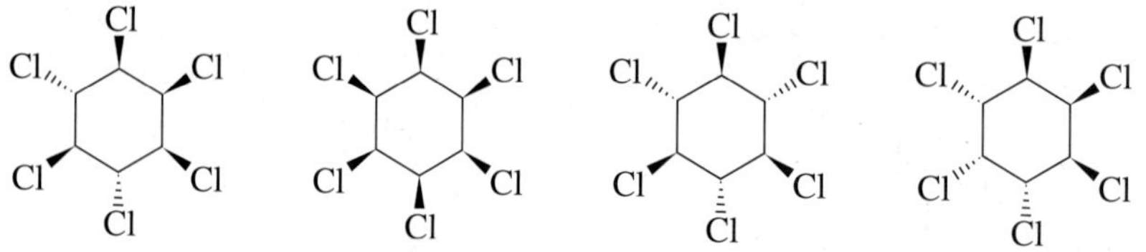

46. The rate of the reaction of 1-bromo-2-butene with ethanol is increased if silver nitrate is added to the reaction mixture. Explain.

47. For each of the following reactions, carried out under $S_N2/E2$ conditions, give the products; if the products can exist as stereoisomers, show which stereoisomers are formed.
    a. (3*S*,4*S*)-3-bromo-4-methylhexane + CH_3O^-
    b. (3*S*,4*R*)-3-bromo-4-methylhexane + CH_3O^-
    c. (3*R*,4*R*)-3-bromo-4-methylhexane + CH_3O^-
    d. (3*R*,4*S*)-3-bromo-4-methylhexane + CH_3O^-

48. Two elimination products are obtained from the following E2 reaction:

$$CH_3CH_2CHDCH_2Br \xrightarrow{HO^-}$$

    a. What are the elimination products?
    b. Which is formed in greater yield? Explain.

49. Three substitution products and three elimination products are obtained from the following reaction:

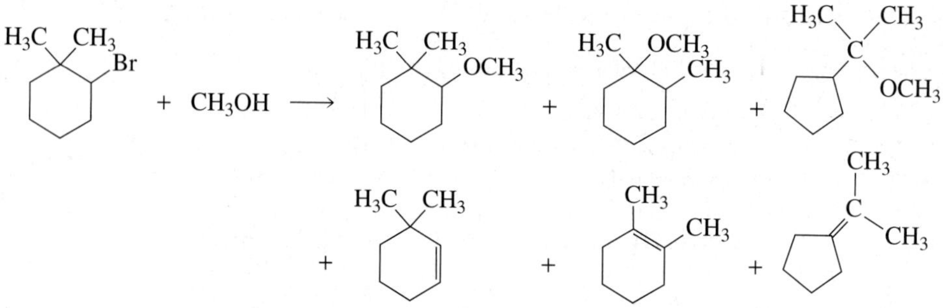

Account for the formation of these products.

50. When the following stereoisomer of 2-chloro-1,3-dimethylcyclohexane reacts with methoxide ion in a solvent that encourages $S_N2/E2$ reactions, only one product is formed:

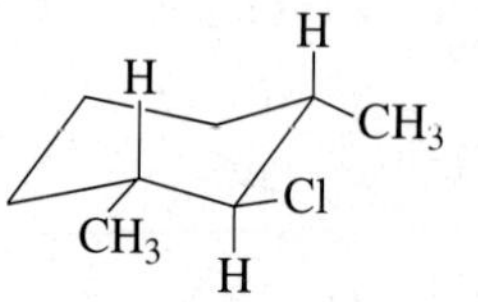

When the same compound reacts with methoxide ion in a solvent that favors $S_N1/E1$ reactions, twelve products are formed. Identify the products that are formed under the two sets of conditions.

51. For each of the following compounds, give the product that will be formed in an E2 reaction and indicate the configuration of the product:
    a. (1*S*,2*S*)-1-bromo-1,2-diphenylpropane
    b. (1*S*,2*R*)-1-bromo-1,2-diphenylpropane

12 Reactions of Alcohols, Ethers, Epoxides, and Sulfur-Containing Compounds • Organometallic Compounds

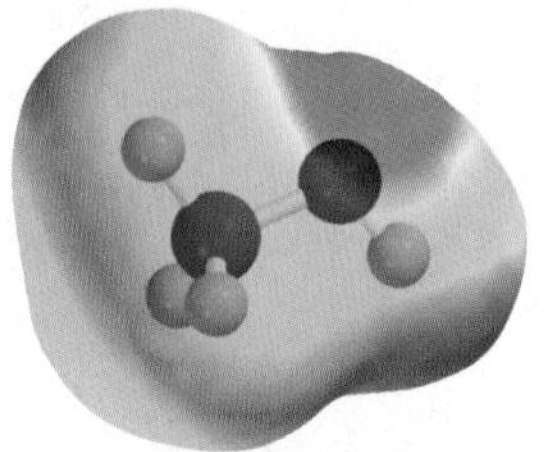

CH_3OH

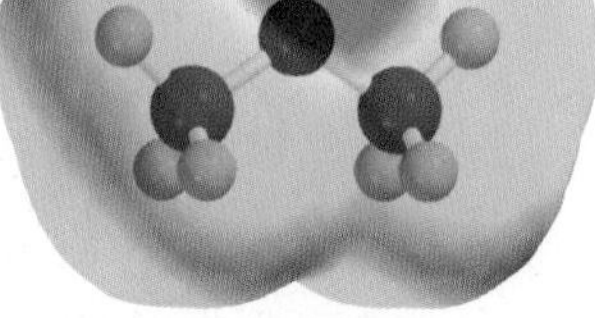

CH_3OCH_3

In Chapters 10 and 11, we saw that alkyl halides undergo substitution and elimination reactions because of their electron-withdrawing halogen atoms. Compounds with other electron-withdrawing groups also undergo substitution and elimination reactions. The relative reactivity of these compounds depends on the electron-withdrawing group.

Alcohols and ethers have leaving groups (^-OH, ^-OR) that are stronger bases than halide ions (^-X). Alcohols and ethers, therefore, are less reactive than alkyl halides in substitution and elimination reactions. We will see that because their leaving groups are strongly basic, alcohols and ethers have to be "activated" before they can undergo a substitution or an elimination reaction. In contrast, sulfonates and sulfonium salts have weakly basic leaving groups, so they undergo substitution reactions with ease.

The weaker the base, the more easily it can be displaced.

The stronger the acid, the weaker is its conjugate base.

R—X	R—O—H	R—O—R	$R—O—\overset{\overset{O}{\|}}{\underset{\underset{O}{\|}}{S}}—R$	$R—\overset{\overset{R}{\|}}{S^+}—R$
an alkyl halide X = F, Cl, Br, I	**an alcohol**	**an ether**	**a sulfonate ester**	**a sulfonium salt**

12.1 Substitution Reactions of Alcohols

An **alcohol** cannot undergo a nucleophilic substitution reaction because it has a strongly basic leaving group (^-OH) that cannot be displaced by a nucleophile.

$$CH_3OH + Br^- \not\rightarrow CH_3Br + \underset{\text{strong base}}{HO^-}$$

An alcohol is able to undergo a nucleophilic substitution reaction if its OH group is converted into a group that is a weaker base (i.e., a better leaving group). One way to convert an OH group into a weaker base is to protonate it. Protonation changes the leaving group from ^{-}OH (a strong base) to H_2O (a weak base), which is weak enough to be displaced by a nucleophile. The substitution reaction is slow and requires heat to take place in a reasonable period of time.

$$CH_3OH + HBr \rightleftharpoons CH_3\overset{+}{O}H_2 \; Br^- \xrightarrow{\Delta} CH_3Br + \underset{\textbf{weak base}}{H_2O}$$

Because the OH group of the alcohol has to be protonated before it can be displaced by a nucleophile, only weakly basic nucleophiles (I^-, Br^-, Cl^-) can be used in the substitution reaction. Moderately and strongly basic nucleophiles (NH_3, RNH_2, CH_3O^-) would be protonated in the acidic solution and, once protonated, would no longer be nucleophiles ($^{+}NH_4$, RNH_3^{+}) or would be poor nucleophiles (CH_3OH).

Primary, secondary, and tertiary alcohols all undergo nucleophilic substitution reactions with HI, HBr, and HCl to form alkyl halides.

$$\underset{\substack{\textbf{1-propanol}\\ \textbf{a primary alcohol}}}{CH_3CH_2CH_2OH} + HI \xrightarrow{\Delta} \underset{\textbf{1-iodopropane}}{CH_3CH_2CH_2I} + H_2O$$

$$\underset{\substack{\textbf{cyclohexanol}\\ \textbf{a secondary alcohol}}}{C_6H_{11}OH} + HBr \xrightarrow{\Delta} \underset{\textbf{bromocyclohexane}}{C_6H_{11}Br} + H_2O$$

$$\underset{\substack{\textbf{2-methyl-2-butanol}\\ \textbf{a tertiary alcohol}}}{CH_3CH_2C(CH_3)_2OH} + HCl \longrightarrow \underset{\textbf{2-chloro-2-methylbutane}}{CH_3CH_2C(CH_3)_2Cl} + H_2O$$

The mechanism of the substitution reaction depends on the structure of the alcohol. Secondary and tertiary alcohols undergo S_N1 reactions. The carbocation intermediate formed in the S_N1 reaction has two possible fates: It can combine with a nucleophile and form a substitution product, or it can lose a proton and form an elimination product. However, only the substitution product is actually obtained, because any alkene formed in an elimination reaction will undergo a subsequent addition reaction with HX to form more of the substitution product.

mechanism of the S_N1 reaction

$$\underset{\substack{\textit{tert}\textbf{-butyl alcohol}\\ \textbf{a tertiary alcohol}}}{(CH_3)_3C{-}\ddot{O}H} + H{-}Br \rightleftharpoons (CH_3)_3C{-}\overset{+}{O}H_2 \rightleftharpoons (CH_3)_3C^{+} + H_2O \xrightarrow{:\ddot{Br}:^-} \underset{\substack{\textbf{substitution}\\ \textbf{product}}}{(CH_3)_3C{-}Br}$$

$$(CH_3)_3C^{+} \rightleftharpoons \underset{\textbf{elimination product}}{CH_3C(CH_3){=}CH_2} + H^+ \xrightarrow{HBr} (CH_3)_3C{-}Br$$

protonation of the most basic atom

formation of a carbocation

reaction of the carbocation with a nucleophile

Tertiary alcohols undergo substitution reactions with hydrogen halides faster than do secondary alcohols because tertiary carbocations are easier to form than secondary carbocations (Section 4.2). Thus, the reaction of a tertiary alcohol with a hydrogen halide proceeds readily at room temperature, whereas the reaction of a secondary alcohol with a hydrogen halide would have to be heated to have the reaction occur at the same rate.

$$CH_3C(CH_3)(OH)CH_2CH_3 + HBr \longrightarrow CH_3C(CH_3)(Br)CH_2CH_3 + H_2O$$

$$CH_3CH(OH)CH_2CH_3 + HBr \xrightarrow{\Delta} CH_3CH(Br)CH_2CH_3 + H_2O$$

Primary alcohols cannot undergo S_N1 reactions because primary carbocations are too unstable to be formed (Section 11.1). Therefore, when a primary alcohol reacts with a hydrogen halide, it must do so via an S_N2 reaction.

mechanism of the S_N2 reaction

$$CH_3CH_2\ddot{O}H + H{-}Br \rightleftharpoons CH_3CH_2{-}\overset{+}{O}H_2 + :\ddot{Br}:^- \longrightarrow CH_3CH_2Br + H_2O$$

ethyl alcohol
a primary alcohol

protonation of the oxygen

back-side attack by the nucleophile

Only the substitution product is obtained. No elimination product is formed because the halide ion, although a good nucleophile, is a weak base, and strong base is needed in an E2 reaction to remove a proton from a β-carbon (Section 11.1). (Remember that a β-carbon is the carbon adjacent to the carbon that is attached to the leaving group.)

Primary alcohols undergo S_N2 reactions with hydrogen halides.

Secondary and tertiary alcohols undergo S_N1 reactions with hydrogen halides.

When HCl is used instead of HBr or HI, the S_N2 reaction is slower because Cl^- is a poorer nucleophile than Br^- or I^- (Section 10.3). The rate of the reaction can be increased by using $ZnCl_2$ as a catalyst.

$$CH_3CH_2CH_2OH + HCl \xrightarrow[\Delta]{ZnCl_2} CH_3CH_2CH_2Cl + H_2O$$

Zn^{2+} is a Lewis acid that complexes strongly with the lone-pair electrons on oxygen. This weakens the C—O bond and creates a better leaving group.

$$CH_3CH_2CH_2\ddot{O}H + ZnCl{-}Cl \longrightarrow CH_3CH_2CH_2{-}\overset{+}{O}H(ZnCl) + :\ddot{Cl}:^- \longrightarrow CH_3CH_2CH_2Cl + HOZnCl$$

THE LUCAS TEST

Whether an alcohol is primary, secondary, or tertiary can be determined by taking advantage of the relative rates at which the three classes of alcohols react with $HCl/ZnCl_2$. This is known as the Lucas test. The alcohol is added to a mixture of HCl and $ZnCl_2$—the Lucas reagent. Low-molecular-weight alcohols are soluble in the Lucas reagent, but the alkyl halide products are not, so the solution turns cloudy as the alkyl halide is formed. When the test is carried out at room temperature, the solution turns cloudy immediately if the alcohol is tertiary, turns cloudy in about five minutes if the alcohol is secondary, and remains clear if the alcohol is primary. Because the test relies on the complete solubility of the alcohol in the Lucas reagent, it is limited to alcohols with fewer than six carbons.

PROBLEM 1◆

The observed relative reactivities of primary, secondary, and tertiary alcohols with a hydrogen halide are 3° > 2° > 1°. If secondary alcohols underwent an S_N2 reaction rather than an S_N1 reaction with a hydrogen halide, what would be the relative reactivities of the three classes of alcohols?

Howard J. Lucas (1885–1963) *was born in Ohio and earned B.S. and M.S. degrees from Ohio State University. He published a description of the Lucas test in 1930. He was a professor of chemistry at the California Institute of Technology.*

Because the reaction of a secondary or a tertiary alcohol with a hydrogen halide is an S_N1 reaction, a carbocation is formed as an intermediate. Therefore, we must check for the possibility of a carbocation rearrangement when predicting the product of the substitution reaction. Remember that a carbocation rearrangement will occur if it leads to formation of a more stable carbocation (Section 4.6). For example, the major product of the reaction of 3-methyl-2-butanol with HBr is 2-bromo-2-methylbutane, because a 1,2-hydride shift converts the initially formed secondary carbocation into a more stable tertiary carbocation.

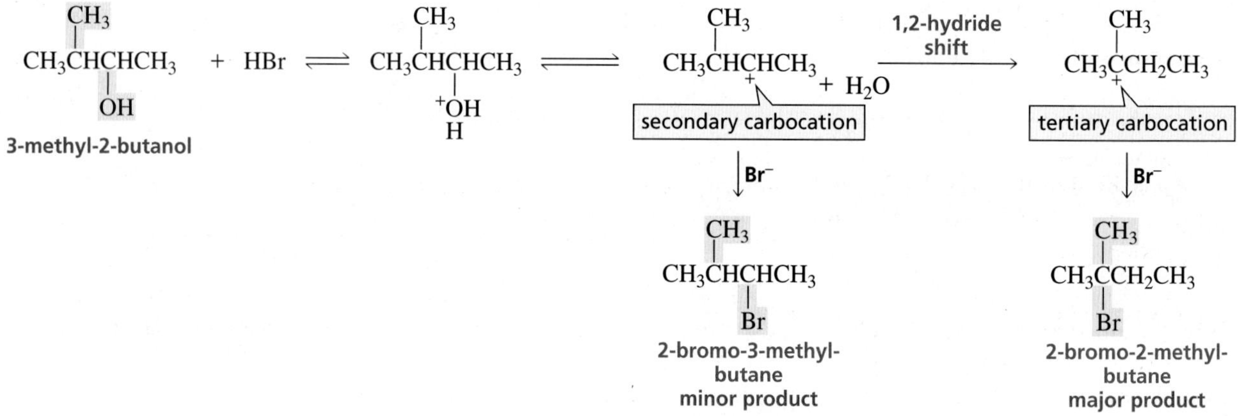

$CH_3CH(CH_3)CH(OH)CH_3$ + HBr ⇌ $CH_3CH(CH_3)CH(\overset{+}{O}H_2)CH_3$ ⇌ $CH_3CH(CH_3)\overset{+}{C}HCH_3$ + H_2O —1,2-hydride shift→ $CH_3\overset{+}{C}(CH_3)CH_2CH_3$

3-methyl-2-butanol

secondary carbocation —Br^-→ $CH_3CH(CH_3)CH(Br)CH_3$ 2-bromo-3-methylbutane, minor product

tertiary carbocation —Br^-→ $CH_3C(CH_3)(Br)CH_2CH_3$ 2-bromo-2-methylbutane, major product

PROBLEM 2◆

Give the major product of each of the following reactions:

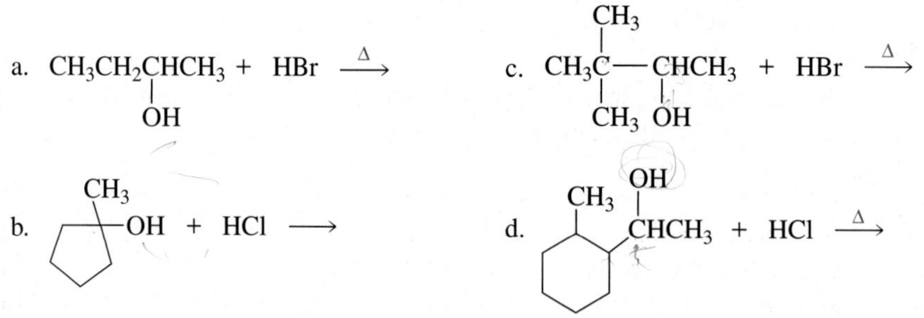

a. $CH_3CH_2CH(OH)CH_3$ + HBr $\xrightarrow{\Delta}$

b. 1-methylcyclopentanol (ring C bearing CH_3 and OH) + HCl →

c. $CH_3C(CH_3)_2—CH(OH)CH_3$ + HBr $\xrightarrow{\Delta}$

d. 2-methylcyclohexyl ring bearing $CH(OH)CH_3$ + HCl $\xrightarrow{\Delta}$

12.2 Amines Do Not Undergo Substitution Reactions

We have just seen that alcohols are much less reactive than alkyl halides in substitution reactions. Amines are even less reactive than alcohols. The relative reactivities of an alkyl fluoride (the least reactive of the alkyl halides), an alcohol, and an amine can be appreciated by comparing the pK_a values of the conjugate acids of their leaving

GRAIN ALCOHOL AND WOOD ALCOHOL

When ethanol is ingested, it acts on the central nervous system. Ingesting moderate amounts of ethanol affects one's judgment and lowers one's inhibitions. Higher concentrations interfere with motor coordination and cause slurred speech and amnesia. Still higher concentrations cause nausea and loss of consciousness. Ingesting very large amounts of ethanol interferes with spontaneous respiration and can be fatal.

The ethanol in alcoholic beverages is produced by the fermentation of glucose, obtained from grapes and from grains such as corn, rye, and wheat, which is why ethanol is also known as grain alcohol. Grains are cooked in the presence of malt (sprouted barley) to convert much of their starch into glucose. Yeast is added to convert glucose into ethanol and carbon dioxide.

$$\underset{\text{glucose}}{C_6H_{12}O_6} \xrightarrow{\text{yeast enzymes}} 2\ \underset{\text{ethanol}}{CH_3CH_2OH} + 2\ CO_2$$

The kind of beverage produced (white or red wine, beer, scotch, bourbon, champagne) depends on the plant species being fermented, whether the CO_2 that is formed is allowed to escape, whether other substances are added, and how the beverage is purified (by sedimentation for wines, by distillation for scotch and bourbon).

The tax on liquor would make ethanol a prohibitively expensive laboratory reagent. Because ethanol is needed in a wide variety of commercial processes, however, laboratory alcohol is not taxed, but it is carefully regulated by the federal government to make certain that it is not used for the preparation of alcoholic beverages. Denatured alcohol—ethanol that has been made undrinkable by adding a denaturant such as benzene or methanol—is not taxed, but the added impurities make it unfit for many laboratory uses.

Methanol, also known as wood alcohol because at one time it was obtained by heating wood in the absence of oxygen, is highly toxic. Ingesting even very small amounts can cause blindness. Ingesting as little as an ounce has been fatal. The antidote to methanol poisoning is discussed in Section 20.11.

groups (recalling that the weaker the acid, the stronger is its conjugate base). The leaving group of an amine ($^-NH_2$) is such a strong base that amines cannot undergo substitution or elimination reactions.

relative reactivities

$$RCH_2F \quad > \quad RCH_2OH \quad > \quad RCH_2NH_2$$

HF	H_2O	NH_3
$pK_a = 3.2$	$pK_a = 15.7$	$pK_a = 36$

Protonation of the amino group makes it a better leaving group, but not nearly as good a leaving group as a protonated alcohol, which is ~13 pK_a units more acidic than a protonated amine.

$$\underset{pK_a = -2.4}{CH_3CH_2\overset{+}{O}H_2} \quad > \quad \underset{pK_a = 11.2}{CH_3CH_2\overset{+}{N}H_3}$$

Therefore, unlike the leaving group of a protonated alcohol, the leaving group of a protonated amine cannot dissociate to form a carbocation or be replaced by a halide ion. Protonated amino groups also cannot be displaced by strongly basic nucleophiles such as HO^- because the nucleophile would react immediately with the acidic hydrogen of the $^+NH_3$ group and thereby be converted to water, a poor nucleophile.

$$CH_3CH_2\overset{+}{N}H_3 + HO^- \rightleftharpoons CH_3CH_2NH_2 + H_2O$$

PROBLEM 3

Why can a halide ion such as Br^- react with a protonated primary alcohol, but not with a protonated primary amine?

12.3 Other Methods for Converting Alcohols into Alkyl Halides

Alcohols are inexpensive and readily available compounds. As we have just seen, they do not undergo nucleophilic substitution because the ^-OH group is too basic to be displaced by a nucleophile. Chemists, therefore, need ways to convert readily available but unreactive alcohols into reactive alkyl halides that can be used as starting materials for the preparation of many organic compounds (Section 10.4).

$$\underset{\text{alcohol}}{R{-}OH} \xrightarrow[\Delta]{HX} \underset{\substack{\text{alkyl halide}\\ X = Cl, Br, I}}{R{-}X} \xrightarrow{^-Nu} R{-}Nu$$

We have just seen that an alcohol can be converted into an alkyl halide by treating it with a hydrogen halide. However, better yields are obtained and carbocation rearrangements can be avoided if a phosphorus trihalide (PCl_3, PBr_3, or PI_3)[1] or thionyl chloride ($SOCl_2$) is used instead. These reagents all act in the same way: They convert an alcohol into an alkyl halide by converting the alcohol into an intermediate with a leaving group that is readily displaced by a halide ion. For example, phosphorus tribromide converts the OH group of an alcohol into a bromophosphite group that can be readily displaced by a bromide ion.

$$CH_3CH_2{-}\ddot{O}H + \underset{\text{phosphorus tribromide}}{PBr_3} \longrightarrow CH_3CH_2{-}\overset{+}{O}(H)PBr_2 \xrightarrow{\text{pyridine}} CH_3CH_2{-}\ddot{O}PBr_2 \xrightarrow{:\ddot{Br}:^-} CH_3CH_2Br + {}^-OPBr_2$$

(pyridine removes H, giving pyridinium ion; $OPBr_2$ = a bromophosphite group)

Thionyl chloride converts an OH group into a chlorosulfite group that can be displaced by Cl^-.

$$CH_3{-}\ddot{O}H + \underset{\text{thionyl chloride}}{Cl{-}\overset{O}{\overset{\|}{S}}{-}Cl} \longrightarrow CH_3{-}\overset{+}{O}(H){-}\overset{O}{\overset{\|}{S}}{-}Cl \xrightarrow{\text{pyridine}} CH_3{-}\ddot{O}{-}\overset{O}{\overset{\|}{S}}{-}Cl \xrightarrow{:\ddot{Cl}:^-} CH_3Cl + SO_2 + Cl^-$$

(pyridine removes H, giving pyridinium ion; $OS(O)Cl$ = a chlorosulfite group)

3-D Molecules: Thionyl chloride; Pyridine; Pyridinium ion

Pyridine is generally used as the solvent in these reactions because it prevents the buildup of HBr or HCl and it is a relatively poor nucleophile.

$$\underset{\text{pyridine}}{C_5H_5N} + HCl \rightleftharpoons C_5H_5\overset{+}{N}H + Cl^-$$

[1]Because of its instability, PI_3 is generated in situ (in the reaction mixture) from the reaction of phosphorus with iodine.

The foregoing reactions work well for primary and secondary alcohols, but tertiary alcohols give poor yields because the intermediate formed by a tertiary alcohol is sterically hindered to attack by the halide ion.

Table 12.1 summarizes some of the commonly used methods for converting alcohols into alkyl halides.

Table 12.1 Commonly Used Methods for Converting Alcohols into Alkyl Halides

$$\mathrm{ROH} + \mathrm{HBr} \xrightarrow{\Delta} \mathrm{RBr}$$

$$\mathrm{ROH} + \mathrm{HI} \xrightarrow{\Delta} \mathrm{RI}$$

$$\mathrm{ROH} + \mathrm{HCl} \xrightarrow{\Delta} \mathrm{RCl}$$

$$\mathrm{ROH} + \mathrm{PBr_3} \xrightarrow[\text{pyridine}]{} \mathrm{RBr}$$

$$\mathrm{ROH} + \mathrm{PCl_3} \xrightarrow[\text{pyridine}]{} \mathrm{RCl}$$

$$\mathrm{ROH} + \mathrm{SOCl_2} \xrightarrow[\text{pyridine}]{} \mathrm{RCl}$$

12.4 Converting Alcohols into Sulfonate Esters

Another way an alcohol can be activated for subsequent reaction with a nucleophile, besides being converted into an alkyl halide, is to be converted into a sulfonate ester. A **sulfonate ester** is formed when an alcohol reacts with a sulfonyl chloride.

$$\underset{\text{a sulfonyl chloride}}{\mathrm{R'{-}SO_2{-}Cl}} + \mathrm{ROH} \xrightarrow[\text{pyridine}]{} \underset{\text{a sulfonate ester}}{\mathrm{R'{-}SO_2{-}OR}} + \mathrm{Cl^-} + \text{pyridine}{-}\mathrm{H^+}$$

The reaction is a nucleophilic substitution reaction. The alcohol displaces the chloride ion. Pyridine is used both as the solvent and to prevent HCl from building up.

$$\mathrm{R'{-}SO_2{-}Cl} + \mathrm{R\ddot{O}H} \longrightarrow \mathrm{R'{-}SO_2{-}\overset{+}{O}(H)R} \xrightarrow{\text{pyridine}} \mathrm{R'{-}SO_2{-}OR} + \mathrm{Cl^-} + \text{pyridine}{-}\mathrm{H^+}$$

A sulfonic acid is a strong acid ($pK_a \sim -1$) because its conjugate base is particularly stable due to delocalization of its negative charge over three oxygen atoms. (Recall from Section 7.6 that electron delocalization stabilizes a charged species.) Since a sulfonic acid is a strong acid, its conjugate base is weak, giving the sulfonate ester an excellent leaving group. (Notice that sulfur has an expanded valence shell—it is surrounded by 12 electrons.)

$$\mathrm{R{-}SO_2{-}OH} \longrightarrow \left[\mathrm{R{-}S(=O)_2{-}O^-} \longleftrightarrow \mathrm{R{-}S(=O)(O^-){=}O} \longleftrightarrow \mathrm{R{-}S(O^-)(=O){=}O}\right] + \mathrm{H^+}$$

resonance contributors

Several sulfonyl chlorides are available to activate OH groups. The most common one is *para*-toluenesulfonyl chloride.

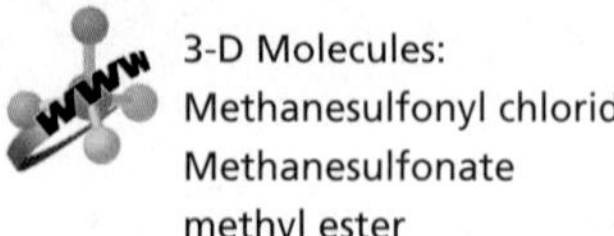

$H_3C-C_6H_4-S(=O)_2-Cl$ **para-toluenesulfonyl chloride**

$H_3C-S(=O)_2-Cl$ **methanesulfonyl chloride**

$F_3C-S(=O)_2-Cl$ **trifluoromethanesulfonyl chloride**

Once the alcohol has been activated by being converted into a sulfonate ester, the appropriate nucleophile is added, generally under conditions that favor S_N2 reactions. The reactions take place readily at room temperature because the leaving group is so good. For example, a *para*-toluenesulfonate ion is about 100 times better than a chloride ion as a leaving group. Sulfonate esters react with a wide variety of nucleophiles, so they can be used to synthesize a wide variety of compounds.

$$CH_3S^- + CH_3CH_2CH_2CH_2-O-S(=O)_2-C_6H_4-CH_3 \longrightarrow CH_3CH_2CH_2CH_2SCH_3 + {}^-O-S(=O)_2-C_6H_4-CH_3$$

ROTs ⁻OTs

$$^-C{\equiv}N + CH_3CH_2CH_2-O-S(=O)_2-C_6H_4-CH_3 \longrightarrow CH_3CH_2CH_2C{\equiv}N + {}^-O-S(=O)_2-C_6H_4-CH_3$$

ROTs ⁻OTs

para-Toluenesulfonyl chloride is called tosyl chloride and is abbreviated TsCl; the product of the reaction of *para*-toluenesulfonyl chloride and an alcohol is called an **alkyl tosylate** and is abbreviated ROTs. The leaving group, therefore, is ⁻OTs. The product of the reaction of trifluoromethanesulfonyl chloride and an alcohol is called an **alkyl triflate** and is abbreviated ROTf.

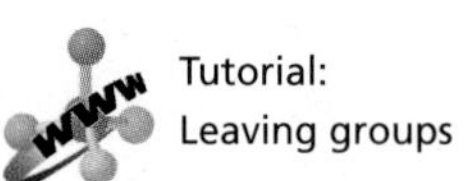

$$CH_3CH_2CH_2OTs + {}^-C{\equiv}N \longrightarrow CH_3CH_2CH_2C{\equiv}N + {}^-OTs$$

an alkyl tosylate

$$CH_3CH_2CH_2OTf + CH_3NH_2 \longrightarrow CH_3CH_2CH_2\overset{+}{N}H_2CH_3 + {}^-OTf$$

an alkyl triflate

PROBLEM 4 SOLVED

Explain why the ether obtained by treating an optically active alcohol with PBr_3 followed by sodium methoxide has the same configuration as the alcohol, whereas the ether obtained by treating the alcohol with tosyl chloride followed by sodium methoxide has a configuration opposite that of the alcohol.

$$\text{R, CH}_3\text{, H, OH (alcohol)} \xrightarrow[\text{2. CH}_3\text{O}^-]{\text{1. PBr}_3\text{/pyridine}} \text{R, CH}_3\text{, H, OCH}_3 \text{ (same configuration)}$$

$$\text{R, CH}_3\text{, H, OH (alcohol)} \xrightarrow[\text{2. CH}_3\text{O}^-]{\text{1. TsCl/pyridine}} \text{CH}_3\text{O, CH}_3\text{, H, R (opposite configuration)}$$

SOLUTION Conversion of the alcohol to the ether via the alkyl halide involves two successive S_N2 reactions: (1) attack of Br^- on the bromophosphite and (2) attack of CH_3O^- on the alkyl halide. Each S_N2 reaction takes place with inversion of configuration, so the final product has the same configuration as the starting material. In contrast, conversion of the alcohol to the ether via the alkyl tosylate involves only one S_N2 reaction: attack of CH_3O^- on the alkyl tosylate. Therefore, the final product and the starting material have opposite configurations.

$$\text{R–C(CH}_3\text{)(H)–OH} \xrightarrow[\text{pyridine}]{\text{PBr}_3} \text{R–C(CH}_3\text{)(H)–OPBr}_2 \xrightarrow{\text{Br}^-} \text{Br–C(CH}_3\text{)(H)–R} \xrightarrow{\text{CH}_3\text{O}^-} \text{R–C(CH}_3\text{)(H)–OCH}_3$$

$$\text{R–C(CH}_3\text{)(H)–OH} \xrightarrow[\text{pyridine}]{\text{TsCl}} \text{R–C(CH}_3\text{)(H)–OTs} \xrightarrow{\text{CH}_3\text{O}^-} \text{CH}_3\text{O–C(CH}_3\text{)(H)–R}$$

PROBLEM 5

Show how 1-butanol can be converted into the following compounds:

a. $CH_3CH_2CH_2CH_2Br$

b. $CH_3CH_2CH_2CH_2OCCH_2CH_3$ (with C=O)

c. $CH_3CH_2CH_2CH_2OCH_3$

d. $CH_3CH_2CH_2CH_2NHCH_2CH_3$

e. $CH_3CH_2CH_2CH_2SH$

f. $CH_3CH_2CH_2CH_2C{\equiv}N$

12.5 Elimination Reactions of Alcohols: Dehydration

An alcohol can undergo an elimination reaction, forming an alkene by losing an OH from one carbon and an H from an adjacent carbon. Overall, this amounts to the elimination of a molecule of water. Loss of water from a molecule is called **dehydration**. Dehydration of an alcohol requires an acid catalyst and heat. Sulfuric acid (H_2SO_4) and phosphoric acid (H_3PO_4) are the most commonly used acid catalysts.

$$\underset{\text{OH}}{\text{CH}_3\text{CH}_2\text{CHCH}_3} \underset{}{\overset{\text{H}_2\text{SO}_4,\ \Delta}{\rightleftharpoons}} \text{CH}_3\text{CH}{=}\text{CHCH}_3 + \text{H}_2\text{O}$$

An acid always reacts with an organic molecule in the same way: It protonates the most basic (electron-rich) atom in the molecule. Thus, in the first step of the dehydration reaction, the acid protonates the oxygen atom of the alcohol. As we saw earlier, protonation converts the very poor leaving group (^-OH) into a good leaving group (H_2O). In the next step, water departs, leaving behind a carbocation. A base in the reaction mixture (water is the base in highest concentration) removes a proton from a β-carbon (a carbon adjacent to the positively charged carbon), forming an alkene and regenerating the acid catalyst. Notice that the dehydration reaction is an E1 reaction of a protonated alcohol.

An acid protonates the most basic atom in a molecule.

mechanism of dehydration (E1)

$$\underset{:\ddot{\text{O}}\text{H}}{\text{CH}_3\text{CHCH}_3} + \text{H–OSO}_3\text{H} \rightleftharpoons \underset{^+\!:\text{OH}\text{ H}}{\text{CH}_3\text{CHCH}_3} + {}^-\text{OSO}_3\text{H} \rightleftharpoons \text{H–CH}_2\text{–}\overset{+}{\text{C}}\text{HCH}_3 + \text{H}_2\ddot{\text{O}}: \rightleftharpoons \text{CH}_2{=}\text{CHCH}_3 + \text{H}_3\text{O}^+$$

protonation of the most basic atom

formation of a carbocation

a base removes a proton from a β-carbon

Movie: Dehydration

When more than one elimination product can be formed, the major product is the more substituted alkene—the one obtained by removing a proton from the β-carbon that is bonded to the fewest hydrogens. (Recall Zaitsev's rule, Section 11.2.) The more substituted alkene is the major product because it is the more stable alkene, so it has the more stable transition state leading to its formation (Figure 12.1).

$$\underset{\text{OH}}{\overset{\text{CH}_3}{\text{CH}_3\text{CCH}_2\text{CH}_3}} \xrightleftharpoons[\Delta]{\text{H}_3\text{PO}_4} \underset{84\%}{\overset{\text{CH}_3}{\text{CH}_3\text{C}{=}\text{CHCH}_3}} + \underset{16\%}{\overset{\text{CH}_3}{\text{CH}_2{=}\text{CCH}_2\text{CH}_3}} + \text{H}_2\text{O}$$

$$\text{1-methylcyclohexanol (H}_3\text{C, OH)} \xrightleftharpoons[\Delta]{\text{H}_2\text{SO}_4} \underset{93\%}{\text{1-methylcyclohexene (CH}_3)} + \underset{7\%}{\text{methylenecyclohexane (CH}_2)} + \text{H}_2\text{O}$$

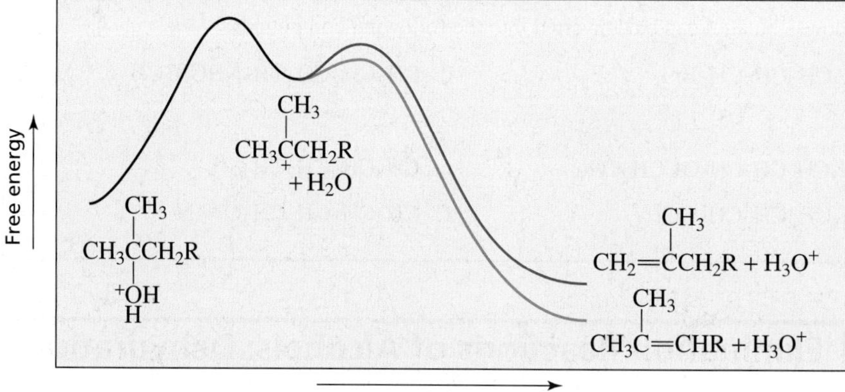

Figure 12.1 ▶ The reaction coordinate diagram for the dehydration of a protonated alcohol. The major product is the more substituted alkene because the transition state leading to its formation is more stable, allowing it to be formed more rapidly.

In Section 4.5, we saw that an alkene is hydrated (adds water) in the presence of an acid catalyst, thereby forming an alcohol. The hydration of an alkene is the reverse of the acid-catalyzed dehydration of an alcohol.

$$\underset{\text{OH}}{\text{RCH}_2\text{CHR}} + \text{H}^+ \xrightleftharpoons[\text{hydration}]{\text{dehydration}} \text{RCH}{=}\text{CHR} + \text{H}_2\text{O} + \text{H}^+$$

To prevent the alkene formed in the dehydration reaction from adding water and reforming the alcohol, the alkene can be removed by distillation as it is formed, because it has a much lower boiling point than the alcohol. Removing a product displaces the reaction to the right. (See Le Châtelier's principle, Section 10.4.)

PROBLEM 6

Explain why the acid-catalyzed dehydration of an alcohol is a reversible reaction, whereas the base-promoted dehydrohalogenation of an alkyl halide is an irreversible reaction.

Because the rate-determining step in the dehydration of a secondary or a tertiary alcohol is formation of a carbocation intermediate, the rate of dehydration parallels the ease with which the carbocation is formed. Tertiary alcohols are the easiest to dehydrate because tertiary carbocations are more stable and therefore are easier to form than secondary and primary carbocations (Section 4.2). In order to undergo dehydration, tertiary alcohols must be heated to about 50 °C in 5% H_2SO_4, secondary alcohols must be heated to about 100 °C in 75% H_2SO_4, and primary alcohols can be dehydrated only under extreme conditions (170 °C in 95% H_2SO_4) and by a different mechanism because primary carbocations are too unstable to be formed (Section 10.5).

Carbocation stability: 3° > 2° > 1°

relative ease of dehydration

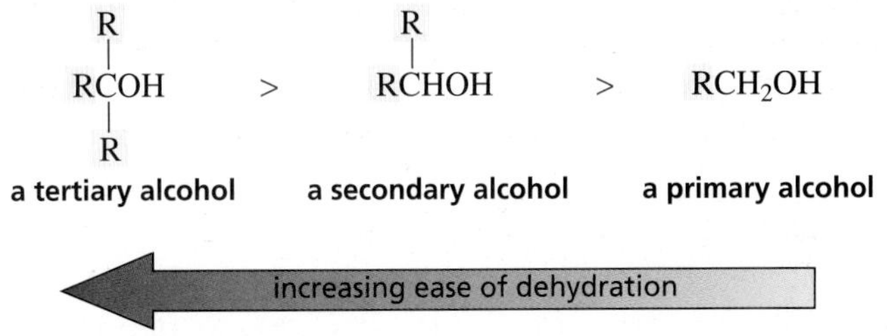

R_3COH > R_2CHOH > RCH_2OH

a tertiary alcohol — a secondary alcohol — a primary alcohol

increasing ease of dehydration (←)

Dehydration of secondary and tertiary alcohols involves the formation of a carbocation intermediate, so be sure to check the structure of the carbocation for the possibility of rearrangement. Remember that a carbocation will rearrange if rearrangement produces a more stable carbocation (Section 4.6). For example, the intitially formed secondary carbocation in the following reaction rearranges to a more stable tertiary carbocation:

3,3-dimethyl-2-butanol $(CH_3)_3CCH(OH)CH_3$ $\xrightarrow[\Delta]{H_3PO_4}$ secondary carbocation $(CH_3)_3C\overset{+}{C}HCH_3$ + H_2O $\xrightarrow{\text{1,2-methyl shift}}$ tertiary carbocation $(CH_3)_2\overset{+}{C}CH(CH_3)CH_3$

secondary carbocation → H^+ + $CH_3C(CH_3)_2CH{=}CH_2$ (3,3-dimethyl-1-butene, 3%)

tertiary carbocation → $CH_3C(CH_3){=}C(CH_3)CH_3$ (2,3-dimethyl-2-butene, 64%) + $CH_2{=}C(CH_3)CH(CH_3)CH_3$ (2,3-dimethyl-1-butene, 33%) + H^+

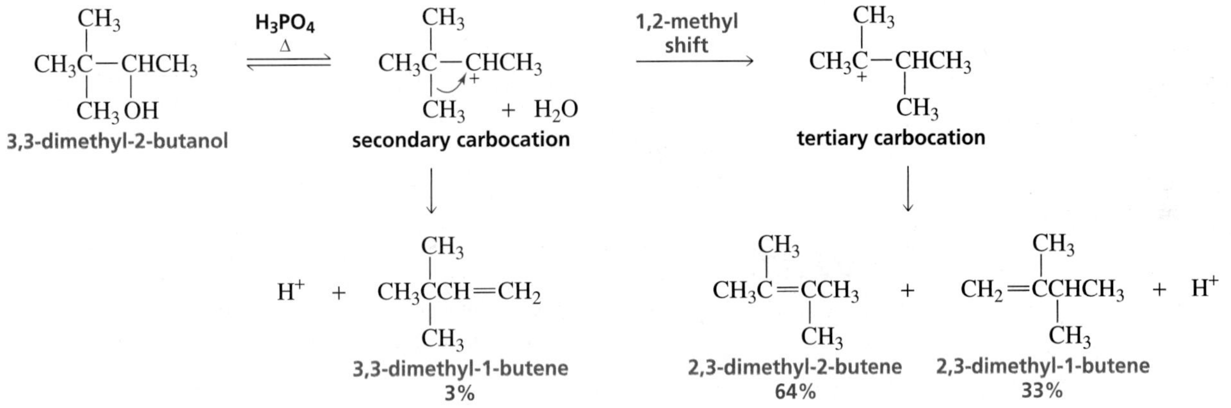

The following is an example of a **ring-expansion rearrangement**. Both the initially formed carbocation and the carbocation to which it rearranges are secondary carbocations, but the initially formed carbocation is less stable because of the strain in its four-membered ring (Section 2.11). Rearrangement relieves this strain. The rearranged secondary carbocation can rearrange by a 1,2-hydride shift to an even more stable tertiary carbocation. (B: is any base present in the solution.)

Tutorial: Carbocation rearrangements

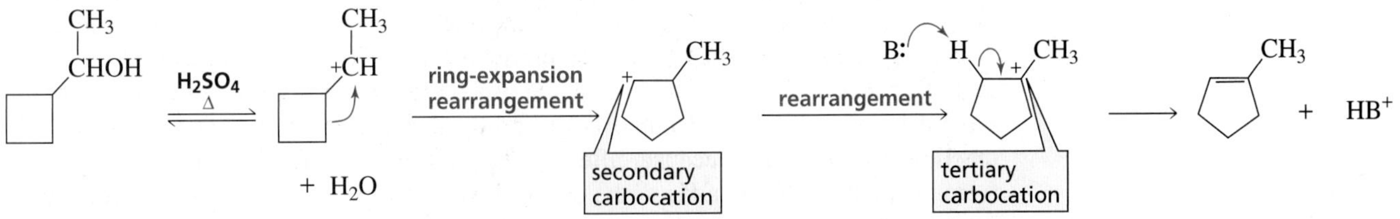

cyclobutyl–CH(OH)CH$_3$ $\xrightarrow[\Delta]{H_2SO_4}$ cyclobutyl–$\overset{+}{C}$HCH$_3$ + H_2O $\xrightarrow{\text{ring-expansion rearrangement}}$ secondary carbocation (2-methylcyclopentyl cation) $\xrightarrow{\text{rearrangement}}$ tertiary carbocation (1-methylcyclopentyl cation; B: removes H) ⟶ 1-methylcyclopentene + HB^+

PROBLEM 7◆

What product would be formed if the preceding alcohol were heated with an equivalent amount of HBr rather than with a catalytic amount of H_2SO_4?

PROBLEM 8

List the following alcohols in order of decreasing rate of dehydration in the presence of acid:

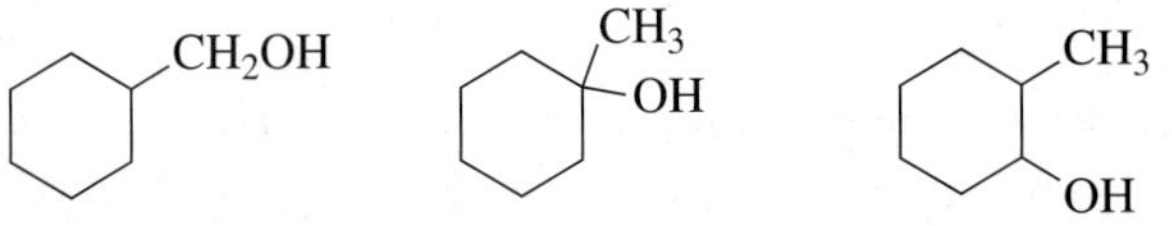

cyclohexyl–CH_2OH 1-methylcyclohexanol (CH_3, OH)   2-methylcyclohexanol (CH_3, OH)

Primary alcohols undergo dehydration by an E2 pathway.

Secondary and tertiary alcohols undergo dehydration by an E1 pathway.

While the dehydration of a tertiary or a secondary alcohol is an E1 reaction, the dehydration of a primary alcohol is an E2 reaction because of the difficulty encountered in forming primary carbocations. Any base (B:) in the reaction mixture (ROH, H_2O, HSO_4^-) can remove the proton in the elimination reaction. An ether is also obtained as the product of a competing S_N2 reaction.

mechanism of dehydration (E2) and competing substitution (S_N2)

$$CH_3CH_2\ddot{O}H + H{-}OSO_3H \rightleftharpoons CH_2{-}CH_2{-}\overset{+}{O}H_2 \xrightarrow{E2} CH_2{=}CH_2 + H_2O + HB^+$$

elimination product

$+ {}^-OSO_3H$

protonation of the most basic atom

removal of a proton from a β-carbon

$$CH_3CH_2\ddot{O}H + CH_3CH_2{-}\overset{+}{O}H_2 \xrightarrow{S_N2} CH_3CH_2\overset{+}{O}(H)CH_2CH_3 \longrightarrow CH_3CH_2OCH_2CH_3 + HB^+$$

substitution product

back-side attack by the nucleophile

B:

PROBLEM 9

Heating an alcohol with sulfuric acid is a good way to prepare a symmetrical ether such as diethyl ether. It is not a good method for preparing an unsymmetrical ether such as ethyl propyl ether.

a. Explain.

b. How would you synthesize ethyl propyl ether?

Although the dehydration of a primary alcohol is an E2 reaction and therefore does not form a carbocation intermediate, the product obtained in most cases is identical to the product that would be obtained if a carbocation had been formed in an E1 reaction and then had rearranged. For example, we would expect 1-butene to be the product of the E2 dehydration of 1-butanol. However, we find that the product is actually 2-butene, which would have been the product if an E1 reaction had occurred and the initially formed primary carbocation intermediate had rearranged to a more stable secondary carbocation. 2-Butene is the product of the reaction, not because an E1 reaction occurred, but because, after the E2 product (1-butene) is formed, a proton from the acidic solution adds to the double bond—adding to the sp^2 carbon bonded to the greater number of hydrogens (in accordance with Markovnikov's rule) to form a carbocation. Loss of a proton from the carbocation—from the β-carbon bonded to the fewest hydrogens (in accordance with Zaitsev's rule)—gives 2-butene, the final product of the reaction.

$$\underset{\text{1-butanol}}{CH_3CH_2CH_2CH_2OH} \underset{\Delta}{\overset{H_2SO_4}{\rightleftharpoons}} \underset{\text{1-butene}}{CH_3CH_2CH{=}CH_2} + H_2O \overset{H^+}{\rightleftharpoons} CH_3CH_2\overset{+}{C}HCH_3 \rightleftharpoons \underset{\text{2-butene}}{CH_3CH{=}CHCH_3} + H^+$$

The stereochemical outcome of the E1 dehydration of an alcohol is identical to the stereochemical outcome of the E1 dehydrohalogenation of an alkyl halide. That is, both the *E* and *Z* isomers are obtained as products. More of the isomer with the bulky groups on opposite sides of the double bond is produced because it is formed more rapidly since, being the more stable isomer, the transition state leading to its formation is more stable (Section 11.5).

$$CH_3CH_2CH(OH)CH_3 \underset{\Delta}{\overset{H_2SO_4}{\rightleftharpoons}} CH_3CH_2\overset{+}{C}HCH_3 + H_2O \longrightarrow \underset{\textit{trans}\text{-2-butene } 74\%}{(H_3C)(H)C{=}C(H)(CH_3)} + \underset{\textit{cis}\text{-2-butene } 23\%}{(H_3C)(H)C{=}C(H)(CH_3)} + \underset{\text{1-butene } 3\%}{CH_3CH_2CH{=}CH_2} + H^+$$

2-butanol

The relatively harsh conditions (acid and heat) required for alcohol dehydration and the structural changes resulting from carbocation rearrangements may result in low yields of the desired alkene. Dehydration, however, can be carried out under milder conditions by using phosphorus oxychloride ($POCl_3$) and pyridine.

$$CH_3CH_2CH(OH)CH_3 \xrightarrow[\text{pyridine, 0 °C}]{POCl_3} CH_3CH{=}CHCH_3$$

Reaction with $POCl_3$ converts the OH group of the alcohol into $OPOCl_2$, a good leaving group. The basic reaction conditions favor an E2 reaction, so a carbocation is not formed and carbocation rearrangements do not occur. Pyridine serves as a base to remove the proton in the elimination reaction and to prevent the buildup of HCl, which would add to the alkene.

$$CH_3CH(CH_3)\text{–}\ddot{O}H + \underset{\text{phosphorus oxychloride}}{P(=O)Cl_3} \longrightarrow CH_3CH(CH_3)\text{–}\overset{+}{O}H\text{–}P(=O)Cl_2 + Cl^- \longrightarrow CH_2(H)\text{–}CH(CH_3)\text{–}\ddot{O}\text{–}P(=O)Cl_2 \longrightarrow CH_2{=}CHCH_3 + 2\,C_5H_5NH^+ + {}^-OPOCl_2$$

BIOLOGICAL DEHYDRATIONS

Dehydration reactions are known to occur in many important biological processes. Instead of being catalyzed by strong acids, which would not be available to a cell, they are catalyzed by enzymes. Fumarase, for example, is an enzyme that catalyzes the dehydration of malate in the citric acid cycle (p. 1038). The citric acid cycle is a series of reactions that oxidize compounds derived from carbohydrates, fatty acids, and amino acids.

Enolase, another enzyme, catalyzes the dehydration of α-phosphoglycerate in glycolysis (p. 1037). Glycolysis is a series of reactions that prepare glucose for entry into the citric acid cycle.

$$\underset{\text{malate}}{{}^-O\text{–}C(=O)\text{–}CH_2\text{–}CH(OH)\text{–}C(=O)\text{–}O^-} \overset{\text{fumarase}}{\rightleftharpoons} \underset{\text{fumarate}}{(H)({}^-O_2C)C{=}C(CO_2^-)(H)} + H_2O$$

$$\underset{\alpha\text{-phosphoglycerate}}{CH_2(OH)\text{–}CH(OPO_3^{2-})\text{–}C(=O)\text{–}O^-} \overset{\text{enolase}}{\rightleftharpoons} \underset{\text{phosphoenolpyruvate}}{CH_2{=}C(OPO_3^{2-})\text{–}C(=O)\text{–}O^-} + H_2O$$

PROBLEM 10◆

What alcohol would you treat with phosphorus oxychloride and pyridine to form each of the following alkenes?

a. $CH_3CH_2C(CH_3){=}CH_2$

b. 1-methylcyclohexene (ring with CH_3)

c. $CH_3CH{=}CHCH_2CH_3$

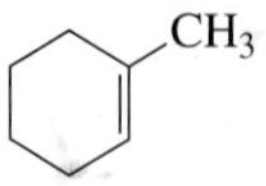

d. 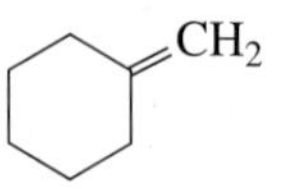 methylenecyclohexane (ring with $=CH_2$)

PROBLEM-SOLVING STRATEGY

Propose a mechanism for the following reaction:

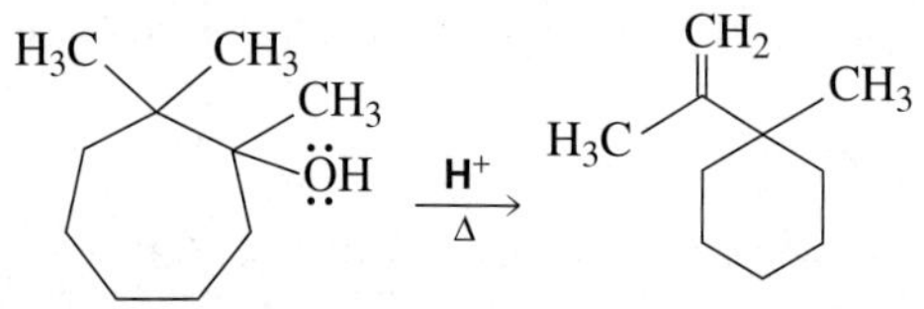

$$\xrightarrow[\Delta]{H^+}$$

Even the most complicated-looking mechanism can be worked out if you proceed one step at a time, keeping in mind the structure of the final product. The oxygen is the only basic atom in the starting material, so that is where protonation occurs. Loss of water forms a tertiary carbocation.

$$\longrightarrow \quad + \ H_2\ddot{O}:$$

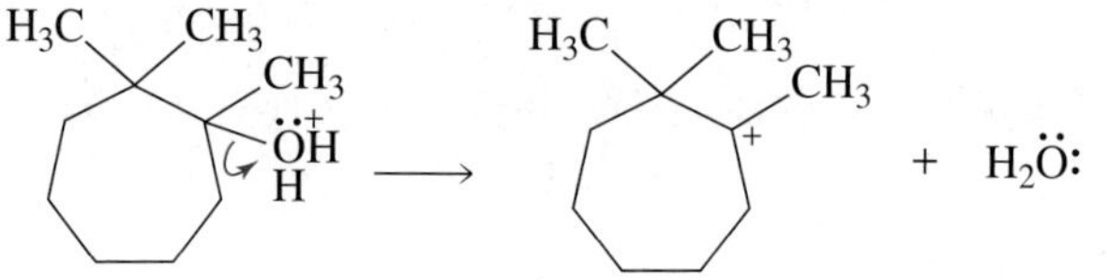

Because the starting material contains a seven-membered ring and the final product has a six-membered ring, a ring-contraction rearrangement must occur. When doing a *ring-contraction* (or a ring-expansion) *rearrangement*, it can be helpful to label the equivalent carbons in the reactant and product. Of the two possible pathways for ring contraction, one leads to a tertiary carbocation and the other leads to a primary carbocation. The correct pathway must be the one that leads to the tertiary carbocation, since it has the same arrangement of atoms as the product, and the primary carbocation would be too unstable to form.

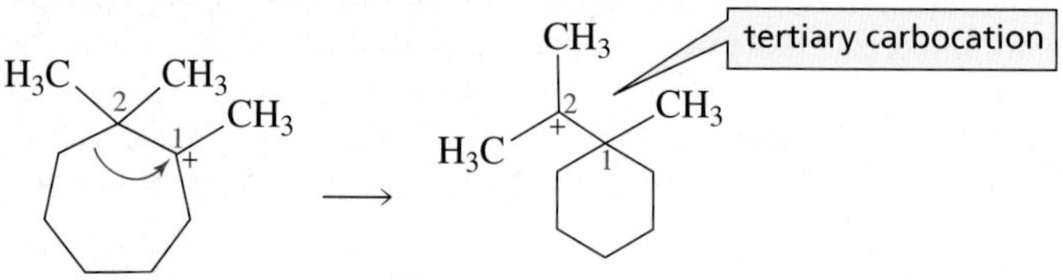

tertiary carbocation

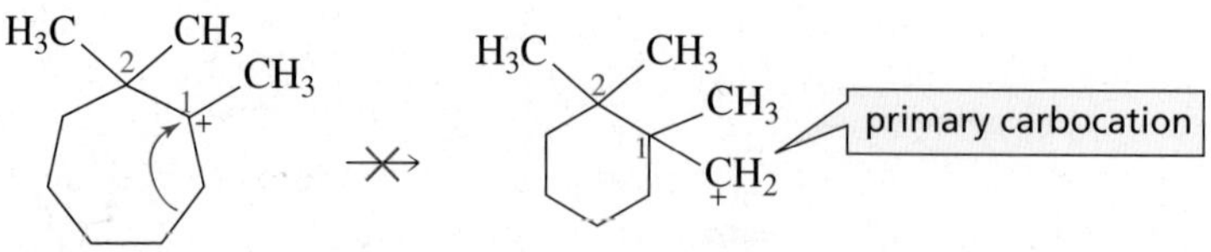

primary carbocation

The final product can now be obtained by removing a proton from the rearranged carbocation.

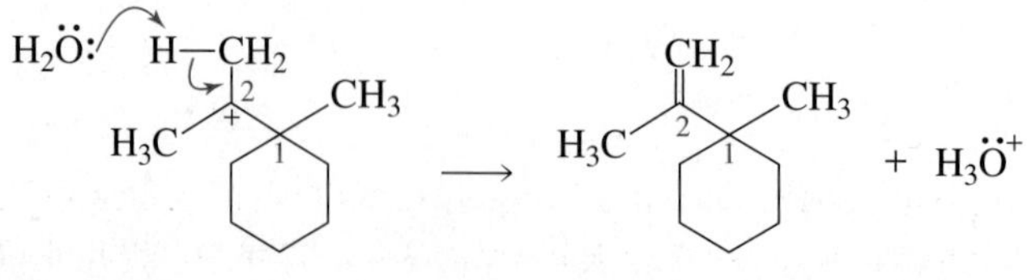

$$\longrightarrow \quad + \ H_3\ddot{O}^+$$

Now continue on to Problem 11.

PROBLEM 11

Propose a mechanism for each of the following reactions:

a. 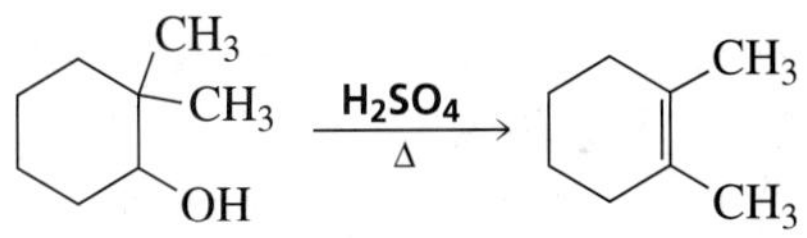 1-(cyclohexyl)-... $\xrightarrow[\Delta]{H_2SO_4}$

b. 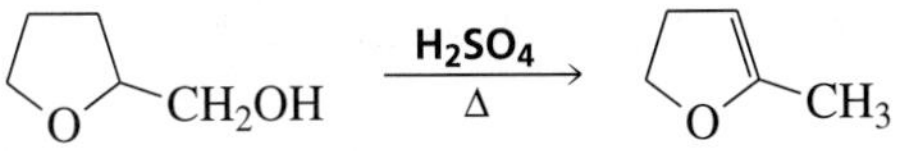 $\xrightarrow[\Delta]{H_2SO_4}$

c. 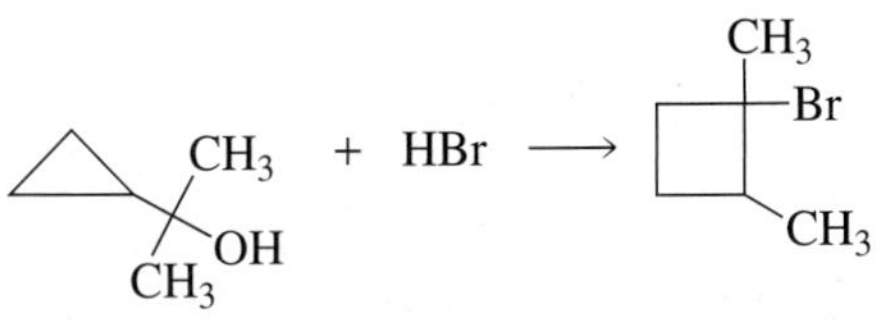 + HBr ⟶

PROBLEM 12◆

Give the major product formed when each of the following alcohols is heated in the presence of H_2SO_4:

a. 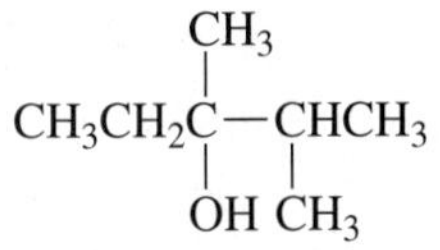 $CH_3CH_2C(CH_3)(OH)-CH(CH_3)CH_3$

b. 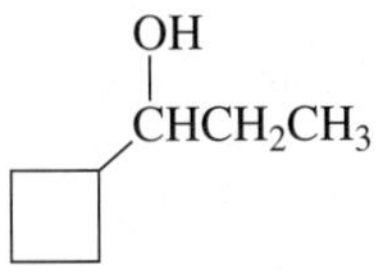 (cyclobutyl)$CH(OH)CH_2CH_3$

c. 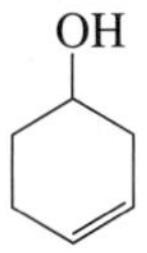 (cyclohexenol)

d. 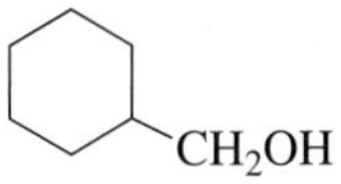 (cyclohexyl)CH_2OH

e. 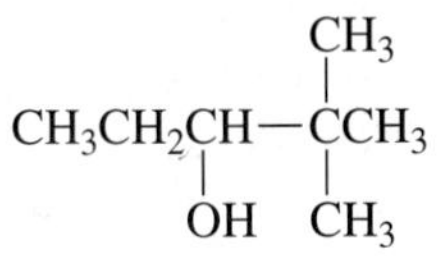 $CH_3CH_2CH(OH)-C(CH_3)_2CH_3$

f. $CH_3CH_2CH_2CH_2CH_2OH$

12.6 Substitution Reactions of Ethers

The —OR group of an **ether** and the —OH group of an alcohol have nearly the same basicity, since their conjugate acids have similar pK_a values. (The pK_a of $CH_3\overset{+}{O}H_2$ is 15.5 and the pK_a of H_3O^+ is 15.7.) Both groups are strong bases, so both are very poor leaving groups. Consequently, alcohols and ethers are equally unreactive toward nucleophilic substitution.

R—O—H an alcohol R—O—R an ether

Many of the reagents that are used to activate alcohols toward nucleophilic substitution (e.g., $SOCl_2$, PCl_3) cannot be used to activate ethers. When an alcohol reacts with an activating agent such as a sulfonyl chloride, a proton dissociates from the intermediate in the second step of the reaction and a stable sulfonate ester results.

$$\underset{\text{an alcohol}}{ROH} + R'-\overset{O}{\underset{O}{\overset{\|}{\underset{\|}{S}}}}-Cl \longrightarrow R\overset{+}{\underset{H}{\underset{|}{O}}}-\overset{O}{\underset{O}{\overset{\|}{\underset{\|}{S}}}}-R' + Cl^- \rightleftharpoons \underset{\text{a sulfonate ester}}{RO-\overset{O}{\underset{O}{\overset{\|}{\underset{\|}{S}}}}-R'} + H^+$$

When an ether reacts with a sulfonyl chloride, however, the oxygen atom does not have a proton that can dissociate. The alkyl group (R) cannot dissociate, so a stable sulfonate ester cannot be formed. Instead, the more stable starting materials are reformed.

$$\underset{\text{an ether}}{ROR} + R'-\overset{O}{\underset{O}{\overset{\|}{\underset{\|}{S}}}}-Cl \rightleftharpoons R\overset{+}{\underset{R}{\underset{|}{O}}}-\overset{O}{\underset{O}{\overset{\|}{\underset{\|}{S}}}}-R' + Cl^-$$

However, like alcohols, ethers can be activated by protonation. Ethers, therefore, can undergo nucleophilic substitution reactions with HBr or HI. As with alcohols, the reaction of ethers with hydrogen halides is slow, and the reaction mixture must be heated in order for the reaction to occur at a reasonable rate.

$$R-O-R' + HI \rightleftharpoons R-\overset{H}{\underset{+}{O}}-R' \;\; I^- \xrightarrow{\Delta} R-I + R'-OH$$

The first step in the cleavage of an ether by HI or HBr is protonation of the ether oxygen. This converts the very basic RO^- leaving group into the less basic ROH leaving group. What happens next in the mechanism depends on the structure of the ether.

If departure of the leaving group creates a relatively stable carbocation (e.g., a tertiary carbocation), an S_N1 reaction occurs—the leaving group departs, and the halide ion combines with the carbocation.

$$CH_3\underset{CH_3}{\overset{CH_3}{C}}-\ddot{O}CH_3 + H^+ \rightleftharpoons CH_3\underset{CH_3}{\overset{CH_3}{C}}-\overset{H}{\overset{+}{O}}CH_3 \xrightarrow{S_N1} CH_3\underset{CH_3}{\overset{CH_3}{C^+}} + CH_3\ddot{O}H \;\; \xrightarrow{:\ddot{I}:^-} CH_3\underset{CH_3}{\overset{CH_3}{C}}-\ddot{I}:$$

protonation

carbocation formation

attack by a nucleophile

However, if departure of the leaving group would create an unstable carbocation (e.g., a methyl, vinyl, aryl, or primary carbocation), the leaving group cannot depart. It has to be displaced by the halide ion. In other words, an S_N2 reaction occurs. The halide ion preferentially attacks the less sterically hindered of the two alkyl groups.

$$CH_3\ddot{O}CH_2CH_2CH_3 + H^+ \rightleftharpoons CH_3-\overset{H}{\overset{+}{O}}CH_2CH_2CH_3 \;\; :\ddot{I}:^- \xrightarrow{S_N2} CH_3\ddot{I}: + CH_3CH_2CH_2OH$$

protonation

nucleophile attacks the less sterically hindered carbon

The cleavage of ethers by HI or HBr occurs more rapidly if the reaction can take place by an S_N1 pathway. If the instability of the carbocation requires the reaction to follow an S_N2 pathway, cleavage will be more rapid with HI than with HBr because I^-

is a better nucleophile than Br^-. Only a substitution product is obtained because the bases present in the reaction mixture (halide ions and H_2O) are too weak to abstract a proton in an E2 reaction, and any alkene formed in an E1 reaction would undergo electrophilic addition with HBr or HI to form the same alkyl halide as would be obtained from the substitution reaction. Concentrated HCl cannot cleave ethers because Cl^- is too poor a nucleophile.

Alcohols and ethers undergo S_N2/E2 reactions unless they would have to form a primary carbocation, in which case they undergo S_N1/E1 reactions.

Because the only reagents with which ethers react are hydrogen halides, ethers are frequently used as solvents. Some common ether solvents are shown in Table 12.2.

Table 12.2 Some Ethers That Are Used as Solvents

$CH_3CH_2OCH_2CH_3$	(five-membered ring with O)	(six-membered ring with O)	(six-membered ring with two O)	$CH_3OCH_2CH_2OCH_3$	$CH_3OC(CH_3)_3$
diethyl ether "ether"	tetrahydrofuran THF	tetrahydropyran	1,4-dioxane	1,2-dimethoxyethane DME	*tert*-butyl methyl ether MTBE

PROBLEM 13

Explain why methyl propyl ether forms both methyl iodide and propyl iodide when it is heated with excess HI.

PROBLEM 14◆

Can HF be used to cleave ethers? Explain.

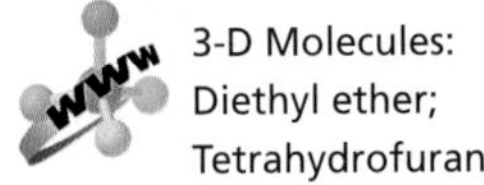

3-D Molecules: Diethyl ether; Tetrahydrofuran

ANESTHETICS

Because diethyl ether (commonly known simply as ether) is a short-lived muscle relaxant, it has been widely used as an inhalation anesthetic. However, because it takes effect slowly and has a slow and unpleasant recovery period, other compounds, such as enflurane, isoflurane, and halothane, have replaced ether as an anesthetic. Diethyl ether is still used where there is a lack of trained anesthesiologists, since it is the safest anesthetic to administer by untrained hands. Anesthetics interact with the nonpolar molecules of cell membranes, causing the membranes to swell, which interferes with their permeability.

$CH_3CH_2OCH_2CH_3$	$CF_3CHClOCHF_2$	$CHClFCF_2OCHF_2$	$CF_3CHClBr$
"ether"	isoflurane	enflurane	halothane

Sodium pentothal (also called thiopental sodium) is commonly used as an intravenous anesthetic. The onset of anesthesia and the loss of consciousness occur within seconds of its administration. Care must be taken when administering sodium pentothal because the dose for effective anesthesia is 75% of the lethal dose. Because of its toxicity, it cannot be used as the sole anesthetic. It is generally used to induce anesthesia before an inhalation anesthetic is administered. Propofol is an anesthetic that has all the properties of the "perfect anesthetic": It can be used as the sole anesthetic by intravenous drip, it has a rapid and pleasant induction period and a wide margin of safety, and recovery from the drug is rapid and pleasant.

Amputation of a leg without anesthetic in 1528.

sodium pentothal **propofol**

PROBLEM 15 SOLVED

Give the major products that would be obtained from heating each of the following ethers with HI:

a. $CH_3CH{=}CHOCH_2CH_3$

b. 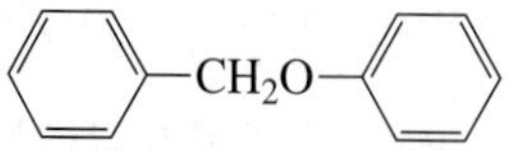 C_6H_5—CH_2O—C_6H_5

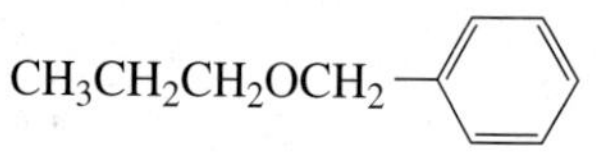

c. $CH_3CH_2CH_2OCH_2$—C_6H_5

d. 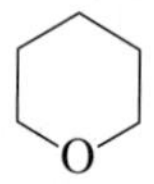 (tetrahydropyran ring)

e. 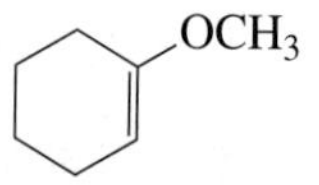 (cyclohexene ring with OCH_3)

f. 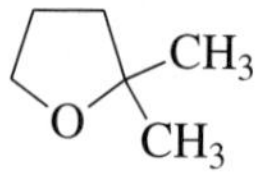 (tetrahydrofuran ring with CH_3, CH_3)

SOLUTION TO 15a The reaction takes place by an S_N2 pathway because neither alkyl group will form a relatively stable carbocation. Iodide ion attacks the carbon of the ethyl group because the other possibility is to attack an sp^2 hybridized carbon, and sp^2 carbons are generally not attacked by nucleophiles (Section 10.8). Thus, the products are ethyl iodide and an enol that immediately rearranges to an aldehyde (Section 6.6).

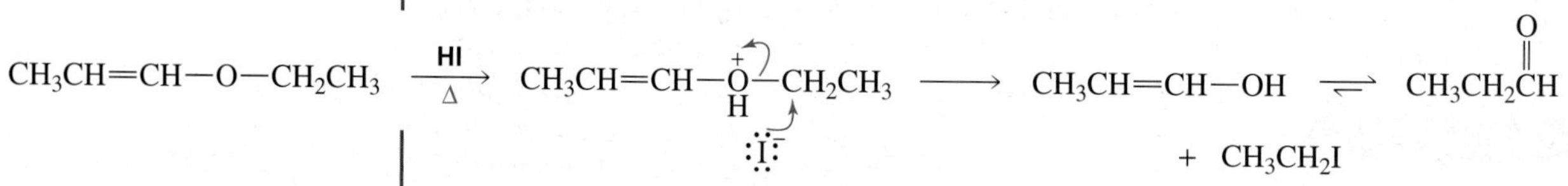

$$CH_3CH{=}CH{-}O{-}CH_2CH_3 \xrightarrow[\Delta]{HI} CH_3CH{=}CH{-}\overset{+}{\underset{H}{O}}{-}CH_2CH_3 \;\; :\ddot{I}:^- \longrightarrow CH_3CH{=}CH{-}OH + CH_3CH_2I \rightleftharpoons CH_3CH_2\overset{O}{\overset{\|}{C}}H$$

12.7 Reactions of Epoxides

Ethers in which the oxygen atom is incorporated into a three-membered ring are called **epoxides** or **oxiranes**. The common name of an epoxide uses the common name of the alkene, followed by "oxide," assuming that the oxygen atom is where the π bond of an alkene would be. The simplest epoxide is ethylene oxide.

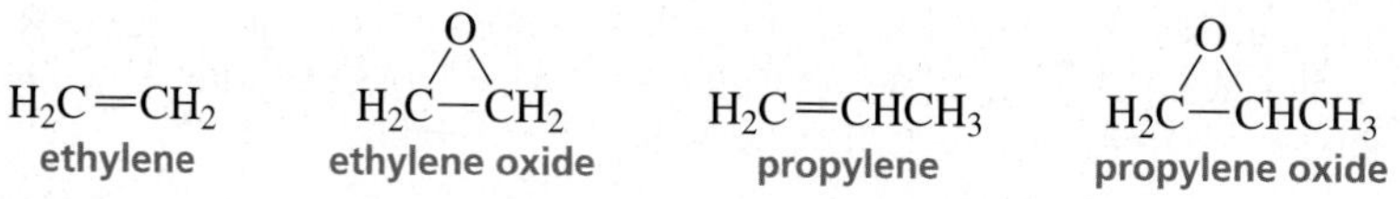

$H_2C{=}CH_2$	$H_2C{-}CH_2$ (O bridge)	$H_2C{=}CHCH_3$	$H_2C{-}CHCH_3$ (O bridge)
ethylene	ethylene oxide	propylene	propylene oxide

There are two systematic ways to name epoxides. One method calls the three-membered oxygen-containing ring "oxirane," with oxygen occupying the 1-position of the ring. Thus, 2-ethyloxirane has an ethyl substituent at the 2-position of the oxirane ring. Alternatively, an epoxide can be named as an alkane, with an "epoxy" prefix that identifies the carbons to which the oxygen is attached.

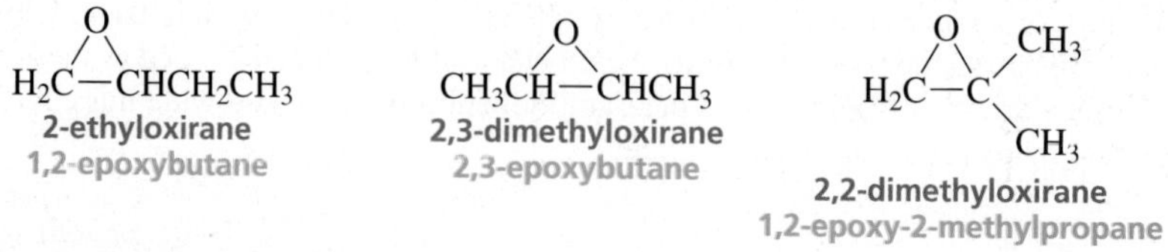

$H_2C{-}CHCH_2CH_3$ (O bridge)	$CH_3CH{-}CHCH_3$ (O bridge)	$H_2C{-}C(CH_3)_2$ (O bridge)
2-ethyloxirane	2,3-dimethyloxirane	2,2-dimethyloxirane
1,2-epoxybutane	2,3-epoxybutane	1,2-epoxy-2-methylpropane

PROBLEM 16◆

Draw the structure of the following compounds:

a. 2-propyloxirane
b. cyclohexene oxide
c. 2,2,3,3-tetramethyloxirane
d. 2,3-epoxy-2-methylpentane

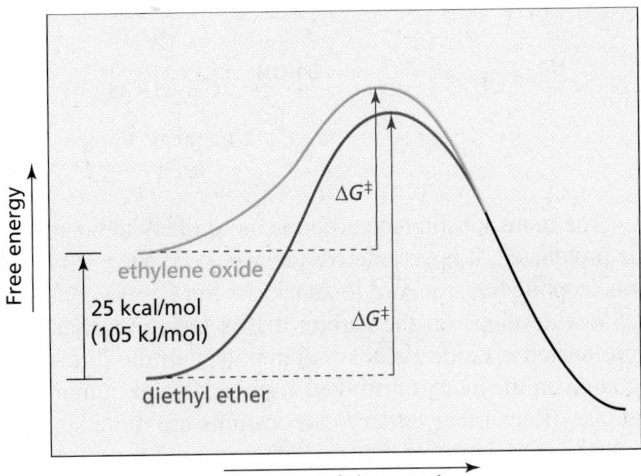

◀ **Figure 12.2**
The reaction coordinate diagrams for nucleophilic attack of hydroxide ion on ethylene oxide and on diethyl ether. The greater reactivity of the epoxide is a result of the strain (ring strain and torsional strain) in the three-membered ring, which increases its free energy.

Although an epoxide and an ether have the same leaving group, epoxides are much more reactive than ethers in nucleophilic substitution reactions because the strain in the three-membered ring is relieved when the ring opens (Figure 12.2). Epoxides, therefore, readily undergo ring-opening reactions with a wide variety of nucleophiles.

Epoxides, like other ethers, react with hydrogen halides. In the first step of the reaction, the nucleophilic oxygen is protonated by the acid. The protonated epoxide is then attacked by the halide ion. Because epoxides are so much more reactive than ethers, the reaction takes place readily at room temperature. (Recall that the reaction of an ether with a hydrogen halide requires heat.)

$$H_2C\overset{O}{—}CH_2 + H—Br: \rightleftharpoons H_2C\overset{\overset{H}{O^+}}{—}CH_2 + :Br:^- \longrightarrow HOCH_2CH_2Br:$$

protonation of the epoxide oxygen atom

back-side attack by the nucleophile

Protonated epoxides are so reactive that they can be opened by poor nucleophiles, such as H_2O and alcohols. (HB^+ is any acid in the reaction mixture; B: is any base.)

$$H_2C\overset{O}{—}CH_2 \overset{H—B^+}{\rightleftharpoons} H_2C\overset{\overset{H}{O^+}}{—}CH_2 \xrightarrow{H_2O:} \underset{\overset{|}{^+OH_2}}{CH_2CH_2OH} \xrightarrow{B:} HOCH_2CH_2OH + HB^+$$

1,2-ethanediol
ethylene glycol

$$CH_3CH\overset{O}{—}CHCH_3 \overset{H—B^+}{\rightleftharpoons} CH_3CH\overset{\overset{H}{O^+}}{—}CHCH_3 \xrightarrow{CH_3OH} CH_3\overset{OH}{C}HCHCH_3 \text{ with } {^+}OCH_3(H) \xrightarrow{B:} CH_3\overset{OH}{C}H\underset{OCH_3}{C}HCH_3 + HB^+$$

3-methoxy-2-butanol

If different substituents are attached to the two carbons of the protonated epoxide (and the nucleophile is something other than H_2O), the product obtained from nucleophilic attack on the 2-position of the oxirane ring will be different from that obtained from nucleophilic attack on the 3-position. The major product is the one resulting from nucleophilic attack on the *more substituted* carbon.

$$CH_3CH\overset{O}{—}CH_2 \underset{}{\overset{H^+}{\rightleftharpoons}} CH_3CH\overset{\overset{H}{O^+}}{—}CH_2 \xrightarrow{CH_3OH} \underset{\text{2-methoxy-1-propanol, major product}}{CH_3\overset{OCH_3}{C}HCH_2OH} + \underset{\text{1-methoxy-2-propanol, minor product}}{CH_3\overset{OH}{C}HCH_2OCH_3} + H^+$$

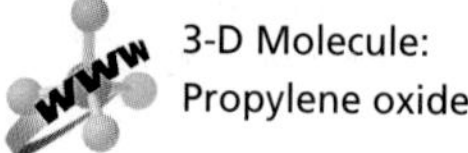

3-D Molecule: Propylene oxide

The more substituted carbon is more likely to be attacked because, after the epoxide is protonated, it is so reactive that one of the C—O bonds begins to break before the nucleophile has a chance to attack. As the C—O bond starts to break, a partial positive charge develops on the carbon that is losing its share of the oxygen's electrons. The protonated epoxide breaks preferentially in the direction that puts the partial positive charge on the more substituted carbon, because a more substituted carbocation is more stable. (Recall that tertiary carbocations are more stable than secondary carbocations, which are more stable than primary carbocations.)

$$CH_3CH\overset{\overset{H}{O^{\delta+}}}{—}CH_2\ (\delta+ \text{ on CH}) \xrightarrow{CH_3\ddot{O}H} CH_3\overset{\overset{H}{^+OCH_3}}{C}HCH_2OH \rightleftharpoons CH_3\overset{OCH_3}{C}HCH_2OH + H^+$$

developing secondary carbocation — major product

$$CH_3CH\overset{\overset{H}{O^{\delta+}}}{—}CH_2\ (\delta+ \text{ on CH}_2) \xrightarrow{CH_3\ddot{O}H} CH_3\overset{OH}{C}HCH_2\overset{+}{\underset{H}{O}}CH_3 \rightleftharpoons CH_3\overset{OH}{C}HCH_2OCH_3 + H^+$$

developing primary carbocation — minor product

The best way to describe the reaction is to say that it occurs by a pathway that is partially S_N1 and partially S_N2. It is not a pure S_N1 reaction because a carbocation intermediate is not fully formed; it is not a pure S_N2 reaction because the leaving group begins to depart before the compound is attacked by the nucleophile.

In Section 12.6, we saw that an ether does not undergo a nucleophilic substitution reaction unless the very basic ^-OR leaving group is converted by protonation into a less basic ROH group. Because of the strain in the three-membered ring, epoxides are reactive enough to open without first being protonated. When a nucleophile attacks an unprotonated epoxide, the reaction is a pure S_N2 reaction. That is, the C—O bond does not begin to break until the carbon is attacked by the nucleophile. In this case, the nucleophile is more likely to attack the *less substituted* carbon because the less substituted carbon is more accessible to attack. (It is less sterically hindered.) Thus, the site of nucleophilic attack on an unsymmetrical epoxide under neutral or basic conditions (when the epoxide *is not* protonated) is different from the site of nucleophilic attack under acidic conditions (when the epoxide *is* protonated).

$$CH_3CH\overset{O}{—}CH_2$$

site of nucleophilic attack under acidic conditions (CH) | site of nucleophilic attack under basic conditions (CH_2)

After the nucleophile has attacked the epoxide, the alkoxide ion can pick up a proton from the solvent or from an acid added after the reaction is over.

$$CH_3CH\overset{O}{—}CH_2 + CH_3\ddot{O}:^- \longrightarrow CH_3\overset{O^-}{C}HCH_2OCH_3 \xrightarrow[\text{or } H^+]{CH_3OH} CH_3\overset{OH}{C}HCH_2OCH_3 + CH_3O^-$$

Epoxides are synthetically useful reagents because they can react with a wide variety of nucleophiles, leading to the formation of a wide variety of products.

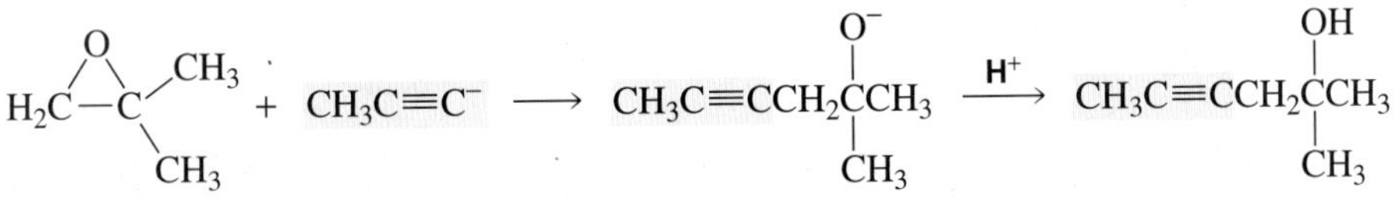

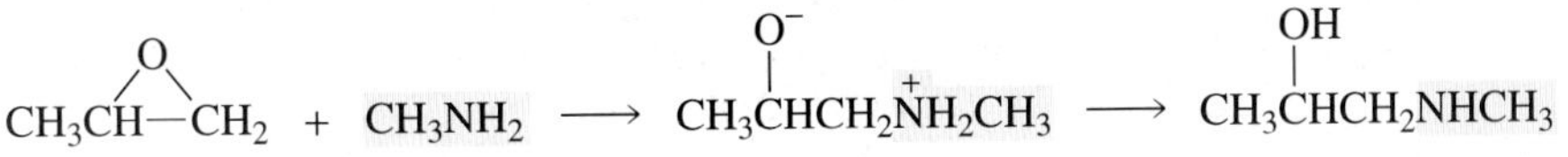

$$\text{H}_2\overset{\text{O}}{\text{C}-\text{C}}(\text{CH}_3)_2 + \text{CH}_3\text{C}{\equiv}\text{C}^- \longrightarrow \text{CH}_3\text{C}{\equiv}\text{CCH}_2\overset{\text{O}^-}{\underset{\text{CH}_3}{\text{C}}}\text{CH}_3 \xrightarrow{\text{H}^+} \text{CH}_3\text{C}{\equiv}\text{CCH}_2\overset{\text{OH}}{\underset{\text{CH}_3}{\text{C}}}\text{CH}_3$$

$$\text{CH}_3\overset{\text{O}}{\text{CH}-\text{CH}_2} + \text{CH}_3\text{NH}_2 \longrightarrow \text{CH}_3\overset{\text{O}^-}{\text{C}}\text{HCH}_2\overset{+}{\text{N}}\text{H}_2\text{CH}_3 \longrightarrow \text{CH}_3\overset{\text{OH}}{\text{C}}\text{HCH}_2\text{NHCH}_3$$

Epoxides also are important in biological processes because they are reactive enough to be attacked by nucleophiles under the conditions found in living systems (Section 12.8).

Notice that the reaction of cyclohexene oxide with a nucleophile leads to trans products because an S_N2 reaction involves back-side nucleophilic attack.

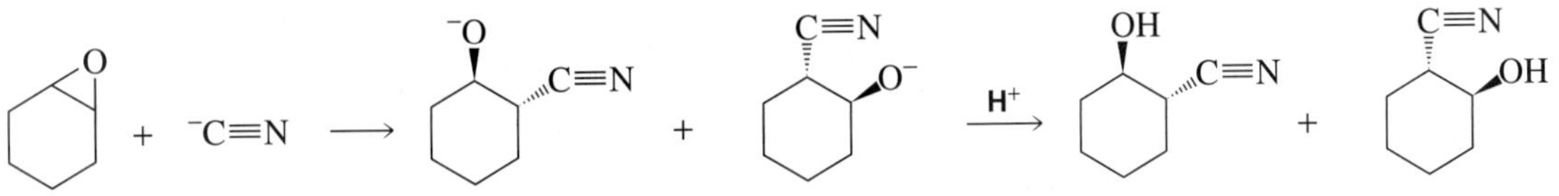

$$\text{cyclohexene oxide} + {}^-\text{C}{\equiv}\text{N} \longrightarrow \text{trans-2-cyanocyclohexanolate (}{}^-\text{O}, \text{C}{\equiv}\text{N)} + \text{enantiomer} \xrightarrow{\text{H}^+} \text{trans-2-cyanocyclohexanol (OH, C}{\equiv}\text{N)} + \text{enantiomer}$$

PROBLEM 17

Why does the preceding reaction form two stereoisomers?

PROBLEM 18◆

Give the major product of each of the following reactions:

a. 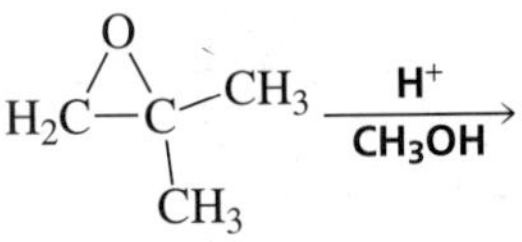 $\text{H}_2\overset{\text{O}}{\text{C}-\text{C}}(\text{CH}_3)_2 \xrightarrow[\text{CH}_3\text{OH}]{\text{H}^+}$

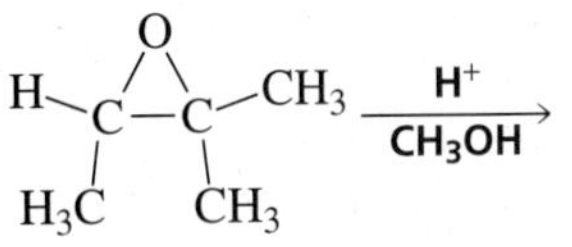

b. 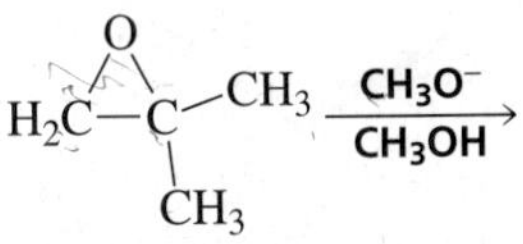 $\text{H}_2\overset{\text{O}}{\text{C}-\text{C}}(\text{CH}_3)_2 \xrightarrow[\text{CH}_3\text{OH}]{\text{CH}_3\text{O}^-}$

c. $\text{H}(\text{H}_3\text{C})\overset{\text{O}}{\text{C}-\text{C}}(\text{CH}_3)_2 \xrightarrow[\text{CH}_3\text{OH}]{\text{H}^+}$

d. 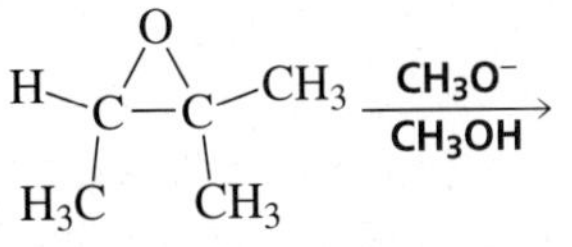 $\text{H}(\text{H}_3\text{C})\overset{\text{O}}{\text{C}-\text{C}}(\text{CH}_3)_2 \xrightarrow[\text{CH}_3\text{OH}]{\text{CH}_3\text{O}^-}$

PROBLEM 19◆

Would you expect the reactivity of a five-membered ring ether such as tetrahydrofuran (Table 12.2) to be more similar to an epoxide or to a noncyclic ether?

12.8 Arene Oxides

When an aromatic hydrocarbon such as benzene is ingested or inhaled, it is enzymatically converted into an *arene oxide*. An **arene oxide** is a compound in which one of the "double bonds" of the aromatic ring has been converted into an epoxide. Formation of an arene oxide is the first step in changing an aromatic compound that enters the body as a foreign substance (e.g., cigarette smoke, drugs, automobile exhaust) into a more water-soluble compound that can eventually be eliminated. The enzyme that detoxifies aromatic hydrocarbons by converting them into arene oxides is called cytochrome P_{450}.

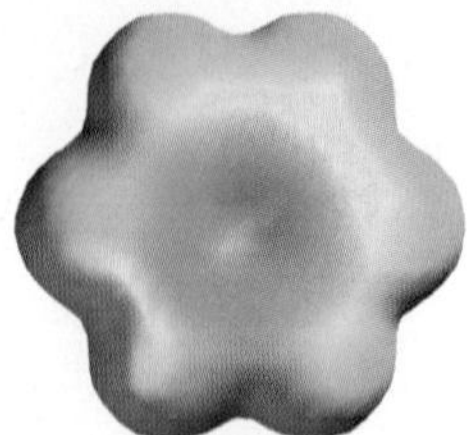
benzene

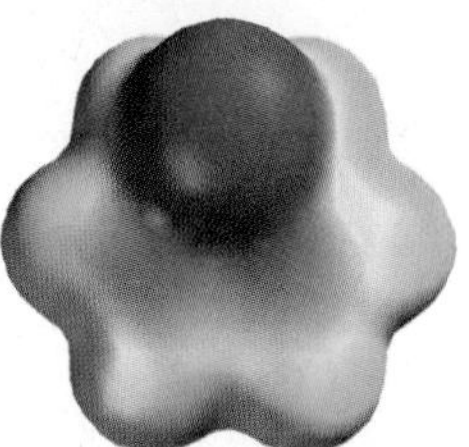
benzene oxide

benzene —cytochrome P_{450}, O_2→ benzene oxide (an arene oxide)

Arene oxides are important intermediates in the biosynthesis of biochemically important phenols such as tyrosine and serotonin.

tyrosine, an amino acid: $HO-C_6H_4-CH_2CH(^+NH_3)COO^-$

serotonin, a vasoconstrictor

An arene oxide can react in two different ways. It can react as a typical epoxide, undergoing attack by a nucleophile to form an addition product (Section 12.7). Alternatively, it can rearrange to form a phenol, something that epoxides such as ethylene oxide cannot do.

3-D Molecule: Benzene oxide

benzene oxide —Y^-→ addition product

benzene oxide → rearranged product

When an arene oxide reacts with a nucleophile, the three-membered ring undergoes back-side nucleophilic attack, forming an addition product.

HO^- + benzene oxide → alkoxide —H_2O→ addition product + HO^-

When an arene oxide undergoes rearrangement, the three-membered epoxide ring opens, picking up a proton from a species in the solution. Instead of immediately losing a proton to form phenol, the carbocation forms an *enone* as a result of a 1,2-hydride shift. This is called an *NIH shift* because it was first observed in a laboratory at the National Institutes of Health. Removal of a proton from the enone forms phenol.

benzene oxide ($H-B^+$) ⇌ a carbocation —NIH shift→ an enone (:B) → phenol + HB^+

a carbocation ↛ phenol + H^+

Because the first step is rate-determining, the rate of formation of phenol depends on the stability of the carbocation. The more stable the carbocation, the easier it is to open the epoxide ring and form the phenol.

Only one arene oxide can be formed from naphthalene because the "double bond" shared by the two rings cannot be epoxidized. Recall from Section 7.11 that benzene rings are particularly stable, so it is much easier to epoxidize a position that will leave one of the benzene rings intact. Naphthalene oxide can rearrange to form either 1-naphthol or 2-naphthol. The carbocation leading to 1-naphthol is more stable because its positive charge can be stabilized by electron delocalization without destroying the aromaticity of the benzene ring on the left of the structure. In contrast, the positive charge on the carbocation leading to 2-naphthol can be stabilized by electron delocalization only if the aromaticity of the benzene ring is destroyed. Consequently, rearrangement leads predominantly to 1-naphthol.

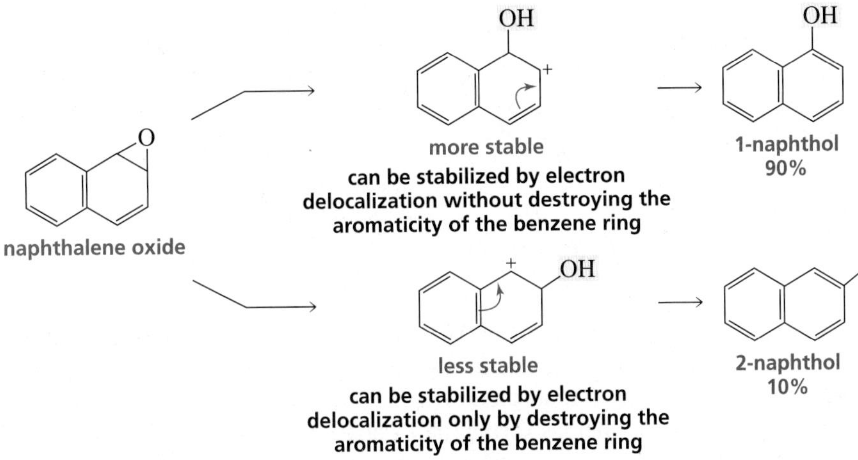

naphthalene oxide

more stable
can be stabilized by electron delocalization without destroying the aromaticity of the benzene ring

1-naphthol
90%

less stable
can be stabilized by electron delocalization only by destroying the aromaticity of the benzene ring

2-naphthol
10%

PROBLEM 20

Draw all possible resonance contributors for the two carbocations in the preceding reaction. Use the resonance contributors to explain why 1-napthol is the major product of the reaction.

PROBLEM 21◆

The existence of the NIH shift was established by determining the major product obtained from rearrangement of the following arene oxide, in which a hydrogen has been replaced by a deuterium:

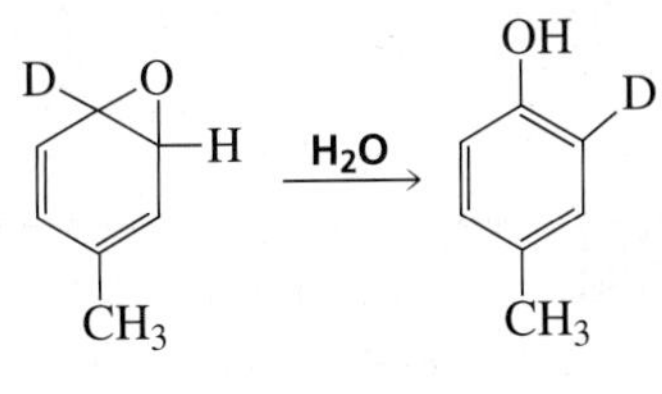

D O H —H_2O→ OH D
CH_3 CH_3

What would be the major product if the NIH shift did not occur? (*Hint:* Recall from Section 11.7 that a C—H bond is easier to break than a C—D bond.)

PROBLEM 22◆

How would the major products obtained from rearrangement of the following arene oxides differ?

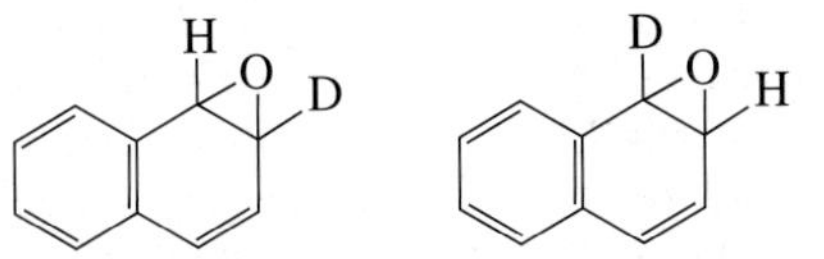

H O D   D O H

A segment of DNA

Some aromatic hydrocarbons are carcinogens—compounds that cause cancer. Investigation has revealed, however, that it is not the hydrocarbon itself that is carcinogenic, but rather the arene oxide into which the hydrocarbon is converted. How do arene oxides cause cancer? We have seen that nucleophiles react with epoxides to form addition products. 2′-Deoxyguanosine, a component of DNA (Section 27.1), has a nucleophilic NH_2 group that is known to react with certain arene oxides. Once 2′-deoxyguanosine becomes covalently attached to an arene oxide, 2′-deoxyguanosine can no longer fit into the DNA double helix. Thus, the genetic code cannot be properly transcribed, which can lead to mutations that cause cancer. Cancer results when cells lose their ability to control their growth and reproduction.

an arene oxide

2′-deoxyguanosine

covalently attached to the arene oxide

Not all arene oxides are carcinogenic. Whether a particular arene oxide is carcinogenic depends on the relative rates of its two reaction pathways—rearrangement or reaction with a nucleophile. Arene oxide rearrangement leads to phenolic products that are not carcinogenic, whereas formation of addition products from nucleophilic attack by DNA can lead to cancer-causing products. Thus, if the rate of arene oxide rearrangement is greater than the rate of nucleophilic attack by DNA, the arene oxide will be harmless. If, however, the rate of nucleophilic attack is greater than the rate of rearrangement, the arene oxide will be a carcinogen.

Because the rate of arene oxide rearrangement depends on the stability of the carbocation formed in the first step of the rearrangement, *an arene oxide's cancer-causing potential depends on the stability of this carbocation*. If the carbocation is relatively stable, rearrangement will be fast and the arene oxide will tend to be noncarcinogenic. If the carbocation is relatively unstable, rearrangement will be slow and the arene oxide will be more likely to undergo nucleophilic attack and be carcinogenic. This means that the more reactive the arene oxide (the more easily it opens to form a carbocation), the less likely it is to be carcinogenic.

PROBLEM 23 SOLVED

Which compound is more likely to be carcinogenic?

a. OCH_3 or NO_2

b. or

SOLUTION TO 23a The nitro-substituted compound is more likely to be carcinogenic. The nitro group destabilizes the carbocation formed when the ring opens by withdrawing electrons from the ring by resonance. In contrast, the methoxy group stabilizes the carbocation by resonance electron donation into the ring. Carbocation formation leads to the harmless product, so the nitro-substituted compound with a less stable (harder-to-form) carbocation will be less likely to undergo rearrangement to a harmless product. In addition, the electron-withdrawing nitro group increases the arene oxide's susceptibility to nucleophilic attack, the cancer-causing pathway.

CHIMNEY SWEEPS AND CANCER

In 1775, a British physician named Percival Potts was the first to recognize that environmental factors can cause cancer, when he became aware that chimney sweeps had a higher incidence of scrotum cancer than the male population as a whole. He theorized that something in the chimney soot was causing cancer. We now know that it was benzo[*a*]pyrene.

Titch Cox, the chimney sweep responsible for cleaning the 800 chimneys at Buckingham Palace.

BENZO[*a*]PYRENE AND CANCER

Benzo[*a*]pyrene is one of the most carcinogenic of the aromatic hydrocarbons. This hydrocarbon is formed whenever an organic compound undergoes incomplete combustion. For example, benzo[*a*]pyrene is found in cigarette smoke, automobile exhaust, and charcoal-broiled meat. Several arene oxides can be formed from benzo[*a*]pyrene. The two most harmful are the 4,5-oxide and the 7,8-oxide. It has been suggested that people who develop lung cancer as a result of smoking may have a higher than normal concentration of cytochrome P_{450} in their lung tissue.

The 4,5-oxide is harmful because it forms a carbocation that cannot be stabilized by electron delocalization without destroying the aromaticity of an adjacent benzene ring. Thus, the carbocation is relatively unstable, so the epoxide will tend not to open until it is attacked by a nucleophile. The 7,8-oxide is harmful because it reacts with water to form a diol, which then forms a diol epoxide. The diol epoxide does not readily undergo rearrangement (the harmless pathway), because it opens to a carbocation that is destabilized by the electron-withdrawing OH groups. Therefore, the diol epoxide can exist long enough to be attacked by nucleophiles (the carcinogenic pathway).

PROBLEM 24

Explain why the two arene oxides in Problem 23a open in opposite directions.

PROBLEM 25

Three arene oxides can be obtained from phenanthrene.

phenanthrene

a. Give the structures of the three phenanthrene oxides.
b. What phenols can be obtained from each phenanthrene oxide?
c. If a phenanthrene oxide can lead to the formation of more than one phenol, which phenol will be obtained in greater yield?
d. Which of the three phenanthrene oxides is most likely to be carcinogenic?

12.9 Crown Ethers

Crown ethers are cyclic compounds that have several ether linkages. A crown ether specifically binds certain metal ions or organic molecules, depending on the size of its cavity. The crown ether is called the "host" and the species it binds is called the "guest." Because the ether linkages are chemically inert, the crown ether can bind the guest without reacting with it. The *crown–guest complex* is called an **inclusion compound**. Crown ethers are named [X]-crown-Y, where X is the total number of atoms in the ring and Y is the number of oxygen atoms in the ring. [15]-Crown-5 selectively binds Na^+ because the crown ether has a cavity diameter of 1.7 to 2.2 Å and Na^+ has an ionic diameter of 1.80 Å. Binding occurs as a result of interaction of the positively charged ion with the lone-pair electrons of the oxygens that point into the cavity of the crown ether.

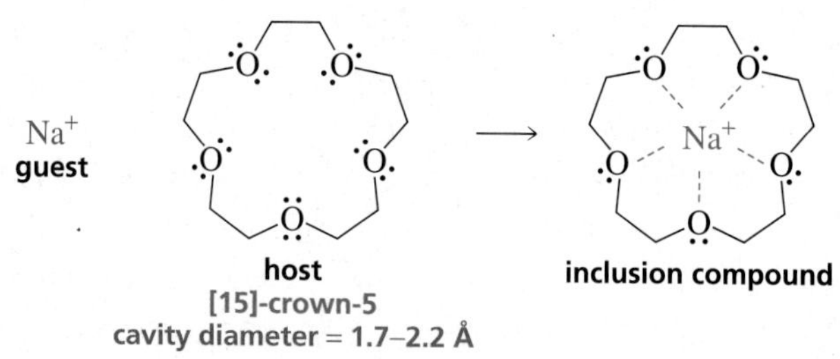

With an ionic diameter of 1.20 Å, Li^+ is too small to be bound by [15]-crown-5, but it binds neatly in [12]-crown-4. On the other hand, K^+, with an ionic diameter of 2.66 Å, is too large to fit into [15]-crown-5, but is bound specifically by [18]-crown-6.

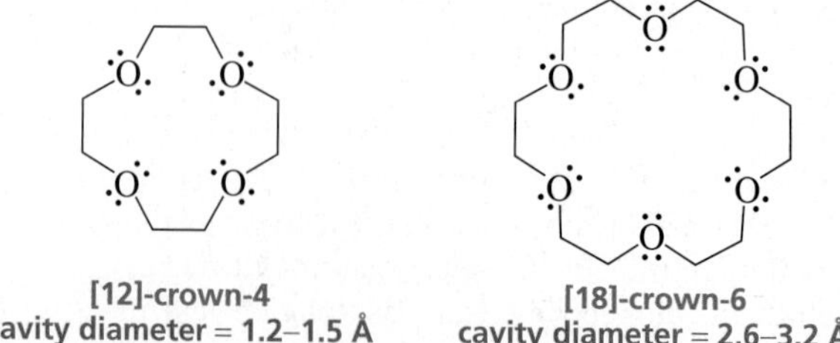

For their work in the field of crown ethers, **Charles J. Pedersen, Donald J. Cram,** *and* **Jean-Marie Lehn** *shared the 1987 Nobel Prize in chemistry.*

Charles J. Pedersen (1904–1989) *was born in Korea to a Korean mother and a Norwegian father. He moved to the United States as a teenager and received a B.S. in chemical engineering from the University of Dayton and an M.S. in organic chemistry from MIT. He joined DuPont in 1927 and retired from there in 1969. He is one of the few people without a Ph.D. to receive a Nobel Prize in the sciences.*

Jean-Marie Lehn *was born in France in 1939. He initially studied philosophy and then switched to chemistry. He received a Ph.D. from the University of Strasbourg. As a postdoctoral fellow, he worked with R. B. Woodward at Harvard on the total synthesis of vitamin* B_{12}. *(See pages 911 and 1177.) He is a professor of chemistry at the Université Louis Pasteur in Strasbourg, France, and at the Collège de France in Paris.*

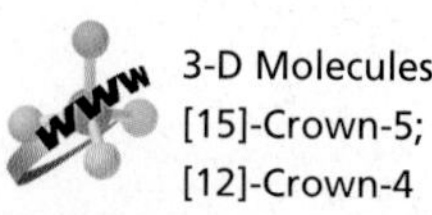

3-D Molecules: [15]-Crown-5; [12]-Crown-4

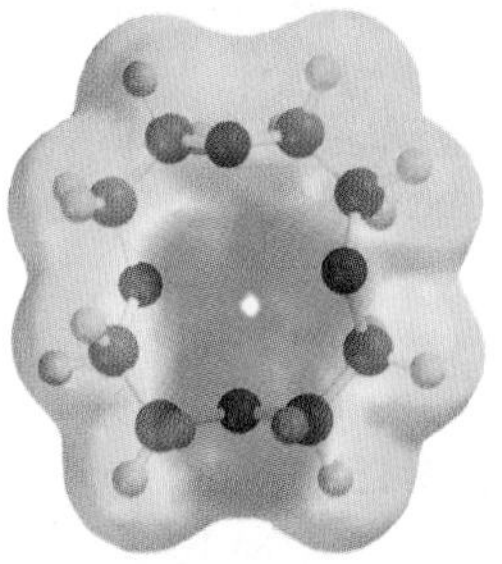

[12]-crown-4

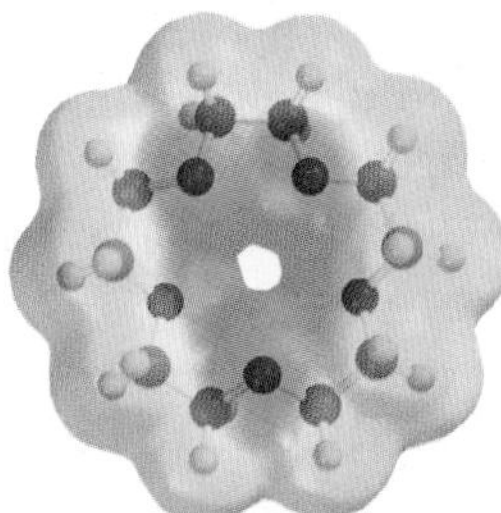
[15]-crown-5

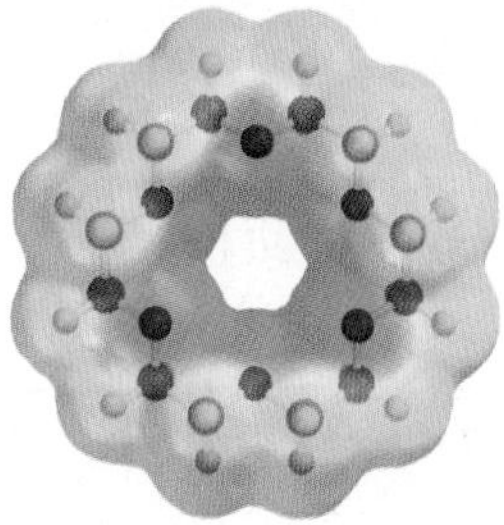
[18]-crown-6

Donald J. Cram (1919–2001) *was born in Vermont. He received a B.S. from Rollins College, an M.S. from the University of Nebraska, and a Ph.D. from Harvard University. He was a professor of chemistry at the University of California, Los Angeles, and was an avid surfer.*

The ability of a host to bind only certain guests is an example of **molecular recognition**. Molecular recognition explains how enzymes recognize their substrates, how antibodies recognize antigens, how drugs recognize receptors, and many other biochemical processes. Only recently have chemists been able to design and synthesize organic molecules that exhibit molecular recognition, although the specificity of these synthetic compounds for their guests is generally less highly developed than the specificity exhibited by biological molecules.

A remarkable property of crown ethers is that they allow inorganic salts to be dissolved in nonpolar organic solvents, thus permitting many reactions to be carried out in nonpolar solvents that otherwise would not be able to take place. For example, the S_N2 reaction of 1-bromohexane with acetate ion poses a problem because potassium acetate is an ionic compound that is soluble only in water, whereas the alkyl halide is insoluble in water. In addition, acetate ion is an extremely poor nucleophile.

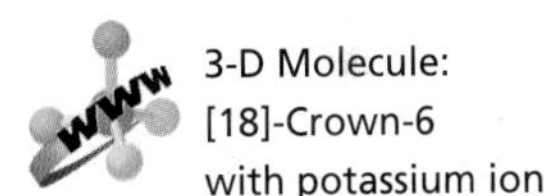

3-D Molecule: [18]-Crown-6 with potassium ion

$$CH_3CH_2CH_2CH_2CH_2CH_2Br + CH_3\overset{O}{\overset{\|}{C}}O^- \xrightarrow{\text{[18]-crown-6}} CH_3CH_2CH_2CH_2CH_2CH_2O\overset{O}{\overset{\|}{C}}CH_3 + Br^-$$

soluble only in a nonpolar sovent

soluble only in water

Potassium acetate can be dissolved in a nonpolar solvent such as benzene if [18]-crown-6 is added to the solution. The crown ether binds potassium in its cavity, and the nonpolar crown ether–potassium complex dissolves in benzene, along with the acetate ion, to maintain electrical neutrality. The crown ether is acting as a **phase-transfer catalyst**—it is bringing a reactant into the phase where it is needed. Because acetate is not solvated in the nonpolar solvent, it is a much better nucleophile than it would be in a

AN IONOPHOROUS ANTIBIOTIC

An antibiotic is a compound that interferes with the growth of microorganisms. Nonactin is a naturally occurring antibiotic that owes its biological activity to its ability to disrupt the carefully maintained electrolyte balance between the inside and outside of a cell. To achieve the gradient between potassium and sodium ions inside and outside the cell that is required for normal cell function, potassium ions are pumped in and sodium ions are pumped out. Nonactin disrupts this gradient by acting like a crown ether. Nonactin's diameter is such that it specifically binds potassium ions. The eight oxygens that point into the cavity and interact with K^+ are highlighted in the structure shown here. The outside of nonactin is nonpolar, so it can easily transport K^+ ions out of the cell through the nonpolar cell membrane. The decreased concentration of K^+ within the cell causes the bacterium to die. Nonactin is an example of an *ionophorous* antibiotic. An *ionophore* is a compound that transports metal ions by binding them tightly.

nonactin

protic polar solvent (Section 10.3). Recall that if a reactant in the rate-determining step is charged, the reaction will be faster in a nonpolar solvent (Section 10.10).

Crown ethers have been used to dissolve many other salts in nonpolar solvents. In each case, the positively charged ion is nestled within the crown ether and is accompanied by a "naked" anion in the nonpolar solvent.

12.10 Thiols, Sulfides, and Sulfonium Salts

Thiols are sulfur analogs of alcohols. Thiols used to be called mercaptans because they form strong complexes with heavy metal cations such as mercury and arsenic (they capture mercury).

$$\underset{\textbf{a thiol}}{2\,CH_3CH_2SH} + \underset{\text{mercuric ion}}{Hg^{2+}} \longrightarrow CH_3CH_2S{-}Hg{-}SCH_2CH_3 + 2\,H^+$$

3-D Molecule: Methanethiol

Thiols are named by adding the suffix "thiol" to the name of the parent hydrocarbon. If there is a second functional group in the molecule, the SH group can be indicated by its substituent name, "mercapto." Like other substituent names, it is cited before the name of the parent hydrocarbon.

$$\underset{\textbf{ethanethiol}}{CH_3CH_2SH} \qquad \underset{\textbf{1-propanethiol}}{CH_3CH_2CH_2SH} \qquad \underset{\textbf{3-methyl-1-butanethiol}}{CH_3\overset{\overset{\displaystyle CH_3}{|}}{C}HCH_2CH_2SH} \qquad \underset{\textbf{2-mercaptoethanol}}{HSCH_2CH_2OH}$$

Low-molecular-weight thiols are noted for their strong and pungent odors, such as those associated with onions, garlic, and skunks. Natural gas is completely odorless and can cause deadly explosions if a leak goes undetected. Therefore, a small amount of a thiol is added to natural gas to give it an odor so that gas leaks can be detected.

Sulfur is larger than oxygen, so the negative charge of the thiolate ion is spread over a larger volume of space, causing it to be more stable than an alkoxide ion. Recall that the more stable the base, the stronger is its conjugate acid (Section 1.18). Thiols, therefore, are stronger acids ($pK_a = 10$) than alcohols, and thiolate ions are weaker bases than alkoxide ions. The larger thiolate ions are less well solvated than alkoxide ions, so in protic solvents thiolate ions are better nucleophiles than alkoxide ions (Section 10.3).

$$CH_3\ddot{\underset{..}{S}}{:}^- + CH_3CH_2{-}Br \xrightarrow[\textbf{CH}_3\textbf{OH}]{} CH_3\underset{..}{\ddot{S}}CH_2CH_3 + Br^-$$

Because sulfur is not as electronegative as oxygen, thiols are not good at hydrogen bonding. Consequently, they have weaker intermolecular attractions and, therefore, considerably lower boiling points than alcohols have (Section 2.9). For example, the boiling point of CH_3CH_2OH is 78 °C, whereas the boiling point of CH_3CH_2SH is 37 °C.

The sulfur analogs of ethers are called **sulfides** or **thioethers**. Sulfur is an excellent nucleophile because its electron cloud is polarizable (Section 10.3). Sulfides, therefore, react readily with alkyl halides to form **sulfonium salts**—a reaction that an ether cannot undergo because oxygen is not as nucleophilic and cannot accommodate a charge as easily.

3-D Molecules: Dimethyl sulfide; Trimethylsulfonium cation

$$\underset{\textbf{dimethyl sulfide}}{CH_3\underset{..}{\ddot{S}}CH_3} + CH_3{-}I \longrightarrow \underset{\substack{\textbf{trimethylsulfonium iodide}\\ \textbf{a sulfonium salt}}}{CH_3\overset{\overset{\displaystyle CH_3}{|}}{\underset{..}{\overset{+}{S}}}CH_3 \;\; I^-}$$

Because it has a weakly basic leaving group, a sulfonium ion readily undergoes nucleophilic substitution reactions. As with other S_N2 reactions, the reaction works best if the group undergoing nucleophilic attack is a methyl group or a primary alkyl group. In Section 10.11, we saw that SAM, a biological methylating agent, is a sulfonium salt.

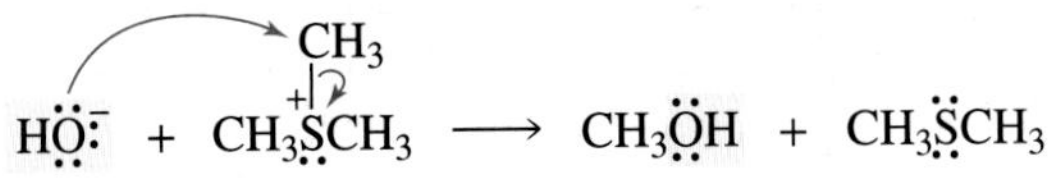

$$HO^- + CH_3\overset{+}{S}(CH_3)CH_3 \longrightarrow CH_3OH + CH_3SCH_3$$

PROBLEM 26

Using an alkyl halide and a thiol as starting materials, how would you prepare the following compounds?

a. $CH_3CH_2SCH_2CH_3$

b. $CH_3SC(CH_3)_2CH_3$

c. $CH_2{=}CHCH(CH_3)SCH_2CH_3$

d. 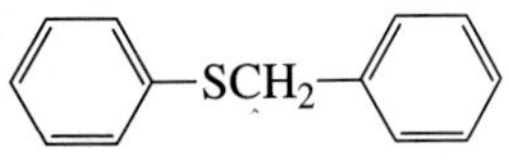 C_6H_5–SCH_2–C_6H_5

MUSTARD GAS

Chemical warfare occurred for the first time in 1915, when Germany released chlorine gas against French and British forces in the battle of Ypres. For the remainder of World War I, both sides used a variety of chemical agents. One of the more common war gases was mustard gas, a reagent that produces blisters over the surface of the body. It is a very reactive compound because the highly nucleophilic sulfur easily displaces a chloride ion by an intramolecular S_N2 reaction, forming a cyclic sulfonium salt that reacts rapidly with a nucleophile. The sulfonium salt is particularly reactive because of the strained three-membered ring.

The blistering caused by mustard gas is due to the high local concentrations of HCl that are produced when water—or any other nucleophile—reacts with the gas when it comes into contact with the skin or lungs. Autopsies of soldiers killed by mustard gas in World War I—estimated to be about 400,000—showed that they had extremely low white blood cell counts, indicating that the gas interfered with bone marrow development. An international treaty in the 1980s banned its use and required that all stockpiles of mustard gas be destroyed.

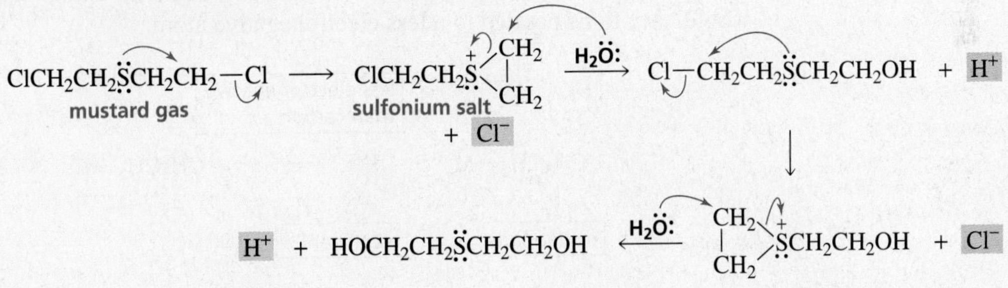

$$\underset{\text{mustard gas}}{ClCH_2CH_2SCH_2CH_2{-}Cl} \longrightarrow \underset{\text{sulfonium salt}}{ClCH_2CH_2\overset{+}{S}\langle(CH_2)_2} + Cl^- \xrightarrow{H_2O:} Cl{-}CH_2CH_2SCH_2CH_2OH + H^+$$

$$\longrightarrow (CH_2)_2\overset{+}{S}CH_2CH_2OH + Cl^- \xrightarrow{H_2O:} H^+ + HOCH_2CH_2SCH_2CH_2OH$$

ANTIDOTE TO A WAR GAS

Lewisite is a war gas developed in 1917 by W. Lee Lewis, an American scientist. The gas rapidly penetrates clothing and skin and is poisonous because it contains arsenic, which combines with thiol groups on enzymes, thereby inactivating the enzymes. During World War II, the Allies were concerned that the Germans would use lewisite, so British scientists developed an antidote to lewisite that the Allies called "British anti-lewisite" (BAL). BAL contains two thiol groups that react with lewisite, thereby preventing it from reacting with the thiol groups of enzymes.

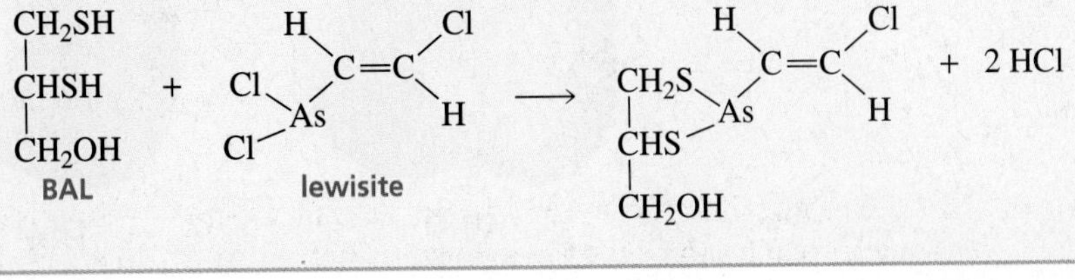

$$\underset{\text{BAL}}{HSCH_2CH(SH)CH_2OH} + \underset{\text{lewisite}}{Cl_2As{-}CH{=}CHCl} \longrightarrow (CH_2S)(CHS)(CH_2OH)As{-}CH{=}CHCl + 2\,HCl$$

PROBLEM 27

Finding that mustard gas interfered with bone marrow development caused chemists to look for less reactive mustard gases that might be used clinically. The following three compounds were studied:

$C_6H_5-N(CH_2CH_2Cl)_2$ $\quad CH_3-N(CH_2CH_2Cl)_2$ $\quad HC(=O)-C_6H_4-N(CH_2CH_2Cl)_2$

One was found to be too reactive, one was found to be too unreactive, and one was found to be too insoluble in water to be injected intravenously. Which is which? (*Hint:* Draw resonance contributors.)

12.11 Organometallic Compounds

Alcohols, ethers, and alkyl halides all contain a carbon atom that is bonded to a more electronegative atom. The carbon atom, therefore, is *electrophilic* and reacts with a nucleophile.

$$\overset{\delta+}{CH_3CH_2}-\overset{\delta-}{Z} + Y^- \longrightarrow CH_3CH_2-Y + Z^-$$

(Z: more electronegative than carbon; CH_3CH_2: electrophile; Y^-: nucleophile)

But what if you wanted a carbon atom to react with an electrophile? For that, you would need a compound with a nucleophilic carbon atom. To be *nucleophilic*, carbon would have to be bonded to a less electronegative atom.

$$\overset{\delta-}{CH_3CH_2}-\overset{\delta+}{M} + E^+ \longrightarrow CH_3CH_2-E + M^+$$

(M: less electronegative than carbon; CH_3CH_2: nucleophile; E^+: electrophile)

A carbon bonded to a metal is nucleophilic because most metals are less electronegative than carbon (Table 12.3). An **organometallic compound** is a compound that contains a carbon–metal bond. *Organolithium compounds* and *organomagnesium compounds* are two of the most common organometallic compounds. The electrostatic potential maps show that the carbon atom attached to the halogen in the alkyl halide is an electrophile (it is blue-green), whereas the carbon atom attached to the metal ion in the organometallic compound is a nucleophile (it is red).

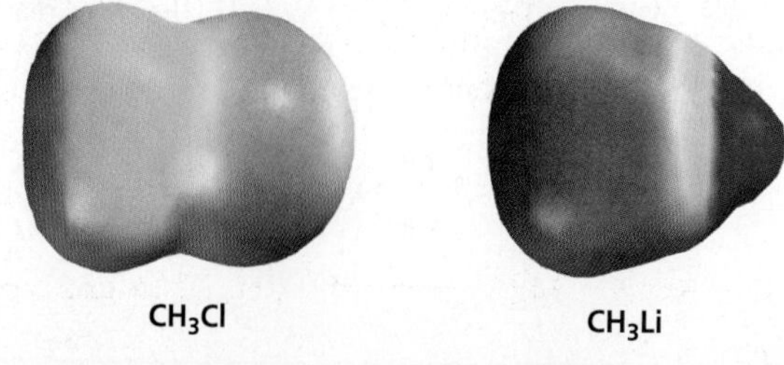

Table 12.3 The Electronegativities of Some of the Elements[a]

IA	IIA	IB	IIB	IIIA	IVA	VA	VIA	VIIA
H 2.1								
Li 1.0	Be 1.5			B 2.0	C 2.5	N 3.0	O 3.5	F 4.0
Na 0.9	Mg 1.2			Al 1.5	Si 1.8	P 2.1	S 2.5	Cl 3.0
K 0.8	Ca 1.0	Cu 1.8	Zn 1.7	Ga 1.8	Ge 2.0			Br 2.8
		Ag 1.4	Cd 1.5		Sn 1.7			I 2.5
			Hg 1.5		Pb 1.6			

[a]From the scale devised by Linus Pauling

Organolithium compounds are prepared by adding lithium to an alkyl halide in a nonpolar solvent such as hexane.

$$CH_3CH_2CH_2CH_2Br + 2\,Li \xrightarrow{\text{hexane}} CH_3CH_2CH_2CH_2Li + LiBr$$

1-bromobutane → butyllithium

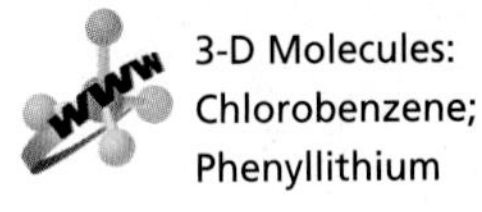

3-D Molecules: Chlorobenzene; Phenyllithium

$$C_6H_5\text{—}Cl + 2\,Li \xrightarrow{\text{hexane}} C_6H_5\text{—}Li + LiCl$$

chlorobenzene → phenyllithium

Organomagnesium compounds, frequently called **Grignard reagents** after their discoverer, are prepared by adding an alkyl halide to magnesium shavings being stirred in anhydrous diethyl ether or THF. The magnesium is inserted between the carbon and the halogen.

$$C_6H_{11}\text{—}Br + Mg \xrightarrow{\text{diethyl ether}} C_6H_{11}\text{—}MgBr$$

cyclohexyl bromide → cyclohexylmagnesium bromide

$$CH_2{=}CHBr + Mg \xrightarrow{\text{THF}} CH_2{=}CHMgBr$$

vinyl bromide → vinylmagnesium bromide

The solvent (usually diethyl ether or tetrahydrofuran) plays a crucial role in the formation of a Grignard reagent. Because the magnesium atom of a Grignard reagent is surrounded by only four electrons, it needs two more pairs of electrons to form an octet. Solvent molecules provide these electrons by coordinating (supplying electron pairs) to the metal. Coordination allows the Grignard reagent to dissolve in the solvent and prevents it from coating the magnesium shavings, which would make them unreactive.

The organometallic compounds form when the metal (Li or Mg) donates its valence electrons to the partially positively charged carbon of the alkyl halide.

$$CH_3\overset{\delta+}{CH_2}\text{—}\overset{\delta-}{Br} + 2\,Li \longrightarrow CH_3\overset{\delta-}{CH_2}\text{:}\overset{\delta+}{Li} + Li^+Br^-$$

$$CH_3\overset{\delta+}{CH_2}\text{—}\overset{\delta-}{Br} + Mg \longrightarrow CH_3\overset{\delta-}{CH_2}\text{:}\overset{\delta+}{MgBr}$$

Reaction with the metal converts the alkyl halide into a compound that will react with electrophiles instead of nucleophiles. Thus, organolithium and organomagnesium compounds react as if they were carbanions.

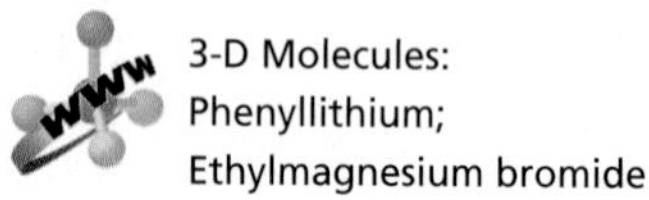

CH_3CH_2MgBr reacts as if it were $CH_3\ddot{C}H_2^{-}\ \overset{+}{M}gBr$
ethylmagnesium bromide

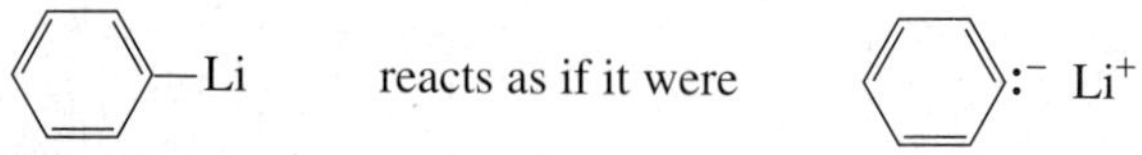

C_6H_5—Li reacts as if it were C_6H_5:⁻ Li^+
phenyllithium

Alkyl halides, vinyl halides, and aryl halides can all be used to form organolithium and organomagnesium compounds. Alkyl bromides are the alkyl halides most often used to form organometallic compounds because they react more readily than alkyl chlorides and are less expensive than alkyl iodides.

The nucleophilic carbon of an organometallic compound reacts with an electrophile. For example, in the following reactions, the organolithium reagent and the Grignard reagent each react with an epoxide.

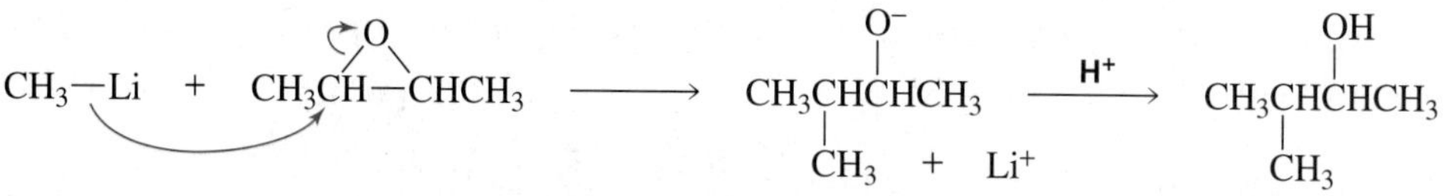

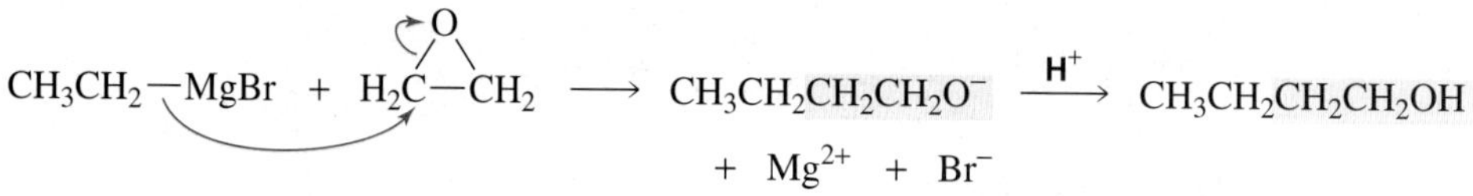

$$CH_3{-}Li + CH_3CH{-}CHCH_3\ (\text{epoxide}) \longrightarrow CH_3CH(O^-)CH(CH_3)CH_3 + Li^+ \xrightarrow{H^+} CH_3CH(OH)CH(CH_3)CH_3$$

$$CH_3CH_2{-}MgBr + H_2C{-}CH_2\ (\text{epoxide}) \longrightarrow CH_3CH_2CH_2CH_2O^- + Mg^{2+} + Br^- \xrightarrow{H^+} CH_3CH_2CH_2CH_2OH$$

Francis August Victor Grignard (1871–1935) *was born in France, the son of a sailmaker. He received a Ph.D. from the University of Lyons in 1901. His synthesis of the first Grignard reagent was announced in 1900. During the next five years, some 200 papers were published about Grignard reagents. He was a professor of chemistry at the University of Nancy and later at the University of Lyons. He shared the Nobel Prize in chemistry in 1912 with Paul Sabatier (p. 173). During World War I, Grignard was drafted into the French army, where he developed a method to detect war gases.*

Notice that when an organometallic compound reacts with ethylene oxide, a primary alcohol containing two more carbons than the organometallic compound is formed.

PROBLEM 28◆

What alcohols would be formed from the reaction of ethylene oxide with the following Grignard reagents?

a. $CH_3CH_2CH_2MgBr$
b. 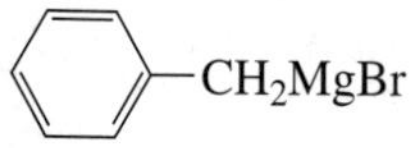 C_6H_5—CH_2MgBr
c. cyclohexyl—MgCl

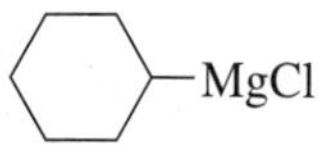

PROBLEM 29

How could the following compounds be prepared, using cyclohexene oxide as a starting material?

a. 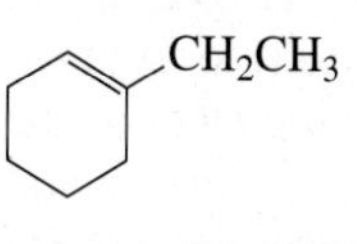 1-ethylcyclohexene (CH_2CH_3)
b. 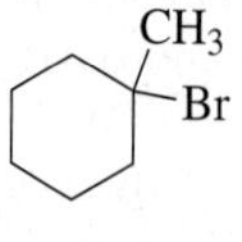 cyclohexane with CH_3 and Br
c. 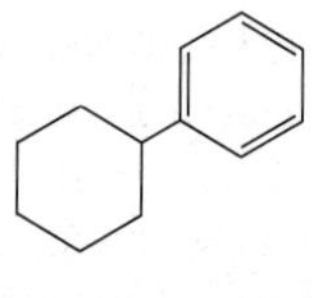 cyclohexylbenzene

Organomagnesium and organolithium compounds are such strong bases that they will react immediately with any acid that is present in the reaction mixture—even with very weak acids such as water and alcohols. When this happens, the organometallic compound is converted into an alkane. If D_2O is used instead of H_2O, a deuterated compound will be obtained.

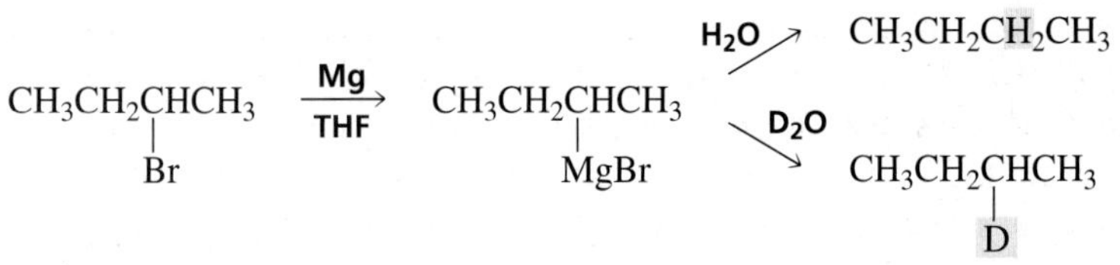

$$CH_3CH_2\underset{\underset{Br}{|}}{C}HCH_3 \xrightarrow[THF]{Mg} CH_3CH_2\underset{\underset{MgBr}{|}}{C}HCH_3 \begin{cases} \xrightarrow{H_2O} CH_3CH_2CH_2CH_3 \\ \xrightarrow{D_2O} CH_3CH_2\underset{\underset{D}{|}}{C}HCH_3 \end{cases}$$

This means that Grignard reagents and organolithium compounds cannot be prepared from compounds that contain acidic groups ($-OH$, $-NH_2$, $-NHR$, $-SH$, $-C\equiv CH$, or $-COOH$ groups). Because even trace amounts of moisture can destroy an organometallic compound, it is important that all reagents be dry when organometallic compounds are being synthesized and when they react with other reagents.

PROBLEM 30 SOLVED

Show, using any necessary reagents, how the following compounds could be prepared with ethylene oxide as one of the reactants:

a. $CH_3CH_2CH_2CH_2OH$
b. $CH_3CH_2CH_2CH_2Br$
c. $CH_3CH_2CH_2CH_2D$
d. $CH_3CH_2CH_2CH_2CH_2CH_2OH$

SOLUTION

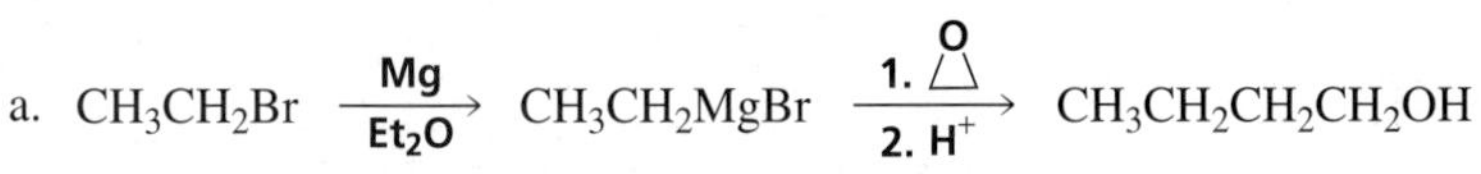

a. $CH_3CH_2Br \xrightarrow[Et_2O]{Mg} CH_3CH_2MgBr \xrightarrow[2.\ H^+]{1.\ \text{ethylene oxide}} CH_3CH_2CH_2CH_2OH$

b. product of a $\xrightarrow{PBr_3} CH_3CH_2CH_2CH_2Br$

c. product of b $\xrightarrow[Et_2O]{Mg} CH_3CH_2CH_2CH_2MgBr \xrightarrow{D_2O} CH_3CH_2CH_2CH_2D$

d. the same reaction sequence as in a, but with butyl bromide in the first step.

PROBLEM 31◆

Which of the following reactions will occur? (For the pK_a values necessary to do this problem, see Appendix II.)

$$CH_3MgBr + H_2O \longrightarrow CH_4 + HOMgBr$$
$$CH_3MgBr + CH_3OH \longrightarrow CH_4 + CH_3OMgBr$$
$$CH_3MgBr + NH_3 \longrightarrow CH_4 + H_2NMgBr$$
$$CH_3MgBr + CH_3NH_2 \longrightarrow CH_4 + CH_3NMgBr$$
$$CH_3MgBr + HC\equiv CH \longrightarrow CH_4 + HC\equiv CMgBr$$

There are many different organometallic compounds. As long as the metal is less electronegative than carbon, the carbon bonded to the metal will be nucleophilic.

$$\overset{\delta-}{C}-\overset{\delta+}{Mg} \quad \overset{\delta-}{C}-\overset{\delta+}{Li} \quad \overset{\delta-}{C}-\overset{\delta+}{Cu} \quad \overset{\delta-}{C}-\overset{\delta+}{Cd} \quad \overset{\delta-}{C}-\overset{\delta+}{Si}$$

$$\overset{\delta-}{C}-\overset{\delta+}{Zn} \quad \overset{\delta-}{C}-\overset{\delta+}{Al} \quad \overset{\delta-}{C}-\overset{\delta+}{Pb} \quad \overset{\delta-}{C}-\overset{\delta+}{Hg} \quad \overset{\delta-}{C}-\overset{\delta+}{Sn}$$

The reactivity of an organometallic compound toward an electrophile depends on the polarity of the carbon–metal bond: The greater the polarity of the bond, the more reactive the compound is as a nucleophile. The polarity of the bond depends on the difference in electronegativity between the metal and carbon (Table 12.3). For example, magnesium has an electronegativity of 1.2, compared with 2.5 for carbon. This large difference in electronegativity makes the carbon–magnesium bond highly polar. (The carbon–magnesium bond is about 52% ionic.) Lithium (1.0) is even less electronegative than magnesium (1.2). Thus, the carbon–lithium bond is more polar than the carbon–magnesium bond. Therefore, an organolithium reagent is a more reactive nucleophilic reagent than a Grignard reagent.

The names of organometallic compounds usually begin with the name of the alkyl group, followed by the name of the metal.

CH_3CH_2MgBr	$CH_3CH_2CH_2CH_2Li$	$(CH_3CH_2CH_2)_2Cd$	$(CH_3CH_2)_4Pb$
ethylmagnesium bromide	butyllithium	dipropylcadmium	tetraethyllead

A Grignard reagent will undergo **transmetallation** (metal exchange) if it is added to a metal halide whose metal is more electronegative than magnesium. In other words, metal exchange will occur if it results in a less polar carbon–metal bond. For example, cadmium (1.5) is more electronegative than magnesium (1.2). Consequently, a carbon–cadmium bond is less polar than a carbon–magnesium bond, so metal exchange occurs.

$$2\ CH_3CH_2MgCl + CdCl_2 \longrightarrow (CH_3CH_2)_2Cd + 2\ MgCl_2$$

ethylmagnesium chloride → diethylcadmium

PROBLEM 32◆

What organometallic compound will be formed from the reaction of methylmagnesium chloride and $SiCl_4$? (*Hint:* See Table 12.3.)

12.12 Coupling Reactions

New carbon–carbon bonds can be made using an organometallic reagent in which the metal ion is a transition metal. Transition metals are indicated by purple in the periodic table on the last page of this book. The reactions are called **coupling reactions** because two groups (any two alkyl, aryl, or vinyl groups) are joined (coupled together).

The first organometallic compounds used in coupling reactions were **Gilman reagents**, also called **organocuprates**. They are prepared by the reaction of an organolithium reagent with cuprous iodide in diethyl ether or THF.

Henry Gilman (1893–1986) *was born in Boston. He received his B.A. and Ph.D. degrees from Harvard University. He joined the faculty at Iowa State University in 1919, where he remained for his entire career. He published over 1000 research papers—more than half after he lost almost all of his sight as the result of a detached retina and glaucoma in 1947. His wife, Ruth, acted as his eyes for 40 years.*

$$2\ CH_3Li + CuI \xrightarrow{\text{THF}} (CH_3)_2CuLi + LiI$$

organolithium reagent → Gilman reagent

Gilman reagents are very useful to synthetic chemists. When a Gilman reagent reacts with an alkyl halide (with the exception of alkyl fluorides, which do not undergo this reaction), one of the alkyl groups of the Gilman reagent replaces the halogen. Notice that this means that an alkane can be formed from two alkyl halides—one alkyl halide is used to form the Gilman reagent, which then reacts with the second alkyl halide. The precise mechanism of the reaction is unknown, but is thought to involve radicals.

$$CH_3CH_2CH_2CH_2Br + (CH_3CH_2CH_2)_2CuLi \xrightarrow{\text{THF}} CH_3CH_2CH_2CH_2CH_2CH_2CH_3 + CH_3CH_2CH_2Cu + LiBr$$

heptane

Gilman reagents can be used to prepare compounds that cannot be prepared by using nucleophilic substitution reactions. For example, S_N2 reactions cannot be used to prepare the following compounds because vinyl and aryl halides cannot undergo nucleophilic attack (Section 10.8):

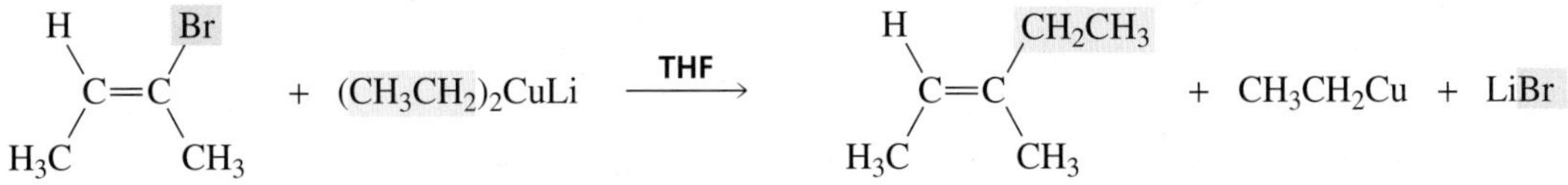

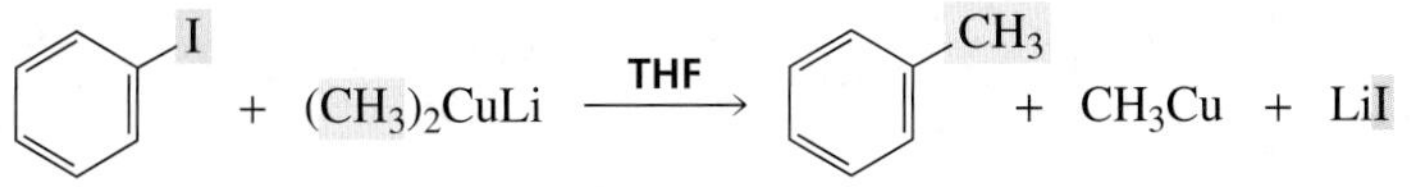

$$(H)(H_3C)C{=}C(Br)(CH_3) + (CH_3CH_2)_2CuLi \xrightarrow{\text{THF}} (H)(H_3C)C{=}C(CH_2CH_3)(CH_3) + CH_3CH_2Cu + LiBr$$

$$C_6H_5I + (CH_3)_2CuLi \xrightarrow{\text{THF}} C_6H_5CH_3 + CH_3Cu + LiI$$

Gilman reagents can even replace halogens in compounds that contain other functional groups.

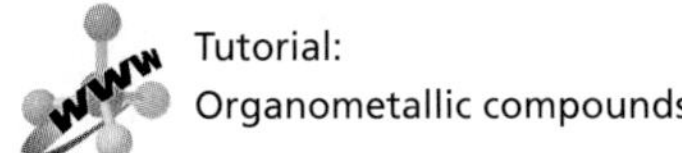

Tutorial:
Organometallic compounds

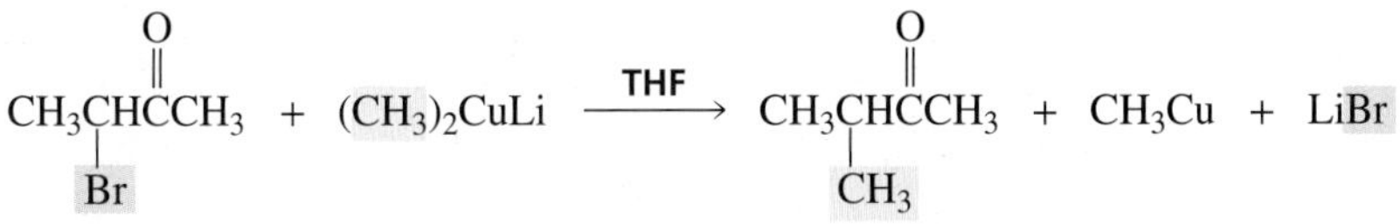

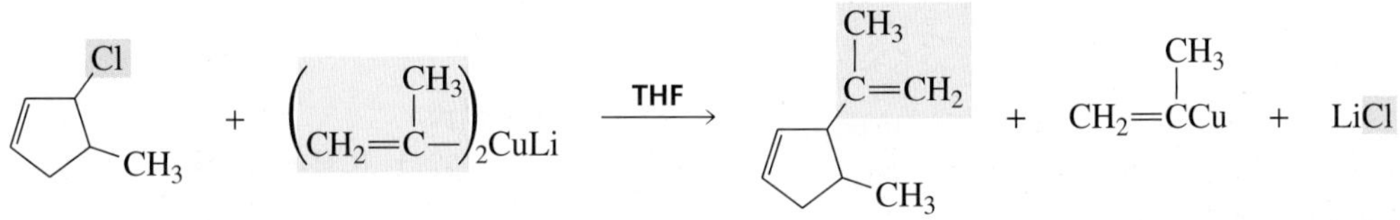

$$CH_3\underset{\substack{|\\ Br}}{C}H\overset{\overset{O}{\|}}{C}CH_3 + (CH_3)_2CuLi \xrightarrow{\text{THF}} CH_3\underset{\substack{|\\ CH_3}}{C}H\overset{\overset{O}{\|}}{C}CH_3 + CH_3Cu + LiBr$$

$$\text{3-chloro-4-methylcyclopentene} + \left(CH_2{=}\overset{\overset{CH_3}{|}}{C}{-}\right)_2CuLi \xrightarrow{\text{THF}} \text{3-}(\overset{\overset{CH_3}{|}}{C}{=}CH_2)\text{-4-methylcyclopentene} + CH_2{=}\overset{\overset{CH_3}{|}}{C}Cu + LiCl$$

PROBLEM 33

Explain why tertiary alkyl halides cannot be used in coupling reactions with Gilman reagents.

PROBLEM 34

Muscalure is the sex attractant of the common housefly. Flies are lured to traps by filling them with fly bait containing both muscalure and an insecticide. Eating the bait is fatal.

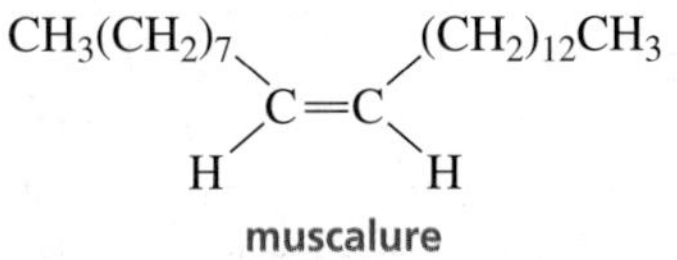

$$(CH_3(CH_2)_7)(H)C{=}C((CH_2)_{12}CH_3)(H)$$

muscalure

How could you synthesize muscalure using the following reagents?

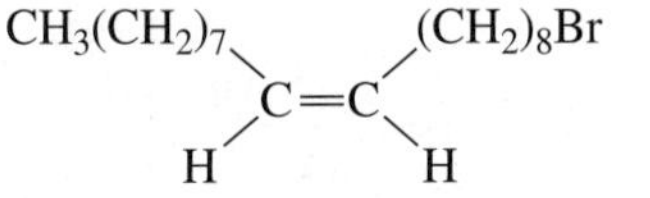

$$(CH_3(CH_2)_7)(H)C{=}C((CH_2)_8Br)(H) \quad \text{and} \quad CH_3(CH_2)_4Br$$

Coupling reactions can be carried out that give high yields of products (80–98%) using a transition-metal catalyst.

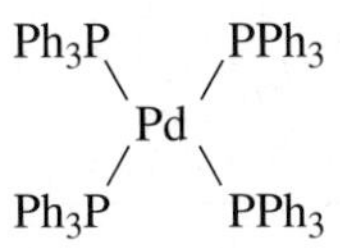

$$Pd(PPh_3)_4$$

transition-metal catalyst

$$PPh_3 = P(C_6H_5)_3$$

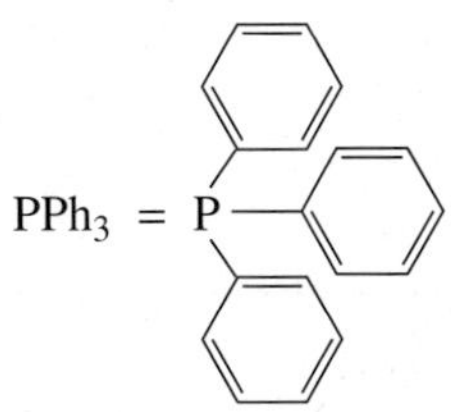

triphenyl phosphine

In each of the following reactions, the catalyst is palladium(0), ligated to four molecules of phenyl phosphine.

The **Heck reaction** couples an aryl, benzyl, or vinyl halide or triflate (Section 12.4) with an alkene in a basic solution.

Heck reaction

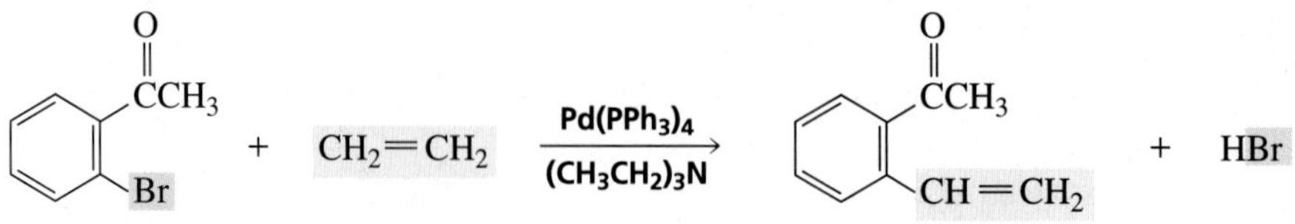

$$\text{(o-Br)C}_6\text{H}_4\text{COCH}_3 + CH_2{=}CH_2 \xrightarrow[(CH_3CH_2)_3N]{Pd(PPh_3)_4} \text{(o-CH=CH}_2\text{)C}_6\text{H}_4\text{COCH}_3 + HBr$$

an aryl halide

$$CH_3O\text{-C}_6\text{H}_4\text{-OTf} + \text{styrene} \xrightarrow[(CH_3CH_2)_3N]{Pd(PPh_3)_4} CH_3O\text{-C}_6\text{H}_4\text{-CH=CH-C}_6\text{H}_5 + HOTf$$

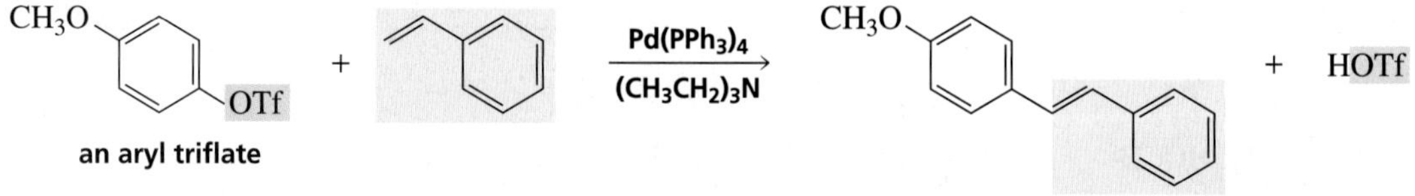

an aryl triflate

The **Stille reaction** couples an aryl, benzyl, or vinyl halide or triflate with a stannane.

Stille reaction

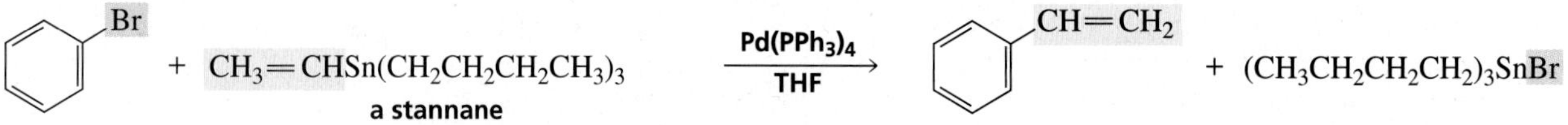

$$\text{C}_6\text{H}_5\text{Br} + CH_3{=}CHSn(CH_2CH_2CH_2CH_3)_3 \xrightarrow[THF]{Pd(PPh_3)_4} \text{C}_6\text{H}_5\text{CH}{=}CH_2 + (CH_3CH_2CH_2CH_2)_3SnBr$$

a stannane

Aryl and vinyl groups are coupled preferentially, but an alkyl group can be coupled if a tetraalkylstannane is used.

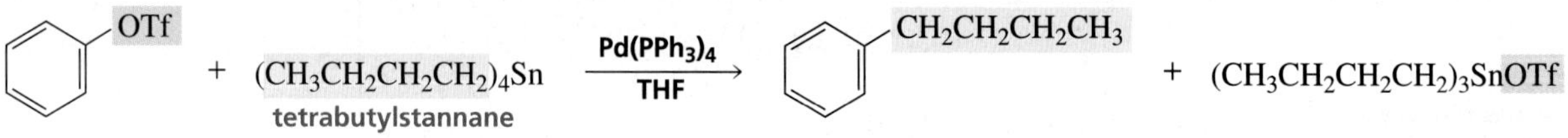

$$\text{C}_6\text{H}_5\text{OTf} + (CH_3CH_2CH_2CH_2)_4Sn \xrightarrow[THF]{Pd(PPh_3)_4} \text{C}_6\text{H}_5\text{CH}_2CH_2CH_2CH_3 + (CH_3CH_2CH_2CH_2)_3SnOTf$$

tetrabutylstannane

The **Suzuki reaction** couples an aryl, benzyl, or vinyl halide with an organoborane in a basic solution.

Suzuki reaction

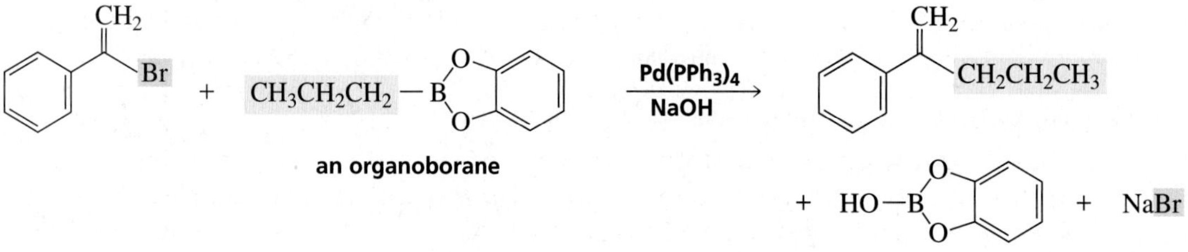

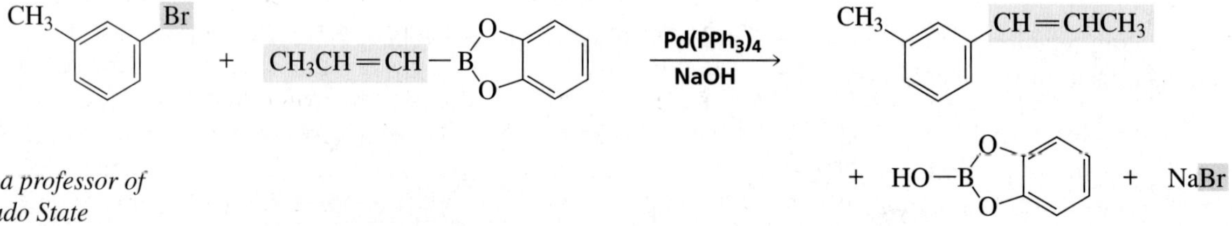

$$\text{C}_6\text{H}_5\text{C(=CH}_2\text{)Br} + CH_3CH_2CH_2-\text{B(catechol)} \xrightarrow[NaOH]{Pd(PPh_3)_4} \text{C}_6\text{H}_5\text{C(=CH}_2\text{)CH}_2CH_2CH_3 + HO-\text{B(catechol)} + NaBr$$

an organoborane

$$\text{m-CH}_3\text{C}_6\text{H}_4\text{Br} + CH_3CH{=}CH-\text{B(catechol)} \xrightarrow[NaOH]{Pd(PPh_3)_4} \text{m-CH}_3\text{C}_6\text{H}_4\text{CH}{=}CHCH_3 + HO-\text{B(catechol)} + NaBr$$

John K. Stille *was a professor of chemistry at Colorado State University until his untimely death in 1989 in a commercial air crash on his way to a scientific meeting in Switzerland.*

PROBLEM 35◆

What alkyl halides would you utilize to synthesize the following compounds, using the organoborane shown?

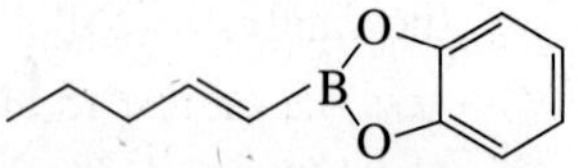

$$CH_3CH_2CH_2CH{=}CH-\text{B(catechol)}$$

a. b. c.

PROBLEM 36◆

The organoborane used in a Suzuki reaction is prepared by the reaction of catecholborane with an alkene or an alkyne.

$$RCH{=}CH_2 + H{-}B(O_2C_6H_4) \longrightarrow RCH_2CH_2{-}B(O_2C_6H_4)$$

catecholborane

What hydrocarbon would you use to prepare the organoborane of Problem 35?

PROBLEM 37◆

Give two sets of alkyl bromide and alkene that could be used in a Heck reaction to prepare the following compound:

$$CH_3\overset{O}{\overset{\|}{C}}{-}C_6H_4{-}CH{=}CH{-}C_6H_4{-}OCH_3$$

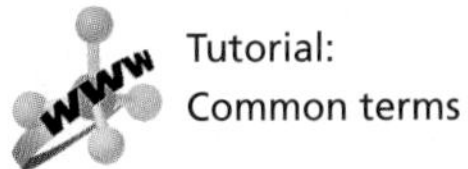

Summary

Alcohols and **ethers** have leaving groups that are stronger bases than halide ions, so alcohols and ethers are less reactive than alkyl halides and have to be "activated" before they can undergo a substitution or an elimination reaction. **Epoxides** do not have to be activated, because ring strain increases their reactivity. **Sulfonate esters** and **sulfonium salts** have weakly basic leaving groups, so they undergo substitution reactions with ease. $^-NH_2$ is such a strong base that amines cannot undergo substitution or elimination reactions.

Primary, secondary, and tertiary alcohols undergo nucleophilic substitution reactions with HI, HBr, and HCl to form alkyl halides. These are S_N2 reactions in the case of primary alcohols and S_N1 reactions in the case of secondary and tertiary alcohols. An alcohol can also be converted into an alkyl halide by phosphorus trihalides or thionyl chloride. These reagents convert the alcohol into an intermediate with a leaving group that is readily displaced by a halide ion.

Conversion to a **sulfonate ester** is another way to activate an alcohol for subsequent reaction with a nucleophile. Because a sulfonic acid is a strong acid, its conjugate base is weak. Activating an alcohol by converting it to a sulfonate ester forms a substitution product with a configuration opposite to that of the alcohol, whereas activating an alcohol by converting it to an alkyl halide forms a substitution product with the same configuration as the alcohol.

An alcohol can be dehydrated if heated with an acid catalyst; **dehydration** is an E2 reaction in the case of primary alcohols and an E1 reaction in the case of secondary and tertiary alcohols. Tertiary alcohols are the easiest to dehydrate and primary alcohols are the hardest. E1 reactions form carbocation intermediates, so carbocation and **ring-expansion rearrangements** can occur. The major product is the more substituted alkene. Both the *E* and *Z* isomers are obtained, but the isomer with the bulky groups on opposite sides of the double bond predominates.

Ethers can undergo nucleophilic substitution reactions with HBr or HI; if departure of the leaving group creates a relatively stable carbocation, an S_N1 reaction occurs; otherwise an S_N2 reaction occurs.

Epoxides undergo ring-opening reactions. Under basic conditions, the least sterically hindered carbon is attacked; under acidic conditions, the most substituted carbon is attacked. **Arene oxides** undergo rearrangment to form phenols or nucleophilic attack to form addition products. An arene oxide's cancer-causing potential depends on the stability of the carbocation formed during rearrangement.

A **crown ether** specifically binds certain metal ions or organic molecules, depending on the size of its cavity, forming an **inclusion compound**. The ability of a host to bind only certain guests is an example of **molecular recognition**. The crown ether can act as a **phase-transfer catalyst**.

Thiols are sulfur analogs of alcohols. They are stronger acids and have lower boiling points than alcohols. Thiolate ions are weaker bases and better nucleophiles in protic solvents than alkoxide ions. Sulfur analogs of ethers are called **sulfides** or **thioethers**. Sulfides react with alkyl halides to form **sulfonium salts**.

Grignard reagents and **organolithium compounds** are the most common **organometallic compounds**—compounds that contain a carbon–metal bond. They cannot be prepared from compounds that contain acidic groups. The carbon atom attached to the halogen in the alkyl halide is an electrophile, whereas the carbon atom attached to the metal ion in the organometallic compound is a nucleophile. The greater the polarity of the carbon–metal bond, the more reactive the organometallic compound is as a nucleophile.

New carbon–carbon bonds can be made using transition-metal organometallic reagents. The reactions are called **coupling reactions** because two carbon-containing groups are joined together. **Gilman reagents** were the first organometallic compounds used in coupling reactions. The **Heck**, **Stille**, and **Suzuki reactions** are coupling reactions.

Summary of Reactions

1. Conversion of an *alcohol* to an *alkyl halide* (Sections 12.1 and 12.3).

$$ROH + HBr \xrightarrow{\Delta} RBr$$

$$ROH + HI \xrightarrow{\Delta} RI$$

$$ROH + HCl \xrightarrow[\Delta]{ZnCl_2} RCl$$

$$ROH + PBr_3 \xrightarrow[\text{pyridine}]{} RBr$$

$$ROH + PCl_3 \xrightarrow[\text{pyridine}]{} RCl$$

$$ROH + SOCl_2 \xrightarrow[\text{pyridine}]{} RCl$$

2. Conversion of an alcohol to a *sulfonate ester* (Section 12.4).

$$ROH + R'SO_2Cl \xrightarrow[\text{pyridine}]{} RO{-}SO_2{-}R' + HCl$$

3. Conversion of an *activated alcohol* (an alkyl halide or a sulfonate ester) to a *compound with a new group bonded to the* sp^3 *carbon* (Section 12.4).

$$RBr + Y^- \longrightarrow RY + Br^-$$

$$RO{-}SO_2{-}R' + Y^- \longrightarrow RY + {}^-O{-}SO_2{-}R'$$

4. Dehydration of alcohols (Section 12.5).

$$-\overset{|}{\underset{H}{C}}-\overset{|}{\underset{OH}{C}}- \underset{\Delta}{\overset{H_2SO_4}{\rightleftharpoons}} \;{>}C{=}C{<} + H_2O$$

$$-\overset{|}{\underset{H}{C}}-\overset{|}{\underset{OH}{C}}- \xrightarrow[\text{pyridine, 0 °C}]{POCl_3} \;{>}C{=}C{<} + H_2O$$

relative rate: tertiary > secondary > primary

5. Cleavage of ethers (Section 12.6).

$$ROR' + HX \xrightarrow{\Delta} ROH + R'X$$

HX = HBr or HI

6. Ring-opening reactions of *epoxides* (Section 12.7)

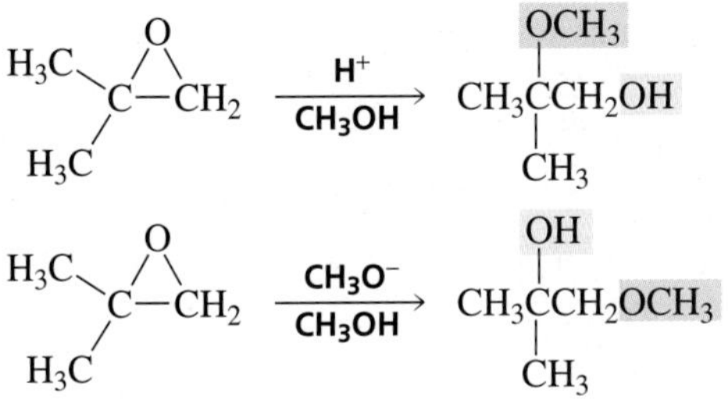

$$(H_3C)_2C\overset{O}{—}CH_2 \xrightarrow[CH_3OH]{H^+} CH_3\underset{CH_3}{\overset{OCH_3}{C}}CH_2OH$$

under acidic conditions, the nucleophile attacks the more substituted ring-carbon

$$(H_3C)_2C\overset{O}{—}CH_2 \xrightarrow[CH_3OH]{CH_3O^-} CH_3\underset{CH_3}{\overset{OH}{C}}CH_2OCH_3$$

under basic conditions, the nucleophile attacks the less sterically hindered ring-carbon

7. Reactions of *arene oxides*: ring opening and rearrangement (Section 12.8).

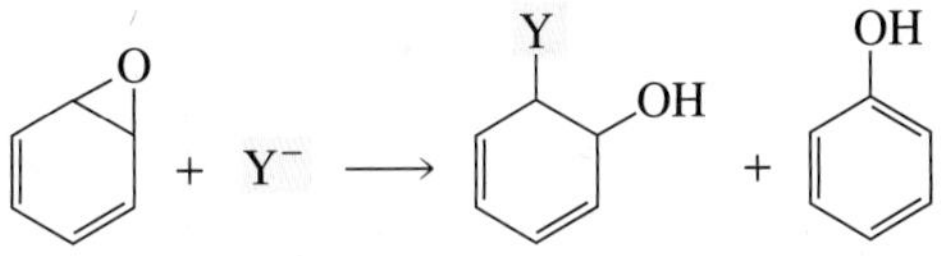

arene oxide + Y^- ⟶ (Y, OH adduct) + phenol (OH)

8. Reactions of thiols, sulfides, and sulfonium salts (Section 12.10).

$$2\ RSH + Hg^{2+} \longrightarrow RS—Hg—SR + 2\ H^+$$

$$RS^- + R'—Br \longrightarrow RSR' + Br^-$$

$$RSR + R'I \longrightarrow R\overset{R'}{\underset{+}{S}}R\ \ I^-$$

$$R\overset{R}{\underset{+}{S}}R + Y^- \longrightarrow RY + RSR$$

9. Reaction of a *Grignard reagent* with an *epoxide* (Section 12.11).

$$RBr \xrightarrow[\text{diethyl ether}]{Mg} RMgBr$$

a Grignard reagent

$$RMgBr + H_2C\overset{O}{—}CH_2 \longrightarrow RCH_2CH_2O^- \xrightarrow{H^+} RCH_2CH_2OH$$

product alcohol contains two more carbons than the Grignard reagent

10. Reaction of a *Gilman reagent* with an *alkyl halide* (Section 12.12).

$$2\ RLi + CuI \xrightarrow{THF} R_2CuLi + LiI$$

Gilman reagent

$$CH_3CH_2CH_2X + R_2CuLi \xrightarrow{THF} CH_3CH_2CH_2R + RCu + LiX$$

X = Cl, Br, or I

11. Reaction of an *aryl*, *benzyl*, or *vinyl halide* or *triflate* with an *alkene*: the Heck reaction (Section 12.12).

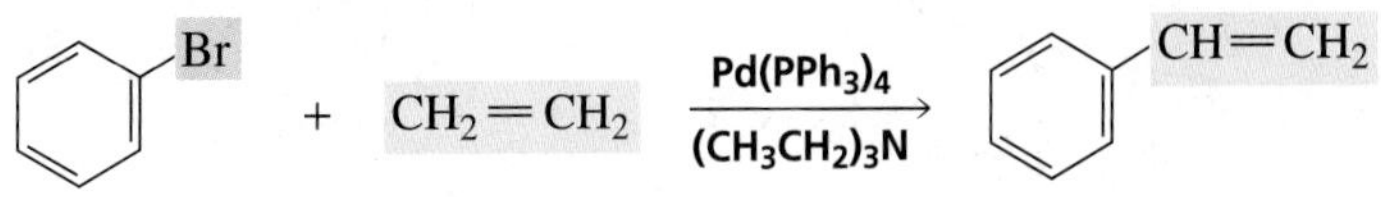

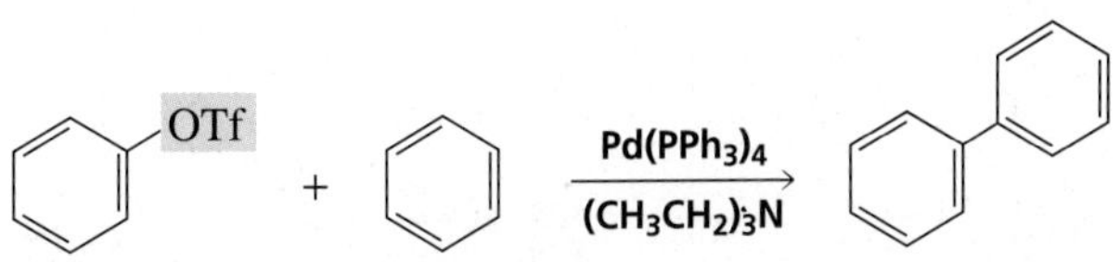

$$C_6H_5Br + CH_2{=}CH_2 \xrightarrow[(CH_3CH_2)_3N]{Pd(PPh_3)_4} C_6H_5CH{=}CH_2$$

$$C_6H_5OTf + C_6H_6 \xrightarrow[(CH_3CH_2)_3N]{Pd(PPh_3)_4} C_6H_5{-}C_6H_5$$

12. Reaction of an *aryl*, *benzyl*, or *vinyl halide* or *triflate* with a *stannane*: the Stille reaction (Section 12.12).

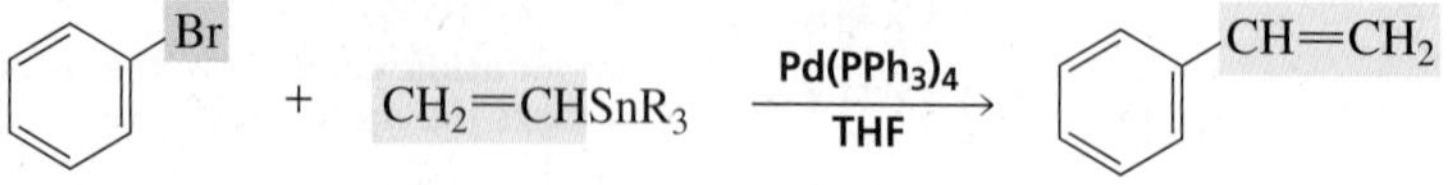

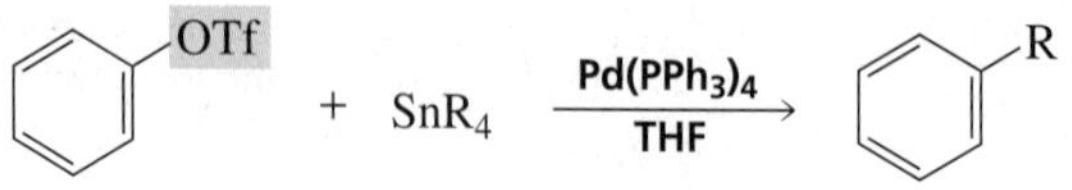

$$C_6H_5Br + CH_2{=}CHSnR_3 \xrightarrow[\text{THF}]{Pd(PPh_3)_4} C_6H_5CH{=}CH_2$$

$$C_6H_5OTf + SnR_4 \xrightarrow[\text{THF}]{Pd(PPh_3)_4} C_6H_5R$$

13. Reaction of an *aryl*, *benzyl*, or *vinyl halide* with an *organoborane*: the Suzuki reaction (Section 12.12).

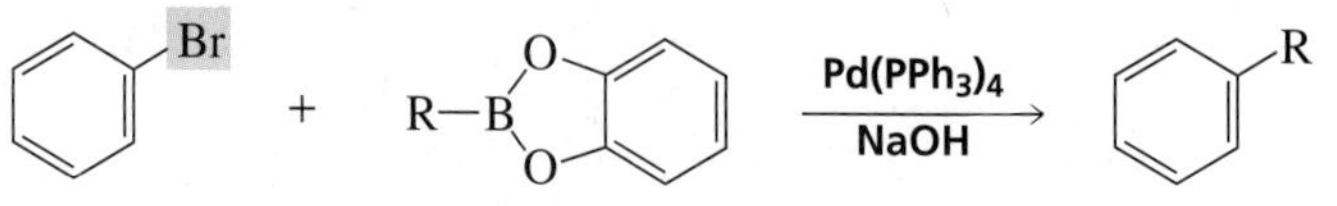

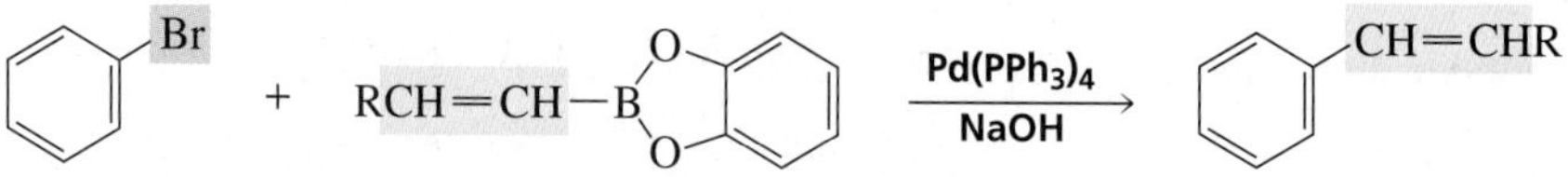

$$C_6H_5Br + R{-}B(O_2C_6H_4) \xrightarrow[\text{NaOH}]{Pd(PPh_3)_4} C_6H_5R$$

$$C_6H_5Br + RCH{=}CH{-}B(O_2C_6H_4) \xrightarrow[\text{NaOH}]{Pd(PPh_3)_4} C_6H_5CH{=}CHR$$

Key Terms

alcohol (p. 437)
alkyl tosylate (p. 444)
alkyl triflate (p. 444)
arene oxide (p. 457)
coupling reaction (p. 470)
crown ether (p. 462)
dehydration (p. 445)
epoxide (p. 454)
ether (p. 451)
Gilman reagent (p. 470)
Grignard reagent (p. 467)
Heck reaction (p. 472)
inclusion compound (p. 462)
molecular recognition (p. 463)
organocuprate (p. 470)
organolithium compound (p. 467)
organomagnesium compound (p. 467)
organometallic compound (p. 466)
oxirane (p. 454)
phase-transfer catalyst (p. 463)
ring-expansion rearrangement (p. 447)
Stille reaction (p. 472)
sulfide (p. 464)
sulfonate ester (p. 443)
sulfonium salt (p. 464)
Suzuki reaction (p. 472)
thioether (p. 464)
thiol (p. 464)
transmetallation (p. 470)

Problems

38. Give the product of each of the following reactions:

a. $CH_3CH_2CH_2OH \xrightarrow[\text{2. } CH_3\overset{O}{\overset{\|}{C}}O^-]{\text{1. methanesulfonyl chloride}}$

b. $CH_3CH_2CH_2CH_2OH + PBr_3 \xrightarrow[\text{pyridine}]{}$

c. 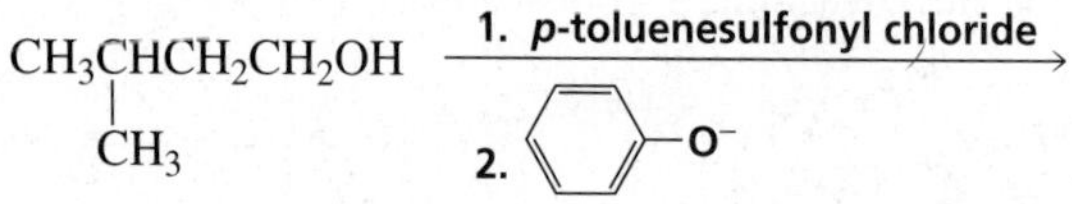 $CH_3CH(CH_3)CH_2CH_2OH \xrightarrow[\text{2. } C_6H_5O^-]{\text{1. } p\text{-toluenesulfonyl chloride}}$

d. 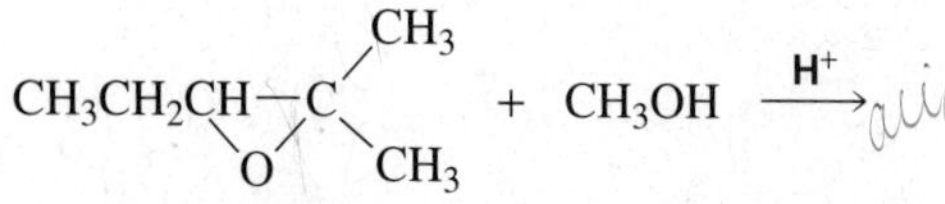 $CH_3CH_2CH{-}C(CH_3)_2$ (epoxide, O bridging) $+ CH_3OH \xrightarrow{H^+}$ acidic

e. 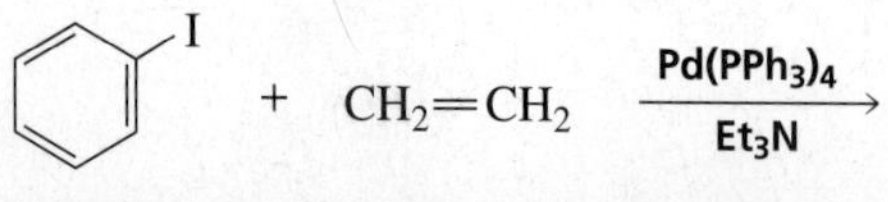 $C_6H_5I + CH_2{=}CH_2 \xrightarrow[\text{Et}_3\text{N}]{Pd(PPh_3)_4}$

f. 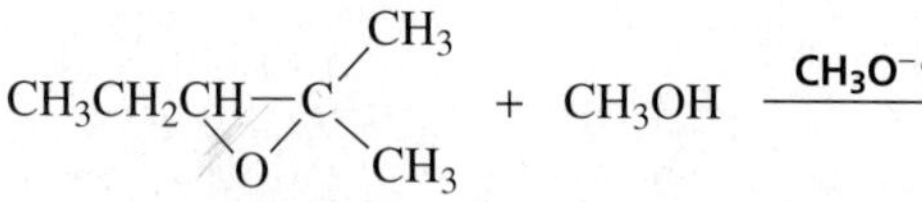 $CH_3CH_2CH{-}C(CH_3)_2$ (epoxide, O bridging) $+ CH_3OH \xrightarrow{CH_3O^-}$ basic

g. $CH_3CH_2CH_2CH_2OH \xrightarrow[\Delta]{H_2SO_4}$?

h. 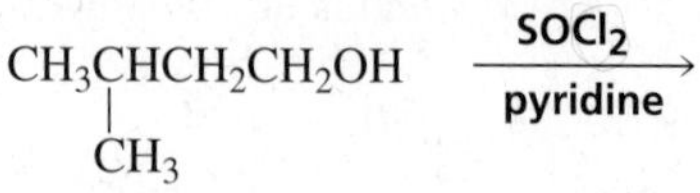 $CH_3CH(CH_3)CH_2CH_2OH \xrightarrow[\text{pyridine}]{SOCl_2}$

i. 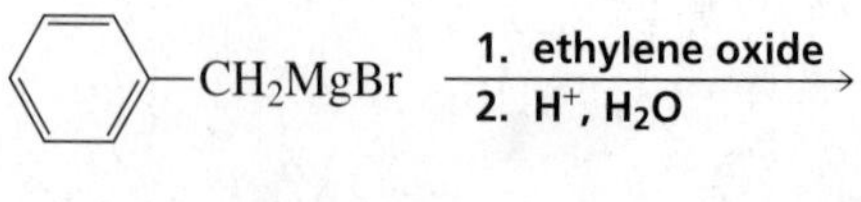 $C_6H_5CH_2MgBr \xrightarrow[\text{2. } H^+,\ H_2O]{\text{1. ethylene oxide}}$

j. 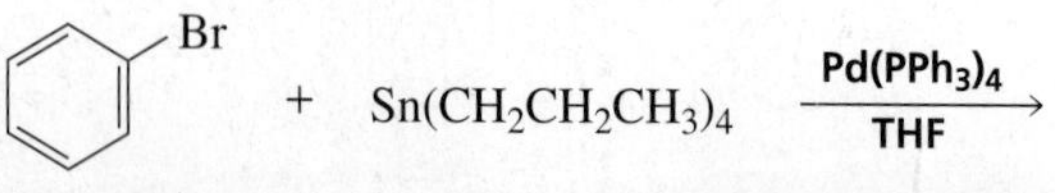 $C_6H_5Br + Sn(CH_2CH_2CH_3)_4 \xrightarrow[\text{THF}]{Pd(PPh_3)_4}$

39. Indicate which alcohol will undergo dehydration more rapidly when heated with H_2SO_4.

a. 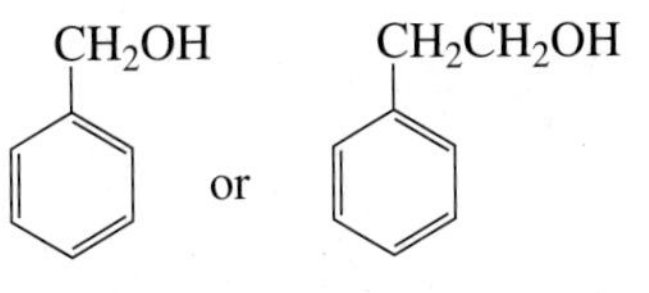 CH_2OH (on benzene ring) or CH_2CH_2OH (on benzene ring)

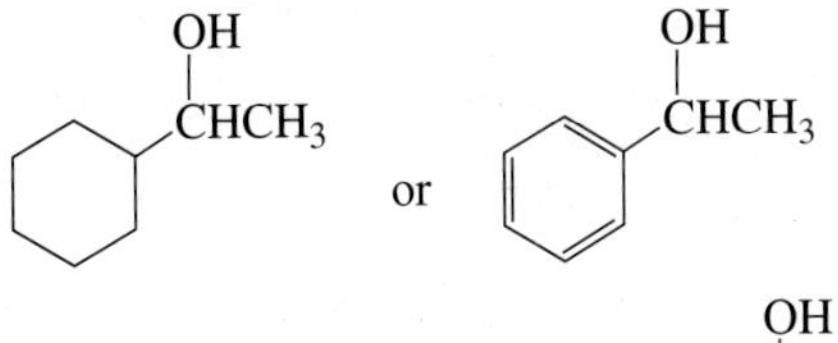

b. 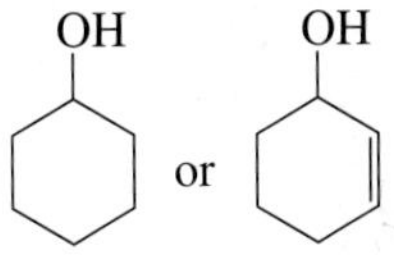 OH (on cyclohexane ring) or OH (on cyclohexene ring)

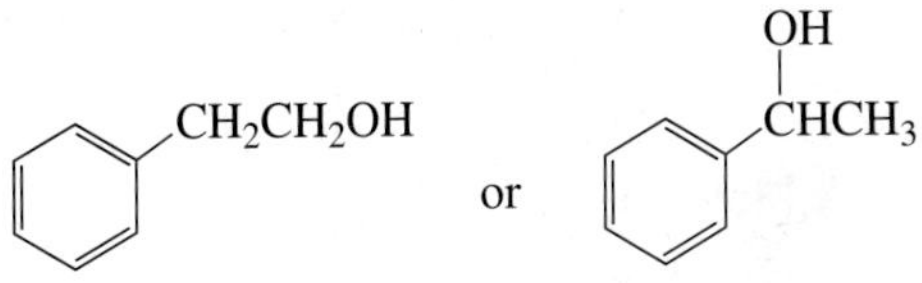

c. 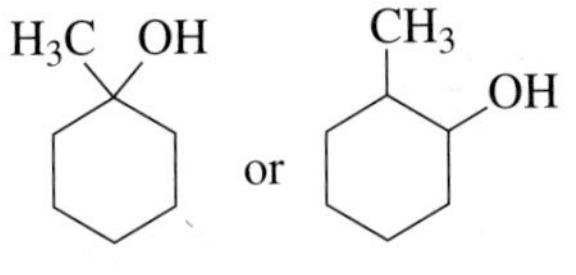 H_3C OH (on cyclohexane ring) or CH_3, OH (on cyclohexane ring)

d. OH–$CHCH_3$ (on cyclohexane ring) or OH–$CHCH_3$ (on benzene ring)

e. CH_2CH_2OH (on benzene ring) or OH–$CHCH_3$ (on benzene ring)

f. 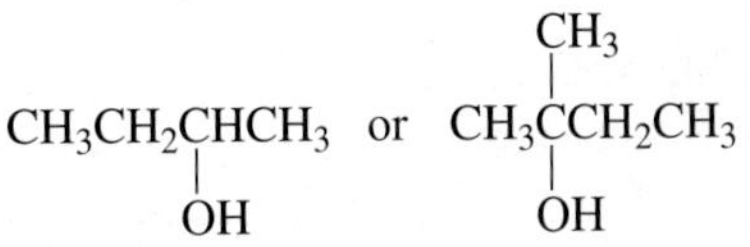 $CH_3CH_2CHCH_3$ (OH) or $CH_3CCH_2CH_3$ (CH_3, OH)

40. Which of the following alkyl halides could be successfully used to form a Grignard reagent?

a. $HOCH_2CH_2CH_2CH_2Br$ b.  $BrCH_2CH_2CH_2\overset{O}{\overset{\|}{C}}OH$ c. $CH_3NCH_2CH_2CH_2Br$ (CH_3) d. $H_2NCH_2CH_2CH_2Br$

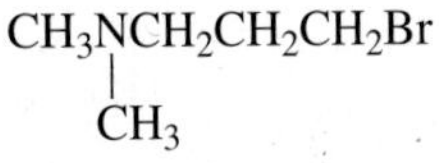

41. Starting with (*R*)-1-deuterio-1-propanol, how could you prepare:
a. (*S*)-1-deuterio-1-propanol? b. (*S*)-1-deuterio-1-methoxypropane? c. (*R*)-1-deuterio-1-methoxypropane?

42. What alkenes would you expect to be obtained from the acid-catalyzed dehydration of 1-hexanol?

43. Give the product of each of the following reactions:

a. 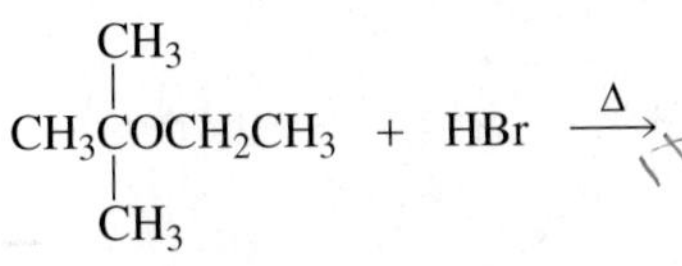 $CH_3COCH_2CH_3$ (CH_3, CH_3) + HBr $\xrightarrow{\Delta}$

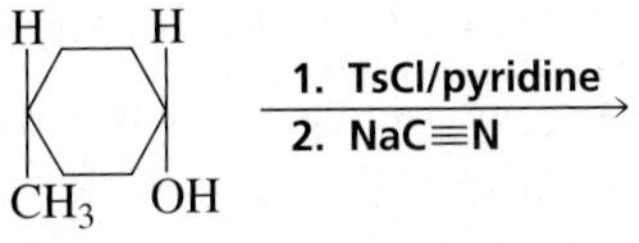

b. 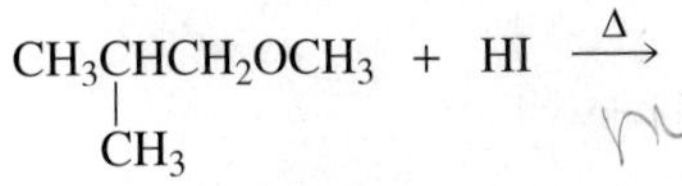 $CH_3CHCH_2OCH_3$ (CH_3) + HI $\xrightarrow{\Delta}$

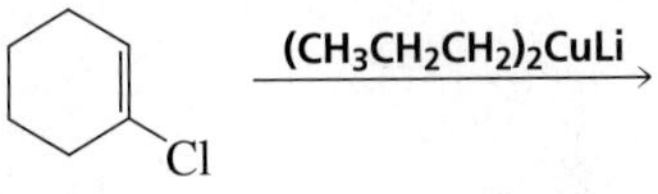

c. 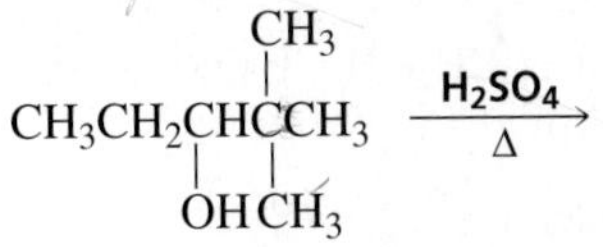 $CH_3CH_2CHCCH_3$ (OH, CH_3, CH_3) $\xrightarrow[\Delta]{H_2SO_4}$

d. $CH_3CH-CCH_3$ (CH_3, CH_3, OH) $\xrightarrow[\Delta]{H_2SO_4}$

e. (cyclohexane spiro-epoxide) $\xrightarrow[CH_3OH]{H^+}$

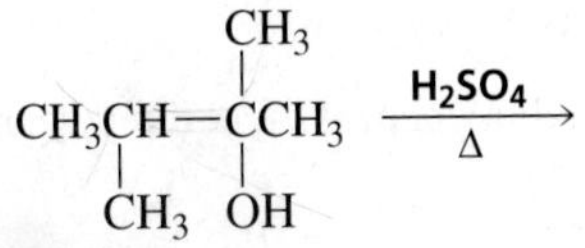

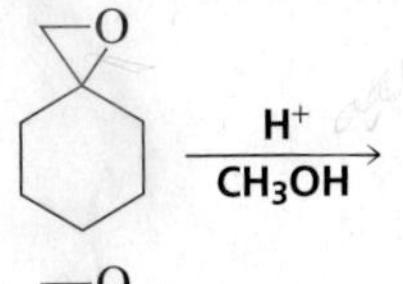

f. 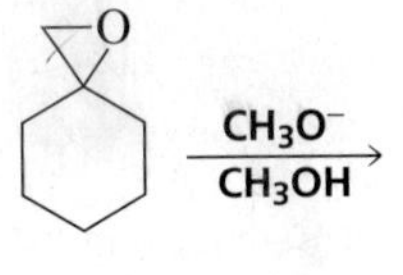 (cyclohexane spiro-epoxide) $\xrightarrow[CH_3OH]{CH_3O^-}$

g. (cyclohexane: H, H, CH_3, OH) $\xrightarrow[\text{2. NaC}\equiv\text{N}]{\text{1. TsCl/pyridine}}$

h. (cyclohexene with Cl) $\xrightarrow{(CH_3CH_2CH_2)_2CuLi}$

44. When heated with H_2SO_4, both 3,3-dimethyl-2-butanol and 2,3-dimethyl-2-butanol are dehydrated to form 2,3-dimethyl-2-butene. Which alcohol dehydrates more rapidly?

45. Using the given starting material, any necessary inorganic reagents, and any carbon-containing compounds with no more than two carbon atoms, indicate how the following syntheses could be carried out:

a. 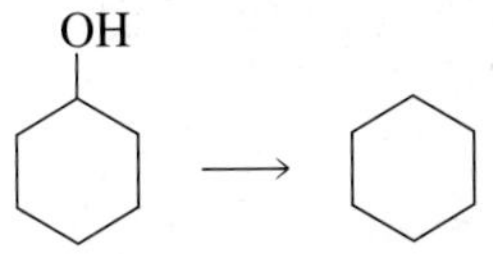 OH (on cyclohexane ring) ⟶ (cyclohexane)

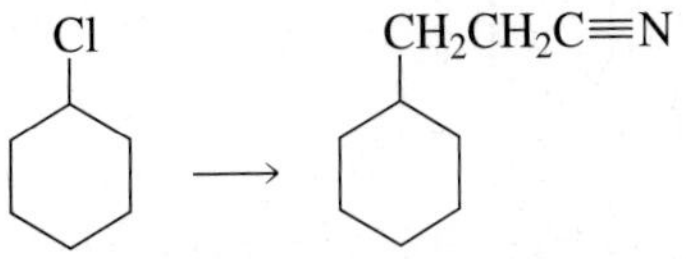

b. 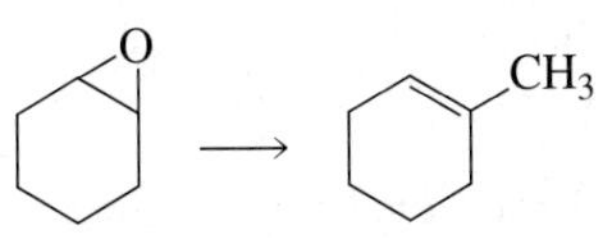 (cyclohexene oxide) ⟶ CH_3 (on cyclohexene ring)

c. Br (on benzene ring) ⟶ CH_2CH_2OH (on benzene ring)

d. Cl (on cyclohexane ring) ⟶ $CH_2CH_2C\equiv N$ (on cyclohexane ring)

e. $CH_3CH_2C\equiv CH$ ⟶ $CH_3CH_2C\equiv CCH_2CH_2OH$

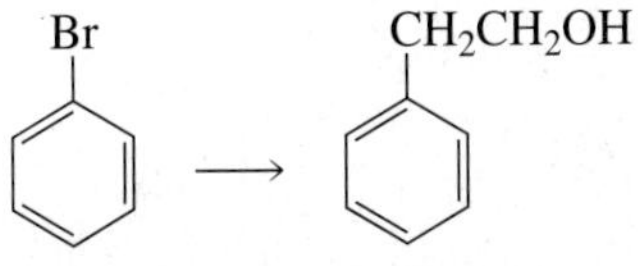

f. 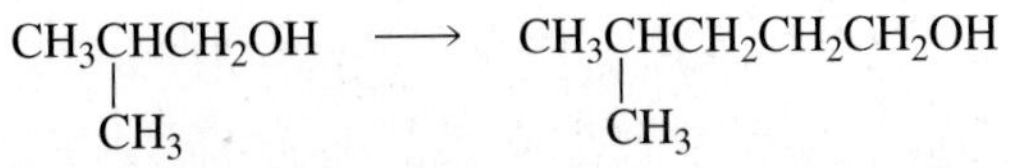 CH_3CHCH_2OH (CH_3) ⟶ $CH_3CHCH_2CH_2CH_2OH$ (CH_3)

46. Propose a mechanism for the following reaction:

$$CH_3\underset{\displaystyle Cl}{CH}CH\overset{O}{-}CH_2 + CH_3O^- \xrightarrow{CH_3OH} CH_3CH\overset{O}{-}CHCH_2OCH_3 + Cl^-$$

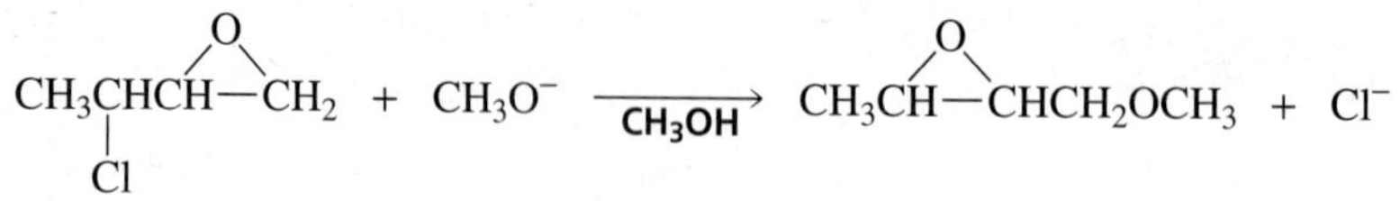

47. When deuterated phenanthrene oxide undergoes an epoxide rearrangement in water, 81% of the deuterium is retained in the product.

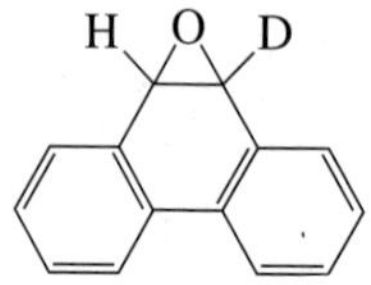

H O D

a. What percentage of the deuterium will be retained if an NIH shift occurs?
b. What percentage of the deuterium will be retained if an NIH shift does not occur?

48. When 3-methyl-2-butanol is heated with concentrated HBr, a rearranged product is obtained. When 2-methyl-1-propanol reacts under the same conditions, a rearranged product is not obtained. Explain.

49. When the following seven-membered ring alcohol is dehydrated, three alkenes are formed:

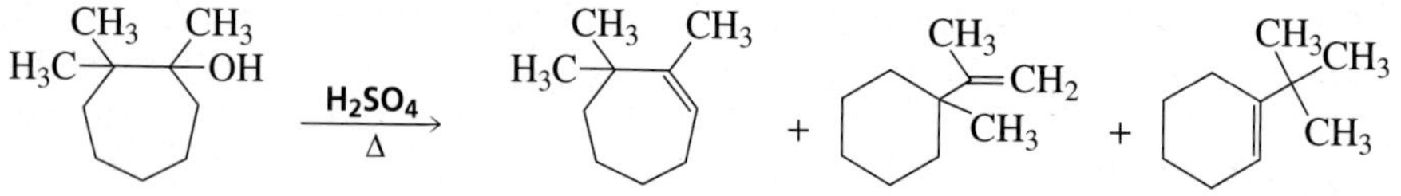

CH_3 CH_3 H_3C OH $\xrightarrow[\Delta]{H_2SO_4}$ CH_3 CH_3 H_3C + CH_3 $=CH_2$ CH_3 + CH_3 CH_3 CH_3

Propose a mechanism for their formation.

50. How could you synthesize isopropyl propyl ether, using isopropyl alcohol as the only carbon-containing reagent?

51. Ethylene oxide reacts readily with HO^- because of the strain in the three-membered ring. Cyclopropane has approximately the same amount of strain, but does not react with HO^-. Explain.

52. Which of the following ethers would be obtained in greatest yield directly from alcohols?

$CH_3OCH_2CH_2CH_3$ $CH_3CH_2OCH_2CH_2CH_3$ $CH_3CH_2OCH_2CH_3$ $CH_3O\underset{\displaystyle CH_3}{\overset{\displaystyle CH_3}{C}}CH_3$

53. Propose a mechanism for each of the following reactions:

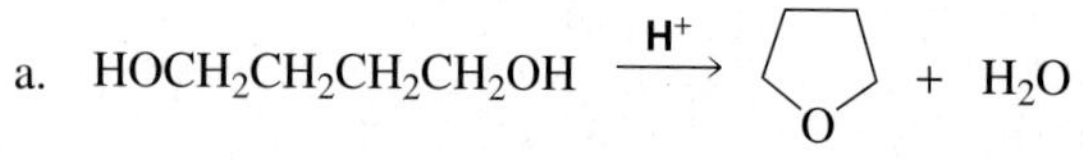

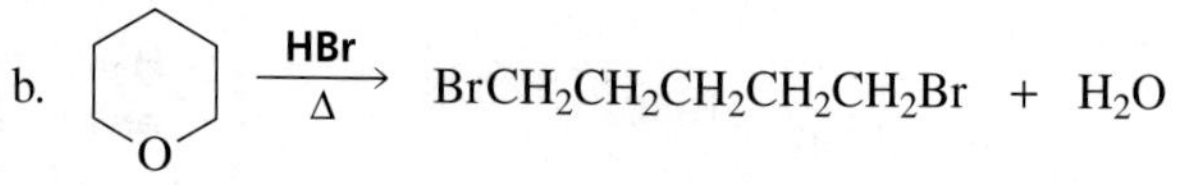

a. $HOCH_2CH_2CH_2CH_2OH \xrightarrow{H^+}$ O + H_2O

b. O $\xrightarrow[\Delta]{HBr}$ $BrCH_2CH_2CH_2CH_2CH_2Br + H_2O$

54. Indicate how each of the following compounds could be prepared, using the given starting material:

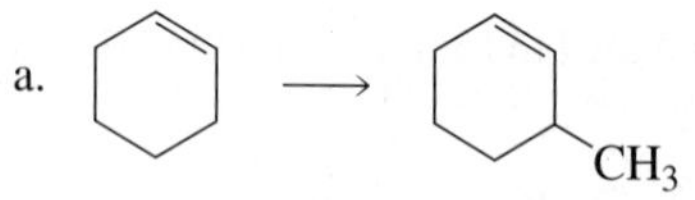

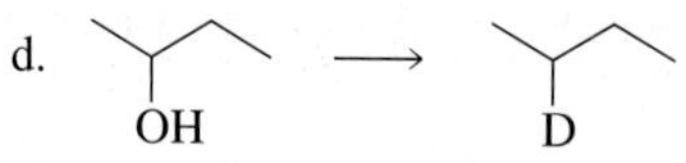

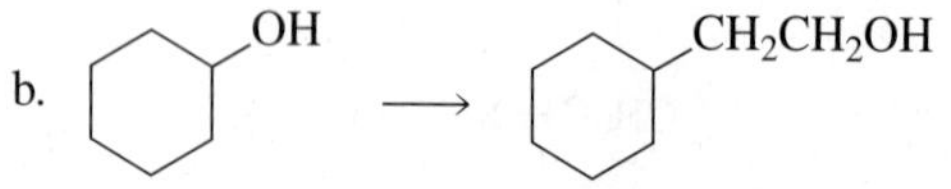

a. ⟶ CH_3

b. OH ⟶ CH_2CH_2OH

c. $CH_3\underset{\displaystyle OH}{\overset{\displaystyle CH_3}{C}}CH_2CH_2CH_3 \longrightarrow CH_3\overset{\displaystyle CH_3}{CH}\underset{\displaystyle Br}{CH}CH_2CH_3$

d. OH ⟶ D

e. $CH_3CH_2CH{=}CH_2 \longrightarrow CH_3CH_2CH_2CH_2CH_2CH_2CH_2CH_3$

55. Triethylene glycol is one of the products obtained from the reaction of ethylene oxide and hydroxide ion. Propose a mechanism for its formation.

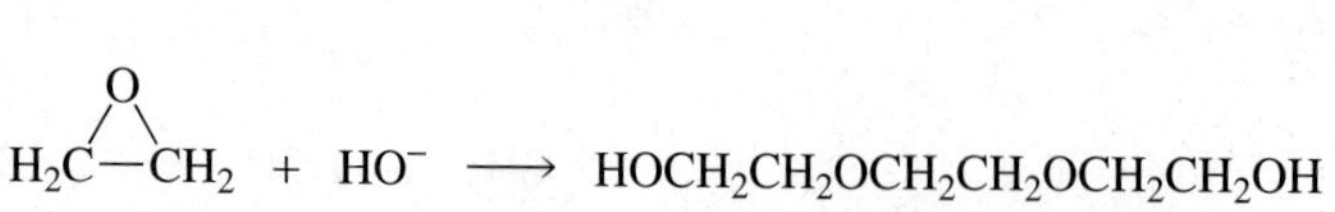

$$H_2C\overset{O}{-}CH_2 + HO^- \longrightarrow \underset{\text{triethylene glycol}}{HOCH_2CH_2OCH_2CH_2OCH_2CH_2OH}$$

56. Give the major product expected from the reaction of 2-ethyloxirane with each of the following reagents:
   a. 0.1 M HCl
   b. CH_3OH/H^+
   c. ethyl magnesium bromide in ether, followed by 0.1 M HCl
   d. 0.1 M NaOH
   e. CH_3OH/CH_3O^-

57. When ethyl ether is heated with excess HI for several hours, the only organic product obtained is ethyl iodide. Explain why ethyl alcohol is not obtained as a product.

58. a. Propose a mechanism for the following reaction:

$CH_2CH_2CH_2Br$ (oxirane) $\xrightarrow{CH_3O^-}$ (tetrahydrofuran)CH_2OCH_3 + Br^-

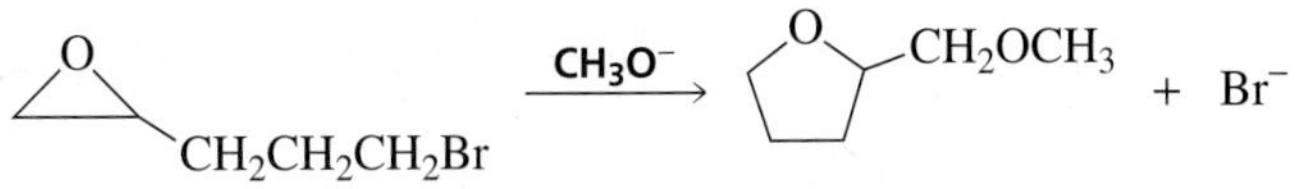

   b. A small amount of a product containing a six-membered ring is also formed. Give the structure of that product.
   c. Why is so little six-membered ring product formed?

59. Identify A through H.

$$CH_3Br \xrightarrow[B]{A} C \xrightarrow[2.\ E]{1.\ D} CH_3CH_2CH_2OH \xrightarrow[2.\ G\ \ 3.\ H]{1.\ F} CH_3CH_2CH_2OCH_2CH_2OH$$

60. Greg Nard added an equivalent of 3,4-epoxy-4-methylcyclohexanol to an ether solution of methyl magnesium bromide and then added dilute hydrochloric acid. He expected that the product would be a diol. He did not get any of the expected product. What product did he get?

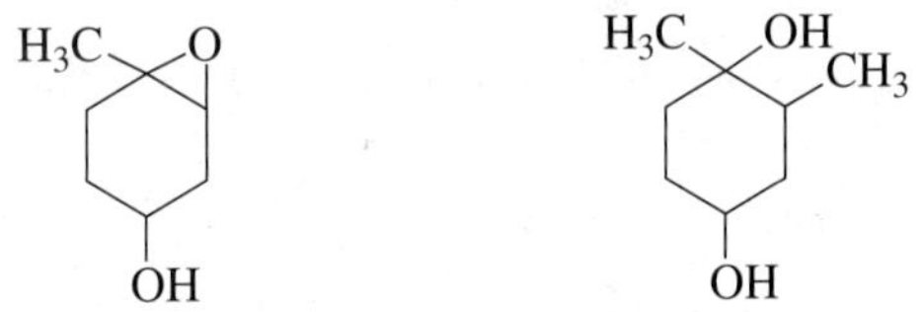

H_3C O OH

3,4-epoxy-4-methyl-cyclohexanol

H_3C OH CH_3 OH

1,2-dimethyl-1,4-cyclohexanediol

61. An ion with a positively charged nitrogen atom in a three-membered ring is called an aziridinium ion. The following aziridinium ion reacts with sodium methoxide to form A and B:

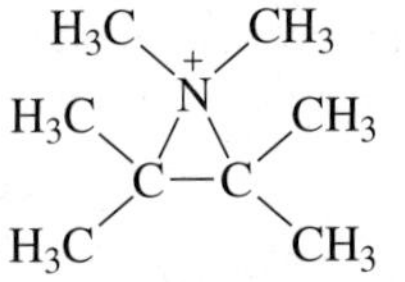

H_3C CH_3 $\overset{+}{N}$ H_3C CH_3 C—C H_3C CH_3

an aziridinium ion

If a small amount of aqueous Br_2 is added to A, the reddish color of Br_2 persists, but the color disappears when Br_2 is added to B. When the aziridinium ion reacts with methanol, only A is formed. Identify A and B.

62. Dimerization is a side reaction that occurs during the preparation of a Grignard reagent. Propose a mechanism that accounts for the formation of the dimer.

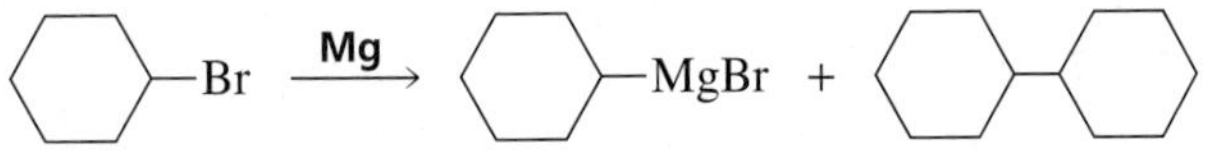

(cyclohexyl)—Br $\xrightarrow{Mg}$ (cyclohexyl)—MgBr + (cyclohexyl)—(cyclohexyl)

a dimer

63. Propose a mechanism for each of the following reactions:

a. O $\xrightarrow[H_2O]{H^+}$ OH HO

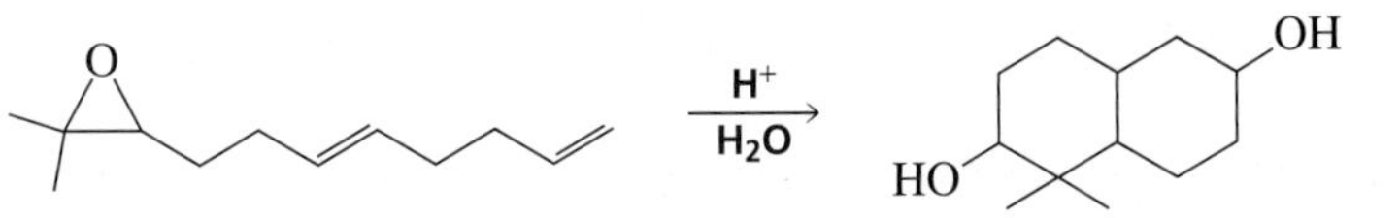

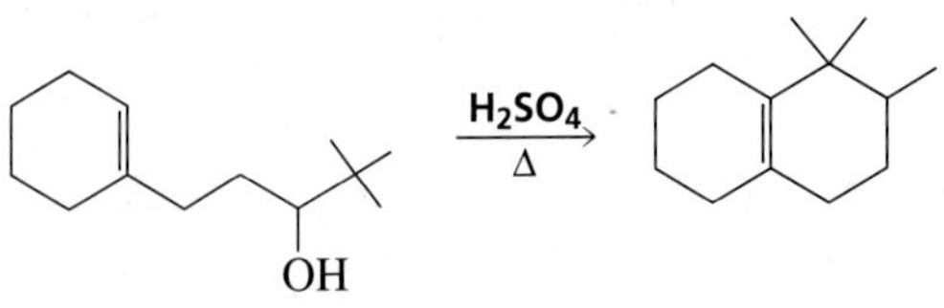

b. OH $\xrightarrow[H_2O]{H^+}$

c. OH $\xrightarrow[\Delta]{H_2SO_4}$

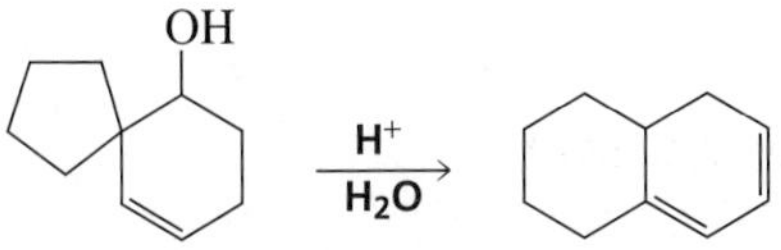

64. One method used to synthesize an epoxide is to treat an alkene with an aqueous solution of Br_2, followed by an aqueous solution of sodium hydroxide.
    a. Propose a mechanism for the conversion of cyclohexene into cyclohexene oxide by this method.
    b. How many products are formed when cyclohexene oxide reacts with methoxide ion in methanol? Draw their structures.

65. Which of the following reactions occurs most rapidly? Why?

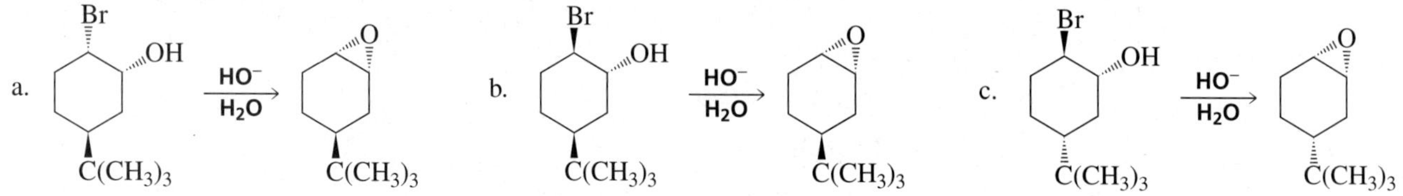

a. Br, OH, $C(CH_3)_3$ $\xrightarrow[H_2O]{HO^-}$ O, $C(CH_3)_3$

b. Br, OH, $C(CH_3)_3$ $\xrightarrow[H_2O]{HO^-}$ O, $C(CH_3)_3$

c. Br, OH, $C(CH_3)_3$ $\xrightarrow[H_2O]{HO^-}$ O, $C(CH_3)_3$

66. When bromobenzene reacts with propene in a Heck reaction, two constitutional isomers are obtained as products. Give the structures of the products and explain why two products are obtained.

67. A vicinal diol has OH groups on adjacent carbons. The dehydration of a vicinal diol is accompanied by a rearrangement called the pinacol rearrangement. Propose a mechanism for this reaction.

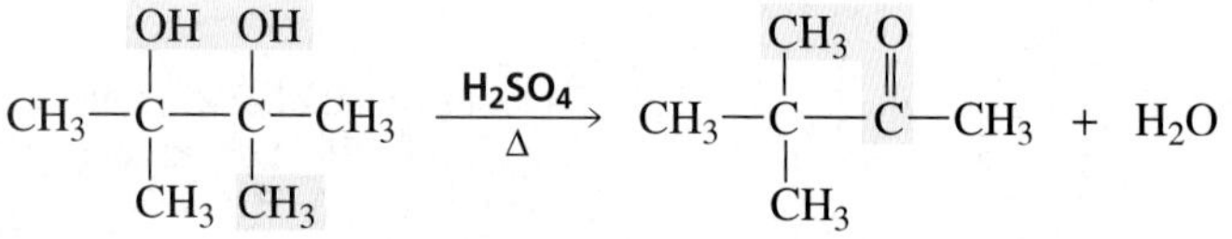

$$CH_3-\underset{CH_3}{\overset{OH}{C}}-\underset{CH_3}{\overset{OH}{C}}-CH_3 \xrightarrow[\Delta]{H_2SO_4} CH_3-\underset{CH_3}{\overset{CH_3}{C}}-\overset{O}{\overset{\|}{C}}-CH_3 + H_2O$$

68. Although 2-methyl-1,2-propanediol is an asymmetrical vicinal diol, only one product is obtained when it is dehydrated in the presence of acid.
    a. What is this product?
    b. Why is only one product formed?

69. What product is obtained when the following vicinal diol is heated in an acidic solution?

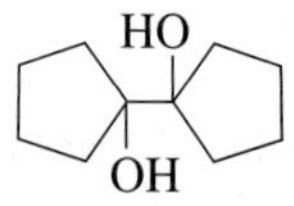

HO, OH

70. Two stereoisomers are obtained from the reaction of cyclopentene oxide and dimethylamine. The *R*,*R*-isomer is used in the manufacture of eclanamine, an antidepressant. What other isomer is obtained?

71. Propose a mechanism for each of the following reactions:

a. HO OH, H_3C, CH_3 $\xrightarrow[\Delta]{H_2SO_4}$ O, CH_3, CH_3 + H_3C, C(=O), CH_3

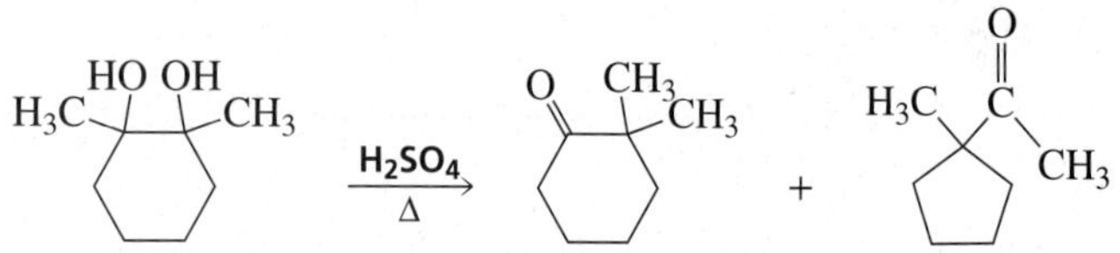

b. OH, CH=CH_2 $\xrightarrow[\Delta]{H_2SO_4}$ O, CH_3

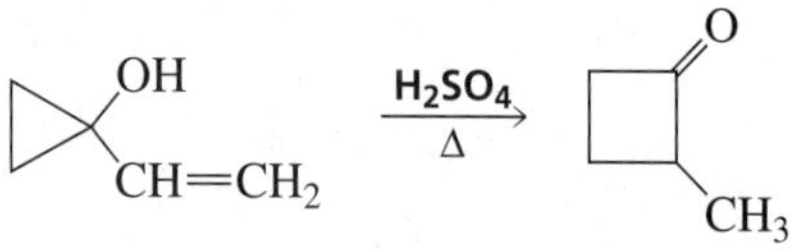

PART FOUR

Identification of Organic Compounds

You have now worked through many problems that asked you to design the synthesis of an organic compound. But if you were actually to go into the laboratory to carry out a synthesis you designed, how would you know that the compound you obtained was the one you had set out to prepare? When a scientist discovers a new compound with physiological activity, its structure must be determined. Only after its structure is known, can methods to synthesize the compound be designed and studies to provide insights into its biological behavior be carried out. Clearly, chemists need to be able to determine the structures of compounds.

In Chapters 13 and 14, you will learn about three instrumental techniques that chemists use to identify compounds.

Mass spectrometry is used to determine the molecular mass and the molecular formula of an organic compound and to identify certain structural features of the compound.

Infrared (IR) spectroscopy is used to determine the kinds of functional groups in an organic compound.

Nuclear magnetic resonance (NMR) spectroscopy helps to identify the carbon–hydrogen framework of an organic compound.

13 Mass Spectrometry and Infrared Spectroscopy

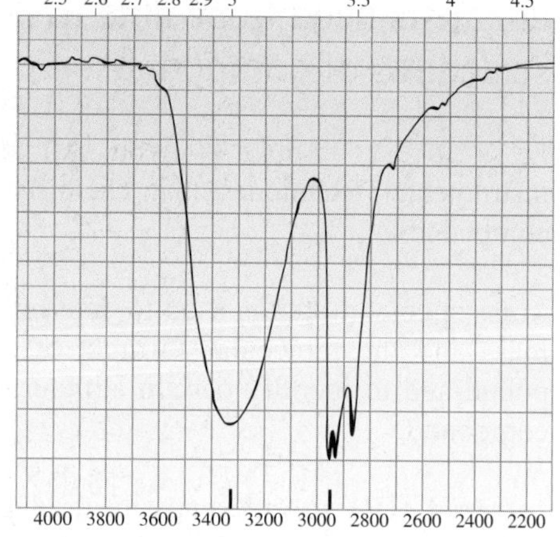

It is essential for chemists to be able to determine the structures of the compounds with which they work. For example, you learned that an aldehyde is formed when a terminal alkyne undergoes hydroboration–oxidation (Section 6.7). But how was it determined that the product of that reaction is actually an aldehyde?

Scientists search the world for new compounds with physiological activity. If a promising compound is found, its structure needs to be determined. Without knowing its structure, chemists cannot design ways to synthesize the compound, nor can they undertake studies to provide insights into its biological behavior.

Before the structure of a compound can be determined, the compound must be isolated. For example, if the product of a reaction carried out in the laboratory is to be identified, it must first be isolated from the solvent and from any unreacted starting materials, as well as from any side products that might have formed. A compound found in nature must be isolated from the organism that manufactures it.

Isolating products and determining their structures used to be daunting tasks. The only tools chemists had to isolate products were distillation (for liquids) and sublimation or fractional recrystallization (for solids). Now a variety of chromatographic techniques allow compounds to be isolated relatively easily. You will learn about these techniques if you take a laboratory course in organic chemistry.

At one time, identifying an organic compound relied upon determining its molecular formula by elemental analysis, determining the compound's physical properties (its melting point, boiling point, etc.), and simple chemical tests that indicated the presence (or absence) of certain functional groups. For example, when an aldehyde is added to a test tube containing a solution of silver oxide in ammonia, a silver mirror is formed on the inside of the test tube. Only aldehydes do this. If a mirror forms, you can conclude that the unknown compound is an aldehyde; if a mirror does not form, you know that the compound is not an aldehyde. Another example of a simple test is the Lucas test, which distinguishes primary, secondary, and tertiary alcohols by how rapidly the test solution turns cloudy after the addition of the Lucas reagent (Section 12.1). These procedures were not sufficient to characterize molecules with complex structures, and because a relatively large sample was needed to carry out all the tests, they were impractical for the analysis of compounds that were difficult to obtain.

Today, a number of different instrumental techniques are used to identify organic compounds. These techniques can be performed quickly on small amounts of a compound and can provide much more information about the compound's structure than simple chemical tests can provide. We have already discussed one such technique: ultraviolet/visible (UV/Vis) spectroscopy, which provides information about organic compounds with conjugated double bonds. In this chapter, we will look at two more instrumental techniques: mass spectrometry and infrared (IR) spectroscopy. **Mass spectrometry** allows us to determine the *molecular mass* and the *molecular formula* of a compound, as well as certain *structural features* of the compound. **Infrared spectroscopy** allows us to determine the *kinds of functional groups* a compound has. In the next chapter, we will look at nuclear magnetic resonance (NMR) spectroscopy, which provides information about the carbon–hydrogen framework of a compound. Of these instrumental techniques, mass spectrometry is the only one that does not involve electromagnetic radiation. Thus, it is called *spectrometry*, whereas the others are called *spectroscopy.*

We will be referring to different classes of organic compounds as we discuss the various instrumental techniques; these classes are listed in Table 13.1. (They are also listed inside the back cover of the book for easy reference.)

Table 13.1 Classes of Organic Compounds

Class	Structure	Note	Class	Structure
Alkane	$-\overset{\mid}{\underset{\mid}{C}}-$	contains only C—C and C—H bonds	Aldehyde	$R\overset{O}{\overset{\|}{C}}H$
Alkene	$\rangle C{=}C\langle$		Ketone	$R\overset{O}{\overset{\|}{C}}R$
Alkyne	$-C{\equiv}C-$		Carboxylic acid	$R\overset{O}{\overset{\|}{C}}OH$
Nitrile	$-C{\equiv}N$		Ester	$R\overset{O}{\overset{\|}{C}}OR$
Alkyl halide	RX	X = F, Cl, Br, or I	Amides	$R\overset{O}{\overset{\|}{C}}NH_2$
Ether	ROR		Amides	$R\overset{O}{\overset{\|}{C}}NHR$
Alcohol	ROH		Amides	$R\overset{O}{\overset{\|}{C}}NR_2$
Phenol	ArOH	Ar = C_6H_5-	Amine (primary)	RNH_2
			Amine (secondary)	R_2NH
Aniline	$ArNH_2$		Amine (tertiary)	R_3N

13.1 Mass Spectrometry

At one time, the molecular weight of a compound was determined by its vapor density or its freezing-point depression, and molecular formulas were determined by elemental analysis, a technique that determined the relative proportions of the elements present in the compound. These were long and tedious techniques that required relatively large amounts of a very pure sample. Today, molecular weights and molecular formulas can be rapidly determined by mass spectrometry from a very small amount of a sample.

In mass spectrometry, a small sample of a compound is introduced into an instrument called a mass spectrometer, where it is vaporized and then ionized as a result of an electron's being removed from each molecule. Ionization can be accomplished in several ways. The most common method bombards the vaporized molecules with a beam of high-energy electrons. The energy of the electron beam can be varied, but a beam of about 70 electron volts (eV) is commonly used. When the electron beam hits a molecule, it knocks out an electron, producing a **molecular ion,** which is a **radical cation**—a species with an unpaired electron and a positive charge.

$$\underset{\textbf{molecule}}{\mathrm{M}} \xrightarrow{\textbf{electron beam}} \underset{\substack{\textbf{molecular ion}\\ \textbf{a radical cation}}}{\mathrm{M}^{\dot{+}}} + \underset{\textbf{electron}}{e^-}$$

Loss of an electron from a molecule weakens the molecule's bonds. Therefore, many of the molecular ions break apart into cations, radicals, neutral molecules, and other radical cations. Not surprisingly, the bonds most likely to break are the weakest ones and those that result in the formation of the most stable products. All the *positively charged fragments* of the molecule pass between two negatively charged plates, which accelerate the fragments into an analyzer tube (Figure 13.1). Neutral

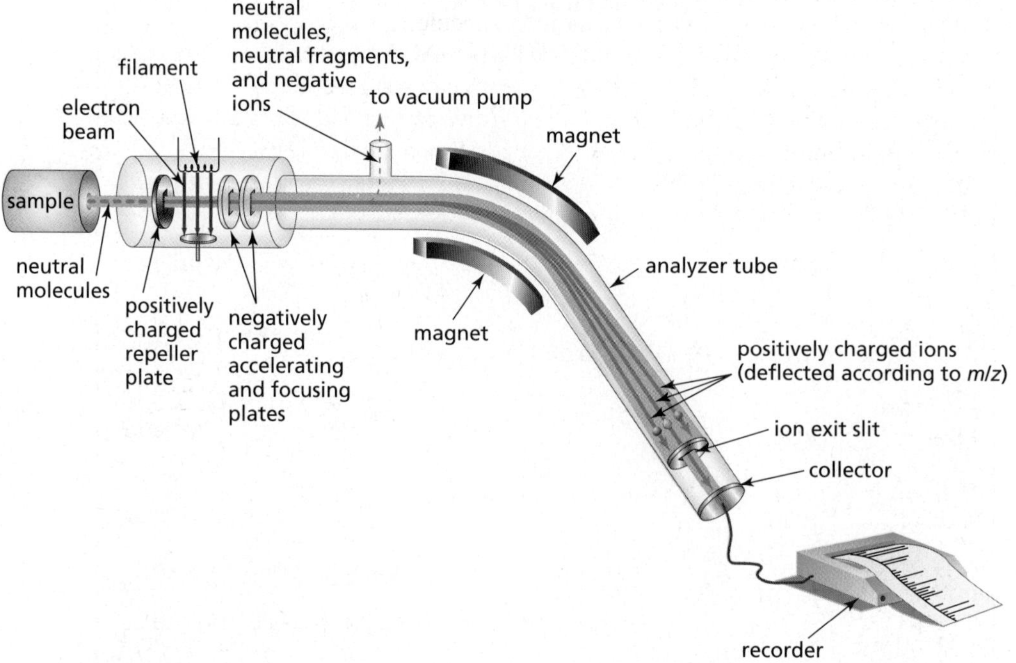

▲ **Figure 13.1**
Schematic of a mass spectrometer. A beam of high-energy electrons causes molecules to ionize and fragment. Positively charged fragments pass through the analyzer tube. Changing the magnetic field strength allows the separation of fragments of varying mass-to-charge ratio.

fragments are not attracted to the negatively charged plates and therefore are not accelerated. They are eventually pumped out of the spectrometer.

The analyzer tube is surrounded by a magnet whose magnetic field deflects the positively charged fragments in a curved path. At a given magnetic field strength, the degree to which the path is curved depends on the mass-to-charge ratio (m/z) of the fragment: The path of a fragment with a smaller m/z value will bend more than that of a heavier fragment. In this way, the particles with the same m/z values can be separated from all the others. If a fragment's path matches the curvature of the analyzer tube, the fragment will pass through the tube and out the ion exit slit. A collector records the relative number of fragments with a particular m/z passing through the slit. The more stable the fragment, the more likely it will make it to the collector. The strength of the magnetic field is gradually increased, so fragments with progressively larger m/z values are guided through the tube and out the exit slit.

The mass spectrometer records a **mass spectrum**—a graph of the relative abundance of each fragment plotted against its m/z value. Because the charge (z) on essentially all the fragments that reach the collector plate is $+1$, m/z is the molecular mass (m) of the fragment. *Remember that only positively charged species reach the collector.*

A mass spectrum records only positively charged fragments.

13.2 The Mass Spectrum • Fragmentation

The mass spectrum of pentane is shown in Figure 13.2. Each m/z value is the **nominal molecular mass** of the fragment—the molecular mass to the nearest whole number. ^{12}C is defined to have a mass of 12.000 atomic mass units (amu), and the masses of other atoms are based on this standard. For example, a proton has a mass of 1.007825 amu. Pentane, therefore, has a *molecular mass* of 72.0939 and a *nominal molecular mass* of 72.

The peak with the highest m/z value in the spectrum—in this case, at $m/z = 72$—is due to the fragment that results when an electron is knocked out of a molecule of the injected sample—in this case, a pentane molecule. In other words, the peak with the highest m/z value represents the molecular ion (M) of pentane. (The tiny peak at $m/z = 73$ will be explained later.) Since it is not known what bond loses the electron, the molecular ion is put in brackets and the positive charge and unpaired electron are assigned to the entire structure. *The m/z value of the molecular ion gives the molecular mass of the compound.* Peaks with smaller m/z values—called **fragment ion peaks**—represent positively charged fragments of the molecule.

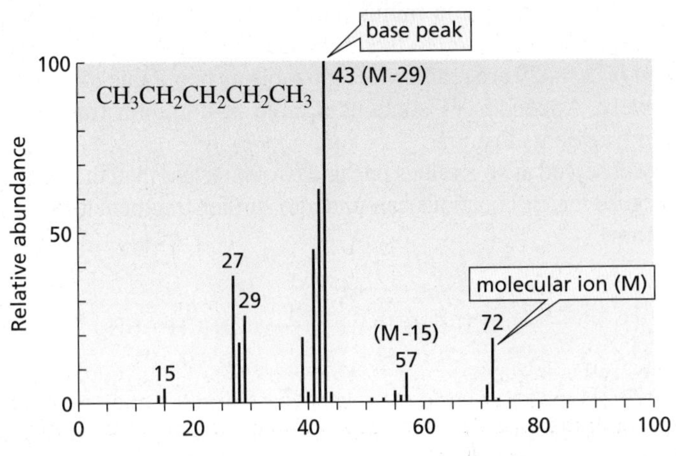

m/z	Relative abundance
73	0.52
72	18.56
71	4.32
57	11.20
43	100.00
42	55.27
41	37.93
39	12.44
29	26.65
28	17.75
27	31.22
15	4.22
14	2.56

Figure 13.2 The mass spectrum of pentane, shown as a bar graph and in tabular form. The base peak represents the fragment that appears in greatest abundance. The m/z value of the molecular ion gives the molecular mass of the compound.

$$CH_3CH_2CH_2CH_2CH_3 \xrightarrow{\text{electron beam}} [CH_3CH_2CH_2CH_2CH_3]^{+\cdot} + e^-$$

molecular ion
$m/z = 72$

The **base peak** is the peak with the greatest intensity, due to its having the greatest relative abundance. The base peak is assigned a relative intensity of 100%, and the relative intensity of each of the other peaks is reported as a percentage of the base peak. Mass spectra can be shown either as bar graphs or in tabular form.

The way a molecular ion fragments depends on the strength of its bonds and the stability of the fragments.

A mass spectrum gives us structural information about the compound because *the m/z values and the relative abundances of the fragments depend on the strength of the molecular ion's bonds and the stability of the fragments.* Weak bonds break in preference to strong bonds, and bonds that break to form more stable fragments break in preference to those that form less stable fragments.

For example, the C—C bonds in the molecular ion formed from pentane have about the same strength. However, the C-2—C-3 bond is more likely to break than the C-1—C-2 bond because C-2—C-3 fragmentation leads to a *primary* carbocation and a *primary* radical, which together are more stable than the *primary* carbocation and *methyl* radical (or *primary* radical and *methyl* cation) obtained from C-1—C-2 fragmentation. C-2—C-3 fragmentation forms ions with $m/z = 43$ or 29, and C-1—C-2 fragmentation forms ions with $m/z = 57$ or 15. The base peak of 43 in the mass spectrum of pentane indicates the preference for C-2—C-3 fragmentation. (See Sections 7.7 and 7.8 to review the relative stabilites of carbocations and radicals.)

$$[\overset{1}{C}H_3\overset{2}{C}H_2\overset{3}{C}H_2\overset{4}{C}H_2\overset{5}{C}H_3]^{+\cdot}$$

molecular ion
$m/z = 72$

$\longrightarrow CH_3\dot{C}H_2 + \overset{+}{C}H_2CH_2CH_3$ ($m/z = 43$)

$\longrightarrow CH_3\overset{+}{C}H_2$ ($m/z = 29$) $+ \dot{C}H_2CH_2CH_3$

$\longrightarrow \dot{C}H_3 + \overset{+}{C}H_2CH_2CH_2CH_3$ ($m/z = 57$)

$\longrightarrow \overset{+}{C}H_3$ ($m/z = 15$) $+ \dot{C}H_2CH_2CH_2CH_3$

A method commonly used to identify fragment ions is to determine the difference between the m/z value of a given fragment ion and that of the molecular ion. For example, the ion with $m/z = 43$ in the mass spectrum of pentane is 29 units smaller than the molecular ion ($M - 29 = 43$). An ethyl radical (CH_3CH_2) has a molecular mass of 29 (because the mass numbers of C and H are 12 and 1, respectively), so the peak at 43 can be attributed to the molecular ion minus an ethyl radical. Similarly, the peak at $m/z = 57$ ($M - 15$) can be attributed to the molecular ion minus a methyl radical. Peaks at $m/z = 15$ and $m/z = 29$ are readily recognizable as being due to methyl and ethyl cations, respectively. Appendix VI contains a table of common fragment ions and a table of common fragments lost.

Peaks are commonly observed at m/z values one and two units less than the m/z values of the carbocations because the carbocations can undergo further fragmentation—losing one or two hydrogen atoms.

$$\underset{m/z = 43}{CH_3CH_2\overset{+}{C}H_2} \xrightarrow{-H\cdot} \underset{m/z = 42}{[CH_3CHCH_2]^{+\cdot}} \xrightarrow{-H\cdot} \underset{m/z = 41}{\overset{+}{C}H_2CH{=}CH_2}$$

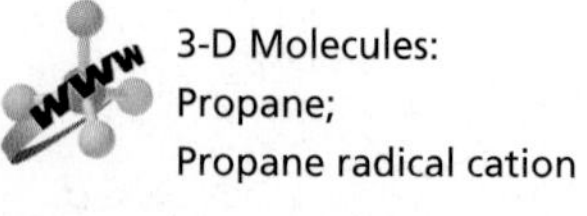
3-D Molecules:
Propane;
Propane radical cation

2-Methylbutane has the same molecular formula as pentane, so it, too, has a molecular ion with $m/z = 72$ (Figure 13.3). Its mass spectrum is similar to that of pentane, with one notable exception: The peak at $m/z = 57$ ($M - 15$) is much more intense.

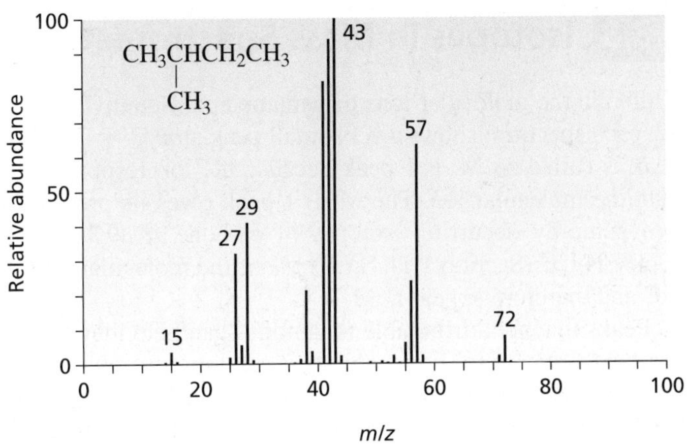

◀ **Figure 13.3**
The mass spectrum of 2-methylbutane.

2-Methylbutane is more likely than pentane to lose a methyl radical, because, when it does, a *secondary* carbocation is formed. In contrast, when pentane loses a methyl radical, a less stable *primary* carbocation is formed.

$$[CH_3\underset{|}{\overset{CH_3}{C}}HCH_2CH_3]^{+\cdot} \longrightarrow CH_3\overset{+}{C}HCH_2CH_3 + \dot{C}H_3$$

molecular ion $m/z = 72$ — $m/z = 57$

PROBLEM 1

How could you distinguish the mass spectrum of 2,2-dimethylpropane from those of pentane and 2-methylbutane?

PROBLEM 2◆

What m/z value is most likely for the base peak in the mass spectrum of 3-methylpentane?

PROBLEM 3 SOLVED

The mass spectra of two very stable cycloalkanes both show a molecular ion peak at $m/z = 98$. One spectrum shows a base peak at $m/z = 69$, the other shows a base peak at $m/z = 83$. Identify the cycloalkanes.

SOLUTION The molecular formula for a cycloalkane is C_nH_{2n}. Because the molecular mass of both cycloalkanes is 98, their molecular formulas must be C_7H_{14} ($7 \times 12 = 84 + 14 = 98$). A base peak of 69 means the loss of an ethyl substituent ($98 - 69 = 29$), whereas a base peak of 83 means the loss of a methyl substituent ($98 - 83 = 15$). Because the cycloalkanes are known to be very stable, we can rule out cycloalkanes with three or four-membered rings. A seven-carbon cycloalkane with a base peak signifying the loss of an ethyl substituent must be ethylcyclopentane. A seven-carbon cycloalkane with a base peak signifying the loss of a methyl substituent must be methylcyclohexane.

ethylcyclopentane (CH_2CH_3 on cyclopentane) $\xrightarrow{\text{electron beam}}$ cyclopentyl cation ($m/z = 69$) + $CH_3\dot{C}H_2$

methylcyclohexane (CH_3 on cyclohexane) $\xrightarrow{\text{electron beam}}$ cyclohexyl cation ($m/z = 83$) + $\dot{C}H_3$

PROBLEM 4

The "nitrogen rule" states that if a compound has an odd-mass molecular ion, the compound contains an odd number of nitrogen atoms.

a. Calculate the m/z value for the molecular ion of the following compounds:
   1. $CH_3CH_2CH_2CH_2NH_2$
   2. $H_2NCH_2CH_2CH_2NH_2$
b. Explain why the nitrogen rule holds.
c. State the rule in terms of an even-mass molecular ion.

13.3 Isotopes in Mass Spectrometry

Although the molecular ions of pentane and 2-methylbutane both have m/z values of 72, each spectrum shows a very small peak at $m/z = 73$ (Figures 13.2 and 13.3). This peak is called an M + 1 peak because the ion responsible for it is one unit heavier than the molecular ion. The M + 1 peak owes its presence to the fact that there are two naturally occurring isotopes of carbon: 98.89% of natural carbon is ^{12}C and 1.11% is ^{13}C (Section 1.1). So 1.11% of the molecular ions contain a ^{13}C instead of a ^{12}C and therefore appear at M + 1.

Peaks that are attributable to isotopes can help identify the compound responsible for a mass spectrum. For example, if a compound contains five carbon atoms, the relative intensity of the M + 1 ion should be $5(1.1\%) = 5(.011)$, multiplied by the relative intensity of the molecular ion. This means that the number of carbon atoms in a compound can be calculated if the relative intensities of both the M and M + 1 peaks are known.

$$\text{number of carbon atoms} = \frac{\text{relative intensity of M + 1 peak}}{.011 \times (\text{relative intensity of M peak})}$$

The isotopic distributions of several elements commonly found in organic compounds are shown in Table 13.2. From the isotopic distributions, we see why the M + 1 peak can be used to determine the number of carbon atoms in a compound: It is because the contributions to the M + 1 peak by isotopes of H, O, and the halogens are very small or nonexistent. This formula does not work as well in predicting the number of carbon atoms in a nitrogen-containing compound because the natural abundance of ^{15}N is relatively high.

Table 13.2 The Natural Abundance of Isotopes Commonly Found in Organic Compounds

Element	Natural abundance			
Carbon	^{12}C 98.89%	^{13}C 1.11%		
Hydrogen	^{1}H 99.99%	^{2}H 0.01%		
Nitrogen	^{14}N 99.64%	^{15}N 0.36%		
Oxygen	^{16}O 99.76%	^{17}O 0.04%	^{18}O 0.20%	
Sulfur	^{32}S 95.0%	^{33}S 0.76%	^{34}S 4.22%	^{36}S 0.02%
Fluorine	^{19}F 100%			
Chlorine	^{35}Cl 75.77%		^{37}Cl 24.23%	
Bromine	^{79}Br 50.69%		^{81}Br 49.31%	
Iodine	^{127}I 100%			

Mass spectra can show M + 2 peaks as a result of a contribution from ^{18}O or from having two heavy isotopes in the same molecule (say, ^{13}C and ^{2}H, or two ^{13}C's). Most of the time, the M + 2 peak is very small. The presence of a large M + 2 peak is evidence of a compound containing either chlorine or bromine, because each of these elements has a high percentage of a naturally occurring isotope that is two units heavier than the most abundant isotope. From the natural abundance of the isotopes of chlorine and bromine in Table 13.2, one can conclude that if the M + 2 peak is one-third the height of the molecular ion peak, then the compound contains one chlorine atom because the natural abundance of ^{37}Cl is one-third that of ^{35}Cl. If the M and M + 2 peaks are about the same height, then the compound contains one bromine atom because the natural abundances of ^{79}Br and ^{81}Br are about the same.

In calculating the molecular masses of molecular ions and fragments, the *atomic mass* of a single isotope of the atom must be used (Cl = 35 or 37, etc.); the *atomic weights* in the periodic table (Cl = 35.453) cannot be used because they are the *weighted averages* of all the naturally occurring isotopes for that element, and mass spectrometry measures the m/z value of an *individual* fragment.

PROBLEM 5◆

The mass spectrum of an unknown compound has a molecular ion peak with a relative intensity of 43.27% and an M + 1 peak with a relative intensity of 3.81%. How many carbon atoms are in the compound?

13.4 Determination of Molecular Formulas: High-Resolution Mass Spectrometry

All the mass spectra shown in this text were determined with a low-resolution mass spectrometer. Such spectrometers give the *nominal molecular mass* of a fragment—the mass to the nearest whole number. High-resolution mass spectrometers can determine the *exact molecular mass* of a fragment to a precision of 0.0001 amu. If we know the exact molecular mass of the molecular ion, we can determine the compound's molecular formula. For example, as the following listing shows, many compounds have a nominal molecular mass of 122 amu, but each of them has a different exact molecular mass.

Some Compounds with a Nominal Molecular Mass of 122 amu and Their Exact Molecular Masses

Molecular formula	C_9H_{14}	$C_7H_{10}N_2$	$C_8H_{10}O$	$C_7H_6O_2$	$C_4H_{10}O_4$	$C_4H_{10}S_2$
Exact molecular mass (amu)	122.1096	122.0845	122.0732	122.0368	122.0579	122.0225

The exact molecular masses of some common isotopes are listed in Table 13.3. Some computer programs can determine the molecular formula of a compound from the compound's exact molecular mass.

Table 13.3 The Exact Masses of Some Common Isotopes

Isotope	Mass	Isotope	Mass
^{1}H	1.007825 amu	^{32}S	31.9721 amu
^{12}C	13.00000 amu	^{35}Cl	34.9689 amu
^{14}N	14.0031 amu	^{79}Br	78.9183 amu
^{16}O	15.9949 amu		

PROBLEM 6◆

Which molecular formula has an exact molecular mass of 86.1096 amu: C_6H_{14}, $C_4H_{10}N_2$, or $C_4H_6O_2$?

13.5 Fragmentation at Functional Groups

Characteristic fragmentation patterns are associated with specific functional groups; these can help identify a substance based on its mass spectrum. The patterns were recognized after the mass spectra of many compounds containing a particular functional group were studied. We will look at the fragmentation patterns of alkyl halides, ethers, alcohols, and ketones as examples.

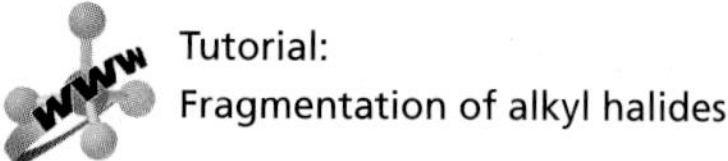

Alkyl Halides

Let's look first at the mass spectrum of 1-bromopropane, shown in Figure 13.4. The relative heights of the M and M + 2 peaks are about equal, so we can conclude that the compound contains a bromine atom. Electron bombardment is most likely to dislodge a lone-pair electron if the molecule has any, because a molecule does not hold onto its lone-pair electrons as tightly as it holds onto its bonding electrons. Thus, electron bombardment dislodges one of bromine's lone-pair electrons.

$$\underset{\textbf{1-bromopropane}}{CH_3CH_2CH_2{-}^{79}\ddot{\underset{..}{Br}}{:} + CH_3CH_2CH_2{-}^{81}\ddot{\underset{..}{Br}}{:}} \xrightarrow{-e^-} \underset{m/z = 122}{CH_3CH_2CH_2{-}^{79}\overset{+}{Br}{:}} + \underset{m/z = 124}{CH_3CH_2CH_2{-}^{81}\overset{+}{Br}{:}} \longrightarrow \underset{m/z = 43}{CH_3CH_2\overset{+}{C}H_2} + {}^{79}{:}\dot{Br}{:} + {}^{81}{:}\dot{Br}{:}$$

The weakest bond in the resulting molecular ion is the C—Br bond (the C—Br bond dissociation energy is 69 kcal/mol; the C—C bond dissociation energy is 85 kcal/mol; see Table 3.1 on p. 129). The bond breaks heterolytically, with both electrons going to the more electronegative of the atoms that were joined by the bond, forming a propyl cation and a bromine atom. As a result, the base peak in the mass spectrum of 1-bromopropane is at $m/z = 43$ [M − 79, or (M + 2) − 81]. The propyl cation has the same fragmentation pattern it exhibited when it was formed from the cleavage of pentane (Figure 13.2).

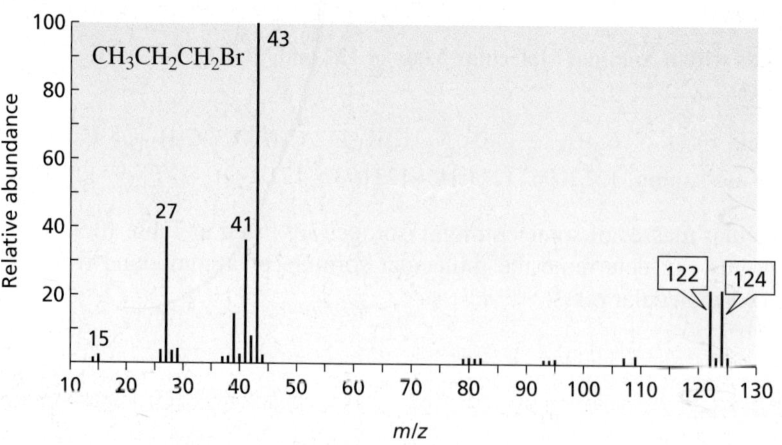

Figure 13.4 ▶ The mass spectrum of 1-bromopropane.

The mass spectrum of 2-chloropropane is shown in Figure 13.5. We know that the compound contains a chlorine atom, because the M + 2 peak is one-third the height of the molecular ion peak. The base peak at $m/z = 43$ results from *heterolytic cleavage* of the C—Cl bond. The peaks at $m/z = 63$ and $m/z = 65$ have a 3 : 1 ratio, indicating that these fragments contain a chlorine atom. They result from *homolytic cleavage* of a C—C bond at the α carbon (the carbon bonded to the chlorine). This cleavage, known as ***α* cleavage**, occurs because the C—Cl (82 kcal/mol) and C—C

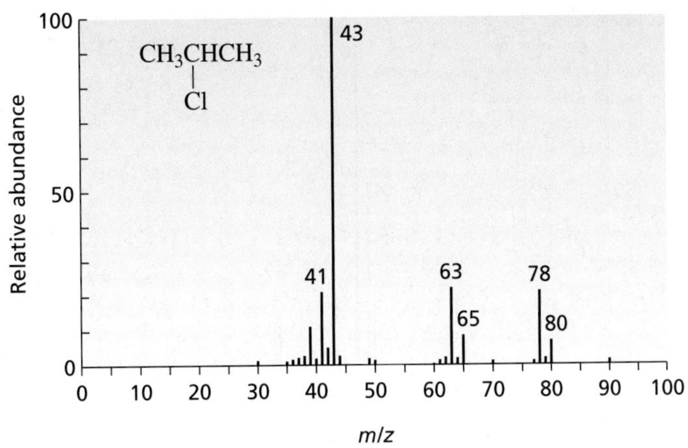

◀ **Figure 13.5**
The mass spectrum of 2-chloropropane.

(85 kcal/mol) bonds have similar strengths, and the species that is formed is a relatively stable cation, since its positive charge is shared by two atoms: $CH_3\overset{+}{C}H—\ddot{C}l\!:$ $\longleftrightarrow$ $CH_3CH{=}\ddot{C}l\!:^+$. Notice that α cleavage does not occur in alkyl bromides because the C—C bond is much weaker than the C—Br bond.

$$CH_3\overset{+}{C}H(CH_3) + {}^{35}\!:\!\ddot{C}l\!: + {}^{37}\!:\!\ddot{C}l\!: \qquad m/z = 43$$

↑ heterolytic cleavage

$$CH_3CH(CH_3)—{}^{35}\dot{C}l\!:^+ + CH_3CH(CH_3)—{}^{37}\dot{C}l\!:^+$$
$$m/z = 78 \qquad m/z = 80$$

$$\xleftarrow{-e^-} \quad CH_3CH(CH_3)—{}^{35}\ddot{C}l\!: + CH_3CH(CH_3)—{}^{37}\ddot{C}l\!: \quad \xrightarrow{-e^-}$$

2-chloropropane

$$CH_3CH(CH_3)—\dot{C}l\!:^{+\,35} + CH_3CH(CH_3)—\dot{C}l\!:^{+\,37}$$
$$m/z = 78 \qquad m/z = 80$$

↓ homolytic cleavage

$$CH_3CH{=}{}^{35}\overset{+}{C}l\!: + CH_3CH{=}{}^{37}\overset{+}{C}l\!: + \cdot CH_3$$
$$m/z = 63 \qquad m/z = 65$$

3-D Molecules: Isopropyl cation; 2-Chloropropane; 2-Chloropropane radical cation

Recall that an arrowhead with one barb represents the movement of one electron.

PROBLEM 7

Sketch the mass spectrum of 1-chloropropane.

Ethers

The mass spectrum of *sec*-butyl isopropyl ether is shown in Figure 13.6. The fragmentation pattern of an ether is similar to that of an alkyl halide.

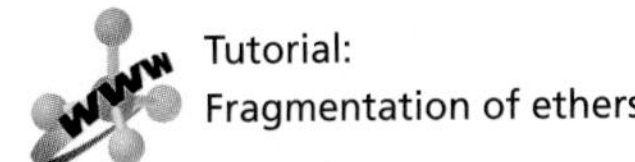

1. Electron bombardment dislodges one of the lone-pair electrons from oxygen.
2. Fragmentation of the resulting molecular ion occurs in two principal ways:
   a. A C—O bond is cleaved heterolytically, with the electrons going to the more electronegative oxygen atom.

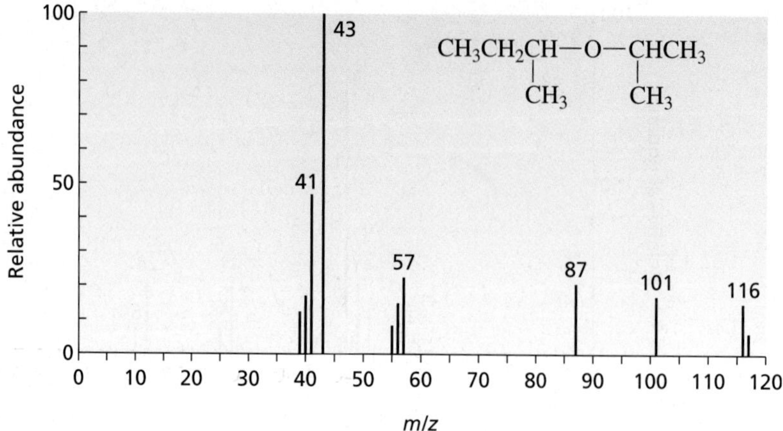

Figure 13.6 ▶ The mass spectrum of *sec*-butyl isopropyl ether.

$CH_3CH_2CH(CH_3){-}\ddot{O}{-}CH(CH_3)CH_3$ (*sec*-butyl isopropyl ether) $\xrightarrow{-e^-}$ $CH_3CH_2CH(CH_3){-}\dot{O}^{+}{-}CH(CH_3)CH_3$ ($m/z = 116$)

$\longrightarrow$ $CH_3CH_2\overset{+}{C}H(CH_3)$ + $:\dot{O}{-}CH(CH_3)CH_3$ ($m/z = 57$)

$\longrightarrow$ $CH_3CH_2CH(CH_3){-}\dot{O}:$ + $\overset{+}{C}H(CH_3)CH_3$ ($m/z = 43$)

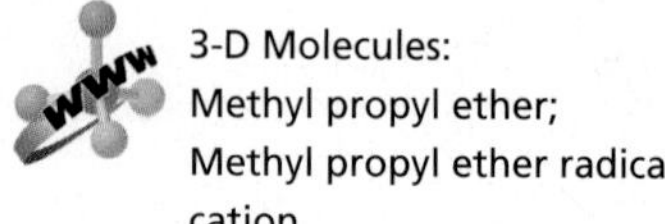

3-D Molecules: Methyl propyl ether; Methyl propyl ether radical cation

b. A C—C bond is cleaved *homolytically* at the α position because it leads to a relatively stable cation in which the positive charge is shared by two atoms (a carbon and an oxygen). The alkyl group leading to the most stable radical is one most easily cleaved. Thus, the peak at $m/z = 87$ is more abundant than the one at $m/z = 101$, even though the compound has three methyl groups bonded to α carbons that can be cleaved to produce a peak at $m/z = 101$, because a primary radical is more stable than a methyl radical.

$CH_3CH_2{-}CH(CH_3){-}\dot{O}^{+}{-}CH(CH_3)CH_3$ $\xrightarrow{\alpha\ \text{cleavage}}$ $CH(CH_3){=}\overset{+}{\ddot{O}}{-}CH(CH_3)CH_3$ + $CH_3\dot{C}H_2$ ($m/z = 87$)

$CH_3CH_2CH(CH_3){-}\dot{O}^{+}{-}CH(CH_3)CH_3$ $\xrightarrow{\alpha\ \text{cleavage}}$ $CH_3CH_2CH{=}\overset{+}{\ddot{O}}{-}CH(CH_3)CH_3$ + $\dot{C}H_3$ ($m/z = 101$)

$CH_3CH_2CH(CH_3){-}\dot{O}^{+}{-}CH(CH_3)CH_3$ $\xrightarrow{\alpha\ \text{cleavage}}$ $CH_3CH_2CH(CH_3){-}\overset{+}{\ddot{O}}{=}CHCH_3$ + $\dot{C}H_3$ ($m/z = 101$)

α carbon

PROBLEM 8◆

The mass spectra of 1-methoxybutane, 2-methoxybutane, and 2-methoxy-2-methylpropane are shown in Figure 13.7. Match the compounds with the spectra.

Alcohols

The molecular ions obtained from alcohols fragment so readily that few of them survive to reach the collector. As a result, the mass spectra of alcohols show small molecular ion peaks. Notice the small molecular ion peak at $m/z = 102$ in the mass spectrum of 2-hexanol (Figure 13.8).

Like alkyl halides and ethers, alcohols undergo α cleavage. Consequently, the mass spectrum of 2-hexanol shows a base peak at $m/z = 45$ (α cleavage leading to a more

Figure 13.7
The mass spectra for Problem 8.

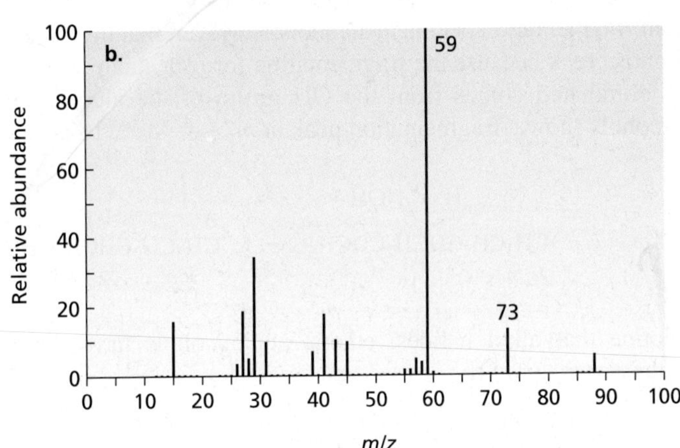

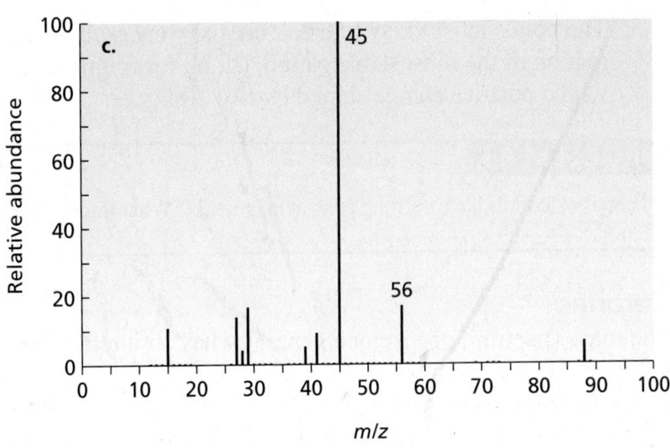

stable butyl radical) and a smaller peak at $m/z = 87$ (α cleavage leading to a less stable methyl radical).

$$\underset{\textbf{2-hexanol}}{CH_3CH_2CH_2CH_2\overset{:\ddot{O}H}{\overset{|}{C}}HCH_3} \xrightarrow{-e^-} \underset{m/z = 102}{CH_3CH_2CH_2CH_2\overset{\dot{+}\ddot{O}H}{\overset{|}{C}}HCH_3}$$

(α carbon)

α **cleavage** $\longrightarrow CH_3CH_2CH_2\dot{C}H_2 + \underset{m/z = 45}{CH_3CH{=}\overset{+}{\ddot{O}}H}$

α **cleavage** $\longrightarrow \underset{m/z = 87}{CH_3CH_2CH_2CH_2CH{=}\overset{+}{\ddot{O}}H} + \dot{C}H_3$

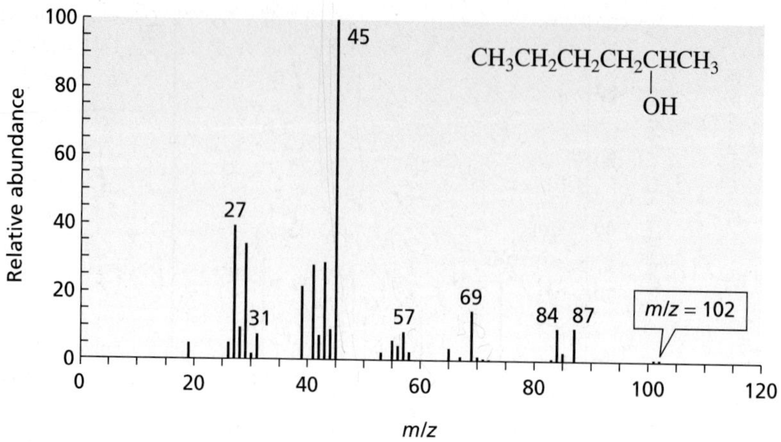

Figure 13.8 ▶ The mass spectrum of 2-hexanol.

In all the fragmentations we have seen so far, only one bond is broken. An important fragmentation occurs in alcohols, however, that involves breaking two bonds. Two bonds break because the fragmentation forms a stable water molecule. The water that is eliminated comes from the OH group of the alcohol and a γ hydrogen. Thus, alcohols show a fragmentation peak at m/z = M − 18 because of loss of water.

$$\underset{\gamma\ \ \beta\ \ \ \alpha}{CH_3CH_2\overset{H}{C}HCH_2\overset{\overset{+\cdot\cdot}{O}H}{C}HCH_3} \longrightarrow CH_3CH_2\dot{C}HCH_2\overset{+}{C}HCH_3 + H_2O$$

$m/z = (102 - 18) = 84$

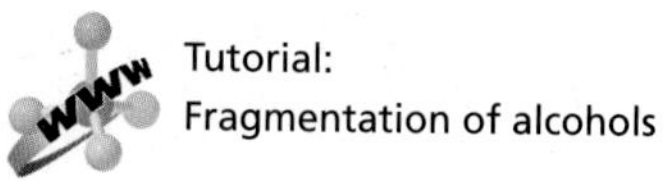

Tutorial: Fragmentation of alcohols

Notice that alkyl halides, ethers, and alcohols have the following fragmentation behavior in common:

1. A bond between carbon and a *more electronegative* atom (a halogen or an oxygen) breaks heterolytically.
2. A bond between carbon and an atom of *similar electronegativity* (a carbon or a hydrogen) breaks homolytically.
3. The bonds most likely to break are the weakest bonds and those that lead to formation of the most stable cation. (Look for fragmentation that results in a cation with a positive charge shared by two atoms.)

PROBLEM 9◆

Primary alcohols have a strong peak at m/z = 31. What fragment is responsible for this peak?

Ketones

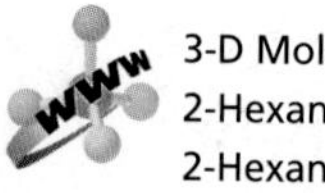

3-D Molecules: 2-Hexanone; 2-Hexanone radical cation; Acetone enol radical cation

The mass spectrum of a ketone generally has an intense molecular ion peak. Ketones fragment homolytically at the C—C bond adjacent to the C=O bond, which results in the formation of a cation with a positive charge shared by two atoms. The alkyl group leading to the more stable radical is the one that is more easily cleaved.

$$\underset{\text{2-pentanone}}{CH_3CH_2CH_2\overset{\overset{\cdot\cdot}{O}:}{\overset{\|}{C}}CH_3} \xrightarrow{-e^-} \underset{m/z = 86}{CH_3CH_2CH_2\overset{\overset{\cdot+}{O}:}{\overset{\|}{C}}CH_3}$$

$$\longrightarrow CH_3CH_2\dot{C}H_2 + \underset{m/z = 43}{CH_3C{\equiv}\overset{+}{O}:}$$

$$\longrightarrow \underset{m/z = 71}{CH_3CH_2CH_2C{\equiv}\overset{+}{O}:} + \dot{C}H_3$$

If one of the alkyl groups attached to the carbonyl carbon has a γ hydrogen, a cleavage known as a **McLafferty rearrangement** may occur. In this rearrangement, the bond between the α carbon and the β carbon breaks homolytically and a hydrogen atom from the γ carbon migrates to the oxygen atom. Again, fragmentation has occurred in a way that produces a cation with a positive charge shared by two atoms.

$$\underset{m/z\,=\,86}{\overset{\gamma}{H_2C}(H)-\overset{\beta}{H_2C}-\overset{\alpha}{CH_2}-\overset{\overset{+}{\ddot{O}}\cdot}{\overset{\|}{C}}-CH_3} \xrightarrow[\text{rearrangement}]{\text{McLafferty}} H_2C{=}CH_2 + \underset{m/z\,=\,58}{\cdot CH_2-\overset{\overset{+}{O}-H}{\overset{\|}{C}}-CH_3}$$

Fred Warren McLafferty *was born in Evanston, Illinois, in 1923. He received a B.S. and an M.S. from the University of Nebraska, and a Ph.D. from Cornell University in 1950. He was a scientist at Dow Chemical Company until he joined the faculty at Purdue University in 1964. He has been a professor of chemistry at Cornell University since 1968.*

PROBLEM 10

How could their mass spectra distinguish the following compounds?

$$CH_3CH_2\overset{O}{\overset{\|}{C}}CH_2CH_3 \qquad CH_3\overset{O}{\overset{\|}{C}}CH_2CH_2CH_3 \qquad CH_3\overset{O}{\overset{\|}{C}}\underset{CH_3}{\underset{|}{C}H}CH_3$$

PROBLEM 11◆

Identify the ketones that are responsible for the mass spectra shown in Figure 13.9.

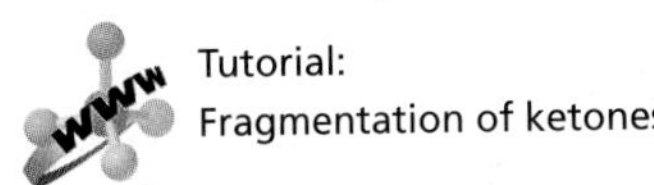

Tutorial:
Fragmentation of ketones

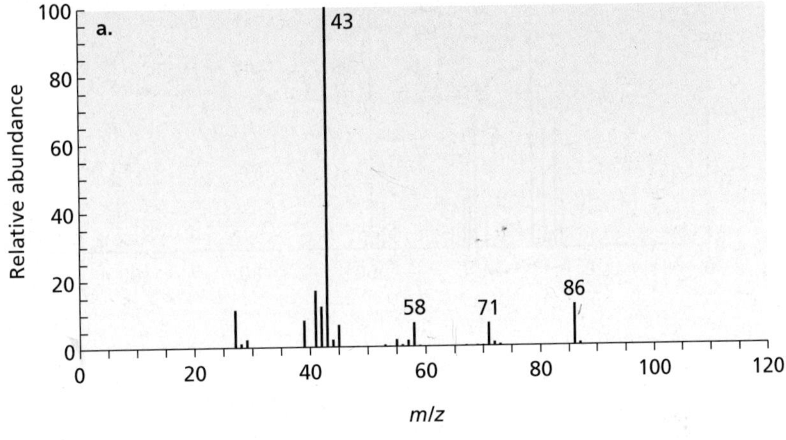

◀ **Figure 13.9**
The mass spectra for Problem 11.

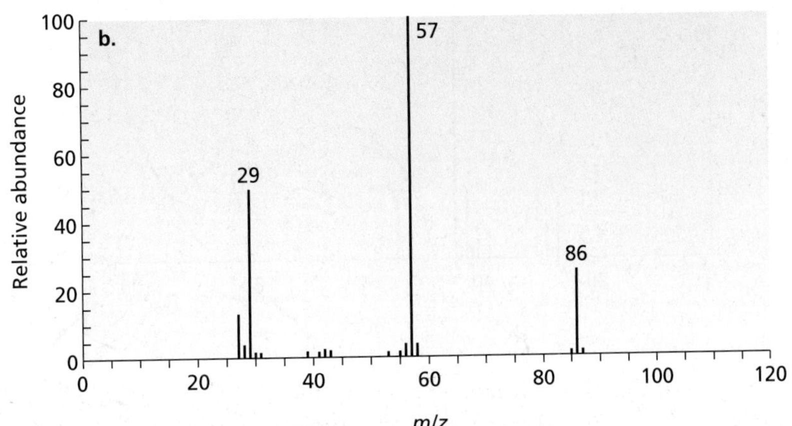

PROBLEM 12

Using curved arrows, show the principal fragments that would be observed in the mass spectrum of each of the following compounds:

a. $CH_3CH_2CH_2CH_2CH_2OH$

b. $CH_3\overset{\overset{\large O}{\|}}{C}CH_2CH_2CH_2CH_3$

c. $CH_3CH_2\underset{\underset{\large OH}{|}}{C}HCH_2CH_2CH_2CH_3$

d. $CH_3CH_2O\overset{\overset{\large CH_2CH_3}{|}}{\underset{\underset{\large CH_3}{|}}{C}}CH_2CH_2CH_3$

e. $CH_3CH_2\underset{\underset{\large CH_3}{|}}{C}HCl$

f. $CH_3\!-\!\overset{\overset{\large CH_3}{|}}{\underset{\underset{\large CH_3}{|}}{C}}\!-\!Br$

PROBLEM 13◆

Two products are obtained from the reaction of (*Z*)-2-pentene with water and a trace of H_2SO_4. The mass spectra of these products are shown in Figure 13.10. Identify the compounds responsible for the spectra.

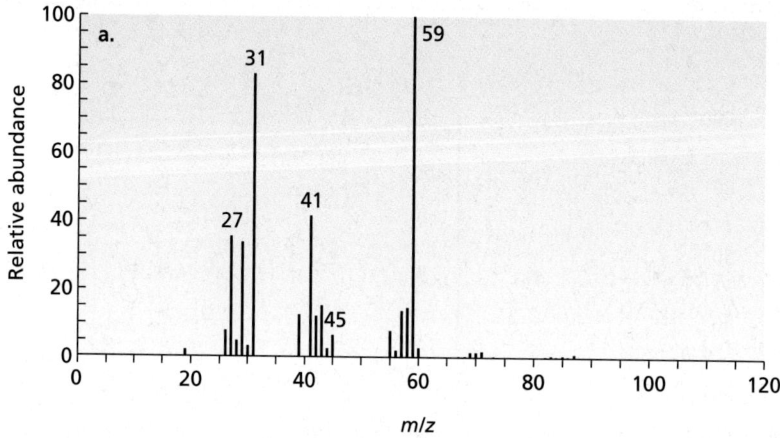

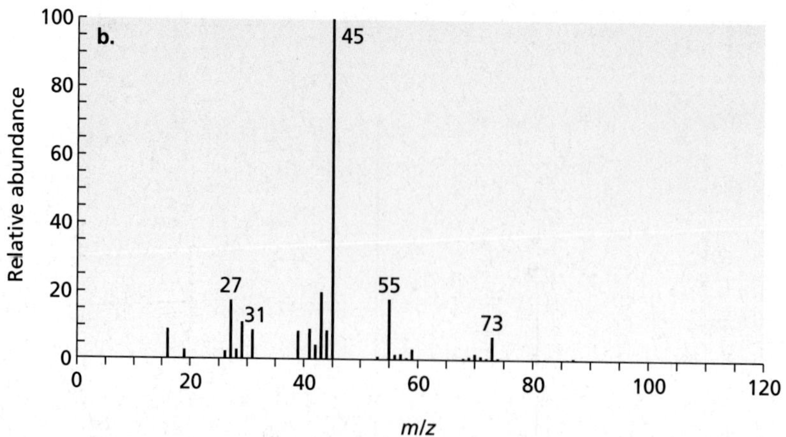

Figure 13.10 ▶
The mass spectra for Problem 13.

The molecular ion and the pattern of fragment ion peaks are unique for each compound. A mass spectrum, therefore, is like a fingerprint of the compound. A positive identification of a compound can be made by comparing its mass spectrum with that of a known sample of the compound.

13.6 Spectroscopy and the Electromagnetic Spectrum

Spectroscopy is the study of the interaction of matter and *electromagnetic radiation.* **Electromagnetic radiation** is radiant energy having the properties of both particles and waves. A continuum of different types of electromagnetic radiation—each type associated with a particular energy range—constitutes the electromagnetic spectrum (Figure 13.11). Visible light is the type of electromagnetic radiation with which we are most familiar, but it represents only a fraction of the range of the entire electromagnetic spectrum. X-rays and radio waves are other types of familiar electromagnetic radiation.

Each of the spectroscopic techniques used to identify compounds that are discussed in this book employs a different type of electromagnetic radiation. You were introduced to ultraviolet/visible (UV/Vis) spectroscopy in Chapter 8. In the current chapter we will look at infrared (IR) spectroscopy, and in the next chapter we will see how compounds can be identified using nuclear magnetic resonance (NMR) spectroscopy.

A particle of electromagnetic radiation is called a *photon*. We may think of electromagnetic radiation as photons traveling at the speed of light. Because electromagnetic radiation has both particle-like and wave-like properties, it can be characterized by either its frequency (ν) or its wavelength (λ). **Frequency** is defined as the number of wave crests that pass by a given point in one second. Frequency has units of hertz (Hz). **Wavelength** is the distance from any point on one wave to the corresponding point on the next wave. Wavelength is generally measured in micrometers or nanometers. One micrometer (μm) is 10^{-6} of a meter; one nanometer (nm) is 10^{-9} of a meter.

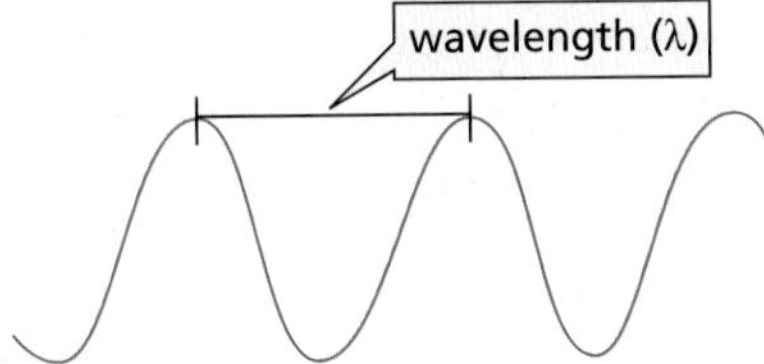

The frequency of electromagnetic radiation, therefore, is equal to the speed of light (c) divided by the radiation's wavelength:

$$\nu = \frac{c}{\lambda} \qquad c = 3 \times 10^{10}\ \text{cm/s}$$

Short wavelenths have high frequencies, and long wavelengths have low frequencies.

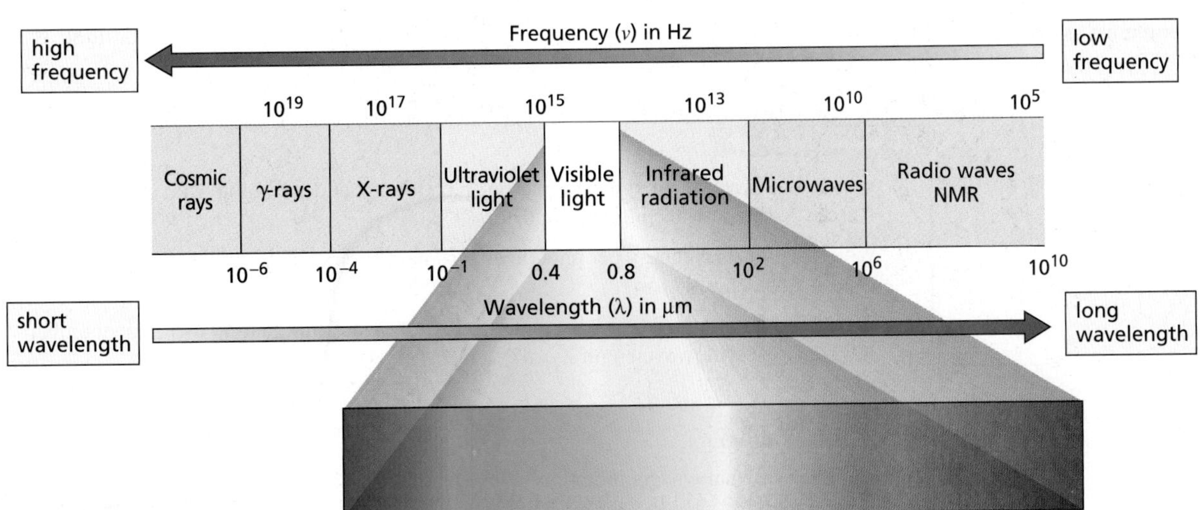

▲ **Figure 13.11**
The electromagnetic spectrum.

The relationship between the energy (*E*) of a photon and the frequency (or the wavelength) of the electromagnetic radiation is described by the equation

$$E = h\nu = \frac{hc}{\lambda}$$

where *h* is the proportionality constant called *Planck's constant*, named after the German physicist who discovered the relationship (Section 3.7). The electromagnetic spectrum is made up of the following components:

- *Cosmic rays*, which consist of radiation discharged by the sun, have the highest energy, the highest frequencies, and the shortest wavelengths.
- *γ-Rays* (gamma rays) are emitted from the nuclei of certain radioactive elements and, because of their high energy, can severely damage biological organisms.
- *X-rays,* somewhat lower in energy than γ-rays, are less harmful, except in high doses. Low-dose X-rays are used to examine the internal structure of organisms. The denser the tissue, the more it blocks X-rays.
- *Ultraviolet (UV) light* is responsible for sunburns, and repeated exposure can cause skin cancer by damaging DNA molecules in skin cells (Section 29.6).
- *Visible light* is the electromagnetic radiation we see.
- We feel *infrared radiation* as heat.
- We cook with *microwaves* and use them in radar.
- *Radio waves* have the lowest energy (lowest frequency). We use them for radio and television communication, digital imaging, remote controls, and wireless linkages for laptop computers. Radio waves are also used in NMR spectroscopy and in magnetic resonance imaging (MRI) (Chapter 14).

Wavenumber ($\tilde{\nu}$) is another way to describe the *frequency* of electromagnetic radiation, and the one most often used in infrared spectroscopy. It is the number of waves in one centimeter, so it has units of reciprocal centimeters (cm^{-1}). Scientists use wavenumbers in preference to wavelengths because, unlike wavelengths, wavenumbers are directly proportional to energy. The relationship between wavenumber (in cm^{-1}) and wavelength (in μm) is given by the equation

$$\tilde{\nu}(cm^{-1}) = \frac{10^4}{\lambda(\mu m)} \qquad (\text{because } 1\ \mu m = 10^{-4}\ cm)$$

High frequencies, large wavenumbers, and short wavelengths are associated with high energy.

So *high frequencies, large wavenumbers,* and *short wavelengths* are associated with *high energy.*

PROBLEM 14◆

a. Which is higher in energy per photon, electromagnetic radiation with wavenumber 100 cm^{-1} or with wavenumber 2000 cm^{-1}?

b. Which is higher in energy per photon, electromagnetic radiation with wavelength 9 μm or with wavelength 8 μm?

c. Which is higher in energy per photon, electromagnetic radiation with wavenumber 3000 cm^{-1} or with wavelength 2 μm?

PROBLEM 15◆

a. Radiation of what wavenumber has a wavelength of 4 μm?

b. Radiation of what wavelength has a wavenumber of 200 cm^{-1}?

13.7 Infrared Spectroscopy

Stretching and Bending Vibrations

The covalent bonds in molecules are constantly vibrating. So when we say that a bond between two atoms has a certain length, we are specifying an average because the bond behaves as if it were a vibrating spring connecting two atoms. A bond vibrates with both stretching and bending motions. A *stretch* is a vibration occurring along the line of the bond that changes the bond length. A *bend* is a vibration that does *not* occur along the line of the bond, but changes the bond angle. A diatomic molecule such as H—Cl can undergo only a **stretching vibration** since it has no bond angles.

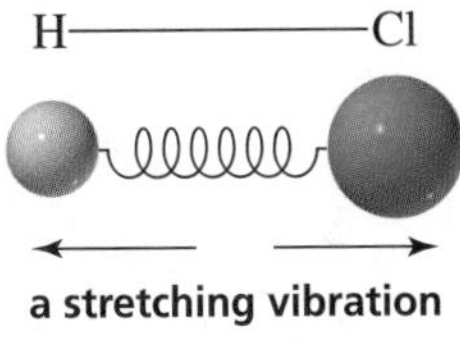

a stretching vibration

The vibrations of a molecule containing three or more atoms are more complex (Figure 13.12). Such molecules can experience symmetric and asymmetric stretches and bends, and their bending vibrations can be either in-plane or out-of-plane. **Bending vibrations** are often referred to by the descriptive terms *rock, scissor, wag,* and *twist.*

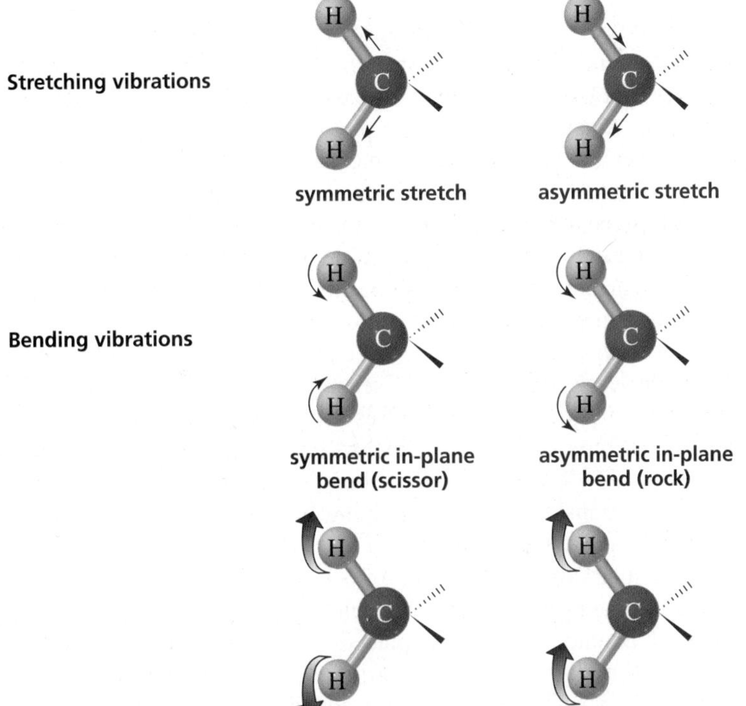

◀ **Figure 13.12**
Stretching and bending vibrations of bonds in organic molecules.

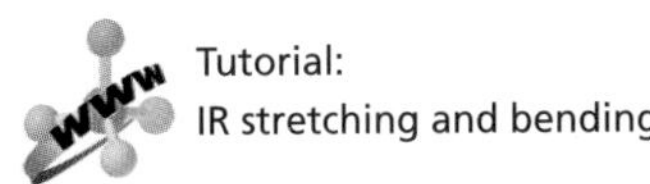
Tutorial:
IR stretching and bending

Each stretching and bending vibration of a bond in a molecule occurs with a characteristic frequency. When a compound is bombarded with radiation of a frequency that exactly matches the frequency of one of its vibrations, the molecule will absorb energy. This allows the bonds to stretch and bend a bit more. Thus, the absorption of energy increases the *amplitude* of the vibration, but does not change its *frequency*. By experimentally determining the wavenumbers of the energy absorbed by a particular compound, we can ascertain what kinds of bonds it has. For example, the stretching vibration of a C=O bond absorbs energy of wavenumber $\sim 1700\ cm^{-1}$, whereas the stretching vibration of an O—H bond absorbs energy of wavenumber $\sim 3450\ cm^{-1}$ (Figure 13.13).

C=O	O—H
~1700 cm^{-1}	~3450 cm^{-1}

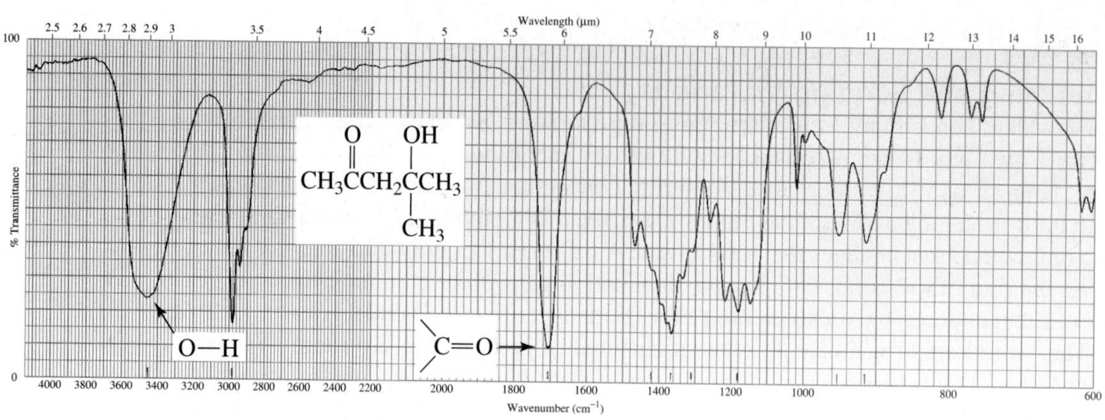

▲ **Figure 13.13**
The infrared spectrum of 4-hydroxy-4-methyl-2-pentanone.

Obtaining an Infrared Spectrum

The instrument used to obtain an *infrared spectrum* is called an *IR spectrometer*. An **infrared spectrum** is obtained by passing infrared radiation through a sample of the compound. A detector generates a plot of percent transmission of radiation versus the wavenumber (or wavelength) of the radiation transmitted (Figure 13.13). At 100% transmission, all the energy of the radiation passes through the molecule. Lower values of percent transmission mean that some of the energy is being absorbed by the compound. Each downward spike in the IR spectrum represents absorption of energy. The spikes are called **absorption bands**. Most chemists report the location of absorption bands using wavenumbers.

A newer type of IR spectrometer, called a Fourier transform IR (FT-IR) spectrometer, has several advantages. Its sensitivity is better because, instead of scanning through the frequences, it measures all frequencies simultaneously. With a conventional IR spectrometer, it can take 2 to 10 minutes to scan through all the frequencies. In contrast, FT-IR spectra can be taken in 1 to 2 seconds. The information is digitized and Fourier transformed by a computer to produce the FT-IR spectrum. The spectra shown in this text are FT-IR spectra.

An IR spectrum can be taken of a gas, a solid, or a liquid sample. Gases are expanded into an evacuated cell (a small container). Solids can be compressed with anhydrous KBr into a disc that is placed in the light beam. Solids can also be examined as mulls. A mull is prepared by grinding a few milligrams of the solid in a mortar. Then a drop or two of mineral oil is added and the grinding continued. In the case of liquid samples, a spectrum can be obtained of the neat (undiluted) liquid by placing a few drops of it between two optically polished plates of NaCl that are placed in the light beam. Alternatively, a small container (called a cell) with optically polished NaCl or AgCl windows is used to hold samples dissolved in solvents. Ionic substances without covalent bonds are used for discs, plates, and cells because they don't absorb IR radiation. (Glass, quartz, and plastics have IR-absorbing covalent bonds.)

When solutions are used, they must be in solvents that have few absorption bands in the region of interest. Commonly used solvents are CH_2Cl_2 and $CHCl_3$. In a double-beam spectrophotometer, the IR radiation is split into two beams—one that passes through the sample cell and the other that passes through a cell containing only the solvent. Any absorptions of the solvent are thus canceled out, so the absorption spectrum is that of the solute alone.

The Functional Group and Fingerprint Regions

Electromagnetic radiation with wavenumbers from 4000 to 600 cm^{-1} has just the right energy to correspond to the stretching and bending vibrations of organic molecules. Electromagnetic radiation in this energy range is known as **infrared radiation** because it is just below the "red region" of visible light. (*Infra* is Latin for "below.")

a.

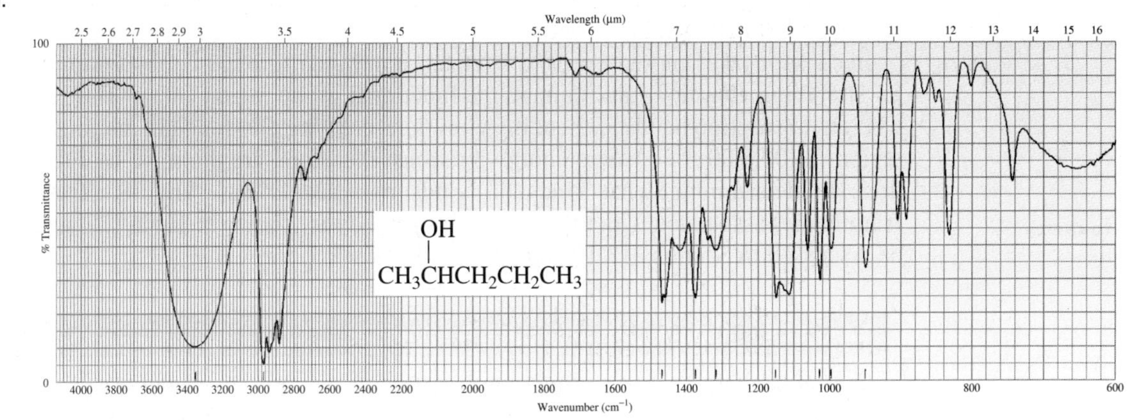

b.

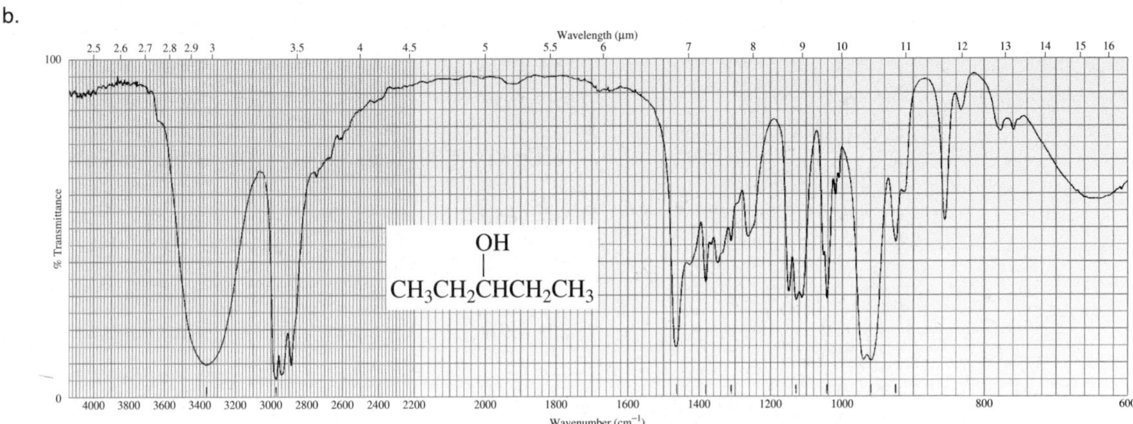

▲ **Figure 13.14**
The IR spectra of (a) 2-pentanol and (b) 3-pentanol. The functional group regions are very similar because the two compounds have the same functional group, but the fingerprint regions are unique for each compound.

An IR spectrum can be divided into two areas. The left-hand two-thirds of an IR spectrum ($4000–1400\ cm^{-1}$) is where most of the functional groups show absorption bands. This region is called the **functional group region**. The right-hand third ($1400–600\ cm^{-1}$) of the IR spectrum is called the **fingerprint region** because it is characteristic of the compound as a whole, just as a fingerprint is characteristic of an individual. Even if two different molecules have the same functional groups, their IR spectra will not be identical, since the functional groups are not in exactly the same environment; this difference is reflected in the pattern of absorption bands in the fingerprint regions. Each compound shows a unique pattern in this region. For example, 2-pentanol and 3-pentanol have the same functional groups, so they show similar absorption bands in the functional group region. Their fingerprint regions are different, however, because the compounds are different (Figure 13.14). Thus a compound can be positively identified by comparing its fingerprint region with the fingerprint region of the spectrum of a known sample of the compound.

13.8 Characteristic Infrared Absorption Bands

The stretching and bending vibrations of each bond in a molecule can give rise to an absorption band, so IR spectra can be quite complex. Organic chemists generally do not try to identify all the absorption bands in an IR spectrum. In this chapter, we will look at some characteristic bands so that you will be able to tell something about the structure of a compound that gives a particular IR spectrum. However, there is a lot more to infrared spectroscopy than we will be able to cover. You can find an extensive table of characteristic group frequencies in Appendix VI. When identifying an unknown compound, one often uses IR spectroscopy in conjunction with information obtained from other spectroscopic techniques. Many of the problems in this chapter and in Chapter 14 give you the opportunity to identify compounds, using information from two or more instrumental methods.

Because it takes more energy to stretch a bond than to bend it, absorption bands for stretching vibrations are found in the functional group region (4000–1400 cm^{-1}), whereas absorption bands for bending vibrations are typically found in the fingerprint region (1400–600 cm^{-1}). Stretching vibrations, therefore, are the most useful vibrations in determining what kinds of bonds a molecule has. The IR **stretching frequencies** associated with different types of bonds are shown in Table 13.4 and will be discussed in Sections 13.10 and 13.11.

It takes more energy to stretch a bond than to bend it.

Table 13.4 Important IR Stretching Frequencies

Type of bond	Wavenumber (cm^{-1})	Intensity
C≡N	2260–2220	medium
C≡C	2260–2100	medium to weak
C=C	1680–1600	medium
C=N	1650–1550	medium
	~1600 and ~1500–1430	strong to weak
C=O	1780–1650	strong
C—O	1250–1050	strong
C—N	1230–1020	medium
O—H (alcohol)	3650–3200	strong, broad
O—H (carboxylic acid)	3300–2500	strong, very broad
N—H	3500–3300	medium, broad
C—H	3300–2700	medium

13.9 The Intensity of Absorption Bands

The intensity of an absorption band depends on the size of the change in dipole moment associated with the vibration: The greater the change in dipole moment, the more intense the absorption. Recall that the dipole moment of a bond is equal to the magnitude of the charge on one of the bonded atoms, multiplied by the distance between the two charges (Section 1.3). When the bond stretches, the increasing distance between the atoms increases the dipole moment. The stretching vibration of an O—H bond will be associated with a greater change in dipole moment than that of an N—H bond because the O—H bond is more polar. Consequently, the stretching vibration of the O—H bond will be more intense. Likewise, the stretching vibration of an N—H bond is more intense than that of a C—H bond because the N—H bond is more polar.

The greater the change in dipole moment, the more intense the absorption.

relative bond polarities
relative intensities of IR absorption

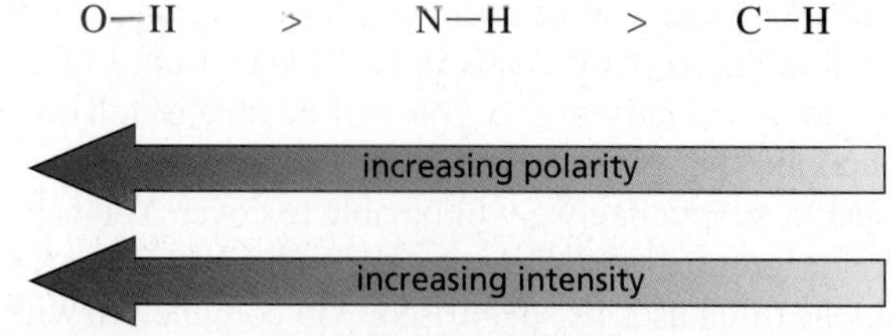

The intensity of an absorption band also depends on the number of bonds responsible for the absorption. For example, the absorption band for the C—H stretch will be more intense for a compound such as octyl iodide, which has 17 C—H bonds, than for methyl iodide, which has only three C—H bonds. The concentration of the sample used to obtain an IR spectrum also affects the intensity of the absorption bands. Concentrated samples have greater numbers of absorbing molecules and, therefore, more intense absorption bands. In the chemical literature, you will find intensities referred to as strong (s), medium (m), weak (w), broad, and sharp.

PROBLEM 16◆

Which would be expected to be more intense, the stretching vibration of a C=O bond or the stretching vibration of a C=C bond?

13.10 The Position of Absorption Bands

Hooke's Law

The amount of energy required to stretch a bond depends on the *strength* of the bond and the *masses* of the bonded atoms. The stronger the bond, the greater the energy required to stretch it, because a stronger bond corresponds to a tighter spring. The frequency of the vibration is inversely related to the mass of the atoms attached to the spring, so heavier atoms vibrate at lower frequencies.

The approximate wavenumber of an absorption can be calculated from the following equation derived from **Hooke's law**, which describes the motion of a vibrating spring:

$$\widetilde{\nu} = \frac{1}{2\pi c}\left[\frac{f(m_1 + m_2)}{m_1 m_2}\right]^{1/2} \qquad c = \text{the speed of light}$$

The equation relates the wavenumber of the stretching vibration ($\widetilde{\nu}$) to the force constant of the bond (f) and the masses of the atoms (in grams) joined by the bond (m_1 and m_2). The force constant is a measure of the strength of the bond. The equation shows that *stronger bonds* and *lighter atoms* give rise to higher frequencies.

Lighter atoms show absorption bands at larger wavenumbers.

C—H ~3000 cm^{-1}
C—D ~2200 cm^{-1}
C—O ~1100 cm^{-1}
C—Cl ~700 cm^{-1}

Stronger bonds show absorption bands at larger wavenumbers.

C≡N ~2200 cm^{-1}
C=N ~1600 cm^{-1}
C—N ~1100 cm^{-1}

THE ORIGINATOR OF HOOKE'S LAW

Robert Hooke (1635–1703) was born on the Isle of Wight off the southern coast of England. A brilliant scientist, he contributed to almost every field of science. He was the first to suggest that light had wave-like properties. He discovered that Gamma Arietis is a double star, and he also discovered Jupiter's Great Red Spot. In a lecture published posthumously, he suggested that earthquakes are caused by the cooling and contracting of the Earth. He examined cork under a microscope and coined the term "cell" to describe what he saw. He wrote about evolutionary development based on his studies of microscopic fossils, and his studies of insects were highly regarded as well. Hooke also invented the balance spring for watches and the universal joint currently used in cars.

Robert Hooke's drawing of a "blue fly" appeared in *Micrographia*, the first book on microscopy, published by Hooke in 1665.

Effect of Bond Order

Bond order affects bond strength, so bond order affects the position of absorption bands. A C≡C bond is stronger than a C=C bond, so a C≡C bond stretches at a higher frequency (~2100 cm^{-1}) than does a C=C bond (~1650 cm^{-1}). C—C bonds show stretching vibrations in the region from 1200 to 800 cm^{-1}, but these vibrations are weak and of little value in identifying compounds. Similarly, a C=O bond stretches at a higher frequency (~1700 cm^{-1}) than does a C—O bond (~1100 cm^{-1}), and a C≡N bond stretches at a higher frequency (~2200 cm^{-1}) than does a C=N bond (~1600 cm^{-1}), which in turn stretches at a higher frequency than does a C—N bond (~1100 cm^{-1}) (Table 13.4).

PROBLEM 17◆

a. Which will occur at a larger wavenumber?
   1. a C≡C stretch or a C=C stretch
   2. a C—H stretch or a C—H bend
   3. a C—N stretch or a C=N stretch

b. Assuming that the force constants are the same, which will occur at a larger wavenumber?
   1. a C—O stretch or a C—Cl stretch
   2. a C—O stretch or a C—C stretch

Resonance and Inductive Electronic Effects

Table 13.4 shows a range of wavenumbers for each stretch because the exact position of the absorption band depends on other structural features of the molecule, such as electron delocalization, the electronic effect of neighboring substituents, and hydrogen bonding. Important details about the structure of a compound can be revealed by the exact position of the absorption band.

For example, the IR spectrum in Figure 13.15 shows that the carbonyl group (C=O) of 2-pentanone absorbs at 1720 cm^{-1}, whereas the IR spectrum in Figure 13.16 shows that the carbonyl group of 2-cyclohexenone absorbs at a lower frequency (1680 cm^{-1}). 2-Cyclohexenone absorbs at a lower frequency because the carbonyl group has less double-bond character due to electron delocalization. Because a single bond is weaker than a double bond, a carbonyl group with significant single-bond character will stretch at a lower frequency than will one with little or no single-bond character.

$CH_3\overset{O}{\overset{\|}{C}}CH_2CH_2CH_3$

2-pentanone

C=O at 1720cm^{-1}

O ⟷ O⁻ +

2-cyclohexenone

C=O at 1680cm^{-1}

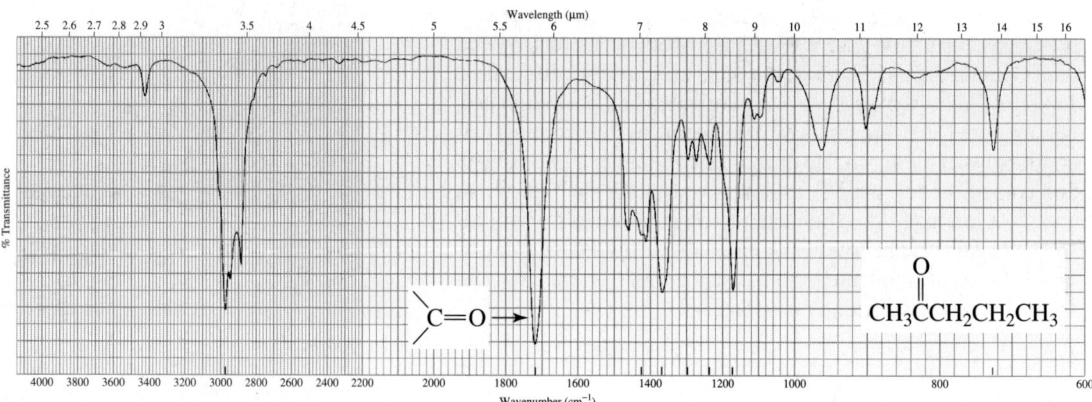

▲ **Figure 13.15**
The IR spectrum of 2-pentanone. The intense absorption band at ~1720 indicates a C=O bond.

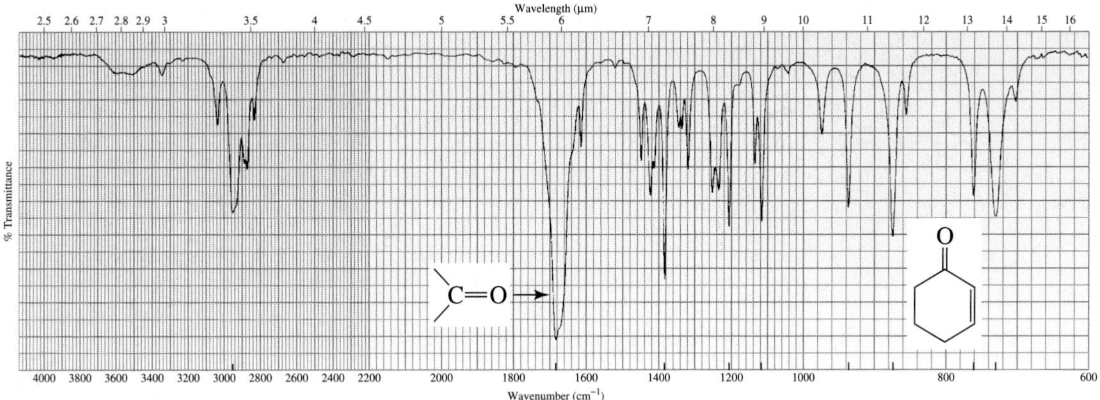

▲ **Figure 13.16**
The IR spectrum of 2-cyclohexenone. Electron delocalization gives its carbonyl group less double-bond character, so it absorbs at a lower frequency (~1680 cm^{-1}) than does a carbonyl group with localized electrons (~1720 cm^{-1}).

Putting an atom other than carbon next to the carbonyl group also causes the position of the carbonyl absorption band to shift. Whether it shifts to a lower or to a higher frequency depends on whether the predominant effect of the atom is to donate electrons by resonance or to withdraw electrons inductively.

$$\mathrm{R{-}C({=}O){-}\ddot{Z}} \longleftrightarrow \mathrm{R{-}C(O^-){=}Z^+} \qquad \mathrm{R{-}C({=}O){-}\ddot{Z}}$$

resonance electron donation — inductive electron withdrawal

The predominant effect of the nitrogen of an amide (Figure 13.17) is electron donation by resonance. Therefore, the carbonyl group of an amide has less double-bond character than does the carbonyl group of a ketone, so it is weaker and stretches more easily (1660 cm^{-1}). In contrast, the predominant effect of the oxygen of an ester is inductive electron withdrawal, so the resonance contributor with the C—O single bond contributes less to the hybrid. The carbonyl group of an ester, therefore, has more double-bond character than does the carbonyl group of a ketone, so the former is stronger and harder to stretch (1740 cm^{-1}) (Figure 13.18).

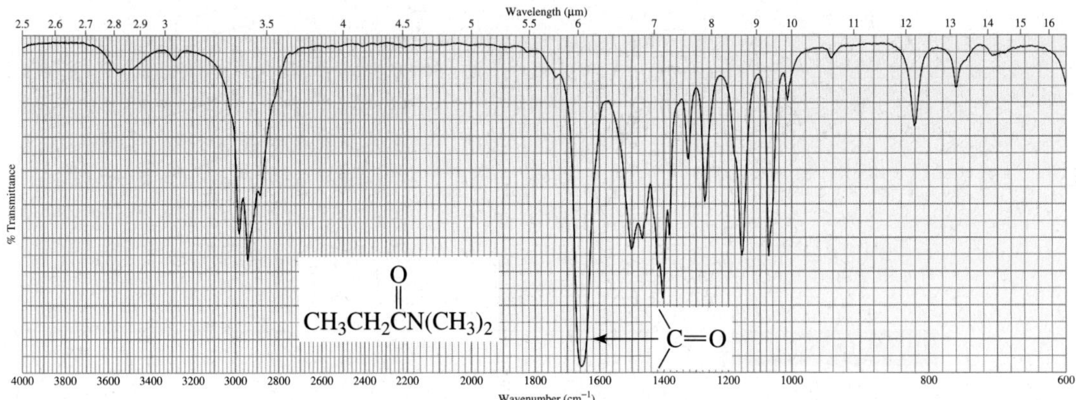

▲ **Figure 13.17**
The IR spectrum of *N*,*N*-dimethylpropanamide. The carbonyl group of an amide has less double-bond character than does the carbonyl group of a ketone, so the former stretches more easily (~1660 cm^{-1}).

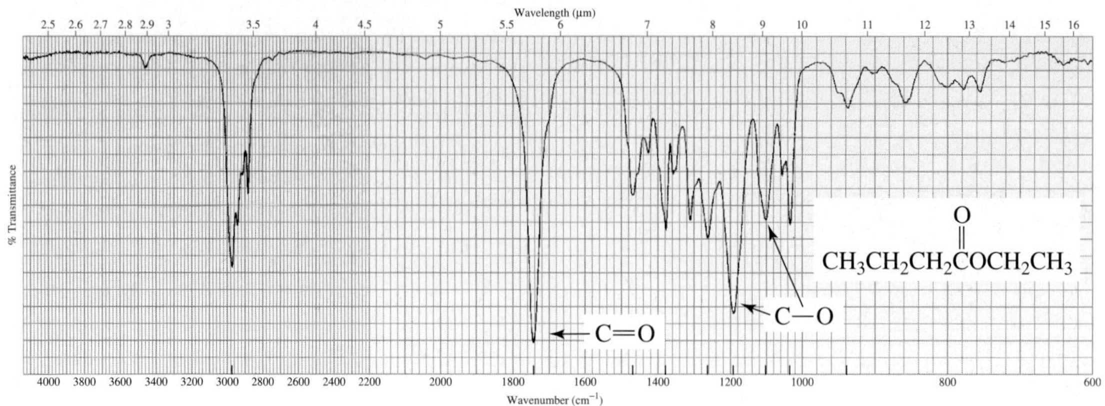

▲ **Figure 13.18**
The IR spectrum of ethyl butanoate. The electron-withdrawing oxygen atom makes the carbonyl group of an ester harder to stretch (~1740 cm^{-1}) than the carbonyl group of a ketone (~1720 cm^{-1}).

A C—O bond shows a stretch between 1250 and 1050 cm^{-1}. If the C—O bond is in an alcohol (Figure 13.19) or an ether, the stretch will occur toward the lower end of the range. If, however, the C—O bond is in a carboxylic acid (Figure 13.20), the stretch will occur at the higher end of the range. The position of the C—O absorption varies because the C—O bond in an alcohol is a pure single bond, whereas the C—O bond in a carboxylic acid has partial double-bond character that is due to resonance electron donation. Esters show C—O stretches at both ends of the range (Figure 13.18) because esters have two C—O single bonds—one that is a pure single bond and one that has partial double-bond character.

$CH_3CH_2—OH$
~1050 cm^{-1}

$CH_3CH_2—O—CH_2CH_3$
~1050 cm^{-1}

$CH_3\overset{O}{\overset{\|}{C}}\ddot{O}H \longleftrightarrow CH_3\overset{O^-}{\overset{|}{C}}{=}\overset{+}{\ddot{O}}H$
~1250 cm^{-1}

$CH_3\overset{O}{\overset{\|}{C}}\ddot{O}—CH_3 \longleftrightarrow CH_3\overset{O^-}{\overset{|}{C}}{=}\overset{+}{\ddot{O}}—CH_3$
~1250 cm^{-1} and ~1050 cm^{-1}

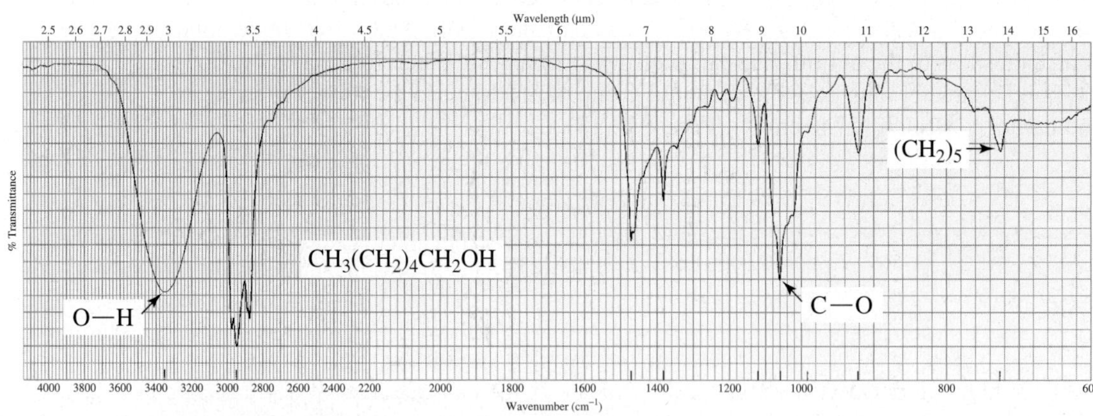

▲ **Figure 13.19**
The IR spectrum of 1-hexanol.

PROBLEM 18◆

Which will occur at a larger wavenumber?

a. the C—N stretch of an amine or the C—N stretch of an amide
b. the C—O stretch of phenol or the C—O stretch of cyclohexanol
c. the C=O stretch of a ketone or the C=O stretch of an amide
d. the stretch or the bend of the C—O bond in ethanol

PROBLEM 19◆

Which would show an absorption band at a larger wavenumber: a carbonyl group bonded to an sp^3 hybridized carbon or a carbonyl group bonded to an sp^2 hybridized carbon?

PROBLEM 20◆

List the following compounds in order of decreasing wavenumber of the C=O absorption band:

a. O O NH

b. O O O

c. $CH_3—\overset{O}{\overset{\|}{C}}—CH_3$ $\quad H—\overset{O}{\overset{\|}{C}}—H$ $\quad CH_3—\overset{O}{\overset{\|}{C}}—H$

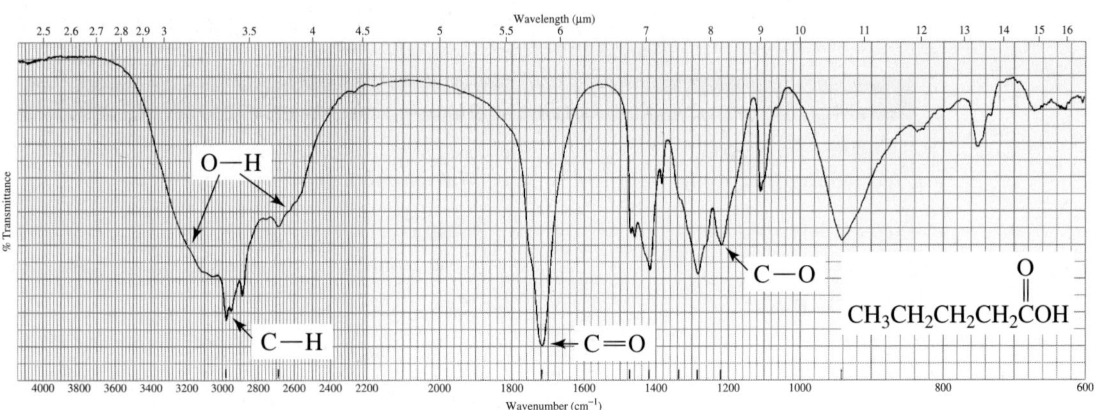

▲ **Figure 13.20**
The IR spectrum of pentanoic acid.

O—H Absorption Bands

O—H absorption bands are easy to detect. Polar O—H bonds show intense absorption bands and the bands are quite broad (Figures 13.19 and 13.20). The position and the breadth of an O—H absorption band depend on the concentration of the solution. The more concentrated the solution, the more likely it is for the OH-containing molecules to form intermolecular hydrogen bonds. It is easier to stretch an O—H bond if it is hydrogen bonded, because the hydrogen is attracted to the oxygen of a neighboring molecule. Therefore, the O—H stretch of a concentrated (hydrogen-bonded) solution of an alcohol occurs at 3550 to 3200 cm^{-1}, whereas the O—H stretch of a dilute solution (with little or no hydrogen bonding) occurs at 3650 to 3590 cm^{-1}. Hydrogen-bonded OH groups also have broader absorption bands because the hydrogen bonds vary in strength (Section 2.9). The absorption bands of non-hydrogen–bonded OH groups are sharper.

hydrogen bond

R—O—H------O(H)—R
concentrated solution
3550–3200 cm^{-1}

R—O—H
dilute solution
3650–3590 cm^{-1}

PROBLEM 21◆

Which will show an O—H stretch at a higher wavenumber, ethanol dissolved in carbon disulfide or an undiluted sample of ethanol?

13.11 C—H Absorption Bands

Stretching Vibrations

The strength of a C—H bond depends on the hybridization of the carbon: The greater the *s* character of the carbon, the stronger the bond it forms (Section 1.14). Therefore, a C—H bond is stronger when the carbon is *sp* hybridized than when it is sp^2 hybridized, which in turn is stronger than when the carbon is sp^3 hybridized. More energy is needed to stretch a stronger bond, and this is reflected in the C—H stretch absorption bands, which occur at ~3300 cm^{-1} if the carbon is *sp* hybridized, at ~3100 cm^{-1} if the carbon is sp^2 hybridized, and at ~2900 cm^{-1} if the carbon is sp^3 hybridized (Table 13.5).

Tutorial: IR spectra

A useful step in the analysis of a spectrum entails looking at the absorption bands in the vicinity of 3000 cm^{-1}. Figures 13.21, 13.22, and 13.23 show the IR spectra for methylcyclohexane, cyclohexene, and ethylbenzene, respectively. The only absorption band in the vicinity of 3000 cm^{-1} in Figure 13.21 is slightly to the right of that value. This tells us that the compound has hydrogens bonded to sp^3 carbons, but none bonded to sp^2 or *sp* carbons. Each of the spectra in Figures 13.22 and 13.23 shows absorption

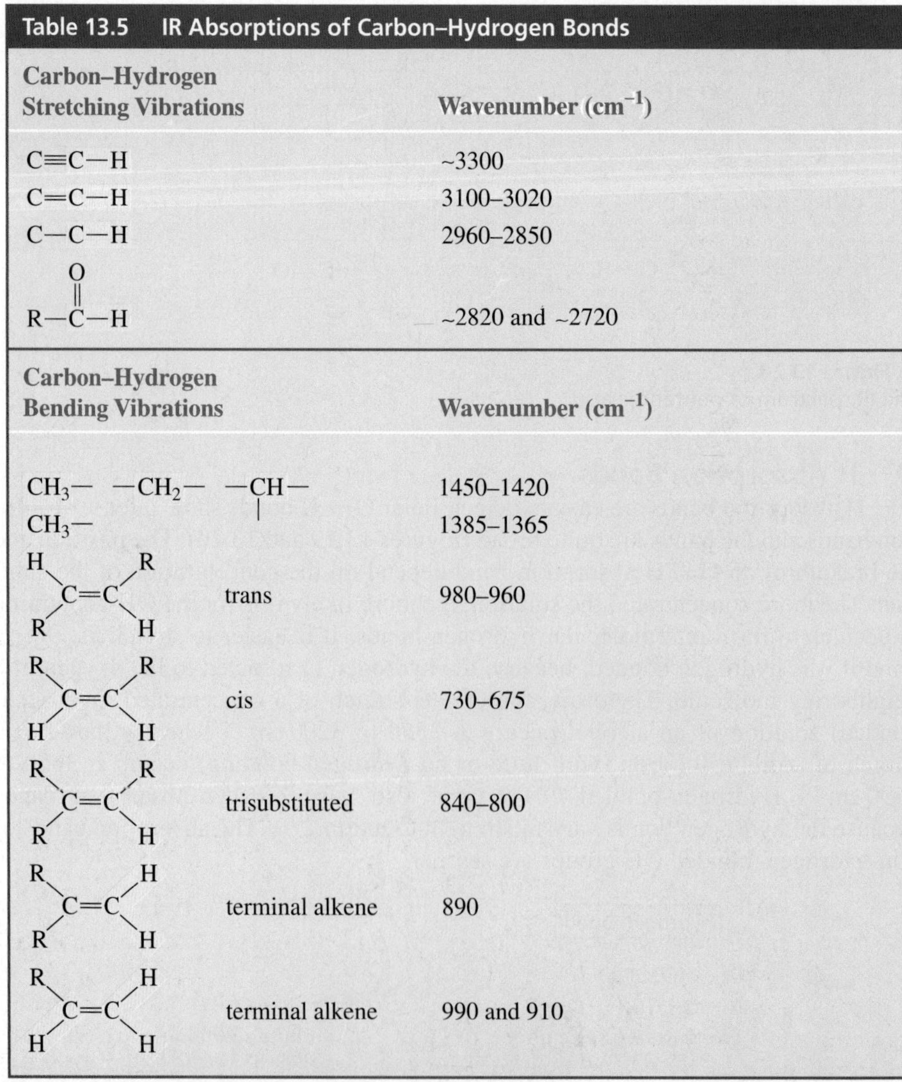

Table 13.5 IR Absorptions of Carbon–Hydrogen Bonds

Carbon–Hydrogen Stretching Vibrations		Wavenumber (cm^{-1})
C≡C—H		~3300
C=C—H		3100–3020
C—C—H		2960–2850
R—C(=O)—H		~2820 and ~2720
Carbon–Hydrogen Bending Vibrations		**Wavenumber (cm^{-1})**
CH_3— —CH_2— —CH—		1450–1420
CH_3—		1385–1365
(H)(R)C=C(R)(H)	trans	980–960
(R)(H)C=C(R)(H)	cis	730–675
(R)(R)C=C(R)(H)	trisubstituted	840–800
(R)(R)C=C(H)(H)	terminal alkene	890
(R)(H)C=C(H)(H)	terminal alkene	990 and 910

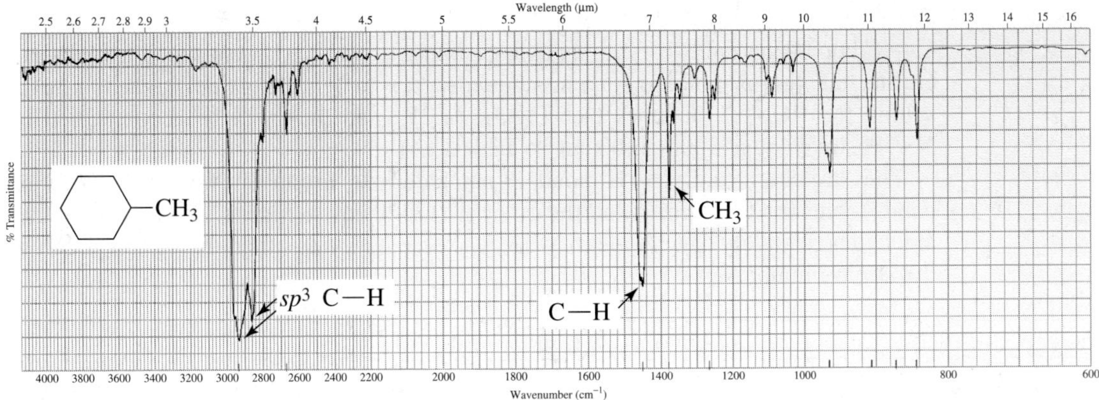

▲ **Figure 13.21**
The IR spectrum of methylcyclohexane.

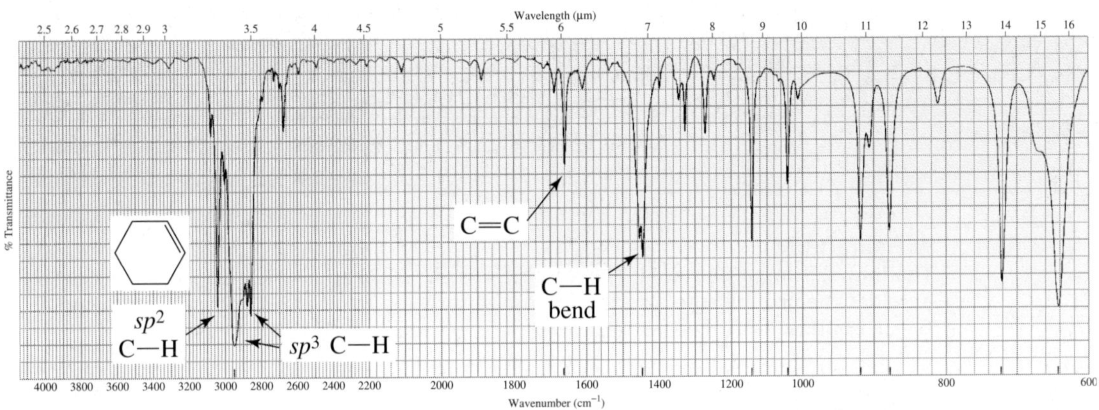

▲ **Figure 13.22**
The IR spectrum of cyclohexene.

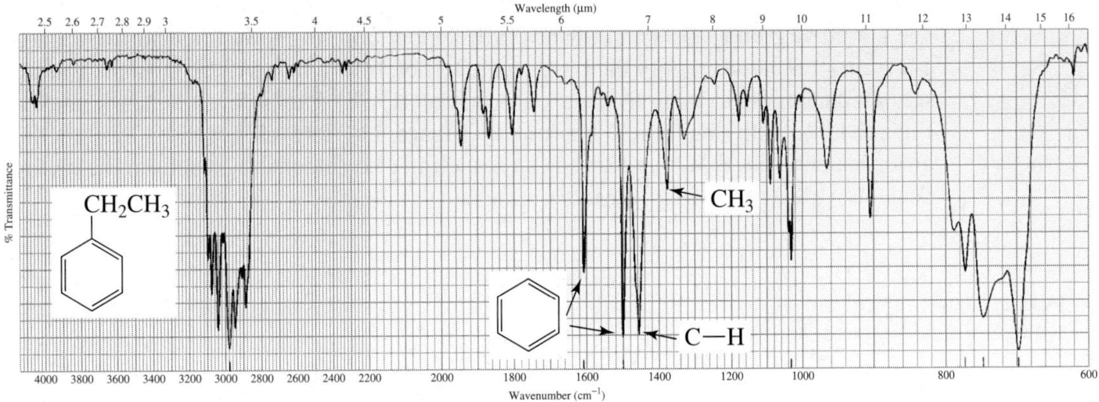

▲ **Figure 13.23**
The IR spectrum of ethylbenzene.

bands slightly to the left and slightly to the right of 3000 cm^{-1}, indicating that the compounds that produced those spectra contain hydrogens bonded to sp^2 and sp^3 carbons.

Once we know that a compound has hydrogens bonded to sp^2 carbons, we need to determine whether those carbons are the sp^2 carbons of an alkene or of a benzene ring. A benzene ring is indicated by sharp absorption bands at ~1600 cm^{-1} and 1500–1430 cm^{-1}, whereas an alkene is indicated by a band at ~1600 cm^{-1} only (Table 13.4). The compound with the spectrum shown in Figure 13.22 is, therefore, an alkene, while that shown in Figure 13.23 has a benzene ring. Be aware that N—H bending vibrations also occur at ~1600 cm^{-1}, so absorption at that wavelength does not always indicate a C=C bond. However, absorption bands resulting from N—H bends tend to be broader (due to hydrogen bonding) and more intense (due to being more polar) than those resulting from C=C stretches (see Figure 13.25), and they will be accompanied by N—H stretching at 3500–3300 cm^{-1} (Table 13.4).

The stretch of the C—H bond in an aldehyde group shows two absorption bands—one at ~2820 cm^{-1} and the other at ~2720 cm^{-1} (Figure 13.24). This makes aldehydes relatively easy to identify because essentially no other absorption occurs at these wavenumbers.

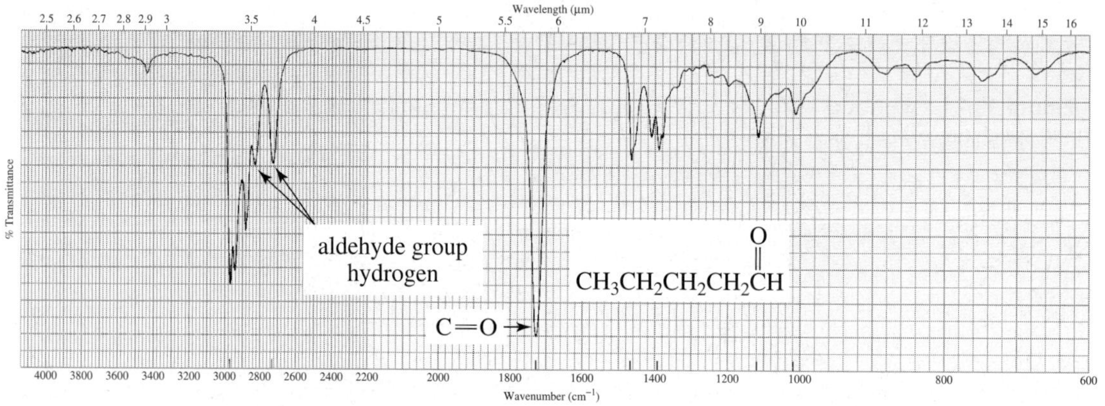

▲ **Figure 13.24**
The IR spectrum of pentanal. The absorptions at ~2820 and ~2720 cm^{-1} readily identify an aldehyde group. Note also the intense absorption band at ~1730 cm^{-1} indicating a C=O bond.

Bending Vibrations

If a compound has sp^3 carbons, a look at ~1400 cm^{-1} will tell you whether the compound has a methyl group. All hydrogens bonded to sp^3 hybridized carbons show a C—H bending vibration slightly to the *left* of 1400 cm^{-1}. Only methyl groups show a C—H bending vibration slightly to the *right* of 1400 cm^{-1}. So if a compound has a methyl group, absorption bands will appear *both* to the left and to the right of 1400 cm^{-1}; otherwise, only the band to the left of 1400 cm^{-1} will be present. You can see evidence of a methyl group in Figure 13.21 (methylcyclohexane) and in Figure 13.23 (ethylbenzene), but not in Figure 13.22 (cyclohexene). Two methyl groups attached to the same carbon can sometimes be detected by a split in the methyl peak at ~1380 cm^{-1} (Figure 13.25).

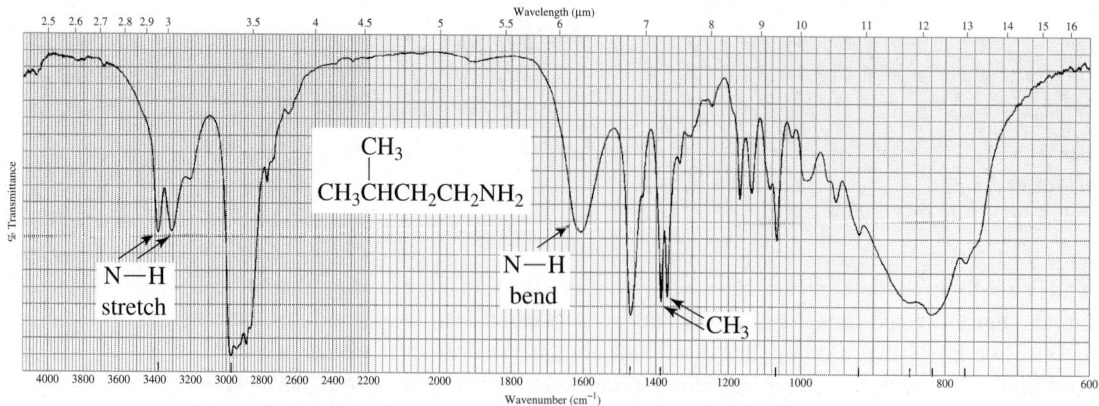

▲ **Figure 13.25**
The IR spectrum of isopentylamine. The double peak at ~1380 cm^{-1} indicates the presence of an isopropyl group. Two N—H bonds are indicated as well.

C—H bending vibrations for hydrogens bonded to sp^2 carbons give rise to absorption bands in the 1000–600 cm^{-1} region (Table 13.5). As the table shows, the frequency of the C—H bending vibrations of an alkene depends on the number of alkyl groups attached to the double bond and the configuration of the alkene. It is important to realize that these absorption bands can be shifted out of the characteristic regions if strongly electron-withdrawing or electron-donating substituents are close to the double bond (Section 13.10). Open-chain compounds with more than four adjacent methylene (CH_2) groups show a characteristic absorption band at 720 cm^{-1} that is due to in-phase rocking of the methylene groups (Figure 13.19).

13.12 The Shape of Absorption Bands

The shape of an absorption band can be helpful in identifying the compound responsible for an IR spectrum. For example, both O—H and N—H bonds stretch at wavenumbers above 3100 cm^{-1}, but the shapes of their stretches are distinctive. Notice the difference in the shape of these absorption bands in the IR spectra of 1-hexanol (Figure 13.19), pentanoic acid (Figure 13.20), and isopentylamine (Figure 13.25). An N—H absorption band (~3300 cm^{-1}) is narrower and less intense than an O—H absorption band (~3300 cm^{-1}), and the O—H absorption band of a carboxylic acid (~3300–2500 cm^{-1}) is broader than the O—H absorption band of an alcohol (Sections 13.9 and 13.10). Notice that two absorption bands are detectable in Figure 13.25 for the N—H stretch because there are two N—H bonds in the compound.

The position, intensity, and shape of an absorption band are helpful in identifying functional groups.

PROBLEM 22

a. Why is an O—H stretch more intense than an N—H stretch?
b. Why is the O—H stretch of a carboxylic acid broader than the O—H stretch of an alcohol?

13.13 Absence of Absorption Bands

The absence of an absorption band can be as useful as the presence of a band in identifying a compound by IR spectroscopy. For example, the spectrum in Figure 13.26 shows a strong absorption at ~1100 cm^{-1}, indicating the presence of a C—O bond. Clearly, the compound is not an alcohol because there is no absorption above 3100 cm^{-1}. Nor is it an ester or any other kind of carbonyl compound because there is no absorption at ~1700 cm^{-1}. The compound has no C≡C, C=C, C≡N, C=N,

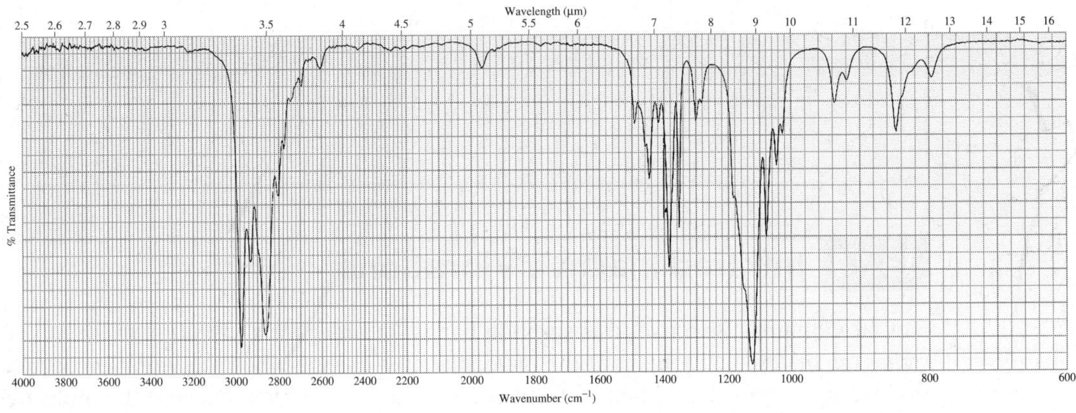

▲ **Figure 13.26**
The IR spectrum of diethyl ether.

or C—N bonds. We may deduce, then, that the compound is an ether. Its C—H absorption bands show that it has only sp^3 hybridized carbons and that it has a methyl group. We also know that the compound has fewer than four adjacent methylene groups, because there is no absorption at ~720 cm^{-1}. The compound is actually diethyl ether.

PROBLEM 23

How does one know that the absorption band at ~1100 cm^{-1} in Figure 13.26 is due to a C—O bond and not to a C—N bond?

PROBLEM 24◆

a. An oxygen-containing compound shows an absorption band at ~1700 cm^{-1} and no absorption bands at ~3300 cm^{-1}, ~2700 cm^{-1}, or ~1100 cm^{-1}. What class of compound is it?

b. A nitrogen-containing compound shows no absorption band at ~3400 cm^{-1} and no absorption bands between 1700 cm^{-1} and 1600 cm^{-1}. What class of compound is it?

PROBLEM 25

How could IR spectroscopy distinguish between the following?

a. a ketone and an aldehyde
b. a cyclic ketone and an open-chain ketone
c. benzene and cyclohexene
d. *cis*-2-hexene and *trans*-2-hexene
e. cyclohexene and cyclohexane
f. a primary amine and a tertiary amine

PROBLEM 26

For each of the following pairs of compounds, give one absorption band that could be used to distinguish between them:

a. $CH_3CH_2CH_2CH_3$ and $CH_3CH_2OCH_3$

b. $CH_3CH_2\overset{O}{\overset{\|}{C}}OCH_3$ and $CH_3CH_2\overset{O}{\overset{\|}{C}}OH$

c. $CH_3CH_2\overset{O}{\overset{\|}{C}}OH$ and $CH_3CH_2CH_2OH$

d. $CH_3CH_2C{\equiv}CCH_3$ and $CH_3CH_2C{\equiv}CH$

e.  and CH_3

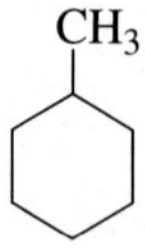

f. 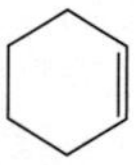 and

13.14 Infrared Inactive Vibrations

Not all vibrations give rise to absorption bands. In order for a vibration to absorb IR radiation, the dipole moment of the molecule must change when the vibration occurs. For example, the C=C bond in 1-butene has a dipole moment because the molecule is not symmetrical about this bond. Recall from Section 1.3 that the dipole moment is equal to the magnitude of the charge on the atoms multiplied by the distance between them. When the C=C bond stretches, the increasing distance between the atoms increases the dipole moment. Because the dipole moment changes when the bond stretches, an absorption band is observed for the C=C stretching vibration.

1-butene	2,3-dimethyl-2-butene	2,3-dimethyl-2-heptene
$(H)(H)C{=}C(H)(CH_2CH_3)$	$(H_3C)(H_3C)C{=}C(CH_3)(CH_3)$	$(H_3C)(H_3C)C{=}C(CH_2CH_2CH_2CH_3)(CH_3)$

2,3-Dimethyl-2-butene, in contrast, is a symmetrical molecule, so its C=C bond has no dipole moment. When the bond stretches, it still has no dipole moment. Since stretching is not accompanied by a change in dipole moment, no absorption band is observed. The vibration is *infrared inactive.* 2,3-Dimethyl-2-heptene experiences a very small change in dipole moment when its C=C bond stretches, so only an extremely weak absorption band (if any) will be detected for the stretching vibration of the bond.

PROBLEM 27◆

Which of the following compounds has a vibration that is infrared inactive: acetone, 1-butyne, 2-butyne, H_2, H_2O, Cl_2, ethene?

PROBLEM 28◆

The mass spectrum and infrared spectrum of an unknown compound are shown in Figures 13.27 and 13.28, respectively. Identify the compound.

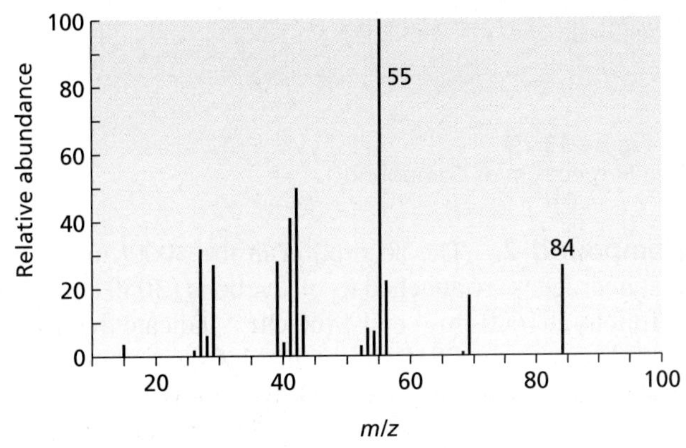

◀ **Figure 13.27**
The mass spectrum for Problem 28.

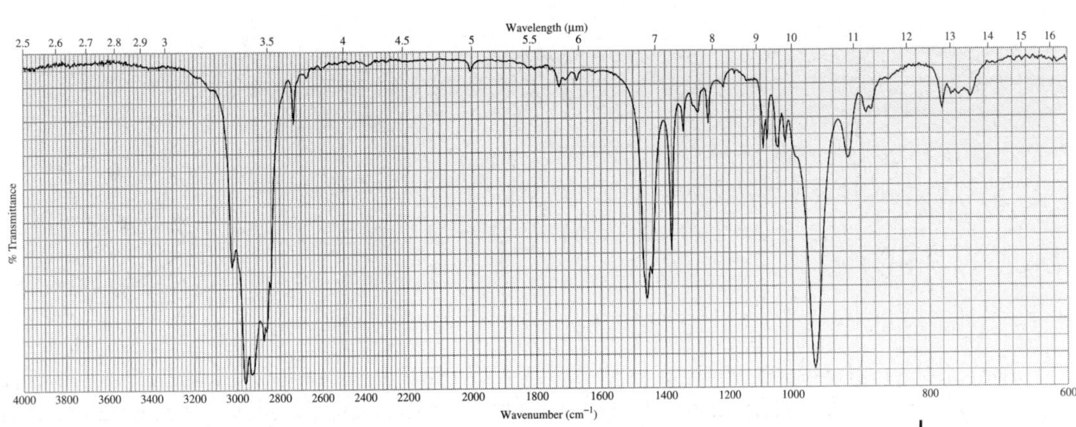

▲ **Figure 13.28**
The IR spectrum for Problem 28.

13.15 Identifying Infrared Spectra

We will now look at some IR spectra and see what we can determine about the structure of the compounds that give rise to the spectra. We might not be able to identify the compound precisely, but when we find out what it is, its structure should fit with our observations.

Compound 1. The absorptions in the 3000 cm^{-1} region in Figure 13.29 indicate that hydrogens are attached to both sp^2 carbons (3075 cm^{-1}) and sp^3 carbons (2950 cm^{-1}). Now let's see if the sp^2 carbons belong to an alkene or to a benzene ring. The absorption at 1650 cm^{-1} and the absorption at ~890 cm^{-1} (Table 13.5) suggest that the compound is a terminal alkene with two alkyl substituents at the 2-position. The lack of absorption at ~720 cm^{-1} indicates that the compound has fewer than four adjacent methylene groups. We are not surprised to find that the compound is 2-methyl-1-pentene.

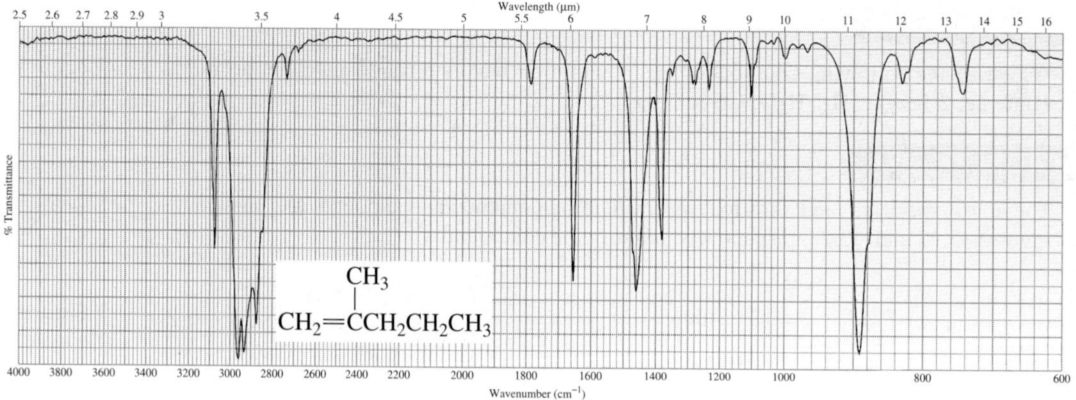

▲ **Figure 13.29**
The IR spectrum of Compound 1.

Compound 2. The absorption in the 3000 cm^{-1} region in Figure 13.30 indicates that hydrogens are attached to sp^2 carbons (3050 cm^{-1}) but not to sp^3 carbons. The absorptions at 1600 cm^{-1} and 1460 cm^{-1} indicate that the compound has a benzene ring. The absorptions at 2810 cm^{-1} and 2730 cm^{-1} show that the compound is an aldehyde. The absorption band for the carbonyl group (C=O) is lower (1700 cm^{-1}) than normal (1720 cm^{-1}), so the carbonyl group has partial single-bond character. Thus, it must be attached directly to the benzene ring. The compound is benzaldehyde.

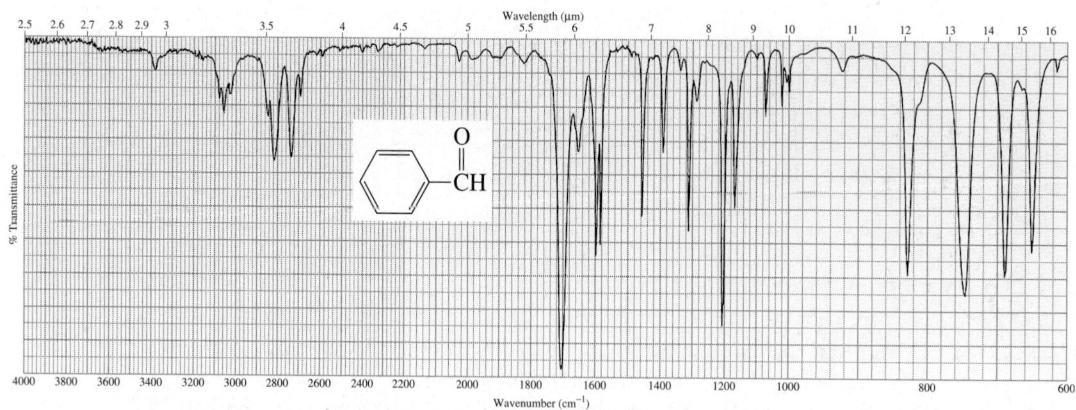

▲ **Figure 13.30**
The IR spectrum of Compound 2.

Compound 3. The absorptions in the 3000 cm^{-1} region in Figure 13.31 indicate that hydrogens are attached to sp^3 carbons (2950 cm^{-1}) but not to sp^2 carbons. The shape of the strong absorption band at 3300 cm^{-1} is characteristic of an O—H group. The absorption at ~2100 cm^{-1} indicates that the compound has a triple bond. The sharp absorption band at 3300 cm^{-1} indicates that the compound is a terminal alkyne. The compound is 2-propyn-1-ol.

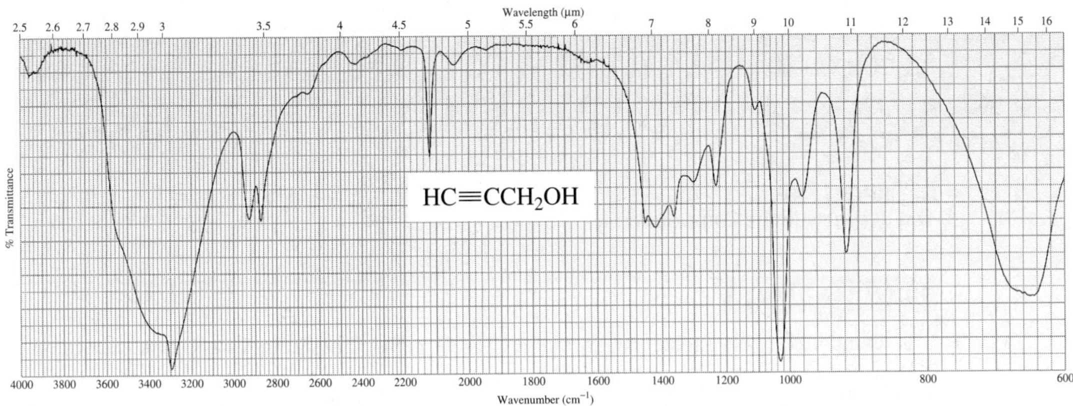

▲ **Figure 13.31**
The IR spectrum of Compound 3.

Compound 4. The absorption in the 3000 cm^{-1} region in Figure 13.32 indicates that hydrogens are attached to sp^3 carbons (2950 cm^{-1}). The relatively strong absorption band at 3300 cm^{-1} suggests that the compound has one N—H bond. The presence of the N—H bond is confirmed by the absorption at 1560 cm^{-1}. The C=O absorption at 1660 cm^{-1} indicates that the compound is an amide. The compound is *N*-methylacetamide.

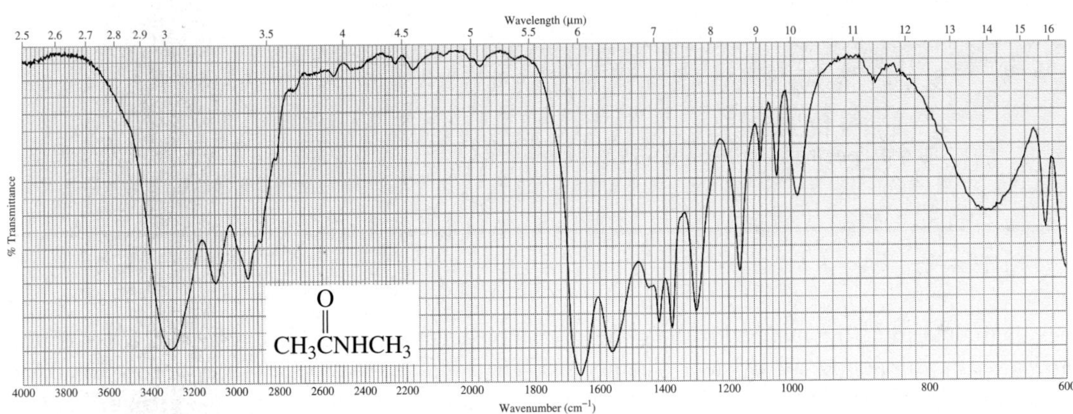

▲ **Figure 13.32**
The IR spectrum of Compound 4.

Compound 5. The absorptions in the 3000 cm^{-1} region in Figure 13.33 indicate that the compound has hydrogens attached to sp^2 carbons (>3000 cm^{-1}) and to sp^3 carbons (<3000 cm^{-1}). The absorptions at 1605 cm^{-1} and 1500 cm^{-1} indicate that the compound contains a benzene ring. The absorption at 1720 cm^{-1} for the carbonyl group indicates that the compound is a ketone and that the carbonyl group is not directly attached to the benzene ring. The absorption at ~1380 cm^{-1} indicates that the compound contains a methyl group. The compound is 1-phenyl-2-butanone.

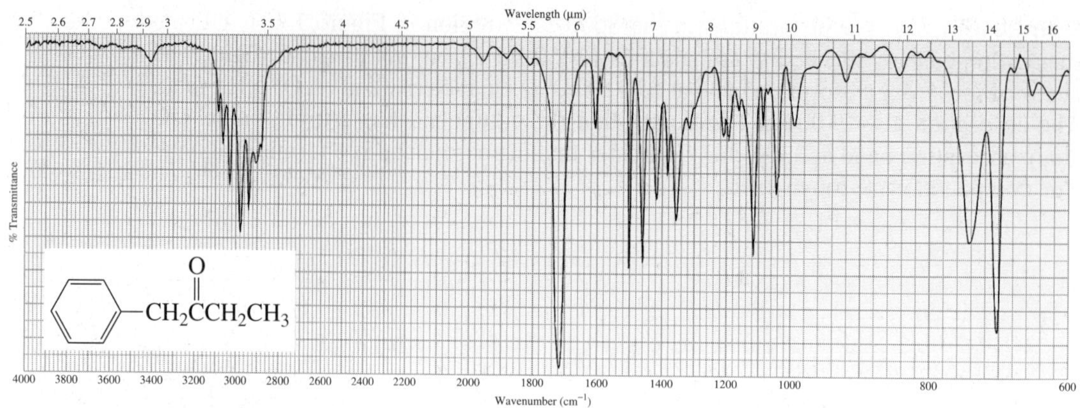

▲ **Figure 13.33**
The IR spectrum of Compound 5.

PROBLEM 29◆

A compound with molecular formula C_4H_6O gives the infrared spectrum shown in Figure 13.34. Identify the compound.

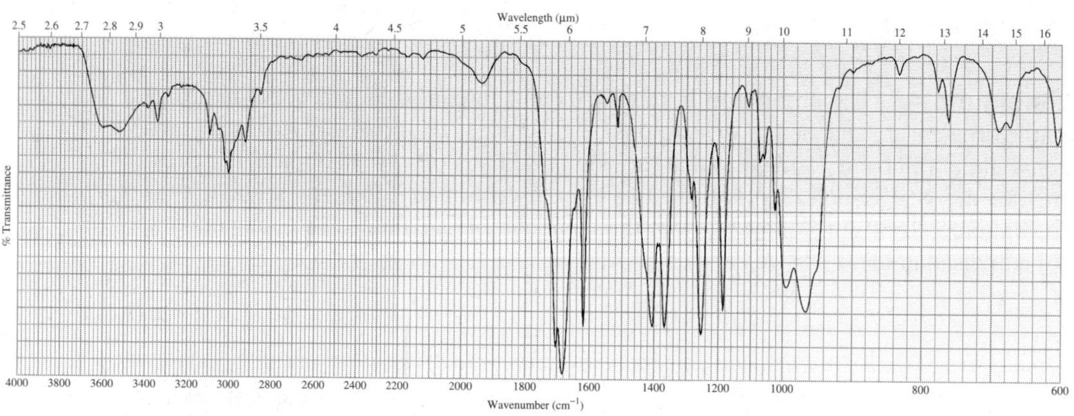

▲ **Figure 13.34**
The IR spectrum for Problem 29.

Summary

Mass spectrometry allows us to determine the *molecular mass* and the *molecular formula* of a compound, as well as certain structural features. In mass spectrometry, a small sample of the compound is vaporized and then ionized as a result of an electron's being removed from each molecule, producing a **molecular ion**—a radical cation. Many of the molecular ions break apart into cations, radicals, neutral molecules, and other radical cations. The bonds most likely to break are the weakest ones and those that result in the formation of the most stable products. The mass spectrometer records a **mass spectrum**—a graph of the relative abundance of each positively charged fragment, plotted against its m/z value.

The molecular ion (M) peak is due to the fragment that results when an electron is knocked out of a molecule; the m/z value of a molecular ion gives the molecular mass of the compound. The "nitrogen rule" states that if a compound has an odd-mass molecular ion, the compound contains an odd number of nitrogen atoms. Peaks with smaller m/z values—**fragment ion peaks**—represent positively charged fragments of the molecule. The **base peak** is the peak with the greatest intensity. High-resolution mass spectrometers determine the exact molecular mass, which allows a compound's molecular formula to be determined.

The M + 1 peak occurs because there are two naturally occurring isotopes of carbon. The number of carbon atoms in a compound can be calculated from the relative intensities of the M and M + 1 peaks. A large M + 2 peak is evidence of a compound containing either chlorine or bromine; if it is one-third the height of the M peak, the compound contains one chlorine atom; if the M and M + 2 peaks are about the same height, the compound contains one bromine atom.

Characteristic fragmentation patterns are associated with specific functional groups. Electron bombardment is most likely to dislodge a lone-pair electron. A bond between carbon and a more electronegative atom breaks *heterolytically*, with the electrons going to the more electronegative atom. A bond between carbon and an atom of similar electronegativity breaks homolytically; **α cleavage** occurs because the species that is formed is a resonance-stabilized cation.

Spectroscopy is the study of the interaction of matter and **electromagnetic radiation**. A continuum of different types of electromagnetic radiation constitutes the electromagnetic spectrum. High-energy radiation is associated with *high frequencies, large wavenumbers*, and *short wavelengths*.

Infrared spectroscopy identifies the kinds of functional groups in a compound. Bonds vibrate with stretching and bending motions. Each stretching and bending vibration occurs with a characteristic frequency. It takes more energy to stretch a bond than to bend it. When a compound is bombarded with radiation of a frequency that exactly matches the frequency of one of its vibrations, the molecule absorbs energy and exhibits an **absorption band**. The **functional group region** of an IR spectrum (4000–1400 cm^{-1}) is where most of the functional groups show absorption bands; the **fingerprint region** (1400–600 cm^{-1}) is characteristic of the compound as a whole.

The position, intensity, and shape of an absorption band help identify functional groups. The amount of energy required to stretch a bond depends on the *strength* of the bond: Stronger bonds show absortion bands at larger wavenumbers. Therefore, the frequency of the absorption depends on bond order, hybridization, electronic, and resonance effects. The frequency is inversely related to the *mass* of the atoms, so heavier atoms vibrate at lower frequencies. The intensity of an absorption band depends on the size of the change in dipole moment associated with the vibration and on the number of bonds responsible for the absorption. In order for a vibration to absorb IR radiation, the dipole moment of the molecule must change when the vibration occurs.

Key Terms

absorption band (p. 500)
base peak (p. 486)
bending vibration (p. 499)
α cleavage (p. 490)
electromagnetic radiation (p. 497)
fingerprint region (p. 501)
fragment ion peak (p. 485)
frequency (p. 497)
functional group region (p. 501)
Hooke's law (p. 503)
infrared radiation (p. 500)
infrared spectroscopy (p. 483)
infrared spectrum (p. 500)
mass spectrometry (p. 483)
mass spectrum (p. 485)
McLafferty rearrangement (p. 494)
molecular ion (p. 484)
nominal molecular mass (p. 485)
radical cation (p. 484)
spectroscopy (p. 497)
stretching frequency (p. 502)
stretching vibration (p. 499)
wavelength (p. 497)
wavenumber (p. 498)

Problems

30. Which peak would be more intense in the mass spectrum of the following compounds—the peak at $m/z = 57$ or the peak at $m/z = 71$?
    a. 3-methylpentane
    b. 2-methylpentane

31. List three factors that influence the intensity of an IR absorption band.

32. For each of the following pairs of compounds, identify one IR absorption band that could be used to distinguish between them:
    a. $CH_3CH_2COCH_3$ (C=O) and $CH_3CH_2CCH_3$ (C=O)
    b. methylcyclohexane and $CH_3CH_2CH_2CH_2CH_2CH_2CH_3$
    c. $CH_3CH_2CH{=}CH_2$ and $CH_3CH_2CH{=}C(CH_3)CH_3$
    d. cyclohexyl–CH_2CH_2OH and cyclohexyl–$CH(OH)CH_3$
    e. $CH_3COCH_2CH_3$ (C=O) and $CH_3CCH_2OCH_3$ (C=O)
    f. $CH_3CH_2CH{=}CHCH_3$ and $CH_3CH_2C{\equiv}CCH_3$

g. $CH_3CH_2\overset{O}{\overset{\|}{C}}H$ and $CH_3CH_2\overset{O}{\overset{\|}{C}}CH_3$

h. 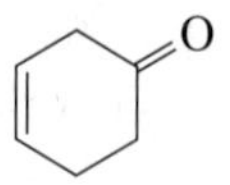 and

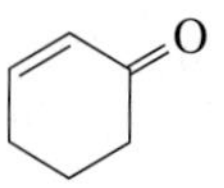

i. *cis*-2-butene and *trans*-2-butene

j. $CH_3CH_2CH_2OH$ and $CH_3CH_2OCH_3$

k. $CH_3CH_2\overset{O}{\overset{\|}{C}}NH_2$ and $CH_3CH_2\overset{O}{\overset{\|}{C}}OCH_3$

l. 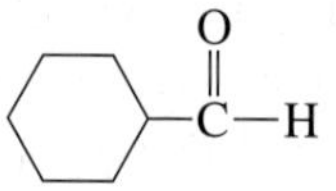 $\overset{O}{\overset{\|}{C}}—H$ and $\overset{O}{\overset{\|}{C}}—H$

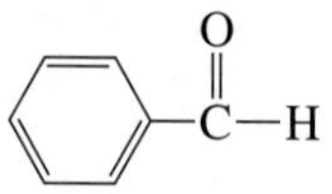

33. a. How could you determine by IR spectroscopy that the following reaction had occurred?

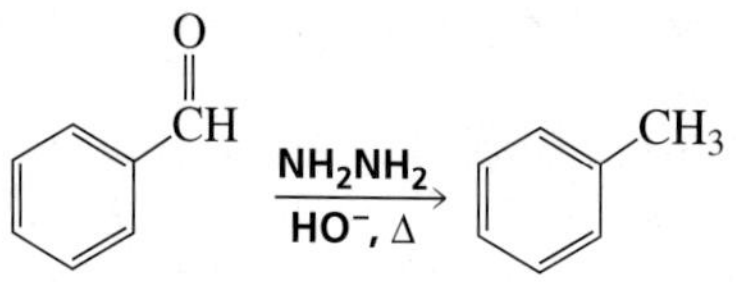

$\overset{O}{\overset{\|}{C}}H \xrightarrow[HO^-,\ \Delta]{NH_2NH_2} CH_3$

b. After purifying the product, how could you determine that all the NH_2NH_2 had been removed?

34. What identifying characteristics would be present in the mass spectrum of a compound containing two bromine atoms?

35. Assuming that the force constant is approximately the same for C—C, C—N, and C—O bonds, predict the relative positions of their stretching vibrations.

36. A mass spectrum shows significant peaks at m/z = 87, 115, 140, and 143. Which of the following compounds is responsible for that mass spectrum: 4,7-dimethyl-1-octanol, 2,6-dimethyl-4-octanol, or 2,2,4-trimethyl-4-heptanol?

37. How could IR spectroscopy distinguish between 1,5-hexadiene and 2,4-hexadiene?

38. A compound gives a mass spectrum with peaks at m/z = 77 (40%), 112 (100%), 114 (33%), and essentially no other peaks. Identify the compound.

39. What hydrocarbons will have a molecular ion peak at m/z = 112?

40. In the following boxes, list the types of bonds and the approximate wavenumber at which each type of bond is expected to show an IR absorption:

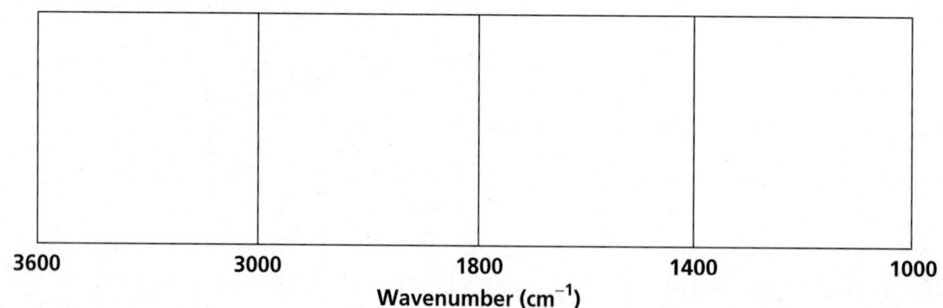

41. For each of the IR spectra in Figures 13.35, 13.36, and 13.37, four compounds are shown. In each case, indicate which of the four compounds is responsible for the spectrum.

a. $CH_3CH_2CH_2C{\equiv}CCH_3$ $CH_3CH_2CH_2CH_2OH$ $CH_3CH_2CH_2CH_2C{\equiv}CH$ $CH_3CH_2CH_2\overset{O}{\overset{\|}{C}}OH$

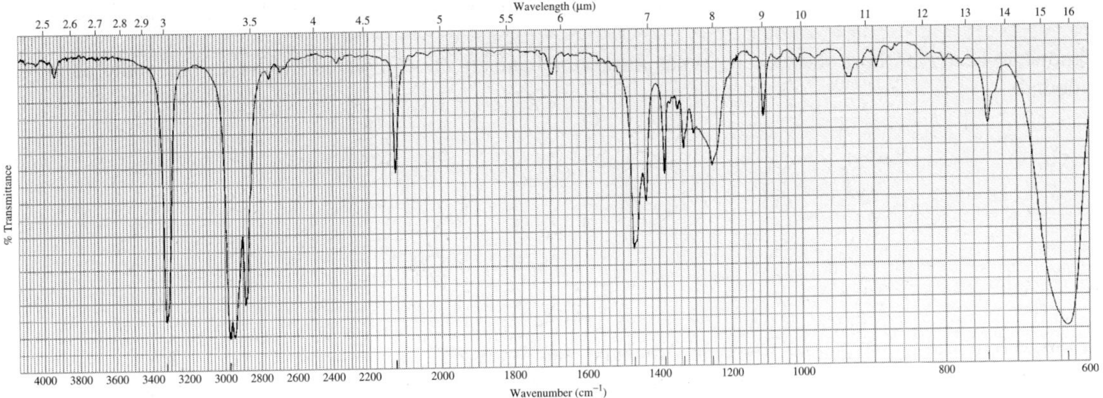

Figure 13.35
The IR spectrum for Problem 41a.

b. $\mathrm{CH_3CH_2\overset{O}{\overset{\|}{C}}OH}$ $\mathrm{CH_3CH_2\overset{O}{\overset{\|}{C}}OCH_2CH_3}$ $\mathrm{CH_3CH_2\overset{O}{\overset{\|}{C}}H}$ $\mathrm{CH_3CH_2\overset{O}{\overset{\|}{C}}CH_3}$

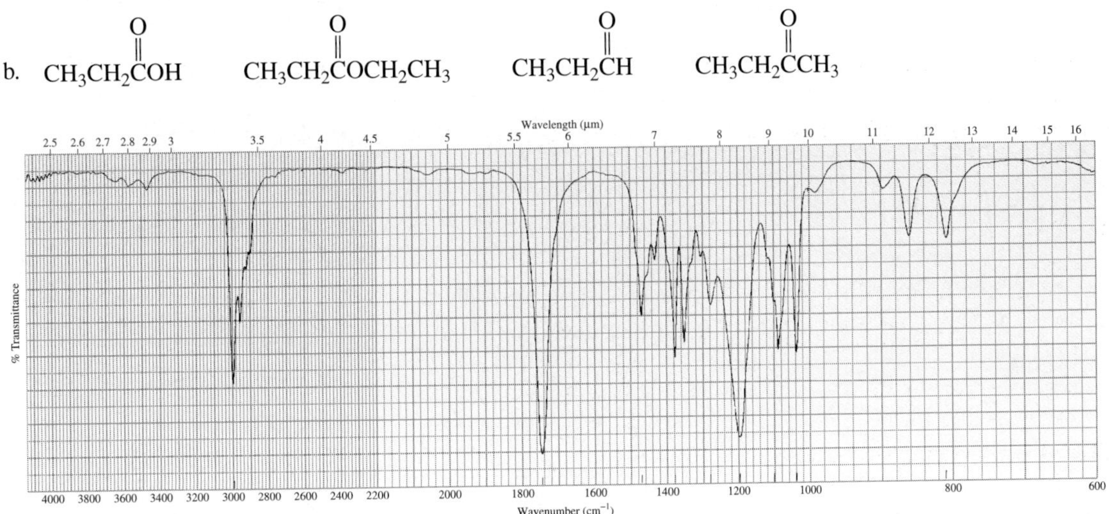

Figure 13.36
The IR spectrum for Problem 41b.

c. $\mathrm{C_6H_5{-}C(CH_3)_3}$ $\mathrm{C_6H_5{-}CH_2CH_2Br}$ $\mathrm{C_6H_5{-}CH{=}CH_2}$ $\mathrm{C_6H_5{-}CH_2OH}$

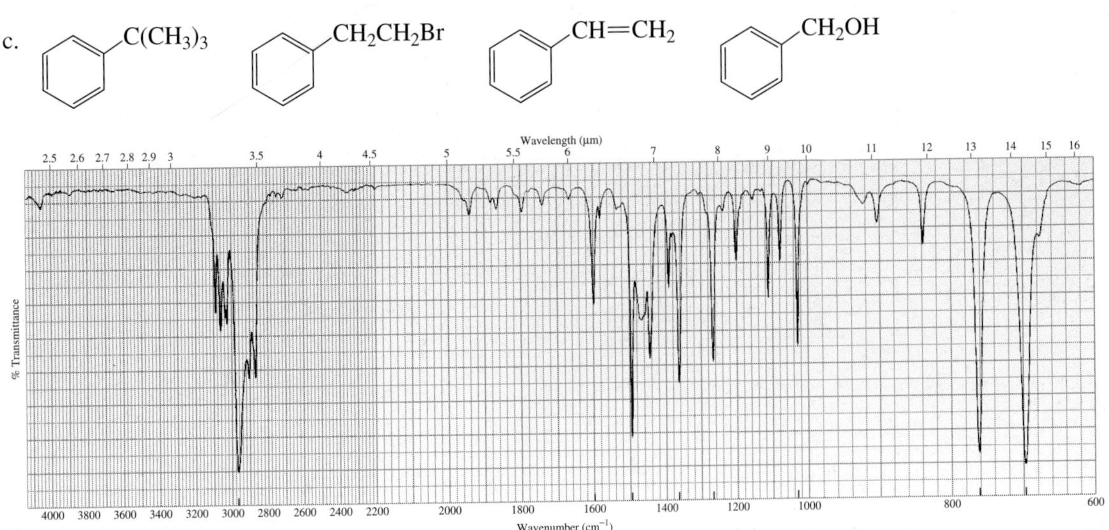

Figure 13.37
The IR spectrum for Problem 41c.

42. What peaks in their mass spectra would be used to distinguish between 4-methyl-2-pentanone and 2-methyl-3-pentanone?

43. A compound is known to be one of those shown here. What absorption bands in the compound's IR spectrum would allow you to identify the compound.

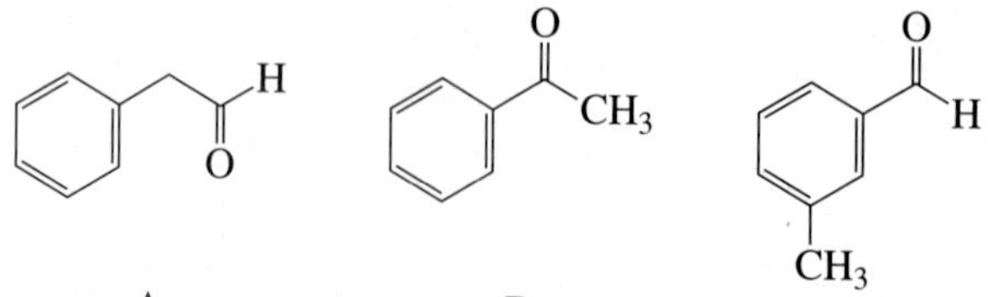

O H CH$_3$ H
CH$_3$
A B C

44. How could IR spectroscopy distinguish among 1-hexyne, 2-hexyne, and 3-hexyne?

45. For each of the IR spectra in Figures 13.38, 13.39, and 13.40, indicate which of the five given compounds is responsible for the spectrum.

a. $CH_3CH_2CH{=}CH_2$ $CH_3CH_2CH_2CH_2OH$ $CH_2{=}CHCH_2CH_2OH$ $CH_3CH_2CH_2OCH_3$ $CH_3CH_2CH_2\overset{O}{\overset{\|}{C}}OH$

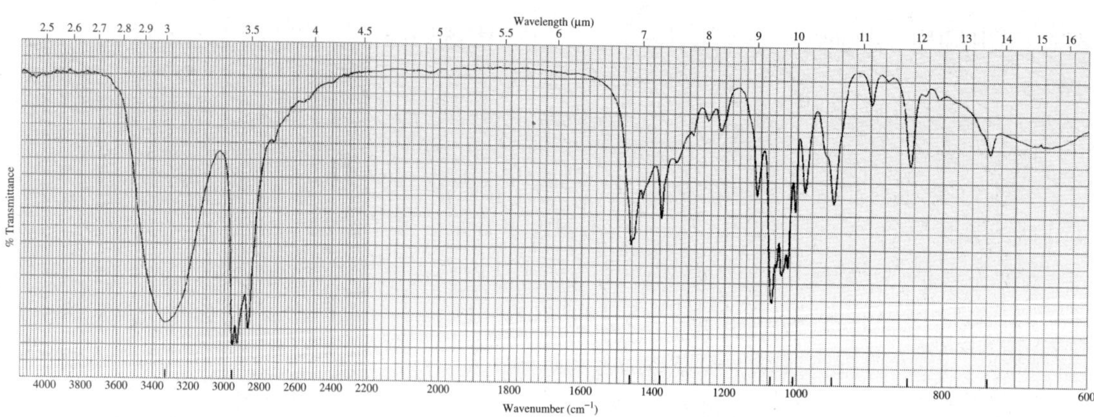

▲ **Figure 13.38**
The IR spectrum for Problem 45a.

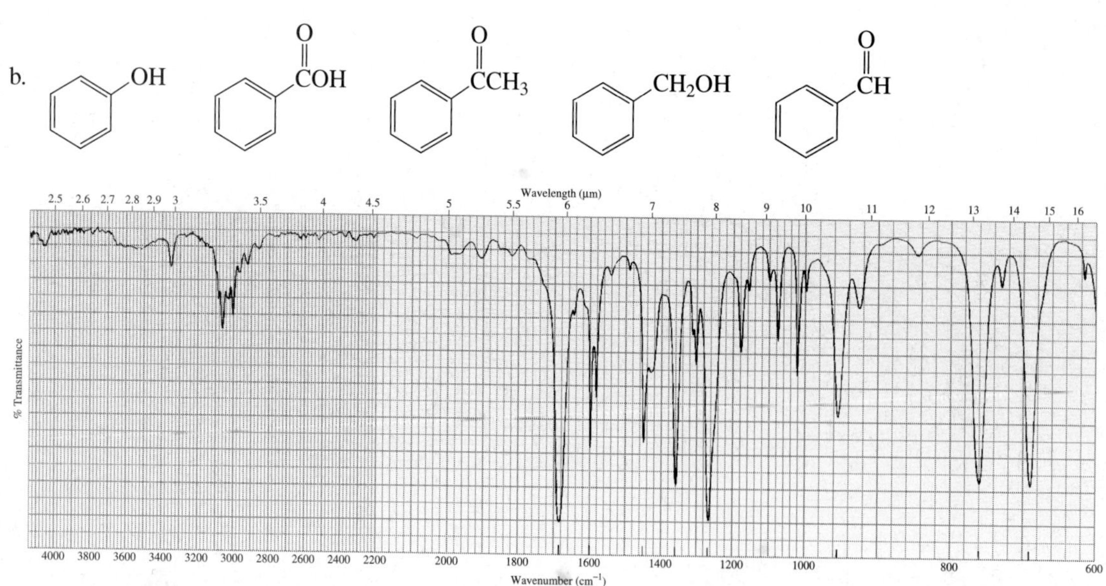

▲ **Figure 13.39**
The IR spectrum for Problem 45b.

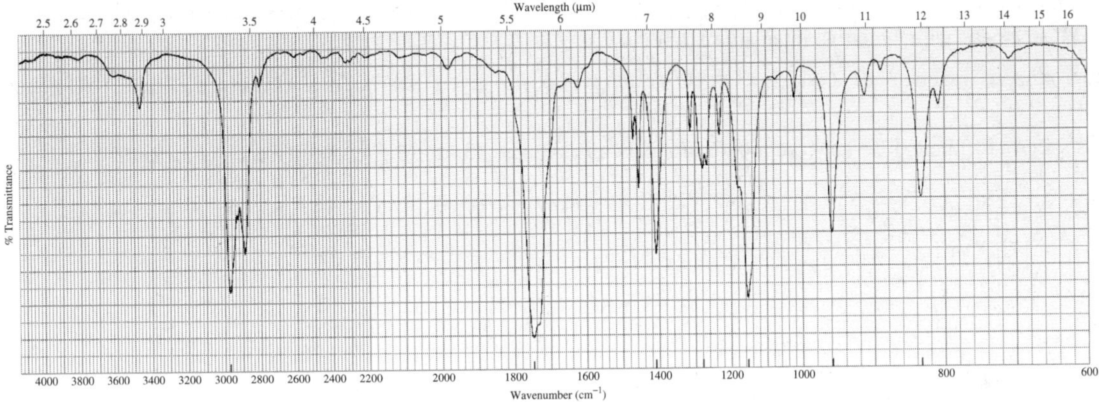

▲ **Figure 13.42**
The IR spectrum for Problem 49.

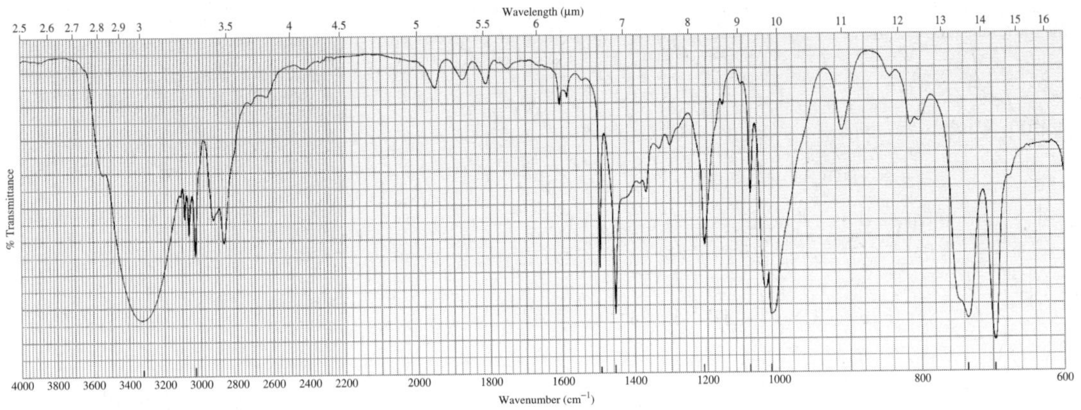

▲ **Figure 13.43**
The IR spectrum for Problem 50.

51. The IR spectrum shown in Figure 13.44 is the spectrum of one of the following compounds. Identify the compound.

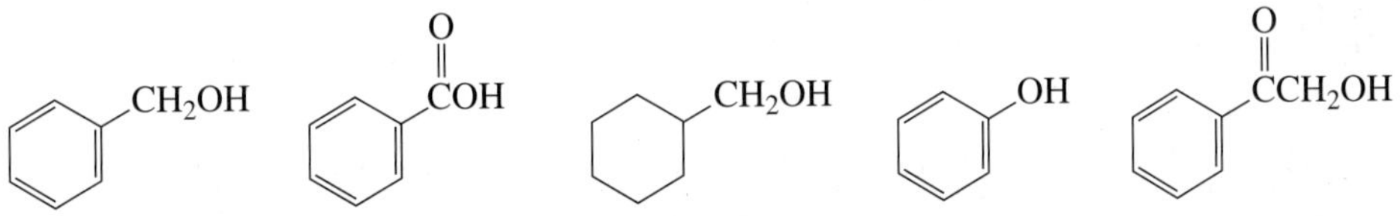

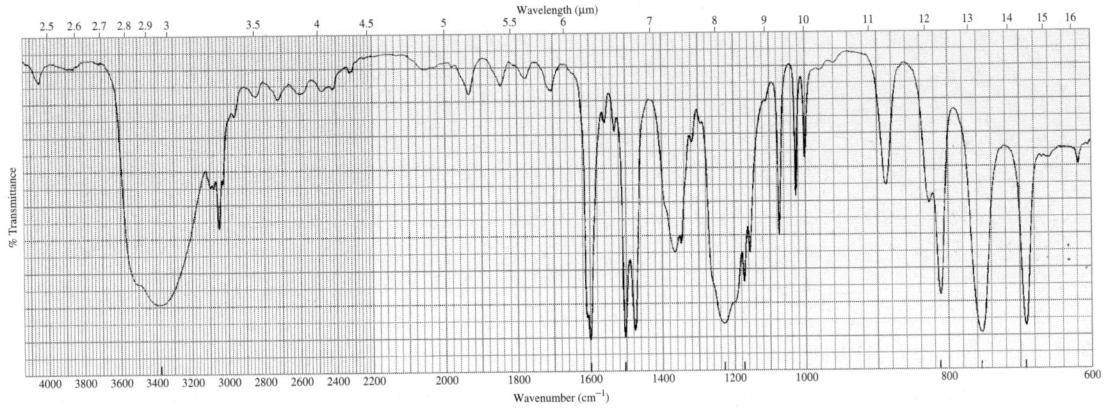

- $C_6H_5CH_2OH$
- C_6H_5COOH (benzene ring–C(=O)OH)
- $C_6H_{11}CH_2OH$ (cyclohexane ring–CH_2OH)
- C_6H_5OH
- $C_6H_5C(=O)CH_2OH$

Wavelength (μm)
2.5 2.6 2.7 2.8 2.9 3 3.5 4 4.5 5 5.5 6 7 8 9 10 11 12 13 14 15 16
% Transmittance
4000 3800 3600 3400 3200 3000 2800 2600 2400 2200 2000 1800 1600 1400 1200 1000 800 600
Wavenumber (cm^{-1})

▲ **Figure 13.44**
The IR spectrum for Problem 51.

52. Determine the molecular formula of a saturated acyclic hydrocarbon with an M peak at $m/z = 100$ with a relative intensity of 27.32%, and an M + 1 peak with a relative intensity of 2.10%.

53. Calculate the approximate wavenumber at which a C=C stretch will occur, given that the force constant for the C=C bond is 10×10^5 gs^{-2}.

54. The IR and mass spectra for three different compounds are shown in Figures 13.45–13.47. Identify each compound.

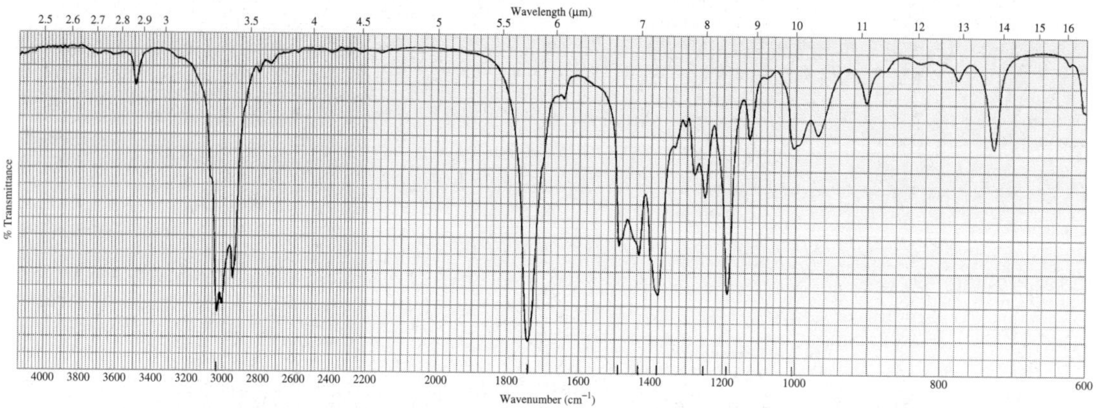

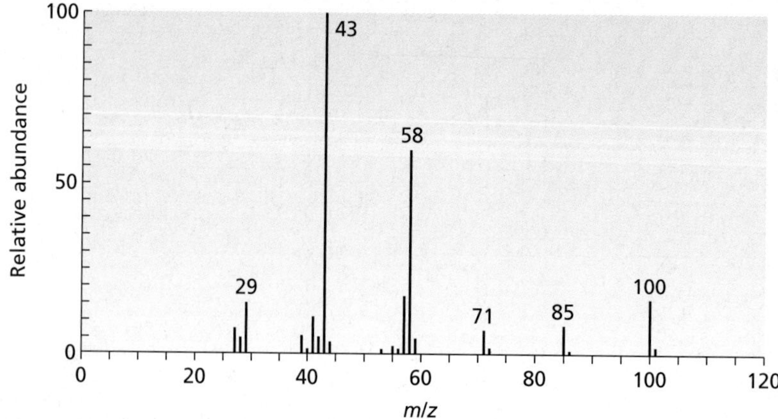

▲ **Figure 13.45**
The IR and mass spectra for Problem 54a.

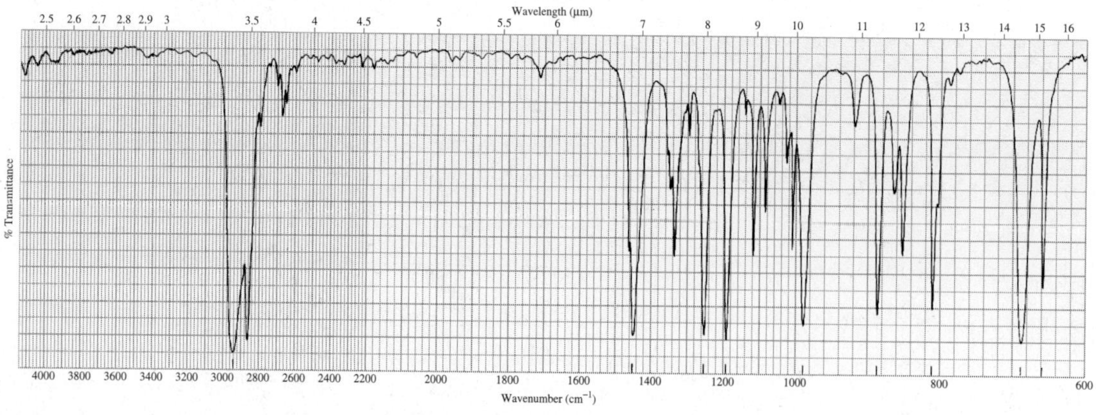

▲ **Figure 13.46**
The IR and mass spectra for Problem 54b.

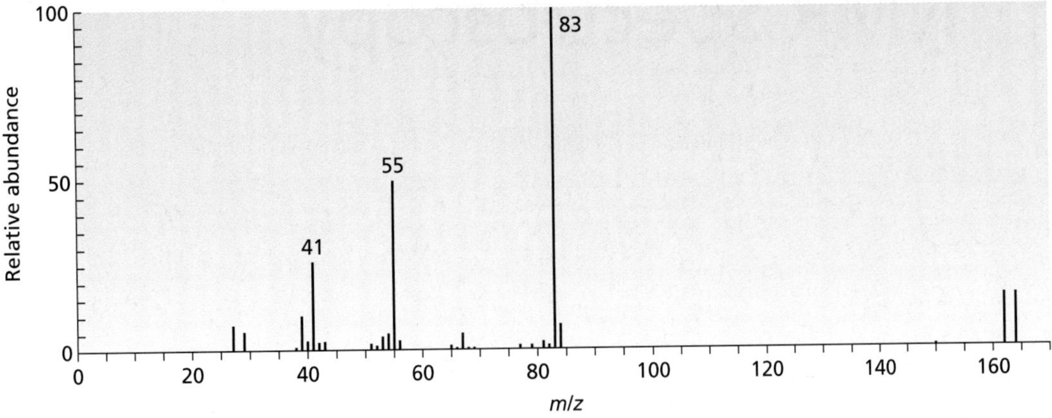

▲ **Figure 13.46 (continued)**

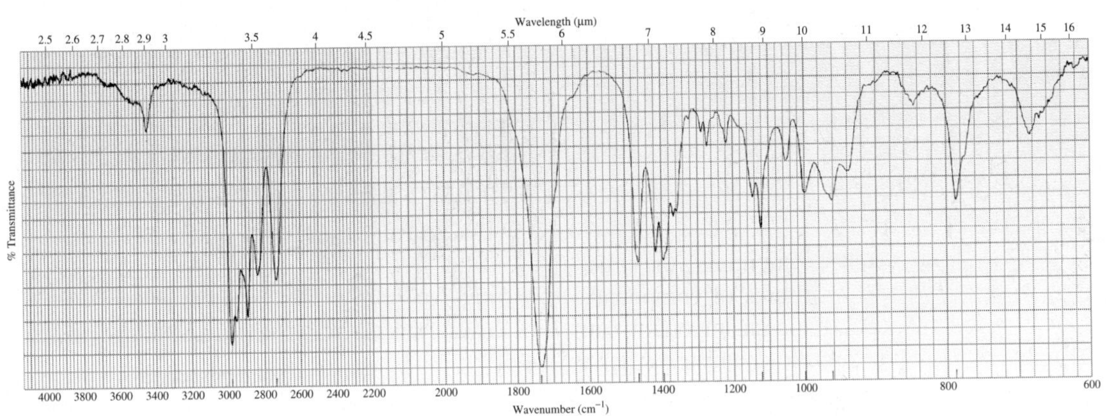

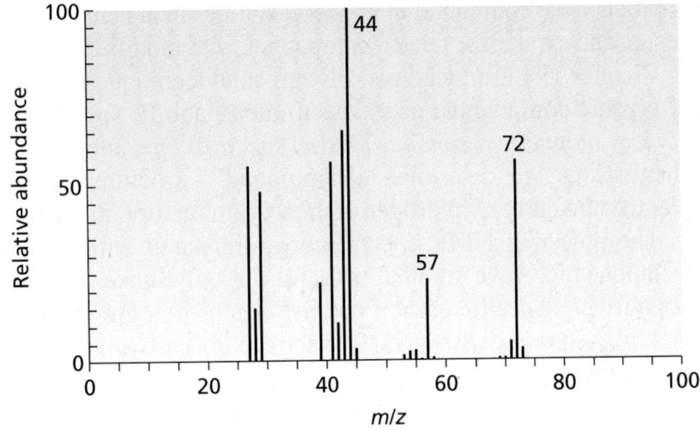

▲ **Figure 13.47**
The IR and mass spectra for Problem 54c.

14 NMR Spectroscopy

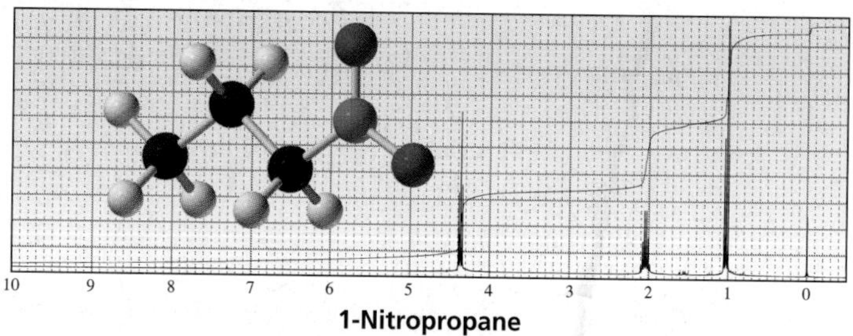

1-Nitropropane

Determining the structures of compounds is an important part of organic chemistry. After a compound has been synthesized, its structure must be confirmed. Chemists who study natural products must determine the structure of a naturally occurring compound before they can design a synthesis to produce the compound in greater quantities than nature can provide or before they can design and synthesize related compounds with modified properties.

Chapter 13 introduced two instrumental techniques used to determine the structure of organic compounds: mass spectrometry and IR spectroscopy. Now we will look at *nuclear magnetic resonance (NMR) spectroscopy*, another instrumental technique that chemists use to determine a compound's structure. **NMR spectroscopy** helps to identify the carbon–hydrogen framework of an organic compound.

The power of NMR spectroscopy, compared with that of the other instrumental techniques we have studied, is that it not only makes it possible to identify the functionality at a specific carbon but also lets us determine what the neighboring carbons look like. In many cases, NMR spectroscopy can be used to determine the entire structure of a molecule.

14.1 Introduction to NMR Spectroscopy

NMR spectroscopy was developed by physical chemists in the late 1940s to study the properties of atomic nuclei. In 1951, chemists realized that NMR spectroscopy could also be used to determine the structures of organic compounds. We have seen that electrons are charged, spinning particles with two allowed spin states: $+1/2$ and $-1/2$ (Section 1.2). Certain nuclei also have allowed spin states of $+1/2$ and $-1/2$, and this property allows them to be studied by NMR. Examples of such nuclei are ^{1}H, ^{13}C, ^{15}N, ^{19}F, and ^{31}P.

Because hydrogen nuclei (protons) were the first nuclei studied by nuclear magnetic resonance, the acronym "NMR" is generally assumed to mean **^{1}H NMR (proton**

magnetic resonance). Spectrometers were later developed for **^{13}C NMR**, **^{15}N NMR**, **^{19}F NMR**, **^{31}P NMR**, and other magnetic nuclei.

Spinning charged nuclei generate a magnetic field, like the field of a small bar magnet. In the absence of an applied magnetic field, the nuclear spins are randomly oriented. However, when a sample is placed in an applied magnetic field (Figure 14.1), the nuclei twist and turn to align themselves *with* or *against* the field of the larger magnet. More energy is needed for a proton to align against the field than with it. Protons that align with the field are in the lower-energy **α-spin state**; protons that align against the field are in the higher-energy **β-spin state**. More nuclei are in the α-spin state than in the β-spin state. The difference in the populations is very small (about 20 out of a million protons), but is sufficient to form the basis of NMR spectroscopy.

Edward Mills Purcell (1912–1997) *and* **Felix Bloch** *did the work on the magnetic properties of nuclei that made the development of NMR spectroscopy possible. They shared the 1952 Nobel Prize in physics. Purcell was born in Illinois. He received a Ph.D. from Harvard University in 1938 and immediately was hired as a faculty member in the physics department.*

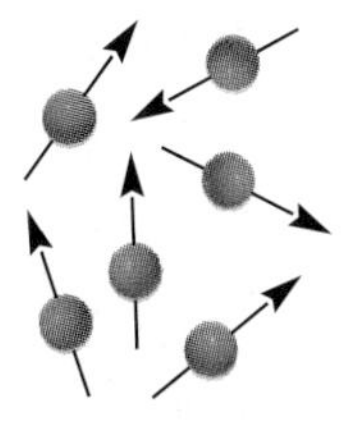

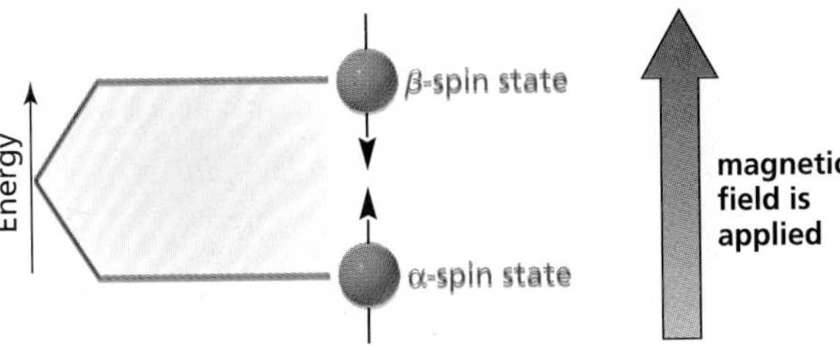

Figure 14.1
In the absence of an applied magnetic field, the spins of the nuclei are randomly oriented. In the presence of an applied magnetic field, the spins of the nuclei line up with or against the field.

The energy difference (ΔE) between the α- and β-spin states depends on the strength of the **applied magnetic field** (B_0). The greater the strength of the magnetic field to which we expose the nucleus, the greater is the difference in energy between the α- and β-spin states (Figure 14.2).

When the sample is subjected to a pulse of radiation whose energy corresponds to the difference in energy (ΔE) between the α- and β-spin states, nuclei in the α-spin state are promoted to the β-spin state. This transition is called "flipping" the spin. Because the energy difference between the α- and β-spin states is so small—for currently available magnets—only a small amount of energy is needed to flip the spin. The radiation required is in the radiofrequency (rf) region of the electromagnetic spectrum and is called **rf radiation**. When the nuclei undergo relaxation (i.e., return to their original state), they emit electromagnetic signals whose frequency depends on

Felix Bloch (1905–1983) *was born in Switzerland. His first academic appointment was at the University of Leipzig. After leaving Germany upon Hitler's rise to power, Bloch worked at universities in Denmark, Holland, and Italy. He eventually came to the United States, becoming a citizen in 1939. He was a professor of physics at Stanford University and worked on the atomic bomb project at Los Alamos, New Mexico, during World War II.*

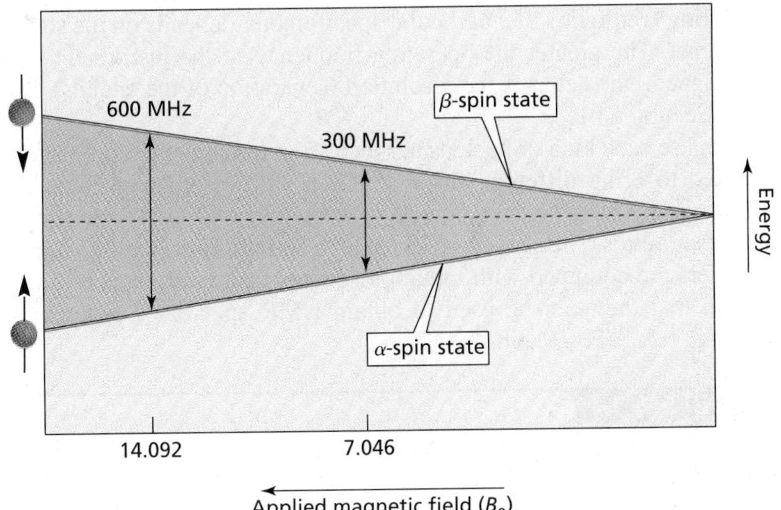

Figure 14.2
The greater the strength of the applied magnetic field, the greater is the difference in energy between the α- and β-spin states.

the difference in energy (ΔE) between the α- and β-spin states. The NMR spectrometer detects these signals and displays them as a plot of signal frequency versus intensity—an NMR spectrum. It is because the nuclei are *in resonance* with the rf radiation that the term "nuclear magnetic resonance" came into being. In this context, "resonance" refers to the flipping back and forth of nuclei between the α- and β-spin states in response to the rf radiation.

Recall that Planck's constant, h, is the proportionality constant relating the difference in energy (ΔE) to the frequency (ν) (Section 13.6) The following equation shows that the energy difference between the spin states (ΔE) depends on the operating frequency of the spectrometer, which in turn depends on the strength of the magnetic field (B_0), measured in tesla (T)[1], and the *gyromagnetic ratio* (γ). The **gyromagnetic ratio** (also called the magnetogyric ratio) is a constant that depends on the magnetic moment of the particular kind of nucleus. In the case of the proton, the value of γ is $2.675 \times 10^8\ \mathrm{T^{-1}\,s^{-1}}$; in the case of a ^{13}C nucleus, it is $6.688 \times 10^7\ \mathrm{T^{-1}\,s^{-1}}$.

$$\Delta E = h\nu = h\frac{\gamma}{2\pi}B_0$$

Canceling Planck's constant on both sides of the equation gives $\nu = \frac{\gamma}{2\pi}B_0$.

The following calculation shows that if an ^1H NMR spectrometer is equipped with a magnet with a magnetic field $(B_0) = 7.046\ \mathrm{T}$, the spectrometer will require an operating frequency of 300 MHz (megahertz):

$$\begin{aligned}\nu &= \frac{\gamma}{2\pi}B_0 \\ &= \frac{2.675 \times 10^8}{2(3.1416)}\mathrm{T^{-1}\,s^{-1}} \times 7.046\ \mathrm{T} \\ &= 300 \times 10^6\ \mathrm{Hz} = 300\ \mathrm{MHz}\end{aligned}$$

Earth's magnetic field is 5×10^5 T, measured at the equator. Its maximum surface magnetic field is 7×10^5 T, measured at the south magnetic pole.

The equation shows that the *magnetic field* (B_0) *is proportional to the operating frequency* (MHz). Therefore, if the spectrometer has a more powerful magnet, it must have a higher operating frequency. For example, a magnetic field of 14.092 T requires an operating frequency of 600 MHz.

The magnetic field is proportional to the operating frequency.

Today's NMR spectrometers operate at frequencies between 60 and 900 MHz. The **operating frequency** of a particular spectrometer depends on the strength of the built-in magnet. The greater the operating frequency of the instrument—and the stronger the magnet—the better is the resolution (separation of the signals) of the NMR spectrum (Section 14.17).

Because each kind of nucleus has its own gyromagnetic ratio, different energies are required to bring different kinds of nuclei into resonance. For example, an NMR spectrometer with a magnet requiring a frequency of 300 MHz to flip the spin of an ^1H nucleus requires a frequency of 75 MHz to flip the spin of a ^{13}C nucleus. NMR spectrometers are equipped with radiation sources that can be tuned to different frequencies so that they can be used to obtain NMR spectra of different kinds of nuclei (^1H, ^{13}C, ^{15}N, ^{19}F, ^{31}P, etc.).

PROBLEM 1◆

What frequency (in MHz) is required to cause a proton to flip its spin when it is exposed to a magnetic field of 1 tesla?

[1]Until recently, the gauss (G) was the unit in which magnetic field strength was commonly measured ($1\ \mathrm{T} = 10^4\ \mathrm{G}$).

NIKOLA TESLA (1856–1943)

Nikola Tesla was born in Croatia, the son of a clergyman. He emigrated to the United States in 1884 and became a citizen in 1891. He was a proponent of alternating current and bitterly fought Edison, who promoted direct current. Although Tesla did not win his dispute with Marconi over which of them invented the radio, Tesla is given credit for developing neon and fluorescent lighting, the electron microscope, the refrigerator motor, and the Tesla coil, a type of transformer.

Nikola Tesla in his laboratory

PROBLEM 2◆

a. Calculate the magnetic field (in tesla) required to flip an ^{1}H nucleus in an NMR spectrometer that operates at 360 MHz.

b. What strength magnetic field is required when a 500-MHz instrument is used?

14.2 Fourier Transform NMR

To obtain an NMR spectrum, one dissolves a small amount of a compound in about 0.5 mL of solvent and puts the solution into a long, thin glass tube, which is placed within a powerful magnetic field (Figure 14.3). The solvents used in NMR are discussed in Section 14.16. Spinning the sample tube about its long axis averages the position of the molecules in the magnetic field and thus greatly increases the resolution of the spectrum.

In modern instruments called *pulsed Fourier transform (FT) spectrometers*, the magnetic field is kept constant and an rf pulse of short duration excites all the protons simultaneously. Because the short rf pulse covers a range of frequencies, the

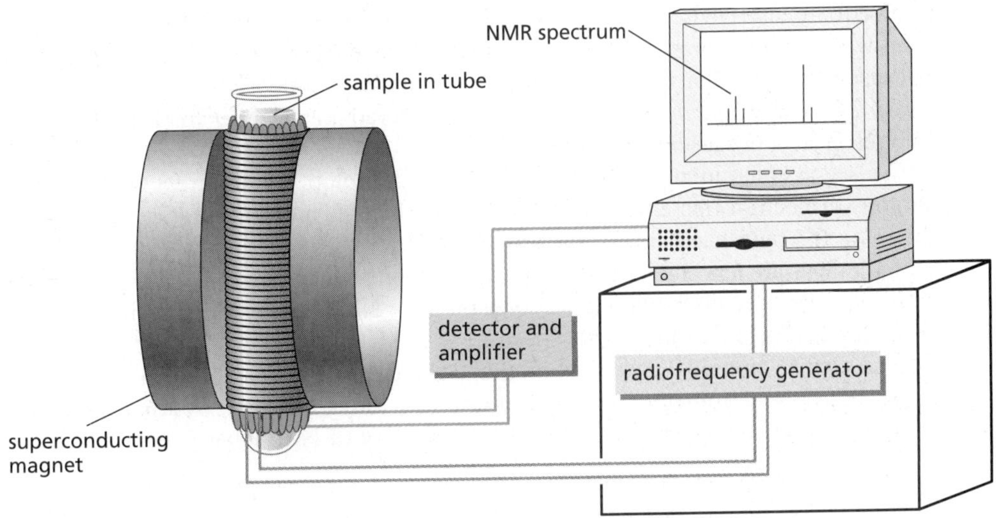

▲ **Figure 14.3**
Schematic of an NMR spectrometer.

The 1991 Nobel Prize in chemistry was awarded to **Richard R. Ernst** *for two important contributions: FT–NMR spectroscopy and an NMR tomography method that forms the basis of magnetic resonance imaging (MRI). Ernst was born in 1933, received a Ph.D. from the Swiss Federal Institute of Technology [Eidgenössische Technische Hochschule (ETH)] in Zurich, and became a research scientist at Varian Associates in Palo Alto, California. In 1968, he returned to the ETH, where he is a professor of chemistry.*

individual protons absorb the frequency required to come into resonance (flip their spin). As the protons relax (i.e., as they return to equilibrium), they produce a complex signal—called a free induction decay (FID)—at a frequency corresponding to ΔE. The intensity of the signal decays as the nuclei lose the energy they gained from the rf pulse. A computer collects and then converts the intensity-versus-time data into intensity-versus-frequency information in a mathematical operation known as a *Fourier transform*, producing a spectrum called a **Fourier transform NMR (FT–NMR) spectrum**. An FT–NMR spectrum can be recorded in about 2 seconds—and many FIDs can be averaged in a few minutes—using less than 5 mg of compound. The NMR spectra in this book are FT–NMR spectra that were taken on a spectrometer with an operating frequency of 300 MHz. This book discusses the theory behind FT–NMR, rather than that behind the older continuous wave (CW) NMR, because FT–NMR is more modern and is easier to understand.

14.3 Shielding

We have seen that when a sample in a magnetic field is irradiated with rf radiation of the proper frequency, each proton[2] in an organic compound gives a signal at a frequency that depends on the energy difference (ΔE) between the α- and β-spin states, where ΔE is determined by the strength of the magnetic field (Figure 14.2). If all the protons in an organic compound were in exactly the same environment, they would all give signals with the same frequency in response to a given applied magnetic field. If this were the case, all NMR spectra would consist of one signal, which would tell us nothing about the structure of the compound, except that it contains protons.

A nucleus, however, is embedded in a cloud of electrons that partly *shields* it from the applied magnetic field. Fortunately for chemists, the shielding varies for different protons within a molecule. In other words, all the protons do not experience the same applied magnetic field.

What causes shielding? In a magnetic field, the electrons circulate about the nuclei and induce a local magnetic field that opposes (i.e., that subtracts from) the applied magnetic field. The **effective magnetic field**, therefore, is what the nuclei "sense" through the surrounding electronic environment:

$$\boldsymbol{B}_{\text{effective}} = \boldsymbol{B}_{\text{applied}} - \boldsymbol{B}_{\text{local}}$$

This means that the greater the electron density of the environment in which the proton is located, the greater B_{local} is and the more the proton is shielded from the applied magnetic field. This type of shielding is called **diamagnetic shielding**. Thus, protons in electron-dense environments sense a *smaller effective magnetic field*. They, therefore, will require a *lower frequency* to come into resonance—that is, flip their spin—because ΔE is smaller (Figure 14.2). Protons in electron-poor environments sense a *larger effective magnetic field* and, therefore, will require a *higher frequency* to come into resonance, because ΔE is larger.

The larger the magnetic field sensed by the proton, the higher is the frequency of the signal.

We see a signal in an NMR spectrum for each proton in a different environment. Protons in electron-rich environments are more shielded and appear at lower frequencies—on the right-hand side of the spectrum (Figure 14.4). Protons in electron-poor environments are less shielded and appear at higher frequencies—on the left-hand side of the spectrum. (Notice that high frequency in an NMR spectrum is on the left-hand side, just as it is in IR and UV/Vis spectra.)

The terms "upfield" and "downfield," which came into use when continuous wave (CW) spectrometers were used (before the advent of Fourier transform spectrometers), are so entrenched in the vocabulary of NMR that you should know

[2]The terms "proton" and "hydrogen" are both used to describe covalently bonded hydrogen in discussions of NMR spectroscopy.

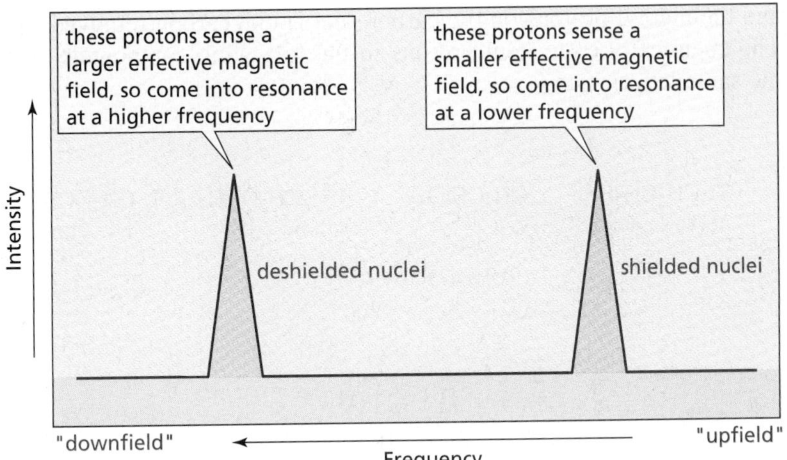
these protons sense a larger effective magnetic field, so come into resonance at a higher frequency
these protons sense a smaller effective magnetic field, so come into resonance at a lower frequency
Intensity
deshielded nuclei
shielded nuclei
"downfield"
"upfield"
Frequency

Figure 14.4
Shielded nuclei come into resonance at lower frequencies than deshielded nuclei.

what they mean. **Upfield** means farther to the right-hand side of the spectrum, and **downfield** means farther to the left-hand side of the spectrum. In contrast to FT–NMR techniques, which hold magnetic field strength constant and vary frequency, continuous-wave techniques hold frequency constant and vary magnetic field. The magnetic field increases from left to right across a spectrum because higher magnetic fields are required for shielded protons to come into resonance at a given frequency (Figure 14.4). Therefore, *upfield* is toward the right and *downfield* is toward the left.

14.4 The Number of Signals in the ¹H NMR Spectrum

Protons in the same environment are called **chemically equivalent protons**. For example, 1-bromopropane has three different sets of chemically equivalent protons. The three methyl protons are chemically equivalent because of rotation about the C—C bond. The two methylene protons on the middle carbon are chemically equivalent, and the two methylene protons on the carbon bonded to the bromine atom make up the third set of chemically equivalent protons.

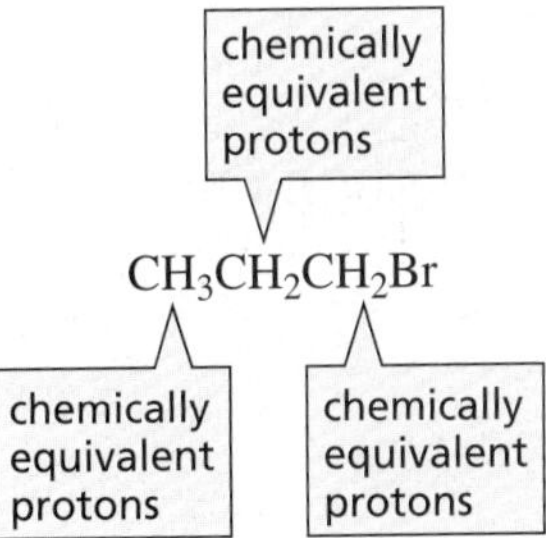
chemically equivalent protons
$CH_3CH_2CH_2Br$
chemically equivalent protons
chemically equivalent protons

Each set of chemically equivalent protons in a compound gives rise to a signal in the ¹H NMR spectrum of that compound. (Sometimes the signals are not sufficiently separated and overlap each other. When this happens, one sees fewer signals than anticipated.) Because 1-bromopropane has three sets of chemically equivalent protons, it has three signals in its ¹H NMR spectrum.

2-Bromopropane has two sets of chemically equivalent protons and, therefore, it has two signals in its ¹H NMR spectrum. The six methyl protons in 2-bromopropane are equivalent, so they give rise to only one signal. Ethyl methyl ether has three sets of chemically equivalent protons: the methyl protons on the carbon adjacent to the oxygen, the methylene protons on the carbon adjacent to the oxygen,

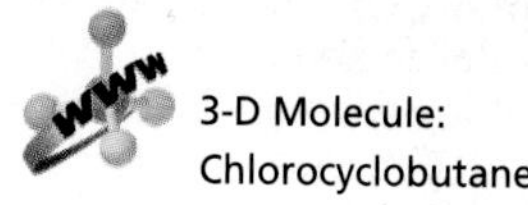

and the methyl protons on the carbon that is one carbon removed from the oxygen. The chemically equivalent protons in the following compounds are designated by the same letter:

$\overset{a}{CH_3}\overset{b}{CH_2}\overset{c}{CH_2}Br$ — three signals

$\overset{a}{CH_3}\overset{b}{CH}(Br)\overset{a}{CH_3}$ — two signals

$\overset{a}{CH_3}\overset{c}{CH_2}O\overset{b}{CH_3}$ — three signals

$\overset{a}{CH_3}O\overset{a}{CH_3}$ — one signal

$\overset{a}{CH_3}C(\overset{a}{CH_3})_2O\overset{b}{CH_3}$ — two signals

$\overset{a}{CH_3}O\overset{b}{CH}Cl_2$ — two signals

$(\overset{a}{H_3C})_2C{=}C\overset{b}{H}_2$ — two signals

$\overset{a}{H}(\overset{b}{H})C{=}C(\overset{c}{H})Br$ — three signals

benzene (all H = a) — one signal

nitrobenzene (ortho H = a, meta H = c, para H = b) — three signals

its ^{1}H NMR spectrum has five signals

chlorocyclobutane
H_a and H_b are not equivalent
H_c and H_d are not equivalent

You can tell how many sets of chemically equivalent protons a compound has from the number of signals in its ^{1}H NMR spectrum.

Sometimes, two protons on the same carbon are not equivalent. For example, the ^{1}H NMR spectrum of chlorocyclobutane has five signals. Even though they are bonded to the same carbon, the H_a and H_b protons are not equivalent because they are not in the same environment: H_a is trans to Cl and H_b is cis to Cl. Similarly, the H_c and H_d protons are not equivalent.

PROBLEM 3◆

How many signals would you expect to see in the ^{1}H NMR spectrum of each of the following compounds?

a. $CH_3CH_2CH_2CH_3$
b. $BrCH_2CH_2Br$
c. $CH_2{=}CHCl$
d. $CH_2{=}CHCH{=}O$ ($CH_2{=}CHCHO$)
e. cyclohexene
f. $CH_3CH_2CH_2C(=O)CH_3$
g. $CH_3CH_2CH(Cl)CH_2CH_3$
h. $CH_3CH(CH_3)CH_2CH(CH_3)CH_3$
i. $CH_3CH(Br)-C_6H_5$
j. $CH_3-C_6H_4-OCH_3$ (para)
k. $CH_3-C_6H_4-CH_3$ (para)
l. 1,3-dibromobenzene (Br, Br meta)
m. $C_6H_5-NO_2$
n. $ClHC{=}CHCl$ (cis: Cl, Cl on same side; H, H on same side)
o. $ClHC{=}CHCH_3$ (cis: Cl, CH_3 on same side; H, H on same side)

PROBLEM 4

How could you distinguish the ^{1}H NMR spectra of the following compounds?

a. $CH_3OCH_2OCH_3$
b. CH_3OCH_3
c. $CH_3OCH_2C(CH_3)_2CH_2OCH_3$

PROBLEM 5◆

There are three isomeric dichlorocyclopropanes. Their ^{1}H NMR spectra show one signal for isomer 1, two signals for isomer 2, and three signals for isomer 3. Draw the structures of isomers 1, 2, and 3.

14.5 The Chemical Shift

A small amount of an inert **reference compound** is added to the sample tube containing the compound whose NMR spectrum is to be taken. The positions of the signals in an NMR spectrum are defined according to how far they are from the signal of the reference compound. The most commonly used reference compound is tetramethylsilane (TMS). Because TMS is a highly volatile compound, it can easily be removed from the sample by evaporation after the NMR spectrum is taken.

$$\begin{array}{c} CH_3 \\ | \\ CH_3-Si-CH_3 \\ | \\ CH_3 \end{array}$$

tetramethylsilane
TMS

The methyl protons of TMS are in a more electron-dense environment than are most protons in organic molecules, because silicon is less electronegative than carbon (electronegativities of 1.8 and 2.5, respectively). Consequently, the signal for the methyl protons of TMS is at a lower frequency than most other signals (i.e., it appears to the right of the other signals).

The position at which a signal occurs in an NMR spectrum is called the *chemical shift.* The **chemical shift** is a measure of how far the signal is from the reference TMS signal. The most common scale for chemical shifts is the δ (delta) scale. The TMS signal is used to define the zero position on this scale. The chemical shift is determined by measuring the distance from the TMS peak (in hertz) and dividing by the operating frequency of the instrument (in megahertz). Because the units are Hz/MHz, a chemical shift has units of parts per million (ppm) of the operating frequency:

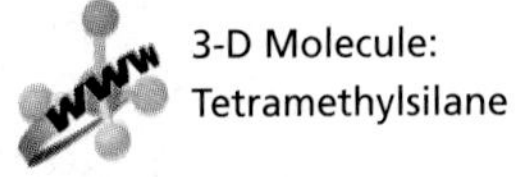

3-D Molecule: Tetramethylsilane

$$\delta = \text{chemical shift (ppm)} = \frac{\text{distance downfield from TMS (Hz)}}{\text{operating frequency of the spectrometer (MHz)}}$$

Most proton chemical shifts fall in the range from 0 to 10 ppm.

The ^{1}H NMR spectrum for 1-bromo-2,2-dimethylpropane in Figure 14.5 shows that the chemical shift of the methyl protons is at 1.05 ppm and the chemical shift of

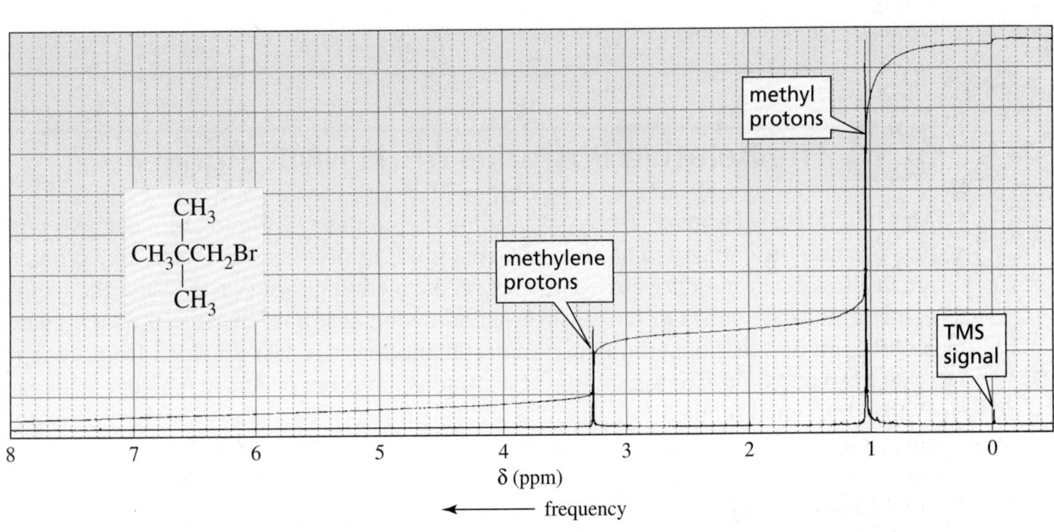

▲ **Figure 14.5**
^{1}H NMR spectrum of 1-bromo-2,2-dimethylpropane. The TMS signal is a reference signal from which chemical shifts are measured; it defines the zero position on the scale.

The greater the chemical shift (δ), the higher the frequency.

The chemical shift (δ) is independent of the operating frequency of the spectrometer.

the methylene protons is at 3.28 ppm. *Notice that low-frequency (upfield, shielded) signals have small δ (ppm) values, whereas high-frequency (downfield, deshielded) signals have large δ values.*

The advantage of the δ scale is that the chemical shift of a given nucleus is *independent of the operating frequency of the NMR spectrometer*. Thus, the chemical shift of the methyl protons of 1-bromo-2,2-dimethylpropane is at 1.05 ppm in both a 60-MHz and a 360-MHz instrument. In contrast, if the chemical shift were reported in hertz, it would be at 63 Hz in a 60-MHz instrument and at 378 Hz in a 360-MHz instrument $(63/60 = 1.05; 378/360 = 1.05)$. The following diagram will help you keep track of the terms associated with NMR spectroscopy:

protons in electron-poor environments	protons in electron-dense environments
deshielded protons	shielded protons
downfield	upfield
high frequency	low frequency
large δ values	small δ values

← δ
← frequency

PROBLEM 6◆

A signal has been reported to occur at 600 Hz downfield from TMS in an NMR spectrometer with a 300-MHz operating frequency.

a. What is the chemical shift of the signal?
b. What would its chemical shift be in an instrument operating at 100 MHz?
c. How many hertz downfield from TMS would the signal be in a 100-MHz spectrometer?

PROBLEM 7◆

a. If two signals differ by 1.5 ppm in a 300-MHz spectrometer, by how much do they differ in a 100-MHz spectrometer?
b. If two signals differ by 90 hertz in a 300-MHz spectrometer, by how much do they differ in a 100-MHz spectrometer?

PROBLEM 8◆

Where would you expect to find the ^1H NMR signal of $(CH_3)_2Mg$ relative to the TMS signal? (*Hint:* See Table 12.3 on p. 467.)

14.6 The Relative Positions of ^1H NMR Signals

The ^1H NMR spectrum of 1-bromo-2,2-dimethylpropane in Figure 14.5 has two signals because the compound has two different kinds of protons. The methylene protons are in a less electron-dense environment than the methyl protons are because the methylene protons are closer to the electron-withdrawing bromine. Because the methylene protons are in a less electron-dense environment, they are less shielded from the applied magnetic field. The signal for these protons therefore occurs at a higher frequency than the signal for the more shielded methyl protons. *Remember that the right-hand side of an NMR spectrum is the low-frequency side, where protons in electron-dense environments (more shielded) show a signal. The left-hand side is the high-frequency side, where less shielded protons show a signal* (Figure 14.4).

We would expect the ^1H NMR spectrum of 1-nitropropane to have three signals because the compound has three different kinds of protons. The closer the protons are to the electron-withdrawing nitro group, the less they are shielded from the applied magnetic field, so the higher the frequency (i.e., the farther downfield) at which their signals will appear. Thus, the protons closest to the nitro group show a signal at the highest frequency (4.37 ppm), and the ones farthest from the nitro group show a signal at the lowest frequency (1.04 ppm).

Electron withdrawal causes NMR signals to appear at higher frequencies (at larger δ values).

$CH_3CH_2CH_2NO_2$ — CH_3: 1.04 ppm; CH_2: 2.07 ppm; CH_2NO_2: 4.37 ppm

Compare the chemical shifts of the methylene protons immediately adjacent to the halogen in each of the following alkyl halides. The position of the signal depends on the electronegativity of the halogen—the more electronegative the halogen, the higher is the frequency of the signal. Thus, the signal for the methylene protons adjacent to fluorine (the most electronegative of the halogens) occurs at the highest frequency, whereas the signal for the methylene protons adjacent to iodine (the least electronegative of the halogens) occurs at the lowest frequency.

$CH_3CH_2CH_2CH_2CH_2F$ — 4.50 ppm

$CH_3CH_2CH_2CH_2CH_2Cl$ — 3.50 ppm

$CH_3CH_2CH_2CH_2CH_2Br$ — 3.40 ppm

$CH_3CH_2CH_2CH_2CH_2I$ — 3.20 ppm

PROBLEM 9♦

a. Which set of protons in each of the following compounds is the least shielded?

1. $CH_3CH_2CH_2Cl$
2. $CH_3CH_2\overset{O}{\overset{\|}{C}}OCH_3$
3. $CH_3CHCHBr$ (with Br on each CH: $CH_3CH(Br)CH(Br)Br$)

b. Which set of protons in each compound is the most shielded?

PROBLEM 10♦

One of the spectra in Figure 14.6 is due to 1-chloropropane, and the other to 1-iodopropane. Which is which?

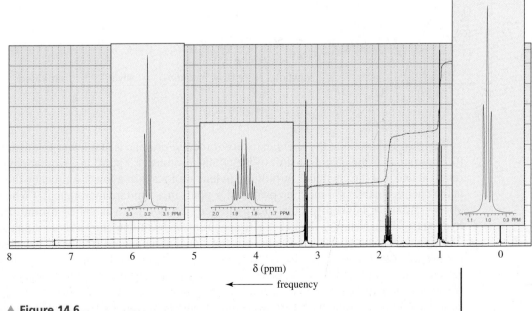

▲ **Figure 14.6**
^{1}H NMR spectra for Problem 10.

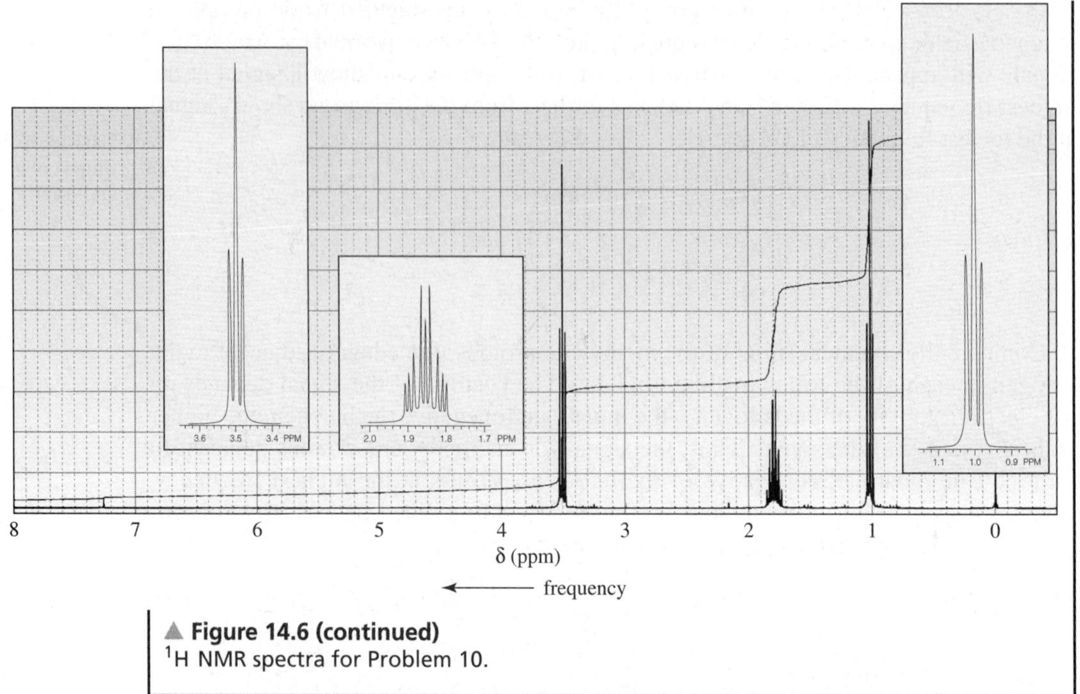

▲ **Figure 14.6 (continued)**
^{1}H NMR spectra for Problem 10.

14.7 Characteristic Values of Chemical Shifts

Approximate values of chemical shifts for different kinds of protons are shown in Table 14.1. (A more extensive compilation of chemical shifts is given in Appendix VI.) An ^{1}H NMR spectrum can be divided into six regions. Rather than memorizing chemical shift values, if you remember the kinds of protons that are in each region, you will be able to tell what kinds of protons a molecule has from a quick look at its NMR spectrum.

12–9.0	9.0–8.0	8.0–6.5	6.5–4.5	4.5–2.5	2.5–1.5	1.5–0
O=C(–)–H; O=C(–)–OH		benzene ring H	C=C–H vinylic	Z–C–H, Z = O, N, halogen	O=C–C–H; C=C–C–H allylic	–C–C–H saturated

δ (ppm)

Table 14.1 shows that the chemical shift of methyl protons is at a lower frequency (0.9 ppm) than is the chemical shift of methylene protons (1.3 ppm) in a similar environment and that the chemical shift of methylene protons is at a lower frequency than is the chemical shift of a methine proton (1.4 ppm) in a similar environment. (When an sp^3 carbon is bonded to only one hydrogen, the hydrogen is called a **methine hydrogen**.) For example, the ^{1}H NMR spectrum of butanone shows three signals. The signal for the ***a*** protons of butanone is the signal at the lowest frequency because the protons are farthest from the electron-withdrawing carbonyl group. (In correlating an NMR spectrum with a structure, the set of protons responsible for the signal at the lowest frequency will be labeled ***a***, the next set will be labeled ***b***, the next set ***c***, etc.) The ***b*** and ***c*** protons are the same distance from the carbonyl group, but the signal

Table 14.1 Approximate Values of Chemical Shifts for ^{1}H NMR[a]

Type of proton	Approximate chemical shift (ppm)	Type of proton	Approximate chemical shift (ppm)
$(CH_3)_4Si$	0	C_6H_5—H (benzene ring—H)	6.5–8
—CH_3	0.9	—C(=O)—H	9.0–10
—CH_2—	1.3	I—C—H	2.5–4
—CH— (methine)	1.4	Br—C—H	2.5–4
—C=C—CH_3	1.7	Cl—C—H	3–4
—C(=O)—CH_3	2.1	F—C—H	4–4.5
C_6H_5—CH_3 (benzene ring—CH_3)	2.3	RNH_2	Variable, 1.5–4
—C≡C—H	2.4	ROH	Variable, 2–5
R—O—CH_3	3.3	ArOH	Variable, 4–7
R—C(R)=CH_2	4.7	—C(=O)—OH	Variable, 10–12
R—C(R)=C(R)—H	5.3	—C(=O)—NH_2	Variable, 5–8

[a]The values are approximate because they are affected by neighboring substituents.

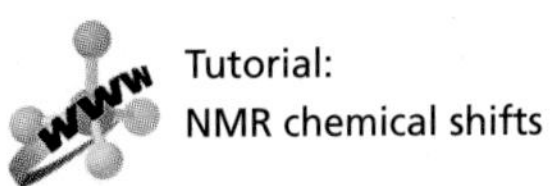

Tutorial: NMR chemical shifts

for the ***b*** protons is at a lower frequency because methyl protons appear at a lower frequency than do methylene protons in a similar environment.

In a similar environment, the signal for methyl protons occurs at a lower frequency than the signal for methylene protons, which in turn occurs at a lower frequency than the signal for a methine proton.

$\underset{a}{CH_3}\underset{c}{CH_2}\overset{O}{\overset{\|}{C}}\underset{b}{CH_3}$
butanone

$\overset{b}{CH_3}O\overset{c}{CH}\overset{a}{CH_3}$ with $\underset{a}{CH_3}$ on the CH
2-methoxypropane

The signal for the ***a*** protons of 2-methoxypropane is the signal at the lowest frequency in the ^{1}H NMR spectrum of this compound because these protons are farthest from the electron-withdrawing oxygen. The ***b*** and ***c*** protons are the same distance from the oxygen, but the signal for the ***b*** protons appears at a lower frequency because, in a similar environment, methyl protons appear at a lower frequency than does a methine proton.

PROBLEM 11◆

In each of the following compounds, which of the underlined protons has the greater chemical shift (i.e., the farther downfield signal or the higher frequency signal)?

a. $CH_3C\underline{H}C\underline{H}Br$
   Br Br

b. $\underline{CH_3}CHOC\underline{H_3}$
   CH_3

c. $CH_3\underline{CH_2}CH\underline{CH_3}$
   Cl

d. $CH_3\underline{CH}C(=O)\underline{CH_2}CH_3$
   CH_3

e. $CH_3\underline{CH_2}CH{=}\underline{CH_2}$

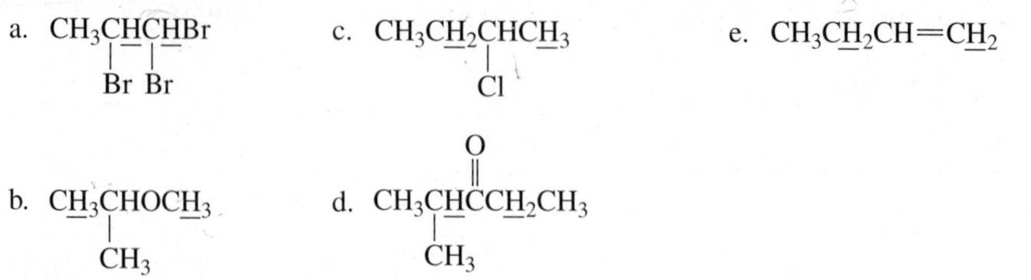

PROBLEM 12◆

In each of the following pairs of compounds, which of the underlined protons has the greater chemical shift (i.e., the farther downfield signal or the higher frequency signal)?

a. $CH_3CH_2\underline{CH_2}Cl$ or $CH_3CH_2\underline{CH_2}Br$

b. $CH_3CH_2\underline{CH_2}Cl$ or $CH_3CH_2\underline{CH}CH_3$
   Cl

c. $CH_3CH_2C(=O)\underline{H}$ or $CH_3CH_2C(=O)O\underline{CH_3}$

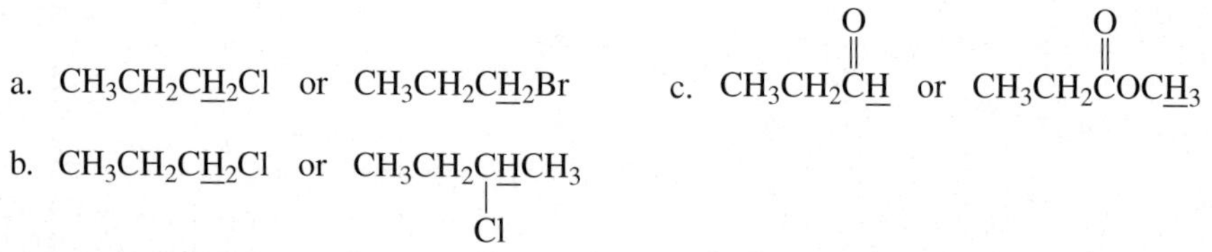

PROBLEM 13

Without referring to Table 14.1, label the protons in the following compounds. The proton that gives the signal at the lowest frequency should be labeled *a*, the next *b*, etc.

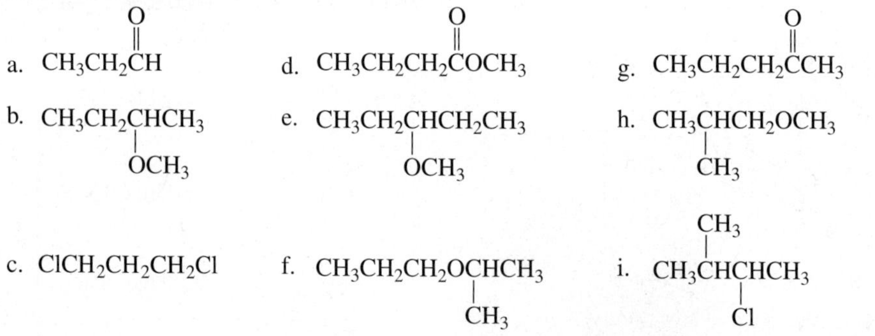

a. $CH_3CH_2C(=O)H$

b. $CH_3CH_2CHCH_3$
   OCH_3

c. $ClCH_2CH_2CH_2Cl$

d. $CH_3CH_2CH_2C(=O)OCH_3$

e. $CH_3CH_2CHCH_2CH_3$
   OCH_3

f. $CH_3CH_2CH_2OCHCH_3$
   CH_3

g. $CH_3CH_2CH_2C(=O)CH_3$

h. $CH_3CHCH_2OCH_3$
   CH_3

i. CH_3
   $CH_3CHCHCH_3$
   Cl

14.8 Integration of NMR Signals

The two signals in the ^1H NMR spectrum of 1-bromo-2,2-dimethylpropane in Figure 14.5 are not the same size because *the area under each signal is proportional to the number of protons that gives rise to the signal.* (The spectrum is shown again in Figure 14.7.) The area under the signal occurring at the lower frequency is larger because the signal is caused by *nine* methyl protons, while the smaller, higher-frequency signal results from *two* methylene protons.

You probably remember from a calculus course that the area under a curve can be determined by integration. An ^1H NMR spectrometer is equipped with a computer that calculates the integrals electronically. Modern spectrometers print out the integrals as numbers on the spectrum. The integrals can also be displayed by a line of integration superimposed on the original spectrum (Figure 14.7). The height of each integration step is proportional to the area under that signal, which, in turn, is proportional to the number of protons giving rise to the signal. By measuring the heights of

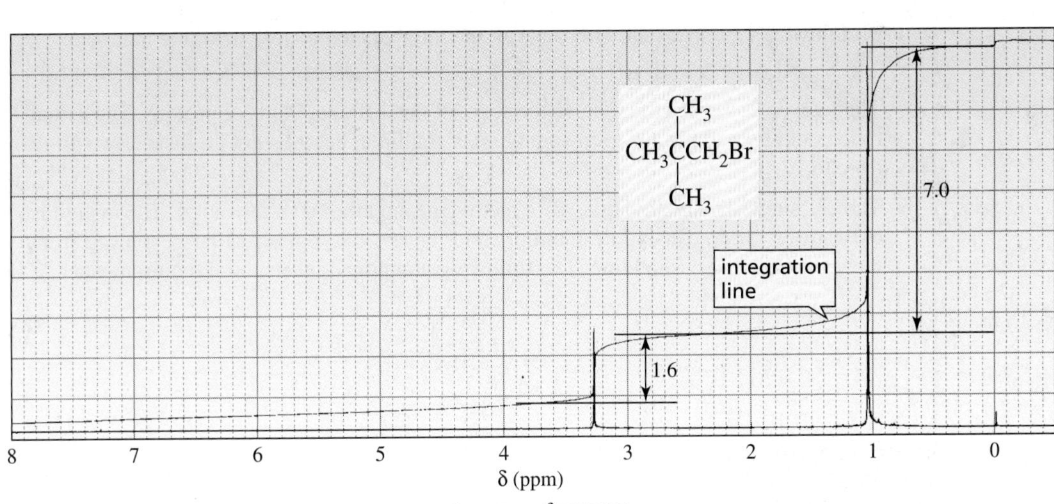

▲ **Figure 14.7**
Analysis of the integration line in the ^{1}H NMR spectrum of 1-bromo-2,2-dimethylpropane.

the integration steps, you can determine that the ratio of the integrals is approximately 1.6 : 7.0 = 1 : 4.4. (The measured integrals are approximate by as much as 10% because of experimental error.) The ratios are multiplied by a number that will cause all the numbers to be close to whole numbers—in this case, we multiply by 2—as there can be only whole numbers of protons. That means that the ratio of protons in the compound is 2 : 8.8, which is rounded to 2 : 9.

The **integration** tells us the *relative* number of protons that give rise to each signal, not the *absolute* number. For example, integration could not distinguish between 1,1-dichloroethane and 1,2-dichloro-2-methylpropane because both compounds would show an integral ratio of 1 : 3.

$CH_3-CH(Cl)-Cl$

1,1-dichloroethane
ratio of protons = 1:3

$CH_3-C(CH_3)(Cl)-CH_2Cl$

1,2-dichloro-2-methylpropane
ratio of protons 2:6 = 1:3

PROBLEM 14◆

How would integration distinguish the ^{1}H NMR spectra of the following compounds?

$CH_3-C(CH_3)_2-CH_2Br$ $\quad$ $CH_3-C(CH_3)(Br)-CH_2Br$ $\quad$ $CH_3-C(CH_2Br)_2-CH_2Br$

PROBLEM 15 SOLVED

a. Calculate the ratios of the different kinds of protons in a compound with an integral ratio of 6 : 4 : 18.4 (going from left to right across the spectrum).

b. Determine the structure of a compound that would give these relative integrals in the observed order.

SOLUTION

a. Divide each by the smallest number:

$$\frac{6}{4} = 1.5 \qquad \frac{4}{4} = 1 \qquad \frac{18.4}{4} = 4.6$$

Multiply by a number that will cause all the numbers to be close to whole numbers:

$$1.5 \times 2 = 3 \qquad 1 \times 2 = 2 \qquad 4.6 \times 2 = 9$$

The ratio 3 : 2 : 9 gives the relative numbers of the different kinds of protons. The actual ratio could be 6 : 4 : 18, or even some higher multiple, but let's not go there if we don't have to.

b. The "3" suggests a methyl, the "2" a methylene, and the "9" a *tert*-butyl. The methyl is closest to a group causing deshielding, and the *tert*-butyl group is farthest away from the group causing deshielding. The following compound meets these requirements:

$$\begin{array}{c} \quad CH_3 \;\; O \\ \quad | \quad\;\; \| \\ CH_3CCH_2COCH_3 \\ | \quad\quad\quad \\ CH_3 \quad\quad\quad \end{array}$$

PROBLEM 16◆

The ^{1}H NMR spectrum shown in Figure 14.8 corresponds to one of the following compounds. Which compound is responsible for this spectrum?

A	B	C	D
HC≡C–C$_6$H$_4$–C≡CH	CH_3–C$_6$H$_4$–CH_3	$ClCH_2$–C$_6$H$_4$–CH_2Cl	Br_2CH–C$_6$H$_4$–$CHBr_2$

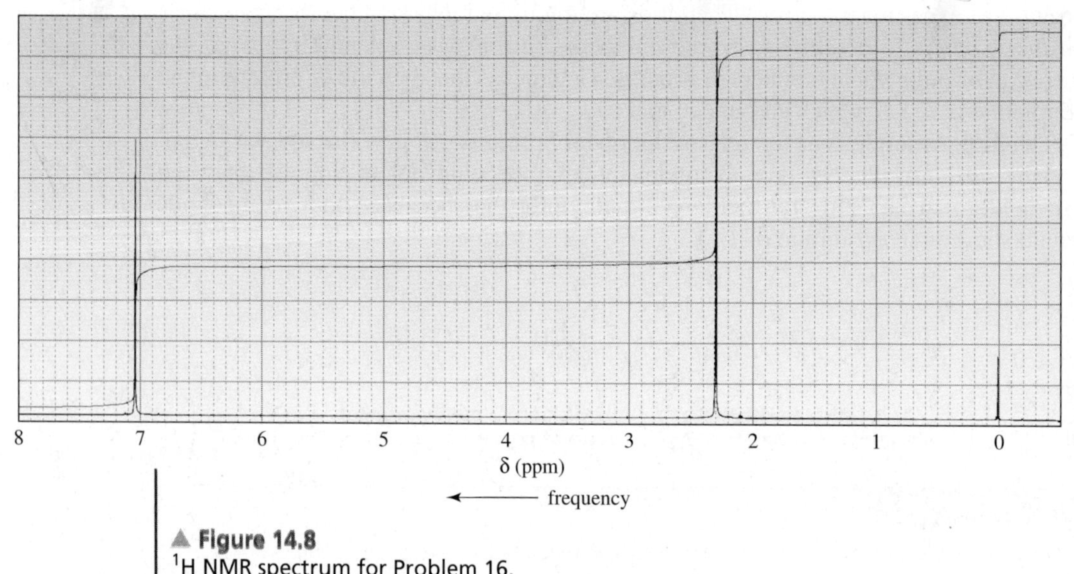

▲ **Figure 14.8**
^{1}H NMR spectrum for Problem 16.

14.9 Diamagnetic Anisotropy

The chemical shifts of hydrogens bonded to sp^2 hybridized carbons are at a higher frequency than one would predict, based on the electronegativity of the sp^2 carbons. For example, a hydrogen bonded to a terminal sp^2 carbon of an alkene appears at 4.7 ppm, a hydrogen bonded to an internal sp^2 carbon appears at 5.3 ppm, and a hydrogen on a benzene ring appears at 6.5–8.0 ppm (Table 14.1).

5.3 ppm

$CH_3CH_2CH{=}CH_2$ C$_6$H$_5$–H

4.7 ppm   7.3 ppm

The unusual chemical shifts associated with hydrogens bonded to carbons that form π bonds are due to *diamagnetic anisotropy.* **Diamagnetic anisotropy** describes an environment in which different magnetic fields are found at different points in space. *Anisotropic* is Greek for "different in different directions." Because π electrons are less tightly held by nuclei than are σ electrons, π electrons are more free to move in response to a magnetic field. When a magnetic field is applied to a compound with π electrons, the π electrons move in a circular path. This electron motion causes an induced magnetic field. How this induced magnetic field affects the chemical shift of a proton depends on the direction of the induced magnetic field—in the region where the proton is located—relative to the direction of the applied magnetic field.

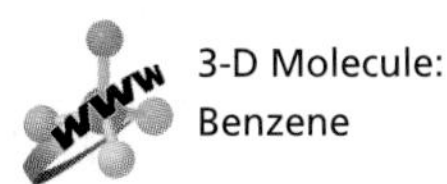

3-D Molecule: Benzene

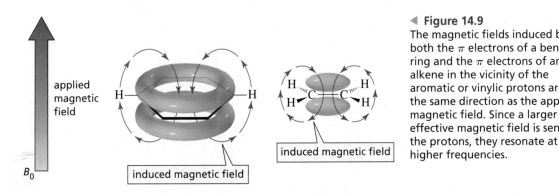

The magnetic field induced by the π electrons of a benzene ring—in the region where benzene's protons are located—is oriented in the same direction as the applied field (Figure 14.9). The magnetic field induced by the π electrons of an alkene—in the region where the protons bonded to the sp^2 carbons of the alkene are located—is also oriented in the same direction as the applied field. Thus, in both cases, a larger effective magnetic field—the sum of the strengths of the applied field and the induced field—is sensed by the protons. Because frequency is proportional to the strength of the magnetic field experienced by the protons, the protons resonate at higher frequencies than they would have if the π electrons had not induced a magnetic field.

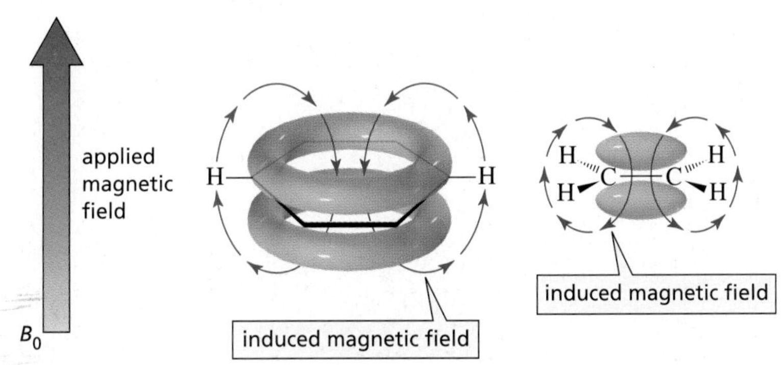

Figure 14.9
The magnetic fields induced by both the π electrons of a benzene ring and the π electrons of an alkene in the vicinity of the aromatic or vinylic protons are in the same direction as the applied magnetic field. Since a larger effective magnetic field is sensed by the protons, they resonate at higher frequencies.

PROBLEM 17◆

[18]-Annulene shows two signals in its ^{1}H NMR spectrum: one at 9.25 ppm and the other very far upfield (beyond TMS) at −2.88 ppm. What hydrogens are responsible for each of the signals? (*Hint:* Notice the direction of the induced magnetic field outside and inside the benzene ring in Figure 14.9.)

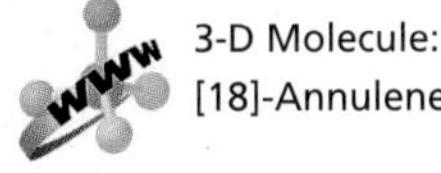

3-D Molecule: [18]-Annulene

14.10 Splitting of the Signals

Notice that the shapes of the signals in the ^{1}H NMR spectrum of 1,1-dichloroethane (Figure 14.10) are different from the shapes of the signals in the ^{1}H NMR spectrum of 1-bromo-2,2-dimethylpropane (Figure 14.5). Both signals in Figure 14.5 are **singlets** (each composed of a single peak), whereas the signal for the methyl protons of

1,1-dichloroethane (the lower-frequency signal) is split into two peaks (a **doublet**), and the signal for the methine proton is split into four peaks (a **quartet**). (Magnifications of the doublet and quartet are shown as insets in Figure 14.10.)

Splitting is caused by protons bonded to adjacent (i.e., directly attached) carbons. The splitting of a signal is described by the $\boldsymbol{N + 1}$ **rule**, where N is the number of *equivalent* protons bonded to *adjacent* carbons. By "equivalent protons," we mean that the protons bonded to an adjacent carbon are equivalent to each other, but not equivalent to the proton giving rise to the signal. Both signals in Figure 14.5 are singlets because neither the carbon adjacent to the methyl groups nor that adjacent to the methylene group in 1-bromo-2,2-dimethylpropane is bonded to any protons $(N + 1 = 0 + 1 = 1)$. In contrast, in Figure 14.10, the carbon adjacent to the methyl group in 1,1-dichloroethane is bonded to one proton, so the signal for the methyl protons is split into a doublet $(N + 1 = 1 + 1 = 2)$. The carbon adjacent to the carbon bonded to the methine proton is bonded to three equivalent protons, so the signal for the methine proton is split into a quartet $(N + 1 = 3 + 1 = 4)$. The number of peaks in a signal is called the **multiplicity** of the signal. Splitting is always mutual: If the ***a*** protons split the ***b*** protons, then the ***b*** protons must split the ***a*** protons. The methine proton and the methyl protons are an example of *coupled protons*. **Coupled protons** split each other's signal.

An ^{1}H NMR signal is split into $N + 1$ peaks, where N is the number of equivalent protons bonded to adjacent carbons.

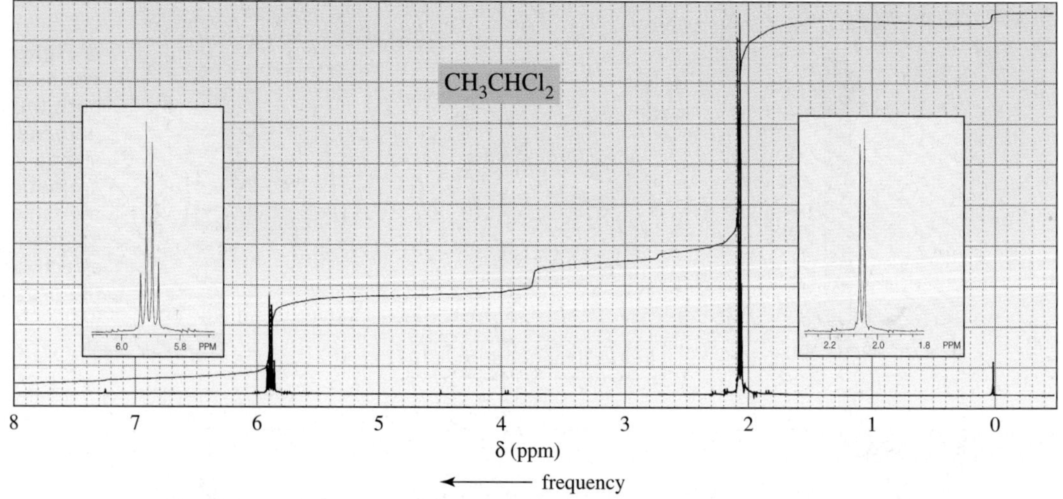

▲ **Figure 14.10**
^{1}H NMR spectrum of 1,1-dichloroethane. The higher-frequency signal is an example of a quartet; the lower-frequency signal is a doublet.

Keep in mind that it is not the number of protons giving rise to a signal that determines the multiplicity of the signal; rather, it is the number of protons bonded to the immediately adjacent carbons that determines the multiplicity. For example, the signal for the ***a*** protons in the following compound will be split into three peaks (a **triplet**) because the adjacent carbon is bonded to two hydrogens. The signal for the ***b*** protons will appear as a quartet because the adjacent carbon is bonded to three hydrogens, and the signal for the ***c*** protons will be a singlet.

$$\underset{a}{CH_3}\underset{b}{CH_2}\overset{O}{\overset{\|}{C}}O\underset{c}{CH_3}$$

More specifically, the splitting of signals occurs when different kinds of protons are close enough for their magnetic fields to influence one another—called **spin–spin coupling**. For example, the frequency at which the methyl protons of 1,1-dichloroethane show a signal is influenced by the magnetic field of the methine proton. If the magnetic field of the methine proton aligns *with* that of the applied magnetic field, it will add to the applied magnetic field, causing the methyl protons to show a signal at a slightly higher frequency. On the other hand, if the magnetic field of the methine proton aligns *against* the applied magnetic field, it will subtract from the applied magnetic field and the methyl protons will show a signal at a lower frequency (Figure 14.11). Therefore, the signal for the methyl protons is split into two peaks, one corresponding to the higher frequency and one corresponding to the lower frequency. Because each spin state has almost the same population, about half the methine protons are lined up with the applied magnetic field and about half are lined up against it. Therefore, the two peaks of the *doublet* have approximately the same height and area.

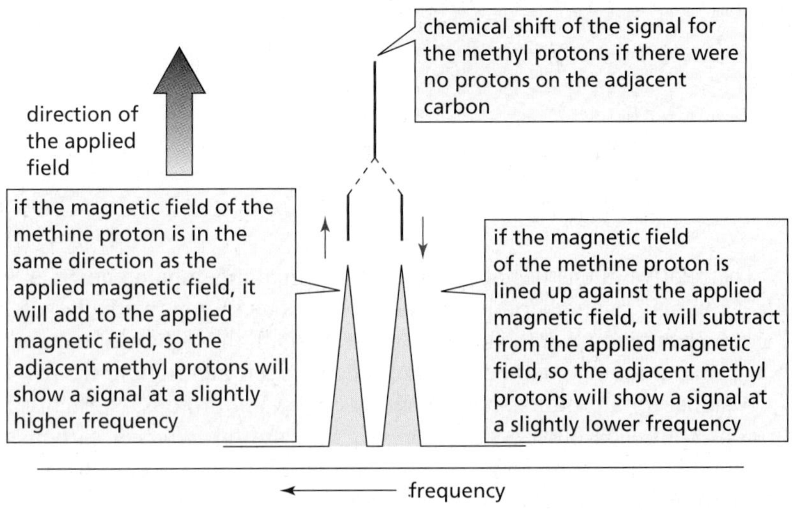
chemical shift of the signal for the methyl protons if there were no protons on the adjacent carbon
direction of the applied field
if the magnetic field of the methine proton is in the same direction as the applied magnetic field, it will add to the applied magnetic field, so the adjacent methyl protons will show a signal at a slightly higher frequency
if the magnetic field of the methine proton is lined up against the applied magnetic field, it will subtract from the applied magnetic field, so the adjacent methyl protons will show a signal at a slightly lower frequency
frequency

Figure 14.11
The signal for the methyl protons of 1,1-dichloroethane is split into a doublet by the methine proton.

Similarly, the frequency at which the methine proton shows a signal is influenced by the magnetic fields of the three protons bonded to the adjacent carbon. The magnetic fields of each of the three methyl protons can align with the applied magnetic field, two can align with the field and one against it, one can align with it and two against it, or all three can align against it. Because the magnetic field that the methine proton senses is affected in four different ways, its signal is a *quartet* (Figure 14.12).

The relative intensities of the peaks in a signal reflect the number of ways the neighboring protons can be aligned relative to the applied magnetic field. For example, a

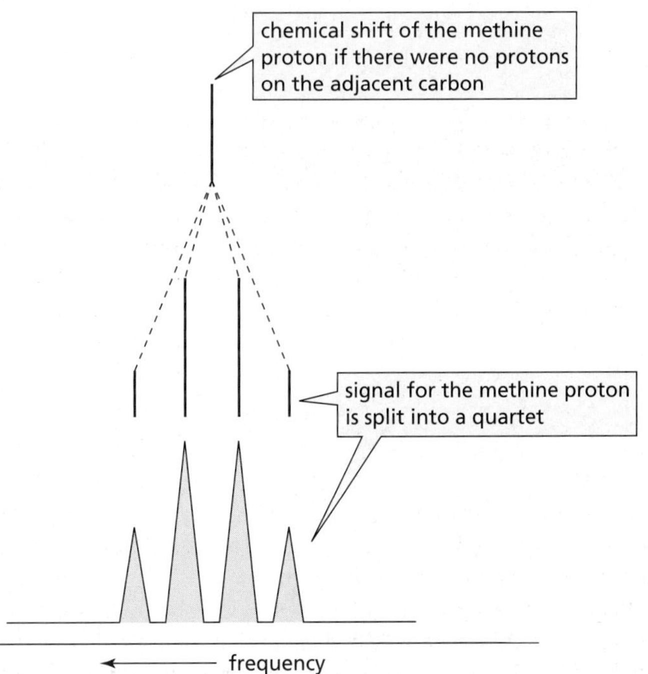
chemical shift of the methine proton if there were no protons on the adjacent carbon
signal for the methine proton is split into a quartet
frequency

Figure 14.12
The signal for the methine proton of 1,1-dichloroethane is split into a quartet by the methyl protons.

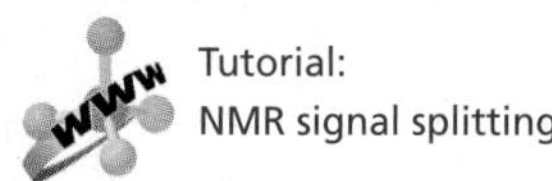

quartet has relative peak intensities of 1 : 3 : 3 : 1 because there is only one way to align the magnetic fields of three protons so that they are all with the field and only one way to align them so that they are all against the field. However, there are three ways to align the magnetic fields of three protons so that two are lined up with the field and one is lined up against the field (Figure 14.13). Likewise, there are three ways to align the magnetic fields of three protons so that one is lined up with the field and two are lined up against it.

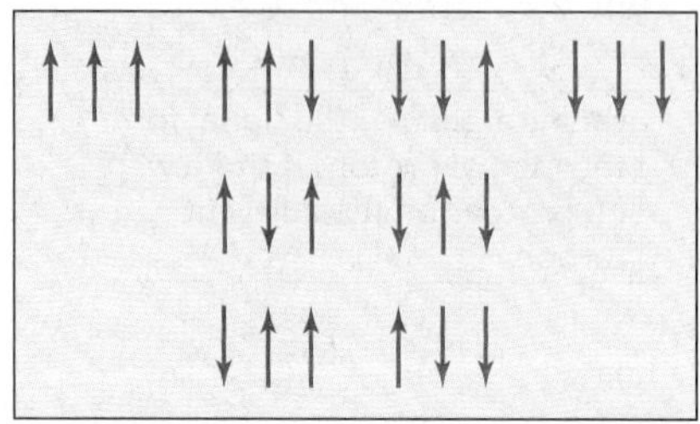

Figure 14.13 ▶ The ways in which the magnetic fields of three protons can be aligned.

The relative intensities obey the mathematical mnemonic known as *Pascal's triangle*. (You may remember this mnemonic from one of your math classes.) According to Pascal, each number at the bottom of a triangle in the rightmost column of Table 14.2 is the sum of the two numbers to its immediate left and right in the row above it.

Equivalent protons do not split each other's signal.

A signal for a proton is never split by *equivalent* protons. Normally, *nonequivalent* protons split each other's signal only if they are on *adjacent* carbons. Splitting is a "through-bond" effect, not a "through-space" effect; it is rarely observed if the protons are separated by more than three σ bonds. If, however, they are separated by more than three bonds and one of the bonds is a double or triple bond, a small splitting is sometimes observed. This is called **long-range coupling**.

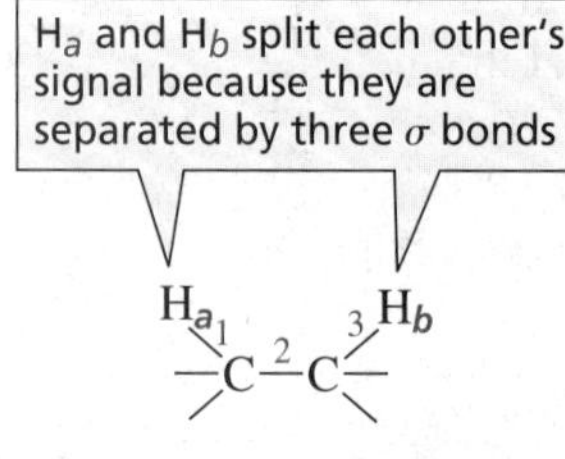

Ha and Hb split each other's signal because they are separated by three σ bonds

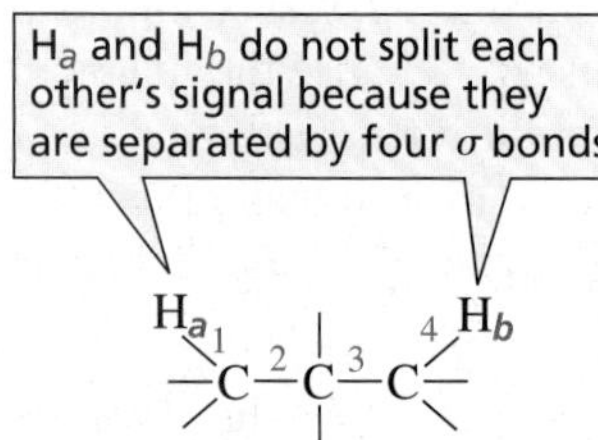

Ha and Hb do not split each other's signal because they are separated by four σ bonds

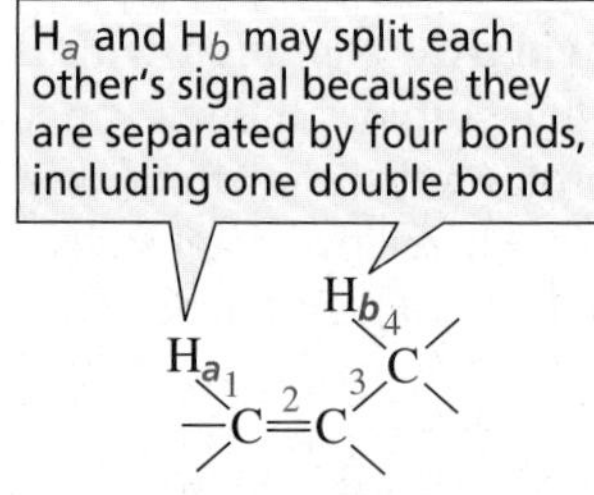

Ha and Hb may split each other's signal because they are separated by four bonds, including one double bond

Blaise Pascal (1623–1662) *was born in France. At age 16, he published a book on geometry, and at 19, he invented a calculating machine. He propounded the modern theory of probability, developed the principle underlying the hydraulic press, and showed that atmospheric pressure decreases as altitude increases. In 1644, he narrowly escaped death when the horses leading a carriage in which he was riding bolted. That scare caused him to devote the rest of his life to meditation and religious writings.*

Table 14.2 Multiplicity of the Signal and Relative Intensities of the Peaks in the Signal

Number of equivalent protons causing splitting	Multiplicity of the signal	Relative peak intensities
0	singlet	1
1	doublet	1:1
2	triplet	1:2:1
3	quartet	1:3:3:1
4	quintet	1:4:6:4:1
5	sextet	1:5:10:10:5:1
6	septet	1:6:15:20:15:6:1

PROBLEM 18

Using a diagram like the one in Figure 14.13, predict

a. the relative intensities of the peaks in a triplet
b. the relative intensities of the peaks in a quintet

PROBLEM 19◆

The ¹H NMR spectra of two carboxylic acids with molecular formula $C_3H_5O_2Cl$ are shown in Figure 14.14. Identify the carboxylic acids. (The "offset" notation means that the signal has been moved to the right by the indicated amount).

a.

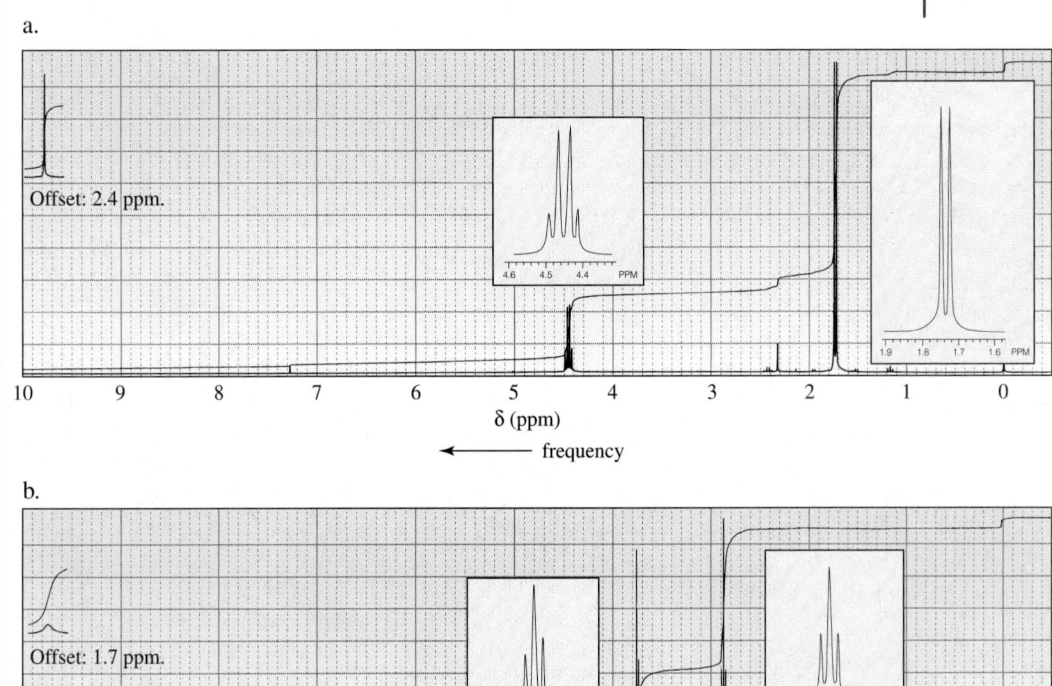

b.

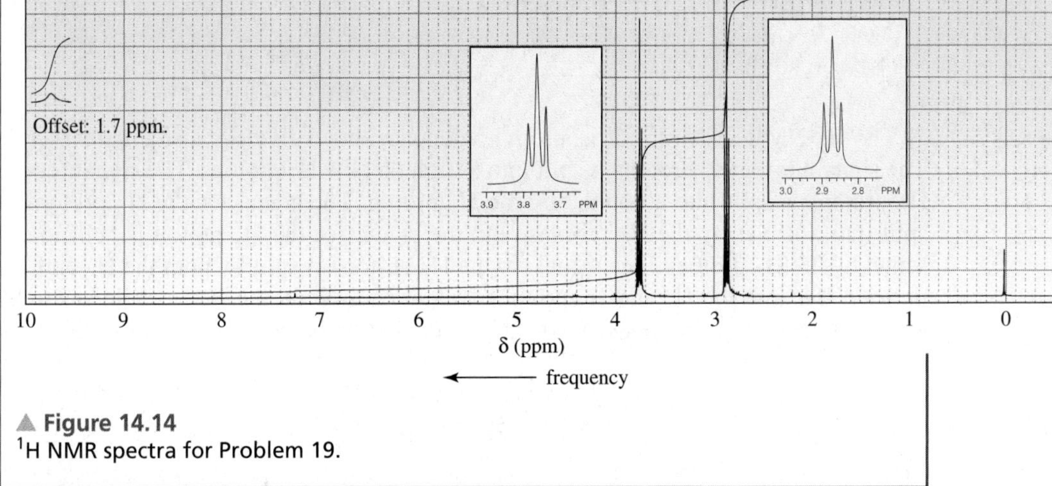

▲ **Figure 14.14**
¹H NMR spectra for Problem 19.

14.11 More Examples of ¹H NMR Spectra

The ¹H NMR spectrum of bromomethane shows one singlet. The three methyl protons are chemically equivalent, and chemically equivalent protons do not split each other's signal. The four protons in 1,2-dichloroethane are also chemically equivalent, so its ¹H NMR spectrum also shows one singlet.

CH_3Br — **bromomethane**

$ClCH_2CH_2Cl$ — **1,2-dichloroethane**

each compound has an NMR spectrum that shows one singlet because equivalent protons do not split each other's signals

There are two signals in the ^{1}H NMR spectrum of 1,3-dibromopropane (Figure 14.15). The signal for the H_b protons is split into a triplet by the two hydrogens on the adjacent carbon. The H_a protons have two adjacent carbons that are bonded to protons. The protons on one adjacent carbon are equivalent to the protons on the other adjacent carbon. Because the two sets of protons are equivalent, the $N + 1$ rule is applied to both sets at the same time. In other words, N is equal to the sum of the equivalent protons on both carbons. So the signal for the H_a protons is split into a quintet $(4 + 1 = 5)$. Integration confirms that two methylene groups contribute to the H_b signal because twice as many protons give rise to the H_b signal as to the H_a signal.

Figure 14.15 ▶ ^{1}H NMR spectrum of 1,3-dibromopropane.

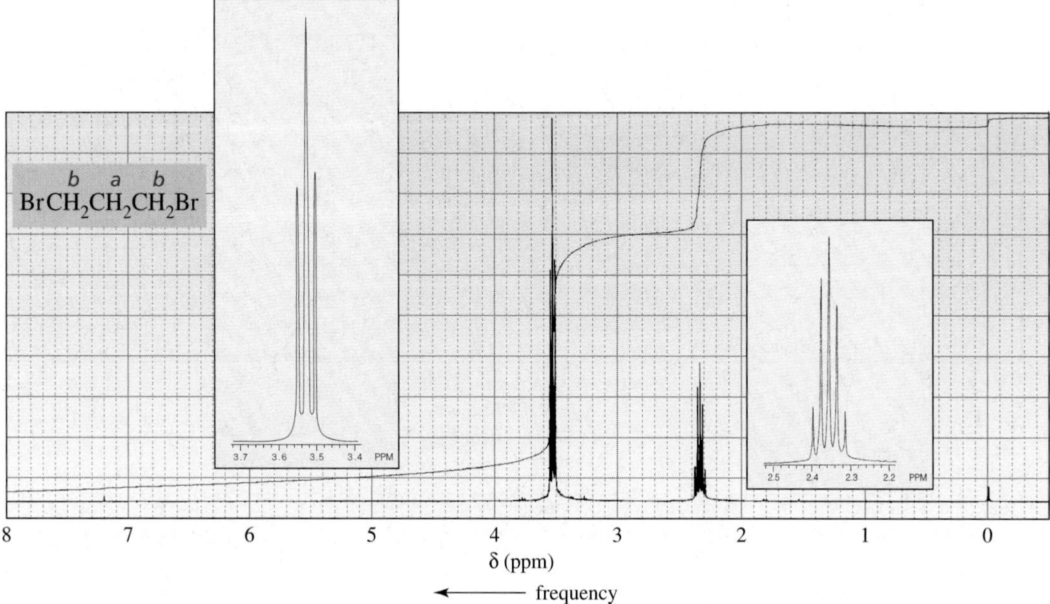

The ^{1}H NMR spectrum of isopropyl butanoate shows five signals (Figure 14.16). The signal for the H_a protons is split into a triplet by the H_c protons. The signal for the H_b protons is split into a doublet by the H_e proton. The signal for the H_d protons is split into a triplet by the H_c protons, and the signal for the H_e proton is split into a septet by the H_b protons. The signal for the H_c protons is split by both the H_a and H_d protons. Because the H_a and H_d protons are not equivalent, the $N + 1$ rule has to be applied separately to each set. Thus, the signal for the H_c protons will be split into a

Figure 14.16 ▶ ^{1}H NMR spectrum of isopropyl butanoate.

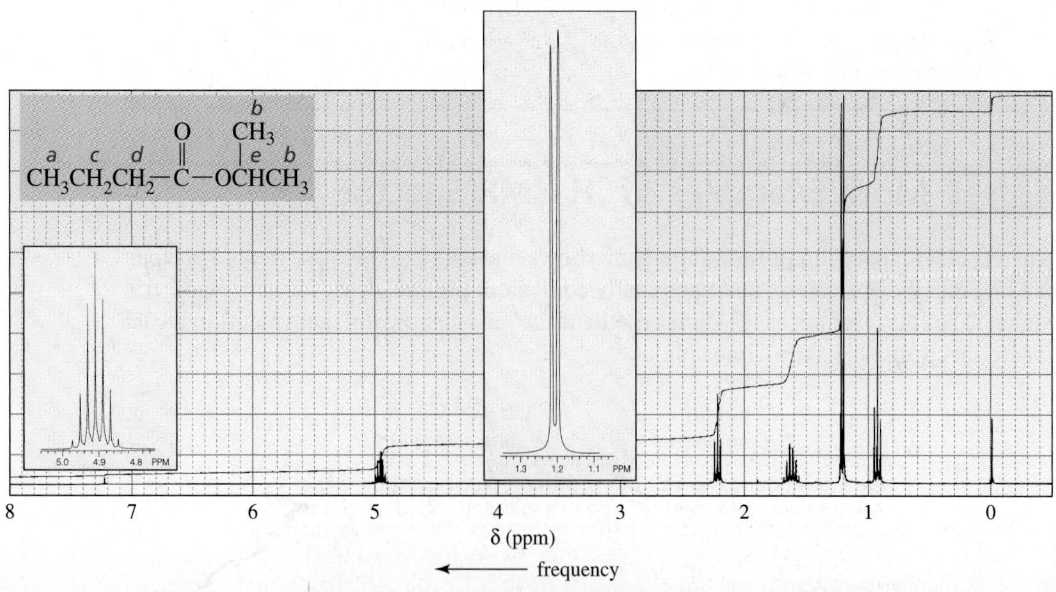

quartet by the H_a protons, and each of these four peaks will be split into a triplet by the H_d protons: $(N_a + 1)(N_d + 1) = (4)(3) = 12$. As a result, the signal for the H_c protons is a **multiplet** (a signal that is more complex that a triplet, quartet, quintet, etc.). The reason we do not see 12 peaks is that some of them overlap (Section 14.13).

PROBLEM 20

Indicate the number of signals and the multiplicity of each signal in the ^{1}H NMR spectrum of each of the following compounds:

a. $ICH_2CH_2CH_2Br$ b. $ClCH_2CH_2CH_2Cl$ c. $ICH_2CH_2CHBr_2$

The ^{1}H NMR spectrum of 3-bromo-1-propene shows four signals (Figure 14.17). Although the H_b and H_c protons are bonded to the same carbon, they are not chemically equivalent (one is cis to the bromomethyl group, the other is trans to the bromomethyl group), so each produces a separate signal. The signal for the H_a protons is split into a doublet by the H_d proton. Notice that the signals for the three vinylic protons are at relatively high frequencies because of diamagnetic anisotropy (Section 14.9). The signal for the H_d proton is a multiplet because it is split separately by the H_a, H_b, and H_c protons.

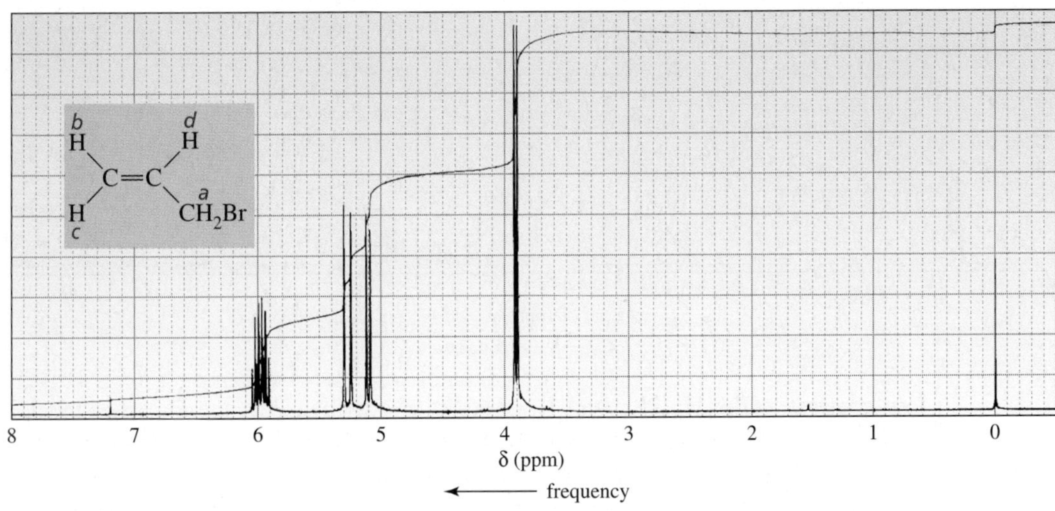

Figure 14.17 ^{1}H NMR spectrum of 3-bromo-1-propene.

Because the H_b and H_c protons are not equivalent, they split one another's signal. This means that the signal for the H_b proton is split into a doublet by the H_d proton and that each of the peaks in the doublet is split into a doublet by the H_c proton. The signal for the H_b proton should be what is called a **doublet of doublets**. The signal for the H_c proton should also be a doublet of doublets. However, the mutual splitting of the signals of two nonidentical protons bonded to the same sp^2 carbon, caused by what is called **geminal coupling**, is often too small to be observed (Section 14.12). Therefore, the signals for the H_b and H_c protons in Figure 14.17 appear as doublets rather than as doublets of doublets. (If the signals were expanded, the doublet of doublets would be observed.)

Notice the difference between a quartet and a doublet of doublets. Both have four peaks. A quartet results from splitting by *three equivalent* adjacent protons; it has relative peak intensities of 1 : 3 : 3 : 1, and the individual peaks are equally spaced. A doublet of doublets, on the other hand, results from splitting by *two nonequivalent* adjacent protons; it has relative peak intensities of 1 : 1 : 1 : 1, and the individual peaks are not necessarily equally spaced (see Figure 14.24).

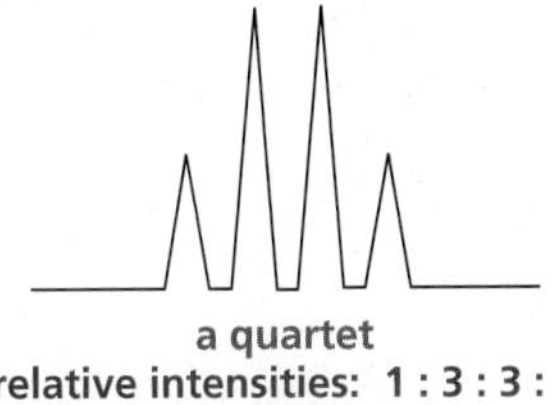

a quartet
relative intensities: 1 : 3 : 3 : 1

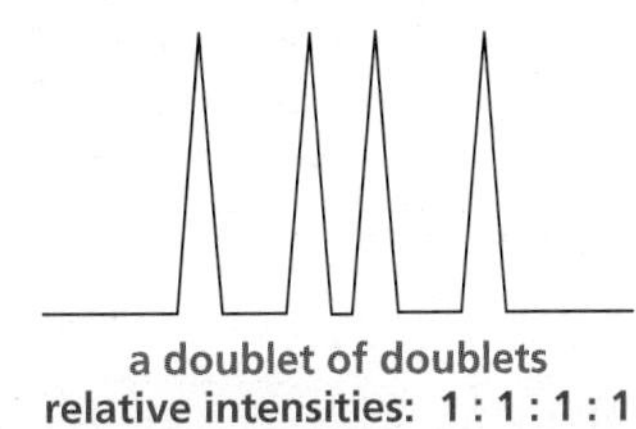

a doublet of doublets
relative intensities: 1 : 1 : 1 : 1

There are five sets of chemically equivalent protons in ethylbenzene (Figure 14.18). We see the expected triplet for the H_a protons and the quartet for the H_b protons. (This is a characteristic pattern for an ethyl group.) We expect the signal for the H_c protons to be a doublet and the signal for the H_e proton to be a triplet. Because the H_c and H_e protons are not equivalent, they must be considered separately in determining the splitting of the signal for the H_d protons. Therefore, we expect the signal for the H_d protons to be split into a doublet by the H_c protons and each peak of the doublet to be split into another doublet by the H_e proton, forming a doublet of doublets. However, we do not see three distinct signals for the H_c, H_d, and H_e protons in Figure 14.18. Instead, we see overlapping signals. Apparently, the electronic effect (i.e., the electron-donating/electron-withdrawing ability) of an ethyl substituent is not sufficiently different from that of a hydrogen to cause a difference in the environments of the H_c, H_d, and H_e protons that is large enough to allow them to appear as separate signals.

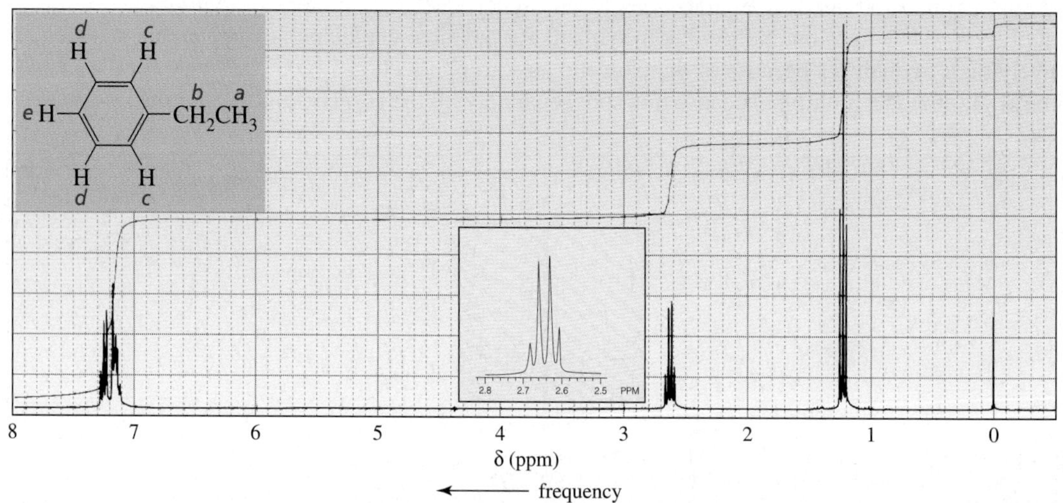

Figure 14.18 ▶ ^{1}H NMR spectrum of ethylbenzene. The signals for the H_c, H_d, and H_e protons overlap.

In contrast to the H_c, H_d, and H_e protons of ethylbenzene, the H_a, H_b, and H_c protons of nitrobenzene show three distinct signals (Figure 14.19), and the multiplicity of each signal is what we predicted for the signals for the benzene ring protons in

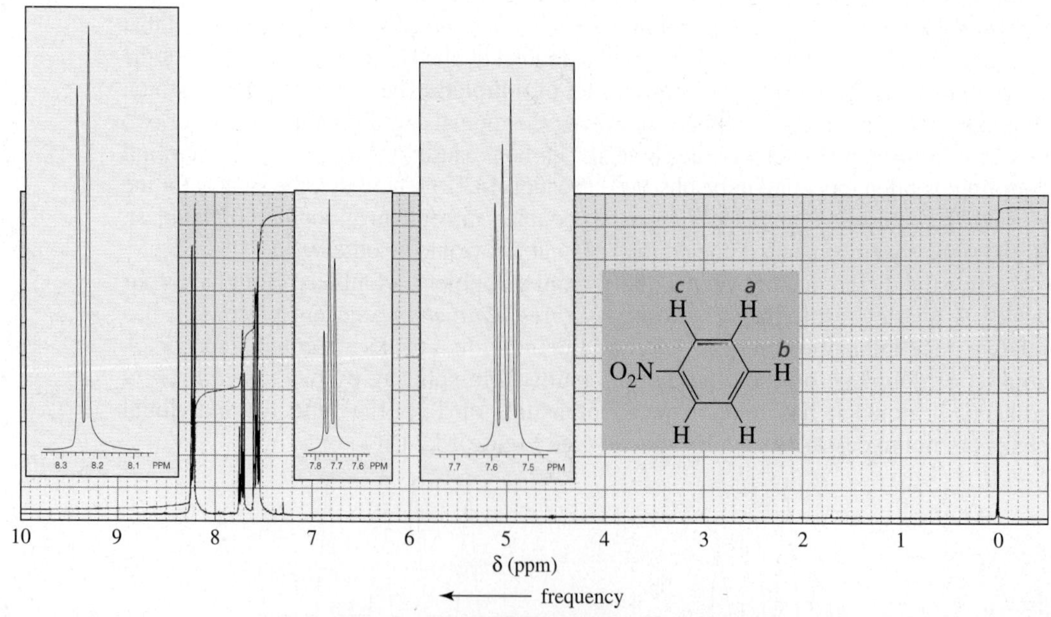

Figure 14.19 ▶ ^{1}H NMR spectrum of nitrobenzene. The signals for the H_a, H_b, and H_c protons do not overlap.

ethylbenzene (H_c is a doublet, H_b is a triplet, and H_a is a doublet of doublets). The nitro group is sufficiently electron withdrawing to cause the H_a, H_b, and H_c protons to be in different enough environments that their signals do not overlap.

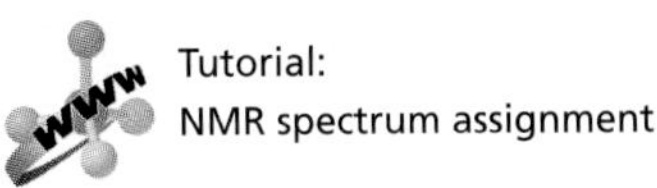

Tutorial:
NMR spectrum assignment

Notice that the signals for the benzene ring protons in Figures 14.18 and 14.19 occur in the 7.0–8.5 ppm region. Other kinds of protons usually do not resonate in this region, so signals in this region of an ^{1}H NMR spectrum indicate that the compound probably contains an aromatic ring.

PROBLEM 21

Explain why the signal for the protons identified as H_a in Figure 14.19 appears at the lowest frequency and the signal for the protons identified as H_c appears at the highest frequency. (*Hint:* Draw the contributing resonance structures.)

PROBLEM 22

How could ^{1}H NMR spectra distinguish the following compounds?

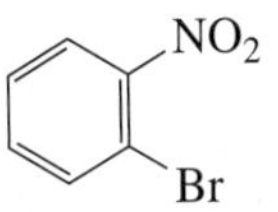

NO_2 / Br (A) NO_2 / Br (B) NO_2 / Br (C)

A 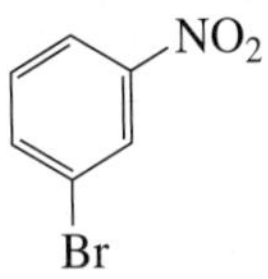 B 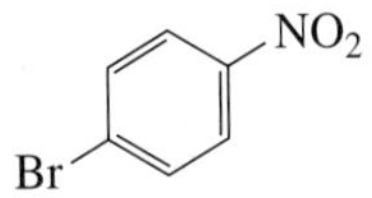 C

PROBLEM 23

How would the ^{1}H NMR spectra for the four compounds with molecular formula $C_3H_6Br_2$ differ?

PROBLEM 24◆

Identify each compound from its molecular formula and its ^{1}H NMR spectrum:

a. C_9H_{12}

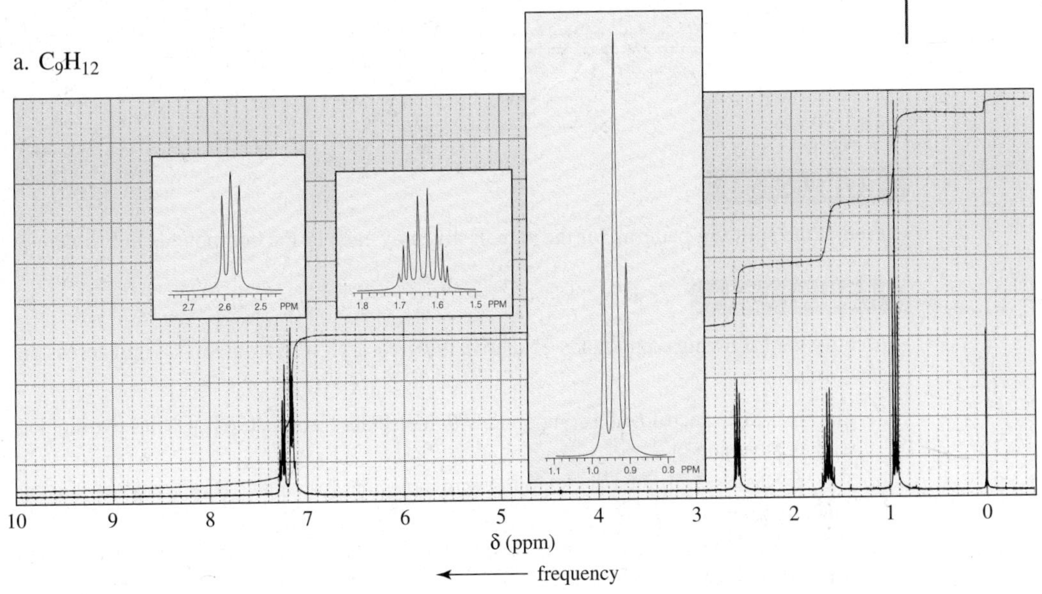

b. $C_5H_{10}O$

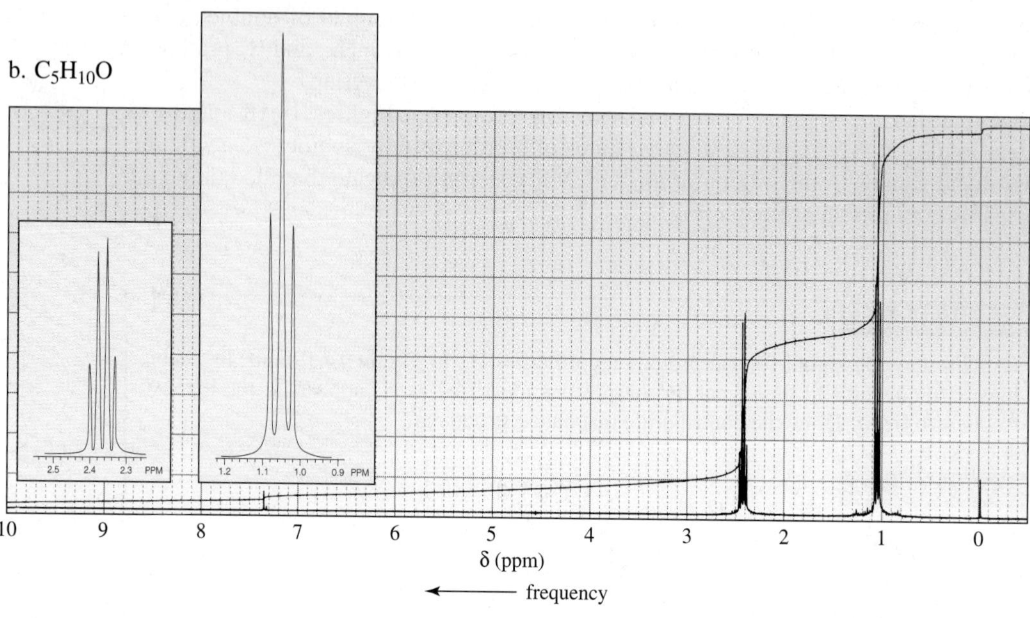

c. $C_9H_{10}O_2$

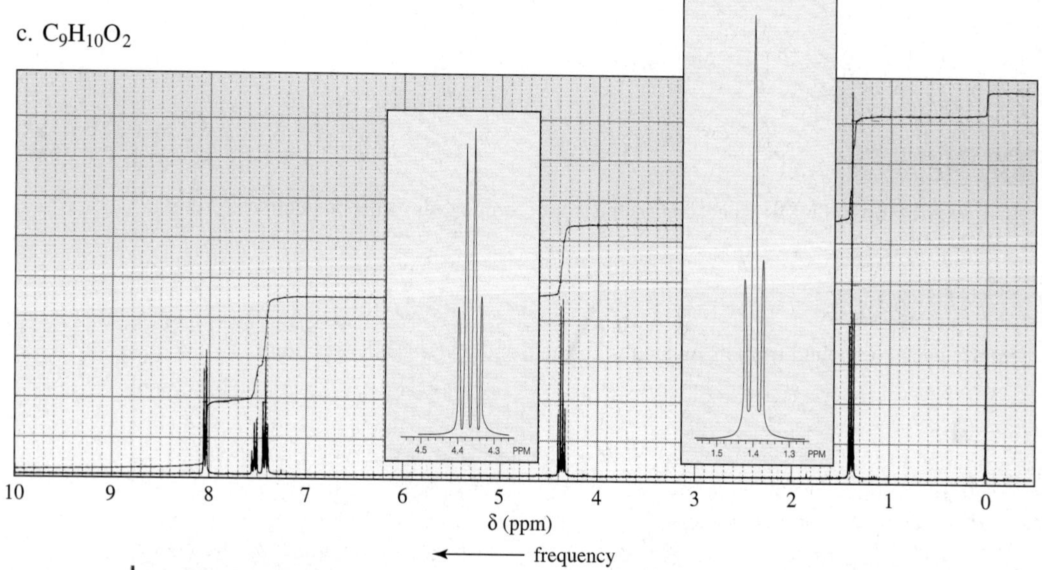

PROBLEM 25

Predict the splitting patterns for the signals given by each of the compounds in Problem 3.

PROBLEM 26◆

Identify the following compounds. (Relative integrals are given from left to right across the spectrum.)

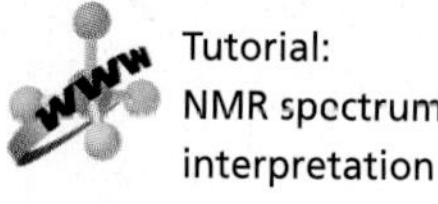
Tutorial: NMR spectrum interpretation

a. The ^{1}H NMR spectrum of a compound with molecular formula $C_4H_{10}O_2$ has two singlets with an area ratio of 2 : 3.
b. The ^{1}H NMR spectrum of a compound with molecular formula $C_6H_{10}O_2$ has two singlets with an area ratio of 2 : 3.
c. The ^{1}H NMR spectrum of a compound with molecular formula $C_8H_6O_2$ has two singlets with an area ratio of 1 : 2.

PROBLEM 27

Describe the ^{1}H NMR spectrum you would expect for each of the following compounds, using relative chemical shifts rather than absolute chemical shifts:

a. $BrCH_2CH_2Br$

b. $CH_3OCH_2CH_2CH_2Br$

c. O= =O

d. $CH_3\overset{CH_3}{\underset{Br}{C}}CH_2CH_3$

e. $CH_3\overset{O}{\overset{\|}{C}}CH_2\overset{O}{\overset{\|}{C}}OCH_3$

f. $\overset{H}{\underset{H}{}}C{=}C\overset{H}{\underset{Cl}{}}$

g. $CH_3CH_2OCH_2CH_3$

h. $CH_3CH_2OCH_2Cl$

i. $CH_3\underset{Cl}{C}HCHCl_2$

j. O

k. $CH_3\overset{CH_3}{C}HCH_2\overset{O}{\overset{\|}{C}}H$

l. $CH_3OCH_2CH_2CH_2OCH_3$

m. $\overset{H}{\underset{H}{}}C{=}C\overset{Cl}{\underset{Cl}{}}$

n. $\overset{Cl}{\underset{H}{}}C{=}C\overset{H}{\underset{Cl}{}}$

o.

14.12 Coupling Constants

The distance, in hertz, between two adjacent peaks of a split NMR signal is called the **coupling constant** (denoted by ***J***). The coupling constant for H_a being split by H_b is denoted by J_{ab}. The signals of coupled protons (protons that split each other's signal) have the same coupling constant; in other words, $J_{ab} = J_{ba}$ (Figure 14.20). Coupling constants are useful in analyzing complex NMR spectra because protons on adjacent carbons can be identified by identical coupling constants.

The magnitude of a coupling constant is independent of the operating frequency of the spectrometer—the same coupling constant is obtained from a 300-MHz instrument as from a 600-MHz instrument. The magnitude of a coupling constant is a measure of how strongly the nuclear spins of the coupled protons influence each other. It, therefore, depends on the number and type of bonds that connect the coupled protons, as well as the geometric relationship of the protons. Characteristic coupling constants are shown in Table 14.3; they range from 0 to 15 Hz.

The coupling constant for two nonequivalent hydrogens on the *same* sp^2 carbon is often too small to be observed (Figure 14.17), but it is large for nonequivalent hydrogens bonded to *adjacent* sp^2 carbons. Apparently, the interaction between the hydrogens is strongly affected by the intervening π electrons. We have seen that π electrons also allow long-range coupling—that is, coupling through four or more bonds (Section 14.10).

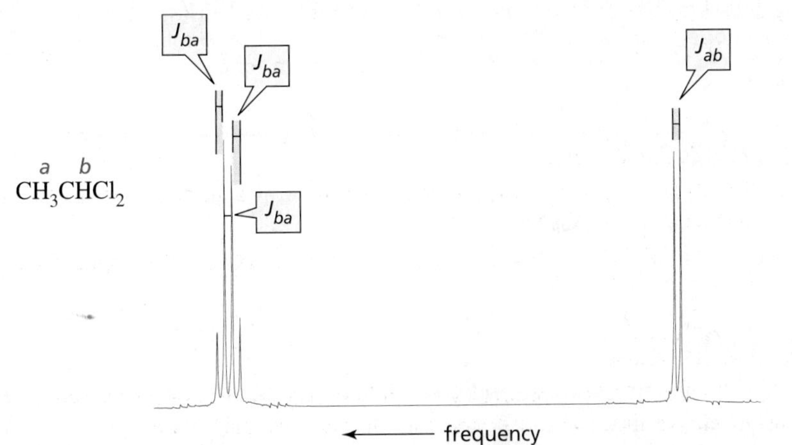

◀ **Figure 14.20**
The H_a and H_b protons of 1,1-dichloroethane are coupled protons, so their signals have the same coupling constant, $J_{ab} = J_{ba}$.

Table 14.3 Approximate Values of Coupling Constants

Approximate value of J_{ab} (Hz)		Approximate value of J_{ab} (Hz)	
H_a–C–C–H_b	7	H_a–C=C–H_b (trans)	15 (trans)
H_a–C–C–C–H_b	0	H_a–C=C–H_b (cis)	10 (cis)
C=C(H_a)(H_b)	2 (geminal coupling)	H_a–C=C–C–H_b	1 (long-range coupling)

The dependence of the coupling constant on the angle between the two C—H bonds is called the Karplus relationship after **Martin Karplus,** *who first observed the relationship. Karplus was born in 1930. He received a B.A. from Harvard University and a Ph.D. from the California Institute of Technology. He is currently a professor of chemistry at Harvard University.*

Coupling constants can be used to distinguish between the ¹H NMR spectra of cis and trans alkenes. The coupling constant of *trans*-vinylic protons is significantly greater than the coupling constant of *cis*-vinylic protons (Figure 14.21), because the coupling constant depends on the dihedral angle between the two C—H bonds in the H—C=C—H unit. The coupling constant is greatest when the angle between the two C—H bonds is 180° (trans) and smaller when it is 0° (cis). Notice the difference between J_{bd} and J_{cd} in the spectrum of 3-bromo-1-propene (Figure 14.17).

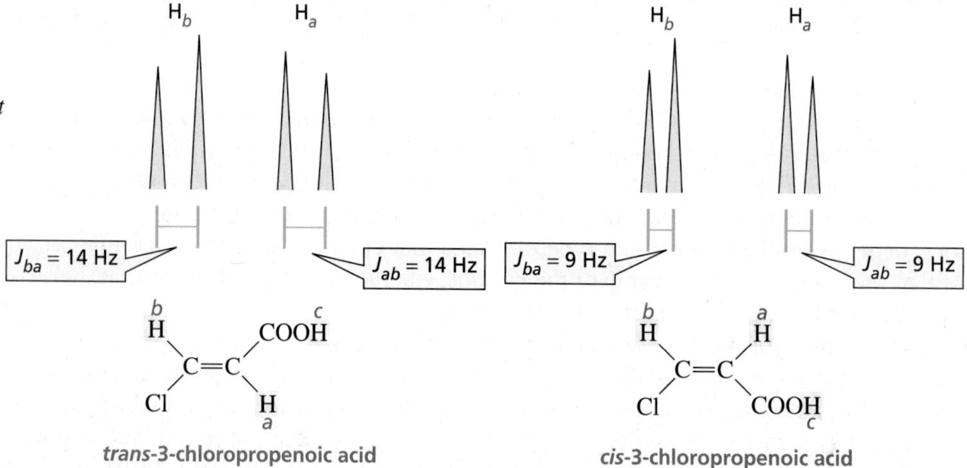

Figure 14.21
The doublets observed for the H_a and H_b protons in the ¹H NMR spectra of *trans*-3-chloropropenoic acid and *cis*-3-chloropropenoic acid. The coupling constant for trans protons (14 Hz) is greater than the coupling constant for cis protons (9 Hz).

The coupling contant for trans protons is greater than the coupling constant for cis protons.

PROBLEM 28

Why is there no coupling between H_a and H_c or between H_b and H_c in *cis*- or *trans*-3-chloropropenoic acid?

Let's now summarize the kind of information that can be obtained from an ¹H NMR spectrum:

1. The number of signals indicates the number of different kinds of protons that are in the compound.

2. The position of a signal indicates the kind of proton(s) responsible for the signal (methyl, methylene, methine, allylic, vinylic, aromatic, etc.) and the kinds of neighboring substituents.
3. The integration of the signal tells the relative number of protons responsible for the signal.
4. The multiplicity of the signal $(N + 1)$ tells the number of protons (N) bonded to adjacent carbons.
5. The coupling constants identify coupled protons.

PROBLEM-SOLVING STRATEGY

Identify the compound with molecular formula $C_9H_{10}O$ that gives the IR and ^{1}H NMR spectra in Figure 14.22.

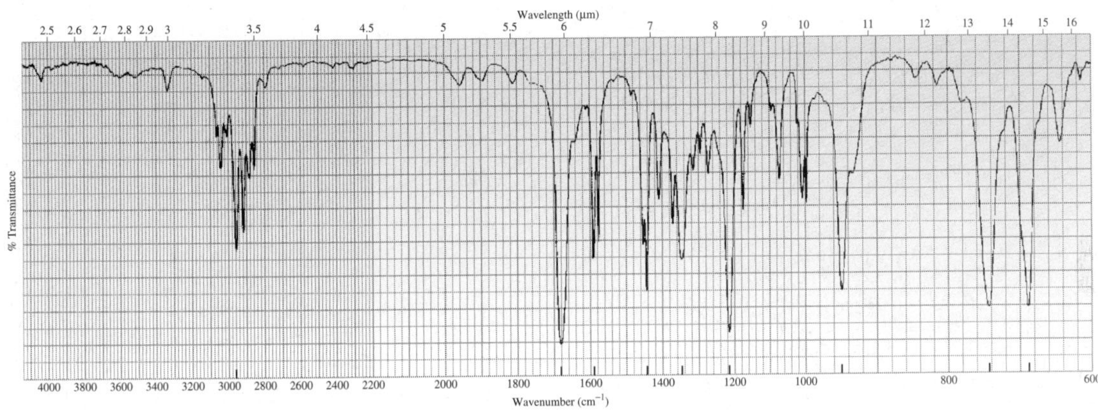

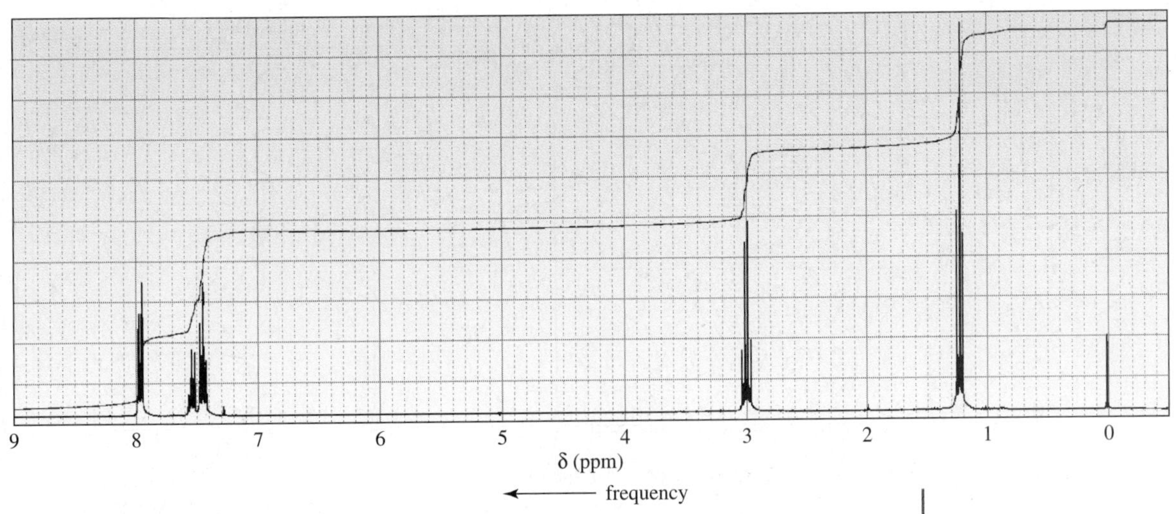

▲ **Figure 14.22**
IR and ^{1}H NMR spectra for this problem-solving strategy.

The best way to approach this kind of problem is to identify whatever structural features you can from the molecular formula and IR spectrum and then use the information from the ^{1}H NMR spectrum to expand on that knowledge. From the molecular formula and IR spectrum, we learn that the compound is a ketone: It has a carbonyl group at ~1680 cm^{-1}, only one oxygen, and no absorption bands at ~2820 and ~2720 cm^{-1} that would indicate an aldehyde. That the carbonyl group absorption is at a lower frequency than normal suggests that it has partial single-bond character as a result of electron delocalization—indicating that it is attached to an sp^2 carbon. The compound contains a benzene ring (>3000 cm^{-1}, ~1600 cm^{-1}, and 1440 cm^{-1}), and it has hydrogens bonded to sp^3 carbons (<3000 cm^{-1}). In the NMR spectrum, the triplet at ~1.2 ppm and the quartet at ~3.0 ppm indicate the presence of an ethyl group that is attached to an electron-withdrawing group. The signals in the 7.4–8.0 ppm region confirm the presence of a benzene ring. From this information, we can conclude that the compound is the following ketone. The integration ratio (5 : 2 : 3) confirms this answer.

$$\text{C}_6\text{H}_5-\overset{\overset{\displaystyle O}{\|}}{C}CH_2CH_3$$

Now continue on to Problem 29.

PROBLEM 29◆

Identify the compound with molecular formula $C_8H_{10}O$ that gives the IR and ^{1}H NMR spectra shown in Figure 14.23.

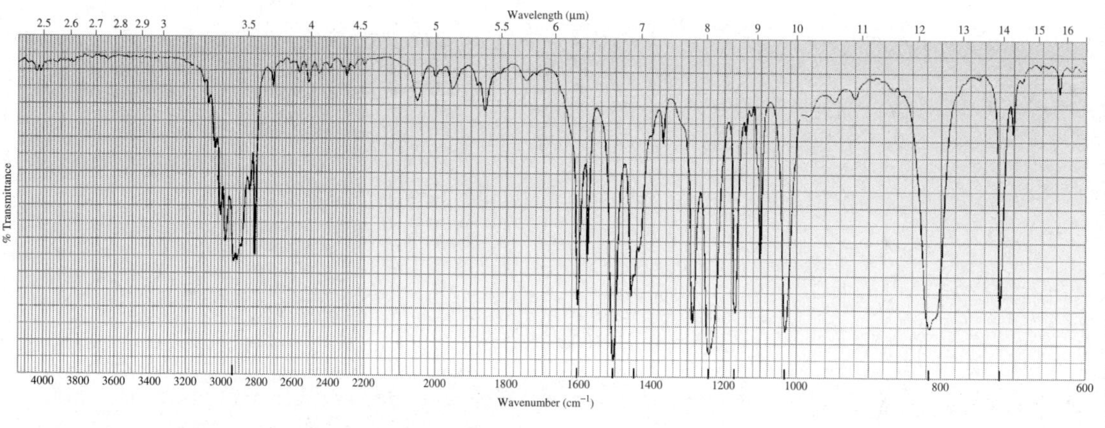

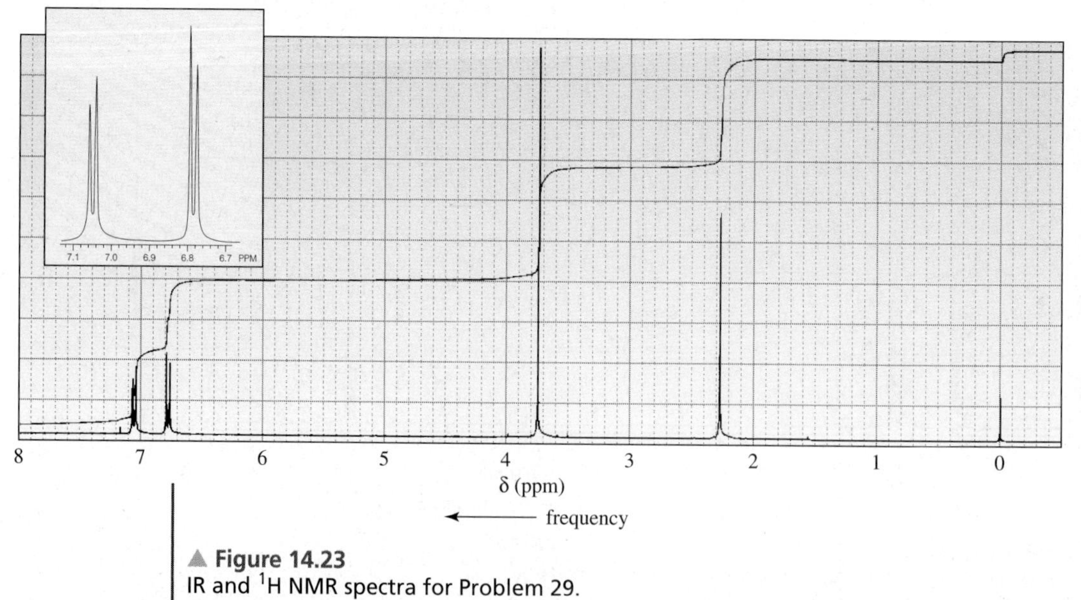

▲ **Figure 14.23**
IR and ^{1}H NMR spectra for Problem 29.

14.13 Splitting Diagrams

The splitting pattern obtained when a signal is split by more than one set of protons can best be understood by using a splitting diagram. In a **splitting diagram** (also called a **splitting tree**), the NMR peaks are shown as vertical lines and the effect of each of the splittings is shown one at a time. For example, a splitting diagram is shown in Figure 14.24 for splitting of the signal for the H_c proton of 1,1,2-trichloro-3-methylbutane into a doublet of doublets by the H_b and H_d protons.

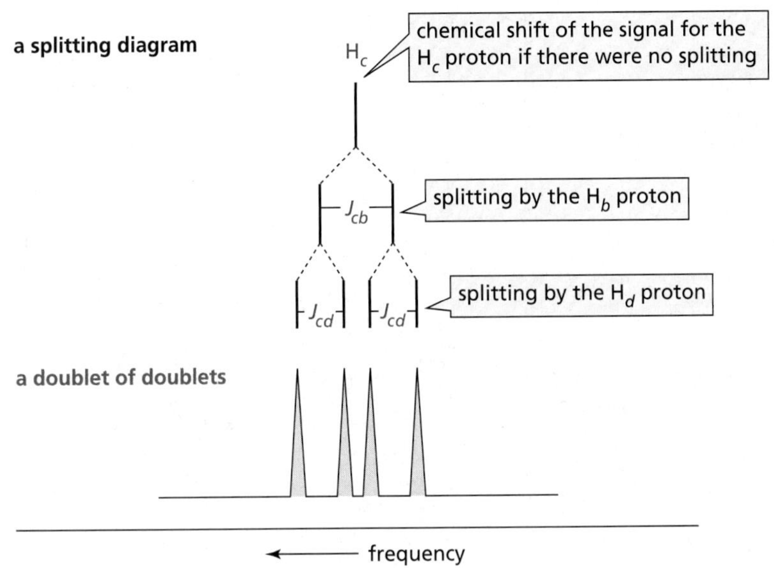

◀ **Figure 14.24**
A splitting diagram for a doublet of doublets.

The signal for the H_b protons of propyl bromide is split into a quartet by the H_a protons, and each of the resulting four peaks is split into a triplet by the H_c protons (Figure 14.25). How many of the 12 peaks are actually seen depends on the relative magnitudes of the two coupling constants, J_{ba} and J_{bc}. For example, the figure shows that there are 12 peaks when J_{ba} is much greater than J_{bc}, 9 peaks when $J_{ba} = 2J_{bc}$, and only 6 peaks when $J_{ba} = J_{bc}$. As you can see, the number of peaks actually observed depends on how many overlap with one another. When peaks overlap, their intensities add together.

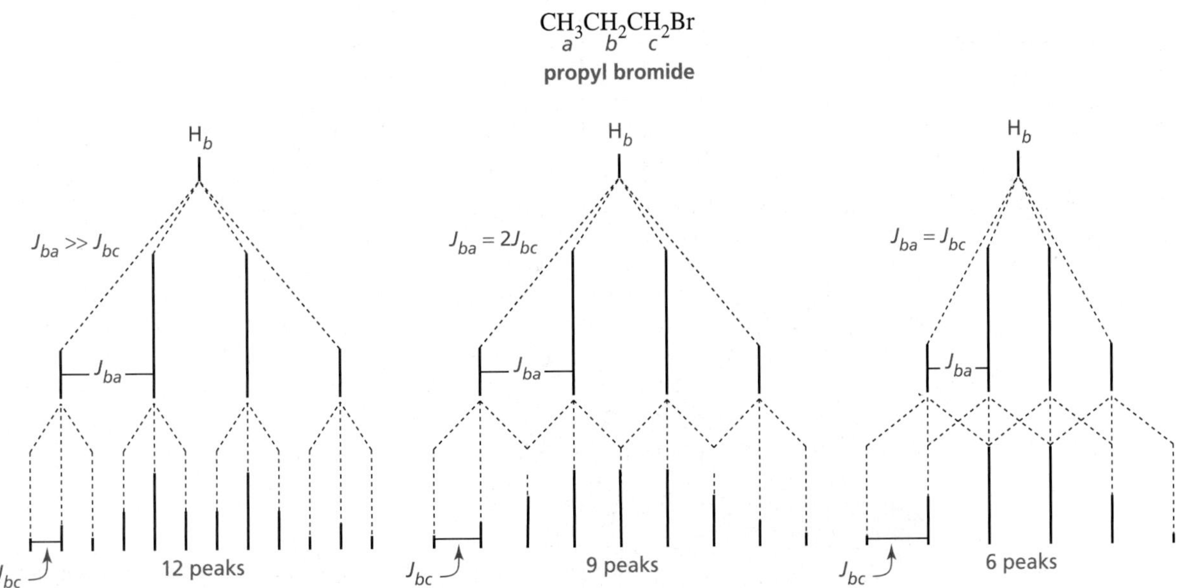

▲ **Figure 14.25**
A splitting diagram for a quartet of triplets. The number of peaks actually observed when a signal is split by two sets of protons depends on the relative magnitudes of the two coupling constants.

We would expect the signal for the H_a protons of 1-chloro-3-iodopropane to be a triplet of triplets (split into nine peaks) because the signal would be split into a triplet by the H_b protons and each of the resulting peaks into a triplet by the H_c protons. The signal, however, is a quintet (Figure 14.26).

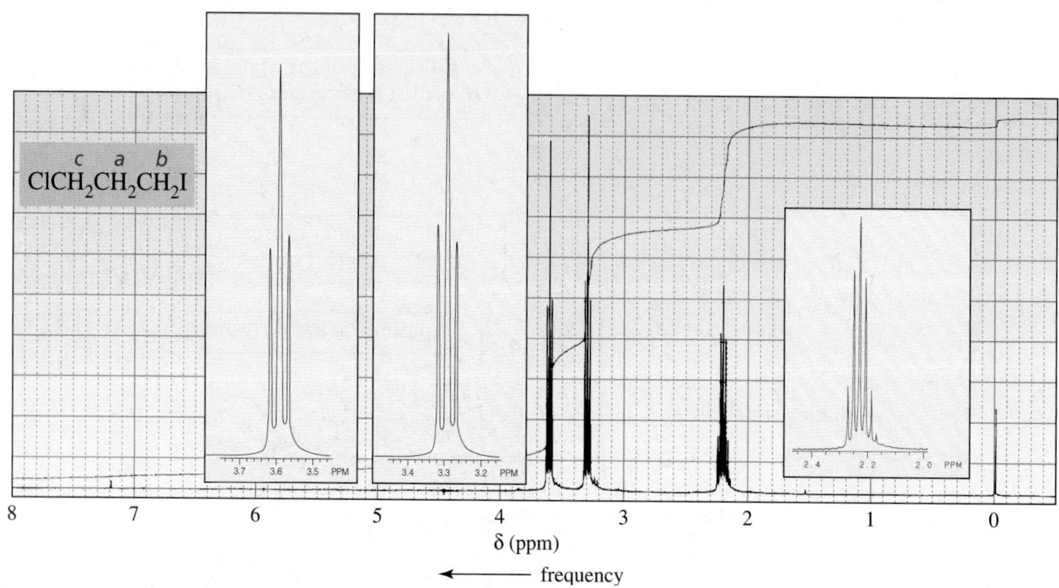

▲ **Figure 14.26**
^{1}H NMR spectrum of 1-chloro-3-iodopropane.

Finding that the signal for the H_a protons of 1-chloro-3-iodopropane is a quintet indicates that J_{ab} and J_{ac} have about the same value. The splitting diagram shows that a quintet results if $J_{ab} = J_{ac}$.

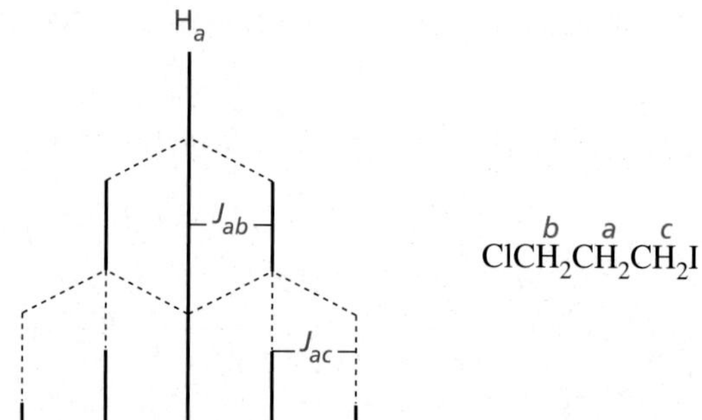

We can conclude that *when two different sets of protons split a signal, the multiplicity of the signal should be determined by using the* N + *1 rule separately for each set of hydrogens when the coupling constants for the two sets are different. When the coupling constants are similar, however, the multiplicity of a signal can be determined by treating both sets of adjacent hydrogens as if they were equivalent.* In other words, the $N + 1$ rule can be applied to both sets of protons simultaneously.

PROBLEM 30 SOLVED

The two hydrogens of a methylene group adjacent to an asymmetric carbon are not equivalent hydrogens because they are in different environments due to the asymmetric carbon. (You can verify this statement by examining molecular models.) Applying the $N + 1$ rule to these two diastereotopic hydrogens (Section 5.16) separately in determining the multiplicity of the signal for the adjacent methyl hydrogens indicates that the signal should be a doublet of doublets. The signal, however, is a triplet. Using a splitting diagram, explain why it is a triplet rather than a doublet of doublets.

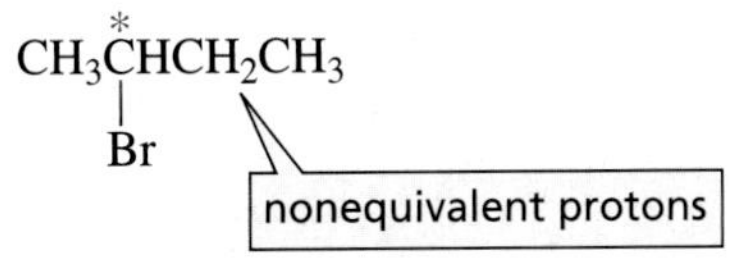

$CH_3\overset{*}{C}HCH_2CH_3$ (Br on the starred carbon)

nonequivalent protons

SOLUTION The observation of a triplet means that the $N + 1$ rule did not have to be applied to the diastereotopic hydrogens separately but could have been applied to the two protons as a set ($N = 2$, so $N + 1 = 3$). This means that the coupling constant for splitting of the methyl signal by one of the methylene hydrogens is similar to the coupling constant for splitting by the other methylene hydrogen.

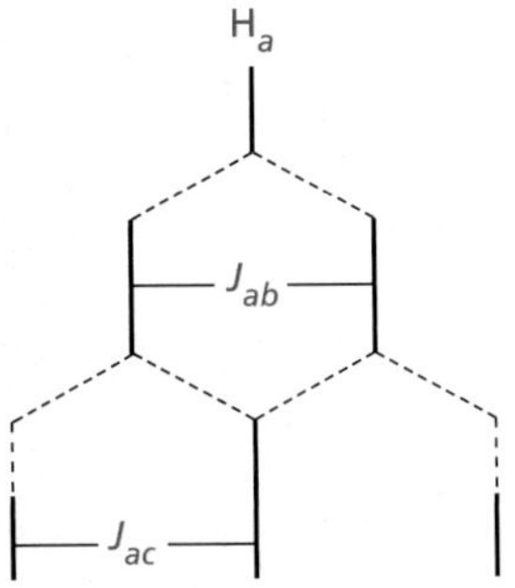

H_a

J_{ab}

J_{ac}

PROBLEM 31

Draw a splitting diagram for H_b, where

a. $J_{ba} = 12$ Hz and $J_{bc} = 6$ Hz

b. $J_{ba} = 12$ Hz and $J_{bc} = 12$ Hz

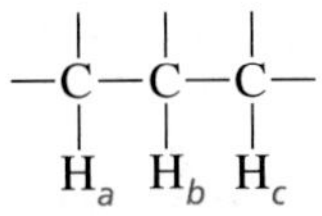

—C—C—C— with H_a H_b H_c

14.14 Time Dependence of NMR Spectroscopy

We have seen that the three methyl hydrogens of ethyl bromide give rise to one signal in the ^1H NMR spectrum because they are chemically equivalent due to rotation about the C—C bond. At any one instant, however, the three hydrogens can be in quite different environments: One can be anti to the bromine, one can be gauche to the bromine, one can be eclipsed with the bromine.

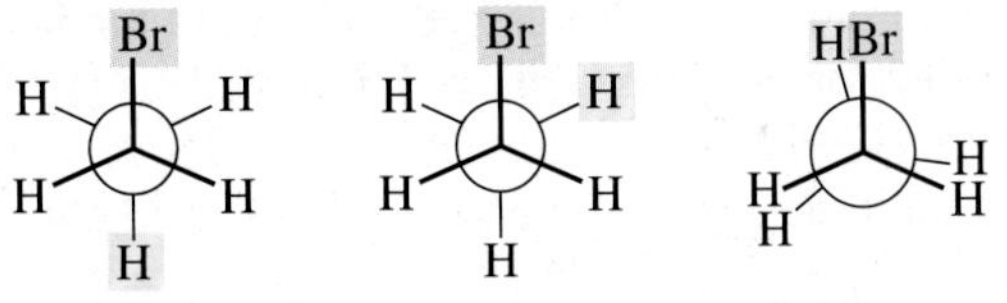

anti gauche eclipsed

An NMR spectrometer is very much like a camera with a slow shutter speed—it is too slow to be able to detect these different environments, so what it sees is an average environment. Because each of the three methyl hydrogens has the same average environment, we see one signal for the methyl group in the ^1H NMR spectrum.

Similarly, the ^1H NMR spectrum of cyclohexane shows only one signal, even though cyclohexane has both axial and equatorial protons. There is only one signal because the interconversion of the chair conformers of cyclohexane occurs too rapidly at

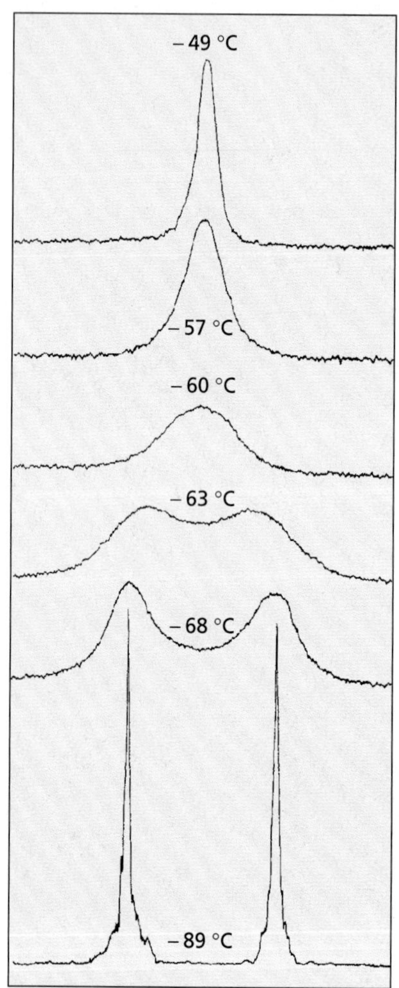

▲ **Figure 14.27**
^{1}H NMR spectra of cyclohexane-d_{11} at various temperatures.

room temperature to be detected by the NMR spectrometer. Because axial protons in one chair conformer are equatorial protons in the other chair conformer, all the protons in cyclohexane have the same average environment on the NMR time scale, so the NMR spectrum shows one signal.

axial
H
H
equatorial
chair–chair interconversion
equatorial
H
H
axial

The rate of chair–chair interconversion is temperature-dependent—the lower the temperature, the slower is the rate of interconversion. Cyclohexane-d_{11} has 11 deuteriums, which means that it has only one hydrogen. ^{1}H NMR spectra of cyclohexane-d_{11} taken at various temperatures are shown in Figure 14.27. Cyclohexane with only one hydrogen was used for this experiment in order to prevent splitting of the signal, which would have complicated the spectrum. Deuterium signals are not detectable in ^{1}H NMR, and splitting by a deuterium on the same or on an adjacent carbon is not normally detectable at the operating frequency of an ^{1}H NMR spectrometer.

At room temperature, the ^{1}H NMR spectrum of cyclohexane-d_{11} shows one sharp signal, which is an average for the axial proton of one chair and the equatorial proton of the other chair. As the temperature decreases, the signal becomes broader and eventually separates into two signals, which are equidistant from the original signal. At −89 °C, two sharp singlets are observed because at that temperature, the rate of chair–chair interconversion has decreased sufficiently to allow the two kinds of protons (axial and equatorial) to be individually detected on the NMR time scale.

14.15 Protons Bonded to Oxygen and Nitrogen

The chemical shift of a proton bonded to an oxygen or a nitrogen depends on the degree to which the proton is hydrogen bonded—the greater the extent of hydrogen bonding, the greater is the chemical shift—because the extent of hydrogen bonding affects the electron density around the proton. For example, the chemical shift of the OH proton of an alcohol ranges from 2 to 5 ppm; the chemical shift of the OH proton of a carboxylic acid, from 10 to 12 ppm; the chemical shift of the NH proton of an amine, from 1.5 to 4 ppm; and the chemical shift of the NH proton of an amide, from 5 to 8 ppm.

The ^{1}H NMR spectrum of pure dry ethanol is shown in Figure 14.28(a), and the ^{1}H NMR spectrum of ethanol with a trace amount of acid is shown in Figure 14.28(b). The spectrum shown in Figure 14.28(a) is what we would predict from what we have learned so far. The signal for the proton bonded to oxygen is farthest downfield and is split into a triplet by the neighboring methylene protons; the signal for the methylene protons is split into a multiplet by the combined effects of the methyl protons and the OH proton.

The spectrum shown in Figure 14.28(b) is the type of spectrum most often obtained for alcohols. The signal for the proton bonded to oxygen is not split, and this proton does not split the signal of the adjacent protons. So the signal for the OH proton is a singlet, and the signal for the methylene protons is a quartet because it is split only by the methyl protons.

The two spectra differ because protons bonded to oxygen undergo **proton exchange**, which means that they are transferred from one molecule to another. Whether the OH proton and the methylene protons split each other's signals depends on how long a particular proton stays on the OH group.

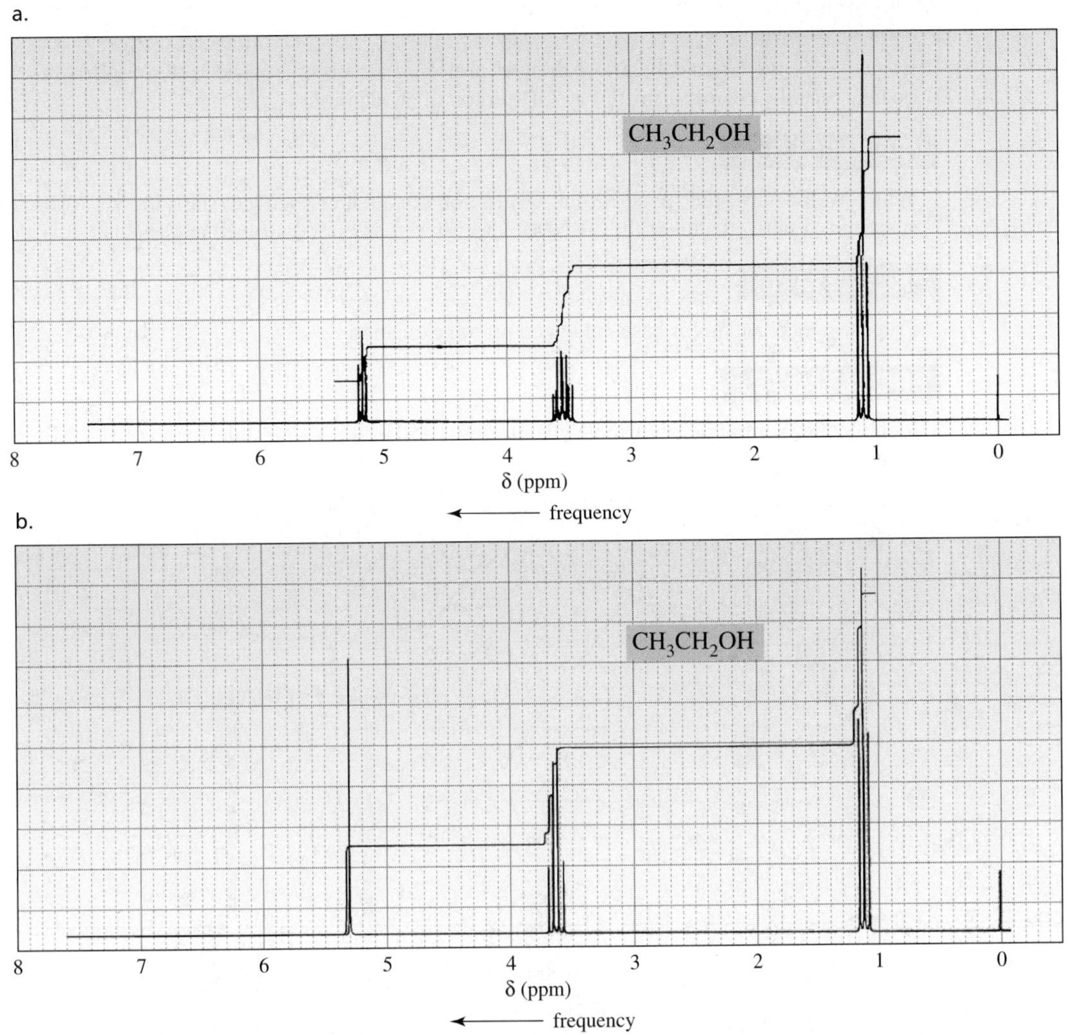

▲ **Figure 14.28**
(a) ¹H NMR spectrum of pure ethanol. (b) ¹H NMR spectrum of ethanol containing a trace amount of acid.

In a sample of pure alcohol, the rate of proton exchange is very slow. This causes the spectrum to look identical to one that would be obtained if proton exchange did not occur. Acids and bases catalyze proton exchange, so if the alcohol is contaminated with just a trace of acid or base, proton exchange becomes rapid. When proton exchange is rapid, the spectrum records only an average of all possible environments. Therefore a rapidly exchanging proton is recorded as a singlet. The effect of a rapidly exchanging proton on adjacent protons is also averaged. Thus, not only is its signal not split by adjacent protons, the rapidly exchanging proton does not cause splitting.

mechanism for acid-catalyzed proton exchange

$$R\ddot{O}{-}H + H\overset{H}{\underset{+}{O}}H \rightleftharpoons R\overset{H}{\underset{+}{\ddot{O}}}{-}H + H\ddot{O}H \rightleftharpoons R\overset{H}{\ddot{O}}: + H\overset{H}{\underset{+}{\ddot{O}}}H$$

The signal for an OH proton is often easy to spot in an ¹H NMR spectrum because it is frequently somewhat broader than other signals (see the signal at δ 4.9 in Figure 14.30(b)). The broadening occurs because the rate of proton exchange is not slow enough to result in a cleanly split signal, as in Figure 14.28(a), or fast enough for a cleanly averaged signal, as in Figure 14.28(b). NH protons also show broad signals, not because of chemical exchange, which is generally quite slow for NH, but because of quadrupolar relaxation.

PROBLEM 32

Explain why the chemical shift of the OH proton of a carboxylic acid is at a higher frequency than the chemical shift of an OH proton of an alcohol.

PROBLEM 33◆

Which would show the signal for the OH proton at a greater chemical shift, the ^1H NMR spectrum of pure ethanol or the ^1H NMR spectrum of ethanol dissolved in CH_2Cl_2?

PROBLEM 34

Propose a mechanism for base-catalyzed proton exchange.

PROBLEM 35◆

Identify the compound with molecular formula C_3H_7NO responsible for the ^1H NMR spectrum in Figure 14.29.

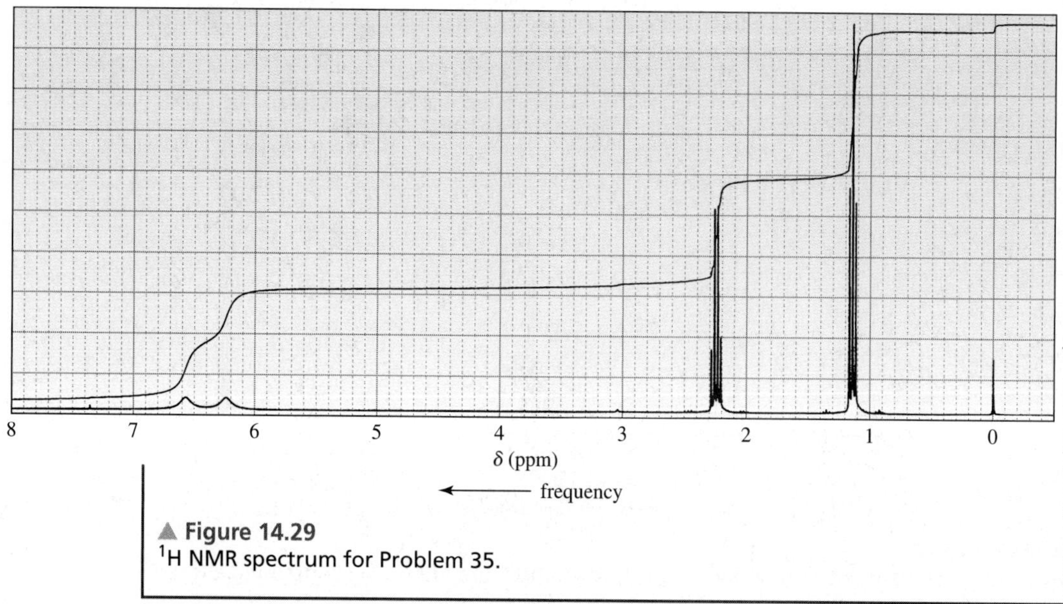

▲ **Figure 14.29**
^1H NMR spectrum for Problem 35.

14.16 Use of Deuterium in ^1H NMR Spectroscopy

Because deuterium signals are not seen in an ^1H NMR spectrum, substituting a deuterium for a hydrogen is a technique used to identify signals and to simplify ^1H NMR spectra (Section 14.14).

If after an alcohol's ^1H NMR spectrum is obtained, a few drops of D_2O are added to the sample and the spectrum is taken again, the OH signal can be identified. It will be the signal that becomes less intense (or disappears) in the second spectrum because of the proton exchange process just discussed. This technique can be used with any proton that undergoes exchange.

$$R—O—H + D—O—D \longrightarrow R—O—D + D—O—H$$

If the ^1H NMR spectrum of $CH_3CH_2OCH_3$ were compared with the ^1H NMR spectrum of $CH_3CD_2OCH_3$, the signal at the highest frequency in the first spectrum would be absent in the second spectrum, indicating that this signal corresponds to the methylene group. Deuterium substitution can be a helpful technique in the analysis of complicated ^1H NMR spectra.

The sample used to obtain an ^{1}H NMR spectrum is made by dissolving the compound in an appropriate solvent. Solvents with protons cannot be used since the signals for solvent protons would be very intense because there is more solvent than compound in a solution. Therefore, deuterated solvents such as $CDCl_3$ and D_2O are commonly used in NMR spectroscopy.

14.17 Resolution of ^{1}H NMR Spectra

The ^{1}H NMR spectrum of 2-*sec*-butylphenol taken on a 60-MHz NMR spectrometer is shown in Figure 14.30(a), and the ^{1}H NMR spectrum of the same compound taken on a 300-MHz instrument is shown in Figure 14.30(b). Why is the resolution (separation of the signals) of the second spectrum so much better?

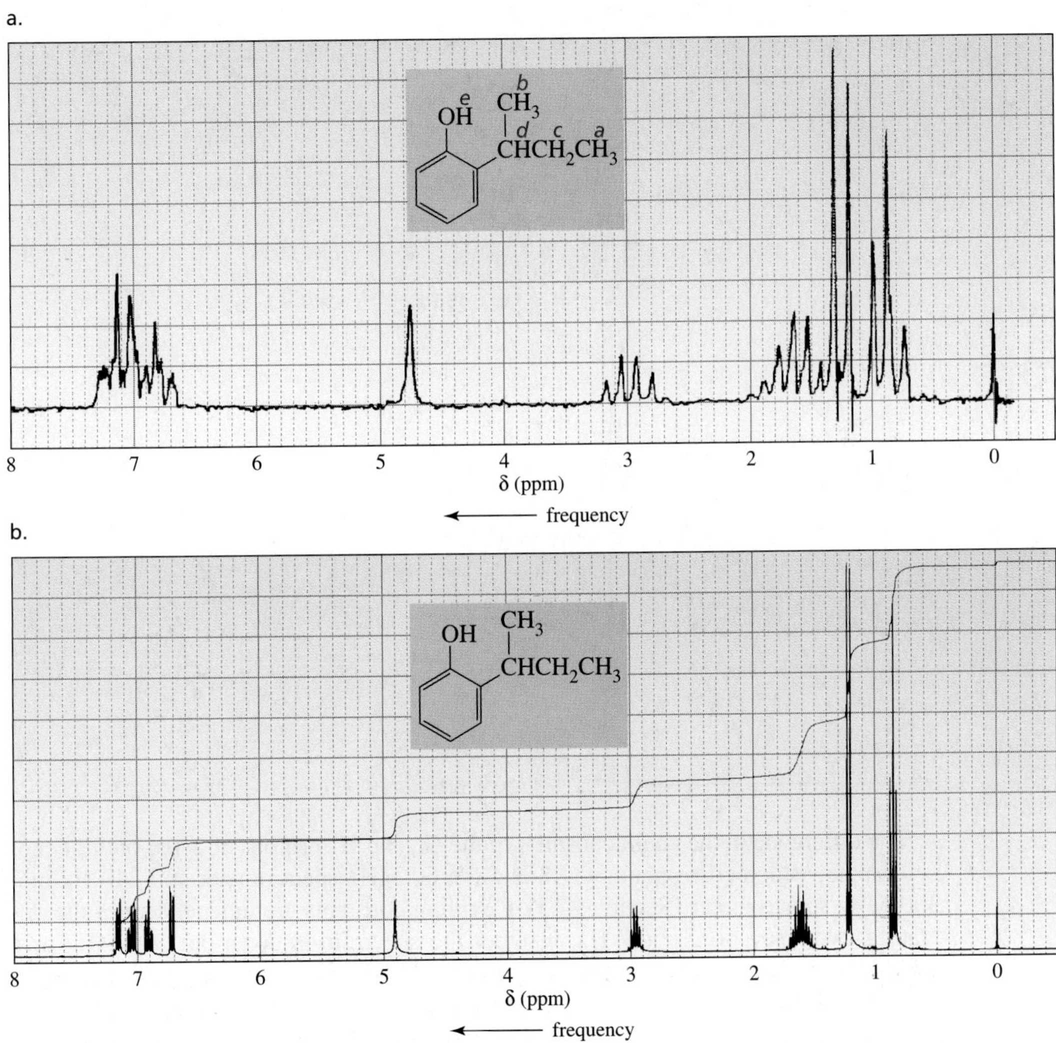

▲ **Figure 14.30**
(a) 60-MHz ^{1}H NMR spectrum of 2-*sec*-butylphenol. (b) 300-MHz ^{1}H NMR spectrum of 2-*sec*-butylphenol.

To observe separate signals with "clean" splitting patterns, the difference in the chemical shifts of two adjacent protons ($\Delta\nu$ in Hz) must be at least 10 times the value of the coupling constant (J). The signals in Figure 14.31 show that as $\Delta\nu/J$ decreases, the two signals appear closer to each other and the outer peaks of the signals become less intense while the inner peaks become more intense. The quartet and triplet of the ethyl group are clearly observed only when $\Delta\nu/J$ is greater than 10.

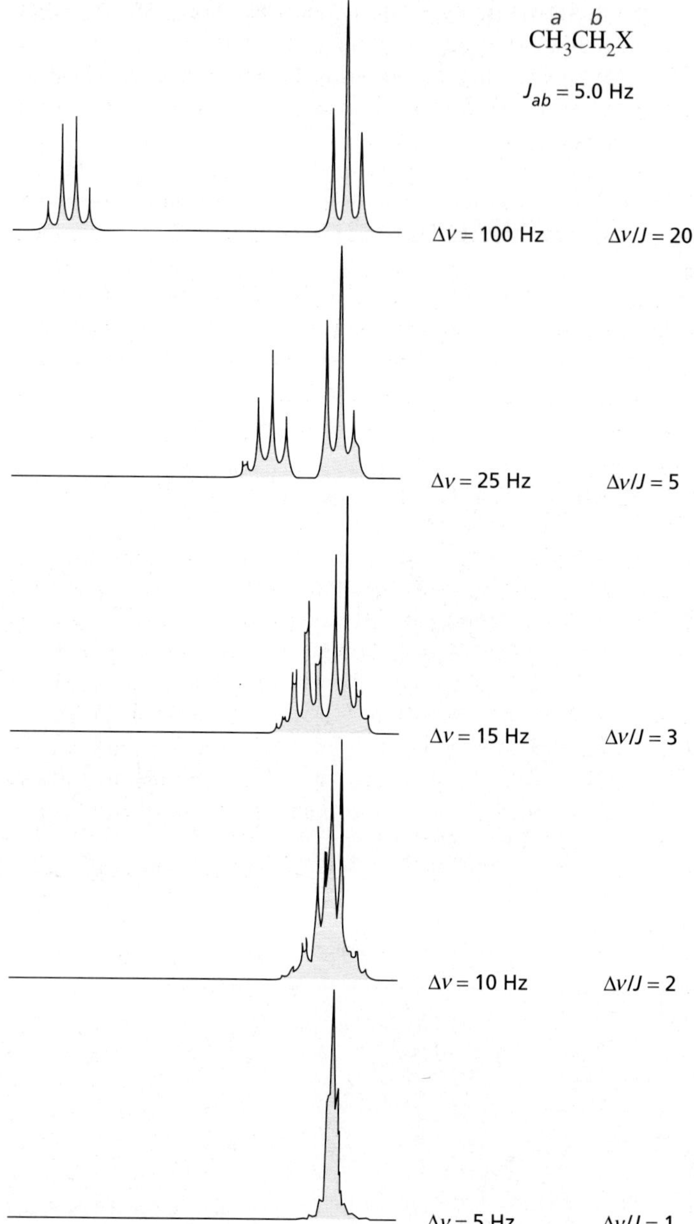

Figure 14.31 ▶ The splitting pattern of an ethyl group as a function of the $\Delta\nu/J$ ratio.

The difference in the chemical shifts of the H_a and H_c protons of 2-*sec*-butylphenol is 0.8 ppm, which corresponds to 48 Hz in a 60-MHz spectrometer and 240 Hz in a 300-MHz spectrometer (Section 14.5). Coupling constants are independent of the operating frequency, so J_{ac} is 7 Hz, whether the spectrum is taken on a 60-MHz or a 300-MHz instrument. Only in the case of the 300-MHz spectrometer is the difference in chemical shift more than 10 times the value of the coupling constant, so only in the 300-MHz spectrum do the signals show clean splitting patterns.

in a 300-MHz spectrometer **in a 60-MHz spectrometer**

$$\frac{\Delta\nu}{J} = \frac{240}{7} = 34 \qquad \frac{\Delta\nu}{J} = \frac{48}{7} = 6.9$$

14.18 ^{13}C NMR Spectroscopy

The number of signals in a ^{13}C NMR spectrum tells you how many different kinds of carbons a compound has—just as the number of signals in an ^1H NMR spectrum tells you how many different kinds of hydrogens a compound has. The principles behind

^{1}H NMR and ^{13}C NMR spectroscopy are essentially the same. There are, however, some differences that make ^{13}C NMR easier to interpret.

The development of ^{13}C NMR spectroscopy as a routine analytical procedure was not possible until computers were available that could carry out a Fourier transform (Section 14.2). ^{13}C NMR requires Fourier transform techniques because the signals obtained from a single scan are too weak to be distinguished from background electronic noise. However, FT–^{13}C NMR scans can be repeated rapidly, so a large number of scans can be recorded and added. ^{13}C signals stand out when hundreds of scans are added, because electronic noise is random, so its sum is close to zero. Without Fourier transform, it could take days to record the number of scans required for a ^{13}C NMR spectrum.

The individual ^{13}C signals are weak because the isotope of carbon (^{13}C) that gives rise to ^{13}C NMR signals constitutes only 1.11% of carbon atoms (Section 13.3). (The most abundant isotope of carbon, ^{12}C, has no nuclear spin and therefore cannot produce an NMR signal.) The low abundance of ^{13}C means that the intensities of the signals in ^{13}C NMR compared with those in ^{1}H NMR are reduced by a factor of approximately 100. In addition, the gyromagnetic ratio (γ) of ^{13}C is about one-fourth that of ^{1}H, and the intensity of a signal is proportional to γ^3. Therefore, the overall intensity of a ^{13}C signal is about 6400 times $(100 \times 4 \times 4 \times 4)$ less than the intensity of an ^{1}H signal.

One advantage to ^{13}C NMR spectroscopy is that the chemical shifts range over about 220 ppm, compared with about 12 ppm for ^{1}H NMR (Table 14.1). This means that signals are less likely to overlap. The ^{13}C NMR chemical shifts of different kinds of carbons are shown in Table 14.4. The reference compound used in ^{13}C NMR is TMS, the reference compound also used in ^{1}H NMR. Notice that ketone and aldehyde carbonyl groups can be easily distinguised from other carbonyl groups.

A disadvantage of ^{13}C NMR spectroscopy is that, unless special techniques are used, the area under a ^{13}C NMR signal is not proportional to the number of atoms

Table 14.4 Approximate Values of Chemical Shifts for ^{13}C NMR

Type of carbon	Approximate chemical shift (ppm)	Type of carbon	Approximate chemical shift (ppm)
$(CH_3)_4Si$	0	C—I	0–40
R—CH_3	8–35	C—Br	25–65
R—CH_2—R	15–50	C—Cl C—N C—O	35–80 40–60 50–80
R—CH(R)—R	20–60	R(—N)C=O	165–175
R—C(R)(R)—R	30–40	R(RO)C=O	165–175
≡C	65–85	R(HO)C=O	175–185
=C	100–150	R(H)C=O	190–200
(benzene ring)C	110–170	R(R)C=O	205–220

giving rise to the signal. Thus, the number of carbons giving rise to a ^{13}C NMR signal cannot routinely be determined by integration.

The ^{13}C NMR spectrum of 2-butanol is shown in Figure 14.32. 2-Butanol has carbons in four different environments, so there are four signals in the spectrum. The relative positions of the signals depend on the same factors that determine the relative positions of the proton signals in ^{1}H NMR. Carbons in electron-dense environments produce low-frequency signals, and carbons close to electron-withdrawing groups produce high-frequency signals. This means that the signals for the carbons of 2-butanol are in the same relative order that we would expect for the signals of the protons on those carbons in the ^{1}H NMR spectrum. Thus, the carbon of the methyl group farther away from the electron-withdrawing OH group gives the lowest-frequency signal. As the frequency increases, the other methyl carbon comes next, followed by the methylene carbon; and the carbon attached to the OH group gives the highest-frequency signal.

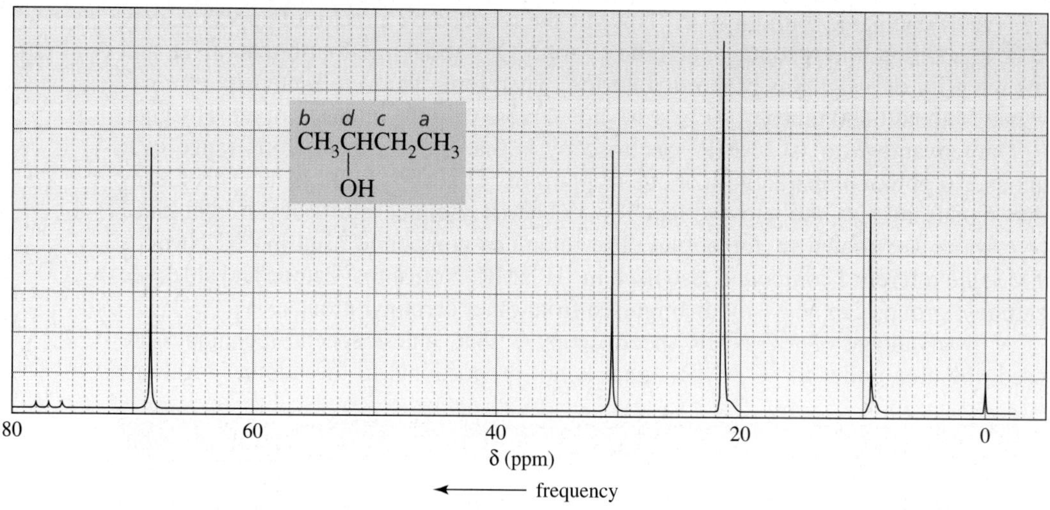

▲ **Figure 14.32**
Proton-decoupled ^{13}C NMR spectrum of 2-butanol.

The signals are not normally split by neighboring carbons because there is little likelihood of an adjacent carbon being a ^{13}C. The probability of two ^{13}C carbons being next to each other is 1.11% × 1.11% (about 1 in 10,000). (Because ^{12}C does not have a magnetic moment, it cannot split the signal of an adjacent ^{13}C.)

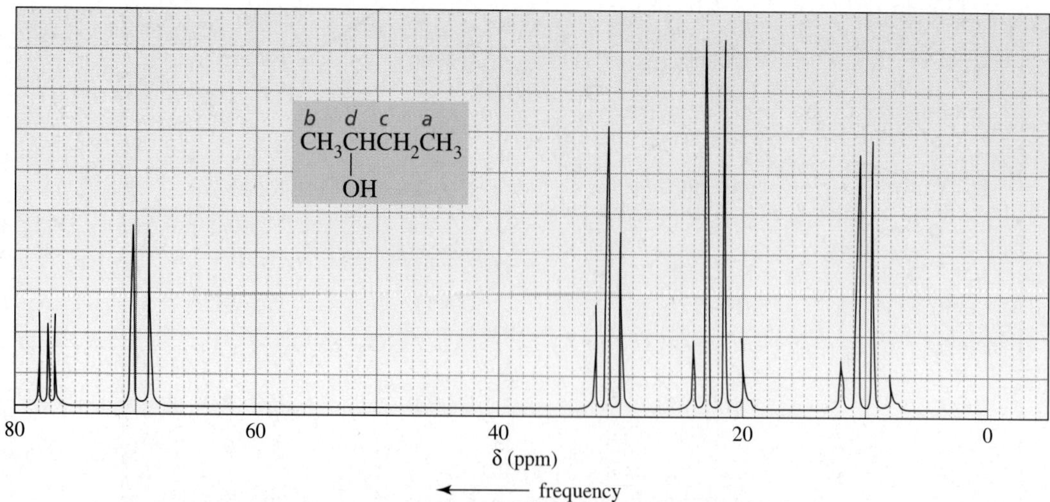

▲ **Figure 14.33**
Proton-coupled ^{13}C NMR spectrum of 2-butanol. If the spectrometer is run in a proton-coupled mode, splitting is observed in a ^{13}C NMR spectrum.

The signals in a ^{13}C NMR spectrum can be split by nearby hydrogens. However, this splitting is not usually observed because the spectra are recorded using spin-decoupling, which obliterates the carbon–proton interactions. Thus, all the signals are singlets in an ordinary ^{13}C NMR spectrum (Figure 14.32).

If the spectrometer is run in a *proton-coupled* mode, the signals show spin–spin splitting. The splitting is not caused by adjacent carbons, but by the hydrogens bonded to the carbon that produces the signal. The multiplicity of the signal is determined by the $N + 1$ rule. The **proton-coupled ^{13}C NMR spectrum** of 2-butanol is shown in Figure 14.33. (The triplet at 78 ppm is produced by the solvent, $CDCl_3$.) The signals for the methyl carbons are each split into a quartet because each methyl carbon is bonded to three hydrogens $(3 + 1 = 4)$. The signal for the methylene carbon is split into a triplet $(2 + 1 = 3)$, and the signal for the carbon bonded to the OH group is split into a doublet $(1 + 1 = 2)$.

The ^{13}C NMR spectrum of 2,2-dimethylbutane is shown in Figure 14.34. The three methyl groups at one end of the molecule are equivalent, so they all appear at the same chemical shift. The intensity of a signal is somewhat related to the number of carbons giving rise to the signal; consequently, the signal for these three methyl groups is the most intense signal in the spectrum. The tiny signal is for the quaternary carbon; carbons that are not attached to hydrogens give very small signals.

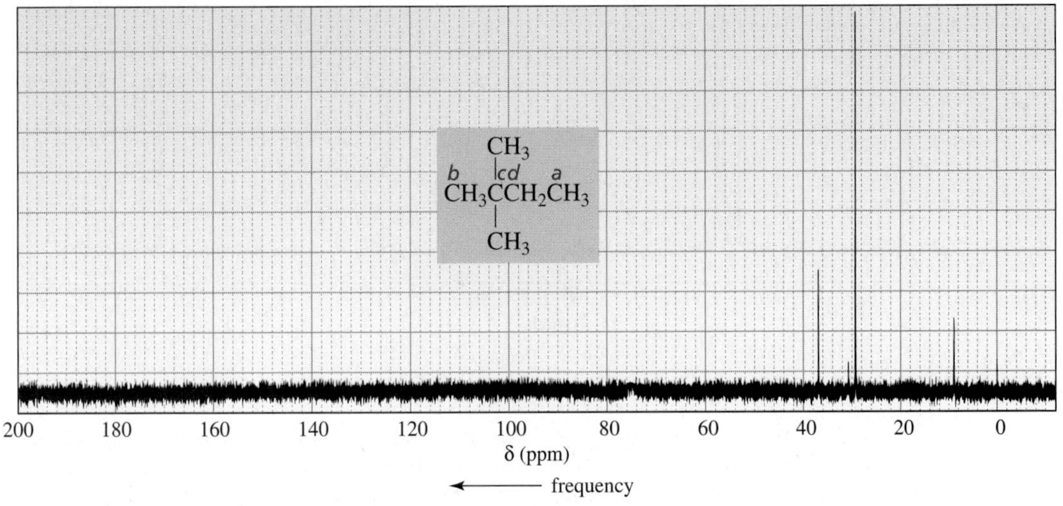

▲ **Figure 14.34**
Proton-decoupled ^{13}C NMR spectrum of 2,2-dimethylbutane.

PROBLEM 36

Answer the following questions for each of the compounds:

a. How many signals are in the ^{13}C NMR spectrum?
b. Which signal is at the lowest frequency?

1. $CH_3CH_2CH_2Br$
2. $(CH_3)_2C{=}CH_2$
3. $CH_3CH_2OCH_3$
4. $CH_2{=}CHBr$
5. $CH_3CH_2\overset{O}{\overset{\|}{C}}OCH_3$
6. $CH_3\underset{CH_3}{\underset{|}{C}}H\overset{O}{\overset{\|}{C}}H$
7. (benzene ring)–Cl
8. $CH_3\overset{O}{\overset{\|}{C}}CH_2CH_2\overset{O}{\overset{\|}{C}}CH_3$
9. $CH_3\overset{CH_3}{\overset{|}{\underset{CH_3}{\underset{|}{C}}}}OCH_3$
10. $CH_3\underset{Br}{\underset{|}{C}}HCH_3$

PROBLEM 37

Describe the proton-coupled ^{13}C NMR spectrum for compounds 1, 3, and 5 in Problem 36, showing relative values (not absolute values) of chemical shifts.

PROBLEM 38

How can 1,2-, 1,3-, and 1,4-dinitrobenzene be distinguished by

a. ^{1}H NMR spectroscopy? b. ^{13}C NMR spectroscopy?

PROBLEM-SOLVING STRATEGY

Identify the compound with molecular formula $C_9H_{10}O_2$ that gives the following ^{13}C NMR spectrum:

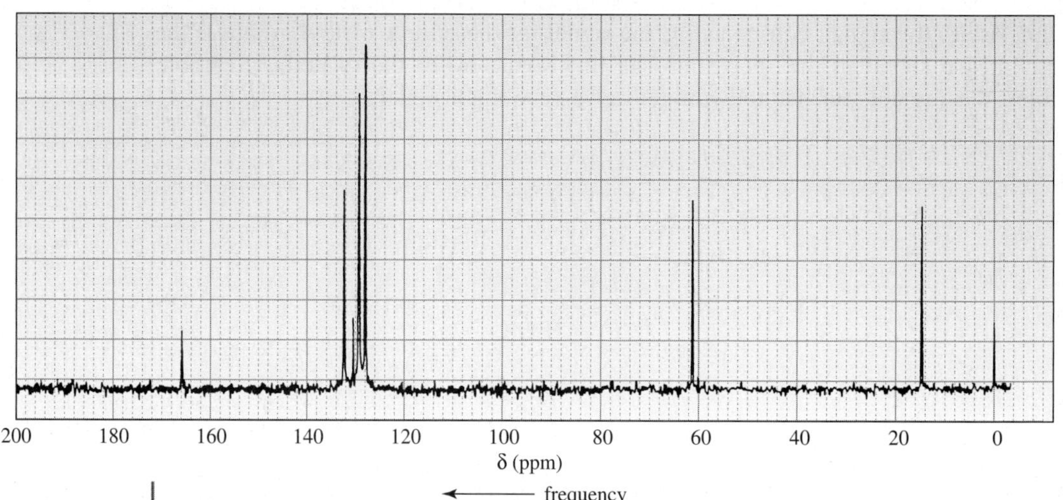

First, pick out the signals that can be absolutely identified. For example, the two oxygen atoms in the molecular formula and the carbonyl carbon signal at 166 ppm indicate that the compound is an ester. The four signals at about 130 ppm suggest that the compound has a benzene ring with a single substituent. (One signal is for the carbon to which the substituent is attached, one is for the ortho carbons, one is for the meta carbons, and one is for the para carbon.) Now subtract the molecular formula of those pieces from the molecular formula of the compound. Subtracting C_6H_5 and CO_2 from the molecular formula leaves C_2H_5, the molecular formula of an ethyl substituent. Therefore, we know that the compound is either ethyl benzoate or phenyl propanoate.

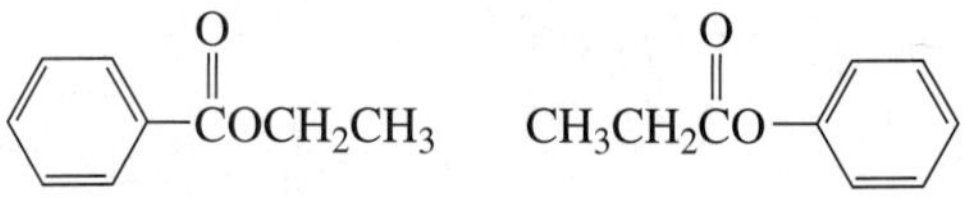

$C_6H_5-\overset{O}{\overset{\|}{C}}OCH_2CH_3$ $\quad$ $CH_3CH_2\overset{O}{\overset{\|}{C}}O-C_6H_5$

ethyl benzoate $\quad$ **phenyl propanoate**

Seeing that the methylene group is at about 60 ppm indicates that it is adjacent to an oxygen. Thus, the compound is ethyl benzoate.

Now continue on to Problem 39.

PROBLEM 39

Identify each compound in Figure 14.35 from its molecular formula and its ^{13}C NMR spectrum.

a. $C_{11}H_{22}O$

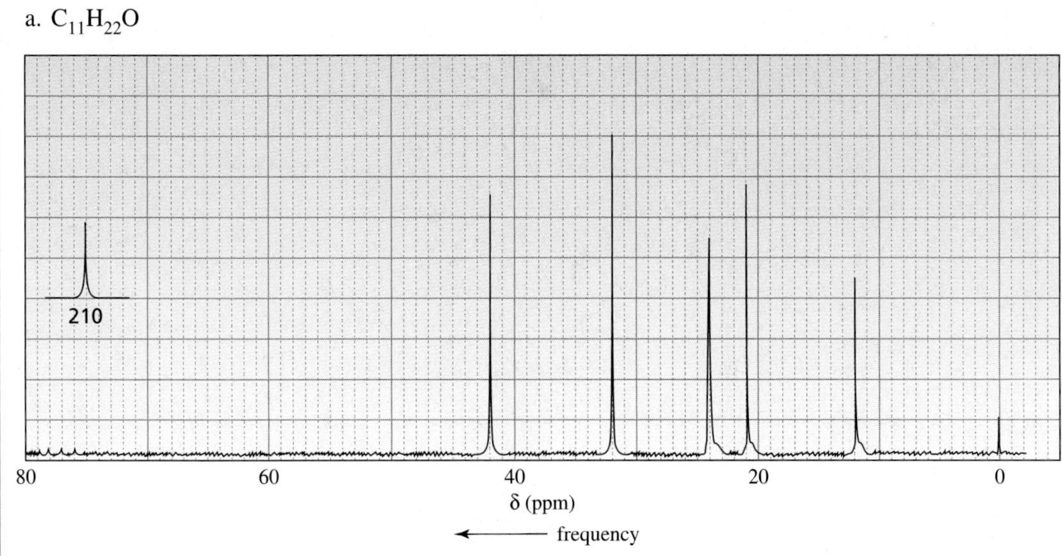

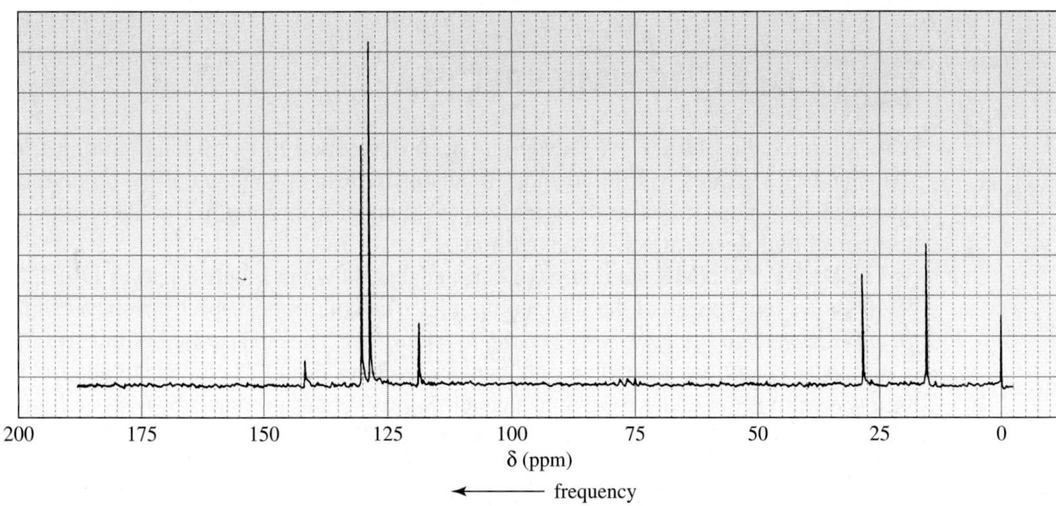

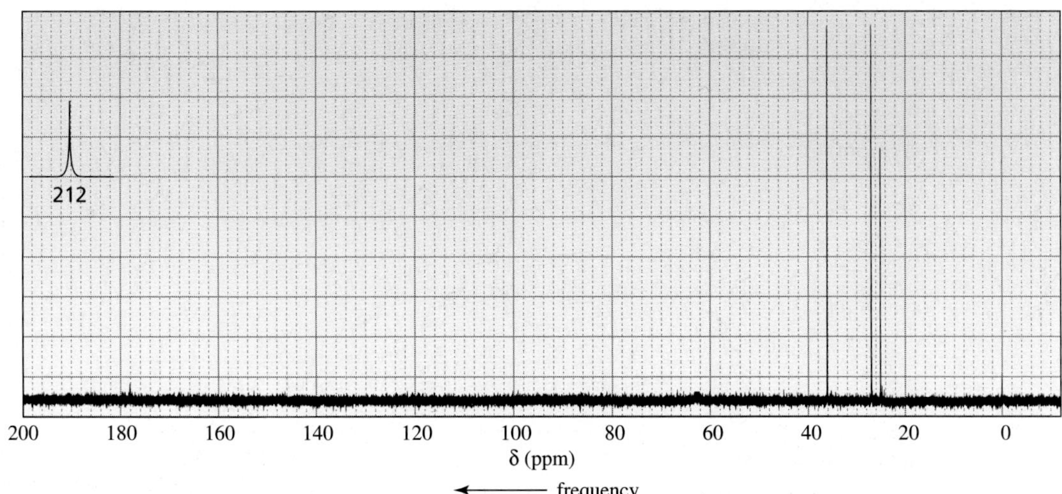

▲ **Figure 14.35**
The ^{13}C NMR spectra for Problem 39.

d. C_6H_{12}

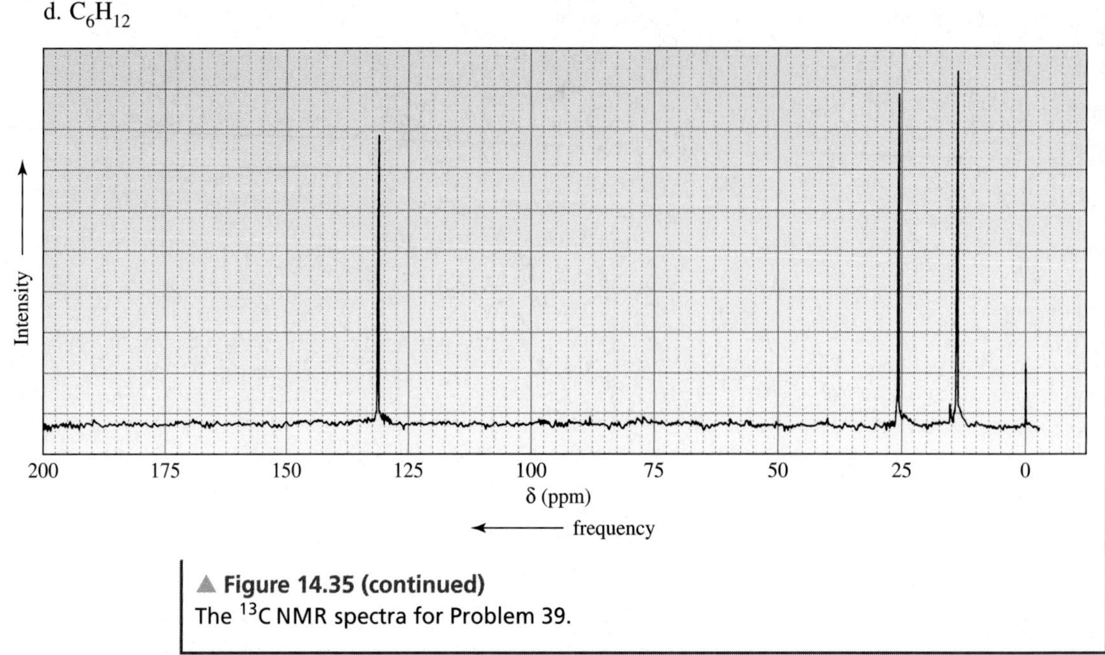

▲ **Figure 14.35 (continued)**
The ^{13}C NMR spectra for Problem 39.

14.19 DEPT ^{13}C NMR SPECTRA

A technique called DEPT ^{13}C NMR has been developed to distinguish among CH_3, CH_2, and CH groups. (DEPT stands for distortionless enhancement by polarization transfer.) It is now much more widely used than proton coupling to determine the number of hydrogens attached to a carbon.

The **DEPT ^{13}C NMR spectrum** of citronellal is shown in Figure 14.36. A DEPT ^{13}C NMR shows four spectra of the same compound. The bottommost spectrum shows signals *for all carbons that are covalently bonded to hydrogens.* The next-to-the-bottom spectrum is run under conditions that allow only signals resulting from CH carbons to appear. The third spectrum is run under conditions that allow only signals produced by CH_2 carbons to appear, and the top spectrum shows only signals produced by CH_3 carbons. Information

Figure 14.36 ▶
DEPT ^{13}C NMR spectrum of citronellal.

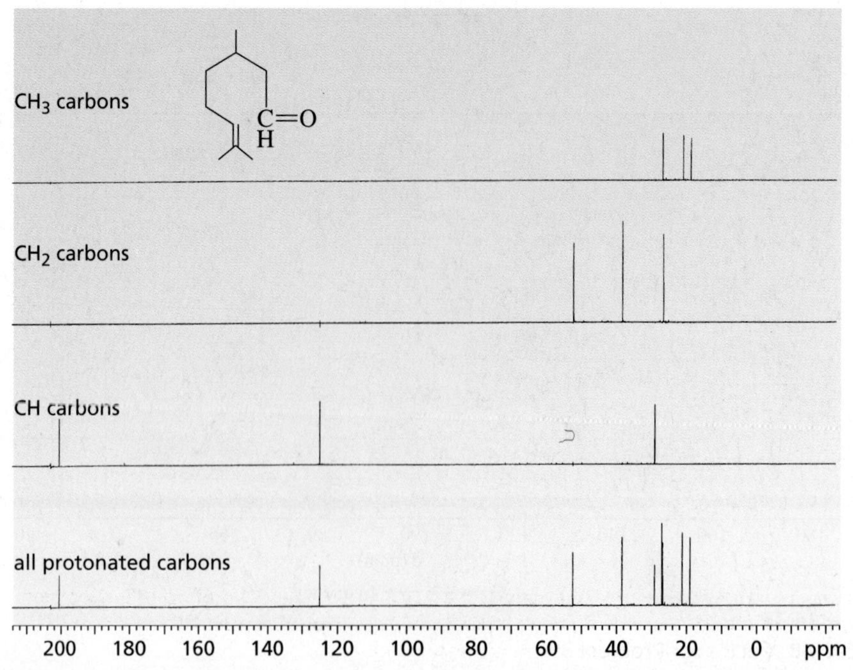

from the upper three spectra lets you determine whether a signal in a ^{13}C NMR spectrum is produced by a CH_3, CH_2, or CH carbon.

Notice that a DEPT ^{13}C NMR spectrum does *not* show a signal for a carbon that is not attached to a hydrogen. For example, the ^{13}C NMR spectrum of 2-butanone shows four signals because it has four nonequivalent carbons, whereas the DEPT ^{13}C NMR spectrum of 2-butanone shows only three signals because the carbonyl carbon is not bonded to a hydrogen, so it will not produce a signal.

14.20 Two-Dimensional NMR Spectroscopy

Complex molecules such as proteins and nucleic acids are difficult to analyze by NMR because the signals in their NMR spectra overlap. Such compounds are now being analyzed by two-dimensional (2-D) NMR. Unlike X-ray crystallography, **2-D NMR** techniques allow scientists to determine the structures of complex molecules in solution. Such a determination is particularly important for biological molecules whose properties depend on how they fold in water. More recently, 3-D and 4-D NMR spectroscopy have been developed and can be used to determine the structures of highly complex molecules. A thorough discussion of 2-D NMR is beyond the scope of this book, but the discussion that follows will give you a brief introduction to this increasingly important spectroscopic technique.

The ^{1}H NMR and ^{13}C NMR spectra discussed in the preceding sections have one frequency axis and one intensity axis; 2-D NMR spectra have two frequency axes and one intensity axis. The most common 2-D spectra involve ^{1}H–^{1}H shift correlations; they identify protons that are coupled (i.e., that split each other's signal). This is called ^{1}H–^{1}H shift-correlated spectroscopy, which is known by the acronym COSY.

A portion of the **COSY spectrum** of ethyl vinyl ether is shown in Figure 14.37(a); it looks like a mountain range viewed from the air because intensity is the third axis. These "mountain-like" spectra (known as *stack plots*) are not the spectra actually used to identify a compound. Instead, the compound is identified using a contour plot (Figure 14.37(b)), where each mountain in Figure 14.37(a) is represented by a large dot (as if its top had been cut off). The two mountains shown in Figure 14.37(a) correspond to the dots labeled B and C in Figure 14.37(b).

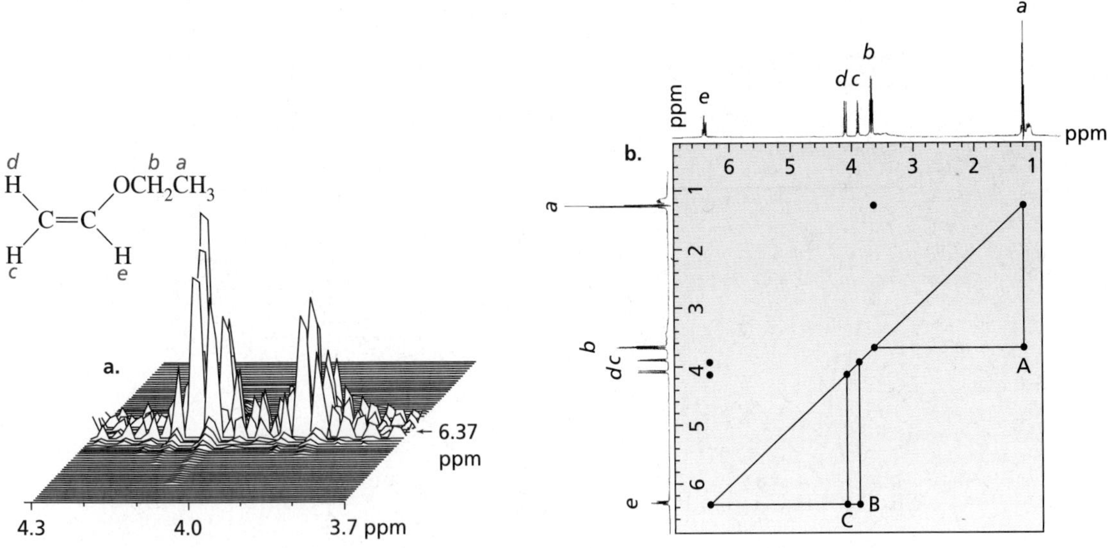

▲ **Figure 14.37**
(a) COSY spectrum of ethyl vinyl ether (stack plot). (b) COSY spectrum of ethyl vinyl ether (contour plot). In a COSY spectrum, an ^{1}H NMR spectrum is plotted on both the *x*- and *y*-axes. Cross peaks *B* and *C* represent the mountains in (a).

In the contour plot (Figure 14.37b), the usual one-dimensional ^1H NMR spectrum is plotted on both the x- and y-axes. To analyze the spectrum, a diagonal line is drawn through the dots that bisect the spectrum. The dots that are *not* on the diagonal (A, B, C) are called *cross peaks.* Cross peaks indicate pairs of protons that are coupled. For example, if we start at the cross peak labeled A and draw a straight line parallel to the y-axis back to the diagonal, we hit the dot on the diagonal at ~1.1 ppm produced by the H_a protons. If we next go back to A and draw a straight line parallel to the x-axis back to the diagonal, we hit the dot on the diagonal at ~3.8 ppm produced by the H_b protons. This means that the H_a and H_b protons are coupled. If we then go to the cross peak labeled B and draw two perpendicular lines back to the diagonal, we see that the H_c and H_e protons are coupled; the cross peak labeled C shows that the H_d and H_e protons are coupled. Notice that we used only cross peaks below the diagonal; the cross peaks above the diagonal give the same information. Notice also that there is no cross peak due to the coupling of H_c and H_d, consistent with the absence of coupling for two protons bonded to an sp^2 carbon in the one-dimensional ^1H NMR spectrum shown in Figure 14.17.

The COSY spectrum of 1-nitropropane is shown in Figure 14.38. Cross peak A shows that the H_a and H_b protons are coupled, and cross peak B shows that the H_b and H_c protons are coupled. Notice that the two triangles in the figure have a common vertex, since the H_b protons are coupled to both the H_a and H_c protons.

Figure 14.38 ▶ COSY spectrum of 1-nitropropane.

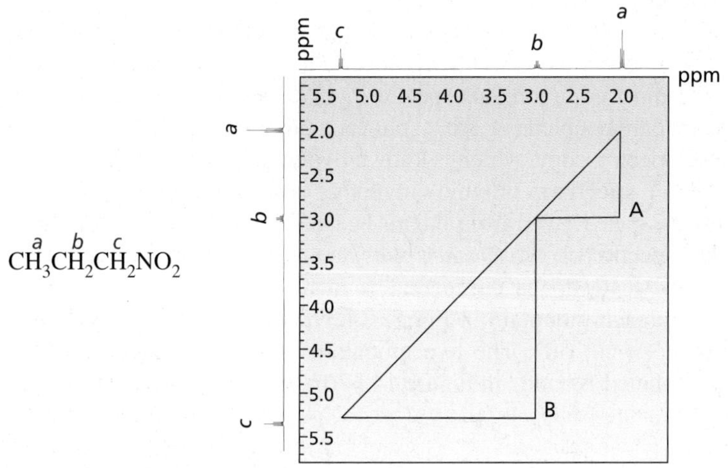

PROBLEM 40

Identify pairs of coupled protons in 2-methyl-3-pentanone, using the COSY spectrum in Figure 14.39.

Figure 14.39 ▶ COSY spectrum for Problem 40.

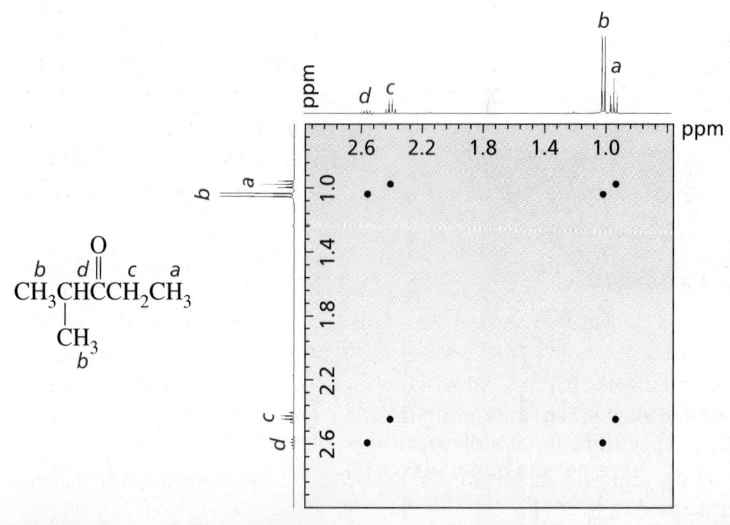

2-D NMR spectra that show ^{13}C–^{1}H shift correlations are called HETCOR (from heteronuclear correlation) spectra. **HETCOR spectra** indicate coupling between protons and the carbon to which they are attached.

The HETCOR spectrum of 2-methyl-3-pentanone is shown in Figure 14.40. The ^{13}C NMR spectrum is shown on the x-axis and the ^{1}H NMR spectrum is shown on the y-axis. The cross peaks in a HETCOR spectrum identify which hydrogens are attached to which carbons. For example, cross peak A indicates that the hydrogens that show a signal at ~0.9 ppm in the ^{1}H NMR spectrum are bonded to the carbon that shows a signal at ~6 ppm in the ^{13}C NMR spectrum. Cross peak C shows that the hydrogens that show a signal at ~2.5 ppm are bonded to the carbon that shows a signal at ~34 ppm.

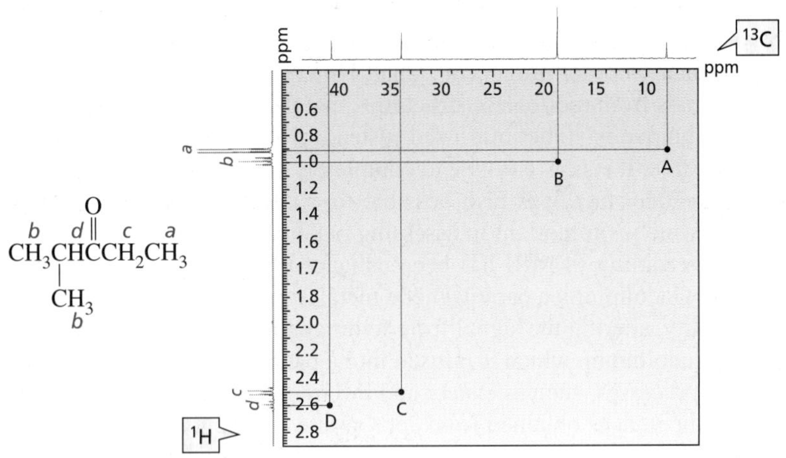

Figure 14.40
HETCOR spectrum of 2-methyl-3-pentanone. A HETCOR spectrum indicates coupling between protons and the carbons to which they are attached.

Clearly, 2-D NMR techniques are not necessary for assigning the signals in the NMR spectrum of a simple compound such as 2-methyl-3-pentanone. However, in the case of many complicated molecules, signals can be assigned only with the aid of 2-D NMR.

Spectra showing ^{13}C–^{13}C shift correlations (called 2-D ^{13}C INADEQUATE spectra) identify directly bonded carbons. There also are spectra in which chemical shifts are plotted on one of the frequency axes and coupling constants on the other. Other 2-D spectra involve the nuclear Overhauser effect (NOESY for very large molecules, ROESY for mid-size molecules).[3] These spectra are used to locate protons that are close together in space.

Albert Warner Overhauser *was born in San Diego in 1925. He was a professor of physics at Cornell University from 1953 to 1958. From 1958 to 1973 he was the director of the Physical Science Laboratory of Ford Motor Company. Since 1973 he has been a professor of physics at Purdue University.*

14.21 Magnetic Resonance Imaging

NMR has become an important tool in medical diagnosis because it allows physicians to probe internal organs and structures without invasive surgical methods or the harmful ionizing radiation of X-rays. When NMR was first introduced into clinical practice in 1981, the selection of an appropriate name was a matter of some debate. Because some members of the public associate nuclear processes with harmful radiation, the "N" was dropped from the medical application of NMR, now known as **magnetic resonance imaging (MRI)**. The spectrometer is called an **MRI scanner**.

An MRI scanner consists of a magnet sufficiently large to accommodate an entire patient, along with additional coils for exciting the nuclei, modifying the magnetic field, and receiving signals. Different tissues yield different signals, which are separated into components by Fourier transform analysis. Each component can be attributed to a specific site of origin in the patient, allowing a cross-sectional image of the patient's body to be constructed.

[3]NOESY and ROESY are the acronyms for nuclear Overhauser effect spectroscopy and rotation-frame Overhauser effect spectroscopy.

An MRI may be taken in any plane, essentially independently of the patient's position, allowing optimum visualization of the anatomical feature of interest. By contrast, the plane of a computed tomography (CT) scan, which uses X-rays, is defined by the position of the patient within the machine and is usually perpendicular to the long axis of the body. CT scans in other planes may be obtained only if the patient is a skilled contortionist.

Most of the signals in an MRI scan originate from the hydrogens of water molecules because these hydrogens are far more abundant in tissues than are the hydrogens of organic compounds. The difference in the way water is bound in different tissues is what produces much of the variation in signal among organs, as well as the variation between healthy and diseased tissue (Figure 14.41). MRI scans, therefore, can sometimes provide much more information than images obtained by other means. For example, MRI can provide detailed images of blood vessels. Flowing fluids such as blood respond differently to excitation in an MRI scanner than do stationary tissues and normally do not produce a signal. However, the data may be processed to eliminate signals from motionless structures, thereby showing signals only from the fluids. This technique is sometimes used instead of more invasive methods to examine the vascular tree. It is now possible to completely suppress the signal from certain types of tissue (usually fat). It is also possible to differentiate intracellular and extracellular edema, which is important in assessing patients suspected of having suffered strokes.

The versatility of MRI has been enhanced by the use of gadolinium as a contrast reagent. Gadolinium, a paramagnetic metal, modifies the magnetic field in its immediate vicinity, altering the signal from hydrogen nuclei in close proximity. The distribution of gadolinium, which is infused into a patient's veins, may be modified by certain disease processes, such as cancer and inflammation. These abnormal distributions appear in the images obtained from appropriate scanning techniques.

NMR spectroscopy using ^{31}P is not yet in routine clinical use, but is being used widely in clinical research. It is of particular interest because ATP and ADP (Sections 17.20 and 27.2) are involved in most metabolic processes, so it will provide a way to investigate cellular metabolism.

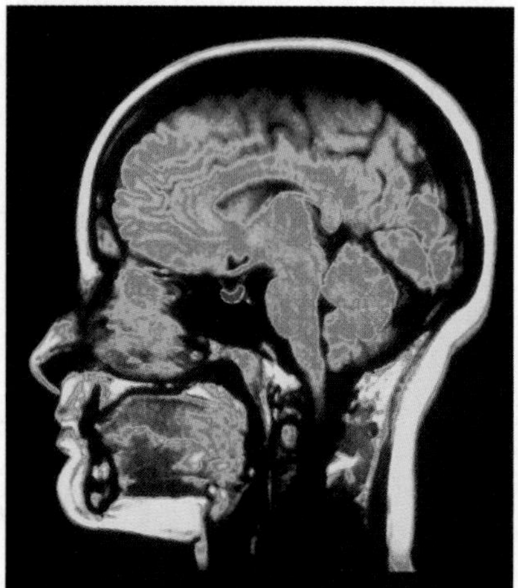

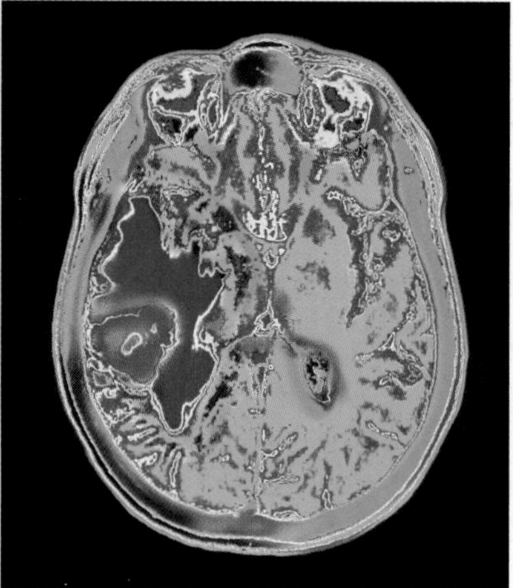

Figure 14.41 ▶
(a) MRI of a normal brain. The pituitary is highlighted (pink).
(b) MRI of an axial section through the brain showing a tumor (purple) surrounded by damaged, fluid-filled tissue (red).

Summary

NMR spectroscopy is used to identify the carbon–hydrogen framework of an organic compound. When a sample is placed in a magnetic field, protons aligning with the field are in the lower-energy **α-spin state**; those aligning against the field are in the higher-energy **β-spin state**. The energy difference between the spin states depends on the strength of the **applied magnetic field**. When subjected to radiation with energy corresponding to the energy difference between the spin states, nuclei in the α-spin state are promoted to the β-spin state. When they return to their original state, they emit signals whose frequency depends on the difference in energy between the spin states. An **NMR spectrometer** detects and displays these signals as a plot of their frequency versus their intensity—an **NMR spectrum**.

Each set of chemically equivalent protons gives rise to a signal, so the number of signals in an ^{1}H NMR spectrum indicates the number of different kinds of protons in a compound. The **chemical shift** is a measure of how far the signal is from the reference TMS signal. The chemical shift (δ) is independent of the **operating frequency** of the spectrometer.

The larger the magnetic field sensed by the proton, the higher is the frequency of the signal. The electron density of the environment in which the proton is located **shields** the proton from the applied magnetic field. Therefore, a proton in an electron-dense environment shows a signal at a lower frequency than a proton near electron-withdrawing groups. Low-frequency (upfield) signals have small δ (ppm) values; high-frequency (downfield) signals have large δ values. Thus, the position of a signal indicates the kind of proton(s) responsible for the signal and the kinds of neighboring substituents. In a similar environment, the chemical shift of methyl protons is at a lower frequency than that of methylene protons, which in turn is at a lower frequency than that of a methine proton. **Diamagnetic anisotropy** causes unusual chemical shifts for hydrogens bonded to carbons that form π bonds. **Integration** tells us the relative number of protons that give rise to each signal.

The **multiplicity** of a signal (the number of peaks in the signal) indicates the number of protons bonded to adjacent carbons. Multiplicity is described by the **N + 1 rule**, where N is the number of equivalent protons bonded to adjacent carbons. A **splitting diagram** can help us understand the splitting pattern obtained when a signal is split by more than one set of protons. Deuterium substitution can be a helpful technique in the analysis of complicated ^{1}H NMR spectra.

The **coupling constant** (***J***) is the distance between two adjacent peaks of a split NMR signal. Coupling constants are independent of the operating frequency of the spectrometer. Coupled protons have the same coupling constant. The coupling constant for trans protons is greater than that for cis protons. When two different sets of protons split a signal, the multiplicity of the signal is determined by using the $N + 1$ rule separately for each set of hydrogens when the coupling constants for the two sets are different. When the coupling constants are similar, the $N + 1$ rule can be applied to both sets simultaneously.

The chemical shift of a proton bonded to an O or an N depends on the degree to which the proton is hydrogen bonded. In the presence of trace amounts of acid or base, protons bonded to oxygen undergo **proton exchange**. In that case, the signal for a proton bonded to an O is not split and does not split the signal of adjacent protons.

The number of signals in a ^{13}C NMR spectrum tells how many different kinds of carbons a compound has. Carbons in electron-dense environments produce low-frequency signals; carbons close to electron-withdrawing groups produce high-frequency signals. Chemical shifts for ^{13}C NMR range over about 220 ppm, compared with about 12 ppm for ^{1}H NMR. ^{13}C NMR signals are not normally split by neighboring carbons, unless the spectrometer is run in a proton-coupled mode.

Key Terms

applied magnetic field (p. 527)
chemically equivalent protons (p. 531)
chemical shift (p. 533)
^{13}C NMR (p. 527)
COSY spectrum (p. 569)
coupled protons (p. 542)
coupling constant (p. 551)
DEPT ^{13}C NMR spectrum (p. 568)
diamagnetic anisotropy (p. 541)
diamagnetic shielding (p. 530)
2-D NMR (p. 569)
doublet (p. 542)
doublet of doublets (p. 547)
downfield (p. 531)
effective magnetic field (p. 530)
Fourier transform NMR (FT–NMR) spectrum (p. 530)
geminal coupling (p. 547)
gyromagnetic ratio (p. 528)
HETCOR spectrum (p. 571)
^{1}H NMR (p. 526)
integration (p. 539)
long-range coupling (p. 544)
magnetic resonance imaging (MRI) (p. 571)
methine hydrogen (p. 536)
MRI scanner (p. 571)
multiplet (p. 547)
multiplicity (p. 542)
$N + 1$ rule (p. 542)
NMR spectroscopy (p. 526)
operating frequency (p. 528)
proton-coupled ^{13}C NMR spectrum (p. 565)
proton exchange (p. 558)
quartet (p. 542)
reference compound (p. 533)
rf radiation (p. 527)
singlet (p. 541)
α-spin state (p. 527)
β-spin state (p. 527)
spin–spin coupling (p. 542)
splitting diagram (p. 554)
splitting tree (p. 554)
triplet (p. 542)
upfield (p. 531)

Problems

41. How many signals are produced by each of the following compounds in its

a. ^{1}H NMR spectrum? b. ^{13}C NMR spectrum?

1. 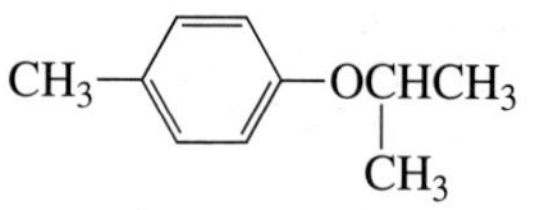 CH_3–C_6H_4–$OCH(CH_3)CH_3$
2. $C_6H_5C(=O)OCH_2CH_3$
3. (cyclic lactone, =O)
4. (oxetane, O)
5. (chlorocyclopropane, Cl)
6. 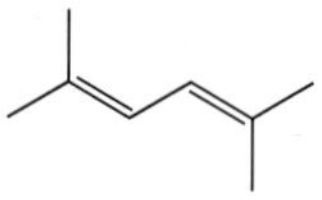 (alkene skeletal structure)

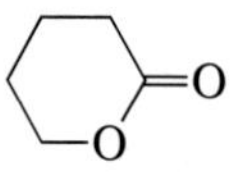

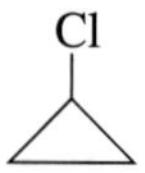

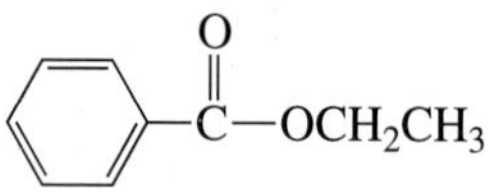

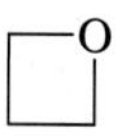

42. Draw a splitting diagram for the H_b proton and indicate its multiplicity if

a. $J_{ba} = J_{bc}$ b. $J_{ba} = 2J_{bc}$

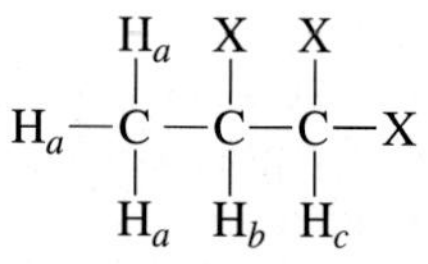

H_a–C(H_a)(H_a)–C(X)(H_b)–C(X)(H_c)–X

43. Label each set of chemically equivalent protons, using *a* for the set that will be at the lowest frequency (farthest upfield) in the ^{1}H NMR spectrum, *b* for the next, etc. Indicate the multiplicity of each signal.

a. 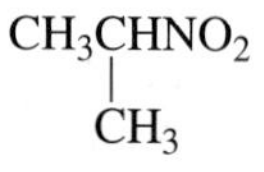 $CH_3CH(CH_3)NO_2$

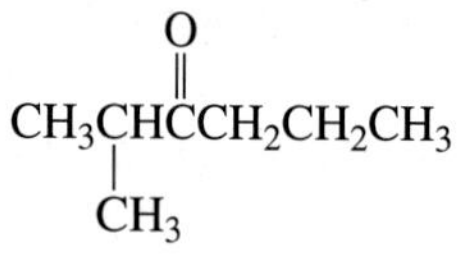

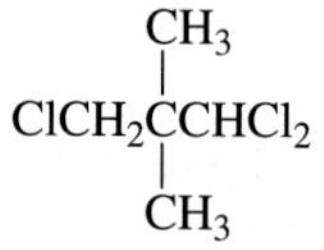

b. $CH_3CH_2CH_2OCH_3$

c. $CH_3CH(CH_3)C(=O)CH_2CH_2CH_3$

d. 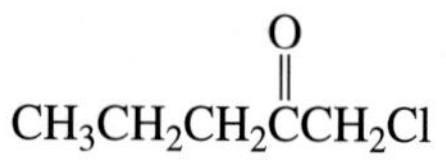 $CH_3CH_2CH_2C(=O)CH_2Cl$

e. $ClCH_2C(CH_3)_2CHCl_2$

f. $ClCH_2CH_2CH_2CH_2CH_2Cl$

44. Match each of the ^{1}H NMR spectra on page 575 with one of the following compounds:

$CH_3CH_2C(=O)CH_3$ $CH_3C(CH_3)_2NO_2$ $CH_3CH_2C(=O)CH_2CH_3$

$CH_3CH_2CH_2NO_2$ $CH_3CH_2NO_2$ $CH_3CH(CH_3)Br$

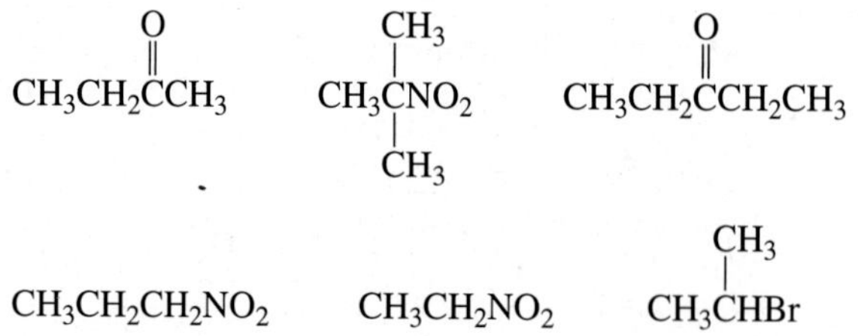

a.

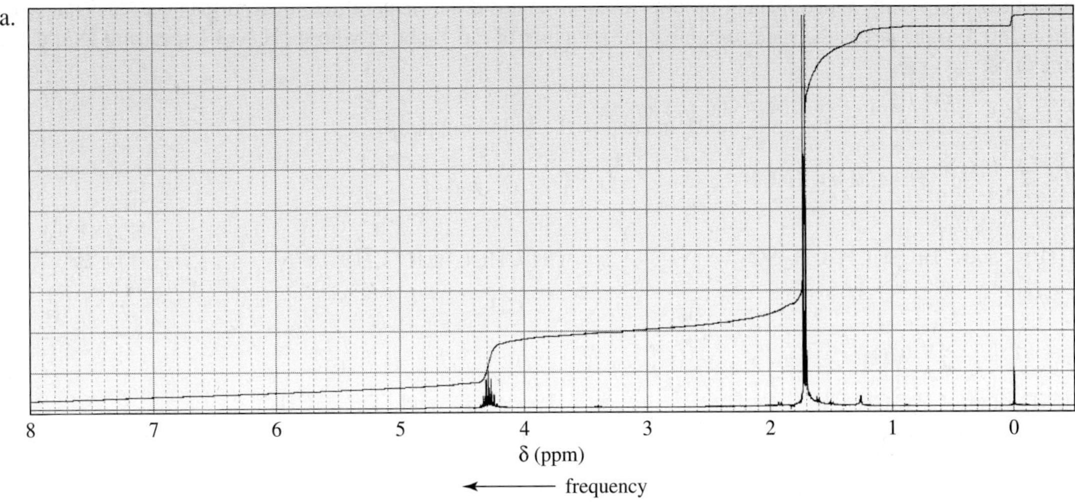
8
7
6
5
4
3
2
1
0
δ (ppm)
frequency

b.

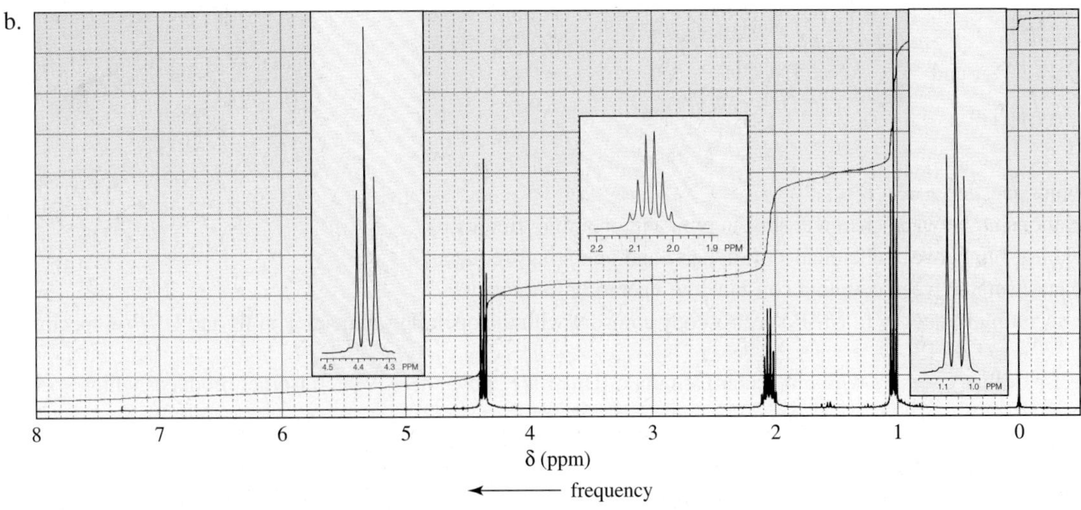
4.5 4.4 4.3 PPM
2.2 2.1 2.0 1.9 PPM
1.1 1.0 PPM
8
7
6
5
4
3
2
1
0
δ (ppm)
frequency

c.

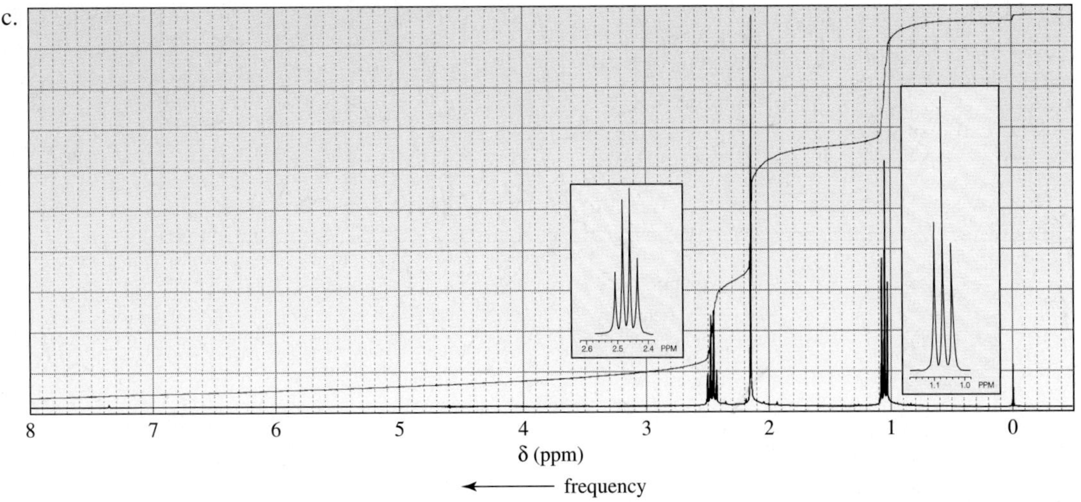
2.6 2.5 2.4 PPM
1.1 1.0 PPM
8
7
6
5
4
3
2
1
0
δ (ppm)
frequency

45. Determine the ratios of the chemically nonequivalent protons in a compound if the steps of the integration curves measure 40.5, 27, 13, and 118 mm, from left to right across the spectrum. Give the structure of a compound whose ^{1}H NMR spectrum would show these integrals in the observed order.

46. How could ^{1}H NMR distinguish between the compounds in each of the following pairs?

a. $CH_3CH_2CH_2OCH_3$ and $CH_3CH_2OCH_2CH_3$

b. $BrCH_2CH_2CH_2Br$ and $BrCH_2CH_2CH_2NO_2$

c. $CH_3CH(CH_3)—CH(CH_3)CH_3$ and $CH_3C(CH_3)_2CH_2CH_3$

d. $CH_3—C(CH_3)_2—C(=O)—OCH_3$ and $CH_3—C(OCH_3)_2—CH_3$

e. $CH_3—C_6H_4—C(CH_3)_2CH_3$ and $C_6H_5—CH_2C(CH_3)_2CH_3$

f. (cyclohexene) and (cyclohexene)

g. $CH_3CH(CH_3)Cl$ and $CH_3CD(CH_3)Cl$

h. Cl above; H—C—CH_3; D—C—H; Cl below and Cl above; D—C—CH_3; H—C—H; Cl below

i. (cyclopropane with H, H above and Cl, Cl below) and (cyclopropane with H, Cl above and Cl, H below)

47. Answer the following questions:

a. What is the relationship between chemical shift in ppm and operating frequency?
b. What is the relationship between chemical shift in hertz and operating frequency?
c. What is the relationship between coupling constant and operating frequency?
d. How does the operating frequency in NMR spectroscopy compare with the operating frequency in IR and UV/Vis spectroscopy?

48. The ^{1}H NMR spectra of three isomers with molecular formula C_4H_9Br are shown here. Which isomer produces which spectrum?

a.

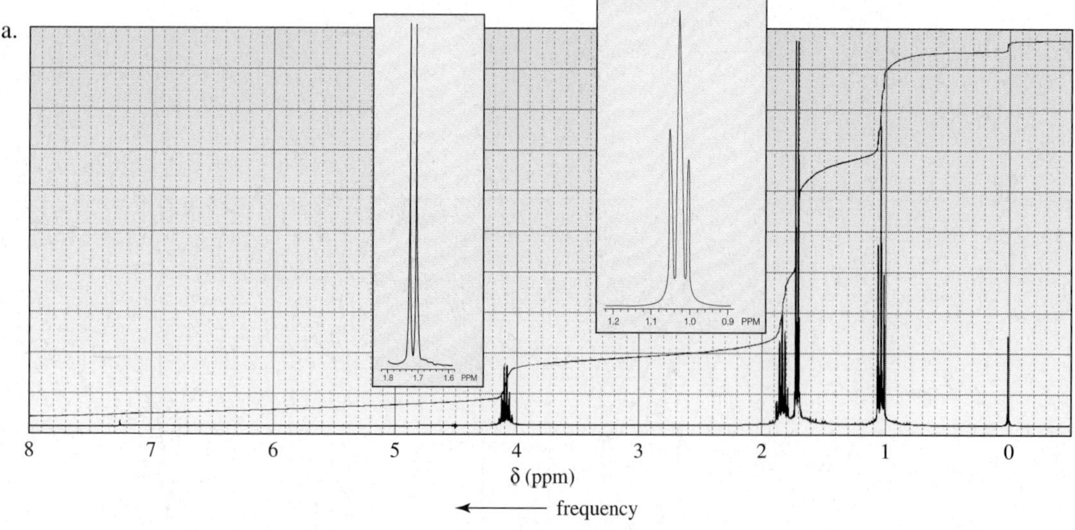

b.

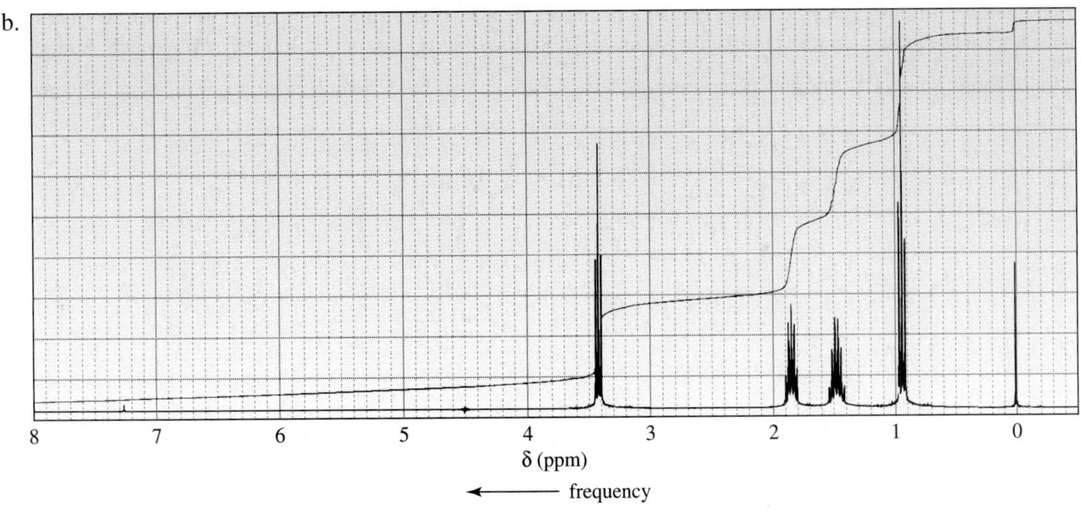

c.

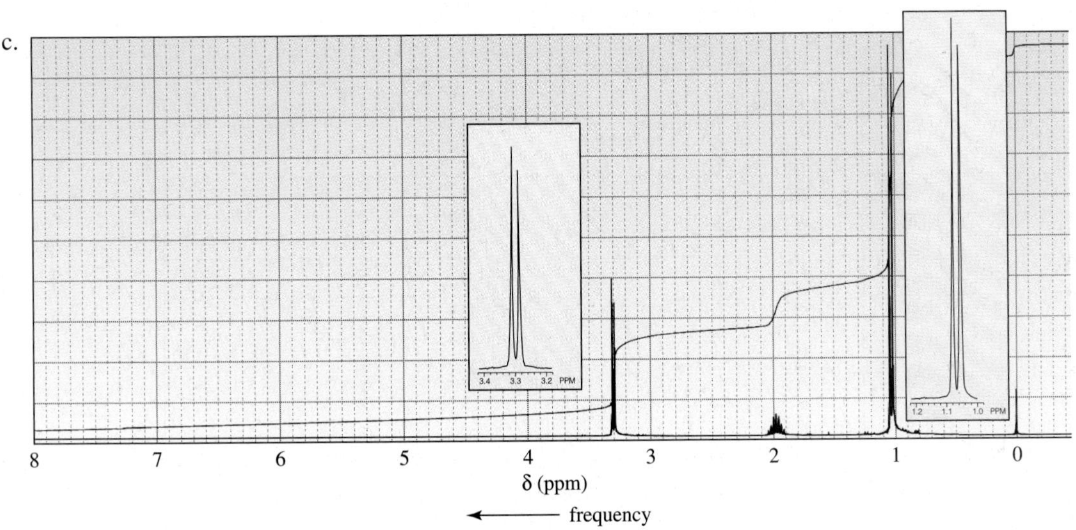

49. Identify each of the following compounds from the ^1H NMR data and molecular formula. The number of hydrogens responsible for each signal is shown in parentheses.

a. $C_4H_8Br_2$
1.97 ppm (6) singlet
3.89 ppm (2) singlet

b. C_8H_9Br
2.01 ppm (3) doublet
5.14 ppm (1) quartet
7.35 ppm (5) broad singlet

c. $C_5H_{10}O_2$
1.15 ppm (3) triplet
1.25 ppm (3) triplet
2.33 ppm (2) quartet
4.13 ppm (2) quartet

50. Identify the compound with molecular formula $C_7H_{14}O$ that gives the following proton-coupled ^{13}C NMR spectrum.

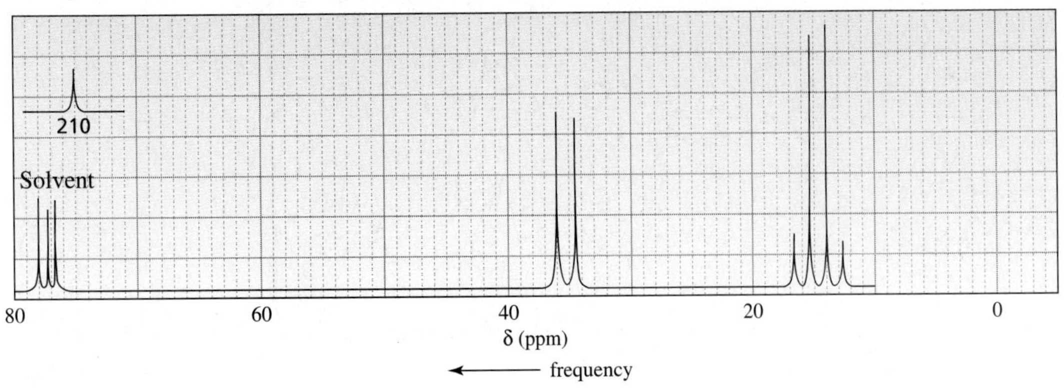

51. Compound A, with molecular formula C_4H_9Cl, shows two signals in its ^{13}C NMR spectrum. Compound B, an isomer of compound A, shows four signals, and in the proton-coupled mode, the signal farthest downfield is a doublet. Identify compounds A and B.

52. The ^{1}H NMR spectra of three isomers with molecular formula $C_7H_{14}O$ are shown here. Which isomer produces which spectrum?

a.

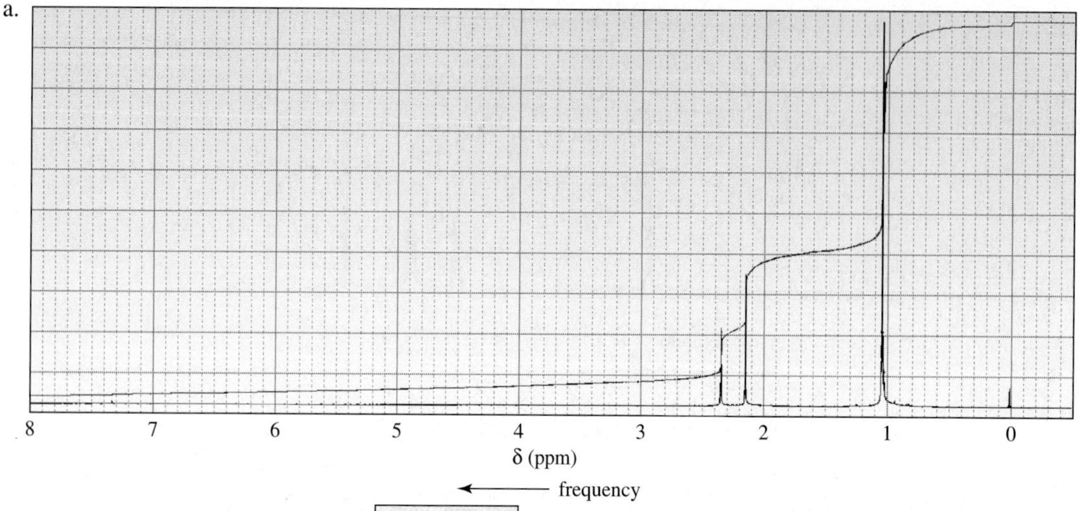

8 7 6 5 4 3 2 1 0

δ (ppm)

← frequency

b.

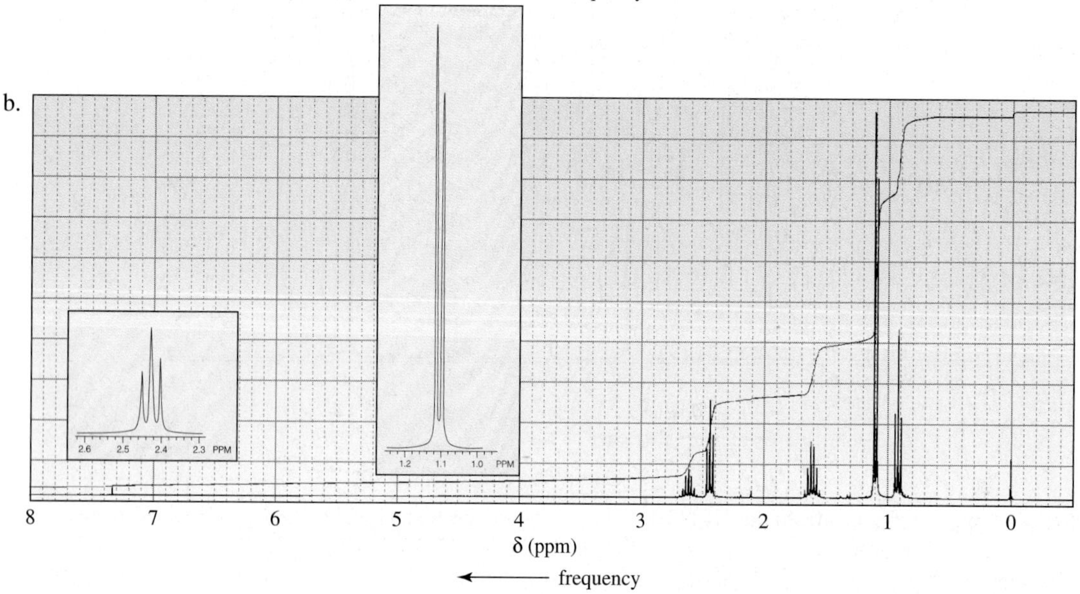

2.6 2.5 2.4 2.3 PPM

1.2 1.1 1.0 PPM

8 7 6 5 4 3 2 1 0

δ (ppm)

← frequency

c.

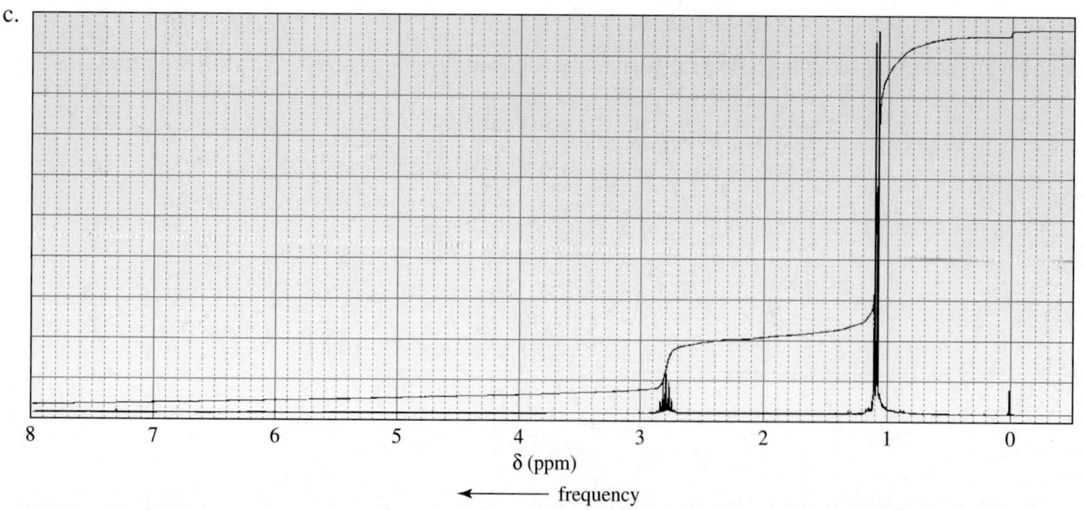

53. Would it be better to use ^{1}H NMR or ^{13}C NMR to distinguish among 1-butene, *cis*-2-butene, and 2-methylpropene? Explain your answer.

54. Determine the structure of each of the following unknown compounds based on its molecular formula and its IR and ^{1}H NMR spectra.

a. $C_5H_{12}O$

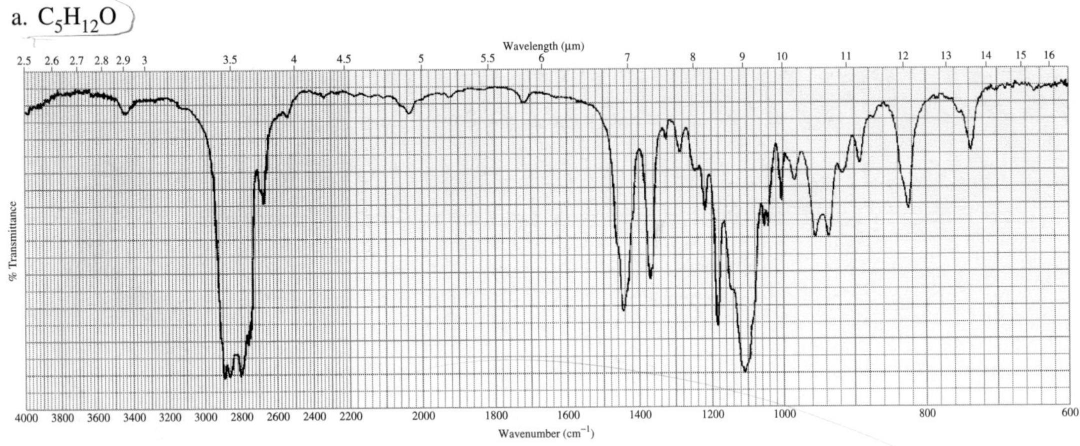

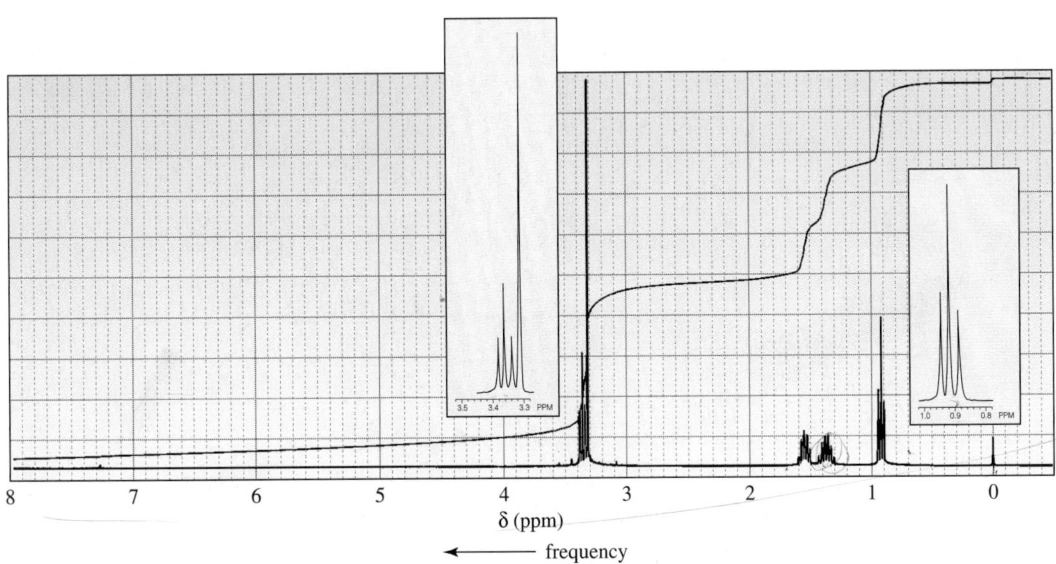

b. $C_6H_{12}O_2$

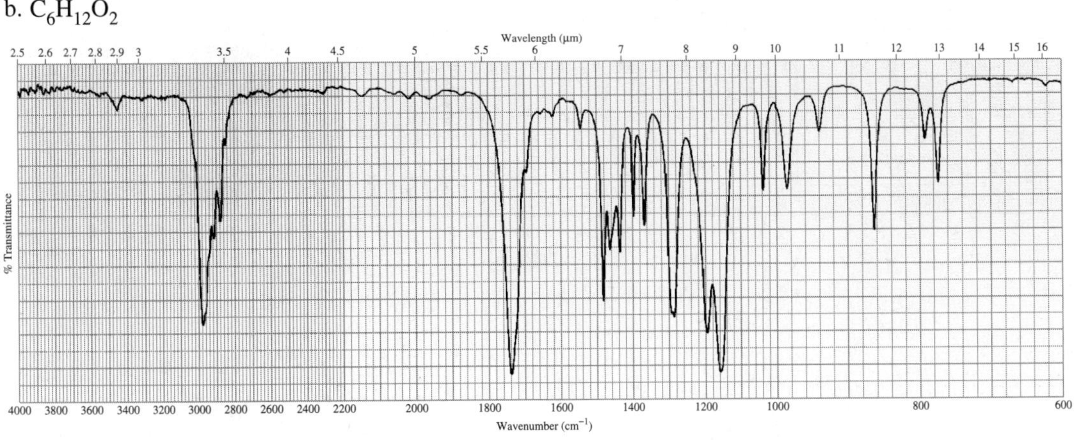

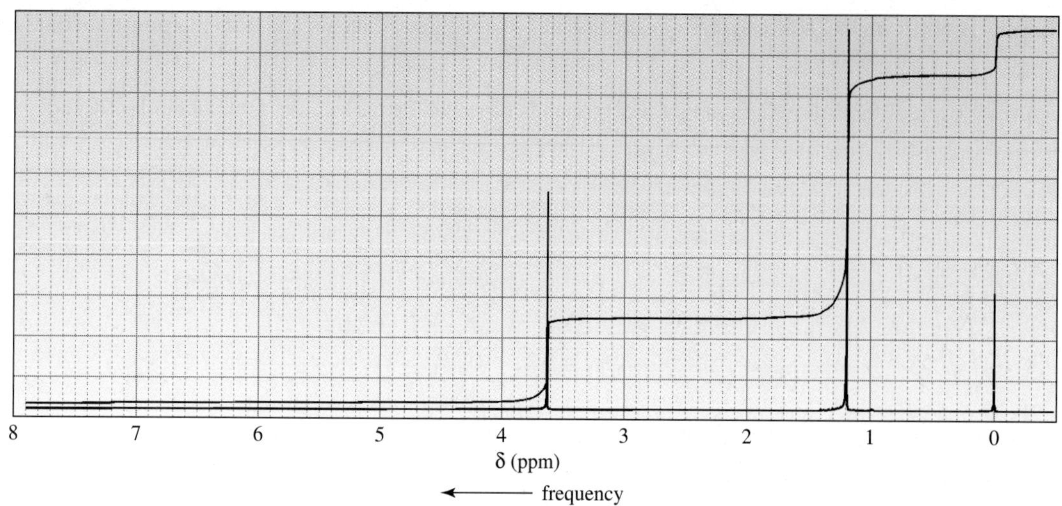

8 7 6 5 4 3 2 1 0
δ (ppm)
frequency

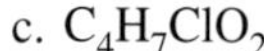

c. $C_4H_7ClO_2$

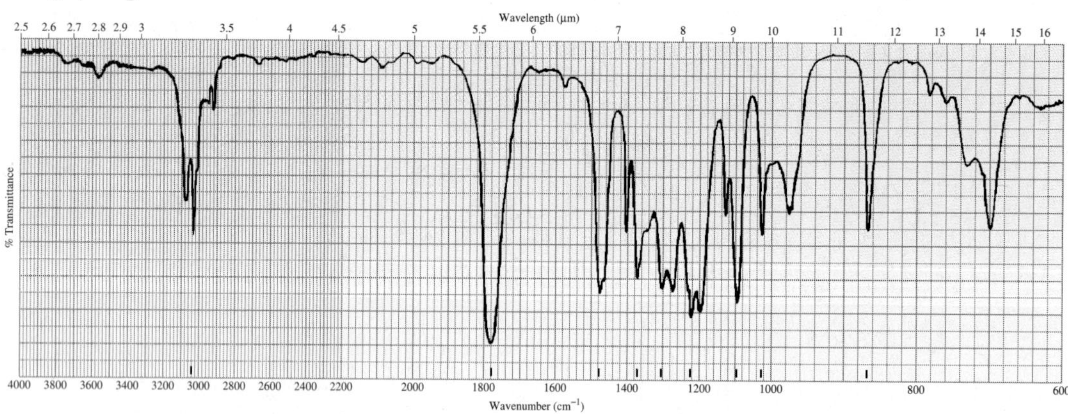

Wavelength (μm)
2.5 2.6 2.7 2.8 2.9 3 3.5 4 4.5 5 5.5 6 7 8 9 10 11 12 13 14 15 16
% Transmittance
4000 3800 3600 3400 3200 3000 2800 2600 2400 2200 2000 1800 1600 1400 1200 1000 800 600
Wavenumber (cm^{-1})

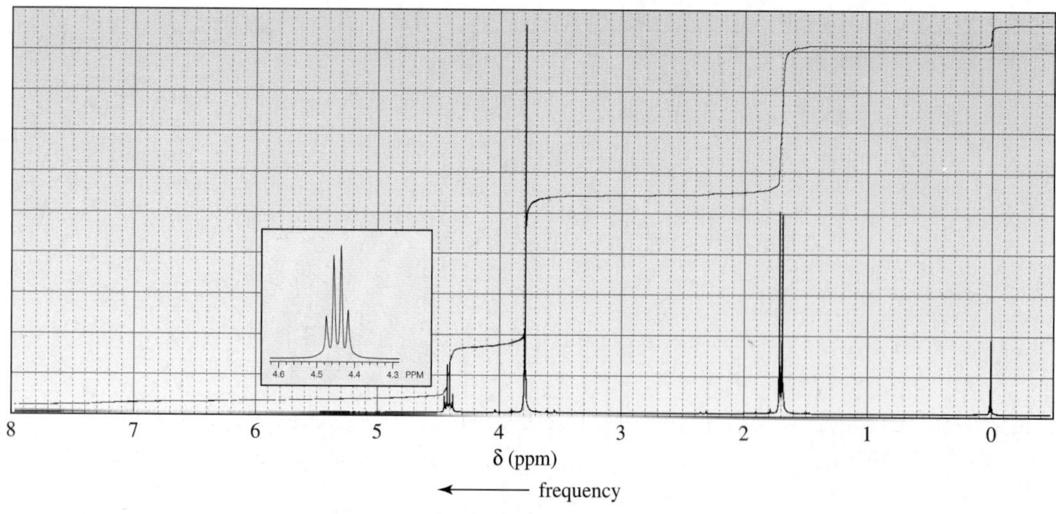

4.6 4.5 4.4 4.3 PPM
8 7 6 5 4 3 2 1 0
δ (ppm)
frequency

d. $C_4H_8O_2$

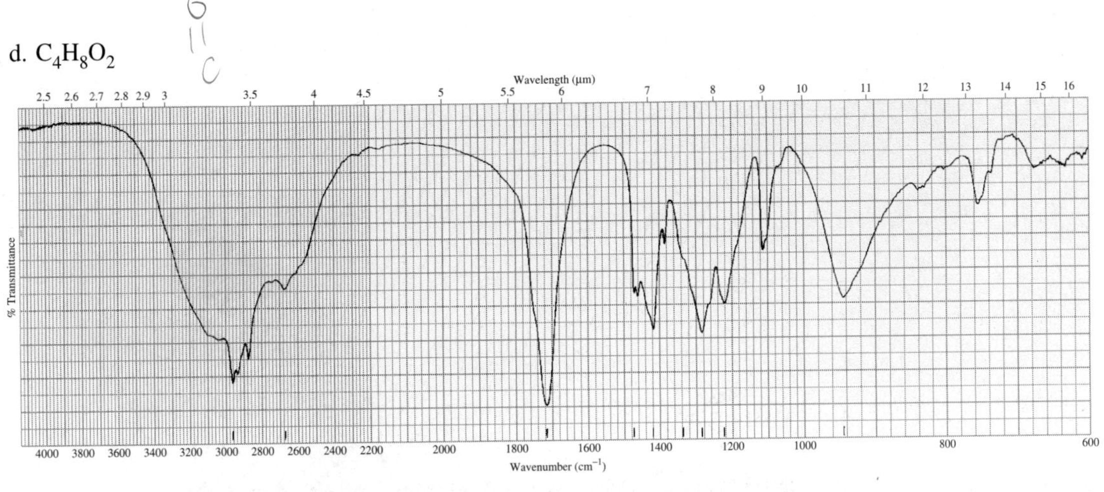

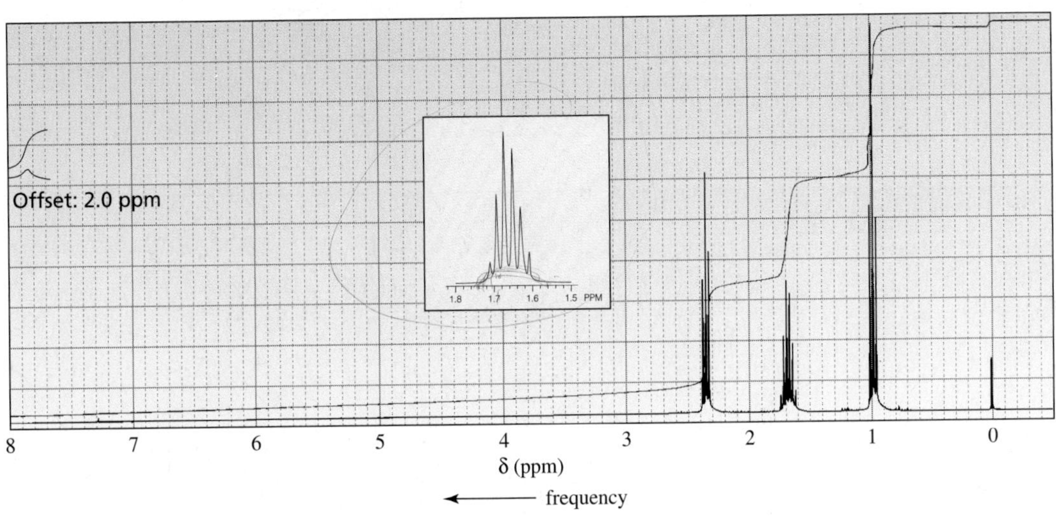

55. There are four esters with molecular formula $C_4H_8O_2$. How could they be distinguished by ^1H NMR?

56. An alkyl halide reacts with an alkoxide ion to form a compound whose ^1H NMR spectrum is shown here. Identify the alkyl halide and the alkoxide ion. (*Hint:* see Section 11.9.)

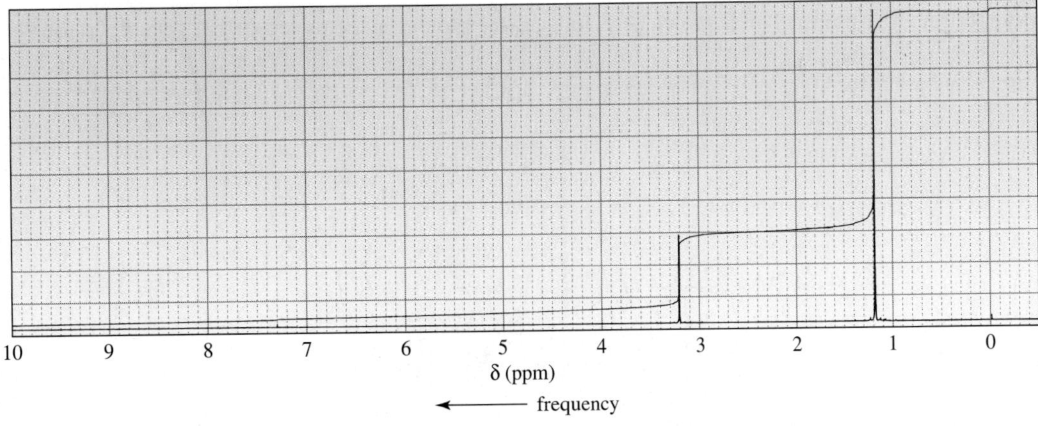

57. Determine the structure of each of the following compounds based on its molecular formula and its ^{13}C NMR spectrum.

a. $C_4H_{10}O$

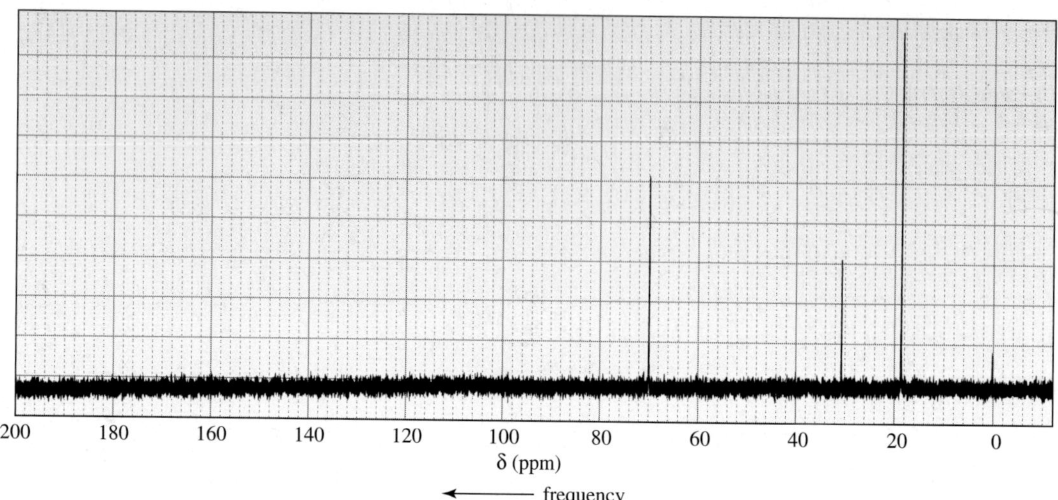

b. $C_6H_{12}O$

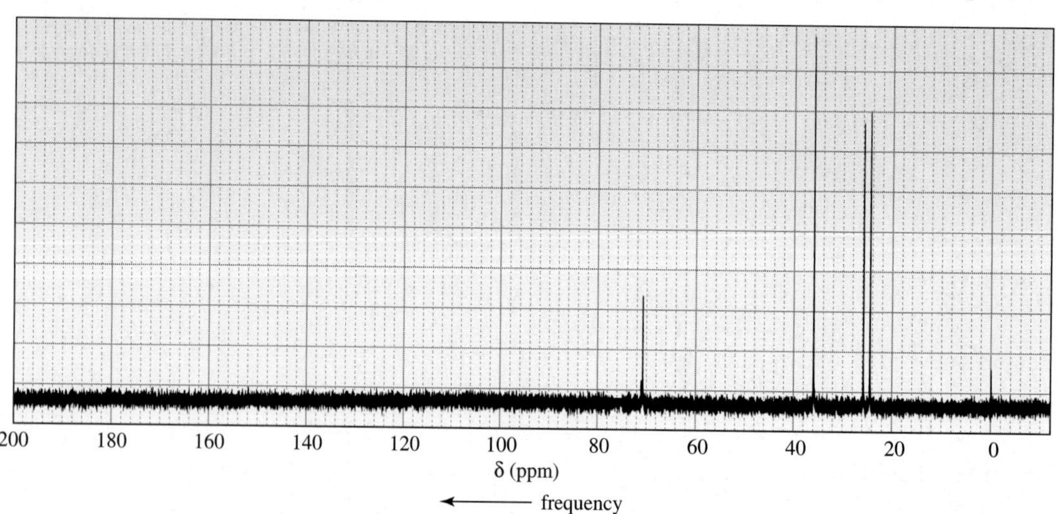

58. The ^1H NMR spectrum of 2-propen-1-ol is shown here. Indicate the protons in the molecule that give rise to each of the signals in the spectrum.

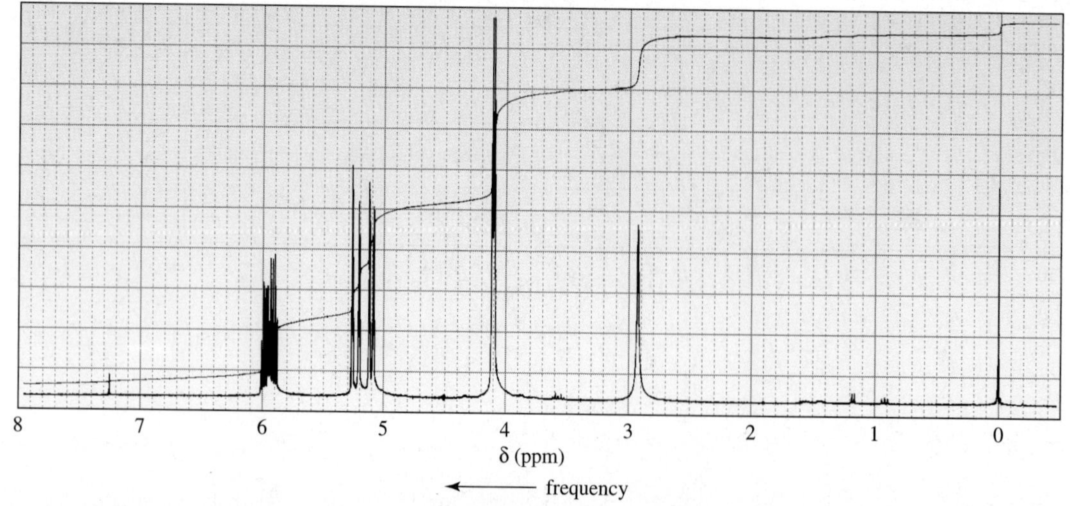

59. How could the signals in the 6.5–8.1-ppm region of their ^1H NMR spectra distinguish among the following compounds?

OCH_3 / NO_2 $\qquad$ OCH_3 / NO_2 $\qquad$ OCH_3 / NO_2

60. The ^1H NMR spectra of two compounds with molecular formula $C_{11}H_{16}$ are shown here. Identify the compounds.

a.

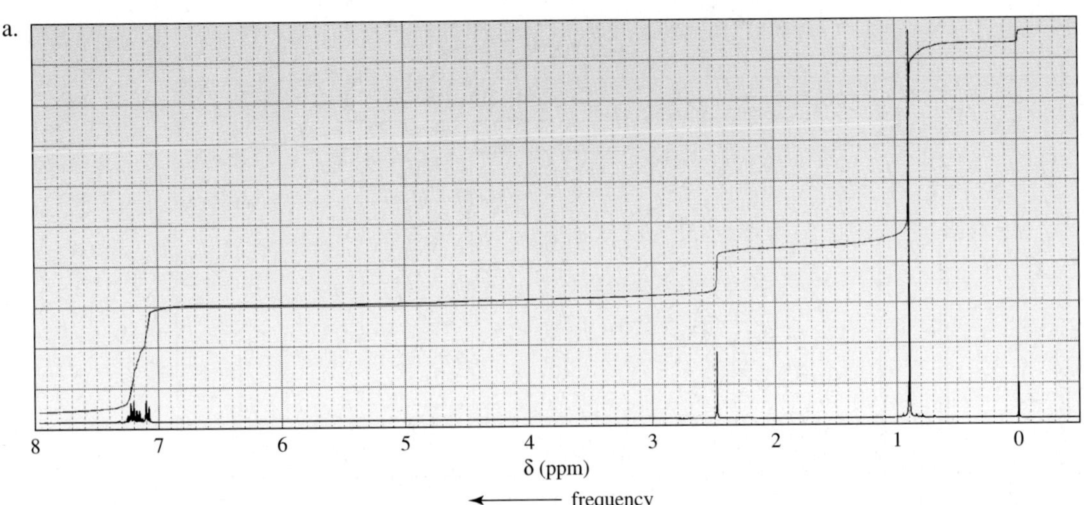

b.

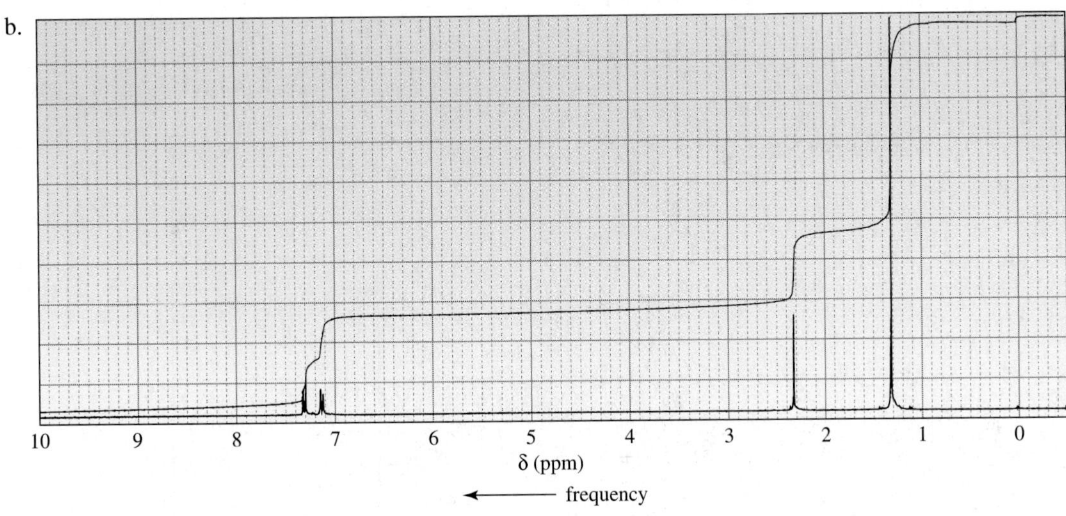

61. Draw a splitting diagram for the H_b proton if $J_{bc} = 10$ and $J_{ba} = 5$.

$$\underset{\substack{H \\ c}}{\overset{Cl}{C}}=\underset{\substack{H \\ b}}{\overset{CH_2Cl\ (a)}{C}}$$

62. Sketch the following spectra that would be obtained for 2-chloroethanol:
    a. The ^{1}H NMR spectrum for a dry sample of the alcohol
    b. The ^{1}H NMR spectrum for a sample of the alcohol that contains a trace amount of acid
    c. The ^{13}C NMR spectrum
    d. The proton-coupled ^{13}C NMR spectrum
    e. The four parts of a DEPT ^{13}C NMR spectrum

63. How could ^{1}H NMR be used to prove that the addition of HBr to propene follows the rule that says that the electrophile adds to the sp^2 carbon bonded to the greater number of hydrogens.

64. Identify each of the following compounds from its molecular formula and its ^{1}H NMR spectrum.

a. C_8H_8

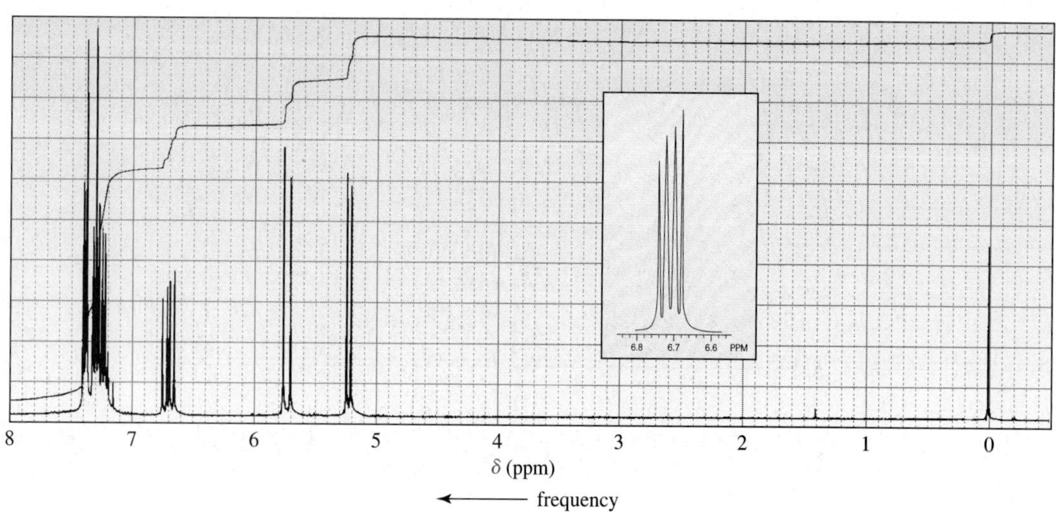

b. $C_6H_{12}O$

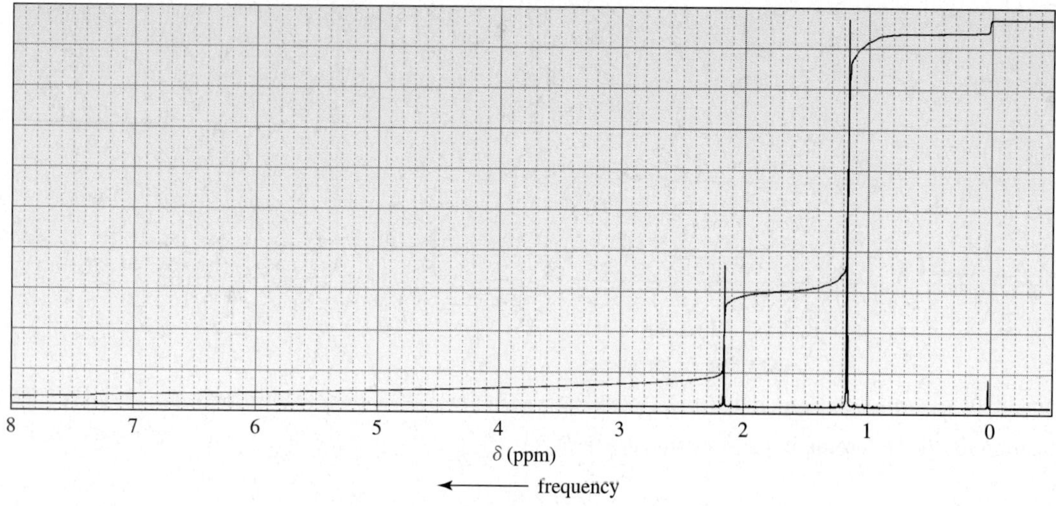

c. $C_9H_{18}O$

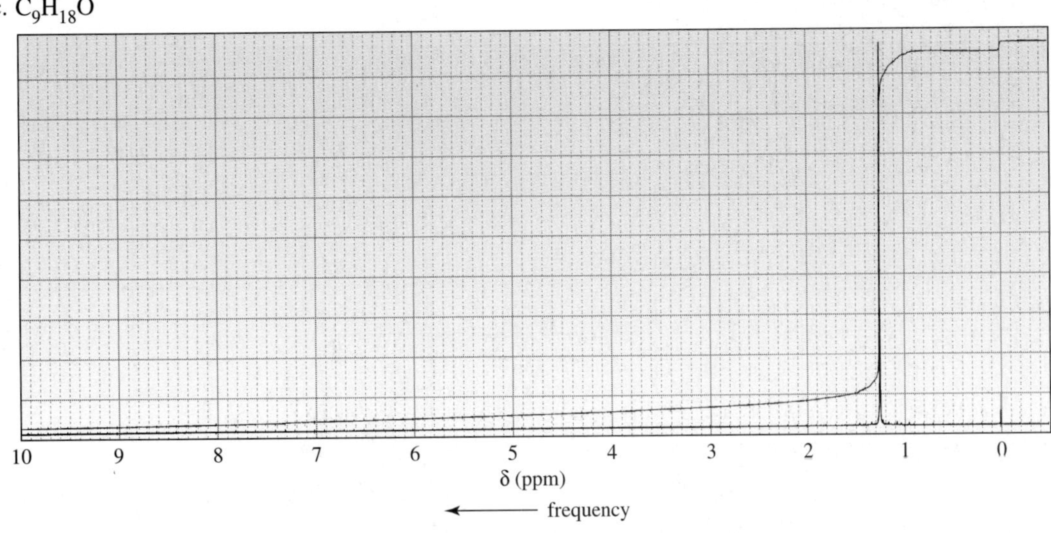

d. C_4H_8O

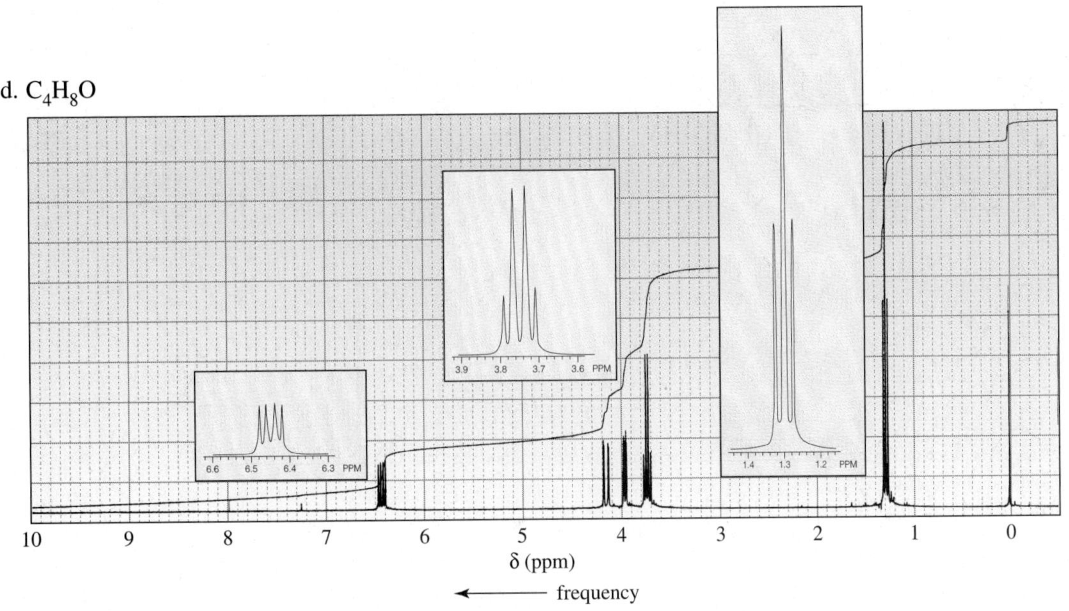

65. Dr. N. M. Arr was called in to help analyze the ^{1}H NMR spectrum of a mixture of compounds known to contain only C, H, and Br. The mixture showed two singlets—one at 1.8 ppm and the other at 2.7 ppm—with relative integrals of 1 : 6, respectively. Dr. Arr determined that the spectrum was that of a mixture of bromomethane and 2-bromo-2-methylpropane. What was the ratio of bromomethane to 2-bromo-2-methylpropane in the mixture?

66. Calculate the amount of energy (in calories) required to flip an ^{1}H nucleus in an NMR spectrometer that operates at 60 MHz.

67. The following ^1H NMR spectra are for four compounds with molecular formula $C_6H_{12}O_2$. Identify the compounds.

a.

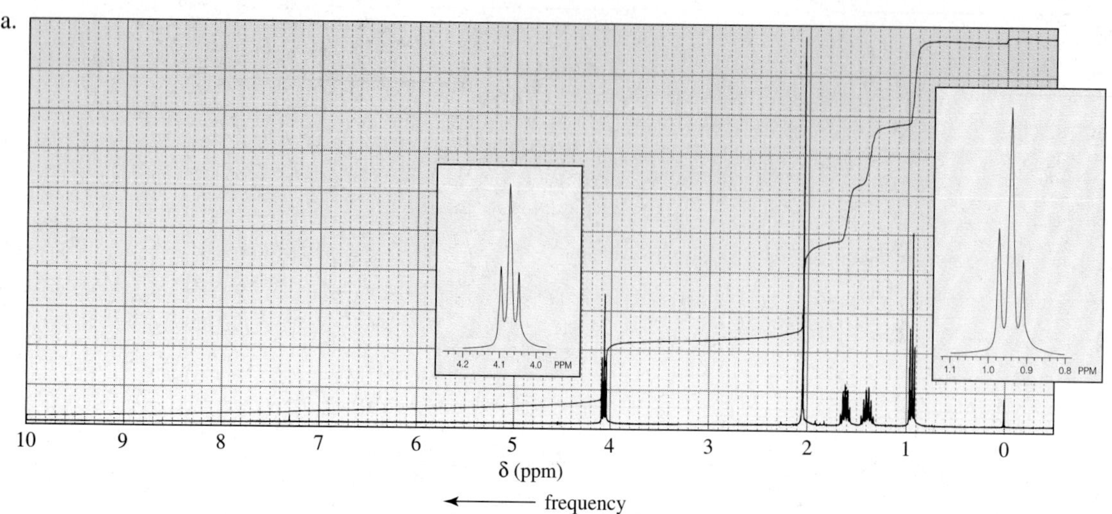

b.

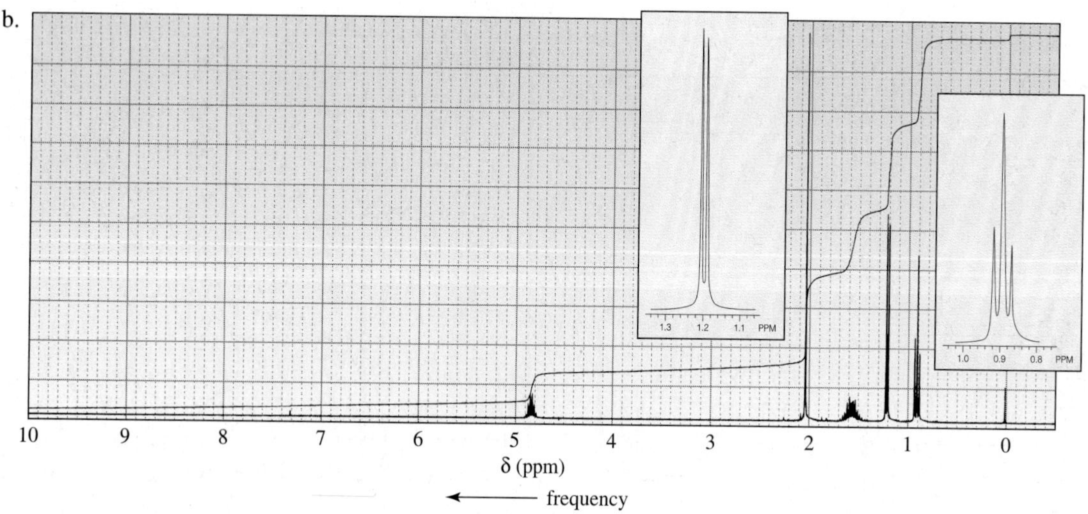

c.

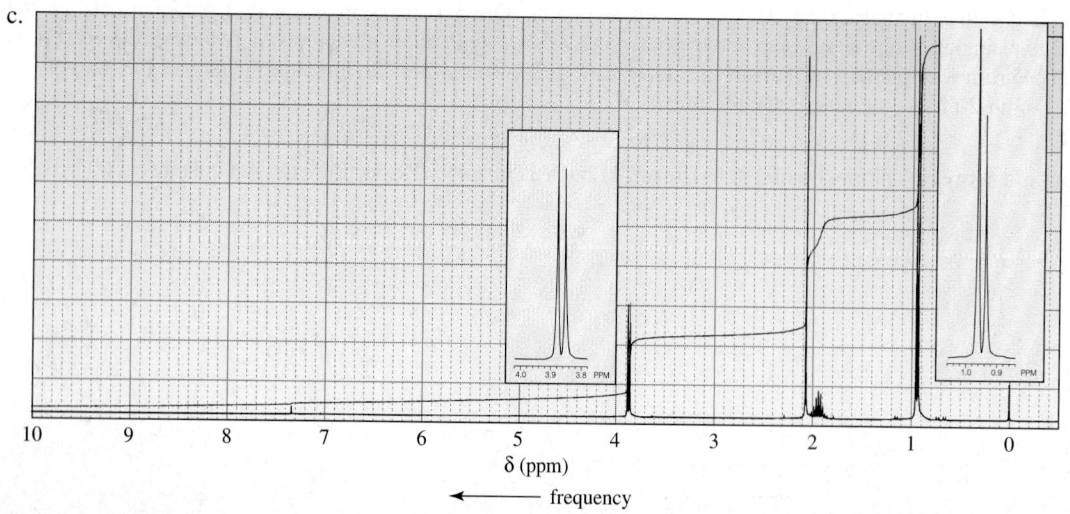

d.

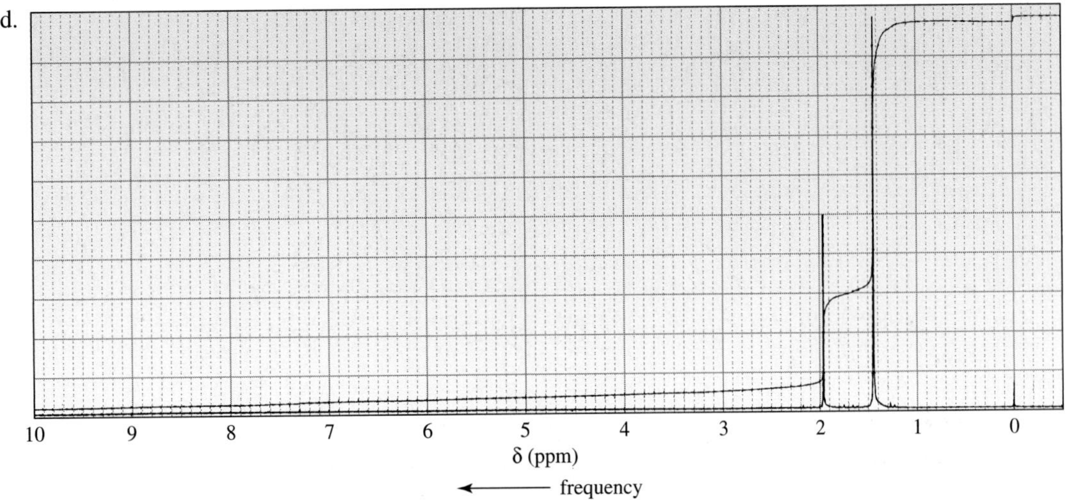

68. When compound A ($C_5H_{12}O$) is treated with HBr, it forms compound B ($C_5H_{11}Br$). The ^1H NMR spectrum of compound A has one singlet (1), two doublets (3, 6), and two multiplets (both 1). (The relative areas of the signals are indicated in parentheses.) The ^1H NMR spectrum of compound B has a singlet (6), a triplet (3), and a quartet (2). Identify compounds A and B.

69. Determine the structure of each of the following compounds based on its molecular formula and its IR and ^1H NMR spectra.

a. $C_6H_{12}O$

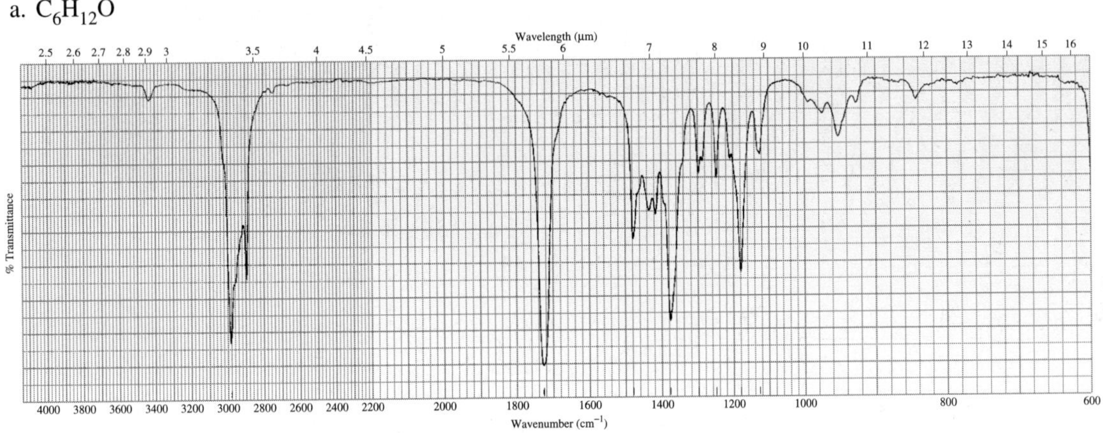

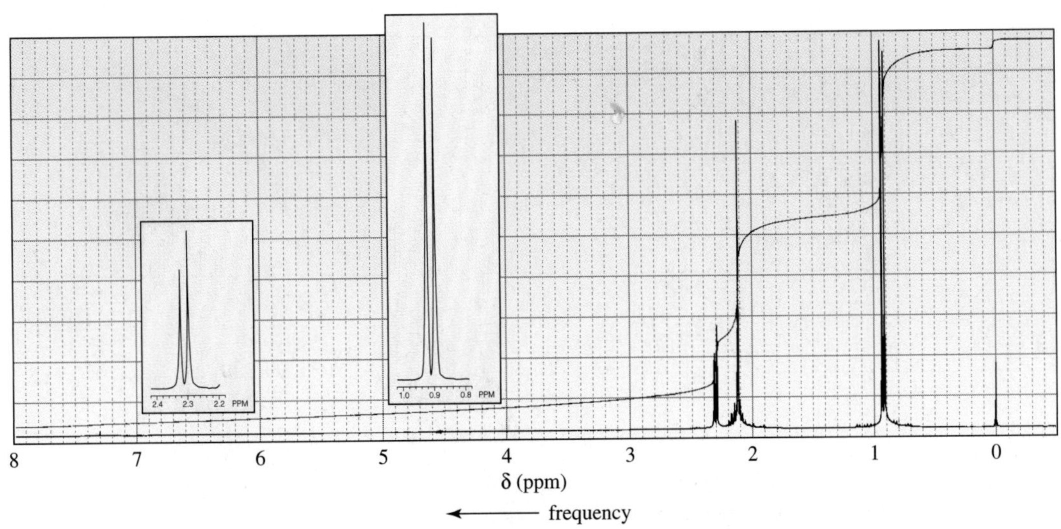

b. $C_6H_{14}O$

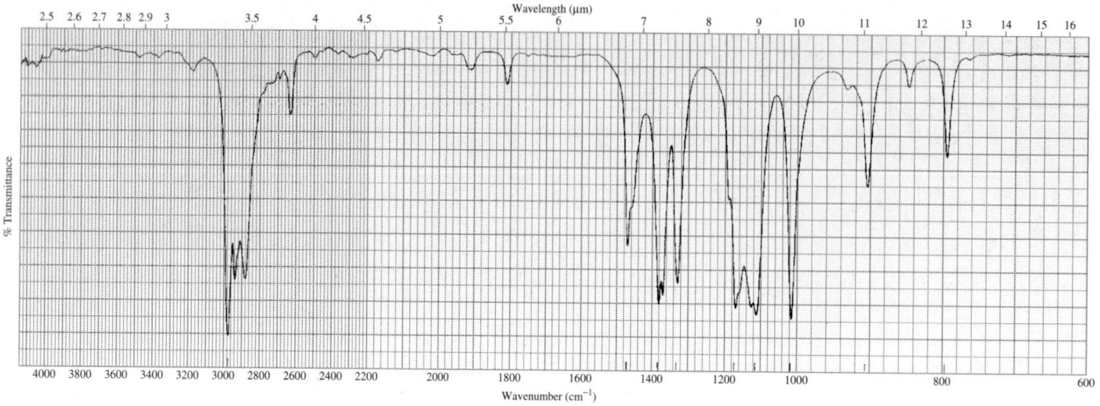

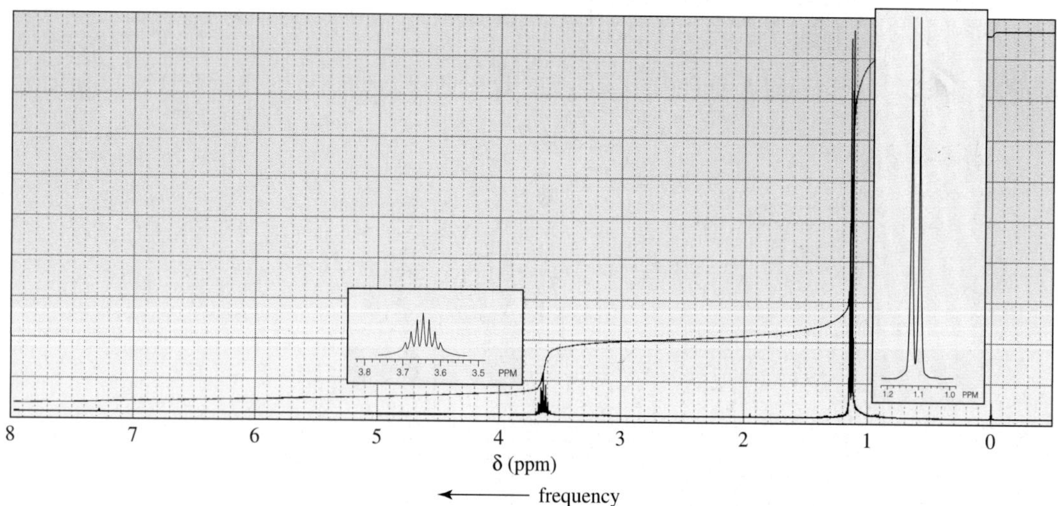

c. $C_{10}H_{13}NO_3$

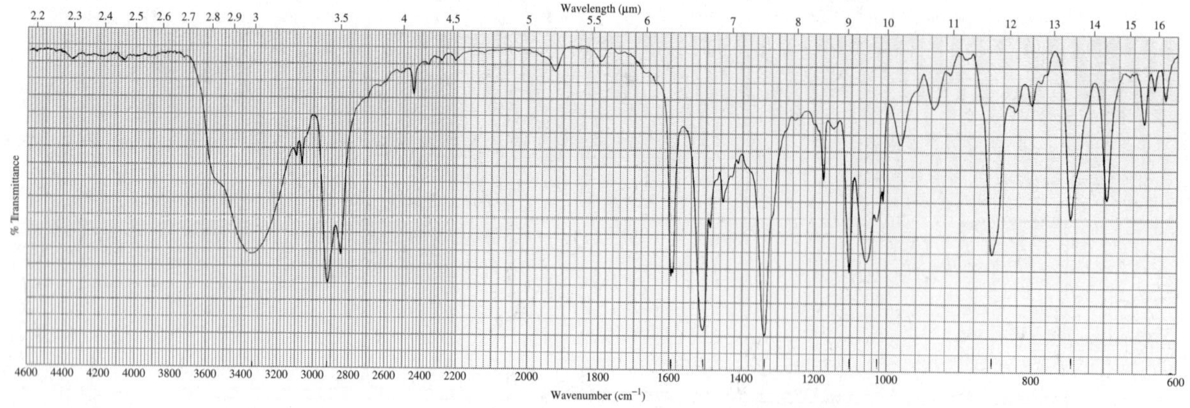

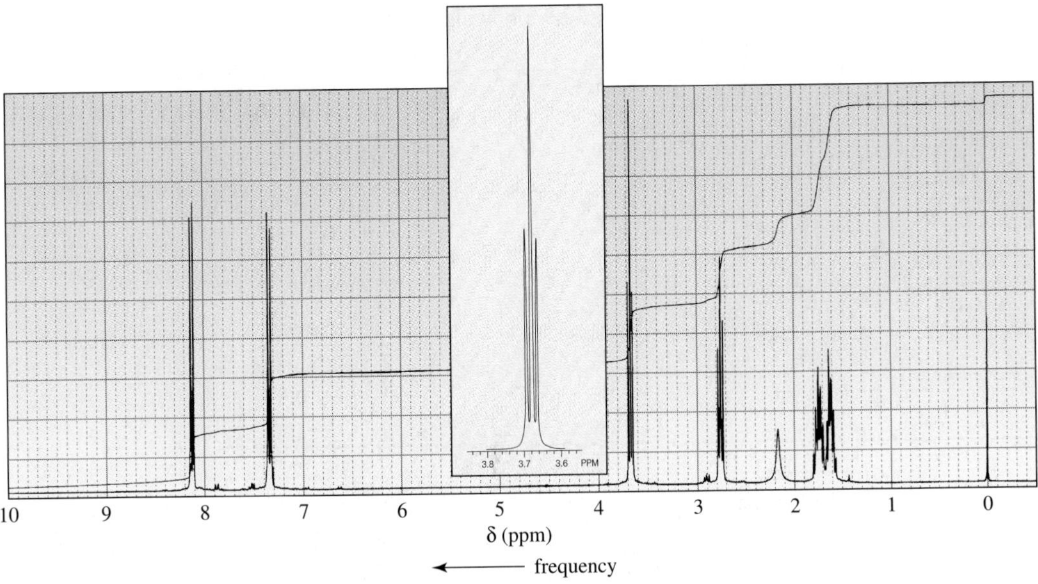
3.8
3.7
3.6
PPM
10
9
8
7
6
5
4
3
2
1
0
δ (ppm)
frequency

d. $C_{11}H_{14}O_2$

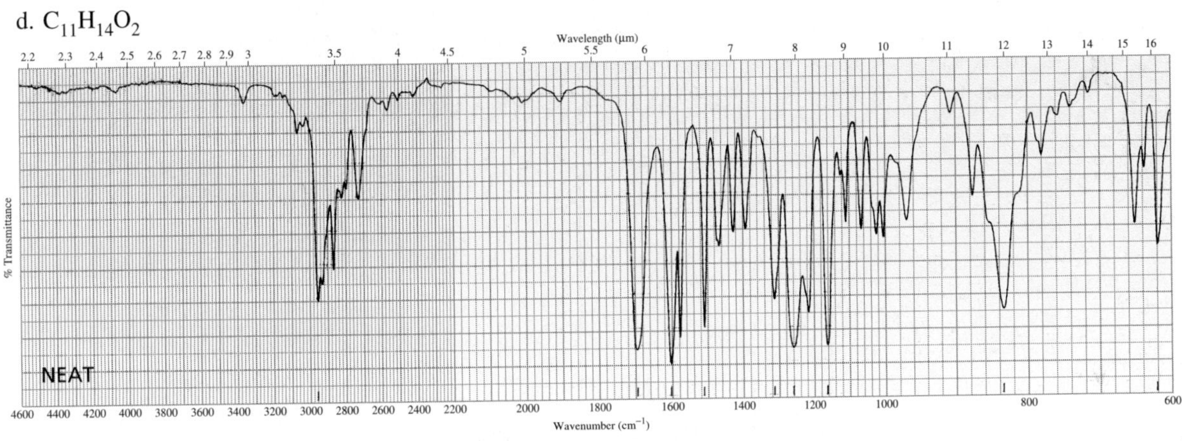
Wavelength (μm)
% Transmittance
NEAT
Wavenumber (cm^{-1})

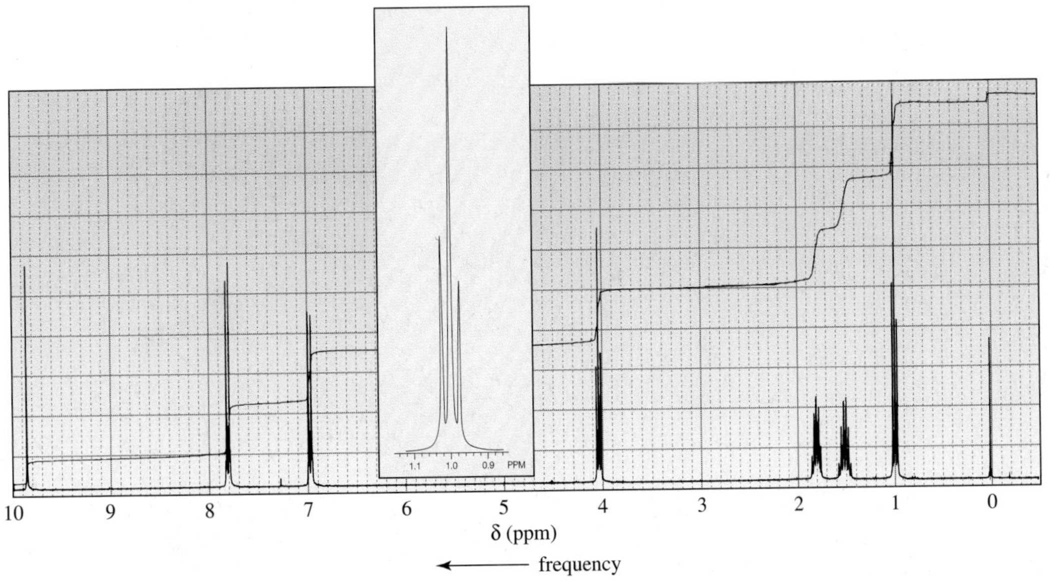
1.1
1.0
0.9
PPM
10
9
8
7
6
5
4
3
2
1
0
δ (ppm)
frequency

70. Identify the compound with molecular formula $C_3H_5Cl_3$ that gives the following ^{13}C NMR spectrum.

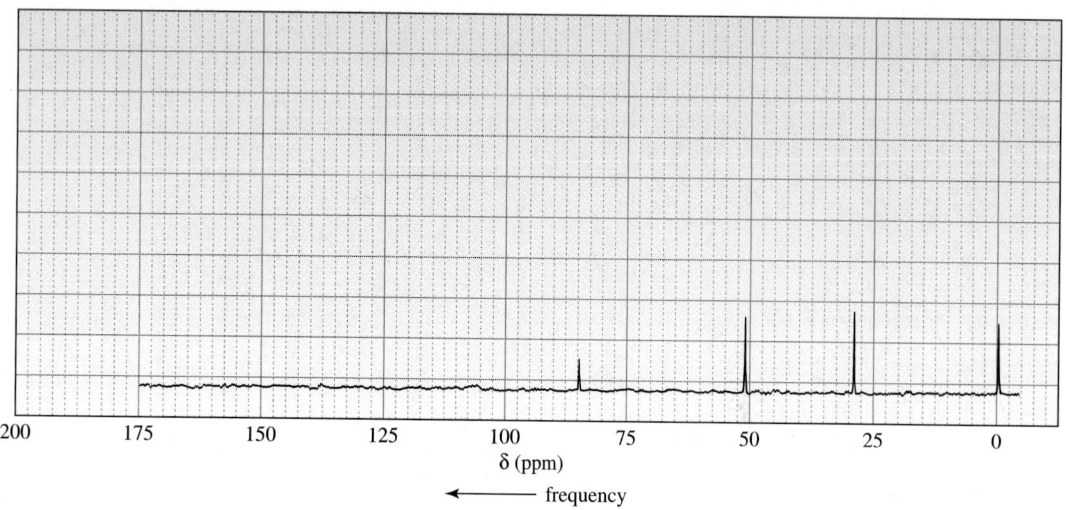

71. Determine the structure of each of the following compounds based on its mass, IR, and ^1H NMR spectra.

a.

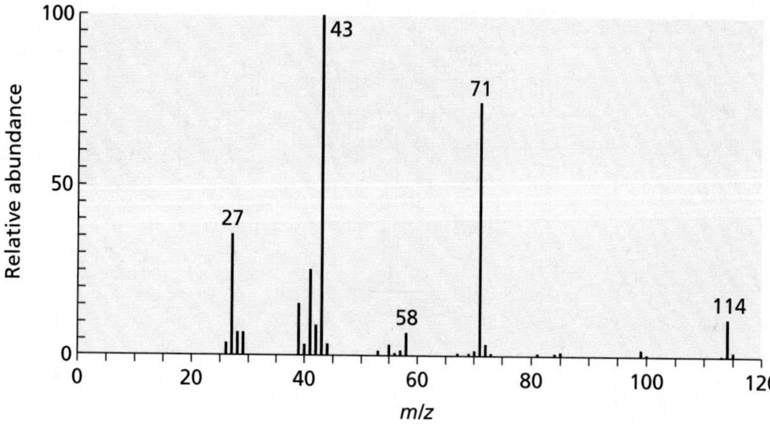

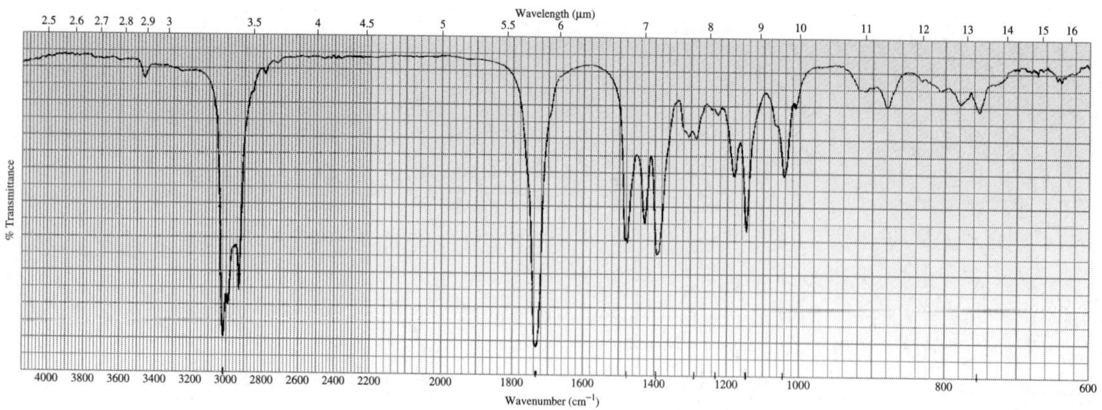

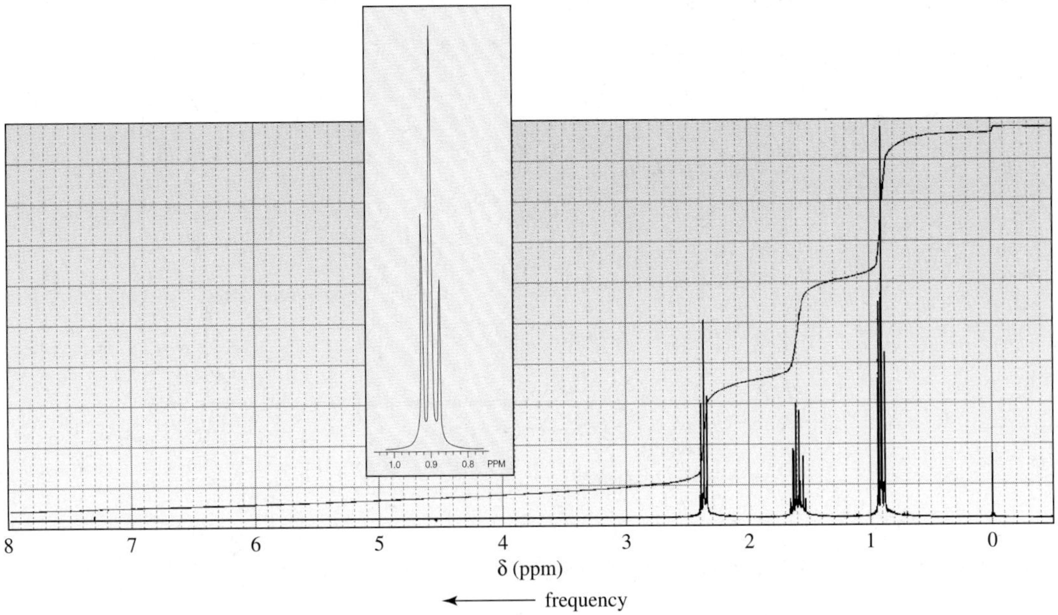
1.0
0.9
0.8
PPM
8
7
6
5
4
3
2
1
0
δ (ppm)
frequency

b.

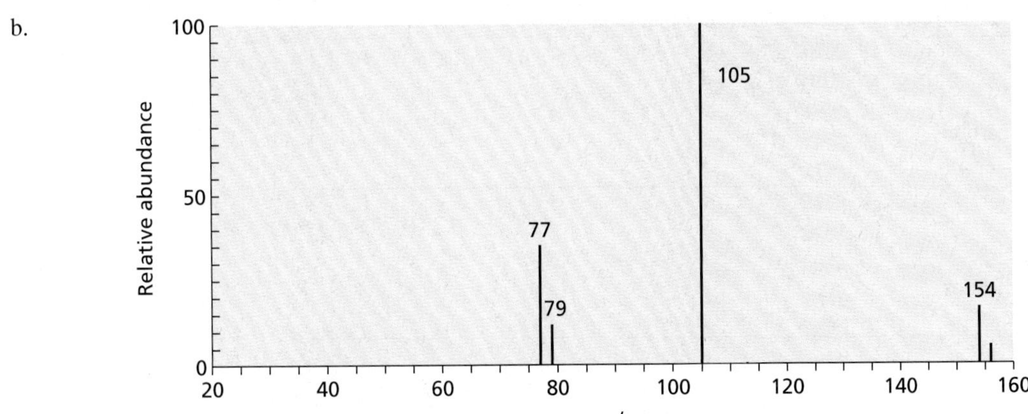
100
50
0
Relative abundance
105
77
79
154
20
40
60
80
100
120
140
160
m/z

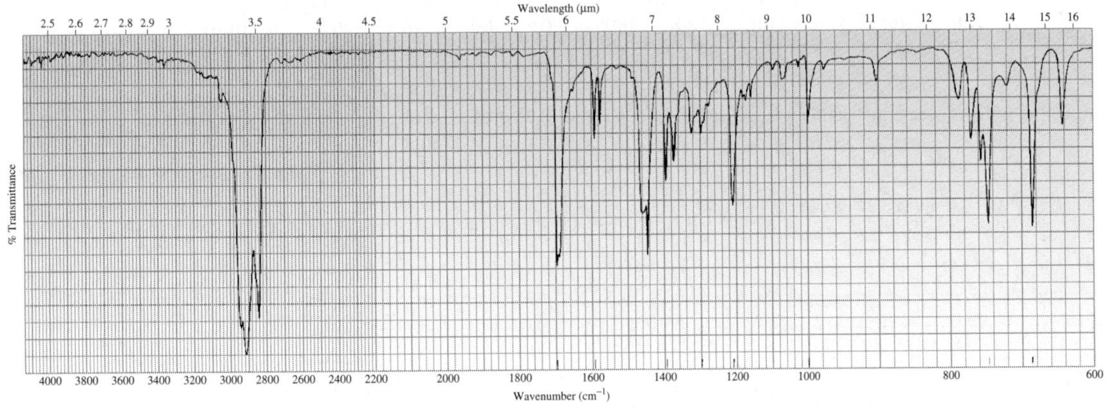
Wavelength (μm)
% Transmittance
Wavenumber (cm⁻¹)

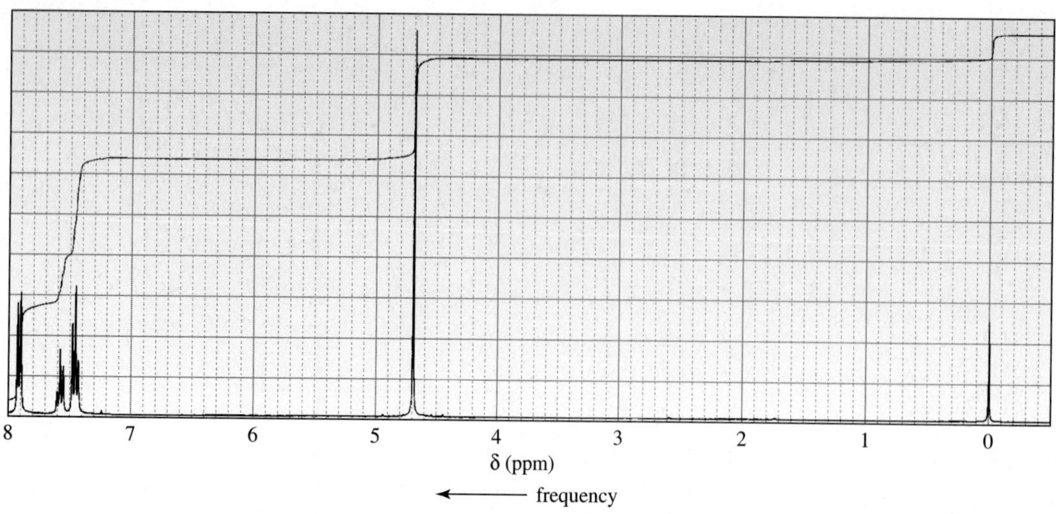

72. Identify the compound with molecular formula $C_6H_{10}O$ that is responsible for the following DEPT ^{13}C NMR spectrum.

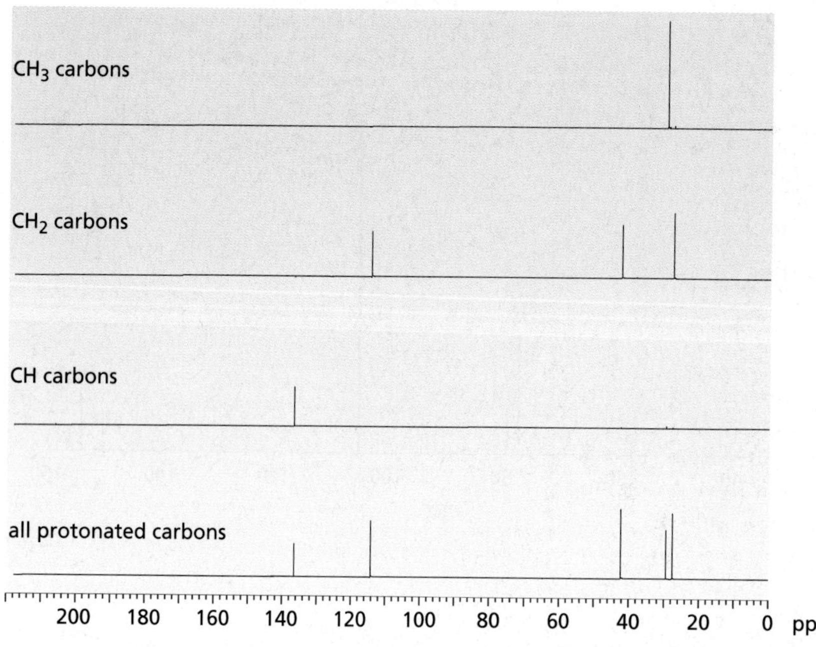

73. Identify the compound with molecular formula C_6H_{14} that is responsible for the following ^1H NMR spectrum.

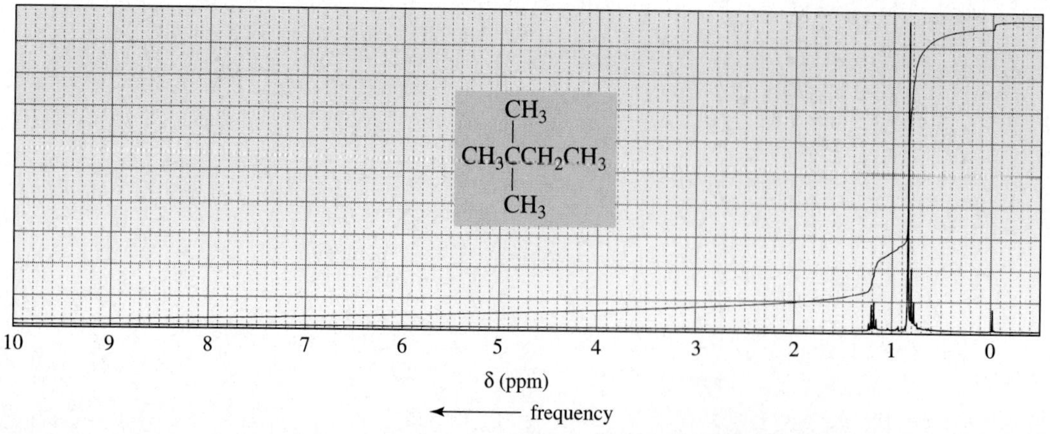

PART FIVE

Aromatic Compounds

The two chapters in Part Five deal with aromaticity and the reactions of aromatic compounds. Aromaticity was first introduced in Chapter 7, where you saw that benzene, a compound with an unusually large resonance energy, is an aromatic compound. We will now look at the criteria that a compound must fulfill in order to be classified as aromatic. Then we will examine the kinds of reactions that aromatic compounds undergo. In Chapter 21, we will return to aromatic compounds when we look at the reactions of aromatic compounds in which one of the ring atoms is an atom other than a carbon.

In **Chapter 15**, we will examine the structural features that cause a compound to be aromatic. We will also look at the features that cause a compound to be antiaromatic. Then we will take a look at the reactions that benzene undergoes. You will see that although benzene, alkenes, and dienes are all nucleophiles (because they all have carbon–carbon π bonds), benzene's aromaticity causes it to undergo reactions that are quite different from the reactions that alkenes and dienes undergo.

In **Chapter 16**, we will look at the reactions of substituted benzenes. First we will study reactions that change the nature of the substituent on the benzene ring; and we will see how the nature of the substituent affects both the reactivity of the ring and the placement of any incoming substituent. Then we will look at three types of reactions that can be used to synthesize substituted benzenes other than those discussed in Chapter 15—reactions of arene diazonium salts, nucleophilic aromatic substitution reactions, and reactions that involve benzyne intermediates. You will then have the opportunity to design syntheses of compounds that contain benzene rings.

15 Aromaticity • Reactions of Benzene

Benzene

Pyrrole

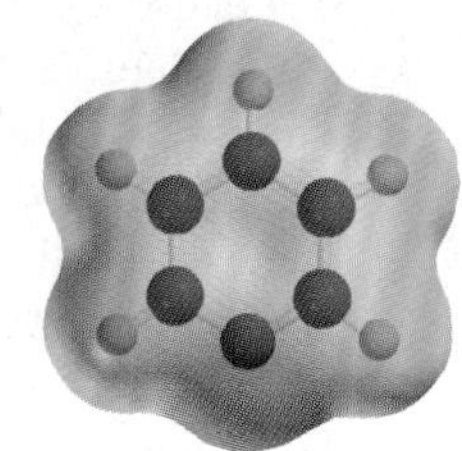

Pyridine

Michael Faraday (1791–1867) *was born in England, a son of a blacksmith. At the age of 14, he was apprenticed to a bookbinder and educated himself by reading the books that he bound. He became an assistant to Sir Humphry Davy in 1812 and taught himself chemistry. In 1825, he became the director of a laboratory at the Royal Institution, and, in 1833, he became a professor of chemistry there. He is best known for his work on electricity and magnetism.*

Eilhardt Mitscherlich (1794–1863) *was born in Germany. He studied oriental languages at the University of Heidelberg and the Sorbonne, where he concentrated on Farsi, hoping that Napoleon would include him in a delegation he intended to send to Persia. That ambition ended with Napoleon's defeat. Mitscherlich returned to Germany to study science, simultaneously receiving a doctorate in Persian studies. He was a professor of chemistry at the University of Berlin.*

The compound we know as benzene was first isolated in 1825 by Michael Faraday, who extracted the compound from a liquid residue obtained after heating whale oil under pressure to produce a gas used to illuminate buildings in London. Because of its origin, chemists suggested that it should be called "pheno" from the Greek word *phainein* ("to shine").

In 1834, Eilhardt Mitscherlich correctly determined benzene's molecular formula (C_6H_6) and decided to call it benzin because of its relationship to benzoic acid, a known substituted form of the compound. Later its name was changed to benzene.

Compounds like benzene, which have relatively few hydrogens in relation to the number of carbons, are typically found in oils produced by trees and other plants. Early chemists called such compounds **aromatic** compounds because of their pleasing fragrances. In this way, they were distinguished from **aliphatic** compounds, with higher hydrogen-to-carbon ratios, that were obtained from the chemical degradation of fats. The chemical meaning of the word "aromatic" now signifies certain kinds of chemical structures. We will now examine the criteria that a compound must satisfy to be classified as aromatic.

15.1 Criteria for Aromaticity

In Chapter 7, we saw that benzene is a planar, cyclic compound with a cyclic cloud of delocalized electrons above and below the plane of the ring (Figure 15.1). Because its π electrons are delocalized, all the C—C bonds have the same length—partway between the length of a typical single and a typical double bond. We also saw that benzene is a particularly stable compound because it has an unusually large resonance energy (36 kcal/mol or 151 kJ/mol). Most compounds with delocalized electrons

a.

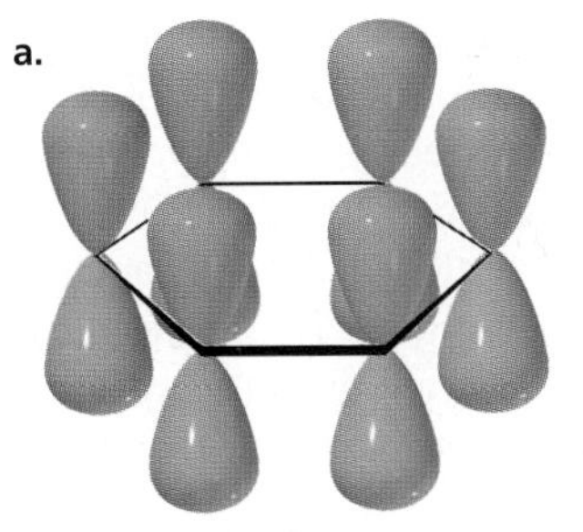

b.

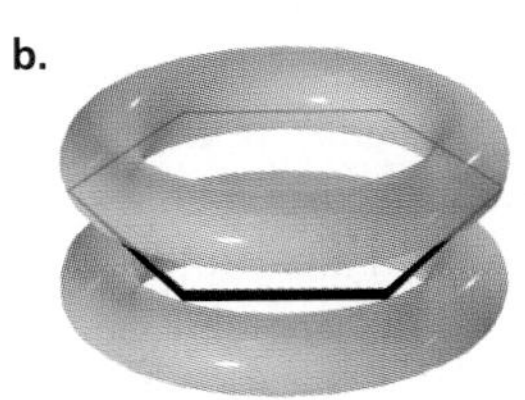

c.

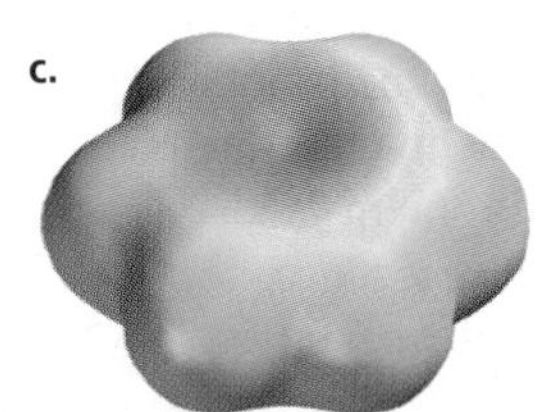

Figure 15.1
(a) Each carbon of benzene has a *p* orbital. (b) The overlap of the *p* orbitals forms a cloud of π electrons above and below the plane of the benzene ring. (c) The electrostatic potential map for benzene shows that all the carbon–carbon bonds have the same electron density.

have much smaller resonance energies. Compounds such as benzene with unusually large resonance energies are called aromatic compounds. How can we tell whether a compound is aromatic by looking at its structure? In other words, what structural features do aromatic compounds have in common?

Aromatic compounds are particularly stable.

To be classified as aromatic, a compound must meet both of the following criteria:

1. *It must have an uninterrupted cyclic cloud of π electrons* (often called a π cloud) *above and below the plane of the molecule.* Let's look a little more closely at what this means:
   For the π cloud to be cyclic, *the molecule must be cyclic.*
   For the π cloud to be uninterrupted, *every atom in the ring must have a* p *orbital.*
   For the π cloud to form, each *p* orbital must overlap with the *p* orbitals on either side of it. Therefore, *the molecule must be planar.*
2. *The π cloud must contain an odd number of pairs of π electrons.*

Benzene is an aromatic compound because it is cyclic and planar, every carbon in the ring has a *p* orbital, and the π cloud contains *three* pairs of π electrons.

For a compound to be aromatic, it must be cyclic and planar and have an uninterrupted cloud of π electrons. The π cloud must contain an odd number of pairs of π electrons.

The German chemist Erich Hückel was the first to recognize that an aromatic compound must have an odd number of pairs of π electrons. In 1931, he described this requirement by what has come to be known as **Hückel's rule**, or the **$4n + 2$ rule**. The rule states that for a planar, cyclic compound to be aromatic, its uninterrupted π cloud must contain $(4n + 2)\pi$ electrons, where n is any whole number. According to Hückel's rule, then, an aromatic compound must have 2 ($n = 0$), 6 ($n = 1$), 10 ($n = 2$), 14 ($n = 3$), 18 ($n = 4$), etc., π electrons. Because there are two electrons in a pair, Hückel's rule requires that an aromatic compound have 1, 3, 5, 7, 9, etc., pairs of π electrons. Thus, Hückel's rule is just a mathematical way of saying that an aromatic compound must have an *odd* number of pairs of π electrons.

Erich Hückel (1896–1980) *was born in Germany. He was a professor of chemistry at the University of Stuttgart and at the University of Marburg.*

PROBLEM 1◆

a. What is the value of n in Hückel's rule when a compound has nine pairs of π electrons?
b. Is such a compound aromatic?

15.2 Aromatic Hydrocarbons

Monocyclic hydrocarbons with alternating single and double bonds are called **annulenes**. A prefix in brackets denotes the number of carbons in the ring. Cyclobutadiene, benzene, and cyclooctatetraene are examples.

cyclobutadiene
[4]-annulene

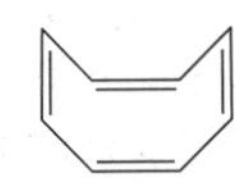

benzene
[6]-annulene

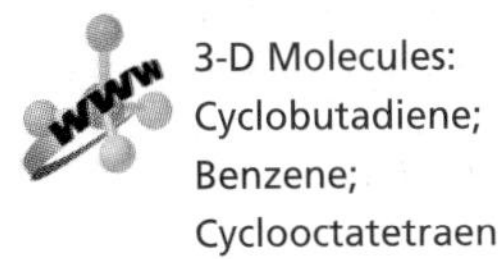

cyclooctatetraene
[8]-annulene

www 3-D Molecules: Cyclobutadiene; Benzene; Cyclooctatetraene

Cyclobutadiene has two pairs of π electrons, and cyclooctatetraene has four pairs of π electrons. Unlike benzene, these compounds are *not* aromatic because they have an *even* number of pairs of π electrons. There is an additional reason why cyclooctatetraene is not aromatic—it is not planar but, instead, tub-shaped. Earlier, we saw that, for an eight-membered ring to be planar, it must have bond angles of 135° (Chapter 2, Problem 28), and we know that sp^2 carbons have 120° bond angles. Therefore, if cyclooctatetraene were planar, it would have considerable angle strain. Because cyclobutadiene and cyclooctatetraene are not aromatic, they do not have the unusual stability of aromatic compounds.

Now let's look at some other compounds and determine whether they are aromatic. Cyclopropene is not aromatic because it does not have an uninterrupted ring of p orbital-bearing atoms. One of its ring atoms is sp^3 hybridized, and only sp^2 and sp hybridized carbons have p orbitals. Therefore, cyclopropene does not fulfill the first criterion for aromaticity.

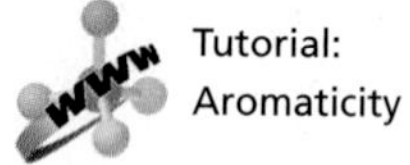

Tutorial: Aromaticity

When drawing resonance contributors, remember that only electrons move, atoms never move.

The cyclopropenyl cation is aromatic because it has an uninterrupted ring of p orbital-bearing atoms and the π cloud contains *one* (an odd number) pair of delocalized π electrons. The cyclopropenyl anion is not aromatic because although it has an uninterrupted ring of p orbital-bearing atoms, its π cloud has *two* (an even number) pairs of π electrons.

resonance contributors of the cyclopropenyl cation

δ+ δ+ δ+

resonance hybrid

Cycloheptatriene is not aromatic. Although it has the correct number of pairs of π electrons (three) to be aromatic, it does not have an uninterrupted ring of p orbital-bearing atoms because one of the ring atoms is sp^3 hybridized. Cyclopentadiene is also not aromatic: It has an even number of pairs of π electrons (two pairs), *and* it does not have an uninterrupted ring of p orbital-bearing atoms. Like cycloheptatriene, cyclopentadiene has an sp^3 hybridized carbon.

sp^3 sp^3

cycloheptatriene cyclopentadiene

3-D Molecules: Phenanthrene; Naphthalene

The criteria for determining whether a monocyclic hydrocarbon compound is aromatic can also be used to determine whether a polycyclic hydrocarbon compound is aromatic. Naphthalene (five pairs of π electrons), phenanthrene (seven pairs of π electrons), and chrysene (nine pairs of π electrons) are aromatic.

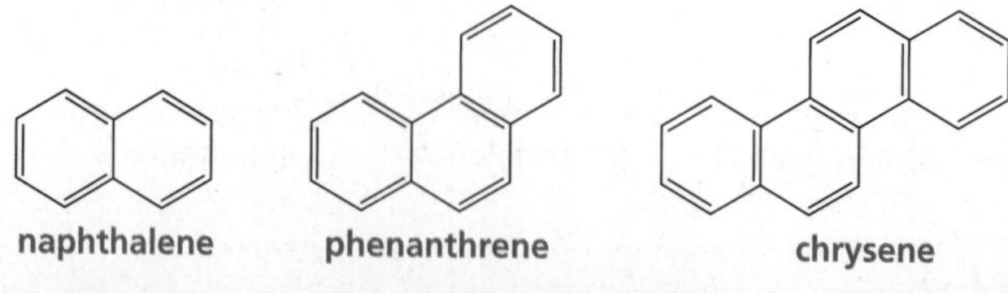

BUCKYBALLS AND AIDS

In addition to diamond and graphite (Section 1.1), a third form of pure carbon was discovered while scientists were conducting experiments designed to understand how long-chain molecules are formed in outer space. R. E. Smalley, R. F. Curl, Jr., and H. W. Kroto, the discoverers of this new form of carbon, shared the 1996 Nobel Prize in chemistry for their discovery. They named this new form buckminsterfullerene (often shortened to fullerene) because it reminded them of the geodesic domes popularized by R. Buckminster Fuller, an American architect and philosopher. The substance is nicknamed "buckyball." Consisting of a hollow cluster of 60 carbons, fullerene is the most symmetrical large molecule known. Like graphite, fullerene has only sp^2 hybridized carbons, but instead of being arranged in layers, the carbons are arranged in rings, forming a hollow cluster of 60 carbons that fit together like the seams of a soccer ball. Each molecule has 32 interlocking rings (20 hexagons and 12 pentagons). At first glance, fullerene would appear to be aromatic because of its benzene-like rings. However, it does not undergo electrophilic substitution reactions; instead, it undergoes electrophilic addition reactions like an alkene. Fullerene's lack of aromaticity is apparently caused by the curvature of the ball, which prevents the molecule from fulfilling the first criterion for aromaticity—that it must be planar.

Buckyballs have extraordinary chemical and physical properties. They are exceedingly rugged and are capable of surviving the extreme temperatures of outer space. Because they are essentially hollow cages, they can be manipulated to make materials never before known. For example, when a buckyball is "doped" by inserting potassium or cesium into its cavity, it becomes an excellent organic superconductor. These molecules are presently being studied for use in many other applications, such as new polymers and catalysts and new drug delivery systems. The discovery of buckyballs is a strong reminder of the technological advances that can be achieved as a result of conducting basic research.

Scientists have even turned their attention to buckyballs in their quest for a cure for AIDS. An enzyme that is required for HIV to reproduce exhibits a nonpolar pocket in its three-dimensional structure. If this pocket is blocked, the production of the virus ceases. Because buckyballs are nonpolar and have approximately the same diameter as the pocket of the enzyme, they are being considered as possible blockers. The first step in pursuing this possibility was to equip the buckyball with polar side chains to make it water soluble so that it could flow through the bloodstream. Scientists have now modified the side chains so that they bind to the enzyme. It's still a long way from a cure for AIDS, but this represents one example of the many and varied approaches that scientists are taking to find a cure for this disease.

A geodesic dome

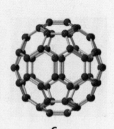

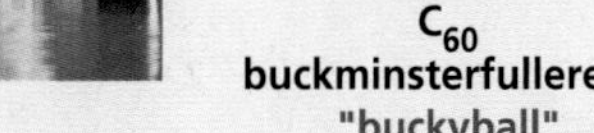

C_{60}
buckminsterfullerene
"buckyball"

PROBLEM 2◆

Which of the following compounds are aromatic?

a. 

b.

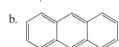

c. cycloheptatrienyl cation

d. 

e. 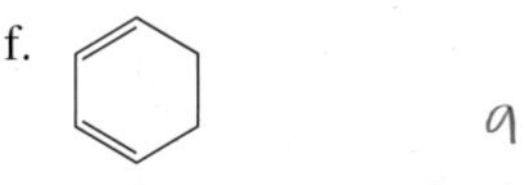

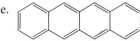

f.

g. cyclononatetraenyl anion

h. $CH_2{=}CHCH{=}CHCH{=}CH_2$

PROBLEM 3 SOLVED

a. How many monobromonaphthalenes are there?
b. How many monobromophenanthrenes are there?

SOLUTION TO 3a There are two monobromonaphthalenes. Substitution cannot occur at either of the carbons shared by both rings, because those carbons are not bonded to a hydrogen. Naphthalene is a flat molecule, so substitution for a hydrogen at any other carbon will result in one of the compounds shown.

Br Br

Richard E. Smalley *was born in 1943 in Akron, Ohio. He received a B.S. from the University of Michigan and a Ph.D. from Princeton University. He is a professor of chemistry at Rice University.*

Robert F. Curl, Jr., *was born in Texas in 1933. He received a B.A. from Rice University and a Ph.D. from the University of California, Berkeley. He is a professor of chemistry at Rice University.*

Sir Harold W. Kroto *was born in 1939 in England and is a professor of chemistry at the University of Sussex.*

3-D Molecules: 1-Chloronaphthalene; 2-Chloronaphthalene

PROBLEM 4

The [10]- and [12]-annulenes have been synthesized, and neither has been found to be aromatic. Explain.

15.3 Aromatic Heterocyclic Compounds

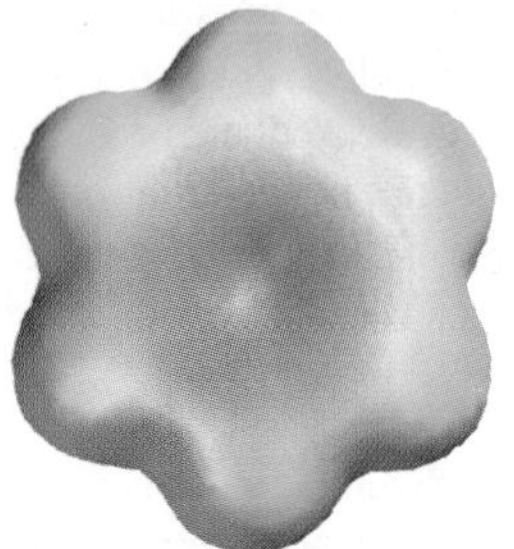
benzene

A compound does not have to be a hydrocarbon to be aromatic. Many *heterocyclic compounds* are aromatic. A **heterocyclic compound** is a cyclic compound in which one or more of the ring atoms is an atom other than carbon. A ring atom that is not carbon is called a **heteroatom**. The name comes from the Greek word *heteros*, which means "different." The most common heteroatoms found in heterocyclic compounds are N, O, and S.

heterocyclic compounds

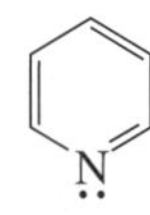

pyridine

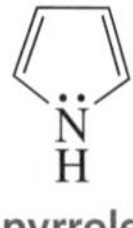

pyrrole

furan

thiophene

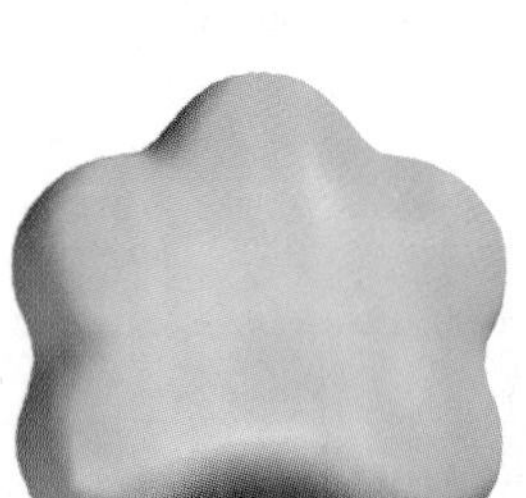
pyridine

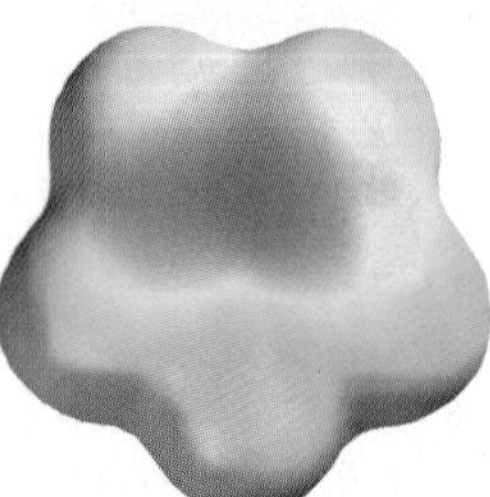
pyrrole

Pyridine is an aromatic heterocyclic compound. Each of the six ring atoms of pyridine is sp^2 hybridized, which means that each has a *p* orbital; and the molecule contains three pairs of π electrons. Don't be confused by the lone-pair electrons on the nitrogen; they are not π electrons. Because nitrogen is sp^2 hybridized, it has three sp^2 orbitals and a *p* orbital. The *p* orbital is used to form the π bond. Two of nitrogen's sp^2 orbitals overlap the sp^2 orbitals of adjacent carbon atoms, and nitrogen's third sp^2 orbital contains the lone pair.

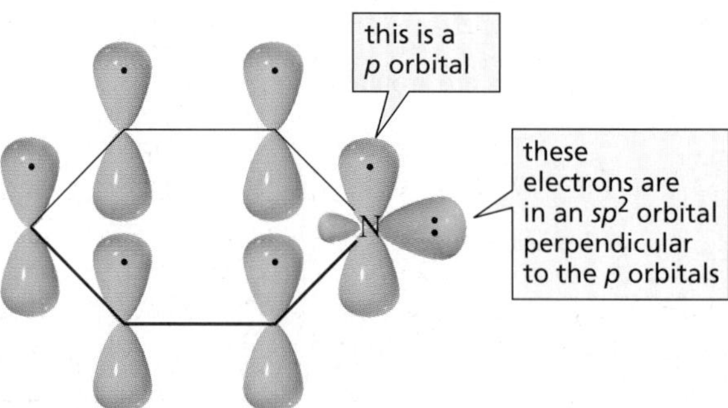

orbital structure of pyridine

It is not immediately apparent that the electrons represented as lone-pair electrons on the nitrogen atom of pyrrole are π electrons. The resonance contributors, however, show that the nitrogen atom is sp^2 hybridized and uses its three sp^2 orbitals to bond to two carbons and one hydrogen. The lone-pair electrons are in a *p* orbital that overlaps the *p* orbitals on adjacent carbons, forming a π bond—thus, they are π electrons. Pyrrole, therefore, has three pairs of π electrons and is aromatic.

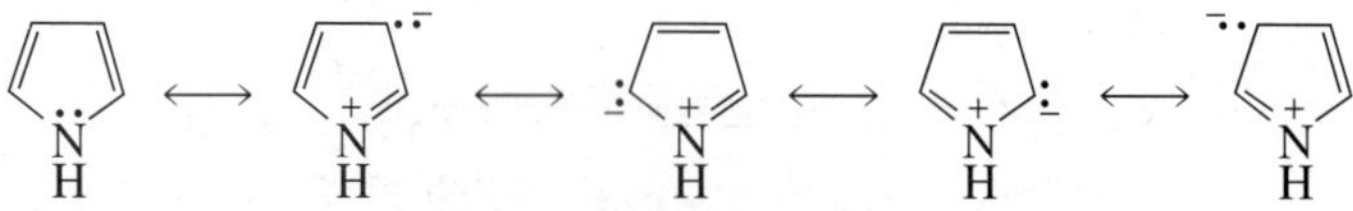

resonance contributors of pyrrole

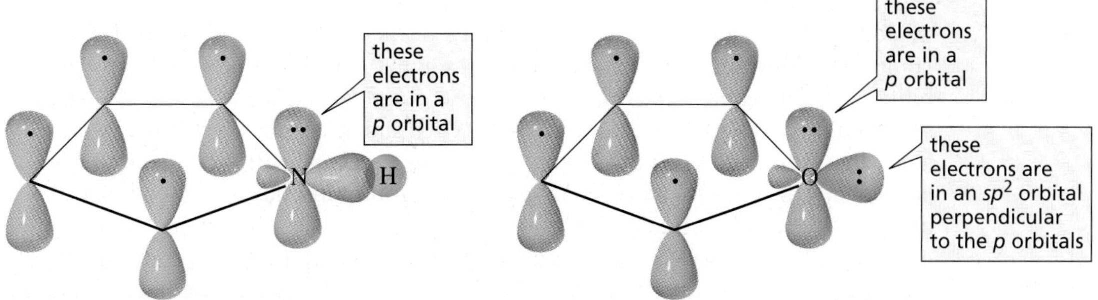
these electrons are in a p orbital
N
H
these electrons are in a p orbital
O
these electrons are in an sp^2 orbital perpendicular to the p orbitals

orbital structure of pyrrole **orbital structure of furan**

Similarly, furan and thiophene are stable aromatic compounds. Both the oxygen in the former and the sulfur in the latter are sp^2 hybridized and have one lone pair in an sp^2 orbital. The second lone pair is in a p orbital that overlaps the p orbitals of adjacent carbons, forming a π bond. Thus, they are π electrons.

resonance contributors of furan

Quinoline, indole, imidazole, purine, and pyrimidine are other examples of heterocyclic aromatic compounds. The heterocyclic compounds discussed in this section are examined in greater detail in Chapter 21.

quinoline indole imidazole purine pyrimidine

PROBLEM 5◆

In what orbitals are the electrons represented as lone pairs when drawing the structures of quinoline, indole, imidazole, purine, and pyrimidine?

PROBLEM 6

Answer the following questions by examining the electrostatic potential maps on p. 598:

a. Why is the bottom part of the electrostatic potential map of pyrrole blue?
b. Why is the bottom part of the electrostatic potential map of pyridine red?
c. Why is the center of the electrostatic potential map of benzene more red than the center of the electrostatic potential map of pyridine?

15.4 Some Chemical Consequences of Aromaticity

The pK_a of cyclopentadiene is 15, which is extraordinarily acidic for a hydrogen that is bonded to an sp^3 hybridized carbon. Ethane, for example, has a pK_a of 50.

$$\text{cyclopentadiene} \rightleftharpoons \text{cyclopentadienyl anion} + H^+$$

cyclopentadiene **pK_a = 15** **cyclopentadienyl anion**

$$CH_3CH_3 \rightleftharpoons CH_3\ddot{C}H_2^- + H^+$$

ethane **pK_a = 50** **ethyl anion**

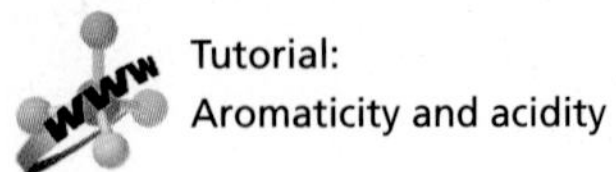
Tutorial:
Aromaticity and acidity

Why is the pK_a of cyclopentadiene so much lower than that of ethane? To answer this question, we must look at the stabilities of the anions that are formed when the compounds lose a proton. (Recall that the strength of an acid is determined by the stability of its conjugate base: The more stable its conjugate base, the stronger is the acid; see Section 1.18.) All the electrons in the ethyl anion are localized. In contrast, the anion that is formed when cyclopentadiene loses a proton fulfills the requirements for aromaticity: It is cyclic and planar, each atom in the ring has a *p* orbital, and the π cloud has three pairs of delocalized π electrons. The negatively charged carbon in the cyclopentadienyl anion is sp^2 hybridized because if it were sp^3 hybridized, the ion would not be aromatic. The resonance hybrid shows that all the carbons in the cyclopentadienyl anion are equivalent. Each carbon has exactly one-fifth of the negative charge associated with the anion.

resonance contributors of the cyclopentadienyl anion

δ– δ– δ– δ– δ–

resonance hybrid

As a result of its aromaticity, the cyclopentadienyl anion is an unusually stable carbanion. This is why cyclopentadiene has an unusually low pK_a. In other words, it is the stability conveyed by the aromaticity of the cyclopentadienyl anion that makes the hydrogen much more acidic than hydrogens bonded to other sp^3 carbons.

PROBLEM 7◆

Predict the relative pK_a values of cyclopentadiene and cycloheptatriene.

PROBLEM 8

a. Draw arrows to show the movement of electrons in going from one resonance contributor to the next in
   1. the cyclopentadienyl anion
   2. pyrrole
b. How many ring atoms share the negative charge in
   1. the cyclopentadienyl anion?
   2. pyrrole?

Another example of the influence of aromaticity on chemical reactivity is the unusual chemical behavior exhibited by cycloheptatrienyl bromide. Recall from Section 2.9 that alkyl halides tend to be relatively nonpolar covalent compounds—they are soluble in nonpolar solvents and insoluble in water. Cycloheptatrienyl bromide, however, is an alkyl halide that behaves like an ionic compound—it is insoluble in nonpolar solvents, but readily soluble in water.

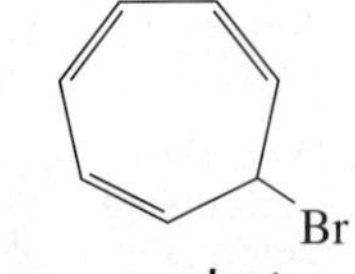

covalent
cycloheptatrienyl bromide

ionic
cycloheptatrienyl bromide
tropylium bromide

Cycloheptatrienyl bromide is an ionic compound because its cation is aromatic. The alkyl halide is *not* aromatic in the covalent form because it has an sp^3 hybridized carbon, so it does *not* have an uninterrupted ring of p orbital-bearing atoms. In the ionic form, however, the cycloheptatrienyl cation (also known as the tropylium cation) *is* aromatic because it is a planar cyclic ion, all the ring atoms are sp^2 hybridized (which means that each ring atom has a p orbital), and it has three pairs of delocalized π electrons. The stability associated with the aromatic cation causes the alkyl halide to exist in the ionic form.

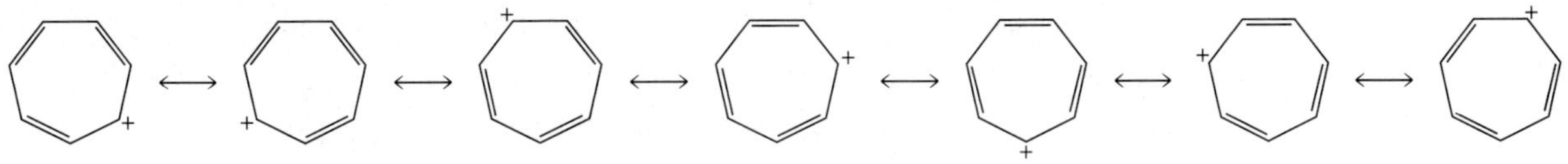

resonance contributors of the cycloheptatrienyl cation

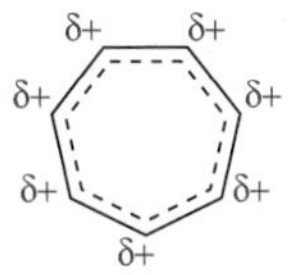

resonance hybrid

PROBLEM-SOLVING STRATEGY

Which of the following compounds has the greater dipole moment?

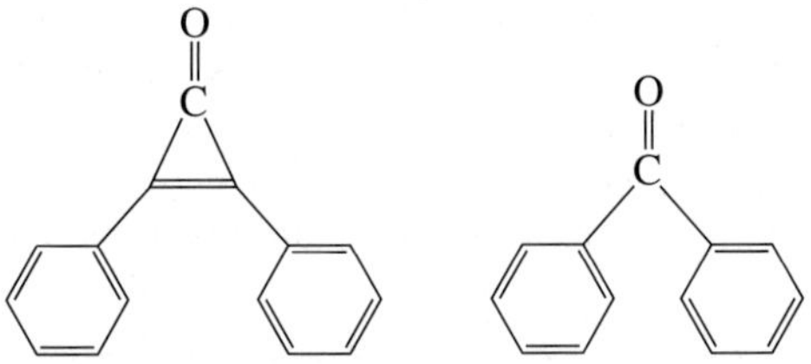

Before attempting to answer this kind of question, make sure that you know exactly what the question is asking. You know that the dipole moment of these compounds results from the unequal sharing of electrons by carbon and oxygen. Therefore, the more unequal the sharing, the greater is the dipole moment. So now the question becomes, which compound has a greater negative charge on its oxygen atom? Draw the structures with separated charges, and determine their relative stabilities. In the case of the compound on the left, the three-membered ring becomes aromatic when the charges are separated. In the case of the compound on the right, the structure with separated charges is not aromatic. Because being aromatic makes a compound more stable, the compound on the left has the greater dipole moment.

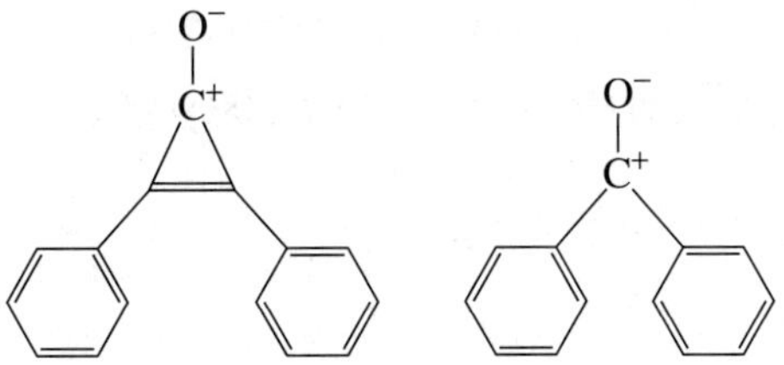

PROBLEM 9

Draw the resonance contributors of the cyclooctatrienyl dianion.

a. Which of the resonance contributors is the least stable?

b. Which of the resonance contributors makes the smallest contribution to the hybrid?

15.5 Antiaromaticity

Antiaromatic compounds are highly unstable.

An aromatic compound is *more stable* than an analogous cyclic compound with localized electrons. In contrast, an **antiaromatic** compound is *less stable* than an analogous cyclic compound with localized electrons. *Aromaticity is characterized by stability, whereas antiaromaticity is characterized by instability.*

relative stabilities

aromatic compound > cyclic compound with localized electrons > antiaromatic compound

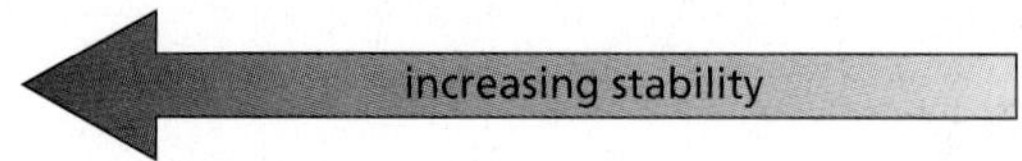

A compound is classified as being antiaromatic if it fulfills the first criterion for aromaticity but does not fulfill the second criterion. In other words, it must be a planar, cyclic compound with an uninterrupted ring of p orbital-bearing atoms, and the π cloud must contain an *even* number of pairs of π electrons. Hückel would state that the π cloud must contain $4n$ π electrons, where n is any whole number—a mathematical way of saying that the cloud must contain an *even* number of pairs of π electrons.

Cyclobutadiene is a planar, cyclic molecule with two pairs of π electrons. Hence, it is expected to be antiaromatic and highly unstable. In fact, it is too unstable to be isolated, although it has been trapped at very cold temperatures. The cyclopentadienyl cation also has two pairs of π electrons, so we can conclude that it is antiaromatic and unstable.

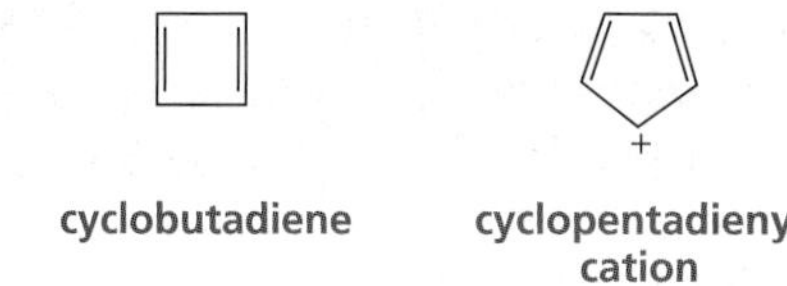

cyclobutadiene cyclopentadienyl cation

PROBLEM 10◆

a. Predict the relative pK_a values of cyclopropene and cyclopropane.
b. Which is more soluble in water, 3-bromocyclopropene or bromocyclopropane?

PROBLEM 11◆

Which of the compounds in Problem 2 are antiaromatic?

15.6 A Molecular Orbital Description of Aromaticity and Antiaromaticity

Why are planar molecules with uninterrupted cyclic π electron clouds highly stable (aromatic) if they have an odd number of pairs of π electrons and highly unstable (antiaromatic) if they have an even number of pairs of π electrons? To answer this question, we must turn to molecular orbital theory.

The relative energies of the π molecular orbitals of a planar molecule with an uninterrupted cyclic π electron cloud can be determined—without having to use any math—by first drawing the cyclic compound with one of its vertices pointed down. The relative energies of the π molecular orbitals correspond to the relative levels of the vertices (Figure 15.2). Molecular orbitals below the midpoint of the cyclic structure are bonding molecular orbitals, those above the midpoint are antibonding molecular orbitals, and any at the midpoint are nonbonding molecular orbitals. This scheme is sometimes called a Frost device (or a Frost circle) in honor of Arthur A. Frost, an

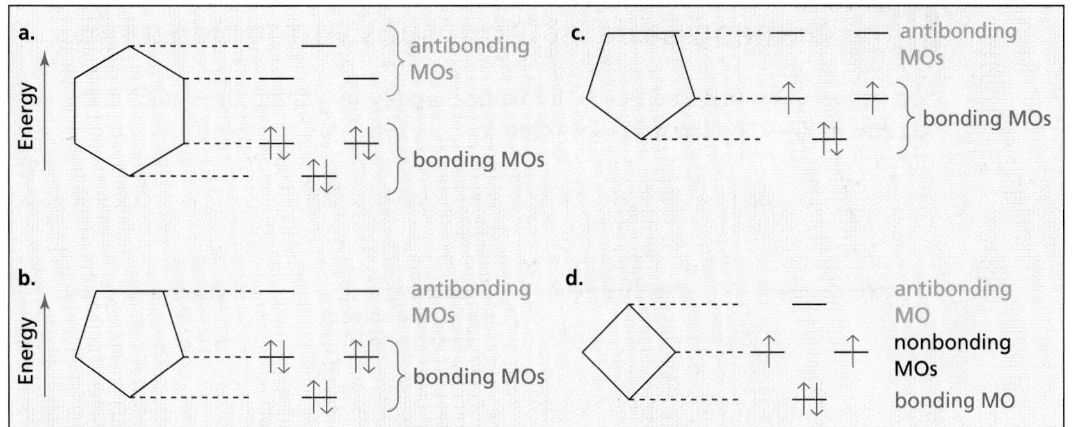

▲ Figure 15.2
The distribution of electrons in the π molecular orbitals of (a) benzene, (b) the cyclopentadienyl anion, (c) the cyclopentadienyl cation, and (d) cyclobutadiene. The relative energies of the π molecular orbitals in a cyclic compound correspond to the relative levels of the vertices. Molecular orbitals below the midpoint of the cyclic structure are bonding, those above the midpoint are antibonding, and those at the midpoint are nonbonding.

American scientist who devised this simple method. Notice that the number of π molecular orbitals is the same as the number of atoms in the ring because each ring atom contributes a *p* orbital. (Recall that orbitals are conserved; Section 7.11.)

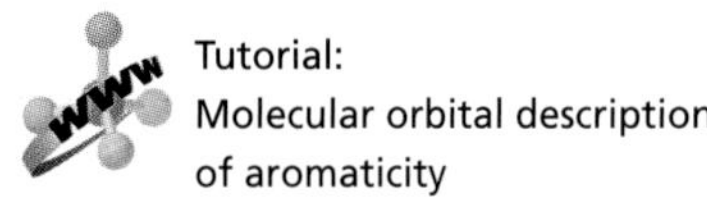

Tutorial:
Molecular orbital description of aromaticity

The six π electrons of benzene occupy its three bonding π molecular orbitals, and the six π electrons of the cyclopentadienyl anion occupy *its* three bonding π molecular orbitals. Notice that there is always an odd number of bonding orbitals because one corresponds to the lowest vertex and the others come in degenerate pairs. This means that aromatic compounds—such as benzene and the cyclopentadienyl anion—with an odd number of pairs of π electrons have completely filled bonding orbitals and no electrons in either nonbonding or antibonding orbitals. This is what gives aromatic molecules their stability. (A more in-depth description of the molecular orbitals in benzene is given in Section 7.11.)

Aromatic compounds are stable because they have filled bonding π molecular orbitals.

Antiaromatic compounds have an even number of pairs of π electrons. Therefore, either they are unable to fill their bonding orbitals (cylopentadienyl cation) or they have a pair of π electrons left over after the bonding orbitals are filled (cyclobutadiene). Hund's rule requires that these two electrons go into two degenerate orbitals (Section 1.2). The unpaired electrons are responsible for the instability of antiaromatic molecules.

PROBLEM 12◆

How many bonding, nonbonding, and antibonding π molecular orbitals does cyclobutadiene have? In which molecular orbitals are the π electrons?

PROBLEM 13◆

Can a radical be aromatic?

PROBLEM 14

Following the instructions for drawing the π molecular orbital energy levels of the compounds shown in Figure 15.2, draw the π molecular orbital energy levels for the cycloheptatrienyl cation, the cycloheptatrienyl anion, and the cyclopropenyl cation. For each compound, show the distribution of the π electrons. Which of the compounds are aromatic? Which are antiaromatic?

15.7 Nomenclature of Monosubstituted Benzenes

Some monosubstituted benzenes are named simply by stating the name of the substituent, followed by the word "benzene."

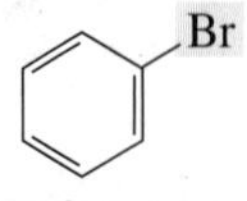

Br — bromobenzene

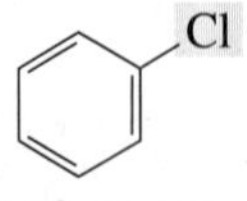

Cl — chlorobenzene

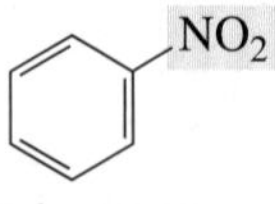

NO_2 — nitrobenzene, used as a solvent in shoe polish

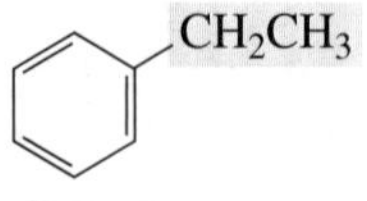

CH_2CH_3 — ethylbenzene

Some monosubstituted benzenes have names that incorporate the name of the substituent. Unfortunately, such names have to be memorized.

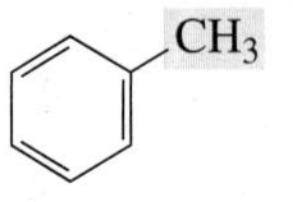

CH_3 — toluene

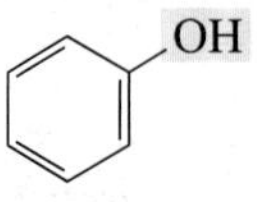

OH — phenol

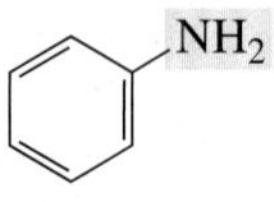

NH_2 — aniline

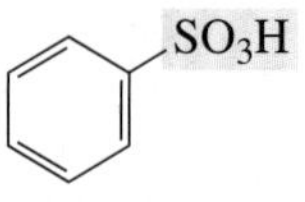

SO_3H — benzenesulfonic acid

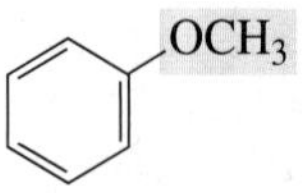

OCH_3 — anisole

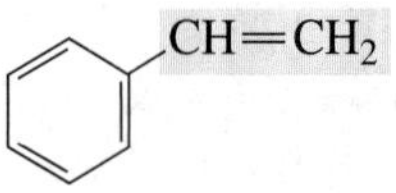

$CH{=}CH_2$ — styrene

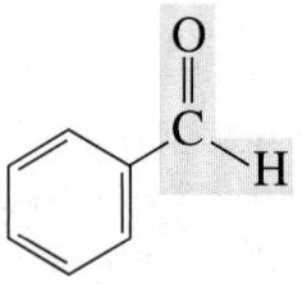

$C(=O)H$ — benzaldehyde

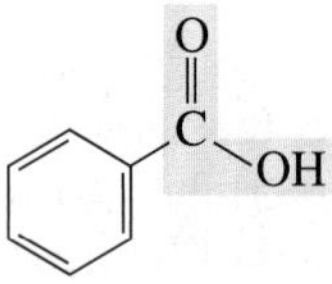

$C(=O)OH$ — benzoic acid

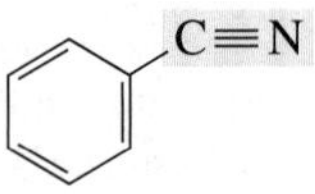

$C{\equiv}N$ — benzonitrile

With the exception of toluene, benzene rings with an alkyl substituent are named as alkyl-substituted benzenes or as phenyl-substituted alkanes.

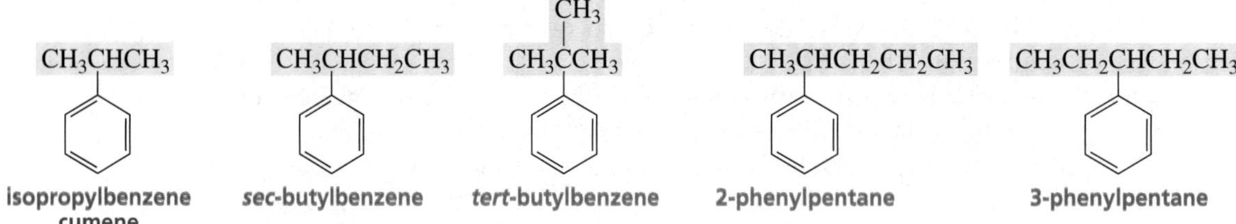

isopropylbenzene (cumene) — *sec*-butylbenzene — *tert*-butylbenzene — 2-phenylpentane — 3-phenylpentane

When a benzene ring is a substituent, it is called a **phenyl group**. A benzene ring with a methylene group is called a **benzyl group**. The phenyl group gets its name from "pheno," the name that was rejected for benzene (Section 15.0).

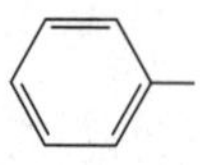

a phenyl group

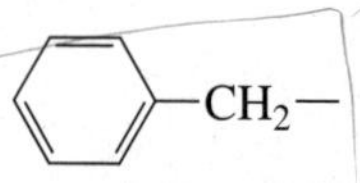

CH_2— — a benzyl group

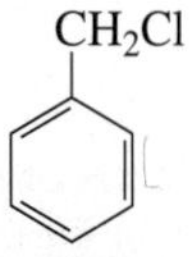

CH_2Cl — chloromethylbenzene (benzyl chloride)

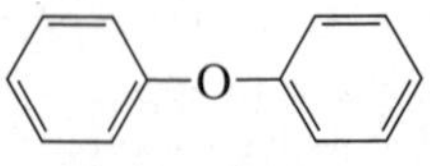

O — diphenyl ether

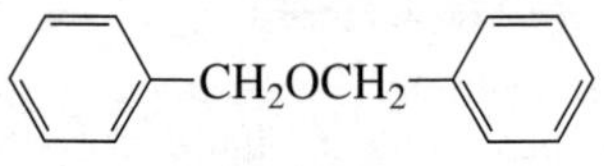

CH_2OCH_2 — dibenzyl ether

An aryl group (Ar) is the general term for either a phenyl group or a substituted phenyl group, just as an alkyl group (R) is the general term for a group derived from an alkane. In other words, ArOH could be used to designate any of the following phenols:

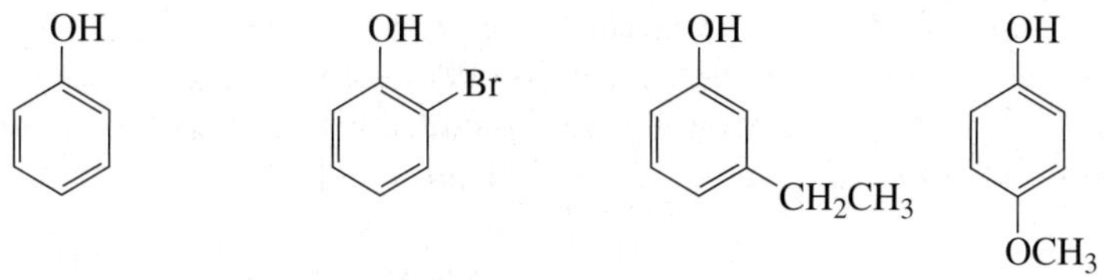

OH OH Br OH CH_2CH_3 OH OCH_3

THE TOXICITY OF BENZENE

Although benzene has been widely used in chemical synthesis and has been frequently used as a solvent, it is toxic. Its major toxic effect is on the central nervous system and on bone marrow. Chronic exposure to benzene causes leukemia and aplastic anemia. A higher-than-average incidence of leukemia has been found in industrial workers with long-term exposure to as little as 1 ppm benzene in the atmosphere. Toluene has replaced benzene as a solvent because, although it is a central nervous system depressant like benzene, it does not cause leukemia or aplastic anemia. "Glue sniffers" seek the narcotic central nervous system effects of solvents such as toluene. This can be a highly dangerous activity.

PROBLEM 15◆

Draw the structure of each of the following compounds:

a. 2-phenylhexane
b. benzyl alcohol
c. 3-benzylpentane
d. bromomethylbenzene

15.8 How Benzene Reacts

As a consequence of the π electrons above and below the plane of its ring, benzene is a nucleophile. It will, therefore, react with an electrophile (Y^+). When an electrophile attaches itself to a benzene ring, a carbocation intermediate is formed.

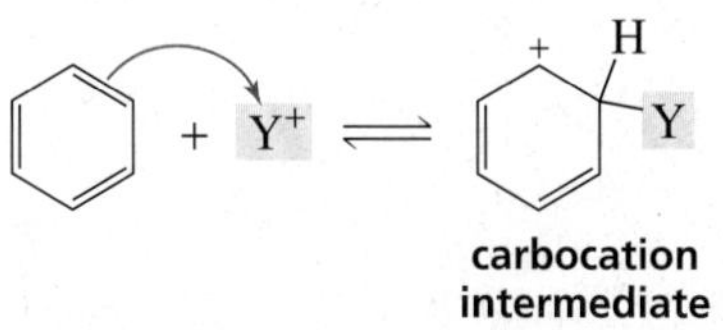

$+ \ Y^+ \rightleftharpoons$ (H, Y, +)

carbocation intermediate

This should remind you of the first step in an electrophilic addition reaction of an alkene: A nucleophilic alkene reacts with an electrophile, thereby forming a carbocation intermediate (Section 3.6). In the second step of an electrophilic addition reaction, the carbocation reacts with a nucleophile (Z^-) to form an addition product.

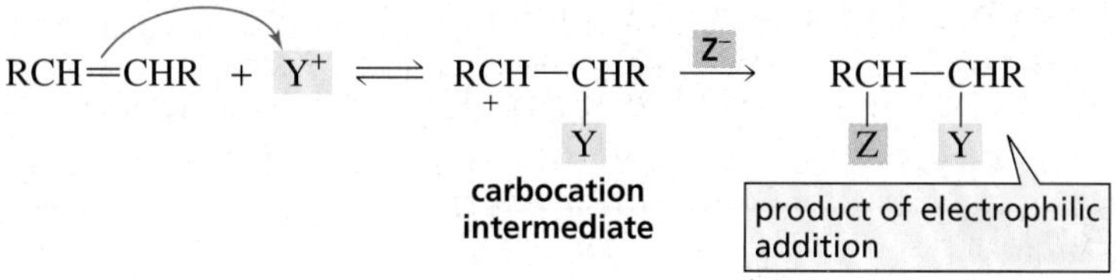

$$\text{RCH}{=}\text{CHR} + Y^+ \rightleftharpoons \underset{+}{\text{RCH}}{-}\underset{\text{Y}}{\text{CHR}} \xrightarrow{Z^-} \underset{\text{Z}}{\text{RCH}}{-}\underset{\text{Y}}{\text{CHR}}$$

carbocation intermediate

product of electrophilic addition

If the carbocation intermediate formed from the reaction of benzene with an electrophile were to react similarly with a nucleophile (depicted as event *b* in Figure 15.3), the addition product would not be aromatic. If, however, the carbocation loses a proton from the site of electrophilic attack (depicted as event *a* in Figure 15.3), the aromaticity of the benzene ring is restored. Because the aromatic product is much more stable than the nonaromatic addition product, the overall reaction is an electrophilic substitution reaction rather than an electrophilic addition reaction. In the substitution reaction, an electrophile substitutes for one of the hydrogens attached to the benzene ring.

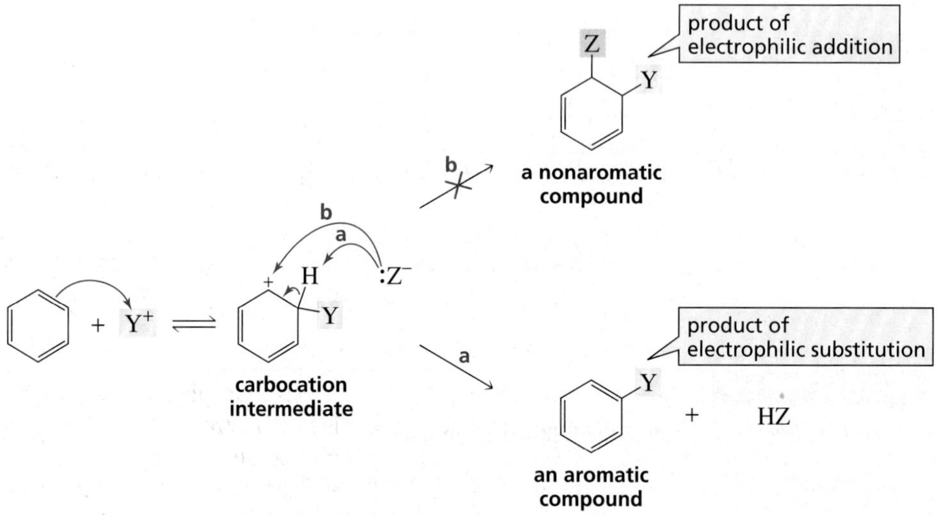

Figure 15.3 ▶
Reaction of benzene with an electrophile. Because the aromatic product is more stable, the reaction proceeds as (a) an electrophilic substitution reaction rather than (b) an electrophilic addition reaction.

product of electrophilic addition
a nonaromatic compound
carbocation intermediate
product of electrophilic substitution
+ HZ
an aromatic compound

The reaction coordinate diagram in Figure 15.4 shows that the reaction of *benzene* to form a *substituted benzene* has a ΔG° close to zero. The reaction of *benzene* to form the much less stable *nonaromatic addition product* would have been a highly endergonic reaction. Consequently, benzene undergoes *electrophilic substitution reactions* that preserve aromaticity, rather than *electrophilic addition reactions* (the reactions characteristic of alkenes), which would destroy aromaticity.

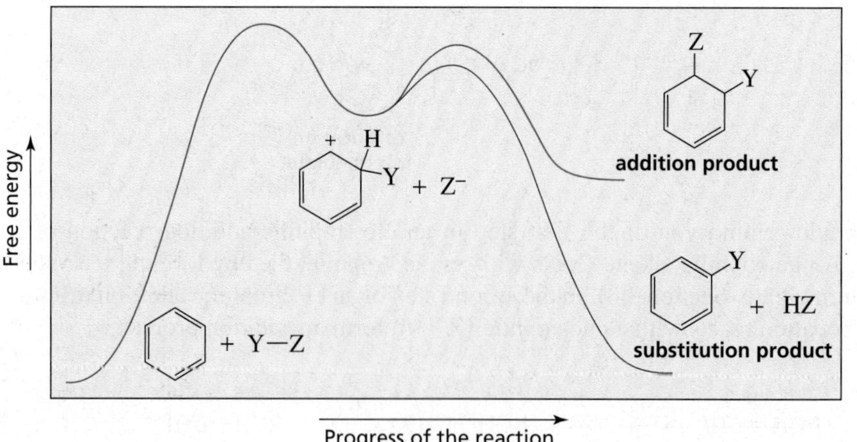

Figure 15.4 ▶
Reaction coordinate diagrams for electrophilic substitution of benzene and electrophilic addition to benzene.

PROBLEM 16

If electrophilic addition to benzene is an endergonic reaction overall, how can electrophilic addition to an alkene be an exergonic reaction overall?

15.9 General Mechanism for Electrophilic Aromatic Substitution Reactions

Because electrophilic substitution of benzene involves the reaction of an electrophile with an aromatic compound, it is more precisely called an electrophilic aromatic substitution reaction. In an **electrophilic aromatic substitution reaction**, an electrophile substitutes for a hydrogen of an aromatic compound.

an electrophilic aromatic substitution reaction

$$C_6H_5\text{–H} + Y^+ \longrightarrow C_6H_5\text{–Y} + H^+$$

The following are the five most common electrophilic aromatic substitution reactions:

1. **Halogenation**: A bromine (Br), a chlorine (Cl), or an iodine (I) substitutes for a hydrogen.
2. **Nitration**: A nitro (NO_2) group substitutes for a hydrogen.
3. **Sulfonation**: A sulfonic acid (SO_3H) group substitutes for a hydrogen.
4. **Friedel–Crafts acylation**: An acyl (RC=O) group substitutes for a hydrogen.
5. **Friedel–Crafts alkylation**: an alkyl (R) group substitutes for a hydrogen.

All of these electrophilic aromatic substitution reactions take place by the same two-step mechanism. In the first step, benzene reacts with an electrophile (Y^+), forming a carbocation intermediate. The structure of the carbocation intermediate can be approximated by three resonance contributors. In the second step of the reaction, a base in the reaction mixture pulls off a proton from the carbocation intermediate, and the electrons that held the proton move into the ring to reestablish its aromaticity. Notice that *the proton is always removed from the carbon that has formed the new bond with the electrophile.*

In an electrophilic aromatic substitution reaction, an electrophile (Y^+) is put on a ring carbon, and the H^+ comes off the same ring carbon.

general mechanism for electrophilic aromatic substitution

benzene + Y^+ $\xrightarrow{\text{slow}}$ [three resonance contributors of the carbocation intermediate bearing H and Y] + :B $\xrightarrow{\text{fast}}$ C_6H_5–Y + HB^+

The first step is relatively slow and endergonic because an aromatic compound is being converted into a much less stable nonaromatic intermediate (Figure 15.4). The second step is fast and strongly exergonic because this step restores the stability-enhancing aromaticity.

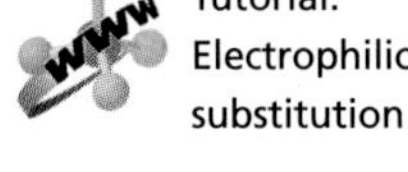

Tutorial: Electrophilic aromatic substitution

We will look at each of these five electrophilic aromatic substitution reactions individually. As you study them, notice that they differ only in how the electrophile (Y^+) needed to start the reaction is generated. Once the electrophile is formed, all five reactions follow the same two-step mechanism for electrophilic aromatic substitution.

PROBLEM 17◆

Which compound will undergo an electrophilic aromatic substitution reaction more rapidly, benzene or hexadeuteriobenzene?

C_6H_6 or C_6D_6

15.10 Halogenation of Benzene

The bromination or chlorination of benzene requires a Lewis acid such as ferric bromide or ferric chloride. Recall that a *Lewis acid* is a compound that accepts a share in a pair of electrons (Section 1.21).

bromination

$$\text{C}_6\text{H}_6 + Br_2 \xrightarrow{FeBr_3} \text{C}_6\text{H}_5\text{Br} + HBr$$

bromobenzene

chlorination

$$\text{C}_6\text{H}_6 + Cl_2 \xrightarrow{FeCl_3} \text{C}_6\text{H}_5\text{Cl} + HCl$$

chlorobenzene

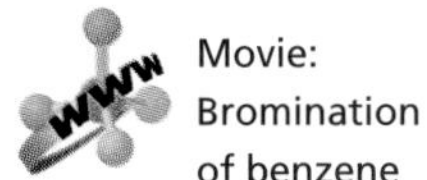
Movie: Bromination of benzene

In the first step of the bromination reaction, bromine donates a lone pair to the Lewis acid. This weakens the Br—Br bond, thereby providing the electrophile necessary for electrophilic aromatic substitution.

mechanism for bromination

$$:\ddot{Br}-\ddot{Br}: + FeBr_3 \longrightarrow :\ddot{Br}-\overset{+}{\ddot{Br}}-\overset{-}{Fe}Br_3$$

$$\text{C}_6\text{H}_6 + :\ddot{Br}-\overset{+}{\ddot{Br}}-\overset{-}{Fe}Br_3 \rightleftharpoons \overset{+}{\text{C}_6\text{H}_6}\text{Br} + \overset{-}{Fe}Br_4 \xrightarrow{:B} \text{C}_6\text{H}_5\text{Br} + HB^+$$

To make the mechanisms easier to understand, only one of the three resonance contributors of the carbocation intermediate is shown in this and subsequent illustrations. Bear in mind, however, that each carbocation intermediate actually has the three resonance contributors shown in Section 15.9. In the last step of the reaction, a base (:B) from the reaction mixture removes a proton from the carbocation intermediate. The following equation shows that the catalyst is regenerated:

$$\overset{-}{Fe}Br_4 + HB^+ \longrightarrow HBr + FeBr_3 + :B$$

Chlorination of benzene occurs by the same mechanism as bromination.

mechanism for chlorination

$$:\ddot{Cl}-\ddot{Cl}: + FeCl_3 \longrightarrow :\ddot{Cl}-\overset{+}{\ddot{Cl}}-\overset{-}{Fe}Cl_3$$

$$\text{C}_6\text{H}_6 + :\ddot{Cl}-\overset{+}{\ddot{Cl}}-\overset{-}{Fe}Cl_3 \rightleftharpoons \overset{+}{\text{C}_6\text{H}_6}\text{Cl} + \overset{-}{Fe}Cl_4 \xrightarrow{:B} \text{C}_6\text{H}_5\text{Cl} + HB^+$$

Ferric bromide and ferric chloride react readily with moisture in the air during handling, which inactivates them as catalysts. Therefore, instead of using the actual salt, ferric bromide or ferric chloride is generated in situ (in the reaction mixture) by adding iron filings and bromine or chlorine to the reaction mixture. Therefore, the halogen in the Lewis acid is the same as the reagent halogen.

$$2\ Fe + 3\ Br_2 \longrightarrow 2\ FeBr_3$$

$$2\ Fe + 3\ Cl_2 \longrightarrow 2\ FeCl_3$$

Unlike the reaction of benzene with Br_2 or Cl_2, the reaction of an alkene with Br_2 or Cl_2 does not require a Lewis acid (Section 4.7). An alkene is more reactive than benzene because an alkene has a smaller activation energy, since carbocation formation is not accompanied by a loss of aromaticity. As a result, the Br—Br or Cl—Cl bond does not have to be weakened to form a better electrophile.

PROBLEM 18

Why does hydration inactivate $FeBr_3$?

Electrophilic iodine (I^+) is obtained by treating I_2 with an oxidizing agent such as nitric acid.

iodination

$$I_2 \xrightarrow{\text{oxidizing agent}} 2\ I^+$$

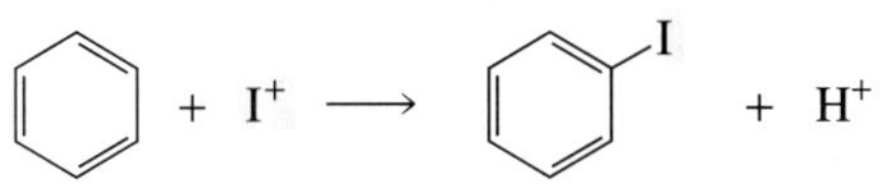

benzene + I^+ ⟶ iodobenzene + H^+

iodobenzene

Once the electrophile is formed, iodination of benzene occurs by the same mechanism as bromination and chlorination.

mechanism for iodination

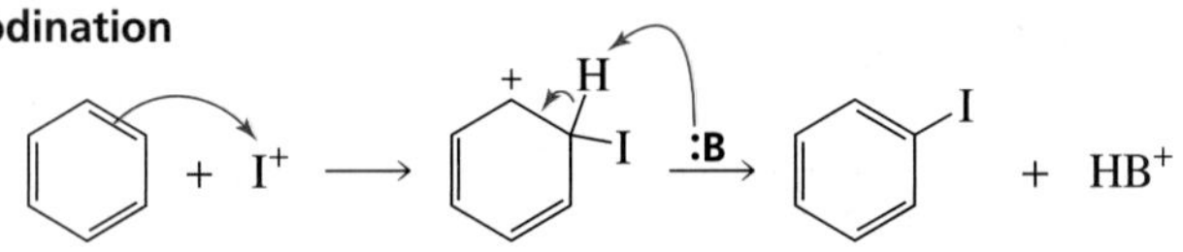

benzene + I^+ ⟶ (carbocation intermediate with H and I) $\xrightarrow{:B}$ iodobenzene + HB^+

THYROXINE

Thyroxine is a hormone that regulates the metabolic rate, causing an increase in the rate at which fats, carbohydrates, and proteins are metabolized. Humans obtain thyroxine from tyrosine (an amino acid) and iodine. We get iodine primarily from the iodized salt in our diet. An enzyme called iodoperoxidase converts the I^- we ingest to I^+, the electrophile needed to place an iodo substituent on the benzene ring. Low thyroxine levels can be corrected by hormone supplements. Chronically low levels of thyroxine cause enlargement of the thyroid gland, a condition known as goiter.

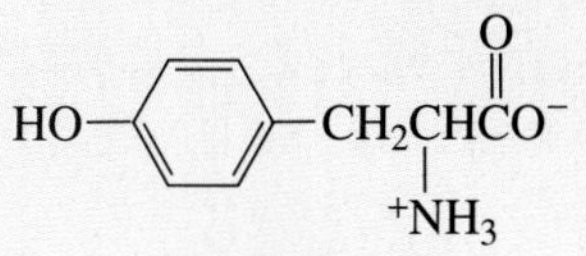

HO–C_6H_4–$CH_2CH(\overset{+}{N}H_3)CO^-$ (C=O)

tyrosine

HO–$C_6H_2I_2$–O–$C_6H_2I_2$–$CH_2CH(\overset{+}{N}H_3)CO^-$ (C=O)

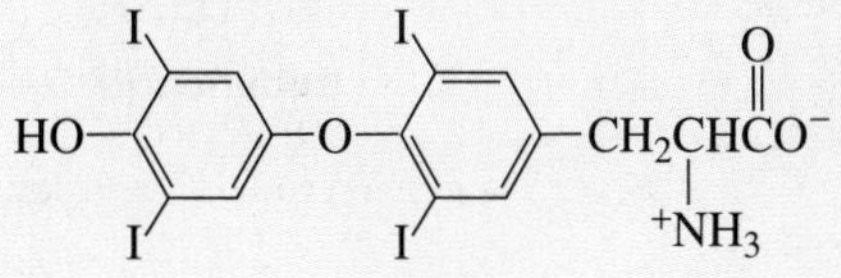

thyroxine

15.11 Nitration of Benzene

Nitration of benzene with nitric acid requires sulfuric acid as a catalyst.

nitration

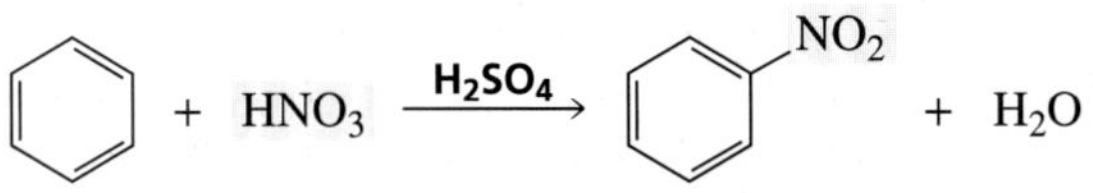

benzene + HNO_3 $\xrightarrow{H_2SO_4}$ nitrobenzene ($C_6H_5NO_2$) + H_2O

nitrobenzene

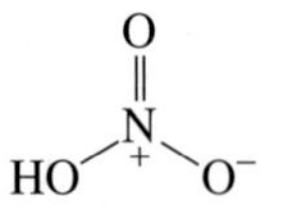

nitric acid

To generate the necessary electrophile, sulfuric acid protonates nitric acid. Loss of water from protonated nitric acid forms a nitronium ion, the electrophile required for nitration. Remember that any base (:B) present in the reaction mixture (H_2O, HSO_4^-, solvent) can remove the proton in the second step of the aromatic substitution reaction.

mechanism for nitration

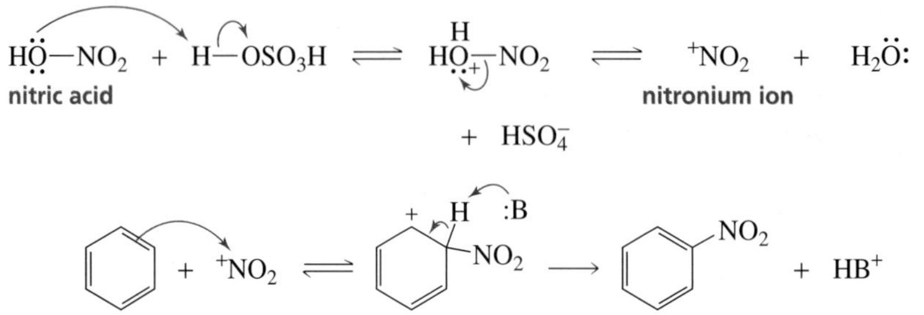

$$HO-NO_2 + H-OSO_3H \rightleftharpoons HO^{+}(H)-NO_2 + HSO_4^- \rightleftharpoons {}^{+}NO_2 + H_2O:$$

nitric acid; nitronium ion

$$C_6H_6 + {}^{+}NO_2 \rightleftharpoons [C_6H_6(H)(NO_2)]^{+} \xrightarrow{:B} C_6H_5NO_2 + HB^{+}$$

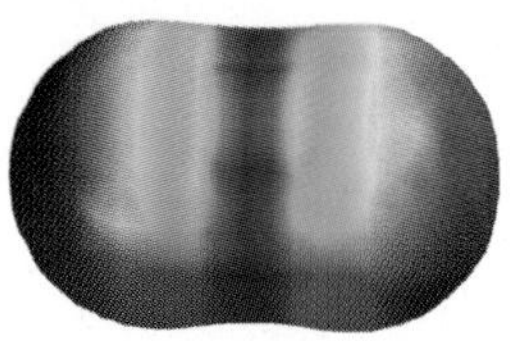

$O=\overset{+}{N}=O$

nitronium ion

15.12 Sulfonation of Benzene

Fuming sulfuric acid (a solution of SO_3 in sulfuric acid) or concentrated sulfuric acid is used to sulfonate aromatic rings.

sulfonation

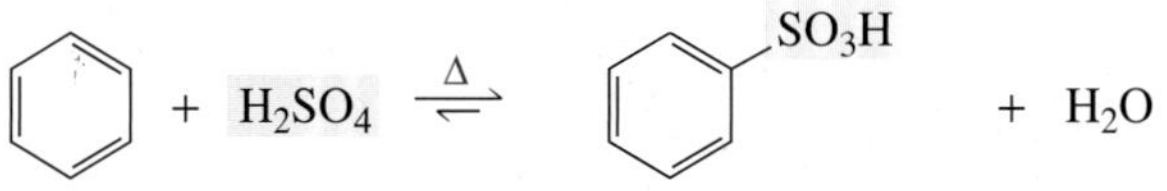

$$C_6H_6 + H_2SO_4 \overset{\Delta}{\rightleftharpoons} C_6H_5SO_3H + H_2O$$

benzenesulfonic acid

As the following mechanism shows, a substantial amount of electrophilic sulfur trioxide (SO_3) is generated when concentrated sulfuric acid is heated, as a result of the $^{+}SO_3H$ electrophile losing a proton. Take a minute to note the similarities in the mechanisms for forming the $^{+}SO_3H$ electrophile for sulfonation and the $^{+}NO_2$ electrophile for nitration.

mechanism for sulfonation

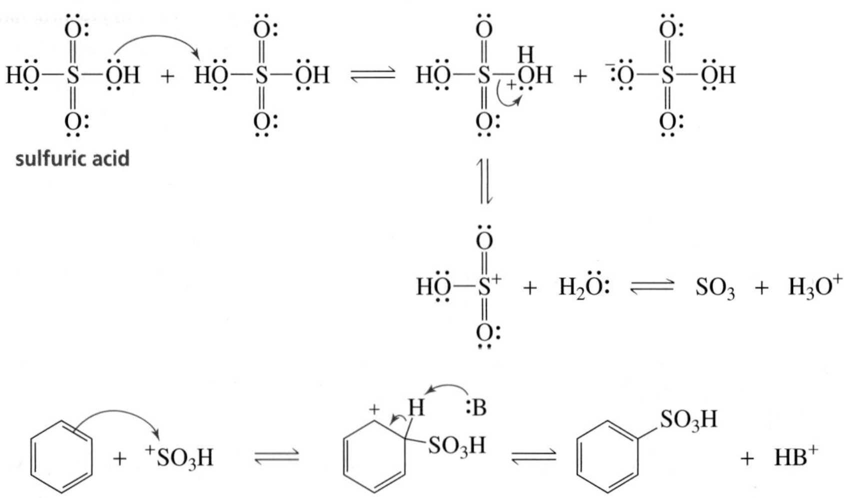

$$HO-SO_2-OH + HO-SO_2-OH \rightleftharpoons HO-SO_2-\overset{+}{O}H_2 + {}^{-}O-SO_2-OH$$

sulfuric acid

$$\rightleftharpoons HO-\overset{+}{S}O_2 + H_2O: \rightleftharpoons SO_3 + H_3O^{+}$$

$$C_6H_6 + {}^{+}SO_3H \rightleftharpoons [C_6H_6(H)(SO_3H)]^{+} \xrightarrow{:B} C_6H_5SO_3H + HB^{+}$$

A sulfonic acid is a strong acid because of the three electron-withdrawing oxygen atoms and the stability of its conjugate base—the electrons left behind when a proton is lost are shared by three oxygen atoms (Section 1.19).

$$C_6H_5SO_3H \underset{}{\overset{pKa = -0.60}{\rightleftharpoons}} C_6H_5SO_3^- + H^+$$

benzenesulfonic acid **benzenesulfonate ion**

Sulfonation of benzene is a reversible reaction. If benzenesulfonic acid is heated in dilute acid, the reaction proceeds in the reverse direction.

$$C_6H_5SO_3H \xrightarrow{H_3O^+ / 100\ °C} C_6H_6 + SO_3 + H^+$$

The **principle of microscopic reversibility** applies to all reactions. It states that the mechanism of a reaction in the reverse direction must retrace each step of the mechanism in the forward direction in microscopic detail. This means that the forward and reverse reactions must have the same intermediates and that the rate-determining "energy hill" must be the same in both directions. For example, sulfonation is described by the reaction coordinate diagram in Figure 15.5, going from left to right. Therefore, desulfonation is described by the same reaction coordinate diagram going from right to left. In sulfonation, the rate-determining step is nucleophilic attack of benzene on the $^+SO_3H$ ion. In desulfonation, the rate-limiting step is loss of the $^+SO_3H$ ion from the benzene ring. An example of the usefulness of desulfonation to synthetic chemists is given in Chapter 16, Problem 19.

mechanism for desulfonation

$$C_6H_5SO_3H + H^+ \rightleftharpoons C_6H_6(H)(SO_3H)^+ \rightleftharpoons C_6H_6 + {}^+SO_3H$$

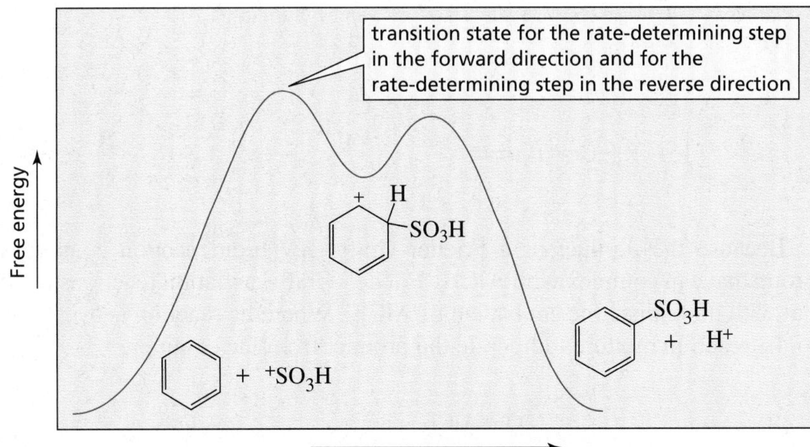

Figure 15.5 Reaction coordinate diagram for the sulfonation of benzene (left to right) and the desulfonation of benzenesulfonic acid (right to left).

PROBLEM 19

The reaction coordinate diagram in Figure 15.5 shows that the rate-determining step for sulfonation is the slower of the two steps, whereas the rate-determining step for desulfonation is the faster of the two steps. Explain how the faster step can be the rate-determining step.

Charles Friedel (1832–1899) *was born in Strasbourg, France. He was a professor of chemistry and director of research at the Sorbonne. At one point, his interest in mineralogy led him to attempt to make synthetic diamonds. He met James Crafts when they both were doing research at L'Ecole de Médicine in Paris. They collaborated scientifically for most of their lives, discovering the Friedel–Crafts reactions in Friedel's laboratory in 1877.*

James Mason Crafts (1839–1917) *was born in Boston, the son of a woolen-goods manufacturer. He graduated from Harvard in 1858 and was a professor of chemistry at Cornell University and the Massachusetts Institute of Technology. He was president of MIT from 1897 to 1900, when he was forced to retire because of chronic poor health.*

15.13 Friedel–Crafts Acylation of Benzene

Two electrophilic substitution reactions bear the names of chemists Charles Friedel and James Crafts. *Friedel–Crafts acylation* places an acyl group on a benzene ring, and *Friedel–Crafts alkylation* places an alkyl group on a benzene ring.

$$\mathrm{R{-}\overset{\overset{\displaystyle O}{\|}}{C}{-}} \qquad\qquad \mathrm{R{-}}$$

an acyl group **an alkyl group**

Either an acyl halide or an acid anhydride can be used for Friedel–Crafts acylation.

Friedel–Crafts acylation

$$\mathrm{C_6H_6} + \underset{\textbf{an acyl chloride}}{\mathrm{RCOCl}} \xrightarrow[\textbf{2. } H_2O]{\textbf{1. } AlCl_3} \mathrm{C_6H_5COR} + \mathrm{HCl}$$

$$\mathrm{C_6H_6} + \underset{\textbf{an acid anhydride}}{\mathrm{RCOOCOR}} \xrightarrow[\textbf{2. } H_2O]{\textbf{1. } AlCl_3} \mathrm{C_6H_5COR} + \mathrm{RCOOH}$$

An acylium ion is the electrophile required for a Friedel–Crafts acylation reaction. This ion is formed by the reaction of an acyl chloride or an acid anhydride with $AlCl_3$, a Lewis acid.

mechanism for Friedel–Crafts acylation

$$\mathrm{RCO\ddot{C}l{:}} + \mathrm{AlCl_3} \longrightarrow \underset{\textbf{an acylium ion}}{\left[\mathrm{R{-}\overset{+}{C}{=}\ddot{O}{:}} \longleftrightarrow \mathrm{R{-}C{\equiv}\overset{+}{O}{:}}\right]} + {}^-\mathrm{AlCl_4}$$

$$\mathrm{C_6H_6} + \mathrm{R{-}\overset{+}{C}{=}\ddot{O}{:}} \rightleftharpoons \text{(arenium ion, H and COR on same carbon)} \xrightarrow{:B} \mathrm{C_6H_5COR} + \mathrm{HB^+}$$

Because the product of a Friedel–Crafts acylation reaction contains a carbonyl group that can complex with $AlCl_3$, Friedel–Crafts acylation reactions must be carried out with more than one equivalent of $AlCl_3$. When the reaction is over, water is added to the reaction mixture to liberate the product from the complex.

$$\mathrm{C_6H_5C(=\ddot{O}{:})R} + \mathrm{AlCl_3} \rightleftharpoons \mathrm{C_6H_5C(={}^+\ddot{O}{-}AlCl_3)R} \xrightarrow{\textbf{3 } H_2O} \mathrm{C_6H_5C(=\ddot{O}{:})R} + \mathrm{Al(OH)_3} + 3\,\mathrm{HCl}$$

PROBLEM 20

Show the mechanism for the generation of the acylium ion if an acid anhydride is used instead of an acyl chloride in a Friedel–Crafts acylation reaction.

PROBLEM 21

Propose a mechanism for the following reaction:

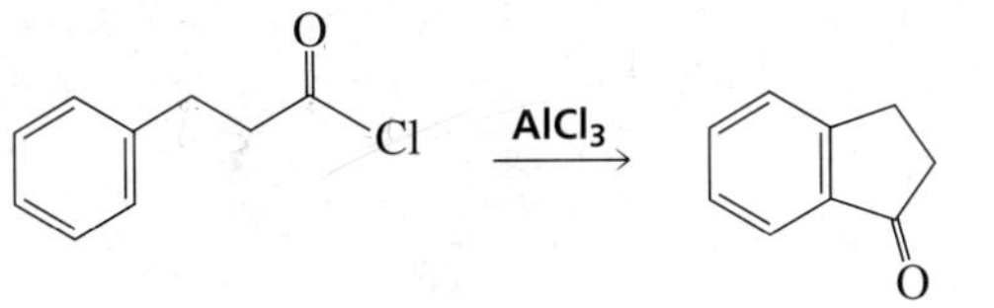

$$\text{PhCH}_2\text{CH}_2\text{COCl} \xrightarrow{\text{AlCl}_3} \text{1-indanone}$$

The synthesis of benzaldehyde from benzene poses a problem because formyl chloride, the acyl halide required for the reaction, is unstable and cannot be purchased. Formyl chloride can be prepared, however, by means of the **Gatterman–Koch** formylation **reaction**. This reaction uses a high-pressure mixture of carbon monoxide and HCl to generate formyl chloride, along with an aluminum chloride–cuprous chloride catalyst to carry out the acylation reaction.

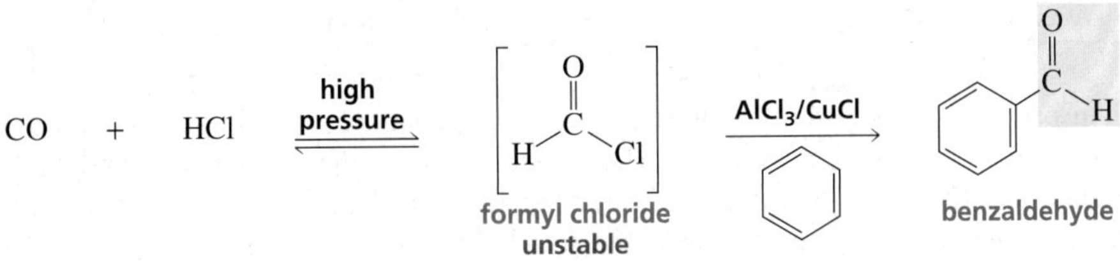

$$\text{CO} + \text{HCl} \underset{}{\overset{\text{high pressure}}{\rightleftharpoons}} \left[\text{HCOCl}\right] \xrightarrow[\text{benzene}]{\text{AlCl}_3/\text{CuCl}} \text{C}_6\text{H}_5\text{CHO}$$

formyl chloride unstable — benzaldehyde

15.14 Friedel–Crafts Alkylation of Benzene

The Friedel–Crafts alkylation reaction substitutes an alkyl group for a hydrogen.

Friedel–Crafts alkylation

$$\text{C}_6\text{H}_6 + \text{RCl} \xrightarrow{\text{AlCl}_3} \text{C}_6\text{H}_5\text{R} + \text{HCl}$$

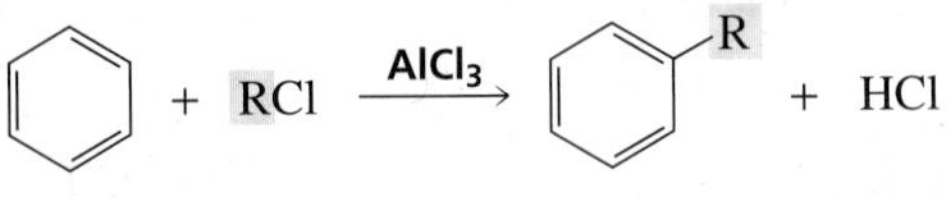

In the first step of the reaction, a carbocation is formed from the reaction of an alkyl halide with $AlCl_3$. Alkyl fluorides, alkyl chlorides, alkyl bromides, and alkyl iodides can all be used. Vinyl halides and aryl halides cannot be used because their carbocations are too unstable to be formed (Section 10.8).

mechanism for Friedel–Crafts alkylation

$$\text{R—}\ddot{\text{Cl}}\text{:} + \text{AlCl}_3 \longrightarrow \text{R}^+ + {}^-\text{AlCl}_4$$

$$\text{C}_6\text{H}_6 + \text{R}^+ \longrightarrow [\text{C}_6\text{H}_6\text{R}]^+ \xrightarrow{:\text{B}} \text{C}_6\text{H}_5\text{R} + \text{HB}^+$$

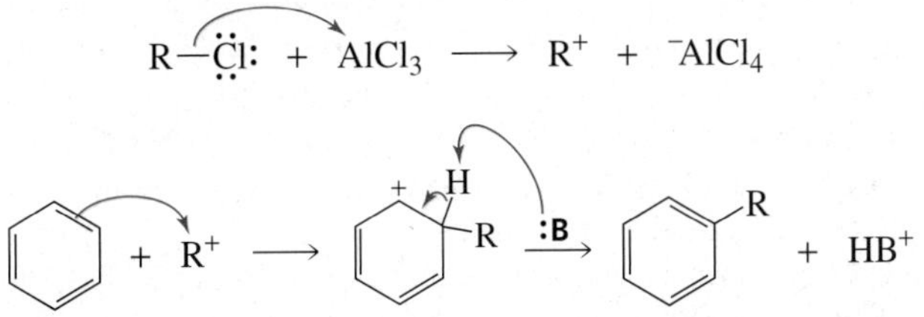

In Section 16.3, we will see that an alkyl-substituted benzene is more reactive than benzene. Therefore, to prevent further alkylation of the alkyl-substituted benzene, a large excess of benzene is used in Friedel–Crafts alkylation reactions. This approach ensures that the electrophile is more likely to encounter a molecule of benzene than a molecule of alkyl-substituted benzene.

Recall that a carbocation will rearrange if rearrangement leads to a more stable carbocation (Section 4.6). When the carbocation can rearrange in a Friedel–Crafts alkylation reaction, the major product will be the product with the rearranged alkyl group on the benzene ring. The relative amounts of rearranged and unrearranged product depend on the increase in carbocation stability achieved as a result of the rearrangement. For example, when benzene reacts with 1-chlorobutane, a primary carbocation rearranges to a secondary carbocation, and 60–80% of the product (the actual percentage depends on the reaction conditions) is the rearranged product.

$$C_6H_6 + CH_3CH_2CH_2CH_2Cl \xrightarrow[0\,°C]{AlCl_3} C_6H_5CH_2CH_2CH_2CH_3 + C_6H_5CH(CH_3)CH_2CH_3$$

1-chlorobutane; 1-phenylbutane 35% (unrearranged alkyl substituent); 2-phenylbutane 65% (rearranged alkyl substituent)

$$CH_3CH_2CH(H)\overset{+}{C}H_2 \xrightarrow{\text{1,2-hydride shift}} CH_3CH_2\overset{+}{C}HCH_3$$

a primary carbocation → a secondary carbocation

When benzene reacts with 1-chloro-2,2-dimethylpropane, a primary carbocation rearranges to a tertiary carbocation. Thus, there is a greater increase in carbocation stability and, therefore, a greater amount of rearranged product—100% of the product (under all reaction conditions) has the rearranged alkyl substituent.

$$C_6H_6 + CH_3C(CH_3)_2CH_2Cl \xrightarrow{AlCl_3} C_6H_5CH_2C(CH_3)_3 + C_6H_5C(CH_3)_2CH_2CH_3$$

1-chloro-2,2-dimethylpropane; 2,2-dimethyl-1-phenylpropane 0% (unrearranged alkyl substituent); 2-methyl-2-phenylbutane 100% (rearranged alkyl substituent)

$$CH_3C(CH_3)_2\overset{+}{C}H_2 \xrightarrow{\text{1,2-methyl shift}} CH_3\overset{+}{C}(CH_3)CH_2CH_3$$

a primary carbocation → a tertiary carbocation

INCIPIENT PRIMARY CARBOCATIONS

For simplicity, we have shown the formation of a primary carbocation in the two preceding reactions. However, as we saw in Section 10.5, primary carbocations are too unstable to be formed in solution. The fact is that a true primary carbocation is never formed in a Friedel–Crafts alkylation reaction. Instead, the carbocation remains complexed with the Lewis acid—it is called an "incipient" carbocation. A carbocation rearrangement occurs because the incipient carbocation has sufficient carbocation character to permit the rearrangement.

$$CH_3CH_2CH_2Cl + AlCl_3 \longrightarrow CH_3CH(H)\overset{\delta+}{C}H_2\text{---}Cl\text{---}\overset{\delta-}{AlCl_3} \xrightarrow{\text{1,2-hydride shift}} CH_3\overset{\delta+}{C}HCH_3\text{---}Cl\text{---}\overset{\delta-}{AlCl_3}$$

incipient primary carbocation

In addition to reacting with carbocations generated from alkyl halides, benzene can react with carbocations generated from the reaction of an alkene (Section 4.1) or an alcohol (Section 12.1) with an acid.

alkylation of benzene by an alkene

$$C_6H_6 + CH_3CH{=}CHCH_3 \xrightarrow{HF} C_6H_5CH(CH_3)CH_2CH_3$$

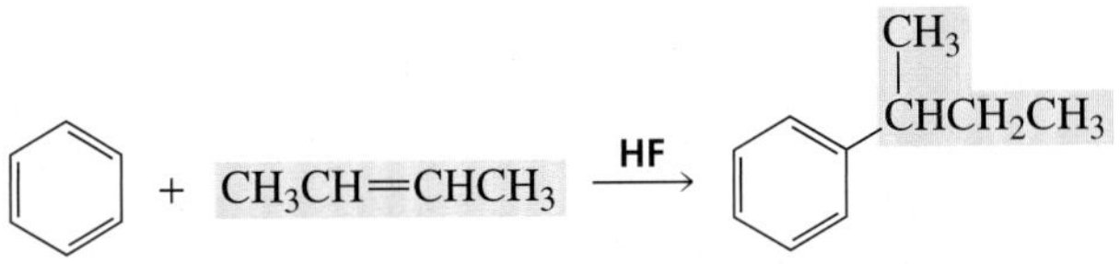

sec-butylbenzene

alkylation of benzene by an alcohol

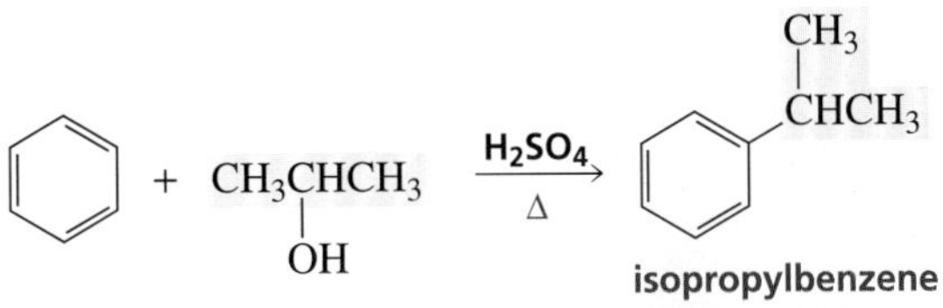

$$C_6H_6 + CH_3CH(OH)CH_3 \xrightarrow[\Delta]{H_2SO_4} C_6H_5CH(CH_3)CH_3$$

isopropylbenzene
cumene

PROBLEM 22

Show the mechanism for alkylation of benzene by an alkene.

PROBLEM 23◆

What would be the major product of a Friedel–Crafts alkylation reaction using the following alkyl halides?

a. CH_3CH_2Cl
b. $CH_3CH_2CH_2Cl$
c. $CH_3CH_2CH(Cl)CH_3$
d. $(CH_3)_3CCH_2Cl$
e. $(CH_3)_2CHCH_2Cl$
f. $CH_2{=}CHCH_2Cl$

15.15 Alkylation of Benzene by Acylation–Reduction

It is not possible to obtain a good yield of an alkylbenzene containing a straight-chain alkyl group via a Friedel–Crafts alkylation reaction, because the incipient primary carbocation will rearrange to a more stable carbocation.

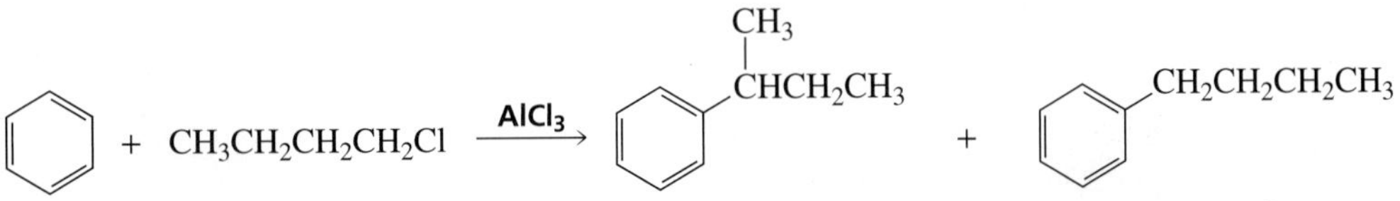

$$C_6H_6 + CH_3CH_2CH_2CH_2Cl \xrightarrow{AlCl_3} C_6H_5CH(CH_3)CH_2CH_3 + C_6H_5CH_2CH_2CH_2CH_3$$

major product minor product

Acylium ions, however, do not rearrange. Consequently, a straight-chain alkyl group can be placed on a benzene ring by means of a Friedel–Crafts acylation reaction, followed by reduction of the carbonyl group to a methylene group. It is called a reduction reaction because the two C—O bonds are replaced by two C—H bonds (Section 4.8). Only a ketone carbonyl group that is adjacent to a benzene ring can be reduced to a methylene group by catalytic hydrogenation (H_2/Pd).

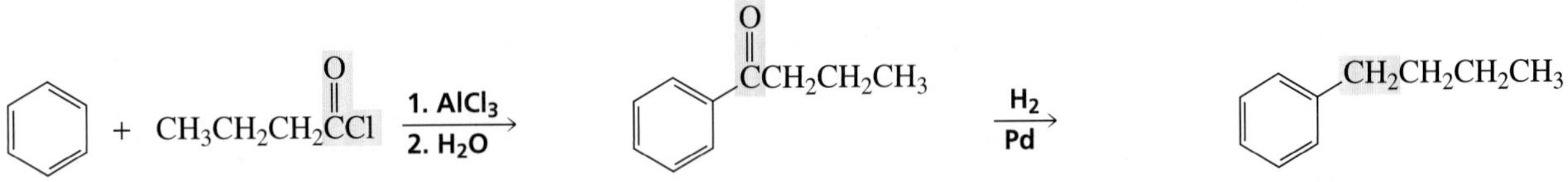

$$C_6H_6 + CH_3CH_2CH_2\overset{O}{\overset{\|}{C}}Cl \xrightarrow[2.\ H_2O]{1.\ AlCl_3} C_6H_5\overset{O}{\overset{\|}{C}}CH_2CH_2CH_3 \xrightarrow[Pd]{H_2} C_6H_5CH_2CH_2CH_2CH_3$$

acyl-substituted benzene alkyl-substituted benzene

E. C. Clemmensen (1876–1941) *was born in Denmark and received a Ph.D. from the University of Copenhagen. He was a scientist at Clemmensen Corp. in Newark, New York.*

Ludwig Wolff (1857–1919) *was born in Germany. He received a Ph.D. from the University of Strasbourg. He was a professor at the University of Jena in Germany.*

N. M. Kishner (1867–1935) *was born in Moscow. He received a Ph.D. from the University of Moscow under the direction of Markovnikov. He was a professor at the University of Tomsk and later at the University of Moscow.*

Besides avoiding carbocation rearrangements, another advantage of preparing alkyl-substituted benzenes by acylation–reduction rather than by direct alkylation is that a large excess of benzene does not have to be used (Section 15.14). Unlike alkyl-substituted benzenes, which are more reactive than benzene (Section 16.3), acyl-substituted benzenes are less reactive than benzene, so they will not undergo additional Friedel–Crafts reactions.

There are more general methods available to reduce a ketone carbonyl group to a methylene group—methods that reduce all ketone carbonyl groups, not just those that are adjacent to benzene rings. Two of the most effective are the Clemmensen reduction and the Wolff–Kishner reduction. The **Clemmensen reduction** uses an acidic solution of zinc dissolved in mercury as the reducing reagent. The **Wolff–Kishner reduction** employs hydrazine (H_2NNH_2) under basic conditions. The mechanism of the Wolff–Kishner reduction is shown in Section 18.6.

$C_6H_5C(=O)CH_2CH_3$ —[Zn(Hg), HCl, Δ; Clemmensen reduction]→ $C_6H_5CH_2CH_2CH_3$

$C_6H_5C(=O)CH_2CH_3$ —[H_2NNH_2, HO^-, Δ; Wolff–Kishner reduction]→ $C_6H_5CH_2CH_2CH_3$

At this point, you may wonder why it is necessary to have more than one way to carry out the same reaction. Alternative methods are useful when there is another functional group in the molecule that could react with the reagents you are using to carry out the desired reaction. For example, heating the following compound with HCl (as required by the Clemmensen reduction) would cause the alcohol to undergo substitution (Section 11.1). Under the basic conditions of the Wolff–Kishner reduction, however, the alcohol group would remain unchanged.

(4-hydroxycyclohexyl)phenyl ketone with CCH_3 (C=O) —[Zn(Hg), HCl, Δ]→ CH_2CH_3-benzene with 4-chlorocyclohexyl (Cl)

—[H_2NNH_2, HO^-, Δ]→ CH_2CH_3-benzene with 4-hydroxycyclohexyl (OH)

Alkylbenzenes with straight-chain alkyl groups can also be prepared by means of the coupling reactions you saw in Section 12.12. One of the alkyl groups of a Gilman reagent can replace the halogen of an aryl halide.

$C_6H_5Br + (CH_3CH_2)_2CuLi \longrightarrow C_6H_5CH_2CH_3 + CH_3CH_2Cu + LiBr$

The **Stille reaction** couples an aryl halide with a stannane.

$C_6H_5Br + (CH_3CH_2CH_2)_4Sn \xrightarrow[\text{THF}]{Pd(PPh_3)_4} C_6H_5CH_2CH_2CH_3 + (CH_3CH_2CH_2)_3SnBr$

tetrapropylstannane

The **Suzuki reaction** couples an aryl halide with an organoborane.

C_6H_5Cl + $CH_3CH_2CH_2{-}B$(catechol) $\xrightarrow[NaOH]{Pd(PPh_3)_4}$ $C_6H_5CH_2CH_2CH_3$ + $HO{-}B$(catechol) + Na

an organoborane — propylbenzene

The required organoborane is obtained from the reaction of an alkene with catecholborane. Because alkenes are readily available, this method can be used to prepare a wide variety of alkyl benzenes.

$CH_3CH{=}CH_2$ + $H{-}B$(catechol) $\longrightarrow$ $CH_3CH_2CH_2{-}B$(catechol)

catecholborane

PROBLEM 24

Describe how the following compounds could be prepared from benzene:

a. $C_6H_5CH(CH_3)CH_2CH_2CH_3$

b. $C_6H_5CH_2CH_2CH_2CH_2CH_3$

Summary

To be classified as **aromatic**, a compound must have an uninterrupted cyclic cloud of π electrons that contains an *odd number of pairs* of π electrons. An **antiaromatic** compound has an uninterrupted cyclic cloud of π electrons with an *even number of pairs* of π electrons. Molecular orbital theory shows that aromatic compounds are stable because their bonding orbitals are completely filled, with no electrons in either nonbonding or antibonding orbitals; in contrast, antiaromatic compounds are unstable because they either are unable to fill their bonding orbitals or they have a pair of π electrons left over after the bonding orbitals are filled. As a result of their aromaticity, the cyclopentadienyl anion and the cycloheptatrienyl cation are unusually stable.

An **annulene** is a monocyclic hydrocarbon with alternating single and double bonds. A **heterocyclic compound** is a cyclic compound in which one or more of the ring atoms is a **heteroatom**—an atom other than carbon. Pyridine, pyrrole, furan, and thiophene are aromatic heterocyclic compounds.

Benzene's aromaticity causes it to undergo **electrophilic aromatic substitution reactions**. The electrophilic addition reactions characteristic of alkenes and dienes would lead to much less stable nonaromatic addition products. The most common electrophilic aromatic substitution reactions are halogenation, nitration, sulfonation, and Friedel–Crafts acylation and alkylation. Once the electrophile is generated, all electrophilic aromatic substitution reactions take place by the same two-step mechanism: (1) The aromatic compound reacts with an electrophile, forming a carbocation intermediate; and (2) a base pulls off a proton from the carbon that formed the bond with the electrophile. The first step is relatively slow and endergonic because an aromatic compound is being converted into a much less stable nonaromatic intermediate; the second step is fast and strongly exergonic because the stability-enhancing aromaticity is being restored.

Some monosubstituted benzenes are named as substituted benzenes (e.g., bromobenzene, nitrobenzene); some have names that incorporate the name of the substituent (e.g., toluene, phenol, aniline). Bromination or chlorination requires a Lewis acid catalyst; iodination requires an oxidizing agent. **Nitration** with nitric acid requires sulfuric acid as a catalyst. Either an acyl halide or an acid anhydride can be used for **Friedel–Crafts acylation**, a reaction that places an acyl group on a benzene ring. If the carbocation formed from the alkyl halide used in a **Friedel–Crafts alkylation** reaction can rearrange, the major product will be the product with the rearranged alkyl group. A straight-chain alkyl group can be placed on a benzene ring via a Friedel–Crafts acylation reaction, followed by reduction of the carbonyl group by catalytic hydrogenation, a **Clemmensen reduction**, or a **Wolff–Kishner reduction**. Alkylbenzenes with straight-chain alkyl groups can also be prepared by means of coupling reactions.

A benzene ring can be sulfonated with fuming or concentrated sulfuric acid. **Sulfonation** is a reversible reaction; heating benzenesulfonic acid in dilute acid removes the sulfonic acid group. The **principle of microscopic reversibility** states that the mechanism of a reaction in the reverse direction must retrace each step of the mechanism in the forward direction in microscopic detail.

Summary of Reactions

1. Electrophilic aromatic substitution reactions:

a. Halogenation (Section 15.10)

$$C_6H_6 + Br_2 \xrightarrow{FeBr_3} C_6H_5Br + HBr$$

$$C_6H_6 + Cl_2 \xrightarrow{FeCl_3} C_6H_5Cl + HCl$$

$$2\,C_6H_6 + I_2 \xrightarrow{HNO_3} 2\,C_6H_5I + 2\,H^+$$

b. Nitration, sulfonation, and desulfonation (Sections 15.11 and 15.12)

$$C_6H_6 + HNO_3 \xrightarrow{H_2SO_4} C_6H_5NO_2 + H_2O$$

$$C_6H_6 + H_2SO_4 \overset{\Delta}{\rightleftharpoons} C_6H_5SO_3H + H_2O$$

c. Friedel–Crafts acylation and alkylation (Sections 15.13 and 15.14)

$$C_6H_6 + RCOCl \xrightarrow[2.\ H_2O]{1.\ AlCl_3} C_6H_5COR + HCl$$

$$\underset{\text{excess}}{C_6H_6} + RCl \xrightarrow{AlCl_3} C_6H_5R + HCl$$

d. Formation of benzaldehyde via a Gatterman–Koch reaction (Section 15.13)

$$CO + HCl + C_6H_6 \xrightarrow[AlCl_3/CuCl]{\text{high pressure}} C_6H_5CHO$$

e. Alkylation with a Gilman reagent (15.15)

$$C_6H_5Br + (R)_2CuLi \longrightarrow C_6H_5R + RCu + LiBr$$

f. Alkylation via a Stille reaction (Section 15.15)

$$C_6H_5Br + R_4Sn \xrightarrow[THF]{Pd(PPh_3)_4} C_6H_5R + R_3SnBr$$

g. Alkylation via a Suzuki reaction (Section 15.15)

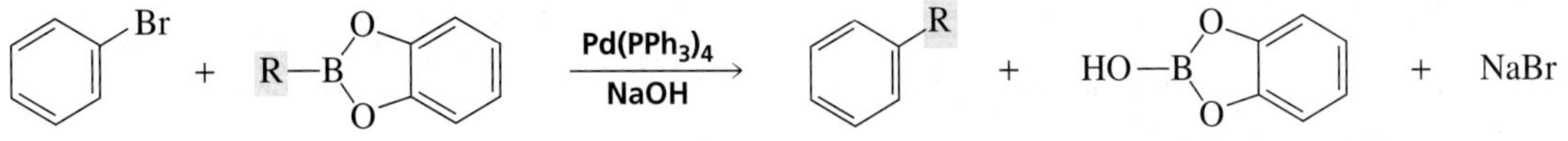

Pd(PPh$_3$)$_4$, NaOH; + NaBr

2. Clemmensen reduction and Wolff–Kishner reduction (Section 15.15)

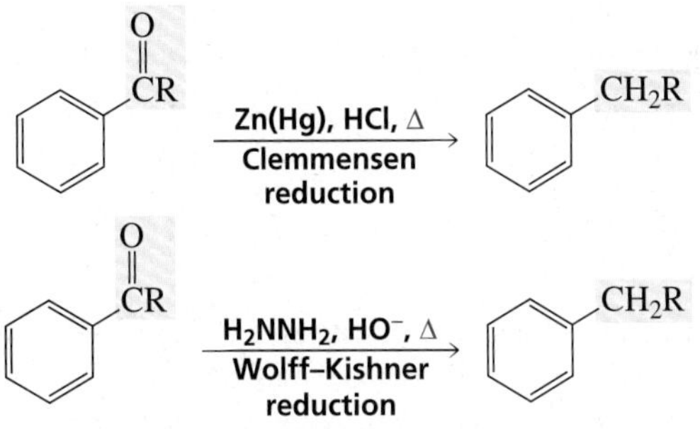

Zn(Hg), HCl, Δ — Clemmensen reduction

H_2NNH_2, HO^-, Δ — Wolff–Kishner reduction

Key Terms

aliphatic (p. 594)
annulene (p. 595)
antiaromatic (p. 602)
aromatic (p. 595)
benzyl group (p. 604)
Clemmensen reduction (p. 616)
electrophilic aromatic substitution reaction (p. 607)
Friedel–Crafts acylation (p. 607)
Friedel–Crafts alkylation (p. 607)
Gatterman–Koch reaction (p. 613)
halogenation (p. 607)
heteroatom (p. 598)
heterocyclic compound (p. 598)
Hückel's rule, or the $4n + 2$ rule (p. 595)
nitration (p. 607)
phenyl group (p. 604)
principle of microscopic reversibility (p. 611)
Stille reaction (p. 616)
sulfonation (p. 607)
Suzuki reaction (p. 617)
Wolff–Kishner reduction (p. 616)

Problems

25. Which of the following compounds are aromatic? Are any antiaromatic? (*Hint:* If possible, a ring will be nonplanar to avoid being antiaromatic.)

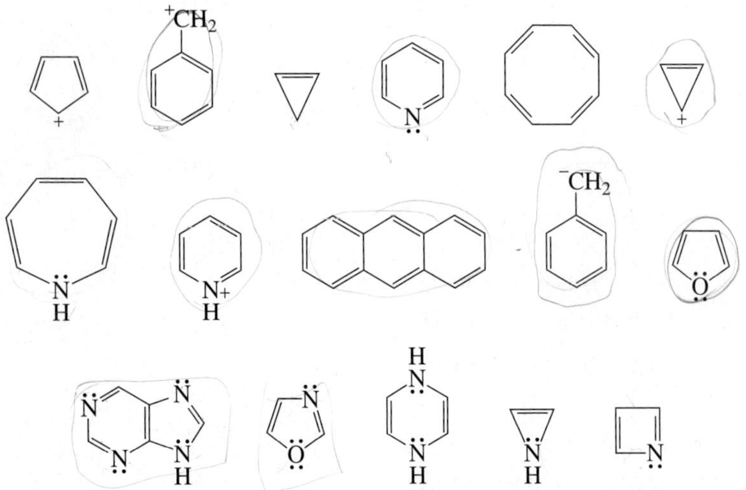

26. Give the product of the reaction of excess benzene with each of the following reagents:

a. isobutyl chloride + $AlCl_3$
b. propene + HF
c. neopentyl chloride + $AlCl_3$
d. dichloromethane + $AlCl_3$

27. Which ion in each of the following pairs is more stable?

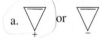

a. (cyclopropenyl cation) or (cyclopropenyl anion)

b. (cycloheptatrienyl cation) or (cycloheptatrienyl anion)

c. (cyclopentadienyl cation) or (cyclopentadienyl anion)

d. or

28. Which can lose a proton more readily, a methyl group bonded to cyclohexane or a methyl group bonded to benzene?

29. How could you prepare the following compounds with benzene as one of the starting materials?

a.

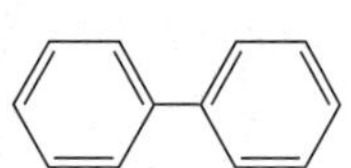

b.

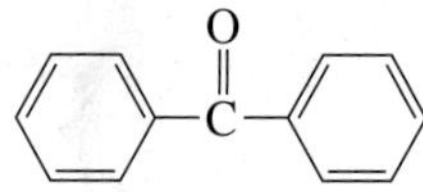

30. Benzene underwent a Friedel–Crafts acylation reaction followed by a Clemmensen reduction. The product gave the following ^{1}H NMR spectrum. What acyl chloride was used in the Friedel–Crafts acylation reaction?

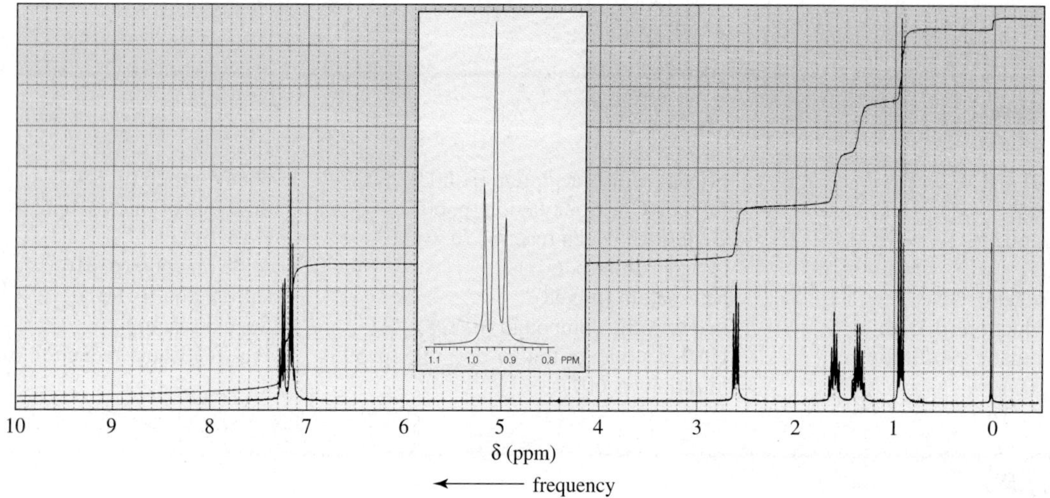

31. Give the products of the following reactions:

a. C_6H_5–$\overset{O}{\overset{\|}{C}}CH_2CH_2CH_2Cl$ $\xrightarrow[\text{2. } H_2O]{\text{1. } AlCl_3}$

b. C_6H_5–$\overset{O}{\overset{\|}{C}}CH_2CH_2Cl$ $\xrightarrow[\text{2. } H_2O]{\text{1. } AlCl_3}$

c. C_6H_5–$CH_2CH_2\overset{O}{\overset{\|}{C}}Cl$ $\xrightarrow[\text{2. } H_2O]{\text{1. } AlCl_3}$

d. C_6H_5–$CH_2CH_2CH_2\overset{O}{\overset{\|}{C}}Cl$ $\xrightarrow[\text{2. } H_2O]{\text{1. } AlCl_3}$

32. Which compound in each of the following pairs is a stronger base? Why?

a.

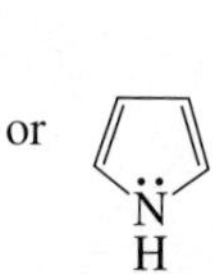

or (pyrrole, N–H)

b. $CH_3\overset{NH_2}{\overset{|}{C}}HCH_3$ or $CH_3\overset{NH}{\overset{\|}{C}}NH_2$

33. a. In what direction is the dipole moment in fulvene? Explain.
b. In what direction is the dipole moment in calicene? Explain.

$=CH_2$

fulvene **calicene**

34. Purine is a heterocyclic compound with four nitrogen atoms.
    a. Which nitrogen is most apt to be protonated?
    b. Which nitrogen is least apt to be protonated?

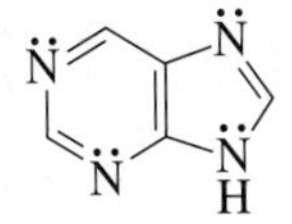

purine

35. Give the product of each of the following reactions:

a. $C_6H_5CH_2CH_2CH_2CH_2Cl \xrightarrow[\Delta]{AlCl_3}$

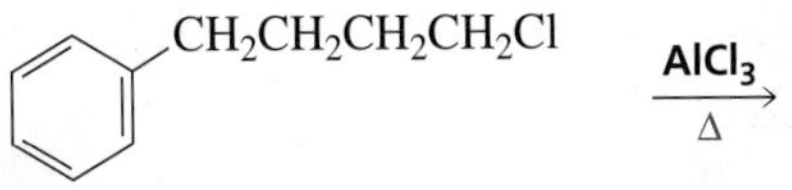

b. benzene + $CH_3CH(Cl)CH_2CH_2CH(Cl)CH_3 \xrightarrow[\Delta]{AlCl_3}$

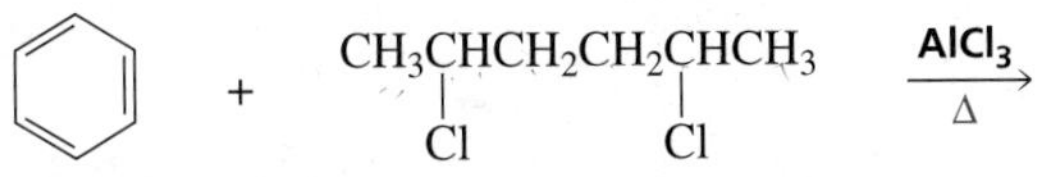

36. Propose a mechanism for each of the following reactions:

a. $C_6H_5CH_2CH_2CH(CH_3)CH{=}CH_2 \xrightarrow{H^+}$ indane bearing H_3C and CH_2CH_3

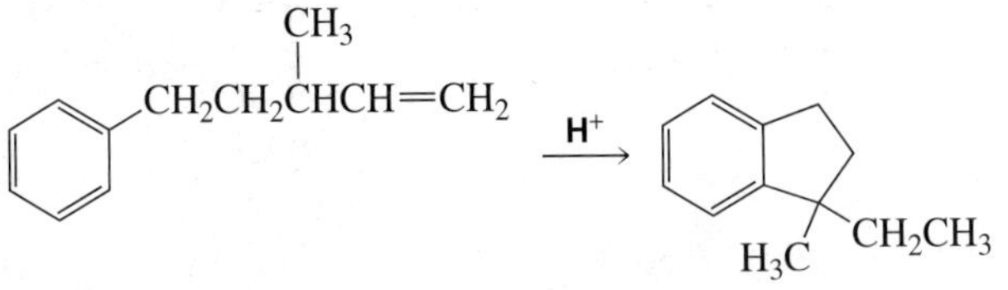

b. $C_6H_5CH{=}CH_2 \xrightarrow{H^+}$ indane bearing CH_3 and phenyl

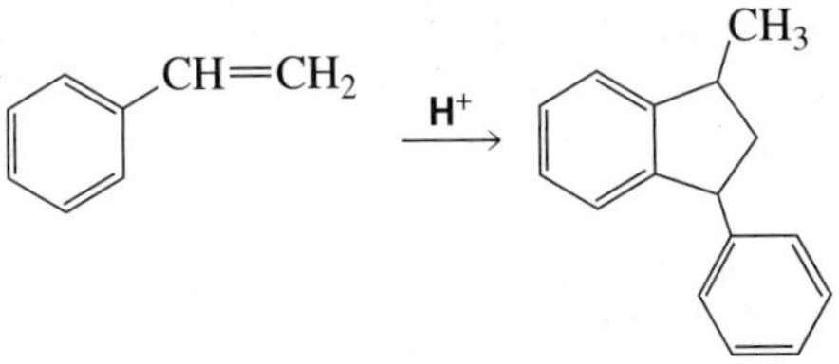

37. Show two ways that the following compound could be synthesized:

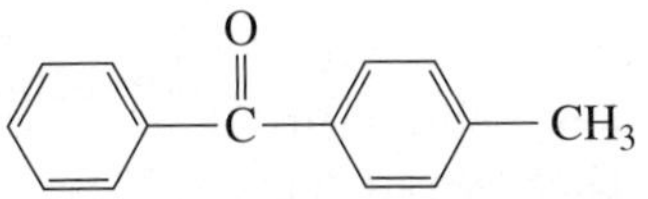

$C_6H_5C(=O)C_6H_4CH_3$

38. In a reaction called the Birch reduction, benzene can be partially reduced to 1,4-cyclohexadiene by an alkali metal (Na, Li, or K) in liquid ammonia and a low-molecular-weight alcohol. Propose a mechanism for this reaction. (*Hint:* See Section 6.8.)

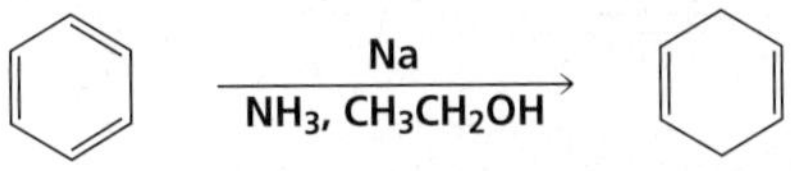

benzene $\xrightarrow[NH_3,\ CH_3CH_2OH]{Na}$ 1,4-cyclohexadiene

39. The *principle of least motion,* which states that the reaction that involves the least change in atomic positions or electronic configuration (all else being equal) is favored, has been suggested to explain why the Birch reduction forms only 1,4-hexadiene. How does this account for the observation that no 1,3-cyclohexadiene is obtained from a Birch reduction?

40. Investigation has shown that cyclobutadiene is actually a rectangular molecule rather than a square molecule. In addition, it has been established that there are two different 1,2-dideuterio-1,3-cyclobutadienes. Explain the reason for these unexpected observations.

cyclobutadiene

16 Reactions of Substituted Benzenes

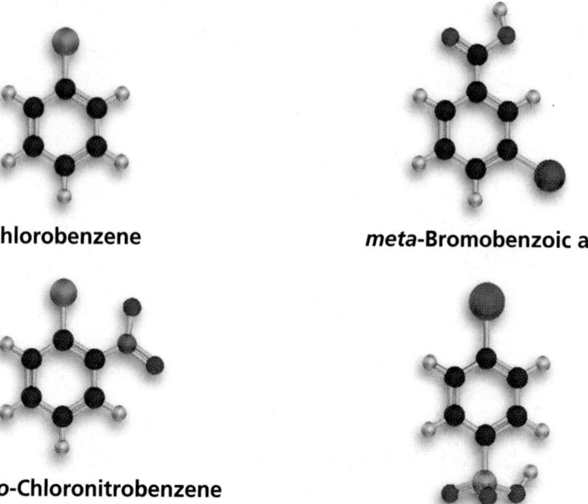

Many substituted benzenes are found in nature. A few that have physiological activity are adrenaline, melanin, ephedrine, chloramphenicol, and mescaline.

O_2N–C_6H_4–$CH(OH)CH(CH_2OH)NHC(=O)CHCl_2$

chloramphenicol
an antibiotic that is particularly effective against typhoid fever

HO, OH–C_6H_3–$CH(OH)CH_2NHCH_3$

adrenaline
epinephrine

C_6H_5–$CH(OH)CH(CH_3)NHCH_3$

ephedrine
a bronchodilator

Many physiologically active substituted benzenes are not found in nature, but exist because chemists have synthesized them. The now-banned diet drug "fen-phen" is a mixture of two synthetic substituted benzenes: fenfluramine and phentermine. Agent Orange, a defoliant widely used in the 1960s during the Vietnam War, is also a mixture of two synthetic substituted benzenes: 2,4-D and 2,4,5-T. The compound TCDD

PEYOTE CULTS

For several centuries, a peyote cult existed among the Aztecs in Mexico, which later spread to many Native North American tribes. By 1880, a religion that combined Christian beliefs with the Native American use of the peyote cactus had developed in southwestern United States, primarily among Native Americans. The followers of this religion believe that the cactus is divinely endowed to shape each person's life. Currently, the only people in the United States who are legally permitted to use peyote are members of the Native American Church—and then only in their religious rites.

$CH_2CH_2NH_2$ / CH_3O, OCH_3, OCH_3 (3,4,5-trimethoxyphenyl)

mescaline
active agent of the peyote cactus

MEASURING TOXICITY

The toxicity of a compound is indicated by its LD_{50} value—the quantity needed to kill 50% of the test animals exposed to the compound. Dioxin, with an LD_{50} value of 0.0006 mg/kg for guinea pigs, is an extremely toxic compound. Compare this with the LD_{50} values of 0.96 mg/kg for strychnine and 15 mg/kg for sodium cyanide—far less toxic compounds. One of the most toxic agents known is the botulism toxin, with an LD_{50} value of about 1×10^{-8} mg/kg.

(known as dioxin) is a contaminant formed during the manufacture of Agent Orange. TCDD has been implicated as the causative agent behind various symptoms suffered by those exposed to Agent Orange during the war.

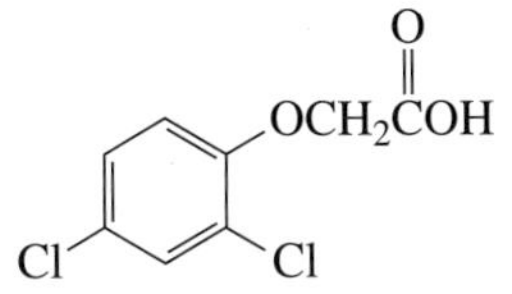

2,4-dichlorophenoxyacetic acid
2,4-D

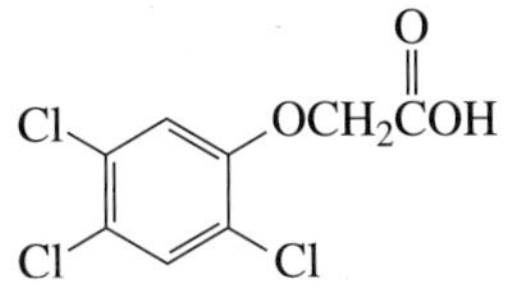

2,4,5-trichlorophenoxyacetic acid
2,4,5-T

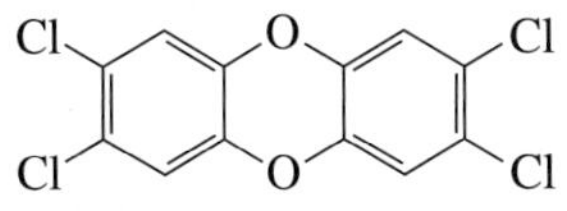

2,3,7,8-tetrachlorodibenzo[*b*,*e*][1,4]dioxin
TCDD

Because of the known physiological activities of adrenaline and mescaline, chemists have synthesized compounds with similar structures. One such compound is amphetamine, a central nervous system stimulant. Amphetamine and a close relative, methamphetamine, are used clinically as appetite suppressants. Methamphetamine is the street drug known as "speed" because of its rapid and intense psychological effects. Two other synthetic substituted benzenes, BHA and BHT, are preservatives (see Section 9.8) found in a wide variety of packaged foods. These compounds represent just a few of the many substituted benzenes that have been synthesized for commercial use by the chemical and pharmaceutical industries.

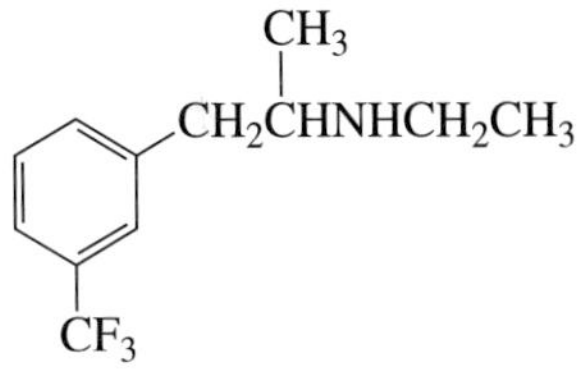

fenfluramine

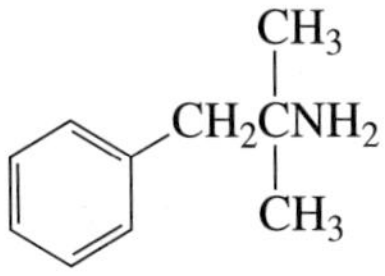

phentermine

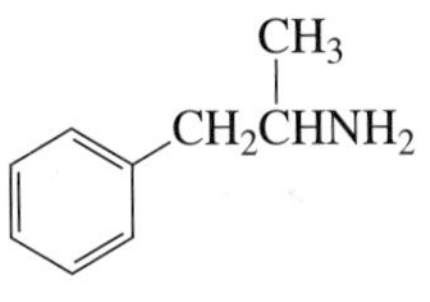

amphetamine

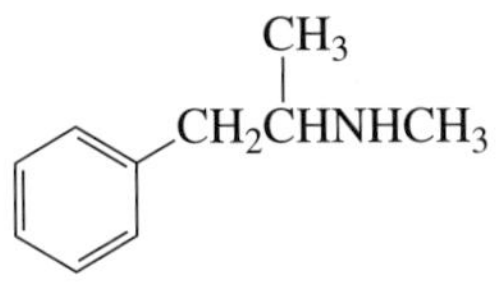

methamphetamine
"speed"

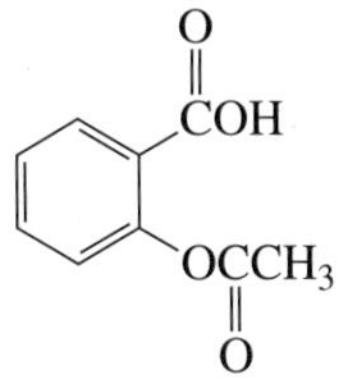

acetylsalicylic acid
aspirin

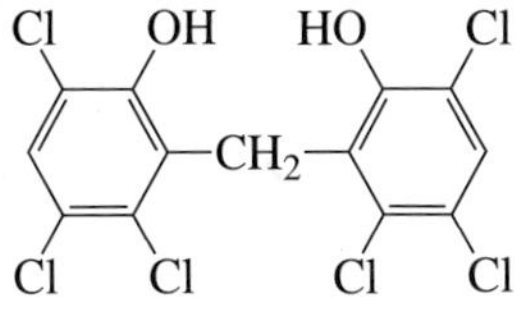

hexachlorophene
a disinfectant

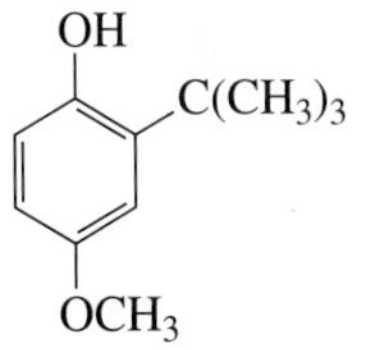

butylated
hydroxyanisole
BHA
a food antioxidant

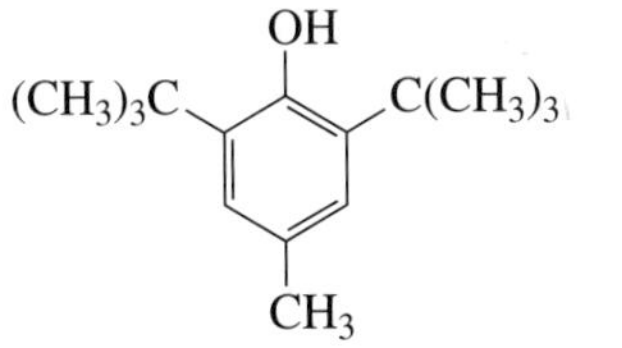

butylated
hydroxytoluene
BHT
a food antioxidant

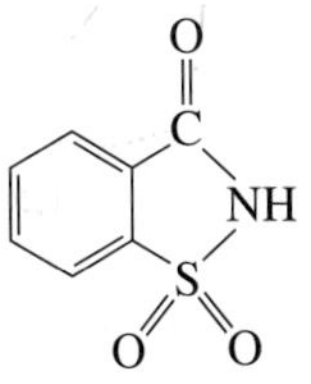

saccharin

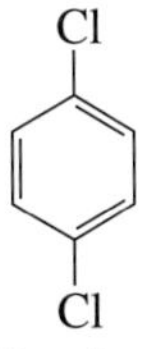

***p*-dichlorobenzene**
mothballs and
air fresheners

In Chapter 15, we looked at the reactions benzene undergoes and we saw how monosubstituted benzenes are named. Now we will see how disubstituted and polysubstituted benzenes are named, and then we will look at the reactions of *substituted benzenes*. The physical properties of several substituted benzenes are given in Appendix I.

16.1 Nomenclature of Disubstituted and Polysubstituted Benzenes

Disubstituted Benzenes

The relative positions of two substituents on a benzene ring can be indicated either by numbers or by the prefixes *ortho, meta*, and *para*. Adjacent substituents are called *ortho*, substituents separated by one carbon are called *meta*, and substituents located opposite one another are designated *para*. Often, only their abbreviations (*o*, *m*, *p*) are used in naming compounds.

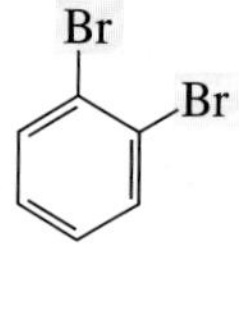

1,2-dibromobenzene
ortho-dibromobenzene
o-dibromobenzene

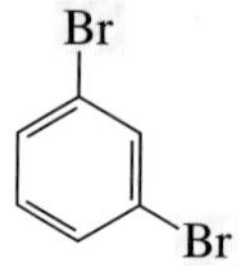

1,3-dibromobenzene
meta-dibromobenzene
m-dibromobenzene

1,4-dibromobenzene
para-dibromobenzene
p-dibromobenzene

If the two substituents are different, they are listed in alphabetical order. The first stated substituent is given the 1-position, and the ring is numbered in the direction that gives the second substituent the lowest possible number.

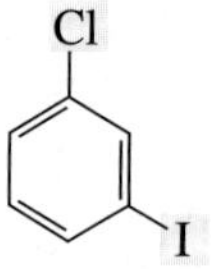

1-chloro-3-iodobenzene
meta-chloroiodobenzene
not
1-iodo-3-chlorobenzene or
meta-iodochlorobenzene

1-bromo-3-nitrobenzene
meta-bromonitrobenzene

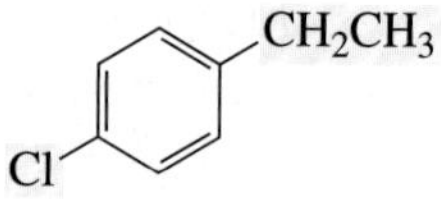

1-chloro-4-ethylbenzene
para-chloroethylbenzene

If one of the substituents can be incorporated into a name (Section 15.7), that name is used and the incorporated substituent is given the 1-position.

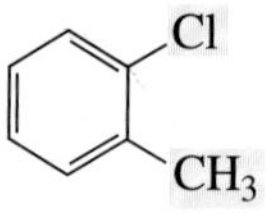

2-chlorotoluene
ortho-chlorotoluene
not
ortho-chloromethylbenzene

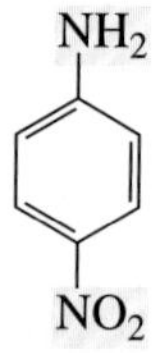

4-nitroaniline
para-nitroaniline
not
para-aminonitrobenzene

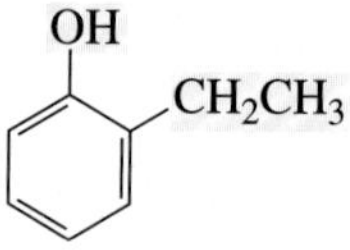

2-ethylphenol
ortho-ethylphenol
not
ortho-ethylhydroxybenzene

A few disubstituted benzenes have names that incorporate both substituents.

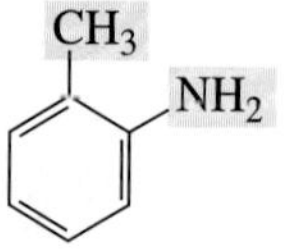

ortho-toluidine

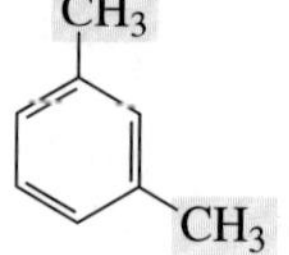

meta-xylene

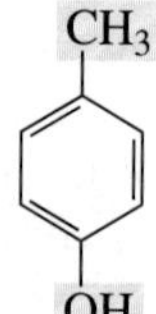

para-cresol
used as a wood preservative until prohibited for environmental reasons

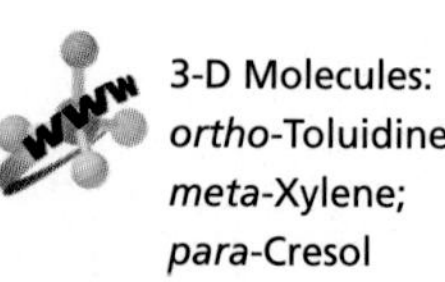

3-D Molecules: *ortho*-Toluidine; *meta*-Xylene; *para*-Cresol

PROBLEM 1◆

Name the following compounds:

a. 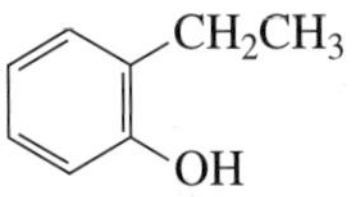 CH_2CH_3, OH b. 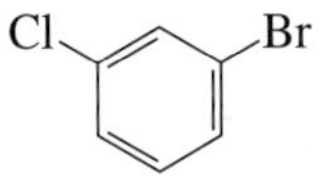 Cl, Br c. 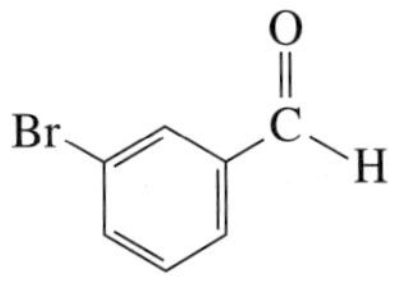 Br, $\overset{O}{\underset{}{C}}$—H d. 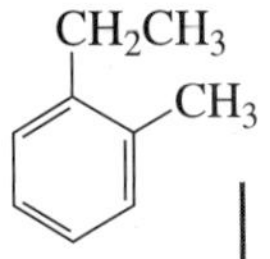 CH_2CH_3, CH_3

PROBLEM 2◆

Draw structures of the following compounds:

a. *para*-toluidine
b. *meta*-cresol
c. *para*-xylene
d. *ortho*-chlorobenzenesulfonic acid

Polysubstituted Benzenes

If the benzene ring has more than two substituents, the substituents are numbered so that the lowest possible numbers are used. The substituents are listed in alphabetical order with their appropriate numbers.

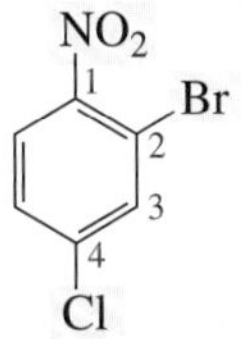

2-bromo-4-chloro-1-nitrobenzene 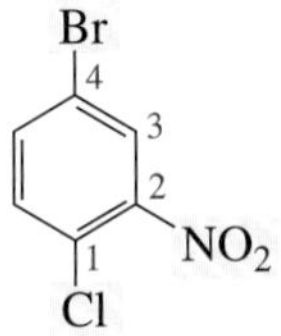 **4-bromo-1-chloro-2-nitrobenzene**

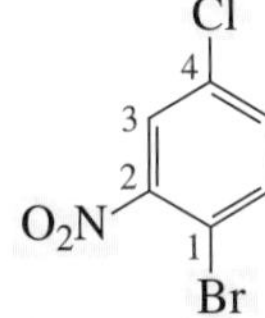

1-bromo-4-chloro-2-nitrobenzene

As with disubstituted benzenes, if one of the substituents can be incorporated into a name, that name is used and the incorporated substituent is given the 1-position. The ring is numbered in the direction that results in the lowest possible numbers in the name of the compound.

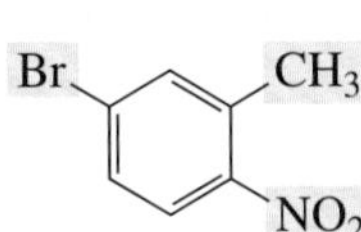

5-bromo-2-nitrotoluene

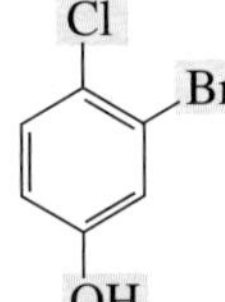

3-bromo-4-chlorophenol

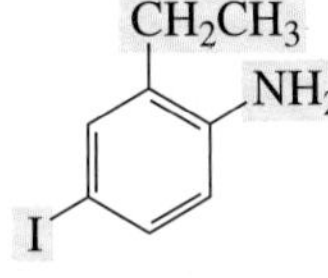

2-ethyl-4-iodoaniline

PROBLEM 3◆

Draw the structure of each of the following compounds:

a. *m*-chlorotoluene
b. *p*-bromophenol
c. *o*-nitroaniline
d. *m*-chlorobenzonitrile
e. 2-bromo-4-iodophenol
f. *m*-dichlorobenzene
g. 2,5-dinitrobenzaldehyde
h. *o*-xylene

PROBLEM 4◆

Correct the following incorrect names:

a. 2,4,6-tribromobenzene
b. 3-hydroxynitrobenzene
c. *para*-methylbromobenzene
d. 1,6-dichlorobenzene

16.2 Reactions of Substituents on Benzene

In Chapter 15, you learned how to prepare benzene rings with halo, nitro, sulfonic acid, alkyl, and acyl substituents.

$$C_6H_6 + Br_2 \xrightarrow{FeBr_3} C_6H_5Br$$

$$C_6H_6 + HNO_3 \xrightarrow{H_2SO_4} C_6H_5NO_2$$

$$C_6H_6 + H_2SO_4 \overset{\Delta}{\rightleftharpoons} C_6H_5SO_3H$$

$$C_6H_6 + RCl \xrightarrow{AlCl_3} C_6H_5R$$

$$C_6H_6 + RCCl(=O) \xrightarrow[2.\ H_2O]{1.\ AlCl_3} C_6H_5C(=O)R$$

Benzene rings with other substituents can be prepared by first synthesizing one of these substituted benzenes and then chemically changing the substituent. Several of these reactions should be familiar.

Reactions of Alkyl Substituents

We have seen that a bromine will selectively substitute for a benzylic hydrogen in a radical substitution reaction. (NBS stands for *N*-bromosuccinimide; Section 9.5.)

$$\underset{\text{propylbenzene}}{C_6H_5CH_2CH_2CH_3} + NBS \xrightarrow[\text{peroxide}]{\Delta} \underset{\text{1-bromo-1-phenylpropane}}{C_6H_5CH(Br)CH_2CH_3} + HBr$$

Once a halogen has been placed in the benzylic position, it can be replaced by a nucleophile by means of an S_N2 or an S_N1 reaction (Section 10.8). A wide variety of substituted benzenes can be prepared this way.

$$\underset{\text{benzyl bromide}}{C_6H_5CH_2Br} \xrightarrow{HO^-} \underset{\text{benzyl alcohol}}{C_6H_5CH_2OH} + Br^-$$

$$\underset{\text{benzyl bromide}}{C_6H_5CH_2Br} \xrightarrow{^-C\equiv N} \underset{\text{phenylacetonitrile}}{C_6H_5CH_2C\equiv N} + Br^-$$

$$\underset{\text{benzyl bromide}}{C_6H_5CH_2Br} \xrightarrow{NH_3} C_6H_5CH_2\overset{+}{N}H_3\ Br^- \xrightarrow{HO^-} \underset{\text{benzylamine}}{C_6H_5CH_2NH_2} + H_2O + Br^-$$

Remember that halo-substituted alkyl groups can also undergo E2 and E1 reactions (Section 11.8). Notice that a bulky base (*tert*-BuO$^-$) is used to encourage elimination over substitution.

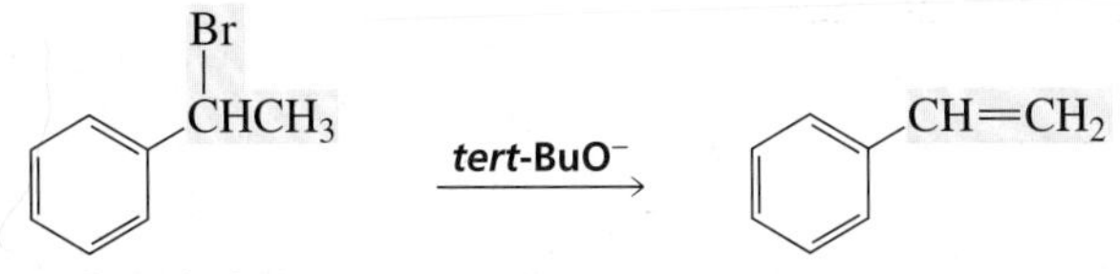

$$C_6H_5CH(Br)CH_3 \xrightarrow{tert\text{-BuO}^-} C_6H_5CH{=}CH_2$$

1-bromo-1-phenylethane → styrene

Substituents with double and triple bonds can undergo catalytic hydrogenation (Section 4.11). Addition of hydrogen to a double or triple bond is an example of a reduction reaction (Section 4.8). When an organic compound is *reduced*, either the number of C—H bonds in the compound increases or the number of C—O, C—N, or C—X (where X denotes a halogen atom) bonds decreases (Section 20.0).

$$C_6H_5CH{=}CH_2 + H_2 \xrightarrow{Pt} C_6H_5CH_2CH_3$$

styrene → ethylbenzene

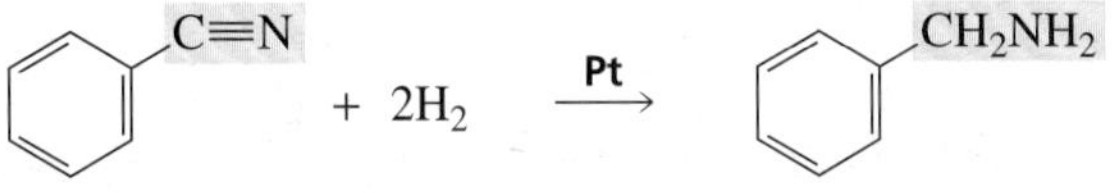

$$C_6H_5C{\equiv}N + 2H_2 \xrightarrow{Pt} C_6H_5CH_2NH_2$$

benzonitrile → benzylamine

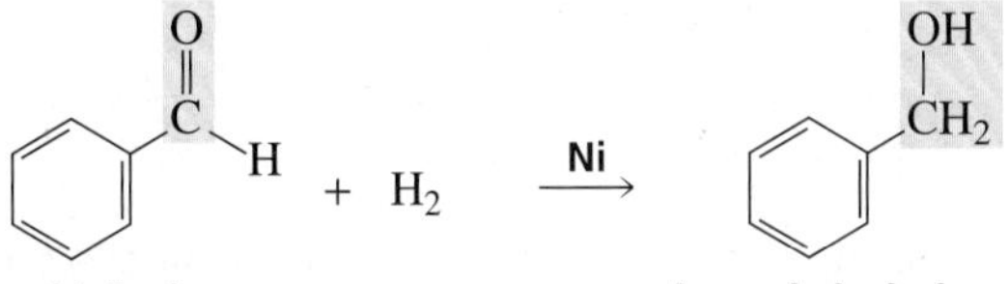

$$C_6H_5CHO + H_2 \xrightarrow{Ni} C_6H_5CH_2OH$$

benzaldehyde → benzyl alcohol

3-D Molecules: Benzyl bromide; Styrene; Benzonitrile; Benzaldehyde

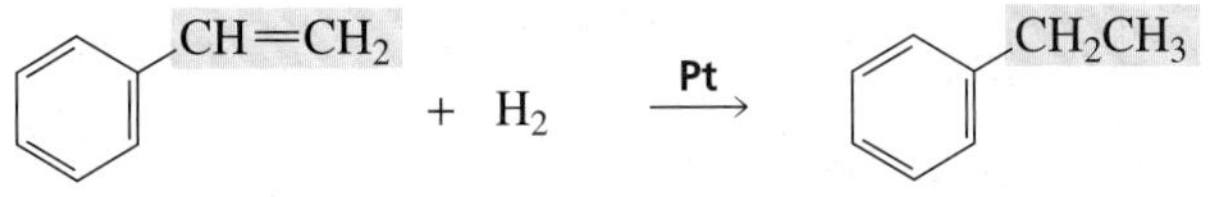

Recall that benzene is an unusually stable compound (Section 7.11). It, therefore, can be reduced only at high temperature and pressure.

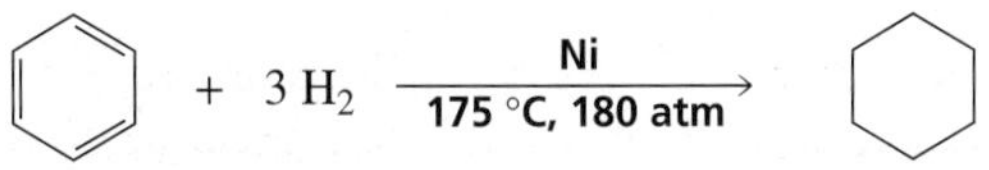

$$C_6H_6 + 3\,H_2 \xrightarrow[175\ ^\circ C,\ 180\ atm]{Ni} C_6H_{12}$$

benzene → cyclohexane

An alkyl group bonded to a benzene ring can be oxidized to a carboxyl group. When an organic compound is *oxidized*, either the number of C—O, C—N, or C—X (where X denotes a halogen atom) bonds increases or the number of C—H bonds decreases (Section 20.0). Commonly used oxidizing agents are potassium permanganate ($KMnO_4$) or acidic solutions of sodium dichromate (H^+, $Na_2Cr_2O_7$). Because the benzene ring is so stable, it will not be oxidized—only the alkyl group is oxidized.

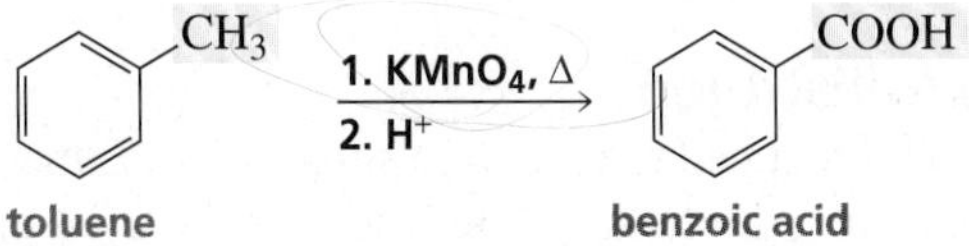

$$C_6H_5CH_3 \xrightarrow[2.\ H^+]{1.\ KMnO_4,\ \Delta} C_6H_5COOH$$

toluene → benzoic acid

Regardless of the length of the alkyl substituent, it will be oxidized to a COOH group, provided that a hydrogen is bonded to the benzylic carbon.

$CH_2CH_2CH_2CH_3$ / $CHCH_3$, CH_3 — $\xrightarrow[\Delta]{Na_2Cr_2O_7,\ H^+}$ — COOH / COOH

m-butylisopropylbenzene → *m*-benzenedicarboxylic acid

If the alkyl group lacks a benzylic hydrogen, the oxidation reaction will not occur because the first step in the oxidation reaction is removal of a hydrogen from the benzylic carbon.

does not have a benzylic hydrogen

$C(CH_3)_3$ $\xrightarrow[\Delta]{Na_2Cr_2O_7,\ H^+}$ no reaction

tert-butylbenzene

The same reagents that oxidize alkyl substituents will oxidize benzylic alcohols to benzoic acid.

1-phenylethanol $\xrightarrow[\Delta]{Na_2Cr_2O_7,\ H^+}$ benzoic acid

If, however, a mild oxidizing agent such as MnO_2 is used, benzylic alcohols are oxidized to aldehydes or ketones.

1-phenylethanol $\xrightarrow[\Delta]{MnO_2}$ acetophenone

phenylmethanol
benzyl alcohol $\xrightarrow[\Delta]{MnO_2}$ benzaldehyde

Reducing a Nitro Substituent

A nitro substituent can be reduced to an amino substituent. Either a metal (tin, iron, or zinc) plus an acid (HCl) or catalytic hydrogenation can be used to carry out the reduction. Recall from Section 1.20 that if acidic conditions are employed, the product will be in its acidic form (anilinium ion). When the reaction is over, base can be added to convert the product into its basic form (aniline).

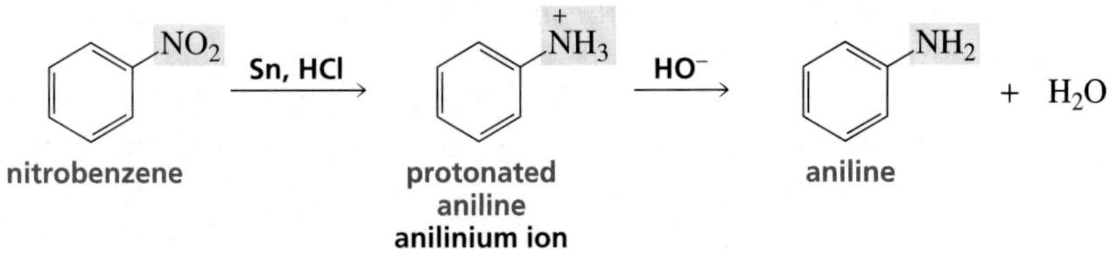

nitrobenzene $\xrightarrow{\text{Sn, HCl}}$ protonated aniline anilinium ion $\xrightarrow{HO^-}$ aniline $+ H_2O$

3-D Molecule: Nitrobenzene

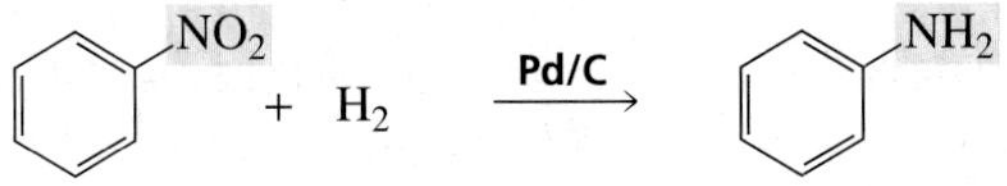

$C_6H_5NO_2 + H_2 \xrightarrow{\text{Pd/C}} C_6H_5NH_2$

It is possible to selectively reduce just one of two nitro groups.

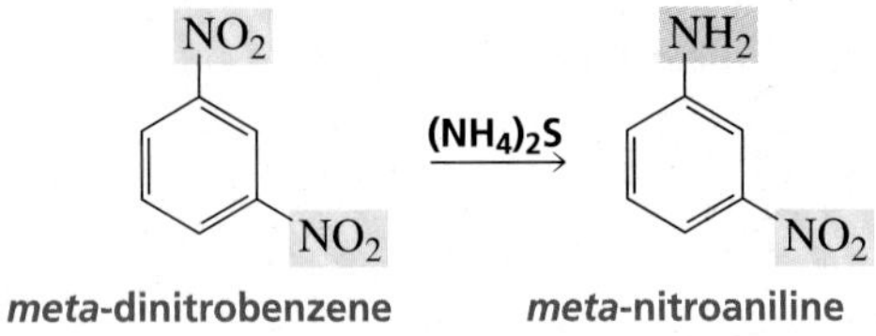

meta-dinitrobenzene $\xrightarrow{(NH_4)_2S}$ *meta*-nitroaniline

PROBLEM 5◆

Give the product of each of the following reactions:

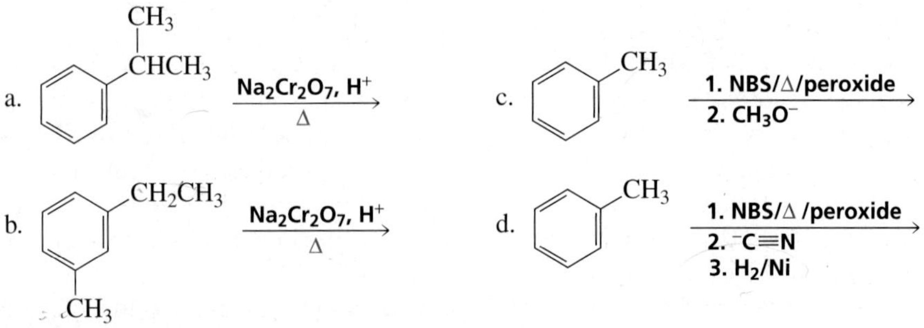

a. $C_6H_5CH(CH_3)CHCH_3$ ... $\xrightarrow[\Delta]{Na_2Cr_2O_7,\ H^+}$

b. CH_2CH_3, CH_3 (meta-substituted benzene) $\xrightarrow[\Delta]{Na_2Cr_2O_7,\ H^+}$

c. $C_6H_5CH_3 \xrightarrow[2.\ CH_3O^-]{1.\ NBS/\Delta/\text{peroxide}}$

d. $C_6H_5CH_3 \xrightarrow[2.\ ^-C{\equiv}N \quad 3.\ H_2/Ni]{1.\ NBS/\Delta\ /\text{peroxide}}$

PROBLEM 6 SOLVED

Show how the following compounds could be prepared from benzene:

a. benzaldehyde
b. styrene
c. 1-bromo-2-phenylethane
d. 2-phenyl-1-ethanol
e. aniline
f. benzoic acid

SOLUTION TO 6a

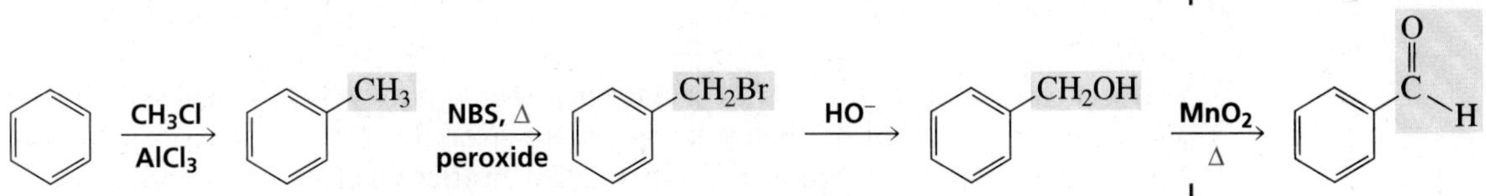

$C_6H_6 \xrightarrow[AlCl_3]{CH_3Cl} C_6H_5CH_3 \xrightarrow[\text{peroxide}]{\text{NBS, }\Delta} C_6H_5CH_2Br \xrightarrow{HO^-} C_6H_5CH_2OH \xrightarrow[\Delta]{MnO_2} C_6H_5CHO$

16.3 The Effect of Substituents on Reactivity

Like benzene, substituted benzenes undergo the five electrophilic aromatic substitution reactions discussed in Chapter 15 and listed in Section 16.2: halogenation, nitration, sulfonation, alkylation, and acylation. Now we need to find out whether a substituted benzene is more reactive or less reactive than benzene itself. The answer depends on the substituent. Some substituents make the ring more reactive and some make it less reactive than benzene toward electrophilic aromatic substitution.

The slow step of an electrophilic aromatic substitution reaction is the addition of an electrophile to the nucleophilic aromatic ring to form a carbocation intermediate (Section 15.9). *Substituents that are capable of donating electrons into the benzene ring will stabilize both the carbocation intermediate and the transition state leading to its formation* (Section 4.3), *thereby increasing the rate of electrophilic aromatic substitution. In contrast, substituents that withdraw electrons from the benzene ring will destabilize the carbocation intermediate and the transition state leading to its formation, thereby decreasing the rate of electrophilic aromatic substitution* (Section 16.4). Before we see how the carbocation intermediate is stabilized by electron donation and destabilized by electron withdrawal, we will look at the ways in which a substituent can donate or withdraw electrons.

Electron-donating substituents increase the reactivity of the benzene ring toward electrophilic aromatic substitution.

Electron-withdrawing substituents decrease the reactivity of the benzene ring toward electrophilic aromatic substitution.

relative rates of electrophilic substitution

Z
Z donates electrons into the benzene ring
>
>
Y
Y withdraws electrons from the benzene ring

There are two ways substituents can donate electrons: *inductive* electron donation and electron donation by *resonance*. There are also two ways substituents can withdraw electrons: *inductive* electron withdrawal and electron withdrawal by *resonance*.

Inductive Electron Donation and Withdrawal

If a substituent that is bonded to a benzene ring is *less electron withdrawing than a hydrogen*, the electrons in the σ bond that attaches the substituent to the benzene ring will move toward the ring more readily than will those in the σ bond that attaches the hydrogen to the ring. Such a substituent donates electrons inductively compared with a hydrogen. Donation of electrons through a σ bond is called **inductive electron donation** (Section 1.18). Alkyl substituents (such as CH_3) donate electrons inductively compared with a hydrogen.

Notice that the electron-donating ability of an alkyl group—not the electron-donating ability of a carbon atom—is compared with that of hydrogen. Carbon is actually slightly less electron donating than hydrogen (because C is more electronegative than H; see Table 1.3), but an alkyl group is more electron donating than hydrogen because of hyperconjugation (Section 4.2).

substituent donates electrons inductively (compared with a hydrogen)
CH_3
substituent withdraws electrons inductively (compared with a hydrogen)
$\overset{+}{N}H_3$

If a substituent is *more electron withdrawing than a hydrogen*, it will withdraw the σ electrons away from the benzene ring more strongly than will a hydrogen. Withdrawal of electrons through a σ bond is called **inductive electron withdrawal**. The $^+NH_3$ group is a substituent that withdraws electrons inductively because it is more electronegative than a hydrogen.

Resonance Electron Donation and Withdrawal

If a substituent has a lone pair on the atom that is directly attached to the benzene ring, the lone pair can be delocalized into the ring; these substituents are said to **donate electrons by resonance**. Substituents such as NH_2, OH, OR, and Cl donate electrons by resonance. These substituents also withdraw electrons inductively because the atom attached to the benzene ring is more electronegative than a hydrogen.

donation of electrons into a benzene ring by resonance

Tutorial:
Donation of electrons into a benzene ring

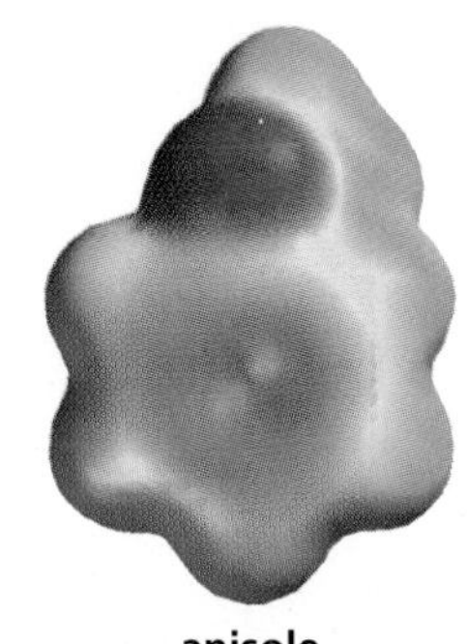

anisole

If a substituent is attached to the benzene ring by an atom that is doubly or triply bonded to a more electronegative atom, the π electrons of the ring can be delocalized onto the substituent; these substituents are said to **withdraw electrons by resonance.** Substituents such as C=O, C≡N, and NO_2 withdraw electrons by resonance. These substituents also withdraw electrons inductively because the atom attached to the benzene ring has a full or partial positive charge and, therefore, is more electronegative than a hydrogen.

withdrawal of electrons from a benzene ring by resonance

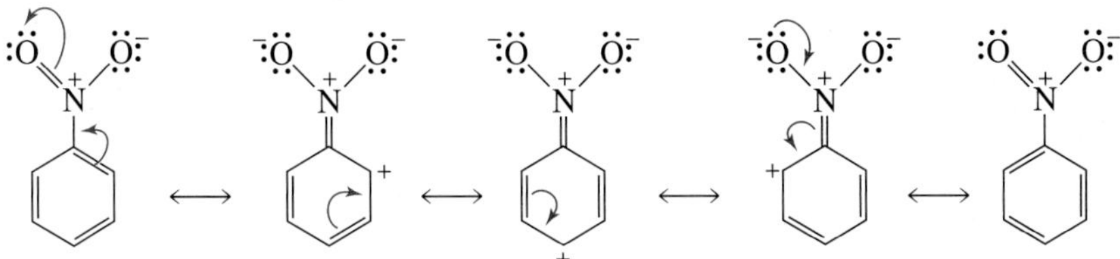

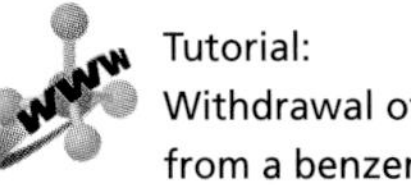

Tutorial:
Withdrawal of electrons from a benzene ring

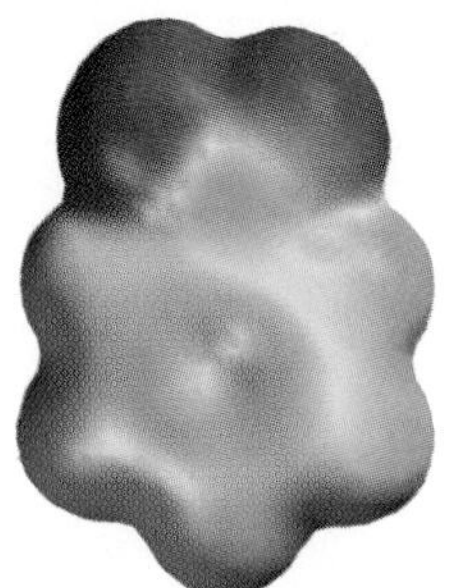

nitrobenzene

PROBLEM 7◆

For each of the following substituents, indicate whether it donates electrons inductively, withdraws electrons inductively, donates electrons by resonance, or withdraws electrons by resonance (inductive effects should be compared with a hydrogen; remember that many substituents can be characterized in more than one way):

a. Br
b. CH_2CH_3
c. $\overset{O}{\overset{\|}{C}}CH_3$
d. $NHCH_3$
e. OCH_3
f. $\overset{+}{N}(CH_3)_3$

Relative Reactivity of Substituted Benzenes

The substituents shown in Table 16.1 are listed according to how they affect the reactivity of the benzene ring toward electrophilic aromatic substitution compared with benzene—in which the substituent is a hydrogen. *The* **activating substituents** *make the benzene ring more reactive toward electrophilic substitution; the* **deactivating substituents** *make the benzene ring less reactive toward electrophilic substitution.* Remember that activating substituents donate electrons into the ring and deactivating substituents withdraw electrons from the ring.

All the *strongly activating substituents* donate electrons into the ring by resonance and withdraw electrons from the ring inductively. The fact that they have been found experimentally to be strong activators indicates that electron donation into the ring by resonance is more significant than inductive electron withdrawal from the ring.

strongly activating substituents

$:NH_2$ $:\ddot{O}H$ $:\ddot{O}R$

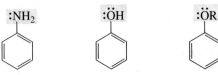

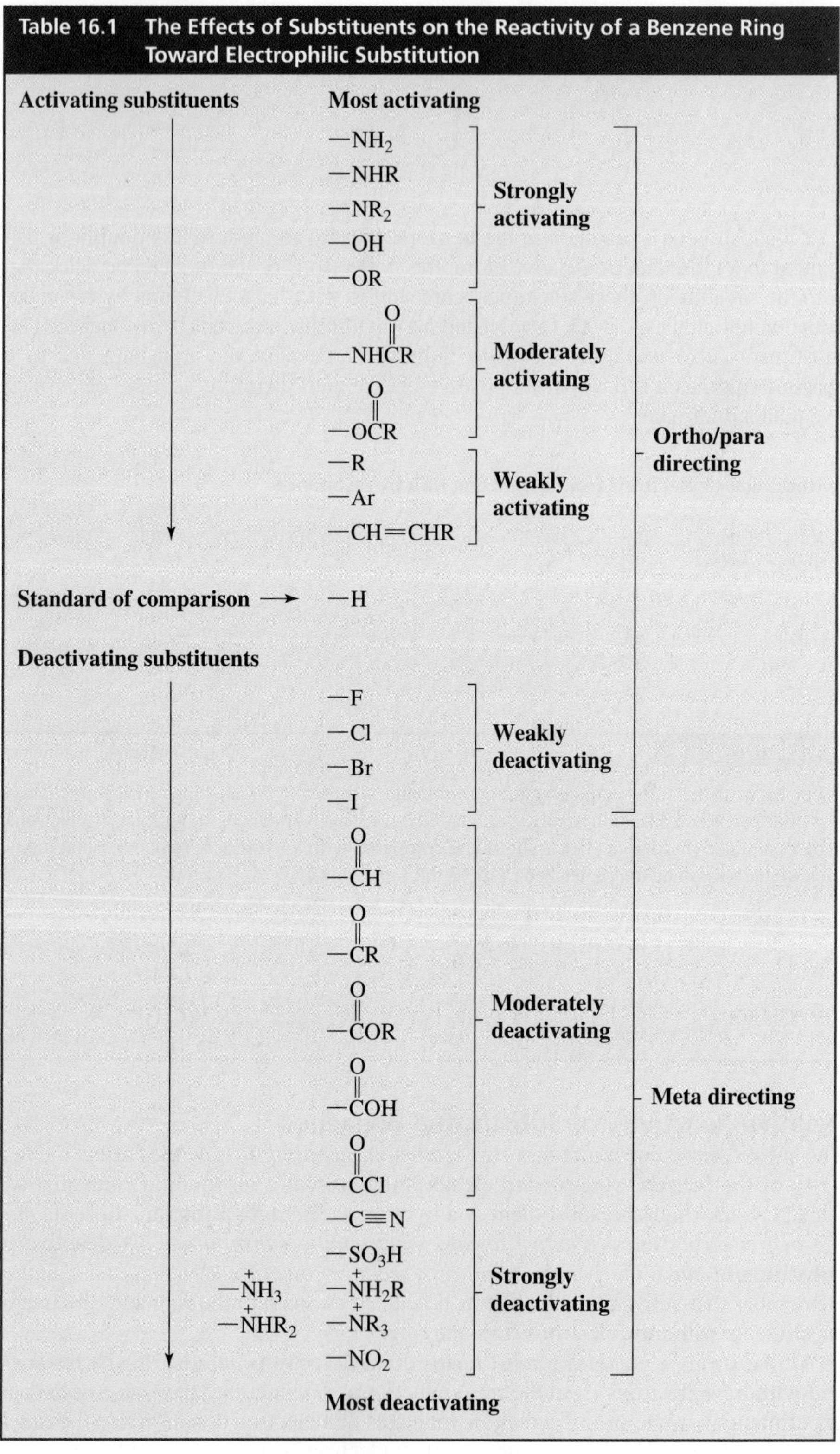

Table 16.1 The Effects of Substituents on the Reactivity of a Benzene Ring Toward Electrophilic Substitution

	Substituent	Effect	Directing
Activating substituents	**Most activating**		
	$-NH_2$	**Strongly activating**	**Ortho/para directing**
	$-NHR$	**Strongly activating**	**Ortho/para directing**
	$-NR_2$	**Strongly activating**	**Ortho/para directing**
	$-OH$	**Strongly activating**	**Ortho/para directing**
	$-OR$	**Strongly activating**	**Ortho/para directing**
	$-NHC(=O)R$	**Moderately activating**	**Ortho/para directing**
	$-OC(=O)R$	**Moderately activating**	**Ortho/para directing**
	$-R$	**Weakly activating**	**Ortho/para directing**
	$-Ar$	**Weakly activating**	**Ortho/para directing**
	$-CH{=}CHR$	**Weakly activating**	**Ortho/para directing**
Standard of comparison →	$-H$		
Deactivating substituents			
	$-F$	**Weakly deactivating**	**Ortho/para directing**
	$-Cl$	**Weakly deactivating**	**Ortho/para directing**
	$-Br$	**Weakly deactivating**	**Ortho/para directing**
	$-I$	**Weakly deactivating**	**Ortho/para directing**
	$-C(=O)H$	**Moderately deactivating**	**Meta directing**
	$-C(=O)R$	**Moderately deactivating**	**Meta directing**
	$-C(=O)OR$	**Moderately deactivating**	**Meta directing**
	$-C(=O)OH$	**Moderately deactivating**	**Meta directing**
	$-C(=O)Cl$	**Moderately deactivating**	**Meta directing**
	$-C{\equiv}N$	**Strongly deactivating**	**Meta directing**
	$-SO_3H$	**Strongly deactivating**	**Meta directing**
$-\overset{+}{N}H_3$	$-\overset{+}{N}H_2R$	**Strongly deactivating**	**Meta directing**
$-\overset{+}{N}HR_2$	$-\overset{+}{N}R_3$	**Strongly deactivating**	**Meta directing**
	$-NO_2$	**Strongly deactivating**	**Meta directing**
	Most deactivating		

The *moderately activating substituents* also donate electrons into the ring by resonance and withdraw electrons from the ring inductively. Because they are only moderately activating, we know that they donate electrons into the ring by resonance less effectively than do the strongly activating substituents.

moderately activating substituents

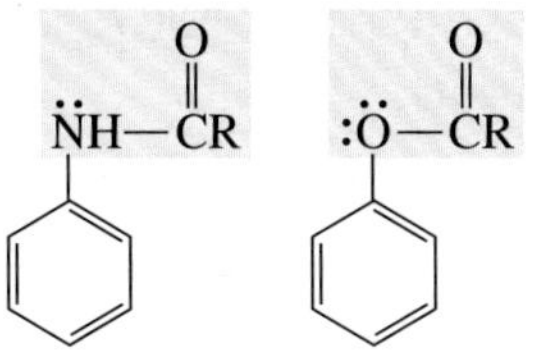

These substituents are less effective at donating electrons into the ring by resonance because, unlike the strongly activating substituents that donate electrons by resonance only *into* the ring, the moderately activating substituents can donate electrons by resonance in two competing directions: *into* the ring and *away from* the ring. The fact that these substituents are activators indicates that, despite their diminished resonance electron donation into the ring, overall they donate electrons by resonance more strongly than they withdraw electrons inductively.

substituent donates electrons by resonance into the benzene ring

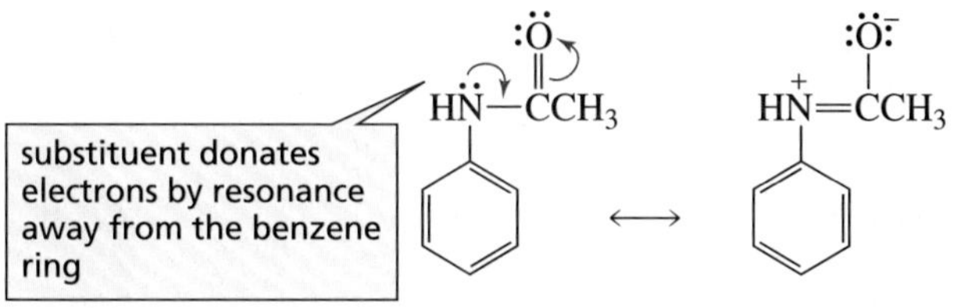

substituent donates electrons by resonance away from the benzene ring

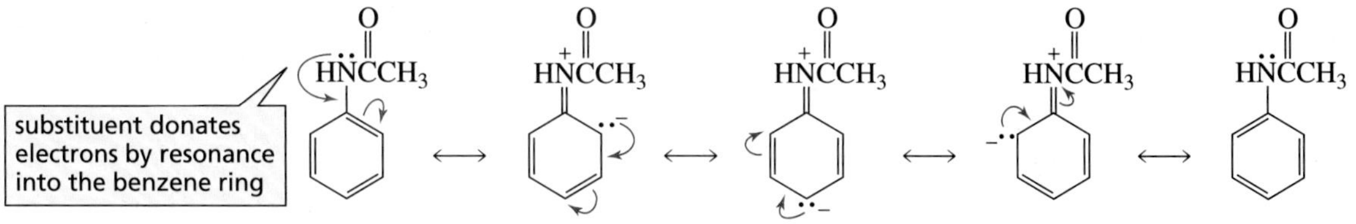

Alkyl, aryl, and CH=CHR groups are *weakly activating substituents*. We have seen that an alkyl substituent, compared with a hydrogen, donates electrons inductively. Aryl and CH=CHR groups can donate electrons into the ring by resonance and can withdraw electrons from the ring by resonance. The fact that they are weak activators indicates that they are slightly more electron donating than they are electron withdrawing.

weakly activating substituents

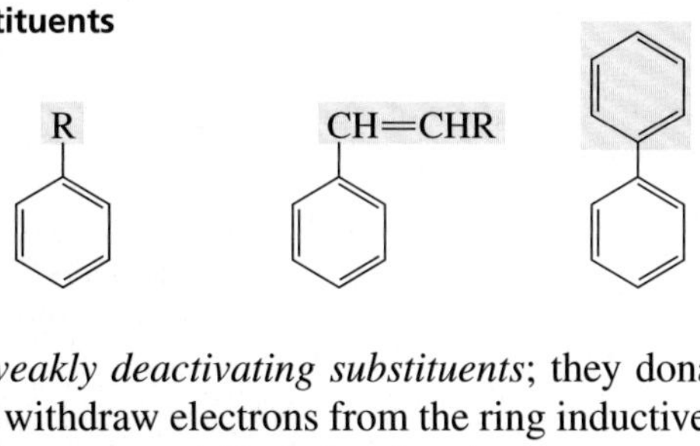

The halogens are *weakly deactivating substituents*; they donate electrons into the ring by resonance and withdraw electrons from the ring inductively. Because the halogens have been found experimentally to be deactivators, we can conclude that they withdraw electrons inductively more strongly than they donate electrons by resonance.

weakly deactivating substituents

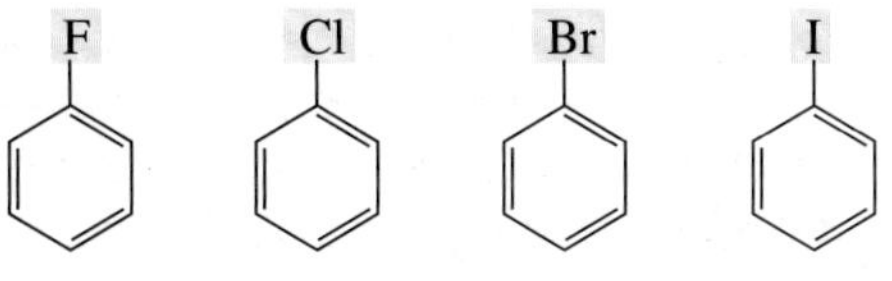

The *moderately deactivating substituents* all have carbonyl groups directly attached to the benzene ring. Carbonyl groups withdraw electrons both inductively and by resonance.

moderately deactivating substituents

$C_6H_5-\overset{O}{\overset{\|}{C}}-H$ $\quad$ $C_6H_5-\overset{O}{\overset{\|}{C}}-R$ $\quad$ $C_6H_5-\overset{O}{\overset{\|}{C}}-OR$ $\quad$ $C_6H_5-\overset{O}{\overset{\|}{C}}-OH$

The *strongly deactivating substituents* are powerful electron withdrawers. Except for the ammonium ions ($^+NH_3$, $^+NH_2R$, $^+NHR_2$, and $^+NR_3$), these substituents withdraw electrons both inductively and by resonance. The ammonium ions have no resonance effect, but the positive charge on the nitrogen atom causes them to strongly withdraw electrons inductively.

strongly deactivating substituents

$C_6H_5-SO_2OH$ (O=S—OH with =O) $\quad$ $C_6H_5-\overset{+}{N}(=O)O^-$ $\quad$ $C_6H_5-C{\equiv}N$ $\quad$ $C_6H_5-\overset{+}{N}R_3$

Take a minute to compare the electrostatic potential maps for anisole, benzene, and nitrobenzene. Notice that an electron-donating substituent (OCH_3) makes the ring more red (more negative), whereas an electron-withdrawing substituent (NO_2) makes the ring less red (less negative).

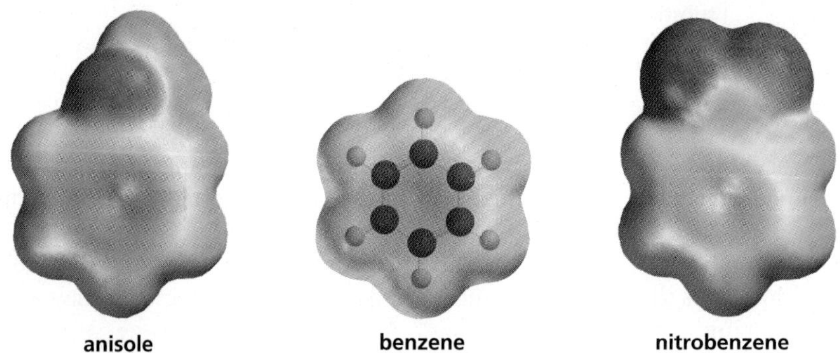

PROBLEM 8◆

List the members of the following sets of compounds in order of decreasing reactivity toward electrophilic aromatic substitution:

a. benzene, phenol, toluene, nitrobenzene, bromobenzene
b. dichloromethylbenzene, difluoromethylbenzene, toluene, chloromethylbenzene

PROBLEM 9 SOLVED

Explain why the halo-substituted benzenes have the relative reactivities shown in Table 16.1.

SOLUTION Table 16.1 shows that fluorine is the least deactivating of the halogen substituents and iodine is the most deactivating. We know that fluorine is the most electronegative of the halogens, which means that it is best at withdrawing electrons inductively. Fluorine is also best at donating electrons by resonance because its $2p$ orbital—compared with the $3p$ orbital of chlorine, the $4p$ orbital of bromine, or the $5p$ orbital of iodine—can better overlap with the $2p$ orbital of carbon. So the fluorine substituent is best both at donating electrons by resonance and at withdrawing electrons inductively. Because the table shows that fluorine is the weakest deactivator of the halogens, we can conclude that electron donation by resonance is the more important factor in determining the relative reactivities of halo-substituted benzenes.

16.4 The Effect of Substituents on Orientation

When a substituted benzene undergoes an electrophilic substitution reaction, where does the new substituent attach itself? Is the product of the reaction the ortho isomer, the meta isomer, or the para isomer?

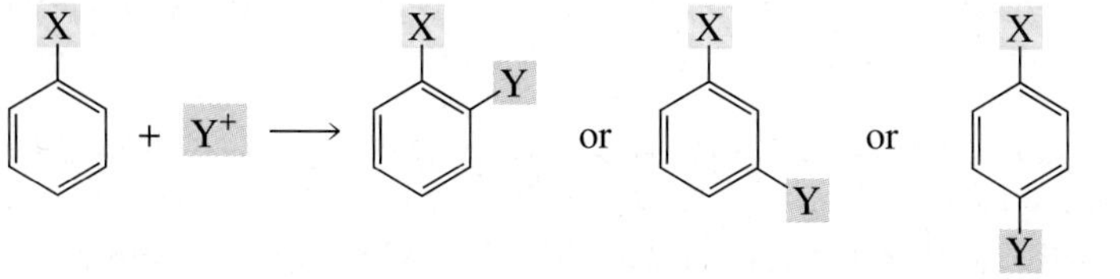

X + Y^+ ⟶ **ortho isomer** or **meta isomer** or **para isomer**

The substituent already attached to the benzene ring determines the location of the new substituent. There are two possibilities: A substituent will direct an incoming substituent either to the ortho *and* para positions, or it will direct an incoming substituent to the meta position. All activating substituents and the weakly deactivating halogens are **ortho–para directors**, and all substituents that are more deactivating than the halogens are **meta directors**. Thus, the substituents can be divided into three groups:

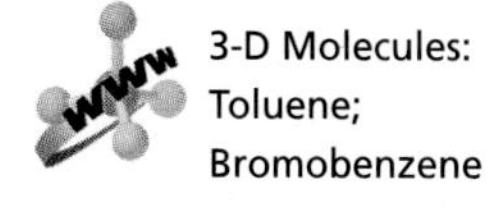
3-D Molecules: Toluene; Bromobenzene

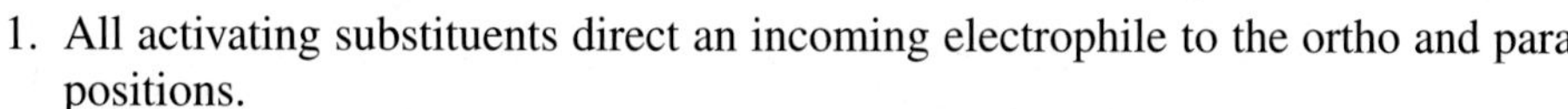

1. All activating substituents direct an incoming electrophile to the ortho and para positions.

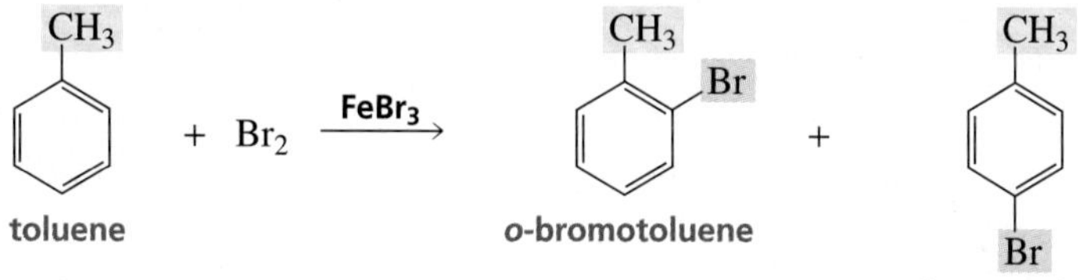

toluene + Br_2 $\xrightarrow{FeBr_3}$ ***o*-bromotoluene** + ***p*-bromotoluene**

All activating substituents are ortho–para directors.

2. The weakly deactivating halogens also direct an incoming electrophile to the ortho and para positions.

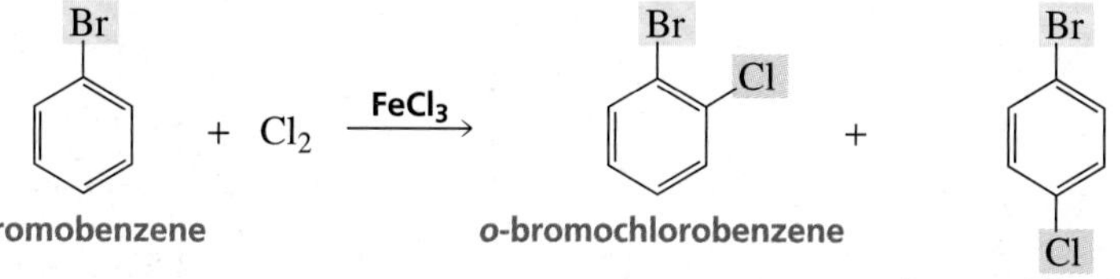

bromobenzene + Cl_2 $\xrightarrow{FeCl_3}$ ***o*-bromochlorobenzene** + ***p*-bromochlorobenzene**

The weakly deactivating halogens are ortho–para directors.

3. All moderately deactivating and strongly deactivating substituents direct an incoming electrophile to the meta position.

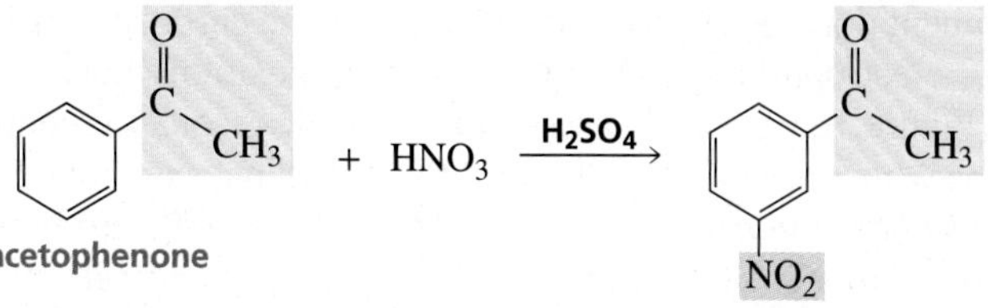

acetophenone + HNO_3 $\xrightarrow{H_2SO_4}$ ***m*-nitroacetophenone**

nitrobenzene + Br_2 $\xrightarrow{FeBr_3}$ ***m*-bromonitrobenzene**

All deactivating substituents (except the halogens) are meta directors.

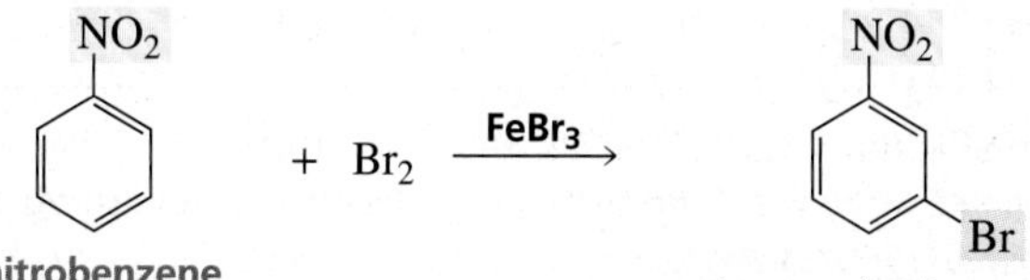

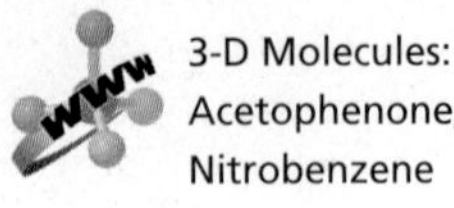
3-D Molecules: Acetophenone; Nitrobenzene

To understand why a substituent directs an incoming electrophile to a particular position, we must look at the stability of the carbocation intermediate that is formed in the rate-determining step. When a substituted benzene undergoes an electrophilic substitution reaction, three different carbocation intermediates can be formed: an *ortho*-substituted carbocation, a *meta*-substituted carbocation, and a *para*-substituted carbocation (Figure 16.1). The relative stabilities of the three carbocations enable us to determine the preferred pathway of the reaction because the more stable the carbocation, the less energy required to make it and the more likely it is that it will be formed (Section 4.3).

3-D Molecule: Anisole

If a substituent donates electrons *inductively*—a methyl group, for example—the indicated resonance contributors in Figure 16.1 are the most stable; the substituent is attached directly to the positively charged carbon, which the substituent can stabilize by inductive electron donation. These relatively stable resonance contributors are obtained only when the incoming group is directed to an ortho or para position. Therefore, the most stable carbocation is obtained by directing the incoming group to the ortho and para positions. Thus, *any substituent that donates electrons inductively is an ortho–para director.*

ortho
most stable
CH_3 + Y^+
toluene
meta
para
most stable

▲ **Figure 16.1**
The structures of the carbocation intermediates formed from the reaction of an electrophile with toluene at the ortho, meta, and para positions.

If a substituent donates electrons by *resonance*, the carbocations formed by putting the incoming electrophile on the ortho and para positions have a fourth resonance contributor (Figure 16.2). This is an especially stable resonance contributor because it is the only one whose atoms (except for hydrogen) all have complete octets. Therefore, *all substituents that donate electrons by resonance are ortho–para directors.*

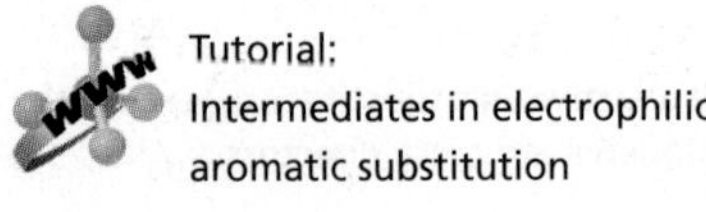
Tutorial: Intermediates in electrophilic aromatic substitution

Substituents with a positive charge or a partial positive charge on the atom attached to the benzene ring, withdraw electrons inductively from the benzene ring, and most withdraw electrons by resonance as well. For all such substituents, the indicated resonance contributors in Figure 16.3 are the least stable because they have a positive charge on each of two adjacent atoms, so the most stable carbocation is formed when the incoming electrophile is directed to the meta position. Thus, *all substituents that withdraw electrons (except for the halogens, which are ortho–para directors because they donate electrons by resonance) are meta directors.*

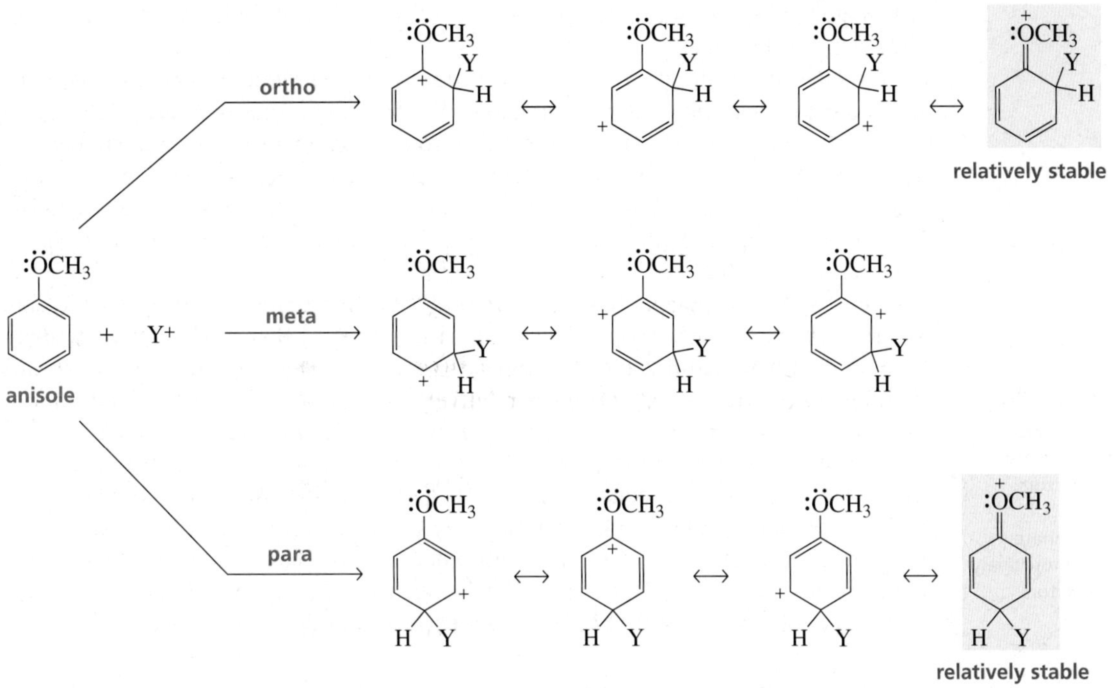

▲ **Figure 16.2**
The structures of the carbocation intermediates formed from the reaction of an electrophile with anisole at the ortho, meta, and para positions.

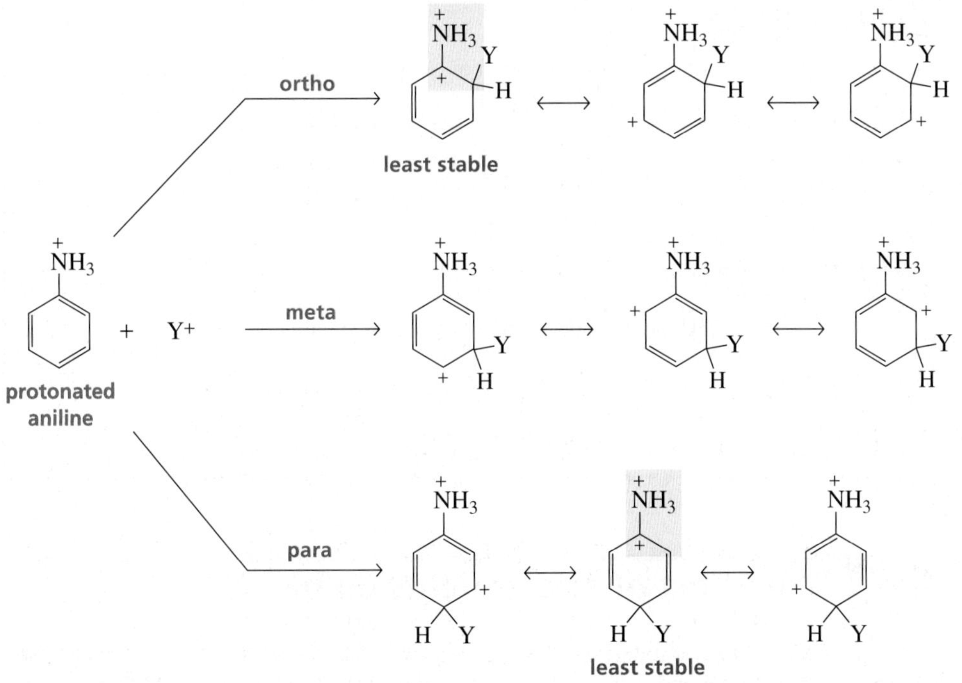

▲ **Figure 16.3**
The structures of the carbocation intermediates formed from the reaction of an electrophile with protonated aniline at the ortho, meta, and para positions.

Notice that the three possible carbocation intermediates in Figures 16.1 and 16.3 are the same, except for the substituent. The nature of the substituent determines whether the resonance contributors with the substituent directly attached to the positively charged carbon are the most stable (electron-donating substituents) or the least stable (electron-withdrawing substituents).

The only deactivating substituents that are ortho–para directors are the halogens, which are the weakest of the deactivators. We have seen that they are deactivators because they inductively withdraw electrons from the ring more strongly than they donate electrons by resonance. The halogens nevertheless are ortho–para directors because of their ability to donate electrons by resonance; they can stabilize the transition states leading to reaction at the ortho and para positions by resonance electron donation (Figure 16.2).

In summary, as shown in Table 16.1, the halogens and all the activating substituents are ortho–para directors. All substituents more deactivating than the halogens are meta directors. Put another way, *all substituents that donate electrons into the ring inductively or by resonance are ortho–para directors, and all substituents that cannot donate electrons into the ring inductively or by resonance are meta directors.*

All substituents that donate electrons into the ring either inductively or by resonance are ortho–para directors.

All substituents that cannot donate electrons into the ring either inductively or by resonance are meta directors.

You don't need to resort to memorization to determine whether a substituent is an ortho–para director or a meta director. It is easy to tell them apart: All ortho–para directors, except for alkyl, aryl, and CH$=$CHR groups, have at least one lone pair on the atom directly attached to the ring; all meta directors have a positive charge or a partial positive charge on the atom attached to the ring. Take a few minutes to examine the substituents listed in Table 16.1 to convince yourself that this is true.

PROBLEM 10

a. Draw the resonance contributors for nitrobenzene.

b. Draw the resonance contributors for chlorobenzene.

PROBLEM 11◆

What product(s) would result from nitration of each of the following compounds?

a. propylbenzene

b. bromobenzene

c. benzaldehyde

d. benzenesulfonic acid

e. cyclohexylbenzene

f. benzonitrile

PROBLEM 12◆

Are the following substituents ortho–para directors or meta directors?

a. CH$=$CHC$\equiv$N

b. NO_2

c. CH_2OH

d. COOH

e. CF_3

f. N$=$O

16.5 The Effect of Substituents on pK_a

We have seen that electron-withdrawing substituents increase the acidity of a compound (Sections 1.18 and 7.10). Therefore, when a substituent either withdraws electrons from or donates electrons into a benzene ring, the pK_a values of substituted phenols, benzoic acids, and protonated anilines will reflect this withdrawal or donation. For example, the pK_a of phenol in H_2O at 25 °C is 9.95. The pK_a of *para*-nitrophenol is lower (7.14) because the nitro substituent withdraws electrons from the ring, whereas the pK_a of *para*-methyl phenol is higher (10.19) because the methyl substituent donates electrons into the ring.

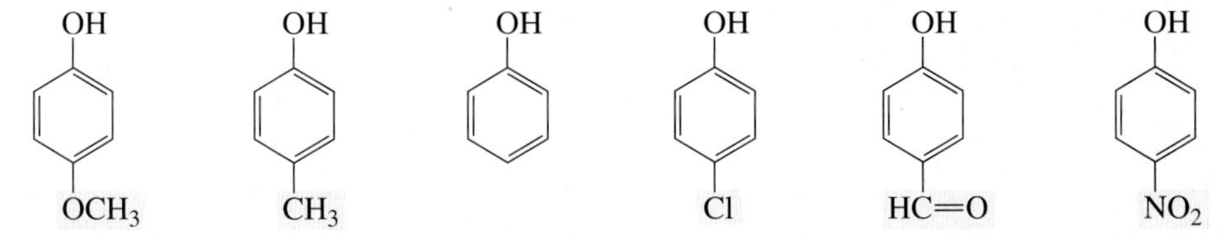

OH, OCH$_3$	OH, CH$_3$	OH	OH, Cl	OH, HC=O	OH, NO$_2$
pK_a = 10.20	pK_a = 10.19	pK_a = 9.95 phenol	pK_a = 9.38	pK_a = 7.66	pK_a = 7.14

Tutorial:
Effect of substituents on pK_a

Take a minute to compare the influence a substituent has on the reactivity of a benzene ring toward electrophilic substitution with its effect on the pK_a of phenol. Notice that the more strongly deactivating the substituent, the lower the pK_a of the phenol; and the more strongly activating the substituent, the higher the pK_a of the phenol. In other words, *electron withdrawal decreases reactivity toward electrophilic substitution and increases acidity, whereas electron donation increases reactivity toward electrophilic substitution and decreases acidity.*

A similar substituent effect on pK_a is observed for substituted benzoic acids and substituted protonated anilines.

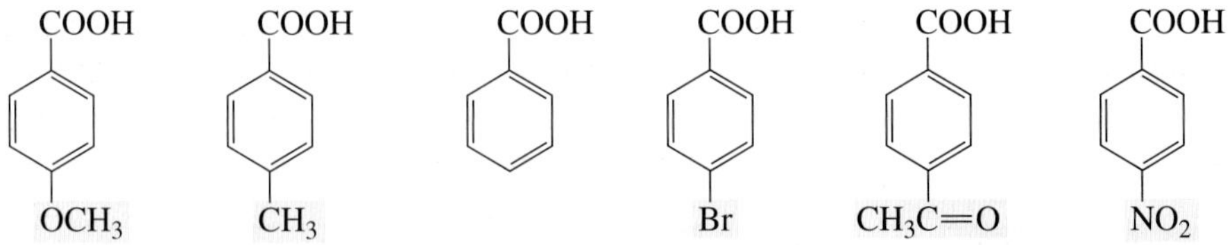

The more deactivating (electron withdrawing) the substituent, the more it increases the acidity of a COOH, an OH, or an $^+$NH$_3$ group attached to a benzene ring.

COOH, OCH$_3$	COOH, CH$_3$	COOH	COOH, Br	COOH, CH$_3$C=O	COOH, NO$_2$
pK_a = 4.47	pK_a = 4.34	pK_a = 4.20	pK_a = 4.00	pK_a = 3.70	pK_a = 3.44

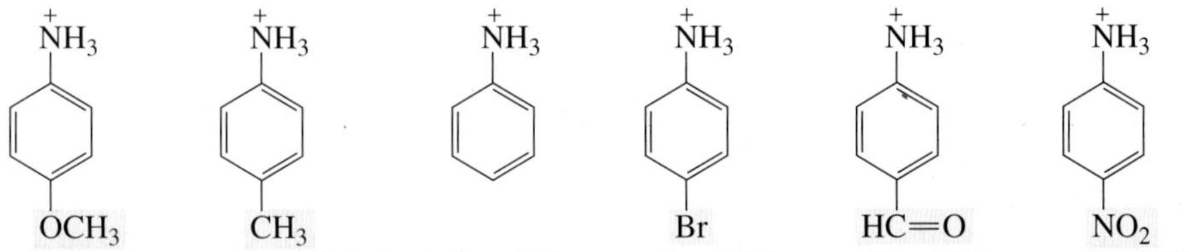

$\overset{+}{N}H_3$, OCH$_3$	$\overset{+}{N}H_3$, CH$_3$	$\overset{+}{N}H_3$	$\overset{+}{N}H_3$, Br	$\overset{+}{N}H_3$, HC=O	$\overset{+}{N}H_3$, NO$_2$
pK_a = 5.29	pK_a = 5.07	pK_a = 4.58	pK_a = 3.91	pK_a = 1.76	pK_a = 0.98

PROBLEM 13◆

Which of the compounds in each of the following pairs is more acidic?

a. CH_3COOH or $ClCH_2COOH$

b. O_2NCH_2COOH or $O_2NCH_2CH_2COOH$

c. CH_3CH_2COOH or $H_3\overset{+}{N}CH_2COOH$

d. 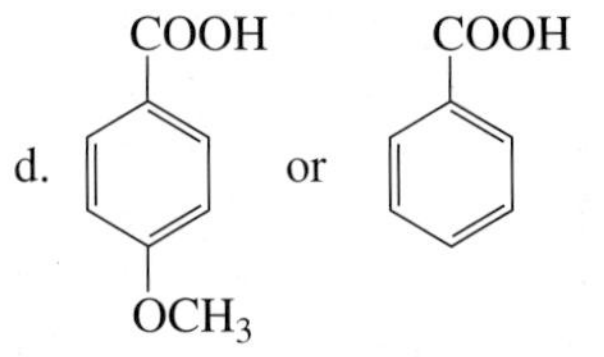 p-methoxybenzoic acid (COOH, OCH$_3$) or benzoic acid (COOH)

e. $HOOCCH_2COOH$ or $^-OOCCH_2COOH$

f. $HCOOH$ or CH_3COOH

g. FCH_2COOH or $ClCH_2COOH$

h. 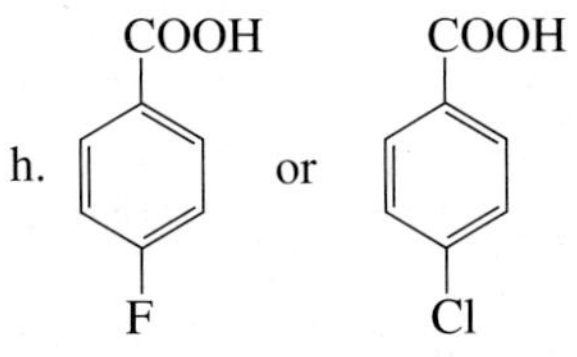 p-fluorobenzoic acid (COOH, F) or p-chlorobenzoic acid (COOH, Cl)

PROBLEM-SOLVING STRATEGY

The *para*-nitroanilinium ion is 3.60 pK_a units more acidic than the anilinium ion (pK_a = 0.98 versus 4.58), but *para*-nitrobenzoic acid is only 0.76 pK_a unit more acidic than benzoic acid (pK_a = 3.44 versus 4.20). Explain why the nitro substituent causes a large change in pK_a in the one case and a small change in pK_a in the other.

Don't expect to be able to solve this kind of problem simply by reading it. First, you need to remember that the acidity of a compound depends on the stability of its conjugate base (Sections 1.18 and 7.10). Next, draw structures of the conjugate bases in question and compare their stabilities.

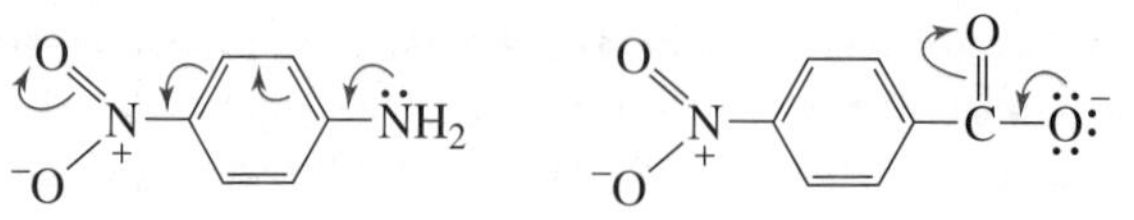

When a proton is lost from the *para*-nitroanilinium ion, the electrons that are left behind are shared by five atoms. (Draw resonance contributors if you want to see which atoms share the electrons.) When a proton is lost from *para*-nitrobenzoic acid, the electrons that are left behind are shared by two atoms. In other words, loss of a proton leads to greater electron delocalization in one base than in the other base. Because electron delocalization stabilizes a compound, we now know why the addition of a nitro substituent has a greater effect on the acidity of an anilinium ion than on benzoic acid. Now continue on to Problem 14.

PROBLEM 14

p-Nitrophenol has a pK_a of 7.14, whereas the pK_a of *m*-nitrophenol is 8.39. Explain.

16.6 The Ortho–Para Ratio

When a benzene ring with an ortho–para-directing substituent undergoes an electrophilic aromatic substitution reaction, what percentage of the product is the ortho isomer and what percentage is the para isomer? Solely on the basis of probability, one would expect more of the ortho product because there are two ortho positions available to the incoming electrophile and only one para position. The ortho position, however, is sterically hindered, whereas the para position is not. Consequently, the para isomer will be formed preferentially if either the substituent on the ring or the incoming electrophile is large. The following nitration reactions illustrate the decrease in the ortho–para ratio with an increase in the size of the alkyl substituent:

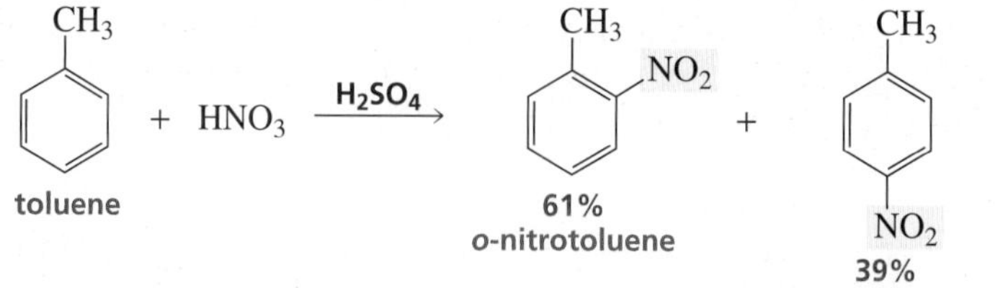

toluene + HNO_3 $\xrightarrow{H_2SO_4}$ 61% *o*-nitrotoluene + 39% *p*-nitrotoluene

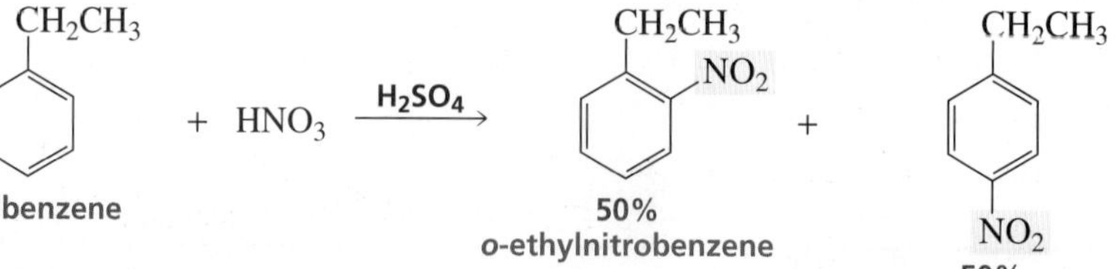

ethylbenzene + HNO_3 $\xrightarrow{H_2SO_4}$ 50% *o*-ethylnitrobenzene + 50% *p*-ethylnitrobenzene

$$\text{tert-butylbenzene} + HNO_3 \xrightarrow{H_2SO_4} \text{o-tert-butylnitrobenzene (18\%)} + \text{p-tert-butylnitrobenzene (82\%)}$$

Fortunately, the difference in the physical properties of the ortho- and para-substituted isomers is sufficient to allow them to be easily separated. Consequently, electrophilic aromatic substitution reactions that lead to both ortho and para isomers are useful in synthetic schemes because the desired product can be easily separated from the reaction mixture.

16.7 Additional Considerations Regarding Substituent Effects

It is important to know whether a substituent is activating or deactivating in determining the conditions needed to carry out a reaction. For example, methoxy and hydroxy substituents are so strongly activating that halogenation is carried out without the Lewis acid ($FeBr_3$ or $FeCl_3$) catalyst.

$$\text{anisole} + Br_2 \longrightarrow \text{p-bromoanisole} + \text{o-bromoanisole}$$

If the Lewis acid catalyst and excess bromine are used, the tribromide is obtained.

$$\text{anisole} + 3\,Br_2 \xrightarrow{FeBr_3} \text{2,4,6-tribromoanisole}$$

All Friedel–Crafts reactions require the Lewis acid catalyst. However, if there is a meta director on the ring (remember that all meta directors are moderate or strong deactivators), the ring will be too unreactive to undergo either Friedel–Crafts acylation or Friedel–Crafts alkylation.

A benzene ring with a meta director cannot undergo a Friedel–Crafts reaction.

$$\text{benzenesulfonic acid} + CH_3CH_2Cl \xrightarrow{AlCl_3} \text{no reaction}$$

$$\text{nitrobenzene} + CH_3\overset{\overset{\displaystyle O}{\|}}{C}Cl \xrightarrow{AlCl_3} \text{no reaction}$$

Aniline and *N*-substituted anilines also do not undergo Friedel–Crafts reactions. The lone pair on the amino group will complex with the Lewis acid ($AlCl_3$) that is needed to carry out the reaction, converting the NH_2 substituent into a deactivating meta director. As we have just seen, Friedel–Crafts reactions do not occur with benzene rings containing meta-directing substituents.

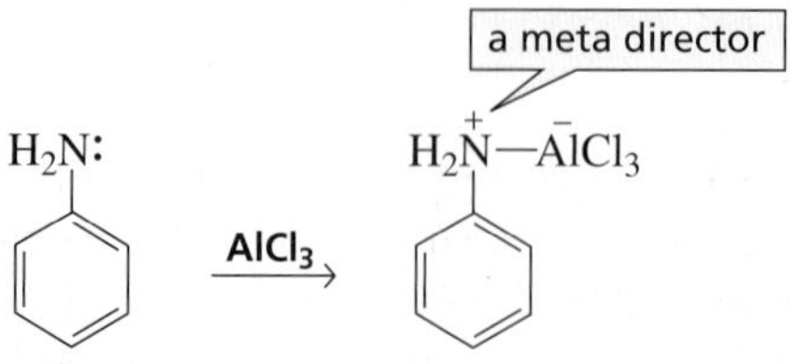

a meta director

$H_2N{:}$ (aniline) $\xrightarrow{AlCl_3}$ $H_2\overset{+}{N}{-}\overset{-}{Al}Cl_3$

aniline

Phenol and anisole undergo Friedel–Crafts reactions—orienting ortho and para—because oxygen, being a weaker base than nitrogen, does not complex with the Lewis acid.

3-D Molecules: *N,N*-Dimethylaniline; *o*-Nitro-*N,N*-dimethylaniline

Aniline also cannot be nitrated because nitric acid is an oxidizing agent and primary amines are easily oxidized. (Nitric acid and aniline can be an explosive combination.) Tertiary aromatic amines, however, can be nitrated. Because the tertiary amino group is a strong activator, nitration is carried out with nitric acid in acetic acid, a milder combination than nitric acid in sulfuric acid. About twice as much para isomer is formed as ortho isomer.

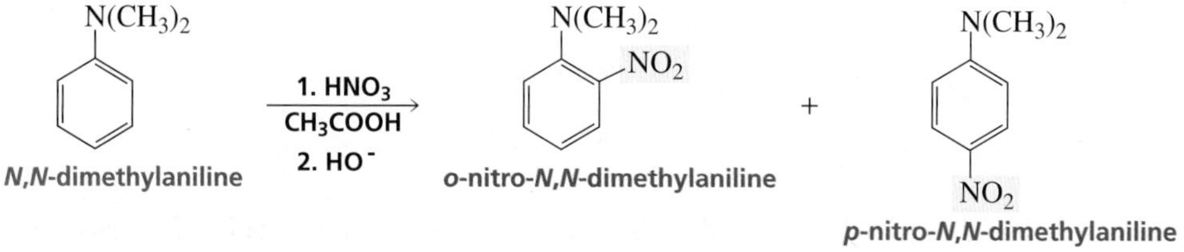

$N(CH_3)_2$ (*N,N*-dimethylaniline) $\xrightarrow[\text{2. HO}^-]{\text{1. HNO}_3,\ \text{CH}_3\text{COOH}}$ $N(CH_3)_2$, NO_2 (*o*-nitro-*N,N*-dimethylaniline) + $N(CH_3)_2$, NO_2 (*p*-nitro-*N,N*-dimethylaniline)

PROBLEM 15◆

Give the products, if any, of each of the following reactions:

a. benzonitrile + methyl chloride + $AlCl_3$
b. aniline + 3 Br_2
c. benzoic acid + CH_3CH_2Cl + $AlCl_3$
d. benzene + 2 CH_3Cl + $AlCl_3$

PROBLEM 16

When *N,N*-dimethylaniline is nitrated, some *meta*-nitro-*N,N*-dimethylaniline is formed. Why does the meta isomer form? (*Hint:* The pK_a of *N,N*-dimethylaniline is 5.07, and more meta isomer is formed if the reaction is carried out at pH = 3.5 than if it is carried out at pH = 4.5.)

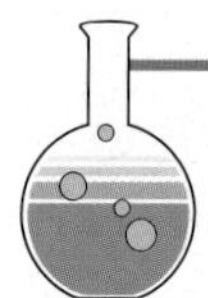

16.8 Designing a Synthesis III: Synthesis of Monosubstituted and Disubstituted Benzenes

As the number of reactions with wich we are familiar increases, we have more reactions to choose from when we design a synthesis. For example, we can now design two very different routes for the synthesis of 2-phenylethanol from benzene.

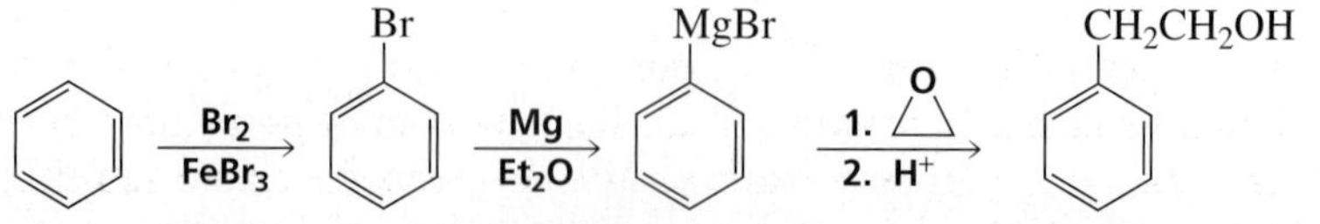

benzene $\xrightarrow[FeBr_3]{Br_2}$ Br-benzene $\xrightarrow[Et_2O]{Mg}$ MgBr-benzene $\xrightarrow[\text{2. H}^+]{\text{1. ethylene oxide}}$ CH_2CH_2OH-benzene

2-phenylethanol

excess benzene $\xrightarrow[AlCl_3]{CH_3CH_2Br}$ $C_6H_5CH_2CH_3$ $\xrightarrow[peroxide]{NBS, \Delta}$ $C_6H_5CHBrCH_3$ $\xrightarrow[\Delta]{tert\text{-}BuO^-}$ $C_6H_5CH{=}CH_2$ $\xrightarrow[2.\ HO^-,\ H_2O_2,\ H_2O]{1.\ BH_3/THF}$ $C_6H_5CH_2CH_2OH$

2-phenylethanol

Which route we choose depends on how easy it is to carry out the required reactions and also on the expected yield of the target molecule (the desired product). For example, the first route shown for the synthesis of 2-phenylethanol is the better procedure. The second route has more steps; it requires excess benzene to prevent polyalkylation;it uses a radical reaction that can produce unwanted side products; the yield of the elimination reaction is not high (because some substitution product is formed as well); and hydroboration–oxidation is not an easy reaction to carry out.

In designing the synthesis of disubstituted benzenes, you must consider carefully the order in which the substituents are to be placed on the ring. For example, if you want to synthesize *meta*-bromobenzenesulfonic acid, the sulfonic acid group has to be placed on the ring first because that group will direct the bromo substituent to the desired meta position.

Tutorial: Multistep synthesis of disubstituted benzenes

benzene $\xrightarrow[\Delta]{H_2SO_4}$ $C_6H_5SO_3H$ $\xrightarrow[FeBr_3]{Br_2}$ *m*-bromobenzenesulfonic acid

However, if the desired product is *para*-bromobenzenesulfonic acid, the order of the two reactions must be reversed because only the bromo substituent is an ortho–para director.

benzene $\xrightarrow[FeBr_3]{Br_2}$ C_6H_5Br $\xrightarrow[\Delta]{H_2SO_4}$ *p*-bromobenzenesulfonic acid + *o*-bromobenzenesulfonic acid

Both substituents of *meta*-nitroacetophenone are meta directors. However, the Friedel–Crafts acylation reaction must be carried out first because the benzene ring of nitrobenzene is too deactivated to undergo a Friedel–Crafts reaction (Section 16.7).

benzene + $CH_3\overset{O}{\overset{\|}{C}}Cl$ $\xrightarrow[2.\ H_2O]{1.\ AlCl_3}$ $C_6H_5COCH_3$ $\xrightarrow[H_2SO_4]{HNO_3}$ *m*-nitroacetophenone

It is also important to determine the point in a reaction sequence at which a substituent should be chemically modified. In the synthesis of *para*-chlorobenzoic acid from toluene, the methyl group is oxidized after it directs the chloro substituent to the para position. (*ortho*-Chlorobenzoic acid is also formed in this reaction.)

$$\text{toluene} \xrightarrow[\text{FeCl}_3]{\text{Cl}_2} p\text{-chlorotoluene} \xrightarrow[\Delta]{\text{KMnO}_4,\ \text{H}^+} \text{COOH / Cl}$$

***para*-chlorobenzoic acid**

In the synthesis of *meta*-chlorobenzoic acid, the methyl group is oxidized before chlorination because a meta director is needed to obtain the desired product.

$$\text{toluene} \xrightarrow[\Delta]{\text{Na}_2\text{Cr}_2\text{O}_7,\ \text{H}^+} \text{benzoic acid} \xrightarrow[\text{FeCl}_3]{\text{Cl}_2} \text{COOH / Cl}$$

***meta*-chlorobenzoic acid**

In the following synthesis of *para*-propylbenzenesulfonic acid, the type of reaction employed, the order in which the substituents are put on the benzene ring, and the point at which a substituent is chemically modified must all be considered: The straight-chain propyl substituent must be put on the ring by a Friedel–Crafts acylation reaction because of the carbocation rearrangement that would occur with a Friedel–Crafts alkylation reaction. The Friedel–Crafts acylation must be carried out before sulfonation because acylation could not be carried out on a ring with a strongly deactivating sulfonic acid substituent, and the sulfonic acid group is a meta director. Finally, the sulfonic acid group must be put on the ring after the carbonyl group is reduced to a methylene group, so that the sulfonic acid group will be directed primarily to the para position by the alkyl group.

$$\text{benzene} \xrightarrow[\text{2. H}_2\text{O}]{\text{1. CH}_3\text{CH}_2\text{CCl (O)},\ \text{AlCl}_3} \text{C}_6\text{H}_5\text{CCH}_2\text{CH}_3\text{(=O)} \xrightarrow[\text{HCl},\ \Delta]{\text{Zn(Hg)}} \text{C}_6\text{H}_5\text{CH}_2\text{CH}_2\text{CH}_3 \xrightarrow[\Delta]{\text{H}_2\text{SO}_4} \text{HO}_3\text{S–C}_6\text{H}_4\text{–CH}_2\text{CH}_2\text{CH}_3$$

***para*-propylbenzenesulfonic acid**

PROBLEM 17

Show how each of the following compounds can be synthesized from benzene:

a. *p*-chloroaniline
b. *m*-chloroaniline
c. *p*-nitrobenzoic acid
d. *m*-nitrobenzoic acid
e. *m*-bromopropylbenzene
f. *o*-bromopropylbenzene
g. 1-phenyl-2-propanol
h. 2-phenylpropene

16.9 Synthesis of Trisubstituted Benzenes

When a disubstituted benzene undergoes an electrophilic aromatic substitution reaction, the directing effect of both substituents has to be considered. If both substituents direct the incoming substituent to the same position, the product of the reaction is easily predicted.

3-D Molecules: *p*-Nitrotoluene; *m*-Chlorotoluene; *p*-Methylphenol; *p*-Chlorotoluene

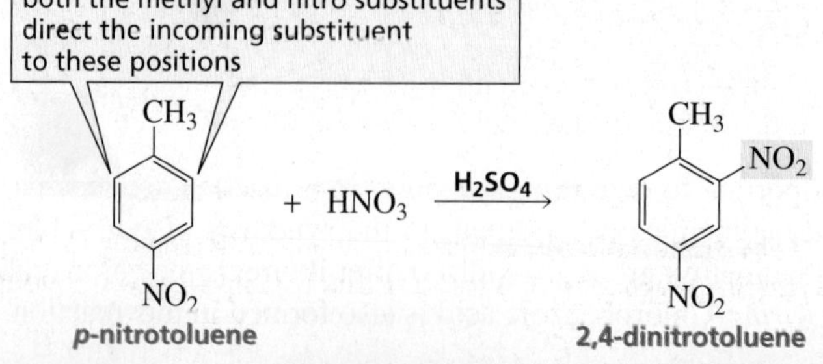

Notice that three positions are activated in the following reaction, but the new substituent ends up on only two of the three positions. Steric hindrance makes the position between the substituents less accessible.

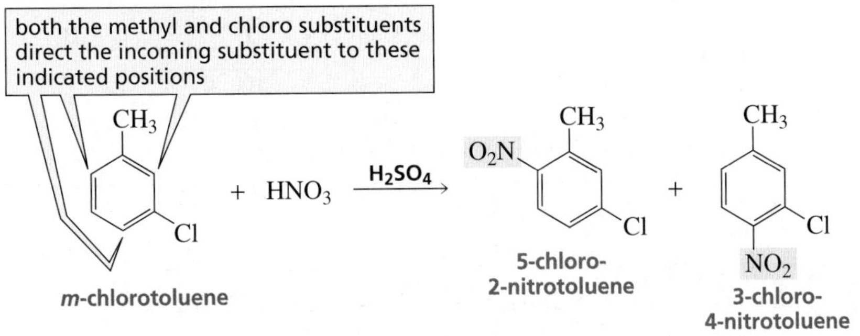

both the methyl and chloro substituents direct the incoming substituent to these indicated positions

m-chlorotoluene + HNO_3 $\xrightarrow{H_2SO_4}$ 5-chloro-2-nitrotoluene + 3-chloro-4-nitrotoluene

If the two substituents direct the new substituent to different positions, a strongly activating substituent will win out over a weakly activating substituent or a deactivating substituent.

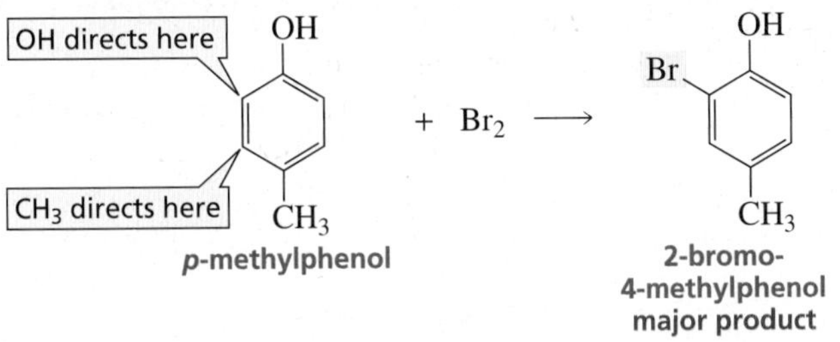

OH directs here

CH_3 directs here

p-methylphenol + Br_2 ⟶ 2-bromo-4-methylphenol major product

If the two substituents have similar activating properties, neither will dominate and a mixture of products will be obtained.

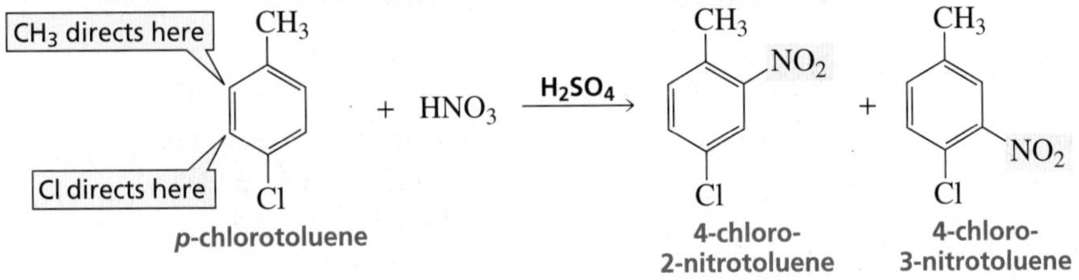

CH_3 directs here

Cl directs here

p-chlorotoluene + HNO_3 $\xrightarrow{H_2SO_4}$ 4-chloro-2-nitrotoluene + 4-chloro-3-nitrotoluene

PROBLEM 18◆

Give the major product(s) of each of the following reactions:

a. nitration of *p*-fluoroanisole
b. chlorination of *o*-benzenedicarboxylic acid
c. bromination of *p*-chlorobenzoic acid

PROBLEM 19 SOLVED

When phenol is treated with Br_2, a mixture of monobromo-, dibromo-, and tribromophenols is obtained. Design a synthesis that would convert phenol primarily to *ortho*-bromophenol.

SOLUTION The bulky sulfonic acid group will add preferentially to the para position. Both the OH and SO_3H groups will direct bromine to the position ortho to the OH group.

Heating in dilute acid removes the sulfonic acid group (Section 15.12). Using a sulfonic acid group to block the para position is a method commonly employed to synthesize high yields of ortho-substituted compounds.

$$\text{C}_6\text{H}_5\text{OH} \xrightarrow[100\ ^\circ\text{C}]{H_2SO_4} p\text{-HOC}_6\text{H}_4\text{SO}_3\text{H} \xrightarrow{Br_2} \text{2-bromo-4-sulfophenol} \xrightarrow[100\ ^\circ\text{C}]{H_3O^+} o\text{-bromophenol}$$

16.10 Synthesis of Substituted Benzenes Using Arenediazonium Salts

So far, we have learned how to place a limited number of different substituents on a benzene ring—those substituents listed in Section 16.2 and those that can be obtained from these substituents by chemical conversion. However, the kinds of substituents that can be placed on benzene rings can be greatly expanded by the use of **arenediazonium salts**.

$$\text{C}_6\text{H}_5\overset{+}{N}{\equiv}N\ \ Cl^-$$

an arenediazonium salt

The drive to form a molecule of stable nitrogen gas (N_2) causes the leaving group of a diazonium ion to be easily displaced by a wide variety of nucleophiles. The mechanism by which a nucleophile displaces the diazonium group depends on the nucleophile: Some displacements involve phenyl cations, while others involve radicals.

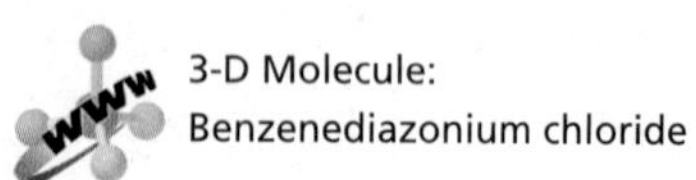

$$\text{C}_6\text{H}_5\overset{+}{N}{\equiv}N\ Cl^- + {}^-Nu \longrightarrow \text{C}_6\text{H}_5Nu + N_2\uparrow + Cl^-$$

benzenediazonium chloride

A primary amine can be converted into a diazonium salt by treatment with nitrous acid (HNO_2). Because nitrous acid is unstable, it is formed in situ, using an aqueous solution of sodium nitrite and HCl or HBr; indeed, N_2 is such a good leaving group that the diazonium salt is synthesized at 0 °C and used immediately without isolation. (The mechanism for conversion of a primary amino group [NH_2] to a diazonium group [$\overset{+}{N}{\equiv}N$] is shown in Section 16.12.)

$$\text{C}_6\text{H}_5NH_2 \xrightarrow[0\ ^\circ\text{C}]{NaNO_2,\ HCl} \text{C}_6\text{H}_5\overset{+}{N}{\equiv}N\ Cl^-$$

Nucleophiles such as ${}^-C{\equiv}N$, Cl^-, and Br^- will replace the diazonium group if the appropriate cuprous salt is added to the solution containing the arenediazonium salt. The reaction of an arenediazonium salt with a cuprous salt is known as a **Sandmeyer reaction**.

Traugott Sandmeyer (1854–1922) *was born in Switzerland. He received a Ph.D. from the University of Heidelberg and discovered the reaction that bears his name in 1884. He was a scientist at Geigy Co. in Basel, Switzerland.*

Sandmeyer reactions

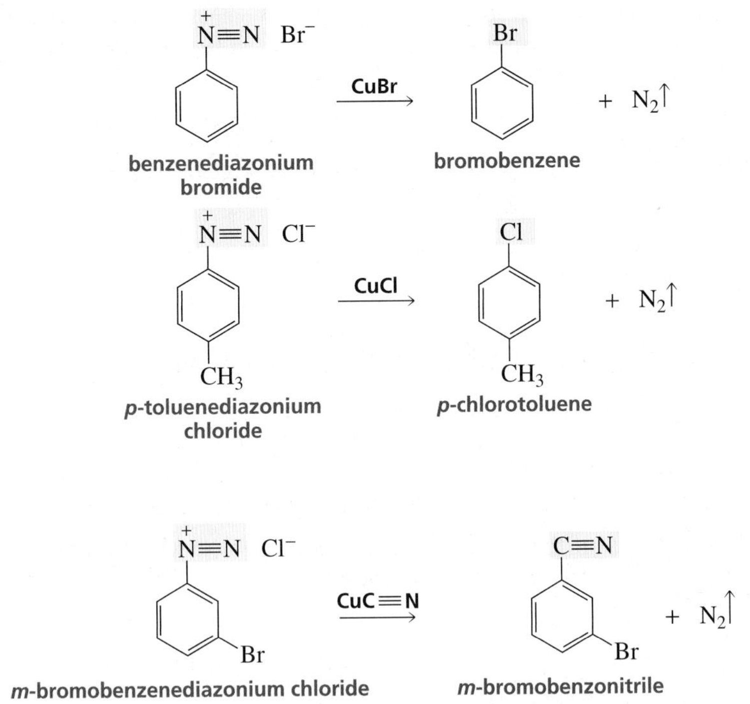

benzenediazonium bromide $\xrightarrow{CuBr}$ bromobenzene + $N_2\uparrow$

p-toluenediazonium chloride $\xrightarrow{CuCl}$ *p*-chlorotoluene + $N_2\uparrow$

m-bromobenzenediazonium chloride $\xrightarrow{CuC\equiv N}$ *m*-bromobenzonitrile + $N_2\uparrow$

KCl and KBr cannot be used in place of CuCl and CuBr in Sandmeyer reactions; the cuprous salts are required. This indicates that the cuprous ion is involved in the reaction, most likely by serving as a radical initiator (Section 4.10).

Although chloro and bromo substituents can be placed directly on a benzene ring by halogenation, the Sandmeyer reaction can be a useful alternative. For example, if you wanted to make *para*-chloroethylbenzene, chlorination of ethylbenzene would lead to a mixture of the ortho and para isomers.

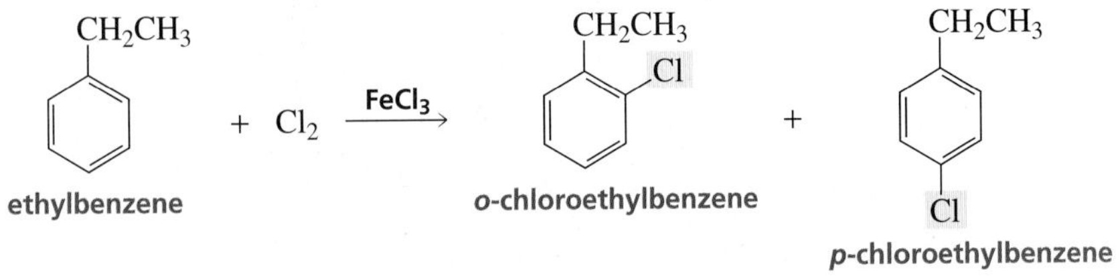

ethylbenzene + Cl_2 $\xrightarrow{FeCl_3}$ *o*-chloroethylbenzene + *p*-chloroethylbenzene

However, if you started with *para*-ethylaniline and used a Sandmeyer reaction for chlorination, only the desired para product would be formed.

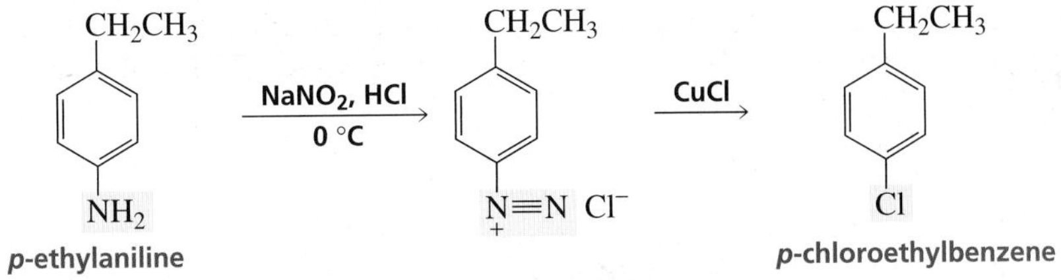

p-ethylaniline $\xrightarrow[0\ °C]{NaNO_2,\ HCl}$ (p-ethylbenzenediazonium chloride) $\xrightarrow{CuCl}$ *p*-chloroethylbenzene

PROBLEM 20

Explain why a diazonium group on a benzene ring cannot be used to direct an incoming substituent to the meta position.

PROBLEM 21

Explain why HBr should be used to generate the benzenediazonium salt if bromobenzene is the desired product of the Sandmeyer reaction and HCl should be used if chlorobenzene is the desired product.

An iodo substituent will replace the diazonium group if potassium iodide is added to the solution containing the diazonium ion.

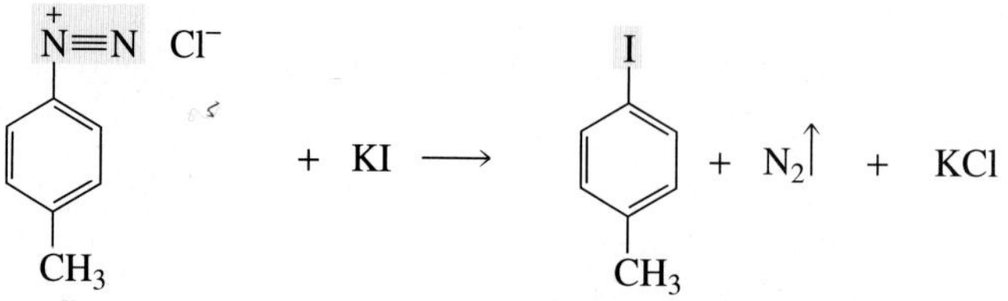

$$p\text{-}CH_3C_6H_4\overset{+}{N}{\equiv}N\ Cl^- + KI \longrightarrow p\text{-}CH_3C_6H_4I + N_2\uparrow + KCl$$

p-toluenediazonium chloride → p-iodotoluene

Fluoro substitution occurs if the arenediazonium salt is heated with fluoroboric acid (HBF_4). This reaction is known as the **Schiemann reaction**.

Günther Schiemann (1899–1969) *was born in Germany. He was a professor of chemistry at the Technologische Hochschule in Hanover, Germany.*

Schiemann reaction

$$C_6H_5\overset{+}{N}{\equiv}N\ Cl^- \xrightarrow{HBF_4} C_6H_5\overset{+}{N}{\equiv}N\ BF_4^- \xrightarrow{\Delta} C_6H_5F + BF_3 + N_2\uparrow + HCl$$

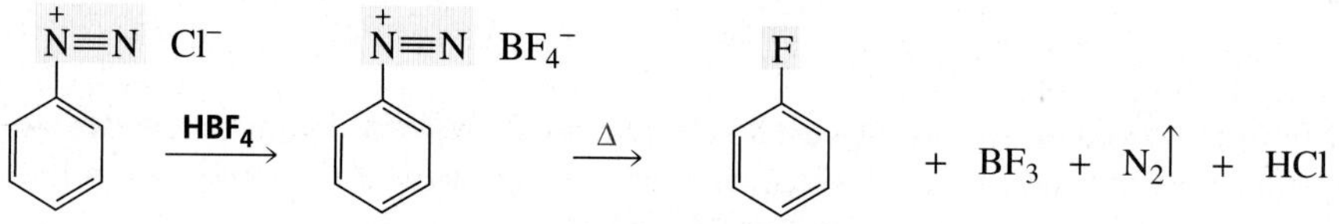

fluorobenzene

If the aqueous solution in which the diazonium salt has been synthesized is acidified and heated, an OH group will replace the diazonium group. (H_2O is the nucleophile.) This is a convenient way to synthesize a phenol.

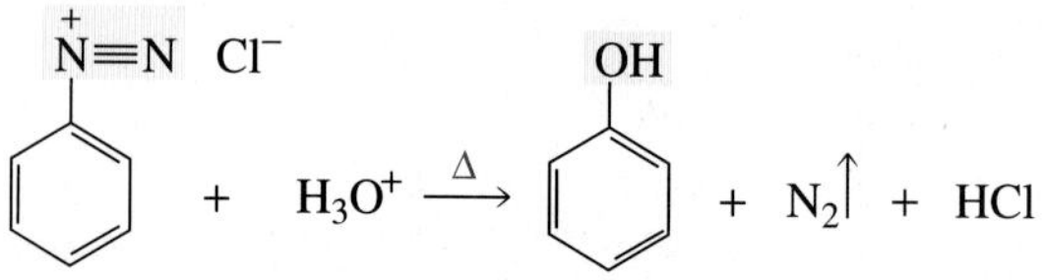

$$C_6H_5\overset{+}{N}{\equiv}N\ Cl^- + H_3O^+ \xrightarrow{\Delta} C_6H_5OH + N_2\uparrow + HCl$$

phenol

Phenols can also be prepared at room temperature using cuprous oxide and aqueous cupric nitrate.

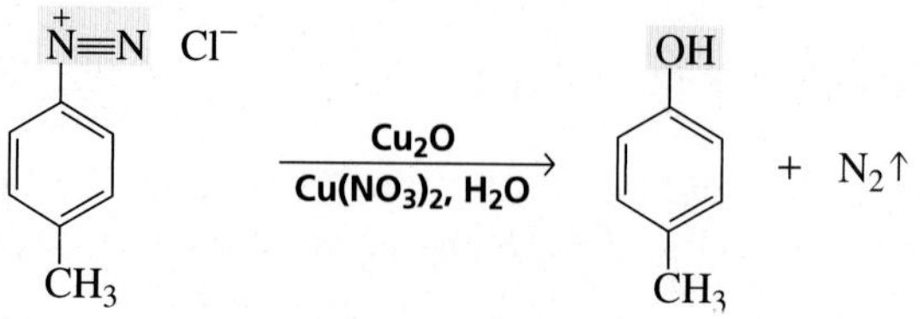

$$p\text{-}CH_3C_6H_4\overset{+}{N}{\equiv}N\ Cl^- \xrightarrow[Cu(NO_3)_2,\ H_2O]{Cu_2O} p\text{-}CH_3C_6H_4OH + N_2\uparrow$$

p-cresol

A hydrogen will replace a diazonium group if the diazonium salt is treated with hypophosphorous acid (H_3PO_2). This is a useful reaction if an amino group or a nitro group is needed for directing purposes and subsequently must be removed. It would be difficult to envision how 1,3,5-tribromobenzene could be synthesized without such a reaction.

$C_6H_5NH_2 + 3\,Br_2 \longrightarrow$ 2,4,6-tribromoaniline $\xrightarrow[0\ °C]{NaNO_2,\ HBr}$ 2,4,6-tribromobenzenediazonium bromide ($\overset{+}{N}{\equiv}N\ Br^-$) $\xrightarrow{H_3PO_2}$ 1,3,5-tribromobenzene $+ N_2\uparrow$

1,3,5-tribromobenzene

PROBLEM 22

Write the sequence of steps involved in the conversion of benzene into benzenediazonium chloride.

PROBLEM 23 SOLVED

Show how the following compounds could be synthesized from benzene:

a. *m*-dibromobenzene
b. *m*-bromophenol
c. *o*-chlorophenol
d. *m*-nitrotoluene
e. *p*-methylbenzonitrile
f. *m*-chlorobenzaldehyde

SOLUTION TO 23a A bromo substituent is an ortho–para director, so halogenation cannot be used to introduce both bromo substituents of *m*-dibromobenzene. Knowing that a bromo substituent can be placed on a benzene ring with a Sandmeyer reaction and that the bromo substituent in a Sandmeyer reaction replaces what originally was a meta-directing nitro substituent, we have a route to the synthesis of the target compound.

benzene $\xrightarrow[H_2SO_4]{HNO_3}$ nitrobenzene (NO_2) $\xrightarrow[FeBr_3]{Br_2}$ *m*-bromonitrobenzene (NO_2, Br) $\xrightarrow[2.\ HO^-]{1.\ Sn/HCl}$ *m*-bromoaniline (NH_2, Br) $\xrightarrow[HBr]{NaNO_2,\ 0\ °C}$ *m*-bromobenzenediazonium bromide ($\overset{+}{N}{\equiv}N\ Br^-$, Br) $\xrightarrow{CuBr}$ *m*-dibromobenzene (Br, Br)

16.11 The Arenediazonium Ion as an Electrophile

In addition to being used to synthesize substituted benzenes, arenediazonium ions can be used as electrophiles in electrophilic aromatic substitution reactions. Because an arenediazonium ion is unstable at room temperature, it can be used as an electrophile only in electrophilic aromatic substitution reactions that can be carried out well below room temperature. In other words, only highly activated benzene rings (phenols, anilines, and *N*-alkylanilines) can undergo electrophilic aromatic substitution reactions with arenediazonium ion electrophiles. The product of the reaction is an *azo compound*. The N=N linkage is called an **azo linkage**. Because the electrophile is so large, substitution takes place preferentially at the less sterically hindered para position.

phenol + *meta*-bromobenzenediazonium chloride ⟶ 3-bromo-4'-hydroxyazobenzene (an azo compound; an azo linkage)

However, if the para position is blocked, substitution will occur at an ortho position.

p-methylphenol + benzenediazonium chloride ⟶ 2-hydroxy-5-methylazobenzene

The mechanism for electrophilic aromatic substitution with an arenediazonium ion electrophile is the same as the mechanism for electrophilic aromatic substitution with any other electrophile.

mechanism for electrophilic aromatic substitution using an arenediazonium ion electrophile

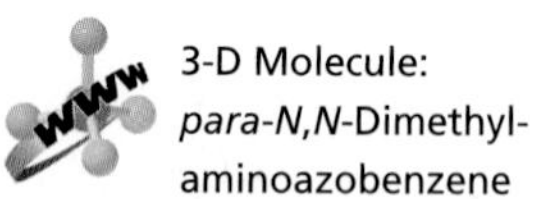
3-D Molecule: *para-N,N*-Dimethylaminoazobenzene

N,N-dimethylaniline ⟶ ⟶ *p-N,N*-dimethylaminoazobenzene + HB

Azo compounds, like alkenes, can exist in cis and trans forms. Because of steric strain, the trans isomer is considerably more stable than the cis isomer (Section 4.11).

trans-azobenzene *cis*-azobenzene

We have seen that azobenzenes are colored compounds because of their extended conjugation and are used commercially as dyes (Section 8.12).

PROBLEM 24

In the mechanism for electrophilic aromatic substitution with a diazonium ion as the electrophile, why does nucleophilic attack occur on the terminal nitrogen atom of the diazonium ion rather than on the nitrogen atom bonded to the benzene ring?

PROBLEM 25

Give the structure of the activated benzene ring and the diazonium ion used in the synthesis of the following compounds:

a. butter yellow b. methyl orange

(The structures of these compounds can be found in Section 8.12.)

16.12 Mechanism for the Reaction of Amines with Nitrous Acid

We have seen that the reaction of a primary amine with nitrous acid produces a diazonium salt. Both aryl amines and alkyl amines undergo this reaction, and both follow the same mechanism.

$$\text{aryl—NH}_2 \xrightarrow[0\,°\text{C}]{\text{NaNO}_2,\ \text{HCl}} \text{aryl—}\overset{+}{\text{N}}\equiv\text{N}\ \ \text{Cl}^-$$

$$\text{alkyl—NH}_2 \xrightarrow[0\,°\text{C}]{\text{NaNO}_2,\ \text{HCl}} \text{alkyl—}\overset{+}{\text{N}}\equiv\text{N}\ \ \text{Cl}^-$$

Conversion of a *primary* amino group to a diazonium group requires a nitrosonium ion that is formed when water is eliminated from protonated nitrous acid.

$$\text{Na}^+\ {}^-\text{:}\ddot{\text{O}}\text{—}\ddot{\text{N}}\text{=}\ddot{\text{O}}\text{:} \rightleftharpoons \text{H}\ddot{\text{O}}\text{—}\ddot{\text{N}}\text{=}\ddot{\text{O}}\text{:} \rightleftharpoons \text{H}\overset{+}{\ddot{\text{O}}}(\text{H})\text{—}\ddot{\text{N}}\text{=}\ddot{\text{O}}\text{:} \rightleftharpoons {}^+\ddot{\text{N}}\text{=}\ddot{\text{O}}\text{:} + \text{H}_2\ddot{\text{O}}\text{:}$$

sodium nitrite (H—Cl; Na⁺Cl⁻) nitrous acid (H—Cl; Cl⁻) nitrosonium ion

The nitrosonium ion accepts a share of the amino nitrogen's lone pair. Loss of a proton from nitrogen forms a **nitrosamine** (also called an ***N*-nitroso compound** because a nitroso substituent is bonded to a nitrogen). Delocalization of nitrogen's lone pair and protonation of oxygen form a protonated *N*-hydroxyazo compound. This compound is in equilibrium with its nonprotonated form, which can be reprotonated on nitrogen (reverse reaction) or protonated on oxygen (forward reaction). Elimination of water forms the diazonium ion.

$$\text{C}_6\text{H}_5\text{NH}_2 + {}^+\text{N=O} \longrightarrow \text{C}_6\text{H}_5\overset{+}{\text{N}}\text{H}_2\text{—N=O} \rightleftharpoons \text{C}_6\text{H}_5\text{NH—N=O} \longrightarrow \text{C}_6\text{H}_5\overset{+}{\text{N}}\text{H=N—OH}$$

aniline, a primary amine; :B; a nitrosamine; H—B; :B

$$\rightleftharpoons \text{C}_6\text{H}_5\ddot{\text{N}}\text{=N—}\ddot{\text{O}}\text{H} \rightleftharpoons \text{C}_6\text{H}_5\ddot{\text{N}}\text{=N—}\overset{+}{\text{O}}\text{H}_2 \longrightarrow \text{C}_6\text{H}_5\overset{+}{\text{N}}\equiv\text{N} + \text{H}_2\text{O}$$

an *N*-hydroxyazo compound; H—B; :B; a diazonium ion

Remember that reactions in which arenediazonium ions are involved must be carried out at 0 °C because they are unstable at higher temperatures. Alkanediazonium ions are even less stable. They lose molecular N_2—even at 0 °C—as they are formed, reacting with whatever nucleophiles are present in the reaction mixture by both $S_N1/E1$ and $S_N2/E2$ mechanisms. Because of the mixture of products obtained, alkanediazonium ions are of limited synthetic use.

PROBLEM 26◆

What products would be formed from the reaction of isopropylamine with sodium nitrite and aqueous HCl?

PROBLEM 27

Diazomethane can be used to convert a carboxylic acid into a methyl ester. Propose a mechanism for this reaction.

$$\text{RCOH (C=O)} + \text{CH}_2\text{N}_2 \longrightarrow \text{RCOCH}_3 \text{ (C=O)} + \text{N}_2\uparrow$$

a carboxylic acid — diazomethane — a methyl ester

Secondary aryl and alkyl amines react with a nitrosonium ion to form nitrosamines rather than diazonium ions. The mechanism of the reaction is similar to that for the reaction of a primary amine with a nitrosonium ion, except that the reaction stops at the nitrosamine stage. The reaction stops because a secondary amine, unlike a primary amine, does not have the second proton that must be lost in order to generate the diazonium ion.

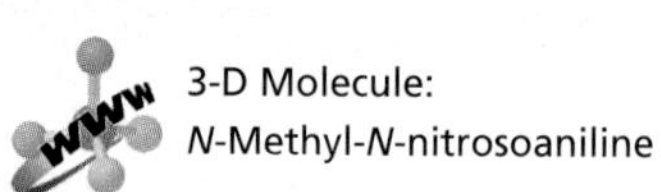

$$\text{C}_6\text{H}_5\text{N(H)CH}_3 + {}^{+}\text{N=O} \longrightarrow \text{C}_6\text{H}_5\overset{+}{\text{N}}\text{(H)(CH}_3\text{)}\text{—N=O} \xrightleftharpoons{:\text{B}} \text{C}_6\text{H}_5\text{N(CH}_3\text{)—N=O} + \text{HB}$$

N-methylaniline
a secondary amine

N-methyl-N-nitrosoaniline
a nitrosamine

The product formed when the nitrogen of a *tertiary* amine shares its lone pair with a nitrosonium ion cannot be stabilized by loss of a proton. A tertiary aryl amine, therefore, can undergo an electrophilic aromatic substitution reaction with a nitrosonium ion. The product of the reaction is primarily the para isomer because the bulky dialkylamino group blocks approach of the nitrosonium ion to the ortho position.

$$\text{C}_6\text{H}_5\text{N(CH}_3\text{)}_2 + {}^{+}\text{N=O} \rightleftharpoons \text{C}_6\text{H}_5\overset{+}{\text{N}}\text{(CH}_3\text{)}_2\text{—N=O}$$

N,N-dimethylaniline
a tertiary amine

$$\downarrow$$

$$(\text{CH}_3)_2\text{N—C}_6\text{H}_4\text{—N=O}$$

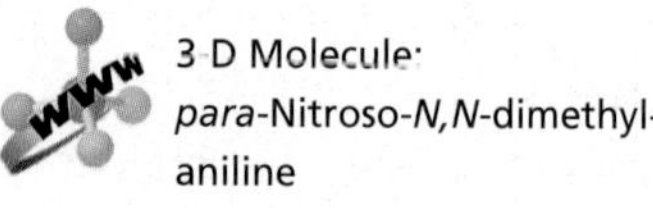

para-nitroso-N,N-dimethylaniline
85%

NITROSAMINES AND CANCER

A 1962 outbreak of food poisoning in sheep in Norway was traced to their ingestion of nitrite-treated fish meal. This incident immediately raised concerns about human consumption of nitrite-treated foods, because sodium nitrite is a commonly used food preservative. It can react with naturally occurring secondary amines present in food, to produce nitrosamines, which are known to be carcinogenic. Smoked fish, cured meats, and beer all contain nitrosamines. Nitrosamines are also found in cheese, which is not surprising because some cheeses are preserved with nitrite and cheese is rich in secondary amines. In the United States, consumer groups asked the Food and Drug Administration to ban the use of sodium nitrite as a preservative. This request was vigorously opposed by the meat-packing industry. Despite extensive investigations, it has not been determined whether the small amounts of nitrosamines present in our food pose a hazard to our health. Until this question can be answered, it will be hard to avoid sodium nitrite in our diet. It is worrisome to note, however, that Japan has both one of the highest gastric cancer rates and the highest average ingestion of sodium nitrite. Some good news is that the concentration of nitrosamines present in bacon has been considerably reduced in recent years by adding ascorbic acid—a nitrosamine inhibitor—to the curing mixture. Also, improvements in the malting process have reduced the level of nitrosamines in beer. Dietary sodium nitrite does have a redeeming feature—there is some evidence that it protects against botulism (a type of severe food poisoning).

16.13 Nucleophilic Aromatic Substitution Reactions

We have seen that aryl halides do not react with nucleophiles under standard reaction conditions because the π electron clouds repel the approach of a nucleophile (Section 10.8).

$\bar{:}Nu$ + chlorobenzene —✕→ no reaction

If, however, the aryl halide has one or more substituents that strongly withdraw electrons from the ring by resonance, **nucleophilic aromatic substitution** reactions can occur without using extreme conditions. The electron-withdrawing groups must be positioned ortho or para to the halogen. The greater the number of electron-withdrawing substituents, the easier it is to carry out the nucleophilic aromatic substitution reaction. Notice the different conditions under which the following reactions occur:

1-chloro-4-nitrobenzene —(1. HO^- (pH 14), 160 °C; 2. H^+)→ 4-nitrophenol + HCl

1-chloro-2,4-dinitrobenzene —(HO^- (pH 10), 100 °C)→ 2,4-dinitrophenol + Cl^-

1-chloro-2,4,6-trinitrobenzene —(H_2O (pH 7), 40 °C)→ 2,4,6-trinitrophenol + Cl^-

3-D Molecule: 1-Chloro-2,4,6-trinitrobenzene

Electron-withdrawing substituents increase the reactivity of the benzene ring toward nucleophilic substitution and decrease the reactivity of the benzene ring toward electrophilic substitution.

Notice also that the strongly electron-withdrawing substituents that *activate* the benzene ring toward *nucleophilic aromatic substitution* reactions are the same substituents that *deactivate* the ring toward *electrophilic aromatic substitution*. In other words, making the ring less electron rich makes it easier for a nucleophile—but more difficult for an electrophile—to approach the ring. Thus, any substituent that deactivates the benzene ring toward electrophilic substitution activates it toward nucleophilic substitution and vice versa.

Nucleophilic aromatic substitution takes place by a two-step reaction known as an **S_NAr reaction** (substitution nucleophilic aromatic). In the first step, the nucleophile attacks the carbon bearing the leaving group from a trajectory that is nearly perpendicular to the aromatic ring. (Recall from Section 10.8 that leaving groups cannot be displaced from sp^2 carbon atoms by back-side attack.) Nucleophilic attack forms a resonance-stabilized carbanion intermediate called a *Meisenheimer complex*, after Jakob Meisenheimer (1876–1934). In the second step of the reaction, the leaving group departs, reestablishing the aromaticity of the ring.

general mechanism for nucleophilic aromatic substitution

In a nucleophilic aromatic substitution reaction, the incoming nucleophile must be a stronger base than the substituent that is being replaced, because the weaker of the two bases will be the one eliminated from the intermediate.

The electron-withdrawing substituent must be ortho or para to the site of nucleophilic attack because the electrons of the attacking nucleophile can be delocalized onto the substituent only if the substituent is in one of those positions.

A variety of substituents can be placed on a benzene ring by means of nucleophilic aromatic substitution reactions. The only requirement is that the incoming group be a stronger base than the group that is being replaced.

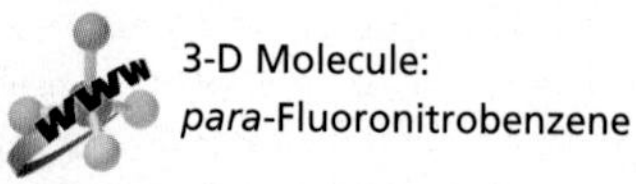

p-fluoronitrobenzene → *p*-nitroanisole

1-bromo-2,4-dinitrobenzene → *N*-ethyl-2,4-dinitroaniline

PROBLEM 28

Draw resonance contributors for the carbanion that would be formed if *meta*-chloronitrobenzene could react with hydroxide ion. Why doesn't it react?

PROBLEM 29◆

a. List the following compounds in order of decreasing reactivity toward nucleophilic aromatic substitution:

chlorobenzene 1-chloro-2,4-dinitrobenzene *p*-chloronitrobenzene

b. List the same compounds in order of decreasing reactivity toward electrophilic aromatic substitution.

PROBLEM 30

Show how each of the following compounds could be synthesized from benzene:

a. *o*-nitrophenol b. *p*-nitroaniline c. *p*-bromoanisole

16.14 Benzyne

An aryl halide such as chlorobenzene can undergo a nucleophilic substitution reaction in the presence of a very strong base such as $^{-}NH_2$. There are two surprising features about this reaction: The aryl halide does not have to contain an electron-withdrawing group, and the incoming substituent does not always end up on the carbon vacated by the leaving group. For example, when chlorobenzene—with the carbon to which the chlorine is attached isotopically labeled with ^{14}C—is treated with amide ion in liquid ammonia, aniline is obtained as the product. Half of the product has the amino group attached to the isotopically labeled carbon (denoted by the asterisk) as expected, but the other half has the amino group attached to the carbon adjacent to the labeled carbon.

Martin D. Kamen (1913–2002) *was born in Toronto. He was the first to isolate ^{14}C, which immediately became the most useful of all the isotopes in chemical and biochemical research. He received a B.S. and a Ph.D. from the University of Chicago and became a U.S. citizen in 1938. He was a professor at the University of California, Berkeley; at Washington University, St. Louis; and at Brandeis University. He was one of the founding professors of the University of California, San Diego. In later years, he became a member of the faculty at the University of Southern California. He received the Fermi medal in 1996.*

Cl, * + $^{-}NH_2$ $\xrightarrow{NH_3\ (liq)}$ NH$_2$, * + *, NH$_2$

chlorobenzene

approximately equal amounts of the two products are obtained

These are the only products formed. Anilines with the amino group two or three carbons removed from the labeled carbon are not formed.

The fact that the two products are formed in approximately equal amounts indicates that the reaction takes place by a mechanism that forms an intermediate in which the two carbons to which the amino group is attached in the product are equivalent. The mechanism that accounts for the experimental observations involves the formation of a **benzyne intermediate**. Benzyne has a triple bond between two adjacent carbon atoms of benzene. In the first step of the mechanism, the strong base ($^{-}NH_2$) removes a proton from the position ortho to the halogen. The resulting anion expels the halide ion, thereby forming benzyne.

The labeling experiment was done by **John D. Roberts**, *who was born in Los Angeles in 1918. He received both his B.A. and Ph.D. degrees from UCLA. He arrived at the California Institute of Technology in 1952 as a Guggenheim fellow and has been a professor there since 1953.*

Cl, *, H + $^{-}:\ddot{N}H_2$ ⟶ Cl, * + NH_3 ⟶ * + Cl^-

benzyne

The incoming nucleophile can attack either of the carbons of the "triple bond" of benzyne. Protonation of the resulting anion forms the substitution product. The overall reaction is an elimination–addition reaction: Benzyne is formed in an elimination reaction and immediately undergoes an addition reaction.

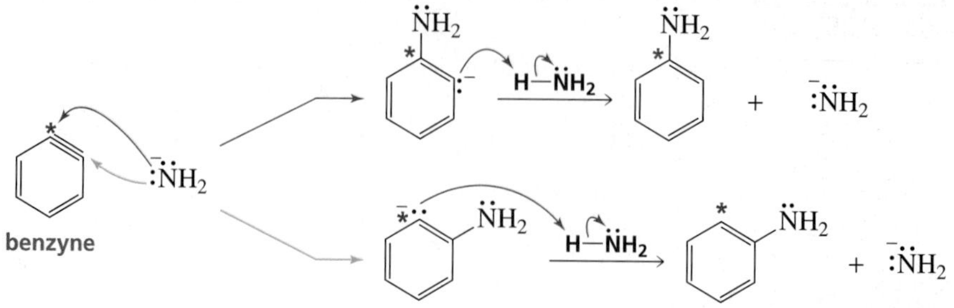

Substitution at the carbon that was attached to the leaving group is called **direct substitution**. Substitution at the adjacent carbon is called **cine substitution** (cine comes from *kinesis*, which is Greek for "movement"). In the following reaction, *o*-toluidine is the direct-substitution product; *m*-toluidine is the cine-substitution product.

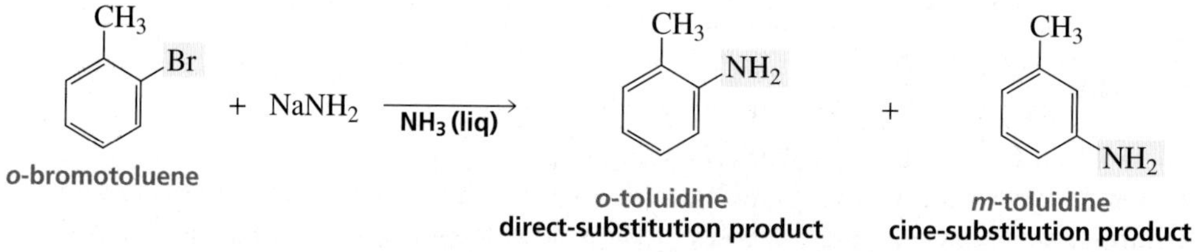

Benzyne is an extremely reactive species. In Section 6.3, we saw that in a molecule with a triple bond, the two *sp* hybridized carbons and the atoms attached to these carbons (C—C≡C—C) are linear because the bond angles are 180°. Four linear atoms cannot be incorporated into a six-membered ring, so the C—C≡C—C system in benzyne is distorted: The original π bond is unchanged, but the sp^2 orbitals that form the new bond are not parallel to one another (Figure 16.4). Therefore, they cannot overlap as well as the *p* orbitals that form a normal π bond, resulting in a much weaker and much more reactive bond.

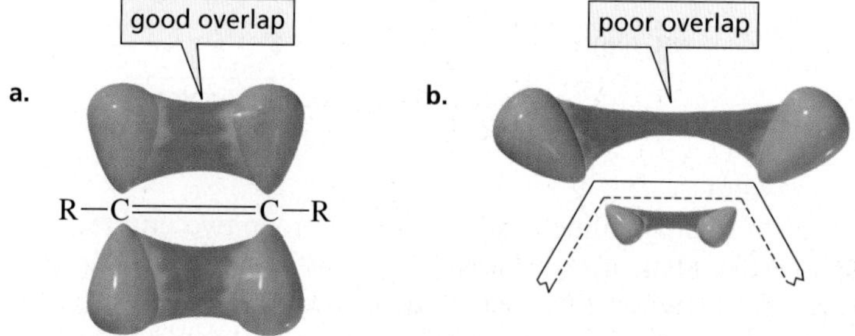

Figure 16.4 ▶
Orbital pictures of the bond formed (a) by overlapping *p* orbitals in a normal triple bond and (b) by overlapping sp^2 orbitals in the distorted "triple bond" in benzyne.

Although benzyne is too unstable to be isolated, evidence that it is formed can be obtained by a trapping experiment. When furan is added to a reaction that forms a benzyne intermediate, furan traps the benzyne intermediate by reacting with it in a Diels–Alder reaction (Section 8.8). The product of the Diels–Alder reaction can be isolated.

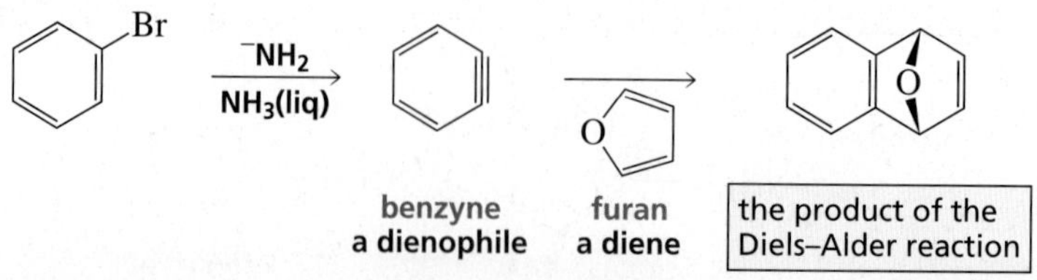

A synthesis employing a benzyne intermediate is a good way to prepare *meta*-amino-substituted aryl ethers (see Problem 31). The amide ion preferentially attacks the meta position because there is less steric hindrance at this position and the resulting negative charge can be stabilized by the adjacent electron-withdrawing oxygen atom.

PROBLEM 31

Starting with benzene, show how *meta*-aminoanisole could be prepared via a synthesis that

a. involves a benzyne intermediate b. does not involve a benzyne intermediate

PROBLEM 32

Give the products that would be obtained from the reaction of the following compounds with sodium amide in liquid ammonia:

a. 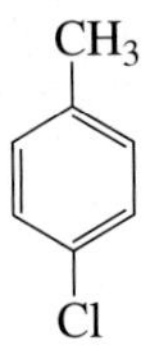 CH_3 / Cl b. 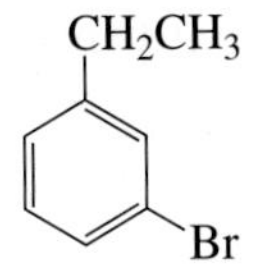 CH_2CH_3 / Br c. 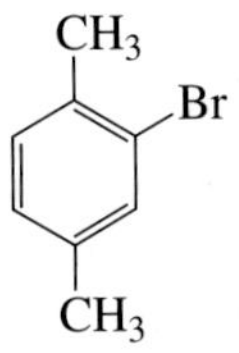 CH_3 / Br / CH_3 d.  OH / Br

16.15 Polycyclic Benzenoid Hydrocarbons

Polycyclic benzenoid hydrocarbons are compounds that contain two or more fused benzene rings. **Fused rings** share two adjacent carbons—naphthalene has two fused rings, anthracene and phenanthrene have three fused rings, and tetracene, triphenylene, pyrene, and chrysene have four fused rings. There are many polycyclic benzenoid hydrocarbons with more than four fused rings.

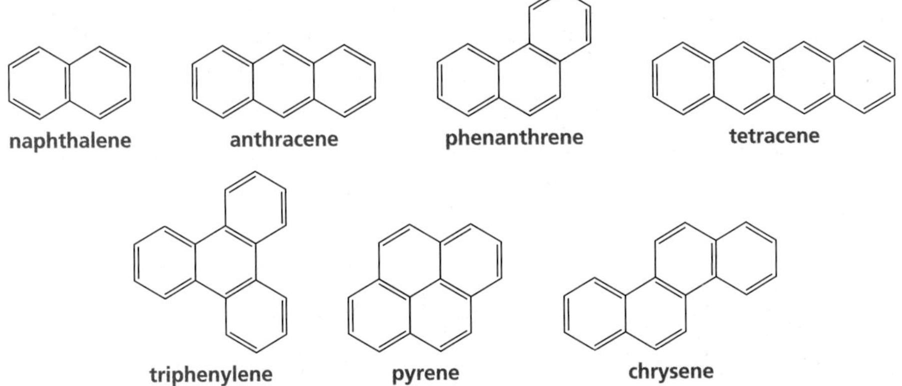

Like benzene, all the larger polycyclic benzenoid hydrocarbons undergo electrophilic substitution reactions. Some of these compounds are well-known carcinogens. The chemical reactions responsible for causing cancer and how you can predict which compounds are carcinogenic were discussed in Section 12.8.

16.16 Electrophilic Substitution Reactions of Naphthalene and Substituted Naphthalenes

Like benzene, naphthalene is an aromatic hydrocarbon. Naphthalene has three resonance contributors.

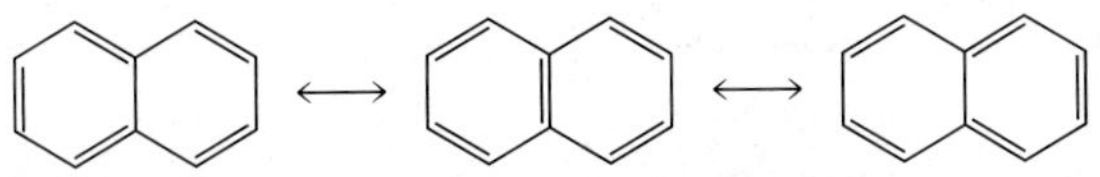

$\longleftrightarrow$ $\longleftrightarrow$

resonance contributors of naphthalene

Also like benzene, naphthalene undergoes electrophilic aromatic substitution reactions. Substitution occurs preferentially at the 1-position. In common nomenclature, the 1-position is called the α-position and the 2-position is called the β-position.

$$\text{naphthalene} + HNO_3 \xrightarrow{H_2SO_4} \text{1-nitronaphthalene} + H_2O$$

1-nitronaphthalene
α-nitronaphthalene

3-D Molecule: Naphthalene

Naphthalene is more reactive than benzene toward electrophilic substitution because the carbocation intermediate is more stable and, therefore, easier to form than the analogous carbocation intermediate formed from benzene. Because of naphthalene's increased reactivity, a Lewis acid is not needed for bromination or chlorination.

$$\text{naphthalene} + Br_2 \longrightarrow \text{1-bromonaphthalene} + HBr$$

1-bromonaphthalene

PROBLEM 33

Use the resonance contributors of naphthalene to predict whether all the carbon–carbon bonds have the same length, as they do in benzene.

PROBLEM 34

Draw the resonance contributors for the carbocation intermediates obtained from electrophilic aromatic substitution at the 1-position and the 2-position of naphthalene. Use the resonance contributors to explain why substitution at the 1-position is preferred.

Sulfonation of naphthalene does not always lead to substitution at the 1-position. If the reaction is carried out under conditions which cause it to be irreversible (80 °C), substitution occurs at the 1-position.

$$\text{naphthalene} + H_2SO_4 \xrightarrow{80\ ^\circ C} \text{naphthalene-1-sulfonic acid} + H_2O$$

naphthalene-1-sulfonic acid

If, however, the reaction is carried out under conditions which cause it to be readily reversible (160 °C), substitution occurs predominantly at the 2-position.

$$\text{naphthalene-1-sulfonic acid} \rightleftharpoons \text{naphthalene} + H_2SO_4 \xrightarrow{160\ ^\circ C} \text{naphthalene-2-sulfonic acid}$$

naphthalene-1-sulfonic acid
kinetic product

naphthalene-2-sulfonic acid
thermodynamic product

3-D Molecules: Naphthalene-1-sulfonic acid; Naphthalene-2-sulfonic acid

This is another example of a reaction whose product composition depends on whether the conditions used in the experiment cause the reaction to be irreversible (under kinetic control) or reversible (under thermodynamic control). (See Section 8.7.) The 1-substituted product is the kinetic product because it is easier to form. It is the predominant product, therefore, when the reaction is carried out under conditions that cause it to be irreversible (mild reaction conditions). The 2-substituted product is the thermodynamic product because it is more stable. It is the predominant product when the reaction is carried out under conditions that cause it to be reversible (higher temperatures).

The 1-substituted product is easier to form because the carbocation leading to its formation is more stable. The 2-substituted product is more stable because there is more room for the bulky sulfonic acid group at the 2-position. In the 1-substituted product, the sulfonic acid group is too close to the hydrogen at the 8-position.

1-Substituted naphthalenes are easier to form; 2-substituted naphthalenes are more stable.

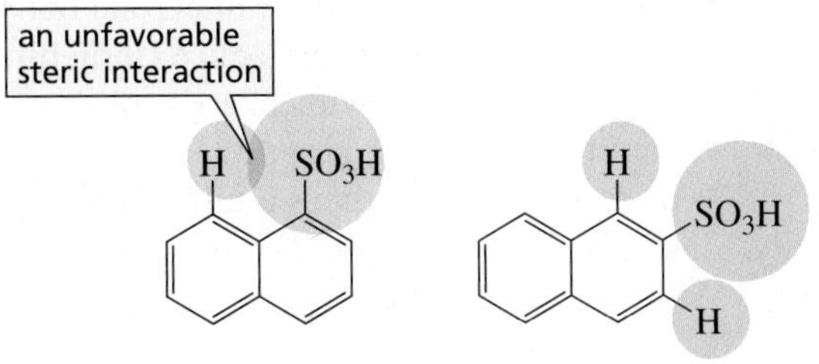

an unfavorable steric interaction

H SO_3H H SO_3H H

In the case of substituted naphthalenes, the nature of the substituent determines which ring will undergo electrophilic substitution. If the substituent is deactivating, the electrophile will attack the 1-position of the ring that lacks the deactivating substituent, because that ring is more reactive than the ring that contains the deactivating substituent.

NO_2 + HNO_3 $\xrightarrow{H_2SO_4}$ NO_2, NO_2

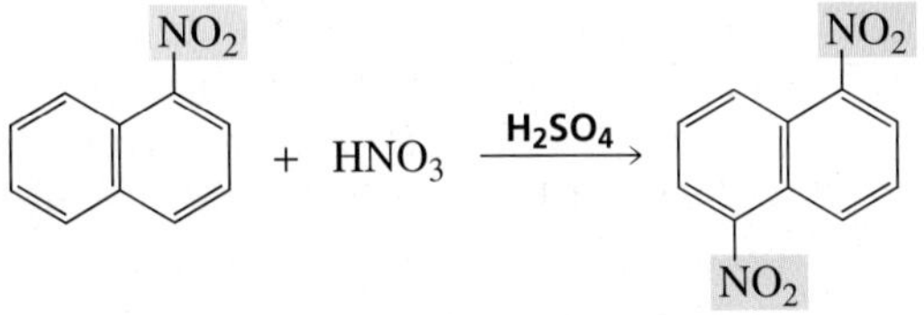

If the substituent is activating, the electrophile will attack the ring that contains the substituent. The incoming electrophile will be directed to a 1-position that is ortho or para to the substituent, because all activating substituents are ortho–para directors (Section 16.4).

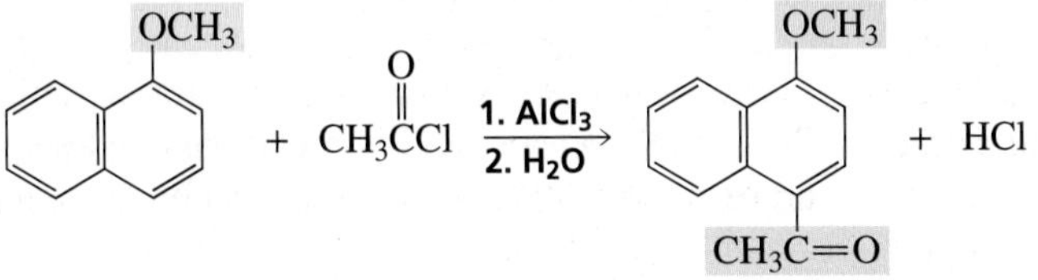

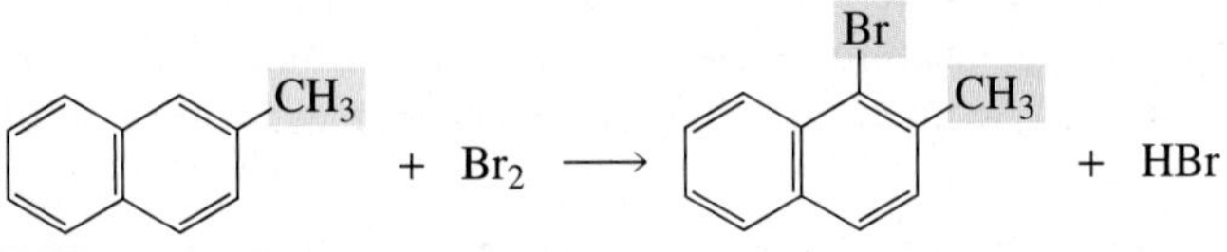

OCH_3 + $CH_3\overset{O}{\overset{\|}{C}}Cl$ $\xrightarrow[2.\ H_2O]{1.\ AlCl_3}$ OCH_3, $CH_3C{=}O$ + HCl

CH_3 + Br_2 $\longrightarrow$ Br, CH_3 + HBr

PROBLEM 35

Give the products that would be obtained from the reaction of the following compounds with Cl_2:

a. 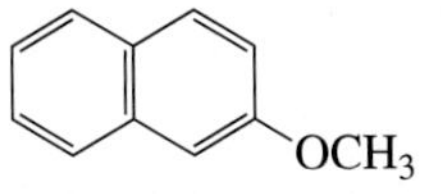 OCH_3

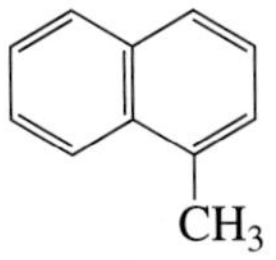

b. 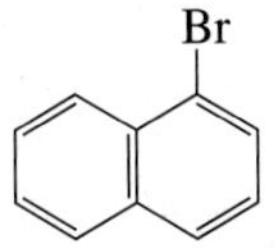 Br

c. CH_3

d. 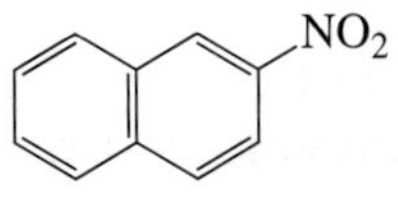 NO_2

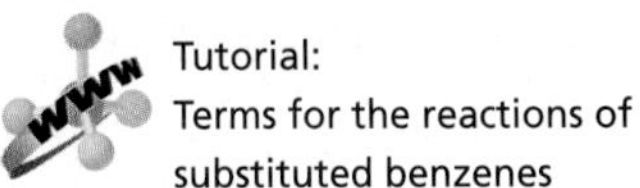

Tutorial:
Terms for the reactions of substituted benzenes

Summary

Benzene rings with substituents other than halo, nitro, sulfonic acid, alkyl, and acyl can be prepared by first synthesizing one of these substituted benzenes and then chemically changing the substituent. The kinds of substituents that can be placed on benzene rings are greatly expanded by reactions of arene diazonium salts, nucleophilic aromatic substitution reactions, and reactions involving a benzyne intermediate. The relative positions of two substituents on a benzene ring are indicated either by numbers or by the prefixes *ortho*, *meta*, and *para*.

The nature of the substituent affects both the reactivity of the benzene ring and the placement of an incoming substituent: The rate of electrophilic aromatic substitution is increased by electron-donating substituents and decreased by electron-withdrawing substituents. Substituents can donate or withdraw electrons **inductively** or by **resonance**.

The stability of the carbocation intermediate determines to which position a substituent directs an incoming electrophile. All activating substituents and the weakly deactivating halogens are **ortho–para directors**; all substituents more deactivating than the halogens are **meta directors**. Ortho–para directors—with the exception of alkyl, aryl, and CH=CHR—have a lone pair on the atom attached to the ring; meta directors have a positive or partial positive charge on the atom attached to the ring. Ortho–para-directing substituents form the para isomer preferentially if either the substituent or the incoming electrophile is large.

In synthesizing disubstituted benzenes, the order in which the substituents are placed on the ring and the point in a reaction sequence at which a substituent is chemically modified are important considerations. When a disubstituted benzene undergoes an electrophilic aromatic substitution reaction, the directing effect of both substituents has to be considered.

RO- and HO-substituted benzenes are halogenated without the Lewis acid. Benzene rings with meta-directing substituents cannot undergo Friedel–Crafts reactions. Aniline and *N*-substituted anilines also cannot undergo Friedel–Crafts reactions.

Aniline and substituted-anilines react with nitrous acid to form **arenediazonium salts**; a diazonium group can be displaced by a nucleophile. Arenediazonium ions can be used as electrophiles with highly activated benzene rings to form **azo compounds** that can exist in cis and trans forms. Secondary amines react with nitrous acid to form **nitrosamines**.

An aryl halide with one or more substituents ortho or para to the leaving group that strongly withdraw electrons by resonance undergoes a **nucleophilic aromatic substitution** (S_NAr) reaction: The nucleophile forms a resonance-stabilized carbanion intermediate, and then the leaving group departs, reestablishing the aromaticity of the ring. The incoming nucleophile must be a stronger base than the substituent being replaced. A substituent that deactivates a benzene ring toward electrophilic substitution activates it toward nucleophilic substitution.

In the presence of a strong base, an aryl halide undergoes a nucleophilic substitution reaction via a **benzyne intermediate**. After a hydrogen halide is eliminated, the nucleophile can attack either of the carbons of the distorted "triple bond" in benzyne. **Direct substitution** is substitution at the carbon that was attached to the leaving group; **cine substitution** is substitution at the adjacent carbon.

The ability of a substituent to withdraw electrons from or donate electrons into a benzene ring is reflected in the pK_a values of substituted phenols, benzoic acids, and protonated anilines: Electron withdrawal increases acidity; electron donation decreases acidity.

Polycyclic benzenoid hydrocarbons contain two or more fused benzene rings; **fused rings** share two adjacent carbons. Polycyclic benzenoid hydrocarbons undergo electrophilic aromatic substitution reactions. Naphthalene undergoes irreversible substitution predominantly at the 1-position and reversible substitution predominantly at the 2-position. The nature of the substituent determines which ring of a substituted-naphthalene undergoes electrophilic substitution.

Summary of Reactions

1. Reactions of substituents on a benzene ring (Section 16.2)

$$C_6H_5CH_3 \xrightarrow[\text{peroxide}]{\text{NBS, }\Delta} C_6H_5CH_2Br \xrightarrow{Z^-} C_6H_5CH_2Z \qquad Z^- = \text{a nucleophile}$$

$$C_6H_5NO_2 \xrightarrow{\text{Sn, HCl}} C_6H_5\overset{+}{N}H_3Cl^- \xrightarrow{HO^-} C_6H_5NH_2$$

$$C_6H_5CH_3 \xrightarrow[\Delta]{Na_2Cr_2O_7,\ H^+} C_6H_5COOH$$

2. Reactions of amines with nitrous acid (Section 16.12)

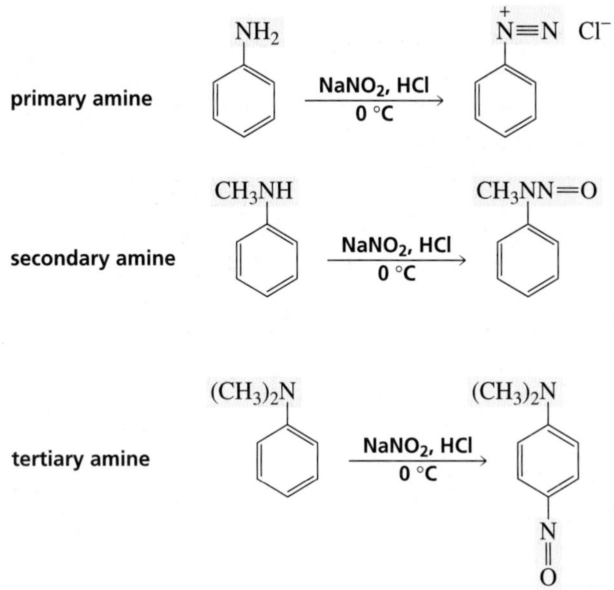

primary amine

$$C_6H_5NH_2 \xrightarrow[0\ °C]{NaNO_2,\ HCl} C_6H_5\overset{+}{N}{\equiv}N\ Cl^-$$

secondary amine

$$C_6H_5NHCH_3 \xrightarrow[0\ °C]{NaNO_2,\ HCl} C_6H_5N(CH_3)N{=}O$$

tertiary amine

$$C_6H_5N(CH_3)_2 \xrightarrow[0\ °C]{NaNO_2,\ HCl} p\text{-}O{=}N{-}C_6H_4N(CH_3)_2$$

3. Replacement of a diazonium group (Section 16.10)

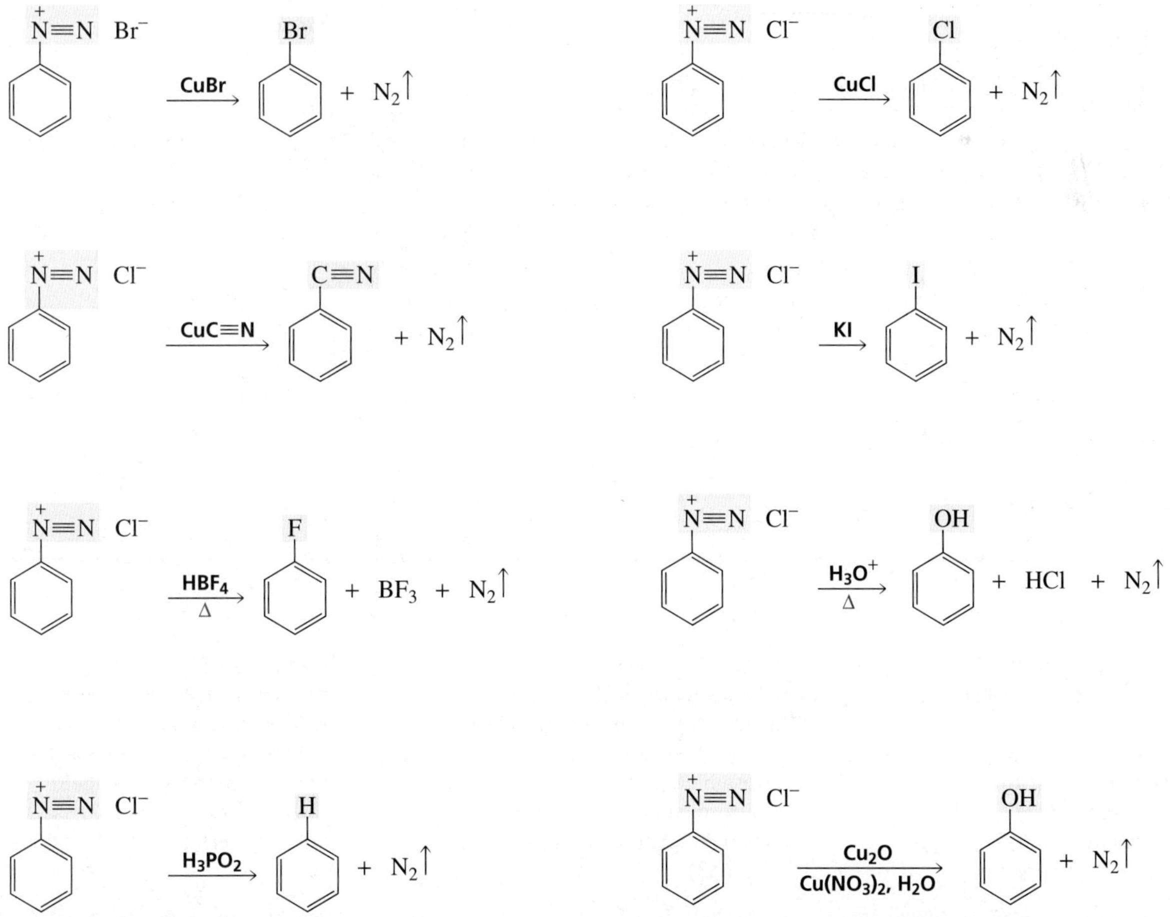

$$C_6H_5\overset{+}{N}{\equiv}N\ Br^- \xrightarrow{CuBr} C_6H_5Br + N_2\uparrow$$

$$C_6H_5\overset{+}{N}{\equiv}N\ Cl^- \xrightarrow{CuCl} C_6H_5Cl + N_2\uparrow$$

$$C_6H_5\overset{+}{N}{\equiv}N\ Cl^- \xrightarrow{CuC{\equiv}N} C_6H_5C{\equiv}N + N_2\uparrow$$

$$C_6H_5\overset{+}{N}{\equiv}N\ Cl^- \xrightarrow{KI} C_6H_5I + N_2\uparrow$$

$$C_6H_5\overset{+}{N}{\equiv}N\ Cl^- \xrightarrow[\Delta]{HBF_4} C_6H_5F + BF_3 + N_2\uparrow$$

$$C_6H_5\overset{+}{N}{\equiv}N\ Cl^- \xrightarrow[\Delta]{H_3O^+} C_6H_5OH + HCl + N_2\uparrow$$

$$C_6H_5\overset{+}{N}{\equiv}N\ Cl^- \xrightarrow{H_3PO_2} C_6H_5H + N_2\uparrow$$

$$C_6H_5\overset{+}{N}{\equiv}N\ Cl^- \xrightarrow[Cu(NO_3)_2,\ H_2O]{Cu_2O} C_6H_5OH + N_2\uparrow$$

4. Formation of an azo compound (Section 16.11)

OH (phenol) + $\overset{+}{N}\equiv N\ Cl^-$ (benzenediazonium chloride) ⟶ HO–C₆H₄–N=N–C₆H₅

5. Nucleophilic aromatic substitution reactions (Section 16.13)

p-bromonitrobenzene + CH_3O^- $\xrightarrow{\Delta}$ p-methoxynitrobenzene (OCH_3, NO_2) + Br^-

2,4-dinitrochlorobenzene (Cl, NO_2, NO_2) + NH_3 $\xrightarrow{\Delta}$ ($^+NH_3Cl^-$, NO_2, NO_2) $\xrightarrow{HO^-}$ (NH_2, NO_2, NO_2) + H_2O

6. Nucleophilic substitution via a benzyne intermediate (Section 16.14)

(CH_3, Br) + $^-NH_2$ $\xrightarrow{NH_3\ (liq)}$ (CH_3, NH_2) + (CH_3, NH_2) + Br^-

direct-substitution product

cine-substitution product

7. Electrophilic aromatic substitution reactions of naphthalene (Section 16.16)

1-methoxynaphthalene (OCH_3) + $CH_3\overset{O}{\overset{\|}{C}}Cl$ $\xrightarrow[2.\ H_2O]{1.\ AlCl_3}$ (OCH_3, $CH_3C{=}O$) + HCl

2-methylnaphthalene (CH_3) + Br_2 ⟶ (Br, CH_3) + HBr

Key Terms

activating substituent (p. 631)
arenediazonium salt (p. 646)
azo linkage (p. 649)
benzyne intermediate (p. 655)
cine substitution (p. 656)
deactivating substituent (p. 631)
direct substitution (p. 656)
donate electrons by resonance (p. 630)
fused rings (p. 657)
inductive electron donation (p. 630)
inductive electron withdrawal (p. 630)
meta director (p. 635)
nitrosamine (p. 651)
N-nitroso compound (p. 651)
nucleophilic aromatic substitution (p. 653)
ortho–para director (p. 635)
resonance electron donation (p. 630)
resonance electron withdrawal (p. 630)
Sandmeyer reaction (p. 646)
Schiemann reaction (p. 648)
S_NAr reaction (p. 654)
withdraw electrons by resonance (p. 631)

Problems

36. Draw the structure of each of the following compounds:
    a. *m*-ethylphenol
    b. *p*-nitrobenzenesulfonic acid
    c. (*E*)-2-phenyl-2-pentene
    d. *o*-bromoaniline
    e. 2-chloroanthracene
    f. *m*-chlorostyrene
    g. *o*-nitroanisole
    h. 2,4-dichlorotoluene

37. Name the following compounds:

a.  COOH, Br
b. 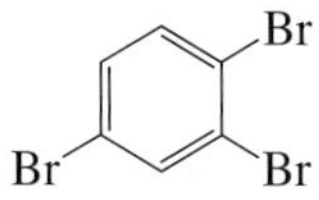 Br, Br, Br
c. OH, H_3C, CH_3
d. $CH{=}CH_2$, NO_2
e. OCH_3, CH_2CH_3
f. SO_3H, Cl, Cl
g. CH_3, Br
h. CH_3
i. CH_2CH_3, Cl, N=N

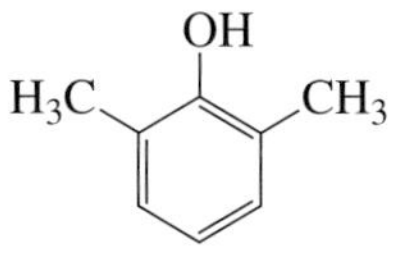

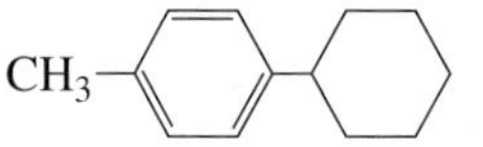

38. Provide the necessary reagents next to the arrows.

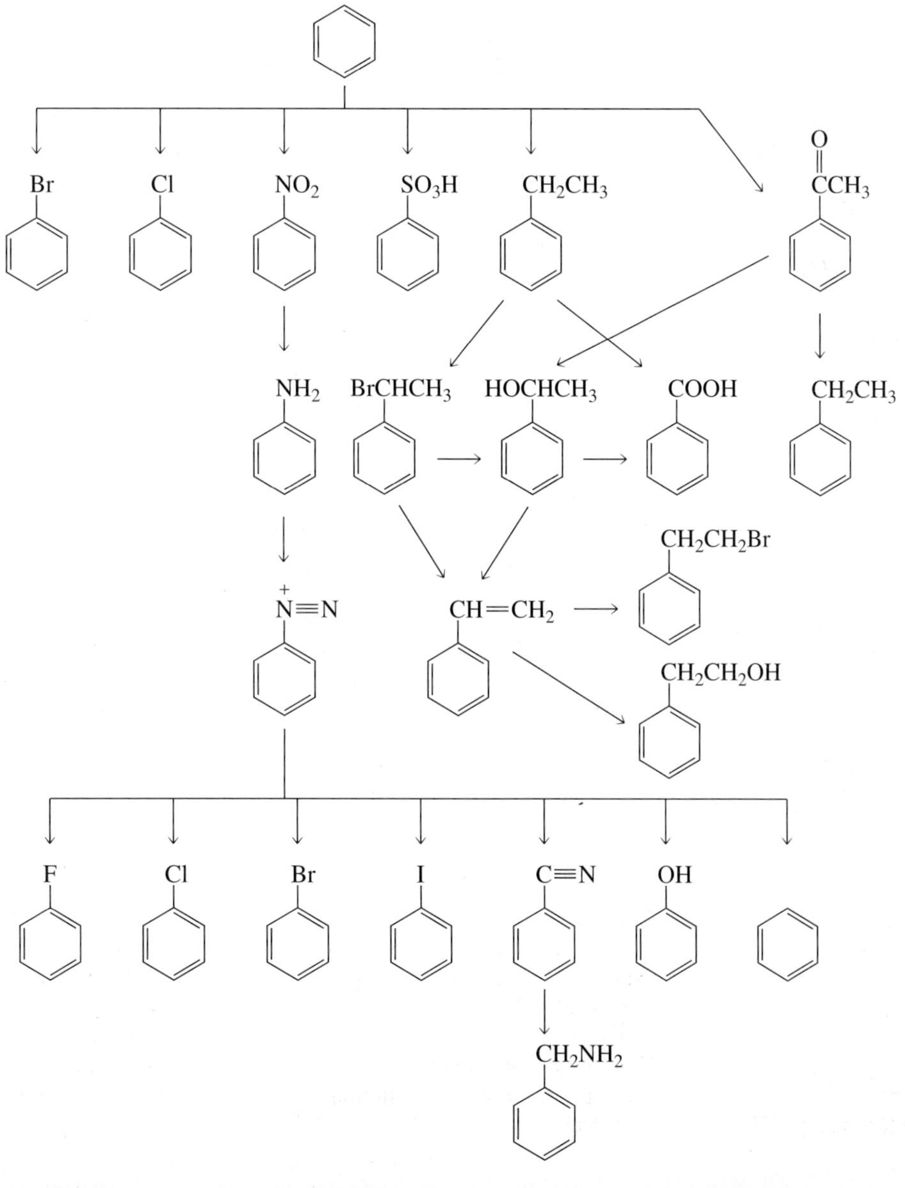

Br, Cl, NO_2, SO_3H, CH_2CH_3, $\overset{O}{\overset{\|}{C}}CH_3$

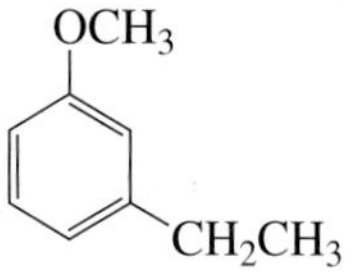

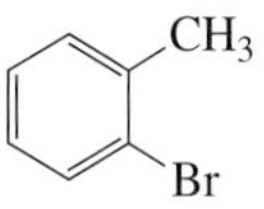

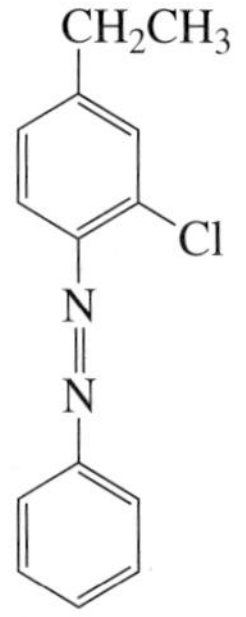

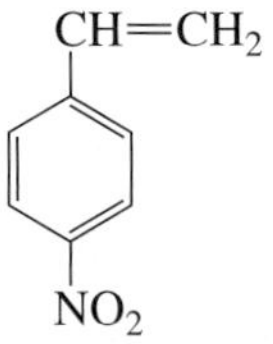

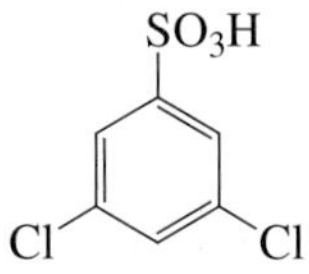

NH_2, $BrCHCH_3$, $HOCHCH_3$, COOH, CH_2CH_3

CH_2CH_2Br

$\overset{+}{N}{\equiv}N$, $CH{=}CH_2$

CH_2CH_2OH

F, Cl, Br, I, $C{\equiv}N$, OH

CH_2NH_2

39. The pK_a values of a few ortho-, meta-, and para-substituted benzoic acids are shown here:

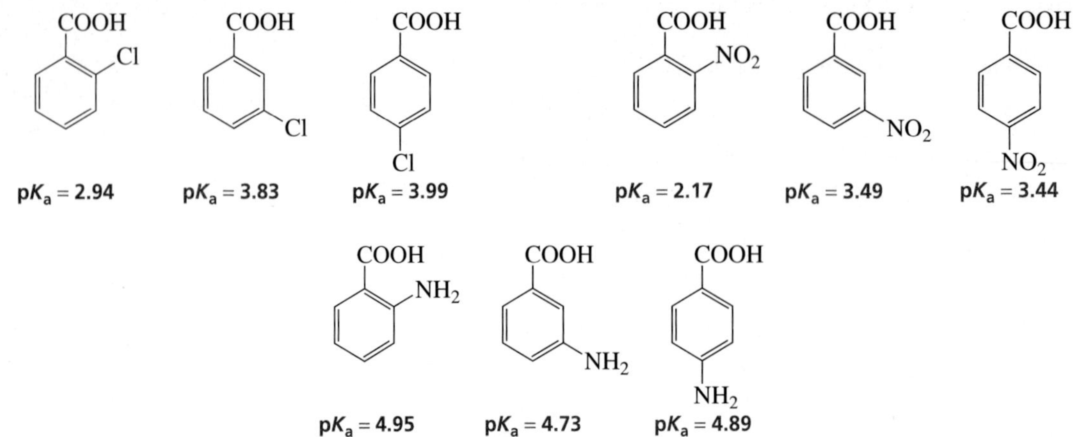

The relative pK_a values depend on the substituent. For chloro-substituted benzoic acids, the ortho isomer is the most acidic and the para isomer is the least acidic; for nitro-substituted benzoic acids, the ortho isomer is the most acidic and the meta isomer is the least acidic; and for amino-substituted benzoic acids, the meta isomer is the most acidic and the ortho isomer is the least acidic. Explain these relative acidities.

Cl: ortho > meta > para $\qquad$ NO_2: ortho > para > meta $\qquad$ NH_2: meta > para > ortho

40. Give the product(s) of each of the following reactions:

a. benzoic acid + HNO_3/H_2SO_4
b. isopropylbenzene + cyclohexene + HF
c. naphthalene + acetyl chloride + $AlCl_3$ followed by H_2O
d. *o*-methylaniline + benzenediazonium chloride
e. cyclohexyl phenyl ether + $Br_2/FeBr_3$
f. phenol + H_2SO_4 + Δ
g. ethylbenzene + $Br_2/FeBr_3$
h. *m*-xylene + $Na_2Cr_2O_7$ + H^+ + Δ

41. Rank the following anions in order of decreasing basicity:

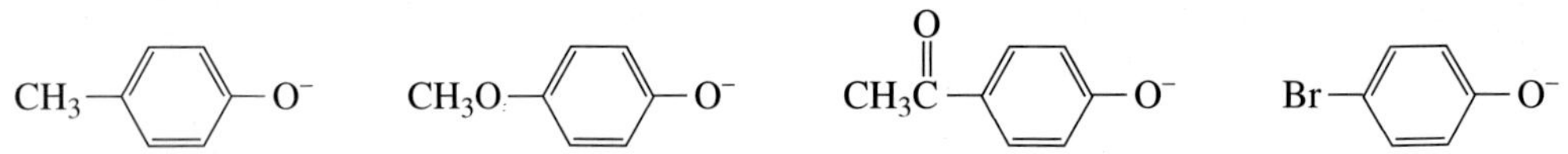

CH_3–C₆H₄–O^- $\qquad$ CH_3O–C₆H₄–O^- $\qquad$ $CH_3C(=O)$–C₆H₄–O^- $\qquad$ Br–C₆H₄–O^-

42. The compound with the following ^{1}H NMR spectrum is known to be highly reactive toward electrophilic aromatic substitution. Identify the compound.

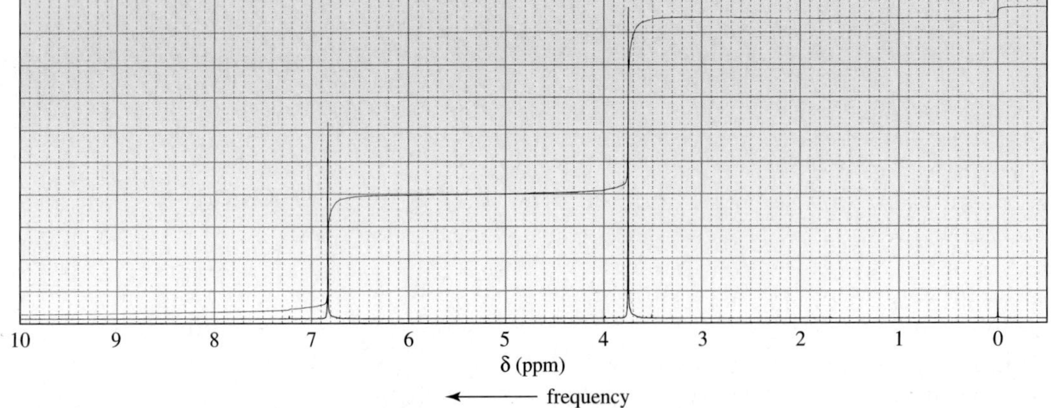

43. Show how the following compounds could be synthesized from benzene:

a. *m*-chlorobenzenesulfonic acid
b. *m*-chloroethylbenzene
c. benzyl alcohol
d. *m*-bromobenzonitrile
e. 1-phenylpentane
f. *m*-hydroxybenzoic acid
g. *m*-bromobenzoic acid
h. *p*-cresol

44. For each of the groups of substituted benzenes, indicate

a. the one that would be the most reactive in an electrophilic aromatic substitution reaction

b. the one that would be the least reactive in an electrophilic aromatic substitution reaction
c. the one that would yield the highest percentage of meta product

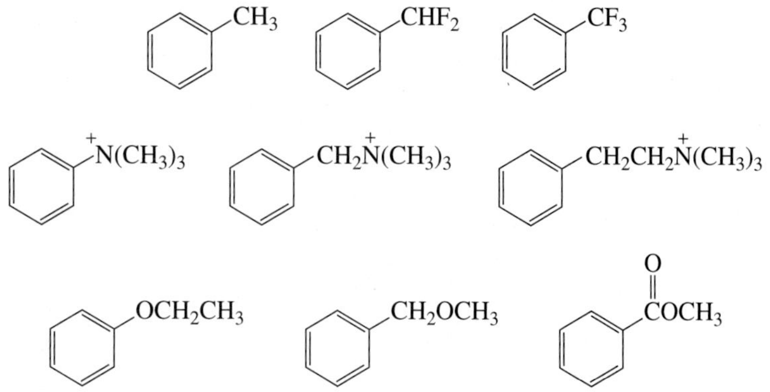

$C_6H_5CH_3$ $C_6H_5CHF_2$ $C_6H_5CF_3$

$C_6H_5\overset{+}{N}(CH_3)_3$ $C_6H_5CH_2\overset{+}{N}(CH_3)_3$ $C_6H_5CH_2CH_2\overset{+}{N}(CH_3)_3$

$C_6H_5OCH_2CH_3$ $C_6H_5CH_2OCH_3$ $C_6H_5COCH_3$

45. Arrange the following groups of compounds in order of decreasing reactivity toward electrophilic aromatic substitution:
a. benzene, ethylbenzene, chlorobenzene, nitrobenzene, anisole
b. 1-chloro-2,4-dinitrobenzene, 2,4-dinitrophenol, 2,4-dinitrotoluene
c. toluene, *p*-cresol, benzene, *p*-xylene
d. benzene, benzoic acid, phenol, propylbenzene
e. *p*-nitrotoluene, 2-chloro-4-nitrotoluene, 2,4-dinitrotoluene, *p*-chlorotoluene
f. bromobenzene, chlorobenzene, fluorobenzene, iodobenzene

46. Give the products of the following reactions:

a. 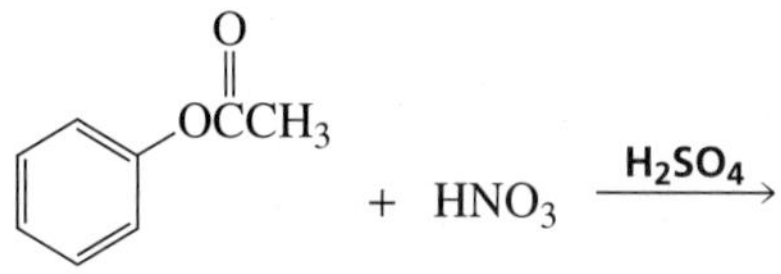 $C_6H_5OCCH_3$ (O) + HNO_3 $\xrightarrow{H_2SO_4}$

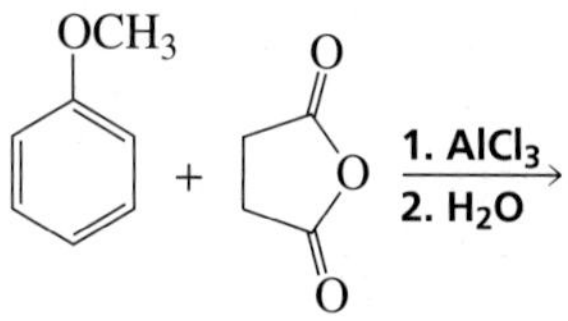

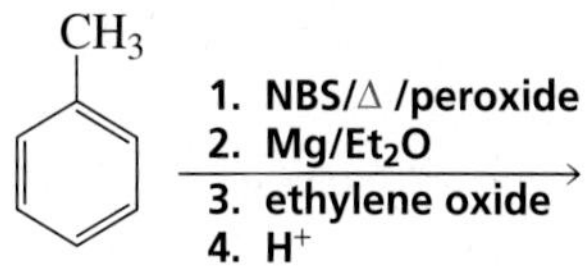

b. 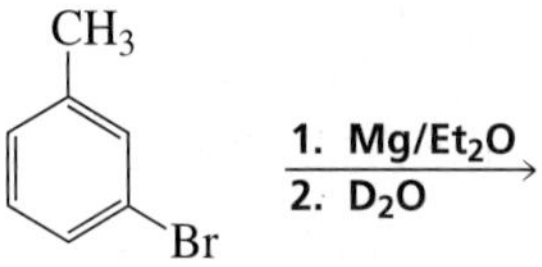 CH_3, Br (meta-substituted benzene) 1. Mg/Et_2O 2. D_2O

c. $C_6H_5OCH_3$ + (succinic anhydride: O, O, O) 1. $AlCl_3$ 2. H_2O

d. (N-phenylpiperidine: N) + Br_2 $\longrightarrow$

e. $C_6H_5CH_3$ 1. NBS/Δ /peroxide 2. Mg/Et_2O 3. ethylene oxide 4. H^+

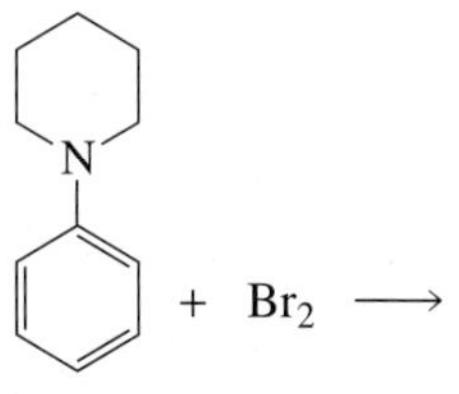

f. 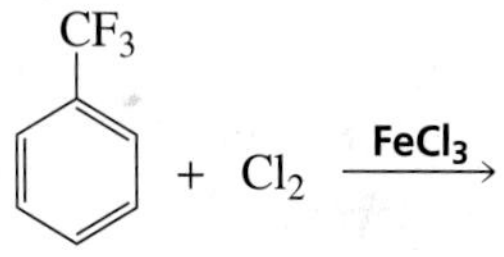 $C_6H_5CF_3$ + Cl_2 $\xrightarrow{FeCl_3}$

47. For each of the statements in Column I, choose a substituent from Column II that fits the description for the compound at the right:

Column I
a. Z donates electrons inductively, but does not donate or withdraw electrons by resonance.
b. Z withdraws electrons inductively and withdraws electrons by resonance.
c. Z deactivates the ring and directs ortho–para.
d. Z withdraws electrons inductively, donates electrons by resonance, and activates the ring.
e. Z withdraws electrons inductively, but does not donate or withdraw electrons by resonance.

Column II
OH
Br
$^+NH_3$
CH_2CH_3
NO_2

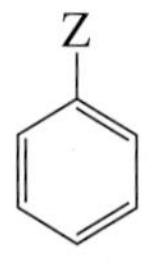

C_6H_5Z

48. For each of the following compounds, indicate the ring carbon that would be nitrated if the compound were treated with HNO_3/H_2SO_4:

a. 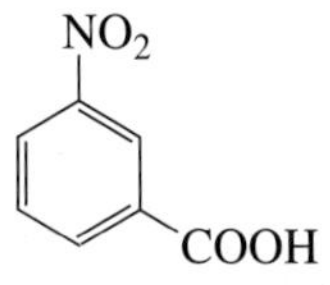 NO_2, COOH (meta-substituted benzene)
b. 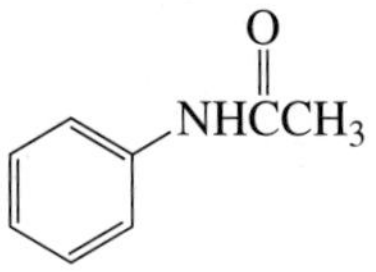 $C_6H_5NHCCH_3$ (O)
c. 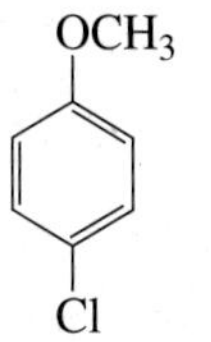 OCH_3, Cl (para-substituted benzene)
d. 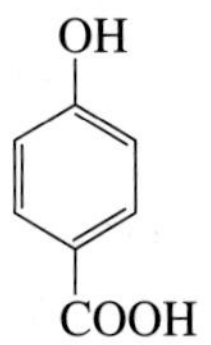 OH, COOH (para-substituted benzene)

e. 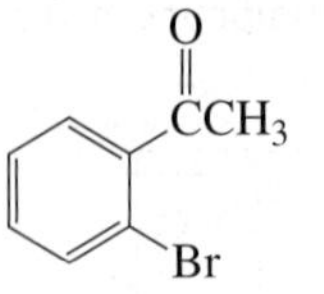 (o-bromophenyl methyl ketone: benzene ring with $\overset{O}{\|}CCH_3$ and Br)

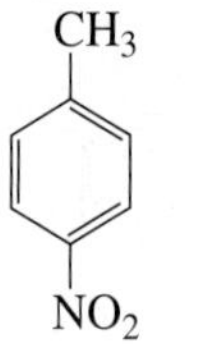

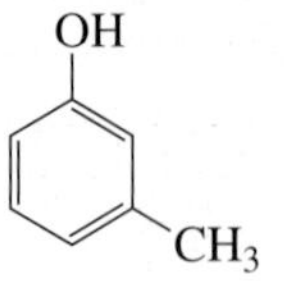

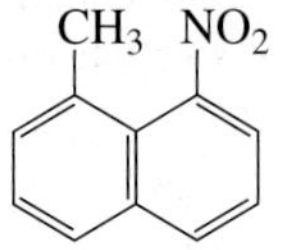

f. 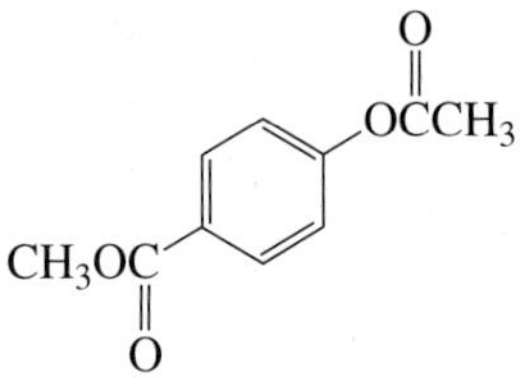 ($CH_3O\overset{O}{\|}C$– benzene ring –$O\overset{O}{\|}CCH_3$)

g. (benzene ring with CH_3 and NO_2 para)

h. 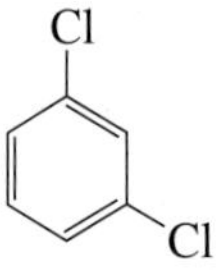 (benzene ring with Cl and Cl meta)

i. (benzene ring with OH and CH_3 meta)

j. 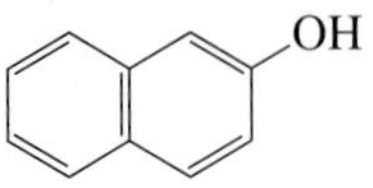 (naphthalene with OH)

k. (naphthalene with CH_3 and NO_2)

l. 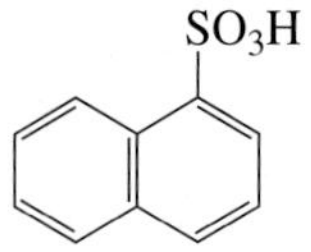 (naphthalene with SO_3H)

49. Show how the following compounds could be synthesized from benzene:

a. *N,N,N*-trimethylanilinium iodide
b. benzyl methyl ether
c. *p*-benzylchlorobenzene
d. 2-methyl-4-nitrophenol
e. *p*-nitroaniline
f. *m*-bromoiodobenzene
g. *p*-dideuteriobenzene
h. *p*-nitro-*N*-methylaniline

50. Which of the following compounds will react with HBr more rapidly?

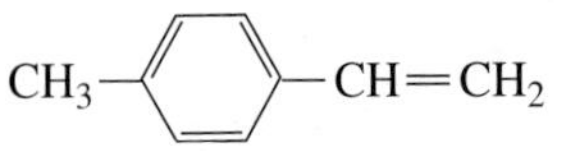

CH_3–C_6H_4–$CH{=}CH_2$ or CH_3O–C_6H_4–$CH{=}CH_2$

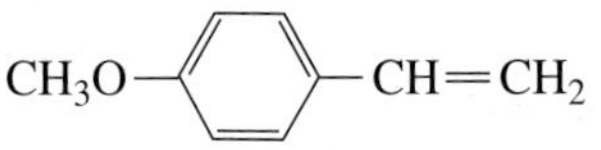

51. Give the product(s) obtained from the reaction of each of the following compounds with $Br_2/FeBr_3$:

a. 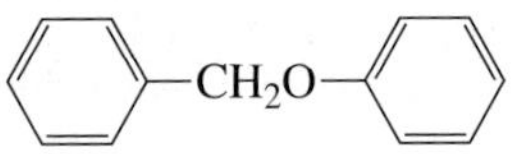 C_6H_5–CH_2O–C_6H_5

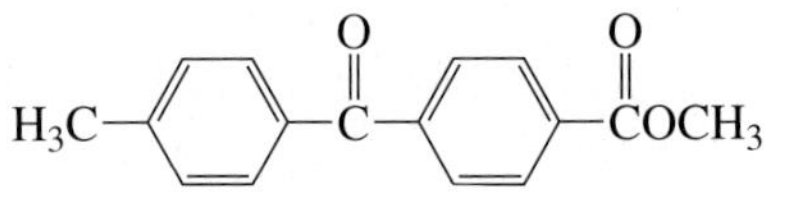

b. 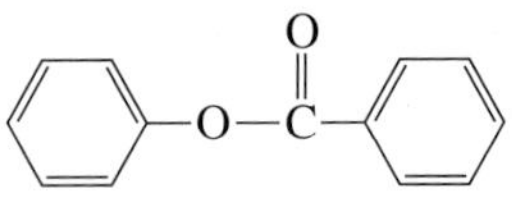 C_6H_5–O–$\overset{O}{\|}C$–C_6H_5

c. H_3C–C_6H_4–$\overset{O}{\|}C$–C_6H_4–$\overset{O}{\|}COCH_3$

d. 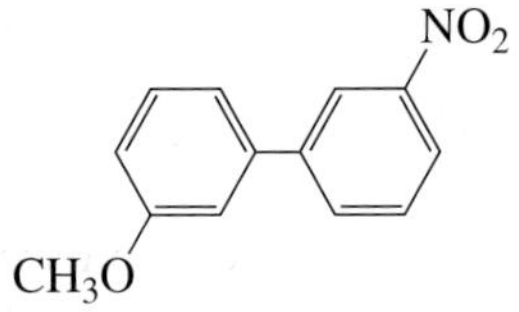 (biphenyl with NO_2 on one ring and CH_3O on the other)

52. Which would react more rapidly with Cl_2 + $FeCl_3$, *m*-xylene or *p*-xylene? Explain.

53. What products would be obtained from the reaction of the following compounds with $Na_2Cr_2O_7 + H^+ + \Delta$?

a. 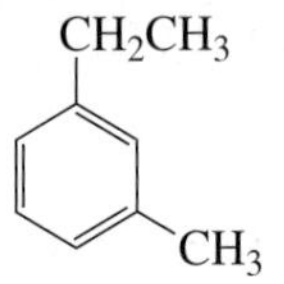 (benzene ring with CH_2CH_3 and CH_3 meta)

b. 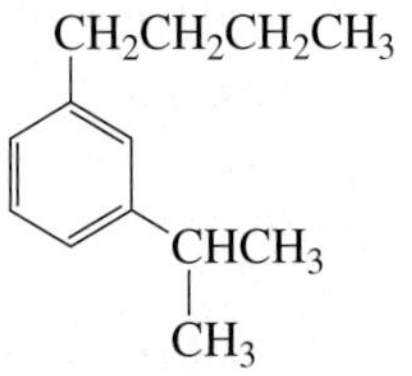 (benzene ring with $CH_2CH_2CH_2CH_3$ and $CHCH_3$ bearing CH_3, meta)

c. 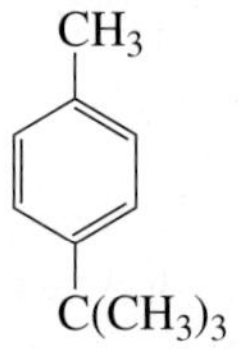 (benzene ring with CH_3 and $C(CH_3)_3$ para)

54. A student had prepared three ethyl-substituted benzaldehydes, but had neglected to label them. The premed student at the next bench said that they could be identified by brominating a sample of each and determining how many bromo-substituted products were formed. Is the premed student's advice sound?

55. Explain, using resonance contributors for the intermediate carbocation, why a phenyl group is an ortho–para director.

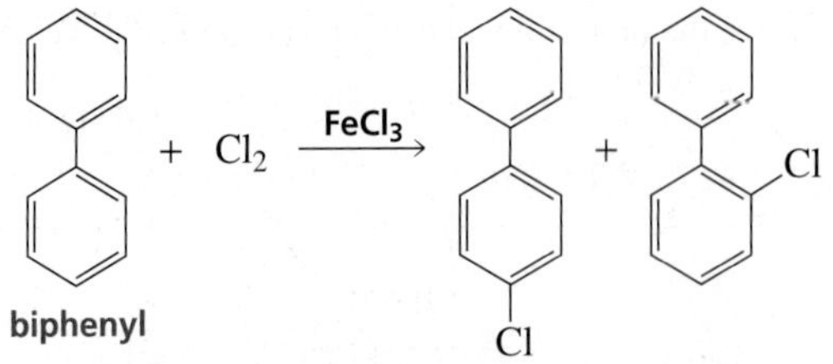

biphenyl + Cl_2 $\xrightarrow{FeCl_3}$ (4-chlorobiphenyl) + (2-chlorobiphenyl)

56. When heated with an acidic solution of sodium dichromate, compound A forms benzoic acid. Identify compound A from its ^{1}H NMR spectrum.

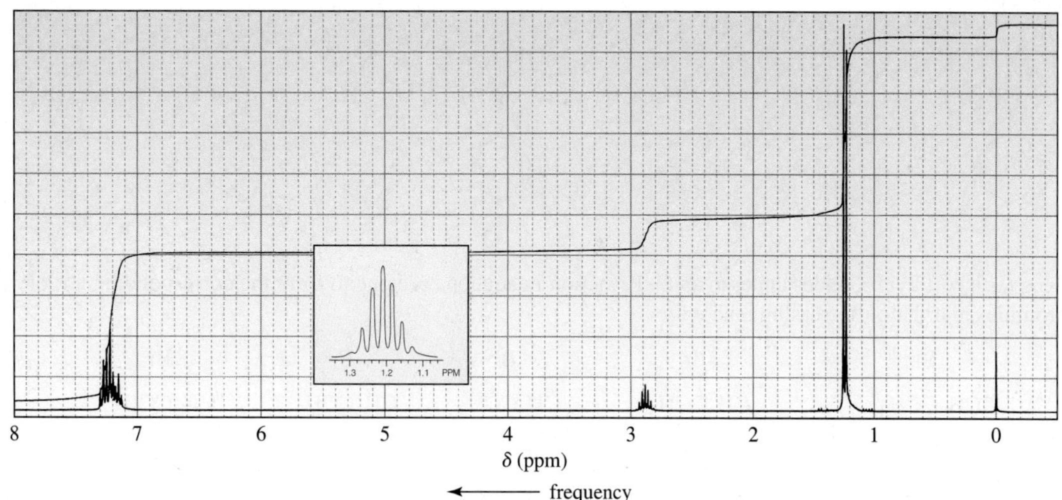

57. Describe two synthetic routes for the preparation of *p*-methoxyaniline with benzene as the starting material.

58. Which is a more stable intermediate?

a. (ring with OH; H, NO_2; +) or (ring with CH_3; H, NO_2; +)

b. (ring with NO_2; HO, Cl; −) or (ring with NO_2; Cl, OH; −)

59. If phenol is allowed to sit in D_2O that contains a small amount of D_2SO_4, what products would be formed?

60. Show how the following compounds could be prepared from benzene:

a. (ring with CH_3 and $CH_3C{=}CH_2$)

b. (ring with OCH_3, Br and NO_2)

c. (ring with $CH_3\overset{O}{\overset{\|}{C}}$, SO_3H and $CH_2CH(CH_3)_2$)

61. An unknown compound reacts with ethyl chloride and aluminum trichloride to form a compound that has the following ^{1}H NMR spectrum. Give the structure of the compound.

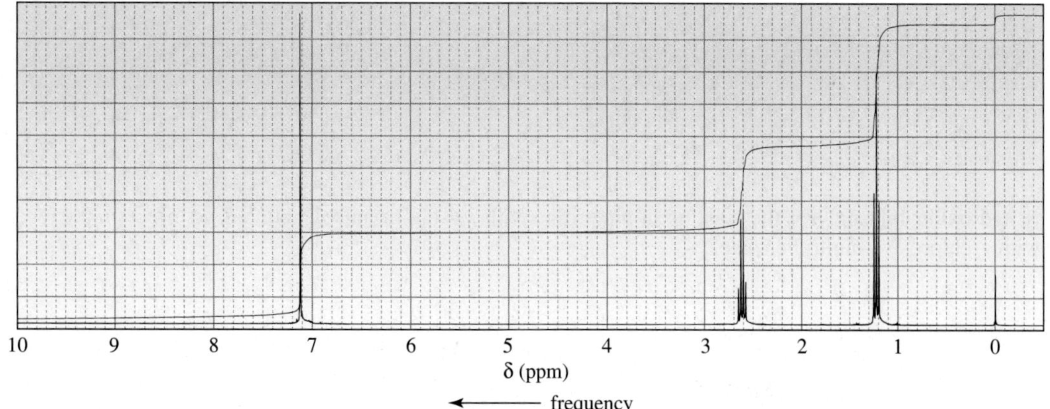

62. How could you distinguish among the following compounds, using
   a. their infrared spectra?
   b. their ^{1}H NMR spectra?

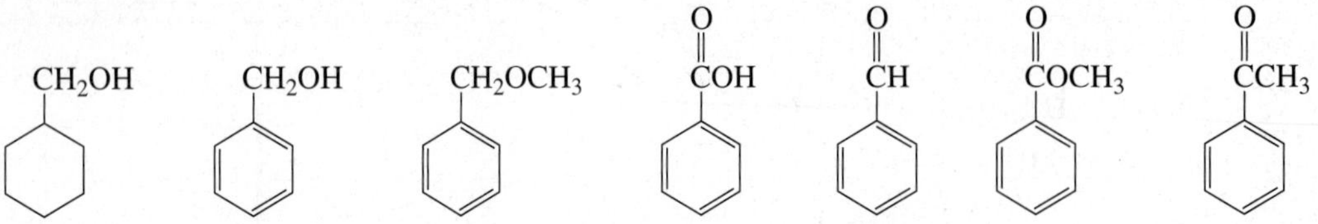

A: CH_2OH (cyclohexyl) B: CH_2OH C: CH_2OCH_3 D: $\overset{O}{\parallel}$COH E: $\overset{O}{\parallel}$CH F: $\overset{O}{\parallel}$COCH$_3$ G: $\overset{O}{\parallel}$CCH$_3$

63. The following tertiary alkyl bromides undergo an S_N1 reaction in aqueous acetone to form the corresponding tertiary alcohols. List the alkyl bromides in order of decreasing reactivity.

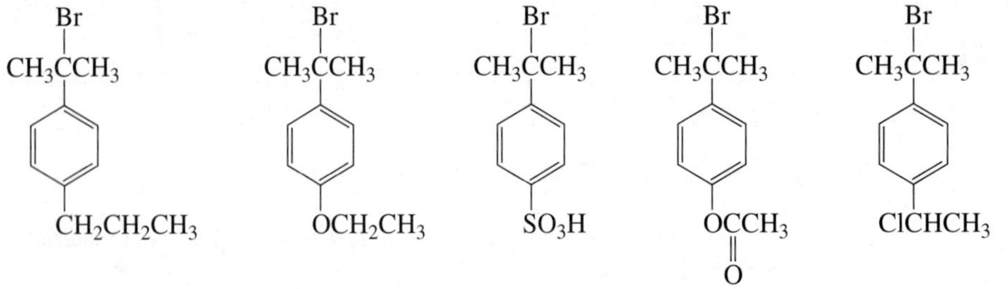

A: CH_3CBrCH_3 / $CH_2CH_2CH_3$ B: CH_3CBrCH_3 / OCH_2CH_3 C: CH_3CBrCH_3 / SO_3H D: CH_3CBrCH_3 / $OCCH_3$ (C=O) E: CH_3CBrCH_3 / $ClCHCH_3$

64. *p*-Fluoronitrobenzene is more reactive than *p*-chloronitrobenzene toward hydroxide ion. What does this tell you about the rate-determining step for nucleophilic aromatic substitution?

65. a. Explain why the following reaction leads to the products shown:

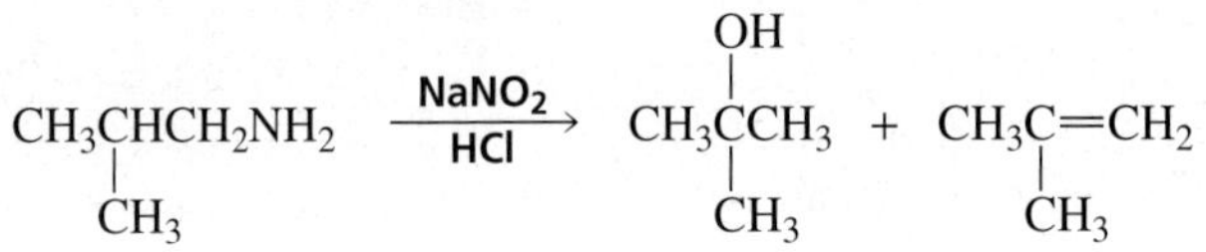

$$CH_3CH(CH_3)CH_2NH_2 \xrightarrow[HCl]{NaNO_2} CH_3C(OH)(CH_3)CH_3 + CH_3C(CH_3){=}CH_2$$

b. What product would be obtained from the following reaction?

$$CH_3-C(OH)(CH_3)-C(NH_2)(CH_3)-CH_3 \xrightarrow[HCl]{NaNO_2}$$

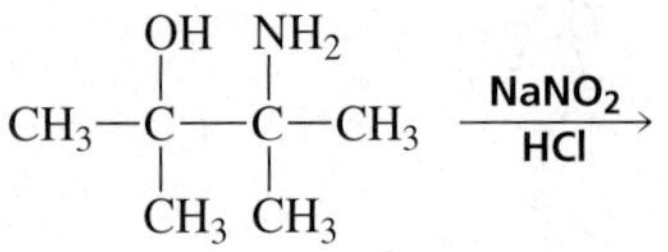

66. Describe how mescaline could be synthesized from benzene. The structure of mescaline is given on p. 622.

67. Propose a mechanism for the following reaction that explains why the configuration of the asymmetric carbon in the reactant is retained in the product:

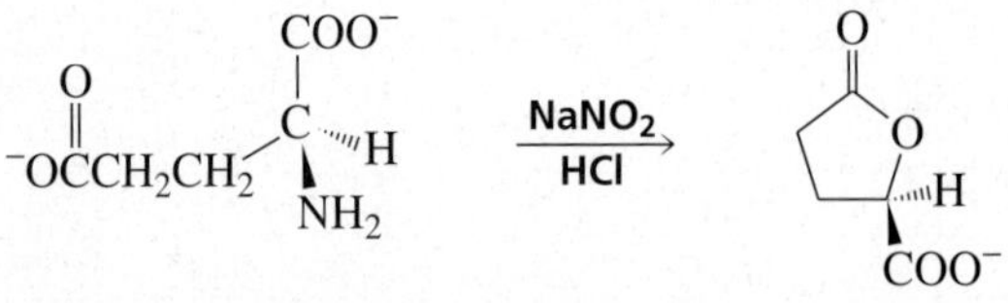

$$^{-}OOCCH_2CH_2C(H)(NH_2)COO^{-} \xrightarrow[HCl]{NaNO_2} \text{lactone (H, } COO^{-})$$

68. Explain why hydroxide ion catalyzes the reaction of piperidine with 2,4-dinitroanisole, but has no effect on the reaction of piperidine with 1-chloro-2,4-dinitrobenzene.

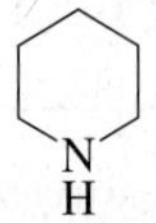

N–H

piperidine

PART SIX

Carbonyl Compounds and Amines

The five chapters in Part Six focus on the reactions of compounds that contain a carbonyl group. Carbonyl compounds can be placed in one of two classes: those that contain a group that can be replaced by another group (Class I) and those that do *not* contain a group that can be replaced by another group (Class II).

Class I carbonyl compounds undergo *nucleophilic acyl substitution* reactions. These reactions are discussed in **Chapter 17**, where you will see that all Class I carbonyl compounds react with nucleophiles in the same way—they form an unstable tetrahedral intermediate that collapses by eliminating the weakest base. So all you need to know to determine the product of a reaction—or even whether a reaction will occur—is the relative basicity of the groups in the tetrahedral intermediate.

The first part of **Chapter 18** compares Class I and Class II carbonyl compounds by discussing their reactions with good nucleophiles—carbon nucleophiles and hydride ion. You will see that while Class I carbonyl compounds undergo *nucleophilic acyl substitution* reactions with these nucleophiles, Class II carbonyl compounds undergo *nucleophilic acyl addition* reactions. The chapter goes on to discuss the reactions of Class II carbonyl compounds with poorer nucleophiles—nitrogen and oxygen nucleophiles. You will see that Class II carbonyl compounds undergo *nucleophilic addition–elimination* reactions with these nucleophiles, forming unstable tetrahedral intermediates that preferentially eliminate the weakest base. In other words, what you learned about the partitioning of tetrahedral intermediates formed by Class I carbonyl compounds is revisited with Class II carbonyl compounds. The reactions of α,β-unsaturated Class I and Class II carbonyl compounds is also discussed.

Many carbonyl compounds have two sites of reactivity: the carbonyl group *and* the α-carbon. Chapter 17 and Chapter 18 discuss the reactions of carbonyl compounds that take place at the carbonyl group. **Chapter 19** examines the reactions of carbonyl compounds that take place at the α-carbon.

You have seen a variety of redox reactions in previous chapters. **Chapter 20** introduces you to more redox reactions. **Chapter 21** extends the coverage of amines that has been presented in earlier chapters and discusses the chemistry of heterocyclic compounds

17 Carbonyl Compounds I

Nucleophilic Acyl Substitution

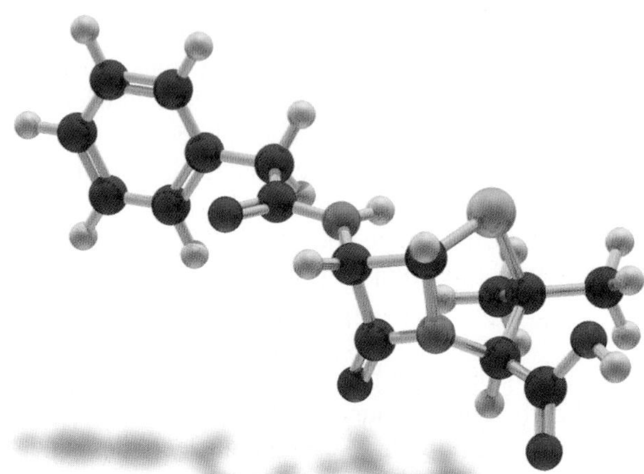
Penicillin G

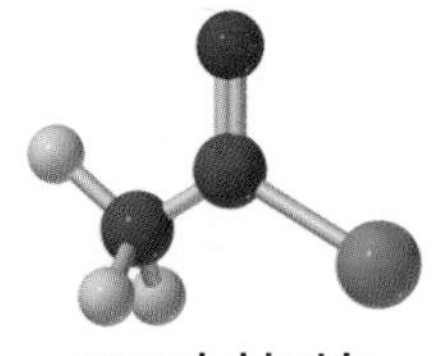
an acyl chloride

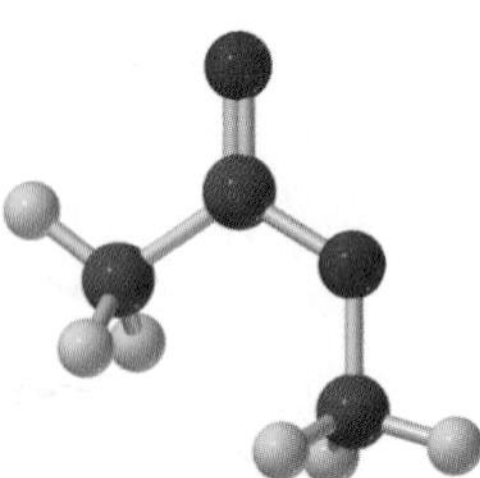
an ester

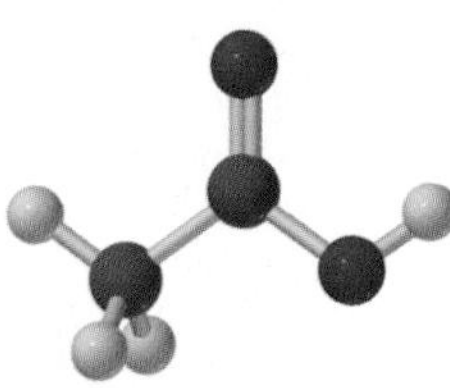
a carboxylic acid

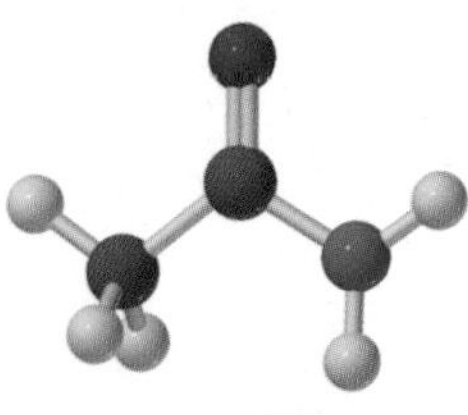
an amide

The **carbonyl group**—a carbon double bonded to an oxygen—is probably the most important functional group found in organic compounds. Compounds containing carbonyl groups— called **carbonyl compounds**— are abundant in nature. Many play important roles in biological processes. Hormones, vitamins, amino acids, drugs, and flavorings are just a few of the carbonyl compounds that affect us daily. An **acyl group** consists of a carbonyl group attached to an alkyl group or to an aryl group.

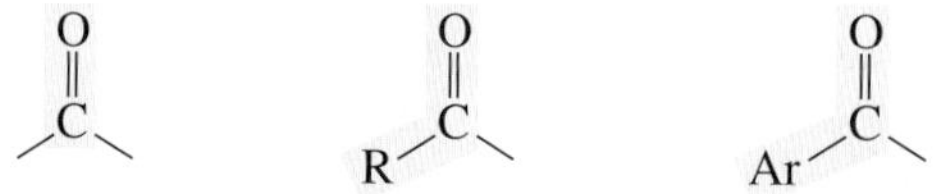

a carbonyl group   acyl groups

The substituents attached to the acyl group strongly affect the reactivity of carbonyl compounds. Carbonyl compounds can be divided into two classes. Class I carbonyl compounds are those in which the acyl group is attached to an atom or a group that *can* be replaced by another group. Carboxylic acids, acyl halides, acid anhydrides, esters, and amides belong to this class. All of these compounds contain a group (—OH, —Cl, —Br, —O(CO)R, —OR, $—NH_2$, —NHR, or $—NR_2$) that can be replaced by a nucleophile. Acyl halides, acid anhydrides, esters, and amides are all called **carboxylic acid derivatives** because they differ from a carboxylic acid only in the nature of the group that has replaced the OH group of the carboxylic acid.

compounds with groups that can be replaced by a nucleophile

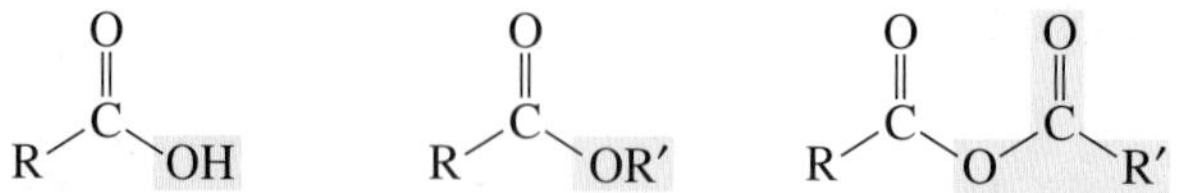

a carboxylic acid

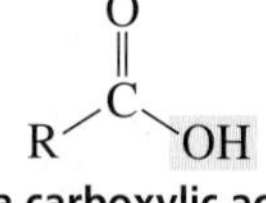
an ester

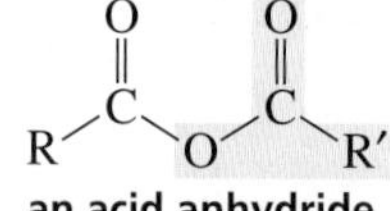
an acid anhydride

compounds with groups that can be replaced by a nucleophile

$R\text{-}C(=O)\text{-}Cl$ an acyl chloride; $R\text{-}C(=O)\text{-}Br$ an acyl bromide — acyl halides

$R\text{-}C(=O)\text{-}NH_2$, $R\text{-}C(=O)\text{-}NHR'$, $R\text{-}C(=O)\text{-}NR'_2$ — amides

Class II carbonyl compounds are those in which the acyl group is attached to a group that *cannot* be readily replaced by another group. Aldehydes and ketones belong to this class. The —H and alkyl or aryl (—R or —Ar) groups of aldehydes and ketones cannot be replaced by a nucleophile.

cannot be replaced by a nucleophile

$R\text{-}C(=O)\text{-}H$ an aldehyde; $R'\text{-}C(=O)\text{-}R$ a ketone

In Chapter 10, we saw that the likelihood of a group's being replaced by another group depends on the relative basicities of the two groups: *The weaker the basicity of a group, the better its leaving ability*. Recall from Section 10.3 that weak bases are good leaving groups because weak bases do not share their electrons as well as strong bases do.

The pK_a values of the conjugate acids of the leaving groups of various carbonyl compounds are listed in Table 17.1. Notice that the acyl groups of Class I carbonyl compounds are attached to weaker bases than are the acyl groups of Class II carbonyl compounds. (Remember that the lower the pK_a, the stronger the acid and the weaker its conjugate base.) The —H of an aldehyde and the alkyl or aryl (—R or —Ar) group of a ketone are too basic to be replaced by another group.

This chapter discusses the reactions of Class I carbonyl compounds. We will see that these compounds undergo substitution reactions because they have an acyl group attached to a group that can be replaced by a nucleophile. The reactions of aldehydes and ketones will be considered in Chapter 18. Because aldehydes and ketones have an acyl group attached to a group that *cannot* be replaced by a nucleophile, we can correctly predict that these compounds *do not* undergo substitution reactions.

17.1 Nomenclature

Carboxylic Acids

In systematic nomenclature, a **carboxylic acid** is named by replacing the terminal "e" of the alkane name with "oic acid." For example, the one-carbon alkane is methan*e*, so the one-carbon carboxylic acid is methan*oic acid*.

	$H\text{-}C(=O)\text{-}OH$	$CH_3\text{-}C(=O)\text{-}OH$	$CH_3CH_2\text{-}C(=O)\text{-}OH$	$CH_3CH_2CH_2\text{-}C(=O)\text{-}OH$
systematic name:	methanoic acid	ethanoic acid	propanoic acid	butanoic acid
common name:	formic acid	acetic acid	propionic acid	butyric acid

$CH_3CH_2CH_2CH_2\text{-}C(=O)\text{-}OH$	$CH_3CH_2CH_2CH_2CH_2\text{-}C(=O)\text{-}OH$	$CH_2{=}CH\text{-}C(=O)\text{-}OH$	$C_6H_5\text{-}C(=O)\text{-}OH$
pentanoic acid	hexanoic acid	propenoic acid	benzenecarboxylic acid
valeric acid	caproic acid	acrylic acid	benzoic acid

Table 17.1 The pK_a Values of the Conjugate Acids of the Leaving Groups of Carbonyl Compounds

Carbonyl compound	Leaving group	Conjugate acid of the leaving group	pK_a
Class I			
$RC(=O)Br$	Br^-	HBr	–9
$RC(=O)Cl$	Cl^-	HCl	–7
$RC(=O)OC(=O)R$	$^-OC(=O)R$	$RC(=O)OH$	~3–5
$RC(=O)OR'$	$^-OR'$	R′OH	~15–16
$RC(=O)OH$	^-OH	H_2O	15.7
$RC(=O)NH_2$	$^-NH_2$	NH_3	36
Class II			
$RC(=O)H$	H^-	H_2	~40
$RC(=O)R$	R^-	RH	~50

Carboxylic acids containing six or fewer carbons are frequently called by their common names. These names were chosen by early chemists to describe some feature of the compound, usually its origin. For example, formic acid is found in ants, bees, and other stinging insects; its name comes from *formica*, which is Latin for "ant." Acetic acid—contained in vinegar—got its name from *acetum*, the Latin word for "vinegar." Propionic acid is the smallest acid that shows some of the characteristics of the larger fatty acids; its name comes from the Greek words *pro* ("the first") and *pion* ("fat"). Butyric acid is found in rancid butter; the Latin word for "butter" is *butyrum*. Caproic acid is found in goat's milk, and if you have the occasion to smell both a goat and caproic acid, you will find that they have similar odors. *Caper* is the Latin word for "goat."

In systematic nomenclature, the position of a substituent is designated by a number. The carbonyl carbon of a carboxylic acid is always the C-1 carbon. In common nomenclature, the position of a substituent is designated by a lowercase Greek letter, and the carbonyl carbon is not given a designation. The carbon adjacent to the carbonyl carbon is the **α-carbon**, the carbon adjacent to the α-carbon is the β-carbon, and so on.

α = alpha
β = beta
γ = gamma
δ = delta
ϵ = epsilon

$$\underset{6}{CH_3}\underset{5}{CH_2}\underset{4}{CH_2}\underset{3}{CH_2}\underset{2}{CH_2}\underset{1}{C}(=O)OH$$

systematic nomenclature

$$\underset{\epsilon}{CH_3}\underset{\delta}{CH_2}\underset{\gamma}{CH_2}\underset{\beta}{CH_2}\underset{\alpha}{CH_2}C(=O)OH$$

common nomenclature

Take a careful look at the following examples to make sure that you understand the difference between systematic (IUPAC) and common nomenclature:

	$CH_3CH_2CH(OCH_3)C(=O)OH$	$CH_3CH_2CH(Br)CH_2C(=O)OH$	$CH_3CH_2CH(Cl)CH_2CH_2C(=O)OH$
systematic name:	**2-methoxybutanoic acid**	**3-bromopentanoic acid**	**4-chlorohexanoic acid**
common name:	**α-methoxybutyric acid**	**β-bromovaleric acid**	**γ-chlorocaproic acid**

The functional group of a carboxylic acid is called a **carboxyl group**.

$—C(=O)OH$
a carboxyl group

$—COOH$ $—CO_2H$
carboxyl groups are frequently shown in abbreviated forms

Carboxylic acids in which a carboxyl group is attached to a ring are named by adding "carboxylic acid" to the name of the cyclic compound.

cyclohexanecarboxylic acid

COOH, CH_3
***trans*-3-methylcyclopentanecarboxylic acid**

COOH, COOH, COOH
1,2,4-benzenetricarboxylic acid

Acyl Halides

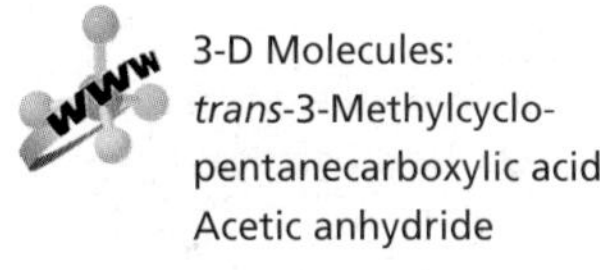

WWW
3-D Molecules:
trans-3-Methylcyclo-
pentanecarboxylic acid;
Acetic anhydride

Acyl halides are compounds that have a halogen atom in place of the OH group of a carboxylic acid. The most common acyl halides are acyl chlorides and acyl bromides. Acyl halides are named by using the acid name and replacing "ic acid" with "yl chloride" (or "yl bromide"). For acids ending with "carboxylic acid," "carboxylic acid" is replaced with "carbonyl chloride" (or "bromide").

	$CH_3C(=O)Cl$	$CH_3CH_2CH(CH_3)CH_2C(=O)Br$	cyclopentyl–$C(=O)Cl$
systematic name:	**ethanoyl chloride**	**3-methylpentanoyl bromide**	**cyclopentanecarbonyl chloride**
common name:	**acetyl chloride**	**β-methylvaleryl bromide**	

Acid Anhydrides

Loss of water from two molecules of a carboxylic acid results in an **acid anhydride**. "Anhydride" means "without water."

$$RC(=O)OH \quad HOC(=O)R \longrightarrow RC(=O)OC(=O)R$$
an acid anhydride

If the two carboxylic acid molecules forming the acid anhydride are the same, the anhydride is a **symmetrical anhydride**. If the two carboxylic acid molecules are different, the anhydride is a **mixed anhydride**. Symmetrical anhydrides are named by using the acid name and replacing "acid" with "anhydride." Mixed anhydrides are named by stating the names of both acids in alphabetical order, followed by "anhydride."

	$CH_3C(=O)OC(=O)CH_3$	$CH_3C(=O)OC(=O)H$
systematic name:	**ethanoic anhydride**	**ethanoic methanoic anhydride**
common name:	**acetic anhydride**	**acetic formic anhydride**
	a symmetrical anhydride	**a mixed anhydride**

Esters

An **ester** is a compound that has an OR′ group in place of the OH group of a carboxylic acid. In naming an ester, the name of the group (R′) attached to the **carboxyl oxygen** is stated first, followed by the name of the acid, with "ic acid" replaced by "ate."

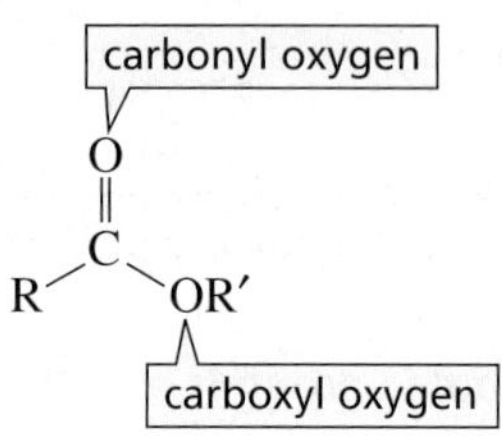

	$CH_3C(=O)OCH_2CH_3$	$CH_3CH_2C(=O)O-C_6H_5$	$CH_3CH(Br)CH_2C(=O)OCH_3$	$C_6H_{11}C(=O)OCH_2CH_3$
systematic name:	**ethyl ethanoate**	**phenyl propanoate**	**methyl 3-bromobutanoate**	**ethyl cyclohexanecarboxylate**
common name:	**ethyl acetate**	**phenyl propionate**	**methyl β-bromobutyrate**	

Salts of carboxylic acids are named in the same way. The cation is named first, followed by the name of the acid, again with "ic acid" replaced by "ate."

Tutorial: Nomenclature of carboxylic acids and their derivatives

	$HC(=O)O^- Na^+$	$CH_3C(=O)O^- K^+$	$C_6H_5C(=O)O^- Na^+$
systematic name:	**sodium methanoate**	**potassium ethanoate**	**sodium benzenecarboxylate**
common name:	**sodium formate**	**potassium acetate**	**sodium benzoate**

Cyclic esters are called **lactones**. In systematic nomenclature, they are named as "2-oxacycloalkanones." Their common names are derived from the common name of the carboxylic acid, which designates the length of the carbon chain, and a Greek letter to indicate the carbon to which the carboxyl oxygen is attached. Thus, four-membered ring lactones are β-lactones (the carboxyl oxygen is on the β-carbon), five-membered ring lactones are γ-lactones, and six-membered ring lactones are δ-lactones.

3-D Molecules: δ-Caprolactone; *cis*-2-Ethylcyclohexane-carboxamide

2-oxacyclopentanone	2-oxacyclohexanone	3-methyl-2-oxacyclohexanone	3-ethyl-2-oxacyclopentanone
γ-butyrolactone	**δ-valerolactone**	**δ-caprolactone**	**γ-caprolactone**

PROBLEM 1

The word "lactone" has its origin in lactic acid, a three-carbon carboxylic acid with an OH group on the α-carbon. Ironically, lactic acid cannot form a lactone. Why not?

Amides

An **amide** has an NH_2, NHR, or NR_2 group in place of the OH group of a carboxylic acid. Amides are named by using the acid name, replacing "oic acid" or "ic acid" with "amide." For acids ending with "carboxylic acid," "ylic acid" is replaced with "amide."

	$CH_3C(=O)NH_2$	$ClCH_2CH_2CH_2C(=O)NH_2$	$C_6H_5C(=O)NH_2$	cyclohexane–$C(=O)NH_2$, 2-CH_2CH_3
systematic name:	ethanamide	4-chlorobutanamide	benzenecarboxamide	*cis*-2-ethylcyclohexanecarboxamide
common name:	acetamide	γ-chlorobutyramide	benzamide	

If a substituent is bonded to the nitrogen, the name of the substituent is stated first (if there is more than one substituent bonded to the nitrogen, they are stated alphabetically), followed by the name of the amide. The name of each substituent is preceded by a capital *N* to indicate that the substituent is bonded to a nitrogen.

$CH_3CH_2C(=O)NH$–cyclohexyl	$CH_3CH_2CH_2CH_2C(=O)N(CH_3)CH_2CH_3$	$CH_3CH_2CH_2C(=O)N(CH_2CH_3)CH_2CH_3$
N-cyclohexylpropanamide	*N*-ethyl-*N*-methylpentanamide	*N*,*N*-diethylbutanamide

Cyclic amides are called **lactams**. Their nomenclature is similar to that of lactones. They are named as "2-azacycloalkanones" in systematic nomenclature ("aza" is used to designate the nitrogen atom). In their common names, the length of the carbon chain is indicated by the common name of the carboxylic acid, and a Greek letter indicates the carbon to which the nitrogen is attached.

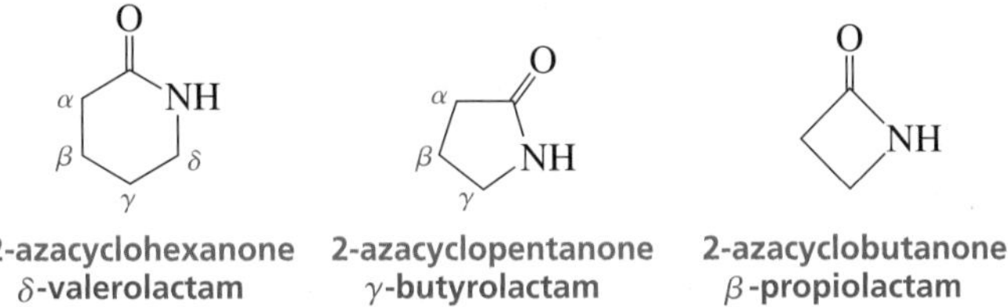

2-azacyclohexanone — δ-valerolactam
2-azacyclopentanone — γ-butyrolactam
2-azacyclobutanone — β-propiolactam

Nitriles

3-D Molecule: Acetonitrile

Nitriles are compounds that contain a C≡N functional group. Nitriles are considered carboxylic acid derivatives because, like all Class I carbonyl compounds, they react with water to form carboxylic acids (Section 17.18). In systematic nomenclature, nitriles are named by adding "nitrile" to the parent alkane name. Notice that the triple-bonded carbon of the nitrile group is counted in the number of carbons in the longest continuous chain. In common nomenclature, nitriles are named by replacing "ic acid" of the carboxylic acid name with "onitrile." They can also be named as alkyl cyanides—stating the name of the alkyl group that is attached to the C≡N group.

	$CH_3C{\equiv}N$	C_6H_5–$C{\equiv}N$	$CH_3CH(CH_3)CH_2CH_2CH_2C{\equiv}N$	$CH_2{=}CHC{\equiv}N$
systematic name:	ethanenitrile	benzenecarbonitrile	5-methylhexanenitrile	propenenitrile
common name:	acetonitrile	benzonitrile	δ-methylcapronitrile	acrylonitrile
	methyl cyanide	**phenyl cyanide**	**isohexyl cyanide**	

PROBLEM 2◆

Name the following compounds:

a. $CH_3CH_2CH_2C{\equiv}N$

b. $CH_3CH_2\overset{\overset{O}{\|}}{C}O\overset{\overset{O}{\|}}{C}CH_3$

c. $CH_3CH_2CH_2\overset{\overset{O}{\|}}{C}O^- \; K^+$

d. $CH_3CH_2CH_2CH_2\overset{\overset{O}{\|}}{C}Cl$

e. $CH_3CH_2CH_2\overset{\overset{O}{\|}}{C}OCH_2\overset{\overset{CH_3}{|}}{C}HCH_3$

f. $CH_3CH_2CH_2CH_2CH_2\overset{\overset{O}{\|}}{C}N(CH_3)_2$

g. (five-membered ring: C=O, NH)

h. (cyclopentane ring with COOH)

i. (six-membered lactone ring: C=O, O, CH_3)

PROBLEM 3

Write a structure for each of the following compounds:

a. phenyl acetate
b. γ-caprolactam
c. butanenitrile
d. *N*-benzylethanamide
e. γ-methylcaproic acid
f. ethyl 2-chloropentanoate
g. β-bromobutyramide
h. propanoic anhydride
i. cyclohexanecarbonyl chloride

17.2 Structures of Carboxylic Acids and Carboxylic Acid Derivatives

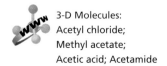

3-D Molecules: Acetyl chloride; Methyl acetate; Acetic acid; Acetamide

The **carbonyl carbon** in carboxylic acids and carboxylic acid derivatives is sp^2 hybridized. It uses its three sp^2 orbitals to form σ bonds to the carbonyl oxygen, the α-carbon, and a substituent (Y). The three atoms attached to the carbonyl carbon are in the same plane, and their bond angles are each approximately 120°.

(~120°, ~120°, ~120° bond angles around the carbonyl carbon C bonded to :O:, the α-carbon, and Y)

The **carbonyl oxygen** is also sp^2 hybridized. One of its sp^2 orbitals forms a σ bond with the carbonyl carbon, and each of the other two sp^2 orbitals contains a lone pair. The remaining p orbital of the carbonyl oxygen overlaps with the remaining p orbital of the carbonyl carbon to form a π bond (Figure 17.1).

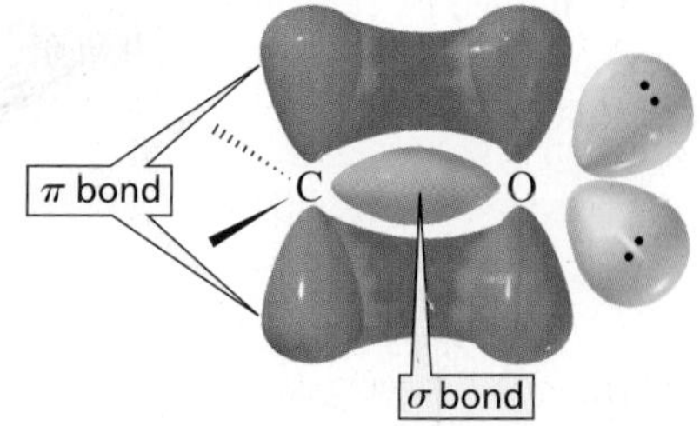

Figure 17.1 ▶ Bonding in a carbonyl group. The π bond is formed by side-to-side overlap of a p orbital of carbon with a p orbital of oxygen.

Esters, carboxylic acids, and amides each have two major resonance contributors. The second resonance contributor for acyl chlorides and acid anyhdrides is much less important.

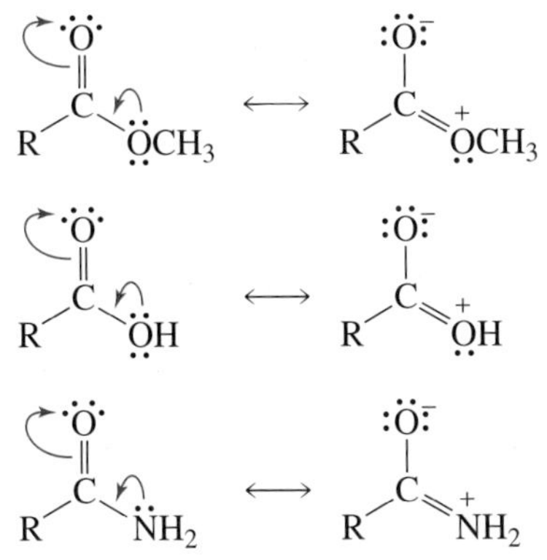

$$RC(=O)OCH_3 \longleftrightarrow RC(-O^-)=\overset{+}{O}CH_3$$

$$RC(=O)OH \longleftrightarrow RC(-O^-)=\overset{+}{O}H$$

$$RC(=O)NH_2 \longleftrightarrow RC(-O^-)=\overset{+}{N}H_2$$

The second resonance contributor (the one on the right) is more important for the amide than it is for the ester or for the carboxylic acid because nitrogen is better than oxygen at sharing its electrons. Nitrogen is less electronegative than oxygen and, therefore, is better able to accommodate a positive charge.

PROBLEM 4◆

Which is longer, the carbon–oxygen single bond in a carboxylic acid or the carbon–oxygen bond in an alcohol? Why?

PROBLEM 5◆

There are three carbon–oxygen bonds in methyl acetate.

a. What are their relative lengths?
b. What are the relative infrared (IR) stretching frequencies of these bonds?

PROBLEM 6

Match the compound with the appropriate carbonyl IR absorption band:

acyl chloride	~1800 and 1750 cm^{-1}
acid anhydride	~1640 cm^{-1}
ester	~1730 cm^{-1}
amide	~1800 cm^{-1}

17.3 Physical Properties of Carbonyl Compounds

The acid properties of carboxylic acids have been discussed previously (Sections 1.18 and 7.10). Recall that carboxylic acids have pK_a values of approximately 3–5 (Appendix II; see also "Special Topics I" in the *Study Guide and Solutions Manual*). The acid properties of dicarboxylic acids will be discussed in Section 17.21. The boiling points and other physical properties of carbonyl compounds are listed in Appendix I. Carbonyl compounds have the following relative boiling points:

relative boiling points

amide > carboxylic acid > nitrile ≫ ester ~ acyl chloride ~ aldehyde ~ ketone

The boiling points of the ester, acyl chloride, ketone, and aldehyde are lower than the boiling point of the alcohol with a comparable molecular weight because the molecules of those carbonyl compounds are unable to form hydrogen bonds with each other. The boiling points of the carbonyl compounds are higher than the boiling point of the ether because of the polar carbonyl group.

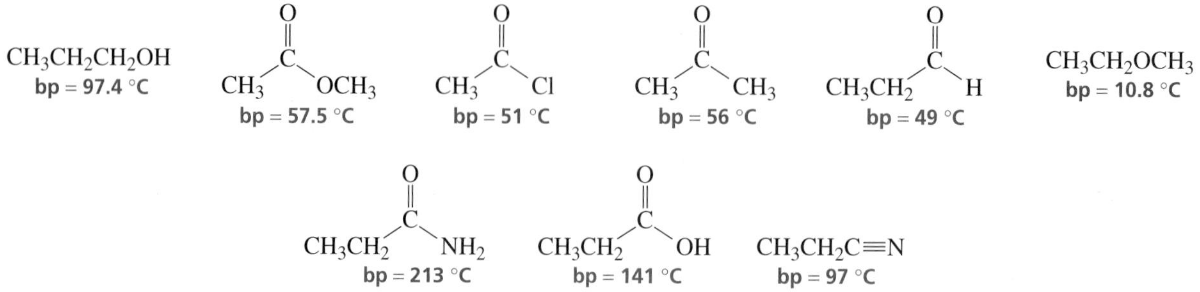

$CH_3CH_2CH_2OH$ bp = 97.4 °C; CH_3COOCH_3 bp = 57.5 °C; CH_3COCl bp = 51 °C; CH_3COCH_3 bp = 56 °C; CH_3CH_2CHO bp = 49 °C; $CH_3CH_2OCH_3$ bp = 10.8 °C

$CH_3CH_2CONH_2$ bp = 213 °C; CH_3CH_2COOH bp = 141 °C; $CH_3CH_2C{\equiv}N$ bp = 97 °C

Carboxylic acids have relatively high boiling points because they form intermolecular hydrogen bonds, giving them larger effective molecular weights.

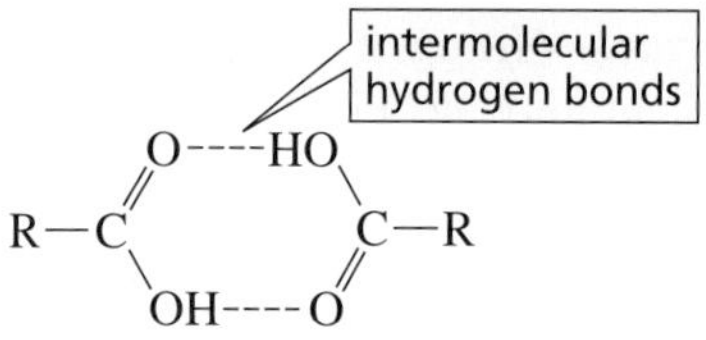

intermolecular hydrogen bonds

Amides have the highest boiling points, because they have strong dipole–dipole interactions since the resonance contributor with separated charges contributes significantly to the overall structure of the compound (Section 17.2). If the nitrogen of an amide is bonded to a hydrogen, hydrogen bonds will form between the molecules. The boiling point of a nitrile is similar to that of an alcohol because a nitrile has strong dipole–dipole interactions (Section 2.9).

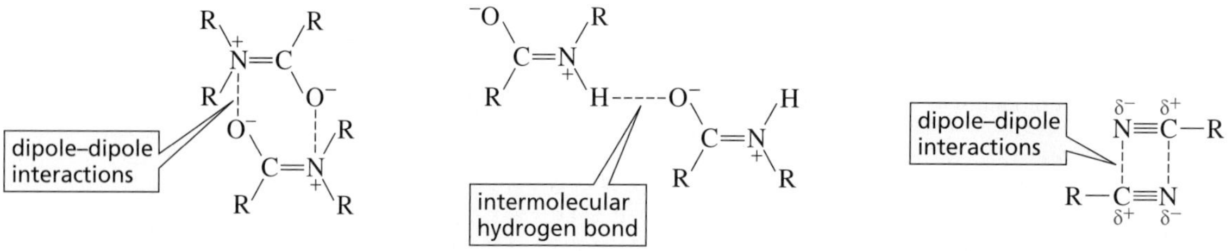

dipole–dipole interactions

intermolecular hydrogen bond

dipole–dipole interactions

Carboxylic acid derivatives are soluble in solvents such as ethers, chlorinated alkanes, and aromatic hydrocarbons. Like alcohols and ethers, carbonyl compounds with fewer than four carbons are soluble in water.

Esters, *N*,*N*-disubstituted amides, and nitriles are often used as solvents because they are polar, but do not have reactive hydroxyl or amino groups. We have seen that dimethylformamide (DMF) is a common aprotic polar solvent (Section 10.3).

17.4 Naturally Occurring Carboxylic Acids and Carboxylic Acid Derivatives

Acyl halides and acid anhydrides are much more reactive than carboxylic acids and esters, which, in turn, are more reactive than amides. We will see the reason for this difference in reactivity in Section 17.5.

Because of their high reactivity, acyl halides and acid anhydrides are not found in nature. Carboxylic acids, on the other hand, are less reactive and *are* found widely in nature. For example, glucose is metabolized to pyruvic acid. (*S*)-(+)-Lactic acid is the compound responsible for the burning sensation felt in muscles during anaerobic exercise, and it is also found in sour milk. Spinach and other leafy green vegetables are rich in oxalic acid. Succinic acid and citric acid are important intermediates in the citric acid cycle (Section 25.1), a series of reactions that oxidize acetyl-CoA to CO_2 in biological systems. Citrus fruits are rich in citric acid; the concentration is greatest in lemons, less in grapefruits, and still less in oranges. (*S*)-(−)-Malic acid is responsible for the sharp taste of unripe apples and pears. As the fruit ripens, the amount of malic acid in the fruit decreases and the amount of sugar increases. The inverse relationship between the levels of malic acid and sugar is important for the propagation of the plant: Animals will not eat the fruit until it becomes ripe—at which time its seeds are mature enough to germinate when they are scattered about. Prostaglandins are locally acting hormones that have several different physiological functions (Sections 17.10 and 26.5), such as stimulating inflammation, causing hypertension, and producing pain and swelling.

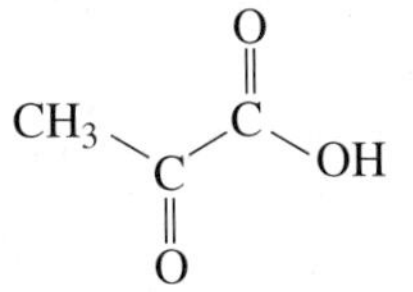

pyruvic acid

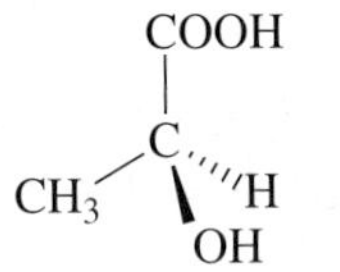

(*S*)-(+)-lactic acid

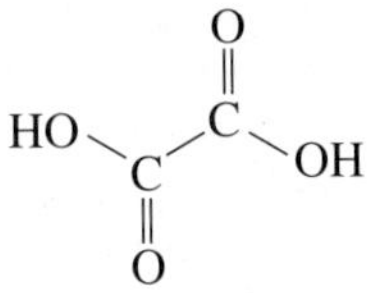

oxalic acid

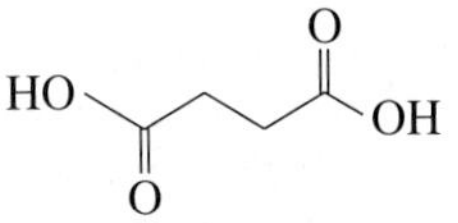

succinic acid

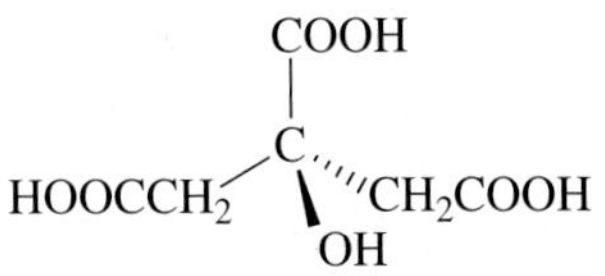

citric acid

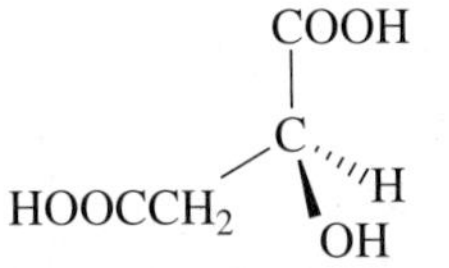

(*S*)-(−)-malic acid

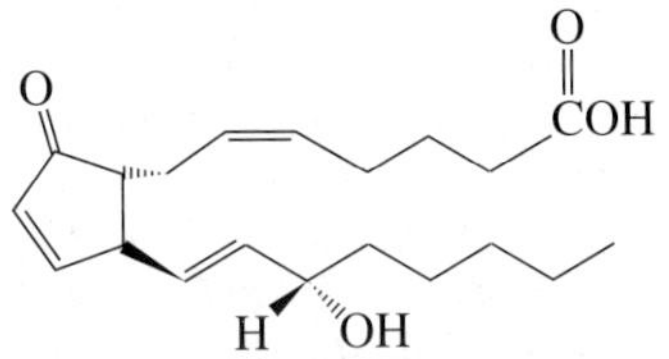

prostaglandin A_2

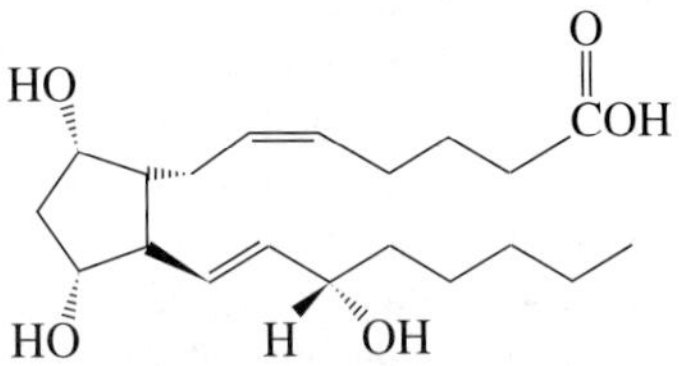

prostaglandin $F_{2\alpha}$

Esters are also commonly found in nature. Many of the fragrances of flowers and fruits are due to esters. (See Problem 27.)

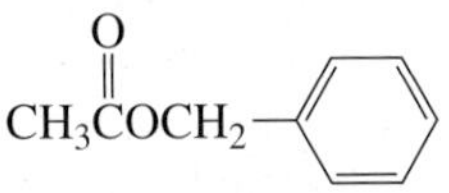

benzyl acetate
jasmine

isopentyl acetate
banana

methyl butyrate
apple

Carboxylic acids with an amino group on the α-carbon are commonly called **amino acids**. Amino acids are linked together by amide bonds to form peptides and proteins (Section 23.7). Caffeine, another naturally occurring amide, is found in cocoa and coffee beans. Penicillin G, a compound with two amide bonds (one of which is in a β-lactam ring), was first isolated from a mold in 1928 by Sir Alexander Fleming.

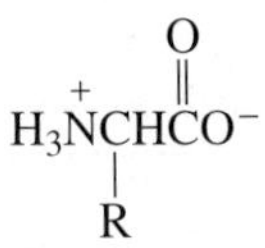

amide bonds

an amino acid

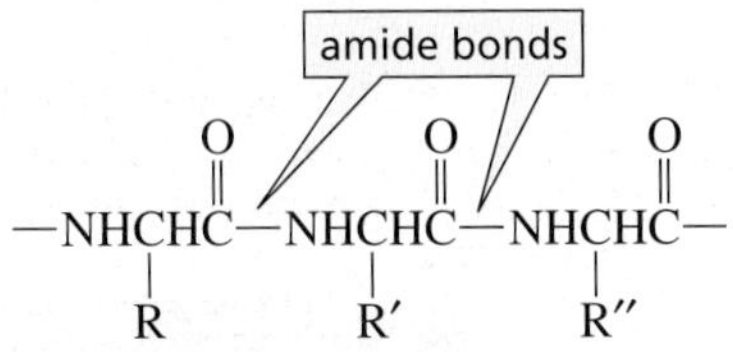

general structure for a peptide or a protein

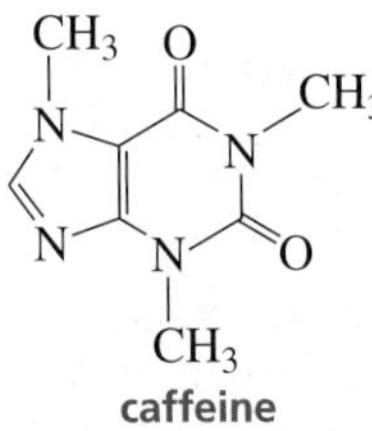

caffeine

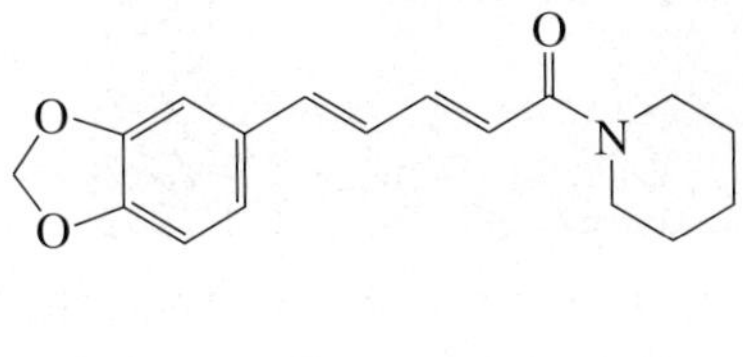

piperine
the major component of black pepper

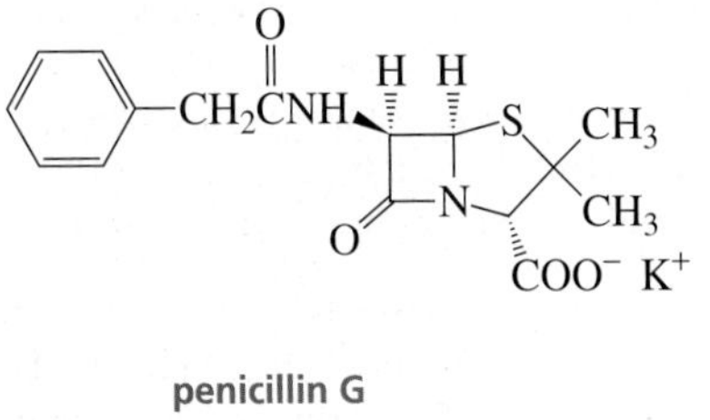

penicillin G

THE DISCOVERY OF PENICILLIN

Sir Alexander Fleming (1881–1955) was born in Scotland. He was a professor of bacteriology at University College, London. The story is told that one day Fleming was about to throw away a culture of staphylococcal bacteria that had been contaminated by a rare strain of the mold *Penicillium notatum.* He noticed that the bacteria had disappeared wherever there was a particle of mold. This suggested to him that the mold must have produced an antibacterial substance. Ten years later, Howard Florey and Ernest Chain isolated the active substance—penicillin G (Section 17.16)—but the delay allowed the sulfa drugs to be the first antibiotics. After penicillin G was found to cure bacterial infections in mice, it was used successfully in 1941 on nine cases of human bacterial infections. By 1943, penicillin G was being produced for the military and was first used for war casualties in Sicily and Tunisia. The drug became available to the civilian population in 1944. The pressure of the war made the determination of penicillin G's structure a priority because once its structure was determined, large quantities of the drug could be produced.

Fleming, Florey, and Chain shared the 1945 Nobel Prize in physiology or medicine. Chain also discovered penicillinase, the enzyme that destroys penicillin (Section 17.16). Although Fleming is generally given credit for the discovery of penicillin, there is clear evidence that the germicidal activity of the mold was recognized in the nineteenth century by Lord Joseph Lister (1827–1912), the English physician renowned for the introduction of aseptic surgery.

Sir Alexander Fleming (1881–1955) *was born in Scotland, the seventh of eight children of a farmer. In 1902, he received a legacy from an uncle that, together with a scholarship, allowed him to study medicine at the University of London. He subsequently became a professor of bacteriology there in 1928. He was knighted in 1944.*

Sir Howard W. Florey (1898–1968) *was born in Australia and received a medical degree from the University of Adelaide. He went to England as a Rhodes Scholar and studied at both Oxford and Cambridge Universities. He became a professor of pathology at the University of Sheffield in 1931 and then at Oxford in 1935. Knighted in 1944, he was given a peerage in 1965 that made him Baron Florey of Adelaide.*

Ernest B. Chain (1906–1979) *was born in Germany and received a Ph.D. from Friedrich-Wilhelm University in Berlin. In 1933, he left Germany for England because Hitler had come to power. He studied at Cambridge, and in 1935 Florey invited him to Oxford. In 1948, he became the director of an institute in Rome, but he returned to England in 1961 to become a professor at the University of London.*

DALMATIANS: DON'T TRY TO FOOL MOTHER NATURE

When amino acids are metabolized, the excess nitrogen is concentrated into uric acid, a compound with five amide bonds. A series of enzyme-catalyzed reactions degrades uric acid to ammonium ion. The extent to which uric acid is degraded in animals depends on the species. Birds, reptiles, and insects excrete excess nitrogen as uric acid. Mammals excrete excess nitrogen as allantoin. Excess nitrogen in aquatic animals is excreted as allantoic acid, urea, or ammonium salts.

uric acid $\xrightarrow{\text{enzyme}}$ allantoin $\xrightarrow{\text{enzyme}}$ allantoic acid $\xrightarrow{\text{enzyme}}$ urea $\xrightarrow{\text{enzyme}}$ $^{+}NH_4X^{-}$

uric acid excreted by: birds, reptiles, insects

allantoin mammals

allantoic acid marine vertebrates

urea cartilaginous fish, amphibia

ammonium salt marine invertebrates

Dalmatians, unlike other mammals, excrete high levels of uric acid. The reason for this is that breeders of Dalmatians select dogs that have no white hairs in their black spots, and the gene that causes the white hairs is linked to the gene that causes uric acid to be converted to allantoin. Dalmatians, therefore, are susceptible to gout (painful deposits of uric acid in joints).

17.5 How Class I Carbonyl Compounds React

The reactivity of carbonyl compounds resides in the polarity of the carbonyl group; oxygen is more electronegative than carbon. The carbonyl carbon, therefore, is an electrophile, so we can safely predict that it will be attacked by nucleophiles.

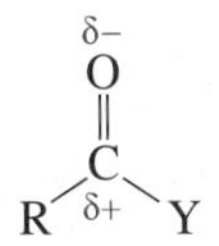

When a nucleophile attacks the carbonyl carbon of a carboxylic acid derivative, the carbon–oxygen π bond breaks and an intermediate is formed. The intermediate is called a **tetrahedral intermediate** because the trigonal (sp^2) carbon in the reactant has become a tetrahedral (sp^3) carbon in the intermediate. Generally, a compound that has an sp^3 carbon bonded to an oxygen atom will be unstable if the sp^3 carbon is bonded to another electronegative atom. The tetrahedral intermediate, therefore, is unstable because Y and Z are both electronegative atoms. A lone pair on the oxygen reforms the π bond, and either Y^- (k_2) or Z^- (k_{-1}) is expelled with its bonding electrons.

A compound that has an sp^3 carbon bonded to an oxygen atom generally will be unstable if the sp^3 carbon is bonded to another electronegative atom.

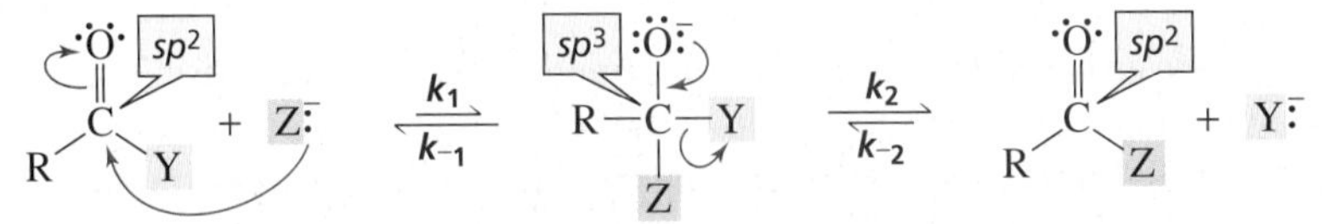

a tetrahedral intermediate

Whether Y^- or Z^- is expelled depends on their relative basicities. The weaker base is expelled preferentially, making this another example of the principle we first saw in Section 10.3: *The weaker the base, the better it is as a leaving group.* Because a weak base does not share its electrons as well as a strong base does, a weaker base forms a weaker bond—one that is easier to break. If Z^- is a much weaker base than Y^-, Z^- will be expelled. In such a case, $k_{-1} \gg k_2$, and the reaction can be written as follows:

The weaker the base, the better it is as a leaving group.

Z^- is a weaker base than Y^-

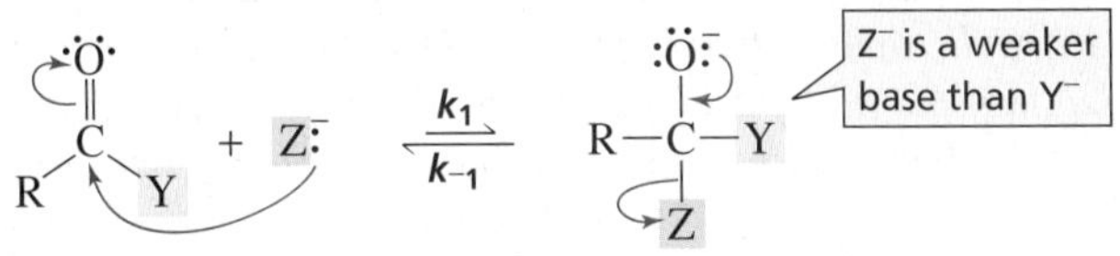

In this case, no new product is formed. The nucleophile attacks the carbonyl carbon, but the tetrahedral intermediate expels the attacking nucleophile and reforms the reactants.

On the other hand, if Y^- is a much weaker base than Z^-, Y^- will be expelled and a new product will be formed. In this case, $k_2 \gg k_{-1}$, and the reaction can be written as follows:

Y^- is a weaker base than Z^-

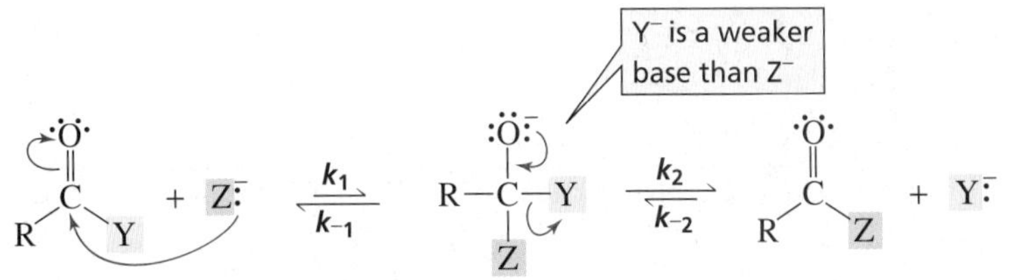

This reaction is called a **nucleophilic acyl substitution reaction** because a nucleophile (Z^-) has replaced the substituent (Y^-) that was attached to the acyl group in the reactant. It is also called an **acyl transfer reaction** because an acyl group has been transferred from one group (Y^-) to another (Z^-).

If the basicities of Y^- and Z^- are similar, the values of k_{-1} and k_2 will be similar. Therefore, some molecules of the tetrahedral intermediate will expel Y^- and others will expel Z^-. When the reaction is over, reactant and product will both be present. The relative amounts of each will depend on the relative basicities of Y^- and Z^-—that is, the relative values of k_2 and k_{-1}—as well as the relative nucleophilicities of Y^- and Z^-—that is, the relative values of k_1 and k_{-2}.

$$\mathrm{R{-}\overset{O}{\overset{\|}{C}}{-}Y} + Z^- \underset{k_{-1}}{\overset{k_1}{\rightleftharpoons}} \mathrm{R{-}\overset{O^-}{\overset{|}{\underset{Z}{\underset{|}{C}}}}{-}Y} \underset{k_{-2}}{\overset{k_2}{\rightleftharpoons}} \mathrm{R{-}\overset{O}{\overset{\|}{C}}{-}Z} + Y^-$$

the basicities of Y^- and Z^- are similar

These three cases are illustrated by the reaction coordinate diagrams shown in Figure 17.2.

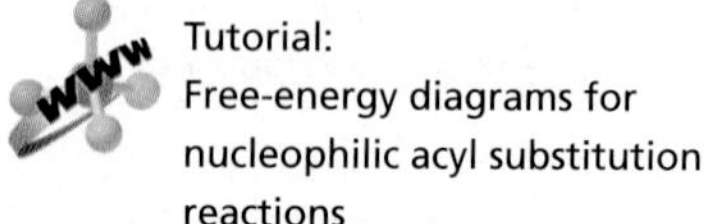
Tutorial: Free-energy diagrams for nucleophilic acyl substitution reactions

1. If the new group in the tetrahedral intermediate is a weaker base than the group that was attached to the acyl group in the reactant, the easier pathway—the lower energy hill—is for the tetrahedral intermediate (TI) to expel the newly added group and reform the reactants, so no reaction takes place (Figure 17.2a).
2. If the new group in the tetrahedral intermediate is a stronger base than the group that was attached to the acyl group in the reactant, the easier pathway is for the tetrahedral intermediate to expel the group that was attached to the acyl group in the reactant and form a substitution product (Figure 17.2b).
3. If both groups in the tetrahedral intermediate have similar basicities, the tetrahedral intermediate can expel either group with similar ease. A mixture of reactant and substitution product will result (Figure 17.2c).

For a carboxylic acid derivative to undergo a nucleophilic acyl substitution reaction, the incoming nucleophile must not be a much weaker base than the group that is to be replaced.

We can make the following general statement about the reactions of carboxylic acid derivatives: *A carboxylic acid derivative will undergo a nucleophilic acyl substitution reaction, provided that the newly added group in the tetrahedral intermediate is not a much weaker base than the group that was attached to the acyl group in the reactant.*

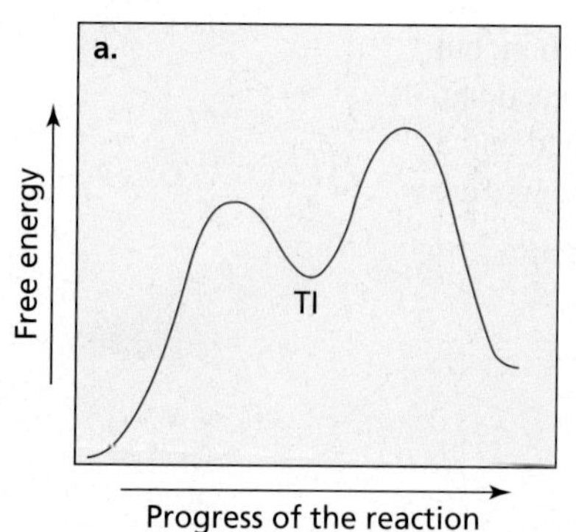

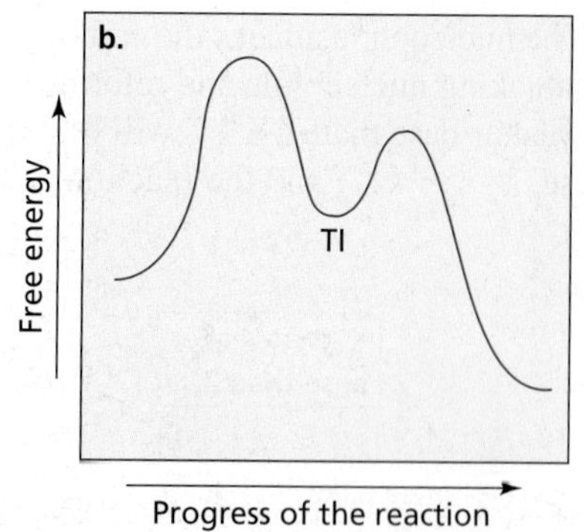

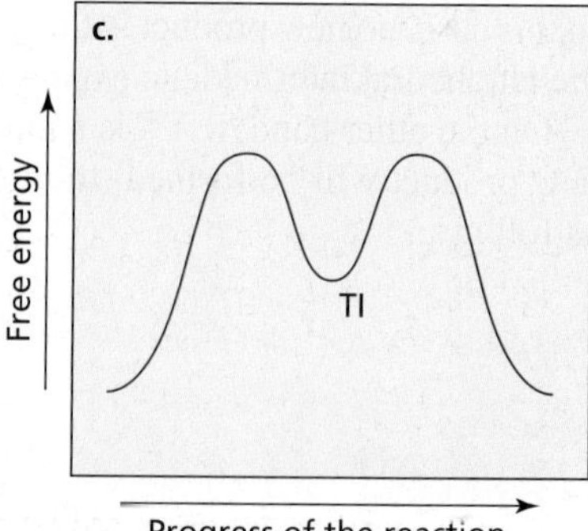

▲ **Figure 17.2**
Reaction coordinate diagrams for nucleophilic acyl substitution reactions in which (a) the nucleophile is a weaker base than the group attached to the acyl group in the reactant, (b) the nucleophile is a stronger base than the group attached to the acyl group in the reactant, and (c) the nucleophile and the group attached to the acyl group in the reactant have similar basicities. TI is the tetrahedral intermediate.

Let's now look at a molecular orbital description of how carbonyl compounds react. In Section 1.6, which first introduced you to molecular orbital theory, you saw that because oxygen is more electronegative than carbon, the $2p$ orbital of oxygen contributes more to the π bonding molecular orbital—it is closer to it in energy—and the $2p$ orbital of carbon contributes more to the π^* antibonding molecular orbital. (See Figure 1.8.) This means that the π^* antibonding orbital is largest at the carbon atom, so that is where the nucleophile's nonbonding orbital—in which the lone pair resides—overlaps. This allows the greatest amount of orbital overlap, and greater overlap means greater stability. When a filled orbital and an empty orbital overlap, the result is a molecular orbital—in this case, a σ molecular orbital—that is more stable than either of the overlapping orbitals (Figure 17.3).

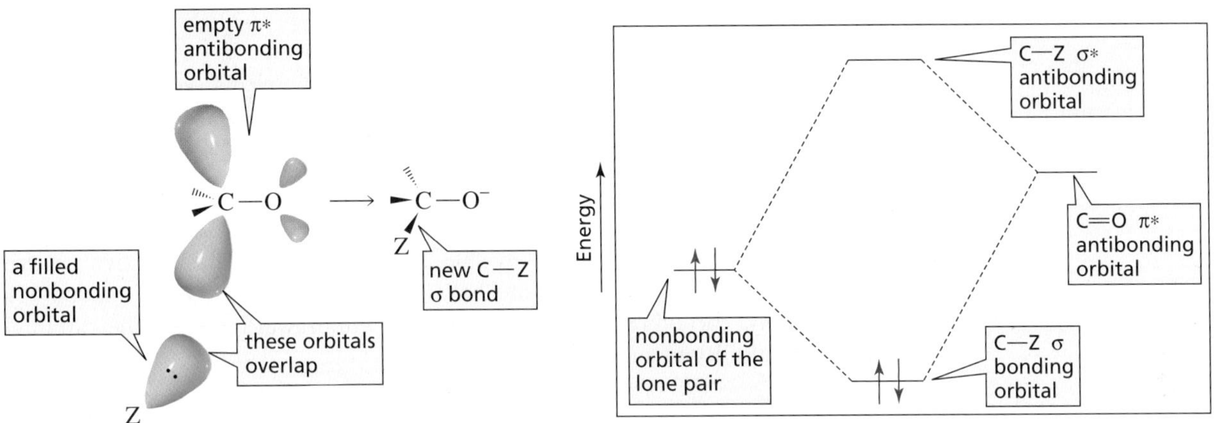

▲ **Figure 17.3**
The filled nonbonding orbital containing the nucleophile's lone pair overlaps the empty π^* antibonding molecular orbital of the carbonyl group, forming the new σ bond in the tetrahedral intermediate.

PROBLEM 7◆

Using the pK_a values in Table 17.1, predict the products of the following reactions:

a. $CH_3C(=O)OCH_3$ + NaCl ⟶

b. $CH_3C(=O)Cl$ + NaOH ⟶

c. $CH_3C(=O)Cl$ + $CH_3C(=O)O^-\ Na^+$ ⟶

d. $CH_3C(=O)OC(=O)CH_3$ + NaCl ⟶

PROBLEM 8◆

Is the following statement true or false?

If the newly added group in the tetrahedral intermediate is a stronger base than the group attached to the acyl group in the reactant, formation of the tetrahedral intermediate is the rate-limiting step of a nucleophilic acyl substitution reaction.

17.6 Relative Reactivities of Carboxylic Acids and Carboxylic Acid Derivatives

We have just seen that there are two steps in a nucleophilic acyl substitution reaction: formation of a tetrahedral intermediate and collapse of the tetrahedral intermediate. The weaker the base attached to the acyl group, the easier it is for *both steps* of the reaction to take place. In other words, the reactivity of a carboxylic acid derivative depends on the basicity of the substituent attached to the acyl group: The less basic the substituent, the more reactive the carboxylic acid derivative.

relative basicities of the leaving groups

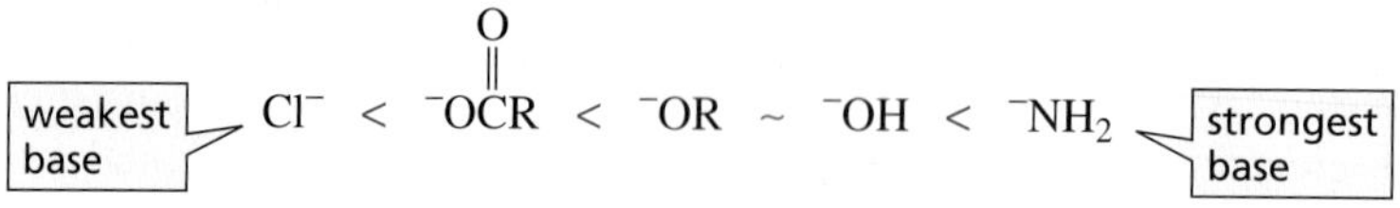

weakest base $Cl^- < {}^-OCR < {}^-OR \sim {}^-OH < {}^-NH_2$ strongest base

relative reactivities of carboxylic acid derivatives

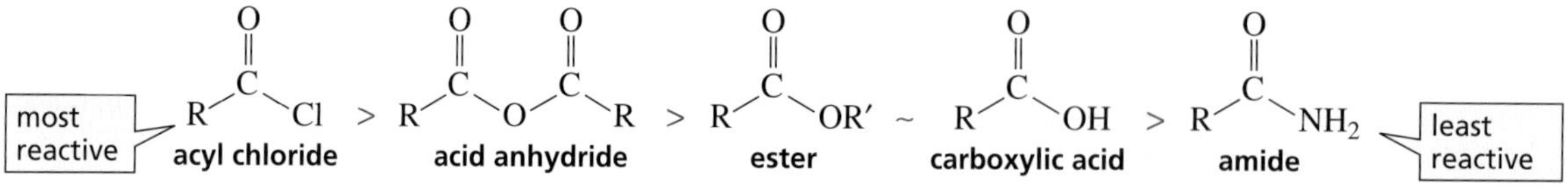

most reactive — acyl chloride (RCOCl) > acid anhydride (RCOOCOR) > ester (RCOOR′) ~ carboxylic acid (RCOOH) > amide ($RCONH_2$) — least reactive

How does having a weak base attached to the acyl group make the *first* step of the nucleophilic substitution reaction easier? First of all, a weaker base is a more electronegative base; that is, it is better able to accommodate its negative charge (Section 1.18). Thus, weaker bases are better at withdrawing electrons inductively from the carbonyl carbon (Section 1.18); electron withdrawal increases the carbonyl carbon's susceptibility to nucleophilic attack.

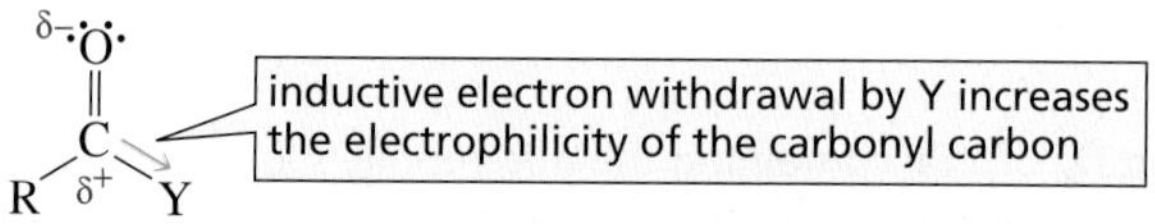

inductive electron withdrawal by Y increases the electrophilicity of the carbonyl carbon

Second, the weaker the basicity of Y, the smaller is the contribution from the resonance contributor with a positive charge on Y (Section 17.2); the less the carboxylic acid derivative is stabilized by electron delocalization, the more reactive it will be.

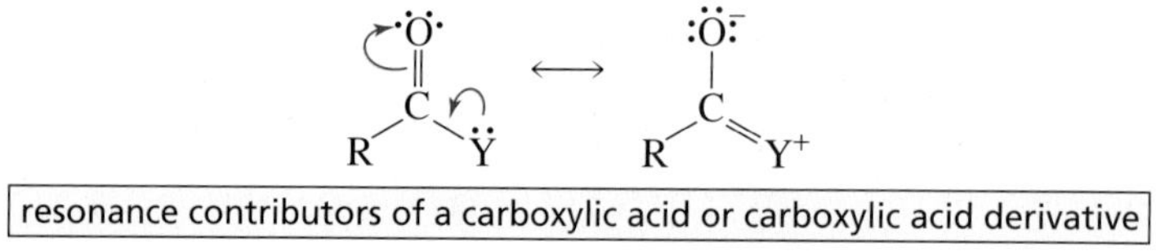

resonance contributors of a carboxylic acid or carboxylic acid derivative

A weak base attached to the acyl group will also make the *second* step of the nucleophilic acyl substitution reaction easier because weak bases are easier to eliminate when the tetrahedral intermediate collapses.

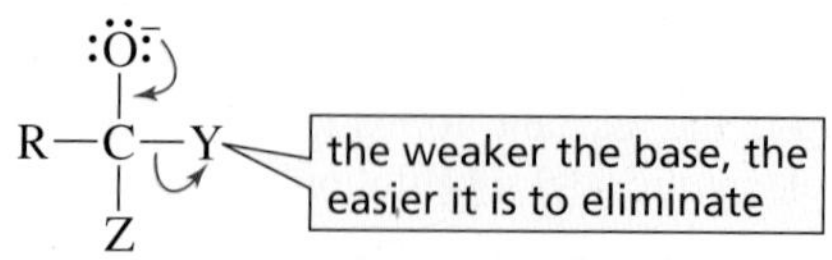

the weaker the base, the easier it is to eliminate

In Section 17.4 we saw that in a nucleophilic acyl substitution reaction, the nucleophile that forms the tetrahedral intermediate must be a stronger base than the base that is already there. This means that a carboxylic acid derivative can be converted into a less reactive carboxylic acid derivative, but not into one that is more reactive. For example, an acyl chloride can be converted into an anhydride because a carboxylate ion is a stronger base than a chloride ion.

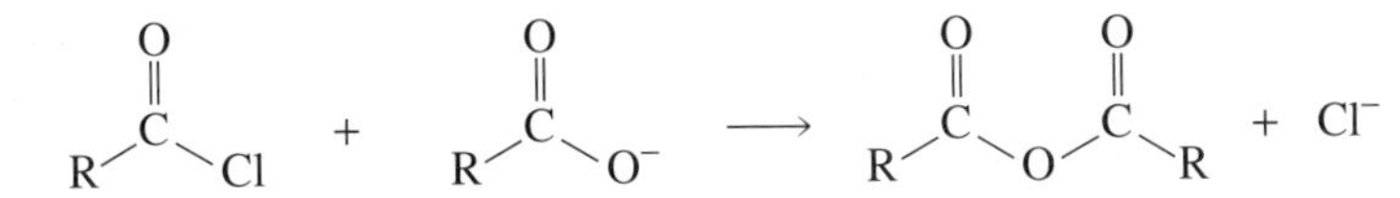

an acyl chloride + carboxylate ion ⟶ an anhydride + Cl^-

An anhydride, however, cannot be converted into an acyl chloride because a chloride ion is a weaker base than a carboxylate ion.

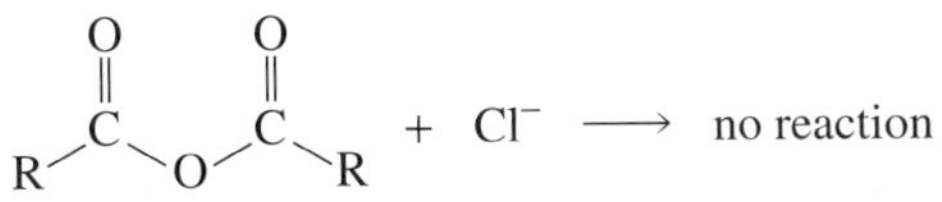

an anhydride + Cl^- ⟶ no reaction

A carboxylic acid derivative will undergo a nucleophilic acyl substitution reaction, provided that the newly added group in the tetrahedral intermediate is not a much weaker base than the group that was attached to the acyl group in the reactant.

17.7 General Mechanism for Nucleophilic Acyl Substitution Reactions

All carboxylic acid derivatives undergo nucleophilic acyl substitution reactions by the same mechanism. If the nucleophile is negatively charged, the mechanism discussed in Section 17.5 is followed: The nucleophile attacks the carbonyl carbon, forming a tetrahedral intermediate. When the tetrahedral intermediate collapses, the weaker base is eliminated.

All carboxylic acid derivatives react by the same mechanism.

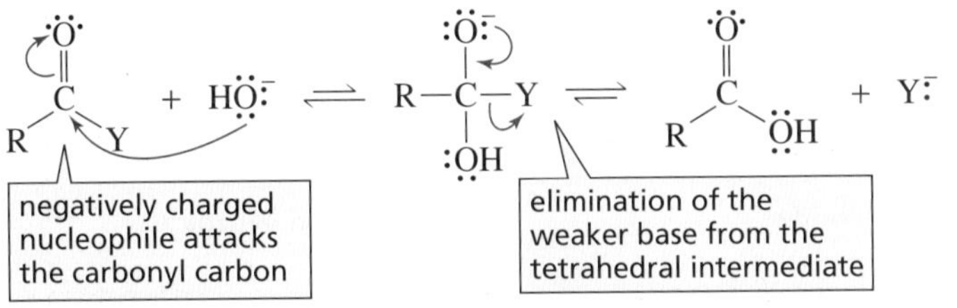

negatively charged nucleophile attacks the carbonyl carbon

elimination of the weaker base from the tetrahedral intermediate

If the nucleophile is neutral, the mechanism has an additional step. A proton is lost from the tetrahedral intermediate formed in the first step, resulting in a tetrahedral intermediate equivalent to the one formed by negatively charged nucleophiles. This tetrahedral intermediate expels the weaker of the two bases—the newly added group after it has lost a proton or the group that was attached to the acyl group in the reactant. (HB^+ represents any species in the solution that is capable of donating a proton, and :B represents any species in the solution that is capable of removing a proton.)

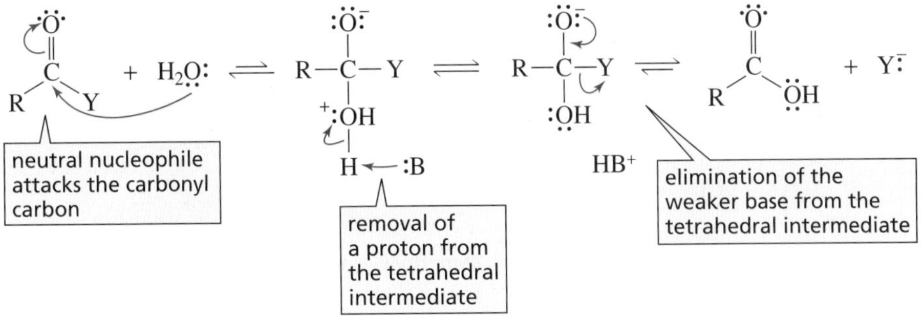

neutral nucleophile attacks the carbonyl carbon

removal of a proton from the tetrahedral intermediate

elimination of the weaker base from the tetrahedral intermediate

Movie: Nucleophilic acyl substitution

The remaining sections of this chapter cover specific examples of these general principles. Keep in mind that *all the reactions follow the same mechanism.* Therefore, you can always determine the outcome of the reactions of carboxylic acids and carboxylic acid derivatives presented in this chapter by examining the tetrahedral intermediate and remembering that the weaker base is preferentially eliminated (Section 17.5).

The tetrahedral intermediate eliminates the weakest base.

PROBLEM 9◆

What will be the product of a nucleophilic acyl substitution reaction—a new carboxylic acid derivative, a mixture of two carboxylic acid derivatives, or no reaction—if the new group in the tetrahedral intermediate is the following?

a. a stronger base than the group that was already there
b. a weaker base than the group that was already there
c. similar in basicity to the group that was already there

17.8 Reactions of Acyl Halides

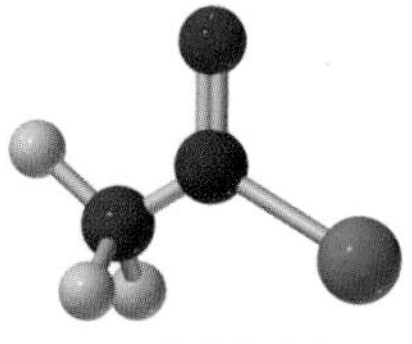
acetyl chloride

Acyl halides react with carboxylate ions to form anhydrides, with alcohols to form esters, with water to form carboxylic acids, and with amines to form amides because in each case the incoming nucleophile is a stronger base than the departing halide ion (Table 17.1). Notice that both alcohols and phenols can be used to prepare esters.

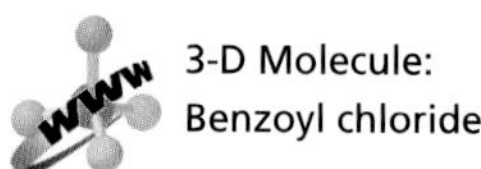
3-D Molecule: Benzoyl chloride

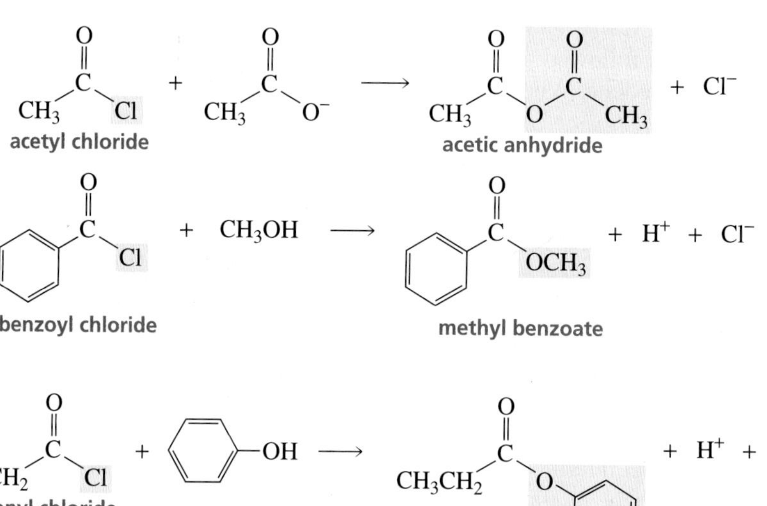

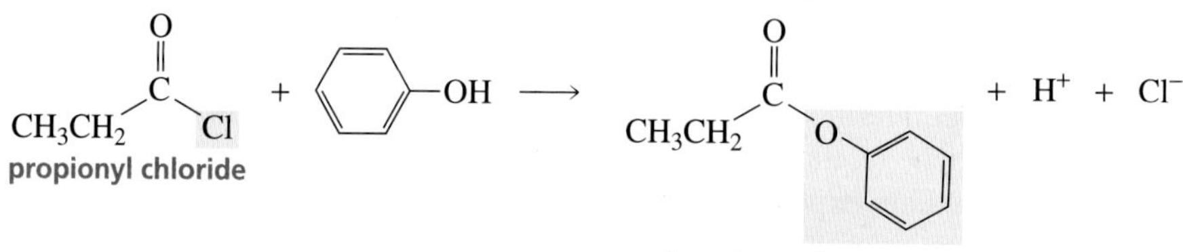

$$CH_3CH_2CH_2C(=O)Cl + H_2O \longrightarrow CH_3CH_2CH_2C(=O)OH + H^+ + Cl^-$$

butyryl chloride → butyric acid

$$C_6H_{11}C(=O)Cl + 2\,CH_3NH_2 \longrightarrow C_6H_{11}C(=O)NHCH_3 + CH_3\overset{+}{N}H_3\,Cl^-$$

cyclohexanecarbonyl chloride → *N*-methylcyclohexanecarboxamide

All the reactions follow the general mechanism described in Section 17.7. In the conversion of an acyl chloride into an acid anhydride, the nucleophilic carboxylate ion attacks the carbonyl carbon of the acyl chloride. Because the resulting tetrahedral intermediate is unstable, the double bond is immediately reformed, expelling chloride ion because it is a weaker base than the carboxylate ion. The final product is an anhydride.

mechanism for the conversion of an acyl chloride into an acid anhydride

$$CH_3COCl + CH_3COO^- \rightleftharpoons CH_3-C(O^-)(Cl)-O-C(=O)CH_3 \longrightarrow CH_3C(=O)-O-C(=O)CH_3 + Cl^-$$

In the conversion of an acyl chloride into an ester, the nucleophilic alcohol attacks the carbonyl carbon of the acyl chloride. Because the protonated ether group is a strong acid (Section 1.17), the tetrahedral intermediate loses a proton. Chloride ion is expelled from the deprotonated tetrahedral intermediate because chloride ion is a weaker base than the alkoxide ion.

mechanism for the conversion of an acyl chloride into an ester

$$CH_3COCl + CH_3\ddot{O}H \rightleftharpoons CH_3-C(O^-)(Cl)-\overset{+}{O}(H)CH_3 \xrightarrow{:B} CH_3-C(O^-)(Cl)-OCH_3 + HB^+ \longrightarrow CH_3C(=O)OCH_3 + Cl^-$$

The reaction of an acyl chloride with ammonia or with a primary or secondary amine forms an amide and HCl. The acid generated in the reaction will protonate unreacted ammonia or unreacted amine; because they are not nucleophiles, the protonated amines cannot react with the acyl chloride. The reaction, therefore, must be carried out with twice as much ammonia or amine as acyl chloride so that there will be enough amine to react with all the acyl halide.

$$CH_3COCl + NH_3 \longrightarrow CH_3CONH_2 + HCl \xrightarrow{NH_3} \overset{+}{N}H_4\,Cl^-$$

$$CH_3CH_2COCl + CH_3NH_2 \longrightarrow CH_3CH_2CONHCH_3 + HCl \xrightarrow{CH_3NH_2} CH_3\overset{+}{N}H_3\,Cl^-$$

$$C_6H_5COCl + 2\,CH_3NHCH_3 \longrightarrow C_6H_5CON(CH_3)CH_3 + CH_3\overset{+}{N}H_2(CH_3)\,Cl^-$$

Because tertiary amines cannot form amides, an equivalent of a tertiary amine such as triethylamine or pyridine can be used instead of excess amine.

$$CH_3COCl + CH_3NH_2 \xrightarrow{\text{pyridine}} CH_3CONHCH_3 + C_5H_5\overset{+}{N}H\;Cl^-$$

PROBLEM 10 SOLVED

a. Two amides are obtained from the reaction of acetyl chloride with a mixture of ethylamine and propylamine. Identify the amides.

b. Only one amide is obtained from the reaction of acetyl chloride with a mixture of ethylamine and pyridine. Why is only one amide obtained?

SOLUTION TO 10a Either of the amines can react with acetyl chloride, so both *N*-ethylacetamide and *N*-propylacetamide are formed.

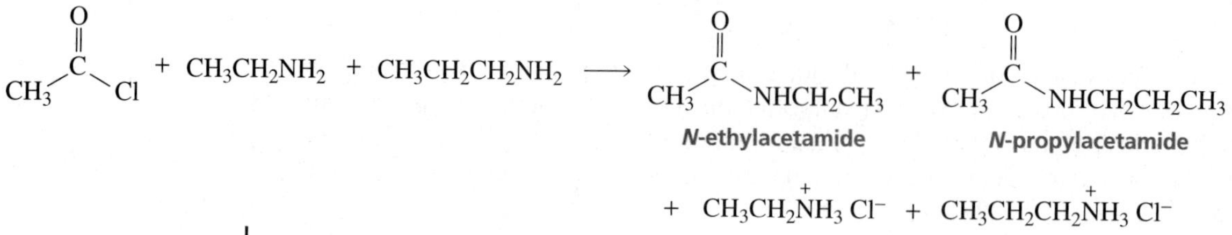

$$CH_3C(=O)Cl + CH_3CH_2NH_2 + CH_3CH_2CH_2NH_2 \longrightarrow CH_3C(=O)NHCH_2CH_3 + CH_3C(=O)NHCH_2CH_2CH_3$$

N-ethylacetamide   *N*-propylacetamide

$$+ CH_3CH_2\overset{+}{N}H_3\ Cl^- + CH_3CH_2CH_2\overset{+}{N}H_3\ Cl^-$$

SOLUTION TO 10b Initially two amides are formed. However, the amide formed by pyridine (a tertiary amine) is very reactive because it has a positively charged nitrogen atom, which makes it an excellent leaving group. Therefore, it will react immediately with unreacted ethylamine, causing *N*-ethylacetamide to be the only amide formed in the reaction.

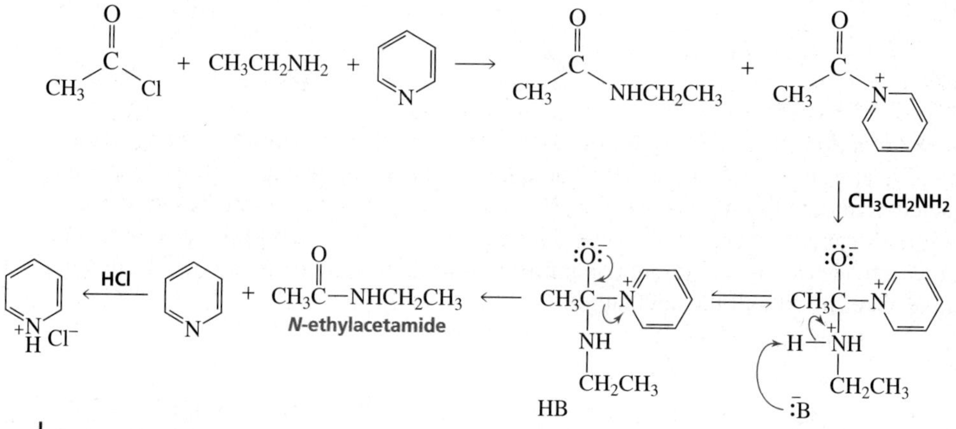

$$CH_3C(=O)Cl + CH_3CH_2NH_2 + \text{pyridine} \longrightarrow CH_3C(=O)NHCH_2CH_3 + CH_3C(=O)\overset{+}{N}C_5H_5$$

$\downarrow CH_3CH_2NH_2$

$$\text{pyridinium } \overset{+}{N}H\ Cl^- \xleftarrow{\text{HCl}} \text{pyridine} + CH_3\overset{O}{\overset{\|}{C}}{-}NHCH_2CH_3 \longleftarrow CH_3C(\ddot{O}:^-)(\overset{+}{N}C_5H_5)(NHCH_2CH_3) \;(\text{HB}) \rightleftharpoons CH_3C(\ddot{O}:^-)(\overset{+}{N}C_5H_5)(H{-}\overset{+}{N}HCH_2CH_3)\;(:\bar{B})$$

N-ethylacetamide

PROBLEM 11

Although excess amine is necessary in the reaction of an acyl chloride with an amine, explain why it is not necessary to use excess alcohol in the reaction of an acyl chloride with an alcohol.

PROBLEM 12

Write the mechanism for the following reactions:

a. the reaction of acetyl chloride with water to form acetic acid

b. the reaction of acetyl bromide with methylamine to form *N*-methylacetamide

PROBLEM 13◆

Starting with acetyl chloride, what nucleophile would you use to make each of the following compounds?

a. $CH_3\overset{O}{\overset{\|}{C}}OCH_2CH_2CH_3$

b. $CH_3\overset{O}{\overset{\|}{C}}NHCH_2CH_3$

c. $CH_3\overset{O}{\overset{\|}{C}}N(CH_3)_2$

d. $CH_3\overset{O}{\overset{\|}{C}}OH$

e. $CH_3\overset{O}{\overset{\|}{C}}O\overset{O}{\overset{\|}{C}}CH_3$

f. 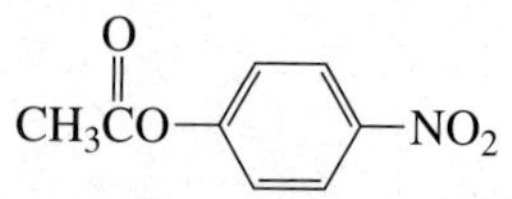 $CH_3\overset{O}{\overset{\|}{C}}O{-}C_6H_4{-}NO_2$

17.9 Reactions of Acid Anhydrides

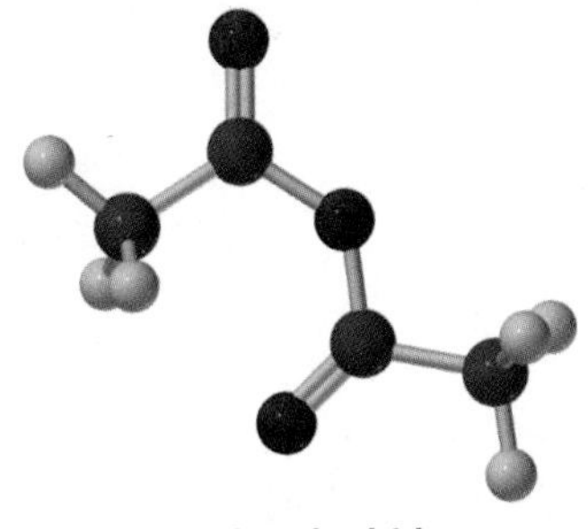
acetic anhydride

Acid anhydrides do not react with sodium chloride or with sodium bromide because the incoming halide ion is a weaker base than the departing carboxylate ion (Table 17.1).

$$CH_3C(=O)OC(=O)CH_3 + Cl^- \longrightarrow \text{no reaction}$$

Because the incoming halide ion is the weaker base, it will be the substituent expelled from the tetrahedral intermediate.

$$CH_3C(=O)OC(=O)CH_3 + :\ddot{Cl}:^- \rightleftharpoons CH_3C(O^-)(Cl)-O-C(=O)CH_3$$

An acid anhydride reacts with an alcohol to form an ester and a carboxylic acid, with water to form two equivalents of a carboxylic acid, and with an amine to form an amide and a carboxylate ion. In each case, the incoming nucleophile—after it loses a proton—is a stronger base than the departing carboxylate ion. In the reaction of an amine with an anhydride, two equivalents of the amine or one equivalent of the amine plus one equivalent of a tertiary amine such as pyridine must be used so that sufficient amine is present to react with the proton produced in the reaction.

$$\underset{\text{acetic anhydride}}{CH_3C(=O)OC(=O)CH_3} + CH_3CH_2OH \longrightarrow \underset{\text{ethyl acetate}}{CH_3C(=O)OCH_2CH_3} + \underset{\text{acetic acid}}{CH_3C(=O)OH}$$

$$\underset{\text{benzoic anhydride}}{C_6H_5C(=O)OC(=O)C_6H_5} + H_2O \longrightarrow 2\ \underset{\text{benzoic acid}}{C_6H_5C(=O)OH}$$

$$\underset{\text{propionic anhydride}}{CH_3CH_2C(=O)OC(=O)CH_2CH_3} + 2\ CH_3NH_2 \longrightarrow \underset{N\text{-methylpropionamide}}{CH_3CH_2C(=O)NHCH_3} + CH_3CH_2C(=O)O^-\ \ H_3\overset{+}{N}CH_3$$

All the reactions follow the general mechanism described in Section 17.7. For example, compare the following mechanism for conversion of an acid anhydride into an ester with the mechanism for conversion of an acyl chloride into an ester presented on p. 687.

mechanism for the conversion of an acid anhydride into an ester (and a carboxylic acid)

$$CH_3C(=O)OC(=O)CH_3 + CH_3\ddot{O}H \rightleftharpoons H_3C-C(O^-)(\overset{+}{O}(H)CH_3)-O-C(=O)CH_3\ \ (:B) \rightleftharpoons H_3C-C(O^-)(\ddot{O}CH_3)-O-C(=O)CH_3\ \ HB^+ \longrightarrow CH_3C(=O)\ddot{O}CH_3 + {}^-:\ddot{O}C(=O)CH_3$$

PROBLEM 14

a. Propose a mechanism for the reaction of acetic anhydride with water.

b. How does this mechanism differ from the mechanism for the reaction of acetic anhydride with an alcohol?

PROBLEM 15

We have seen that acid anhydrides react with alcohols, water, and amines. In which one of these three reactions does the tetrahedral intermediate not have to lose a proton before it eliminates the carboxylate ion? Explain.

methyl acetate

17.10 Reactions of Esters

Esters do not react with halide ions or with carboxylate ions because these nucleophiles are much weaker bases than the RO^- leaving group of the ester (Table 17.1).

An ester reacts with water to form a carboxylic acid and an alcohol. This is an example of a **hydrolysis** reaction—a reaction with water that converts one compound into two compounds (*lysis* is Greek for "breaking down").

a hydrolysis reaction

$$CH_3C(=O)OCH_3 + H_2O \underset{}{\overset{HCl}{\rightleftharpoons}} CH_3C(=O)OH + CH_3OH$$

methyl acetate → acetic acid

An ester reacts with an alcohol to form a new ester and a new alcohol. This is an example of an **alcoholysis** reaction. This particular alcoholysis reaction is also called a **transesterification reaction** because one ester is converted to another ester.

a transesterification reaction

$$C_6H_5C(=O)OCH_3 + CH_3CH_2OH \overset{HCl}{\rightleftharpoons} C_6H_5C(=O)OCH_2CH_3 + CH_3OH$$

methyl benzoate → ethyl benzoate

Both the hydrolysis and the alcoholysis of an ester are very slow reactions because water and alcohols are poor nucleophiles and esters have very basic leaving groups. These reactions, therefore, are always catalyzed when carried out in the laboratory. Both hydrolysis and alcoholysis of an ester can be catalyzed by acids. (See Section 17.11.) The rate of hydrolysis can also be increased by hydroxide ion, and the rate of alcoholysis can be increased by the conjugate base (RO^-) of the reactant alcohol (Section 17.12).

Esters also react with amines to form amides. A reaction with an amine that converts one compound into two compounds is called **aminolysis**. Notice that the aminolysis of an ester requires only one equivalent of amine, unlike the aminolysis of an acyl halide or an acid anhydride, which requires two equivalents (Sections 17.8 and 17.9). This is because the leaving group of an ester (RO^-) is more basic than the amine, so the alkoxide ion—rather than unreacted amine—picks up the proton generated in the reaction.

an aminolysis reaction

$$CH_3CH_2C(=O)OCH_2CH_3 + CH_3NH_2 \longrightarrow CH_3CH_2C(=O)NHCH_3 + CH_3CH_2OH$$

ethyl propionate → *N*-methylpropionamide

The reaction of an ester with an amine is not as slow as the reaction of an ester with water or an alcohol, because an amine is a better nucleophile. This is fortunate, because the rate of the reaction of an ester with an amine cannot be increased by acid or by HO^- or RO^- (Problem 20). The aminolysis of an ester can be driven to completion by using excess amine or by distilling off the alcohol as it is formed.

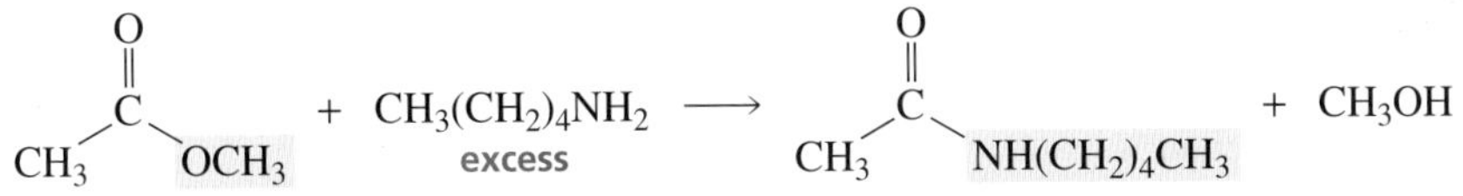

$$CH_3C(=O)OCH_3 + \underset{\text{excess}}{CH_3(CH_2)_4NH_2} \longrightarrow CH_3C(=O)NH(CH_2)_4CH_3 + CH_3OH$$

In Section 7.10, we saw that phenols are stronger acids than alcohols. Therefore, phenoxide ions (ArO^-) are weaker bases than alkoxide ions (RO^-), which means that phenyl esters are more reactive than alkyl esters.

3-D Molecule: Phenyl acetate

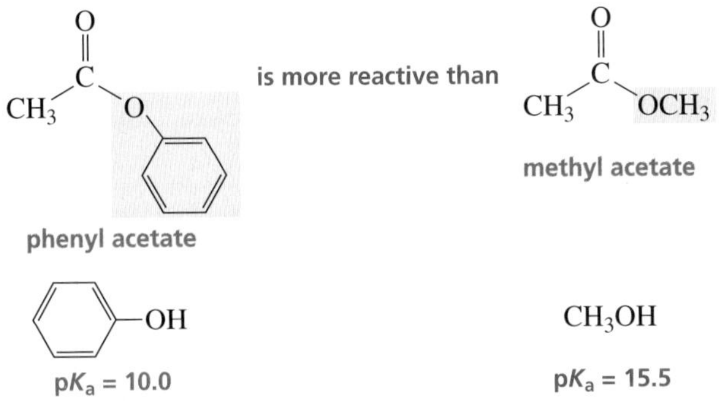

$CH_3C(=O)OC_6H_5$ (phenyl acetate) **is more reactive than** $CH_3C(=O)OCH_3$ (methyl acetate)

C_6H_5OH pK_a = 10.0 $\quad$ CH_3OH pK_a = 15.5

ASPIRIN

A transesterification reaction that blocks prostaglandin synthesis is responsible for aspirin's activity as an anti-inflammatory agent. Prostaglandins have several different biological functions, one of which is to stimulate inflammation. The enzyme prostaglandin synthase catalyzes the conversion of arachidonic acid into PGH_2, a precursor of prostaglandins and the related thromboxanes (Section 26.5).

$$\text{arachidonic acid} \xrightarrow{\textbf{prostaglandin synthase}} PGH_2 \rightarrow \text{prostaglandins},\ \text{thromboxanes}$$

Prostaglandin synthase is composed of two enzymes. One of the enzymes—cyclooxygenase—has a CH_2OH group (called a serine hydroxyl group because it is part of the amino acid called serine) that is necessary for enzyme activity. The CH_2OH group reacts with aspirin (acetylsalicylic acid) in a transesterification reaction and becomes acetylated. This inactivates the enzyme. Prostaglandin therefore cannot be synthesized, and inflammation is suppressed.

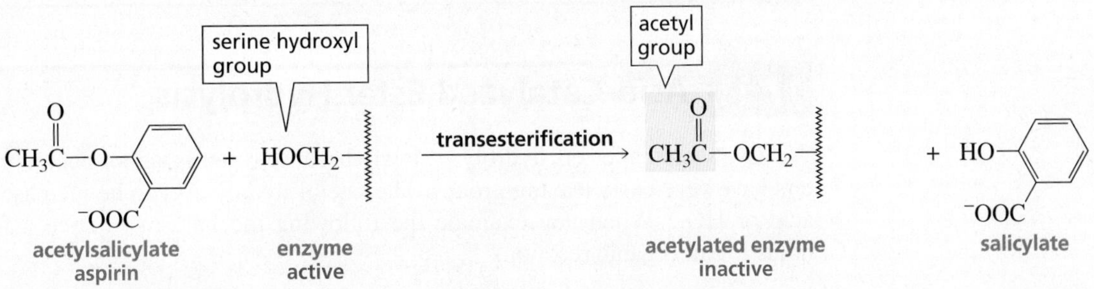

$$\underset{\substack{\text{acetylsalicylate}\\\text{aspirin}}}{CH_3C(=O)-O-C_6H_4COO^-} + \underset{\substack{\text{enzyme}\\\text{active}}}{HOCH_2-} \xrightarrow{\text{transesterification}} \underset{\substack{\text{acetylated enzyme}\\\text{inactive}}}{CH_3C(=O)-OCH_2-} + \underset{\text{salicylate}}{HO-C_6H_4COO^-}$$

(serine hydroxyl group; acetyl group)

Thromboxanes stimulate platelet aggregation. Because aspirin inhibits the formation of PGH_2, it inhibits thromboxane production and, therefore, platelet aggregation. Presumably, this is why low levels of aspirin have been reported to reduce the incidence of strokes and heart attacks that result from blood clot formation. Aspirin's activity as an anticoagulant is why doctors caution patients not to take aspirin for several days before surgery.

PROBLEM 16

Write a mechanism for the following reactions:

a. the noncatalyzed hydrolysis of methyl propionate
b. the aminolysis of phenyl formate, using methylamine

PROBLEM 17◆

a. State three factors that contribute to the fact that the noncatalyzed hydrolysis of an ester is a slow reaction.
b. Which is faster, hydrolysis of an ester or aminolysis of the same ester?

PROBLEM 18 SOLVED

a. List the following esters in order of decreasing reactivity toward hydrolysis:

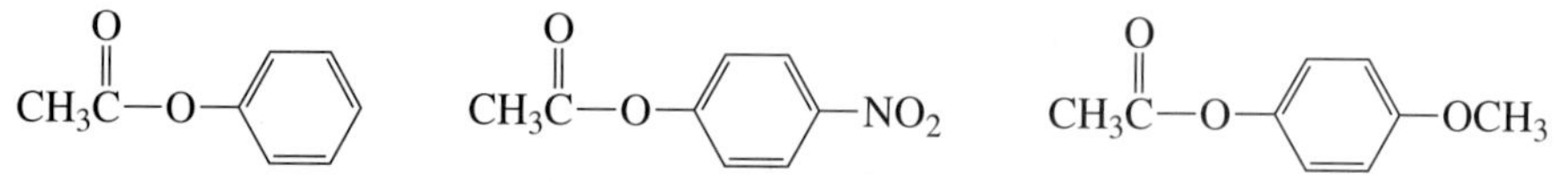

$$CH_3\overset{O}{\overset{\|}{C}}-O-C_6H_5 \qquad CH_3\overset{O}{\overset{\|}{C}}-O-C_6H_4-NO_2 \qquad CH_3\overset{O}{\overset{\|}{C}}-O-C_6H_4-OCH_3$$

b. How would the rate of hydrolysis of the *para*-methylphenyl ester compare with the rate of hydrolysis of these three esters?

SOLUTION TO 18a Because the nitro group withdraws electrons from the benzene ring and the methoxy group donates electrons into the benzene ring, the nitro-substituted ester will be the most susceptible, and the methoxy-substituted ester the least susceptible, to nucleophilic attack. We know that electron withdrawal increases acidity and electron donation decreases acidity, so *para*-nitrophenol is a stronger acid than phenol, which is a stronger acid than *para*-methoxyphenol. Therefore, the *para*-nitrophenoxide ion is the weakest base and the best leaving group of the three, whereas the *para*-methoxyphenoxide ion is the strongest base and the worst leaving group. Thus, both relatively slow steps of the hydrolysis reaction are fastest for the ester with the electron-withdrawing nitro substituent and slowest for the ester with the electron-donating methoxy substituent.

$$CH_3\overset{O}{\overset{\|}{C}}-O-C_6H_4-NO_2 > CH_3\overset{O}{\overset{\|}{C}}-O-C_6H_5 > CH_3\overset{O}{\overset{\|}{C}}-O-C_6H_4-OCH_3$$

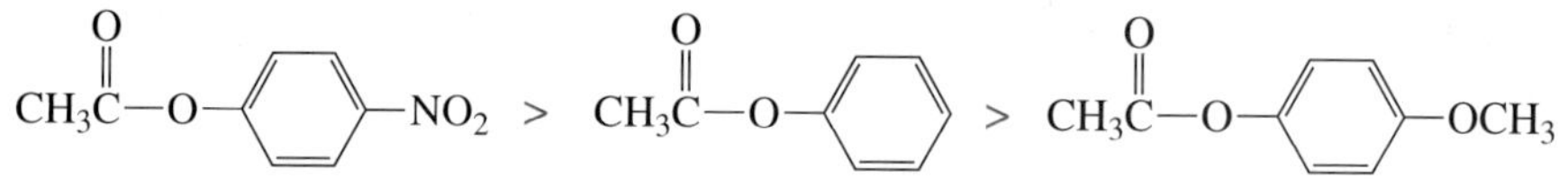

SOLUTION TO 18b The methyl substituent donates electrons inductively into the benzene ring, but donates electrons to a lesser extent than does the resonance-donating methoxy substituent. Therefore, the rate of hydrolysis of the methyl-substituted ester is slower than the rate of hydrolysis of the unsubstituted ester, but faster than the rate of hydrolysis of the methoxy-substituted ester.

$$CH_3\overset{O}{\overset{\|}{C}}-O-C_6H_5 > CH_3\overset{O}{\overset{\|}{C}}-O-C_6H_4-CH_3 > CH_3\overset{O}{\overset{\|}{C}}-O-C_6H_4-OCH_3$$

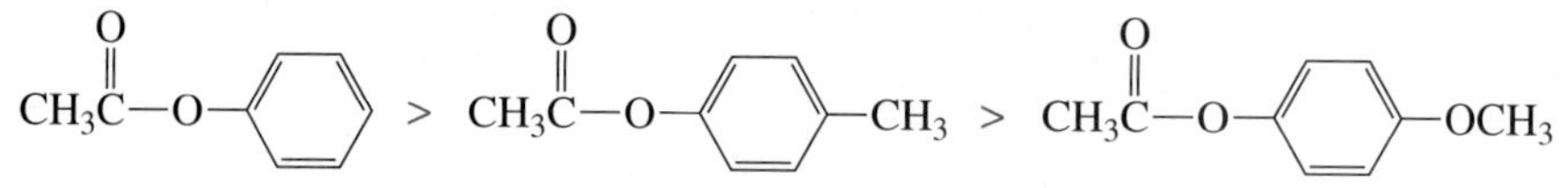

17.11 Acid-Catalyzed Ester Hydrolysis

We have seen that esters hydrolyze slowly because water is a poor nucleophile and esters have very basic leaving groups. The rate of hydrolysis can be increased by either acid or HO^-. When you examine the following mechanisms, notice a feature that holds for all organic reactions:

1. In an acid-catalyzed reaction, all organic intermediates and products are positively charged or neutral; *negatively charged organic intermediates and products are not formed in acidic solutions.*
2. In a reaction in which HO^- is used to increase the rate of the reaction, all organic intermediates and products are negatively charged or neutral; *positively charged organic intermediates and products are not formed in basic solutions.*

Normal Ester Hydrolysis

The first step in the mechanism for acid-catalyzed ester hydrolysis is protonation of the carbonyl oxygen by the acid. Recall that HB^+ represents any species in the solution that is capable of donating a proton and :B represents any species in the solution that is capable of removing a proton.

The carbonyl oxygen is protonated because it is the atom with the greatest electron density, as shown by the resonance contributors.

resonance contributors of an ester

HB^+ represents any species in the solution that is capable of donating a proton and :B represents any species in the solution that is capable of removing a proton.

In the second step of the mechanism, the nucleophile (H_2O) attacks the protonated carbonyl group. The resulting protonated tetrahedral intermediate (tetrahedral intermediate I) is in equilibrium with its nonprotonated form (tetrahedral intermediate II). Either the OH or the OR group of tetrahedral intermediate II (in this case, OR = OCH_3) can be protonated. Because the OH and OR groups have approximately the same basicity, both tetrahedral intermediate I (OH is protonated) and tetrahedral intermediate III (OR is protonated) are formed. (From Section 1.20, we know that the relative amounts of the three tetrahedral intermediates depend on the pH of the solution and the pK_a values of the protonated intermediates.) When tetrahedral intermediate I collapses, it expels H_2O in preference to CH_3O^- because H_2O is a weaker base, thereby reforming the ester. When tetrahedral intermediate III collapses, it expels CH_3OH rather than HO^- because CH_3OH is a weaker base, thereby forming the carboxylic acid. (Tetrahedral intermediate II is less likely to collapse, because both HO^- and CH_3O^- are strong bases.) Acid-catalyzed ester hydrolysis is discussed in greater detail in Section 24.3.

mechanism for acid-catalyzed ester hydrolysis

protonation of the carbonyl group

formation of a tetrahedral intermediate

tetrahedral intermediate I

equilibration of the three tetrahedral intermediates

tetrahedral intermediate II

removal of a proton from the carbonyl group

elimination of the weaker base

tetrahedral intermediate III

Because H_2O and CH_3OH have approximately the same basicity, it will be as likely for tetrahedral intermediate I to collapse to reform the ester as it will for tetrahedral intermediate III to collapse to form the carboxylic acid. Consequently, when the reaction has reached equilibrium, both ester and carboxylic acid will be present in approximately equal amounts.

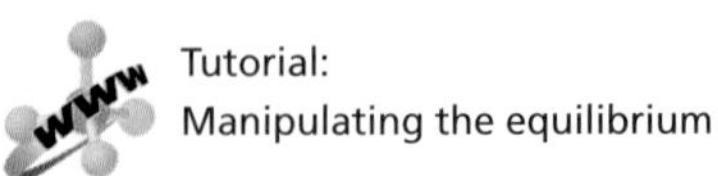

Tutorial: Manipulating the equilibrium

$$CH_3C(=O)OCH_3 + H_2O \overset{HCl}{\rightleftharpoons} CH_3C(=O)OH + CH_3OH$$

both ester and carboxylic acid will be present in approximately equal amounts when the reaction has reached equilibrium

Excess water will force the equilibrium to the right (Le Châtelier's principle; Section 10.4). Or, if the boiling point of the product alcohol is significantly lower than the boiling points of the other components of the reaction, the reaction can be driven to the right by distilling off the product alcohol as it is formed.

$$CH_3C(=O)OCH_3 + \underset{\text{excess}}{H_2O} \overset{HCl}{\rightleftharpoons} CH_3C(=O)OH + CH_3OH$$

The mechanism for the acid-catalyzed reaction of a carboxylic acid and an alcohol to form an ester and water is the exact reverse of the mechanism for the acid-catalyzed hydrolysis of an ester to form a carboxylic acid and an alcohol. If the ester is the desired product, the reaction should be carried out under conditions that will drive the equilibrium to the left—using excess alcohol or removing water as it is formed (Section 17.14).

$$CH_3C(=O)OCH_3 + H_2O \overset{HCl}{\rightleftharpoons} CH_3C(=O)OH + \underset{\text{excess}}{CH_3OH}$$

PROBLEM 19◆

Referring to the mechanism for the acid-catalyzed hydrolysis of methyl acetate:

a. What species could be represented by HB^+?
b. What species could be represented by :B?
c. What species is HB^+ most likely to be in a hydrolysis reaction?
d. What species is HB^+ most likely to be in the reverse reaction?

PROBLEM 20

Referring to the mechanism for the acid-catalyzed hydrolysis of methyl acetate, write the mechanism—showing all the curved arrows—for the acid-catalyzed reaction of acetic acid and methanol to form methyl acetate. Use HB^+ and :B to represent proton-donating and proton-removing species, respectively.

Now let's see how the acid increases the rate of ester hydrolysis. The acid is a catalyst. Recall that **catalyst** is a substance that increases the rate of a reaction without being consumed or changed in the overall reaction (Section 4.5). For a catalyst to increase the rate of a reaction, it must increase the rate of the slow step of the reaction. Changing the rate of a fast step will not affect the rate of the overall reaction. Four of the six steps in the mechanism for acid-catalyzed ester hydrolysis are proton transfer steps. Proton transfer to or from an electronegative atom such as oxygen or nitrogen is a fast step. So there are two relatively slow steps in the mechanism: formation of a tetrahedral intermediate and collapse of a tetrahedral intermediate. The acid increases the rates of both of these steps.

The acid increases the rate of formation of the tetrahedral intermediate by protonating the carbonyl oxygen. Protonated carbonyl groups are more susceptible than nonprotonated carbonyl groups to nucleophilic attack because a positively charged oxygen is more electron withdrawing than a neutral oxygen. Increased electron withdrawal by the oxygen increases the electron deficiency of the carbonyl carbon, which increases its attractiveness to nucleophiles.

protonation of the carbonyl oxygen increases the susceptibility of the carbonyl carbon to nucleophilic attack

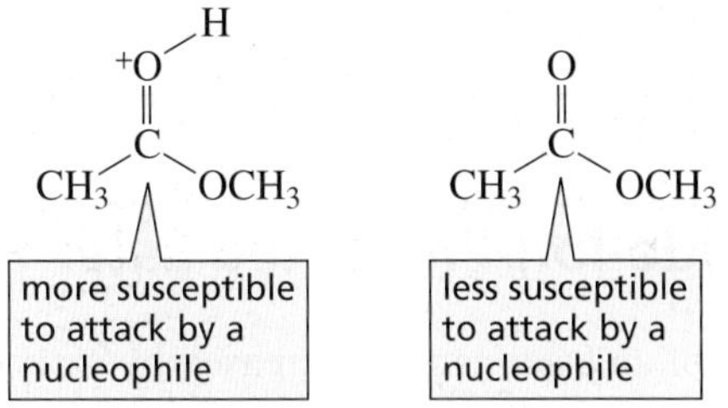

more susceptible to attack by a nucleophile

less susceptible to attack by a nucleophile

The acid increases the rate of collapse of the tetrahedral intermediate by decreasing the basicity of the leaving group, which makes it easier to eliminate. In the acid-catalyzed hydrolysis of an ester, the leaving group is ROH, a weaker base than the leaving group (RO^-) in the uncatalyzed reaction.

An acid catalyst increases the electrophilicity of the carbonyl carbon atom and decreases the basicity of the leaving group.

tetrahedral intermediate in acid-catalyzed ester hydrolysis

tetrahedral intermediate in uncatalyzed ester hydrolysis

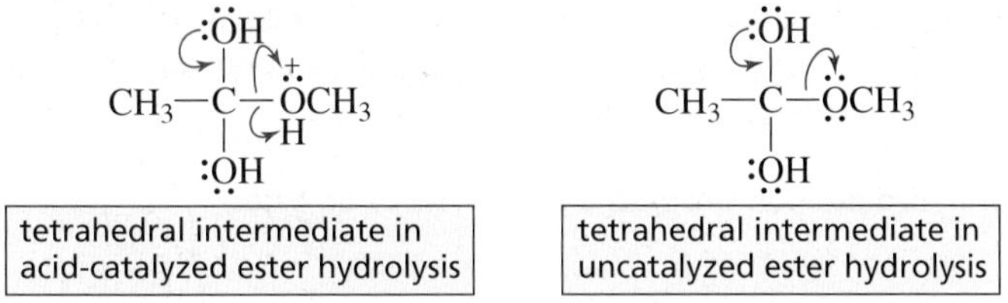

Hydrolysis of Esters with Tertiary Alkyl Groups

Esters with tertiary alkyl groups undergo hydrolysis much more rapidly than do other esters because they hydrolyze by a completely different mechanism—one that does not involve formation of a tetrahedral intermediate. The hydrolysis of an ester with a tertiary alkyl group is an S_N1 reaction because when the carboxylic acid leaves, it leaves behind a relatively stable tertiary carbocation.

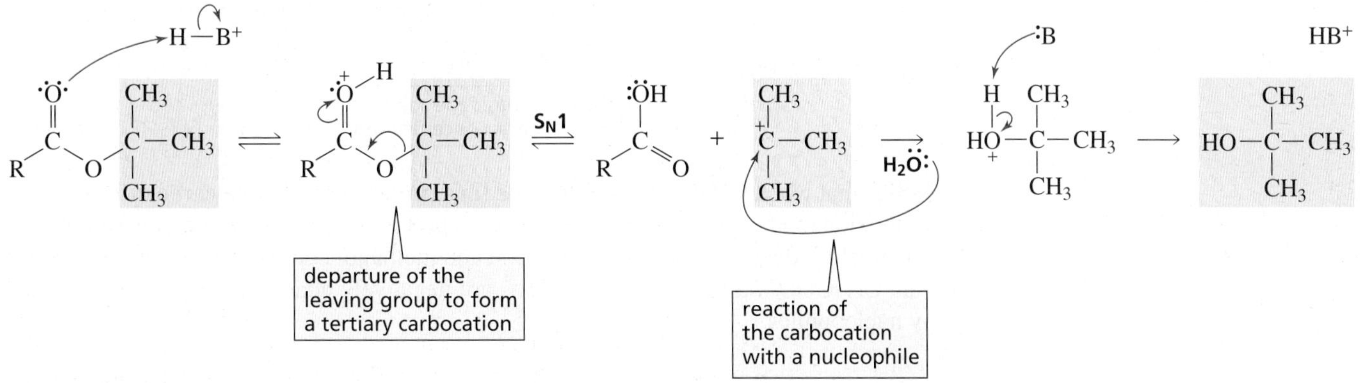

departure of the leaving group to form a tertiary carbocation

reaction of the carbocation with a nucleophile

Transesterification

Transesterification—the reaction of an ester with an alcohol—is also catalyzed by acid. The mechanism for transesterification is identical to the mechanism for normal ester hydrolysis, except that the nucleophile is ROH rather than H_2O. As in hydrolysis, the leaving groups in the tetrahedral intermediate formed in transesterification have approximately the same basicity. Consequently, an excess of the reactant alcohol must be used to produce more of the desired product.

$$CH_3COOCH_3 + CH_3CH_2CH_2OH \overset{HCl}{\rightleftharpoons} CH_3COOCH_2CH_2CH_3 + CH_3OH$$

methyl acetate — propyl alcohol — propyl acetate — methyl alcohol

PROBLEM 21

Write the mechanism for the acid-catalyzed transesterification reaction of methyl acetate with ethanol.

17.12 Hydroxide-Ion-Promoted Ester Hydrolysis

The rate of hydrolysis of an ester can be increased by carrying out the reaction in a basic solution. Like an acid, hydroxide ion increases the rates of both slow steps of the reaction.

Hydroxide ion increases the rate of formation of the tetrahedral intermediate because HO^- is a better nucleophile than H_2O, so HO^- more readily attacks the carbonyl carbon. Hydroxide ion increases the rate of collapse of the tetrahedral intermediate because a smaller fraction of the negatively charged tetrahedral intermediate becomes protonated in a basic solution. A negatively charged oxygen can more readily expel the very basic leaving group (RO^-) because the oxygen does not develop a partial positive charge in the transition state.

Hydroxide ion is a better nucleophile than water.

mechanism for hydroxide-ion-promoted hydrolysis of an ester

$$CH_3COOCH_3 + HO^- \rightleftharpoons CH_3{-}C(O^-)(OH){-}OCH_3 \rightleftharpoons CH_3COOH + CH_3O^-$$

$$CH_3{-}C(O^-)(OH){-}OCH_3 \underset{H_2O}{\overset{HO^-}{\rightleftharpoons}} CH_3{-}C(OH)_2{-}OCH_3$$

$$CH_3COOH + CH_3O^- \rightarrow CH_3COO^- + CH_3OH$$

the more basic the solution, the lower its concentration

Notice that when CH_3O^- is expelled, the final products are not the carboxylic acid and methoxide ion because if only one species is protonated, it will be the more basic one. The final products are the carboxylate ion and methanol because CH_3O^- is more basic than CH_3COO^-. Since carboxylate ions are negatively charged, they are not attacked by nucleophiles.

$$CH_3COO^- + :Nu \;\not\rightarrow$$

Therefore, the hydroxide-ion-promoted hydrolysis of an ester, unlike the acid-catalyzed hydrolysis of an ester, is *not* a reversible reaction.

$$RCOOCH_3 + H_2O \underset{}{\overset{HCl}{\rightleftharpoons}} RCOOH + CH_3OH$$

reversible reaction

$$RCOOCH_3 + H_2O \xrightarrow{NaOH} RCOO^- + CH_3OH$$

irreversible reaction

This reaction is called a hydroxide-ion-promoted reaction rather than a base-catalyzed reaction because hydroxide ion increases the rate of the first step of the reaction by being a better nucleophile than water—not by being a stronger base than water—and because hydroxide ion is consumed in the overall reaction. To be a catalyst, a species must not be changed by or consumed in the reaction (Section 17.11). So hydroxide ion is actually a reagent rather than a catalyst. Therefore, it is more accurate to call the reaction a hydroxide-ion-*promoted* reaction than a hydroxide-ion-*catalyzed* reaction.

Hydroxide ion promotes only hydrolysis reactions, not transesterification reactions or aminolysis reactions. Hydroxide ion cannot promote reactions of carboxylic acid derivatives with alcohols or with amines because one function of hydroxide ion is to provide a strong nucleophile for the first step of the reaction. Thus, when the nucleophile is supposed to be an alcohol or an amine, nucleophilic attack by hydroxide ion would form a product different from the product that would be formed from nucleophilic attack by an alcohol or amine. Hydroxide can be used to promote a hydrolysis reaction because the same product is formed, regardless of whether the attacking nucleophile is H_2O or HO^-.

Reactions in which the nucleophile is an alcohol can be promoted by the conjugate base of the alcohol. The function of the alkoxide ion is to provide a strong nucleophile for the reaction, so only reactions in which the nucleophile is an alcohol can be promoted by its conjugate base.

$$C_6H_5COOCH_3 + \underset{\text{excess}}{CH_3CH_2OH} \overset{CH_3CH_2O^-}{\rightleftharpoons} C_6H_5COOCH_2CH_3 + CH_3OH$$

PROBLEM 22◆

a. What species other than an acid can be used to increase the rate of a transesterification reaction that converts methyl acetate to propyl acetate?

b. Explain why the rate of aminolysis of an ester cannot be increased by H^+, HO^-, or RO^-.

You have seen that nucleophilic acyl substitution reactions take place by a mechanism in which a tetrahedral intermediate is formed and subsequently collapses. The tetrahedral intermediate, however, is too unstable to be isolated. How, then, do we know that it is formed? How do we know that the reaction doesn't take place by a one-step direct-displacement mechanism (similar to the mechanism of an S_N2 reaction) in which the incoming nucleophile attacks the carbonyl carbon and displaces the leaving group—a mechanism that would not form a tetrahedral intermediate?

$$\overset{\delta-}{HO}\cdots\underset{R}{\overset{O}{C}}\cdots\overset{\delta-}{OR}$$

transition state for a hypothetical one-step direct-displacement mechanism

Myron L. Bender (1924–1988) *was born in St. Louis. He was a professor of chemistry at the Illinois Institute of Technology and at Northwestern University.*

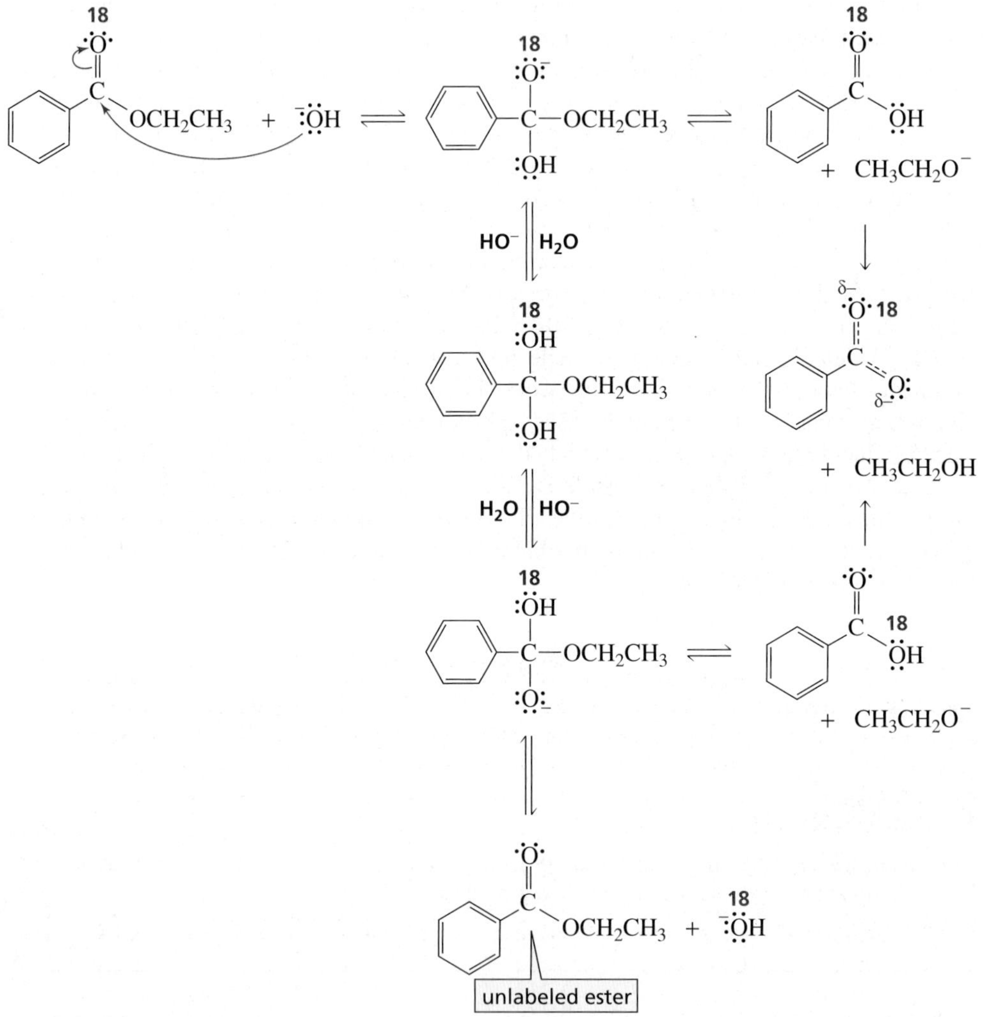

To answer this question, Myron Bender investigated the hydroxide-ion-promoted hydrolysis of ethyl benzoate, with the carbonyl oxygen of ethyl benzoate labeled with ^{18}O. When he isolated ethyl benzoate from an incomplete reaction mixture, he found that some of the ester was no longer labeled. If the reaction had taken place by a one-step direct-displacement mechanism, all the isolated ester would have remained labeled because the carbonyl group would not have participated in the reaction. On the other hand, if the mechanism involved a tetrahedral intermediate, some of the isolated ester would no longer be labeled because some of the label would have been transferred to the hydroxide ion. By this experiment, Bender provided evidence for the reversible formation of a tetrahedral intermediate.

18 O; C; OCH_2CH_3 + $^-$:OH ⇌ 18 :O:$^-$; C—OCH_2CH_3; :OH ⇌ 18 O; C; OH + $CH_3CH_2O^-$

HO$^-$ | H_2O

18 :OH; C—OCH_2CH_3; :OH

δ− O 18; C; O: δ− + CH_3CH_2OH

H_2O | HO$^-$

18 :OH; C—OCH_2CH_3; :O:$^-$ ⇌ O; C; 18 OH + $CH_3CH_2O^-$

O; C; OCH_2CH_3 + 18 $^-$:OH

unlabeled ester

PROBLEM 23◆

If butanoic acid and ^{18}O labeled methanol are allowed to react under acidic conditions, what compounds will be labeled when the reaction has reached equilibrium?

PROBLEM 24◆

D. N. Kursanov, a Russian chemist, proved that the bond that is broken in the hydroxide-ion-promoted hydrolysis of an ester is the acyl C—O bond, rather than the alkyl C—O bond, by studying the reaction of the following ester with HO^-/H_2O:

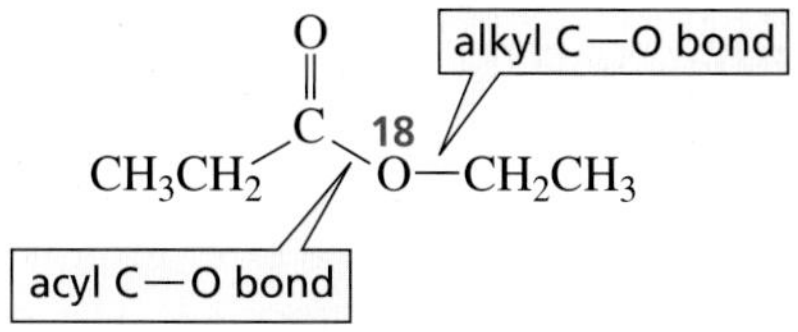

$$CH_3CH_2\overset{O}{\overset{\|}{C}}{-}{}^{18}O{-}CH_2CH_3$$

alkyl C—O bond

acyl C—O bond

a. Which of the products contained the ^{18}O label?

b. What product would have contained the ^{18}O label if the alkyl C—O bond had broken?

PROBLEM 25 SOLVED

Early chemists could envision several possible mechanisms for hydroxide-ion-promoted ester hydrolysis:

1. a nucleophilic acyl substitution reaction

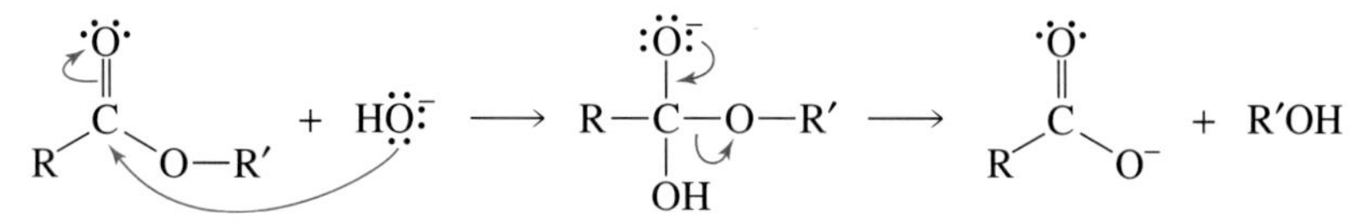

$$RC(=O)OR' + HO^- \longrightarrow R{-}C(O^-)(OH){-}O{-}R' \longrightarrow RC(=O)O^- + R'OH$$

2. an S_N2 reaction

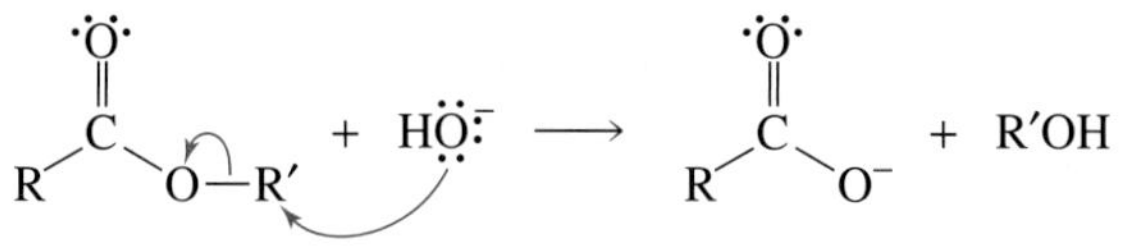

$$RC(=O)O{-}R' + HO^- \longrightarrow RC(=O)O^- + R'OH$$

3. an S_N1 reaction

$$RC(=O)O{-}R' \longrightarrow RC(=O)O^- + R'^+ \xrightarrow{HO^-} R'OH$$

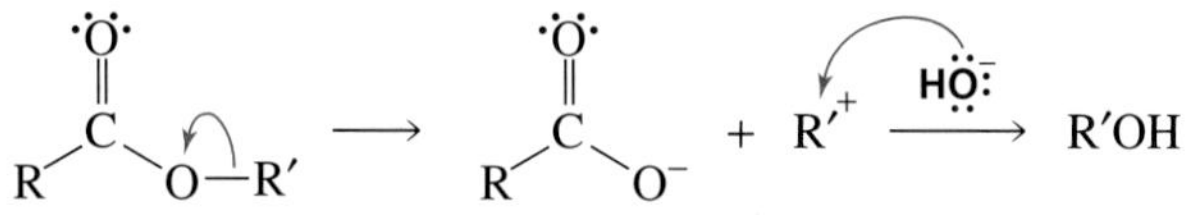

Devise an experiment that would distinguish among these three reactions.

SOLUTION Start with a single stereoisomer of an alcohol whose OH group is bonded to an asymmetric carbon, and determine the specific rotation of the alcohol. Then convert the alcohol into an ester, using a method that does not break any bonds to the asymmetric carbon. Next, hydrolyze the ester, isolate the alcohol obtained from hydrolysis, and determine its specific rotation.

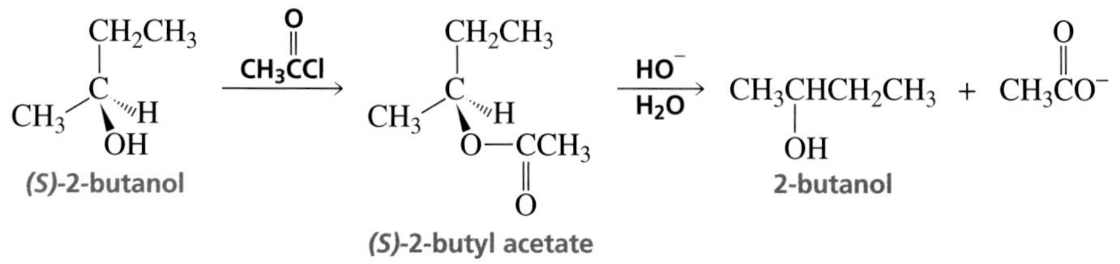

$$\underset{\textbf{(S)-2-butanol}}{CH_3CH(OH)CH_2CH_3} \xrightarrow{CH_3\overset{O}{\overset{\|}{C}}Cl} \underset{\textbf{(S)-2-butyl acetate}}{CH_3CH(OCCH_3)CH_2CH_3} \xrightarrow[H_2O]{HO^-} \underset{\textbf{2-butanol}}{CH_3\underset{OH}{CH}CH_2CH_3} + CH_3\overset{O}{\overset{\|}{C}}O^-$$

If the reaction is a nucleophilic acyl substitution reaction, the product alcohol will have the same specific rotation as the reactant alcohol because no bonds to the asymmetric carbon are broken (Section 5.12).

If the reaction is an S_N2 reaction, the product alcohol and the reactant alcohol will have opposite specific rotations because the mechanism requires back-side attack on the asymmetric carbon (Section 10.2).

If the reaction is an S_N1 reaction, the product alcohol will have a small (or zero) specific rotation because the mechanism requires carbocation formation, which leads to racemization of the alcohol (Section 10.7).

17.13 Soaps, Detergents, and Micelles

Fats and **oils** are triesters of glycerol. Glycerol contains three alcohol groups and therefore can form three ester groups. When the ester groups are hydrolyzed in a basic solution, glycerol and carboxylate ions are formed. The carboxylic acids that are bonded to glycerol in fats and oils have long, unbranched R groups. Because they are obtained from fats, long-chain unbranched carboxylic acids are called **fatty acids**. In Section 26.3, we will see that the difference between a fat and an oil resides in the structure of the fatty acids.

$$\begin{array}{l} CH_2O-\overset{O}{\overset{\|}{C}}-R^1 \\ |\\ CHO-\overset{O}{\overset{\|}{C}}-R^2 \\ | \\ CH_2O-\overset{O}{\overset{\|}{C}}-R^3 \end{array} + H_2O \xrightarrow{\text{NaOH}} \begin{array}{l} CH_2OH \\ | \\ CHOH \\ | \\ CH_2OH \end{array} + \begin{array}{l} R^1-\overset{O}{\overset{\|}{C}}-O^- Na^+ \\ R^2-\overset{O}{\overset{\|}{C}}-O^- Na^+ \\ R^3-\overset{O}{\overset{\|}{C}}-O^- Na^+ \end{array}$$

a fat or an oil — glycerol — sodium salts of fatty acids soap

Soaps are sodium or potassium salts of fatty acids. Thus, soaps are obtained when fats or oils are hydrolyzed under basic conditions. The hydrolysis of an ester in a basic solution is called **saponification**—the Latin word for "soap" is *sapo*. The following compounds are three of the most common soaps:

$$CH_3(CH_2)_{16}\overset{O}{\overset{\|}{C}}O^- Na^+$$
sodium stearate

$$CH_3(CH_2)_7CH{=}CH(CH_2)_7\overset{O}{\overset{\|}{C}}O^- Na^+$$
sodium oleate

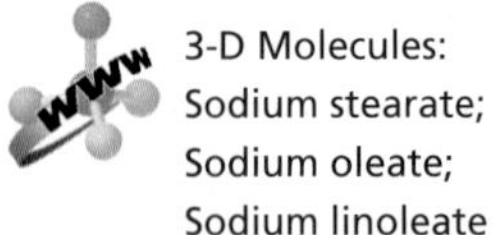
3-D Molecules: Sodium stearate; Sodium oleate; Sodium linoleate

$$CH_3(CH_2)_4CH{=}CHCH_2CH{=}CH(CH_2)_7\overset{O}{\overset{\|}{C}}O^- Na^+$$
sodium linoleate

PROBLEM 26 SOLVED

An oil obtained from coconuts is unusual in that all three fatty acid components are identical. The molecular formula of the oil is $C_{45}H_{86}O_6$. What is the molecular formula of the carboxylate ion obtained when the oil is saponified?

SOLUTION When the oil is saponified, it forms glycerol and three equivalents of carboxylate ion. In losing glycerol, the fat loses three carbons and five hydrogens. Thus, the three equivalents of carboxylate ion have a combined molecular formula of $C_{42}H_{81}O_6$. Dividing by three gives a molecular formula of $C_{14}H_{27}O_2$ for the carboxylate ion.

Long-chain carboxylate ions do not exist as individual ions in aqueous solution. Instead, they arrange themselves in spherical clusters called **micelles**, as shown in Figure 17.4. Each micelle contains 50 to 100 long-chain carboxylate ions. A micelle resembles a large ball, with the polar head of each carboxylate ion and its counterion on the outside of the ball because of their attraction for water and the nonpolar tail buried in the interior of the ball to minimize its contact with water. The attractive forces of hydrocarbon chains for each other in water are called **hydrophobic interactions** (Section 23.14). Soap has cleansing ability because nonpolar oil molecules, which carry dirt, dissolve in the nonpolar interior of the micelle and are carried away with the soap during rinsing.

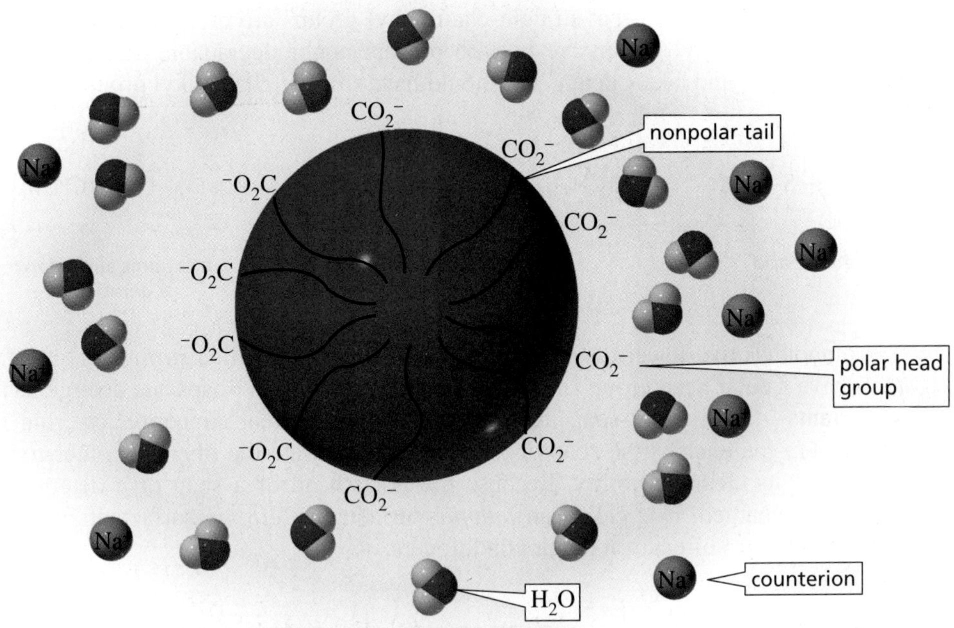

▲ **Figure 17.4**
In aqueous solution, soap forms micelles, with the polar heads (carboxylate groups) on the surface and the nonpolar tails (fatty acid R groups) in the interior.

Because the surface of the micelle is negatively charged, the individual micelles repel each other instead of clustering to form larger aggregates. As river water flows over and around rocks, it leaches out calcium and magnesium ions. The concentration of calcium and magnesium ions in water is described by its "hardness." Hard water contains high concentrations of these ions; soft water contains few, if any, calcium and magnesium ions. While micelles with sodium and potassium cations are dispersed in water, micelles with calcium and magnesium cations form aggregates. In hard water, therefore, soaps form a precipitate that we know as "bathtub ring" or "soap scum."

MAKING SOAP

For thousands of years, soap was prepared by heating animal fat with wood ashes. Wood ashes contain potassium carbonate, which makes the solution basic. The modern commercial method of making soap involves boiling fats or oils in aqueous sodium hydroxide and adding sodium chloride to precipitate the soap, which is then dried and pressed into bars. Perfume can be added for scented soaps, dyes can be added for colored soaps, sand can be added for scouring soaps, and air can be blown into the soap to make it float in water.

Making soap

The formation of soap scum in hard water led to a search for synthetic materials that would have the cleansing properties of soap, but would not form scum when they encountered calcium and magnesium ions. The synthetic "soaps" that were developed, known as **detergents**, are salts of benzene sulfonic acids. Calcium and magnesium sulfonate salts do not form aggregates. "Detergent" comes from the Latin *detergere*, which means "to wipe off." After the initial introduction of detergents into the marketplace, it was discovered that straight-chain alkyl groups are biodegradable, whereas branched-chain alkyl groups are not. To prevent nonbiodegradable detergents from polluting rivers and lakes, detergents should have straight-chain alkyl groups.

$R-C_6H_4-SO_2-OH$ — a sulfonic acid

$CH_3(CH_2)_{11}-C_6H_4-SO_2-O^- Na^+$ — an anionic surfactant, a detergent

$C_6H_5-CH_2\overset{+}{N}(CH_3)_2(CH_2)_{15}CH_3 \; Cl^-$ — a cationic surfactant, a germicide

Compounds that lower the surface tension of water are called *surfactants*. Surfactants have a polar head group and a long-chain nonpolar tail. Soaps and detergents are surfactants—that is why soap and detergent solutions feel slippery. Lowering the surface tension enables the soap or detergent to penetrate the fibers of a fabric, thus enhancing its cleaning ability. Because the polar head of a soap or a detergent is negatively charged, it is called an *anionic* surfactant. *Cationic* surfactants are used widely as fabric softeners and hair conditioners.

17.14 Reactions of Carboxylic Acids

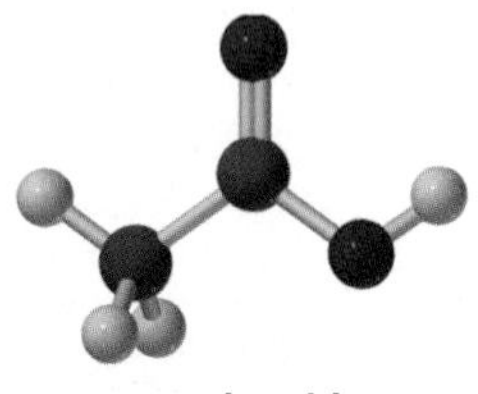
acetic acid

Carboxylic acids can undergo nucleophilic acyl substitution reactions only when they are in their acidic forms. The basic form of a carboxylic acid does not undergo nucleophilic acyl substitution reactions because the negatively charged carboxylate ion is resistant to nucleophilic attack (Section 17.12). Thus, carboxylate ions are even less reactive toward nucleophilic acyl substitution reactions than are amides.

relative reactivities toward nucleophilic acyl substitution

$$RCOOH > RCONH_2 > RCOO^-$$

most reactive (RCOOH) ... least reactive ($RCOO^-$)

Carboxylic acids have approximately the same reactivity as esters—the HO^- leaving group of a carboxylic acid has approximately the same basicity as the RO^- leaving group of an ester. Therefore, like esters, carboxylic acids do not react with halide ions or with carboxylate ions.

Carboxylic acids react with alcohols to form esters. The reaction must be carried out in an acidic solution, not only to catalyze the reaction but also to keep the carboxylic acid in its acid form so that it will react with the nucleophile. Because the tetrahedral intermediate formed in this reaction has two potential leaving groups of approximately the same basicity, the reaction must be carried out with excess alcohol to drive it toward products. Emil Fischer (Section 5.5) was the first to discover that an ester could be prepared by treating a carboxylic acid with excess alcohol in the presence of an acid catalyst, so the reaction is called a **Fischer esterification**.

$$CH_3COOH + CH_3OH \overset{HCl}{\rightleftharpoons} CH_3COOCH_3 + H_2O$$

acetic acid; methyl alcohol excess; methyl acetate

Carboxylic acids do not undergo nucleophilic acyl substitution reactions with amines. Because a carboxylic acid has a lower pK_a than a protonated amine, the carboxylic acid immediately donates a proton to the amine when the two compounds are mixed. The ammonium carboxylate salt is the final product of the reaction; the carboxylate ion is unreactive and the protonated amine is not a nucleophile.

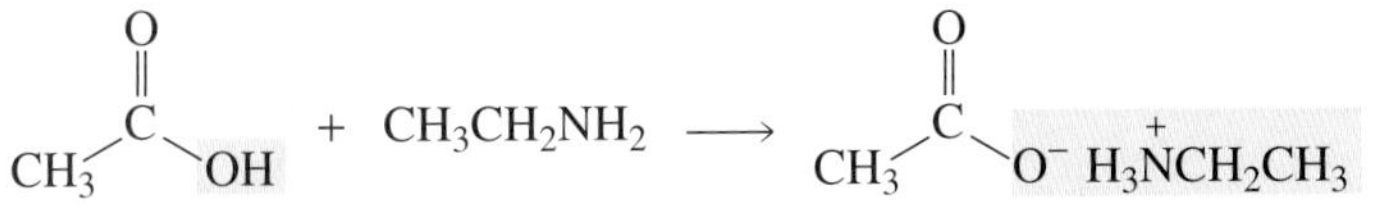

$$CH_3\overset{O}{\overset{\|}{C}}OH + CH_3CH_2NH_2 \longrightarrow CH_3\overset{O}{\overset{\|}{C}}O^- \; H_3\overset{+}{N}CH_2CH_3$$

an ammonium carboxylate salt

$$CH_3CH_2\overset{O}{\overset{\|}{C}}OH + NH_3 \longrightarrow CH_3CH_2\overset{O}{\overset{\|}{C}}O^- \; \overset{+}{N}H_4$$

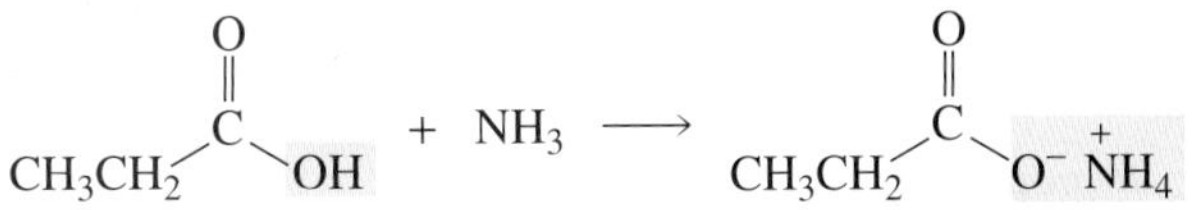

PROBLEM 27

Using an alcohol for one method and an alkyl halide for the other, show two ways to make each of the following esters:

a. methyl butyrate (odor of apples)
b. propyl acetate (odor of pears)
c. ethyl butyrate (odor of pineapple)
d. octyl acetate (odor of oranges)
e. isopentyl acetate (odor of bananas)
f. methyl phenylethanoate (odor of honey)

PROBLEM-SOLVING STRATEGY

Propose a mechanism for the following reaction:

$$\xrightarrow[CH_2Cl_2]{Br_2} \quad + \; HBr$$

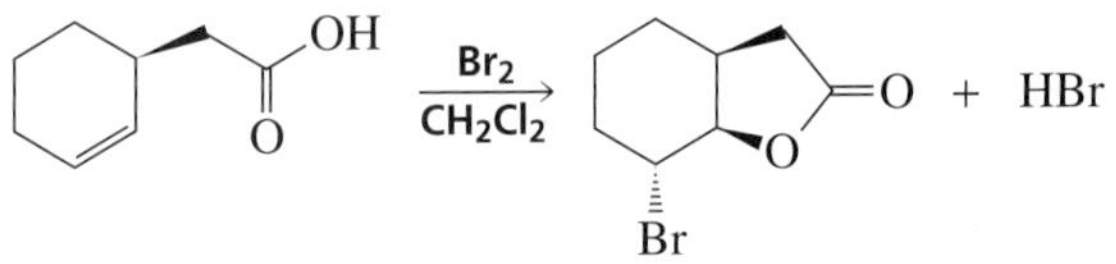

When you are asked to propose a mechanism, look carefully at the reagents to determine the first step of the mechanism. One of the reagents has two functional groups: a carboxyl group and a carbon–carbon double bond. The other reagent, Br_2, does not react with carboxylic acids, but does react with alkenes (Section 4.7). One side of the alkene is sterically hindered by the carboxyl group. Therefore, Br_2 will add to the other side of the double bond, forming a bromonium ion. We know that in the second step of the addition reaction, a nucleophile will attack the bromonium ion. Of the two nucleophiles present, the carbonyl oxygen is positioned to attack the back side of the bromonium ion, resulting in a compound with the observed configuration. Loss of a proton gives the final product of the reaction.

$$:\ddot{Br}-\ddot{Br}: \longrightarrow \; + \; Br^- \longrightarrow \; \rightleftharpoons \; + \; HBr$$

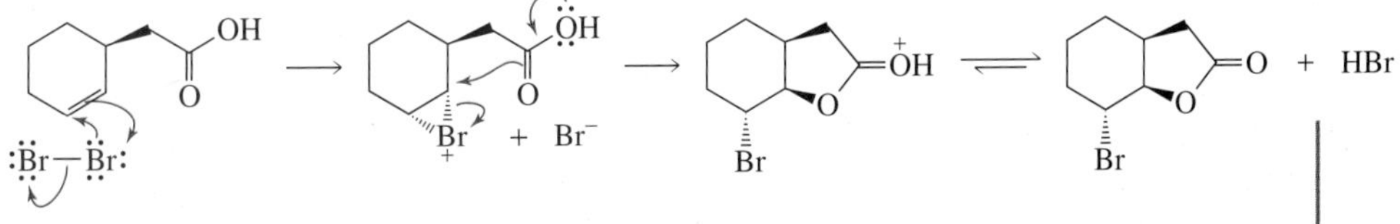

Now continue on to Problem 28.

PROBLEM 28

Propose a mechanism for the following reaction. (*Hint:* Number the carbons to help you see where they end up in the product.)

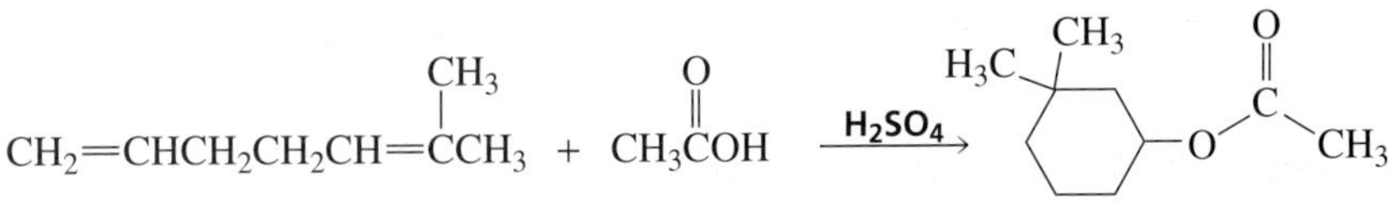

$$CH_2{=}CHCH_2CH_2CH{=}\overset{CH_3}{\overset{|}{C}}CH_3 + CH_3\overset{O}{\overset{\|}{C}}OH \xrightarrow{H_2SO_4}$$

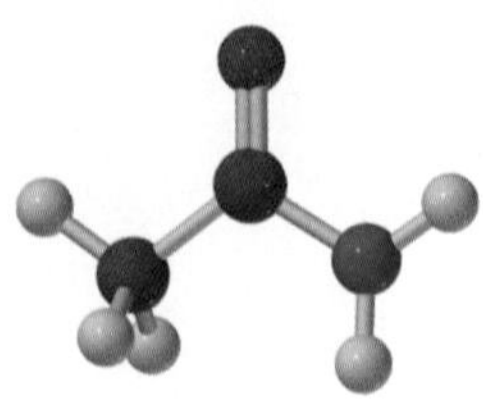
acetamide

17.15 Reactions of Amides

Amides are very unreactive compounds, which is comforting, since proteins are composed of amino acids linked together by amide bonds (Section 23.11). Amides do not react with halide ions, carboxylate ions, alcohols, or water because, in each case, the incoming nucleophile is a weaker base than the leaving group of the amide (Table 17.1).

$$CH_3C(=O)NHCH_2CH_2CH_3 + Cl^- \longrightarrow \text{no reaction}$$
N-propylacetamide

$$CH_3CH_2C(=O)N(CH_3)_2 + CH_3C(=O)O^- \longrightarrow \text{no reaction}$$
N,N-dimethylpropionamide

$$C_6H_5C(=O)NHCH_3 + CH_3OH \longrightarrow \text{no reaction}$$
N-methylbenzamide

$$CH_3CH_2C(=O)NHCH_2CH_3 + H_2O \longrightarrow \text{no reaction}$$
N-ethylpropanamide

Amides do, however, react with water and alcohols if the reaction mixture is heated in the presence of an acid. The reason for this will be explained in Section 17.16.

$$CH_3C(=O)NHCH_2CH_3 + H_2O \xrightarrow[\Delta]{HCl} CH_3C(=O)OH + CH_3CH_2\overset{+}{N}H_3$$
N-ethylacetamide

$$C_6H_5C(=O)NHCH_3 + CH_3CH_2OH \xrightarrow[\Delta]{HCl} C_6H_5C(=O)OCH_2CH_3 + CH_3\overset{+}{N}H_3$$
N-methylbenzamide

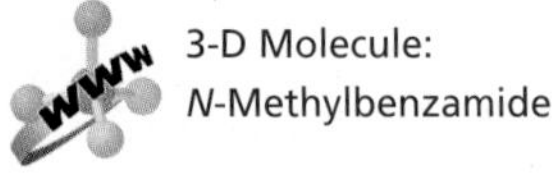
3-D Molecule: *N*-Methylbenzamide

Molecular orbital theory can explain why amides are unreactive. In Section 17.3, we saw that amides have an important resonance contributor in which nitrogen shares its lone pair with the carbonyl carbon—the orbital that contains the lone pair overlaps the empty π^* orbital of the carbonyl group (Figure 17.3). This overlap lowers the energy of the lone pair—it is neither basic nor nucleophilic—and it raises the energy of the π^* orbital of the carbonyl group, making it less reactive to nucleophiles (Figure 17.5).

An amide with an NH_2 group can be dehydrated to a nitrile. Dehydrating reagents commonly employed for this purpose are P_2O_5, $POCl_3$, and $SOCl_2$.

$$CH_3CH_2C(=O)NH_2 \xrightarrow[80\,°C]{P_2O_5} CH_3CH_2C{\equiv}N$$

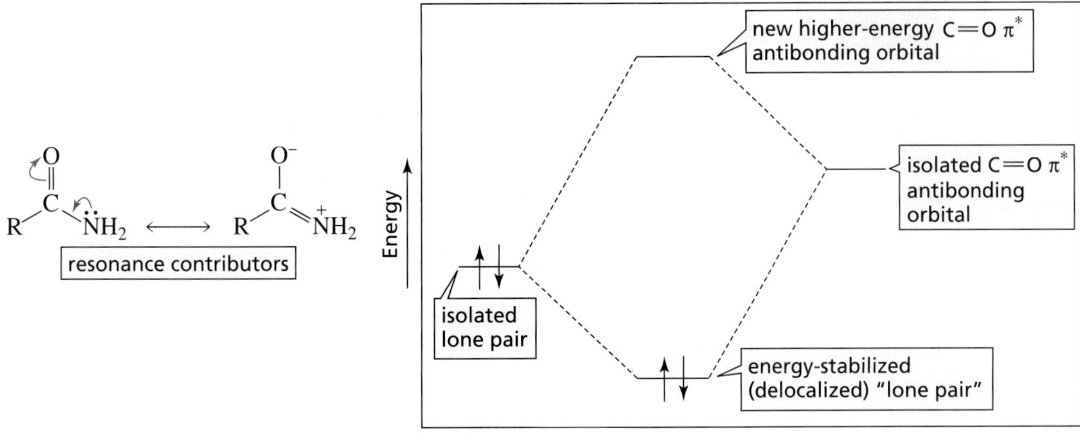

▲ **Figure 17.5**
The filled nonbonding orbital containing nitrogen's lone pair overlaps the empty π^* antibonding molecular orbital of the carbonyl group. This stabilizes the lone pair, making it less reactive, and raises the energy of the π^* orbital of the carbonyl group, making it less able to react with nucleophiles.

PROBLEM 29

a. Which of the following reactions would lead to the formation of an amide?

1. $RC(=O)OH + CH_3NH_2$
2. $RC(=O)OCH_3 + CH_3NH_2$
3. $RC(=O)OCH_3 + CH_3NH_2 \xrightarrow{CH_3O^-}$
4. $RC(=O)O^- + CH_3NH_2$
5. $RC(=O)Cl + CH_3NH_2$
6. $RC(=O)OCH_3 + CH_3NH_2 \xrightarrow{HO^-}$

b. For those reactions that do form amides, what could you do to improve either the rate of amide formation or the yield of the amide product?

PROBLEM 30

Propose a mechanism for the reaction of an amide with thionyl chloride to form a nitrile. (*Hint:* In the first step of the reaction, the amide is the nucleophile and thionyl chloride is the electrophile.)

NATURE'S SLEEPING PILL

Melatonin, a naturally occurring amide, is a hormone that is synthesized by the pineal gland from the amino acid tryptophan. Melatonin regulates the dark–light clock that governs such things as the sleep–wake cycle, body temperature, and hormone production.

tryptophan
an amino acid

melatonin

Melatonin levels increase from evening to night and then decrease as morning approaches. People with high levels of melatonin sleep longer and more soundly than those with low levels. The concentration of the hormone in the blood varies with age—6-year-olds have more than five times the concentration that 80-year-olds have—which is one of the reasons why young people have less trouble sleeping than older people. Melatonin supplements are used to treat insomnia, jet lag, and seasonal affective disorder.

17.16 Acid-Catalyzed Hydrolysis of Amides

When an amide is hydrolyzed under acidic conditions, the acid protonates the carbonyl oxygen, increasing the susceptibility of the carbonyl carbon to nucleophilic attack. Nucleophilic attack by water on the carbonyl carbon leads to tetrahedral intermediate I, which is in equilibrium with its nonprotonated form, tetrahedral intermediate II. Reprotonation can occur either on oxygen to reform tetrahedral intermediate I or on nitrogen to form tetrahedral intermediate III. Protonation on nitrogen is favored because the NH_2 group is a stronger base than the OH group. Of the two possible leaving groups in tetrahedral intermediate III (HO^- and NH_3), NH_3 is the weaker base, so it is expelled, forming the carboxylic acid as the final product. Since the reaction is carried out in an acidic solution, NH_3 will be protonated after it is expelled from the tetrahedral intermediate. This prevents the reverse reaction from occurring.

An acid catalyst increases the reactivity of a carbonyl group.

An acid catalyst can make a group a better leaving group.

mechanism for acid-catalyzed hydrolysis of an amide

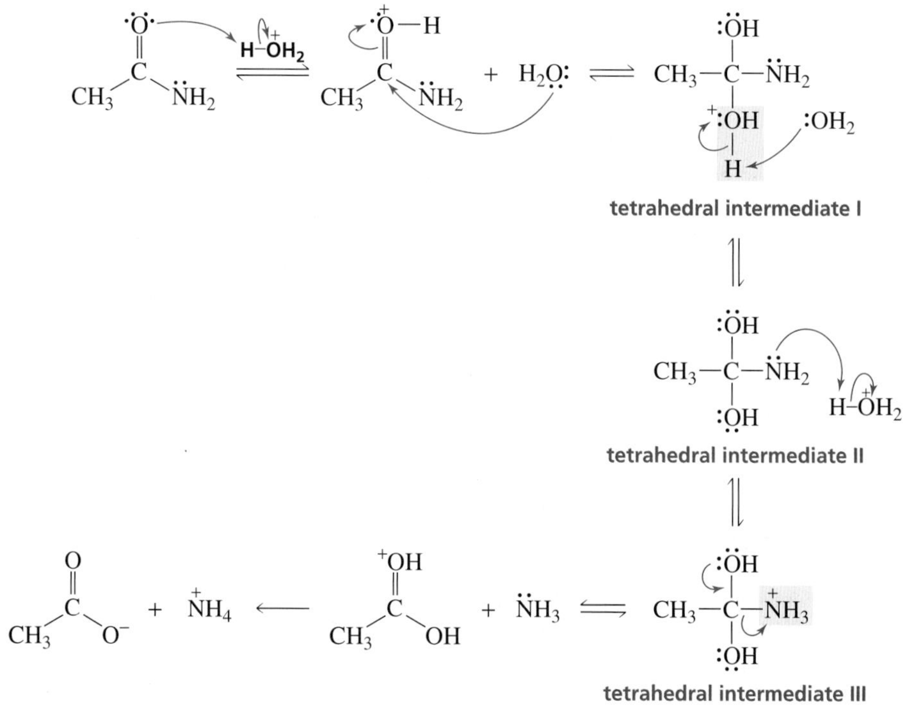

tetrahedral intermediate I

tetrahedral intermediate II

tetrahedral intermediate III

Let's take a minute to see why an amide cannot be hydrolyzed without a catalyst. In the uncatalyzed reaction, the amide is not protonated. Therefore, water, a very poor nucleophile, must attack a neutral amide that is much less susceptible to nucleophilic attack than a protonated amide would be. In addition, the NH_2 group of the tetrahedral intermediate is not protonated in the uncatalyzed reaction. Therefore, HO^- is the group expelled from the tetrahedral intermediate—because HO^- is a weaker base than $^-NH_2$—which reforms the amide.

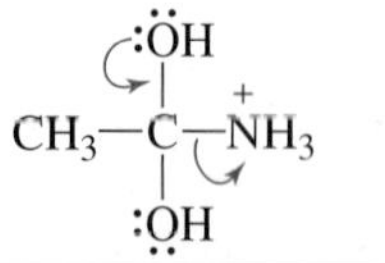

tetrahedral intermediate in acid-catalyzed amide hydrolysis

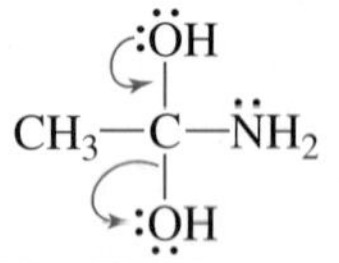

tetrahedral intermediate in uncatalyzed amide hydrolysis

Tutorial:
Conversions between carboxylic acid derivatives

An amide reacts with an alcohol in the presence of acid for the same reason that it reacts with water in the presence of acid.

PENICILLIN AND DRUG RESISTANCE

Penicillin contains a strained β-lactam ring. The strain in the four-membered ring increases the amide's reactivity. It is thought that the antibiotic activity of penicillin results from its ability to acylate (put an acyl group on) a CH_2OH group of an enzyme that is involved in the synthesis of bacterial cell walls. Acylation inactivates the enzyme, and actively growing bacteria die because they are unable to produce functional cell walls. Penicillin has no effect on mammalian cells because mammalian cells are not enclosed by cell walls. To minimize hydrolysis of the β-lactam ring during storage, penicillins are refrigerated.

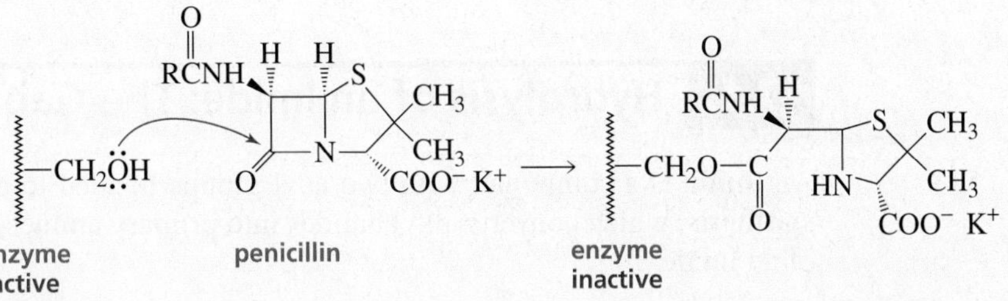

enzyme active —$CH_2\ddot{O}H$ + penicillin ⟶ enzyme inactive —CH_2O—C(=O)—

Bacteria that are resistant to penicillin secrete penicillinase, an enzyme that catalyzes the hydrolysis of the β-lactam ring of penicillin. The ring-opened product has no antibacterial activity.

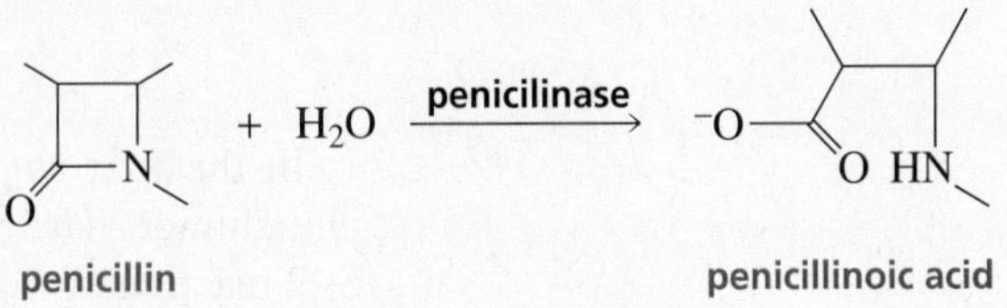

penicillin + H_2O —penicilinase→ penicillinoic acid

PENICILLINS IN CLINICAL USE

More than 10 different penicillins are currently in clinical use. They differ only in the group (R) attached to the carbonyl group. Some of these penicillins are shown here. In addition to their structural differences, the penicillins differ in the organisms against which they are most effective. They also differ in their resistance to penicillinase. For example, ampicillin, a synthetic penicillin, is clinically effective against bacteria that are resistant to penicillin G, a naturally occurring penicillin. Almost 19% of humans are allergic to penicillin G.

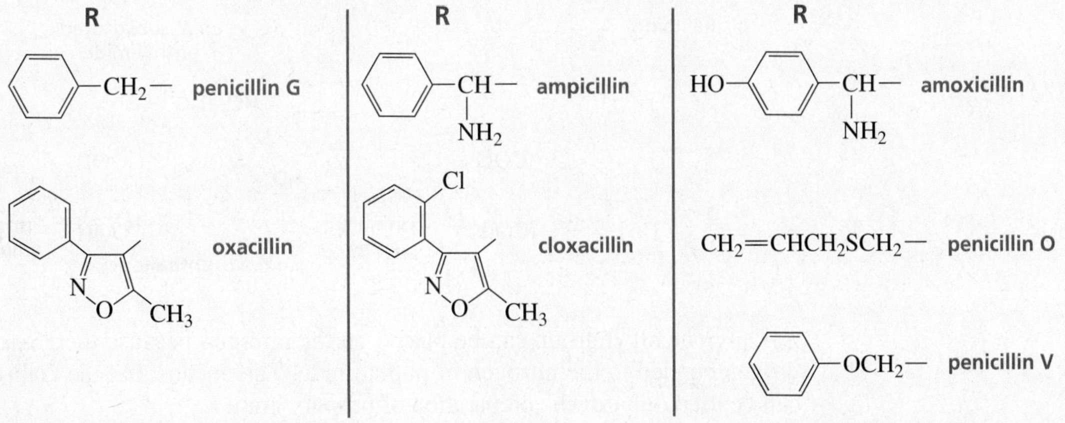

R	R	R
$C_6H_5CH_2$— penicillin G	$C_6H_5CH(NH_2)$— ampicillin	HO—C_6H_4—$CH(NH_2)$— amoxicillin
oxacillin	cloxacillin	$CH_2{=}CHCH_2SCH_2$— penicillin O
		$C_6H_5OCH_2$— penicillin V

Penicillin V is a semisynthetic penicillin that is in clinical use. It is not a naturally occurring penicillin; nor is it a true synthetic penicillin because chemists don't synthesize it. The *Penicillium* mold synthesizes it after the mold is fed 2-phenoxyethanol, the compound it needs for the side chain.

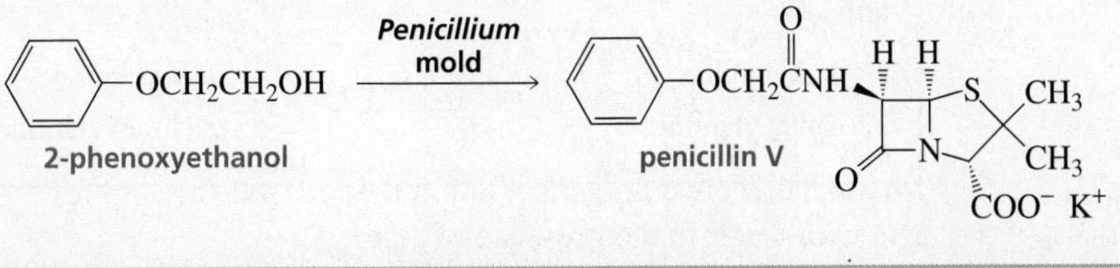

2-phenoxyethanol $C_6H_5OCH_2CH_2OH$ —*Penicillium* mold→ penicillin V

PROBLEM 31◆

List the following amides in order of decreasing reactivity toward acid-catalyzed hydrolysis:

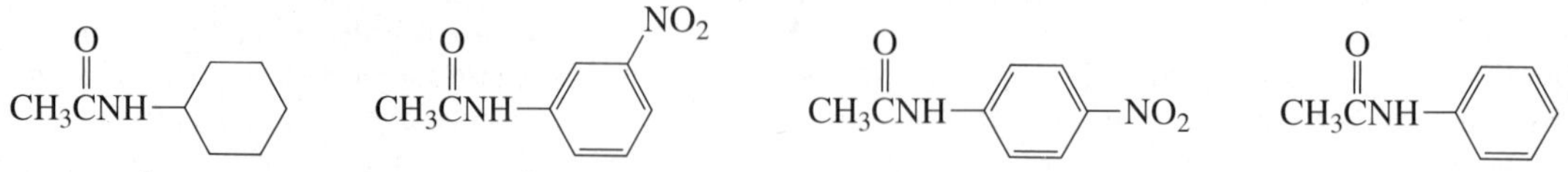

A: CH_3CNH–cyclohexyl (C=O)

B: CH_3CNH–phenyl with NO_2 (meta)

C: CH_3CNH–phenyl–NO_2 (para)

D: CH_3CNH–phenyl

17.17 Hydrolysis of an Imide: The Gabriel Synthesis

An **imide** is a compound with two acyl groups bonded to a nitrogen. The **Gabriel synthesis**, which converts alkyl halides into primary amines, involves the hydrolysis of an imide.

$$RCH_2Br \xrightarrow{\text{Gabriel synthesis}} RCH_2NH_2$$

In the first step of the reaction, a base removes a proton from the nitrogen of phthalimide. The resulting nucleophile reacts with an alkyl halide. Because this is an S_N2 reaction, it works best with primary alkyl halides (Section 10.2). Hydrolysis of the *N*-substituted phthalimide is catalyzed by acid. Because the solution is acidic, the final products are a primary alkyl ammonium ion and phthalic acid. Neutralization of the ammonium ion with base forms the primary amine. Notice that the alkyl group of the primary amine is identical to the alkyl group of the alkyl halide.

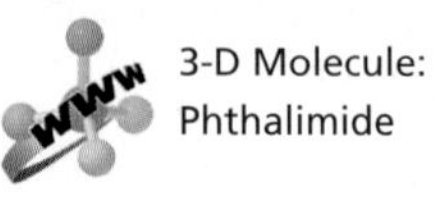

3-D Molecule: Phthalimide

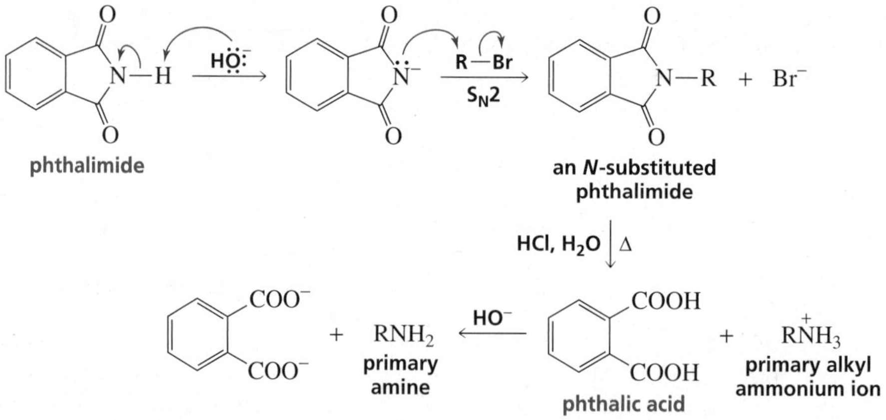

phthalimide (N–H) $\xrightarrow{HO^-}$ N$^-$ $\xrightarrow[S_N2]{R-Br}$ an *N*-substituted phthalimide (N–R) + Br^-

$\downarrow$ HCl, H_2O, Δ

phthalic acid (COOH, COOH) + $R\overset{+}{N}H_3$ primary alkyl ammonium ion $\xrightarrow{HO^-}$ phthalate (COO^-, COO^-) + RNH_2 primary amine

Only one alkyl group can be placed on the nitrogen because there is only one hydrogen bonded to the nitrogen of phthalimide. This means that the Gabriel synthesis can be used only for the preparation of primary amines.

PROBLEM 32◆

What alkyl bromide would you use in a Gabriel synthesis to prepare each of the following amines?

a. pentylamine
b. isohexylamine
c. benzylamine
d. cyclohexylamine

PROBLEM 33

Primary amines can also be prepared by the reaction of an alkyl halide with azide ion, followed by catalytic hydrogenation. What advantage do this method and the Gabriel synthesis have over the synthesis of a primary amine using an alkyl halide and ammonia?

$$CH_3CH_2CH_2Br \xrightarrow{^-N_3} CH_3CH_2CH_2N{=}\overset{+}{N}{=}\overset{-}{N} \xrightarrow[Pt]{H_2} CH_3CH_2CH_2NH_2 + N_2$$

17.18 Hydrolysis of Nitriles

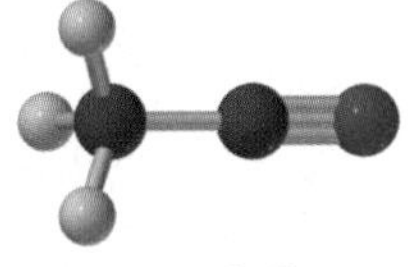

acetonitrile

Nitriles are even harder to hydrolyze than amides. Nitriles are slowly hydrolyzed to carboxylic acids when heated with water and an acid.

$$CH_3CH_2C{\equiv}N + H_2O \xrightarrow[\Delta]{HCl} CH_3CH_2C(=O)OH + \overset{+}{N}H_4$$

In the first step of the acid-catalyzed hydrolysis of a nitrile, the acid protonates the nitrogen of the cyano group, making it easier for water to attack the carbon of the cyano group in the next step. Attack on the cyano group by water is analogous to attack on a carbonyl group by water. Because nitrogen is a stronger base than oxygen, oxygen loses a proton and nitrogen gains a proton, resulting in a product that is a protonated amide (whose two resonance contributors are shown). The amide is immediately hydrolyzed to a carboxylic acid—because an amide is easier to hydrolyze than a nitrile—following the acid-catalyzed mechanism shown in Section 17.15.

mechanism for acid-catalyzed hydrolysis of a nitrile

$$R{-}C{\equiv}N: \xrightleftharpoons{H-\overset{+}{O}H_2} R{-}C{\equiv}\overset{+}{N}H + H_2\ddot{O}: \rightleftharpoons R{-}C(=\ddot{N}H){-}\overset{+}{O}H{-}H \;\; (H_2O:) \rightleftharpoons R{-}C(=\ddot{N}H){-}\ddot{O}H \;\; (H{-}\overset{+}{O}H_2)$$

$$R{-}C(=\ddot{O}:){-}OH \xleftarrow[\text{(several steps)}]{H_2O} R{-}C(=\overset{+}{O}H){-}\ddot{N}H_2 \longleftrightarrow R{-}C(=\overset{+}{N}H_2){-}\ddot{O}H$$

a carboxylic acid **a protonated amide**

Because nitriles can be prepared from the reaction of an alkyl halide with cyanide ion (Section 10.4), you now know how to convert an alkyl halide into a carboxylic acid. Notice that the carboxylic acid has one more carbon than the alkyl halide.

$$CH_3CH_2Br \xrightarrow[DMF]{^-C{\equiv}N} CH_3CH_2C{\equiv}N \xrightarrow[\Delta]{HCl,\ H_2O} CH_3CH_2C(=O)OH$$

PROBLEM 34◆

Which alkyl halides form the following carboxylic acids after reacting with sodium cyanide and the product heated in an acidic aqueous solution?

a. butyric acid

b. isovaleric acid

c. cyclohexanecarboxylic acid

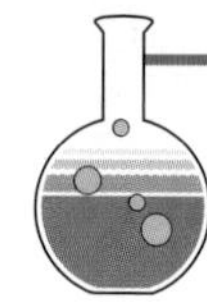

17.19 Designing a Synthesis IV: The Synthesis of Cyclic Compounds

Most of the reactions we have studied are intermolecular reactions: The two reacting groups are in different molecules. *Cyclic compounds are formed from intramolecular reactions*: The two reacting groups are in the same molecule. We have seen that intramolecular reactions are particularly favorable if the reaction forms a compound with a five- or a six-membered ring (Section 11.11).

In designing the synthesis of a cyclic compound, we must examine the target molecule to determine what kinds of reactive groups will be necessary for a successful synthesis. For example, we know that an ester is formed from the acid-catalyzed reaction of a carboxylic acid with an alcohol. Therefore, a cyclic ester (lactone) can be prepared from a reactant that has both a carboxylic acid group and an alcohol group in the same molecule. The size of the lactone ring will be determined by the number of carbon atoms between the carboxylic acid group and the alcohol group.

$$HOCH_2CH_2CH_2CH_2\overset{O}{\overset{\|}{C}}OH \xrightarrow{HCl}$$ (six-membered lactone)

four intervening carbon atoms

$$HOCH_2CH_2CH_2\overset{O}{\overset{\|}{C}}OH \xrightarrow{HCl}$$ (five-membered lactone)

three intervening carbon atoms

A compound with a ketone group attached to a benzene ring can be prepared using a Friedel–Crafts acylation reaction (Section 15.13). Therefore, a cyclic ketone will result if a Lewis acid ($AlCl_3$) is added to a compound that contains both a benzene ring and an acyl chloride group separated by the appropriate number of carbon atoms.

$$C_6H_5CH_2CH_2\overset{O}{\overset{\|}{C}}Cl \xrightarrow[\text{2. } H_2O]{\text{1. } AlCl_3}$$ (1-indanone)

$$C_6H_5CH_2CH_2CH_2\overset{O}{\overset{\|}{C}}Cl \xrightarrow[\text{2. } H_2O]{\text{1. } AlCl_3}$$ (1-tetralone)

A cyclic ether can be prepared by an intramolecular Williamson ether synthesis (Section 11.9).

$$ClCH_2CH_2CH_2\underset{:\ddot{O}H}{\overset{CH_3}{C}}CH_3 \xrightarrow{Na} Cl{-}CH_2CH_2CH_2\underset{:\ddot{O}:^-}{\overset{CH_3}{C}}CH_3 \longrightarrow$$ (2,2-dimethyltetrahydrofuran)

A cyclic ether can also be prepared by an intramolecular electrophilic addition reaction.

$$CH_2{=}CHCH_2CH_2CH_2CH(OH)CH_3 \xrightarrow{HCl} CH_3\overset{+}{C}HCH_2CH_2CH_2CH(\ddot{O}H)CH_3 \longrightarrow \text{2,6-dimethyl-oxanium (O–H}^+) \longrightarrow \text{2,6-dimethyltetrahydropyran} + H^+$$

The product obtained from the intramolecular reaction can undergo further reactions allowing for the synthesis of many different compounds. For example, the alkyl bromide could undergo an elimination reaction, could undergo substitution with a wide variety of nucleophiles, or could be converted into a Grignard reagent that could react with many different electrophiles.

$$C_6H_5CH_2CH_2CH_2CH_2Cl \xrightarrow{AlCl_3} \text{tetralin} \xrightarrow[\Delta/\text{peroxide}]{NBS} \text{1-bromotetralin (Br)}$$

PROBLEM 35

Design a synthesis for each of the following compounds, using an intramolecular reaction:

a. 2-ethyltetrahydrofuran (CH_2CH_3)

b. 1-methyltetralin (CH_3)

c. 1,2-dihydronaphthalene

d. tetralin-1-carboxylic acid (COOH)

e. 2-allyltetrahydrofuran ($CH_2CH{=}CH_2$)

f. tetrahydrofuran-2-yl $CH_2CH_2CH_2OH$

17.20 Synthesis of Carboxylic Acid Derivatives

Of the various classes of carbonyl compounds discussed in this chapter—acyl halides, acid anhydrides, esters, carboxylic acids, and amides—carboxylic acids are the most commonly available both in the laboratory and in biological systems. This means that carboxylic acids are the reagents most likely to be available when a chemist or a cell needs to synthesize a carboxylic acid derivative. However, we have seen that carboxylic acids are relatively unreactive toward nucleophilic acyl substitution reactions because the OH group of a carboxylic acid is a strong base and therefore a poor leaving group. In neutral solutions (physiological pH $= 7.3$), a carboxylic acid is even more resistant to nucleophilic acyl substitution reactions because it exists predominantly in its unreactive negatively charged basic form (Sections 1.20 and 17.14). Therefore, both organic chemists and cells need a way to activate carboxylic acids so that they can readily undergo nucleophilic acyl substitution reactions.

Activation of Carboxylic Acids for Nucleophilic Acyl Substitution Reactions in the Laboratory

Because acyl halides are the most reactive of the carboxylic acid derivatives, the easiest way to synthesize any other carboxylic acid derivative is to add the appropriate nucleophile to an acyl halide. Consequently, organic chemists activate carboxylic acids by converting them into acyl halides.

A carboxylic acid can be converted into an acyl chloride by heating it either with thionyl chloride ($SOCl_2$) or with phosphorus trichloride (PCl_3). Acyl bromides can be synthesized by using phosphorus tribromide (PBr_3).

$$CH_3COOH \text{ (acetic acid)} + SOCl_2 \text{ (thionyl chloride)} \xrightarrow{\Delta} CH_3COCl \text{ (acetyl chloride)} + SO_2 + HCl$$

$$3\ CH_3CH_2CH_2COOH \text{ (butanoic acid)} + PCl_3 \text{ (phosphorus trichloride)} \xrightarrow{\Delta} 3\ CH_3CH_2CH_2COCl \text{ (butanoyl chloride)} + H_3PO_3$$

$$3\ C_6H_5COOH \text{ (benzoic acid)} + PBr_3 \text{ (phosphorus tribromide)} \xrightarrow{\Delta} 3\ C_6H_5COBr \text{ (benzoyl bromide)} + H_3PO_3 \text{ (phosphorous acid)}$$

All these reagents convert the OH group of a carboxylic acid into a better leaving group than the halide ion.

$$RC(=O)-O-S(=O)Cl \qquad RC(=O)-O-PCl_2 \qquad RC(=O)-O-PBr_2$$

good leaving groups

Therefore, when the halide ion subsequently attacks the carbonyl carbon and forms a tetrahedral intermediate, the halide ion is *not* the group that is eliminated.

$$RC(=O)-O-S(=O)Cl + :\ddot{Cl}:^- \longrightarrow R-C(O^-)(Cl)-O-S(=O)Cl \longrightarrow RC(=O)Cl + SO_2 + Cl^-$$

Notice that the reagents that cause the OH group of a carboxylic acid to be replaced by a halogen are the same reagents that cause the OH group of an alcohol to be replaced by a halogen (Section 12.3).

Once the acyl halide has been prepared, a wide variety of carboxylic acid derivatives can be synthesized by adding the appropriate nucleophile (Section 17.8).

$$RC(=O)Cl + RC(=O)O^- \longrightarrow RC(=O)OC(=O)R \text{ (an anhydride)} + Cl^-$$

$$RC(=O)Cl + ROH \longrightarrow RC(=O)OR \text{ (an ester)} + HCl$$

$$RC(=O)Cl + 2\ RNH_2 \longrightarrow RC(=O)NHR \text{ (an amide)} + R\overset{+}{N}H_3\ Cl^-$$

Carboxylic acids can also be activated for nucleophilic acyl substitution reactions by being converted into anhydrides. Treating a carboxylic acid with a strong dehydrating agent such as P_2O_5 yields an anhydride.

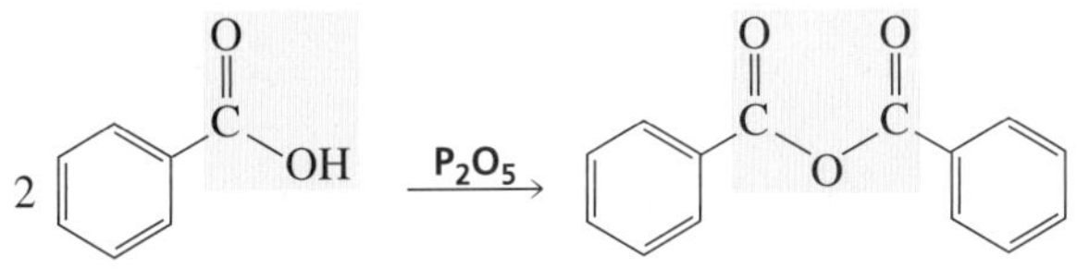

$$2\ C_6H_5COOH \xrightarrow{P_2O_5} C_6H_5C(=O)OC(=O)C_6H_5$$

Carboxylic acids and carboxylic acid derivatives can also be prepared by methods other than nucleophilic acyl substitution reactions. A summary of the methods used to synthesize these compounds is given in Appendix IV.

Activation of Carboxylate Ions for Nucleophilic Acyl Substitution Reactions in Biological Systems

The synthesis of compounds by biological organisms is called **biosynthesis**. Acyl halides and acid anhydrides are too reactive to be used as reagents in biological systems. Cells live in a predominantly aqueous environment, and acyl halides and acid anhydrides are rapidly hydrolyzed in water. So living organisms must activate carboxylic acids in a different way.

One way living organisms activate carboxylic acids is to convert them into acyl phosphates, acyl pyrophosphates, and acyl adenylates.

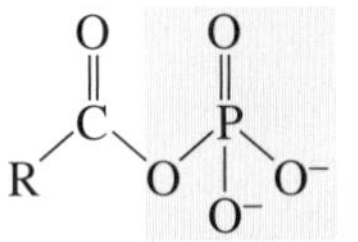

$RC(=O)OP(=O)(O^-)O^-$ — **an acyl phosphate**

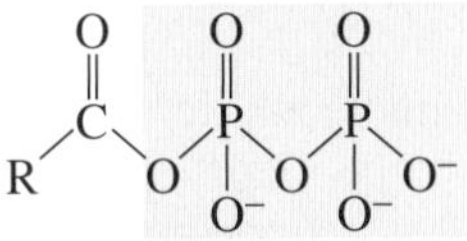

$RC(=O)OP(=O)(O^-)OP(=O)(O^-)O^-$ — **an acyl pyrophosphate**

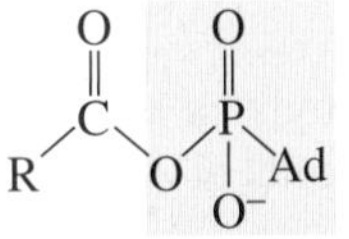

$RC(=O)OP(=O)(O^-)Ad$ — **an acyl adenylate**

An **acyl phosphate** is a mixed anhydride of a carboxylic acid and phosphoric acid; an **acyl pyrophosphate** is a mixed anhydride of a carboxylic acid and pyrophosphoric acid; an **acyl adenylate** is a mixed anhydride of a carboxylic acid and adenosine monophosphate (AMP).

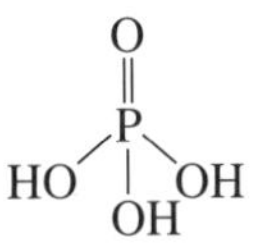

$HOP(=O)(OH)OH$ — **phosphoric acid**

$HOP(=O)(OH)OP(=O)(OH)OH$ — **pyrophosphoric acid**

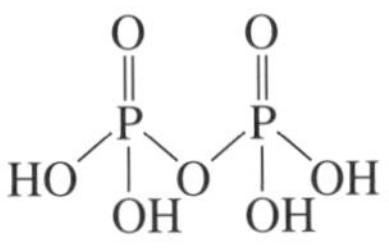

The structure of adenosine triphosphate (ATP) is shown below with "Ad" in place of the adenosyl group; adenosine monophosphate has two fewer phosphate groups.

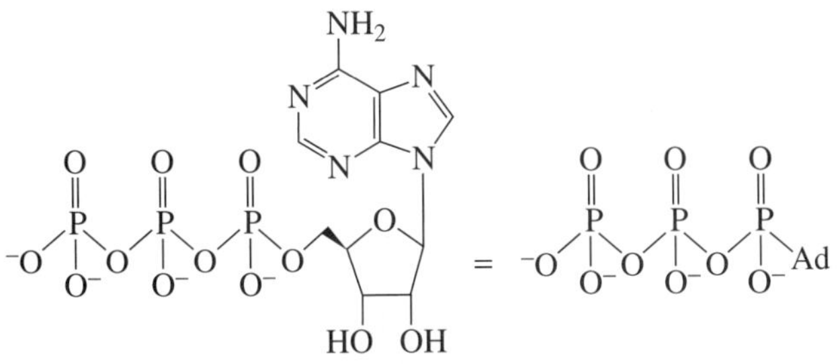

$$^-OP(=O)(O^-)OP(=O)(O^-)OP(=O)(O^-)O\text{-CH}_2\text{-ribose(HO, OH)-adenine}(NH_2) = {}^-OP(=O)(O^-)OP(=O)(O^-)OP(=O)(O^-)Ad$$

adenosine triphosphate
ATP

3-D Molecule: Adenosine triphosphate

Acyl phosphates are formed by nucleophilic attack of a carboxylate ion on the γ-phosphorus (the terminal phosphorus) of ATP. Attack of a nucleophile on the P=O group breaks a **phosphoanhydride bond** (rather than the π bond), so an intermediate is not formed. Essentially, it is an S_N2 reaction with an adenosine pyrophosphate leaving group. This reaction and the ones that follow will be discussed in greater detail in Sections 27.3 and 27.4.

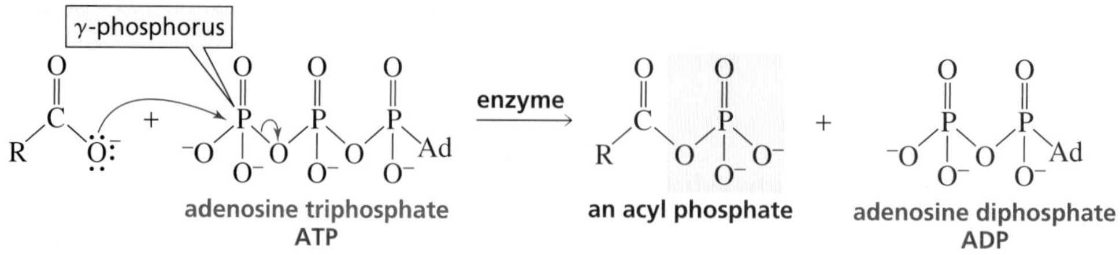

γ-phosphorus

adenosine triphosphate ATP —enzyme→ an acyl phosphate + adenosine diphosphate ADP

Acyl pyrophosphates are formed by nucleophilic attack of a carboxylate ion on the β-phosphorus of ATP.

β-phosphorus

adenosine triphosphate ATP —enzyme→ an acyl pyrophosphate + adenosine monophosphate AMP

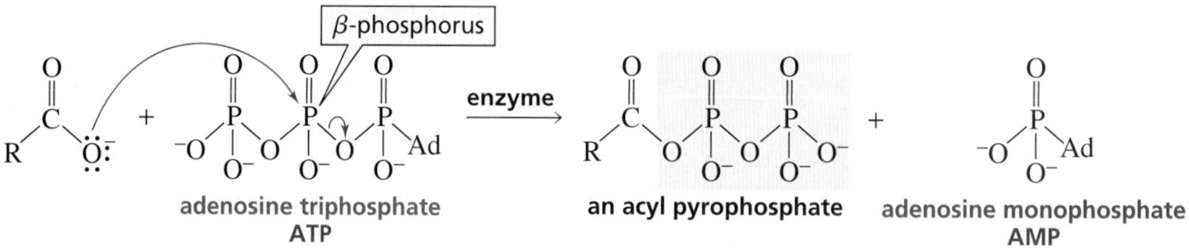

Acyl adenylates are formed by nucleophilic attack of a carboxylate ion on the α-phosphorus of ATP.

α-phosphorus

adenosine triphosphate ATP —enzyme→ an acyl adenylate + pyrophosphate

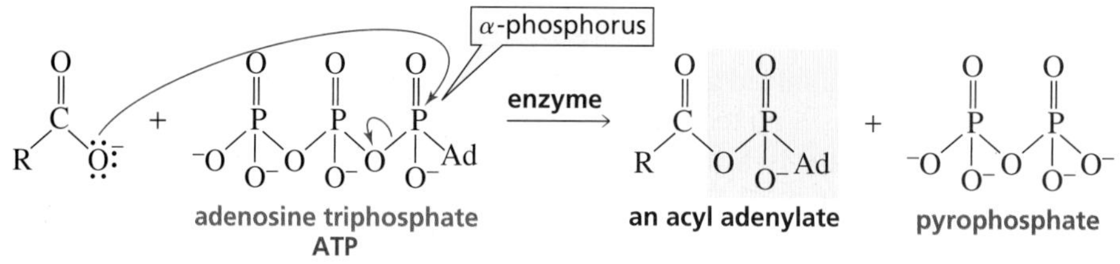

Because these mixed anhydrides are negatively charged, they are not readily approached by nucleophiles. Thus, they are used only in enzyme-catalyzed reactions. One of the functions of enzymes that catalyze biological nucleophilic acyl substitution reactions is to neutralize the negative charges of the mixed anhydride (Section 27.5). Another function of the enzyme is to exclude water from the site where the reaction takes place. Otherwise hydrolysis of the mixed anhydride would compete with the desired nucleophilic acyl substitution reaction.

A **thioester** is an ester with a sulfur atom in place of the oxygen atom between the acyl and alkyl groups.

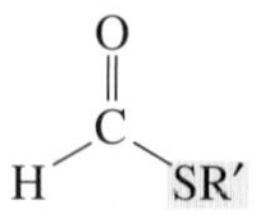

a thioester

Thioesters are the most common forms of activated carboxylic acids in a cell. Although thioesters hydrolyze at about the same rate as oxygen esters, they are much more reactive than oxygen esters toward attack by nitrogen and carbon nucleophiles. This allows a thioester to survive in the aqueous environment of the cell—without being hydrolyzed—waiting to be a substrate in a nucleophilic acyl substitution reaction.

The carbonyl carbon of a thioester is more susceptible to nucleophilic attack than is the carbonyl carbon of an oxygen ester because there is less electron delocalization onto the carbonyl oxygen when Y is S than when Y is O. There is less electron delocalization because there is less overlap between the 3*p* orbital of sulfur and the 2*p* orbital of carbon, compared with the amount of overlap between the 2*p* orbital of oxygen and the 2*p* orbital of carbon (Section 17.2). In addition, a thiolate ion is a weaker base and therefore a better leaving group than an alkoxide ion.

$R-C(=O)-\ddot{Y}R \longleftrightarrow R-C(-O^-)=\overset{+}{Y}R$

CH_3CH_2SH pK_a = 10.5 CH_3CH_2OH pK_a = 15.9

The thiol used in biological systems for the formation of thioesters is coenzyme A. The compound is written "CoASH" to emphasize that the thiol group is the reactive part of the molecule.

Coenzyme A was discovered by **Fritz A. Lipmann (1899–1986)**. *He also was the first to recognize its importance in intermediary metabolism. Lipmann was born in Germany. To escape the Nazis, he moved to Denmark in 1932 and to the United States in 1939, becoming a U.S. citizen in 1944. For his work on coenzyme A, he received the Nobel Prize in physiology or medicine in 1953, sharing it with Hans Krebs.*

$HS-CH_2CH_2NH\overset{O}{\overset{\|}{C}}CH_2CH_2NH\overset{O}{\overset{\|}{C}}CH(OH)-C(CH_3)_2CH_2O-P(=O)(O^-)-O-P(=O)(O^-)-O$–ribose(3'-$OPO_3^{2-}$, 2'-OH)–adenine($NH_2$)

coenzyme A
CoASH

The first step in converting a carboxylic acid into a thioester is to convert the carboxylic acid into an acyl adenylate. The acyl adenylate then reacts with CoASH to form the thioester. The most common thioester in cells is acetyl-CoA.

$CH_3C(=O)O^-$ + ATP ($^-O-P(=O)(O^-)-O-P(=O)(O^-)-O-P(=O)(O^-)-Ad$) $\xrightarrow{\text{enzyme}}$ $CH_3C(=O)-O-P(=O)(O^-)-Ad$ + pyrophosphate $\xrightarrow{\text{CoASH}}$ $CH_3C(=O)SCoA$ (acetyl-CoA) + AMP

Acetylcholine, an ester, is one example of a carboxylic acid derivative that cells synthesize using acetyl-CoA. Acetylcholine is a *neurotransmitter*—a compound that transmits nerve impulses across the synapses between nerve cells.

$CH_3C(=O)SCoA$ (acetyl-CoA) + $HOCH_2CH_2\overset{+}{N}(CH_3)_3$ (choline) $\xrightarrow{\text{enzyme}}$ $CH_3C(=O)OCH_2CH_2\overset{+}{N}(CH_3)_3$ (acetylcholine) + CoASH

It is believed that one way genes are activated is by amide formation between acetyl-CoA and a lysine residue (an amine) of a DNA-bound protein.

$CH_3C(=O)SCoA$ + $-CH(CH_2CH_2CH_2CH_2NH_2)-C(=O)NH-$ (a lysine residue) $\xrightarrow{\text{enzyme}}$ $-CH(CH_2CH_2CH_2CH_2NH-C(=O)CH_3)-C(=O)NH-$ (an amide) + CoASH

NERVE IMPULSES, PARALYSIS, AND INSECTICIDES

After a nerve impulse is transmitted between cells, acetylcholine must be rapidly hydrolyzed to enable the recipient cell to receive another impulse.

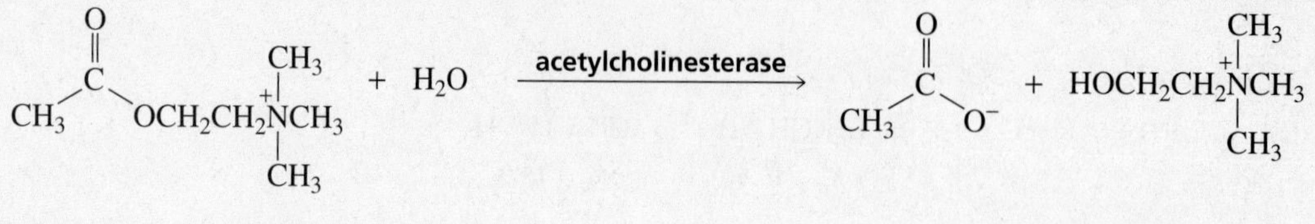

$$CH_3C(=O)OCH_2CH_2\overset{+}{N}(CH_3)_3 + H_2O \xrightarrow{\text{acetylcholinesterase}} CH_3C(=O)O^- + HOCH_2CH_2\overset{+}{N}(CH_3)_3$$

Acetylcholinesterase, the enzyme that catalyzes this hydrolysis, has a CH_2OH group that is necessary for its catalytic activity. Diisopropyl fluorophosphate (DFP), a military nerve gas, inhibits acetylcholinesterase by reacting with the CH_2OH group. When the enzyme is inhibited, paralysis occurs because the nerve impulses cannot be transmitted properly. DFP is extremely toxic: Its LD_{50} (the lethal dose for 50% of the test animals) is only 0.5 mg/kg of body weight.

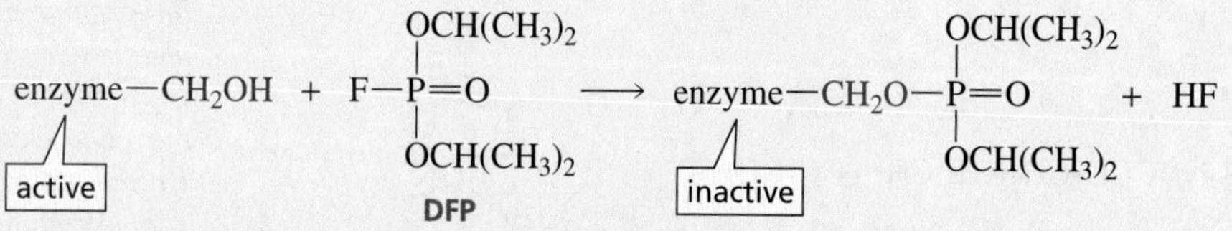

$$\underset{\text{active}}{\text{enzyme}}-CH_2OH + \underset{\text{DFP}}{F-P(=O)(OCH(CH_3)_2)_2} \longrightarrow \underset{\text{inactive}}{\text{enzyme}}-CH_2O-P(=O)(OCH(CH_3)_2)_2 + HF$$

Malathion and parathion, compounds related to DFP, are used as insecticides. The LD_{50} of malathion is 2800 mg/kg. Parathion is more toxic, with an LD_{50} of 2 mg/kg.

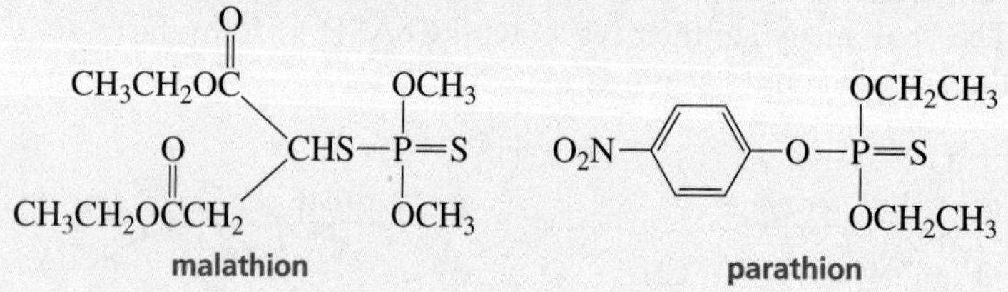

$$\underset{\text{malathion}}{(CH_3CH_2OC(=O))(CH_3CH_2OC(=O)CH_2)CHS-P(=S)(OCH_3)_2} \qquad \underset{\text{parathion}}{O_2N-C_6H_4-O-P(=S)(OCH_2CH_3)_2}$$

17.21 Dicarboxylic Acids and Their Derivatives

The structures of some common dicarboxylic acids and their pK_a values are shown in Table 17.2.

Although the two carboxyl groups of a dicarboxylic acid are identical, the two pK_a values are different because the protons are lost one at a time and therefore leave from different species. The first proton is lost from a neutral molecule, whereas the second proton is lost from a negatively charged ion.

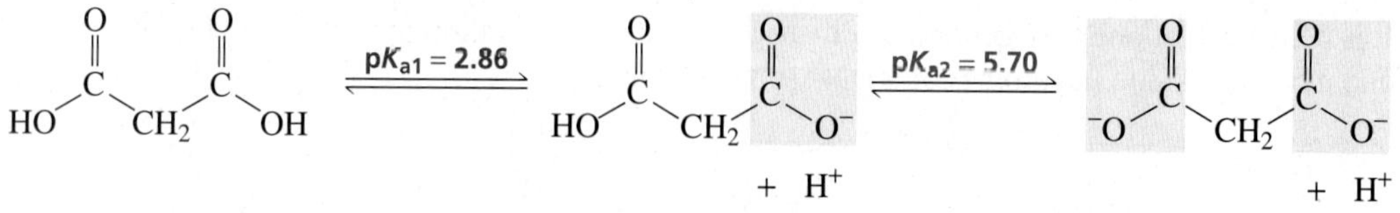

$$HOC(=O)CH_2C(=O)OH \underset{}{\overset{pK_{a1} = 2.86}{\rightleftharpoons}} HOC(=O)CH_2C(=O)O^- + H^+ \underset{}{\overset{pK_{a2} = 5.70}{\rightleftharpoons}} {}^-OC(=O)CH_2C(=O)O^- + H^+$$

A COOH group withdraws electrons (more strongly than does an H) and therefore increases the stability of the conjugate base that is formed when the first COOH group loses a proton—thereby increasing its acidity. The pK_a values of the dicarboxylic acids

Table 17.2 Structures, Names, and pK_a Values of Some Simple Dicarboxylic Acids

Dicarboxylic acid	Common name	pK_{a1}	pK_{a2}
$HOC(=O)OH$	Carbonic acid	3.58	6.35
$HOC(=O)-C(=O)OH$	Oxalic acid	1.27	4.27
$HOC(=O)CH_2C(=O)OH$	Malonic acid	2.86	5.70
$HOC(=O)CH_2CH_2C(=O)OH$	Succinic acid	4.21	5.64
$HOC(=O)CH_2CH_2CH_2C(=O)OH$	Glutaric acid	4.34	5.27
$HOC(=O)CH_2CH_2CH_2CH_2C(=O)OH$	Adipic acid	4.41	5.28
$C_6H_4(C(=O)OH)_2$ (ortho)	Phthalic acid	2.95	5.41

show that the acid-strengthening effect of the COOH group decreases as the separation between the two carboxyl groups increases.

Dicarboxylic acids readily lose water when heated if they can form a cyclic anhydride with a five- or a six-membered ring.

$$HOC(=O)CH_2CH_2CH_2C(=O)OH \underset{}{\overset{\Delta}{\rightleftharpoons}} \text{glutaric anhydride} + H_2O$$

glutaric acid glutaric anhydride

$$C_6H_4(C(=O)OH)_2 \overset{\Delta}{\rightleftharpoons} \text{phthalic anhydride} + H_2O$$

phthalic acid phthalic anhydride

Cyclic anhydrides are more easily prepared if the dicarboxylic acid is heated in the presence of acetyl chloride or acetic anhydride or if it is treated with a strong dehydrating agent such as P_2O_5.

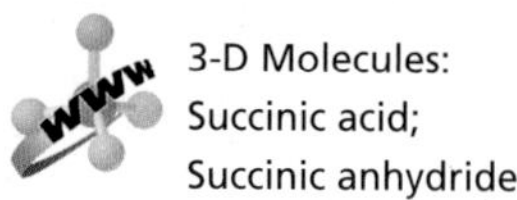

succinic acid + acetic anhydride $\xrightarrow{\Delta}$ succinic anhydride + 2 CH_3COOH

glutaric acid $\xrightarrow{P_2O_5}$ glutaric anhydride

PROBLEM 36

a. Propose a mechanism for the formation of succinic anhydride in the presence of acetic anhydride.
b. How does acetic anhydride help in the formation of succinic anhydride?

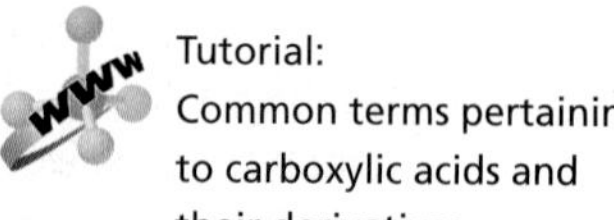

Tutorial:
Common terms pertaining to carboxylic acids and their derivatives

Carbonic acid—a compound with two OH groups bonded to the carbonyl carbon—is unstable, readily breaking down to CO_2 and H_2O. The reaction is reversible, so carbonic acid is formed when CO_2 is bubbled into water (Section 1.20).

$$\text{HO–C(=O)–OH} \rightleftharpoons CO_2 + H_2O$$

carbonic acid

SYNTHETIC POLYMERS

Synthetic polymers play important roles in our daily lives. Polymers are compounds that are made by linking together many small molecules called monomers. In many synthetic polymers, the monomers are held together by ester and amide bonds. For example, Dacron® is a polyester and nylon is a polyamide.

$$\left[\!-\text{OCH}_2\text{CH}_2\text{O}-\overset{\text{O}}{\overset{\|}{\text{C}}}-\text{C}_6\text{H}_4-\overset{\text{O}}{\overset{\|}{\text{C}}}\right]_n\!\!-\text{OCH}_2\text{CH}_2\text{O}-$$

Dacron®

$$-\text{NH(CH}_2)_5\overset{\text{O}}{\overset{\|}{\text{C}}}\left[\text{NH(CH}_2)_5\overset{\text{O}}{\overset{\|}{\text{C}}}\right]_n\text{NH(CH}_2)_5\overset{\text{O}}{\overset{\|}{\text{C}}}-$$

nylon 6

Synthetic polymers have taken the place of metals, fabrics, glass, ceramics, wood, and paper, allowing us to have a greater variety and larger quantities of materials than nature could have provided. New polymers are continually being designed to fit human needs. For example, Kevlar® has a tensile strength greater than steel. It is used for high-performance skis and bulletproof vests. Lexan® is a strong and transparent polymer used for such things as traffic light lenses and compact disks.

$$-\overset{\text{O}}{\overset{\|}{\text{C}}}-\text{C}_6\text{H}_4-\overset{\text{O}}{\overset{\|}{\text{C}}}\left[\text{NH}-\text{C}_6\text{H}_4-\text{NH}-\overset{\text{O}}{\overset{\|}{\text{C}}}-\text{C}_6\text{H}_4-\overset{\text{O}}{\overset{\|}{\text{C}}}\right]_n\text{NH}-\text{C}_6\text{H}_4-\text{NH}-$$

Kevlar®

$$\text{C}_6\text{H}_5-\text{C(CH}_3)_2-\text{C}_6\text{H}_4-\text{O}\left[\overset{\text{O}}{\overset{\|}{\text{C}}}-\text{O}-\text{C}_6\text{H}_4-\text{C(CH}_3)_2-\text{C}_6\text{H}_4-\text{O}\right]_n\overset{\text{O}}{\overset{\|}{\text{C}}}-\text{O}-\text{C}_6\text{H}_4-\text{C(CH}_3)_2-\text{C}_6\text{H}_4-$$

Lexan®

These and other synthetic polymers are discussed in detail in Chapter 28.

We have seen that the OH group of a carboxylic acid can be substituted to give a variety of carboxylic acid derivatives. Similarly, the OH groups of carbonic acid can be substituted by other groups.

$Cl-C(=O)-Cl$ phosgene; $CH_3O-C(=O)-OCH_3$ dimethyl carbonate; $H_2N-C(=O)-NH_2$ urea; $H_2N-C(=O)-OH$ carbamic acid; $H_2N-C(=O)-OCH_3$ methyl carbamate

PROBLEM 37◆

What products would you expect to obtain from the following reactions?

a. phosgene + excess diethylamine
b. malonic acid + 2 acetyl chloride
c. methyl carbamate + methylamine
d. urea + water
e. urea + water + H^+
f. β-ethylglutaric acid + acetyl chloride + Δ

Summary

A **carbonyl group** is a carbon double-bonded to an oxygen; an **acyl group** is a carbonyl group attached to an alkyl or aryl group. **Acyl halides, acid anhydrides, esters**, and **amides** are called **carboxylic acid derivatives** because they differ from a carboxylic acid only in the nature of the group that has replaced the OH group of the carboxylic acid. Cyclic esters are called **lactones**; cyclic amides are **lactams**. There are **symmetrical anhydrides** and **mixed anhydrides**.

Carbonyl compounds can be placed in one of two classes. Class I carbonyl compounds contain a group that can be replaced by another group; carboxylic acids and carboxylic acid derivatives belong to this class. Class II carbonyl compounds do not contain a group that can be replaced by another group; aldehydes and ketones belong to this class.

The reactivity of carbonyl compounds resides in the polarity of the carbonyl group; the carbonyl carbon has a partial positive charge that is attractive to nucleophiles. Class I carbonyl compounds undergo **nucleophilic acyl substitution reactions**: a nucleophile replaces the substituent that was attached to the acyl group in the reactant. All Class I carbonyl compounds react with nucleophiles in the same way: the nucleophile attacks the carbonyl carbon, forming an unstable **tetrahedral intermediate**. (Generally, a compound with an sp^3 carbon bonded to an oxygen is unstable if the sp^3 carbon is bonded to another electronegative atom.) The tetrahedral intermediate reforms a carbonyl compound by eliminating the weakest base.

A carboxylic acid derivative will undergo a nucleophilic acyl substitution reaction provided that the newly added group in the tetrahedral intermediate is not a much weaker base than the group that was attached to the acyl group in the reactant. The weaker the base attached to the acyl group, the easier it is for both steps of the nucleophilic acyl substitution reaction to take place. The relative reactivities toward nucleophilic acyl substitution: acyl halides > acid anhydrides > carboxylic acids and esters > amides > carboxylate ions.

Hydrolysis, **alcoholysis**, and **aminolysis** are reactions in which water, alcohols, and amines, respectively, convert one compound into two compounds. A **transesterification reaction** converts one ester to another ester. Treating a carboxylic acid with excess alcohol and an acid catalyst is called a **Fischer esterification**. An ester with a tertiary alkyl group hydrolyzes via an S_N1 reaction.

The rate of hydrolysis can be increased by either acid or HO^-; the rate of alcoholysis can be increased by either acid or RO^-. An acid increases the rate of formation of the tetrahedral intermediate by protonating the carbonyl oxygen, which increases the electrophilicity of the carbonyl group, and by decreasing the basicity of the leaving group, which makes it easier to eliminate. Hydroxide (or alkoxide) ion increases the rate of formation of the tetrahedral intermediate—it is a better nucleophile than water (or an alcohol)—and increases the rate of collapse of the tetrahedral intermediate. Hydroxide ion promotes only hydrolysis reactions; alkoxide ion promotes only alcoholysis reactions. In an acid-catalyzed reaction, all organic reactants, intermediates, and products are positively charged or neutral; in hydroxide-ion- or alkoxide-ion-promoted reactions, all organic reactants, intermediates, and products are negatively charged or neutral.

Fats and **oils** are triesters of glycerol. Hydrolyzing the ester groups in a basic solution (**saponification**) forms glycerol and fatty acid salts (soaps). Long-chain carboxylate ions arrange themselves in spherical clusters called **micelles**. The attractive forces of hydrocarbon chains for each other in water are called **hydrophobic interactions**.

Amides are unreactive compounds but do react with water and alcohols if the reaction mixture is heated in the presence of an acid. Nitriles are harder to hydrolyze than amides. The **Gabriel synthesis**, which converts alkyl halides into primary amines, involves the hydrolysis of an **imide**.

Organic chemists activate carboxylic acids by converting them into acyl halides or acid anhydrides. Cells activate carboxylic acids by converting them into **acyl phosphates, acyl pyrophosphates, acyl adenylates**, and **thioesters**.

Summary of Reactions

1. Reactions of acyl halides (Section 17.8)

$$RCOCl + CH_3COO^- \longrightarrow RCOOCOCH_3 + Cl^-$$

$$RCOCl + CH_3OH \longrightarrow RCOOCH_3 + HCl$$

$$RCOCl + H_2O \longrightarrow RCOOH + HCl$$

$$RCOCl + 2\,CH_3NH_2 \longrightarrow RCONHCH_3 + CH_3\overset{+}{N}H_3\,Cl^-$$

2. Reactions of acid anhydrides (Section 17.9)

$$RCOOCOR + CH_3OH \longrightarrow RCOOCH_3 + RCOOH$$

$$RCOOCOR + H_2O \longrightarrow 2\,RCOOH$$

$$RCOOCOR + 2\,CH_3NH_2 \longrightarrow RCONHCH_3 + RCOO^-\,H_3\overset{+}{N}CH_3$$

3. Reactions of esters (Sections 17.10–17.13)

$$RCOOR + CH_3OH \underset{}{\overset{HCl}{\rightleftharpoons}} RCOOCH_3 + ROH$$

$$RCOOR + H_2O \underset{}{\overset{HCl}{\rightleftharpoons}} RCOOH + ROH$$

$$RCOOR + H_2O \xrightarrow{HO^-} RCOO^- + ROH$$

$$RCOOR + CH_3NH_2 \longrightarrow RCONHCH_3 + ROH$$

4. Reactions of carboxylic acids (Section 17.14)

$$RCOOH + CH_3OH \underset{}{\overset{HCl}{\rightleftharpoons}} RCOOCH_3 + H_2O$$

$$RCOOH + CH_3NH_2 \longrightarrow RCOO^- \ H_3\overset{+}{N}CH_3$$

5. Reactions of amides (Sections 17.15 and 17.16)

$$RCONH_2 + H_2O \xrightarrow[\Delta]{HCl} RCOOH + \overset{+}{N}H_4$$

$$RCONH_2 \xrightarrow[\Delta]{P_2O_5} RC{\equiv}N$$

6. Gabriel synthesis of primary amines (Section 17.17)

$$RCH_2Br \xrightarrow[\text{2. HCl, } H_2O, \Delta \quad \text{3. } HO^-]{\text{1. phthalimide, } HO^-} RCH_2NH_2$$

7. Hydrolysis of nitriles (Section 17.18)

$$RC{\equiv}N + H_2O \xrightarrow[\Delta]{HCl} RCOOH + \overset{+}{N}H_4$$

8. Activation of carboxylic acids by chemists (Section 17.20)

$$RCOOH + SOCl_2 \xrightarrow[\Delta]{} RCOCl + SO_2 + HCl$$

$$3\ RCOOH + PCl_3 \xrightarrow[\Delta]{} 3\ RCOCl + H_3PO_3$$

$$2\ RCOOH \xrightarrow{P_2O_5} RCOOCOR$$

9. Activation of carboxylic acids by cells (Section 17.20)

$$RCOO^- + {}^-O{-}P(O)(O^-){-}O{-}P(O)(O^-){-}O{-}P(O)(O^-){-}Ad \xrightarrow{\text{enzyme}} RC(O){-}O{-}P(O)(O^-){-}O^- + {}^-O{-}P(O)(O^-){-}O{-}P(O)(O^-){-}Ad$$

$$RCOO^- + {}^-O{-}P(O)(O^-){-}O{-}P(O)(O^-){-}O{-}P(O)(O^-){-}Ad \xrightarrow{\text{enzyme}} RC(O){-}O{-}P(O)(O^-){-}O{-}P(O)(O^-){-}O^- + {}^-O{-}P(O)(O^-){-}Ad$$

$$RCOO^- + {}^-O\text{-}P(O)(O^-)\text{-}O\text{-}P(O)(O^-)\text{-}O\text{-}P(O)(O^-)\text{-}Ad \xrightarrow{\text{enzyme}} RC(O)\text{-}O\text{-}P(O)(O^-)\text{-}Ad + {}^-O\text{-}P(O)(O^-)\text{-}O\text{-}P(O)(O^-)\text{-}O^-$$

$$CH_3COO^- + {}^-O\text{-}P(O)(O^-)\text{-}O\text{-}P(O)(O^-)\text{-}O\text{-}P(O)(O^-)\text{-}Ad \xrightarrow{\text{enzyme}} CH_3C(O)\text{-}O\text{-}P(O)(O^-)\text{-}Ad \xrightarrow{\text{CoASH}} CH_3\text{—}\overset{O}{\overset{\|}{C}}\text{—}SCoA + AMP$$

acetyl-CoA

10. Dehydration of dicarboxylic acids (Section 17.21)

$$HOOC(CH_2)_2COOH \underset{}{\overset{\Delta}{\rightleftharpoons}} \text{succinic anhydride} + H_2O$$

$$HOOC(CH_2)_2COOH + CH_3C(O)OC(O)CH_3 \longrightarrow \text{succinic anhydride} + 2\ CH_3COOH$$

$$HOOC(CH_2)_2COOH \xrightarrow{P_2O_5} \text{succinic anhydride}$$

Key Terms

acid anhydride (p. 673)
acyl adenylate (p. 713)
acyl group (p. 670)
acyl halide (p. 673)
acyl phosphate (p. 713)
acyl pyrophosphate (p. 713)
acyl transfer reaction (p. 682)
alcoholysis (p. 690)
α-carbon (p. 672)
amide (p. 675)
amino acid (p. 679)
aminolysis (p. 690)
biosynthesis (p. 713)
carbonyl carbon (p. 676)
carbonyl compound (p. 670)
carbonyl group (p. 670)
carbonyl oxygen (p. 676)
carboxyl group (p. 673)
carboxylic acid (p. 671)
carboxylic acid derivative (p. 670)
carboxyl oxygen (p. 674)
catalyst (p. 694)
detergent (p. 701)
ester (p. 674)
fatty acid (p. 700)
fats (p. 700)
Fischer esterification (p. 702)
Gabriel synthesis (p. 708)
hydrolysis (p. 690)
hydrophobic interactions (p. 700)
imide (p. 708)
lactam (p. 675)
lactone (p. 674)
micelle (p. 700)
mixed anhydride (p. 673)
neurotransmitter (p. 715)
nitriles (p. 675)
nucleophilic acyl substitution reaction (p. 682)
oils (p. 700)
phoshoanhydride bond (p. 714)
saponification (p. 700)
soap (p. 700)
symmetrical anhydride (p. 673)
tetrahedral intermediate (p. 681)
thioester (p. 714)
transesterification reaction (p. 690)

Problems

38. Write a structure for each of the following compounds:
   a. *N,N*-dimethylhexanamide
   b. 3,3-dimethylhexanamide
   c. cyclohexanecarbonyl chloride
   d. propanenitrile
   e. propionyl bromide
   f. sodium acetate
   g. benzoic anhydride
   h. β-valerolactone
   i. 3-methylbutanenitrile
   j. cycloheptanecarboxylic acid

39. Name the following compounds:

a. 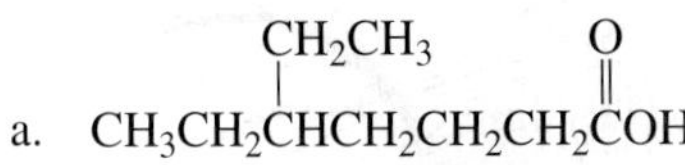

e. $CH_3CH_2CH_2C(=O)N(CH_3)_2$

h. $CH_2{=}CHCH_2C(=O)NHCH_3$

b. $CH_3CH_2C(=O)OCH_2CH_2CH_3$

f. $CH_3CH_2CH_2CH_2C(=O)Cl$

i. CH_2CH_3, C, H, CH_3, CH_2COOH

c. $CH_3CH_2CH_2CH_2C{\equiv}N$

g. (benzene ring)$-C(=O)OC(=O)CH_3$

j. $CH_2C{\equiv}N$, C, H, CH_3, $CH_2CH_2CH_3$

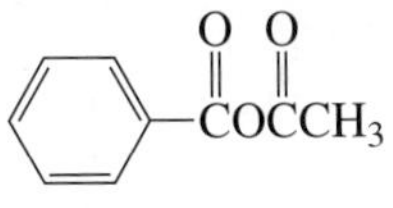

d. $CH_3CH_2C(=O)OC(=O)CH_2CH_3$

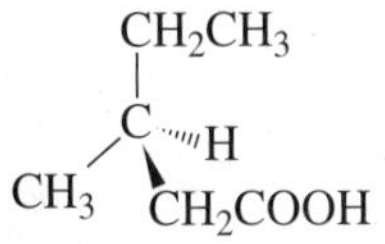

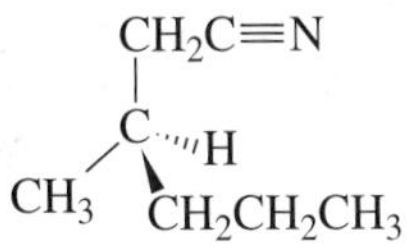

40. What products would be formed from the reaction of benzoyl chloride with the following reagents?

a. sodium acetate
b. water
c. dimethylamine
d. aqueous HCl
e. aqueous NaOH
f. cyclohexanol
g. benzylamine
h. 4-chlorophenol
i. isopropyl alcohol
j. aniline

41. a. List the following esters in order of decreasing reactivity in the first step of a nucleophilic acyl substitution reaction (formation of the tetrahedral intermediate):

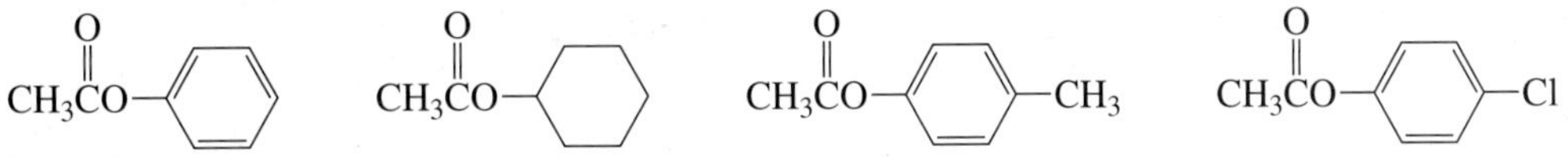

A: $CH_3C(=O)O-$(benzene ring)
B: $CH_3C(=O)O-$(cyclohexane ring)
C: $CH_3C(=O)O-$(benzene ring)$-CH_3$
D: $CH_3C(=O)O-$(benzene ring)$-Cl$

b. List the same esters in order of decreasing reactivity in the second step of a nucleophilic acyl substitution reaction (collapse of the tetrahedral intermediate).

42. a. Which compound would you expect to have a higher dipole moment, methyl acetate or butanone?

$CH_3C(=O)OCH_3$ **methyl acetate** $\quad$ $CH_3C(=O)CH_2CH_3$ **butanone**

b. Which would you expect to have a higher boiling point?

43. How could you use ^1H NMR spectroscopy to distinguish among the following esters?

A: $CH_3C(=O)OCH_2CH_3$
B: $HC(=O)OCH_2CH_2CH_3$
C: $CH_3CH_2C(=O)OCH_3$
D: $HC(=O)OCH(CH_3)CH_3$

44. If propionyl chloride is added to one equivalent of methylamine, only a 50% yield of *N*-methylpropanamide is obtained. If, however, the acyl chloride is added to two equivalents of methylamine, the yield of *N*-methylpropanamide is almost 100%. Explain these observations.

45. a. When a carboxylic acid is dissolved in isotopically labeled water (H_2O^{18}), the label is incorporated into both oxygens of the acid. Propose a mechanism to account for this.

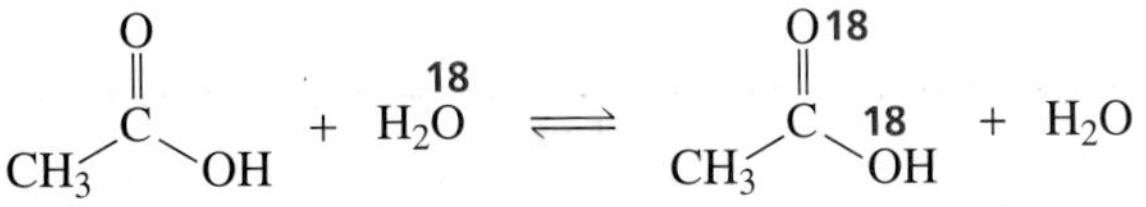

$$CH_3C(=O)OH + H_2\overset{18}{O} \rightleftharpoons CH_3C(=O^{18})\overset{18}{O}H + H_2O$$

b. If a carboxylic acid is dissolved in isotopically labeled methanol ($CH_3{}^{18}OH$) and an acid catalyst is added, where will the label reside in the product?

46. What reagents would you use to convert methyl propanoate into the following compounds?

a. isopropyl propanoate
b. sodium propanoate
c. *N*-ethylpropanamide
d. propanoic acid

47. A compound with molecular formula $C_5H_{10}O_2$ gives the following IR spectrum. When it undergoes acid-catalyzed hydrolysis, the compound with the following ^1H NMR spectrum is formed. Identify the compounds.

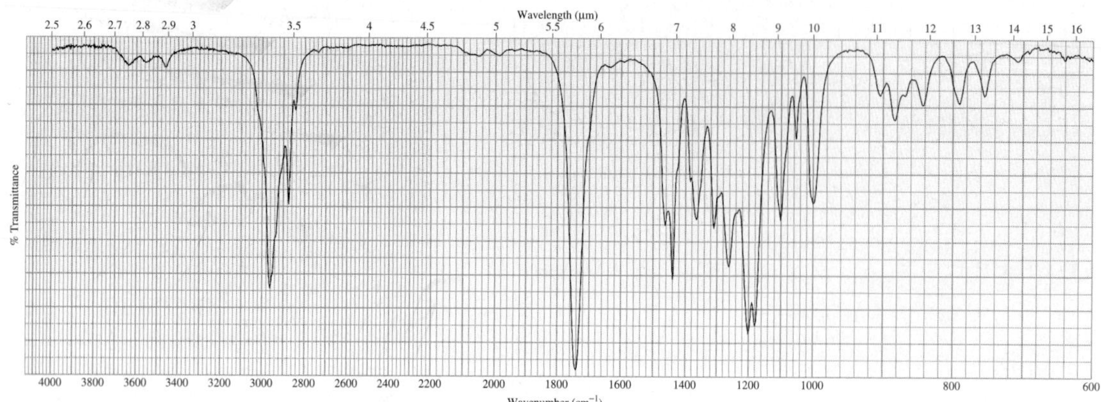

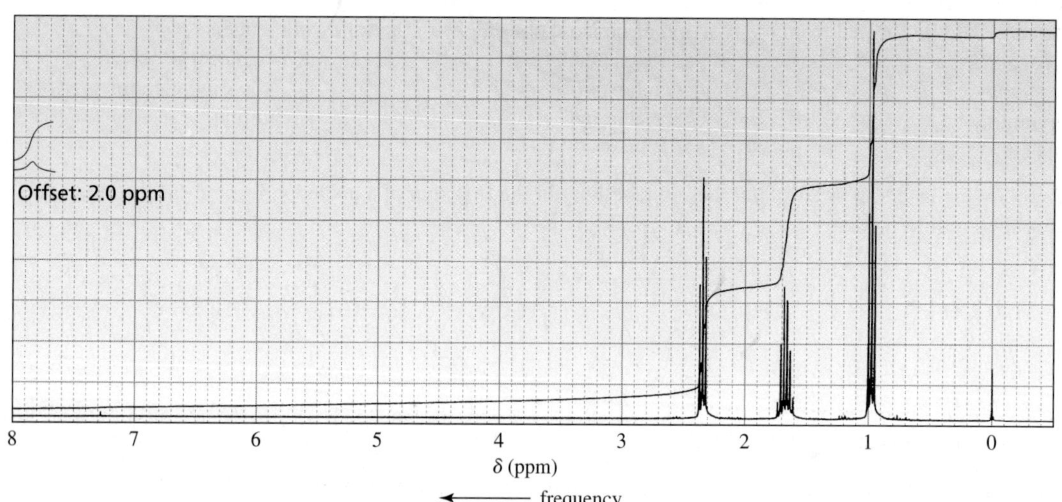

48. Aspartame, the sweetener used in the commercial products NutraSweet® and Equal®, is 160 times sweeter than sucrose. What products would be obtained if aspartame were hydrolyzed completely in an aqueous solution of HCl?

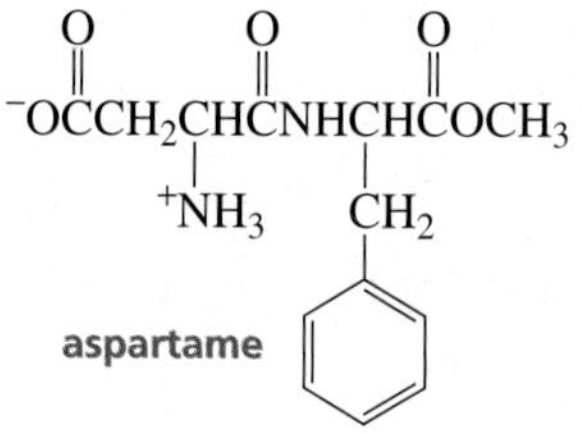

$$^{-}O\overset{O}{\overset{\|}{C}}CH_2\underset{^{+}NH_3}{\underset{|}{C}H}\overset{O}{\overset{\|}{C}}NH\underset{CH_2C_6H_5}{\underset{|}{C}H}\overset{O}{\overset{\|}{C}}OCH_3$$

aspartame

49. a. Which of the following reactions will not give the carbonyl product shown?

1. $CH_3\overset{O}{\overset{\|}{C}}OH + CH_3\overset{O}{\overset{\|}{C}}O^- \longrightarrow CH_3\overset{O}{\overset{\|}{C}}O\overset{O}{\overset{\|}{C}}CH_3$

2. $CH_3\overset{O}{\overset{\|}{C}}Cl + CH_3\overset{O}{\overset{\|}{C}}O^- \longrightarrow CH_3\overset{O}{\overset{\|}{C}}O\overset{O}{\overset{\|}{C}}CH_3$

3. $CH_3\overset{O}{\overset{\|}{C}}NH_2 + Cl^- \longrightarrow CH_3\overset{O}{\overset{\|}{C}}Cl$

4. $CH_3\overset{O}{\overset{\|}{C}}OH + CH_3NH_2 \longrightarrow CH_3\overset{O}{\overset{\|}{C}}NHCH_3$

5. $CH_3\overset{O}{\overset{\|}{C}}OCH_3 + CH_3NH_2 \longrightarrow CH_3\overset{O}{\overset{\|}{C}}NHCH_3$

6. $CH_3\overset{O}{\overset{\|}{C}}OCH_3 + Cl^- \longrightarrow CH_3\overset{O}{\overset{\|}{C}}Cl$

7. $CH_3\overset{O}{\overset{\|}{C}}NHCH_3 + CH_3\overset{O}{\overset{\|}{C}}O^- \longrightarrow CH_3\overset{O}{\overset{\|}{C}}O\overset{O}{\overset{\|}{C}}CH_3$

8. $CH_3\overset{O}{\overset{\|}{C}}Cl + H_2O \longrightarrow CH_3\overset{O}{\overset{\|}{C}}OH$

9. $CH_3\overset{O}{\overset{\|}{C}}NHCH_3 + H_2O \longrightarrow CH_3\overset{O}{\overset{\|}{C}}OH$

10. $CH_3\overset{O}{\overset{\|}{C}}O\overset{O}{\overset{\|}{C}}CH_3 + CH_3OH \longrightarrow CH_3\overset{O}{\overset{\|}{C}}OCH_3$

b. Which of the reactions that do not occur can be made to occur if an acid catalyst is added to the reaction mixture?

50. 1,4-Diazabicyclo[2.2.2]octane (abbreviated DABCO) is a tertiary amine that catalyzes transesterification reactions. Propose a mechanism to show how it does this.

1,4-diazabicyclo[2.2.2]octane
DABCO

51. Identify the major and minor products of the following reaction:

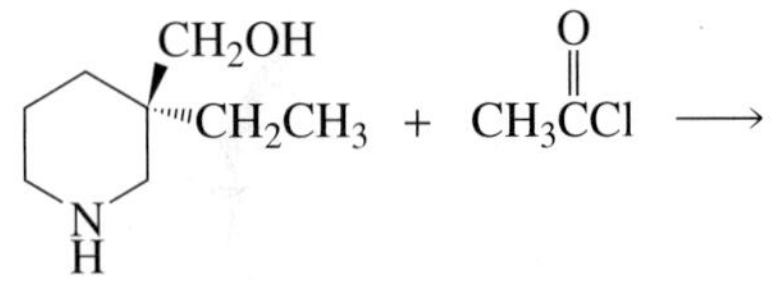

(2-piperidine with CH_2OH and CH_2CH_3 substituents) $+ CH_3\overset{O}{\overset{\|}{C}}Cl \longrightarrow$

52. Two products, A and B, are obtained from the reaction of 1-bromobutane with NH_3. Compound A reacts with acetyl chloride to form C, and B reacts with acetyl chloride to form D. The IR spectra of C and D are shown. Identify A, B, C, and D.

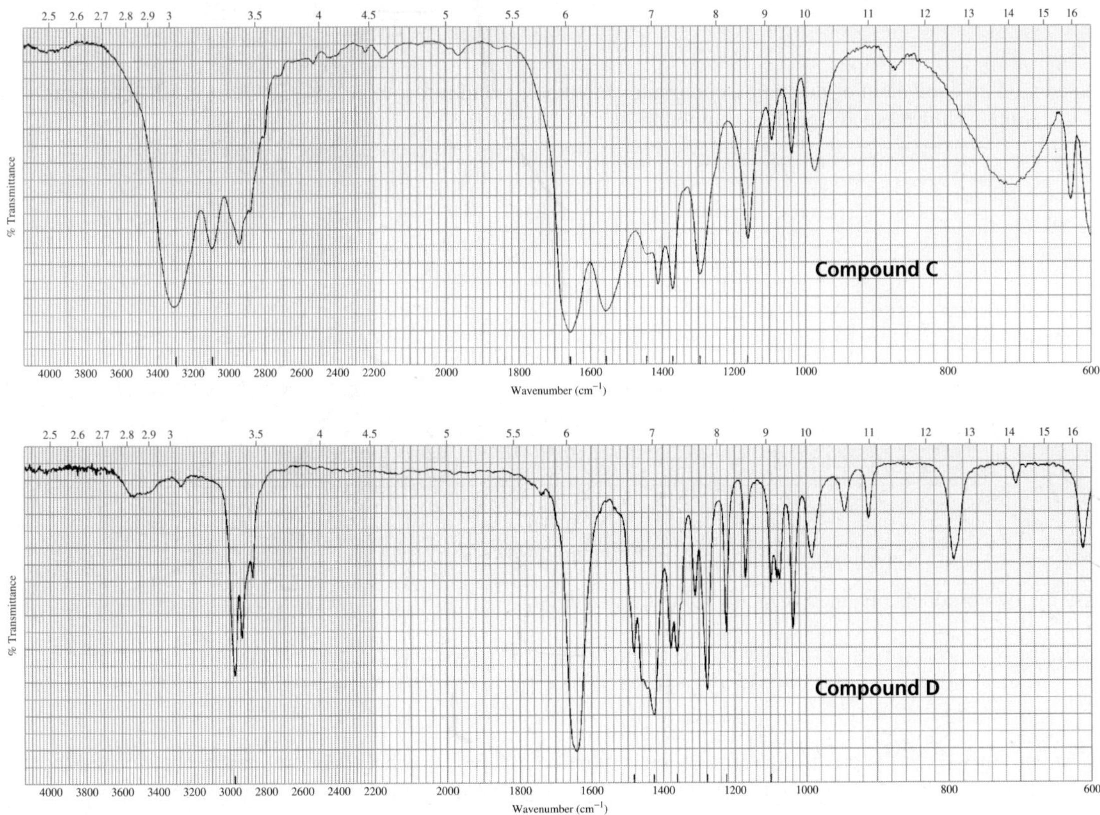

53. Phosgene ($COCl_2$) was used as a poison gas in World War I. Give the product that would be formed from the reaction of phosgene with each of the following reagents:
1. one equivalent of methanol
2. excess methanol
3. excess propylamine
4. one equivalent of ethanol followed by one equivalent of methylamine

54. When Ethyl Ester treated butanedioic acid with thionyl chloride, she was surprised to find that the product she obtained was an anhydride rather than an acyl chloride. Propose a mechanism to explain why she obtained an anhydride.

55. Give the products of the following reactions:

a. $CH_3\overset{O}{\overset{\|}{C}}Cl + KF \longrightarrow$

b. (2-pyrrolidinone) $+ H_2O \xrightarrow{HCl}$

c. $C_6H_5\overset{O}{\overset{\|}{C}}OH \xrightarrow[2.\ 2\ CH_3NH_2]{1.\ SOCl_2}$

d. (succinic anhydride) $+ H_2O \longrightarrow$

e. $Cl\overset{O}{\overset{\|}{C}}Cl$ + (1,2-benzenediol) $\longrightarrow$

f. (γ-butyrolactone) $+ \underset{\text{excess}}{H_2O} \xrightarrow{HCl}$

g. $CH_3\overset{O}{\overset{\|}{C}}CH_2O\overset{O}{\overset{\|}{C}}CH_3 + \underset{\text{excess}}{CH_3OH} \xrightarrow{CH_3O^-}$

h. (2-carboxyphenylacetic acid, $CH_2\overset{O}{\overset{\|}{C}}OH$ and $\overset{O}{\overset{\|}{C}}OH$ on a benzene ring) $\xrightarrow[\Delta]{(CH_3\overset{O}{\overset{\|}{C}})_2O}$

i. (phthalic anhydride) $+ \underset{\text{excess}}{NH_3} \longrightarrow$

j. (homophthalic anhydride) $+ \underset{\text{excess}}{CH_3OH} \xrightarrow{HCl}$

56. When treated with an equivalent of methanol, compound A, with molecular formula $C_4H_6Cl_2O$, forms the compound whose ^1H NMR spectrum is shown below. Identify compound A.

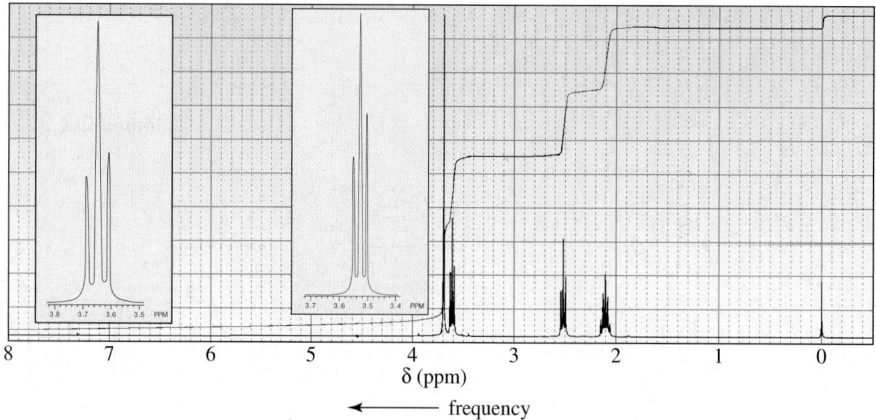

57. a. Identify the two products obtained from the following reaction:

$$\underset{\text{excess}}{CH_3\overset{O}{\overset{\|}{C}}O\overset{O}{\overset{\|}{C}}CH_3} + CH_3\overset{NH_2}{\overset{|}{C}}HCH_2CH_2OH \longrightarrow$$

b. Eddie Amine carried out the preceding reaction, but stopped it before it was half over, whereupon he isolated the major product. He was surprised to find that the product he isolated was neither of the products obtained when the reaction was allowed to go to completion. What product did he isolate?

58. An aqueous solution of a primary or secondary amine reacts with an acyl chloride to form an amide as the major product. However, if the amine is tertiary, an amide is not formed. What product *is* formed? Explain.

59. a. Ann Hydride did not obtain any ester when she added 2,4,6-trimethylbenzoic acid to an acidic solution of methanol. Why? (*Hint:* Build models.)

b. Would Ann have encountered the same problem if she had tried to synthesize the methyl ester of *p*-methylbenzoic acid in the same way?

c. How could she prepare the methyl ester of 2,4,6-trimethylbenzoic acid? (*Hint:* See Section 16.12.)

60. When a compound with molecular formula $C_{11}H_{14}O_2$ undergoes acid-catalyzed hydrolysis, one of the products that is isolated gives the following ^{1}H NMR spectrum. Identify the compound.

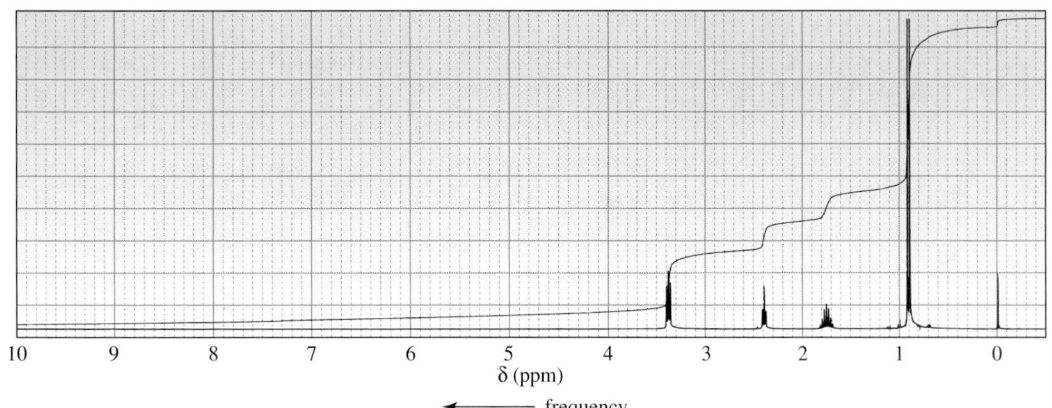

61. List the following compounds in order of decreasing frequency of the carbon–oxygen double-bond stretch:

$$CH_3\overset{\overset{\displaystyle O}{\|}}{C}OCH_3 \qquad CH_3\overset{\overset{\displaystyle O}{\|}}{C}Cl \qquad CH_3\overset{\overset{\displaystyle O}{\|}}{C}H \qquad CH_3\overset{\overset{\displaystyle O}{\|}}{C}NH_2$$

62. a. If the equilibrium constant for the reaction of acetic acid and ethanol to form ethyl acetate is 4.02, what will be the concentration of ethyl acetate at equilibrium if the reaction is carried out with equal amounts of acetic acid and ethanol?

b. What will be the concentration of ethyl acetate at equilibrium if the reaction is carried out with 10 times more ethanol than acetic acid? *Hint:* Recall the quadratic equation: For $ax^2 + bx + c = 0$,

$$x = \frac{-b \pm (b^2 - 4ac)^{1/2}}{2a}$$

c. What will be the concentration of ethyl acetate at equilibrium if the reaction is carried out with 100 times more ethanol than acetic acid?

63. The ^{1}H NMR spectra for two esters with molecular formula $C_8H_8O_2$ are shown below. If each of the esters is added to an aqueous solution with a pH of 10, which of the esters will be hydrolyzed more completely when the hydrolysis reactions have reached equilibrium?

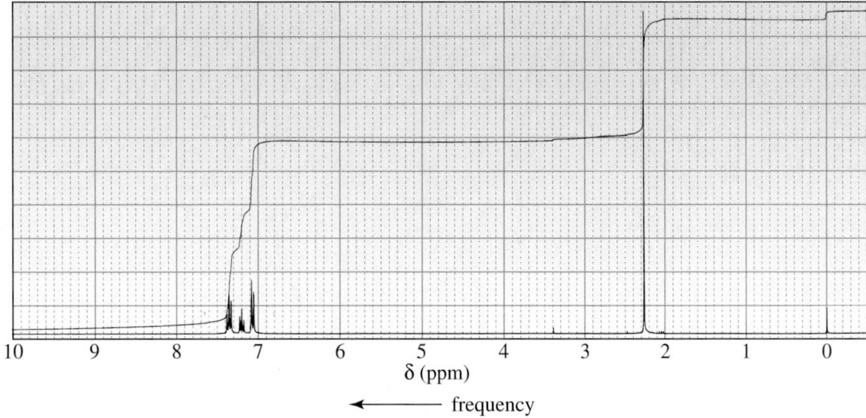

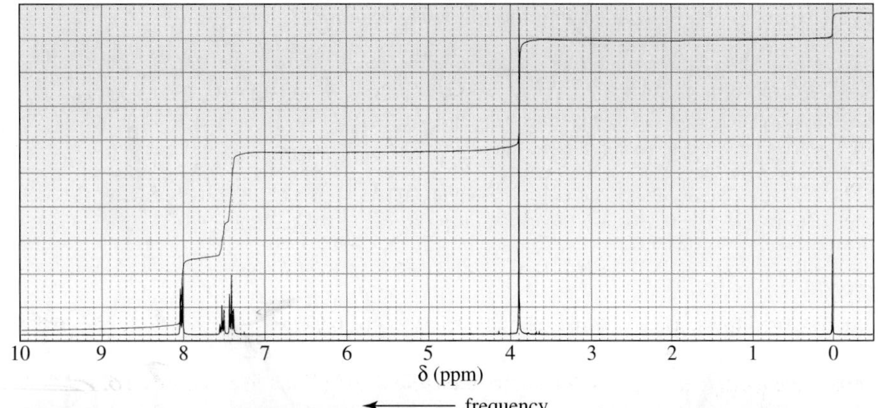

64. Show how the following compounds could be prepared from the given starting materials. You can use any necessary organic or inorganic reagents.

a. $CH_3CH_2CONH_2 \longrightarrow CH_3CH_2COCl$

b. $CH_3CH_2CH_2CH_2OH \longrightarrow CH_3CH_2CH_2CH_2COOH$

c. $CH_3(CH_2)_{10}COOH \longrightarrow CH_3(CH_2)_{11}$–$C_6H_4$–$SO_3^-\ Na^+$

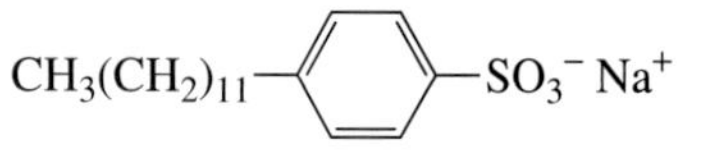

a detergent

d. 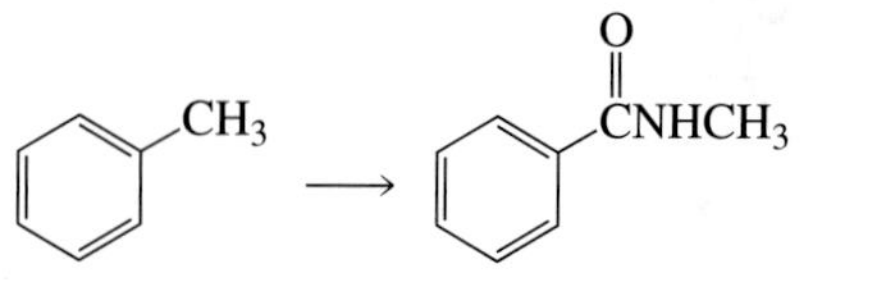 $C_6H_5CH_3 \longrightarrow C_6H_5CONHCH_3$

e. 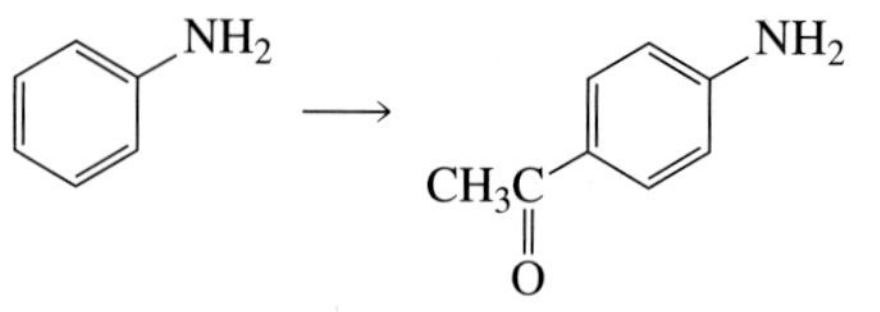 $C_6H_5NH_2 \longrightarrow$ p-CH_3CO–C_6H_4–NH_2

f. 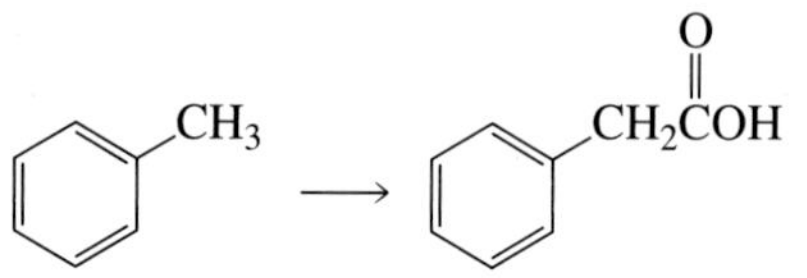 $C_6H_5CH_3 \longrightarrow C_6H_5CH_2COOH$

65. Is the acid-catalyzed hydrolysis of acetamide a reversible or an irreversible reaction? Explain.

66. The reaction of a nitrile with an alcohol in the presence of a strong acid forms a secondary amide. This reaction is known as the *Ritter reaction*. The Ritter reaction does not work with primary alcohols.

$$RC{\equiv}N + R'OH \xrightarrow{H^+} RCONHR'$$

the Ritter reaction

a. Propose a mechanism for the Ritter reaction.
b. Why does the Ritter reaction not work with primary alcohols?
c. How does the Ritter reaction differ from the acid-catalyzed hydrolysis of a nitrile to form a primary amide?

67. The intermediate shown here is formed during the hydroxide-ion-promoted hydrolysis of the ester group. Propose a mechanism for the reaction.

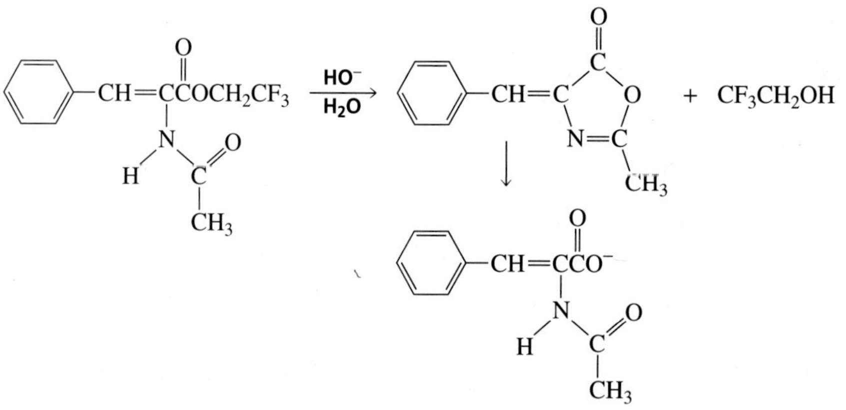

$$C_6H_5CH{=}C(NHCOCH_3)COOCH_2CF_3 \xrightarrow[H_2O]{HO^-} \text{(oxazolone intermediate: } C_6H_5CH{=}C\text{ in ring with } C{=}O,\ O,\ N{=}C(CH_3)) + CF_3CH_2OH$$

$$\downarrow$$

$$C_6H_5CH{=}C(NHCOCH_3)COO^-$$

68. What product would you expect to obtain from each of the following reactions?

a. $CH_3CH_2CH(OH)CH_2CH_2CH_2COOH \xrightarrow{HCl}$

b. 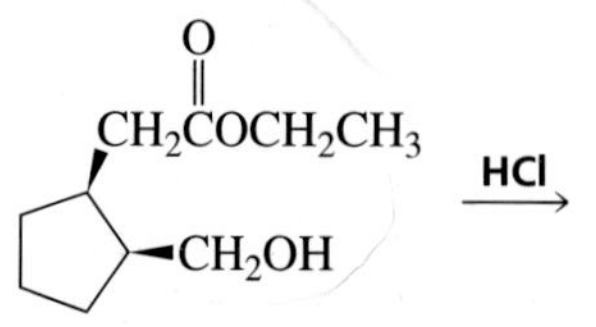 cyclopentane bearing $CH_2COOCH_2CH_3$ and CH_2OH (wedge bonds) $\xrightarrow{HCl}$

c. 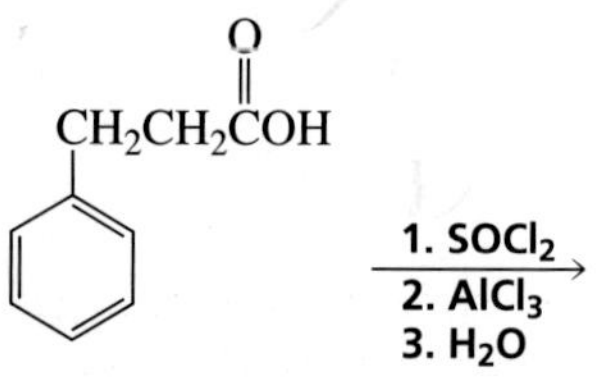 $C_6H_5CH_2CH_2COOH \xrightarrow{1.\ SOCl_2\quad 2.\ AlCl_3\quad 3.\ H_2O}$

69. Sulfonamides, the first antibiotics, were introduced clinically in 1934 (Sections 25.8 and 30.4). Show how a sulfonamide can be prepared from benzene.

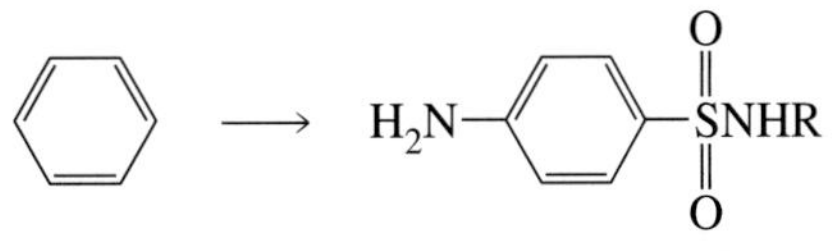

a sulfonamide

70. a. How could aspirin be synthesized, starting with benzene?
    b. Ibuprofen is the active ingredient in pain relievers such as Advil®, Motrin®, and Nuprin®. How could ibuprofen be synthesized, starting with benzene?

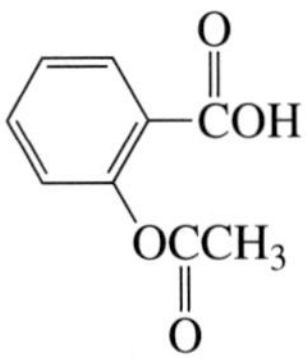

aspirin

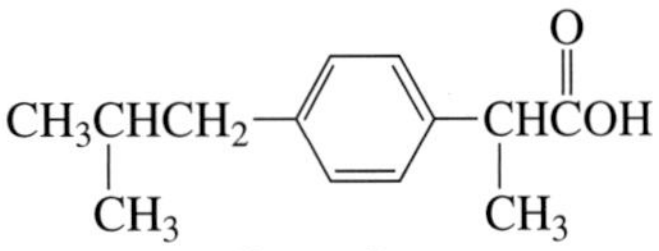

ibuprofen

71. The following compound has been found to be an inhibitor of penicillinase. The enzyme can be reactivated by hydroxylamine (NH_2OH). Propose a mechanism to account for the inhibition and for the reactivation.

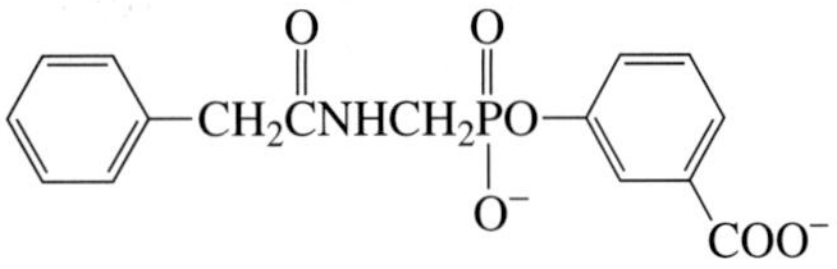

72. For each of the following reactions, propose a mechanism that will account for the formation of the product:

a. HCl

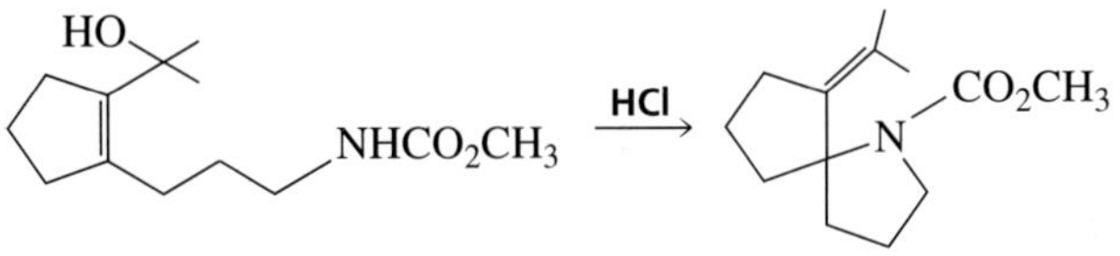

b. CH_3O^- Na^+

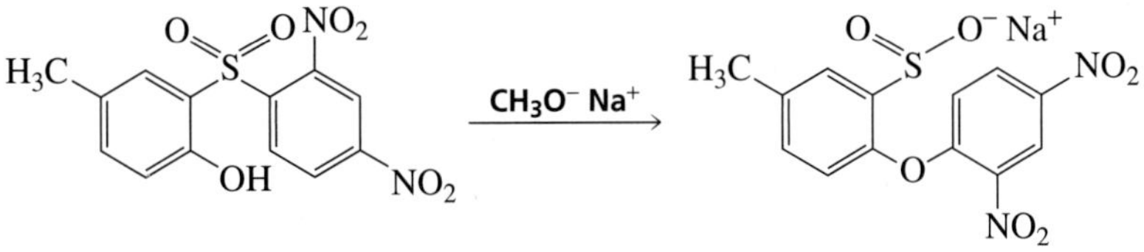

73. Show how Novocain®, a painkiller used frequently by dentists (Section 30.3), can be prepared from benzene.

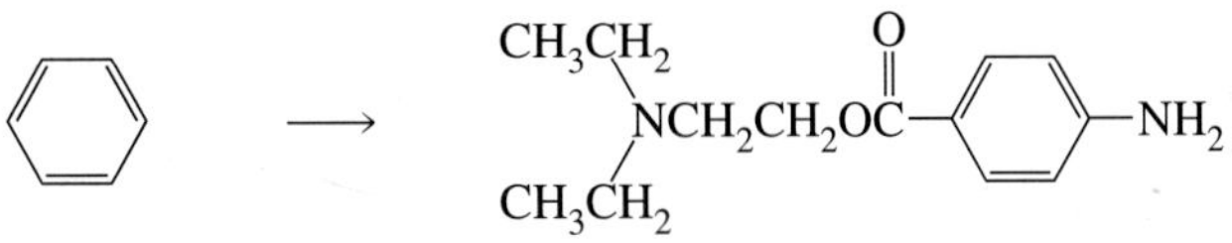

Novocain®

74. Catalytic antibodies catalyze a reaction by binding to the transition state, thereby stabilizing it. As a result, the energy of activation is lowered and the reaction goes faster. The synthesis of the antibody is carried out in the presence of a transition state analog—a

stable molecule that structurally resembles the transition state. This causes an antibody to be generated that will recognize and bind to the transition state. For example, the following transition state analog has been used to generate a catalytic antibody that catalyzes the hydrolysis of the structurally similar ester:

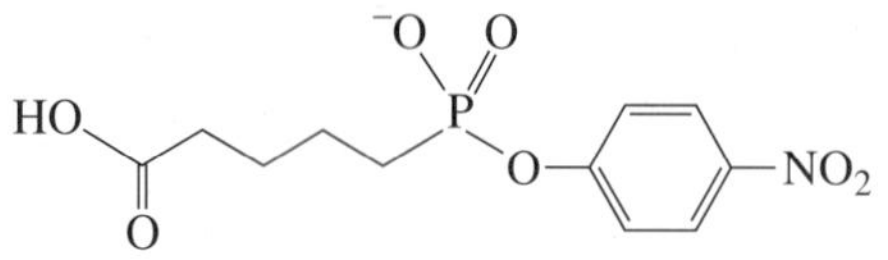

transition state analog

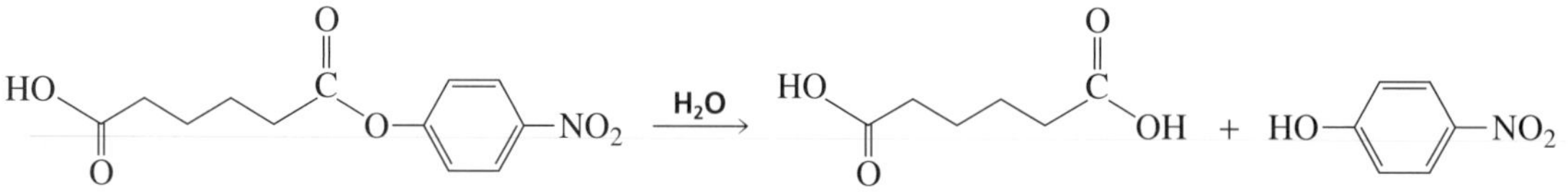

a. Draw the transition state for the hydrolysis reaction.
b. The following transition state analog is used to generate a catalytic antibody for the catalysis of ester hydrolysis. Give the structure of an ester whose rate of hydrolysis would be increased by this catalytic antibody.

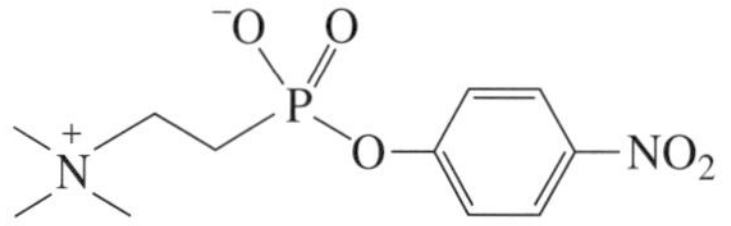

c. Design a transition state analog that would catalyze amide hydrolysis at the amide group indicated.

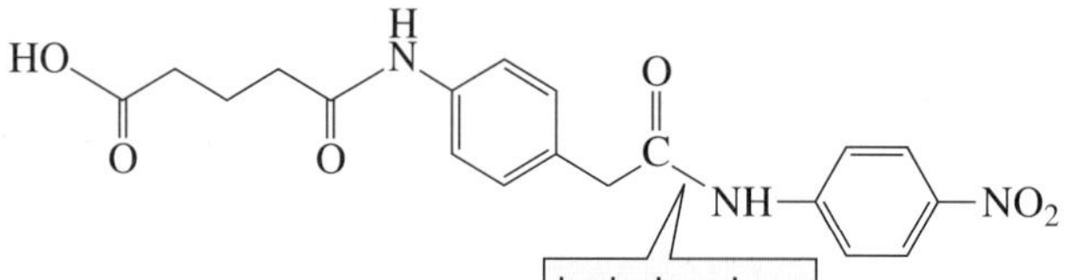

hydrolyze here

75. Saccharin, an artificial sweetener, is about 300 times sweeter than sucrose. Describe how saccharin could be prepared, using benzene as the starting material.

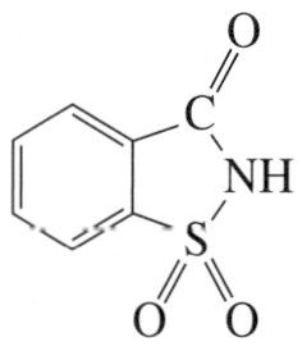

saccharin

76. Information about the mechanism of reaction of a series of substituted benzenes can be obtained by plotting the logarithm of the observed rate constant obtained at a particular pH against the Hammett substituent constant (σ) for the particular substituent. The σ value for hydrogen is 0. Electron-donating substituents have negative σ values; electron-withdrawing substituents have positive σ values. The more strongly electron donating the substituent, the more negative its σ value will be; the more strongly electron withdrawing the substituent, the more positive value its σ value will be. The slope of a plot of the logarithm of the rate constant versus σ is called the ρ (rho) value. The ρ value for the hydroxide-ion-promoted hydrolysis of a series of meta- and para-substituted ethyl benzoates is $+2.46$; the ρ value for amide formation for the reaction of a series of meta- and para-substituted anilines with benzoyl chloride is -2.78.
a. Why does one set of experiments give a positive ρ value while the other set of experiments gives a negative ρ value?
b. Why do you think that ortho-substituted compounds were not included in the experiment?
c. What would you predict the sign of the ρ value to be for the ionization of a series of meta- and para-substituted benzoic acids?

18 Carbonyl Compounds II

Nucleophilic Acyl Addition, Nucleophilic Acyl Substitution, and Nucleophilic Addition–Elimination • Reactions of α,β-Unsaturated Carbonyl Compounds

formaldehyde acetaldehyde

acetone

In Section 17.0, we saw that carbonyl compounds—compounds that possess a carbonyl group (C=O)—can be divided into two classes: Class I carbonyl compounds, which have a group that can be replaced by a nucleophile, and Class II carbonyl compounds, which comprise aldehydes and ketones. Unlike Class I carbonyl compounds, Class II carbonyl compounds do not have a group that can be replaced by a nucleophile.

The carbonyl carbon of the simplest aldehyde, formaldehyde, is bonded to two hydrogens. The carbonyl carbon in all other **aldehydes** is bonded to a hydrogen and to an alkyl (or an aryl) group. The carbonyl carbon of a **ketone** is bonded to two alkyl (or aryl) groups. Aldehydes and ketones *do not have* a group that can be replaced by another group, because hydride ions (H^-) and carbanions (R^-) and are too basic to be displaced by nucleophiles under normal conditions.

formaldehyde

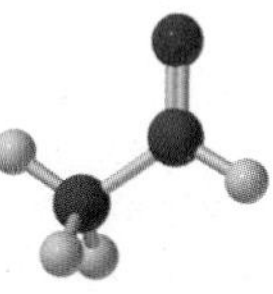

acetaldehyde

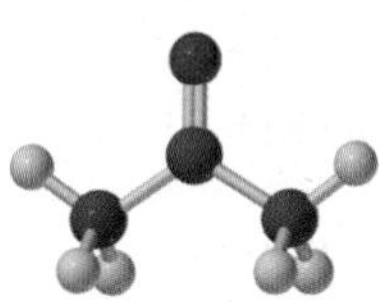

acetone

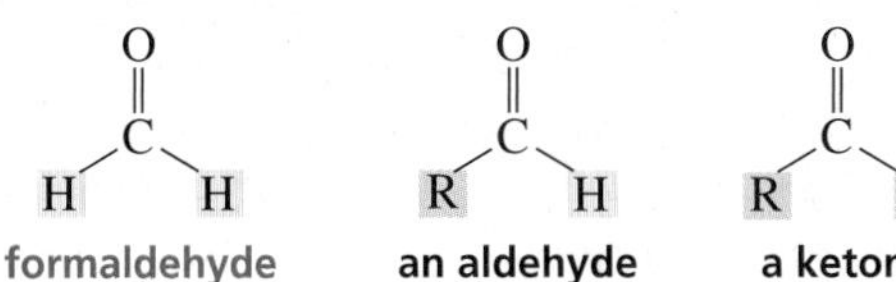

formaldehyde an aldehyde a ketone

The physical properties of aldehydes and ketones are discussed in Section 17.3 (see also Appendix I), and the methods used to prepare aldehydes and ketones are summarized in Appendix IV.

Many compounds found in nature have aldehyde or ketone functional groups. Aldehydes have pungent odors, whereas ketones tend to smell sweet. Vanillin and cinnamaldehyde are examples of naturally occurring aldehydes. A whiff of vanilla extract will allow you to appreciate the pungent odor of vanilla. The ketones carvone and camphor are responsible for the characteristic sweet odors of spearmint leaves, caraway seeds, and the camphor tree.

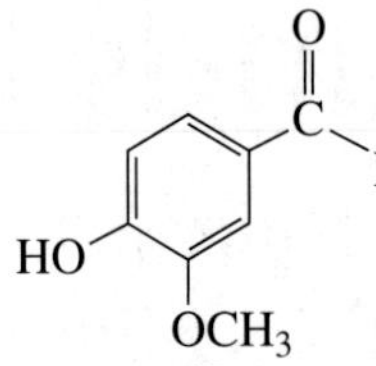

vanillin
vanilla flavoring

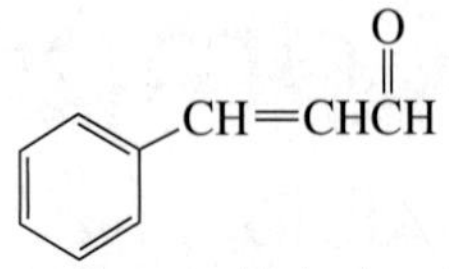

cinnamaldehyde
cinnamon flavoring

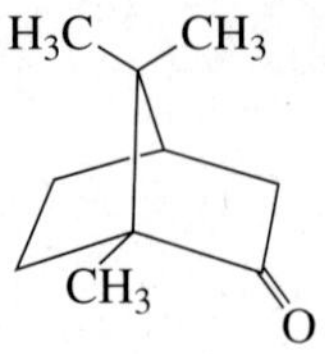

camphor

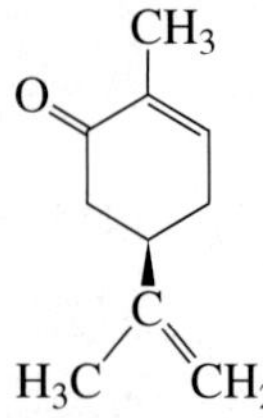

(R)-(–)-carvone
spearmint oil

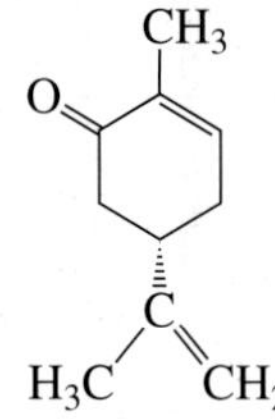

(S)-(+)-carvone
caraway seed oil

In ketosis, a pathological condition that can occur in people with diabetes, the body produces more acetoacetate than can be metabolized. The excess acetoacetate breaks down to acetone—a ketone—and CO_2. Ketosis can be recognized by the smell of acetone on a person's breath.

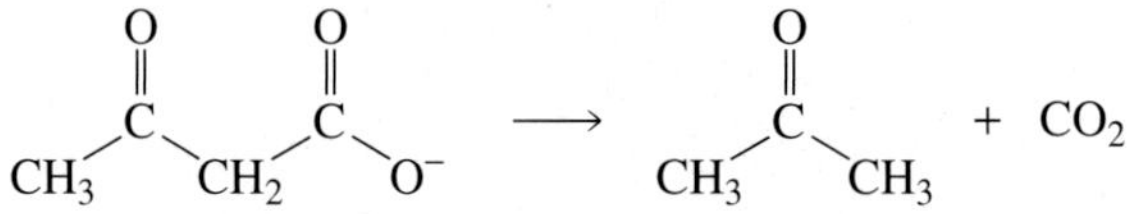

$$CH_3C(=O)CH_2C(=O)O^- \longrightarrow CH_3C(=O)CH_3 + CO_2$$

acetoacetate — acetone

Two ketones that are of biological importance illustrate how a small difference in structure can be responsible for a large difference in biological activity: Progesterone is a female sex hormone synthesized primarily in the ovaries, whereas testosterone is a male sex hormone synthesized primarily in the testes.

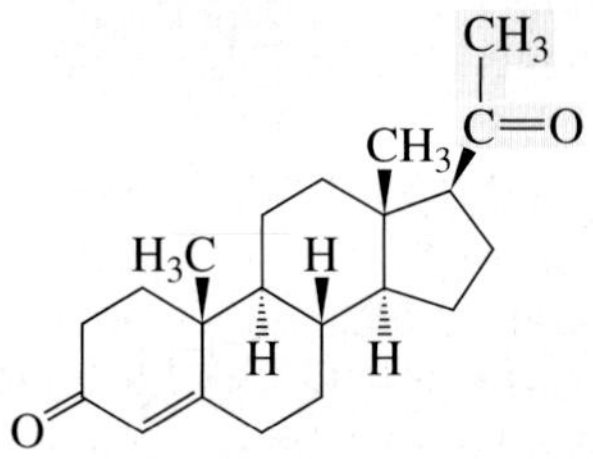

progesterone
a female sex hormone

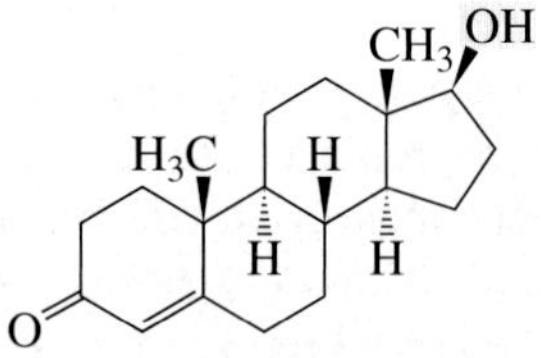

testosterone
a male sex hormone

18.1 Nomenclature

Aldehydes

The systematic name of an aldehyde is obtained by replacing the terminal "e" from the name of the parent hydrocarbon with "al." For example, a one-carbon aldehyde is methan*al*; a two-carbon aldehyde is ethan*al*. The position of the carbonyl carbon does not have to be designated, because it is always at the end of the parent hydrocarbon and therefore always has the 1-position.

The common name of an aldehyde is the same as the common name of the corresponding carboxylic acid (Section 17.1), except that "aldehyde" is substituted for "ic acid" (or "oic acid"). When common names are used, the position of a substituent is designated by a lowercase Greek letter. The carbonyl carbon is not designated; the carbon adjacent to the carbonyl carbon is the α-carbon.

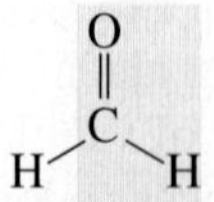

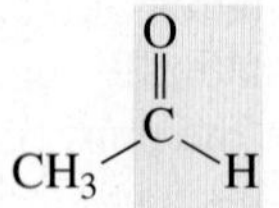

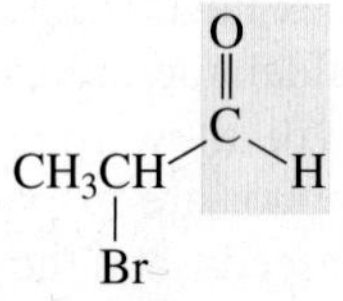

	HCHO	CH_3CHO	$CH_3CHBrCHO$
systematic name:	methanal	ethanal	2-bromopropanal
common name:	formaldehyde	acetaldehyde	α-bromopropionaldehyde

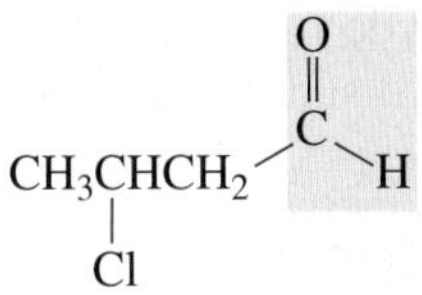

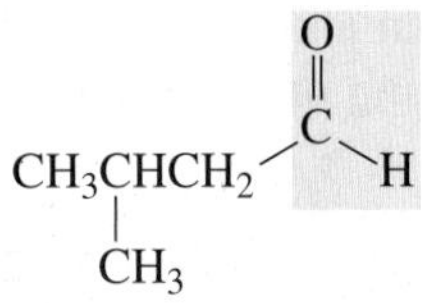

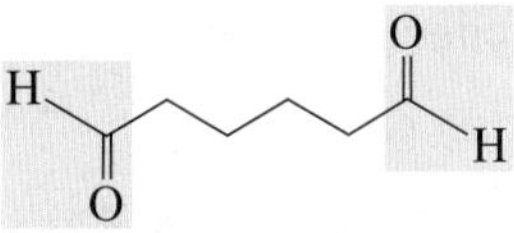

CH_3CHCH_2CHO (Cl on C-3) — systematic name: **3-chlorobutanal**; common name: **β-chlorobutyraldehyde**

CH_3CHCH_2CHO (CH_3 on C-3) — systematic name: **3-methylbutanal**; common name: **isovaleraldehyde**

systematic name: **hexanedial**

Notice that the terminal "e" is not removed in hexanedial; the "e" is removed only to avoid two successive vowels.

If the aldehyde group is attached to a ring, the aldehyde is named by adding "carbaldehyde" to the name of the cyclic compound.

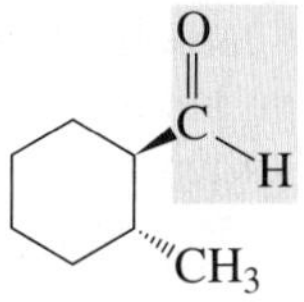

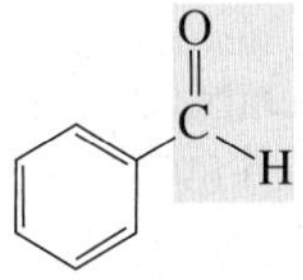

systematic name: ***trans*-2-methylcyclohexanecarbaldehyde**

systematic name: **benzenecarbaldehyde**; common name: **benzaldehyde**

In Section 8.1, we saw that a carbonyl group has a higher nomenclature priority than an alcohol or an amine group. However, all carbonyl compounds do not have the same priority. The nomenclature priorities of the various carbonyl groups are shown in Table 18.1.

If a compound has two functional groups, the one with the lower priority is indicated by its prefix. The prefix of an aldehyde oxygen that is part of the parent hydrocarbon is "oxo." The prefix of a one-carbon aldehyde group that is not part of the parent hydrocarbon is "formyl."

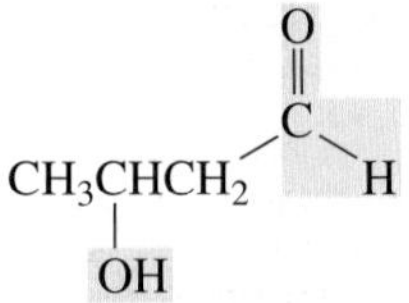

CH_3CHCH_2CHO (OH on C-3) — **3-hydroxybutanal**

$HC(=O)CH_2CH_2CH_2C(=O)OCH_3$ — **methyl 5-oxopentanoate**

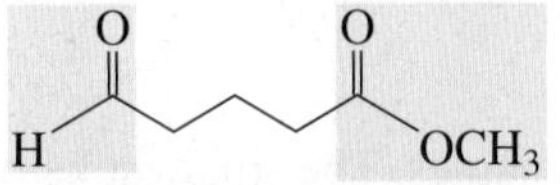

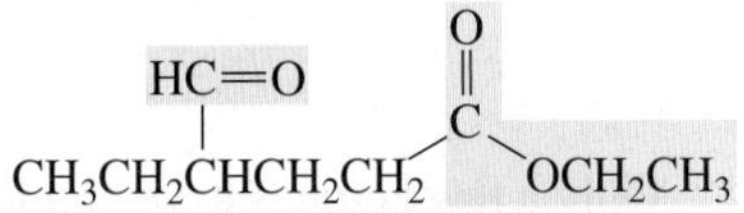

$CH_3CH_2CH(HC{=}O)CH_2CH_2C(=O)OCH_2CH_3$ — **ethyl 4-formylhexanoate**

Table 18.1 Summary of Functional Group Nomenclature

Class	Suffix name	Prefix name
Carboxylic acid	-oic acid	Carboxy
Ester	-oate	Alkoxycarbonyl
Amide	-amide	Amido
Nitrile	-nitrile	Cyano
Aldehyde	-al	Oxo (=O)
Aldehyde	-al	Formyl (—CH=O)
Ketone	-one	Oxo (=O)
Alcohol	-ol	Hydroxy
Amine	-amine	Amino
Alkene	-ene	Alkenyl
Alkyne	-yne	Alkynyl
Alkane	-ane	Alkyl
Ether	—	Alkoxy
Alkyl halide	—	Halo

(arrow: increasing priority, from bottom to top)

If the compound has both an alkene and an aldehyde functional group, the alkene is cited first, with the "e" ending omitted to avoid two successive vowels (Section 8.1).

$CH_3CH{=}CHCH_2-\overset{O}{\overset{\|}{C}}-H$

3-pentenal

Ketones

The systematic name of a ketone is obtained by removing the "e" from the name of the parent hydrocarbon and adding "one." The chain is numbered in the direction that gives the carbonyl carbon the smaller number. In the case of cyclic ketones, a number is not necessary because the carbonyl carbon is assumed to be at the 1-position. Frequently, derived names are used for ketones—the substituents attached to the carbonyl group are cited in alphabetical order, followed by "ketone."

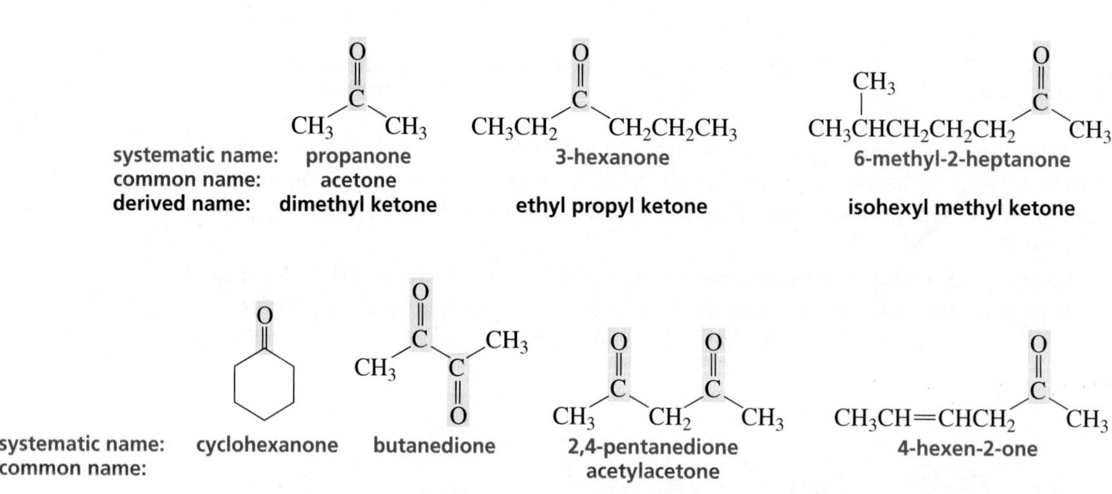

Only a few ketones have common names. The smallest ketone, propanone, is usually referred to by its common name, acetone. Acetone is a common laboratory solvent. Common names are also used for some phenyl-substituted ketones; the number of carbons (other than those of the phenyl group) is indicated by the common name of the corresponding carboxylic acid, substituting "ophenone" for "ic acid."

common name:	acetophenone	butyrophenone	benzophenone
derived name:	methyl phenyl ketone	phenyl propyl ketone	diphenyl ketone

If the ketone has a second functional group of higher naming priority, the ketone oxygen is indicated by the prefix "oxo."

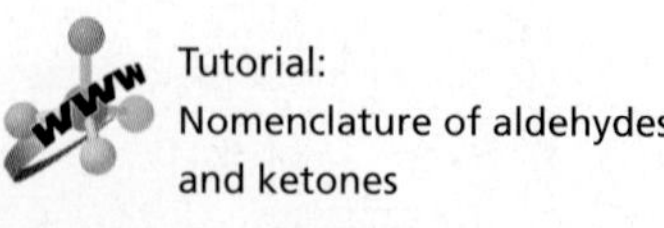

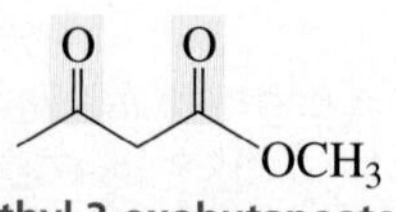

systematic name: 4-oxopentanal — methyl 3-oxobutanoate — 2-(3-oxopentyl)-cyclohexanone

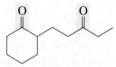

BUTANEDIONE: AN UNPLEASANT COMPOUND

Fresh perspiration is odorless. Bacteria that are always present on our skin produce lactic acid, thereby creating an acidic environment that allows other bacteria to break down the components of perspiration, producing compounds with the unappealing odors we associate with armpits and sweaty feet. One such compound is butanedione.

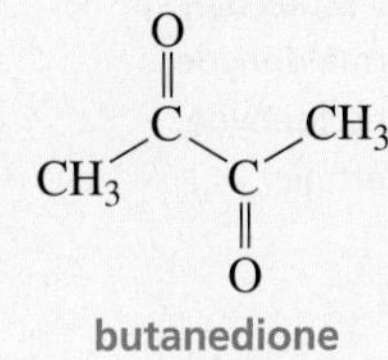

butanedione

PROBLEM 1

Why are numbers not used to designate the positions of the functional groups in propanone and butanedione?

PROBLEM 2◆

Give two names for each of the following compounds:

a. 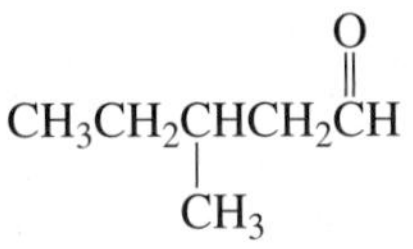 $CH_3CH_2CH(CH_3)CH_2CHO$

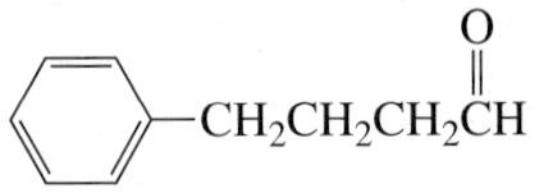

b. 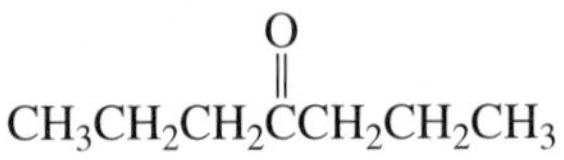 $CH_3CH_2CH_2C(=O)CH_2CH_2CH_3$

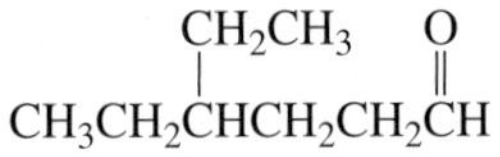

c. 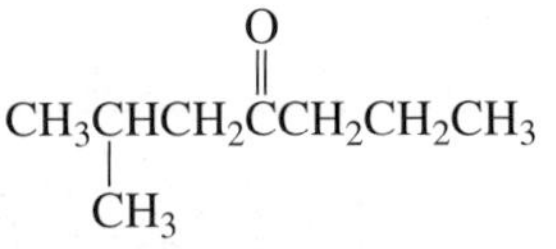 $CH_3CH(CH_3)CH_2C(=O)CH_2CH_2CH_3$

d. $C_6H_5CH_2CH_2CH_2CHO$

e. $CH_3CH_2CH(CH_2CH_3)CH_2CH_2CHO$

f.  $CH_2{=}CHC(=O)CH_2CH_2CH_2CH_3$

PROBLEM 3◆

Name the following compounds:

a. 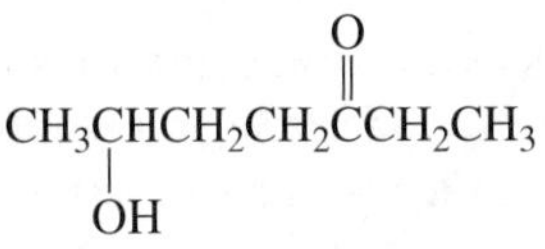 $CH_3CH(OH)CH_2CH_2C(=O)CH_2CH_3$

b. 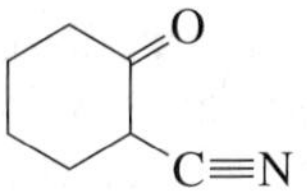 2-oxocyclohexane ring bearing C≡N (cyclohexanone with C≡N on the adjacent carbon)

c. $CH_3CH_2CH(HC{=}O)CH_2C(=O)NH_2$

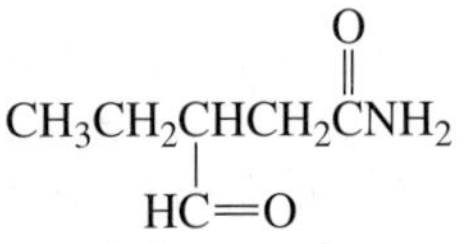

18.2 Relative Reactivities of Carbonyl Compounds

We have seen that the carbonyl group is polar because oxygen, being more electronegative than carbon, has a greater share of the electrons of the double bond (Section 17.5). The partial positive charge on the carbonyl carbon causes carbonyl compounds to be attacked by nucleophiles. The electron deficiency of the carbonyl carbon is indicated by the blue areas in the electrostatic potential maps.

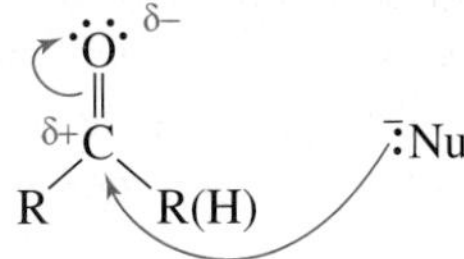

An aldehyde has a greater partial positive charge on its carbonyl carbon than does a ketone because a hydrogen is electron withdrawing compared with an alkyl group (Section 4.2). An aldehyde, therefore, is less stable than a ketone, which makes it more reactive toward nucleophilic attack.

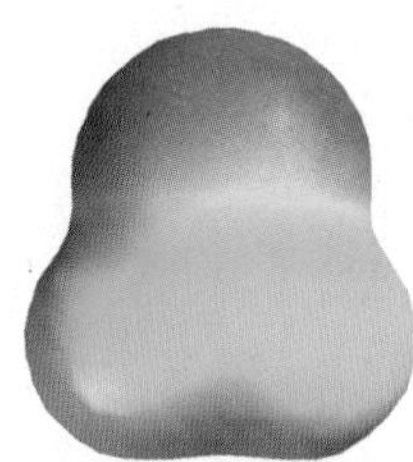
formaldehyde

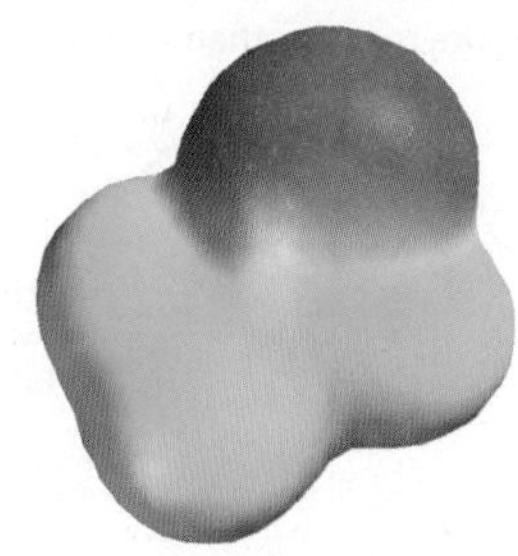
acetaldehyde

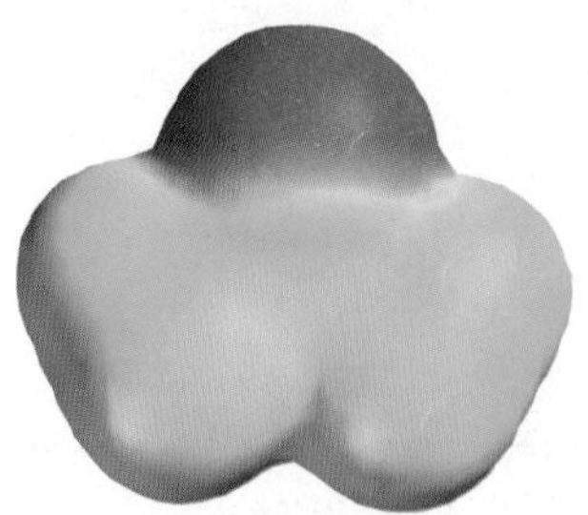
acetone

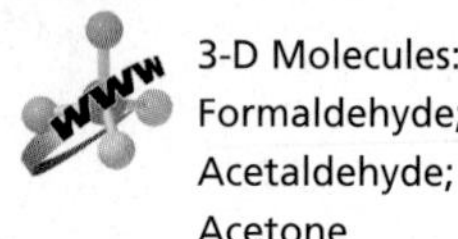

relative reactivities

most reactive → HCHO (formaldehyde) > RCHO (an aldehyde) > RCOR' (a ketone) ← least reactive

Steric factors also contribute to the greater reactivity of an aldehyde. The carbonyl carbon of an aldehyde is more accessible to the nucleophile than is the carbonyl carbon of a ketone because the hydrogen attached to the carbonyl carbon of an aldehyde is smaller than the alkyl group attached to the carbonyl carbon of a ketone. Steric factors also become important in the tetrahedral transition state because the bond angles are 109.5°; therefore, the alkyl groups are closer to one another than they are in the carbonyl compound, in which the bond angles are 120°. Ketones have greater steric crowding in their transition states, so they have less stable transition states than aldehydes have.

For the same reason, ketones with small alkyl groups bonded to the carbonyl carbon are more reactive than ketones with large alkyl groups.

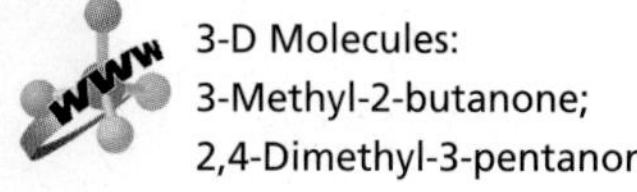

relative reactivities

most reactive → $H_3C-CO-CH_3$ > $H_3C-CO-CH(CH_3)CH_3$ > $CH_3CH(CH_3)-CO-CH(CH_3)CH_3$ ← least reactive

PROBLEM 4◆

Which ketone is more reactive?

a. 2-heptanone or 4-heptanone
b. *p*-nitroacetophenone or *p*-methoxyacetophenone

Aldehydes and ketones are less reactive than acyl chlorides and acid anhydrides, but are more reactive than esters and amides.

How does the reactivity of an aldehyde or a ketone toward nucleophiles compare with the reactivity of the carbonyl compounds whose reactions you studied in Chapter 17? Aldehydes and ketones are right in the middle—less reactive than acyl halides and acid anhydrides, but more reactive than esters, carboxylic acids, and amides.

relative reactivities of carbonyl compounds toward nucleophiles

acyl halide > acid anhydride > aldehyde > ketone > ester ~ carboxylic acid > amide > carboxylate ion

(acyl halide: most reactive; carboxylate ion: least reactive)

The carbonyl compounds discussed in Chapter 17 have a lone pair on an atom attached to the carbonyl group that can be shared with the carbonyl carbon by resonance, which makes the carbonyl carbon less electron deficient. We have seen that the reactivity of these carbonyl compounds is related to the basicity of Y^- (Section 17.5). The weaker the basicity of Y^-, the more reactive is the carbonyl group because weak bases are less able to donate electrons by resonance to the carbonyl carbon and are better able to withdraw electrons inductively from the carbonyl carbon.

$$R-C(=\ddot{O})-\ddot{Y} \longleftrightarrow R-C(-\ddot{O}:^-)=Y^+$$

Consequently, aldehydes and ketones are not as reactive as carbonyl compounds in which Y^- is a very weak base (acyl halides and acid anhydrides), but are more reactive than carbonyl compounds in which Y^- is a relatively strong base (carboxylic acids, esters, and amides). A molecular orbital explanation of why resonance electron donation decreases the reactivity of the carbonyl group is given in Section 17.15.

18.3 How Aldehydes and Ketones React

In Section 17.5, we saw that the carbonyl group of a carboxylic acid or a carboxylic acid derivative is attached to a group that can be replaced by another group. These compounds therefore react with nucleophiles to form substitution products.

$$\mathrm{RC(=O)Y} + Z^- \rightleftharpoons \mathrm{R{-}C(O^-)(Z){-}Y} \rightleftharpoons \mathrm{RC(=O)Z} + Y^-$$

product of nucleophilic substitution

Carboxylic acid derivatives undergo nucleophilic substitution reactions with nucleophiles.

In contrast, the carbonyl group of an aldehyde or a ketone is attached to a group that is too strong a base (H^- or R^-) to be eliminated under normal conditions, so it cannot be replaced by another group. Consequently, aldehydes and ketones react with nucleophiles to form addition products, not substitution products. Thus, aldehydes and ketones undergo **nucleophilic addition** reactions, whereas carboxylic acid derivatives undergo **nucleophilic acyl substitution** reactions.

$$\mathrm{RC(=O)R'} + \mathrm{HZ} \longrightarrow \mathrm{R{-}C(OH)(Z){-}R'}$$

product of nucleophilic addition

When a nucleophile adds to a carbonyl group, the hybridization of the carbonyl carbon changes from sp^2 in the carbonyl compound to sp^3 in the addition product. In Section 17.5, we saw that a compound that has an sp^3 carbon bonded to an oxygen atom generally will be unstable if the sp^3 carbon is also bonded to another electronegative atom. Therefore, if the nucleophile that adds to the aldehyde or ketone is one in which Z *is not* electronegative (Z is an H or a C nucleophile), the tetrahedral addition product will be stable. It will be an alkoxide ion that can be protonated either by the solvent or by added acid. (HB^+ is any species that provides a proton; :B is any species that removes a proton.)

$$\underset{sp^2}{\mathrm{RC(=O)R'}} + Z^- \longrightarrow \underset{sp^3}{\mathrm{R{-}C(O^-)(Z){-}R'}} \underset{:B}{\overset{HB^+}{\rightleftharpoons}} \mathrm{R{-}C(OH)(Z){-}R'}$$

product of nucleophilic addition is stable; Z = C or H

Aldehydes and ketones undergo nucleophilic addition reactions with hydride ion and with carbon nucleophiles.

If, on the other hand, the nucleophile that adds to the aldehyde or ketone is one in which Z *is* electronegative (Z is an O or an N nucleophile), the tetrahedral addition

Aldehydes and ketones undergo nucleophilic addition–elimination reactions with oxygen and nitrogen nucleophiles.

product will *not* be stable. Water will be eliminated from the addition product. This is called a **nucleophilic addition–elimination reaction**. We will see that the fate of the dehydrated product depends on the identity of Z.

$$R-\overset{OH}{\underset{:Z}{C}}-R' \underset{:B}{\overset{HB^+}{\rightleftharpoons}} R-\overset{\overset{H}{+OH}}{\underset{:Z}{C}}-R' \rightleftharpoons \underset{+Z}{\overset{R\quad R'}{C}} + H_2O$$

product of nucleophilic addition is not stable; Z = O or N

product of nucleophilic addition–elimination

18.4 Reactions of Carbonyl Compounds with Carbon Nucleophiles

Few reactions in organic chemistry result in the formation of new C—C bonds. Consequently, those reactions that do are very important to synthetic organic chemists when they need to synthesize larger organic molecules from smaller molecules. The addition of a carbon nucleophile to a carbonyl compound is an example of a reaction that forms a new C—C bond and therefore forms a product with more carbon atoms than the starting material.

Reaction with Grignard Reagents

Addition of a Grignard reagent to a carbonyl compound is a versatile reaction that leads to the formation of a new C—C bond. The reaction can produce compounds with a variety of structures because both the structure of the carbonyl compound and the structure of the Grignard reagent can be varied. In Section 12.11, we saw that a Grignard reagent can be prepared by adding an alkyl halide to magnesium shavings in diethyl ether. We also saw that a Grignard reagent reacts as if it were a carbanion.

$$CH_3CH_2Br \xrightarrow[Et_2O]{Mg} CH_3CH_2MgBr$$

$$CH_3CH_2MgBr \text{ reacts as if it were } CH_3\ddot{C}H_2^- \ \overset{+}{Mg}Br$$

Attack of a Grignard reagent on a carbonyl carbon forms an alkoxide ion that is complexed with magnesium ion. Addition of water or dilute acid breaks up the complex. When a Grignard reagent reacts with formaldehyde, the addition product is a primary alcohol.

$$\underset{\text{formaldehyde}}{H_2C{=}\ddot{O}} + \underset{\text{butylmagnesium bromide}}{CH_3CH_2CH_2CH_2-MgBr} \longrightarrow \underset{\text{an alkoxide ion}}{CH_3CH_2CH_2CH_2CH_2\ddot{O}:^- \ \overset{+}{Mg}Br} \xrightarrow{H_3O^+} \underset{\text{1-pentanol, a primary alcohol}}{CH_3CH_2CH_2CH_2CH_2\ddot{O}H}$$

When a Grignard reagent reacts with an aldehyde other than formaldehyde, the addition product is a secondary alcohol.

$$\underset{\text{propanal}}{CH_3CH_2CH{=}\ddot{O}} + \underset{\text{propylmagnesium bromide}}{CH_3CH_2CH_2-MgBr} \longrightarrow CH_3CH_2\overset{:\ddot{O}:^-\ \overset{+}{Mg}Br}{C}HCH_2CH_2CH_3 \xrightarrow{H_3O^+} \underset{\text{3-hexanol, a secondary alcohol}}{CH_3CH_2\overset{:\ddot{O}H}{C}HCH_2CH_2CH_3}$$

When a Grignard reagent reacts with a ketone, the addition product is a tertiary alcohol.

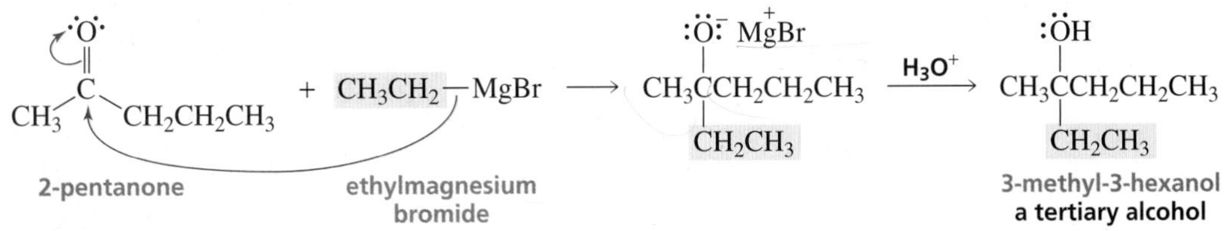

$$\underset{\text{2-pentanone}}{CH_3\overset{O}{\overset{\|}{C}}CH_2CH_2CH_3} + \underset{\text{ethylmagnesium bromide}}{CH_3CH_2-MgBr} \longrightarrow CH_3\overset{:\ddot{O}:^-\ \overset{+}{Mg}Br}{\overset{|}{C}}(CH_2CH_3)CH_2CH_2CH_3 \xrightarrow{H_3O^+} \underset{\substack{\text{3-methyl-3-hexanol}\\ \text{a tertiary alcohol}}}{CH_3\overset{:\ddot{O}H}{\overset{|}{C}}(CH_2CH_3)CH_2CH_2CH_3}$$

In the following reactions, numbers are used with the reagents to indicate that the acid is not added until the reaction with the Grignard reagent is complete:

Movie: Reactions of Grignard reagents with ketones

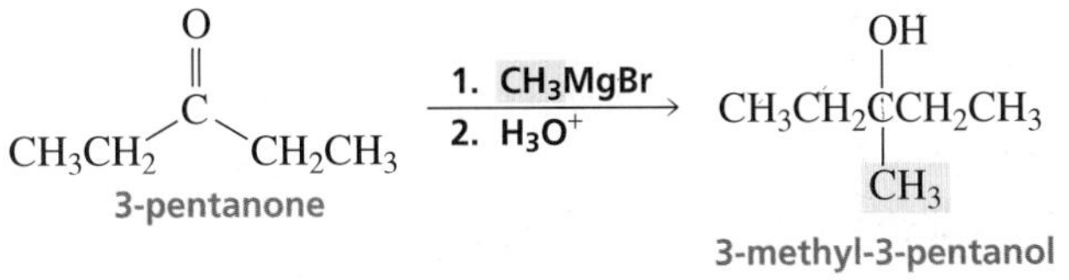

$$\underset{\text{3-pentanone}}{CH_3CH_2\overset{O}{\overset{\|}{C}}CH_2CH_3} \xrightarrow[\text{2. } H_3O^+]{\text{1. } CH_3MgBr} \underset{\text{3-methyl-3-pentanol}}{CH_3CH_2\overset{OH}{\overset{|}{C}}(CH_3)CH_2CH_3}$$

$$\underset{\text{butanal}}{CH_3CH_2CH_2\overset{O}{\overset{\|}{C}}H} \xrightarrow[\text{2. } H_3O^+]{\text{1. } C_6H_5-MgBr} \underset{\text{1-phenyl-1-butanol}}{CH_3CH_2CH_2\overset{OH}{\overset{|}{C}}H-C_6H_5}$$

A Grignard reagent can also react with carbon dioxide. The product of the reaction is a carboxylic acid with one more carbon atom than the Grignard reagent has.

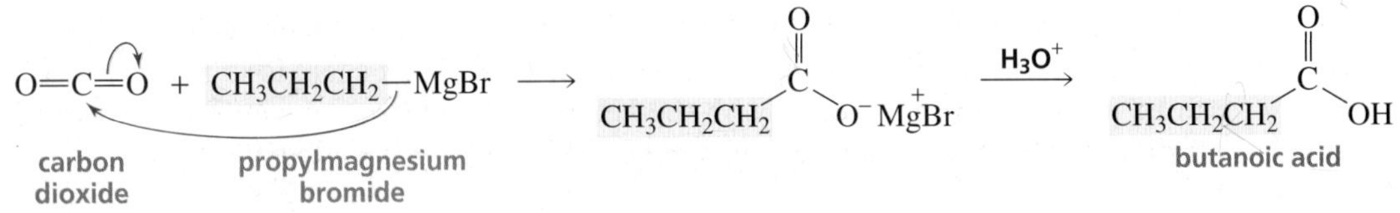

$$\underset{\text{carbon dioxide}}{O{=}C{=}O} + \underset{\text{propylmagnesium bromide}}{CH_3CH_2CH_2-MgBr} \longrightarrow CH_3CH_2CH_2\overset{O}{\overset{\|}{C}}O^-\ \overset{+}{Mg}Br \xrightarrow{H_3O^+} \underset{\text{butanoic acid}}{CH_3CH_2CH_2\overset{O}{\overset{\|}{C}}OH}$$

PROBLEM 5◆

a. How many isomers are obtained from the reaction of 2-pentanone with ethylmagnesium bromide followed by treatment with aqueous acid?
b. How many isomers are obtained from the reaction of 2-pentanone with methylmagnesium bromide followed by treatment with aqueous acid?

PROBLEM 6◆

We saw that 3-methyl-3-hexanol can be synthesized from the reaction of 2-pentanone with ethylmagnesium bromide. What two other combinations of ketone and Grignard reagent could be used to prepare the same tertiary alcohol?

In addition to reacting with aldehydes and ketones—Class II carbonyl compounds—Grignard reagents react with Class I carbonyl compounds—carbonyl compounds that have groups that can be replaced by another group.

Class I carbonyl compounds undergo two successive reactions with the Grignard reagent. For example, when an ester reacts with a Grignard reagent, the first reaction is a *nucleophilic acyl substitution reaction* because an ester, unlike an aldehyde or a ketone, has a group that can be replaced by the Grignard reagent. The product of the reaction is a ketone. The reaction does not stop at the ketone stage, however, because ketones are more reactive than esters toward nucleophilic attack (Section 18.2). Reaction of the ketone with a second molecule of the Grignard reagent forms a tertiary alcohol. Because the tertiary alcohol is formed as a result of two successive reactions with a Grignard reagent, the alcohol has two identical groups bonded to the tertiary carbon.

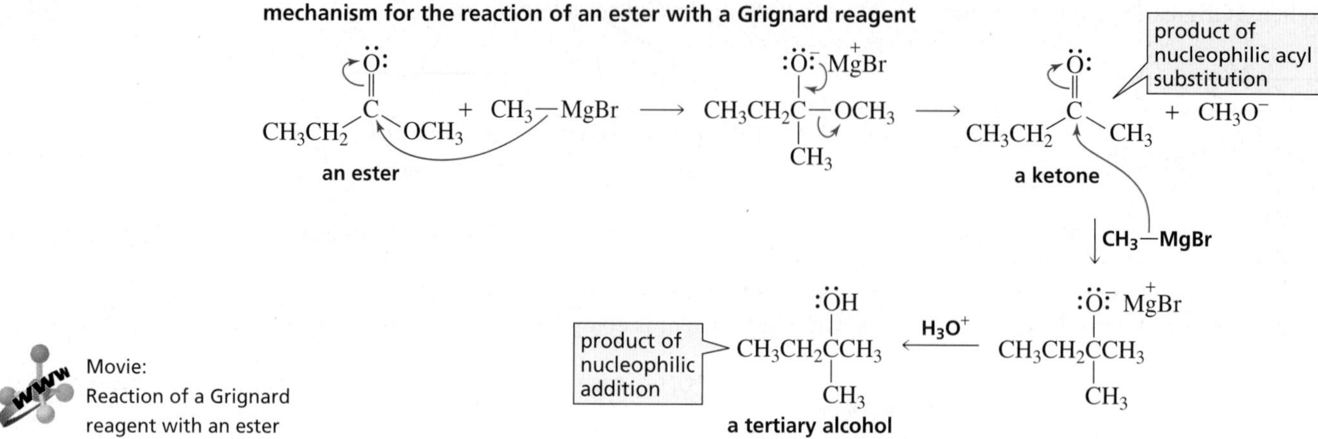

Movie: Reaction of a Grignard reagent with an ester

Tertiary alcohols are also formed from the reaction of two equivalents of a Grignard reagent with an acyl halide.

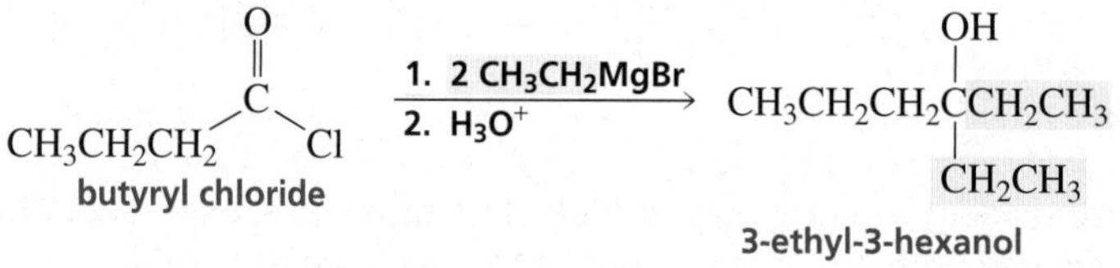

$$\underset{\text{butyryl chloride}}{CH_3CH_2CH_2C(=O)Cl} \xrightarrow[\text{2. } H_3O^+]{\text{1. 2 } CH_3CH_2MgBr} \underset{\text{3-ethyl-3-hexanol}}{CH_3CH_2CH_2C(OH)(CH_2CH_3)CH_2CH_3}$$

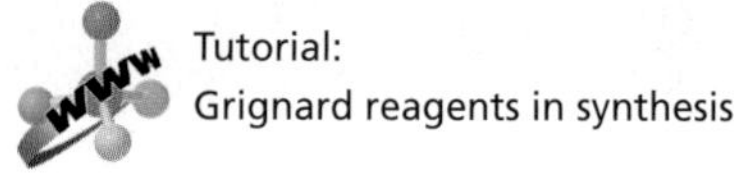

Tutorial: Grignard reagents in synthesis

In theory, we should be able to stop this reaction at the ketone stage because a ketone is less reactive than an acyl halide. However, the Grignard reagent is so reactive that it can be prevented from reacting with the ketone only under very carefully controlled conditions. There are better ways to synthesize ketones (Appendix IV).

PROBLEM 7◆

What product would be obtained from the reaction of one equivalent of a carboxylic acid with one equivalent of a Grignard reagent?

PROBLEM 8 SOLVED

a. Which of the following tertiary alcohols cannot be prepared from the reaction of an ester with excess Grignard reagent?

1. $CH_3C(OH)(CH_3)CH_3$
2. $CH_3C(OH)(CH_3)CH_2CH_3$
3. $CH_3CH_2C(OH)(CH_3)CH_2CH_2CH_3$
4. $CH_3CH_2C(OH)(CH_3)CH_2CH_3$
5. $CH_3C(OH)(CH_2CH_3)CH_2CH_2CH_2CH_3$
6. $C_6H_5C(OH)(CH_3)C_6H_5$

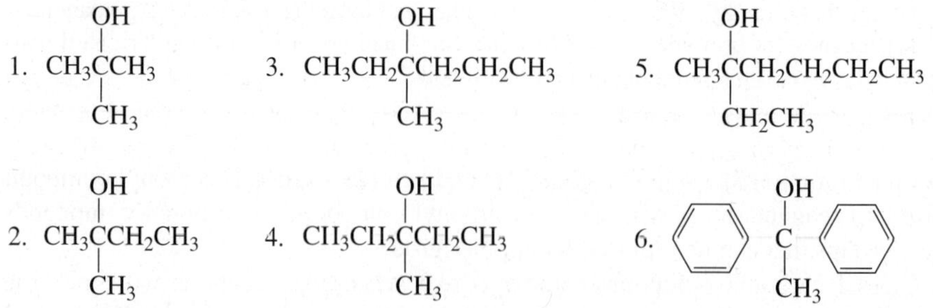

b. For those alcohols that can be prepared by the reaction of an ester with excess Grignard reagent, what ester and what Grignard reagent should be used?

SOLUTION TO 8a A tertiary alcohol is obtained from the reaction of an ester with two equivalents of a Grignard reagent. Therefore, tertiary alcohols prepared in this way must have two identical substituents on the carbon to which the OH is bonded, because two substituents come from the Grignard reagent. Alcohols (3) and (5) cannot be prepared in this way because they do not have two identical substituents.

SOLUTION TO 8b(2) Methyl propanoate and excess methylmagnesium bromide.

PROBLEM 9◆

Which of the following secondary alcohols can be prepared from the reaction of methyl formate with excess Grignard reagent?

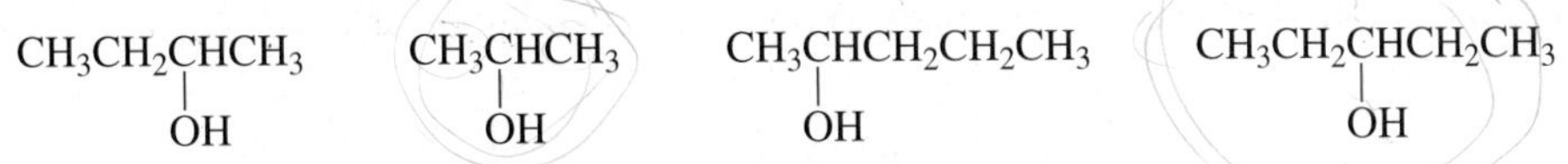

$CH_3CH_2CH(OH)CH_3$ $CH_3CH(OH)CH_3$ $CH_3CH(OH)CH_2CH_2CH_3$ $CH_3CH_2CH(OH)CH_2CH_3$

Reaction with Acetylide Ions

We have seen that a terminal alkyne can be converted into an acetylide ion by a strong base (Section 6.9).

$$CH_3C{\equiv}CH \xrightarrow[NH_3]{NaNH_2} CH_3C{\equiv}C{:}^-$$

An acetylide ion is another example of a carbon nucleophile that reacts with carbonyl compounds. When the reaction is over, a weak acid (one that will not react with the triple bond, such as pyridinium ion), is added to the reaction mixture to protonate the alkoxide ion.

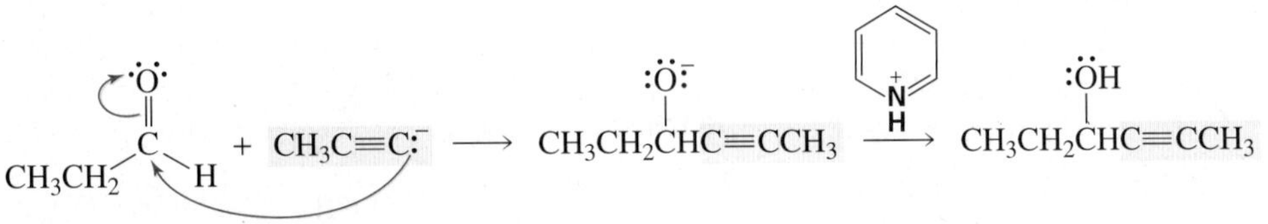

$$CH_3CH_2CH{=}O + CH_3C{\equiv}C{:}^- \longrightarrow CH_3CH_2CH(O^-)C{\equiv}CCH_3 \xrightarrow{\text{pyridinium } (C_5H_5NH^+)} CH_3CH_2CH(OH)C{\equiv}CCH_3$$

PROBLEM 10

Show how the following compounds could be prepared, using ethyne as one of the starting materials. Explain why ethyne should be alkylated before, rather than after, nucleophilic addition.

a. 1-pentyn-3-ol
b. 1-phenyl-2-butyn-1-ol
c. 2-methyl-3-hexyn-2-ol

Reaction with Hydrogen Cyanide

Hydrogen cyanide adds to aldehydes and ketones to form **cyanohydrins**. This reaction forms a product with one more carbon atom than the reactant. In the first step of the reaction, the cyanide ion attacks the carbonyl carbon. The alkoxide ion then accepts a proton from an undissociated molecule of hydrogen cyanide.

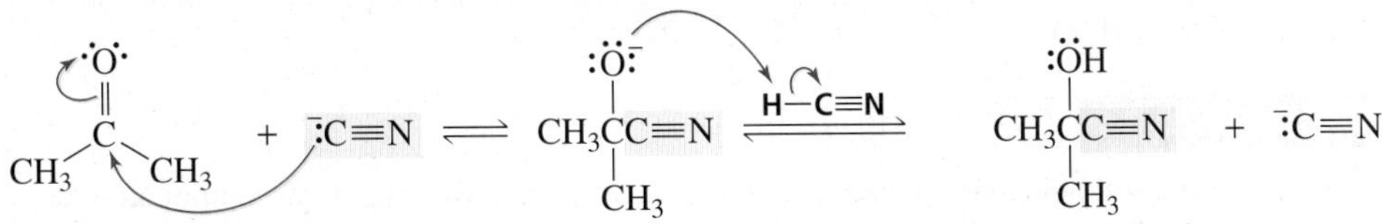

$$\underset{\text{acetone}}{(CH_3)_2C{=}O} + {:}^-C{\equiv}N \rightleftharpoons CH_3C(O^-)(CH_3)C{\equiv}N \xrightleftharpoons{H{-}C{\equiv}N} \underset{\text{acetone cyanohydrin}}{CH_3C(OH)(CH_3)C{\equiv}N} + {:}^-C{\equiv}N$$

Because hydrogen cyanide is a toxic gas, the best way to carry out this reaction is to generate hydrogen cyanide during the reaction by adding HCl to a mixture of the aldehyde or ketone and excess sodium cyanide. Excess sodium cyanide is used in order to make sure that some cyanide ion is available to act as a nucleophile.

Compared with other carbon nucleophiles, cyanide ion is a relatively weak base (the pK_a of HC≡N is 9.14, the pK_a of HC≡CH is 25, the pK_a of CH_3CH_3 is 50), which means that the cyano group is the most easily eliminated of the carbon nucleophiles from the addition product. Cyanohydrins, however, are stable because the OH group will not eliminate the cyano group; the transition state for the elimination reaction would be relatively unstable since the oxygen atom would bear a partial positive charge. If the OH group loses its proton, however, the cyano group will be eliminated because the oxygen atom would have a partial negative charge instead of a partial positive charge in the transition state of the elimination reaction. Therefore, in basic solutions, a cyanohydrin is converted back to the carbonyl compound.

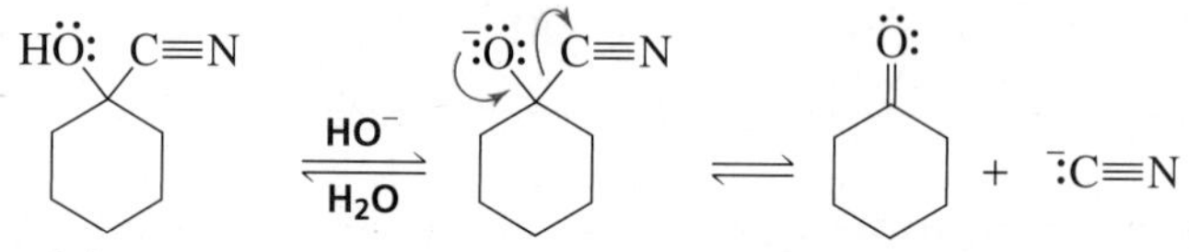

HÖ: C≡N (cyclohexane) $\xrightarrow[H_2O]{HO^-}$:Ö:⁻ C≡N (cyclohexane) ⇌ Ö: (cyclohexanone) + ⁻:C≡N

cyclohexanone cyanohydrin

Cyanide ion does not react with esters because the cyanide ion is a weaker base than an alkoxide ion, so the cyanide ion would be eliminated from the tetrahedral intermediate.

The addition of hydrogen cyanide to aldehydes and ketones is a synthetically useful reaction because of the subsequent reactions that can be carried out on the cyanohydrin. For example, the acid-catalyzed hydrolysis of a cyanohydrin forms an α-hydroxycarboxylic acid (Section 17.18).

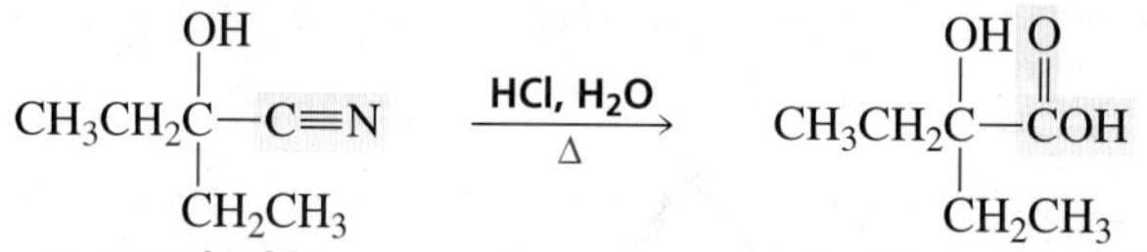

$$CH_3CH_2C(OH)(CH_2CH_3)—C≡N \xrightarrow[\Delta]{HCl,\ H_2O} CH_3CH_2C(OH)(CH_2CH_3)—COH(=O)$$

a cyanohydrin → an α-hydroxy carboxylic acid

The catalytic addition of hydrogen to a cyanohydrin produces a primary amine with an OH group on the β-carbon.

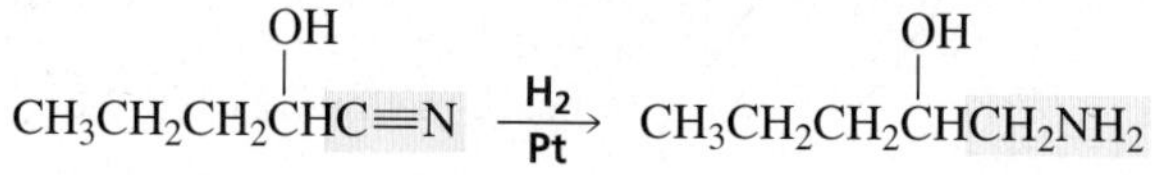

$$CH_3CH_2CH_2CH(OH)C≡N \xrightarrow[Pt]{H_2} CH_3CH_2CH_2CH(OH)CH_2NH_2$$

PROBLEM 11

Can a cyanohydrin be prepared by treating a ketone with sodium cyanide?

PROBLEM 12

Explain why aldehydes and ketones react with a weak acid such as hydrogen cyanide in the presence of $^-$C≡N, but do not react with strong acids such as HCl or H_2SO_4 in the presence of Cl^- or HSO_4^-.

PROBLEM 13 SOLVED

How can the following compounds be prepared, starting with a carbonyl compound with one fewer carbon atoms than the desired product?

a. $HOCH_2CH_2NH_2$

b. 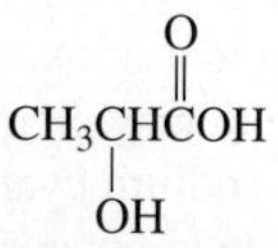 $CH_3CH(OH)COH(=O)$

SOLUTION TO 13a The starting material for the synthesis of the two-carbon compound must be formaldehyde. Addition of hydrogen cyanide followed by addition of H_2 to the triple bond of the cyanohydrin forms the desired compound.

$$\overset{\displaystyle O}{\overset{\|}{HCH}} \xrightarrow[\text{HCl}]{\text{NaC}\equiv\text{N}} HOCH_2C\equiv N \xrightarrow[\text{Pt}]{H_2} HOCH_2CH_2NH_2$$

SOLUTION TO 13b The starting material for the synthesis of the three-carbon α-hydroxycarboxylic acid must be ethanal. Addition of hydrogen cyanide, followed by hydrolysis of the cyanohydrin, forms the target molecule.

$$CH_3\overset{\displaystyle O}{\overset{\|}{C}}H \xrightarrow[\text{HCl}]{\text{NaC}\equiv\text{N}} CH_3\underset{\displaystyle OH}{\underset{|}{C}}HC\equiv N \xrightarrow[\Delta]{\text{HCl, } H_2O} CH_3\underset{\displaystyle OH}{\underset{|}{C}}H\overset{\displaystyle O}{\overset{\|}{C}}OH$$

18.5 Reactions of Carbonyl Compounds with Hydride Ion

Addition of hydride ion to an aldehyde or ketone forms an alkoxide ion. Subsequent protonation by an acid produces an alcohol. The overall reaction adds H_2 to the carbonyl group. Recall that the addition of hydrogen to an organic compound is a **reduction reaction** (Section 4.8).

$$R\overset{\displaystyle O}{\overset{\|}{C}}R' + :H^- \longrightarrow R-\underset{\displaystyle H}{\underset{|}{\overset{\displaystyle O^-}{\overset{|}{C}}}}-R' \underset{:B}{\overset{HB^+}{\rightleftharpoons}} R-\underset{\displaystyle H}{\underset{|}{\overset{\displaystyle OH}{\overset{|}{C}}}}-R'$$

Aldehydes and ketones are generally reduced using sodium borohydride ($NaBH_4$) as the source of hydride ion. Aldehydes are reduced to primary alcohols, and ketones are reduced to secondary alcohols. Notice that the acid is not added to the reaction mixture until the reaction with the hydride donor is complete.

$$CH_3CH_2CH_2\overset{\displaystyle O}{\overset{\|}{C}}H \xrightarrow[\text{2. } H_3O^+]{\text{1. NaBH}_4} CH_3CH_2CH_2CH_2OH$$

butanal
an aldehyde

1-butanol
a primary alcohol

$$CH_3CH_2CH_2\overset{\displaystyle O}{\overset{\|}{C}}CH_3 \xrightarrow[\text{2. } H_3O^+]{\text{1. NaBH}_4} CH_3CH_2CH_2\overset{\displaystyle OH}{\overset{|}{C}}HCH_3$$

2-pentanone
a ketone

2-pentanol
a secondary alcohol

PROBLEM 14◆

What alcohols are obtained from the reduction of the following compounds with sodium borohydride?

a. 2-methylpropanal
b. cyclohexanone
c. benzaldehyde
d. acetophenone

The reaction of a Class I carbonyl compound (i.e., a carbonyl compound with a group that can be replaced by another group) with hydride ion involves two successive reactions with the nucleophile. (Recall that Class I carbonyl compounds also undergo two successive reactions with a Grignard reagent; see Section 18.4.) Sodium borohydride ($NaBH_4$) is not a sufficiently strong hydride donor to react with the less reactive (compared with aldehydes and ketones) esters, carboxylic acids, and amides, so esters, carboxylic acids, and amides must be reduced with lithium aluminum hydride ($LiAlH_4$), a more reactive hydride donor.

Esters undergo two successive reactions with hydride ion and with Grignard reagents.

Because lithium aluminum hydride is more reactive than sodium borohydride, it is not as safe or as easy to use. Since it reacts violently with protic solvents, lithium aluminum hydride must be used in a dry, aprotic solvent.

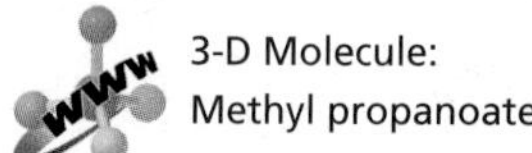

3-D Molecule: Methyl propanoate

The reaction of an ester with $LiAlH_4$ produces two alcohols, one corresponding to the acyl portion of the ester and one corresponding to the alkyl portion.

$$CH_3CH_2C(=O)OCH_3 \xrightarrow[2.\ H_3O^+]{1.\ LiAlH_4} CH_3CH_2CH_2OH + CH_3OH$$

methyl propanoate (an ester) → 1-propanol + methanol

When an ester reacts with hydride ion, the first reaction is a nucleophilic acyl substitution reaction because an ester has a group that can be substituted by hydride ion. The product of this reaction is an aldehyde. The aldehyde then undergoes a nucleophilic addition reaction with a second equivalent of hydride ion, forming an alkoxide ion, which when protonated gives a primary alcohol. The reaction cannot be stopped at the aldehyde stage because an aldehyde is more reactive than an ester toward nucleophilic attack.

mechanism for the reaction of an ester with hydride ion

$$CH_3CH_2C(=O)OCH_3 + H{-}\bar{A}lH_3 \longrightarrow CH_3CH_2CH(O^-)OCH_3 \longrightarrow CH_3CH_2C(=O)H + CH_3O^-$$

an ester → an aldehyde (product of nucleophilic acyl substitution)

$$CH_3CH_2C(=O)H \xrightarrow{H{-}\bar{A}lH_3} CH_3CH_2CH_2O^- \xrightarrow{H_3O^+} CH_3CH_2CH_2OH$$

a primary alcohol (product of nucleophilic addition)

Chemists have found that if diisobutylaluminum hydride (DIBALH) is used as the hydride donor at a low temperature, the reaction can be stopped after the addition of one equivalent of hydride ion. This reagent, therefore, makes it possible to convert esters into aldehydes, which is initially surprising, since aldehydes are more reactive than esters toward hydride ion.

$$(CH_3)_2CHCH_2{-}AlH{-}CH_2CH(CH_3)_2$$

diisobutylaluminum hydride
DIBALH

$$CH_3CH_2CH_2CH_2C(=O)OCH_3 \xrightarrow[2.\ H_2O]{1.\ [(CH_3)_2CHCH_2]_2AlH,\ -78\ ^\circ C} CH_3CH_2CH_2CH_2C(=O)H + CH_3OH$$

methyl pentanoate → pentanal

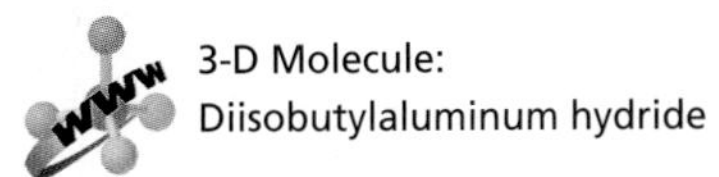

The reaction is carried out at −78° C (the temperature of a dry ice–acetone bath). At this cold temperature, the initially formed tetrahedral intermediate is stable, so it does not eliminate the alkoxide ion. All of the unreacted hydride donor is removed from the solution before the solution warms up. Therefore, when the tetrahedral intermediate eliminates the alkoxide ion, there is no reducing agent to react with the aldehyde. If, however, the reaction is carried out at room temperature, a primary alcohol is obtained—the same product obtained when an ester reacts with $LiAlH_4$.

The reaction of a carboxylic acid with $LiAlH_4$ forms a single primary alcohol.

CH_3COOH (acetic acid) $\xrightarrow[2.\ H_3O^+]{1.\ LiAlH_4}$ CH_3CH_2OH (ethanol)

C_6H_5COOH (benzoic acid) $\xrightarrow[2.\ H_3O^+]{1.\ LiAlH_4}$ $C_6H_5CH_2OH$ (benzyl alcohol)

In the first step of the reaction, a hydride ion reacts with the acidic hydrogen of the carboxylic acid, forming H_2 and a carboxylate ion. We have seen that nucleophiles do not react with carboxylate ions because of their negative charge. However, in this case, an electrophile (AlH_3) is present that accepts a pair of electrons from the carboxylate ion and forms a new hydride donor. Then, analogous to the reduction of an ester by $LiAlH_4$, two successive additions of hydride ion take place, with an aldehyde being formed as an intermediate on the way to the primary alcohol.

mechanism for the reaction of a carboxylic acid with hydride ion

$CH_3C(=O)-O-H + H-\bar{A}lH_3 \longrightarrow CH_3C(=O)-O^- + H_2 \xrightarrow{AlH_3} CH_3C(=O)-O-\bar{A}lH_3 \longrightarrow CH_3CH(O-AlH_2)-O^- \longrightarrow CH_3CH=O + AlH_2O^-$

a carboxylic acid

hydride ion removes an acidic proton

new hydride donor

second addition of hydride ion

$CH_3CH=O$ (an aldehyde) $\xrightarrow{H-\bar{A}lH_3}$ $CH_3CH_2O^-$ $\xrightarrow{H_3O^+}$ CH_3CH_2OH (a primary alcohol)

Acyl chlorides, like esters and carboxylic acids, undergo two successive additions of hydride ion when treated with $LiAlH_4$.

$CH_3CH_2CH_2COCl$ (butanoyl chloride) $\xrightarrow[2.\ H_3O^+]{1.\ LiAlH_4}$ $CH_3CH_2CH_2CH_2OH$ (1-butanol)

Amides also undergo two successive additions of hydride ion when they react with $LiAlH_4$. The product of the reaction is an amine. Primary, secondary, or tertiary amines can be formed, depending on the number of substituents bonded to the nitrogen of the amide. Overall, the reaction converts a carbonyl group into a methylene group. (Notice that H_2O rather than H_3O^+ is used in the second step of the reaction. The product, therefore, is an amine rather than an ammonium ion.)

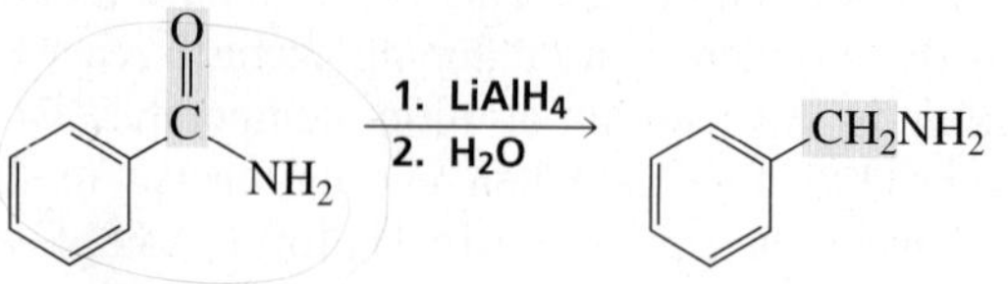

benzamide —(1. $LiAlH_4$ 2. H_2O)→ $C_6H_5CH_2NH_2$ benzylamine, a primary amine

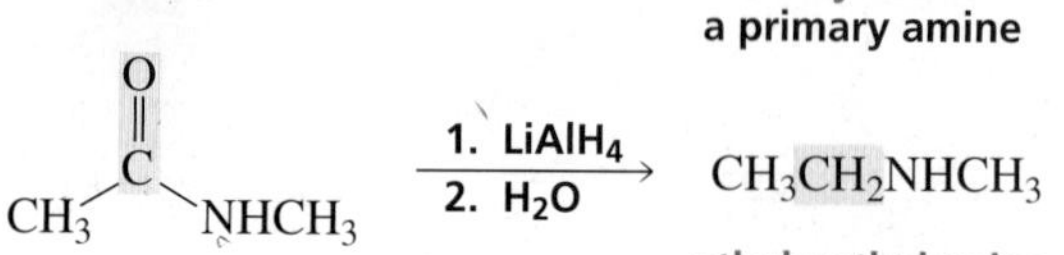

$CH_3C(=O)NHCH_3$ N-methylacetamide —(1. $LiAlH_4$ 2. H_2O)→ $CH_3CH_2NHCH_3$ ethylmethylamine, a secondary amine

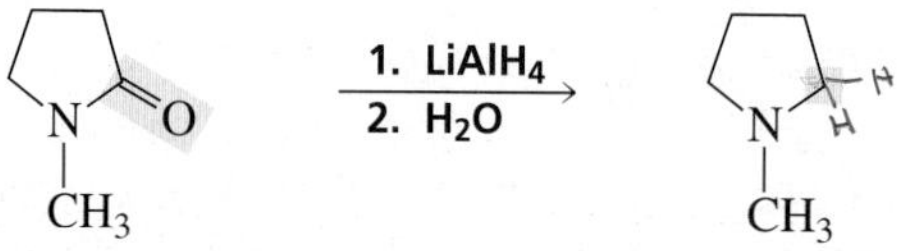

N-methyl-γ-butyrolactam —(1. $LiAlH_4$ 2. H_2O)→ N-methylpyrrolidine, a tertiary amine

The mechanism of the reaction shows why the product of the reaction is an amine. Take a minute to note the similarities between the mechanisms for the reaction of hydride ion with an *N*-substituted amide and with a carboxylic acid.

mechanism for the reaction of an *N*-substituted amide with hydride ion

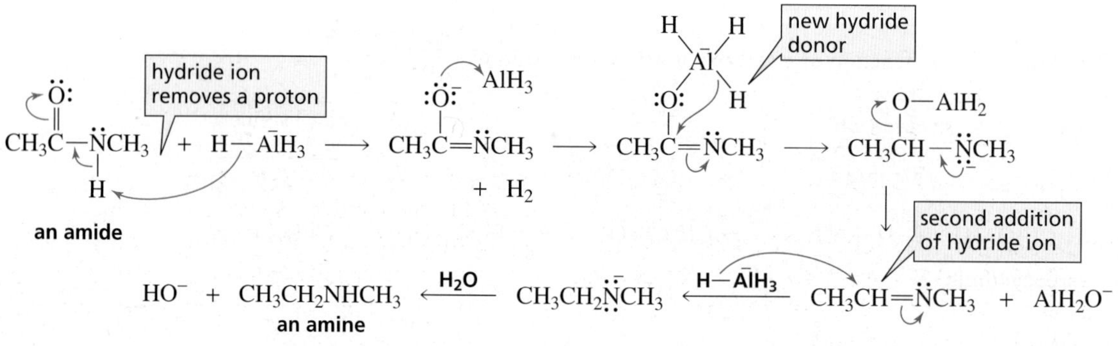

$CH_3C(=O)NHCH_3$ (an amide) + $H-\bar{A}lH_3$ (hydride ion removes a proton) ⟶ $CH_3C(-O^-)=NCH_3$ + H_2 ⟶ $CH_3C(-O-AlH_3)=NCH_3$ (new hydride donor) ⟶ $CH_3CH(O-AlH_2)-\bar{N}CH_3$ ⟶ $CH_3CH=NCH_3$ + AlH_2O^- —($H-\bar{A}lH_3$, second addition of hydride ion)→ $CH_3CH_2\bar{N}CH_3$ —(H_2O)→ HO^- + $CH_3CH_2NHCH_3$ (an amine)

The mechanisms for the reaction of $LiAlH_4$ with unsubstituted and *N,N*-disubstituted amines are somewhat different, but have the same result: the conversion of a carbonyl group into a methylene group.

PROBLEM 15◆

What amides would you treat with $LiAlH_4$ in order to prepare the following amines?

a. benzylmethylamine
b. ethylamine
c. diethylamine
d. triethylamine

PROBLEM 16

Starting with *N*-benzylbenzamide, how would you make the following compounds?

a. dibenzylamine
b. benzoic acid
c. benzaldehyde
d. benzyl alcohol

18.6 Reactions of Aldehydes and Ketones with Nitrogen Nucleophiles

Aldehydes and ketones react with a *primary* amine (RNH_2) to form an imine. An **imine** is a compound with a carbon–nitrogen double bond. The imine obtained from the reaction of a carbonyl compound and a primary amine is often called a **Schiff base**.

$$\underset{\text{an aldehyde or a ketone}}{\text{H–C–C=O}} \; + \; \underset{\text{a primary amine}}{R{-}NH_2} \; \rightleftharpoons \; \underset{\text{an imine, a Schiff base}}{\text{H–C–C=N–R}} \; + \; H_2O$$

The orbital model of a C=N group (Figure 18.1) is similar to the orbital model of a C=O group (Figure 17.1 on p. 676). The imine nitrogen is sp^2 hybridized. One of its sp^2 orbitals forms a σ bond with the imine carbon, one forms a σ bond with a substituent, and the third contains a lone pair. The p orbital of nitrogen and the p orbital of carbon overlap to form a π bond.

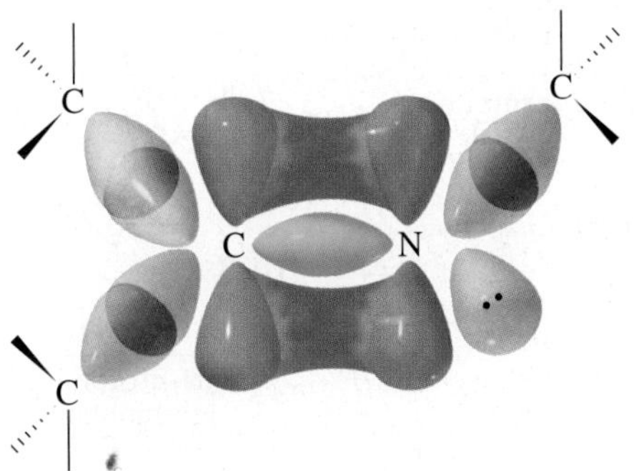

Figure 18.1 Bonding in an imine

Aldehydes and ketones react with a *secondary amine* to form an enamine (pronounced "ENE-amine"). An **enamine** is an α,β-unsaturated tertiary amine—a tertiary amine with a double bond in the α,β-position relative to the nitrogen atom. Notice that the double bond is in the part of the molecule that comes from the aldehyde or ketone. The name "enamine" comes from "ene" + "amine," with the "e" omitted in order to avoid two successive vowels.

$$\underset{\text{an aldehyde or a ketone}}{\text{H–C–C=O}} \; + \; \underset{\text{a secondary amine}}{R_2NH} \; \rightleftharpoons \; \underset{\text{an enamine}}{\text{C=C–NR}_2} \; + \; H_2O$$

When you first look at the products of imine and enamine formation, they appear to be quite different. However, when you look at the mechanisms for the reactions, you will see that the mechanisms are exactly the same except for the site from which a proton is lost in the last step of the reaction.

Addition of Primary Amines

Aldehydes and ketones react with primary amines to form imines. The reaction requires a catalytic (small) amount of acid—we will see that the pH of the reaction mixture must be carefully controlled.

$$C_6H_5CH{=}O + CH_3CH_2NH_2 \underset{}{\overset{\text{catalytic } H^+}{\rightleftharpoons}} C_6H_5CH{=}NCH_2CH_3 + H_2O$$

benzaldehyde
an aldehyde

ethylamine
a primary amine

an imine

$$(CH_3CH_2)_2C{=}O + H_2NCH_2C_6H_5 \overset{\text{catalytic } H^+}{\rightleftharpoons} (CH_3CH_2)_2C{=}NCH_2C_6H_5 + H_2O$$

3-pentanone
a ketone

benzylamine
a primary amine

an imine

In the first step of the mechanism for imine formation, the amine attacks the carbonyl carbon. Gain of a proton by the alkoxide ion and loss of a proton by the ammonium ion forms a neutral tetrahedral intermediate. The neutral tetrahedral intermediate, called a *carbinolamine*, is in equilibrium with two protonated forms. Protonation can take place on either the nitrogen or the oxygen atom. Elimination of water from the oxygen-protonated intermediate forms a protonated imine that loses a proton to yield the imine.

HB^+ represents any species in the solution that is capable of donating a proton, and :B represents any species in the solution that is capable of removing a proton.

mechanism for imine formation

cyclohexanone + RNH_2 ⇌ O^- / $\overset{+}{N}H_2R$ ⇌ OH / $\overset{+}{N}H_2R$ (H :B) ⇌ OH / NHR

N-protonated carbinolamine

neutral tetrahedral intermediate
a carbinolamine

⇌ $\overset{+}{O}H_2$ / NHR — O-protonated carbinolamine; elimination of water ⇌ $=\overset{+}{N}R$ + H_2O (H :B) — a protonated imine ⇌ $=NR$ + HB^+ — an imine

The equilibrium favors the nitrogen-protonated tetrahedral intermediate because nitrogen is more basic than oxygen. The equilibrium can be forced toward the imine by removing water as it is formed or by precipitation of the imine product.

Overall, the addition of a nitrogen nucleophile to an aldehyde or a ketone is a *nucleophilic addition–elimination reaction*: nucleophilic addition of an amine to form an unstable tetrahedral intermediate, followed by elimination of water. The tetrahedral intermediates are unstable because the newly formed sp^3 carbon is bonded to an oxygen and to a nitrogen—another electronegative atom. Water is eliminated, and loss of a proton from the resulting protonated imine forms a stable imine.

A compound with an sp^3 carbon bonded to an oxygen and to another electronegative atom is unstable.

In contrast, the reaction of an aldehyde or a ketone with a carbon or hydrogen nucleophile forms a stable tetrahedral compound because the newly formed sp^3 carbon is *not* bonded to a second electronegative atom. Thus, aldehydes and ketones undergo *nucleophilic addition reactions* with carbon and hydrogen nucleophiles, whereas they undergo *nucleophilic addition–elimination reactions* with nitrogen nucleophiles.

The pH at which imine formation is carried out must be carefully controlled. There must be sufficient acid present to protonate the tetrahedral intermediate so that H_2O rather than the much more basic HO^- is the leaving group. However, if too much acid is present, it will protonate all of the reactant amine. Protonated amines are not nucleophiles, so they cannot react with carbonyl groups. Therefore, unlike the acid-catalyzed reactions we have seen previously (Section 17.11), there is not sufficient acid present to protonate the carbonyl group in the first step of the reaction (see Problem 17).

A plot of the observed rate constant for the reaction of acetone with hydroxylamine as a function of the pH of the reaction mixture is shown in Figure 18.2. This type of plot is called a **pH-rate profile**. The pH-rate profile in the figure is a bell-shaped curve with the maximum rate occurring at about pH 4.5, 1.5 pH units below the pK_a of hydroxylamine ($pK_a = 6.0$). As the acidity increases below pH 4.5, the rate of the reaction decreases because more and more of the amine becomes protonated. As a result, less and less of the amine is present in the nucleophilic nonprotonated form. As the acidity decreases above pH 4.5, the rate decreases because less and less of the tetrahedral intermediate is present in the reactive protonated form.

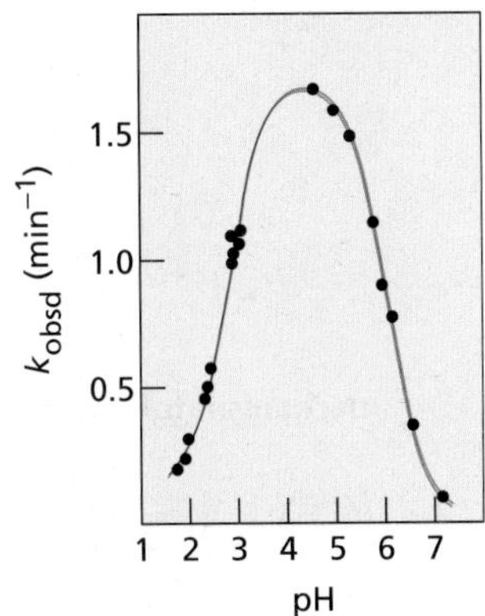

▲ **Figure 18.2**
Dependence of the rate of the reaction of acetone with hydroxylamine on the pH of the reaction mixture.

Imine formation is reversible: In acidic aqueous solutions, imines are hydrolyzed back to the carbonyl compound and amine.

$$C_6H_5\text{-}CH{=}NCH_2CH_3 + H_2O \xrightarrow{HCl} C_6H_5\text{-}CH{=}O + CH_3CH_2\overset{+}{N}H_3$$

In an acidic solution, the amine is protonated and, therefore, is unable to react with the carbonyl compound to reform the imine.

Imine formation and hydrolysis are important reactions in biological systems (Sections 19.21, 24.9, and 25.6). Imine hydrolysis is the reason DNA contains A, G, C, and T nucleotides, whereas RNA contains A, G, C, and U nucleotides (Section 27.14).

PROBLEM 17◆

The pK_a of protonated acetone is about −7.5 and the pK_a of protonated hydroxylamine is 6.0.

a. In its reaction with hydroxylamine at pH = 4.5 (Figure 18.2), what fraction of acetone will be present in its acidic, protonated form? (*Hint*: See Section 1.20.)
b. In its reaction with hydroxylamine at pH = 1.5, what fraction of acetone will be present in its acidic, protonated form?
c. In its reaction with acetone at pH = 1.5 (Figure 18.2), what fraction of hydroxylamine will be present in its reactive basic form?

PROBLEM 18

A ketone can be prepared from the reaction of a nitrile with a Grignard reagent. Describe the intermediate that is formed in this reaction, and explain how it can be converted to a ketone.

Aldehydes and ketones react with primary amines to form imines.

An imine undergoes acid-catalyzed hydrolysis to form a carbonyl compound and a primary amine.

Addition of Secondary Amines

Aldehydes and ketones react with secondary amines to form enamines. Like imine formation, the reaction requires a trace amount of an acid catalyst.

3-D Molecule: *N,N*-Diethyl-1-cyclopentenamine

$$\text{cyclopentanone} + (CH_3CH_2)_2NH \underset{}{\overset{\text{catalytic } H^+}{\rightleftharpoons}} \text{C}_5\text{H}_7\text{-}N(CH_2CH_3)_2 + H_2O$$

cyclopentanone; diethylamine, a secondary amine; an enamine

$$\text{cyclohexanone} + \text{pyrrolidine (NH)} \overset{\text{catalytic } H^+}{\rightleftharpoons} \text{cyclohexenyl-N(pyrrolidine)} + H_2O$$

cyclohexanone; pyrrolidine, a secondary amine; an enamine

The mechanism for enamine formation is exactly the same as that for imine formation, until the last step of the reaction. When a primary amine reacts with an aldehyde or a ketone, the protonated imine loses a proton from nitrogen in the last step of the reaction, forming a neutral imine. However, when the amine is secondary, the positively charged nitrogen is not bonded to a hydrogen. A stable neutral molecule is obtained by removing a proton from the α-carbon of the compound derived from the carbonyl compound. An enamine is the result.

mechanism for enamine formation

N-protonated carbinolamine

neutral tetrahedral intermediate
a carbinolamine

O-protonated carbinolamine

elimination of water

an enamine

this intermediate cannot lose a proton from N, so it loses a proton from an α-carbon

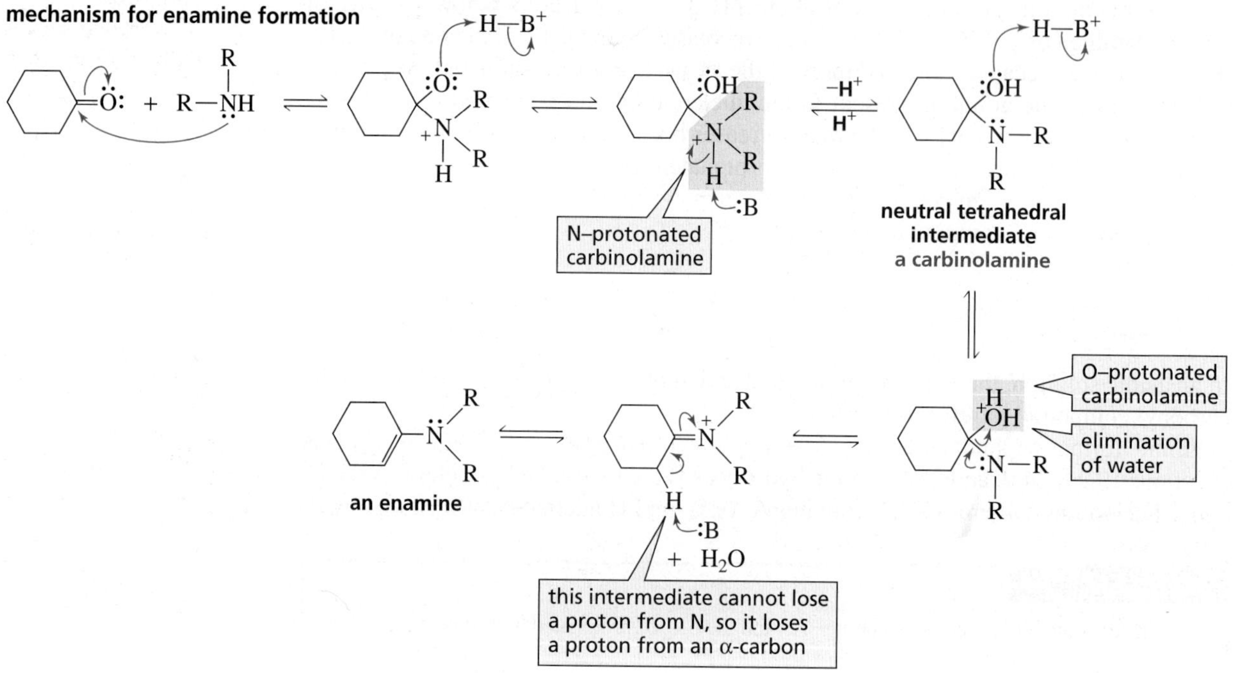

In aqueous acidic solutions, an enamine is hydrolyzed back to the carbonyl compound and secondary amine, a reaction that is similar to the acid-catalyzed hydrolysis of an imine back to the carbonyl compound and a primary amine.

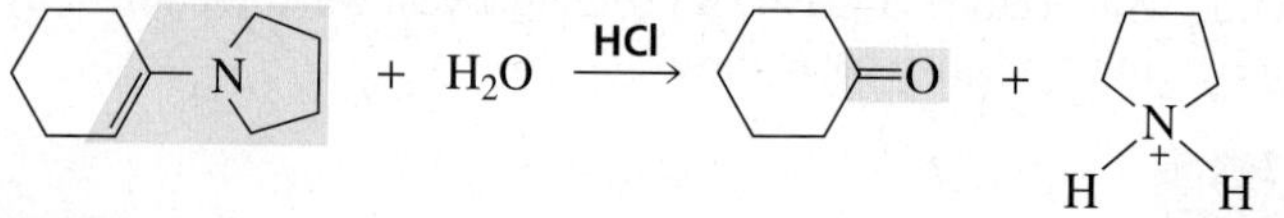

PROBLEM 19

a. Write the mechanism for the following reactions:
   1. the acid-catalyzed hydrolysis of an imine to a carbonyl compound and a primary amine
   2. the acid-catalyzed hydrolysis of an enamine to a carbonyl compound and a secondary amine

b. How do these mechanisms differ?

PROBLEM 20◆

Give the products of the following reactions. (A catalytic amount of acid is present in each reaction.)

a. cyclopentanone + ethylamine
b. cyclopentanone + diethylamine
c. acetophenone + hexylamine
d. acetophenone + cyclohexylamine

Formation of Imine Derivatives

Compounds such as hydroxylamine (NH_2OH), hydrazine (NH_2NH_2), and semicarbazide ($NH_2NHCONH_2$) are similar to primary amines in that they all have an NH_2 group. Thus, like primary amines, they react with aldehydes and ketones to form imines—often called *imine derivatives* because the substituent attached to the imine nitrogen is not an R group. The imine obtained from the reaction with hydroxylamine is called an **oxime**, the imine obtained from the reaction with hydrazine is called a **hydrazone**, and the imine obtained from the reaction with semicarbazide is called a **semicarbazone**.

3-D Molecules:
The *N*-methylimine of *acetone*;
The oxime of *acetone*;
The hydrazone of acetone;
The semicarbazone of acetone

$$C_6H_5\text{-}CH{=}O + \underset{\text{hydroxylamine}}{H_2NOH} \underset{}{\overset{\text{catalytic } H^+}{\rightleftharpoons}} \underset{\text{an oxime}}{C_6H_5\text{-}CH{=}NOH} + H_2O$$

$$C_6H_5(CH_3)C{=}O + \underset{\text{hydrazine}}{H_2NNH_2} \overset{\text{catalytic } H^+}{\rightleftharpoons} \underset{\text{a hydrazone}}{C_6H_5(CH_3)C{=}NNH_2} + H_2O$$

$$\text{cyclohexanone} + \underset{\text{semicarbazide}}{H_2NNH\overset{O}{\overset{\|}{C}}NH_2} \overset{\text{catalytic } H^+}{\rightleftharpoons} \underset{\text{a semicarbazone}}{C_6H_{10}{=}NNH\overset{O}{\overset{\|}{C}}NH_2} + H_2O$$

Tutorial:
Imine and oxime formation

Phenyl-substituted hydrazines react with aldehydes and ketones to form **phenylhydrazones**.

$$\text{cyclopentanone} + \underset{\text{phenylhydrazine}}{H_2NNH\text{-}C_6H_5} \overset{\text{catalytic } H^+}{\rightleftharpoons} \underset{\text{a phenylhydrazone}}{C_5H_8{=}NNH\text{-}C_6H_5} + H_2O$$

$$CH_3CH_2CH{=}O + \underset{\text{2,4-dinitrophenylhydrazine}}{H_2NNH\text{-}C_6H_3(O_2N)\text{-}NO_2} \overset{\text{catalytic } H^+}{\rightleftharpoons} \underset{\text{a 2,4-dinitrophenylhydrazone}}{CH_3CH_2CH{=}NNH\text{-}C_6H_3(O_2N)\text{-}NO_2} + H_2O$$

PROBLEM 21

Imines can exist as stereoisomers. The isomers are named by the *E*, *Z* system of nomenclature. (The lone pair has the lowest priority.)

R, R′ C=N X (lone pair) R, R′ C=N (lone pair) X

Draw the structure of each of the following compounds:

a. (*E*)-benzaldehyde semicarbazone
b. (*Z*)-propiophenone oxime
c. cyclohexanone 2,4-dinitrophenylhydrazone

PROBLEM 22

Semicarbazide has two NH_2 groups. Explain why only one of them forms an imine.

NONSPECTROPHOTOMETRIC IDENTIFICATION OF ALDEHYDES AND KETONES

Before spectrophotometric techniques were available, unknown aldehydes and ketones were identified by preparing imine derivatives. For example, suppose you have an unknown ketone whose boiling point you have determined to be 140 °C. This allows you to narrow the possibilities to the five ketones (A to E) listed in the following table, based on their boiling points (ketones boiling at 139 °C and 141 °C cannot be excluded, unless your thermometer is calibrated perfectly and your laboratory technique is sensational).

Ketone	bp (°C)	2,4-Dinitrophenylhydrazone mp (°C)	Oxime mp (°C)	Semicarbazone mp (°C)
A	140	94	57	98
B	140	102	68	123
C	139	121	79	121
D	140	101	69	112
E	141	90	61	101

Adding 2,4-dinitrophenylhydrazine to a sample of the unknown ketone produces crystals of a 2,4-dinitrophenylhydrazone that melt at 102 °C. You can now narrow the choice to two ketones: B and D. Preparing the oxime of the unknown ketone will not distinguish between B and D because the oximes of B and D have similar melting points, but preparing the semicarbazone will allow you to identify the ketone. Finding that the semicarbazone of the unknown ketone has a melting point of 112 °C establishes that the unknown ketone is D.

The Wolff–Kishner Reduction

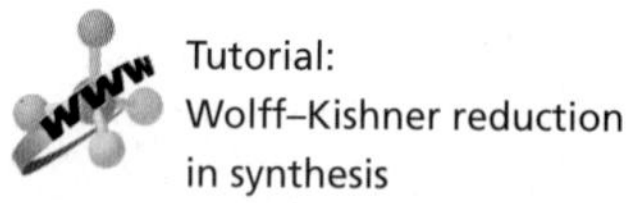

In Section 15.15, we saw that when a ketone or an aldehyde is heated in a basic solution of hydrazine, the carbonyl group is converted into a methylene group. This process is called **deoxygenation** because an oxygen is removed from the reactant. The reaction is known as the *Wolff–Kishner reduction.*

$$C_6H_5COCH_3 \xrightarrow[HO^-,\ \Delta]{NH_2NH_2} C_6H_5CH_2CH_3$$

Hydroxide ion and heat differentiate the Wolff–Kishner reduction from ordinary hydrazone formation. Initially, the ketone reacts with hydrazine to form a hydrazone. After the hydrazone is formed, hydroxide ion removes a proton from the NH_2 group. Heat is required because this proton is not easily removed. The negative charge can be delocalized onto carbon, which abstracts a proton from water. The last two steps are repeated to form the deoxygenated product and nitrogen gas.

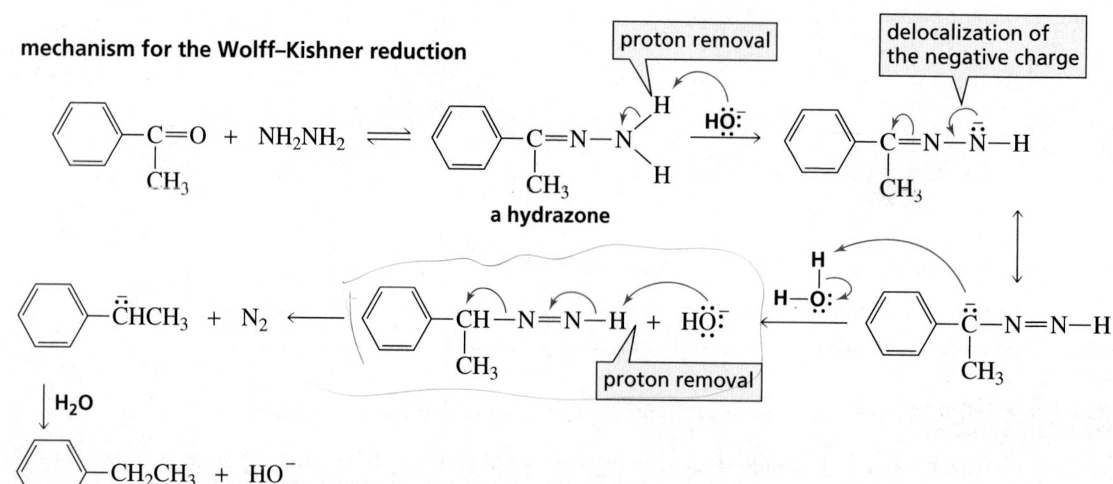

18.7 Reactions of Aldehydes and Ketones with Oxygen Nucleophiles

Addition of Water

Water adds to an aldehyde or a ketone to form a *hydrate*. A **hydrate** is a molecule with two OH groups on the same carbon. Hydrates are also called **gem-diols** (*gem* comes from *geminus,* Latin for "twin"). Hydrates of aldehydes or ketones are generally too unstable to be isolated because the tetrahedral carbon is attached to two oxygen atoms.

Most hydrates are too unstable to be isolated.

$$\underset{\text{an aldehyde or a ketone}}{RC(=O)R\,(H)} + H_2O \rightleftharpoons \underset{\text{a } gem\text{-diol, a hydrate}}{R-C(OH)_2-R\,(H)}$$

Water is a poor nucleophile and therefore adds relatively slowly to a carbonyl group. The rate of the reaction can be increased by an acid catalyst (Figure 18.3). Keep in mind that a catalyst has no effect on the position of the equilibrium. A catalyst affects the *rate* at which the equilibrium is achieved. In other words, the catalyst affects the rate at which an aldehyde or a ketone is converted to a hydrate; it has no effect on the *amount* of aldehyde or ketone converted to hydrate (Section 24.0).

mechanism for acid-catalyzed hydrate formation

$$\underset{42\%}{CH_3CH{=}O} + H{-}\overset{+}{O}H_2 \rightleftharpoons CH_3CH{=}\overset{+}{O}H + H_2O: \rightleftharpoons CH_3-CH(OH)-\overset{+}{O}H_2 \;(+\,H_2O) \rightleftharpoons \underset{58\%}{CH_3-CH(OH)_2} + H_3O^+$$

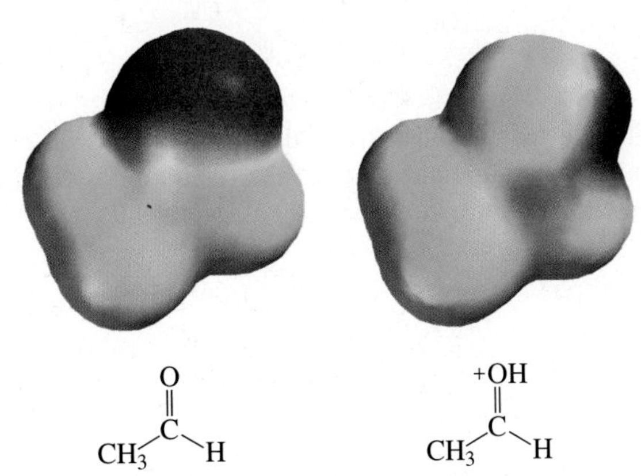

Figure 18.3 The electrostatic potential maps show that the carbonyl carbon of the protonated aldehyde is more susceptible to nucleophilic attack (the blue is more intense) than the carbonyl carbon of the unprotonated aldehyde.

PROBLEM 23

Hydration of an aldehyde can also be catalyzed by hydroxide ion. Propose a mechanism for hydroxide-ion-catalyzed hydration.

The extent to which an aldehyde or a ketone is hydrated in an aqueous solution depends on the aldehyde or ketone. For example, only 0.2% of acetone is hydrated at equilibrium, but 99.9% of formaldehyde is hydrated. Why is there such a difference?

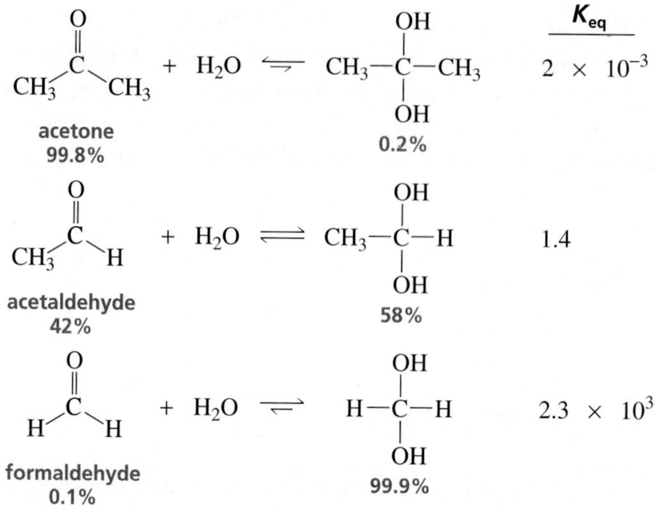

$$CH_3COCH_3 + H_2O \rightleftharpoons CH_3C(OH)_2CH_3 \qquad K_{eq} = 2 \times 10^{-3}$$

acetone 99.8% ⇌ 0.2%

$$CH_3CHO + H_2O \rightleftharpoons CH_3CH(OH)_2 \qquad K_{eq} = 1.4$$

acetaldehyde 42% ⇌ 58%

$$HCHO + H_2O \rightleftharpoons H_2C(OH)_2 \qquad K_{eq} = 2.3 \times 10^{3}$$

formaldehyde 0.1% ⇌ 99.9%

The equilibrium constant for a reaction depends on the relative stabilities of the reactants and products. The equilibrium constant for hydrate formation, therefore, depends on the relative stabilities of the carbonyl compound and the hydrate. We have seen that electron-donating alkyl groups make a carbonyl compound *more stable* (less reactive) (Section 18.2).

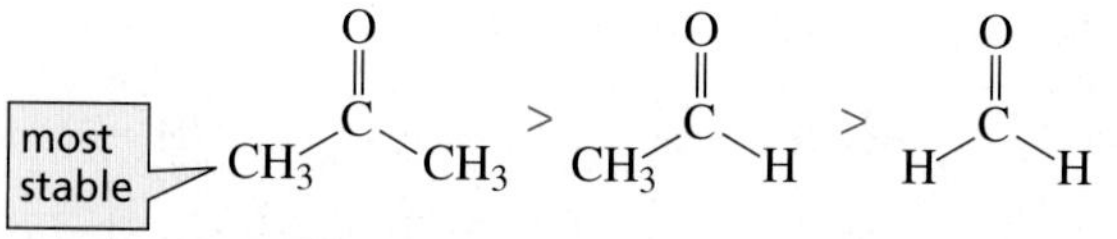

most stable: $CH_3COCH_3 > CH_3CHO > HCHO$

In contrast, alkyl groups make the hydrate *less stable* because of steric interactions between the alkyl groups.

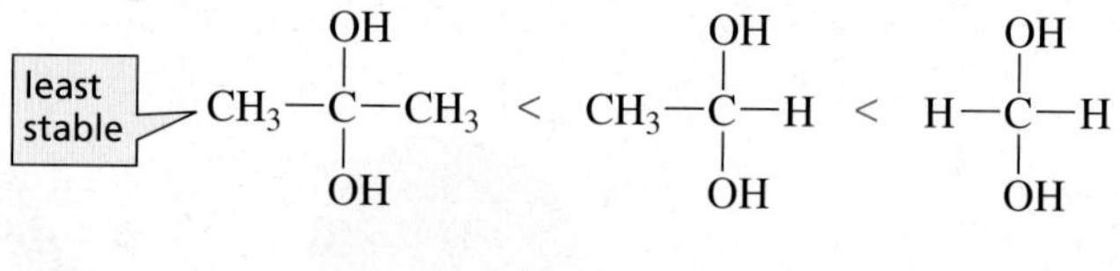

least stable: $CH_3C(OH)_2CH_3 < CH_3CH(OH)_2 < H_2C(OH)_2$

The electron clouds of the alkyl substituents do not interfere with each other in the carbonyl compound because the bond angles are 120°. However, the bond angles in the tetrahedral hydrate are 109.5°, so the alkyl groups are closer to one another.

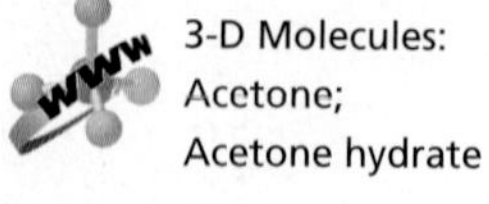
3-D Molecules: Acetone; Acetone hydrate

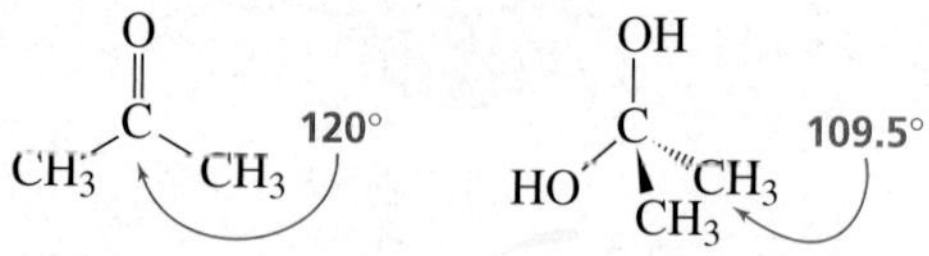

CH_3COCH_3 (120°) $\qquad$ $(HO)_2C(CH_3)_2$ (109.5°)

Alkyl groups, therefore, shift the equilibrium to the left—toward reactants—because they stabilize the carbonyl compound and destabilize the hydrate. As a result, less acetone than formaldehyde is hydrated at equilibrium.

In conclusion, the percentage of hydrate present in solution at equilibrium depends on both electronic and steric effects. Electron donation and bulky substituents *decrease* the percentage of hydrate present at equilibrium, whereas electron withdrawal and small substituents *increase* it.

PRESERVING BIOLOGICAL SPECIMENS

A 37% solution of formaldehyde in water is known as *formalin*—commonly used in the past to preserve biological specimens. Because formaldehyde is an eye and skin irritant, it has been replaced in most biology laboratories by other preservatives. One preservative frequently used is a solution of 2 to 5% phenol in ethanol with added antimicrobial agents.

If the amount of hydrate formed from the reaction of water with a ketone is too small to detect, how do we know that the reaction has even occurred? We can prove that it occurs by adding the ketone to ^{18}O-labeled water and isolating the ketone after equilibrium has been established. Finding that the label has been incorporated into the ketone indicates that hydration has occurred.

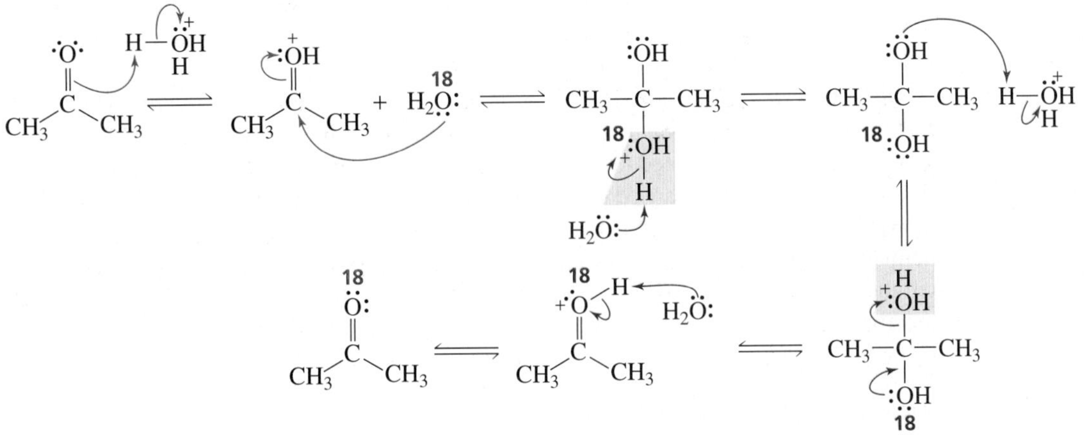

PROBLEM 24

Trichloroacetaldehyde has such a large equilibrium constant for its reaction with water that the reaction is essentially irreversible. Therefore, chloral hydrate, the product of the reaction, is one of the few hydrates that can be isolated. Chloral hydrate is a sedative that can be lethal. A cocktail laced with it is commonly known—in detective novels, at least—as a "Mickey Finn." Explain the favorable equilibrium constant.

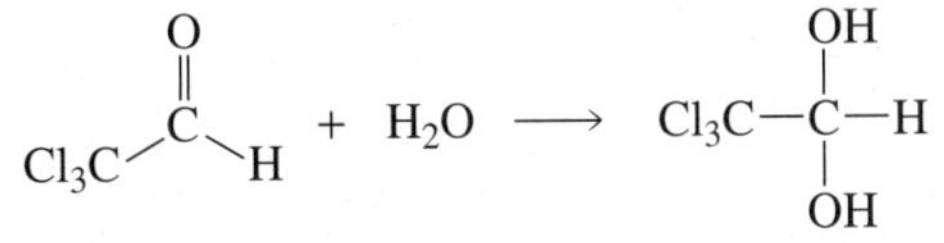

$$Cl_3C{-}C({=}O){-}H + H_2O \longrightarrow Cl_3C{-}C(OH)_2{-}H$$

trichloroacetaldehyde — chloral hydrate

PROBLEM 25◆

Which of the following ketones has the largest equilibrium constant for the addition of water?

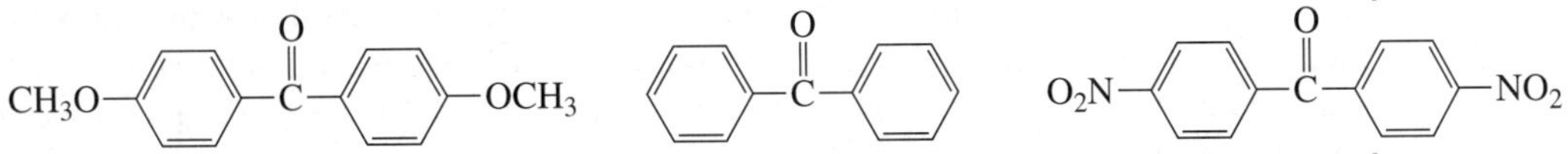

$CH_3O-C_6H_4-C(=O)-C_6H_4-OCH_3$ $C_6H_5-C(=O)-C_6H_5$ $O_2N-C_6H_4-C(=O)-C_6H_4-NO_2$

Addition of Alcohol

The product formed when one equivalent of an alcohol adds to an *aldehyde* is called a **hemiacetal**. The product formed when a second equivalent of alcohol is added is called an **acetal**. Like water, an alcohol is a poor nucleophile, so an acid catalyst is required for the reaction to take place at a reasonable rate.

$$CH_3CHO + CH_3OH \underset{}{\overset{HCl}{\rightleftharpoons}} CH_3-CH(OH)-OCH_3 \underset{}{\overset{CH_3OH,\ HCl}{\rightleftharpoons}} CH_3-CH(OCH_3)_2 + H_2O$$

an aldehyde — a hemiacetal — an acetal

When the carbonyl compound is a *ketone* instead of an aldehyde, the addition products are called a **hemiketal** and a **ketal**, respectively.

$$CH_3COCH_3 + CH_3OH \underset{}{\overset{HCl}{\rightleftharpoons}} CH_3-C(OH)(OCH_3)-CH_3 \underset{}{\overset{CH_3OH,\ HCl}{\rightleftharpoons}} CH_3-C(OCH_3)_2-CH_3 + H_2O$$

a ketone — a hemiketal — a ketal

Hemi is the Greek word for "half." When one equivalent of alcohol has added to an aldehyde or a ketone, the compound is halfway to the final acetal or ketal, which contains groups from two equivalents of alcohol.

In the first step of acetal (or ketal) formation, the acid protonates the carbonyl oxygen, making the carbonyl carbon more susceptible to nucleophilic attack (Figure 18.3). Loss of a proton from the protonated tetrahedral intermediate gives the hemiacetal (or hemiketal). Because the reaction is carried out in an acidic solution, the hemiacetal (or hemiketal) is in equilibrium with its protonated form. The two oxygen atoms of the hemiacetal (or hemiketal) are equally basic, so either one can be protonated. Loss of water from the tetrahedral intermediate with a protonated OH group forms a compound that is very reactive because of its electron-deficient carbon. Nucleophilic attack on this compound by a second molecule of alcohol, followed by loss of a proton, forms the acetal (or ketal).

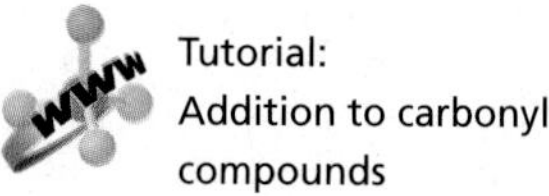
Tutorial: Addition to carbonyl compounds

mechanism for acid-catalyzed acetal or ketal formation

$$CH_3COCH_3 \overset{H-B^+}{\rightleftharpoons} CH_3C(=\overset{+}{O}H)CH_3 + CH_3\ddot{O}H \rightleftharpoons CH_3-C(OH)(\overset{+}{O}HCH_3)-CH_3 \overset{:B}{\rightleftharpoons} CH_3-C(OH)(OCH_3)-CH_3 \text{ (a hemiketal)}$$

$$\overset{H-B^+}{\rightleftharpoons} CH_3-C(\overset{+}{O}H_2)(OCH_3)-CH_3 \rightleftharpoons CH_3C(=\overset{+}{O}CH_3)CH_3 + H_2O \overset{CH_3\ddot{O}H}{\rightleftharpoons} CH_3-C(\overset{+}{O}HCH_3)(OCH_3)-CH_3 \overset{:B}{\rightleftharpoons} CH_3-C(OCH_3)_2-CH_3 \text{ (a ketal)} + H-B^+$$

Although the tetrahedral carbon of an acetal or ketal is bonded to two oxygen atoms, causing us to predict that the acetal or ketal is not stable, the acetal or ketal can be isolated if the water eliminated from the hemiacetal (or hemiketal) is removed from the reaction mixture. This is because, if water is not available, the only compound the acetal or ketal can form is an O-methylated carbonyl compound, which is less stable than the acetal or ketal.

The acetal or ketal can be transformed back to the aldehyde or ketone in an acidic aqueous solution.

$$CH_3CH_2-C(OCH_2CH_3)_2-CH_3 + \underset{\text{excess}}{H_2O} \overset{H_3O^+}{\rightleftharpoons} CH_3CH_2COCH_3 + 2\ CH_3CH_2OH$$

PROBLEM 26

Show the mechanism for the acid-catalyzed hydrolysis of an acetal.

PROBLEM 27

Explain mechanistically why an acetal or ketal can be isolated but most hydrates can't be isolated.

Notice that the mechanisms for imine, enamine, hydrate, and acetal (or ketal) formation are similar. After the nucleophile (a primary amine in the case of imine formation, a secondary amine in the case of enamine formation, water in the case of hydrate formation, and an alcohol in the case of acetal or ketal formation) has added to the carbonyl group, water is eliminated from the protonated tetrahedral intermediate, forming a positively charged intermediate. In imine and hydrate formation, a neutral product is achieved by losing a proton from a nitrogen and an oxygen, respectively. In enamine formation, a neutral product is achieved by losing a proton from an α-carbon. In acetal formation, a neutral compound is achieved by adding a second equivalent of alcohol. Also notice that, because the nucleophile in hydrate formation is water, elimination of water gives back the original aldehyde or ketone.

Tutorial:
Common terms, the addition of nucleophiles to carbonyl compounds

PROBLEM-SOLVING STRATEGY

Explain why acetals and ketals are hydrolyzed back to the aldehyde or ketone in acidic aqueous solutions, but are stable in basic aqueous solutions.

The best way to approach this kind of question is to write out the structures and the mechanism that describe what the question is asking. When the mechanism is written, the answer should become apparent. In an acidic solution, the acid protonates an oxygen of the acetal. This creates a weak base (CH_3OH) that can be expelled by the other CH_3O group. When the group is expelled, water can attack the reactive intermediate, and you are then on your way back to the ketone (or aldehyde).

$$R_2C(OCH_3)_2 \xrightarrow{H_3O^+} R_2C(OCH_3)(\overset{+}{O}HCH_3) \rightleftharpoons R_2C{=}\overset{+}{O}CH_3 \xrightarrow{H_2O} R_2C(\overset{+}{O}H_2)(OCH_3) \rightleftharpoons R_2C(OH)(OCH_3) \rightleftharpoons R_2C(OH)(\overset{+}{O}HCH_3) \rightleftharpoons R_2C{=}\overset{+}{O}H \rightleftharpoons R_2C{=}O$$

In a basic solution, the CH_3O group cannot be protonated. Therefore, the group that would have to be eliminated to reform the ketone (or aldehyde) would be the very basic CH_3O^- group. A CH_3O^- group is too basic to be eliminated by the other CH_3O group, which has little driving force because of the positive charge that would be placed on its oxygen atom if elimination were to occur.

$$R_2C(OCH_3)_2 \not\rightarrow$$

Now continue on to Problem 28.

PROBLEM 28

a. Would you expect hemiacetals to be stable in basic solutions? Explain your answer.
b. Acetal formation must be catalyzed by an acid. Explain why it cannot be catalyzed by CH_3O^-.
c. Can the rate of hydrate formation be increased by hydroxide ion as well as by acid? Explain.

18.8 Protecting Groups

Ketones (or aldehydes) react with 1,2-diols to form five-membered ring ketals (or acetals) and with 1,3-diols to form six-membered ring ketals (or acetals). Recall that five- and six-membered rings are formed relatively easily (Section 11.11). The mechanism is the same as that shown in Section 18.7 for acetal formation, except that instead of reacting with two separate molecules of alcohol, the carbonyl compound reacts with the two alcohol groups of a single molecule of the diol.

$$CH_3CH_2\overset{O}{\overset{\|}{C}}CH_2CH_3 + \underset{\text{1,2-ethanediol}}{HOCH_2CH_2OH} \overset{HCl}{\rightleftharpoons} \text{(1,3-dioxolane ketal with } CH_3CH_2 \text{ and } CH_2CH_3 \text{ on C)} + H_2O$$

$$\text{cyclohexanone} + \underset{\text{1,3-propanediol}}{HOCH_2CH_2CH_2OH} \overset{HCl}{\rightleftharpoons} \text{(spiro 1,3-dioxane ketal)} + H_2O$$

If a compound has two functional groups that will react with a given reagent and you want only one of them to react, it is necessary to protect the other functional group from the reagent. A group that protects a functional group from a synthetic operation that it would not otherwise survive is called a **protecting group**.

If you have ever painted a room with a spray gun, you may have taped over the things you do not want to paint, such as baseboards and window frames. In a similar way, 1,2-diols and 1,3-diols are used to protect the carbonyl group of aldehydes and ketones. The diol is like the tape. For example, in the synthesis of the following hydroxyketone from the keto ester, $LiAlH_4$ can reduce both functional groups of a keto ester, and the one that you don't want to react—the keto group—is the more reactive of the two.

$$\text{2-oxocyclohexyl-}\overset{O}{\overset{\|}{C}}OCH_3 \xrightarrow{?} \text{2-oxocyclohexyl-}CH_2OH$$

If the keto group is converted to a ketal, only the ester group will react with $LiAlH_4$. The protecting group can be removed by acid-catalyzed hydrolysis after the ester has been reduced. It is critical that the conditions used to remove a protecting group do not affect other groups in the molecule. Acetals and ketals are good protecting groups because, being ethers, they do not react with bases, reducing agents, or oxidizing agents.

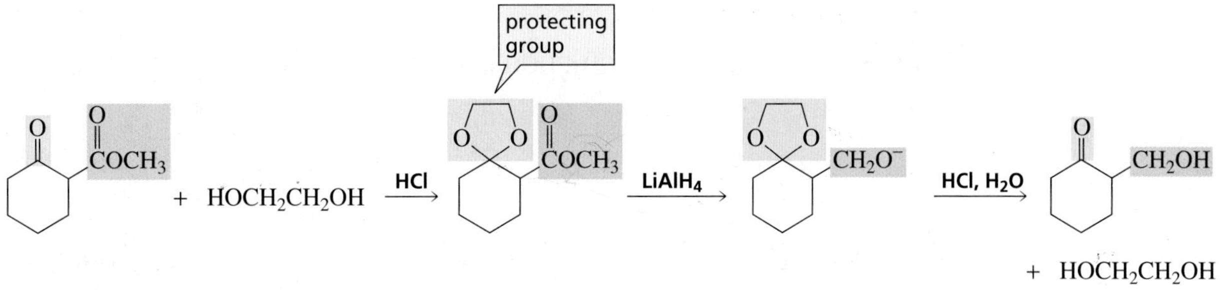

protecting group

$+ HOCH_2CH_2OH \xrightarrow{HCl}$ $\xrightarrow{LiAlH_4}$ $\xrightarrow{HCl,\ H_2O}$ $+ HOCH_2CH_2OH$

PROBLEM 29◆

a. What would have been the product of the preceding reaction with $LiAlH_4$ if the keto group had not been protected?

b. What reagent could you use to reduce only the keto group?

PROBLEM 30

Why don't acetals react with nucleophiles?

In the following reaction, the aldehyde reacts with the diol because aldehydes are more reactive than ketones. The Grignard reagent will now react only with the keto group. The protecting group can be removed by acid-catalyzed hydrolysis.

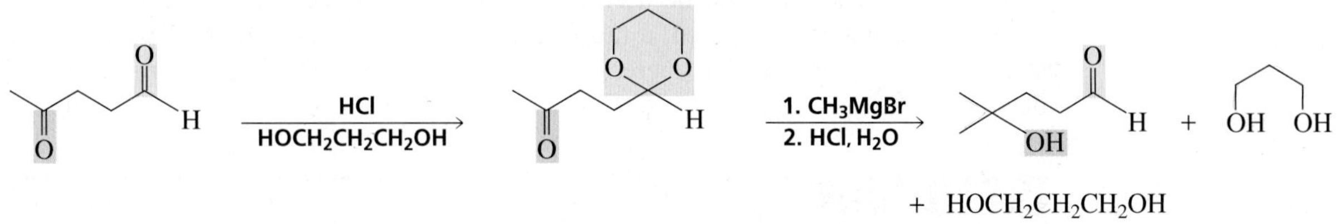

$\xrightarrow[HOCH_2CH_2CH_2OH]{HCl}$ $\xrightarrow[2.\ HCl,\ H_2O]{1.\ CH_3MgBr}$ $+ HOCH_2CH_2CH_2OH$

One of the best ways to protect an OH group of an alcohol is to convert it to a trimethylsilyl (TMS) ether by treating the alcohol with chlorotrimethylsilane and a tertiary amine. The ether is formed by an S_N2 reaction. Although a tertiary alkyl halide does not undergo an S_N2 reaction, the tertiary silyl compound does because Si—C bonds are longer than C—C bonds, reducing steric hindrance at the site of nucleophilic attack. The amine prevents the solution from becoming acidic by reacting with the HCl generated in the reaction. The TMS ether, which is stable in neutral and basic solutions, can be removed with aqueous acid under mild conditions.

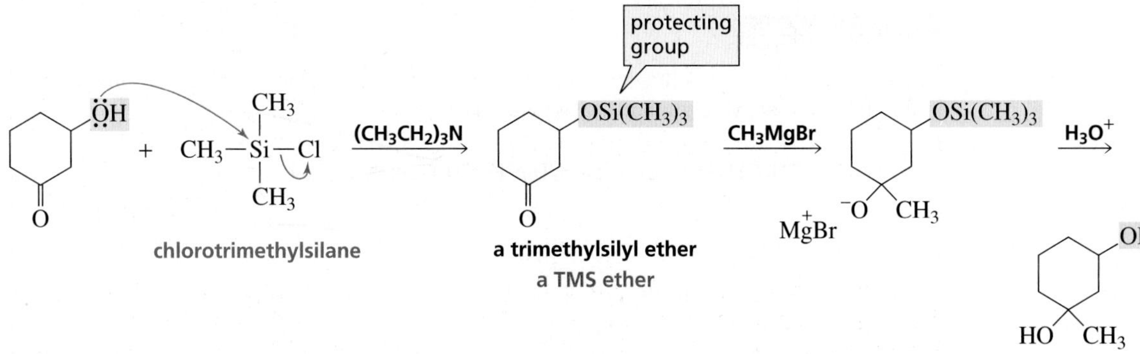

protecting group

$+ CH_3-Si(CH_3)_2-Cl \xrightarrow{(CH_3CH_2)_3N}$ $OSi(CH_3)_3$ $\xrightarrow{CH_3MgBr}$ $\xrightarrow{H_3O^+}$

chlorotrimethylsilane

a trimethylsilyl ether

a TMS ether

The OH group of a carboxylic acid group can be protected by converting the carboxylic acid into an ester.

$$\underset{\text{OH}}{\text{CH}_3\text{CHCH}_2\text{COH}}\ (\text{C=O}) \xrightarrow[\text{excess}]{\text{HCl},\ \text{CH}_3\text{CH}_2\text{OH}} \underset{\text{OH}}{\text{CH}_3\text{CHCH}_2\text{COCH}_2\text{CH}_3} \xrightarrow{\text{SOCl}_2} \underset{\text{Cl}}{\text{CH}_3\text{CHCH}_2\text{COCH}_2\text{CH}_3}$$

$$\downarrow {}^-\text{C}{\equiv}\text{N}$$

$$\text{CH}_3\text{CH}_2\text{OH} + \underset{\text{COOH}}{\text{CH}_3\text{CHCH}_2\text{COH}} \xleftarrow[\text{H}_2\text{O},\ \Delta]{\text{HCl}} \underset{\text{C}\equiv\text{N}}{\text{CH}_3\text{CHCH}_2\text{COCH}_2\text{CH}_3}$$

An amino group can be protected by being converted into an amide (Section 17.8). The acetyl group can subsequently be removed by acid-catalyzed hydrolysis.

$$\underset{\text{aniline}}{\text{C}_6\text{H}_5\text{NH}_2} \xrightarrow{\text{CH}_3\text{CCl (C=O)}} \underset{\text{acetanilide}}{\text{C}_6\text{H}_5\text{NHCCH}_3} \xrightarrow[\text{H}_2\text{SO}_4]{\text{HNO}_3} \underset{p\text{-nitroacetanilide}}{p\text{-O}_2\text{NC}_6\text{H}_4\text{NHCCH}_3} \xrightarrow[\text{2. HO}^-]{\text{1. HCl, H}_2\text{O}, \Delta} \underset{p\text{-nitroaniline}}{p\text{-O}_2\text{NC}_6\text{H}_4\text{NH}_2} + \text{CH}_3\text{CO}^-\ (\text{C=O})$$

Protecting groups should be used only when absolutely necessary, because each time a protecting group is used, it must be attached and then taken off. This adds two steps to the synthesis, which decreases the overall yield of the target compound (the desired product).

PROBLEM 31

What products would be formed from the preceding reaction if aniline's amino group were not protected?

PROBLEM 32◆

a. In a six-step synthesis, what will be the yield of the target compound if each of the reactions employed gives an 80% yield? (An 80% yield is a relatively high laboratory yield.)

b. What would be the yield if two more steps were added to the synthesis?

PROBLEM 33

Show how each of the following compounds can be prepared from the given starting material. In each case, you will need to use a protecting group.

a. $\text{HOCH}_2\text{CH}_2\text{CH}_2\text{Br} \longrightarrow \underset{\text{OH}}{\text{HOCH}_2\text{CH}_2\text{CH}_2\text{CHCH}_3}$

b. $m\text{-BrC}_6\text{H}_4\text{CH (C=O)} \longrightarrow m\text{-HOOCC}_6\text{H}_4\text{CH (C=O)}$

18.9 Addition of Sulfur Nucleophiles

Aldehydes and ketones react with thiols to form thioacetals and thioketals. The mechanism for addition of a thiol is the same as that for addition of an alcohol. Recall that thiols are sulfur analogs of alcohols (Section 12.10).

$$CH_3CH_2C(=O)CH_2CH_3 + \underset{\text{methanethiol}}{2\,CH_3SH} \overset{\text{HCl}}{\rightleftharpoons} \underset{\text{a thioketal}}{CH_3CH_2C(SCH_3)_2CH_2CH_3} + H_2O$$

$$\text{cyclohexanone} + \underset{\text{1,3-propanedithiol}}{HSCH_2CH_2CH_2SH} \overset{\text{HCl}}{\rightleftharpoons} \text{a thioketal} + H_2O$$

Thioacetal (or thioketal) formation is a synthetically useful reaction because a thioacetal (or thioketal) is desulfurized when it reacts with H_2 and Raney nickel. Desulfurization replaces the C—S bonds with C—H bonds.

$$\text{(cyclic thioketal of cyclohexanone)} \xrightarrow[\text{Raney Ni}]{H_2} \text{cyclohexane}$$

$$\text{(cyclic thioketal, } CH_3CH_2\ CH_2CH_3) \xrightarrow[\text{Raney Ni}]{H_2} CH_3CH_2CH_2CH_2CH_3$$

Thioketal formation followed by desulfurization provides us with a third method that can be used to convert the carbonyl group of a ketone into a methylene group. We have already seen the other two methods—the Clemmensen reduction and the Wolff–Kishner reduction (Sections 15.15 and 18.6).

18.10 The Wittig Reaction

An aldehyde or a ketone reacts with a phosphonium ylide (pronounced "ILL-id") to form an alkene. An **ylide** is a compound that has opposite charges on adjacent covalently bonded atoms with complete octets. The ylide can also be written in the double-bonded form because phosphorus can have more than eight valence electrons.

$$\underset{\text{a phosphonium ylide}}{(C_6H_5)_3\overset{+}{P}-\overset{-}{C}H_2} \longleftrightarrow (C_6H_5)_3P{=}CH_2$$

The reaction of an aldehyde or a ketone with a phosphonium ylide to form an alkene is called a **Wittig reaction**. The overall reaction amounts to interchanging the double-bonded oxygen of the carbonyl compound and the double-bonded carbon group of the phosphonium ylide.

Georg Friedrich Karl Wittig (1897–1987) *was born in Germany. He received a Ph.D. from the University of Marburg in 1926. He was a professor of chemistry at the Universities of Braunschweig, Freiberg, Tübingen, and Heidelberg, where he studied phosphorus-containing organic compounds. He received the Nobel Prize in chemistry in 1979, sharing it with H.C. Brown (Section 4.9).*

$$(CH_3)_2C{=}O + (C_6H_5)_3P{=}CHCH_3 \longrightarrow (CH_3)_2C{=}CHCH_3 + (C_6H_5)_3P{=}O$$

a phosphonium ylide; triphenylphosphine oxide

$$\text{cyclohexanone} + (C_6H_5)_3P{=}C(CH_3)_2 \longrightarrow \text{cyclohexylidene}{=}C(CH_3)_2 + (C_6H_5)_3P{=}O$$

Evidence has accumulated that the Wittig reaction is a concerted [2 + 2] cycloaddition reaction, with the nucleophilic carbon of the ylide attacking the electrophilic carbon of the carbonyl compound. It is called a [2 + 2] cycloaddition reaction because, of the four π electrons involved in the cyclic transition state, two come from the carbonyl group and two come from the ylide (Section 29.4). Elimination of triphenylphosphine oxide forms the alkene product.

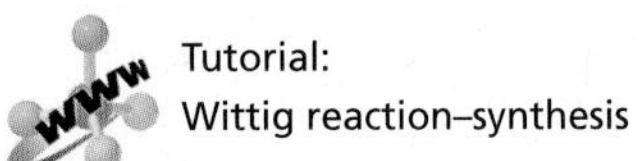

Tutorial: Wittig reaction–synthesis

$$R_2C{=}O + {:}\overset{-}{C}H_2{-}\overset{+}{P}(C_6H_5)_3 \longrightarrow R_2C(O{-}P(C_6H_5)_3){-}CH_2 \longrightarrow O{=}P(C_6H_5)_3 + R_2C{=}CH_2$$

a [2 + 2] cycloaddition reaction

The phosphonium ylide needed for a particular synthesis is obtained by an S_N2 reaction between triphenylphosphine and an alkyl halide with the appropriate number of carbon atoms. A proton on the carbon adjacent to the positively charged phosphorus atom is sufficiently acidic ($pK_a = 35$) to be removed by a strong base such as butyllithium (Section 12.11).

$$(C_6H_5)_3P{:} + CH_3CH_2{-}Br \xrightarrow{S_N2} (C_6H_5)_3\overset{+}{P}{-}CH_2CH_3\ Br^- \xrightarrow{CH_3CH_2CH_2\overset{-}{C}H_2\ \overset{+}{Li}} (C_6H_5)_3\overset{+}{P}{-}\overset{-}{C}HCH_3$$

triphenylphosphine; a phosphonium ylide

If two sets of reagents are available for the synthesis of an alkene, it is better to use the one that requires the less sterically hindered alkyl halide for synthesis of the ylide. Recall that the more sterically hindered the alkyl halide, the less reactive it is in an S_N2 reaction (Section 10.2). For example, it is better to use a three-carbon alkyl halide and a five-carbon carbonyl compound than a five-carbon alkyl halide and a three-carbon carbonyl compound for the synthesis of 3-ethyl-3-hexene because it would be easier to form an ylide from 1-bromopropane than from 3-bromopentane.

$$CH_3CH_2C(=O)CH_2CH_3 + (C_5H_6)_3P{=}CHCH_2CH_3 \longrightarrow CH_3CH_2C(CH_2CH_3){=}CHCH_2CH_3 \quad \text{(preferred method)}$$

$$CH_3CH_2C(=O)H + (C_5H_6)_3P{=}C(CH_2CH_3)CH_2CH_3 \longrightarrow CH_3CH_2C(CH_2CH_3){=}CHCH_2CH_3$$

3-ethyl-3-hexene

The Wittig reaction is a very powerful way to make an alkene because the reaction is completely regioselective—the double bond will be in only one place.

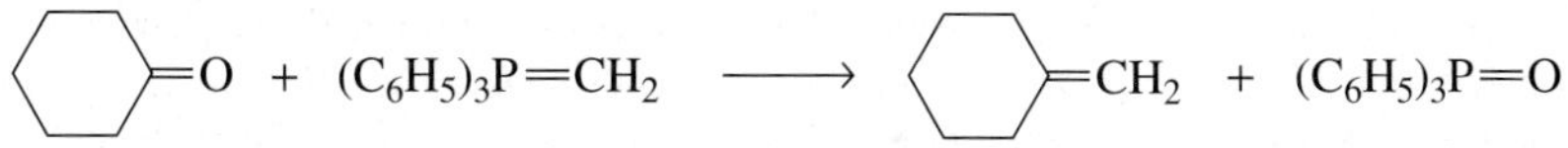

$$\text{cyclohexanone} + (C_6H_5)_3P{=}CH_2 \longrightarrow \text{methylenecyclohexane} + (C_6H_5)_3P{=}O$$

methylenecyclohexane

The Wittig reaction also is the best way to make a terminal alkene such as methylenecyclohexane because other methods would form a terminal alkene only as a minor product if at all.

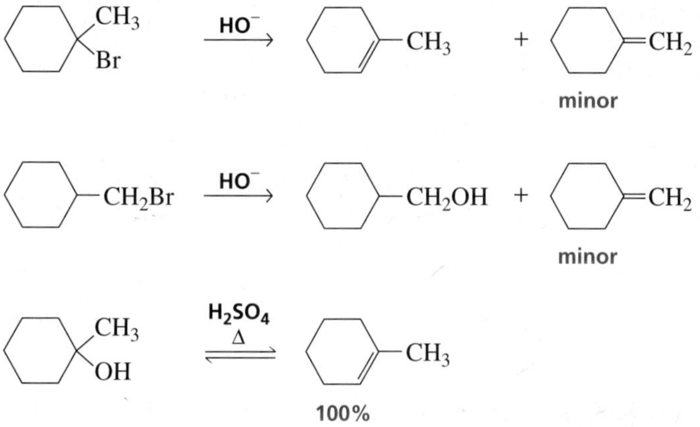

$$\text{1-bromo-1-methylcyclohexane} \xrightarrow{HO^-} \text{1-methylcyclohexene} + \text{methylenecyclohexane (minor)}$$

$$\text{C}_6\text{H}_{11}\text{-}CH_2Br \xrightarrow{HO^-} \text{C}_6\text{H}_{11}\text{-}CH_2OH + \text{methylenecyclohexane (minor)}$$

$$\text{1-methylcyclohexanol} \underset{\Delta}{\overset{H_2SO_4}{\rightleftharpoons}} \text{1-methylcyclohexene (100\%)}$$

The stereoselectivity of the Wittig reaction depends on the structure of the ylide. Ylides can be divided into two types: *Stabilized ylides* have a group, such as a carbonyl group, that can share the carbanion's negative charge; *unstabilized ylides* do not have such a group.

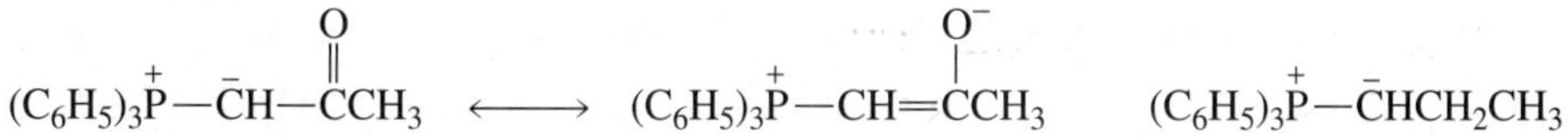

$$(C_6H_5)_3\overset{+}{P}-\overset{-}{C}H-\overset{O}{\overset{\|}{C}}CH_3 \longleftrightarrow (C_6H_5)_3\overset{+}{P}-CH{=}\overset{O^-}{\overset{|}{C}}CH_3 \qquad (C_6H_5)_3\overset{+}{P}-\overset{-}{C}HCH_2CH_3$$

a stabilized ylide — an unstabilized ylide

Stabilized ylides form primarily E isomers, and unstabilized ylides form primarily Z isomers.

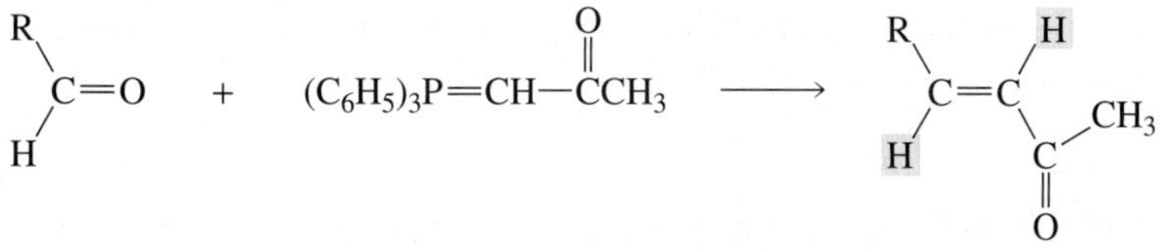

$$RHC{=}O + (C_6H_5)_3P{=}CH-\overset{O}{\overset{\|}{C}}CH_3 \longrightarrow (E)\text{-}RCH{=}CHC(=O)CH_3$$

E alkene

$$RHC{=}O + (C_6H_5)_3P{=}CHCH_2CH_3 \longrightarrow (Z)\text{-}RCH{=}CHCH_2CH_3$$

Z alkene

β-CAROTENE

β-Carotene is found in yellow-orange fruits and vegetables such as apricots, mangoes, carrots, and sweet potatoes. The synthesis of β-carotene from vitamin A is an important example of the use of the Wittig reaction in industry. Notice that the ylide is a stabilized ylide and the product has the *E* configuration at the reaction site.

β-Carotene is used in the food industry to color margarine. Many people take β-carotene as a dietary supplement because there is some evidence that high levels of β-carotene are associated with a low incidence of cancer. More recent evidence, however, suggests that β-carotene taken in pill form does not have the cancer-preventing effects of β-carotene obtained from vegetables.

vitamin A aldehyde + ylide $=P(C_6H_5)_3$ → β-carotene

PROBLEM 34 SOLVED

a. What carbonyl compound and what phosphonium ylide are required for the synthesis of the following alkenes?

1. $CH_3CH_2CH_2CH{=}C(CH_3)CH_3$
2. cyclohexylidene $=CHCH_2CH_3$
3. $(C_6H_5)_2C{=}CHCH_3$
4. $C_6H_5{-}CH{=}CH_2$

b. What alkyl halide is required to prepare each of the phosphonium ylides?

SOLUTION TO 34a (1) The atoms on either side of the double bond can come from the carbonyl compound, so there are two pairs of compounds that can be used.

$$CH_3\overset{O}{\overset{\|}{C}}CH_3 + (C_6H_5)_3P{=}CHCH_2CH_2CH_3 \quad \textbf{or} \quad CH_3CH_2CH_2\overset{O}{\overset{\|}{C}}H + (C_6H_5)_3P{=}\underset{CH_3}{\underset{|}{C}}CH_3$$

SOLUTION TO 34b (1) The alkyl halide required depends on which phosphonium ylide is used; it would be either 1-bromobutane or 2-bromopropane.

$$CH_3CH_2CH_2CH_2Br \quad \textbf{or} \quad CH_3\underset{Br}{\underset{|}{C}}HCH_3$$

The primary alkyl halide would be more reactive in the S_N2 reaction required to make the ylide, so the best method would be to use acetone and the ylide obtained from 1-bromobutane.

18.11 Stereochemistry of Nucleophilic Addition Reactions: *Re* and *Si* Faces

A carbonyl carbon bonded to two different substituents is a **prochiral carbonyl carbon** because it will become a chirality center (asymmetric carbon) if it adds a group unlike either of the groups already bonded to it. The addition product will be a pair of enantiomers.

a prochiral carbonyl carbon

$$X\!-\!C(=O)\!-\!Y \xrightarrow{HZ} \text{X–C(OH)(Y)(Z)} \quad \text{Y–C(OH)(X)(Z)}$$

a pair of enantiomers

The carbonyl carbon and the three atoms attached to it define a plane. The nucleophile can approach either side of the plane. One side of the carbonyl compound is called the *Re* (pronounced "ree") face, and the other side is called the *Si* (pronounced "sigh") face; *Re* is for *rectus* and *Si* is for *sinister*—similar to *R* and *S*. To distinguish between the ***Re* and *Si* faces**, the three groups attached to the carbonyl carbon are assigned priorities using the Cahn–Ingold–Prelog system of priorities used in *E*, *Z* and *R*, *S* nomenclature (Sections 3.5 and 5.6 respectively). The *Re* face is the face closest to the observer when decreasing priorities ($1 > 2 > 3$) are in a clockwise direction, and the *Si* face is the opposite face—the one closest to the observer when decreasing priorities are in a counterclockwise direction.

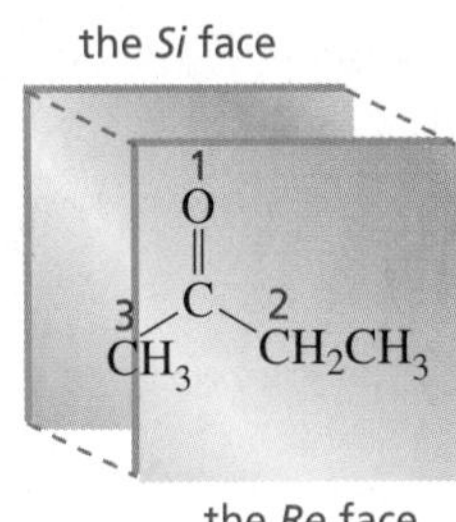

Attack by a nucleophile on the *Re* face forms one enantiomer, whereas attack on the *Si* face forms the other enantiomer. For example, attack by hydride ion on the *Re* face of butanone forms (*S*)-2-butanol, and attack on the *Si* face forms (*R*)-2-butanol.

$CH_3C(=O)CH_2CH_3$ (O = 1, CH_2CH_3 = 2, CH_3 = 3) — the *Re* face is closest to the observer

H^- attack on the *Re* face followed by H^+ → (*S*)-2-butanol

H^- attack on the *Si* face followed by H^+ → (*R*)-2-butanol

3-D Molecules: 2-Butanone; (*S*)-2-Butanol; (*R*)-2-Butanol

Whether attack on the *Re* face forms the *R* or *S* enantiomer depends on the priority of the attacking nucleophile relative to the priorities of the groups attached to the carbonyl carbon. For example, we just saw that attack by hydride ion on the *Re* face of butanone forms (*S*)-2-butanol, but attack by a methyl Grignard reagent on the *Re* face of propanal forms (*R*)-2-butanol.

$CH_3CH_2C(=O)H$ (O = 1, CH_3CH_2 = 2, H = 3) — the *Si* face is closest to the observer

CH_3MgBr attack on the *Re* face followed by H^+ → (*R*)-2-butanol

CH_3MgBr attack on the *Si* face followed by H^+ → (*S*)-2-butanol

Because the carbonyl carbon and the three atoms attached to it define a plane, the *Re* and *Si* faces have an equal probability of being attacked. Consequently, an addition reaction forms equal amounts of the two enantiomers.

ENZYME-CATALYZED CARBONYL ADDITIONS

In an enzyme-catalyzed addition to a carbonyl compound, only one of the enantiomers is formed. The enzyme can block one face of the carbonyl compound so that it cannot be attacked, or it can position the nucleophile so that it is able to attack the carbonyl group from only one side of the molecule.

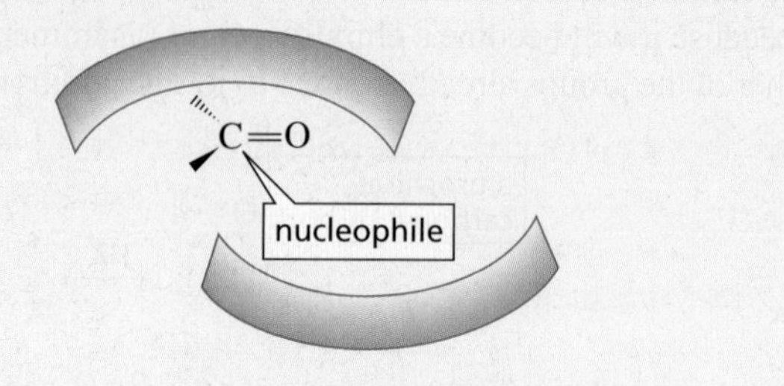

PROBLEM 35◆

Which enantiomer is formed when a methyl Grignard reagent attacks the *Re* face of each of the following carbonyl compounds?

a. propiophenone
b. benzaldehyde
c. 2-pentanone
d. 3-hexanone

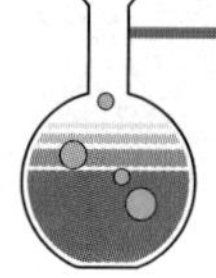

18.12 Designing a Synthesis V: Disconnections, Synthons, and Synthetic Equivalents

The synthesis of a complicated molecule from simple starting materials is not always obvious. We have seen that it is often easier to work backward from the desired product to available starting materials—a process called *retrosynthetic analysis* (Section 6.11). In a retrosynthetic analysis, the chemist dissects a molecule into smaller and smaller pieces until readily available starting materials are obtained.

retrosynthetic analysis

target molecule ⟹ Y ⟹ X ⟹ W ⟹ starting materials

A useful step in a retrosynthetic analysis is a **disconnection**—breaking a bond to produce two fragments. Typically, one fragment is positively charged and one is negatively charged. The fragments of a disconnection are called **synthons**. Synthons are often not real compounds—they can be imaginary species. For example, if we consider the retrosynthetic analysis of cyclohexanol, we see that a disconnection gives two synthons—an α-hydroxycarbocation and a hydride ion.

retrosynthetic analysis

HO H (cyclohexanol) —a disconnection⟹ OH (cyclohexyl cation, +) + H^-

synthons

open arrow represents a retrosynthetic operation

A **synthetic equivalent** is the reagent that is actually used as the source of the synthon. Cyclohexanone is the synthetic equivalent for the α-hydroxycarbocation, and sodium borohydride is the synthetic equivalent for hydride ion. Thus, cyclohexanol, the target molecule, can be prepared by treating cyclohexanone with sodium borohydride.

synthesis

cyclohexanone $\xrightarrow[\text{2. } H^+,\ H_2O]{\text{1. } NaBH_4}$ cyclohexanol

When carrying out a disconnection, you must decide, after breaking the bond, which fragment gets the positive charge and which gets the negative charge. In the retrosynthetic analysis of cyclohexanol, we could have given the positive charge to the hydrogen, and many acids (HCl, HBr, etc.) could have been used for the synthetic equivalent for H^+. However, we would have been at a loss to find a synthetic equivalent for an α-hydroxycarbanion. Therefore, when we carry out the disconnection, we assign the positive charge to the carbon and the negative charge to the hydrogen.

Cyclohexanol can also be disconnected by breaking the C—O bond instead of the C—H bond, forming a carbocation and hydroxide ion.

retrosynthetic analysis

cyclohexanol $\Longrightarrow$ cyclohexyl cation + HO^-

The problem now becomes choosing a synthetic equivalent for the carbocation. A synthetic equivalent for a positively charged synthon needs an electron-withdrawing group at just the right place. Cyclohexyl bromide, with an electron-withdrawing bromine, is a synthetic equivalent for the cyclohexyl carbocation. Cyclohexanol, therefore, can be prepared by treating cyclohexyl bromide with hydroxide ion. This method, however, is not as good as the first synthesis we proposed—reduction of cyclohexanone—because some of the alkyl halide is converted into an alkene, so the overall yield of the target compound is lower.

synthesis

cyclohexyl bromide + HO^- $\longrightarrow$ cyclohexanol + cyclohexene

Retrosynthetic analysis shows that 1-methylcyclohexanol can be formed from the reaction of cyclohexanone, the synthetic equivalent for the α-hydroxycarbocation, and methylmagnesium bromide, the synthetic equivalent for the methyl anion (Section 18.4).

retrosynthetic analysis

1-methylcyclohexanol (HO, CH_3) $\Longrightarrow$ α-hydroxycarbocation (OH, +) + $^-CH_3$

synthesis

cyclohexanone $\xrightarrow[\text{2. } H_3O^+]{\text{1. } CH_3MgBr}$ 1-methylcyclohexanol (HO, CH_3)

Other disconnections of 1-methylcyclohexanol are possible because any bond to carbon can serve as a disconnection site. For example, one of the ring C—C bonds could be broken. However, these are not useful disconnections, because the synthetic equivalents of the synthons they produce are not easily prepared. A retrosynthetic step must lead to readily obtainable starting materials.

retrosynthetic analysis

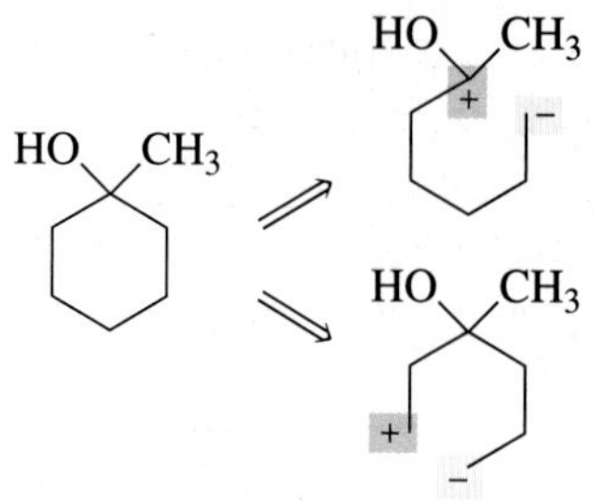

HO CH$_3$ (1-methylcyclohexanol) ⟹ HO CH$_3$ (synthon with + at carbinol carbon, − at chain end) ; HO CH$_3$ (synthon with + at ring carbon, − at chain end)

PROBLEM 36

Using bromocyclohexane as a starting material, how could you synthesize the following compounds?

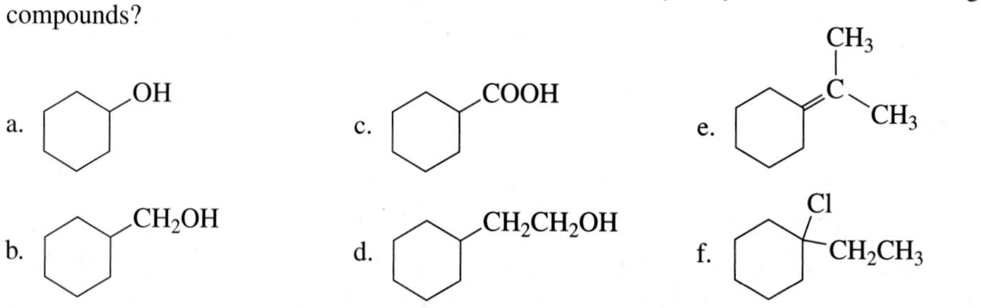

a. cyclohexyl–OH

b. cyclohexyl–CH_2OH

c. cyclohexyl–COOH

d. cyclohexyl–CH_2CH_2OH

e. cyclohexylidene=$C(CH_3)CH_3$

f. 1-chloro-1-ethylcyclohexane (Cl, CH_2CH_3)

SYNTHESIZING ORGANIC COMPOUNDS

Organic chemists synthesize compounds for many reasons: to study their properties or to answer a variety of chemical questions, or because they have unusual shapes or other unusual structural features or useful properties. One reason chemists synthesize natural products is to provide us with greater supplies of these compounds than nature can produce. For example, Taxol®—a compound that has been successful in treating ovarian and breast cancer—is extracted from the bark of *Taxus*, the yew tree found in the Pacific Northwest. The supply of natural Taxol® is limited because yew trees are uncommon and grow very slowly and stripping the bark kills the tree. The bark of one tree provides only one dose of the drug. In addition, *Taxus* forests serve as habitats for the spotted owl, an endangered species, so harvesting the trees would accelerate the owl's demise. Once chemists were successful in determining the structure of Taxol®, efforts could be undertaken to synthesize it in order to make it more widely available as an anticancer drug. Several syntheses have been successful.

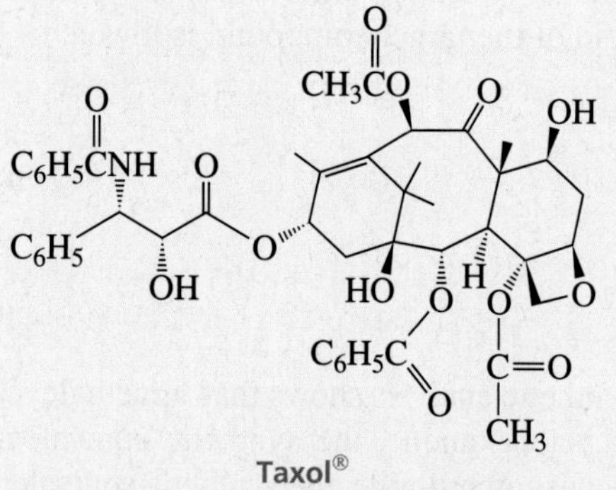

(Structure: CH_3CO, O, OH, C_6H_5CNH, O, C_6H_5, OH, HO, H, O, O, O, C_6H_5C, O, C=O, CH_3)

Taxol®

Once a compound has been synthesized, chemists can study its properties to learn how it works; then they can design and synthesize safer or more potent analogs. For example, chemists have found that the anticancer activity of Taxol® is substantially reduced if its four ester groups are hydrolyzed. This gives one small clue as to how the molecule functions.

SEMISYNTHETIC DRUGS

Taxol® is a difficult molecule to synthesize because of its complicated structure. Chemists have made the synthesis a lot easier by allowing the yew tree to carry out the first part of the synthesis. Chemists extract the drug precursor from the needles of the tree, and the precursor is converted to Taxol® in the laboratory. Thus, the precursor is isolated from a renewable resource, while the drug itself could be obtained only by killing the tree. In this manner, chemists have learned to synthesize compounds jointly with nature.

18.13 Nucleophilic Addition to α,β-Unsaturated Aldehydes and Ketones

The resonance contributors for an α,β-unsaturated carbonyl compound show that the molecule has two electrophilic sites: the carbonyl carbon and the β-carbon.

$$\overset{\beta}{\text{RCH}}{=}\overset{\alpha}{\text{CH}}{-}\overset{\overset{\text{:}\ddot{\text{O}}}{\|}}{\text{CR}} \longleftrightarrow \text{RCH}{=}\text{CH}{-}\overset{\overset{\text{:}\ddot{\text{O}}\text{:}^-}{|}}{\underset{+}{\text{CR}}} \longleftrightarrow \underset{+}{\text{RCH}}{-}\text{CH}{=}\overset{\overset{\text{:}\ddot{\text{O}}\text{:}^-}{|}}{\text{CR}}$$

an α,β-unsaturated carbonyl compound

electrophilic site | electrophilic site

This means that if an aldehyde or a ketone has a double bond in the α,β-position, a nucleophile can add either to the carbonyl carbon or to the β-carbon.

Nucleophilic addition to the carbonyl carbon is called **direct addition** or 1,2-addition.

direct addition

$$\text{Y}^{:-} + \text{RCH}{=}\text{CH}{-}\overset{\overset{\text{:}\ddot{\text{O}}}{\|}}{\text{CR}} \longrightarrow \text{RCH}{=}\text{CH}{-}\underset{\underset{\text{Y}}{|}}{\overset{\overset{\text{:}\ddot{\text{O}}\text{:}^-}{|}}{\text{CR}}} \xrightarrow{\text{H}_3\text{O}^+} \text{RCH}{=}\text{CH}{-}\underset{\underset{\text{Y}}{|}}{\overset{\overset{\text{OH}}{|}}{\text{CR}}}$$

Nucleophilic addition to the β-carbon is called **conjugate addition** or 1,4-addition, because addition occurs at the 1- and 4-positions (i.e., across the conjugated system). After 1,4-addition has occurred, the product—an enol—tautomerizes to a ketone (or to an aldehyde, Section 6.6), so the overall reaction amounts to addition to the carbon–carbon double bond, with the nucleophile adding to the β-carbon and a proton from the reaction mixture adding to the α-carbon. Compare these reactions with the 1,2- and 1,4-addition reactions you studied in Section 8.7.

conjugate addition

$$\text{Y}^{:-} + \text{RCH}{=}\text{CH}{-}\overset{\overset{\text{O}}{\|}}{\text{CR}} \longrightarrow \left[\underset{\underset{\text{Y}}{|}}{\text{RCH}}{-}\ddot{\text{C}}\text{H}^-{-}\overset{\overset{\text{O}}{\|}}{\text{CR}} \longleftrightarrow \underset{\underset{\text{Y}}{|}}{\text{RCH}}{-}\text{CH}{=}\overset{\overset{\text{O}^-}{|}}{\text{CR}}\right]$$

resonance contributors

$$\downarrow \text{H}_3\text{O}^+$$

$$\underset{\underset{\text{Y}}{|}}{\text{RCH}}{-}\text{CH}_2{-}\overset{\overset{\text{O}}{\|}}{\text{CR}} \rightleftharpoons \underset{\underset{\text{Y}}{|}}{\text{RCH}}{-}\text{CH}{=}\overset{\overset{\text{OH}}{|}}{\text{CR}}$$

keto tautomer | **enol tautomer**

Whether the product obtained from nucleophilic addition to an α,β-unsaturated aldehyde or ketone is the direct addition product or the conjugate addition product depends on the nature of the nucleophile, the structure of the carbonyl compound, and the conditions under which the reaction is carried out.

Nucleophiles that form *unstable addition products*—that is, nucleophiles that are weak bases, allowing direct addition to be reversible—form conjugate addition products because conjugate addition is *not* reversible, and the conjugate addition product is more stable. Nucleophiles in this group include halide ions, cyanide ion, thiols, alcohols, and amines.

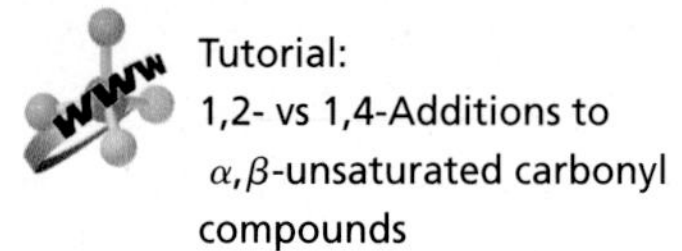

Tutorial:
1,2- vs 1,4-Additions to α,β-unsaturated carbonyl compounds

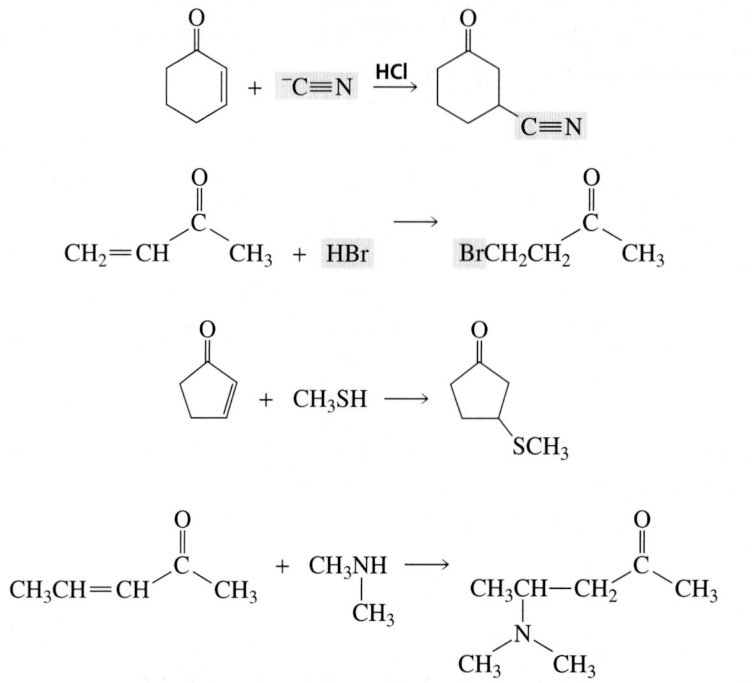

Nucleophiles that form *stable addition products*—that is, nucleophiles that are strong bases, thereby making direct addition irreversible—can form *either* direct addition products or conjugate addition products. Nucleophiles in this group include hydride ion and carbanions. The reaction that prevails is the one that is faster, so the product that is formed will depend on the reactivity of the carbonyl group. Compounds with reactive carbonyl groups form primarily direct addition products because for those compounds, direct addition is faster, whereas compounds with less reactive carbonyl groups form primarily conjugate addition products because for *those* compounds, conjugate addition is faster. For example, aldehydes have more reactive carbonyl groups than do ketones, so sodium borohydride forms primarily direct addition products with aldehydes. Compared with aldehydes, ketones form less of the direct addition product and more of the conjugate addition product.

product of direct addition
97%

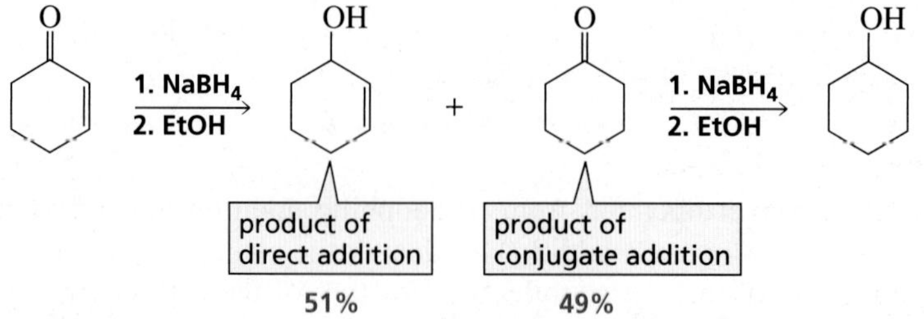

product of direct addition
51%

product of conjugate addition
49%

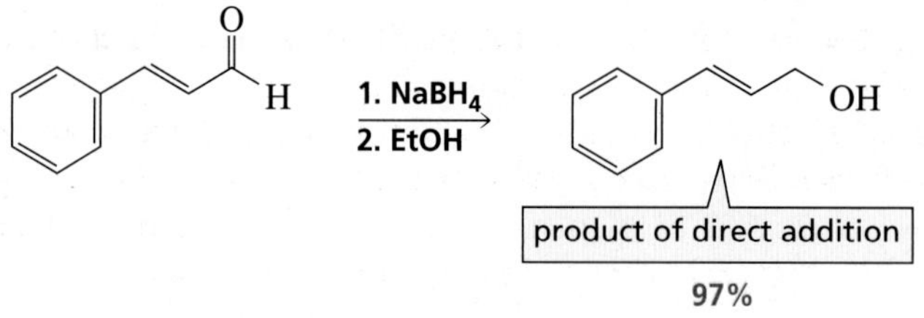

Notice that a saturated alcohol is the final product of conjugate addition in the preceding reaction because the carbonyl group of the ketone will react with a second equivalent of hydride ion.

If direct addition is the desired outcome of hydride addition, it can be achieved by carrying out the reaction in the presence of cerium chloride, a Lewis acid that activates the carbonyl group toward nucleophilic attack by complexing with the carbonyl oxygen.

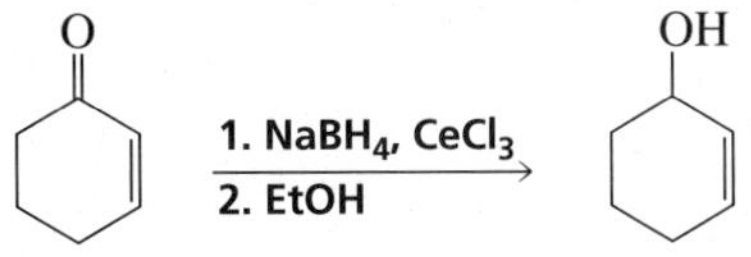

1. $NaBH_4$, $CeCl_3$
2. EtOH

99%

Like hydride ions, Grignard reagents add irreversibly to carbonyl groups. Therefore, Grignard reagents react with α,β-unsaturated aldehydes and unhindered α,β-unsaturated ketones to form direct addition products.

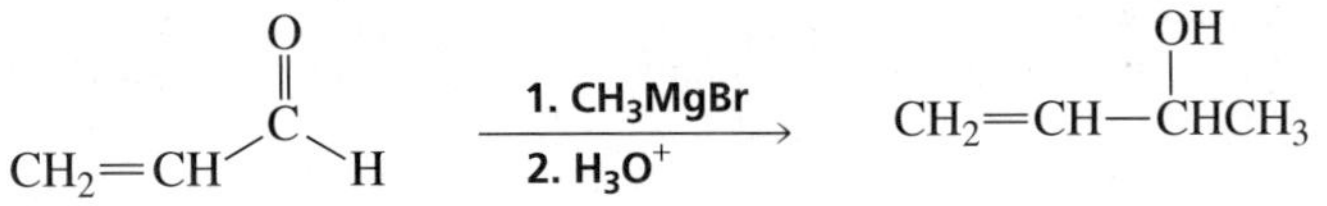

$$CH_2{=}CH{-}\overset{O}{\overset{\|}{C}}{-}H \xrightarrow[2.\ H_3O^+]{1.\ CH_3MgBr} CH_2{=}CH{-}\overset{OH}{\overset{|}{C}}HCH_3$$

If, however, the rate of direct addition is slowed down by steric hindrance, a Grignard reagent will form a conjugate addition product because conjugate addition then becomes the faster reaction.

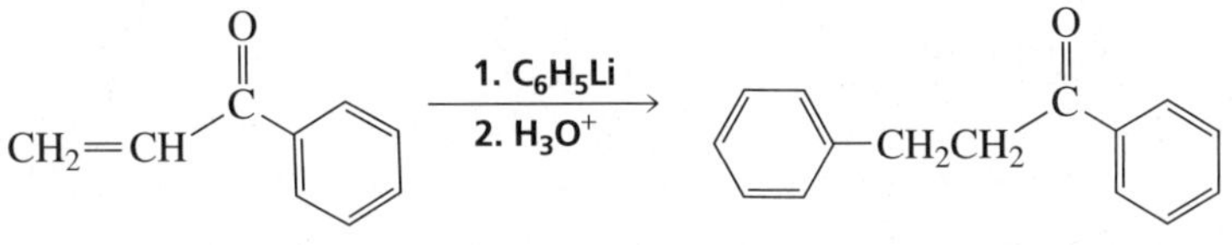

$$CH_2{=}CH{-}\overset{O}{\overset{\|}{C}}{-}C_6H_5 \xrightarrow[2.\ H_3O^+]{1.\ C_6H_5Li} C_6H_5{-}CH_2CH_2{-}\overset{O}{\overset{\|}{C}}{-}C_6H_5$$

Only conjugate addition occurs when Gilman reagents (lithium dialkylcuprates, Section 12.12) react with α,β-unsaturated aldehydes and ketones. Therefore, Grignard reagents should be used when you want to add an alkyl group to the carbonyl carbon, whereas Gilman reagents should be used when you want to add an alkyl group to the β-carbon.

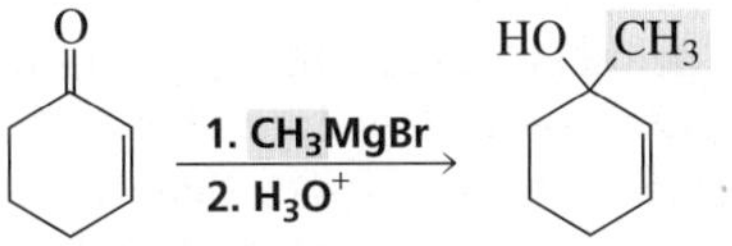

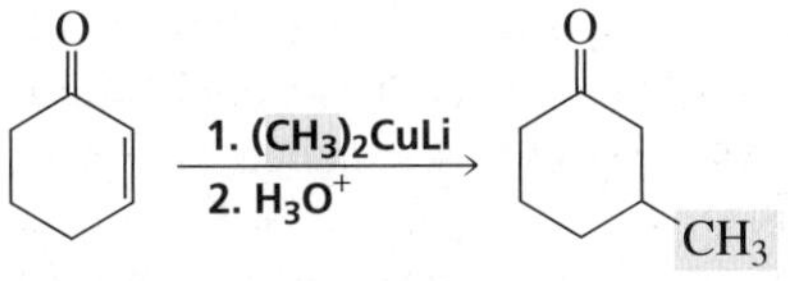

1. CH_3MgBr
2. H_3O^+

HO CH$_3$

1. $(CH_3)_2CuLi$
2. H_3O^+

CH$_3$

Electrophiles and nucleophiles can be classified as either *hard* or *soft*. Hard electrophiles and nucleophiles are more polarized than soft ones. Hard nucleophiles prefer to react with hard electrophiles, and soft nucleophiles prefer to react with soft electrophiles. Therefore, a Grignard reagent with a highly polarized C—Mg bond prefers to react with the harder C=O bond, whereas a Gilman reagent with a much less polarized C—Cu bond prefers to react with the softer C=C bond.

PROBLEM 37

Give the major product of each of the following reactions:

a. 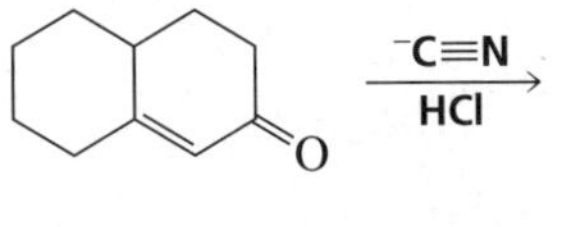 $\xrightarrow[HCl]{^-C{\equiv}N}$

b. $\xrightarrow[2.\ H_3O^+]{1.\ NaBH_4}$

c. $CH_3\overset{CH_3}{\overset{|}{C}}{=}CH\overset{O}{\overset{\|}{C}}CH_3 \xrightarrow[2.\ H_3O^+]{1.\ CH_3MgBr}$

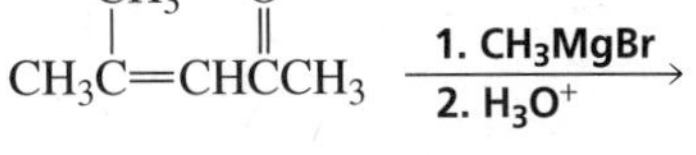

d. 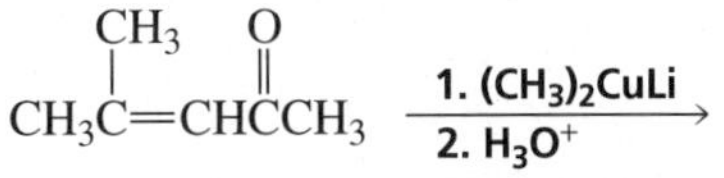 $CH_3\overset{CH_3}{\overset{|}{C}}{=}CH\overset{O}{\overset{\|}{C}}CH_3 \xrightarrow[2.\ H_3O^+]{1.\ (CH_3)_2CuLi}$

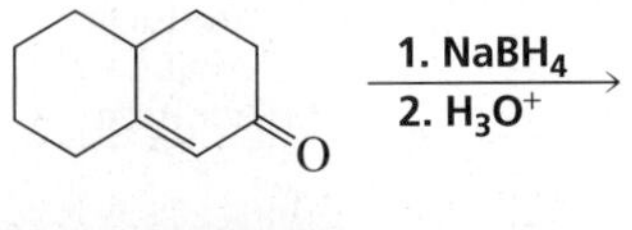

PROBLEM 38◆

Which would give a higher yield of an unsaturated alcohol when treated with sodium borohydride, a sterically hindered ketone or a nonsterically hindered ketone?

CANCER CHEMOTHERAPY

Two compounds—vernolepin and helinalin—owe their effectiveness as anticancer drugs to conjugate addition reactions.

Cancer cells are cells that have lost their ability to control their growth; therefore, they proliferate rapidly. DNA polymerase is an enzyme that a cell needs to make a copy of its DNA for a new cell. DNA polymerse has an SH group and each of these drugs has two α,β-unsaturated carbonyl groups. Irreversible conjugate addition of the enzyme to an α,β-unsaturated carbonyl group inactivates the enzyme.

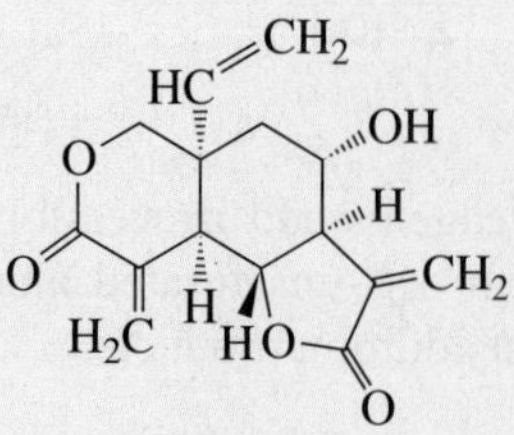

vernolepin

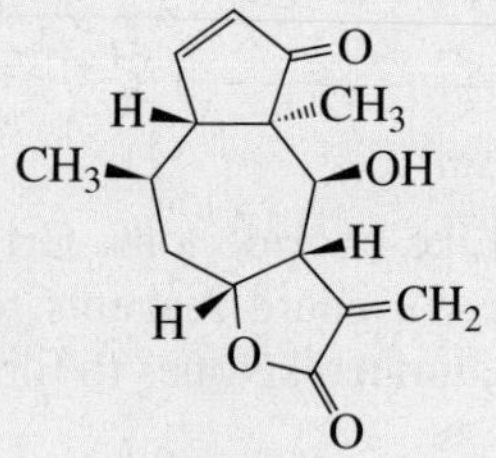

helenalin

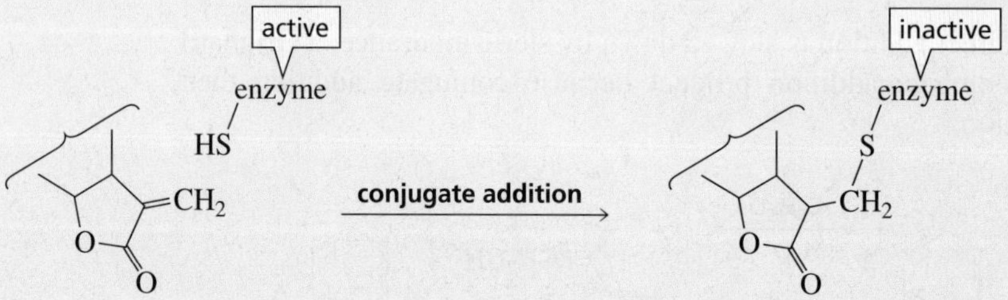

active enzyme — HS — CH_2 — **conjugate addition** → inactive enzyme — S — CH_2

18.14 Nucleophilic Addition to α,β-Unsaturated Carboxylic Acid Derivatives

α,β-Unsaturated carboxylic acid derivatives, like α,β-unsaturated aldehydes and ketones, have two electrophilic sites for nucleophilic attack: They can undergo *conjugate addition* or *nucleophilic acyl substitution*. Notice that they undergo *nucleophilic acyl substitution* rather than *direct addition* because the α,β-unsaturated carbonyl compound had a group that can be replaced by a nucleophile. In other words, as with non-conjugated carbonyl compounds, nucleophilic acyl addition becomes nucleophilic acyl substitution if the carbonyl group is attached to a group that can be replaced by another group (Section 18.3).

Nucleophiles react with α,β-unsaturated carboxylic acid derivatives with reactive carbonyl groups, such as acyl chlorides, to form nucleophilic acyl substitution products. Conjugate addition products are formed from the reaction of nucleophiles with less reactive carbonyl groups, such as esters and amides.

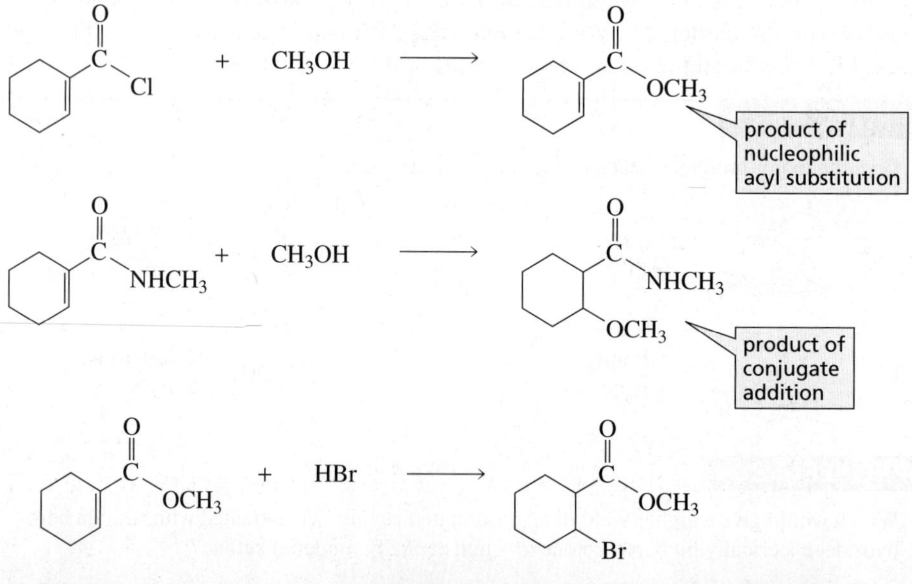

(cyclohexenyl)C(=O)Cl + CH_3OH ⟶ (cyclohexenyl)C(=O)OCH_3 — product of nucleophilic acyl substitution

(cyclohexenyl)C(=O)$NHCH_3$ + CH_3OH ⟶ (2-methoxycyclohexyl)C(=O)$NHCH_3$ — product of conjugate addition

(cyclohexenyl)C(=O)OCH_3 + HBr ⟶ (2-bromocyclohexyl)C(=O)OCH_3

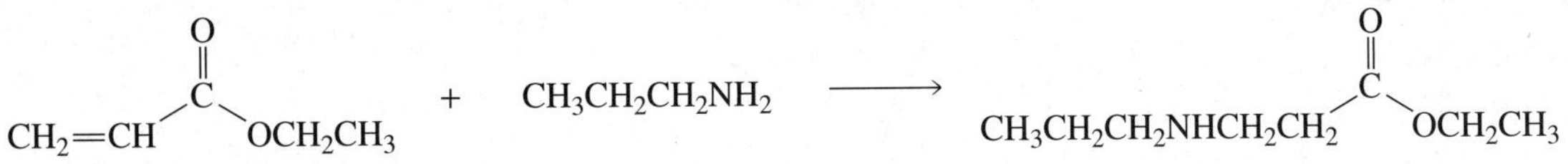

$$CH_2{=}CH{-}C({=}O){-}OCH_2CH_3 + CH_3CH_2CH_2NH_2 \longrightarrow CH_3CH_2CH_2NHCH_2CH_2{-}C({=}O){-}OCH_2CH_3$$

PROBLEM 39

Give the major product of each of the following reactions:

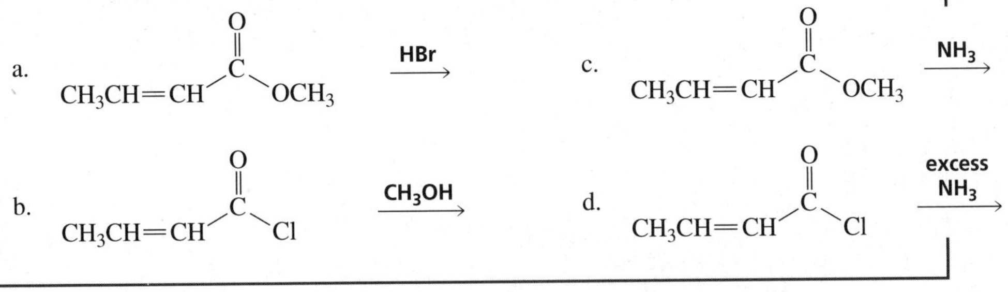

a. $CH_3CH{=}CH{-}C({=}O){-}OCH_3 \xrightarrow{\text{HBr}}$

b. $CH_3CH{=}CH{-}C({=}O){-}Cl \xrightarrow{\text{CH}_3\text{OH}}$

c. $CH_3CH{=}CH{-}C({=}O){-}OCH_3 \xrightarrow{\text{NH}_3}$

d. $CH_3CH{=}CH{-}C({=}O){-}Cl \xrightarrow{\text{excess NH}_3}$

18.15 Enzyme-Catalyzed Additions to α,β-Unsaturated Carbonyl Compounds

Several reactions in biological systems involve addition to α,β-unsaturated carbonyl compounds. The following are examples of conjugate addition reactions that occur in biological systems. Notice that the carbonyl groups are either unreactive (COO^-) or have low reactivity (the CoA ester) toward the nucleophile, so conjugate addition occurs in each case. The last reaction is an important step in the biosynthesis of fatty acids (Section 19.21).

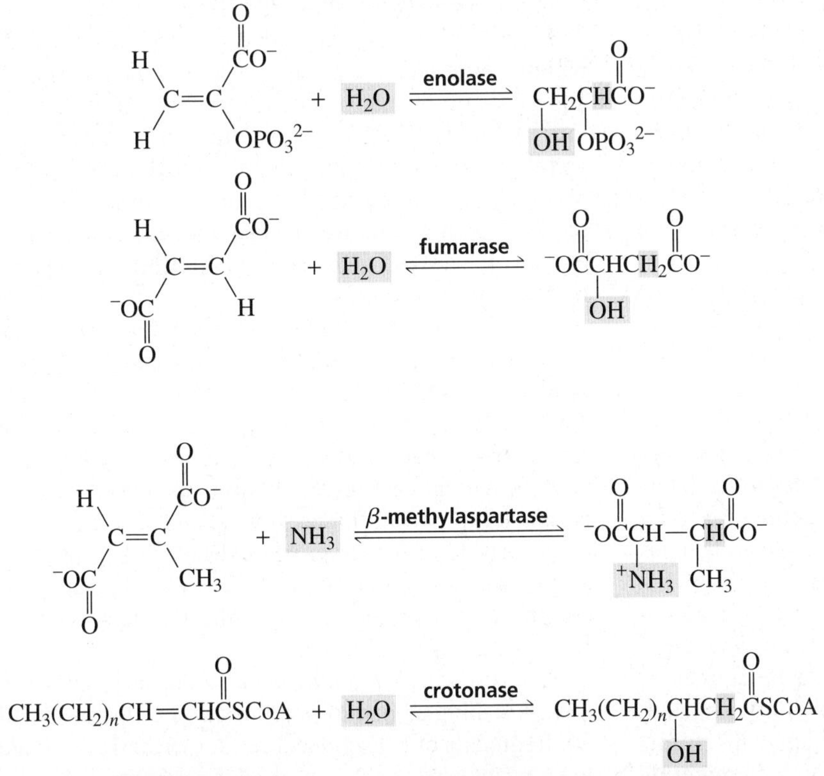

$$H_2C{=}C(OPO_3^{2-}){-}C({=}O)O^- + H_2O \xrightleftharpoons{\text{enolase}} CH_2(OH){-}CH(OPO_3^{2-}){-}C({=}O)O^-$$

$$^-OC({=}O){-}CH{=}CH{-}C({=}O)O^- + H_2O \xrightleftharpoons{\text{fumarase}} {}^-OC({=}O)CH(OH)CH_2C({=}O)O^-$$

$$^-OC({=}O){-}CH{=}C(CH_3){-}C({=}O)O^- + NH_3 \xrightleftharpoons{\beta\text{-methylaspartase}} {}^-OC({=}O)CH(\overset{+}{N}H_3){-}CH(CH_3)C({=}O)O^-$$

$$CH_3(CH_2)_nCH{=}CHC({=}O)SCoA + H_2O \xrightleftharpoons{\text{crotonase}} CH_3(CH_2)_nCH(OH)CH_2C({=}O)SCoA$$

ENZYME-CATALYZED CIS–TRANS INTERCONVERSION

Enzymes that catalyze the interconversion of cis and trans isomers are called cis–trans isomerases. These isomerases are all known to contain thiol (SH) groups. Thiols are weak bases and therefore add to the β-carbon of an α,β-unsaturated carbonyl compound (conjugate addition). The resulting carbon–carbon single bond rotates before the enol is able to tautomerize to the ketone. When tautomerization occurs, the thiol is eliminated. Rotation results in cis–trans interconversion.

cis double bond

enzyme–SH

enzyme–S

rotation

trans double bond

enzyme–SH

Summary

Aldehydes and **ketones** are Class II carbonyl compounds; they have an acyl group attached to a group (—H, —R, or —Ar) that cannot be readily replaced by another group. Aldehydes and ketones undergo **nucleophilic addition reactions** with C and H nucleophiles and **nucleophilic addition–elimination reactions** with O and N nucleophiles. With the exception of amides, carboxylic acid derivatives (Class I carbonyl compounds) undergo **nucleophilic acyl substitution** reactions with C and H nucleophiles to form a Class II carbonyl compound, which then undergoes a **nucleophilic addition** reaction with a second equivalent of the C or H nucleophile. Notice that the tetrahedral intermediate formed by attack of a nucleophile on a carbonyl compound is stable if the newly formed tetrahedral carbon is not bonded to a second electronegative atom or group and is generally unstable if it is.

Electronic and steric factors cause an aldehyde to be more reactive than a ketone toward nucleophilic attack. Aldehydes and ketones are less reactive than acyl halides and acid anhydrides and are more reactive than esters, carboxylic acids, and amides.

Grignard reagents react with aldehydes to form secondary alcohols, with ketones and acyl halides to form tertiary alcohols, and with carbon dioxide to form carboxylic acids. Aldehydes are reduced to primary alcohols, ketones to secondary alcohols, and amides to amines.

Aldehydes and ketones undergo acid-catalyzed addition of water to form hydrates. Electron donation and bulky substituents decrease the percentage of hydrate present at equilibrium. Most hydrates are too unstable to be isolated. Acid-catalyzed addition of alcohol to aldehydes forms **hemiacetals** and **acetals**, and to ketones forms **hemiketals** and **ketals**. Acetal and ketal formation are reversible. Cyclic acetals and ketals serve as **protecting groups** for aldehyde and ketone functional groups. Aldehydes and ketones react with thiols to form thioacetals and thioketals; desulfurization replaces the C—S bonds with C—H bonds.

Aldehydes and ketones react with primary amines to form **imines** and with secondary amines to form **enamines**. The mechanisms are the same, except for the site from which a proton is lost in the last step of the reaction. Imine and enamine formation are reversible; imines and enamines are hydrolyzed under acidic conditions back to the carbonyl compound and amine. A **pH-rate profile** is a plot of the observed rate constant as a function of the pH of the reaction mixture. Hydroxide ion and heat differentiate the **Wolff–Kishner reduction** from ordinary **hydrazone** formation.

An aldehyde or a ketone reacts with a phosphonium ylide in a **Wittig reaction** to form an alkene. A Wittig reaction is a concerted [2 + 2] cycloaddition reaction; it is completely regioselective. Stabilized ylides form primarily *E* isomers; unstabilized ylides form primarily *Z* isomers.

A **prochiral carbonyl carbon** is a carbonyl carbon that is bonded to two different substituents. The *Re* face is the one closest to the observer when decreasing priorities are in a clockwise direction; the *Si* face is the opposite face. Attack by a nucleophile on either the *Re* face or the *Si* face forms a pair of enantiomers.

A useful step in a retrosynthetic analysis is a **disconnection**—breaking a bond to produce two fragments. **Synthons** are fragments of a disconnection. A **synthetic equivalent** is the reagent used as the source of the synthon.

Nucleophilic addition to the carbonyl carbon of an α,β-unsaturated Class II carbonyl compound is called **direct addition**; addition to the β-carbon is called **conjugate addition**. Whether direct or conjugate addition occurs depends on the nature of the nucleophile, the structure of the carbonyl compound, and the reaction conditions. Nucleophiles that form unstable direct addition products—halide ions, cyanide ion, thiols, alcohols, and amines—form conjugate addition products. Nucleophiles that form stable addition products—hydride ion and carbanions—form direct addition products with reactive carbonyl groups and conjugate addition products with less reactive carbonyl groups. A Grignard reagent with a highly polarized C—Mg bond reacts with the harder C=O bond; a Gilman reagent with a less polarized C—Cu bond reacts with the softer C=C bond.

Nucleophiles form nucleophilic acyl substitution products with α,β-unsaturated Class I carbonyl compounds that have reactive carbonyl groups and conjugate addition products with compounds with less reactive carbonyl groups.

Summary of Reactions

1. Reaction of *carbonyl compounds* with carbon nucleophiles (Section 18.4).

   a. Reaction of *formaldehyde* with a Grignard reagent forms a primary alcohol:

$$\text{HC(=O)H} \xrightarrow[\text{2. } H_3O^+]{\text{1. } CH_3MgBr} CH_3CH_2OH$$

   b. Reaction of an *aldehyde* (other than formaldehyde) with a Grignard reagent forms a secondary alcohol:

$$\text{RC(=O)H} \xrightarrow[\text{2. } H_3O^+]{\text{1. } CH_3MgBr} \text{R—C(OH)(CH}_3\text{)—H}$$

   c. Reaction of a *ketone* with a Grignard reagent forms a tertiary alcohol:

$$\text{RC(=O)R}' \xrightarrow[\text{2. } H_3O^+]{\text{1. } CH_3MgBr} \text{R—C(OH)(CH}_3\text{)—R}'$$

   d. Reaction of an *ester* with a Grignard reagent forms a tertiary alcohol with two identical substituents:

$$\text{RC(=O)OR}' \xrightarrow[\text{2. } H_3O^+]{\text{1. 2 } CH_3MgBr} \text{R—C(OH)(CH}_3\text{)—CH}_3$$

   e. Reaction of an *acyl chloride* with a Grignard reagent forms a tertiary alcohol with two identical substituents:

$$\text{RC(=O)Cl} \xrightarrow[\text{2. } H_3O^+]{\text{1. 2 } CH_3MgBr} \text{R—C(OH)(CH}_3\text{)—CH}_3$$

   f. Reaction of CO_2 with a Grignard reagent forms a carboxylic acid:

$$O{=}C{=}O \xrightarrow[\text{2. } H_3O^+]{\text{1. } CH_3MgBr} CH_3\text{—C(=O)—OH}$$

g. Reaction with acetylide ions:

$$R-\overset{O}{\overset{\|}{C}}-R \xrightarrow[\text{2. } H_3O^+]{\text{1. } RC\equiv C^-} R-\underset{R}{\underset{|}{\overset{OH}{\overset{|}{C}}}}-C\equiv CR$$

h. Reaction with cyanide ion:

$$R-\overset{O}{\overset{\|}{C}}-R \xrightarrow[\text{HCl}]{^-C\equiv N} R-\underset{R}{\underset{|}{\overset{OH}{\overset{|}{C}}}}-C\equiv N$$

2. Reactions of *carbonyl compounds* with hydride ion donors (Section 18.5).

a. Reaction of an *aldehyde* with sodium borohydride forms a primary alcohol:

$$R-\overset{O}{\overset{\|}{C}}-H \xrightarrow[\text{2. } H_3O^+]{\text{1. } NaBH_4} RCH_2OH$$

b. Reaction of a *ketone* with sodium borohydride forms a secondary alcohol:

$$R-\overset{O}{\overset{\|}{C}}-R \xrightarrow[\text{2. } H_3O^+]{\text{1. } NaBH_4} R-\overset{OH}{\overset{|}{CH}}-R$$

c. Reaction of an *ester* with lithium aluminum hydride forms two alcohols:

$$R-\overset{O}{\overset{\|}{C}}-OR' \xrightarrow[\text{2. } H_3O^+]{\text{1. } LiAlH_4} RCH_2OH + R'OH$$

d. Reaction of an *ester* with diisobutylaluminum hydride forms an aldehyde:

$$R-\overset{O}{\overset{\|}{C}}-OR' \xrightarrow[\text{2. } H_2O]{\text{1. } [(CH_3)_2CHCH_2]_2AlH,\ -78\ ^\circ C} R-\overset{O}{\overset{\|}{C}}-H$$

e. Reaction of a *carboxylic acid* with lithium aluminum hydride forms a primary alcohol:

$$R-\overset{O}{\overset{\|}{C}}-OH \xrightarrow[\text{2. } H_3O^+]{\text{1. } LiAlH_4} R-CH_2-OH$$

f. Reaction of an *acyl chloride* with lithium aluminum hydride forms a primary alcohol:

$$R-\overset{O}{\overset{\|}{C}}-Cl \xrightarrow[\text{2. } H_3O^+]{\text{1. } LiAlH_4} R-CH_2-OH$$

g. Reaction of an *amide* with lithium aluminum hydride forms an amine:

$$\text{RC(=O)NH}_2 \xrightarrow[\text{2. H}_2\text{O}]{\text{1. LiAlH}_4} \text{R—CH}_2\text{—NH}_2$$

$$\text{RC(=O)NHR}' \xrightarrow[\text{2. H}_2\text{O}]{\text{1. LiAlH}_4} \text{R—CH}_2\text{—NHR}'$$

$$\text{RC(=O)NR}'\text{R}'' \xrightarrow[\text{2. H}_2\text{O}]{\text{1. LiAlH}_4} \text{R—CH}_2\text{—N(R}'')\text{—R}'$$

3. Reactions of *aldehydes* and *ketones* with amines (Section 18.6).

a. Reaction with a *primary amine* forms an imine:

$$\text{R}_2\text{C=O} + \text{H}_2\text{NZ} \underset{}{\overset{\text{catalytic H}^+}{\rightleftharpoons}} \text{R}_2\text{C=NZ} + \text{H}_2\text{O}$$

When Z = R, the product is a Schiff base; Z can also be OH, NH_2, NHC_6H_5, $NHC_6H_3(NO_2)_2$, or $NHCONH_2$.

b. Reaction with a *secondary amine* forms an enamine:

$$\text{R(—CH<)C=O} + \text{RNHR} \overset{\text{catalytic H}^+}{\rightleftharpoons} \text{R(—C<)=C—NR}_2 + \text{H}_2\text{O}$$

c. The Wolff–Kishner reduction converts a carbonyl group to a methylene group:

$$\text{RC(=O)R}' \xrightarrow[\text{HO}^-,\,\Delta]{\text{NH}_2\text{NH}_2} \text{R—CH}_2\text{—R}'$$

4. Reactions of an *aldehyde* or a *ketone* with oxygen nucleophiles.

a. Reaction of an *aldehyde* or a *ketone* with water forms a hydrate (Section 18.7):

$$\text{RC(=O)R}' + \text{H}_2\text{O} \overset{\text{HCl}}{\rightleftharpoons} \text{R—C(OH)}_2\text{—R}'$$

b. Reaction of an *aldehyde* or a *ketone* with excess alcohol forms an acetal or a ketal (Section 18.8):

$$\text{RC(=O)R}' + 2\,\text{R}''\text{OH} \overset{\text{HCl}}{\rightleftharpoons} \text{R—C(OH)(OR}'')\text{—R}' \rightleftharpoons \text{R—C(OR}'')_2\text{—R}' + \text{H}_2\text{O}$$

5. Protecting groups (Section 18.8).

a. *Aldehydes* and *ketones* can be protected by being converted to acetals:

$$\text{RC(=O)R} + \text{HOCH}_2\text{CH}_2\text{OH} \overset{\text{HCl}}{\rightleftharpoons} \text{R}_2\text{C(OCH}_2\text{CH}_2\text{O)} + \text{H}_2\text{O}$$

b. The OH group of an *alcohol* can be protected by being converted to a TMS ether:

$$\text{R—OH} + (\text{CH}_3)_3\text{SiCl} \xrightarrow{(\text{CH}_3\text{CH}_2)_3\text{N}} \text{R—OSi(CH}_3)_3$$

c. The OH group of a *carboxylic acid* can be protected by being converted to an ester:

$$\text{RC(=O)OH} + \underset{\text{excess}}{\text{CH}_3\text{OH}} \overset{\text{HCl}}{\rightleftharpoons} \text{RC(=O)OCH}_3 + \text{H}_2\text{O}$$

d. An *amino group* can be protected by being converted to an amide:

$$\text{R—NH}_2 + \text{RC(=O)Cl} \longrightarrow \text{RC(=O)NHR} + \text{HCl}$$

6. Reaction of an *aldehyde* or a *ketone* with a thiol forms a thioacetal or a thioketal (Section 18.9):

$$\text{RC(=O)R}' + 2\,\text{R}''\text{SH} \overset{\text{HCl}}{\rightleftharpoons} \text{R—C(SR}'')_2\text{—R}' + \text{H}_2\text{O}$$

7. Desulfurization of *thioacetals* and *thioketals* forms alkanes (Section 18.9):

$$\text{R—C(SR}'')_2\text{—R}' \xrightarrow[\text{Raney Ni}]{\text{H}_2} \text{R—CH}_2\text{—R}'$$

8. Reaction of an *aldehyde* or a *ketone* with a phosphonium ylide (a Wittig reaction) forms an alkene (Section 18.10):

$$\text{RC(=O)R}' + (\text{C}_6\text{H}_5)_3\text{P}{=}\text{CR}_2 \longrightarrow \text{R}_2\text{C}{=}\text{CRR}' + (\text{C}_6\text{H}_5)_3\text{P}{=}\text{O}$$

9. Reactions of *α,β-unsaturated aldehydes* and *ketones* with nucleophiles (Section 18.13):

$$\text{RCH}{=}\text{CHC(=O)R}' + \text{NuH} \nearrow \text{RCH}{=}\text{CHC(OH)(Nu)R}' \text{ (direct addition)}$$

$$\text{RCH}{=}\text{CHC(=O)R}' + \text{NuH} \searrow \text{RCH(Nu)CH}_2\text{C(=O)R}' \text{ (conjugate addition)}$$

Nucleophiles that are weak bases ($^-$CN, RSH, RNH_2, Br^-) and R_2CuLi form conjugate addition products. Nucleophiles that are strong bases (RLi, RMgBr, and H^-) form direct addition products with reactive carbonyl groups and conjugate addition products with less reactive carbonyl groups.

10. Reactions of *α,β-unsaturated carboxylic acid derivatives* with nucleophiles (Section 18.14):

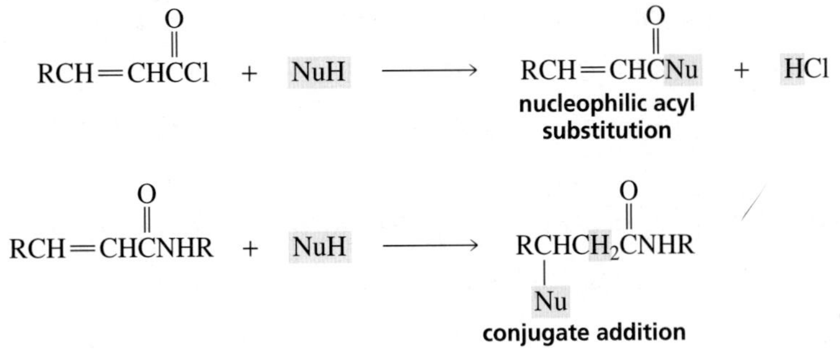

$$\mathrm{RCH{=}CH\overset{O}{\overset{\|}{C}}Cl} + \mathrm{NuH} \longrightarrow \mathrm{RCH{=}CH\overset{O}{\overset{\|}{C}}Nu} + \mathrm{HCl}$$

nucleophilic acyl substitution

$$\mathrm{RCH{=}CH\overset{O}{\overset{\|}{C}}NHR} + \mathrm{NuH} \longrightarrow \mathrm{R\underset{Nu}{\underset{|}{C}}HCH_2\overset{O}{\overset{\|}{C}}NHR}$$

conjugate addition

Nucleophiles form nucleophilic acyl substitution products with reactive carbonyl groups and conjugate addition products with less reactive carbonyl groups.

Key Terms

acetal (p. 755)
aldehyde (p. 731)
conjugate addition (p. 769)
cyanohydrin (p. 741)
deoxygenation (p. 752)
disconnection (p. 766)
direct addition (p. 769)
enamine (p. 747)
gem-diol (p. 753)
hemiacetal (p. 755)
hemiketal (p. 756)
hydrate (p. 753)
hydrazone (p. 751)
imine (p. 747)
ketal (p. 756)
ketone (p. 731)
nucleophilic acyl substitution (p. 737)
nucleophilic addition (p. 737)
nucleophilic addition–elimination reaction (p. 738)
oxime (p. 751)
phenylhydrazone (p. 751)
pH–rate profile (p. 749)
prochiral carbonyl carbon (p. 765)
protecting group (p. 758)
Re and *Si* faces (p. 765)
reduction reaction (p. 743)
Schiff base (p. 747)
semicarbazone (p. 751)
synthetic equivalent (p. 766)
synthon (p. 766)
Wittig reaction (p. 761)
ylide (p. 761)

Problems

40. Draw the structure for each of the following compounds:
 a. isobutyraldehyde
 b. 4-hexenal
 c. diisopentyl ketone
 d. 3-methylcyclohexanone
 e. 2,4-pentanedione
 f. 4-bromo-3-heptanone
 g. γ-bromocaproaldehyde
 h. 2-ethylcyclopentanecarbaldehyde
 i. 4-methyl-5-oxohexanal
 j. benzene-1,3-dicarbaldehyde

41. Give the products of each of the following reactions:

 a. 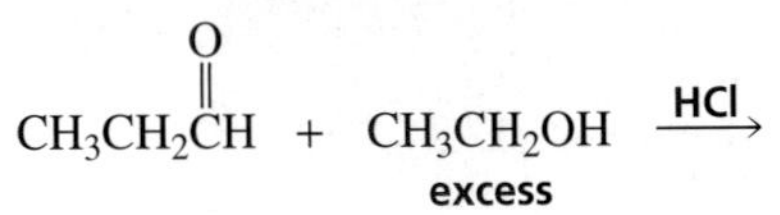 $\mathrm{CH_3CH_2\overset{O}{\overset{\|}{C}}H} + \underset{\text{excess}}{\mathrm{CH_3CH_2OH}} \xrightarrow{\text{HCl}}$

 b. $\mathrm{C_6H_5\overset{O}{\overset{\|}{C}}CH_2CH_3} + \mathrm{NH_2NH_2} \xrightarrow[\text{H}^+]{\text{catalytic}}$

 c. $\mathrm{C_6H_5\overset{O}{\overset{\|}{C}}CH_2CH_3} + \mathrm{NH_2NH_2} \xrightarrow[\Delta]{\text{HO}^-}$

 d. $\mathrm{CH_3CH_2\overset{O}{\overset{\|}{C}}CH_3} \xrightarrow[\text{2. H}_3\text{O}^+]{\text{1. NaBH}_4}$

 e. $\mathrm{CH_3CH_2\overset{O}{\overset{\|}{C}}CH_2CH_3} + \underset{\text{excess}}{\mathrm{NaC{\equiv}N}} \xrightarrow{\text{HCl}}$

 f. $\mathrm{CH_3CH_2CH_2\overset{O}{\overset{\|}{C}}OCH_2CH_3} \xrightarrow[\text{2. H}_3\text{O}^+]{\text{1. LiAlH}_4}$

 g. $\mathrm{CH_3CH_2CH_2\overset{O}{\overset{\|}{C}}CH_3} + \mathrm{HOCH_2CH_2OH} \xrightarrow{\text{HCl}}$

 h. 3-methyl-2-cyclohexenone (ring structure: C=O, ring C=C, CH_3) $+ \underset{\text{excess}}{\mathrm{NaC{\equiv}N}} \xrightarrow{\text{HCl}}$

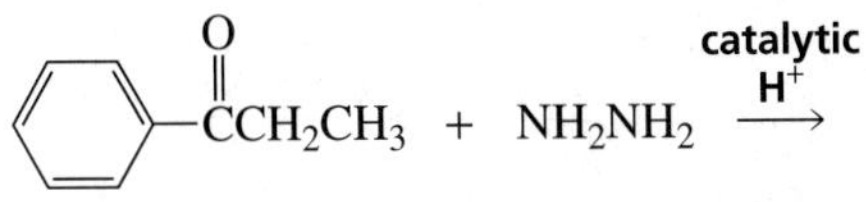

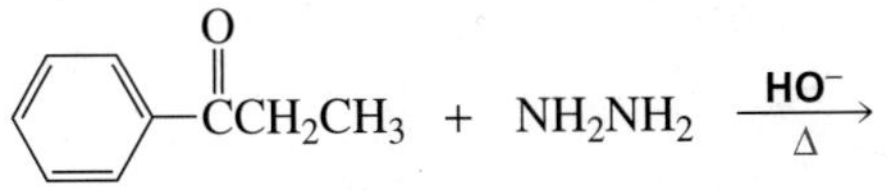

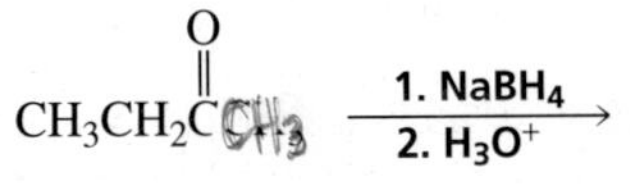

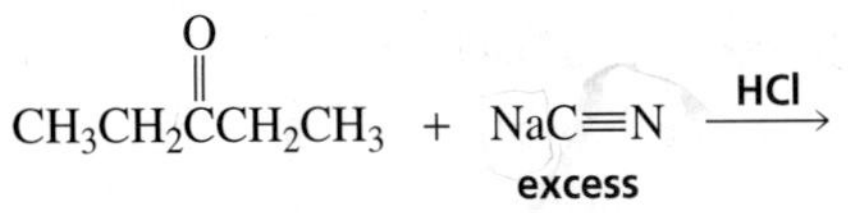

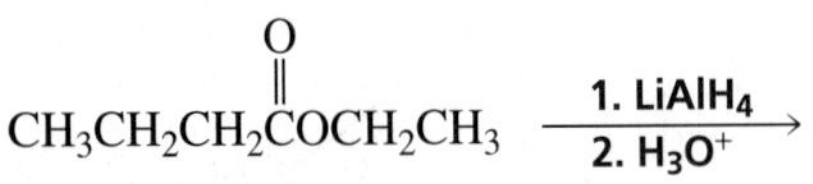

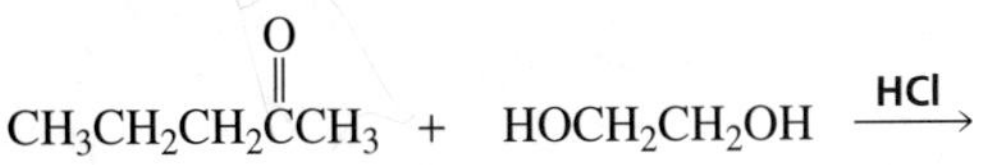

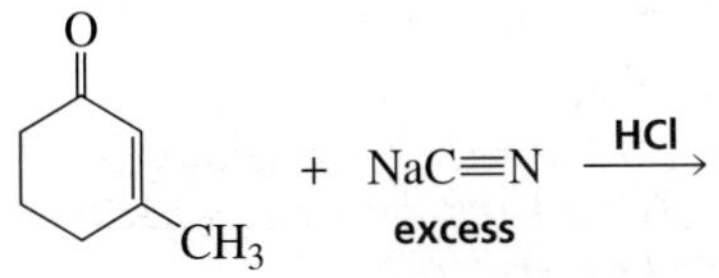

42. List the following compounds in order of decreasing reactivity toward nucleophilic attack:

$CH_3CH_2CH(CH_3)C(=O)CH_2CH_3$ $CH_3CH_2C(=O)H$ $CH_3CH_2CH(CH_3)C(=O)C(CH_3)HCH_2CH_3$

$CH_3CH_2C(=O)CH_2CH_3$ $CH_3CH_2CH(CH_3)C(=O)CH(CH_3)CH_2CH_3$ $CH_3CH(CH_3)CH_2C(=O)CH_2CH_3$

43. a. Show the reagents required to form the primary alcohol.

$R{-}C(=O){-}H \rightarrow RCH_2OH$

$R'CH{=}CH_2 \rightarrow RCH_2OH$

$R{-}C(=O){-}OH \rightarrow RCH_2OH$

$RCH_2Br \rightarrow RCH_2OH$

$R{-}C(=O){-}OR \rightarrow RCH_2OH$

$RCH_2OCH_3 \xrightarrow{\Delta} RCH_2OH$

$R{-}C(=O){-}Cl \rightarrow RCH_2OH$

$R{-}C(=O){-}O{-}C(=O){-}R \rightarrow RCH_2OH$

$H_2C{=}O \rightarrow RCH_2OH$

ethylene oxide (O) $\rightarrow RCH_2OH$

b. Which of the reactions cannot be used for the synthesis of isobutyl alcohol?

c. Which of the reactions changes the carbon skeleton of the starting material?

44. Using cyclohexanone as the starting material, describe how each of the following compounds could be synthesized:

a. OH   d. NH_2 g. $CH{=}CH_2$

b.   e. CH_2NH_2 h. (show two methods)

c. Br   f. $N(CH_3)_2$ i. CH_2CH_3 (show two methods)

45. Propose a mechanism for the following reaction:

$$HOCH_2CH_2CH_2CH_2C(=O)H \xrightarrow[CH_3OH]{HCl} \text{O}\ \ \text{OCH}_3$$

46. List the following compounds in order of decreasing K_{eq} for hydrate formation:

$C_6H_5C(=O)CH_3$ $Cl{-}C_6H_4C(=O)CH_3$ $O_2N{-}C_6H_4C(=O)CH_3$ $CH_3O{-}C_6H_4C(=O)CH_3$

47. Fill in the boxes:

a. $CH_3OH \xrightarrow{\square} CH_3Br \xrightarrow[\square]{\square} \square \xrightarrow[2.\ \square]{1.\ \square} CH_3CH_2OH$

b. $CH_4 \xrightarrow{\square} CH_3Br \xrightarrow[\square]{\square} \square \xrightarrow[2.\ \square]{1.\ \square} CH_3CH_2CH_2OH$

48. Give the products of each of the following reactions:

a. $C_6H_5-C(CH_2CH_3){=}NCH_2CH_3 + H_2O \xrightarrow{HCl}$

b. $CH_3CH_2\overset{O}{\overset{\|}{C}}CH_3 \xrightarrow[2.\ H_3O^+]{1.\ CH_3CH_2MgBr}$

c. cyclopentanone $+ (C_6H_5)_3P{=}CHCH_3 \longrightarrow$

d. $CH_3CH_2\overset{O}{\overset{\|}{C}}OCH_3 \xrightarrow[2.\ H_3O^+]{1.\ CH_3CH_2MgBr\ \text{excess}}$

e. $C_6H_5COCH{=}CHCH_3 + CH_3OH \xrightarrow{HCl}$

f. 2-pyrrolidinone (N–H lactam) $\xrightarrow[2.\ H_2O]{1.\ LiAlH_4}$

g. cyclohexanone $+ CH_3CH_2NH_2 \xrightarrow{\text{catalytic } H^+}$

h. cyclohexanone $+ (CH_3CH_2)_2NH \xrightarrow{\text{catalytic } H^+}$

i. $CH_3\overset{CH_3}{\overset{|}{C}}{=}CH\overset{O}{\overset{\|}{C}}CH_3 + HBr \longrightarrow$

j. $2\ CH_2{=}CH{-}\overset{O}{\overset{\|}{C}}OCH_3 + CH_3NH_2 \longrightarrow$

49. Thiols can be prepared from the reaction of thiourea with an alkyl halide, followed by hydroxide-ion-promoted hydrolysis.

$$\underset{\text{thiourea}}{H_2N\overset{S}{\overset{\|}{C}}NH_2} \xrightarrow[2.\ HO^-,\ H_2O]{1.\ CH_3CH_2Br} \underset{\text{urea}}{H_2N\overset{O}{\overset{\|}{C}}NH_2} + \underset{\text{ethanethiol}}{CH_3CH_2SH}$$

a. Propose a mechanism for the reaction.
b. What thiol would be formed if the alkyl halide employed were pentyl bromide?

50. The only organic compound obtained when compound Z undergoes the following sequence of reactions gives the ^1H NMR spectrum shown. Identify compound Z.

$$\text{Compound Z} \xrightarrow[2.\ H_3O^+]{1.\ \text{phenylmagnesium bromide}} \xrightarrow[\Delta]{MnO_2}$$

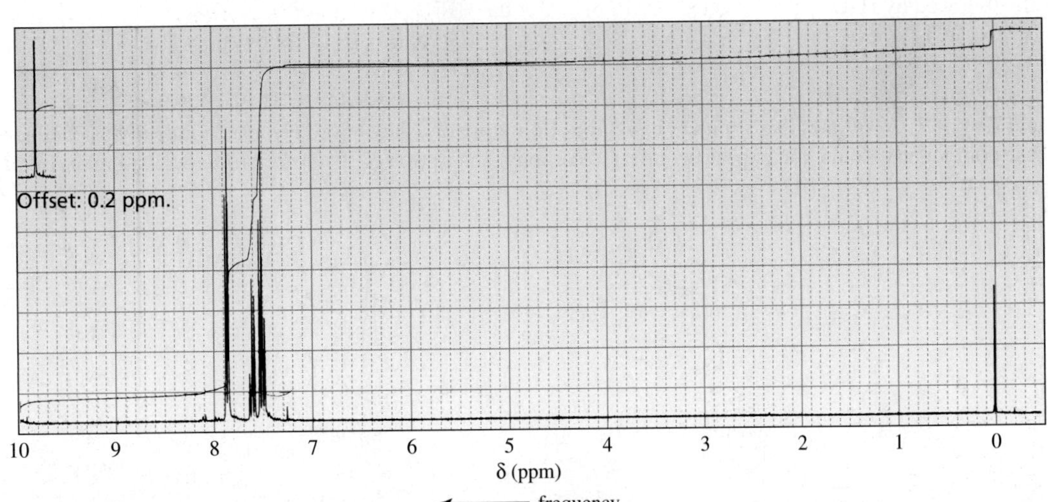

51. Propose a mechanism for each of the following reactions:

a. 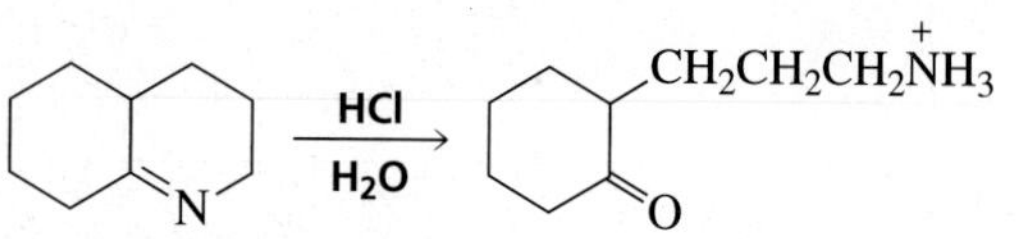 [bicyclic imine] $\xrightarrow[H_2O]{HCl}$ 2-substituted cyclohexanone with $CH_2CH_2CH_2\overset{+}{N}H_3$

c. 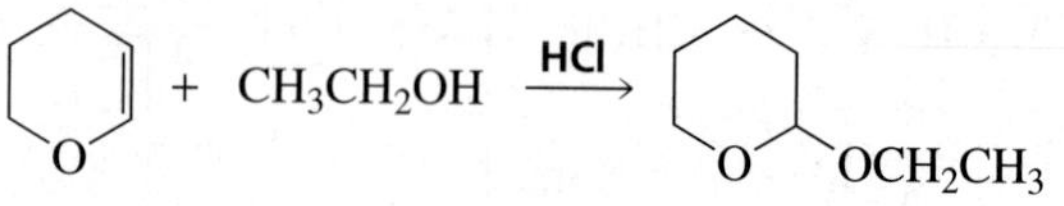 [dihydropyran] + CH_3CH_2OH $\xrightarrow{HCl}$ [tetrahydropyran] OCH_2CH_3

b. 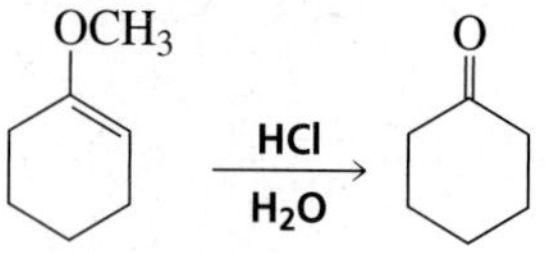 [1-methoxycyclohexene] (OCH_3) $\xrightarrow[H_2O]{HCl}$ [cyclohexanone] (O)

52. How many signals would the product of the following reaction show in these spectra

a. its ^{1}HNMR spectrum

b. its ^{13}C NMR spectrum

$$CH_3\overset{O}{\overset{\|}{C}}CH_2CH_2\overset{O}{\overset{\|}{C}}OCH_3 \xrightarrow[2.\ H_3O^+]{1.\ \text{excess } CH_3MgBr}$$

53. Give the products of the following rections. Show all stereoisomers that are formed.

a. 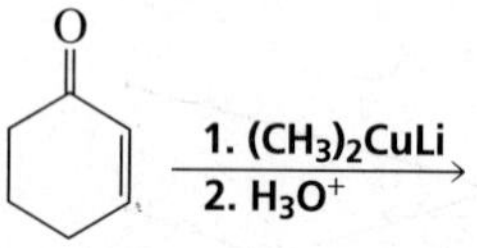 [2-cyclohexenone] $\xrightarrow[2.\ H_3O^+]{1.\ (CH_3)_2CuLi}$

c. 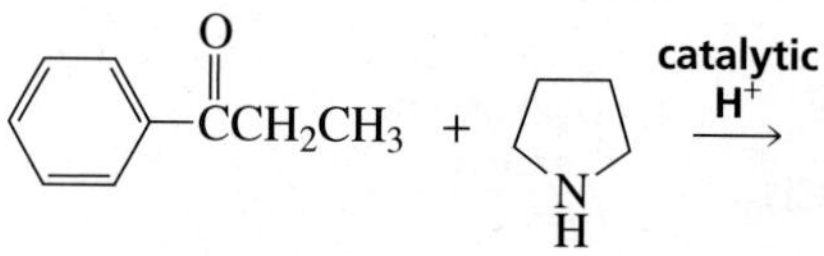 [benzene]$-\overset{O}{\overset{\|}{C}}CH_2CH_3$ + [pyrrolidine, N–H] $\xrightarrow{\text{catalytic } H^+}$

b. 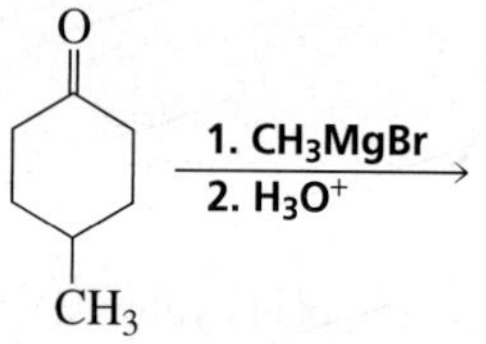 [4-methylcyclohexanone] (CH_3) $\xrightarrow[2.\ H_3O^+]{1.\ CH_3MgBr}$

d. $CH_3CH_2\overset{O}{\overset{\|}{C}}CH_2CH_2CH_2CH_3 \xrightarrow[2.\ H_3O^+]{1.\ LiAlH_4}$

54. List three different sets of reagents (a carbonyl compound and a Grignard reagent) that could be used to prepare each of the following tertiary alcohols:

a. 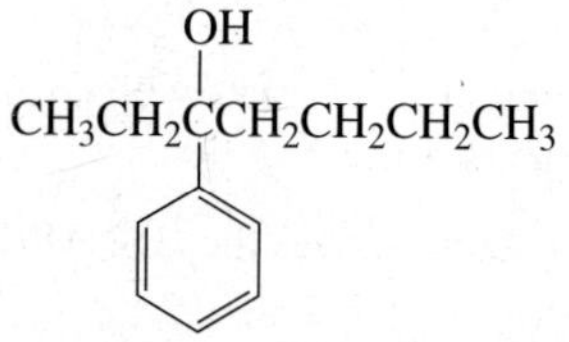 $CH_3CH_2\overset{OH}{\overset{|}{C}}CH_2CH_2CH_2CH_3$ (with phenyl on the carbinol carbon)

b. $CH_3CH_2\overset{OH}{\overset{|}{C}}CH_2CH_2CH_3$ (with CH_2CH_3 on the carbinol carbon)

55. Give the product of the reaction of 3-methyl-2-cyclohexenone with each of the following reagents:

a. CH_3MgBr followed by H_3O^+

b. excess NaCN, HCl

c. H_2, Pd

d. HBr

e. $(CH_3CH_2)_2CuLi$ followed by H_3O^+

f. CH_3CH_2SH

56. Norlutin® and Enovid® are ketones that suppress ovulation. Consequently, they have been used clinically as contraceptives. For which of these compounds would you expect the infrared carbonyl absorption (C=O stretch) to be at a higher frequency? Explain.

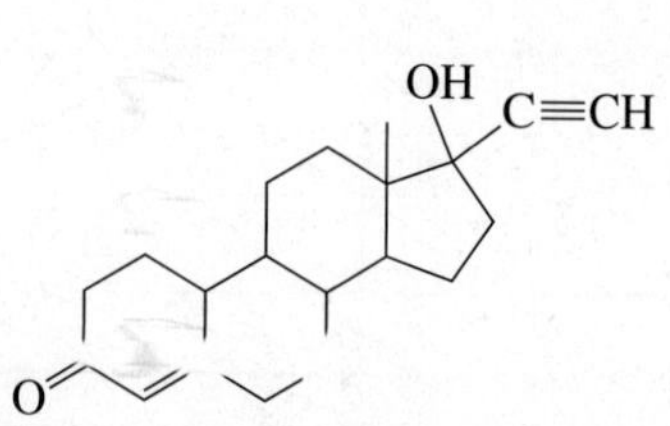

[Structure: OH, C≡CH, O] Norlutin®

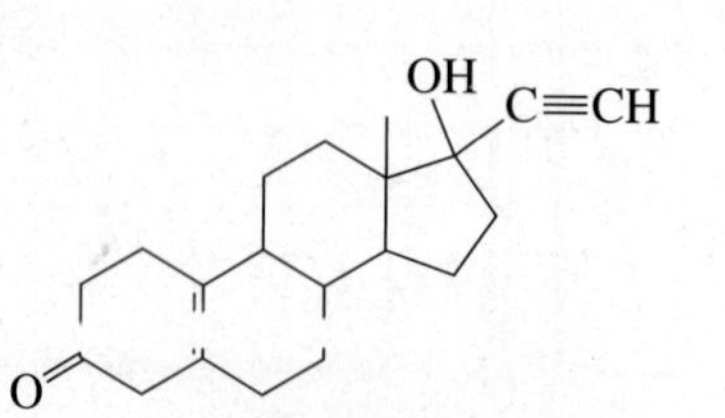

[Structure: OH, C≡CH, O] Enovid®

57. A compound gives the following IR spectrum. Upon reaction with sodium borohydride followed by acidification, the product with the following ^{1}HNMR spectrum is formed. Identify the compounds.

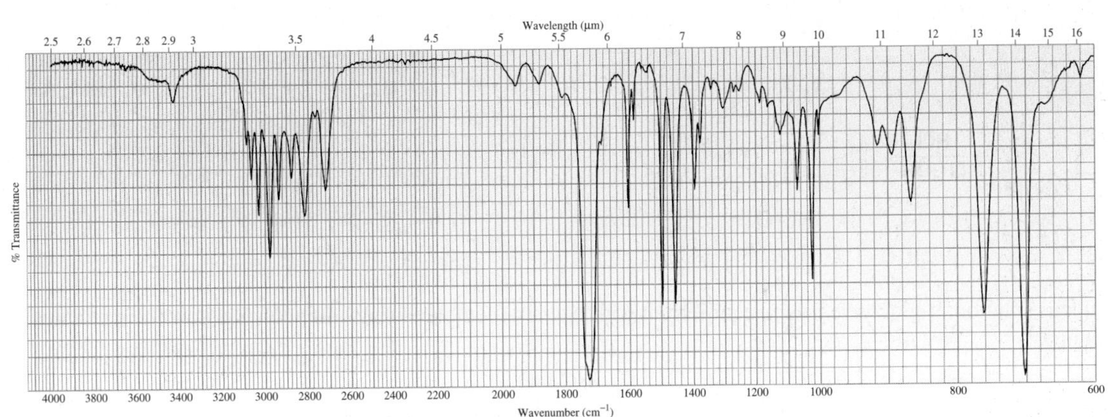

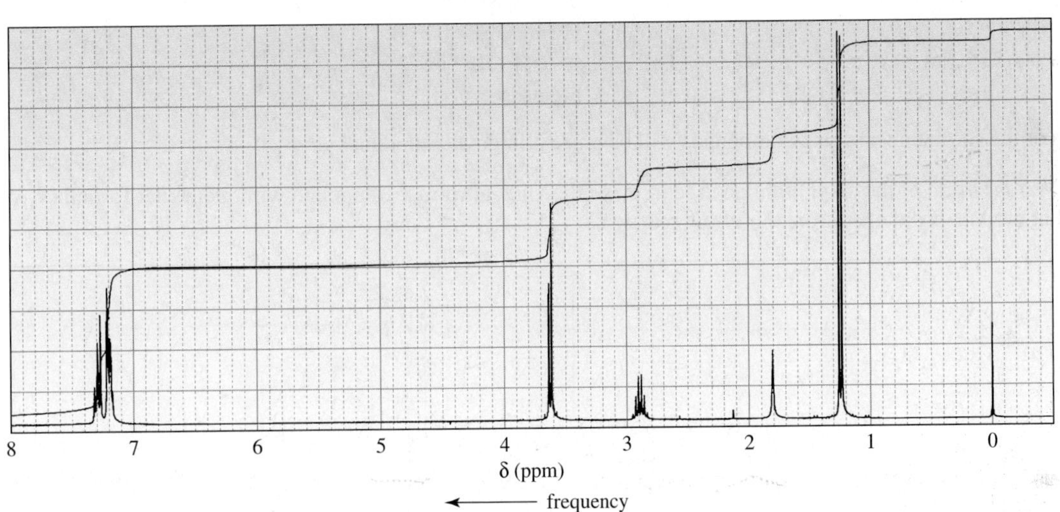

58. Unlike a phosphonium ylide, which reacts with an aldehyde or ketone to form an alkene, a sulfonium ylide reacts with an aldehyde or ketone to form an epoxide. Explain why one ylide forms an alkene, whereas the other forms an epoxide.

$$CH_3CH_2\overset{\overset{\displaystyle O}{\|}}{C}H + (CH_3)_2S{=}CH_2 \longrightarrow CH_3CH_2\overset{\overset{\displaystyle O}{\diagup\ \diagdown}}{CH{-}CH_2} + CH_3SCH_3$$

59. Indicate how the following compounds could be prepared from the given starting materials:

a. $C_6H_5\overset{\overset{\displaystyle O}{\|}}{C}OCH_3 \longrightarrow C_6H_5C(OH)(CH_3)CH_3$

b. $C_6H_5\overset{\overset{\displaystyle O}{\|}}{C}OCH_3 \longrightarrow C_6H_5\overset{\overset{\displaystyle O}{\|}}{C}H$

c. $CH_3CH_2CH_2CH_2Br \longrightarrow CH_3CH_2CH_2CH_2\overset{\overset{\displaystyle O}{\|}}{C}OH$

d. 6-methyl-2-piperidinone (O= on ring carbon next to N–H, CH_3 on other carbon next to N) $\longrightarrow$ 2-methylpiperidine (N–H, CH_3)

e. cyclohexanol (OH) $\longrightarrow$ N-methylcyclohexylamine ($NHCH_3$)

60. Propose a reasonable mechanism for each of the following reactions:

a. 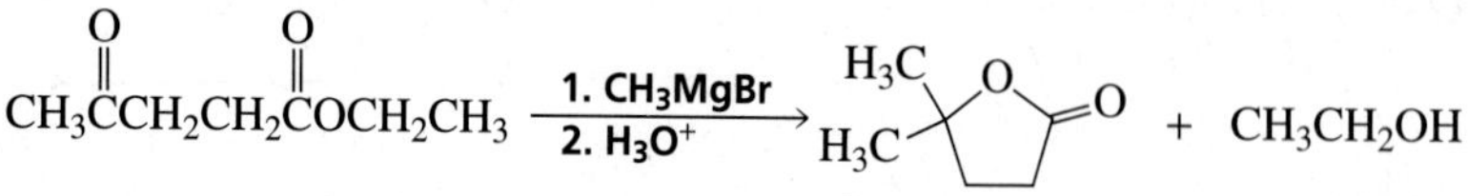 $CH_3\overset{O}{\overset{\|}{C}}CH_2CH_2\overset{O}{\overset{\|}{C}}OCH_2CH_3 \xrightarrow[2.\ H_3O^+]{1.\ CH_3MgBr}$ (H₃C)(H₃C) lactone $+ CH_3CH_2OH$

b. 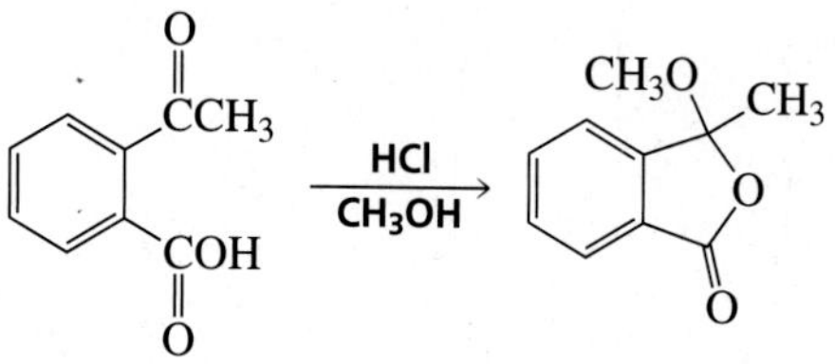 o-$C_6H_4(\overset{O}{\overset{\|}{C}}CH_3)(\overset{O}{\overset{\|}{C}}OH) \xrightarrow[CH_3OH]{HCl}$ cyclic product with CH_3O, CH_3, O, =O

61. a. In aqueous solution, D-glucose exists in equilibrium with two six-membered ring compounds. Draw the structures of these compounds.

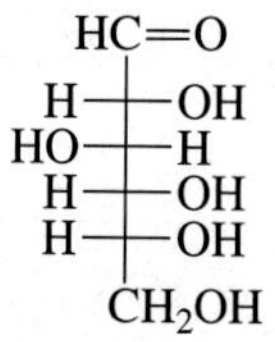

$HC{=}O$; H—OH; HO—H; H—OH; H—OH; CH_2OH

D-glucose

b. Which of the six-membered ring compounds will be present in greater amount?

62. The ^1H NMR spectrum of the alkyl bromide used to make the ylide to form a compound with molecular formula $C_{11}H_{14}$ is shown below. What product is obtained from the Wittig reaction?

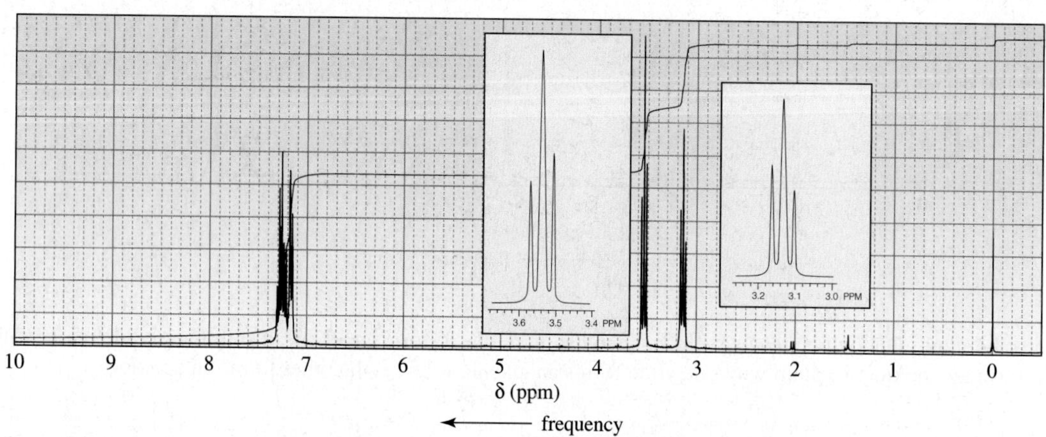

63. In the presence of an acid catalyst, acetaldehyde forms a trimer known as paraldehyde. Because it induces sleep when it is administered to animals in large doses, paraldehyde is used as a sedative or hypnotic. Propose a mechanism for the formation of paraldehyde.

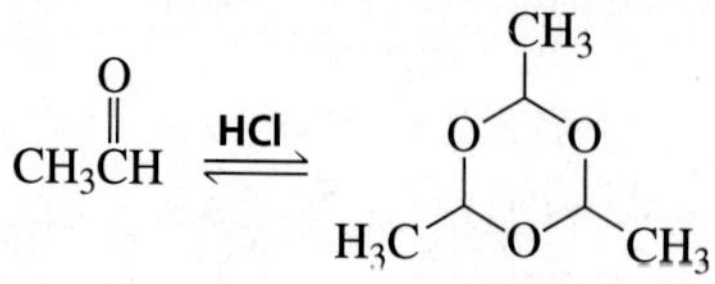

$CH_3\overset{O}{\overset{\|}{C}}H \overset{HCl}{\rightleftharpoons}$ paraldehyde (trimethyl-1,3,5-trioxane: CH_3, H_3C, CH_3)

paraldehyde

64. The addition of hydrogen cyanide to benzaldehyde forms a compound called mandelonitrile. (*R*)-Mandelonitrile is formed from the hydrolysis of amygdalin, a compound found in the pits of peaches and apricots. Amygdalin is the principal constituent of laetrile, a compound that was once highly touted as a treatment for cancer. The drug was subsequently found to be ineffective. Is (*R*)-mandelonitrile formed by attack of cyanide ion on the *Re* face or the *Si* face of benzaldehyde?

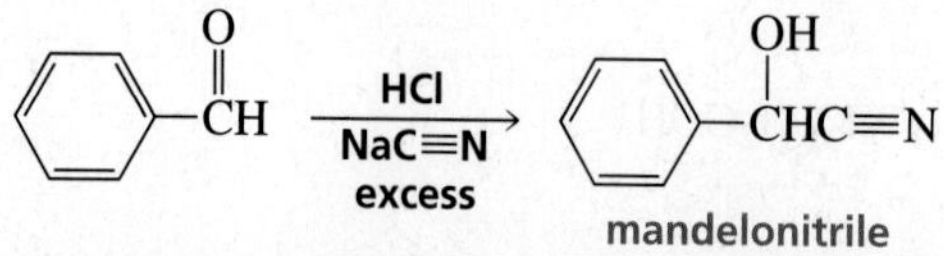

$C_6H_5\overset{O}{\overset{\|}{C}}H \xrightarrow[NaC{\equiv}N\ \text{excess}]{HCl} C_6H_5\overset{OH}{\overset{|}{C}}HC{\equiv}N$

mandelonitrile

65. What carbonyl compound and what phosphonium ylide are needed to synthesize the following compounds?

a. 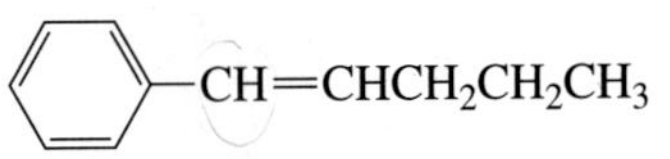 C_6H_5–CH=CHCH$_2$CH$_2$CH$_3$ b. (cyclopentylidene)=CHCH$_2$CH$_3$ c. C_6H_5–CH=CH–C_6H_5 d. 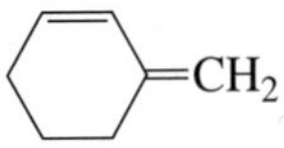 (cyclohexylidene)=CH$_2$

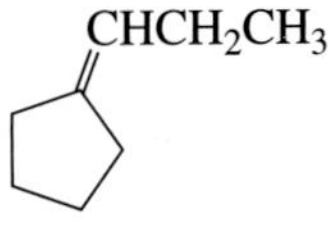

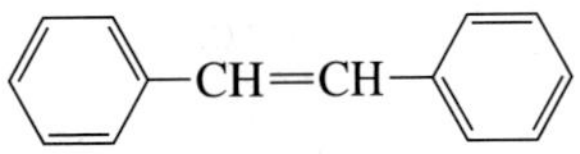

66. Identify compounds A and B:

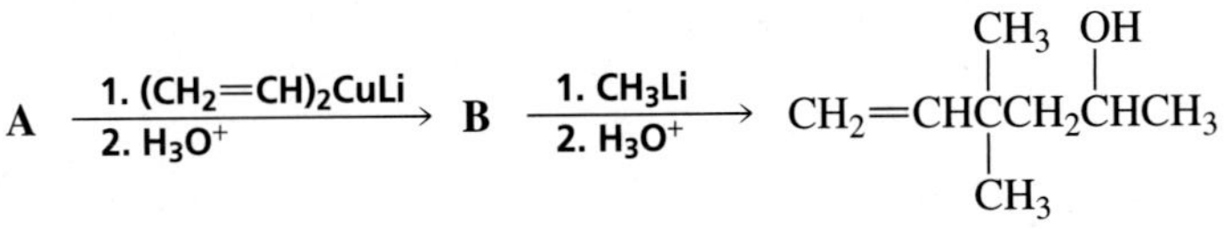

$$\text{A} \xrightarrow[\text{2. } H_3O^+]{\text{1. } (CH_2{=}CH)_2CuLi} \text{B} \xrightarrow[\text{2. } H_3O^+]{\text{1. } CH_3Li} CH_2{=}CH\underset{CH_3}{\overset{CH_3}{C}}CH_2\overset{OH}{C}HCH_3$$

67. Propose a reasonable mechanism for each of the following reactions:

a. 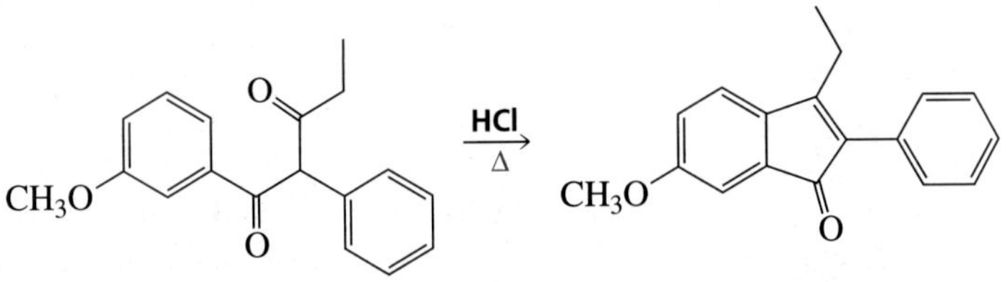 CH$_3$O ... $\xrightarrow[\Delta]{\text{HCl}}$ CH$_3$O ...

b. 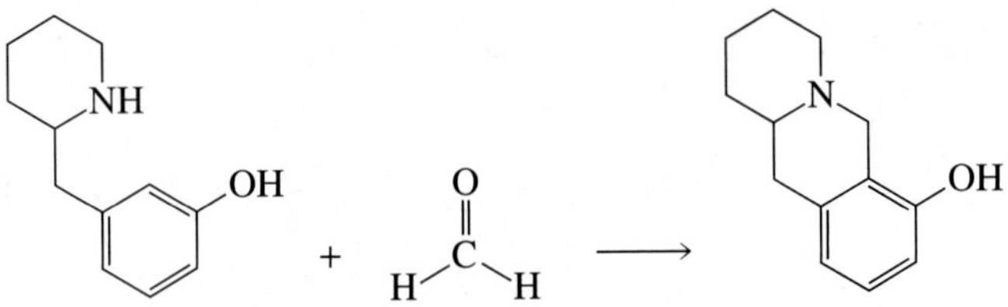 NH ... OH + $H{-}\overset{O}{\overset{\|}{C}}{-}H$ ⟶ N ... OH

68. A compound reacts with methylmagnesium bromide followed by acidification to form the product with the following ^1H NMR spectrum. Identify the compound.

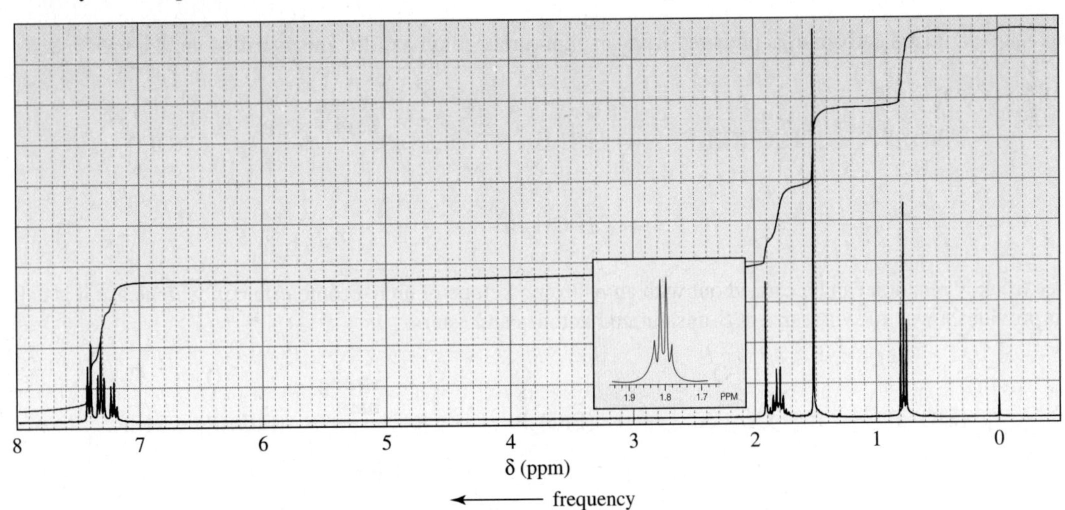

69. Show how each of the following compounds can be prepared from the given starting material. In each case, you will need to use a protecting group.

a. 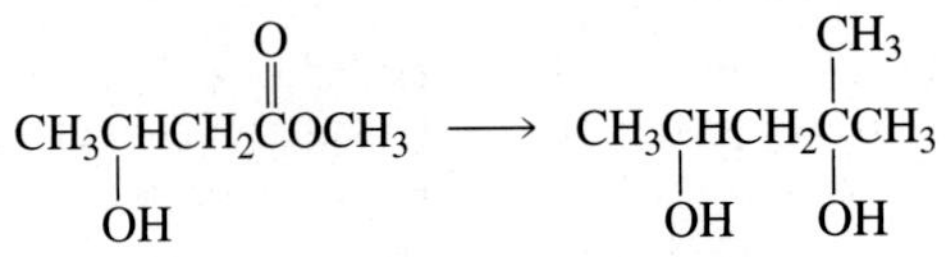 $CH_3\underset{OH}{C}HCH_2\overset{O}{\overset{\|}{C}}OCH_3$ ⟶ $CH_3\underset{OH}{C}HCH_2\underset{OH}{\overset{CH_3}{C}}CH_3$

b. 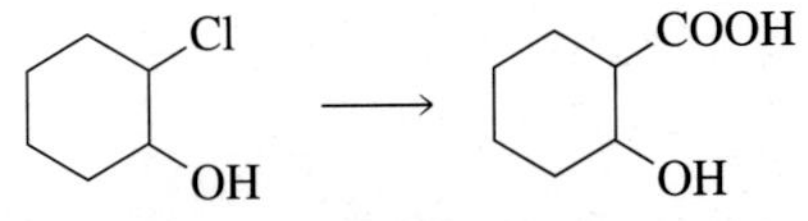 (2-chlorocyclohexanol: Cl, OH) ⟶ (2-hydroxycyclohexanecarboxylic acid: COOH, OH)

c. (3-bromoacetophenone: $\overset{O}{\overset{\|}{C}}$–CH$_3$, Br) ⟶ ($\overset{O}{\overset{\|}{C}}$–CH$_3$, CH$_2CH_2$OH)

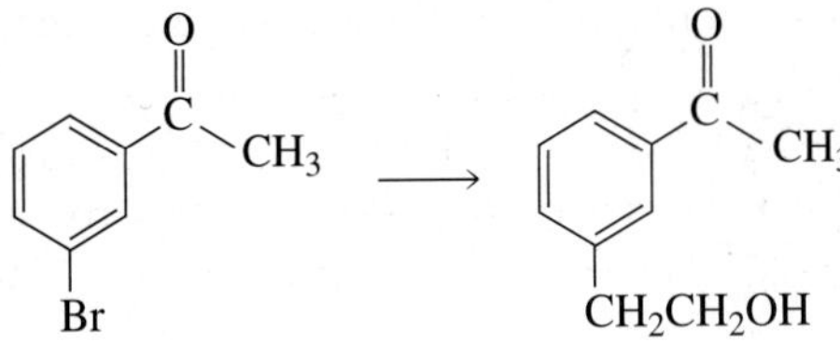

70. When a cyclic ketone reacts with diazomethane, the next larger cyclic ketone is formed. This is called a ring expansion reaction. Provide a mechanism for this reaction.

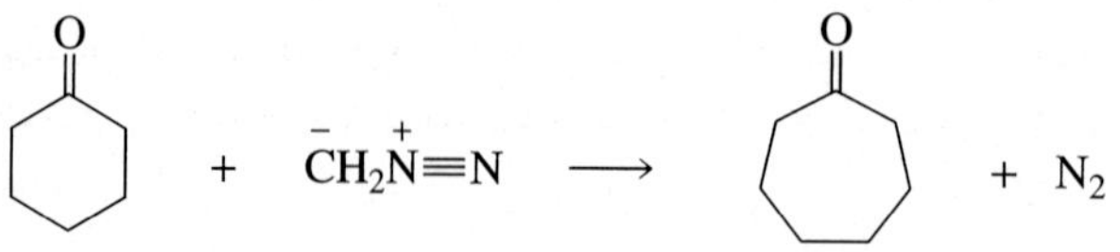

$$\text{cyclohexanone} + \overset{-}{C}H_2\overset{+}{N}\equiv N \longrightarrow \text{cycloheptanone} + N_2$$

cyclohexanone diazomethane cycloheptanone

71. The pK_a values of oxaloacetic acid are 2.22 and 3.98.

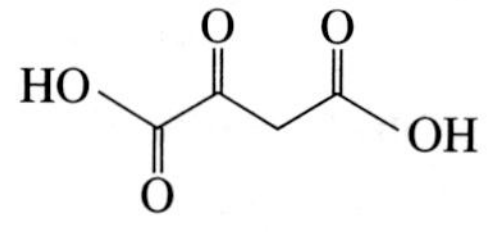

$$HOOC{-}CO{-}CH_2{-}COOH$$

oxaloacetic acid

a. Which carboxyl group is more acidic?
b. The amount of hydrate present in an aqueous solution of oxaloacetic acid depends on the pH of the solution: 95% at pH = 0, 81% at pH = 1.3, 35% at pH = 3.1, 13% at pH = 4.7, 6% at pH = 6.7, and 6% at pH = 12.7. Explain this pH dependence.

72. The *Horner–Emmons modification* is a variation of a Wittig reaction in which a phosphonate-stabilized carbanion in used instead of a phosphonium ylide.

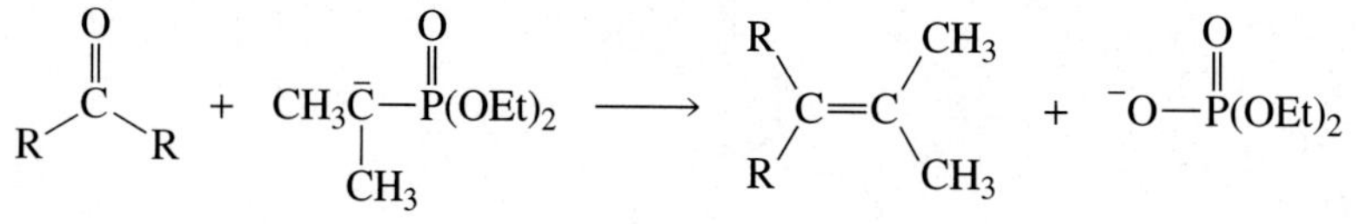

$$R{-}\overset{O}{\overset{\|}{C}}{-}R + CH_3\overset{-}{C}(CH_3){-}\overset{O}{\overset{\|}{P}}(OEt)_2 \longrightarrow R_2C{=}C(CH_3)_2 + {}^{-}O{-}\overset{O}{\overset{\|}{P}}(OEt)_2$$

$Et = CH_3CH_2$

The phosphonate-stabilized carbanion is prepared from an appropriate alkyl halide. This is called the *Arbuzov reaction.*

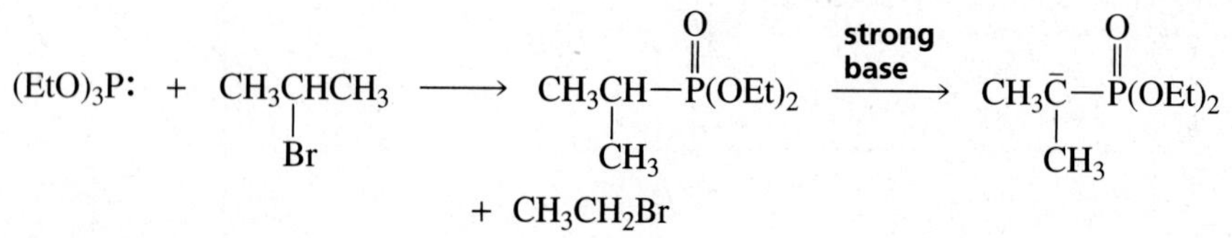

$$(EtO)_3P\!: + CH_3\underset{Br}{\underset{|}{C}}HCH_3 \longrightarrow CH_3\underset{CH_3}{\underset{|}{C}}H{-}\overset{O}{\overset{\|}{P}}(OEt)_2 + CH_3CH_2Br \xrightarrow{\text{strong base}} CH_3\underset{CH_3}{\underset{|}{\overset{-}{C}}}{-}\overset{O}{\overset{\|}{P}}(OEt)_2$$

Because the Arbuzov reaction can be carried out with an α-bromo ketone or an α-bromo ester (in which case it is called a *Perkow reaction*), it provides a way to synthesize α,β-unsaturated ketones and esters.

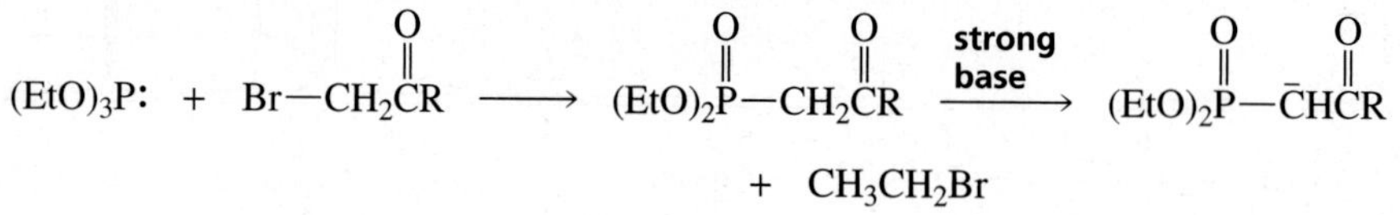

$$(EtO)_3P\!: + Br{-}CH_2\overset{O}{\overset{\|}{C}}R \longrightarrow (EtO)_2\overset{O}{\overset{\|}{P}}{-}CH_2\overset{O}{\overset{\|}{C}}R + CH_3CH_2Br \xrightarrow{\text{strong base}} (EtO)_2\overset{O}{\overset{\|}{P}}{-}\overset{-}{C}H\overset{O}{\overset{\|}{C}}R$$

a. Propose a mechanism for the Arbuzov reaction.
b. Propose a mechanism for the Horner–Emmons modification.
c. Show how the following compounds can be prepared from the given starting material:

1. $CH_3CH_2\overset{O}{\overset{\|}{C}}H \longrightarrow CH_3CH_2CH{=}CH\overset{O}{\overset{\|}{C}}CH_3$

2. 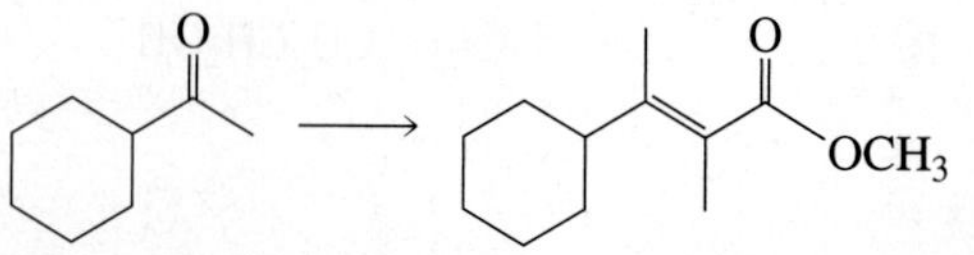 $C_6H_{11}{-}\overset{O}{\overset{\|}{C}}{-}CH_3 \longrightarrow C_6H_{11}{-}CH{=}C(CH_3){-}\overset{O}{\overset{\|}{C}}{-}OCH_3$

73. In order to solve this problem, you must read the description of the Hammett σ, ρ treatment given in Chapter 17, Problem 76. When the rate constants for the hydrolysis of several morpholine enamines of para-substituted propiophenones are determined at pH = 4.7, the ρ value is positive; however, when the rates of hydrolysis are determined at pH = 10.4, the ρ value is negative.

   a. What is the rate-determining step of the hydrolysis reaction when it is carried out in a basic solution?

   b. What is the rate-determining step of the reaction when it is carried out in an acidic solution?

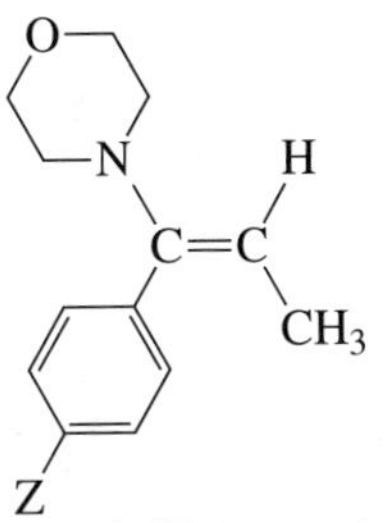

a morpholine enamine of a para-substituted propiophenone

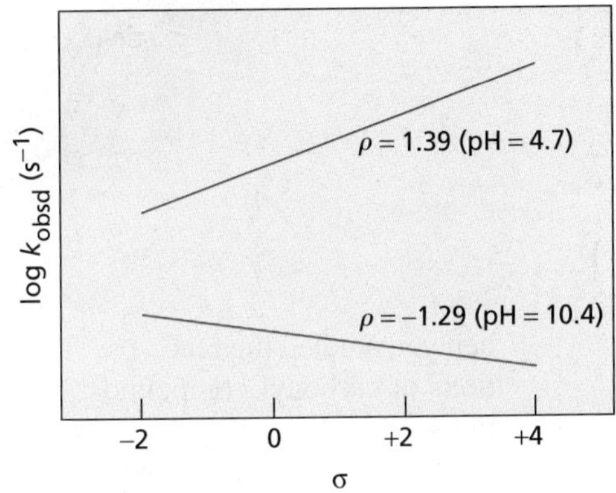

74. Propose a mechanism for each of the following reactions:

   a. 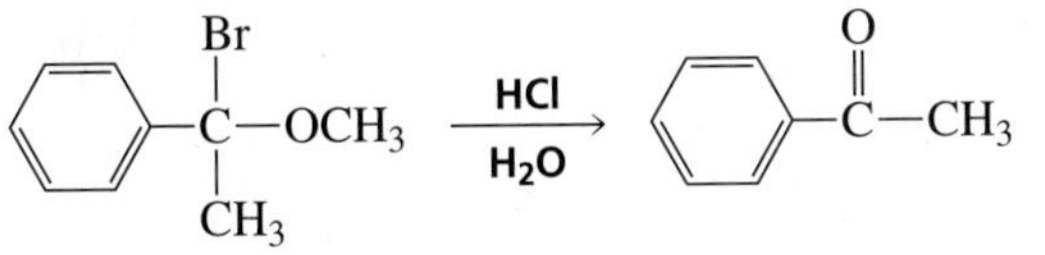 $C_6H_5C(Br)(CH_3)OCH_3 \xrightarrow[H_2O]{HCl} C_6H_5COCH_3$

   b. $\xrightarrow[H_2O]{HCl}$ $(HOCH_2CH_2CH_2)_2CHCH{=}O$

   c. + $CH_3CH_2SH \longrightarrow$ (OH, OH, SCH_2CH_3)

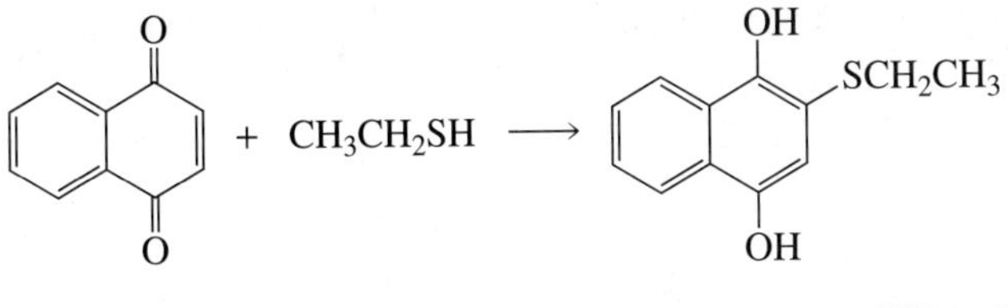

   d. 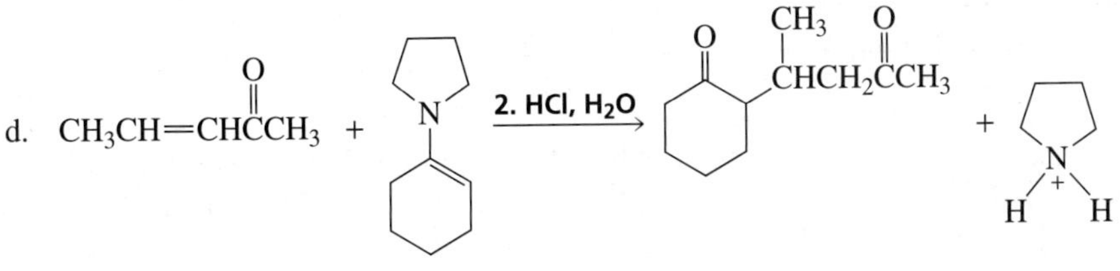 $CH_3CH{=}CHCCH_3$ (C=O) + $\xrightarrow{\text{2. HCl, } H_2O}$ $CHCH_2CCH_3$ (CH_3, O) +

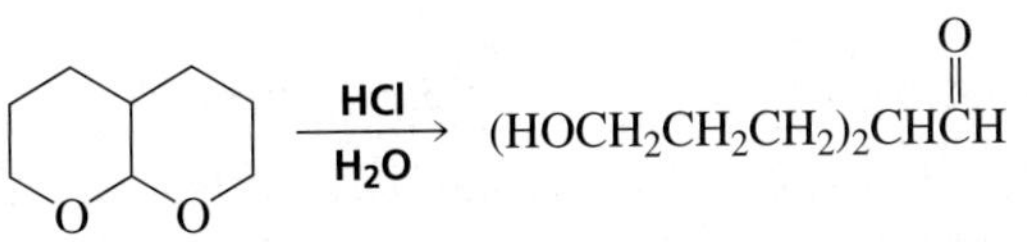

19 Carbonyl Compounds III

Reactions at the α-Carbon

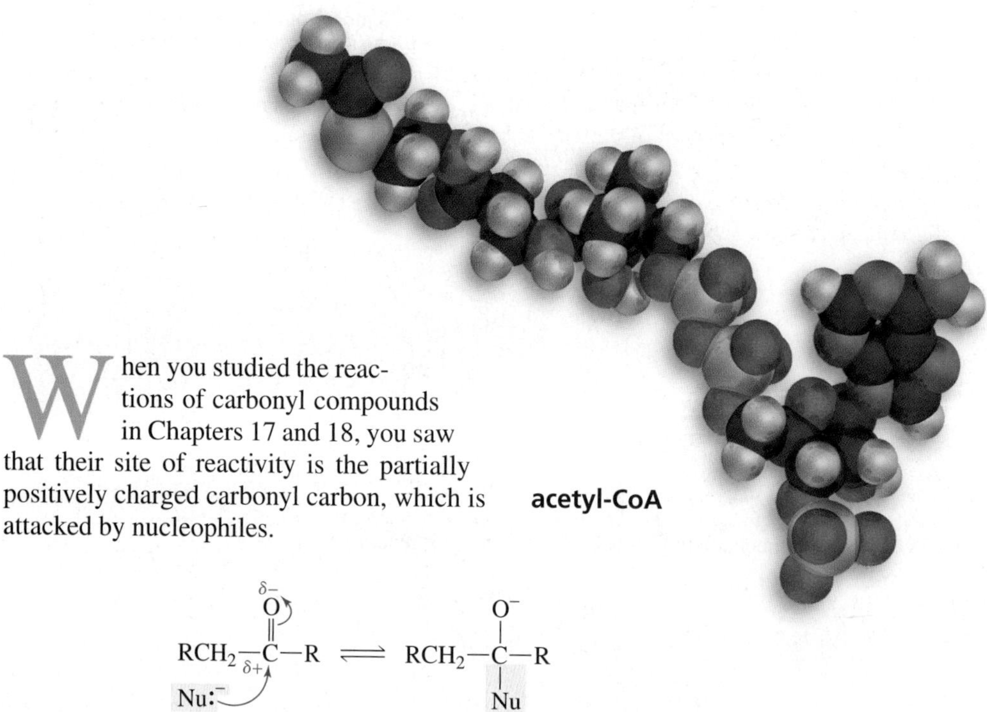

When you studied the reactions of carbonyl compounds in Chapters 17 and 18, you saw that their site of reactivity is the partially positively charged carbonyl carbon, which is attacked by nucleophiles.

$$\mathrm{RCH_2}-\overset{\overset{\delta-}{\mathrm{O}}}{\overset{\|}{\underset{\delta+}{\mathrm{C}}}}-\mathrm{R} \;\; + \;\; \mathrm{Nu}:^- \;\rightleftharpoons\; \mathrm{RCH_2}-\overset{\mathrm{O^-}}{\overset{|}{\underset{\underset{\mathrm{Nu}}{|}}{\mathrm{C}}}}-\mathrm{R}$$

Aldehydes, ketones, esters, and *N,N*-disubstituted amides have a second site of reactivity. A hydrogen bonded to *a carbon adjacent to a carbonyl carbon* is sufficiently acidic to be removed by a strong base. The carbon adjacent to a carbonyl carbon is called an ***α*-carbon**. A hydrogen bonded to an α-carbon is called an ***α*-hydrogen**.

$$\mathrm{R}-\underset{\underset{\mathrm{H}}{|}}{\mathrm{CH}}-\overset{\overset{\mathrm{O}}{\|}}{\mathrm{C}}-\mathrm{R} \;\; + \;\; :\mathrm{base}^- \;\rightleftharpoons\; \mathrm{R}-\underset{..}{\overset{-}{\mathrm{C}}\mathrm{H}}-\overset{\overset{\mathrm{O}}{\|}}{\mathrm{C}}-\mathrm{R} \;\; + \;\; \mathrm{baseH}$$

an α-carbon

an α-hydrogen

In Section 19.1, you will find out why a hydrogen bonded to a carbon adjacent to a carbonyl carbon is more acidic than other hydrogens attached to sp^3 hybridized carbons, and you will look at some reactions that result from this acidity. At the end of this chapter, you will see that a proton is not the only substituent that can be removed from an α-carbon: A carboxyl group bonded to an α-carbon can be removed as CO_2. Finally, you will be introduced to some synthetic schemes that rely on being able to remove protons and carboxyl groups from α-carbons.

19.1 Acidity of α-Hydrogens

Hydrogen and carbon have similar electronegativities, which means that the electrons binding them together are shared almost equally by the two atoms. Consequently, a hydrogen bonded to a carbon is usually not acidic. This is particularly true for hydrogens bonded to sp^3 hybridized carbons, because these carbons are the most similar to hydrogen in electronegativity (Section 6.9). The high pK_a of ethane is evidence of the low acidity of hydrogens bonded to sp^3 hybridized carbons.

CH_3CH_3 — pK_a = 50

A hydrogen bonded to an sp^3 hybridized carbon adjacent to a carbonyl carbon is much more acidic than hydrogens bonded to other sp^3 hybridized carbons. For example, the pK_a for dissociation of an α-hydrogen from an aldehyde or a ketone ranges from 16 to 20, and the pK_a for dissociation of an α-hydrogen from an ester is about 25 (Table 19.1). Notice that, although an α-hydrogen is more acidic than most other carbon-bound hydrogens, it is less acidic than a hydrogen of water (pK_a = 15.7). A compound that contains a relatively acidic hydrogen bonded to an sp^3 hybridized carbon is called a **carbon acid**.

RCH_2CHO (RCH₂C(=O)H), RCH_2COR (RCH₂C(=O)R) — pK_a ~ 16–20

RCH_2CO_2R (RCH₂C(=O)OR) — pK_a ~ 25

Table 19.1 The pK_a Values of Some Carbon Acids

	pK_a		pK_a
$\underset{H}{C}H_2CON(CH_3)_2$ (CH₂C(=O)N(CH₃)₂, H on CH₂)	30	N≡C$\underset{H}{C}$HC≡N	11.8
$\underset{H}{C}H_2COOCH_2CH_3$ (CH₂C(=O)OCH₂CH₃)	25	$CH_3C(=O)\underset{H}{C}HC(=O)OCH_2CH_3$	10.7
$\underset{H}{C}H_2C{\equiv}N$	25	$C_6H_5C(=O)\underset{H}{C}HC(=O)CH_3$	9.4
$\underset{H}{C}H_2COCH_3$ (CH₂C(=O)CH₃)	20	$CH_3C(=O)\underset{H}{C}HC(=O)CH_3$	8.9
$\underset{H}{C}H_2CHO$ (CH₂C(=O)H)	17	$CH_3C(=O)\underset{H}{C}HC(=O)H$	5.9
$CH_3\underset{H}{C}HNO_2$	8.6	$O_2N\underset{H}{C}HNO_2$	3.6

Why is a hydrogen bonded to an sp^3 hybridized carbon that is adjacent to a carbonyl carbon so much more acidic than hydrogens bonded to other sp^3 hybridized carbons? An α-hydrogen is more acidic because the base formed when the proton is removed from the α-carbon is more stable than the base formed when a proton is removed from other sp^3 hybridized carbons, and acid strength is determined by the stability of the conjugate base that is formed when the acid gives up a proton (Section 1.17).

Why is the base more stable? When a proton is removed from ethane, the electrons left behind reside solely on a carbon atom. Because carbon is not very electronegative, a carbanion is relatively unstable and therefore difficult to form. As a result, the pK_a of its conjugate acid is very high.

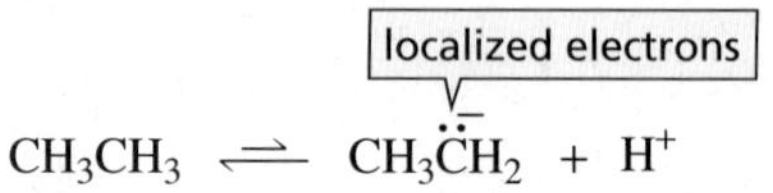

localized electrons

$$CH_3CH_3 \rightleftharpoons CH_3\ddot{C}H_2^- + H^+$$

When a proton is removed from a carbon adjacent to a carbonyl carbon, two factors combine to increase the stability of the base that is formed. First, the electrons left behind when the proton is removed are delocalized, and electron delocalization increases the stability of a compound (Section 7.6). More important, the electrons are delocalized onto an oxygen, an atom that is better able to accommodate the electrons because it is more electronegative than carbon.

electrons are better accomodated on O than on C

$$RCH(H)-C(=O)-R \rightleftharpoons R\ddot{C}H^--C(=O)-R \longleftrightarrow RCH=C(O^-)-R \quad + H^+$$

delocalized electrons

resonance contributors

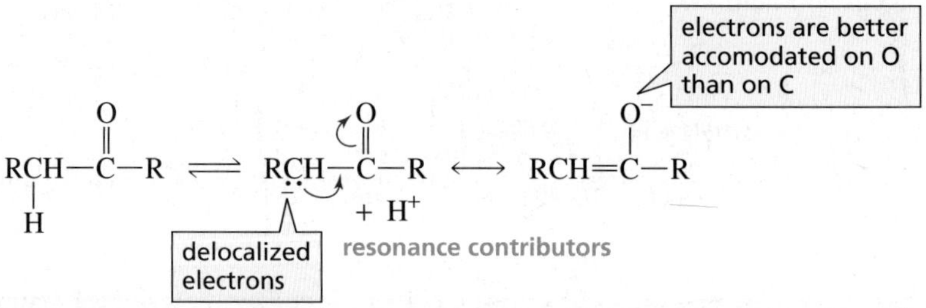

PROBLEM 1

The pK_a of propene is 42, which is greater than the pK_a of the carbon acids listed in Table 19.1, but less than the pK_a of an alkane. Explain.

Now we can understand why aldehydes and ketones ($pK_a = 16–20$) are more acidic than esters ($pK_a = 25$). The electrons left behind when an α-hydrogen is removed from an ester are not as readily delocalized onto the carbonyl oxygen as are the electrons left behind when an α-hydrogen is removed from an aldehyde or a ketone. Because a lone pair on the oxygen of the OR group of the ester can also be delocalized onto the carbonyl oxygen, the two pairs of electrons compete for delocalization onto oxygen.

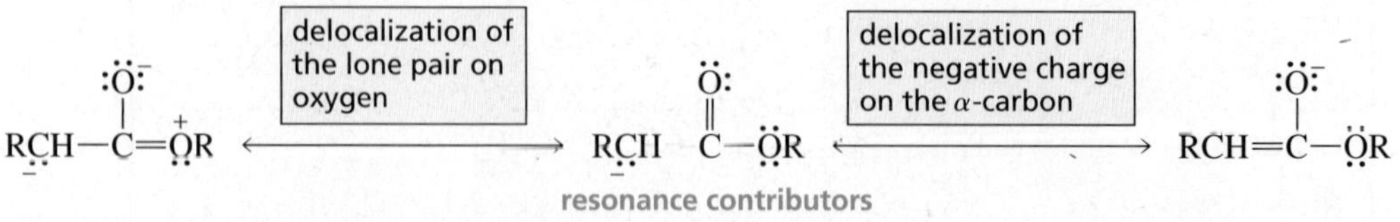

delocalization of the lone pair on oxygen

delocalization of the negative charge on the α-carbon

$$R\ddot{C}H^--C(\ddot{O}:^-)=\overset{+}{\ddot{O}}R \longleftrightarrow R\ddot{C}H^--C(=\ddot{O}:)-\ddot{O}R \longleftrightarrow RCH=C(\ddot{O}:^-)-\ddot{O}R$$

resonance contributors

Nitroalkanes, nitriles, and *N,N*-disubstituted amides also have relatively acidic α-hydrogens (Table 19.1) because in each case the electrons left behind when the proton is removed can be delocalized onto an atom that is more electronegative than carbon.

$CH_3CH_2NO_2$	$CH_3CH_2C{\equiv}N$	$CH_3C(=O)N(CH_3)_2$
nitroethane	propanenitrile	*N,N*-dimethylacetamide
$pK_a = 8.6$	$pK_a = 26$	$pK_a = 30$

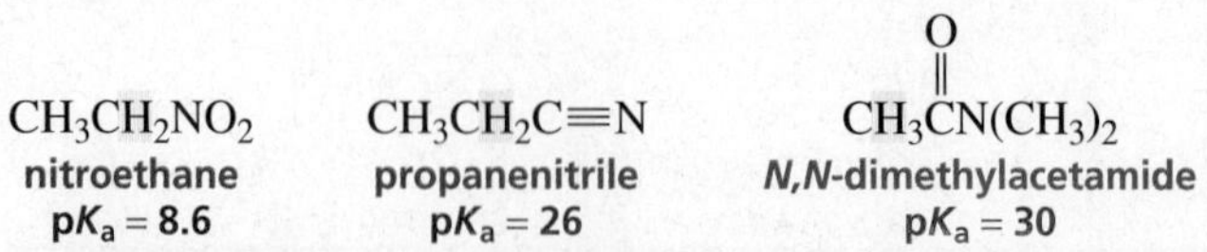

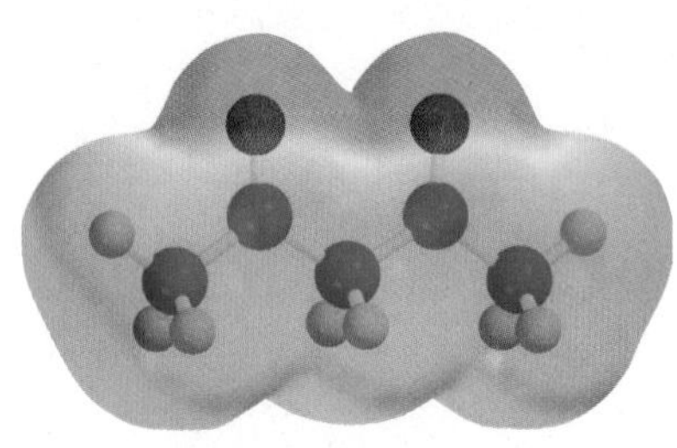

2,4-pentanedione

If the α-carbon is *between* two carbonyl groups, the acidity of an α-hydrogen is even greater (Table 19.1). For example, an α-hydrogen of ethyl 3-oxobutyrate, a compound with an α-carbon between a ketone carbonyl group and an ester carbonyl group, has a pK_a of 10.7. An α-hydrogen of 2,4-pentanedione, a compound with an α-carbon between two ketone carbonyl groups, has a pK_a of 8.9. Ethyl 3-oxobutyrate is classified as a **β-keto ester** because the ester has a carbonyl group at the β-position. 2,4-Pentanedione is a **β-diketone**.

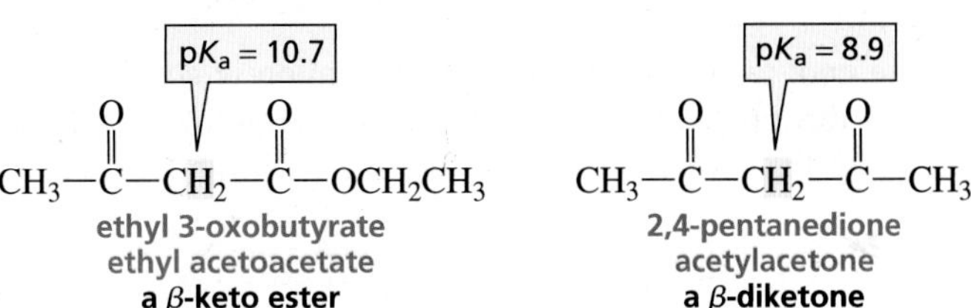

ethyl 3-oxobutyrate
ethyl acetoacetate
a β-keto ester

2,4-pentanedione
acetylacetone
a β-diketone

The acidity of α-hydrogens bonded to carbons flanked by two carbonyl groups increases because the electrons left behind when the proton is removed can be delocalized onto *two* oxygen atoms. β-Diketones have lower pK_a values than β-keto esters because electrons are more readily delocalized onto ketone carbonyl groups than they are onto ester carbonyl groups.

$$CH_3-\overset{\overset{\ddot{O}:}{\|}}{C}-CH=\overset{\overset{:\ddot{O}:^-}{|}}{C}-CH_3 \longleftrightarrow CH_3-\overset{\overset{:\ddot{O}}{\|}}{C}-\underset{\cdot\cdot}{\overset{-}{C}}H-\overset{\overset{:\ddot{O}}{\|}}{C}-CH_3 \longleftrightarrow CH_3-\overset{\overset{:\ddot{O}:^-}{|}}{C}=CH-\overset{\overset{:\ddot{O}}{\|}}{C}-CH_3$$

resonance contributors for the 2,4-pentanedione anion

PROBLEM 2◆

Give an example of

a. a β-keto nitrile

b. a β-diester

PROBLEM 3◆

List the compounds in each of the following groups in order of decreasing acidity:

a. $CH_2{=}CH_2$ $\quad CH_3CH_3 \quad CH_3\overset{O}{\overset{\|}{C}}H \quad HC{\equiv}CH$

b. $CH_3\overset{O}{\overset{\|}{C}}CH_2\overset{O}{\overset{\|}{C}}CH_3 \quad CH_3O\overset{O}{\overset{\|}{C}}CH_2\overset{O}{\overset{\|}{C}}OCH_3 \quad CH_3\overset{O}{\overset{\|}{C}}CH_2\overset{O}{\overset{\|}{C}}OCH_3 \quad CH_3\overset{O}{\overset{\|}{C}}CH_3$

c. N-methyl-2-piperidinone (ring C=O with NCH_3) $\quad$ δ-valerolactone (ring C=O with O) $\quad$ cyclohexanone

19.2 Keto–Enol Tautomerism

A ketone exists in equilibrium with its enol tautomer. You were introduced to tautomers in Section 6.6. Recall that **tautomers** are isomers that are in rapid equilibrium. Keto–enol tautomers differ in the location of a double bond and a hydrogen.

$$RCH_2-\overset{O}{\overset{\|}{C}}-R \rightleftharpoons RCH=\overset{OH}{\overset{|}{C}}-R$$

keto tautomer $\quad$ **enol tautomer**

For most ketones, the enol tautomer is much less stable than the keto tautomer. For example, an aqueous solution of acetone exists as an equilibrium mixture of more than 99.9% keto tautomer and less than 0.1% enol tautomer.

$$CH_3-\overset{\overset{\displaystyle O}{\|}}{C}-CH_3 \rightleftharpoons CH_2=\overset{\overset{\displaystyle OH}{|}}{C}-CH_3$$

> 99.9% **keto tautomer**   < 0.1% **enol tautomer**

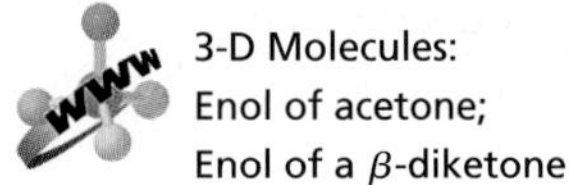
3-D Molecules: Enol of acetone; Enol of a β-diketone

The fraction of the enol tautomer in an aqueous solution is considerably greater for a β-diketone because the enol tautomer is stabilized by intramolecular hydrogen bonding and by conjugation of the carbon–carbon double bond with the second carbonyl group.

a hydrogen bond

85% **keto tautomer**   15% **enol tautomer**

Phenol is unusual in that its enol tautomer is *more* stable than its keto tautomer because the enol tautomer is aromatic, but the keto tautomer is not.

enol tautomer   **keto tautomer**

PROBLEM 4

Only 15% of 2,4-pentanedione exists as the enol tautomer in water, but 92% exists as the enol tautomer in hexane. Explain why this is so.

Now that we know that a hydrogen on a carbon adjacent to a carbonyl carbon is somewhat acidic, we can understand why keto and enol tautomers interconvert as we first saw in Chapter 6. **Keto–enol interconversion** is also called **keto–enol tautomerization** or **enolization**. The interconversion of the tautomers can be catalyzed by either acids or bases.

In a basic solution, hydroxide ion removes a proton from the α-carbon of the keto tautomer. The anion that is formed has two resonance contributors: a carbanion and an enolate ion. The enolate ion contributes more to the resonance hybrid because the negative charge is better accommodated by oxygen than by carbon. Protonation on oxygen forms the enol tautomer, whereas protonation on the α-carbon reforms the keto tautomer.

base-catalyzed keto–enol interconversion

removal of a proton from the α-carbon

protonation of oxygen

$$RCH(H)-\overset{\overset{\displaystyle O}{\|}}{C}-R + HO^- \rightleftharpoons R\overset{-}{C}H-\overset{\overset{\displaystyle O}{\|}}{C}-R \longleftrightarrow RCH=\overset{\overset{\displaystyle O^-}{|}}{C}-R \xrightarrow{H_2O} RCH=\overset{\overset{\displaystyle OH}{|}}{C}-R + HO^-$$

keto tautomer   **enolate ion**   **enol tautomer**

In an acidic solution, the carbonyl oxygen of the keto tautomer is protonated and water removes a proton from the α-carbon, forming the enol.

acid-catalyzed keto–enol interconversion

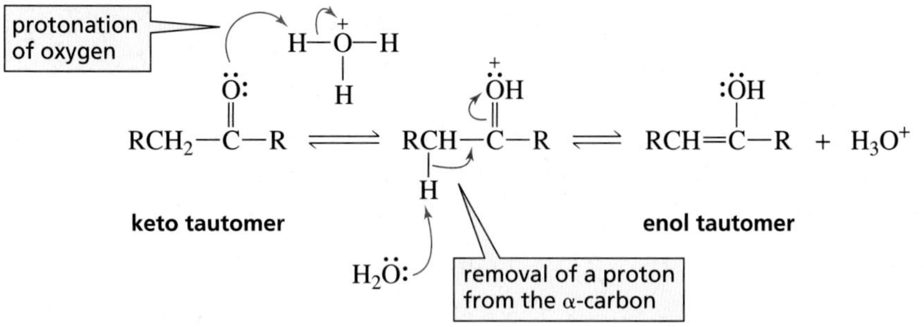

$$RCH_2-\overset{\overset{\large O}{\|}}{C}-R \rightleftharpoons RCH-\overset{\overset{\large {}^{+}OH}{\|}}{C}-R \rightleftharpoons RCH{=}\overset{\overset{\large OH}{|}}{C}-R + H_3O^+$$

protonation of oxygen (H_3O^+); keto tautomer; removal of a proton from the α-carbon (H_2O:); enol tautomer

Notice that the steps are reversed in the base- and acid-catalyzed reactions. In the base-catalyzed reaction, the base removes the α-proton in the first step and the oxygen is protonated in the second step. In the acid-catalyzed reaction, the acid protonates the oxygen in the first step and the α-proton is removed in the second step. Notice also how the catalyst is regenerated in both the acid- and base-catalyzed mechanisms.

PROBLEM 5◆

Draw the enol tautomers for each of the following compounds. For those compounds that have more than one enol tautomer, indicate which is more stable.

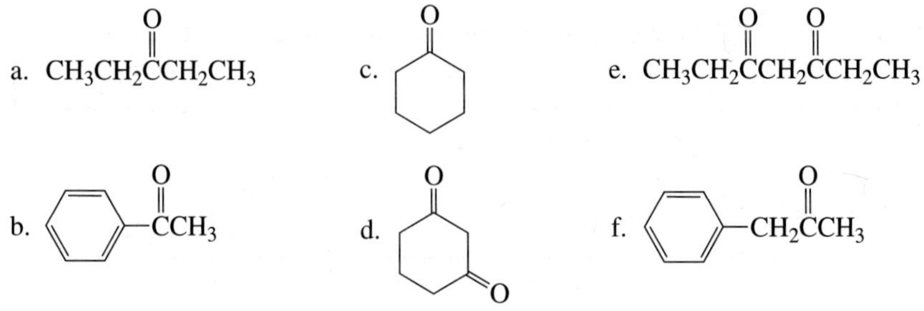

a. $CH_3CH_2\overset{\overset{O}{\|}}{C}CH_2CH_3$

b. $C_6H_5\overset{\overset{O}{\|}}{C}CH_3$

c. cyclohexanone

d. 1,3-cyclohexanedione

e. $CH_3CH_2\overset{\overset{O}{\|}}{C}CH_2\overset{\overset{O}{\|}}{C}CH_2CH_3$

f. $C_6H_5CH_2\overset{\overset{O}{\|}}{C}CH_3$

19.3 How Enols and Enolate Ions React

The carbon–carbon double bond of an enol suggests that it is a nucleophile—like an alkene. An enol is more electron rich than an alkene because the oxygen atom donates electrons by resonance. An enol, therefore, is a better nucleophile than an alkene.

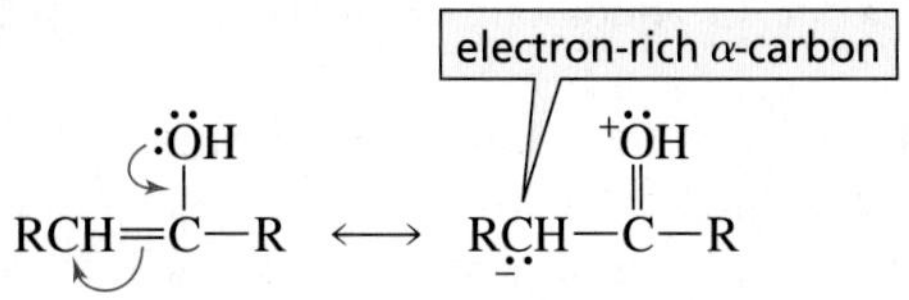

$$RCH{=}\overset{\overset{\large :\ddot{O}H}{|}}{C}-R \longleftrightarrow R\overset{-}{\ddot{C}}H-\overset{\overset{\large {}^{+}\ddot{O}H}{\|}}{C}-R$$

electron-rich α-carbon

resonance contributors of an enol

Carbonyl compounds that form enols undergo substitution reactions at the α-carbon. When an α-substitution reaction takes place under acidic conditions, water removes a proton from the α-carbon of the protonated carbonyl compound. The nucleophilic enol then reacts with an electrophile. The overall reaction is an **α-substitution reaction**—one electrophile (E^+) is substituted for another (H^+).

acid-catalyzed α-substitution reaction

$$RCH_2\text{—}C(=O)\text{—}R \overset{H_3O^+}{\rightleftharpoons} RCH_2\text{—}C(=\overset{+}{O}H)\text{—}R \overset{H_2O}{\rightleftharpoons} RCH{=}C(OH)\text{—}R + H_3O^+ \xrightarrow{E^+} RCH(E)\text{—}C(=\overset{+}{O}H)\text{—}R \overset{H_2O}{\rightleftharpoons} RCH(E)\text{—}C(=O)\text{—}R + H_3O^+$$

enol

α-substituted product

E = Electrophile

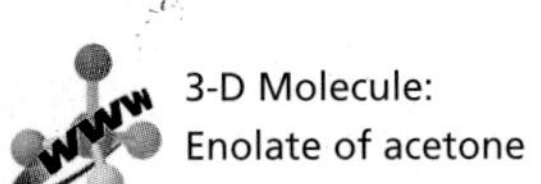
3-D Molecule: Enolate of acetone

When an α-substitution reaction takes place under basic conditions, a base removes a proton from the α-carbon and the nucleophilic enolate ion then reacts with an electrophile. Enolate ions are much better nucleophiles than enols because they are negatively charged.

base-catalyzed α-substitution reaction

$$RCH_2\text{—}C(=O)\text{—}R \overset{HO^-}{\rightleftharpoons} R\overset{-}{C}H\text{—}C(=O)\text{—}R \longleftrightarrow RCH{=}C(O^-)\text{—}R \;+\; H_2O \xrightarrow{E^+} RCH(E)\text{—}C(=O)\text{—}R$$

enolate ion

α-substituted product

The resonance contributors of the enolate ion show that it has two electron-rich sites: the α-carbon and the oxygen. The enolate ion is an example of an *ambident nucleophile* (*ambi* is Latin for "both"; *dent* is Latin for "teeth"). An **ambident nucleophile** is a nucleophile with two nucleophilic sites ("two teeth").

electron-rich oxygen

electron-rich α-carbon

$$RCH{=}C(O^-)\text{—}R \longleftrightarrow R\overset{-}{C}H\text{—}C(=O)\text{—}R$$

resonance contributors of an enolate ion

Which nucleophilic site (C or O) reacts with the electrophile depends on the electrophile and on the reaction conditions. Protonation occurs preferentially on oxygen because of the greater concentration of negative charge on the more electronegative oxygen atom. However, when the electrophile is something other than a proton, carbon is more likely to be the nucleophile because carbon is a better nucleophile than oxygen.

Notice the similarity between keto–enol interconversion and α-substitution. Actually, keto–enol interconversion is an α-substitution reaction in which hydrogen serves as both the electrophile that is removed from the α-carbon and the electrophile that is added to the α-carbon (when the enol or enolate reverts back to the keto tautomer).

As various α-substitution reactions are discussed in this chapter, notice that they all follow basically the same mechanism: A base removes a proton from an α-carbon and the resulting enol or enolate reacts with an electrophile. The reactions differ only in the nature of the base and the electrophile—and in whether the reactions are carried out under acidic or basic conditions.

PROBLEM 6

Explain why the aldehydic hydrogen (the one attached to the carbonyl carbon) is not exchanged with deuterium.

$$CH_3\overset{O}{\overset{\|}{C}}H \xrightleftharpoons[D_2O]{^-OD} CD_3\overset{O}{\overset{\|}{C}}H$$

19.4 Halogenation of the α-Carbon of Aldehydes and Ketones

Acid-Catalyzed Halogenation

When Br_2, Cl_2, or I_2 is added to an *acidic* solution of an aldehyde or a ketone, a halogen replaces *one* of the α-hydrogens of the carbonyl compound.

Under acidic conditions, one α-hydrogen is substituted for a halogen.

$$\text{cyclohexanone} + Cl_2 \xrightarrow{H_3O^+} \text{2-chlorocyclohexanone} + HCl$$

$$C_6H_5\overset{O}{\overset{\|}{C}}CH_2CH_3 + I_2 \xrightarrow{H_3O^+} C_6H_5\overset{O}{\overset{\|}{C}}\underset{I}{\underset{|}{C}}HCH_3 + HI$$

In the first step of this acid-catalyzed reaction, the carbonyl oxygen is protonated. Water is the base that removes a proton from the α-carbon, forming an enol that reacts with an electrophilic halogen.

acid-catalyzed halogenation

$$CH_3-\overset{\ddot{O}:}{\overset{\|}{C}}-C_6H_5 \xrightleftharpoons{H_3O^+} \underset{H}{\underset{|}{C}}H_2-\overset{\overset{+}{\ddot{O}}H}{\overset{\|}{C}}-C_6H_5 \xrightleftharpoons{H_2\ddot{O}:} CH_2=\overset{:\ddot{O}H}{\overset{|}{C}}-C_6H_5 + H_3O^+$$

$$CH_2=\overset{:\ddot{O}H}{\overset{|}{C}}-C_6H_5 + Br-Br \longrightarrow \underset{Br}{\underset{|}{C}}H_2-\overset{+\ddot{O}H}{\overset{\|}{C}}-C_6H_5 + Br^-$$

$$\underset{Br}{\underset{|}{C}}H_2-\overset{+\ddot{O}H}{\overset{\|}{C}}-C_6H_5 \xrightleftharpoons{H_2\ddot{O}:} \underset{Br}{\underset{|}{C}}H_2-\overset{\ddot{O}:}{\overset{\|}{C}}-C_6H_5 + H_3O^+$$

Base-Promoted Halogenation

When excess Br_2, Cl_2, or I_2 is added to a *basic* solution of an aldehyde or a ketone, the halogen replaces *all* the α-hydrogens.

3-D Molecule: α-Bromoacetone

$$R-CH_2-\overset{O}{\overset{\|}{C}}-R + \underset{\text{excess}}{Br_2} \xrightarrow{HO^-} R-\overset{Br}{\underset{Br}{C}}-\overset{O}{\overset{\|}{C}}-R + 2\,Br^-$$

Under basic conditions, all the α-hydrogens are substituted for halogens.

In the first step of this base-promoted reaction, hydroxide ion removes a proton from the α-carbon. The enolate ion then reacts with the electrophilic bromine. These two steps are repeated until all the α-hydrogens are replaced by bromine.

base-promoted halogenation

$$\mathrm{RCH_2-\overset{O}{\overset{\|}{C}}-R} \rightleftharpoons \mathrm{R\overset{-}{C}H-\overset{O}{\overset{\|}{C}}-R} \longleftrightarrow \mathrm{RCH=\overset{O^-}{\overset{|}{C}}-R} \xrightarrow{Br-Br} \mathrm{R\underset{Br}{\underset{|}{C}}H-\overset{O}{\overset{\|}{C}}-R} \xrightarrow{\text{repeat first two steps}} \mathrm{R\overset{Br}{\overset{|}{\underset{Br}{\underset{|}{C}}}}-\overset{O}{\overset{\|}{C}}-R}$$

$\mathrm{HO^-}$ $\quad + \mathrm{H_2O} \quad + \mathrm{Br^-}$

Each successive halogenation is *more rapid* than the previous one because the electron-withdrawing bromine increases the acidity of the remaining α-hydrogens. This is why *all* the α-hydrogens are replaced by bromines. Under acidic conditions, on the other hand, each successive halogenation is *slower* than the previous one because the electron-withdrawing bromine decreases the basicity of the carbonyl oxygen, thereby making protonation of the carbonyl oxygen less favorable.

The Haloform Reaction

In the presence of excess base and excess halogen, a methyl ketone is first converted into a trihalo-substituted ketone. Then hydroxide ion attacks the carbonyl carbon of the trihalo-substituted ketone. Because the trihalomethyl ion is a weaker base than hydroxide ion (the pK_a of CHI_3 is 14; the pK_a of H_2O is 15.7), the trihalomethyl ion is the group more easily expelled from the tetrahedral intermediate, so the final product is a carboxylic acid. The conversion of a methyl ketone to a carboxylic acid is called a **haloform reaction** because one of the products is haloform—$CHCl_3$ (chloroform), $CHBr_3$ (bromoform), or CHI_3 (iodoform). Before spectroscopy became a routine analytical tool, the haloform reaction served as a test for methyl ketones. The presence of a methyl ketone was indicated by the formation of iodoform, a bright yellow compound.

the haloform reaction

$$\mathrm{R-\overset{O}{\overset{\|}{C}}-CH_3} \xrightarrow[\text{I}_2\ \text{excess}]{\text{HO}^-\ \text{excess}} \underset{\text{a trihalo-substituted ketone}}{\mathrm{R-\overset{O}{\overset{\|}{C}}-CI_3}} \xrightarrow{\mathrm{HO^-}} \mathrm{R-\overset{O^-}{\overset{|}{\underset{OH}{\underset{|}{C}}}}-CI_3} \longrightarrow \mathrm{R-\overset{O}{\overset{\|}{C}}-OH} + {}^-\mathrm{CI_3} \rightleftharpoons \mathrm{R-\overset{O}{\overset{\|}{C}}-O^-} + \mathrm{CHI_3}$$

PROBLEM 7

Why do only methyl ketones undergo the haloform reaction?

PROBLEM 8◆

A ketone undergoes acid-catalyzed bromination, acid-catalyzed chlorination, and acid-catalyzed deuterium exchange at the α-carbon, all at about the same rate. What does this tell you about the mechanism of the reactions?

19.5 Halogenation of the α-Carbon of Carboxylic Acids: The Hell–Volhard–Zelinski Reaction

Carboxylic acids do not undergo substitution reactions at the α-carbon because a base will remove a proton from the OH group rather than from the α-carbon, since the OH group is more acidic. If, however, a carboxylic acid is treated with PBr_3 and Br_2, then the α-carbon can be brominated. (Red phosphorus can be used in place of PBr_3, since P and excess Br_2 react to form PBr_3.) This halogenation reaction is called the **Hell–Volhard–Zelinski**

reaction or, more simply, the **HVZ reaction**. You will see when you examine the mechanism of the HVZ reaction that α-substitution occurs because an acyl bromide, rather than a carboxylic acid, is the compound that undergoes α-substitution.

the HVZ reaction

$$\mathrm{RCH_2COH} \xrightarrow[\text{2. } H_2O]{\text{1. } PBr_3 \text{ (or P)}, Br_2} \mathrm{RCHCOH}$$

In the first step of the HVZ reaction, PBr_3 converts the carboxylic acid into an acyl bromide by a mechanism similar to the one by which PBr_3 converts an alcohol into an alkyl bromide (Section 12.3). (Notice that in both reactions PBr_3 replaces an OH with a Br.) The acyl bromide is in equilibrium with its enol. Bromination of the enol forms the α-brominated acyl bromide, which is hydrolyzed to the α-brominated carboxylic acid.

mechanism for the Hell–Volhard–Zelinski reaction

$$\underset{\text{a carboxylic acid}}{\mathrm{CH_3CH_2{-}\overset{O}{\overset{\|}{C}}{-}OH}} \xrightarrow{PBr_3} \underset{\text{an acyl bromide}}{\mathrm{CH_3CH_2{-}\overset{O}{\overset{\|}{C}}{-}Br}} \rightleftharpoons \underset{\text{enol}}{\mathrm{CH_3CH{=}\overset{OH}{\overset{|}{C}}{-}Br}} \quad + \ \mathrm{Br{-}Br}$$

$$\downarrow$$

$$\underset{\text{an }\alpha\text{-brominated carboxylic acid}}{\mathrm{CH_3\underset{Br}{\underset{|}{C}}H{-}\overset{O}{\overset{\|}{C}}{-}OH}} \xleftarrow{H_2O} \underset{\text{an }\alpha\text{-brominated acyl bromide}}{\mathrm{CH_3\underset{Br}{\underset{|}{C}}H{-}\overset{O}{\overset{\|}{C}}{-}Br}} + \mathrm{HBr} \Longleftarrow \mathrm{CH_3\underset{Br}{\underset{|}{C}}H{-}\overset{^+OH}{\overset{\|}{C}}{-}Br} + :\ddot{Br}:^-$$

Carl Magnus von Hell (1849–1926) *was born in Germany. He studied with Hermann von Fehling at the University of Stuttgart and with Richard Erlenmeyer (1825–1909) at the University of Munich. Von Hell reported the HVZ reaction in 1881, and the reaction was independently confirmed by both Volhard and Zelinski in 1887.*

Jacob Volhard (1834–1910) *was also born in Germany. Brilliant, but lacking direction, he was sent by his parents to England to be with August von Hofmann (Section 21.5), a family friend. After working with Hofmann, Volhard became a professor of chemistry, first at the University of Munich, then at the University of Erlangen, and later at the University of Halle. He was the first to synthesize sarcosine and creatine.*

Nikolai Dimitrievich Zelinski (1861–1953) *was born in Moldavia. He was a professor of chemistry at the University of Moscow. In 1911, he left the university to protest the firing of the entire administration by the Ministry of Education. He went to St. Petersburg, where he directed the laboratory of the Ministry of Finances. In 1917, after the Russian Revolution, he returned to the University of Moscow.*

19.6 α-Halogenated Carbonyl Compounds in Synthesis

You have seen that when a base removes a proton from an α-carbon of an aldehyde or a ketone (Section 19.2), the α-carbon becomes *nucleophilic*—it reacts with electrophiles.

cyclohexanone $\xrightleftharpoons{\text{base}}$ enolate anion ⟷ enolate (O⁻) $\xrightarrow{E^+}$ 2-E-cyclohexanone

α-carbon reacts with electrophiles

However, when the α-position is halogenated, the α-carbon becomes *electrophilic*—it reacts with nucleophiles. Therefore, both electrophiles and nucleophiles can be placed on α-carbons.

cyclohexanone $\xrightarrow[Br_2]{H_3O^+}$ 2-bromocyclohexanone $\xrightarrow{\ddot{N}u^-}$ 2-Nu-cyclohexanone + Br^-

α-carbon reacts with electrophiles

Tutorial: α-Halogenated carbonyl compounds in synthesis

α-Brominated carbonyl compounds are also useful to synthetic chemists because once a bromine has been introduced into the α-position of a carbonyl compound, an α,β-unsaturated carbonyl compound can be prepared by means of an E2 elimination reaction, using a strong and bulky base to encourage elimination over substitution (Section 11.8).

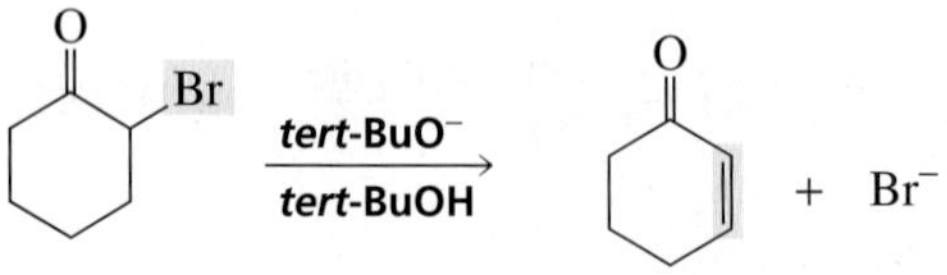

$$\text{2-bromocyclohexanone} \xrightarrow[\textit{tert}\text{-BuOH}]{\textit{tert}\text{-BuO}^-} \text{2-cyclohexenone} + Br^-$$

an α,β-unsaturated carbonyl compound

PROBLEM 9

How would you prepare the following compounds from the given starting materials?

a. $CH_3CH_2CHO \longrightarrow CH_3CH(N(CH_3)_2)CHO$

b. $CH_3CH_2CHO \longrightarrow CH_3CH(OH)CHO$

c. cyclohexanone $\longrightarrow$ 2-methoxycyclohexanone (OCH_3)

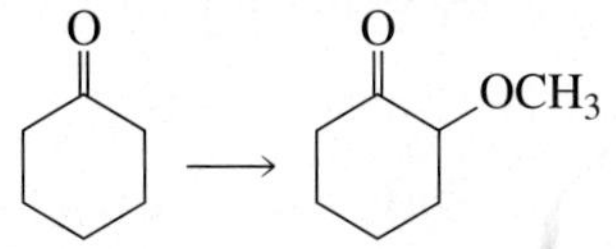

d. 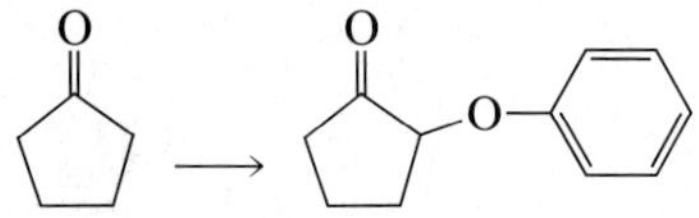 cyclopentanone $\longrightarrow$ 2-phenoxycyclopentanone

PROBLEM 10

How could the following compounds be prepared from a carbonyl compound with no carbon–carbon double bonds?

a. $CH_3CH{=}CHCCH_2CH_2CH_3$ (C=O on the fourth carbon)

b. 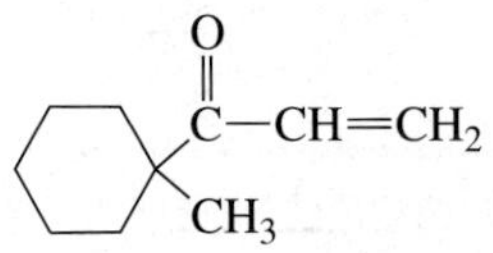 1-methylcyclohexyl vinyl ketone (cyclohexane ring bearing CH_3 and $\overset{O}{\overset{\|}{C}}{-}CH{=}CH_2$)

PROBLEM 11

How could the following compounds be prepared from cyclohexanone?

a. 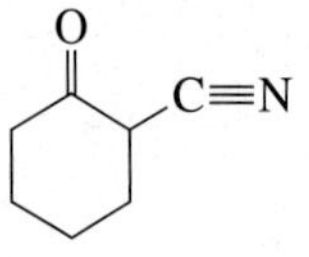 2-cyanocyclohexanone ($C{\equiv}N$)

b. 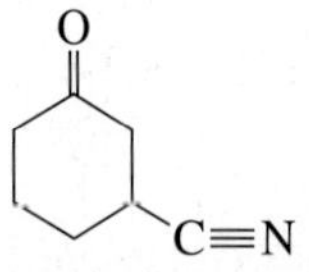 3-cyanocyclohexanone ($C{\equiv}N$)

c. 3-(methylthio)cyclohexanone (SCH_3)

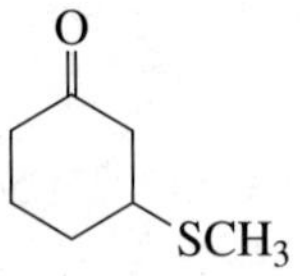

d. 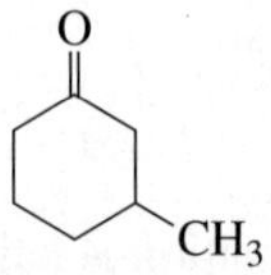 3-methylcyclohexanone (CH_3)

19.7 Using LDA to Form an Enolate

The amount of carbonyl compound converted to enolate depends on the pK_a of the carbonyl compound and the particular base used to remove the α-hydrogen. For example, when hydroxide ion (the pK_a of its conjugate acid is 15.7) is used to remove an α-hydrogen from cyclohexanone (pK_a = 17), only a small amount of the carbonyl compound is converted into the enolate because hydroxide ion is a weaker base than the base being formed. (Recall that the equilibrium of an acid–base reaction favors reaction of the strong acid and formation of the weak acid, Section 1.17.)

cyclohexanone + HO^- ⇌ enolate (carbanion ⟷ O^- form) + H_2O

pK_a = 17 < 0.1% pK_a = 15.7

In contrast, when lithium diisopropylamide (LDA) is used to remove the α-proton (the pK_a of LDA's conjugate acid is about 35), essentially all the carbonyl compound is converted to enolate because LDA is a much stronger base than the base being formed (Section 1.17). Therefore, LDA is the base of choice for those reactions that require the carbonyl compound to be completely converted to enolate before it reacts with an electrophile.

cyclohexanone + LDA ⟶ enolate (carbanion ⟷ O^- form) + DIA

pK_a = 17 ~100% pK_a = 35

Using a nitrogen base to form an enolate can be a problem because a nitrogen base can also react as a nucleophile and attack the carbonyl carbon (Section 18.6). However, the two bulky alkyl substituents bonded to the nitrogen of LDA make it difficult for the nitrogen to get close enough to the carbonyl carbon to react with it. Consequently, LDA is a strong base but a poor nucleophile, so it removes an α-hydrogen much faster than it attacks a carbonyl carbon. LDA is easily prepared by adding butyllithium to diisopropylamine (DIA) in THF at −78 °C.

3-D Molecule: Lithium diisopropylamide (LDA)

$$\underset{\substack{\text{diisopropylamine} \\ pK_a = 35}}{\mathrm{CH_3\overset{CH_3}{C}HNH\overset{CH_3}{C}HCH_3}} + \underset{\text{butyllithium}}{\mathrm{CH_3CH_2CH_2\overset{-}{C}H_2\overset{+}{Li}}} \xrightarrow[-78\,°\mathrm{C}]{\mathrm{THF}} \underset{\substack{\text{lithium diisopropylamide} \\ \text{LDA}}}{\mathrm{CH_3\overset{CH_3}{C}H\overset{-}{N}\overset{CH_3}{C}HCH_3 \; Li^+}} + \underset{\substack{\text{butane} \\ pK_a \sim 50}}{\mathrm{CH_3CH_2CH_2CH_3}}$$

19.8 Alkylation of the α-Carbon of Carbonyl Compounds

Alkylation of the α-carbon of a carbonyl compound is an important reaction because it gives us another way to form a carbon–carbon bond. Alkylation is carried out by first removing a proton from the α-carbon with a strong base such as LDA and then adding the appropriate alkyl halide. Because the alkylation is an S_N2 reaction, it works best with methyl halides and primary alkyl halides (Section 10.2).

cyclopentanone $\xrightarrow{\text{LDA/THF}}$ enolate (carbanion ⟷ :Ö:⁻ form) $\xrightarrow{\mathrm{CH_3CH_2{-}Br}}$ 2-ethylcyclopentanone (CH_2CH_3) + Br^-

Ketones, esters, and nitriles can be alkylated at the α-carbon in this way. Aldehydes, however, give poor yields of α-alkylated products (Section 19.11).

$C_6H_5CH_2COCH_3 \xrightarrow[\text{2. } CH_3I]{\text{1. LDA/THF}} C_6H_5CH(CH_3)COCH_3$

$CH_3CH_2CH_2C{\equiv}N \xrightarrow[\text{2. } CH_3CH_2I]{\text{1. LDA/THF}} CH_3CH_2CH(CH_2CH_3)C{\equiv}N$

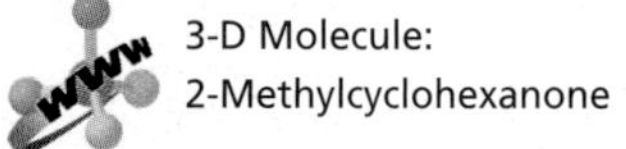

Two different products can be formed when the ketone is not symmetrical, because either α-carbon can be alkylated. For example, methylation of 2-methylcyclohexanone with one equivalent of methyl iodide forms both 2,6-dimethylcyclohexanone and 2,2-dimethylcyclohexanone. The relative amounts of the two products depend on the reaction conditions.

2-methylcyclohexanone → (LDA) two enolates; each reacts with $CH_3{-}I$ → 2,6-dimethylcyclohexanone + 2,2-dimethylcyclohexanone

The enolate leading to 2,6-dimethylcyclohexanone is the *kinetic* enolate because the α-hydrogen that is removed to make this enolate is more accessible (particularly if a hindered base like LDA is used) and slightly more acidic. So 2,6-dimethylcyclohexanone is formed faster and is the major product if the reaction is carried out at −78 °C.

The enolate leading to 2,2-dimethylcyclohexanone is the *thermodynamic* enolate because it has the more substituted double bond, making it the more stable enolate. (Alkyl substitution increases enolate stability for the same reason that alkyl substitution increases alkene stability, Section 4.11.) Therefore, 2,2-dimethylcyclohexanone is the major product if the reaction is carried out under conditions that cause enolate formation to be reversible and if a less hindered base (KH) is used.

The less substituted α-carbon can be alkylated—without having to control the conditions to make certain that the reaction does not become reversible—by first making the *N,N*-dimethylhydrazone of the ketone.

2-methylcyclohexanone $\xrightarrow{(CH_3)_2NNH_2}$ an *N,N*-dimethylhydrazone $\xrightarrow{Bu^-Li^+}$ (Li coordinated to N, Bu⁻ removes α-H) → carbanion $\xrightarrow{CH_3CH_2Br}$ 2-ethyl-6-methyl hydrazone $\xrightarrow[H_2O]{HCl}$ 2-ethyl-6-methylcyclohexanone + $(CH_3)_2\overset{+}{N}HNH_3$...

The *N,N*-dimethylhydrazone will form so that the dimethylamino group is pointing away from the more substituted α-carbon. The nitrogen of the dimethylamino group directs the base to the less substituted carbon by coordinating with the lithium ion of butyllithium (Bu^-Li^+), the base generally employed in this reaction. Hydrolysis of the hydrazone reforms the ketone (Section 18.6).

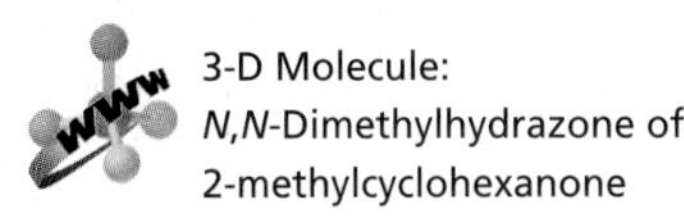

3-D Molecule: *N,N*-Dimethylhydrazone of 2-methylcyclohexanone

THE SYNTHESIS OF ASPIRIN

The first step in the industrial synthesis of aspirin is known as the **Kolbe–Schmitt carboxylation reaction**. The phenolate ion reacts with carbon dioxide under pressure to form *o*-hydroxybenzoic acid, also known as salicylic acid. Acetylation of salicylic acid with acetic acid forms acetylsalicylic acid (aspirin).

During World War I, the American subsidiary of the Bayer Company bought as much phenol as it could from the international market, knowing that eventually all the phenol could be converted into aspirin. This left little phenol available for other countries to purchase for the synthesis of 2,4,6-trinitrophenol (TNT), a common explosive.

[Reaction scheme: phenolate + O=C=O → intermediate → salicylate; HCl → salicylic acid; $(CH_3C)_2O$ → acetylsalicylic acid]

salicylic acid
o-hydroxybenzoic acid

acetylsalicylic acid
aspirin

Hermann Kolbe (1818–1884) *and* **Rudolph Schmitt (1830–1898)** *were born in Germany. Kolbe was a professor at the Universities of Marburg and Leipzig. Schmitt received a Ph.D. from the University of Marburg and was a professor at the University of Dresden. Kolbe discovered how to prepare aspirin in 1859. Schmitt modified the synthesis in 1885, making aspirin available in large quantities at a low price.*

PROBLEM 12◆

What compound is formed when cyclohexanone is shaken with NaOD in D_2O for several hours?

PROBLEM 13

Explain why alkylation of an α-carbon works best if the alkyl halide used in the reaction is a primary alkyl halide and why it does not work at all if it is a tertiary alkyl halide.

PROBLEM-SOLVING STRATEGY

How could you prepare 4-methyl-3-hexanone from a ketone containing no more than six carbon atoms?

$$CH_3CH_2\overset{O}{\overset{\|}{C}}\underset{CH_3}{\underset{|}{C}}HCH_2CH_3$$

4-methyl-3-hexanone

Two sets of ketone and alkyl halide could be used for the synthesis: 3-hexanone and a methyl halide or 3-pentanone and an ethyl halide.

$$\underset{\text{3-hexanone}}{CH_3CH_2\overset{O}{\overset{\|}{C}}CH_2CH_2CH_3} + CH_3Br \quad \text{or} \quad \underset{\text{3-pentanone}}{CH_3CH_2\overset{O}{\overset{\|}{C}}CH_2CH_3} + CH_3CH_2Br$$

3-Pentanone and an ethyl halide are the preferred starting materials. Because 3-pentanone is symmetrical, only one α-substituted ketone will be formed. In contrast, 3-hexanone can form two different enolates, so two α-substituted ketones can be formed.

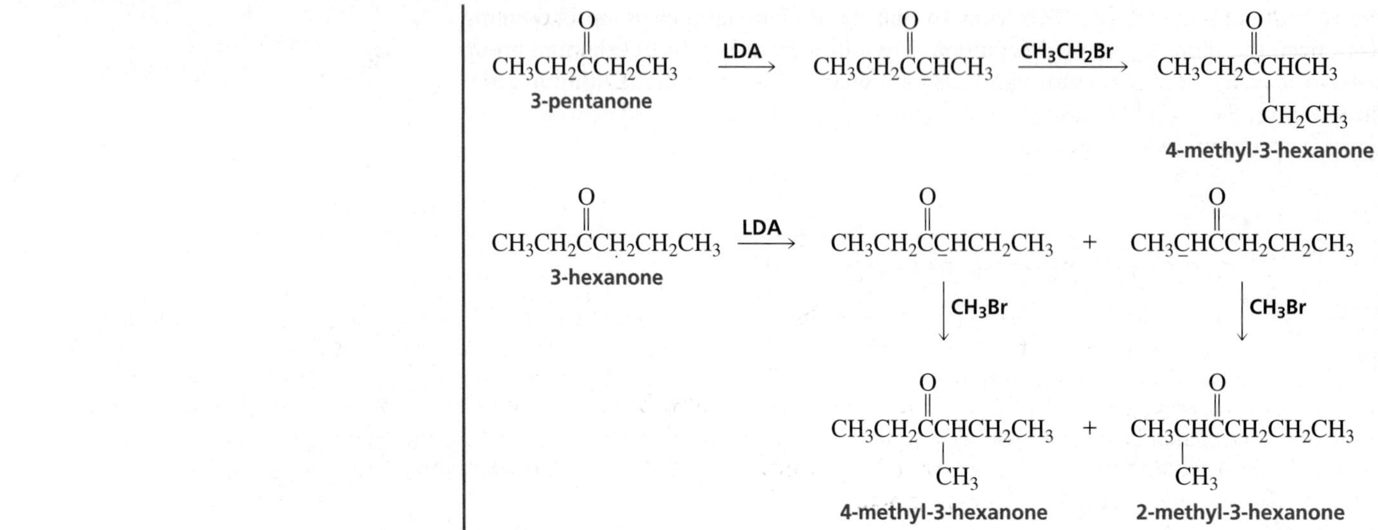

$$CH_3CH_2\overset{O}{\overset{\|}{C}}CH_2CH_3 \xrightarrow{\text{LDA}} CH_3CH_2\overset{O}{\overset{\|}{C}}\bar{C}HCH_3 \xrightarrow{CH_3CH_2Br} CH_3CH_2\overset{O}{\overset{\|}{C}}\underset{CH_2CH_3}{\underset{|}{C}H}CH_3$$

3-pentanone → 4-methyl-3-hexanone

$$CH_3CH_2\overset{O}{\overset{\|}{C}}CH_2CH_2CH_3 \xrightarrow{\text{LDA}} CH_3CH_2\overset{O}{\overset{\|}{C}}\bar{C}HCH_2CH_3 + CH_3\bar{C}H\overset{O}{\overset{\|}{C}}CH_2CH_2CH_3$$

3-hexanone

$\downarrow CH_3Br$ (each)

$$CH_3CH_2\overset{O}{\overset{\|}{C}}\underset{CH_3}{\underset{|}{C}H}CH_2CH_3 + CH_3\underset{CH_3}{\underset{|}{C}H}\overset{O}{\overset{\|}{C}}CH_2CH_2CH_3$$

4-methyl-3-hexanone, 2-methyl-3-hexanone

Now continue on to Problem 14.

PROBLEM 14

How could each of the following compounds be prepared from a ketone and an alkyl halide?

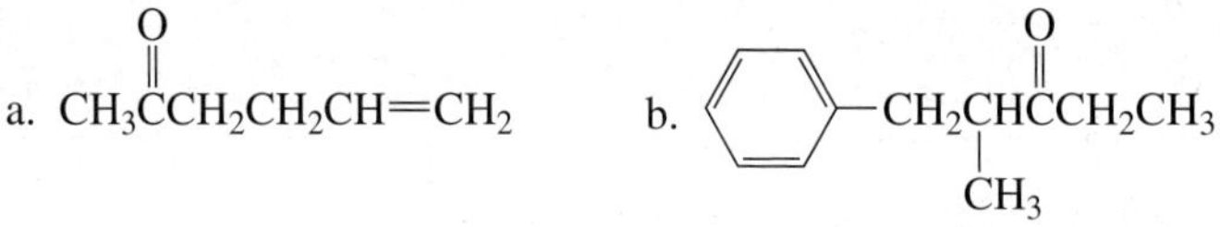

a. $CH_3\overset{O}{\overset{\|}{C}}CH_2CH_2CH{=}CH_2$

b. $C_6H_5{-}CH_2\underset{CH_3}{\underset{|}{C}H}\overset{O}{\overset{\|}{C}}CH_2CH_3$

19.9 Alkylation and Acylation of the α-Carbon via an Enamine Intermediate

We have seen that an enamine is formed when an aldehyde or a ketone reacts with a secondary amine (Section 18.6).

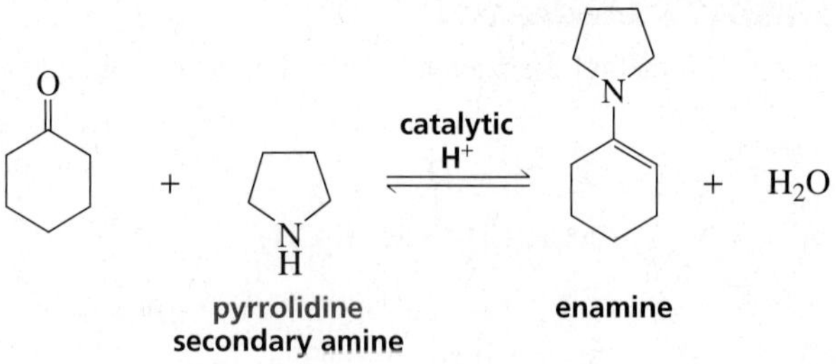

cyclohexanone + pyrrolidine (secondary amine) $\underset{\text{catalytic } H^+}{\rightleftharpoons}$ enamine + H_2O

Enamines react with electrophiles in the same way that enolates do.

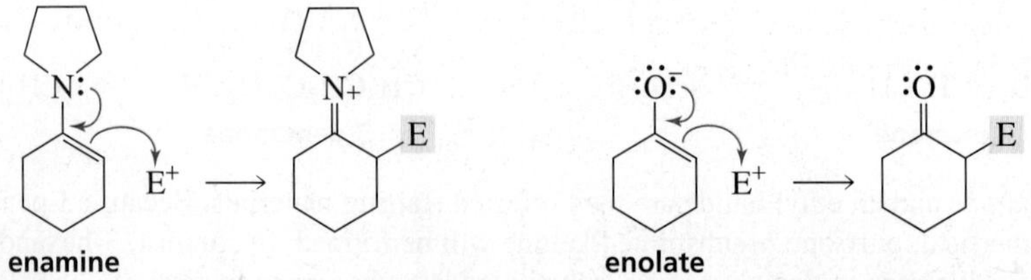

enamine + E^+ → iminium ion (α-E)

enolate + E^+ → ketone (α-E)

This means that electrophiles can be added to the α-carbon of an aldehyde or a ketone by first converting the carbonyl compound to an enamine (by treating the carbonyl compound with a secondary amine), adding the electrophile, and then hydrolyzing the imine back to the ketone.

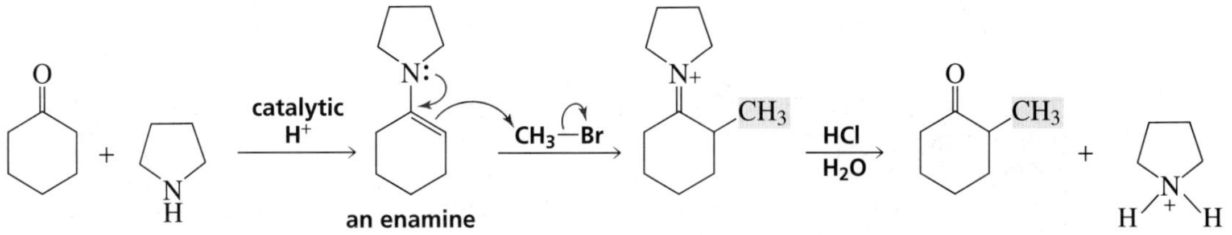

catalytic H^+ — an enamine — CH_3—Br — HCl, H_2O

Because the alkylation step is an S_N2 reaction, only primary alkyl halides or methyl halides should be used (Section 10.2).

3-D Molecule: Pyrrolidine enamine of cyclohexanone

One advantage to using an enamine intermediate to alkylate an aldehyde or a ketone is that only the monoalkylated product is formed.

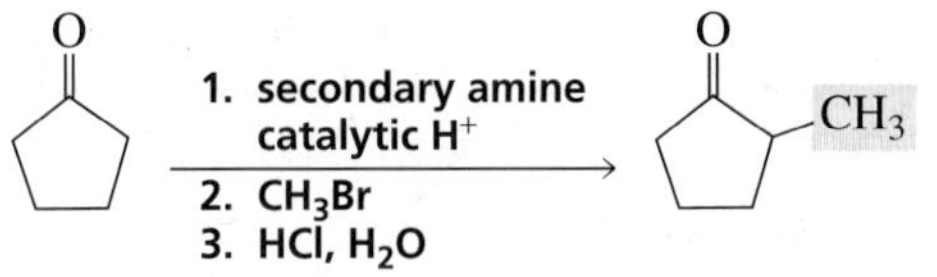

1. secondary amine catalytic H^+
2. CH_3Br
3. HCl, H_2O

In contrast, when a carbonyl compound is alkylated directly, dialkylated and O-alkylated products can also be formed.

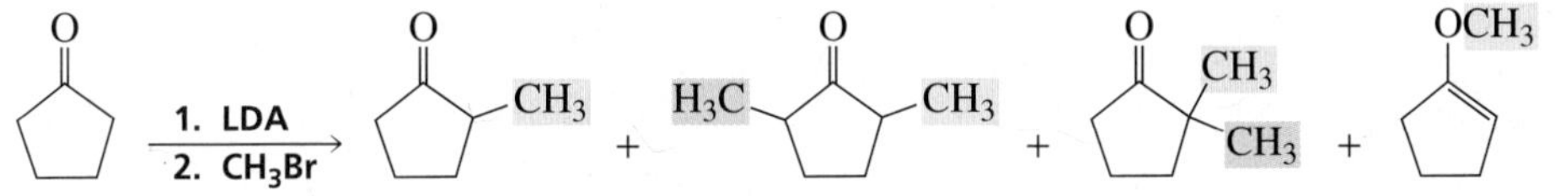

1. LDA
2. CH_3Br

In addition to being able to be alkylated, aldehydes and ketones can also be acylated via an enamine intermediate.

catalytic H^+ — CH_3CCl — HCl, H_2O

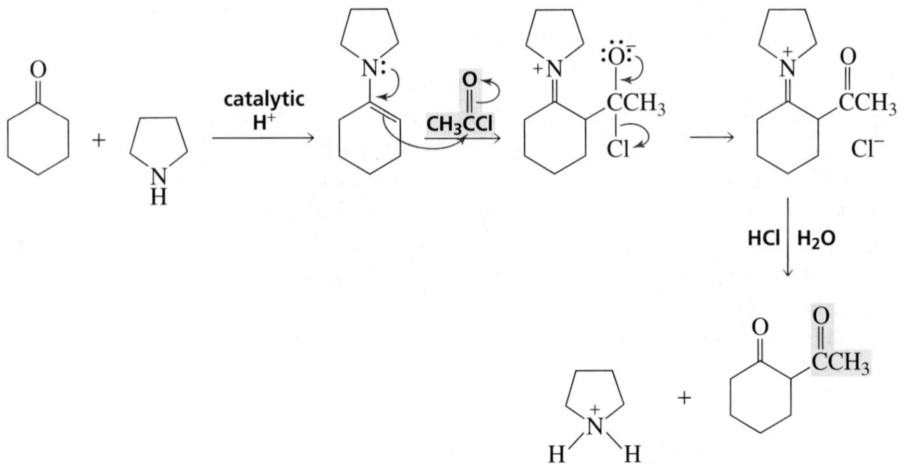

Notice that the α-carbon of an aldehyde or a ketone can be made to react with an *electrophile* either by first treating the carbonyl compound with LDA or by converting the carbonyl compound to an enamine.

LDA — E^+

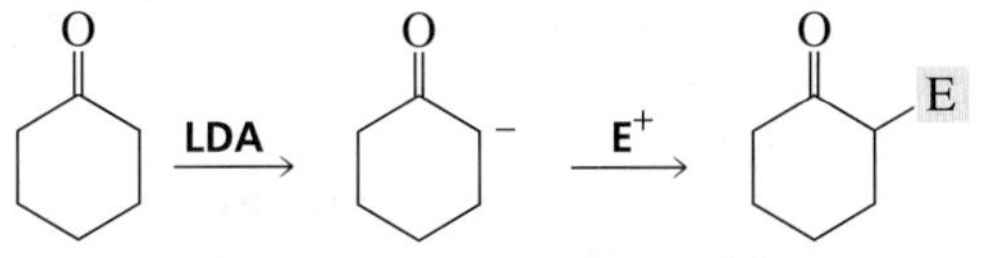

Alternatively, the α-carbon of an aldehyde or a ketone can be made to react with a *nucleophile* by first brominating the α-position of the carbonyl compound.

cyclohexanone $\xrightarrow[H_3O^+]{Br_2}$ 2-bromocyclohexanone $\xrightarrow{^-Nu}$ 2-Nu-cyclohexanone

PROBLEM 15

Describe how the following compounds could be prepared using an enamine intermediate:

a. 2-($CH_2CH_2CH_3$)cyclohexanone   b. 2-($\overset{O}{\|}{C}CH_2CH_3$)cyclohexanone

19.10 Alkylation of the β-Carbon: The Michael Reaction

In Section 18.13, you saw that nucleophiles react with α,β-unsaturated aldehydes and ketones, forming either direct addition products or conjugate addition products; in Section 18.14, you saw that nucleophiles react with α,β-unsaturated carboxylic acid derivatives, forming either nucleophilic acyl substitution products or conjugate addition products.

$$RCH{=}CH\overset{O}{\overset{\|}{C}}R \xrightarrow[2.\ H_3O^+]{1.\ ^-Nu} RCH{=}CH\underset{Nu}{\overset{OH}{C}}R + R\underset{Nu}{CH}CH_2\overset{O}{\overset{\|}{C}}R$$

direct addition   conjugate addition

Michael reactions form 1,5-dicarbonyl compounds.

When the nucleophile is an enolate, the addition reaction has a special name—it is called a **Michael reaction**. The enolates that work best in Michael reactions are those that are flanked by two electron-withdrawing groups: enolates of β-diketones, β-diesters, β-keto esters, and β-keto nitriles. Because these enolates are relatively weak bases, addition occurs at the β-carbon of α,β-unsaturated aldehydes and ketones. The enolates also add to the β-carbon of α,β-unsaturated esters and amides because of the low reactivity of the carbonyl group. Notice that Michael reactions form 1,5-dicarbonyl compounds—if one carbonyl carbon is given the 1-position, the other carbonyl carbon is at the 5-position.

Arthur Michael (1853–1942) *was born in Buffalo, New York. He studied at the University of Heidelberg, the University of Berlin, and L'École de Médicine in Paris. He was a professor of chemistry at Tufts and Harvard Universities, retiring from Harvard when he was 83.*

$$\underset{\text{an }\alpha,\beta\text{-unsaturated aldehyde}}{CH_2{=}CH\overset{O}{\overset{\|}{C}}H} + \underset{\text{a }\beta\text{-diketone}}{CH_3\overset{O}{\overset{\|}{C}}CH_2\overset{O}{\overset{\|}{C}}CH_3} \xrightarrow{HO^-} \underset{CH_3\underset{O}{\underset{\|}{C}}CH\underset{O}{\underset{\|}{C}}CH_3}{CH_2CH_2\overset{O}{\overset{\|}{C}}H}$$

$$\underset{\text{an }\alpha,\beta\text{-unsaturated ketone}}{CH_3CH{=}CH\overset{O}{\overset{\|}{C}}CH_3} + \underset{\text{a }\beta\text{-diester}}{CH_3CH_2O\overset{O}{\overset{\|}{C}}CH_2\overset{O}{\overset{\|}{C}}OCH_2CH_3} \xrightarrow{CH_3CH_2O^-} CH_3\underset{CH(\underset{O}{\underset{\|}{C}}OCH_2CH_3)_2}{CH}CH_2\overset{O}{\overset{\|}{C}}CH_3$$

$$CH_3CH{=}CH\overset{O}{\overset{\|}{C}}NH_2 + CH_3CH_2\overset{O}{\overset{\|}{C}}CH_2\overset{O}{\overset{\|}{C}}OCH_3 \xrightarrow{CH_3O^-} CH_3CHCH_2\overset{O}{\overset{\|}{C}}NH_2 \text{ (with } CH_3CH_2\underset{O}{\underset{\|}{C}}\underset{O}{\underset{\|}{CHC}}OCH_3 \text{ on C3)}$$

an α,β-unsaturated amide; a β-keto ester

$$CH_3CH_2CH{=}CH\overset{O}{\overset{\|}{C}}OCH_3 + CH_3\overset{O}{\overset{\|}{C}}CH_2C{\equiv}N \xrightarrow{CH_3O^-} CH_3CH_2CHCH_2\overset{O}{\overset{\|}{C}}OCH_3 \text{ (with } CH_3\underset{O}{\underset{\|}{C}}CHC{\equiv}N \text{ on C3)}$$

an α,β-unsaturated ester; a β-keto nitrile

All these reactions take place by the same mechanism: A base removes a proton from the α-carbon of the carbon acid, the enolate adds to the β-carbon of an α,β-unsaturated carbonyl compound, and the α-carbon obtains a proton from the solvent.

mechanism of the Michael reaction

$$R\overset{O}{\overset{\|}{C}}CH_2\overset{O}{\overset{\|}{C}}R \underset{}{\overset{HO^-}{\rightleftharpoons}} R\overset{O}{\overset{\|}{C}}\ddot{C}H\overset{O}{\overset{\|}{C}}R \xrightarrow{RCH{=}CH{-}\overset{O}{\overset{\|}{C}}R} RCH{-}CH{=}\overset{:\ddot{O}:^-}{C}R \text{ (RCCHCR)} \xrightarrow{H_2O} RCH{-}CH_2{-}\overset{:\ddot{O}}{\overset{\|}{C}}R \text{ (RCCHCR)} + HO^-$$

removal of a proton from the α-carbon

addition of the enolate to the β-carbon

protonation of the α-carbon

Notice that if either of the reactants in a Michael reaction has an ester group, the base used to remove the α-proton is the same as the leaving group of the ester. This is done because the base, in addition to being able to remove an α-proton, can react as a nucleophile and attack the carbonyl group of the ester. If the nucleophile is identical to the OR group of the ester, nucleophilic attack on the carbonyl group will not change the reactant.

$$CH_3{-}\overset{\ddot{O}:}{\overset{\|}{C}}{-}\ddot{O}CH_3 + CH_3\ddot{O}:^- \rightleftharpoons CH_3{-}\overset{:\ddot{O}:^-}{\underset{:\ddot{O}CH_3}{C}}{-}\ddot{O}CH_3$$

Gilbert Stork *was born in Belgium in 1921. He graduated from the University of Florida and received a Ph.D. from the University of Wisconsin. He was a professor of chemistry at Harvard University and has been a professor at Columbia University since 1953. He is responsible for the development of many new synthetic procedures in addition to the one bearing his name.*

Enamines can be used in place of enolates in Michael reactions. When an enamine is used as a nucleophile in a Michael reaction, the reaction is called a **Stork enamine reaction**.

mechanism of the Stork enamine reaction

$$\text{enamine} \xrightarrow{CH_2{=}CH{-}\overset{O}{\overset{\|}{C}}H} \text{iminium}{-}CH_2{-}CH{=}CH{-}\ddot{O}:^- \xrightarrow{H_2O} \text{iminium}{-}CH_2CH_2\overset{O}{\overset{\|}{C}}H + HO^-$$

$$\xrightarrow{HCl,\ H_2O} \text{2-oxocyclohexyl}{-}CH_2CH_2\overset{O}{\overset{\|}{C}}H + \text{pyrrolidinium } (H_2N^+)$$

PROBLEM 16◆

What reagents would you use to prepare the following compounds?

a. (2-oxocyclohexyl) substituted $CH(COCH_3)_2$: cyclohexanone bearing $CH_3C(=O)$—CH—$CCH_3(=O)$ at the α-carbon

b. $CH_3\overset{O}{\overset{\|}{C}}CH_2CH_2CH(\overset{O}{\overset{\|}{C}}OCH_2CH_3)_2$

19.11 The Aldol Addition

In Chapter 18, we saw that aldehydes and ketones are electrophiles and therefore react with nucleophiles. In the preceding sections of this chapter, we have seen that when a proton is removed from the α-carbon of an aldehyde or a ketone, the resulting anion is a nucleophile and therefore reacts with electrophiles. An **aldol addition** is a reaction in which *both* of these activities are observed: One molecule of a carbonyl compound—after a proton is removed from an α-carbon—reacts as a *nucleophile* and attacks the *electrophilic* carbonyl carbon of a second molecule of the carbonyl compound.

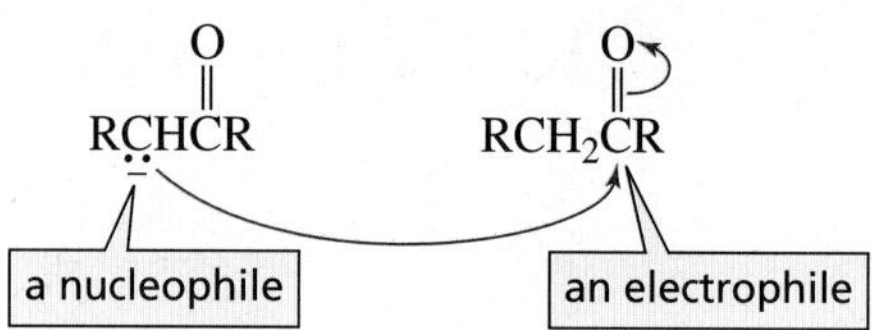

3-D Molecules: β-Hydroxyaldehyde; β-Hydroxyketone

An aldol addition is a reaction between two molecules of an *aldehyde* or two molecules of a *ketone*. When the reactant is an aldehyde, the addition product is a β-hydroxyaldehyde, which is why the reaction is called an aldol addition ("ald" for aldehyde, "ol" for alcohol). When the reactant is a ketone, the addition product is a β-hydroxyketone. Because the addition reaction is reversible, good yields of the addition product are obtained only if it is removed from the solution as it is formed.

aldol additions

$$2\ CH_3CH_2\overset{O}{\overset{\|}{C}}H \underset{}{\overset{HO^-}{\rightleftharpoons}} CH_3CH_2\overset{OH}{\overset{|}{C}}H-\overset{O}{\overset{\|}{C}}HCH\ (\text{with } CH_3 \text{ on the } \alpha\text{-carbon})$$

a β-hydroxyaldehyde

$$2\ CH_3\overset{O}{\overset{\|}{C}}CH_3 \overset{HO^-}{\rightleftharpoons} CH_3\overset{OH}{\overset{|}{C}}(CH_3)-CH_2\overset{O}{\overset{\|}{C}}CH_3$$

a β-hydroxyketone

The new C—C bond formed in an aldol addition is between the α-carbon of one molecule and the carbon that formerly was the carbonyl carbon of the other molecule.

In the first step of an aldol addition, a base removes an α-proton from the carbonyl compound, creating an enolate. The enolate adds to the carbonyl carbon of a second molecule of the carbonyl compound, and the resulting negatively charged oxygen is protonated by the solvent.

mechanism for the aldol addition

$$CH_3CH_2\overset{O}{\overset{\|}{C}}H \overset{HO^-}{\rightleftharpoons} CH_3\ddot{\underset{-}{C}}H\overset{O}{\overset{\|}{C}}H \overset{CH_3CH_2\overset{O}{\overset{\|}{C}}H}{\rightleftharpoons} CH_3CH_2\overset{O^-}{\overset{|}{C}}H-\underset{CH_3}{\underset{|}{C}}H\overset{O}{\overset{\|}{C}}H \underset{HO^-}{\overset{H_2O}{\rightleftharpoons}} CH_3CH_2\overset{OH}{\overset{|}{C}}H-\underset{CH_3}{\underset{|}{C}}H\overset{O}{\overset{\|}{C}}H$$

a β-hydroxyaldehyde

Ketones are less susceptible than aldehydes to attack by nucleophiles (Section 18.2), so aldol additions occur more slowly with ketones. The relatively high reactivity of aldehydes in competing aldol addition reactions is what causes them to give low yields of α-alkylation products (Section 19.8).

$$CH_3\overset{O}{\overset{\|}{C}}CH_3 \underset{}{\overset{HO^-}{\rightleftharpoons}} \overset{-}{C}H_2\overset{O}{\overset{\|}{C}}CH_3 \overset{CH_3\overset{O}{\overset{\|}{C}}CH_3}{\rightleftharpoons} CH_3\underset{CH_3}{\underset{|}{\overset{O^-}{\overset{|}{C}}}}-CH_2\overset{O}{\overset{\|}{C}}CH_3 \underset{HO^-}{\overset{H_2O}{\rightleftharpoons}} CH_3\underset{CH_3}{\underset{|}{\overset{OH}{\overset{|}{C}}}}-CH_2\overset{O}{\overset{\|}{C}}CH_3$$

a β-hydroxyketone

Because an aldol addition reaction occurs between two molecules of the same carbonyl compound, the product has twice as many carbons as the reacting aldehyde or ketone.

PROBLEM 17

Show the aldol addition product that would be formed from each of the following compounds:

a. $CH_3CH_2CH_2CH_2\overset{O}{\overset{\|}{C}}H$

b. $CH_3\underset{CH_3}{\underset{|}{C}}HCH_2CH_2\overset{O}{\overset{\|}{C}}H$

c. $CH_3CH_2\overset{O}{\overset{\|}{C}}CH_2CH_3$

d. cyclohexanone

PROBLEM 18◆

For each of the following compounds, indicate the aldehyde or ketone from which it would be formed by an aldol addition:

a. 2-ethyl-3-hydroxyhexanal
b. 4-hydroxy-4-methyl-2-pentanone
c. 2,4-dicyclohexyl-3-hydroxybutanal
d. 5-ethyl-5-hydroxy-4-methyl-3-heptanone

PROBLEM 19

An aldol addition can be catalyzed by acids as well as by bases. Propose a mechanism for the acid-catalyzed aldol addition of propanal.

19.12 Dehydration of Aldol Addition Products: Formation of α,β-Unsaturated Aldehydes and Ketones

You have seen that alcohols are dehydrated when they are heated with acid (Section 12.5). The β-hydroxyaldehyde and β-hydroxyketone products of aldol addition reactions are easier to dehydrate than many other alcohols because the double bond formed as the result of dehydration is conjugated with a carbonyl group. Conjugation increases the stability of the product (Section 8.3) and, therefore, makes it easier to form. If the product of an aldol addition is dehydrated, the overall reaction is called an **aldol condensation**. A **condensation reaction** is a reaction that combines two molecules while removing a small molecule (usually water or an alcohol).

An aldol addition product loses water to form an aldol condensation product.

$$2\ CH_3CH_2\overset{O}{\overset{\|}{C}}H \overset{HO^-}{\rightleftharpoons} CH_3CH_2\overset{OH}{\overset{|}{C}}H-\underset{CH_3}{\underset{|}{C}}H\overset{O}{\overset{\|}{C}}H \xrightarrow[\Delta]{H_3O^+} CH_3CH_2CH=\underset{CH_3}{\underset{|}{C}}\overset{O}{\overset{\|}{C}}H + H_2O$$

a β-hydroxyaldehyde — an α,β-unsaturated aldehyde

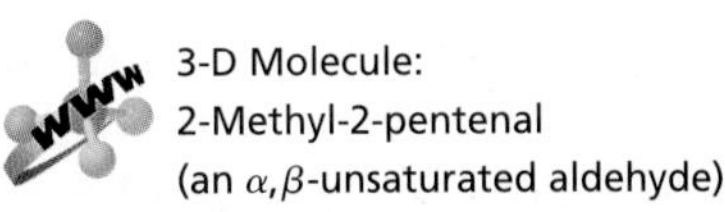

3-D Molecule: 2-Methyl-2-pentenal (an α,β-unsaturated aldehyde)

β-Hydroxyaldehydes and β-hydroxyketones can also be dehydrated under basic conditions, so heating the aldol addition product in either acid or base leads to dehydration. The product of dehydration is called an enone—"ene" for the double bond and "one" for the carbonyl group.

$$2\,CH_3\overset{O}{\overset{\|}{C}}CH_3 \overset{HO^-}{\rightleftharpoons} CH_3\underset{CH_3}{\underset{|}{\overset{OH}{\overset{|}{C}}}}{-}CH_2\overset{O}{\overset{\|}{C}}CH_3 \xrightarrow[\Delta]{HO^-} CH_3\underset{CH_3}{\underset{|}{C}}{=}CH\overset{O}{\overset{\|}{C}}CH_3 + H_2O$$

a β-hydroxyketone; an α,β-unsaturated ketone, an enone

Dehydration sometimes occurs under the conditions in which the aldol addition is carried out, without additional heating. In such cases, the β-hydroxycarbonyl compound is an intermediate and the enone is the final product of the reaction. For example, the β-hydroxyketone formed from the aldol addition of acetophenone loses water as soon as it is formed, because the double bond formed by loss of water is conjugated not only with the carbonyl group, but also with the benzene ring. Conjugation stabilizes the dehydrated product and therefore makes it relatively easy to form.

$$2\,C_6H_5\overset{O}{\overset{\|}{C}}{-}CH_3 \overset{HO^-}{\rightleftharpoons} \left[C_6H_5\underset{CH_3}{\underset{|}{\overset{OH}{\overset{|}{C}}}}{-}CH_2{-}\overset{O}{\overset{\|}{C}}{-}C_6H_5\right] \longrightarrow C_6H_5{-}\underset{CH_3}{\underset{|}{C}}{=}CH{-}\overset{O}{\overset{\|}{C}}{-}C_6H_5 + H_2O$$

PROBLEM 20 SOLVED

How could you prepare the following compounds using a starting material containing no more than three carbon atoms?

a. $CH_3CH_2CH_2\overset{O}{\overset{\|}{C}}OH$

b. $CH_3CH_2\underset{Br}{\underset{|}{C}}H\underset{CH_3}{\underset{|}{C}}H\overset{O}{\overset{\|}{C}}H$

c. $CH_3CH_2CH_2\overset{O}{\overset{\|}{C}}CH_3$

SOLUTION TO 20a A compound with the correct four-carbon skeleton can be obtained if a two-carbon aldehyde undergoes an aldol addition. Dehydration of the addition product forms an α,β-unsaturated aldehyde. Catalytic hydrogenation forms an aldehyde. Some of the α,β-unsaturated aldehyde might be reduced to an alcohol, but that's all right because both the aldehyde and the alcohol can be oxidized by an acidic solution of CrO_3 to the target compound (Section 20.2).

$$CH_3\overset{O}{\overset{\|}{C}}H \xrightarrow{HO^-} CH_3\overset{OH}{\overset{|}{C}}HCH_2\overset{O}{\overset{\|}{C}}H \xrightarrow[\Delta]{H_3O^+\text{ or }HO^-} CH_3CH{=}CH\overset{O}{\overset{\|}{C}}H$$

$$\downarrow H_2/Pt$$

$$CH_3CH_2CH_2\overset{O}{\overset{\|}{C}}OH \xleftarrow[H_2SO_4]{CrO_3} \underset{\text{major}}{CH_3CH_2CH_2\overset{O}{\overset{\|}{C}}H} + \underset{\text{minor}}{CH_3CH_2CH_2CH_2OH}$$

19.13 The Mixed Aldol Addition

If two different carbonyl compounds are used in an aldol addition, four products can be formed because each enolate can react both with another molecule of the carbonyl compound from which the enolate was formed and with the other carbonyl compound. In the following example, both carbonyl compound A and carbonyl compound B can lose a proton from an α-carbon to form enolates A^- and B^-; A^- can react with either A or B, and B^- can react with either A or B:

$$CH_3CH_2CHO\ (A) + CH_3CH(CH_3)CH_2CHO\ (B) \underset{H_2O}{\overset{HO^-}{\rightleftharpoons}} CH_3\overset{-}{C}HCHO\ (A^-)$$

$A^- + A \rightarrow CH_3CH_2CH(O^-)-CH(CH_3)CHO \underset{HO^-}{\overset{H_2O}{\rightleftharpoons}} CH_3CH_2CH(OH)-CH(CH_3)CHO$

$A^- + B \rightarrow CH_3CH(CH_3)CH_2CH(O^-)-CH(CH_3)CHO \underset{HO^-}{\overset{H_2O}{\rightleftharpoons}} CH_3CH(CH_3)CH_2CH(OH)-CH(CH_3)CHO$

$$A + B \underset{H_2O}{\overset{HO^-}{\rightleftharpoons}} CH_3CH(CH_3)\overset{-}{C}HCHO\ (B^-)$$

$B^- + A \rightarrow CH_3CH_2CH(O^-)-CH(CH(CH_3)CH_3)CHO \underset{HO^-}{\overset{H_2O}{\rightleftharpoons}} CH_3CH_2CH(OH)-CH(CH(CH_3)CH_3)CHO$

$B^- + B \rightarrow CH_3CH(CH_3)CH_2CH(O^-)-CH(CH(CH_3)CH_3)CHO \underset{HO^-}{\overset{H_2O}{\rightleftharpoons}} CH_3CH(CH_3)CH_2CH(OH)-CH(CH(CH_3)CH_3)CHO$

The preceding reaction is called a **mixed aldol addition** or a **crossed aldol addition**. The four products have similar physical properties, making them difficult to separate. Consequently, a mixed aldol addition that forms four products is not a synthetically useful reaction.

Under certain conditions, a mixed aldol addition can lead primarily to one product. If one of the carbonyl compounds does not have any α-hydrogens, it cannot form an enolate. This reduces the number of possible products from four to two. A greater amount of one of the two products will be formed if the compound *without* α-hydrogens is always present in excess, because the enolate will be more likely to react with it, rather than with another molecule of the carbonyl compound from which the enolate was formed, if there is more of the compound without α-hydrogens in solution. Therefore, the compound with α-hydrogens should be added slowly to a basic solution of the compound without α-hydrogens.

Tutorial: Aldol reactions—synthesis

$$C_6H_5CHO\ (\text{excess}) + CH_3CH_2CH_2CHO\ (\text{add slowly}) \overset{HO^-}{\rightleftharpoons} C_6H_5CH(OH)-CH(CH_2CH_3)CHO \longrightarrow C_6H_5CH{=}C(CH_2CH_3)CHO + H_2O$$

If both carbonyl compounds have α-hydrogens, primarily one aldol product can be obtained if LDA is used to remove the α-hydrogen that creates the enolate. Because LDA is a strong base (Section 19.7), all the carbonyl compound will be converted into an enolate, so there will be no carbonyl compound left with which the enolate could react in an aldol addition. An aldol addition will not be able to occur until the second carbonyl compound is added to the reaction mixture. If the second carbonyl compound is added slowly, the chance that it will form an enolate and then react with another molecule of its parent carbonyl compound will be minimized.

$$\text{cyclohexanone} \xrightarrow{LDA} \text{enolate} \xrightarrow[2.\ H_2O]{1.\ CH_3CH_2CH_2CHO\ \text{add slowly}} \text{2-}(CH(OH)CH_2CH_2CH_3)\text{cyclohexanone}$$

PROBLEM 21

Give the products obtained from mixed aldol additions of the following compounds:

a. $CH_3CH_2CH_2CH(=O)$ + $CH_3CH_2CH_2CH_2CH(=O)$

b. $CH_3C(=O)CH_3$ + $CH_3CH_2C(=O)CH_2CH_3$

c. 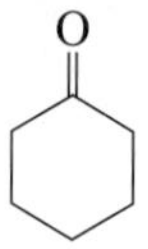 cyclohexanone + $CH_3CH_2CH(=O)$

d. $HCH(=O)$ (excess) + $CH_3CH_2CH(=O)$

PROBLEM 22

Describe how the following compounds could be prepared using an aldol addition in the first step of the synthesis:

a. $CH_3CH_2C(=O)CH(CH_3)CH_2OH$

b. 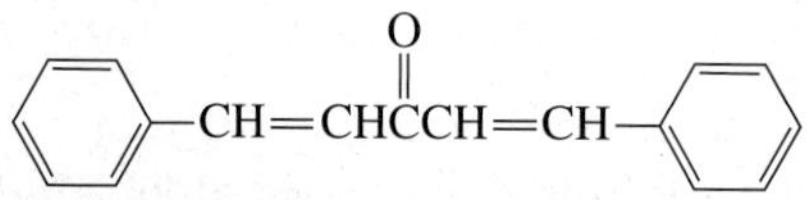 C_6H_5—CH=CHC(=O)CH=CH—C_6H_5

c. 2-benzylidenecyclohexanone

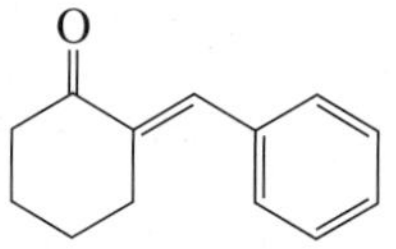

PROBLEM 23

Propose a mechanism for the following reaction:

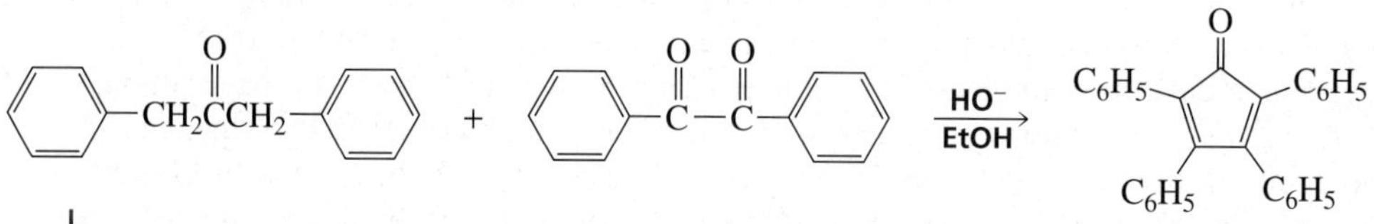

$$C_6H_5CH_2C(=O)CH_2C_6H_5 + C_6H_5C(=O)C(=O)C_6H_5 \xrightarrow[\text{EtOH}]{\text{HO}^-} \text{tetraphenylcyclopentadienone}$$

Product: cyclopentadienone with four C_6H_5 substituents.

Ludwig Claisen (1851–1930) *was born in Germany and received a Ph.D. from the University of Bonn, studying under Kekulé. He was a professor of chemistry at the University of Bonn, Owens College (Manchester, England), the University of Munich, the University of Aachen, the University of Kiel, and the University of Berlin.*

19.14 The Claisen Condensation

When two molecules of an *ester* undergo a condensation reaction, the reaction is called a **Claisen condensation**. The product of a Claisen condensation is a β-keto ester.

$$2\ CH_3CH_2\overset{O}{\overset{\|}{C}}OCH_2CH_3 \xrightarrow[\text{2. HCl}]{\text{1. } CH_3CH_2O^-} CH_3CH_2\overset{O}{\overset{\|}{C}}-\underset{CH_3}{\underset{|}{CH}}\overset{O}{\overset{\|}{C}}OCH_2CH_3 + CH_3CH_2OH$$

a β-keto ester

3-D Molecule: β Keto ester

As in an aldol addition, in a Claisen condensation one molecule of carbonyl compound is converted into an enolate when an α-hydrogen is removed by a strong base. The enolate attacks the carbonyl carbon of a second molecule of ester. The base employed corresponds to the leaving group of the ester so that the reactant is not changed if the base acts as a nucleophile and attacks the carbonyl group (Section 19.10).

mechanism for the Claisen condensation

$$CH_3CH_2\overset{O}{\overset{\|}{C}}OCH_3 \xrightleftharpoons{CH_3\ddot{O}^-} CH_3\overset{-}{C}H\overset{O}{\overset{\|}{C}}OCH_3 + CH_3OH \xrightleftharpoons{CH_3CH_2\overset{O}{\overset{\|}{C}}OCH_3} CH_3CH_2\overset{O^-}{\overset{|}{C}}(OCH_3)-\overset{CH_3}{CH}\overset{O}{\overset{\|}{C}}OCH_3 \rightleftharpoons CH_3CH_2\overset{O}{\overset{\|}{C}}-\underset{CH_3}{\underset{|}{C}H}\overset{O}{\overset{\|}{C}}OCH_3 + CH_3O^-$$

After nucleophilic attack, the Claisen condensation and the aldol addition differ. In the Claisen condensation, the negatively charged oxygen reforms the carbon–oxygen π bond and expels the $^-$OR group. In the aldol addition, the negatively charged oxygen obtains a proton from the solvent.

Claisen condensation:

formation of a π bond by expulsion of RO$^-$

$$RCH_2\overset{:\ddot{O}:^-}{\overset{|}{C}}(OR)-\underset{R}{\underset{|}{C}H}\overset{O}{\overset{\|}{C}}OR \rightleftharpoons RCH_2\overset{:\ddot{O}}{\overset{\|}{C}}\underset{R}{\underset{|}{C}H}\overset{O}{\overset{\|}{C}}OR + RO^-$$

aldol addition

protonation of O$^-$

$$RCH_2\overset{:\ddot{O}:^-}{\overset{|}{C}H}-\underset{R}{\underset{|}{C}H}\overset{O}{\overset{\|}{C}}H \xrightarrow{H_2O} RCH_2\overset{:\ddot{O}H}{\overset{|}{C}H}\underset{R}{\underset{|}{C}H}\overset{O}{\overset{\|}{C}}H + HO^-$$

The difference between the last step of the Claisen condensation and the last step of the aldol addition arises from the difference between esters and aldehydes or ketones. With esters, the carbon to which the negatively charged oxygen is bonded is also bonded to a group that can be expelled. With aldehydes or ketones, the carbon to which the negatively charged oxygen is bonded is not bonded to a group that can be expelled. Thus, the Claisen condensation is a substitution reaction, whereas the aldol addition is an addition reaction.

Expulsion of the alkoxide ion is reversible because the alkoxide ion can readily reform the tetrahedral intermediate by reacting with the β-keto ester. The condensation reaction can be driven to completion, however, if a proton is removed from the β-keto ester. Removing a proton prevents the reverse reaction from occurring, because the negatively charged alkoxide ion will not react with the negatively charged β-keto ester anion. It is easy to remove a proton from the β-keto ester because its central α-carbon is flanked by two carbonyl groups, making its α-hydrogen much more acidic than the α-hydrogens of the ester starting material.

$$RCH_2\overset{O^-}{\overset{|}{C}}(CH_3O)-\underset{R}{\underset{|}{C}H}\overset{O}{\overset{\|}{C}}OCH_3 \rightleftharpoons \underset{\beta\text{-keto ester}}{RCH_2\overset{O}{\overset{\|}{C}}-\underset{R}{\underset{|}{C}H}\overset{O}{\overset{\|}{C}}OCH_3} + CH_3O^- \longrightarrow \underset{\beta\text{-keto ester anion}}{RCH_2\overset{O}{\overset{\|}{C}}-\underset{R}{\underset{|}{\overset{-}{C}}}=\overset{O}{\overset{|}{C}}OCH_3} + CH_3OH$$

Consequently, a successful Claisen condensation requires an ester with two α-hydrogens and an equivalent amount of base rather than a catalytic amount of base. When the reaction is over, addition of acid to the reaction mixture reprotonates the β-keto ester anion. Any remaining alkoxide ion that could cause the reaction to reverse would also be protonated.

$$RCH_2\overset{O}{\overset{\|}{C}}-\underset{R}{\underset{|}{\bar{C}}}-\overset{O}{\overset{\|}{C}}OCH_3 + CH_3CH_2O^- \xrightarrow{HCl} RCH_2\overset{O}{\overset{\|}{C}}-\underset{R}{\underset{|}{CH}}-\overset{O}{\overset{\|}{C}}OCH_3 + CH_3CH_2OH$$

PROBLEM 24◆

Give the products of the following reactions:

a. $CH_3CH_2CH_2\overset{O}{\overset{\|}{C}}OCH_3 \xrightarrow[\text{2. HCl}]{\text{1. } CH_3O^-}$

b. $CH_3\underset{CH_3}{\underset{|}{C}}HCH_2\overset{O}{\overset{\|}{C}}OCH_2CH_3 \xrightarrow[\text{2. HCl}]{\text{1. } CH_3CH_2O^-}$

PROBLEM 25

Explain why a Claisen condensation product is not obtained from esters such as ethyl benzoate and ethyl 2-methylbutanoate.

19.15 The Mixed Claisen Condensation

A **mixed Claisen condensation** is a condensation reaction between two different esters. Like a mixed aldol addition, a mixed Claisen condensation is a useful reaction only if it is carried out under conditions that foster the formation of primarily one product. Otherwise, a mixture of products that are difficult to separate will be formed. Primarily one product will be formed if one of the esters has no α-hydrogens (and therefore cannot form an enolate) and the other ester is added slowly so that the ester without α-hydrogens is always in excess.

a mixed Claisen condensation

$$\underset{\text{add slowly}}{CH_3CH_2CH_2\overset{O}{\overset{\|}{C}}OCH_2CH_3} + \underset{\text{excess}}{C_6H_5\overset{O}{\overset{\|}{C}}-OCH_2CH_3} \xrightarrow[\text{2. HCl}]{\text{1. } CH_3CH_2O^-} C_6H_5\overset{O}{\overset{\|}{C}}-\underset{CH_2CH_3}{\underset{|}{C}}H\overset{O}{\overset{\|}{C}}OCH_2CH_3 + CH_3CH_2OH$$

A reaction similar to a mixed Claisen condensation is the condensation of a ketone and an ester. Because the α-hydrogens of a ketone are more acidic than those of an ester, primarily one product is formed if the ketone and the base are each added slowly to the ester. The product is a β-diketone. Because of the difference in acidities of the α-hydrogens, primarily one condensation product is obtained even if both reagents have α-hydrogens.

condensation of a ketone and an ester

$$\underset{\text{add slowly}}{\text{cyclohexanone}} + \underset{\substack{\text{ethyl acetate} \\ \text{excess}}}{CH_3\overset{O}{\overset{\|}{C}}-OCH_2CH_3} \xrightarrow[\text{2. HCl}]{\text{1. } CH_3CH_2O^-} \underset{\text{a } \beta\text{-diketone}}{\text{2-(}\overset{O}{\overset{\|}{C}}CH_3\text{)cyclohexanone}} + CH_3CH_2OH$$

A β-keto aldehyde is formed when a ketone condenses with a formate ester.

$$\text{cyclohexanone (add slowly)} + \underset{\text{ethyl formate, excess}}{HC(=O){-}OCH_2CH_3} \xrightarrow[\text{2. HCl}]{\text{1. } CH_3CH_2O^-} \underset{\text{a }\beta\text{-keto aldehyde}}{\text{2-(CH=O)cyclohexanone}} + CH_3CH_2OH$$

A β-keto ester is formed when a ketone condenses with diethyl carbonate.

$$\text{cyclohexanone (add slowly)} + \underset{\text{diethyl carbonate, excess}}{CH_3CH_2OC(=O){-}OCH_2CH_3} \xrightarrow[\text{2. HCl}]{\text{1. } CH_3CH_2O^-} \underset{\text{a }\beta\text{-keto ester}}{\text{2-(COCH}_2\text{CH}_3\text{)cyclohexanone}} + CH_3CH_2OH$$

PROBLEM 26

Give the product of each of the following reactions:

a. $CH_3CH_2\overset{O}{\overset{\|}{C}}OCH_3 + CH_3\overset{O}{\overset{\|}{C}}OCH_3 \xrightarrow[\text{2. HCl}]{\text{1. } CH_3O^-}$

b. $C_6H_5\overset{O}{\overset{\|}{C}}CH_3 + \underset{\text{excess}}{C_6H_5\overset{O}{\overset{\|}{C}}OCH_2CH_3} \xrightarrow[\text{2. HCl}]{\text{1. } CH_3CH_2O^-}$

c. $\underset{\text{excess}}{H\overset{O}{\overset{\|}{C}}OCH_3} + CH_3CH_2CH_2\overset{O}{\overset{\|}{C}}OCH_3 \xrightarrow[\text{2. HCl}]{\text{1. } CH_3O^-}$

d. $\underset{\text{excess}}{CH_3CH_2O\overset{O}{\overset{\|}{C}}OCH_2CH_3} + CH_3CH_2\overset{O}{\overset{\|}{C}}OCH_2CH_3 \xrightarrow[\text{2. HCl}]{\text{1. } CH_3CH_2O^-}$

PROBLEM 27 SOLVED

Show how the following compounds could be prepared, starting with 3-cyanocyclohexanone:

a. 2-($\overset{O}{\overset{\|}{C}}CH_2CH_3$)-3-($C{\equiv}N$)cyclohexanone

b. 2-($CH_2CH_2\overset{O}{\overset{\|}{C}}CH_3$)-3-($C{\equiv}N$)cyclohexanone

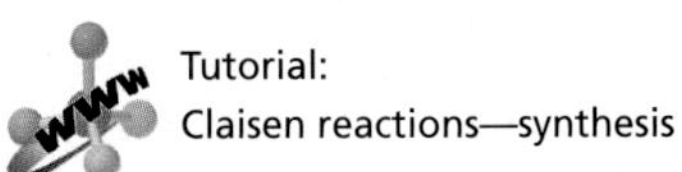

Tutorial:
Claisen reactions—synthesis

SOLUTION TO 27a Because the desired compound is a 1,3-dicarbonyl compound, it can be prepared by treating an enolate with an ester.

$$\text{3-cyanocyclohexanone} \xrightarrow{CH_3O^-} \text{enolate (carbanion at C-2)} \xrightarrow[\text{2. } H_3O^+]{\text{1. } CH_3CH_2\overset{O}{\overset{\|}{C}}OCH_3} \text{2-(}\overset{O}{\overset{\|}{C}}CH_2CH_3\text{)-3-(C}{\equiv}\text{N)cyclohexanone}$$

SOLUTION TO 27b Because the desired compound is a 1,5-dicarbonyl compound, it can be prepared by a Michael reaction—treating an enolate with an α,β-unsaturated carbonyl compound.

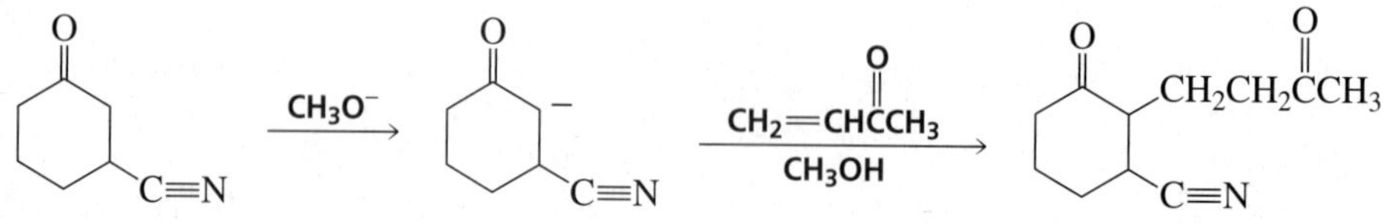

$$\text{2-cyanocyclohexanone} \xrightarrow{CH_3O^-} \text{enolate} \xrightarrow[CH_3OH]{CH_2{=}CHCCH_3} \text{2-(CH}_2\text{CH}_2\text{CCH}_3\text{)-2-cyanocyclohexanone}$$

19.16 Intramolecular Condensation and Addition Reactions

We have seen that if a compound has two functional groups that can react with each other, an intramolecular reaction readily occurs if the reaction leads to the formation of five- or six-membered rings (Section 11.11). Consequently, if base is added to a compound that contains two carbonyl groups, an intramolecular reaction occurs if a product with a five- or six-membered ring can be formed. Thus, a compound with two ester groups undergoes an intramolecular Claisen condensation, and a compound with two aldehyde or ketone groups undergoes an intramolecular aldol addition.

Intramolecular Claisen Condensations

The addition of base to a 1,6-diester causes the diester to undergo an intramolecular Claisen condensation, thereby forming a five-membered ring β-keto ester. An intramolecular Claisen condensation is called a **Dieckmann condensation.**

Walter Dieckmann (1869–1925) *was born in Germany. He received a Ph.D. from the University of Munich and became a professor of chemistry there.*

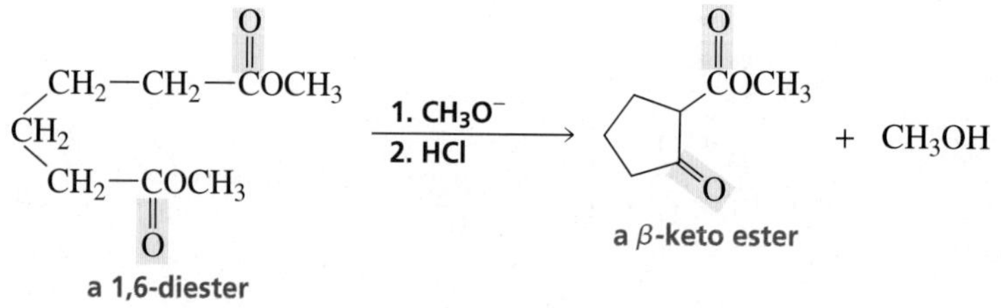

$$CH_2(CH_2CH_2COCH_3)(CH_2COCH_3) \xrightarrow[2.\ HCl]{1.\ CH_3O^-} \text{cyclopentanone-2-COCH}_3 + CH_3OH$$

a 1,6-diester — a β-keto ester

A six-membered ring β-keto ester is formed from a Dieckmann condensation of a 1,7-diester.

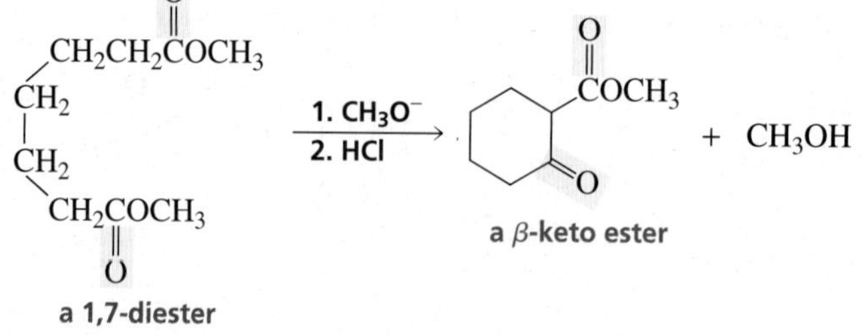

$$CH_2CH_2COCH_3\text{–}CH_2\text{–}CH_2\text{–}CH_2COCH_3 \xrightarrow[2.\ HCl]{1.\ CH_3O^-} \text{cyclohexanone-2-COCH}_3 + CH_3OH$$

a 1,7-diester — a β-keto ester

The mechanism of the Dieckmann condensation is the same as the mechanism of the Claisen condensation. The only difference between the two reactions is that the attacking enolate and the carbonyl group undergoing nucleophilic attack are in different molecules in the Claisen condensation, but are in the same molecule in the Dieckmann condensation. The Dieckmann condensation, like the Claisen condensation, is driven to completion by carrying out the reaction with enough base to remove a proton from the α-carbon of the β-keto ester product. When the reaction is over, acid is added to reprotonate the condensation product.

mechanism for the Dieckmann condensation

$CH_2CH_2COCH_3$ / CH_2 / CH_2COCH_3 $\xrightleftharpoons{CH_3O^-}$ $CH_2\ddot{C}HCOCH_3$ / CH_2 / CH_2COCH_3 + CH_3OH ⇌ (cyclic tetrahedral intermediate with $COCH_3$, OCH_3, $:\ddot{O}:^-$) ⇌ (cyclopentanone with $COCH_3$) + CH_3O^-

PROBLEM 28

Write the mechanism for the base-catalyzed formation of a cyclic β-keto ester from a 1,7-diester.

Intramolecular Aldol Additions

Because a 1,4-diketone has two different sets of α-hydrogens, two different intramolecular addition products can potentially form—one with a five-membered ring, the other with a three-membered ring. The greater stability of five- and six-membered rings causes them to be formed preferentially (Section 2.11). In fact, the five-membered ring product is the only product formed from the intramolecular aldol addition of a 1,4-diketone.

$CH_3CCH_2CH_2CCH_3$ (2,5-hexanedione, a 1,4-diketone) $\xrightleftharpoons[H_2O]{HO^-}$ $CH_3CCH_2CH_2C\ddot{C}H_2^-$ ⟶ (cyclopentanone alkoxide, ^-O, CH_3) $\xrightarrow{H_2O}$ (HO, CH_3 cyclopentanone)

$\xrightleftharpoons[H_2O]{HO^-}$ $CH_3CCH_2\ddot{C}HCCH_3$ ⟶✕ (cyclopropane, ^-O, CH_3, $C(=O)-CH_3$) $\xrightarrow{H_2O}$ (HO, CH_3, $C(=O)-CH_3$ cyclopropane) **not formed**

The intramolecular aldol addition of a 1,6-diketone potentially can lead to either a seven- or a five-membered ring product. Again, the more stable product—the one with the five-membered ring—is the only product of the reaction.

$CH_3CCH_2CH_2CH_2CH_2CCH_3$ (2,7-octanedione, a 1,6-diketone) $\xrightleftharpoons[H_2O]{HO^-}$ $CH_3CCH_2CH_2CH_2CH_2C\ddot{C}H_2^-$ ⟶✕ (cycloheptanone, O^-, CH_3) $\xrightarrow{H_2O}$ (cycloheptanone, OH, CH_3) **not formed**

$\xrightleftharpoons[H_2O]{HO^-}$ $CH_3CCH_2CH_2CH_2\ddot{C}HCCH_3$ ⟶ (cyclopentane, ^-O, CH_3, CCH_3=O) $\xrightarrow{H_2O}$ (cyclopentane, HO, CH_3, CCH_3=O)

1,5-Diketones and 1,7-diketones undergo intramolecular aldol additions to form six-membered ring products.

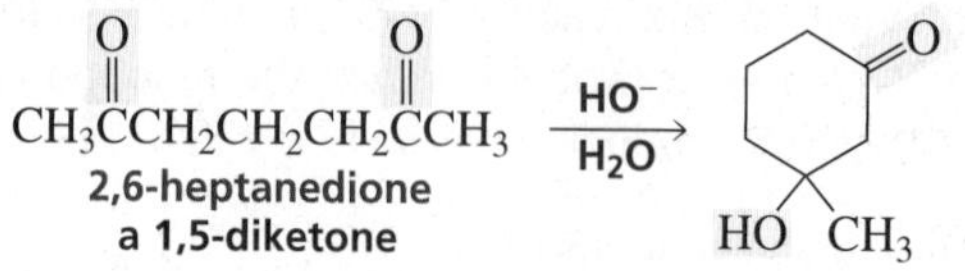

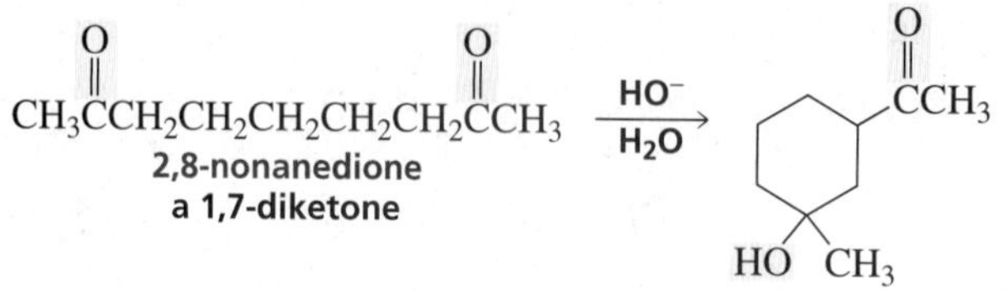

$$\underset{\text{2,6-heptanedione}\\\text{a 1,5-diketone}}{CH_3\overset{O}{\overset{\|}{C}}CH_2CH_2CH_2\overset{O}{\overset{\|}{C}}CH_3} \xrightarrow[H_2O]{HO^-} \text{(3-hydroxy-3-methylcyclohexanone)}$$

2,6-heptanedione
a 1,5-diketone

$$\underset{\text{2,8-nonanedione}\\\text{a 1,7-diketone}}{CH_3\overset{O}{\overset{\|}{C}}CH_2CH_2CH_2CH_2CH_2\overset{O}{\overset{\|}{C}}CH_3} \xrightarrow[H_2O]{HO^-} \text{(ring with } \overset{O}{\overset{\|}{C}}CH_3 \text{ and HO, } CH_3\text{)}$$

2,8-nonanedione
a 1,7-diketone

PROBLEM 29◆

If the preference for formation of a six-membered ring were not so great, what other cyclic product would be formed from the intramolecular aldol addition of

a. 2,6-heptanedione?
b. 2,8-nonanedione?

PROBLEM 30

Can 2,4-pentanedione undergo an intramolecular aldol addition? If so, why? If not, why not?

PROBLEM 31 SOLVED

What products can be obtained if the following keto aldehyde is treated with a base? Which would you expect to be the major product?

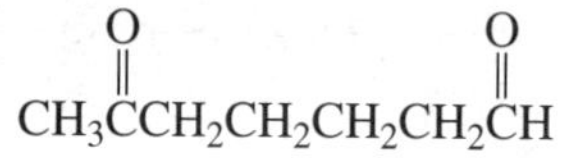

$$CH_3\overset{O}{\overset{\|}{C}}CH_2CH_2CH_2CH_2\overset{O}{\overset{\|}{C}}H$$

SOLUTION Three products are possible because there are three different sets of α-hydrogens. More B and C are formed than A because a five-membered ring is formed in preference to a seven-membered ring. The major product depends on the reaction conditions. B is the thermodynamic product because it is formed from the more stable enolate. C is the kinetic product because the α-hydrogen of the aldehyde is more acidic than the α-hydrogen of the ketone (Table 19.1).

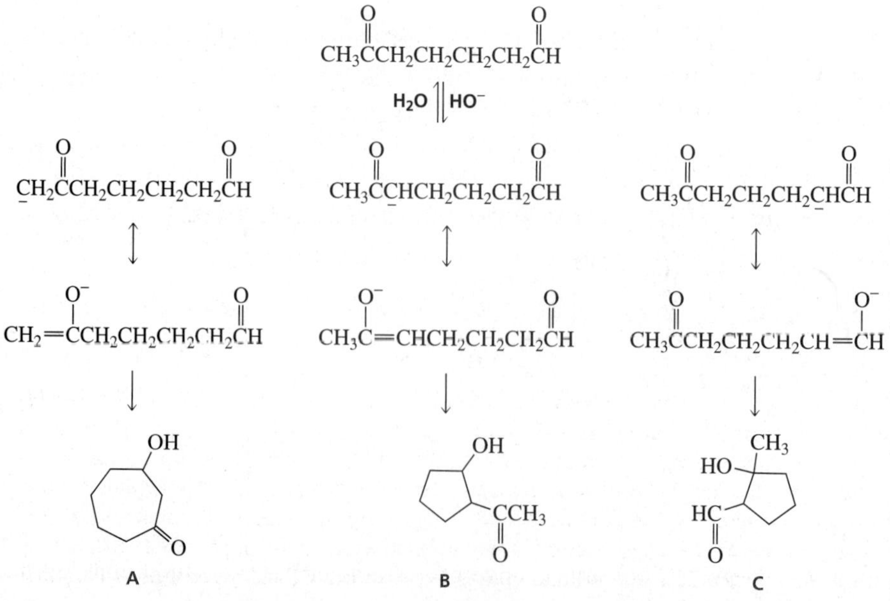

$$CH_3\overset{O}{\overset{\|}{C}}CH_2CH_2CH_2CH_2\overset{O}{\overset{\|}{C}}H \underset{H_2O}{\overset{HO^-}{\rightleftharpoons}}$$

$$\overset{-}{C}H_2\overset{O}{\overset{\|}{C}}CH_2CH_2CH_2CH_2\overset{O}{\overset{\|}{C}}H \longleftrightarrow CH_2{=}\overset{O^-}{\overset{|}{C}}CH_2CH_2CH_2CH_2\overset{O}{\overset{\|}{C}}H \longrightarrow \text{A}$$

$$CH_3\overset{O}{\overset{\|}{C}}\overset{-}{C}HCH_2CH_2CH_2\overset{O}{\overset{\|}{C}}H \longleftrightarrow CH_3\overset{O^-}{\overset{|}{C}}{=}CHCH_2CH_2CH_2\overset{O}{\overset{\|}{C}}H \longrightarrow \text{B}$$

$$CH_3\overset{O}{\overset{\|}{C}}CH_2CH_2CH_2\overset{-}{C}H\overset{O}{\overset{\|}{C}}H \longleftrightarrow CH_3\overset{O}{\overset{\|}{C}}CH_2CH_2CH_2CH{=}\overset{O^-}{\overset{|}{C}}H \longrightarrow \text{C}$$

A: OH, O (cycloheptanone ring)
B: OH, $\overset{O}{\overset{\|}{C}}CH_3$ (cyclopentane ring)
C: CH_3, HO, HC=O (cyclopentane ring)

PROBLEM 32◆

Give the product of the reaction of each of the following compounds with a base:

a. 2-($CH_2CH_2CCH_3$, C=O)cyclohexanone

c. $HCCH_2CH_2CH_2CH_2CH_2CH$ (O over each terminal C)

b. cyclodecane-1,6-dione

d. 2-($CH_2CH_2CH_2CCH_3$, C=O)cyclohexanone

The Robinson Annulation

Reactions that form carbon–carbon bonds are important to synthetic chemists. Without such reactions, large organic molecules could not be prepared from smaller ones. We have seen that Michael reactions and aldol additions form carbon–carbon bonds. The **Robinson annulation** is a reaction that puts these two carbon–carbon bond-forming reactions together, providing a route to the synthesis of many complicated organic molecules. "Annulation" comes from *annulus*, Latin for "ring." Thus, an **annulation reaction** is a ring-forming reaction.

The first stage of a Robinson annulation is a Michael reaction that forms a 1,5-diketone. You just saw that a 1,5-diketone undergoes an intramolecular aldol addition when treated with base—this is the second stage of the Robinson annulation. Notice that a Robinson annulation results in a product that has a 2-cyclohexenone ring.

Sir Robert Robinson (1886–1975) *was born in England, the son of a manufacturer. After receiving a Ph.D. from the University of Manchester, he joined the faculty at the University of Sydney in Australia. He returned to England three years later, becoming a professor of chemistry at Oxford in 1929. Robinson was an accomplished mountain climber. Knighted in 1939, he received the 1947 Nobel Prize in chemistry for his work on alkaloids.*

the Robinson annulation

$CH_2{=}CHCCH_3$ (C=O) + cyclohexanone $\xrightarrow[\text{a Michael reaction}]{HO^-}$ 2-(3-oxobutyl)cyclohexanone (C=O, CH_3) $\xrightarrow[\text{an intramolecular aldol addition}]{HO^-}$ bicyclic keto alcohol (OH) $\xrightarrow{\Delta}$ bicyclic enone + H_2O

PROBLEM-SOLVING STRATEGY

Propose a synthesis for each of the following compounds, using a Robinson annulation:

a. 4-methylcyclohex-2-enone (CH_3)

b. 3-methylcyclohex-2-enone (CH_3)

By analyzing a Robinson annulation, we will be able to determine which part of the molecule comes from which reactant. This, then, will allow us to choose appropriate reactants for any other Robinson annulation. Analysis shows that the keto group of the cyclohexenone comes from the α,β-unsaturated carbonyl compound and the double bond results from attack of the enolate of the α,β-unsaturated carbonyl compound on the carbonyl group of the other reactant. Thus, we can arrive at the appropriate reactants by cutting through the double bond and cutting between the β- and γ-carbons on the other side of the carbonyl group.

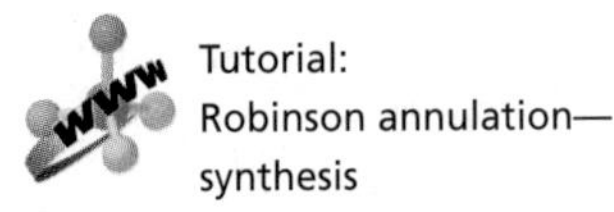

Tutorial: Robinson annulation—synthesis

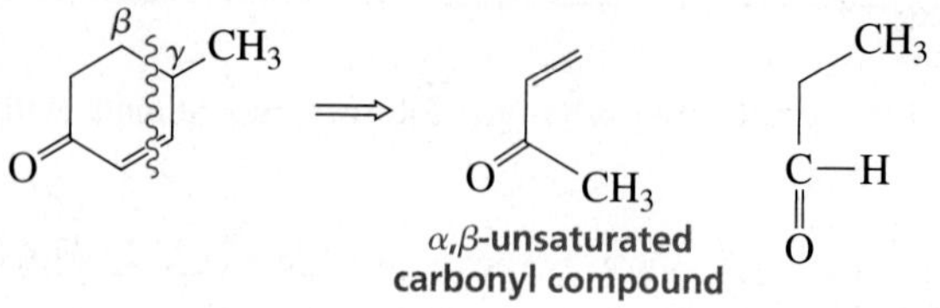

β γ CH_3 $\Longrightarrow$ α,β-unsaturated carbonyl compound + $CH_3CH_2C(=O)H$

α,β-unsaturated carbonyl compound

Therefore, the required reactants for (a) are:

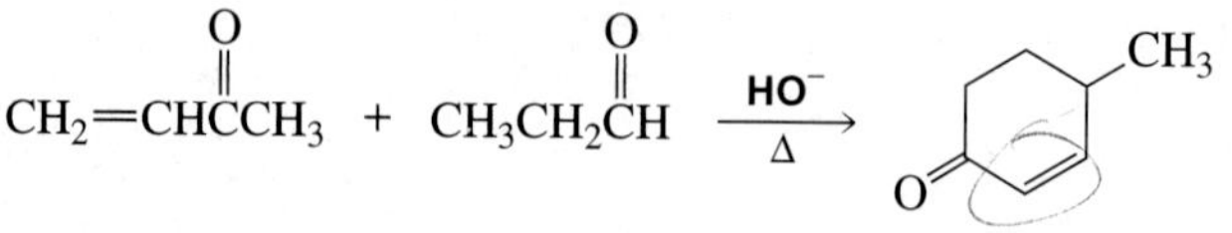

$$CH_2{=}CH\overset{O}{\overset{\|}{C}}CH_3 + CH_3CH_2\overset{O}{\overset{\|}{C}}H \xrightarrow[\Delta]{HO^-}$$

By cutting through (b), we can determine the required reactants for its synthesis:

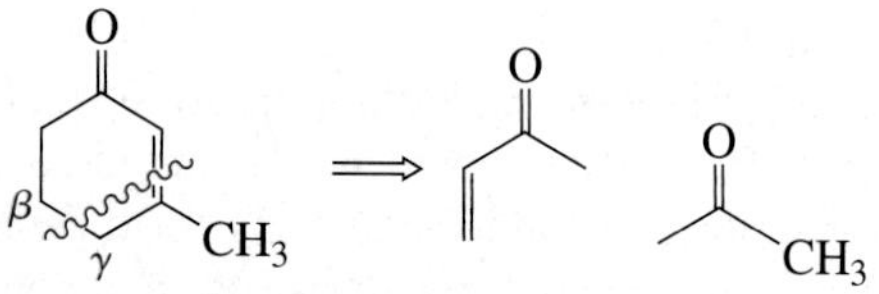

Therefore, the required reactants for (b) are:

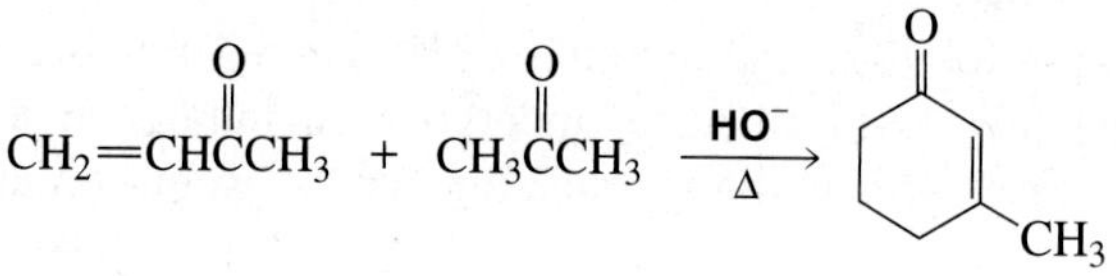

$$CH_2{=}CH\overset{O}{\overset{\|}{C}}CH_3 + CH_3\overset{O}{\overset{\|}{C}}CH_3 \xrightarrow[\Delta]{HO^-}$$

Now continue on to Problem 33.

PROBLEM 33

Propose a synthesis for each of the following compounds, using a Robinson annulation:

a. 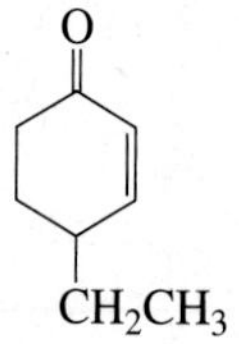 CH_2CH_3

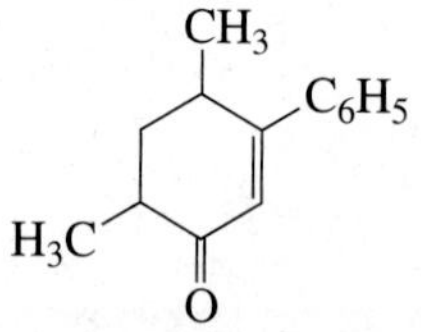

b. 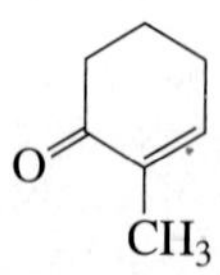 CH_3

c. CH_3, C_6H_5, H_3C, O

d. 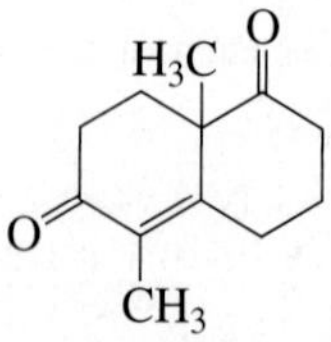 H_3C, O, CH_3

19.17 Decarboxylation of 3-Oxocarboxylic Acids

Carboxylate ions do not lose CO_2, for the same reason that alkanes such as ethane do not lose a proton—because the leaving group would be a carbanion. Carbanions are very strong bases and therefore are very poor leaving groups.

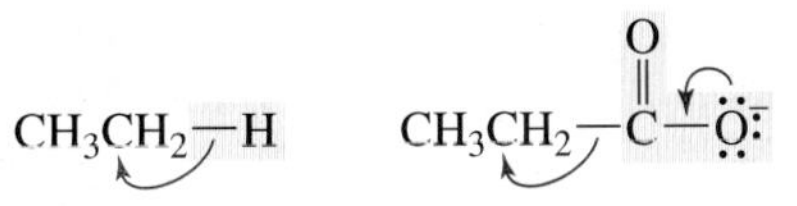

If, however, the CO_2 group is bonded to a carbon that is adjacent to a carbonyl carbon, the CO_2 group can be removed because the electrons left behind can be delocalized onto the carbonyl oxygen. Consequently, 3-oxocarboxylate ions (carboxylate ions with a keto group at the 3-position) lose CO_2 when they are heated. Loss of CO_2 from a molecule is called **decarboxylation**.

removing CO_2 from an α-carbon

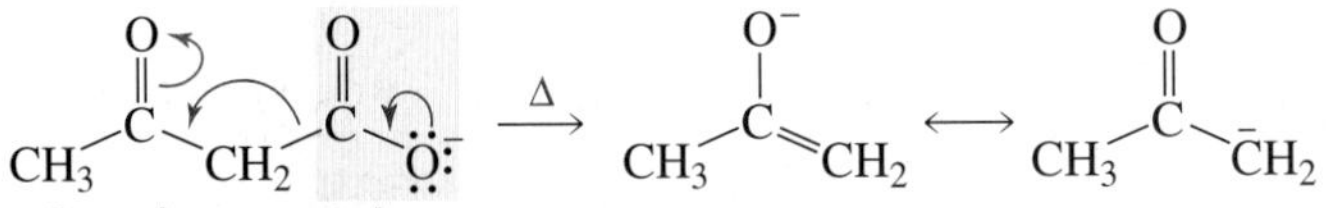

3-oxobutanoate ion
acetoacetate ion

$+ \; CO_2$

Notice the similarity between removal of CO_2 from a 3-oxocarboxylate ion and removal of a proton from an α-carbon. In both reactions, a substituent—CO_2 in one case, H^+ in the other—is removed from an α-carbon and its bonding electrons are delocalized onto an oxygen.

removing a proton from an α-carbon

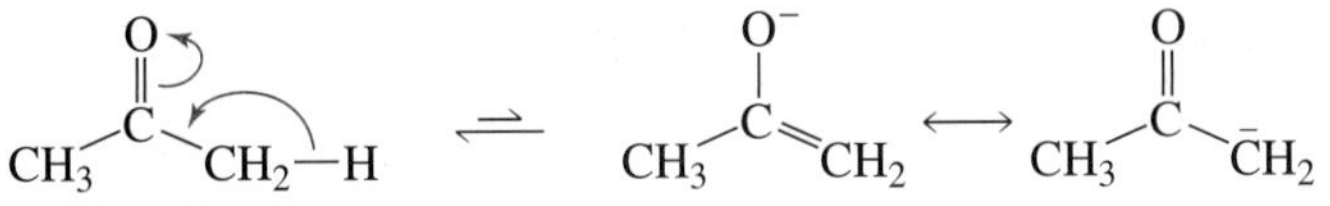

propanone
acetone

$+ \; H^+$

Decarboxylation is even easier if the reaction is carried out under acidic conditions, because the reaction is catalyzed by an intramolecular transfer of a proton from the carboxyl group to the carbonyl oxygen. The enol that is formed immediately tautomerizes to a ketone.

3-Oxocarboxylic acids decarboxylate when heated.

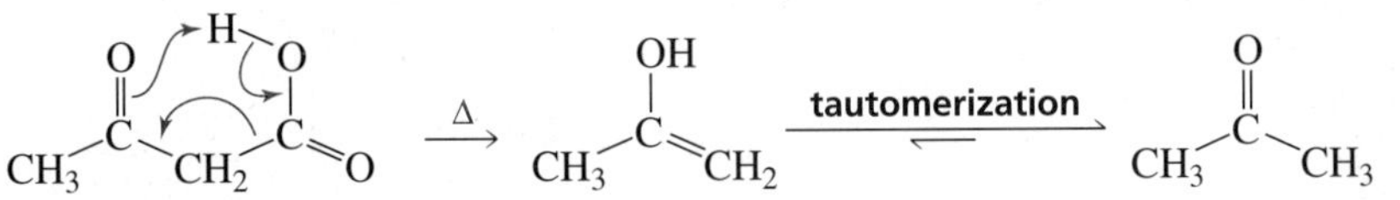

tautomerization

3-oxobutanoic acid
acetoacetic acid
a β-keto acid

$+ \; CO_2$

We saw in Section 19.1 that it is harder to remove a proton from an α-carbon if the electrons are delocalized onto the carbonyl group of an ester rather than onto the carbonyl group of a ketone. For the same reason, a higher temperature is required to decarboxylate a β-dicarboxylic acid such as malonic acid than to decarboxylate a β-keto acid.

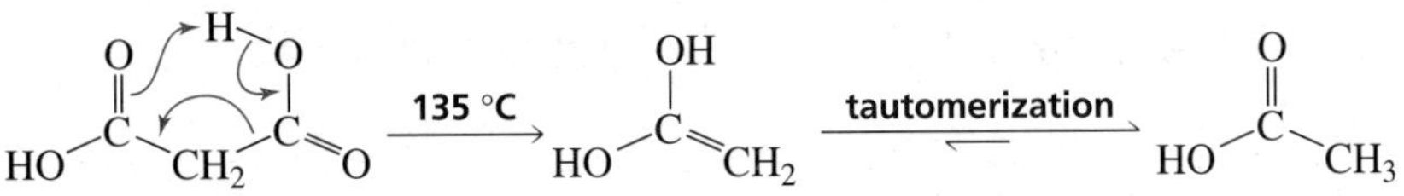

135 °C

tautomerization

malonic acid

$+ \; CO_2$

In summary, carboxylic acids with a carbonyl group at the 3-position (both β-ketocarboxylic acids and β-dicarboxylic acids) lose CO_2 when they are heated.

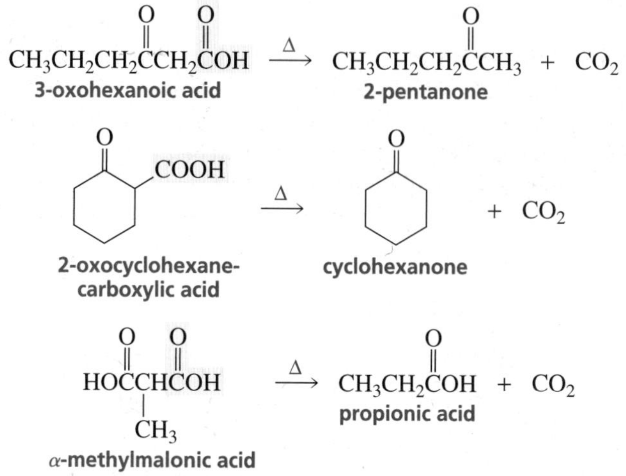

$$CH_3CH_2CH_2\overset{O}{\overset{\|}{C}}CH_2\overset{O}{\overset{\|}{C}}OH \xrightarrow{\Delta} CH_3CH_2CH_2\overset{O}{\overset{\|}{C}}CH_3 + CO_2$$

3-oxohexanoic acid → 2-pentanone

2-oxocyclohexane-carboxylic acid (COOH) $\xrightarrow{\Delta}$ cyclohexanone $+ CO_2$

$$HO\overset{O}{\overset{\|}{C}}\underset{CH_3}{\underset{|}{C}H}\overset{O}{\overset{\|}{C}}OH \xrightarrow{\Delta} CH_3CH_2\overset{O}{\overset{\|}{C}}OH + CO_2$$

α-methylmalonic acid → propionic acid

THE HUNSDIECKER REACTION

Heinz and Clare Hunsdiecker found that a carboxylic acid can be decarboxylated if a heavy metal salt of the carboxylic acid is heated with bromine or iodine. The product is an alkyl halide with one less carbon atom than the starting carboxylic acid. The heavy metal can be silver ion, mercuric ion, or lead(IV). The reaction is now known as the **Hunsdiecker reaction**.

$$CH_3CH_2CH_2CH_2\overset{O}{\overset{\|}{C}}OH \xrightarrow[\text{2. } Br_2,\ \Delta]{\text{1. } Ag_2O} CH_3CH_2CH_2CH_2Br + CO_2 + AgBr$$

pentanoic acid → 1-bromobutane

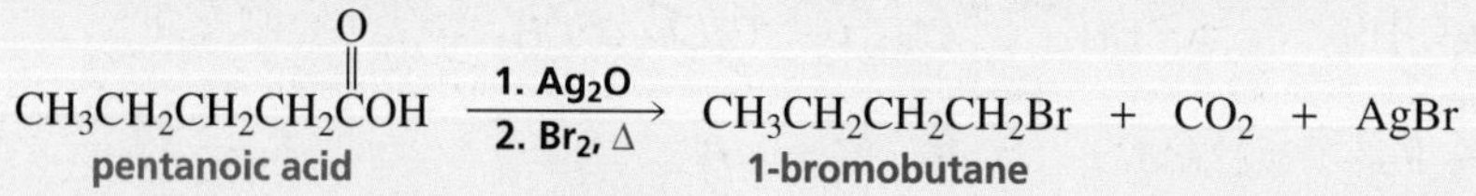

The reaction involves formation of a hypobromite as a result of precipitation of AgBr. A radical reaction is initiated by homolytic cleavage of the O—Br bond of the hypobromite. The carboxyl radical loses CO_2, and the alkyl radical thus formed abstracts a bromine radical from the hypobromite to propagate the reaction.

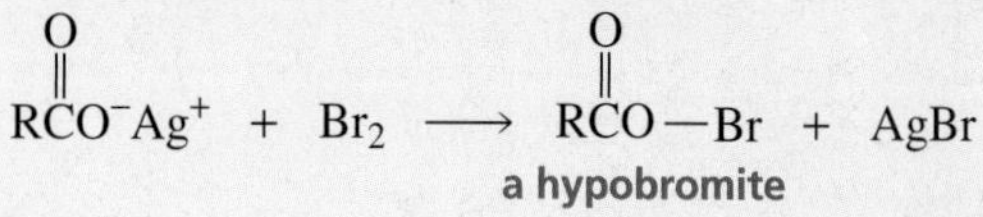

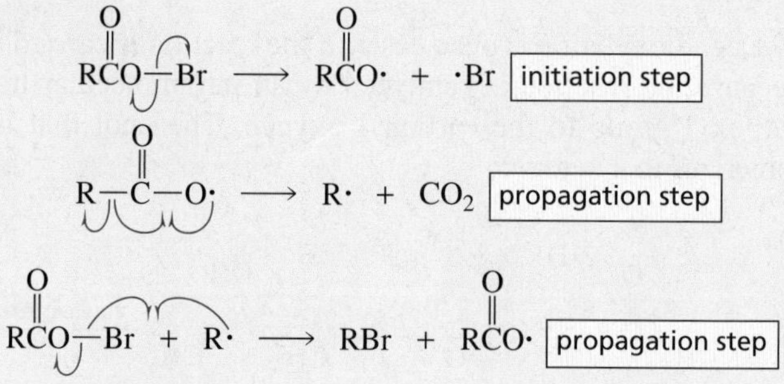

$$R\overset{O}{\overset{\|}{C}}O^-Ag^+ + Br_2 \longrightarrow R\overset{O}{\overset{\|}{C}}O{-}Br + AgBr$$

a hypobromite

$$R\overset{O}{\overset{\|}{C}}O{-}Br \longrightarrow R\overset{O}{\overset{\|}{C}}O\cdot + \cdot Br$$ initiation step

$$R{-}\overset{O}{\overset{\|}{C}}{-}O\cdot \longrightarrow R\cdot + CO_2$$ propagation step

$$R\overset{O}{\overset{\|}{C}}O{-}Br + R\cdot \longrightarrow RBr + R\overset{O}{\overset{\|}{C}}O\cdot$$ propagation step

Heinz (1904–1981) *and* **Clare (1903–1995) Hunsdiecker** *were both born in Germany, Heinz in Cologne and Clare in Kiel. They both received Ph.D.s from the University of Cologne and spent their careers working in a private laboratory in Cologne.*

PROBLEM 34◆

Which of the following compounds would be expected to decarboxylate when heated?

a. 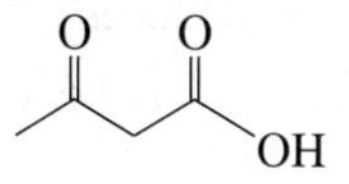 [structure: CH_3COCH_2COOH]

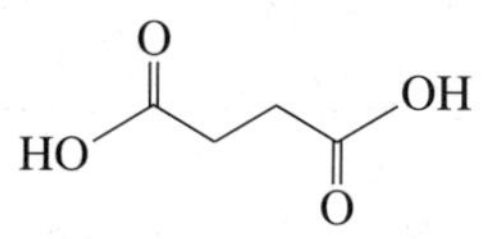

b. 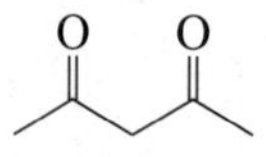 [structure: $CH_3COCH_2COCH_3$]

c. [structure: $HOOCCH_2CH_2COOH$]

d. 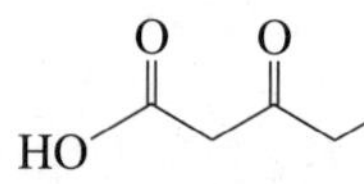 [structure: $HOOCCH_2COCH_2CH_3$]

19.18 The Malonic Ester Synthesis: Synthesis of Carboxylic Acids

A combination of two of the reactions discussed in this chapter—alkylation of an α-carbon and decarboxylation of a β-dicarboxylic acid—can be used to prepare carboxylic acids of any desired chain length. The procedure is called the **malonic ester synthesis** because the starting material for the synthesis is the diethyl ester of malonic acid. The first two carbons of the carboxylic acid come from malonic ester, and the rest of the carboxylic acid comes from the alkyl halide used in the second step of the reaction.

A malonic ester synthesis forms a carboyxlic acid with two more carbon atoms than the alkyl halide.

malonic ester synthesis

$$C_2H_5O\overset{O}{\overset{\|}{C}}-CH_2-\overset{O}{\overset{\|}{C}}OC_2H_5 \xrightarrow[\text{2. RBr} \quad \text{3. HCl, } H_2O,\ \Delta]{\text{1. } CH_3CH_2O^-} R-CH_2\overset{O}{\overset{\|}{C}}OH$$

diethyl malonate
malonic ester

from malonic ester

from the alkyl halide

In the first part of the malonic ester synthesis, the α-carbon of the diester is alkylated (Section 19.8). A proton is easily removed from the α-carbon because it is flanked by two ester groups ($pK_a = 13$). The resulting α-carbanion reacts with an alkyl halide, forming an α-substituted malonic ester. Because alkylation is an S_N2 reaction, it works best with methyl halides and primary alkyl halides (Section 10.2). Heating the α-substituted malonic ester in an acidic aqueous solution hydrolyzes it to an α-substituted malonic acid, which, upon further heating, loses CO_2, forming a carboxylic acid with two more carbons than the alkyl halide.

Tutorial: Malonic ester synthesis

$$C_2H_5O\overset{O}{\overset{\|}{C}}-CH_2-\overset{O}{\overset{\|}{C}}OC_2H_5 \xrightarrow{CH_3CH_2O^-} C_2H_5O\overset{O}{\overset{\|}{C}}-\ddot{C}H-\overset{O}{\overset{\|}{C}}OC_2H_5 \xrightarrow{R-Br} C_2H_5O\overset{O}{\overset{\|}{C}}-\underset{R}{\underset{|}{C}H}-\overset{O}{\overset{\|}{C}}OC_2H_5 + Br^-$$

removal of a proton from the α-carbon

akylation of the α-carbon

an α-substituted malonic ester

HCl, H_2O, Δ — hydrolysis

$$R-CH_2-\overset{O}{\overset{\|}{C}}OH + CO_2 \xleftarrow{\Delta} HO\overset{O}{\overset{\|}{C}}-\underset{R}{\underset{|}{C}H}-\overset{O}{\overset{\|}{C}}OH + 2\ CH_3CH_2OH$$

decarboxylation

an α-substituted malonic acid

PROBLEM 35◆

What alkyl bromide(s) should be used in the malonic ester synthesis of each of the following carboxylic acids?

a. propanoic acid
b. 2-methylpropanoic acid
c. 3-phenylpentanoic acid
d. 4-methylpentanoic acid

Carboxylic acids with two substituents bonded to the α-carbon can be prepared by carrying out two successive α-carbon alkylations.

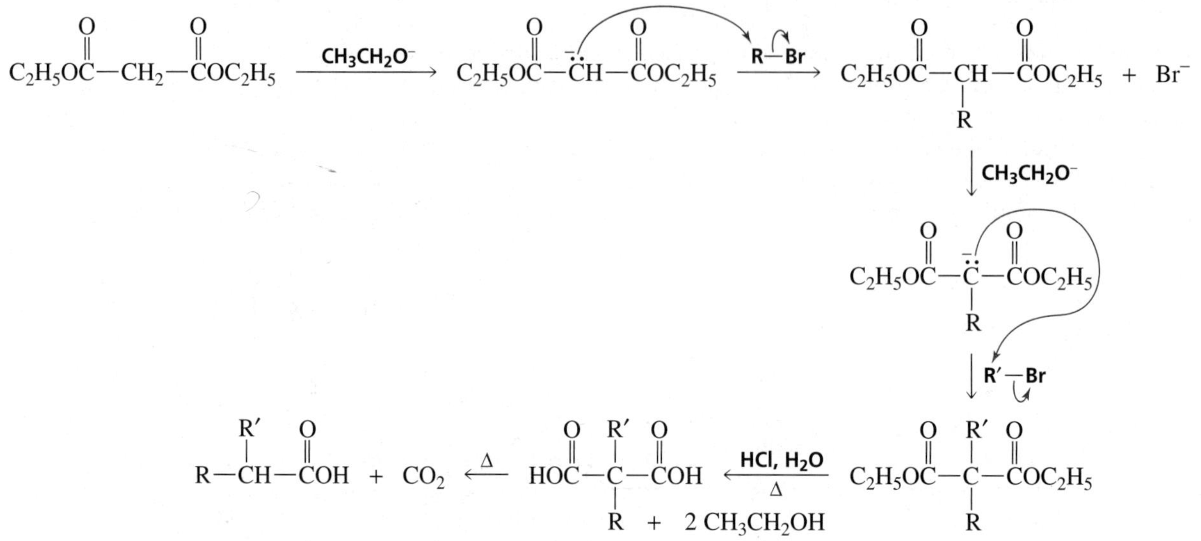

$$C_2H_5OC(=O)-CH_2-C(=O)OC_2H_5 \xrightarrow{CH_3CH_2O^-} C_2H_5OC(=O)-\ddot{C}H^{-}-C(=O)OC_2H_5 \xrightarrow{R-Br} C_2H_5OC(=O)-CH(R)-C(=O)OC_2H_5 + Br^-$$

$$\xrightarrow{CH_3CH_2O^-} C_2H_5OC(=O)-\ddot{C}^{-}(R)-C(=O)OC_2H_5 \xrightarrow{R'-Br} C_2H_5OC(=O)-C(R)(R')-C(=O)OC_2H_5$$

$$\xrightarrow[\Delta]{HCl,\ H_2O} HOC(=O)-C(R)(R')-C(=O)OH + 2\ CH_3CH_2OH \xrightarrow{\Delta} R-CH(R')-C(=O)OH + CO_2$$

PROBLEM 36

Explain why the following carboxylic acids cannot be prepared by the malonic ester synthesis:

a. 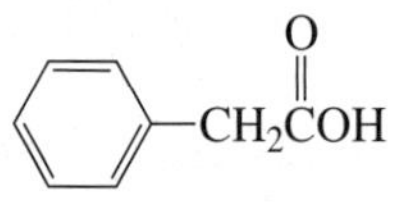 $C_6H_5-CH_2C(=O)OH$

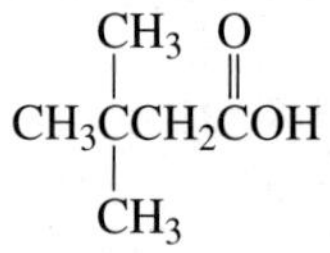

b. 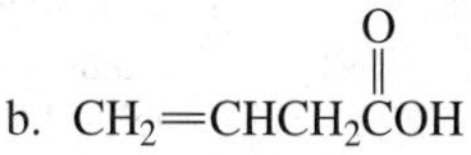 $CH_2=CHCH_2C(=O)OH$

c. $CH_3C(CH_3)_2CH_2C(=O)OH$

19.19 The Acetoacetic Ester Synthesis: Synthesis of Methyl Ketones

The only difference between the acetoacetic ester synthesis and the malonic ester synthesis is the use of acetoacetic ester rather than malonic ester as the starting material. The difference in starting material causes the product of the **acetoacetic ester synthesis** to be a *methyl ketone* rather than a *carboxylic acid.* The carbonyl group of the methyl ketone and the carbon atoms on either side of it come from acetoacetic ester, and the rest of the ketone comes from the alkyl halide used in the second step of the reaction.

acetoacetic ester synthesis

$$CH_3C(=O)-CH_2-C(=O)OC_2H_5 \xrightarrow[\text{2. RBr} \quad \text{3. HCl, } H_2O,\ \Delta]{\text{1. } CH_3CH_2O^-} R-CH_2C(=O)CH_3$$

ethyl 3-oxobutanoate
ethyl acetoacetate
"acetoacetic ester"

from acetoacetic ester

from the alkyl halide

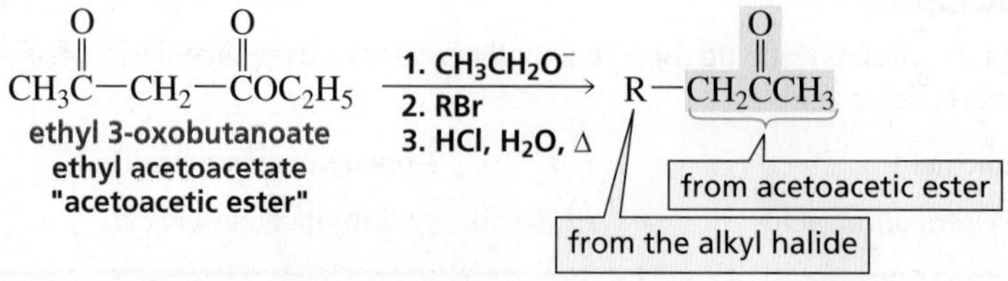

The mechanisms for the acetoacetic ester synthesis and the malonic ester synthesis are similar. The last step in the acetoacetic ester synthesis is the decarboxylation of a substituted acetoacetic acid rather than a substituted malonic acid.

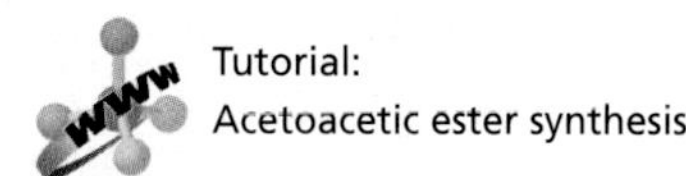

$$CH_3\overset{O}{\overset{\|}{C}}-CH_2-\overset{O}{\overset{\|}{C}}OC_2H_5 \xrightarrow{CH_3CH_2O^-} CH_3\overset{O}{\overset{\|}{C}}-\ddot{\bar{C}}H-\overset{O}{\overset{\|}{C}}OC_2H_5 \xrightarrow{R-Br} CH_3\overset{O}{\overset{\|}{C}}-\underset{R}{\underset{|}{C}H}-\overset{O}{\overset{\|}{C}}OC_2H_5 + Br^-$$

removal of a proton from the α-carbon

alkylation of the α-carbon

$\downarrow$ HCl, H_2O, Δ — hydrolysis

$$CH_3\overset{O}{\overset{\|}{C}}-CH_2-R + CO_2 \xleftarrow{\Delta} CH_3\overset{O}{\overset{\|}{C}}-\underset{R}{\underset{|}{C}H}-\overset{O}{\overset{\|}{C}}OH + CH_3CH_2OH$$

decarboxylation

PROBLEM 37 SOLVED

Starting with methyl propanoate, how could you prepare 4-methyl-3-heptanone?

$$\underset{\text{methyl propanoate}}{CH_3CH_2\overset{O}{\overset{\|}{C}}OCH_3} \xrightarrow{?} \underset{\text{4-methyl-3-heptanone}}{CH_3CH_2\overset{O}{\overset{\|}{C}}\underset{CH_3}{\underset{|}{C}H}CH_2CH_2CH_3}$$

Because the target molecule has several more carbon atoms than the starting material, a Claisen condensation appears to be a good way to start this synthesis. The Claisen condensation forms a β-keto ester that can be alkylated at the desired carbon because it is flanked by two carbonyl groups. Acid-catalyzed hydrolysis will form a 3-oxocarboxylic acid that will decarboxylate when heated.

$$CH_3CH_2\overset{O}{\overset{\|}{C}}OCH_3 \xrightarrow[2.\ H_3O^+]{1.\ CH_3O^-} CH_3CH_2\overset{O}{\overset{\|}{C}}-\underset{CH_3}{\underset{|}{C}H}-\overset{O}{\overset{\|}{C}}OCH_3 \xrightarrow[2.\ CH_3CH_2CH_2Br]{1.\ CH_3O^-} CH_3CH_2\overset{O}{\overset{\|}{C}}-\underset{CH_2CH_2CH_3}{\underset{|}{\overset{CH_3}{\overset{|}{C}}}}-\overset{O}{\overset{\|}{C}}OCH_3$$

$\downarrow$ Δ HCl, H_2O

$$CH_3CH_2\overset{O}{\overset{\|}{C}}\underset{CH_3}{\underset{|}{C}H}CH_2CH_2CH_3$$

PROBLEM 38◆

What alkyl bromide should be used in the acetoacetic ester synthesis of each of the following methyl ketones?

a. 2-pentanone
b. 2-octanone
c. 4-phenyl-2-butanone

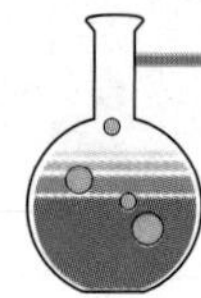

19.20 Designing a Synthesis VI: Making New Carbon–Carbon Bonds

When you are planning the synthesis of a compound that requires the formation of a new carbon–carbon bond, first locate the new bond that must be made. For example, in the synthesis of the following β-diketone, the new bond is the one that makes the second five-membered ring:

new bond

Next, determine which of the atoms that form the bond should be the nucleophile and which should be the electrophile. In this case, it is easy to choose between the two possibilities because you know that a carbonyl carbon is an electrophile.

nucleophile / electrophile or electrophile / nucleophile

Now you need to determine what compound you could use that would give you the desired electrophilic and nucleophilic sites. If the starting material is indicated, use it as a clue to arrive at the desired compound. For example, an ester carbonyl group would be a good electrophile for this synthesis because it has a group that would be eliminated. Furthermore, the α-hydrogens of the ketone are more acidic than the α-hydrogens of the ester, so it would be easy to obtain the desired nucleophile. The ester can be easily prepared from the carboxylic acid starting material.

1. $SOCl_2$ 2. CH_3OH ; 1. CH_3O^- 2. H_3O^+

In the next synthesis, two new carbon–carbon bonds must be formed.

After determining the electrophilic and nucleophilic sites, you can see that two successive alkylations of a diester of malonic acid, using 1,5-dibromopentane for the alkyl halide, will produce the desired compound.

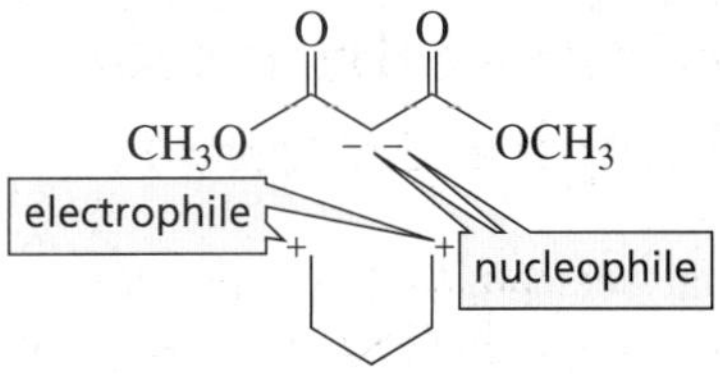

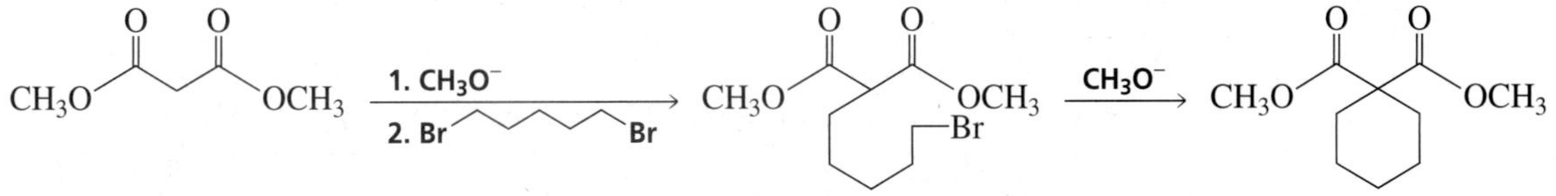

electrophile
nucleophile

In planning the following synthesis, the diester that is given as the starting material suggests that you should use a Dieckmann condensation to obtain the cyclic compound:

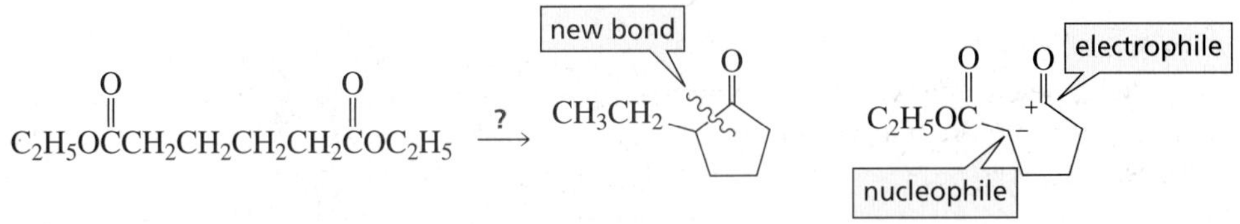

$$C_2H_5OCCH_2CH_2CH_2CH_2COC_2H_5 \xrightarrow{?}$$

new bond
electrophile
nucleophile

After the cyclopentanone ring is formed from a Dieckmann condensation of the starting material, alkylation of the α-carbon followed by hydrolysis of the β-keto ester and decarboxylation forms the desired product.

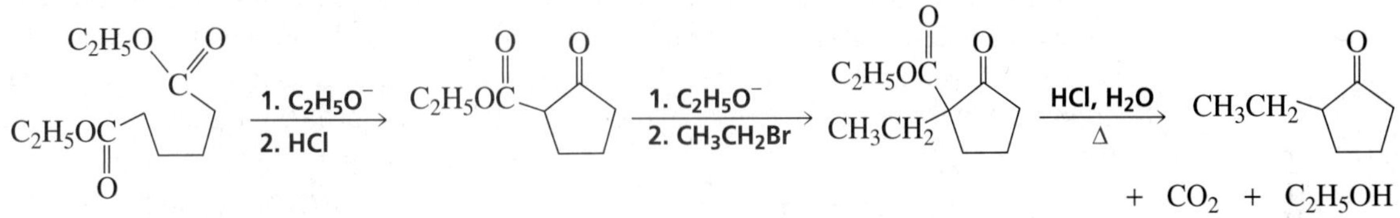

1. $C_2H_5O^-$ 2. HCl; 1. $C_2H_5O^-$ 2. CH_3CH_2Br; HCl, H_2O, Δ → + CO_2 + C_2H_5OH

PROBLEM 39

Design a synthesis for each of the following compounds using the given starting material:

a. 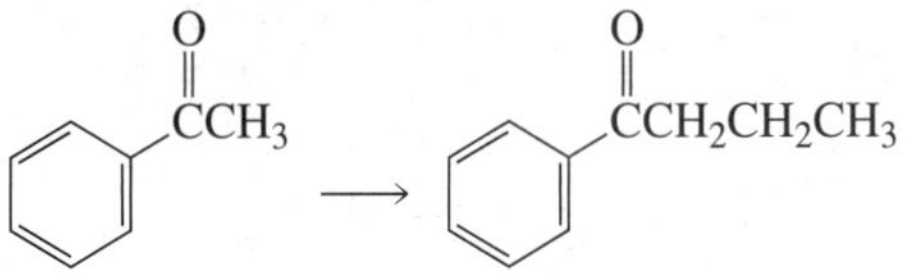 (phenyl)CCH_3 → (phenyl)$CCH_2CH_2CH_3$

b. 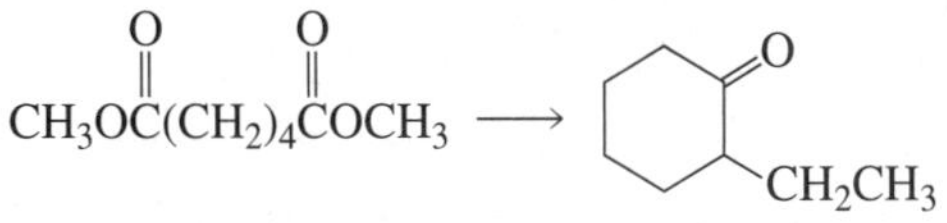 $CH_3OC(CH_2)_4COCH_3$ →

c. 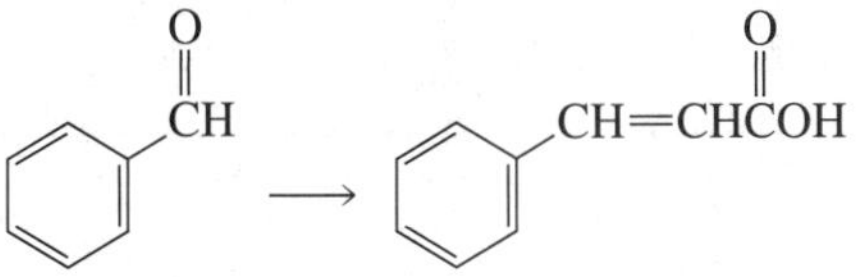 (phenyl)CH → (phenyl)CH=CHCOH

d. 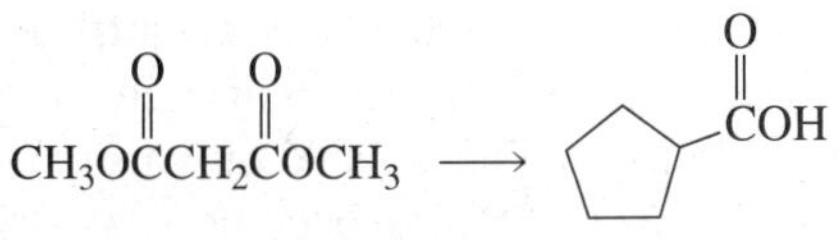 $CH_3OCCH_2COCH_3$ →

19.21 Reactions at the α-Carbon in Biological Systems

Many reactions that occur in biological systems involve reactions at the α-carbon—the kinds of reactions you have studied in this chapter. We will now look at a few examples.

A Biological Aldol Addition

D-Glucose, the most abundant sugar found in nature, is synthesized by biological systems from two molecules of pyruvate. The series of reactions that convert two molecules of pyruvate into D-glucose is called **gluconeogenesis**. The reverse process—the breakdown of D-glucose into two molecules of pyruvate—is called **glycolysis** (Section 25.1).

$$2\,CH_3\overset{O}{\overset{\|}{C}}-\overset{O}{\overset{\|}{C}}O^- \;\rightleftharpoons\;\rightleftharpoons\;\xrightleftharpoons{\text{several steps}}\;\rightleftharpoons\; \text{D-glucose}$$

pyruvate → (gluconeogenesis) / ← (glycolysis)

D-glucose (Fischer projection):

```
      HC=O
   H ─┼─ OH
  HO ─┼─ H
   H ─┼─ OH
   H ─┼─ OH
      CH2OH
```

D-glucose

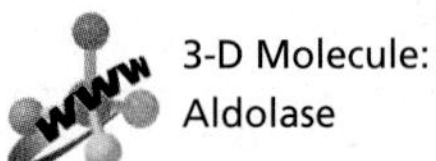
3-D Molecule: Aldolase

Because D-glucose has twice as many carbons as pyruvate, it should not be surprising that one of the steps in the biosynthesis of D-glucose is an aldol addition. An enzyme called aldolase catalyzes an aldol addition between dihydroxyacetone phosphate and D-glyceraldehyde-3-phosphate. The product is D-fructose-1,6-diphosphate, which is subsequently converted to D-glucose. The mechanism of this reaction is discussed in Section 24.9.

```
  CH2OPO3^2-
  |
  C=O
  |
  CH2OH
```

dihydroxyacetone phosphate

```
      O
      ‖
      C─H
   H ─┼─ OH
      CH2OPO3^2-
```

D-glyceraldehyde-3-phosphate

$\xrightleftharpoons{\text{aldolase}}$

```
      CH2OPO3^2-
      |
      C=O
  HO ─┼─ H
   H ─┼─ OH
   H ─┼─ OH
      CH2OPO3^2-
```

D-fructose-1,6-diphosphate

PROBLEM 40

Propose a mechanism for the formation of D-fructose-1,6-diphosphate from dihydroxyacetone phosphate and D-glyceraldehyde-3-phosphate, using HO^- as the catalyst.

A Biological Aldol Condensation

Collagen is the most abundant protein in mammals, amounting to about one-fourth of the total protein. It is the major fibrous component of bone, teeth, skin, cartilage, and tendons. It also is responsible for holding groups of cells together in discrete units. Individual collagen molecules—called tropocollagen—can be isolated only from tissues of young animals. As animals age, the individual molecules become cross-linked, which is why meat from older animals is tougher than meat from younger ones. Collagen cross-linking is an example of an aldol condensation.

Before collagen molecules can cross-link, the primary amino groups of the lysine residues of collagen must be converted to aldehyde groups. The enzyme that catalyzes this reaction is called lysyl oxidase. An aldol condensation between two aldehyde residues results in a cross-linked protein.

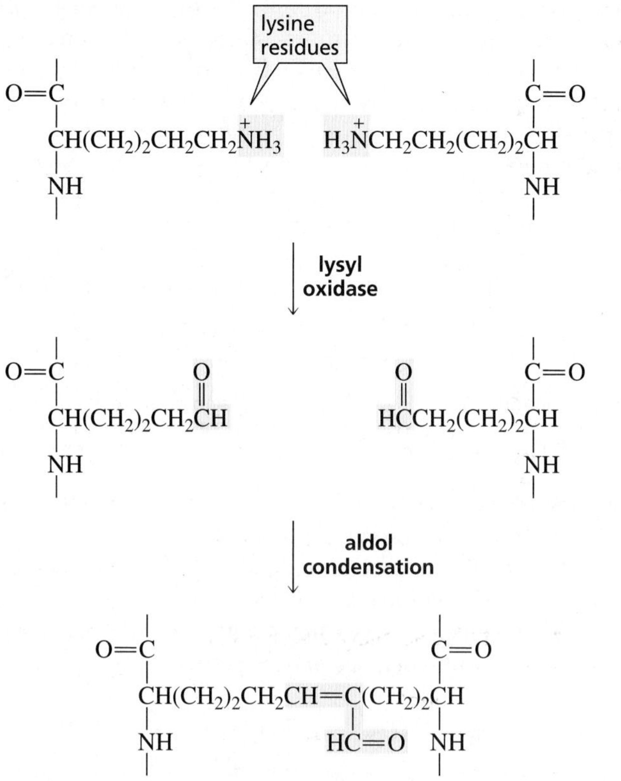

lysine residues

$$\text{O=C(–CH(CH}_2)_2\text{CH}_2\text{CH}_2\overset{+}{\text{N}}\text{H}_3\text{)(–NH–)} \qquad \text{H}_3\overset{+}{\text{N}}\text{CH}_2\text{CH}_2(\text{CH}_2)_2\text{CH(–C=O)(–NH–)}$$

lysyl oxidase ↓

$$\text{O=C(–CH(CH}_2)_2\text{CH}_2\text{CH=O)(–NH–)} \qquad \text{O=HCCH}_2(\text{CH}_2)_2\text{CH(–C=O)(–NH–)}$$

aldol condensation ↓

$$\text{O=C(–CH(CH}_2)_2\text{CH}_2\text{CH=C(HC=O)(CH}_2)_2\text{CH(–C=O)(–NH–))(–NH–)}$$

cross-linked collagen

A Biological Claisen Condensation

Fatty acids are long-chain, unbranched carboxylic acids (Sections 17.13 and 26.1). Most naturally occurring fatty acids contain an even number of carbons because they are synthesized from acetate, which has two carbon atoms.

In Section 17.20, you saw that carboxylic acids can be activated in biological systems by being converted to thioesters of coenzyme A.

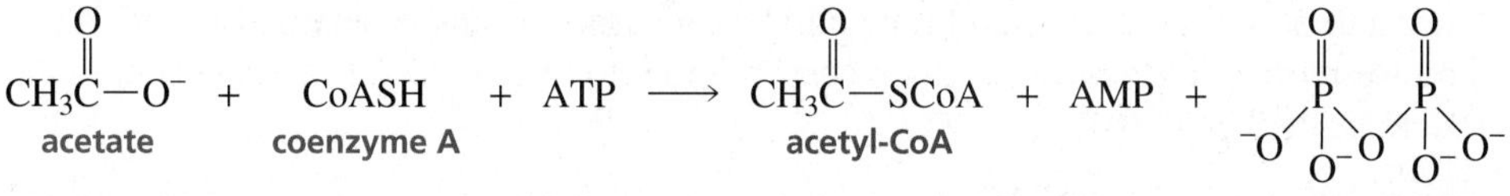

$$\underset{\text{acetate}}{\text{CH}_3\overset{\text{O}}{\overset{\|}{\text{C}}}\text{–O}^-} + \underset{\text{coenzyme A}}{\text{CoASH}} + \text{ATP} \longrightarrow \underset{\text{acetyl-CoA}}{\text{CH}_3\overset{\text{O}}{\overset{\|}{\text{C}}}\text{–SCoA}} + \text{AMP} + \underset{\text{pyrophosphate}}{{}^-\text{O–P(=O)(O}^-\text{)–O–P(=O)(O}^-\text{)–O}^-}$$

One of the necessary reactants for fatty acid synthesis is malonyl-CoA, which is obtained by carboxylation of acetyl-CoA. The mechanism for this reaction is discussed in Section 25.5.

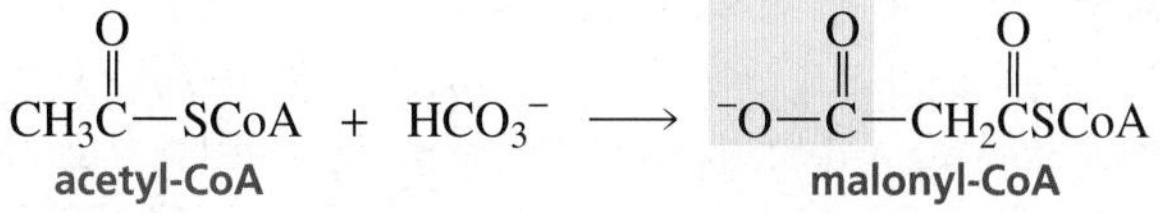

$$\underset{\text{acetyl-CoA}}{\text{CH}_3\overset{\text{O}}{\overset{\|}{\text{C}}}\text{–SCoA}} + \text{HCO}_3^- \longrightarrow \underset{\text{malonyl-CoA}}{{}^-\text{O–}\overset{\text{O}}{\overset{\|}{\text{C}}}\text{–CH}_2\overset{\text{O}}{\overset{\|}{\text{C}}}\text{SCoA}}$$

Before fatty acid synthesis can occur, however, the acyl groups of acetyl-CoA and malonyl-CoA are transferred to other thiols by means of a transesterification reaction.

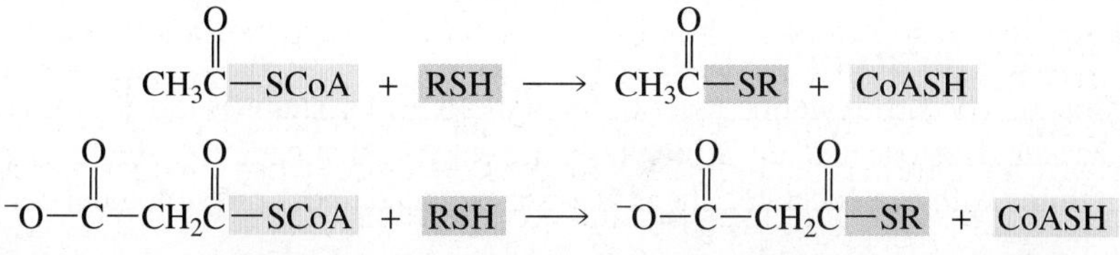

$$\text{CH}_3\overset{\text{O}}{\overset{\|}{\text{C}}}\text{–SCoA} + \text{RSH} \longrightarrow \text{CH}_3\overset{\text{O}}{\overset{\|}{\text{C}}}\text{–SR} + \text{CoASH}$$

$${}^-\text{O–}\overset{\text{O}}{\overset{\|}{\text{C}}}\text{–CH}_2\overset{\text{O}}{\overset{\|}{\text{C}}}\text{–SCoA} + \text{RSH} \longrightarrow {}^-\text{O–}\overset{\text{O}}{\overset{\|}{\text{C}}}\text{–CH}_2\overset{\text{O}}{\overset{\|}{\text{C}}}\text{–SR} + \text{CoASH}$$

The first step in the biosynthesis of a fatty acid is a Claisen condensation between a molecule of acetyl thioester and a molecule of malonyl thioester. You have seen that, in the laboratory the nucleophile needed for a Claisen condensation is obtained by using a strong base to remove an α-hydrogen. Strong bases are not available for biological reactions because they take place at neutral pH. So the required nucleophile is generated by removing CO_2—rather than a proton—from the α-carbon of malonyl thioester. (Recall that 3-oxocarboxylic acids are easily decarboxylated; Section 19.17.) Loss of CO_2 also serves to drive the condensation reaction to completion. The product of the condensation reaction undergoes a reduction, a dehydration, and a second reduction to give a four-carbon thioester.

$$CH_3\overset{O}{\overset{\|}{C}}-SR \;\; {}^{-}\ddot{O}-\overset{O}{\overset{\|}{C}}-CH_2\overset{O}{\overset{\|}{C}}-SR \longrightarrow CH_3\overset{:\ddot{O}:^{-}}{\overset{|}{C}}(SR)-CH_2\overset{O}{\overset{\|}{C}}-SR + CO_2 \longrightarrow CH_3\overset{O}{\overset{\|}{C}}-CH_2\overset{O}{\overset{\|}{C}}-SR$$

$$\xrightarrow{\text{reduction}} CH_3\overset{OH}{\overset{|}{C}H}-CH_2\overset{O}{\overset{\|}{C}}-SR \xrightarrow{\text{dehydration}} CH_3CH{=}CH\overset{O}{\overset{\|}{C}}-SR \xrightarrow{\text{reduction}} CH_3CH_2CH_2\overset{O}{\overset{\|}{C}}SR$$

The four-carbon thioester undergoes a Claisen condensation with another molecule of malonyl thioester. Again, the product of the condensation reaction undergoes a reduction, a dehydration, and a second reduction, this time to form a six-carbon thioester. The sequence of reactions is repeated, and each time two more carbons are added to the chain. This mechanism explains why naturally occurring fatty acids are unbranched and generally contain an even number of carbons.

$$CH_3CH_2CH_2\overset{O}{\overset{\|}{C}}-SR + {}^{-}O-\overset{O}{\overset{\|}{C}}-CH_2\overset{O}{\overset{\|}{C}}-SR \xrightarrow{\text{Claisen condensation}} CH_3CH_2CH_2\overset{O}{\overset{\|}{C}}-CH_2\overset{O}{\overset{\|}{C}}-SR$$

$$\xrightarrow{\text{1. reduction; 2. dehydration; 3. reduction}} CH_3CH_2CH_2CH_2CH_2\overset{O}{\overset{\|}{C}}-SR$$

Once a thioester with the appropriate number of carbon atoms is obtained, it undergoes a transesterification reaction with glycerol in order to form fats, oils, and phospholipids (Sections 26.3 and 26.4).

PROBLEM 41◆

Palmitic acid (hexadecanoic acid) is a saturated 16-carbon fatty acid. How many moles of malonyl-CoA are required for the synthesis of one mole of palmitic acid?

PROBLEM 42◆

a. If the biosynthesis of palmitic acid were carried out with CD_3COSR and nondeuterated malonyl thioester, how many deuteriums would be incorporated into palmitic acid?
b. If the biosynthesis of palmitic acid were carried out with $^{-}OOCCD_2COSR$ and nondeuterated acetyl thioester, how many deuteriums would be incorporated into palmitic acid?

A Biological Decarboxylation

An example of a decarboxylation that occurs in biological systems is the decarboxylation of acetoacetate. Acetoacetate decarboxylase, the enzyme that catalyzes the reaction, first forms an imine with acetoacetate. The imine is protonated under physiological conditions and therefore readily accepts the pair of electrons left behind when the substrate loses CO_2. Decarboxylation forms an enamine. Hydrolysis of the enamine produces the decarboxylated product (acetone) and regenerates the enzyme (Section 18.6).

$$\text{E—NH}_2 + \text{acetoacetate} \rightleftharpoons \text{E—}\overset{+}{\text{N}}\text{H=C(CH}_3\text{)CH}_2\text{COO}^- \;\; (\text{a protonated imine})$$

acetoacetate decarboxylase · acetoacetate · a protonated imine

$$\text{E—}\ddot{\text{N}}\text{H—C(=CH}_2\text{)CH}_3 + CO_2 \;\; (\text{an enamine})$$

$$\text{E—NH}_2 + \text{O=C(CH}_3)_2 \xleftarrow{H_2O} \text{E—}\overset{+}{\text{N}}\text{H=C(CH}_3)_2 \xleftarrow{H_3O^+} \text{E—}\ddot{\text{N}}\text{H—C(=CH}_2\text{)CH}_3$$

acetone · an enamine

Tutorial: Common terms: reactions at the α-carbon

PROBLEM 43

When the enzymatic decarboxylation of acetoacetate is carried out in $H_2{}^{18}O$, the acetone that is formed contains ^{18}O. What does this tell you about the mechanism of the reaction?

Summary

A hydrogen bonded to an **α-carbon** of an aldehyde, ketone, ester, or *N,N*-disubstituted amide is sufficiently acidic to be removed by a strong base because the base that is formed when the proton is removed is stabilized by delocalization of its negative change onto an oxygen. A **carbon acid** is a compound with a relatively acidic hydrogen bonded to an sp^3 hybridized carbon. Aldehydes and ketones ($pK_a \sim 16–20$) are more acidic than esters ($pK_a \sim 25$). **β-Diketones** ($pK_a \sim 9$) and **β-keto esters** ($pK_a \sim 11$) are even more acidic.

Keto–enol interconversion can be catalyzed by acids or by bases. Generally, the **keto tautomer** is more stable. When an **α-substitution reaction** takes place under acidic conditions, an enol reacts with an electrophile; when the reaction takes place under basic conditions, an enolate ion reacts with an electrophile. Whether C or O reacts with the electrophile depends on the electrophile and the reaction conditions.

Aldehydes and ketones react with Br_2, Cl_2, or I_2: Under acidic conditions, a halogen replaces one of the α-hydrogens of the carbonyl compound; under basic conditions, halogens replace all the α-hydrogens. The **HVZ** reaction brominates the α-carbon of a carboxylic acid. When the α-position is halogenated, the α-carbon reacts with nucleophiles.

LDA is used to form an enolate in reactions that require the carbonyl compound to be completely converted to enolate before it reacts with an electrophile. If the electrophile is an alkyl halide, the enolate is alkylated. The less substituted α-carbon is alkylated when the reaction is under kinetic control; the more substituted α-carbon is alkylated when the reaction is under thermodynamic control. Aldehydes and ketones can be alkylated or acylated via an **enamine intermediate**. Enolates of β-diketones, β-diesters, β-keto esters, and β-keto nitriles undergo **Michael reactions** with α,β-unsaturated carbonyl compounds. Michael reactions form 1,5-dicarbonyl compounds.

In an **aldol addition**, the enolate of an aldehyde or a ketone reacts with the carbonyl carbon of a second molecule of aldehyde or ketone, forming a β-hydroxyaldehyde or a β-hydroxyketone. The new C—C bond forms between the α-carbon of one molecule and the carbon that formerly was the carbonyl carbon of the other molecule. The product of an aldol addition can be dehydrated to give an **aldol condensation** product. In a **Claisen condensation**, the enolate of an ester reacts with a second molecule of ester, eliminating an $^-$OR group to form a β-keto ester. A **Dieckmann condensation** is an intramolecular Claisen condensation. A **Robinson annulation** is a ring-forming reaction in which a Michael reaction and an intramolecular aldol addition occur sequentially.

Carboxylic acids with a carbonyl group at the 3-position **decarboxylate** when they are heated. Carboxylic acids can be prepared by a **malonic ester synthesis**; the α-carbon of the diester is alkylated and the α-substituted malonic ester undergoes acid-catalyzed hydrolysis and decarboxylation;

the resulting carboxylic acid has two more carbon atoms than the alkyl halide. Similarly, methyl ketones can be prepared by an **acetoacetic ester synthesis**; the carbonyl group and the carbon atoms on either side of it come from acetoacetic ester, and the rest of the methyl ketone comes from the alkyl halide.

When planning the synthesis of a compound that requires the formation of a new carbon–carbon bond, first locate the new bond that must be made, and then determine which of the atoms that form the bond should be the nucleophile and which should be the electrophile.

Summary of Reactions

1. Halogenation of the α-carbon of aldehydes and ketones (Section 19.4).

$$RCH_2\overset{O}{\overset{\|}{C}}H + X_2 \xrightarrow{H_3O^+} R\underset{X}{\underset{|}{C}}H\overset{O}{\overset{\|}{C}}H + HX$$

$$RCH_2\overset{O}{\overset{\|}{C}}R + \underset{\text{excess}}{X_2} \xrightarrow{HO^-} R\overset{X}{\overset{|}{\underset{X}{\underset{|}{C}}}}-\overset{O}{\overset{\|}{C}}R + 2X^-$$

$$X_2 = Cl_2, Br_2, \text{ or } I_2$$

2. Halogenation of the α-carbon of carboxylic acids: the Hell–Volhard–Zelinski reaction (Section 19.5).

$$RCH_2\overset{O}{\overset{\|}{C}}OH \xrightarrow[\text{2. } H_2O]{\text{1. } PBr_3 \text{ (or P)}, Br_2} R\underset{Br}{\underset{|}{C}}H\overset{O}{\overset{\|}{C}}OH$$

3. The α-carbon can serve as a nucleophile and react with an electrophile (Section 19.6).

$$R\underset{H}{\underset{|}{C}}H\overset{O}{\overset{\|}{C}}R \overset{\text{base}}{\rightleftharpoons} R\overset{-}{C}H\overset{O}{\overset{\|}{C}}R \xrightarrow{E^+} R\underset{E}{\underset{|}{C}}H\overset{O}{\overset{\|}{C}}R$$

4. The α-carbon can serve as an electrophile and react with a nucleophile (Section 19.6).

$$R\underset{H}{\underset{|}{C}}H\overset{O}{\overset{\|}{C}}R \xrightarrow[Br_2]{H_3O^+} R\underset{Br}{\underset{|}{C}}H\overset{O}{\overset{\|}{C}}R \xrightarrow{\bar{N}u} R\underset{Nu}{\underset{|}{C}}H\overset{O}{\overset{\|}{C}}R$$

5. Compounds with halogenated α-carbons can form α,β-unsaturated carbonyl compounds (Section 19.6).

$$RCH_2\underset{Br}{\underset{|}{C}}H\overset{O}{\overset{\|}{C}}R' \xrightarrow{\text{base}} RCH{=}CH\overset{O}{\overset{\|}{C}}R'$$

6. Alkylation of the α-carbon of carbonyl compounds (Section 19.8).

$$RCH_2\overset{O}{\overset{\|}{C}}R' \xrightarrow[2.\ RCH_2X]{1.\ LDA/THF} R\underset{CH_2R}{\underset{|}{C}}H\overset{O}{\overset{\|}{C}}R' \qquad X = \text{halogen}$$

$$RCH_2\overset{O}{\overset{\|}{C}}OR' \xrightarrow[2.\ RCH_2X]{1.\ LDA/THF} R\underset{CH_2R}{\underset{|}{C}}H\overset{O}{\overset{\|}{C}}OR'$$

$$RCH_2C{\equiv}N \xrightarrow[2.\ RCH_2X]{1.\ LDA/THF} R\underset{CH_2R}{\underset{|}{C}}HC{\equiv}N$$

7. Alkylation and acylation of the α-carbon of aldehydes and ketones by means of an enamine intermediate (Sections 19.9 and 19.10).

$$\text{cyclohexanone} \xrightarrow[\text{catalytic } H^+]{\text{pyrrolidine (N–H)}} \text{1-(pyrrolidin-1-yl)cyclohexene}$$

$$\xrightarrow[2.\ HCl,\ H_2O]{1.\ RCH_2Br} \text{2-}(CH_2R)\text{cyclohexanone}$$

$$\xrightarrow[2.\ HCl,\ H_2O]{1.\ CH_3CH_2\overset{O}{\overset{\|}{C}}Cl} \text{2-}(\overset{O}{\overset{\|}{C}}CH_2CH_3)\text{cyclohexanone}$$

$$\xrightarrow[2.\ HCl,\ H_2O]{1.\ CH_2{=}CH\overset{O}{\overset{\|}{C}}H} \text{2-}(CH_2CH_2\overset{O}{\overset{\|}{C}}H)\text{cyclohexanone}$$

8. Michael reaction: attack of an enolate on an α,β-unsaturated carbonyl compound (Section 19.10).

$$RCH{=}CH\overset{O}{\overset{\|}{C}}R + R\overset{O}{\overset{\|}{C}}CH_2\overset{O}{\overset{\|}{C}}R \xrightarrow{HO^-} R\underset{R\underset{\|}{\underset{O}{C}}\underset{}{C}H\underset{\|}{\underset{O}{C}}R}{\underset{|}{C}}HCH_2\overset{O}{\overset{\|}{C}}R$$

9. Aldol addition of two aldehydes, two ketones, or an aldehyde and a ketone (Sections 19.11 and 19.13).

$$2\ RCH_2\overset{O}{\overset{\|}{C}}H \xrightarrow{HO^-} RCH_2\overset{OH}{\overset{|}{C}}H\underset{R}{\underset{|}{C}}H\overset{O}{\overset{\|}{C}}H$$

10. Aldol condensation: dehydration of the product of an aldol addition (Section 19.12).

$$RCH_2\overset{OH}{\overset{|}{C}}H\underset{R}{\underset{|}{C}}H\overset{O}{\overset{\|}{C}}H \xrightarrow[\Delta]{H_3O^+ \text{ or } HO^-} RCH_2CH{=}\underset{R}{\underset{|}{C}}\overset{O}{\overset{\|}{C}}H + H_2O$$

11. Claisen condensation of two esters (Sections 19.14 and 19.15).

$$2\ RCH_2\overset{O}{\overset{\|}{C}}OCH_3 \xrightarrow[2.\ HCl]{1.\ CH_3O^-} RCH_2\overset{O}{\overset{\|}{C}}\underset{R}{\underset{|}{C}}H\overset{O}{\overset{\|}{C}}OCH_3 + CH_3OH$$

12. Condensation of a ketone and an ester (Section 19.15).

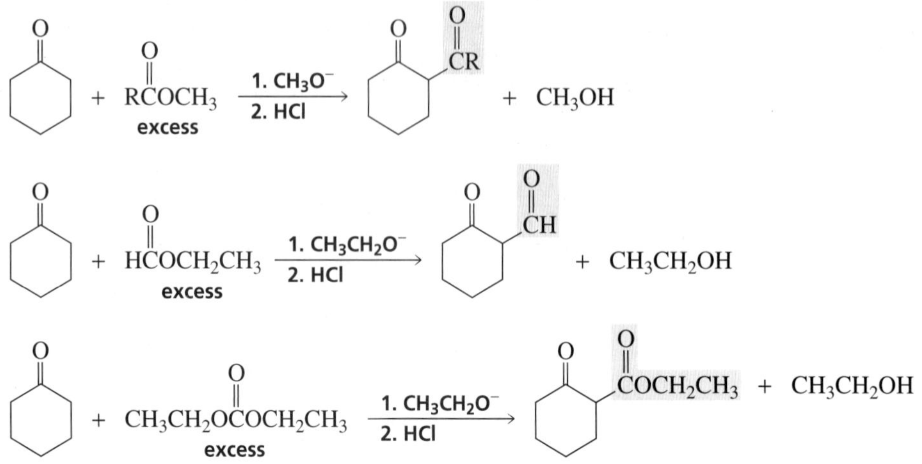

cyclohexanone + RCOCH_3 (C=O) (excess) $\xrightarrow[\text{2. HCl}]{\text{1. CH}_3\text{O}^-}$ 2-acylcyclohexanone (CR, C=O) + CH_3OH

cyclohexanone + $\text{HCOCH}_2\text{CH}_3$ (C=O) (excess) $\xrightarrow[\text{2. HCl}]{\text{1. CH}_3\text{CH}_2\text{O}^-}$ 2-formylcyclohexanone (CH, C=O) + CH_3CH_2OH

cyclohexanone + $\text{CH}_3\text{CH}_2\text{OCOCH}_2\text{CH}_3$ (C=O) (excess) $\xrightarrow[\text{2. HCl}]{\text{1. CH}_3\text{CH}_2\text{O}^-}$ 2-(COCH_2CH_3)cyclohexanone + CH_3CH_2OH

13. Robinson annulation (Section 19.16).

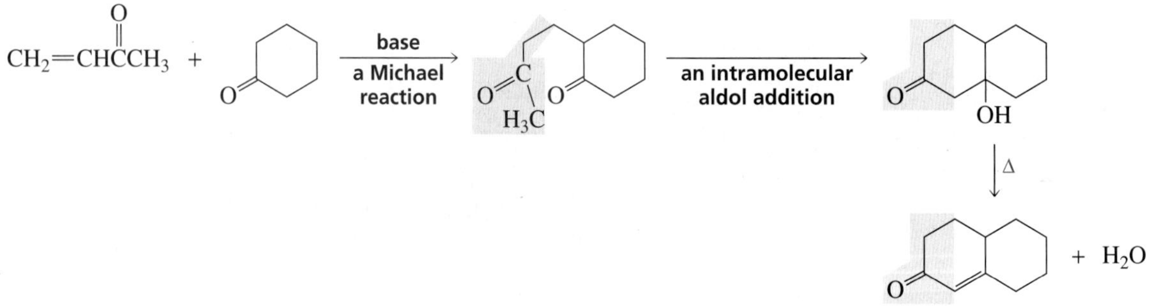

$\text{CH}_2{=}\text{CHCCH}_3$ (C=O) + cyclohexanone $\xrightarrow[\text{a Michael reaction}]{\text{base}}$ 2-(3-oxobutyl)cyclohexanone ($\text{O}{=}\text{C}$, H_3C) $\xrightarrow{\text{an intramolecular aldol addition}}$ bicyclic β-hydroxy ketone (OH) $\xrightarrow{\Delta}$ bicyclic α,β-unsaturated ketone + H_2O

14. Decarboxylation of 3-oxocarboxylic acids (Section 19.17).

$$\text{RCCH}_2\text{COH} \xrightarrow{\Delta} \text{RCCH}_3 + \text{CO}_2$$

15. Malonic ester synthesis: preparation of carboxylic acids (Section 19.18).

$$\text{C}_2\text{H}_5\text{OCCH}_2\text{COC}_2\text{H}_5 \xrightarrow[\text{2. RBr} \quad \text{3. HCl, H}_2\text{O}, \Delta]{\text{1. CH}_3\text{CH}_2\text{O}^-} \text{RCH}_2\text{COH} + \text{CO}_2$$

16. Acetoacetic ester synthesis: preparation of methyl ketones (Section 19.19).

$$\text{CH}_3\text{CCH}_2\text{COC}_2\text{H}_5 \xrightarrow[\text{2. RBr} \quad \text{3. HCl, H}_2\text{O}, \Delta]{\text{1. CH}_3\text{CH}_2\text{O}^-} \text{RCH}_2\text{CCH}_3 + \text{CO}_2$$

Key Terms

acetoacetic ester synthesis (p. 822)
aldol addition (p. 806)
aldol condensation (p. 807)
ambident nucleophile (p. 794)
annulation reaction (p. 817)
α-carbon (p. 788)
carbon acid (p. 789)
Claisen condensation (p. 810)
condensation reaction (p. 807)
crossed aldol addition (p. 809)
decarboxylation (p. 819)
Dieckmann condensation (p. 814)
β-diketone (p. 791)
enolization (p. 792)
gluconeogenesis (p. 826)

glycolysis (p. 826)
haloform reaction (p. 796)
Hell–Volhard–Zelinski (HVZ) reaction (p. 796)
Hunsdiecker reaction (p. 820)
α-hydrogen (p. 788)
keto–enol interconversion (p. 792)
keto–enol tautomerization (p. 792)
β-keto ester (p. 791)
Kolbe–Schmitt carboxylation reaction (p. 801)
malonic ester synthesis (p. 821)
Michael reaction (p. 804)
mixed aldol addition (p. 809)
mixed Claisen condensation (p. 812)
Robinson annulation (p. 817)
α-substitution reaction (p. 793)
Stork enamine reaction (p. 805)
tautomers (p. 791)

Problems

44. Write a structure for each of the following:
    a. ethyl acetoacetate
    b. α-methylmalonic acid
    c. a β-keto ester
    d. the enol tautomer of cyclopentanone
    e. the carboxylic acid obtained from the malonic ester synthesis when the alkyl halide is propyl bromide

45. Give the products of the following reactions:
    a. diethyl heptanedioate: (1) sodium ethoxide; (2) HCl
    b. pentanoic acid + PBr_3 + Br_2, followed by hydrolysis
    c. acetone + ethyl acetate: (1) sodium ethoxide; (2) HCl
    d. diethyl 2-ethylhexanedioate: (1) sodium ethoxide; (2) HCl
    e. diethyl malonate: (1) sodium ethoxide; (2) isobutyl bromide; (3) HCl, H_2O + Δ
    f. acetophenone + diethyl carbonate: (1) sodium ethoxide; (2) HCl
    g. 1,3-cyclohexanedione + allyl bromide + sodium hydroxide
    h. dibenzyl ketone + methyl vinyl ketone + excess sodium hydroxide
    i. cyclopentanone: (1) pyrrolidine + catalytic H^+; (2) ethyl bromide; (3) HCl, H_2O
    j. γ-butyrolactone + LDA in THF followed by methyl iodide
    k. 2,7-octanedione + sodium hydroxide
    l. cyclohexanone + NaOD in D_2O
    m. diethyl 1,2-benzenedicarboxylate + ethyl acetate: (1) excess sodium ethoxide; (2) HCl

46. The chemical shifts of nitromethane, dinitromethane, and trinitromethane are at δ 6.10, δ 4.33, and δ 7.52. Match each chemical shift with the compound. Explain how chemical shift correlates with pK_a.

47. a. Explain why a racemic mixture of 2-methyl-1-phenyl-1-butanone is formed when (*R*)-2-methyl-1-phenyl-1-butanone is dissolved in an acidic or basic aqueous solution.
    b. Give an example of another ketone that would undergo acid- or base-catalyzed racemization.

48. Identify A–L. (*Hint:* A shows three singlets in its ^1H NMR spectrum with integral ratios 3 : 2 : 3 and gives a positive iodoform test; see Section 19.4)

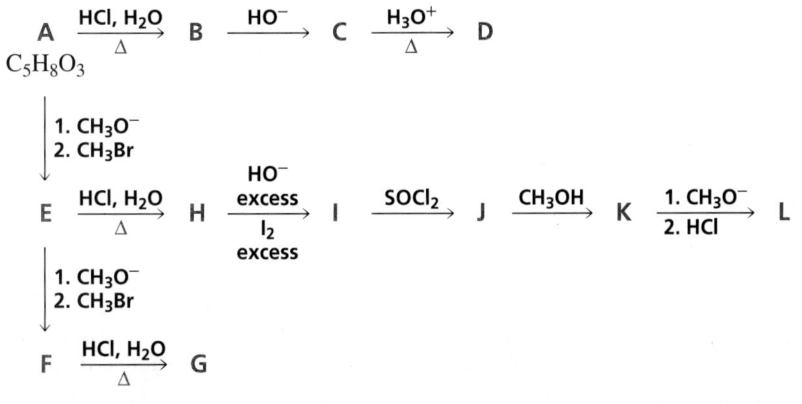

$$\underset{C_5H_8O_3}{A} \xrightarrow[\Delta]{HCl,\ H_2O} B \xrightarrow{HO^-} C \xrightarrow[\Delta]{H_3O^+} D$$

A $\xrightarrow{1.\ CH_3O^-;\ 2.\ CH_3Br}$ E

$$E \xrightarrow[\Delta]{HCl,\ H_2O} H \xrightarrow[I_2\ \text{excess}]{HO^-\ \text{excess}} I \xrightarrow{SOCl_2} J \xrightarrow{CH_3OH} K \xrightarrow[2.\ HCl]{1.\ CH_3O^-} L$$

E $\xrightarrow{1.\ CH_3O^-;\ 2.\ CH_3Br}$ F

$$F \xrightarrow[\Delta]{HCl,\ H_2O} G$$

49. Show how the following compounds could be prepared from cyclohexanone:

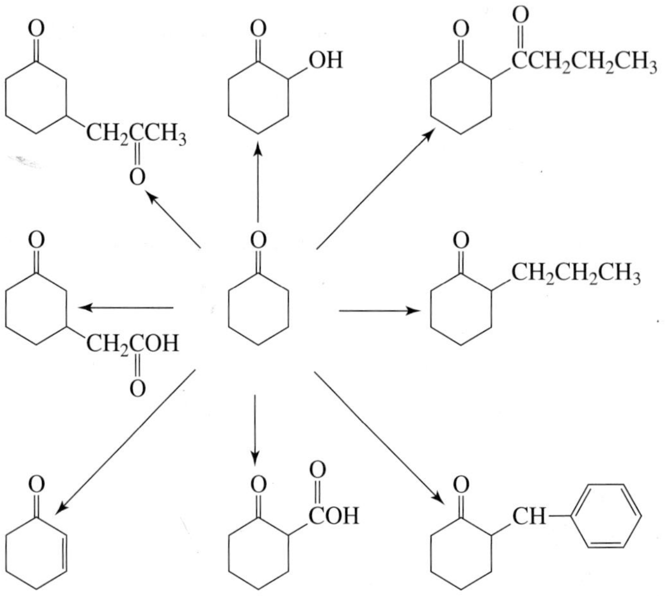

50. A β,γ-unsaturated carbonyl compound rearranges to a more stable conjugated α,β-unsaturated compound in the presence of either acid or base.

a. Propose a mechanism for the base-catalyzed rearrangement.

b. Propose a mechanism for the acid-catalyzed rearrangement.

$\xrightarrow{H_3O^+ \text{ or } HO^-}$

a β,γ-unsaturated carbonyl compound | **an α,β-unsaturated carbonyl compound**

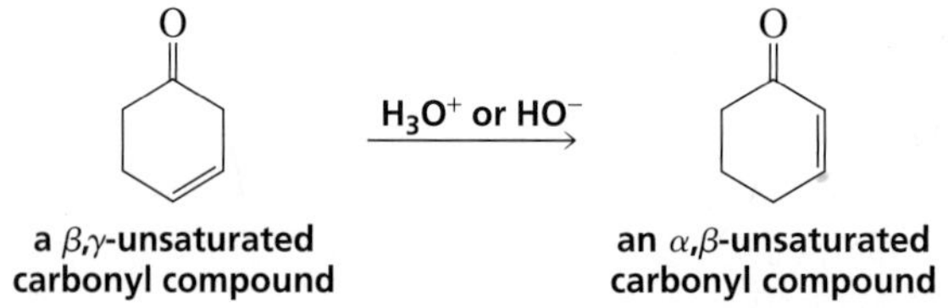

51. There are other condensation reactions similar to the aldol and Claisen condensations:

a. The *Perkin condensation* is the condensation of an aromatic aldehyde and acetic anhydride. Give the product obtained from the following Perkin condensation:

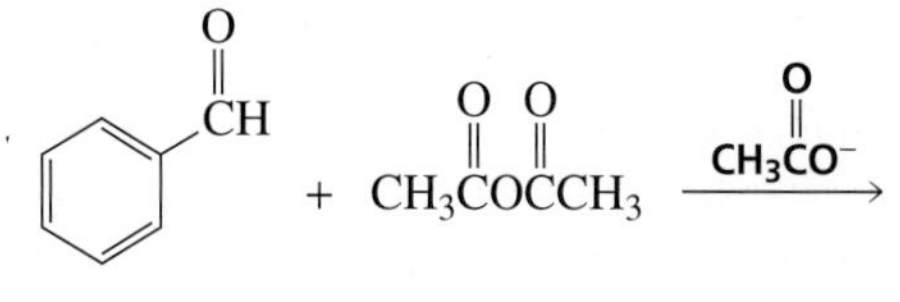

$$C_6H_5CHO + CH_3COCCH_3 \xrightarrow{CH_3CO^-}$$

b. What compound would result if water were added to the product of the Perkin condensation?

c. The *Knoevenagel condensation* is the condensation of an aldehyde or a ketone that has no α-hydrogens and a compound such as diethyl malonate that has an α-carbon flanked by two electron-withdrawing groups. Give the product obtained from the following Knoevenagel condensation:

$$C_6H_5CHO + C_2H_5OCCH_2COC_2H_5 \xrightarrow{CH_3CH_2O^-}$$

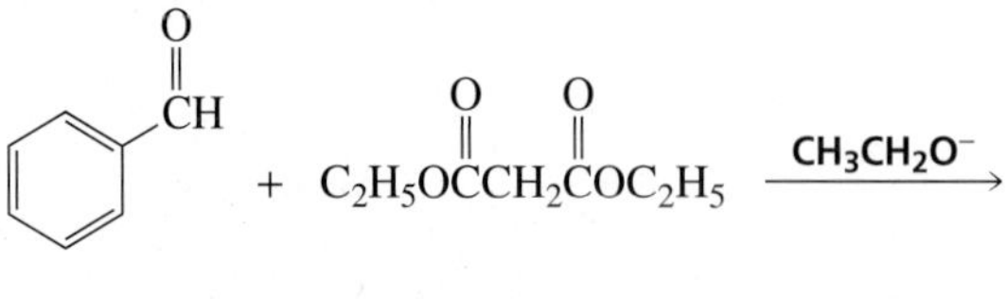

d. What product would be obtained if the product of the Knoevenagel condensation were heated in an aqueous acidic solution?

52. The *Reformatsky reaction* is an addition reaction in which an organozinc reagent is used instead of a Grignard reagent to attack the carbonyl group of an aldehyde or a ketone. Because the organozinc reagent is less reactive than a Grignard reagent, a nucleophilic addition to the ester group does not occur. The organozinc reagent is prepared by treating an a-bromo ester with zinc.

$$CH_3CH_2\overset{O}{\overset{\|}{C}}H + CH_3\underset{ZnBr}{\underset{|}{C}}H\overset{O}{\overset{\|}{C}}OCH_3 \longrightarrow CH_3CH_2\overset{^{+}ZnBr\ \ O^-}{\overset{|}{C}}H\underset{CH_3}{\underset{|}{C}}H\overset{O}{\overset{\|}{C}}OCH_3 \xrightarrow{H_2O} CH_3CH_2\overset{OH}{\overset{|}{C}}H\underset{CH_3}{\underset{|}{C}}H\overset{O}{\overset{\|}{C}}OCH_3$$

an organozinc reagent **a β-hydroxy ester**

Describe how each of the following compounds could be prepared, using a Reformatsky reaction:

a. $CH_3CH_2CH_2\overset{OH}{\overset{|}{C}}HCH_2\overset{O}{\overset{\|}{C}}OCH_3$

b. $CH_3CH_2\overset{OH}{\overset{|}{C}}H\underset{CH_2CH_3}{\underset{|}{C}}H\overset{O}{\overset{\|}{C}}OH$

c. $CH_3CH_2CH{=}\underset{CH_3}{\underset{|}{C}}\overset{O}{\overset{\|}{C}}OH$

d. $CH_3CH_2\overset{OH}{\overset{|}{\underset{CH_2CH_3}{\underset{|}{C}}}}CH_2\overset{O}{\overset{\|}{C}}OCH_3$

53. The ketone whose ^{1}H NMR spectrum is shown here was obtained as the product of an acetoacetic ester synthesis. What alkyl halide was used in the synthesis?

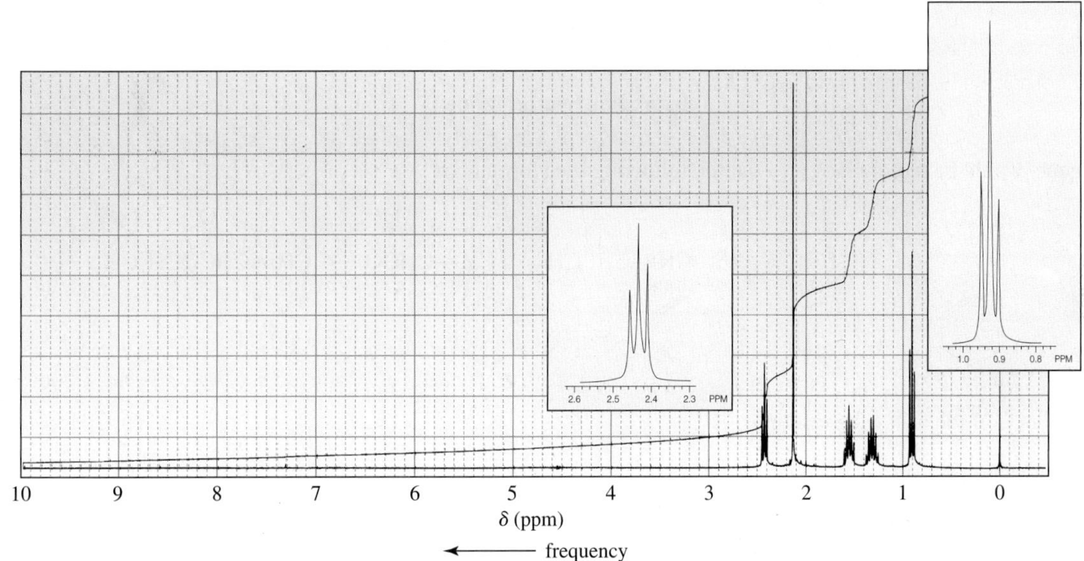

54. Indicate how the following compounds could be synthesized from cyclohexanone and any other necessary reagents:

a. 2-($CH_2CH_2CH_2CH_3$)cyclohexanone

b. 2-($\overset{O}{\overset{\|}{C}}CH_2CH_2CH_3$)cyclohexanone **(two ways)**

c. 2-($CH_2CH_2\overset{O}{\overset{\|}{C}}CH_3$)cyclohexanone

d. 2-($\overset{O}{\overset{\|}{C}}OCH_2CH_3$)cyclohexanone

e. 2-($\overset{O}{\overset{\|}{C}}H$)cyclohexanone

f. bicyclic ketone (O= on fused bicyclic ring)

55. Compound A with molecular formula C_6H_{10} has two peaks in its ^{1}H NMR spectrum, both of which are singlets (with ratio 9 : 1). Compound A reacts with an acidic aqueous solution containing mercuric sulfate to form compound B that gives a positive iodoform test (Section 19.4) and that has an ^{1}H NMR spectrum that shows two singlets (with ratio 3 : 1). Identify A and B.

56. Indicate how each of the following compounds could be synthesized from the given starting material and any other necessary reagents:

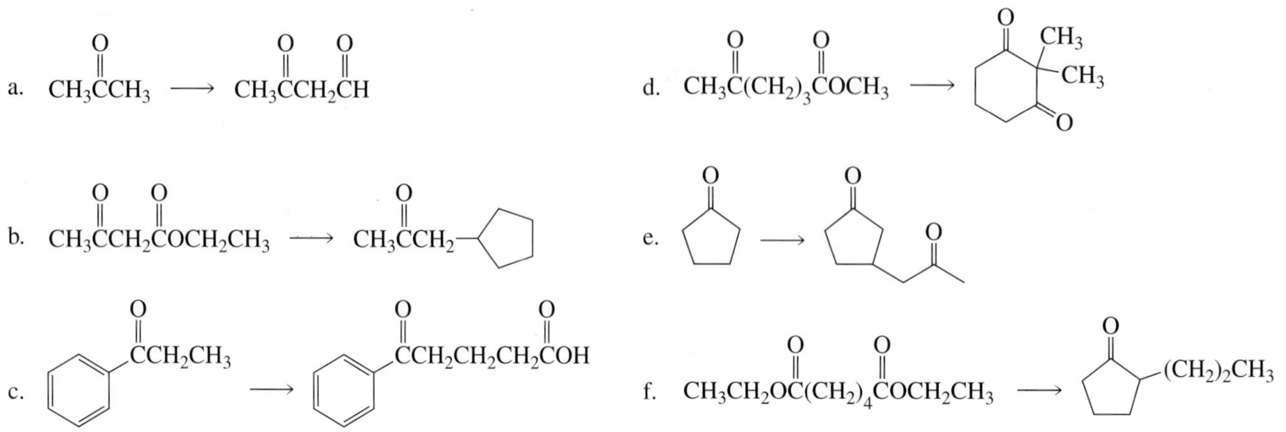

a. $CH_3\overset{O}{\overset{\|}{C}}CH_3 \longrightarrow CH_3\overset{O}{\overset{\|}{C}}CH_2\overset{O}{\overset{\|}{C}}H$

b. $CH_3\overset{O}{\overset{\|}{C}}CH_2\overset{O}{\overset{\|}{C}}OCH_2CH_3 \longrightarrow CH_3\overset{O}{\overset{\|}{C}}CH_2\text{–cyclopentyl}$

c. $C_6H_5\overset{O}{\overset{\|}{C}}CH_2CH_3 \longrightarrow C_6H_5\overset{O}{\overset{\|}{C}}CH_2CH_2CH_2\overset{O}{\overset{\|}{C}}OH$

d. $CH_3\overset{O}{\overset{\|}{C}}(CH_2)_3\overset{O}{\overset{\|}{C}}OCH_3 \longrightarrow$ 2,2-dimethylcyclohexane-1,3-dione

e. cyclopentanone $\longrightarrow$ 2-(2-oxopropyl)cyclopentanone

f. $CH_3CH_2O\overset{O}{\overset{\|}{C}}(CH_2)_4\overset{O}{\overset{\|}{C}}OCH_2CH_3 \longrightarrow$ 2-$(CH_2)_2CH_3$-cyclopentanone

57. Bupropion hydrochloride is an antidepressant marketed under the trade name Wellbutrin®. Propose a synthesis of bupropion hydrochloride, starting with benzene.

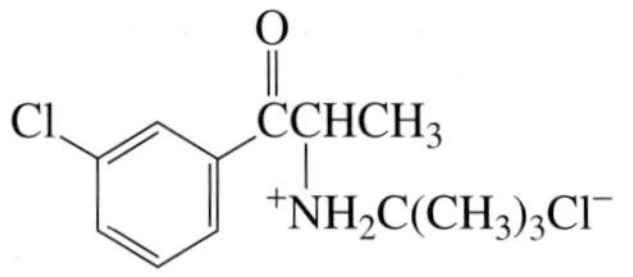

3-Cl-$C_6H_4\overset{O}{\overset{\|}{C}}CHCH_3$ with $^+NH_2C(CH_3)_3Cl^-$ on the CH

bupropion hydrochloride

58. What reagents would be required to carry out the following transformations?

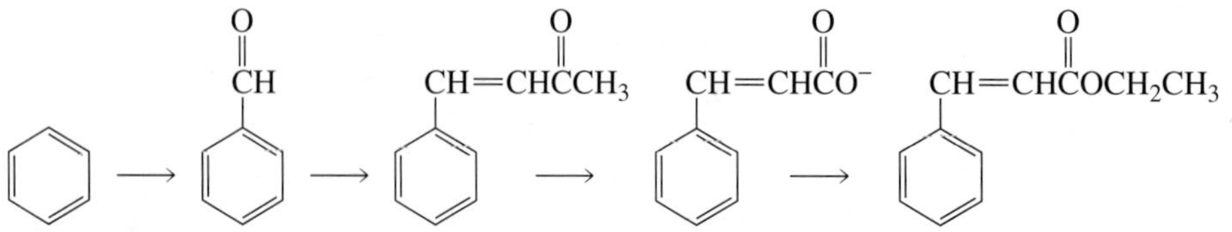

$C_6H_6 \longrightarrow C_6H_5\overset{O}{\overset{\|}{C}}H \longrightarrow C_6H_5CH{=}CH\overset{O}{\overset{\|}{C}}CH_3 \longrightarrow C_6H_5CH{=}CH\overset{O}{\overset{\|}{C}}O^- \longrightarrow C_6H_5CH{=}CH\overset{O}{\overset{\|}{C}}OCH_2CH_3$

59. Give the products of the following reactions:

a. 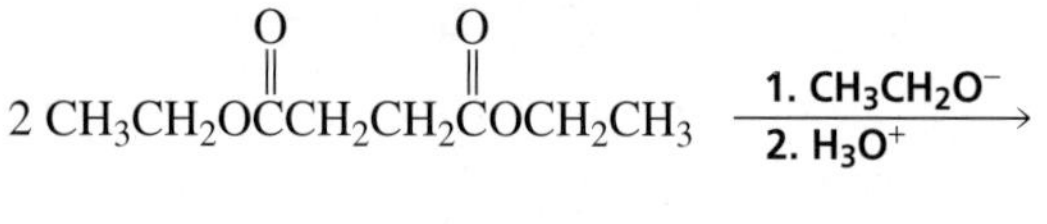 $2\ CH_3CH_2O\overset{O}{\overset{\|}{C}}CH_2CH_2\overset{O}{\overset{\|}{C}}OCH_2CH_3 \xrightarrow[\text{2. } H_3O^+]{\text{1. } CH_3CH_2O^-}$

b. 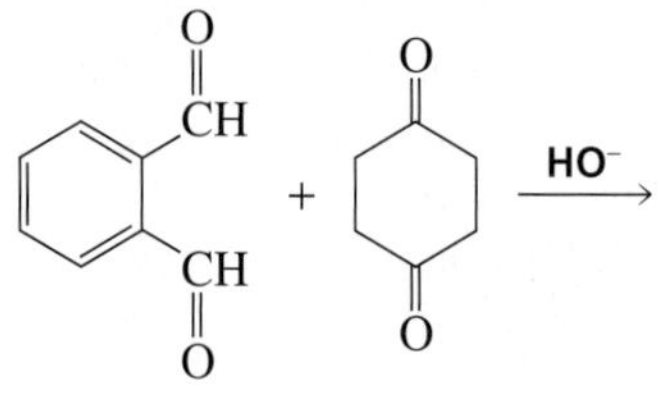 benzene-1,2-dicarbaldehyde ($\overset{O}{\overset{\|}{C}}H$, $\overset{O}{\overset{\|}{C}}H$) + cyclohexane-1,4-dione $\xrightarrow{HO^-}$

60. a. Show how the amino acid alanine can be synthesized from propanoic acid.
b. Show how the amino acid glycine can be synthesized from phthalimide and diethyl 2-bromomalonate.

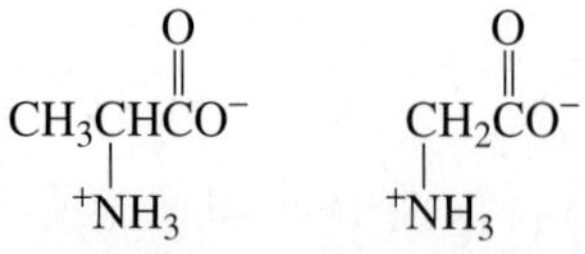

$CH_3\underset{^+NH_3}{\underset{|}{C}H}\overset{O}{\overset{\|}{C}}O^-$ **alanine**

$\underset{^+NH_3}{\underset{|}{C}H_2}\overset{O}{\overset{\|}{C}}O^-$ **glycine**

61. Cindy Synthon tried to prepare the following compounds using aldol condensations. Which of these compounds was she successful in synthesizing? Explain why the other syntheses were not successful.

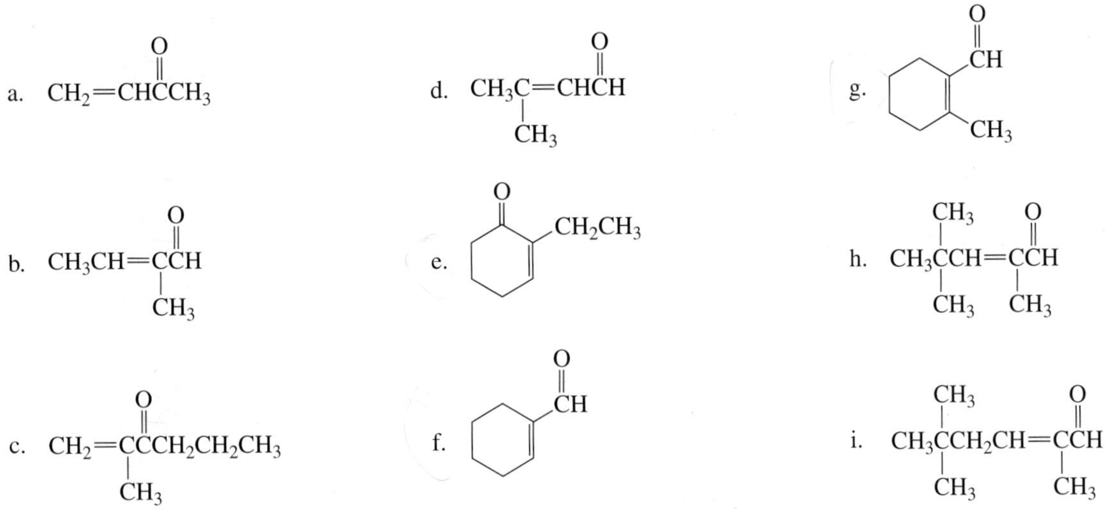

a. $CH_2{=}CHCCH_3$ (C=O)

b. $CH_3CH{=}C(CH_3)CH$ (C=O)

c. $CH_2{=}C(CH_3)CCH_2CH_2CH_3$ (C=O)

d. $CH_3C(CH_3){=}CHCH$ (C=O)

e. 2-cyclohexenone with CH_2CH_3

f. cyclohexene with CH(=O)

g. cyclohexane with CH(=O) and CH_3

h. $CH_3C(CH_3)_2CH{=}C(CH_3)CH$ (C=O)

i. $CH_3C(CH_3)_2CH_2CH{=}C(CH_3)CH$ (C=O)

62. Explain why the following bromoketone forms different bicyclic compounds under different reaction conditions:

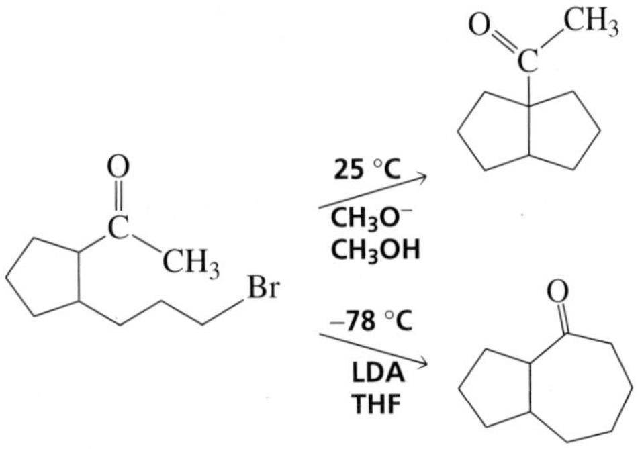

25 °C, CH_3O^-, CH_3OH

−78 °C, LDA, THF

63. Explain why the product obtained in the following reactions depends on the number of equivalents of base used in the reaction:

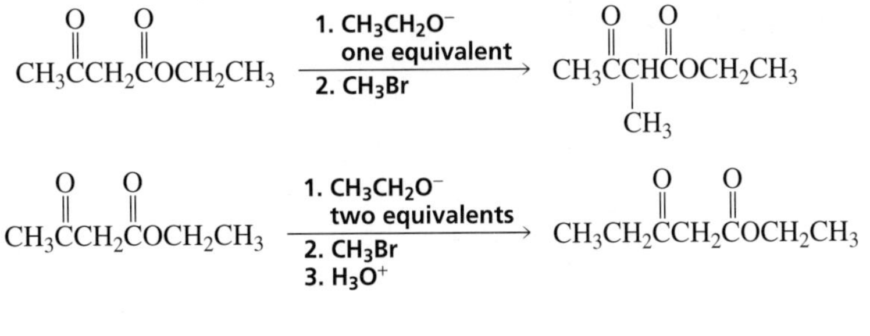

$$CH_3\overset{O}{\overset{\|}{C}}CH_2\overset{O}{\overset{\|}{C}}OCH_2CH_3 \xrightarrow[\text{2. } CH_3Br]{\text{1. } CH_3CH_2O^- \text{ one equivalent}} CH_3\overset{O}{\overset{\|}{C}}\underset{CH_3}{\underset{|}{C}H}\overset{O}{\overset{\|}{C}}OCH_2CH_3$$

$$CH_3\overset{O}{\overset{\|}{C}}CH_2\overset{O}{\overset{\|}{C}}OCH_2CH_3 \xrightarrow[\text{2. } CH_3Br \text{ 3. } H_3O^+]{\text{1. } CH_3CH_2O^- \text{ two equivalents}} CH_3CH_2\overset{O}{\overset{\|}{C}}CH_2\overset{O}{\overset{\|}{C}}OCH_2CH_3$$

64. A *Mannich reaction* puts an $>NCH_2-$ group on the α-carbon of a carbon acid. Propose a mechanism for the reaction.

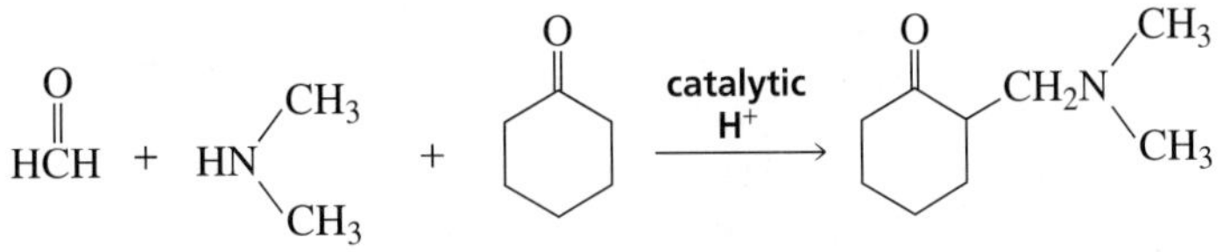

$$\overset{O}{\overset{\|}{HCH}} + HN(CH_3)_2 + \text{cyclohexanone} \xrightarrow{\text{catalytic } H^+} \text{2-}(CH_2N(CH_3)_2)\text{cyclohexanone}$$

65. What carbonyl compounds are required to prepare a compound with molecular formula $C_{10}H_{10}O$ whose ^{1}H NMR spectrum is shown?

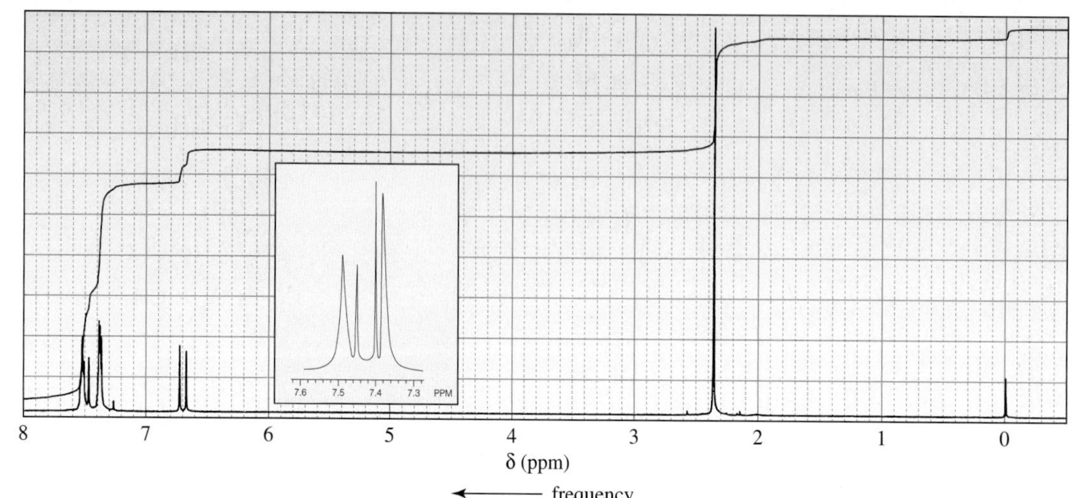

66. Ninhydrin reacts with an amino acid to form a purple-colored compound. Propose a mechanism to account for the formation of the colored compound.

2 ninhydrin + H_2NCHCO^- (R) (an amino acid) $\xrightarrow{\text{catalytic } H^+}$ purple-colored compound + CO_2 + RCH(=O) + 3 H_2O

67. A carboxylic acid is formed when an α-haloketone reacts with hydroxide ion. This reaction is called a *Favorskii reaction*. Propose a mechanism for the following Favorskii reaction. (*Hint:* In the first step, HO^- removes a proton from the α-carbon that is not bonded to Br; a three-membered ring is formed in the second step; and HO^- is a nucleophile in the third step.)

$$CH_3CH(Br)C(=O)CH_2CH_3 \xrightarrow[H_2O]{HO^-} CH_3CH_2CH(CH_3)C(=O)O^-$$

68. Give the products of the following reactions. (*Hint:* See Problem 67.)

a. 2-bromocyclohexanone $\xrightarrow[H_2O]{HO^-}$

b. bicyclic ketone with $OCCH_3$ (=O) $\xrightarrow[H_2O]{HO^-}$

69. An α,β-unsaturated carbonyl compound can be prepared by a reaction known as a selenenylation–oxidation reaction. A selenoxide is formed as an intermediate. Propose a mechanism for the reaction.

cyclohexanone $\xrightarrow{\text{1. LDA/THF; 2. } C_6H_5SeBr; \text{ 3. } H_2O_2}$ 2-cyclohexenone

a selenoxide: 2-($\overset{+}{Se}C_6H_5$, O^-)cyclohexanone

70. a. What carboxylic acid would be formed if the malonic ester synthesis were carried out with one equivalent of malonic ester, one equivalent of 1,5-dibromopentane, and two equivalents of base?
    b. What carboxylic acid would be formed if the malonic ester synthesis were carried out with two equivalents of malonic ester, one equivalent of 1,5-dibromopentane, and two equivalents of base?

71. A *Cannizzaro reaction* is the reaction of an aldehyde that has no α-hydrogens with concentrated aqueous sodium hydroxide. In this reaction, half the aldehyde is converted to a carboxylic acid and the other half is converted to an alcohol. Propose a reasonable mechanism for the following Cannizzaro reaction:

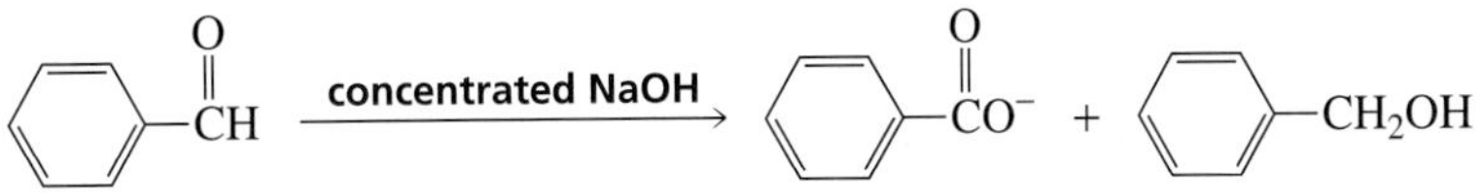

$$C_6H_5CHO \xrightarrow{\text{concentrated NaOH}} C_6H_5CO^- (O) + C_6H_5CH_2OH$$

72. Propose a reasonable mechanism for each of the following reactions:

a. $\xrightarrow{H_3O^+}$

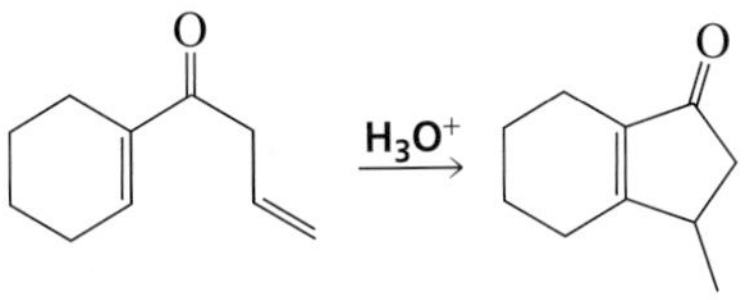

b. $\xrightarrow{H_3O^+}$ HO ... OH

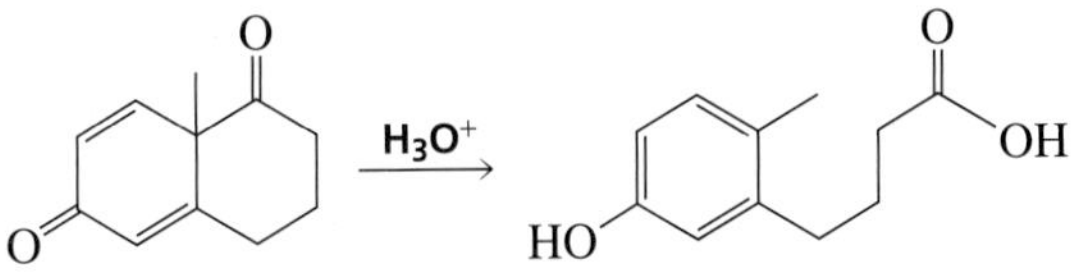

73. The following reaction is known as the *benzoin condensation*. The reaction will not take place if sodium hydroxide is used instead of sodium cyanide. Propose a mechanism for the reaction.

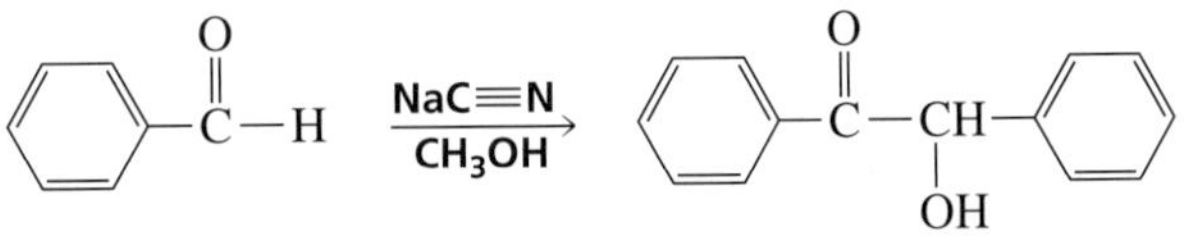

$$C_6H_5\overset{O}{\overset{\|}{C}}-H \xrightarrow[CH_3OH]{NaC{\equiv}N} C_6H_5\overset{O}{\overset{\|}{C}}-\underset{OH}{\underset{|}{CH}}-C_6H_5$$

benzoin

74. Orsellinic acid, a common constituent of lichens, is synthesized from the condensation of acetyl thioester and malonyl thioester. If a lichen were grown on a medium containing acetate that was radioactively labeled with ^{14}C at the carbonyl carbon, which carbons would be labeled in orsellinic acid?

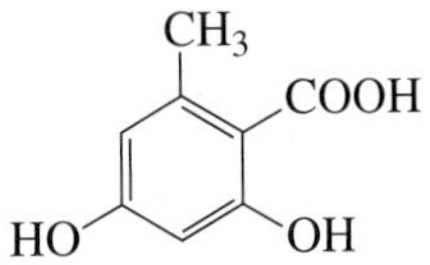

CH3, COOH, HO, OH

orsellinic acid

75. A compound known as *Hagemann's ester* can be prepared by treating a mixture of formaldehyde and ethyl acetoacetate first with base and then with acid and heat. Write the structure for the product of each of the steps.
    a. The first step is an aldol-like condensation.
    b. The second step is a Michael addition.
    c. The third step is an intramolecular aldol condensation.
    d. The fourth step is a hydrolysis followed by a decarboxylation.

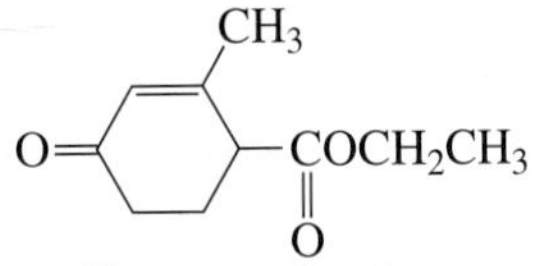

CH3, O=, COCH2CH3, O

Hagemann's ester

76. Amobarbital is a sedative marketed under the trade name Amytal®. Propose a synthesis of amobarbital, using diethyl malonate and urea as two of the starting materials.

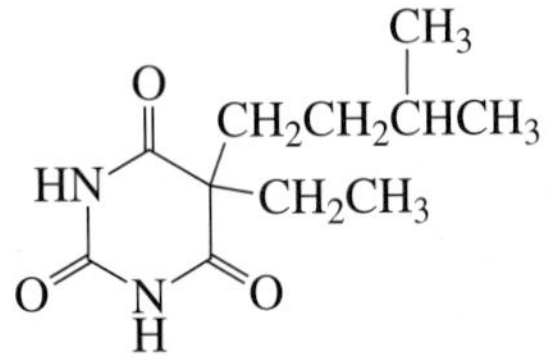

$CH_2CH_2CH(CH_3)CH_3$, CH_2CH_3, HN, O, N, H

Amytal®

77. Propose a reasonable mechanism for each of the following reactions:

a. 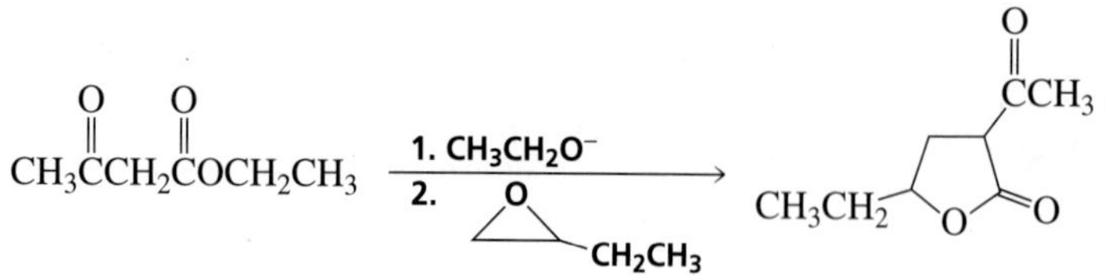 $CH_3\overset{O}{\overset{\|}{C}}CH_2\overset{O}{\overset{\|}{C}}OCH_2CH_3 \xrightarrow[\text{2. epoxide-}CH_2CH_3]{\text{1. }CH_3CH_2O^-}$ (lactone product bearing CH_3CH_2 and $\overset{O}{\overset{\|}{C}}CH_3$ groups)

b. 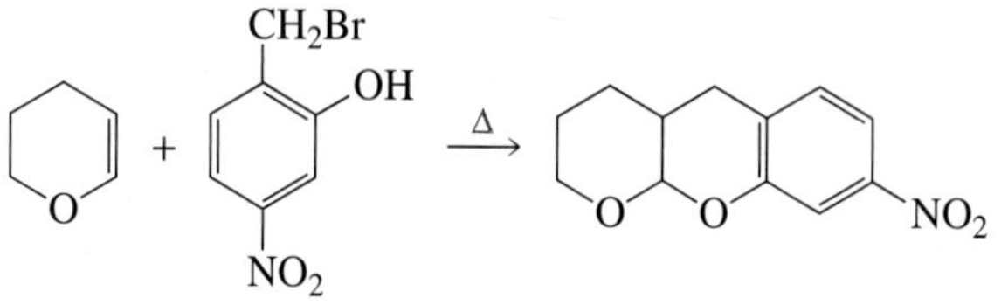 dihydropyran + CH_2Br, OH, NO_2-substituted benzene $\xrightarrow{\Delta}$ tricyclic product with NO_2

78. Tyramine is an alkaloid found in mistletoe, ripe cheese, and putrefied animal tissue. Dopamine is a neurotransmitter involved in the regulation of the central nervous system.

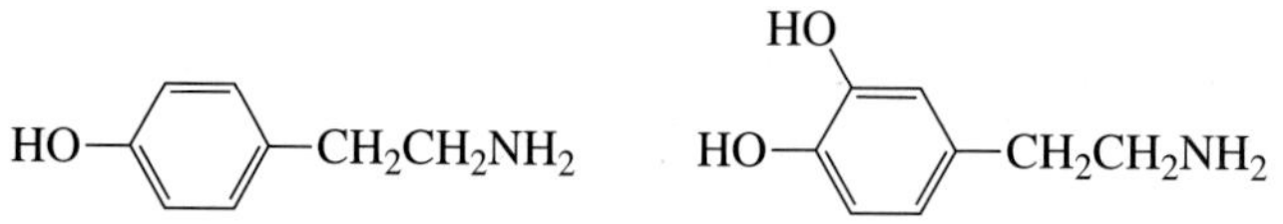

HO–C6H4–$CH_2CH_2NH_2$
tyramine

HO, HO–C6H3–$CH_2CH_2NH_2$
dopamine

a. Give two ways to prepare β-phenylethylamine from β-phenylethyl chloride.
b. How can β-phenylethylamine be prepared from benzyl chloride?
c. How can β-phenylethylamine be prepared from benzaldehyde?
d. How can tyramine be prepared from β-phenylethylamine?
e. How can dopamine be prepared from tyramine?

79. Show how estrone, a steroid hormone, can be prepared from the given starting materials. (*Hint:* Start with a Robinson annulation.)

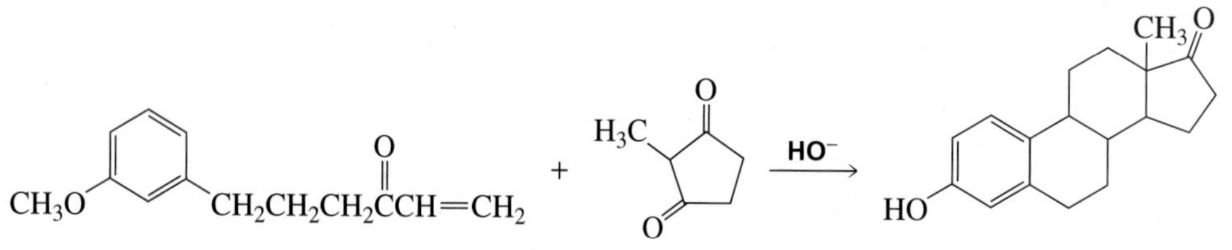

CH_3O–C6H4–$CH_2CH_2CH_2\overset{O}{\overset{\|}{C}}CH{=}CH_2$ + 2-methylcyclopentane-1,3-dione (H_3C) $\xrightarrow{HO^-}$ HO–, CH_3, O (estrone)

estrone

80. a. Ketoprofen, like ibuprofen, is an anti-inflammatory analgesic. How could ketoprofen be synthesized from the given starting material?

(benzophenone bearing CH_3) $\longrightarrow$ (benzophenone bearing $\overset{O}{\overset{\|}{C}}HCOH$ with CH_3)

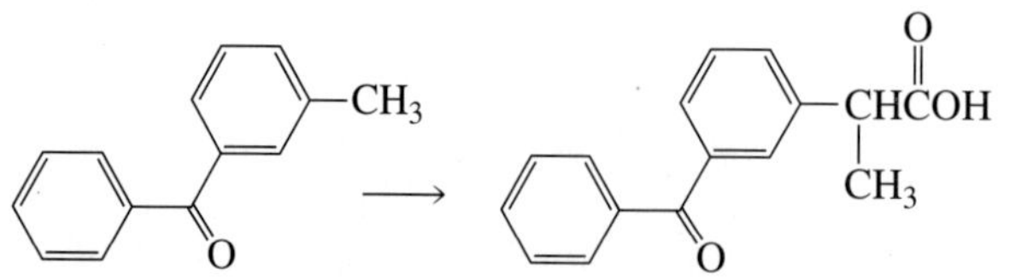

ketoprofen

b. Ketoprofen and ibuprofen both have a propanoic acid substituent (see Problem 70 in Chapter 17). Explain why the identical subunits are synthesized in different ways.

20 More About Oxidation–Reduction Reactions

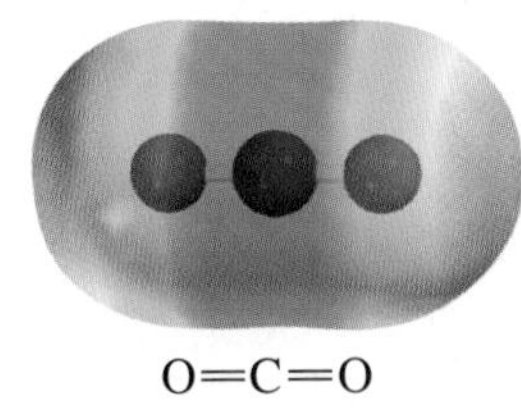

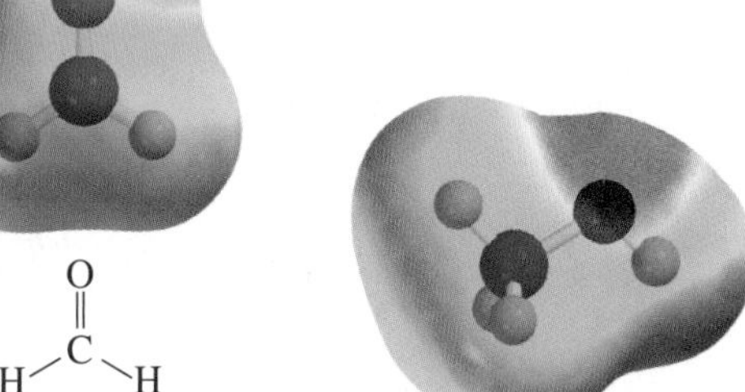

An important group of organic reactions consists of those that involve the transfer of electrons from one molecule to another. Organic chemists use these reactions—called **oxidation–reduction reactions** or **redox reactions**—to synthesize a large variety of compounds. Redox reactions are also important in biological systems because many of these reactions produce energy. You have seen a number of oxidation and reduction reactions in other chapters, but discussing them as a group will give you the opportunity to compare them.

In an oxidation–reduction reaction, one compound loses electrons and one compound gains electrons. The compound that loses electrons is oxidized, and the one that gains electrons is reduced. One way to remember the difference between oxidation and reduction is with the phrase "LEO the lion says GER": *Loss of Electrons is Oxidation; Gain of Electrons is Reduction.*

The following is an example of an oxidation–reduction reaction involving inorganic reagents:

$$Cu^{+} + Fe^{3+} \longrightarrow Cu^{2+} + Fe^{2+}$$

In this reaction, Cu^{+} loses an electron, so Cu^{+} is oxidized. Fe^{3+} gains an electron, so Fe^{3+} is reduced. The reaction demonstrates two important points about oxidation–reduction reactions. First, *oxidation is always coupled with reduction.* In other words, a compound cannot gain electrons (be reduced) unless another compound in the reaction simultaneously loses electrons (is oxidized). Second, the compound that is oxidized (Cu^{+}) is called the **reducing agent** because it loses the electrons that are used to reduce the other compound (Fe^{3+}). Similarly, the compound that is reduced (Fe^{3+}) is called the **oxidizing agent** because it gains the electrons given up by the other compound (Cu^{+}) when it is oxidized.

It is easy to tell whether an organic compound has been oxidized or reduced simply by looking at the change in the structure of the compound. We will be looking primarily at reactions where **oxidation** or **reduction** has taken place on carbon: If the reaction increases the number of C—H bonds or decreases the number of C—O, C—N, or C—X bonds (where X denotes a halogen), the compound has been reduced. If the reaction decreases the number of C—H bonds or increases the number of C—O, C—N, or C—X bonds, the compound has been oxidized. Notice that the **oxidation state** of a carbon atom equals the total number of its C—O, C—N, and C—X bonds.

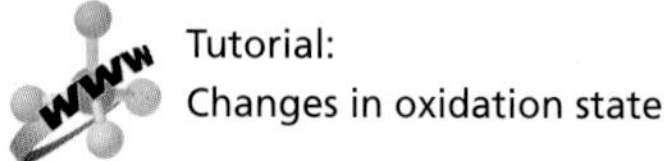

Tutorial: Changes in oxidation state

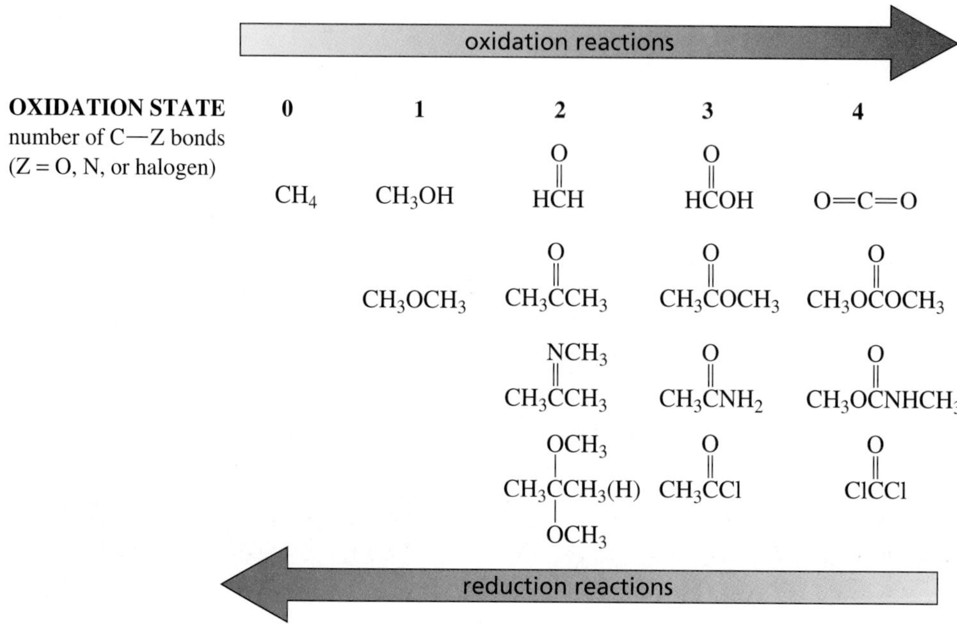

Let's now take a look at some examples of oxidation–reduction reactions that take place on carbon. You have seen these reactions in previous chapters. Notice that in each of the following reactions, the product has more C—H bonds than the reactant has: The alkene, aldehyde, and ketone, therefore, are being reduced (Sections 4.11, 18.5, and 15.15). Hydrogen, sodium borohydride, and hydrazine are the reducing agents.

Reduction at carbon increases the number of C—H bonds or decreases the number of C—O, C—N, or C—X bonds.

Oxidation at carbon decreases the number of C—H bonds or increases the number of C—O, C—N, or C—X bonds.

$$\underset{\text{an alkene}}{\mathrm{RCH{=}CHR}} \xrightarrow[\mathrm{Pt}]{\mathrm{H_2}} \mathrm{RCH_2CH_2R}$$

$$\underset{\text{an aldehyde}}{\mathrm{R{-}C({=}O){-}H}} \xrightarrow[\text{2. } \mathrm{H_3O^+}]{\text{1. } \mathrm{NaBH_4}} \mathrm{RCH_2OH}$$

$$\underset{\text{a ketone}}{\mathrm{R{-}C({=}O){-}R}} \xrightarrow[\mathrm{HO^-},\ \Delta]{\mathrm{H_2NNH_2}} \mathrm{RCH_2R}$$

In the next group of reactions, the number of C—Br bonds increases in the first reaction. In the second and third reactions, the number of C—H bonds decreases and the number of C—O bonds increases. This means that the alkene, the aldehyde, and the alcohol are being oxidized. Bromine and chromic acid (H_2CrO_4) are the oxidizing agents. Notice that the increase in the number of C—O bonds in the third reaction results from a carbon–oxygen single bond becoming a carbon–oxygen double bond.

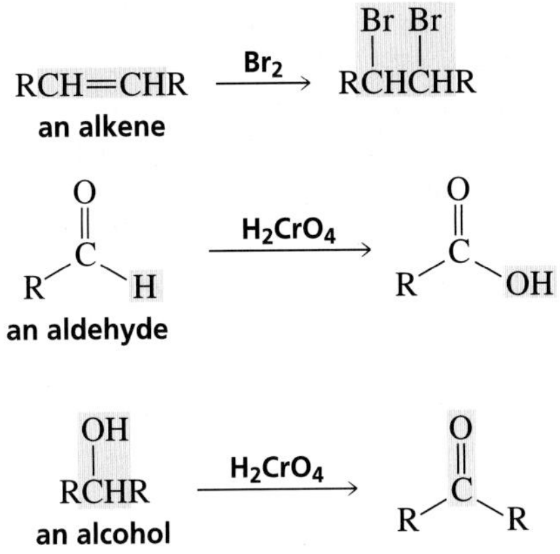

$$RCH{=}CHR \xrightarrow{Br_2} \underset{}{RCH(Br)CH(Br)R}$$
an alkene

$$RCHO \xrightarrow{H_2CrO_4} RCOOH$$
an aldehyde

$$RCH(OH)R \xrightarrow{H_2CrO_4} RC(=O)R$$
an alcohol

If water is added to an alkene, the product has one more C—H bond than the reactant, but it also has one more C—O bond. In this reaction, one carbon is reduced and another is oxidized. The two processes cancel each other as far as the overall molecule is concerned, so the overall reaction is neither an oxidation nor a reduction.

$$RCH{=}CHR \xrightarrow[H_2O]{H^+} RCH_2CH(OH)R$$

Oxidation–reduction reactions that take place on nitrogen or sulfur show similar structural changes. The number of N—H or S—H bonds increases in reduction reactions, and the number of N—O or S—O bonds increases in oxidation reactions. In the following reactions, nitrobenzene and the disulfide are being reduced (Sections 16.2 and 23.7), and the thiol is being oxidized to a sulfonic acid:

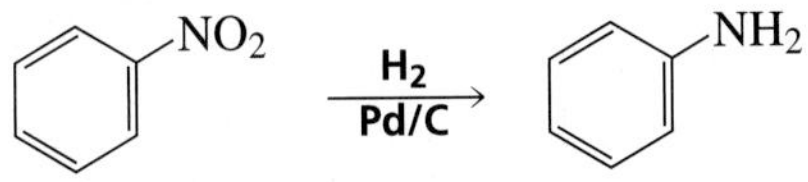

$$C_6H_5NO_2 \xrightarrow[Pd/C]{H_2} C_6H_5NH_2$$
nitrobenzene

$$CH_3CH_2S{-}SCH_2CH_3 \xrightarrow[Zn]{HCl} 2\ CH_3CH_2SH$$
a disulfide → a thiol

$$CH_3CH_2SH \xrightarrow{HNO_3} CH_3CH_2SO_3H$$
a thiol → a sulfonic acid

Many oxidizing reagents and many reducing reagents are available to organic chemists. This chapter highlights only a small fraction of the available reagents. The ones selected are some of the more common reagents that illustrate the types of transformations caused by oxidation and reduction.

PROBLEM 1◆

Indicate whether each of the following reactions is an oxidation reaction, a reduction reaction, or neither:

a. $$CH_3C(=O)Cl \xrightarrow[\text{partially deactivated Pd}]{H_2} CH_3C(=O)H$$

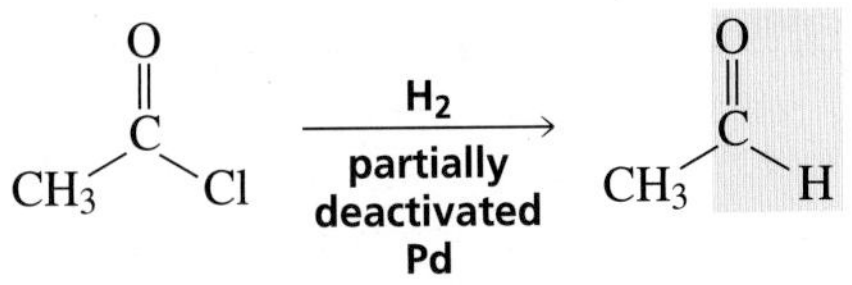

b. $RCH{=}CHR \xrightarrow{HBr} RCH_2CHBrR$

c. cyclohexane $\xrightarrow[h\nu]{Br_2}$ bromocyclohexane

d. $CH_3CH_2OH \xrightarrow{H_2CrO_4} CH_3COOH$

e. $CH_3C{\equiv}N \xrightarrow[Pt]{H_2} CH_3CH_2NH_2$

f. $CH_3CH_2CH_2Br \xrightarrow{HO^-} CH_3CH_2CH_2OH$

20.1 Reduction Reactions

An organic compound is reduced when hydrogen (H_2) is added to it. A molecule of H_2 can be thought of as being composed of (1) two hydrogen atoms, (2) two electrons and two protons, or (3) a hydride ion and a proton. In the sections that follow, you will see that these three ways to describe H_2 correspond to the three mechanisms by which H_2 is added to an organic compound.

components of H:H

H· ·H	·⁻ H⁺ ·⁻ H⁺	H:⁻ H⁺
two hydrogen atoms	two electrons and two protons	a hydride ion and a proton

Reduction by Addition of Two Hydrogen Atoms

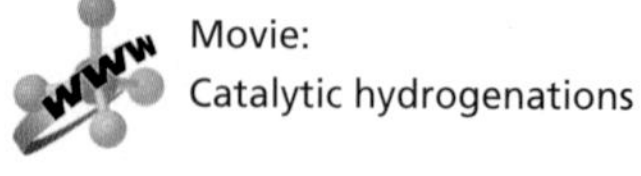

Movie: Catalytic hydrogenations

You have already seen that hydrogen can be added to carbon–carbon double and triple bonds in the presence of a metal catalyst (Sections 4.11 and 6.8). These reactions, called **catalytic hydrogenations**, are reduction reactions because there are more C—H bonds in the products than in the reactants. Alkenes and alkynes are both reduced to alkanes.

$$\underset{\text{1-butene}}{CH_3CH_2CH{=}CH_2} + H_2 \xrightarrow{\text{Pt, Pd, or Ni}} \underset{\text{butane}}{CH_3CH_2CH_2CH_3}$$

$$\underset{\text{1-pentyne}}{CH_3CH_2CH_2C{\equiv}CH} + 2\,H_2 \xrightarrow{\text{Pt, Pd, or Ni}} \underset{\text{pentane}}{CH_3CH_2CH_2CH_2CH_3}$$

In a catalytic hydrogenation, the H—H bond breaks homolytically (Section 4.11). This means that the reduction reaction involves the addition of two hydrogen atoms to the organic molecule.

We have seen that the catalytic hydrogenation of an alkyne can be stopped at a cis alkene if a partially deactivated catalyst is used (Section 6.8).

$$\underset{\text{2-butyne}}{CH_3C{\equiv}CCH_3} + H_2 \xrightarrow{\text{Lindlar catalyst}} \underset{\textit{cis}\text{-2-butene}}{(CH_3)HC{=}CH(CH_3)}$$

Only the alkene substituent is reduced in the following reaction. The very stable benzene ring can be reduced only under special conditions.

$$C_6H_5\text{–}CH{=}CH_2 \xrightarrow[\text{Pd/C}]{H_2} C_6H_5\text{–}CH_2CH_3$$

3-D Molecules: Styrene; Ethyl benzene

Catalytic hydrogenation can also be used to reduce carbon–nitrogen double and triple bonds. The reaction products are amines.

$$CH_3CH_2CH{=}NCH_3 + H_2 \xrightarrow{\text{Pd/C}} \underset{\text{methylpropylamine}}{CH_3CH_2CH_2NHCH_3}$$

$$CH_3CH_2CH_2C{\equiv}N + 2\,H_2 \xrightarrow{\text{Pd/C}} \underset{\text{butylamine}}{CH_3CH_2CH_2CH_2NH_2}$$

The carbonyl group of ketones and aldehydes can be reduced by catalytic hydrogenation, with Raney nickel as the metal catalyst. (Raney nickel is finely dispersed nickel with adsorbed hydrogen, so an external source of H_2 is not needed.) Aldehydes are reduced to primary alcohols, and ketones are reduced to secondary alcohols.

Murray Raney (1885–1966) *was born in Kentucky. He received a B.A. from the University of Kentucky in 1909, and in 1951 the university awarded him an honorary Doctor of Science. He worked at the Gilman Paint and Varnish Co. in Chattanooga, Tennessee, were he patented several chemical and metallurgical processes. In 1963, the company was sold and renamed W. R. Grace & Co., Raney Catalyst Division.*

$$\underset{\text{an aldehyde}}{CH_3CH_2CH_2C(=O)H} \xrightarrow[\text{Raney Ni}]{H_2} \underset{\text{a primary alcohol}}{CH_3CH_2CH_2CH_2OH}$$

$$\underset{\text{a ketone}}{CH_3CH_2C(=O)CH_3} \xrightarrow[\text{Raney Ni}]{H_2} \underset{\text{a secondary alcohol}}{CH_3CH_2CH(OH)CH_3}$$

The reduction of an acyl chloride can be stopped at an aldehyde if a partially deactivated catalyst is used. This reaction is known as the **Rosenmund reduction**. The catalyst for the Rosenmund reduction is similar to the partially deactivated palladium catalyst used to stop the reduction of an alkyne at a cis alkene (Section 6.8).

Karl W. Rosenmund (1884–1964) *was born in Berlin. He was a professor of chemistry at the University of Kiel.*

$$\underset{\text{an acyl chloride}}{CH_3CH_2C(=O)Cl} \xrightarrow[\text{partially deactivated Pd}]{H_2} \underset{\text{an aldehyde}}{CH_3CH_2C(=O)H}$$

The carbonyl groups of carboxylic acids, esters, and amides are less reactive, so they are harder to reduce than the carbonyl groups of aldehydes and ketones (Section 18.5). They cannot be reduced by catalytic hydrogenation (except under extreme conditions). They can, however, be reduced by a method we will discuss later in this section.

$$\underset{\text{a carboxylic acid}}{CH_3CH_2C(=O)OH} \xrightarrow[\text{Raney Ni}]{H_2} \text{no reaction}$$

$$\underset{\text{an ester}}{CH_3CH_2C(=O)OCH_3} \xrightarrow[\text{Raney Ni}]{H_2} \text{no reaction}$$

$$\underset{\text{an amide}}{CH_3CH_2C(=O)NHCH_3} \xrightarrow[\text{Raney Ni}]{H_2} \text{no reaction}$$

PROBLEM 2◆

Give the products of the following reactions:

a. $CH_3CH_2CH_2CH_2\overset{O}{\overset{\|}{C}}H \xrightarrow[\text{Raney Ni}]{H_2}$

b. $CH_3CH_2CH_2C{\equiv}N \xrightarrow[\text{Pd/C}]{H_2}$

c. $CH_3CH_2CH_2C{\equiv}CCH_3 \xrightarrow[\text{Lindlar catalyst}]{H_2}$

d. $CH_3\overset{O}{\overset{\|}{C}}OCH_3 \xrightarrow[\text{Raney Ni}]{H_2}$

e. $CH_3\overset{O}{\overset{\|}{C}}Cl \xrightarrow[\text{partially deactivated Pd}]{H_2}$

f. $CH_3\overset{O}{\overset{\|}{C}}Cl \xrightarrow[\text{Raney Ni}]{H_2}$

g. 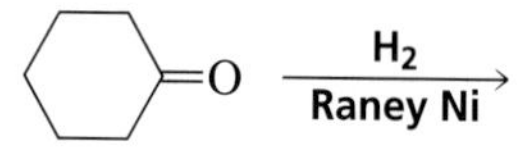 cyclohexanone (ring)=O $\xrightarrow[\text{Raney Ni}]{H_2}$

h. (cyclohexane ring)=NCH_3 $\xrightarrow[\text{Pd/C}]{H_2}$

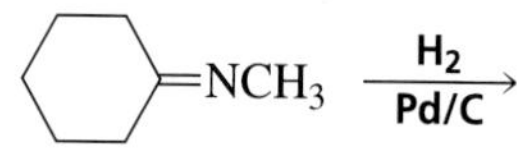

Reduction by Addition of an Electron, a Proton, an Electron, and a Proton

When a compound is reduced using sodium in liquid ammonia, sodium donates an electron to the compound and ammonia donates a proton. This sequence is then repeated, so the overall reaction adds two electrons and two protons to the compound. Such a reaction is known as a **dissolving-metal reduction**.

In Section 6.8, you saw the mechanism for the dissolving-metal reduction that converts an alkyne to a trans alkene.

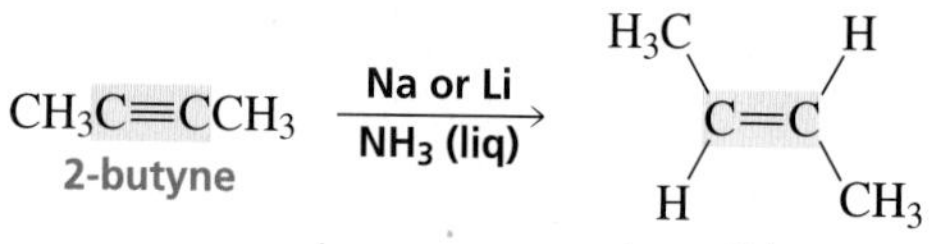

$$\underset{\text{2-butyne}}{CH_3C{\equiv}CCH_3} \xrightarrow[NH_3\ (liq)]{\text{Na or Li}} \underset{trans\text{-2-butene}}{(H_3C)(H)C{=}C(H)(CH_3)}$$

Sodium (or lithium) in liquid ammonia cannot reduce a carbon–carbon double bond. This makes it a useful reagent for reducing a triple bond in a compound that also contains a double bond.

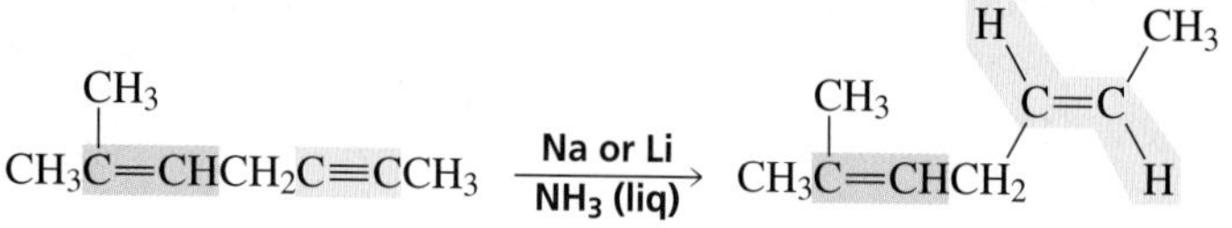

$$CH_3\overset{CH_3}{\overset{|}{C}}{=}CHCH_2C{\equiv}CCH_3 \xrightarrow[NH_3\ (liq)]{\text{Na or Li}} CH_3\overset{CH_3}{\overset{|}{C}}{=}CHCH_2(H)C{=}C(CH_3)(H)$$

Reduction by Addition of a Hydride Ion and a Proton

Carbonyl groups are easily reduced by metal hydrides such as sodium borohydride ($NaBH_4$) or lithium aluminum hydride. The actual reducing agent in **metal-hydride reductions** is hydride ion (H^-). Hydride ion adds to the carbonyl carbon, and the alkoxide ion that is formed is subsequently protonated. In other words, the carbonyl group is reduced by adding an H^- followed by an H^+. The mechanisms for reduction by these reagents are discussed in Section 18.5.

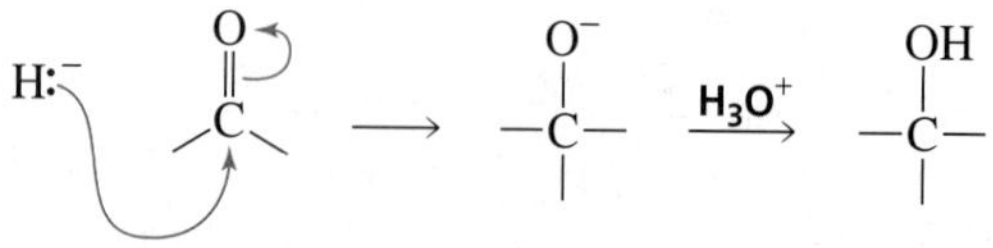

$$H{:}^- \; + \; \overset{O}{\overset{\|}{C}} \longrightarrow -\overset{O^-}{\overset{|}{\underset{|}{C}}}- \xrightarrow{H_3O^+} -\overset{OH}{\overset{|}{\underset{|}{C}}}-$$

Aldehydes, ketones, and acyl halides can be reduced by sodium borohydride.

Remember, the numbers in front of the reagents above or below a reaction arrow indicate that the second reagent is not added until reaction with the first reagent is completed.

$$CH_3CH_2CH_2C(=O)H \xrightarrow[\text{2. } H_3O^+]{\text{1. NaBH}_4} CH_3CH_2CH_2CH_2OH$$

an aldehyde → a primary alcohol

$$CH_3CH_2CH_2C(=O)CH_3 \xrightarrow[\text{2. } H_3O^+]{\text{1. NaBH}_4} CH_3CH_2CH_2CH(OH)CH_3$$

a ketone → a secondary alcohol

$$CH_3CH_2CH_2C(=O)Cl \xrightarrow[\text{2. } H_3O^+]{\text{1. NaBH}_4} CH_3CH_2CH_2CH_2OH$$

an acyl chloride → a primary alcohol

The metal–hydrogen bonds in lithium aluminum hydride are more polar than the metal–hydrogen bonds in sodium borohydride. As a result, $LiAlH_4$ is a stronger reducing agent than $NaBH_4$. Consequently, both $LiAlH_4$ and $NaBH_4$ reduce aldehydes, ketones, and acyl halides, but $LiAlH_4$ is not generally used for this purpose since $NaBH_4$ is safer and easier to use. $LiAlH_4$ is generally used to reduce only compounds—such as carboxylic acids, esters, and amides—that cannot be reduced by the milder reagent.

$$CH_3CH_2CH_2C(=O)OH \xrightarrow[\text{2. } H_3O^+]{\text{1. LiAlH}_4} CH_3CH_2CH_2CH_2OH + H_2O$$

a carboxylic acid → a primary alcohol

$$CH_3CH_2C(=O)OCH_3 \xrightarrow[\text{2. } H_3O^+]{\text{1. LiAlH}_4} CH_3CH_2CH_2OH + CH_3OH$$

an ester → a primary alcohol

If diisobutylaluminum hydride (DIBALH) is used as the hydride donor at a low temperature instead of $LiAlH_4$, the reduction of the ester can be stopped after the addition of one equivalent of hydride ion. Therefore, the final products of the reaction are an aldehyde and an alcohol (Section 18.5).

$$CH_3CH_2CH_2C(=O)OCH_3 \xrightarrow[\text{2. } H_2O]{\text{1. } [(CH_3)_2CHCH_2]_2AlH,\ -78\ ^\circ C} CH_3CH_2CH_2C(=O)H + CH_3OH$$

an ester → an aldehyde

Replacing some of the hydrogens of $LiAlH_4$ with OR groups decreases the reactivity of the metal hydride. For example, lithium tri-*tert*-butoxyaluminum hydride reduces an acyl chloride to an aldehyde, whereas $LiAlH_4$ reduces the acyl chloride all the way to an alcohol.

$$CH_3CH_2CH_2CH_2C(=O)Cl \xrightarrow[\text{2. } H_2O]{\text{1. LiAl}[OC(CH_3)_3]_3H,\ -78\ ^\circ C} CH_3CH_2CH_2CH_2C(=O)H$$

an acyl chloride → an aldehyde

The carbonyl group of an amide is reduced to a methylene group (CH_2) by lithium aluminum hydride (Section 18.5). Primary, secondary, and tertiary amines are formed, depending on the number of substituents bonded to the nitrogen of the amide. To obtain the amine in its neutral basic form, acid is not used in the second step of the reaction.

$$CH_3CH_2CH_2C(=O)NH_2 \xrightarrow[2.\ H_2O]{1.\ LiAlH_4} CH_3CH_2CH_2CH_2NH_2$$
a primary amine

$$CH_3CH_2CH_2C(=O)NHCH_3 \xrightarrow[2.\ H_2O]{1.\ LiAlH_4} CH_3CH_2CH_2CH_2NHCH_3$$
a secondary amine

$$CH_3CH_2CH_2C(=O)N(CH_3)CH_3 \xrightarrow[2.\ H_2O]{1.\ LiAlH_4} CH_3CH_2CH_2CH_2N(CH_3)CH_3$$
a tertiary amine

Because sodium borohydride cannot reduce an ester, an amide, or a carboxylic acid, it can be used to selectively reduce an aldehyde or a ketone group in a compound that also contains a less reactive group. Acid is not used in the second step of the following reaction, in order to avoid hydrolyzing the ester:

$$CH_3C(=O)CH_2CH_2C(=O)OCH_3 \xrightarrow[2.\ H_2O]{1.\ NaBH_4} CH_3CH(OH)CH_2CH_2C(=O)OCH_3$$

The multiply bonded carbon atoms of alkenes and alkynes do not possess a partial positive charge and therefore will not react with reagents that reduce compounds by donating a hydride ion.

$$CH_3CH_2CH{=}CH_2 \xrightarrow{NaBH_4} \text{no reduction reaction}$$

$$CH_3CH_2C{\equiv}CH \xrightarrow{NaBH_4} \text{no reduction reaction}$$

Because sodium borohydride cannot reduce carbon–carbon double bonds, a carbonyl group in a compound that also has an alkene functional group can be selectively reduced, as long as the double bonds are not conjugated (Section 18.13). Acid is not used in the second step of the reaction, in order to avoid addition to the double bond.

$$CH_3CH{=}CHCH_2C(=O)CH_3 \xrightarrow[2.\ H_2O]{1.\ NaBH_4} CH_3CH{=}CHCH_2CH(OH)CH_3$$

A **chemoselective reaction** is a reaction in which a reagent reacts with one functional group in preference to another. For example, $NaBH_4$ in isopropyl alcohol reduces aldehydes faster than it reduces ketones.

$$CH_3C(=O)CH_2CH_2C(=O)H \xrightarrow[\text{isopropyl alcohol}]{NaBH_4} CH_3C(=O)CH_2CH_2CH_2OH$$

In contrast, $NaBH_4$ in aqueous ethanol at $-15\ °C$ in the presence of cerium trichloride reduces ketones faster than it reduces aldehydes. There are many reducing reagents—and conditions under which those reagents should be used—available to the synthetic chemist. We can cover only a fraction of these in this chapter.

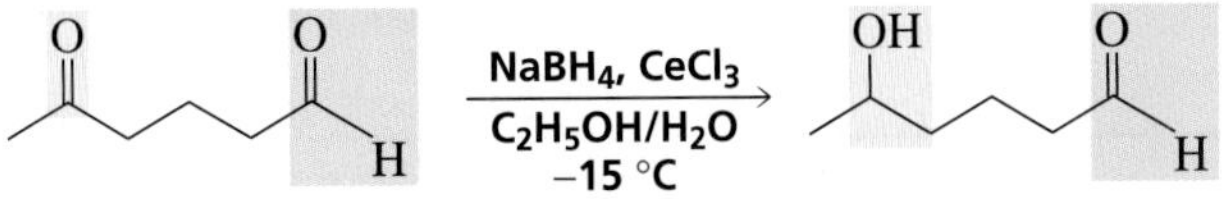

$$\xrightarrow[\substack{C_2H_5OH/H_2O \\ -15\ °C}]{NaBH_4,\ CeCl_3}$$

PROBLEM 3

Explain why terminal alkynes cannot be reduced by Na in liquid NH_3.

PROBLEM 4◆

Give the products of the following reactions:

a. 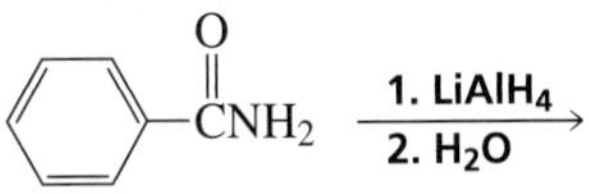 $C_6H_5CNH_2$ (C=O) $\xrightarrow[2.\ H_2O]{1.\ LiAlH_4}$

b. C_6H_5COH (C=O) $\xrightarrow[2.\ H_3O^+]{1.\ LiAlH_4}$

c. $CH_3CH_2\overset{O}{\overset{\|}{C}}CH_2CH_3 \xrightarrow[2.\ H_3O^+]{1.\ NaBH_4}$

d. $C_6H_{11}\overset{O}{\overset{\|}{C}}OCH_2CH_3 \xrightarrow[2.\ H_3O^+]{1.\ LiAlH_4}$

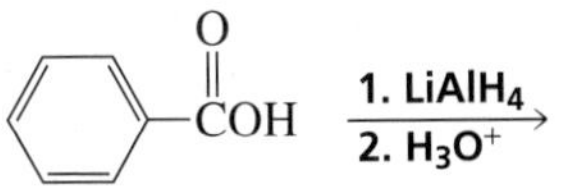

e. $CH_3CH_2\overset{O}{\overset{\|}{C}}NHCH_2CH_3 \xrightarrow[2.\ H_2O]{1.\ LiAlH_4}$

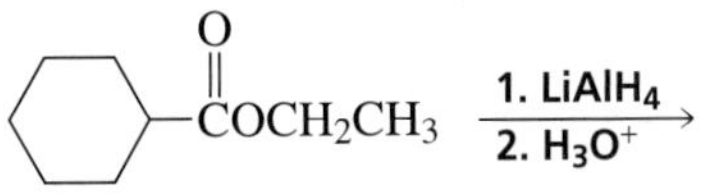

f. $CH_3CH_2CH_2\overset{O}{\overset{\|}{C}}OH \xrightarrow[2.\ H_3O^+]{1.\ LiAlH_4}$

PROBLEM 5

Can carbon–nitrogen double and triple bonds be reduced by lithium aluminum hydride? Explain your answer.

PROBLEM 6

Give the products of the following reactions (assume that excess reducing agent is used in d):

a. 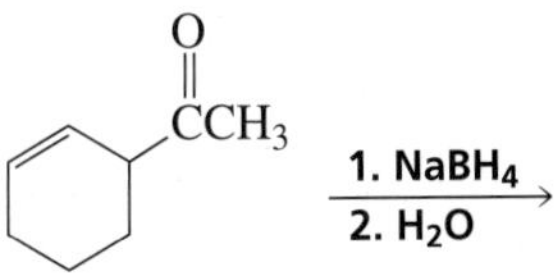 (cyclohexenyl)$\overset{O}{\overset{\|}{C}}CH_3 \xrightarrow[2.\ H_2O]{1.\ NaBH_4}$

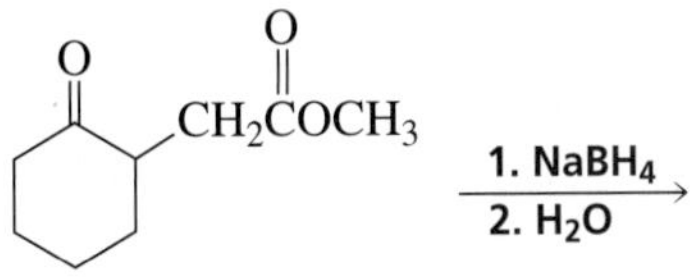

b. 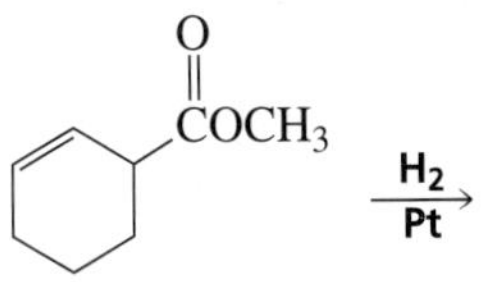 (cyclohexenyl)$\overset{O}{\overset{\|}{C}}OCH_3 \xrightarrow[Pt]{H_2}$

c. (2-oxocyclohexyl)$CH_2\overset{O}{\overset{\|}{C}}OCH_3 \xrightarrow[2.\ H_2O]{1.\ NaBH_4}$

d. 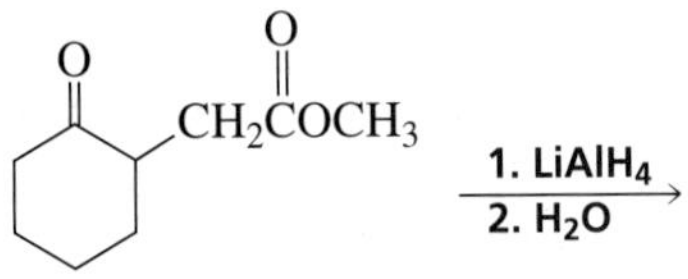 (2-oxocyclohexyl)$CH_2\overset{O}{\overset{\|}{C}}OCH_3 \xrightarrow[2.\ H_2O]{1.\ LiAlH_4}$

PROBLEM 7 SOLVED

How could you synthesize the following compounds from starting materials containing no more than four carbons?

a. 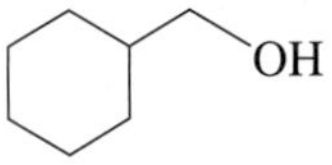 (cyclohexyl)CH_2CH_2OH

b. 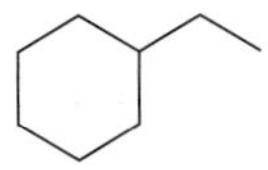 (cyclohexyl)CH_2CH_3

SOLUTION TO 7a The six-membered ring indicates that the compound can be synthesized by means of a Diels–Alder reaction.

$$\text{1,3-butadiene} + \text{CH}_2\text{=CHCHO} \xrightarrow{\Delta} \text{cyclohexenyl-CHO} \xrightarrow[\text{Raney Ni}]{H_2} \text{cyclohexyl-CH}_2\text{OH}$$

20.2 Oxidation of Alcohols

Oxidation is the reverse of reduction. For example, a ketone is *reduced* to a secondary alcohol, and the reverse reaction is the *oxidation* of a secondary alcohol to a ketone.

$$\text{ketone} \underset{\text{oxidation}}{\overset{\text{reduction}}{\rightleftharpoons}} \text{secondary alcohol}$$

3-D Molecules: Chromic acid; Chromium trioxide; Sodium dichromate

A reagent that is often used to oxidize alcohols is chromic acid (H_2CrO_4), which is formed when chromium trioxide (CrO_3) or sodium dichromate ($Na_2Cr_2O_7$) is dissolved in aqueous acid. These reactions are easily recognized as oxidations because the number of C—H bonds in the reactant decreases and the number of C—O bonds increases.

$$CH_3CH_2CH(OH)CH_3 \xrightarrow[H_2SO_4]{CrO_3} CH_3CH_2C(=O)CH_3$$

$$\text{cyclohexanol} \xrightarrow[H_2SO_4]{Na_2Cr_2O_7} \text{cyclohexanone}$$

$$\text{cyclopentyl-CH(OH)CH}_2\text{CH}_3 \xrightarrow{H_2CrO_4} \text{cyclopentyl-C(=O)CH}_2\text{CH}_3$$

secondary alcohols **ketones**

Primary alcohols are initially oxidized to aldehydes by these reagents. The reaction, however, does not stop at the aldehyde. Instead, the aldehyde is further oxidized to a carboxylic acid.

$$\underset{\text{a primary alcohol}}{CH_3CH_2CH_2CH_2OH} \xrightarrow{H_2CrO_4} \underset{\text{an aldehyde}}{\left[CH_3CH_2CH_2CH{=}O\right]} \xrightarrow[\text{oxidation}]{\text{further}} \underset{\text{a carboxylic acid}}{CH_3CH_2CH_2C(=O)OH}$$

Notice that the oxidation of either a primary or a secondary alcohol involves removal of a hydrogen from the carbon to which the OH is attached. The carbon bearing the OH group in a tertiary alcohol is not bonded to a hydrogen, so the OH group cannot be oxidized to a carbonyl group.

Chromic acid and the other chromium-containing oxidizing reagents oxidize an alcohol by first forming a chromate ester. The carbonyl compound is formed when the chromate ester undergoes an E2 elimination (Section 11.1).

mechanism for alcohol oxidation by chromic acid

$$HO{-}Cr(=O)_2{-}\ddot{O}H \xrightleftharpoons{H-B^+} HO{-}Cr(=O)_2{-}\overset{+}{O}H_2 \xrightarrow{RCH_2\ddot{O}H} RCH_2{-}\overset{+}{O}H{-}Cr(=O)_2{-}OH \xrightleftharpoons{H_2\ddot{O}:} \underset{\text{a chromate ester}}{RCH(H){-}O{-}Cr(=O)_2{-}OH} \xrightarrow[:B]{\text{an E2 reaction}} RCH{=}O + H_2CrO_3$$

$$HB^+ + H_2O$$

BLOOD ALCOHOL CONTENT

As blood passes through the arteries in the lungs, an equilibrium is established between the alcohol in one's blood and the alcohol in one's breath. So if the concentration of one is known, the concentration of the other can be estimated. The test that law enforcement agencies use to approximate a person's blood alcohol level is based on the oxidation of breath ethanol by sodium dichromate. The test employs a sealed glass tube that contains the oxidizing agent impregnated onto an inert material. The ends of the tube are broken off, and one end of the tube is attached to a mouthpiece and the other to a balloon-type bag. The person undergoing the test blows into the mouthpiece until the bag is filled with air.

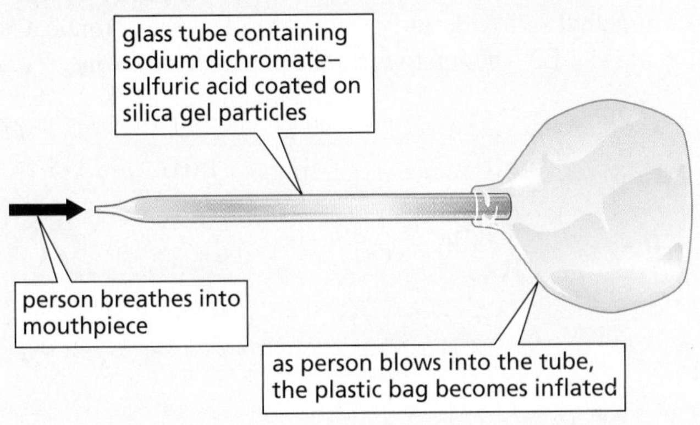

Any ethanol in the breath is oxidized as it passes through the column. When ethanol is oxidized, the red-orange oxidizing agent ($Cr_2O_7^{2-}$) is reduced to green chromic ion. The greater the concentration of alcohol in the breath, the farther the green color spreads through the tube.

$$\underset{}{CH_3CH_2OH} + \underset{\textbf{red orange}}{Cr_2O_7^{2-}} \xrightarrow{H^+} CH_3\overset{O}{\overset{\|}{C}}OH + \underset{\textbf{green}}{Cr^{3+}}$$

If the person fails this test—determined by the extent to which the green color spreads through the tube—a more accurate Breathalyzer® test is administered. The Breathalyzer® test also depends on the oxidation of breath ethanol by sodium dichromate, but it provides more accurate results because it is quantitative. In the test, a known volume of breath is bubbled through an acidic solution of sodium dichromate, and the concentration of chromic ion is measured precisely with a spectrophotometer.

The oxidation of a primary alcohol can be easily stopped at the aldehyde if pyridinium chlorochromate (PCC) is used as the oxidizing agent and the reaction is carried out in an anhydrous solvent such as dichloromethane, as explained in the following box:

$$\underset{\textbf{a primary alcohol}}{CH_3CH_2CH_2CH_2OH} \xrightarrow[CH_2Cl_2]{PCC} \underset{\textbf{an aldehyde}}{CH_3CH_2CH_2\overset{O}{\overset{\|}{C}}H}$$

THE ROLE OF HYDRATES IN THE OXIDATION OF PRIMARY ALCOHOLS

When a primary alcohol is oxidized to a carboxylic acid, the alcohol is initially oxidized to an aldehyde, which is in equilibrium with its hydrate (Section 18.7). It is the hydrate that is subsequently oxidized to a carboxylic acid.

The oxidation reaction can be stopped at the aldehyde if the reaction is carried out with pyridinium chlorochromate (PCC), because PCC is used in an anhydrous solvent. If water is not present, the hydrate cannot be formed.

$$CH_3CH_2OH \xrightarrow{H_2CrO_4} CH_3\overset{O}{\overset{\|}{C}}H \underset{}{\overset{H^+}{\underset{H_2O}{\rightleftharpoons}}} CH_3\overset{OH}{\overset{|}{\underset{OH}{\underset{|}{C}}}}H \xrightarrow{H_2CrO_4} CH_3\overset{O}{\overset{\|}{C}}OH$$

Because of the toxicity of chromium-based reagents, other methods for the oxidation of alcohols have been developed. One of the most widely employed methods, called the **Swern oxidation**, uses dimethyl sulfoxide [$(CH_3)_2SO$], oxalyl chloride [$(COCl)_2$], and triethylamine. Since the reaction is *not* carried out in an

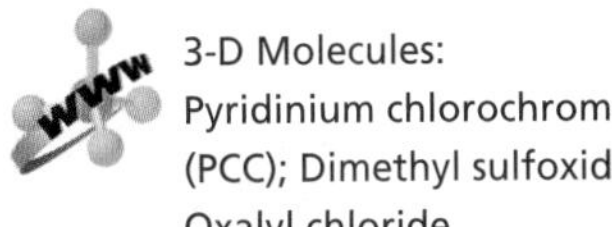

3-D Molecules: Pyridinium chlorochromate (PCC); Dimethyl sulfoxide; Oxalyl chloride

aqueous solution, the oxidation of a primary alcohol (like PCC oxidation) stops at the aldehyde. Secondary alcohols are oxidized to ketones.

$$\underset{\text{a primary alcohol}}{CH_3CH_2CH_2OH} \xrightarrow[\text{2. triethylamine}]{\text{1. } CH_3\overset{O}{\overset{\|}{S}}CH_3,\ Cl-\overset{O}{\overset{\|}{C}}-\overset{O}{\overset{\|}{C}}-Cl,\ -60\ °C} \underset{\text{an aldehyde}}{CH_3CH_2\overset{O}{\overset{\|}{C}}H}$$

$$\underset{\text{a secondary alcohol}}{CH_3CH_2\overset{OH}{\overset{|}{C}}HCH_3} \xrightarrow[\text{2. triethylamine}]{\text{1. } CH_3\overset{O}{\overset{\|}{S}}CH_3,\ Cl-\overset{O}{\overset{\|}{C}}-\overset{O}{\overset{\|}{C}}-Cl,\ -60\ °C} \underset{\text{a ketone}}{CH_3CH_2\overset{O}{\overset{\|}{C}}CH_3}$$

The actual oxidizing agent in the Swern oxidation is dimethylchlorosulfonium ion, which is formed from the reaction of dimethyl sulfoxide and oxalyl chloride. Like chromic acid oxidation, the Swern oxidation uses an E2 reaction to form the aldehyde or ketone.

mechanism of the Swern oxidation

$$\underset{\text{alcohol}}{R-\overset{R(H)}{\underset{H}{C}}-\ddot{O}H} + \underset{\text{dimethylchlorosulfonium ion}}{(CH_3)_2\overset{+}{S}-Cl} \longrightarrow R-\overset{R(H)}{\underset{H}{C}}-\overset{+}{\underset{H}{\ddot{O}}}-\overset{+}{S}(CH_3)_2 + Cl^- \rightleftharpoons R-\overset{R(H)}{\underset{H}{C}}-\ddot{O}-\overset{+}{S}(CH_3)_2 + HCl \xrightarrow{\text{an E2 reaction}} \underset{\text{aldehyde or ketone}}{R-\overset{R(H)}{\overset{|}{C}}=O} + CH_3SCH_3$$

$\underset{\text{triethylamine}}{(CH_3CH_2)_3\ddot{N}}$

To understand how dimethyl sulfoxide and oxalyl chloride react to form the dimethylchlorosulfonium ion, see Problem 64.

PROBLEM 8◆

Give the product formed from the reaction of each of the following alcohols with

a. an acidic solution of sodium dichromate:
b. the reagents required for a Swern oxidation:

1. 3-pentanol
2. 1-pentanol
3. 2-methyl-2-pentanol
4. 2,4-hexanediol
5. cyclohexanol
6. 1,4-butanediol

PROBLEM 9

Propose a mechanism for the chromic acid oxidation of 1-propanol to propanal.

PROBLEM 10 SOLVED

How could butanone be prepared from butane?

$$\underset{\text{butane}}{CH_3CH_2CH_2CH_3} \xrightarrow{?} \underset{\text{butanone}}{CH_3\overset{O}{\overset{\|}{C}}CH_2CH_3}$$

SOLUTION We know that the first reaction has to be a radical halogenation because that is the only reaction that an alkane undergoes. Bromination will lead to a greater yield of the desired 2-halo-substituted compound than will chlorination because the bromine radical is more selective than a chlorine radical. To maximize the yield of substitution product (Section 11.8), the alkyl bromide is treated with acetate ion and the ester is then hydrolyzed to the alcohol, which forms the target compound when it is oxidized.

$$CH_3CH_2CH_2CH_3 \xrightarrow[h\nu]{Br_2} CH_3\overset{Br}{\overset{|}{C}}HCH_2CH_3 \xrightarrow[\text{2. HCl, } H_2O]{\text{1. } CH_3COO^-} CH_3\overset{OH}{\overset{|}{C}}HCH_2CH_3 \xrightarrow[H_2SO_4]{Na_2Cr_2O_7} CH_3\overset{O}{\overset{\|}{C}}CH_2CH_3$$

20.3 Oxidation of Aldehydes and Ketones

Aldehydes are oxidized to carboxylic acids. Because aldehydes are generally easier to oxidize than primary alcohols, any of the reagents described in the preceding section for oxidizing primary alcohols to carboxylic acids can be used to oxidize aldehydes to carboxylic acids.

$$CH_3CH_2CHO \xrightarrow[H_2SO_4]{Na_2Cr_2O_7} CH_3CH_2COOH$$

$$C_6H_{11}CHO \xrightarrow{H_2CrO_4} C_6H_{11}COOH$$

aldehydes carboxylic acids

Silver oxide is a mild oxidizing agent. A dilute solution of silver oxide in aqueous ammonia (*Tollens reagent*) will oxidize an aldehyde, but it is too weak to oxidize an alcohol or any other functional group. An advantage to using Tollens reagent to oxidize an aldehyde is that the reaction occurs under basic conditions. Therefore, you do not have to worry about harming other functional groups in the molecule that may undergo a reaction in an acidic solution.

$$CH_3CH_2CHO \xrightarrow[2.\ H_3O^+]{1.\ Ag_2O,\ NH_3} CH_3CH_2COOH + \underset{\text{metallic silver}}{Ag}$$

Bernhard Tollens (1841–1918) *was born in Germany. He was a professor of chemistry at the University of Göttingen, the same university from which he received a Ph.D.*

The oxidizing agent in Tollens reagent is Ag^+, which is reduced to metallic silver. The **Tollens test** is based on this reaction: If Tollens reagent is added to a small amount of an aldehyde in a test tube, the inside of the test tube becomes coated with a shiny mirror of metallic silver. Consequently, if a mirror is not formed when Tollens reagent is added to a compound, it can be concluded that the compound does not have an aldehyde functional group.

Ketones do not react with most of the reagents used to oxidize aldehydes. However, both aldehydes *and* ketones can be oxidized by a peroxyacid. Aldehydes are oxidized to carboxylic acids and ketones are oxidized to esters. A **peroxyacid** (also called a percarboxylic acid or an acyl hydroperoxide) contains one more oxygen than a carboxylic acid, and it is this oxygen that is inserted between the carbonyl carbon and the H of an aldehyde or the R of a ketone. The reaction is called a **Baeyer–Villiger oxidation**.

Baeyer-Villiger oxidations

$$\underset{\text{an aldehyde}}{CH_3CH_2CH_2CHO} + \underset{\text{a peroxyacid}}{RCOOOH} \longrightarrow \underset{\text{a carboxylic acid}}{CH_3CH_2CH_2COOH} + RCOOH$$

$$\underset{\text{a ketone}}{CH_3CH_2COCH_2CH_3} + \underset{\text{a peroxyacid}}{RCOOOH} \longrightarrow \underset{\text{an ester}}{CH_3CH_2COOCH_2CH_3} + RCOOH$$

Johann Friedrich Wilhelm Adolf von Baeyer (1835–1917) *started his study of chemistry under Bunsen and Kekulé at the University of Heidelberg and received a Ph.D. from the University of Berlin, studying under Hofmann. (See also Section 2.11.)*

Victor Villiger (1868–1934) *was Baeyer's student. The two published the first paper on the Baeyer–Villiger oxidation in* Chemische Berichte *in 1899.*

If the two alkyl substituents attached to the carbonyl group of the ketone are not the same, on which side of the carbonyl carbon is the oxygen inserted? For example, does the oxidation of cyclohexyl methyl ketone form methyl cyclohexanecarboxylate or cyclohexyl acetate?

cyclohexyl methyl ketone $\xrightarrow{RCOOH}$ methyl cyclohexanecarboxylate or ? cyclohexyl acetate

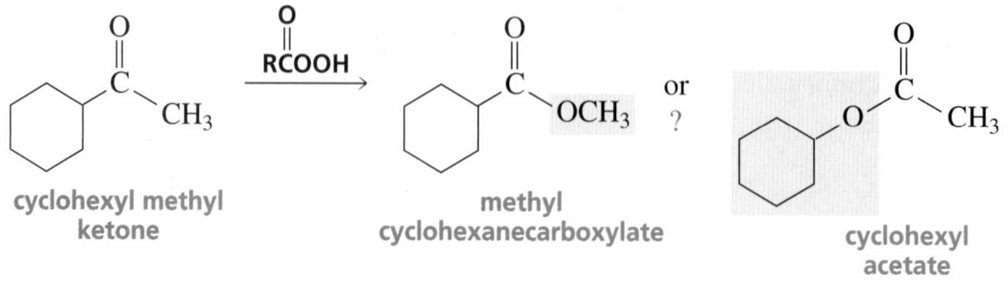

To answer this question, we must look at the mechanism of the reaction. The ketone and the peroxyacid react to form an unstable tetrahedral intermediate with a very weak O—O bond. As the O—O bond breaks heterolytically, one of the alkyl groups migrates to an oxygen. This rearrangement is similar to the 1,2-shifts that occur when carbocations rearrange (Section 4.6).

mechanism of the Baeyer–Villiger oxidation

$RC(=O)R' + CH_3CO_3H \rightleftharpoons R{-}C(O^-)(R'){-}HO^+{-}OCCH_3$ (B:) $\rightleftharpoons R{-}C(O^-)(R'){-}O{-}OCCH_3$ (an unstable intermediate) $\longrightarrow R'C(=O)OR + CH_3CO^-$ (HB$^+$)

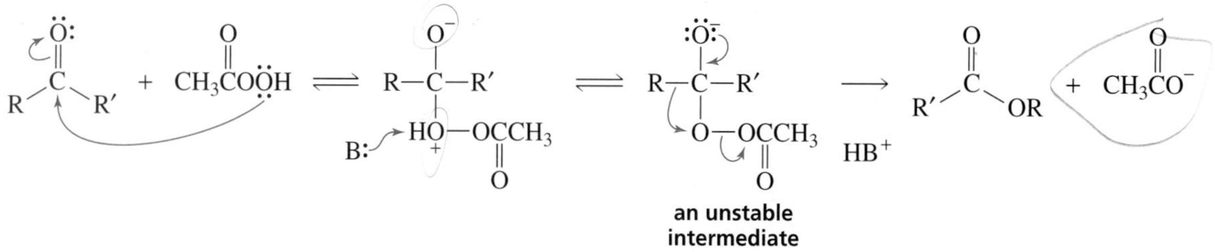

Several studies have established the following order of group migration tendencies:

relative migration tendencies

most likely to migrate H > *tert*-alkyl > *sec*-alkyl = phenyl > primary alkyl > methyl least likely to migrate

Therefore, the product of the Baeyer–Villiger oxidation of cyclohexyl methyl ketone will be cyclohexyl acetate because a secondary alkyl group (the cyclohexyl group) is more likely to migrate than a methyl group. Aldehydes are always oxidized to carboxylic acids, since H has the greatest tendency to migrate.

PROBLEM 11

Give the products of the following reactions:

a. 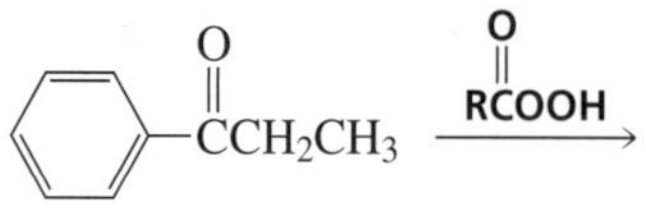 $C_6H_5CCH_2CH_3$ (C=O) $\xrightarrow{RCOOH}$

b. C_6H_5CH (C=O) $\xrightarrow{RCOOH}$

c. 2-methylcyclopentanone $\xrightarrow{RCOOH}$

d. $CH_3CH(CH_3){-}C(=O){-}C(CH_3)_2CH_3$ $\xrightarrow{RCOOH}$

e. $CH_3CH_2CH_2CH$ (C=O) $\xrightarrow[2.\ H_3O^+]{1.\ Ag_2O,\ NH_3}$

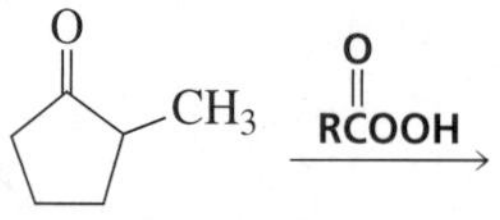

f. 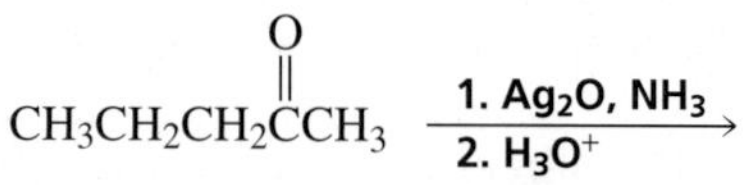 $CH_3CH_2CH_2CCH_3$ (C=O) $\xrightarrow[2.\ H_3O^+]{1.\ Ag_2O,\ NH_3}$

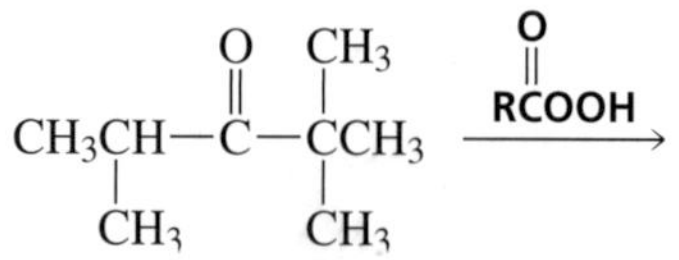

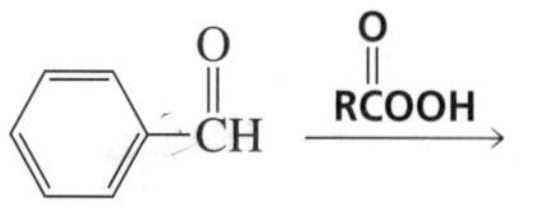

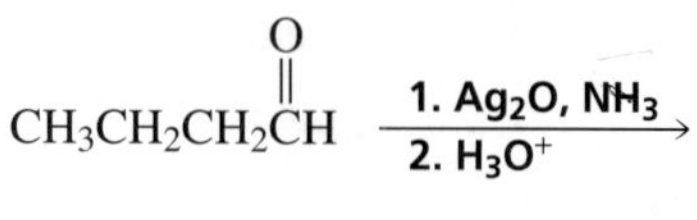

20.4 Oxidation of Alkenes with Peroxyacids

An alkene can be oxidized to an epoxide by a peroxyacid. The overall reaction amounts to the transfer of an oxygen atom from the peroxyacid to the alkene.

$$RCH{=}CH_2 + RCOOH \longrightarrow RCH{-}CH_2 + RCOH$$

an alkene + a peroxyacid ⟶ an epoxide + a carboxylic acid

Recall than an O—O bond is weak and easily broken (Section 20.3).

R—C(=O)—O—O—H (a weak bond: the O—O bond)

The oxygen atom of the OH group of the peroxyacid accepts a pair of electrons from the π bond of the alkene, causing the weak O—O bond to break heterolytically. The electrons from the O—O bond are delocalized onto the carbonyl group. The electrons left behind as the O—H bond breaks add to the carbon of the alkene that becomes electron deficient when the π bond breaks. Notice that **epoxidation** of an alkene is a concerted reaction: All the bond-forming and bond-breaking processes take place in a single step.

3-D Molecules: Peroxyacetic acid; *m*-Chloroperoxybenzoic acid

mechanism for epoxidation of an alkene

The mechanism for the addition of oxygen to a double bond to form an epoxide is analogous to the mechanism described in Section 4.7 for the addition of bromine to a double bond to form a cyclic bromonium ion. In one case the electrophile is oxygen, and in the other it is bromine. So the reaction of an alkene with a peroxyacid, like the reaction of an alkene with Br_2, is an electrophilic addition reaction.

The addition of oxygen to an alkene is a stereospecific reaction. Because the reaction is concerted, the C—C bond cannot rotate, so there is no opportunity for the relative positions of the groups bonded to the sp^2 carbons of the alkene to change. Therefore, a cis alkene forms a cis epoxide. Similarly, a trans alkene forms a trans epoxide.

3-D Molecules: *cis*-2,3-Dimethyloxirane; *trans*-2,3-Dimethyloxirane

cis-2-butene —RCOOH→ *cis*-2,3-dimethyloxirane

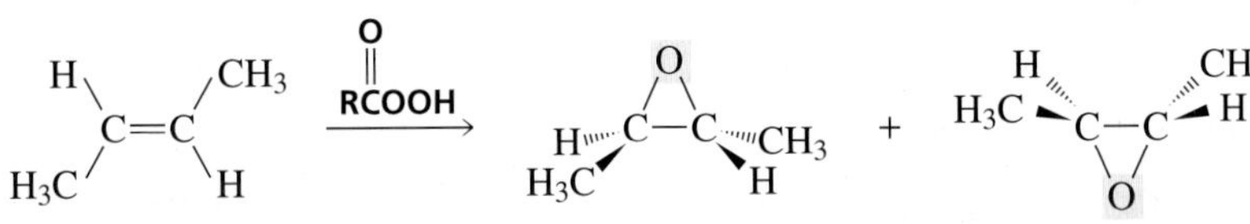

trans-2-butene *trans*-2,3-dimethyloxirane

Because the oxygen can add from the top or the bottom of the plane containing the double bond, *trans*-2-butene forms a pair of enantiomers; *cis*-2-butene forms a meso compound—it and its mirror image are identical (Section 5.10).

Increasing the electron density of the double bond increases the rate of epoxidation because it makes the double bond more nucleophilic. Alkyl substituents increase the electron density of the double bond. Therefore, if a diene is treated with only enough peroxyacid to react with one of the double bonds, it will be the most substituted double bond that is epoxidized.

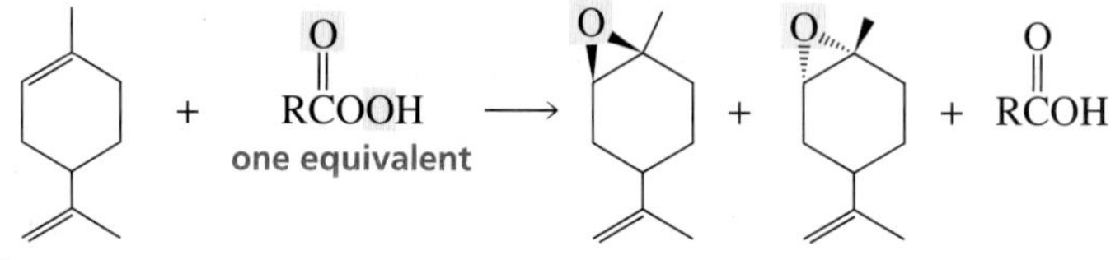

limonene + $RCOOH$ (one equivalent) ⟶ [epoxide] + [epoxide] + $RCOH$ (with C=O)

PROBLEM 12◆

What alkene would you treat with a peroxyacid in order to obtain each of the following epoxides?

a. 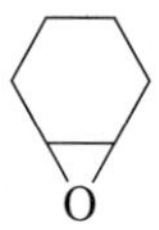 [cyclohexene oxide]

b.  $H_2C-CHCH_2CH_3$ (epoxide, O bridging)

c. 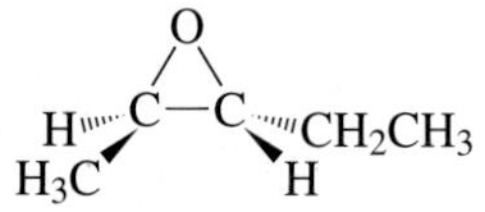 H, H_3C on C; CH_2CH_3, H on C (epoxide)

d. 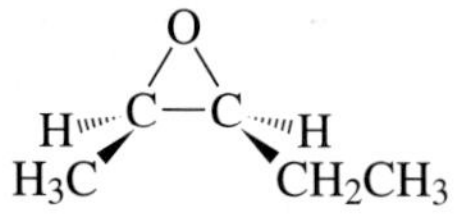 H, H_3C on C; H, CH_2CH_3 on C (epoxide)

PROBLEM 13

Give the major product of the reaction of each of the following compounds with one equivalent of a peroxyacid. Indicate the configuration of the product.

a. 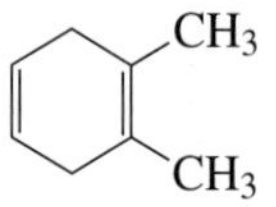 [1,2-dimethylcyclohexa-1,4-diene with CH_3, CH_3]

b. 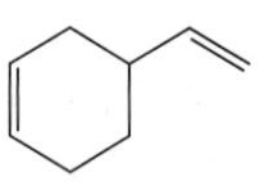 [cyclohexene with vinyl substituent]

c. 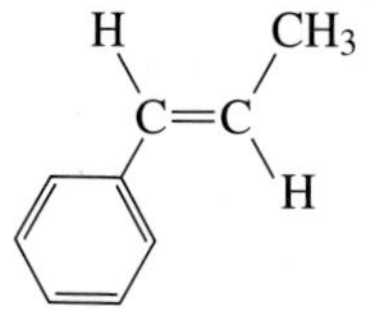 H, phenyl on C=C; CH_3, H

d. 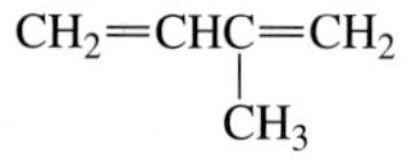 $CH_2=CHC=CH_2$ with CH_3 on C-2

PROBLEM 14

Show how the following target molecules could be synthesized from propene:

a. 1-methoxy-2-propanol

b. 2-butanol

c. 2-butanone

PROBLEM 15

Explain why an epoxide is a relatively stable product, whereas a bromonium ion is a reactive intermediate.

20.5 Designing a Synthesis VII: Controlling Stereochemistry

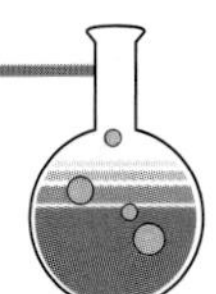

The target molecule of a synthesis may be one of several stereoisomers. The actual number of stereoisomers depends on the number of double bonds and asymmetric carbons in the molecule because each double bond can exist in an *E* or *Z* configuration (Section 3.5) and each asymmetric carbon can have an *R* or *S* configuration (Section 5.6). In addition, if the target molecule has rings with a common bond, the rings can be either trans fused or cis fused (Section 2.15). In designing a synthesis, care must be taken to make sure that each double bond, each asymmetric carbon, and each ring fusion in the target molecule has the appropriate configuration. If the stereochemistry of the reactions is not controlled, the resulting mixture of stereoisomers may be difficult or even impossible to separate. Therefore, in planning a synthesis, an organic chemist must consider the stereochemical outcomes of all reactions and must use highly stereoselective reactions to achieve the desired configurations. Some stereoselective reactions are also *enantioselective*; an **enantioselective reaction** forms more of one enantiomer than of another.

We have seen that an enantiomerically pure target molecule can be obtained if an enzyme is used to catalyze the reaction that forms the target molecule. Enzyme-catalyzed reactions result in the exclusive formation of one enantiomer since enzymes are chiral (Section 5.20). For example, ketones are enzymatically reduced to alcohols by enzymes called alcohol dehydrogenases. Whether the *R* or the *S* enantiomer is formed depends on the particular alcohol dehydrogenase used: Alcohol dehydrogenase from the bacterium *Lactobacillus kefir* forms *R* alcohols, whereas alcohol dehydrogenases from yeast, horse liver, and the bacterium *Thermoanaerobium brocki* form *S* alcohols. The alcohol dehydrogenases use NADPH to carry out the reduction (Section 25.2). Using an enzyme-catalyzed reaction to control the configuration of a target molecule is not a universally useful method because enzymes require substrates of very specific size and shape (Section 24.8).

$$C_6H_5COCF_3 + NADPH + H^+ \xrightarrow[\text{alcohol dehydrogenase}]{\textit{Lactobacillus kefir}} C_6H_5CH(OH)CF_3 + NADP^+$$

(*R*)-2,2,2-trifluoro-1-phenyl-1-ethanol

$$C_6H_5COCF_3 + NADPH + H^+ \xrightarrow[\text{alcohol dehydrogenase}]{\textit{Thermoanaerobium brocki}} C_6H_5CH(OH)CF_3 + NADP^+$$

(*S*)-2,2,2-trifluoro-1-phenyl-1-ethanol

Alternatively, an enantiomerically pure catalyst that is not an enzyme can be used to obtain an enantiomerically pure target molecule. For example, an enantiomerically pure epoxide of an allylic alcohol can be prepared by treating the alcohol with *tert*-butyl hydroperoxide, titanium isopropoxide, and enantiomerically pure diethyl tartrate (DET). The structure of the epoxide depends on the enantiomer of diethyl tartrate used.

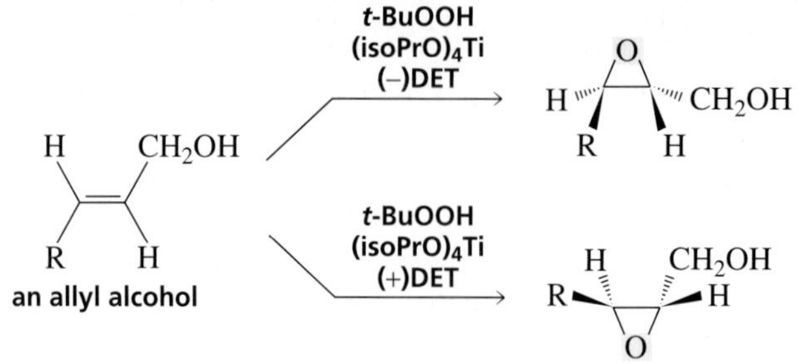

K. Barry Sharpless was born in Philadelphia in 1941. He received a B.A. from Dartmouth in 1963 and a Ph.D. in chemistry from Stanford in 1968. He served as a professor at MIT and Stanford. Currently, he is at the Scripps Research Institute in La Jolla, California. He received the 2001 Nobel Prize in chemistry for his work on chirally catalyzed oxidation reactions. (See also Section 24.3.)

an allyl alcohol

t-BuOOH, (isoPrO)$_4$Ti, (−)DET

t-BuOOH, (isoPrO)$_4$Ti, (+)DET

This method, developed in 1980 by Barry Sharpless, has proven to be useful for the synthesis of a wide variety of enantiomerically pure compounds, because an epoxide can easily be converted into a compound with two adjacent asymmetric carbons, since epoxides are very susceptible to attack by nucleophiles. In the following example, an allylic alcohol is converted into an enantiomerically pure epoxide, which is used to form an enantiomerically pure diol.

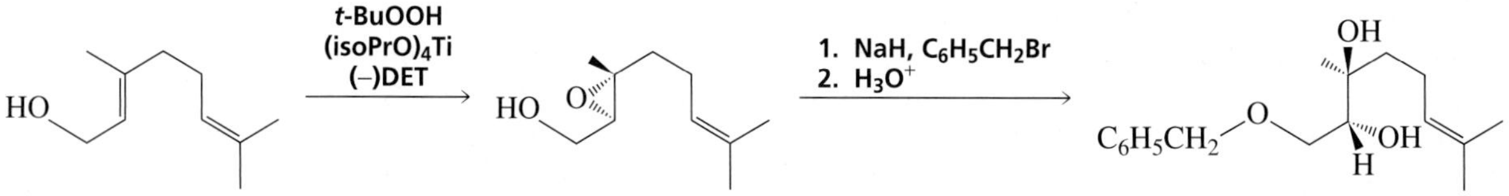

t-BuOOH, (isoPrO)$_4$Ti, (−)DET

1. NaH, $C_6H_5CH_2Br$
2. H_3O^+

PROBLEM 16

What is the product of the reaction of methylmagnesium bromide with either of the enantiomerically pure epoxides that can be prepared from (*E*)-3-methyl-2-pentene by the preceding method? Assign *R* or *S* configurations to the asymmetric carbons of each product.

PROBLEM 17◆

Is the addition of Br_2 to an alkene such as *trans*-2-pentene a stereoselective reaction? Is it a stereospecific reaction? Is it an enantioselective reaction?

20.6 Hydroxylation of Alkenes

An alkene can be oxidized to a 1,2-diol either by potassium permanganate ($KMnO_4$) in a cold basic solution or by osmium tetroxide (OsO_4). The solution of potassium permanganate must be basic, and the oxidation must be carried out at room temperature or below. If the solution is heated or if it is acidic, the diol will be oxidized further (Section 20.8). A diol is also called a **glycol**. The OH groups are on adjacent carbons in 1,2-diols, so 1,2-diols are also known as **vicinal diols** or **vicinal glycols**.

$$CH_3CH{=}CHCH_3 \xrightarrow[\text{cold}]{KMnO_4,\ HO^-,\ H_2O} CH_3\underset{OH}{CH}-\underset{OH}{CH}CH_3$$

a vicinal diol

$$CH_3CH_2CH{=}CH_2 \xrightarrow[2.\ H_2O_2]{1.\ OsO_4} CH_3CH_2\underset{OH}{CH}CH_2OH$$

a vicinal diol

Tutorial: Hydroxylation reactions—synthesis

Both $KMnO_4$ and OsO_4 form a cyclic intermediate when they react with an alkene. The reactions occur because manganese and osmium are in a highly positive oxidation state and, therefore, attract electrons. (Since the oxidation state is given by the number of bonds to oxygen, magnesium and osmium have oxidations states of +7 and +8, respectively.) Formation of the cyclic intermediate is a syn addition because both oxygens are delivered to the same side of the double bond. Therefore, the oxidation reaction is stereospecific—a cis cycloalkene forms only a cis diol.

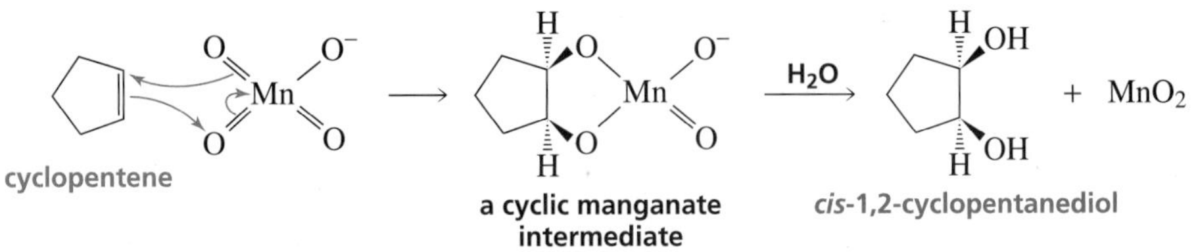

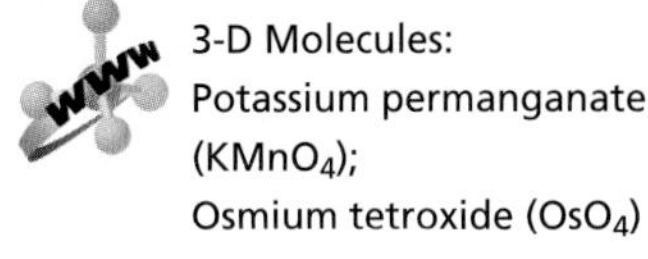

3-D Molecules: Potassium permanganate ($KMnO_4$); Osmium tetroxide (OsO_4)

mechanism for cis glycol formation

cyclopentene → a cyclic manganate intermediate $\xrightarrow{H_2O}$ *cis*-1,2-cyclopentanediol + MnO_2

The cyclic osmate intermediate is hydrolyzed with hydrogen peroxide that reoxidizes osmium to osmium tetroxide.

mechanism for cis glycol formation

cyclohexene → a cyclic osmate intermediate $\xrightarrow{H_2O_2}$ *cis*-1,2-cyclohexanediol + OsO_3

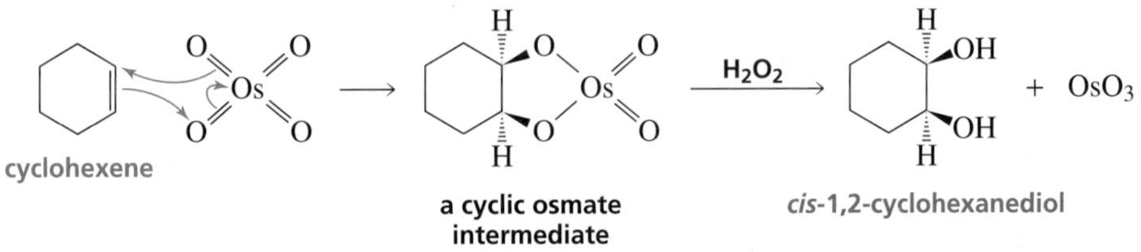

Higher yields of the diol are obtained with osmium tetroxide because the cyclic osmate intermediate is less likely to undergo side reactions.

PROBLEM 18◆

Give the products that would be formed from the reaction of each of the following alkenes with OsO_4, followed by aqueous H_2O_2:

a. $CH_3C{=}CHCH_2CH_3$ (with CH_3 on C-2)

b. 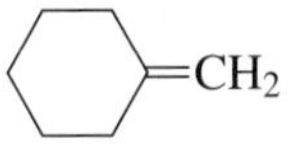 (cyclohexane)$=CH_2$

PROBLEM 19

What stereoisomers would be formed from the reaction of each of the following alkenes with OsO_4 followed by H_2O_2?

a. *trans*-2-butene

b. *cis*-2-butene

c. *cis*-2-pentene

d. *trans*-2-pentene

20.7 Oxidative Cleavage of 1,2-Diols

1,2-Diols are oxidized to ketones and/or aldehydes by periodic acid (HIO_4). Periodic acid reacts with the diol to form a cyclic intermediate. The reaction takes place because iodine is in a highly positive oxidation state (+7), so it readily accepts electrons. When the intermediate breaks down, the bond between the two carbons bonded to the OH groups breaks. If the carbon that is bonded to an OH group is also bonded to two

R groups, the product will be a ketone. If the carbon is bonded to an R and an H, the product will be an aldehyde. Because this oxidation reaction cuts the reactant into two pieces, it is called an **oxidative cleavage**.

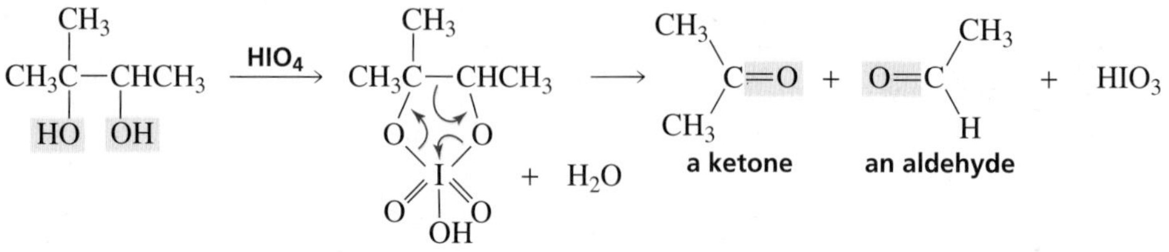

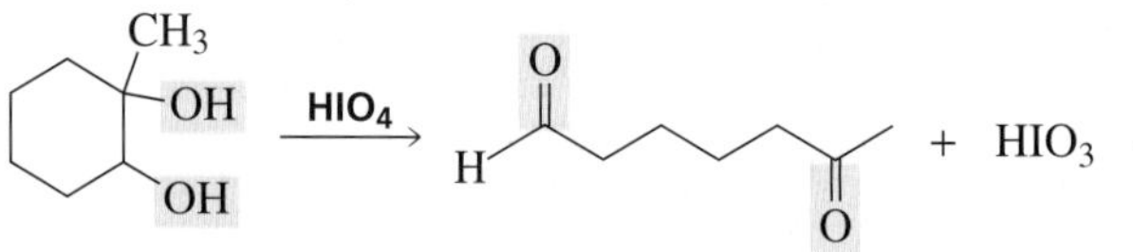

$$CH_3C(CH_3)(OH)-CH(OH)CH_3 \xrightarrow{HIO_4} \text{cyclic periodate ester} + H_2O \longrightarrow (CH_3)_2C{=}O + O{=}C(CH_3)H + HIO_3$$

a ketone   an aldehyde

$$\text{1-methyl-1,2-cyclohexanediol} \xrightarrow{HIO_4} HC(=O)CH_2CH_2CH_2CH_2C(=O)CH_3 + HIO_3$$

PROBLEM 20◆

An alkene is treated with OsO_4 followed by H_2O_2. When the resulting diol is treated with HIO_4, the only product obtained is an unsubstituted cyclic ketone with molecular formula $C_6H_{10}O$. What is the structure of the alkene?

PROBLEM-SOLVING STRATEGY

Of the following five compounds, explain why only **D** cannot be cleaved by periodic acid.

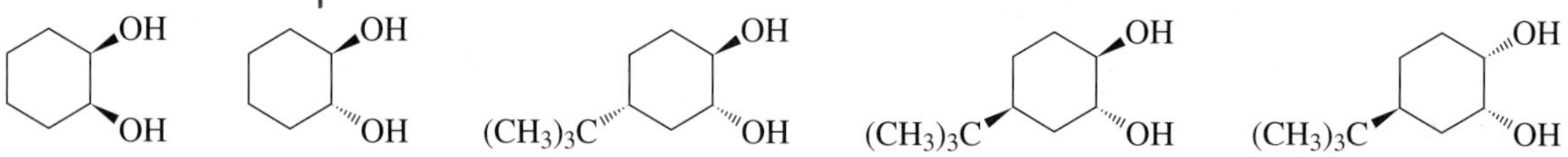

A B C D E

To figure out why one of a series of similar compounds is unreactive, we first need to consider what kinds of compounds undergo the reaction and any stereochemical requirements of the reaction. We know that periodic acid cleaves 1,2-diols. Because the reaction forms a cyclic intermediate, the two OH groups of the diol must be positioned so that they can form the intermediate.

The two OH groups of a 1,2-cyclohexanediol can both be equatorial, they can both be axial, or one can be equatorial and the other axial.

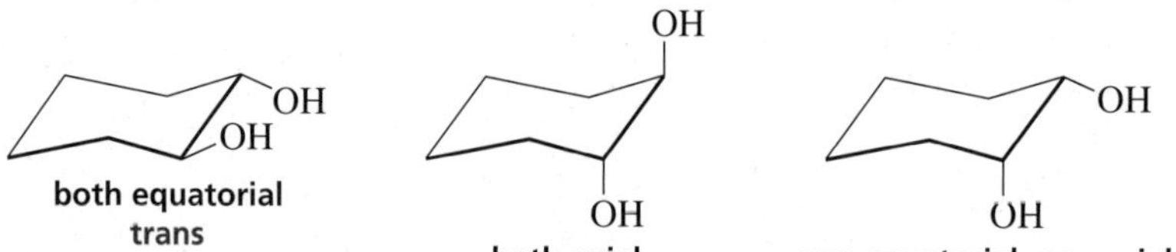

both equatorial trans   both axial trans   one equatorial, one axial cis

In a cis 1,2-cyclohexanediol, one OH is equatorial and the other is axial. Because both cis 1,2-diols (A and E) are cleaved, we know that the cyclic intermediate can be formed when the OH groups are in these positions. In a trans 1,2-diol, both OH groups are equatorial *or* both are axial (Section 2.14). Two of the trans 1,2-diols can be cleaved (B and C), and one cannot (D). We can conclude that the one that cannot be cleaved must have both OH groups in axial positions because they would be too far from each other to form a cyclic intermediate. Now we need to draw the most stable conformers of B, C, and D to see why only D has both OH groups in axial positions.

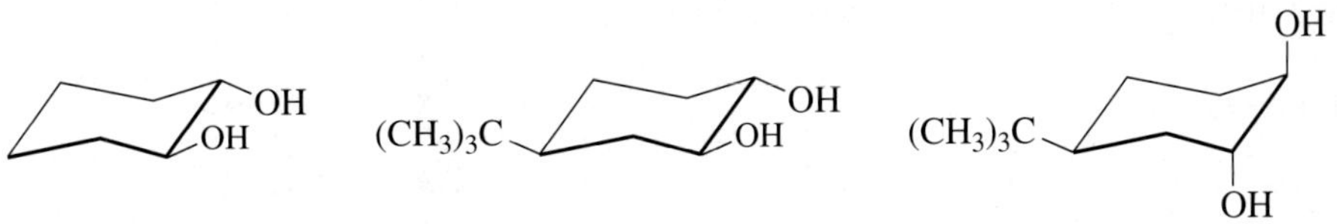

B C D

The most stable conformer of B is the one with both OH groups in equatorial positions. The steric requirements of the bulky *tert*-butyl group force it into an equatorial position, where there is more room for such a large substituent. This causes both OH groups in compound C to be in equatorial positions and both OH groups in compound D to be in axial positions. Therefore, C can be cleaved by periodic acid, but D cannot.

Now continue on to Problem 21.

PROBLEM 21◆

Which of each pair of diols is cleaved more rapidly by periodic acid?

a. 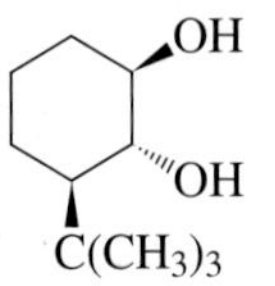 or 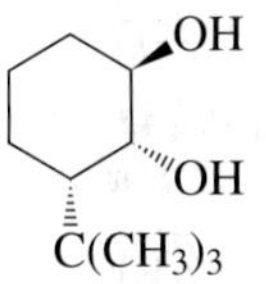 b. 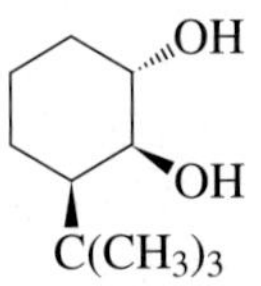 or

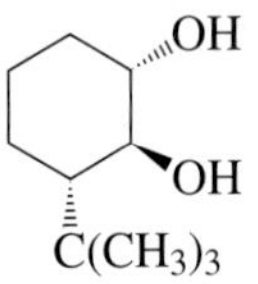

20.8 Oxidative Cleavage of Alkenes

Ozonolysis

We have seen that alkenes can be oxidized to 1,2-diols and that 1,2-diols can be further oxidized to aldehydes and ketones (Sections 20.6 and 20.7, respectively). Alternatively, alkenes can be directly oxidized to aldehydes and ketones by ozone (O_3). When an alkene is treated with ozone at low temperatures, the double bond breaks and the carbons that were doubly bonded to each other find themselves doubly bonded to oxygens instead. This oxidation reaction is known as **ozonolysis**.

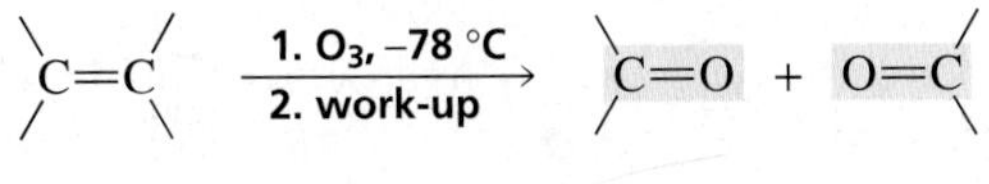

$$\text{C=C} \xrightarrow[\text{2. work-up}]{\text{1. } O_3,\ -78\ ^\circ\text{C}} \text{C=O} + \text{O=C}$$

Ozone is produced by passing oxygen gas through an electric discharge. The structure of ozone can be represented by the following resonance contributors:

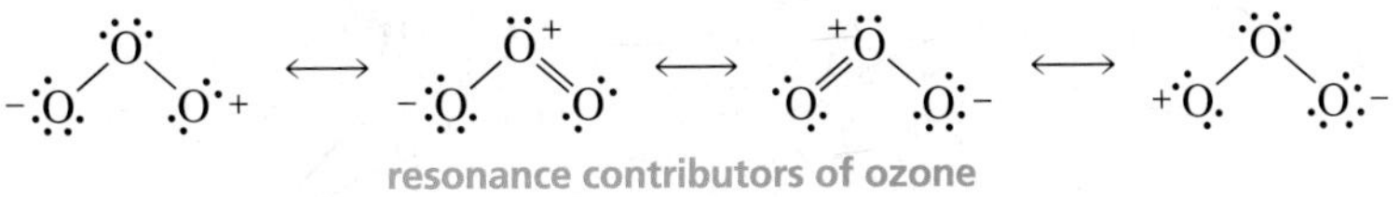

resonance contributors of ozone

Ozone and the alkene undergo a concerted cycloaddition reaction—the oxygen atoms add to the two sp^2 carbons in a single step. The addition of ozone to the alkene should remind you of the electrophilic addition reactions of alkenes discussed in Chapter 4. An electrophile adds to one of the sp^2 carbons, and a nucleophile adds to the other. The electrophile is the oxygen at one end of the ozone molecule, and the nucleophile is the oxygen at the other end. The product of ozone addition to an alkene is a

molozonide. (The name "molozonide" indicates that one mole of ozone has added to the alkene.) The molozonide is unstable because it has two O—O bonds; it immediately rearranges to a more stable **ozonide**.

mechanism for ozonide formation

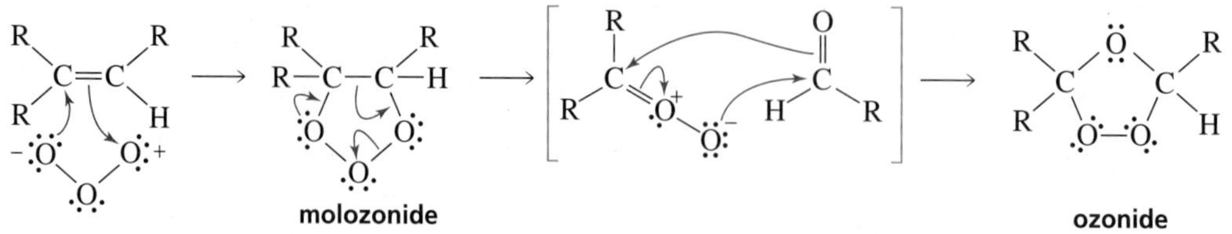

molozonide

ozonide

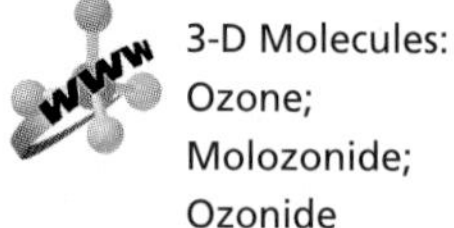
3-D Molecules: Ozone; Molozonide; Ozonide

Ozonides are explosive, so they are seldom isolated. In solution, they are easily cleaved to carbonyl compounds. If the ozonide is cleaved in the presence of a reducing agent such as zinc or dimethyl sulfide, the products will be ketones and/or aldehydes. (The product will be a ketone if the sp^2 carbon of the alkene is bonded to two carbon-containing substituents; the product will be an aldehyde if at least one of the substituents bonded to the sp^2 carbon is a hydrogen.) The reducing agent prevents aldehydes from being oxidized to carboxylic acids. Cleaving the ozonide in the presence of zinc or dimethyl sulfide is referred to as "working up the ozonide under reducing conditions."

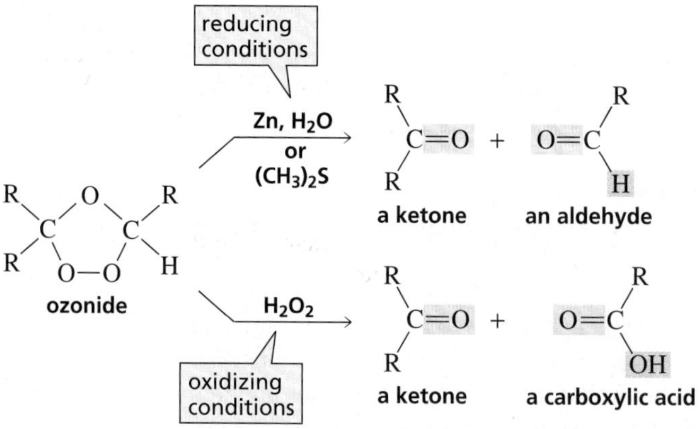

reducing conditions

Zn, H_2O or $(CH_3)_2S$

ozonide

a ketone

an aldehyde

H_2O_2

oxidizing conditions

a ketone

a carboxylic acid

If the ozonide is cleaved in the presence of an oxidizing agent such as hydrogen peroxide (H_2O_2), the products will be ketones and/or carboxylic acids. Carboxylic acids are formed instead of aldehydes because any aldehyde that is initially formed will be immediately oxidized to a carboxylic acid by hydrogen peroxide. Cleavage in the presence of H_2O_2 is referred to as "working up the ozonide under oxidizing conditions."

The following reactions are examples of the oxidative cleavage of alkenes by ozonolysis:

To determine the product of ozonolysis, replace C=C with C=O O=C. If work-up is done under oxidizing conditions, convert any aldehyde products to carboxylic acids.

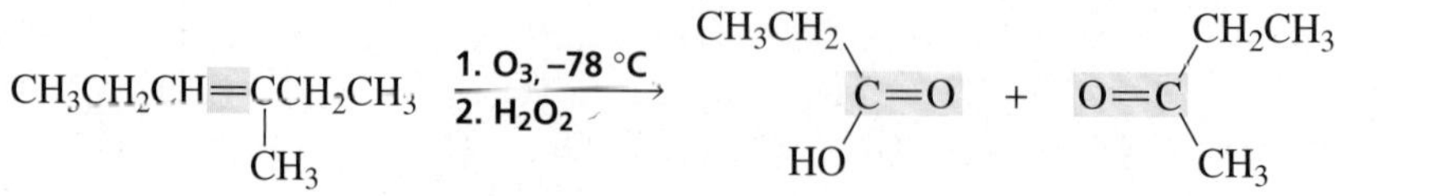

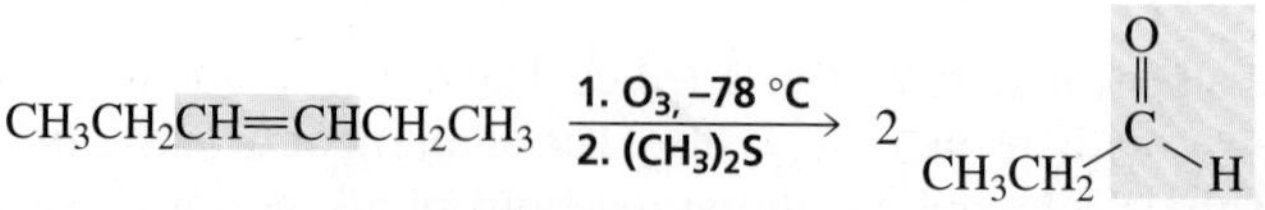

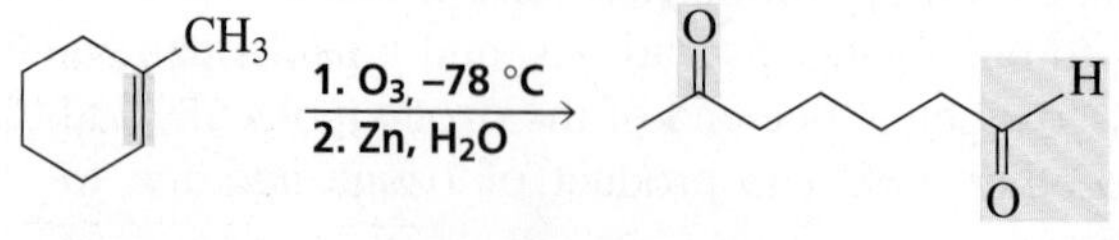

$CH_3CH_2CH{=}C(CH_3)CH_2CH_3$ → (1. O_3, −78 °C; 2. H_2O_2) → $CH_3CH_2C(=O)OH$ + $O{=}C(CH_2CH_3)CH_3$

$CH_3CH_2CH{=}CHCH_2CH_3$ → (1. O_3, −78 °C; 2. $(CH_3)_2S$) → 2 CH_3CH_2CHO

1-methylcyclohexene → (1. O_3, −78 °C; 2. Zn, H_2O) → $CH_3CO(CH_2)_4CHO$

The one-carbon fragment obtained from the reaction of a terminal alkene with ozone will be oxidized to formaldehyde if the ozonide is worked up under reducing conditions and to formic acid if it is worked up under oxidizing conditions.

$$CH_3CH_2CH_2CH{=}CH_2 \xrightarrow[\text{2. Zn, } H_2O]{\text{1. } O_3, -78\ °C} (CH_3CH_2CH_2)(H)C{=}O + O{=}C(H)(H)$$

$$CH_3CH_2CH_2CH{=}CH_2 \xrightarrow[\text{2. } H_2O_2]{\text{1. } O_3, -78\ °C} (CH_3CH_2CH_2)(HO)C{=}O + O{=}C(H)(OH)$$

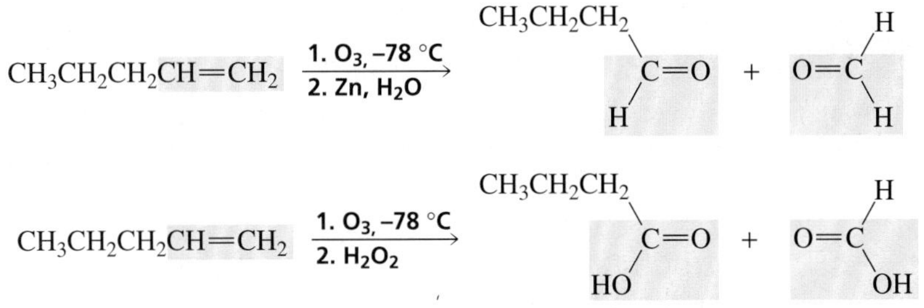

Only the side-chain double bond will be oxidized in the following reaction because the stable benzene ring is oxidized only under prolonged exposure to ozone.

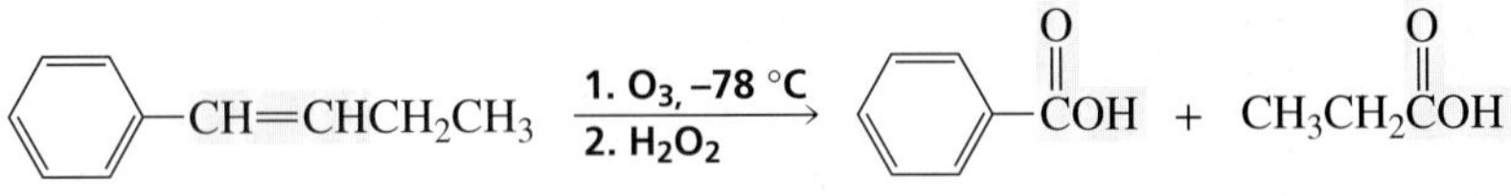

$$C_6H_5\text{–}CH{=}CHCH_2CH_3 \xrightarrow[\text{2. } H_2O_2]{\text{1. } O_3, -78\ °C} C_6H_5\text{–}\overset{O}{\overset{\|}{C}}OH + CH_3CH_2\overset{O}{\overset{\|}{C}}OH$$

PROBLEM 22

Give an example of an alkene that will form the same ozonolysis products, regardless of whether the ozonide is worked up under reducing conditions (Zn, H_2O) or oxidizing conditions (H_2O_2).

PROBLEM 23

Give the products that you would expect to obtain when the following compounds are treated with ozone, followed by work-up with

a. Zn, H_2O: b. H_2O_2:

1. $CH_3CH_2CH_2\overset{CH_3}{\overset{|}{C}}{=}CHCH_3$
2. $CH_2{=}CHCH_2CH_2CH_2CH_3$
3. 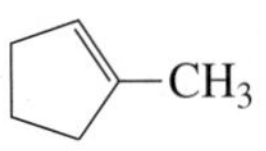 1-methylcyclopentene (cyclopentene ring–CH_3)
4. 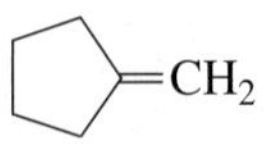 methylenecyclopentane (cyclopentane ring$=CH_2$)
5. $CH_3CH_2CH_2CH{=}CHCH_2CH_2CH_3$
6. 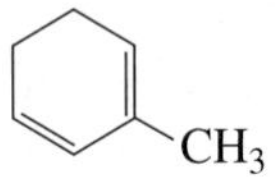 1-methylcyclohexene (cyclohexene ring–CH_3)

Ozonolysis can be used to determine the structure of an unknown alkene. If you know what carbonyl compounds are formed by ozonolysis, you can mentally work backward to deduce the structure of the alkene. For example, if ozonolysis of an alkene followed by a work-up under reducing conditions forms acetone and butanal as products, you can conclude that the alkene was 2-methyl-2-hexene.

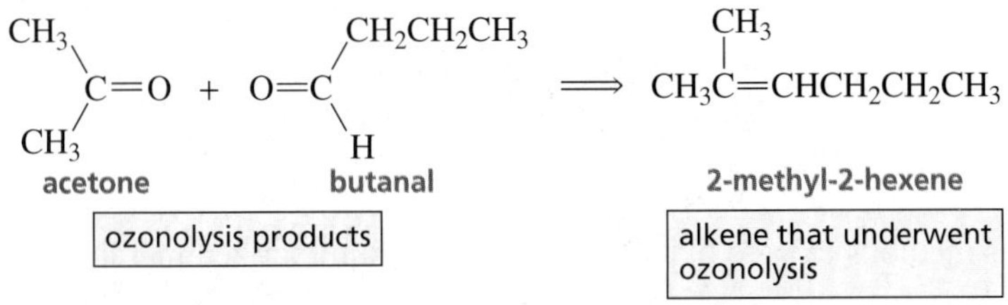

$$\underset{\text{acetone}}{(CH_3)_2C{=}O} + \underset{\text{butanal}}{O{=}C(H)CH_2CH_2CH_3} \Longrightarrow \underset{\text{2-methyl-2-hexene}}{CH_3\overset{CH_3}{\overset{|}{C}}{=}CHCH_2CH_2CH_3}$$

ozonolysis products alkene that underwent ozonolysis

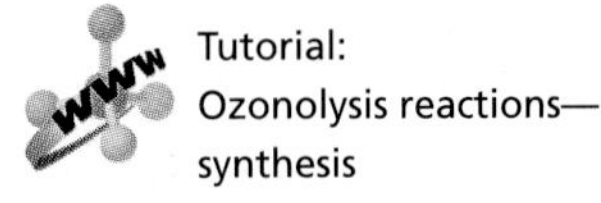

Tutorial: Ozonolysis reactions—synthesis

PROBLEM 24◆

a. What alkene would give only acetone as a product of ozonolysis?
b. What alkenes would give only butanal as a product of ozonolysis?

PROBLEM 25 SOLVED

The following products were obtained from ozonolysis of a diene followed by work-up under reducing conditions. Give the structure of the diene.

$$\overset{\displaystyle O}{\overset{\|}{H C}}CH_2CH_2CH_2\overset{\displaystyle O}{\overset{\|}{C}}H \; + \; H\overset{\displaystyle O}{\overset{\|}{C}}H \; + \; CH_3CH_2\overset{\displaystyle O}{\overset{\|}{C}}H$$

SOLUTION The five-carbon dicarbonyl compound indicates that the diene must contain five carbons flanked by two double bonds.

$$\overset{\displaystyle O}{\overset{\|}{H C}}CH_2CH_2CH_2\overset{\displaystyle O}{\overset{\|}{C}}H \Longrightarrow =CHCH_2CH_2CH_2CH=$$

One of the carbonyl compounds obtained from ozonolysis has one carbon atom, and the other has three carbon atoms. Therefore, one carbon has to be added to one end of the diene, and three carbons have to be added to the other end.

$$CH_2{=}CHCH_2CH_2CH_2CH{=}CHCH_2CH_3$$

PROBLEM 26◆

What aspect of the structure of the alkene does ozonolysis not tell you?

Permanganate Cleavage

We have seen that alkenes are oxidized to 1,2-diols by a basic solution of potassium permanganate at room temperature or below, and the 1,2-diols can subsequently be cleaved by periodic acid to form aldehydes and/or ketones (Sections 20.6 and 20.7). If, however, the basic solution of potassium permanganate is heated or if the solution is acidic, the reaction will not stop at the diol. Instead, the alkene will be cleaved, and the reaction products will be ketones and carboxylic acids. If the reaction is carried out under basic conditions, any carboxylic acid product will be in its basic form ($RCOO^-$); if the reaction is carried out under acidic conditions, any carboxylic acid product will be in its acidic form (RCOOH) (Section 1.20). Terminal alkenes form CO_2 as a product.

A peroxyacid, OsO_4, and (cold basic) $KMnO_4$ break only the π bond of the alkene. Ozone and acidic (or hot basic) $KMnO_4$ break both the π bond and the σ bond.

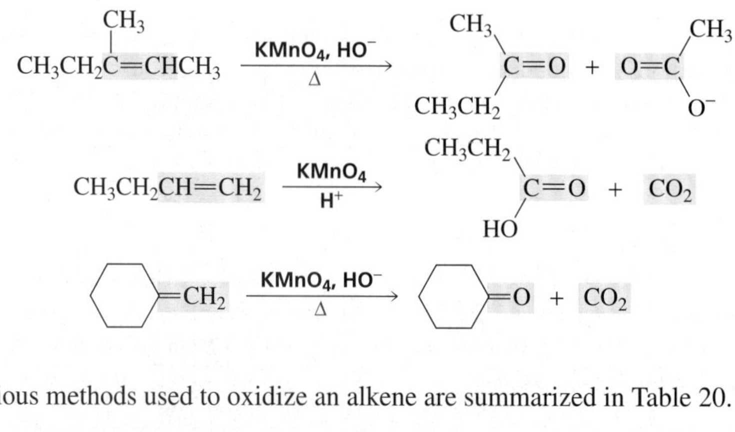

$$CH_3CH_2\overset{\displaystyle CH_3}{\overset{|}{C}}{=}CHCH_3 \xrightarrow[\Delta]{KMnO_4,\ HO^-} (CH_3)(CH_3CH_2)C{=}O + O{=}C(CH_3)O^-$$

$$CH_3CH_2CH{=}CH_2 \xrightarrow[H^+]{KMnO_4} (CH_3CH_2)(HO)C{=}O + CO_2$$

$$\text{cyclohexylidene}{=}CH_2 \xrightarrow[\Delta]{KMnO_4,\ HO^-} \text{cyclohexanone}{=}O + CO_2$$

The various methods used to oxidize an alkene are summarized in Table 20.1.

Table 20.1 Summary of the Methods Used to Oxidize an Alkene

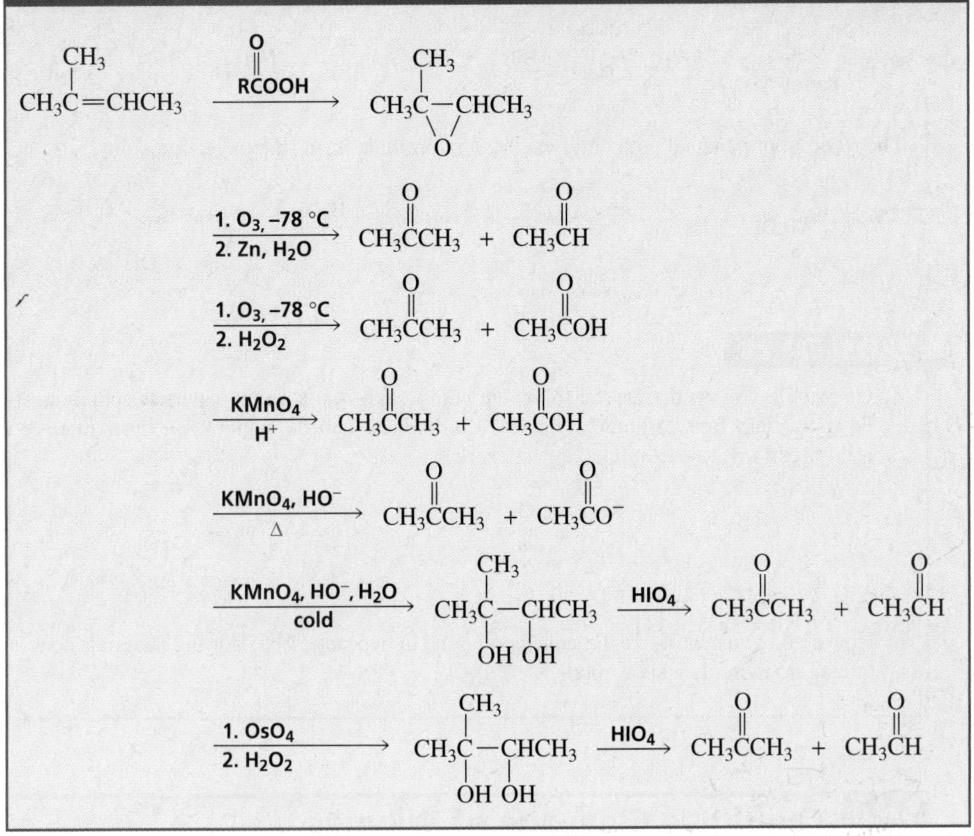

$$CH_3\overset{CH_3}{\overset{|}{C}}{=}CHCH_3 \xrightarrow{RCOOH} CH_3\overset{CH_3}{\overset{|}{C}}{-}CHCH_3 \text{ (epoxide, bridged by O)}$$

$$\xrightarrow[2.\ Zn,\ H_2O]{1.\ O_3,\ -78\ ^\circ C} CH_3\overset{O}{\overset{\|}{C}}CH_3 + CH_3\overset{O}{\overset{\|}{C}}H$$

$$\xrightarrow[2.\ H_2O_2]{1.\ O_3,\ -78\ ^\circ C} CH_3\overset{O}{\overset{\|}{C}}CH_3 + CH_3\overset{O}{\overset{\|}{C}}OH$$

$$\xrightarrow[H^+]{KMnO_4} CH_3\overset{O}{\overset{\|}{C}}CH_3 + CH_3\overset{O}{\overset{\|}{C}}OH$$

$$\xrightarrow[\Delta]{KMnO_4,\ HO^-} CH_3\overset{O}{\overset{\|}{C}}CH_3 + CH_3\overset{O}{\overset{\|}{C}}O^-$$

$$\xrightarrow[cold]{KMnO_4,\ HO^-,\ H_2O} CH_3\overset{CH_3}{\overset{|}{C}}{-}CHCH_3 \text{ (OH OH)} \xrightarrow{HIO_4} CH_3\overset{O}{\overset{\|}{C}}CH_3 + CH_3\overset{O}{\overset{\|}{C}}H$$

$$\xrightarrow[2.\ H_2O_2]{1.\ OsO_4} CH_3\overset{CH_3}{\overset{|}{C}}{-}CHCH_3 \text{ (OH OH)} \xrightarrow{HIO_4} CH_3\overset{O}{\overset{\|}{C}}CH_3 + CH_3\overset{O}{\overset{\|}{C}}H$$

PROBLEM 27 SOLVED

Describe how the following compound could be prepared, using the given starting materials (perform a retrosynthetic analysis to help you arrive at your answer):

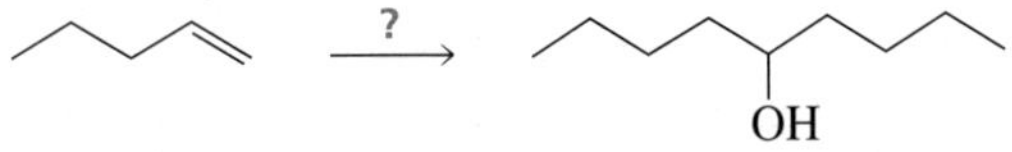

$\xrightarrow{?}$ OH

SOLUTION Retrosynthetic analysis shows that the target molecule can be disconnected to give a five-carbon positively charged fragment and a four-carbon negatively charged fragment (Section 18.12). Pentanal and a butyl magnesium bromide are the synthetic equivalents for the two fragments. The five-carbon starting material can be converted to the required four-carbon compound by ozonolysis.

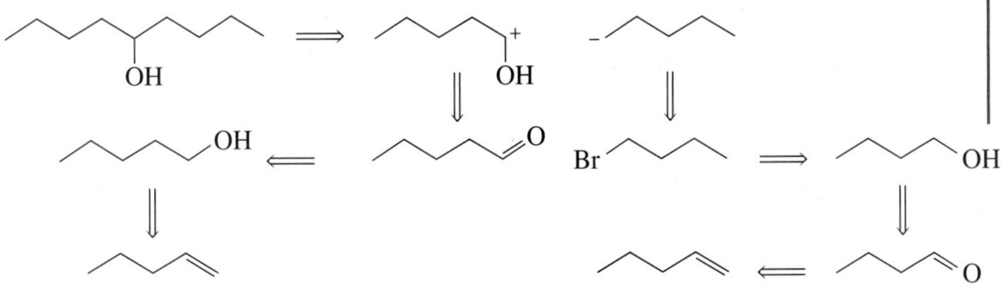

OH ⟹ + OH –

OH ⟸ O Br ⟹ OH

⟸ O

Now the synthesis of pentanal and butyl magnesium bromide from the given starting material can be written in the forward direction, including the reagents necessary for the transformations.

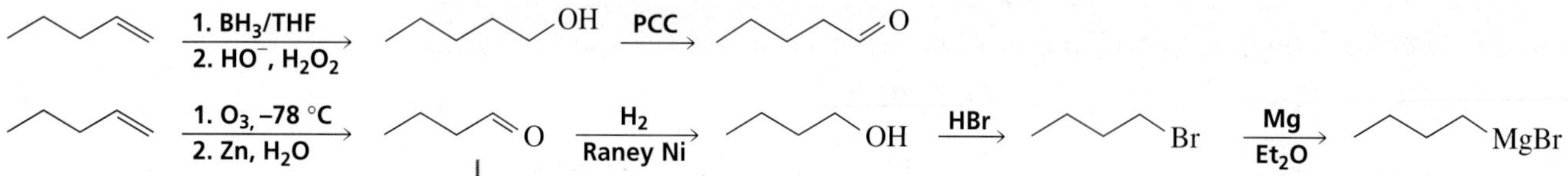

1. BH_3/THF 2. HO^-, H_2O_2 → OH; PCC → O

1. O_3, −78 °C 2. Zn, H_2O → O; H_2/Raney Ni → OH; HBr → Br; Mg/Et_2O → MgBr

The reaction of pentanal with butyl magnesium bromide forms the target compound.

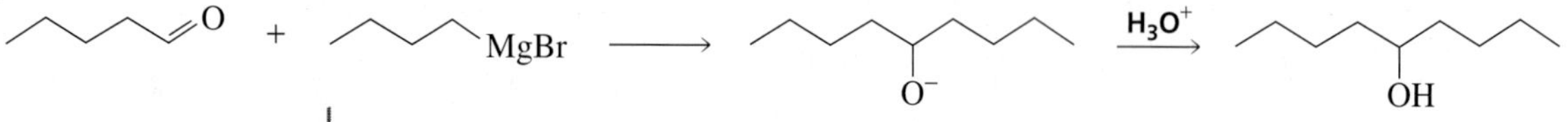

O + MgBr → O^-; H_3O^+ → OH

PROBLEM 28

a. How could you synthesize the following compound from starting materials containing no more than four carbons? (*Hint:* A 1,6-diketone can be synthesized by oxidative cleavage of a 1,2-disubstituted cyclohexene.)

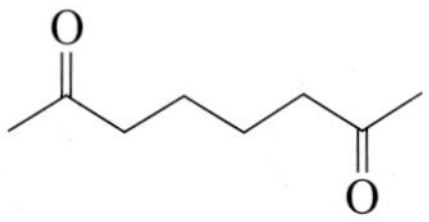

b. How could you synthesize the same compound in two steps from starting materials containing no more than six carbons?

20.9 Oxidative Cleavage of Alkynes

The same reagents that oxidize alkenes also oxidize alkynes. Alkynes are oxidized to diketones by a basic solution of $KMnO_4$ at room temperature and are cleaved by ozonolysis to carboxylic acids. Ozonolysis requires neither oxidative nor reductive work-up—it is followed only by hydrolysis. Carbon dioxide is obtained from the CH group of a terminal alkyne.

$$\underset{\text{2-pentyne}}{CH_3C{\equiv}CCH_2CH_3} \xrightarrow[HO^-]{KMnO_4} CH_3\overset{O}{\overset{\|}{C}}{-}\overset{O}{\overset{\|}{C}}CH_2CH_3$$

$$\underset{\text{2-pentyne}}{CH_3C{\equiv}CCH_2CH_3} \xrightarrow[2.\ H_2O]{1.\ O_3,\ -78\ °C} CH_3\overset{O}{\overset{\|}{C}}OH + CH_3CH_2\overset{O}{\overset{\|}{C}}OH$$

$$\underset{\text{1-pentyne}}{CH_3CH_2CH_2C{\equiv}CH} \xrightarrow[2.\ H_2O]{1.\ O_3,\ -78\ °C} CH_3CH_2CH_2\overset{O}{\overset{\|}{C}}OH + CO_2$$

PROBLEM 29♦

What is the structure of the alkyne that gives each of the following sets of products upon ozonolysis followed by hydrolysis?

a. 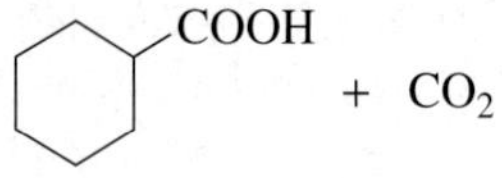 (cyclohexyl)COOH + CO_2

b. 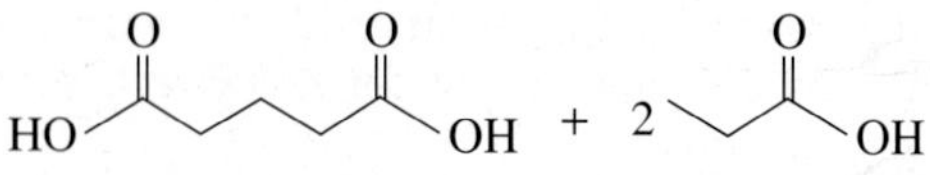 HO(C=O)CH₂CH₂CH₂(C=O)OH + 2 CH₃CH₂(C=O)OH

20.10 Designing a Synthesis VIII: Functional Group Interconversion

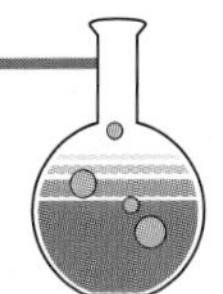

Converting one functional group into another is called **functional group interconversion**. Our knowledge of oxidation–reduction reactions has greatly expanded our ability to carry out functional group interconversions. For example, an aldehyde can be converted into a primary alcohol, an alkene, a secondary alcohol, a ketone, a carboxylic acid, an acyl chloride, an ester, an amide, or an amine.

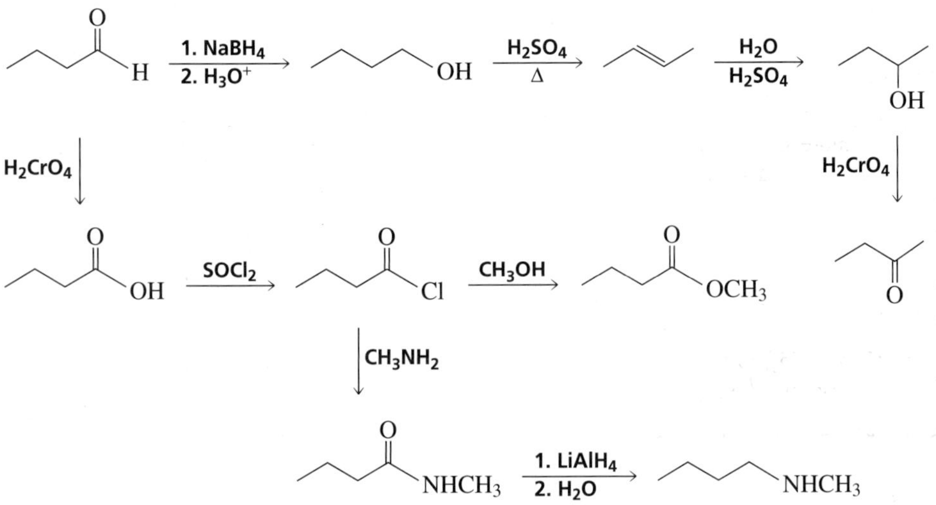

1. $NaBH_4$ 2. H_3O^+ ; H_2SO_4, Δ ; H_2O, H_2SO_4 ; H_2CrO_4 ; H_2CrO_4 ; $SOCl_2$; CH_3OH ; CH_3NH_2 ; 1. $LiAlH_4$ 2. H_2O

Tutorial: Multistep synthesis

A ketone can be converted into an ester or an alcohol.

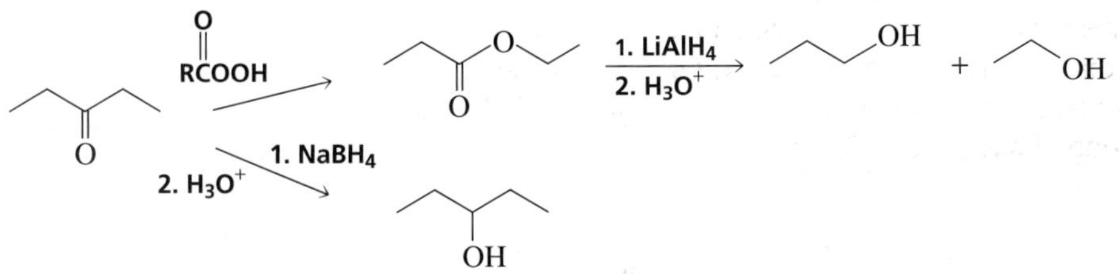

RCOOH ; 1. $LiAlH_4$ 2. H_3O^+ ; 1. $NaBH_4$ 2. H_3O^+

As the number of reactions you learn grows, so will the number of functional group interconversions you can perform. You will also find that you have more than one route available when you design a synthesis. The route you actually decide to use will depend on the availability and cost of the starting materials and on the ease with which the reactions in the synthetic pathway can be carried out.

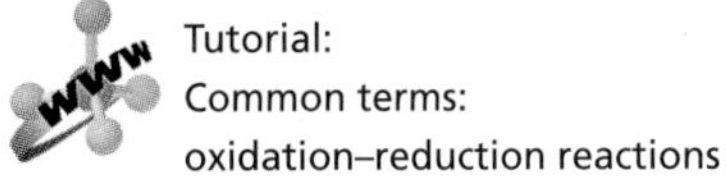

Tutorial: Common terms: oxidation–reduction reactions

PROBLEM 30

Add the necessary reagents over the reaction arrows.

a.

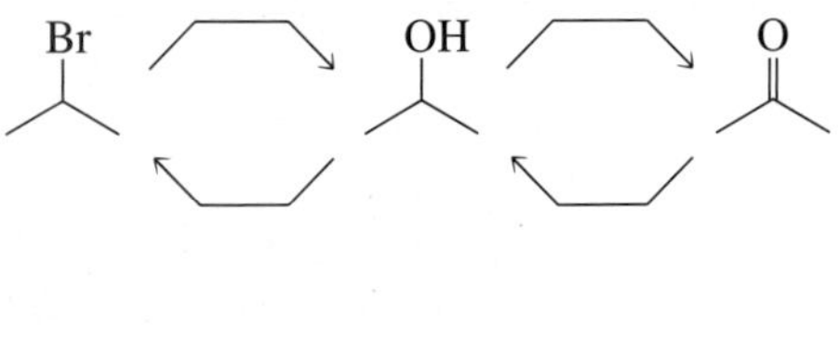

b.

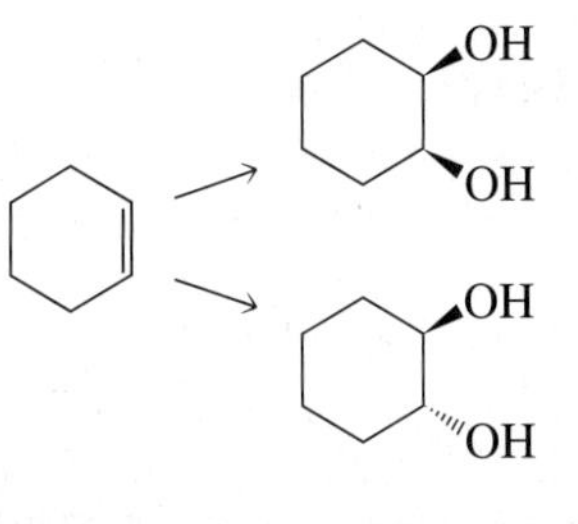

PROBLEM 31

a. Show two ways to convert an alkyl halide into an alcohol that contains one additional carbon atom.

b. Show how a primary alkyl halide can be converted into an amine that contains one additional carbon atom.

c. Show how a primary alkyl halide can be converted into an amine that contains one less carbon atom.

PROBLEM 32

Show how each of the following compounds could be synthesized from the given starting material:

a. 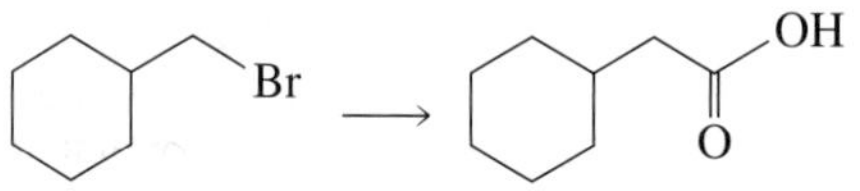 (cyclohexyl)ethyl bromide (Br) ⟶ cyclohexylpropanoic acid (C(=O)OH)

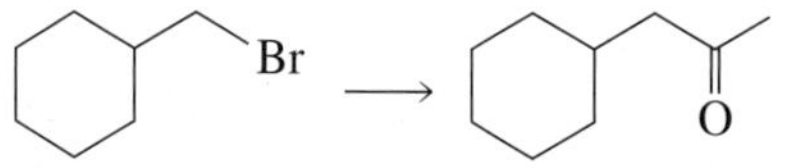

b. (cyclohexyl)ethyl bromide (Br) ⟶ cyclohexyl butan-2-one (C=O)

c. (cyclohexyl)ethyl bromide (Br) ⟶ cyclohexanecarbaldehyde (O)

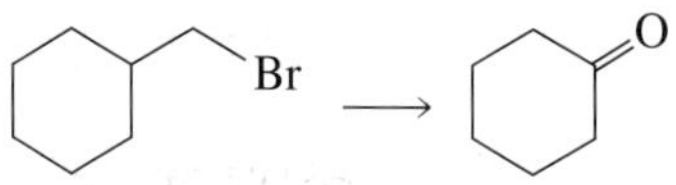

d. 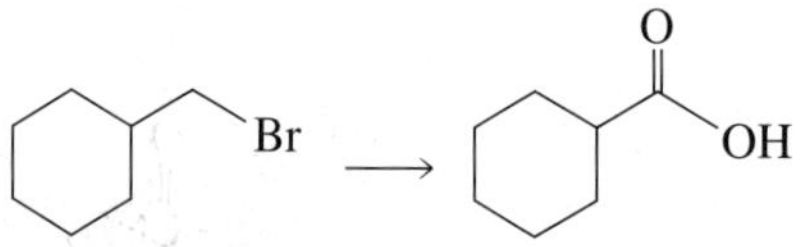 (cyclohexyl)ethyl bromide (Br) ⟶ cyclohexylacetic acid (O, OH)

PROBLEM 33

How many different functional groups can you use to synthesize a primary alcohol?

20.11 Biological Oxidation–Reduction Reactions

Both oxidation reactions and reduction reactions are important in living systems. An example of an oxidation reaction that takes place in animal cells is the oxidation of ethanol to acetaldehyde, a reaction catalyzed by the enzyme alcohol dehydrogenase. Ingestion of a moderate amount of ethanol lowers inhibitions and causes a light-headed feeling, but the physiological effects of acetaldehyde are not as pleasant. Acetaldehyde is responsible for the feeling known as a hangover. (In Section 25.4, we will see how vitamin B_1 can help cure a hangover.)

$$\underset{\text{ethanol}}{CH_3CH_2OH} + NAD^+ \xrightarrow[\text{dehydrogenase}]{\text{alcohol}} \underset{\text{acetaldehyde}}{CH_3\overset{O}{\overset{\|}{C}}H} + NADH + H^+$$

The enzyme cannot oxidize ethanol to acetaldehyde unless an oxidizing agent is present. Oxidizing agents used by organic chemists, such as chromate and permanganate salts, are not present in living systems. NAD^+ (nicotinamide adenine dinucleotide), the most common oxidizing agent available in living systems, is used by cells to oxidize alcohols to aldehydes (Section 25.2). Notice that NAD^+ is written with a positive charge to reflect the positive charge on the nitrogen atom of the pyridine ring.

NAD^+ is reduced to NADH when it oxidizes a compound. NADH is used by the cell as a reducing agent. When NADH reduces a compound, it is oxidized back to NAD^+, which can then be used for another oxidation. Although NAD^+ and NADH are complicated-looking molecules, the structural changes that occur when they act as oxidizing and reducing agents take place on a relatively small part of the molecule. The

TREATING ALCOHOLICS WITH ANTABUSE

Disulfiram, most commonly known by one of its trade names, Antabuse®, is used to treat alcoholics. The drug causes violently unpleasant effects when ethanol is consumed, even when it is consumed a day or two after Antabuse® is taken.

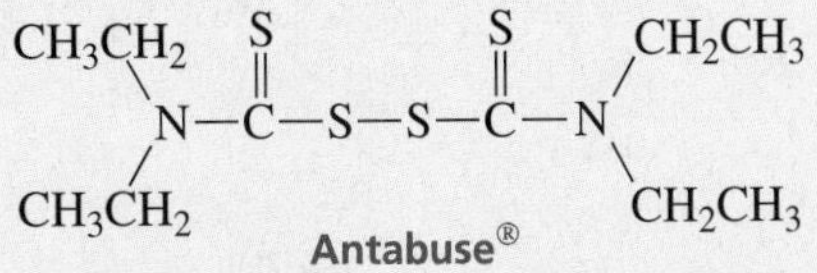

$$(CH_3CH_2)_2N{-}\overset{S}{\overset{\|}{C}}{-}S{-}S{-}\overset{S}{\overset{\|}{C}}{-}N(CH_2CH_3)_2$$

Antabuse®

Antabuse® inhibits aldehyde dehydrogenase, the enzyme responsible for oxidizing acetaldehyde to acetic acid, resulting in a buildup of acetaldehyde. Acetaldehyde causes the unpleasant physiological effects of intoxication: intense flushing, nausea, dizziness, sweating, throbbing headaches, decreased blood pressure, and, ultimately, shock. Consequently, Antabuse® should be taken only under strict medical supervision. In some people, aldehyde dehydrogenase does not function properly. Their symptoms in response to ingesting alcohol are nearly the same as those of individuals who are medicated with Antabuse®.

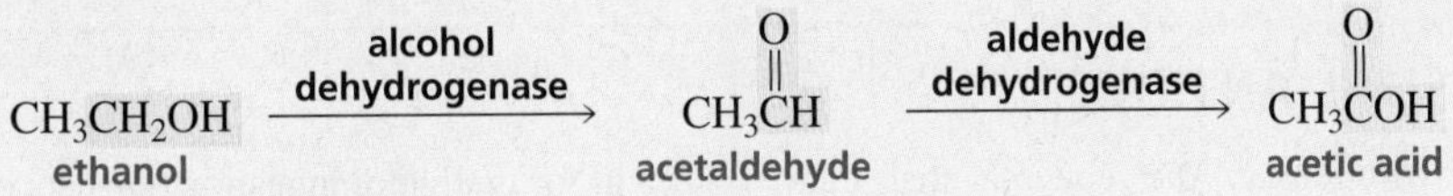

$$\underset{\text{ethanol}}{CH_3CH_2OH} \xrightarrow{\text{alcohol dehydrogenase}} \underset{\text{acetaldehyde}}{CH_3\overset{O}{\overset{\|}{C}}H} \xrightarrow{\text{aldehyde dehydrogenase}} \underset{\text{acetic acid}}{CH_3\overset{O}{\overset{\|}{C}}OH}$$

rest of the molecule is used to bind NAD^+ or NADH to the proper site on the enzyme that catalyzes the reaction.

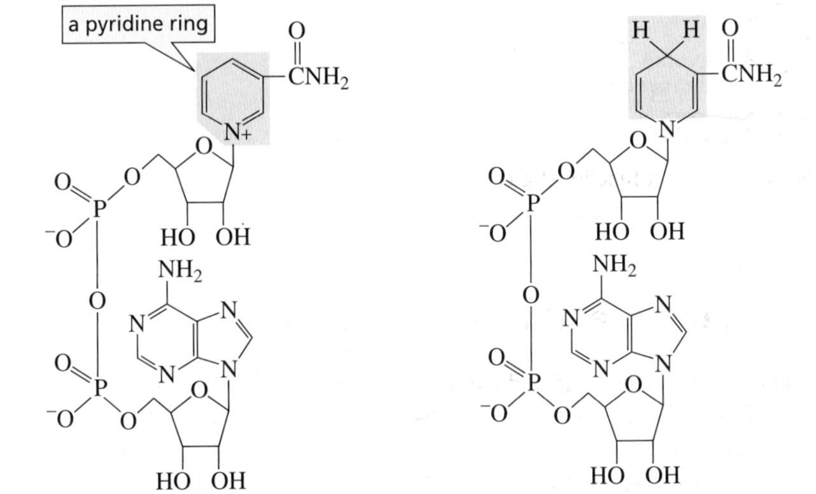

a pyridine ring

nicotinamide adenine dinucleotide
NAD^+

reduced nicotinamide adenine dinucleotide
NADH

3-D Molecules: NAD^+; NADH

NAD^+ oxidizes a compound by accepting a hydride ion from it. In this way, the number of carbon–hydrogen bonds in the compound decreases (the compound is oxidized) and the number of carbon–hydrogen bonds in NAD^+ increases (NAD^+ is reduced). NAD^+ can accept a hydride ion at the 4-position of the pyridine ring because the electrons can be delocalized onto the positively charged nitrogen atom. Although NAD^+ could also accept a hydride ion at the 2-position, the hydride ion is always delivered to the 4-position in enzyme-catalyzed reactions.

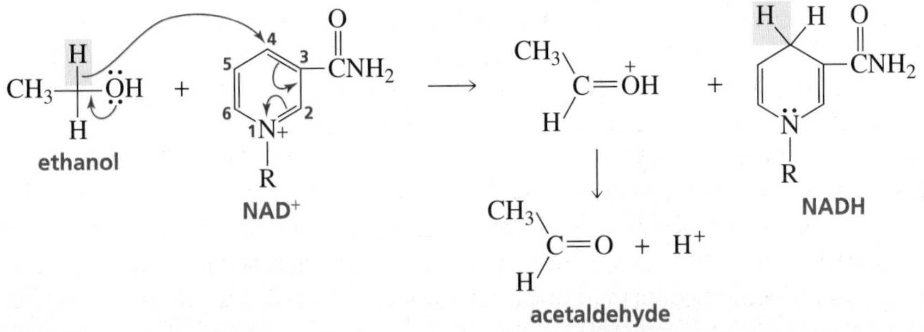

ethanol + NAD^+ ⟶ $CH_3(H)C{=}\overset{+}{O}H$ + NADH

$\downarrow$

$CH_3(H)C{=}O + H^+$

acetaldehyde

NADH reduces a compound by donating a hydride ion from the 4-position of the six-membered ring. Thus, NADH, $NaBH_4$, and $LiAlH_4$ all act as reducing agents in the same way—they donate a hydride ion.

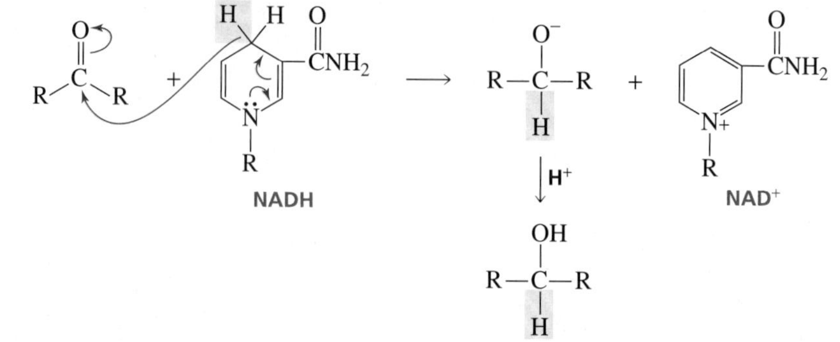

NADH

NAD^+

AN UNUSUAL ANTIDOTE

Alcohol dehydrogenase, the enzyme that catalyzes the oxidation of ethanol to acetaldehyde, catalyzes the oxidation of other alcohols as well. For example, it catalyzes the oxidation of methanol to formaldehyde.

Methanol itself is not harmful, but ingestion of methanol can be fatal because formaldehyde is extremely toxic. The treatment for methanol ingestion consists of giving the patient intravenous injections of ethanol. Alcohol dehydrogenase has 25 times the affinity for ethanol that it has for methanol. Thus, if the enzyme can be kept loaded with ethanol, methanol will be excreted before it has the opportunity to be oxidized.

$$CH_3OH + NAD^+ \xrightarrow{\text{alcohol dehydrogenase}} HCH(=O) + NADH + H^+$$

methanol → formaldehyde

FETAL ALCOHOL SYNDROME

The damage done to a human fetus when the mother drinks alcohol during her pregnancy is known as *fetal alcohol syndrome*. It has been shown that the harmful effects—growth retardation, decreased mental functioning, and facial and limb abnormalities—are attributable to the acetaldehyde that is formed from the oxidation of ethanol, which crosses the placenta and accumulates in the liver of the fetus.

20.12 Oxidation of Hydroquinones and Reduction of Quinones

Hydroquinone, a *para*-benzenediol, is easily oxidized to *para*-benzoquinone. Although a wide variety of oxidizing agents can be used, Fremy's salt (dipotassium nitrosodisulfonate) is the preferred oxidizing agent. The quinone can easily be reduced back to hydroquinone.

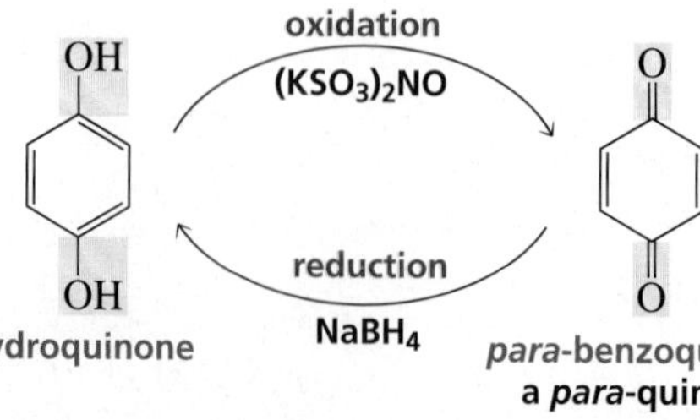

oxidation $(KSO_3)_2NO$

reduction $NaBH_4$

hydroquinone

para-benzoquinone
a *para*-quinone

Similarly, *ortho*-benzenediols are oxidized to *ortho*-quinones.

oxidation $(KSO_3)_2NO$; reduction $NaBH_4$

ortho-benzoquinone
an *ortho*-quinone

Overall, the oxidation reaction involves the loss of two hydrogen atoms and the reduction reaction involves the gain of two hydrogen atoms. In Section 9.8, we saw that phenols are used as radical inhibitors because of their ability to lose a hydrogen atom.

mechanism for hydroquinone oxidation–quinone reduction

hydroquinone ⇌ semiquinone ⇌ *para*-benzoquinone

Coenzyme Q (CoQ) is a quinone found in the cells of all aerobic organisms. It is also called ubiquinone because it is ubiquitous (found everywhere) in nature. Its function is to carry electrons in the electron-transport chain. The oxidized form of CoQ accepts a pair of electrons from a biological reducing agent such as NADH and ultimately transfers them to O_2.

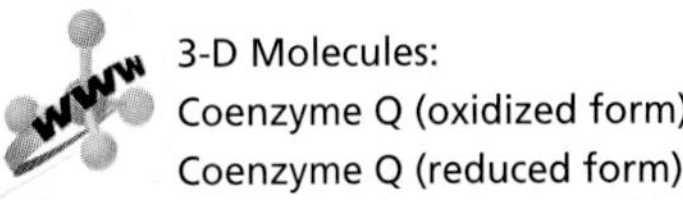

3-D Molecules:
Coenzyme Q (oxidized form);
Coenzyme Q (reduced form)

NADH + H^+ + coenzyme Q oxidized form ⟶ coenzyme Q reduced form + NAD^+

$R = (CH_2CH{=}C(CH_3)CH_2)_nH$
$n = 1–10$

$\frac{1}{2}$ O_2 + coenzyme Q reduced form ⟶ coenzyme Q oxidized form + H_2O

In this way, biological oxidizing agents are recycled: NAD^+ oxidizes a compound, thereby forming NADH, which is oxidized back to NAD^+ by oxygen, via coenzyme Q, which is unchanged in the overall reaction. Biological redox reagents and their recycling are discussed further in Sections 25.2 and 25.3.

NAD^+ + Substrate$_{\text{reduced}}$ ⟶ Substrate$_{\text{oxidized}}$ + NADH + H^+

NADH + H^+ + $\frac{1}{2}$ O_2 ⟶ NAD^+ + H_2O

THE CHEMISTRY OF PHOTOGRAPHY

Black-and-white photography depends on the fact that hydroquinone is easily oxidized. Photographic film is covered by an emulsion of silver bromide. When light hits the film, the silver bromide is sensitized. Sensitized silver bromide is a better oxidizing agent than silver bromide that has not been exposed to light. When the exposed film is put into a solution of hydroquinone (a common photographic developer), hydroquinone is oxidized to quinone by sensitized silver ion and the silver ion is reduced to silver metal, which remains in the emulsion. The exposed film is "fixed" by washing away unsensitized silver bromide with $Na_2S_2O_3/H_2O$. Black silver deposits are left in regions where light has struck the film. This is the black, opaque part of a photographic negative.

Summary

Oxidation is coupled with reduction: A **reducing agent** is oxidized and an **oxidizing agent** is reduced. For reactions in which oxidation or reduction has taken place on carbon, if the reaction increases the number of C—H bonds or decreases the number of C—O, C—N, or C—X bonds (where X d notes a halogen), the compound has been reduced; if the reaction decreases the number of C—H bonds or increases the number of C—O, C—N, or C—X bonds, the compound has been oxidized. Similarly, the number of N—H or S—H bonds increases in reduction reactions, and the number of N—O or S—O bonds increases in oxidation reactions. The **oxidation state** of a carbon atom equals the total number of its C—O, C—N, and C—X bonds.

An organic compound is reduced by the addition of H_2 by one of three mechanisms: **Catalytic hydrogenations** add two hydrogen atoms, **dissolving metal reductions** add two electrons and two protons, and **metal hydride reductions** involve the addition of a hydride ion followed by a proton. Carbon–carbon, carbon–nitrogen, and some carbon–oxygen multiple bonds can be reduced by catalytic hydrogenation. An alkyne is reduced by sodium and liquid ammonia to a trans alkene. $LiAlH_4$ is a stronger reducing agent than $NaBH_4$. $NaBH_4$ is used to reduce aldehydes, ketones, and acyl halides; $LiAlH_4$ is used to reduce carboxylic acids, esters, and amides. Replacing some of the hydrogens of $LiAlH_4$ with OR groups decreases the reactivity of the metal hydride. Multiply bonded carbon atoms cannot be reduced by metal hydrides.

Primary alcohols are oxidized to carboxylic acids by chromium-containing reagents and to aldehydes by PCC or a **Swern oxidation**. Secondary alcohols are oxidized to ketones. Tollens reagent can oxidize only aldehydes. A **peroxyacid** oxidizes an aldehyde to a carboxylic acid, a ketone to an ester (in a **Baeyer–Villiger oxidation**), and an alkene to an epoxide. Alkenes are oxidized to 1,2-diols by potassium permanganate ($KMnO_4$) in a cold basic solution or by osmium tetroxide (OsO_4).

1,2-Diols are **oxidatively cleaved** to ketones and/or aldehydes by periodic acid (HIO_4). Ozonolysis oxidatively cleaves alkenes to ketones and/or aldehydes when worked up under reducing conditions and to ketones and/or carboxylic acids when worked up under oxidizing conditions. Acidic solutions and hot basic solutions of potassium permanganate also oxidatively cleave alkenes to ketones and/or carboxylic acids.

A **chemoselective reaction** is a reaction in which a reagent reacts with one functional group in preference to another. An **enantioselective reaction** forms more of one enantiomer than of another. Converting one functional group into another is called **functional group interconversion**.

NAD^+ and NADH are the most common redox reagents in living systems. NAD^+ oxidizes a compound by accepting a hydride ion from it; NADH reduces a compound by donating a hydride ion to it.

Summary of Reactions

1. Catalytic hydrogenation of double and triple bonds (Section 20.1).

a.
$$RCH{=}CHR + H_2 \xrightarrow{\text{Pt, Pd, or Ni}} RCH_2CH_2R$$
$$RC{\equiv}CR + 2\,H_2 \xrightarrow{\text{Pt, Pd, or Ni}} RCH_2CH_2R$$
$$RCH{=}NR + H_2 \xrightarrow{\text{Pt, Pd, or Ni}} RCH_2NHR$$
$$RC{\equiv}N + 2\,H_2 \xrightarrow{\text{Pt, Pd, or Ni}} RCH_2NH_2$$

b.
$$RC(=O)H + H_2 \xrightarrow{\text{Raney Ni}} RCH_2OH$$
$$RC(=O)R + H_2 \xrightarrow{\text{Raney Ni}} RCH(OH)R$$

c. $RCCl$ (C=O) + H_2 $\xrightarrow{\text{partially deactivated Pd}}$ RCH (C=O)

$$\text{RCCl(=O)} + H_2 \xrightarrow{\text{partially deactivated Pd}} \text{RCH(=O)}$$

2. Reduction of alkynes to alkenes (Section 20.1).

$$RC{\equiv}CR \xrightarrow[\text{Lindlar catalyst}]{H_2} \text{(H)(R)C=C(H)(R)} \text{ (cis)}$$

$$RC{\equiv}CR \xrightarrow[NH_3\text{ (liq)}]{\text{Na or Li}} \text{(H)(R)C=C(R)(H)} \text{ (trans)}$$

3. Reduction of carbonyl compounds with reagents that donate hydride ion (Section 20.1).

a. $$\text{RCH(=O)} \xrightarrow[2.\ H_3O^+]{1.\ NaBH_4} RCH_2OH$$

b. $$\text{RCR(=O)} \xrightarrow[2.\ H_3O^+]{1.\ NaBH_4} \text{RCH(OH)R}$$

c. $$\text{RCCl(=O)} \xrightarrow[2.\ H_3O^+]{1.\ NaBH_4} RCH_2OH$$

d. $$\text{RCOR}'\text{(=O)} \xrightarrow[2.\ H_2O]{1.\ [(CH_3)_2CHCH_2]_2AlH,\ -78\ ^\circ C} \text{RCH(=O)} + R'OH$$

e. $$\text{RCOH(=O)} \xrightarrow[2.\ H_3O^+]{1.\ LiAlH_4} RCH_2OH$$

f. $$\text{RCOR}'\text{(=O)} \xrightarrow[2.\ H_3O^+]{1.\ LiAlH_4} RCH_2OH + R'OH$$

g. $$\text{RCNHR}'\text{(=O)} \xrightarrow[2.\ H_2O]{1.\ LiAlH_4} RCH_2NHR'$$

h. $$\text{RCCl(=O)} \xrightarrow[2.\ H_2O]{1.\ LiAl[OC(CH_3)_3]_3H,\ -78\ ^\circ C} \text{RCH(=O)}$$

4. Oxidation of alcohols (Section 20.2).

primary alcohols $$RCH_2OH \xrightarrow{H_2CrO_4} \left[\text{RCH(=O)}\right] \xrightarrow[\text{oxidation}]{\text{further}} \text{RCOH(=O)}$$

$$RCH_2OH \xrightarrow[CH_2Cl_2]{PCC} \text{RCH(=O)}$$

$$RCH_2OH \xrightarrow[2.\ \text{triethylamine}]{1.\ CH_3\overset{O}{S}CH_3,\ ClC(=O){-}C(=O)Cl,\ -60\ ^\circ C} \text{RCH(=O)}$$

secondary alcohols $$\text{RCH(OH)R} \xrightarrow[H_2SO_4]{Na_2Cr_2O_7} \text{RCR(=O)}$$

$$\text{RCH(OH)R} \xrightarrow[2.\ \text{triethylamine}]{1.\ CH_3\overset{O}{S}CH_3,\ ClC(=O){-}C(=O)Cl,\ -60\ ^\circ C} \text{RCR(=O)}$$

5. Oxidation of aldehydes and ketones (Section 20.3).

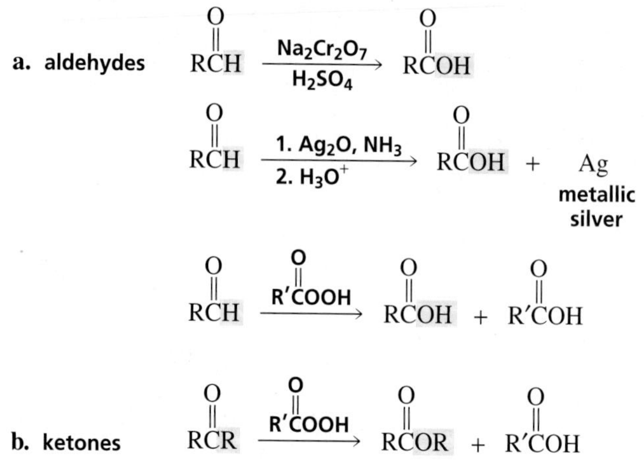

a. aldehydes

$$\text{RCH=O} \xrightarrow[\text{H}_2\text{SO}_4]{\text{Na}_2\text{Cr}_2\text{O}_7} \text{RC(=O)OH}$$

$$\text{RCH=O} \xrightarrow[\text{2. H}_3\text{O}^+]{\text{1. Ag}_2\text{O, NH}_3} \text{RC(=O)OH} + \underset{\text{metallic silver}}{\text{Ag}}$$

$$\text{RCH=O} \xrightarrow{\text{R'C(=O)OOH}} \text{RC(=O)OH} + \text{R'C(=O)OH}$$

b. ketones

$$\text{RC(=O)R} \xrightarrow{\text{R'C(=O)OOH}} \text{RC(=O)OR} + \text{R'C(=O)OH}$$

6. Oxidation of alkenes (Sections 20.4, 20.6, 20.8).

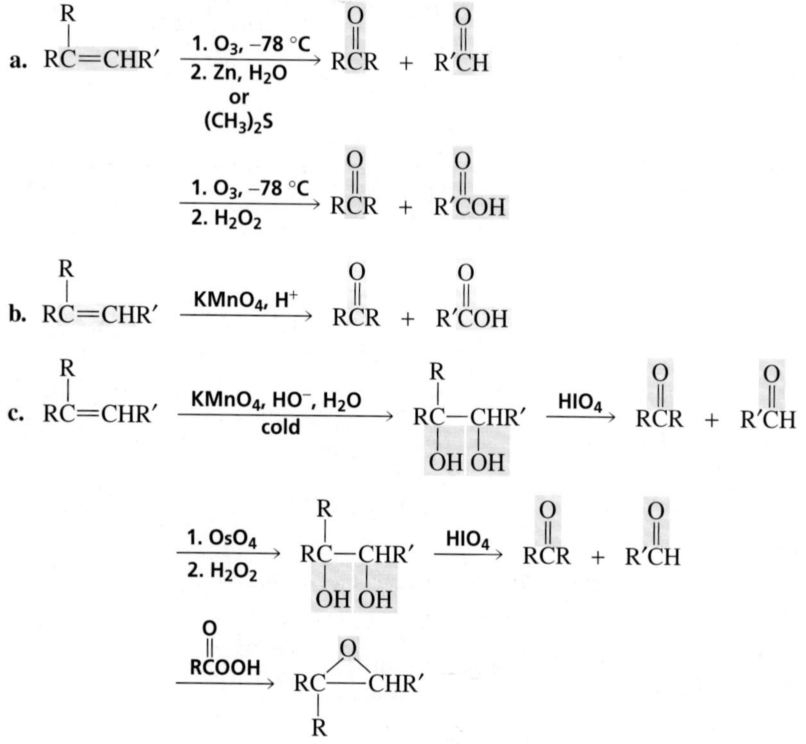

a.

$$\text{RC(R)=CHR'} \xrightarrow[\text{2. Zn, H}_2\text{O or (CH}_3)_2\text{S}]{\text{1. O}_3, -78\ ^\circ\text{C}} \text{RC(=O)R} + \text{R'CH=O}$$

$$\xrightarrow[\text{2. H}_2\text{O}_2]{\text{1. O}_3, -78\ ^\circ\text{C}} \text{RC(=O)R} + \text{R'C(=O)OH}$$

b.

$$\text{RC(R)=CHR'} \xrightarrow{\text{KMnO}_4,\ \text{H}^+} \text{RC(=O)R} + \text{R'C(=O)OH}$$

c.

$$\text{RC(R)=CHR'} \xrightarrow[\text{cold}]{\text{KMnO}_4,\ \text{HO}^-,\ \text{H}_2\text{O}} \text{RC(R)(OH)–CH(OH)R'} \xrightarrow{\text{HIO}_4} \text{RC(=O)R} + \text{R'CH=O}$$

$$\xrightarrow[\text{2. H}_2\text{O}_2]{\text{1. OsO}_4} \text{RC(R)(OH)–CH(OH)R'} \xrightarrow{\text{HIO}_4} \text{RC(=O)R} + \text{R'CH=O}$$

$$\xrightarrow{\text{RC(=O)OOH}} \text{RC(R)–CHR'}\ (\text{epoxide, bridged by O})$$

7. Oxidation of 1,2-diols (Section 20.7).

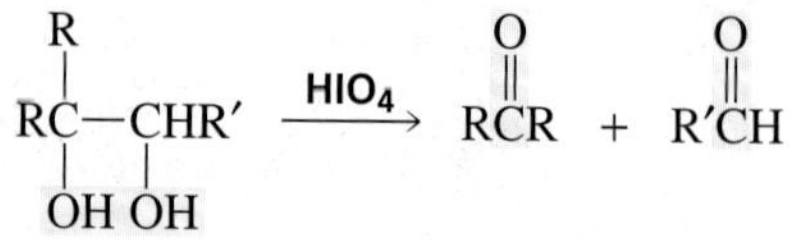

$$\text{RC(R)(OH)–CH(OH)R'} \xrightarrow{\text{HIO}_4} \text{RC(=O)R} + \text{R'CH=O}$$

8. Oxidation of alkynes (Section 20.9).

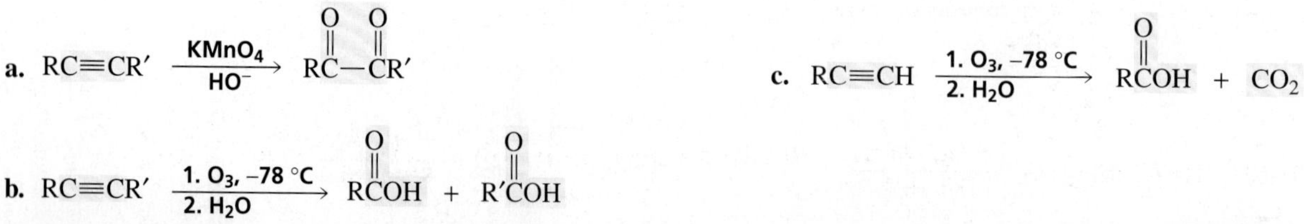

a.

$$\text{RC≡CR'} \xrightarrow[\text{HO}^-]{\text{KMnO}_4} \text{RC(=O)–C(=O)R'}$$

b.

$$\text{RC≡CR'} \xrightarrow[\text{2. H}_2\text{O}]{\text{1. O}_3, -78\ ^\circ\text{C}} \text{RC(=O)OH} + \text{R'C(=O)OH}$$

c.

$$\text{RC≡CH} \xrightarrow[\text{2. H}_2\text{O}]{\text{1. O}_3, -78\ ^\circ\text{C}} \text{RC(=O)OH} + \text{CO}_2$$

9. Oxidation of hydroquinones and reduction of quinones (Section 20.12).

$$\text{hydroquinone (OH, OH)} \underset{\text{NaBH}_4,\ \text{reduction}}{\overset{(\text{KSO}_3)_2\text{NO},\ \text{oxidation}}{\rightleftharpoons}} \text{quinone (O, O)}$$

Key Terms

Baeyer–Villiger oxidation (p. 853)
catalytic hydrogenation (p. 844)
chemoselective reaction (p. 848)
dissolving-metal reduction (p. 846)
enantioselective reaction (p. 857)
epoxidation (p. 855)
functional group interconversion (p. 867)
glycol (p. 858)
metal-hydride reduction (p. 846)
molozonide (p. 862)
oxidation (p. 842)
oxidation–reduction reaction (p. 841)
oxidation state (p. 842)
oxidative cleavage (p. 860)
oxidizing agent (p. 841)
ozonide (p. 862)
ozonolysis (p. 861)
peroxyacid (p. 853)
redox reaction (p. 841)
reducing agent (p. 841)
reduction (p. 842)
Rosenmund reduction (p. 845)
Swern oxidation (p. 851)
Tollens test (p. 853)
vicinal diol (p. 858)
vicinal glycol (p. 858)

Problems

34. Fill in the blank with "oxidized" or "reduced."

a. Secondary alcohols are ______ to ketones.
b. Acyl halides are ______ to aldehydes.
c. Aldehydes are ______ to primary alcohols.
d. Alkenes are ______ to aldehydes and/or ketones.
e. Aldehydes are ______ to carboxylic acids.
f. Alkenes are ______ to 1,2-diols.
g. Alkenes are ______ to alkanes.

35. Give the products of the following reactions. Indicate whether each reaction is an oxidation or a reduction:

a. $CH_3CH_2CH_2CH_2CH_2OH \xrightarrow[H_2SO_4]{Na_2Cr_2O_7}$

b. $C_6H_5CH{=}CH_2 \xrightarrow[HO^-,\ \Delta]{KMnO_4}$

c. $CH_3CH_2CH_2\overset{O}{\overset{\|}{C}}Cl \xrightarrow[\text{partially deactivated Pd}]{H_2}$

d. $CH_3CH_2C{\equiv}CH \xrightarrow{\text{1. disiamylborane; 2. } H_2O_2,\ HO^-,\ H_2O\text{; 3. } LiAlH_4\text{; 4. } H_3O^+}$

e. $CH_3CH_2CH{=}CHCH_2CH_3 \xrightarrow[\text{2. Zn, } H_2O]{\text{1. } O_3,\ -78\ °C}$

f. $CH_3CH_2CH_2\overset{O}{\overset{\|}{C}}NHCH_3 \xrightarrow[\text{2. } H_2O]{\text{1. } LiAlH_4}$

g. $C_6H_5\overset{O}{\overset{\|}{C}}H \xrightarrow{R\overset{O}{\overset{\|}{C}}OOH}$

h. $C_6H_5\overset{O}{\overset{\|}{C}}OCH(CH_3)CH_3 \xrightarrow[\text{2. } H_3O^+]{\text{1. } LiAlH_4}$

i. $(H_3C)(H)C{=}C(H)(CH_3)$ (trans) $\xrightarrow{R\overset{O}{\overset{\|}{C}}OOH}$

j. $C_6H_5\overset{O}{\overset{\|}{C}}H \xrightarrow[\text{Raney Ni}]{H_2}$

k. $CH_3CH_2CH_2C{\equiv}CCH_3 \xrightarrow[NH_3\ (liq)]{Na}$

l. $CH_3CH_2CH_2C{\equiv}CCH_3 \xrightarrow[2.\ H_2O]{1.\ O_3,\ -78\ °C}$

m. 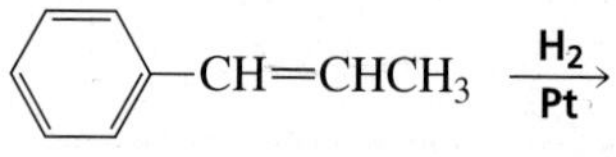 $C_6H_5{-}CH{=}CHCH_3 \xrightarrow[Pt]{H_2}$

n. 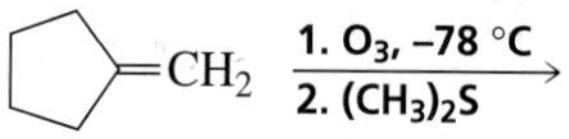 (cyclopentylidene)$=CH_2 \xrightarrow[2.\ (CH_3)_2S]{1.\ O_3,\ -78\ °C}$

o. 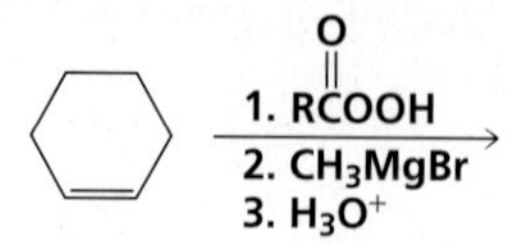 cyclohexene $\xrightarrow{1.\ RCOOH\ (peroxyacid,\ R\overset{O}{\overset{\|}{C}}OOH)\quad 2.\ CH_3MgBr\quad 3.\ H_3O^+}$

p. 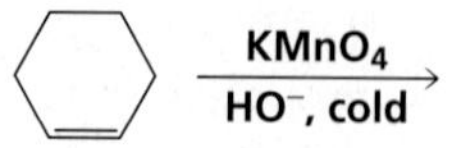 cyclohexene $\xrightarrow[HO^-,\ cold]{KMnO_4}$

q. 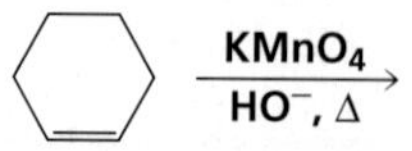 cyclohexene $\xrightarrow[HO^-,\ \Delta]{KMnO_4}$

r. cyclohexene $\xrightarrow[2.\ H_2O_2]{1.\ O_3,\ -78\ °C}$

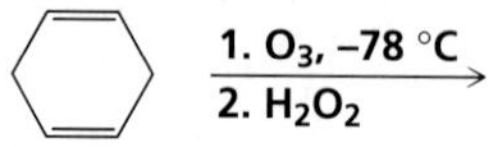

36. How could each of the following compounds be converted to $CH_3CH_2CH_2\overset{O}{\overset{\|}{C}}OH$?

a. $CH_3CH_2CH_2\overset{O}{\overset{\|}{C}}H$

b. $CH_3CH_2CH_2CH_2OH$

c. $CH_3CH_2CH_2CH_2Br$

d. $CH_3CH_2CH{=}CH_2$

37. Identify A–G:

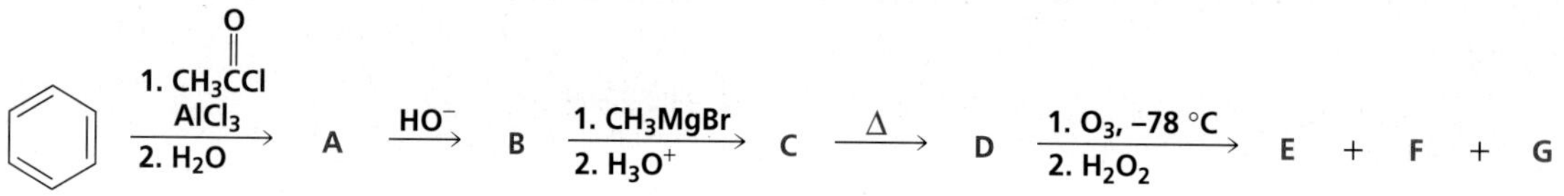

benzene $\xrightarrow[2.\ H_2O]{1.\ CH_3\overset{O}{\overset{\|}{C}}Cl,\ AlCl_3}$ A $\xrightarrow{HO^-}$ B $\xrightarrow[2.\ H_3O^+]{1.\ CH_3MgBr}$ C $\xrightarrow{\Delta}$ D $\xrightarrow[2.\ H_2O_2]{1.\ O_3,\ -78\ °C}$ E + F + G

38. Identify the alkene that would give each of the following products upon ozonolysis followed by treatment with hydrogen peroxide:

a. $CH_3CH_2CH_2\overset{O}{\overset{\|}{C}}OH + CH_3\overset{O}{\overset{\|}{C}}CH_3$

b. $CH_3\overset{O}{\overset{\|}{C}}CH_2CH_2CH_2CH_2\overset{O}{\overset{\|}{C}}CH_2CH_3$

c. 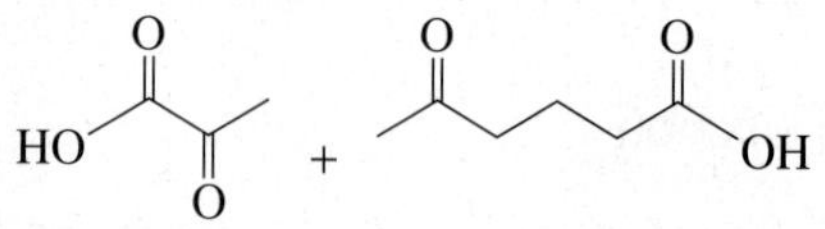 $HO\overset{O}{\overset{\|}{C}}CH_2\overset{O}{\overset{\|}{C}}H$ (HOOC–CH(CH₃)–C(=O)– skeletal) + $CH_3CH_2\overset{O}{\overset{\|}{C}}CH_2CH_2\overset{O}{\overset{\|}{C}}OH$

d. 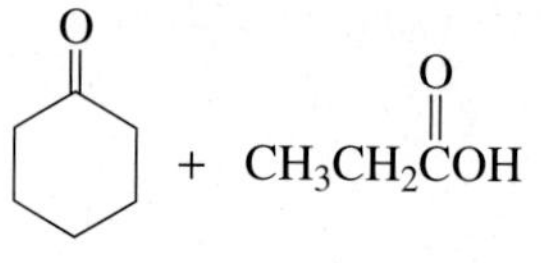 cyclohexanone + $CH_3CH_2\overset{O}{\overset{\|}{C}}OH$

e. 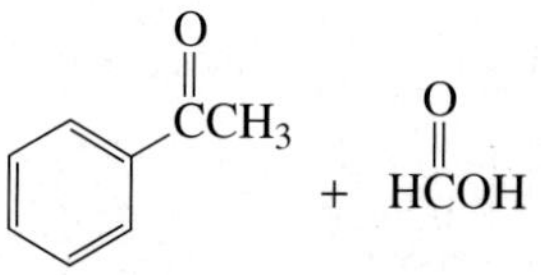 $C_6H_5\overset{O}{\overset{\|}{C}}CH_3 + H\overset{O}{\overset{\|}{C}}OH$

f. 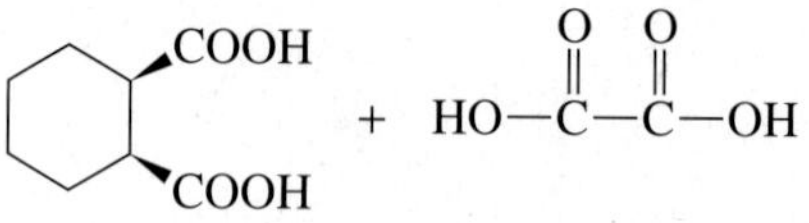 cis-cyclohexane-1,2-dicarboxylic acid (COOH, COOH) $+ HO{-}\overset{O}{\overset{\|}{C}}{-}\overset{O}{\overset{\|}{C}}{-}OH$

39. Fill in each box with the appropriate reagent:

a. $CH_3CH_2CH{=}CH_2 \xrightarrow[2.\ \square]{1.\ \square} CH_3CH_2CH_2CH_2OH \xrightarrow{\square} CH_3CH{=}CHCH_3 \xrightarrow{\square} CH_3\overset{O}{\overset{\|}{C}}OH$

b. $CH_3CH_2Br \xrightarrow{\square} \square \xrightarrow[2.\ \square]{1.\ \square} CH_3CH_2CH_2CH_2OH \xrightarrow{\square} CH_3CH_2CH_2\overset{O}{\overset{\|}{C}}H$

c. 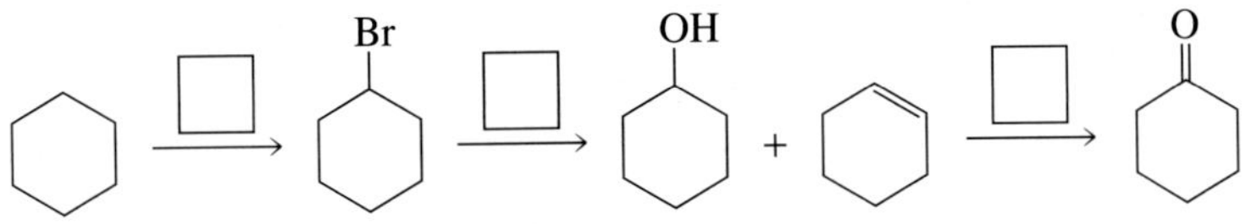 [cyclohexane] $\xrightarrow{\square}$ [bromocyclohexane] $\xrightarrow{\square}$ [cyclohexanol] + [cyclohexene] $\xrightarrow{\square}$ [cyclohexanone] + $HO\overset{O}{\overset{\|}{C}}CH_2CH_2CH_2CH_2\overset{O}{\overset{\|}{C}}OH$

40. Describe how 1-butyne can be converted into each of the following compounds:

a. 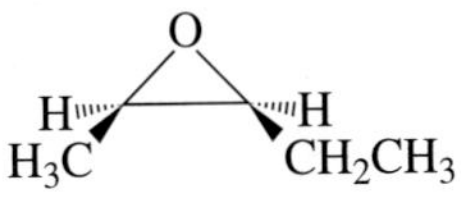 [epoxide: H, H_3C / H, CH_2CH_3]

b. 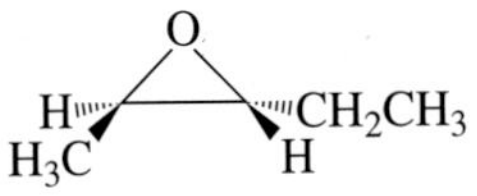 [epoxide: H, H_3C / CH_2CH_3, H]

41. a. Give the products obtained from ozonolysis of each of the following compounds, followed by work-up under oxidizing conditions:

1. 2. 3. 4. 5.

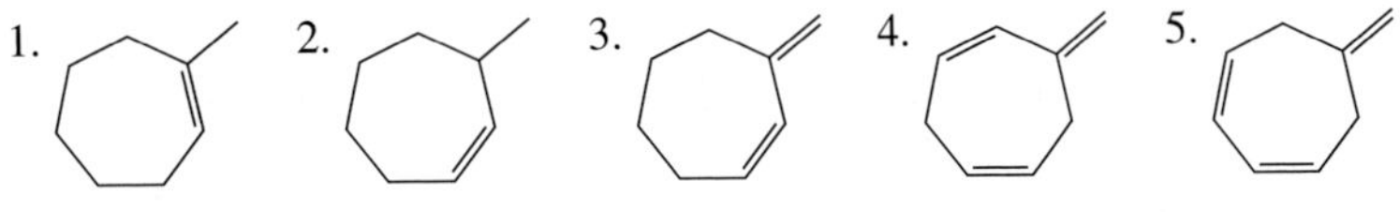

b. What compound would form the following products upon reaction with ozone, followed by work-up under reducing conditions?

$H\overset{O}{\overset{\|}{C}}CH_2CH_2\overset{O}{\overset{\|}{C}}-\overset{O}{\overset{\|}{C}}H$ + $H-\overset{O}{\overset{\|}{C}}-\overset{O}{\overset{\|}{C}}-H$ + $H\overset{O}{\overset{\|}{C}}H$

42. Show how each of the following compounds can be prepared from cyclohexene:

a. 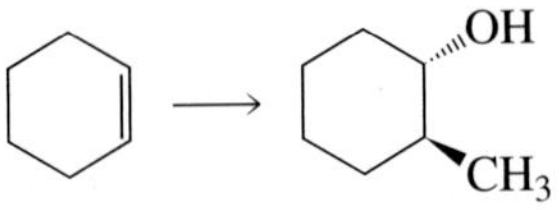 [cyclohexene] → [cyclohexanol with OH and CH_3]

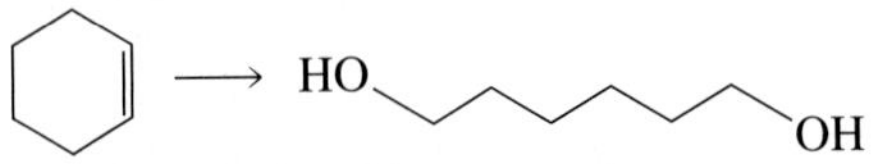

b. 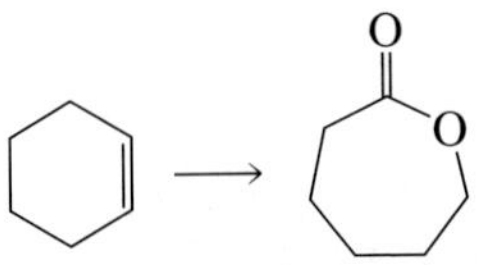 [cyclohexene] → [lactone]

c. [cyclohexene] → HO~~~~~~OH

d. 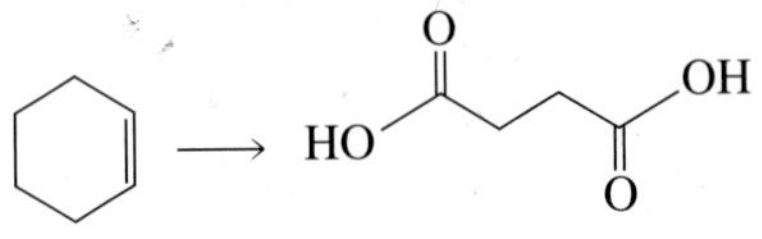 [cyclohexene] → HO(C=O)CH₂CH₂(C=O)CH₂OH

43. The ^1H NMR spectrum of the product obtained when an unknown alkene reacts with ozone and the ozonolysis product is worked up under oxidizing conditions is shown. Identify the alkene.

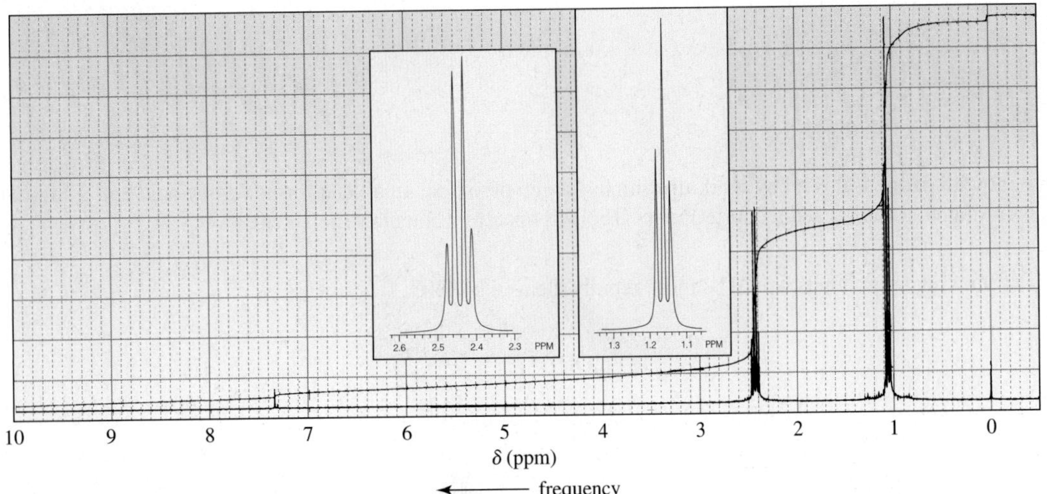

44. Identify A–N:

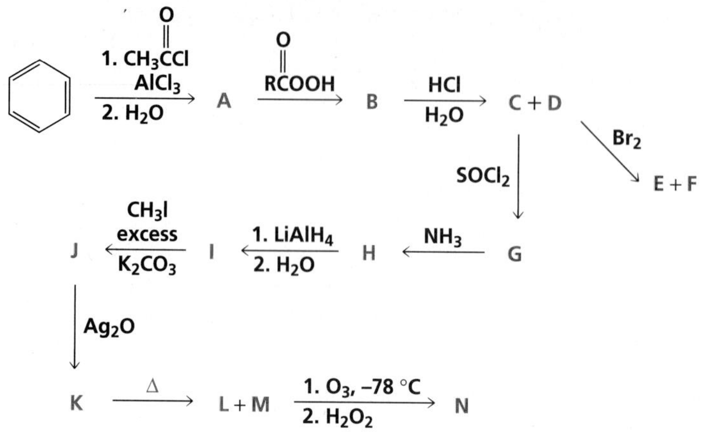

Benzene $\xrightarrow[\text{2. } H_2O]{\text{1. } CH_3\overset{O}{\overset{\|}{C}}Cl,\ AlCl_3}$ A $\xrightarrow{RC\overset{O}{\overset{\|}{O}}OH}$ B $\xrightarrow[H_2O]{HCl}$ C + D

C + D $\xrightarrow{Br_2}$ E + F

C + D $\xrightarrow{SOCl_2}$ G $\xrightarrow{NH_3}$ H $\xrightarrow[\text{2. } H_2O]{\text{1. } LiAlH_4}$ I $\xrightarrow[K_2CO_3]{CH_3I \text{ excess}}$ J

J $\xrightarrow{Ag_2O}$ K $\xrightarrow{\Delta}$ L + M $\xrightarrow[\text{2. } H_2O_2]{\text{1. } O_3,\ -78\ °C}$ N

45. Chromic acid oxidizes 2-propanol six times faster than it oxidizes 2-deuterio-2-propanol. Explain. (*Hint:* See Section 11.7.)

46. Fill in each box with the appropriate reagent:

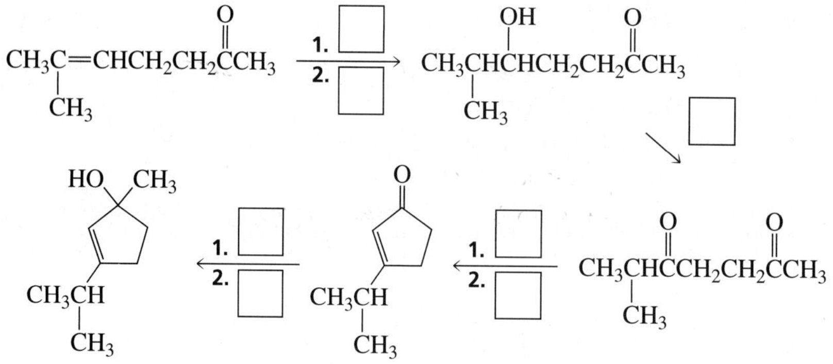

$CH_3\underset{CH_3}{\underset{|}{C}}{=}CHCH_2CH_2\overset{O}{\overset{\|}{C}}CH_3$ $\xrightarrow[\text{2. } \square]{\text{1. } \square}$ $CH_3\underset{CH_3}{\underset{|}{C}H}\overset{OH}{\overset{|}{C}H}CH_2CH_2\overset{O}{\overset{\|}{C}}CH_3$ $\xrightarrow{\square}$ $CH_3\underset{CH_3}{\underset{|}{C}H}\overset{O}{\overset{\|}{C}}CH_2CH_2\overset{O}{\overset{\|}{C}}CH_3$ $\xrightarrow[\text{2. } \square]{\text{1. } \square}$ 2-(CH₃CHCH₃)-cyclopent-2-enone $\xrightarrow[\text{2. } \square]{\text{1. } \square}$ 1-methyl-2-(CH₃CHCH₃)-cyclopent-2-en-1-ol (HO, CH₃)

47. Show how each of the following compounds could be prepared, using the given starting material:

a. $CH_3CH_2\overset{O}{\overset{\|}{C}}H \longrightarrow CH_3CH_2\overset{O}{\overset{\|}{C}}OCH_2CH_2CH_3$

b. $CH_3CH_2CH_2CH_2OH \longrightarrow CH_3CH_2CH_2\overset{O}{\overset{\|}{C}}CH_2CH_3$

c. 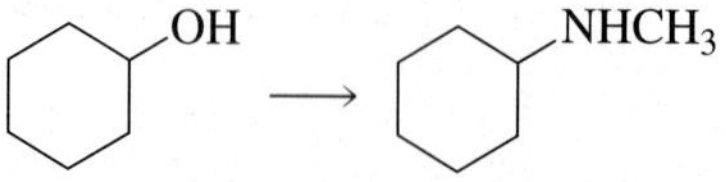 cyclohexyl–OH $\longrightarrow$ cyclohexyl–NHCH₃

d. 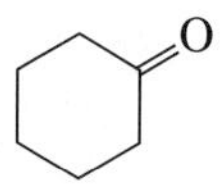 cyclohexanone $\longrightarrow HO\overset{O}{\overset{\|}{C}}CH_2CH_2CH_2CH_2\overset{O}{\overset{\|}{C}}OH$

e. 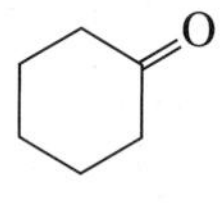 cyclohexanone $\longrightarrow HO\overset{O}{\overset{\|}{C}}CH_2CH_2CH_2CH_2\overset{O}{\overset{\|}{C}}CH_3$

48. Upon treatment with ozone followed by work-up with hydrogen peroxide, an alkene forms formic acid and a compound that shows three signals (a singlet, a triplet, and a quartet) in its ^1H NMR spectrum. Identify the alkene.

49. Which of the following compounds would be more rapidly cleaved by HIO_4?

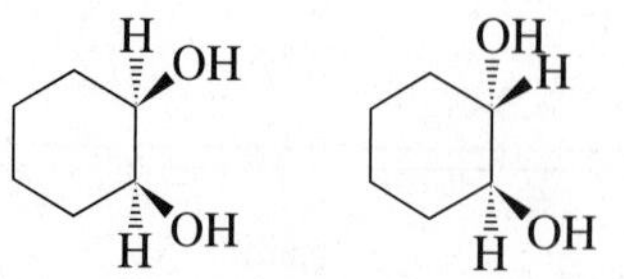

A: cyclohexane-1,2-diol (H, OH; H, OH) B: cyclohexane-1,2-diol (OH, H; H, OH)

50. Show how cyclohexylacetylene can be converted into each of the following compounds:

a. cyclohexyl–$\overset{O}{\overset{\|}{C}}OH$

b. cyclohexyl–$CH_2\overset{O}{\overset{\|}{C}}OH$

51. Show how the following compounds could be synthesized. The only carbon-containing reagents that are available for each synthesis are shown.

a. $CH_3CH_2CH_2OH \longrightarrow CH_3CH_2CH_2\underset{CH_3}{\underset{|}{C}}HCH_2OH$

b. $CH_3\underset{CH_3}{\underset{|}{C}}HOH \longrightarrow CH_3\underset{CH_3}{\underset{|}{C}}HCH_2CH_2CH_3$

c. $CH_3CH_2OH + CH_3\underset{CH_3}{\underset{|}{C}}HOH \longrightarrow$ (2,2,4-trimethyl-1,3-dioxane)

52. The catalytic hydrogenation of 0.5 g of a hydrocarbon at 25 °C consumed about 200 mL of H_2 under 1 atm of pressure. Reaction of the hydrocarbon with ozone, followed by treatment with hydrogen peroxide gave one product, which was found to be a four-carbon carboxylic acid. Identify the hydrocarbon.

53. Tom Thumbs was asked to prepare the following compounds from the given starting materials. The reagents he chose to use for each synthesis are shown.

a. Which of his syntheses were successful?
b. What products did he obtain from the other syntheses?
c. In his unsuccessful syntheses, what reagents should he have used to obtain the desired product?

1. $CH_3CH_2\overset{CH_3}{\overset{|}{C}}{=}CHCH_3 \xrightarrow[H_2SO_4]{KMnO_4} CH_3CH_2\overset{CH_3}{\overset{|}{C}}(OH){-}CH(OH)CH_3$

2. $CH_3CH_2\overset{O}{\overset{\|}{C}}OCH_3 \xrightarrow[2.\ H_3O^+]{1.\ NaBH_4} CH_3CH_2CH_2OH + CH_3OH$

3. 1,2-dimethylcyclohexene (with H_3C, CH_3) $\xrightarrow[2.\ HO^-]{1.\ R\overset{O}{\overset{\|}{C}}OOH}$ 1,2-dimethylcyclohexane-1,2-diol (OH OH; H_3C CH_3)

54. The catalytic hydrogenation of compound A formed compound B. The IR spectrum of compound A and the ¹H NMR spectrum of compound B are shown. Identify the compounds.

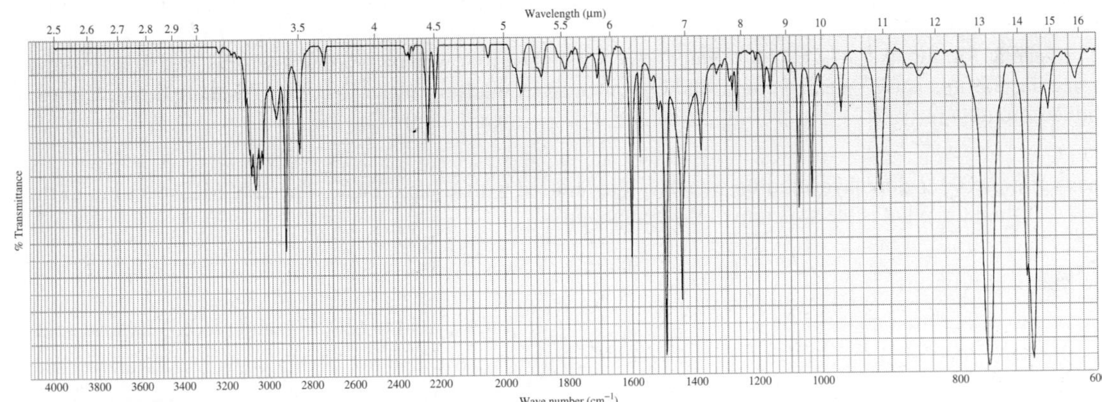

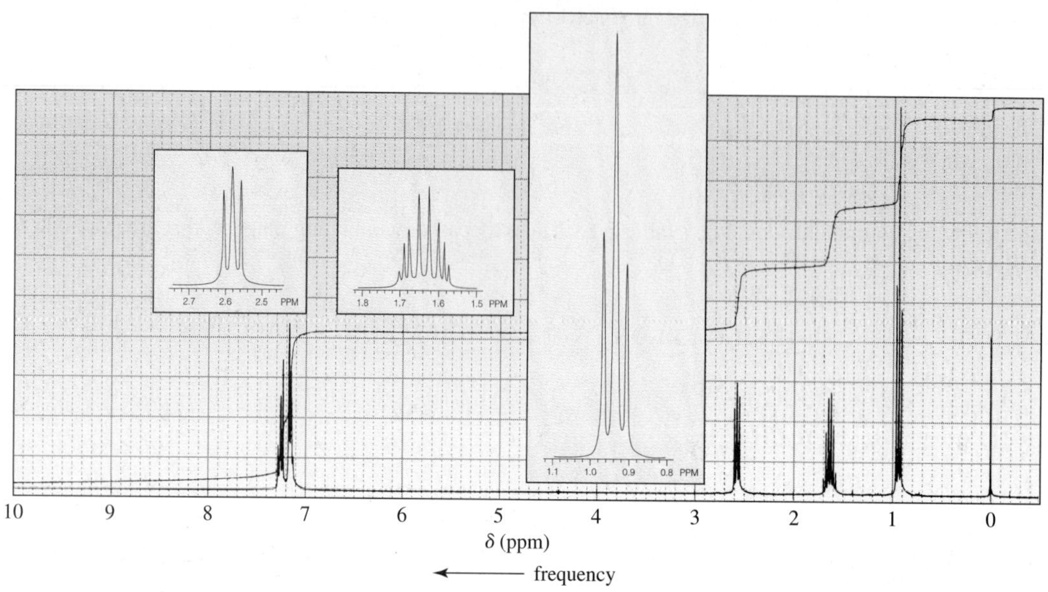

55. Diane Diol worked for several days to prepare the following compounds:

CH_3	CH_3	CH_3	CH_3	CH_3	CH_2CH_3
H—OH	H—OH	H—OH	H—OH	H—OH	H—OH
CH_2	H—OH	HO—H	HO—H	CH_2	H—OH
H—OH	CH_3	CH_3	CH_2CH_3	HO—H	CH_2CH_3
CH_3				CH_3	

She labeled them carefully and went to lunch. To her horror, when she returned, she found that the labels had fallen off the bottles and onto the floor. Gladys Glycol, the student at the next bench, told her that the diols could be easily distinguished by two experiments. All Diane had to do was determine which ones were optically active and how many products were obtained when each was treated with periodic acid. Diane did what Gladys suggested and found the following to be true:

1. Compounds A, E, and F are optically active, and B, C, and D are optically inactive.
2. One product is obtained from the reaction of A, B, and D with periodic acid.
3. Two products are obtained from the reaction of F with periodic acid.
4. C and E do not react with periodic acid.

Will Diane be able to distinguish among the six diols and label them from A to F with only the preceding information? Label the structures.

56. Show how propyl propionate could be prepared, using allyl alcohol as the only source of carbon.

57. Compound A has a molecular formula of $C_5H_{12}O$ and is oxidized by an acidic solution of sodium dichromate to give compound B, whose molecular formula is $C_5H_{10}O$. When compound A is heated with H_2SO_4, C and D are obtained. Considerably more D is obtained than C. Compound C reacts with O_3, followed by treatment with H_2O_2, to give two products: formic acid and compound E, whose molecular formula is C_4H_8O. Compound D reacts with O_3, followed by treatment with H_2O_2, to give compound F, whose molecular formula is C_3H_6O, and compound G, whose molecular formula is $C_2H_4O_2$. What are the structures of compounds A through G?

58. A compound forms *cis*-1,2-dimethycyclopropane when it is reduced with H_2 and Pd/C. The ^1H NMR spectrum of the compound shows only two singlets. What is the structure of the compound?

59. Show how you could convert
   a. maleic acid to (2*R*,3*S*)-tartaric acid
   b. fumaric acid to (2*R*,3*S*)-tartaric acid
   c. maleic acid to (2*R*,3*R*)- and (2*S*,3*S*)-tartaric acid
   d. fumaric acid to (2*R*,3*R*)- and (2*S*,3*S*)-tartaric acid

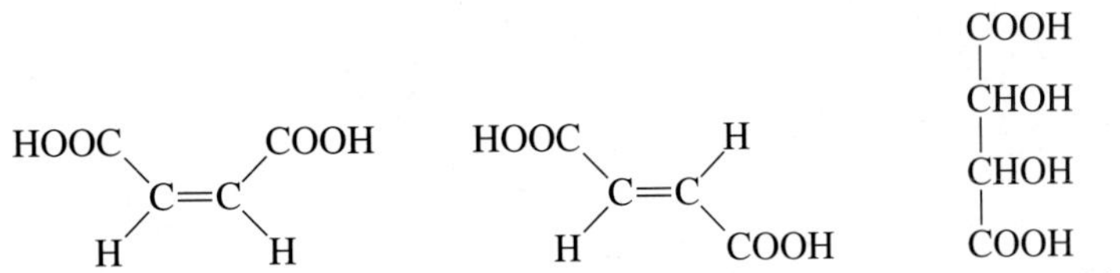

HOOC(H)C=C(H)COOH — maleic acid; HOOC(H)C=C(H)COOH (trans) — fumaric acid; HOOC–CHOH–CHOH–COOH — tartaric acid

60. Identify A through O:

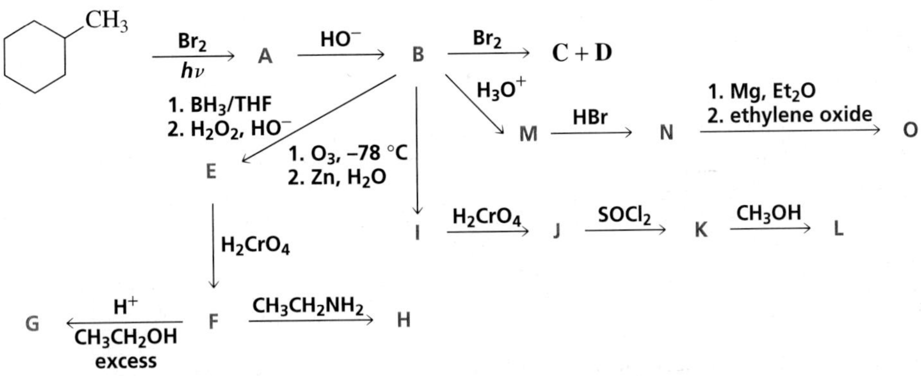

Cyclohexyl-CH_3 $\xrightarrow[h\nu]{Br_2}$ A $\xrightarrow{HO^-}$ B $\xrightarrow{Br_2}$ C + D

B $\xrightarrow{1.\ BH_3/THF;\ 2.\ H_2O_2,\ HO^-}$ E $\xrightarrow{H_2CrO_4}$ F $\xrightarrow{CH_3CH_2NH_2}$ H

F $\xrightarrow[CH_3CH_2OH\ excess]{H^+}$ G

B $\xrightarrow{1.\ O_3,\ -78\ °C;\ 2.\ Zn,\ H_2O}$ I $\xrightarrow{H_2CrO_4}$ J $\xrightarrow{SOCl_2}$ K $\xrightarrow{CH_3OH}$ L

B $\xrightarrow{H_3O^+}$ M $\xrightarrow{HBr}$ N $\xrightarrow{1.\ Mg,\ Et_2O;\ 2.\ ethylene\ oxide}$ O

61. Show how the following compounds could be prepared, using only the indicated starting material as the source of carbon:

a. $CH_3\overset{O}{\overset{\|}{C}}CH_3$ from $CH_3\overset{CH_3}{\overset{|}{C}}HCH_3$

b. $CH_3CH{=}\overset{CH_3}{\overset{|}{C}}CH_3$ using propane as the only source of carbon

c. $CH_3\overset{O}{\overset{\|}{C}}-\overset{CH_3}{\overset{|}{C}}HCH_3$ from propane and any molecule with two carbon atoms

d. $CH_3CH_2\overset{O}{\overset{\|}{C}}H$ from two molecules of ethane

62. A primary alcohol can be oxidized only as far as the aldehyde stage if the alcohol is first treated with tosyl chloride (TsCl) and the resulting tosylate is allowed to react with dimethyl sulfoxide (DMSO). Propose a mechanism for this reaction. (*Hint:* See Section 20.2.)

$$CH_3CH_2CH_2CH_2OH \xrightarrow[pyridine]{TsCl} CH_3CH_2CH_2CH_2OTs \xrightarrow{DMSO} CH_3CH_2CH_2\overset{O}{\overset{\|}{C}}H$$

63. Identify the alkene that gives each of the following products upon ozonolysis, followed by treatment with dimethyl sulfide:

a. 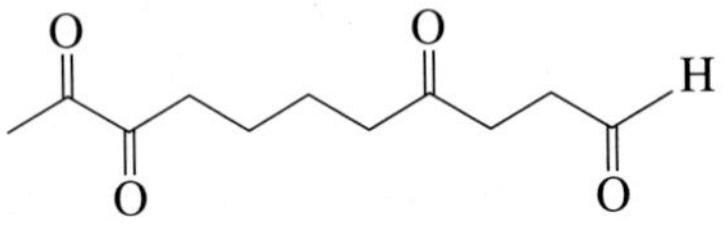 [skeletal structure: polycarbonyl product with O labels and terminal CHO (H)]

b. 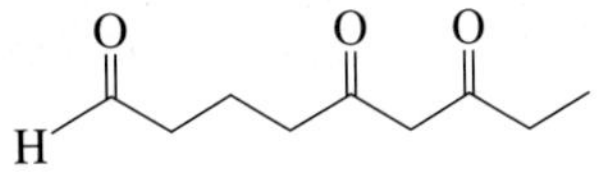 [skeletal structure: H–C(=O)– chain with two ketone O's] + $H\overset{O}{\overset{\|}{C}}H$

64. Propose a mechanism to explain how dimethyl sulfoxide and oxalyl chloride react to form the dimethylchlorosulfonium ion used as the oxidizing agent in the Swern oxidation.

$$\underset{\text{dimethyl sulfoxide}}{CH_3-\overset{O}{\overset{\|}{S}}-CH_3} + \underset{\text{oxalyl chloride}}{Cl-\overset{O}{\overset{\|}{C}}-\overset{O}{\overset{\|}{C}}-Cl} \longrightarrow \underset{\text{dimethylchloro-sulfonium ion}}{CH_3-\overset{Cl}{\overset{|}{\underset{+}{S}}}-CH_3} + CO_2 + CO + Cl^-$$

65. Show how the following compounds could be prepared, using only the indicated starting material as the source of carbon:

a. 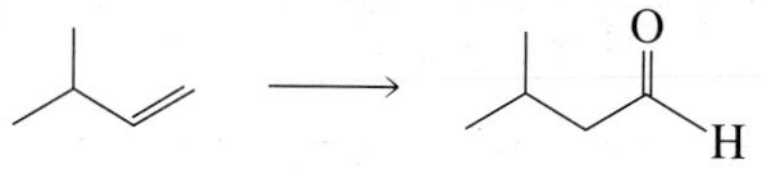

b. 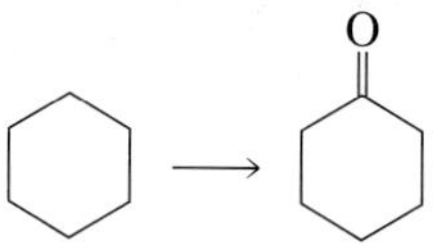

c. 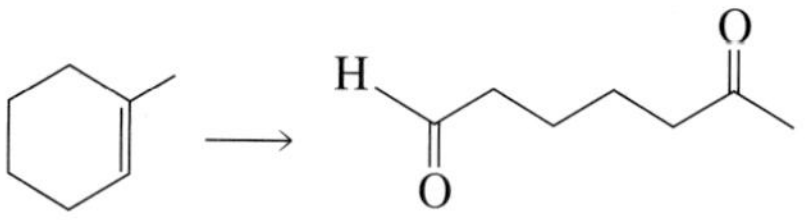

d. 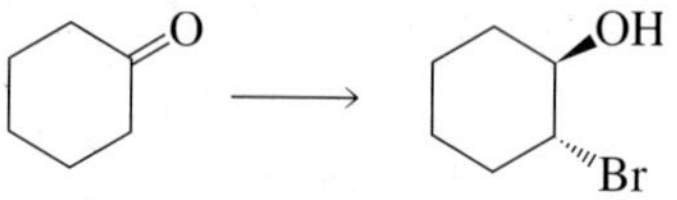

e. 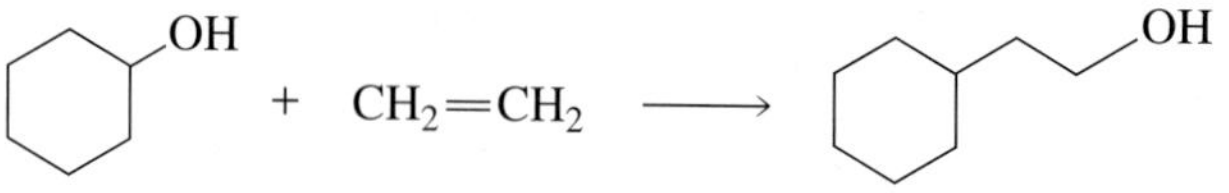

f. 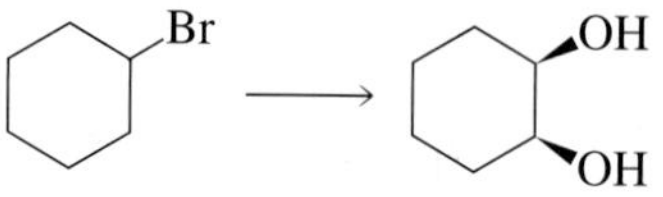

66. Terpineol ($C_{10}H_{18}O$) is an optically active compound with one asymmetric carbon. It is used as an antiseptic. Reaction of terpineol with H_2/Pt forms an optically inactive compound ($C_{10}H_{20}O$). Heating the reduced compound in acid followed by ozonolysis and work-up under reducing conditions produces acetone and a compound whose ^1H NMR and ^{13}C NMR are shown. What is the structure of terpineol?

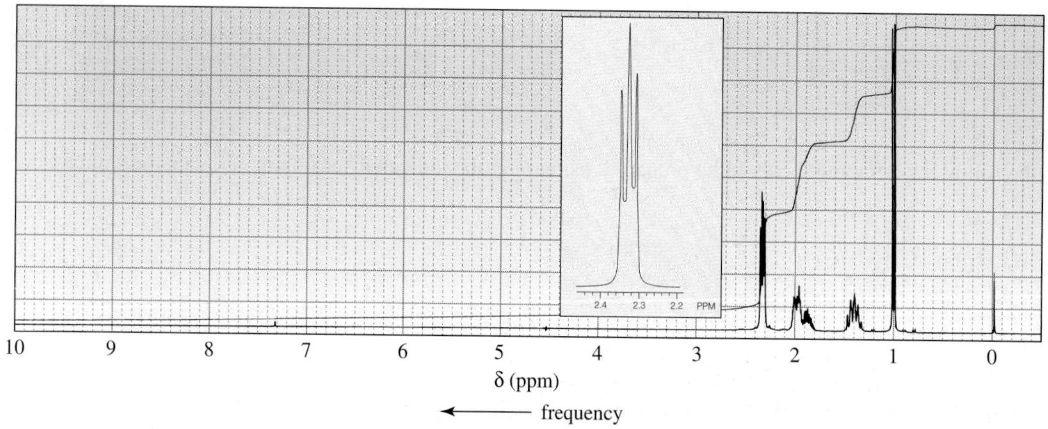

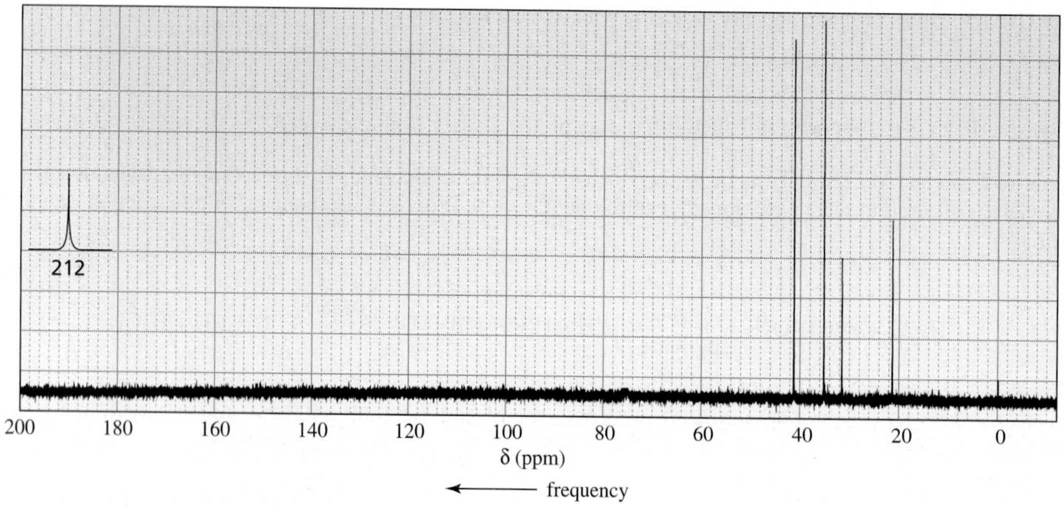

67. Propose a mechanism for the following enzyme-catalyzed reaction. (*Hint:* Notice that Br is attached to the more substituted carbon.)

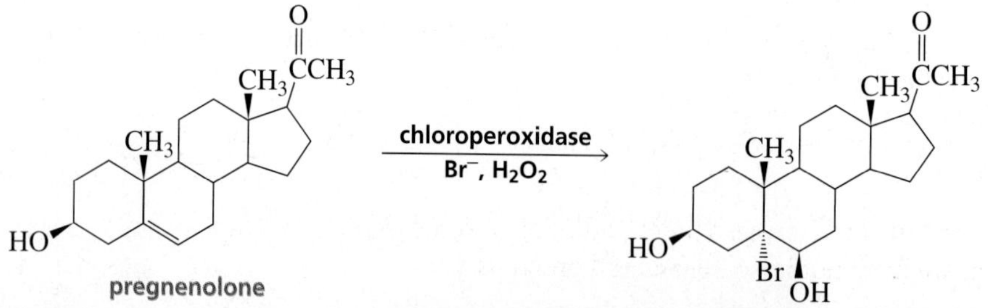

pregnenolone

21 More About Amines • Heterocyclic Compounds

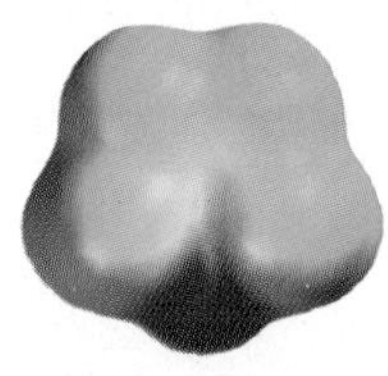
pyrrolidine

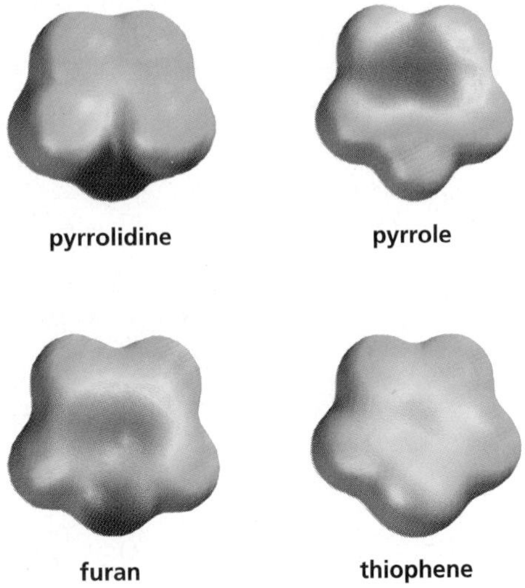

pyrrole

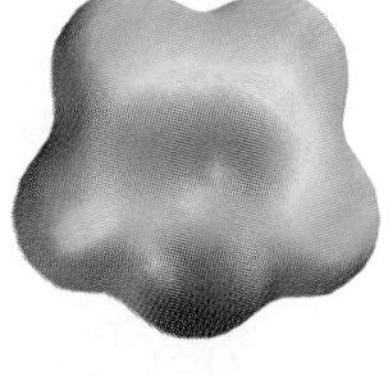
furan

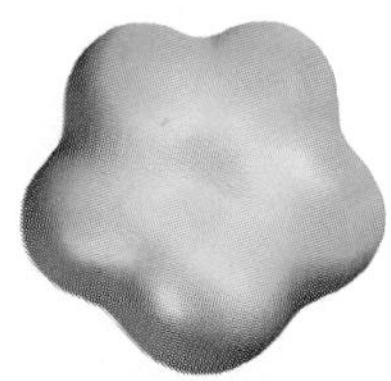
thiophene

Amines are compounds in which one or more of the hydrogens of ammonia (NH_3) have been replaced by an alkyl group. Amines are among some of the most abundant compounds in the biological world. We will appreciate their importance in Chapter 23, when we look at amino acids and proteins; in Chapter 24, when we study how enzymes catalyze chemical reactions; in Chapter 25, when we investigate the ways in which coenzymes—compounds derived from vitamins—help enzymes catalyze chemical reactions; in Chapter 27, when we study nucleic acids (DNA and RNA); and in Chapter 30, when we take a look at how drugs are discovered and designed.

Amines are also exceedingly important compounds to organic chemists, far too important to leave until the end of a course in organic chemistry. We have, therefore, already studied many aspects of amines and their chemistry. For example, we have seen that the nitrogen in amines is sp^3 hybridized and the lone pair resides in an empty sp^3 orbital (Section 2.8). We also have examined the physical properties of amines—their hydrogen bonding properties, boiling points, and solubilities (Section 2.9). In Section 2.7, we learned how amines are named. Most important, we have seen that the lone-pair electrons of the nitrogen atom cause amines to react as bases, sharing their lone pair with a proton, and to react as nucleophiles, sharing their lone pair with an atom other than a proton.

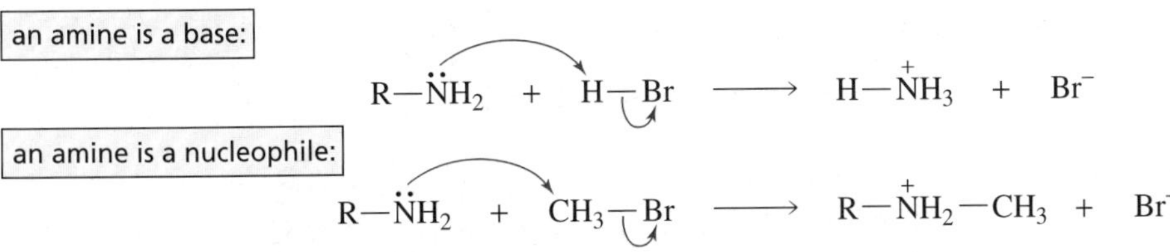

In this chapter, we will revisit some of these topics and look at some aspects of amines and their chemistry that we have not been considered previously.

Some amines are **heterocyclic compounds** (or **heterocycles**)—cyclic compounds in which one or more of the atoms of the ring are heteroatoms. A **heteroatom** is an atom other than carbon. The name comes from the Greek word *heteros*, which means "different." A variety of atoms, such as N, O, S, Se, P, Si, B, and As, can be incorporated into ring structures.

Heterocycles are an extraordinarily important class of compounds, making up more than half of all known organic compounds. Almost all the compounds we know as drugs, most vitamins, and many other natural products are heterocycles. In this chapter, we will consider the most prevalent heterocyclic compounds—the ones that contain the heteroatoms N, O, and S.

A **natural product** is a compound synthesized by a plant or an animal. **Alkaloids** are natural products that contain one or more nitrogen heteroatoms and are found in the leaves, bark, roots, or seeds of plants. Examples include caffeine (found in tea leaves, coffee beans, and cola nuts) and nicotine (found in tobacco leaves). Morphine is an alkaloid obtained from opium, the juice derived from a species of poppy. Morphine is 50 times stronger than aspirin as an analgesic, but it is addictive and suppresses respiration. Heroin is a synthetic compound that is made by acetylating morphine (Section 30.3).

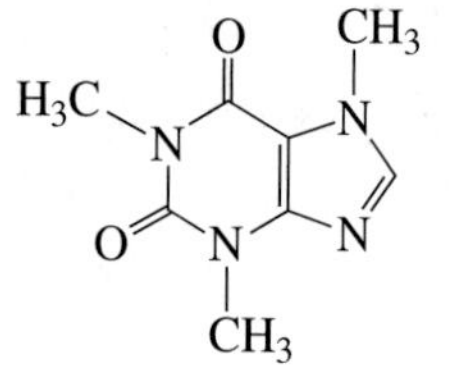

caffeine 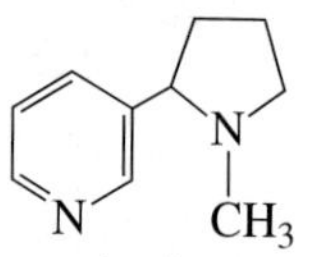 nicotine 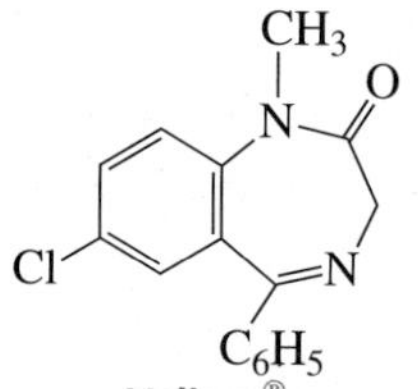 Valium® 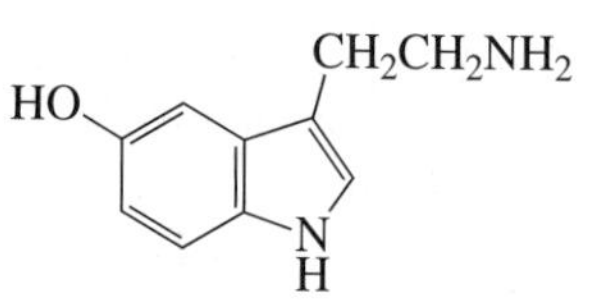 serotonin

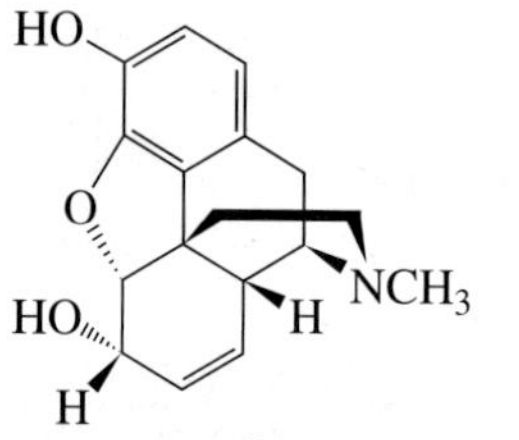

morphine 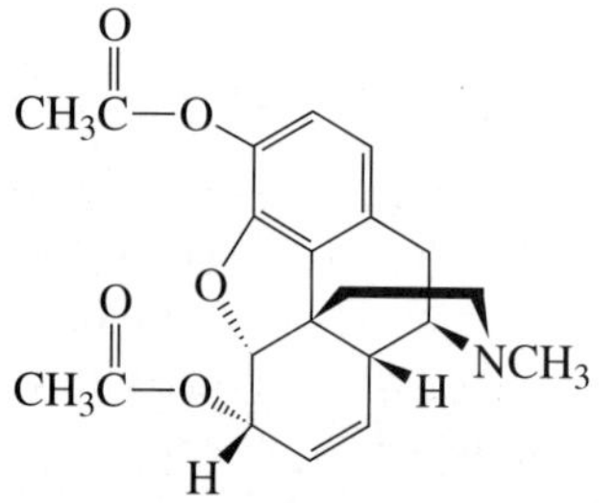 heroin

Two other heterocycles are Valium®, a synthetic tranquilizer, and serotonin, a neurotransmitter. Serotonin is responsible for, among other things, the feeling of having had enough to eat. When food is ingested, brain neurons are signaled to release serotonin. A once widely used diet drug (actually a combination of two drugs, fenfluramine and phentermine), popularly known as fen/phen, works by causing brain neurons to release extra serotonin (Chapter 16, p. 622). After finding that many of those who took fenfluramine had abnormal echocardiograms due to heart valve problems, the Food and Drug Administration asked the manufacturer of these diet drugs to withdraw the products. There is some evidence that faulty metabolism of serotonin plays a role in bipolar affective disorder.

21.1 More About Nomenclature

In Section 2.7, we saw that amines are classified as primary, secondary, or tertiary, depending on whether one, two, or three hydrogens of ammonia, respectively, have been replaced by an alkyl group. We also saw that amines have both common and systematic names. Common names are obtained by citing the names of the alkyl subsitutents (in alphabetical order) that have replaced the hydrogens of ammonia. Systematic names employ "amine" as a functional group suffix.

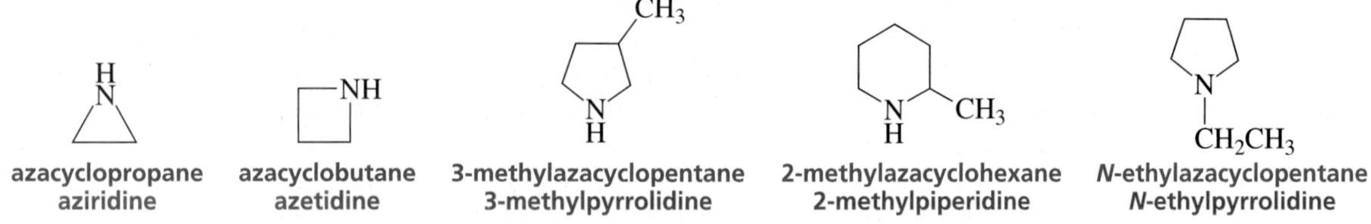

A saturated cyclic amine—a cyclic amine without any double bonds—can be named as a cycloalkane, using the prefix "aza" to denote the nitrogen atom. There are, however, other acceptable names. Some of the more commonly used names are shown here. Notice that heterocyclic rings are numbered so that the heteroatom has the lowest possible number.

azacyclopropane aziridine · azacyclobutane azetidine · 3-methylazacyclopentane 3-methylpyrrolidine · 2-methylazacyclohexane 2-methylpiperidine · N-ethylazacyclopentane N-ethylpyrrolidine

Heterocycles with oxygen and sulfur heteroatoms are named similarly. The prefix for oxygen is "oxa" and that for sulfur is "thia."

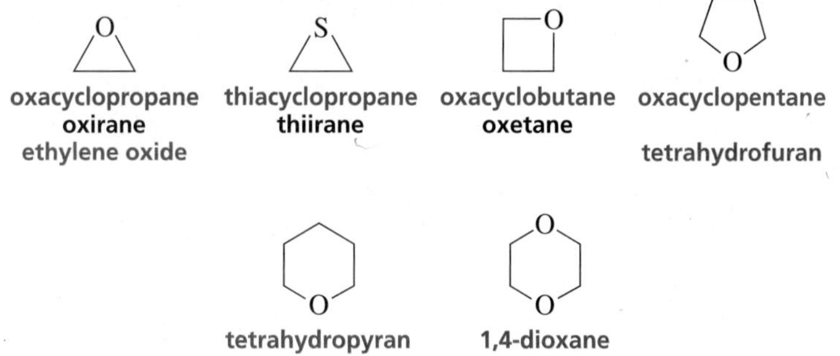

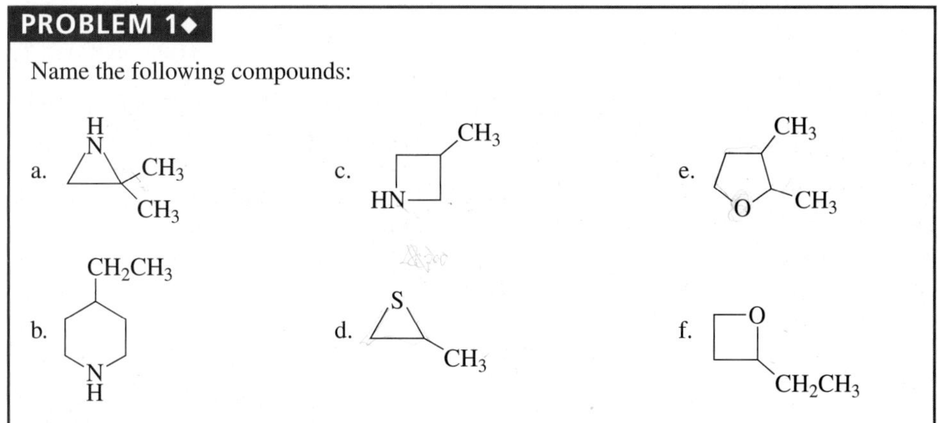

21.2 Amine Inversion

The lone-pair electrons on nitrogen allow an amine to turn "inside out" rapidly at room temperature. This is called **amine inversion**. One way to picture amine inversion is to compare it to an umbrella that turns inside out in a windstorm.

amine inversion

The lone pair is required for inversion: Quaternary ammonium ions—ions with four bonds to nitrogen and hence no lone pair—do not invert.

Notice that amine inversion takes place through a transition state in which the sp^3 nitrogen becomes an sp^2 nitrogen. The three groups bonded to the sp^2 nitrogen are coplanar in the transition state with bond angles of 120°, and the lone pair is in a *p* orbital. The "inverted" and "non-inverted" amine molecules are enantiomers, but they cannot be separated because amine inversion is rapid. The energy required for amine inversion is approximately 6 kcal/mol (or 25 kJ/mol), about twice the amount of energy required for rotation about a carbon–carbon single bond, but still low enough to allow the enantiomers to interconvert rapidly at room temperature.

21.3 More About the Acid–Base Properties of Amines

Amines are the most common organic bases. We have seen that ammonium ions have pK_a values of about 11 (Section 1.17) and that anilinium ions have pK_a values of about 5 (Sections 7.10 and 16.5). The greater acidity of anilinium ions compared with ammonium ions is due to the greater stability of the conjugate bases of the anilinium ions as a result of electron delocalization. Amines have very high pK_a values. For example, the pK_a of methylamine is 40.

$CH_3CH_2CH_2\overset{+}{N}H_3$ $pK_a = 10.8$

$CH_3\overset{+}{N}H_2CH_3$ $pK_a = 10.9$

$CH_3CH_2\overset{+}{N}H(CH_2CH_3)_2$ $pK_a = 11.1$

$C_6H_5\overset{+}{N}H_3$ $pK_a = 4.58$

$CH_3C_6H_4\overset{+}{N}H_3$ $pK_a = 5.07$

CH_3NH_2 $pK_a = 40$

Saturated heterocycles containing five or more atoms have physical and chemical properties typical of acyclic compounds that contain the same heteroatom. For example, pyrrolidine, piperidine, and morpholine are typical secondary amines, and *N*-methylpyrrolidine and quinuclidine are typical tertiary amines. The conjugate acids of these amines have pK_a values expected for ammonium ions. We have seen that the basicity of amines allows them to be easily separated from other organic compounds (Chapter 1, Problems 70 and 71).

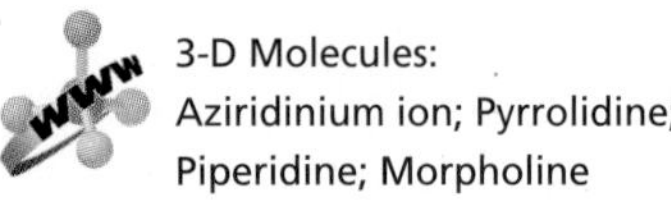
3-D Molecules: Aziridinium ion; Pyrrolidine; Piperidine; Morpholine

the ammonium ions of:

pyrrolidine	piperidine	morpholine	*N*-methylpyrrolidine	quinuclidine
$pK_a = 11.27$	$pK_a = 11.12$	$pK_a = 9.28$	$pK_a = 10.32$	$pK_a = 11.38$

PROBLEM 2◆

Why is the pK_a of the conjugate acid of morpholine significantly lower than the pK_a of the conjugate acid of piperidine?

PROBLEM 3◆

a. Draw the structure of 3-quinuclidinone.

b. What is the approximate pK_a of its conjugate acid?

c. Which has a lower pK_a, the conjugate acid of 3-bromoquinuclidine or the conjugate acid of 3-chloroquinuclidine?

21.4 Reactions of Amines

The lone pair on the nitrogen of an amine causes it to be nucleophilic as well as basic. We have seen amines act as nucleophiles in a number of different kinds of reactions: in nucleophilic substitution reactions—reactions that *alkylate* the amine (Section 10.4)—such as

$$CH_3CH_2Br + \underset{\text{methylamine}}{CH_3NH_2} \longrightarrow \underset{Br^-}{CH_3CH_2\overset{+}{N}H_2CH_3} \rightleftharpoons \underset{\text{ethylmethylamine}}{CH_3CH_2NHCH_3} + HBr$$

in nucleophilic acyl substitution reactions—reactions that *acylate* the amine (Sections 17.8, 17.9, and 17.10)—for example,

$$CH_3CH_2COCl + \underset{\text{methylamine}}{2\ CH_3NH_2} \longrightarrow \underset{\text{an amide}}{CH_3CH_2CONHCH_3} + CH_3\overset{+}{N}H_3Cl^-$$

$$CH_3COOCOCH_3 + 2\ \underset{\text{piperidine}}{\text{piperidine}} \longrightarrow \underset{\text{an amide}}{CH_3CO\text{-}N(\text{piperidine})} + \text{piperidinium } (H_2N^+)\ {}^-OCCH_3$$

in nucleophilic addition–elimination reactions—the reactions of aldehydes and ketones with primary amines to form imines and with secondary amines to form enamines (Section 18.6)—such as

$$\text{cyclopentanone}{=}O + \underset{\text{benzylamine}}{H_2NCH_2C_6H_5} \underset{H^+}{\overset{\text{catalytic}}{\rightleftharpoons}} \underset{\text{an imine}}{\text{cyclopentylidene}{=}NCH_2C_6H_5} + H_2O$$

$$\text{cyclohexanone}{=}O + \underset{\text{pyrrolidine}}{HN(CH_2)_4} \underset{H^+}{\overset{\text{catalytic}}{\rightleftharpoons}} \underset{\text{an enamine}}{\text{cyclohexenyl}\text{-}N(CH_2)_4} + H_2O$$

and in conjugate addition reactions (Section 18.13)—for instance,

$$CH_3\overset{CH_3}{\overset{|}{C}}{=}CH\overset{O}{\overset{\|}{C}}H + CH_3\underset{CH_3}{\underset{|}{N}}H \longrightarrow CH_3\overset{CH_3}{\overset{|}{C}}(NCH_3CH_3)\text{—}CH_2\overset{O}{\overset{\|}{C}}H$$

We have seen that primary arylamines react with nitrous acid to form stable arenediazonium salts (Section 16.12). Arenediazonium salts are useful to synthetic chemists because the diazonium group can be replaced by a wide variety of nucleophiles. This reaction allows a wider variety of substituted benzenes to be prepared than can be prepared solely from electrophilic aromatic substitution reactions.

$$C_6H_5\text{–}NH_2 \xrightarrow[NaNO_2]{HCl} C_6H_5\text{–}\overset{+}{N}{\equiv}N\ Cl^- \xrightarrow{Nu^-} C_6H_5\text{–}Nu + N_2 + Cl^-$$

an arenediazonium salt

Amines are much less reactive than other compounds with electron-withdrawing groups bonded to sp^3 hybridized carbons, such as alkyl halides, alcohols, and ethers. The relative reactivities of an alkyl fluoride (the least reactive of the alkyl halides), an alcohol, an ether, and an amine can be appreciated by comparing the pK_a values of the conjugate acids of their leaving groups, recalling that the weaker the acid, the stronger its conjugate base and the poorer the base is as a leaving group. The leaving group of an amine ($^-NH_2$) is such a strong base that amines cannot undergo the substitution and elimination reactions that alkyl halides undergo.

relative reactivities

most reactive	RCH_2F >	RCH_2OH ~	RCH_2OCH_3 >	RCH_2NH_2	least reactive
strongest acid, weakest conjugate base	HF pK_a = 3.2	H_2O pK_a = 15.7	RCH_2OH pK_a = 15.5	NH_3 pK_a = 36	weakest acid, strongest conjugate base

Protonation of the amino group makes it a weaker base and therefore a better leaving group, but it still is not nearly as good a leaving group as a protonated alcohol. Recall that protonated ethanol is more than 13 pK_a units more acidic than protonated ethylamine.

$$CH_3CH_2\overset{+}{O}H_2 \quad pK_a = -2.4 \qquad CH_3CH_2\overset{+}{N}H_3 \quad pK_a = 11.2$$

So, unlike the leaving group of a protonated alcohol, the leaving group of a protonated amine cannot dissociate to form a carbocation or be replaced by a halide ion. Protonated amino groups also cannot be displaced by strongly basic nucleophiles such as HO^- because the base would react immediately with the acidic hydrogen, and protonation would convert it into a poor nucleophile.

$$CH_3CH_2\overset{+}{N}H_3 + HO^- \rightleftharpoons CH_3CH_2NH_2 + H_2O$$

PROBLEM 4

Why is it that a halide ion such as Br^- can react with a protonated primary alcohol, but cannot react with a protonated primary amine?

PROBLEM 5

Give the product of each of the following reactions:

a. 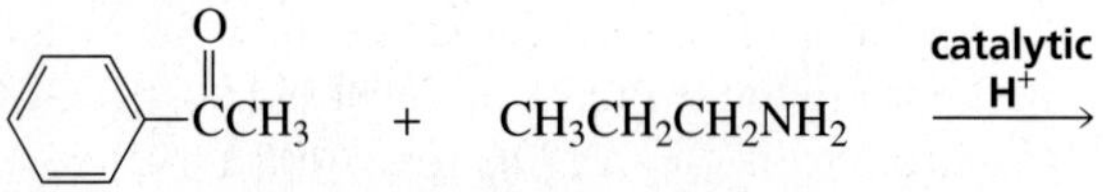 $C_6H_5\text{–}\overset{O}{\overset{\|}{C}}CH_3 + CH_3CH_2CH_2NH_2 \xrightarrow{\text{catalytic } H^+}$

b. $CH_3\overset{\overset{O}{\|}}{C}Cl$ + 2 (pyrrolidine, N–H) $\longrightarrow$

c. (benzene)–NH_2 $\xrightarrow[\text{2. } H_2O,\ Cu_2O,\ Cu(NO_3)_2]{\text{1. HCl, } NaNO_2,\ 0\ ^\circ C}$

d. (benzene)–$\overset{\overset{O}{\|}}{C}CH_3$ + $CH_3CH_2NHCH_2CH_3$ $\xrightarrow[H^+]{\text{catalytic}}$

21.5 Reactions of Quaternary Ammonium Hydroxides

The leaving group of a **quaternary ammonium ion** has about the same leaving tendency as a protonated amino group, but it does not have an acidic hydrogen that would protonate a basic reactant. A quaternary ammonium ion, therefore, can undergo a reaction with a strong base. The reaction of a quaternary ammonium ion with hydroxide ion is known as a **Hofmann elimination reaction**. The leaving group in a Hofmann elimination reaction is a tertiary amine. Because a tertiary amine is a relatively poor leaving group, the reaction requires heat.

$$CH_3CH_2CH_2\overset{+}{N}(CH_3)_3\ HO^- \xrightarrow{\Delta} CH_3CH{=}CH_2 + N(CH_3)_3 + H_2O$$

August Wilhelm von Hofmann (1818–1892) *was born in Germany. He first studied law and then changed to chemistry. He founded the German Chemical Society. Hofmann taught at the Royal College of Chemistry in London for 20 years and then returned to Germany to teach at the University of Berlin. He was one of the founders of the German dye industry. Married four times—he was left a widower three times—he had 11 children.*

A Hofmann elimination reaction is an E2 reaction. Recall that an E2 reaction is a concerted, one-step reaction—the proton and the tertiary amine are removed in the same step (Section 11.1). Very little substitution product is formed.

mechanism of the Hofmann elimination

$$CH_3\underset{\underset{H}{|}}{CH}-CH_2-\overset{+}{N}(CH_3)_3 \longrightarrow CH_3CH{=}CH_2 + N(CH_3)_3 + H_2O$$

($H\ddot{O}:^-$ removes the β-hydrogen)

PROBLEM 6

What is the difference between the reaction that occurs when isopropyltrimethylammonium hydroxide is heated and the reaction that occurs when 2-bromopropane is treated with hydroxide ion?

The carbon to which the tertiary amine is attached is designated as the α-carbon, so the adjacent carbon, from which the proton is removed, is called the β-carbon. (Recall that E2 reactions are also called β-elimination reactions, since elimination is initiated by removing a proton from the β-carbon; Section 11.1.) If the quaternary ammonium ion has more than one β-carbon, the major alkene product is the one obtained by removing a proton from the β-carbon bonded to the greater number of hydrogens. In the following reaction, the major alkene product is obtained by removing a hydrogen from

In a Hofmann elimination reaction, the hydrogen is removed from the β-carbon bonded to the most hydrogens.

the β-carbon bonded to three hydrogens, and the minor alkene product results from removing a hydrogen from the β-carbon bonded to two hydrogens.

β-carbon β-carbon

$$\underset{\underset{CH_3\ \ HO^-}{\overset{|+}{CH_3NCH_3}}}{CH_3CHCH_2CH_2CH_3} \xrightarrow{\Delta} \underset{\substack{\text{1-pentene}\\\text{major product}}}{CH_2{=}CHCH_2CH_2CH_3} + \underset{\substack{\text{2-pentene}\\\text{minor product}}}{CH_3CH{=}CHCH_2CH_3} + \underset{\text{trimethylamine}}{\underset{CH_3}{CH_3\overset{}{N}CH_3}} + H_2O$$

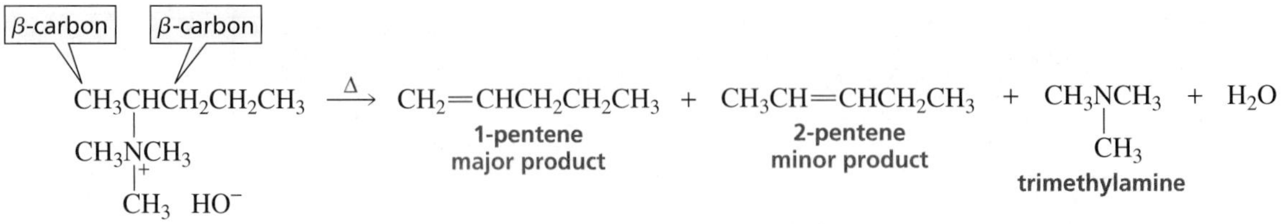

In the next reaction, the major alkene product comes from removing a hydrogen from the β-carbon bonded to two hydrogens, because the other β-carbon is bonded to only one hydrogen.

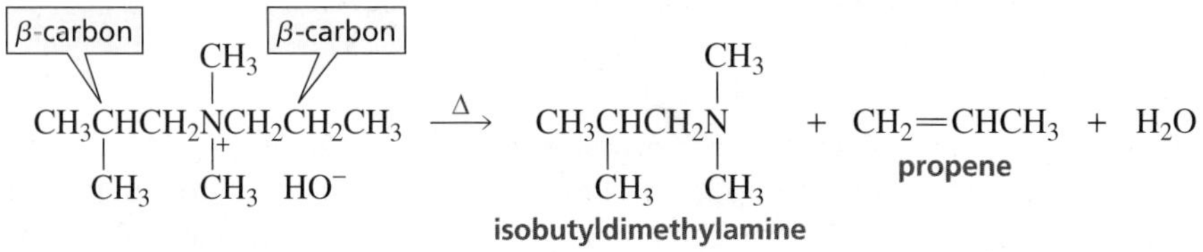

β-carbon β-carbon

$$\underset{CH_3\quad CH_3\ HO^-}{CH_3CHCH_2\overset{CH_3}{\overset{|}{N}}{}^{+}CH_2CH_2CH_3} \xrightarrow{\Delta} \underset{\text{isobutyldimethylamine}}{\underset{CH_3\quad CH_3}{CH_3CHCH_2\overset{CH_3}{N}}} + \underset{\text{propene}}{CH_2{=}CHCH_3} + H_2O$$

PROBLEM 7◆

What are the minor products in the preceding Hofmann elimination reaction?

We saw that in an E2 reaction of an alkyl chloride, alkyl bromide, or alkyl iodide, a hydrogen is removed from the β-carbon bonded to the *fewest* hydrogens (*Zaitsev's rule*; Section 11.2). Now we see that in an E2 reaction of a quaternary ammonium ion, the hydrogen is removed from the β-carbon bonded to the *most* hydrogens (*anti-Zaitsev elimination*).

Why do alkyl halides follow Zaitsev's rule, while quaternary amines violate the rule? When hydroxide ion starts to remove a proton from the alkyl bromide, the bromide ion immediately begins to depart and a transition state with an *alkene-like* structure results. The proton is removed from the β-carbon bonded to the fewest hydrogens in order to achieve the most stable alkene-like transition state.

Zaitsev elimination
alkene-like transition state

anti-Zaitsev elimination
carbanion-like transition state

$\overset{\delta-}{OH}$ ⋮ H ⋮ $CH_3CH{\cdots}CHCH_3$ ⋮ $\overset{\delta-}{Br}$	$\overset{\delta-}{OH}$ ⋮ H ⋮ $CH_3CH_2C{\cdots}CH_2$ ⋮ $\overset{\delta-}{Br}$	$\overset{\delta-}{OH}$ ⋮ H ⋮ $\overset{\delta-}{C}H_2CHCH_2CH_2CH_3$ | $^+N(CH_3)_3$	$\overset{\delta-}{OH}$ ⋮ H ⋮ $CH_3CH\overset{\delta-}{C}HCH_2CH_3$ | $^+N(CH_3)_3$
more stable	less stable	more stable	less stable

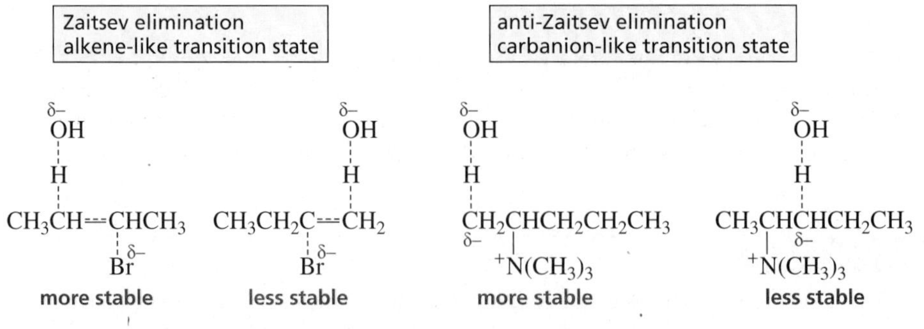

When, however, hydroxide ion starts to remove a proton from a quaternary ammonium ion, the leaving group does not immediately begin to leave, because a tertiary amine is not as good a leaving group as Cl^-, Br^-, or I^-. As a result, a partial negative charge builds up on the carbon from which the proton is being removed. This gives the transition state a *carbanion-like* structure rather than an alkene-like structure. By removing a proton from the β-carbon bonded to the most hydrogens, the most stable carbanion-like transition state is achieved. (Recall from Section 11.2 that primary carbanions are more stable than secondary carbanions, which are more stable than tertiary carbanions.) Steric factors in the Hofmann reaction also favor anti-Zaitsev elimination.

Because the Hofmann elimination reaction occurs in an anti-Zaitsev manner, anti-Zaitsev elimination is also referred to as Hofmann elimination. We have previously seen anti-Zaitsev elimination in the E2 reactions of alkyl fluorides as a result of fluoride ion being a poorer leaving group than chloride, bromide, or iodide ions. As in a Hofmann elimination reaction, the poor leaving group results in a carbanion-like transition state rather than an alkene-like transition state (Section 11.2).

PROBLEM 8◆

Give the major products of each of the following reactions:

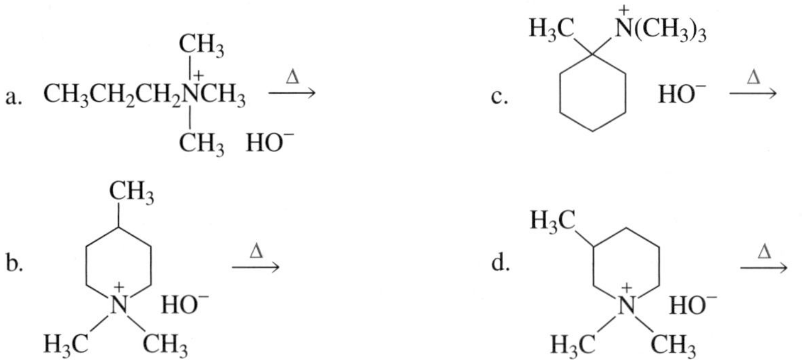

a. $CH_3CH_2CH_2\overset{+}{N}(CH_3)_3 \; HO^- \xrightarrow{\Delta}$

b. 1,1-dimethyl-4-methylpiperidinium ($\overset{+}{N}(CH_3)_2$, CH_3 at C-4) $HO^- \xrightarrow{\Delta}$

c. 2-methylcyclohexyl($\overset{+}{N}(CH_3)_3$) $HO^- \xrightarrow{\Delta}$

d. 1,1-dimethyl-3-methylpiperidinium ($\overset{+}{N}(CH_3)_2$, H_3C at C-3) $HO^- \xrightarrow{\Delta}$

For a quaternary ammonium ion to undergo an elimination reaction, the counterion must be hydroxide ion because a strong base is needed to start the reaction by removing a proton from a β-carbon. Since halide ions are weak bases, quaternary ammonium *halides* cannot undergo a Hofmann elimination reaction. However, a quaternary ammonium *halide* can be converted into a quaternary ammonium *hydroxide* by treating it with silver oxide and water. The silver halide precipitates, and the halide ion is replaced by hydroxide ion. The compound can now undergo an elimination reaction.

$$2\,R_4N^+\,I^- + Ag_2O + H_2O \longrightarrow 2\,R_4N^+\,HO^- + 2\,AgI\downarrow$$

The reaction of an amine with sufficient methyl iodide to convert the amine into a quaternary ammonium iodide is called **exhaustive methylation**. (See Chapter 10, Problem 8.) The reaction is carried out in a basic solution of potassium carbonate, so that the amines formed as intermediates will be predominantly in their basic forms.

exhaustive methylation

$$CH_3CH_2CH_2NH_2 + \underset{\text{excess}}{CH_3I} \xrightarrow{K_2CO_3} CH_3CH_2CH_2\overset{+}{N}(CH_3)_3\; I^-$$

The Hofmann elimination reaction was used by early organic chemists as the last step of a process known as a Hofmann degradation—a method used to identify amines. In a *Hofmann degradation*, an amine is exhaustively methylated with methyl iodide, treated with silver oxide to convert the quaternary ammonium iodide to a quaternary ammonium hydroxide, and then heated to allow it to undergo a Hofmann elimination. Once the alkene is identified, working backwards gives the structure of the amine.

A USEFUL BAD-TASTING COMPOUND

Several practical uses have been found for Bitrex®, a quaternary ammonium salt, because it is one of the most bitter-tasting substances known and is nontoxic. Bitrex® is put on bait to encourage deer to look elsewhere for their food, it is put on the backs of animals to keep them from biting one another, it is put on children's' fingers to persuade them to stop sucking their thumbs or biting their fingernails, and it is added to toxic substances to keep them from being ingested accidentally.

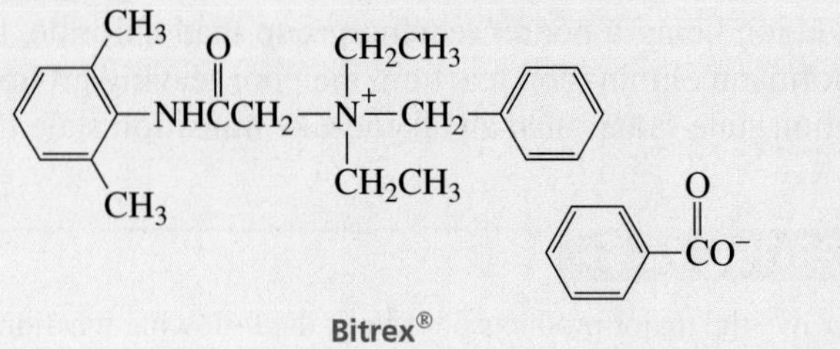

$$\text{2,6-}(CH_3)_2C_6H_3{-}NHC(=O)CH_2{-}\overset{+}{N}(CH_2CH_3)_2{-}CH_2C_6H_5 \quad C_6H_5CO^-\text{(=O)}$$

Bitrex®

PROBLEM 9

Identify the amine in each case.

a. 4-Methyl-2-pentene is obtained from the Hofmann degradation of a primary amine.
b. 2-Methyl-1-3-butadiene is obtained from two successive Hofmann degradations of a secondary amine.

PROBLEM 10 SOLVED

Describe a synthesis for each of the following compounds, using the given starting material and any necessary reagents:

a. $CH_3CH_2CH_2CH_2NH_2 \longrightarrow CH_3CH_2CH{=}CH_2$

b. $CH_3CH_2CH_2CH(Br)CH_3 \longrightarrow CH_3CH_2CH_2CH{=}CH_2$

c. 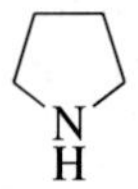 pyrrolidine (five-membered ring, NH) $\longrightarrow CH_2{=}CH{-}CH{=}CH_2$

SOLUTION TO 10a Although an amine cannot undergo an elimination reaction, a quaternary ammonium hydroxide can. The amine, therefore, must first be converted into a quaternary ammonium hydroxide. Reaction with excess methyl iodide converts the amine into a quaternary ammonium iodide, and treatment with aqueous silver oxide forms the quaternary ammonium hydroxide. Heat is required for the elimination reaction.

$$CH_3CH_2CH_2CH_2NH_2 \xrightarrow[K_2CO_3]{CH_3I\ \text{excess}} CH_3CH_2CH_2CH_2\overset{+}{N}(CH_3)_3\ I^- \xrightarrow[H_2O]{Ag_2O} CH_3CH_2CH_2CH_2\overset{+}{N}(CH_3)_3\ HO^- \xrightarrow{\Delta} CH_3CH_2CH{=}CH_2 + H_2O$$

21.6 Phase-Transfer Catalysis

A problem organic chemists face in the laboratory is finding a solvent that will dissolve all the reactants needed for a given reaction. For example, if we want cyanide ion to react with 1-bromohexane, we encounter a problem: Sodium cyanide is an ionic compound that is soluble only in water, whereas the alkyl halide is insoluble in water. Therefore, if we mix an aqueous solution of sodium cyanide with a solution of 1-bromohexane in a nonpolar solvent, there will be two distinct phases—an aqueous phase and an organic phase—because the solutions are immiscible. How, then, can sodium cyanide react with the alkyl halide?

$$\underset{\text{1-bromohexane}}{CH_3CH_2CH_2CH_2CH_2CH_2Br} + {}^{-}C{\equiv}N \xrightarrow{?} CH_3CH_2CH_2CH_2CH_2CH_2C{\equiv}N + Br^-$$

The two compounds will be able to react with each other if a catalytic amount of a **phase-transfer catalyst** is added to the reaction mixture.

phase-transfer catalyst

$$CH_3CH_2CH_2CH_2CH_2CH_2Br + {}^{-}C{\equiv}N \xrightarrow{R_4\overset{+}{N}\ HSO_4^-} CH_3CH_2CH_2CH_2CH_2CH_2C{\equiv}N + Br^-$$

Quaternary ammonium salts are the most common phase-transfer catalysts. However, we saw in Section 12.9 that crown ethers can also be used as phase-transfer catalysts.

phase-transfer catalysts

$CH_3CH_2CH_2CH_2\overset{+}{N}(CH_2CH_2CH_2CH_3)_3\ HSO_4^-$	$CH_3(CH_2)_{14}CH_2\overset{+}{N}(CH_3)_3\ HSO_4^-$	$C_6H_5CH_2\overset{+}{N}(CH_2CH_3)_3\ HSO_4^-$
tetrabutylammonium hydrogen sulfate	**hexadecyltrimethylammonium hydrogen sulfate**	**benzyltriethylammonium hydrogen sulfate**

How does the addition of a phase-transfer catalyst allow the reaction of cyanide ion with 1-bromohexane to take place? Because of its nonpolar alkyl groups, the quaternary ammonium salt is soluble in nonpolar solvents, but because of its charge, it is also soluble in water. This means that the quaternary ammonium salt can act as a mediator between the two immiscible phases. When a phase-transfer catalyst such as tetrabutylammonium hydrogen sulfate passes into the nonpolar, organic phase, it must carry a counterion with it to balance its positive charge. The counterion can be either its original counterion (hydrogen sulfate) or another ion that is present in the solution (in the reaction under discussion, it will be cyanide ion). Because there is more cyanide ion than hydrogen sulfate ion in the aqueous phase, cyanide ion will more often be the accompanying ion. Once in the organic phase, cyanide ion can react with the alkyl halide. (When hydrogen sulfate is transported into the oganic phase, it is unreactive because it is both a weak base and a poor nucleophile.) The quaternary ammonium ion will pass back into the aqueous phase carrying with it either hydrogen sulfate or bromide ion as a counterion. The reaction continues with the phase-transfer catalyst shuttling back and forth between the two phases. **Phase-transfer catalysis** has been successfully used in a wide variety of organic reactions.

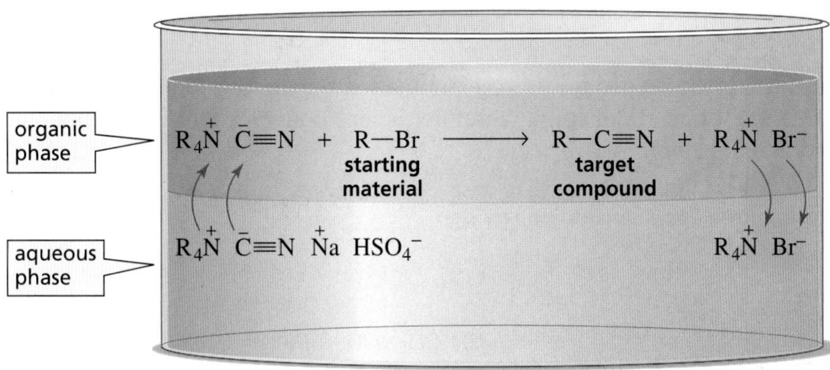

21.7 Oxidation of Amines; The Cope Elimination Reaction

Amines are easily oxidized, sometimes just by being exposed to air. Amines, therefore, are stored as salts (e.g., as amine hydrochlorides), and drugs that contain amino groups are often sold as salts.

Primary amines are oxidized to hydroxylamines, which in turn are oxidized to nitroso compounds, which are oxidized to nitro compounds. Hydrogen peroxide, peroxyacids, and other common oxidizing agents are used to oxidize amines. The oxidation reactions generally take place by mechanisms that involve radicals, so they are not well characterized.

$$\underset{\text{a primary amine}}{R-NH_2} \xrightarrow{\text{oxidation}} \underset{\text{a hydroxylamine}}{R-NH-OH} \xrightarrow{\text{oxidation}} \underset{\text{a nitroso compound}}{R-N=O} \xrightarrow{\text{oxidation}} \underset{\text{a nitro compound}}{R-\overset{+}{N}(=O)O^-}$$

Secondary amines are oxidized to secondary hydroxylamines, and tertiary amines are oxidized to tertiary amine oxides.

$$\underset{\text{a secondary amine}}{R_2\ddot{N}H} + HO-OH \longrightarrow R_2\overset{+}{N}H(OH) + HO^- \longrightarrow \underset{\text{a secondary hydroxylamine}}{R_2N-OH} + H_2O$$

$$\underset{\text{a tertiary amine}}{R_3\ddot{N}} + HO-OH \longrightarrow R_3\overset{+}{N}-OH + HO^- \longrightarrow \underset{\text{an tertiary amine oxide}}{R_3\overset{+}{N}-O^-} + H_2O$$

Arthur C. Cope (1909–1966) *was born in Indiana. He received a Ph.D. from the University of Wisconsin and was a professor of chemistry at Bryn Mawr College, Columbia University, and MIT.*

Tertiary amine oxides undergo a reaction similar to the Hofmann elimination reaction, called a *Cope elimination reaction.* In a **Cope elimination reaction**, a tertiary amine oxide rather than a quaternary ammonium ion undergoes elimination. The Cope elimination reaction occurs under milder conditions than does a Hofmann elimination reaction.

$$\underset{\text{a tertiary amine oxide}}{CH_3CH_2CH_2\overset{+}{N}(CH_3)_2O^-} \xrightarrow{\Delta} CH_3CH=CH_2 + \underset{\text{a hydroxylamine}}{(CH_3)_2N-OH}$$

A strong base is not needed for a Cope elimination because the amine oxide acts as its own base. The Cope elimination, therefore, is an intramolecular E2 reaction and involves syn elimination.

mechanism of the Cope elimination reaction

$$CH_3CH(H)-CH_2-\overset{+}{N}(CH_3)_2-\ddot{O}:^- \xrightarrow{\Delta} CH_3CH=CH_2 + (CH_3)_2N-\ddot{O}H$$

The major product of the Cope elimination, like that of the Hofmann elimination, is the one obtained by removing a hydrogen from the β-carbon bonded to the greater number of hydrogens.

In a Cope elimination, the hydrogen is removed from the β-carbon bonded to the most hydrogens.

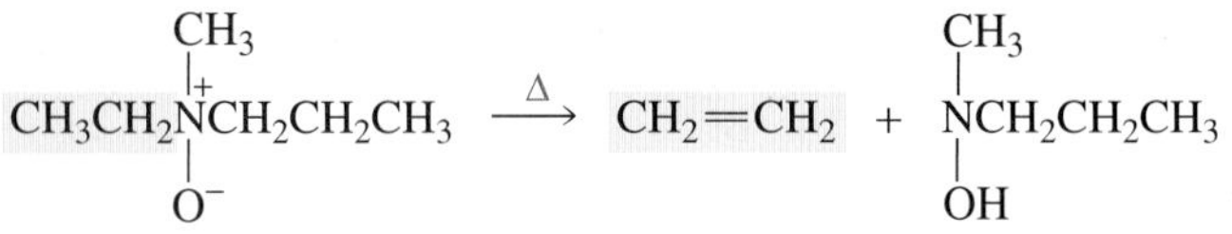

$$CH_3CH_2\overset{+}{N}(CH_3)(O^-)CH_2CH_2CH_3 \xrightarrow{\Delta} CH_2{=}CH_2 + N(CH_3)(OH)CH_2CH_2CH_3$$

PROBLEM 11◆

Does the Cope elimination have an alkene-like transition state or a carbanion-like transition state?

PROBLEM 12◆

Give the products that would be obtained by treating the following tertiary amines with hydrogen peroxide followed by heat:

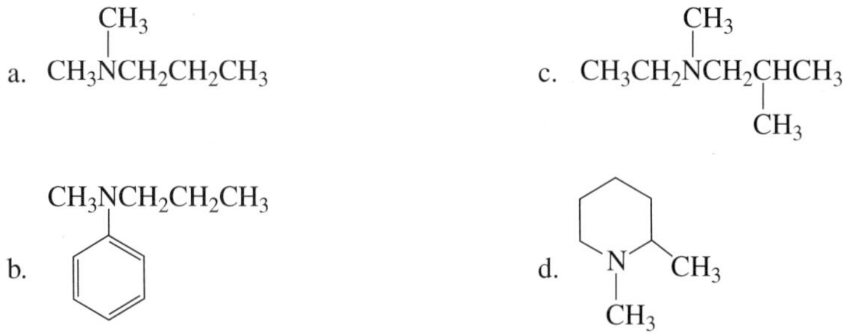

a. $CH_3N(CH_3)CH_2CH_2CH_3$

b. $CH_3NCH_2CH_2CH_3$ with a phenyl group on N

c. $CH_3CH_2N(CH_3)CH_2CH(CH_3)CH_3$

d. a piperidine ring with CH_3 on N and CH_3 on the carbon next to N

21.8 Synthesis of Amines

Because ammonia and amines are good nucleophiles, they readily undergo S_N2 reactions with alkyl halides. (X denotes a halogen.)

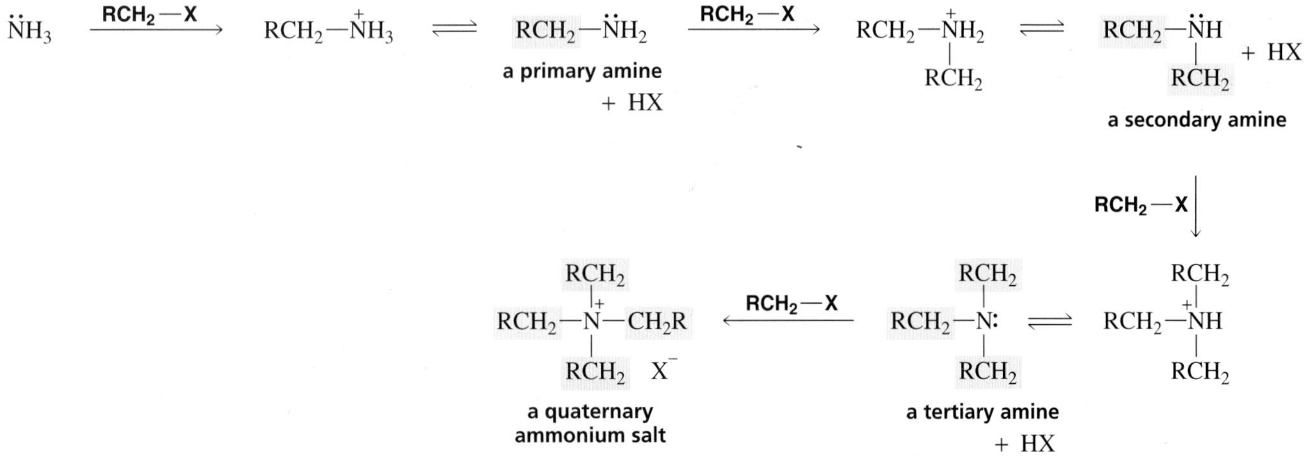

$$\ddot{N}H_3 \xrightarrow{RCH_2-X} RCH_2-\overset{+}{N}H_3 \rightleftharpoons RCH_2-\ddot{N}H_2 + HX$$

a primary amine

$$RCH_2-\ddot{N}H_2 \xrightarrow{RCH_2-X} RCH_2-\overset{+}{N}H_2(RCH_2) \rightleftharpoons RCH_2-\ddot{N}H(RCH_2) + HX$$

a secondary amine

$$RCH_2-\ddot{N}H(RCH_2) \xrightarrow{RCH_2-X} RCH_2-\overset{+}{N}H(RCH_2)_2 \rightleftharpoons RCH_2-\ddot{N}(RCH_2)_2 + HX$$

a tertiary amine

$$RCH_2-\ddot{N}(RCH_2)_2 \xrightarrow{RCH_2-X} RCH_2-\overset{+}{N}(RCH_2)_2-CH_2R \quad X^-$$

a quaternary ammonium salt

Although these S_N2 reactions can be used to synthesize amines, the yields are poor because it is difficult to stop the reaction when the desired number of alkyl substituents have been placed on the nitrogen since ammonia and primary, secondary, and tertiary amines have similar reactivities.

A much better way to prepare a primary amine is by means of a Gabriel synthesis (Section 17.17). This reaction involves alkylating phthalimide and then hydrolyzing the *N*-substituted phthalimide.

Gabriel synthesis

$$\text{phthalimide} \xrightarrow[\text{2. } CH_3CH_2Br]{\text{1. } HO^-} \text{N}-CH_2CH_3 \text{ (N-ethylphthalimide)} + Br^- \xrightarrow[\Delta]{H_3O^+} \text{phthalic acid (COOH, COOH)} + CH_3CH_2\overset{+}{N}H_3$$

phthalimide; a protonated primary amine

Primary amines also can be prepared in good yields if azide ion ($^-N_3$) is used as the nucleophile in an S_N2 reaction. The product of the reaction is an alkyl azide, which can be reduced to a primary amine. (See Chapter 10, Problem 9.)

$$\underset{\text{butyl bromide}}{CH_3CH_2CH_2CH_2Br} \xrightarrow{^-N_3} \underset{\text{butyl azide}}{CH_3CH_2CH_2CH_2N{=}\overset{+}{N}{=}N^-} \xrightarrow[Pd/C]{H_2} \underset{\text{butylamine}}{CH_3CH_2CH_2CH_2CH_2NH_2}$$

Other reduction reactions also result in the formation of primary amines. For example, the catalytic reduction of a nitrile forms a primary amine. (Recall that a nitrile can be obtained from the reaction of cyanide ion with an alkyl halide.)

$$\underset{\text{butyl bromide}}{CH_3CH_2CH_2CH_2Br} \xrightarrow[HCl]{NaC{\equiv}N} \underset{\text{pentanenitrile}}{CH_3CH_2CH_2CH_2C{\equiv}N} \xrightarrow[Pd/C]{H_2} \underset{\text{pentylamine}}{CH_3CH_2CH_2CH_2CH_2NH_2}$$

Amines are obtained from the reduction of amides with $LiAlH_4$ (Sections 18.5 and 20.1). This method can be used to synthesize primary, secondary, and tertiary amines. The class of amine obtained depends on the number of substituents on the nitrogen atom of the amide.

$$R-\overset{O}{\overset{\|}{C}}-NH_2 \xrightarrow[\text{2. } H_2O]{\text{1. } LiAlH_4} \underset{\text{a primary amine}}{RCH_2NH_2}$$

$$R-\overset{O}{\overset{\|}{C}}-NHCH_3 \xrightarrow[\text{2. } H_2O]{\text{1. } LiAlH_4} \underset{\text{a secondary amine}}{RCH_2NHCH_3}$$

$$R-\overset{O}{\overset{\|}{C}}-N(CH_3)CH_3 \xrightarrow[\text{2. } H_2O]{\text{1. } LiAlH_4} \underset{\text{a tertiary amine}}{RCH_2N(CH_3)CH_3}$$

A primary amine can be obtained from the reaction of an aldehyde or a ketone with excess ammonia in the presence of H_2 and Raney nickel. Because the imine does not have a substituent other than a hydrogen bonded to the nitrogen, it is relatively unstable, so the amine is obtained by adding H_2 to the C=N bond as it is formed. This is called **reductive amination**.

$$(CH_3CH_2)_2C{=}O + \underset{\text{excess}}{NH_3} \longrightarrow \underset{\text{unstable}}{\left[(CH_3CH_2)_2C{=}NH\right]} \xrightarrow[Pd/C]{H_2} (CH_3CH_2)_2CHNH_2$$

Secondary and tertiary amines can be prepared from imines and enamines by reducing the imine or enamine. Sodium triacetoxyborohydride is a commonly used reducing agent for this reaction.

$$C_6H_5\text{-}CH{=}O + CH_3CH_2NH_2 \underset{}{\overset{\text{catalytic } H^+}{\rightleftharpoons}} C_6H_5\text{-}CH{=}NCH_2CH_3 \xrightarrow{NaBH(OCCH_3)_3} C_6H_5\text{-}CH_2NHCH_2CH_3$$

an imine

$$\text{cyclohexanone} + CH_3NHCH_3 \overset{\text{catalytic } H^+}{\rightleftharpoons} \text{cyclohexenyl-}N(CH_3)_2 \xrightarrow{NaBH(OCCH_3)_3} \text{cyclohexyl-}N(CH_3)_2$$

an enamine

A primary amine is obtained from the reduction of a nitroalkane, and an arylamine is obtained from the reduction of nitrobenzene.

$$\underset{\text{nitroethane}}{CH_3CH_2NO_2} + H_2 \xrightarrow{Pd/C} \underset{\text{ethylamine}}{CH_3CH_2NH_2}$$

$$\underset{\text{nitrobenzene}}{C_6H_5\text{-}NO_2} + H_2 \xrightarrow{Pd/C} \underset{\text{aniline}}{C_6H_5\text{-}NH_2}$$

PROBLEM 13◆

Excess ammonia must be used when a primary amine is synthesized by reductive amination. What product will be obtained if the reaction is carried out with an excess of the carbonyl compound instead?

21.9 Aromatic Five-Membered-Ring Heterocycles

Pyrrole, Furan, and Thiophene

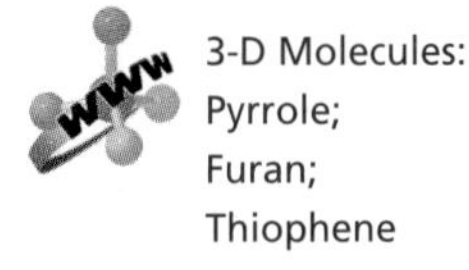

3-D Molecules: Pyrrole; Furan; Thiophene

Pyrrole, **furan**, and **thiophene** are five-membered-ring heterocycles. Each has three pairs of delocalized π electrons: Two of the pairs are shown as π bonds, and one pair is shown as a lone pair on the heteroatom. Furan and thiophene have a second lone pair that is not part of the π cloud. These electrons are in an sp^2 orbital perpendicular to the p orbitals. Pyrrole, furan, and thiophene are aromatic because they are cyclic and planar, every carbon in the ring has a p orbital, and the π cloud contains *three* pairs of π electrons (Sections 15.1 and 15.3).

pyrrole furan thiophene

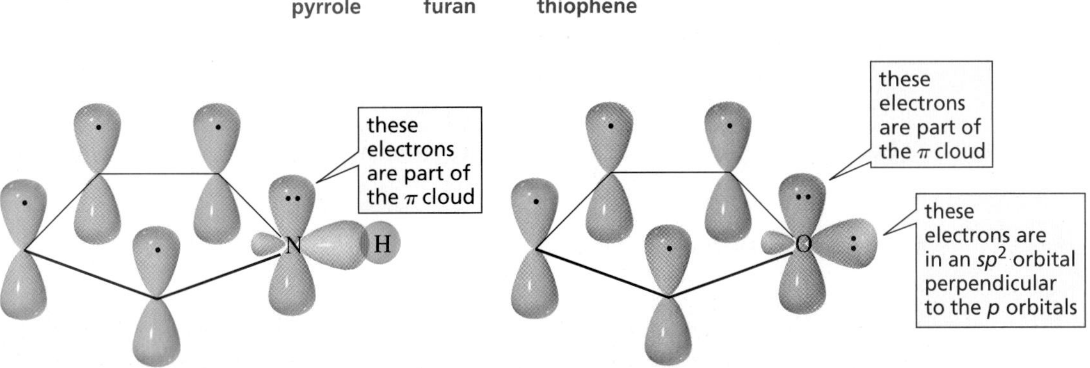

orbital structure of pyrrole orbital structure of furan

Pyrrole is an extremely weak base because the electrons shown as a lone pair are part of the π cloud. Therefore, when pyrrole is protonated, its aromaticity is destroyed. Consequently, the conjugate acid of pyrrole is a very strong acid ($pK_a = -3.8$); that is, it has a strong tendency to lose a proton.

The resonance contributors of pyrrole show that nitrogen donates the electrons depicted as a lone pair into the five-membered ring.

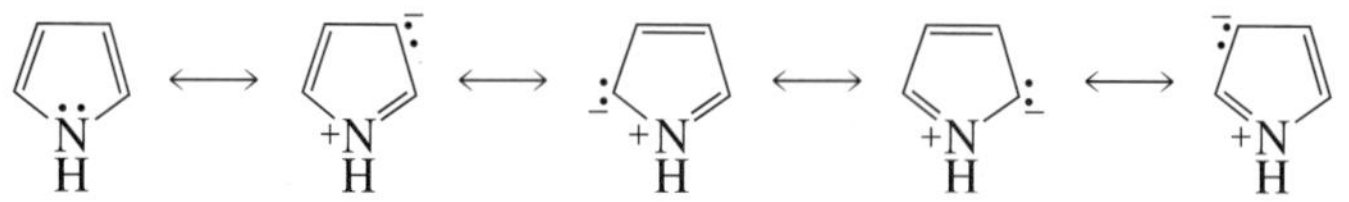

resonance contributors of pyrrole

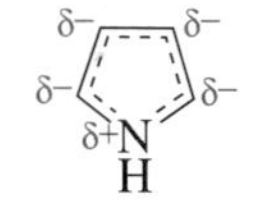

resonance hybrid

Pyrrolidine—a saturated five-membered-ring heterocyclic amine—has a dipole moment of 1.57 D because the nitrogen atom is electron withdrawing. Pyrrole—an unsaturated five-membered-ring heterocyclic amine—has a slightly larger dipole moment (1.80 D), but as we see from the electrostatic potential maps, the two dipole moments are in opposite directions. (The red areas are on opposite sides of the two molecules.) Apparently, the ability of pyrrole's nitrogen to donate electrons into the ring by resonance more than makes up for its inductive electron withdrawal (Section 16.3).

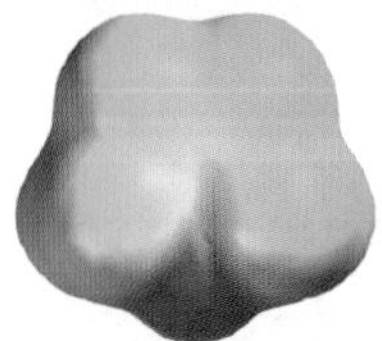

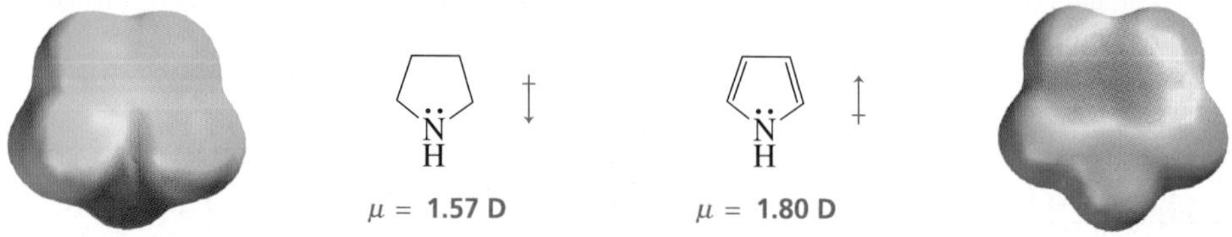

μ = 1.57 D

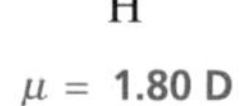

μ = 1.80 D

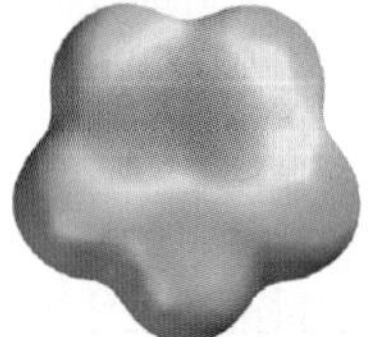

pyrrolidine **pyrrole**

In Section 7.6, we saw that the more stable and more nearly equivalent the resonance contributors, the greater is the resonance energy. The resonance energies of pyrrole, furan, and thiophene are not as great as the resonance energies of benzene and the cyclopentadienyl anion, compounds for which the resonance contributors are all equivalent. Thiophene, with the least electronegative heteroatom, has the greatest resonance energy of these five-membered heterocycles; and furan, with the most electronegative heteroatom, has the smallest resonance energy.

relative resonance energies of some aromatic compounds

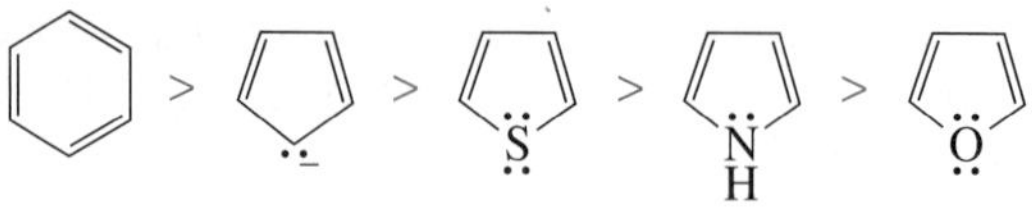

Pyrrole, furan, and thiophene undergo electrophilic substitution preferentially at C-2.

Because pyrrole, furan, and thiophene are aromatic, they undergo electrophilic aromatic substitution reactions.

mechanism for electrophilic aromatic substitution

$+ \; Y^+ \xrightarrow{\text{slow}} \; \xrightarrow{\text{fast}} \; + \; HB^+$

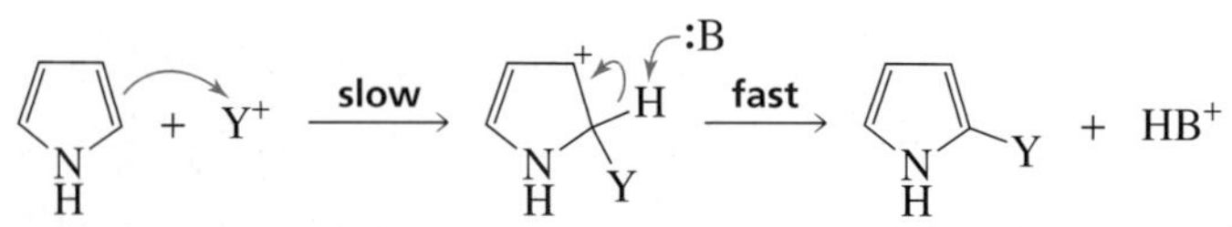

furan + Br_2 ⟶ 2-bromofuran + HBr

2-bromofuran

2-methylpyrrole + HNO_3 $\xrightarrow{(CH_3CO)_2O}$ 2-methyl-5-nitropyrrole + H_2O

2-methyl-5-nitropyrrole

Substitution occurs preferentially at C-2 because the intermediate obtained by attaching a substituent at this position is more stable than the intermediate obtained by attaching a substituent at C-3 (Figure 21.1). Both intermediates have a relatively stable resonance contributor in which all the atoms (except H) have complete octets. The intermediate resulting from C-2 substitution of pyrrole has *two* additional resonance contributors, each with a positive charge on a *secondary allylic* carbon. The intermediate resulting from C-3 substitution, however, has only *one* additional resonance contributor, which has a positive charge on a *secondary* carbon. This resonance contributor is further destabilized by being adjacent to an electron-withdrawing nitrogen atom. If both positions adjacent to the heteroatom are occupied, electrophilic substitution will take place at C-3.

Pyrrole, furan, and thiophene are more reactive than benzene toward electrophilic aromatic substitution.

2,5-dimethylfuran + Br_2 ⟶ 3-bromo-2,5-dimethylfuran + HBr

3-bromo-2,5-dimethylfuran

Pyrrole, furan, and thiophene are all more reactive than benzene toward electrophilic substitution because they are better able to stabilize the positive charge on the carbocation intermediate, since the lone pair on the hetereoatom can donate electrons into the ring by resonance (Figure 21.1).

relative reactivity toward electrophilic aromatic substitution

pyrrole > furan > thiophene > benzene

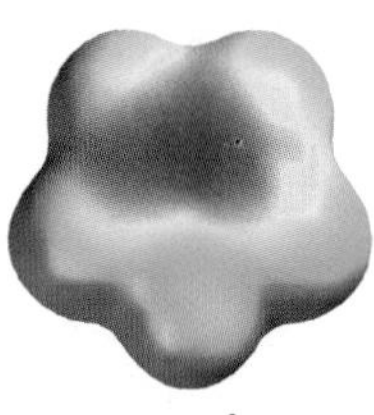

pyrrole

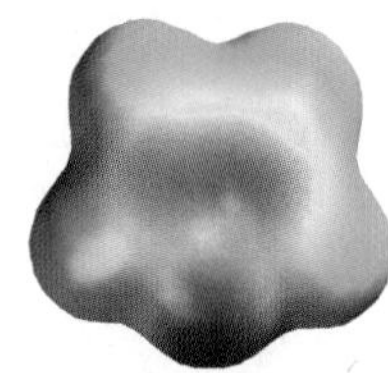

furan

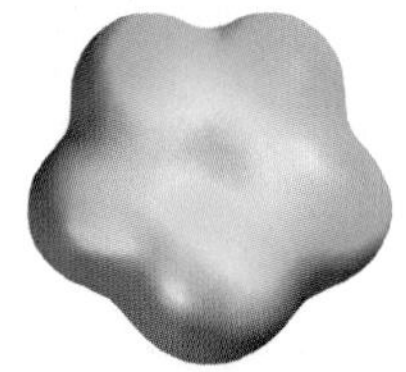

thiophene

Furan is not as reactive as pyrrole in electrophilic aromatic substitution reactions. The oxygen of furan is more electronegative than the nitrogen of pyrrole, so the oxygen is not as effective as nitrogen in stabilizing the carbocation. Thiophene is less reactive than furan toward electrophilic substitution because sulfur's π electrons are in a $3p$ orbital, which overlaps less effectively than the $2p$ orbital of nitrogen or oxygen with the $2p$ orbital of carbon. The electrostatic potential maps illustrate the different electron densities of the three rings.

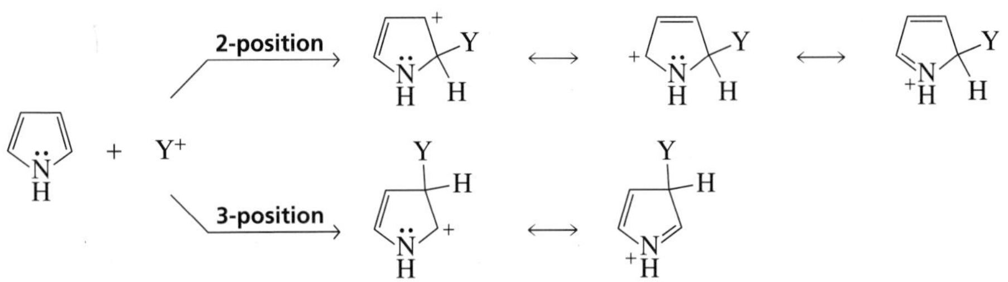

Figure 21.1 Structures of the intermediates that can be formed from the reaction of an electrophile with pyrrole at C-2 and C-3.

The relative reactivities of the five-membered-ring heterocycles are reflected in the Lewis acid required to catalyze a Friedel–Crafts acylation reaction (Section 15.13). Benzene requires $AlCl_3$, a relatively strong Lewis acid. Thiophene is more reactive than benzene, so it can undergo a Friedel–Crafts reaction using $SnCl_4$, a weaker Lewis acid. An even weaker Lewis acid, BF_3, can be used when the substrate is furan. Pyrrole is so reactive that an anhydride is used instead of a more reactive acyl chloride, and no catalyst is necessary.

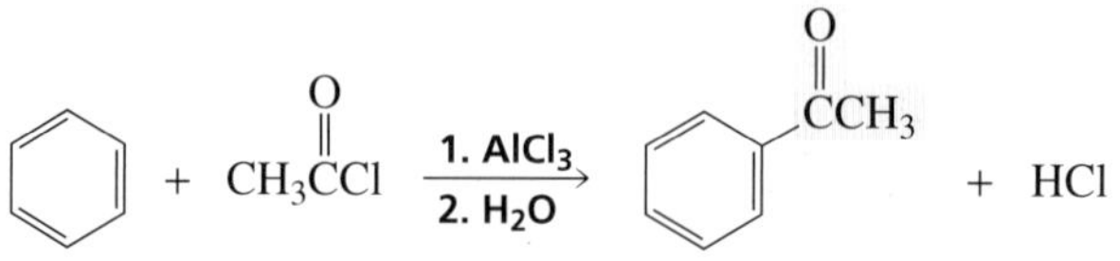

benzene + $CH_3\overset{O}{\overset{\|}{C}}Cl$ $\xrightarrow[2.\ H_2O]{1.\ AlCl_3}$ $C_6H_5\overset{O}{\overset{\|}{C}}CH_3$ + HCl

phenylethanone

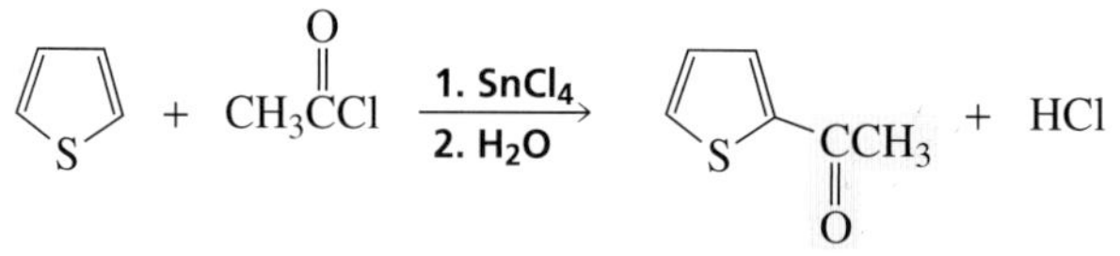

thiophene + $CH_3\overset{O}{\overset{\|}{C}}Cl$ $\xrightarrow[2.\ H_2O]{1.\ SnCl_4}$ 2-thienyl–$\overset{O}{\overset{\|}{C}}CH_3$ + HCl

2-acetylthiophene

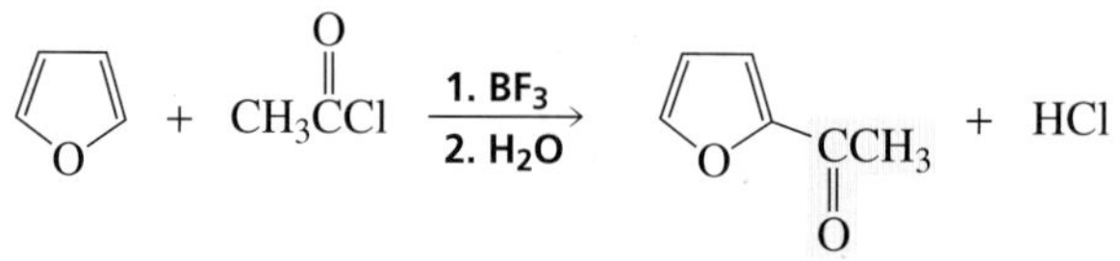

furan + $CH_3\overset{O}{\overset{\|}{C}}Cl$ $\xrightarrow[2.\ H_2O]{1.\ BF_3}$ 2-furyl–$\overset{O}{\overset{\|}{C}}CH_3$ + HCl

2-acetylfuran

pyrrole + $CH_3\overset{O}{\overset{\|}{C}}O\overset{O}{\overset{\|}{C}}CH_3$ $\longrightarrow$ 2-pyrrolyl–$\overset{O}{\overset{\|}{C}}CH_3$ + $CH_3\overset{O}{\overset{\|}{C}}OH$

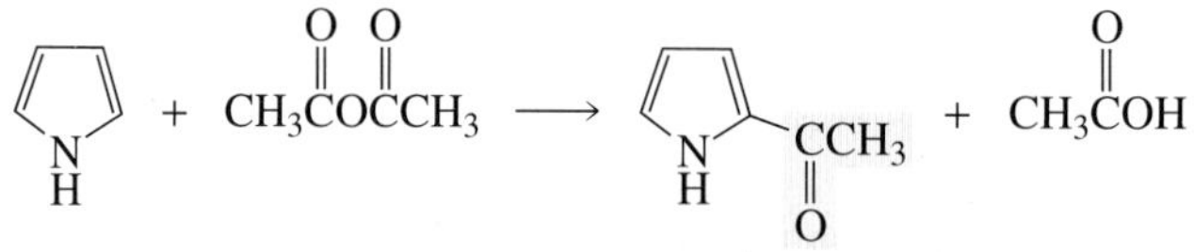

2-acetylpyrrole

The resonance hybrid of pyrrole indicates that there is a partial positive charge on the nitrogen. Therefore, pyrrole is protonated on C-2 rather than on nitrogen. Remember, a proton is an electrophile and, like other electrophiles, attaches to the C-2 position of pyrrole.

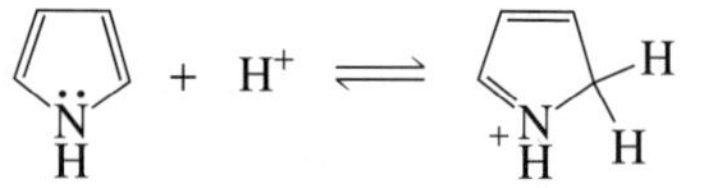

pyrrole + H^+ $\rightleftharpoons$ 2H-pyrrolium ion

$pK_a = -3.8$

Pyrrole is unstable in strongly acidic solutions because once protonated, it can readily polymerize.

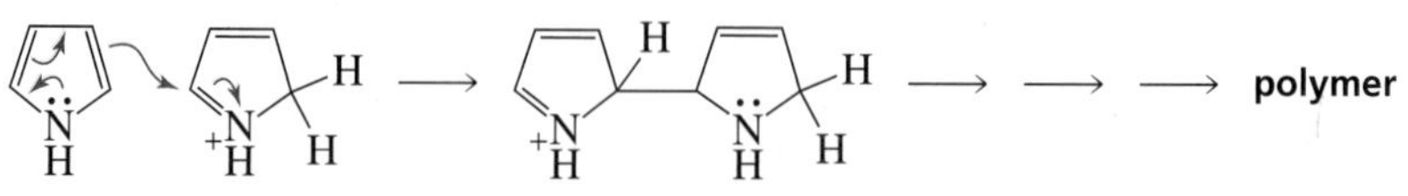

pyrrole + protonated pyrrole $\longrightarrow$ dimer $\longrightarrow \longrightarrow \longrightarrow$ polymer

Pyrrole is more acidic ($pK_a = \sim 17$) than the analogous saturated amine ($pK_a = \sim 36$), because the nitrogen in pyrrole is sp^2 hybridized and is, therefore, more electronegative than the sp^3 nitrogen of a saturated amine (Table 21.1). Pyrrole's acidity also is increased as a result of its conjugate base being stabilized by electron

TABLE 21.1 The pK_a Values of Several Nitrogen Heterocycles

pK_a = −3.8	pK_a = −2.4	pK_a = 1.0	pK_a = 2.5	pK_a = 4.85	pK_a = 5.16
pK_a = 6.8	pK_a = 8.0	pK_a = 11.1	pK_a = 14.4	pK_a = ~17	pK_a = ~36

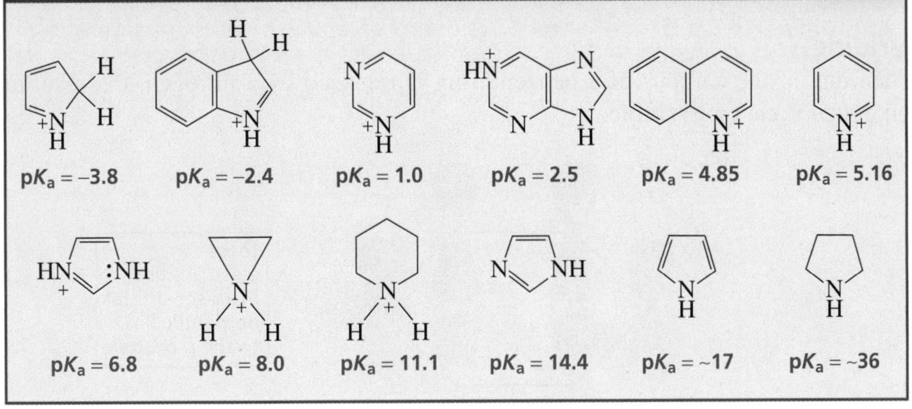

Tutorial: Basic sites of nitrogen heterocycles

delocalization. (Recall that the more stable the base, the stronger is its conjugate acid; Section 1.17).

pK_a = ~17 + H^+

pK_a = ~36 + H^+

PROBLEM 14

When pyrrole is added to a dilute solution of D_2SO_4 in D_2O, 2-deuteriopyrrole is formed. Propose a mechanism to account for the formation of this compound.

PROBLEM 15

Use resonance contributors to explain why pyrrole is protonated on C-2 rather than on nitrogen.

PROBLEM 16

Explain why pyrrole (pK_a ~ 17) is less acidic than cyclopentadiene (pK_a = 15), even though nitrogen is considerably more electronegative than carbon.

Indole, Benzofuran, and Benzothiophene

Indole, benzofuran, and benzothiophene contain a five-membered aromatic ring fused to a benzene ring. The rings are numbered in a way that gives the heteroatom the lowest possible number. Indole, benzofuran, and benzothiophene are aromatic because they are cyclic and planar, every carbon in the ring has a *p* orbital, and the π cloud of each compound contains *five* pairs of π electrons (Section 15.1). Notice that the electrons shown as a lone pair on the indole nitrogen are part of the π cloud; therefore, the conjugate acid of indole, like the conjugate acid of pyrrole, is a strong acid (pK_a = −2.4). In other words, indole is an extremely weak base.

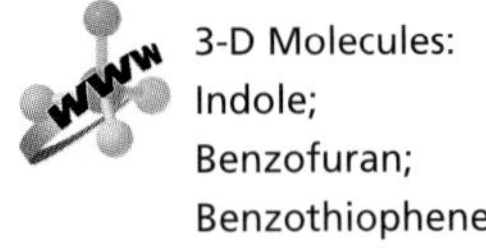

3-D Molecules: Indole; Benzofuran; Benzothiophene

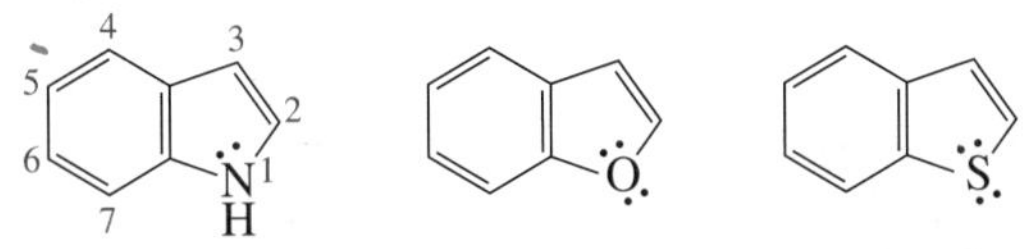

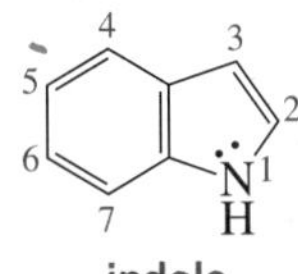

indole

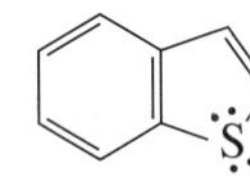

benzofuran

benzothiophene

21.10 Aromatic Six-Membered-Ring Heterocycles

Pyridine

When one of the carbons of a benzene ring is replaced by a nitrogen, the resulting compound is called **pyridine**.

3-D Molecule: Pyridine

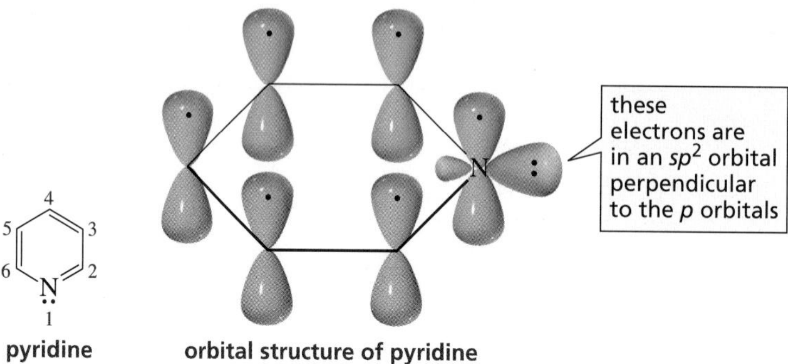

pyridine orbital structure of pyridine

The pyridinium ion is a stronger acid than a typical ammonium ion because the acidic hydrogen of a pyridinium ion is attached to an sp^2 hybridized nitrogen, which is more electronegative than an sp^3 hybridized nitrogen (Section 6.9).

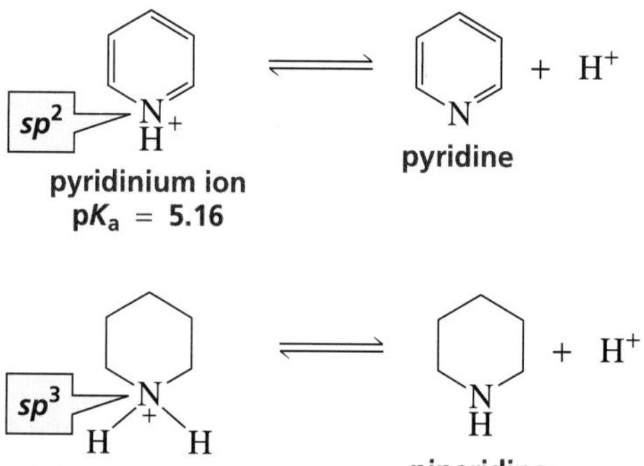

pyridinium ion $pK_a = 5.16$ ⇌ pyridine + H^+

Tutorial: Lone-pair electrons on nitrogen heterocycles

piperidinium ion $pK_a = 11.12$ ⇌ piperidine + H^+

Pyridine is a tertiary amine, so it undergoes reactions characteristic of tertiary amines. For example, pyridine undergoes S_N2 reactions with alkyl halides (Section 10.4), and it reacts with hydrogen peroxide to form an *N*-oxide (Section 21.7).

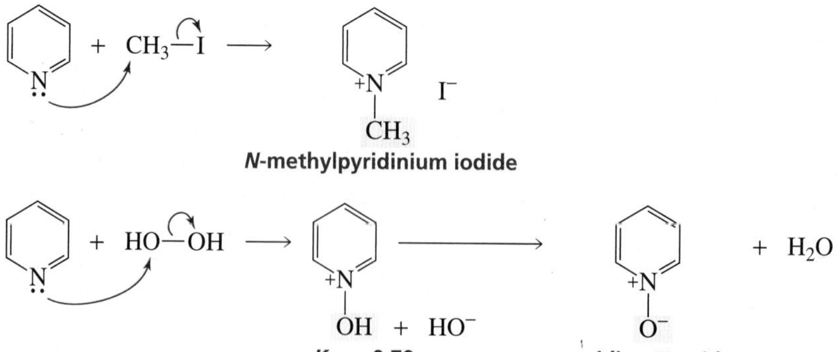

pyridine + CH_3—I ⟶ *N*-methylpyridinium iodide

pyridine + HO—OH ⟶ (OH + HO^-, $pK_a = 0.79$) ⟶ pyridine-*N*-oxide + H_2O

PROBLEM 17 SOLVED

Will an amide be formed from the reaction of an acyl chloride with an aqueous solution of pyridine? Explain your answer.

SOLUTION An amide will *not* be formed because the positively charged nitrogen causes pyridine to be an excellent leaving group. Therefore, the final product of the reaction will be a carboxylic acid. (If the final pH of the solution is greater than the pK_a of the carboxylic acid, the carboxylic acid will be predominantly in its basic form.)

$$RCOCl + \text{pyridine} \longrightarrow RCO\text{-}\overset{+}{N}C_5H_5 \xrightarrow{H_2O} RCO^- + \text{pyridine}$$

Pyridine is aromatic. Like benzene, it has two uncharged resonance contributors. Because of the electron-withdrawing nitrogen, pyridine has three charged resonance contributors that benzene does not have.

resonance contributors of pyridine

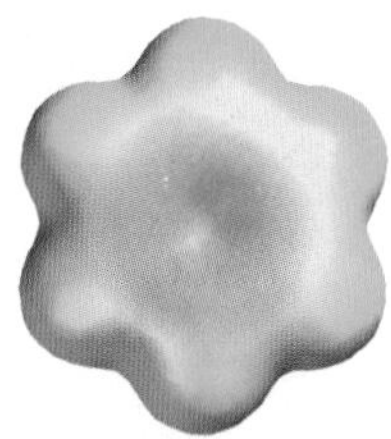
benzene

The dipole moment of pyridine is 1.57 D. As the resonance contributors and the electrostatic potential map indicate, the electron-withdrawing nitrogen is the negative end of the dipole.

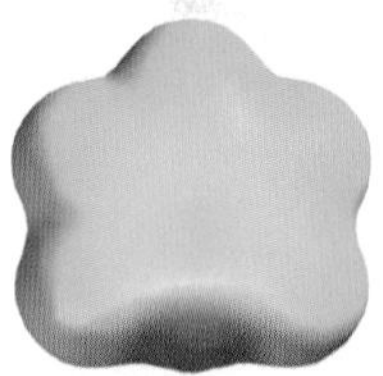
pyridine

μ = 1.57 D

Because it is aromatic, pyridine (like benzene) undergoes electrophilic aromatic substitution reactions (:B is any base in the solution).

mechanism for electrophilic aromatic substitution

$$\text{pyridine} + Y^+ \xrightarrow{\text{slow}} \text{intermediate} \xrightarrow{\text{fast}} \text{3-Y-pyridine} + HB^+$$

Pyridine undergoes electrophilic aromatic substitution at C-3.

Electrophilic aromatic substitution of pyridine takes place at C-3 because the most stable intermediate is obtained by placing an electrophilic substituent at that position (Figure 21.2). When the substituent is placed at C-2 or C-4, one of the resulting resonance contributors is particularly unstable because its nitrogen atom has an incomplete octet *and* a positive charge. The electron-withdrawing nitrogen atom makes the intermediate obtained from electrophilic aromatic substitution of pyridine less stable than the carbocation intermediate obtained from electrophilic aromatic substitution of benzene. Pyridine, therefore, is less reactive than benzene. Indeed, it is even less reactive than nitrobenzene. (Recall from Section 16.3 that an electron-withdrawing nitro group strongly deactivates a benzene ring toward electrophilic aromatic substitution.)

relative reactivity toward electrophilic aromatic substitution

benzene > nitrobenzene (NO_2) > pyridine > 1,3-dinitrobenzene (NO_2, NO_2)

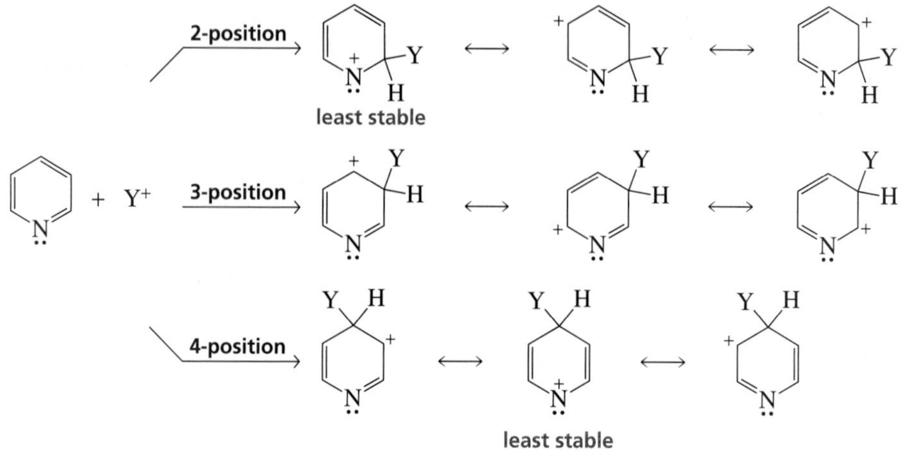

Figure 21.2 ▶
Structures of the intermediates that can be formed from the reaction of an electrophile with pyridine.

2-position
least stable
3-position
4-position
least stable

Pyridine, therefore, undergoes electrophilic aromatic substitution reactions only under vigorous conditions, and the yields of these reactions are often quite low. If the nitrogen becomes protonated under the reaction conditions, the reactivity is further decreased because a positively charged nitrogen would make the carbocation intermediate even less stable.

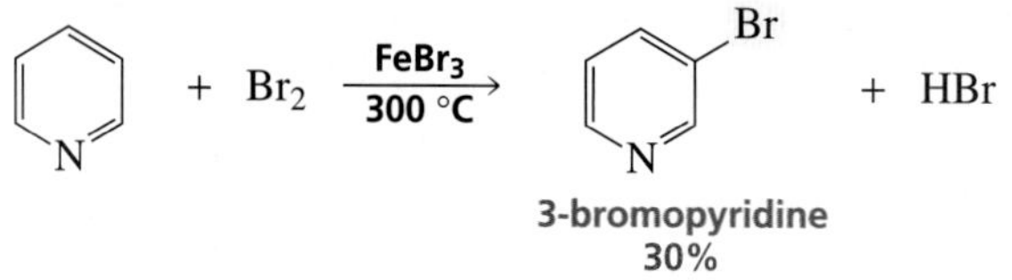

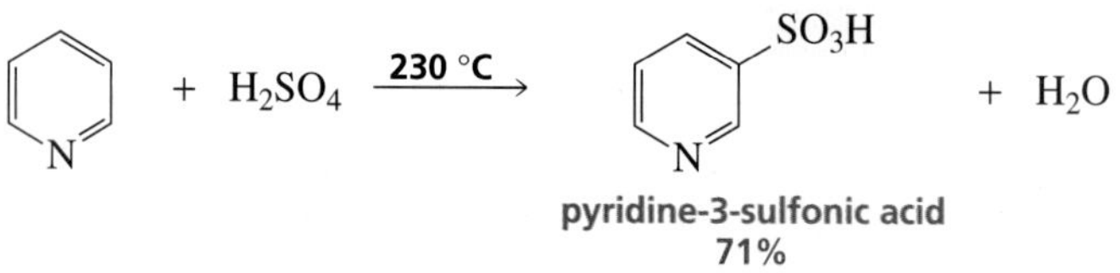

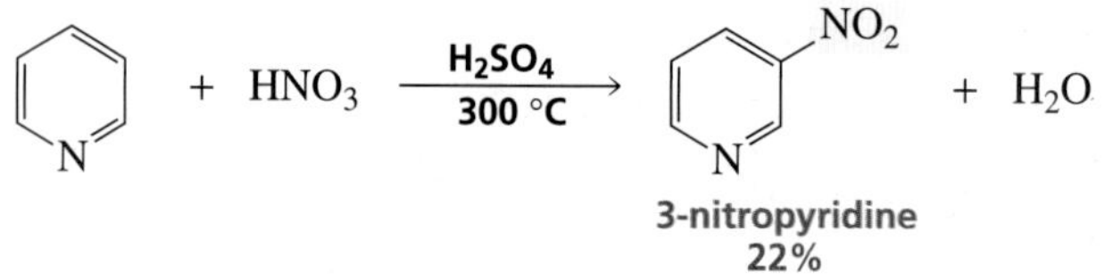

$$\text{pyridine} + Br_2 \xrightarrow[300\ °C]{FeBr_3} \text{3-bromopyridine (30\%)} + HBr$$

$$\text{pyridine} + H_2SO_4 \xrightarrow{230\ °C} \text{pyridine-3-sulfonic acid (71\%)} + H_2O$$

$$\text{pyridine} + HNO_3 \xrightarrow[300\ °C]{H_2SO_4} \text{3-nitropyridine (22\%)} + H_2O$$

We have seen that highly deactivated benzene rings do not undergo Friedel–Crafts alkylation or acylation reactions. Therefore, pyridine, whose reactivity is similar to that of a highly deactivated benzene, does not undergo these reactions either.

$$\text{pyridine} + CH_3CH_2Cl \xrightarrow{AlCl_3} \text{no electrophilic aromatic substitution reaction}$$

PROBLEM 18

Give the product of the following reaction:

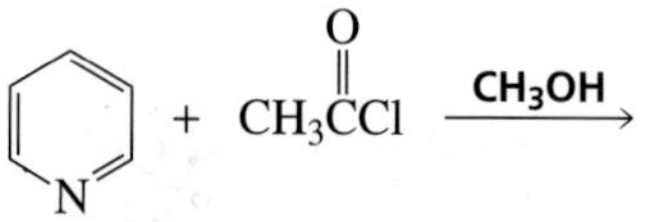

$$\text{pyridine} + CH_3\overset{O}{\overset{\|}{C}}Cl \xrightarrow{CH_3OH}$$

Since pyridine is *less* reactive than benzene toward *electrophilic* aromatic substitution, it is not surprising that pyridine is *more* reactive than benzene toward *nucleophilic* aromatic substitution. The electron-withdrawing nitrogen atom that destabilizes the intermediate in electrophilic aromatic substitution stabilizes it in nucleophilic aromatic substitution.

mechanism for nucleophilic aromatic substitution

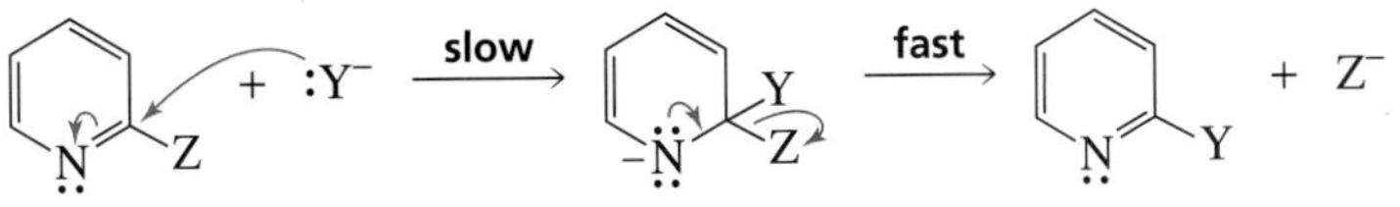

slow fast + Z^-

Nucleophilic aromatic substitution of pyridine takes place at C-2 and C-4, because attack at these positions leads to the most stable intermediate. Only when nucleophilic attack occurs at these positions is a resonance contributor obtained that has the greatest electron density on nitrogen, the most electronegative of the ring atoms (Figure 21.3).

Pyridine undergoes nucleophilic aromatic substitution at C-2 and C-4.

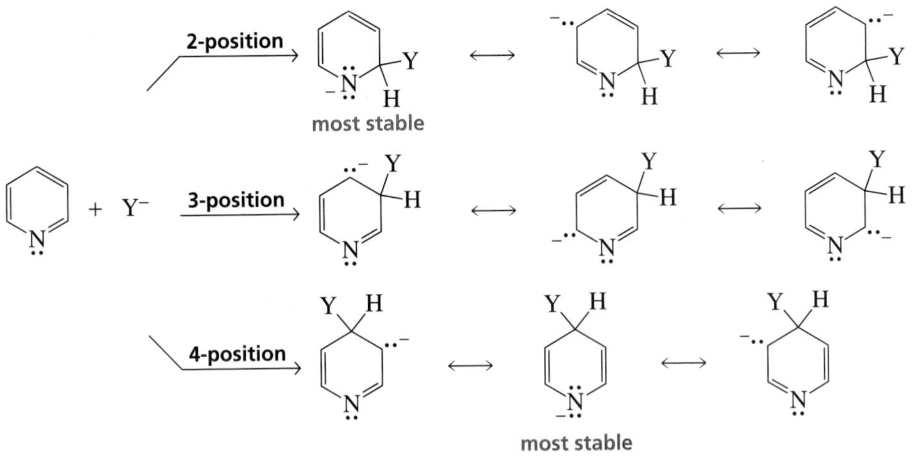

2-position

most stable

+ Y^- 3-position

4-position

most stable

Figure 21.3
Structures of the intermediates that can be formed from the reaction of a nucleophile with pyridine.

If the leaving groups at C-2 and C-4 are different, the incoming nucleophile will preferentially substitute for the weaker base (the better leaving group).

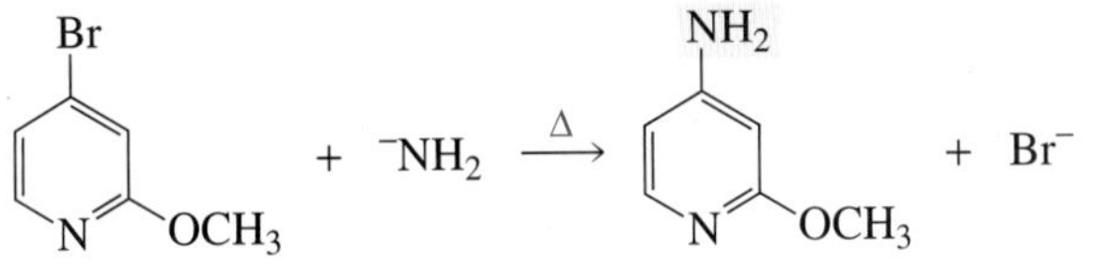

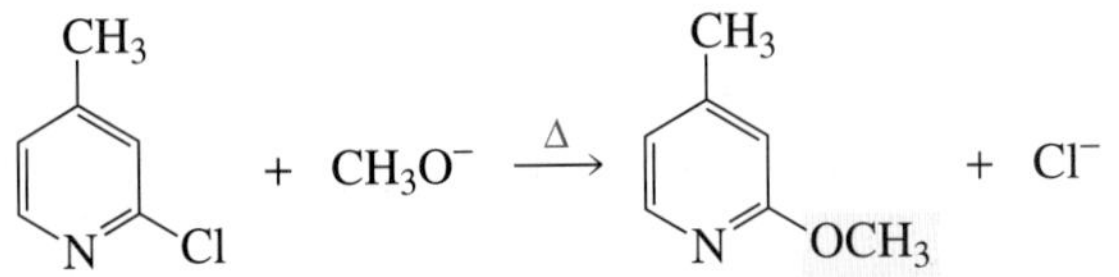

$+ {}^-NH_2 \xrightarrow{\Delta} + Br^-$

$+ CH_3O^- \xrightarrow{\Delta} + Cl^-$

Pyridine is *less* reactive than benzene toward electrophilic aromatic substitution and *more* reactive than benzene toward nucleophilic aromatic substitution.

PROBLEM 19

Compare the mechanisms of the following reactions:

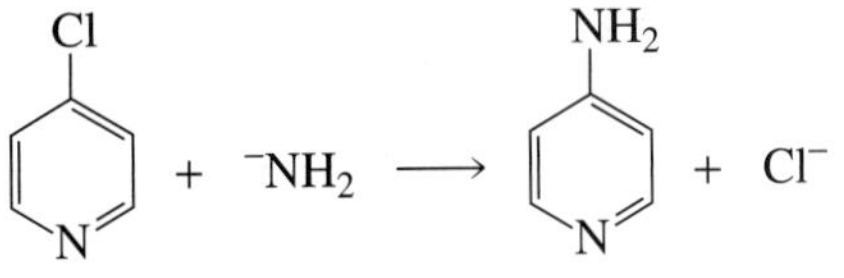

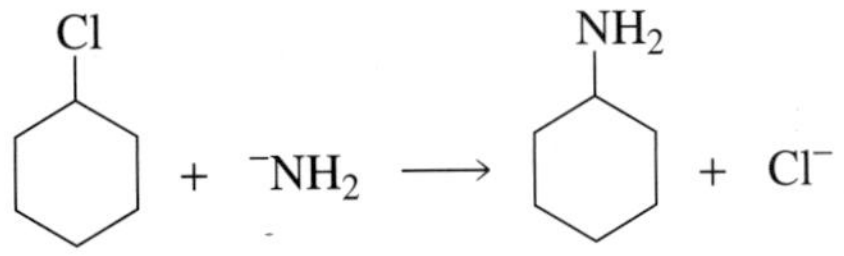

$+ {}^-NH_2 \longrightarrow + Cl^-$

$+ {}^-NH_2 \longrightarrow + Cl^-$

PROBLEM 20

a. Propose a mechanism for the following reaction:

$\xrightarrow[\Delta]{KOH/H_2O}$

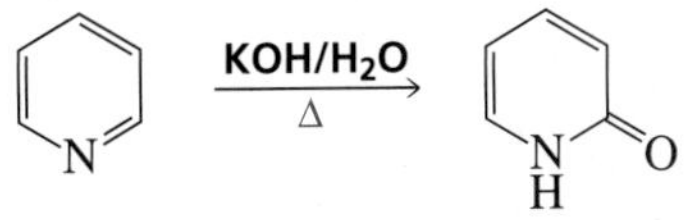

b. What other product is formed?

Substituted pyridines undergo many of the side-chain reactions that substituted benzenes undergo. For example, alkyl-substituted pyridines can be brominated and oxidized.

NBS, Δ/peroxide: 2-ethylpyridine (CH_2CH_3) → 2-(1-bromoethyl)pyridine ($CHCH_3$, Br)

$Na_2Cr_2O_7$, H_2SO_4, Δ: 4-methylpyridine (CH_3) → pyridine-4-carboxylic acid (COOH)

When 2- or 4-aminopyridine is diazotized, α-pyridone or γ-pyridone is formed. Apparently, the diazonium salt reacts immediately with water to form a hydroxypyridine (Section 16.10). The product of the reaction is a pyridone because the keto form of a hydroxypyridine is more stable than the enol form. (The mechanism for the conversion of a primary amino group into a diazonium group is shown in Section 16.12).

2-aminopyridine (NH_2) —$NaNO_2$, HCl, 0 °C→ [$\overset{+}{N}{\equiv}N\ Cl^-$] —$H_2O$→ 2-hydroxypyridine enol form (OH) ⇌ α-pyridone keto form (N–H, O)

4-aminopyridine (NH_2) —$NaNO_2$, HCl, 0 °C→ [$\overset{+}{N}{\equiv}N\ Cl^-$] —$H_2O$→ 4-hydroxypyridine enol form (OH) ⇌ γ-pyridone keto form (O, N–H)

The electron-withdrawing nitrogen causes the α-hydrogens of alkyl groups attached to the 2- and 4-positions of the pyridine ring to have about the same acidity as the α-hydrogens of ketones (Section 19.1).

4-methylpyridine (CH_3) ⇌ (HO⁻) $\overset{-}{C}H_2$ ⟷ CH_2 ⟷ CH_2 ⟷ CH_2 ⟷ $\overset{-}{C}H_2$ (resonance contributors of the carbanion)

Consequently, the α-hydrogens of alkyl substituents can be removed by base, and the resulting carbanions can react as nucleophiles.

4-methylpyridine (CH_3) ⇌ (HO⁻) $\overset{-}{C}H_2$ —(benzaldehyde, O=CH; an aldol condensation)→ 4-styrylpyridine (CH=CH–Ph) + H_2O

2,3-dimethylpyridine (CH_3, CH_3) —$^-NH_2$→ (CH_3, $\overset{-}{C}H_2$) —CH_3Br, an S_N2 reaction→ (CH_3, CH_2CH_3) + Br^-

PROBLEM 24◆

Imidazole boils at 257 °C, whereas *N*-methylimidazole boils at 199 °C. Explain this difference in boiling points.

PROBLEM 25◆

What percent of imidazole will be protonated at physiological pH (7.3)?

Purine and Pyrimidine

Nucleic acids (DNA and RNA) contain substituted **purines** and substituted **pyrimidines** (Section 27.1); DNA contains A, G, C, and T, and RNA contains A, G, C, and U. (Why DNA contains T instead of U is explained in Section 27.14.) Unsubstituted purine and pyrimidine are not found in nature. Notice that hydroxypurines and hydroxypyrimidines are more stable in the keto form. We will see that the preference for the keto form is crucial for proper base pairing in DNA (Section 27.7).

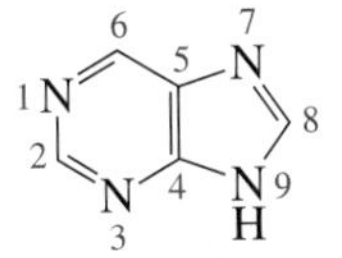

purine 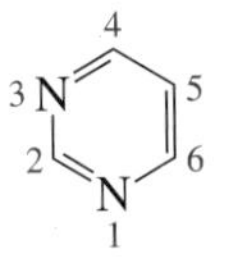 **pyrimidine**

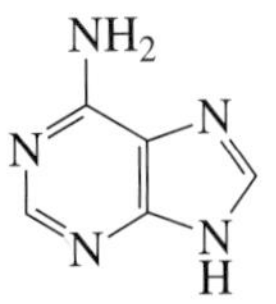

adenine 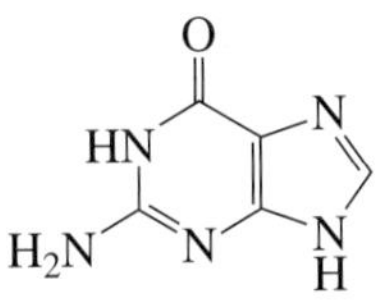 **guanine** 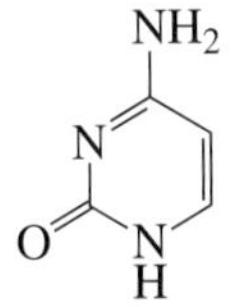 **cytosine** 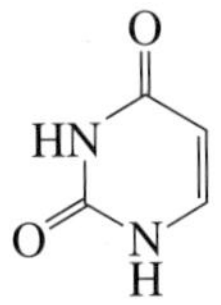 **uracil** 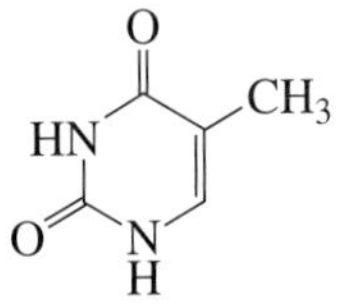 **thymine**

Porphyrin

Substituted *porphyrins* are important naturally occurring heterocyclic compounds. A **porphyrin ring system** consists of four pyrrole rings joined by one-carbon bridges. Heme, which is found in hemoglobin and myoglobin, contains an iron ion (Fe^{2+}) ligated by the four nitrogens of a porphyrin ring system. **Ligation** is the sharing of nonbonding electrons with a metal ion. The porphyrin ring system of heme is known as **protoporphyrin IX**; the ring system plus the iron atom is called *iron protoporphyrin IX*.

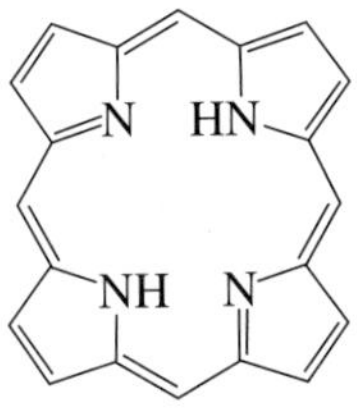

a porphyrin ring system

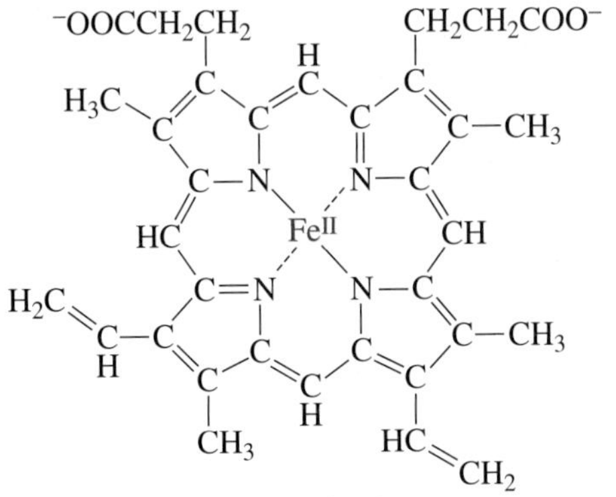

iron protoporphyrin IX
heme

Hemoglobin is responsible for transporting oxygen to cells and carbon dioxide away from cells, whereas myoglobin is responsible for storing oxygen in cells. Hemoglobin has four polypeptide chains and four heme groups; myoglobin has one polypeptide chain and one heme group. The iron atoms in hemoglobin and myoglobin, in addition to being ligated to the four nitrogens of the porphyrin ring, are also ligated to a histidine of the protein component (globin), and the sixth ligand is oxygen or carbon dioxide. Carbon monoxide is about the same size and shape as O_2, but CO binds

3-D Molecule: Heme

more tightly than O_2 to Fe^{2+}. Consequently, breathing carbon monoxide can be fatal because it prevents the transport of oxygen in the bloodstream.

The extensive conjugated system of porphyrin gives blood its characteristic red color. Its high molar absorptivity (about 160,000) allows concentrations as low as 1×10^{-8} M to be detected by UV spectroscopy (Section 8.10).

The biosynthesis of porphyrin involves the formation of porphobilinogen from two molecules of δ-aminolevulinic acid. The precise mechanism for this biosynthesis is as yet unknown. A possible mechanism starts with the formation of an imine between the enzyme that catalyzes the reaction and one of the molecules of δ-aminolevulinic acid. An aldol-type condensation occurs between the imine and a free molecule of δ-aminolevulinic acid. Nucleophilic attack by the amino group on the imine closes the ring. The enzyme is then eliminated, and removal of a proton creates the aromatic ring.

a mechanism for the biosynthesis of porphyrin

E—NH₂ enzyme; δ-aminolevulinic acid; base; H_2O; H^+; porphobilinogen

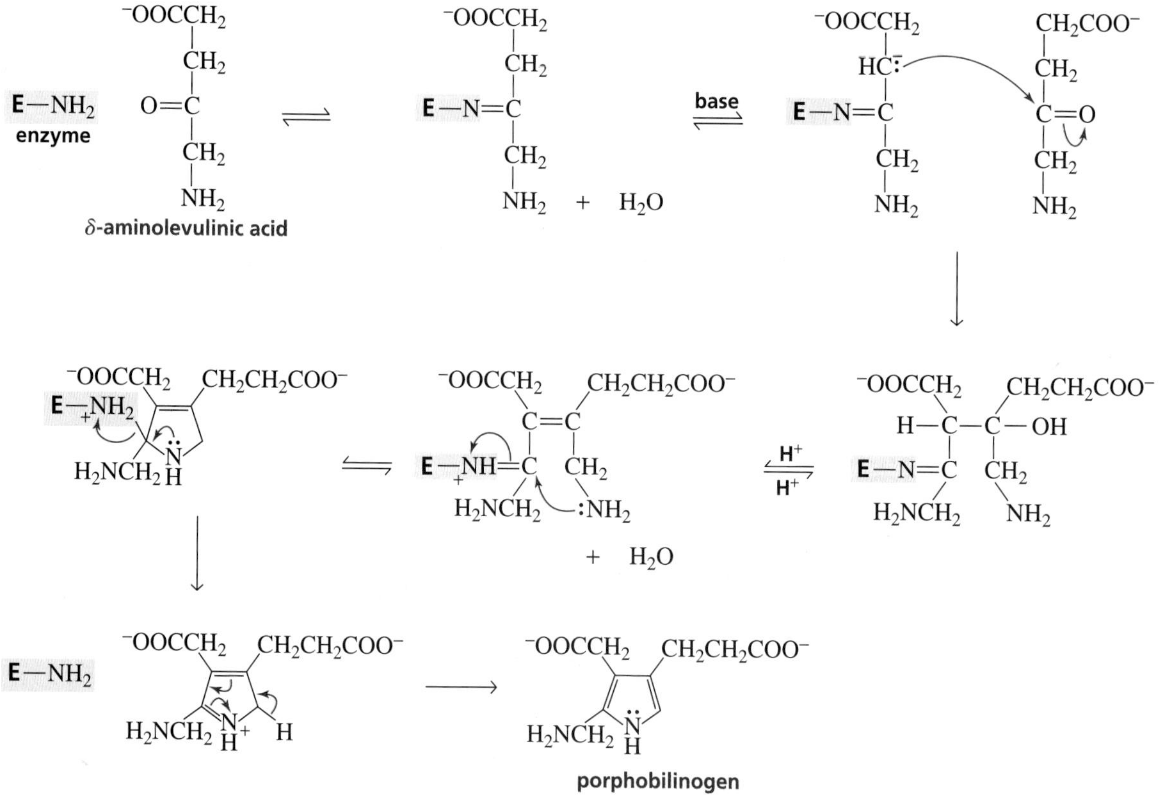

Four porphobilinogen molecules react to form porphyrin.

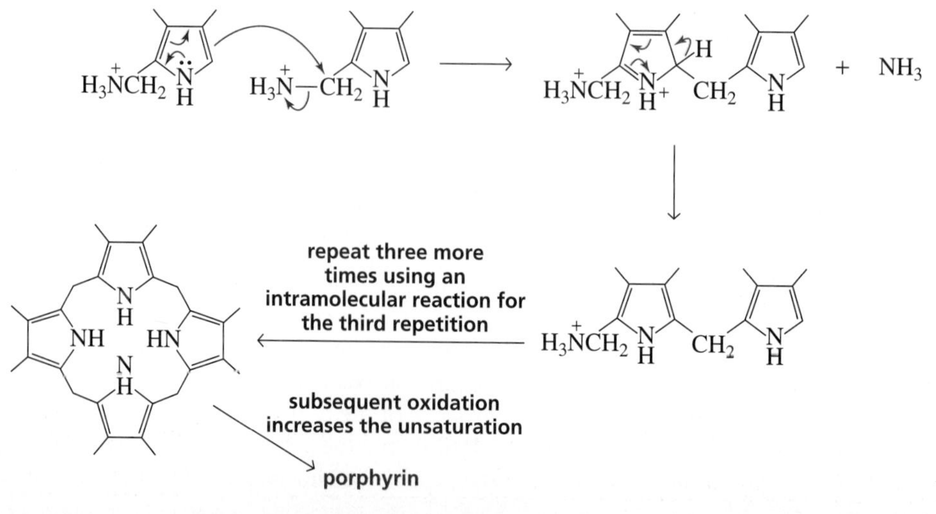

repeat three more times using an intramolecular reaction for the third repetition

subsequent oxidation increases the unsaturation

porphyrin

3-D Molecule: Chlorophyll *a*

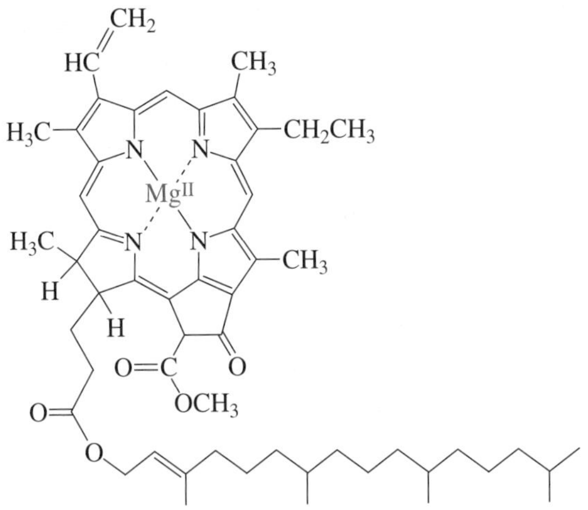

The ring system in chlorophyll *a*, the substance responsible for the green color of plants, is similar to porphyrin but contains a cyclopentanone ring, and one of its pyrrole rings is partially reduced. The metal atom in chlorophyll *a* is magnesium (Mg^{2+})

chlorophyll *a*

Vitamin B_{12} also has a ring system similar to porphyrin, but one of the methine bridges is missing. The ring system of vitamin B_{12} is known as a corrin ring system. The metal atom in vitamin B_{12} is cobalt (Co^{3+}). The chemistry of vitamin B_{12} is discussed in Section 25.7.

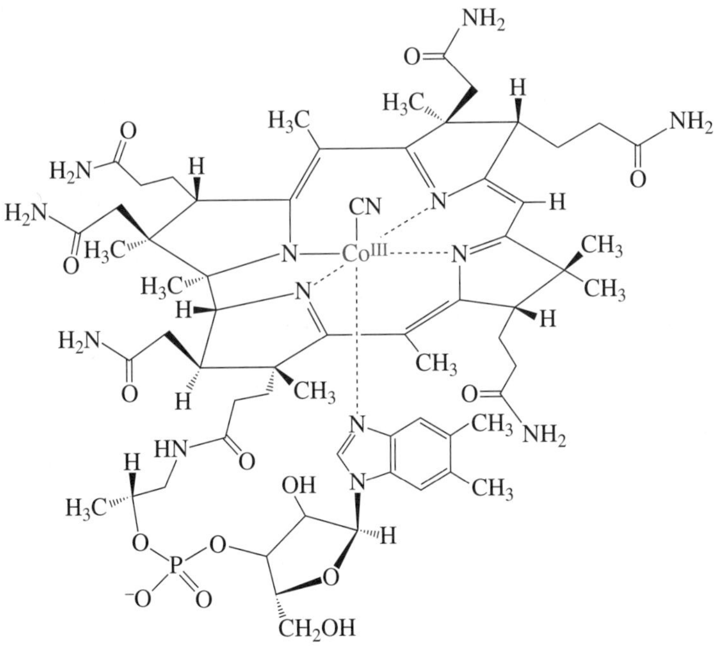

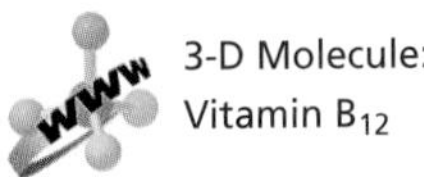

3-D Molecule: Vitamin B_{12}

vitamin B_{12}

PROBLEM 26◆

Is porphyrin aromatic?

PROBLEM 27

Show how the last two porphobilinogen molecules are incorporated into the porphyrin ring.

PORPHYRIN, BILIRUBIN, AND JAUNDICE

The average human turns over about 6 g of hemoglobin each day. The protein portion (globin) and the iron are reutilized, but the porphyrin ring is broken down. First it is reduced to biliverdin, a green compound, which is subsequently reduced to bilirubin, a yellow compound. If more bilirubin is formed than can be excreted by the liver, bilirubin accumulates in the blood. When the concentration of bilirubin in the blood reaches a certain level, bilirubin diffuses into the tissues, causing them to become yellow. This condition is known as jaundice.

Summary

Amines are compounds in which one or more of the hydrogens of ammonia have been replaced by R groups. Amines are classified as primary, secondary, or tertiary, depending on whether one, two, or three hydrogens of ammonia have been replaced. Amines undergo **amine inversion** through a transition state in which the sp^3 nitrogen becomes an sp^2 nitrogen.

Some amines are **heterocyclic compounds**—cyclic compounds in which one or more of the atoms of the ring is an atom other than carbon. Heterocyclic rings are numbered so that the **heteroatom** has the lowest possible number. A **natural product** is a compound synthesized by a plant or an animal. **Alkaloids** are natural products containing one or more nitrogen heteroatoms and are found in the leaves, bark, roots, or seeds of plants.

Because of the lone pair on the nitrogen, amines are both bases and nucleophiles. Amines react as nucleophiles in nucleophilic substitution reactions, in nucleophilic acyl substitution reactions, in nucleophilic addition–elimination reactions, and in conjugate addition reactions.

Amines cannot undergo the substitution and elimination reactions that alkyl halides undergo, because the leaving groups of amines are too basic. Protonated amines also cannot undergo the reactions that protonated alcohols and protonated ethers undergo. Amines are easily oxidized. Saturated heterocycles containing five or more atoms have physical and chemical properties typical of acyclic compounds that contain the same heteroatom.

Quaternary ammonium hydroxides and amine oxides undergo E2 elimination reactions known as Hofmann elimination reactions and Cope elimination reactions, respectively. In both reactions, the proton from the β-carbon bonded to the greater number of hydrogens is removed. Quaternary ammonium salts are the most common **phase-transfer catalysts**.

Primary amines can be synthesized by means of a Gabriel synthesis, by reduction of an alkyl azide or a nitrile, by **reductive amination**, and by reduction of an amide.

Pyrrole, **furan**, and **thiophene** are aromatic compounds that undergo electrophilic aromatic substitution reactions preferentially at C-2. These compounds are more reactive than benzene toward electrophilic aromatic substitution. When pyrrole is protonated, its aromaticity is destroyed. Pyrrole polymerizes in strongly acidic solutions. Indole, benzofuran, and benzothiophene are aromatic compounds that contain a five-membered aromatic ring fused to a benzene ring.

Replacing one of benzene's carbons with a nitrogen forms **pyridine**, an aromatic compound that undergoes electrophilic aromatic substitution reactions at C-3 and nucleophilic aromatic substitution reactions at C-2 and C-4. Pyridine is less reactive than benzene toward electrophilic aromatic substitution and more reactive toward nucleophilic aromatic substitution. Quinoline and isoquinoline are aromatic compounds with both a benzene ring and a pyridine ring.

Imidazole is the heterocyclic ring of the amino acid histidine. The conjugate acid of imidazole has a pK_a of 6.8, allowing it to exist in both the protonated and unprotonated forms at physiological pH (pH = 7.3). Nucleic acids (DNA and RNA) contain substituted **purines** and substituted **pyrimidines**. Hydroxypurines and hydroxypyrimidines are more stable in the keto form. A **porphyrin ring system** consists of four pyrrole rings joined by one-carbon bridges; in hemoglobin and myoglobin, the four nitrogen atoms are ligated to Fe^{2+}. The metal atom in chlorophyll *a* is Mg^{2+} and the metal atom in vitamin B_{12} is Co^{2+}.

Summary of Reactions

1. Reaction of amines as nucleophiles (Section 21.4).
   a. In alkylation reactions:

$$R'-\ddot{N}H_2 \xrightarrow{RBr} R'-\overset{+}{N}H_2(R)\ Br^- \rightleftharpoons R'-\ddot{N}H(R) + HBr \xrightarrow{RBr} R-\overset{+}{N}H(R)(R)\ Br^- \rightleftharpoons R'-N{:}(R)(R) + HBr \xrightarrow{RBr} R'-\overset{+}{N}(R)(R)-R\ Br^-$$

b. In acylation reactions:

$$RC(=O)Cl + 2\ R'NH_2 \longrightarrow RC(=O)NHR' + R'NH_3^+\ Cl^-$$

c. In nucleophilic addition–elimination reactions:

i Reaction of a primary amine with an aldehyde or ketone to form an imine:

$$R_2C{=}O + R{-}NH_2 \underset{}{\overset{\text{catalytic } H^+}{\rightleftharpoons}} R_2C{=}N{-}R + H_2O$$

ii Reaction of a secondary amine with an aldehyde or ketone to form an enamine:

$$RCH_2C(R){=}O + R{-}NH{-}R \overset{\text{catalytic } H^+}{\rightleftharpoons} RCH{=}C(R){-}NR_2 + H_2O$$

d. In conjugate addition reactions:

$$RCH{=}CHC(=O)R + R'NH_2 \longrightarrow RCH(NHR'){-}CH_2C(=O)R$$

2. Primary arylamines react with nitrous acid to form stable arenediazonium salts (Section 21.4).

$$C_6H_5{-}NH_2 \xrightarrow[NaNO_2]{HCl} C_6H_5{-}\overset{+}{N}{\equiv}N\ Cl^-$$

3. Oxidation of amines: Primary amines are oxidized to nitro compounds, secondary amines to hydroxylamines, and tertiary amines to amine oxides (Section 21.7).

$$R{-}NH_2 \xrightarrow{\text{oxidation}} R{-}NH{-}OH \xrightarrow{\text{oxidation}} R{-}N{=}O \xrightarrow{\text{oxidation}} R{-}\overset{+}{N}(=O){-}O^-$$

$$R_2\ddot{N}H \xrightarrow{\text{oxidation}} R_2N{-}OH + H_2O$$

a secondary amine — a secondary hydroxylamine

$$R_3\ddot{N} \xrightarrow{\text{oxidation}} R_3\overset{+}{N}{-}O^- + H_2O$$

a tertiary amine — a tertiary amine oxide

4. Elimination reactions of *quaternary ammonium hydroxides* or *tertiary amine oxides* (Sections 21.5 and 21.7).

$$RCH_2CH_2\overset{+}{N}(CH_3)_3\ HO^- \xrightarrow[\text{Hofmann elimination}]{\Delta} RCH{=}CH_2 + N(CH_3)_3 + H_2O$$

$$RCH_2CH_2N(CH_3)_2 \xrightarrow{H_2O_2} RCH_2CH_2\overset{+}{N}(CH_3)_2{-}O^- \xrightarrow[\text{Cope elimination}]{\Delta} RCH{=}CH_2 + (CH_3)_2N{-}OH$$

in both eliminations, the proton is removed from the β-carbon bonded to the most hydrogens

5. Synthesis of amines (Section 21.8).

a. Gabriel synthesis of primary amines:

$$\text{phthalimide (NH)} \xrightarrow[\text{2. R—Br}]{\text{1. HO}^-} \text{N-alkylphthalimide (N—R)} + Br^- \xrightarrow[\Delta]{H_3O^+} \text{phthalic acid (COOH, COOH)} + R\overset{+}{N}H_3$$

b. Reduction of an alkyl azide or a nitrile:

$$R—Br \xrightarrow{^-N_3} R—N{=}\overset{+}{N}{=}N^- \xrightarrow[\text{Pd/C}]{H_2} R—NH_2$$

$$R—C{\equiv}N \xrightarrow[\text{Pd/C}]{H_2} R—CH_2NH_2$$

c. Reduction of a nitroalkane or nitrobenzene:

$$CH_3CH_2CH_2NO_2 + H_2 \xrightarrow{\text{Pd/C}} CH_3CH_2CH_2NH_2$$

$$C_6H_5—NO_2 + H_2 \xrightarrow{\text{Pd/C}} C_6H_5—NH_2$$

d. Aldehydes and ketones react (1) with excess *ammonia* plus H_2/metal catalyst to form primary amines, (2) with a *primary amine* followed by reduction with sodium triacetoxyborohydride to form secondary amines, and (3) with a *secondary amine* followed by reduction with sodium triacetoxyborohydride to form tertiary amines:

$$R_2C{=}O \xrightarrow[H_2,\ \text{Pd/C}]{\text{excess } NH_3} R_2CH—NH_2 + H_2O$$

$$R_2C{=}O + CH_3NH_2 \overset{\text{catalytic } H^+}{\rightleftharpoons} R_2C{=}NCH_3 \xrightarrow{NaBH(OCCH_3)_3} R_2CH—NHCH_3$$

$$\text{cyclohexanone} + CH_3NHCH_3 \overset{\text{catalytic } H^+}{\rightleftharpoons} \text{1-cyclohexenyl—}N(CH_3)_2 \xrightarrow{NaBH(OCCH_3)_3} \text{cyclohexyl—}N(CH_3)_2$$

6. Electrophilic aromatic substitution reactions.

a. Pyrrole, furan, and thiophene (Section 21.9):

$$\text{pyrrole} + HNO_3 \xrightarrow{(CH_3CO)_2O} \text{2-nitropyrrole} + H_2O$$

$$\text{furan} + Br_2 \longrightarrow \text{2-bromofuran} + HBr$$

$$\text{thiophene} + CH_3\overset{O}{\overset{\|}{C}}Cl \xrightarrow[\text{2. } H_2O]{\text{1. } SnCl_4} \text{2-thienyl—}\overset{O}{\overset{\|}{C}}CH_3 + HCl$$

b. Pyridine (Section 21.10):

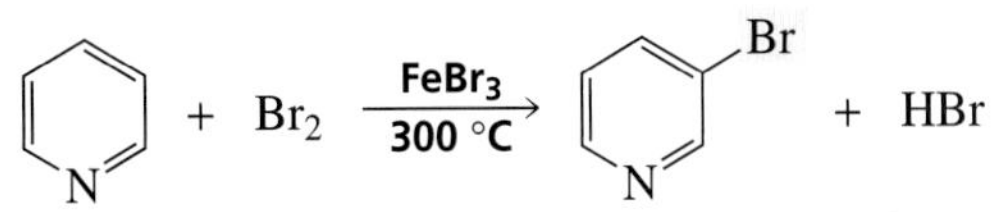

$$\text{pyridine} + Br_2 \xrightarrow[300\ °C]{FeBr_3} \text{3-bromopyridine} + HBr$$

7. Nucleophilic aromatic substitution reactions of pyridine (Section 21.10).

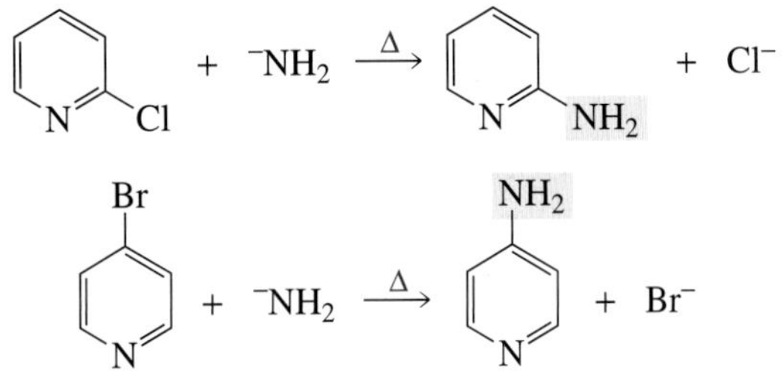

$$\text{2-chloropyridine} + {}^{-}NH_2 \xrightarrow{\Delta} \text{2-aminopyridine} + Cl^-$$

$$\text{4-bromopyridine} + {}^{-}NH_2 \xrightarrow{\Delta} \text{4-aminopyridine} + Br^-$$

Key Terms

alkaloid (p. 884)
amine inversion (p. 885)
amines (p. 883)
Cope elimination reaction (p. 894)
exhaustive methylation (p. 891)
furan (p. 897)
heteroatom (p. 884)
heterocycle (p. 884)
heterocyclic compound (p. 884)
Hofmann elimination reaction (p. 889)
imidazole (p. 907)
ligation (p. 909)
natural product (p. 884)
phase-transfer catalysis (p. 893)
phase-transfer catalyst (p. 893)
porphyrin ring system (p. 909)
protoporphyrin IX (p. 909)
purine (p. 909)
pyridine (p. 902)
pyrimidine (p. 909)
pyrrole (p. 897)
quaternary ammonium ion (p. 889)
reductive amination (p. 896)
thiophene (p. 897)

Problems

28. Name the following compounds:

a. 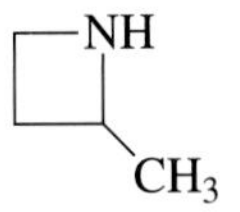 (azetidine with CH_3 substituent)

b. 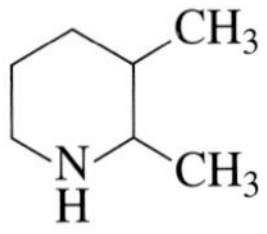 (piperidine with two CH_3 substituents, N–H)

c. 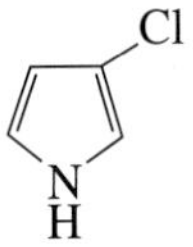 (pyrrole with Cl substituent, N–H)

d. 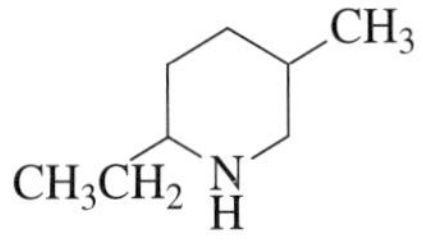 (piperidine with CH_3CH_2 and CH_3 substituents, N–H)

29. Give the product of each of the following reactions:

a. 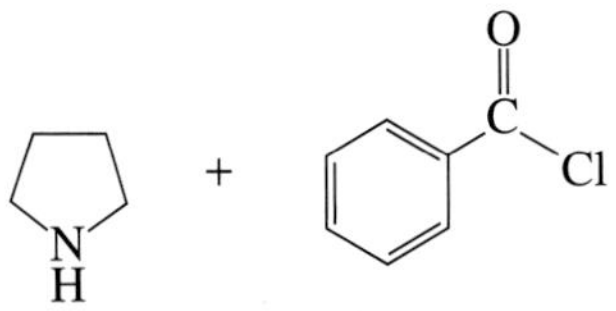 pyrrolidine + C_6H_5COCl

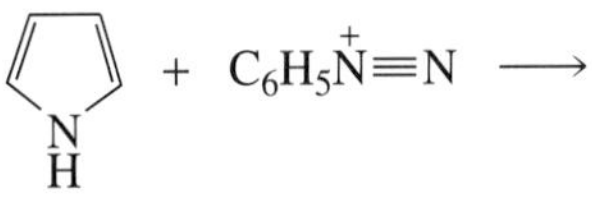

b. 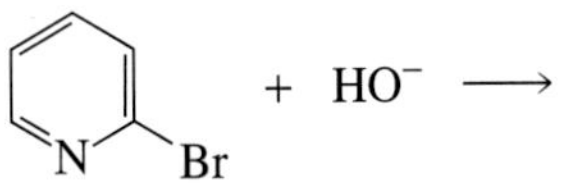 2-bromopyridine + $HO^- \longrightarrow$

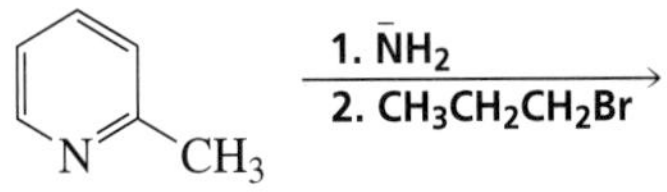

c. $CH_3CH_2CH_2CH_2Br \xrightarrow[2.\ H_2/Pd]{1.\ {}^{-}C{\equiv}N}$

d. 2-methylfuran + $CH_3\overset{O}{\overset{\|}{C}}Cl \longrightarrow$

e. 2-chloropyridine $\xrightarrow{C_6H_5Li}$

f. 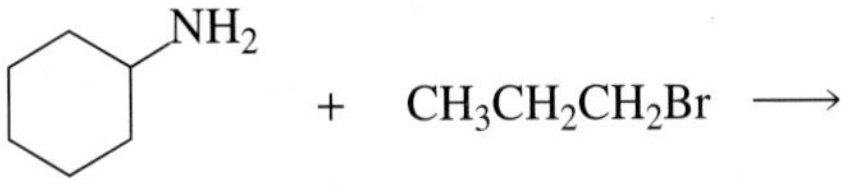 cyclohexylamine (NH_2) + $CH_3CH_2CH_2Br \longrightarrow$

g. pyrrole + $C_6H_5\overset{+}{N}{\equiv}N \longrightarrow$

h. 2-methylpyridine (CH_3) $\xrightarrow[2.\ CH_3CH_2CH_2Br]{1.\ {}^{-}NH_2}$

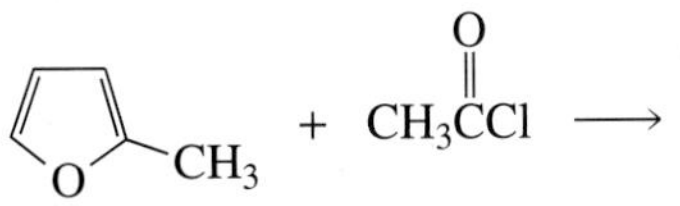

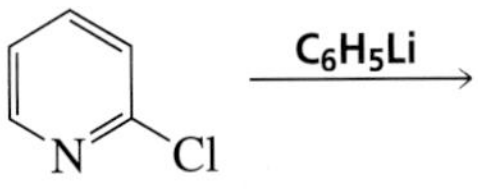

i. 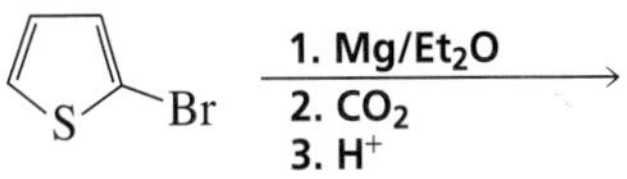 2-bromothiophene $\xrightarrow[2.\ CO_2\quad 3.\ H^+]{1.\ Mg/Et_2O}$

30. List the following compounds in order of decreasing acidity:

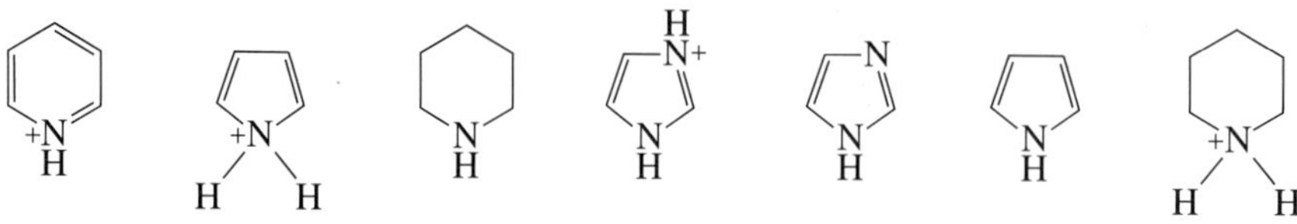

31. Which of the following compounds is easier to decarboxylate?

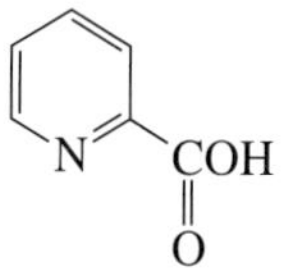

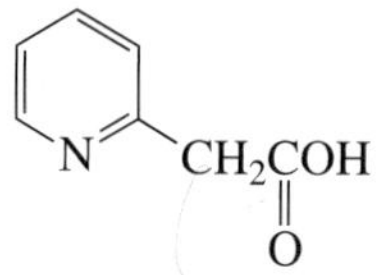

32. Rank the following compounds in order of decreasing reactivity in an electrophilic aromatic substitution reaction:

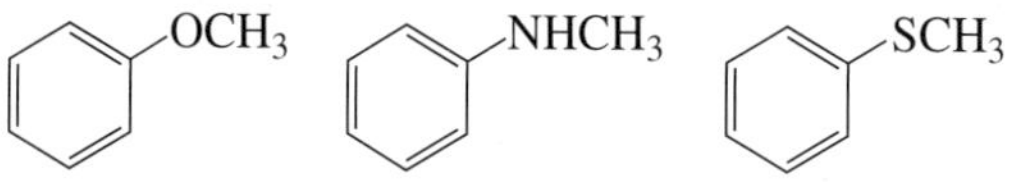

33. One of the following compounds undergoes electrophilic aromatic substitution predominantly at C-3, and one undergoes electrophilic aromatic substitution predominantly at C-4. Which is which?

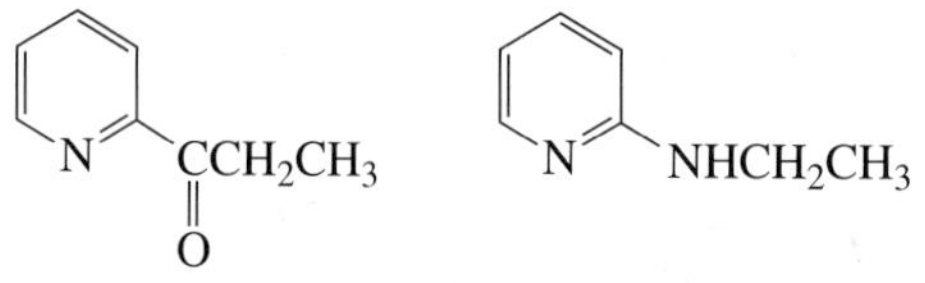

34. Benzene undergoes electrophilic aromatic substitution reactions with aziridines in the presence of a Lewis acid such as $AlCl_3$.
   a. What are the major and minor products of the following reaction?

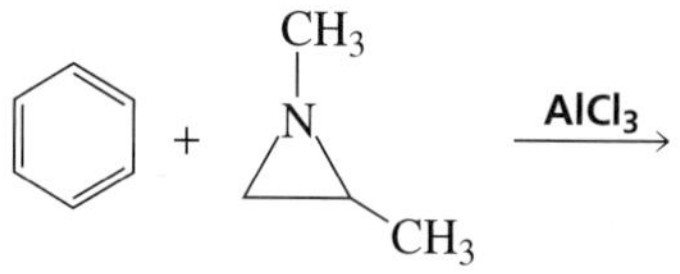

   b. Would you expect epoxides to undergo similar reactions?

35. A Hofmann degradation of a primary amine forms an alkene that gives butanal and 2-methylpropanal upon ozonolysis and work-up under reducing conditions. Identify the amine.

36. The dipole moments of furan and tetrahydrofuran are in the same direction. One compound has a dipole moment of 0.70 D, and the other has a dipole moment of 1.73 D. Which is which?

37. Show how the vitamin niacin can be synthesized from nicotine.

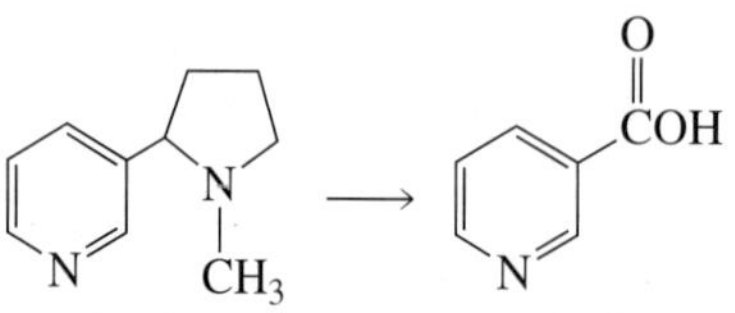

nicotine niacin

38. The chemical shifts of the C-2 hydrogen in the ^{1}H NMR spectra of pyrrole, pyridine, and pyrrolidine are δ2.82, δ6.42, and δ8.50, Match each chemical shift with its heterocycle.

39. Explain why protonation of aniline has a dramatic effect on the compound's UV spectrum, whereas protonation of pyridine has only a small effect on that compound's UV spectrum.

40. Explain why pyrrole (pK_a ~ 17) is a much stronger acid than ammonia (pK_a = 36).

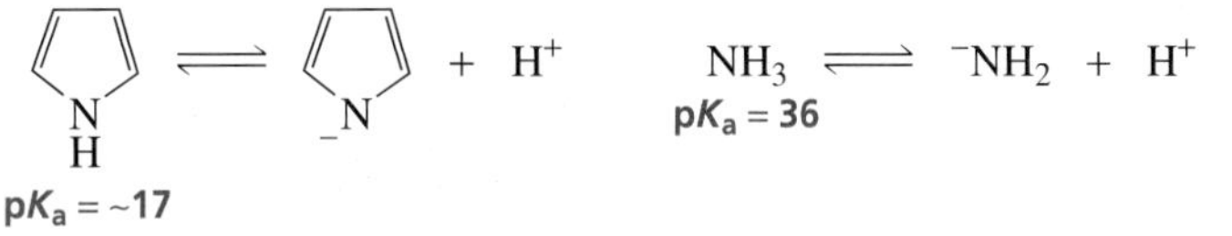

pK_a = ~17

$NH_3 \rightleftharpoons {}^-NH_2 + H^+$

pK_a = 36

41. Propose a mechanism for the following reaction:

2 pyrrole + $H_2C{=}O$ —trace H^+→ (pyrrole)$-CH_2-$(pyrrole)

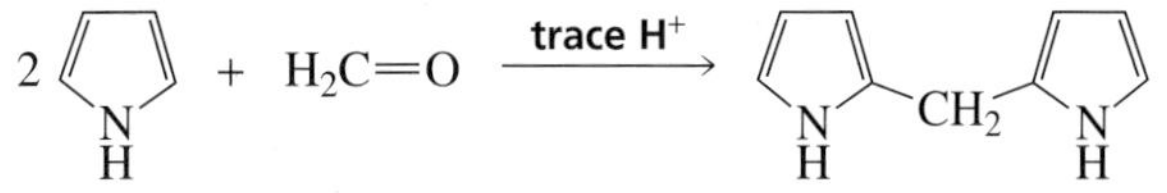

42. Quinolines are commonly synthesized by a method known as the Skraup synthesis, which involves the reaction of aniline with glycerol under acidic conditions. Nitrobenzene is added to the reaction mixture to serve as an oxidizing agent. The first step in the synthesis is the dehydration of glycerol to propenal.

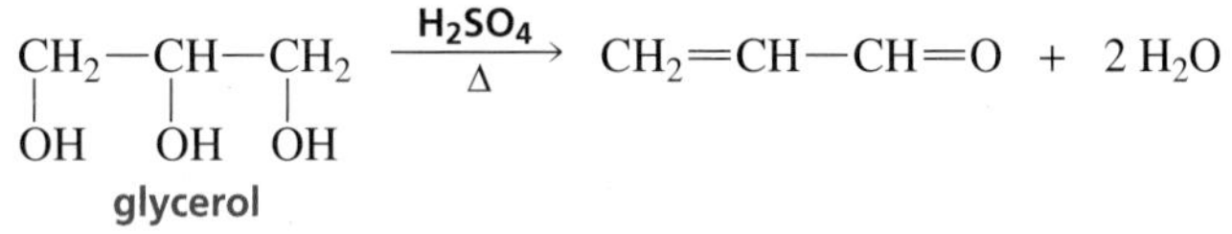

$$\underset{\text{glycerol}}{\underset{OH\ \ \ \ \ OH\ \ \ \ OH}{CH_2-CH-CH_2}} \xrightarrow[\Delta]{H_2SO_4} CH_2{=}CH-CH{=}O + 2\,H_2O$$

a. What product would be obtained if *para*-ethylaniline were used instead of aniline?
b. What product would be obtained if 3-hexen-2-one were used instead of glycerol?
c. What starting materials are needed for the synthesis of 2,7-diethyl-3-methylquinoline?

43. Propose a mechanism for each of the following reactions:

a. 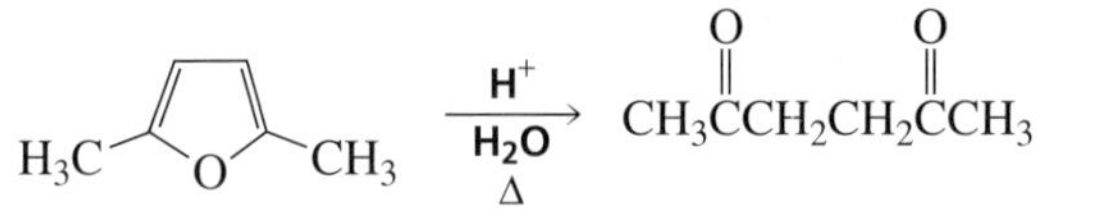 2,5-dimethylfuran (H_3C, CH_3) $\xrightarrow[H_2O,\ \Delta]{H^+}$ $CH_3\overset{O}{\overset{\|}{C}}CH_2CH_2\overset{O}{\overset{\|}{C}}CH_3$

b. 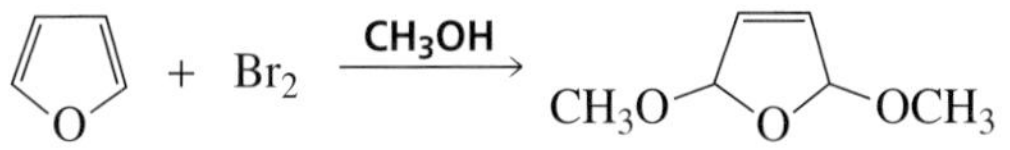 furan + Br_2 $\xrightarrow{CH_3OH}$ CH_3O–(2,5-dihydrofuran)–OCH_3

44. Give the major product of each of the following reactions:

a. 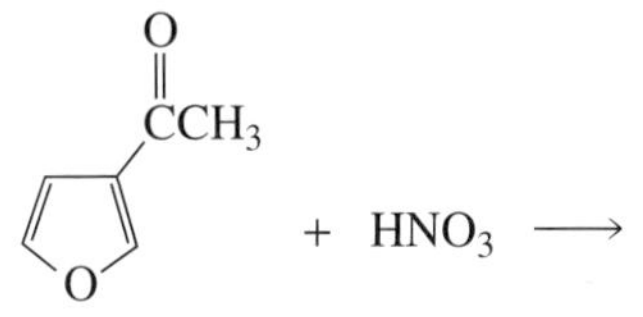 3-acetylfuran ($\overset{O}{\overset{\|}{C}}CH_3$) + HNO_3 ⟶

b. 2-nitrothiophene (NO_2) + Br_2 ⟶

c. $CH_3CHCH_2\overset{+}{N}CH_2CH_3$ (with CH_3 on C2, two CH_3 on N) HO^- $\xrightarrow{\Delta}$

d. 4-(dimethylamino)pyridine (CH_3NCH_3) + CH_3I ⟶

e. 2-pyridone (N–H, =O) + PCl_5 ⟶

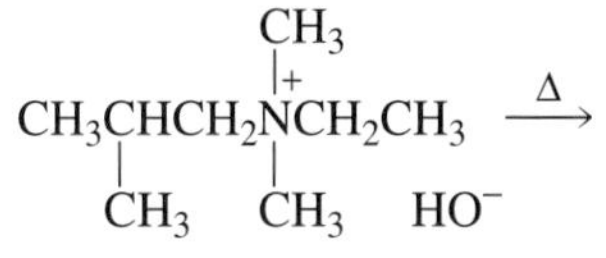

f. 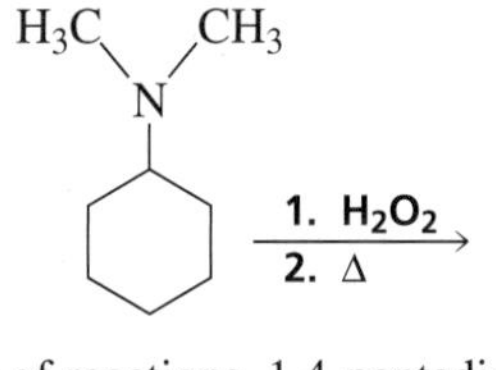 N,N-dimethylcyclohexylamine (H_3C, CH_3, N) $\xrightarrow{1.\ H_2O_2\ \ 2.\ \Delta}$

g. 4-methyl-N-methylpyridinium (CH_3, ^+N–CH_3) $\xrightarrow{1.\ HO^-\ \ 2.\ H_2C{=}O\ \ 3.\ H^+}$

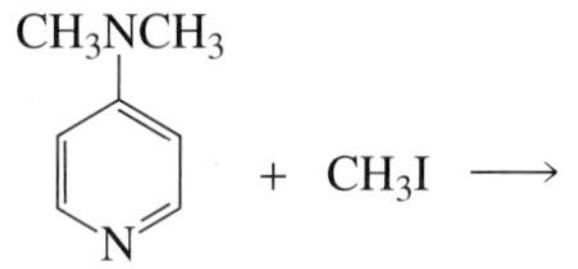

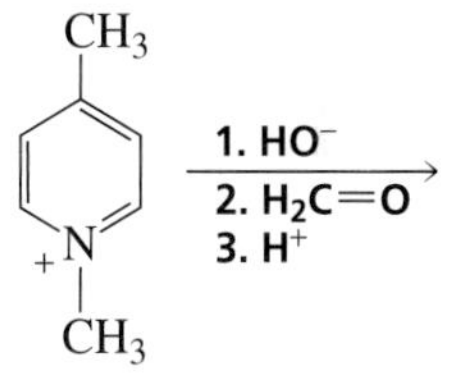

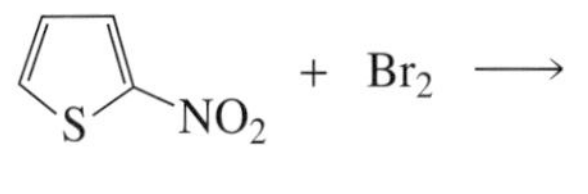

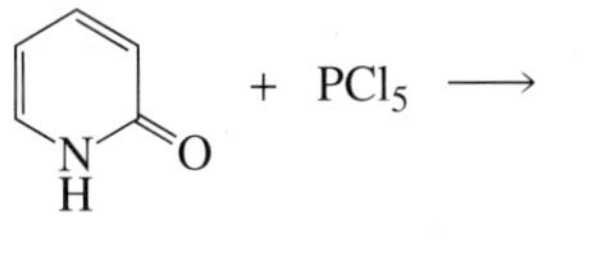

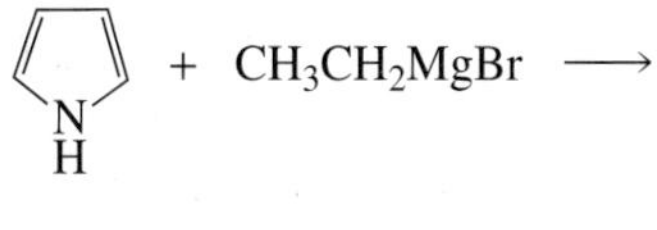

h. pyrrole (N–H) + CH_3CH_2MgBr ⟶

i. 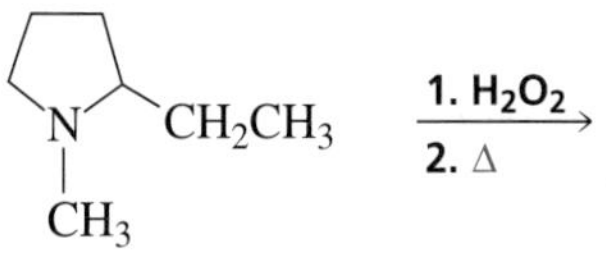 1-methyl-2-ethylpyrrolidine (CH_2CH_3, N–CH_3) $\xrightarrow{1.\ H_2O_2\ \ 2.\ \Delta}$

45. When piperidine undergoes the indicated series of reactions, 1,4-pentadiene is obtained as the product. When the four different methyl-substituted piperidines undergo the same series of reactions, each forms a different diene: 1,5-hexadiene, 1,4-pentadiene, 2-methyl-1,4-pentadiene, and 3-methyl-1,4-pentadiene. Which methyl-substituted piperidine yields which diene?

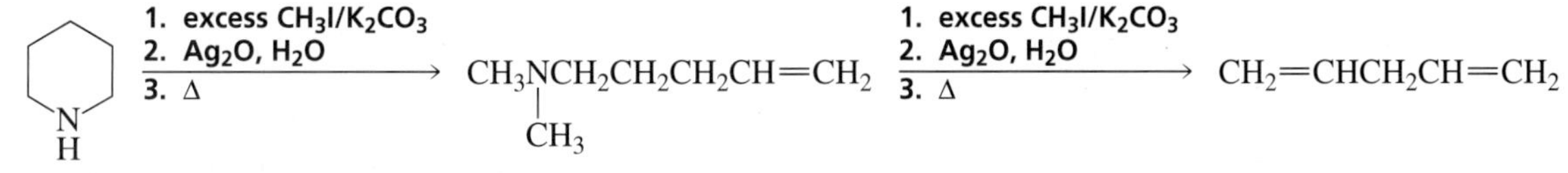

piperidine $\xrightarrow{1.\ \text{excess}\ CH_3I/K_2CO_3\ \ 2.\ Ag_2O,\ H_2O\ \ 3.\ \Delta}$ $CH_3\underset{CH_3}{\underset{|}{N}}CH_2CH_2CH_2CH{=}CH_2$ $\xrightarrow{1.\ \text{excess}\ CH_3I/K_2CO_3\ \ 2.\ Ag_2O,\ H_2O\ \ 3.\ \Delta}$ $CH_2{=}CHCH_2CH{=}CH_2$

46. a. Draw resonance contributors to show why pyridine-*N*-oxide is more reactive than pyridine toward electrophilic aromatic substitution
    b. At what position does pyridine-*N*-oxide undergo electrophilic aromatic substitution?

47. Propose a mechanism for the following reaction:

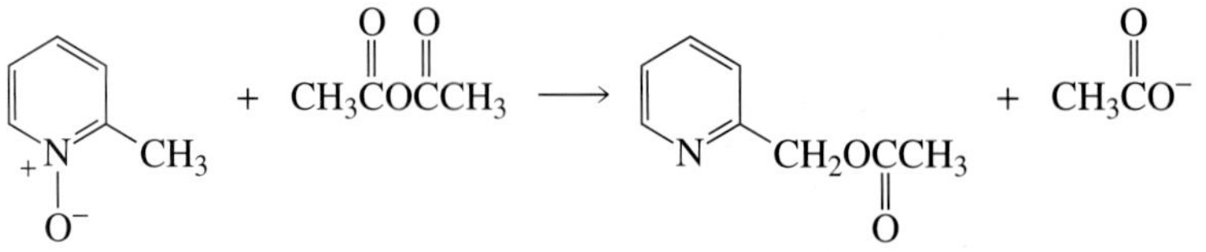

$$\text{2-methylpyridine-}N\text{-oxide} + CH_3COCCH_3 \text{ (O, O)} \longrightarrow \text{2-pyridyl-}CH_2OCCH_3 \text{ (O)} + CH_3CO^-$$

48. Explain why the aziridinium ion has a considerably lower pK_a (8.0) than that of a typical secondary ammonium ion (~11). (*Hint:* Recall that the larger the bond angle, the greater the *s* character, and the greater the *s* character, the more electronegative the atom.)

aziridinium ion ⇌ aziridine + H^+

aziridinium ion
pK_a = 8.04

49. Pyrrole reacts with excess *para*-(*N*,*N*-dimethylamino)benzaldehyde to form a highly colored compound. Draw the structure of the colored compound.

50. 2-Phenylindole is prepared from the reaction of acetophenone and phenylhydrazine, a method known as the Fischer indole synthesis. Propose a mechanism for this reaction. (*Hint:* The reactive species is the enamine tautomer of the phenylhydrazone.)

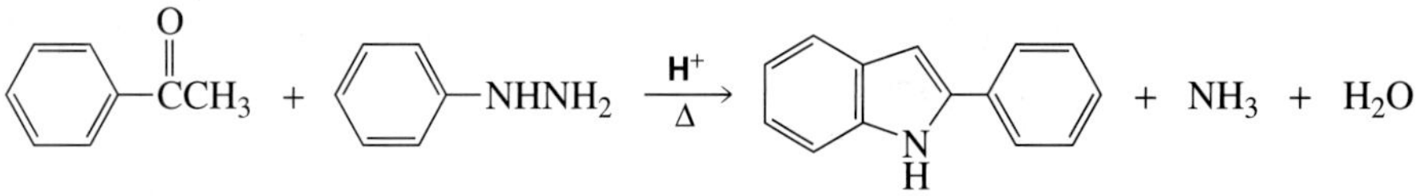

$$C_6H_5CCH_3 \text{ (O)} + C_6H_5NHNH_2 \xrightarrow[\Delta]{H^+} \text{2-phenylindole} + NH_3 + H_2O$$

51. What starting materials are required for the synthesis of the following compounds, using the Fischer indole synthesis? (*Hint:* See Problem 50.)

a. 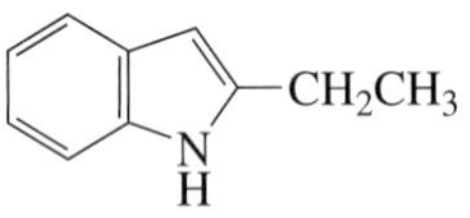 2-ethylindole (CH_2CH_3, NH)
b. 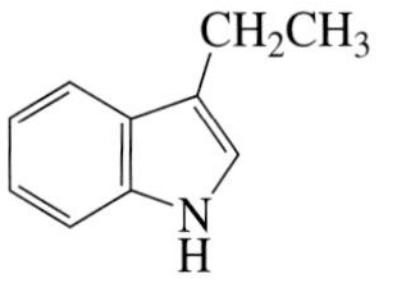 3-ethylindole (CH_2CH_3, NH)
c. 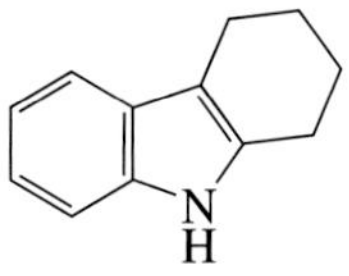 tetrahydrocarbazole (NH)

52. Organic chemists work with tetraphenylporphyrins rather than with porphyrins because tetraphenylporphyrins are much more resistant to air oxidation. Tetraphenylporphyrin can be prepared by the reaction of benzaldehyde with pyrrole. Propose a mechanism for the formation of the ring system shown here:

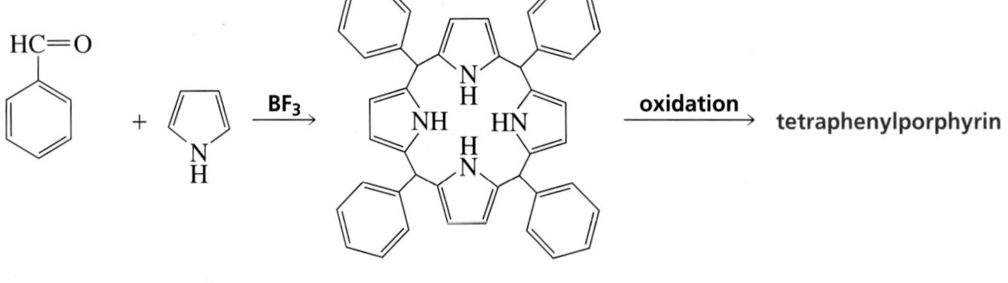

$$\text{benzaldehyde (HC=O)} + \text{pyrrole} \xrightarrow{\mathbf{BF_3}} \text{ring system (NH, HN)} \xrightarrow{\textbf{oxidation}} \textbf{tetraphenylporphyrin}$$

53. Propose a mechanism different from the one shown in Section 21.11 for the biosynthesis of porphobilinogen.

PART SEVEN

Bioorganic Compounds

Chapters 22 through 27 discuss the chemistry of organic compounds found in biological systems. Many of these compounds are larger than the organic compounds you have seen up to this point, and they may have more than one functional group, but the principles that govern their structure and reactivity are essentially the same as those that govern the structure and reactivity of the compounds that you have been studying. These chapters, therefore, will give you the opportunity to review much of the organic chemistry you have learned as you apply your knowledge of organic reactions to compounds found in the biological world.

Chapter 22 introduces you to the chemistry of carbohydrates, the most abundant class of compounds in the biological world. First you will learn about the structures and reactions of monosaccharides. Then you will see how they are linked to form disaccharides and polysaccharides. A wide variety of carbohydrates found in nature will be discussed.

Chapter 23 starts by looking at the physical properties of amino acids. Then you will see how amino acids are linked to form peptides and proteins. You will also see how proteins are made in the laboratory, and you will be able to compare this with how they are made in nature when you read Chapter 27. The structure of proteins will be examined to prepare you for understanding how enzymes catalyze chemical reactions, which is covered in Chapter 24.

Chapter 24 first describes the various ways that organic reactions are catalyzed and then shows how enzymes employ the same principles in their catalysis of reactions that occur in biological systems.

Chapter 25 describes the chemistry of the coenzymes—organic compounds that some enzymes need to catalyze biological reactions. Coenzymes play a variety of chemical roles: Some function as oxidizing and reducing agents, some allow electrons to be delocalized, some activate groups for further reaction, and some provide good nucleophiles or strong bases needed for reactions. Because coenzymes are derived from vitamins, you will see why vitamins are necessary for many of the organic reactions that occur in biological systems.

Chapter 26 discusses the chemistry of lipids. Lipids are water-insoluble compounds found in animals and plants. First you will first study the structure and function of different kinds of lipids. You will then be able to understand such things as how aspirin prevents inflammation and what causes butter to become rancid. How cholesterol and other terpenes are synthesized in nature will be explained as well.

Chapter 27 covers the chemistry and structures of nucleosides, nucleotides, and nucleic acids (RNA and DNA). You will see—mechanistically—why ATP is the universal carrier of chemical energy, how nucleotides are linked to form nucleic acids, why DNA contains thymines instead of uracils, and how the genetic messages encoded in DNA are transcribed into mRNA and then translated into proteins. Also explained are how the sequence of bases in DNA is determined and how DNA with specific base sequences can be synthesized.

22 Carbohydrates

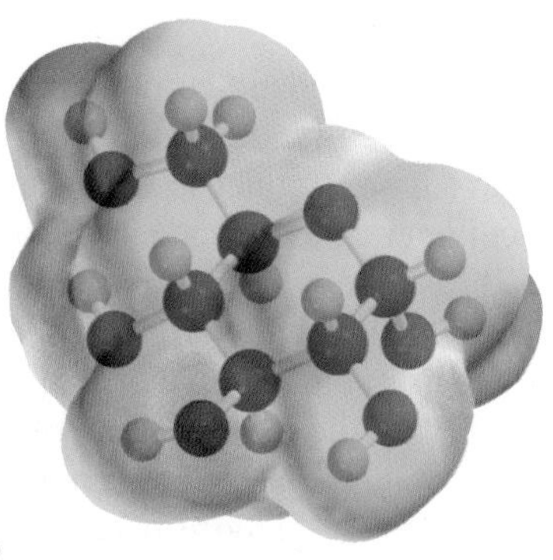
α-D-glucose

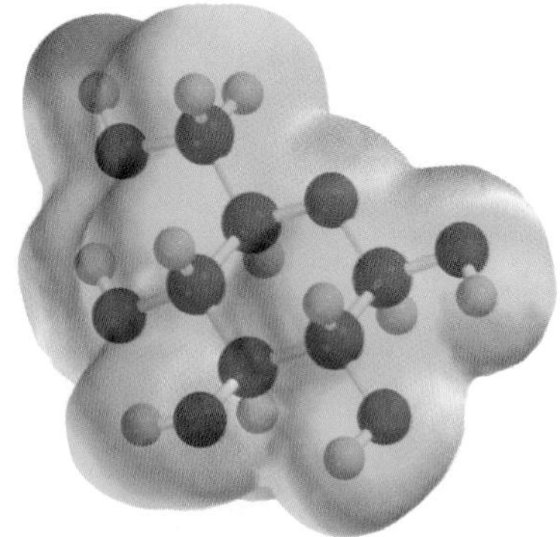
β-D-glucose

Bioorganic compounds are organic compounds found in biological systems. Bioorganic compounds follow the same principles of structure and reactivity as the organic molecules we have discussed so far. There is great similarity between the organic reactions chemists carry out in the laboratory and those performed by nature inside a living cell. In other words, bioorganic reactions can be thought of as organic reactions that take place in tiny flasks called cells.

Most bioorganic compounds have more complicated structures than those of the organic compounds that you are used to seeing, but do not let the structures fool you into thinking that their chemistry is equally complicated. One reason the structures of bioorganic compounds are more complicated is that bioorganic compounds must be able to recognize each other, and much of their structure is for that purpose—a function called **molecular recognition**.

D-glucose

D-fructose

The first group of bioorganic compounds we will study are *carbohydrates*—the most abundant class of compounds in the biological world, making up more than 50% of the dry weight of the Earth's biomass. Carbohydrates are important constituents of all living organisms and have a variety of different functions. Some are important structural components of cells; others act as recognition sites on cell surfaces. For example, the first event in all our lives was a sperm recognizing a carbohydrate on the surface of an egg's wall. Other carbohydrates serve as a major source of metabolic energy. For example, the leaves, fruits, seeds, stems, and roots of plants contain carbohydrates that plants use for their own metabolic needs and that then serve the metabolic needs of the animals that eat the plants.

Early chemists noted that carbohydrates have molecular formulas that make them appear to be hydrates of carbon, $C_n(H_2O)_n$—hence the name. Later structural studies revealed that these compounds were *not* hydrates because they did not contain intact water molecules, but the term "carbohydrate" persists. **Carbohydrates** are polyhydroxy aldehydes such as D-glucose, polyhydroxy ketones such as D-fructose, and compounds such as sucrose that can be hydrolyzed to polyhydroxy aldehydes or polyhydroxy ketones (Section 22.17). The chemical structures of carbohydrates are commonly represented by wedge-and-dash structures or by Fischer projections. Notice that both D-glucose and D-fructose have the molecular formula $C_6H_{12}O_6$, consistent with the general formula $C_6(H_2O)_6$ that made early chemists think that those compounds were hydrates of carbon.

Recall from Section 5.4 that horizontal bonds point toward the viewer and vertical bonds point away from the viewer in Fischer projections.

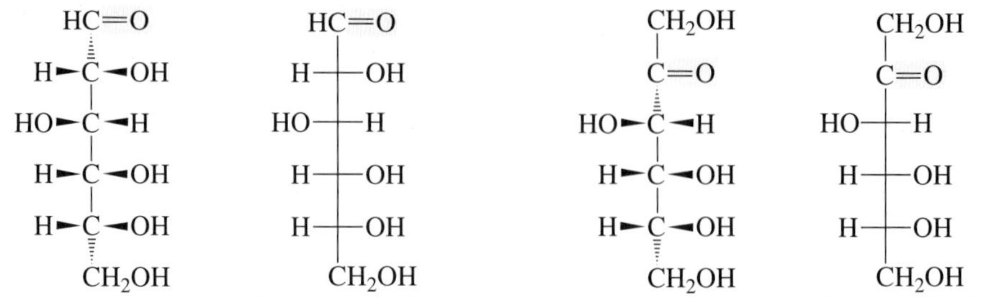

HC=O
H–C–OH
HO–C–H
H–C–OH
H–C–OH
CH_2OH
wedge-and-dash structure

HC=O
H–OH
HO–H
H–OH
H–OH
CH_2OH
Fischer projection

D-glucose
a polyhydroxy aldehyde

CH_2OH
C=O
HO–C–H
H–C–OH
H–C–OH
CH_2OH
wedge-and-dash structure

CH_2OH
C=O
HO–H
H–OH
H–OH
CH_2OH
Fischer projection

D-fructose
a polyhydroxy ketone

3-D Molecules: D-Glucose; D-Fructose

The most abundant carbohydrate in nature is D-glucose. Living cells oxidize D-glucose in the first of a series of processes that provide them with energy. When animals have more D-glucose than they need for energy, they convert excess D-glucose into a polymer called glycogen (Section 22.18). When an animal needs energy, glycogen is broken down into individual D-glucose molecules. Plants convert excess D-glucose into a polymer known as starch. Cellulose—the major structural component of plants—is another polymer of D-glucose. Chitin, a carbohydrate similar to cellulose, makes up the exoskeletons of crustaceans, insects, and other arthropods and is also the structural material of fungi.

Animals obtain glucose from food—such as plants—that contains glucose. Plants produce glucose by *photosynthesis*. During photosynthesis, plants take up water through their roots and use carbon dioxide from the air to synthesize glucose and oxygen. Because photosynthesis is the reverse of the process used by organisms to obtain energy—the oxidation of glucose to carbon dioxide and water—plants require energy to carry out photosynthesis. Plants obtain the energy they need for photosynthesis from sunlight, captured by chlorophyll molecules in green plants. Photosynthesis uses the CO_2 that animals exhale as waste and generates the O_2 that animals inhale to sustain life. Nearly all the oxygen in the atmosphere has been released by photosynthetic processes.

$$\underset{\text{glucose}}{C_6H_{12}O_6} + 6\,O_2 \underset{\text{photosynthesis}}{\overset{\text{oxidation}}{\rightleftharpoons}} 6\,CO_2 + 6\,H_2O + \text{energy}$$

22.1 Classification of Carbohydrates

The terms "carbohydrate," "saccharide," and "sugar" are often used interchangeably. "Saccharide" comes from the word for "sugar" in several early languages (*sarkara* in Sanskrit, *sakcharon* in Greek, and *saccharum* in Latin).

There are two classes of carbohydrates: *simple carbohydrates* and *complex carbohydrates*. **Simple carbohydrates** are **monosaccharides** (single sugars), whereas **complex carbohydrates** contain two or more sugar subunits linked together. **Disaccharides** have two sugar subunits linked together, **oligosaccharides** have three to 10 sugar subunits (*oligos* is Greek for "few") linked together, and **polysaccharides** have more than 10 sugar subunits linked together. Disaccharides, oligosaccharides, and polysaccharides can be broken down to monosaccharide subunits by hydrolysis.

a sugar subunit

—M—M—M—M—M—M—M—M—M— $\xrightarrow{\text{hydrolysis}}$ x M

polysaccharide **monosaccharide**

A *monosaccharide* can be a polyhydroxy aldehyde such as D-glucose or a polyhydroxy ketone such as D-fructose. Polyhydroxy aldehydes are called **aldoses** ("ald" is for aldehyde; "ose" is the suffix for a sugar), whereas polyhydroxy ketones are called **ketoses**. Monosaccharides are also classified according to the number of carbons they contain: Monosaccharides with three carbons are **trioses**, those with four carbons are **tetroses**, those with five carbons are **pentoses**, and those with six and seven carbons are **hexoses** and **heptoses**, respectively. A six-carbon polyhydroxy aldehyde such as D-glucose is an aldohexose, whereas a six-carbon polyhydroxy ketone such as D-fructose is a ketohexose.

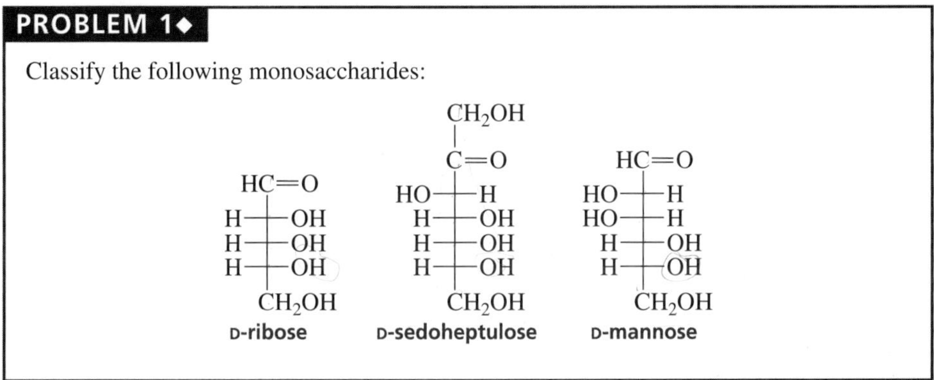

22.2 The D and L Notation

The smallest aldose, and the only one whose name does not end in "ose," is glyceraldehyde, an aldotriose.

$$\text{HOCH}_2\text{CH(OH)CH=O}$$

glyceraldehyde

A carbon to which four different groups are attached is an asymmetric carbon.

Because glyceraldehyde has an asymmetric carbon, it can exist as a pair of enantiomers.

(R)-(+)-glyceraldehyde **(S)-(−)-glyceraldehyde**

perspective formulas

(R)-(+)-glyceraldehyde **(S)-(−)-glyceraldehyde**

Fischer projections

Emil Fischer and his colleagues studied carbohydrates in the late nineteenth century, when techniques for determining the configurations of compounds were not available. Fischer arbitrarily assigned the *R*-configuration to the dextrorotatory isomer of glyceraldehyde that we call D-glyceraldehyde. He turned out to be correct: D-Glyceraldehyde is (*R*)-(+)-glyceraldehyde, and L-glyceraldehyde is (*S*)-(−)-glyceraldehyde (Section 5.13).

D-glyceraldehyde **L-glyceraldehyde**

The notations D and L are used to describe the configurations of carbohydrates and amino acids (Section 23.2), so it is important to learn what D and L signify. In Fischer projections of monosaccharides, the carbonyl group is always placed on top (in the case of aldoses) or as close to the top as possible (in the case of ketoses). From its structure, you can see that galactose has four asymmetric carbons (C-2, C-3, C-4, and C-5). *If the OH group attached to the bottom-most asymmetric carbon (the carbon that is second from the bottom) is on the right, then the compound is a D-sugar. If the OH group is on the left, then the compound is an L-sugar.* Almost all sugars found in nature are D-sugars. Notice that the mirror image of a D-sugar is an L-sugar.

HC=O
H—OH
HO—H
HO—H
H—OH
CH_2OH
D-galactose

the OH group is on the right

HC=O
HO—H
H—OH
H—OH
HO—H
CH_2OH
L-galactose
mirror image of D-galactose

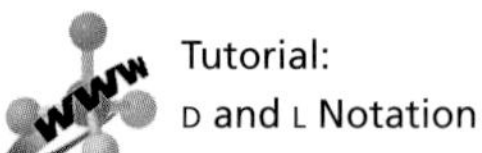

Tutorial:
D and L Notation

Like *R* and *S*, D and L indicate the configuration of an asymmetric carbon, but they do not indicate whether the compound rotates polarized light to the right (+) or to the left (−) (Section 5.7). For example, D-glyceraldehyde is dextrorotatory, whereas D-lactic acid is levorotatory. In other words, optical rotation, like melting or boiling points, is a physical property of a compound, whereas "*R*, *S*, D, and L" are conventions humans use to indicate the configuration of a molecule.

HC=O
H—OH
CH_2OH
D-(+)-glyceraldehyde

COOH
H—OH
CH_3
D-(−)-lactic acid

The common name of the monosaccharide, together with the D or L designation, completely defines its structure because the relative configurations of all the asymmetric carbons are implicit in the common name.

PROBLEM 2

Draw Fischer projections of L-glucose and L-fructose.

PROBLEM 3◆

Indicate whether each of the following is D-glyceraldehyde or L-glyceraldehyde, assuming that the horizontal bonds point toward you and the vertical bonds point away from you (Section 5.6):

a. HC=O / $HOCH_2$—OH / H

b. H / HO—CH_2OH / HC=O

c. CH_2OH / HO—H / HC=O

22.3 Configurations of Aldoses

Aldotetroses have two asymmetric carbons and therefore four stereoisomers. Two of the stereoisomers are D-sugars and two are L-sugars. The names of the aldotetroses—erythrose and threose—were used to name the erythro and threo pairs of enantiomers described in Section 5.9.

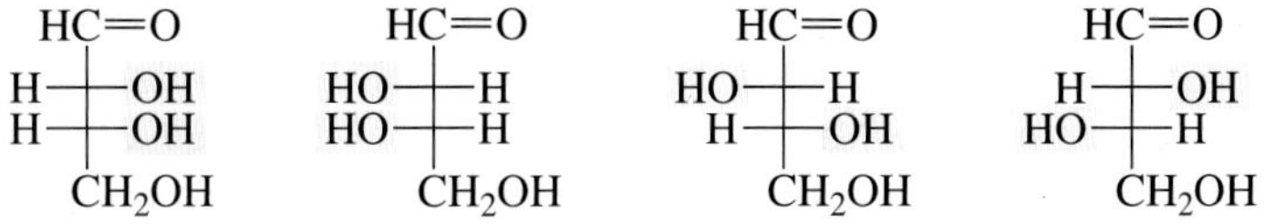

D-erythrose	L-erythrose	D-threose	L-threose
HC=O	HC=O	HC=O	HC=O
H—OH	HO—H	HO—H	H—OH
H—OH	HO—H	H—OH	HO—H
CH_2OH	CH_2OH	CH_2OH	CH_2OH

Aldopentoses have three asymmetric carbons and therefore eight stereoisomers (four pairs of enantiomers), while aldohexoses have four asymmetric carbons and 16 stereoisomers (eight pairs of enantiomers). The four D-aldopentoses and the eight D-aldohexoses are shown in Table 22.1.

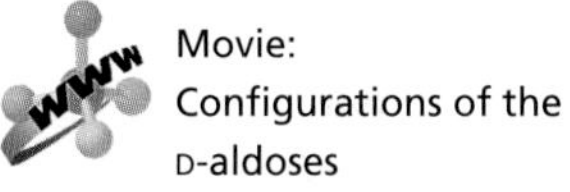

Movie: Configurations of the D-aldoses

Diastereomers that differ in configuration at only one asymmetric carbon are called **epimers**. For example, D-ribose and D-arabinose are C-2 epimers (they differ in configuration only at C-2), and D-idose and D-talose are C-3 epimers.

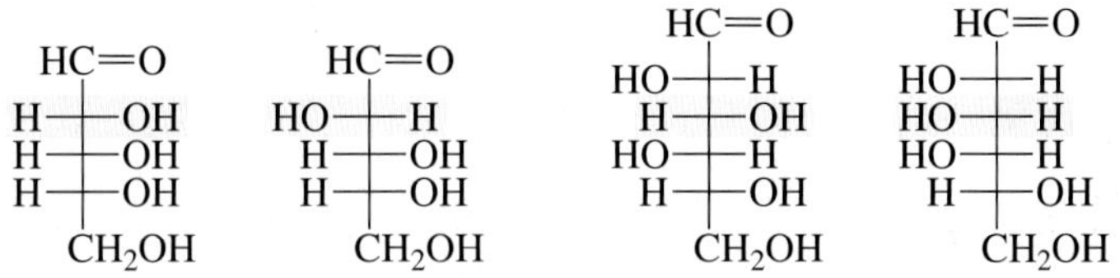

D-ribose	D-arabinose	D-idose	D-talose
HC=O	HC=O	HC=O	HC=O
H—OH	HO—H	HO—H	HO—H
H—OH	H—OH	H—OH	HO—H
H—OH	H—OH	HO—H	HO—H
		H—OH	H—OH
CH_2OH	CH_2OH	CH_2OH	CH_2OH
C-2 epimers		C-3 epimers	

TABLE 22.1 Configurations of the D-Aldoses

D-glyceraldehyde: HC=O; H—OH; CH_2OH

↙ D-erythrose ↘ D-threose

D-erythrose	D-threose
HC=O	HC=O
H—OH	HO—H
H—OH	H—OH
CH_2OH	CH_2OH

D-erythrose → D-ribose, D-arabinose; D-threose → D-xylose, D-lyxose

D-ribose	D-arabinose	D-xylose	D-lyxose
HC=O	HC=O	HC=O	HC=O
H—OH	HO—H	H—OH	HO—H
H—OH	H—OH	HO—H	HO—H
H—OH	H—OH	H—OH	H—OH
CH_2OH	CH_2OH	CH_2OH	CH_2OH

D-ribose → D-allose, D-altrose; D-arabinose → D-glucose, D-mannose; D-xylose → D-gulose, D-idose; D-lyxose → D-galactose, D-talose

D-allose	D-altrose	D-glucose	D-mannose	D-gulose	D-idose	D-galactose	D-talose
HC=O	HC=O	HC=O	HC=O	HC=O	HC=O	HC=O	HC=O
H—OH	HO—H	H—OH	HO—H	H—OH	HO—H	H—OH	HO—H
H—OH	H—OH	HO—H	HO—H	H—OH	H—OH	HO—H	HO—H
H—OH	H—OH	H—OH	H—OH	HO—H	HO—H	HO—H	HO—H
H—OH	H—OH	H—OH	H—OH	H—OH	H—OH	H—OH	H—OH
CH_2OH	CH_2OH	CH_2OH	CH_2OH	CH_2OH	CH_2OH	CH_2OH	CH_2OH

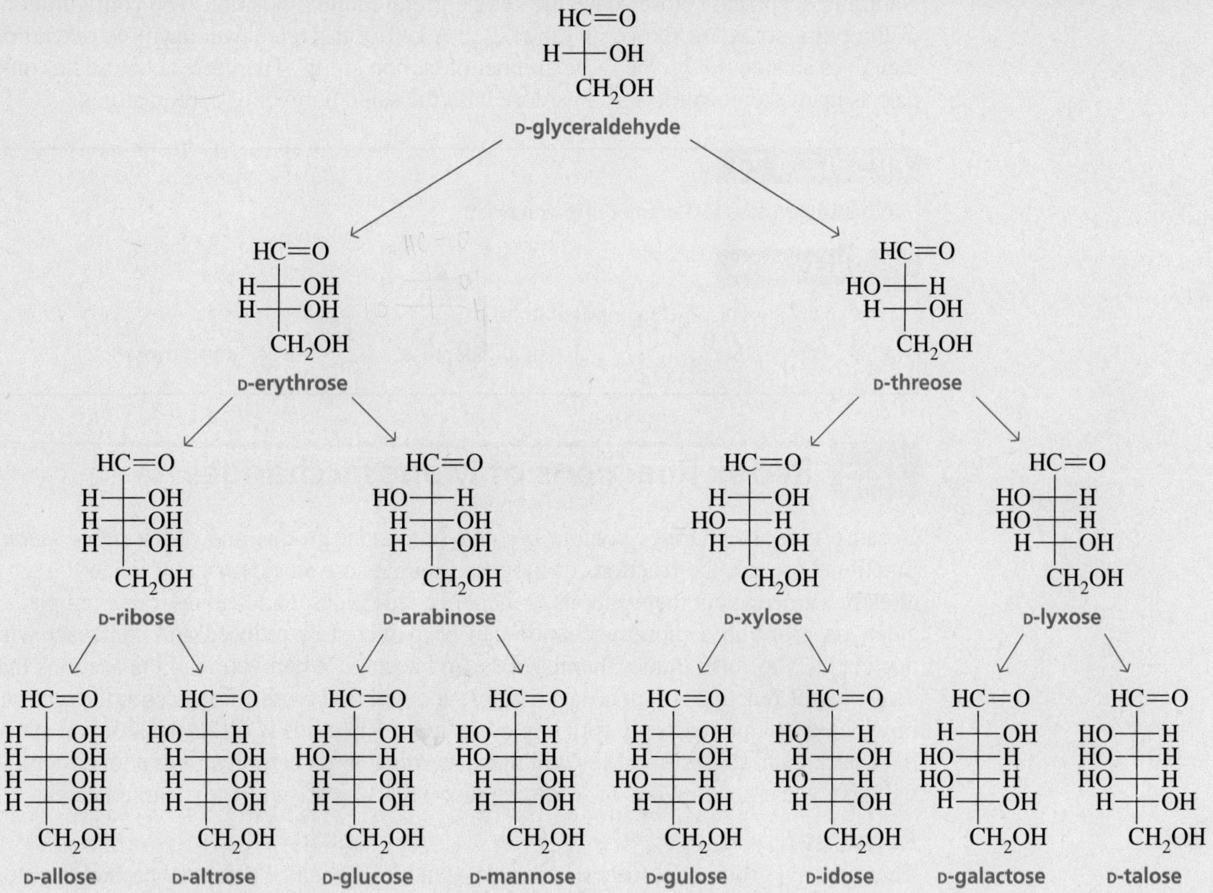

D-Mannose is the C-2 epimer of D-glucose.

D-Galactose is the C-4 epimer of D-glucose.

Diastereomers are configurational isomers that are not enantiomers.

D-Glucose, D-mannose, and D-galactose are the most common aldohexoses in biological systems. An easy way to learn their structures is to memorize the structure of D-glucose and then remember that D-mannose is the C-2 epimer of D-glucose and D-galactose is the C-4 epimer of D-glucose. Sugars such as D-glucose and D-galactose are also diastereomers because they are stereoisomers that are not enantiomers (Section 5.9). An epimer is a particular kind of diastereomer.

PROBLEM 4◆

a. Are D-erythrose and L-erythrose enantiomers or diastereomers?
b. Are L-erythrose and L-threose enantiomers or diastereomers?

PROBLEM 5◆

a. What sugar is the C-3 epimer of D-xylose?
b. What sugar is the C-5 epimer of D-allose?
c. What sugar is the C-4 epimer of L-gulose?

PROBLEM 6◆

Give systematic names to the following compounds. Indicate the configuration (*R* or *S*) of each asymmetric carbon:

a. D-glucose b. L-glucose c. D-galactose

22.4 Configurations of Ketoses

Naturally occurring ketoses have the ketone group in the 2-position. The configurations of the D-2-ketoses are shown in Table 22.2. A ketose has one fewer asymmetric carbon than does an aldose with the same number of carbon atoms. Therefore, a ketose has only half as many stereoisomers as an aldose with the same number of carbon atoms.

PROBLEM 7◆

What sugar is the C-3 epimer of D-fructose?

PROBLEM 8◆

How many stereoisomers are possible for

a. a 2-ketoheptose? b. an aldoheptose? c. a ketotriose?

22.5 Redox Reactions of Monosaccharides

Because monosaccharides contain *alcohol* functional groups and *aldehyde* or *ketone* functional groups, the reactions of monosaccharides are an extension of what you have already learned about the reactions of alcohols, aldehydes, and ketones. For example, an aldehyde group in a monosaccharide can be oxidized or reduced and can react with nucleophiles to form imines, hemiacetals, and acetals. When you read the sections that deal with the reactions of monosaccharides, you will find cross-references to the sections in which the same reactivity for simple organic compounds is discussed. As you study, refer back to these sections; they will make learning about carbohydrates a lot easier and will give you a good review of some chemistry that you have already learned about.

Reduction

The carbonyl group of aldoses and ketoses can be reduced by the usual carbonyl-group reducing agents (e.g., $NaBH_4$; Section 20.1). The product of the reduction is a polyalcohol, known as an **alditol**. Reduction of an aldose forms one alditol. Reduction of a ketose forms two alditols because the reaction creates a new asymmetric carbon in the

TABLE 22.2 Configurations of the D-Ketoses

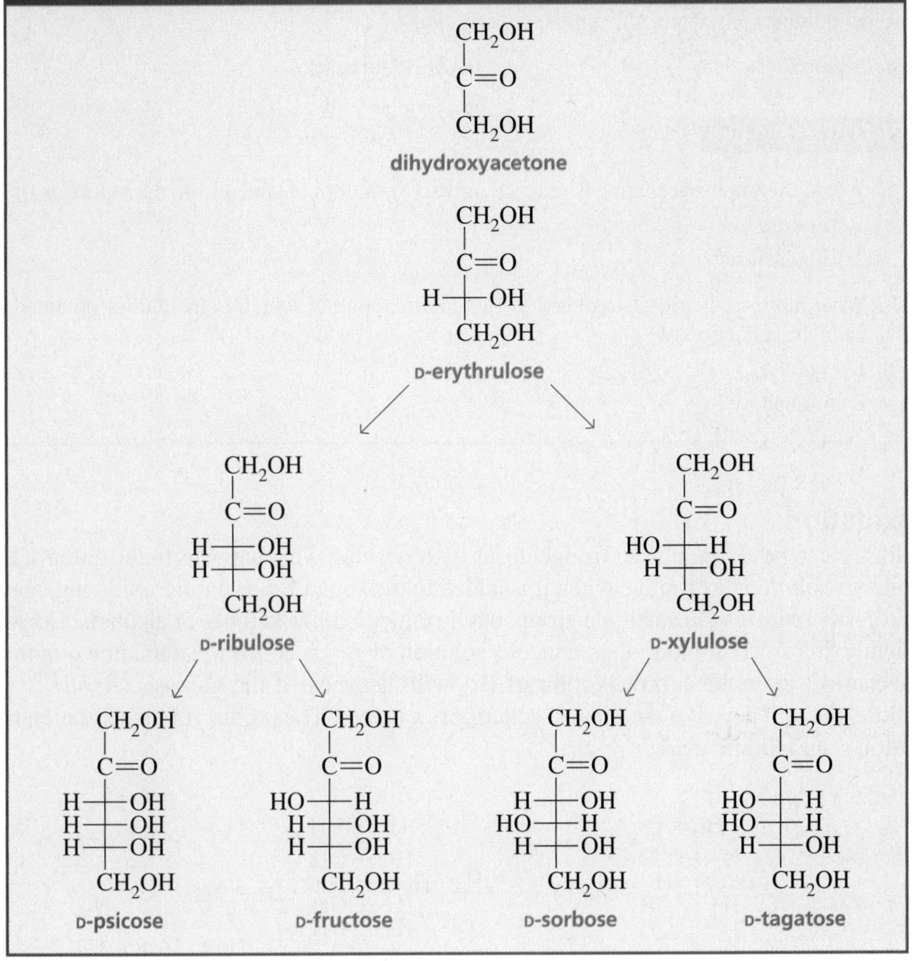

CH_2OH
C=O
CH_2OH
dihydroxyacetone

CH_2OH
C=O
H—OH
CH_2OH
D-erythrulose

CH_2OH
C=O
H—OH
H—OH
CH_2OH
D-ribulose

CH_2OH
C=O
HO—H
H—OH
CH_2OH
D-xylulose

CH_2OH
C=O
H—OH
H—OH
H—OH
CH_2OH
D-psicose

CH_2OH
C=O
HO—H
H—OH
H—OH
CH_2OH
D-fructose

CH_2OH
C=O
H—OH
HO—H
H—OH
CH_2OH
D-sorbose

CH_2OH
C=O
HO—H
HO—H
H—OH
CH_2OH
D-tagatose

product. D-Mannitol, the alditol formed from the reduction of D-mannose, is found in mushrooms, olives, and onions. The reduction of D-fructose forms D-mannitol and D-glucitol, the C-2 epimer of D-mannitol. D-Glucitol—also called sorbitol—is about 60% as sweet as sucrose. It is found in plums, pears, cherries, and berries and is used as a sugar substitute in the manufacture of candy.

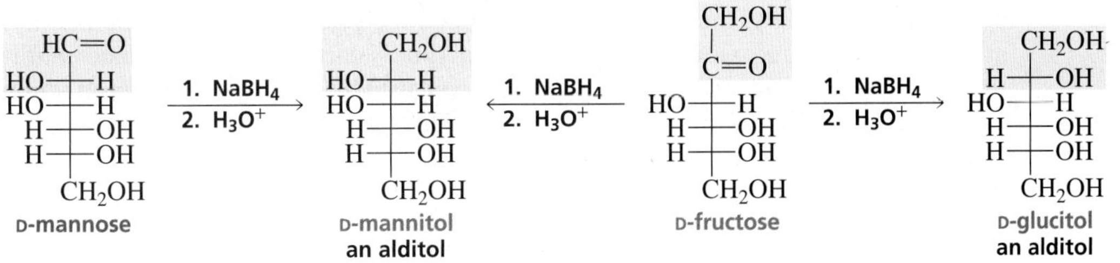

HC=O
HO—H
HO—H
H—OH
H—OH
CH_2OH
D-mannose

→ 1. $NaBH_4$ 2. H_3O^+

CH_2OH
HO—H
HO—H
H—OH
H—OH
CH_2OH
D-mannitol
an alditol

← 1. $NaBH_4$ 2. H_3O^+

CH_2OH
C=O
HO—H
H—OH
H—OH
CH_2OH
D-fructose

→ 1. $NaBH_4$ 2. H_3O^+

CH_2OH
H—OH
HO—H
H—OH
H—OH
CH_2OH
D-glucitol
an alditol

D-Glucitol is also obtained from the reduction of either D-glucose or L-gulose.

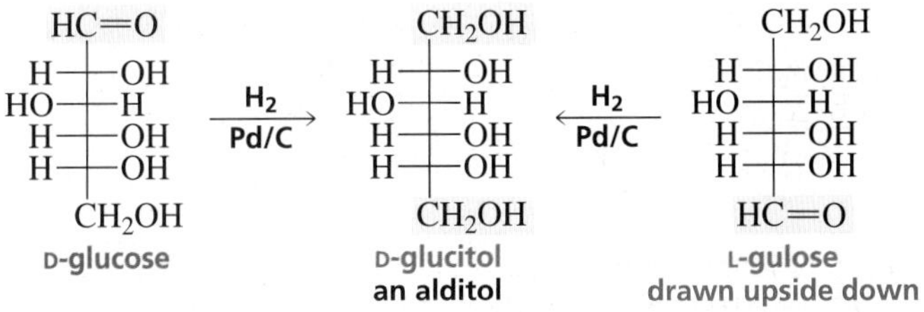

HC=O
H—OH
HO—H
H—OH
H—OH
CH_2OH
D-glucose

→ H_2 Pd/C

CH_2OH
H—OH
HO—H
H—OH
H—OH
CH_2OH
D-glucitol
an alditol

← H_2 Pd/C

CH_2OH
H—OH
HO—H
H—OH
H—OH
HC=O
L-gulose
drawn upside down

D-Xylitol—obtained from the reduction of D-xylose—is used as a sweetening agent in cereals and in "sugarless" gum.

PROBLEM 9◆

What products are obtained from the reduction of

a. D-idose?

b. D-sorbose?

PROBLEM 10◆

a. What other monosaccharide is reduced only to the alditol obtained from the reduction of
   1. D-talose?
   2. D-galactose?

b. What monosaccharide is reduced to two alditols, one of which is the alditol obtained from the reduction of
   1. D-talose?
   2. D-mannose?

Oxidation

Aldoses can be distinguished from ketoses by observing what happens to the color of an aqueous solution of bromine when it is added to the sugar. Br_2 is a mild oxidizing agent and easily oxidizes the aldehyde group, but it cannot oxidize ketones or alcohols. Consequently, if a small amount of an aqueous solution of Br_2 is added to an unknown monosaccharide, the reddish-brown color of Br_2 will disappear if the monosaccharide is an aldose, but will persist if the monosaccharide is a ketose. The product of the oxidation reaction is an **aldonic acid**.

HC=O / H—OH / HO—H / H—OH / H—OH / CH_2OH (D-glucose) + Br_2 (red) $\xrightarrow{H_2O}$ COOH / H—OH / HO—H / H—OH / H—OH / CH_2OH (D-gluconic acid, an aldonic acid) + 2 Br^- (colorless)

Both aldoses and ketoses are oxidized to aldonic acids by Tollens reagent (Ag^+, NH_3, HO^-), so that reagent cannot be used to distinguish between aldoses and ketoses. Recall from Section 20.3, however, that Tollens reagent oxidizes aldehydes but not ketones. Why, then, are ketoses oxidized by Tollens reagent, while ketones are not? Ketoses are oxidized because the reaction is carried out under basic conditions, and in a basic solution, ketoses are converted into aldoses by enolization (Section 19.2). For example, the ketose D-fructose is in equilibrium with its enol. However, the enol of D-fructose is also the enol of D-glucose, as well as the enol of D-mannose. Therefore, when the enol reketonizes, all three carbonyl compounds are formed.

CH_2OH / C=O / HO—H / H—OH / H—OH / CH_2OH (D-fructose, a ketose) $\underset{HO^-, H_2O}{\overset{HO^-, H_2O}{\rightleftharpoons}}$ HC—OH ‖ C—OH / HO—H / H—OH / H—OH / CH_2OH (enol of D-fructose, enol of D-glucose, enol of D-mannose) $\underset{HO^-, H_2O}{\overset{HO^-, H_2O}{\rightleftharpoons}}$ HC=O / H—OH / HO—H / H—OH / H—OH / CH_2OH (D-glucose, an aldose) + HC=O / HO—H / HO—H / H—OH / H—OH / CH_2OH (D-mannose, an aldose)

PROBLEM 11

Write the mechanism for the base-catalyzed conversion of D-fructose into D-glucose and D-mannose.

PROBLEM 12◆

When D-tagatose is added to a basic aqueous solution, an equilibrium mixture of three monosaccharides is obtained. What are these monosaccharides?

If an oxidizing agent stronger than those discussed previously is used (such as HNO_3), one or more of the alcohol groups can be oxidized in addition to the aldehyde group. A primary alcohol is the one most easily oxidized. The product that is obtained when both the aldehyde and the primary alcohol groups of an aldose are oxidized is called an **aldaric acid**. (In an ald**on**ic acid, **one** end is oxidized. In an ald**ar**ic acid, both ends **ar**e oxidized.)

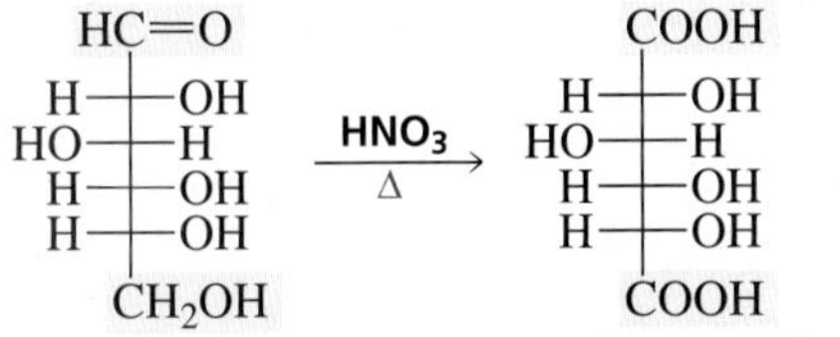

```
    HC=O                        COOH
 H ─┼─ OH                    H ─┼─ OH
HO ─┼─ H       HNO3         HO ─┼─ H
 H ─┼─ OH     ──────→        H ─┼─ OH
 H ─┼─ OH        Δ           H ─┼─ OH
   CH2OH                        COOH
 D-glucose                 D-glucaric acid
                           an aldaric acid
```

PROBLEM 13◆

a. Name an aldohexose other than D-glucose that is oxidized to D-glucaric acid by nitric acid.
b. What is another name for D-glucaric acid?
c. Name another pair of aldohexoses that are oxidized to identical aldaric acids.

22.6 Osazone Formation

The tendency of monosaccharides to form syrups that do not crystallize made the purification and isolation of monosaccharides difficult. Emil Fischer found that when phenylhydrazine is added to an aldose or a ketose, a yellow crystalline solid that is insoluble in water is formed. He called this derivative an **osazone** ("ose" for sugar; "azone" for hydrazone). Osazones are easily isolated and purified and were once used extensively to identify monosaccharides.

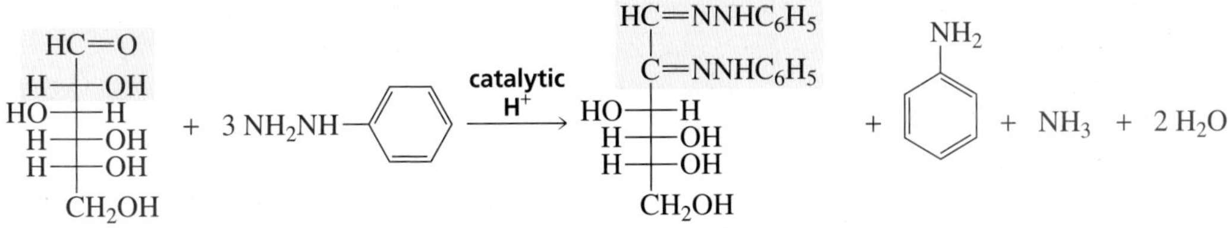

```
    HC=O                                              HC=NNHC6H5                NH2
 H ─┼─ OH                                             |                          |
HO ─┼─ H                             catalytic         C=NNHC6H5
 H ─┼─ OH   + 3 NH2NH─C6H5            H+          HO ─┼─ H          +   C6H5      + NH3 + 2 H2O
 H ─┼─ OH                          ──────→         H ─┼─ OH
   CH2OH                                           H ─┼─ OH
 D-glucose                                           CH2OH
                                              the osazone of D-glucose
```

Aldehydes and ketones react with one equivalent of phenylhydrazine, forming phenylhydrazones (Section 18.6). Aldoses and ketoses, in contrast, react with three equivalents of phenylhydrazine, forming osazones. One equivalent functions as an oxidizing agent and is reduced to aniline and ammonia. Two equivalents form imines with carbonyl groups. The reaction stops at this point, regardless of how much phenylhydrazine is present. (Recall that the pH at which imine formation is carried out must be carefully controlled; Section 18.6.)

C-2 epimers form identical osazones.

Because the configuration of the number-2 carbon is lost during osazone formation, C-2 epimers form identical osazones. For example, D-idose and D-gulose, which are C-2 epimers, both form the same osazone.

D-idose: HC=O; HO–H; H–OH; HO–H; H–OH; CH_2OH —(3 NH_2NH–C_6H_5, catalytic H^+)→ the osazone of D-idose and of D-gulose: $HC{=}NNHC_6H_5$; $C{=}NNHC_6H_5$; H–OH; HO–H; H–OH; CH_2OH ←(3 NH_2NH–C_6H_5, catalytic H^+)— D-gulose: HC=O; H–OH; H–OH; HO–H; H–OH; CH_2OH

The number-1 and number-2 carbons of ketoses react with phenylhydrazine, too. Consequently, D-fructose, D-glucose, and D-mannose all form the same osazone.

D-glucose: HC=O; H–OH; HO–H; H–OH; H–OH; CH_2OH —(3 H_2NNH–C_6H_5, catalytic H^+)→ the osazone of D-glucose and/of D-fructose: $HC{=}NNHC_6H_5$; $C{=}NNHC_6H_5$; HO–H; H–OH; H–OH; CH_2OH ←(3 H_2NNH–C_6H_5, catalytic H^+)— D-fructose: CH_2OH; C=O; HO–H; H–OH; H–OH; CH_2OH

PROBLEM 14◆

Name a ketose and another aldose that form the same osazone as

a. D-ribose
b. D-altrose
c. L-idose
d. D-galactose

PROBLEM 15◆

What monosaccharides form the same osazone as D-sorbose?

MEASURING THE BLOOD GLUCOSE LEVELS OF DIABETICS

Glucose reacts with an NH_2 group of hemoglobin to form an imine (Section 18.6) that subsequently undergoes an irreversible rearrangement to a more stable α-aminoketone known as hemoglobin-A_{Ic}.

D-glucose: HC=O; H–OH; HO–H; H–OH; H–OH; CH_2OH —(NH_2–hemoglobin, catalytic H^+)→ HC=N–hemoglobin; H–OH; HO–H; H–OH; H–OH; CH_2OH —(rearrangement)→ hemoglobin-A_{Ic}: CH_2NH–hemoglobin; C=O; HO–H; H–OH; H–OH; CH_2OH

Diabetes results when the body does not produce sufficient insulin or when the insulin it produces does not properly stimulate its target cells. Because insulin is the hormone that maintains the proper level of glucose in the blood, diabetics have increased blood glucose levels. The amount of hemoglobin-A_{Ic} formed is proportional to the concentration of glucose in the blood, so diabetics have a higher concentration of hemoglobin-A_{Ic} than nondiabetics. Thus, measuring the hemoglobin-A_{Ic} level is a way to determine whether the blood glucose level of a diabetic is being controlled.

Cataracts, a common complication in diabetics, are caused by the reaction of glucose with the NH_2 group of proteins in the lens of the eye. It is thought that the arterial rigidity common in old age may be attributable to a similar reaction of glucose with the NH_2 group of proteins.

22.7 Chain Elongation: The Kiliani–Fischer Synthesis

The carbon chain of an aldose can be increased by one carbon in a **Kiliani–Fischer synthesis**. In other words, tetroses can be converted into pentoses, and pentoses can be converted into hexoses.

In the first step of the synthesis (the Kiliani portion), the aldose is treated with sodium cyanide and HCl (Section 18.4). Addition of cyanide ion to the carbonyl group creates a new asymmetric carbon. Consequently, two cyanohydrins that differ only in configuration at C-2 are formed. The configurations of the other asymmetric carbons do not change, because no bond to any of the asymmetric carbons is broken during the course of the reaction (Section 5.12). Kiliani went on to hydrolyze the cyanohydrins to aldonic acids (Section 17.18), and Fischer had previously developed a method to convert aldonic acids to aldoses. This reaction sequence was used for many years, but the method currently employed to convert the cyanohydrins to aldoses was developed by Serianni and Barker in 1979; it is referred to as the modified Kiliani–Fischer synthesis. Serianni and Barker reduced the cyanohydrins to imines, using a partially deactivated palladium (on barium sulfate) catalyst so that the imines would not be further reduced to amines. The imines could then be hydrolyzed to aldoses (Section 18.6).

Heinrich Kiliani (1855–1945) *was born in Germany. He received a Ph.D. from the University of Munich, studying under Professor Emil Erlenmeyer. Kiliani became a professor of chemistry at the University of Freiburg.*

the modified Kiliani–Fischer synthesis

D-erythrose (HC=O; H–OH; H–OH; CH_2OH) + $^-C{\equiv}N$ →(HCl)

- C≡N; H–OH; H–OH; H–OH; CH_2OH →(H_2, $Pd/BaSO_4$) HC=NH; H–OH; H–OH; H–OH; CH_2OH →(HCl, H_2O) HC=O; H–OH; H–OH; H–OH; CH_2OH (D-ribose) + $^+NH_4$
- C≡N; HO–H; H–OH; H–OH; CH_2OH →(H_2, $Pd/BaSO_4$) HC=NH; HO–H; H–OH; H–OH; CH_2OH →(HCl, H_2O) HC=O; HO–H; H–OH; H–OH; CH_2OH (D-arabinose) + $^+NH_4$

Notice that the synthesis leads to a pair of C-2 epimers because the first step of the reaction converts the carbonyl carbon in the starting material to an asymmetric carbon. Therefore, the OH bonded to C-2 in the product can be on the right or on the left in the Fischer projection. The two epimers are not obtained in equal amounts, however, because the first step of the reaction produces a pair of diastereomers and diastereomers are generally formed in unequal amounts (Section 5.19).

The Kiliani–Fischer synthesis leads to a pair of C-2 epimers.

PROBLEM 16◆

What monosaccharides would be formed in a Kiliani–Fischer synthesis starting with one of these?

a. D-xylose b. L-threose

22.8 Chain Shortening: The Ruff Degradation

The *Ruff degradation* is the opposite of the Kiliani–Fischer synthesis. Thus, the **Ruff degradation** shortens an aldose chain by one carbon: Hexoses are converted into pentoses, and pentoses are converted into tetroses. In the Ruff degradation, the calcium salt of an aldonic acid is oxidized with hydrogen peroxide. Ferric ion catalyzes the oxidation reaction, which cleaves the bond between C-1 and C-2, forming CO_2 and an aldehyde. It is known that the reaction involves the formation of radicals, but the precise mechanism is not well understood.

Otto Ruff (1871–1939) *was born in Germany. He received a Ph.D. from the University of Berlin. He was a professor of chemistry at the University of Danzig and later at the University of Breslau.*

the Ruff degradation

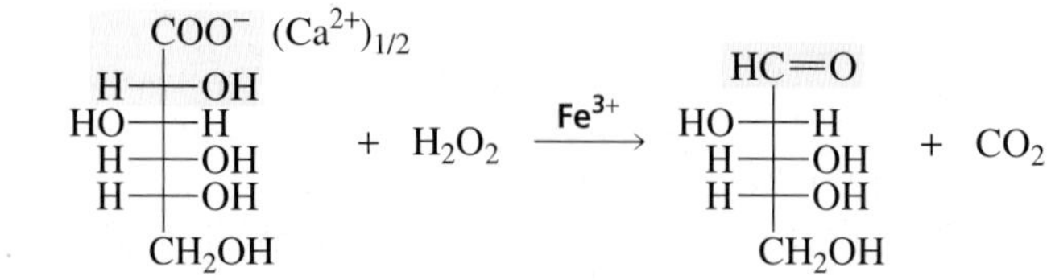

COO^- $(Ca^{2+})_{1/2}$
H—OH
HO—H
H—OH
H—OH
CH_2OH
calcium D-gluconate

$+ \; H_2O_2 \xrightarrow{Fe^{3+}}$

HC=O
HO—H
H—OH
H—OH
CH_2OH
D-arabinose

$+ \; CO_2$

The calcium salt of the aldonic acid necessary for the Ruff degradation is easily obtained by oxidizing an aldose with an aqueous solution of bromine and then adding calcium hydroxide to the reaction mixture.

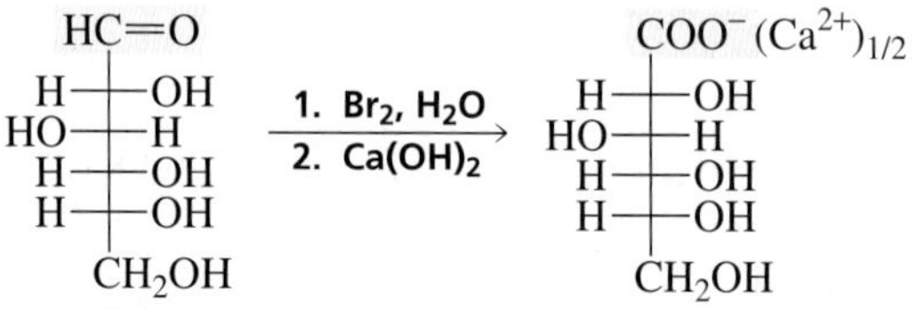

HC=O
H—OH
HO—H
H—OH
H—OH
CH_2OH
D-glucose

$\xrightarrow[\text{2. } Ca(OH)_2]{\text{1. } Br_2, \, H_2O}$

COO^- $(Ca^{2+})_{1/2}$
H—OH
HO—H
H—OH
H—OH
CH_2OH
calcium D-gluconate

PROBLEM 17◆

What two monosaccharides can be degraded to

a. D-arabinose?
b. D-glyceraldehyde?
c. L-ribose?

22.9 Stereochemistry of Glucose: The Fischer Proof

In 1891, Emil Fischer determined the stereochemistry of glucose using one of the most brilliant examples of reasoning in the history of chemistry. He chose (+)-glucose for his study because it was the most common monosaccharide found in nature.

Fischer knew that (+)-glucose was an aldohexose, but 16 different structures can be written for an aldohexose. Which of them represents the structure of (+)-glucose? The 16 stereoisomers of the aldohexoses are actually eight pairs of enantiomers, so if you know the structures of one set, you automatically know the structures of the other set. Therefore, Fischer needed to consider only one set of eight. He considered the eight stereoisomers that had the C-5 OH group on the right in the Fischer projection (the stereoisomers shown below that we now call the D-sugars). One of these would be one enantiomer of glucose, and its mirror image would be the other enantiomer. It was not possible to determine whether (+)-glucose was D-glucose or L-glucose until 1951 (Section 5.13). Fischer used the following information to determine glucose's stereochemistry—that is, to determine the configuration of each of its asymmetric carbons.

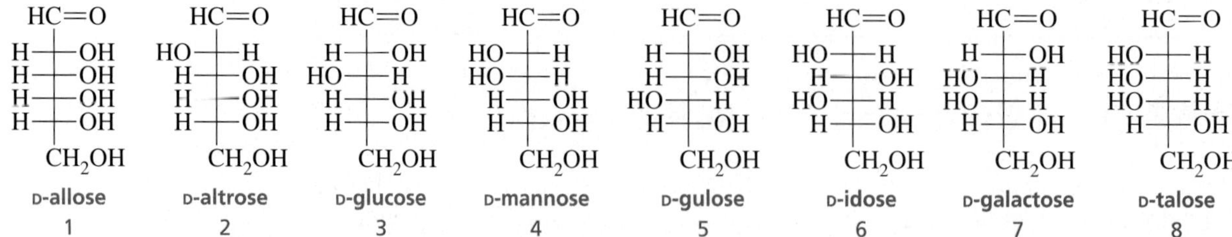

1. When the Kiliani–Fischer synthesis is performed on the sugar known as (−)-arabinose, the two sugars known as (+)-glucose and (+)-mannose are obtained. This means that (+)-glucose and (+)-mannose are C-2 epimers; in other words, they have the same configuration at C-3, C-4, and C-5. Consequently, (+)-glucose and (+)-mannose have to be one of the following pairs: sugars 1 and 2, 3 and 4, 5 and 6, or 7 and 8.

2. (+)-Glucose and (+)-mannose are both oxidized by nitric acid to optically active aldaric acids. The aldaric acids of sugars 1 and 7 would not be optically active because they have a plane of symmetry. (A compound with a plane of symmetry is achiral—it has a superimposable mirror image; Section 5.10.) Excluding sugars 1 and 7 means that (+)-glucose and (+)-mannose must be sugars 3 and 4 or 5 and 6.

3. Because (+)-glucose and (+)-mannose are the products obtained when the Kiliani–Fischer synthesis is carried out on (−)-arabinose, there are only two possibilities for the structure of (−)-arabinose. That is, if (+)-glucose and (+)-mannose are sugars 3 and 4, then (−)-arabinose has the structure shown below on the left; on the other hand, if (+)-glucose and (+)-mannose are sugars 5 and 6, then (−)-arabinose has the structure shown on the right:

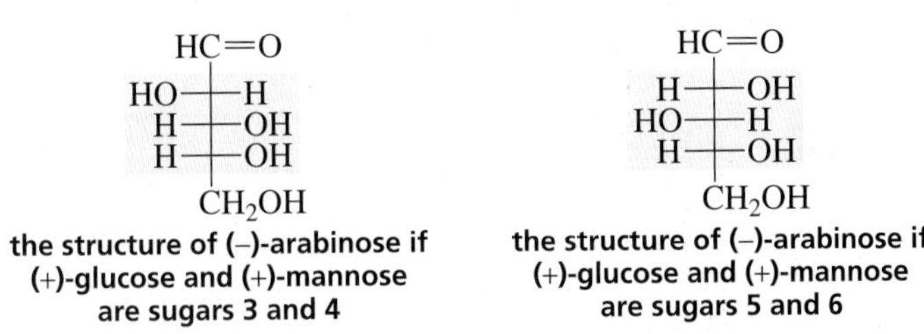

When (−)-arabinose is oxidized with nitric acid, the aldaric acid that is obtained is optically active. This means that the aldaric acid does *not* have a plane of symmetry. Therefore, (−)-arabinose must have the structure shown on the left because the aldaric acid of the sugar on the right has a plane of symmetry. Thus, (+)-glucose and (+)-mannose are represented by sugars 3 and 4.

4. The last step in the Fischer proof was to determine whether (+)-glucose is sugar 3 or sugar 4. To answer this question, Fischer had to develop a chemical method that would interchange the aldehyde and primary alcohol groups of an aldohexose. When he chemically interchanged the aldehyde and primary alcohol groups of the sugar known as (+)-glucose, he obtained an aldohexose that was different from (+)-glucose. When he chemically interchanged the aldehyde and primary alcohol groups of (+)-mannose, he still had (+)-mannose. Therefore, he concluded that (+)-glucose is sugar 3 because reversing the aldehyde and alcohol groups of sugar 3 leads to a different sugar (L-gulose).

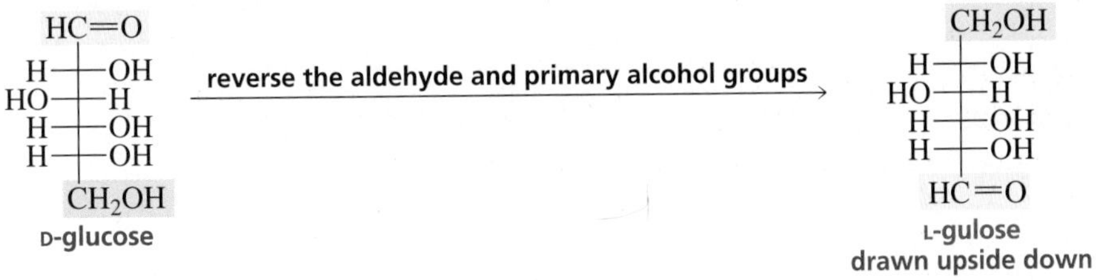

If (+)-glucose is sugar 3, (+)-mannose must be sugar 4. As predicted, when the aldehyde and primary alcohol groups of sugar 4 are reversed, the same sugar is obtained.

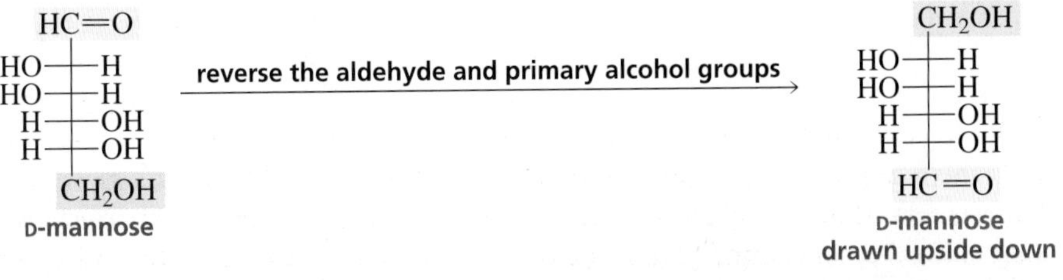

Using similar reasoning, Fischer went on to determine the stereochemistry of 14 of the 16 aldohexoses. He received the Nobel Prize in chemistry in 1902 for this achievement. His original guess that (+)-glucose is a D-sugar was later shown to be correct, so all of his structures are correct (Section 5.13). If he had been wrong and (+)-glucose had been an L-sugar, his contribution to the stereochemistry of aldoses would still have had the same significance, but all his stereochemical assignments would have been reversed.

GLUCOSE/DEXTROSE

André Dumas first used the term "glucose" in 1838 to refer to the sweet compound that comes from honey and grapes. Later, Kekulé (Section 7.1) decided that it should be called dextrose because it was dextrorotatory. When Fischer studied the sugar, he called it glucose, and chemists have called it glucose ever since, although dextrose is often found on food labels.

Jean-Baptiste-André Dumas (1800–1884) *was born in France. Apprenticed to an apothecary, he left to study chemistry in Switzerland. He became a professor of chemistry at the University of Paris and at the Collège de France. He was the first French chemist to teach laboratory courses. In 1848, he left science for a political career. He became a senator, master of the French mint, and mayor of Paris.*

PROBLEM 18 SOLVED

Aldohexoses A and B form the same osazone. A is oxidized by nitric acid to an optically active aldaric acid, and B is oxidized to an optically inactive aldaric acid. Ruff degradation of either A or B forms aldopentose C, which is oxidized by nitric acid to an optically active aldaric acid. Ruff degradation of C forms D, which is oxidized by nitric acid to an optically active aldaric acid. Ruff degradation of D forms (+)-glyceraldehyde. Identify A, B, C, and D.

SOLUTION This is the kind of problem that should be solved by working backwards. The bottom-most asymmetric carbon in D must have the OH group on the right because D is degraded to (+)-glyceraldehyde. D must be D-threose, since D is oxidized to an optically active aldaric acid. The two bottom-most asymmetric carbons in C and D have the same configuration because C is degraded to D. C must be D-lyxose, since it is oxidized to an optically active aldaric acid. A and B, therefore, must be D-galactose and D-talose. Because A is oxidized to an optically active aldaric acid, it must be D-talose and B must be D-galactose.

PROBLEM 19◆

Identify A, B, C, and D in the preceding problem if D is oxidized to an optically *inactive* aldaric acid, A, B, and C are oxidized to optically active aldaric acids, and interchanging the aldehyde and alcohol groups of A leads to a different sugar.

22.10 Cyclic Structure of Monosaccharides: Hemiacetal Formation

D-Glucose exists in three different forms: the open-chain form of D-glucose that we have been discussing and two cyclic forms—α-D-glucose and β-D-glucose. We know that the two cyclic forms are different, because they have different physical properties: α-D-Glucose melts at 146 °C, whereas β-D-glucose melts at 150 °C; α-D-glucose has a specific rotation of +112.2°, whereas β-D-glucose has a specific rotation of +18.7°.

How can D-glucose exist in a cyclic form? In Section 18.7, we saw that an aldehyde reacts with an equivalent of an alcohol to form a hemiacetal. A monosaccharide such as D-glucose has an aldehyde group and several alcohol groups. The alcohol group bonded to C-5 of D-glucose reacts intramolecularly with the aldehyde group, forming a six-membered-ring hemiacetal. Why are there two different cyclic forms? Two different hemiacetals are formed because the carbonyl carbon of the open-chain sugar becomes a new asymmetric carbon in the hemiacetal. If the OH group bonded to the new asymmetric carbon is on the right, the hemiacetal is α-D-glucose; if the OH group is on the left, the hemiacetal is β-D-glucose. The mechanism for cyclic hemiacetal formation is the same as the mechanism for hemiacetal formation between individual aldehyde and alcohol molecules (Section 18.7).

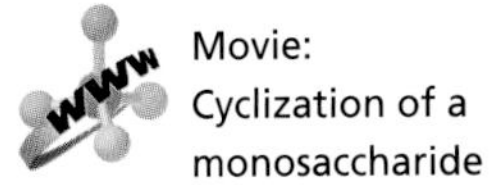

Movie: Cyclization of a monosaccharide

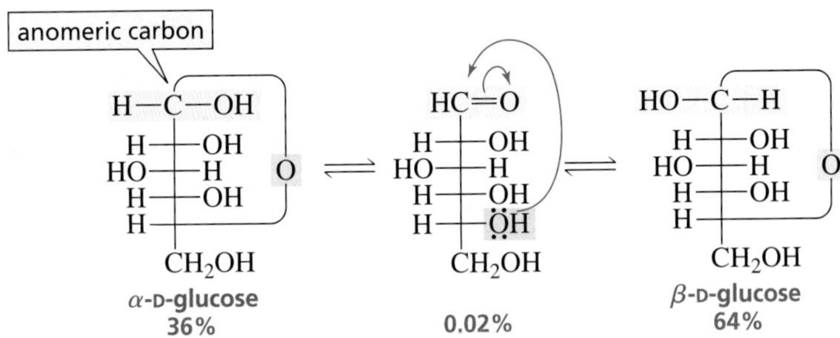

α-D-Glucose and β-D-glucose are called anomers. **Anomers** are two sugars that differ in configuration only at the carbon that was the carbonyl carbon in the open-chain form. This carbon is called the **anomeric carbon**. *Ano* is Greek for "upper"; thus, anomers differ in configuration at the upper-most asymmetric carbon. The anomeric carbon is the only carbon in the molecule that is bonded to two oxygens. The prefixes α- and β- denote the configuration about the anomeric carbon. Anomers, like epimers, differ in configuration at only one carbon atom. Anomers are a particular kind of diastereomers.

In an aqueous solution, the open-chain compound is in equilibrium with the two cyclic hemiacetals. Formation of the cyclic hemiacetals proceeds nearly to completion (unlike formation of acyclic hemiacetals), so very little glucose exists in the open-chain form (about 0.02%). At equilibrium, there is almost twice as much β-D-glucose (64%) as α-D-glucose (36%). The sugar still undergoes the reactions discussed in previous sections (oxidation, reduction, and osazone formation) because the reagents react with the small amount of open-chain aldehyde that is present. As the aldehyde reacts, the equilibrium shifts to form more open-chain aldehyde, which can then undergo reaction. Eventually, all the glucose molecules react by way of the open-chain aldehyde.

When crystals of pure α-D-glucose are dissolved in water, the specific rotation gradually changes from +112.2° to +52.7°. When crystals of pure β-D-glucose are dissolved in water, the specific rotation gradually changes from +18.7° to +52.7°. This change in rotation occurs because, in water, the hemiacetal opens to form the aldehyde and, when the aldehyde recyclizes, both α-D-glucose and β-D-glucose can be formed. Eventually, the three forms of glucose reach equilibrium concentrations. The specific rotation of the equilibrium mixture is +52.7°—this is why the same specific rotation results whether the crystals originally dissolved in water are α-D-glucose or β-D-glucose. A slow change in optical rotation to an equilibrium value is known as **mutarotation**.

If an aldose can form a five- or a six-membered-ring, it will exist predominantly as a cyclic hemiacetal in solution. Whether a five- or a six-membered ring is formed depends on their relative stabilities. Six-membered-ring sugars are called **pyranoses**, and five-membered-ring sugars are called **furanoses**. These names come from *pyran* and *furan*, the names of the five- and six-membered-ring cyclic ethers shown in the margin. Consequently, α-D-glucose is also called α-D-glucopyranose. The prefix α- indicates the configuration about the anomeric carbon, and "pyranose" indicates that the sugar exists as a six-membered-ring cyclic hemiacetal.

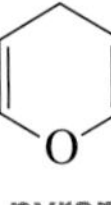

pyran

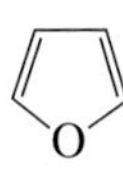

furan

Fischer projections are not the best way to show the structure of a cyclic sugar, because of how the C—O—C bond is represented. A somewhat more satisfactory representation is given by a **Haworth projection**.

In a Haworth projection of a D-pyranose, the six-membered ring is represented as flat and is viewed edge on. The ring oxygen is always placed in the back right-hand corner of the ring, with the anomeric carbon (C-1) on the right-hand side and the primary alcohol group drawn *up* from the back left-hand corner (C-5). Groups on the *right* in a Fischer projection are *down* in a Haworth projection, whereas groups on the *left* in a Fischer projection are *up* in a Haworth projection.

Groups on the *right* in a Fischer projection are *down* in a Haworth projection.

Groups on the *left* in a Fischer projection are *up* in a Haworth projection.

Sir Walter Norman Haworth (1883–1950) *was born in England. He received a Ph.D. in Germany from the University of Göttingen and later was a professor of chemistry at the Universities of Durham and Birmingham in Great Britain. He was the first to synthesize vitamin C and was the one who named it ascorbic acid. During World War II, he worked on the atomic bomb project. He received the Nobel Prize in Chemistry in 1937 and was knighted in 1947.*

The Haworth projection of a D-furanose is viewed edge on, with the ring oxygen away from the viewer. The anomeric carbon is on the right-hand side of the molecule, and the primary alcohol group is drawn *up* from the back left-hand corner.

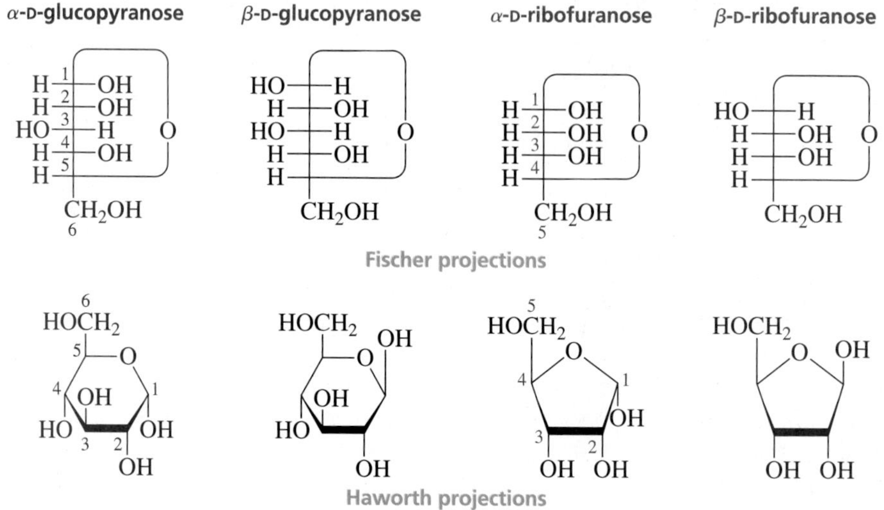

Ketoses also exist predominantly in cyclic forms. D-Fructose forms a five-membered-ring-hemiketal as a consequence of the C-5 OH group reacting with the ketone carbonyl group (Section 18.7). If the new asymmetric carbon has the OH group on the right in a Fischer projection, the compound is α-D-fructose; if the OH group is on the left, the compound is β-D-fructose. These sugars can also be called α-D-fructofuranose and β-D-fructofuranose. Notice that in fructose the anomeric carbon is C-2, not C-1 as in aldoses. D-Fructose can also form a six-membered ring by using the C-6 OH group. The pyranose form predominates in the monosaccharide, whereas the furanose form predominates when the sugar is part of a disaccharide. (See the structure of sucrose in Section 22.17.)

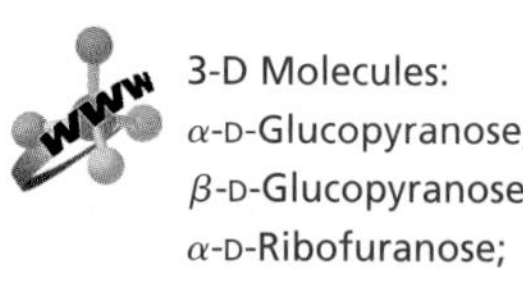

3-D Molecules: α-D-Glucopyranose; β-D-Glucopyranose; α-D-Ribofuranose; β-D-Ribofuranose

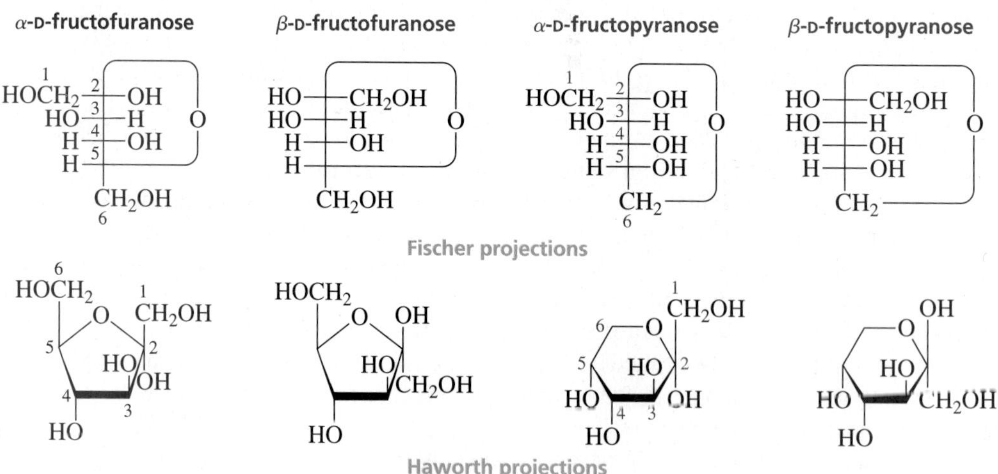

Haworth projections are useful because they allow us to see easily whether the OH groups on the ring are cis or trans to each other. Because five-membered rings are close to planar, furanoses are well represented by Haworth projections. However, Haworth projections are structurally misleading for pyranoses because a six-membered ring is not flat but exists preferentially in a chair conformation (Section 2.12).

PROBLEM 20

4-Hydroxy- and 5-hydroxyaldehydes exist primarily in the cyclic hemiacetal form. Give the structure of the cyclic hemiacetal formed by each of the following:

a. 4-hydroxybutanal
b. 5-hydroxypentanal
c. 4-hydroxypentanal
d. 4-hydroxyheptanal

PROBLEM 21

Draw the following sugars using Haworth projections:

a. β-D-galactopyranose
b. α-D-tagatopyranose
c. α-L-glucopyranose

PROBLEM 22

D-Glucose most often exists as a pyranose, but it can also exist as a furanose. Draw the Haworth projection of α-D-glucofuranose.

22.11 Stability of Glucose

Drawing D-glucose in its chair conformation shows why it is the most common D-aldohexose in nature. To convert the Haworth projection of D-glucose into a chair conformation, start by drawing the chair so that the back is on the left and the footrest is on the right. Then place the ring oxygen at the back right-hand corner and the primary alcohol group in the equatorial position. (It would be helpful here to build a molecular model.) The primary alcohol group is the largest of all the substituents, and large substituents are more stable in the equatorial position because there is less steric strain in that position (Section 2.13). Because the OH group bonded to C-4 is trans to the primary alcohol group (this is easily seen in the Haworth projection), the C-4 OH group is also in the equatorial position. (Recall from Section 2.14 that 1,2-diequatorial substituents are trans to one another.) The C-3 OH group is trans to the C-4 OH group, so the C-3 OH group is also in the equatorial position. As you move around the ring, you will find that all the OH substituents in β-D-glucose are in equatorial positions. The axial positions are all occupied by hydrogens, which require little space and therefore experience little steric strain. No other aldohexose exists in such a strain-free conformation. This means that β-D-glucose is the most stable of all the aldohexoses, so it is not surprising that it is the most prevalent aldohexose in nature.

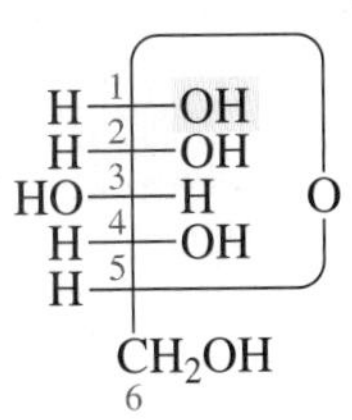

Fischer projection

Haworth projection

chair conformation

α-D-glucose

The α-position is to the right in a Fischer projection, down in a Haworth projection, and axial in a chair conformation.

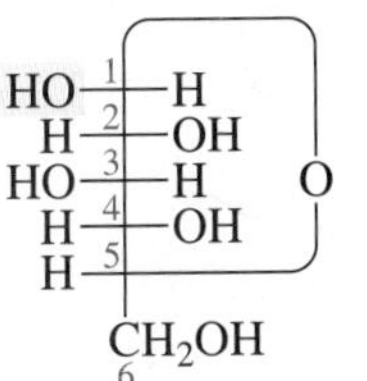

Fischer projection

Haworth projection

chair conformation

β-D-glucose

The β-position is to the left in a Fischer projection, up in a Haworth projection, and equatorial in a chair conformation.

Why is there more β-D-glucose than α-D-glucose in an aqueous solution at equilibrium? The OH group bonded to the anomeric carbon is in the equatorial position in β-D-glucose, whereas it is in the axial position in α-D-glucose. Therefore, β-D-glucose is more stable than α-D-glucose, so β-D-glucose predominates at equilibrium in an aqueous solution.

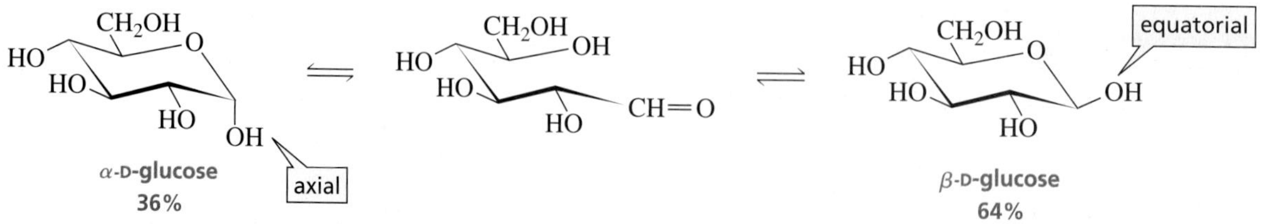

CH_2OH HO HO HO OH axial ⇌ CH_2OH OH HO HO HO CH=O ⇌ CH_2OH HO HO HO OH equatorial

α-D-glucose 36%   β-D-glucose 64%

3-D Molecules: α-D-Galactose; β-D-Gulose; β-L-Gulose

If you remember that all the OH groups in β-D-glucose are in equatorial positions, it is easy to draw the chair conformation of any other pyranose. For example, if you want to draw α-D-galactose, you would put all the OH groups in equatorial positions, except the OH groups at C-4 (because galactose is a C-4 epimer of glucose) and at C-1 (because it is the α-anomer). You would put these two OH groups in axial positions.

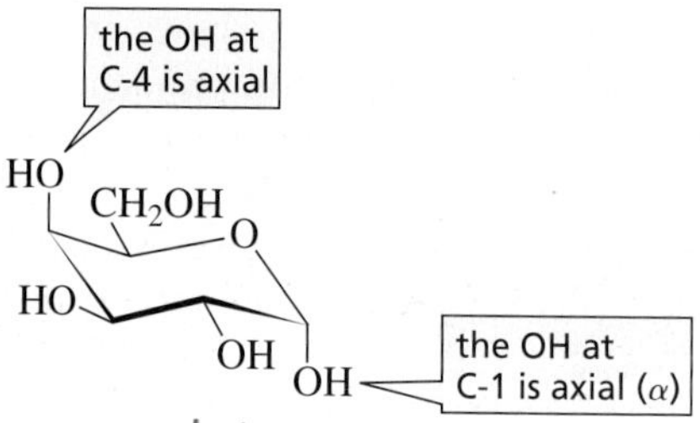

the OH at C-4 is axial
HO CH_2OH HO OH OH
the OH at C-1 is axial (α)

α-D-galactose

To draw an L-pyranose, draw the D-pyranose first, and then draw its mirror image. For example, to draw β-L-gulose, first draw β-D-gulose. (Gulose differs from glucose at C-3 and C-4, so the OH groups at these positions are in axial positions.) Then draw the mirror image of β-D-gulose to get β-L-gulose.

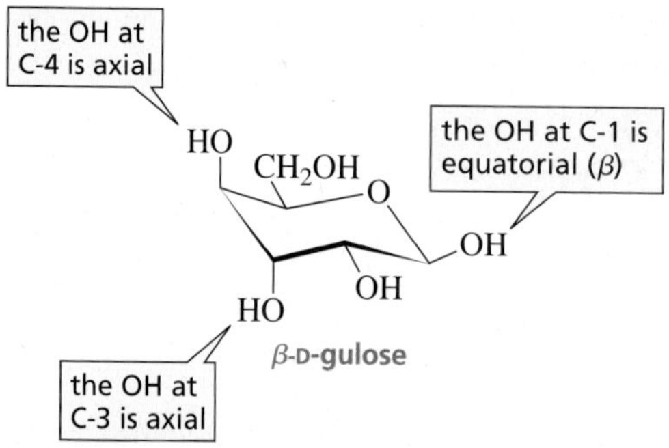

the OH at C-4 is axial
HO CH_2OH OH OH HO
the OH at C-1 is equatorial (β)
the OH at C-3 is axial

β-D-gulose

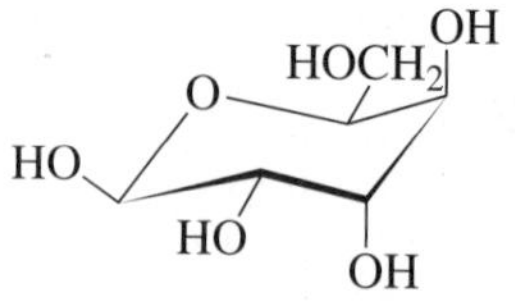

OH $HOCH_2$ HO HO OH

β-L-gulose

PROBLEM 23◆

Which OH groups are in the axial position in

a. β-D-mannopyranose?

b. β-D-idopyranose?

c. α-D-allopyranose?

22.12 Acylation and Alkylation of Monosaccharides

The OH groups of monosaccharides show the chemistry typical of alcohols. For example, they react with acetyl chloride or acetic anhydride to form esters (Sections 17.8 and 17.9).

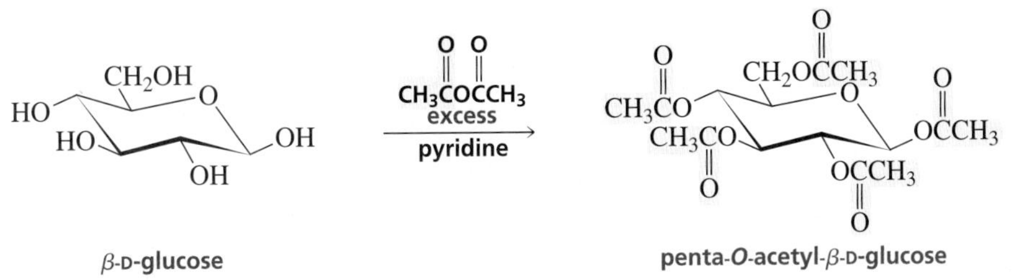

β-D-glucose $\xrightarrow[\text{pyridine}]{\text{CH}_3\text{COCCH}_3 \text{ excess}}$ penta-*O*-acetyl-β-D-glucose

The OH groups also react with methyl iodide/silver oxide to form ethers (Section 10.4). The OH group is a relatively poor nucleophile, so silver oxide is used to increase the leaving tendency of the iodide ion in the S_N2 reaction.

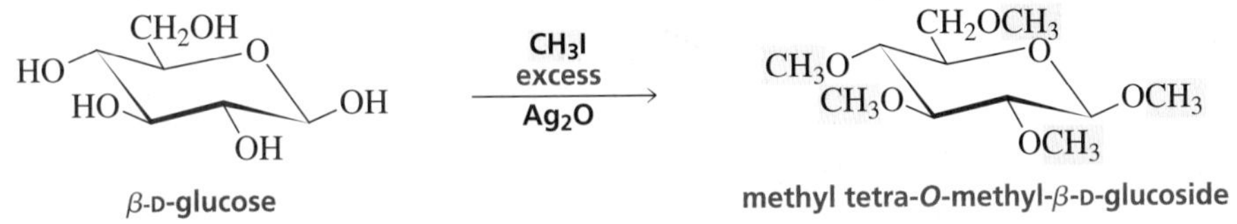

β-D-glucose $\xrightarrow[\text{Ag}_2\text{O}]{\text{CH}_3\text{I excess}}$ methyl tetra-*O*-methyl-β-D-glucoside

22.13 Formation of Glycosides

In Section 18.7, we saw that after an aldehyde reacts with an equivalent of an alcohol to form a hemiacetal, the hemiacetal reacts with a second equivalent of alcohol to form an acetal. Similarly, the cyclic hemiacetal (or hemiketal) formed by a monosaccharide can react with an alcohol to form an acetal (or ketal). The acetal (or ketal) of a sugar is called a **glycoside**, and the bond between the anomeric carbon and the alkoxy oxygen is called a **glycosidic bond**. Glycosides are named by replacing the "e" ending of the sugar's name with "ide." Thus, a glycoside of glucose is a glucoside, a glycoside of galactose is a galactoside, etc. If the pyranose or furanose name is used, the acetal is called a **pyranoside** or a **furanoside**.

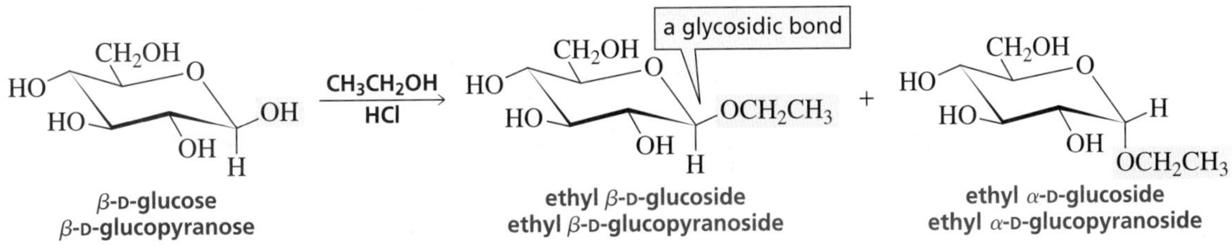

β-D-glucose (β-D-glucopyranose) $\xrightarrow[\text{HCl}]{\text{CH}_3\text{CH}_2\text{OH}}$ ethyl β-D-glucoside (ethyl β-D-glucopyranoside) + ethyl α-D-glucoside (ethyl α-D-glucopyranoside)

a glycosidic bond

Notice that the reaction of a single anomer with an alcohol leads to the formation of both the α- and β-glycosides. The mechanism of the reaction shows why both glycosides are formed. The OH group bonded to the anomeric carbon becomes protonated in the acidic solution, and a lone pair on the ring oxygen helps expel a molecule of water. The anomeric carbon in the resulting oxocarbenium ion is sp^2 hybridized, so that part of the molecule is planar. (An **oxocarbenium ion** has a positive charge that is shared by a carbon and an oxygen.) When the alcohol comes in from the top of the plane, the β-glycoside is formed; when the alcohol comes in from the bottom of the plane, the α-glycoside is formed. Notice that the mechanism is the same as that shown for acetal formation in Section 18.7.

mechanism of glycoside formation

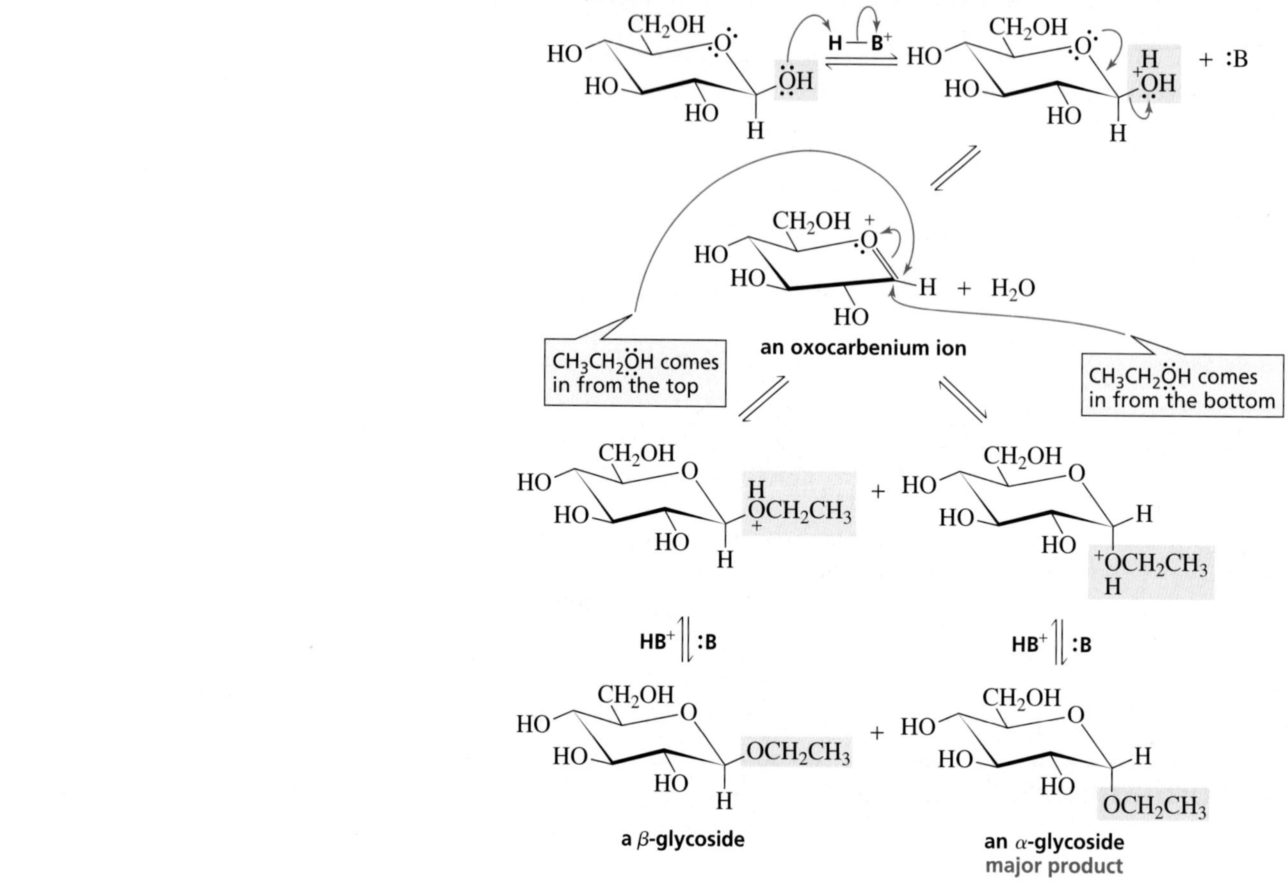

HO HO HO CH_2OH O ÖH H H—B⁺ HO HO HO CH_2OH O H ⁺OH H + :B

$CH_3CH_2\ddot{O}H$ comes in from the top

HO HO HO CH_2OH ⁺O H + H_2O

an oxocarbenium ion

$CH_3CH_2\ddot{O}H$ comes in from the bottom

HO HO HO CH_2OH O H ⁺OCH$_2$CH$_3$ H + HO HO HO CH_2OH O H ⁺OCH$_2$CH$_3$ H

HB^+ :B HB^+ :B

HO HO HO CH_2OH O OCH_2CH_3 H + HO HO HO CH_2OH O H OCH_2CH_3

a β-glycoside **an α-glycoside** **major product**

Surprisingly, D-glucose forms more of the α-glycoside than the β-glycoside. The reason for this is explained in the next section.

Similar to the reaction of a monosaccharide with an alcohol is the reaction of a monosaccharide with an amine in the presence of a trace amount of acid. The product of the reaction is an *N*-glycoside. An ***N*-glycoside** has a nitrogen in place of the oxygen at the glycosidic linkage. The subunits of DNA and RNA are β-*N*-glycosides (Section 27.1).

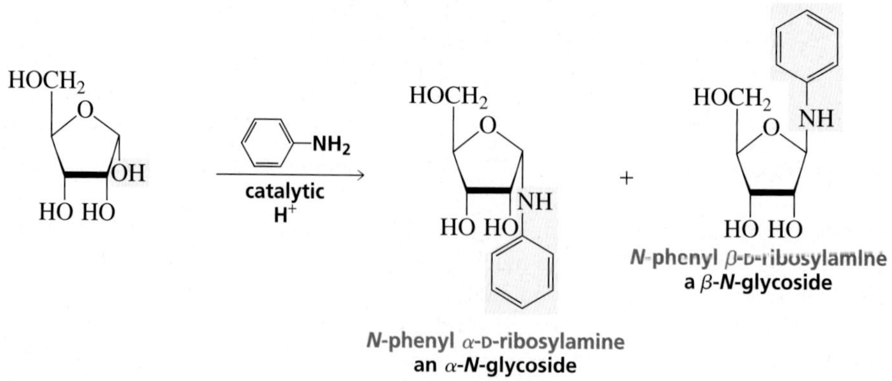

$HOCH_2$ O OH HO HO —NH$_2$ catalytic H^+ → $HOCH_2$ O NH HO HO + $HOCH_2$ O NH HO HO

N-phenyl α-D-ribosylamine
an α-N-glycoside

N-phenyl β-D-ribosylamine
a β-N-glycoside

PROBLEM 24◆

Why is only a trace amount of acid used in the formation of an *N*-glycoside?

22.14 The Anomeric Effect

We have seen that β-D-glucose is more stable than α-D-glucose because there is more room for a substituent in the equatorial position. The relative amounts of β-D-glucose and α-D-glucose are only 2 : 1, however, so the preference of the OH group for the equatorial position is surprisingly small (Section 22.10). Contrast this to the preference of the OH group for the equatorial position in cyclohexane, which is 5.4 : 1 (Table 2.10 on p. 99).

When glucose reacts with an alcohol to form a glucoside, the major product is the α-glucoside. Since acetal formation is reversible, the α-glucoside must be more stable than the β-glucoside. The preference of certain substituents bonded to the anomeric carbon for the axial position is called the **anomeric effect**.

What is responsible for the anomeric effect? One clue is that substituents that prefer the axial position have a lone pair on the atom (Z) bonded to the ring. The C—Z bond has a σ^* antibonding orbital. If one of the ring oxygen's lone pairs is in an orbital that is parallel to the σ^* antibonding orbital, the molecule can be stabilized by electron density from oxygen moving into the σ^* orbital. The orbital containing the axial lone pair of the ring oxygen can overlap the σ^* orbital only if the substituent is axial. If the substituent is equatorial, neither of the orbitals that contain a lone pair is aligned correctly for overlap. As a result of overlap between the lone pair and the σ^* orbital, the C—Z bond is longer and weaker than normal, and the C—O bond within the ring is shorter and stronger than normal.

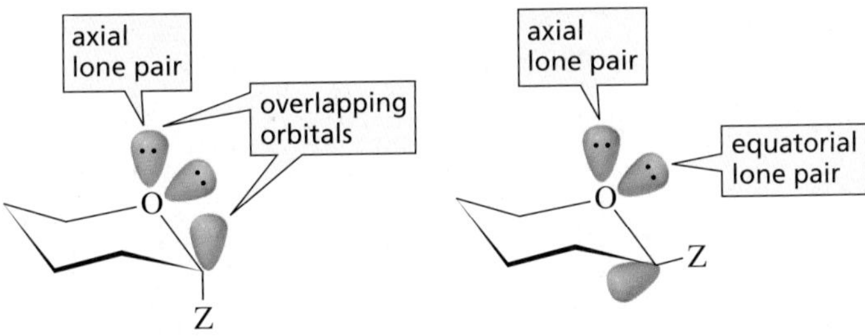

22.15 Reducing and Nonreducing Sugars

Because glycosides are acetals (or ketals), they are not in equilibrium with the open-chain aldehyde (or ketone) in neutral or basic aqueous solutions. Because they are not in equilibrium with a compound with a carbonyl group, they cannot be oxidized by reagents such as Ag^+ or Br_2. Glycosides, therefore, are nonreducing sugars—they cannot reduce Ag^+ or Br_2.

Hemiacetals (or hemiketals) are in equilibrium with the open-chain sugars in aqueous solution. So as long as a sugar has an aldehyde, a ketone, a hemiacetal, or a hemiketal group, it is able to reduce an oxidizing agent and therefore is classified as a **reducing sugar**. Without one of these groups, it is a **nonreducing sugar**.

A sugar with an aldehyde, a ketone, a hemiacetal, or a hemiketal group is a reducing sugar. A sugar without one of these groups is a nonreducing sugar.

PROBLEM 25 SOLVED

Name the following compounds, and indicate whether each is a reducing sugar or a nonreducing sugar:

a. 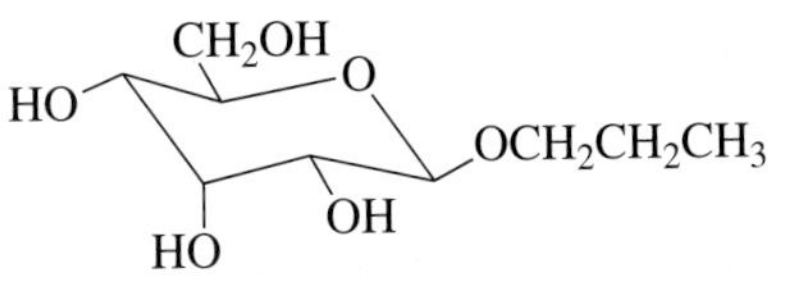 CH_2OH, HO, HO, OH, O, $OCH_2CH_2CH_3$

b. 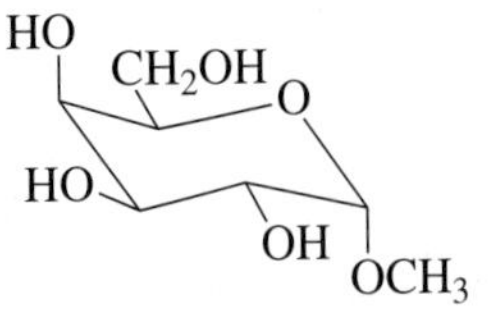 HO, CH_2OH, O, HO, OH, OCH_3

c. 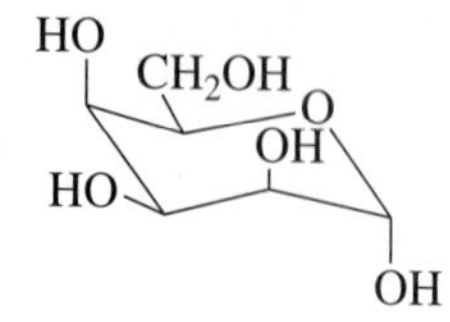

d. 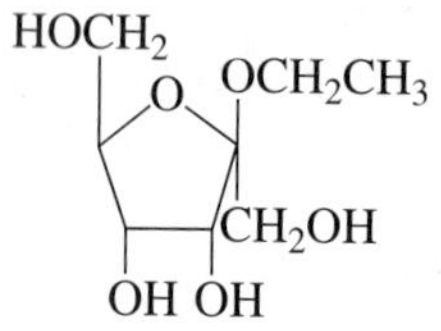

SOLUTION TO 25a The only OH group in an axial position in (a) is the one at C-3. Therefore, this sugar is the C-3 epimer of D-glucose, which is D-allose. The substituent at the anomeric carbon is in the β-position. Thus, the sugar's name is propyl β-D-alloside or propyl β-D-allopyranoside. Because the sugar is an acetal, it is a nonreducing sugar.

22.16 Determination of Ring Size

Two different procedures can be used to determine what size ring a monosaccharide forms. In the first procedure, treatment of the monosaccharide with excess methyl iodide and silver oxide converts all the OH groups to OCH_3 groups (Section 22.12). Acid-catalyzed hydrolysis of the acetal then forms a hemiacetal, which is in equilibrium with its open-chain form. The size of the ring can be determined from the structure of the open-chain form because the sole OH group is the one that had formed the cyclic hemiacetal.

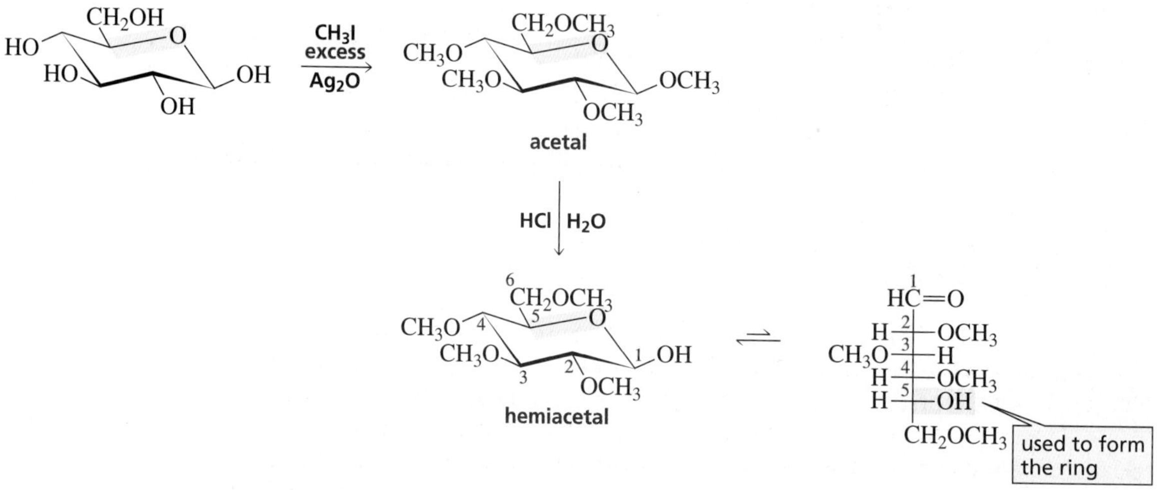

In the second procedure used to determine a monosaccharide's ring size, an acetal of the monosaccharide is oxidized with excess periodic acid. (Recall from Section 20.7 that periodic acid cleaves 1,2-diols.)

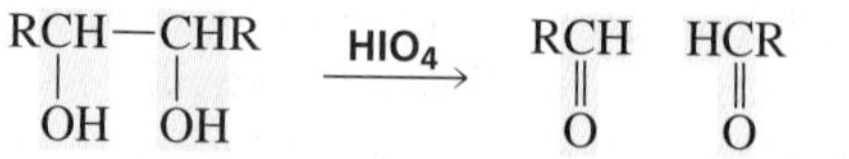

The α-hydroxyaldehyde formed when periodic acid cleaves a 1,2,3-diol is further oxidized to formic acid and another aldehyde.

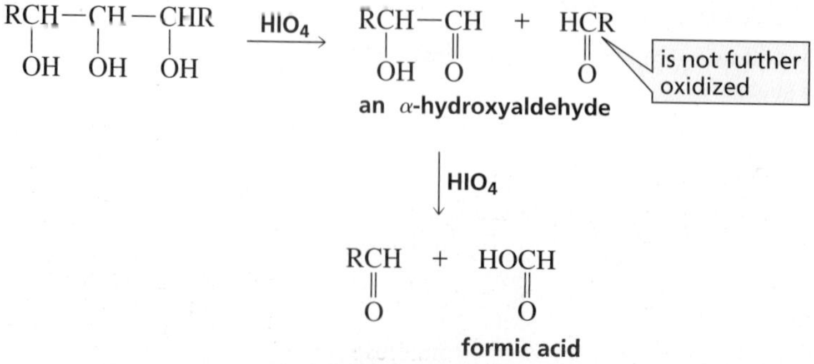

The products obtained from periodate cleavage of a six-membered-ring acetal are different from those obtained from cleavage of a five-membered-ring acetal.

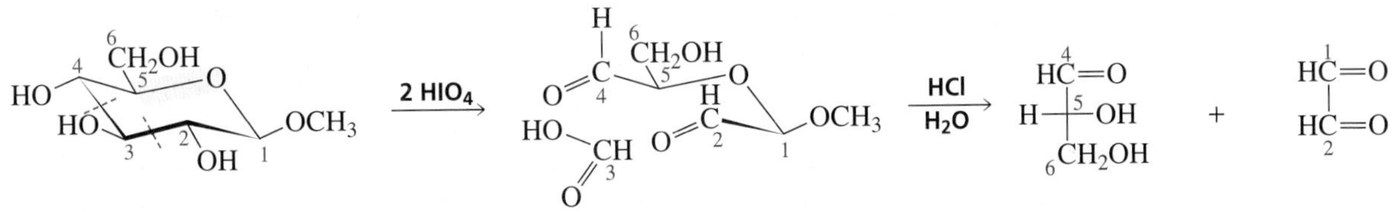

six-membered ring $\xrightarrow{2\ HIO_4}$ formic acid $\xrightarrow[H_2O]{HCl}$ D-glyceraldehyde + $HC{=}O$–$HC{=}O$

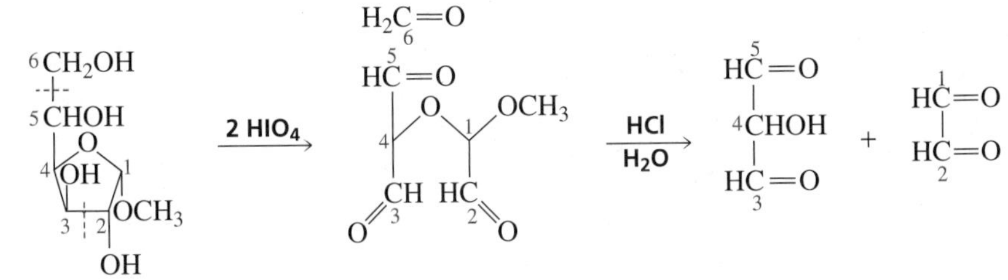

five-membered ring $\xrightarrow{2\ HIO_4}$ $\xrightarrow[H_2O]{HCl}$ $HC{=}O$–CHOH–$HC{=}O$ + $HC{=}O$–$HC{=}O$

PROBLEM 26◆

What kind of aldohexose would form L-glyceraldehyde when its acetal is oxidized with periodic acid?

22.17 Disaccharides

If the hemiacetal group of a monosaccharide forms an acetal by reacting with an alcohol group of another monosaccharide, the glycoside that is formed is a disaccharide. **Disaccharides** are compounds consisting of two monosaccharide subunits hooked together by an acetal linkage. For example, maltose is a disaccharide obtained from the hydrolysis of starch. It contains two D-glucose subunits hooked together by an acetal linkage. This particular acetal linkage is called an **α-1,4′-glycosidic linkage**. The linkage is between C-1 of one sugar subunit and C-4 of the other. The "prime" superscript indicates that C-4 is not in the same ring as C-1. The linkage is an α-1,4′-glycosidic linkage because the oxygen atom involved in the glycosidic linkage is in the α-position. *Remember that the α-position is axial when a sugar is shown in a chair conformation and is down when the sugar is shown in a Haworth projection; the β-position is equatorial when a sugar is shown in a chair conformation and is up when the sugar is shown in a Haworth projection.*

an α-1,4′-glycosidic linkage

the configuration of this carbon is not specified

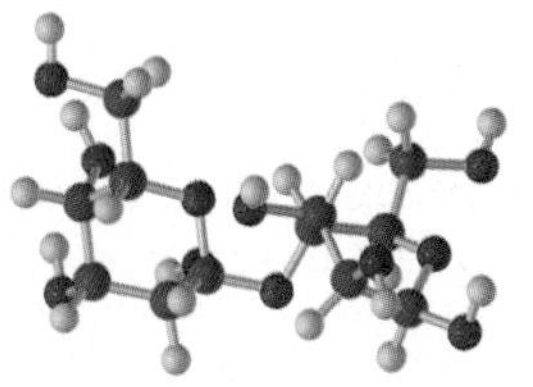

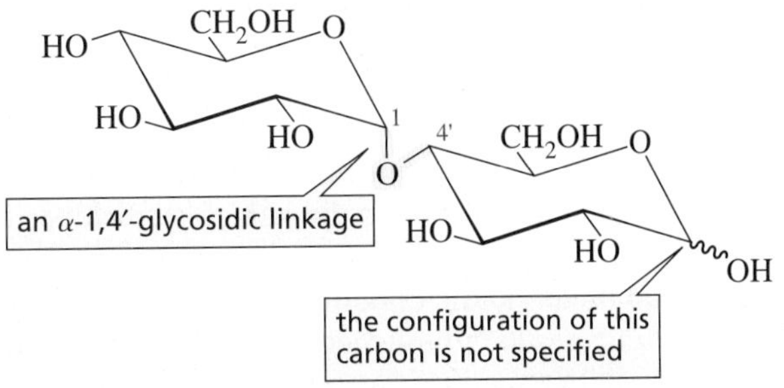

maltose

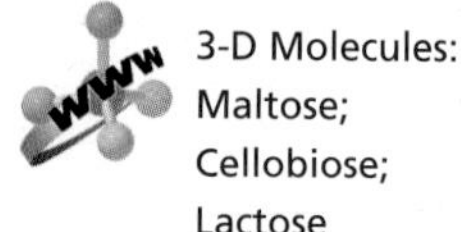

3-D Molecules: Maltose; Cellobiose; Lactose

Note that the structure of maltose is shown without specifying the configuration of the anomeric carbon that is not an acetal (the anomeric carbon of the subunit on the right marked with a wavy line), because maltose can exist in both the α and β forms. In α-maltose, the OH group bonded to this anomeric carbon is in the axial position. In β-maltose, the OH group is in the equatorial position. Because maltose can exist in both α and β forms, mutarotation occurs when crystals of one form are dissolved in a solvent. Maltose is a reducing sugar because the right-hand subunit is a hemiacetal and therefore is in equilibrium with the open-chain aldehyde that is easily oxidized.

Cellobiose, a disaccharide obtained from the hydrolysis of cellulose, also contains two D-glucose subunits. Cellobiose differs from maltose in that the two glucose subunits are hooked together by a **β-1,4′-glycosidic linkage**. Thus, the only difference in the structures of maltose and cellobiose is the configuration of the glycosidic linkage. Like maltose, cellobiose exists in both α and β forms because the OH group bonded to the anomeric carbon not involved in acetal formation can be in either the axial position (in α-cellobiose) or the equatorial position (in β-cellobiose). Cellobiose is a reducing sugar because the subunit on the right is a hemiacetal.

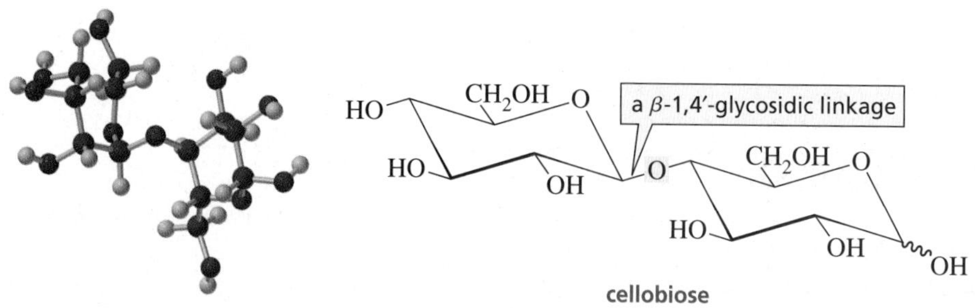

cellobiose

Lactose is a disaccharide found in milk. Lactose constitutes 4.5% of cow's milk by weight and 6.5% of human milk. One of the subunits of lactose is D-galactose, and the other is D-glucose. The D-galactose subunit is an acetal, and the D-glucose subunit is a hemiacetal. The subunits are joined by a β-1,4′-glycosidic linkage. Because one of the subunits is a hemiacetal, lactose is a reducing sugar and undergoes mutarotation.

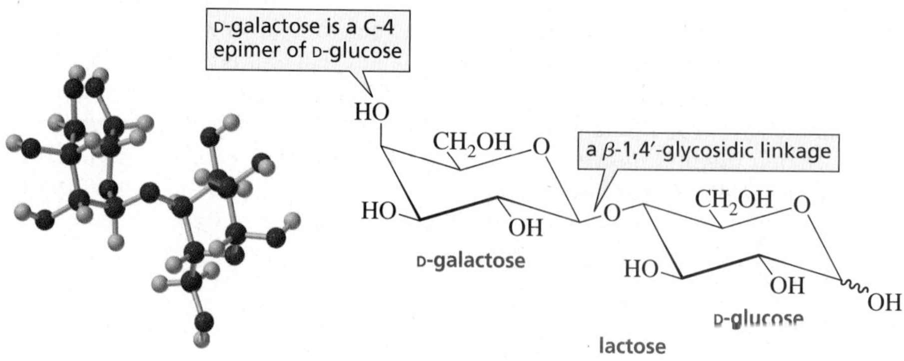

lactose

A simple experiment can prove that the hemiacetal linkage in lactose is in the glucose residue rather than in the galactose residue. The disaccharide is treated with excess methyl iodide in the presence of Ag_2O (Section 22.12), and the product is hydrolyzed under acidic conditions. The two acetal linkages are hydrolyzed, but all the ether linkages are untouched. Identification of the products shows that the galactose residue contained the acetal linkage in the disaccharide because it was able to react with methyl iodide at C-4. The glucose residue, on the other hand, was unable to react with methyl iodide at C-4 because the OH group at that position was used to form the acetal with galactose.

acetal linkage

hemiacetal linkage

$\xrightarrow[\text{Ag}_2\text{O}]{\text{CH}_3\text{I excess}}$

acetal is hydrolyzed

acetal is hydrolyzed

$\downarrow \text{H}_2\text{O}$

reacted with CH_3I

did not react with CH_3I

2,3,4,6-tetra-*O*-methylgalactose

2,3,6-tri-*O*-methylglucose

LACTOSE INTOLERANCE

Lactase is an enzyme that specifically breaks the β-1,4′-glycosidic linkage of lactose. Cats and dogs lose their intestinal lactase when they become adults; they are then no longer able to digest lactose. Consequently, when they are fed milk or milk products, the undegraded lactose causes digestive problems such as bloating, abdominal pain, and diarrhea. These problems occur because only monosaccharides can pass into the bloodstream, so lactose has to pass undigested into the large intestine. When humans have stomach flu or other intestinal disturbances, they can temporarily lose their lactase, thereby becoming lactose intolerant. Some humans lose their lactase permanently as they mature. Approximately 10% of the adult Caucasian population of the United States has lost its lactase. Lactose intolerance is much more common in people whose ancestors came from nondairy-producing countries. For example, only 3% of Danes but 97% of Thais, are lactose intolerant.

GALACTOSEMIA

After lactose is degraded into glucose and galactose, the galactose must be converted into glucose before it can be used by cells. Individuals who do not have the enzyme that converts galactose into glucose have the genetic disease known as galactosemia. Without this enzyme, galactose accumulates in the bloodstream. This can cause mental retardation and even death in infants. Galactosemia is treated by excluding galactose from the diet.

The most common disaccharide is sucrose (table sugar). Sucrose is obtained from sugar beets and sugarcane. About 90 million tons of sucrose are produced in the world each year. Sucrose consists of a D-glucose subunit and a D-fructose subunit linked by a glycosidic bond between C-1 of glucose (in the α-position) and C-2 of fructose (in the β-position).

Unlike the other disaccharides that have been discussed, sucrose is not a reducing sugar and does not exhibit mutarotation because the glycosidic bond is between the anomeric carbon of glucose and the anomeric carbon of fructose. Sucrose, therefore, does not have a hemiacetal or hemiketal group, so it is not in equilibrium with the readily oxidized open-chain aldehyde or ketone form in aqueous solution.

α linkage at glucose

β linkage at fructose

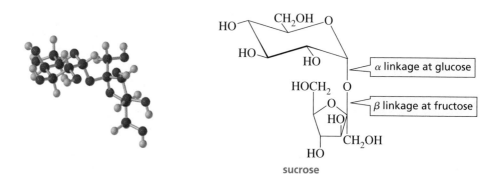

sucrose

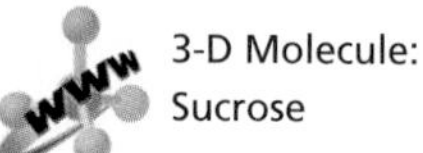

3-D Molecule: Sucrose

Sucrose has a specific rotation of $+66.5°$. When it is hydrolyzed, the resulting equimolar mixture of glucose and fructose has a specific rotation of $-22.0°$. Because the sign of the rotation changes when sucrose is hydrolyzed, a 1 : 1 mixture of glucose and fructose is called *invert sugar*. The enzyme that catalyzes the hydrolysis of sucrose is called *invertase*. Honeybees have invertase, so the honey they produce is a mixture of sucrose, glucose, and fructose. Because fructose is sweeter than sucrose, invert sugar is sweeter than sucrose. Some foods are advertised as containing fructose instead of sucrose, which means that they achieve the same level of sweetness with a lower sugar content.

22.18 Polysaccharides

Polysaccharides contain as few as 10 or as many as several thousand monosaccharide units joined together by glycosidic linkages. The molecular weight of the individual polysaccharide chains is variable. The most common polysaccharides are starch and cellulose.

Starch is the major component of flour, potatoes, rice, beans, corn, and peas. Starch is a mixture of two different polysaccharides: amylose (about 20%) and amylopectin (about 80%). Amylose is composed of unbranched chains of D-glucose units joined by α-1,4′-glycosidic linkages.

an α-1,4′-glycosidic linkage

three subunits of amylose

an α-1,4'-glycosidic bond

an α-1,6'-glycosidic bond

Figure 22.1
Branching in amylopectin. The hexagons represent glucose units. They are joined by α-1,4′- and α-1,6′-glyclosidic bonds.

Amylopectin is a branched polysaccharide. Like amylose, it is composed of chains of D-glucose units joined by α-1,4′-glycosidic linkages. Unlike amylose, however, amylopectin also contains **α-1,6′-glycosidic linkages**. These linkages create the branches in the polysaccharide (Figure 22.1). Amylopectin can contain up to 10^6 glucose units, making it one of the largest molecules found in nature.

Tutorial: Identifying glycosidic linkages and numbering pyranose and furanose rings

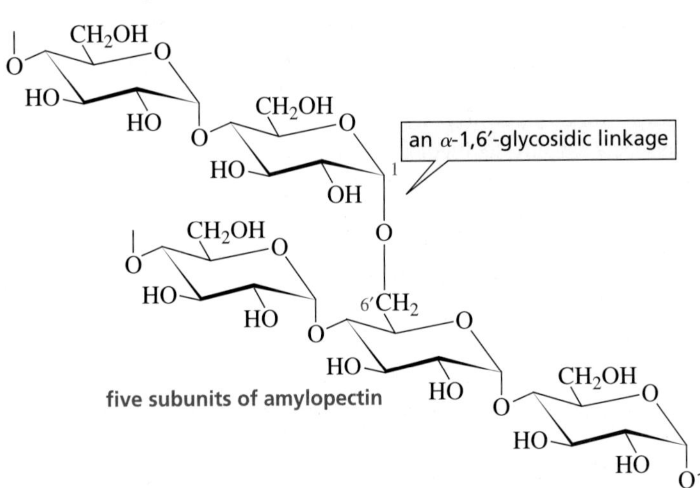

Animals store their excess glucose in a polysaccharide known as glycogen. Glycogen has a structure similar to that of amylopectin, but glycogen has more branches (Figure 22.2). The branch points in glycogen occur about every 10 residues, whereas those in amylopectin occur about every 25 residues. The high degree of branching in glycogen has important physiological effects. When the body needs energy, many individual glucose units can be simultaneously removed from the ends of many branches.

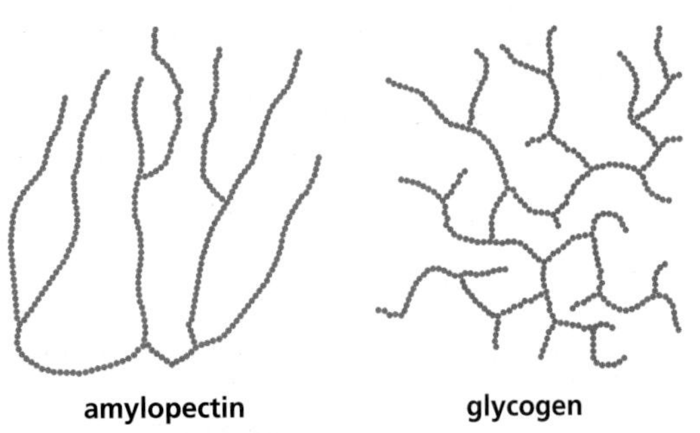

Figure 22.2
A comparison of the branching in amylopectin and glycogen.

WHY THE DENTIST IS RIGHT

Bacteria found in the mouth have an enzyme that converts sucrose into a polysaccharide called dextran. Dextran is made up of glucose units joined mainly through α-1,3′- and α-1,6′-glycosidic linkages. About 10% of dental plaque is composed of dextran. This is the chemical basis of why your dentist cautions you not to eat candy.

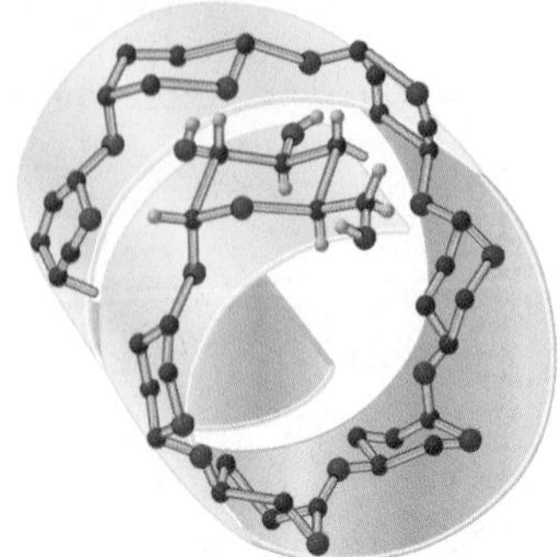

▲ **Figure 22.3**
The α-1,4′-glycosidic linkages in amylose cause it to form a left-handed helix. Many of its OH groups form hydrogen bonds with water molecules.

Cellulose is the structural material of higher plants. Cotton, for example, is composed of about 90% cellulose, and wood is about 50% cellulose. Like amylose, cellulose is composed of unbranched chains of D-glucose units. Unlike amylose, however, the glucose units in cellulose are joined by β-1,4′-glycosidic linkages rather than by α-1,4′-glycosidic linkages.

a β-1,4′-glycosidic linkage

three subunits of cellulose

α-1,4′-Glycosidic linkages are easier to hydrolyze than β-1,4′-glycosidic linkages because of the anomeric effect that weakens the bond to the anomeric carbon (Section 22.14). All mammals have the enzyme (α-glucosidase) that hydrolyzes the α-1,4′-glycosidic linkages that join glucose units, but they do not have the enzyme (β-glucosidase) that hydrolyzes β-1,4′-glycosidic linkages. As a result, mammals *cannot* obtain the glucose they need by eating cellulose. However, bacteria that possess β-glucosidase inhabit the digestive tracts of grazing animals, so cows can eat grass and horses can eat hay to meet their nutritional requirements for glucose. Termites also harbor bacteria that break down the cellulose in the wood they eat.

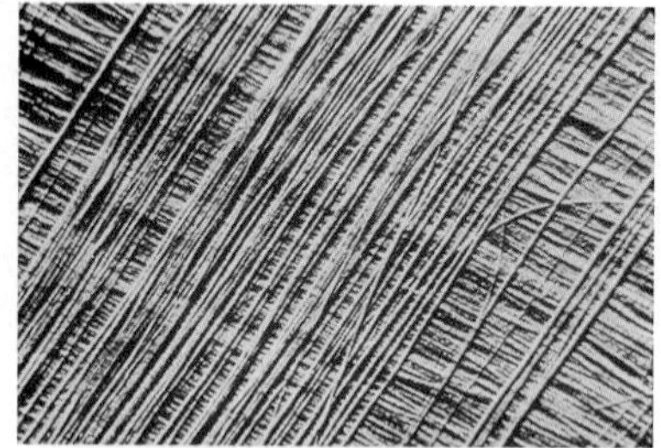

▲ Strands of cellulose in a plant fiber

The different glycosidic linkages in starch and cellulose give these compounds very different physical properties. The α-linkages in starch cause amylose to form a helix that promotes hydrogen bonding of its OH groups to water molecules (Figure 22.3). As a result, starch is soluble in water.

On the other hand, the β-linkages in cellulose promote the formation of intramolecular hydrogen bonds. Consequently, these molecules line up in linear arrays (Figure 22.4), and intermolecular hydrogen bonds form between adjacent chains. These large aggregates cause cellulose to be insoluble in water. The strength of these bundles of polymer chains makes cellulose an effective structural material. Processed cellulose is also used for the production of paper and cellophane.

Chitin is a polysaccharide that is structurally similar to cellulose. It is the major structural component of the shells of crustaceans (e.g., lobsters, crabs, and shrimps) and the exoskeletons of insects. Like cellulose, chitin has β-1,4′-glycosidic linkages. It differs from cellulose, though, in that it has an *N*-acetylamino group instead of an OH group at the C-2 position. The β-1,4′-glycosidic linkages give chitin its structural rigidity.

▲ The shell of this bright orange crab from Australia is composed largely of chitin.

three subunits of chitin

Figure 22.4 ▶
The β-1,4′-glycosidic linkages in cellulose form intramolecular hydrogen bonds, which cause the molecules to line up in linear arrays.

intermolecular hydrogen bond

PROBLEM 27

What is the main structural difference between

a. amylose and cellulose?
b. amylose and amylopectin?
c. amylopectin and glycogen?
d. cellulose and chitin?

CONTROLLING FLEAS

Several different drugs have been developed to help pet owners control fleas. One of these drugs is lufenuron, the active ingredient in Program®. Lufenuron interferes with the production of chitin. Since the exoskeleton of a flea is composed primarily of chitin, a flea cannot live if it cannot make chitin.

lufenuron

22.19 Some Naturally Occurring Products Derived from Carbohydrates

Deoxy sugars are sugars in which one of the OH groups is replaced by a hydrogen (*deoxy* means "without oxygen"). 2-Deoxyribose—it is missing the oxygen at the C-2 position—is an important example of a deoxy sugar. Ribose is the sugar component of ribonucleic acid (RNA), whereas 2-deoxyribose is the sugar component of deoxyribonucleic acid (DNA). RNA and DNA are *N*-glycosides—their subunits consist of an amine bonded to the β-position of the anomeric carbon of ribose or 2-deoxyribose. The subunits are linked by a phosphate group between C-3 of one sugar and C-5 of the next sugar (Section 27.1).[1]

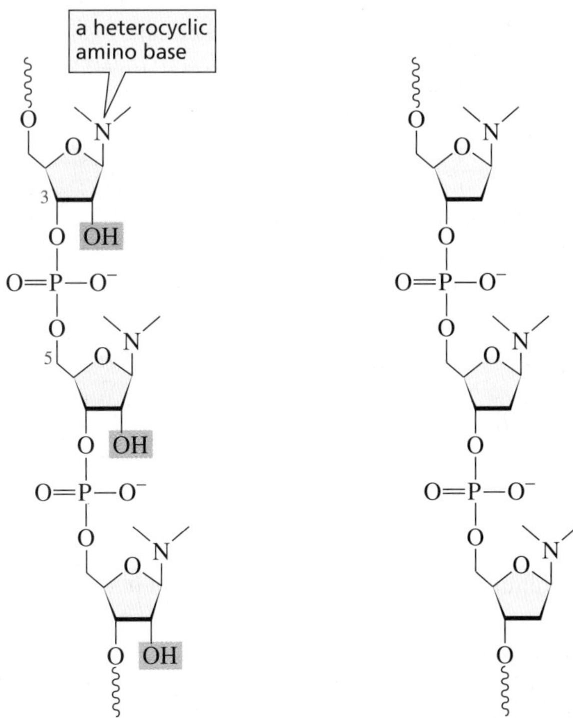

a short segment of RNA
the sugar component is D-ribose

a short segment of DNA
the sugar component is D-2′-deoxyribose

[1]In referring to the sugar found in DNA, it is called 2-deoxyribose. In numbering the sugar in a DNA molecule, it is 2′-deoxyribose because the nonprimed numbers refer to positions on the heterocyclic amine components.

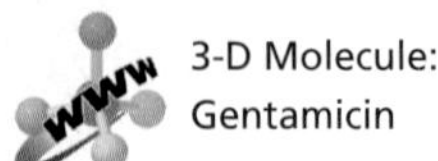

3-D Molecule: Gentamicin

In **amino sugars**, one of the OH groups is replaced by an amino group. *N*-Acetylglucosamine—the subunit of chitin and one of the subunits of certain bacterial cell walls—is an example of an amino sugar (Sections 22.18 and 24.9). Some important antibiotics contain amino sugars. For example, the three subunits of the antibiotic gentamicin are deoxyamino sugars. Notice that the middle subunit is missing the ring oxygen, so it really isn't a sugar.

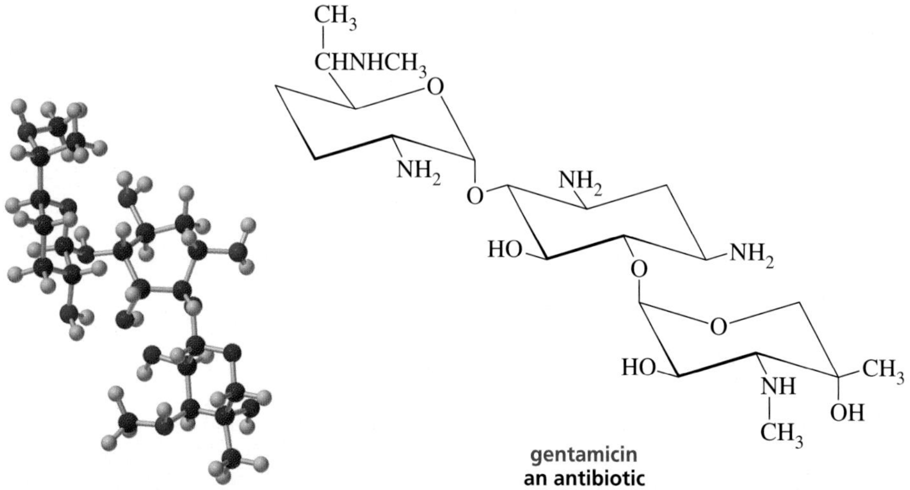

gentamicin
an antibiotic

HEPARIN

Heparin is an anticoagulant that is released to prevent excessive blood clot formation when an injury occurs. Heparin is a polysaccharide made up of glucosamine, glucuronic acid, and iduronic acid subunits. The C-6 OH groups of the glucosamine subunits and the C-2 OH groups of the iduronic acid subunits are sulfonated. Some of the amino groups are sulfonated and some are acetylated. Thus, heparin is a highly negatively charged molecule, found principally in cells that line arterial walls. Heparin is widely used clinically as an anticoagulant.

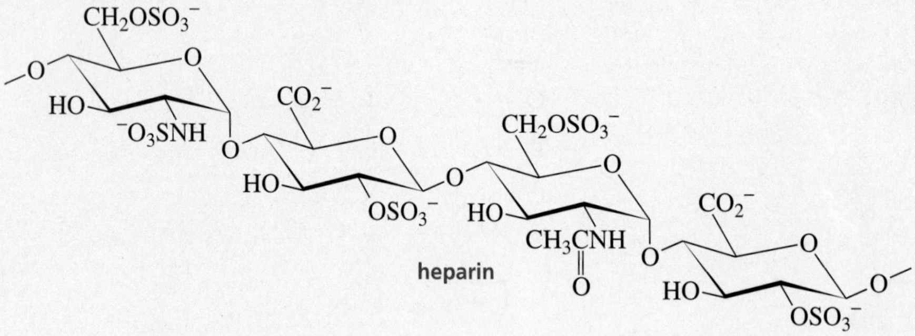

heparin

L-Ascorbic acid (vitamin C) is synthesized in plants and in the livers of most vertebrates. Humans, monkeys, and guinea pigs do not have the enzymes necessary for the biosynthesis of vitamin C, so they must obtain the vitamin in their diets. The biosynthesis of vitamin C involves the enzymatic conversion of D-glucose into L-gulonic acid—reminiscent of the last step in the Fischer proof. L-Gulonic acid is converted into a γ-lactone by the enzyme lactonase, and then an enzyme called oxidase oxidizes the lactone to L-ascorbic acid. The L-configuration of ascorbic acid refers to the configuration at C-5, which was C-2 in D-glucose.

the synthesis of L-ascorbic acid

D-glucose $\xrightarrow{\text{oxidizing enzyme}}$ $\xrightarrow{\text{reducing enzyme}}$ rotate 180° = L-gulonic acid $\xrightarrow{\text{lactonase}}$ a γ-lactone $\xrightarrow{\text{oxidase}}$ L-ascorbic acid vitamin C $\xrightarrow{\text{oxidation}}$ L-dehydroascorbic acid

L-configuration

$pK_a = 4.17$

Although L-ascorbic acid does not have a carboxylic acid group, it is an acidic compound because the pK_a of the C-3 OH group is 4.17. L-Ascorbic acid is readily oxidized to L-dehydroascorbic acid, which is also physiologically active. If the lactone ring is opened by hydrolysis, all vitamin C activity is lost. Therefore, not much intact vitamin C survives in food that has been thoroughly cooked. Worse, if the food is cooked in water and then drained, the water-soluble vitamin is thrown out with the water!

VITAMIN C

Vitamin C traps radicals formed in aqueous environments (Section 9.8). It is an antioxidant because it prevents oxidation reactions by radicals. Not all the physiological functions of vitamin C are known. What is known, though, is that vitamin C is required for the synthesis of collagen, which is the structural protein of skin, tendons, connective tissue, and bone. If vitamin C is not present in the diet (it is abundant in citrus fruits and tomatoes), lesions appear on the skin, severe bleeding occurs about the gums, in the joints, and under the skin, and wounds heal slowly. The disease caused by a deficiency of vitamin C is known as *scurvy*. British sailors who shipped out to sea after the late 1700s were required to eat limes to prevent scurvy. This is how they came to be called "limeys." Scurvy was the first disease to be treated by adjusting the diet. *Scorbutus* is Latin for "scurvy"; *ascorbic*, therefore, means "no scurvy."

PROBLEM 28

Explain why the C-3 OH group of vitamin C is more acidic than the C-2 OH group.

22.20 Carbohydrates on Cell Surfaces

The surfaces of many cells contain short polysaccharide chains that allow the cells to interact with each other, as well as to interact with invading viruses and bacteria. These polysaccharides are linked to the cell surface by the reaction of an OH or an NH_2 group of a protein with the anomeric carbon of a cyclic sugar. Proteins bonded to polysaccharides are called **glycoproteins**. The percentage of carbohydrate in glycoproteins is variable; some glycoproteins contain as little as 1% carbohydrate by weight, whereas others contain as much as 80%.

Many different types of proteins are glycoproteins. For example, structural proteins such as collagen, proteins found in mucous secretions, immunoglobulins, follicle-stimulating hormone and thyroid-stimulating hormone, interferon (an antiviral protein), and blood plasma proteins are all glycoproteins. One of the functions of the polysaccharide chain is to act as a receptor site on the cell surface in order to transmit signals from hormones and other molecules across the cell membrane into the cell. The carbohydrates on the surfaces of cells also serve as points of attachment for other cells, viruses, and toxins.

Carbohydrates on the surfaces of cells provide a way for cells to recognize one another. The interaction between surface carbohydrates has been found to play a role in many diverse activites, such as infection, the prevention of infection, fertilization, inflammatory diseases like rheumatoid arthritis and septic shock, and blood clotting. For example, the goal of the HIV protease inhibitor drugs is to prevent HIV from recognizing and penetrating cells. The fact that several known antibiotics contain amino sugars suggests that they function by recognizing target cells. Carbohydrate interactions also are involved in the regulation of cell growth, so changes in membrane glycoproteins are thought to be correlated with malignant transformations.

Blood type (A, B, or O) is determined by the nature of the sugar bound to the protein on the outer surfaces of red blood cells. Each type of blood is associated with a different carbohydrate structure (Figure 22.5). Type AB blood has the carbohydrate structure of both type A and type B.

Antibodies are proteins that are synthesized by the body in response to a foreign substance, called an *antigen*. Interaction with the antibody either causes the antigen to precipitate or flags it for destruction by immune system cells. This is why, for example, blood cannot be transferred from one person to another unless the carbohydrate

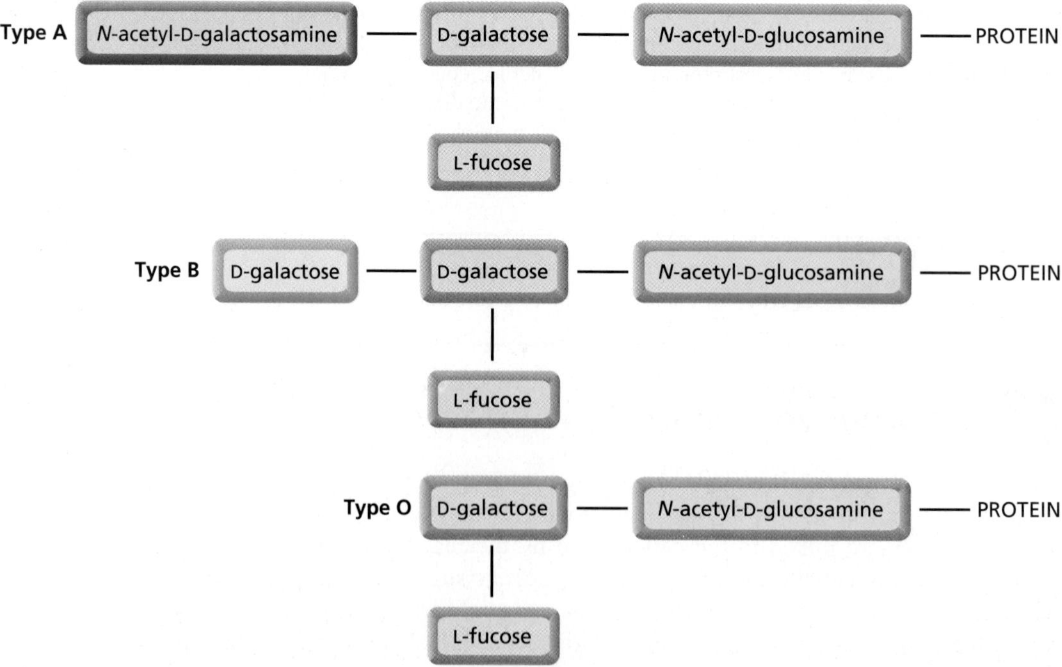

Figure 22.5
Blood type is determined by the nature of the sugar on the surfaces of red blood cells. Fucose is 6-deoxygalactose.

portions of the donor and acceptor are compatible. Otherwise the donated blood will be considered a foreign substance.

Looking at Figure 22.5, we can see why the immune system of type A people recognizes type B blood as foreign and vice versa. The immune system of people with type A, B, or AB blood does not, however, recognize type O blood as foreign, because the carbohydrate in type O blood is also a component of types A, B, and AB blood. Thus, anyone can accept type O blood, so people with type O blood are called universal donors. Type AB people can accept types AB, A, B, and O blood, so people with type AB blood are referred to as universal acceptors.

PROBLEM 29◆

From the nature of the carbohydrate bound to red blood cells, answer the following questions:

a. People with type O blood can donate blood to anyone, but they cannot receive blood from everyone. From whom can they *not* receive blood?

b. People with type AB blood can receive blood from anyone, but they cannot give blood to everyone. To whom can they *not* give blood?

22.21 Synthetic Sweeteners

For a molecule to taste sweet, it must bind to a receptor on a taste bud cell of the tongue. The binding of this molecule causes a nerve impulse to pass from the taste bud to the brain, where the molecule is interpreted as being sweet. Sugars differ in their degree of "sweetness." The relative sweetness of glucose is 1.00, that of sucrose is 1.45, and that of fructose, the sweetest of all sugars, is 1.65. Developers of synthetic sweeteners must consider several factors—such as toxicity, stability, and cost—in addition to taste.

Saccharin, the first synthetic sweetener, was discovered by Ira Remsen and his student Constantine Fahlberg at Johns Hopkins University in 1878. Fahlberg was studying the oxidation of ortho-substituted toluenes in Remsen's laboratory when he found that one of his newly synthesized compounds had an extremely sweet taste. (As strange as it may seem today, at one time it was common for chemists to taste compounds in order to characterize them.) He called this compound saccharin, and it was eventually found to be about 300 times sweeter than glucose. Notice that, in spite of its name, saccharin is *not* a saccharide.

Ira Remsen (1846–1927) *was born in New York. After receiving an M.D. from Columbia University, he decided to become a chemist. He earned a Ph.D. in Germany and then returned to the United States in 1872 to accept a faculty position at Williams College. In 1876, he became a professor of chemistry at the newly established Johns Hopkins University, where he initiated the first center for chemical research in the United States. He later became the second president of Johns Hopkins.*

saccharin (benzene ring fused to a five-membered ring: C=O, NH, SO_2)

CH_3CH_2O–C_6H_4–$NH\overset{O}{\overset{\|}{C}}NH_2$
dulcin

C_6H_{11}–$NHSO_3^-\ Na^+$
sodium cyclamate

$^-O\overset{O}{\overset{\|}{C}}CH_2CH(\overset{+}{N}H_3)\overset{O}{\overset{\|}{C}}NHCH(CH_2C_6H_5)\overset{O}{\overset{\|}{C}}OCH_3$
aspartame

Because it has no caloric value, when it became commercially available in 1885, saccharin became an important substitute for sucrose. The chief nutritional problem in the West was—and still is—the overconsumption of sugar and its consequences: obesity, heart disease, and dental decay. Saccharin is also important to diabetics, who must limit their consumption of sucrose and glucose. Although the toxicity of saccharin was not studied carefully when the compound first became available to the public (our current concern with toxicity is a fairly recent development), extensive studies done since then have shown saccharin to be a safe sugar substitute. In 1912, saccharin was temporarily banned in the United States, not because of any concern about its toxicity, but because of a concern that people would miss out on the nutritional benefits of sugar.

THE WONDER OF DISCOVERY

Ira Remsen gave the following account of why he became a scientist:[2] He was working as a physician and came across the statement "Nitric acid acts upon copper" while reading a chemistry book. He decided to see what "acts on" meant. He poured nitric acid on a penny sitting on a table. "But what was this wonderful thing which I beheld? The cent had already changed and it was not small change either. A greenish blue liquid foamed and fumed over the cent and over the table. The air in the neighborhood of the performance became dark red. A great colored cloud arose. This was disagreeable and suffocating—how should I stop this? I tried to get rid of the objectionable mess by picking it up and throwing it out of the window, which I had meanwhile opened. I learned another fact—nitric acid not only acts upon copper but it acts upon my fingers. The pain led to another unpremeditated experiment. I drew my fingers across my trousers and another fact was discovered. Nitric acid also acts upon trousers. Taking everything else into consideration, that was the most impressive experiment, and, relatively, probably the most expensive experiment I ever performed. I tell of it even now with interest. It was a revelation to me. It resulted in a desire on my part to learn even more about that remarkable kind of action. Plainly the only way to learn about it was to see its results, to experiment, to work in a laboratory."

[2] L. R. Summerlin, C. L. Bordford, and J. B. Ealy, *Chemical Demonstrations,* 2nd ed. (Washington, DC: American Chemical Society, 1988).

Dulcin® was the second synthetic sweetener to be discovered (in 1884). Even though it did not have the bitter, metallic aftertaste associated with saccharin, it never achieved much popularity. Dulcin® was taken off the market in 1951 in response to questions about its toxicity.

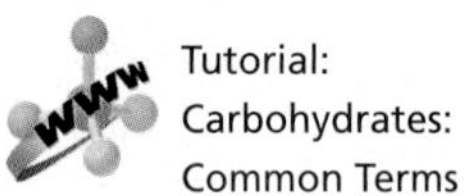

Tutorial: Carbohydrates: Common Terms

Sodium cyclamate became a widely used nonnutritive sweetener in the 1950s, but was banned in the United States some 20 years later in response to two studies that appeared to show that large amounts of sodium cyclamate cause liver cancer in mice.

Aspartame was approved by the U.S. Food and Drug Administration (FDA) in 1981. About 200 times sweeter than sucrose, aspartame is sold under the trade name NutraSweet® (Section 23.8). Because NutraSweet® contains a phenylalanine subunit, it should not be used by people with the genetic disease known as PKU (Section 25.6).

The fact that these four synthetic sweeteners have such different structures, all of which are very different from those of monosaccharides, indicates that the sensation of sweetness is not induced by a single molecular shape.

Summary

Bioorganic compounds—organic compounds found in biological systems—obey the same principles of structure and reactivity as do small organic molecules. Much of the structure of bioorganic compounds is for **molecular recognition**.

Carbohydrates are the most abundant class of compounds in the biological world. They are polyhydroxy aldehydes (**aldoses**) and polyhydroxy ketones (**ketoses**) or compounds formed by linking up aldoses and ketoses. D and L notations describe the configuration of the bottom-most asymmetric carbon of a **monosaccharide**; the configurations of the other carbons are implicit in the common name. Most naturally occurring sugars are D-sugars. Naturally occurring ketoses have the ketone group in the 2-position. **Epimers** differ in configuration at only one asymmetric carbon: D-mannose is the C-2 epimer of D-glucose and D-galactose is the C-4 epimer of D-glucose.

Reduction of an aldose forms one **alditol**; reduction of a ketose forms two alditols. Br_2 oxidizes aldoses, but not ketoses; Tollens reagent oxidizes both. Aldoses are oxidized to **aldonic acids** or **aldaric acids**. Aldoses and ketoses react with three equivalents of phenylhydrazine, forming **osazones**. C-2 epimers form identical osazones. The **Kiliani–Fischer synthesis** increases the carbon chain of an aldose by one carbon—it forms C-2 epimers. The **Ruff degradation** decreases the carbon chain by one carbon. The OH groups of monosaccharides react with acetyl chloride to form esters and with methyl iodide/silver oxide to form ethers.

The aldehyde or keto group of a **monosaaccharide** reacts with one of its OH groups to form cyclic hemiacetals or hemiketals: Glucose forms α-D-glucose and β-D-glucose. The α-position is axial when a sugar is shown in a chair conformation and down when the sugar is shown in a Haworth projection; the β-position is equatorial when a sugar is shown in a chair conformation and up when the sugar is shown in a Haworth projection. At equilibrium, there is more β-D-glucose than α-D-glucose. α-D-Glucose and β-D-glucose are **anomers**—they differ in configuration only at the carbon (**anomeric carbon**) that was the carbonyl carbon in the open-chain form. Anomers have different physical properties. Six-membered-ring sugars are **pyranoses**; five-membered-ring sugars are **furanoses**. The most abundant monosaccharide in nature is D-glucose. All the OH groups in β-D-glucose are in equatorial positions. A slow change in optical rotation to an equilibrium value is known as **mutarotation**.

The cyclic hemiacetal (or hemiketal) can react with an alcohol to form an acetal (or ketal), called a **glycoside**. If the name "pyranose" or "furanose" is used, the acetal is called a **pyranoside** or a **furanoside**. The bond between the anomeric carbon and the alkoxy oxygen is called a **glycosidic bond**. The preference for the axial position by certain substituents bonded to the anomeric carbon is called the **anomeric effect**. If a sugar has an aldehyde, ketone, hemiacetal, or hemiketal group, it is a reducing sugar.

Disaccharides consist of two monosaccharide subunits hooked together by an acetal linkage. Maltose has an **α-1,4′-glycosidic linkage**; cellobiose has a **β-1,4′-glycosidic linkage**. The most common disaccharide is sucrose; it consists of a D-glucose subunit and a D-fructose subunit linked by their anomeric carbons.

Polysaccharides contain as few as 10 or as many as several thousand monosaccharide units joined together by glycosidic linkages. Starch is composed of amylose and amylopectin. Amylose has unbranched chains of D-glucose units joined by α-1,4′-glycosidic linkages. Amylopectin, too, has chains of D-glucose units joined by α-1,4′-glycosidic linkages, but it also has α-1,6′-glycosidic linkages that create branches. Glycogen is similar to amylopectin, but has more branches. Cellulose has unbranched chains of D-glucose units joined by β-1,4′-glycosidic linkages. The α-linkages cause amylose to form a helix; the β-linkages allow the molecules of cellulose to form intramolecular hydrogen bonds.

The surfaces of many cells contain short polysaccharide chains that allow the cells to interact with each other. These polysaccharides are linked to the cell surface by protein groups. Proteins bonded to polysaccharides are called **glycoproteins**.

Summary of Reactions

1. Reduction (Section 22.5).

$$\text{HC=O}-(\text{CHOH})_n-\text{CH}_2\text{OH} \xrightarrow[\text{Pd/C}]{\text{H}_2} \text{CH}_2\text{OH}-(\text{CHOH})_n-\text{CH}_2\text{OH}$$

$$\text{CH}_2\text{OH}-\text{C=O}-(\text{CHOH})_n-\text{CH}_2\text{OH} \xrightarrow[\text{2. H}_3\text{O}^+]{\text{1. NaBH}_4} \text{CH}_2\text{OH}-\text{CHOH}-(\text{CHOH})_n-\text{CH}_2\text{OH}$$

2. Oxidation (Section 22.5).

a. $$\text{HC=O}-(\text{CHOH})_n-\text{CH}_2\text{OH} \xrightarrow[\text{HO}^-]{\text{Ag}^+,\ \text{NH}_3} \text{COO}^- -(\text{CHOH})_n-\text{CH}_2\text{OH} + \text{Ag}$$

b. $$\text{CH}_2\text{OH}-\text{C=O}-(\text{CHOH})_n-\text{CH}_2\text{OH} \xrightarrow[\text{HO}^-]{\text{Ag}^+,\ \text{NH}_3} \text{COO}^- -\text{CHOH}-(\text{CHOH})_n-\text{CH}_2\text{OH} + \text{Ag}$$

c. $$\text{HC=O}-(\text{CHOH})_n-\text{CH}_2\text{OH} \xrightarrow[\text{H}_2\text{O}]{\text{Br}_2} \text{COOH}-(\text{CHOH})_n-\text{CH}_2\text{OH} + 2\ \text{Br}^-$$

d. $$\text{HC=O}-(\text{CHOH})_n-\text{CH}_2\text{OH} \xrightarrow[\Delta]{\text{HNO}_3} \text{COOH}-(\text{CHOH})_n-\text{COOH}$$

3. Enolization (Section 22.5).

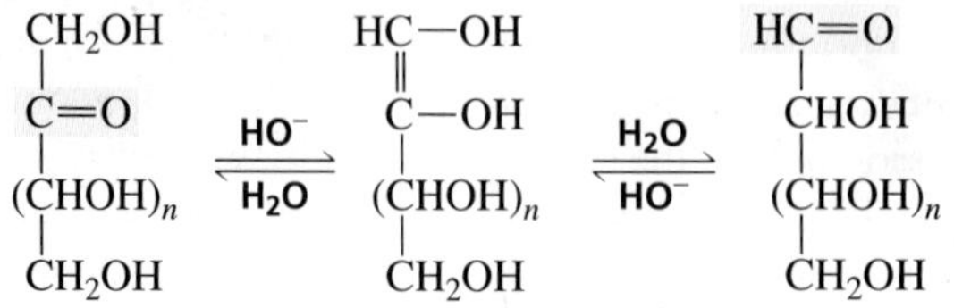

$$\begin{matrix} CH_2OH \\ | \\ C{=}O \\ | \\ (CHOH)_n \\ | \\ CH_2OH \end{matrix} \underset{H_2O}{\overset{HO^-}{\rightleftharpoons}} \begin{matrix} HC{-}OH \\ \| \\ C{-}OH \\ | \\ (CHOH)_n \\ | \\ CH_2OH \end{matrix} \underset{HO^-}{\overset{H_2O}{\rightleftharpoons}} \begin{matrix} HC{=}O \\ | \\ CHOH \\ | \\ (CHOH)_n \\ | \\ CH_2OH \end{matrix}$$

4. Osazone formation (Section 22.6).

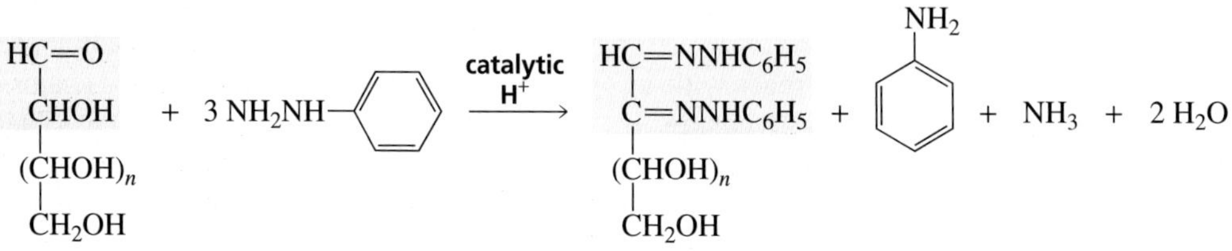

$$\begin{matrix} HC{=}O \\ | \\ CHOH \\ | \\ (CHOH)_n \\ | \\ CH_2OH \end{matrix} + 3\,NH_2NH{-}C_6H_5 \xrightarrow[H^+]{\text{catalytic}} \begin{matrix} HC{=}NNHC_6H_5 \\ | \\ C{=}NNHC_6H_5 \\ | \\ (CHOH)_n \\ | \\ CH_2OH \end{matrix} + C_6H_5NH_2 + NH_3 + 2\,H_2O$$

5. Chain elongation: the Kiliani–Fischer synthesis (Section 22.7).

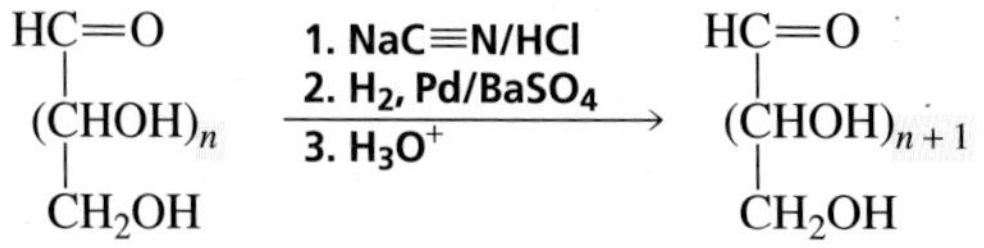

$$\begin{matrix} HC{=}O \\ | \\ (CHOH)_n \\ | \\ CH_2OH \end{matrix} \xrightarrow[3.\ H_3O^+]{1.\ NaC{\equiv}N/HCl \quad 2.\ H_2,\ Pd/BaSO_4} \begin{matrix} HC{=}O \\ | \\ (CHOH)_{n+1} \\ | \\ CH_2OH \end{matrix}$$

6. Chain shortening: the Ruff degradation (Section 22.8).

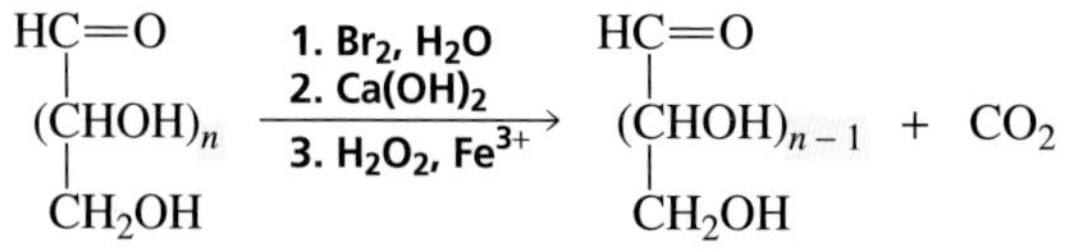

$$\begin{matrix} HC{=}O \\ | \\ (CHOH)_n \\ | \\ CH_2OH \end{matrix} \xrightarrow[3.\ H_2O_2,\ Fe^{3+}]{1.\ Br_2,\ H_2O \quad 2.\ Ca(OH)_2} \begin{matrix} HC{=}O \\ | \\ (CHOH)_{n-1} \\ | \\ CH_2OH \end{matrix} + CO_2$$

7. Acylation (Section 22.12).

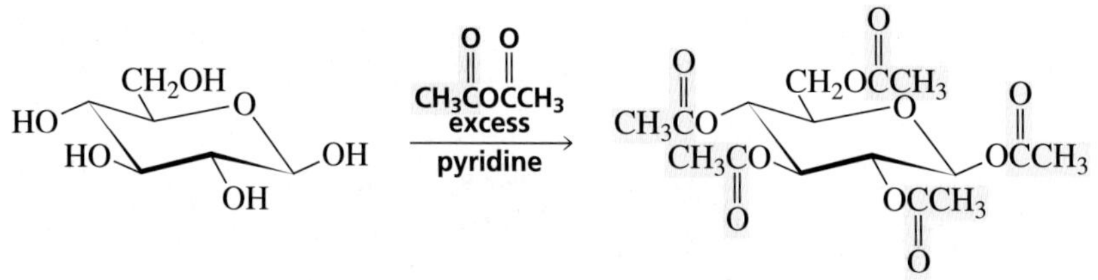

Glucopyranose (HO, HO, OH, OH, CH_2OH) $\xrightarrow[\text{pyridine}]{CH_3COCCH_3\ (\text{O O})\ \text{excess}}$ penta-acetate (CH_2OCCH_3, CH_3CO, CH_3CO, $OCCH_3$, $OCCH_3$; each with C=O)

8. Alkylation (Section 22.12).

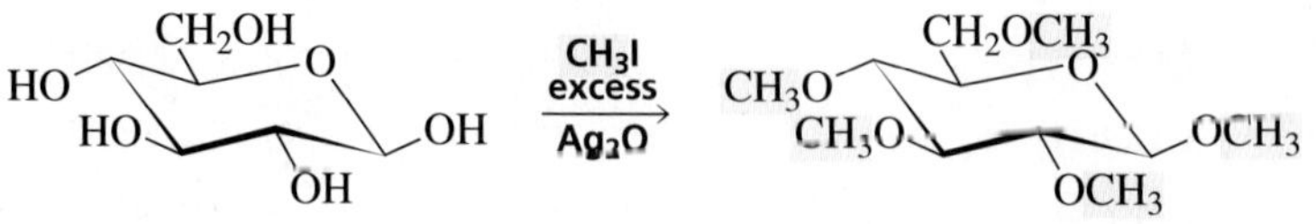

Glucopyranose (HO, HO, OH, OH, CH_2OH) $\xrightarrow[Ag_2O]{CH_3I\ \text{excess}}$ (CH_2OCH_3, CH_3O, CH_3O, OCH_3, OCH_3)

9. Acetal (and ketal) formation (Section 22.13).

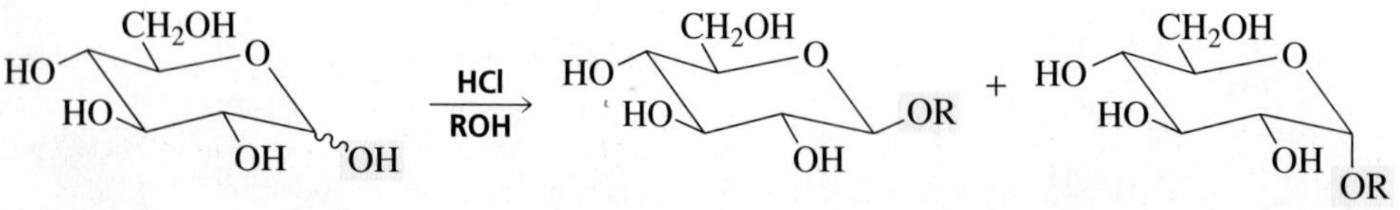

Glucopyranose (HO, HO, OH, OH, CH_2OH) $\xrightarrow[ROH]{HCl}$ β-glycoside (HO, HO, OH, OR, CH_2OH) + α-glycoside (HO, HO, OH, OR, CH_2OH)

Key Terms

aldaric acid (p. 929)
alditol (p. 926)
aldonic acid (p. 928)
aldose (p. 923)
amino sugar (p. 950)
anomeric carbon (p. 935)
anomeric effect (p. 941)
anomers (p. 935)
bioorganic compound (p. 921)
carbohydrate (p. 921)
complex carbohydrate (p. 922)
deoxy sugar (p. 949)
disaccharide (p. 922)
epimers (p. 925)
furanose (p. 935)
furanoside (p. 939)
glycoprotein (p. 951)
glycoside (p. 939)
N-glycoside (p. 940)
glycosidic bond (p. 939)
α-1,4′-glycosidic linkage (p. 943)
α-1,6′-glucosidic linkage (p. 947)
β-1,4′-glycosidic linkage (p. 944)
Haworth projection (p. 935)
heptose (p. 923)
hexose (p. 923)
ketose (p. 923)
Kiliani–Fischer synthesis (p. 931)
molecular recognition (p. 921)
monosaccharide (p. 922)
mutarotation (p. 935)
nonreducing sugar (p. 941)
oligosaccharide (p. 922)
osazone (p. 929)
oxocarbenium ion (p. 939)
pentose (p. 923)
polysaccharide (p. 922)
pyranose (p. 935)
pyranoside (p. 939)
reducing sugar (p. 941)
Ruff degradation (p. 931)
simple carbohydrate (p. 922)
tetrose (p. 923)
triose (p. 923)

Problems

30. Give the product or products that are obtained when D-galactose reacts with the following:
    a. nitric acid
    b. Tollens reagent
    c. $H_2/Pd/C$
    d. three equivalents of phenylhydrazine
    e. Br_2 in water
    f. ethanol + HCl
    g. product of reaction e (above) + $Ca(OH)_2$, Fe^{3+}, H_2O_2

31. Identify the following sugars:
    a. An aldopentose that is not D-arabinose forms D-arabinitol when it is reduced with $NaBH_4$.
    b. A sugar forms the same osazone as D-galactose with phenylhydrazine, but it is not oxidized by an aqueous solution of Br_2.
    c. A sugar that is not D-altrose forms D-altraric acid when it reacts with nitric acid.
    d. A ketose, when reduced with H_2 + Pd/C, forms D-altritol and D-allitol.

32. Answer the following questions about the eight aldopentoses:
    a. Which are enantiomers?
    b. Which give identical osazones?
    c. Which form an optically active compound when oxidized with nitric acid?

33. The reaction of D-ribose with one equivalent of methanol plus HCl forms four products. Give the structures of the products.

34. Determine the structure of D-galactose, using arguments similar to those used by Fischer to prove the structure of D-glucose.

35. Dr. Isent T. Sweet isolated a monosaccharide and determined that it had a molecular weight of 150. Much to his surprise, he found that it was not optically active. What is the structure of the monosaccharide?

36. The ^1H NMR spectrum of D-glucose in D_2O exhibits two high-frequency (low-field) doublets. What is responsible for these doublets?

37. Treatment with sodium borohydride converts aldose A into an optically inactive alditol. Ruff degradation of A forms B, whose alditol is optically inactive. Ruff degradation of B forms D-glyceraldehyde. Identify A and B.

38. D-Glucuronic acid is found widely in plants and animals. One of its functions is to detoxify poisonous HO-containing compounds by reacting with them in the liver to form glucuronides. Glucuronides are water soluble and therefore readily excreted. After ingestion of a poison such as turpentine, morphine, or phenol, the glucuronides of these compounds are found in the urine. Give the structure of the glucuronide formed by the reaction of β-D-glucuronic acid and phenol.

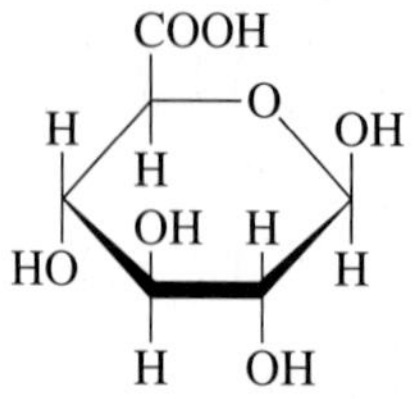

COOH H O OH H OH H HO H H OH

β-D-glucuronic acid

39. Hyaluronic acid, a component of connective tissue, is the fluid that lubricates the joints. It is an alternating polymer of *N*-acetyl-D-glucosamine and D-glucuronic acid joined by β-1,3′-glycosidic linkages. Draw a short segment of hyaluronic acid.

40. In order to synthesize D-galactose, Professor Amy Losse went to the stockroom to get some D-lyxose to use as a starting material. She found that the labels had fallen off the bottles containing D-lyxose and D-xylose. How could she determine which bottle contains D-lyxose?

41. When D-fructose is dissolved in D_2O and the solution is made basic, the D-fructose recovered from the solution has an average of 1.7 deuterium atoms attached to carbon per molecule. Show the mechanism that accounts for the incorporation of these deuterium atoms into D-fructose.

42. A D-aldopentose is oxidized by nitric acid to an optically active aldaric acid. A Ruff degradation of the aldopentose leads to a monosaccharide that is oxidized by nitric acid to an optically inactive aldaric acid. Identify the D-aldopentose.

43. How many aldaric acids are obtained from the 16 aldohexoses?

44. Calculate the percentages of α-D-glucose and β-D-glucose present at equilibrium from the specific rotations of α-D-glucose, β-D-glucose, and the equilibrium mixture. Compare your values with those given in Section 22.10. (*Hint:* The specific rotation of the mixture equals the specific rotation of α-D-glucose times the fraction of glucose present in the α-form plus the specific rotation of β-D-glucose times the fraction of glucose present in the β-form.)

45. Predict whether D-altrose exists preferentially as a pyranose or a furanose. (*Hint:* The most stable arrangement for a five-membered ring is for all the adjacent substituents to be trans.)

46. Propose a mechanism for the rearrangement that converts an α-hydroxyimine into an α-aminoketone in the presence of a trace amount of acid (Section 22.6).

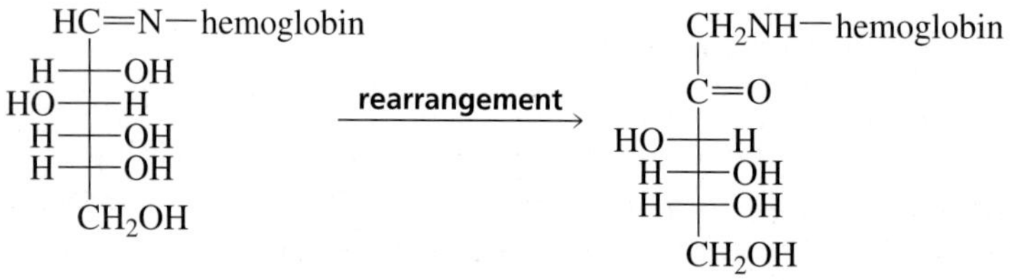

HC=N—hemoglobin H—OH HO—H H—OH H—OH CH_2OH **rearrangement** CH_2NH—hemoglobin C=O HO—H H—OH H—OH CH_2OH

47. A disaccharide forms a silver mirror with Tollens reagent and is hydrolyzed by a β-glycosidase. When the disaccharide is treated with excess methyl iodide in the presence of Ag_2O and then hydrolyzed with water under acidic conditions, 2,3,4-tri-*O*-methylmannose and 2,3,4,6-tetra-*O*-methylgalactose are formed.
   a. Draw the structure of the disaccharide.
   b. What is the function of Ag_2O?

48. All the glucose units in dextran have six-membered rings. When a sample of dextran is treated with methyl iodide and silver oxide and the product is hydrolyzed under acidic conditions, the products obtained are 2,3,4,6-tetra-*O*-methyl-D-glucose, 2,4,6-tri-*O*-methyl-D-glucose, 2,3,4-tri-*O*-methyl-D-glucose, and 2,4-di-*O*-methyl-D-glucose. Draw a short segment of dextran.

49. When a pyranose is in the chair conformation in which the CH_2OH group and the C-1 OH group are both in the axial position, the two groups can react to form an acetal. This is called the anhydro form of the sugar. (It has lost water.) The anhydro form of D-idose is shown here. In an aqueous solution at 100 °C, about 80% of D-idose exists in the anhydro form. Under the same conditions, only about 0.1% of D-glucose exists in the anhydro form. Explain.

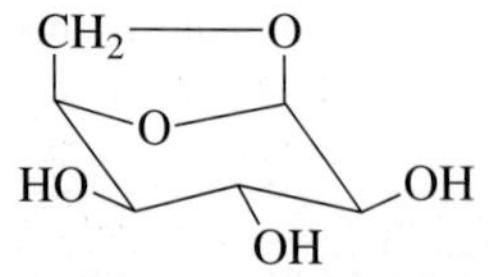

CH_2—O O HO OH OH

anhydro form of D-idose

50. Devise a method to convert D-glucose into D-allose.

23 Amino Acids, Peptides, and Proteins

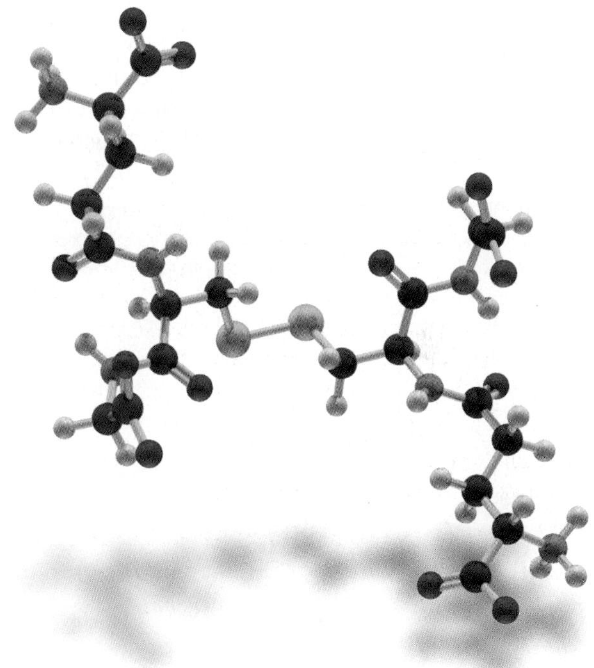

oxidized glutathione

The three kinds of polymers that are prevalent in nature are polysaccharides, proteins, and nucleic acids. You have already learned about polysaccharides, which are naturally occurring polymers of sugar subunits (Section 22.18), and nucleic acids are covered in Chapter 27. We will now look at proteins and the structurally similar, but shorter, peptides. **Peptides** and **proteins** are polymers of **amino acids** linked together by amide bonds. The repeating units are called **amino acid residues**.

Amino acid polymers can be composed of any number of monomers. A **dipeptide** contains two amino acid residues, a **tripeptide** contains three, an **oligopeptide** contains three to 10, and a **polypeptide** contains many amino acid residues. Proteins are naturally occurring polypeptides that are made up of 40 to 4000 amino acid residues.

From the structure of an amino acid, we can see that the name is not very precise. The compounds commonly called amino acids are more precisely called α-aminocarboxylic acids.

$$\mathrm{R{-}CH(\overset{+}{N}H_3){-}\overset{\overset{\large O}{\|}}{C}{-}OH}$$

α-aminocarboxylic acid
an amino acid

$$\mathrm{{-}NHCH(R)\overset{\overset{\large O}{\|}}{C}{-}NHCH(R')\overset{\overset{\large O}{\|}}{C}{-}NHCH(R'')\overset{\overset{\large O}{\|}}{C}{-}}$$

amino acids are linked together by amide bonds

Proteins and peptides serve many functions in biological systems. Some protect organisms from their environment or impart strength to certain biological structures. Hair, horns, hooves, feathers, fur, and the tough outer layer of skin are all composed largely of a **structural protein** called keratin. Collagen, another structural protein, is a major component of bones, muscles, and tendons. Some proteins have other protective functions. Snake venoms and plant toxins, for example, protect their owners from other species, blood-clotting proteins protect the vascular system when it is injured, and antibodies and peptide antibiotics protect us from disease. A group of proteins called **enzymes** catalyzes the chemical reactions that occur in living systems, and some of the hormones that regulate these reactions are peptides. Proteins are also responsible for many physiological functions, such as the transport and storage of oxygen in the body and the contraction of muscles.

23.1 Classification and Nomenclature of Amino Acids

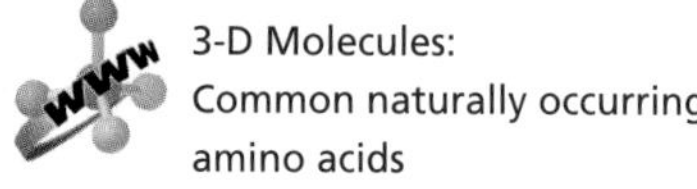

3-D Molecules: Common naturally occurring amino acids

The structures of the 20 most common naturally occurring amino acids and the frequency with which each occurs in proteins are shown in Table 23.1. Other amino acids occur in nature, but only infrequently. All amino acids except proline contain a primary amino group. Proline contains a secondary amino group incorporated into a five-membered ring. The amino acids differ only in the substituent (R) attached to the α-carbon. The wide variation in these substituents (called side chains) is what gives proteins their great structural diversity and, as a consequence, their great functional diversity.

Table 23.1 The Most Common Naturally Occurring Amino Acids
The amino acids are shown in the form that predominates at physiological pH (7.3).

	Formula	Name	Abbreviations		Average relative abundance in proteins
Aliphatic side chain amino acids	$H{-}CH(\overset{+}{N}H_3)C(=O)O^-$	Glycine	Gly	G	7.5%
	$CH_3{-}CH(\overset{+}{N}H_3)C(=O)O^-$	Alanine	Ala	A	9.0%
	$CH_3CH(CH_3){-}CH(\overset{+}{N}H_3)C(=O)O^-$	Valine*	Val	V	6.9%
	$CH_3CH(CH_3)CH_2{-}CH(\overset{+}{N}H_3)C(=O)O^-$	Leucine*	Leu	L	7.5%
	$CH_3CH_2CH(CH_3){-}CH(\overset{+}{N}H_3)C(=O)O^-$	Isoleucine*	Ile	I	4.6%
Hydroxy-containing amino acids	$HOCH_2{-}CH(\overset{+}{N}H_3)C(=O)O^-$	Serine	Ser	S	7.1%
	$CH_3CH(OH){-}CH(\overset{+}{N}H_3)C(=O)O^-$	Threonine*	Thr	T	6.0%

* Essential amino acids

Table 23.1 (continued)

	Formula	Name	Abbreviations		Average relative abundance in proteins
Sulfur-containing amino acids	$HSCH_2{-}CH(NH_3^+)CO_2^-$	Cysteine	Cys	C	2.8%
	$CH_3SCH_2CH_2{-}CH(NH_3^+)CO_2^-$	Methionine*	Met	M	1.7%
Acidic amino acids	$^-O_2CCH_2{-}CH(NH_3^+)CO_2^-$	Aspartate (aspartic acid)	Asp	D	5.5%
	$^-O_2CCH_2CH_2{-}CH(NH_3^+)CO_2^-$	Glutamate (glutamic acid)	Glu	E	6.2%
Amides of acidic amino acids	$H_2NC(O)CH_2{-}CH(NH_3^+)CO_2^-$	Asparagine	Asn	N	4.4%
	$H_2NC(O)CH_2CH_2{-}CH(NH_3^+)CO_2^-$	Glutamine	Gln	Q	3.9%
Basic amino acids	$H_3\overset{+}{N}CH_2CH_2CH_2CH_2{-}CH(NH_3^+)CO_2^-$	Lysine*	Lys	K	7.0%
	$H_2NC(=\overset{+}{N}H_2)NHCH_2CH_2CH_2{-}CH(NH_3^+)CO_2^-$	Arginine*	Arg	R	4.7%
Benzene-containing amino acids	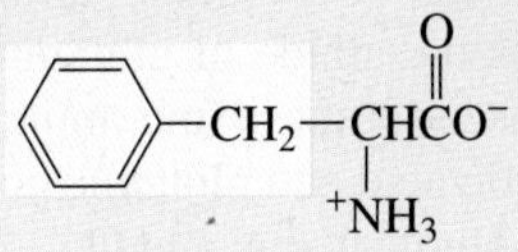 $C_6H_5{-}CH_2{-}CH(NH_3^+)CO_2^-$	Phenylalanine*	Phe	F	3.5%
	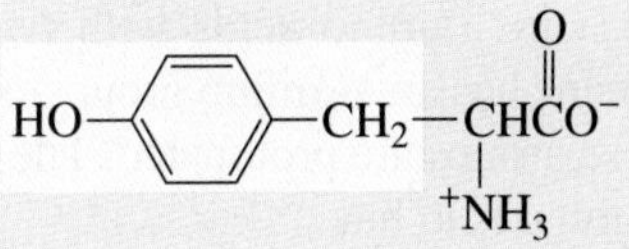 $HO{-}C_6H_4{-}CH_2{-}CH(NH_3^+)CO_2^-$	Tyrosine	Tyr	Y	3.5%
Heterocylic amino acids	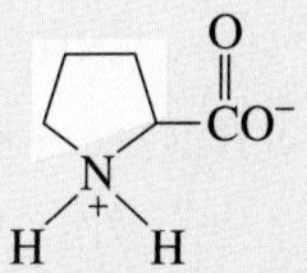 pyrrolidine ring with $\overset{+}{N}H_2$ and CO_2^-	Proline	Pro	P	4.6%

* Essential amino acids

Table 23.1 (continued)

	Formula	Name	Abbreviations		Average relative abundance in proteins
Heterocyclic amino acids (continued)	(imidazolyl)$CH_2-\overset{\overset{\displaystyle O}{\|}}{C}HCO^-$ with $^+NH_3$ on α-carbon	Histidine*	His	H	2.1%
	(3-indolyl)$CH_2-\overset{\overset{\displaystyle O}{\|}}{C}HCO^-$ with $^+NH_3$ on α-carbon	Tryptophan*	Trp	W	1.1%

* Essential amino acids

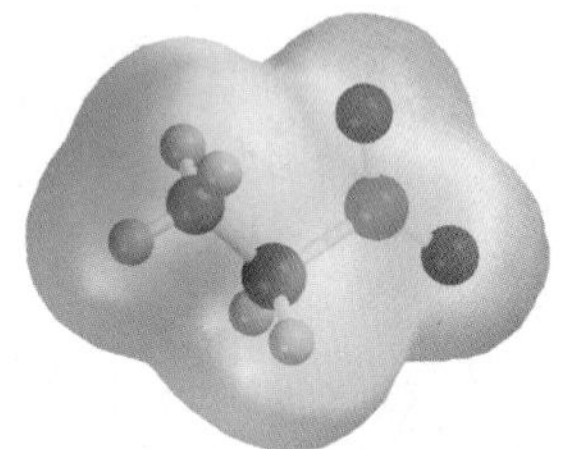

glycine

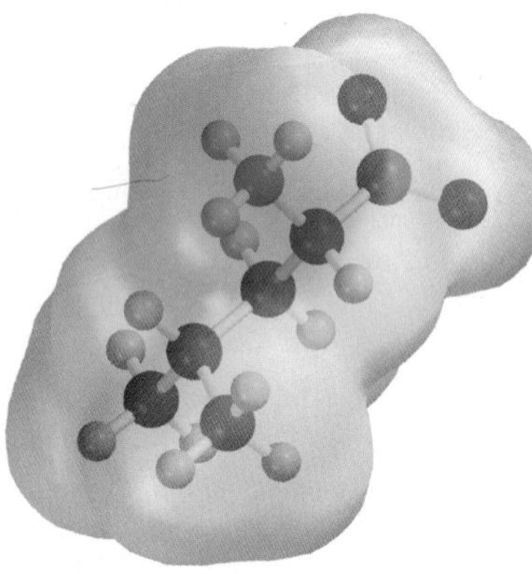

leucine

The amino acids are almost always called by their common names. Often, the name tells you something about the amino acid. For example, glycine got its name because of its sweet taste (*glykos* is Greek for "sweet"), and valine, like valeric acid, has five carbon atoms. Asparagine was first found in asparagus, and tyrosine was isolated from cheese (*tyros* is Greek for "cheese").

Dividing the amino acids into classes makes them easier to learn. The aliphatic side chain amino acids include glycine, the amino acid in which R = H, and four amino acids with alkyl side chains. Alanine is the amino acid with a methyl side chain, and valine has an isopropyl side chain. Can you guess which amino acid—leucine or isoleucine—has an isobutyl side chain? If you gave the obvious answer, you guessed incorrectly. Isoleucine does *not* have an "iso" group; it is leucine that has an isobutyl substituent—isoleucine has a *sec*-butyl substituent. Each of the amino acids has both a three-letter abbreviation (the first three letters of the name in most cases) and a single-letter abbreviation.

Two amino acid side chains—serine and threonine—contain alcohol groups. Serine is an HO-substituted alanine and threonine has a branched ethanol substituent. There are also two sulfur-containing amino acids: Cysteine is an HS-substituted alanine and methionine has a 2-methylthioethyl substituent.

There are two acidic amino acids (amino acids with two carboxylic acid groups): aspartate and glutamate. Aspartate is a carboxy-substituted alanine and glutamate has one more methylene group than aspartate. (If their carboxyl groups are protonated, they are called aspartic acid and glutamic acid, respectively.) Two amino acids—asparagine and glutamine—are amides of the acidic amino acids; asparagine is the amide of aspartate and glutamine is the amide of glutamate. Notice that the obvious one-letter abbreviations cannot be used for these four amino acids because A and G are used for alanine and glycine. Aspartic acid and glutamic acid are abbreviated D and E, and asparagine and glutamine are abbreviated N and Q.

There are two basic amino acids (amino acids with two basic nitrogen-containing groups): lysine and arginine. Lysine has an ϵ-amino group and arginine has a δ-guanidino group. At physiological pH, these groups are protonated. The ϵ and δ can remind you how many methylene groups each amino acid has.

$$H_3\overset{+}{N}-\overset{\epsilon}{C}H_2\overset{\delta}{C}H_2\overset{\gamma}{C}H_2\overset{\beta}{C}H_2\overset{\alpha}{C}H\overset{\overset{\displaystyle O}{\|}}{C}O^- \quad (\alpha\text{-}C\text{ bears } {}^+NH_3)$$

an ϵ-amino group

lysine

$$H_2N-\overset{\overset{\displaystyle {}^+NH_2}{\|}}{C}-NH-\overset{\delta}{C}H_2\overset{\gamma}{C}H_2\overset{\beta}{C}H_2\overset{\alpha}{C}H\overset{\overset{\displaystyle O}{\|}}{C}O^- \quad (\alpha\text{-}C\text{ bears } {}^+NH_3)$$

a δ-guanidino group

arginine

Two amino acids—phenylalanine and tyrosine—contain benzene rings. As its name indicates, phenylalanine is phenyl-substituted alanine. Tyrosine is phenylalanine with a *para*-hydroxy substituent.

Proline, histidine, and tryptophan are heterocyclic amino acids. Proline has its nitrogen incorporated into a five-membered ring—it is the only amino acid that contains a secondary amino group. Histidine is an imidazole-substituted alanine. Imidazole is an aromatic compound because it is cyclic and planar and has three pairs of delocalized π electrons (Section 21.11). The pK_a of a protonated imidazole ring is 6.0, so the ring will be protonated in acidic solutions and nonprotonated in basic solutions (Section 23.3).

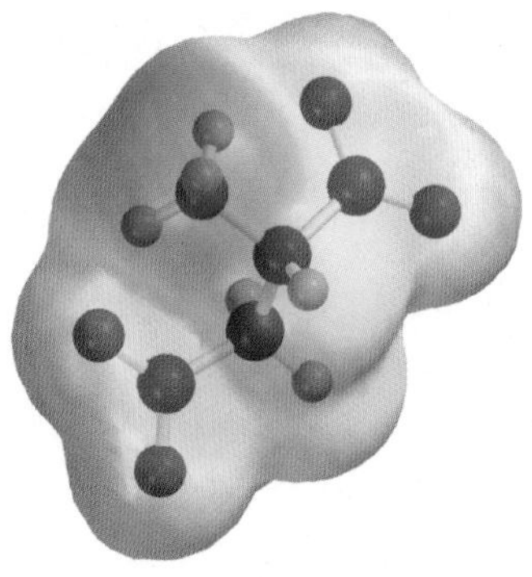
aspartate

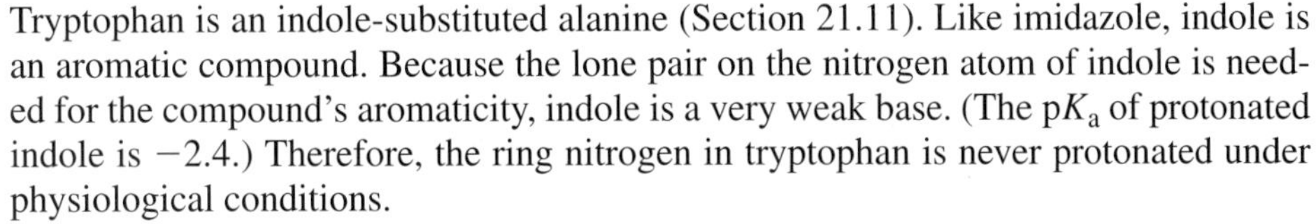

Tryptophan is an indole-substituted alanine (Section 21.11). Like imidazole, indole is an aromatic compound. Because the lone pair on the nitrogen atom of indole is needed for the compound's aromaticity, indole is a very weak base. (The pK_a of protonated indole is −2.4.) Therefore, the ring nitrogen in tryptophan is never protonated under physiological conditions.

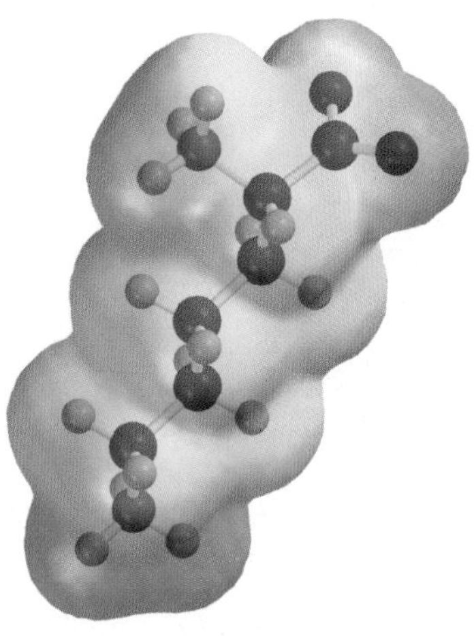
lysine

Ten amino acids are *essential amino acids*. We humans must obtain these 10 **essential amino acids** from our diets because we either cannot synthesize them at all or cannot synthesize them in adequate amounts. For example, we must have a dietary source of phenylalanine because we cannot synthesize benzene rings. However, we do not need tyrosine in our diets, because we can synthesize the necessary amounts from phenylalanine. The essential amino acids are denoted by red asterisks (*) in Table 23.1. Although humans can synthesize arginine, it is needed for growth in greater amounts than can be synthesized. So arginine is an essential amino acid for children, but a nonessential amino acid for adults. Not all proteins contain the same amino acids. Bean protein is deficient in methionine, for example, and wheat protein is deficient in lysine. They are *incomplete* proteins: They contain too little of one or more essential amino acids to support growth. Therefore, a balanced diet must contain proteins from different sources.

Dietary protein is hydrolyzed in the body to individual amino acids. Some of these amino acids are used to synthesize proteins needed by the body, some are broken down further to supply energy to the body, and some are used as starting materials for the synthesis of nonprotein compounds the body needs, such as adrenaline, thyroxine, and melanin (Section 25.6).

N̈
H
indole

PROBLEM 1

a. Explain why, when the imidazole ring of histidine is protonated, the double-bonded nitrogen is the nitrogen that accepts the proton.

$$\text{:N}\diagdown\!\!\text{:NH (ring)}-\text{CH}_2\text{CHCOO}^-\ (\text{NH}_2) + 2\,\text{H}^+ \rightleftharpoons \overset{+}{\text{H}}\text{N}\diagdown\!\!\text{:NH (ring)}-\text{CH}_2\text{CHCOO}^-\ (\overset{+}{\text{N}}\text{H}_3)$$

b. Explain why, when the guanidino group of arginine is protonated, the double-bonded nitrogen is the nitrogen that accepts the proton.

$$\text{H}_2\ddot{\text{N}}\overset{\ddot{\text{N}}\text{H}}{\overset{\|}{\text{C}}}\ddot{\text{N}}\text{HCH}_2\text{CH}_2\text{CH}_2\underset{\text{NH}_2}{\text{CH}}\overset{\text{O}}{\overset{\|}{\text{C}}}\text{O}^- + 2\,\text{H}^+ \rightleftharpoons \text{H}_2\ddot{\text{N}}\overset{^+\text{NH}_2}{\overset{\|}{\text{C}}}\ddot{\text{N}}\text{HCH}_2\text{CH}_2\text{CH}_2\underset{^+\text{NH}_3}{\text{CH}}\overset{\text{O}}{\overset{\|}{\text{C}}}\text{O}^-$$

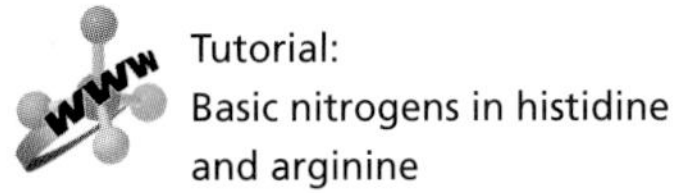

Tutorial:
Basic nitrogens in histidine and arginine

23.2 Configuration of Amino Acids

alanine
an amino acid

The α-carbon of all the naturally occurring amino acids except glycine is an asymmetric carbon. Therefore, 19 of the 20 amino acids listed in Table 23.1 can exist as enantiomers. The D and L notation used for monosaccharides (Section 22.2) is also used for amino acids. The D and L isomers of monosaccharides and amino acids are defined the same way. Thus, an amino acid drawn in a Fischer projection with the carboxyl group on the top and the R group on the bottom of the vertical axis is a **D-amino acid** if the amino group is on the right and an **L-amino acid** if the amino group is on the left. Unlike monosaccharides, where the D isomer is the one found in nature, most amino acids found in nature have the L configuration. To date, D-amino acid residues have been found only in a few peptide antibiotics and in some small peptides attached to the cell walls of bacteria.

D-glyceraldehyde: O=C—H on top; H—C—OH; CH_2OH at bottom

L-glyceraldehyde: O=C—H on top; HO—C—H; CH_2OH at bottom

D-amino acid: O=C—O^- on top; H—C—$\overset{+}{N}H_3$; R at bottom

L-amino acid: O=C—O^- on top; $H_3\overset{+}{N}$—C—H; R at bottom

Why D-sugars and L-amino acids? While it makes no difference which isomer nature "selected" to be synthesized, it is important that the same isomer be synthesized by all organisms. For example, if mammals ended up having L-amino acids, then L-amino acids would need to be the isomers synthesized by the organisms upon which mammals depend for food.

AMINO ACIDS AND DISEASE

The Chamorro people of Guam have a high incidence of a syndrome that resembles amyotrophic lateral sclerosis (ALS or Lou Gehrig's disease) with elements of Parkinson's disease and dementia. This syndrome developed during World War II when, as a result of food shortages, the tribe ate large quantities of *Cycas circinalis* seeds. These seeds contain β-methylamino-L-alanine, an amino acid that binds to glutamate receptors. When monkeys are given β-methylamino-L-alanine, they develop some of the features of this syndrome. There is hope that, by studying the mechanism of action of β-methylamino-L-alanine, we may gain an understanding of how ALS and Parkinson's disease arise.

PROBLEM 2◆

a. Which isomer—(*R*)-alanine or (*S*)-alanine—is D-alanine?
b. Which isomer—(*R*)-aspartate or (*S*)-aspartate—is D-aspartate?
c. Can a general statement be made relating *R* and *S* to D and L?

PROBLEM 3◆

Which amino acids in Table 23.1 have more than one asymmetric carbon?

23.3 Acid–Base Properties of Amino Acids

Every amino acid has a carboxyl group and an amino group, and each group can exist in an acidic form or a basic form, depending on the pH of the solution in which the amino acid is dissolved. The carboxyl groups of the amino acids have pK_a values of approximately 2, and the protonated amino groups have pK_a values near 9 (Table 23.2). Both groups, therefore, will be in their acidic forms in a very acidic solution (pH ~ 0). At pH = 7, the pH of the solution is greater than the pK_a of the carboxyl group, but less than the pK_a of the protonated amino group. The carboxyl group, therefore, will be in its basic form and the amino group will be in its acidic form. In a strongly basic solution (pH ~ 11), both groups will be in their basic forms.

Recall from the Henderson–Hasselbalch equation (Section 1.20) that the acidic form predominates if the pH of the solution is less than the pK_a of the compound and the basic form predominates if the pH of the solution is greater than the pK_a of the compound.

$$\underset{\text{pH} = 0}{R-\underset{\underset{^{+}NH_3}{|}}{CH}-\overset{\overset{O}{\|}}{C}-OH} \rightleftharpoons \underset{\substack{\text{a zwitterion} \\ \text{pH} = 7}}{R-\underset{\underset{^{+}NH_3}{|}}{CH}-\overset{\overset{O}{\|}}{C}-O^- + H^+} \rightleftharpoons \underset{\text{pH} = 11}{R-\underset{\underset{NH_2}{|}}{CH}-\overset{\overset{O}{\|}}{C}-O^- + H^+}$$

Notice that an amino acid can never exist as an uncharged compound, regardless of the pH of the solution. To be uncharged, an amino acid would have to lose a proton from an $^{+}NH_3$ group with a pK_a of about 9 before it would lose a proton from a COOH group with a pK_a of about 2. This clearly is impossible: A weak acid cannot be more acidic than a strong acid. Therefore, at physiological pH (7.3) an amino acid exists as a dipolar ion, called a *zwitterion*. A **zwitterion** is a compound that has a negative charge

Table 23.2 The pK_a Values of Amino Acids

Amino acid	pK_a α-COOH	pK_a α-NH_3^+	pK_a side chain
Alanine	2.34	9.69	—
Arginine	2.17	9.04	12.48
Asparagine	2.02	8.84	—
Aspartic acid	2.09	9.82	3.86
Cysteine	1.92	10.46	8.35
Glutamic acid	2.19	9.67	4.25
Glutamine	2.17	9.13	—
Glycine	2.34	9.60	—
Histidine	1.82	9.17	6.04
Isoleucine	2.36	9.68	—
Leucine	2.36	9.60	—
Lysine	2.18	8.95	10.79
Methionine	2.28	9.21	—
Phenylalanine	2.16	9.18	—
Proline	1.99	10.60	—
Serine	2.21	9.15	—
Threonine	2.63	9.10	—
Tryptophan	2.38	9.39	—
Tyrosine	2.20	9.11	10.07
Valine	2.32	9.62	—

on one atom and a positive charge on a nonadjacent atom. (The name comes from *zwitter*, German for "hermaphrodite" or "hybrid.")

A few amino acids have side chains with ionizable hydrogens (Table 23.2). The protonated imidazole side chain of histidine, for example, has a pK_a of 6.04. Histidine, therefore, can exist in four different forms, and the form that predominates depends on the pH of the solution.

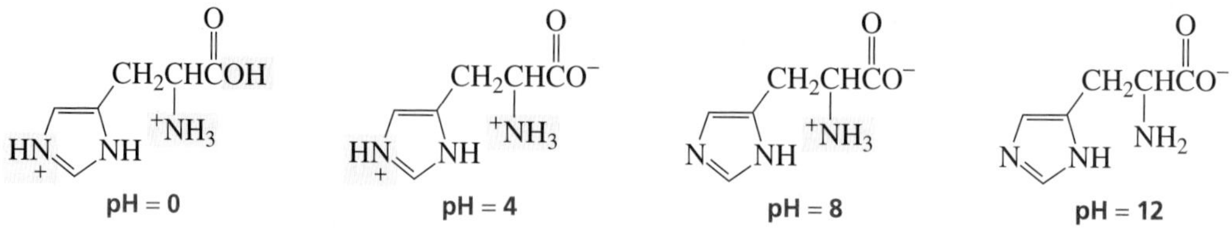

pH = 0 pH = 4 pH = 8 pH = 12

histidine

PROBLEM 4◆

Why are the carboxylic acid groups of the amino acids so much more acidic ($pK_a \sim 2$) than a carboxylic acid such as acetic acid ($pK_a = 4.76$)?

PROBLEM 5 SOLVED

Draw the form in which each of the following amino acids predominantly exists at physiological pH (7.3):

a. aspartic acid
b. histidine
c. glutamine
d. lysine
e. arginine
f. tyrosine

SOLUTION TO 5a Both carboxyl groups are in their basic forms because the pH is greater than their pK_a's. The protonated amino group is in its acidic form because the pH is less than its pK_a.

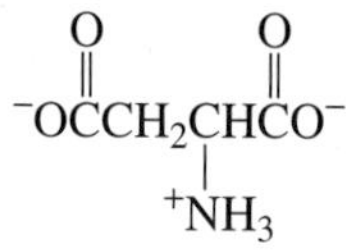

PROBLEM 6◆

Draw the form in which glutamic acid predominantly exists in a solution with the following pH:

a. pH = 0
b. pH = 3
c. pH = 6
d. pH = 11

PROBLEM 7

a. Why is the pK_a of the glutamic acid side chain greater than the pK_a of the aspartic acid side chain?
b. Why is the pK_a of the arginine side chain greater than the pK_a of the lysine side chain?

23.4 The Isoelectric Point

The **isoelectric point** (pI) of an amino acid is the pH at which it has no net charge. In other words, it is the pH at which the amount of positive charge on an amino acid exactly balances the amount of negative charge:

pI (isoelectric point) = pH at which there is no net charge

The pI of an amino acid that does *not* have an ionizable side chain—such as alanine—is midway between its two pK_a values. This is because at pH = 2.34, half the molecules have a negatively charged carboxyl group and half have an uncharged carboxyl group, and at pH = 9.69, half the molecules have a positively charged amino group and half have an uncharged amino group. As the pH increases from 2.34, the carboxyl group of more molecules becomes negatively charged; as the pH decreases from 9.69, the amino group of more molecules becomes positively charged. Therefore, at the average of the two pK_a values, the number of negatively charged groups equals the number of positively charged groups.

Recall from the Henderson–Hasselbalch equation that when pH = pK_a, half the group is in its acidic form and half is in its basic form (Section 1.20).

An amino acid will be positively charged if the pH of the solution is less than the pI of the amino acid and will be negatively charged if the pH of the solution is greater than the pI of the amino acid.

$$CH_3CH(\overset{+}{N}H_3)COOH$$

$pK_a = 2.34$ (COOH); $pK_a = 9.69$ ($^+NH_3$)

alanine

$$pI = \frac{2.34 + 9.69}{2} = \frac{12.03}{2} = 6.02$$

The pI of an amino acid that *has* an ionizable side chain is the average of the pK_a values of the similarly ionizing groups (a positively charged group ionizing to an uncharged group or an uncharged group ionizing to a negatively charged group). For example, the pI of lysine is the average of the pK_a values of the two groups that are positively charged in their acidic form and uncharged in their basic form. The pI of glutamate, on the other hand, is the average of the pK_a values of the two groups that are uncharged in their acidic form and negatively charged in their basic form.

$$H_3\overset{+}{N}CH_2CH_2CH_2CH_2CH(\overset{+}{N}H_3)COOH$$

$pK_a = 10.79$ (side-chain $^+NH_3$); $pK_a = 2.18$ (COOH); $pK_a = 8.95$ (α-$^+NH_3$)

lysine

$$pI = \frac{8.95 + 10.79}{2} = \frac{19.74}{2} = 9.87$$

$$HOOCCH_2CH_2CH(\overset{+}{N}H_3)COOH$$

$pK_a = 4.25$ (side-chain COOH); $pK_a = 2.19$ (COOH); $pK_a = 9.67$ ($^+NH_3$)

glutamic acid

$$pI = \frac{2.19 + 4.25}{2} = \frac{6.44}{2} = 3.22$$

PROBLEM 8

Explain why the pI of lysine is the average of the pK_a values of its two protonated amino groups.

PROBLEM 9◆

Calculate the pI of each of the following amino acids:

a. asparagine b. arginine c. serine d. aspartic acid

PROBLEM 10◆

a. Which amino acid has the lowest pI value?

b. Which amino acid has the highest pI value?

c. Which amino acid has the greatest amount of negative charge at pH = 6.20?

d. Which amino acid—glycine or methionine—has a greater negative charge at pH = 6.20?

PROBLEM 11

Explain why the pI values of tyrosine and cysteine cannot be determined by the method just described.

23.5 Separation of Amino Acids

Electrophoresis

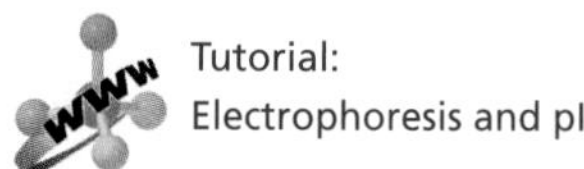

A mixture of amino acids can be separated by several different techniques. **Electrophoresis** separates amino acids on the basis of their pI values. A few drops of a solution of an amino acid mixture are applied to the middle of a piece of filter paper or to a gel. When the paper or the gel is placed in a buffered solution between two electrodes and an electric field is applied, an amino acid with a pI greater than the pH of the solution will have an overall positive charge and will migrate toward the cathode (the negative electrode). The farther the amino acid's pI is from the pH of the buffer, the more positive the amino acid will be and the farther it will migrate toward the cathode in a given amount of time. An amino acid with a pI less than the pH of the buffer will have an overall negative charge and will migrate toward the anode (the positive electrode). If two molecules have the same charge, the larger one will move more slowly during electrophoresis because the same charge has to move a greater mass.

Since amino acids are colorless, how can we detect them after they have been separated? When amino acids are heated with ninhydrin, they form a colored product. After electrophoretic separation of the amino acids, the filter paper is sprayed with ninhydrin and dried in a warm oven. Most amino acids form a purple product. The number of different kinds of amino acids in the mixture is determined by the number of colored spots on the filter paper (Figure 23.1). The individual amino acids are identified by their location on the paper compared with a standard.

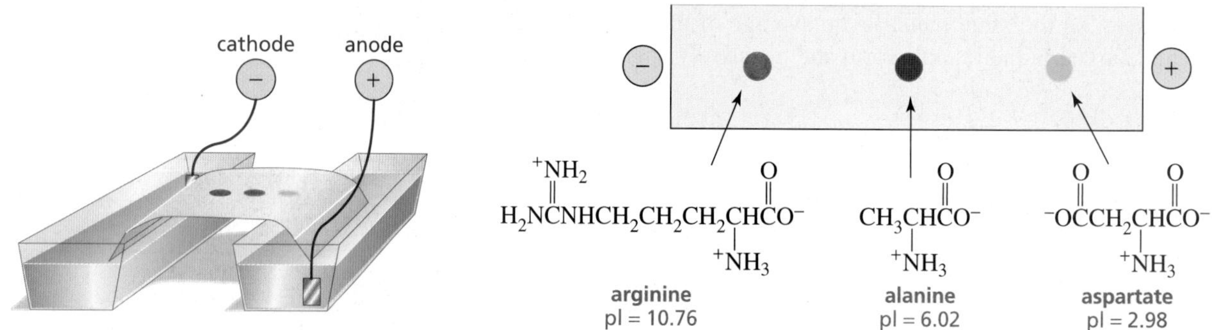

▲ **Figure 23.1**
Arginine, alanine, and aspartic acid separated by electrophoresis at pH = 5.

The mechanism for formation of the colored product is as shown, omitting the mechanisms for the steps involving dehydration, imine formation, and imine hydrolysis. (These mechanisms are shown in Sections 18.6 and 18.7.)

mechanism for the reaction of an amino acid with ninhydrin to form a colored product

ninhydrin ⇌ (indane-1,2,3-trione) + H_2O

+ H_2NCHCO^- (C=O, R) **an amino acid** ⟶ (indanedione)=N−CH(R)−C(=O)−$\ddot{O}$:$^-$ + H_2O

↓

(indanedione enolate, $\ddot{O}$:$^-$)−N=CHR + H−O−H + CO_2 ⟶ (indanedione)−N=CHR + HO^- ⟶ (H_2O) (indanedione)−NH_2 + RCH(=O)

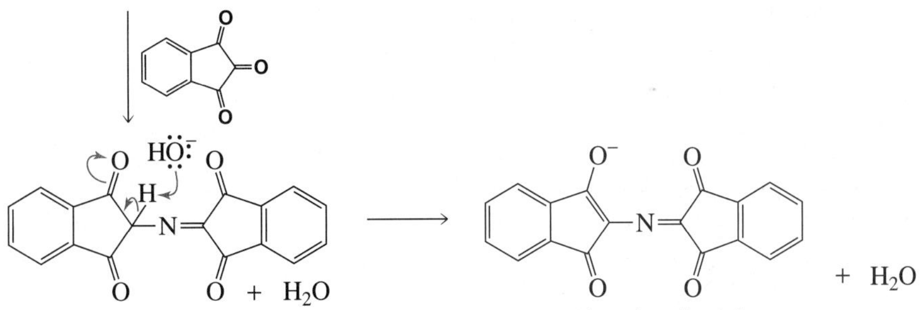

$+ H_2O$ $\longrightarrow$ $+ H_2O$

purple-colored product

Paper Chromatography and Thin-Layer Chromatography

Paper chromatography once played an important role in biochemical analysis because it provided a method for separating amino acids using very simple equipment. Although more modern techniques are now more commonly used, we will describe the principles behind paper chromatography because many of the same principles are employed in modern separation techniques.

The technique of paper chromatography separates amino acids on the basis of polarity. A few drops of a solution of an amino acid mixture are applied to the bottom of a strip of filter paper. The edge of the paper is then placed in a solvent (typically a mixture of water, acetic acid, and butanol). The solvent moves up the paper by capillary action, carrying the amino acids with it. Depending on their polarities, the amino acids have different affinities for the mobile (solvent) and stationary (paper) phases and therefore travel up the paper at different rates. The more polar the amino acid, the more strongly it is adsorbed onto the relatively polar paper. The less polar amino acids travel up the paper more rapidly, since they have a greater affinity for the mobile phase. Therefore, when the paper is developed with ninhydrin, the colored spot closest to the origin is the most polar amino acid and the spot farthest away from the origin is the least polar amino acid (Figure 23.2).

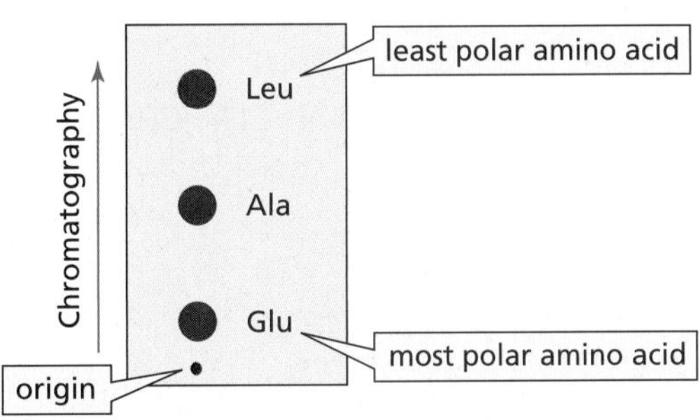

Figure 23.2 Separation of glutamate, alanine, and leucine by paper chromatography.

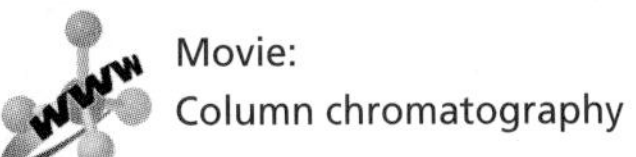

The most polar amino acids are those with charged side chains, the next most polar are those with side chains that can form hydrogen bonds, and the least polar are those with hydrocarbon side chains. For amino acids with hydrocarbon side chains, the larger the alkyl group, the less polar the amino acid. In other words, leucine is less polar than valine.

Paper chromatography has largely been replaced by **thin-layer chromatography** (TLC). TLC differs from paper chromatography in that TLC uses a plate with a coating of solid material instead of filter paper. The physical property on which the separation is based depends on the solid material and the solvent chosen for the mobile phase.

PROBLEM 12◆

A mixture of seven amino acids (glycine, glutamate, leucine, lysine, alanine, isoleucine, and aspartate) is separated by TLC. Explain why only six spots show up when the chromatographic plate is sprayed with ninhydrin and heated.

Ion-Exchange Chromatography

Electrophoresis and thin-layer chromatography are analytical separations—small amounts of amino acids are separated for analysis. Preparative separation, in which larger amounts of amino acids are separated for use in subsequent processes, can be achieved using **ion-exchange chromatography**. This technique employs a column packed with an insoluble resin. A solution of a mixture of amino acids is loaded onto the top of the column and eluted with a buffer. The amino acids separate because they flow through the column at different rates, as explained below.

The resin is a chemically inert material with charged side chains. One commonly used resin is a copolymer of styrene and divinylbenzene (see Chapter 28, Problem 32) with negatively charged sulfonic acid groups on some of the benzene rings (Figure 23.3). If a mixture of lysine and glutamate in a solution with a pH of 6 were loaded onto the column, glutamate would travel down the column rapidly because its negatively charged side chain would be repelled by the negatively charged sulfonic acid groups of the resin. The positively charged side chain of lysine, on the other hand, would cause that amino acid to be retained on the column. This kind of resin is called a **cation-exchange resin** because it exchanges the Na^+ counterions of the SO_3^- groups for the positively charged species that are added to the column. In addition, the relatively nonpolar nature of the column causes it to retain nonpolar amino acids longer than polar amino acids. Resins with positively charged groups are called **anion-exchange resins** because they impede the flow of anions by exchanging their negatively charged counterions for negatively charged species that are added to the column. A common anion-exchange resin (Dowex® 1) has $CH_2N^+(CH_3)_3Cl^-$ groups in place of the $SO_3^-Na^+$ groups in Figure 23.3.

Cations bind most strongly to cation-exchange resins.

Anions bind most strongly to anion-exchange resins.

$SO_3^-Na^+$ $SO_3^-Na^+$ $SO_3^-Na^+$

$-CH_2-CH-CH_2-CH-CH-CH_2-CH-CH_2-CH-CH_2-CH-CH_2-CH-$

$-CH_2-CH-CH_2-CH-CH_2-CH-CH_2-CH-CH_2-CH-CH_2-CH-$

$SO_3^-Na^+$ $SO_3^-Na^+$

Figure 23.3 ▶ A section of a cation-exchange resin. This particular resin is called Dowex® 50.

An **amino acid analyzer** is an instrument that automates ion-exchange chromatography. When a solution of an amino acid mixture passes through the column of an amino acid analyzer containing a cation-exchange resin, the amino acids move through the column at different rates, depending on their overall charge. The effluent (the solution leaving the column) is collected in fractions, which are collected often enough that a different amino acid ends up in each fraction (Figure 23.4). If ninhydrin is added to each of the fractions, the concentration of the amino acid in each fraction

WATER SOFTENERS: EXAMPLES OF CATION-EXCHANGE CHROMATOGRAPHY

Water softeners contain a column with a cation-exchange resin that has been flushed with concentrated sodium chloride. In Section 17.13, we saw that the presence of calcium and magnesium ions in water is what causes the water to be "hard." When water passes through the column, the resin binds magnesium and calcium ions more tightly than it binds sodium ions. In this way, the water softener removes magnesium and calcium ions from water, replacing them with sodium ions. The resin must be recharged from time to time by flushing it with concentrated sodium chloride to replace the bound magnesium and calcium ions with sodium ions.

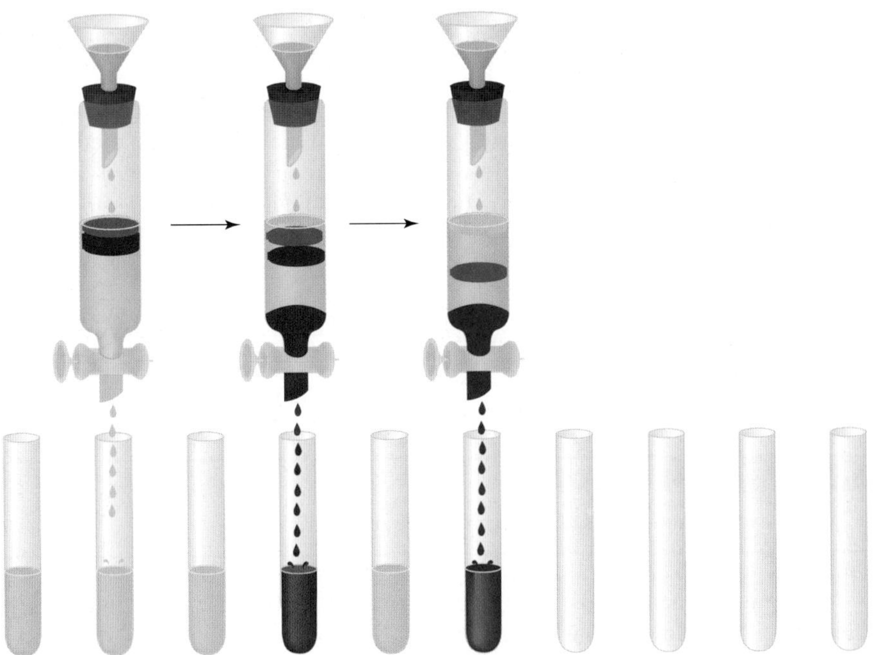

Figure 23.4
Separation of amino acids by ion-exchange chromatography.

can be determined by the amount of absorption at 570 nm—because the colored compound formed by the reaction of an amino acid with ninhydrin has a λ_{max} of 570 (Section 8.11). In this way, the identity and the relative amount of each amino acid can be determined (Figure 23.5).

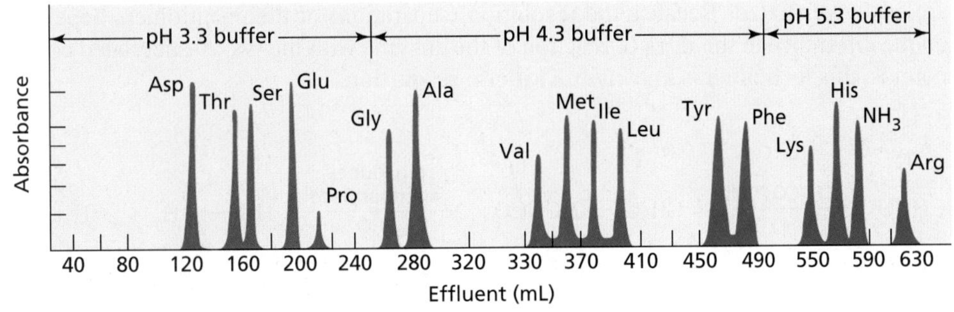

Figure 23.5
A typical chromatogram obtained from the separation of a mixture of amino acids using an automated amino acid analyzer.

PROBLEM 13

Why are buffer solutions of increasingly higher pH used to elute the column that generates the chromatogram shown in Figure 23.5?

PROBLEM 14

Explain the order of elution (with a buffer of pH 4) of each of the following pairs of amino acids on a column packed with Dowex® 50 (Figure 23.3):

a. aspartate before serine
b. glycine before alanine
c. valine before leucine
d. tyrosine before phenylalanine

PROBLEM 15◆

In what order would the following amino acids be eluted with a buffer of pH 4 from a column containing an anion-exchange resin (Dowex® 1)?

histidine, serine, aspartate, valine

23.6 Resolution of Racemic Mixtures of Amino Acids

Chemists do not have to rely on nature to produce amino acids; they can synthesize them in the laboratory, using a variety of methods. One of the oldest methods replaces an α-hydrogen of a carboxylic acid with a bromine in a Hell–Volhard–Zelinski reaction (Section 19.5). The resulting α-bromocarboxylic acid then undergoes an S_N2 reaction with ammonia to form the amino acid (Section 10.4).

$$\underset{\text{a carboxylic acid}}{RCH_2\overset{O}{\overset{\|}{C}}OH} \xrightarrow[\text{2. } H_2O]{\text{1. } Br_2,\ PBr_3} R\underset{Br}{\underset{|}{C}}H\overset{O}{\overset{\|}{C}}OH \xrightarrow{\text{excess } NH_3} \underset{\text{an amino acid}}{R\underset{^+NH_3}{\underset{|}{C}}H\overset{O}{\overset{\|}{C}}O^-} + \overset{+}{N}H_4Br^-$$

PROBLEM 16

Why is excess ammonia used in the preceding reaction?

When amino acids are synthesized in nature, only the L-enantiomer is formed (Section 5.20). However, when amino acids are synthesized in the laboratory, the product is usually a racemic mixture of D and L enantiomers. If only one isomer is desired, the enantiomers must be separated. They can be separated by means of an enzyme-catalyzed reaction. Because an enzyme is chiral, it will react at a different rate with each of the enantiomers (Section 5.20). For example, pig kidney aminoacylase is an enzyme that catalyzes the hydrolysis of *N*-acetyl-L-amino acids, but not *N*-acetyl-D-amino acids. Therefore, if the racemic amino acid is converted into a pair of *N*-acetylamino acids and the *N*-acetylated mixture is hydrolyzed with pig kidney aminoacylase, the products will be the L-amino acid and *N*-acetyl-D-amino acid, which are easily separated. Because the resolution (separation) of the enantiomers depends on the difference in the rates of reaction of the enzyme with the two *N*-acetylated compounds, this technique is known as a **kinetic resolution**.

$$\underset{\substack{\text{D-amino acid}\\+\\\text{L-amino acid}}}{H_2N\underset{R}{\underset{|}{C}}H\overset{O}{\overset{\|}{C}}O^-} \xrightarrow{CH_3\overset{O}{\overset{\|}{C}}O\overset{O}{\overset{\|}{C}}CH_3} \underset{\substack{N\text{-acetyl-D-amino acid}\\+\\N\text{-acetyl-L-amino acid}}}{CH_3\overset{O}{\overset{\|}{C}}-NH\underset{R}{\underset{|}{C}}H\overset{O}{\overset{\|}{C}}O^-} \xrightarrow[H_2O]{\text{pig kidney aminoacylase}} \underset{\text{L-amino acid}}{H_2N-\overset{CO^-}{\underset{R}{C}}-H} + CH_3\overset{O}{\overset{\|}{C}}O^- + \underset{N\text{-acetyl-D-amino acid}}{CH_3\overset{O}{\overset{\|}{C}}-NH\underset{R}{\underset{|}{C}}H\overset{O}{\overset{\|}{C}}O^-}$$

PROBLEM 17

Pig liver esterase is an enzyme that catalyzes the hydrolysis of esters. It hydrolyzes esters of L-amino acids more rapidly than esters of D-amino acids. How can this enzyme be used to separate a racemic mixture of amino acids?

PROBLEM 18◆

Amino acids can be synthesized by reductive amination of α-keto acids (Section 21.8).

$$R\underset{O}{\underset{\|}{C}}-\overset{O}{\overset{\|}{C}}-OH \xrightarrow[H_2/\text{Raney Ni}]{\text{excess ammonia}} R\underset{^+NH_3}{\underset{|}{C}}H-\overset{O}{\overset{\|}{C}}-O^-$$

Biological organisms can also convert α-keto acids into amino acids, but because H_2 and metal catalysts are not available to the cell, they do so by a different mechanism (Section 25.6.)

a. What amino acid is obtained from the reductive amination of each of the following metabolic intermediates in the cell?

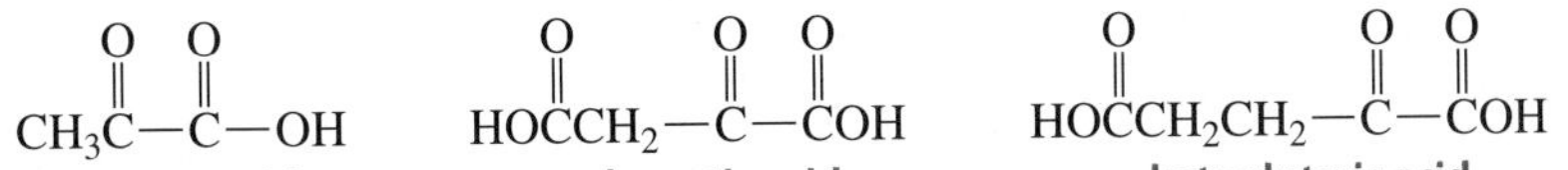

$$\underset{\text{pyruvic acid}}{CH_3\overset{O}{\overset{\|}{C}}-\overset{O}{\overset{\|}{C}}-OH} \qquad \underset{\text{oxaloacetic acid}}{HO\overset{O}{\overset{\|}{C}}CH_2-\overset{O}{\overset{\|}{C}}-\overset{O}{\overset{\|}{C}}OH} \qquad \underset{\alpha\text{-ketoglutaric acid}}{HO\overset{O}{\overset{\|}{C}}CH_2CH_2-\overset{O}{\overset{\|}{C}}-\overset{O}{\overset{\|}{C}}OH}$$

b. What amino acids are obtained from the same metabolic intermediates when they are synthesized in the laboratory?

23.7 Peptide Bonds and Disulfide Bonds

Peptide bonds and disulfide bonds are the only covalent bonds that hold amino acid residues together in a peptide or a protein.

Peptide Bonds

The amide bonds that link amino acid residues are called **peptide bonds**. By convention, peptides and proteins are written with the free amino group (the **N-terminal amino acid**) on the left and the free carboxyl group (the **C-terminal amino acid**) on the right.

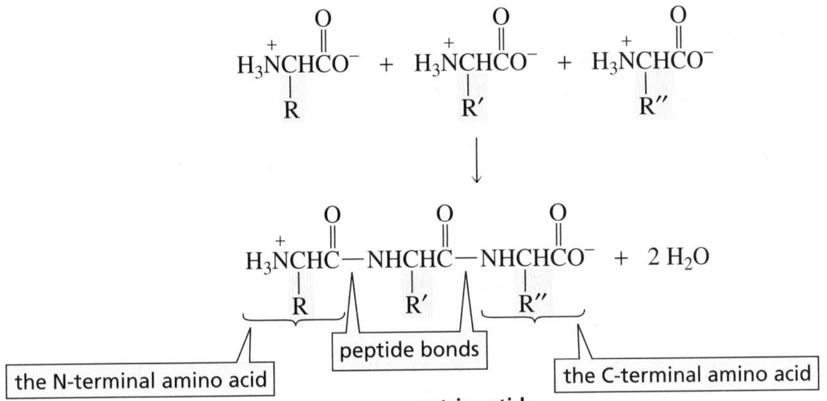

$$H_3\overset{+}{N}\underset{\substack{|\\R}}{C}H\overset{O}{\overset{\|}{C}}O^- + H_3\overset{+}{N}\underset{\substack{|\\R'}}{C}H\overset{O}{\overset{\|}{C}}O^- + H_3\overset{+}{N}\underset{\substack{|\\R''}}{C}H\overset{O}{\overset{\|}{C}}O^-$$

$$\downarrow$$

$$H_3\overset{+}{N}\underset{\substack{|\\R}}{C}H\overset{O}{\overset{\|}{C}}-NH\underset{\substack{|\\R'}}{C}H\overset{O}{\overset{\|}{C}}-NH\underset{\substack{|\\R''}}{C}H\overset{O}{\overset{\|}{C}}O^- + 2\,H_2O$$

the N-terminal amino acid — peptide bonds — the C-terminal amino acid

a tripeptide

When the identities of the amino acids in a peptide are known but their sequence is not known, the amino acids are written separated by commas. When the sequence of amino acids is known, the amino acids are written separated by hyphens. In the following pentapeptide shown on the right, valine is the N-terminal amino acid and histidine is the C-terminal amino acid. The amino acids are numbered starting with the N-terminal end. The glutamate residue is referred to as Glu 4 because it is the fourth amino acid from the N-terminal end. In naming the peptide, adjective names (ending in "yl") are used for all the amino acids except the C-terminal amino acid. Thus, this pentapeptide is named valylcysteylalanylglutamylhistidine.

Glu, Cys, His, Val, Ala
the pentapeptide contains the indicated amino acids, but their sequence is not known

Val-Cys-Ala-Glu-His
the amino acids in the pentapeptide have the indicated sequence

A peptide bond has about 40% double-bond character because of electron delocalization (Section 17.2). Steric hindrance causes the trans configuration to be more stable than the cis configuration, so the α-carbons of adjacent amino acids are trans to each other (Section 4.11).

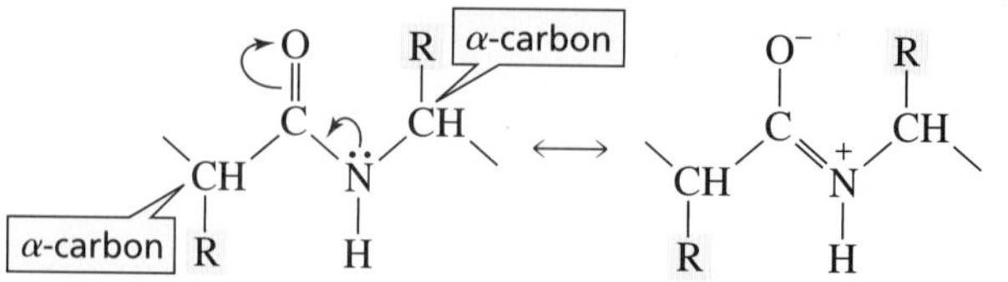

α-carbon α-carbon

trans configuration

Free rotation about the peptide bond is not possible because of its partial double-bond character. The carbon and nitrogen atoms of the peptide bond and the two atoms to which each is attached are held rigidly in a plane (Figure 23.6). This regional planarity affects the way a chain of amino acids can fold, so it has important implications for the three-dimensional shapes of peptides and proteins (Section 23.13).

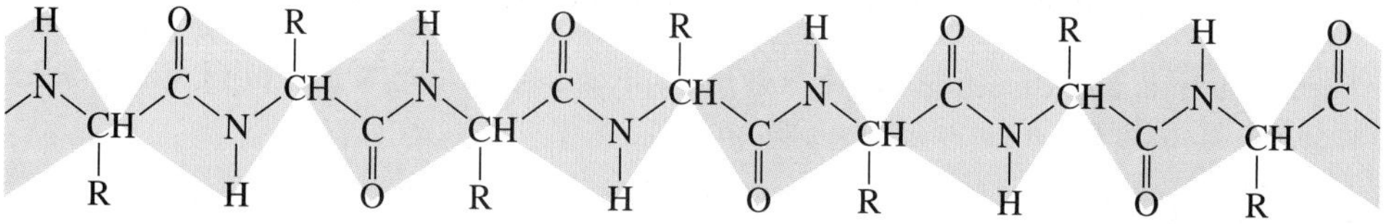

Figure 23.6 A segment of a polypeptide chain. The plane defined by each peptide bond is indicated. Notice that the R groups bonded to the α-carbons are on alternate sides of the peptide backbone.

PROBLEM 19

Draw a peptide bond in a cis configuration.

Disulfide Bonds

When thiols are oxidized under mild conditions, they form disulfides. A **disulfide** is a compound with an S—S bond.

$$2\,R{-}SH \xrightarrow{\text{mild oxidation}} RS{-}SR$$

a thiol → a disulfide

An oxidizing agent commonly used for this reaction is Br_2 (or I_2) in a basic solution.

mechanism for oxidation of a thiol to a disulfide

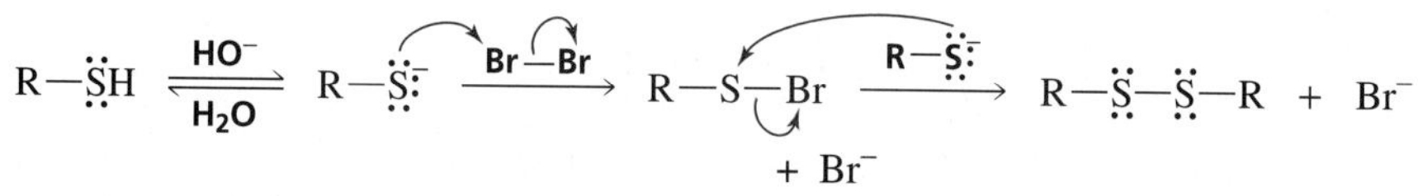

$$R{-}\ddot{S}H \underset{H_2O}{\overset{HO^-}{\rightleftharpoons}} R{-}\ddot{S}{:}^- \xrightarrow{Br{-}Br} R{-}S{-}Br + Br^- \xrightarrow{R{-}\ddot{S}{:}^-} R{-}\ddot{S}{-}\ddot{S}{-}R + Br^-$$

Because thiols can be oxidized to disulfides, disulfides can be reduced to thiols.

$$RS{-}SR \xrightarrow{\text{reduction}} 2\,R{-}SH$$

a disulfide → a thiol

Cysteine is an amino acid that contains a thiol group. Two cysteine molecules therefore can be oxidized to a disulfide. This disulfide is called cystine.

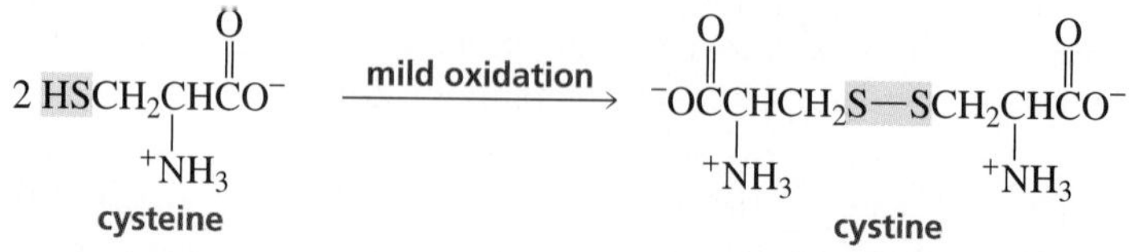

$$2\,HSCH_2CH(\overset{+}{N}H_3)CO_2^- \xrightarrow{\text{mild oxidation}} {}^-O_2CCH(\overset{+}{N}H_3)CH_2S{-}SCH_2CH(\overset{+}{N}H_3)CO_2^-$$

cysteine → cystine

Two cysteine residues in a protein can be oxidized to a disulfide. This is known as a **disulfide bridge**. Disulfide bridges are the only covalent bonds that are found between nonadjacent amino acids in peptides and proteins. They contribute to the overall shape of a protein by holding the cysteine residues in close proximity, as shown in Figure 23.7.

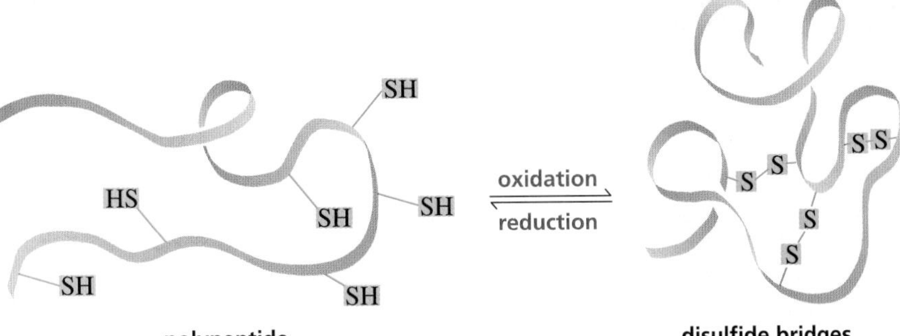

Figure 23.7 Disulfide bridges cross-linking portions of a peptide.

Insulin, a hormone secreted by the pancreas, controls the level of glucose in the blood by regulating glucose metabolism. Insulin is a polypeptide with two peptide chains. The short chain (the A-chain) contains 21 amino acids and the long chain (the B-chain) contains 30 amino acids. The two chains are held together by two disulfide bridges. These are **interchain disulfide bridges** (between the A- and B-chains). Insulin also has an **intrachain disulfide bridge** (within the A-chain).

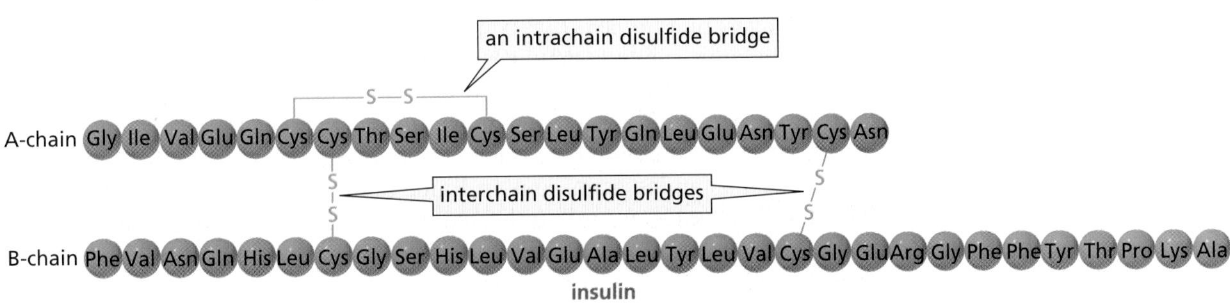

HAIR: STRAIGHT OR CURLY?

Hair is made up of a protein known as keratin. Keratin contains an unusually large number of cysteine residues (about 8% of the amino acids), which give it many disulfide bridges to maintain its three-dimensional structure. People can alter the structure of their hair (if they feel that it is either too straight or too curly) by changing the location of these disulfide bridges. This is accomplished by first applying a reducing agent to the hair to reduce all the disulfide bridges in the protein strands. Then the hair is given the desired shape (using curlers to curl it or combing it straight to uncurl it), and an oxidizing agent is applied that forms new disulfide bridges. The new disulfide bridges maintain the hair's new shape. When this treatment is applied to straight hair, it is called a "permanent." When it is applied to curly hair, it is called "hair straightening."

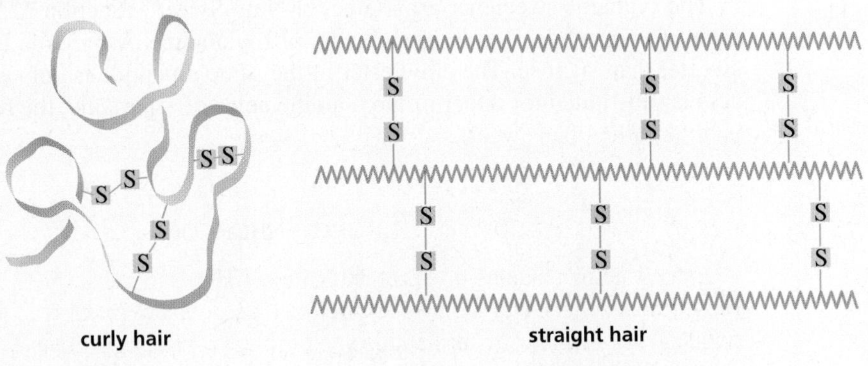

PROBLEM 20◆

a. How many different octapeptides can be made from the 20 naturally occurring amino acids?

b. How many different proteins containing 100 amino acids can be made from the 20 naturally occurring amino acids?

PROBLEM 21◆

Which bonds in the backbone of a peptide can rotate freely?

23.8 Some Interesting Peptides

Oxytocin was the first small peptide to be synthesized. Its synthesis was achieved in 1953 by **Vincent du Vigneaud (1901–1978)***, who later synthesized vasopressin. Du Vigneaud was born in Chicago and was a professor at George Washington University Medical School and later at Cornell University Medical College. For synthesizing these nonapeptides, he received the Nobel Prize in chemistry in 1955.*

Enkephalins are pentapeptides that are synthesized by the body to control pain. They decrease the body's sensitivity to pain by binding to receptors in certain brain cells. Part of the three-dimensional structures of enkephalins must be similar to those of morphine and painkillers such as Demerol® because they bind to the same receptors (Sections 30.3 and 30.6).

Tyr-Gly-Gly-Phe-Leu	Tyr-Gly-Gly-Phe-Met
leucine enkephalin	**methionine enkephalin**

Bradykinin, vasopressin, and oxytocin are peptide hormones. They are all nonapeptides. Bradykinin inhibits the inflammation of tissues. Vasopressin controls blood pressure by regulating the contraction of smooth muscle. It is also an antidiuretic. Oxytocin induces labor in pregnant women and stimulates milk production in nursing mothers. Vasopressin and oxytocin both have an intrachain disulfide bond, and their C-terminal amino acids contain amide rather than carboxyl groups. Notice that the C-terminal amide group is indicated by writing "NH_2" after the name of the C-terminal amino acid. In spite of their very different physiological effects, vasopressin and oxytocin differ only by two amino acids.

bradykinin Arg-Pro-Pro-Gly-Phe-Ser-Pro-Phe-Arg

vasopressin Cys-Tyr-Phe-Gln-Asn-Cys-Pro-Arg-Gly-NH_2 (S—S bond between the two Cys)

oxytocin Cys-Tyr-Ile-Gln-Asn-Cys-Pro-Leu-Gly-NH_2 (S—S bond between the two Cys)

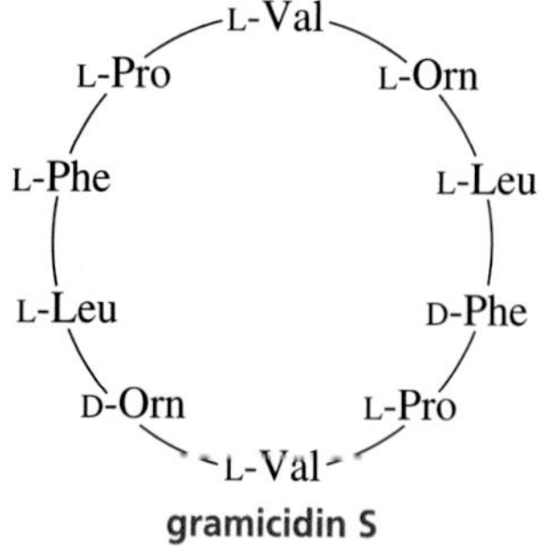

gramicidin S

Gramicidin S is an antibiotic produced by a strain of bacteria. It is a cyclic decapeptide. Notice that it contains the amino acids L-ornithine (L-Orn), D-ornithine (D-Orn), and also D-phenylalanine. Ornithine is not listed in Table 23.1 because it occurs rarely in nature. Ornithine resembles lysine, but has one less methylene group in its side chain.

The synthetic sweetener aspartame, or NutraSweet® (Section 22.21), is the methyl ester of a dipeptide of L-aspartate and L-phenylalanine. Aspartame is about 200 times sweeter than sucrose. The ethyl ester of the same dipeptide is not sweet. If a D-amino acid is substituted for either of the L-amino acids of aspartame, the resulting dipeptide is bitter rather than sweet.

$$H_3\overset{+}{N}CH_2CH_2CH_2\underset{\overset{|}{^+NH_3}}{CH}\overset{\overset{O}{\|}}{C}O^-$$

ornithine

$$H_3\overset{+}{N}\underset{\underset{COO^-}{|}}{\underset{|}{CH_2}}\!\!\!\!CH\overset{\overset{O}{\|}}{C}—NH\underset{\underset{C_6H_5}{|}}{\underset{|}{CH_2}}\!\!\!\!CH\overset{\overset{O}{\|}}{C}OCH_3$$

aspartame
NutraSweet®

Glutathione is a tripeptide of glutamate, cysteine, and glycine. Its function is to destroy harmful oxidizing agents in the body. Oxidizing agents are thought to be responsible for some of the effects of aging and are believed to play a role in cancer (Section 9.8). Glutathione removes oxidizing agents by reducing them. Consequently, glutathione is oxidized, forming a disulfide bond between two glutathione molecules. An enzyme subsequently reduces the disulfide bond, allowing glutathione to react with more oxidizing agents.

$$2\,H_3\overset{+}{N}CH(COO^-)CH_2CH_2\overset{O}{\overset{\|}{C}}-NHCH(CH_2SH)\overset{O}{\overset{\|}{C}}-NHCH_2\overset{O}{\overset{\|}{C}}O^-$$

glutathione

reducing agent ⇅ oxidizing agent

$$H_3\overset{+}{N}CH(COO^-)CH_2CH_2\overset{O}{\overset{\|}{C}}-NHCH(CH_2-S-S-CH_2)\overset{O}{\overset{\|}{C}}-NHCH_2\overset{O}{\overset{\|}{C}}O^-$$

$$H_3\overset{+}{N}CH(COO^-)CH_2CH_2\underset{O}{\underset{\|}{C}}-NHCH\underset{O}{\underset{\|}{C}}-NHCH_2\underset{O}{\underset{\|}{C}}O^-$$

oxidized glutathione

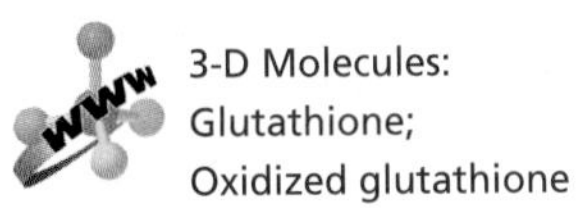

3-D Molecules: Glutathione; Oxidized glutathione

PROBLEM 22

What is unusual about glutathione's structure? (If you can't answer this question, draw the structure you would expect for glu-cys-gly, and compare your structure with the structure of glutathione.)

23.9 Strategy of Peptide Bond Synthesis: N-Protection and C-Activation

Because amino acids have two functional groups, a problem arises when one attempts to make a particular peptide bond. For example, suppose you wanted to make the dipeptide Gly-Ala. That dipeptide is only one of four possible dipeptides that could be formed from alanine and glycine.

$$H_3\overset{+}{N}CH_2\overset{O}{\overset{\|}{C}}-NHCH(CH_3)\overset{O}{\overset{\|}{C}}O^-$$
Gly-Ala

$$H_3\overset{+}{N}CH(CH_3)\overset{O}{\overset{\|}{C}}-NHCH(CH_3)\overset{O}{\overset{\|}{C}}O^-$$
Ala-Ala

$$H_3\overset{+}{N}CH_2\overset{O}{\overset{\|}{C}}-NHCH_2\overset{O}{\overset{\|}{C}}O^-$$
Gly-Gly

$$H_3\overset{+}{N}CH(CH_3)\overset{O}{\overset{\|}{C}}-NHCH_2\overset{O}{\overset{\|}{C}}O^-$$
Ala-Gly

If the amino group of the amino acid that is to be on the N-terminal end (in this case, Gly) is protected, it will not be available to form a peptide bond. If the carboxyl group of this same amino acid is activated before the second amino acid is added, the amino group of the added amino acid (in this case, Ala) will react with the activated carboxyl group of glycine in preference to reacting with a nonactivated carboxyl group of another alanine molecule.

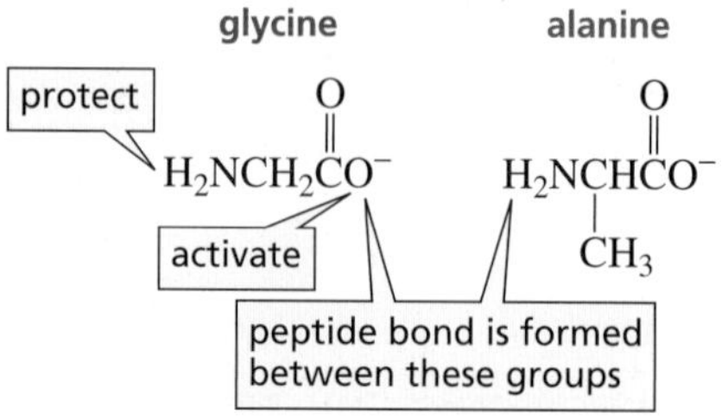

glycine: $H_2NCH_2CO^-$ (with C=O); alanine: H_2NCHCO^- (with C=O and CH_3)

protect

activate

peptide bond is formed between these groups

The reagent that is most often used to protect the amino group of an amino acid is di-*tert*-butyl dicarbonate. Its popularity is due to the ease with which the protecting group can be removed when the need for protection is over. The protecting group is known by the acronym *t*-BOC (pronounced "tee-bok").

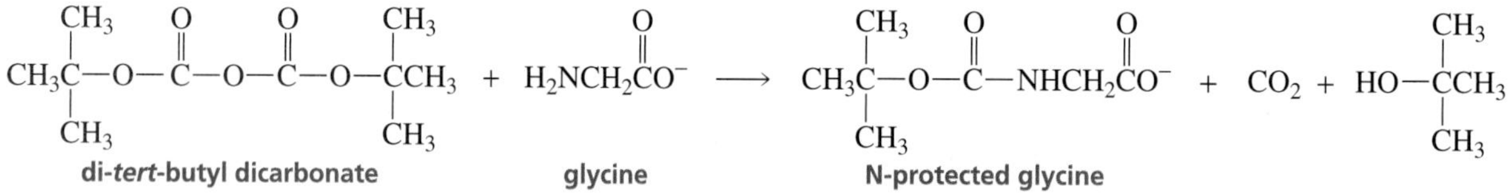

$$(CH_3)_3C{-}O{-}\overset{O}{\overset{\|}{C}}{-}O{-}\overset{O}{\overset{\|}{C}}{-}O{-}C(CH_3)_3 + H_2NCH_2\overset{O}{\overset{\|}{C}}O^- \longrightarrow (CH_3)_3C{-}O{-}\overset{O}{\overset{\|}{C}}{-}NHCH_2\overset{O}{\overset{\|}{C}}O^- + CO_2 + HO{-}C(CH_3)_3$$

di-*tert*-butyl dicarbonate; glycine; N-protected glycine

Carboxylic acids are generally activated by being converted into acyl chlorides (Section 17.20). Acyl chlorides, however, are so reactive that they can readily react with the side chains of some of the amino acids during peptide synthesis, creating unwanted products. The preferred method for activating the carboxyl group of an N-protected amino acid is to convert it into an imidate using dicyclohexylcarbodiimide (DCC). (By now, you have probably noticed that biochemists are even more fond of acronyms than organic chemists are.) DCC activates a carboxyl group by putting a good leaving group on the carbonyl carbon.

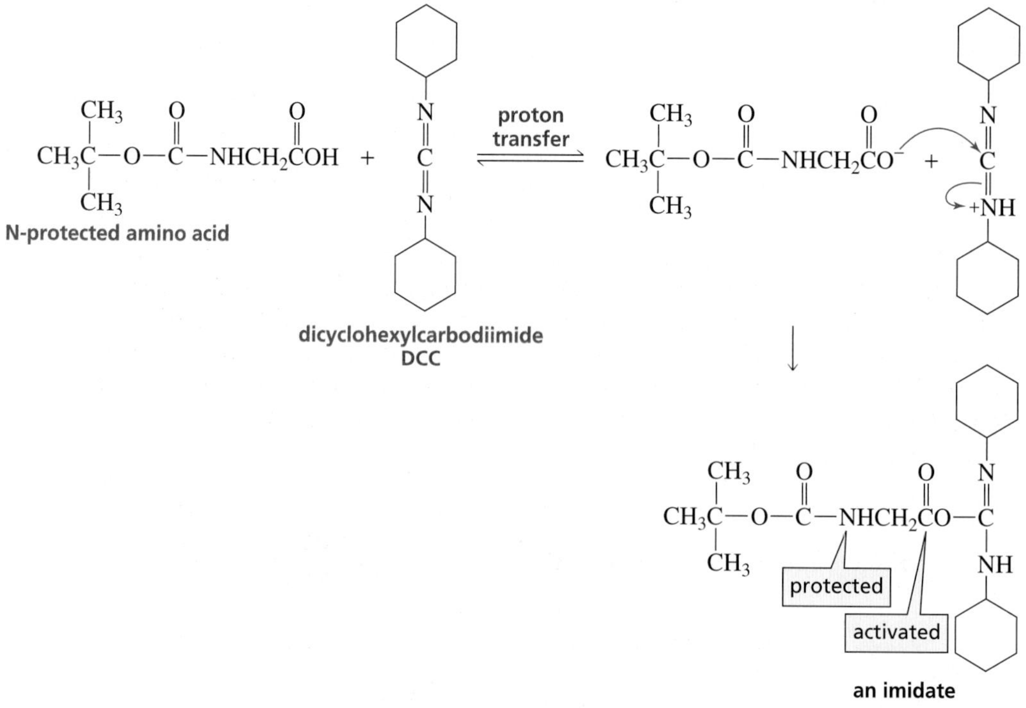

$$(CH_3)_3C{-}O{-}\overset{O}{\overset{\|}{C}}{-}NHCH_2\overset{O}{\overset{\|}{C}}OH + C_6H_{11}{-}N{=}C{=}N{-}C_6H_{11} \underset{}{\overset{\text{proton transfer}}{\rightleftharpoons}} (CH_3)_3C{-}O{-}\overset{O}{\overset{\|}{C}}{-}NHCH_2\overset{O}{\overset{\|}{C}}O^- + C_6H_{11}{-}N{=}C{=}\overset{+}{N}H{-}C_6H_{11}$$

N-protected amino acid; dicyclohexylcarbodiimide DCC

$$\downarrow$$

$$(CH_3)_3C{-}O{-}\overset{O}{\overset{\|}{C}}{-}NHCH_2\overset{O}{\overset{\|}{C}}O{-}C({=}N{-}C_6H_{11}){-}NH{-}C_6H_{11}$$

protected

activated

an imidate

After the amino acid has its N-terminal group protected and its C-terminal group activated, the second amino acid is added. The unprotected amino group of the second amino acid attacks the activated carboxyl group, forming a tetrahedral intermediate.

The C—O bond of the tetrahedral intermediate is easily broken because the bonding electrons are delocalized, forming dicyclohexylurea, a stable diamide. [Recall that the weaker (more stable) the base, the better it is as a leaving group; see Section 17.5.]

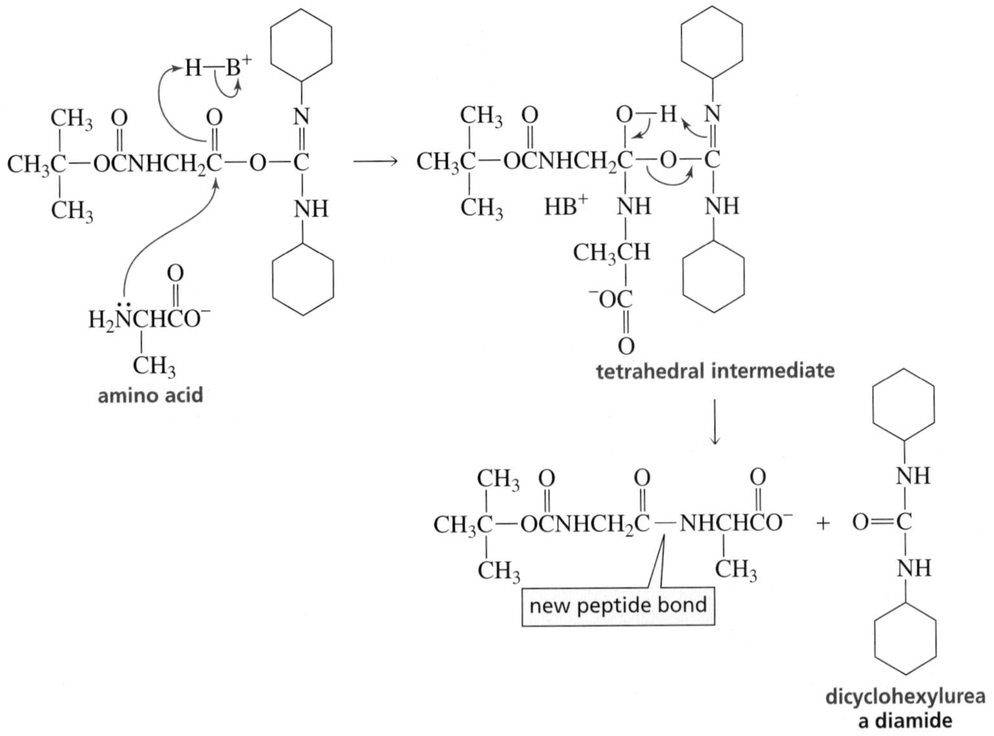

amino acid

tetrahedral intermediate

new peptide bond

dicyclohexylurea
a diamide

Amino acids can be added to the growing C-terminal end by repeating these two steps: activating the carboxyl group of the C-terminal amino acid of the peptide by treating it with DCC and then adding a new amino acid.

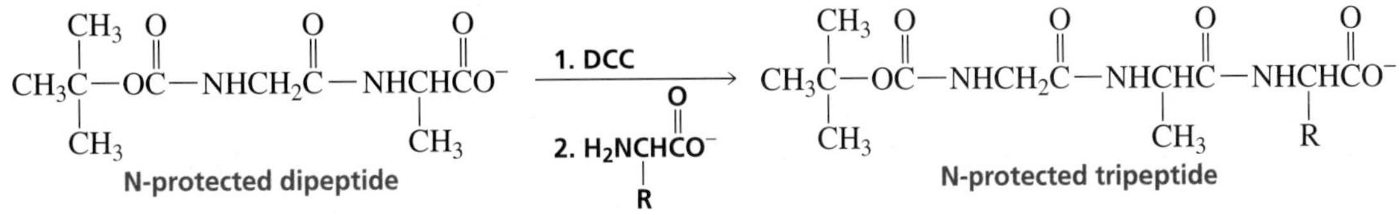

N-protected dipeptide

1. DCC

2. H_2NCHCO^- (with R)

N-protected tripeptide

When the desired number of amino acids has been added to the chain, the protecting group on the N-terminal amino acid is removed. *t*-BOC is an ideal protecting group because it can be removed by washing with trifluoroacetic acid and methylene chloride, reagents that will not break any other covalent bonds. The protecting group is removed by an elimination reaction, forming isobutylene and carbon dioxide. Because these products are gases, they escape, driving the reaction to completion.

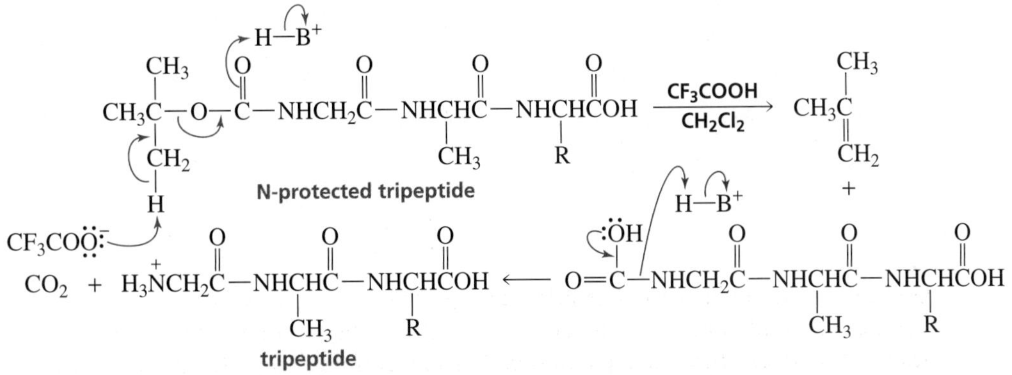

N-protected tripeptide

CF_3COOH
CH_2Cl_2

tripeptide

Theoretically, one should be able to make as long a peptide as desired with this technique. Reactions do not produce 100% yields, however, and the yields are further decreased during the purification process. After each step of the synthesis, the peptide must be purified to prevent subsequent unwanted reactions with leftover reagents. Assuming that each amino acid can be added to the growing end of the peptide chain in an 80% yield (a relatively high yield, as you can probably appreciate from your own experience in the laboratory), the overall yield of a nonapeptide such as bradykinin would be only 17%. It is clear that large polypeptides could never be synthesized in this way.

Number of amino acids	**2**	**3**	**4**	**5**	**6**	**7**	**8**	**9**
Overall yield	80%	64%	51%	41%	33%	26%	21%	17%

PROBLEM 23

What dipeptides would be formed by heating a mixture of valine and N-protected leucine?

PROBLEM 24

Suppose you are trying to synthesize the dipeptide Val-Ser. Compare the product that would be obtained if the carboxyl group of N-protected valine were activated with thionyl chloride with the product that would be obtained if the carboxyl group were activated with DCC.

PROBLEM 25

Show the steps in the synthesis of the tetrapeptide Leu-Phe-Lys-Val.

PROBLEM 26◆

a. Calculate the overall yield of bradykinin if the yield for the addition of each amino acid to the chain is 70%.
b. What would be the overall yield of a peptide containing 15 amino acid residues if the yield for the incorporation of each is 80%?

23.10 Automated Peptide Synthesis

In addition to producing low overall yields, the method of peptide synthesis described in Section 23.9 is extremely time-consuming because the product must be purified at each step of the synthesis. In 1969, Bruce Merrifield described a method that revolutionized the synthesis of peptides because it provided a much faster way to produce peptides in much higher yields. Furthermore, because it is automated, the synthesis requires fewer hours of direct attention. Using this technique, bradykinin was synthesized with an 85% yield in 27 hours. Subsequent refinements in the technique now allow a reasonable yield of a peptide containing 100 amino acids to be synthesized in four days.

In the Merrifield method, the C-terminal amino acid is covalently attached to a solid support contained in a column. Each N-terminal blocked amino acid is added one at a time, along with other needed reagents, so the protein is synthesized from the C-terminal end to the N-terminal end. Notice that this is opposite to the way proteins are synthesized in nature (from the N-terminal end to the C-terminal end; Section 27.13). Because it uses a solid support and is automated, Merrifield's method of protein synthesis is called **automated solid-phase peptide synthesis**.

R. Bruce Merrifield *was born in 1921 and received a B.S. and a Ph.D. from the University of California, Los Angeles. He is a professor of chemistry at Rockefeller University. Merrifield received the 1984 Nobel Prize in chemistry for developing automated solid-phase peptide synthesis.*

Merrifield automated solid-phase synthesis of a tripeptide

$(CH_3)_3C{-}O{-}C({=}O){-}NHCH(R)CO^-$ + $ClCH_2$–C₆H₄–(resin)

N-protected amino acid

resin

↓

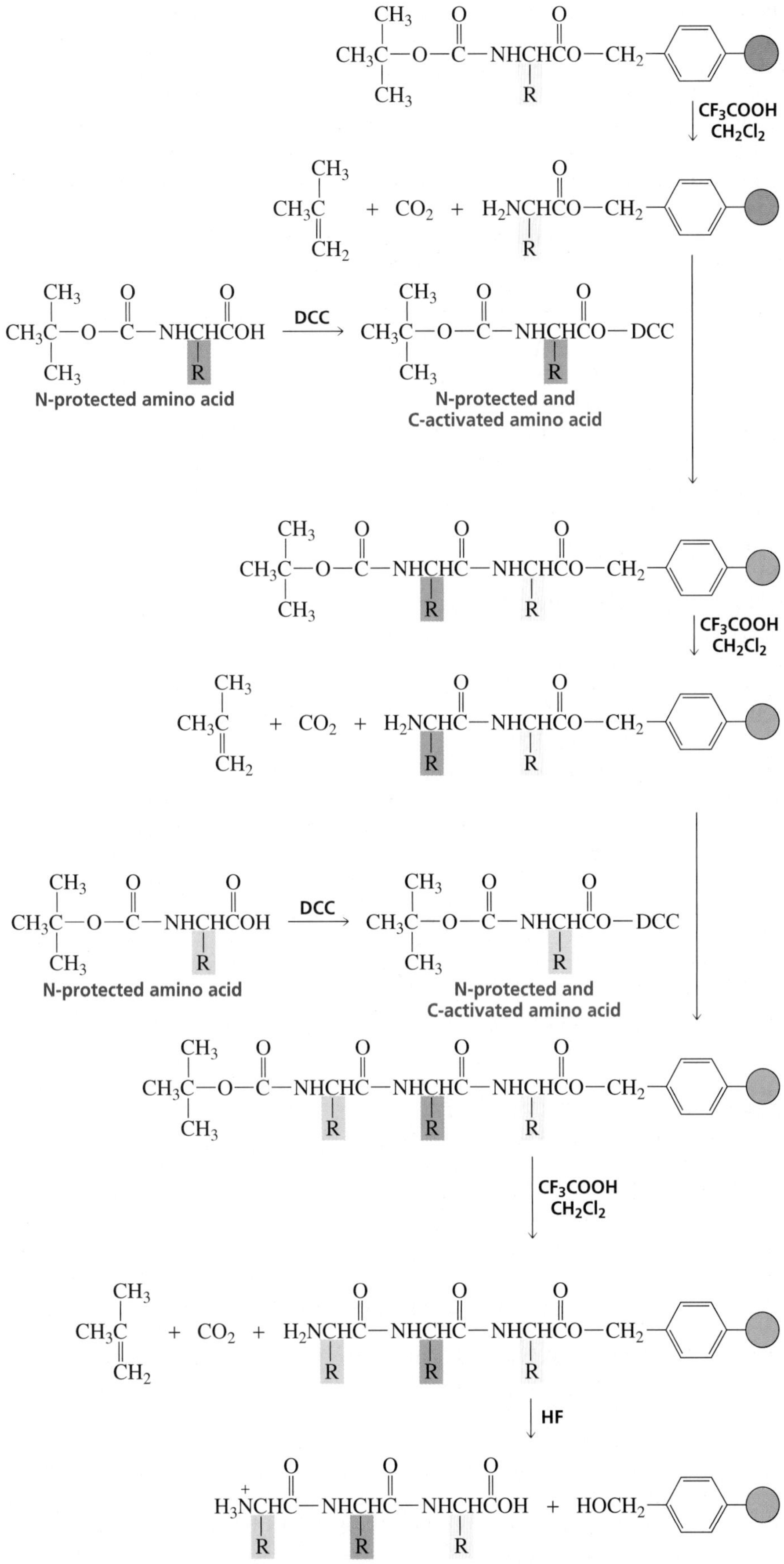

Tutorial: Merrifield automated solid-phase peptide synthesis

$$(CH_3)_3C{-}O{-}\overset{O}{\overset{\|}{C}}{-}NH\underset{R}{CH}\overset{O}{\overset{\|}{C}}O{-}CH_2{-}C_6H_4{-}\text{resin}$$

$\downarrow$ CF_3COOH, CH_2Cl_2

$$CH_3\overset{CH_3}{C}{=}CH_2 + CO_2 + H_2N\underset{R}{CH}\overset{O}{\overset{\|}{C}}O{-}CH_2{-}C_6H_4{-}\text{resin}$$

$$(CH_3)_3C{-}O{-}\overset{O}{\overset{\|}{C}}{-}NH\underset{R}{CH}\overset{O}{\overset{\|}{C}}OH \xrightarrow{\text{DCC}} (CH_3)_3C{-}O{-}\overset{O}{\overset{\|}{C}}{-}NH\underset{R}{CH}\overset{O}{\overset{\|}{C}}O{-}DCC$$

N-protected amino acid — N-protected and C-activated amino acid

$\downarrow$

$$(CH_3)_3C{-}O{-}\overset{O}{\overset{\|}{C}}{-}NH\underset{R}{CH}\overset{O}{\overset{\|}{C}}{-}NH\underset{R}{CH}\overset{O}{\overset{\|}{C}}O{-}CH_2{-}C_6H_4{-}\text{resin}$$

$\downarrow$ CF_3COOH, CH_2Cl_2

$$CH_3\overset{CH_3}{C}{=}CH_2 + CO_2 + H_2N\underset{R}{CH}\overset{O}{\overset{\|}{C}}{-}NH\underset{R}{CH}\overset{O}{\overset{\|}{C}}O{-}CH_2{-}C_6H_4{-}\text{resin}$$

$$(CH_3)_3C{-}O{-}\overset{O}{\overset{\|}{C}}{-}NH\underset{R}{CH}\overset{O}{\overset{\|}{C}}OH \xrightarrow{\text{DCC}} (CH_3)_3C{-}O{-}\overset{O}{\overset{\|}{C}}{-}NH\underset{R}{CH}\overset{O}{\overset{\|}{C}}O{-}DCC$$

N-protected amino acid — N-protected and C-activated amino acid

$\downarrow$

$$(CH_3)_3C{-}O{-}\overset{O}{\overset{\|}{C}}{-}NH\underset{R}{CH}\overset{O}{\overset{\|}{C}}{-}NH\underset{R}{CH}\overset{O}{\overset{\|}{C}}{-}NH\underset{R}{CH}\overset{O}{\overset{\|}{C}}O{-}CH_2{-}C_6H_4{-}\text{resin}$$

$\downarrow$ CF_3COOH, CH_2Cl_2

$$CH_3\overset{CH_3}{C}{=}CH_2 + CO_2 + H_2N\underset{R}{CH}\overset{O}{\overset{\|}{C}}{-}NH\underset{R}{CH}\overset{O}{\overset{\|}{C}}{-}NH\underset{R}{CH}\overset{O}{\overset{\|}{C}}O{-}CH_2{-}C_6H_4{-}\text{resin}$$

$\downarrow$ HF

$$H_3\overset{+}{N}\underset{R}{CH}\overset{O}{\overset{\|}{C}}{-}NH\underset{R}{CH}\overset{O}{\overset{\|}{C}}{-}NH\underset{R}{CH}\overset{O}{\overset{\|}{C}}OH + HOCH_2{-}C_6H_4{-}\text{resin}$$

The solid support to which the C-terminal amino acid is attached is a resin similar to the one used in ion-exchange chromatography (Section 23.5), except that the benzene rings have chloromethyl substituents instead of sulfonic acid substituents. Before the C-terminal amino acid is attached to the resin, its amino group is protected with *t*-BOC to prevent the amino group from reacting with the resin. The C-terminal amino acid is attached to the resin by means of an S_N2 reaction—its carboxyl group attacks a benzyl carbon of the resin, displacing a chloride ion (Section 10.4).

After the C-terminal amino acid is attached to the resin, the *t*-BOC protecting group is removed (Section 23.9). The next amino acid, with its amino group protected with *t*-BOC and its carboxyl group activated with DCC, is added to the column.

A huge advantage of the Merrifield method of peptide synthesis is that the growing peptide can be purified by washing the column with an appropriate solvent after each step of the procedure. The impurities are washed out of the column because they are not attached to the solid support. Since the peptide is covalently attached to the resin, none of it is lost in the purification step, leading to high yields of purified product.

After the required amino acids have been added one by one, the peptide can be removed from the resin by treatment with HF under mild conditions that do not break the peptide bonds.

Merrifield's technique is constantly being improved so that peptides can be made more rapidly and more efficiently. However, it still cannot begin to compare with nature: A bacterial cell is able to synthesize a protein containing thousands of amino acids in seconds and can simultaneously synthesize thousands of different proteins with no mistakes.

Since the early 1980s, it has been possible to synthesize proteins by genetic engineering techniques. Strands of DNA can be introduced into bacterial cells, causing the cells to produce large amounts of a desired protein (Section 27.13). For example, mass quantities of human insulin are produced from genetically modified *E. coli*. Genetic engineering techniques also have been useful in synthesizing proteins that differ in one or a few amino acids from the natural protein. Such synthetic proteins have been used, for example, to learn how a change in a single amino acid affects the properties of a protein (Section 24.9).

PROBLEM 27

Show the steps in the synthesis of the peptide in Problem 25, using Merrifield's method.

23.11 Protein Structure

Protein molecules are described by several levels of structure. The **primary structure** of a protein is the sequence of amino acids in the chain and the location of all the disulfide bridges. The **secondary structure** describes the regular conformation assumed by segments of the protein's backbone. In other words, the secondary structure describes how local regions of the backbone fold. The **tertiary structure** describes the three-dimensional structure of the entire polypeptide. If a protein has more than one polypeptide chain, it has quaternary structure. The **quaternary structure** of a protein is the way the individual protein chains are arranged with respect to each other.

Proteins can be divided roughly into two classes. **Fibrous proteins** contain long chains of polypeptides that occur in bundles. These proteins are insoluble in water. All the structural proteins described at the beginning of this chapter, such as keratin and collagen, are fibrous proteins. **Globular proteins** are soluble in water and tend to have roughly spherical shapes. Essentially all enzymes are globular proteins.

PRIMARY STRUCTURE AND EVOLUTION

When we examine the primary structures of proteins that carry out the same function in different organisms, we can relate the number of amino acid differences between the proteins to the taxonomic differences between the species. For example, cytochrome *c*, a protein that transfers electrons in biological oxidations, has about 100 amino acid residues. Yeast cytochrome *c* differs by 48 amino acids from horse cytochrome *c*, while duck cytochrome *c* differs by only two amino acids from chicken cytochrome *c*. Chickens and turkeys have cytochrome *c*'s with identical primary structures. Humans and chimpanzees also have identical cytochrome *c*'s, differing by one amino acid from the cytochrome *c* of the rhesus monkey.

23.12 Determining the Primary Structure of a Protein

The first step in determining the sequence of amino acids in a peptide or a protein is to reduce any disulfide bridges in the peptide or protein. A commonly used reducing agent is 2-mercaptoethanol, which is oxidized to a disulfide. Reaction of the protein thiol groups with iodoacetic acid prevents the disulfide bridges from reforming as a result of oxidation by O_2.

cleaving disulfide bridges

$$\text{—NHCH(CH}_2\text{S—SCH}_2\text{)CHNH— (two —NHCH—C(=O)— chains)} + 2\,HSCH_2CH_2OH \longrightarrow \text{—NHCH(CH}_2\text{SH)C(=O)— (two chains)} + \begin{matrix} SCH_2CH_2OH \\ | \\ SCH_2CH_2OH \end{matrix}$$

2-mercaptoethanol

$$\xrightarrow[\text{iodoacetic acid}]{ICH_2COOH} \text{—NHCH(CH}_2SCH_2COOH\text{)C(=O)— (two chains)} + 2\,HI$$

PROBLEM 28

Write the mechanism for the reaction of a cysteine residue with iodoacetic acid.

Insulin was the first protein for which the primary sequence was determined. This was done in 1953 by **Frederick Sanger**, *who received the 1958 Nobel Prize in chemistry for his work. Sanger was born in England in 1918 and received a Ph.D. from Cambridge University, where he has worked for his entire career. He also received a share of the 1980 Nobel Prize in chemistry (Section 27.15) for being the first to sequence a DNA molecule (with 5375 nucleotide pairs).*

The next step is to determine the number and kinds of amino acids in the peptide or protein. To do this, a sample of the peptide or protein is dissolved in 6 N HCl and heated at 100 °C for 24 hours. This treatment hydrolyzes all the amide bonds in the protein, including the amide bonds of asparagine and glutamine.

$$\text{protein} \xrightarrow[\text{100 °C, 24 h}]{\text{6 N HCl}} \text{amino acids}$$

The mixture of amino acids is then passed through an amino acid analyzer to determine the number and kind of each amino acid in the peptide or protein (Section 23.5).

Because all the asparagine and glutamine residues have been hydrolyzed to aspartate and glutamate residues, the number of aspartate or glutamate residues in the amino acid mixture tells us the number of aspartate plus asparagine or glutamate plus glutamine residues in the original protein. A separate technique must be used to distinguish between aspartate and asparagine or between glutamate and glutamine in the original protein.

The strongly acidic conditions used for hydrolysis destroy all the tryptophan residues because the indole ring is unstable in acid (Section 21.9). The tryptophan content of the protein can be determined by hydroxide-ion-promoted hydrolysis of the protein. This is not a general method for peptide bond hydrolysis because the strongly basic conditions destroy several other amino acid residues.

There are several ways to identify the N-terminal amino acid of a peptide or protein. One of the most widely used methods is to treat the protein with phenyl isothiocyanate (PITC), more commonly known as **Edman's reagent**. This reagent reacts with the N-terminal amino group, and the resulting thiazolinone derivative is cleaved from the protein under mildly acidic conditions. The thiazolinone derivative is extracted into an organic solvent and in the presence of acid, rearranges to a more stable phenylthiohydantoin (PTH).

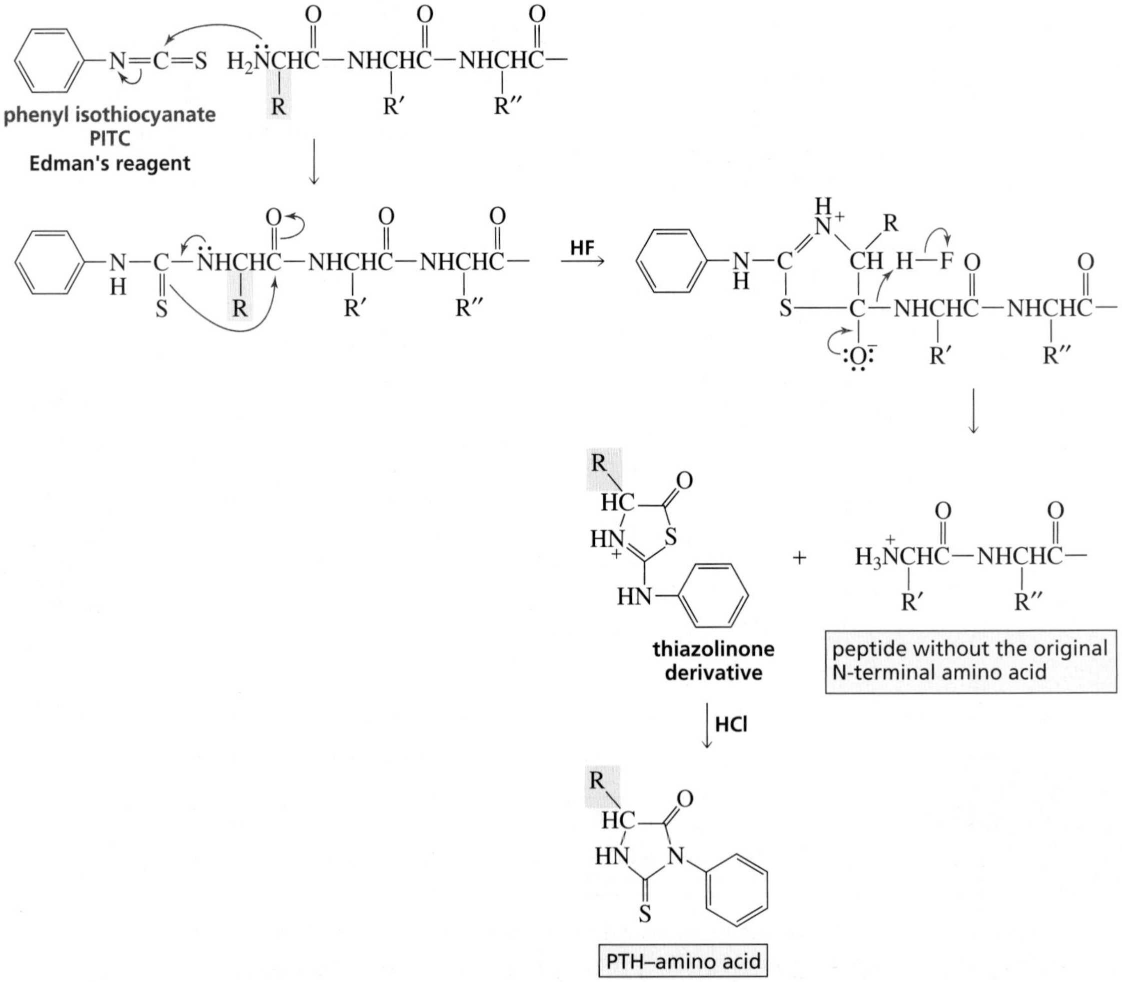

phenyl isothiocyanate
PITC
Edman's reagent

HF

thiazolinone derivative

peptide without the original N-terminal amino acid

HCl

PTH–amino acid

Because each amino acid has a different substituent (R), each amino acid forms a different PTH–amino acid. The particular PTH–amino acid can be identified by chromatography using known standards. Several successive Edman degradations can be carried out on a protein. The entire primary sequence cannot be determined in this way, however, because side products accumulate that interfere with the results. An automated instrument known as a *sequenator* allows about 50 successive Edman degradations to be carried out on a protein.

The C-terminal amino acid of the peptide or protein can be identified by treating the protein with carboxypeptidase A. Carboxypeptidase A is an enzyme that cleaves off the C-terminal amino acid as long as it is *not* arginine or lysine (Section 24.9). On the other hand, carboxypeptidase B cleaves off the C-terminal amino acid *only* if it is arginine or lysine. Carboxypeptidases are exopeptidases. An **exopeptidase** is an enzyme that catalyzes the hydrolysis of a peptide bond at the end of a peptide chain.

site where carboxypeptidase cleaves

$$\text{—NHCH(R)C(=O)—NHCH(R')C(=O)—NHCH(R'')CO}^-$$

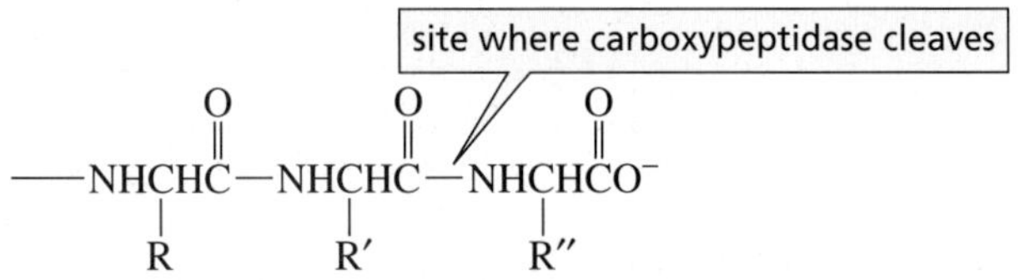

Once the N-terminal and C-terminal amino acids have been identified, a sample of the protein is hydrolyzed with dilute acid. This treatment, called **partial hydrolysis**, hydrolyzes only some of the peptide bonds. The resulting fragments are separated, and the amino acid composition of each is determined. The N-terminal and C-terminal amino acids of each fragment can also be identified. The sequence of the original protein can then be determined by lining up the peptides and looking for points of overlap.

PROBLEM-SOLVING STRATEGY

A nonapeptide undergoes partial hydrolysis to give peptides whose amino acid compositions are shown. Reaction of the intact nonapeptide with Edman's reagent releases PTH-Leu. What is the sequence of the nonapeptide?

a. Pro, Ser
b. Gly, Glu
c. Met, Ala, Leu
d. Gly, Ala
e. Glu, Ser, Val, Pro
f. Glu, Pro, Gly
g. Met, Leu
h. His, Val

Let's start with the N-terminal amino acid. We know that it is Leu. Now we need to look for a fragment that contains Leu. Fragment (g) tells us that Met is next to Leu and fragment (c) tells us that Ala is next to Met. Now we look for a fragment that contains Ala. Fragment (d) contains Ala and tells us that Gly is next to Ala. From fragment (b), we know that Glu comes next. Glu is in both fragments (e) and (f). Fragment (e) has three amino acids we have yet to place in the growing peptide (Ser, Val, Pro), but fragment (f) has only one, so from fragment (f), we know that Pro is the next amino acid. Fragment (a) tells us that the next amino acid is Ser. Now we can use fragment (e). Fragment (e) tells us that the next amino acid is Val, and fragment (h) tells us that His is the last (C-terminal) amino acid. Thus, the amino acid sequence of the nonapeptide is

Leu-Met-Ala-Gly-Glu-Pro-Ser-Val-His

Now continue on to Problem 29.

PROBLEM 29◆

A decapeptide undergoes partial hydrolysis to give peptides whose amino acid compositions are shown. Reaction of the intact decapeptide with Edman's reagent releases PTH-Gly. What is the sequence of the decapeptide?

a. Ala, Trp
b. Val, Pro, Asp
c. Pro, Val
d. Ala, Glu
e. Trp, Ala, Arg
f. Arg, Gly
g. Glu, Ala, Leu
h. Met, Pro, Leu, Glu

The peptide or protein can also be partially hydrolyzed using endopeptidases. An **endopeptidase** is an enzyme that catalyzes the hydrolysis of a peptide bond that is not at the end of a peptide chain. Trypsin, chymotrypsin, and elastase are endopeptidases

Table 23.3 Specificity of Peptide or Protein Cleavage

Reagent	Specificity
Chemical reagents	
Edman's reagent	removes the N-terminal amino acid
Cyanogen bromide	hydrolyzes on the C-side of Met
Exopeptidases*	
Carboxypeptidase A	removes the C-terminal amino acid (not Arg or Lys)
Carboxypeptidase B	removes the C-terminal amino acid (only Arg or Lys)
Endopeptidases*	
Trypsin	hydrolyzes on the C-side of Arg and Lys
Chymotrypsin	hydrolyzes on the C-side of amino acids that contain aromatic six-membered rings (Phe, Tyr, Trp)
Elastase	hydrolyzes on the C-side of small amino acids (Gly and Ala)

*Cleavage will not occur if Pro is on either side of the bond to be hydrolyzed.

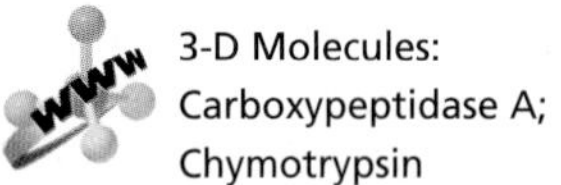
3-D Molecules: Carboxypeptidase A; Chymotrypsin

that catalyze the hydrolysis of only the specific peptide bonds listed in Table 23.3. Trypsin, for example, catalyzes the hydrolysis of the peptide bond on the C-side of only arginine or lysine residues.

C-side of lysine | C-side of arginine

$$-NHCH(R)C(=O)-NHCH(CH_2CH_2CH_2CH_2\overset{+}{N}H_3)C(=O)-NHCH(R')C(=O)-NHCH(R'')C(=O)-NHCH(CH_2CH_2CH_2NHC(=\overset{+}{N}H_2)NH_2)C(=O)-NHCH(R''')C(=O)-$$

Thus, trypsin will catalyze the hydrolysis of three peptide bonds in the following peptide, creating a hexapeptide, a dipeptide, and two tripeptides.

Ala-Lys-Phe-Gly-Asp-Trp-Ser-Arg-Met-Val-Arg-Tyr-Leu-His

cleavage by trypsin

Chymotrypsin catalyzes the hydrolysis of the peptide bond on the C-side of amino acids that contain aromatic six-membered rings (Phe, Tyr, Trp).

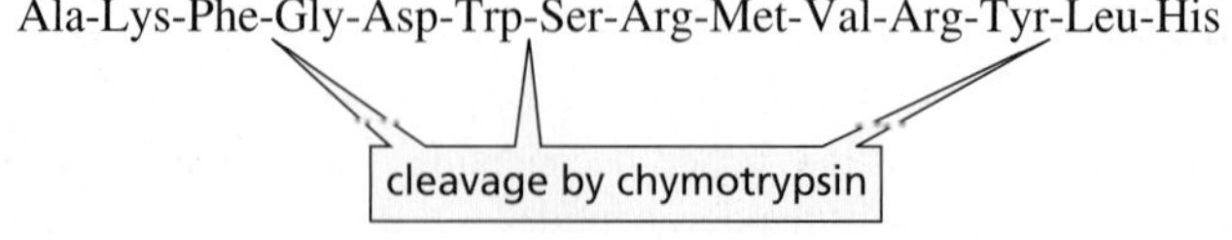

Elastase catalyzes the hydrolysis of peptide bonds on the C-side of the two smallest amino acids (Gly, Ala). Chymotrypsin and elastase are much less specific than trypsin. (An explanation for the specificity of these enzymes is given in Section 24.9.)

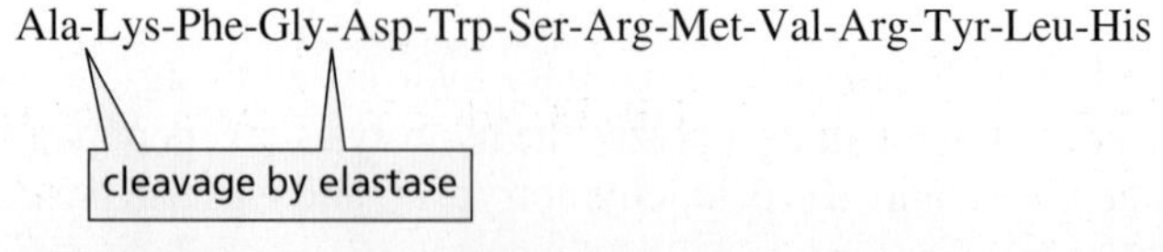

None of the exopeptidases or endopeptidases that we have mentioned will catalyze the hydrolysis of an amide bond if proline is at the hydrolysis site. These enzymes recognize the appropriate hydrolysis site by its shape and charge, and the cyclic structure of proline causes the hydrolysis site to have an unrecognizable three-dimensional shape.

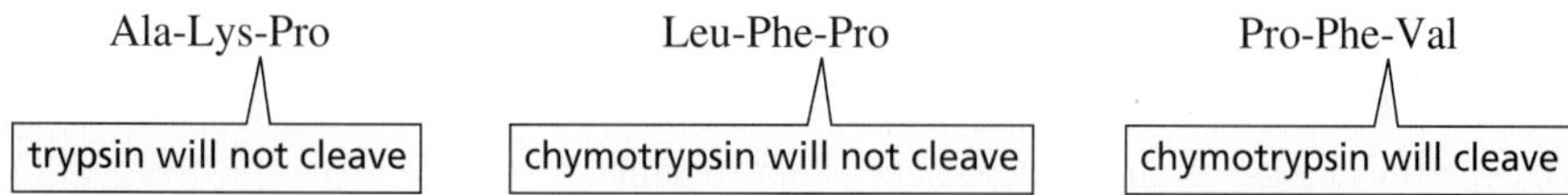

Cyanogen bromide (BrC≡N) causes the hydrolysis of the amide bond on the C-side of a methionine residue. Cyanogen bromide is more specific than the endopeptidases about what peptide bonds it cleaves, so it provides more reliable information about the primary structure (the sequence of amino acids). Because cyanogen bromide is not a protein and therefore does not recognize the substrate by its shape, cyanogen bromide will still cleave the peptide bond if proline is at the cleavage site.

Ala-Lys-Phe-Gly-Asp-Trp-Ser-Arg-Met-Val-Arg-Tyr-Leu-His

cleavage by cyanogen bromide

The first step in the mechanism for cleavage of a peptide bond by cyanogen bromide is attack by the highly nucleophilic sulfur of methionine on cyanogen bromide. Formation of a five-membered ring with departure of the weakly basic leaving group is followed by acid-catalyzed hydrolysis, which cleaves the protein (Section 18.6). Further hydrolysis can cause the lactone (a cyclic ester) to open to a carboxyl group and an alcohol group (Section 17.11).

mechanism for the cleavage of a peptide bond by cyanogen bromide

Br—C≡N

$-NHCH(R)C(=O)NHCH(CH_2CH_2\ddot{S}CH_3)-C(=O)-NHCH(R')C(=O)-$ ⟶ $-NHCH(R)C(=O)NHCH(CH_2CH_2\overset{+}{S}(CH_3)-C\equiv N)-C(=O)-\ddot{N}HCH(R')C(=O)-$ + Br^-

↓

$-NHCH(R)C(=O)NHCH-C(=\overset{+}{N}HCH(R')C(=O)-)$ (five-membered ring: CH_2, CH_2, O) + $CH_3SC\equiv N$

HCl | H_2O ↓

$-NHCH(R)C(=O)NHCH-C=O$ (lactone ring: CH_2, CH_2, O) + $H_3\overset{+}{N}CH(R')C(=O)-$

HCl, H_2O ⟵

$-NHCH(R)C(=O)NHCH(CH_2CH_2OH)-COH(=O)$

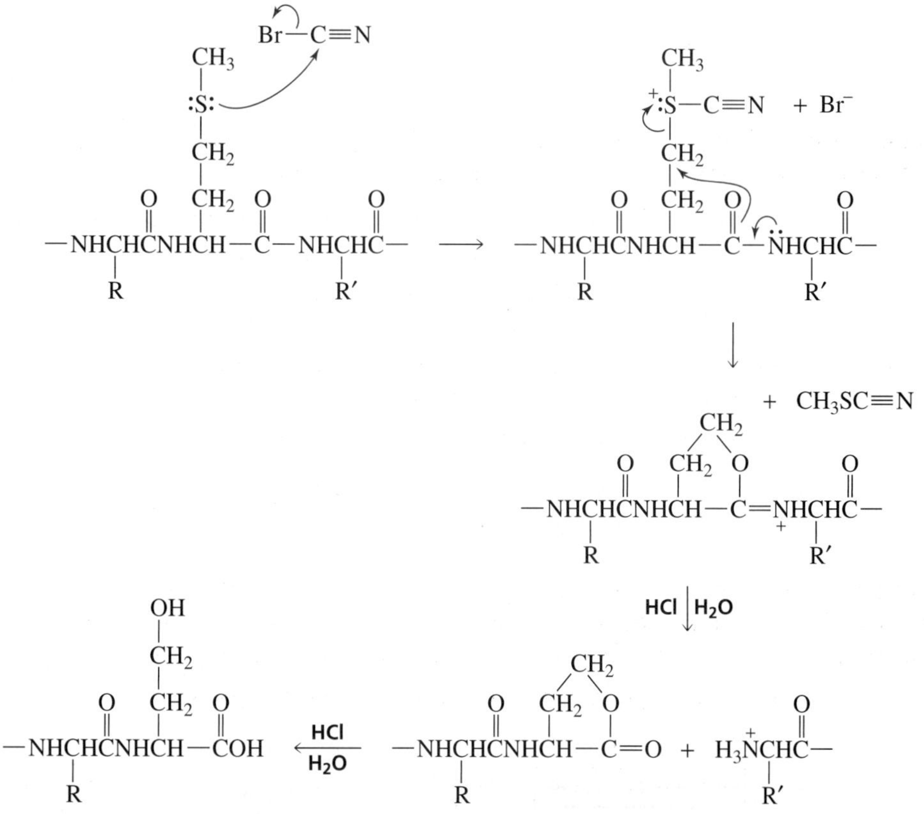

The last step in determining the primary structure of a protein is to figure out the location of any disulfide bonds. This is done by hydrolyzing a sample of the protein that has intact disulfide bonds. From a determination of the amino acids in the cysteine-containing fragments, the locations of the disulfide bonds in the protein can be established (Problem 47).

PROBLEM 30

Why won't cyanogen bromide cleave at cysteine residues?

PROBLEM 31◆

In determining the primary structure of insulin, what would lead you to conclude that it had more than one polypeptide chain?

PROBLEM 32 SOLVED

Determine the amino acid sequence of a polypeptide from the following results:

Acid hydrolysis gives Ala, Arg, His, 2 Lys, Leu, 2 Met, Pro, 2 Ser, Thr, Val.

Carboxypeptidase A releases Val.

Edman's reagent releases PTH-Leu.

Cleavage with cyanogen bromide gives three peptides with the following amino acid compositions:

1. His, Lys, Met, Pro, Ser
2. Thr, Val
3. Ala, Arg, Leu, Lys, Met, Ser

Trypsin-catalyzed hydrolysis gives three peptides and a single amino acid:

1. Arg, Leu, Ser
2. Met, Pro, Ser, Thr, Val
3. Lys
4. Ala, His, Lys, Met

SOLUTION Acid hydrolysis shows that the polypeptide has 13 amino acids. The N-terminal amino acid is Leu (Edman's reagent), and the C-terminal amino acid is Val (carboxypeptidase A).

Leu ___ ___ ___ ___ ___ ___ ___ ___ ___ ___ ___ Val

Because cyanogen bromide cleaves on the C-side of Met, any peptide containing Met must have Met as its C-terminal amino acid. The peptide that does not contain Met must be the C-terminal peptide. We know that peptide 3 is the N-terminal peptide because it contains Leu. Since it is a hexapeptide, we know that the 6th amino acid in the 13-amino acid peptide is Met. We also know that the eleventh amino acid is Met because cyanogen bromide cleavage gave the dipeptide Thr, Val.

Ala, Arg, Lys, Ser His, Lys, Pro, Ser

Leu ___ ___ ___ ___ Met ___ ___ ___ ___ Met Thr Val

Because trypsin cleaves on the C-side of Arg and Lys, any peptide containing Arg or Lys must have that amino acid as its C-terminal amino acid. Therefore, Arg is the C-terminal amino acid of peptide 1, so we now know that the first three amino acids are Leu-Ser-Arg. We also know that the next two are Lys-Ala because if they were Ala-Lys, trypsin cleavage would give an Ala, Lys dipeptide. The trypsin data also identify the positions of His and Lys.

Pro, Ser

Leu Ser Arg Lys Ala Met His Lys ___ ___ Met Thr Val

Finally, because trypsin successfully cleaves on the C-side of Lys, Pro cannot be adjacent to Lys. Thus, the amino acid sequence of the given polypeptide is

Leu Ser Arg Lys Ala Met His Lys Ser Pro Met Thr Val

PROBLEM 33◆

Determine the primary structure of an octapeptide from the following data:

Acid hydrolysis gives 2 Arg, Leu, Lys, Met, Phe, Ser, Tyr.
Carboxypeptidase A releases Ser.
Edman's reagent releases Leu.
Cyanogen bromide forms two peptides with the following amino acid compositions:

1. Arg, Phe, Ser
2. Arg, Leu, Lys, Met, Tyr

Trypsin forms the following two amino acids and two peptides:

1. Arg
2. Ser
3. Arg, Met, Phe
4. Leu, Lys, Tyr

23.13 Secondary Structure of Proteins

Secondary structure describes the conformation of segments of the backbone chain of a peptide or protein. To minimize energy, a polypeptide chain tends to fold in a repeating geometric structure such as an α-helix or a β-pleated sheet. Three factors determine the choice of secondary structure:

- the regional planarity about each peptide bond (as a result of the partial double-bond character of the amide bond), which limits the possible conformations of the peptide chain (Section 23.7)
- maximization of the number of peptide groups that engage in hydrogen bonding (i.e., hydrogen bonding between the carbonyl oxygen of one amino acid residue and the amide hydrogen of another)
- adequate separation between nearby R groups to avoid steric hindrance and repulsion of like charges

hydrogen bonding between peptide groups

α-Helix

One type of secondary structure is the **α-helix**. In an α-helix, the backbone of the polypeptide coils around the long axis of the protein molecule (Figure 23.8). The helix is stabilized by hydrogen bonds: Each hydrogen attached to an amide nitrogen is

a.
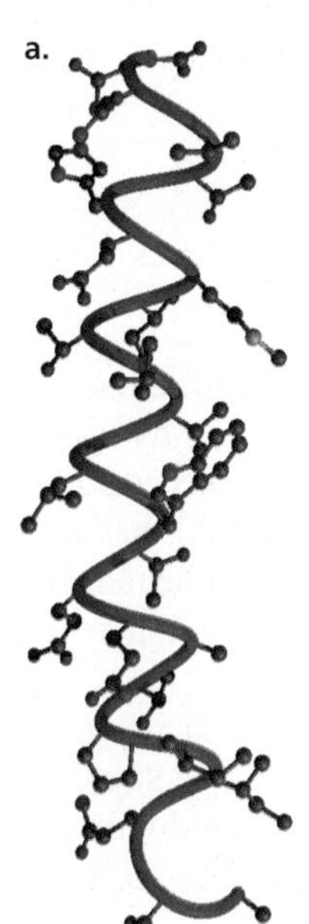

b.
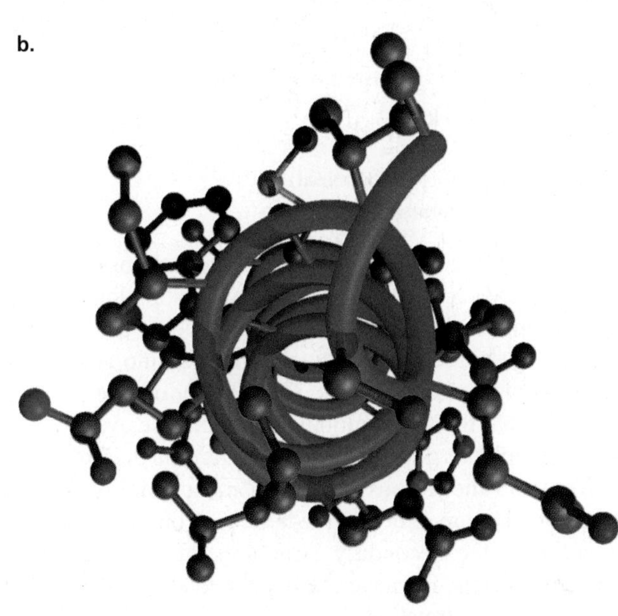

Figure 23.8
(a) A segment of a protein in an α-helix. (b) Looking up the longitudinal axis of an α-helix.

hydrogen bonded to a carbonyl oxygen of an amino acid four residues away. The substituents on the α-carbons of the amino acids protrude outward from the helix, thereby minimizing steric hindrance. Because the amino acids have the L-configuration, the α-helix is a right-handed helix; that is, it rotates in a clockwise direction as it spirals down. Each turn of the helix contains 3.6 amino acid residues, and the repeat distance of the helix is 5.4 Å.

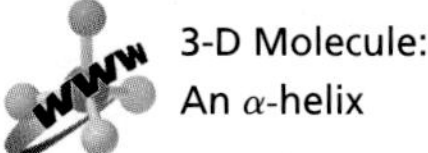

3-D Molecule: An α-helix

Not all amino acids are able to fit into an α-helix. A proline residue, for example, forces a bend in a helix because the bond between the proline nitrogen and the α-carbon cannot rotate to enable it to fit readily into a helix. Similarly, two adjacent amino acids that have more than one substituent on a β-carbon (valine, isoleucine, or threonine) cannot fit into a helix because of steric crowding between the R groups. Finally, two adjacent amino acids with like-charged substituents cannot fit into a helix because of electrostatic repulsion between the R groups. The percentage of amino acid residues coiled into an α-helix varies from protein to protein, but, on average, about 25% of the residues in globular proteins are in α-helices.

β-Pleated Sheet

The second type of secondary structure is the **β-pleated sheet**. In a β-pleated sheet, the polypeptide backbone is extended in a zigzag structure resembling a series of pleats. A β-pleated sheet is almost fully extended—the average two-residue repeat distance is 7.0 Å. The hydrogen bonding in a β-pleated sheet occurs between neighboring peptide chains. The adjacent hydrogen-bonded peptide chains can run in the same direction or in opposite directions. In a **parallel β-pleated sheet**, the adjacent chains run in the same direction. In an **antiparallel β-pleated sheet**, the adjacent chains run in opposite directions (Figure 23.9).

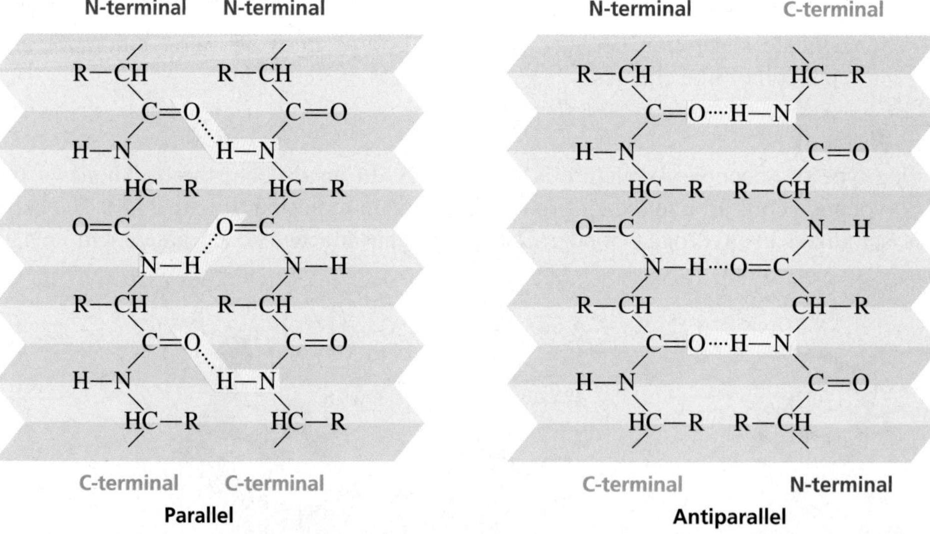

Figure 23.9 ▶ Segment of a β-pleated sheet drawn to illustrate its pleated character. Note that the first is parallel and the second is antiparallel.

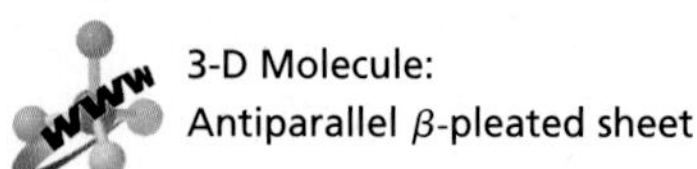

3-D Molecule: Antiparallel β-pleated sheet

Because the substituents (R) on the α-carbons of the amino acids on adjacent chains are close to each other, the chains can nestle closely together to maximize hydrogen-bonding interactions only if the substituents are small. Silk, for example, a protein with a large number of relatively small amino acids (glycine and alanine), has large segments of β-pleated sheets. The number of side-by-side strands in a β-pleated sheet ranges from 2 to 15 in a globular protein. The average strand in a β-pleated sheet section of a globular protein contains six amino acid residues.

Wool and the fibrous protein of muscle are examples of proteins with secondary structures that are almost all α-helices. Consequently, these proteins can be stretched. In contrast, the secondary structures of silk and spider webs are predominantly β-pleated sheets. Because the β-pleated sheet is a fully extended structure, these proteins cannot be stretched.

Coil Conformation

Generally, less than half of a globular protein is in an α-helix or β-pleated sheet (Figure 23.10). Most of the rest of the protein is still highly ordered but is difficult to describe. These polypeptide fragments are said to be in a **coil conformation** or a **loop conformation**.

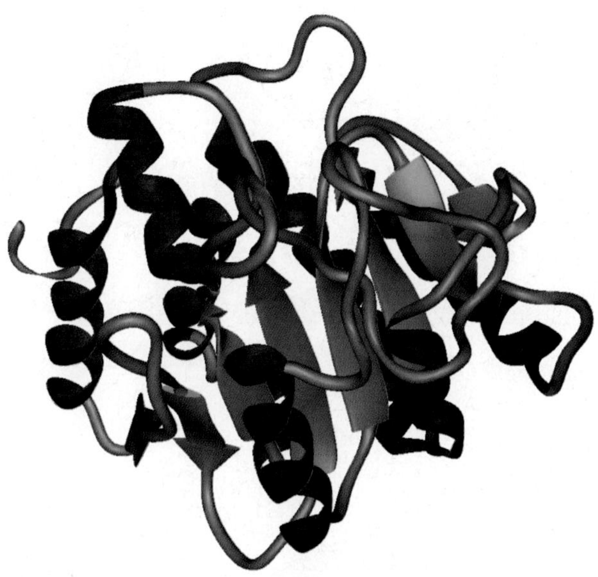

Figure 23.10
The backbone structure of carboxypeptidase A: α-helical segments are purple; β-pleated sheets are indicated by flat green arrows pointing in the N ⟶ C direction.

PROBLEM 34◆

How long is an α-helix that contains 74 amino acids? Compare the length of this α-helix with the length of a fully extended peptide chain containing the same number of amino acids. (The distance between consecutive amino acids in a fully extended chain is 3.5 Å.)

β-PEPTIDES: AN ATTEMPT TO IMPROVE ON NATURE

Chemists are currently studying β-peptides, which are polymers of β-amino acids. These peptides have backbones one carbon longer than the peptides nature synthesizes using α-amino acids. Therefore, each β-amino acid residue has two carbons to which side chains can be attached.

Like α-polypeptides, β-polypeptides fold into relatively stable helical and pleated sheet conformations, causing scientists to wonder whether biological activity might be possible with such peptides. Recently, a β-peptide with biological activity has been synthesized—one that mimics the activity of the hormone somatostatin. There is hope that β-polypeptides will provide a source of new drugs and catalysts. Surprisingly, the peptide bonds in β-polypeptides are resistant to the enzymes that catalyze the hydrolysis of peptide bonds in α-polypeptides. This resistance to hydrolysis means that a β-polypeptide drug will have a longer duration of action in the bloodstream.

$$H_3\overset{+}{N}-CH(R)-C(=O)-O^-$$
α-amino acid

$$H_3\overset{+}{N}-CH(R)-CH(R')-C(=O)-O^-$$
β-amino acid

23.14 Tertiary Structure of Proteins

The *tertiary structure* of a protein is the three-dimensional arrangement of all the atoms in the protein. Proteins fold spontaneously in solution in order to maximize their stability. Every time there is a stabilizing interaction between two atoms, free energy is released. The more free energy released (the more negative the $\Delta G°$), the more stable the protein. So a protein tends to fold in a way that maximizes the number of stabilizing interactions (Figure 23.11).

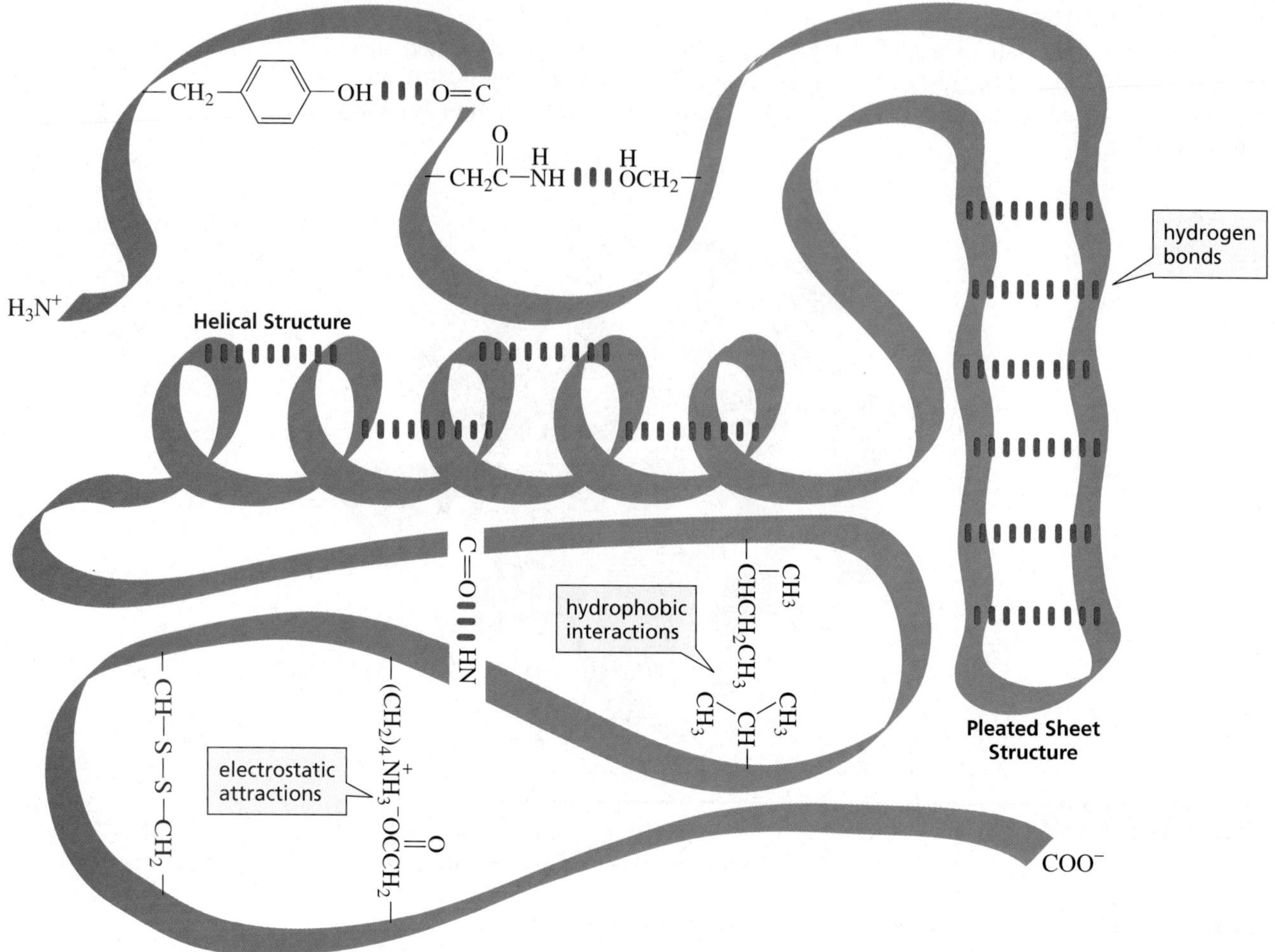

▲ **Figure 23.11**
Stabilizing interactions responsible for the tertiary structure of a protein.

Max Ferdinand Perutz *and* **John Cowdery Kendrew** *were the first to determine the tertiary structure of a protein. Using X-ray diffraction, they determined the tertiary structure of myoglobin (1957) and hemoglobin (1959). For this work, they shared the 1962 Nobel Prize in chemistry.*

Max Perutz *was born in Austria in 1914. In 1936, because of the rise of Nazism, he moved to England. He received a Ph.D. from, and became a professor at, Cambridge University. He worked on the three-dimensional structure of hemoglobin and assigned the work on myoglobin (a smaller protein) to* **John Kendrew (1917–1997)**. *Kendrew was born in England and was educated at Cambridge University.*

The stabilizing interactions include disulfide bonds, hydrogen bonds, electrostatic attractions (attractions between opposite charges), and hydrophobic (van der Waals) interactions. Stabilizing interactions can occur between peptide groups (atoms in the backbone of the protein), between side-chain groups (α-substituents), and between peptide and side-chain groups. Because the side-chain groups help determine how a protein folds, the tertiary structure of a protein is determined by its primary structure.

Disulfide bonds are the only covalent bonds that can form when a protein folds. The other bonding interactions that occur in folding are much weaker, but because there are so many of them (Figure 23.12), they are the important interactions in determining how a protein folds.

Most proteins exist in aqueous environments. Therefore, they tend to fold in a way that exposes the maximum number of polar groups to the aqueous environment and that buries the nonpolar groups in the interior of the protein, away from water.

The interactions between nonpolar groups are known as **hydrophobic interactions**. These interactions increase the stability of a protein by increasing the entropy of water molecules. Water molecules that surround nonpolar groups are highly structured. When two nonpolar groups come together, the surface area in contact with water decreases, decreasing the amount of structured water. Decreasing structure increases entropy, which in turn decreases the free energy, which increases the stability of the protein. (Recall that $\Delta G° = \Delta H° - T\Delta S°$.)

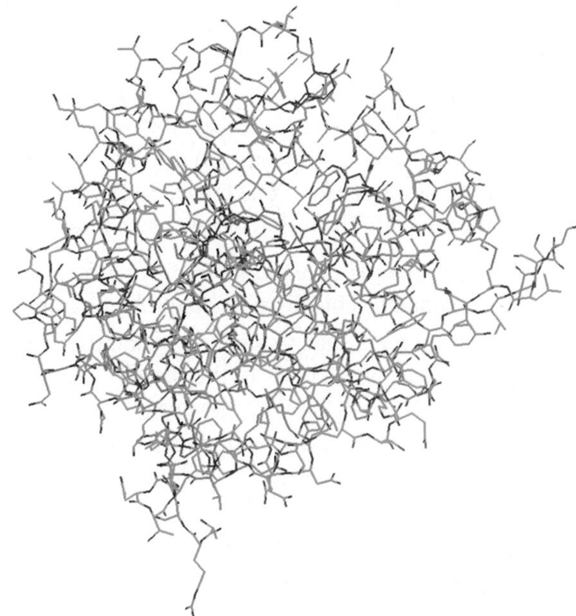

◀ **Figure 23.12**
The three-dimensional structure of carboxypeptidase A.

PROBLEM 35◆

How would a protein that resides in the interior of a membrane fold, compared with the water-soluble protein just discussed? (*Hint:* See Section 26.4.)

23.15 Quaternary Structure of Proteins

Proteins that have more than one peptide chain are called **oligomers**. The individual chains are called **subunits**. A protein with a single subunit is called a *monomer*, one with two subunits is called a *dimer*; one with three subunits is called a *trimer*, and one with four subunits is called a *tetramer*. Hemoglobin is an example of a tetramer. It has two different kinds of subunits and two of each kind. The quaternary structure of hemoglobin is shown in Figure 23.13.

The subunits are held together by the same kinds of interactions that hold the individual protein chains in a particular three-dimensional conformation: hydrophobic interactions, hydrogen bonding, and electrostatic attractions. The quaternary structure of a protein describes the way the subunits are arranged in space. Some of the possible arrangements of the six subunits of a hexamer are shown here:

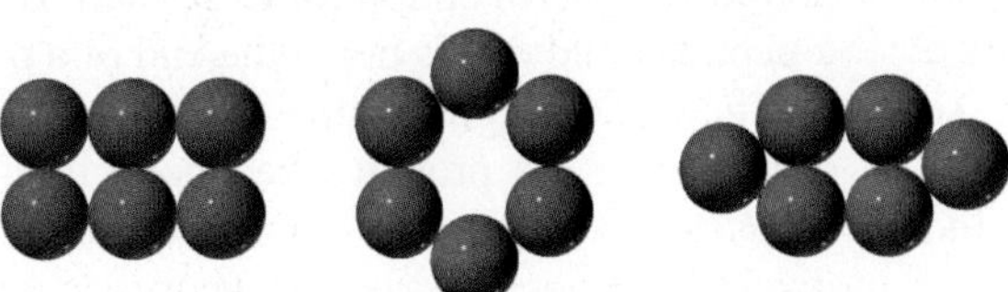

possible quaternary structures for a hexamer

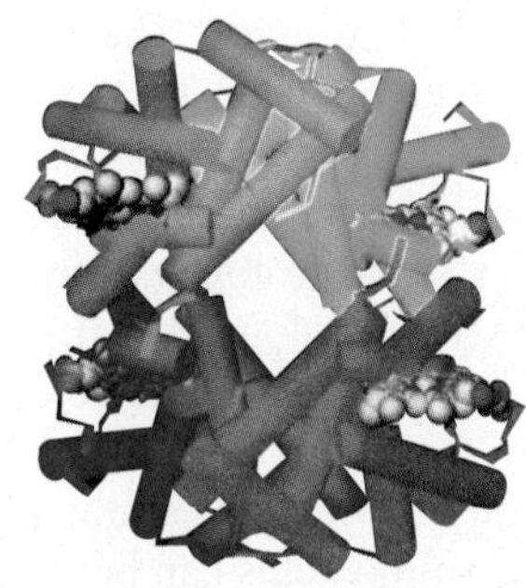

▲ **Figure 23.13**
Computer graphic representation of the quaternary structure of hemoglobin. The orange and pink subunits are identical, as are the green and purple subunits. The cylindrical tubes represent the polypeptide chains, while the beads represent the iron-containing porphyrin rings (Section 21.11).

PROBLEM 36◆

a. Which of the following water-soluble proteins would have the greatest percentage of polar amino acids—a spherical protein, a cigar-shaped protein, or a subunit of a hexamer?
b. Which of these would have the smallest percentage of polar amino acids?

23.16 Protein Denaturation

Destroying the highly organized tertiary structure of a protein is called **denaturation**. Anything that breaks the bonds responsible for maintaining the three-dimensional shape of the protein will cause the protein to denature (unfold). Because these bonds are weak, proteins are easily denatured. The totally random conformation of a denatured protein is called a **random coil**. The following are some of the ways that proteins can be denatured:

- Changing the pH denatures proteins because it changes the charges on many of the side chains. This disrupts electrostatic attractions and hydrogen bonds.
- Certain reagents such as urea and guanidine hydrochloride denature proteins by forming hydrogen bonds to the protein groups that are stronger than the hydrogen bonds formed between the groups.
- Detergents such as sodium dodecyl sulfate denature proteins by associating with the nonpolar groups of the protein, thus interfering with the normal hydrophobic interactions.
- Organic solvents denature proteins by disrupting hydrophobic interactions.
- Proteins can also be denatured by heat or by agitation. Both increase molecular motion, which can disrupt the attractive forces. A well-known example is the change that occurs to the white of an egg when it is heated or whipped.

Summary

Peptides and **proteins** are polymers of **amino acids** linked together by **peptide** (amide) **bonds**. A **dipeptide** contains two amino acid residues, a **tripeptide** contains three, an **oligopeptide** contains three to 10, and a **polypeptide** contains many amino acid residues. Proteins have 40 to 4000 amino acid residues. The **amino acids** differ only in the substituent attached to the α-carbon. Most amino acids found in nature have the L configuration.

The carboxyl groups of the amino acids have pK_a values of ~2 and the protonated amino groups have pK_a values of ~9. At physiological pH, an amino acid exists as a **zwitterion**. A few amino acids have side chains with ionizable hydrogens. The **isoelectric point** (pI) of an amino acid is the pH at which the amino acid has no net charge. A mixture of amino acids can be separated based on their pI's by **electrophoresis** or based on their polarities by **paper chromatography** or **thin-layer chromatography**. Preparative separation can be achieved using **ion-exchange chromatography** employing a **cation-exchange resin**. An **amino acid analyzer** is an instrument that automates ion-exchange chromatography. A racemic mixture of amino acids can be separated by a **kinetic resolution**.

The amide bonds that link amino acid residues are called **peptide bonds**. A peptide bond has about 40% double-bond character. Two cysteine residues can be oxidized to a **disulfide bridge**. Disulfide bridges are the only covalent bonds that form between nonadjacent amino acids. By convention, peptides and proteins are written with the free amino group (the **N-terminal amino acid**) on the left and the free carboxyl group (the **C-terminal amino acid**) on the right.

To synthesize a peptide bond, the amino group of the first amino acid must be protected (by *t*-BOC) and its carboxyl group activated (with DCC). The second amino acid is added to form a dipeptide. Amino acids can be added to the growing C-terminal end by repeating these two steps: activating the carboxyl group of the C-terminal amino acid with DCC and adding a new amino acid. **Automated solid-phase peptide synthesis** allows peptides to be made more rapidly and in higher yields.

The **primary structure** of a protein is the sequence of its amino acids and the location of all its disulfide bridges. The N-terminal amino acid of a peptide or protein can be determined with **Edman's reagent**. The C-terminal amino acid can be identified with carboxypeptidase. **Partial hydrolysis** hydrolyzes only some of the peptide bonds. An **exopeptidase** catalyzes the hydrolysis of a peptide bond at the end of a peptide chain. An **endopeptidase** catalyzes the hydrolysis of a peptide bond that is not at the end of a peptide chain.

The **secondary structure** of a protein describes how local segments of the protein's backbone folds. A protein folds so as to maximize the number of stabilizing interactions: disulfide bonds, hydrogen bonds, electrostatic attractions (attraction between opposite charges), and **hydrophobic interactions** (interactions between nonpolar groups). An **α-helix**, a **β-pleated sheet**, and a **coil conformation** are types of secondary structure. The **tertiary structure** of a protein is the three-dimensional arrangement of all the atoms in the protein. Proteins with more than one peptide chain are called **oligomers**. The individual chains are called **subunits**. The **quaternary structure** of a protein describes the way the **subunits** are arranged with respect to each other in space.

Key Terms

amino acid (p. 959)
D-amino acid (p. 964)
L-amino acid (p. 964)
amino acid analyzer (p. 970)
amino acid residue (p. 959)
anion-exchange resin (p. 970)
antiparallel β-pleated sheet (p. 990)
automated solid-phase peptide synthesis (p. 980)
cation-exchange resin (p. 970)
coil conformation (p. 991)
C-terminal amino acid (p. 973)
denaturation (p. 994)
dipeptide (p. 959)
disulfide (p. 974)
disulfide bridge (p. 974)
Edman's reagent (p. 984)
electrophoresis (p. 968)
endopeptidase (p. 985)
enzyme (p. 960)
essential amino acid (p. 963)
exopeptidase (p. 985)
fibrous protein (p. 982)
globular protein (p. 982)
α-helix (p. 989)
hydrophobic interactions (p. 992)
interchain disulfide bridge (p. 975)
intrachain disulfide bridge (p. 975)
ion-exchange chromatography (p. 970)
isoelectric point (p. 966)
kinetic resolution (p. 972)
loop conformation (p. 991)
N-terminal amino acid (p. 973)
oligomer (p. 993)
oligopeptide (p. 959)
paper chromatography (p. 969)
parallel β-pleated sheet (p. 990)
partial hydrolysis (p. 985)
peptide (p. 959)
peptide bond (p. 973)
β-pleated sheet (p. 990)
polypeptide (p. 959)
primary structure (p. 982)
protein (p. 959)
quaternary structure (p. 982)
random coil (p. 994)
secondary structure (p. 982)
structural protein (p. 959)
subunit (p. 993)
tertiary structure (p. 982)
thin-layer chromatography (p. 969)
tripeptide (p. 959)
zwitterion (p. 965)

Problems

37. Unlike most amines and carboxylic acids, amino acids are insoluble in diethyl ether. Explain.

38. Indicate the peptides that would result from cleavage by the indicated reagent:
    a. His-Lys-Leu-Val-Glu-Pro-Arg-Ala-Gly-Ala by trypsin
    b. Leu-Gly-Ser-Met-Phe-Pro-Tyr-Gly-Val by chymotrypsin
    c. Val-Arg-Gly-Met-Arg-Ala-Ser by carboxypeptidase A
    d. Ser-Phe-Lys-Met-Pro-Ser-Ala-Asp by cyanogen bromide
    e. Arg-Ser-Pro-Lys-Lys-Ser-Glu-Gly by trypsin

39. Aspartame has a pI of 5.9. Draw its most prevalent form at physiological pH.

40. Draw the form of aspartic acid that predominates at
    a. pH = 1.0 b. pH = 2.6 c. pH = 6.0 d. pH = 11.0

41. Dr. Kim S. Tree was preparing a manuscript for publication in which she reported that the pI of the tripeptide Lys-Lys-Lys was 10.6. One of her students pointed out that there must be an error in her calculations because the pK_a of the ε-amino group of lysine is 10.8 and the pI of the tripeptide has to be greater than any of its individual pK_a values. Was the student correct?

42. A mixture of amino acids that do not separate sufficiently when a single technique is used can often be separated by two-dimensional chromatography. In this technique, the mixture of amino acids is applied to a piece of filter paper and separated by chromatographic techniques. The paper is then rotated 90°, and the amino acids are further separated by electrophoresis, producing a type of chromatogram called a *fingerprint*. Identify the spots in the fingerprint obtained from a mixture of Ser, Glu, Leu, His, Met, and Thr.

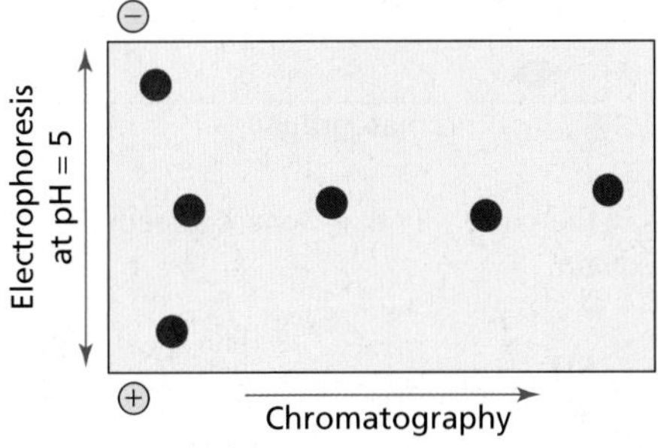

43. Explain the difference in the pK_a values of the carboxyl groups of alanine, serine, and cysteine.

44. Which would be a more effective buffer at physiological pH, a solution of 0.1 M glycylglycylglycylglycine or a solution of 0.2 M glycine?

45. Identify the location and type of charge on the hexapeptide Lys-Ser-Asp-Cys-His-Tyr at
   a. pH = 7
   b. pH = 5
   c. pH = 9

46. The following polypeptide was treated with 2-mercaptoethanol and then with iodoacetic acid. After reacting with maleic anhydride, the peptide was hydrolyzed by trypsin. (Treatment with maleic anhydride causes trypsin to cleave a peptide only at arginine residues.)

   Gly-Ser-Asp-Ala-Leu-Pro-Gly-Ile-Thr-Ser-Arg-Asp-Val-Ser-Lys-Val-Glu-Tyr-Phe-Glu-Ala-Gly-Arg-Ser-Glu-Phe-Lys-Glu-Pro-Arg-Leu-Tyr-Met-Lys-Val-Glu-Gly-Arg-Pro-Val-Ser-Ala-Gly-Leu-Trp

   a. Why, after a peptide is treated with maleic anhydride, does trypsin no longer cleave it at lysine residues?
   b. How many fragments are obtained from the peptide?
   c. In what order would the fragments be eluted from an anion-exchange column using a buffer of pH = 5?

47. Treatment of a polypeptide with 2-mercaptoethanol yields two polypeptides with the following primary sequences:

   Val-Met-Tyr-Ala-Cys-Ser-Phe-Ala-Glu-Ser
   Ser-Cys-Phe-Lys-Cys-Trp-Lys-Tyr-Cys-Phe-Arg-Cys-Ser

   Treatment of the original intact polypeptide with chymotrypsin yields the following peptides:
   a. Ala, Glu, Ser
   b. 2 Phe, 2 Cys, Ser
   c. Tyr, Val, Met
   d. Arg, Ser, Cys
   e. Ser, Phe, 2 Cys, Lys, Ala, Trp
   f. Tyr, Lys

   Determine the positions of the disulfide bridges in the original polypeptide.

48. Show how aspartame can be synthesized using DCC.

49. Reaction of a polypeptide with carboxypeptidase A releases Met. The polypeptide undergoes partial hydrolysis to give the following peptides. What is the sequence of the polypeptide?
   a. Ser, Lys, Trp
   b. Gly, His, Ala
   c. Glu, Val, Ser
   d. Leu, Glu, Ser
   e. Met, Ala, Gly
   f. Ser, Lys, Val
   g. Glu, His
   h. Leu, Lys, Trp
   i. Lys, Ser
   j. Glu, His, Val
   k. Trp, Leu, Glu
   l. Ala, Met

50. Glycine has pK_a values of 2.3 and 9.6. Would you expect the pK_a values of glycylglycine to be higher or lower than these values?

51. A mixture of 15 amino acids gave the fingerprint shown below (see also Problem 42). Identify the spots. (*Hint 1:* Pro reacts with ninhydrin to form a yellow color; Phe and Tyr form a yellow-green color. *Hint 2:* Count the number of spots before you start.)

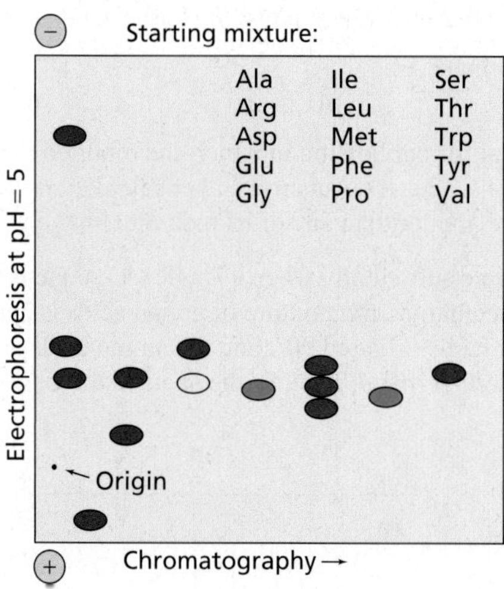

52. Dithiothreitol reacts with disulfide bridges in the same way that 2-mercaptoethanol does. With dithiothreitol, however, the equilibrium lies much more to the right. Explain.

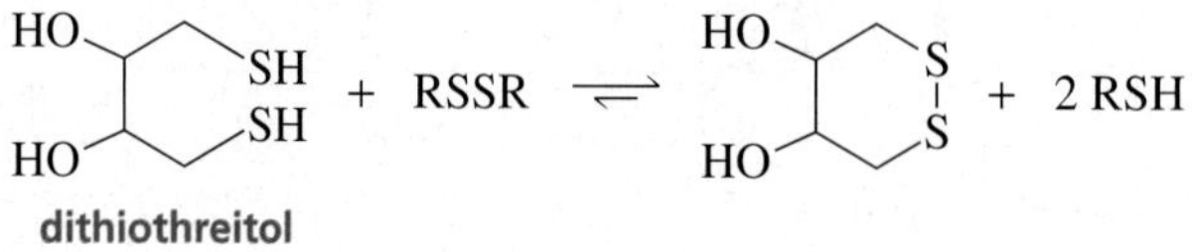

HO, HO, SH, SH (dithiothreitol) + RSSR ⇌ HO, HO, S–S (cyclic disulfide) + 2 RSH

dithiothreitol

53. α-Amino acids can be prepared by treating an aldehyde with ammonia and hydrogen cyanide, followed by acid-catalyzed hydrolysis.
    a. Give the structures of the two intermediates formed in this reaction.
    b. What amino acid is formed when the aldehyde that is used is 3-methylbutanal?
    c. What aldehyde would be needed to prepare valine?

54. The UV spectra of tryptophan, tyrosine, and phenylalanine are shown here. Each spectrum is that of a 1×10^{-3} M solution of the amino acid, buffered at pH = 6.0. Calculate the approximate molar absorptivity of each of the three amino acids at 280 nm.

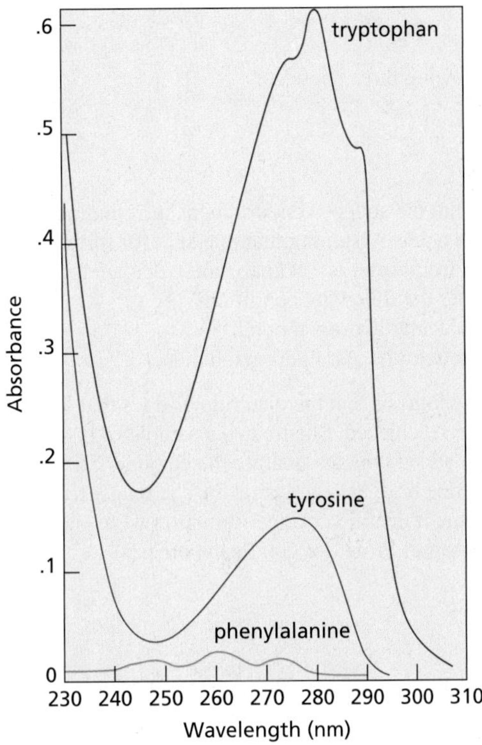

55. A normal polypeptide and a mutant of the polypeptide were hydrolyzed by an endopeptidase under the same conditions. The normal and mutant differ by one amino acid residue. The fingerprints of the peptides obtained from the normal and mutant polypeptides are as shown. What kind of amino acid substitution occurred as a result of the mutation (That is, the substituted amino acid more or less polar than the original amino acid? Is its pI lower or higher?)

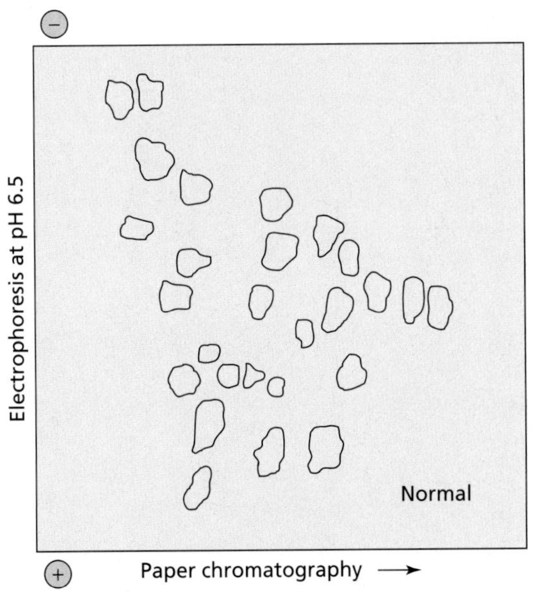

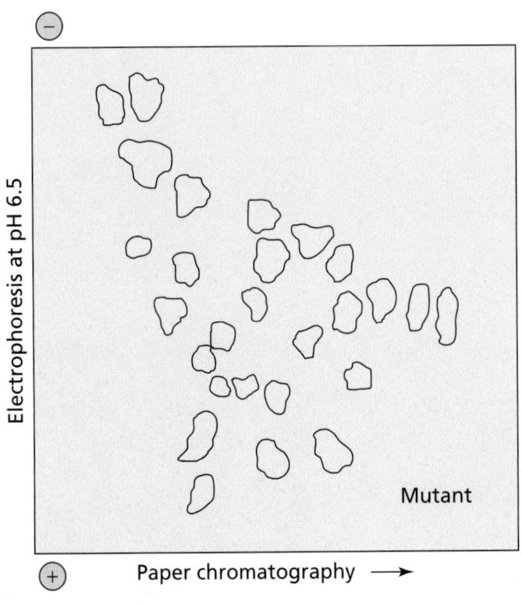

56. Determine the amino acid sequence of a polypeptide from the following results:
    a. Complete hydrolysis of the peptide yields the following amino acids: Ala, Arg, Gly, 2 Lys, Met, Phe, Pro, 2 Ser, Tyr, Val.
    b. Treatment with Edman's reagent gives PTH-Val.
    c. Carboxypeptidase A releases Ala.
    d. Treatment with cyanogen bromide yields the following two peptides:
        1. Ala, 2 Lys, Phe, Pro, Ser, Tyr
        2. Arg, Gly, Met, Ser, Val
    e. Treatment with chymotrypsin yields the following three peptides:
        1. 2 Lys, Phe, Pro
        2. Arg, Gly, Met, Ser, Tyr, Val
        3. Ala, Ser
    f. Treatment with trypsin yields the following three peptides:
        1. Gly, Lys, Met, Tyr
        2. Ala, Lys, Phe, Pro, Ser
        3. Arg, Ser, Val

57. The C-terminal end of a protein extends into the aqueous environment surrounding the protein. The C-terminal amino acids are Gln, Asp, 2 Ser, and three nonpolar amino acids. Assuming that the $\Delta G°$ for formation of a hydrogen bond is -3 kcal/mol and the $\Delta G°$ for removal of a hydrophobic group from water is -4 kcal/mol, calculate the $\Delta G°$ for folding the C-terminal end of the protein into the interior of the protein under the following conditions:
    a. All the polar groups form one intramolecular hydrogen bond.
    b. All but two of the polar groups form intramolecular hydrogen bonds.

58. Professor Mary Gold wanted to test her hypothesis that the disulfide bridges that form in many proteins do so after the minimum energy conformation of the protein has been achieved. She treated a sample of lysozyme, an enzyme containing four disulfide bridges, with 2-mercaptoethanol and then added urea to denature the enzyme. She slowly removed these reagents so that the enzyme could refold and reform the disulfide bridges. The lysozyme she recovered had 80% of its original activity. What would be the percent activity in the recovered enzyme if disulfide bridge formation were entirely random rather than determined by the tertiary structure? Does this experiment support Professor Gold's hypothesis?

24 Catalysis

A **catalyst** is a substance that *increases the rate of a chemical reaction without itself being consumed or changed in the reaction*. We have seen that the rate of a chemical reaction depends on the energy barrier that must be overcome in the process of converting reactants into products (Section 3.7). The height of the "energy hill" is indicated by the free energy of activation ($\Delta G^{\ddagger}$). A catalyst increases the rate of a chemical reaction by providing a pathway with a lower $\Delta G^{\ddagger}$.

A catalyst can decrease $\Delta G^{\ddagger}$ in one of three ways:

1. The catalyzed and uncatalyzed reactions can follow different, but similar, mechanisms, with the catalyst providing a way to convert the reactant into a *less stable species* (Figure 24.1a).
2. The catalyzed and uncatalyzed reactions can follow different, but similar, mechanisms, with the catalyst providing a way to make *the transition state more stable* (Figure 24.1b).

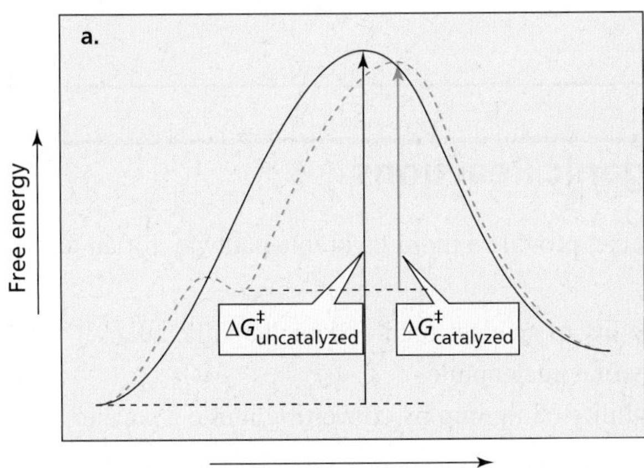

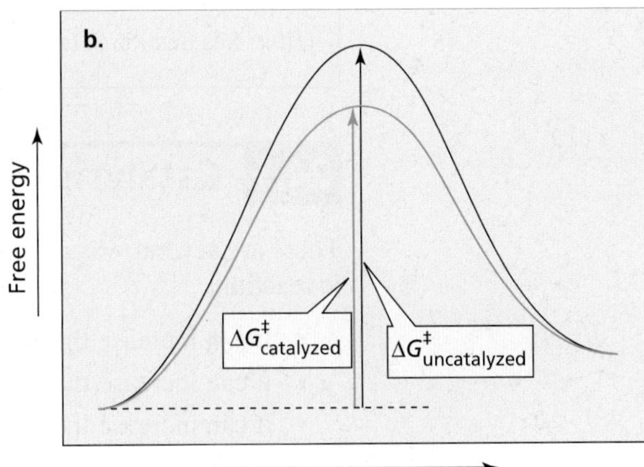

▲ **Figure 24.1**
Reaction coordinate diagrams for an uncatalyzed reaction and for a catalyzed reaction. (a) The catalyst converts the reactant to a less stable species. (b) The catalyst stabilizes the transition state.

3. The catalyst can completely *change the mechanism* of the reaction, providing an alternative pathway with a smaller $\Delta G^\ddagger$ than that of the uncatalyzed reaction (Figure 24.2).

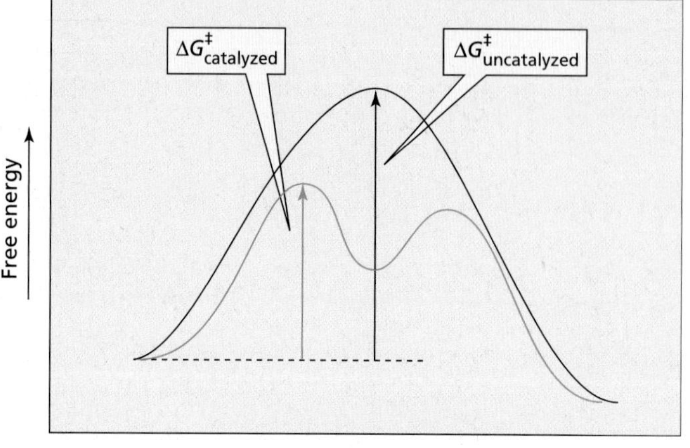

Figure 24.2 ▶
Reaction coordinate diagrams for an uncatalyzed reaction and for a catalyzed reaction. The catalyzed reaction takes place by an alternative and energetically more favorable pathway.

When we say that a catalyst is neither consumed nor changed by a reaction, we do not mean that it does not participate in the reaction. A catalyst *must* participate in the reaction if it is going to make it go faster. What we mean is that a catalyst has the same form after the reaction that it had before the reaction. Because the catalyst is not used up during the reaction—if it is used up in one step of the reaction, it must be regenerated in a subsequent step—only a small amount of the catalyst is needed. Therefore, a catalyst is added to a reaction mixture in small *catalytic* amounts, much less than the number of moles of reactant.

Notice that the stability of the original reactants and final products is the same in both the catalyzed and uncatalyzed reactions. In other words, the catalyst does not change the equilibrium constant of the reaction. (Observe that $\Delta G°$ is the same for the catalyzed and uncatalyzed reactions in Figures 24.1a, 24.1b, and 24.2.) Because the catalyst does not change the equilibrium constant, it does not change the *amount* of product formed during the reaction. It changes only the *rate* at which the product is formed.

PROBLEM 1◆

Which of the following parameters would be different for a reaction carried out in the presence of a catalyst, compared with the same reaction carried out in the absence of a catalyst?

$$\Delta G°, \Delta H^\ddagger, E_a, \Delta S^\ddagger, \Delta H°, K_{eq}, \Delta G^\ddagger, \Delta S°, k_{rate}$$

(*Hint:* See Section 3.7.)

24.1 Catalysis in Organic Reactions

There are several ways a catalyst can provide a more favorable pathway for an organic reaction:

- It can increase the susceptibility of an electrophile to nucleophilic attack.
- It can increase the reactivity of a nucleophile.
- It can increase the leaving ability of a group by converting it into a weaker base.

In this chapter, we will look at some of the most common catalysts—nucleophilic catalysts, acid catalysts, base catalysts, and metal-ion catalysts—and the ways in which they provide an energetically more favorable pathway for an organic reaction. We will then see how the same modes of catalysis are used in enzyme-catalyzed reactions.

THE NOBEL PRIZE

Throughout this text, you have found biographical sketches that give you some information about the men and women who created the science you are studying. You have seen that many of these people are Nobel Prize winners. The Nobel Prize is considered by many to be the most coveted award a scientist can receive. The awards were established by **Alfred Bernhard Nobel (1833–1896)**. The first prizes were awarded in 1901. Nobel was born in Stockholm, Sweden. When he was nine, he moved with his parents to St. Petersburg, where his father manufactured the torpedoes and submarine mines he had invented for the Russian government. As a young man, Alfred did research on explosives in a factory his father owned near Stockholm. In 1864, an explosion in the factory killed his younger brother, causing Alfred to look for ways to make it easier to handle and transport explosives. The Swedish government would not allow the factory to be rebuilt because of the many accidents that had occurred there. Nobel, therefore, established an explosives factory in Germany, where, in 1867, he discovered that if nitroglycerin were mixed with diatomaceous earth, the mixture could be molded into sticks that could not be set off without a detonating cap. Thus, Nobel invented dynamite. He also invented blasting gelatin and smokeless powder. Although he was the inventor of explosives used by the military, he was a strong supporter of peace movements.

The 355 patents Nobel held made him a wealthy man. He never married, and when he died, his will stipulated that the bulk of his estate ($9,200,000) be used to establish prizes to be awarded to those who "have conferred the greatest benefit on mankind." He instructed that the money be invested and the interest earned each year be divided into five equal portions "to be awarded to the persons having made the most important contributions in the fields of chemistry, physics, physiology or medicine, literature, and to the one who had done the most or the best work toward fostering fraternity among nations, the abolition of standing armies, and the holding and promotion of peace congresses." Nobel also stipulated that in awarding the prizes, no consideration be given to the nationality of the candidate, that each prize be shared by no more than three persons, and that no prize be awarded posthumously.

Nobel gave instructions that the prizes for chemistry and physics were to be awarded by the Royal Swedish Academy of Sciences, for physiology or medicine by the Karolinska Institute in Stockholm, for literature by the Swedish Academy, and for peace by a five-person committee elected by the Norwegian Parliament. The deliberations are secret, and the decisions cannot be appealed. In 1969, the Swedish Central Bank established a prize in economics in Nobel's honor. The recipient of this prize is selected by the Royal Swedish Academy of Sciences. On December 10—the anniversary of Nobel's death—the prizes are awarded in Stockholm, except for the peace prize, which is awarded in Oslo.

Alfred Bernhard Nobel

24.2 Nucleophilic Catalysis

A **nucleophilic catalyst** increases the rate of a reaction by acting as a nucleophile; it forms an intermediate by forming a covalent bond with one of the reactants. **Nucleophilic catalysis**, therefore, is also called **covalent catalysis**. A nucleophilic catalyst increases the reaction rate by completely changing the mechanism of the reaction.

A nucleophilic catalyst forms an intermediate by forming a covalent bond with a reactant.

In the following reaction, iodide ion increases the rate of conversion of ethyl chloride into ethyl alcohol by acting as a nucleophilic catalyst:

a nucleophilic catalyst

$$CH_3CH_2Cl + HO^- \xrightarrow[H_2O]{I^-} CH_3CH_2OH + Cl^-$$

To understand how iodide ion catalyzes this reaction, we need to look at the mechanism of the reaction both with and without the catalyst. In the absence of iodide ion, ethyl chloride is converted into ethyl alcohol in a one-step S_N2 reaction.

mechanism of the uncatalyzed reaction

$$HO^- + CH_3CH_2{-}Cl \longrightarrow CH_3CH_2OH + Cl^-$$

If iodide ion is present in the reaction mixture, the reaction takes place by two successive S_N2 reactions.

The 2001 Nobel Prize in chemistry was awarded to **K. Barry Sharpless, William S. Knowles,** *and* **Royji Noyroi** *for their work in the field of catalysis.*

William Knowles, *was born in 1917. He received a Ph.D. from Columbia University in 1942 and is a scientist at Monsanto Company, St. Louis, Missouri.*

Royji Noyroi *was born in 1938 in Kobe, Japan. He received a Ph.D. from Kyoto University and is a Professor of materials science at Nagoya University in Nagoya, Japan. Knowles and Noyroi were cited for their work on chirally catalyzed hydrogenation.*

K. Barry Sharpless *was cited for his work on chirally catalyzed oxidation reactions (Section 20.5).*

mechanism of the iodide-ion-catalyzed reaction

$$I^- + CH_3CH_2{-}Cl \longrightarrow CH_3CH_2I + Cl^-$$
$$HO^- + CH_3CH_2{-}I \longrightarrow CH_3CH_2OH + I^-$$

The first S_N2 reaction in the catalyzed reaction is faster than the uncatalyzed reaction because in a protic solvent, iodide ion is a better nucleophile than hydroxide ion, which is the nucleophile in the uncatalyzed reaction (Section 10.3). The second S_N2 reaction in the catalyzed reaction is also faster than the uncatalyzed reaction because iodide ion is a weaker base and therefore a better leaving group than chloride ion, the leaving group in the uncatalyzed reaction. Thus, iodide ion increases the rate of formation of ethanol by changing a relatively slow one-step reaction into a reaction with two relatively fast steps (Figure 24.2).

Iodide ion is a nucleophilic catalyst because it reacts as a nucleophile, forming a covalent bond with the reactant. The iodide ion that is consumed in the first reaction is regenerated in the second, so it comes out of the reaction unchanged.

Another reaction in which a nucleophilic catalyst provides a more favorable pathway by changing the mechanism of the reaction is the imidazole-catalyzed hydrolysis of an ester.

a nucleophilic catalyst

$$CH_3COO{-}C_6H_5 + H_2O \xrightarrow{\text{imidazole}} CH_3COOH + HO{-}C_6H_5$$

phenyl acetate — acetic acid — phenol

Imidazole is a better nucleophile than water, so imidazole reacts faster with the ester than water does. The acyl imidazole that is formed is particularly reactive because the positively charged nitrogen makes imidazole a very good leaving group. Therefore, it is hydrolyzed much more rapidly than the ester would have been. Because formation of the acyl imidazole and its subsequent hydrolysis are both faster than ester hydrolysis, imidazole increases the rate of ester hydrolysis.

an ester + imidazole ⇌ tetrahedral intermediate ⇌ an acyl imidazole + $^-O{-}C_6H_5$

an acyl imidazole $\xrightarrow{H_2O}$ CH_3CO^- (O) + $H{:}N^+$ imidazolium + $HO{-}C_6H_5$

24.3 Acid Catalysis

An **acid catalyst** increases the rate of a reaction by donating a proton to a reactant. For example, the rate of hydrolysis of an ester is markedly increased by an acid catalyst (HB^+).

A proton is donated to the reactant in an acid-catalyzed reaction.

an acid catalyst

$$CH_3\overset{O}{\overset{\|}{C}}OCH_3 + H_2O \overset{HB^+}{\rightleftharpoons} CH_3\overset{O}{\overset{\|}{C}}OH + CH_3OH$$

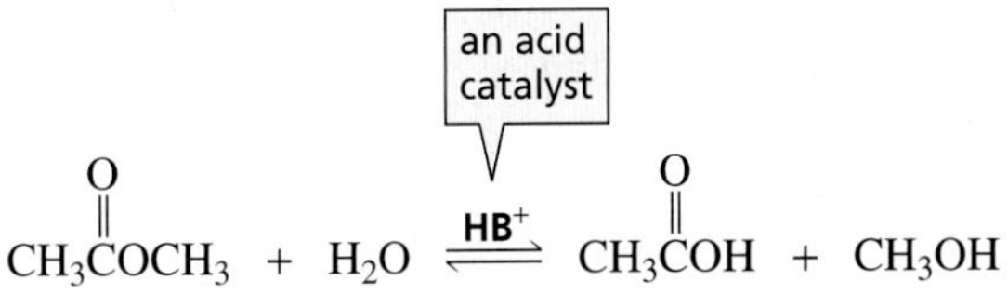

How an acid catalyzes the hydrolysis of an ester can be understood by examining the mechanism for the acid-catalyzed reaction. *Donation of a proton to, and removal of a proton from, an electronegative atom such as oxygen are fast steps.* Therefore, the reaction has two slow steps: formation of the tetrahedral intermediate and collapse of the tetrahedral intermediate. A catalyst must increase the rate of a slow step because increasing the rate of a fast step will not increase the rate of the overall reaction.

A catalyst must increase the rate of a slow step. Increasing the rate of a fast step will not increase the rate of the overall reaction.

mechanism for acid-catalyzed ester hydrolysis

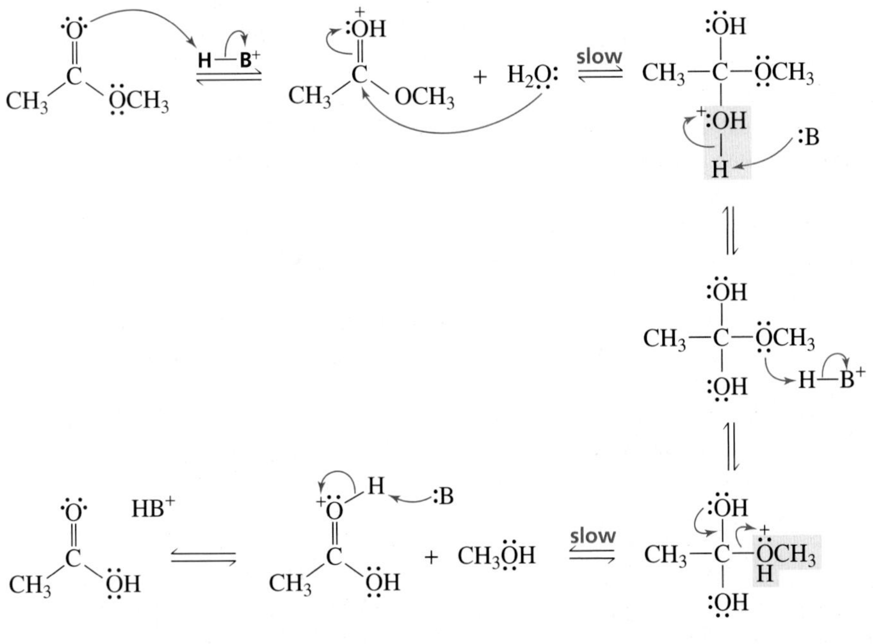

$CH_3C(=O)OCH_3$ + H—B$^+$ ⇌ $CH_3C(=\overset{+}{O}H)OCH_3$ + H_2O: $\underset{\text{slow}}{\rightleftharpoons}$ $CH_3—C(OH)(OCH_3)—\overset{+}{O}H_2$ + :B ⇌ $CH_3—C(OH)(OCH_3)—OH$ + H—B$^+$ ⇌ $CH_3—C(OH)(OH)—\overset{+}{O}HCH_3$ $\underset{\text{slow}}{\rightleftharpoons}$ $CH_3C(=\overset{+}{O}H)OH$ + CH_3OH + :B ⇌ $CH_3C(=O)OH$ + HB^+

The acid increases the rates of both slow steps of this reaction. It increases the rate of formation of the tetrahedral intermediate by protonating the carbonyl oxygen, thereby increasing the reactivity of the carbonyl group. We have seen that a protonated carbonyl group is more susceptible to nucleophilic attack than an unprotonated carbonyl group because the protonated carbonyl group is more electrophilic. In other words, the protonated carbonyl group is more susceptible to nucleophilic attack (Section 17.11). Increasing the reactivity of the carbonyl group by protonating it is an example of providing a way to convert the reactant into a less stable (i.e., more reactive) species (Figure 24.1a).

acid-catalyzed first slow step

$CH_3C(=\overset{+}{O}H)OCH_3 + H_2O$:

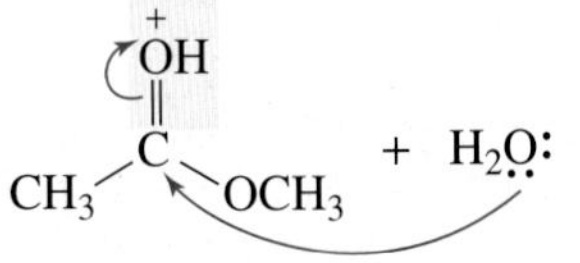

uncatalyzed first slow step

$CH_3C(=O)OCH_3 + H_2O$:

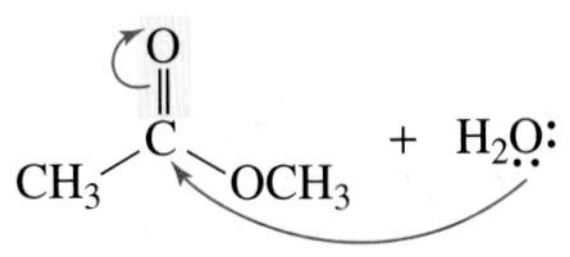

The acid increases the rate of the second slow step by changing the basicity of the group that is eliminated when the tetrahedral intermediate collapses. In the presence of an acid, methanol (pK_a of $CH_3\overset{+}{O}H_2$ = −2.5) is eliminated; in the absence of an acid, methoxide ion (pK_a of CH_3OH = 15.7) is eliminated. Methanol is a weaker base than methoxide ion, so it is more easily eliminated.

acid-catalyzed second slow step **uncatalyzed second slow step**

$$CH_3-\underset{OH}{\overset{:\ddot{O}H}{C}}-\overset{+}{O}CH_3\ (H) \qquad CH_3-\underset{OH}{\overset{:\ddot{O}H}{C}}-OCH_3$$

The mechanism for the acid-catalyzed hydrolysis of an ester shows that the reaction can be divided into two distinct phases: formation of a tetrahedral intermediate and collapse of a tetrahedral intermediate. There are three steps in each phase. Notice that in each phase, the first step is a fast protonation step, the second step is a slow catalyzed step that involves either breaking a π bond or forming a π bond, and the last step is a fast deprotonation step (to regenerate the catalyst).

PROBLEM 2

Compare each of the following mechanisms with the mechanism for each phase of the acid-catalyzed hydrolysis of an ester, indicating

a. similarities b. differences

1. acid-catalyzed formation of a hydrate (Section 18.7)
2. acid-catalyzed conversion of an aldehyde into a hemiacetal (Section 18.7)
3. acid-catalyzed conversion of a hemiacetal into an acetal (Section 18.7)
4. acid-catalyzed hydrolysis of an amide (Section 17.16)

There are two types of acid catalysis: *specific-acid catalysis* and *general-acid catalysis*. In **specific-acid catalysis**, the proton is fully transferred to the reactant *before* the slow step of the reaction begins (Figure 24.3a). In **general-acid catalysis**, the proton is transferred to the reactant *during* the slow step of the reaction (Figure 24.3b). The mechanism for acid-catalyzed hydrolysis on p. 1003 shows that the slow steps of the reaction are specific-acid catalyzed.

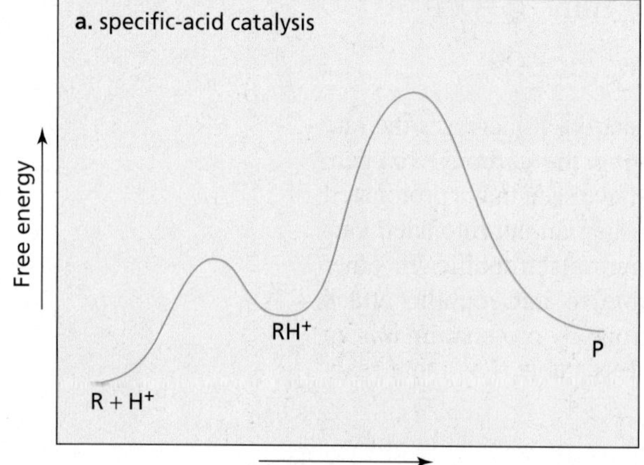

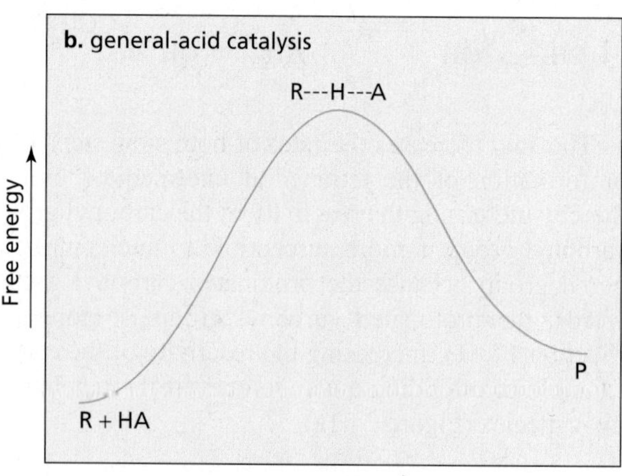

▲ **Figure 24.3**
(a) Reaction coordinate diagram for a specific-acid-catalyzed reaction. (The proton is completely transferred to the reactant before the slow step of the reaction begins.)
(b) Reaction coordinate diagram for a general-acid-catalyzed reaction. (The proton is partially transferred to the reactant in the transition state of the slow step of the reaction.)

Specific-acid and general-acid catalysts speed up a reaction in the same way—by donating a proton in order to make bond making and bond breaking easier. The two types of acid catalysis differ only in the extent to which the proton is transferred in the transition state of the slow step of the reaction. In a specific-acid-catalyzed reaction, the transition state has a fully transferred proton, whereas in a general-acid-catalyzed reaction, the transition state has a partially transferred proton (Figure 24.3).

In the examples that follow, notice the difference in the extent of proton transfer that has occurred when the nucleophile attacks the reactant. In specific-acid-catalyzed attack by water on a carbonyl group, the nucleophile attacks a fully protonated carbonyl group. In general-acid-catalyzed attack by water on a carbonyl group, the carbonyl group becomes protonated as the nucleophile attacks it.

specific-acid-catalyzed attack by water

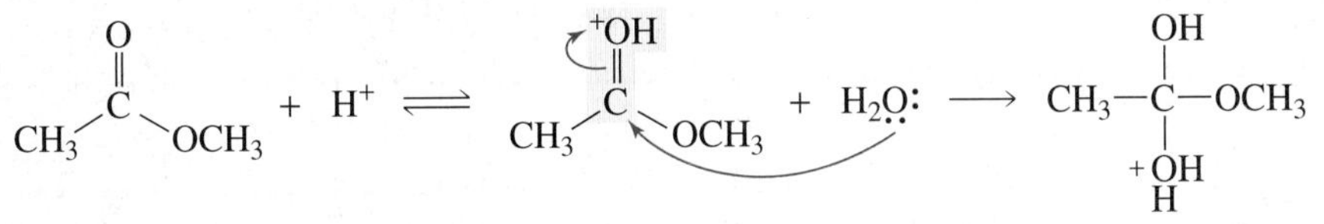

$$CH_3C(=O)OCH_3 + H^+ \rightleftharpoons CH_3C(={}^+OH)OCH_3 + H_2\ddot{O}: \longrightarrow CH_3-C(OH)(\overset{+}{O}H_2)-OCH_3$$

general-acid-catalyzed attack by water

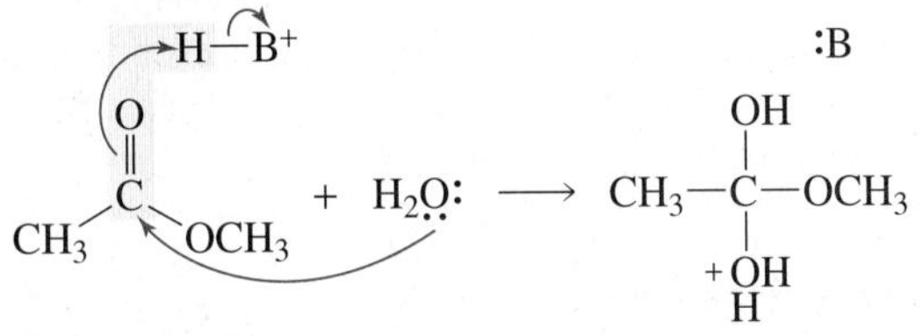

$$CH_3C(=O)OCH_3 \;\; H-B^+ + H_2\ddot{O}: \longrightarrow CH_3-C(OH)(\overset{+}{O}H_2)-OCH_3 \;\; :B$$

In specific-acid-catalyzed collapse of a tetrahedral intermediate, a fully protonated leaving group is eliminated, whereas in general-acid-catalyzed collapse of a tetrahedral intermediate, the leaving group picks up a proton as the group is eliminated.

specific-acid-catalyzed elimination of the leaving group

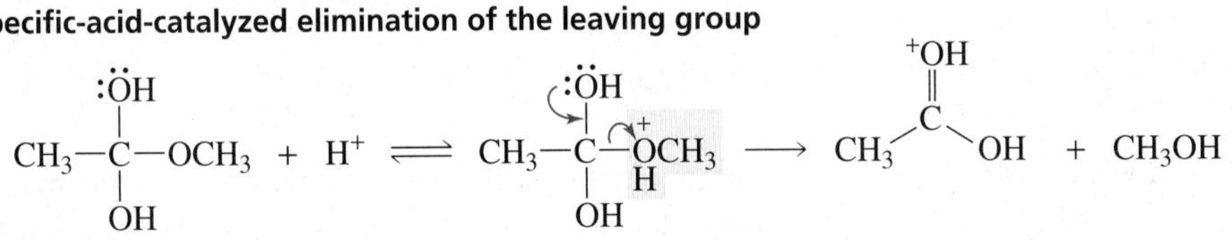

$$CH_3-C(:\ddot{O}H)(OH)-OCH_3 + H^+ \rightleftharpoons CH_3-C(:\ddot{O}H)(OH)-\overset{+}{O}(H)CH_3 \longrightarrow CH_3C(={}^+OH)OH + CH_3OH$$

general-acid-catalyzed elimination of the leaving group

$$CH_3-C(:\ddot{O}H)(OH)-OCH_3 \;\; H-B^+ \longrightarrow CH_3C(={}^+OH)OH + CH_3OH \;\; :B$$

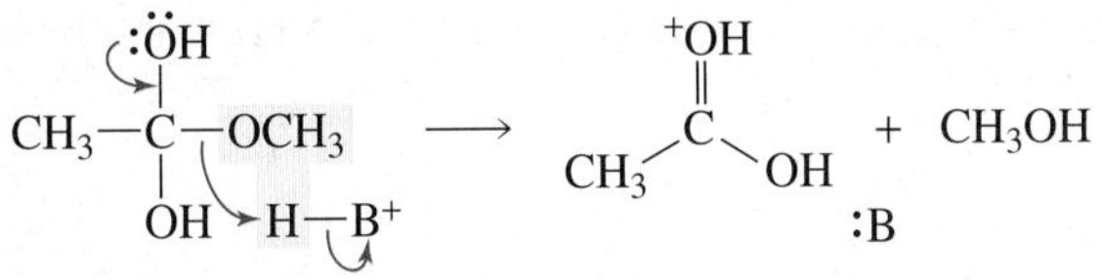

A specific-acid catalyst must be an acid that is strong enough to protonate the reactant fully before the slow step begins. A general-acid catalyst can be a weaker acid because it only partially transfers a proton in the transition state of the slow step.

PROBLEM 3

The following reaction occurs by a general-acid-catalyzed mechanism:

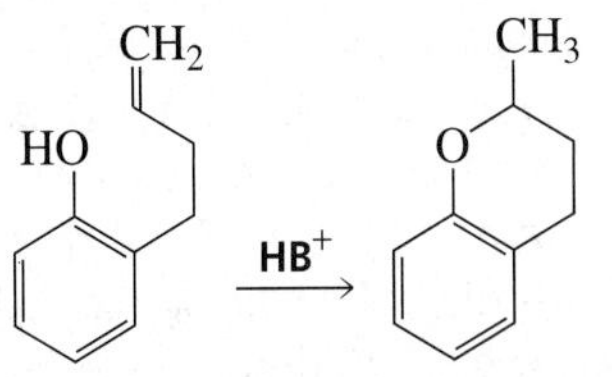

2-(3-butenyl)phenol (HO, CH_2=) $\xrightarrow{HB^+}$ 2-methylchromane (CH_3, O)

Propose a mechanism for this reaction.

PROBLEM 4 SOLVED

An alcohol will not react with aziridine unless an acid is present. Why is the acid necessary?

$$\text{aziridine} + CH_3OH \xrightarrow{HCl} H_3\overset{+}{N}CH_2CH_2OCH_3 \quad {}^{-}Cl$$

SOLUTION Although relief of ring strain is sufficient by itself to cause an epoxide to undergo a ring-opening reaction (Section 12.7), it is not sufficient to cause an aziridine to undergo a ring-opening reaction. A negatively charged nitrogen is a stronger base, and therefore a poorer leaving group, than a negatively charged oxygen. An acid, therefore, is needed to make the ring nitrogen a better leaving group by protonating it.

24.4 Base Catalysis

A proton is removed from the reactant in a base-catalyzed reaction.

A **base catalyst** increases the rate of a reaction by removing a proton from the reactant. For example, dehydration of a hydrate in the presence of hydroxide ion is a base-catalyzed reaction. Hydroxide ion (the base) increases the rate of the reaction by removing a proton from the neutral hydrate.

specific-base catalyzed dehydration

$$\underset{\text{a hydrate}}{ClCH_2C(OH)_2CH_2Cl} \underset{}{\overset{^{-}OH}{\rightleftharpoons}} ClCH_2C(O^{-})(OH)CH_2Cl + H_2O \overset{\text{slow}}{\rightleftharpoons} ClCH_2C(=O)CH_2Cl + {}^{-}OH$$

Removing a proton from the hydrate increases the rate of dehydration by providing a pathway with a more stable transition state. The transition state for elimination of ^{-}OH from a negatively charged tetrahedral intermediate is more stable because a positive charge does not develop on the electronegative oxygen atom, as it does in the transition state for elimination of ^{-}OH from a neutral tetrahedral intermediate.

$\delta-$O ‖ C ¦ $\delta-$OH — transition state for elimination of ^{-}OH from a negatively charged tetrahedral intermediate

$\delta+$OH ‖ C ¦ $\delta-$OH — transition state for elimination of ^{-}OH from a neutral tetrahedral intermediate

The foregoing base-catalyzed dehydration of a hydrate is an example of specific-base catalysis. In **specific-base catalysis**, the proton is completely removed from the reactant *before* the slow step of the reaction begins. In **general-base catalysis**, on the other hand, the proton is removed from the reactant *during* the slow step of the reaction. Compare the extent of proton transfer in the slow step of the preceding specific-base-catalyzed dehydration with the extent of proton transfer in the slow step of the following general-base-catalyzed dehydration:

general-base-catalyzed dehydration

$$\underset{\text{a hydrate}}{ClCH_2\overset{O-H}{\underset{OH}{C}}CH_2Cl} \xrightleftharpoons[]{\text{slow}} ClCH_2\overset{O}{\overset{\|}{C}}CH_2Cl + {}^-OH + HB \rightleftharpoons H_2O + {:}B^-$$

In specific-base catalysis, the base has to be strong enough to remove a proton from the reactant completely before the slow step begins. In general-base catalysis, the base can be weaker because the proton is only partially transferred to the base in the transition state of the slow step. We will see that enzymes catalyze reactions using general-acid and general-base catalytic groups because at physiological pH (7.3), too small a concentration ($\sim 1 \times 10^{-7}$ M) of H^+ for specific-acid catalysis or HO^- for specific-base catalysis is available.

PROBLEM 5◆

The mechanism for hydroxide-ion-promoted ester hydrolysis is shown in Section 17.12. What catalytic role does hydroxide ion play in this mechanism?

PROBLEM 6

The following reaction occurs by a mechanism involving general-base catalysis:

$$C_6H_4(COOCH_2CH_3)(CH_2OH) \xrightarrow{\text{base}} \text{phthalide (lactone)} + CH_3CH_2OH$$

Propose a mechanism for this reaction.

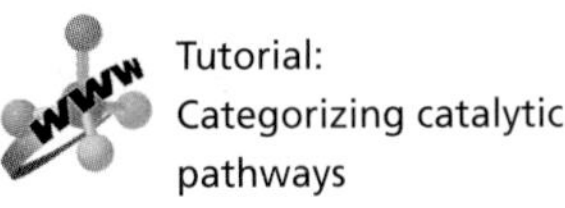

Tutorial:
Categorizing catalytic pathways

24.5 Metal-Ion Catalysis

Metal ions exert their catalytic effect by coordinating (i.e., complexing) with atoms that have lone-pair electrons. In other words, metal ions are Lewis acids (Section 1.21). A *metal ion* can increase the rate of a reaction in several ways.

- It can make a reaction center more susceptible to receiving electrons, as in A in the following diagram:

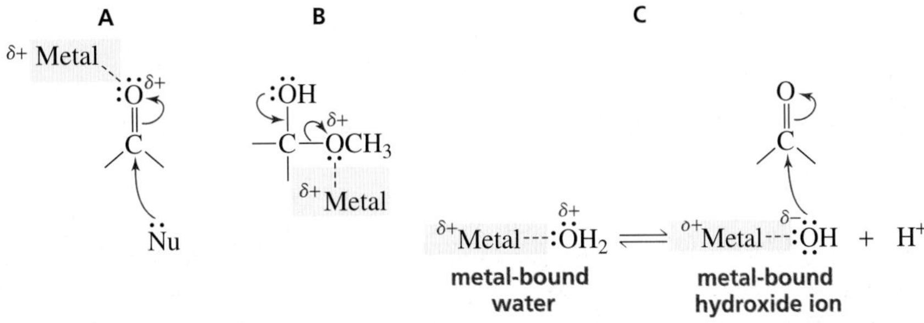

- It can make a leaving group a weaker base, and therefore a better leaving group, as in B.
- It can increase the rate of a hydrolysis reaction by increasing the nucleophilicity of water, as in C.

In cases A and B, the metal ion exerts the same kind of catalytic effect as a proton does. In a reaction in which the metal ion has the same catalytic effect as a proton (increasing the electrophilicity of a reaction center or decreasing the basicity of a leaving group), the metal ion is often called an **electrophilic catalyst**.

In case C, metal-ion complexation of water increases its nucleophilicity by converting it to metal bound hydroxide ion. The pK_a of water is 15.7. When a metal ion complexes with water, it increases its tendency to lose a proton: The pK_a of metal-bound water depends on the metal atom (Table 24.1). When metal-bound water loses a proton, metal-bound hydroxide ion is formed. Metal bound hydroxide ion, while not as good a nucleophile as hydroxide ion, is a better nucleophile than water.

Table 24.1 The pK_a of Metal-Bound Water

M^{2+}	pK_a	M^{2+}	pK_a
Ca^{2+}	12.7	Co^{2+}	8.9
Mg^{2+}	11.8	Zn^{2+}	8.7
Cd^{2+}	11.6	Fe^{2+}	7.2
Mn^{2+}	10.6	Cu^{2+}	6.8
Ni^{2+}	9.4	Be^{2+}	5.7

Now we will look at some examples of metal-ion-catalyzed reactions. Co^{2+} catalyzes the condensation of two molecules of the ethyl ester of glycine to form the ethyl ester of glycylglycine. The actual catalyst is a cobalt complex, $[Co(ethylenediamine)_2]^{2+}$.

$$2\ H_2NCH_2\overset{O}{\overset{\|}{C}}OCH_2CH_3 \xrightarrow{Co^{2+}} H_2NCH_2\overset{O}{\overset{\|}{C}}NHCH_2\overset{O}{\overset{\|}{C}}OCH_2CH_3 + CH_3CH_2OH$$

Coordination of the metal ion with the carbonyl oxygen makes the carbonyl group more susceptible to nucleophilic attack by stabilizing the developing negative charge on the oxygen in the transition state.

$\overset{+}{Co}$ ---- $\overset{\delta+}{:NH_2}$; $\delta+$; $:\ddot{O}$; CH_2 ; C ; OCH_2CH_3 ; $H_2\ddot{N}CH_2\overset{O}{\overset{\|}{C}}OCH_2CH_3$

The decarboxylation of dimethyloxaloacetate can be catalyzed by either Cu^{2+} or Al^{3+}.

$$^{-}O-\overset{O}{\overset{\|}{C}}-\overset{O}{\overset{\|}{C}}-\underset{CH_3}{\overset{CH_3}{\overset{|}{\underset{|}{C}}}}-\overset{O}{\overset{\|}{C}}-O^{-} \xrightarrow{Cu^{2+}\text{ or }Al^{3+}} {}^{-}O-\overset{O}{\overset{\|}{C}}-\overset{O}{\overset{\|}{C}}-\underset{CH_3}{\underset{|}{C}}HCH_3 + CO_2$$

dimethyloxaloacetate

In this reaction, the metal ion complexes with two oxygen atoms of the reactant. Complexation increases the rate of decarboxylation by making the carbonyl group more susceptible to receiving the electrons left behind when CO_2 is eliminated.

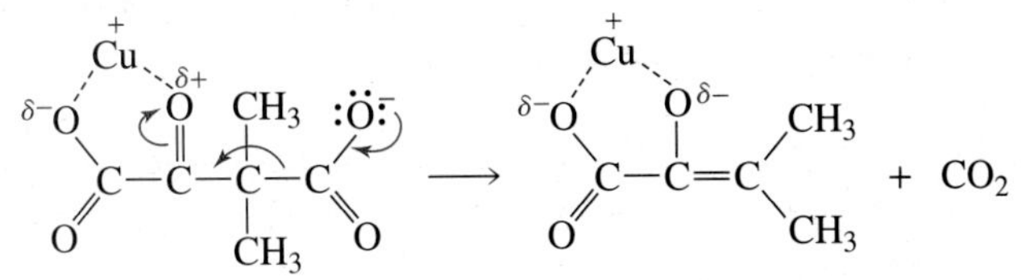

$$\text{Cu}^{+}\text{-chelated dimethyloxaloacetate} \longrightarrow \text{Cu}^{+}\text{-chelated enolate} + CO_2$$

The hydrolysis of methyl trifluoroacetate has two slow steps. Zn^{2+} increases the rate of the first slow step by providing metal-bound hydroxide ion, a better nucleophile than water. Zn^{2+} increases the rate of the second slow step by decreasing the basicity of the group that is eliminated from the tetrahedral intermediate.

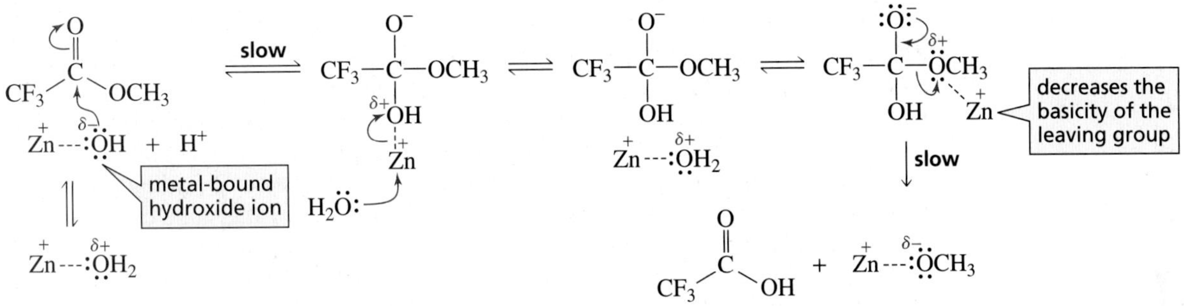

$$CF_3C(=O)OCH_3 + \overset{+}{Zn}\text{---}:\ddot{O}H \xrightarrow{\text{slow}} CF_3-\underset{OH}{\overset{O^-}{C}}-OCH_3 \rightleftharpoons CF_3-\underset{OH}{\overset{O^-}{C}}-OCH_3 \rightleftharpoons CF_3-\underset{OH}{\overset{:\ddot{O}:^-}{C}}-\ddot{O}CH_3\cdots\overset{+}{Zn} \xrightarrow{\text{slow}} CF_3C(=O)OH + \overset{+}{Zn}\text{---}:\ddot{O}CH_3$$

metal-bound hydroxide ion

decreases the basicity of the leaving group

PROBLEM 7◆

The rate constant for the uncatalyzed reaction of two molecules of glycine ethyl ester to form glycylglycine ethyl ester is $0.6\ s^{-1}\ M^{-1}$. In the presence of $[Co(ethylenediamine)_2]^{2+}$, the rate constant is $1.5 \times 10^6\ s^{-1}\ M^{-1}$. What rate enhancement does the catalyst provide?

PROBLEM 8

Although metal ions increase the rate of decarboxylation of dimethyloxaloacetate, they have no effect on the rate of decarboxylation either of the monoethyl ester of dimethyloxaloacetate or of acetoacetate. Explain why this is so.

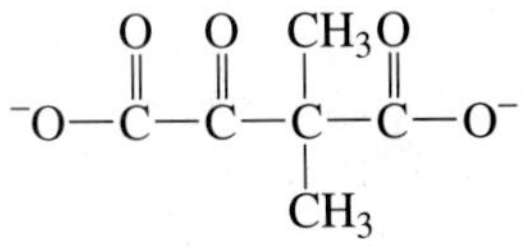

$$^-O-\overset{O}{\overset{\|}{C}}-\overset{O}{\overset{\|}{C}}-\underset{CH_3}{\overset{CH_3}{C}}-\overset{O}{\overset{\|}{C}}-O^-$$
dimethyloxaloacetate

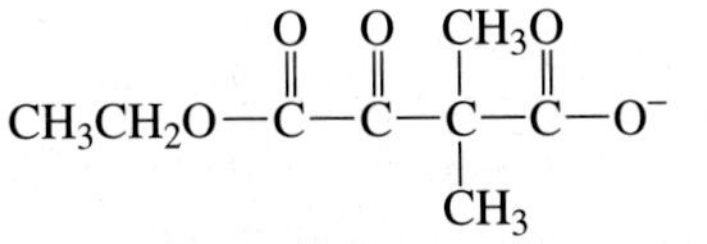

$$CH_3CH_2O-\overset{O}{\overset{\|}{C}}-\overset{O}{\overset{\|}{C}}-\underset{CH_3}{\overset{CH_3}{C}}-\overset{O}{\overset{\|}{C}}-O^-$$
monoethyl ester of dimethyloxaloacetate

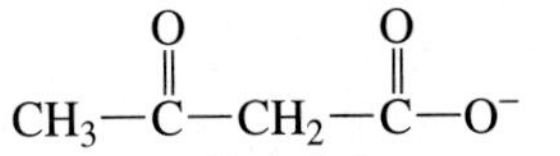

$$CH_3-\overset{O}{\overset{\|}{C}}-CH_2-\overset{O}{\overset{\|}{C}}-O^-$$
acetoacetate

PROBLEM 9

The hydrolysis of glycinamide is catalyzed by $[Co(ethylenediamine)_2]^{2+}$. Propose a mechanism for this reaction.

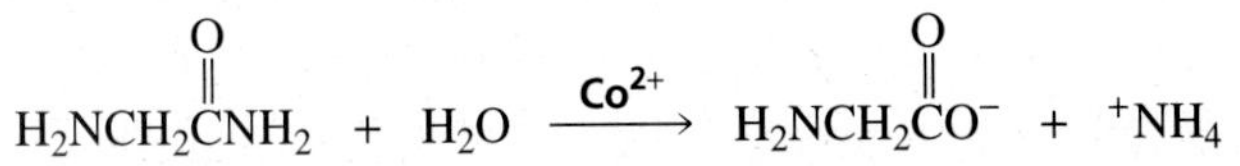

$$H_2NCH_2\overset{O}{\overset{\|}{C}}NH_2 + H_2O \xrightarrow{Co^{2+}} H_2NCH_2\overset{O}{\overset{\|}{C}}O^- + {}^+NH_4$$

24.6 Intramolecular Reactions

The rate of a chemical reaction is determined by the number of molecular collisions with sufficient energy *and* with the proper orientation in a given period of time (Section 3.7):

$$\textbf{rate of reaction} = \frac{\textbf{number of collisions}}{\textbf{unit of time}} \times \begin{array}{c}\textbf{fraction with}\\ \textbf{sufficient energy}\end{array} \times \begin{array}{c}\textbf{fraction with}\\ \textbf{proper orientation}\end{array}$$

Because a catalyst decreases the energy barrier of a reaction, it increases the reaction rate by increasing the number of collisions that occur with sufficient energy to overcome the barrier.

The rate of a reaction can also be increased by increasing the frequency of the collisions *and* the number of collisions that take place with the proper orientation. We have seen that an intramolecular reaction that results in formation of a five- or a six-membered ring occurs more readily than the analogous intermolecular reaction. This is because an intramolecular reaction has the advantage that the reacting groups are tied together in the same molecule, giving them a better chance of finding each other than if they were in two different molecules in a solution of the same concentration (Section 11.11). As a result, the frequency of the collisions increases.

If, in addition to being in the same molecule, the reacting groups are juxtaposed in a way that increases the probability that they will collide with each other in the proper orientation, the rate of the reaction is further increased. The relative rates shown in Table 24.2 demonstrate the enormous increase that occurs in the rate of a reaction when the reacting groups are properly juxtaposed.

Rate constants for a series of reactions are generally compared in terms of relative rates because relative rates allow one to see immediately how much faster one reaction is than another. **Relative rates** are obtained by dividing the rate constant for each of the reactions by the rate constant of the slowest reaction in the series. Because an intramolecular reaction is a first-order reaction (it has units of time^{-1}) and an intermolecular reaction is a second-order reaction (it has units of time^{-1} M^{-1}), the relative rates in Table 24.2 have units of molarity (M) (Section 3.7).

$$\textbf{relative rate} = \frac{\textbf{first-order rate constant}}{\textbf{second-order rate constant}} = \frac{\textbf{time}^{-1}}{\textbf{time}^{-1}\ \textbf{M}^{-1}} = \textbf{M}$$

The relative rates shown in Table 24.2 are also called *effective molarities*. **Effective molarity** is the concentration of the reactant that would be required in an *intermolecular* reaction for it to have the same rate as the *intramolecular* reaction. In other words, the effective molarity is the advantage given to a reaction by having the reacting groups in the same molecule. In some cases, juxtaposing the reacting groups provides such an enormous increase in rate that the effective molarity is greater than the concentration of the reactant in its solid state!

The first reaction shown in Table 24.2 (A) is an intermolecular reaction between an ester and a carboxylate ion. The second reaction (B) has the same reacting groups in a single molecule. The rate of the intramolecular reaction is 1000 times faster than the rate of the intermolecular reaction.

The reactant in B has four C—C bonds that are free to rotate, whereas the reactant in D has only three such bonds. Conformers in which the large groups are rotated away from each other are more stable. However, when these groups are pointed away from each other, they are in an unfavorable conformation for reaction. Because the reactant in D has fewer bonds that are free to rotate, the groups are less apt to be in a conformation that is unfavorable for a reaction. Therefore, reaction D is faster than reaction B. The relative rate constants for the reactions shown in Table 24.2 are quantitatively

Table 24.2 Relative Rates of an Intermolecular Reaction and Five Intramolecular Reactions

	Reaction	Relative reaction rate
A	$CH_3\overset{O}{\overset{\|}{C}}-O-C_6H_4-Br + CH_3\overset{O}{\overset{\|}{C}}-O^- \longrightarrow CH_3\overset{O}{\overset{\|}{C}}-O-\overset{O}{\overset{\|}{C}}CH_3 + {}^-O-C_6H_4-Br$	1.0
B	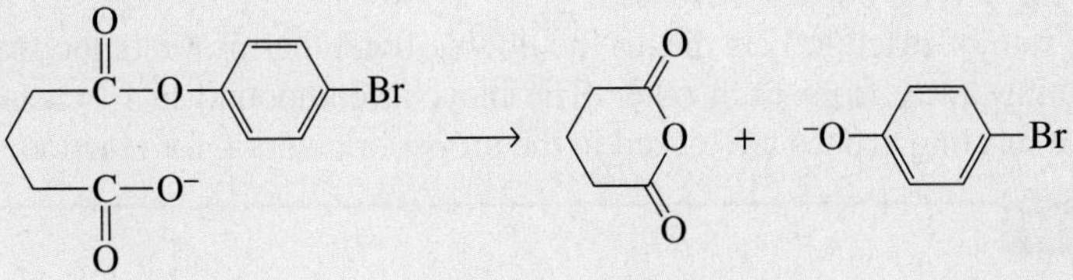 (glutarate monoester) $\longrightarrow$ (six-membered cyclic anhydride) $+ {}^-O-C_6H_4-Br$	1×10^3 M
C	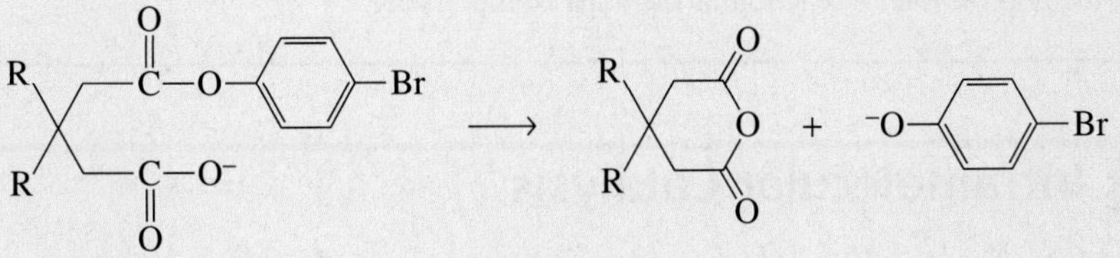 (R-disubstituted glutarate monoester) $\longrightarrow$ (R-disubstituted cyclic anhydride) $+ {}^-O-C_6H_4-Br$	2.3×10^4 M R = CH_3 1.3×10^6 M R = *iso*-C_3H_7
D	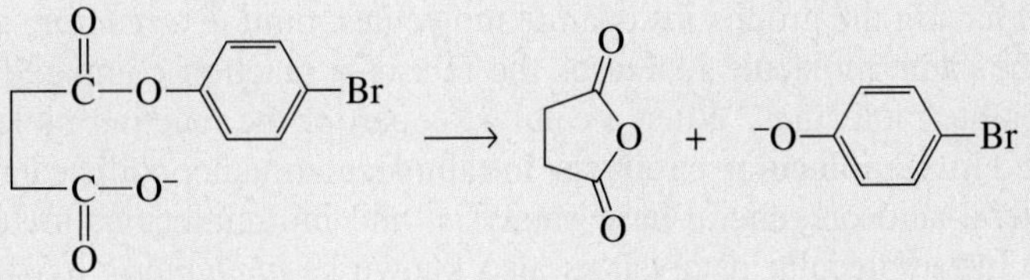 (succinate monoester) $\longrightarrow$ (five-membered cyclic anhydride) $+ {}^-O-C_6H_4-Br$	2.2×10^5 M
E	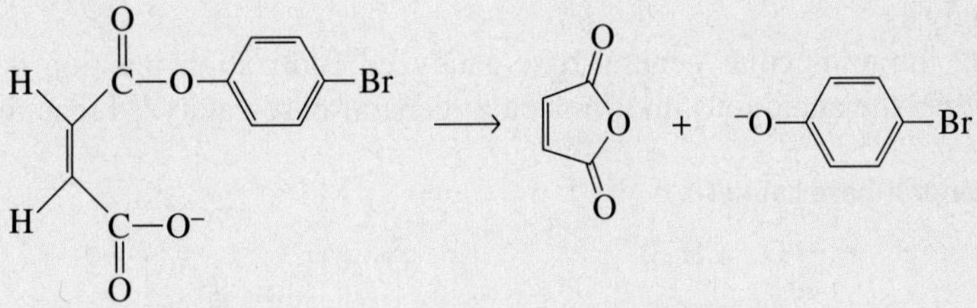 (maleate monoester) $\longrightarrow$ (maleic anhydride) $+ {}^-O-C_6H_4-Br$	1×10^7 M
F	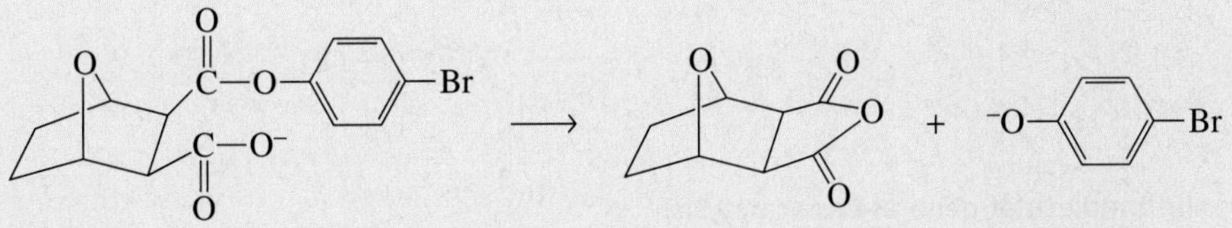 (bicyclic oxa-bridged dicarboxylate monoester) $\longrightarrow$ (bicyclic anhydride) $+ {}^-O-C_6H_4-Br$	5×10^7 M

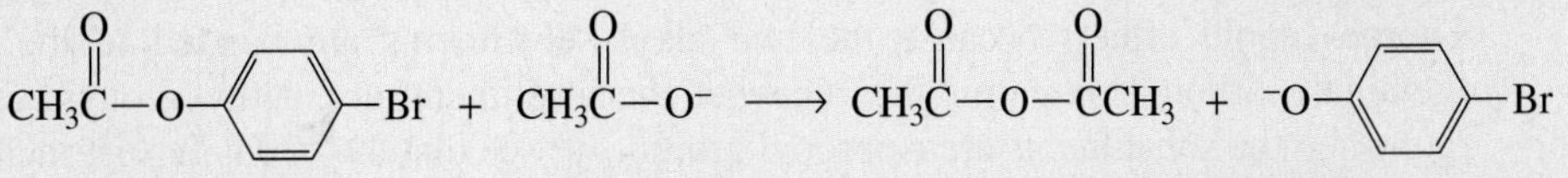

related to the calculated probability of forming the conformation in which the carboxylate ion is in position to attack the carbonyl carbon.

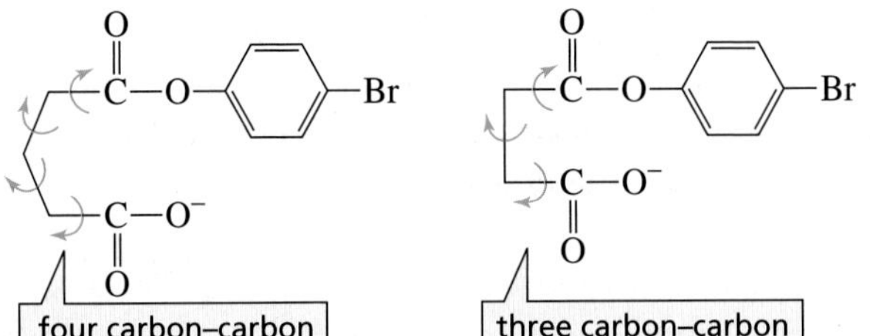

four carbon–carbon bonds can rotate

three carbon–carbon bonds can rotate

Reaction C is faster than reaction B because the alkyl substituents of the reactant in C decrease the available volume for rotation of the reactive groups away from each other. Thus, there is a greater probability that the molecule will be in a conformation with the reacting groups positioned for ring closure. This is called the *gem-dialkyl effect* (or Thorpe–Ingold effect) because the two alkyl substituents are bonded to the same (geminal) carbon. Comparing the rate when the substituents are methyl groups with the rate when the substituents are isopropyl groups, we see that the rate is further increased when the size of the alkyl groups is increased.

The increased rate of reaction E is due to the double bond that prevents the reacting groups from rotating away from each other. The bicyclic compound in F reacts even faster, because the reacting groups are locked in the proper orientation for reaction.

PROBLEM 10◆

The relative rate of reaction of the cis alkene (E) is given in Table 24.2. What would you expect the relative rate of reaction of the trans isomer to be?

24.7 Intramolecular Catalysis

Just as putting two reacting groups in the same molecule increases the rate of a reaction compared with having the groups in separate molecules, putting a *reacting group* and a *catalyst* in the same molecule increases the rate of a reaction compared with having them in separate molecules. When a catalyst is part of the reacting molecule, the catalysis is called **intramolecular catalysis**. Intramolecular nucleophilic catalysis, intramolecular general-acid or general-base catalysis, and intramolecular metal-ion catalysis all occur. Intramolecular catalysis is also known as *anchimeric assistance* (*anchimeric* means "adjacent parts" in Greek). Let's now look at some examples of intramolecular catalysis.

The following intramolecular general-base-catalyzed enolization reaction is considerably faster than the analogous intermolecular general-base-catalyzed reaction:

intramolecular general-base catalysis

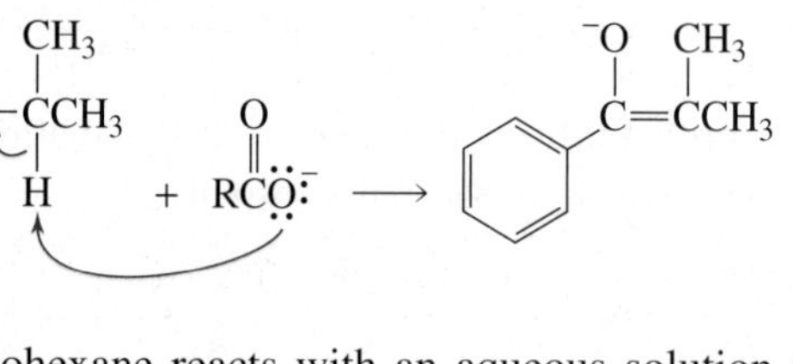

intermolecular general-base catalysis

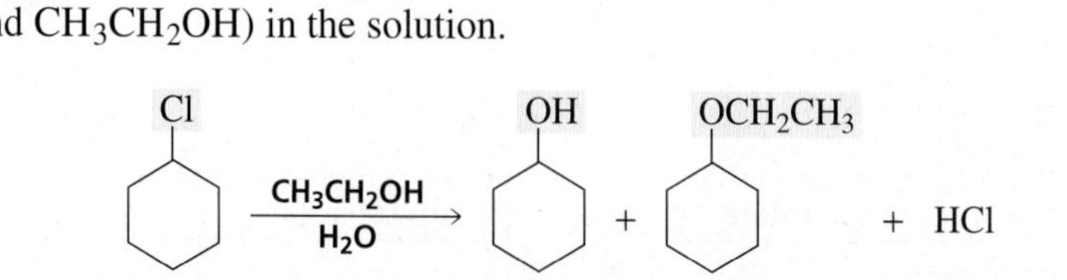

When chlorocyclohexane reacts with an aqueous solution of ethanol, an alcohol and an ether are formed. Two products are formed because there are two nucleophiles (H_2O and CH_3CH_2OH) in the solution.

Cl — $\xrightarrow[H_2O]{CH_3CH_2OH}$ OH + OCH_2CH_3 + HCl

A 2-thio-substituted chlorocyclohexane undergoes the same reaction. However, the rate of the reaction depends on whether the thio substituent is cis or trans to the chloro substituent. If it is trans, the 2-thio-substituted compound reacts about 70,000 times faster than the unsubstituted compound. But if it is cis, the 2-thio-substituted compound reacts a little more slowly than the unsubstituted compound.

2-(SC_6H_5)-chlorocyclohexane $\xrightarrow[H_2O]{CH_3CH_2OH}$ 2-(SC_6H_5)-cyclohexanol + 2-(SC_6H_5)-1-(OCH_2CH_3)cyclohexane + HCl

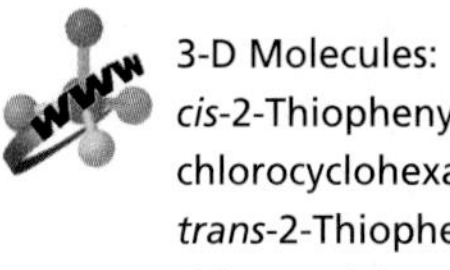
3-D Molecules: *cis*-2-Thiophenylchlorocyclohexane; *trans*-2-Thiophenylchlorocyclohexane

What accounts for the much faster reaction of the trans-substituted compound? In this reaction, the thio substituent is an intramolecular nucleophilic catalyst. It displaces the chloro substituent by attacking the back side of the carbon to which the chloro substituent is attached. Back-side attack requires both substituents to be in axial positions, and only the trans isomer can have both of its substituents in axial positions (Section 2.14). Subsequent attack by water or ethanol on the sulfonium ion is rapid because the positively charged sulfur is an excellent leaving group and breaking the three-membered ring releases strain.

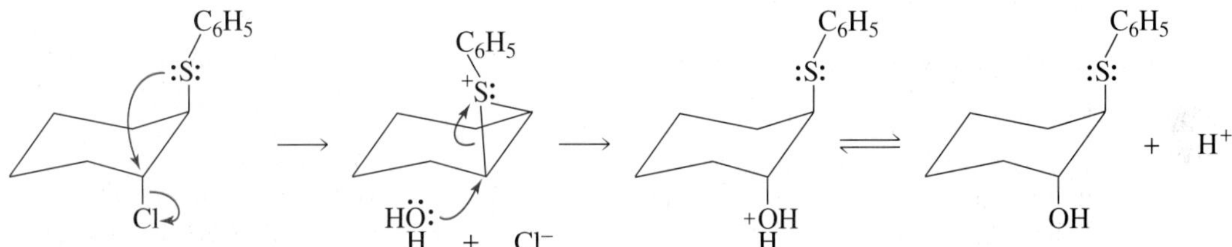

PROBLEM 11◆

Show all the products, and their configurations, that would be obtained from solvolysis of the trans-substituted compound illustrated in the preceding diagram.

The rate of hydrolysis of phenyl acetate is increased about 150-fold at neutral pH by the presence of a carboxylate ion in the ortho position. The *ortho*-carboxyl-substituted ester is commonly known as aspirin (Section 19.9). In the following reactions, the reactants and products are shown in the form that predominates at physiological pH (7.3).

$CH_3C(=O)–O–C_6H_5 + H_2O \xrightarrow{\text{relative rate} = 1} CH_3CO^- (=O) + HO–C_6H_5$

$CH_3C(=O)–O–C_6H_4–COO^- + H_2O \xrightarrow{\text{relative rate} \sim 150} CH_3CO^- (=O) + HO–C_6H_4–COO^-$

The *ortho*-carboxylate group is an intramolecular general-base catalyst that increases the nucleophilicity of water, thereby increasing the rate of formation of the tetrahedral intermediate.

3-D Molecule: Aspirin

$CH_3C(=O)–O–C_6H_4–C(=O)O^-$ with H–Ö:–H attacking the carbonyl carbon and the carboxylate oxygen removing H from water

If nitro groups are put on the benzene ring, the *ortho*-carboxyl substituent acts as an intramolecular *nucleophilic catalyst* instead of an intramolecular *general-base catalyst*. It increases the rate of the hydrolysis reaction by converting the ester into an anhydride, and an anhydride is more rapidly hydrolyzed than an ester.

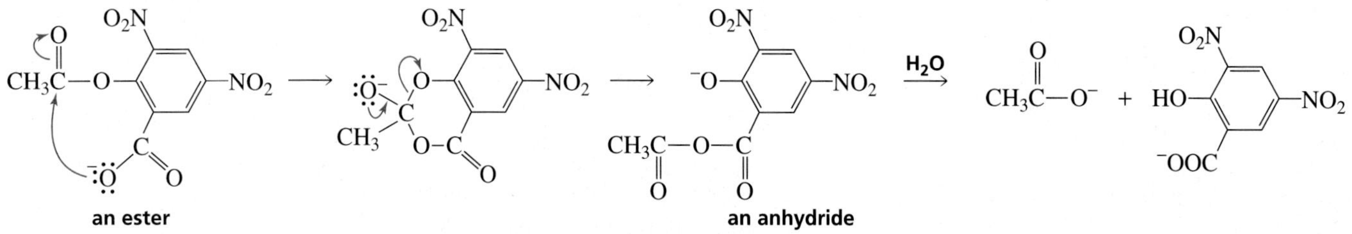

an ester

an anhydride

PROBLEM 12 SOLVED

What causes the mode of catalysis to change from general base to nucleophilic in the hydrolysis of an *ortho*-carboxyl-substituted phenyl acetate?

SOLUTION The *ortho*-carboxyl substituent is in position to form a tetrahedral intermediate. If the carboxyl group in the tetrahedral intermediate is a better leaving group than the phenoxy group, the carboxyl group will be eliminated preferentially from the intermediate. This will reform the starting material, which will be hydrolyzed by a general-base-catalyzed mechanism (path A). However, if the phenoxy group is a better leaving group than the carboxyl group, the phenoxy group will be eliminated, thereby forming the anhydride, and the reaction will have occurred by a mechanism involving nucleophilic catalysis (path B).

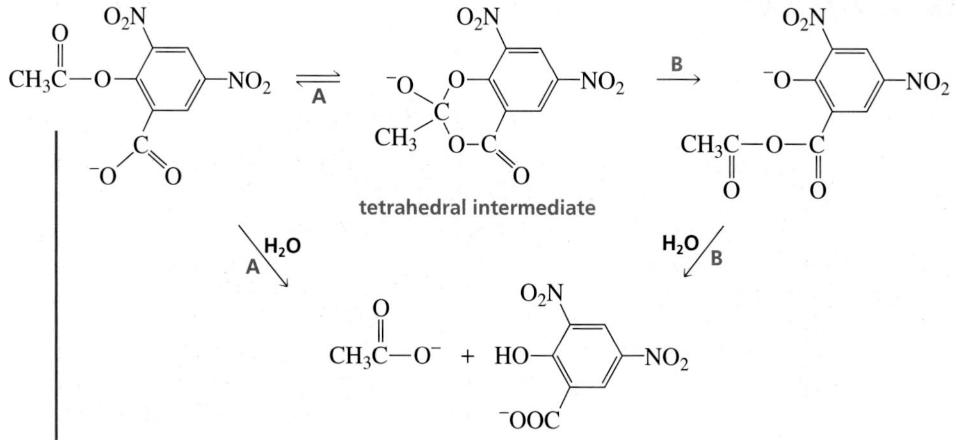

tetrahedral intermediate

PROBLEM 13◆

Why do the nitro groups change the relative leaving tendencies of the carboxyl and phenyl groups in the tetrahedral intermediate in Problem 12?

PROBLEM 14

Whether the *ortho*-carboxyl substituent acts as an intramolecular general-base catalyst or as an intramolecular nucleophilic catalyst can be determined by carrying out the hydrolysis of aspirin with ^{18}O-labeled water and determining whether ^{18}O is incorporated into *ortho*-carboxyl-substituted phenol. Explain the results that would be obtained with the two types of catalysis.

The following reaction, in which Ni^{2+} catalyzes the hydrolysis of the ester, is an example of intramolecular metal-ion catalysis:

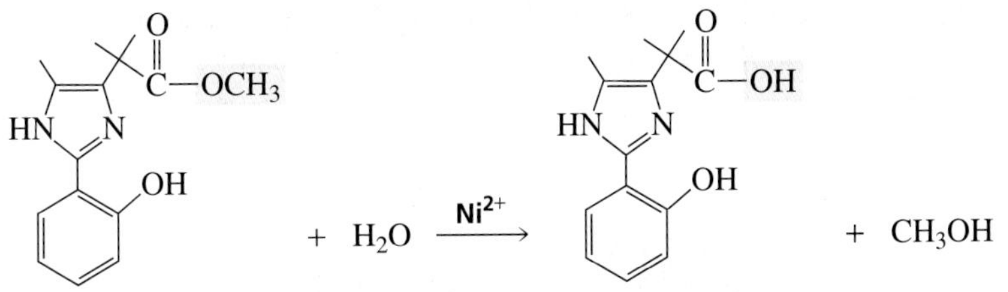

$$\text{(imidazole–}C(=O)OCH_3\text{ ester)} + H_2O \xrightarrow{Ni^{2+}} \text{(imidazole–}C(=O)OH\text{ acid)} + CH_3OH$$

The metal ion complexes with an oxygen and a nitrogen of the reactant, as well as with a molecule of water. The metal ion increases the rate of the reaction by positioning the water molecule and increasing its nucleophilicity by converting it to metal-bound hydroxide.

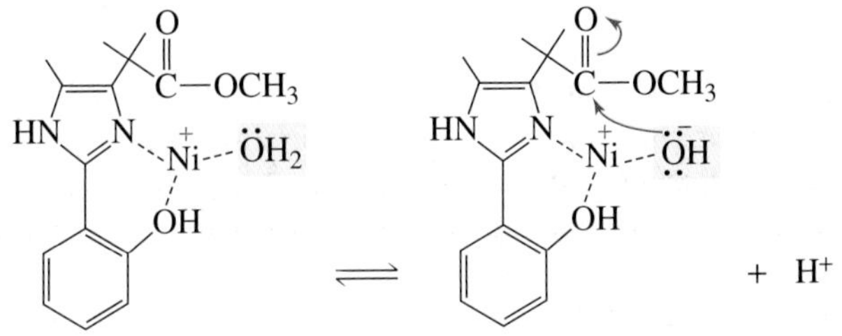

$$\text{(ester–}N\cdots Ni^{+}\cdots \ddot{O}H_2\text{)} \rightleftharpoons \text{(ester–}N\cdots Ni^{+}\cdots \ddot{O}H^{-}\text{)} + H^+$$

24.8 Catalysis in Biological Reactions

Essentially all organic reactions that occur in biological systems require a catalyst. Most biological catalysts are **enzymes**, which are globular proteins (Section 23.11). Each biological reaction is catalyzed by a different enzyme. Enzymes are extraordinarily good catalysts—they can increase the rate of an intermolecular reaction by as much as 10^{16}. In contrast, rate enhancements achieved by nonbiological catalysts in intermolecular reactions are seldom greater than 10,000-fold.

The reactant of an enzyme-catalyzed reaction is called a **substrate**. The enzyme has a pocket or cleft known as an active site. The substrate specifically fits and binds to the **active site**, and all the bond-making and bond-breaking steps of the reaction occur while the substrate is at that site. Enzymes differ from nonbiological catalysts in that they are specific for the reactant whose reaction they catalyze (Section 5.20). Not all enzymes have the same degree of specificity. Some are specific for a single compound and will not tolerate even the slightest variation in structure, whereas some catalyze the reaction of an entire family of compounds with related structures. The specificity of an enzyme for its substrate is an example of the phenomenon known as **molecular recognition**—the ability of one molecule to recognize another.

The specificity of an enzyme results from its conformation and the particular amino acid side chains that make up the active site. For example, a negatively charged side chain of an amino acid at the active site of an enzyme can associate with a positively charged group on the substrate, a hydrogen-bond donor on the enzyme can associate with a hydrogen-bond acceptor on the substrate, and hydrophobic groups on the enzyme associate with hydrophobic groups on the substrate. The specificity of an enzyme for its substrate has been described by the lock-and-key model. In the **lock-and-key model**, the substrate is said to fit the enzyme much as a key fits a lock.

The energy released as a result of binding the substrate to the enzyme can be used to induce a change in the conformation of the enzyme, leading to more precise binding between the substrate and the active site. This change in conformation of the enzyme is known as induced fit. In the **induced-fit model**, the shape of the active site does not become completely complementary to the shape of the substrate until the enzyme has bound the substrate.

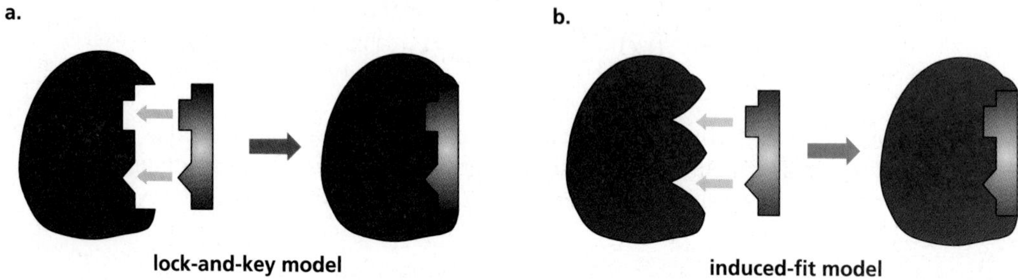

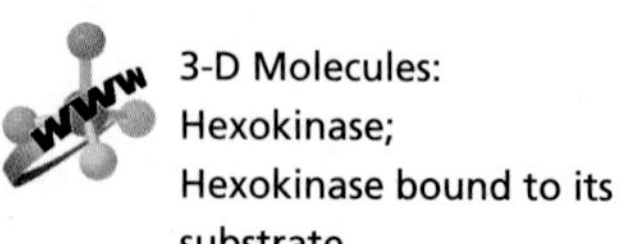

3-D Molecules: Hexokinase; Hexokinase bound to its substrate

An example of induced fit is shown in Figure 24.4. The three-dimensional structure of the enzyme hexokinase is shown before and after binding glucose, its substrate. Notice the change in conformation that occurs upon binding the substrate.

The following are some of the most important factors that contribute to the remarkable catalytic ability of enzymes:

- Reacting groups are brought together at the active site in the proper orientation for reaction. This is analogous to the way that proper positioning of reacting groups increases the rate of intramolecular reactions (Section 24.6).
- Some of the amino acid side chains of the enzyme serve as catalytic groups, and many enzymes have metal ions at their active site that act as catalysts. These species are positioned in orientations relative to the substrate needed for catalysis. This is analogous to the rate enhancements observed for intramolecular catalysis by acids, bases, and metal ions (Section 24.7).
- Groups on the enzyme can stabilize transition states and intermediates by van der Waals interactions, electrostatic interactions, and hydrogen bonding (Figure 24.1b).

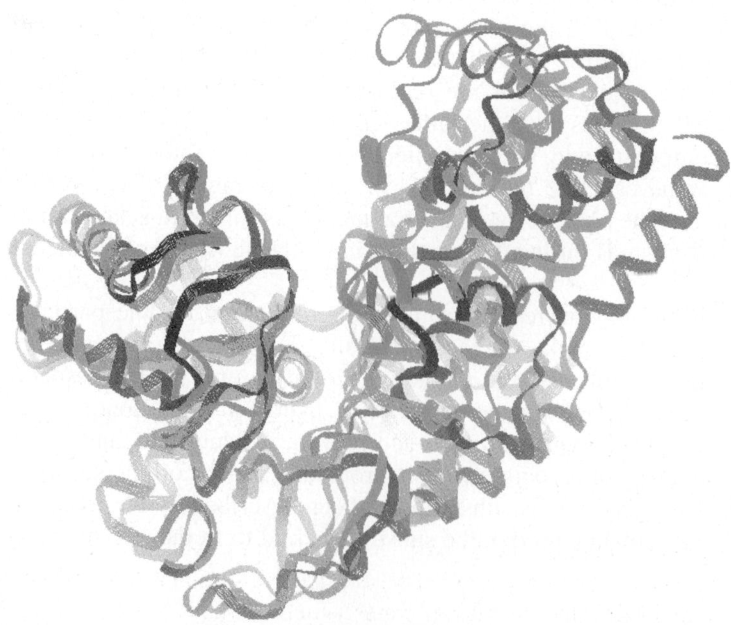

Figure 24.4 ▶
The structure of hexokinase before binding its substrate is shown in red. The structure of hexokinase after binding its substrate is shown in green.

As we look at some examples of enzyme-catalyzed reactions, notice that the functional groups on the enzyme side chains are the same functional groups you are used to seeing in simple organic compounds, and the modes of catalysis used by enzymes are the same as the modes of catalysis used in organic reactions. The remarkable catalytic ability of enzymes stems in part from their ability to use several modes of catalysis in the same reaction. Factors other than those listed can contribute to the increased rate of enzyme-catalyzed reactions, but not all factors are employed by every enzyme. We will consider some of these factors when we discuss individual enzymes. Now let's look at the mechanisms for five enzyme-catalyzed reactions.

24.9 Enzyme-Catalyzed Reactions

Mechanism for Carboxypeptidase A

Carboxypeptidase A is an *exopeptidase*—an enzyme that catalyzes the hydrolysis of the C-terminal peptide bond in peptides and proteins, releasing the C-terminal amino acid (Section 23.12).

$$\text{—NHCH(R)C(=O)—NHCH(R')C(=O)—NHCH(R'')CO}^- + H_2O$$

$$\big\downarrow \text{ carboxypeptidase A}$$

$$\text{—NHCH(R)C(=O)—NHCH(R')CO}^- + H_3\overset{+}{N}CH(R'')CO^-$$

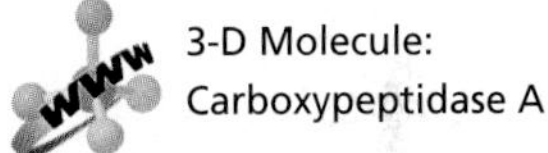
3-D Molecule: Carboxypeptidase A

Carboxypeptidase A is a *metalloenzyme*—an enzyme that contains a tightly bound metal ion. The metal ion in carboxypeptidase A is Zn^{2+}. Carboxypeptidase A is one of several hundred enzymes known to contain zinc. In bovine pancreatic carboxypeptidase A, Zn^{2+} is bound to the enzyme at its active site by forming a complex with Glu 72, His 196, and His 69, as well as with a water molecule (Figure 24.5). (The source of the enzyme is specified because, although carboxypeptidase A's from different sources follow the same mechanism, they have slightly different primary structures.)

Several groups at the active site of carboxypeptidase A participate in binding the substrate in the optimum position for reaction (Figure 24.5). Arg 145 forms two hydrogen bonds and Tyr 248 forms one hydrogen bond with the C-terminal carboxyl group of the substrate. The side chain of the C-terminal amino acid is positioned in a hydrophobic pocket, which is why carboxypeptidase A is not active if the C-terminal amino acid is arginine or lysine. Apparently, the long, positively charged side chains of these amino acid residues (Table 23.1) cannot fit into the nonpolar pocket. The reaction proceeds as follows:

- When the substrate binds to the active site, Zn^{2+} partially complexes with the oxygen of the carbonyl group of the amide that will be hydrolyzed (Figure 24.5). Zn^{2+} polarizes the carbon–oxygen double bond, making the carbonyl carbon more susceptible to nucleophilic attack and stabilizing the negative charge that develops on the oxygen atom in the transition state that leads to the tetrahedral intermediate. Arg 127 also increases the carbonyl group's electrophilicity and stabilizes the developing negative charge on the oxygen atom in the transition state. Zn^{2+} also complexes with water, thereby making it a better nucleophile. Glu 270 functions as a general-base catalyst, further increasing water's nucleophilicity.

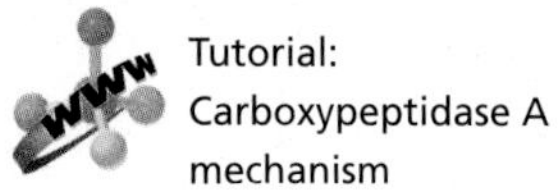
Tutorial: Carboxypeptidase A mechanism

- In the second step of the reaction, Glu 270 functions as a general-acid catalyst, increasing the leaving tendency of the amino group. When the reaction is over, the amino acid (phenylalanine in this example) and the peptide with one less amino acid residue dissociate from the enzyme, and another molecule of

Overall Reaction

$$\sim\!\overset{O}{\overset{\|}{C}}-NH\underset{CH_2C_6H_5}{\underset{|}{C}H}\overset{O}{\overset{\|}{C}}O^- + H_2O \xrightarrow{\text{carboxypeptidase A}} \sim\!\overset{O}{\overset{\|}{C}}O^- + H_3\overset{+}{N}\underset{CH_2C_6H_5}{\underset{|}{C}H}\overset{O}{\overset{\|}{C}}O^-$$

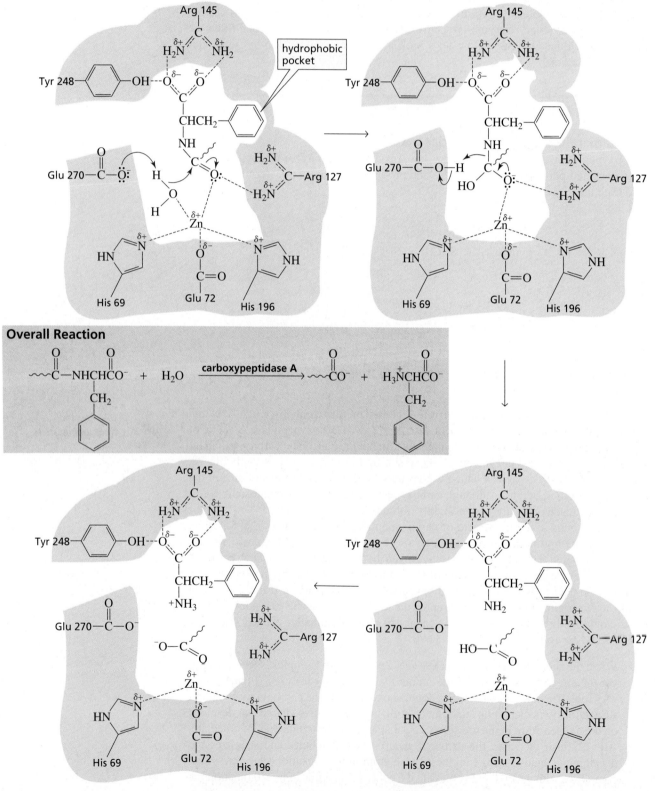

▲ **Figure 24.5**
Proposed mechanism for the carboxypeptidase A–catalyzed hydrolysis of a peptide bond.

substrate binds to the active site. It has been suggested that the unfavorable electrostatic interaction between the negatively charged carboxyl group of the peptide product and the negatively charged Glu 270 residue facilitates the release of the product from the enzyme.

PROBLEM 15◆

Which of the following C-terminal peptide bonds would be more readily cleaved by carboxypeptidase A?

Ser-Ala-Phe or Ser-Ala-Asp

Explain.

PROBLEM 16

Carboxypeptidase A has esterase activity as well as peptidase activity. In other words, the compound can hydrolyze ester bonds as well as peptide bonds. When carboxypeptidase A hydrolyzes ester bonds, Glu 270 acts as a nucleophilic catalyst instead of a general-base catalyst. Propose a mechanism for the carboxypeptidase A-catalyzed hydrolysis of an ester bond.

Mechanism for the Serine Proteases

Trypsin, chymotrypsin, and elastase are members of a large group of *endopeptidases* known collectively as serine proteases. Recall that an endopeptidase cleaves a peptide bond that is not at the end of a peptide chain (Section 23.12). They are called *proteases* because they catalyze the hydrolysis of protein peptide bonds. They are called *serine proteases* because they all have a serine residue at the active site that participates in the catalysis.

The various serine proteases have similar primary structures, suggesting that they are evolutionarily related. They all have the same three catalytic residues at the active site: an aspartate, a histidine, and a serine. But they have one important difference—the composition of the pocket at the active site that binds the side chain of the amino acid residue undergoing hydrolysis (Figure 24.6). This pocket is what gives the serine proteases their different specificities (Section 23.12).

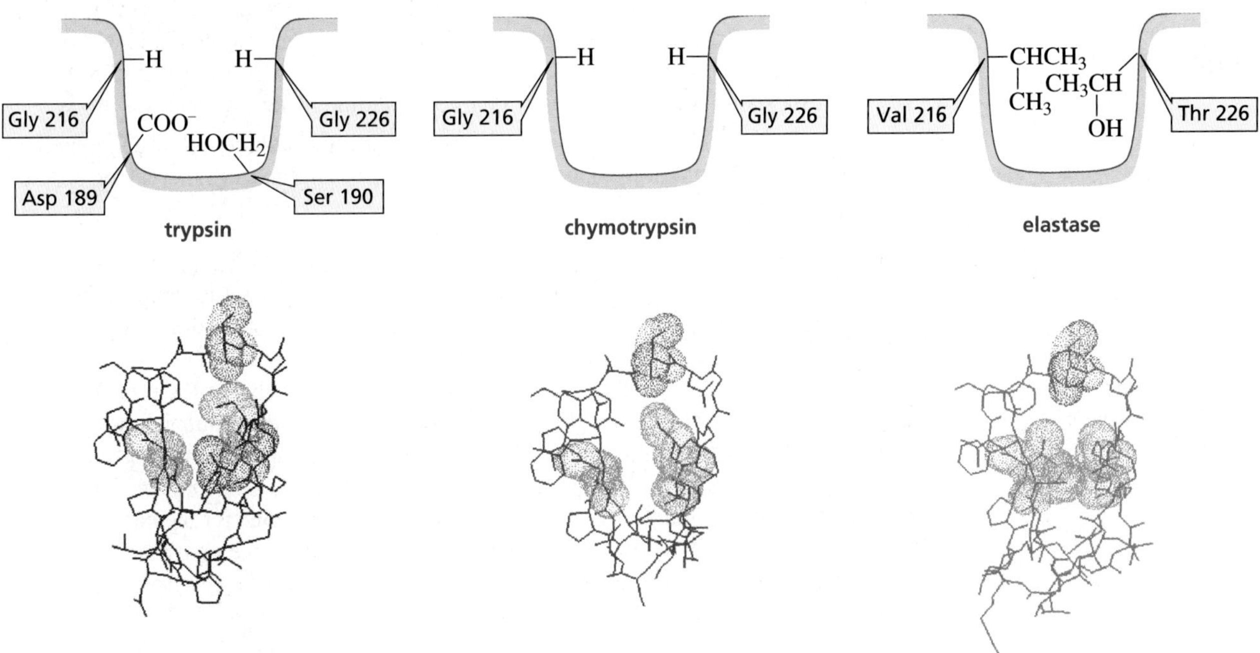

▲ **Figure 24.6**
The binding pockets in trypsin, chymotrypsin, and elastase. The negatively charged aspartate is shown in red, and the relatively nonpolar amino acids are shown in green. The structures of the binding pockets explain why trypsin binds long, positively charged amino acids; chymotrypsin binds flat, nonpolar amino acids; and elastase binds only small amino acids.

The pocket in trypsin is narrow and has a serine and a negatively charged aspartate carboxyl group at its bottom. The shape and charge of the binding pocket cause it to bind long, positively charged amino acid side chains (Lys and Arg). This is why trypsin hydrolyzes peptide bonds on the C-side of arginine and lysine residues. The pocket in chymotrypsin is narrow and is lined with nonpolar amino acids, so chymotrypsin cleaves on the C-side of amino acids with flat, nonpolar side chains (Phe, Tyr, Trp). In elastase, two glycines on the sides of the pocket in trypsin and in chymotrypsin are replaced by relatively bulky valine and threonine residues. Consequently, only small amino acids can fit into the pocket. Elastase, therefore, hydrolyzes peptide bonds on the C-side of small amino acids (Gly and Ala).

The mechanism for bovine chymotrypsin-catalyzed hydrolysis of a peptide bond is shown in Figure 24.7. The other serine proteases follow the same mechanism. The reaction proceeds as follows:

- As a result of binding the flat, nonpolar side chain in the pocket, the amide linkage that is to be hydrolyzed is positioned very close to Ser 195. His 57 functions as a general-base catalyst, increasing the nucleophilicity of serine, which attacks the carbonyl group. This process is helped by Asp 102, which uses its negative charge to stabilize the resulting positive charge on His 57 and to position the five-membered ring so that its basic N atom is close to the OH of serine. The stabilization of a charge by an opposite charge is called **electrostatic catalysis**. Formation of the tetrahedral intermediate causes a slight change in the conformation of the protein that allows the negatively charged oxygen to slip into a previously unoccupied area of the active site known as the *oxyanion hole*. Once in the oxyanion hole, the negatively charged oxygen can hydrogen bond with two peptide groups (Gly 193 and Ser 195), which stabilizes the tetrahedral intermediate.
- In the next step, the tetrahedral intermediate collapses, expelling the amino group. This is a strongly basic group that cannot be expelled without the participation of His 57, which acts as a general-acid catalyst. The product of the second step is an **acyl-enzyme intermediate** because the serine group of the enzyme has been acylated. (An acyl group has been put on it.)
- The third step is just like the first step, except that water instead of serine is the nucleophile. Water attacks the acyl group of the acyl-enzyme intermediate, with His 57 functioning as a general-base catalyst to increase water's nucleophilicity and Asp 102 stabilizing the positively charged histidine residue.
- In the final step of the reaction, the tetrahedral intermediate collapses, expelling serine. His 57 functions as a general-acid catalyst in this step, increasing serine's leaving tendency.

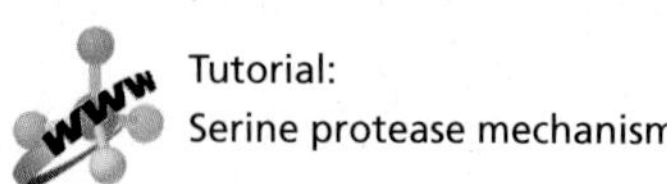

The mechanism for chymotrypsin-catalyzed hydrolysis shows the importance of histidine as a catalytic group. Because the pK_a of the imidazole ring of histidine ($pK_a = 6.0$) is close to neutrality, histidine can act both as a general-acid catalyst and as a general-base catalyst at physiological pH.

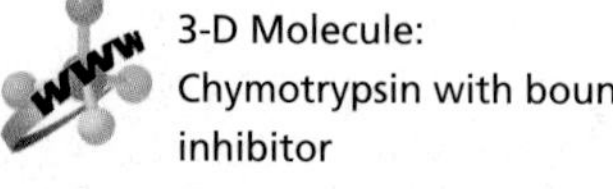

Much information about the relationship between the structure of a protein and its function has been determined by **site-specific mutagenesis**, a technique that replaces one amino acid of a protein with another. For example, when Asp 102 of chymotrypsin is replaced with Asn 102, the enzyme's ability to bind the substrate is unchanged, but its ability to catalyze the reaction decreases to less than 0.05% of the value for the native enzyme. Clearly, Asp 102 must be involved in the catalytic process. We have seen that its role is to position histidine and use its negative charge to stabilize histidine's positive charge.

side chain of an aspartate (Asp) residue — side chain of an asparagine (Asn) residue

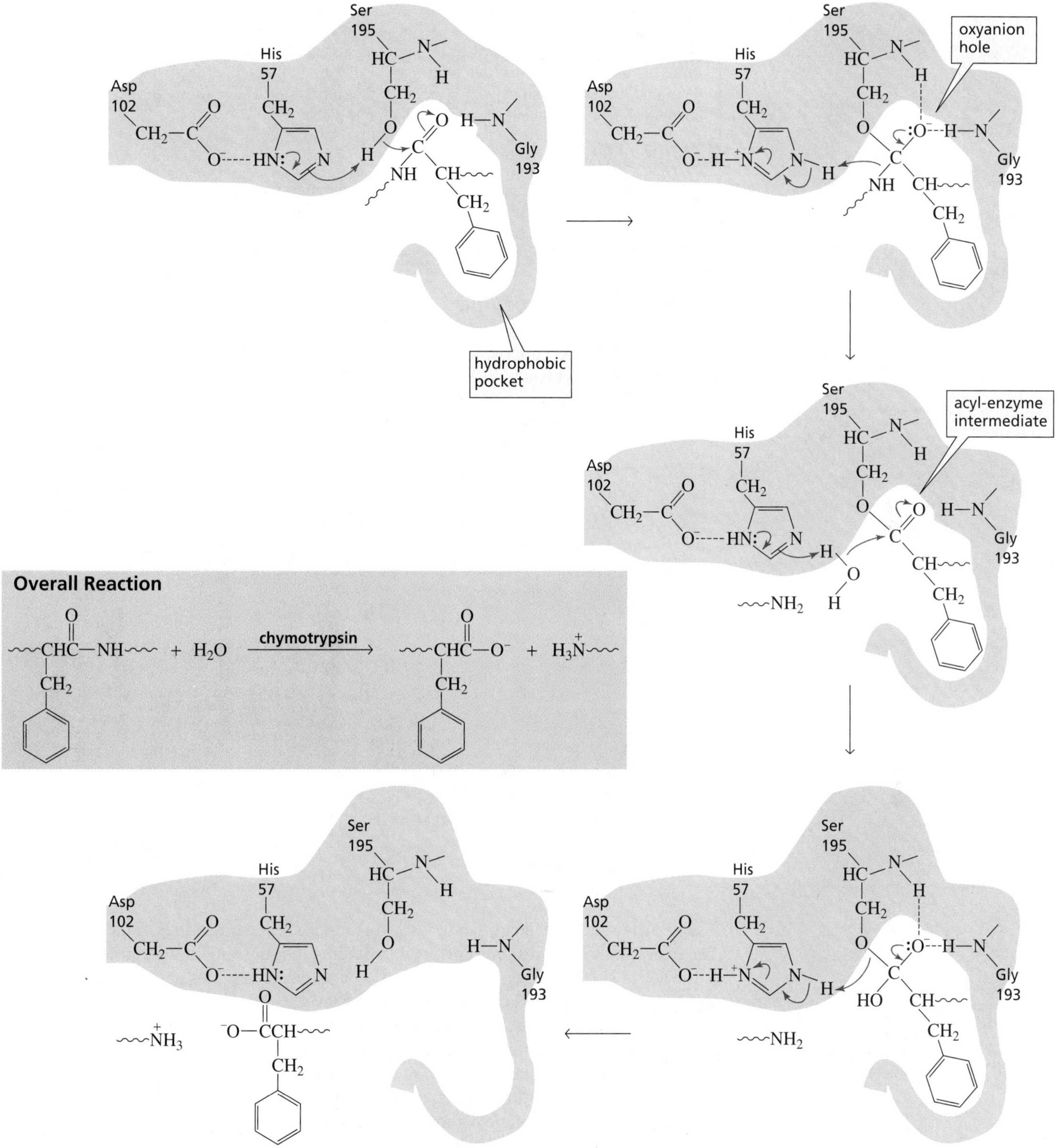

Overall Reaction

$$\sim\!\!\sim CH(CH_2C_6H_5)C(=O)-NH\sim\!\!\sim \; + \; H_2O \xrightarrow{\text{chymotrypsin}} \sim\!\!\sim CH(CH_2C_6H_5)C(=O)-O^- \; + \; H_3\overset{+}{N}\sim\!\!\sim$$

▲ **Figure 24.7**
Proposed mechanism for the chymotrypsin-catalyzed hydrolysis of a peptide bond.

PROBLEM 17◆

Arginine and lysine side chains fit into trypsin's binding pocket. One of these side chains forms a direct hydrogen bond with serine and an indirect hydrogen bond (mediated through a water molecule) with aspartate. The other side chain forms direct hydrogen bonds with both serine and aspartate. Which is which?

PROBLEM 18

Serine proteases do not catalyze hydrolysis if the amino acid at the hydrolysis site is a D-amino acid. Trypsin, for example, cleaves on the C-side of L-Arg and L-Lys, but not on the C-side of D-Arg and D-Lys. Explain.

Mechanism for Lysozyme

Lysozyme is an enzyme that destroys bacterial cell walls. These cell walls are composed of alternating *N*-acetylmuramic acid (NAM) and *N*-acetylglucosamine (NAG) units. Lysozyme destroys the cell wall by catalyzing the hydrolysis of the NAM–NAG bond.

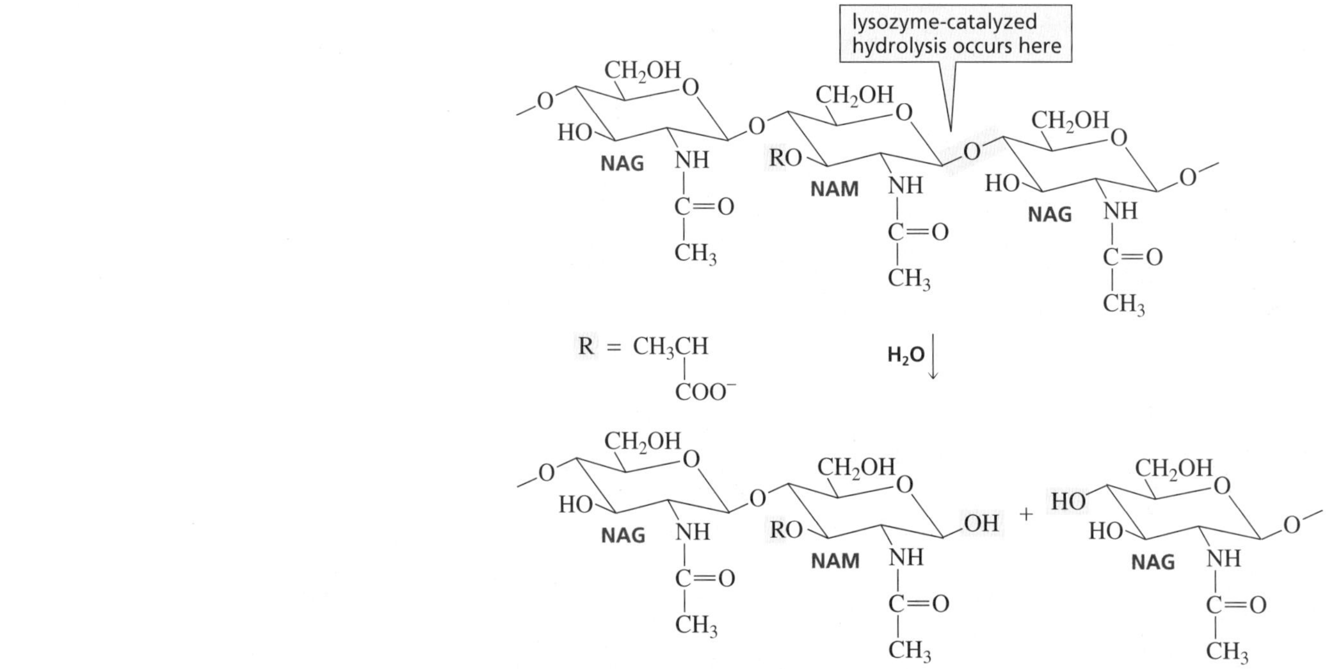

lysozyme-catalyzed hydrolysis occurs here

NAG — NAM — NAG

$R = CH_3CH(COO^-)$

$\xrightarrow{H_2O}$

NAG — NAM–OH + HO–NAG

The active site of hen egg-white lysozyme binds six sugar residues of the substrate. The many amino acid residues involved in binding the substrate in the correct position in the active site are shown in Figure 24.8. The six sugar residues are labeled A, B, C, D, E, and F. The carboxylic acid substituent of NAM cannot fit into the binding site for C or E. This means that NAM units must be in the sites for B, D, and F. Hydrolysis occurs between D and E.

Lysozyme has two catalytic groups at the active site: Glu 35 and Asp 52 (Figure 24.9). Once it was determined that the enzyme-catalyzed reaction takes place with retention of configuration at the anomeric carbon, it could be concluded that it cannot be a one-step S_N2 reaction; the reaction must involve two sequential S_N2 reactions or an S_N1 reaction with the enzyme blocking one face of the oxocarbenium ion intermediate from nucleophilic attack. Although lysozyme was the first enzyme to have its mechanism studied—and the mechanism has been studied extensively for almost 40 years—only recently have data been obtained that support the mechanism involving two sequential S_N2 reactions shown in Figure 24.9:

- In the first step of the reaction, Asp 52 acts as a nucleophilic catalyst and attacks C-1 of the NAM residue, displacing the leaving group. Glu 35 acts as a general-acid

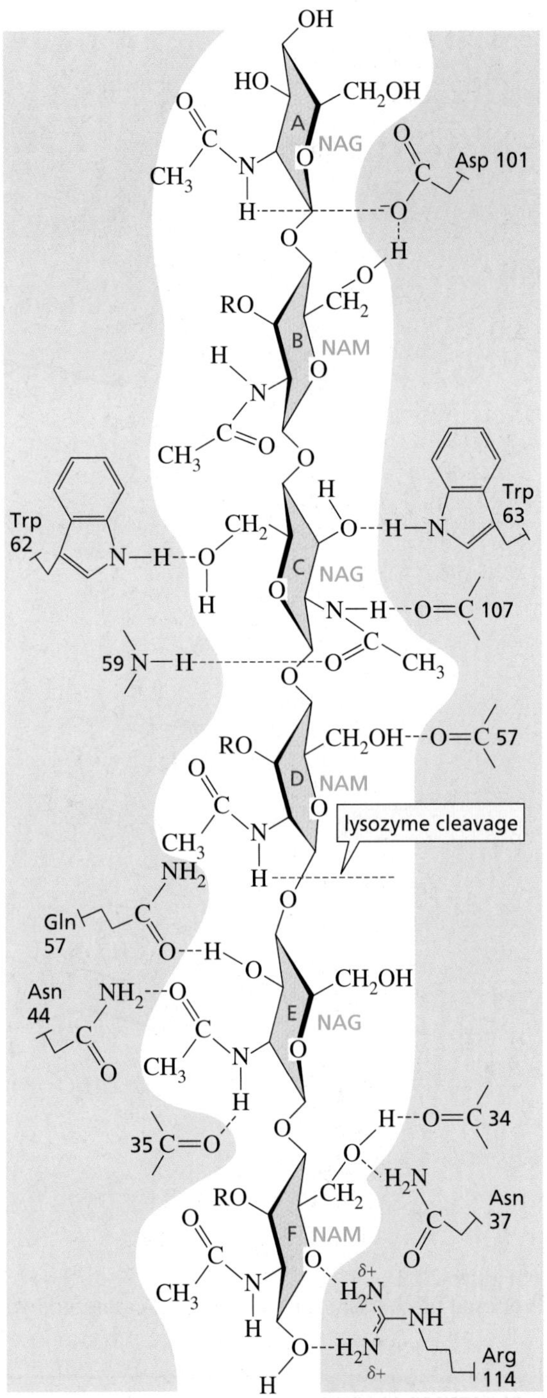

Figure 24.8
The amino acids at the active site of lysozyme that are involved in binding the substrate.

catalyst, protonating the leaving group and thereby making it a weaker base and a better leaving group. The lone pair on the ring oxygen can help displace the leaving group. Site-specific mutagenesis studies show that when Glu 35 is replaced by Asp, the enzyme has only weak activity. Apparently, Asp does not have the optimal distance and angle to the oxygen atom that needs to be protonated. When Glu 35 is replaced by Ala, an amino acid that cannot act as an acid catalyst, the activity of the enzyme is completely lost.

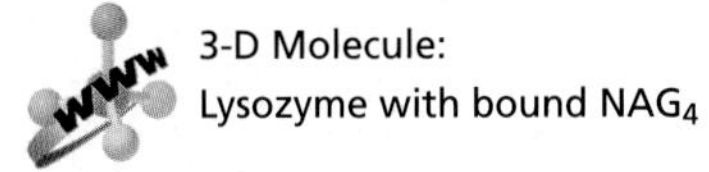

3-D Molecule: Lysozyme with bound NAG_4

- In the second step of the reaction, Glu 35 acts as a general-base catalyst to increase water's nucleophilicity.

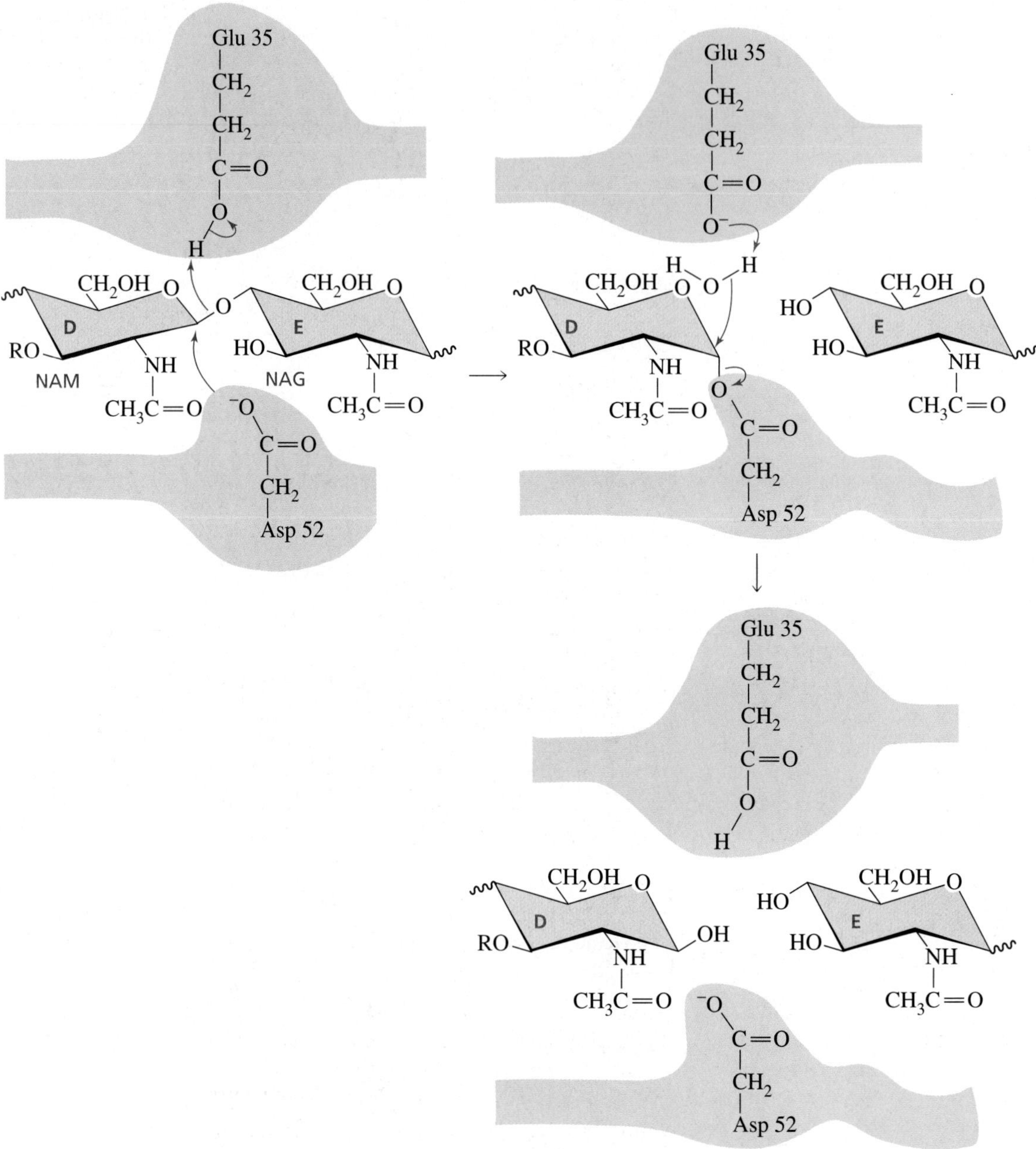

▲ **Figure 24.9**
Proposed mechanism for the lysozyme-catalyzed hydrolysis of a cell wall.

PROBLEM 19◆

If H_2O^{18} were used to hydrolyze lysozyme, which ring would contain the label, NAM or NAG?

A plot of the activity of an enzyme as a function of the pH of the reaction mixture is called a **pH–activity profile** or a **pH–rate profile** (Section 18.6). The pH–activity profile for lysozyme is shown in Figure 24.10. It is a bell-shaped curve with the maximum rate occurring at about pH 5.3. The pH at which the enzyme is 50% active is 3.8 on the ascending leg of the curve and 6.7 on the descending leg. These pH values correspond to the pK_a values of the enzyme's catalytic groups. (This is true for all bell-shaped pH–rate profiles, provided that the pK_a values are at least two pK_a units apart. If the difference between them is less than two pK_a units, the precise pK_a values of the catalytic groups must be determined in other ways.)

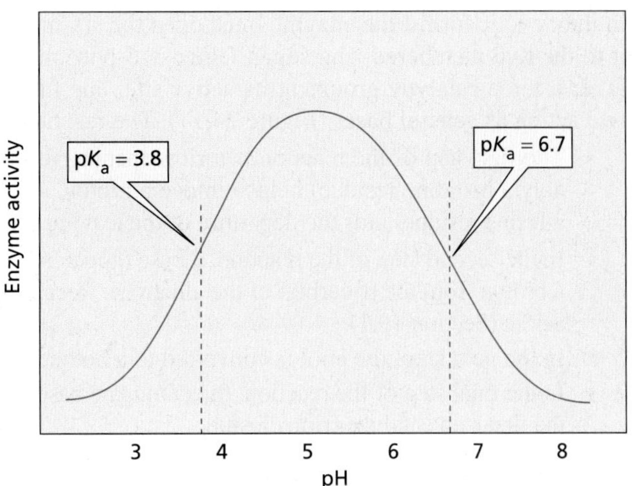

◀ **Figure 24.10**
Dependence of lysozyme activity on the pH of the reaction mixture.

The pK_a given by the ascending leg is the pK_a of a group that is catalytically active in its basic form. When that group is fully protonated, the enzyme is not active. As the pH of the reaction mixture increases, a larger fraction of the group is present in its basic form, and as a result, the enzyme shows increasing activity. Similarly, the pK_a given by the descending leg is the pK_a of a group that is catalytically active in its acidic form. Maximum catalytic activity occurs when the group is fully protonated; activity decreases with increasing pH because a greater fraction of the group lacks a proton.

From the lysozyme mechanism shown in Figure 24.9, we can conclude that Asp 52 is the group with a pK_a of 3.8 and Glu 35 is the group with a pK_a of 6.7. The pH–activity profile indicates that lysozyme is maximally active when Asp 52 is in its basic form and Glu 35 is in its acidic form.

Table 23.2 shows that the pK_a of aspartic acid is 3.86 and the pK_a of glutamic acid is 4.25. The pK_a of Asp 52 agrees with the pK_a of aspartic acid, but the pK_a of Glu 35 is much greater than the pK_a of glutamic acid. Why is the pK_a of the glutamic acid residue at the active site of the enzyme so much greater than the pK_a given for glutamic acid in the table? The pK_a values in the table were determined in water. In the enzyme, Asp 52 is surrounded by polar groups, which means that its pK_a should be close to the pK_a determined in water, a polar solvent. Glu 35, however, is in a predominantly nonpolar microenvironment, so its pK_a should be greater than the pK_a determined in water. We have seen that the pK_a of a carboxylic acid is greater in a nonpolar solvent because there is less tendency to form charged species in nonpolar solvents (Section 10.10).

Part of the catalytic efficiency of lysozyme results from its ability to provide different solvent environments at the active site. This allows one catalytic group to exist in its acidic form at the same surrounding pH at which a second catalytic group exists in its basic form. This property is unique to enzymes; chemists cannot provide different solvent environments for different parts of nonenzymatic systems.

PROBLEM 20◆

When apples that have been cut are exposed to oxygen, an enzyme-catalyzed reaction causes them to turn brown. They can be prevented from turning brown by coating them with lemon juice (pH ~ 3.5). Explain why this is so.

Mechanism for Glucose-6-phosphate Isomerase

Glycolysis is the name given to the series of enzyme-catalyzed reactions responsible for converting D-glucose into two molecules of pyruvate (Sections 19.21 and 25.1). The second reaction in glycolysis is an isomerization reaction that converts D-glucose-6-phosphate to D-fructose-6-phosphate. Recall that the open-chain form of glucose is an aldohexose, whereas the open-chain form of fructose is a ketohexose. Therefore, the enzyme that catalyzes this reaction—glucose-6-phosphate isomerase—converts an aldose to a ketose (Section 22.1). Because, in solution, the sugars exist predominantly in their cyclic forms, the enzyme must open the six-membered-ring sugar and convert it to the five-membered-ring sugar. Glucose-6-phosphate isomerase is known to have at least three catalytic groups at its active site, one functioning as a general acid and two acting as general bases (Figure 24.11). The reaction proceeds as follows:

- The first step of the reaction is a ring-opening reaction. A general base (presumably a histidine residue) helps remove a proton, and a general acid (thought to be a lysine residue) aids the departure of the leaving group (Section 18.7).
- In the second step of the reaction, a base (apparently a glutamate residue) removes a proton from the α-carbon of the aldehyde. Recall that α-hydrogens are relatively acidic (Section 19.1).
- In the next step, the enol is converted to a ketone (Section 19.2).
- In the final step of the reaction, the conjugate base of the general acid employed in the first step catalyzes ring closure.

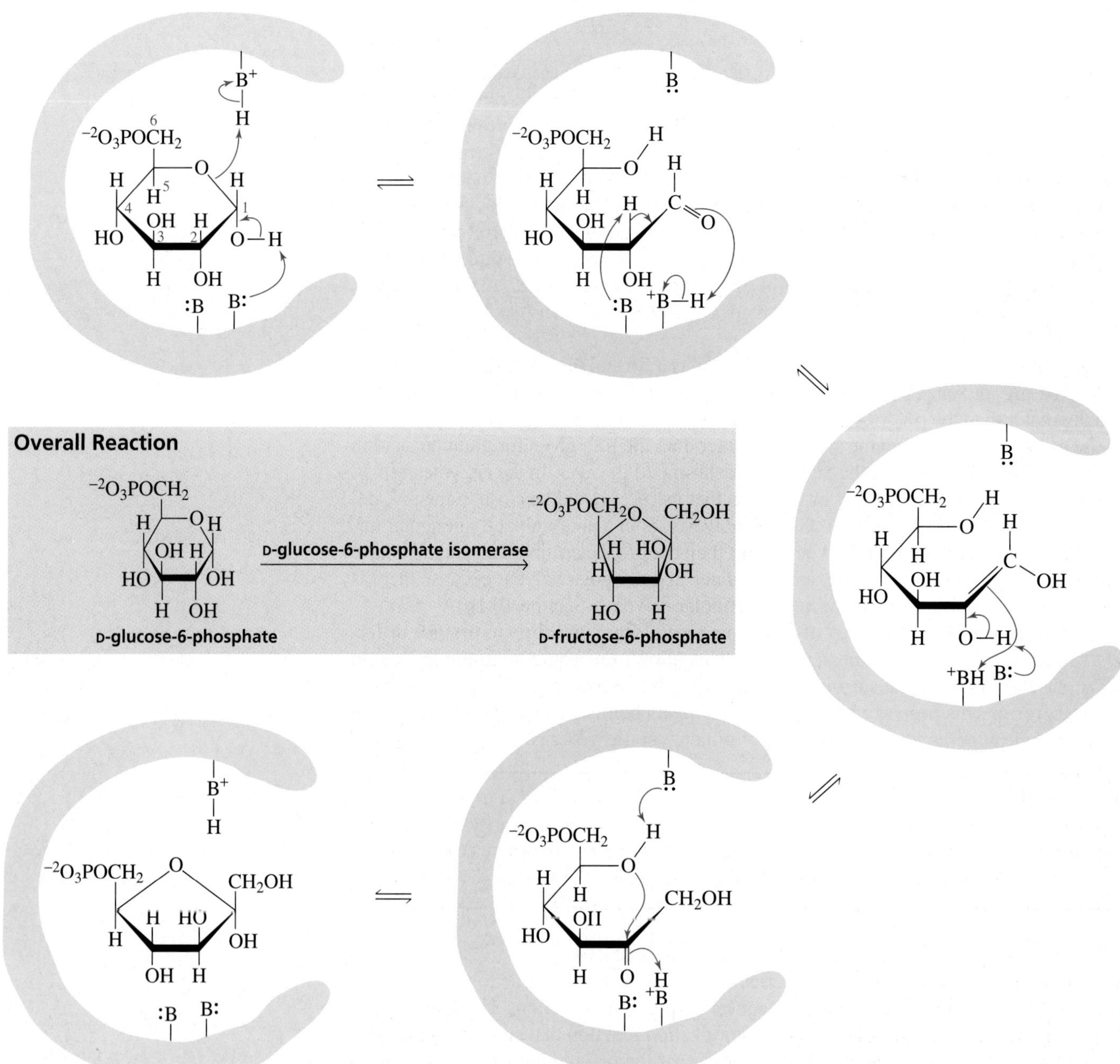

▲ **Figure 24.11**
Proposed mechanism for the isomerization of D-glucose-6-phosphate to D-fructose-6-diphosphate.

PROBLEM 21

When D-glucose undergoes isomerization in the absence of the enzyme, three products result: D-glucose, D-fructose, and D-mannose (Section 22.5). Why is D-mannose not formed in the enzyme-catalyzed reaction?

PROBLEM 22◆

The descending leg of the pH–rate profile for glucose-6-phosphate isomerase indicates that one of the amino acid side chains at the active site of the enzyme has a pK_a value of 9.3. Identify the amino acid side chain.

Mechanism for Aldolase

The substrate for the first enzyme-catalyzed reaction of glycolysis is a six-carbon compound (D-glucose). The final product of glycolysis is two molecules of a three-carbon compound (pyruvate). Therefore, at some point in the series of enzyme-catalyzed reactions, a six-carbon compound must be cleaved into two three-carbon compounds. The enzyme *aldolase* catalyzes this cleavage (Figure 24.12). Aldolase converts D-fructose-1,6-diphosphate into D-glyceraldehyde-3-phosphate and dihydroxyacetone phosphate. The enzyme is called aldolase because the reverse reaction is an aldol addition reaction (Section 19.13). The reaction proceeds as follows:

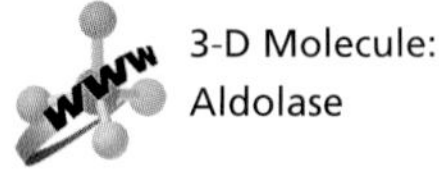

3-D Molecule: Aldolase

- In the first step of the aldolase-catalyzed reaction, D-fructose-1,6-diphosphate forms an imine, with a lysine residue at the active site of the enzyme (Section 18.6).
- A tyrosine residue functions as a general-base catalyst in the step that cleaves the bond between C-3 and C-4. The molecule of D-glyceraldehyde-3-phosphate formed in this step dissociates from the enzyme.
- The enamine intermediate rearranges to an imine, with the tyrosine residue now functioning as a general-acid catalyst.
- Hydrolysis of the imine releases dihydroxyacetone phosphate, the other three-carbon product.

PROBLEM 23

Propose a mechanism for the aldolase-catalyzed cleavage of D-fructose-1,6-diphosphate if it did not form an imine with the substrate. What is the advantage gained by imine formation?

PROBLEM 24

In glycolysis, why must D-glucose-6-phosphate isomerize to D-fructose-6-phosphate before the cleavage reaction with aldolase occurs? (See p. 1037.)

PROBLEM 25◆

Aldolase shows no activity if it is incubated with iodoacetic acid before D-fructose-1,6-diphosphate is added to the reaction mixture. Suggest what could cause the loss of activity.

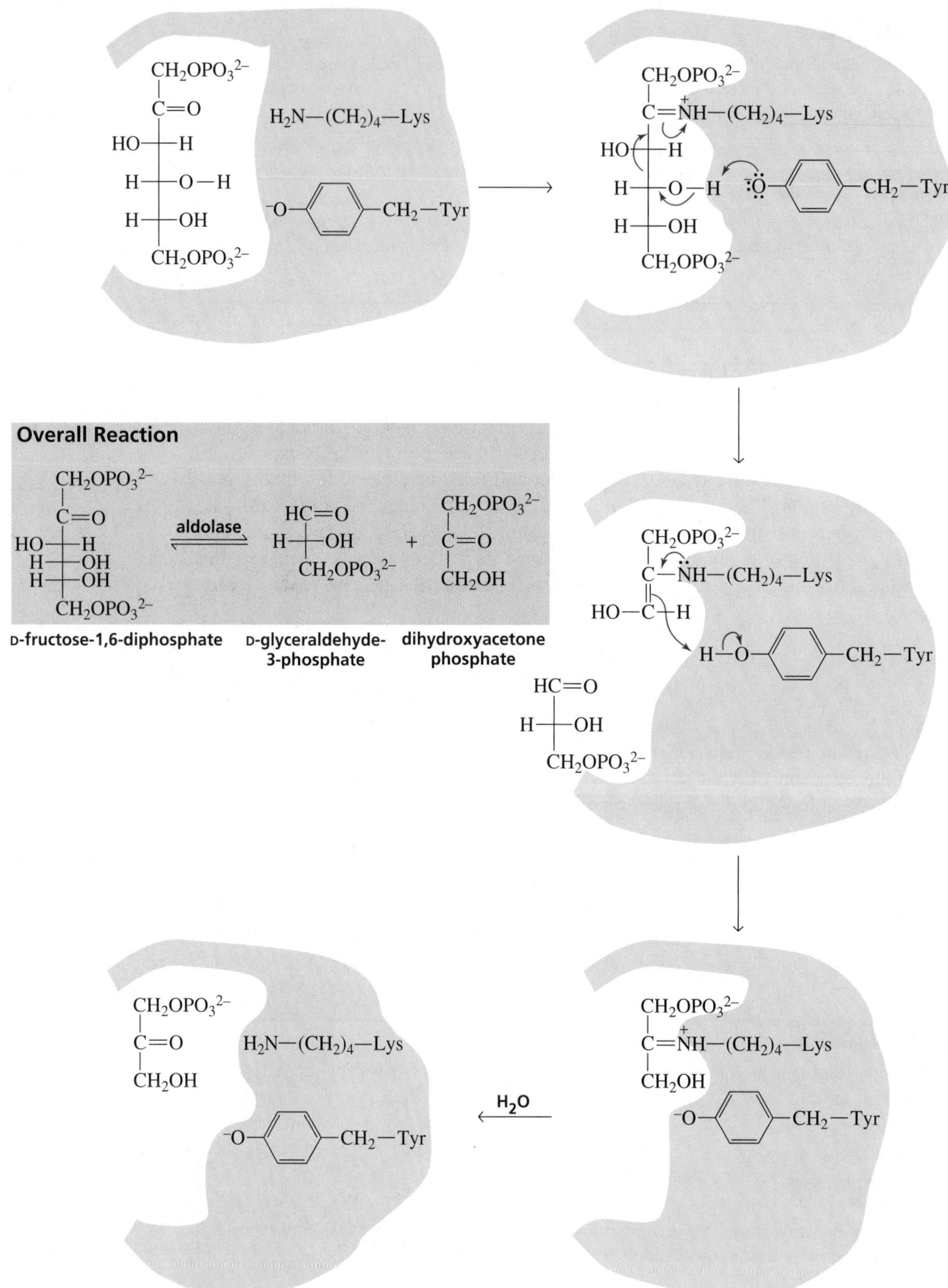

▲ **Figure 24.12**
Proposed mechanism for the aldolase-catalyzed cleavage of D-fructose-1,6-diphosphate.

Summary

A **catalyst** increases the rate of a chemical reaction but is not consumed or changed in the reaction. A catalyst changes the rate at which a product is formed, not the amount of product formed. A catalyst must increase the rate of a slow step. It does this by providing a pathway with a lower $\Delta G^{\ddagger}$. To provide a lower $\Delta G^{\ddagger}$, a catalyst can convert the reactant into a less stable species, make the transition state more stable, or completely change the mechanism of the reaction. Some of the ways a catalyst provides a more favorable pathway for a reaction is by increasing the susceptibility of an electrophile to nucleophilic attack, increasing the reactivity of a nucleophile, or increasing the leaving ability of a group.

A **nucleophilic catalyst** increases the rate of a reaction by acting as a nucleophile: It forms an intermediate by forming a covalent bond with a reactant. Stabilization of a charge by an opposite charge is called **electrostatic catalysis**. An acid catalyst increases the rate of a reaction by donating a proton to a reactant. There are two types of acid catalysis: In **specific-acid catalysis**, the proton is fully transferred to the reactant before the slow step of the reaction; in **general-acid catalysis**, the proton is transferred during the slow step. A **base catalyst** increases the rate of a reaction by removing a proton from the reactant. There are two types of base catalysis: In **specific-base catalysis**, the proton is completely removed from the reactant before the slow step of the reaction; in **general-base catalysis**, the proton is removed during the slow step.

A **metal ion** can increase the rate of a reaction by making a reaction center more susceptible to receiving electrons, by making a leaving group a weaker base, or by increasing the nucleophilicity of water. An **electrophilic catalyst** is a metal ion with the same catalytic effect as a proton.

The rate of a chemical reaction is determined by the number of collisions between two molecules or between two intramolecular constituents with sufficient energy and with the proper orientation in a given period of time. An **intramolecular reaction** that forms a five- or a six-membered ring occurs more readily than the analogous **intermolecular reaction** because both the frequency of the collisions and the probability that collisions will occur in the proper orientation increases. **Effective molarity** is the concentration of the reactant that would be required in an intermolecular reaction for it to have the same rate as the corresponding intramolecular reaction. When a catalyst is part of the reacting molecule, the catalysis is called **intramolecular catalysis**; intramolecular nucleophilic catalysis, intramolecular general-acid or general-base catalysis, and intramolecular metal-ion catalysis all are possible.

Essentially all organic reactions that occur in biological systems require a catalyst. Most biological catalysts are **enzymes**. The reactant of an enzyme-catalyzed reaction is called a **substrate**. The substrate specifically binds to the **active site** of the enzyme, and all the bond-making and bond-breaking steps of the reaction occur while it is at that site. The specificity of an enzyme for its substrate is an example of **molecular recognition**. The change in conformation of the enzyme when it binds the substrate is known as **induced fit**.

Two important factors contributing to the remarkable catalytic ability of enzymes are that reacting groups are brought together at the active site in the proper orientation for reaction and that the amino acid side chains and a metal ion in the case of some enzymes are in the proper position relative to the substrate needed for catalysis. Information about the relationship between the structure of a protein and its function has been determined by **site-specific mutagenesis**. A **pH–rate profile** is a plot of the activity of an enzyme as a function of the pH of the reaction mixture.

Key Terms

acid catalyst (p. 1003)
active site (p. 1015)
acyl-enzyme intermediate (p. 1020)
base catalyst (p. 1006)
catalyst (p. 999)
covalent catalysis (p. 1001)
effective molarity (p. 1010)
electrophilic catalyst (p. 1008)
electrostatic catalysis (p. 1020)
enzyme (p. 1015)
general-acid catalysis (p. 1004)
general-base catalysis (p. 1006)
induced-fit model (p. 1016)
intramolecular catalysis (p. 1012)
lock-and-key model (p. 1015)
metal-ion catalysis (p. 1007)
molecular recognition (p. 1015)
nucleophilic catalysis (p. 1001)
nucleophilic catalyst (p. 1001)
pH–activity profile (p. 1024)
pH–rate profile (p. 1024)
relative rate (p. 1010)
site-specific mutagenesis (p. 1020)
specific-acid catalysis (p. 1004)
specific-base catalysis (p. 1006)
substrate (p. 1015)

Problems

26. Which of the following two compounds would eliminate HBr more rapidly in basic solutions?

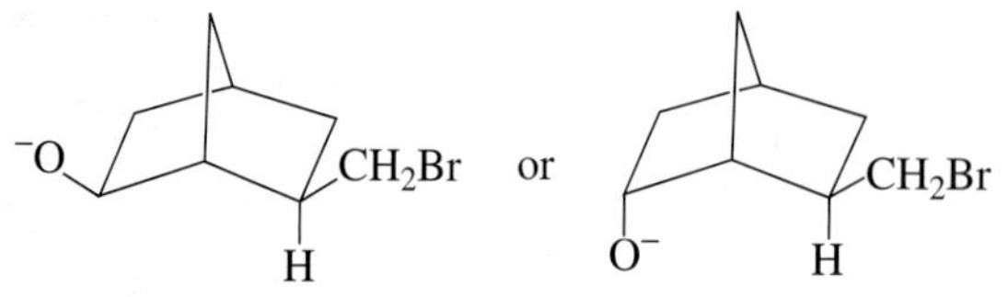

27. Which compound would form a lactone more rapidly?

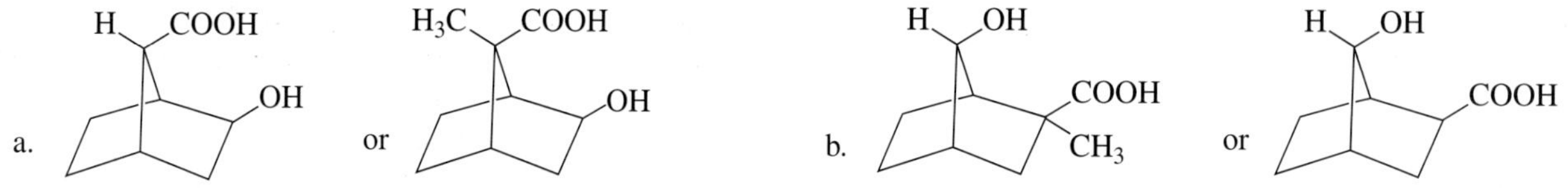

a. or

b. or

28. Which compound would form an anhydride more rapidly?

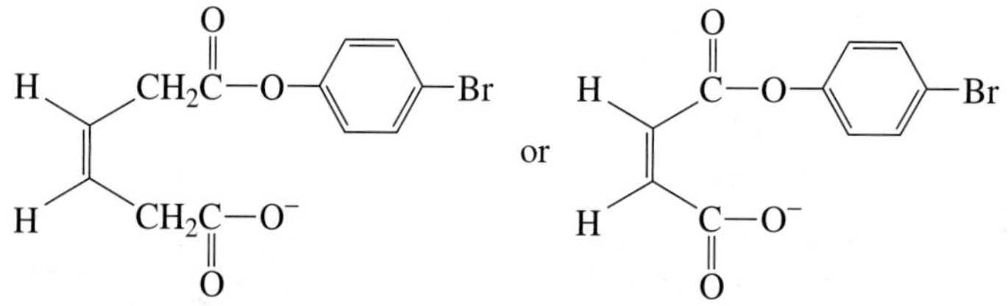

or

29. Which compound has the greatest rate of hydrolysis: benzamide, *o*-carboxybenzamide, *o*-formylbenzamide, or *o*-hydroxybenzamide?

30. Indicate the type of catalysis that is occurring in the slow step in each of the following reaction sequences:

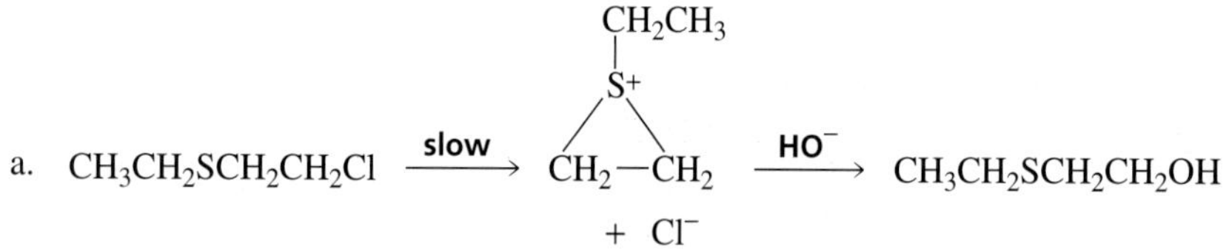

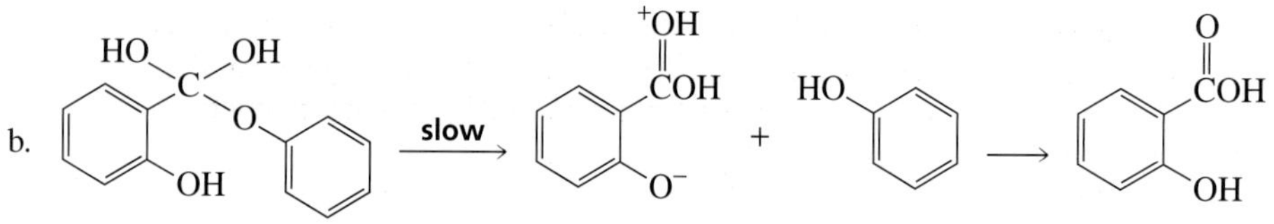

a. $CH_3CH_2SCH_2CH_2Cl \xrightarrow{\text{slow}}$ (episulfonium ion, S^+–CH_2CH_3, $CH_2—CH_2$) $+ Cl^- \xrightarrow{HO^-} CH_3CH_2SCH_2CH_2OH$

b. $\xrightarrow{\text{slow}}$ + $\longrightarrow$

31. The deuterium kinetic isotope effect (k_{H_2O}/k_{D_2O}) for the hydrolysis of aspirin is 2.2. What does this tell you about the kind of catalysis exerted by the *ortho*-carboxyl substituent? (*Hint:* It is easier to break an O—H bond than an O—D bond.)

32. Draw the pH–activity profile for an enzyme with one catalytic group at the active site. The catalytic group is a general-acid catalyst with a pK_a of 5.6.

33. A Co^{2+} complex catalyzes the following hydrolysis of the lactam shown:

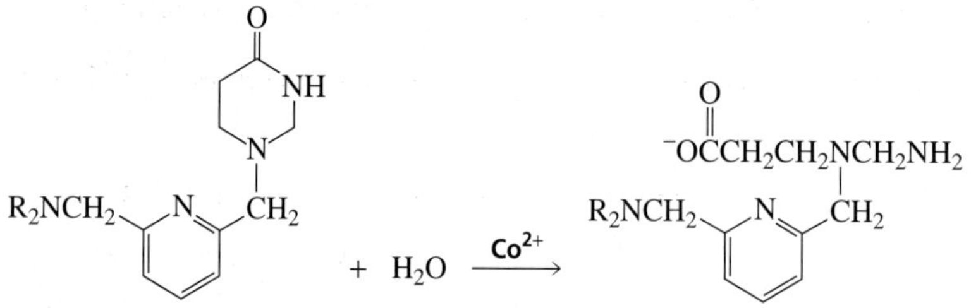

$+ H_2O \xrightarrow{Co^{2+}}$ $^-OCCH_2CH_2NCH_2NH_2$

Propose a mechanism for the metal-ion catalyzed reaction.

34. There are two kinds of aldolases. Class I aldolases are found in animals and plants; Class II aldolases are found in fungi, algae, and some bacteria. Only Class I aldolases form an imine. Class II aldolases have a metal ion (Zn^{2+}) at the active site. The mechanism for catalysis by Class I aldolases was given in Section 24.9. Propose a mechanism for catalysis by Class II aldolases.

35. Propose a mechanism for the following reaction. (*Hint*: The rate of the reaction is much slower if the nitrogen atom is replaced by CH.)

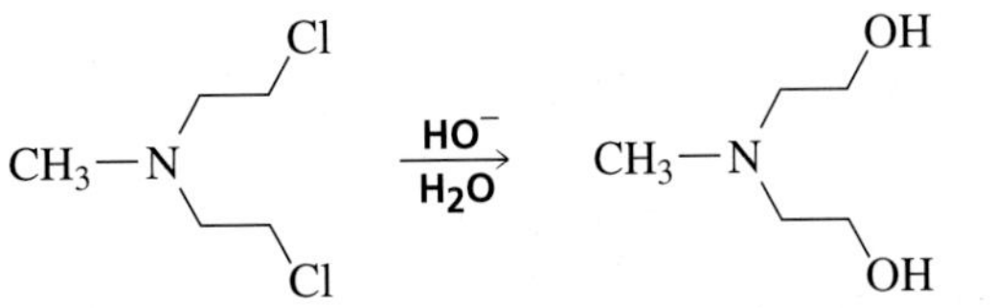

$CH_3—N(CH_2CH_2Cl)_2 \xrightarrow[H_2O]{HO^-} CH_3—N(CH_2CH_2OH)_2$

36. The hydrolysis of the ester shown here is catalyzed by morpholine, a secondary amine. Propose a mechanism for this reaction. (*Hint:* The pK_a of the conjugate acid of morpholine is 9.3, so morpholine is too weak a base to function as a general-base catalyst in this reaction.)

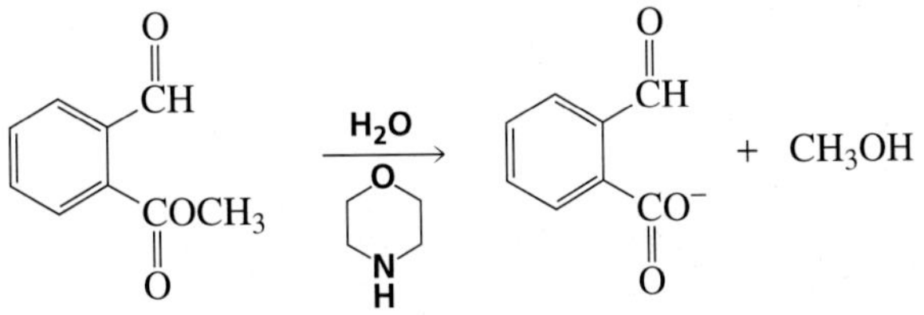

(o-CHO)C_6H_4COCH$_3$ (ester) $\xrightarrow[\text{morpholine}]{H_2O}$ (o-CHO)C_6H_4CO$^-$ (carboxylate) + CH_3OH

morpholine

37. The enzyme carbonic anhydrase catalyzes the conversion of carbon dioxide into bicarbonate ion (Section 1.20). It is a metalloenzyme, with Zn^{2+} coordinated at the active site by three histidine residues. Propose a mechanism for this reaction.

$$CO_2 + H_2O \xrightarrow{\text{carbonic anhydrase}} HCO_3^- + H^+$$

38. At pH = 12, the rate of hydrolysis of ester A is greater than the rate of hydrolysis of ester B. At pH = 8, the relative rates reverse. (Ester B hydrolyzes faster than ester A.) Explain these observations.

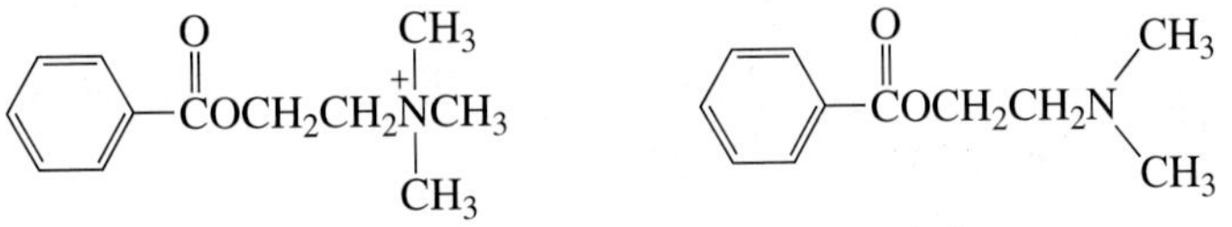

A: $C_6H_5COCH_2CH_2\overset{+}{N}(CH_3)_3$ (benzoate ester, C=O)

B: $C_6H_5COCH_2CH_2N(CH_3)_2$ (benzoate ester, C=O)

39. 2-Acetoxycyclohexyl tosylate reacts with acetate ion to form 1,2-cyclohexanediol diacetate. The reaction is stereospecific; the stereoisomers obtained as products depend on the stereoisomer used as a reactant. Explain the following observations:
   a. Both cis reactants form an optically active trans product, but each cis reactant forms a different trans product.
   b. Both trans reactants form the same racemic mixture.
   c. A trans reactant is more reactive than a cis reactant.

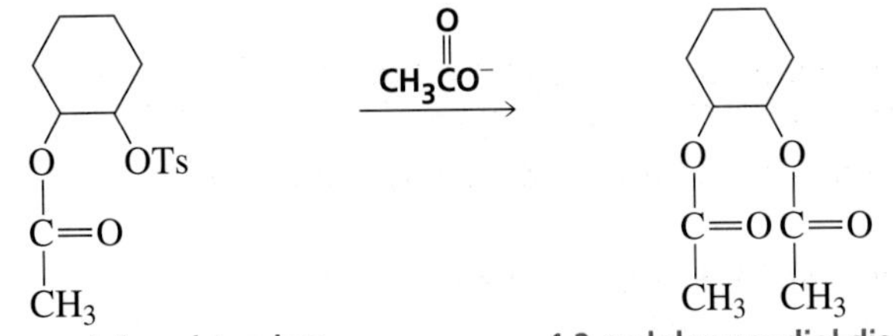

2-acetoxycyclohexyl tosylate (OCOCH$_3$, OTs) $\xrightarrow{CH_3CO_2^-}$ 1,2-cyclohexanediol diacetate (OCOCH$_3$, OCOCH$_3$)

40. *Staphylococcus* nuclease is an enzyme that catalyzes the hydrolysis of DNA. The overall hydrolysis reaction is as follows:

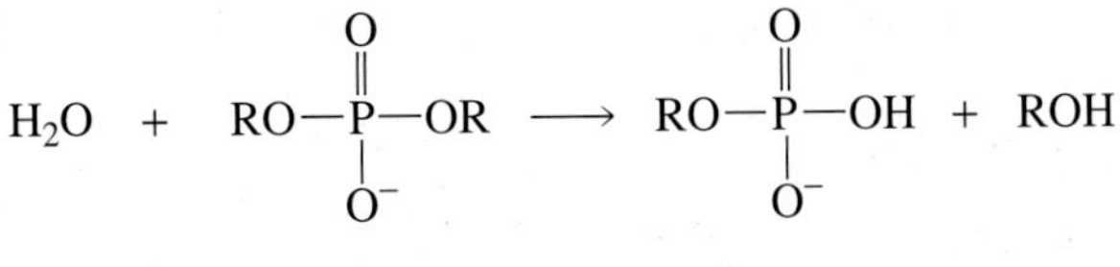

$$H_2O + RO—\underset{O^-}{\overset{O}{\overset{\|}{P}}}—OR \longrightarrow RO—\underset{O^-}{\overset{O}{\overset{\|}{P}}}—OH + ROH$$

Recall that the nucleotides in DNA have phosphodiester linkages. The reaction is catalyzed by Ca^{2+}, Glu 43, and Arg 87. Propose a mechanism for this reaction.

41. Proof that an imine was formed between aldolase and its substrate was obtained by using D-fructose-1,6-diphosphate, labeled at the C-2 position with ^{14}C, as the substrate. $NaBH_4$ was added to the reaction mixture. A radioactive product was isolated from the reaction mixture and hydrolyzed in an acidic solution. Draw the structure of the radioactive product obtained from the acidic solution. (*Hint:* $NaBH_4$ reduces an imine linkage.)

42. 3-Amino-2-oxindole catalyzes the decarboxylation of α-keto acids.
   a. Propose a mechanism for the catalyzed reaction.
   b. Would 3-aminoindole be equally effective as a catalyst?

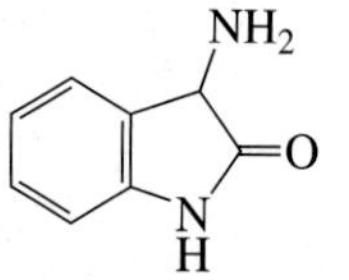

NH$_2$ / =O / N H

3-amino-2-oxindole

43. a. Explain why the alkyl halide shown here reacts much more rapidly than do primary alkyl halides, such as butyl chloride and pentyl chloride, with guanine residues.

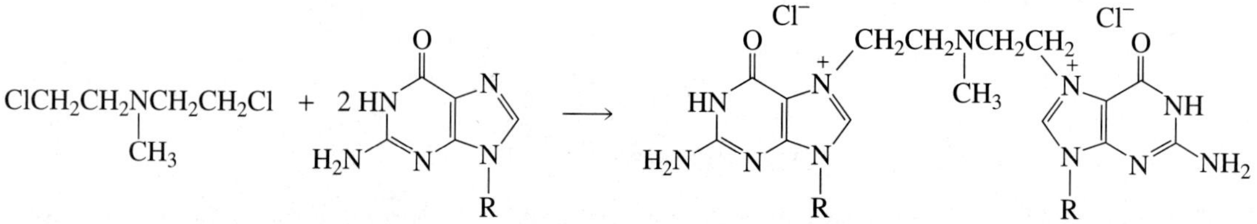

$ClCH_2CH_2NCH_2CH_2Cl$ (N–CH_3) + 2 guanine (HN, H_2N, N–R) ⟶ guanine–$CH_2CH_2NCH_2CH_2$–guanine (N–CH_3), 2 Cl^-

   b. The alkyl halide can react with two guanine residues in two different chains, thereby cross-linking the chains. Propose a mechanism for the cross-linking reaction.

44. Triosephosphate isomerase catalyzes the conversion of dihydroxyacetone phosphate to glyceraldehyde-3-phosphate. The enzyme's catalytic groups are Glu 165 and His 95. In the first step of the reaction, these catalytic groups function as a general-base and a general-acid catalyst, respectively. Propose a mechanism for the reaction.

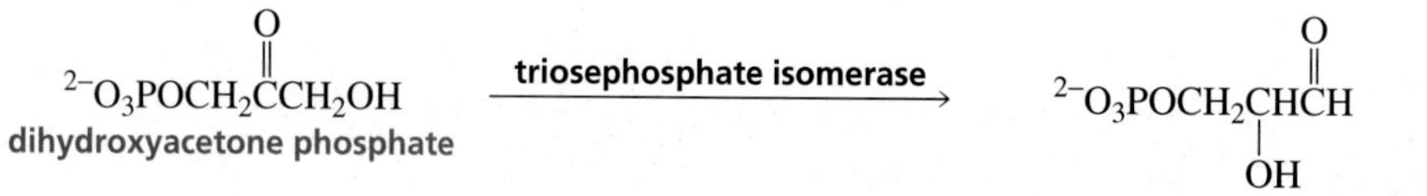

$^{2-}O_3POCH_2C(=O)CH_2OH$ (dihydroxyacetone phosphate) $\xrightarrow{\text{triosephosphate isomerase}}$ $^{2-}O_3POCH_2CH(OH)CH(=O)$ (glyceraldehyde-3-phosphate)

25 The Organic Mechanisms of the Coenzymes • Metabolism

N^5,N^{10}-methylenetetrahydrofolate

Many enzymes cannot catalyze a reaction without the help of a cofactor. **Cofactors** assist enzymes in catalyzing a variety of reactions that cannot be catalyzed solely by the amino acid side chains of the protein. Some cofactors are *metal ions*, while others are *organic molecules*.

A metal-ion cofactor acts as a Lewis acid in a variety of ways to help an enzyme catalyze a reaction. It can coordinate with groups on the enzyme, causing them to align in a geometry advantageous for reaction; it can help bind the substrate to the active site of the enzyme; it can form a coordination complex with the substrate to increase its reactivity; or it can increase the nucleophilicity of water at the active site (Section 24.5). An enzyme that has a tightly bound metal ion (Co^{2+}, Cu^{2+}, Cu^{2+}, Cu^{2+}, Fe^{2+}, Mo^{2+}, Zn^{2+}) is called a **metalloenzyme**. Carboxypeptidase A is an example of a metalloenzyme (Section 24.9).

PROBLEM 1◆

How does the metal ion in carboxypeptidase A increase its catalytic activity?

Cofactors that are organic molecules are called **coenzymes**. Coenzymes are derived from organic compounds commonly known as *vitamins*. Table 25.1 lists the vitamins and their biochemically active coenzyme forms.

A **vitamin** is a substance the body cannot synthesize that is needed in small amounts for normal body function. Sir Frederick Hopkins was the first to suggest that diseases such as rickets and scurvy might result from the absence of substances in the diet that are needed only in very small quantities. Because the first such compound recognized to be essential in the diet was an amine, Casimir Funk incorrectly concluded that all such compounds were amines and called them vitamines ("life-amines"). The *e* was later dropped from the name.

Casimir Funk (1884–1967) *was born in Poland, received his medical degree from the University of Bern, and became a U.S. citizen in 1920. In 1923, he returned to Poland to direct the State Institute of Hygiene. He returned to the United States permanently when World War II broke out.*

We have seen that enzymes catalyze reactions following the principles of organic chemistry (Section 24.9). Coenzymes use these same principles. We will see that

Table 25.1 The Vitamins, Their Coenzymes, and Their Chemical Functions

Vitamin	Coenzyme	Reaction catalyzed	Human deficiency disease
Water-Soluble Vitamins			
Niacin (niacinate)	NAD^+, $NADP^+$	Oxidation	Pellagra
	NADH, NADPH	Reduction	
Riboflavin (vitamin B_2)	FAD, FMN	Oxidation	Skin inflammation
	$FADH_2$, $FMNH_2$	Reduction	
Thiamine (vitamin B_1)	Thiamine pyrophosphate (TPP)	Two-carbon transfer	Beriberi
Lipoic acid (lipoate)	Lipoate	Oxidation	—
	Dihydrolipoate	Reduction	
Pantothenic acid (pantothenate)	Coenzyme A (CoASH)	Acyl transfer	—
Biotin (vitamin H)	Biotin	Carboxylation	—
Pyridoxine (vitamin B_6)	Pyridoxal phosphate (PLP)	Decarboxylation	Anemia
		Transamination	
		Racemization	
		$C_\alpha—C_\beta$ bond cleavage	
		α,β-Elimination	
		β-Substitution	
Vitamin B_{12}	Coenzyme B_{12}	Isomerization	Pernicious anemia
Folic acid (folate)	Tetrahydrofolate (THF)	One-carbon transfer	Megaloblastic anemia
Ascorbic acid (vitamin C)	—	—	Scurvy
Water-Insoluble (lipid-soluble) Vitamins			
Vitamin A	—	—	—
Vitamin D	—	—	Rickets
Vitamin E	—	—	—
Vitamin K	Vitamin KH_2	Carboxylation	—

coenzymes play a variety of chemical roles that the amino acid side chains of enzymes cannot play. Some coenzymes function as oxidizing and reducing agents, some allow electrons to be delocalized, some activate groups for further reaction, and some provide good nucleophiles or strong bases needed for a reaction. Because it would be highly inefficient for the body to use a compound only once and then discard it, coenzymes are recycled. Therefore, we will see that any coenzyme that is changed during the course of a reaction is subsequently converted back to its original form.

An enzyme plus the cofactor it requires to catalyze a reaction is called a **holoenzyme**. An enzyme that has had its cofactor removed is called an **apoenzyme**. Holoenzymes are catalytically active, whereas apoenzymes are catalytically inactive because they have lost their cofactors.

Early nutritional studies divided vitamins into two classes: water-soluble vitamins and water-insoluble vitamins (Table 25.1). Vitamins A, D, E, and K are water insoluble. Vitamin K is the only water-insoluble vitamin currently known to function as a coenzyme. Vitamin A is required for proper vision, vitamin D regulates calcium and phosphate metabolism, and vitamin E is an antioxidant. Because they do not function as coenzymes, vitamins A, D, and E are not discussed in this chapter. Vitamins A and E are discussed in Sections 9.8 and 26.7, and vitamin D is discussed in Section 29.6.

All the water-soluble vitamins except vitamin C function as coenzymes. In spite of its name, vitamin C is not actually a vitamin because it is required in fairly high amounts and most mammals are able to synthesize it (Section 22.19). Humans and guinea pigs cannot synthesize it, however, so it must be included in their diets. We have seen that vitamin C and vitamin E are radical inhibitors and therefore are antioxidants. Vitamin C traps radicals formed in aqueous environments, whereas vitamin E traps radicals formed in nonpolar environments (Section 9.8).

It is impossible to overdose on water-soluble vitamins because the body can readily eliminate any excess. You can, however, overdose on water-insoluble vitamins because they are *not* easily eliminated by the body and can accumulate in cell membranes and other nonpolar components of the body. For example, excess vitamin D causes calcification of soft tissues. The kidneys are particularly susceptible to calcification, which eventually leads to kidney failure. Vitamin D is formed in the skin as a result of a photochemical reaction caused by the ultraviolet rays from the sun (Section 29.6).

Sir Frederick G. Hopkins (1861–1947) *was born in England. His finding that one sample of a protein supported life while another did not led him to conclude that the former contained a trace amount of a substance essential to life. His hypothesis later became known as the "vitamin concept," for which he received a share of the 1929 Nobel Prize in physiology or medicine. He also originated the concept of essential amino acids.*

VITAMIN B_1

Christiaan Eijkman (1858–1930) was a member of a medical team that was sent to the East Indies to study beriberi in 1886. At that time, all diseases were thought to be caused by microorganisms. When the microorganism that caused beriberi could not be found, the team left the East Indies. Eijkman stayed behind to become the director of a new bacteriological laboratory. In 1896, he accidentally discovered the cause of beriberi when he noticed that chickens used in the laboratory had developed symptoms characteristic of the disease. He found that the symptoms had developed when a cook had started feeding the chickens rice meant for hospital patients. The symptoms disappeared when a new cook resumed feeding chicken feed to the chickens. Later it was recognized that thiamine (vitamin B_1) is present in rice hulls but not in polished rice. For this work, Eijkman shared the 1929 Nobel Prize in physiology or medicine with Frederick Hopkins.

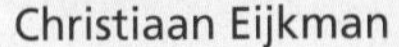
Christiaan Eijkman

Sir Frederick Hopkins

25.1 Overall View of Metabolism

The reactions that living organisms carry out to obtain the energy they need and to synthesize the compounds they require are collectively known as **metabolism**. The process of metabolism can be divided into two parts: **catabolism** and **anabolism**. *Catabolic reactions* break down complex nutrient molecules to provide energy and simple precursor molecules for synthesis. *Anabolic reactions* require energy and result in the synthesis of complex biomolecules from simpler precursor molecules.

catabolism: complex molecules $\longrightarrow$ simple molecules + energy

anabolism: simple molecules + energy $\longrightarrow$ complex molecules

It is important to remember that almost every reaction that occurs in a living system is catalyzed by an enzyme. The enzyme holds the reactants and any necessary cofactors in place, so that the reacting functional groups and catalytic groups are oriented in a way that will cause a specific well-defined chemical reaction (Section 24.8). In some cases, it may be helpful to see where a particular enzyme-catalyzed reaction discussed in this chapter fits into the overall metabolic scheme, so we will start by taking an overall view of metabolism.

Catabolism can be divided into four stages. The *first stage of catabolism* is called digestion. The reactants required for all life processes ultimately come from our diet. In that sense, we really are what we eat. In this first stage, fats, carbohydrates, and proteins are hydrolyzed into fatty acids, monosaccharides, and amino acids, respectively (Figure 25.1).

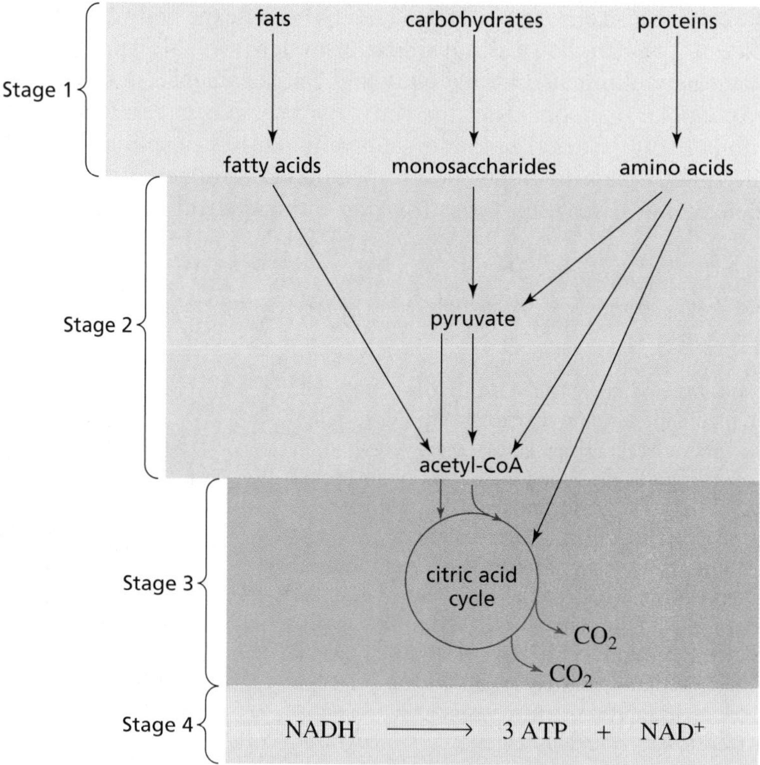

Figure 25.1 ▶
The four stages of catabolism:
1. digestion (hydrolysis of polymers to monomers)
2. conversion of monomers to compounds that can enter the citric acid cycle
3. the citric acid cycle
4. oxidative phosphorylation

In the *second stage of catabolism*, these hydrolysis products—fatty acids, monosaccharides, and amino acids—are converted to compounds that can enter the citric acid cycle. To enter the citric acid cycle, a compound must be either one of the compounds in the cycle itself (called a citric acid cycle intermediate), acetyl-CoA, or pyruvate. Acetyl CoA is the only non–citric acid cycle intermediate that can enter the cycle (Section 17.20); pyruvate can enter the citric acid cycle only because it can be converted to acetyl-CoA or to oxaloacetate, a citric acid cycle intermediate (Sections 25.4 and 25.5). Fatty acids are converted to acetyl-CoA, monosaccharides are converted to pyruvate via glycolysis (Figure 25.2), and amino acids are converted to acetyl-CoA, pyruvate, and/or citric acid cycle intermediates, depending on the amino acid.

The *third stage of catabolism* is the citric acid cycle (also known as the Krebs cycle or the tricarboxylic acid (TCA) cycle; Figure 25.3). For every acetyl-CoA that enters the cycle, two molecules of CO_2 are formed.

Metabolic energy is measured in terms of adenosine triphosphate (ATP). How the body uses ATP is described in Sections 27.2 and 27.3. Very little ATP is formed in the first three stages of catabolism. However, in the *fourth stage of catabolism*, every NADH (see Section 25.2) that is formed in the process of carrying out oxidation reactions in the earlier stages is converted into three ATPs in a process known as *oxidative phosphorylation*. Thus, the bulk of the energy provided by fats, carbohydrates, and proteins is obtained in this fourth stage.

Anabolism can be thought of as the reverse of catabolism. In anabolism, acetyl-CoA, pyruvate, and citric acid cycle intermediates are the starting materials for the synthesis of fatty acids, monosaccharides, and amino acids. These compounds are then used to form fats, carbohydrates, and proteins. The mechanisms utilized by biological systems to synthesize fats and proteins are discussed in Sections 19.21 and 27.13.

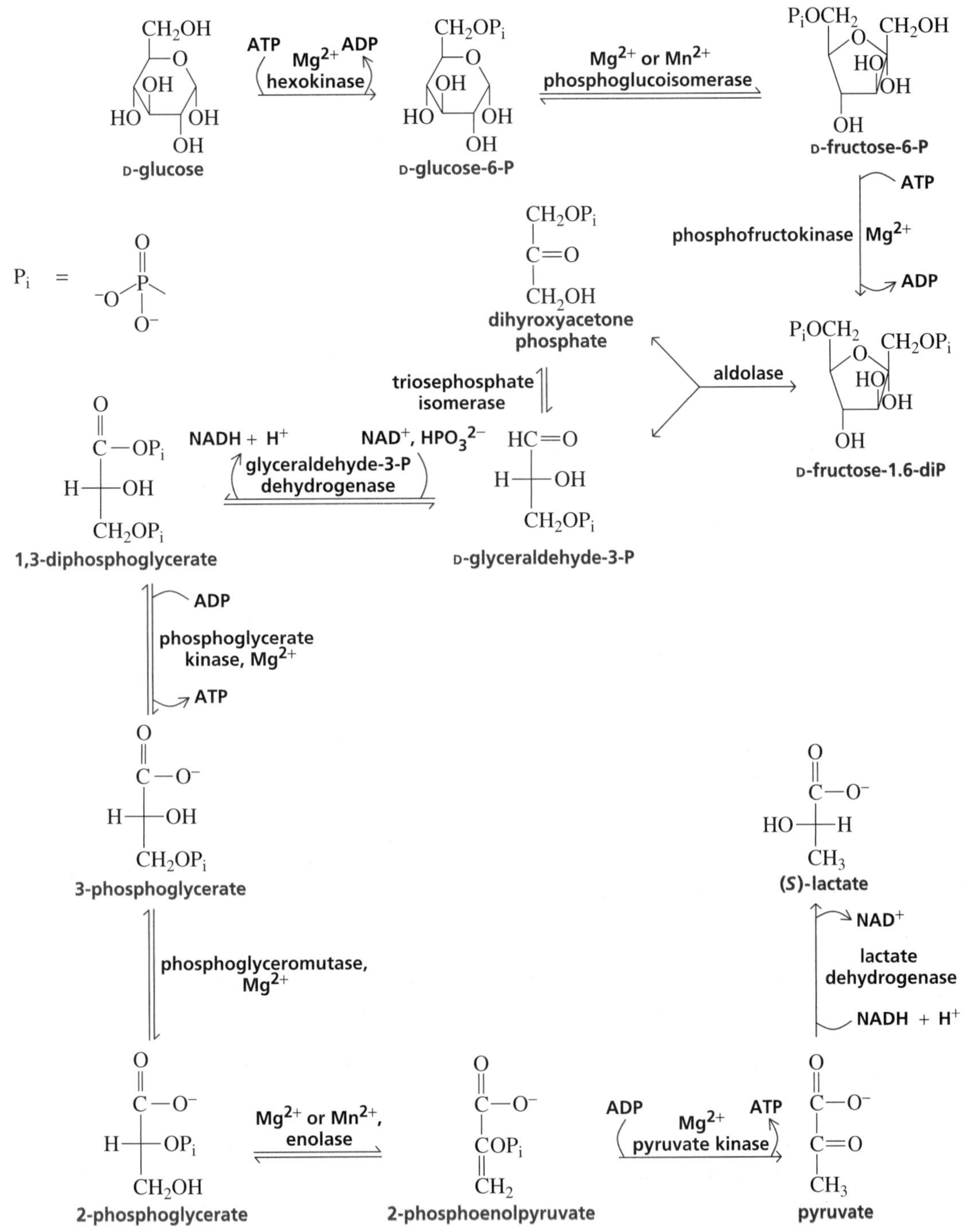

▲ **Figure 25.2**
Glycolysis—the series of enzyme-catalyzed reactions responsible for the conversion of 1 mole of D-glucose into 2 moles of pyruvate.

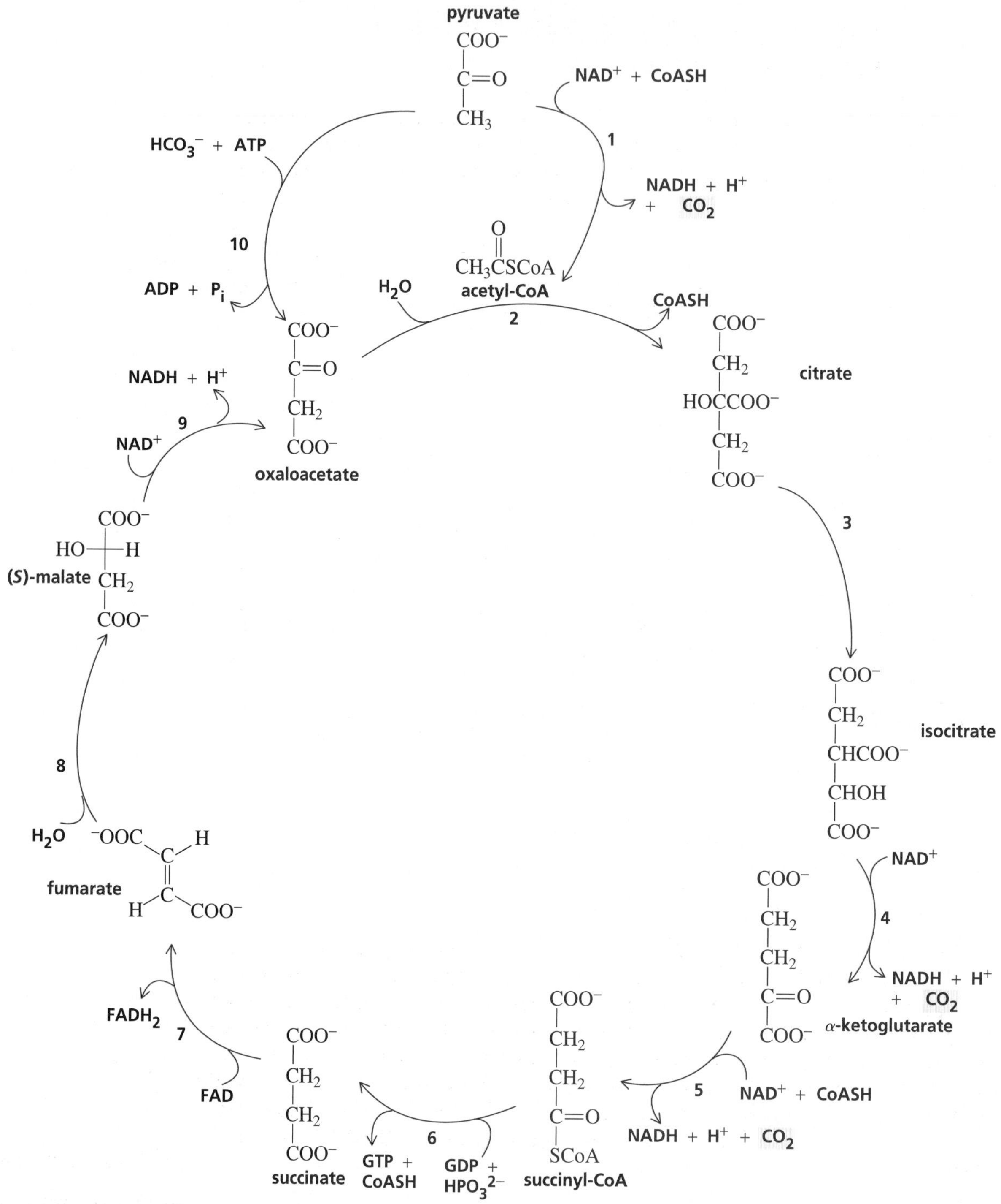

pyruvate COO^- $C{=}O$ CH_3

NAD^+ + CoASH

1

NADH + H^+ + CO_2

CH_3CSCoA (O above C)
acetyl-CoA

HCO_3^- + ATP

10

ADP + P_i

H_2O

2

CoASH

citrate COO^- CH_2 $HOCCOO^-$ CH_2 COO^-

oxaloacetate COO^- $C{=}O$ CH_2 COO^-

NADH + H^+

9

NAD^+

(S)-malate COO^- HO—C—H CH_2 COO^-

3

isocitrate COO^- CH_2 $CHCOO^-$ CHOH COO^-

NAD^+

4

NADH + H^+ + CO_2

8

H_2O

fumarate ^-OOC—CH=CH—COO^-

α-ketoglutarate COO^- CH_2 CH_2 $C{=}O$ COO^-

5

NAD^+ + CoASH

NADH + H^+ + CO_2

succinyl-CoA COO^- CH_2 CH_2 $C{=}O$ SCoA

6

GDP + HPO_3^{2-}

GTP + CoASH

succinate COO^- CH_2 CH_2 COO^-

7

FAD

$FADH_2$

▲ **Figure 25.3**
The citric acid cycle (also known as the Krebs cycle or the tricarboxylic acid (TCA) cycle)—the series of enzyme-catalyzed reactions responsible for the oxidation of acetyl-CoA to CO_2 and for the entry of pyruvate and acetyl-CoA into the cycle.

1 pyruvate dehydrogenase
2 citrate synthase
3 aconitase
4 isocitrate dehydrogenase
5 α-ketogutarate dehydrogenase
6 succinyl-CoA synthetase
7 succinate dehydrogenase
8 fumarase
9 malate dehydrogenase
10 pyruvate carboxylase

25.2 Niacin: The Vitamin Needed for Many Redox Reactions

An enzyme that catalyzes an oxidation reaction or a reduction reaction requires a coenzyme because none of the amino acid side chains can act as oxidizing or reducing agents. The coenzyme is the oxidizing or reducing agent. The enzyme's role is to hold the substrate and coenzyme together so that the oxidation or reduction reaction can take place.

The coenzymes most commonly used by enzymes to catalyze oxidation reactions are **nicotinamide adenine dinucleotide (NAD^+)** and **nicotinamide adenine dinucleotide phosphate ($NADP^+$)**. (You were introduced to NAD^+ in Section 20.11.)

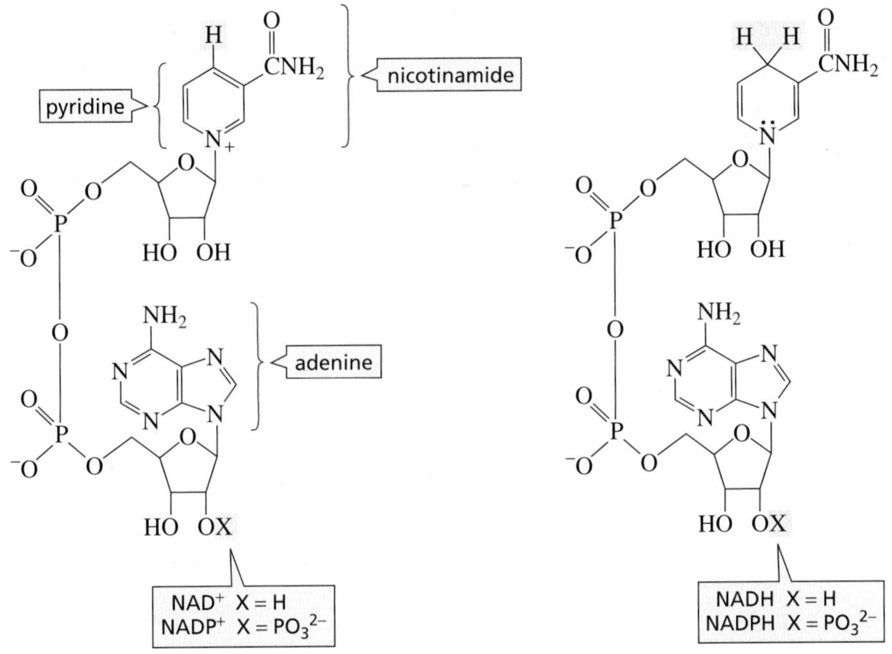

pyridine
nicotinamide
adenine
NAD^+ X = H
$NADP^+$ X = PO_3^{2-}

NADH X = H
NADPH X = PO_3^{2-}

When NAD^+ (or $NADP^+$) oxidizes a substrate, the coenzyme is reduced to NADH (or NADPH). NADH and NADPH are reducing agents; they are used as coenzymes by enzymes that catalyze reduction reactions. Enzymes that catalyze oxidation reactions bind NAD^+ (or $NADP^+$) more tightly than they bind NADH (or NADPH). When the oxidation reaction is over, the relatively loosely bound NADH (or NADPH) dissociates from the enzyme. Likewise, enzymes that catalyze reduction reactions bind NADH (or NADPH) more tightly than they bind NAD^+ (or $NADP^+$). When the reduction reaction is over, the relatively loosely bound NAD^+ (or $NADP^+$) dissociates from the enzyme.

NAD^+ and $NADP^+$ are oxidizing agents.

NADH and NADPH are reducing agents.

$$\text{substrate}_{\text{reduced}} + NAD^+ \xrightleftharpoons{\text{enzyme}} \text{substrate}_{\text{oxidized}} + NADH + H^+$$

$$\text{substrate}_{\text{reduced}} + NADP^+ \xrightleftharpoons{\text{enzyme}} \text{substrate}_{\text{oxidized}} + NADPH + H^+$$

NAD^+ is composed of two nucleotides linked together through their phosphate groups. A **nucleotide** consists of a heterocycle attached, in a β-configuration, to C-1 of a phosphorylated ribose (Section 27.1). A **heterocycle** is a cyclic compound in which one or more of the ring atoms is an atom other than carbon (see p. 884).

a nucleotide (phosphate–ribose–heterocycle, with HO OH on the ribose)   adenine   niacinamide / nicotinamide   niacin / nicotinic acid

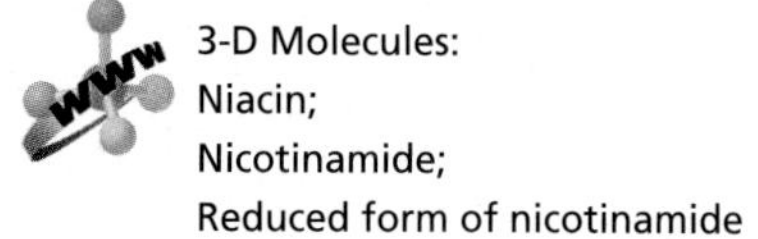

The heterocyclic component of one of the nucleotides of NAD^+ is nicotinamide, and the heterocyclic component of the other is adenine. This accounts for the coenzyme's name (**n**icotinamide **a**denine **d**inucleotide). The positive charge in the NAD^+ abbreviation indicates the positively charged nitrogen of the substituted pyridine ring.

The only way in which $NADP^+$ differs structurally from NAD^+ is in the phosphate group bonded to the 2′-OH group of the ribose of the adenine nucleotide; this explains the addition of "P" to the name. NAD^+ and NADH are generally used as coenzymes in catabolic reactions; $NADP^+$ and NADPH are generally used as coenzymes in anabolic reactions.

The adenine nucleotide for the coenzymes is provided by ATP. The nicotinamide nucleotide is derived from the vitamin known as niacin. Humans can synthesize a small amount of niacin from the amino acid tryptophan, but not in sufficient quantity to meet the body's metabolic needs.

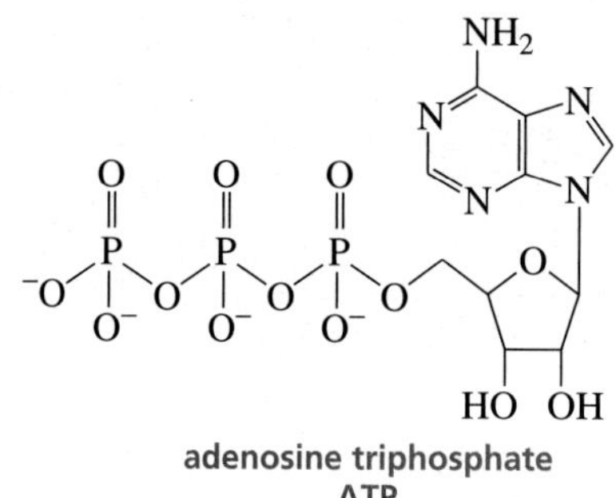

adenosine triphosphate
ATP

NIACIN DEFICIENCY

A deficiency in niacin causes pellagra, a disease that begins with dermatitis and ultimately causes insanity and death. More than 120,000 cases of pellagra were reported in the United States in 1927, mainly among poor people with unvaried diets. A factor known to be present in preparations of vitamin B prevented pellagra, but it was not until 1937 that the factor was identified as nicotinic acid.

When bread companies started adding nicotinic acid to their bread, they insisted that its name be changed to niacin because they thought that nicotinic acid sounded too much like nicotine and they did not want their vitamin-enriched bread to be associated with a harmful substance. Niacinamide is a nutritionally equivalent form of the vitamin.

The oxidation of the *secondary alcohol group* of malate to a *ketone group* is one of the reactions in the citric acid cycle (Figure 25.2). NAD^+ is the oxidizing reagent in this reaction. Many enzymes that catalyze oxidation reactions are called **dehydrogenases**. Recall that the number of C—H bonds decreases in an oxidation reaction (Section 20.0). In other words, dehydrogenases remove hydrogen.

$$\underset{\text{malate}}{^-OOCCH_2CH(OH)COO^-} + NAD^+ \xrightleftharpoons{\text{malate dehydrogenase}} \underset{\text{oxaloacetate}}{^-OOCCH_2COCOO^-} + NADH + H^+$$

β-Aspartate-semialdehyde is reduced to homoserine in an anabolic pathway, with NADPH as the reducing agent.

$$\underset{\beta\text{-aspartate-semialdehyde}}{\overset{\overset{O}{\|}\quad\overset{O}{\|}}{HCCH_2\underset{\underset{^{+}NH_3}{|}}{C}HCO^-}} + NADPH + H^+ \xrightarrow{\text{homoserine dehydrogenase}} \underset{\text{homoserine}}{HOCH_2CH_2\underset{\underset{^{+}NH_3}{|}}{C}H\overset{\overset{O}{\|}}{C}O^-} + NADP^+$$

The differentiation between the coenzymes used in catabolism and anabolism is maintained because the enzymes that catalyze these oxidation–reduction reactions exhibit strong specificity for a particular coenzyme. For example, an enzyme that catalyzes an oxidation reaction can readily tell the difference between NAD^+ and $NADP^+$; if the enzyme is in a catabolic pathway, it will bind NAD^+, but not $NADP^+$. In addition, the relative concentrations of the coenzymes in the cell encourage binding of the appropriate coenzyme. For example, because NAD^+ and NADH are catabolic coenzymes and catabolic reactions are most often oxidation reactions, the NAD^+ concentration in the cell is much greater than the NADH concentration. (The cell maintains its $[NAD^+]/[NADH]$ ratio near 1000.) Because $NADP^+$ and NADPH are anabolic coenzymes and anabolic pathways are predominantly reduction reactions, the concentration of NADPH in the cell is greater than the concentration of $NADP^+$. (The ratio of $[NADP^+]/[NADPH]$ is maintained at about 0.01.)

Mechanisms for Pyridine Nucleotide Coenzymes

How do these oxidation–reduction reactions take place? All the chemistry of the pyridine nucleotide coenzymes (NAD^+, $NADP^+$, NADH, and NADPH) takes place at the 4-position of the pyridine ring. The rest of the molecule is important for binding the coenzyme to the proper site on the enzyme. If a substrate is being *oxidized*, it donates a hydride ion (H^-) to the 4-position of the pyridine ring. In the following reaction, the primary alcohol is oxidized to an aldehyde. A basic amino acid side chain of the enzyme can help the reaction by removing a proton from the oxygen in the substrate.

oxidation of substrate
reduction of coenzyme

Glyceraldehyde-3-phosphate dehydrogenase is an example of an enzyme that uses NAD^+ as an oxidizing coenzyme. The enzyme catalyzes the oxidation of the aldehyde group of glyceraldehyde-3-phosphate (GAP) to an anhydride of a carboxylic acid and phosphoric acid. This is a reaction that occurs in glycolysis (Figure 25.3).

D-glyceraldehyde-3-phosphate + NAD^+ + $HOPO_3^{2-}$ $\xrightarrow{\text{glyceraldehyde-3-phosphate dehydrogenase}}$ D-1,3-diphosphoglycerate + NADH + H^+

In the first step of the mechanism for this reaction, an SH group of a cysteine side chain at the active site of the enzyme reacts with glyceraldehyde-3-phosphate to form a tetrahedral intermediate. A side chain of the enzyme increases cysteine's nucleophilicity by acting as a general-base catalyst. The tetrahedral intermediate expels a hydride ion, transferring it to the 4-position of the pyridine ring of an NAD^+ that is bonded to the enzyme at an adjacent site. NADH dissociates from the enzyme, and the enzyme binds a new NAD^+. Phosphate reacts with the thioester, forming the anhydride product

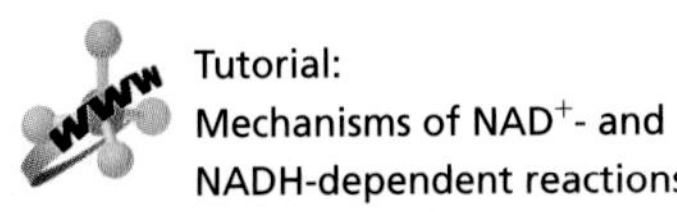

Tutorial: Mechanisms of NAD^+- and NADH-dependent reactions

and releasing cysteine. Notice that at the end of the reaction the holoenzyme is exactly as it was at the beginning, so the catalytic cycle can be repeated. The NADH produced in the reaction is reoxidized to NAD^+ in the fourth stage of catabolism (Figure 25.1).

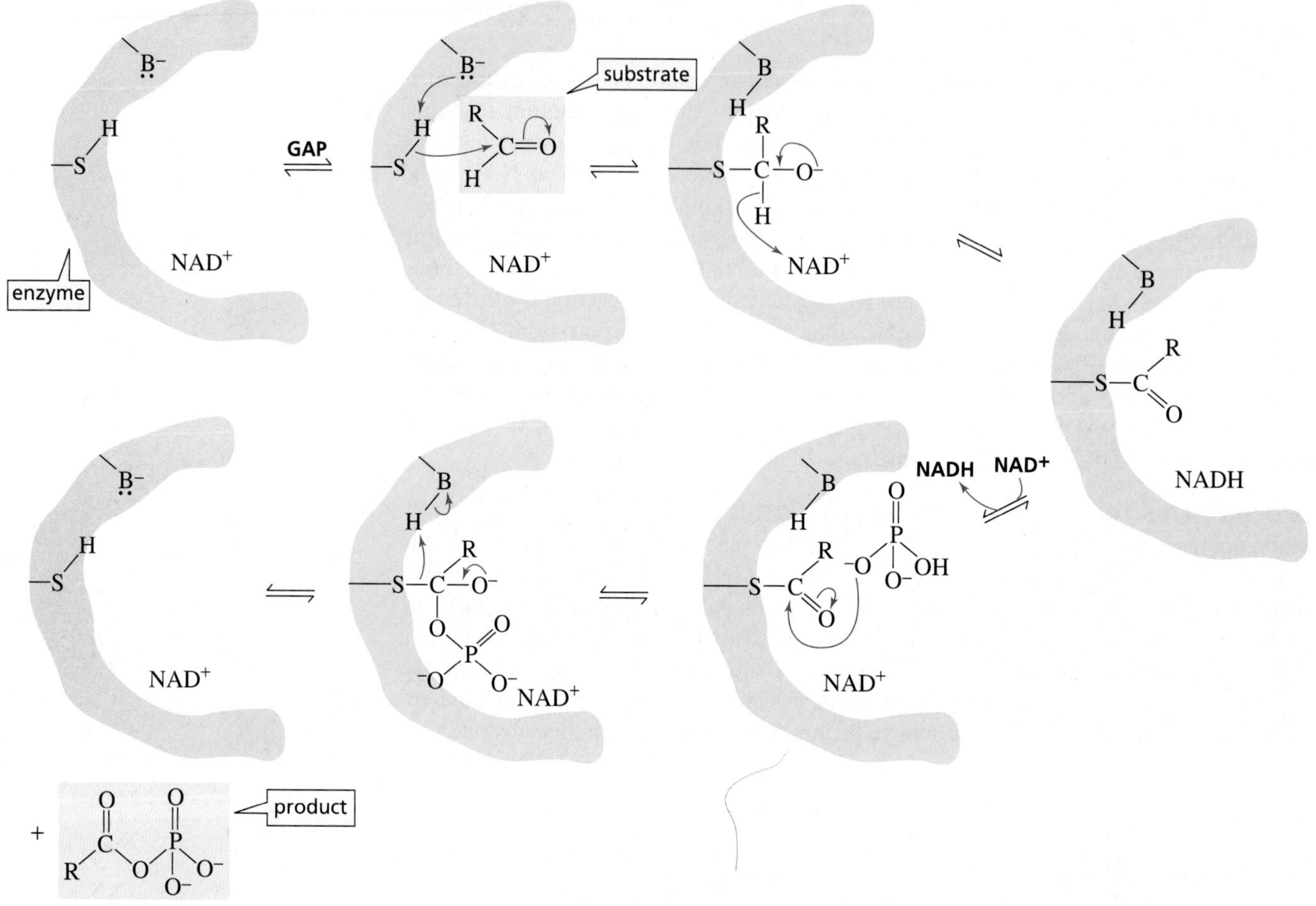

enzyme
GAP
substrate
NAD^+
NADH
product

The mechanism for reduction by NADH (or by NADPH) is the reverse of the mechanism for oxidation by NAD^+ (or by $NADP^+$). If a substrate is being *reduced*, the dihydropyridine ring donates a hydride ion from its 4-position to the substrate. An acidic side chain of the enzyme aids the reaction by donating a proton to the substrate.

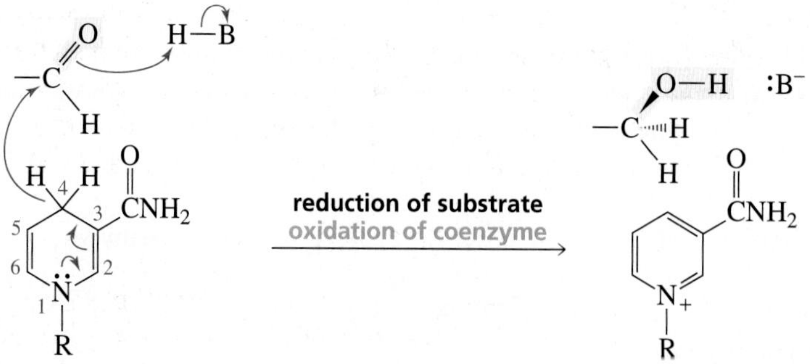

reduction of substrate
oxidation of coenzyme

Because NADH and NADPH reduce compounds by donating a hydride ion, they can be considered biological equivalents of $NaBH_4$ or $LiAlH_4$—hydride donors used as reducing reagents in nonbiological reactions (Section 20.1).

Why are the structures of biological redox (reducing and oxidizing) reagents so much more complicated than the structures of the redox reagents used to carry out the same reactions in the laboratory? NADH is certainly more complicated than $LiAlH_4$, although both reagents reduce compounds by donating a hydride ion. Much of the structural complexity of a coenzyme is for molecular recognition—to allow it to be recognized by the enzyme. *Molecular recognition* allows the enzyme to bind the substrate and the coenzyme in the proper orientation for reaction.

In addition, a redox reagent found in a biological system must be less reactive than a laboratory redox reagent because it must be selective. For example, a biological reducing agent cannot reduce just any reducible compound with which it comes into contact. Biological reactions are much more carefully controlled than that. Because the coenzymes are relatively unreactive compared with nonbiological redox agents, the reaction between the substrate and the coenzyme does not occur at all or takes place very slowly without the enzyme. For example, NADH will reduce an aldehyde or a ketone only in the presence of an enzyme. $NaBH_4$ and $LiAlH_4$ are more reactive hydride donors—much too reactive to exist in the aqueous environment of the cell. Similarly, NAD^+ is a much more selective oxidizing agent than the typical oxidizing agents used in the laboratory; for example, NAD^+ will oxidize an alcohol only in the presence of an enzyme.

Because a biological reducing agent must be recycled (rather than having its oxidized form thrown away as would be the case for a reducing agent used in a laboratory), the equilibrium constant between the oxidized and reduced forms is generally close to unity. Therefore, biological redox reactions are not highly exergonic; rather, they are equilibrium reactions that are driven in the appropriate direction by the removal of reaction products as a result of participation in subsequent reactions.

As you study the coenzymes in this chapter, don't let the complexity of their structures deter you. Notice that only a small part of the coenzyme is actually involved in the chemical reaction. Notice also that the coenzymes follow the same rules of organic chemistry as do the simple organic molecules with which you are familiar.

We saw in Section 5.16 that an oxidizing enzyme can distinguish between the two hydrogens on the carbon from which it catalyzes the removal of a hydride ion. For example, alcohol dehydrogenase removes only the pro-*R* hydrogen, H_a, of ethanol. (H_b is the pro-*S* hydrogen.)

$$CH_3C(H_a)(H_b)OH + NAD^+ \xrightarrow{\text{alcohol dehydrogenase}} CH_3C(H_b){=}O + NADH_a + H^+$$

ethanol → acetaldehyde

Similarly, a reducing enzyme can distinguish between the two hydrogens at the 4-position of the nicotinamide ring of NADH. An enzyme has a specific binding site for the coenzyme, and when it binds the coenzyme, it blocks one of its sides. If the enzyme blocks the B-side of NADH, the substrate will bind to the A-side and the H_a hydride ion will be transferred to the substrate. If the enzyme blocks the A-side of the coenzyme, the substrate will have to bind to the B-side and the H_b hydride ion will be transferred. Currently, 156 dehydrogenases are known to transfer H_a, and 121 are known to transfer H_b.

substrate
A-side
enzyme
B-side

The enzyme blocks the B-side of the coenzyme, so the substrate binds to the A-side.

The enzyme blocks the A-side of the coenzyme, so the substrate binds to the B-side.

25.3 Flavin Adenine Dinucleotide and Flavin Mononucleotide: Vitamin B_2

A *flavoprotein* is an enzyme that contains either **flavin adenine dinucleotide (FAD)** or **flavin mononucleotide (FMN)** as a coenzyme. FAD and FMN, like NAD^+ and $NADP^+$, are coenzymes used in oxidation reactions. As its name indicates, FAD is a dinucleotide in which one of the heterocyclic components is flavin and the other is adenine. FMN contains flavin but not adenine—it is a mononucleotide. (Flavin is a bright yellow compound; *flavus* is Latin for "yellow.") Notice that instead of ribose, the flavin nucleotide has a ribitol group (a reduced ribose). Flavin plus ribitol is called *riboflavin*. Riboflavin is also known as vitamin B_2. A vitamin B_2 deficiency causes inflammation of the skin.

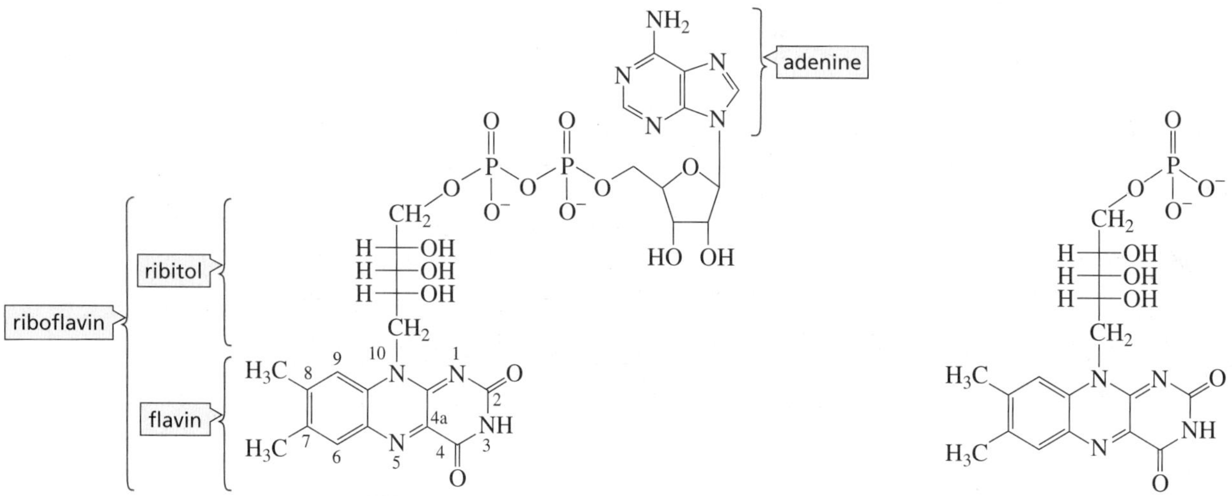

riboflavin ribitol flavin adenine

FAD

FMN

3-D Molecules:
Flavin adenine dinucleotide (FAD);
Flavin mononucleotide (FMN)

In most flavoproteins, FAD (or FMN) is bound quite tightly. Tight binding allows the enzyme to control the oxidation potential of the coenzyme. (The more positive the oxidation potential, the stronger is the oxidizing agent.) Consequently, some flavoproteins are stronger oxidizing agents than others.

PROBLEM 2◆

FAD is obtained by an enzyme-catalyzed reaction that uses FMN and ATP as substrates. What is the other product of this reaction?

How can we tell which enzymes use FAD (or FMN) rather than NAD^+ (or $NADP^+$) as the oxidizing coenzyme? A rough guideline is that NAD^+ and $NADP^+$ are the coenzymes used in enzyme-catalyzed oxidation reactions involving carbonyl compounds (alcohols being oxidized to ketones, aldehydes, or carboxylic acids), while FAD and FMN are the coenzymes used in other types of oxidations. For example, in the following reactions, FAD oxidizes a dithiol to a disulfide, an amine to an imine, and a saturated alkyl group to an unsaturated alkene, and FMN oxidizes NADH to NAD^+. (This is only an approximate guideline, however, because FAD is involved in some oxidations that involve carbonyl compounds, and NAD^+ and $NADP^+$ are involved in some oxidations that do not involve carbonyl compounds.)

$$\text{dihydrolipoate} + FAD \xrightarrow{\text{dihydrolipoyl dehydrogenase}} \text{lipoate} + FADH_2$$

$$\underset{\text{D-amino acid or L-amino acid}}{RCH(NH_2)CO^-} + FAD \xrightarrow{\text{D-amino acid oxidase or L-amino acid oxidase}} RC(=NH)CO^- + FADH_2$$

$$\underset{\text{succinate}}{{}^-OCCH_2CH_2CO^-} + FAD \xrightarrow{\text{succinate dehydrogenase}} \underset{\text{fumarate}}{{}^-OOCCH{=}CHCOO^-} + FADH_2$$

$$NADH + H^+ + FMN \xrightarrow{\text{NADH dehydrogenase}} NAD^+ + FMNH_2$$

Mechanisms for Flavin Nucleotide Coenzymes

When FAD (or FMN) oxidizes a substrate (S), the coenzyme is reduced to $FADH_2$ (or $FMNH_2$). $FADH_2$ and $FMNH_2$, like NADH and NADPH, are reducing agents. All the oxidation–reduction chemistry takes place on the flavin ring. Reduction of the flavin ring disrupts the conjugated system, so the reduced coenzymes are less colored than their oxidized forms.

FAD and FMN are oxidizing agents.

$FADH_2$ and $FMNH_2$ are reducing agents.

$$\underset{\text{FAD / FMN}}{\text{flavin}} + S_{red} \longrightarrow \underset{FADH_2 / FMNH_2}{\text{reduced flavin}} + S_{ox}$$

PROBLEM 3◆

How many conjugated double bonds are there in

a. FAD?

b. $FADH_2$?

In the first step of the mechanism for the FAD-catalyzed oxidation of dihydrolipoate to lipoate, the thiolate ion attacks the C-4a position of the flavin ring. This reaction is general-acid catalyzed—as the thiolate ion attacks the ring, a proton is donated to the N-5 nitrogen. A second nucleophilic attack by a thiolate ion, this time on the sulfur that is covalently attached to the coenzyme, generates the oxidized product and $FADH_2$. Section 25.4 discusses where this FAD-catalyzed reaction fits into metabolism.

mechanism for dihydrolipoyl dehydrogenase

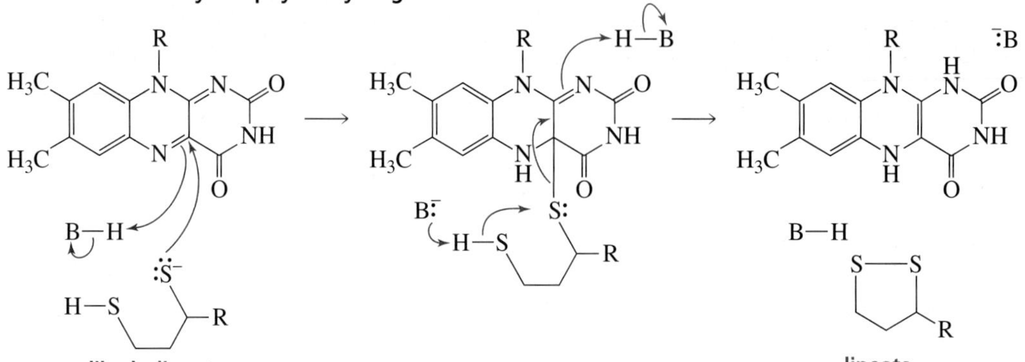

dihydrolipoate

lipoate

In the first step of the flavin-catalyzed oxidation of an amino acid to an imino acid, a basic side chain at the active site of the enzyme removes a proton from the α-carbon of the amino acid. The carbanion that is formed attacks the N-5 position of the flavin ring. Collapse of the resulting tetrahedral intermediate gives the oxidized amino acid (an imino acid) and the reduced coenzyme ($FADH_2$).

mechanism for D- or L-amino acid oxidase

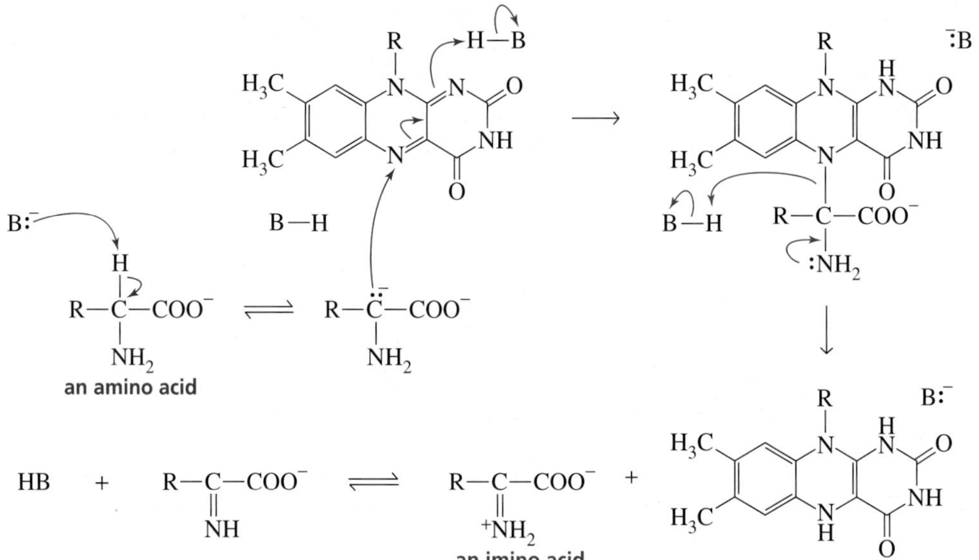

an amino acid

an imino acid

Notice that FAD is a more versatile coenzyme than NAD^+. Unlike NAD^+, which always uses the same mechanism, flavin coenzymes can use several different mechanisms to carry out an oxidation. For example, we have just seen that when FAD oxidizes dihydrolipoate, nucleophilic attack occurs on the C-4a position of the flavin ring, but when it oxidizes an amino acid, nucleophilic attack occurs on the N-5 position.

Cells contain very low concentrations of FAD and much higher concentrations of NAD^+. This difference in concentration is responsible for a significant difference in the enzymes (**E**) that use NAD^+ as an oxidizing agent and those that use FAD. Generally, FAD is covalently bound to its enzyme and remains bound after being reduced to $FADH_2$. Unlike NADH, $FADH_2$ does not dissociate from the enzyme. It, therefore, must be reoxidized to FAD before the enzyme can enter another round of catalysis. The oxidizing agent used for this reaction is NAD^+ or O_2. Therefore, an enzyme that uses an oxidizing coenzyme other than NAD^+ may still require NAD^+ to reoxidize the reduced coenzyme. For this reason, NAD^+ has been called the "common currency" of biological oxidation–reduction reactions.

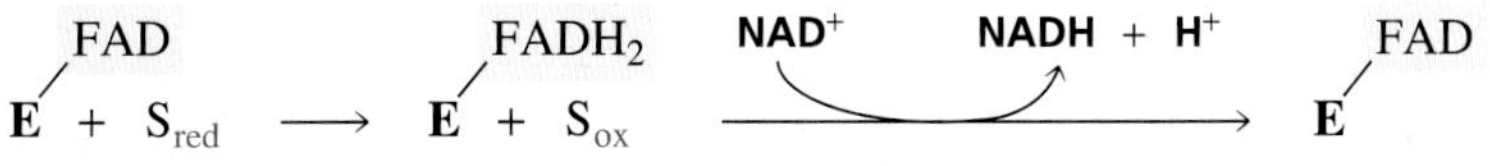

$$\overset{\text{FAD}}{\text{E}} + \text{S}_{\text{red}} \longrightarrow \overset{\text{FADH}_2}{\text{E}} + \text{S}_{\text{ox}} \xrightarrow{\text{NAD}^+ \;\; \text{NADH} + \text{H}^+} \overset{\text{FAD}}{\text{E}}$$

PROBLEM 4

Propose a mechanism for the reduction of lipoate by $FADH_2$.

PROBLEM 5 SOLVED

A common way for FAD to become covalently bound to its enzyme is by having a proton removed from the C-8 methyl group and a proton donated to N-1. Then a cysteine or other nucleophilic side chain of the enzyme attacks the methylene carbon at C-8 as a proton is donated to N-5. Describe these events mechanistically.

SOLUTION

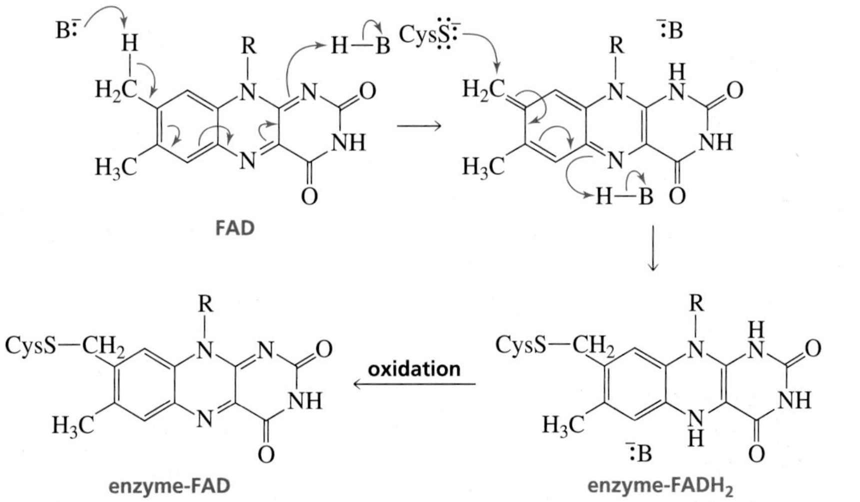

FAD

enzyme-FAD

oxidation

enzyme-$FADH_2$

Notice that during the process of attaching FAD to the enzyme, FAD is reduced to $FADH_2$. It is subsequently oxidized back to FAD. Once the coenzyme is attached to the enzyme, it does not come off.

PROBLEM 6

Explain why the hydrogens of the methyl group bonded to flavin at C-8 are more acidic than those of the methyl group bonded at C-7.

25.4 Thiamine Pyrophosphate: Vitamin B_1

Thiamine was the first of the B vitamins to be identified, so it became known as vitamin B_1. The absence of thiamine in the diet causes a disease called beriberi, which damages the heart and impairs nerve reflexes. In extreme cases, it causes paralysis. One major dietary source of vitamin B_1 is the hulls of rice kernels (p. 1035). A deficiency is therefore most likely to occur when highly polished rice is a major component of the diet. A deficiency is also seen in alcoholics who are severely malnourished.

Thiamine pyrophosphate (TPP) is required by enzymes that catalyze the transfer of a two-carbon fragment from one species to another.

The coenzyme form of vitamin B_1 is **thiamine pyrophosphate (TPP)**. TPP is the coenzyme required by enzymes that catalyze the transfer of a two-carbon fragment from one species to another.

thiamine pyrophosphate
TPP

Pyruvate decarboxylase is an enzyme that requires thiamine pyrophosphate. Pyruvate decarboxylate catalyzes the decarboxylation of pyruvate and transfers the resulting two-carbon fragment to a proton, resulting in the formation of acetaldehyde.

$$CH_3-\overset{O}{\overset{\|}{C}}-\overset{O}{\overset{\|}{C}}-O^- + H^+ \xrightarrow{\text{pyruvate decarboxylase}} CH_3-\overset{O}{\overset{\|}{C}}-H + CO_2$$

pyruvate
an α-keto acid

acetaldehyde

You may wonder why an α-keto acid such as pyruvate can be decarboxylated, since the electrons left behind when CO_2 is removed cannot be delocalized onto the carbonyl oxygen. We will see that the thiazolium ring of the coenzyme provides a site for the delocalization of the electrons. A site to which electrons can be delocalized is called an **electron sink**.

Mechanism for Pyruvate Decarboxylase

The hydrogen bonded to the imine carbon of TPP is relatively acidic ($pK_a = 12.7$) because the ylide carbanion formed when the proton is removed is stabilized by the adjacent positively charged nitrogen. The ylide carbanion is a good nucleophile.

$$\text{thiazolium ring} \xrightleftharpoons{pK_a = 12.7} \text{ylide carbanion} + H^+$$

thiazolium ring **ylide carbanion**

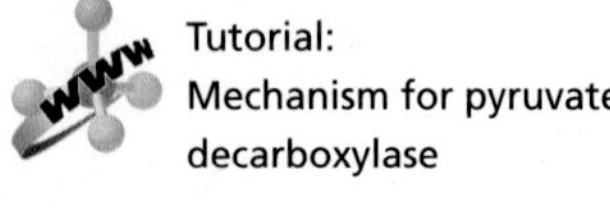
Tutorial:
Mechanism for pyruvate decarboxylase

In the first step of the mechanism for pyruvate decarboxylase, the nucleophilic carbanion attacks the electrophilic ketone group of the α-keto acid. The resulting intermediate can easily undergo decarboxylation because the electrons left behind when CO_2 is removed can be delocalized onto the positively charged nitrogen. The positively charged nitrogen of TPP is a more effective *electron sink* than the β-keto group of a β-keto acid, a class of compounds that are fairly easily decarboxylated (Section 19.17). The product of decarboxylation is stabilized by electron delocalization. One of the resonance contributors is neutral and the other has separated charges—we will call it a *resonance-stabilized carbanion*. Protonation of the resonance-stabilized carbanion and a subsequent elimination reaction form acetaldehyde and regenerate the coenzyme.

mechanism for pyruvate decarboxylase

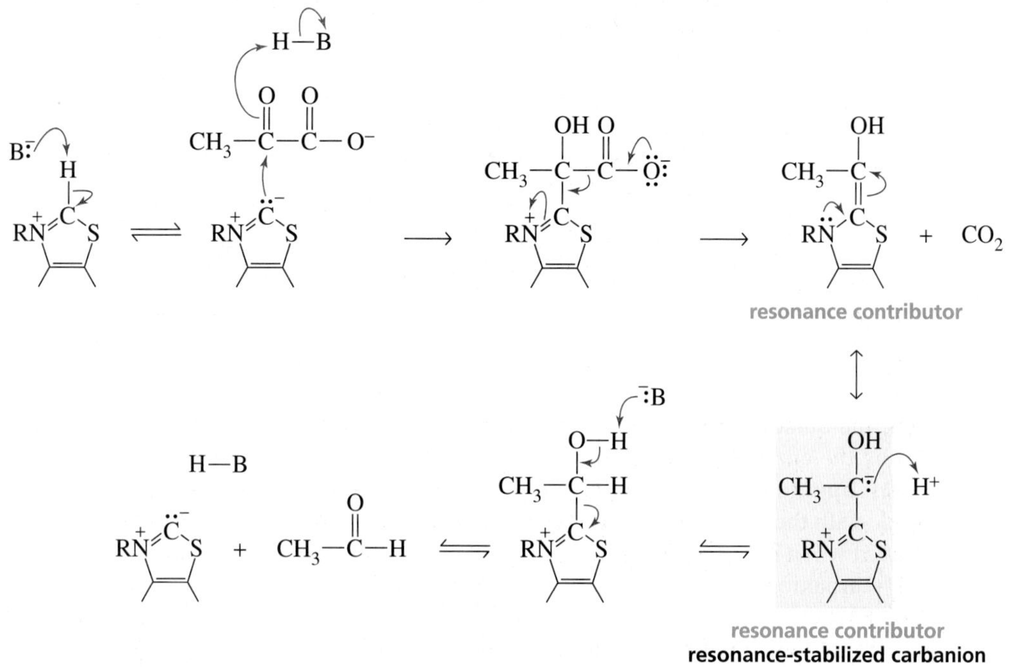

resonance contributor

resonance contributor

resonance-stabilized carbanion

PROBLEM 7

Draw structures that show the similarity between decarboxylation of the pyruvate–TPP intermediate and decarboxylation of a β-keto acid.

PROBLEM 8

Acetolactate synthase is another TPP-requiring enzyme. It also catalyzes the decarboxylation of pyruvate but transfers the resulting two-carbon fragment to another molecule of pyruvate, forming acetolactate. This is the first step in the biosynthesis of the amino acids valine and leucine. Propose a mechanism for acetolactate synthase.

$$2\,CH_3-\overset{O}{\overset{\|}{C}}-\overset{O}{\overset{\|}{C}}-O^- \xrightarrow[\text{TPP}]{\text{acetolactate synthase}} CH_3-\overset{O}{\overset{\|}{C}}-\underset{CH_3}{\underset{|}{\overset{OH}{\overset{|}{C}}}}-\overset{O}{\overset{\|}{C}}-O^- + CO_2$$

pyruvate — acetolactate

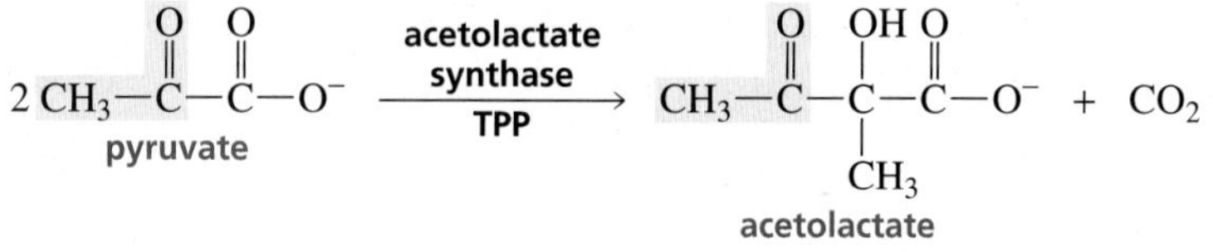

PROBLEM 9

Acetolactate synthase can also transfer the two-carbon fragment from pyruvate to α-ketobutyrate, forming α-aceto-α-hydroxybutyrate. This is the first step in the formation of isoleucine. Propose a mechanism for this reaction.

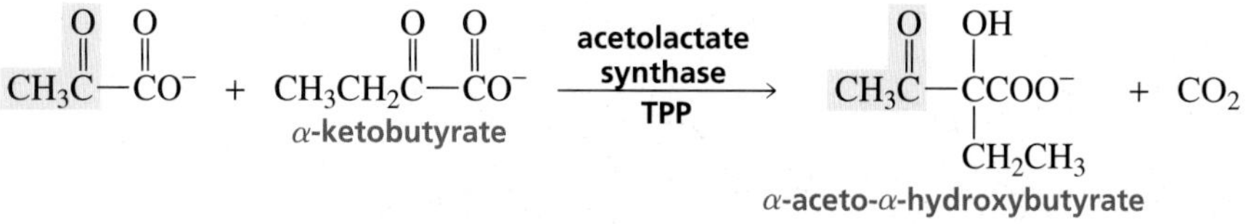

$$CH_3\overset{O}{\overset{\|}{C}}-\overset{O}{\overset{\|}{C}}O^- + CH_3CH_2\overset{O}{\overset{\|}{C}}-\overset{O}{\overset{\|}{C}}O^- \xrightarrow[\text{TPP}]{\text{acetolactate synthase}} CH_3\overset{O}{\overset{\|}{C}}-\underset{CH_2CH_3}{\underset{|}{\overset{OH}{\overset{|}{C}}}}COO^- + CO_2$$

α-ketobutyrate — α-aceto-α-hydroxybutyrate

Mechanism for the Pyruvate Dehydrogenase System

A compound must enter the citric acid cycle to be completely metabolized (Figure 25.1). Thus, fats, carbohydrates, and proteins must be converted into compounds that are part of the citric acid cycle or that can enter the cycle. Acetyl-CoA is the only compound capable of entering the citric acid cycle (Figure 25.3). The final product of carbohydrate metabolism is pyruvate (Figure 25.2). To enter the carboxylic acid cycle, pyruvate must be converted either to acetyl-CoA or to oxaloacetate (Section 25.5)—a citric acid cycle intermediate.

The *pyruvate dehydrogenase system* is a group of three enzymes responsible for the conversion of pyruvate to acetyl-CoA. The pyruvate dehydrogenase system requires TPP and four other coenzymes: lipoate, coenzyme A, FAD, and NAD^+.

$$\underset{\text{pyruvate}}{CH_3-\overset{O}{\overset{\|}{C}}-\overset{O}{\overset{\|}{C}}-O^-} + CoASH \xrightarrow[\text{system}]{\text{pyruvate dehydrogenase}} \underset{\text{acetyl-CoA}}{CH_3-\overset{O}{\overset{\|}{C}}-SCoA} + CO_2$$

The first enzyme in the system catalyzes the reaction of TPP with pyruvate to form the same resonance-stabilized carbanion formed by pyruvate decarboxylase and by the enzyme in Problems 8 and 9. The second enzyme of the system (**E_2**) requires **lipoate**, a coenzyme that becomes attached to its enzyme by forming an amide with the amino group of a lysine side chain. The disulfide linkage of lipoate is cleaved when it undergoes nucleophilic attack by the carbanion. In the next step, the TPP carbanion is eliminated from the tetrahedral intermediate. **Coenzyme A (CoASH)** reacts with the thioester in a transthioesterification reaction (one thioester is converted into another), substituting coenzyme A for dihydrolipoate. At this point, the final reaction product (acetyl-CoA) has been formed. However, before another catalytic cycle can occur, dihydrolipoate must be oxidized back to lipoate. This is done by the third enzyme (**E_3**), an FAD-requiring enzyme (Section 25.3). Oxidation of dihydrolipoate by FAD forms enzyme-bound $FADH_2$. NAD^+ then oxidizes $FADH_2$ back to FAD.

mechanism for the pyruvate dehydrogenase system

resonance-stabilized carbanion + lipoate–$C(=O)NH(CH_2)_4\mathbf{E_2}$ ⟶ tetrahedral intermediate (H–B, :B)

⟶ $CH_3-C(=O)-S$ … SH–$C(=O)NH(CH_2)_4\mathbf{E_2}$ + TPP carbanion

$\xrightarrow{\text{CoASH}}$ $CH_3-C(=O)-SCoA$ (acetyl-CoA) + SH SH dihydrolipoate–$C(=O)NH(CH_2)_4\mathbf{E_2}$

$\xrightarrow{\text{FAD}-\mathbf{E_3}}$ S–S lipoate–$C(=O)NH(CH_2)_4\mathbf{E_2}$ + $FADH_2-\mathbf{E_3}$ $\xrightarrow{NAD^+ \;\; NADH + H^+}$ $FAD-\mathbf{E_3}$

3-D Molecule: Coenzyme A

The vitamin precursor of CoASH is pantothenate. We have seen that CoASH is used in biological systems to activate carboxylic acids by converting them into thioesters, which are much more reactive toward nucleophilic acyl substitution reactions than are carboxylic acids (Section 17.20). At physiological pH (7.3), a carboxylic acid would be present in its negatively charged basic form, which could not be approached by a nucleophile.

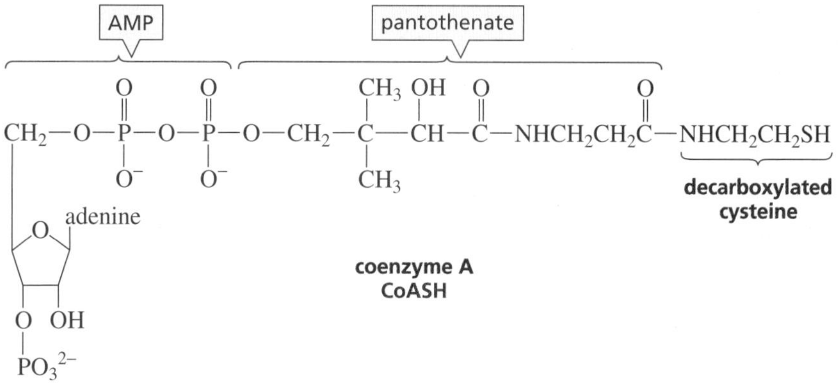

AMP | pantothenate

$CH_2{-}O{-}P(=O)(O^-){-}O{-}P(=O)(O^-){-}O{-}CH_2{-}C(CH_3)_2{-}CH(OH){-}C(=O){-}NHCH_2CH_2C(=O){-}NHCH_2CH_2SH$

adenine

O OH

PO_3^{2-}

decarboxylated cysteine

coenzyme A
CoASH

PROBLEM 10 SOLVED

TPP is a coenzyme for transketolase, the enzyme that catalyzes the conversion of a ketopentose (xylulose-5-phosphate) and an aldopentose (ribose-5-phosphate) into an aldotriose (glyceraldehyde-3-phosphate) and a ketoheptose (sedoheptulose-7-phosphate). Notice that the total number of carbon atoms in the reactants and products does not change (5 + 5 = 3 + 7). Propose a mechanism for this reaction.

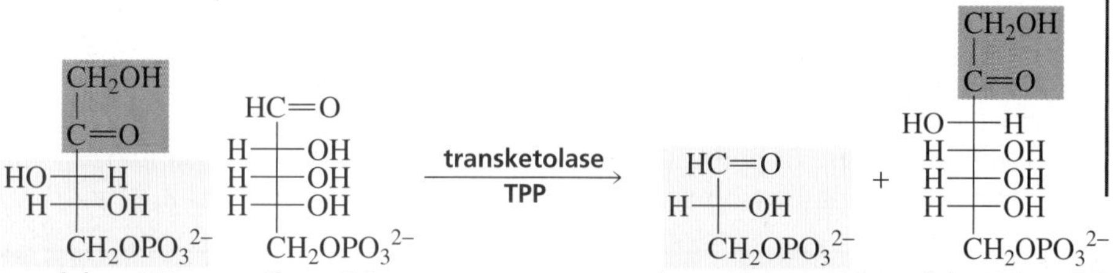

xylulose-5-P + **ribose-5-P** $\xrightarrow[\text{TPP}]{\text{transketolase}}$ **glyceraldehyde-3-P** + **sedoheptulose-7-P**

SOLUTION The reaction shows that a two-carbon fragment is transferred from xylulose-5-phosphate to ribose-5-phosphate. Because TPP transfers two-carbon fragments, we know that TPP must remove the two-carbon fragment that is to be transferred from xylulose-5-phosphate. Thus, the reaction must start by TPP attacking the carbonyl group of xylulose-5-phosphate. We can add an acid group to accept the electrons from the carbonyl group and a basic group to aid in the removal of the two-carbon fragment. The two-carbon fragment, that becomes attached to TPP, is a resonance-stabilized carbanion that adds to the carbonyl group of ribose-5-phosphate. Again, an acid group accepts the electrons from the carbonyl group, and a basic group aids in the elimination of TPP.

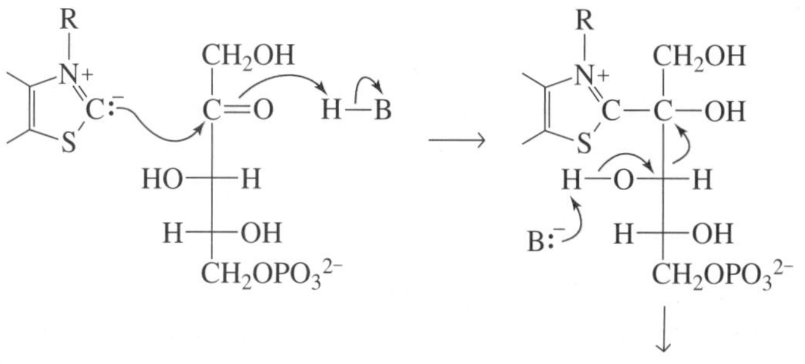

H—B B:

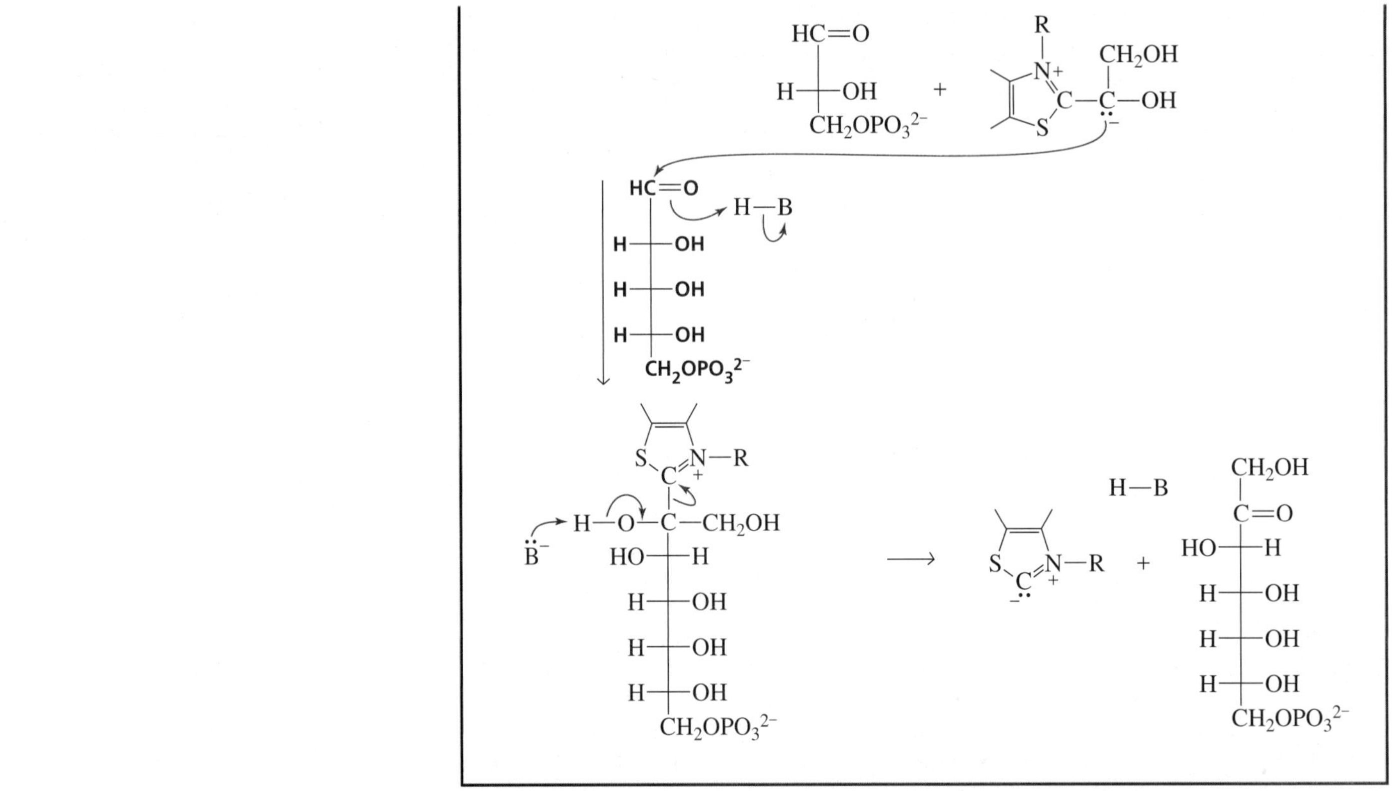

Notice the similar function of TPP in all TPP-requiring enzymes. In each reaction, the nucleophilic coenzyme attacks a carbonyl group of the substrate and allows a bond to that carbonyl group to be broken because the electrons left behind can be delocalized into the thiazolium ring. The resulting two-carbon fragment is then transferred—to a proton in the case of pyruvate decarboxylase, to coenzyme A (via lipoate) in the pyruvate dehydrogenase system, and to a carbonyl group in Problems 8, 9, and 10.

PROBLEM 11

An unfortunate effect of drinking too much alcohol—a hangover—is attributable to the acetaldehyde formed when ethanol is oxidized. There is some evidence that vitamin B_1 can cure a hangover. How can the vitamin do this?

25.5 Biotin: Vitamin H

Biotin (vitamin H) is an unusual vitamin because it can be synthesized by bacteria that live in the intestine. Consequently, biotin does not have to be included in our diet and deficiencies are rare. Biotin deficiencies, however, can be found in people who maintain a diet high in raw eggs. Egg whites contain a protein (avidin) that binds biotin tightly and thereby prevents it from acting as a coenzyme. When eggs are cooked, avidin is denatured, and the denatured protein does not bind biotin. Biotin is attached to its enzyme by forming an amide with the amino group of a lysine side chain.

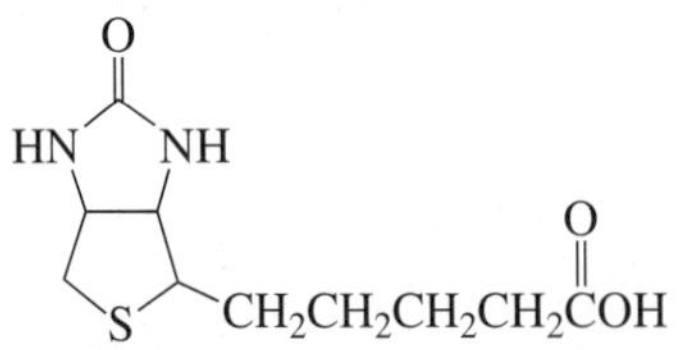

$CH_2CH_2CH_2CH_2COOH$

biotin

$CH_2CH_2CH_2CH_2CONH(CH_2)_4$**E**

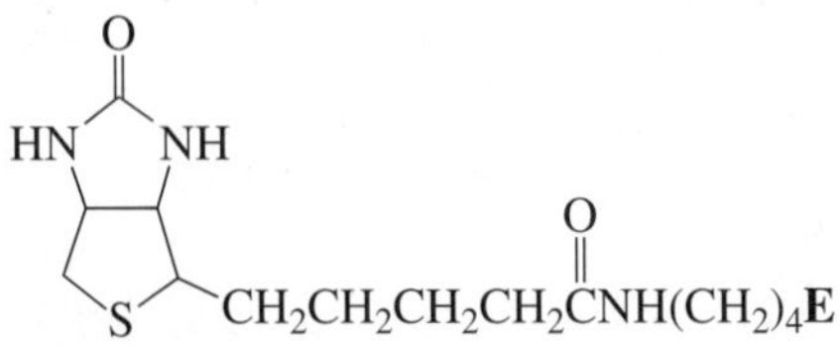

enzyme-bound biotin

Biotin is the coenzyme required by enzymes that catalyze carboxylation of a carbon adjacent to a carbonyl group. For example, pyruvate carboxylase converts pyruvate—the end product of carbohydrate metabolism—to oxaloacetate, a citric acid cycle intermediate (Figure 25.2). Acetyl-CoA carboxylase converts acetyl-CoA into malonyl-CoA, one of the reactions in the anabolic pathway that converts acetyl-CoA into fatty acids (Section 19.21). Biotin-requiring enzymes use bicarbonate (HCO_3^-) for the source of the carboxyl group that becomes attached to the substrate.

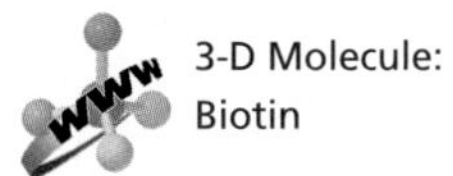

Biotin is required by enzymes that catalyze the carboxylation of a carbon adjacent to a carbonyl group.

$$CH_3COCOO^- + HCO_3^- + ATP \xrightarrow[\text{Mg}^{2+},\ \text{biotin}]{\text{pyruvate carboxylase}} {}^-OOCCH_2COCOO^- + ADP + HPO_4^{2-}$$

pyruvate → oxaloacetate

$$CH_3COSCoA + HCO_3^- + ATP \xrightarrow[\text{Mg}^{2+},\ \text{biotin}]{\text{acetyl-CoA carboxylase}} {}^-OOCCH_2COSCoA + ADP + HPO_4^{2-}$$

acetyl-CoA → malonyl-CoA

Mechanism for Biotin

In addition to requiring bicarbonate, biotin-requiring enzymes require Mg^{2+} and ATP. The function of Mg^{2+} is to decrease the overall negative charge on ATP by complexing with two of its negatively charged oxygens (Section 27.5). Unless its negative charge is reduced, ATP cannot be approached by a nucleophile. The function of ATP is to increase the reactivity of bicarbonate by converting it to "activated bicarbonate," a compound with a good leaving group (Section 27.2). Notice that "activated bicarbonate" is a mixed anhydride of carbonic acid and phosphoric acid (Section 17.1).

bicarbonate + ATP (Mg^{2+} complexed, adenosine) → activated bicarbonate (Mg^{2+} complexed) + ADP

Nucleophilic attack by biotin on activated bicarbonate forms carboxybiotin. Because the nitrogen of an amide is not nucleophilic, it is likely that the active form of biotin has an enolate-like structure. Nucleophilic attack by the substrate (in this case, the enolate form of acetyl-CoA) on carboxybiotin transfers the carboxyl group from biotin to the substrate.

activated bicarbonate + "enolate-like" structure of enzyme-bound biotin [S, $(CH_2)_4CNH(CH_2)_4$**E**] → tetrahedral intermediate (PO_3^{2-}, Mg^{2+})

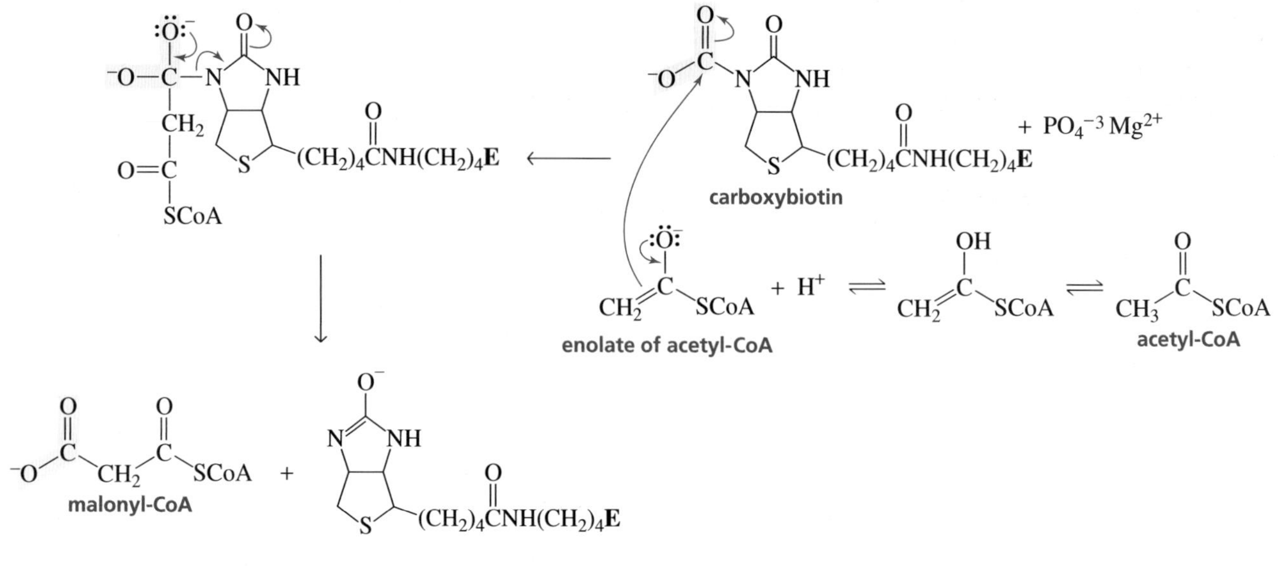

carboxybiotin

enolate of acetyl-CoA

acetyl-CoA

malonyl-CoA

All biotin-requiring enzymes follow the same three steps: activation of bicarbonate by ATP, reaction of activated bicarbonate with biotin to form carboxybiotin, and transfer of the carboxyl group from carboxybiotin to the substrate.

25.6 Pyridoxal Phosphate: Vitamin B_6

Pyridoxal phosphate (PLP) is the coenzyme required by enzymes that catalyze certain transformations of amino acids. The coenzyme is derived from pyridoxine, the vitamin known as vitamin B_6. The name "pyridox*al*" indicates that the coenzyme is a pyridine aldehyde. A deficiency in vitamin B_6 causes anemia; severe deficiencies can cause seizures and death.

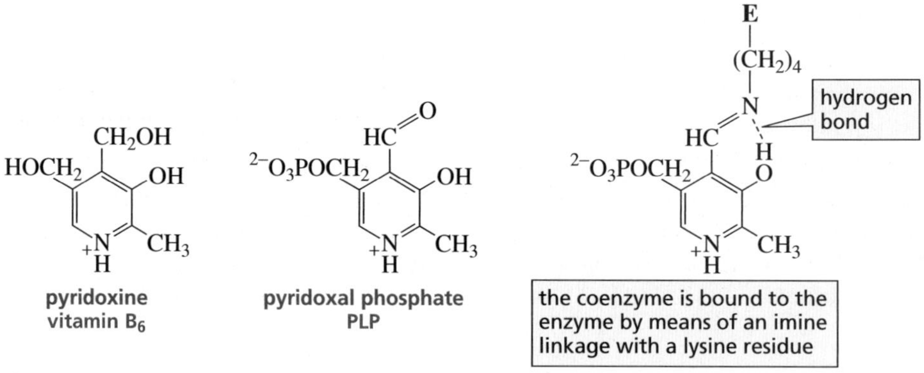

pyridoxine
vitamin B_6

pyridoxal phosphate
PLP

hydrogen bond

the coenzyme is bound to the enzyme by means of an imine linkage with a lysine residue

Pyridoxal phosphate (PLP) is required by enzymes that catalyze certain transformations of amino acids.

Several different kinds of amino acid transformations are catalyzed by PLP-requiring enzymes. The most common are decarboxylation, transamination, racemization (the interconversion of L and D amino acids), C_α C_β bond cleavage, and α,β-elimination.

decarboxylation

$$RCH(\overset{+}{N}H_3)-\overset{O}{\overset{\|}{C}}-O^- \xrightarrow[\text{PLP}]{\text{E}} RCH_2\overset{+}{N}H_3 + CO_2$$

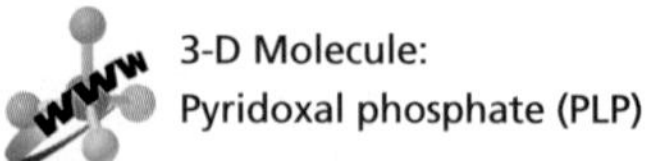

3-D Molecule:
Pyridoxal phosphate (PLP)

transamination

$$\underset{{}^{+}NH_3}{RCHCO^-}(C{=}O) + {}^-OCCH_2CH_2CCO^- \xrightarrow[\text{PLP}]{\text{E}} RCCO^- + {}^-OCCH_2CH_2CHCO^-$$

$$\text{RCH(}{}^{+}\text{NH}_3\text{)C(=O)O}^- + {}^-\text{O(O=)CCH}_2\text{CH}_2\text{C(=O)C(=O)O}^- \xrightarrow[\text{PLP}]{\text{E}} \text{RC(=O)C(=O)O}^- + {}^-\text{O(O=)CCH}_2\text{CH}_2\text{CH(}{}^{+}\text{NH}_3\text{)C(=O)O}^-$$

α-ketoglutarate — glutamate

racemization

$$\text{R–C(H)(}{}^{+}\text{NH}_3\text{)–C(=O)–O}^- \xrightarrow[\text{PLP}]{\text{E}} \text{R–C(H)(}{}^{+}\text{NH}_3\text{)–C(=O)–O}^- + \text{R–C(}{}^{+}\text{NH}_3\text{)(H)–C(=O)–O}^-$$

L-amino acid — L-amino acid — D-amino acid

C_α—C_β bond cleavage

$$\text{HOCH(R)CH(}{}^{+}\text{NH}_3\text{)C(=O)O}^- \xrightarrow[\text{PLP}]{\text{E}} \text{O=CH(R)} + \text{CH}_2\text{(}{}^{+}\text{NH}_3\text{)C(=O)O}^-$$

***α*,*β*-elimination**

$$\text{XCH}_2\text{CH(}{}^{+}\text{NH}_3\text{)C(=O)O}^- \xrightarrow[\text{PLP}]{\text{E}} \text{CH}_3\text{C(=O)C(=O)O}^- + \text{X}^- + \overset{+}{\text{N}}\text{H}_4$$

In each of these transformations, one of the bonds to the *α*-carbon of the amino acid substrate is broken in the first step of the reaction. Decarboxylation breaks the bond joining the carboxyl group to the *α*-carbon; transamination, racemization, and *α*,*β*-elimination break the bond joining the hydrogen to the *α*-carbon; and C_α—C_β bond cleavage breaks the bond joining the R group to the *α*-carbon.

$$\text{NH}_2\text{–C(COO}^-\text{)(R)–H}$$

bond broken in decarboxylation

bond broken in C_α—C_β bond cleavage

bond broken in transamination, racemization, and *α*,*β*-elimination

PLP becomes attached to its enzyme by forming an imine with the amino group of a lysine side chain. The first step in all PLP-requiring enzymes is a **transimination** reaction, a reaction in which one imine is converted into another imine. In the transimination reaction, the amino acid substrate reacts with the imine formed by *PLP and the enzyme*, forming a tetrahedral intermediate. The lysine group of the enzyme is expelled, forming a new imine between *PLP and the amino acid*.

transimination

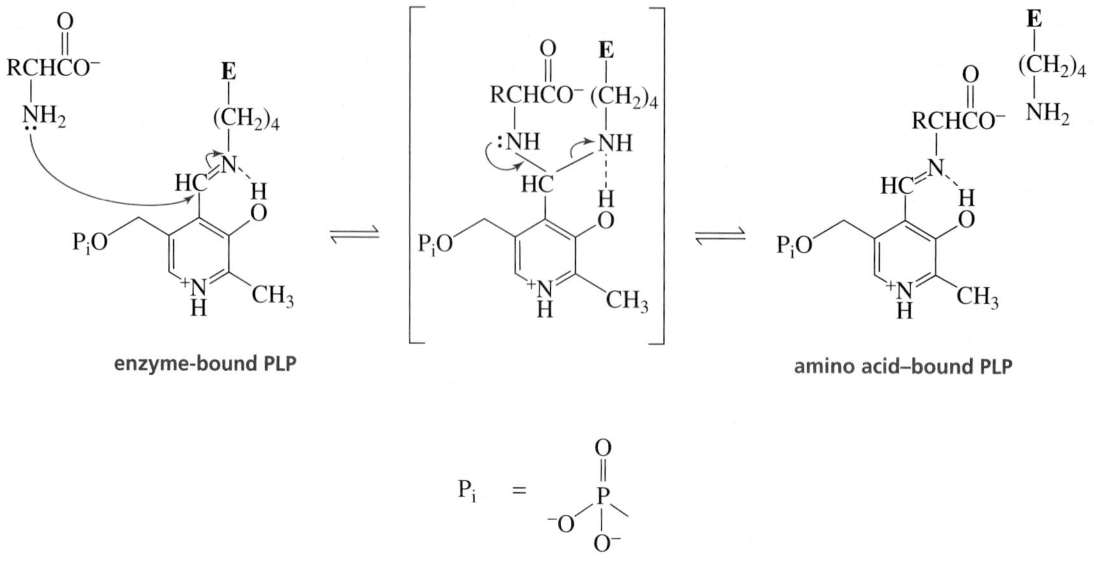

enzyme-bound PLP

amino acid–bound PLP

P_i = $^{-}O-P(=O)(O^{-})-$

Once the amino acid has formed an imine with PLP, a bond to the α-carbon can be broken because the electrons left behind when the bond breaks can be delocalized onto the positively charged protonated nitrogen of the pyridine ring. In other words, the protonated nitrogen of the pyridine ring is an electron sink. If the OH group is removed from the pyridine ring, the cofactor loses much of its activity. Apparently, the hydrogen bond formed by the OH group helps weaken the bond to the α-carbon.

Mechanism for Decarboxylation

If the PLP-catalyzed reaction is a decarboxylation, the carboxyl group is removed from the α-carbon of the amino acid. Electron rearrangement and protonation of the α-carbon of the decarboxylated intermediate by a protonated amino group of a lysine side chain or by some other acid group then reestablishes the aromaticity of the pyridine ring. The last step in all PLP-requiring enzymes is a transimination reaction with a lysine side chain, in order to release the product of the enzyme-catalyzed reaction and regenerate enzyme-bound PLP.

mechanism for PLP-catalyzed decarboxylation of an amino acid

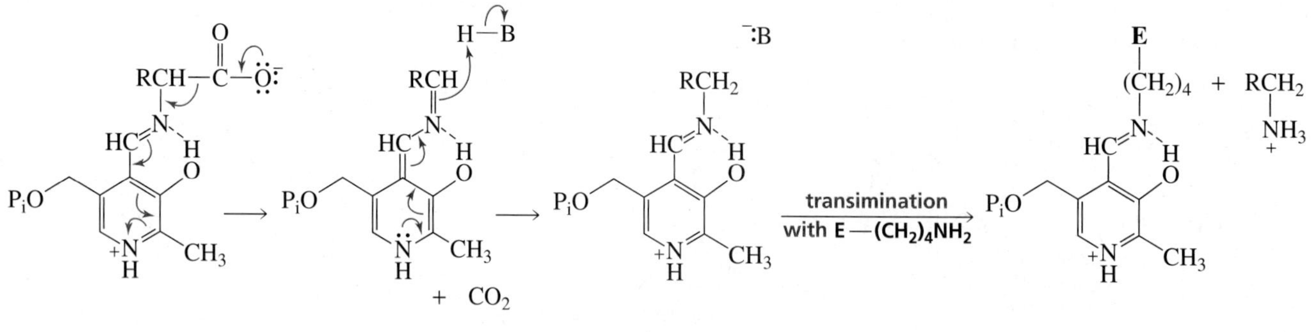

+ CO_2

transimination with E—$(CH_2)_4NH_2$

$(CH_2)_4$ + $RCH_2NH_3^+$

Mechanism for Transamination

The first reaction in the catabolism of most amino acids is replacement of the amino group of the amino acid by a ketone group. This is called a **transamination** reaction because the amino group removed from the amino acid is not lost, but is *transferred* to the ketone group of α-ketoglutarate, thereby forming glutamate. The enzymes that catalyze transamination reactions are called *aminotransferases*. Transamination allows the amino groups of the various amino acids to be collected into a single amino acid (glutamate) so that excess nitrogen can be easily excreted. (Do not confuse *transamination* with *transimination*, discussed previously.)

In the first step of transamination, a proton is removed from the α-carbon of the amino acid bound to PLP. Rearrangement of the electrons and protonation of the carbon attached to the pyridine ring, followed by hydrolysis of the imine, forms the α-keto acid and pyridoxamine. At this point, the amino group has been removed from the amino acid, but pyridoxamine has to be converted back to enzyme-bound PLP before another round of catalysis can occur. Pyridoxamine forms an imine with α-ketoglutarate, the second substrate of the reaction. Removal of a proton from the carbon attached to the pyridine ring, followed by rearrangement of the electrons and donation of a proton to the α-carbon of the substrate forms an imine that, when transiminated with a lysine side chain, releases glutamate and reforms enzyme-bound PLP.

Notice that the proton transfer steps are reversed in the two phases of the reaction. Transfer of the amino group of the amino acid to pyridoxal requires removal of the proton from the α-carbon and donation of a proton to the carbon bonded to the pyridine ring. Transfer of the amino group of pyridoxamine to α-ketoglutarate requires removal of the proton from the carbon bonded to the pyridine ring and donation of a proton to the α-carbon.

mechanism for PLP-catalyzed transamination of an amino acid

H_2O

pyridoxamine

transaminated amino acid
an α-keto acid

$^-OCCH_2CH_2CCO^-$

α-ketoglutarate

$+ H_2O$

transimination with $E—(CH_2)_4NH_2$

glutamate

HEART ATTACKS: ASSESSING THE DAMAGE

Damage to heart muscle after a myocardial infarction (a heart attack) allows aminotransferases and other enzymes to leak from the damaged cells of the heart into the bloodstream. Following a heart attack, the severity of damage done to the heart can be determined from the concentrations of alanine aminotransferase and aspartate aminotransferase in the bloodstream.

PHENYLKETONURIA: AN INBORN ERROR OF METABOLISM

Tyrosine is a nonessential amino acid because the body can make it by hydroxylating phenylalanine. About 1 in every 20,000 babies, however, is born without phenylalanine hydroxylase, the enzyme that converts phenylalanine into tyrosine. This genetic disease is called phenylketonuria (PKU). Without phenylalanine hydroxylase, the level of phenylalanine builds up, and when it reaches a high concentration, it is transaminated to phenylpyruvate. The high level of phenylpyruvate found in urine gave the disease its name.

phenylalanine ($C_6H_5CH_2CH(^+NH_3)COO^-$) —phenylalanine hydroxylase→ tyrosine ($HO-C_6H_4-CH_2CH(^+NH_3)COO^-$) —tyrosine aminotransferase→ *para*-hydroxyphenylpyruvate ($HO-C_6H_4-CH_2C(=O)COO^-$)

phenylalanine → phenylpyruvate ($C_6H_5CH_2C(=O)COO^-$)

tyrosine → dihydroxyphenylalanine, dopa ($(HO)_2C_6H_3-CH_2CH(^+NH_3)COO^-$)

dopa → melanin

dopa → dopamine ($(HO)_2C_6H_3-CH_2CH_2\overset{+}{N}H_3$) → norepinephrine, noradrenaline ($(HO)_2C_6H_3-CH(OH)CH_2\overset{+}{N}H_3$) → epinephrine, adrenaline ($(HO)_2C_6H_3-CH(OH)CH_2\overset{+}{N}H_2CH_3$)

para-hydroxyphenylpyruvate —*para*-hydroxyphenylpyruvate dioxygenase→ homogentisate ($(HO)_2C_6H_3-CH_2COO^-$) —homogentisate dioxygenase→ $H-C(COO^-)=C(H)-C(=O)CH_2C(=O)CH_2COO^-$ → → fumarate + acetyl-CoA

About 3 days after they begin drinking milk, all babies born in the United States are tested for high serum phenylalanine levels, which would indicate a buildup of phenylalanine. If a baby is found to lack phenylalanine hydroxylase, he or she is immediately put on a diet low in phenylalanine and high in tyrosine. As long as the phenylalanine level is kept under careful control for the first 5 to 10 years of life, the baby will experience no adverse effects.

If the diet is not controlled, however, the baby will be severely mentally retarded by the time he or she is a few months old. Untreated children have paler skin and fairer hair than other members of their family because they don't synthesize tyrosine, so their melanin levels are low. Melanin is the precursor of the compound responsible for skin pigmentation. Melanin is formed from dopa, which in turn is formed from the transamination of tyrosine. Half of untreated phenylketonurics are dead by age 20. When a woman with PKU becomes pregnant, she must return to the low phenylalanine diet she had as a child, because a high level of phenylalanine can cause abnormal development of the fetus.

Another genetic disease that results from a deficiency of an enzyme in the pathway for phenylalanine degradation is alcaptonuria, which is caused by lack of homogentisate dioxygenase. The only ill effect of this enzyme deficiency is black urine. The urine of those afflicted with alcaptonuria turns black because the homogentisate they excrete immediately oxidizes in the air.

PROBLEM 12◆

Taking into account the pathway for phenylalanine degradation, answer the following questions:

a. What coenzyme and what other organic compound are needed by tyrosine aminotransferase?
b. What bond in homogentisate is oxidized by homogentisate dioxygenase? (*Hint:* Keto-enol tautomerism occurs after oxidation.)
c. What compound is used to supply the methyl group needed to convert noradrenaline into adrenaline? (*Hint:* See Section 10.11.)

PROBLEM 13◆

α-Keto acids other than α-ketoglutarate can be used to accept the amino group from pyridoxamine in enzyme-catalyzed transaminations. What amino acids are formed from the following α-keto acids?

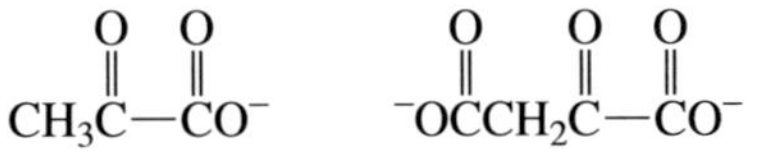

$$CH_3\overset{O}{\overset{\|}{C}}-\overset{O}{\overset{\|}{C}}O^- \qquad {}^-O\overset{O}{\overset{\|}{C}}CH_2\overset{O}{\overset{\|}{C}}-\overset{O}{\overset{\|}{C}}O^-$$

pyruvate oxaloacetate

Mechanism for Racemization

The first step in the PLP-catalyzed racemization of an L-amino acid is the same as the first step in the PLP-catalyzed transamination of an amino acid—removal of a proton from the α-carbon of the amino acid bound to PLP. In the second step of racemization, reprotonation occurs on the α-carbon. The proton can be donated to the sp^2 hybridized α-carbon from either side of the plane defined by the double bond. Consequently, both D- and L-amino acids are formed. In other words, the L-amino acid is racemized.

mechanism for PLP-catalyzed racemization of an L-amino acid

B:⁻ H O R—C—CO⁻ → B—H RC—CO⁻ → B:⁻ RCH—CO⁻ —transimination with E—$(CH_2)_4NH_2$→ E $(CH_2)_4$ + $RCHCO^-$ $^+NH_3$

racemized amino acid

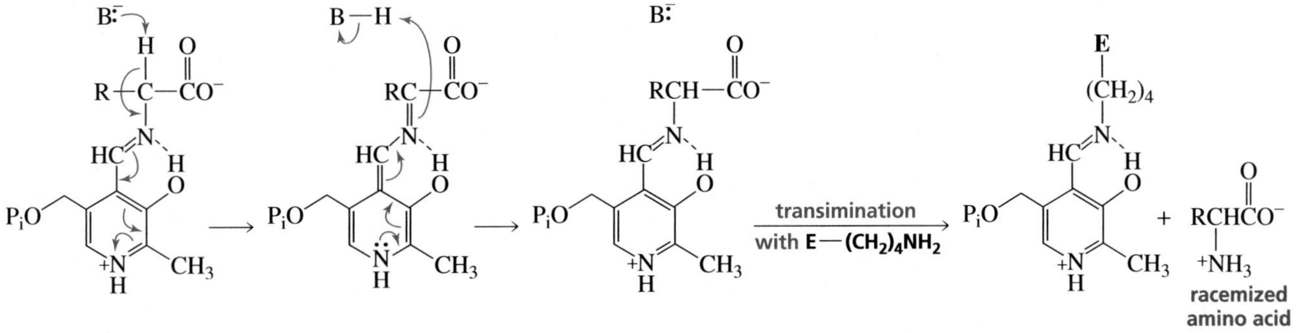

Compare the second step in a PLP-catalyzed transamination with the second step in a PLP-catalyzed racemization. In an enzyme that catalyzes transamination, an acidic group at the active site of the enzyme is in position to donate a proton to the carbon attached to the pyridine ring. The enzyme that catalyzes racemization does not have this acidic group, so the substrate is reprotonated at the α-carbon. In other words, the *coenzyme* carries out the chemical reaction, but the *enzyme* determines the course of the reaction.

Mechanism for C_α—C_β Bond Cleavage

In the first step of the mechanism for PLP-catalyzed C_α—C_β bond cleavage, a basic group at the active site of the enzyme removes a proton from an OH group bonded to the β-carbon of the amino acid. This causes the C_α—C_β bond to be cleaved. Serine and threonine are the only two amino acids that can serve as substrates for the reaction because they are the only amino acids with an OH group bonded to their β-carbon. When serine is the substrate, the product of the cleavage reaction is formaldehyde (R = H); when threonine is the substrate, the product of the cleavage reaction is acetaldehyde (R = CH_3). Electron rearrangement and protonation of the α-carbon of the amino acid, followed by transimination with a lysine side chain, releases glycine.

mechanism for PLP-catalyzed C_α—C_β bond cleavage

transimination with E—$(CH_2)_4NH_2$

+ RCH(=O)

glycine

The formaldehyde formed when serine undergoes C_α—C_β bond cleavage never leaves the active site of the enzyme; it is immediately transferred to tetrahydrofolate (Section 25.8).

PROBLEM 14

Propose a mechanism for a PLP-catalyzed α,β-elimination.

Choosing the Bond to Be Cleaved

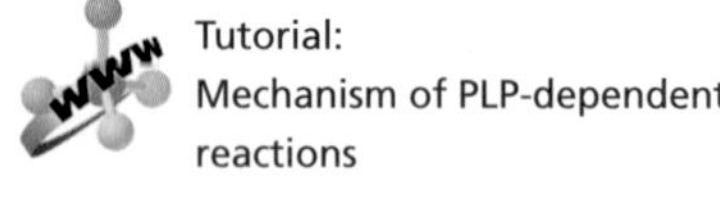

If all PLP-requiring enzymes start with the same substrate—an amino acid bound to pyridoxal phosphate by an imine linkage—how can three different bonds be cleaved in the first step of the reaction? The bond cleaved in the first step depends on the conformation of the amino acid that the enzyme binds. There is free rotation about the C_α—N bond of the amino acid, and an enzyme can bind any of the possible conformations about this bond. The enzyme will bind the conformation in which the overlapping orbitals of the bond to be broken in the first step of the reaction lie parallel to the *p* orbitals of the conjugated system. In this way, the orbital containing the electrons left behind when the bond is broken can overlap with the orbitals of the conjugated system. If such overlap cannot occur, the electrons cannot be delocalized into the conjugated system and the carbanion intermediate cannot be stabilized.

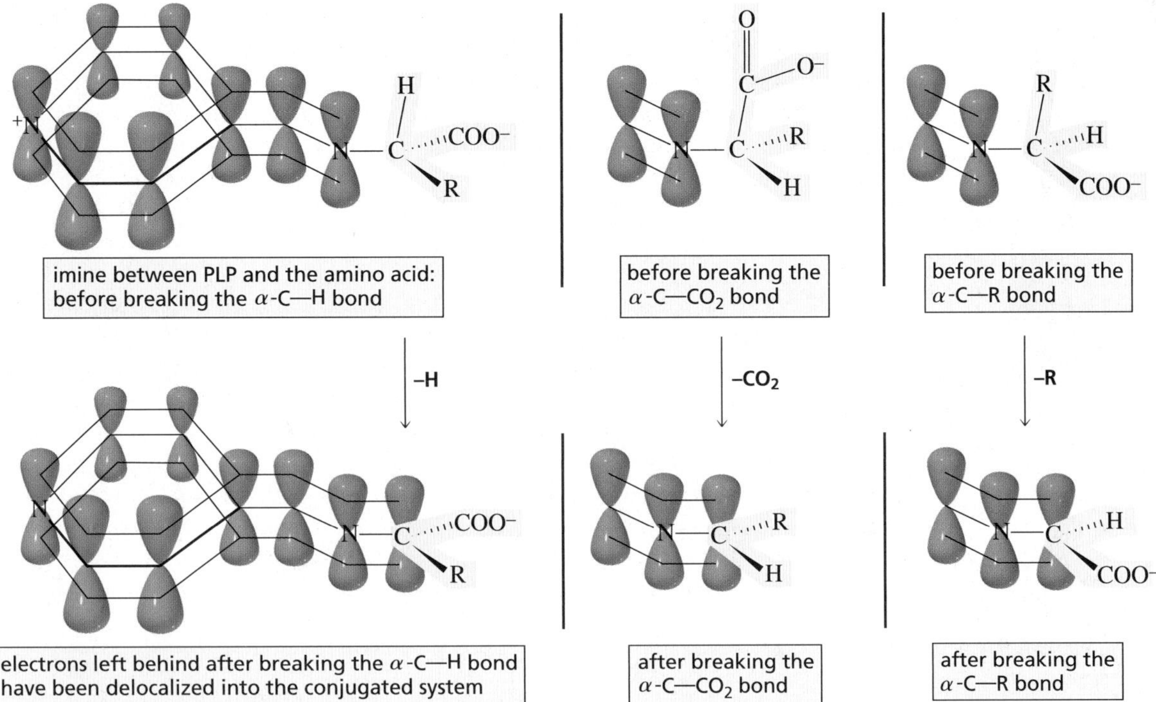

PROBLEM 15

Which of the following compounds is more easily decarboxylated?

$$\text{(4-pyridinium)}-CH_2CH_2-\overset{O}{\overset{\|}{C}}-O^- \quad \text{or} \quad \text{(4-pyridinium)}-CH_2-\overset{O}{\overset{\|}{C}}-O^-$$

PROBLEM 16

Explain why the ability of PLP to catalyze an amino acid transformation is greatly reduced if a PLP-requiring enzymatic reaction is carried out at a pH at which the pyridine nitrogen is not protonated.

PROBLEM 17

Explain why the ability of PLP to catalyze an amino acid transformation is greatly reduced if the OH substituent of pyridoxal phosphate is replaced by an OCH_3.

25.7 Coenzyme B_{12}: Vitamin B_{12}

Enzymes that catalyze certain rearrangement reactions require **coenzyme B_{12}**, a coenzyme derived from vitamin B_{12}. The structure of vitamin B_{12} was determined by Dorothy Crowfoot Hodgkin, using X-ray crystallography. The vitamin has a cyano group (or HO^- or H_2O) coordinated with cobalt (Section 21.11). In coenzyme B_{12}, the cyano group is replaced by a 5′-deoxyadenosyl group.

Dorothy Crowfoot Hodgkin (1910–1994) *was born in Egypt to English parents. She received an undergraduate degree from Somerville College, Oxford University, and earned a Ph.D. from Cambridge University. She determined the structures of penicillin, insulin, and vitamin B_{12}. For her work on vitamin B_{12}, she received the 1964 Nobel Prize in chemistry. She was a professor of chemistry at Somerville, where one of her graduate students was former Prime Minister of England Margaret Thatcher. Hodgkin was a founding member of Pugwash, an organization whose purpose was to further communication between scientists on both sides of the Iron Curtain.*

coenzyme B_{12}

Animals and plants cannot synthesize vitamin B_{12}. In fact, only a few microorganisms can synthesize it. Humans must obtain all their vitamin B_{12} from their diet, particularly from meat. Because vitamin B_{12} is needed in only very small amounts, deficiencies caused by consumption of insufficient amounts of the vitamin are rare, but have been found in vegetarians who eat no animal products. Deficiencies are most commonly caused by an inability to absorb the vitamin in the intestine. The deficiency causes pernicious anemia. The following are examples of enzyme-catalyzed reactions that require coenzyme B_{12}:

$$\underset{\beta\text{-methylaspartate}}{CH_3CH(COO^-)CH(^+NH_3)COO^-} \xrightleftharpoons[\text{coenzyme } B_{12}]{\text{glutamate mutase}} \underset{\text{glutamate}}{^-OOCCH_2CH_2CH(^+NH_3)COO^-}$$

$$\underset{\text{methylmalonyl-CoA}}{CH_3CH(COO^-)CSCoA} \xrightleftharpoons[\text{coenzyme } B_{12}]{\text{methylmalonyl-CoA mutase}} \underset{\text{succinyl-CoA}}{^-OOCCH_2CH_2CSCoA}$$

$$CH_3\underset{\displaystyle OH}{\underset{|}{C}}HCH_2OH \xrightarrow[\text{coenzyme } B_{12}]{\text{dioldehydrase}} \left[CH_3CH_2\underset{\displaystyle OH}{\underset{|}{C}}HOH\right] \xrightarrow{-H_2O} CH_3CH_2\overset{\displaystyle O}{\overset{\|}{C}}H$$

1,2-propanediol — a hydrate — propanal

3-D Molecule: Coenzyme B_{12}

In each of these coenzyme B_{12}–requiring reactions, a group (Y) bonded to one carbon changes places with a hydrogen bonded to an adjacent carbon.

Coenzyme B_{12} is required by enzymes that catalyze the exchange of a hydrogen bonded to one carbon with a group bonded to an adjacent carbon.

$$-\underset{\displaystyle Y}{\underset{|}{\overset{|}{C}}}-\underset{\displaystyle H}{\underset{|}{\overset{|}{C}}}- \xrightleftharpoons[]{\text{a coenzyme } B_{12}\text{-requiring enzyme}} -\underset{\displaystyle H}{\underset{|}{\overset{|}{C}}}-\underset{\displaystyle Y}{\underset{|}{\overset{|}{C}}}-$$

For example, glutamate mutase and methylmalonyl-CoA mutase both catalyze a reaction in which the COO^- group bonded to one carbon changes places with a hydrogen of an adjacent methyl group. In the reaction catalyzed by dioldehydrase, an OH group changes places with a methylene hydrogen. The resulting product is a hydrate that loses water to form propanal.

Mechanism for Coenzyme B_{12}

The chemistry of coenzyme B_{12} takes place at the bond joining the cobalt and the 5′-deoxyadenosyl group. The currently accepted mechanism for dioldehydrase involves an initial homolytic cleavage of this unusually weak bond (26 kcal/mol or 109 kJ/mol, compared with 99 kcal/mol or 414 kJ/mol for a C—H bond). Breaking the bond forms a 5′-deoxyadenosyl radical and reduces Co(III) to Co(II). The 5′-deoxyadenosyl radical abstracts a hydrogen atom from the C-1 carbon of the substrate, thereby becoming 5′-deoxyadenosine. A hydroxyl radical (·OH) migrates from C-2 to C-1, creating a radical at C-2. This radical abstracts a hydrogen atom from 5′-deoxyadenosine, forming the rearranged product and regenerating the 5′-deoxyadenosyl radical that recombines with Co(II) to regenerate the coenzyme. The enzyme–coenzyme complex is then ready for another catalytic cycle. The initial product is a hydrate that loses water to form propanal, the final product of the reaction.

the role of 5′-deoxyadenosylcobalamin in a coenzyme B_{12}-requiring enzyme-catalyzed reaction

Ad—CH₂—Co(III) + CH_3–CH(OH)–CH_2OH ⇌ Ad—ĊH₂ Co(II) + CH_3–CH(OH)–CH_2OH ⇌ Ad—CH₃ Co(II) + CH_3–CH(OH)–ĊH(OH) ⇌ Ad—CH₃ Co(II) + CH_3–ĊH–CH(OH)OH ⇌ Ad—ĊH₂ Co(II) + CH_3–CH₂–CH(OH)OH (a hydrate) ⇌ Ad—CH₂—Co(III) + $CH_3CH_2CH{=}O$ + H_2O (propanal)

It is likely that all coenzyme B_{12}–requiring enzymes catalyze reactions by means of the same general mechanism. The role of the coenzyme is to provide a way to remove a hydrogen atom from the substrate. Once the hydrogen atom has been removed, an adjacent group can migrate to take its place. The coenzyme then gives back the hydrogen atom, delivering it to the carbon that lost the migrating group.

PROBLEM 18

Ethanolamine ammonia lyase, a coenzyme B_{12}–requiring enzyme, catalyzes the following reaction:

$$HOCH_2CH_2NH_2 \xrightarrow{\text{ethanolamine ammonia lyase}} CH_3\overset{\overset{\displaystyle O}{\|}}{C}H + NH_3$$

Propose a mechanism for this reaction.

PROBLEM 19◆

A fatty acid with an even number of carbon atoms is metabolized to acetyl-CoA, which can enter the citric acid cycle. A fatty acid with an odd number of carbon atoms is metabolized to acetyl-CoA and one equivalent of propionyl-CoA. Two coenzyme-requiring enzymes are needed to convert propionyl-CoA into succinyl-CoA, a citric acid cycle intermediate. Write the two enzyme-catalyzed reactions and indicate the required coenzymes.

25.8 Tetrahydrofolate: Folic Acid

Tetrahydrofolate (THF) is the coenzyme required by enzymes that catalyze the transfer of a group containing one carbon to their substrates.

Tetrahydrofolate (THF) is the coenzyme used by enzymes catalyzing reactions that transfer a group containing a single carbon to their substrates. The one-carbon group can be a methyl group (CH_3), a methylene group (CH_2), or a formyl group (HC=O). Tetrahydrofolate results from the reduction of two double bonds of folic acid (folate), its precursor vitamin. Bacteria synthesize folate, but mammals cannot.

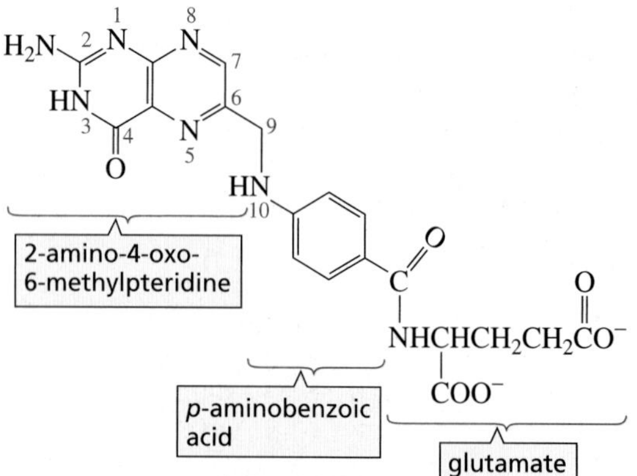

2-amino-4-oxo-6-methylpteridine

p-aminobenzoic acid

glutamate

folic acid (folate)

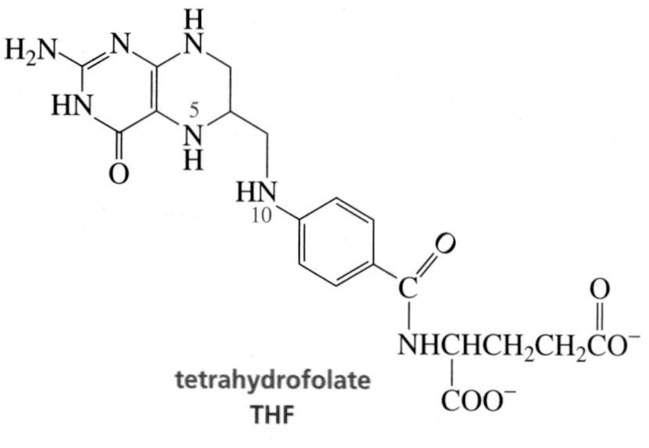

tetrahydrofolate THF

There are six different THF-coenzymes. N^5-Methyl-THF transfers a methyl group (CH_3), N^5,N^{10}-methylene-THF transfers a methylene group (CH_2), and the others transfer a formyl group (HC=O).

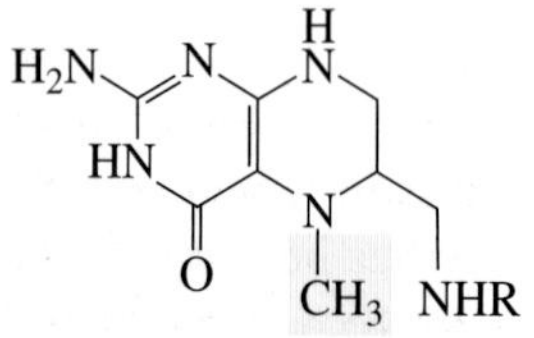

N^5-methyl-THF

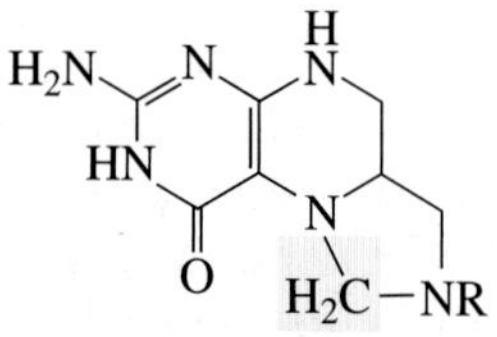

N^5,N^{10}-methylene-THF

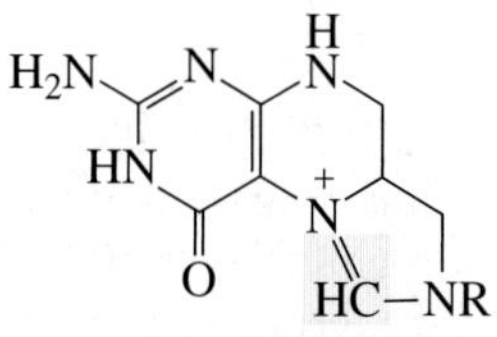

N^5,N^{10}-methenyl-THF

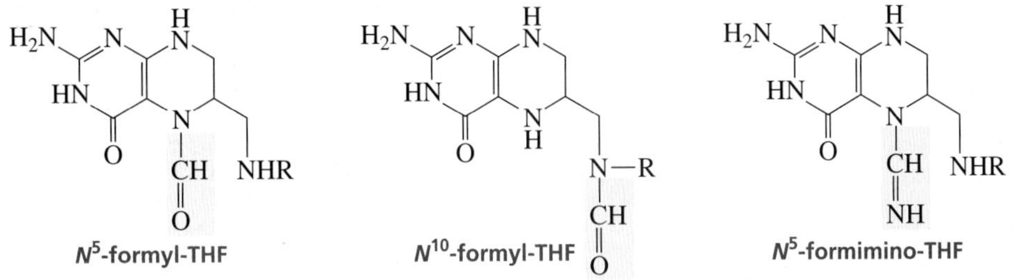

N^5-formyl-THF **N^{10}-formyl-THF** **N^5-formimino-THF**

3-D Molecule: Tetrahydrofolate (THF)

Homocysteine methyl transferase and glycinamide ribonucleotide (GAR) transformylase are examples of enzymes that require THF-coenzymes.

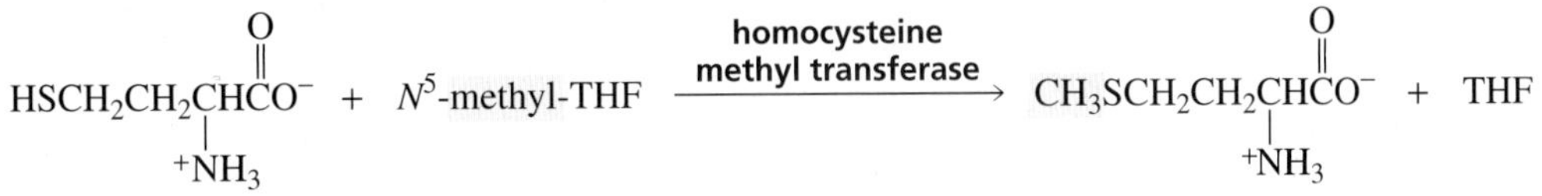

$$HSCH_2CH_2CH(^+NH_3)CO_2^- + N^5\text{-methyl-THF} \xrightarrow{\text{homocysteine methyl transferase}} CH_3SCH_2CH_2CH(^+NH_3)CO_2^- + \text{THF}$$

homocysteine **methionine**

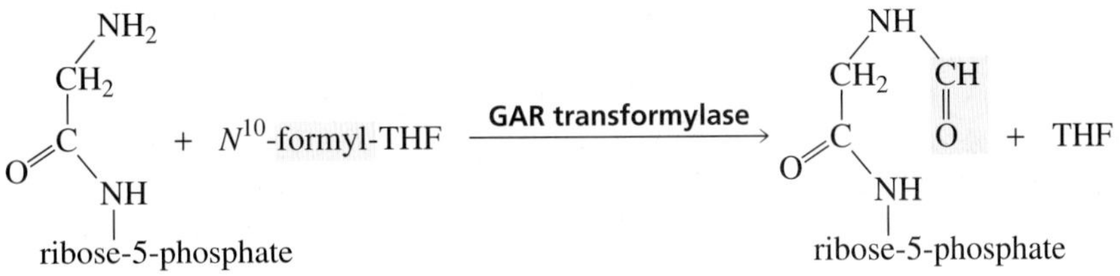

GAR transformylase

$$\text{ribose-5-phosphate-NHC(=O)CH}_2\text{NH}_2 + N^{10}\text{-formyl-THF} \longrightarrow \text{ribose-5-phosphate-NHC(=O)CH}_2\text{NHCH=O} + \text{THF}$$

Thymidylate Synthase: The Enzyme That Converts U's into T's

The heterocyclic bases in RNA are adenine, guanine, cytosine, and uracil (A, G, C, and U); the heterocyclic bases in DNA are adenine, guanine, cytosine, and thymine (A, G, C, and T). In other words, the heterocyclic bases in RNA and DNA are the same, except that RNA contains U's, whereas DNA contains T's (Sections 27.1 and 27.14). The T's used for the biosynthesis of DNA are synthesized from U's by thymidylate synthase, an enzyme that requires N^5,N^{10}-methylene-THF as a coenzyme. Even though the only structural difference between a U and a T is a *methyl* group, a T is synthesized by first transferring a *methylene* group to a U.

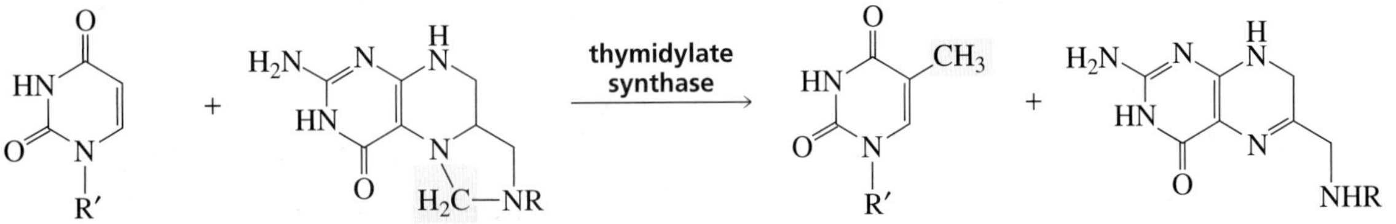

thymidylate synthase

2′-deoxyuridine 5′-monophosphate dUMP + **N^5,N^{10}-methylene-THF** → **2′-deoxythymidine 5′-monophosphate dTMP** + **dihydrofolate DHF**

R′ = 2′-deoxyribose-5-phosphate

In the first step of the reaction catalyzed by thymidylate synthase, a nucleophilic cysteine group at the active site of the enzyme attacks the β-carbon of uridine. (This is an example of conjugate addition; see Section 18.13.) A subsequent nucleophilic attack by the α-carbon of uridine on the methylene group of N^5,N^{10}-methylene-THF forms a covalent bond between uridine and the coenzyme. A proton on the α-carbon of uridine is removed with the help of a basic group at the active site of the enzyme, eliminating the coenzyme. Transfer of a hydride ion from the coenzyme to uridine and elimination of the enzyme forms thymidine and dihydrofolate (DHF).

mechanism for catalysis by thymidylate synthase

oxidized coenzyme

Notice that the coenzyme initially transfers a methylene group to the substrate. The methylene group is subsequently reduced to a methyl group. Because the coenzyme is the reducing agent, it is simultaneously oxidized. The oxidized coenzyme is dihydrofolate.

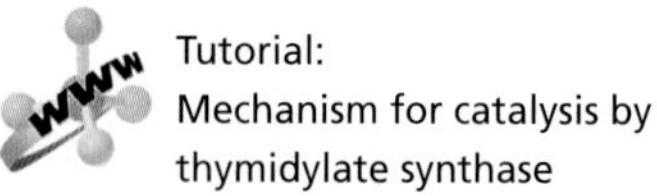

Tutorial: Mechanism for catalysis by thymidylate synthase

When the reaction is over, dihydrofolate must be converted back to N^5,N^{10}-methylene-THF so that the coenzyme can undergo another catalytic cycle. Dihydrofolate is first reduced to tetrahydrofolate. Then serine hydroxymethyl transferase—the PLP-requiring enzyme that cleaves the $C_\alpha—C_\beta$ bond of serine to form glycine and formaldehyde—transfers formaldehyde to the coenzyme (Section 25.6). In other words, the formaldehyde that is cleaved off serine is immediately transferred to THF to form N^5,N^{10}-methylene-THF, which is fortunate because formaldehyde is cytotoxic. (It kills cells.)

$$\text{dihydrofolate} + \text{NADPH} + H^+ \xrightarrow{\textbf{dihydrofolate reductase}} \text{tetrahydrofolate} + \text{NADP}^+$$

$$\text{tetrahydrofolate} + \underset{\textbf{serine}}{HOCH_2\underset{^+NH_3}{CH}COO^-} \xrightarrow[\textbf{PLP}]{\textbf{serine hydroxymethyl transferase}} N^5,N^{10}\text{-methylene-THF} + \underset{\textbf{glycine}}{\underset{^+NH_3}{CH_2}COO^-}$$

Cancer Chemotherapy

Cancer is associated with rapidly growing and proliferating cells. Because cells cannot multiply if they cannot synthesize DNA, several cancer chemotherapeutic agents have been developed to inhibit thymidylate synthase and dihydrofolate reductase. If a cell cannot make thymidine, it cannot synthesize DNA. Inhibiting dihydrofolate reductase also prevents the synthesis of thymidine because cells have a limited amount of tetrahydrofolate. If they cannot convert dihydrofolate back to tetrahydrofolate, they cannot continue to synthesize thymidine.

A common anticancer drug that inhibits thymidylate synthase is 5-fluorouracil. 5-Fluorouracil and uracil react with thymidylate synthase in the same way. However, the fluorine at the 5-position cannot be removed by the base in the third step of the reaction because fluorine is too electronegative to come off as F^+. Because fluorine cannot be removed, the reaction stops at this point, leaving the enzyme permanently attached to the substrate. The active site of the enzyme is now blocked with 5-fluorouracil, so the enzyme cannot bind uracil, which means that thymidine can no longer be synthesized. Consequently, the synthesis of DNA is also stopped. Unfortunately, most anticancer drugs cannot discriminate between diseased and normal cells. As a result, cancer chemotherapy is accompanied by terrible side effects. However, cancer cells undergo uncontrolled cell division so, because they are dividing more rapidly than normal cells, they are harder hit by cancer-fighting chemotherapeutic agents.

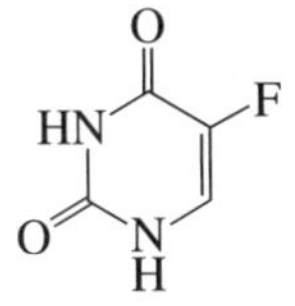

5-fluorouracil
5-FU

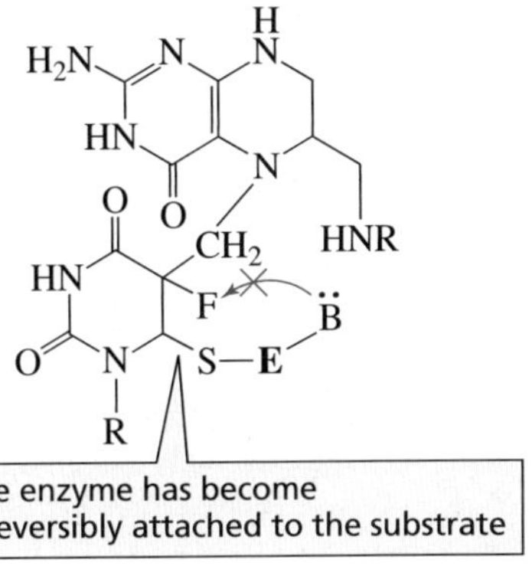

the enzyme has become irreversibly attached to the substrate

5-Fluorouracil is a **mechanism-based inhibitor**—it inactivates the enzyme by taking part in the normal catalytic mechanism. It is also called a **suicide inhibitor** because when the enzyme reacts with it, the enzyme "commits suicide." The use of 5-fluorouracil illustrates the importance of knowing the mechanism for an enzyme-catalyzed reaction. If you know the mechanism, you may be able to design an inhibitor to turn the reaction off at a certain step.

Aminopterin and methotrexate are anticancer drugs that are inhibitors of dihydrofolate reductase. Because their structures are similar to that of dihydrofolate, they compete with it for binding at the active site of the enzyme. Since they bind 1000 times more tightly to the enzyme than does dihydrofolate, they inhibit the enzyme. These two compounds are examples of **competitive inhibitors**.

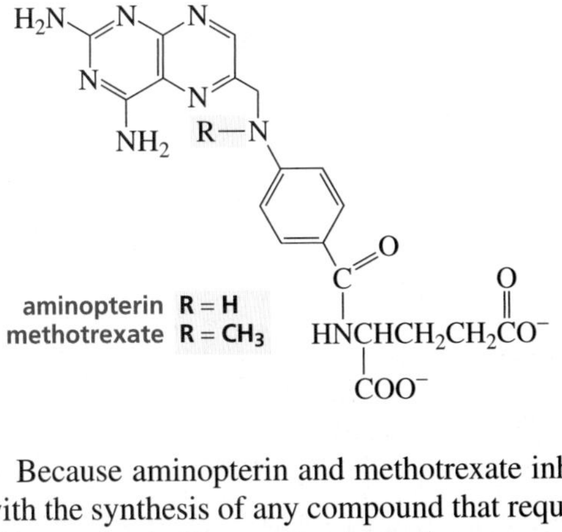

aminopterin R = H
methotrexate R = CH_3

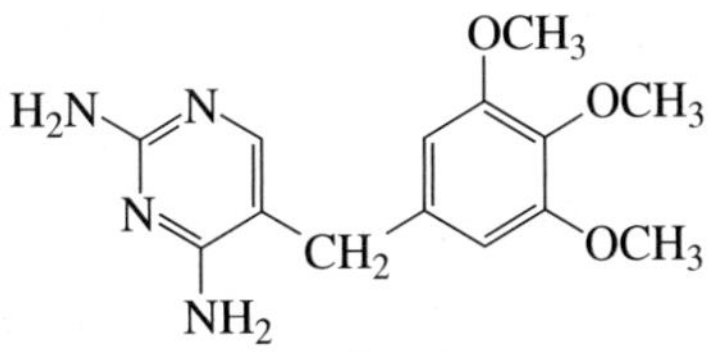

trimethoprim

Because aminopterin and methotrexate inhibit the synthesis of THF, they interfere with the synthesis of any compound that requires a THF-coenzyme in one of the steps of its synthesis. Thus, not only do they prevent the synthesis of thymidine but they also inhibit the synthesis of adenine and guanine—other heterocyclic compounds needed

Donald D. Woods (1912–1964) *was born in Ipswich, England, and received a B.A. and a Ph.D. from Cambridge University. He worked in Paul Flores's laboratory at the London Hospital Medical College.*

for the synthesis of DNA—because their synthesis also requires a THF-coenzyme. One clinical technique used in chemotherapy to fight cancer calls for the patient to be given a lethal dose of methotrexate and then to "save" him or her by administering N^5-formyl-THF.

Trimethoprim is used as an antibiotic because it binds to bacterial dihydrofolate reductase much more tightly than to mammalian dihydrofolate reductase.

THE FIRST ANTIBACTERIAL DRUGS

Sulfonamides—commonly known as sulfa drugs—were introduced clinically in 1934 as the first effective antibacterial drugs (Section 30.4). Donald Woods, a British bacteriologist, noticed that sulfanilamide, initially the most widely used sulfonamide, was structurally similar to *p*-aminobenzoic acid, a compound necessary for bacterial growth. He proposed that sulfanilamide's antibacterial properties were due to its being able to block the normal utilization of *p*-aminobenzoic acid.

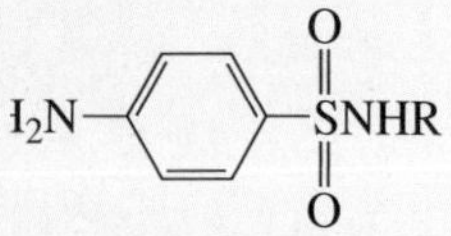

a sulfonamide

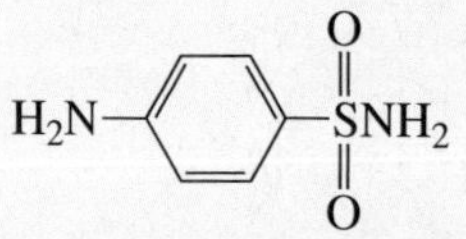

sulfanilamide

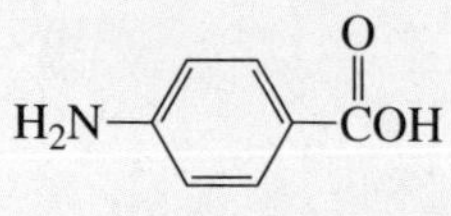

p-aminobenzoic acid

Woods and Paul Flores suggested that sulfanilamide acts by inhibiting the enzyme that incorporates *p*-aminobenzoic acid into folic acid. Because the enzyme cannot tell the difference between sulfanilamide and *p*-aminobenzoic acid, both compounds compete for the active site of the enzyme. Humans are not adversely affected by the drug because they do not synthesize folate—they get all their folate from their diets.

Paul B. Flores (1882–1971) *was born in London. He moved his laboratory to Middlesex Hospital Medical School when a bacterial chemistry unit was established there. He was knighted in 1946.*

PROBLEM 20◆

What is the source of the methyl group in thymidine?

25.9 Vitamin KH_2: Vitamin K

Vitamin K is required for proper clotting of blood. The letter K comes from *koagulation*, which is German for "clotting." A series of reactions utilizing six proteins is involved in blood clotting. In order for blood to clot, the blood-clotting proteins must bind Ca^{2+}. Vitamin K is required for proper Ca^{2+} binding. Vitamin K deficiencies are rare because the vitamin is synthesized by intestinal bacteria. Vitamin K is also found in the leaves of green plants. **Vitamin KH_2** (the hydroquinone of vitamin K) is the coenzyme form of the vitamin (Section 20.12).

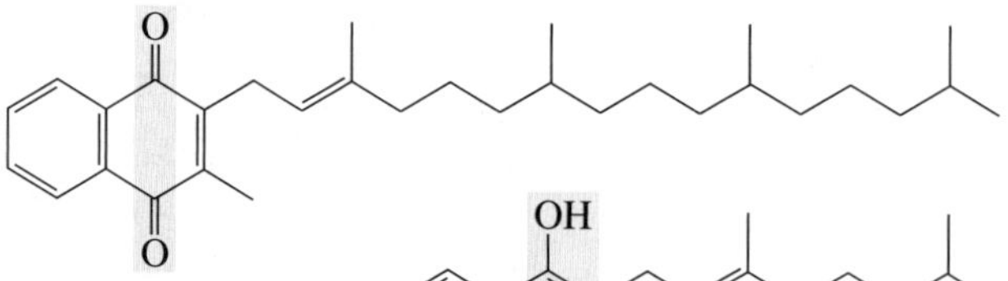

vitamin K
a quinone

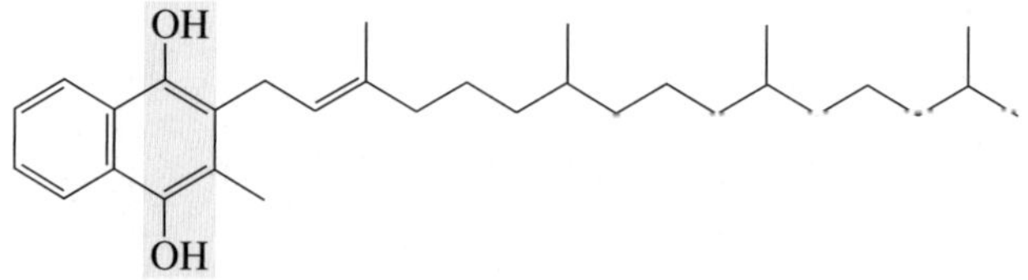

vitamin KH_2
a hydroquinone

3-D Molecule: Vitamin K

Vitamin KH_2 is the coenzyme for the enzyme that catalyzes the carboxylation of the γ-carbon of glutamate side chains in proteins, forming γ-carboxyglutamates. γ-Carboxyglutamates complex Ca^{2+} much more effectively than glutamates do. The proteins involved in blood clotting all have several glutamates near their N-terminal ends. For example, prothrombin has glutamates at positions 7, 8, 15, 17, 20, 21, 26, 27, 30, and 33.

Vitamin KH_2 is required by the enzyme that catalyzes the carboxylation of the γ-carbon of a glutamate side chain in a protein.

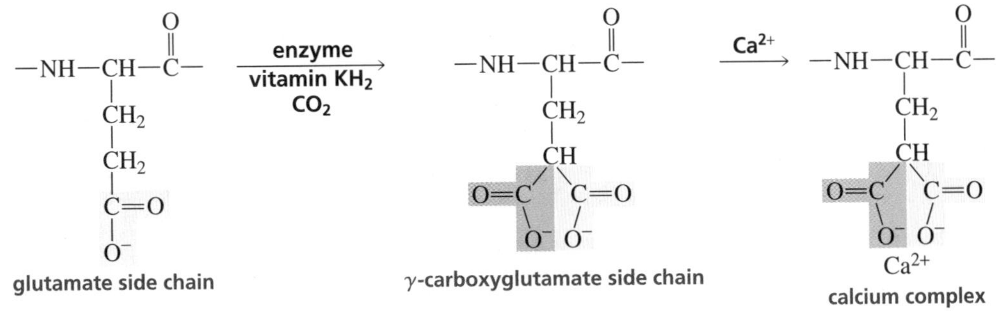

enzyme
vitamin KH_2
CO_2

Ca^{2+}

glutamate side chain

γ-carboxyglutamate side chain

calcium complex

The mechanism for the vitamin KH_2–catalyzed carboxylation of glutamate had puzzled chemists because the γ-proton that must be removed from glutamate before it can attack CO_2 is not very acidic. The mechanism, therefore, must involve the creation of a strong base. The following mechanism has been proposed by Paul Dowd: The vitamin loses a proton from a phenolic OH group, and the base that is thereby formed attacks molecular oxygen. A dioxetane is formed and then collapses to give a vitamin K base that is strong enough to remove a proton from the γ-carbon of glutamate. The glutamate carbanion attacks CO_2 to form γ-carboxyglutamate, and the protonated vitamin K base (a hydrate) loses water, forming vitamin K epoxide.

Paul Dowd (1936–1996) *was born in Brockton, Massachusetts. He did his undergraduate work at Harvard University and received a Ph.D. from Columbia University. He was a professor of chemistry at Harvard University and was a professor of chemistry from 1970 to 1996 at the University of Pittsburgh.*

mechanism for the vitamin KH_2-dependent carboxylation of glutamate

a dioxetane

vitamin K base

$-H_2O$

vitamin K epoxide

γ-carboxyglutamate

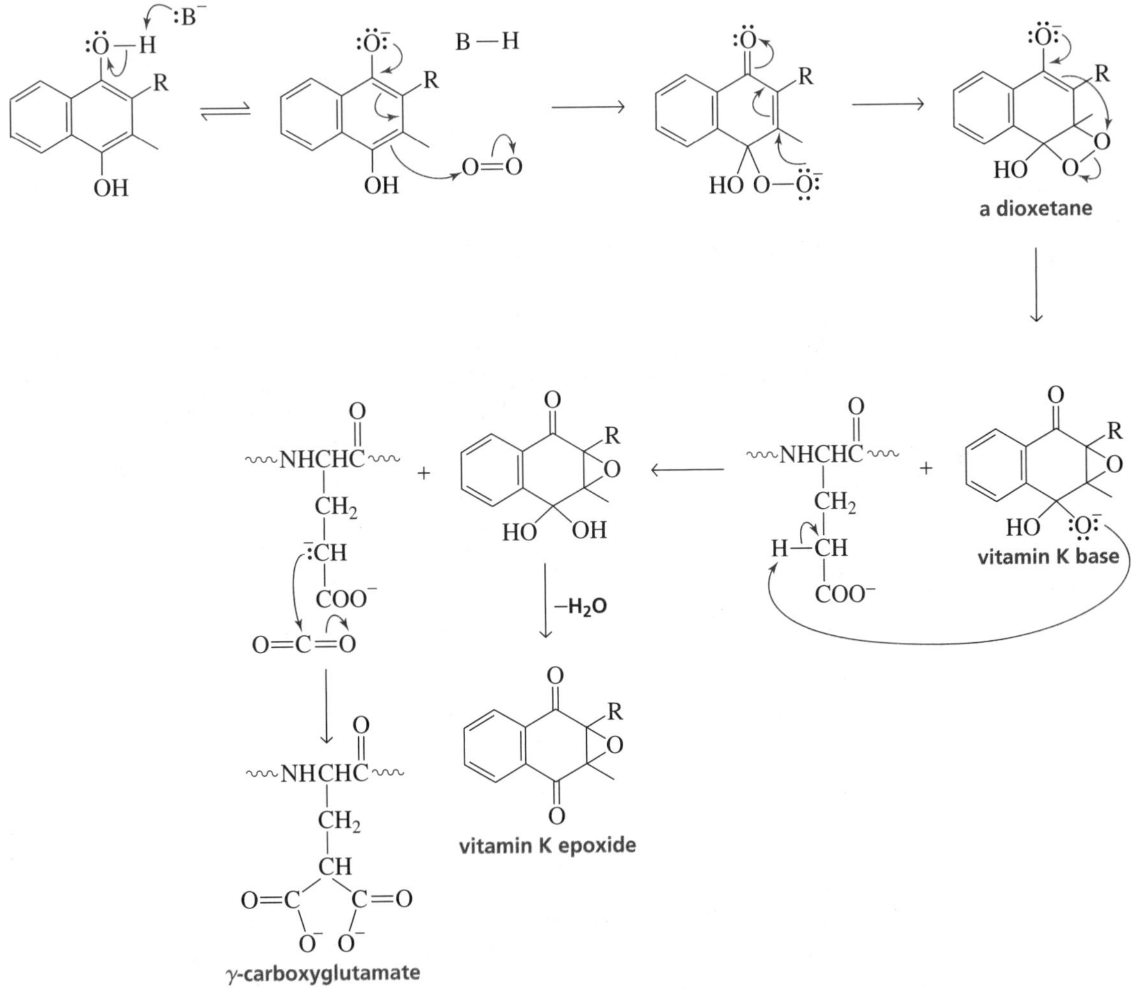

Vitamin K epoxide is reduced back to vitamin KH_2 by an enzyme that uses the coenzyme dihydrolipoate as the reducing agent. The epoxide is first reduced to vitamin K, which is then further reduced to vitamin KH_2.

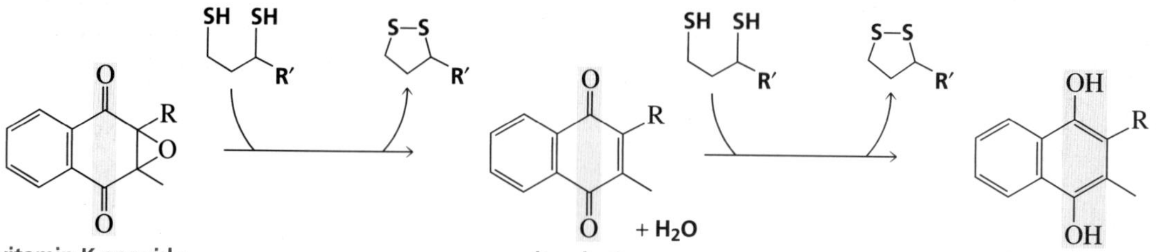

vitamin K epoxide → vitamin K + H_2O → vitamin KH_2

Warfarin and dicoumarol are used clinically as anticoagulants. Warfarin is also a common rat poison, causing death by internal bleeding. These compounds prevent clotting by inhibiting the enzyme that reduces vitamin K epoxide to vitamin KH_2, thereby preventing the carboxylation of glutamate. The enzyme cannot tell the difference between these two compounds and vitamin K epoxide, so the compounds act as *competitive inhibitors*.

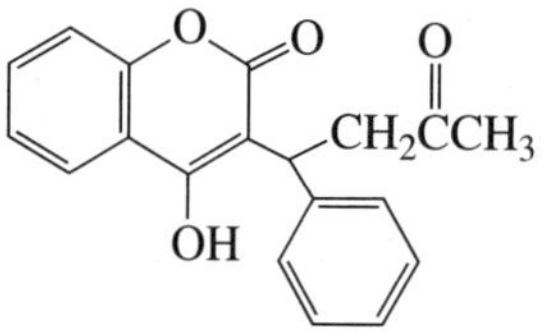

warfarin

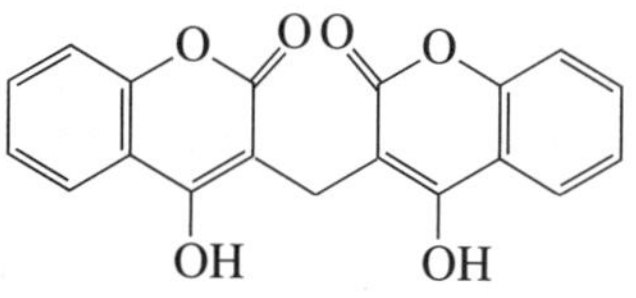

dicoumarol

Vitamin E has recently been found to be an anticoagulant. It directly inhibits the enzyme that carboxylates glutamate residues.

TOO MUCH BROCCOLI

An article describing two women with diseases characterized by abnormal blood clotting reported that they did not improve when they were given warfarin. When questioned about their diets, one woman said that she ate at least a pound (0.45 kg) of broccoli every day, and the other ate broccoli soup and a broccoli salad every day. When broccoli was removed from their diets, warfarin became effective in preventing the abnormal clotting of their blood. Because broccoli is high in vitamin K, these patients had been getting enough dietary vitamin K to compete with the drug, thereby making the drug ineffective.

PROBLEM 21

Thiols such as ethanethiol and propanethiol can be used to reduce vitamin K epoxide back to vitamin KH_2, but they react much more slowly than dihydrolipoate. Explain.

Summary

Cofactors assist enzymes in catalyzing a variety of reactions that cannot be catalyzed solely by their amino acid side chains. Cofactors can be metal ions or organic molecules. An enzyme with a tightly bound metal ion is called a **metalloenzyme**. Cofactors that are organic molecules are called **coenzymes** and these are derived from **vitamins**. A vitamin is a substance the body cannot synthesize that is needed in small amounts for normal body function. All the water-soluble vitamins except vitamin C function as coenzymes. Vitamin K is the only water-insoluble vitamin currently recognized to function as a coenzyme.

Coenzymes play a variety of chemical roles that the amino acid side chains of enzymes cannot play: Some function as oxidizing and reducing agents; some allow electrons to be delocalized; some activate groups for further reaction; and some provide good nucleophiles or strong bases needed for a reaction. **Molecular recognition** allows the enzyme to bind the substrate and the coenzyme in the proper orientation for reaction. Coenzymes are recycled. An enzyme plus its cofactor is called a **holoenzyme**. An enzyme that has had its cofactor removed is called an **apoenzyme**.

Metabolism—the set of reactions that living organisms carry out to obtain energy and to synthesize the compounds they require—can be divided into **catabolism** and **anabolism**. **Catabolic reactions** break down complex molecules to provide energy and simple molecules. **Anabolic reactions** require energy and lead to the synthesis of complex biomolecules.

The coenzymes used by enzymes to catalyze oxidation reactions are **NAD$^+$**, **NADP$^+$**, **FAD**, and **FMN**; those used to catalyze reduction reactions are **NADH**, **NADPH**, **FADH$_2$**, and **FMNH$_2$**. Many enzymes that catalyze oxidation reactions are called **dehydrogenases**. All the redox chemistry of the **pyridine nucleotide coenzymes** takes place at the 4-position of the pyridine ring. All the redox chemistry of the **flavin coenzymes** takes place on the flavin ring.

Thiamine pyrophosphate (TPP) is the coenzyme required by enzymes that catalyze the transfer of a two-carbon fragment. **Biotin** is the coenzyme required by enzymes that catalyze carboxylation of a carbon adjacent to a carbonyl group. **Pyridoxal phosphate (PLP)** is the coenzyme required by enzymes that catalyze certain transformations of amino acids: decarboxylation, transamination, racemization, $C_\alpha - C_\beta$ bond cleavage, and α,β-elimination. In a **transimination reaction**, one imine is converted into another imine; in a **transamination reaction**, the amino group is removed from a substrate and transferred to another molecule.

In a **coenzyme B$_{12}$**–requiring reaction, a group bonded to one carbon changes places with a hydrogen bonded to an adjacent carbon. **Tetrahydrofolate (THF)** is the coenzyme used by enzymes catalyzing reactions that transfer a group containing a single carbon—methyl, methylene, or formyl—to their substrates. **Vitamin KH$_2$** is the coenzyme for the enzyme that catalyzes the carboxylation of the γ-carbon of glutamate side chains—a reaction required for blood clotting. A **suicide inhibitor** inactivates an enzyme by taking part in the normal catalytic mechanism. **Competitive inhibitors** compete with the substrate for binding at the active site of the enzyme.

Key Terms

anabolism (p. 1035)
apoenzyme (p. 1034)
biotin (p. 1052)
catabolism (p. 1035)
coenzyme (p. 1033)
coenzyme A (CoASH) (p. 1050)
coenzyme B_{12} (p. 1061)
cofactor (p. 1033)
competitive inhibitor (p. 1067)
dehydrogenase (p. 1040)
electron sink (p. 1048)
flavin adenine dinucleotide (FAD) (p. 1044)
flavin mononucleotide (FMN) (p. 1044)
heterocycle (p. 1039)
holoenzyme (p. 1034)
lipoate (p. 1050)
mechanism-based inhibitor (p. 1067)
metabolism (p. 1035)
metalloenzyme (p. 1033)
nicotinamide adenine dinucleotide (NAD^+) (p. 1039)
nicotinamide adenine dinucleotide phosphate ($NADP^+$) (p. 1039)
nucleotide (p. 1039)
pyridoxal phosphate (PLP) (p. 1054)
suicide inhibitor (p. 1067)
tetrahydrofolate (THF) (p. 1064)
thiamine pyrophosphate (TPP) (p. 1048)
transamination (p. 1056)
transimination (p. 1055)
vitamin (p. 1033)
vitamin KH_2 (p. 1068)

Problems

22. Answer the following questions:
    a. What six cofactors act as oxidizing agents?
    b. What are the cofactors that donate one-carbon groups?
    c. What three one-carbon groups are various tetrahydrofolates capable of donating to substrates?

d. What is the function of FAD in the pyruvate dehydrogenase complex?
e. What is the function of NAD^+ in the pyruvate dehydrogenase complex?
f. What is the reaction necessary for proper blood clotting catalyzed by vitamin KH_2?
g. What coenzymes are used for decarboxylation reactions?
h. What kinds of substrates do the decarboxylating coenzymes work on?
i. What coenzymes are used for carboxylation reactions?
j. What kinds of substrates do the carboxylating coenzymes work on?

23. Name the coenzymes that
   a. allow electrons to be delocalized
   b. activate groups for further reaction
   c. provide a good nucleophile
   d. provide a strong base

24. For each of the following reactions, name the enzyme that catalyzes the reaction and name the required coenzyme:

a. 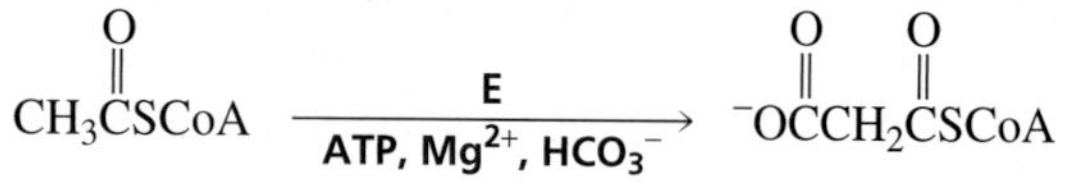 $$CH_3\overset{O}{\overset{\|}{C}}SCoA \xrightarrow[\text{ATP, }Mg^{2+},\ HCO_3^-]{E} {}^-O\overset{O}{\overset{\|}{C}}CH_2\overset{O}{\overset{\|}{C}}SCoA$$

b. $$\text{(dithiol: ring chain with SH, SH)}-(CH_2)_4\overset{O}{\overset{\|}{C}}O^- \xrightarrow{E} \text{(S—S ring)}-(CH_2)_4\overset{O}{\overset{\|}{C}}O^-$$

c. $$^-O\overset{O}{\overset{\|}{C}}\underset{CH_3}{\underset{|}{C}H}\overset{O}{\overset{\|}{C}}SCoA \xrightarrow{E} {}^-O\overset{O}{\overset{\|}{C}}CH_2CH_2\overset{O}{\overset{\|}{C}}SCoA$$

d. $$CH_3\overset{O}{\overset{\|}{C}}-\overset{O}{\overset{\|}{C}}O^- \xrightarrow[\text{a catabolic reaction}]{E} CH_3\overset{OH}{\overset{|}{C}H}-\overset{O}{\overset{\|}{C}}O^-$$

e. $$^-O\overset{O}{\overset{\|}{C}}CH_2\underset{^+NH_3}{\underset{|}{C}H}\overset{O}{\overset{\|}{C}}O^- + {}^-O\overset{O}{\overset{\|}{C}}CH_2CH_2\underset{O}{\underset{\|}{C}}\overset{O}{\overset{\|}{C}}O^- \xrightarrow{E} {}^-O\overset{O}{\overset{\|}{C}}CH_2\underset{O}{\underset{\|}{C}}\overset{O}{\overset{\|}{C}}O^- + {}^-O\overset{O}{\overset{\|}{C}}CH_2CH_2\underset{^+NH_3}{\underset{|}{C}H}\overset{O}{\overset{\|}{C}}O^-$$

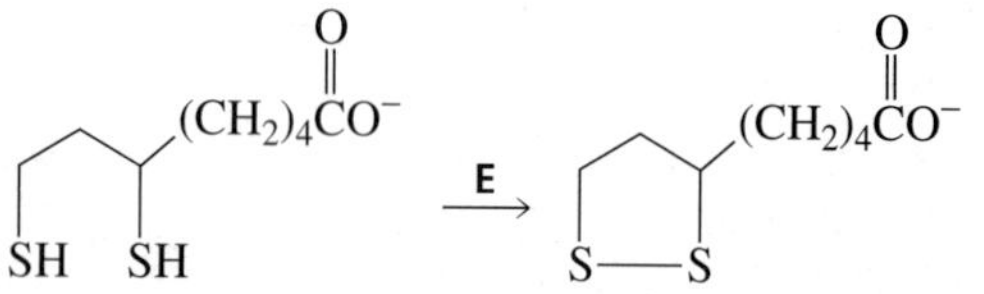

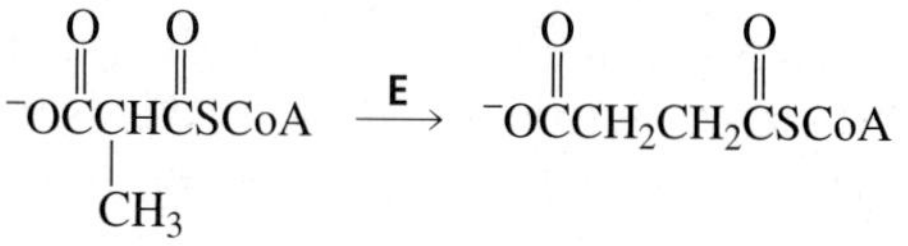

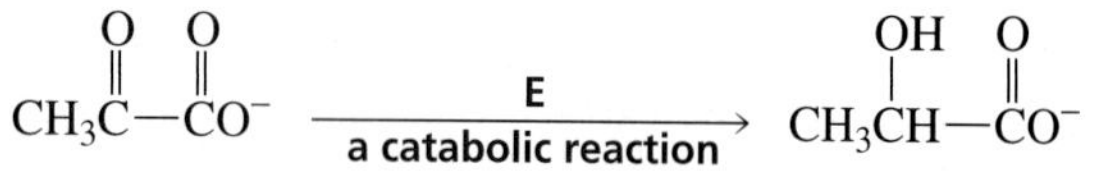

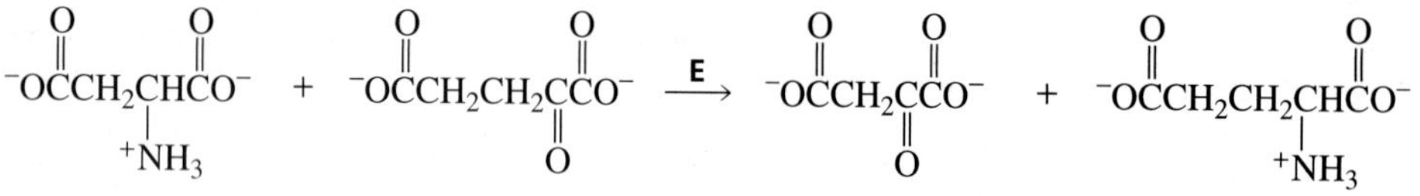

f. 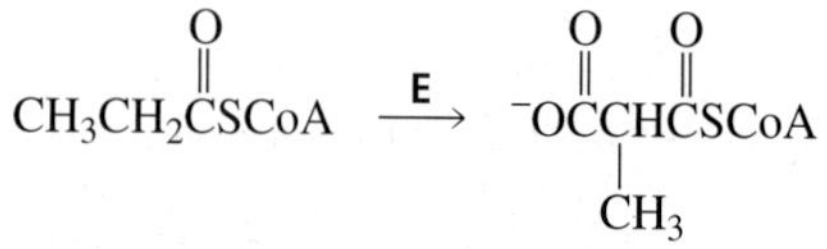 $$CH_3CH_2\overset{O}{\overset{\|}{C}}SCoA \xrightarrow{E} {}^-O\overset{O}{\overset{\|}{C}}\underset{CH_3}{\underset{|}{C}H}\overset{O}{\overset{\|}{C}}SCoA$$

25. *S*-Adenosylmethionine (SAM) is formed from the reaction between ATP (Section 10.11) and methionine. The other product of the reaction is triphosphate. Propose a mechanism for this reaction.

26. Five coenzymes are required by α-ketoglutarate dehydrogenase, the enzyme in the citric acid cycle that converts α-ketoglutarate to succinyl-CoA.
   a. Identify the coenzymes.
   b. Propose a mechanism for this reaction.

$$\underset{\alpha\text{-ketoglutarate}}{^-O\overset{O}{\overset{\|}{C}}CH_2CH_2\overset{O}{\overset{\|}{C}}-\overset{O}{\overset{\|}{C}}O^-} \xrightarrow{\alpha\text{-ketoglutarate dehydrogenase}} \underset{\text{succinyl-CoA}}{^-O\overset{O}{\overset{\|}{C}}CH_2CH_2\overset{O}{\overset{\|}{C}}SCoA} + CO_2$$

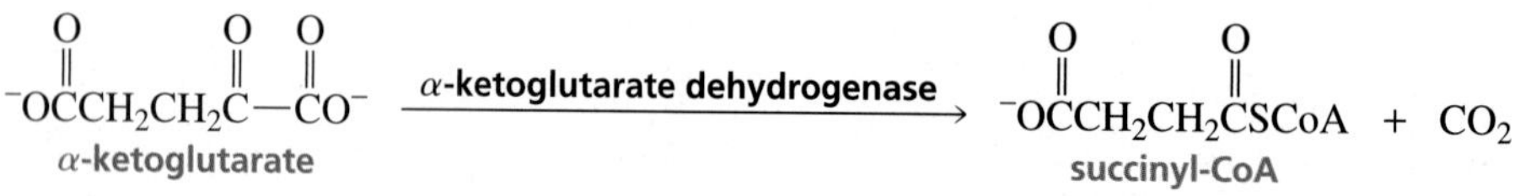

27. Give the products of the following reaction, where T is tritium:

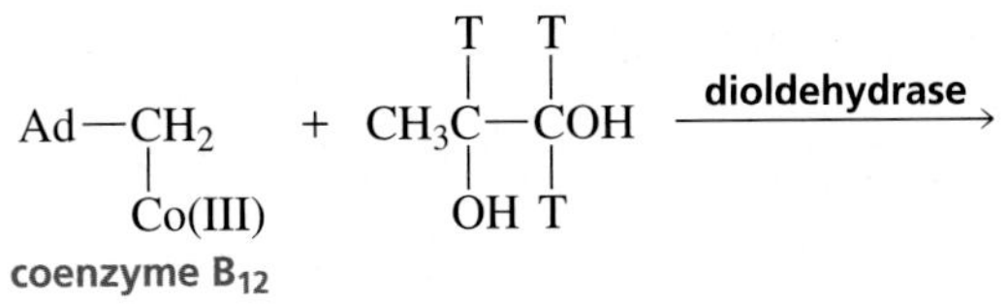

$$\underset{\text{coenzyme } B_{12}}{\underset{|}{Ad-CH_2} \atop Co(III)} + CH_3\overset{T}{\underset{OH}{C}}-\overset{T}{\underset{T}{C}}OH \xrightarrow{\text{dioldehydrase}}$$

(*Hint:* Tritium is a hydrogen atom with two neutrons. Although a C—T bond breaks four times more slowly than a C—H bond, it is still the first bond in the substrate that breaks.)

28. Propose a mechanism for methylmalonyl-CoA mutase, the enzyme that converts methylmalonyl-CoA into succinyl-CoA.

29. When transaminated, the three branched-chain amino acids (valine, leucine, and isoleucine) form compounds that have the characteristic odor of maple syrup. An enzyme known as branched-chain α-keto acid dehydrogenase converts these compounds into CoA esters. People who do not have this enzyme have the genetic disease known as maple syrup urine disease—so-called because their urine smells like maple syrup.
   a. Give the structures of the compounds that smell like maple syrup.
   b. Give the structures of the CoA esters.
   c. Branched-chain α-keto acid dehydrogenase has five coenzymes. Identify them.
   d. Suggest a way to treat maple syrup urine disease.

30. When UMP is dissolved in T_2O (T = tritium; see Problem 27), exchange of T for H occurs at the 5-position. Propose a mechanism for this exchange.

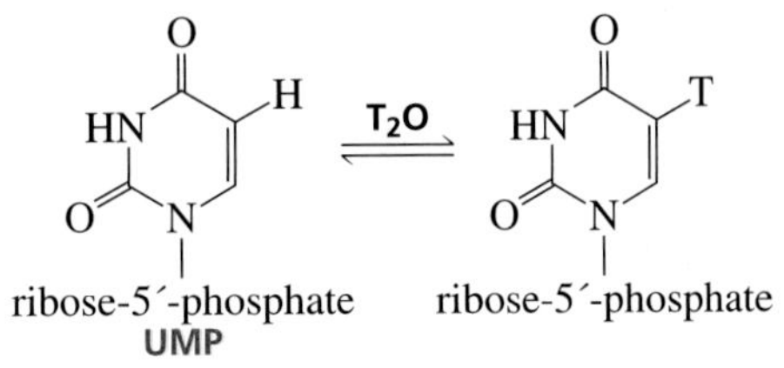

$$\text{UMP (5-H, ribose-5'-phosphate)} \underset{}{\overset{T_2O}{\rightleftharpoons}} \text{5-T uracil (ribose-5'-phosphate)}$$

ribose-5′-phosphate
UMP
ribose-5′-phosphate

31. Dehydratase is a pyridoxal-requiring enzyme that catalyzes an α,β-elimination reaction. Propose a mechanism for this reaction.

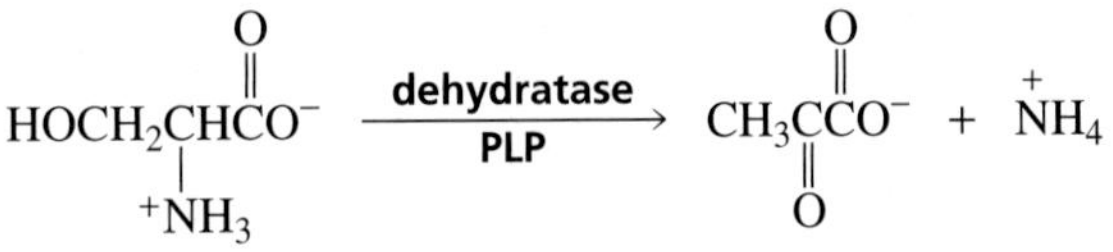

$$HOCH_2\underset{^+NH_3}{CH}\overset{O}{\overset{\|}{C}}O^- \xrightarrow[\text{PLP}]{\text{dehydratase}} CH_3\underset{O}{\underset{\|}{C}}\overset{O}{\overset{\|}{C}}O^- + \overset{+}{N}H_4$$

32. In addition to the reactions mentioned in Section 25.6, PLP can catalyze β-substitution reactions. Propose a mechanism for the following PLP-catalyzed β-substitution reaction:

$$XCH_2\underset{^+NH_3}{CH}\overset{O}{\overset{\|}{C}}O^- + Y^- \xrightarrow[\text{PLP}]{\text{E}} YCH_2\underset{^+NH_3}{CH}\overset{O}{\overset{\|}{C}}O^- + X^-$$

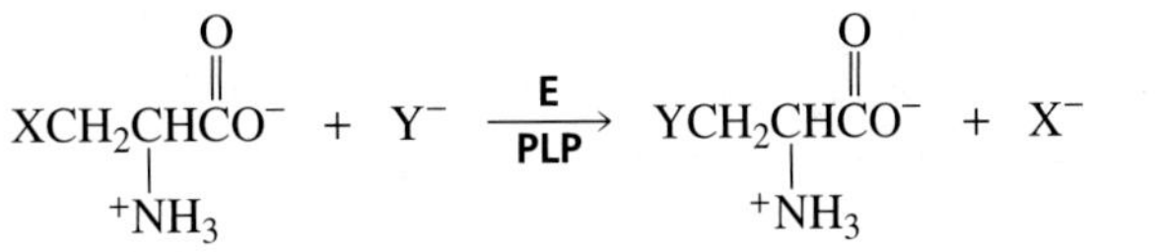

33. PLP can catalyze both α,β-elimination reactions (Problem 31) and β,γ-elimination reactions. Propose a mechanism for the following PLP-catalyzed β,γ-elimination:

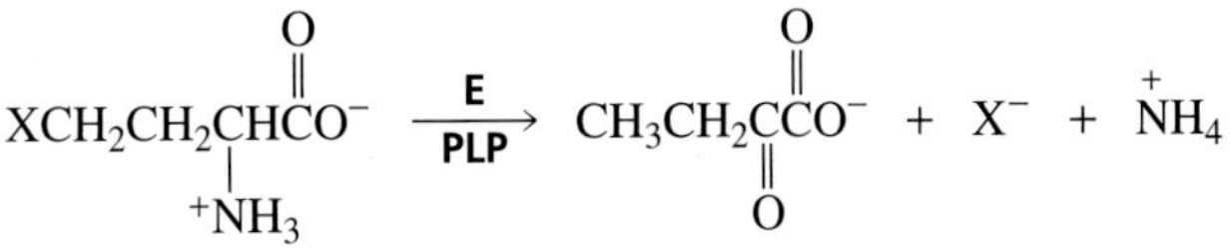

$$XCH_2CH_2\underset{^+NH_3}{CH}\overset{O}{\overset{\|}{C}}O^- \xrightarrow[\text{PLP}]{\text{E}} CH_3CH_2\underset{O}{\underset{\|}{C}}\overset{O}{\overset{\|}{C}}O^- + X^- + \overset{+}{N}H_4$$

34. The glycine cleavage system is a group of four enzymes that together catalyze the following reaction:

$$\text{glycine} + \text{THF} \xrightarrow{\text{glycine cleavage system}} N^5,N^{10}\text{-methylene-THF} + CO_2$$

Use the following information to determine the sequence of reactions involved in the glycine cleavage system:

a. The first enzyme involved in the reaction is a PLP-requiring decarboxylase.
b. The second enzyme is aminomethyltransferase. This enzyme has a lipoate coenzyme.
c. The third enzyme is an N^5,N^{10}-methylene-THF synthesizing enzyme. It catalyzes a reaction that forms $^{+}NH_4$ as one of the products.
d. The fourth enzyme is an FAD-requiring enzyme.
e. The cleavage system also requires NAD^+.

35. Nonenzyme-bound FAD is a stronger oxidizing agent than NAD^+. How, then, can NAD^+ oxidize the reduced flavoenzyme in the pyruvate dehydrogenase system?

36. $FADH_2$ reduces α,β-unsaturated thioesters to saturated thioesters. The reaction is thought to take place by a mechanism that involves radicals. Propose a mechanism for this reaction.

$$RCH{=}CH\overset{\overset{\displaystyle O}{\|}}{C}SR + FADH_2 \longrightarrow RCH_2CH_2\overset{\overset{\displaystyle O}{\|}}{C}SR + FAD$$

26 Lipids

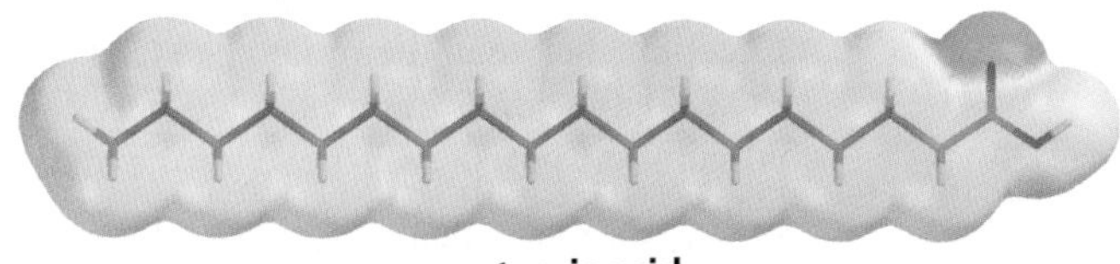

stearic acid

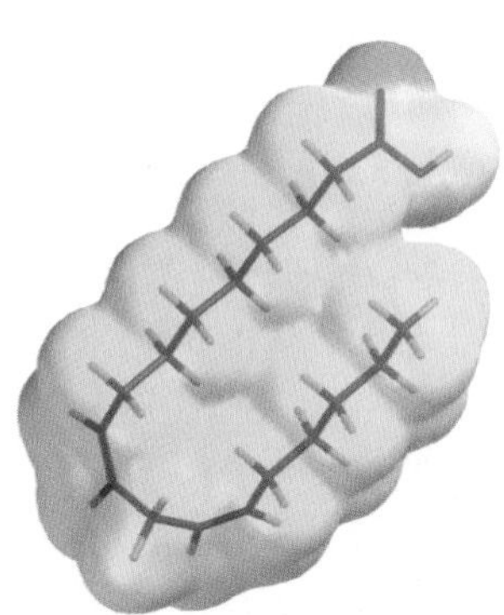

linoleic acid

Lipids are organic compounds, found in living organisms, that are soluble in nonpolar organic solvents. Because compounds are classified as lipids on the basis of a physical property—their solubility in an organic solvent—rather than as a result of their structures, lipids have a variety of structures and functions, as the following examples illustrate:

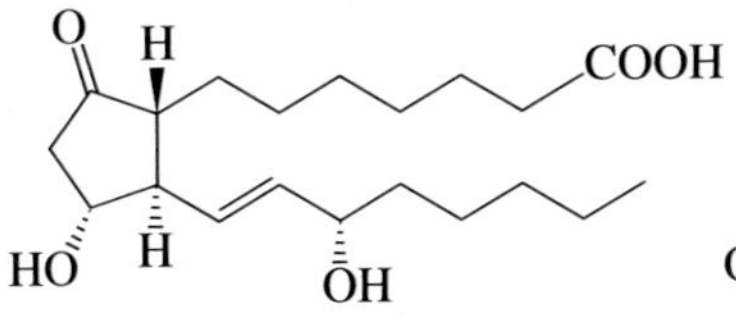

PGE_1
a vasodilator

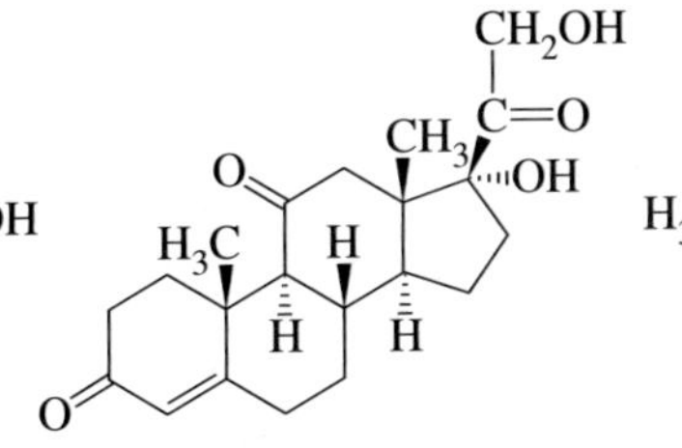

cortisone
a hormone

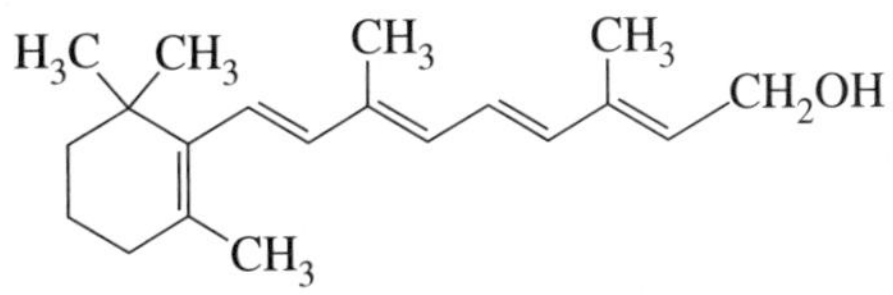

vitamin A
a vitamin

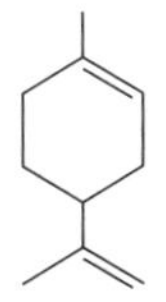

limonene
in orange and lemon oils

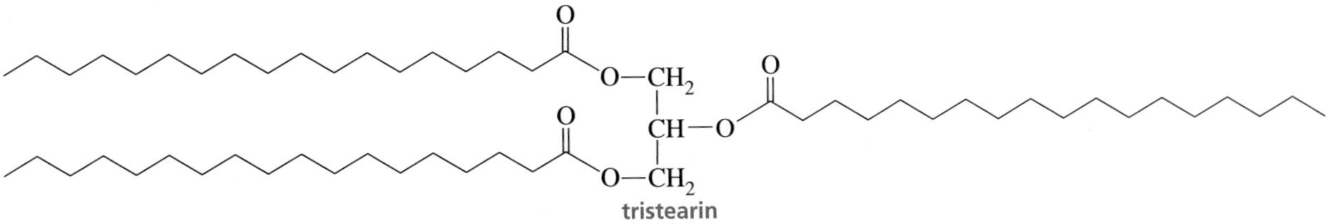

tristearin
a fat

The solubility of lipids in nonpolar organic solvents results from their significant hydrocarbon component. The hydrocarbon portion of the compound is responsible for its "oiliness" or "fattiness." The word *lipid* comes from the Greek *lipos*, which means "fat."

26.1 Fatty Acids

Fatty acids are carboxylic acids with long hydrocarbon chains. The fatty acids most frequently found in nature are shown in Table 26.1. Because they are synthesized from acetate, a compound with two carbon atoms, most naturally occurring fatty acids contain an even number of carbon atoms and are unbranched. The mechanism for the biosynthesis of fatty acids is discussed in Section 19.21. Fatty acids can be saturated with hydrogen (and therefore have no carbon–carbon double bonds) or unsaturated (have carbon–carbon double bonds). Fatty acids with more than one double bond are called **polyunsaturated fatty acids**. Double bonds in naturally occurring unsaturated fatty acids are never conjugated—they are always separated by one methylene group.

The physical properties of a fatty acid depend on the length of the hydrocarbon chain and the degree of unsaturation. As expected, the melting points of saturated fatty acids increase with increasing molecular weight because of increased van der Waals interactions between the molecules (Section 2.9).

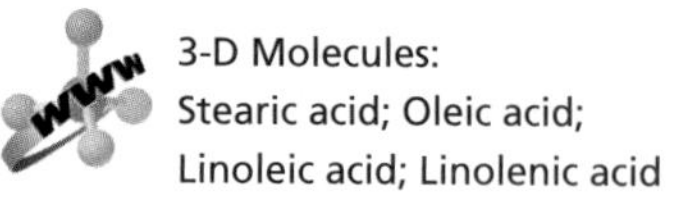
3-D Molecules: Stearic acid; Oleic acid; Linoleic acid; Linolenic acid

The double bonds in unsaturated fatty acids generally have the cis configuration. This configuration produces a bend in the molecules, which prevents them from packing together as tightly as fully saturated fatty acids. As a result, unsaturated fatty acids have

Table 26.1 Common Naturally Occurring Fatty Acids

Number of carbons	Common name	Systematic name	Structure	Melting point °C
Saturated				
12	lauric acid	dodecanoic acid	COOH	44
14	myristic acid	tetradecanoic acid	COOH	58
16	palmitic acid	hexadecanoic acid	COOH	63
18	stearic acid	octadecanoic acid	COOH	69
20	arachidic acid	eicosanoic acid	COOH	77
Unsaturated				
16	palmitoleic acid	(9Z)-hexadecenoic acid	COOH	0
18	oleic acid	(9Z)-octadecenoic acid	COOH	13
18	linoleic acid	(9Z,12Z)-octadecadienoic acid	COOH	−5
18	linolenic acid	(9Z,12Z,15Z)-octadecatrienoic acid	COOH	−11
20	arachidonic acid	(5Z,8Z,11Z,14Z)-eicosatetraenoic acid	COOH	−50
20	EPA	(5Z,8Z,11Z,14Z,17Z)-eicosapentaenoic acid	COOH	−50

fewer intermolecular interactions and, therefore, lower melting points than saturated fatty acids with comparable molecular weights (Table 26.1). The melting points of the unsaturated fatty acids decrease as the number of double bonds increases. For example, an 18-carbon fatty acid melts at 69 °C if it is saturated, at 13 °C if it has one double bond, at −5 °C if it has two double bonds, and at −11 °C if it has three double bonds.

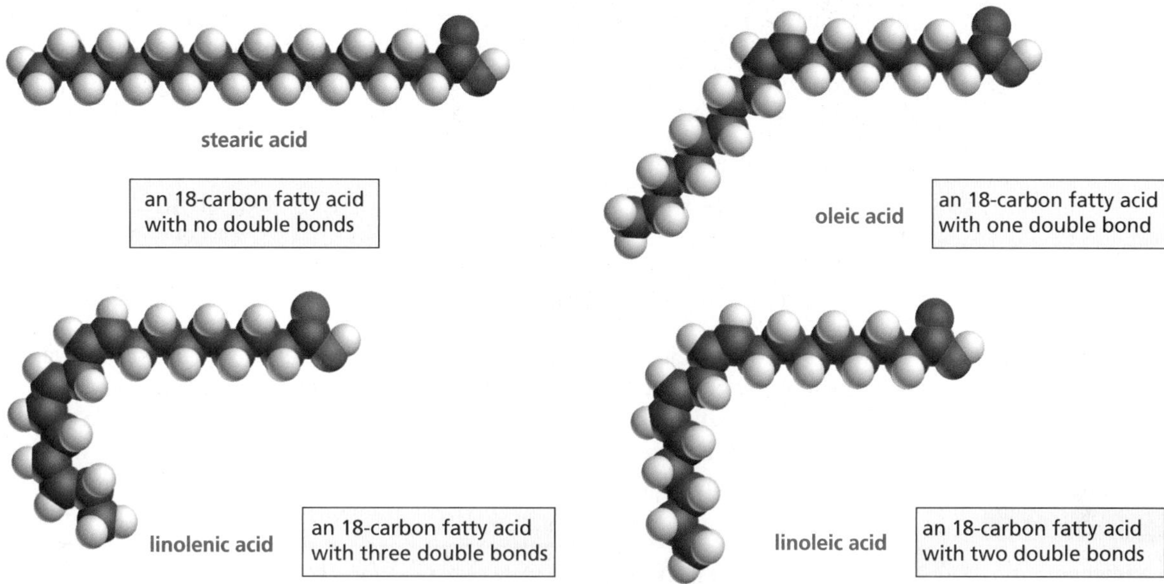

PROBLEM 1

Explain the difference in the melting points of the following fatty acids:

a. palmitic acid and stearic acid
b. palmitic acid and palmitoleic acid
c. oleic acid and linoleic acid

PROBLEM 2◆

What products are formed when arachidonic acid reacts with excess ozone followed by treatment with H_2O_2? (*Hint:* See Section 20.8.)

OMEGA FATTY ACIDS

Omega is a term used to indicate the position of the first double bond—from the methyl end—in an unsaturated fatty acid. For example, linoleic acid is called omega-6 fatty acid because its first double bond is after the sixth carbon, and linolenic acid is called omega-3 fatty acid because its first double bond is after the third carbon. Mammals lack the enzyme that introduces a double bond beyond C-9 (the carboxyl carbon is C-1). Linoleic acid and linolenic acids, therefore, are essential fatty acids for mammals. In other words, the acids must be included in the diet because, although they cannot be synthesized, they are required for normal body function.

COOH

omega-6 fatty acid

linoleic acid

omega-3 fatty acid

COOH

linolenic acid

▲ Layers of honeycomb in a beehive

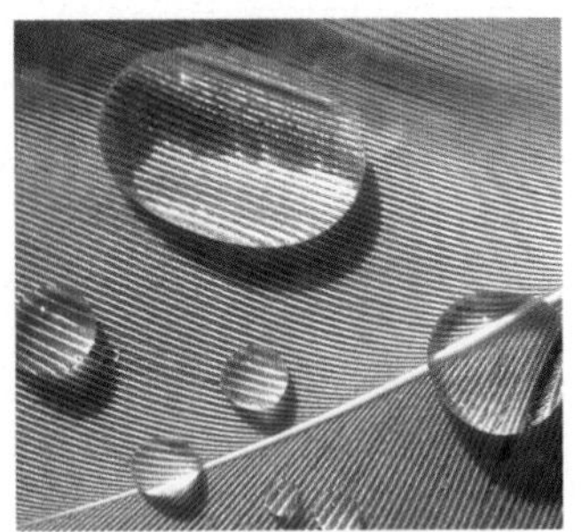
▲ Raindrops on a feather

26.2 Waxes

Waxes are esters formed from long-chain carboxylic acids and long-chain alcohols. For example, beeswax, the structural material of beehives, has a 26-carbon carboxylic acid component and a 30-carbon alcohol component. The word *wax* comes from the Old English *weax,* meaning "material of the honeycomb." Carnauba wax is a particularly hard wax because of its relatively high molecular weight, arising from a 32-carbon carboxylic acid component and a 34-carbon alcohol component. Carnauba wax is widely used as a car wax and in floor polishes.

$$CH_3(CH_2)_{24}\overset{\overset{\displaystyle O}{\|}}{C}O(CH_2)_{29}CH_3$$
a major component of beeswax
structural material of beehives

$$CH_3(CH_2)_{30}\overset{\overset{\displaystyle O}{\|}}{C}O(CH_2)_{33}CH_3$$
a major component of carnauba wax
coating on the leaves of a Brazilian palm

$$CH_3(CH_2)_{14}\overset{\overset{\displaystyle O}{\|}}{C}O(CH_2)_{15}CH_3$$
a major component of spermaceti wax
from the heads of sperm whales

Waxes are common in living organisms. The feathers of birds are coated with wax to make them water repellent. Some vertebrates secrete wax in order to keep their fur lubricated and water repellent. Insects secrete a waterproof, waxy layer on the outside of their exoskeletons. Wax is also found on the surfaces of certain leaves and fruits, where it serves as a protectant against parasites and minimizes the evaporation of water.

26.3 Fats and Oils

Triacylglycerols, also called triglycerides, are compounds in which the three OH groups of glycerol are esterified with fatty acids. If the three fatty acid components of a triacylglycerol are the same, the compound is called a **simple triacylglycerol**. **Mixed triacylglycerols**, on the other hand, contain two or three different fatty acid components and are more common than simple triacylglycerols. Not all triacylglycerol molecules from a single source are necessarily identical; substances such as lard and olive oil, for example, are mixtures of several different triacylglycerols (Table 26.2).

Table 26.2 Approximate Percentage of Fatty Acids in Some Common Fats and Oils

		Saturated fatty acids				Unsaturated fatty acids		
	mp (°C)	lauric C_{12}	myristic C_{14}	palmitic C_{16}	stearic C_{18}	oleic C_{18}	linoleic C_{18}	linolenic C_{18}
Animal fats								
Butter	32	2	11	29	9	27	4	—
Lard	30	—	1	28	12	48	6	—
Human fat	15	1	3	25	8	46	10	—
Whale blubber	24	—	8	12	3	35	10	—
Plant oils								
Corn	20	—	1	10	3	50	34	—
Cottonseed	−1	—	1	23	1	23	48	—
Linseed	−24	—	—	6	3	19	24	47
Olive	−6	—	—	7	2	84	5	—
Peanut	3	—	—	8	3	56	26	—
Safflower	−15	—	—	3	3	19	70	3
Sesame	−6	—	—	10	4	45	40	—
Soybean	−16	—	—	10	2	29	51	7

$$\begin{array}{l} CH_2-OH \\ | \\ CH-OH \\ | \\ CH_2-OH \end{array} \qquad \begin{array}{l} R^1-\overset{\overset{\displaystyle O}{\|}}{C}-OH \\ R^2-\overset{\overset{\displaystyle O}{\|}}{C}-OH \\ R^3-\overset{\overset{\displaystyle O}{\|}}{C}-OH \end{array} \qquad \begin{array}{l} CH_2-O-\overset{\overset{\displaystyle O}{\|}}{C}-R^1 \\ | \\ CH-O-\overset{\overset{\displaystyle O}{\|}}{C}-R^2 \\ | \\ CH_2-O-\overset{\overset{\displaystyle O}{\|}}{C}-R^3 \end{array}$$

glycerol fatty acids a triacylglycerol a fat or an oil

Triacylglycerols that are solids or semisolids at room temperature are called **fats**. Fats are usually obtained from animals and are composed largely of triacylglycerols with either saturated fatty acids or fatty acids with only one double bond. The saturated fatty acid tails pack closely together, giving the triacylglycerols relatively high melting points, causing them to be solids at room temperature.

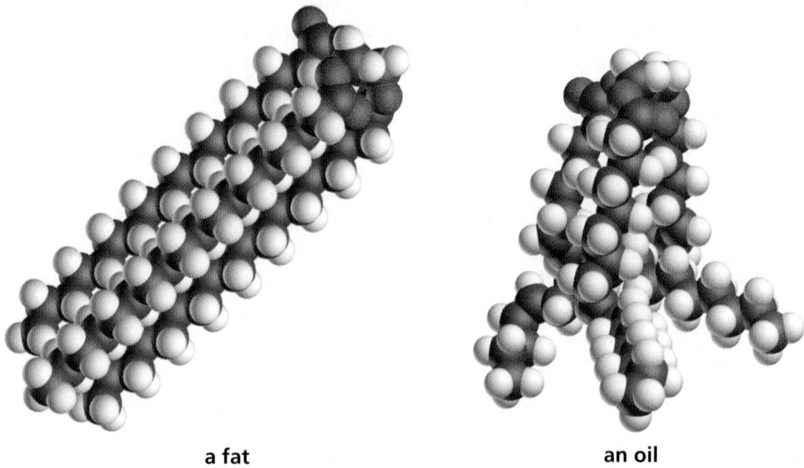

a fat an oil

Liquid triacylglycerols are called **oils**. Oils typically come from plant products such as corn, soybeans, olives, and peanuts. They are composed primarily of triacylglycerols with unsaturated fatty acids that cannot pack tightly together. Consequently, they have relatively low melting points, causing them to be liquids at room temperature. The approximate fatty acid compositions of some common fats and oils are shown in Table 26.2.

Some or all of the double bonds of polyunsaturated oils can be reduced by catalytic hydrogenation (Section 4.11). Margarine and shortening are prepared by hydrogenating vegetable oils such as soybean oil and safflower oil until they have the desired consistency. This process is called "hardening of oils." The hydrogenation reaction must be carefully controlled, however, because reducing all the carbon–carbon double bonds would produce a hard fat with the consistency of beef tallow.

$$RCH{=}CHCH_2CH{=}CHCH_2CH{=}CH- \xrightarrow[Pt]{H_2} RCH_2CH_2CH_2CH{=}CHCH_2CH_2CH_2-$$

Vegetable oils have become popular in food preparation because some studies have linked the consumption of saturated fats with heart disease. Recent studies have shown that unsaturated fats may also be implicated in heart disease. However, one unsaturated fatty acid—a 20-carbon fatty acid with five double bonds, known as EPA and found in high concentrations in fish oils—is thought to lower the chance of developing certain forms of heart disease. Once consumed, dietary fat is hydrolyzed in the intestine, regenerating glycerol and fatty acids. We have seen that the hydrolysis of fats under basic conditions forms glycerol and salts of fatty acids that are commonly known as *soap* (Section 17.13).

▲ This puffin's diet is high in fish oil.

PROBLEM 3♦

Do all triacylglycerols have the same number of asymmetric carbons?

PROBLEM 4♦

Which has a higher melting point, glyceryl tripalmitoleate or glyceryl tripalmitate?

Organisms store energy in the form of triacylglycerols. A fat provides about six times as much metabolic energy as an equal weight of hydrated glycogen because fats are less oxidized than carbohydrates and, since fats are nonpolar, they do not bind water. In contrast, two-thirds of the weight of stored glycogen is water (Section 22.18).

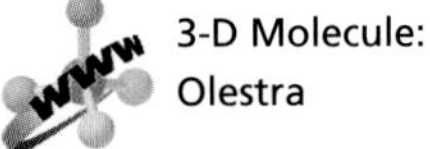
3-D Molecule: Olestra

Animals have a subcutaneous layer of fat cells that serves as both an energy source and an insulator. The fat content of the average man is about 21%, whereas the fat content of the average woman is about 25%. Humans can store sufficient fat to provide for the body's metabolic needs for two to three months, but can store only enough carbohydrate to provide for its metabolic needs for less than 24 hours. Carbohydrates, therefore, are used primarily as a quick, short-term energy source.

Polyunsaturated fats and oils are easily oxidized by O_2 by means of a radical chain reaction. In the initiation step, a radical removes a hydrogen from a methylene group that is flanked by two double bonds. This hydrogen is the one most easily removed

OLESTRA: NONFAT WITH FLAVOR

Chemists have been searching for ways to reduce the caloric content of foods without decreasing their flavor. Many people who believe that "no fat" is synonymous with "no flavor" can understand this problem. The federal Food and Drug Administration (Section 30.13) approved the limited use of Olestra as a substitute for dietary fat in snack foods. Procter and Gamble spent 30 years and more than $2 billion to develop this compound. Its approval was based on the results of more than 150 studies.

Olestra is a semisynthetic compound. That is, Olestra does not exist in nature, but its components do. Developing a compound that can be made from units that are a normal part of our diet decreases the potential toxic effects of the new compound. Olestra is made by esterifying all the OH groups of sucrose with fatty acids obtained from cottonseed oil and soybean oil. Therefore, its component parts are table sugar and vegetable oil. Olestra works as a fat substitute because its ester linkages are too hindered to be hydrolyzed by digestive enzymes. As a result, Olestra tastes like fat, but it has no caloric value since it cannot be digested.

Courtesy of Procter & Gamble Company

Olestra

because the resulting radical is resonance stabilized by both double bonds. This radical reacts with O_2, forming a peroxy radical with conjugated double bonds. The peroxy radical removes a hydrogen from a methylene group of another molecule of fatty acid, forming an alkyl hydroperoxide. The two propagating steps are repeated over and over.

RCH=CH—CH(H)—CH=CH— + X· —initiation→ RCH=CH—ĊH—CH=CH— + HX

resonance contributor with isolated double bonds

↕

RCH(·)—CH=CH—CH=CH—

resonance contributor with conjugated double bonds

+ ·O—O· —propagation→

RCH(O—O·)—CH=CH—CH=CH—

a peroxy radical

RCH=CH—CH_2—CH=CH— —propagation→

RCH=CH—ĊH—CH=CH— + RCH(O—OH)—CH=CH—CH=CH—

an alkyl hydroperoxide

The reaction of fatty acids with O_2 causes them to become rancid. The unpleasant taste and smell associated with rancidity are the results of further oxidation of the alkyl hydroperoxide to shorter chain carboxylic acids such as butyric acid that have strong odors. The same process contributes to the odor associated with sour milk.

PROBLEM 5

Draw the resonance contributors for the radical formed when a hydrogen atom is removed from C-10 of arachidonic acid.

WHALES AND ECHOLOCATION

Whales have enormous heads, accounting for 33% of their total weight. They have large deposits of fat in their heads and lower jaws. This fat is very different from both the whale's normal body fat and its dietary fat. Because major anatomical modifications were necessary to accommodate this fat, it must have some important function for the animal. It is now believed that the fat is used for echolocation—emitting sounds in pulses and gaining information by analyzing the returning echoes. The fat in the whale's head focuses the emitted sound waves in a directional beam, and the echoes are received by the fat organ in the lower jaw. This organ transmits the sound to the brain for processing and interpretation, providing the whale with information about the depth of the water, changes in the seafloor, and the position of the coastline. The fat deposits in the whale's head and jaw therefore give the animal a unique acoustic sensory system and allow it to compete successfully for survival with the shark, which also has a well-developed sense of sound direction.

Humpback whale in Alaska

26.4 Membranes

For biological systems to operate, some parts of organisms must be separated from other parts. On a cellular level, the outside of the cell must be separated from the inside. "Greasy" lipid **membranes** serve as the barrier. In addition to isolating the cell's contents, these membranes allow the selective transport of ions and organic molecules into and out of the cell.

Phospholipids

Phosphoacylglycerols (also called **phosphoglycerides**) are the major components of cell membranes. They are similar to triacylglycerols except that a terminal OH group of glycerol is esterified with phosphoric acid rather than with a fatty acid, forming a **phosphatidic acid**. Because phosphoacylglycerols are lipids that contain a phosphate group, they are classified as **phospholipids**. The C-2 carbon of glycerol in phosphoacylglycerols has the *R* configuration.

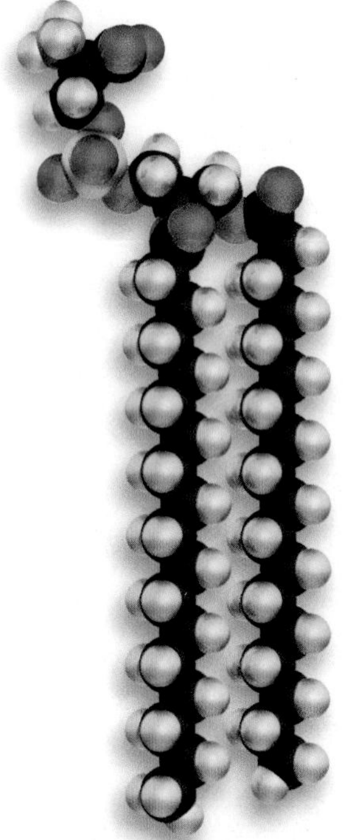

phosphatidylserine

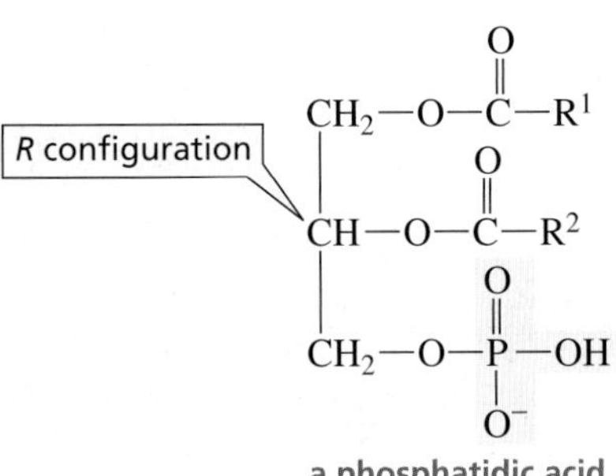

a phosphatidic acid

Phosphatidic acids are the simplest phosphoacylglycerols and are present only in small amounts in membranes. The most common phosphoacylglycerols in membranes have a second phosphate ester linkage. The alcohols most commonly used to form this second ester group are ethanolamine, choline, and serine. Phosphatidylethanolamines are also called *cephalins*, and phosphatidylcholines are called *lecithins*. Used as emulsifying agents, lecithins are added to foods such as mayonnaise to prevent the aqueous and fat components from separating.

$CH_2—O—C(=O)—R^1$
$CH—O—C(=O)—R^2$
$CH_2—O—P(=O)(O^-)—OCH_2CH_2\overset{+}{N}H_3$

a phosphatidylethanolamine
a cephalin

$CH_2—O—C(=O)—R^1$
$CH—O—C(=O)—R^2$
$CH_2—O—P(=O)(O^-)—OCH_2CH_2\overset{+}{N}(CH_3)_3$

a phosphatidylcholine
a lecithin

$CH_2—O—C(=O)—R^1$
$CH—O—C(=O)—R^2$
$CH_2—O—P(=O)(O^-)—OCH_2CH(\overset{+}{N}H_3)COO^-$

a phosphatidylserine

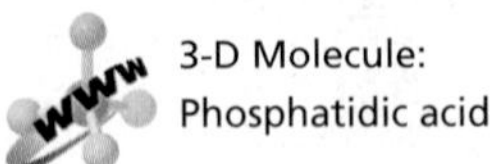

3-D Molecule:
Phosphatidic acid

Phosphoacylglycerols form membranes by arranging themselves in a **lipid bilayer**. The polar heads of the phosphoacylglycerols are on the outside of the bilayer, and the fatty acid chains form the interior of the bilayer. Cholesterol—a membrane lipid discussed in Section 26.9—is also found in the interior of the bilayer (Figure 26.1). A typical bilayer is about 50 Å thick. [Compare the bilayer with the micelles formed by soap in aqueous solution (Section 17.13).]

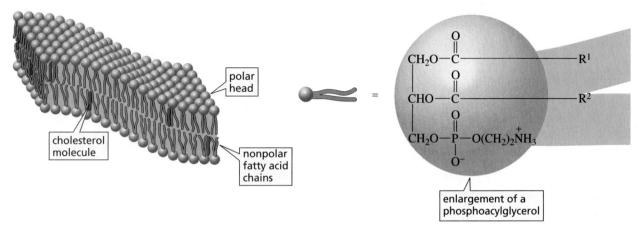

▲ **Figure 26.1**
A lipid bilayer.

The fluidity of a membrane is controlled by the fatty acid components of the phosphoacylglycerols. Saturated fatty acids decrease membrane fluidity because their hydrocarbon chains can pack closely together. Unsaturated fatty acids increase fluidity because they pack less closely together. Cholesterol also decreases fluidity (Section 26.9). Only animal membranes contain cholesterol, so they are more rigid than plant membranes.

The unsaturated fatty acid chains of phosphoacylglycerols are susceptible to reaction with O_2, similar to the reaction described on p. 1081 for fats and oils. Oxidation of phosphoacylglycerols can lead to the degradation of membranes. Vitamin E is an important antioxidant that protects fatty acid chains from degradation via oxidation. Vitamin E, also called α-tocopherol, is classified as a lipid because it is soluble in nonpolar organic solvents. Because vitamin E reacts more rapidly with oxygen than triacylglycerols do, the vitamin prevents biological membranes from reacting with oxygen (Section 9.8). There are some who believe that vitamin E slows the aging process. Because vitamin E also reacts with oxygen more rapidly than fats do, it is added to many foods to prevent spoilage.

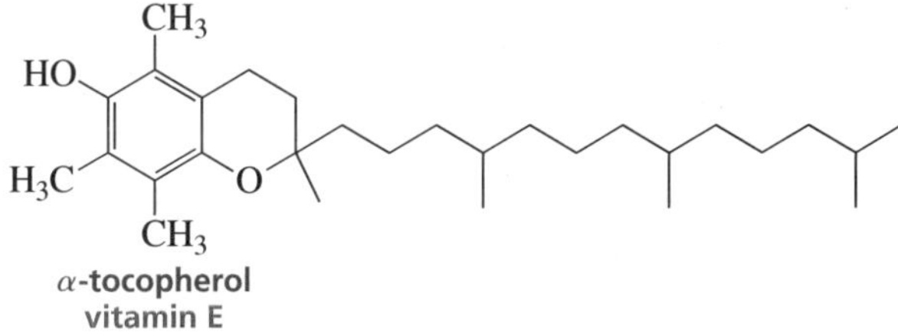

α-tocopherol
vitamin E

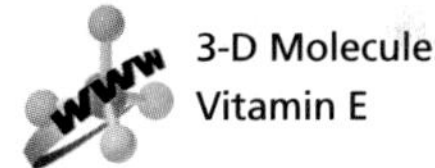
3-D Molecule:
Vitamin E

IS CHOCOLATE A HEALTH FOOD?

We have long been told that our diets should include lots of fruits and vegetables because they are good sources of antioxidants. Antioxidants protect against cardiovascular disease, cancer, and cataracts, and they are thought to slow the effects of aging. Recent studies show that chocolate has high levels of antioxidants—complex mixtures of phenolic compounds (Section 9.8). On a weight basis, the concentration of antioxidants in chocolate is higher than the concentration in red wine or green tea and 20 times higher than the concentration in tomatoes. Dark chocolate contains more than twice the level of antioxidants as milk chocolate. Unfortunately, white chocolate contains no antioxidants. Another piece of good news is that stearic acid, the main fatty acid in chocolate, does not appear to raise blood cholesterol levels the way other saturated fatty acids do.

PROBLEM 6◆

Membranes contain proteins. Integral membrane proteins extend partly or completely through the membrane, whereas peripheral membrane proteins are found on the inner or outer surfaces of the membrane. What is the likely difference in the overall amino acid composition of integral and peripheral membrane proteins?

PROBLEM 7◆

A colony of bacteria accustomed to an environment at 25 °C was moved to an identical environment, at 35 °C. The higher temperature increased the fluidity of the bacterial membranes. What could the bacteria do to regain their original membrane fluidity?

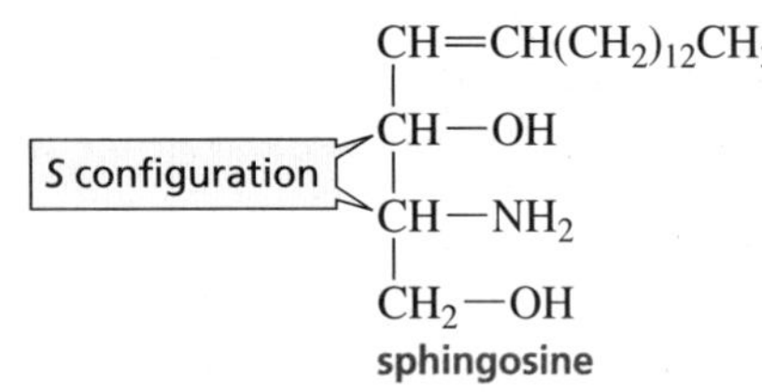

sphingosine

3-D Molecule: Sphingosine

Sphingolipids

Sphingolipids are also found in membranes. They are the major lipid components in the myelin sheaths of nerve fibers. Sphingolipids contain sphingosine instead of glycerol. In sphingolipids, the amino group of sphingosine is bonded to the acyl group of a fatty acid. Both asymmetric carbons in sphingosine have the *S* configuration.

Two of the most common kinds of sphingolipids are *sphingomyelins* and *cerebrosides*. In sphingomyelins, the primary OH group of sphingosine is bonded to phosphocholine or phosphoethanolamine, similar to the bonding in lecithins and cephalins. In cerebrosides, the primary OH group of sphingosine is bonded to a sugar residue through a β-glycosidic linkage (Section 22.13). Sphingomyelins are phospholipids because they contain a phosphate group. Cerebrosides, on the other hand, are not phospholipids.

a sphingomyelin

a glucocerebroside

MULTIPLE SCLEROSIS AND THE MYELIN SHEATH

The myelin sheath is a lipid-rich material that is wrapped around the axons of nerve cells. Composed largely of sphingomyelins and cerebrosides, the sheath functions so as to increase the velocity of nerve impulses. Multiple sclerosis is a disease characterized by loss of the myelin sheath, a consequent slowing of nerve impulses, and eventual paralysis.

PROBLEM 8

a. Draw the structures of three different sphingomyelins.
b. Draw the structure of a galactocerebroside.

PROBLEM 9

The membrane phospholipids in animals such as deer and elk have a higher degree of unsaturation in cells closer to the hoof than in cells closer to the body. Explain how this trait can be important for survival.

26.5 Prostaglandins

Prostaglandins are found in all body tissues and are responsible for regulating a variety of physiological responses, such as inflammation, blood pressure, blood clotting, fever, pain, the induction of labor, and the sleep–wake cycle. All prostaglandins have a five-membered ring with a seven-carbon carboxylic acid substituent and an eight-carbon hydrocarbon substituent. The two substituents are trans to each other.

prostaglandin skeleton

Prostaglandins are named in accordance with the format PGX, where X designates the functional groups of the five-membered ring. PGAs, PGBs, and PGCs all contain a carbonyl group and a double bond in the five-membered ring. The location of the double bond determines whether a prostaglandin is a PGA, PGB, or PGC. PGDs and PGEs are β-hydroxy ketones, and PGFs are 1,3-diols. A subscript indicates the total number of double bonds in the side chains, and "α" and "β" indicate the configuration of the two OH groups in a PGF: "α" indicates a cis diol and "β" indicates a trans diol.

PGAs PGBs PGCs PGDs

PGE_1 PGE_2

$PGF_{2\alpha}$

Ulf Svante von Euler (1905–1983) *first identified prostaglandins—from semen—in the early 1930s. He named them for their source, the prostate gland. By the time it was realized that all cells except red blood cells synthesize prostaglandins, their name had become entrenched. Von Euler was born in Stockholm and received an M.D. from the Karolinska Institute, where he remained as a member of the faculty. He discovered noradrenaline and identified its function as a chemical intermediate in nerve transmission. For this work, he shared the 1970 Nobel Prize in physiology or medicine with Julius Axelrod and Sir Bernard Katz.*

Prostaglandins are synthesized from arachidonic acid, a 20-carbon fatty acid with four cis double bonds. In the cell, arachidonic acid is found esterified to the 2-position of glycerol in many phospholipids. Arachidonic acid is synthesized from linoleic acid. Because linoleic acid cannot be synthesized by mammals, it must be included in the diet.

An enzyme called prostaglandin endoperoxide synthase catalyzes the conversion of arachidonic acid to PGH_2, the precursor of all prostaglandins. There are two forms of this enzyme; one carries out the normal physiological production of prostaglandin, and the other synthesizes additional prostaglandin in response to inflammation. The enzyme has two activities: a cyclooxygenase activity and a hydroperoxidase activity. It uses its cyclooxygenase activity to form the five-membered ring. In the first step of this transformation, a hydrogen atom is removed from a carbon flanked by two double bonds. This hydrogen is removed relatively easily because the resulting radical is

For their work on prostaglandins, **Sune Bergström, Bengt Ingemar Samuelsson,** *and* **John Robert Vane** *shared the 1982 Nobel Prize in physiology or medicine. Bergström and Samuelsson were born in Sweden—Bergström in 1916 and Samuelsson in 1934. They are both at the Karolinska Institute. Vane was born in England in 1927 and is at the Wellcome Foundation in Beckenham, England.*

stabilized by electron delocalization. The radical reacts with oxygen to form a peroxy radical. Notice that these two steps are the same as the first two steps in the reaction that causes fats to become rancid (Section 26.3). The peroxy radical rearranges and reacts with a second molecule of oxygen. The enzyme then uses its hydroperoxidase activity to convert the OOH group into an OH group, forming PGH_2, which rearranges to form PGE_2, a prostaglandin.

biosynthesis of prostaglandins, thromboxanes, and prostacyclins

arachidonic acid

cyclooxygenase

a peroxy radical

hydroperoxidase

PGH_2

several steps

a thromboxane

several steps

a prostacyclin

PGE_2
a prostaglandin

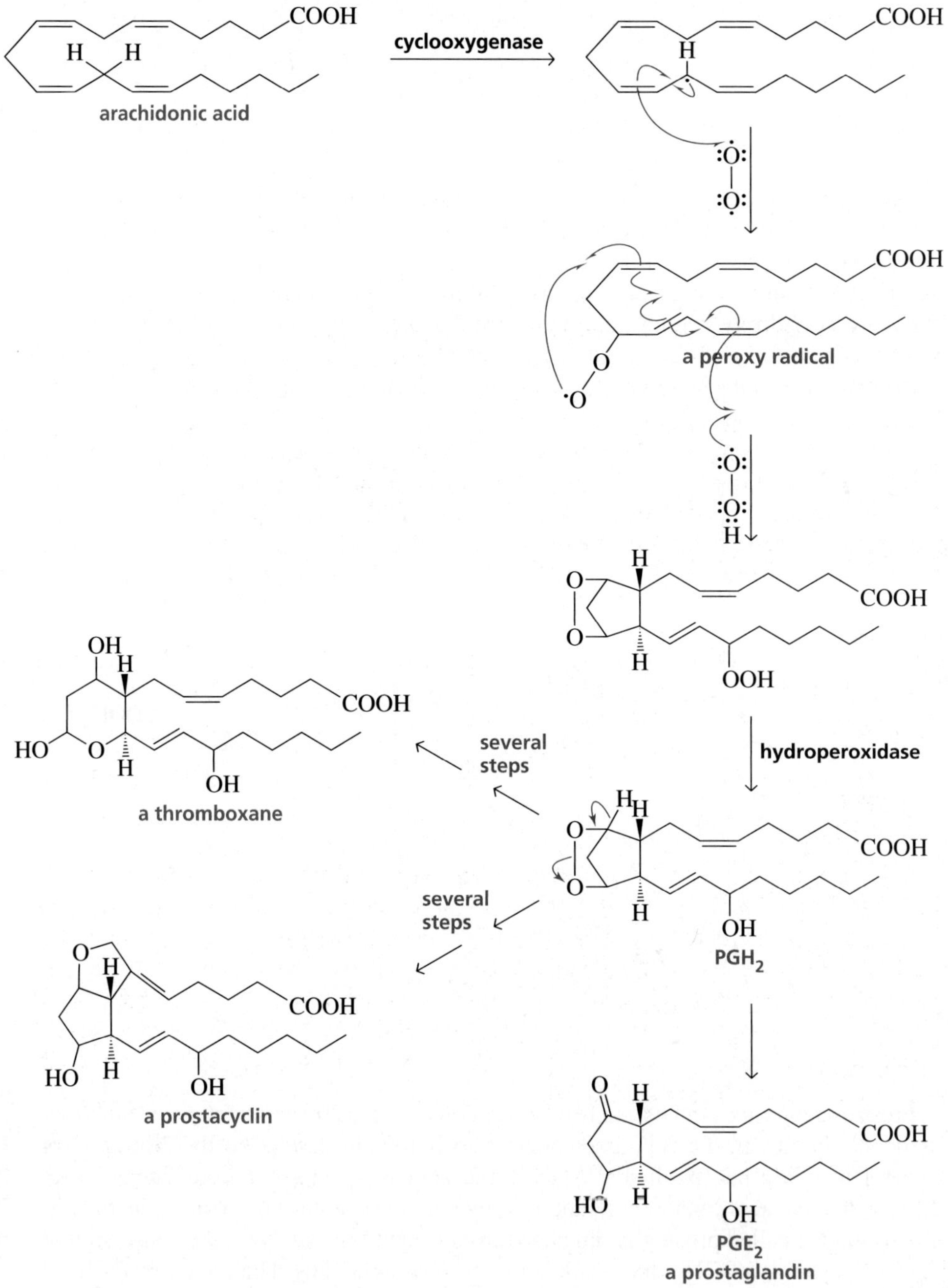

In addition to serving as a precursor for the synthesis of prostaglandins, PGH_2 is a precursor for the synthesis of *thromboxanes* and *prostacyclins*. Thromboxanes constrict blood vessels and stimulate the aggregation of platelets, the first step in blood clotting. Prostacyclins have the opposite effect, dilating blood vessels and inhibiting the aggregation of platelets. The levels of these two compounds must be carefully controlled to maintain the proper balance in the blood.

Aspirin (acetylsalicylic acid) inhibits the cyclooxygenase activity of prostaglandin endoperoxide synthase. It does this by transferring an acetyl group to a serine hydroxyl group of the enzyme (Section 17.10). Aspirin, therefore, inhibits the synthesis of prostaglandins and, in that way, decreases the inflammation produced by these compounds. Aspirin also inhibits the synthesis of thromboxanes and prostacyclins. Overall, this causes a slight decrease in the rate of blood clotting, which is why some doctors recommend one aspirin tablet every other day to reduce the chance of a heart attack or stroke caused by clotting in blood vessels.

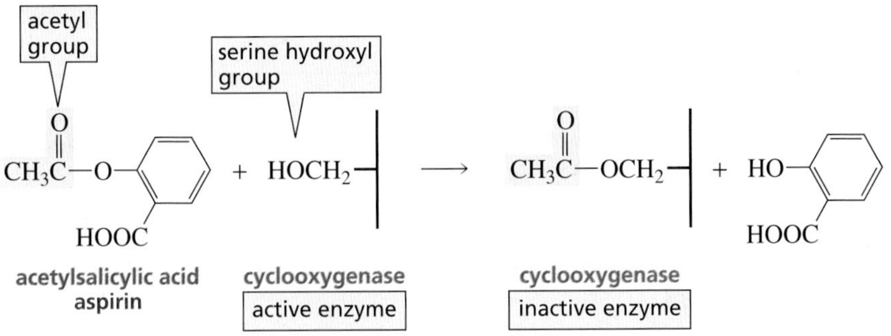

acetyl group — serine hydroxyl group

$CH_3C(=O)-O-C_6H_4-COOH$ + $HOCH_2$–[enzyme] ⟶ $CH_3C(=O)-OCH_2$–[enzyme] + $HO-C_6H_4-COOH$

acetylsalicylic acid aspirin — cyclooxygenase active enzyme — cyclooxygenase inactive enzyme

Other anti-inflammatory drugs, such as ibuprofen (the active ingredient in Advil®, Motrin®, and Nuprin®) and naproxen (the active ingredient in Aleve®), also inhibit the synthesis of prostaglandins. They compete with either arachidonic acid or the peroxy radical for the enzyme's binding site.

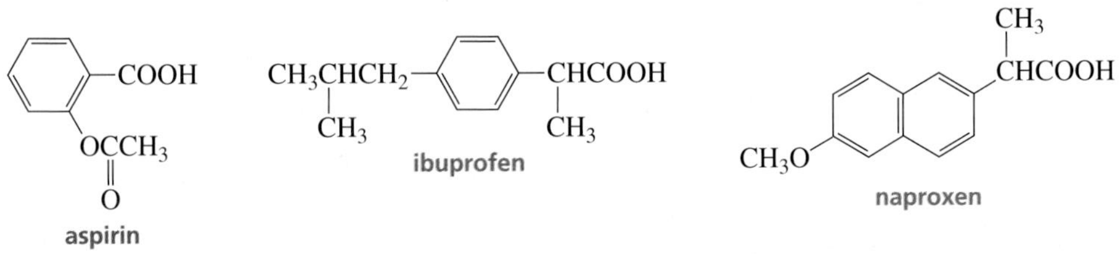

aspirin — ibuprofen — naproxen

Both aspirin and these other nonsteroidal anti-inflammatory drugs (NSAIDs) inhibit the synthesis of all prostaglandins—those produced under normal physiological conditions and those produced in response to inflammation. The production of acid in the stomach is regulated by a prostaglandin. When prostaglandin synthesis stops, therefore, the acidity of the stomach can rise above normal levels. New drugs—Celebrex® and Vioxx®—that recently have become available inhibit only the enzyme that produces prostaglandin in response to stress. Thus, inflammatory conditions now can be treated without some of the harmful side effects.

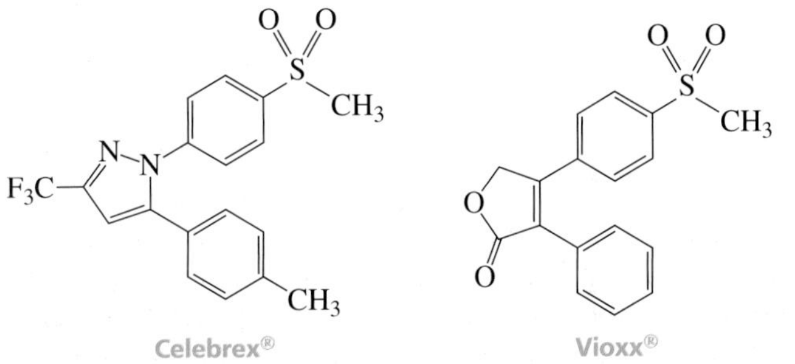

Celebrex® — Vioxx®

Arachidonic acid can also be converted into a *leukotriene*. Because they induce contraction of the muscle that lines the airways to the lungs, leukotrienes are implicated in allergic reactions, inflammatory reactions, and heart attacks. Leukotrienes also bring on the symptoms of asthma and are implicated in anaphylactic shock, a potentially fatal allergic reaction. There are several antileukotriene agents available for the treatment of asthma.

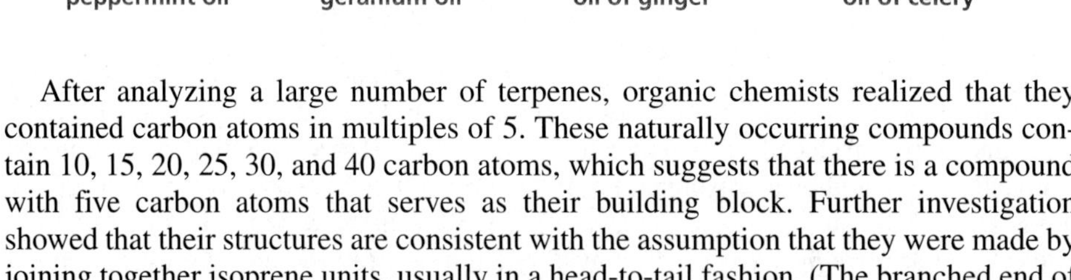

arachidonic acid → a leukotriene

PROBLEM 10

Treating PGA_2 with a strong base such as sodium *tert*-butoxide followed by addition of acid converts it to PGC_2. Propose a mechanism for this reaction.

26.6 Terpenes

Terpenes are a diverse class of lipids. More than 20,000 terpenes are known. They can be hydrocarbons, or they can contain oxygen and be alcohols, ketones, or aldehydes. Oxygen-containing terpenes are sometimes called **terpenoids**. Certain terpenes and terpenoids have been used as spices, perfumes, and medicines for many thousands of years.

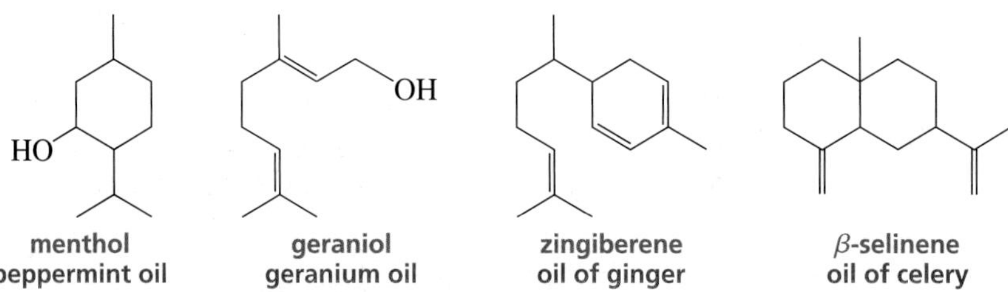

menthol peppermint oil | geraniol geranium oil | zingiberene oil of ginger | β-selinene oil of celery

Leopold Stephen Ružička (1887–1976) *was the first to recognize that many organic compounds contain multiples of five carbons. A Croatian, Ružička attended college in Switzerland and became a Swiss citizen in 1917. He was a professor of chemistry at the University of Utrecht in the Netherlands and later at the Federal Institute of Technology in Zürich. For his work on terpenes, he shared the 1939 Nobel Prize in chemistry with Adolph Butenandt (p. 1099).*

After analyzing a large number of terpenes, organic chemists realized that they contained carbon atoms in multiples of 5. These naturally occurring compounds contain 10, 15, 20, 25, 30, and 40 carbon atoms, which suggests that there is a compound with five carbon atoms that serves as their building block. Further investigation showed that their structures are consistent with the assumption that they were made by joining together isoprene units, usually in a head-to-tail fashion. (The branched end of isoprene is called the head, and the unbranched end is called the tail.) Isoprene is the common name for 2-methyl-1,3-butadiene, a compound containing five carbon atoms.

That isoprene units are linked in a head-to-tail fashion to form terpenes is known as the **isoprene rule**.

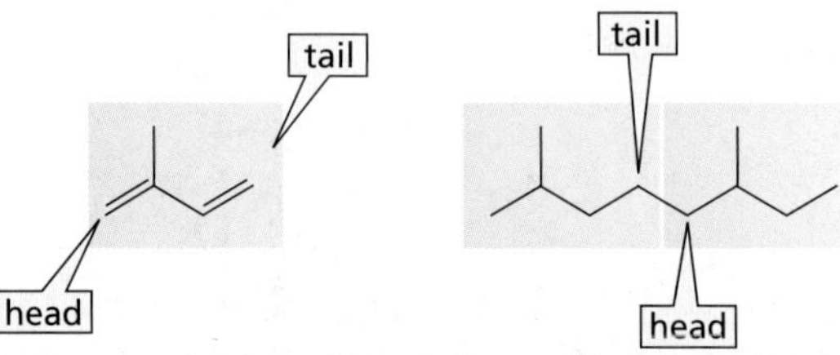

carbon skeleton of two isoprene units with a bond between the tail of one and the head of another

In the case of cyclic compounds, the linkage of the head of one isoprene unit to the tail of another is followed by an additional linkage to form the ring. The second linkage is not necessarily head-to-tail, but is whatever is necessary to form a stable five- or six-membered ring.

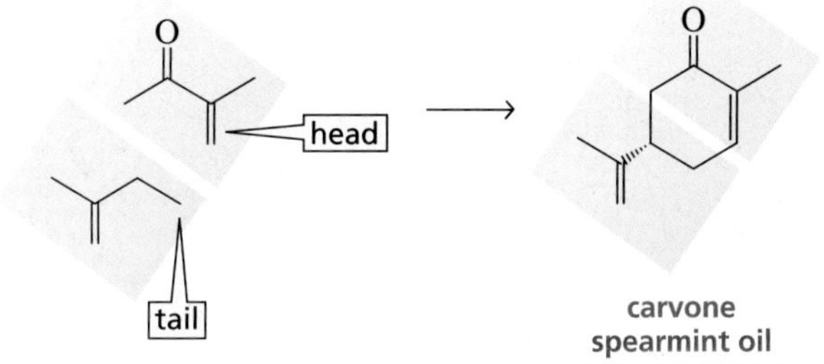

head

tail

carvone
spearmint oil
a monoterpene

In Section 26.8, we will see that the compound actually used in the biosynthesis of terpenes is not isoprene, but isopentenyl pyrophosphate, a compound that has the same carbon skeleton as isoprene. We will also look at the mechanism by which isopentenyl pyrophosphate units are joined together in a head-to-tail fashion.

Terpenes are classified according to the number of carbons they contain (Table 26.3). **Monoterpenes** are composed of two isoprene units, so they have 10 carbons. **Sesquiterpenes**, with 15 carbons, are composed of three isoprene units. Many fragrances and flavorings found in plants are monoterpenes and sesquiterpenes. These compounds are known as *essential oils*.

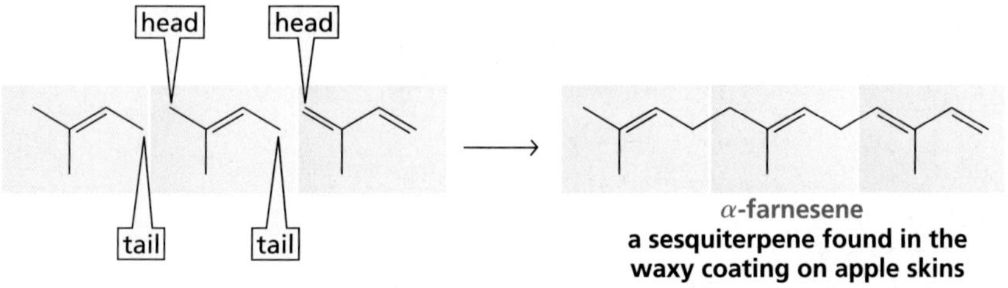

head head

tail tail

α-farnesene
a sesquiterpene found in the waxy coating on apple skins

Table 26.3 Classification of Terpenes

Carbon atoms	Classification	Carbon atoms	Classification
10	monoterpenes	25	sesterterpenes
15	sesquiterpenes	30	triterpenes
20	diterpenes	40	tetraterpenes

Triterpenes (six isoprene units) and **tetraterpenes** (eight isoprene units) have important biological roles. For example, **squalene**, a triterpene, is a precursor of steroid molecules (Section 26.9).

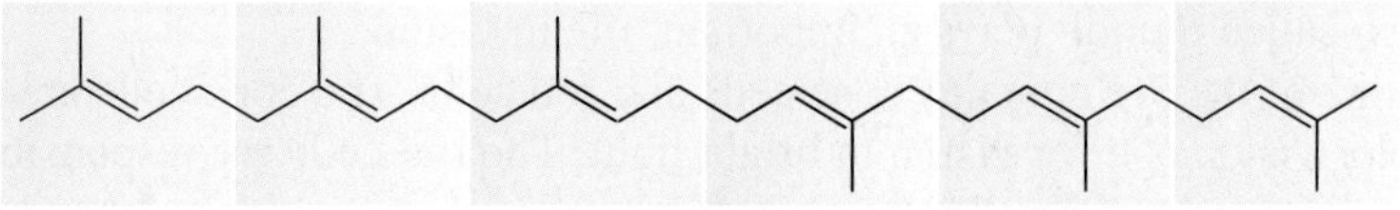

squalene

Carotenoids are tetraterpenes. Lycopene, the compound responsible for the red coloring of tomatoes and watermelon, and β-carotene, the compound that causes carrots and apricots to be orange, are examples of carotenoids. β-Carotene is also the coloring agent used in margarine. β-Carotene and other colored compounds are found in the leaves of trees, but their characteristic colors are usually obscured by the green color of chlorophyll. In the fall, when chlorophyll degrades, the colors become apparent. The many conjugated double bonds in lycopene and β-carotene cause the compounds to be colored (Section 8.13).

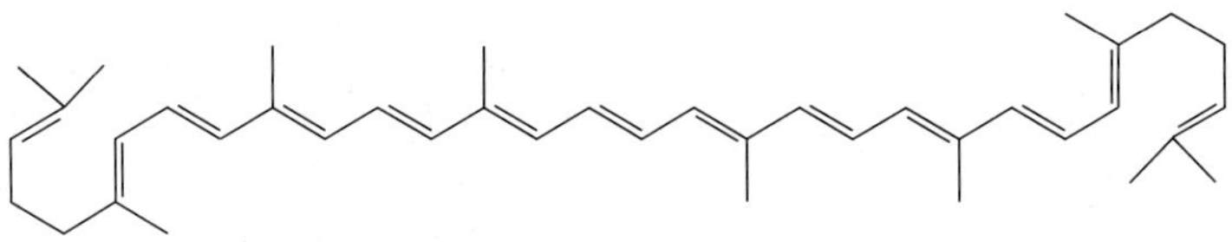

lycopene

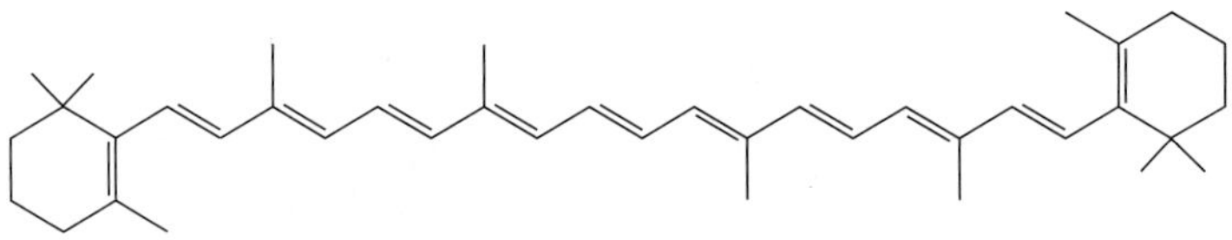

β-carotene

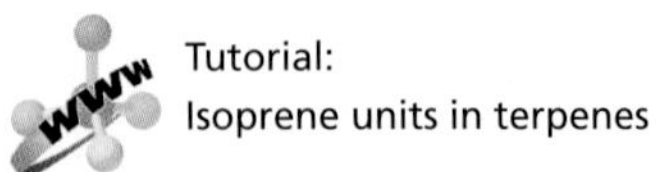
Tutorial:
Isoprene units in terpenes

PROBLEM 11 SOLVED

Mark off the isoprene units in menthol, zingiberene, β-selinene, and squalene.

SOLUTION For zingiberene, we have

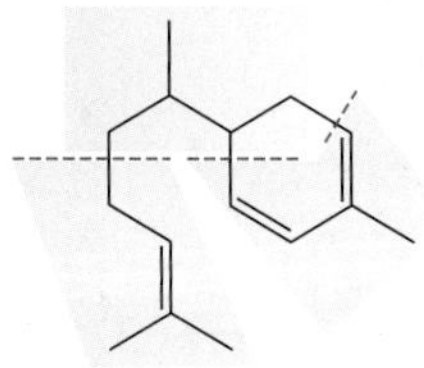

PROBLEM 12◆

One of the linkages in squalene is tail-to-tail, not head-to-tail. What does this suggest about how squalene is synthesized in nature? (*Hint*: Locate the position of the tail-to-tail linkage.)

PROBLEM 13

Mark off the isoprene units in lycopene and β-carotene. Can you detect a similarity in the way in which squalene, lycopene, and β-carotene are biosynthesized?

26.7 Vitamin A

Vitamins A, D, E, and K are lipids (Sections 25.9 and 29.6). Vitamin A is the only water-insoluble vitamin we have not already discussed. β-Carotene, which is cleaved to form two molecules of vitamin A, is the major dietary source of the vitamin. Vitamin A, also called retinol, plays an important role in vision.

The retina of the eye contains cone cells and rod cells. The cone cells are responsible for color vision and for vision in bright light. The rod cells are responsible for vision in dim light. In rod cells, vitamin A is oxidized to an aldehyde and the trans double bond at C-11 is isomerized to a cis double bond. The mechanism for the enzyme-catalyzed interconversion of cis and trans double bonds is discussed in Section 18.15. The protein *opsin* uses a lysine side chain (Lys 216) to form an imine with (11*Z*)-retinal, resulting in a complex known as *rhodopsin.* When rhodopsin absorbs visible light, it isomerizes to the trans isomer. This change in molecular geometry causes an electrical signal to be sent to the brain, where it is perceived as a visual image. The trans isomer of rhodopsin is not stable and is hydrolyzed to (11*E*)-retinal and opsin in a reaction referred to as bleaching of the visual pigment. (11*E*)-Retinal is then converted back to (11*Z*)-retinal to complete the vision cycle.

the chemistry of vision

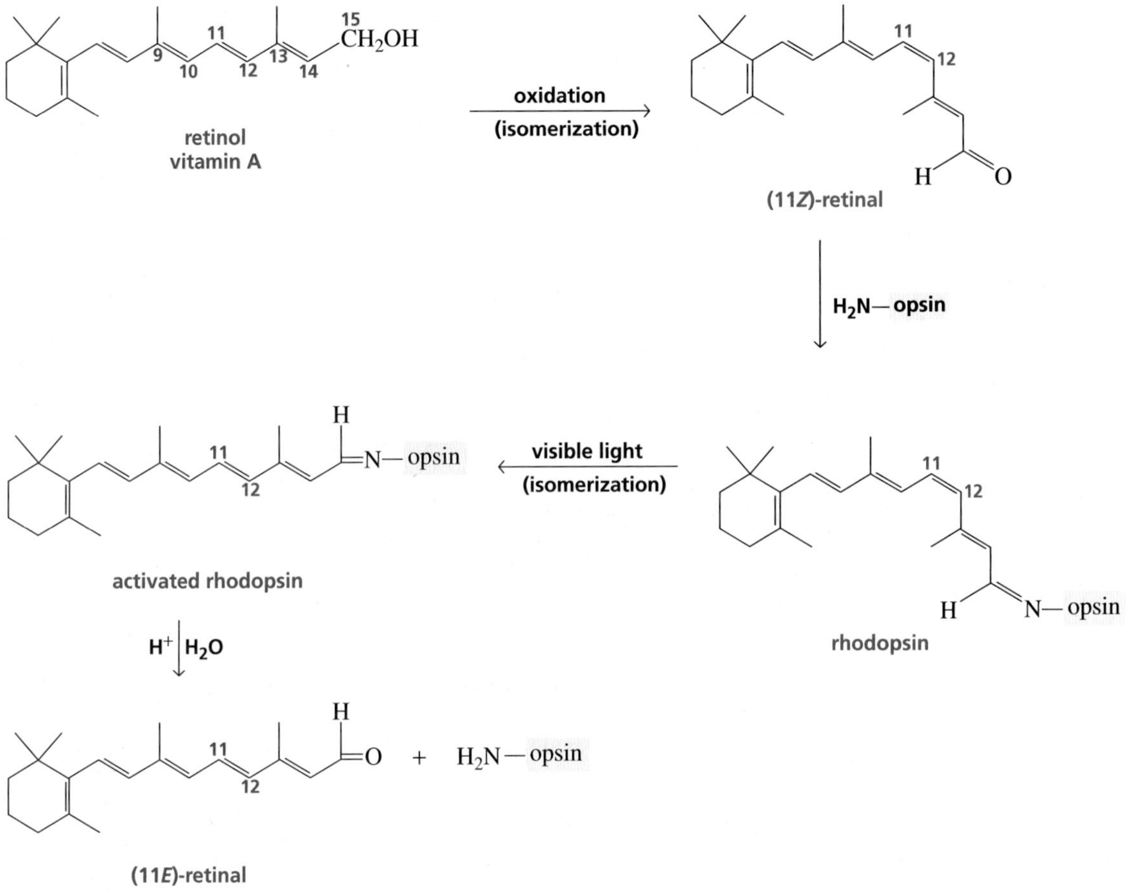

retinol
vitamin A

oxidation
(isomerization)

(11*Z*)-retinal

H_2N— opsin

rhodopsin

visible light
(isomerization)

activated rhodopsin

H^+ H_2O

(11*E*)-retinal + H_2N— opsin

The details of how the foregoing sequence of reactions creates a visual image are not clearly understood. The fact that a simple change in configuration can be responsible for initiating a process as complicated as vision, though, is remarkable.

26.8 Biosynthesis of Terpenes

Biosynthesis of Isopentenyl Pyrophosphate

The five-carbon compound used for the biosynthesis of terpenes is 3-methyl-3-butenylpyrophosphate, loosely called isopentenyl pyrophosphate by biochemists. Each step in its biosynthesis is catalyzed by a different enzyme. The first step is the same Claisen condensation that occurs in the first step of the biosynthesis of fatty acids, except that the acetyl and malonyl groups remain attached to coenzyme A rather than being transferred to the acyl carrier protein (Section 19.21). The Claisen condensation is followed by an aldol addition with a second molecule of malonyl-CoA. The resulting thioester is reduced with two equivalents of NADPH to form mevalonic acid (Section 25.2). A pyrophosphate group is added by means of two successive phosphorylations with ATP. Decarboxylation and loss of the OH group result in isopentenyl pyrophosphate.

biosynthesis of isopentenyl pyrophosphate

acetyl-CoA + malonyl-CoA —(Claisen condensation)→ acetoacetyl-CoA + CO_2 + CoASH

acetoacetyl-CoA —(1. malonyl-CoA; 2. H_2O)→ hydroxymethylglutaryl-CoA + CoASH + CO_2

hydroxymethylglutaryl-CoA —(2 NADPH → 2 NADP$^+$)→ mevalonic acid —(ATP → ADP)→ mevalonyl phosphate

mevalonyl phosphate —(ATP → ADP)→ mevalonyl pyrophosphate —(ATP → ADP)→ isopentenyl pyrophosphate + CO_2 + HPO_4^{2-}

The mechanism for converting mevalonic acid into mevalonyl phosphate is essentially an S_N2 reaction with an adenosyl pyrophosphate leaving group (Section 27.3). A second S_N2 reaction converts mevalonyl phosphate to mevalonyl pyrophosphate. ATP is an excellent phosphorylating reagent for nucleophiles because its phosphoanhydride bonds are easily broken. The reason that phosphoanhydride bonds are so easily broken is discussed in Section 27.4.

mevalonic acid + ATP → mevalonyl phosphate + ADP + H$^+$

mevalonyl phosphate + ATP → mevalonyl pyrophosphate + ADP

PROBLEM 14 SOLVED

Give the mechanism for the last step in the biosynthesis of isopentenyl pyrophosphate, showing why ATP is required.

SOLUTION In the last step of the biosynthesis of isopentenyl pyrophosphate, elimination of CO_2 is accompanied by elimination of an ^{-}OH group, which is a strong base and therefore a poor leaving group. ATP is used to convert the OH group into a phosphate group, which is easily eliminated because it is a weak base.

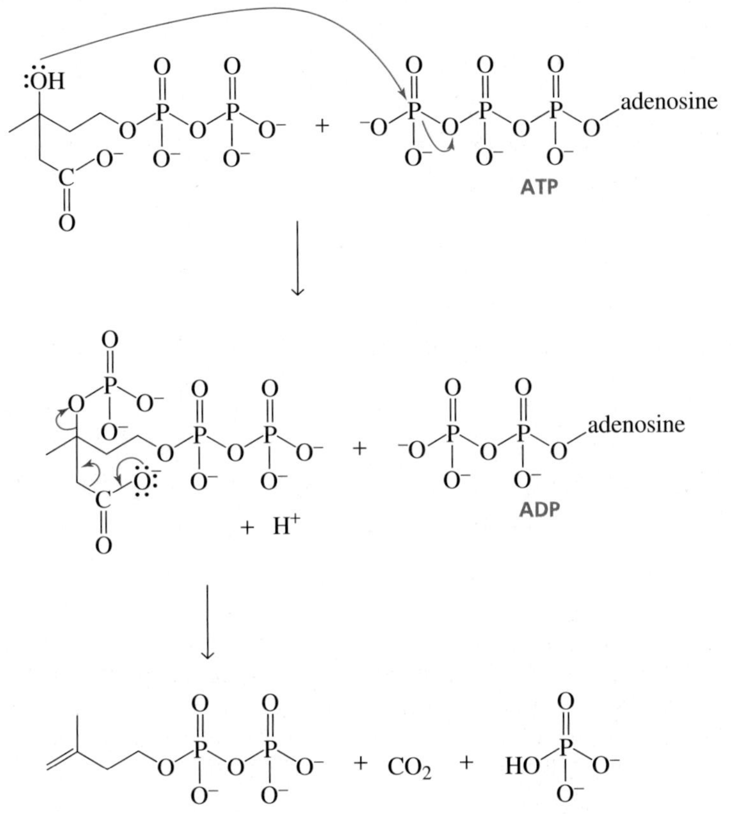

ATP

adenosine

ADP

$+ \ H^+$

$+ \ CO_2 \ +$

PROBLEM 15

Give the mechanisms for the Claisen condensation and aldol addition that occur in the first two steps of the biosynthesis of isopentenyl pyrophosphate.

Biosynthesis of Dimethylallyl Pyrophosphate

Both **isopentenyl pyrophosphate** and **dimethylallyl pyrophosphate** are needed for the biosynthesis of terpenes. Therefore, some isopentenyl pyrophosphate is converted to dimethylallyl pyrophosphate by an enzyme-catalyzed isomerization reaction. The isomerization involves addition of a proton to the sp^2 carbon of isopentenyl pyrophosphate that is bonded to the greater number of hydrogens (Section 4.4) and elimination of a proton from the carbocation intermediate in accordance with Zaitsev's rule (Section 11.2).

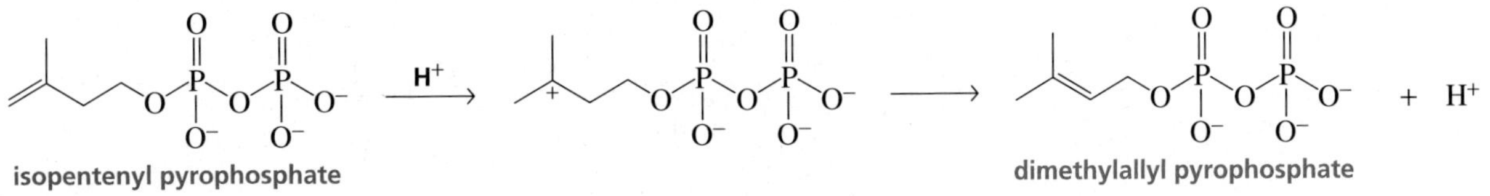

H^+ → → $+ \ H^+$

isopentenyl pyrophosphate

dimethylallyl pyrophosphate

Terpene Biosynthesis

The reaction of dimethylallyl pyrophosphate with isopentenyl pyrophosphate forms geranyl pyrophosphate, a 10-carbon compound. In the first step of the reaction, isopentenyl pyrophosphate acts as a nucleophile and displaces a pyrophosphate group from dimethylallyl pyrophosphate. Pyrophosphate is an excellent leaving group: Its four OH groups have pK_a values of 0.9, 2.0, 6.6, and 9.4. Therefore, three of the four groups will be primarily in their basic forms at physiological pH (pH = 7.3). A proton is removed in the next step, resulting in the formation of geranyl pyrophosphate.

dimethylallyl pyrophosphate + isopentenyl pyrophosphate → (carbocation intermediate) + pyrophosphate → geranyl pyrophosphate + H^+

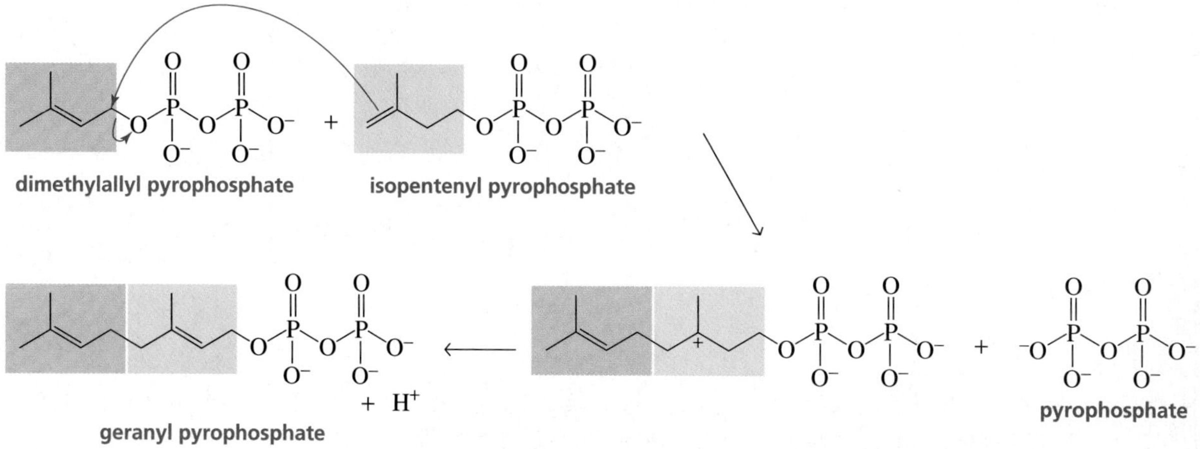

The following scheme shows how some of the many monoterpenes could be synthesized from geranyl pyrophosphate:

geranyl pyrophosphate —H_2O→ geraniol (in rose and geranium oils) —reduction→ citronellol (in rose and geranium oils) —oxidation→ citronellal (in lemon oil)

geranyl pyrophosphate → (cyclic carbocation) —H_2O→ α-terpineol (in juniper oil) —H^+, H_2O→ terpin hydrate (a common constituent of cough medicine)

(cyclic carbocation) → limonene (in orange and lemon oils) —oxidation→ (ketone) —reduction→ menthol (in peppermint oil)

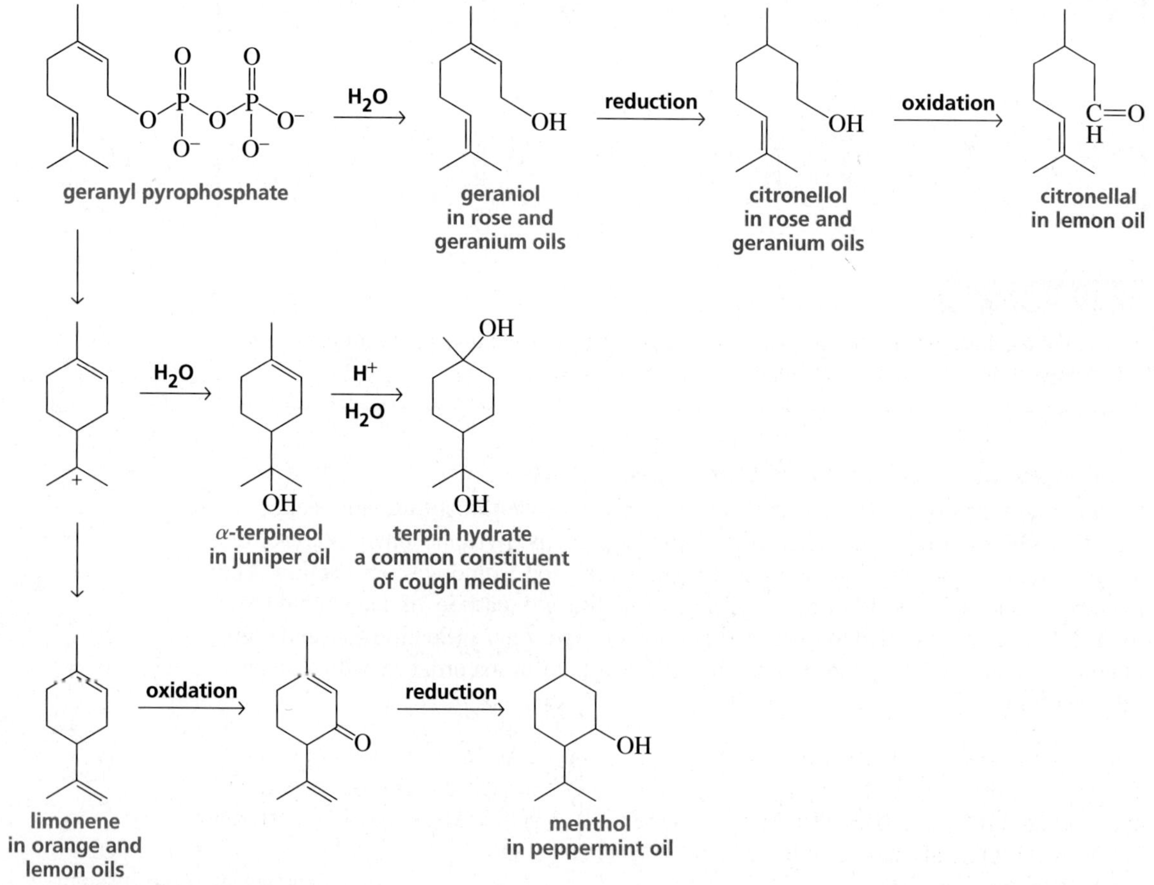

PROBLEM 16

Propose a mechanism for the conversion of the *E* isomer of geranyl pyrophosphate to the *Z* isomer.

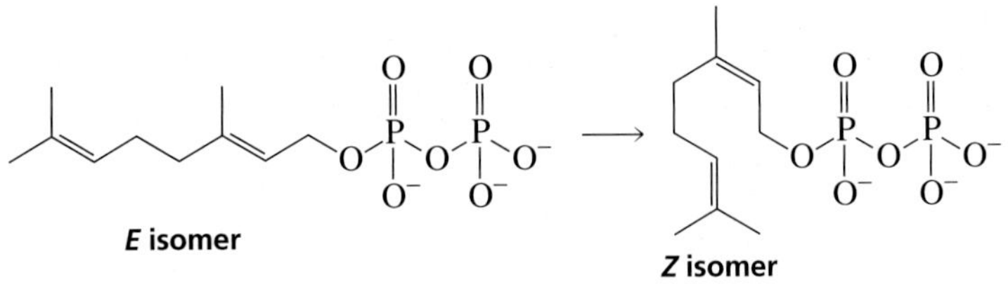

E isomer

Z isomer

PROBLEM 17

Propose mechanisms for the formation of α-terpineol and limonene from geranyl pyrophosphate.

Geranyl pyrophosphate can react with another molecule of isopentenyl pyrophosphate to form farnesyl pyrophosphate, a 15-carbon compound.

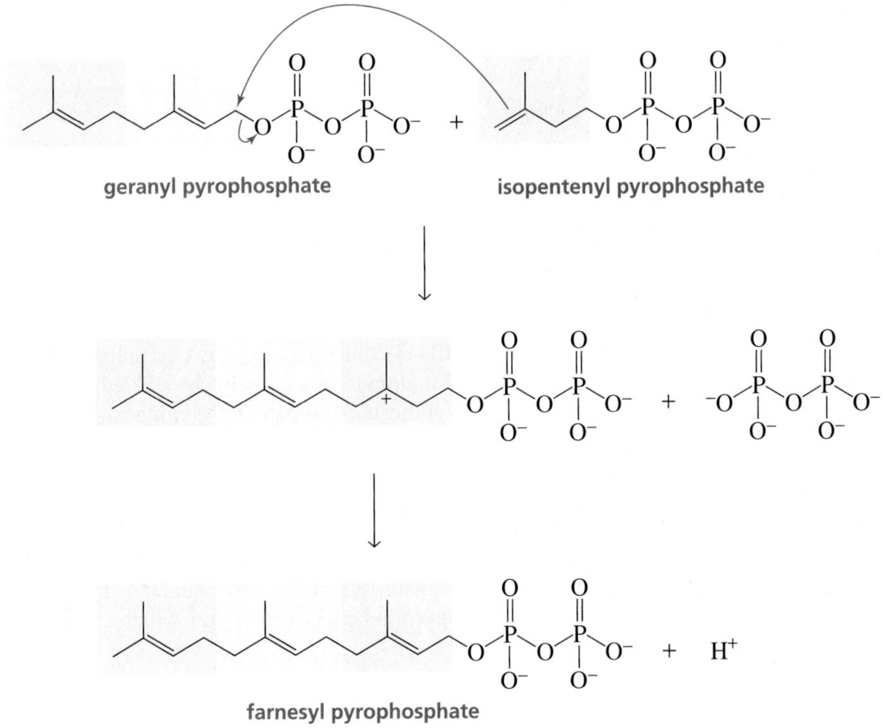

geranyl pyrophosphate

isopentenyl pyrophosphate

farnesyl pyrophosphate

Two molecules of farnesyl pyrophosphate form squalene, a 30-carbon compound. The reaction is catalyzed by the enzyme squalene synthase, which joins the two molecules in a tail-to-tail linkage. Squalene is the precursor of cholesterol, and cholesterol is the precursor of all other steroids.

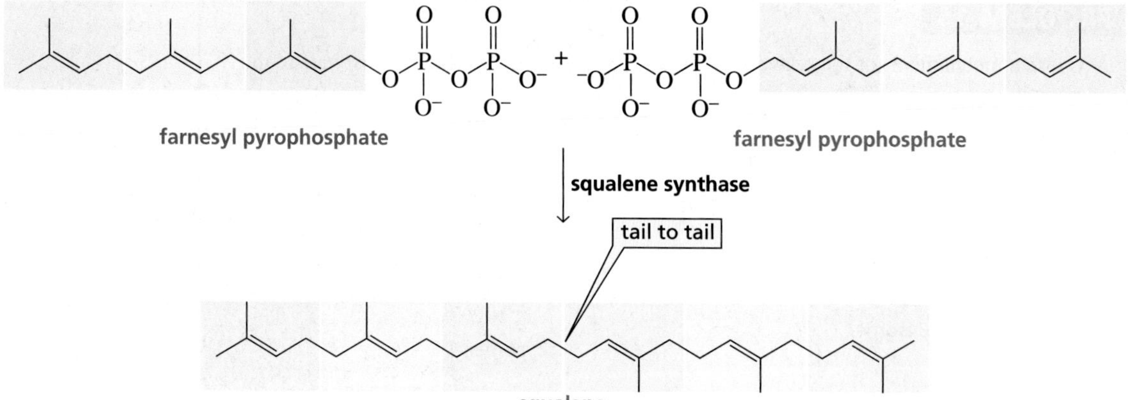

Farnesyl pyrophosphate can react with another molecule of isopentenyl pyrophosphate to form geranylgeranyl pyrophosphate, a 20-carbon compound. Two geranylgeranyl pyrophosphates can join to form phytoene, a 40-carbon compound. Phytoene is the precursor of the carotenoid (tetraterpene) pigments in plants.

PROBLEM 18

In aqueous acidic solution, farnesyl pyrophosphate forms the following sesquiterpene:

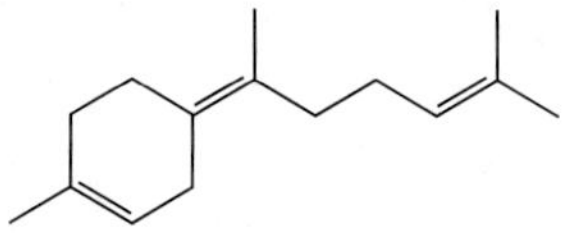

Propose a mechanism for this reaction.

PROBLEM 19 SOLVED

If squalene were synthesized in a medium containing acetate whose carbonyl carbon were labeled with radioactive ^{14}C, which carbons in squalene would be labeled?

SOLUTION Acetate reacts with ATP to form acetyl adenylate, which then reacts with CoASH to form acetyl-CoA (Section 17.20). Because malonyl-CoA is prepared from acetyl-CoA, the thioester carbonyl carbon of malonyl-CoA will also be labeled. Examining each step of the mechanism for the biosynthesis of isopentenyl pyrophosphate from acetyl-CoA and malonyl-CoA allows you to determine the locations of the radioactively labeled carbons in isopentenyl pyrophosphate. Similarly, the locations of the radioactively labeled carbons in geranyl pyrophosphate can be determined from the mechanism for its biosynthesis from isopentenyl pyrophosphate. And the locations of the radioactively labeled carbons in farnesyl pyrophosphate can be determined from the mechanism for its biosynthesis from geranyl pyrophosphate. Knowing that squalene is obtained from a tail-to-tail linkage of two farnesyl pyrophosphates tells you which carbons in squalene will be labeled.

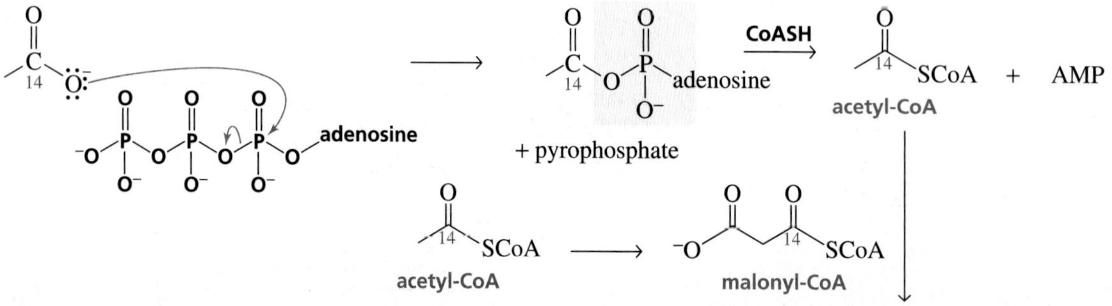

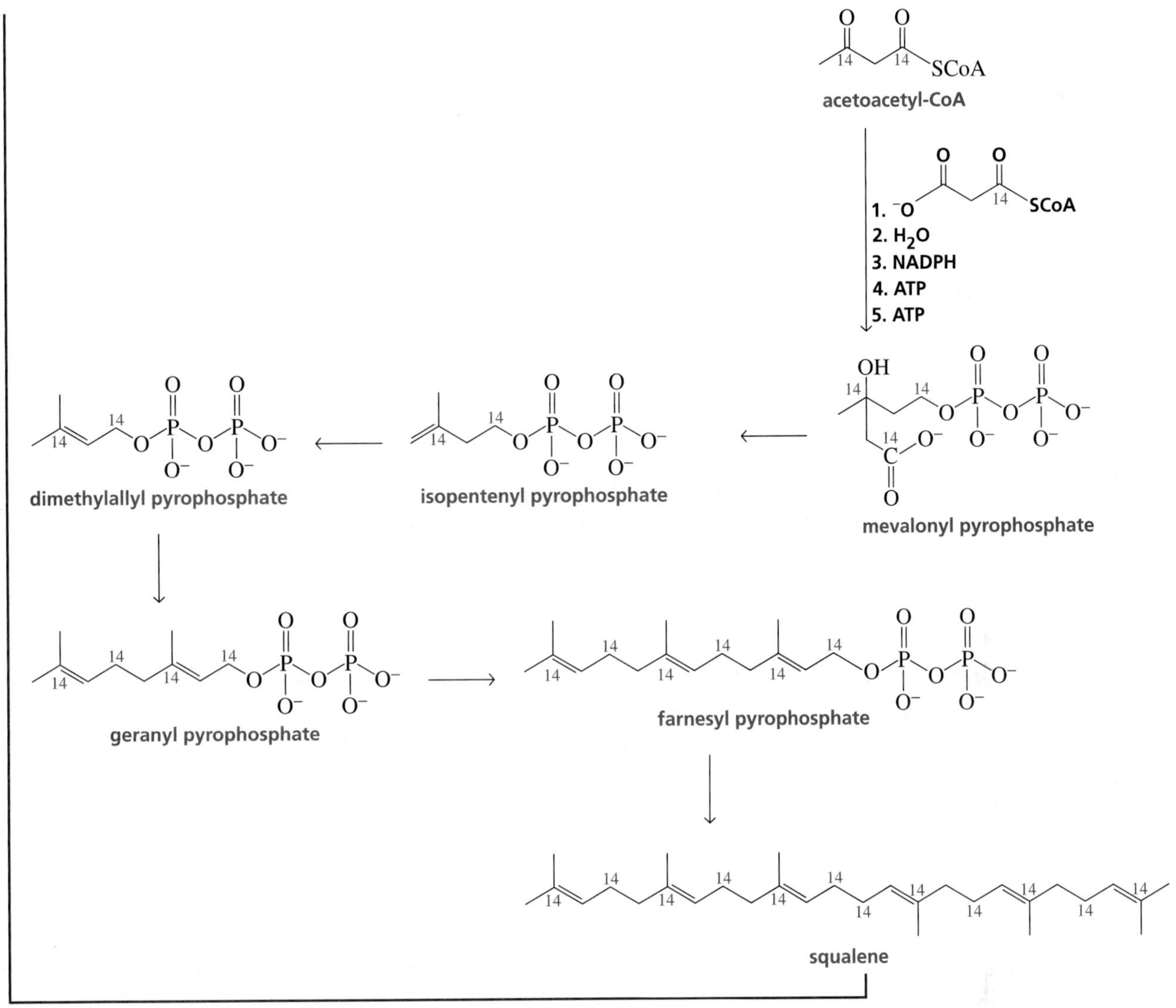

acetoacetyl-CoA

1. ^-O … SCoA
2. H_2O
3. NADPH
4. ATP
5. ATP

mevalonyl pyrophosphate

isopentenyl pyrophosphate

dimethylallyl pyrophosphate

geranyl pyrophosphate

farnesyl pyrophosphate

squalene

26.9 Steroids

Hormones are chemical messengers—organic compounds synthesized in glands and delivered by the bloodstream to target tissues in order to stimulate or inhibit some process. Many hormones are **steroids**. Because steroids are nonpolar compounds, they are lipids. Their nonpolar character allows them to cross cell membranes, so they can leave the cells in which they are synthesized and enter their target cells.

All steroids contain a tetracyclic ring system. The four rings are designated A, B, C, and D. A, B, and C are six-membered rings and D is a five-membered ring. The carbons in the steroid ring system are numbered as shown.

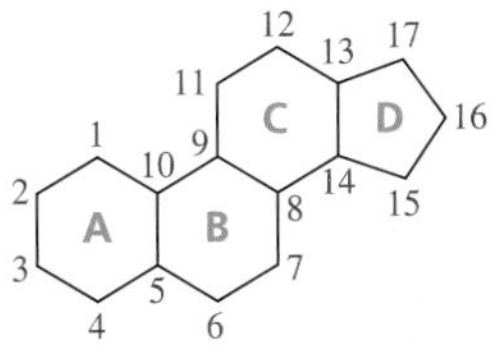

the steroid ring system

We have seen that rings can be **trans fused** or **cis fused** and that trans fused rings are more stable (Section 2.15). In steroids, the B, C, and D rings are all trans fused. In most naturally occurring steroids, the A and B rings are also trans fused.

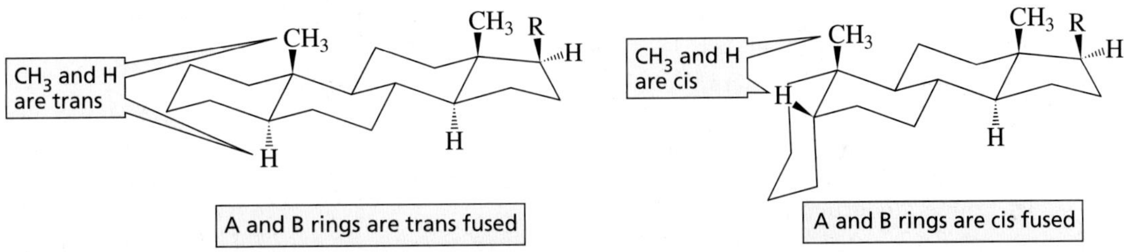

CH_3 and H are trans

A and B rings are trans fused

CH_3 and H are cis

A and B rings are cis fused

Two German chemists, **Heinrich Otto Wieland (1877–1957)** *and* **Adolf Windaus (1876–1959)**, *each received a Nobel Prize in chemistry (Wieland in 1927 and Windaus in 1928) for work that led to the determination of the structure of cholesterol.*

Heinrich Wieland, *the son of a chemist, was a professor at the University of Munich, where he showed that bile acids were steroids and determined their individual structures. During World War II, he remained in Germany but was openly anti-Nazi.*

Adolf Windaus *originally intended to be a physician, but the experience of working with Emil Fischer for a year changed his mind. He discovered that vitamin D was a steroid, and he was the first to recognize that vitamin B_1 contained sulfur.*

Many steroids have methyl groups at the 10- and 13-positions. These are called **angular methyl groups**. When steroids are drawn, both angular methyl groups are shown to be above the plane of the steroid ring system. Substituents on the same side of the steroid ring system as the angular methyl groups are designated **β-substituents** (indicated by a solid wedge). Those on the opposite side of the plane of the ring system are **α-substituents** (indicated by a hatched wedge).

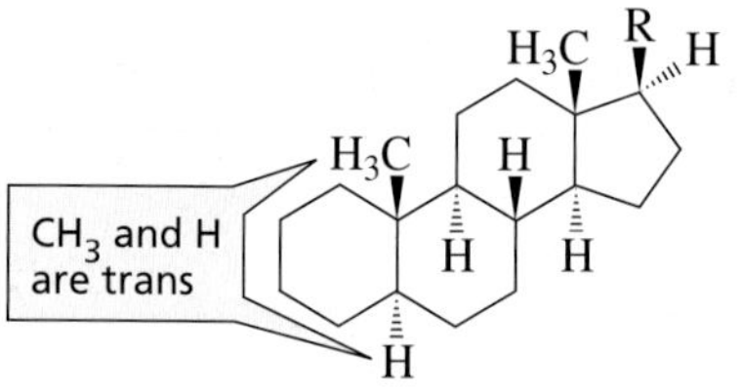

CH_3 and H are trans

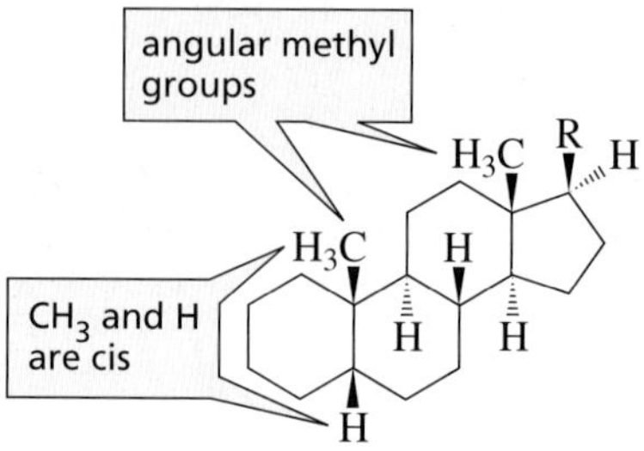

angular methyl groups

CH_3 and H are cis

PROBLEM 20◆

A β-hydrogen at C-5 means that the A and B rings are _____ fused; an α-hydrogen at C-5 means that they are _____ fused.

The most abundant member of the steroid family in animals is **cholesterol**, the precursor of all other steroids. Cholesterol is biosynthesized from squalene, a triterpene (Section 26.6). Cholesterol is an important component of cell membranes (Figure 26.1). Its ring structure makes it more rigid than other membrane lipids. Because cholesterol has eight asymmetric carbons, 256 stereoisomers are possible, but only one exists in nature (Chapter 5, Problem 20).

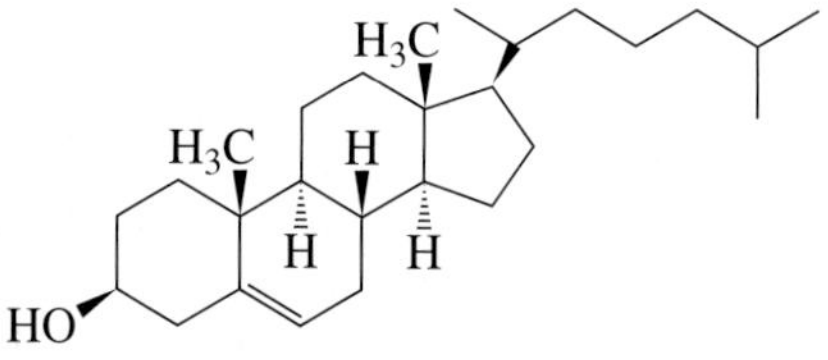

cholesterol

The steroid hormones can be divided into five classes: glucocorticoids, mineralocorticoids, androgens, estrogens, and progestins. Glucocorticoids and mineralocorticoids are synthesized in the adrenal cortex and are collectively known as *adrenal cortical steroids*. All adrenal cortical steroids have an oxygen at C-11.

Glucocorticoids, as their name suggests, are involved in glucose metabolism, as well as in the metabolism of proteins and fatty acids. Cortisone is an example of a glucocorticoid. Because of its anti-inflammatory effect, it is used clinically to treat arthritis and other inflammatory conditions.

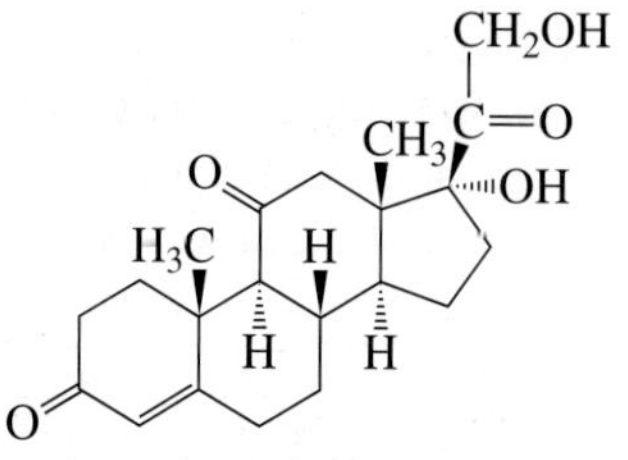

cortisone

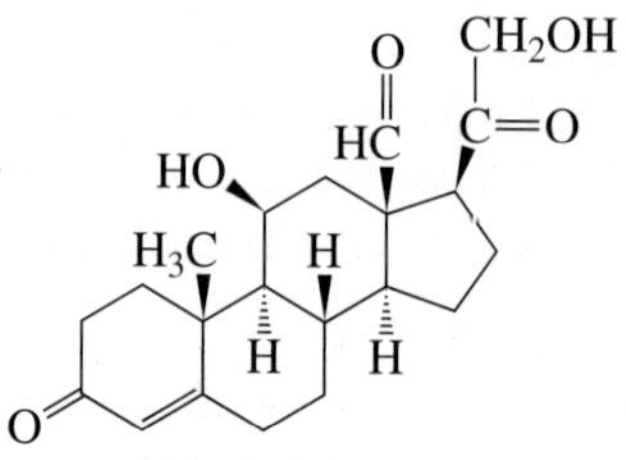

aldosterone

Mineralocorticoids cause increased reabsorption of Na^+, Cl^-, and HCO_3^- by the kidneys, leading to an increase in blood pressure. Aldosterone is an example of a mineralocorticoid.

PROBLEM 21◆

Is the OH substituent of the A ring of cholesterol an α-substituent or a β-substituent?

PROBLEM 22

Aldosterone is in equilibrium with its cyclic hemiacetal. Draw the hemiacetal form of aldosterone.

The male sex hormones, known as *androgens*, are secreted by the testes. They are responsible for the development of male secondary sex characteristics during puberty. They also promote muscle growth. Testosterone and 5α-dihydrotestosterone are androgens.

testosterone

5α-dihydrotestosterone

Adolf Friedrich Johann Butenandt (1903–1995) *was born in Germany. He shared the 1939 Nobel Prize in chemistry (with Ružička; see p. 1088) for isolating and determining the structures of estrone, androsterone, and progesterone. Forced by the Nazi government to refuse the prize, he accepted it after World War II. He was the director of the Kaiser Wilhelm Institute in Berlin and later was a professor at the Universities of Tübingen and Munich.*

Estradiol and estrone are female sex hormones known as *estrogens*. They are secreted by the ovaries and are responsible for the development of female secondary sex characteristics. They also regulate the menstrual cycle. Progesterone is the hormone that prepares the lining of the uterus for implantation of an ovum and is essential for the maintenance of pregnancy. It also prevents ovulation during pregnancy.

estradiol

estrone

progesterone

Michael S. Brown *and* **Joseph Leonard Goldstein** *shared the 1985 Nobel Prize in physiology or medicine for their work on the regulation of cholesterol metabolism and the treatment of disease caused by elevated cholesterol levels in the blood. Brown was born in New York in 1941; Goldstein, in South Carolina in 1940. They are both professors of medicine at the University of Texas Southwestern Medical Center.*

Although the various steroid hormones have remarkably different physiological effects, their structures are quite similar. For example, the only difference between testosterone and progesterone is the substituent at C-17, and the only difference between 5α-dihydrotestosterone and estradiol is one carbon and six hydrogens, but these compounds make the difference between being male and being female. These examples illustrate the extreme specificity of biochemical reactions.

PROBLEM 23

The acid component of a cholesterol ester is a fatty acid such as linoleic acid. Draw the structure of a cholesterol ester.

In addition to being the precursor of all the steroid hormones in animals, cholesterol is the precursor of the *bile acids*. In fact, the word *cholesterol* is derived from the Greek words *chole* meaning "bile" and *stereos* meaning "solid." The bile acids—cholic acid and chenodeoxycholic acid—are synthesized in the liver, stored in the gallbladder, and secreted into the small intestine, where they act as emulsifying agents so that fats and oils can be digested by water-soluble digestive enzymes. Cholesterol is also the precursor of vitamin D (Section 29.6).

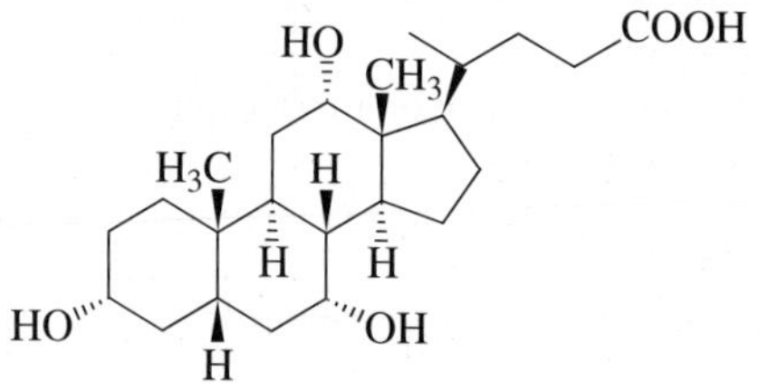

cholic acid

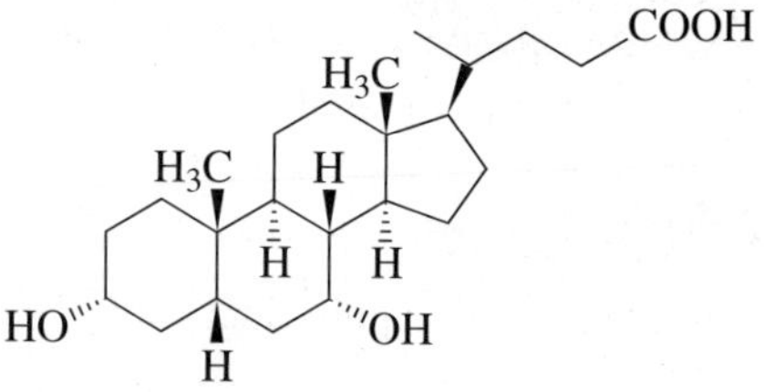

chenodeoxycholic acid

PROBLEM 24◆

Are the three OH groups of cholic acid axial or equatorial?

CHOLESTEROL AND HEART DISEASE

Cholesterol is probably the best-known lipid because of the correlation between cholesterol levels in the blood and heart disease. Cholesterol is synthesized in the liver and is also found in almost all body tissues. Cholesterol is also found in many foods, but we do not require it in our diet because the body can synthesize all we need. A diet high in cholesterol can lead to high levels of cholesterol in the bloodstream, and the excess can accumulate on the walls of arteries, restricting the flow of blood. This disease of the circulatory system is known as *atherosclerosis* and is a primary cause of heart disease. Cholesterol travels through the bloodstream packaged in particles that also contain cholesterol esters, phospholipids, and proteins. The particles are classified according to their density. LDL (low-density lipoprotein) particles transport cholesterol from the liver to other tissues. Receptors on the surfaces of cells bind LDL particles, allowing them to be brought into the cell so that it can use the cholesterol. HDL (high-density lipoprotein) is a cholesterol scavenger, removing cholesterol from the surfaces of membranes and delivering it back to the liver, where it is converted into bile acids. LDL is the so-called bad cholesterol, whereas HDL is the "good" cholesterol. The more cholesterol we eat, the less the body synthesizes. But this does not mean that the presence of dietary cholesterol has no effect on the total amount of cholesterol in the bloodstream, because dietary cholesterol also inhibits the synthesis of the LDL receptors. So the more cholesterol we eat, the less the body synthesizes, but also, the less the body can get rid of by bringing it into target cells.

CLINICAL TREATMENT OF HIGH CHOLESTEROL

Statins are the newest class of cholesterol-reducing drugs. Statins reduce serum cholesterol levels by inhibiting the enzyme that catalyzes the reduction of hydroxymethylglutaryl-CoA to mevalonic acid (Section 26.8). Decreasing the mevalonic acid concentration decreases the isopentenyl pyrophosphate concentration, so the biosynthesis of all terpenes, including cholesterol, is diminished. As a consequence of diminished cholesterol synthesis in the liver, the liver expresses more LDL receptors—the receptors that help clear LDL from the bloodstream. Studies show that for every 10% that cholesterol is reduced, deaths from coronary heart disease are reduced by 15% and total death risk is reduced by 11%.

Compactin and lovastatin are natural statins used clinically under the trade names Zocor® and Mevacor®. Atorvastatin (Lipitor)®, a synthetic statin, is now the most popular statin. Lipitor® has greater potency and a longer half-life than natural statins have, because its metabolites are as active as the parent drug in reducing cholesterol levels. Therefore, smaller doses of the drug may be administered. The required dose is reduced further because Lipitor® is marketed as a single enantiomer. In addition, it is more lipophilic than compactin and lovastatin, so it has a greater tendency to remain in the endoplasmic reticulum of the liver cells, where it is needed.

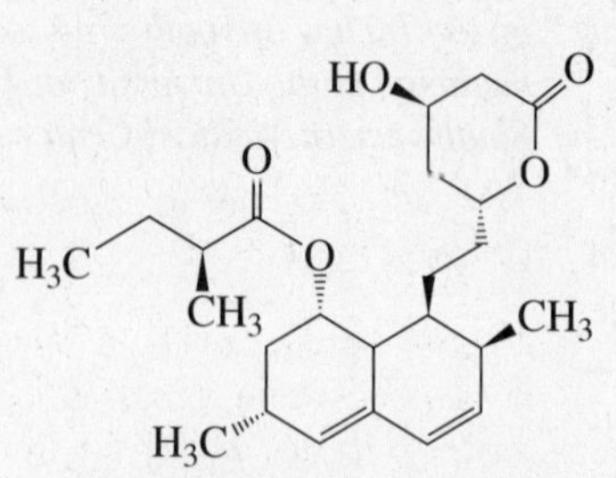

lovastatin
Mevacor®

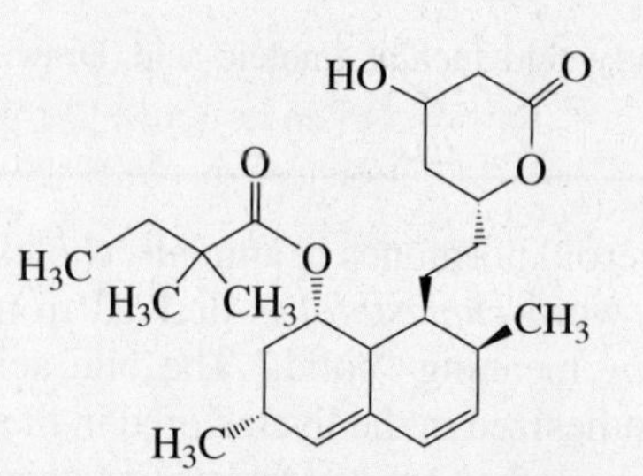

simvastatin
Zocor®

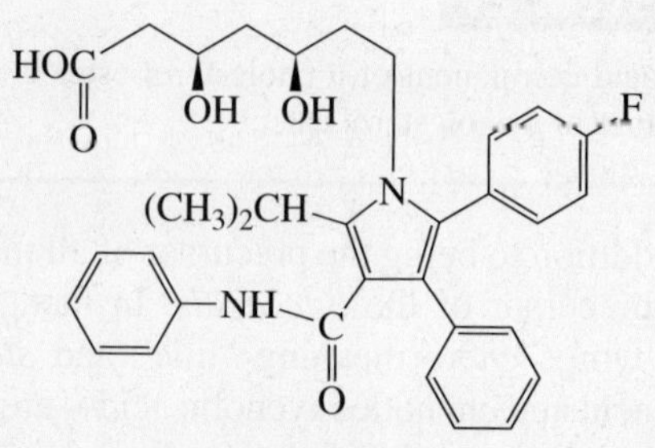

atorvastatin
Lipitor®

26.10 Biosynthesis of Cholesterol

How is cholesterol, the precursor of all the steroid hormones, biosynthesized? The starting material for the biosynthesis is the triterpene squalene, which must first be converted to lanosterol. Lanosterol is converted to cholesterol in a series of 19 steps.

The first step in the conversion of squalene to lanosterol is epoxidation of the 2,3-double bond of squalene. Acid-catalyzed opening of the epoxide initiates a series of cyclizations resulting in the protosterol cation. Elimination of a C-9 proton from the cation initiates a series of 1,2-hydride and 1,2-methyl shifts, resulting in lanosterol.

biosynthesis of lanosterol and cholesterol

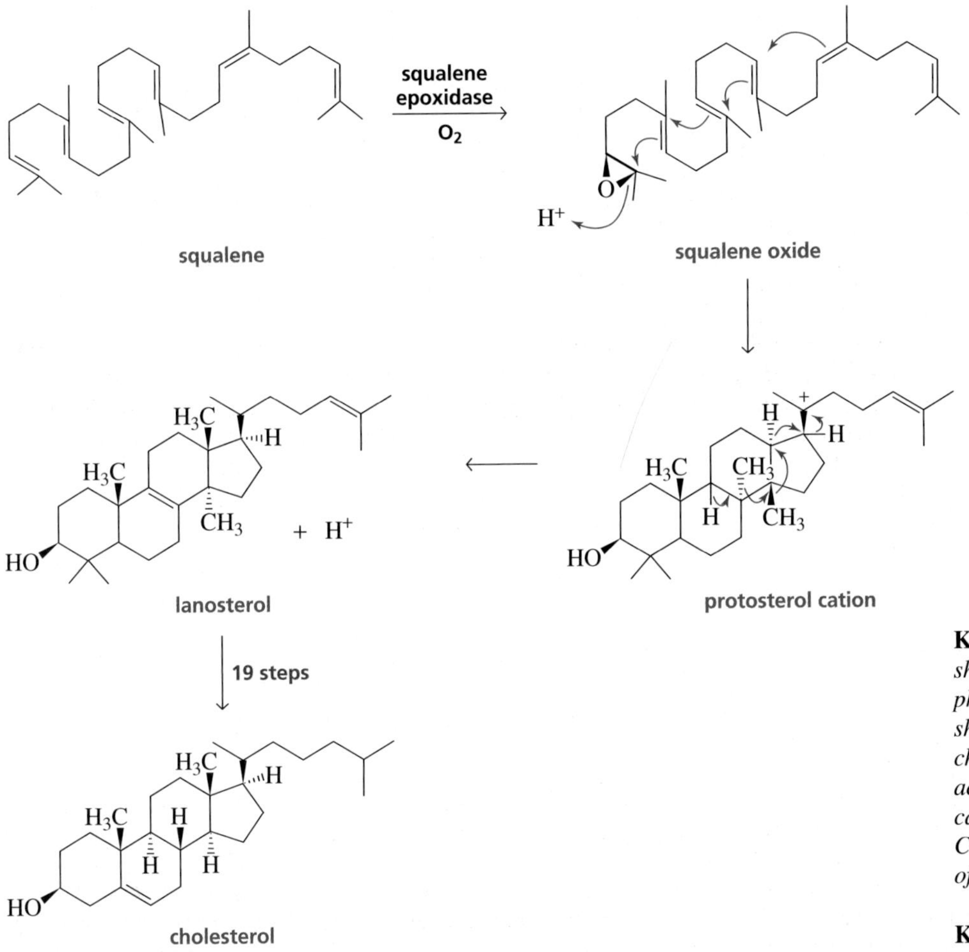

squalene → (squalene epoxidase, O_2) → squalene oxide (H^+) → protosterol cation → lanosterol + H^+ → (19 steps) → cholesterol

Converting lanosterol to cholesterol requires removing three methyl groups from lanosterol (in addition to reducing two double bonds and creating a new double bond). Removing methyl groups from carbon atoms is not easy: Many different enzymes are required to carry out the 19 steps. So why does nature bother? Why not just use lanosterol instead of cholesterol? Konrad Bloch answered that question when he found that membranes containing lanosterol instead of cholesterol are much more permeable. Small molecules are able to pass easily through lanosterol-containing membranes. As each methyl group is removed from lanosterol, the membrane becomes less and less permeable.

PROBLEM 25

Draw the individual 1,2-hydride and 1,2-methyl shifts responsible for conversion of the protosterol cation to lanosterol. How many hydride shifts are involved? How many methyl shifts?

Konrad Bloch *and* **Feodor Lynen** *shared the 1964 Nobel Prize in physiology or medicine. Bloch showed how fatty acids and cholesterol are biosynthesized from acetate. Lynen showed that the two-carbon acetate unit is actually acetyl-CoA, and he determined the structure of coenzyme A.*

Konrad Emil Bloch *(1912–2000) was born in Upper Silesia (then a part of Germany), left Nazi Germany for Switzerland in 1934, and came to the United States in 1936, becoming a U.S. citizen in 1944. He received a Ph.D. from Columbia in 1938, taught at the University of Chicago, and became a professor of biochemistry at Harvard in 1954.*

Feodor Lynen (1911–1979) *was born in Germany, received a Ph.D. under Heinrich Wieland, and married Wieland's daughter. He was head of the Institute of Cell Chemistry at the University of Munich.*

26.11 Synthetic Steroids

The potent physiological effects of steroids led scientists, in their search for new drugs, to synthesize steroids that are not available in nature and to investigate their physiological effects. Stanozolol and Dianabol are drugs developed in this way. They have the same muscle-building effect as testosterone. Steroids that aid in the development of muscle are called *anabolic steroids*. These drugs are available by prescription and are used to treat people suffering from traumas accompanied by muscle deterioration. The same drugs have been administered to athletes and racehorses to increase their muscle mass. Stanozolol was the drug detected in several athletes in the 1988 Olympics. Anabolic steroids, when taken in relatively high dosages, have been found to cause liver tumors, personality disorders, and testicular atrophy.

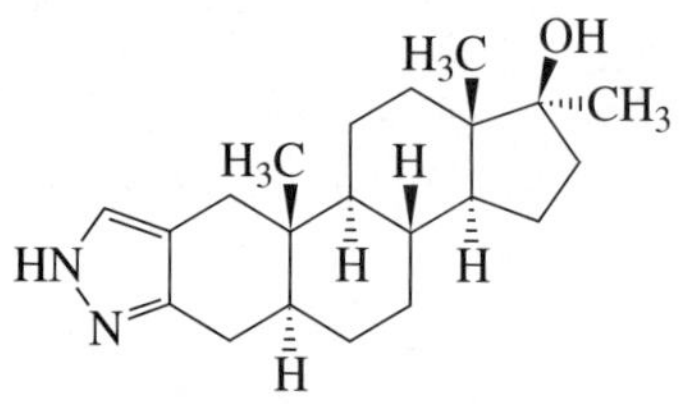

stanozolol 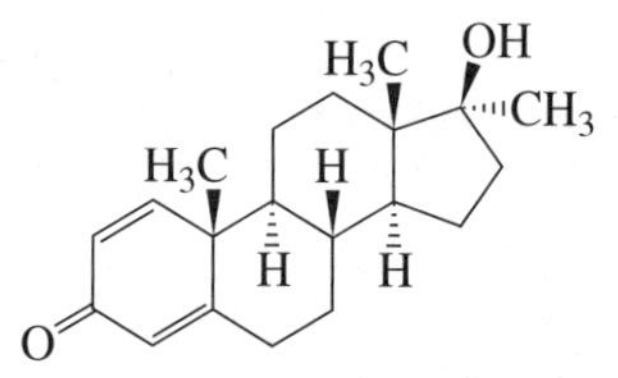 **Dianabol®**

Many synthetic steroids have been found to be much more potent than natural steroids. Norethindrone, for example, is better than progesterone in arresting ovulation. Another synthetic steroid, RU 486, when taken along with prostaglandins, terminates pregnancy within the first nine weeks of gestation. Notice that these oral contraceptives have structures similar to that of progesterone.

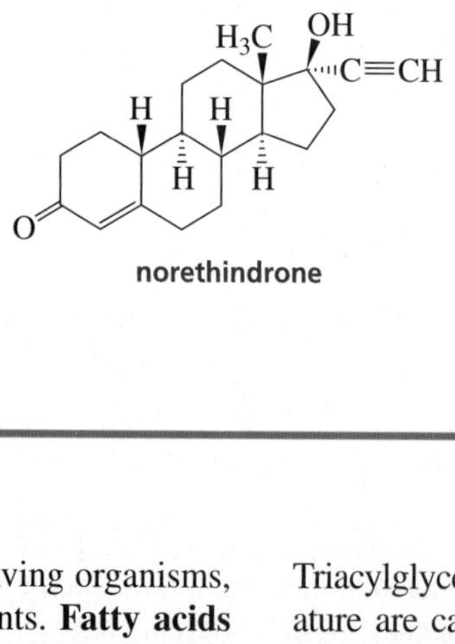

norethindrone 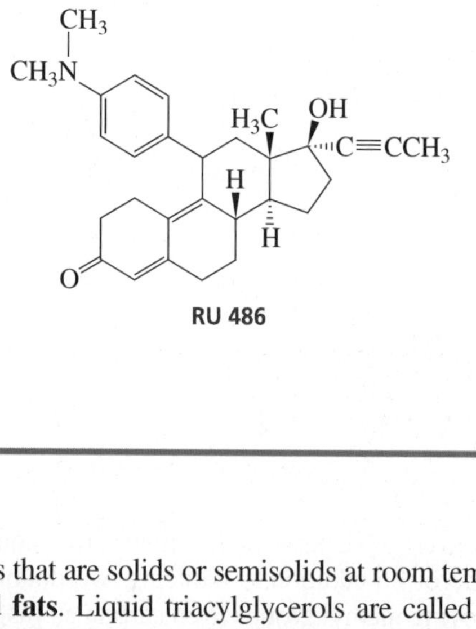 **RU 486**

Summary

Lipids are organic compounds, found in living organisms, that are soluble in nonpolar organic solvents. **Fatty acids** are carboxylic acids with long hydrocarbon chains. Double bonds in fatty acids have the cis configuration. Fatty acids with more than one double bond are called **polyunsaturated fatty acids**. Double bonds in naturally occurring unsaturated fatty acids are separated by one methylene group. **Waxes** are esters formed from long-chain carboxylic acids and long-chain alcohols. **Prostaglandins** are synthesized from arachidonic acid and are responsible for regulating a variety of physiological responses.

Triacylglycerols (triglycerides) are compounds in which the three OH groups of glycerol are esterified with fatty acids. Triacylglycerols that are solids or semisolids at room temperature are called **fats**. Liquid triacylglycerols are called **oils**. Some or all of the double bonds of polyunsaturated oils can be reduced by catalytic hydrogenation. **Phosphoacylglycerols** differ from triacylglycerols in that the terminal OH group of glycerol is esterified with phosphoric acid instead of a fatty acid. Phosphoacylglycerols form membranes by arranging themselves in a **lipid bilayer**. **Phospholipids** are lipids that contain a phosphate group. **Sphingolipids**, also found in membranes, contain sphingosine instead of glycerol.

Terpenes contain carbon atoms in multiples of 5. They are made by joining together five-carbon isoprene units, usually in a head-to-tail fashion—the **isoprene rule**.

Monoterpenes—terpenes with two isoprene units—have 10 carbons; **sesquiterpenes** have 15. **Squalene**, a **triterpene** (a terpene with six isoprene units), is a precursor of steroid molecules. Lycopene and β-carotene are **tetraterpenes** called **carotenoids**. β-Carotene is cleaved to form two molecules of vitamin A.

The five-carbon compound used for the synthesis of terpenes is isopentenyl pyrophosphate. The reaction of **dimethylallyl pyrophosphate** (formed from isopentenyl pyrophosphate) with **isopentenyl pyrophosphate** forms geranyl pyrophosphate, a 10-carbon compound. Geranyl pyrophosphate can react with another molecule of isopentenyl pyrophosphate to form farnesyl pyrophosphate, a 15-carbon compound. Two molecules of farnesyl pyrophosphate form **squalene**, a 30-carbon compound. Squalene is the precursor of **cholesterol**. Farnesyl pyrophosphate can react with another molecule of isopentenyl pyrophosphate to form geranylgeranyl pyrophosphate, a 20-carbon compound. Two geranylgeranyl pyrophosphates join to form phytoene, a 40-carbon compound. Phytoene is the precursor of the **carotenoids**.

Hormones are chemical messengers. Many hormones are **steroids**. All steroids contain a tetracyclic ring system. The B, C, and D rings are **trans fused**. In most naturally occurring steroids, the A and B rings are also trans fused. Methyl groups at C-10 and C-13 are called **angular methyl groups**. **β-Substituents** are on the same side of the steroid ring system as the angular methyl groups; **α-substituents** are on the opposite side. Synthetic steroids are steroids that are not found in nature.

The most abundant member of the steroid family in animals is **cholesterol**, the precursor of all other steroids. Cholesterol is an important component of cell membranes; its ring structure causes it to be more rigid than other membrane lipids. In the biosynthesis of cholesterol, squalene is converted to lanosterol, which is converted to cholesterol.

Key Terms

angular methyl group (p. 1098)
carotenoid (p. 1089)
cholesterol (p. 1098)
cis fused (p. 1097)
dimethylallyl pyrophosphate (p. 1093)
fat (p. 1079)
fatty acid (p. 1076)
hormone (p. 1097)
isopentenyl pyrophosphate (p. 1093)
isoprene rule (p. 1088)
lipid (p. 1075)
lipid bilayer (p. 1082)
membrane (p. 1082)
mixed triacylglycerol (p. 1078)
monoterpene (p. 1089)
oil (p. 1079)
phosphatidic acid (p. 1082)
phosphoacylglycerol (p. 1082)
phosphoglycerides (p. 1082)
phospholipid (p. 1082)
polyunsaturated fatty acid (p. 1076)
prostaglandin (p. 1085)
sesquiterpene (p. 1089)
simple triacylglycerol (p. 1078)
sphingolipid (p. 1084)
squalene (p. 1089)
steroid (p. 1097)
α-substituent (p. 1098)
β-substituent (p. 1098)
terpene (p. 1088)
terpenoid (p. 1088)
tetraterpene (p. 1089)
trans fused (p. 1097)
triacylglycerol (p. 1078)
triterpene (p. 1089)
wax (p. 1078)

Problems

26. An optically active fat, when completely hydrolyzed, yields twice as much stearic acid as palmitic acid. Draw the structure of the fat.

27. a. How many different triacylglycerols are there in which one of the fatty acid components is lauric acid and two are myristic acid?
    b. How many different triacylglycerols are there in which one of the fatty acid components is lauric acid, one is myristic acid, and one is palmitic acid?

28. Cardiolipins are found in heart muscles. Give the products formed when a cardiolipin undergoes complete acid-catalyzed hydrolysis.

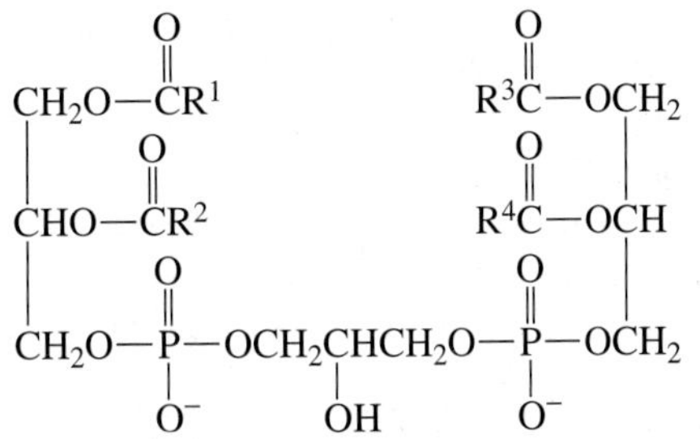

```
           O                            O
           ‖                            ‖
CH2O—CR1                     R3C—OCH2
  |        O                    O     |
  |        ‖                    ‖     |
CHO—CR2                      R4C—OCH
  |        O                    O     |
  |        ‖                    ‖     |
CH2O—P—OCH2CHCH2O—P—OCH2
           |        |           |
           O−       OH          O−
```

a cardiolipin

29. Nutmeg contains a simple, fully saturated triacylglycerol with a molecular weight of 722. Draw its structure.

30. Give the product that would be obtained from the reaction of cholesterol with each of the following reagents: (*Hint:* Because of steric hindrance from the angular methyl groups, the α-face is more susceptible to attack by reagents than the β-face.)
    a. H_2O, H^+
    b. BH_3 in THF, followed by $H_2O_2 + HO^-$
    c. $H_2, Pd/C$
    d. $Br_2 + H_2O$
    e. peroxyacetic acid
    f. the product of part e + CH_3O^-

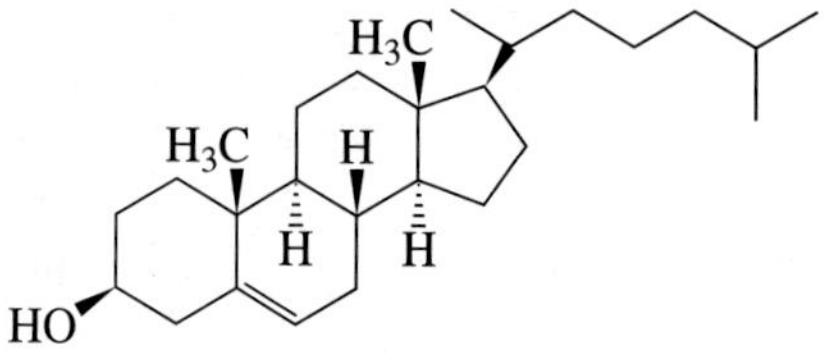

cholesterol

31. Dr. Cole S. Terol synthesized the following samples of mevalonic acid and fed them to a group of lemon trees:

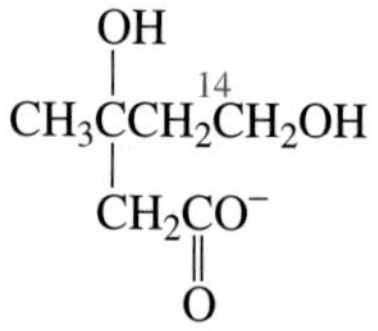

$CH_3C(OH)(CH_2CO_2^-)CH_2\overset{14}{C}H_2OH$ — sample A

$CH_3C(OH)(CH_2\overset{14}{C}O_2^-)CH_2CH_2OH$ — sample B

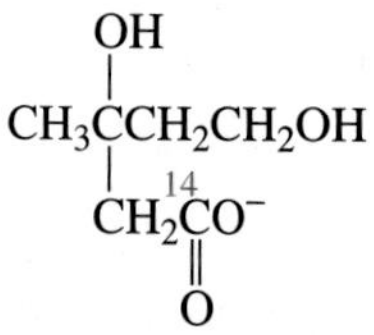

$CH_3C(OH)(\overset{14}{C}H_2CO_2^-)CH_2CH_2OH$ — sample C

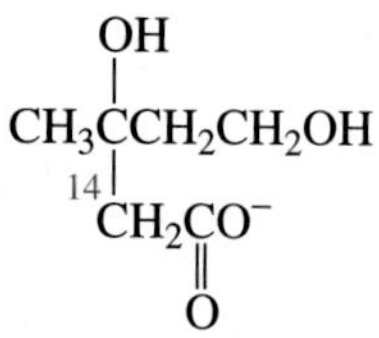

Which carbons in citronellal, which is isolated from lemon oil, will be labeled in trees that were fed the following?
    a. sample A
    b. sample B
    c. sample C

32. An optically active monoterpene (compound A) with molecular formula $C_{10}H_{18}O$ undergoes catalytic hydrogenation to form an optically inactive compound with molecular formula $C_{10}H_{20}O$ (compound B). When compound B is heated with acid, followed by reaction with O_3, and workup under reducing conditions (Zn, H_2O), one of the products obtained is 4-methylcyclohexanone. Give possible structures for compound A.

33. If junipers were allowed to grow in a medium containing acetate in which the methyl carbon was labeled with ^{14}C, which carbons in α-terpineol would be labeled?

34. a. Propose a mechanism for the following reaction:

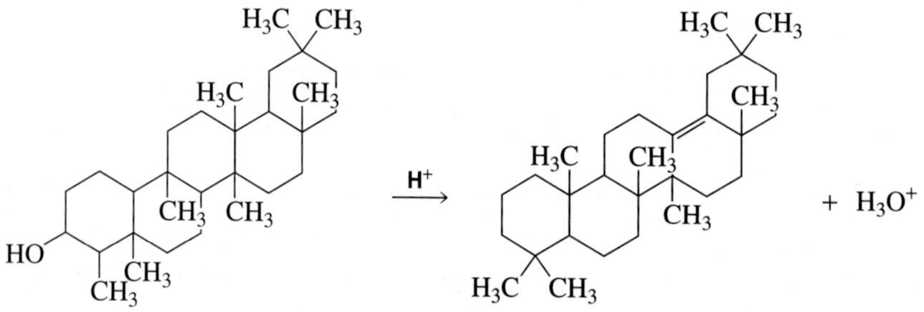

$\xrightarrow{H^+}$ + H_3O^+

    b. To what class of terpene does the starting material belong? Mark off the isoprene units in the starting material.

35. 5-Androstene-3,17-dione is isomerized to 4-androstene-3,17-dione by hydroxide ion. Propose a mechanism for this reaction.

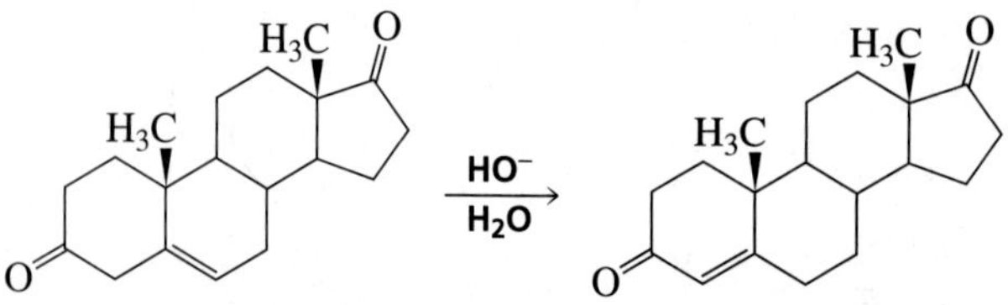

$\xrightarrow[H_2O]{HO^-}$

5-androstene-3,17-dione — 4-androstene-3,17-dione

36. Both OH groups of one of the following steroid diols react with excess ethyl chloroformate, but only one OH group of the other steroid diol reacts under the same conditions:

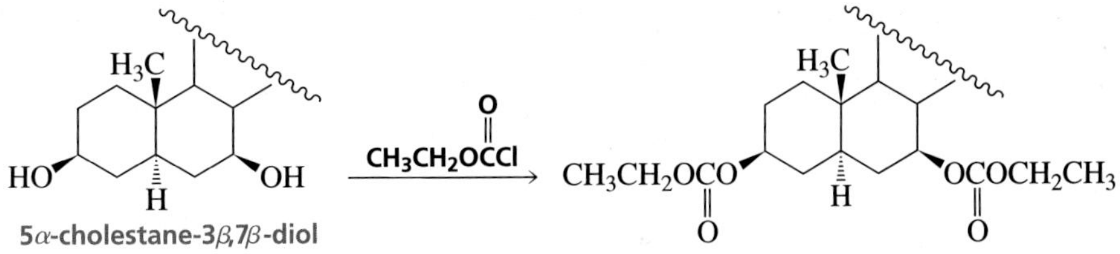

5α-cholestane-3β,7β-diol $\xrightarrow{CH_3CH_2OCCl}$ (both OH converted to $CH_3CH_2OC(=O)O$– groups)

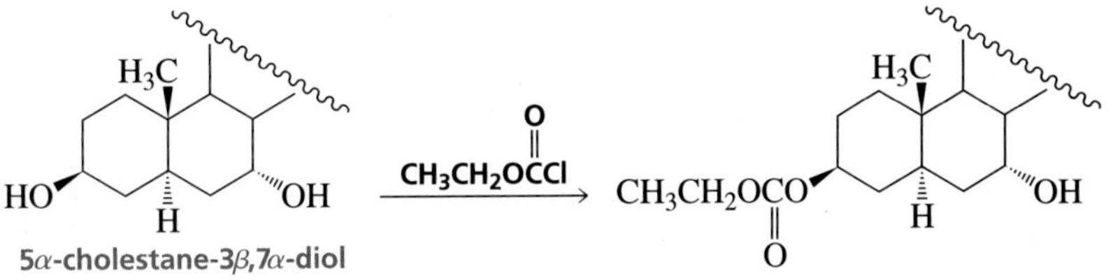

5α-cholestane-3β,7α-diol $\xrightarrow{CH_3CH_2OCCl}$ (only the 3-OH converted to $CH_3CH_2OC(=O)O$–; 7-OH remains)

Explain the difference in reactivity.

37. The acid-catalyzed dehydration of an alcohol to a rearranged alkene is known as a Wagner–Meerwein rearrangement. Propose a mechanism for the following Wagner–Meerwein rearrangement:

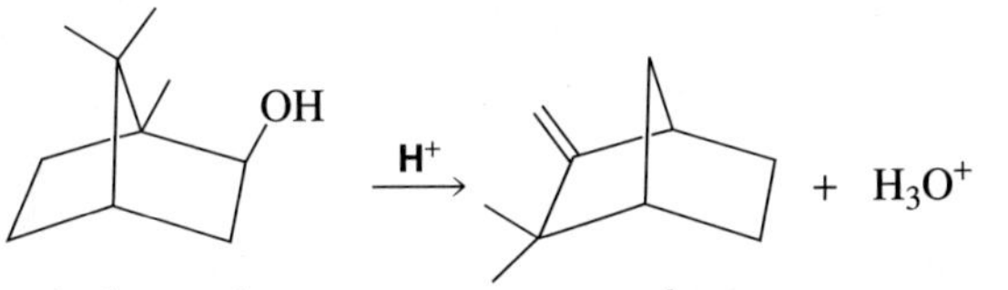

isoborneol $\xrightarrow{H^+}$ camphene + H_3O^+

38. Diethylstilbestrol (DES) was given to pregnant women to prevent miscarriage, until it was found that the drug caused cancer in both the mothers and their female children. DES has estradiol activity even though it is not a steroid. Draw DES in a way that shows that it is structurally similar to estradiol.

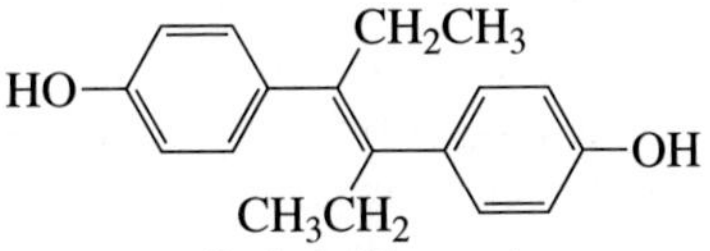

diethylstilbestrol
DES

27 Nucleosides, Nucleotides, and Nucleic Acids

an RNA catalyst

In previous chapters, we studied two of the three major kinds of biopolymers—polysaccharides and proteins. Now we will look at the third—nucleic acids. There are two types of nucleic acids—**deoxyribonucleic acid (DNA)** and **ribonucleic acid (RNA)**. DNA encodes an organism's entire hereditary information and controls the growth and division of cells. In most organisms, the genetic information stored in DNA is transcribed into RNA. This information can then be translated for the synthesis of all the proteins needed for cellular structure and function.

Studies that determined the structures of the nucleic acids and paved the way for the discovery of the DNA double helix were carried out by **Phoebus Levene** *and elaborated by* **Sir Alexander Todd**.

Phoebus Aaron Theodor Levene (1869–1940) *was born in Russia. When he immigrated to the United States with his family in 1891, his Russian name Fishel was changed to Phoebus. Because his medical school education had been interrupted, he returned to Russia to complete his studies. When he returned to the United States, he took chemistry courses at Columbia University. Deciding to forgo medicine for a career in chemistry, he went to Germany to study under Emil Fischer. He was a professor of chemistry at the Rockefeller Institute (now Rockefeller University).*

DNA was first isolated in 1869 from the nuclei of white blood cells. Because this material was found in the nucleus and was acidic, it was called *nucleic acid*. Eventually, scientists found that the nuclei of all cells contain DNA, but it wasn't until 1944 that they realized that nucleic acids are the carriers of genetic information. In 1953, James Watson and Francis Crick described the three-dimensional structure of DNA—the famed double helix.

27.1 Nucleosides and Nucleotides

Nucleic acids are chains of five-membered-ring sugars linked by phosphate groups (Figure 27.1). The anomeric carbon of each sugar is bonded to a nitrogen of a heterocyclic compound in a β-glycosidic linkage. (Recall from Section 22.10 that a β-linkage is one in which the substituents at C-1 and C-4 are on the same side of the furanose ring.) Because the heterocyclic compounds are amines, they are commonly referred to as **bases**. In RNA the five-membered-ring sugar is D-ribose. In DNA it is 2-deoxy-D-ribose (D-ribose without an OH group in the 2-position).

Phosphoric acid links the sugars in both RNA and DNA. The acid has three dissociable OH groups with pK_a values of 1.9, 6.7, and 12.4. Each of the OH groups can react with an alcohol to form a *phosphomonoester*, a *phosphodiester*, or a *phosphotriester*. In nucleic acids the phosphate group is a **phosphodiester**.

THE STRUCTURE OF DNA: WATSON, CRICK, FRANKLIN, AND WILKINS

James D. Watson was born in Chicago in 1928. He graduated from the University of Chicago at the age of 19 and received a Ph.D. three years later from Indiana University. In 1951, as a postdoctoral fellow at Cambridge University, Watson worked on determining the three-dimensional structure of DNA.

Francis H. C. Crick was born in Northampton, England, in 1916. Originally trained as a physicist, Crick was involved in radar research during World War II. After the war, he entered Cambridge University to study for a Ph.D. in chemistry, which he received in 1953. He was a graduate student when he carried out his portion of the work that led to the proposal of the double helical structure of DNA.

Rosalind Franklin was born in London in 1920. She graduated from Cambridge University and in 1942 quit her graduate studies to accept a position as a research officer in the British Coal Utilisation Research Association. After the war, she studied X-ray diffraction techniques in Paris. In 1951 she returned to England, accepting a position to develop an X-ray diffraction unit in the biophysics department at King's College. Her X-ray studies showed that DNA was a helix with phosphate groups on the outside of the molecule. Franklin died in 1958 without knowing the significance her work had played in determining the structure of DNA.

Watson and Crick shared the 1962 Nobel Prize in medicine or physiology with Maurice Wilkins for determining the double helical structure of DNA. Wilkins contributed X-ray studies that confirmed the double helical structure. Wilkins was born in New Zealand in 1916 and moved to England six years later with his parents. During World War II he joined other British scientists who were working with American scientists on the development of the atomic bomb.

Francis Crick (left) and James Watson (right)

Rosalind Franklin

By courtesy of the National Portrait Gallery, London

RNA

DNA

▲ **Figure 27.1**
Nucleic acids consist of a chain of five-membered-ring sugars linked by phosphate groups. Each sugar (D-ribose in RNA, 2′-deoxy-D-ribose in DNA) is bonded to a heterocyclic amine in a β-glycosidic linkage.

Alexander R. Todd (1907–1997) *was born in Scotland. He received two Ph.D. degrees, one from Johann Wolfgang Goethe University in Frankfurt (1931) and one from Oxford University (1933). He was a professor of chemistry at the University of Edinburgh, at the University of Manchester, and from 1944 to 1971 at Cambridge University. He was knighted in 1954 and was made a baron in 1962 (Baron Todd of Trumpington). For his work on nucleotides, he was awarded the 1957 Nobel Prize in chemistry.*

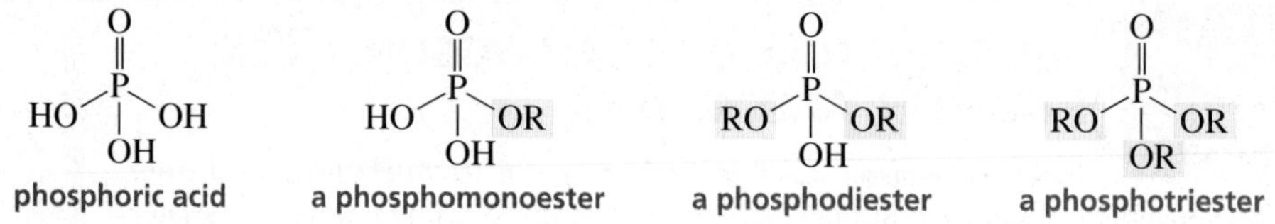

phosphoric acid · a phosphomonoester · a phosphodiester · a phosphotriester

The vast differences in heredity among species and among members of the same species are determined by the sequence of the bases in DNA. Surprisingly, there are only four bases in DNA—two are substituted purines (adenine and guanine), and two are substituted pyrimidines (cytosine and thymine).

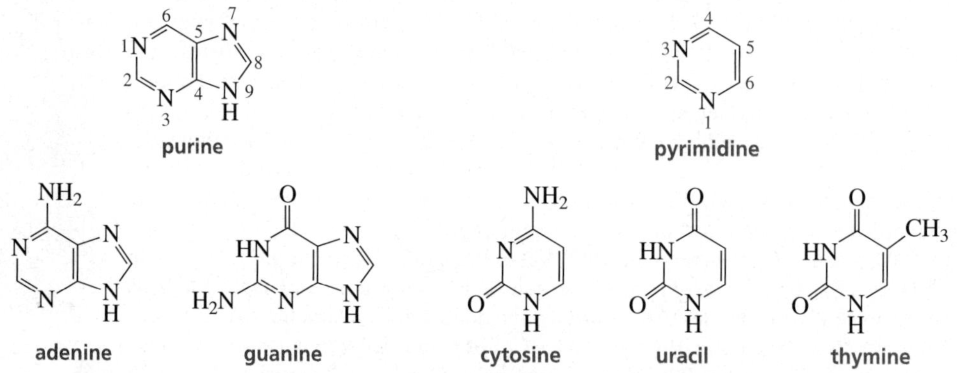

purine · pyrimidine

adenine · guanine · cytosine · uracil · thymine

RNA also contains only four bases. Three (adenine, guanine, and cytosine) are the same as those in DNA, but the fourth base in RNA is uracil instead of thymine. Notice that thymine and uracil differ only by a methyl group—thymine is 5-methyluracil. The reason DNA contains thymine instead of uracil is explained in Section 27.14.

nucleoside = base + sugar

The purines and pyrimidines are bonded to the anomeric carbon of the furanose ring—purines at N-9 and pyrimidines at N-1—in a β-glycosidic linkage. A compound containing a base bonded to D-ribose or to 2-deoxy-D-ribose is called a **nucleoside**. In a nucleoside the ring positions of the sugar are indicated by primed numbers to distinguish them from the ring positions of the base. This is why the sugar component of DNA is referred to as 2′-deoxy-D-ribose. Notice the difference in the base names and their corresponding nucleoside names in Table 27.1. For example, adenine is the base, whereas adenosine is the nucleoside. Similarly, cytosine is the base, whereas cytidine is the nucleoside, and so forth. Because uracil is found only in RNA, it is shown attached to D-ribose but not to 2-deoxy-D-ribose; because thymine is found only in DNA, it is shown attached to 2-deoxy-D-ribose but not to D-ribose.

nucleosides

adenosine · guanosine · cytidine · uridine

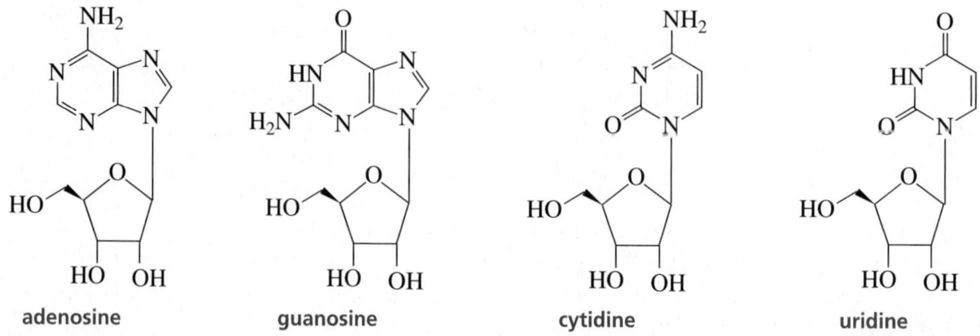

2′-deoxyadenosine **2′-deoxyguanosine** **2′-deoxycytidine** **thymidine**

Table 27.1 The Names of the Bases, the Nucleosides, and the Nucleotides

Base	Nucleoside		Ribonucleotide	Deoxyribonucleotide
Adenine	Adenosine	2′-Deoxyadenosine	Adenosine 5′-phosphate	2′-Deoxyadenosine 5′-phosphate
Guanine	Guanosine	2′-Deoxyguanosine	Guanosine 5′-phosphate	2′-Deoxyguanosine 5′-phosphate
Cytosine	Cytidine	2′-Deoxycytidine	Cytidine 5′-phosphate	2′-Deoxycytidine 5′-phosphate
Thymine	—	Thymidine	—	Thymidine 5′-phosphate
Uracil	Uridine	—	Uridine 5′-phosphate	

PROBLEM 1

In acidic solutions, nucleosides are hydrolyzed to a sugar and a heterocyclic base. Propose a mechanism for this reaction.

A **nucleotide** is a nucleoside with either the 5′- or the 3′-OH group bonded in an ester linkage to phosphoric acid. The nucleotides of RNA—where the sugar is D-ribose—are more precisely called **ribonucleotides**, whereas the nucleotides of DNA—where the sugar is 2-deoxy-D-ribose—are called **deoxyribonucleotides**.

nucleotide = base + sugar + phosphate

nucleotides

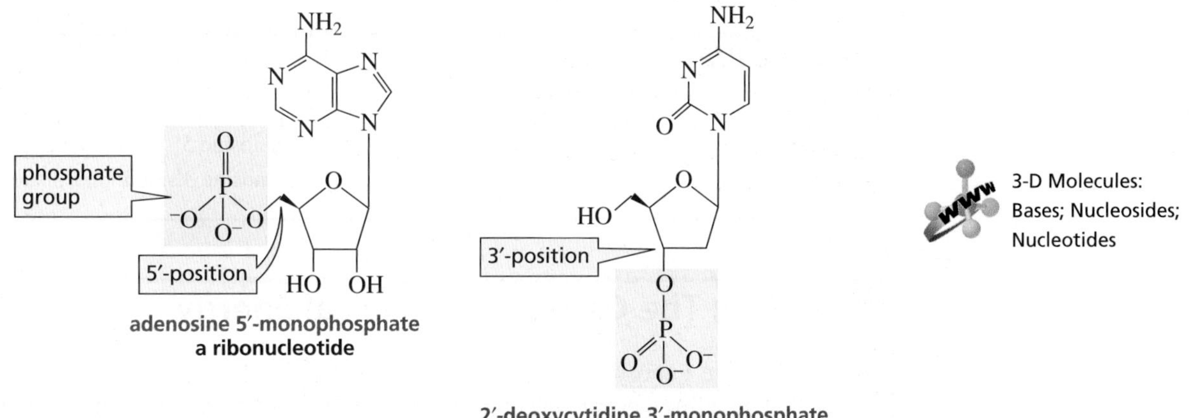

adenosine 5′-monophosphate
a ribonucleotide

2′-deoxycytidine 3′-monophosphate
a deoxyribonucleotide

3-D Molecules: Bases; Nucleosides; Nucleotides

When phosphoric acid is heated with P_2O_5 it loses water, forming a phosphoanhydride called pyrophosphoric acid. Its name comes from *pyr*, the Greek word for "fire." Thus, pyrophosphoric acid is prepared by "fire"—that is, by heating. Triphosphoric acid and higher polyphosphoric acids are also formed.

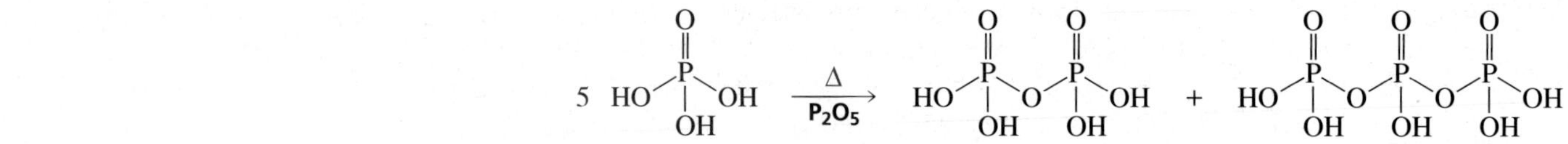

5 phosphoric acid $\xrightarrow[P_2O_5]{\Delta}$ pyrophosphoric acid + triphosphoric acid

Because phosphoric acid can form an anhydride, nucleotides can exist as monophosphates, diphosphates, and triphosphates. They are named by adding *monophosphate* or *diphosphate* or *triphosphate* to the name of the nucleoside.

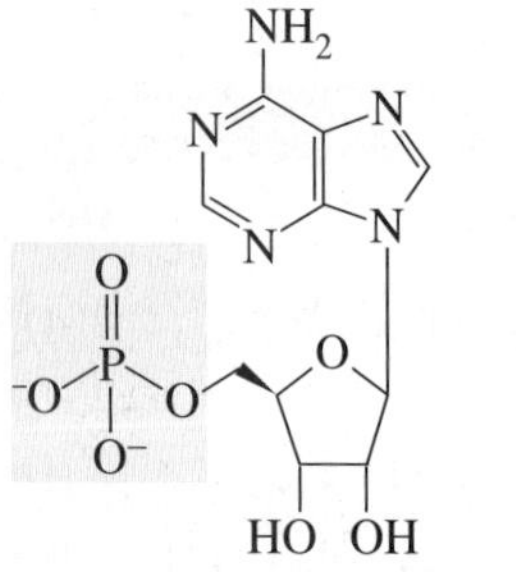

adenosine 5′-monophosphate AMP

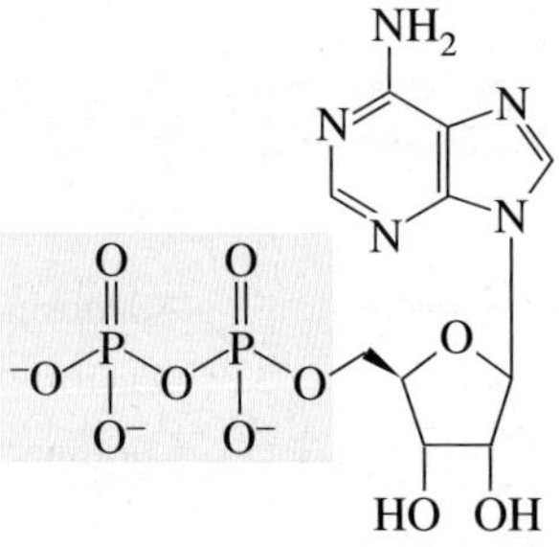

adenosine 5′-diphosphate ADP

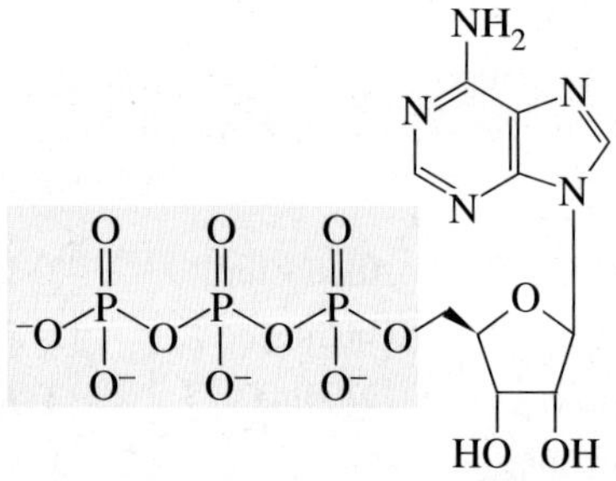

adenosine 5′-triphosphate ATP

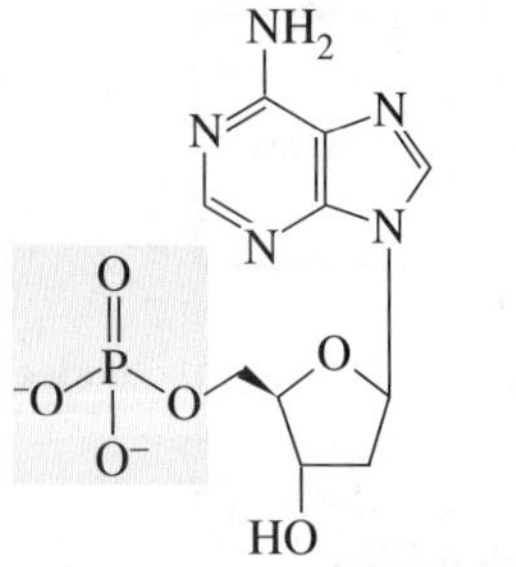

2′-deoxyadenosine 5′-monophosphate dAMP

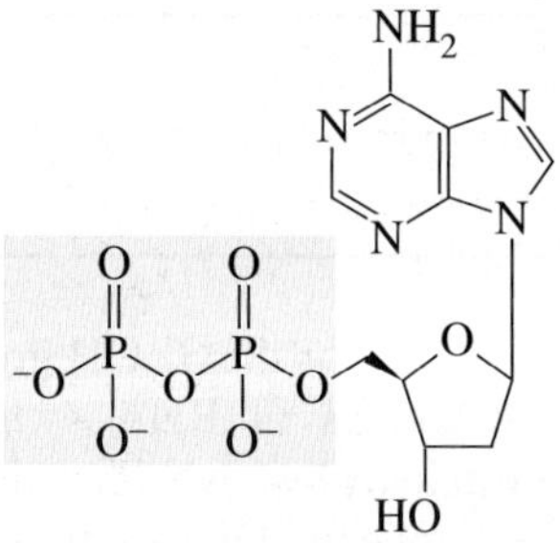

2′-deoxyadenosine 5′-diphosphate dADP

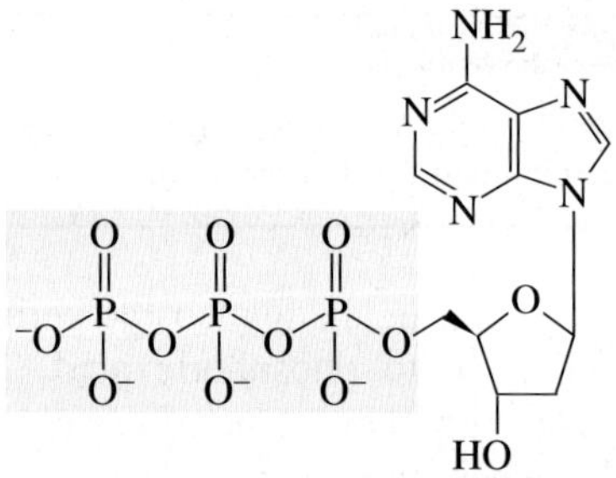

2′-deoxyadenosine 5′-triphosphate dATP

PROBLEM 2

Draw the structure for each of the following:

a. dCDP
b. dTTP
c. dUMP
d. UDP
e. guanosine 5′-triphosphate
f. adenosine 3′-monophosphate

27.2 ATP: The Carrier of Chemical Energy

All cells require energy to ensure their survival and reproduction. They get the energy they need by converting nutrients into a chemically useful form of energy. The most important form of chemical energy is **adenosine 5′-triphosphate (ATP)**. The importance of ATP to biological reactions is shown by its turnover rate in humans—each day, a person uses an amount of ATP equivalent to his or her body weight. ATP is known as the universal carrier of chemical energy because, as it is commonly stated, "the energy of hydrolysis of ATP converts endergonic reactions into exergonic reactions."

In other words, the ability of ATP to enable otherwise unfavorable reactions to occur is attributed to the large amount of energy released when ATP is hydrolyzed, which can be used to drive an endergonic reaction. For example, the reaction of

D-glucose with hydrogen phosphate to form D-glucose-6-phosphate is endergonic ($\Delta G^{\circ\prime} = +3.3$ kcal/mol or $+13.8$ kJ/mol).[1] The hydrolysis of ATP, on the other hand, is highly exergonic ($\Delta G^{\circ\prime} = -7.3$ kcal/mol or -30.5 kJ/mol). When the two reactions are added together (the species occurring on both sides of the reaction arrow cancel), the net reaction is exergonic ($\Delta G^{\circ\prime} = -4.0$ kcal/mol or -16.7 kJ/mol). Thus, the energy released from the hydrolysis of ATP is more than enough to drive the phosphorylation of D-glucose. Two reactions in which the energy of one is used to drive the other are known as *coupled reactions*.

Reaction	$\Delta G^{\circ\prime}$
D-glucose + hydrogen phosphate ⟶ D-glucose-6-phosphate + H_2O	+ 3.3 kcal/mol or + 13.8 kJ/mol
ATP + H_2O ⟶ ADP + hydrogen phosphate	− 7.3 kcal/mol or − 30.5 kJ/mol
D-glucose + ATP ⟶ D-glucose-6-phosphate + ADP	− 4.0 kcal/mol or − 16.7 kJ/mol

This nonmechanistic description of ATP's power makes ATP look like a magical source of energy. Let's look at the mechanism of the reaction to see what really happens. The reaction is a simple one-step nucleophilic substitution reaction. The 6-OH group of glucose attacks the terminal phosphate of ATP, breaking a **phosphoanhydride bond** without forming an intermediate. Essentially it is an S_N2 reaction with an adenosine pyrophosphate leaving group.

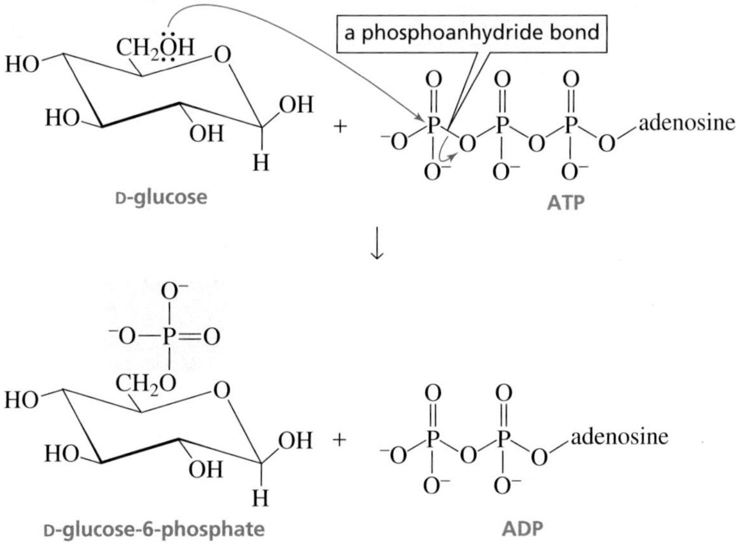

a phosphoanhydride bond

D-glucose + ATP ⟶ D-glucose-6-phosphate + ADP

3-D Molecules: Adenosine 5′-triphosphate (ATP)

Now we have a chemical understanding of why the phosphorylation of glucose requires ATP. Without ATP, the 6-OH group of D-glucose would have to displace a very basic $^{-}$OH group from hydrogen phosphate. With ATP, the 6-OH group of D-glucose displaces the weakly basic ADP.

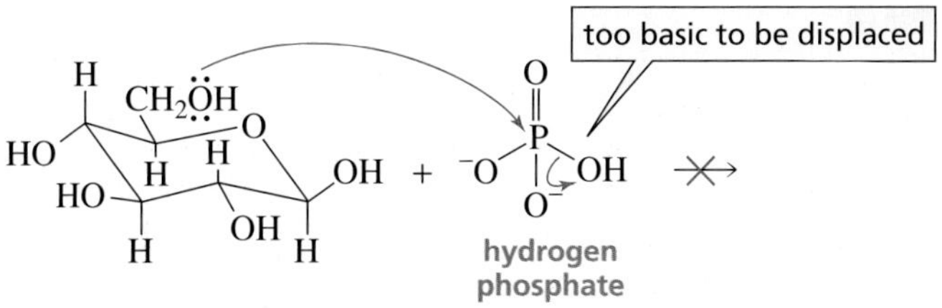

too basic to be displaced

hydrogen phosphate

[1]The prime in $\Delta G^{\circ\prime}$ indicates that two additional parameters have been added to the ΔG° defined in Section 3.7—the reaction occurs in aqueous solution at pH = 7 and the concentration of water is assumed to be constant.

Although the phosphorylation of glucose is described as being driven by the "hydrolysis" of ATP, you can see from the mechanism that glucose does not react with hydrogen phosphate and that ATP is not hydrolyzed because it does not react with water. In other words, neither of the coupled reactions actually occurs. Instead, the phosphate group of ATP is transferred directly to D-glucose.

The transfer of a phosphate group from ATP to D-glucose is an example of a **phosphoryl transfer reaction**. There are many phosphoryl transfer reactions in biological systems. In all of these reactions, the electrophilic phosphate group is transferred to a nucleophile as a result of breaking a phosphoanhydride bond. This example of a phosphoryl transfer reaction demonstrates the actual chemical function of ATP—*it provides a reaction pathway involving a good leaving group for a reaction that cannot occur (or would occur very slowly) because of a poor leaving group.*

ATP provides a reaction pathway with a good leaving group for a reaction that cannot occur because of a poor leaving group.

PROBLEM 3 SOLVED

The hydrolysis of phosphoenolpyruvate is so highly exergonic ($\Delta G^{\circ\prime} = -14.8$ kcal/mol or -61.9 kJ/mol) that it can be used to "drive the formation" of ATP from ADP and hydrogen phosphate ($\Delta G^{\circ\prime} = +7.3$ kcal/mol or $+30.5$ kJ/mol). Propose a mechanism for this reaction.

SOLUTION As we saw in the example with ATP, neither of the coupled reactions actually occurs: phosphoenolpyruvate does not react with water and ADP does not react with hydrogen phosphate. Just as ATP "drives the formation" of glucose-6-phosphate by supplying glucose (a nucleophile) with a phosphate that has a good leaving group (ADP), phosphoenolpyruvate "drives the formation" of ATP by supplying ADP (a nucleophile) with a phosphate that has a good leaving group (pyruvate).

$CH_2{=}C(OPO_3^{2-})C(=O)O^-$ (phosphoenolpyruvate) + ADP (^-O–P(=O)(O$^-$)–O–P(=O)(O$^-$)–O–adenosine) $\longrightarrow$ $CH_2{=}C(O^-)C(=O)O^-$ + ATP (^-O–P(=O)(O$^-$)–O–P(=O)(O$^-$)–O–P(=O)(O$^-$)–O–adenosine)

phosphoenolpyruvate ADP ATP

$\downarrow H_2O$

$CH_2{=}C(OH)C(=O)O^-$ $\xrightarrow{\text{tautomerization}}$ $CH_3C(=O)C(=O)O^-$

pyruvate

PROBLEM 4◆

Why is pyruvate a good leaving group?

PROBLEM 5◆

Several important biomolecules and the $\Delta G^{\circ\prime}$ values for their hydrolysis are listed here. Which of them "hydrolyzes" with sufficient energy to "drive the formation" of ATP?

glycerol-1-phosphate:
-2.2 kcal/mol (-9.2 kJ/mol)

phosphocreatine:
-11.8 kcal/mol (-49.4 kJ/mol)

fructose-6-phosphate:
-3.8 kcal/mol3 (-15.9 kJ/mol)

glucose-6-phosphate:
-3.3 kcal/mol (13.8 kJ/mol).

27.3 Three Mechanisms for Phosphoryl Transfer Reactions

There are three possible mechanisms for a phosphoryl transfer reaction. We will illustrate them using the following nucleophilic acyl substitution reaction.

$$CH_3C(=O)O^- + RSH \rightleftharpoons CH_3-C(OH)(O^-)-SR \not\rightarrow CH_3C(=O)SR + HO^-$$

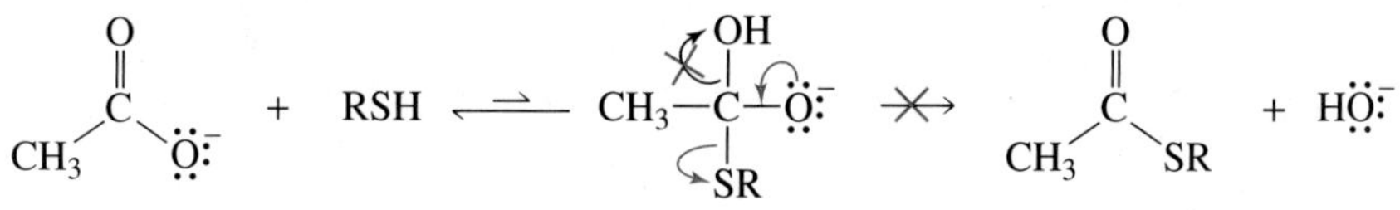

This reaction does not occur without ATP because the negatively charged carboxylate ion resists nucleophilic attack and, if the tetrahedral intermediate could be formed, the incoming nucleophile is a weaker base than the base that would have to be expelled from the tetrahedral intermediate to form the thioester. In other words, the thiol would be expelled from the tetrahedral intermediate, reforming the carboxylate ion (Section 17.5).

If ATP is added to the reaction mixture, the reaction occurs. The carboxylate ion attacks one of the phosphate groups of ATP, breaking a phosphoanhydride bond. This puts a leaving group on the carboxyl group that can be displaced by the thiol. There are three possible mechanisms for the reaction of a nucleophile with ATP because each of the three phosphorus atoms of ATP can undergo nucleophilic attack. Each mechanism puts a different phosphate leaving group on the nucleophile.

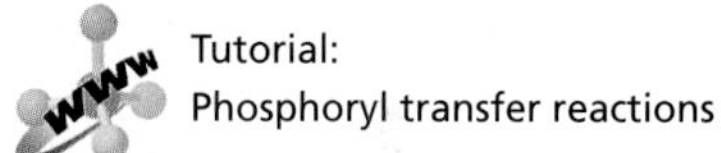

Tutorial:
Phosphoryl transfer reactions

$$^-O-\overset{\gamma}{P}(=O)(O^-)-O-\overset{\beta}{P}(=O)(O^-)-O-\overset{\alpha}{P}(=O)(O^-)-O-\text{adenosine}$$

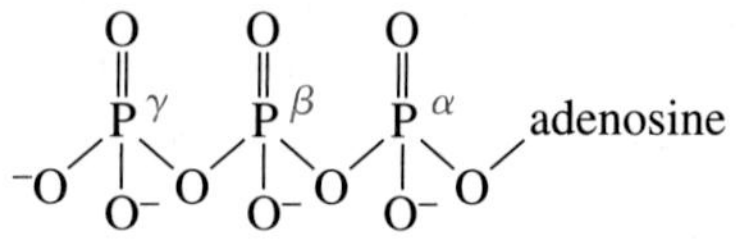

ATP

If the carboxylate ion attacks the γ-phosphorus of ATP, an **acyl phosphate** is formed. The acyl phosphate then reacts with the thiol in a nucleophilic acyl substitution reaction (Section 17.5) to form the thioester.

nucleophilic attack on the γ-phosphorus

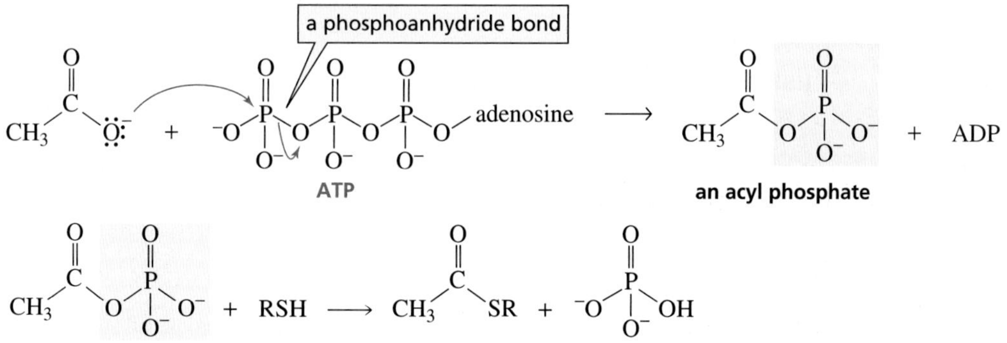

a phosphoanhydride bond

$$CH_3C(=O)O^- + \underset{\text{ATP}}{^-O-P(=O)(O^-)-O-P(=O)(O^-)-O-P(=O)(O^-)-O-\text{adenosine}} \longrightarrow \underset{\text{an acyl phosphate}}{CH_3C(=O)-O-P(=O)(O^-)-O^-} + \text{ADP}$$

$$CH_3C(=O)-O-P(=O)(O^-)-O^- + RSH \longrightarrow CH_3C(=O)SR + {^-O}-P(=O)(O^-)-OH$$

overall reaction

$$CH_3C(=O)O^- + RSH + ATP \longrightarrow CH_3C(=O)SR + ADP + {^-O}-P(=O)(O^-)-OH$$

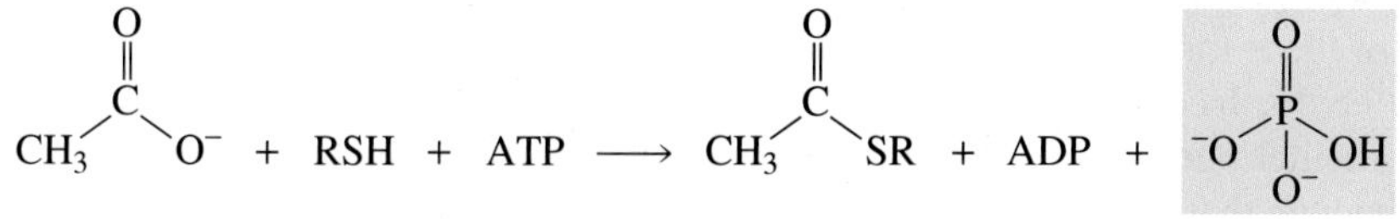

If the carboxylate ion attacks the β-phosphorus of ATP, an **acyl pyrophosphate** is formed. The acyl phosphate then reacts with the thiol in a nucleophilic acyl substitution reaction to form the thioester.

nucleophilic attack on the β-phosphorus

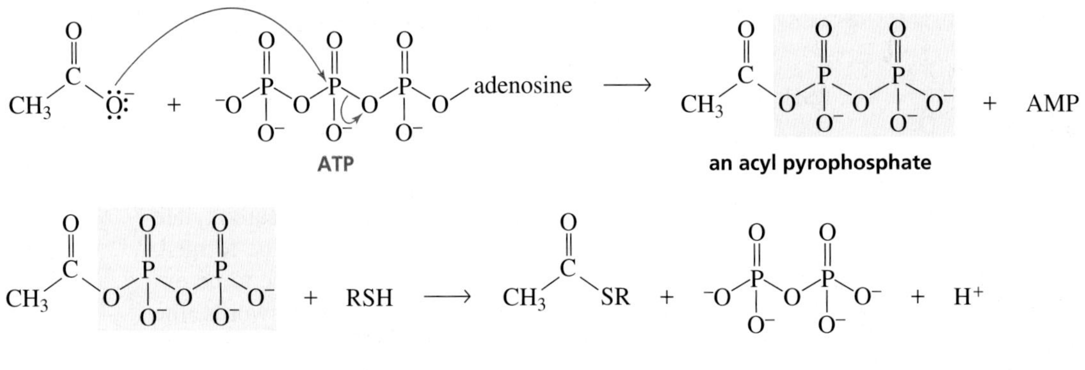

$$CH_3COO^- + \underset{\textbf{ATP}}{{}^-O-P(=O)(O^-)-O-P(=O)(O^-)-O-P(=O)(O^-)-O\text{-adenosine}} \longrightarrow \underset{\textbf{an acyl pyrophosphate}}{CH_3C(=O)-O-P(=O)(O^-)-O-P(=O)(O^-)-O^-} + AMP$$

$$CH_3C(=O)-O-P(=O)(O^-)-O-P(=O)(O^-)-O^- + RSH \longrightarrow CH_3C(=O)SR + {}^-O-P(=O)(O^-)-O-P(=O)(O^-)-O^- + H^+$$

overall reaction

$$CH_3C(=O)O^- + RSH + ATP \longrightarrow CH_3C(=O)SR + AMP + {}^-O-P(=O)(O^-)-O-P(=O)(O^-)-O^- + H^+$$

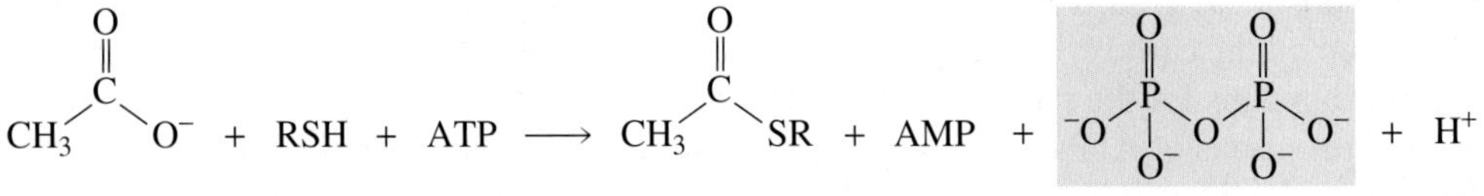

In the third possible mechanism, the carboxylate ion attacks the α-phosphorus of ATP, forming an **acyl adenylate**. The acyl phosphate then reacts with the thiol in a nucleophilic acyl substitution reaction to form the thioester.

nucleophilic attack on the α-phosphorus

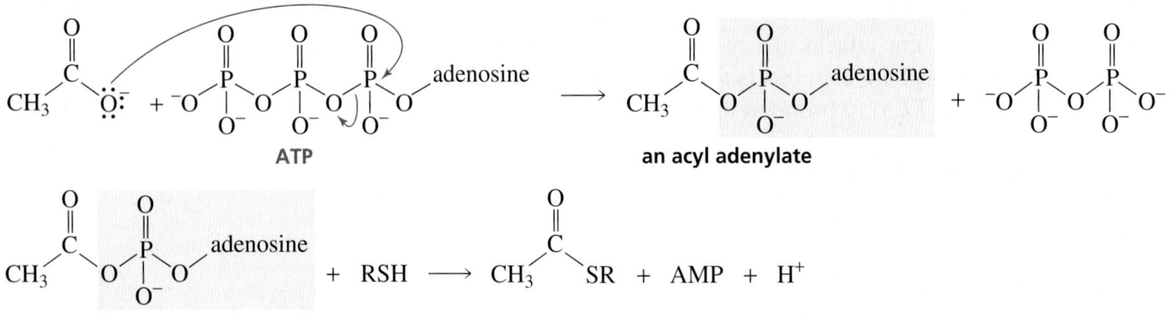

$$CH_3COO^- + \underset{\textbf{ATP}}{{}^-O-P(=O)(O^-)-O-P(=O)(O^-)-O-P(=O)(O^-)-O\text{-adenosine}} \longrightarrow \underset{\textbf{an acyl adenylate}}{CH_3C(=O)-O-P(=O)(O^-)-O\text{-adenosine}} + {}^-O-P(=O)(O^-)-O-P(=O)(O^-)-O^-$$

$$CH_3C(=O)-O-P(=O)(O^-)-O\text{-adenosine} + RSH \longrightarrow CH_3C(=O)SR + AMP + H^+$$

overall reaction

$$CH_3C(=O)O^- + RSH + ATP \longrightarrow CH_3C(=O)SR + AMP + {}^-O-P(=O)(O^-)-O-P(=O)(O^-)-O^- + H^+$$

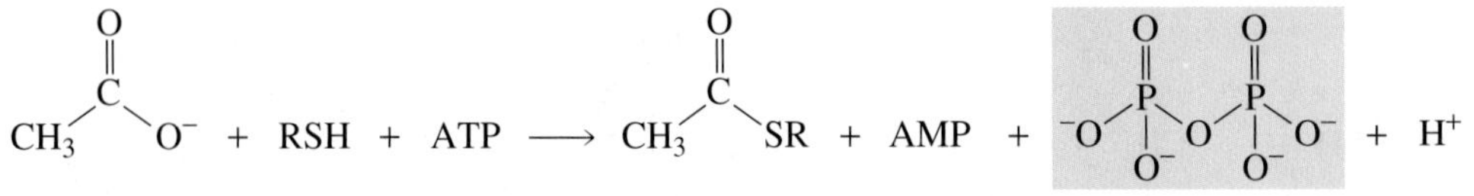

In Section 17.20 we saw that carboxylic acids in biological systems can be activated by being converted into acyl phosphates, acyl pyrophosphates, and acyl adenylates. Each of these three mechanisms puts a leaving group on the carboxylic acid that can easily be displaced by a nucleophile. The only difference in the three mechanisms is the particular phosphate atom that is attacked by the nucleophile and the nature of the intermediate that is formed.

Many different nucleophiles react with ATP in biological systems. Whether nucleophilic attack occurs on the α-, β-, or γ-phosphorus in any particular reaction depends on the enzyme catalyzing the reaction (Section 27.5). Mechanisms involving nucleophilic attack on the γ-phosphorus form ADP and phosphate as side products, whereas mechanisms involving nucleophilic attack on the α- or β-phosphorus form AMP and pyrophosphate as side products.

When pyrophosphate is one of the side products, it is subsequently hydrolyzed to two equivalents of phosphate. Consequently, in reactions in which pyrophosphate is formed as a product, its subsequent hydrolysis drives the reaction to the right, ensuring its irreversibility.

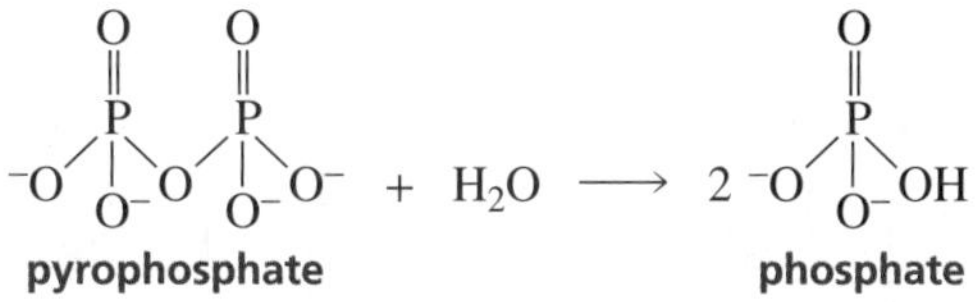

$$\underset{\textbf{pyrophosphate}}{{}^{-}O-P(=O)(O^{-})-O-P(=O)(O^{-})-O^{-}} + H_2O \longrightarrow 2\ \underset{\textbf{phosphate}}{{}^{-}O-P(=O)(O^{-})-OH}$$

Therefore, enzyme-catalyzed reactions in which irreversibility is important take place by one of the mechanisms that form pyrophosphate as a product (attack on the α- or β-phosphorus of ATP). For example, both the reaction that links nucleotide subunits to form nucleic acids (Section 27.7) and the reaction that binds an amino acid to a tRNA (the first step in translating RNA into a protein; Section 27.12) involve nucleophilic attack on the α-phosphorus of ATP.

PROBLEM 6

The β-phosphorus of ATP has two phosphoanhydride linkages, but only the one linking the β-phosphorus to the α-phosphorus is broken in phosphoryl transfer reactions. Explain why the one linking the β-phosphorus to the γ-phosphorus is never broken.

27.4 The "High-Energy" Character of Phosphoanhydride Bonds

Because the hydrolysis of a phosphoanhydride bond is a highly exergonic reaction, phosphoanhydride bonds are called **"high-energy bonds."** The term "high-energy" in this context means that a lot of energy is released when a reaction occurs that causes the bond to break. Do not confuse it with "bond energy," the term chemists use to describe how difficult it is to break a bond. A bond with a *high bond energy* is hard to break, whereas a *high-energy bond* breaks readily.

Why is the hydrolysis of a phosphoanhydride bond so exergonic? In other words, why is the $\Delta G^{\circ\prime}$ value for its hydrolysis large and negative? A large negative $\Delta G^{\circ\prime}$ means that the products of the reaction are much more stable than the reactants. Let's look at ATP and its hydrolysis products, phosphate and ADP, to see why this is so.

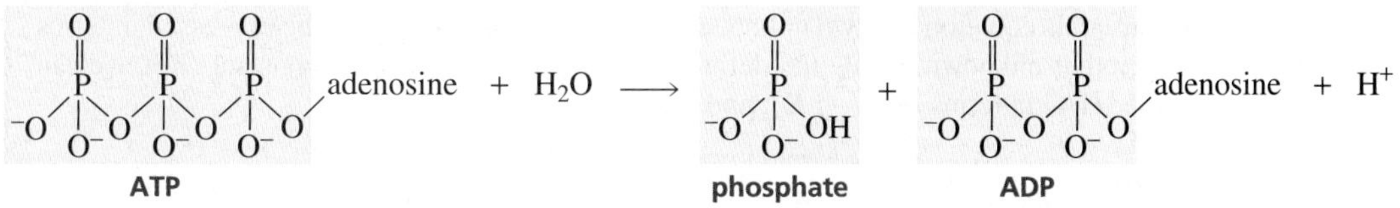

$$\underset{\textbf{ATP}}{{}^{-}O-P(=O)(O^{-})-O-P(=O)(O^{-})-O-P(=O)(O^{-})-O}\text{adenosine} + H_2O \longrightarrow \underset{\textbf{phosphate}}{{}^{-}O-P(=O)(O^{-})-OH} + \underset{\textbf{ADP}}{{}^{-}O-P(=O)(O^{-})-O-P(=O)(O^{-})-O}\text{adenosine} + H^{+}$$

Three factors contribute to the greater stability of ADP and phosphate compared to ATP:

1. **Greater electrostatic repulsion in ATP.** At physiological pH (pH = 7.3), ATP has 3.3 negative charges, ADP has 2.8 negative charges, and phosphate has 1.1 negative charges (Section 1.20). Because of ATP's greater negative charge, more electrostatic repulsions are present in ATP than in ADP or phosphate. Electrostatic repulsions destabilize a molecule.

2. **More solvation in the products.** Negatively charged ions are stabilized in an aqueous solution by solvation. Because the reactant has 3.3 negative charges, while the sum of the negative charges on the products is 3.9 $(2.8 + 1.1)$, there is more solvation in the products than in the reactant.
3. **Greater resonance stabilization in the products.** Delocalization of a lone pair on the oxygen joining the two phosphorus atoms is not very effective because it puts a positive charge on an oxygen that is next to a partially positively charged phosphorus atom. When the phosphoanhydride bond breaks, one additional lone pair can be effectively delocalized.

three lone pairs can be effectively delocalized

four lone pairs can be effectively delocalized

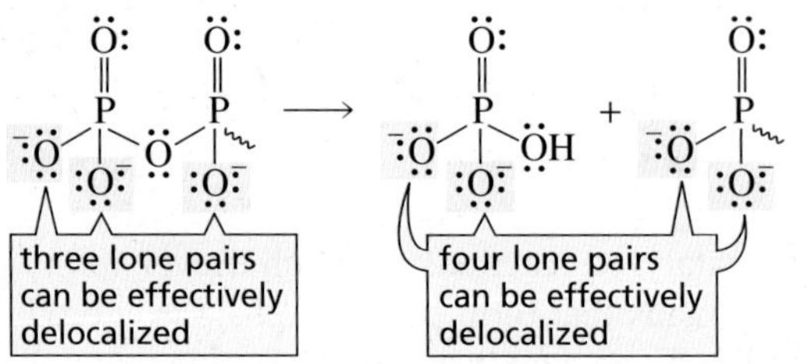

Similar factors explain the large negative $\Delta G^{\circ\prime}$ when ATP is hydrolyzed to AMP and pyrophosphate, and when pyrophosphate is hydrolyzed to two equivalents of phosphate.

PROBLEM 7 SOLVED

ATP has pK_a values of 0.9, 1.5, 2.3, and 7.7; ADP has pK_a values of 0.9, 2.8, and 6.8; and phosphoric acid has pK_a values of 1.9, 6.7, and 12.4.
Do the calculation showing that at pH 7.3:

a. the charge on ATP is -3.3

b. the charge on ADP is -2.8

c. the charge on phosphate is -1.1

SOLUTION TO 7a Because pH 7.3 is much more basic than the pK_a values of the first three ionizations of ATP, we know that these three groups will be entirely in their basic forms at that pH, giving ATP three negative charges. We need to determine what fraction of the group with pK_a 7.7 will be in its basic form at pH 7.3.

$$\frac{\text{concentration in the basic form}}{\text{total concentration}} = \frac{[A^-]}{[A^-] + [HA]}$$

$$[A^-] = \text{concentration of the basic form}$$

$$[HA] = \text{concentration of the acidic form}$$

Because this equation has two unknowns, one of the unknowns must be expressed in terms of the other unknown. Using the definition of the acid dissociation constant (K_a), we can define $[HA]$ in terms of $[A^-]$, K_a, and $[H^+]$.

$$K_a = \frac{[A^-][H^+]}{[HA]}$$

$$[HA] = \frac{[A^-][H^+]}{K_a}$$

$$\frac{[A^-]}{[A^-] + [HA]} = \frac{[A^-]}{[A^-] + \dfrac{[A^-][H^+]}{K_a}} = \frac{K_a}{K_a + [H^+]}$$

Now we can calculate the fraction of the group with pK_a 7.7 that will be in the basic form.

$$\frac{K_a}{K_a + [H^+]} = \frac{2.0 \times 10^{-8}}{2.0 \times 10^{-8} + 5.0 \times 10^{-8}} = 0.3$$

$$\text{total negative charge on ATP} = 3.0 + 0.3 = 3.3$$

27.5 Kinetic Stability of ATP in the Cell

Although ATP reacts readily in enzyme-catalyzed reactions, it reacts quite slowly in the absence of an enzyme. For example, carboxylic acid anhydrides hydrolyze in a matter of minutes, but ATP takes several weeks to hydrolyze. The low rate of ATP hydrolysis is important because it allows ATP to exist in the cell until it is needed for an enzyme-catalyzed reaction.

The negative charges on ATP are what make it relatively unreactive. These negative charges repel the approach of nucleophiles. When ATP is bound at an active site of an enzyme, it complexes with magnesium (Mg^{2+}), which decreases the overall negative charge on ATP. (This is why ATP-requiring enzymes also require metal ions; Section 25.5.) The other two negative charges can be stabilized by positively charged groups such as arginine or lysine residues at the active site, as shown in Figure 27.2. In this form, ATP is readily approached by nucleophiles, so ATP reacts rapidly in an enzyme-catalyzed reaction, but only very slowly in the absence of the enzyme.

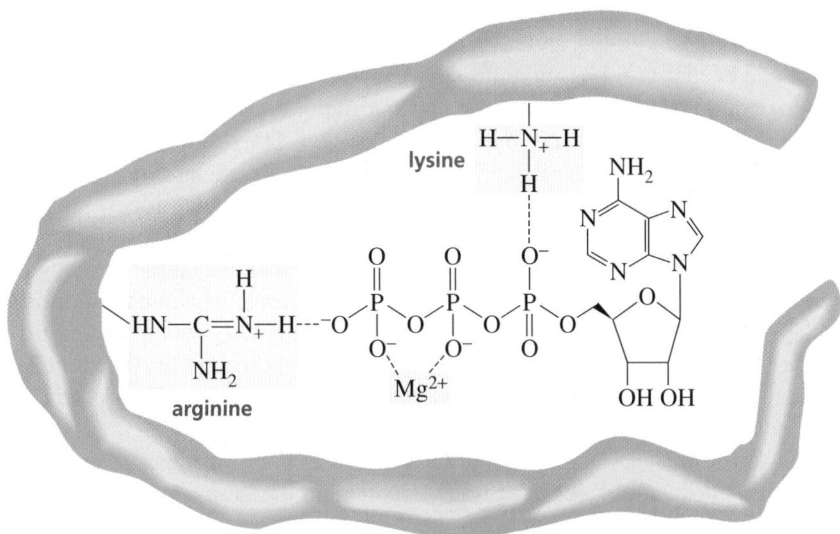

◀ **Figure 27.2**
The interactions between ATP, Mg^{2+}, and arginine and lysine residues at the active site of an enzyme.

27.6 Other Important Nucleotides

ATP is not the only biologically important nucleotide. Guanosine 5′-triphosphate (GTP) is used in place of ATP in some phosphoryl transfer reactions. We have also seen in Sections 25.2 and 25.3 that dinucleotides are used as oxidizing agents (NAD^+, $NADP^+$, FAD, FMN) and reducing agents (NADH, NADPH, $FADH_2$, $FMNH_2$).

Another important nucleotide is adenosine 3′,5′-monophosphate, commonly known as cyclic AMP. Cyclic AMP is called a "second messenger" because it serves as a link between several hormones (the first messengers) and certain enzymes that regulate cellular function. Secretion of certain hormones, such as adrenaline, activates adenylate cyclase, the enzyme responsible for the synthesis of cyclic AMP from ATP. Cyclic AMP then activates an enzyme, generally by phosphorylating it. Cyclic nucleotides are so important in regulating cellular reactions that an entire scientific journal is devoted to these processes.

ATP —adenylate cyclase→ cyclic AMP + pyrophosphate (Mg^{2+}) + H^+

3-D Molecule: Cyclic AMP

PROBLEM 8

What products would be obtained from the hydrolysis of cyclic AMP?

27.7 The Nucleic Acids

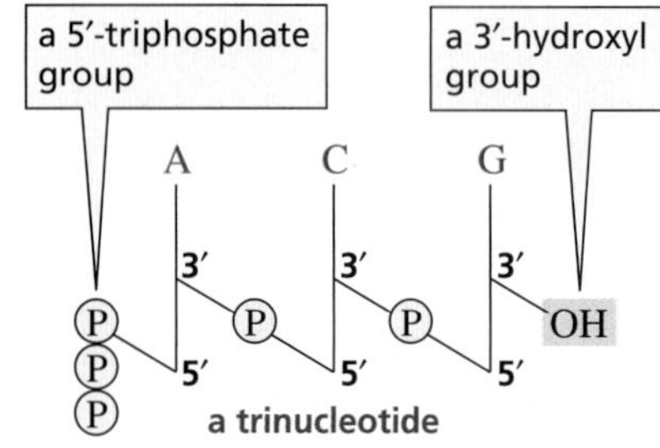

Nucleic acids are composed of long strands of nucleotide subunits linked by phosphodiester bonds. These linkages join the 3′-OH group of one nucleotide to the 5′-OH group of the next nucleotide (Figure 27.1). A **dinucleotide** contains two nucleotide subunits, an **oligonucleotide** contains three to ten subunits, and a **polynucleotide** contains many subunits. DNA and RNA are polynucleotides. Notice that the nucleotide at one end of the strand has an unlinked 5′-triphosphate group, and the nucleotide at the other end of the strand has an unlinked 3′-hydroxyl group.

Nucleotide triphosphates are the starting materials for the biosynthesis of nucleic acids. DNA is synthesized by enzymes called *DNA polymerases*, and RNA is synthesized by enzymes called *RNA polymerases*. The nucleotide strand is formed as a result of nucleophilic attack by a 3′-OH group of one nucleotide triphosphate on the α-phosphorus of another nucleotide triphosphate, breaking a phosphoanhydride bond and eliminating pyrophosphate (Figure 27.3). This means that the growing polymer is synthesized in the $5' \longrightarrow 3'$ direction; in other words, new nucleotides are added to the 3′-end. Pyrophosphate is subsequently hydrolyzed, which makes the reaction irreversible (Section 27.3). RNA strands are biosynthesized in the same way, using ribonucleotides instead of 2′-deoxyribonucleotides. The **primary structure** of a nucleic acid is the sequence of bases in the strand.

Watson and Crick concluded that DNA consists of two strands of nucleic acids with the sugar–phosphate backbone on the outside and the bases on the inside. The chains are held together by hydrogen bonds between the bases on one strand and the bases on the other strand (Figure 27.4). The width of the double-stranded molecule is relatively constant, so a purine must pair with a pyrimidine. If the larger purines paired, the strand would bulge; if the smaller pyrimidines paired, the strands would have to contract to bring the two pyrimidines close enough to form hydrogen bonds.

Erwin Chargaff (1905-2002) *was born in Austria and received a Ph.D. from the University of Vienna. To escape Hitler, he came to the United States in 1935, becoming a professor at Columbia University College of Physicians and Surgeons. He modified paper chromatography, a technique developed to identify amino acids (Section 23.5), so that it could be used to quantify the different bases in a sample of DNA.*

Critical to Watson and Crick's proposal for the secondary structure of DNA were experiments carried out by Erwin Chargaff. These experiments showed that the number of adenines in DNA equals the number of thymines and the number of guanines equals the number of cytosines. Chargaff also noted that the number of adenines and thymines relative to the number of guanines and cytosines is characteristic of a given species but varies from species to species. In human DNA, for example, 60.4% of the bases are adenines and thymines, whereas 74.2% of them are adenines and thymines in the DNA of the bacterium *Sarcina lutea*.

Chargaff's data showing that [adenine] = [thymine] and [guanine] = [cytosine] could be explained if adenine (A) always paired with thymine (T) and guanine (G) always paired with cytosine (C). This means the two strands are *complementary*—where there is an A in one strand, there is a T in the opposing strand, and where there is a G in

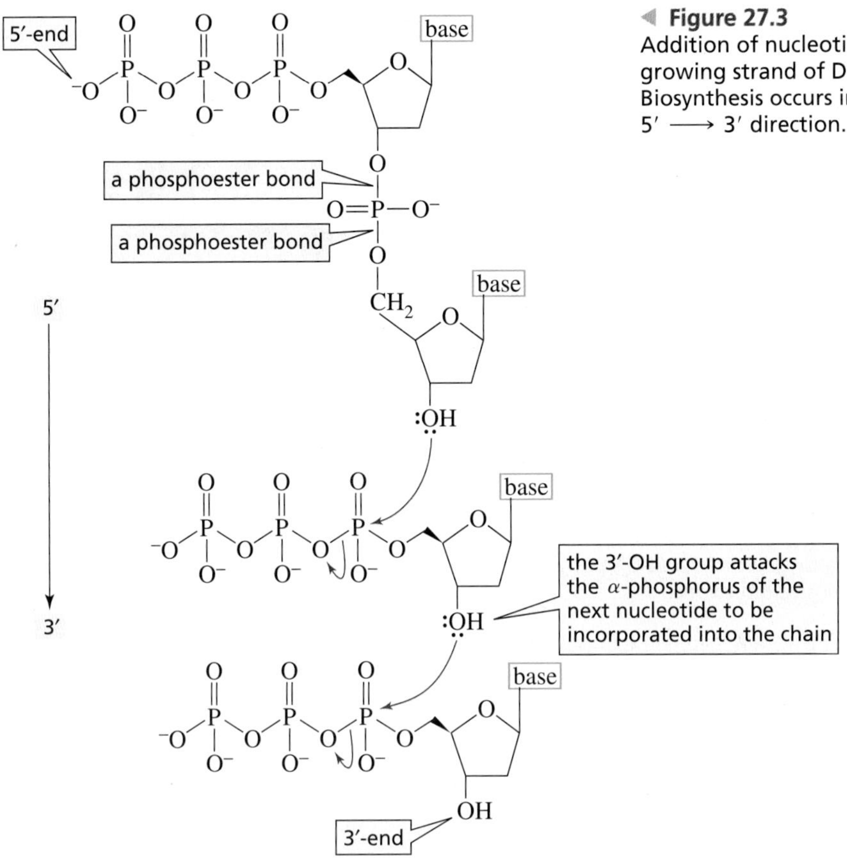

Figure 27.3
Addition of nucleotides to a growing strand of DNA. Biosynthesis occurs in the 5′ ⟶ 3′ direction.

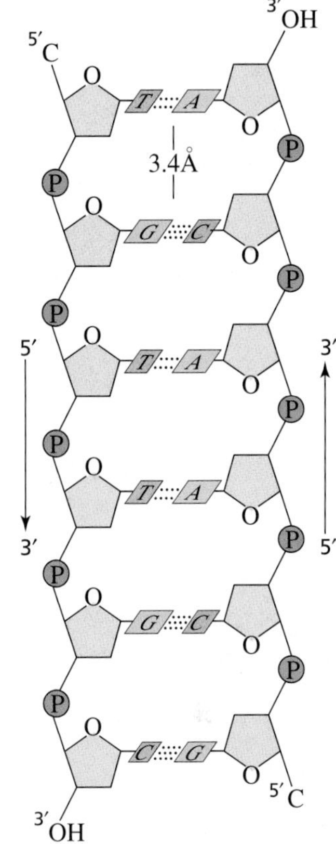

Figure 27.4
Complementary base pairing in DNA. Adenine (a purine) always pairs with thymine (a pyrimidine); guanine (a purine) always pairs with cytosine (a pyrimidine).

[A] = [T]
[G] = [C]

one strand there is a C in the other strand (Figure 27.4). Thus, if you know the sequence of bases in one strand, you can figure out the sequence of bases in the other strand.

What causes adenine to pair with thymine rather than with cytosine (the other pyrimidine)? The base pairing is dictated by hydrogen bonding. Learning that the bases exist in the keto form (Section 19.2) allowed Watson to explain the pairing.[2] Adenine forms two hydrogen bonds with thymine but would form only one hydrogen bond with cytosine. Guanine forms three hydrogen bonds with cytosine but would form only one hydrogen bond with thymine (Figure 27.5). The N—H----N and

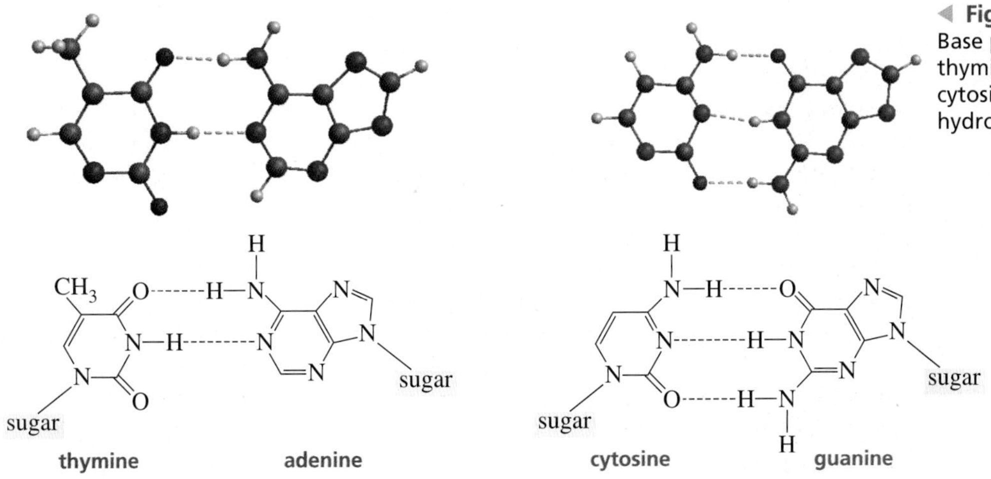

Figure 27.5
Base pairing in DNA: Adenine and thymine form two hydrogen bonds; cytosine and guanine form three hydrogen bonds.

thymine adenine cytosine guanine

[2]Watson was having difficulty understanding the base pairing in DNA because he thought the bases existed in the enol form (see Problem 10). When Jerry Donohue, an American crystallographer, informed him that the bases more likely existed in the keto form, Chargaff's data could easily be explained by hydrogen bonding between adenine and thymine and between guanine and cytosine.

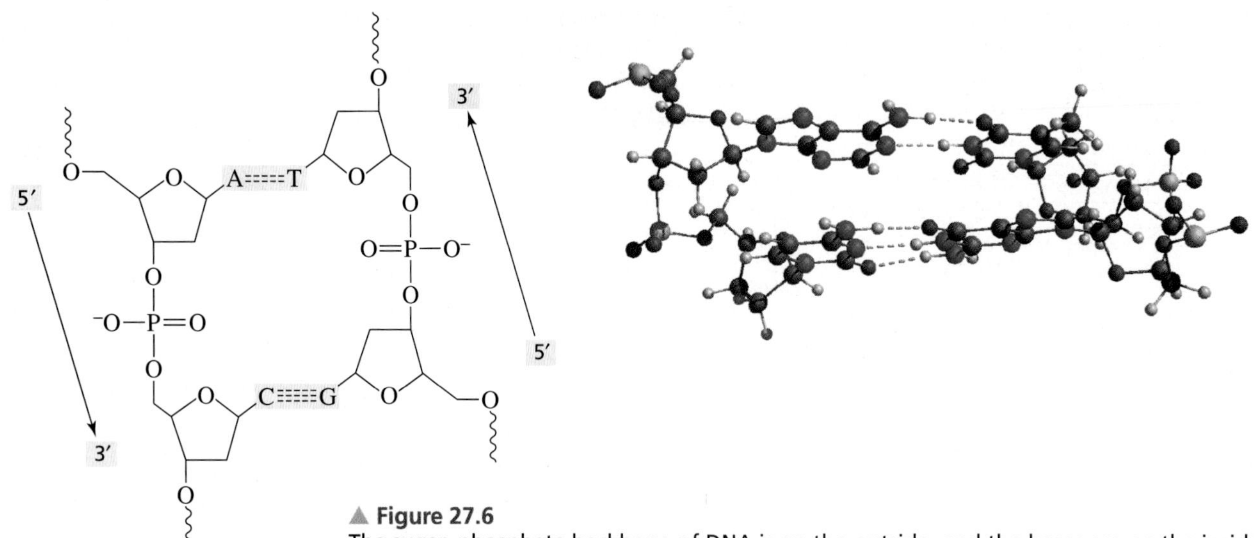

▲ **Figure 27.6**
The sugar–phosphate backbone of DNA is on the outside, and the bases are on the inside, with A's pairing with T's and G's pairing with C's. The two strands are antiparallel—they run in opposite directions.

N—H----O bonds that hold the bases together are all about the same length (2.9 ± 0.1 Å).

The two DNA strands are antiparallel—they run in opposite directions, with the sugar–phosphate backbone on the outside and the bases on the inside (Figures 27.4 and 27.6). By convention, the sequence of bases in a polynucleotide is written in the $5' \longrightarrow 3'$ direction (the 5′-end is on the left).

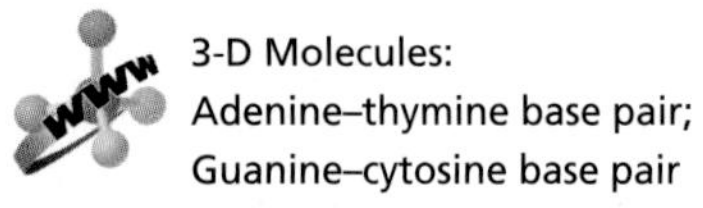

3-D Molecules:
Adenine–thymine base pair;
Guanine–cytosine base pair

The DNA strands are not linear but are twisted into a helix around a common axis (see Figure 27.7a). The base pairs are planar and parallel to each other on the inside of the helix (Figures 27.7b and c). The secondary structure is therefore known as a

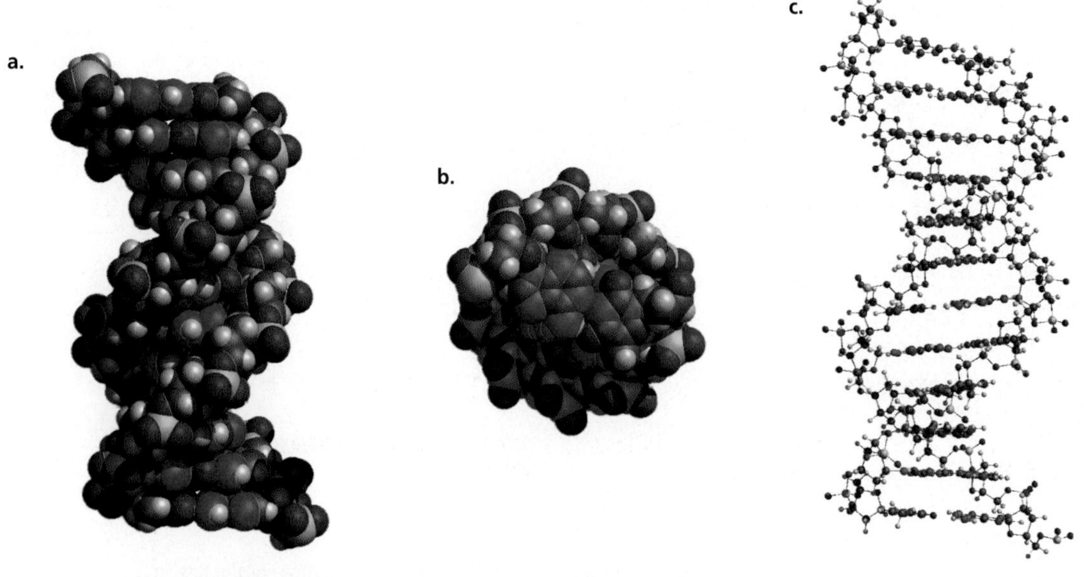

▲ **Figure 27.7**
(a) The DNA double helix. (b) View looking down the long axis of the helix. (c) The bases are planar and parallel on the inside of the helix.

double helix. The double helix resembles a ladder (the base pairs are the rungs) twisted around an axis running down through its rungs (Figures 27.4 and 27.7c). The sugar–phosphate backbone is wrapped around the bases. The phosphate OH group has a pK_a of about 2, so it is in its basic form (negatively charged) at physiological pH. The negatively charged backbone repels nucleophiles, thereby preventing cleavage of the phosphodiester bonds.

Unlike DNA, RNA is easily cleaved because the 2′-OH group of ribose can act as the nucleophile that cleaves the strand (Figure 27.8). This explains why the 2′-OH group is absent in DNA. To preserve the genetic information, DNA must remain intact throughout the life span of a cell. Cleavage of DNA would have disastrous consequences for the cell and for life itself. RNA, in contrast, is synthesized as it is needed and is degraded once it has served its purpose.

Hydrogen bonding between base pairs is just one of the forces holding the two strands of the DNA double helix together. The bases are planar aromatic molecules that stack on top of one another. Each pair is slightly rotated with respect to the next pair, like a partially spread-out hand of cards. There are favorable van der Waals interactions between the mutually induced dipoles of adjacent pairs of bases. These interactions, known as **stacking interactions**, are weak attractive forces, but when added together they contribute significantly to the stability of the double helix. Stacking interactions are strongest between two purines and weakest between two pyrimidines. Confining the bases to the inside of the helix has an additional stabilizing effect—it reduces the surface area of the relatively nonpolar residues exposed to water. This increases the entropy of the surrounding water molecules (Section 23.14).

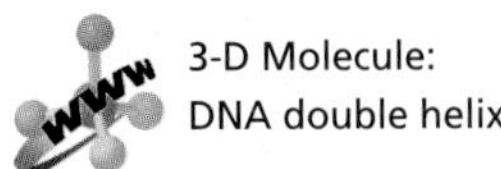

3-D Molecule: DNA double helix

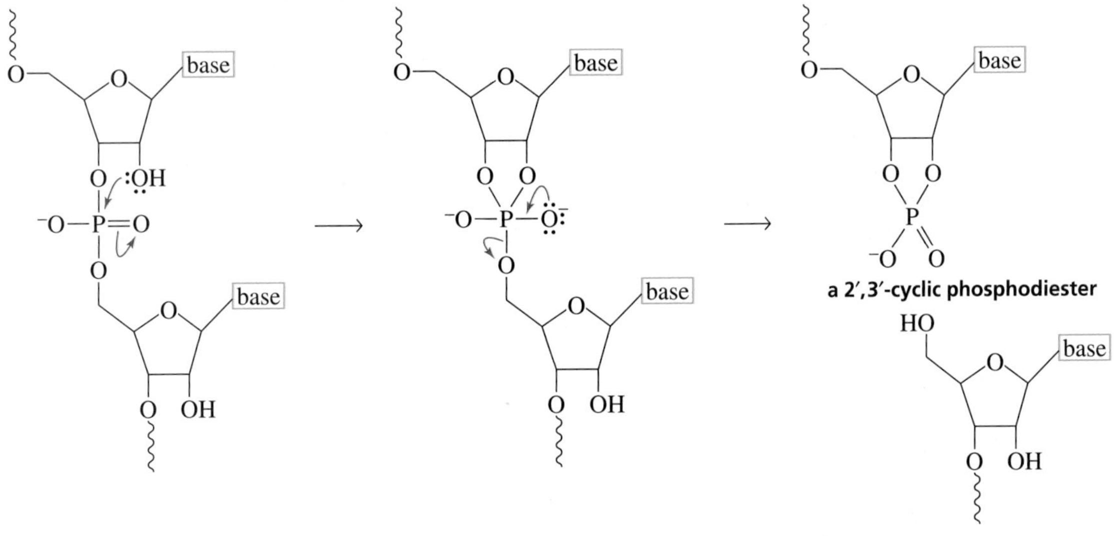

a 2′,3′-cyclic phosphodiester

▲ **Figure 27.8**
Hydrolysis of RNA. The 2′-OH group acts as an intramolecular nucleophilic catalyst. It has been estimated that RNA is hydrolyzed 3 billion times faster than DNA.

PROBLEM 9

Indicate whether each functional group of the five heterocyclic bases in nucleic acids can function as a hydrogen bond acceptor (A), a hydrogen bond donor (D), or both (D/A).

PROBLEM 10

Using the D, A, and D/A designations in Problem 9, explain how base pairing would be affected if the bases existed in the enol form.

PROBLEM 11

The 2′,3′-cyclic phosphodiester, which is formed when RNA is hydrolyzed (Figure 27.8), reacts with water, forming a mixture of nucleotide 2′- and 3′-phosphates. Propose a mechanism for this reaction.

PROBLEM 12◆

If one of the strands of DNA has the following sequence of bases running in the 5′ ⟶ 3′ direction,

5′—G—G—A—C—A—A—T—C—T—G—C—3′

a. What is the sequence of bases in the complementary strand?

b. What base is closest to the 5′-end in the complementary strand?

27.8 Helical Forms of DNA

Naturally occurring DNA can exist in the three different helical forms shown in Figure 27.9. The B- and A-helices are both right-handed. The B-helix is the predominant form in aqueous solution, while the A-helix is the predominant form in nonpolar solvents. Nearly all the DNA in living organisms is in a B-helix. The Z-helix is a left-handed helix. It occurs in regions where there is a high content of G—C base pairs. The A-helix is shorter (for a given number of base pairs) and about 3% broader than the B-helix, which is shorter and broader than the Z-helix.

Helices are characterized by the number of bases per 360° turn and the distance (the rise) between adjacent base pairs. A-DNA has 11 base pairs per turn and a 2.3 Å rise; B-DNA has 10 base pairs per turn and a 3.4 Å rise; and Z-DNA has 12 base pairs per turn and a 3.8 Å rise.

If you examine Figure 27.9, you will see that there are two kinds of alternating grooves in DNA. In B-DNA the **major groove** is wider than the **minor groove**. Cross sections of the double helix show that one side of each base pair faces into the major groove and the other side faces into the minor groove (Figure 27.10).

Proteins and other molecules can bind to the grooves. The hydrogen-bonding properties of the functional groups facing into each groove determine what kind of molecules will bind to the groove. Mitomycin is a naturally occurring compound that has been found to have both antibacterial activity and anticancer activity. It works by binding to the minor groove of DNA. It binds at regions rich in A's and T's (Section 30.10).

Figure 27.9 ▶ The three helical forms of DNA.

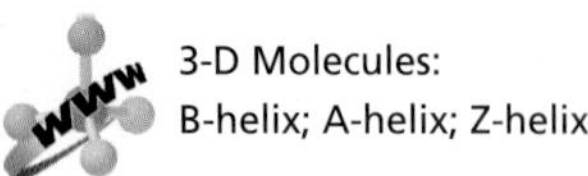

3-D Molecules: B-helix; A-helix; Z-helix

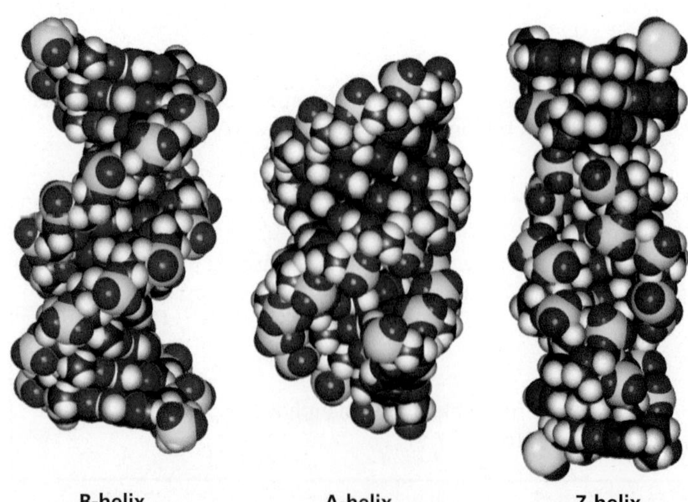

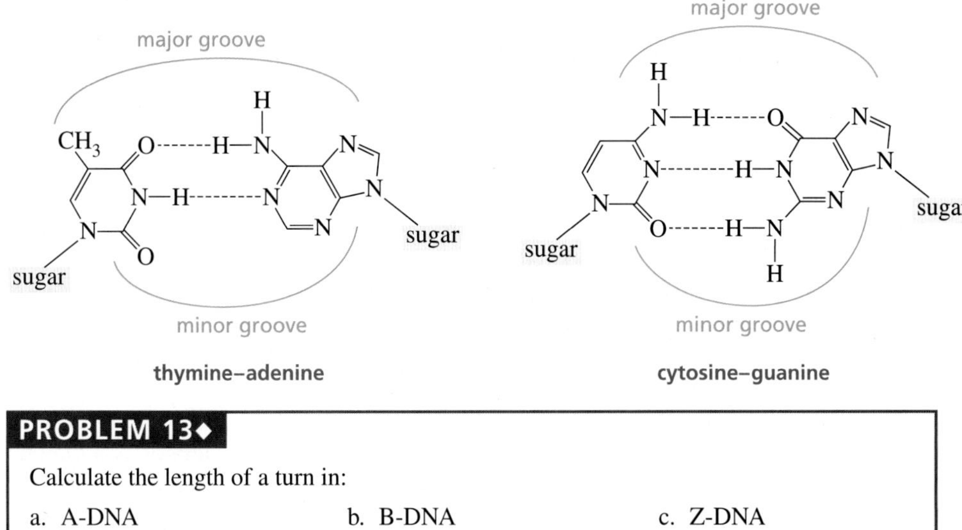

◀ **Figure 27.10**
One side of each base pair faces into the major groove, and the other side faces into the minor groove.

PROBLEM 13◆

Calculate the length of a turn in:

a. A-DNA b. B-DNA c. Z-DNA

27.9 Biosynthesis of DNA: Replication

Watson and Crick's proposal for the structure of DNA was an exciting development because the structure immediately suggested how DNA is able to pass on genetic information to succeeding generations. Because the two strands are complementary, both carry the same genetic information. Both strands serve as templates for the synthesis of complementary new strands (Figure 27.11). The new (daughter) DNA

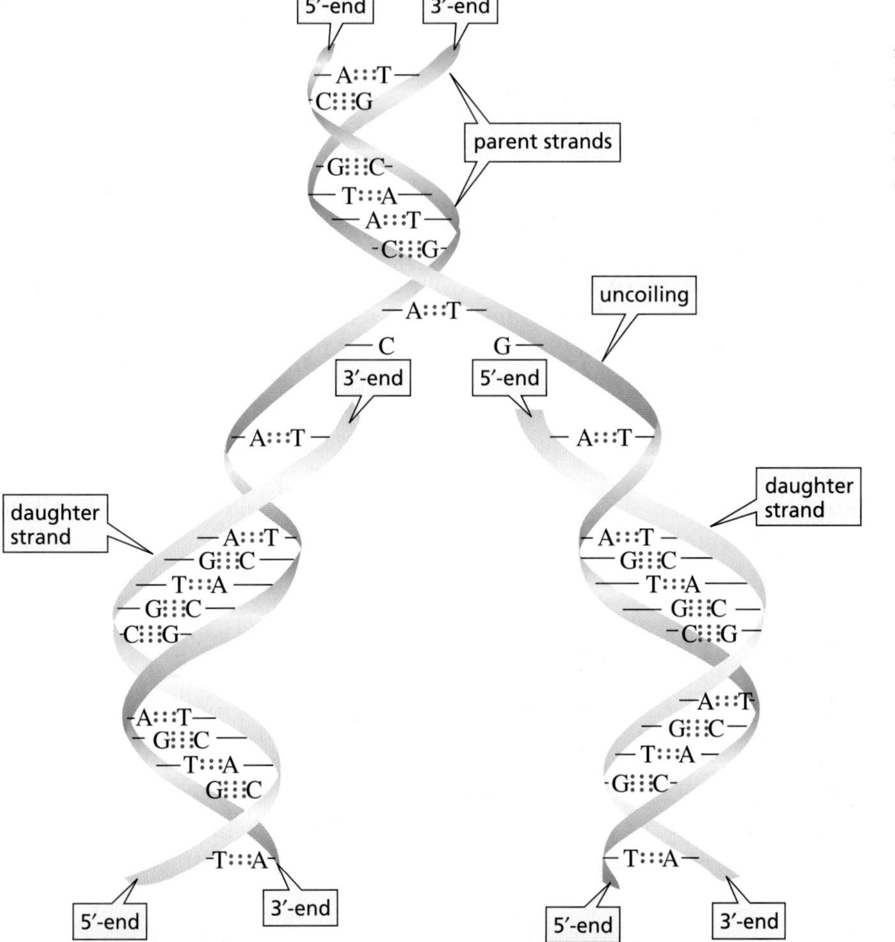

◀ **Figure 27.11**
Replication of DNA. The daughter strand on the left is synthesized continuously in the 5′ ⟶ 3′ direction; the daughter strand on the right is synthesized discontinuously in the 5′ ⟶ 3′ direction.

molecules are identical to the original (parent) molecule—they contain all the original genetic information. The synthesis of identical copies of DNA is called **replication**.

All reactions involved in nucleic acid synthesis are catalyzed by enzymes. The synthesis of DNA takes place in a region of the molecule where the strands have started to separate, called a **replication fork**. Because a nucleic acid can be synthesized only in the 5′ ⟶ 3′ direction, only the daughter strand on the left in Figure 27.11 is synthesized continuously in a single piece (because it is synthesized in the 5′ ⟶ 3′ direction). The other daughter strand needs to grow in the 3′ ⟶ 5′ direction, so it is synthesized discontinuously in small pieces. Each piece is synthesized in the 5′ ⟶ 3′ direction and the fragments are joined together by an enzyme called DNA ligase. Each of the two resulting daughter molecules of DNA that result contains one of the original strands (blue strand) plus a newly synthesized strand (green strand). This process is called **semiconservative replication**.

The genetic information of a human cell is contained in 23 pairs of chromosomes. Each chromosome is composed of several thousand **genes** (segments of DNA). The total DNA of a human cell—the **human genome**—contains 3.1 billion base pairs.

PROBLEM 14

Using a dark line for parental DNA and wavy lines for DNA synthesized from parental DNA, show what the population of DNA molecules would look like in the fourth generation.

PROBLEM 15◆

Assuming that the human genome, with its 3.1 billion base pairs, is entirely in a B-helix, how long is the DNA in a human cell?

PROBLEM 16

Why doesn't DNA unravel completely before replication begins?

27.10 Biosynthesis of RNA: Transcription

Severo Ochoa *was the first to prepare synthetic strands of RNA by incubating nucleotides in the presence of enzymes that are involved in the biosynthesis of RNA.* **Arthur Kornberg** *prepared synthetic strands of DNA in a similar manner. For this work, they shared the 1959 Nobel Prize in physiology or medicine.*

Severo Ochoa (1905–1993) *was born in Spain. He graduated from the University of Malaga in 1921 and received an M.D. from the University of Madrid. He spent the next four years studying in Germany and England and then joined the faculty at New York University College of Medicine. He became a U.S. citizen in 1956.*

Arthur Kornberg *was born in New York in 1918. He graduated from the College of the City of New York and received an M.D. from the University of Rochester. He is a member of the faculty of the biochemistry department at Stanford University.*

The sequence of DNA bases provides the blueprint for the synthesis of RNA. The synthesis of RNA from a DNA blueprint, called **transcription**, takes place in the nucleus of the cell. This initial RNA is the precursor to all RNA: messenger RNA, ribosomal RNA, and transfer RNA. The newly synthesized RNA leaves the nucleus, carrying the genetic information into the cytoplasm (the cell material outside the nucleus), where translation of this information into proteins takes place (see Figure 27.17).

DNA contains sequences of bases known as *promoter sites*. The **promoter sites** mark the beginning of genes. An enzyme recognizes a promoter site and binds to it, initiating RNA synthesis. The DNA at a promoter site unwinds to give two single strands, exposing the bases. One of the strands is called the **sense strand** or **informational strand**. The complementary strand is called the **template strand** or **antisense strand**. The template strand is read in the 3′ ⟶ 5′ direction, so that RNA can be synthesized in the 5′ ⟶ 3′ direction (Figure 27.12). The bases in the template strand specify the bases that need to be incorporated into RNA, following the same base pairing found in DNA. For example, each guanine in the template strand specifies the incorporation of a cytosine into RNA, and each adenine in the template strand specifies the incorporation of a uracil into RNA. (Recall that in RNA, uracil is used instead of thymine.). Because both the sense strand and RNA are complementary to the template strand, the sense strand and RNA have the same base sequence, except that RNA has a uracil wherever the sense strand has a thymine. Just as there are promoter sites that signal the places to start RNA synthesis, there are sites in DNA that signal that no more bases should be added to the growing strand of RNA, at which point synthesis stops.

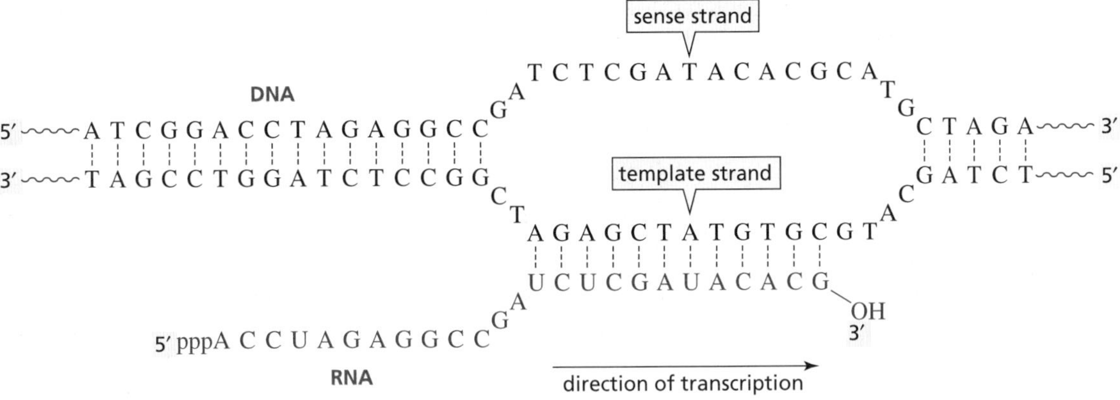

▲ **Figure 27.12**
Transcription: using DNA as a blueprint for RNA.

Surprisingly, a gene is not necessarily a continuous sequence of bases. Often the bases of a gene are interrupted by bases that appear to have no informational content. A stretch of bases representing a portion of a gene is called an **exon**, while a stretch of bases that contains no genetic information is called an **intron**. The RNA that is synthesized is complementary to the entire sequence of DNA bases—exons and introns. So after the RNA is synthesized, but before it leaves the nucleus, the so-called nonsense bases (encoded by the introns) are cut out and the informational fragments are spliced together, resulting in a much shorter RNA molecule. This RNA processing step is known as **RNA splicing**. Scientists have found that only about 2% of DNA contains genetic information, while 98% consists of introns.

It has been suggested that the purpose of introns is to make RNA more versatile. The originally synthesized long strand of RNA can be spliced in different ways to create a variety of shorter RNAs.

PROBLEM 17

Why do both thymine and uracil specify the incorporation of adenine?

27.11 Ribosomal RNA

RNA is much shorter than DNA and is generally single-stranded. Although DNA molecules can have billions of base pairs, RNA molecules rarely have more than 10,000 nucleotides. There are three kinds of RNA—**messenger RNA (mRNA)** whose sequence of bases determines the sequence of animo acids in a protein, **ribosomal RNA (rRNA)**, a structural component of ribosomes, and **transfer RNA (tRNA)**, the carriers of amino acids for protein synthesis.

The biosynthesis of proteins takes place on particles known as **ribosomes**. A ribosome is composed of about 40% protein and about 60% rRNA. There is increasing evidence that protein synthesis is catalyzed by rRNA molecules rather than by enzymes. RNA molecules—found in ribosomes—that act as catalysts are known as **ribozymes**. The protein molecules in the ribosome enhance the functioning of the rRNA molecules.

Ribosomes are made up of two subunits. The size of the subunits depends on whether they are found in prokaryotic organisms or eukaryotic organisms. **Prokaryotic organisms** (*pro*, Greek for "before"; *karyon*, Greek for "kernel" or "nut") are the earliest organisms. They are unicellular and do not have nuclei. A **eukaryotic organism** (*eu*, Greek for "well") is much more complicated. Eukaryotic organisms can be unicellular or multicellular and their cells have nuclei. A prokaryotic ribosome is composed of a 50S subunit and a smaller 30S subunit; together they form a 70S ribosome. A eukaryotic ribosome has a 60S subunit and a 40S subunit; together they form an 80S ribosome.

Sidney Altman *and* **Thomas R. Cech** *received the 1989 Nobel Prize in chemistry for their discovery of the catalytic properties of RNA.*

Sidney Altman *was born in Montreal in 1939. He received a B.S. from MIT and a Ph.D. from the University of Colorado, Boulder. He was a postdoctoral fellow in Francis Crick's laboratory at Cambridge University. He is a professor of biology at Yale University.*

Thomas Cech *was born in Chicago in 1947. He received a B.A. from Grinnell College and a Ph.D. from the University of California, Berkeley. He was a postdoctoral fellow at MIT. He is a professor of chemistry at the University of Colorado, Boulder.*

The S stands for the **sedimentation constant**, which designates where a given component sediments during centrifugation.[3]

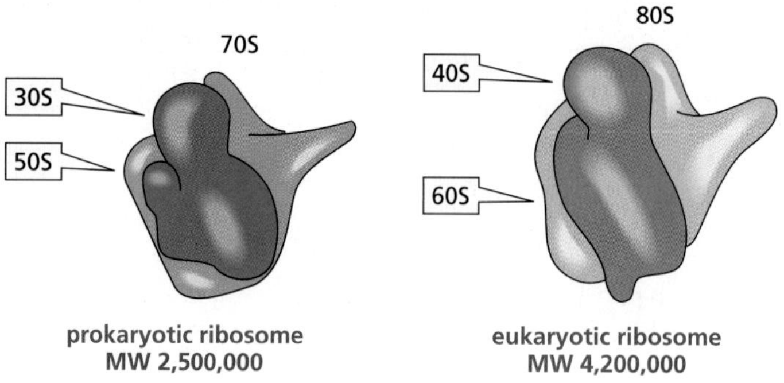

27.12 Transfer RNA

Transfer RNA (tRNA) is much smaller than mRNA or rRNA. It contains only 70 to 90 nucleotides. The single strand of tRNA is folded into a characteristic cloverleaf structure strung out with three loops and a little bulge next to the right-hand loop (Figure 27.13a). There are at least four regions with complementary base pairing. All tRNAs have a CCA sequence at the 3′-end. The three bases at the bottom of the loop directly opposite the 5′- and 3′-ends are called an **anticodon** (Figures 27.13a and b).

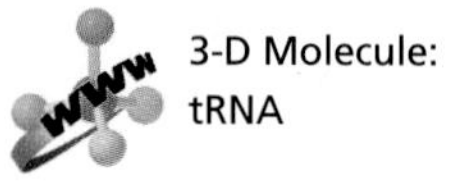

Each tRNA can carry an amino acid bound as an ester to its terminal 3′-OH group. The amino acid will be inserted into a protein during protein biosynthesis. Each tRNA can carry only one particular amino acid. A tRNA that carries alanine is designated as $tRNA^{Ala}$.

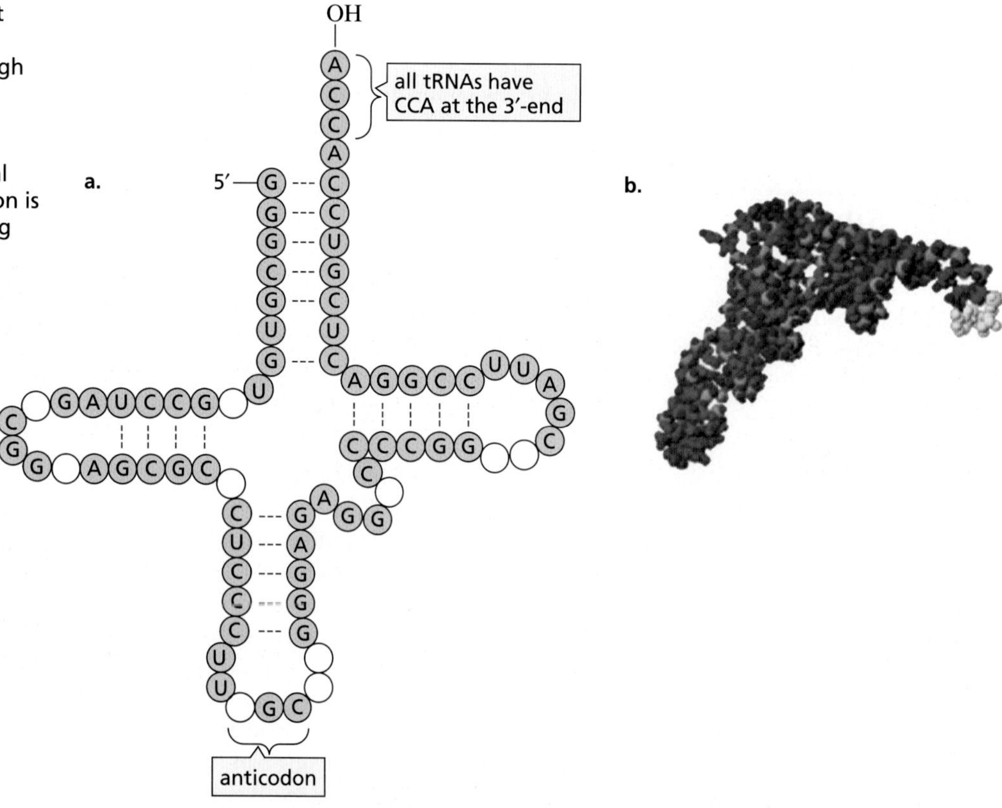

Figure 27.13 ▶
(a) $tRNA^{Ala}$, a transfer RNA that carries alanine. Compared with other RNAs, tRNA contains a high percentage of unusual bases (shown as empty circles). These bases result from enzymatic modification of the four normal bases. (b) $tRNA^{Ser}$: The anticodon is shown in red; the serine binding site is shown in yellow.

[3]Sedimentation constants are not additive, which is why a 50S and a 30S can combine to form a 70S.

How does an amino acid become attached to a tRNA? Attachment of the amino acid is catalyzed by an enzyme called aminoacyl-tRNA synthetase. In the first step of the enzyme-catalyzed reaction (Figure 27.14), the carboxyl group of the amino acid attacks the α-phosphorus of ATP, activating the carboxylate group by forming an acyl adenylate. The pyrophosphate that is expelled is subsequently hydrolyzed, ensuring the irreversibility of the phosphoryl transfer reaction (Section 27.3). Then a nucleophilic acyl substitution reaction occurs—the 3′-OH group of tRNA attacks the carbonyl carbon of the acyl adenylate, forming a tetrahedral intermediate. The amino acyl tRNA is formed when AMP is expelled from the tetrahedral intermediate. All the steps take place at the active site of the enzyme. Each amino acid has its own aminoacyl-tRNA synthetase. Each synthetase has two specific binding sites, one for the amino acid and one for the tRNA that carries that amino acid (Figure 27.15).

Elizabeth Keller (1918–1997) *was the first to recognize that tRNA had a cloverleaf structure. She received a B.S. from the University of Chicago in 1940 and a Ph.D. from Cornell University Medical College in 1948. She worked at the Huntington Memorial Laboratory of Massachusetts General Hospital and at the United States Public Health Service. Later she became a professor at MIT and then at Cornell University.*

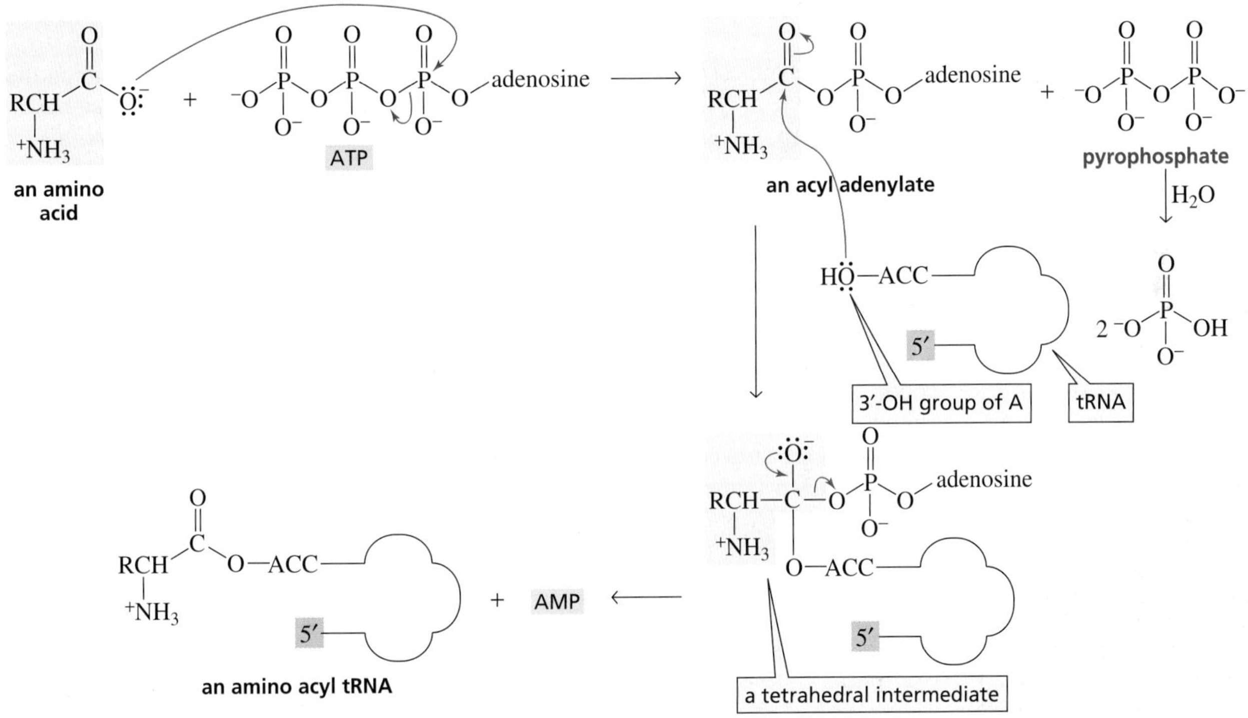

▲ **Figure 27.14**
The proposed mechanism for aminoacyl-tRNA synthetase—the enzyme that catalyzes the attachment of an amino acid to a tRNA.

It is critical that the correct amino acid is attached to the tRNA. Otherwise, the correct protein will not be synthesized. Fortunately, the synthetases correct their own mistakes. For example, valine and threonine are approximately the same size—threonine has an OH group in place of a CH_3 group of valine. Both amino acids, therefore, can bind at the amino acid binding site of the aminoacyl-tRNA synthetase for valine, and both can then be activated by reacting with ATP to form an acyl adenylate. The aminoacyl-tRNA synthetase for valine has two adjacent catalytic sites, one for

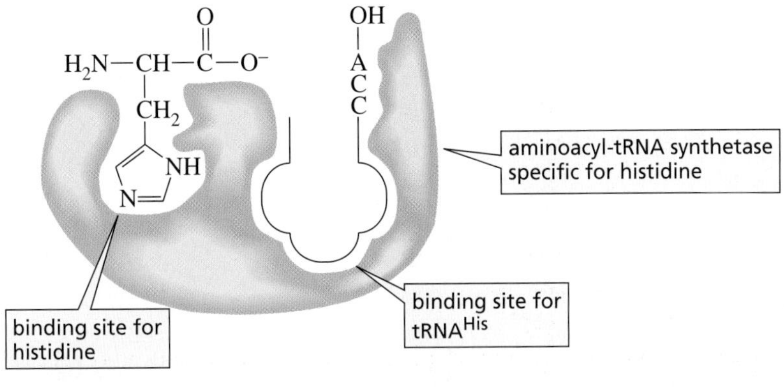

◀ **Figure 27.15**
An aminoacyl-tRNA synthetase has a binding site for tRNA and a binding site for the particular amino acid that is to be attached to that tRNA. Histidine is the amino acid and $tRNA^{His}$ is the tRNA molecule in this example.

attaching the acyl adenylate to tRNA and one for hydrolyzing the acyl adenylate. The acylation site is hydrophobic, so valine is preferred over threonine for the tRNA acylation reaction. The hydrolytic site is polar, so threonine is preferred over valine for the hydrolysis reaction. Thus, if threonine is activated by the aminoacyl-tRNA synthetase for valine, it will be hydrolyzed rather than transferred to the tRNA.

$$\underset{\text{valine}}{CH_3\underset{\displaystyle CH_3}{\underset{|}{C}}H-\underset{\displaystyle {}^{+}NH_3}{\underset{|}{C}}H\overset{\displaystyle O}{\overset{\|}{C}}O^-} \qquad \underset{\text{threonine}}{CH_3\underset{\displaystyle OH}{\underset{|}{C}}H-\underset{\displaystyle {}^{+}NH_3}{\underset{|}{C}}H\overset{\displaystyle O}{\overset{\|}{C}}O^-}$$

27.13 Biosynthesis of Proteins: Translation

A protein is synthesized from its N-terminal end to its C-terminal end by reading the bases along the mRNA strand in the 5′ $\longrightarrow$ 3′ direction. A sequence of three bases, called a **codon**, specifies a particular amino acid that is to be incorporated into a protein. The bases are read consecutively and are never skipped. A codon is written with the 5′-nucleotide on the left. Each amino acid is specified by a three-base sequence known as the **genetic code** (Table 27.2). For example, UCA on mRNA codes for the amino acid serine, whereas CAG codes for glutamine.

The genetic code was worked out independently by **Marshall Nirenberg** *and* **Har Gobind Khorana**, *for which they shared the 1968 Nobel Prize in physiology or medicine.* **Robert Holley**, *who worked on the structure of tRNA molecules, also shared that year's prize.*

Because there are four bases and the codons are triplets, $4^3 = 64$ different codons are possible. This is many more than are needed to specify the 20 different amino acids, so all the amino acids—except methionine and tryptophan—have more than one codon. It is not surprising, therefore, that methionine and tryptophan are the least abundant amino acids in proteins. Actually, 61 of the bases specify amino acids, and three bases are stop codons. **Stop codons** tell the cell to "stop protein synthesis here."

Marshall Nirenberg *was born in New York in 1927. He received a bachelor's degree from the University of Florida and a Ph.D. from the University of Michigan. He is a scientist at the National Institutes of Health.*

Translation is the process by which the genetic message in DNA that has been passed to mRNA is decoded and used to build proteins. Each of the approximately 100,000 proteins in the human body is synthesized from a different mRNA. Don't

Har Gobind Khorana *was born in India in 1922. He received a bachelor's and a master's degree from Punjab University and a Ph.D. from the University of Liverpool. In 1960 he joined the faculty at the University of Wisconsin and later became a professor at MIT.*

Table 27.2 The Genetic Code

5′-Position	Middle position				3′-Position
	U	C	A	G	
U	Phe	Ser	Tyr	Cys	U
	Phe	Ser	Tyr	Cys	C
	Leu	Ser	Stop	Stop	A
	Leu	Ser	Stop	Trp	G
C	Leu	Pro	His	Arg	U
	Leu	Pro	His	Arg	C
	Leu	Pro	Gln	Arg	A
	Leu	Pro	Gln	Arg	G
A	Ile	Thr	Asn	Ser	U
	Ile	Thr	Asn	Ser	C
	Ile	Thr	Lys	Arg	A
	Met	Thr	Lys	Arg	G
G	Val	Ala	Asp	Gly	U
	Val	Ala	Asp	Gly	C
	Val	Ala	Glu	Gly	A
	Val	Ala	Glu	Gly	G

confuse transcription and translation—these words are used just as they are used in English. Transcription (DNA to RNA) is copying *within the same language* of nucleotides. Translation (RNA to protein) is *changing to another language*—the language of amino acids.

Transcription: DNA ⟶ RNA

Translation: mRNA ⟶ protein

How the information in mRNA is translated into a polypeptide is shown in Figure 27.16. In this figure, serine was the last amino acid incorporated into the

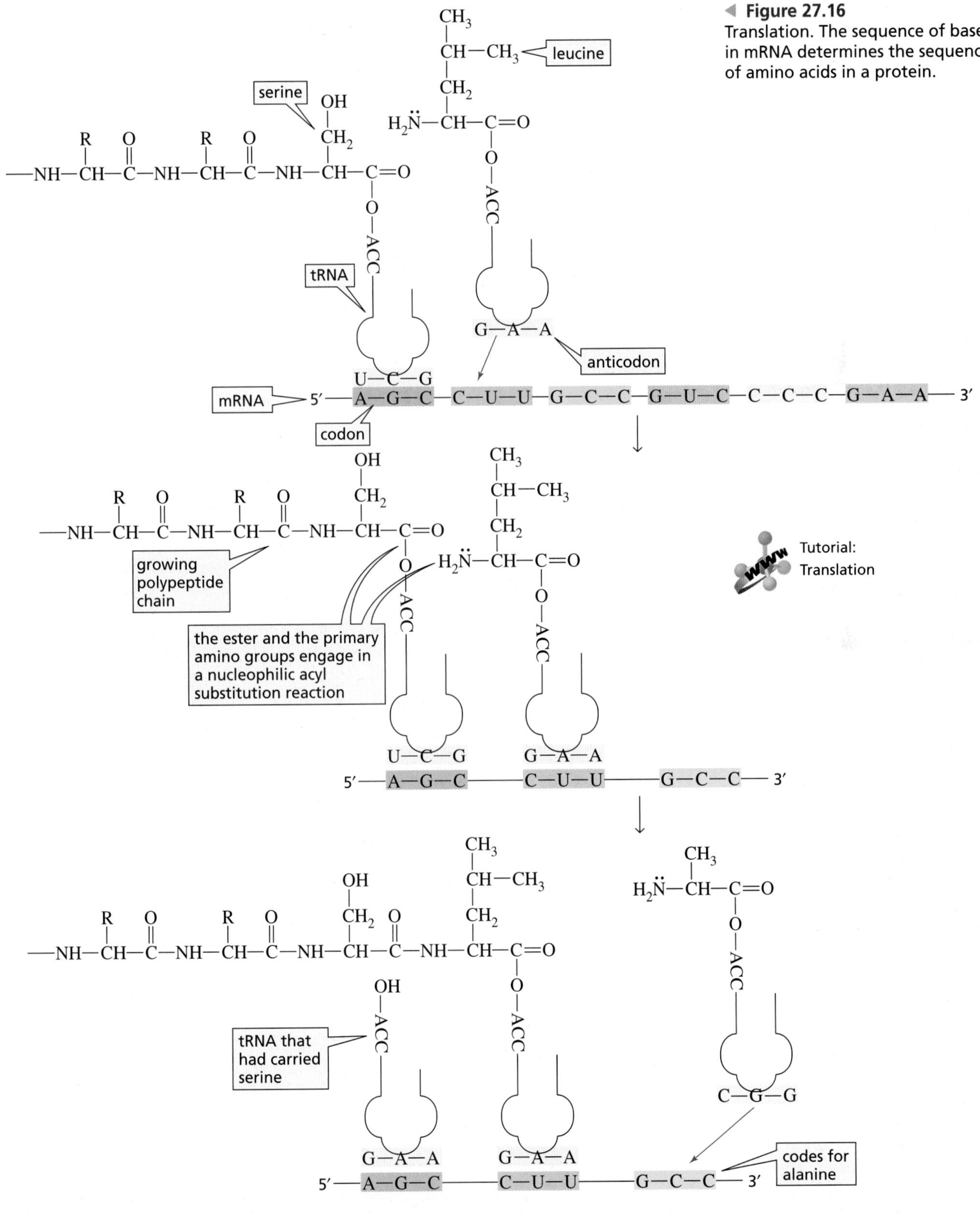

Figure 27.16
Translation. The sequence of bases in mRNA determines the sequence of amino acids in a protein.

Tutorial: Translation

growing polypeptide chain. Serine was specified by the AGC codon because the anticodon of the tRNA that carries serine is GCU (3′-UCG-5′). (Remember that a base sequence is read in the 5′ ⟶ 3′ direction, so the sequence of bases in an anticodon must be read from right to left.) The next codon is CUU, signaling for a tRNA with an anticodon of AAG (3′-GAA-5′). That particular tRNA carries leucine. The amino group of leucine reacts in an enzyme-catalyzed nucleophilic acyl substitution reaction with the ester on the adjacent tRNA, displacing the tRNA. The next codon (GCC) brings in a tRNA carrying alanine. The amino group of alanine displaces the tRNA that brought in leucine. Subsequent amino acids are brought in one at a time in the same way, with the codon in mRNA specifying the amino acid to be incorporated by complementary base pairing with the anticodon of the tRNA that carries that amino acid.

Protein synthesis takes place on the ribosomes (Figure 27.17). The smaller subunit of the ribosome (30S in prokaryotic cells) has three binding sites for RNA molecules. It binds the mRNA whose base sequence is to be read, the tRNA carrying the growing peptide chain, and the tRNA carrying the next amino acid to be incorporated into the protein. The larger subunit of the ribosome (50S in prokaryotic cells) catalyzes peptide bond formation.

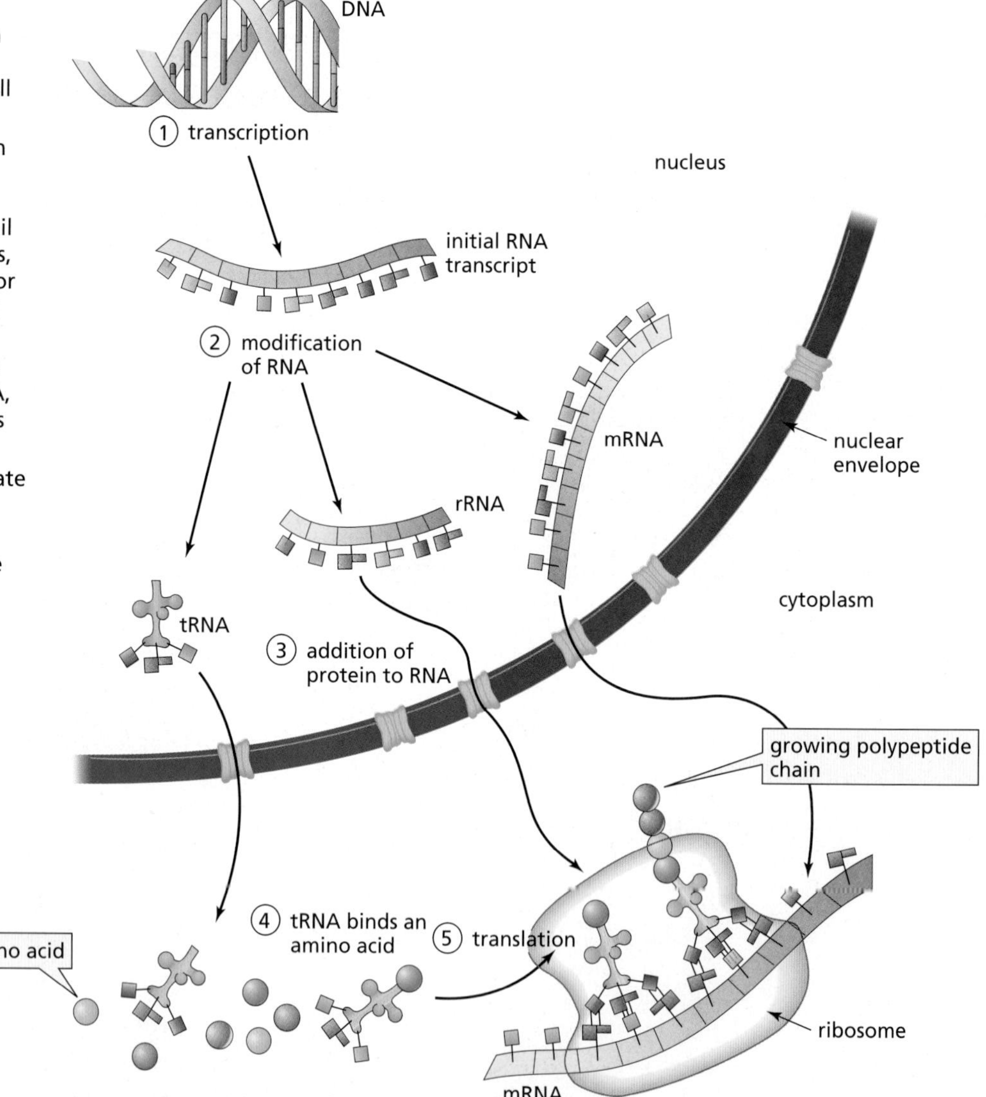

Figure 27.17 ▶

1. Transcription of DNA occurs in the nucleus. The initial RNA transcript is the precursor of all RNA: tRNA, rRNA, and mRNA.
2. The initially formed RNA often must be chemically modified before it acquires biological activity. Modification can entail removing nucleotide segments, adding nucleotides to the 5′- or 3′-ends, or chemically altering certain nucleotides.
3. Proteins are added to rRNA to form ribosomal subunits. tRNA, mRNA, and ribosomal subunits leave the nucleus.
4. Each tRNA binds the appropriate amino acid.
5. tRNA, mRNA, and a ribosome work together to translate the mRNA information into a protein.

PROBLEM 18◆

If methionine is the first amino acid incorporated into a heptapeptide, what is the sequence of the amino acids encoded for by the following stretch of mRNA?

5′—G—C—A—U—G—G—A—C—C—C—C—G—U—U—A—U—U—A—A—A—C—A—C—3′

PROBLEM 19◆

Four C's occur in a row in the segment of mRNA in Problem 18. What polypeptide would be formed from the mRNA if one of the four C's were cut out of the strand?

PROBLEM 20

UAA is a stop codon. Why does the UAA sequence in mRNA in Problem 18 not cause protein synthesis to stop?

PROBLEM 21◆

Write the sequences of bases in the sense strand of DNA that resulted in the mRNA in Problem 18.

PROBLEM 22

List the possible codons on mRNA that specify each amino acid in Problem 18 and the anticodon on the tRNA that carries that amino acid.

3-D Molecules: Chloramphenicol complexed to acetyl-transferase; Tetracycline

A sculpture done by Robert Holley.

Robert W. Holley (1922–1993) *was born in Illinois and received a bachelor's degree from the University of Illinois and a Ph.D. from Cornell University. During World War II he worked on the synthesis of penicillin at Cornell Medical School. He was a professor at Cornell and later at the University of California, San Diego. He was also a noted sculptor.*

SICKLE CELL ANEMIA

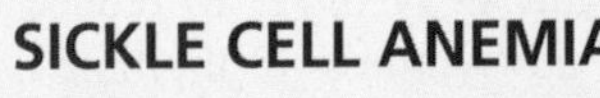

Sickle cell anemia is an example of the damage that can be caused by a change in a single base of DNA (Problem 55 in Chapter 23). It is a hereditary disease caused when a GAG triplet becomes a GTG triplet in the sense strand of a section of DNA that codes for the β-subunit of hemoglobin. As a consequence, the mRNA codon becomes GUG—which signals for incorporation of valine—rather than GAG, which would have signaled for incorporation of glutamic acid. The change from a polar glutamic acid to a nonpolar valine is sufficient to change the shape of the deoxyhemoglobin molecule and induce aggregation, causing it to precipitate in red blood cells. This stiffens the cells, making it difficult for them to squeeze through a capillary. Blocked capillaries cause severe pain and can be fatal.

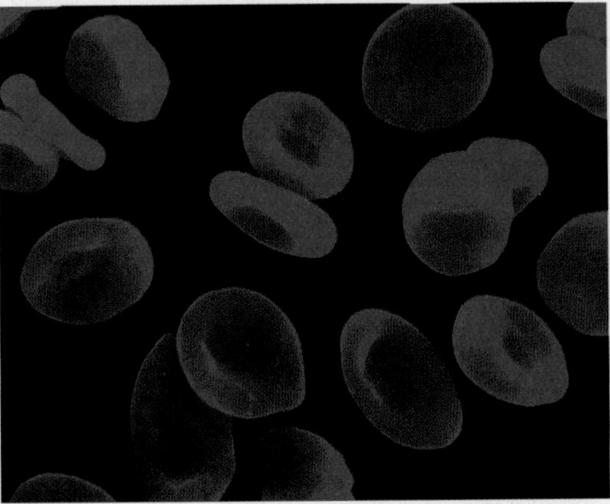

Normal red blood cells

Sickle red blood cells

ANTIBIOTICS THAT ACT BY INHIBITING TRANSLATION

Puromycin is a naturally occurring antibiotic. It is one of several antibiotics that act by inhibiting translation. Puromycin mimics the 3′-CCA-aminoacyl portion of a tRNA. If, during translation, the enzyme is fooled into transferring the growing peptide chain to the amino group of puromycin rather than to the amino group of the incoming 3′-CCA-aminoacyl tRNA, protein synthesis stops. Because puromycin blocks protein synthesis in eukaryotes as well as in prokaryotes, it is poisonous to humans and therefore is not a clinically useful antibiotic. To be clinically useful, an antibiotic must affect protein synthesis only in prokaryotic cells.

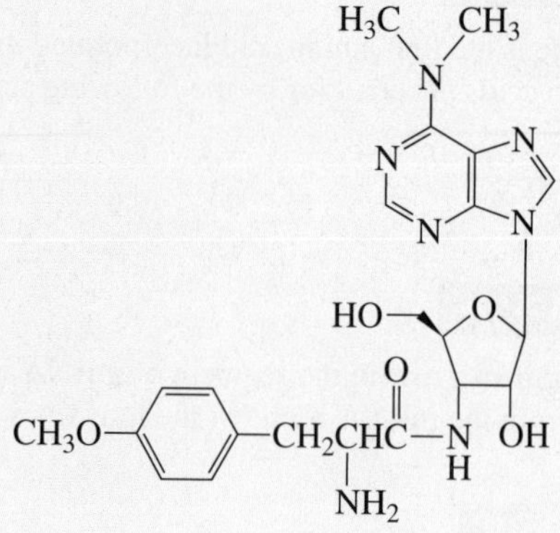

puromycin

Clinically useful antibiotics	Mode of action
Tetracycline	Prevents the aminoacyl-tRNA from binding to the ribosome
Erythromycin	Prevents the incorporation of new amino acids into the protein
Streptomycin	Inhibits the initiation of protein synthesis
Chloramphenicol	Prevents the new peptide bond from being formed

27.14 Why DNA Contains Thymine Instead of Uracil

In Section 25.8 we saw that dTMP is formed when dUMP is methylated, with coenzyme N^5,N^{10}-methylenetetrahydrofolate supplying the methyl group. Because the incorporation of the methyl group into uracil oxidizes tetrahydrofolate to dihydrofolate, dihydrofolate must be reduced back to tetrahydrofolate to prepare the cofactor for another catalytic cycle. The reducing agent is NADPH. Every NADPH formed in a biological organism can drive the formation of three ATPs, so using an NADPH to reduce dihydrofolate comes at the expense of ATP. This means that the synthesis of thymine is energetically expensive, so there must be a good reason for DNA to contain thymine instead of uracil.

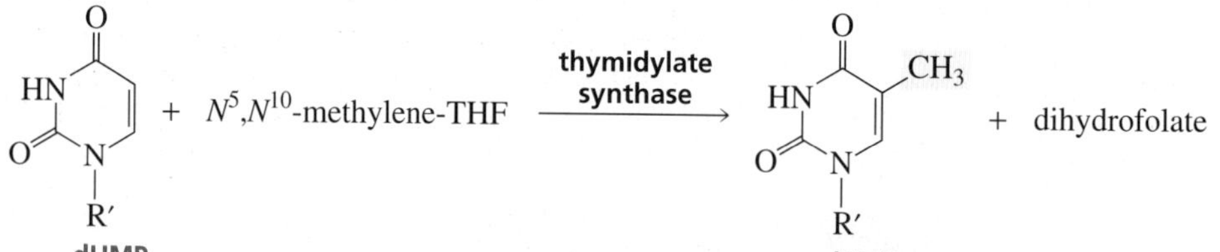

$$\text{dUMP} + N^5,N^{10}\text{-methylene-THF} \xrightarrow{\text{thymidylate synthase}} \text{dTMP} + \text{dihydrofolate}$$

R′ = 2′-deoxyribose-5-P

$$\text{dihydrofolate} + \text{NADPH} + \text{H}^+ \xrightarrow{\text{dihydrofolate reductase}} \text{tetrahydrofolate} + \text{NADP}^+$$

The presence of thymine instead of uracil in DNA prevents potentially lethal mutations. Cytosine can tautomerize to form an imine, which can be hydrolyzed to uracil (Section 18.6). The overall reaction is called a **deamination** since it removes an amino group.

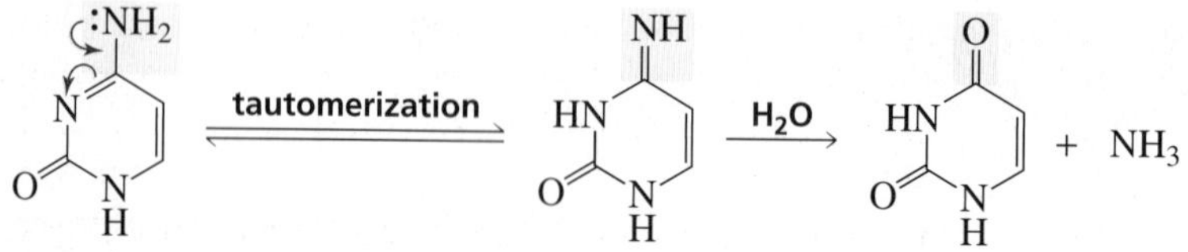

$$\text{cytosine} \xrightleftharpoons{\text{tautomerization}} \text{imine} \xrightarrow{H_2O} \text{uracil} + NH_3$$

If a cytosine in DNA is deaminated to a uracil, uracil will specify incorporation of an adenine into the daughter strand during replication instead of the guanine that would have been specified by cytosine. Fortunately, a U in DNA is recognized as a "mistake" by cell enzymes before an incorrect base can be inserted into the daughter strand. These enzymes cut out the U and replace it with a C. If U's were normally found in DNA, the enzymes could not distinguish between a normal U and a U formed by deamination of cytosine. Having T's in place of U's in DNA allows the U's that are found in DNA to be recognized as mistakes.

Unlike DNA, which replicates itself, any mistake in RNA does not survive for long because RNA is constantly being degraded and then resynthesized from the DNA template. Therefore, it is not worth spending the extra energy to incorporate T's into RNA.

PROBLEM 23◆

Adenine can be deaminated to hypoxanthine, and guanine can be deaminated to xanthine. Draw structures for hypoxanthine and xanthine.

PROBLEM 24

Explain why thymine cannot be deaminated.

27.15 Determining the Base Sequence of DNA

In June 2000, two teams of scientists (one from a private biotechnology company and one from the publicly funded Human Genome Project) announced that they had completed the first draft of the sequence of the 3.1 billion base pairs in human DNA. This is an enormous accomplishment. For example, if the sequence of 1 million base pairs were determined each day, it would take more than 10 years to complete the sequence of the human genome.

DNA molecules are too large to sequence as a unit, so DNA is first cleaved at specific base sequences and the resulting DNA fragments are sequenced. The enzymes that cleave DNA at specific base sequences are called **restriction endonucleases**, and the DNA fragments that are formed are called **restriction fragments**. Several hundred restriction enzymes are now known. A few examples of restriction enzymes, the base sequence each recognizes, and the point of cleavage in that base sequence are shown here.

restriction enzyme	recognition sequence
*Alu*I	AG|CT TC|GA
*Fnu*DI	GG|CC CC|GG
*Pst*I	CTGCA|G G|ACGTC

The base sequences that most restriction enzymes recognize are *palindromes*. A palindrome is a word or a group of words that reads the same forward and backward. "Toot" and "race car" are examples of palindromes.[4] A restriction enzyme recognizes

[4]Some other palindromes are "Mom," "Dad," "Bob," "Lil," "radar," "noon," "wow," "poor Dan in a droop", "a man, a plan, a canal, Panama," "Sex at noon taxes," and "He lived as a devil, eh?"

a piece of DNA in which *the template strand is a palindrome of the sense strand.* In other words, the sequence of bases in the template strand (reading from right to left) is identical to the sequence of bases in the sense strand (reading from left to right).

PROBLEM 25◆

Which of the following base sequences would most likely be recognized by a restriction endonuclease?

a. ACGCGT
b. ACGGGT
c. ACGGCA
d. ACACGT
e. ACATCGT
f. CCAACC

The restriction fragments can be sequenced using a chain-terminator procedure developed by Frederick Sanger known as the **dideoxy method**. This method involves generating fragments whose length depends on the last base added to the fragment. Because of its simplicity, it has superceded alternative methods.

In the dideoxy method a small piece of DNA called a primer, labeled at the 5′-end with ^{32}P, is added to the restriction fragment whose sequence is to be determined. Next, the four 2′-deoxyribonucleoside triphosphates are added as well as DNA polymerase, the enzyme that adds nucleotides to a strand of DNA. In addition, a small amount of the 2′,3′-dideoxynucleoside triphosphate of one of the bases is added to the reaction mixture. A 2′,3′-dideoxynucleoside triphosphate has no OH groups at the 2′- and 3′-positions.

Frederick Sanger (Section 23.12) and **Walter Gilbert** *shared half of the 1980 Nobel Prize in chemistry for their work on DNA sequencing. The other half went to* **Paul Berg**, *who had developed a method of cutting nucleic acids at specific sites and recombining the fragments in new ways, a technique known as recombinant DNA technology.*

Walter Gilbert *was born in Boston in 1932. He received a master's degree in physics from Harvard and a Ph.D. in mathematics from Cambridge University. In 1958 he joined the faculty at Harvard, where he became interested in molecular biology.*

Paul Berg *was born in New York in 1926. He received a Ph.D. from Western Reserve University (now Case Western Reserve University). He joined the faculty at Washington University in St. Louis in 1955 and became a professor of biochemistry at Stanford in 1959.*

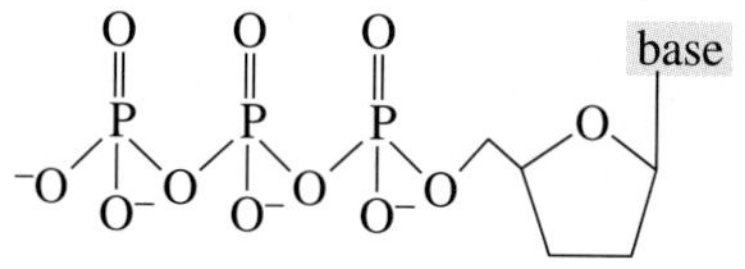

a 2′,3′-dideoxynucleoside triphosphate

Nucleotides will be added to the primer by base pairing with the restriction fragment. Synthesis will stop if the 2′,3′-dideoxy analog of dATP is added instead of dATP, because the 2′,3′-dideoxy analog does not have a 3′-OH to which additional nucleotides can be added. Therefore, three different chain-terminated fragments will be obtained from the DNA restriction fragment shown here.

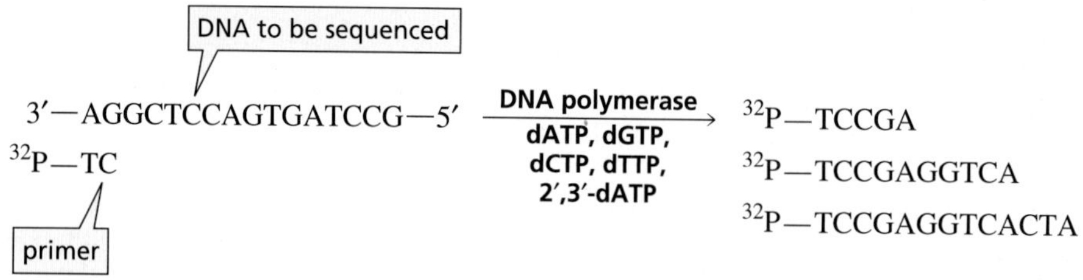

The procedure is repeated three more times using a 2′,3′-dideoxy analog of dGTP, then a 2′,3′-dideoxy analog of dCTP, and then a 2′,3′-dideoxy analog of dTTP.

PROBLEM 26

What labeled fragments would be obtained from the segment of DNA shown above if a 2′,3′-dideoxy analog of dGTP had been added to the reaction mixture instead of a 2′,3′-dideoxy analog of dATP?

The chain-terminated fragments obtained from each of the four experiments are loaded onto separate lanes of a buffered polyacrylamide gel—the fragments obtained from using a 2′,3′-dideoxy analog of dATP are loaded onto one lane, the fragments obtained from using a 2′,3′-dideoxy analog of dGTP onto another lane, and so on. An electric field is applied across the ends of the gel, causing the negatively charged fragments to travel toward the positively charged electrode (the anode). The smaller fragments fit through the spaces in the gel relatively easily and therefore travel through the gel faster, while the larger fragments pass through the gel more slowly.

After the fragments have been separated, the gel is placed in contact with a photographic plate. Radiation from ^{32}P causes a dark spot to appear on the plate opposite the location of each labeled fragment in the gel. This technique is called autoradiography, and the exposed photographic plate is known as an **autoradiograph** (Figure 27.18).

The sequence of bases in the original restriction fragment can be read directly from the autoradiograph. The identity of each base is determined by noting the column where each successive dark spot (larger piece of labeled fragment) appears, starting at the bottom of the gel to identify the base at the 5′-end. The sequence of the fragment of DNA responsible for the autoradiograph in Figure 27.18 is shown on the left-hand side of the figure.

Once the sequence of bases in a restriction fragment is determined, the results can be checked by determining the base sequence of the complementary strand. The base sequence in the original piece of DNA can be determined by repeating the entire procedure with a different restriction endonuclease and noting overlapping fragments.

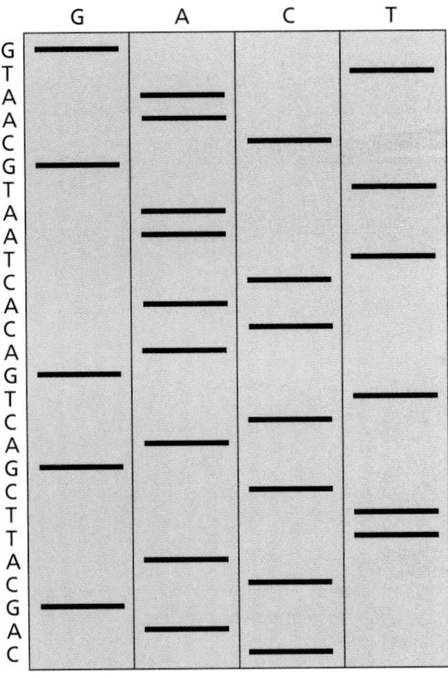

▲ **Figure 27.18**
An autoradiograph.

DNA FINGERPRINTING

The base sequence of the human genome varies from individual to individual, generally by a single base change every few hundred base pairs. Because some of these changes occur in base sequences recognized by restriction endonucleases, the fragments formed when human DNA reacts with a particular restriction endonuclease vary in size depending on the individual. It is this variation that forms the basis of DNA fingerprinting (also called DNA profiling or DNA typing). This technique is used by forensic chemists to compare DNA samples collected at the scene of a crime with the DNA of the suspected perpetrator. The most powerful technique for DNA identification analyzes restriction fragment length polymorphisms (RFLPs) obtained from regions of DNA in which individual variations are most common. This technique takes four to six weeks and requires a blood stain about the size of a dime. The chance of identical results from two different persons is thought to be one in a million. The second type of DNA profiling uses a polymerase chain reaction (PCR), which amplifies a specific region of DNA and compares differences at that site among individuals. This technique can be done in less than a week and requires only 1% of the amount required for RFLP, but does not discriminate as well among individuals. The chance of identical results from two different people is 1 in 500 to 1 in 2000. DNA fingerprinting is also being used to establish paternity, accounting for about 100,000 DNA profiles a year.

27.16 Laboratory Synthesis of DNA Strands

There is a great deal of interest in the synthesis of oligonucleotides with specific base sequences. This would allow scientists to synthesize genes that could be inserted into the DNA of microorganisms, causing the organisms to synthesize a particular protein. Alternatively, a synthetic gene could be inserted into the DNA of an organism defective in that gene—a process known as **gene therapy**.

Synthesizing an oligonucleotide with a particular base sequence is an even more challenging task than synthesizing a polypeptide with a specific amino acid sequence because each nucleotide has several groups that must be protected and then deprotected at the proper times. The approach taken was to develop an automated method similar to automated peptide synthesis (Section 23.10). The growing nucleotide is attached to a solid support so that it can be purified by flushing the reaction container with an appropriate solvent. Therefore, none of the synthesized product will be lost during purification.

Phosphoramidite Monomers

One current method synthesizes oligonucleotides using phosphoramidite monomers. The 5′-OH group of each phosphoramidite monomer is attached to a *para*-dimethoxytrityl (DMTr) protecting group. The particular group (Pr) used to protect the base depends on the base.

a phosphoramidite

a phosphoramidite monomer used for oligonucleotide synthesis

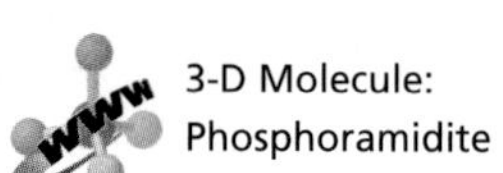

The 3′-nucleoside of the oligonucleotide to be synthesized is attached to a controlled-pore glass solid support and the oligonucleotide is synthesized from the 3′-end. When a monomer is added to the nucleoside attached to the solid support, the only nucleophile in the reaction mixture is the 5′-OH group of the sugar bonded to the solid support. This nucleophile attacks the phosphorus of the phosphoramidite, displacing the amine and forming a phosphite. The amine is too strong a base to be expelled without being protonated. Protonated tetrazole is the acid used for protonation because it is strong enough to protonate the diisopropylamine leaving group, but not strong enough to remove the DMTr protecting group. The phosphite is oxidized to a phosphate using I_2 or *tert*-butylhydroperoxide. The DMTr protecting group on the 5′-end of the dinucleotide is removed with mild acid. The cycle of (1) monomer addition, (2) oxidation, and (3) deprotection with acid is repeated over and over until a polymer of the desired length and sequence is obtained. Notice that the DNA polymer is synthesized in the 3′ ⟶ 5′ direction, which is opposite to the direction (5′ ⟶ 3′) DNA is synthesized in nature.

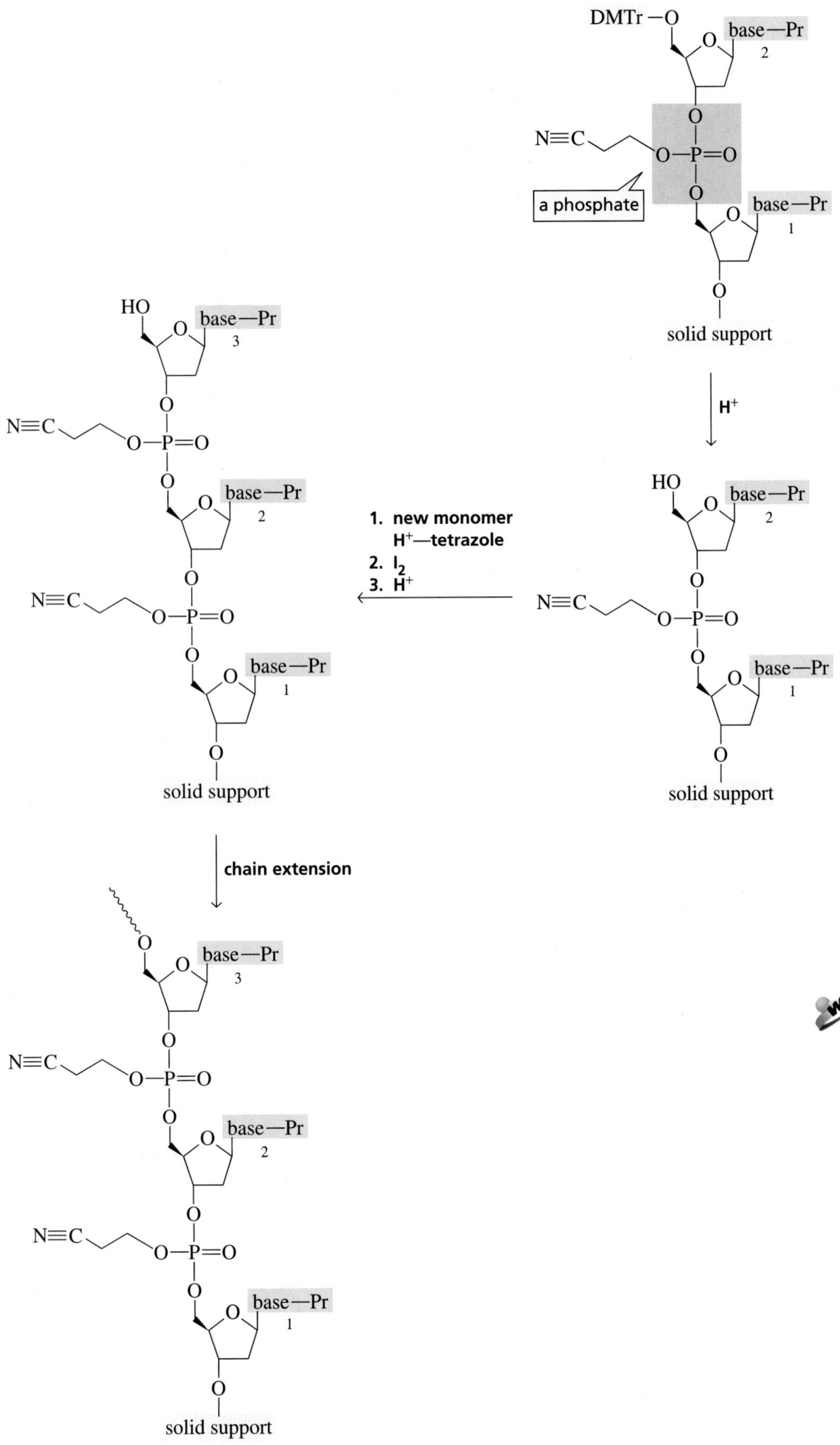

DMTr—O base—Pr 2

N≡C O—P=O

a phosphate

base—Pr 1

solid support

H^+

HO base—Pr 2

N≡C O—P=O

base—Pr 1

solid support

1. new monomer H^+—tetrazole
2. I_2
3. H^+

HO base—Pr 3

N≡C O—P=O

base—Pr 2

N≡C O—P=O

base—Pr 1

solid support

chain extension

base—Pr 3

N≡C O—P=O

base—Pr 2

N≡C O—P=O

base—Pr 1

solid support

Tutorial:
Oligonucleotide synthesis
with phosphoramidites

The NH_2 groups of cytosine, adenine, and guanine are nucleophiles and therefore must be protected to prevent them from reacting with a newly added monomer. The NH_2 groups of cytosine and adenine are protected as benzamides and the NH_2 group of guanine is protected as a butanamide. Thymine does not contain any nucleophilic groups, so it does not have to be protected.

unprotected bases **protected bases**

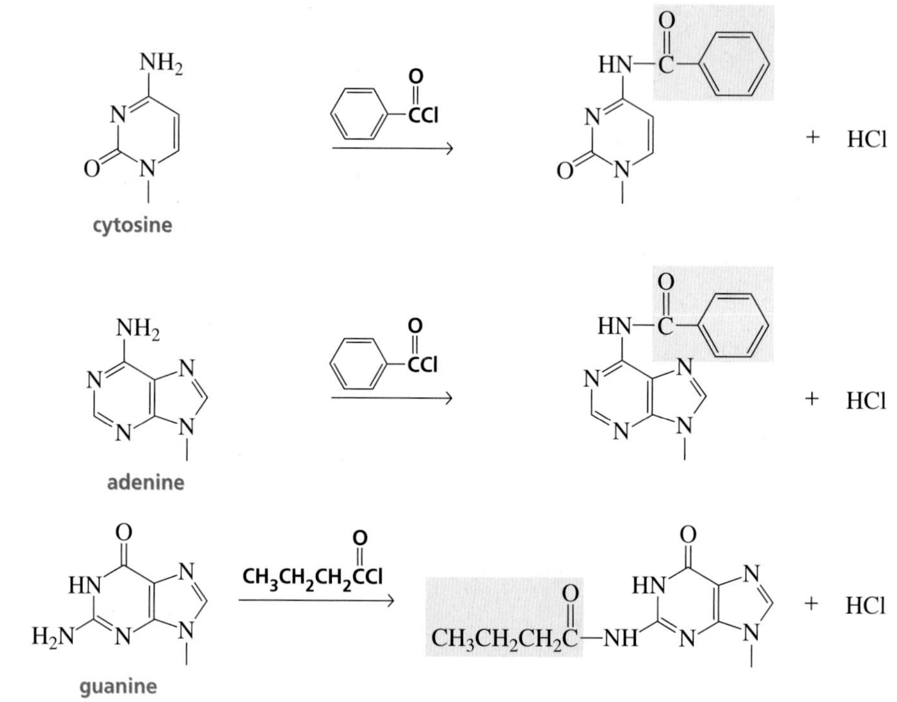

cytosine + HCl

adenine + HCl

$CH_3CH_2CH_2CCl$ → $CH_3CH_2CH_2C(=O)$—NH + HCl

guanine

After the oligonucleotide is synthesized, the protecting groups on the phosphates and the protecting groups on the bases must be removed, and the oligonucleotide has to be detached from the solid support. This can all be done in a single step using aqueous ammonia. Because a hydrogen bonded to a carbon adjacent to a cyano group is relatively acidic (Section 19.1), ammonia can be used as the base in the elimination reaction that removes the phosphate protecting group.

$N{\equiv}C$—CH(H)—CH$_2$—O—P(=O)(OR)(OR) + $\ddot{N}H_3$ ⟶ $N{\equiv}C$—CH=CH$_2$ + ^-O—P(=O)(OR)(OR) + $^+NH_4$

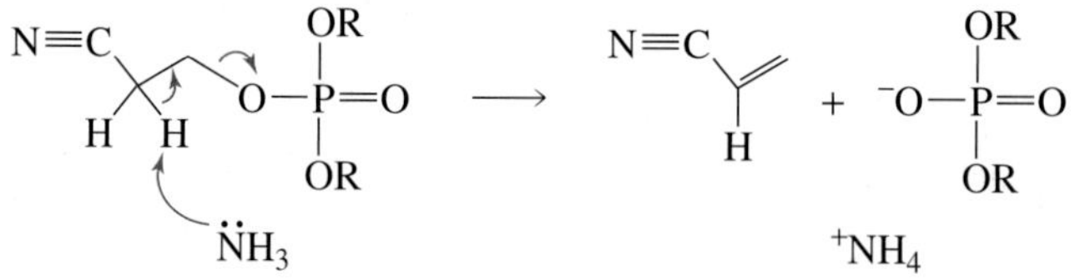

Currently, automated DNA synthesizers can synthesize nucleic acids containing as many as 130 nucleotides in acceptable yields, adding one nucleotide every 2 to 3 minutes. Longer nucleic acids can be prepared by splicing together two or more individually prepared strands. To ensure a good overall yield of oligonucleotide, the addition of each monomer must occur in high yield (>98%). This can be accomplished by using a large excess of monomer. This, however, makes oligonucleotide synthesis very expensive because the unreacted monomers are wasted.

H-Phosphonate Monomers

A second method for synthesizing oligonucleotides with specific base sequences, using H-phosphonate monomers, has the advantage over the phosphoramidite method in that the monomers are easier to handle and phosphate protecting groups are not needed. However, the yields are not as good. The H-phosphonate monomers are activated by reacting with an acyl chloride, which converts them into phosphoanhydrides. The 5′-OH group of the nucleoside attached to the solid support reacts with the phosphoanhydride, forming a dimer. The DMTr protecting group is removed with acid under mild conditions, and a second activated monomer is added. Monomers are added one at a time in this way until the strand is complete. Oxidation with I_2 (or *tert*-butyl hydroperoxide) converts the H-phosphonate groups to phosphate groups. The base-protecting groups are removed by aqueous ammonia as in the phosphoramidite method.

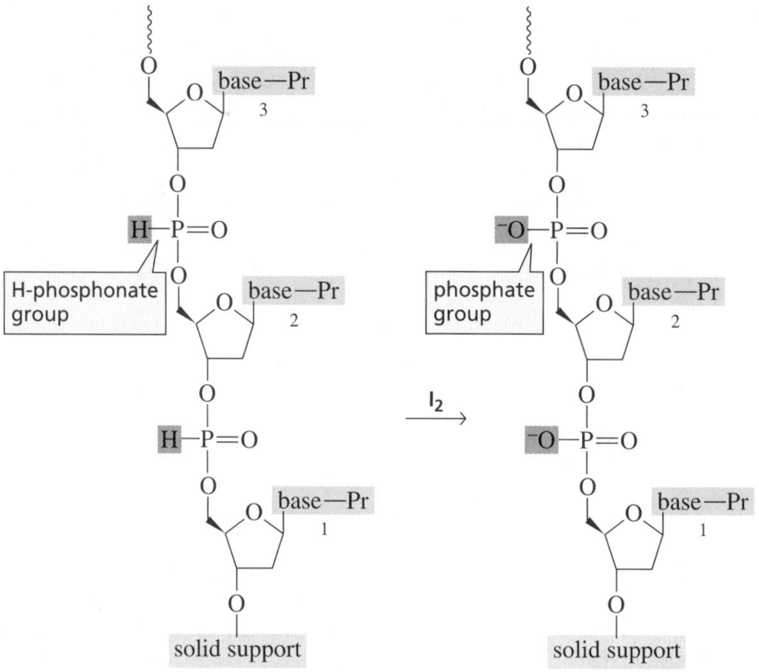

H-phosphonate group

base—Pr 3

base—Pr 2

base—Pr 1

solid support

$\xrightarrow{I_2}$

phosphate group

base—Pr 3

base—Pr 2

base—Pr 1

solid support

Notice that in the phosphoramidite method an oxidation is done each time a monomer is added, whereas in the H-phosphonate method a single oxidation is carried out after the entire strand has been synthesized.

PROBLEM 27

Propose a mechanism for removal of the DMTr protecting group by treatment with acid.

27.17 Rational Drug Design

Certain diseases such as acquired immunodeficiency syndrome (AIDS) and herpes are caused by **retroviruses**. The genetic information of a retrovirus is contained in its RNA. The retrovirus uses the sequence of bases in RNA as a template to synthesize DNA. It is called a retrovirus because its genetic information flows from RNA to DNA instead of the more typical flow from DNA to RNA.

Drugs that interfere with the synthesis of DNA by retroviruses have been designed and developed. If the retrovirus cannot synthesize DNA, it cannot take over the genetic machinery of the cell to produce more retroviral RNA and retroviral proteins. Designing drugs with particular structures to achieve specific purposes is called **rational drug design**. AZT is perhaps the best known of the drugs designed to interfere with retroviral DNA synthesis. AZT is taken up by the T lymphocytes, cells that are particularly susceptible to human immunodeficiency virus (HIV), the retrovirus that causes AIDS. Once in the cell, AZT is converted to AZT-triphosphate. The retroviral enzyme (reverse transcriptase) that catalyzes DNA formation from RNA binds AZT-triphosphate more tightly than it binds dTTP. Therefore, AZT rather than T is added to the growing DNA chain. Because AZT does not have a 3′-OH group, no additional nucleotides can be added to the chain and DNA synthesis comes to a sudden halt. Fortunately, the concentration of AZT required to affect reverse transcription is too low to affect most cellular DNA replication. A newer drug, 2′,3′-dideoxyinosine (ddI), has a similar mechanism of action. Rational drug design is further discussed in Chapter 30.

HN, O, CH_3, O, N, O, HO, $^-N{=}\overset{+}{N}{=}N$

3′-azido-2′-deoxythymidine
AZT

HN, O, N, N, N, O, HO

2′,3′-dideoxyinosine
ddI

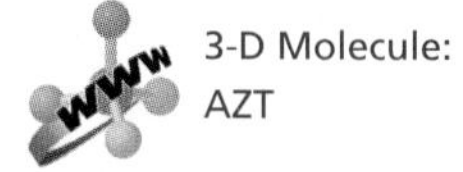

Because viruses, bacteria, and cancer cells all require DNA to grow and reproduce, chemists are trying to design compounds that will bind to the DNA of invading organisms and interfere with their reproduction. Chemists are also attempting to design polymers that will bind to specific sequences of human DNA. Such compounds could disrupt the expression of a gene (interfere with its transcription into RNA). For example, there is hope that compounds can be designed that will interfere with the expression of genes that contribute to the development of cancer. Polymers that bind to DNA are called *antigene agents*; those that bind to mRNA are called *antisense agents*.

For a compound to target a particular gene, the compound must be able to recognize a specific sequence of 15 to 20 bases. A sequence that long might occur only once in the human genome, so the compound would be specific for a particular site on DNA. In contrast, if the compound recognizes a sequence of only six bases, it could affect the human genome at more than a million locations because that sequence could occur once in every 2000 bases. However, since only 10% of the genes are expressed in most cells, a compound that recognizes a specific sequence of 10 to 12 bases may confer a gene-specific effect.

One approach to **site-specific recognition** uses a synthetic strand of oligonucleotides. When a strand of oligonucleotides is added to natural double-stranded DNA, the strand wraps around the major groove of DNA, forming a triple helix (Figure 27.19). The hope is that if the DNA sequence of a particular gene is known, a deoxyribonucleotide can be synthesized that will bind to that gene.

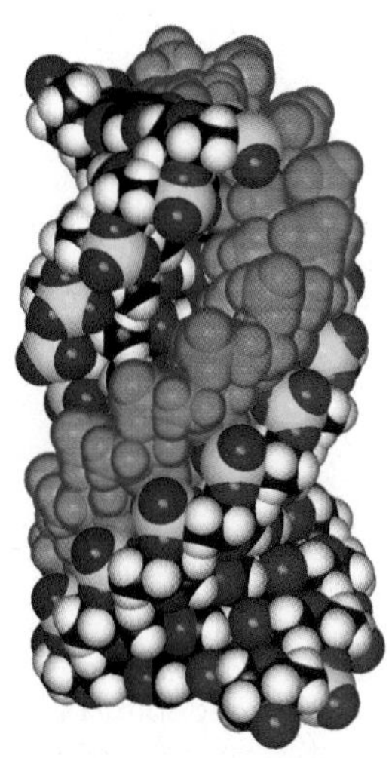

▲ **Figure 27.19**
A triple helix. A synthetic strand of oligonucleotides is wrapped around the major groove of double-stranded DNA.

A triple helix is formed through *Hoogsteen base pairing* between the existing base pairs in DNA and bases in the third synthetic strand. In **Hoogsteen base pairing**, a T in the synthetic strand binds to an A of an AT base pair, and a protonated cytosine in the synthetic strand binds to a G of a GC base pair (Figure 27.20). Thus, oligonucleotides can be prepared with sequences that will base-pair to the sense strand of

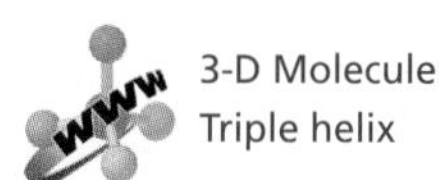

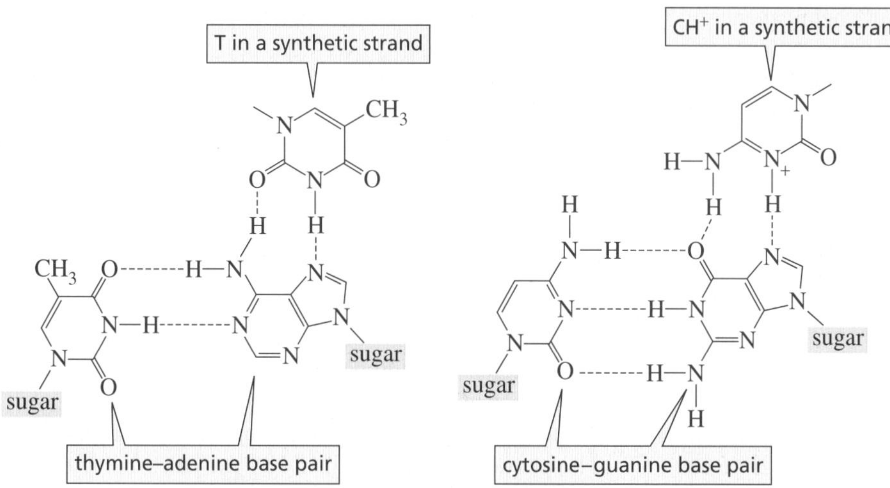

◀ **Figure 27.20**
Hoogsteen base pairing: A T in a synthetic strand of oligonucleotides binds to the A of an A—T base pair in double-stranded DNA; a protonated C (^+CH) in the synthetic strand binds to the G of a G—C base pair in DNA.

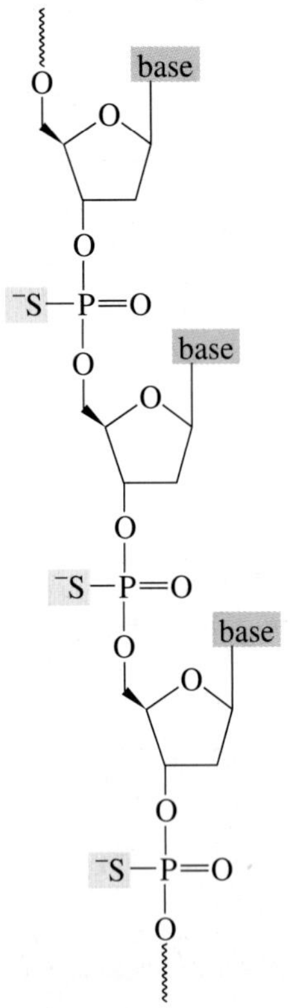

DNA at the desired location. Several methods are being investigated that will cut out the piece of targeted DNA after the synthetic strand binds to the double helix.

A problem with using synthetic oligonucleotides to target DNA is that the synthetic strands are susceptible to enzymes, such as restriction endonucleases, that catalyze the hydrolysis of DNA. Consequently, other polymers are being designed that will recognize specific DNA sequences but will not be enzymatically hydrolyzed. A compound that has shown some promise is a polymer of phosphorothioates. The polymer differs from DNA in that the negatively charged oxygen bonded to the phosphorus is replaced by a negatively charged sulfur (Figure 27.21). The polymer binds to both DNA and RNA with complementary base pairing. Oligonucleotide phosphorothioates consisting of various lengths of deoxycytidine residues have recently been found to be effective in protecting T cells from HIV infection.

▲ **Figure 27.21**
A synthetic oligonucleotide with negatively charged sulfurs in place of negatively charged oxygens.

PROBLEM 28

5-Bromouracil, a highly mutagenic compound, is used in cancer chemotherapy. When administered to a patient, it is converted to the triphosphate and incorporated into DNA in place of thymine, which it resembles sterically. Why does it cause mutations? (*Hint*: The bromo substituent increases the stability of the enol tautomer.)

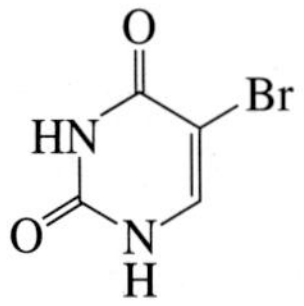

5-bromouracil 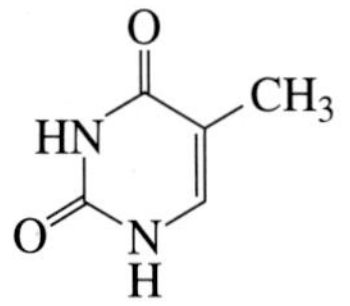   thymine

Summary

There are two types of nucleic acids—**deoxyribonucleic acid (DNA)** and **ribonucleic acid (RNA)**. DNA encodes an organism's hereditary information and controls the growth and division of cells. In most organisms the genetic information stored in DNA is **transcribed** into RNA. This information can then be **translated** for the synthesis of all the proteins needed for cellular structure and function.

ATP is a cell's most important source of chemical energy; ATP provides a reaction pathway involving a good leaving group for a reaction that would not otherwise occur because of a poor leaving group. This occurs by way of a **phosphoryl transfer reaction** in which a phosphate-containing group of ATP is transferred to a nucleophile as a result of breaking a **phosphoanhydride bond**. The reaction involves one of three intermediates—an **acyl phosphate**, an **acyl pyrophosphate**, or an **acyl adenylate**. Cleavage of a phosphoanhydride bond is highly exergonic because of electrostatic repulsions, solvation, and electron delocalization

A **nucleoside** contains a base bonded to D-ribose or to 2-deoxy-D-ribose. A **nucleotide** is a nucleoside with either the 5′- or the 3′-OH group bonded to phosphoric acid by an ester linkage. **Nucleic acids** are composed of long strands of nucleotide subunits linked by phosphodiester bonds. These linkages join the 3′-OH group of one nucleotide to the 5′-OH group of the next nucleotide. A **dinucleotide** contains two nucleotide subunits, an **oligonucleotide** contains three to ten subunits, and a **polynucleotide** contains many subunits. DNA contains 2′-deoxy-D-ribose while RNA contains D-ribose. The difference in the sugars causes DNA to be stable and RNA to be easily cleaved.

The **primary structure** of a nucleic acid is the sequence of bases in its strand. DNA contains **A**, **G**, **C**, and **T**; RNA contains **A**, **G**, **C**, and **U**. RNA has thymine instead of uracil to prevent mutations caused by imine hydrolysis of C to form U. DNA is a double-stranded helix with a major and a minor groove; the strands run in opposite directions and are twisted into a helix. The bases are confined to the inside of the helix and the sugar and phosphate groups are on the outside. The strands are held together by hydrogen bonds between bases of opposing strands as well as by **stacking interactions**—van der Waals attractions between adjacent bases on the same strand. The two strands—one is called a **sense strand** and the other a **template strand**—are complementary: **A** pairs with **T**, and **G** pairs with **C**. DNA is synthesized in the 5′ ⟶ 3′ direction by a process called **semiconservative replication**.

The sequence of bases in DNA provides the blueprint for the synthesis (**transcription**) of RNA. RNA is synthesized in the 5′ ⟶ 3′ direction by transcribing the DNA template strand in the 3′ ⟶ 5′ direction. There are three kinds of RNA: messenger RNA, ribosomal RNA, and transfer RNA. Protein synthesis (**translation**) takes place from the N-terminal end to the C-terminal end by reading the bases along the mRNA strand in the 5′ ⟶ 3′ direction. Each three-base sequence—**a codon**—specifies the particular amino to be incorporated into a protein. A tRNA carries the amino acid bound as an ester to its 3′-terminal position. The codons and the amino acids they specify are known as the **genetic code**.

Restriction endonucleases cleave DNA at specific palindromes forming **restriction fragments**. The **dideoxy method** is the preferred method to determine the sequence of bases in the restriction fragments. Oligonucleotides with specific base sequences can be synthesized using phosphoramidite monomers or H-phosphonate monomers. Polymers that bind to DNA are called antigene agents; those that bind to RNA are called antisense agents.

Key Terms

acyl adenylate (p. 1114)
acyl phosphate (p. 1113)
acyl pyrophosphate (p. 1114)
adenosine 5′-triphosphate (ATP) (p. 1110)
anticodon (p. 1126)
antisense strand (p. 1124)
autoradiograph (p. 1135)
base (p. 1106)
codon (p. 1128)
deamination (p. 1132)
deoxyribonucleic acid (DNA) (p. 1106)
deoxyribonucleotide (p. 1109)
dideoxy method (p. 1134)
dinucleotide (p. 1118)
double helix (p. 1121)
eukaryotic organism (p. 1125)
exon (p. 1125)
gene (p. 1124)
gene therapy (p. 1135)
genetic code (p. 1128)
high-energy bond (p. 1115)
Hoogsteen base pairing (p. 1141)
human genome (p. 1124)
informational strand (p. 1124)
intron (p. 1125)
major groove (p. 1122)
messenger RNA (mRNA) (p. 1125)
minor groove (p. 1122)
nucleic acid (p. 1106)
nucleoside (p. 1108)
nucleotide (p. 1109)
oligonucleotide (p. 1118)
phosphoanhydride bond (p. 1111)
phosphodiester (p. 1106)
phosphoryl transfer reaction (p. 1112)
polynucleotide (p. 1118)
primary structure (p. 1118)
prokaryotic organism (p. 1125)
promoter site (p. 1124)
rational drug design (p. 1140)
replication (p. 1124)
replication fork (p. 1124)
restriction endonuclease (p. 1133)
restriction fragment (p. 1133)
retrovirus (p. 1140)
ribonucleic acid (RNA) (p. 1106)
ribonucleotide (p. 1109)
ribosomal RNA (rRNA) (p. 1125)
ribosome (p. 1125)
ribozyme (p. 1125)
RNA splicing (p. 1125)
sedimentation constant (p. 1126)
semiconservative replication (p. 1124)
sense strand (p. 1124)
site-specific recognition (p. 1141)
stacking interactions (p. 1121)
stop codon (p. 1128)
template strand (p. 1124)
transcription (p. 1124)
transfer RNA (tRNA) (p. 1125)
translation (p. 1128)

Problems

29. Name the following compounds:

a. [structure: guanine nucleoside with 3′-phosphate; labels O, HN, N, H_2N, N, N, O, HO, O, OH, $O{=}P{-}O^-$, O^-]

b. [structure: cytidine 5′-diphosphate; labels NH_2, N, O, N, O, O, P, P, ^-O, O^-, O, O^-, O, O, HO, OH]

c. [structure: 2′-deoxyadenosine 5′-monophosphate; labels NH_2, N, N, N, N, O, P, ^-O, O^-, O, O, HO]

d. [structure: thymidine; labels O, CH_3, HN, O, N, O, HO, HO]

30. What nonapeptide is coded for by the following piece of mRNA?

5′—AAA—GUU—GGC—UAC—CCC—GGA—AUG—GUG—GUC—3′

31. What would be the base sequence of the segment of DNA that is responsible for the biosynthesis of the following hexapeptide?

Gly-Ser-Arg-Val-His-Glu

32. Propose a mechanism for the following reaction:

$$^{-}OCCH_2CH_2CHCO^{-}(\text{both C=O}; {}^{+}NH_3 \text{ on CH}) + NH_3 + ATP \longrightarrow H_2NCCH_2CH_2CHCO^{-}(\text{both C=O}; {}^{+}NH_3 \text{ on CH}) + ADP + {}^{-}O{-}P(=O)(OH){-}O^{-}$$

33. Match the codon with the anticodon:

Codon	Anticodon
AAA	ACC
GCA	CCU
CUU	UUU
AGG	AGG
CCU	UGA
GGU	AAG
UCA	GUC
GAC	UGC

34. Using the single-letter abbreviations for the amino acids in Table 23.1, write the sequence of amino acids in a tetrapeptide represented by the first four letters in your first name. Do not use any letter twice. (Because not all letters are assigned to amino acids, you might have to use one or two letters in your last name.) Write the sequence of bases in mRNA that would result in the synthesis of that polypeptide. Write the sequence of bases in the sense strand of DNA that would result in formation of the appropriate mRNA.

35. Which of the following pairs of dinucleotides are present in equal amounts in DNA?
   a. CC and GG
   b. CG and GT
   c. CA and TG
   d. CG and AT
   e. GT and CA
   f. TA and AT

36. Why is the codon a triplet rather than a doublet or a quartet?

37. RNAase, the enzyme that catalyzes the hydrolysis of RNA, has two catalytically active histidine residues at its active site. One of the histidine residues is catalytically active in its acidic form and the other is catalytically active in its basic form. Propose a mechanism for RNAase.

38. The amino acid sequences of peptide fragments obtained from a normal protein and from the same protein synthesized by a defective gene were compared. They were found to differ in only one peptide fragment. The primary sequences of the fragments are shown here.

Normal: Gln-Tyr-Gly-Thr-Arg-Tyr-Val
Mutant: Gln-Ser-Glu-Pro-Gly-Thr

   a. What is the defect in DNA?
   b. It was later determined that the normal peptide fragment is an octapeptide with a C-terminal Val-Leu. What is the C-terminal amino acid of the mutant peptide?

39. Whether the mechanism requiring activation of a carboxylate ion by ATP involves attack of the carboxylate ion on the α-phosphorus or the β-phosphorus of ATP cannot be determined from the reaction products because AMP and pyrophosphate are obtained as products in both mechanisms. The mechanisms, however, can be distinguished by a labeling experiment in which the enzyme, the carboxylate ion, ATP, and radioactively labeled pyrophosphate are incubated, and the ATP is isolated. If the isolated ATP is radioactive, the mechanism involves attack on the α-phosphorus. If it is not radioactive, the mechanism involves attack on the β-phosphorus. Explain these conclusions.

40. What would be the results of the experiment in Problem 39 if radioactive AMP were added to the incubation mixture instead of radioactive pyrophosphate?

41. Which cytosine in the following sense strand of DNA could cause the most damage to the organism if it were deaminated?

$$5'{-}A{-}T{-}G{-}T{-}C{-}G{-}C{-}T{-}A{-}A{-}T{-}C{-}3'$$

42. Sodium nitrite, a common food preservative (p. 653), is capable of causing mutations in an acidic environment by converting cytosines to uracils. Explain how this occurs.

43. The first amino acid incorporated into a polypeptide chain during its biosynthesis in prokaryotes is *N*-formylmethionine. Explain the purpose of the formyl group.

PART EIGHT

Special Topics in Organic Chemistry

In previous chapters, the polymers synthesized by biological systems—proteins, carbohydrates, and nucleic acids—were discussed.

Chapter 28 discusses polymers synthesized by chemists. These synthetic polymers have physical properties that make them useful in everyday life, in applications such as fabrics, bottles, food wrap, automobile parts, and compact discs.

Chapter 29 discusses pericyclic reactions—reactions that occur as a result of a cyclic reorganization of electrons. In this chapter, you will learn how the conservation of orbital symmetry theory explains the relationships among reactant, product, and reaction conditions in a pericyclic reaction.

Chapter 30 introduces you to medicinal chemistry. Here you will see how many of our commonly used drugs were discovered, and you will learn about some of the techniques used to develop new drugs.

28 Synthetic Polymers

Super glue

Probably no group of synthetic compounds is more important to modern life than synthetic polymers. A **polymer** is a large molecule made by linking together repeating units of small molecules called **monomers**. The process of linking them together is called **polymerization**.

$$n\text{M} \xrightarrow{\text{polymerization}} \text{—M—M—M—M—M—M—M—M—M—}$$

monomer → polymer

ethylene monomers → polyethylene

Unlike small organic molecules, which are of interest because of their *chemical* properties, these giant molecules are interesting because of their *physical* properties, which make them useful in everyday life. Some synthetic polymers resemble natural substances, but most are quite different from those found in nature. Such diverse items as photographic film, compact discs, rugs, food wrap, artificial joints, Super glue, toys, bottles, weather stripping, automobile body parts, shoe soles, and condoms are made of synthetic polymers.

Polymers can be divided into two broad groups: **synthetic polymers** and **biopolymers** (natural polymers). Synthetic polymers are synthesized by scientists, whereas biopolymers are synthesized by organisms. Examples of biopolymers are

DNA, the storage molecule for genetic information—the molecule that determines whether a fertilized egg becomes a human or a honeybee; RNA and proteins, the molecules that induce biochemical transformations; and polysaccharides. The structures and properties of these molecules are presented in other chapters. In this chapter, we will explore synthetic polymers.

Humans first relied on *natural polymers* for clothing, wrapping themselves in animal skins and furs. Later, they learned to spin natural fibers into thread and to weave the thread into cloth. Today, much of our clothing is made of *synthetic polymers* (e.g., nylon, polyester, polyacrylonitrile). Many people prefer clothing made of natural polymers (e.g., cotton, wool, silk), but it has been estimated that if synthetic polymers were not available, all the arable land in the United States would have to be used for the production of cotton and wool for clothing.

The first **plastic**—a polymer capable of being molded—was celluloid. Invented in 1856 by Alexander Parke, it was a mixture of nitrocellulose and camphor. Celluloid was used in the manufacture of billiard balls and piano keys, replacing scarce ivory. The invention of celluloid provided a reprieve for many elephants, but caused some moments of consternation in billiard parlors because nitrocellulose is flammable and explosive. Celluloid was used for motion picture film until it was replaced by cellulose acetate, a less dangerous polymer.

The first synthetic fiber was rayon. In 1865, the French silk industry was threatened by an epidemic that killed many silkworms, highlighting the need for an artificial silk substitute. Louis Chardonnet accidentally discovered the starting material for a synthetic fiber when, while wiping up some spilled nitrocellulose from a table, he noticed long silklike strands adhering to both the cloth and the table. "Chardonnet silk" was introduced at the Paris Exposition in 1891. It was called *rayon* because it was so shiny that it appeared to give off rays of light.

The first synthetic rubber was synthesized by German chemists in 1917. Their efforts were in response to a severe shortage of raw materials as a result of blockading during World War I.

Hermann Staudinger was the first to recognize that the various polymers being produced were not disorderly conglomerates of monomers, but were made up of chains of monomers joined together. Today, the synthesis of polymers has grown from a process carried out with little chemical understanding to a sophisticated science in which molecules are engineered with predetermined specifications in order to produce new materials tailored to fit human needs. Recent examples of the many new polymers that are constantly being designed include Lycra®, a fabric with elastic properties, and Dyneema®, the strongest fabric commercially available.

Polymer chemistry is part of the larger discipline of **materials science**, which involves the creation of new materials to replace metals, glass, ceramics, fabrics, wood, cardboard, and paper. Polymer chemistry has evolved into a multibillion-dollar industry. Currently, there are approximately 30,000 patented polymers in the United States. More than 2.5×10^{13} kilograms of synthetic polymers are produced in the United States each year, and we can expect many more new materials to be developed by scientists in the years to come.

Alexander Parke (1813–1890) *was born in Birmingham, England. He called the polymer that he invented "pyroxylin." He was, however, unable to market it.*

The inventor **John Wesley Hyatt (1837–1920)** *was born in New York. When a New York firm offered a prize of $10,000 for a substitute for ivory billiard balls, Hyatt improved the synthesis of pyroxylin. He changed its name to "celluloid" and patented a method for making billiard balls. He did not, however, win the prize.*

Louis-Marie-Hilaire Bernigaud, Comte de Chardonnet (1839–1924) *was born in France. In the early stages of his career, he was an assistant to Louis Pasteur. Because the rayon he initially produced was made from nitrocellulose, it was dangerously flammable. Eventually, chemists learned to remove some of the nitro groups after the fiber was formed, which made the fiber much less flammable but not as strong.*

Hermann Staudinger (1881–1965), *the son of a professor, was born in Germany. He became a professor at the Technical Institute of Karlsruhe and at the University of Freiburg. He received the Nobel Prize in chemistry in 1953 for his contributions to polymer chemistry.*

28.1 General Classes of Synthetic Polymers

Synthetic polymers can be divided into two major classes, depending on their method of preparation. **Chain-growth polymers**, also known as **addition polymers**, are made by **chain reactions**—the addition of monomers to the end of a growing chain. The end of the chain is reactive because it is a radical, a cation, or an anion. Polystyrene—used for disposable food containers, insulation, and toothbrush handles, among other things—is an example of a chain-growth polymer. Polystyrene is pumped full of air to produce the material known as Styrofoam®.

Chain-growth polymers are also called addition polymers.

Chain-growth polymers are made by chain reactions.

$$CH_2{=}CH(C_6H_5) \quad CH_2{=}CH(C_6H_5) \quad CH_2{=}CH(C_6H_5) \longrightarrow -CH_2-CH(C_6H_5)-[CH_2-CH(C_6H_5)]_n-CH_2-CH(C_6H_5)-$$

styrene — polystyrene, a chain-growth polymer (repeating unit)

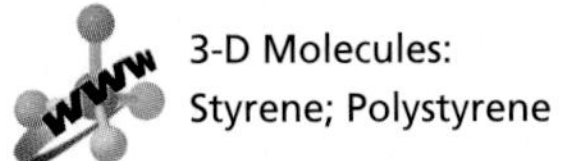

3-D Molecules: Styrene; Polystyrene

Step-growth polymers are also called condensation polymers.

Step-growth polymers, also called **condensation polymers**, are made by combining two molecules while, in most cases, removing a small molecule, generally water or an alcohol. The reacting molecules have reactive functional groups at each end. Unlike chain-growth polymerization, which requires the individual molecules to add to the end of a growing chain, step-growth polymerization allows any two reactive molecules to combine. Dacron® is an example of a step-growth polymer.

$$CH_3O-\overset{O}{\overset{\|}{C}}-C_6H_4-\overset{O}{\overset{\|}{C}}-OCH_3 + HOCH_2CH_2OH \xrightarrow{\Delta} \left[OCH_2CH_2O-\overset{O}{\overset{\|}{C}}-C_6H_4-\overset{O}{\overset{\|}{C}}\right]_n-OCH_2CH_2O- \; + \; 2n\,CH_3OH$$

dimethyl terephthalate — 1,2-ethanediol — poly(ethylene terephthalate), Dacron®, a step-growth polymer (repeating unit)

Dacron® is the most common of the group of polymers known as **polyesters**—polymers with many ester groups. Polyesters are used for clothing and are responsible for the wrinkle-resistant behavior of many fabrics. Polyester is also used to make the plastic film called Mylar®, needed in the manufacture of magnetic recording tape. This film is tear-resistant and, when processed, has a tensile strength nearly as great as that of steel. Aluminized Mylar® was used to make the Echo satellite that was put into orbit around the Earth as a giant reflector. The polymer used to make soft drink bottles is also a polyester.

28.2 Chain-Growth Polymers

The monomers used most commonly in chain-growth polymerization are ethylene (ethene) and substituted ethylenes. In the chemical industry, monosubstituted ethylenes are known as *alpha olefins*. Polymers formed from ethylene or substituted ethylenes are called **vinyl polymers**. Some of the many vinyl polymers synthesized by chain-growth polymerization are listed in Table 28.1.

Chain-growth polymerization proceeds by one of three mechanisms: **radical polymerization**, **cationic polymerization**, or **anionic polymerization**. Each mechanism has three distinct phases: an *initiation step* that starts the polymerization, *propagation steps* that allow the chain to grow, and *termination steps* that stop the growth of the chain. We will see that the choice of mechanism depends on the structure of the monomer *and* the initiator used to activate the monomer.

Radical Polymerization

For chain-growth polymerization to occur by a radical mechanism, a radical initiator must be added to the monomer to convert some of the monomer molecules into radicals. The initiator breaks homolytically into radicals, and each radical adds to an alkene monomer, converting it into a radical. This radical reacts with another monomer, adding a new subunit that propagates the chain. The radical site is now at the end of the most recent unit added to the end of the chain. This is called the **propagating site**.

Table 28.1 Some Important Chain-Growth Polymers and Their Uses

Monomer	Repeating unit	Polymer name	Uses
$CH_2{=}CH_2$	$-CH_2-CH_2-$	polyethylene	film, toys, bottles, plastic bags
$CH_2{=}CH(Cl)$	$-CH_2-CH(Cl)-$	poly(vinyl chloride)	"squeeze" bottles, pipe, siding, flooring
$CH_2{=}CH-CH_3$	$-CH_2-CH(CH_3)-$	polypropylene	molded caps, margarine tubs, indoor/outdoor carpeting, upholstery
$CH_2{=}CH(C_6H_5)$	$-CH_2-CH(C_6H_5)-$	polystyrene	packaging, toys, clear cups, egg cartons, hot drink cups
$CF_2{=}CF_2$	$-CF_2-CF_2-$	poly(tetrafluoroethylene) Teflon®	nonsticking surfaces, liners, cable insulation
$CH_2{=}CH(C{\equiv}N)$	$-CH_2-CH(C{\equiv}N)-$	poly(acrylonitrile) Orlon®, Acrilan®	rugs, blankets, yarn, apparel, simulated fur
$CH_2{=}C(CH_3)-C({=}O)OCH_3$	$-CH_2-C(CH_3)(C({=}O)OCH_3)-$	poly(methyl methacrylate) Plexiglas®, Lucite®	lighting fixtures, signs, solar panels, skylights
$CH_2{=}CH(OC({=}O)CH_3)$	$-CH_2-CH(OC({=}O)CH_3)-$	poly(vinyl acetate)	latex paints, adhesives

RECYCLING SYMBOLS

When plastics are recycled, the various types must be separated from one another. To aid in the separation, many states require manufacturers to include a recycling symbol on their products to indicate the type of plastic. You are probably familiar with these symbols, which are found on the bottom of plastic containers. The symbols consist of three arrows around one of seven numbers; an abbreviation below the symbol indicates the type of polymer from which the container is made. The lower the number in the middle of the symbol, the greater is the ease with which the material can be recycled: 1 (PET) stands for poly(ethylene terephthalate), 2 (HDPE) for high-density polyethylene, 3 (V) for poly(vinyl chloride), 4 (LDPE) for low-density polyethylene, 5 (PP) for polypropylene, 6 (PS) for polystyrene, and 7 for all other plastics.

Recycling labels

chain-initiating steps

$$\underset{\text{a radical initiator}}{RO{-}OR} \xrightarrow{\Delta} \underset{\text{radicals}}{2\ RO\cdot}$$

$$RO\cdot + CH_2{=}\underset{\substack{|\\ Z}}{CH} \longrightarrow ROCH_2\underset{\substack{|\\ Z}}{\dot{C}H}$$

the alkene monomer reacts with a radical

chain-propagating steps

propagating sites

$$ROCH_2\underset{\substack{|\\ Z}}{\dot{C}H} + CH_2{=}\underset{\substack{|\\ Z}}{CH} \longrightarrow ROCH_2\underset{\substack{|\\ Z}}{CH}CH_2\underset{\substack{|\\ Z}}{\dot{C}H}$$

$$ROCH_2\underset{\substack{|\\ Z}}{CH}CH_2\underset{\substack{|\\ Z}}{\dot{C}H} + CH_2{=}\underset{\substack{|\\ Z}}{CH} \longrightarrow ROCH_2\underset{\substack{|\\ Z}}{CH}CH_2\underset{\substack{|\\ Z}}{CH}CH_2\underset{\substack{|\\ Z}}{\dot{C}H} \xrightarrow{\text{etc.}}$$

This process is repeated over and over. Hundreds or even thousands of alkene monomers can add one at a time to the growing chain. Eventually, the chain reaction stops because the propagating sites are destroyed. Propagating sites can be destroyed when two chains combine at their propagating sites; when two chains undergo *disproportionation*, with one chain being oxidized to an alkene and the other being reduced to an alkane; or when a chain reacts with an impurity that consumes the radical.

three ways to terminate the chain

chain combination

$$2\ RO{+}\!\Big[CH_2\underset{\substack{|\\ Z}}{CH}\Big]_n\!{-}CH_2\underset{\substack{|\\ Z}}{\dot{C}H} \longrightarrow RO{+}\!\Big[CH_2\underset{\substack{|\\ Z}}{CH}\Big]_n\!{-}CH_2\underset{\substack{|\\ Z}}{CH}\underset{\substack{|\\ Z}}{CH}CH_2{-}\Big[\underset{\substack{|\\ Z}}{CH}CH_2\Big]_n\!{-}OR$$

disproportionation

$$2\ RO{-}\Big[CH_2\underset{\substack{|\\ Z}}{CH}\Big]_n\!{-}CH_2\underset{\substack{|\\ Z}}{\dot{C}H} \longrightarrow RO{-}\Big[CH_2\underset{\substack{|\\ Z}}{CH}\Big]_n\!{-}CH{=}\underset{\substack{|\\ Z}}{CH} + RO{-}\Big[CH_2\underset{\substack{|\\ Z}}{CH}\Big]_n\!{-}CH_2\underset{\substack{|\\ Z}}{CH_2}$$

reaction with an impurity

$$RO{-}\Big[CH_2\underset{\substack{|\\ Z}}{CH}\Big]_n\!{-}CH_2\underset{\substack{|\\ Z}}{\dot{C}H} + \text{impurity} \longrightarrow RO{-}\Big[CH_2\underset{\substack{|\\ Z}}{CH}\Big]_n\!{-}CH_2\underset{\substack{|\\ Z}}{CH}{-}\text{impurity}$$

Thus, *radical polymerizations* have chain-initiating, chain-propagating, and chain-terminating steps similar to the steps that take place in the radical reactions discussed in Sections 4.10 and 9.2.

As long as the polymer has a high molecular weight, the groups at the ends of the polymer are relatively unimportant in determining its physical properties and are generally not even specified; it is the rest of the molecule that determines the properties of the polymer.

The molecular weight of the polymer can be controlled by a process known as **chain transfer**. In chain transfer, the growing chain reacts with a molecule XY in a manner that allows X· to terminate the chain, leaving behind Y· to initiate a new chain. XY can be a solvent, a radical initiator, or any molecule with a bond that can be cleaved homolytically.

$$-CH_2{-}\Big[CH_2\underset{\substack{|\\ Z}}{CH}\Big]_n\!{-}CH_2\underset{\substack{|\\ Z}}{\dot{C}H} + XY \longrightarrow -CH_2{-}\Big[CH_2\underset{\substack{|\\ Z}}{CH}\Big]_n\!{-}CH_2\underset{\substack{|\\ Z}}{CH}X + Y\cdot$$

Chain-growth polymerization of monosubstituted ethylenes exhibits a marked preference for **head-to-tail addition**, where the head of one monomer is attached to the tail of another.

tail head

$CH_2{=}CH(Z)$

$-CH_2CH(Z)CH_2CH(Z)-$ **head-to-tail**

$-CH_2CH(Z)CH(Z)CH_2-$ **head-to-head**

$-CH(Z)CH_2CH_2CH(Z)-$ **tail-to-tail**

Head-to-tail addition of a substituted ethylene results in a polymer in which every other carbon bears a substituent.

$CH_2{=}CH(Cl)$ (**vinyl chloride**) $\longrightarrow$ $-CH_2CH(Cl)CH_2CH(Cl)CH_2CH(Cl)CH_2CH(Cl)CH_2CH(Cl)CH_2CH(Cl)-$ (**poly(vinyl chloride)**)

3-D Molecules: Vinyl chloride; Poly(vinyl chloride)

Head-to-tail addition is favored for steric reasons because the propagating site preferentially attacks the less sterically hindered unsubstituted sp^2 carbon of the alkene. Groups that stabilize radicals also favor head-to-tail addition. For example, when Z is a phenyl substituent, the benzene ring stabilizes the radical by electron delocalization, so the propagating site is the carbon that bears the phenyl substituent.

$-CH_2\dot{C}H(C_6H_5) \longleftrightarrow -CH_2CH{=}(C_6H_5) \longleftrightarrow -CH_2CH{=}(C_6H_5) \longleftrightarrow -CH_2CH{=}(C_6H_5) \longleftrightarrow -CH_2\dot{C}H(C_6H_5)$

In cases where Z is small—which makes steric considerations less important—and is less able to stabilize the growing end of the chain by electron delocalization, some head-to-head addition and some tail-to-tail addition also occur. This has been observed primarily in situations where Z is fluorine. Abnormal addition, however, has never been found to constitute more than 10% of the overall chain.

Monomers that most readily undergo chain-growth polymerization by a radical mechanism are those in which the substituent Z is able to stabilize the growing radical species by electron delocalization. Examples of monomers that undergo radical polymerization are shown in Table 28.2.

Table 28.2 Examples of Alkenes That Undergo Radical Polymerization

$CH_2{=}CH(C_6H_5)$ **styrene**	$CH_2{=}CH\text{–}OC(=O)CH_3$ **vinyl acetate**	$CH_2{=}C(CH_3)\text{–}C(=O)OCH_3$ **methyl methacrylate**
$CH_2{=}CH(Cl)$ **vinyl chloride**	$CH_2{=}CH\text{–}C{\equiv}N$ **acrylonitrile**	$CH_2{=}CH\text{–}CH{=}CH_2$ **1,3-butadiene**

Any compound that readily undergoes homolytic cleavage to form radicals that are sufficiently energetic to convert an alkene into a radical can serve as an initiator for radical polymerization. Several radical initiators are shown in Table 28.3.

Table 28.3 Some Radical Initiators

$$CH_3\underset{CH_3}{\overset{CH_3}{C}}O{-}OH \longrightarrow CH_3\underset{CH_3}{\overset{CH_3}{C}}O\cdot \;+\; \cdot OH$$

$$KO\underset{O}{\overset{O}{S}}O{-}O\underset{O}{\overset{O}{S}}OK \longrightarrow 2\,KO\underset{O}{\overset{O}{S}}O\cdot$$

$$CH_3\underset{CH_3}{\overset{CH_3}{C}}O{-}O\underset{CH_3}{\overset{CH_3}{C}}CH_3 \longrightarrow 2\,CH_3\underset{CH_3}{\overset{CH_3}{C}}O\cdot$$

$$C_6H_5{-}\overset{O}{\overset{\|}{C}}O{-}O\overset{O}{\overset{\|}{C}}{-}C_6H_5 \longrightarrow 2\,C_6H_5{-}\overset{O}{\overset{\|}{C}}O\cdot$$

$$CH_3\underset{C\equiv N}{\overset{CH_3}{C}}{-}N{=}N{-}\underset{C\equiv N}{\overset{CH_3}{C}}CH_3 \longrightarrow 2\,CH_3\underset{C\equiv N}{\overset{CH_3}{C}}\cdot \;+\; N_2$$

A common feature of all radical initiators is a relatively weak bond that readily undergoes homolytic cleavage. In all but one of the radical initiators shown in Table 28.3, the weak bond is an oxygen–oxygen bond. Two factors enter into the choice of radical initiator for a particular chain-growth polymerization. The first is the desired solubility of the initiator. For example, potassium persulfate is often used if the initiator needs to be soluble in water, whereas an initiator with several carbons is chosen if the initiator must be soluble in a nonpolar solvent. The second factor is the temperature at which the polymerization reaction is to be carried out. For example, a *tert*-butoxy radical is relatively stable, so an initiator that forms a *tert*-butoxy radical is used for polymerizations carried out at relatively high temperatures.

PROBLEM 1◆

What monomer would you use to form each of the following polymers?

a. $-CH_2\underset{Cl}{CH}CH_2\underset{Cl}{CH}CH_2\underset{Cl}{CH}CH_2\underset{Cl}{CH}CH_2\underset{Cl}{CH}-$

b. $-CH_2\underset{\underset{\underset{CH_3}{O}}{C=O}}{\overset{CH_3}{C}}CH_2\underset{\underset{\underset{CH_3}{O}}{C=O}}{\overset{CH_3}{C}}CH_2\underset{\underset{\underset{CH_3}{O}}{C=O}}{\overset{CH_3}{C}}CH_2\underset{\underset{\underset{CH_3}{O}}{C=O}}{\overset{CH_3}{C}}CH_2\underset{\underset{\underset{CH_3}{O}}{C=O}}{\overset{CH_3}{C}}CH_2\underset{\underset{\underset{CH_3}{O}}{C=O}}{\overset{CH_3}{C}}-$

c. $-CF_2CF_2CF_2CF_2CF_2CF_2CF_2CF_2CF_2CF_2-$

PROBLEM 2◆

Which polymer would be more apt to contain abnormal head-to-head linkages: poly(vinyl chloride) or polystyrene?

PROBLEM 3

Draw a segment of polystyrene that contains abnormal head-to-head and tail-to-tail linkages.

PROBLEM 4

Show the mechanism for the formation of a segment of poly(vinyl chloride) containing three units of vinyl chloride and initiated by hydrogen peroxide.

Branching of the Polymer Chain

If the propagating site abstracts a hydrogen atom from a chain, a branch can grow off the chain at that point.

$$-CH_2CH_2CH_2\dot{C}H_2 + -CH_2CH_2\underset{}{\overset{H}{\overset{|}{C}}}HCH_2CH_2CH_2-$$

$$\downarrow$$

$$-CH_2CH_2CH_2\overset{H}{\overset{|}{C}}H_2 + -CH_2CH_2\dot{C}HCH_2CH_2CH_2- \xrightarrow{CH_2=CH_2} -CH_2CH_2\overset{\dot{C}H_2-CH_2}{\overset{|}{C}}HCH_2CH_2CH_2-$$

Abstraction of a hydrogen atom from a carbon near the end of a chain leads to short branches, whereas abstraction of a hydrogen atom from a carbon near the middle of a chain results in long branches. Short branches are more likely to be formed than long ones because the ends of the chain are more accessible.

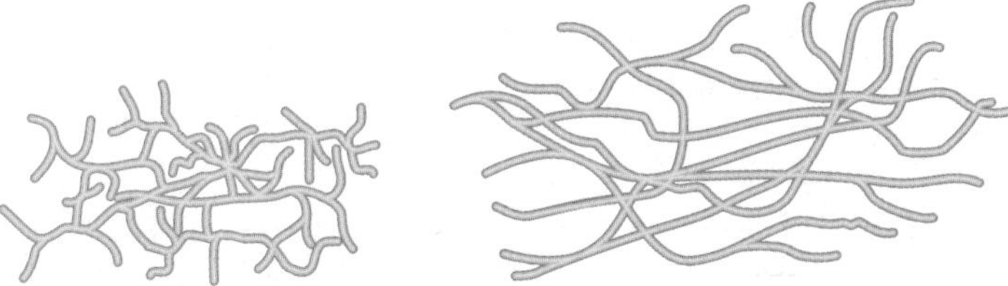

chain with short branches **chain with long branches**

Branching greatly affects the physical properties of the polymer. Linear unbranched chains can pack together more closely than branched chains can. Consequently, linear polyethylene (known as high-density polyethylene) is a relatively hard plastic, used for the production of such things as artificial hip joints, while branched polyethylene (low-density polyethylene) is a much more flexible polymer, used for trash bags and dry-cleaning bags.

Branched polymers are more flexible.

PROBLEM 5◆

Polyethylene can be used for the production of beach chairs and beach balls. Which of these items is made from more highly branched polyethylene?

PROBLEM 6

Draw a short segment of branched polystyrene that shows the linkages at the branch point.

Cationic Polymerization

In cationic polymerization, the initiator is an electrophile that adds to the alkene, causing it to become a cation. The initiator most often used in cationic polymerization is a Lewis acid, such as BF_3 or $AlCl_3$. The advantage of such an initiator is that it does not have an accompanying nucleophile that could act as a chain terminator, as would be the case with a proton-donating acid such as HCl. The cation formed in the initiation step reacts with a second monomer, forming a new cation that reacts in turn with a third monomer. As each subsequent monomer adds to the chain, the positively charged propagating site always ends up on the last unit added.

chain-initiating step

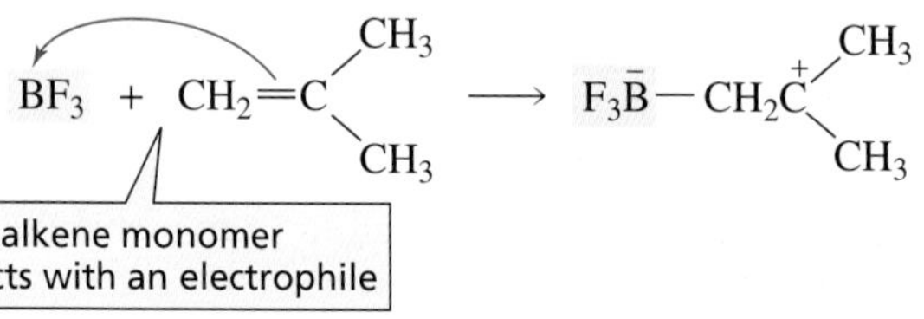

$$BF_3 + CH_2{=}C(CH_3)_2 \longrightarrow F_3\bar{B}{-}CH_2\overset{+}{C}(CH_3)_2$$

the alkene monomer reacts with an electrophile

chain-propagating steps

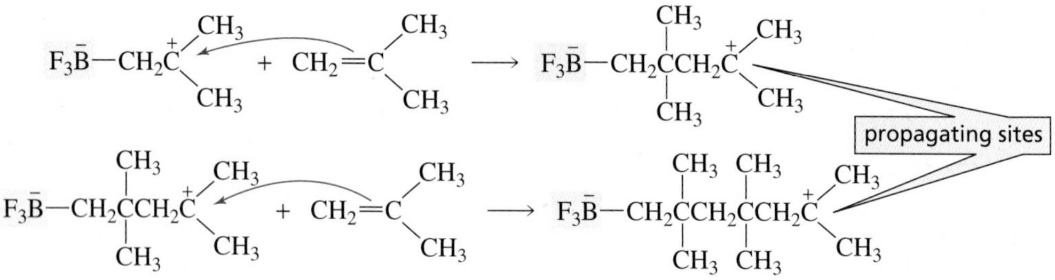

$$F_3\bar{B}{-}CH_2\overset{+}{C}(CH_3)_2 + CH_2{=}C(CH_3)_2 \longrightarrow F_3\bar{B}{-}CH_2C(CH_3)_2CH_2\overset{+}{C}(CH_3)_2$$

$$F_3\bar{B}{-}CH_2C(CH_3)_2CH_2\overset{+}{C}(CH_3)_2 + CH_2{=}C(CH_3)_2 \longrightarrow F_3\bar{B}{-}CH_2C(CH_3)_2CH_2C(CH_3)_2CH_2\overset{+}{C}(CH_3)_2$$

propagating sites

Cationic polymerization can be terminated by loss of a proton or by addition of a nucleophile that reacts with the propagating site. The chain can also be terminated by a chain-transfer reaction with the solvent (XY).

three ways to terminate the chain

loss of a proton

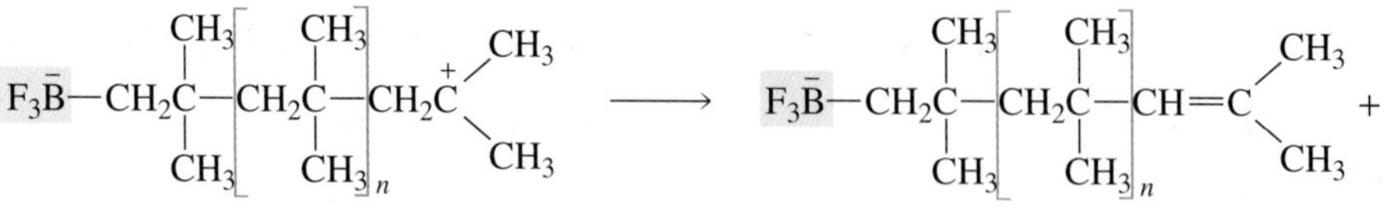

$$F_3\bar{B}{-}CH_2C(CH_3)_2{-}[CH_2C(CH_3)_2]_n{-}CH_2\overset{+}{C}(CH_3)_2 \longrightarrow F_3\bar{B}{-}CH_2C(CH_3)_2{-}[CH_2C(CH_3)_2]_n{-}CH{=}C(CH_3)_2 + H^+$$

reaction with a nucleophile

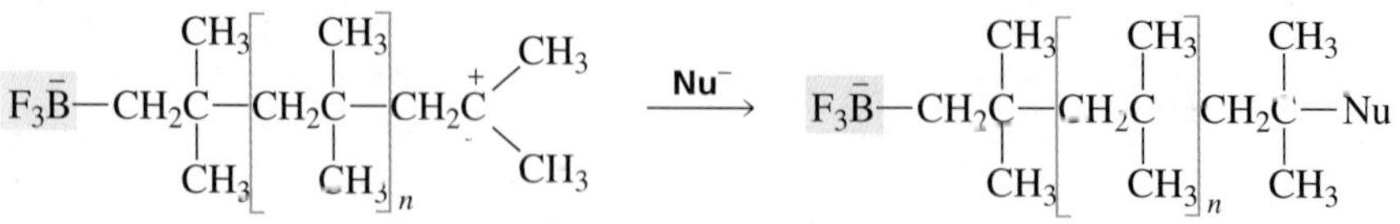

$$F_3\bar{B}{-}CH_2C(CH_3)_2{-}[CH_2C(CH_3)_2]_n{-}CH_2\overset{+}{C}(CH_3)_2 \xrightarrow{Nu^-} F_3\bar{B}{-}CH_2C(CH_3)_2{-}[CH_2C(CH_3)_2]_n{-}CH_2C(CH_3)_2{-}Nu$$

chain-transfer reaction with the solvent

$$F_3\bar{B}{-}CH_2C(CH_3)_2{-}[CH_2C(CH_3)_2]_n{-}CH_2\overset{+}{C}(CH_3)_2 \xrightarrow{XY} F_3\bar{B}{-}CH_2C(CH_3)_2{-}[CH_2C(CH_3)_2]_n{-}CH_2C(CH_3)_2{-}X + Y^+$$

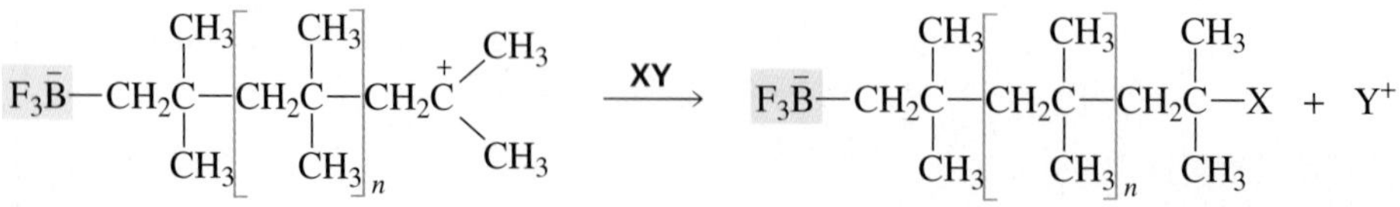

The carbocation intermediates formed during cationic polymerization, like any other carbocations, can undergo rearrangement by either a 1,2-hydride shift or a 1,2-methyl shift if rearrangement leads to a more stable carbocation (Section 4.6). For example, the polymer formed from the cationic polymerization of 3-methyl-1-butene contains both unrearranged and rearranged units. The unrearranged propagating site is a secondary carbocation, whereas the rearranged propagating site—obtained by a 1,2-hydride shift—is a more stable tertiary carbocation. The extent of rearrangement depends on the reaction temperature.

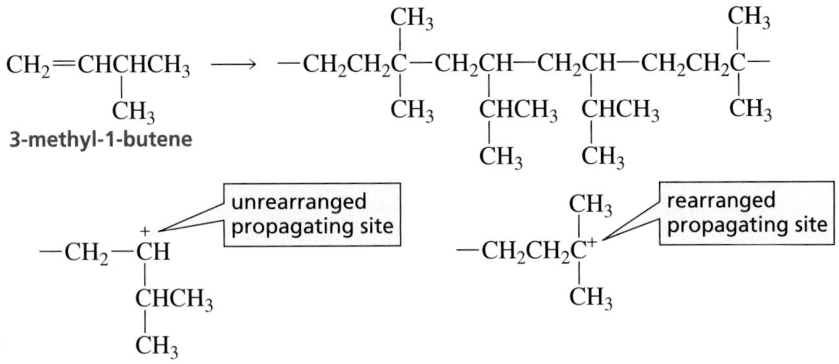

$$CH_2{=}CHCH(CH_3)CH_3 \longrightarrow -CH_2CH_2C(CH_3)_2-CH_2CH(CHCH_3CH_3)-CH_2CH(CHCH_3CH_3)-CH_2CH_2C(CH_3)_2-$$

3-methyl-1-butene

$$-CH_2-\overset{+}{C}H(CHCH_3CH_3)$$ unrearranged propagating site

$$-CH_2CH_2\overset{+}{C}(CH_3)_2$$ rearranged propagating site

Monomers that are best able to undergo polymerization by a cationic mechanism are those with substituents that can stabilize the positive charge at the propagating site by donating electrons inductively or by resonance. Examples of monomers that undergo cationic polymerization are given in Table 28.4.

Table 28.4 Examples of Alkenes That Undergo Cationic Polymerization

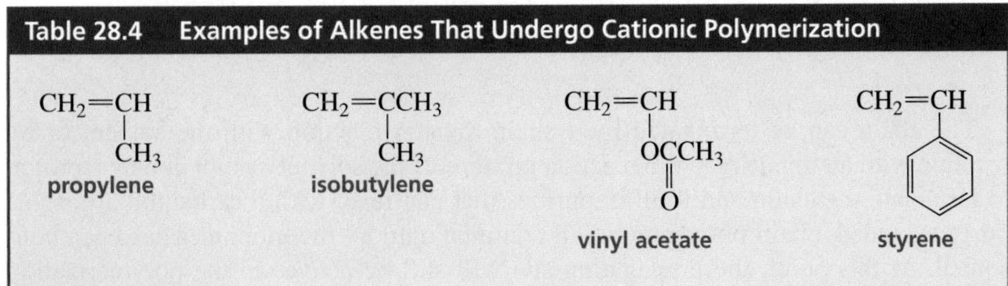

PROBLEM 7◆

List the following groups of monomers in order of decreasing ability to undergo cationic polymerization:

a. $CH_2{=}CH-C_6H_4-NO_2$ (para) $\quad CH_2{=}CH-C_6H_4-CH_3$ (para) $\quad CH_2{=}CH-C_6H_4-OCH_3$ (para)

b. $CH_2{=}CHCH_3 \quad CH_2{=}CHO\overset{O}{\overset{\|}{C}}CH_3 \quad CH_2{=}CH\overset{O}{\overset{\|}{C}}OCH_3$

c. $CH_2{=}CH-C_6H_5 \quad CH_2{=}C(CH_3)-C_6H_5$

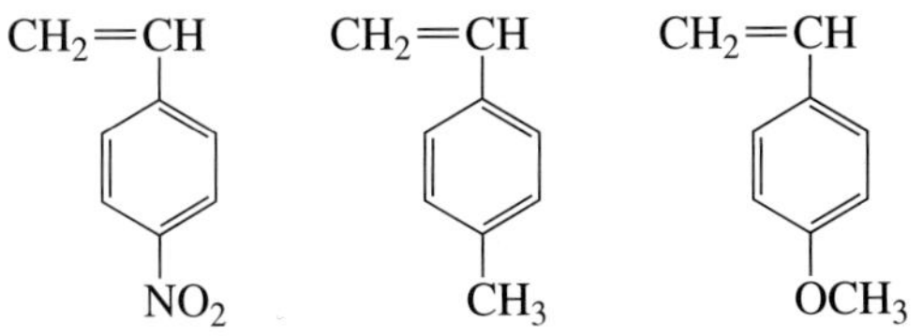

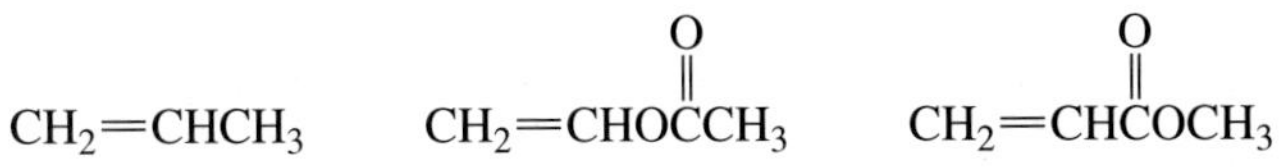

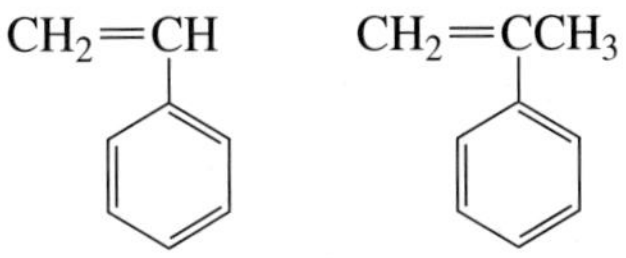

Anionic Polymerization

In anionic polymerization, the initiator is a nucleophile that reacts with the alkene to form a propagating site that is an anion. Nucleophilic attack on an alkene does not occur readily because alkenes are themselves electron rich. Therefore, the initiator must be a very good nucleophile, such as sodium amide or butyllithium, and the alkene must contain an electron-withdrawing substituent to decrease its electron density. Some alkenes that undergo polymerization by an anionic mechanism are shown in Table 28.5.

chain-initiating step

$$\overset{-}{\ddot{\mathrm{B}}}\mathrm{u}\ \mathrm{Li^+} + \mathrm{CH_2{=}CH(C{\equiv}N)} \longrightarrow \mathrm{Bu{-}CH_2}\overset{-}{\ddot{\mathrm{C}}}\mathrm{H(C{\equiv}N)}$$

the alkene monomer reacts with a nucleophile

chain-propagating steps

$$\mathrm{Bu{-}CH_2}\overset{-}{\ddot{\mathrm{C}}}\mathrm{H(C{\equiv}N)} + \mathrm{CH_2{=}CH(C{\equiv}N)} \longrightarrow \mathrm{Bu{-}CH_2CH(C{\equiv}N){-}CH_2}\overset{-}{\ddot{\mathrm{C}}}\mathrm{H(C{\equiv}N)}$$

$$\mathrm{Bu{-}CH_2CH(C{\equiv}N){-}CH_2}\overset{-}{\ddot{\mathrm{C}}}\mathrm{H(C{\equiv}N)} + \mathrm{CH_2{=}CH(C{\equiv}N)} \longrightarrow \mathrm{Bu{-}CH_2CH(C{\equiv}N){-}CH_2CH(C{\equiv}N){-}CH_2}\overset{-}{\ddot{\mathrm{C}}}\mathrm{H(C{\equiv}N)}$$

propagating sites

The chain can be terminated by a chain transfer reaction with the solvent or by reaction with an impurity in the reaction mixture. If the solvent cannot donate a proton to terminate the chain and if all impurities that can react with a carbanion are rigorously excluded, chain propagation will continue until all the monomer has been consumed. At this point, the propagating site will still be active, so the polymerization reaction will continue if more monomer is added to the system. Such nonterminated chains are called **living polymers** because the chains remain active until they are "killed." Living polymers usually result from anionic polymerization because the chains cannot be terminated by proton loss from the polymer, as they can in cationic polymerization, or by disproportionation or radical recombination, as they can in radical polymerization.

Super glue is a polymer of methyl α-cyanoacrylate. Because the monomer has two electron-withdrawing groups, it requires only a moderately good nucleophile to initiate anionic polymerization. An OH group of cellulose or a nucleophilic group of a protein can act as an initiator. You may well have experienced this reaction if you have ever spilled a drop of Super glue on your fingers. A nucleophilic group of the

Table 28.5 Examples of Alkenes That Undergo Anionic Polymerization

$CH_2{=}CH(Cl)$	$CH_2{=}CH(C{\equiv}N)$	$CH_2{=}C(CH_3)C({=}O)OCH_3$	$CH_2{=}CH(C_6H_5)$
vinyl chloride	acrylonitrile	methyl methacrylate	styrene

protein on the surface of the skin initiates the polymerization reaction, with the result that two fingers can become firmly glued together. The ability to form covalent bonds with groups on the surfaces of the objects to be glued together is what gives Super glue its amazing strength. Polymers similar to Super glue (they are butyl, isobutyl, or octyl esters rather than methyl esters) are used by surgeons to close wounds.

$$CH_2{=}C(C{\equiv}N)(C{=}O)OCH_3 \longrightarrow -CH_2C(C{\equiv}N)(COOCH_3)-CH_2C(C{\equiv}N)(COOCH_3)-CH_2C(C{\equiv}N)(COOCH_3)-CH_2-\left[C(C{\equiv}N)(COOCH_3)-CH_2\right]_n-C(C{\equiv}N)(COOCH_3)-CH_2C(C{\equiv}N)(COOCH_3)-$$

methyl α-cyanoacrylate — Super glue

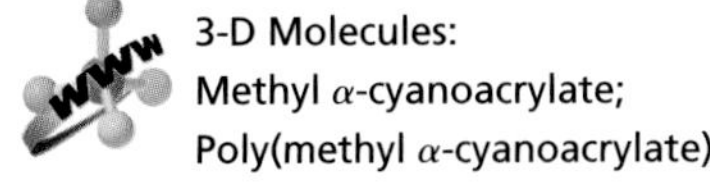
3-D Molecules: Methyl α-cyanoacrylate; Poly(methyl α-cyanoacrylate)

PROBLEM 8◆

List the following groups of monomers in order of decreasing ability to undergo anionic polymerization:

a. $CH_2{=}CH-C_6H_4-NO_2$ $\quad$ $CH_2{=}CH-C_6H_4-CH_3$ $\quad$ $CH_2{=}CH-C_6H_4-OCH_3$

b. $CH_2{=}CHCH_3$ $\quad$ $CH_2{=}CHCl$ $\quad$ $CH_2{=}CHC{\equiv}N$

What Determines the Mechanism?

We have seen that the substituent on the alkene determines the best mechanism for chain-growth polymerization. Alkenes with substituents that can stabilize radicals readily undergo radical polymerization, alkenes with electron-donating substituents that can stabilize cations undergo cationic polymerization, and alkenes with electron-withdrawing substituents that can stabilize anions undergo anionic polymerizations.

Some alkenes undergo polymerization by more than one mechanism. For example, styrene can undergo polymerization by radical, cationic, and anionic mechanisms because the phenyl group can stabilize benzylic radicals, benzylic cations, and benzylic anions. The particular mechanism followed for the polymerization of styrene depends on the nature of the initiator chosen to start the reaction.

Ring-Opening Polymerizations

Although ethylene and substituted ethylenes are the monomers most commonly used for chain-growth polymerization reactions, other compounds can polymerize as well. For example, epoxides undergo chain-growth polymerization reactions. If the initiator is a nucleophile such as HO^- or RO^-, polymerization occurs by an anionic mechanism.

$$R\ddot{O}^{-} + \text{(propylene oxide)} \longrightarrow RO-CH_2CH(CH_3)O^-$$

propylene oxide

$$RO-CH_2CH(CH_3)\ddot{O}^{-} + \text{(propylene oxide)} \longrightarrow RO-CH_2CH(CH_3)OCH_2CH(CH_3)O^-$$

If the initiator is a Lewis acid or a proton-donating acid, epoxides are polymerized by a cationic mechanism. Polymerization reactions that involve ring-opening reactions, such as the polymerization of propylene oxide, are called **ring-opening polymerizations**.

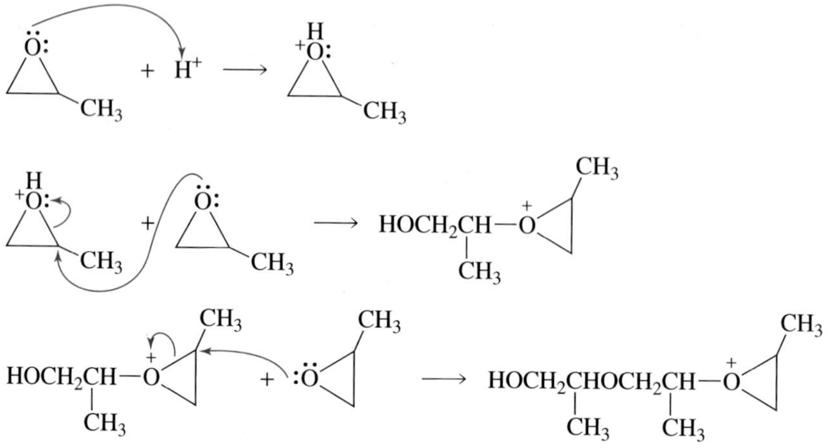

$$\text{propylene oxide} + H^+ \longrightarrow \text{protonated propylene oxide } (\overset{+}{O}H)$$

$$\text{protonated propylene oxide} + \text{propylene oxide} \longrightarrow HOCH_2\underset{CH_3}{CH}{-}\overset{+}{O}\text{(ring, CH}_3\text{)}$$

$$HOCH_2\underset{CH_3}{CH}{-}\overset{+}{O}\text{(ring, CH}_3\text{)} + \text{propylene oxide} \longrightarrow HOCH_2\underset{CH_3}{CH}OCH_2\underset{CH_3}{CH}{-}\overset{+}{O}\text{(ring, CH}_3\text{)}$$

PROBLEM 9

Explain why when propylene oxide undergoes anionic polymerization, nucleophilic attack occurs at the less substituted carbon of the epoxide, but when it undergoes cationic polymerization, nucleophilic attack occurs at the more substituted carbon.

PROBLEM 10

Describe the polymerization of 2,2-dimethyloxirane

a. by an anionic mechanism

b. by a cationic mechanism

PROBLEM 11◆

Which monomer and which type initiator would you use to synthesize each of the following polymers?

a. 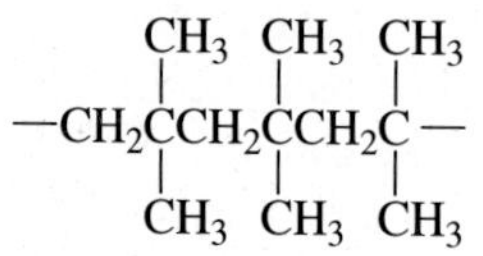 $-CH_2\underset{CH_3}{\overset{CH_3}{C}}CH_2\underset{CH_3}{\overset{CH_3}{C}}CH_2\underset{CH_3}{\overset{CH_3}{C}}-$

b. 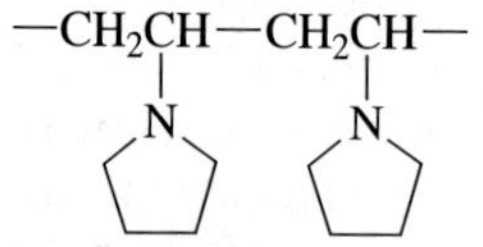 $-CH_2\underset{N}{CH}-CH_2\underset{N}{CH}-$ (N = pyrrolidin-1-yl)

c. $-CH_2CH_2OCH_2CH_2O-$

d. 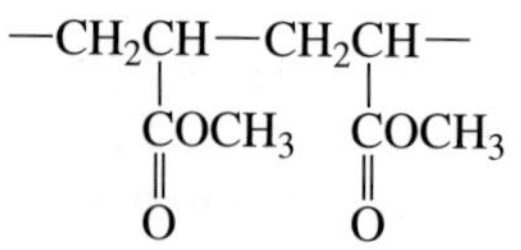 $-CH_2\underset{COCH_3}{CH}-CH_2\underset{COCH_3}{CH}-$ (C=O)

PROBLEM 12◆

Draw a short segment of the polymer formed from cationic polymerization of 3,3-dimethyloxacyclobutane.

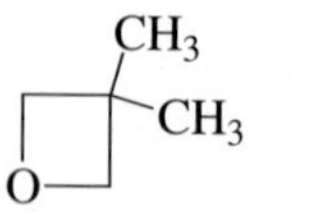

3,3-dimethyloxacyclobutane

28.3 Stereochemistry of Polymerization • Ziegler–Natta Catalysts

Polymers formed from monosubstituted ethylenes can exist in three configurations: isotactic, syndiotactic, and atactic. An **isotactic polymer** has all of its substituents on the same side of the fully extended carbon chain. (*Iso* and *taxis* are Greek for "the same" and "order," respectively.) In a **syndiotactic polymer** (*syndio* means "alternating"), the substituents regularly alternate on both sides of the carbon chain. The substituents in an **atactic polymer** are randomly oriented.

isotactic configuration (same side)

H CH_3 H CH_3 H CH_3 H CH_3 H CH_3 H CH_3

syndiotactic configuration (both sides)

H CH_3 H_3C H H CH_3 H_3C H H CH_3 H_3C H

The configuration of the polymer affects its physical properties. Polymers in the isotactic or syndiotactic configuration are more likely to be crystalline solids because positioning the substituents in a regular order allows for a more regular packing arrangement. Polymers in the atactic configuration are more disordered and cannot pack together as well, so these polymers are less rigid and, therefore, softer.

The configuration of the polymer depends on the mechanism by which polymerization occurs. In general, radical polymerization leads primarily to branched polymers in the atactic configuration. Cationic polymerization produces polymers with a considerable fraction of the chains in the isotactic or syndiotactic configuration. Anionic polymerization produces polymers with the most stereoregularity. The percentage of chains in the isotactic or syndiotactic configuration increases as the polymerization temperature decreases.

In 1953, Karl Ziegler and Giulio Natta found that the structure of a polymer could be controlled if the growing end of the chain and the incoming monomer were coordinated with an aluminum–titanium initiator. These initiators are now called **Ziegler–Natta catalysts**. Long, unbranched polymers with either the isotactic or the syndiotactic configuration can be prepared using Ziegler–Natta catalysts. Whether the chain is isotactic or syndiotactic depends on the particular Ziegler–Natta catalyst used. These catalysts revolutionized the field of polymer chemistry because they allow the synthesis of stronger and stiffer polymers that have greater resistance to cracking and heat. High-density polyethylene is prepared using a Ziegler–Natta process.

The mechanism of the Ziegler–Natta-catalyzed polymerization of a substituted ethylene is shown in Figure 28.1. The monomer forms a π complex (Section 6.5) with titanium at an open coordination site (i.e., a site available to accept electrons) and the coordinated alkene is inserted between the titanium and the growing polymer, thereby extending the polymer chain. Because a new coordination site opens up during insertion of the monomer, the process can be repeated over and over.

Polyacetylene is another polymer prepared by a Ziegler–Natta process. It is a **conducting polymer** because the conjugated double bonds in polyacetylene make it possible to conduct electricity down its backbone after several electrons are removed from or added to the backbone.

$$HC{\equiv}CH \xrightarrow{\text{a Ziegler–Natta catalyst}} -CH{=}CH{+}CH{=}CH{+}_n CH{=}CH-$$

acetylene — polyacetylene

Karl Ziegler (1898–1973), *the son of a minister, was born in Germany. He was a professor at the University of Frankfurt and then at the University of Heidelberg.*

Karl Ziegler *and* **Giulio Natta** *did not work together, but each independently developed the catalyst system used in polymerization. They shared the 1963 Nobel Prize in chemistry.*

Giulio Natta (1903–1979) *was the son of an Italian judge. He was a professor at the Polytechnic Institute in Milan, where he became the director of the Industrial Chemistry Research Center.*

▲ **Figure 28.1**
The mechanism of the Ziegler–Natta-catalyzed polymerization of a substituted ethylene. A monomer forms a π complex with an open coordination site of titanium and then is inserted between the titanium and the growing polymer.

28.4 Polymerization of Dienes • The Manufacture of Rubber

When the bark of a rubber tree is cut, a sticky white liquid oozes out. This is the same liquid found inside the stalks of dandelions and milkweed. The sticky material is *latex*, a suspension of rubber particles in water. Latex protects the tree after an injury, by covering the wound like a bandage.

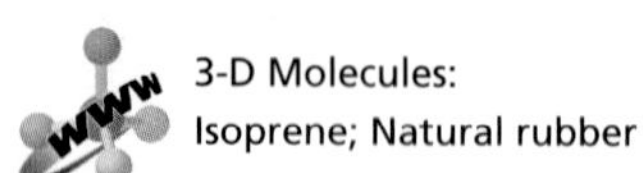
3-D Molecules: Isoprene; Natural rubber

Natural rubber is a polymer of 2-methyl-1,3-butadiene (isoprene; Section 26.6). On average, a molecule of rubber contains 5000 isoprene units. All the double bonds in natural rubber are cis. Rubber is a waterproof material because it consists of a tangle of hydrocarbon chains that have no affinity for water. Charles Macintosh, a Scotsman, was the first to use rubber as a waterproof coating for raincoats.

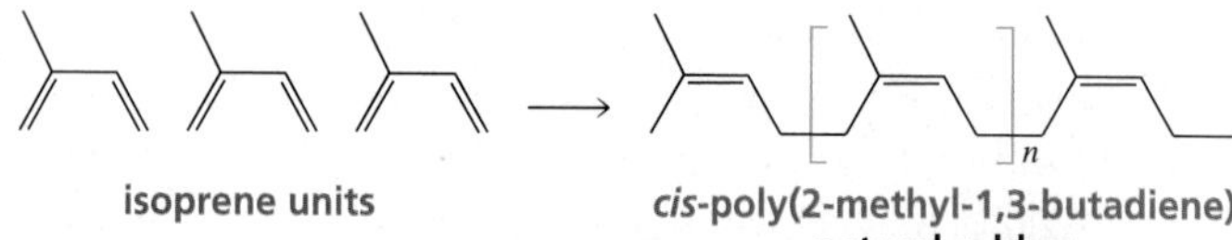
isoprene units → *cis*-poly(2-methyl-1,3-butadiene) **natural rubber**

▲ Latex being collected from a rubber tree.

Gutta-percha (from the Malaysian words *getah*, meaning "gum," and *percha*, meaning "tree") is a naturally occurring isomer of rubber in which all the double bonds are trans. Like rubber, gutta-percha is exuded by certain trees, but it is much less common. It is also harder and more brittle than rubber. Gutta-percha is the filling material that dentists use in root canals and the material used for the casing of golf balls.

PROBLEM 13

Draw a short segment of gutta-percha.

By mimicking nature, scientists have learned to make synthetic rubbers with properties tailored to meet human needs. These materials have some of the properties of natural rubber, including being waterproof and elastic, but they have some improved properties as well—they are tougher, more flexible, and more durable than natural rubber.

Synthetic rubbers have been made by polymerizing dienes other than isoprene. One synthetic rubber is a polymer of 1,3-butadiene in which all the double bonds are cis. Polymerization is carried out in the presence of a Ziegler–Natta catalyst so that the configuration of the double bonds in the polymer can be controlled.

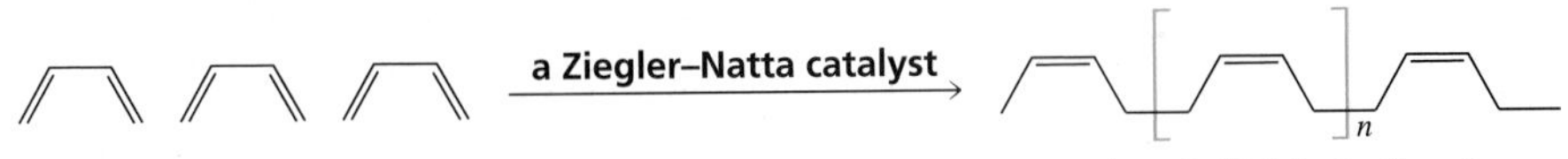

1,3-butadiene monomers —a Ziegler–Natta catalyst→ *cis*-poly(1,3-butadiene) a synthetic rubber

Neoprene is a synthetic rubber made by polymerizing 2-chloro-1,3-butadiene in the presence of a Ziegler–Natta catalyst that causes all the double bonds in the polymer to have the trans configuration. Neoprene is used to make wet suits, shoe soles, tires, hoses, and coated fabrics.

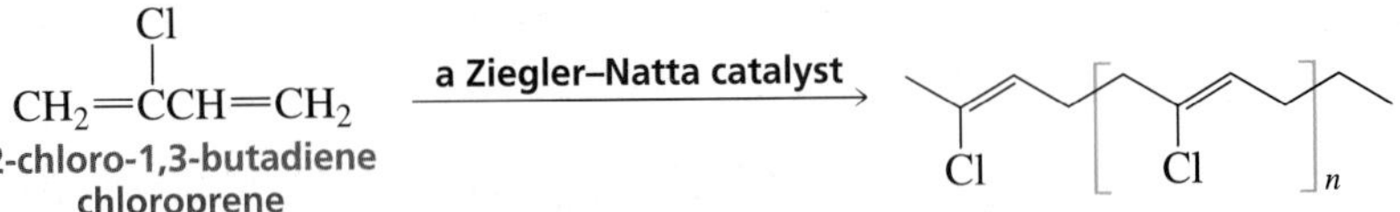

$CH_2{=}C(Cl)CH{=}CH_2$ —a Ziegler–Natta catalyst→ neoprene

2-chloro-1,3-butadiene
chloroprene

A problem common to both natural and most synthetic rubbers is that the polymers are very soft and sticky. They can be hardened by a process known as *vulcanization*. Charles Goodyear discovered this process while trying to improve the properties of rubber. He accidentally spilled a mixture of rubber and sulfur on a hot stove. To his surprise, the mixture became hard but flexible. He called the heating of rubber with sulfur **vulcanization**, after Vulcan, the Roman god of fire.

Heating rubber with sulfur causes **cross-linking** of separate polymer chains through disulfide bonds (Figure 28.2). Instead of the individual chains just being entangled together, the vulcanized chains are covalently bonded together in one giant molecule. Because the polymer has double bonds, the chains have bends and kinks that prevent them from forming a tightly packed crystalline polymer. When rubber is stretched, the chains straighten out along the direction of the pull. Cross-linking prevents the polymer from being torn when it is stretched, and the cross-links provide a reference framework for the material to return to when the stretching force is removed.

Charles Goodyear (1800–1860), *the son of an inventor of farm implements, was born in Connecticut. He patented the process of vulcanization in 1844. The process was so simple, however, that it could be easily copied, so he spent many years contesting infringements on his patent. In 1852, with Daniel Webster as his lawyer, he obtained the right to the patent.*

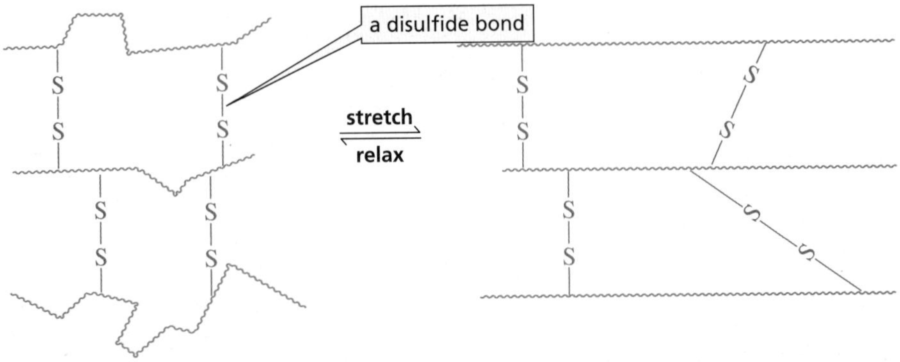

a disulfide bond
stretch
relax

◀ **Figure 28.2**
The rigidity of rubber is increased by cross-linking the polymer chains with disulfide bonds. When rubber is stretched, the randomly coiled chains straighten out and orient themselves along the direction of the stretch.

The physical properties of rubber can be controlled by regulating the amount of sulfur used in the vulcanization process. Rubber made with 1–3% sulfur is soft and stretchy and is used to make rubber bands. Rubber made with 3–10% sulfur is more rigid and is used in the manufacture of tires. Goodyear's name can be found on many tires sold today. The story of rubber is an example of a scientist taking a natural material and finding ways to improve its useful properties.

The greater the degree of cross-linking, the more rigid is the polymer.

PROBLEM 14

The polymer formed from a diene such as 1,3-butadiene contains vinyl branches. Propose an anionic polymerization mechanism to account for the formation of these branches.

$$CH_2{=}CHCH{=}CH_2 \longrightarrow -CH_2CH{=}CHCH_2CH(CH{=}CH_2)CH_2CH_2CH{=}CHCH_2-$$

28.5 Copolymers

The polymers we have discussed so far are formed from only one type of monomer and are called **homopolymers**. Often, two or more different monomers are used to form a polymer. The resulting product is called a **copolymer**. Increasing the number of different monomers used to form the copolymer dramatically increases the number of different copolymers that can be formed. Even if only two kinds of monomers are used, copolymers with very different properties can be prepared by varying the amounts of each monomer. Both chain-growth polymers and step-growth polymers can be copolymers. Many of the synthetic polymers used today are copolymers. Table 28.6 shows some common copolymers and the monomers from which they are synthesized.

Table 28.6 Some Examples of Copolymers and Their Uses

Monomer	Copolymer name	Uses
 $CH_2{=}CHCl$ (vinyl chloride) + $CH_2{=}CCl_2$ (vinylidene chloride)	Saran	film for wrapping food
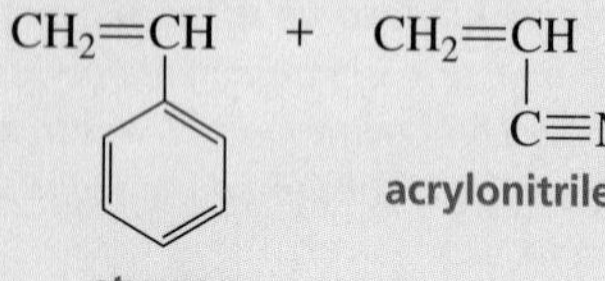 $CH_2{=}CHC_6H_5$ (styrene) + $CH_2{=}CHC{\equiv}N$ (acrylonitrile)	SAN	dishwasher-safe objects, vaccum cleaner parts
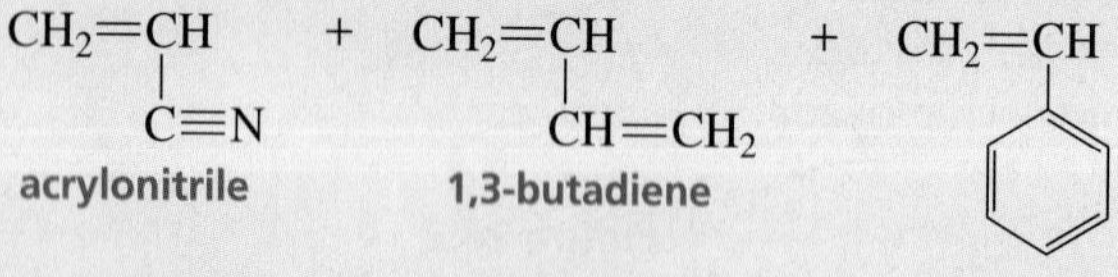 $CH_2{=}CHC{\equiv}N$ (acrylonitrile) + $CH_2{=}CHCH{=}CH_2$ (1,3-butadiene) + $CH_2{=}CHC_6H_5$ (styrene)	ABS	bumpers, crash helmets, telephones, luggage
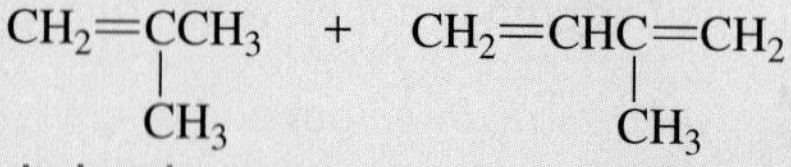 $CH_2{=}C(CH_3)CH_3$ (isobutylene) + $CH_2{=}CHC(CH_3){=}CH_2$ (isoprene)	butyl rubber	inner tubes, balls, inflatable sporting goods

There are four types of copolymers. In an **alternating copolymer**, the two monomers alternate. In a **block copolymer**, there are blocks of each kind of monomer. In a **random copolymer**, the distribution of monomers is random. A **graft copolymer** contains branches derived from one monomer grafted onto a backbone derived from another monomer. These structural differences extend the range of physical properties available to the scientist designing the copolymer.

an alternating copolymer	ABABABABABABABABABABABA
a block copolymer	AAAAABBBBBAAAAABBBBBAAA
a random copolymer	AABABABBABAABBABABBAAAB
a graft copolymer	AAAAAAAAAAAAAAAAAAAAAAA (with BBBBBB branches)

```
AAAAAAAAAAAAAAAAAAAAAAA
  B         B         B
  B         B         B
  B         B         B
  B         B         B
  B         B         B
  B         B         B
```

28.6 Step-Growth Polymers

Step-growth polymers are formed by the intermolecular reaction of bifunctional molecules (molecules with two functional groups). When the functional groups react, in most cases a small molecule such as H_2O, alcohol, or HCl is lost. This is why these polymers are also called *condensation polymers.*

There are two types of step-growth polymers. One type is formed by the reaction of a single monomer that possesses two different functional groups A and B. Functional group A of one monomer reacts with functional group B of another monomer.

$$\text{A—B} \quad \text{A—B} \longrightarrow \text{A—X—B}$$

The other type of step-growth polymer is formed by the reaction of two different bifunctional monomers. One monomer contains two A functional groups and the other monomer contains two B functional groups.

Step-growth polymers are made by combining molecules with reactive groups at each end.

$$\text{A—A} \quad \text{B—B} \longrightarrow \text{A—X—B}$$

The formation of step-growth polymers, unlike the formation of chain-growth polymers, does not involve chain reactions. Any two monomers (or short chains) can react. The progress of a typical step-growth polymerization is shown schematically in Figure 28.3. When the reaction is 50% complete (12 bonds have formed between 25 monomers), the reaction products are primarily dimers and trimers. Even at 75% completion, no long chains have been formed. This means that if step-growth polymerization is to lead to long-chain polymers, very high yields must be achieved.

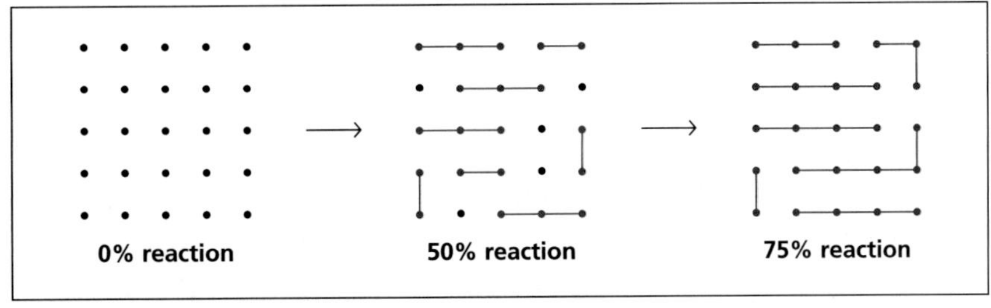

Figure 28.3 Progress of a step-growth polymerization.

Polyamides

Nylon is the common name of a synthetic **polyamide**. Nylon 6 is an example of a step-growth polymer formed by a monomer with two different functional groups. The carboxylic acid group of one monomer reacts with the amino group of another monomer, resulting in the formation of amide groups. Structurally, the reaction is similar to the polymerization of α-amino acids to form proteins (Section 23.7). This particular nylon is called nylon 6 because it is formed from the polymerization of 6-aminohexanoic acid, a compound that contains six carbons.

$$\mathrm{H_3\overset{+}{N}(CH_2)_5\overset{\overset{\large O}{\|}}{C}O^-} \xrightarrow[-\mathrm{H_2O}]{\Delta} \mathrm{—NH(CH_2)_5\overset{\overset{\large O}{\|}}{C}\left[NH(CH_2)_5\overset{\overset{\large O}{\|}}{C}\right]_n NH(CH_2)_5\overset{\overset{\large O}{\|}}{C}—}$$

6-aminohexanoic acid

nylon 6
a polyamide

Nylon was first synthesized in 1931 by **Wallace Carothers (1896–1937)**. *He was born in Iowa and received a Ph.D. from the University of Illinois. He taught there and at Harvard before being hired by DuPont to head its program in basic science. Nylon was introduced to the public in 1939, but its widespread use was delayed until after World War II because all nylon produced during the war was used by the military. Carothers died unaware of the era of synthetic fibers that dawned after the war.*

The starting material for nylon 6 is ϵ-caprolactam. The lactam is opened by hydrolysis.

$$\text{ϵ-caprolactam} \xrightarrow[\Delta]{H^+,\ H_2O} H_3\overset{+}{N}CH_2CH_2CH_2CH_2CH_2\overset{O}{\overset{\|}{C}}OH$$

ϵ-caprolactam

ϵ-aminocaproic acid
6-aminohexanoic acid

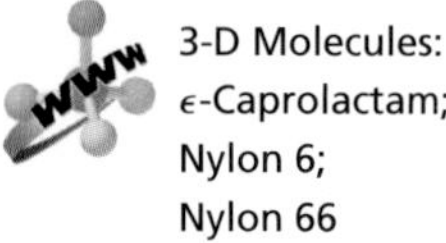
3-D Molecules: ϵ-Caprolactam; Nylon 6; Nylon 66

Nylon 66 is an example of a step-growth polymer formed by two different bifunctional monomers: adipic acid and 1,6-hexanediamine. It is called nylon 66 because it is a polyamide formed from a six-carbon diacid and a six-carbon diamine.

$$\underset{\text{adipic acid}}{HO\overset{O}{\overset{\|}{C}}(CH_2)_4\overset{O}{\overset{\|}{C}}OH} + \underset{\text{1,6-hexanediamine}}{H_2N(CH_2)_6NH_2} \xrightarrow[-H_2O]{\Delta} \underset{\text{nylon 66}}{-\overset{O}{\overset{\|}{C}}(CH_2)_4\overset{O}{\overset{\|}{C}}\left[NH(CH_2)_6NH\overset{O}{\overset{\|}{C}}(CH_2)_4\overset{O}{\overset{\|}{C}}\right]_n NH(CH_2)_6NH-}$$

Nylon first found wide use in textiles and carpets. Because it is resistant to stress, it is now used in many other applications, such as mountaineering ropes, tire cords and fishing lines, and as a substitute for metal in bearings and gears. The extended applications of nylon precipitated a search for new "super fibers" with super strength and super heat resistance.

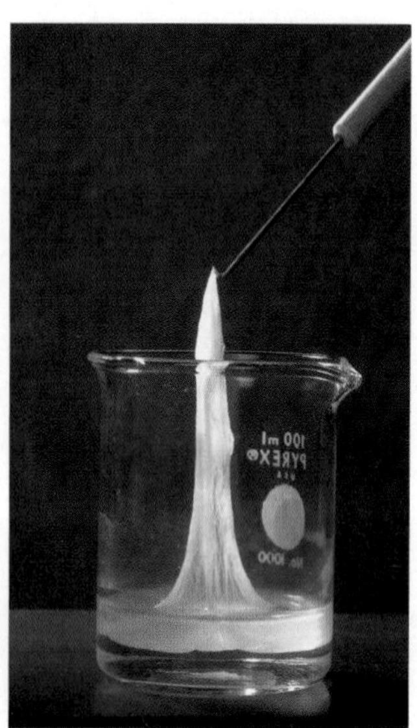
▲ Nylon is pulled from a beaker of adipoyl chloride and 1,6-hexanediamine.

PROBLEM 15◆

a. Draw a short segment of nylon 4.
b. From what lactam is nylon 4 synthesized?
c. Draw a short segment of nylon 44.

PROBLEM 16

Write an equation that explains what will happen if a scientist working in the laboratory spills sulfuric acid on her nylon 66 hose.

One super fiber is Kevlar®, a polymer of 1,4-benzenedicarboxylic acid and 1,4-diaminobenzene. The incorporation of aromatic rings into polymers has been found to result in polymers with great physical strength. Aromatic polyamides are called **aramides**. Kevlar® is an aramide with a tensile strength greater than that of steel. Army helmets are made of Kevlar®, which is also used for lightweight bulletproof vests and high-performance skis. Because it is stable at very high temperatures, it is used in the protective clothing worn by firefighters.

$$\underset{\text{1,4-benzenedicarboxylic acid}}{HO-\overset{O}{\overset{\|}{C}}-C_6H_4-\overset{O}{\overset{\|}{C}}-OH} + \underset{\text{1,4-diaminobenzene}}{H_2N-C_6H_4-NH_2} \xrightarrow[-H_2O]{\Delta}$$

$$-\overset{O}{\overset{\|}{C}}-C_6H_4-\overset{O}{\overset{\|}{C}}\left[NH-C_6H_4-NH-\overset{O}{\overset{\|}{C}}-C_6H_4-\overset{O}{\overset{\|}{C}}\right]_n NH-C_6H_4-NH-$$

Kevlar®
an aramide

Kevlar® owes its strength to the way in which the individual polymer chains interact with each other. The chains are hydrogen bonded, causing them to form a sheetlike structure.

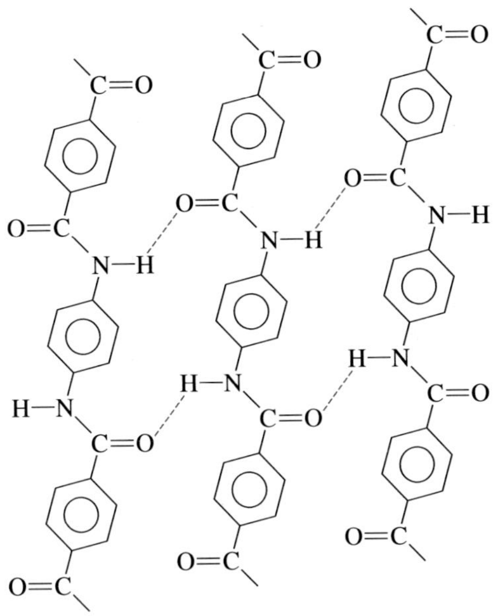

Polyesters

Polyesters are step-growth polymers in which the monomer units are joined together by ester groups. They have found wide commercial use as fibers, plastics, and coatings. The most common polyester is known by the trade name Dacron® and is made by the transesterification (Section 17.10) of dimethyl terephthalate with ethylene glycol. High resilience, durability, and moisture resistance are the properties of this polymer that contribute to its "wash-and-wear" characteristics.

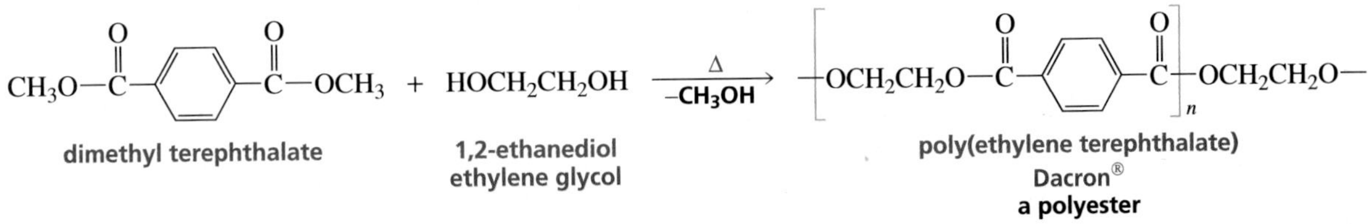

$$CH_3O{-}\overset{O}{\overset{\|}{C}}{-}C_6H_4{-}\overset{O}{\overset{\|}{C}}{-}OCH_3 + HOCH_2CH_2OH \xrightarrow[-CH_3OH]{\Delta} {-}\!\left[OCH_2CH_2O{-}\overset{O}{\overset{\|}{C}}{-}C_6H_4{-}\overset{O}{\overset{\|}{C}}\right]_n\!{-}OCH_2CH_2O{-}$$

dimethyl terephthalate — 1,2-ethanediol ethylene glycol — poly(ethylene terephthalate) Dacron® **a polyester**

Kodel® polyester is formed by the transesterification of dimethyl terephthalate with 1,4-di(hydroxymethyl)cyclohexane. The stiff polyester chain causes the fiber to have a harsh feel that can be softened by blending it with wool or cotton.

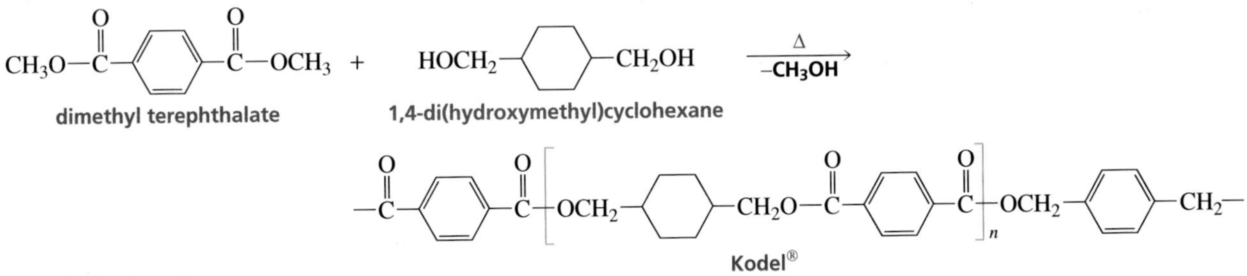

$$CH_3O{-}\overset{O}{\overset{\|}{C}}{-}C_6H_4{-}\overset{O}{\overset{\|}{C}}{-}OCH_3 + HOCH_2{-}C_6H_{10}{-}CH_2OH \xrightarrow[-CH_3OH]{\Delta}$$

dimethyl terephthalate — 1,4-di(hydroxymethyl)cyclohexane

$${-}\overset{O}{\overset{\|}{C}}{-}C_6H_4{-}\overset{O}{\overset{\|}{C}}\!\left[OCH_2{-}C_6H_{10}{-}CH_2O{-}\overset{O}{\overset{\|}{C}}{-}C_6H_4{-}\overset{O}{\overset{\|}{C}}\right]_n\!OCH_2{-}C_6H_{10}{-}CH_2{-}$$

Kodel®

PROBLEM 17

What happens to polyester slacks if aqueous NaOH is spilled on them?

Polyesters with two ester groups bonded to the same carbon are known as **polycarbonates**. Lexan® is produced by the reaction of phosgene with bisphenol A. Lexan® is a strong, transparent polymer used for bulletproof windows and traffic-light lenses. In recent years, polycarbonates have become important polymers in the automobile industry as well as in the manufacture of compact discs.

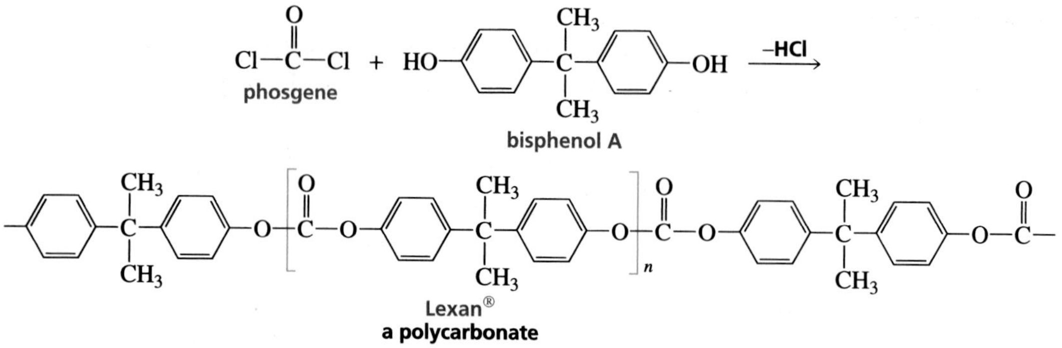

$$\mathrm{Cl{-}C({=}O){-}Cl} + \mathrm{HO{-}C_6H_4{-}C(CH_3)_2{-}C_6H_4{-}OH} \xrightarrow{-\mathrm{HCl}}$$

phosgene, bisphenol A

$$\left[\mathrm{{-}O{-}C({=}O){-}O{-}C_6H_4{-}C(CH_3)_2{-}C_6H_4{-}}\right]_n$$

Lexan®
a polycarbonate

Epoxy Resins

Tutorial: Repeating units in step-growth polymers

Epoxy resins are the strongest adhesives known. They can adhere to almost any kind of surface and are resistant to solvents and to extremes of temperature. When an epoxy cement is used, a low-molecular-weight *prepolymer* (the most common is a polymer of bisphenol A and epichlorohydrin) is mixed with a *hardener*—a compound that will react with the prepolymer to form a cross-linked polymer.

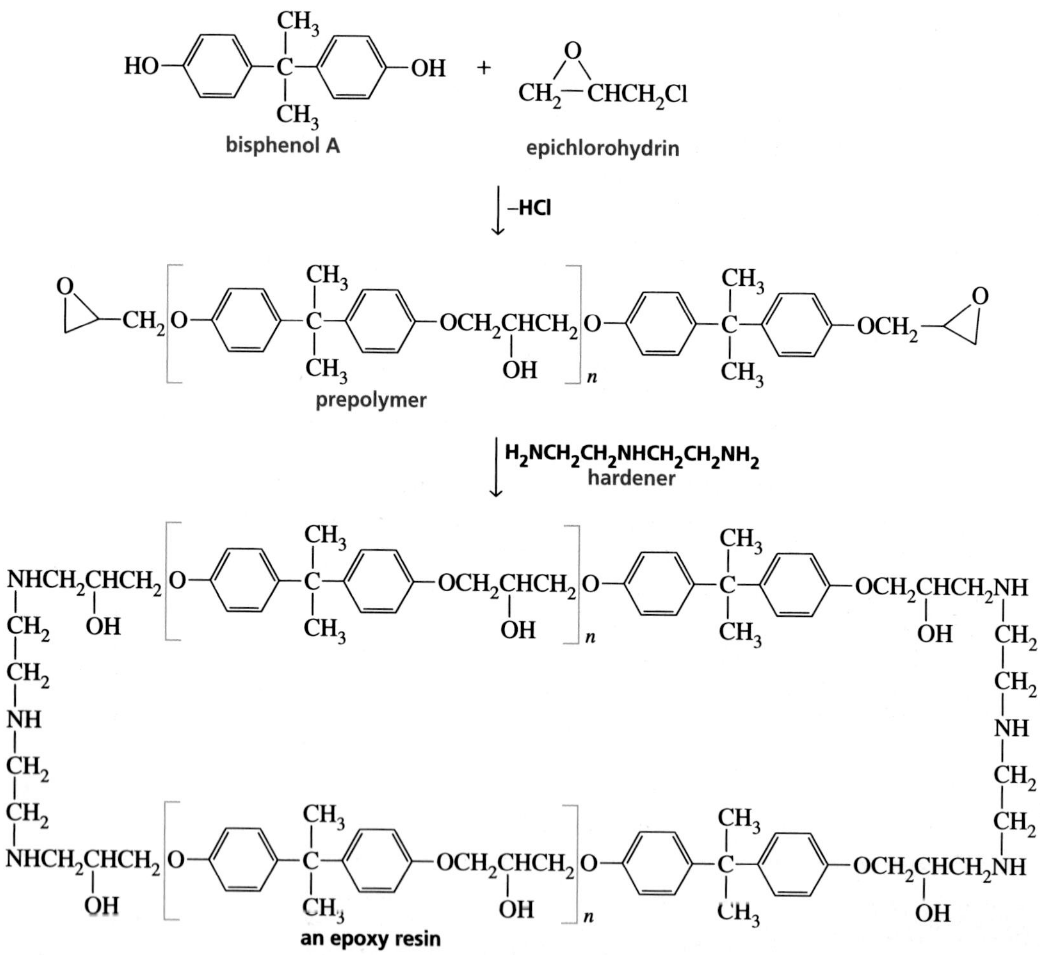

$$\mathrm{HO{-}C_6H_4{-}C(CH_3)_2{-}C_6H_4{-}OH} + \mathrm{\overset{O}{CH_2{-}CHCH_2Cl}}$$

bisphenol A, epichlorohydrin

$$\downarrow -\mathrm{HCl}$$

$$\mathrm{(epoxide){-}CH_2O}\left[\mathrm{{-}C_6H_4{-}C(CH_3)_2{-}C_6H_4{-}OCH_2CH(OH)CH_2O}\right]_n\mathrm{{-}C_6H_4{-}C(CH_3)_2{-}C_6H_4{-}OCH_2{-}(epoxide)}$$

prepolymer

$$\downarrow \mathrm{H_2NCH_2CH_2NHCH_2CH_2NH_2}\ \text{(hardener)}$$

$$\mathrm{NHCH_2CH(OH)CH_2O}\left[\mathrm{{-}C_6H_4{-}C(CH_3)_2{-}C_6H_4{-}OCH_2CH(OH)CH_2O}\right]_n\mathrm{{-}C_6H_4{-}C(CH_3)_2{-}C_6H_4{-}OCH_2CH(OH)CH_2NH}$$

(the terminal NH groups of two such chains are linked through $\mathrm{CH_2CH_2NHCH_2CH_2}$ bridges)

an epoxy resin

PROBLEM 18

a. Propose a mechanism for the formation of the prepolymer formed by bisphenol A and epichlorohydrin.

b. Propose a mechanism for the reaction of the prepolymer with the hardener.

Polyurethanes

A **urethane**—also called a carbamate—is a compound that has an OR group and an NHR group bonded to the same carbonyl carbon. Urethanes can be prepared by treating an isocyanate with an alcohol.

$$\underset{\text{an isocyanate}}{RN{=}C{=}O} + \underset{\text{an alcohol}}{ROH} \longrightarrow \underset{\text{a urethane}}{RNH{-}\overset{\overset{\displaystyle O}{\|}}{C}{-}OR}$$

Polyurethanes are polymers that contain urethane groups. One of the most common polyurethanes is prepared by the polymerization of toluene-2,6-diisocyanate and ethylene glycol. If the reaction is carried out in the presence of a blowing agent, the product is a polyurethane foam. Blowing agents are gases such as nitrogen or carbon dioxide. At one time, chlorofluorocarbons—low-boiling liquids that vaporize on heating—were used, but they have been banned for environmental reasons (Section 9.9). Polyurethane foams are used for furniture stuffing, carpet backings, and insulation. Notice that polyurethanes prepared from diisocyanates and diols are the only step-growth polymers that we have seen in which a small molecule is *not* lost during polymerization.

toluene-2,6-diisocyanate (O=C=N–, CH$_3$, –N=C=O on benzene ring) + $HOCH_2CH_2OH$ (ethylene glycol) ⟶

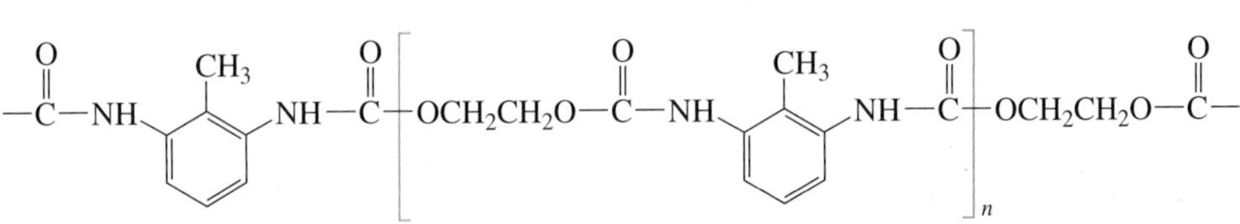

a polyurethane

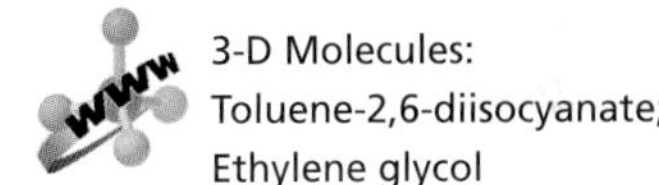

3-D Molecules: Toluene-2,6-diisocyanate; Ethylene glycol

One of the most important uses of polyurethanes is in fabrics with elastic properties, such as spandex (Lycra®). These materials are block copolymers in which some of the polymer segments are polyurethanes, some are polyesters, and some are polyamides. The blocks of polyurethane are soft, amorphous segments that become crystalline on stretching (Section 28.7). When the tension is released, they revert to the amorphous state.

PROBLEM 19

If a small amount of glycerol is added to the reaction mixture of toluene-2,6-diisocyanate and ethylene glycol during the synthesis of polyurethane foam, a much stiffer foam is obtained. Explain.

$$\underset{\text{glycerol}}{\underset{\displaystyle OH}{CH_2}{-}\underset{\displaystyle OH}{CH}{-}\underset{\displaystyle OH}{CH_2}}$$

28.7 Physical Properties of Polymers

The individual chains of a polymer such as polyethylene are held together by van der Waals forces. Because these forces operate only at small distances, they are strongest if the polymer chains can line up in an ordered, closely packed array. The regions of the polymer in which the chains are highly ordered with respect to one another are called **crystallites** (Figure 28.4). Between the crystallites are amorphous, noncrystalline regions in which the chains are randomly oriented. The more crystalline—the more

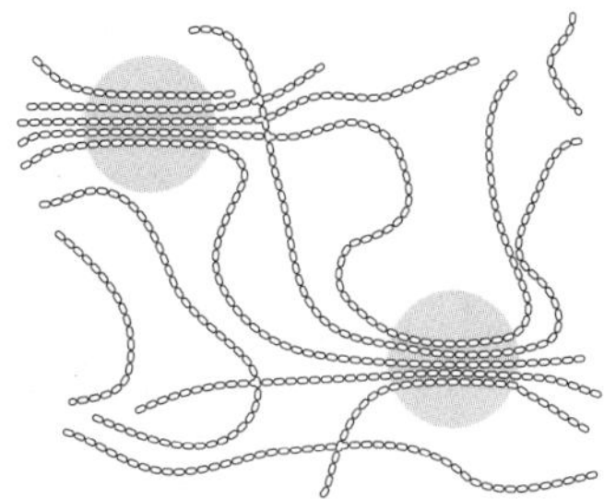
▲ **Figure 28.4**
In the circled regions, called crystallites, the polymer chains are highly ordered, similar to the ordering found in crystals. Between the circles are noncrystalline regions in which the polymer chains are randomly oriented.

ordered—the polymer is, the denser, harder, and more resistant to heat it is (Table 28.7). If the polymer chains possess substituents (as does poly[methyl methacrylate], for example) or have branches that prevent them from packing closely together, the density of the polymer is reduced.

Table 28.7 Properties of Polyethylene as a Function of Crystallinity

Crystallinity (%)	55	62	70	77	85
Density (g/cm^3)	0.92	0.93	0.94	0.95	0.96
Melting point (°C)	109	116	125	130	133

Thermoplastic Polymers

Plastics can be classified according to the physical properties imparted to them by the way in which their individual chains are arranged. **Thermoplastic polymers** have both ordered crystalline regions and amorphous noncrystalline regions. Thermoplastic polymers are hard at room temperature, but soft enough to be molded when heated, because the individual chains can slip past one another at elevated temperatures. Thermoplastic polymers are the plastics we encounter most often in our daily lives—in combs, toys, switch plates, and telephone casings, for example. They are the plastics that are easily cracked.

Thermosetting Polymers

Very strong and rigid materials can be obtained if polymer chains are cross-linked. The greater the degree of cross-linking, the more rigid is the polymer. Such cross-linked polymers are called **thermosetting polymers**. After they are hardened, they cannot be remelted by heating, because the cross-links are covalent bonds, not intermolecular van der Waals forces. Cross-linking reduces the mobility of the polymer chains, causing the polymer to be relatively brittle. Because thermosetting polymers do not have the wide range of properties characteristic of thermoplastic polymers, they are less widely used.

Melmac®, a highly cross-linked thermosetting polymer of melamine and formaldehyde, is a hard, moisture-resistant material. Because it is colorless, Melmac® can be made into materials with pastel colors. It is used to make lightweight dishes and counter surfaces.

H_2N, N, NH_2, N, N, NH_2 + $H_2C{=}O$ $\xrightarrow{-H_2O}$ —NH, N, $NHCH_2NH$, N, NH—, N, N, N, N, NH, NH, CH_2, CH_2, —NH, N, $NHCH_2NH$, N, NH—, N, N, N, N, NH, NH

melamine

formaldehyde

Melmac®

Leo Hendrik Baekeland (1863–1944) *discovered Bakelite while looking for a substitute for shellac in his home laboratory. He was born in Belgium and became a professor of chemistry at the University of Ghent. A fellowship brought him to the United States in 1889, and he decided to stay. His hobby was photography, and he invented photographic paper that could be developed under artificial light. He sold the patent to Eastman-Kodak.*

PROBLEM 20

Propose a mechanism for the formation of Melmac®.

PROBLEM 21

Bakelite was the first of the thermosetting polymers. It is a highly cross-linked polymer formed from the acid-catalyzed polymerization of phenol and formaldehyde. It is a much darker polymer than Melmac®, so the range of colors of products made from Bakelite is limited. Propose a structure for Bakelite.

DESIGNING A POLYMER

A polymer used for making dental impressions must be soft enough initially to be molded around the teeth, but must become hard enough later to maintain a fixed shape. The polymer commonly used for dental impressions contains three-membered aziridine rings that react to cross-link the chains. Because aziridine rings are not very reactive, cross-linking occurs relatively slowly, so most of the hardening of the polymer does not occur until the polymer is removed from the patient's mouth.

polymer used to make dental impressions

A polymer used for making contact lenses must be sufficiently hydrophilic to allow lubrication of the eye. Such a polymer, therefore, has many OH groups.

$$-CH_2-\underset{\underset{CH_3}{|}}{\underset{C=O}{|}}{CH}-CH_2-\underset{\underset{CH_2OH}{|}}{\underset{\underset{CH_2}{|}}{\underset{C=O}{|}}}{CH}-CH_2-\underset{\underset{CH_3}{|}}{\underset{C=O}{|}}{CH}-CH_2-\underset{\underset{CH_2OH}{|}}{\underset{\underset{CH_2}{|}}{\underset{C=O}{|}}}{CH}-$$

polymer used to make contact lenses

Elastomers

An **elastomer** is a polymer that stretches and then reverts to its original shape. It is a randomly oriented amorphous polymer, but it must have some cross-linking so that the chains do not slip over one another. When elastomers are stretched, the random chains stretch out. The van der Waals forces are not strong enough to maintain them in that arrangement; therefore, when the stretching force is removed, the chains go back to their random shapes. Rubber is an example of an elastomer.

Oriented Polymers

Polymers that are stronger than steel or that conduct electricity almost as well as copper can be made by taking the polymer chains obtained by conventional polymerization, stretching them out, and putting them back together in a parallel fashion (Figure 28.5). Such polymers are called **oriented polymers**. Converting conventional polymers into oriented polymers has been compared to "uncooking" spaghetti. The conventional polymer is disordered cooked spaghetti, whereas the oriented polymer is ordered raw spaghetti.

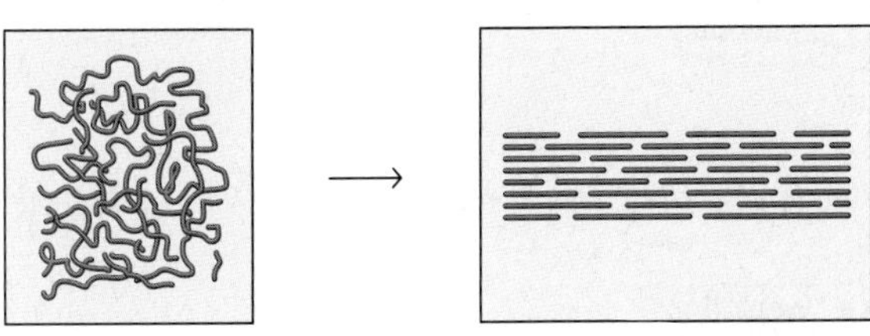

Figure 28.5 The creation of an oriented polymer.

Dyneema®, the strongest commercially available fabric, is an oriented polyethylene polymer. Its molecular weight is 100 times greater than that of high-density polyethylene. It is lighter than Kevlar® and at least 40% stronger. A rope made of Dyneema® can lift almost 119,000 pounds, whereas a steel rope of similar size fails before the weight reaches 13,000 pounds! It is astounding that a chain of carbon atoms can be stretched and properly oriented to produce a material stronger than steel. Dyneema is used to make full-face crash helmets, protective fencing suits, and hang gliders.

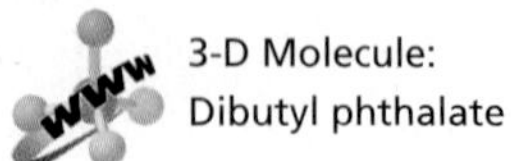

Plasticizers

A plasticizer can be added to a polymer to make it more flexible. A **plasticizer** is an organic compound that dissolves in the polymer, lowering the attractions between the polymer chains, which allows them to slide past one another. Dibutyl phthalate is a commonly used plasticizer. It is added to poly(vinyl chloride)—normally a brittle polymer—to make products such as vinyl raincoats, shower curtains, and garden hoses.

$$C_6H_4(COOCH_2CH_2CH_2CH_3)_2$$

dibutyl phthalate
a plasticizer

An important criterion to consider in choosing a plasticizer is its permanence—how well the plasticizer remains in the polymer. The "new-car smell" appreciated by car owners is the odor of the plasticizer that has vaporized from the vinyl upholstery. When a significant amount of the plasticizer has evaporated, the upholstery becomes brittle and cracks. Phthalates with higher molecular weights and lower vapor pressures than those of dibutyl phthalate are now commonly used for car interiors.

Biodegradable Polymers

Biodegradable polymers are polymers that can be broken into small segments by enzyme-catalyzed reactions. The enzymes are produced by microorganisms. The carbon–carbon bonds of chain-growth polymers are inert to enzyme-catalyzed reactions, so they are nonbiodegradable unless bonds that *can* be broken by enzymes are inserted into the polymer. Then, when the polymer is buried as waste, microorganisms present in the ground can degrade the polymer. One method used to make a polymer biodegradable involves inserting hydrolyzable ester groups into it. For example, if the acetal shown below is added to an alkene undergoing radical polymerization, ester groups will be inserted into the polymer, forming "weak links" that are susceptible to enzyme-catalyzed hydrolysis.

$$-CH_2\dot{C}H(R) + CH_2{=}C(OCH_2CH_2CH_2CH_2O) \longrightarrow -CH_2CH(R)CH_2-\dot{C}(OCH_2CH_2CH_2CH_2O)$$

an acetal

$$\downarrow$$

$$-CH_2CH(R)CH_2-C(=O)-OCH_2CH_2CH_2CH_2CH_2\dot{C}H(R) \xleftarrow{CH_2=CHR} -CH_2CH(R)CH_2-C(=O)-OCH_2CH_2CH_2\dot{C}H_2$$

a "weak link"

Summary

A **polymer** is a large molecule made by linking together repeating units of small molecules called **monomers**. The process of linking them is called **polymerization**. **Polymer chemistry** is part of the larger discipline of **materials science**. Polymers can be divided into two groups: **synthetic polymers**, which are synthesized by scientists, and **biopolymers**, which are synthesized by organisms. Synthetic polymers can be divided into two classes, depending on their method of preparation: **chain-growth polymers** (also called **addition polymers**) and **step-growth polymers** (also known as **condensation polymers**).

Chain-growth polymers are made by **chain reactions**—by the addition of monomers to the end of a growing chain. These reactions take place by one of three mechanisms: **radical polymerization**, **cationic polymerization**, or **anionic polymerization**. Each mechanism has an initiation step that starts the polymerization, propagation steps that allow the chain to grow at the **propagating site**, and termination steps that stop the growth of the chain. The choice of mechanism depends on the structure of the **monomer** and the initiator used to activate the monomer. In radical polymerization, the initiator is a radical; in cationic polymerization, it is an electrophile; and in cationic polymerization, it is a nucleophile. Nonterminated polymer chains are called **living polymers**.

Chain-growth polymerization exhibits a preference for **head-to-tail addition**. Branching affects the physical properties of the polymer because linear unbranched chains can pack together more closely than branched chains can. The substituents are on the same side of the carbon chain in an **isotactic polymer**, alternate on both sides of the chain in a **syndiotactic polymer**, and are randomly oriented in an **atactic polymer**. The structure of a polymer can be controlled with **Ziegler–Natta catalysts**. Natural rubber is a polymer of 2-methyl-1,3-butadiene. Synthetic rubbers have been made by polymerizing dienes other than isoprene. Heating rubber with sulfur to cross-link the chains is called **vulcanization**.

Homopolymers are made of one kind of monomer, whereas **copolymers** are made of more than one kind. In an **alternating copolymer**, the two monomers alternate. In a **block copolymer**, there are blocks of each kind of monomer. In a **random copolymer**, the distribution of monomers is random. A **graft copolymer** contains branches derived from one monomer grafted onto a backbone derived from another.

Step-growth polymers are made by combining two molecules with reactive functional groups at each end. There are two types of step-growth polymers. One type is formed by using a single monomer with two different functional groups, A and B. The other is formed by using two different bifunctional monomers, one containing two A functional groups and the other containing two B functional groups. The formation of step-growth polymers does not involve chain reactions.

Nylon is a **polyamide**. Aromatic polyamides are called **aramides**. Dacron® is a polyester. **Polycarbonates** are polyesters with two ester groups bonded to the same carbon. A **urethane** is a compound that has an ester and an amide group bonded to the same carbon.

Crystallites are highly ordered regions of a polymer. The more crystalline the polymer is, the denser, harder, and more resistant to heat it is. **Thermoplastic polymers** have crystalline and noncrystalline regions. **Thermosetting polymers** have cross-linked polymer chains. The greater the degree of cross-linking, the more rigid is the polymer. An **elastomer** is a plastic that stretches and then reverts to its original shape. A **plasticizer** is an organic compound that dissolves in the polymer and allows the chains to slide past one another. **Biodegradable polymers** can be broken into small segments by enzyme-catalyzed reactions.

Key Terms

addition polymer (p. 1147)
alternating copolymer (p. 1162)
anionic polymerization (p. 1148)
aramide (p. 1164)
atactic polymer (p. 1159)
biodegradable polymer (p. 1170)
biopolymer (p. 1146)
block copolymer (p. 1162)
cationic polymerization (p. 1148)
chain-growth polymer (p. 1147)
chain reaction (p. 1147)
chain transfer (p. 1150)
condensation polymer (p. 1148)
conducting polymer (p. 1159)
copolymer (p. 1162)
cross-linking (p. 1161)
crystallites (p. 1167)
elastomer (p. 1169)
epoxy resin (p. 1166)
graft copolymer (p. 1162)
head-to-tail addition (p. 1151)
homopolymer (p. 1162)
isotactic polymer (p. 1159)
living polymer (p. 1156)
materials science (p. 1147)
monomer (p. 1146)
oriented polymer (p. 1169)
plastic (p. 1147)
plasticizer (p. 1170)
polyamide (p. 1163)
polycarbonate (p. 1165)
polyester (p. 1165)
polymer (p. 1146)
polymer chemistry (p. 1147)
polymerization (p. 1146)
polyurethane (p. 1167)
propagating site (p. 1148)
radical polymerization (p. 1148)
random copolymer (p. 1162)
ring-opening polymerization (p. 1158)
step-growth polymer (p. 1148)
syndiotactic polymer (p. 1159)
synthetic polymer (p. 1146)
thermoplastic polymer (p. 1168)
thermosetting polymer (p. 1168)
urethane (p. 1167)
vinyl polymer (p. 1148)
vulcanization (p. 1161)
Ziegler–Natta catalyst (p. 1159)

Problems

22. Draw short segments of the polymers obtained from the following monomers:

a. $CH_2{=}CHF$

b. $CH_2{=}CHCO_2H$

c. $HO(CH_2)_5\overset{O}{\overset{\|}{C}}OH$

d. $Cl\overset{O}{\overset{\|}{C}}(CH_2)_5\overset{O}{\overset{\|}{C}}Cl + H_2N(CH_2)_5NH_2$

e. 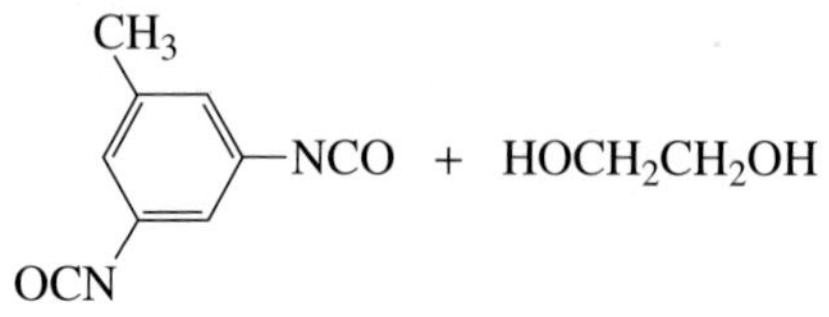 (benzene ring bearing CH_3, $-NCO$, and OCN) $+ HOCH_2CH_2OH$

For each polymer, indicate whether the polymerization is a chain-growth or a step-growth polymerization.

23. Draw the repeating unit of the step-growth polymer that will be formed from each of the following reactions:

a. 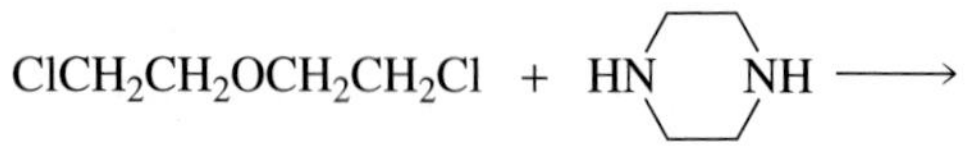 $ClCH_2CH_2OCH_2CH_2Cl + HN\langle\rangle NH \longrightarrow$ (piperazine)

b. 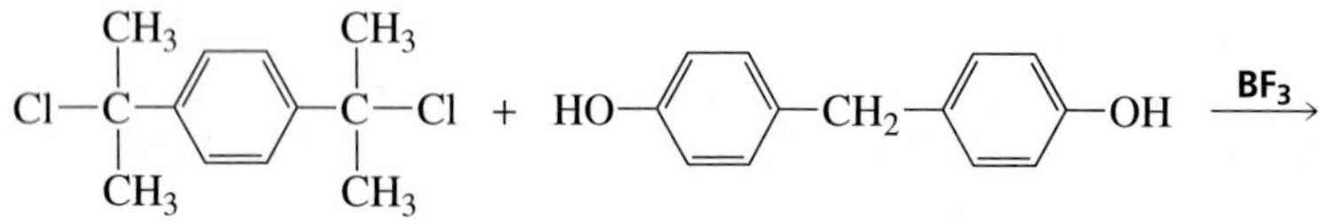 $Cl{-}C(CH_3)_2{-}C_6H_4{-}C(CH_3)_2{-}Cl + HO{-}C_6H_4{-}CH_2{-}C_6H_4{-}OH \xrightarrow{BF_3}$

c. $H_2N{-}C_6H_4{-}OCH_2CH_2CH_2O{-}C_6H_4{-}NH_2 + H\overset{O}{\overset{\|}{C}}{-}\overset{O}{\overset{\|}{C}}H \longrightarrow$

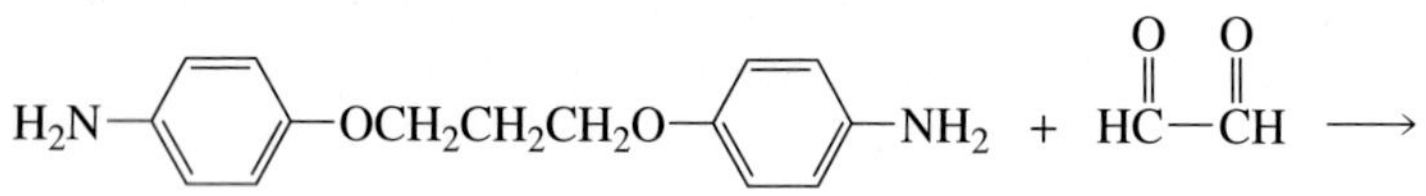

d. 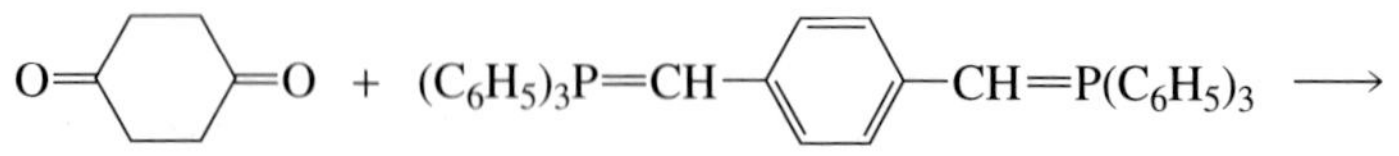 $O{=}C_6H_8{=}O + (C_6H_5)_3P{=}CH{-}C_6H_4{-}CH{=}P(C_6H_5)_3 \longrightarrow$ (cyclohexane-1,4-dione)

24. Draw the structure of the monomer or monomers used to synthesize the following polymers:

a. $-CH_2\underset{CH_2CH_3}{\underset{|}{C}H}-$

b. $-CH_2\underset{CH_3}{\underset{|}{C}H}O-$

c. $-SO_2{-}C_6H_4{-}SO_2NH(CH_2)_6NH-$

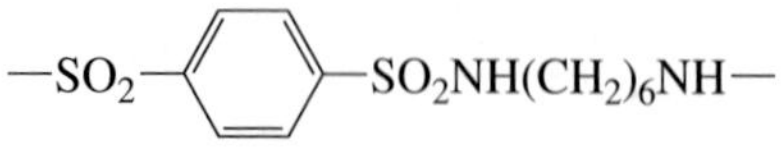

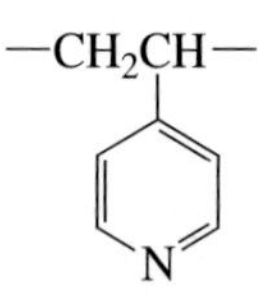

d. $-CH_2CH-$ (CH bearing a 4-pyridyl ring)

e. $-CH_2\overset{CH_3}{\overset{|}{C}}{=}CHCH_2-$

f. $-CH_2CH_2CH_2CH_2\overset{O}{\overset{\|}{C}}O-$

g. 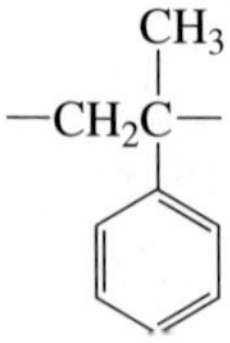 $-CH_2\overset{CH_3}{\overset{|}{C}}-$ (C bearing a phenyl ring)

h. 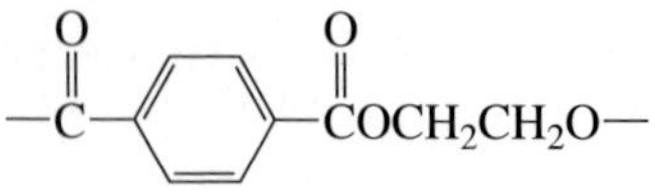 $-\overset{O}{\overset{\|}{C}}{-}C_6H_4{-}\overset{O}{\overset{\|}{C}}OCH_2CH_2O-$

For each polymer, indicate whether it is a chain-growth polymer or a step-growth polymer.

25. Explain why the configuration of a polymer of isobutylene is not isotactic, syndiotactic, or atactic.

26. Draw short segments of the polymers obtained from the following compounds under the given reaction conditions:

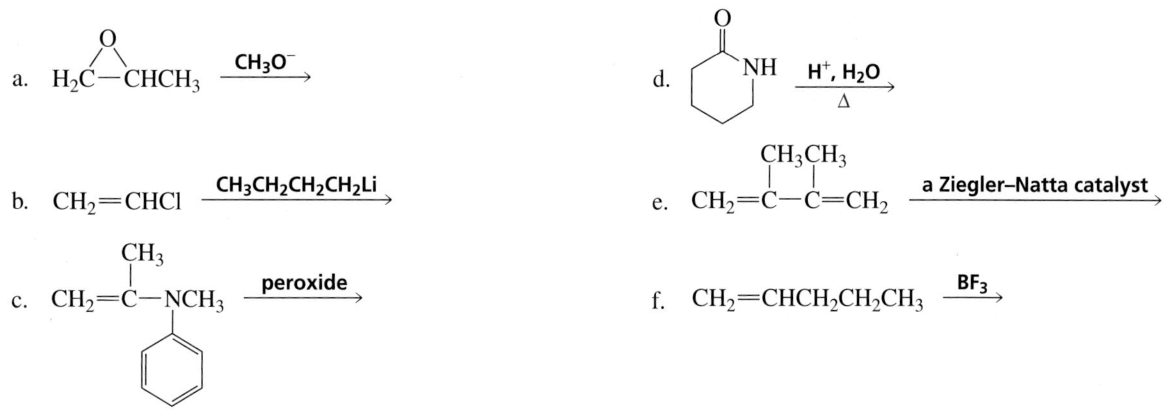

a. $H_2C\overset{O}{—}CHCH_3 \xrightarrow{CH_3O^-}$

b. $CH_2{=}CHCl \xrightarrow{CH_3CH_2CH_2CH_2Li}$

c. $CH_2{=}C(CH_3){-}N(CH_3)C_6H_5 \xrightarrow{peroxide}$

d. (six-membered cyclic lactam, C=O adjacent to NH) $\xrightarrow[\Delta]{H^+,\ H_2O}$

e. $CH_2{=}C(CH_3){-}C(CH_3){=}CH_2 \xrightarrow{\text{a Ziegler–Natta catalyst}}$

f. $CH_2{=}CHCH_2CH_2CH_3 \xrightarrow{BF_3}$

27. Quiana® is a synthetic fabric that feels very much like silk.
a. Is Quiana® a nylon or a polyester?
b. What monomers are used to synthesize Quiana®?

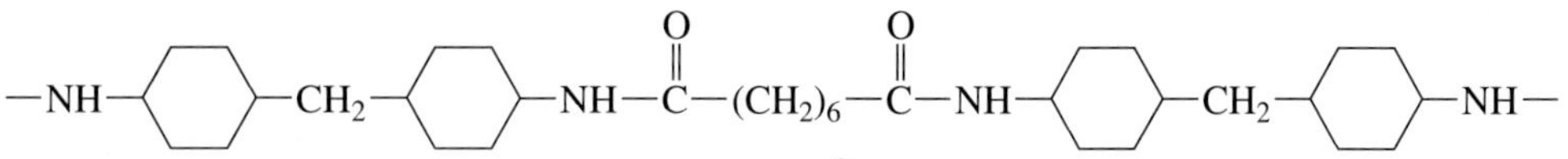

$$—NH—C_6H_{10}—CH_2—C_6H_{10}—NH—\overset{O}{\overset{\|}{C}}—(CH_2)_6—\overset{O}{\overset{\|}{C}}—NH—C_6H_{10}—CH_2—C_6H_{10}—NH—$$

Quiana®

28. Explain why a random copolymer is obtained when 3,3-dimethyl-1-butene undergoes cationic polymerization.

$$CH_2{=}CH—C(CH_3)_2—CH_3 \longrightarrow —CH_2—CH(C(CH_3)_3)—CH_2—CH(CH_3)—C(CH_3)_2—CH_2—CH(CH_3)—C(CH_3)_2—CH_2—CH(C(CH_3)_3)—$$

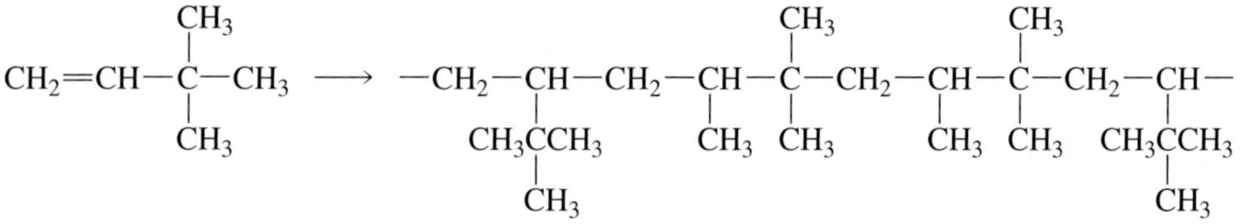

29. Polly Propylene has started two polymerization reactions. One flask contains a monomer that polymerizes by a chain-growth mechanism, and the other flask contains a monomer that polymerizes by a step-growth mechanism. When the reactions are terminated and the contents of the flasks analyzed, one flask contains a high-molecular-weight polymer and some monomer, but very little material of intermediate molecular weight. The other flask contains mainly material of intermediate molecular weight and very little monomer or high-molecular-weight material. Which flask contains which product? Explain.

30. Poly(vinyl alcohol) is a polymer used to make fibers and adhesives. It is synthesized by hydrolysis or alcoholysis of the polymer obtained from polymerization of vinyl acetate.
a. Why is poly(vinyl alcohol) not prepared by polymerizing vinyl alcohol?
b. Is poly(vinyl acetate) a polyester?

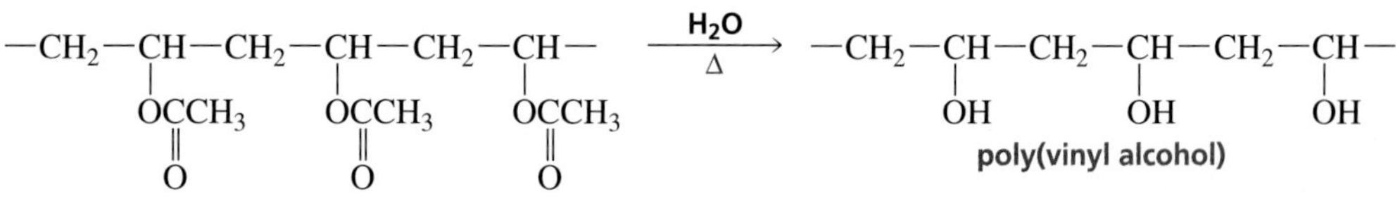

$$—CH_2—CH(OCCH_3)—CH_2—CH(OCCH_3)—CH_2—CH(OCCH_3)— \xrightarrow[\Delta]{H_2O} —CH_2—CH(OH)—CH_2—CH(OH)—CH_2—CH(OH)—$$

(each $OCCH_3$ bears C=O)

poly(vinyl acetate) → **poly(vinyl alcohol)**

31. Five different repeating units are found in the polymer obtained by cationic polymerization of 4-methyl-1-pentene. Identify these repeating units.

32. If a peroxide is added to styrene, the polymer known as polystyrene is formed. If a small amount of 1,4-divinylbenzene is added to the reaction mixture, a stronger and more rigid polymer is formed. Draw a short section of this more rigid polymer.

$$CH_2{=}CH—C_6H_4—CH{=}CH_2$$

1,4-divinylbenzene

33. A particularly strong and rigid polyester used for electronic parts is marketed under the trade name Glyptal®. It is a polymer of terephthalic acid and glycerol. Draw a segment of the polymer, and explain why it is so strong.

34. Draw a short section of the polymer obtained from anionic polymerization of β-propiolactone.

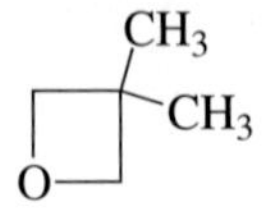

CH_3 CH_3 O

3,3-dimethyloxacyclobutane

35. Which monomer would give a greater yield of polymer, 5-hydroxypentanoic acid or 6-hydroxyhexanoic acid? Explain your choice.

36. When rubber balls and other objects made of natural rubber are exposed to the air for long periods, they turn brittle and crack. This does not happen to objects made of polyethylene. Explain.

37. Why do vinyl raincoats become brittle as they get old, even if they are not exposed to air or to any pollutants?

38. The polymer shown here is synthesized by hydroxide ion–promoted hydrolysis of an alternating copolymer of *para*-nitrophenyl methacrylate and acrylate.
   a. Propose a mechanism for the formation of the alternating copolymer.
   b. Explain why hydrolysis of the copolymer to form the polymer occurs much more rapidly than hydrolysis of *para*-nitrophenyl acetate.

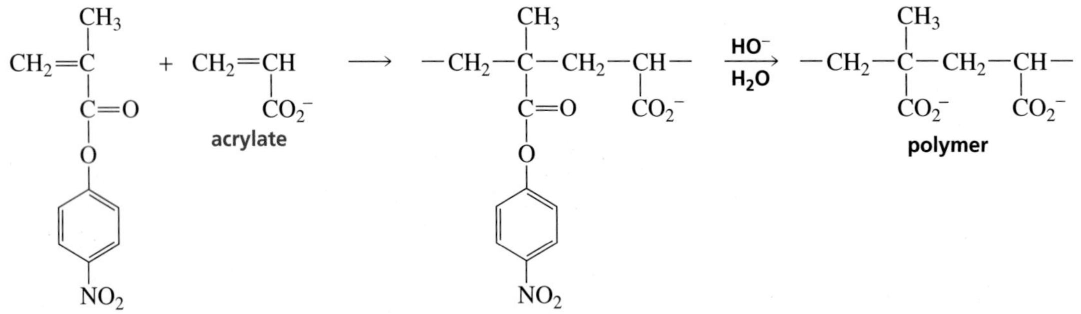

$CH_2{=}C(CH_3)C(=O)O{-}C_6H_4{-}NO_2$ + $CH_2{=}CHCO_2^-$ $\longrightarrow$ $-CH_2{-}C(CH_3)(C(=O)O{-}C_6H_4{-}NO_2){-}CH_2{-}CH(CO_2^-){-}$ $\xrightarrow[H_2O]{HO^-}$ $-CH_2{-}C(CH_3)(CO_2^-){-}CH_2{-}CH(CO_2^-){-}$

para-nitrophenyl methacrylate **acrylate** **alternating copolymer** **polymer**

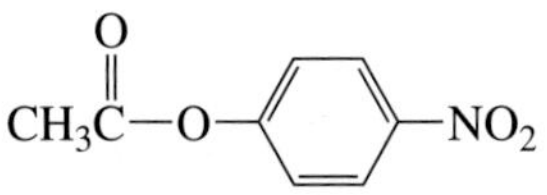

$CH_3C(=O){-}O{-}C_6H_4{-}NO_2$

***para*-nitrophenyl acetate**

39. An alternating copolymer of styrene and vinyl acetate can be turned into a graft copolymer by hydrolyzing it and then adding ethylene oxide. Draw the structure of the graft copolymer.

40. How could head-to-head poly(vinyl bromide) be synthesized?

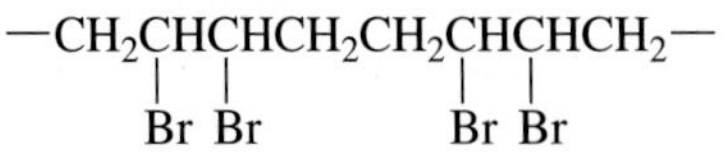

$-CH_2CH(Br)CH(Br)CH_2CH_2CH(Br)CH(Br)CH_2-$

head-to-head poly(vinyl bromide)

41. Delrin® (polyoxymethylene) is a tough self-lubricating polymer used in gear wheels. It is made by polymerizing formaldehyde in the presence of an acid catalyst.
   a. Propose a mechanism for formation of a segment of the polymer.
   b. Is Delrin® a chain-growth polymer or a step-growth polymer?

29 Pericyclic Reactions

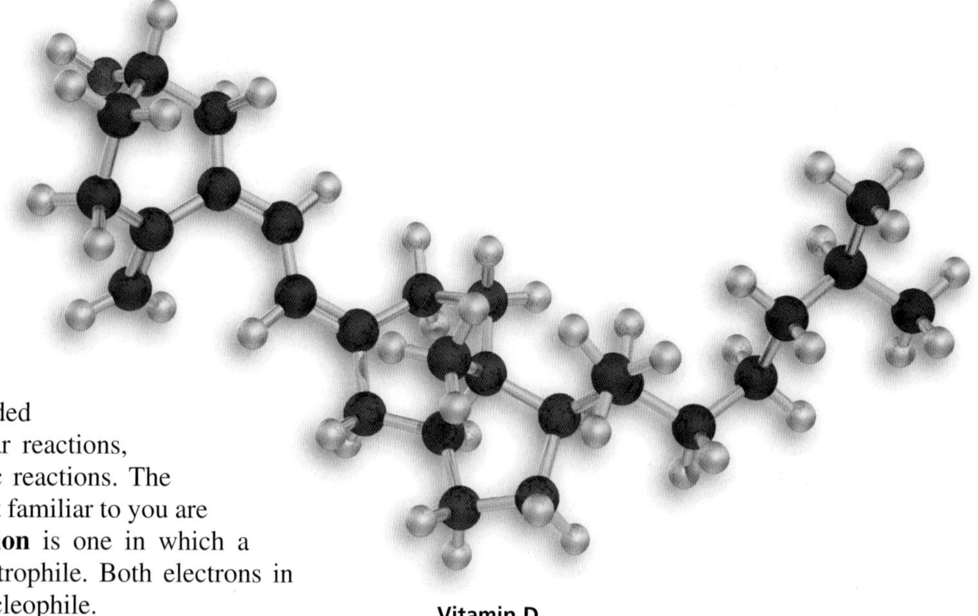
Vitamin D

Reactions of organic compounds can be divided into three classes—polar reactions, radical reactions, and pericyclic reactions. The most common and the ones most familiar to you are polar reactions. A **polar reaction** is one in which a nucleophile reacts with an electrophile. Both electrons in the new bond come from the nucleophile.

a polar reaction

$$H\ddot{\underset{..}{O}}{:}^{-} + \overset{\delta+}{C}H_3-\overset{\delta-}{Br} \longrightarrow CH_3OH + Br^{-}$$

A **radical reaction** is one in which a new bond is formed using one electron from each of the reactants.

a radical reaction

$$CH_3\dot{C}H_2 + Cl-Cl \longrightarrow CH_3CH_2Cl + \cdot Cl$$

A **pericyclic reaction** occurs as a result of reorganizing the electrons in the reactant(s). In this chapter we will look at the three most common types of pericyclic reactions—electrocyclic reactions, cycloaddition reactions, and sigmatropic rearrangements.

29.1 Three Kinds of Pericyclic Reactions

An **electrocyclic reaction** is an intramolecular reaction in which a new σ (sigma) bond is formed between the ends of a conjugated π (pi) system. This reaction is easy to recognize—the product is a *cyclic* compound that has one more ring and one fewer π bond than the reactant.

an electrocyclic reaction

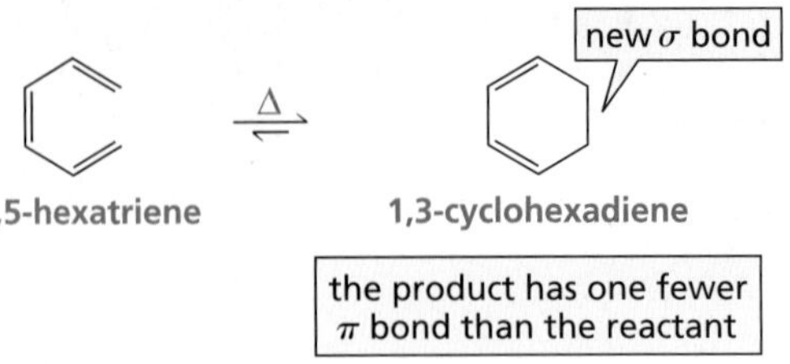

new σ bond

1,3,5-hexatriene ⇌(Δ) 1,3-cyclohexadiene

the product has one fewer π bond than the reactant

Electrocyclic reactions are reversible. In the reverse direction, an electrocyclic reaction is one in which a σ bond in a cyclic reactant breaks, forming a conjugated π system that has one *more* π bond than the cyclic reactant.

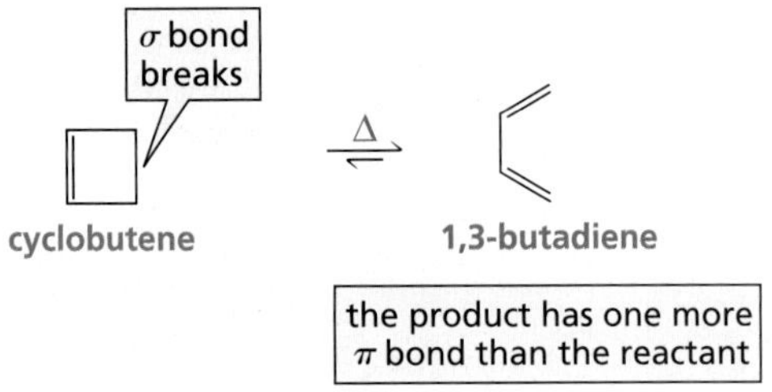

σ bond breaks

cyclobutene ⇌(Δ) 1,3-butadiene

the product has one more π bond than the reactant

In a **cycloaddition reaction**, two different π bond–containing molecules react to form a cyclic compound. Each of the reactants loses a π bond, and the resulting cyclic product has two new σ bonds. The Diels–Alder reaction is a familiar example of a cycloaddition reaction (Section 8.8).

a cycloaddition reaction

1,3-butadiene + ethene ⟶ cyclohexene

new σ bond

new σ bond

the product has two fewer π bonds than the sum of the π bonds in the reactants

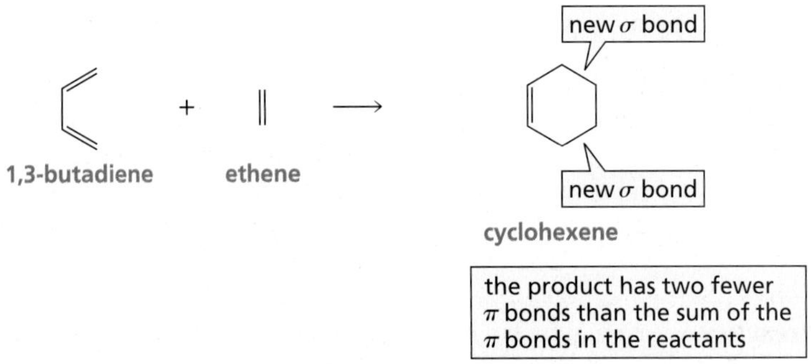

In a **sigmatropic rearrangement**, a σ bond is broken in the reactant, a new σ bond is formed in the product, and the π bonds rearrange. The number of π bonds does not change (the reactant and the product have the same number of π bonds). The σ bond that is broken can be in the middle of the π system or at the end of the π system. The π system consists of the double-bonded carbons and the carbons immediately adjacent to them.

sigmatropic rearrangements

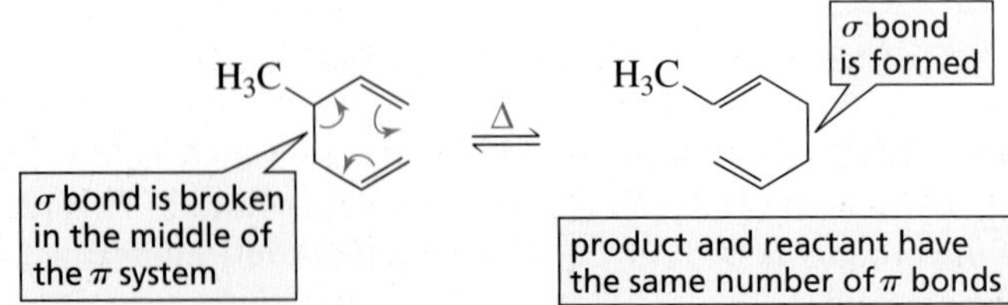
H_3C
Δ
H_3C
σ bond is formed
σ bond is broken in the middle of the π system
product and reactant have the same number of π bonds

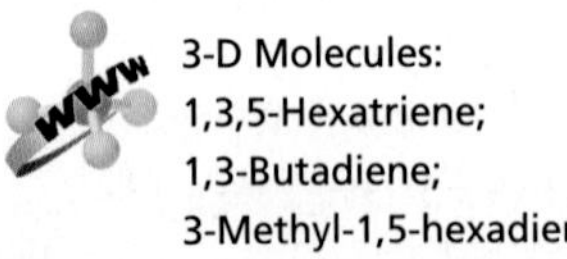
WWW

3-D Molecules: 1,3,5-Hexatriene; 1,3-Butadiene; 3-Methyl-1,5-hexadiene

H_3C CH_2 H → H_3C CH_2 H

σ bond is broken at the end of the π system

σ bond is formed

Notice that electrocyclic reactions and sigmatropic rearrangements occur within a single π system—they are *intra*molecular reactions. In contrast, cycloaddition reactions involve the interaction of two different π systems—they are usually *inter*molecular reactions. The three kinds of pericyclic reactions share the following common features:

- They are all concerted reactions. This means that all the electron reorganization takes place in a single step. Therefore, there is one transition state and no intermediate.
- Because the reactions are concerted, they are highly stereoselective.
- The reactions are generally not affected by catalysts.

We will see that the configuration of the product formed in a pericyclic reaction depends on

- the configuration of the reactant
- the number of conjugated double bonds or pairs of electrons in the reacting system
- whether the reaction is a thermal reaction or a photochemical reaction

A **photochemical reaction** is one that takes place when a reactant absorbs light. A **thermal reaction** takes place *without* the absorption of light. Despite its name, a thermal reaction does not necessarily require more heat than what is available at room temperature. Some thermal reactions do require additional heat in order to take place at a reasonable rate, but others readily occur at, or even below, room temperature.

For many years, pericyclic reactions puzzled chemists. Why did some pericyclic reactions take place only under thermal conditions, whereas others took place only under photochemical conditions, and yet others were successfully carried out under both thermal and photochemical conditions? Another puzzling aspect of pericyclic reactions was the configurations of the products that were formed. After many pericyclic reactions had been investigated, it became apparent that if a pericyclic reaction could take place under both thermal and photochemical conditions, the configuration of the product formed under one set of conditions was different from the configuration of the product formed under the other set of conditions. For example, if the cis isomer was obtained under thermal conditions, the trans isomer was obtained under photochemical conditions and vice versa.

It took two very talented chemists, each bringing his own expertise to the problem, to explain the puzzling behavior of pericyclic reactions. In 1965, R. B. Woodward, an experimentalist, and Roald Hoffmann, a theorist, developed the **conservation of orbital symmetry theory** to explain the relationship among the structure and configuration of the reactant, the conditions (thermal and/or photochemical) under which the reaction takes place, and the configuration of the product. Because the behavior of pericyclic reactions is so precise, it is not surprising that everything about their behavior can be explained by one simple theory. The difficult part was having the insight to arrive at the theory.

The conservation of orbital symmetry theory states that *in-phase orbitals overlap during the course of a pericyclic reaction*. The conservation of orbital symmetry theory was based on the **frontier orbital theory** put forth by Kenichi Fukui in 1954. Although Fukui's theory was more than 10 years old, it had been overlooked because of its mathematical complexity and Fukui's failure to apply it to stereoselective reactions.

Roald Hoffmann *and* **Kenichi Fukui** *shared the 1981 Nobel Prize in chemistry for the conservation of orbital symmetry theory and the frontier orbital theory.* **R. B. Woodward** *did not receive a share in the prize because he died two years before it was awarded, and Alfred Nobel's will stipulates that the prize cannot be awarded posthumously. Woodward, however, had received the 1965 Nobel Prize in chemistry for his work in organic synthesis.*

Roald Hoffmann *was born in Poland in 1937 and came to the United States when he was 12. He received a B.S. from Columbia and a Ph.D. from Harvard. When the conservation of orbital symmetry theory was proposed, he and Woodward were both on the faculty at Harvard. Hoffmann is presently a professor of chemistry at Cornell.*

Kenichi Fukui (1918–1998) *was born in Japan. He was a professor at Kyoto University until 1982 when he became president of the Kyoto Institute of Technology. He was the first Japanese to receive the Nobel Prize in chemistry.*

Robert B. Woodward (1917–1979) *was born in Boston and first became acquainted with chemistry in his home laboratory. He entered MIT at the age of 16 and received a Ph.D. the same year that those who entered MIT with him received their B.A.'s. He went to Harvard as a postdoctoral fellow and remained there for his entire career. Cholesterol, cortisone, strychnine, reserpine (the first tranquilizing drug), chlorophyll, tetracycline, and vitamin B_{12} are just some of the complicated organic molecules Woodward synthesized. He received the Nobel Prize in chemistry in 1965.*

According to the conservation of orbital symmetry theory, whether a compound will undergo a pericyclic reaction under particular conditions *and* what product will be formed both depend on molecular orbital symmetry. To understand pericyclic reactions, therefore, we must now review molecular orbital theory. We will then be able to understand how the symmetry of a molecular orbital controls both the conditions under which a pericyclic reaction takes place and the configuration of the product that is formed.

PROBLEM 1◆

Examine the following pericyclic reactions. For each reaction, indicate whether it is an electrocyclic reaction, a cycloaddition reaction, or a sigmatropic rearrangement.

a. $\xrightarrow{\Delta}$

b. H CH_3 $\xrightarrow{\Delta}$ CH_3

c. CH_2 CH_2 + $CHOCH_3$ ‖ CH_2 $\xrightarrow{\Delta}$ OCH_3

d. $CHCH_3$ CH_2 + $HC{\equiv}CH$ $\xrightarrow{\Delta}$ CH_3

29.2 Molecular Orbitals and Orbital Symmetry

The overlap of *p* atomic orbitals to form π molecular orbitals can be described mathematically using quantum mechanics. The result of the mathematical treatment can be described simply in nonmathematical terms by **molecular orbital (MO) theory**. You were introduced to molecular orbital theory in Sections 1.6 and 7.11. Take a few minutes to review the following key points raised in these sections.

- The two lobes of a *p* orbital have opposite phases. When two in-phase atomic orbitals interact, a covalent bond is formed. When two out-of-phase atomic orbitals interact, a node is created between the two nuclei.
- Electrons fill molecular orbitals according to the same rules—the aufbau principle, the Pauli exclusion principle, Hund's rule—that govern how they fill atomic orbitals: An electron goes into the available molecular orbital with the lowest energy, and only two electrons can occupy a particular molecular orbital (Section 1.2).
- Because the π-bonding portion of a molecule is perpendicular to the framework of the σ bonds, the π bonds can be treated independently. Each carbon atom that forms a π bond has a *p* orbital, and the *p* orbitals of the carbon atoms combine to produce a π molecular orbital. Thus, a molecular orbital can be described by the **linear combination of atomic orbitals (LCAO)**. In a π molecular orbital, each electron that previously occupied a *p* atomic orbital surrounding an individual carbon nucleus now surrounds the entire part of the molecule that is included in the interacting *p* orbitals.

A molecular orbital description of ethene is shown in Figure 29.1. (To show the different phases of the two lobes of a *p* orbital, one phase is represented by a blue lobe and the other phase by a green lobe.)[1] Because ethene has one π bond, it has two *p* atomic orbitals that combine to produce two π molecular orbitals. The in-phase interaction of the two *p* atomic orbitals gives a **bonding π molecular orbital**, designated by ψ_1 (ψ is the Greek letter psi). The bonding molecular orbital is of lower energy than the isolated *p* atomic orbitals. The two *p* atomic orbitals of ethene can also interact out-of-phase. Interaction of out-of-phase orbitals gives an **antibonding π^* molecular orbital,** ψ_2, which is of higher energy than the *p* atomic orbitals. The bonding molecular orbital results from additive interaction of the atomic orbitals, whereas the antibonding molecular orbital results from subtractive interaction. In other words, the interaction of in-phase orbitals holds atoms together, while the interaction of out-of-phase orbitals pushes atoms apart. Because electrons reside in the available molecular orbitals with the lowest energy and two electrons can occupy a molecular orbital, the two π electrons of ethene reside in the bonding π molecular orbital. This molecular orbital picture describes all molecules with one carbon–carbon double bond.

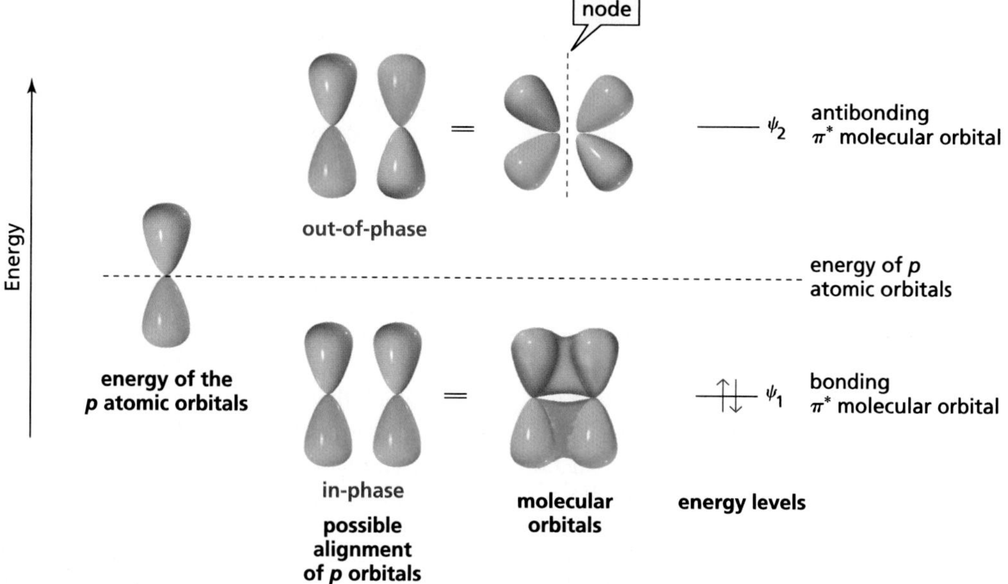

▲ **Figure 29.1**
Interaction of in-phase *p* atomic orbitals gives a bonding π molecular orbital that is lower in energy than the *p* atomic orbitals. Interaction of out-of-phase π atomic orbitals gives an antibonding π^* molecular orbital that is higher in energy than the *p* atomic orbitals.

1,3-Butadiene has two conjugated π bonds, so it has four *p* atomic orbitals (Figure 29.2). Four atomic orbitals can combine linearly in four different ways. Consequently, there are four π molecular orbitals: ψ_1, ψ_2, ψ_3, and ψ_4. Notice that orbitals are conserved: Four atomic orbitals combine to produce four molecular orbitals. Half are bonding molecular orbitals (ψ_1 and ψ_2) and the other half are antibonding molecular orbitals (ψ_3 and ψ_4). Also notice that the bonding MOs are lower in energy and the antibonding MOs higher in energy than the *p* atomic orbitals. Because the four π electrons will reside in the available molecular orbitals with the lowest energy, two electrons are in ψ_1 and two are in ψ_2. Remember that although the molecular orbitals have different energies, they all coexist. This molecular orbital picture describes all molecules with two conjugated carbon–carbon double bonds.

Orbitals are conserved—two atomic orbitals combine to produce two molecular orbitals, four atomic orbitals combine to produce four molecular orbitals, six atomic orbitals combine to produce six molecular orbitals, etc.

[1]Because the different phases of the *p* orbital result from the different mathematical signs (+ and −) of the wave function of the electron, some chemists represent the different phases by a (+) and a (−).

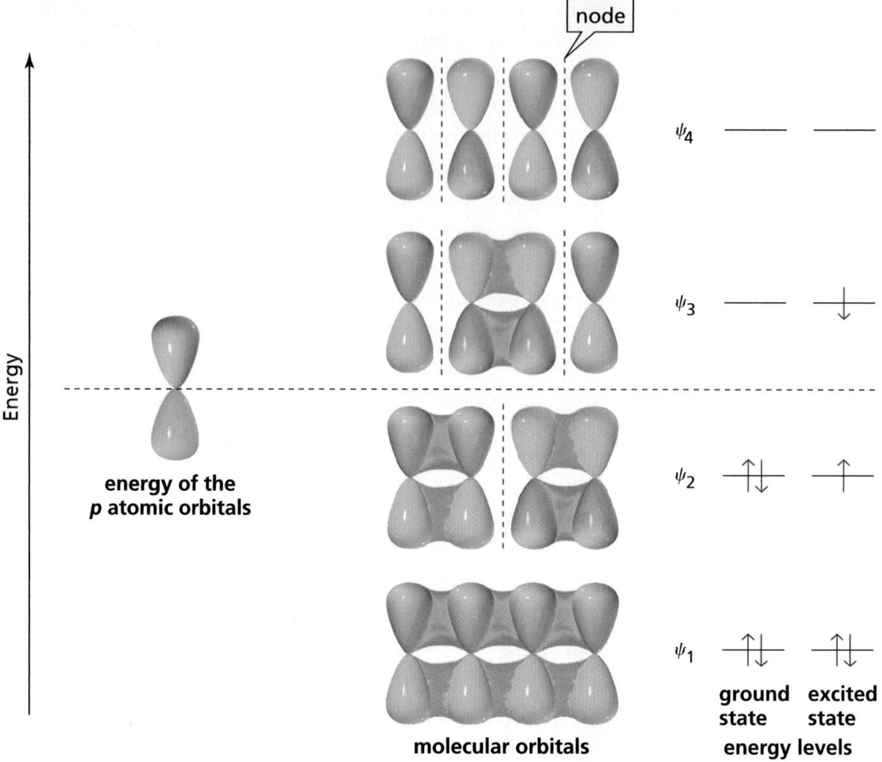

▲ **Figure 29.2**
Four p atomic orbitals interact to give the four π molecular orbitals of 1,3-butadiene.

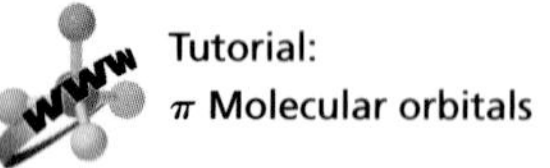

If you examine the interacting orbitals in Figure 29.2, you will see that in-phase orbitals interact to give a bonding interaction and out-of-phase orbitals interact to create a node. Recall that a node is a place in which there is zero probability of finding an electron (Section 1.5). You will also see that as the energy of the molecular orbital increases, the number of bonding interactions decreases and the number of nodes *between* the nuclei increases. For example, ψ_1 has three bonding interactions and zero nodes between the nuclei, ψ_2 has two bonding interactions and one node between the nuclei, ψ_3 has one bonding interaction and two nodes between the nuclei, and ψ_4 has zero bonding interactions and three nodes between the nuclei. *Notice that a molecular orbital is bonding if the number of bonding interactions is greater than the number of nodes between the nuclei, and a molecular orbital is antibonding if the number of bonding interactions is fewer than the number of nodes between the nuclei.*

The normal electronic configuration of a molecule is known as its **ground state**. In the ground state of 1,3-butadiene, the *highest occupied molecular orbital (HOMO)* is ψ_2, and the *lowest unoccupied molecular orbital (LUMO)* is ψ_3. If a molecule absorbs light of an appropriate wavelength, the light will promote an electron from its ground-state HOMO to its LUMO (from ψ_2 to ψ_3). The molecule is then in an **excited state.** In the excited state, the HOMO is ψ_3 and the LUMO is ψ_4. *In a thermal reaction the reactant is in its ground state; in a photochemical reaction the reactant is in an excited state.*

Some molecular orbitals are *symmetric* and some are *asymmetric* (also called *dissymetric*), and they are easy to distinguish. If the p orbitals at the ends of the molecular orbital are in-phase (both have blue lobes on the top and green lobes on the bottom), the molecular orbital is symmetric. If the two end p orbitals are out-of-phase, the molecular orbital is asymmetric. In Figure 29.2, ψ_1 and ψ_3 are **symmetric molecular orbitals** and ψ_2 and ψ_4 are **asymmetric molecular orbitals**. Notice that as the molecular orbitals increase in energy, they alternate in being symmetric and asymmetric. Therefore, *the ground-state HOMO and the excited-state HOMO always have opposite symmetries—one is symmetric and the other is asymmetric.* A molecular orbital description of 1,3,5-hexatriene, a compound with three conjugated double bonds, is shown in Figure 29.3. As a review, examine the figure and note

The ground-state HOMO and the excited-state HOMO have opposite symmetries.

- the distribution of electrons in the ground and excited states
- that the number of bonding interactions decreases and the number of nodes increases as the molecular orbitals increase in energy
- that the molecular orbitals alternate from symmetric to asymmetric as the molecular orbitals increase in energy

Although the chemistry of a compound is determined by all its molecular orbitals, we can learn a great deal about the chemistry of a compound by looking at only the **highest occupied molecular orbital (HOMO)** and the **lowest unoccupied molecular orbital (LUMO)**. These two molecular orbitals are known as the **frontier orbitals**. We will now see that simply by evaluating *one* of the frontier molecular orbitals of the reactant(s) in a pericyclic reaction, we can predict the conditions under which the reaction will occur (thermal or photochemical, or both) and the products that will be formed.

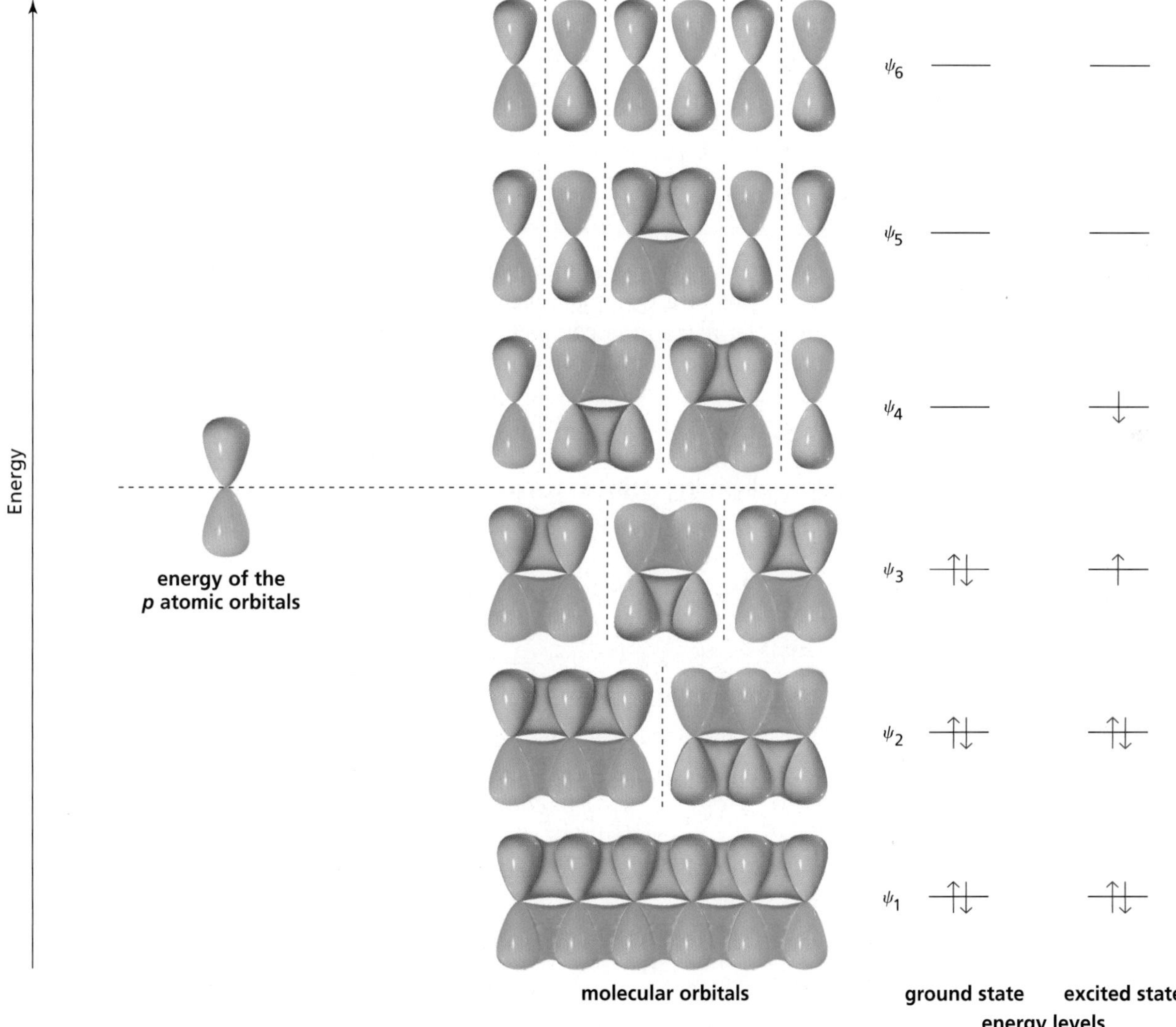

▲ **Figure 29.3**
Six *p* atomic orbitals interact to give the six π molecular orbitals of 1,3,5-hexatriene.

PROBLEM 2◆

Answer the following questions for the π molecular orbitals of 1,3,5-hexatriene:

a. Which are the bonding orbitals, and which are the antibonding orbitals?
b. Which orbitals are the HOMO and the LUMO in the ground state?
c. Which orbitals are the HOMO and the LUMO in the excited state?
d. Which orbitals are symmetric, and which are asymmetric?
e. What is the relationship between HOMO and LUMO and symmetric and asymmetric orbitals?

PROBLEM 3◆

a. How many π molecular orbitals does 1,3,5,7-octatetraene have?
b. What is the designation of its HOMO (ψ_1, ψ_2, etc.)?
c. How many nodes does its highest-energy π molecular orbital have between the nuclei?

PROBLEM 4

Give a molecular orbital description for each of the following:

a. 1,3-pentadiene
b. 1,4-pentadiene
c. 1,3,5-heptatriene
d. 1,3,5,8-nonatetraene

29.3 Electrocyclic Reactions

An *electrocyclic reaction* is an intramolecular reaction in which the rearrangement of π electrons leads to a cyclic product that has one fewer π bond than the reactant. An electrocyclic reaction is completely stereoselective—it preferentially forms one stereoisomer; an electrocyclic reaction is also stereoselective. For example, when (2*E*,4*Z*,6*E*)-octatriene undergoes an electrocyclic reaction under thermal conditions, only the cis product is formed; when (2*E*,4*Z*,6*Z*)-octatriene undergoes an electrocyclic reaction under thermal conditions, only the trans product is formed. Recall that *E* means the high-priority groups are on opposite sides of the double bond, and *Z* means the high-priority groups are on the same side of the double bond (Section 3.5).

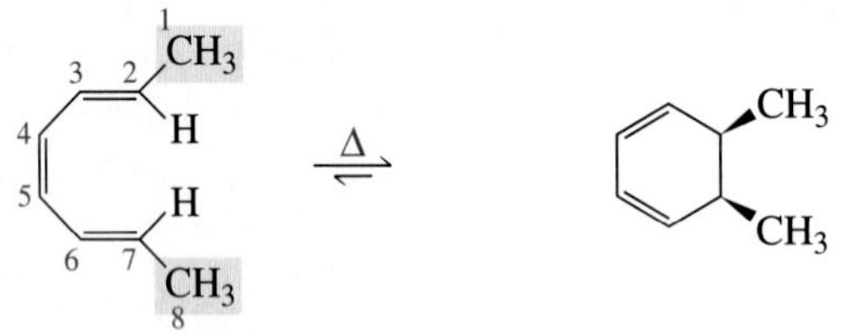

1 CH_3 3 2 H 4 5 H 6 7 CH_3 8 $\xrightleftharpoons{\Delta}$ CH_3 CH_3

(2*E*,4*Z*,6*E*)-octatriene **cis-5,6-dimethyl-1,3-cyclohexadiene**

CH_3 H CH_3 H $\xrightleftharpoons{\Delta}$ CH_3 CH_3

(2*E*,4*Z*,6*Z*)-octatriene ***trans*-5,6-dimethyl-1,3-cyclohexadiene**

However, when the reactions are carried out under photochemical conditions, the products have opposite configurations: The compound that forms the cis isomer under thermal conditions forms the trans isomer under photochemical conditions, and the compound that forms the trans isomer under thermal conditions forms the cis isomer under photochemical conditions.

CH_3 H H CH_3 $\underset{}{\overset{h\nu}{\rightleftharpoons}}$ CH_3 CH_3

(2*E*,4*Z*,6*E*)-octatriene **trans-5,6-dimethyl-1,3-cyclohexadiene**

CH_3 H CH_3 H $\underset{}{\overset{h\nu}{\rightleftharpoons}}$ CH_3 CH_3

(2*E*,4*Z*,6*Z*)-octatriene ***cis*-5,6-dimethyl-1,3-cyclohexadiene**

Under thermal conditions, (2*E*,4*Z*)-hexadiene cyclizes to *cis*-3,4-dimethylcyclobutene, and (2*E*,4*E*)-hexadiene cyclizes to *trans*-3,4-dimethylcyclobutene.

1 CH_3 2 3 H 4 CH_3 5 6 H $\underset{}{\overset{\Delta}{\rightleftharpoons}}$ CH_3 H CH_3 H

(2*E*,4*Z*)-hexadiene ***cis*-3,4-dimethylcyclobutene**

CH_3 H H CH_3 $\underset{}{\overset{\Delta}{\rightleftharpoons}}$ CH_3 H H CH_3

(2*E*,4*E*)-hexadiene ***trans*-3,4-dimethylcyclobutene**

As we saw with the octatrienes, the configuration of the product changes if the reactions are carried out under photochemical conditions: The trans isomer is obtained from (2*E*,4*Z*)-hexadiene instead of the cis isomer; the cis isomer is obtained from (2*E*,4*E*)-hexadiene instead of the trans isomer.

CH_3 H CH_3 H $\underset{}{\overset{h\nu}{\rightleftharpoons}}$ H CH_3 CH_3 H

(2*E*,4*Z*)-hexadiene ***trans*-3,4-dimethylcyclobutene**

CH_3 H H CH_3 $\underset{}{\overset{h\nu}{\rightleftharpoons}}$ CH_3 H CH_3 H

(2*E*,4*E*)-hexadiene ***cis*-3,4-dimethylcyclobutene**

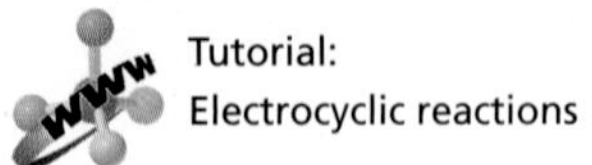

Electrocyclic reactions are reversible. The cyclic compound is favored for electrocyclic reactions that form six-membered rings, whereas the open-chain compound is favored for electrocyclic reactions that form four-membered rings because of the angle strain and torsional strain associated with four-membered rings (Section 2.11).

Now we will use what we have learned about molecular orbitals to explain the configuration of the products of the previous reactions. We will then be able to predict the configuration of the product of any other electrocyclic reaction.

The product of an electrocyclic reaction results from the formation of a new σ bond. For this bond to form, the p orbitals at the ends of the conjugated system must rotate so they overlap head-to-head (and rehybridize to sp^3). Rotation can occur in two ways. If both orbitals rotate in the same direction (both clockwise or both counterclockwise), ring closure is **conrotatory.**

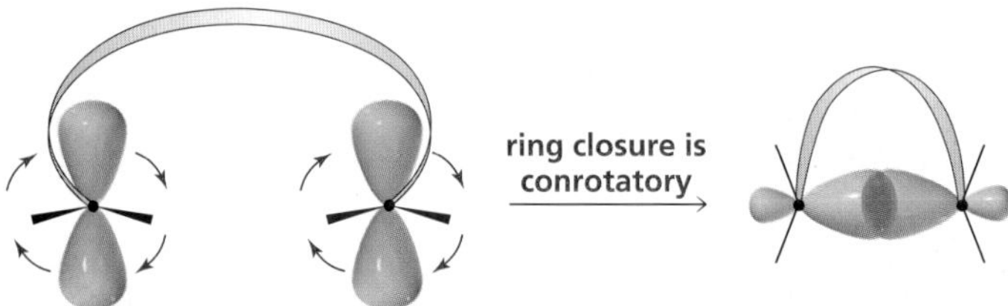

If the orbitals rotate in opposite directions (one clockwise, the other counterclockwise), ring closure is **disrotatory**.

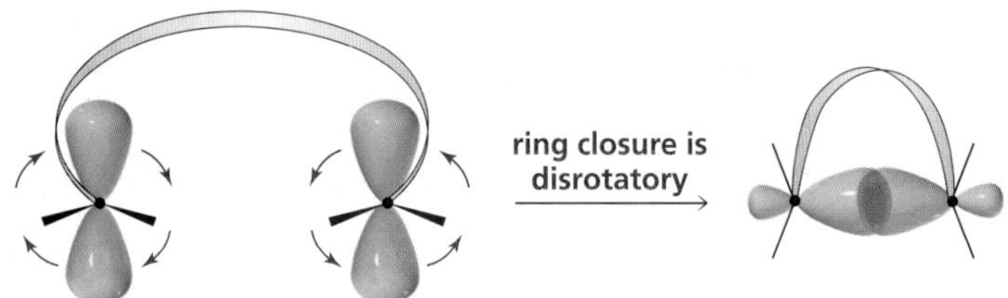

The mode of ring closure depends on the symmetry of the HOMO of the compound undergoing ring closure. Only the symmetry of the HOMO is important in determining the course of the reaction because this is where the highest energy electrons are. These are the most loosely held electrons and therefore the ones most easily moved during a reaction.

To form the new σ bond, the orbitals must rotate so that in-phase p orbitals overlap, because in-phase overlap is a bonding interaction. Out-of-phase overlap would be an antibonding interaction. If the HOMO is symmetric (the end orbitals are identical), rotation will have to be disrotatory to achieve in-phase overlap. In other words, disrotatory ring closure is symmetry-allowed, whereas conrotatory ring closure is symmetry-forbidden.

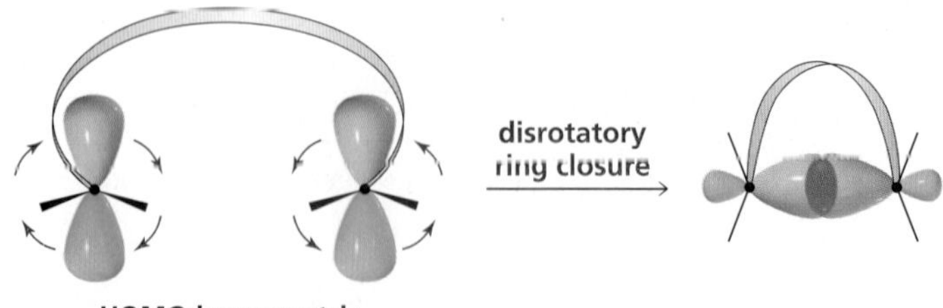

If the HOMO is asymmetric, rotation has to be conrotatory in order to achieve in-phase overlap. In other words, conrotatory ring closure is symmetry-allowed, whereas disrotatory ring closure is symmetry-forbidden.

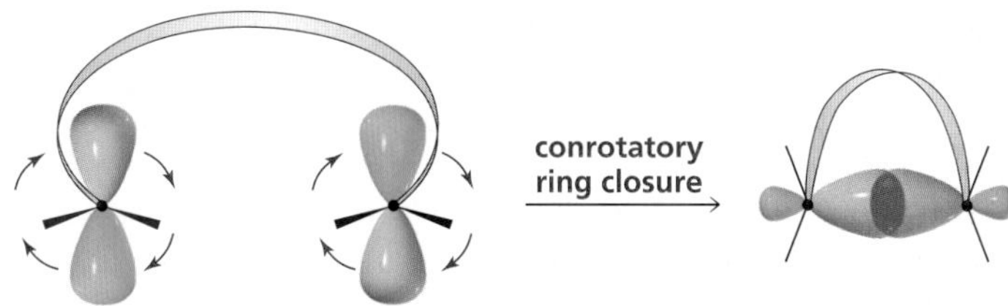

HOMO is asymmetric

Notice that a **symmetry-allowed pathway** is one in which in-phase orbitals overlap; a **symmetry-forbidden pathway** is one in which out-of-phase orbitals would overlap. A symmetry-allowed reaction can take place under relatively mild conditions. If a reaction is symmetry-forbidden, it cannot take place by a concerted pathway. If a symmetry-forbidden reaction takes place at all, it must do so by a nonconcerted mechanism.

A symmetry-allowed pathway requires in-phase orbital overlap.

Now we are ready to learn why the electrocyclic reactions discussed at the beginning of this section form the indicated products, and why the configuration of the product changes if the reaction is carried out under photochemical conditions.

The ground-state HOMO (ψ_3) of a compound with three conjugated π bonds, such as (2*E*,4*Z*,6*E*)-octatriene, is symmetric (Figure 29.3). This means that ring closure under *thermal conditions* is disrotatory. In disrotatory ring closure of (2*E*,4*Z*,6*E*)-octatriene, the methyl groups are both pushed up (or down), which results in formation of the cis product.

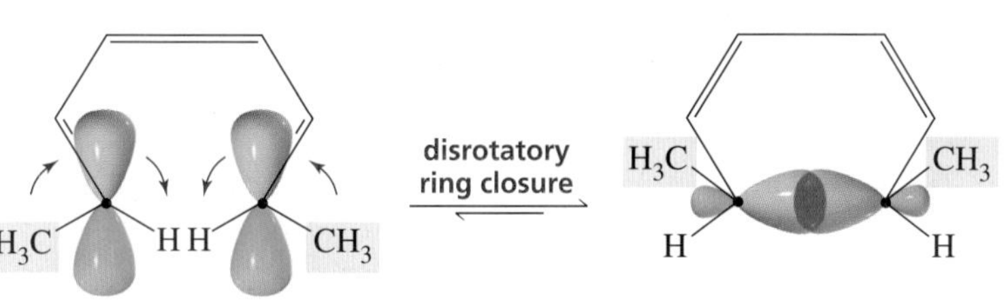

(2*E*,4*Z*,6*E*)-octatriene   *cis*-5,6-dimethyl-1,3-cyclohexadiene

In disrotatory ring closure of (2*E*,4*Z*,6*Z*)-octatriene, one methyl group is pushed up and the other is pushed down, which results in formation of the trans product.

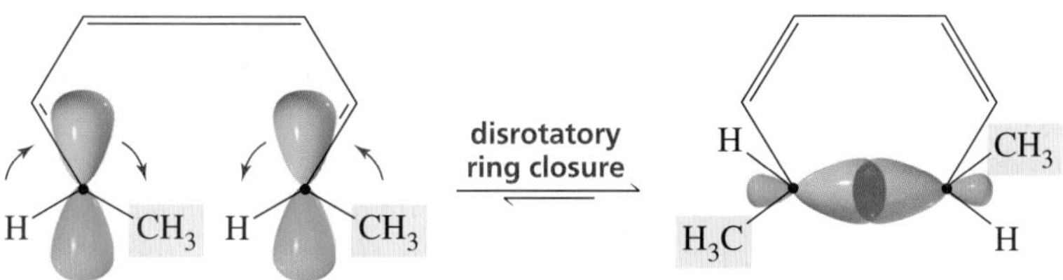

(2*E*,4*Z*,6*Z*)-octatriene   *trans*-5,6-dimethyl-1,3-cyclohexadiene

If the reaction takes place under *photochemical conditions,* we must consider the excited-state HOMO rather than the ground-state HOMO. The excited-state HOMO (ψ_4) of a compound with three π bonds is asymmetric (Figure 29.3). Therefore, under photochemical conditions, (2*E*,4*Z*,6*Z*)-octatriene undergoes conrotatory ring closure, so both methyl groups are pushed down (or up) and the cis product is formed.

The symmetry of the HOMO of the compound undergoing ring closure controls the stereochemical outcome of an electrocyclic reaction.

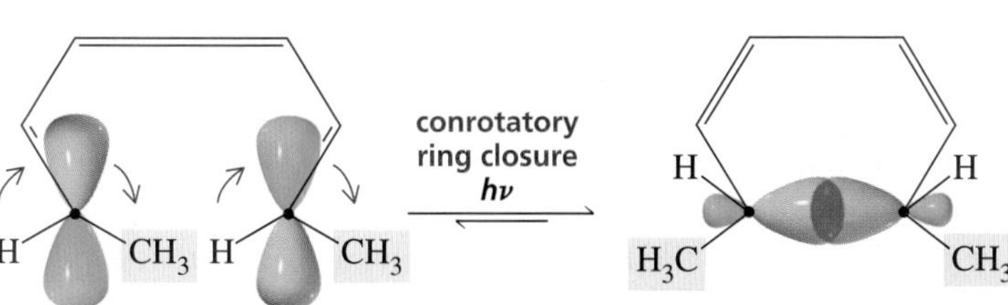

(2*E*,4*Z*,6*Z*)-octatriene   *cis*-5,6-dimethyl-1,3-cyclohexadiene

We have just seen why the configuration of the product formed under photochemical conditions is the opposite of the configuration of the product formed under thermal conditions: The ground-state HOMO is symmetric—so disrotatory ring closure occurs, whereas the excited-state HOMO is asymmetric—so conrotatory ring closure occurs. Thus, the stereochemical outcome of an electrocyclic reaction depends on the symmetry of the HOMO of the compound undergoing ring closure.

Now let's see why ring closure of (2*E*,4*Z*)-hexadiene forms *cis*-3,4-dimethylcyclobutene. The compound undergoing ring closure has two conjugated π bonds. The ground-state HOMO of a compound with two conjugated π bonds is asymmetric (Figure 29.2), so ring closure is conrotatory. Conrotatory ring closure of (2*E*,4*Z*)-hexadiene leads to the cis product.

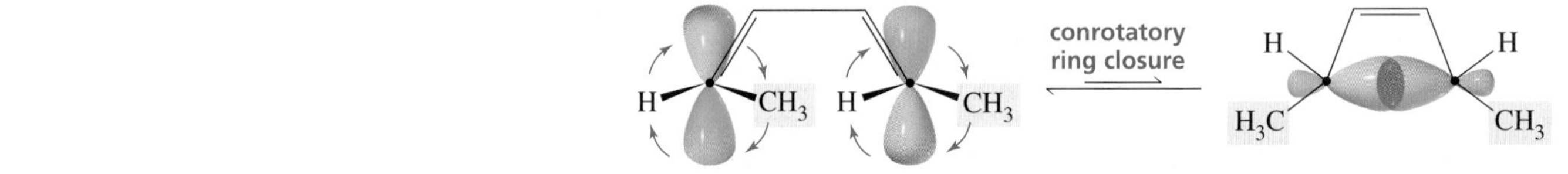

conrotatory ring closure

H, CH_3, H, CH_3 — H, H, H_3C, CH_3

(2*E*,4*Z*)-hexadiene **cis-3,4-dimethylcyclobutene**

Similarly, conrotatory ring closure of (2*E*,4*E*)-hexadiene leads to the trans product.

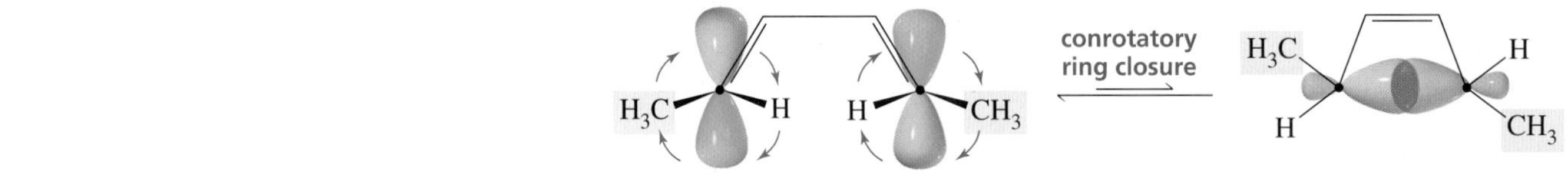

conrotatory ring closure

H_3C, H, H, CH_3 — H_3C, H, H, CH_3

(2*E*,4*E*)-hexadiene ***trans*-3,4-dimethylcyclobutene**

If the reaction is carried out under photochemical conditions, however, the excited-state HOMO of a compound with two conjugated π bonds is symmetric. (Recall that the ground-state HOMO and the excited-state HOMO have opposite symmetries.) So (2*E*,4*Z*)-hexadiene will undergo disrotatory ring closure, resulting in the trans product, whereas (2*E*,4*E*)-hexadiene will undergo disrotatory ring closure and form the cis product.

We have seen that the ground-state HOMO of a compound with two conjugated double bonds is asymmetric, whereas the ground-state HOMO of a compound with three conjugated double bonds is symmetric. If we examine molecular orbital diagrams for compounds with four, five, six, and more conjugated double bonds, we can conclude that *the ground-state HOMO of a compound with an even number of conjugated double bonds is asymmetric, whereas the ground-state HOMO of a compound with an odd number of conjugated double bonds is symmetric.* Therefore, from the number of conjugated double bonds in a compound, we can immediately tell whether ring closure will be conrotatory (an even number of conjugated double bonds) or disrotatory (an odd number of conjugated double bonds) under thermal conditions. However, if the reaction takes place under photochemical conditions, everything is reversed since the ground-state and excited-state HOMOs have opposite symmetries; if the ground-state HOMO is symmetric, the excited-state HOMO is asymmetric and vice versa.

The ground-state HOMO of a compound with an even number of conjugated double bonds is asymmetric.

The ground-state HOMO of a compound with an odd number of conjugated double bonds is symmetric.

We have seen that the stereochemistry of an electrocyclic reaction depends on the mode of ring closure, and the mode of ring closure depends on the number of conjugated π bonds in the reactant *and* on whether the reaction is carried out under thermal or photochemical conditions. What we have learned about electrocyclic reactions can be summarized by the **selection rules** listed in Table 29.1. These are also known as the **Woodward–Hoffmann rules** for electrocyclic reactions.

The rules in Table 29.1 are for determining whether a given electrocyclic reaction is "allowed by orbital symmetry." There are also selection rules to determine whether cycloaddition reactions (Table 29.3) and sigmatropic rearrangements (Table 29.4) are

Table 29.1 Woodward–Hoffmann Rules for Electrocyclic Reactions

Number of conjugated π bonds	Reaction conditions	Allowed mode of ring closure
Even number	Thermal	Conrotatory
	Photochemical	Disrotatory
Odd number	Thermal	Disrotatory
	Photochemical	Conrotatory

"allowed by orbital symmetry." It can be rather burdensome to memorize these rules (and worrisome if they are forgotten during an exam), but all the rules can be summarized by the mnemonic "TE-AC." How to use "TE-AC" is explained in Section 29.7.

PROBLEM 5

a. For conjugated systems with two, three, four, five, six, and seven conjugated π bonds, construct quick molecular orbitals (just draw the *p* orbitals at the ends of the conjugated system as they are drawn on pp. 1185–1186) to show whether the HOMO is symmetric or asymmetric).

b. Using these drawings, convince yourself that the Woodward–Hoffmann rules in Table 29.1 are valid.

c. Using these drawings, convince yourself that the "TE-AC" shortcut method for learning the information in Table 29.1 is valid (see Section 29.7).

PROBLEM 6◆

a. Under thermal conditions, will ring closure of (2*E*,4*Z*,6*Z*,8*E*)-decatetraene be conrotatory or disrotatory?

b. Will the product have the cis or the trans configuration?

c. Under photochemical conditions, will ring closure be conrotatory or disrotatory?

d. Will the product have the cis or the trans configuration?

The series of reactions in Figure 29.4 illustrates just how easy it is to determine the mode of ring closure and therefore the product of an electrocyclic reaction. The reactant of the first reaction has three conjugated double bonds and is undergoing ring closure under thermal conditions. Ring closure, therefore, is disrotatory (Table 29.1). Disrotatory ring closure of this reactant causes the hydrogens to be cis in the ring-closed product. To determine the relative positions of the hydrogens, draw them in the reactant and then draw arrows showing disrotatory ring closure (Figure 29.4a).

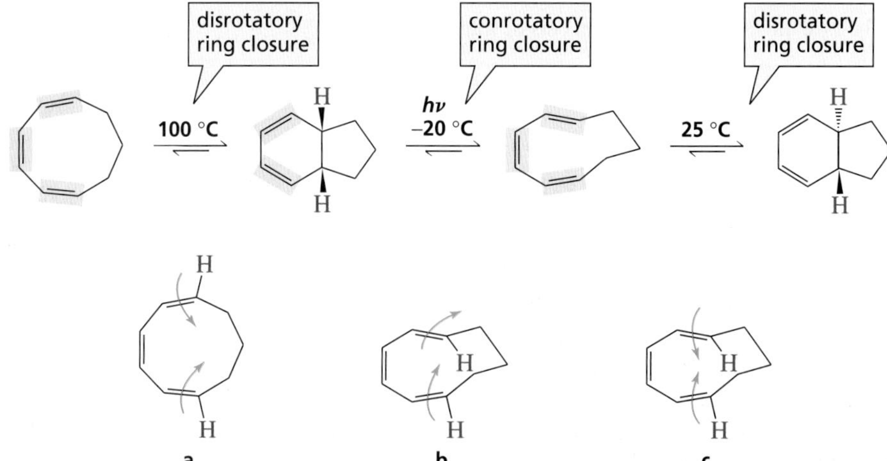

◄ **Figure 29.4**
Determining the stereochemistry of the product of an electrocyclic reaction.

The second step is a ring-opening electrocyclic reaction that takes place under photochemical conditions. Because of the principle of microscopic reversibility (Section 15.12), the orbital symmetry rules used for a ring-closure reaction also apply to the reverse ring-opening reaction. The compound undergoing reversible ring closure has three conjugated double bonds. Because the reaction occurs under photochemical conditions, ring opening (and ring closure) is conrotatory. (Notice that the number of conjugated double bonds used to determine the mode of ring opening/closure in reversible electrocyclic reactions is the number in the compound undergoing ring closure.) If conrotatory rotation is to result in a product with cis hydrogens, the hydrogens in the compound undergoing ring closure must point in the same direction (Figure 29.4b).

The third step is a thermal ring closure of a compound with three conjugated double bonds, so ring closure is disrotatory. Drawing the hydrogens and the arrows (Figure 29.4c) allows you to determine the relative positions of the hydrogens in the ring-closed product.

Notice that in all these electrocyclic reactions, if the bonds to the substituents in the reactant point in *opposite directions* (as in Figure 29.4a), the substituents will be cis in the product if ring closure is disrotatory and trans if ring closure is conrotatory. On the other hand, if they point in the *same direction* (as in Figure 29.4b or 29.4c), they will be trans in the product if ring closure is disrotatory and cis if ring closure is conrotatory (Table 29.2).

Table 29.2 Configuration of the Product of an Electrocyclic Reaction

Substituents in the reactant	Mode of ring closure	Configuration of the product
Point in opposite directions	Disrotatory	cis
	Conrotatory	trans
Point in the same direction	Disrotatory	trans
	Conrotatory	cis

PROBLEM 7◆

Which of the following are correct? Correct any false statements.

a. A conjugated diene with an even number of double bonds undergoes conrotatory ring closure under thermal conditions.

b. A conjugated diene with an asymmetric HOMO undergoes conrotatory ring closure under thermal conditions.

c. A conjugated diene with an odd number of double bonds has a symmetric HOMO.

PROBLEM 8◆

a. Identify the mode of ring closure for each of the following electrocyclic reactions.

b. Are the indicated hydrogens cis or trans?

1. [cyclooctatetraene-type ring] $\xrightarrow{\Delta}$ [bicyclic product with H, H]

2. [cyclic triene] $\xrightarrow{\Delta}$ [bicyclic product with H, H]

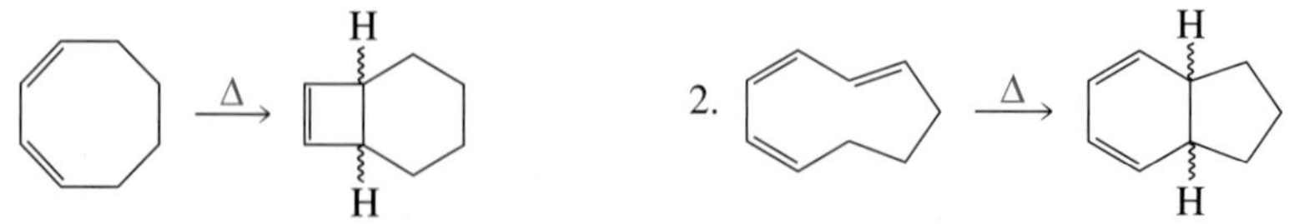

29.4 Cycloaddition Reactions

In a *cycloaddition reaction,* two different π bond–containing molecules react to form a cyclic molecule by rearranging the π electrons and forming two new σ bonds. The Diels–Alder reaction is one of the best known examples of a cycloaddition reaction (Section 8.8). Cycloaddition reactions are classified according to the number of π electrons that interact in the reaction. The Diels–Alder reaction is a [4 + 2] cycloaddition reaction because one reactant has four interacting π electrons and the other reactant has two interacting π electrons. Only the π electrons participating in electron rearrangement are counted.

[4 + 2] cycloaddition (a Diels–Alder reaction)

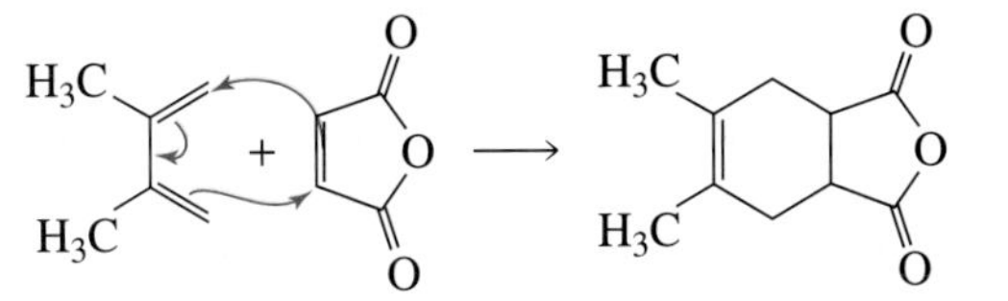

[2 + 2] cycloaddition

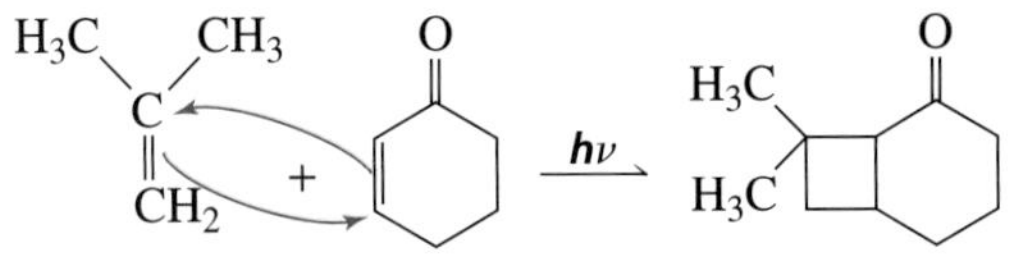

[8 + 2] cycloaddition

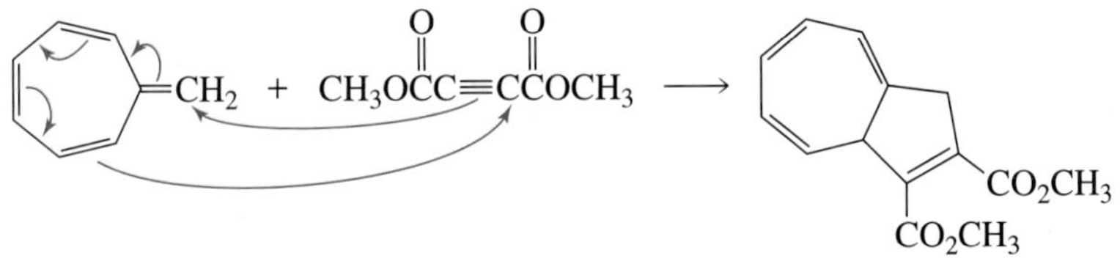

In a cycloaddition reaction, the orbitals of one molecule must overlap with the orbitals of the second molecule. Therefore, the frontier molecular orbitals of both reactants must be evaluated to determine the outcome of the reaction. Because the new σ bonds in the product are formed by donation of electron density from one reactant to the other reactant, we must consider the HOMO of one of the molecules and the LUMO of the other because only an empty orbital can accept electrons. It does not matter which reacting molecule's HOMO is used. It is required only that we use the HOMO of one and the LUMO of the other.

There are two modes of orbital overlap for the simultaneous formation of two σ bonds—*suprafacial* and *antarafacial*. Bond formation is **suprafacial** if both σ bonds form on the same side of the π system. Bond formation is **antarafacial** if the two σ bonds form on opposite sides of the π system. Suprafacial bond formation is similar to syn addition, whereas antarafacial bond formation resembles anti addition (Section 5.19).

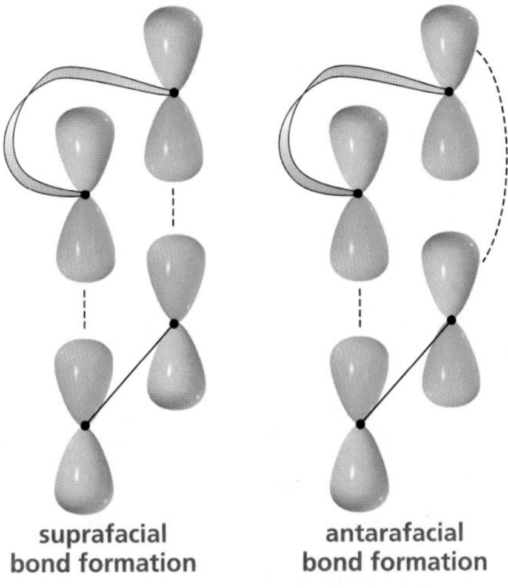

suprafacial bond formation

antarafacial bond formation

A cycloaddition reaction that forms a four-, five-, or six-membered ring must involve suprafacial bond formation. The geometric constraints of these small rings make the antarafacial approach highly unlikely even if it is symmetry-allowed. (Remember that symmetry-allowed means the overlapping orbitals are in-phase.) Antarafacial bond formation is more likely in cycloaddition reactions that form larger rings.

Frontier orbital analysis of a [4 + 2] cycloaddition reaction shows that overlap of in-phase orbitals to form the two new σ bonds requires suprafacial orbital overlap (Figure 29.5). This is true whether we use the LUMO of the dienophile (a system with one π bond; Figure 29.1) and the HOMO of the diene (a system with two conjugated π bonds; Figure 29.2) or the HOMO of the dienophile and the LUMO of the diene. Now we can understand why Diels–Alder reactions occur with relative ease (Section 8.8).

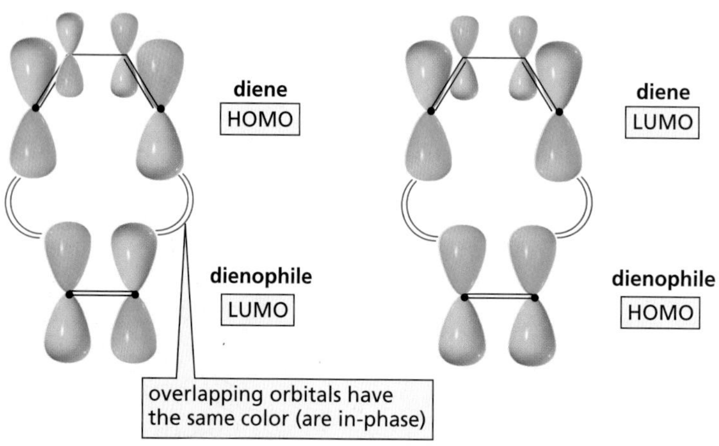

Figure 29.5 ▶
Frontier molecular orbital analysis of a [4 + 2] cycloaddition reaction. The HOMO of either of the reactants can be used with the LUMO of the other. Both situations require suprafacial overlap for bond formation.

A [2 + 2] cycloaddition reaction does not occur under thermal conditions but does take place under photochemical conditions.

$\xrightarrow{\Delta}$ no reaction

$\xrightarrow{h\nu}$

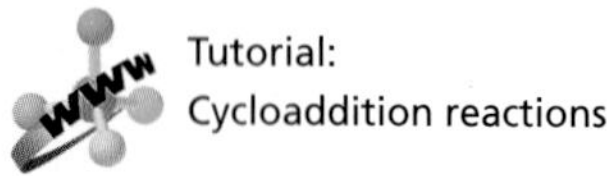

Tutorial:
Cycloaddition reactions

The frontier molecular orbitals in Figure 29.6 show why this is so. Under thermal conditions, suprafacial overlap is not symmetry-allowed (the overlapping orbitals are out-of-phase). Antarafacial overlap is symmetry-allowed but is not possible because of the small size of the ring. Under photochemical conditions, however, the reaction can take place because the symmetry of the excited-state HOMO is opposite that of the ground-state HOMO. Therefore, overlap of the excited-state HOMO of one alkene with the LUMO of the second alkene involves symmetry-allowed suprafacial bond formation.

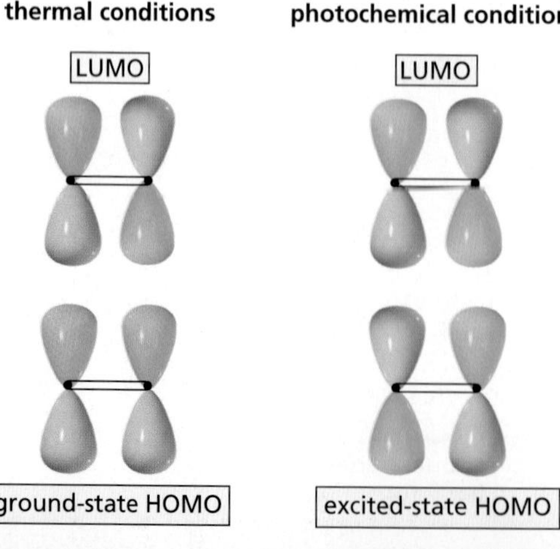

Figure 29.6 ▶
Frontier molecular orbital analysis of a [2 + 2] cycloaddition reaction under thermal and photochemical conditions.

Notice in the photochemical reaction that only one of the reactants is in an excited state. Because of the very short lifetimes of excited states, it is unlikely that two reactants in their excited states would find one another to interact. The selection rules for cycloaddition reactions are summarized in Table 29.3.

Table 29.3 Woodward–Hoffmann Rules for Cycloaddition Reactions

Sum of the number of π bonds in the reacting systems of both reagents	Reaction conditions	Allowed mode of ring closure
Even number	Thermal	Antarafacial[a]
	Photochemical	Suprafacial
Odd number	Thermal	Suprafacial
	Photochemical	Antarafacial[a]

[a] Although antarafacial ring closure is symmetry-allowed, it can occur only with large rings.

COLD LIGHT

A reverse [2 + 2] cycloaddition reaction is responsible for the cold light given off by light sticks. A light stick contains a thin glass vial that holds a mixture of sodium hydroxide and hydrogen peroxide suspended in a solution of diphenyloxalate and a dye. When the vial breaks, two nucleophilic acyl substitution reactions occur that form a compound with an unstable four-membered ring. Recall that phenoxide ion is a relatively good leaving group (Section 17.10).

$$\text{HO–OH} \xrightarrow{\text{HO}^-} \text{:}\ddot{\text{O}}\text{–OH} + H_2O$$

Suprafacial overlap to form a four-membered ring can take place only under photochemical conditions, so one of the reactants must be in an excited state. Therefore, one of the two carbon dioxide molecules formed when the four-membered ring breaks is in an excited state (indicated by an asterisk). When the electron in the excited state drops down to the ground state, a photon of ultraviolet light is released—which is *not* visible to the human eye. However, in the presence of a dye, the excited carbon dioxide molecule can transfer some of its energy to the dye molecule, which causes an electron in the dye to be promoted to an excited state. When the electron of the dye drops down to the ground state, a photon of visible light is released—which *is* visible to the human eye. In Section 29.6 you will see that a similar reaction is responsible for the light given off by fireflies.

has an electron in an excited state

dye + visible light

PROBLEM 9

Explain why maleic anhydride reacts rapidly with 1,3-butadiene but does not react at all with ethene under thermal conditions.

maleic anhydride

PROBLEM 10 SOLVED

Compare the reaction of 2,4,6-cycloheptatrienone with cyclopentadiene to that with ethene. Why does 2,4,6-cycloheptatrienone use two π electrons in one reaction and four π electrons in the other?

a. [2,4,6-cycloheptatrienone + cyclopentadiene →] b. [2,4,6-cycloheptatrienone] + $CH_2{=}CH_2$ →

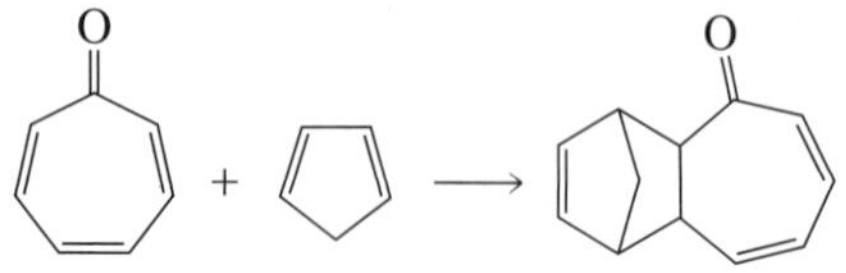

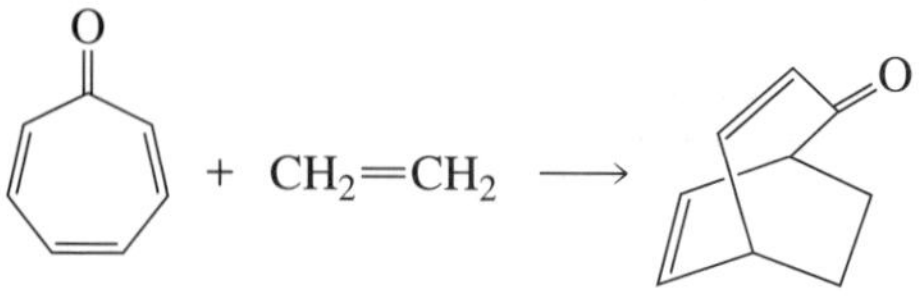

SOLUTION Both reactions are [4 + 2] cycloaddition reactions. When 2,4,6-cycloheptatrienone reacts with cyclopentadiene, it uses two of its π electrons because cyclopentadiene is the four-π-electron reactant. When 2,4,6-cycloheptatrienone reacts with ethene, it uses four of its π electrons because ethene is the two-π-electron reactant.

a.

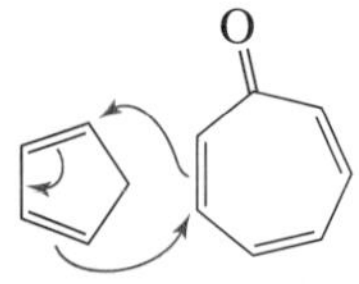

b. $CH_2{=}CH_2$

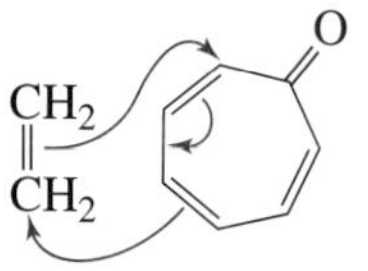

PROBLEM 11◆

Will a concerted reaction take place between 1,3-butadiene and 2-cyclohexenone in the presence of ultraviolet light?

29.5 Sigmatropic Rearrangements

The last class of concerted pericyclic reactions that we will consider is the group of reactions known as *sigmatropic rearrangements.* In a **sigmatropic rearrangement**, a σ bond in the reactant is broken, a new σ bond is formed, and the π electrons rearrange. The σ bond that breaks is a bond to an allylic carbon. It can be a σ bond between a carbon and a hydrogen, between a carbon and another carbon, or between a carbon and an oxygen, nitrogen, or sulfur. "Sigmatropic" comes from the Greek word *tropos,* which means "change," so sigmatropic means "sigma-change."

The numbering system used to describe a sigmatropic rearrangement differs from any numbering system you have seen previously. First, mentally break the σ bond in the reactant and give a number-1 label to both atoms that were attached by the bond. Then look at the new σ bond in the product. Count the number of atoms in each of the fragments that connect the broken σ bond and the new σ bond. The two numbers are put in brackets with the smaller number stated first. For example, in the following [2,3] sigmatropic rearrangement, two atoms (N, N) connect the old and new σ bonds in one fragment and three atoms (C, C, C) connect the old and new σ bonds in the other fragment.

a [2,3] sigmatropic rearrangement

bond broken; $R{-}\overset{+}{N}{=}\overset{-}{N}$ (N labels 1, 2; carbons 1, 2, 3) with CH_3 $\xrightarrow{\Delta}$ $R{-}N{=}N$ with CH_3; new bond formed

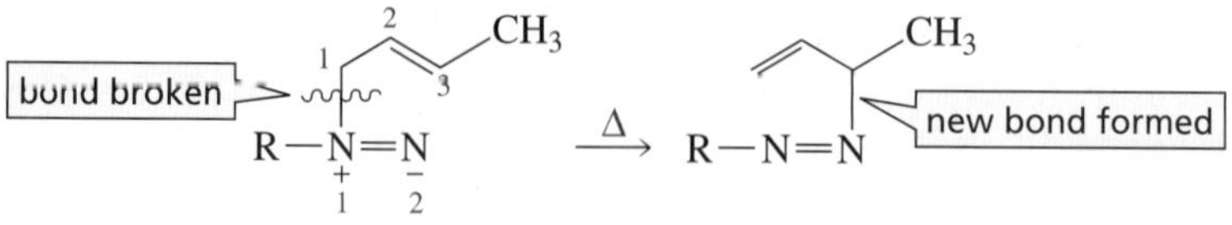

a [1,5] sigmatropic rearrangement

$$CH_3\underset{\underset{1}{H}}{\overset{1}{C}H}{-}\overset{2}{C}H{=}\overset{3}{C}H{-}\overset{4}{C}H{=}\overset{5}{C}H_2 \xrightarrow{\Delta} CH_3CH{=}CH{-}CH{=}CH{-}\underset{H}{CH_2}$$

bond broken; new bond formed

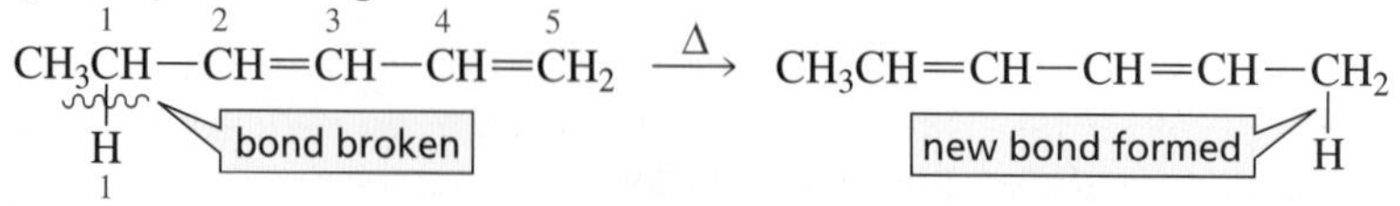

a [1,3] sigmatropic rearrangement

$$CH_3\overset{CH_3}{\underset{CH_3}{\overset{|}{\underset{|}{C}}}}-CH=CH_2 \xrightarrow{\Delta} CH_3\overset{CH_3}{\overset{|}{C}}=CH-\underset{CH_3}{\underset{|}{CH_2}}$$

bond broken

new bond formed

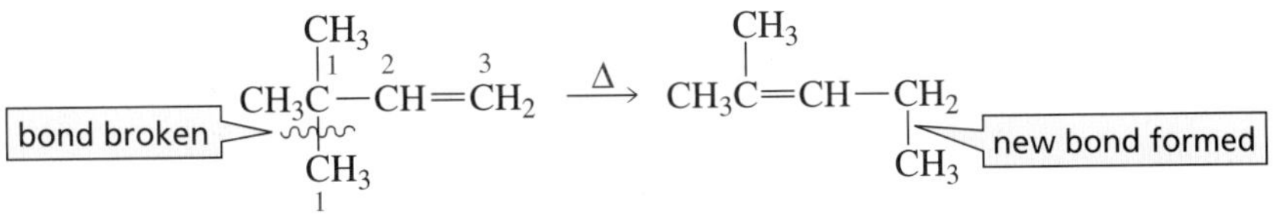

a [3,3] sigmatropic rearrangement

bond broken

$\xrightarrow{\Delta}$

new bond formed

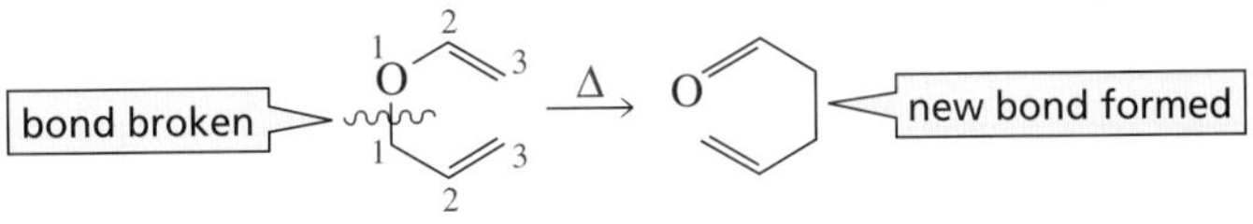

PROBLEM 12

a. Name the kind of sigmatropic rearrangement that occurs in each of the following reactions.
b. Using arrows, show the electron rearrangement that takes place in each of the reactions.

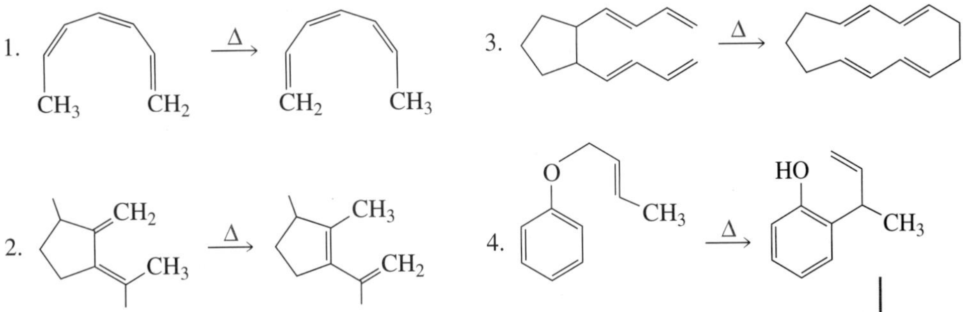

1. CH_3 CH_2 $\xrightarrow{\Delta}$ CH_2 CH_3

2. CH_2 CH_3 $\xrightarrow{\Delta}$ CH_3 CH_2

3. $\xrightarrow{\Delta}$

4. O CH_3 $\xrightarrow{\Delta}$ HO CH_3

In the transition state of a sigmatropic rearrangement, the group that migrates is partially bonded to the migration origin and partially bonded to the migration terminus. There are two possible modes for rearrangement. If the migrating group remains on the same face of the π system, the rearrangement is **suprafacial**. If the migrating group moves to the opposite face of the π system, the rearrangement is **antarafacial**.

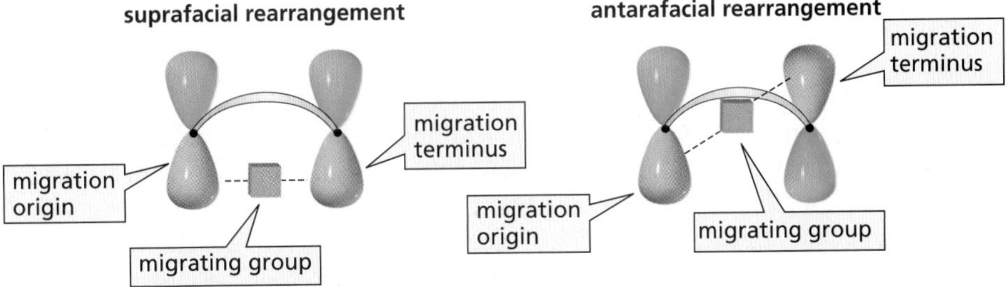

Sigmatropic rearrangements have cyclic transition states. If the transition state has six or fewer atoms in the ring, rearrangement must be suprafacial because of the geometric constraints of small rings.

A [1,3] sigmatropic rearrangement involves a π bond and a pair of σ electrons, or we can say that it involves two pairs of electrons. A [1,5] sigmatropic rearrangement involves two π bonds and a pair of σ electrons (three pairs of electrons), and a [1,7] sigmatropic rearrangement involves four pairs of electrons. The symmetry rules for sigmatropic rearrangements are nearly the same as those for cycloaddition reactions—the only difference is that we count the number of pairs of electrons rather than the number of π bonds. (Compare Tables 29.3 and 29.4.) *Recall that the ground-state HOMO of a compound with an even number of conjugated double bonds is asymmetric, whereas the ground-state HOMO of a compound with an odd number of conjugated double bonds is symmetric.*

Table 29.4 Woodward–Hoffmann Rules for Sigmatropic Rearrangements

Number of pairs of electrons in the reacting system	Reaction conditions	Allowed mode of ring closure
Even number	Thermal	Antarafacial[a]
	Photochemical	Suprafacial
Odd number	Thermal	Suprafacial
	Photochemical	Antarafacial[a]

[a] Although antarafacial ring closure is symmetry-allowed, it can occur only with large rings.

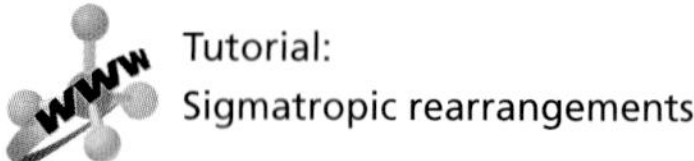

Tutorial:
Sigmatropic rearrangements

A **Cope[2] rearrangement** is a [3,3] sigmatropic rearrangement of a 1,5-diene. A **Claisen[3] rearrangement** is a [3,3] sigmatropic rearrangement of an allyl vinyl ether. Both rearrangements form six-membered-ring transition states. The reactions, therefore, must be able to take place by a suprafacial pathway. Whether or not a suprafacial pathway is symmetry-allowed depends on the number of pairs of electrons involved in the rearrangement (Table 29.4). Because [3,3] sigmatropic rearrangements involve three pairs of electrons, they occur by a suprafacial pathway under thermal conditions. Therefore, both Cope and Claisen rearrangements readily take place under thermal conditions.

a Cope rearrangement

a Claisen rearrangement

PROBLEM 13◆

a. Give the product of the following reaction:

b. If the terminal sp^2 carbon of the substituent bonded to the benzene ring is labeled with ^{14}C, where will the label be in the product?

Migration of Hydrogen

When a hydrogen migrates in a sigmatropic rearrangement, the *s* orbital of hydrogen is partially bonded to both the migration origin and the migration terminus in the transition state. Therefore, a [1,3] sigmatropic migration of hydrogen involves a four-membered-ring transition state. Because two pairs of electrons are involved, the HOMO is asymmetric. The selection rules, therefore, require an antarafacial rearrangement for a 1,3-hydrogen shift under thermal conditions (Table 29.4). Consequently, 1,3-hydrogen shifts do not occur under thermal conditions because the four-membered-ring transition state does not allow the required antarafacial rearrangement.

migration of hydrogen

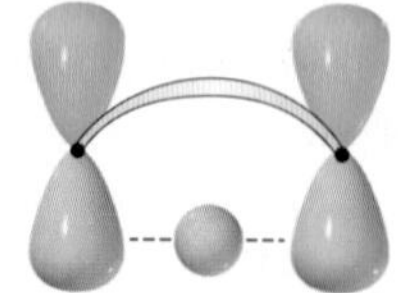

suprafacial rearrangement

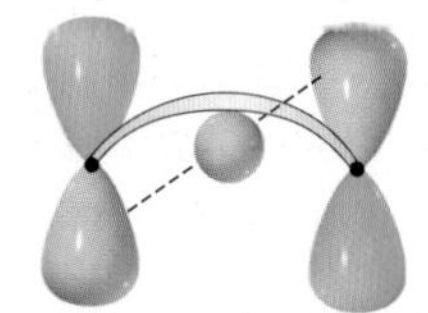

antarafacial rearrangement

[2] Cope also discovered the Cope elimination (p. 894).

[3] Claisen also discovered the Claisen condensation (p. 810).

1,3-Hydrogen shifts can take place if the reaction is carried out under photochemical conditions because the HOMO is symmetric under photochemical conditions, which means that hydrogen can migrate by a suprafacial pathway (Table 29.4).

1,3-hydrogen shifts

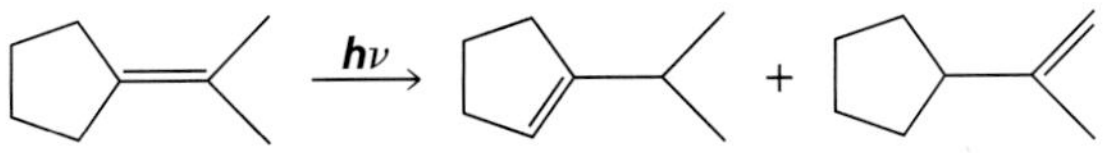

$h\nu$ +

Two products are obtained in the previous reaction because two different allylic hydrogens can undergo a 1,3-hydrogen shift.

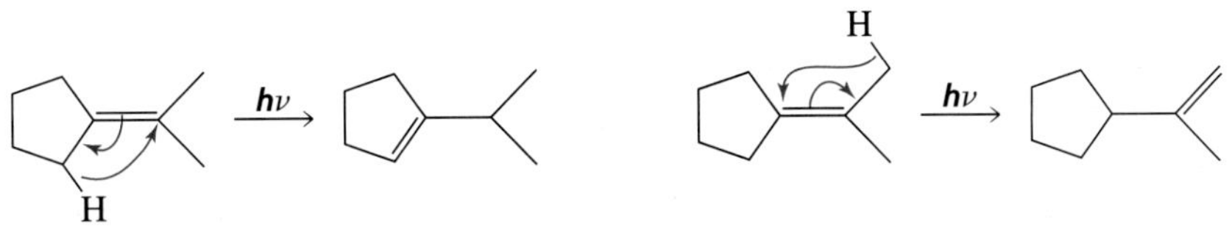

H $h\nu$ H $h\nu$

[1,5] Sigmatropic migrations of hydrogen are well known. They involve three pairs of electrons, so they take place by a suprafacial pathway under thermal conditions.

1,5-hydrogen shifts

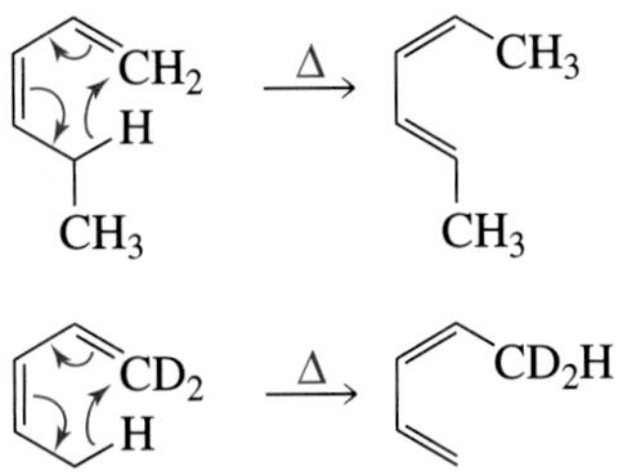

CH_2 H CH_3 $\xrightarrow{\Delta}$ CH_3 CH_3

CD_2 H $\xrightarrow{\Delta}$ CD_2H

PROBLEM 14

Why was a deuterated compound used in the preceding example?

PROBLEM 15

Account for the difference in the products obtained under photochemical and thermal conditions.

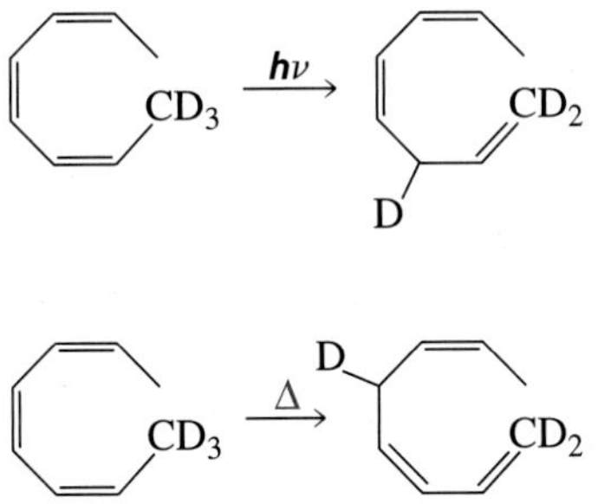

CD_3 $\xrightarrow{h\nu}$ CD_2 D

CD_3 $\xrightarrow{\Delta}$ D CD_2

[1,7] Sigmatropic hydrogen migrations involve four pairs of electrons. They can take place under thermal conditions because the eight-membered-ring transition state allows the required antarafacial rearrangement.

1,7-hydrogen shift

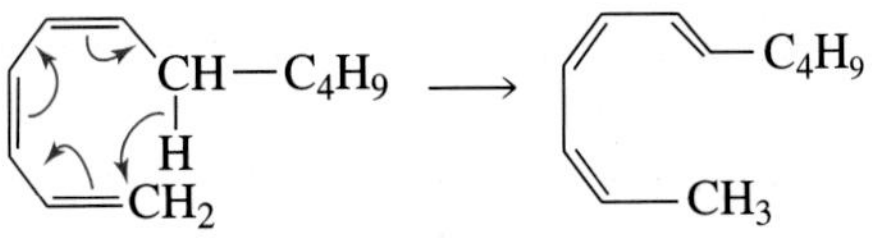

$CH-C_4H_9$ H CH_2 $\longrightarrow$ C_4H_9 CH_3

PROBLEM 16 SOLVED

5-Methyl-1,3-cyclopentadiene rearranges to give a mixture of 5-methyl-1,3-cyclopentadiene, 1-methyl-1,3-cyclopentadiene, and 2-methyl-1,3-cyclopentadiene. Show how these products are formed.

SOLUTION Notice that both equilibria involve [1,5] sigmatropic rearrangements. Although a hydride moves from one carbon to an adjacent carbon, the rearrangements are not called 1,2-shifts because this would not account for all the atoms involved in the rearranged π electron system.

H CH$_3$ ⇌ H H CH$_3$ ⇌ H H CH$_3$

5-methyl-1,3-cyclopentadiene **1-methyl-1,3-cyclopentadiene** **2-methyl-1,3-cyclopentadiene**

Migration of Carbon

Unlike hydrogen, which can migrate in only one way because of its spherical *s* orbital, carbon has two ways to migrate because it has a two-lobed *p* orbital. Carbon can simultaneously interact with the migration origin and the migration terminus using one of its lobes.

carbon migrating with one of its lobes interacting

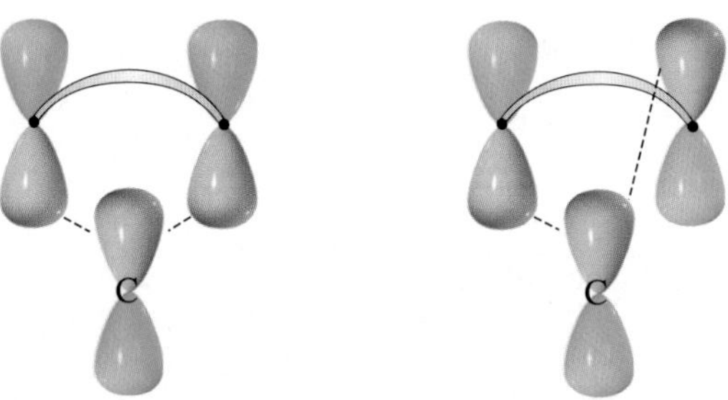

suprafacial rearrangement **antarafacial rearrangement**

Carbon can also simultaneously interact with the migration source and the migration terminus using both lobes of its *p* orbital.

carbon migrating with both of its lobes interacting

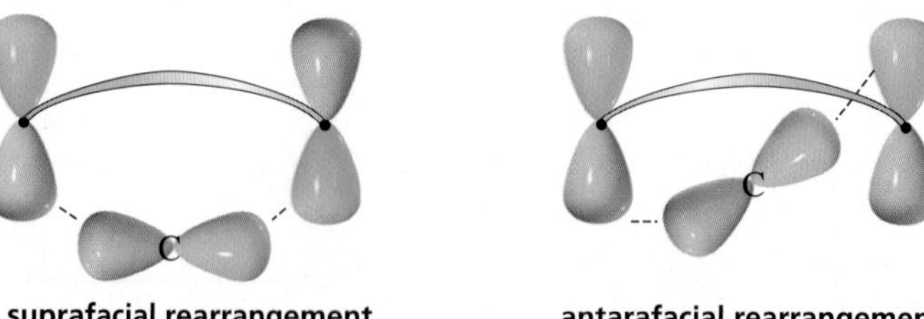

suprafacial rearrangement **antarafacial rearrangement**

If the reaction requires a suprafacial rearrangement, carbon will migrate using one of its lobes if the HOMO is symmetric and will migrate using both of its lobes if the HOMO is asymmetric.

When carbon migrates with only one of its *p* lobes interacting with the migration source and migration terminus, the migrating group retains its configuration because bonding is always to the same lobe. When carbon migrates with both of its *p* lobes interacting, bonding in the reactant and bonding in the product involve different lobes. Therefore, migration occurs with inversion of configuration.

The following [1,3] sigmatropic rearrangement has a four-membered-ring transition state that requires a suprafacial pathway. The reacting system has two pairs of electrons, so its HOMO is asymmetric. Therefore, the migrating carbon interacts with the migration source and the migration terminus using both of its lobes, so it undergoes inversion of configuration.

inverted configuration

PROBLEM 17

[1,3] Sigmatropic migrations of hydrogen cannot occur under thermal conditions, but [1,3] sigmatropic migrations of carbon can occur under thermal conditions. Explain.

PROBLEM 18◆

a. Will thermal 1,3-migrations of carbon occur with retention or inversion of configuration?

b. Will thermal 1,5-migrations of carbon occur with retention or inversion of configuration?

29.6 Pericyclic Reactions in Biological Systems

Biological Cycloaddition Reactions

Exposure to ultraviolet light may cause skin cancer. This is one of the reasons why many scientists are concerned about the thinning ozone layer. The ozone layer absorbs ultraviolet radiation high in the atmosphere, protecting organisms on Earth's surface (Section 9.9). One cause of skin cancer is the formation of *thymine dimers.* At any point in DNA where there are two adjacent thymine residues (Section 27.1), a [2 + 2] cycloaddition reaction can occur, resulting in the formation of a thymine dimer. Because [2 + 2] cycloaddition reactions take place only under photochemical conditions, the reaction takes place only in the presence of ultraviolet light.

two adjacent thymine residues on DNA $\xrightarrow{h\nu}$ mutation-causing thymine dimer

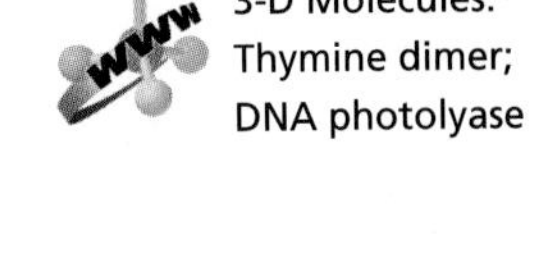

Thymine dimers can cause cancer because they interfere with the structural integrity of DNA. Any modification of DNA structure can lead to mutations and possibly to cancer.

Fortunately, there are enzymes that repair damaged DNA. When a repair enzyme recognizes a thymine dimer, it reverses the [2 + 2] cycloaddition reaction to regenerate the original T–T sequence. Repair enzymes, however, are not perfect, and some damage always remains uncorrected. People who do not have the repair enzyme (called DNA photolyase) that removes thymine dimers rarely live beyond the age of 20. Fortunately, this genetic defect is rare.

Fireflies are one of several species that emit light as a result of a retro (i.e., reverse) [2 + 2] cycloaddition reaction, similar to the reaction that produces the cold light in light sticks (Section 29.4). Fireflies have an enzyme (luciferase) that catalyzes the reaction between luciferin, ATP, and molecular oxygen that forms a compound with an unstable four-membered ring. When the four-membered ring breaks, an electron in

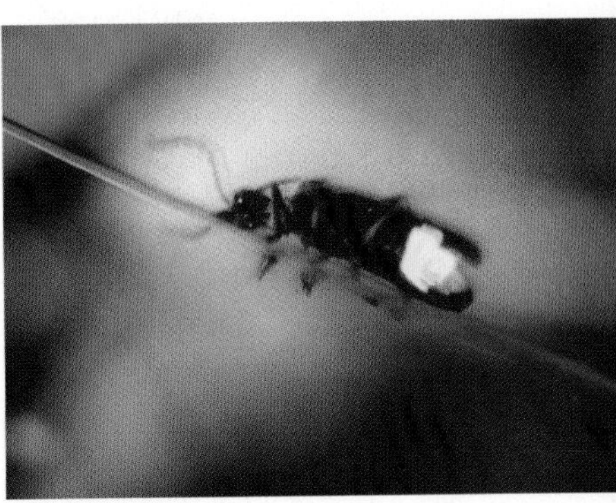

▲ A firefly with its stomach aglow

oxyluciferin is promoted to an excited state because suprafacial overlap can occur only under photochemical conditions. When the electron in the excited state drops down to the ground state, a photon of light is released. In this case, the luciferin molecule acts as both the source of the unstable four-membered ring and as the dye molecule that had to be added to the cold-light reaction.

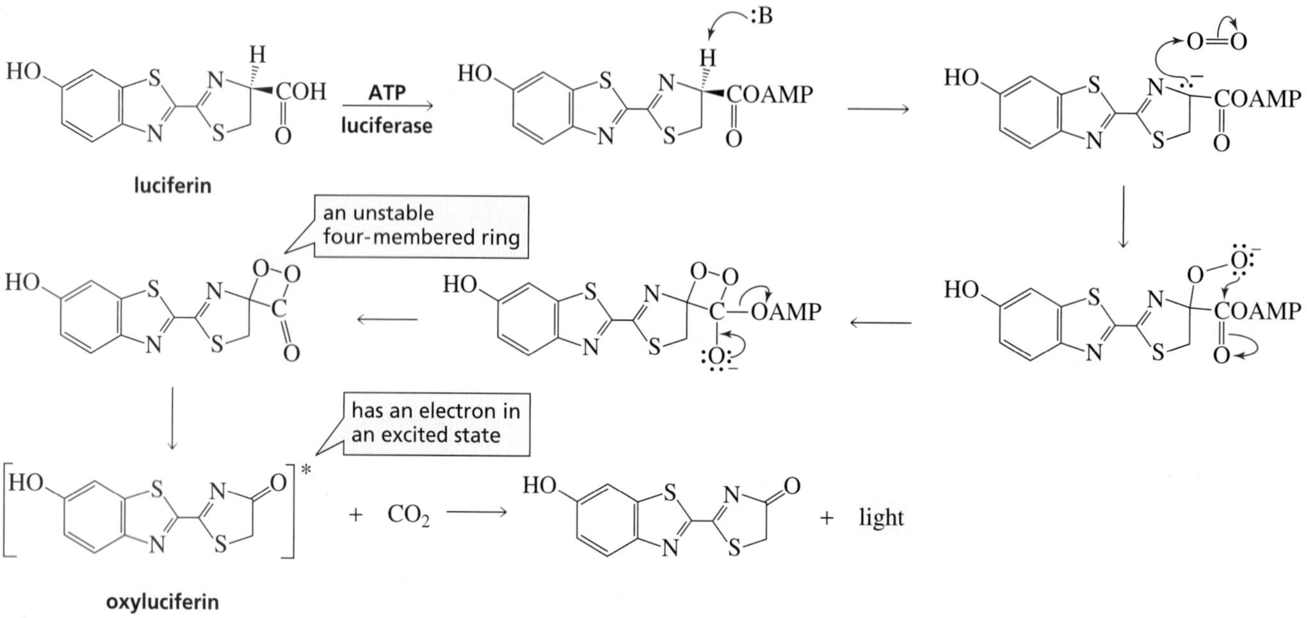

luciferin

ATP
luciferase

an unstable four-membered ring

has an electron in an excited state

oxyluciferin

$+ CO_2$

+ light

A Biological Reaction Involving an Electrocyclic Reaction and a Sigmatropic Rearrangement

7-Dehydrocholesterol, a steroid found in skin, is converted into vitamin D_3 by two pericyclic reactions. The first is an electrocyclic reaction that opens one of the six-membered rings in the starting material to form provitamin D_3. This reaction occurs under photochemical conditions. The provitamin then undergoes a [1,7] sigmatropic rearrangement to form vitamin D_3. The sigmatropic rearrangement takes place under thermal conditions and is slower than the electrocyclic reaction, so vitamin D_3 continues to be synthesized for several days after exposure to sunlight. Vitamin D_3 is not the active form of the vitamin. The active form requires two successive hydroxylations of vitamin D_3—the first occurs in the liver and the second in the kidneys.

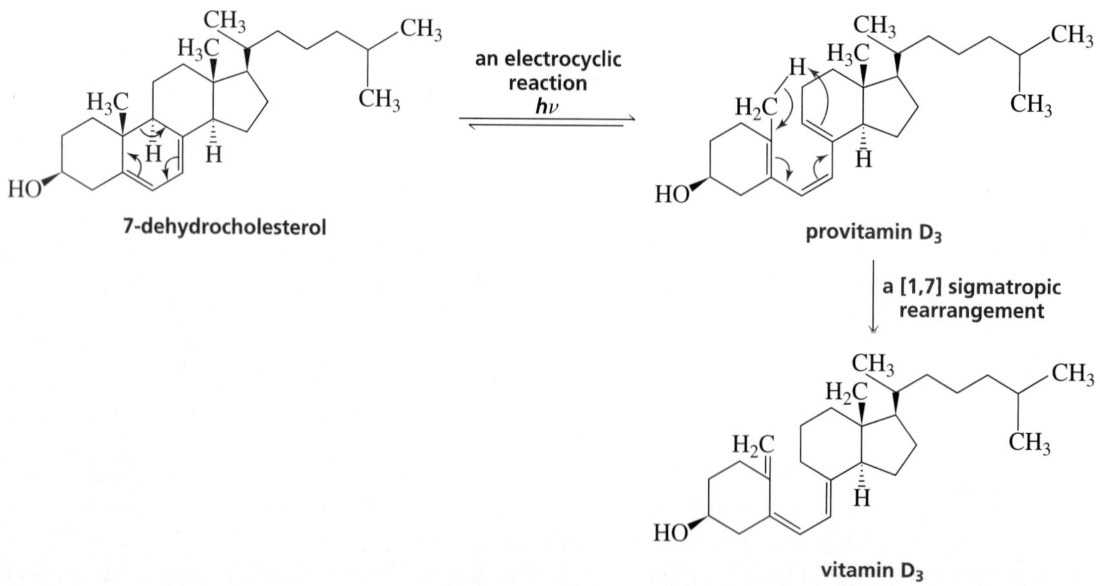

7-dehydrocholesterol

an electrocyclic reaction
$h\nu$

provitamin D_3

a [1,7] sigmatropic rearrangement

vitamin D_3

A deficiency in vitamin D, which can be prevented by getting enough sun, causes a disease known as rickets. Rickets is characterized by deformed bones and stunted growth. Too much vitamin D is also harmful because it causes the calcification of soft tissues. It is thought that skin pigmentation evolved to protect the skin from the sun's ultraviolet rays in order to prevent the synthesis of too much vitamin D_3. This agrees with the observation that peoples who are indigenous to countries close to the equator have greater skin pigmentation.

PROBLEM 19◆

Does the [1,7] sigmatropic rearrangement that converts provitamin D_3 to vitamin D_3 involve suprafacial or antarafacial rearrangement?

PROBLEM 20

Explain why photochemical ring closure of provitamin D_3 to form 7-dehydrocholesterol results in the hydrogen and methyl substituents being trans to one another.

29.7 Summary of the Selection Rules for Pericyclic Reactions

The selection rules that determine the outcome of electrocyclic reactions, cycloaddition reactions, and sigmatropic rearrangements are summarized in Tables 29.1, 29.3, and 29.4, respectively. This is still a lot to remember. Fortunately, the selection rules for all pericyclic reactions can be summarized in one word: "**TE-AC**."

- If **TE** (**T**hermal/**E**ven) describes the reaction, the outcome *is given* by **AC** (**A**ntarafacial or **C**onrotatory).
- If one of the letters of **TE** is incorrect (it is not **T**hermal/**E**ven but is **T**hermal/**O**dd or **P**hotochemical/**E**ven), the outcome *is not given* by **AC** (the outcome is **S**uprafacial or **D**isrotatory).
- If both of the letters of **TE** are incorrect (**P**hotochemical/**O**dd), the outcome *is given* by **AC** (**A**ntarafacial or **C**onrotatory)—"two negatives make a positive."

Summary

A **pericyclic reaction** occurs when the electrons in the reactant(s) are reorganized in a cyclic manner. Pericyclic reactions are concerted, highly stereoselective reactions that are generally not affected by catalysts or by a change in solvent. The three most common types of pericyclic reactions are *electrocyclic reactions, cycloaddition reactions*, and *sigmatropic rearrangements*. The configuration of the product of a pericyclic reaction depends on the configuration of the reactant, the number of conjugated double bonds or pairs of electrons in the reacting system, and whether the reaction is **thermal** or **photochemical**. The outcome of pericyclic reactions is given by a set of **selection rules**, which can be summarized by **TE-AC**.

The two lobes of a *p* orbital have opposite phases. When two in-phase atomic orbitals interact, a covalent bond is formed; two out-of-phase orbitals interact to create a node. The conservation of orbital symmetry theory states that in-phase orbitals overlap during the course of a pericyclic reaction; a **symmetry-allowed pathway** is one in which in-phase orbitals overlap. If the *p* orbitals at the ends of the molecular orbital are in-phase, the molecular orbital is **symmetric**. If the two end *p* orbitals are out-of-phase, the molecular orbital is **asymmetric**.

The ground-state **HOMO** of a compound with an even number of conjugated double bonds or an even number of pairs of electrons is **asymmetric**; the ground-state HOMO of a compound with an odd number of conjugated double bonds or an odd number of pairs of electrons is **symmetric**. The **ground-state** HOMO and the **excited-state** HOMO have opposite symmetries.

An **electrocyclic reaction** is an intramolecular reaction in which a new σ (sigma) bond is formed between the ends of a conjugated π (pi) system. To form this new σ bond, *p* orbitals at the ends of the conjugated system rotate so they can engage in in-phase overlap. If both orbitals rotate in the same direction, ring closure is **conrotatory**; if they rotate in opposite directions, it is **disrotatory**. If the HOMO is asymmetric, conrotatory ring closure occurs: if it is symmetric, disrotatory ring closure occurs.

In a **cycloaddition reaction** two different π bond-containing molecules react to form a cyclic compound by rearranging the π electrons and forming two new σ bonds. Because a cycloaddition reaction involves two molecules, the HOMO of one molecule and the **LUMO** of the other must be considered. Bond formation is **suprafacial** if both σ bonds form on the same side of the π system; it is **antarafacial** if the two σ bonds form on opposite sides of the π system. Formation of small rings occurs only by suprafacial overlap.

In a **sigmatropic rearrangement**, a σ bond is broken in the reactant, a new σ bond is formed in the product, and the π bonds rearrange. If the migrating group remains on the same face of the π system, the rearrangement is **suprafacial**; if it moves to the opposite face of the π system, it is **antarafacial**.

Key Terms

antarafacial bond formation (p. 1189)
antarafacial rearrangement (p. 1193)
antibonding π^* molecular orbital etc. (p. 1179)
asymmetric molecular orbital (p. 1180)
bonding π molecular orbital (p. 1179)
Claisen rearrangement (p. 1194)
conrotatory ring closure (p. 1184)
conservation of orbital symmetry theory (p. 1177)
Cope rearrangement (p. 1194)
cycloaddition reaction (p. 1176)
disrotatory ring closure (p. 1184)
electrocyclic reaction (p. 1175)
excited state (p. 1180)
frontier orbital analysis (p. 1190)
frontier orbitals (p. 1181)
frontier orbital theory (p. 1177)
ground state (p. 1180)
highest occupied molecular orbital (HOMO) (p. 1181)
linear combination of atomic orbitals (LCAO) (p. 1178)
lowest unoccupied molecular orbital (LUMO) (p. 1181)
molecular orbital (MO) theory (p. 1178)
pericyclic reaction (p. 1175)
photochemical reaction (p. 1177)
polar reaction (p. 1175)
radical reaction (p. 1175)
selection rules (p. 1186)
sigmatropic rearrangement (p. 1176)
suprafacial bond formation (p. 1189)
suprafacial rearrangement (p. 1193)
symmetric molecular orbital (p. 1180)
symmetry-allowed pathway (p. 1185)
symmetry-forbidden pathway (p. 1185)
thermal reaction (p. 1177)
Woodward–Hoffmann rules (p. 1186)

Problems

21. Give the product of each of the following reactions:

a. 2 [butadiene] $\xrightarrow{\Delta}$

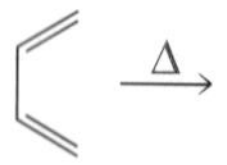

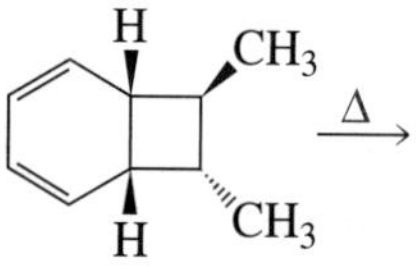

b. 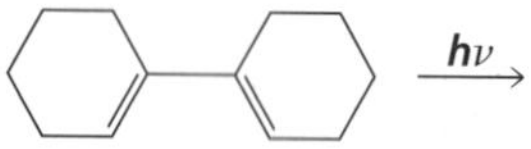 [bicyclohexenyl] $\xrightarrow{h\nu}$

c. 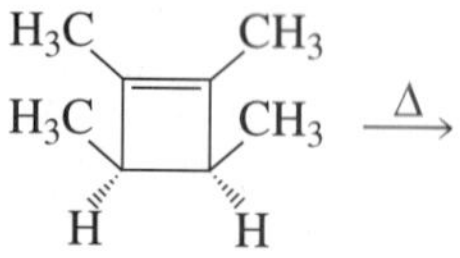 [H_3C, CH_3, H_3C, CH_3, H, H substituted bicyclic cyclobutene] $\xrightarrow{\Delta}$

d. 2 [cyclopentadiene] $\xrightarrow{\Delta}$

e. 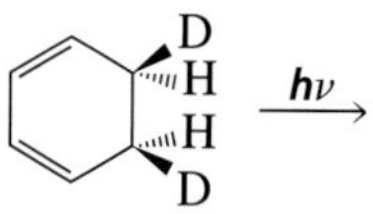 [D, H, H, D substituted cyclohexadiene] $\xrightarrow{h\nu}$

f. 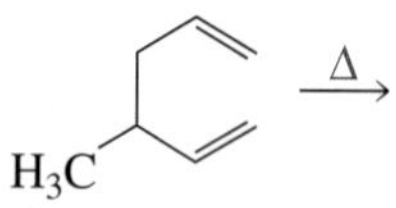 [H_3C substituted diene] $\xrightarrow{\Delta}$

g. [H, CH_3, H, CH_3 substituted bicyclic compound] $\xrightarrow{\Delta}$

h. [cyclodecatriene] $\xrightarrow{\Delta}$

22. Give the product of each of the following reactions:

a. 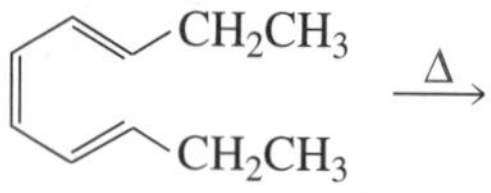 [triene with CH_2CH_3, CH_2CH_3] $\xrightarrow{\Delta}$

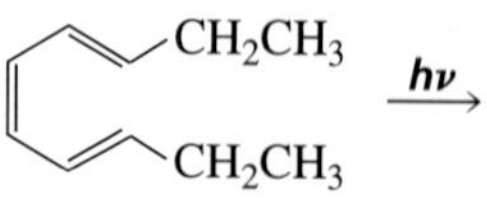

b. 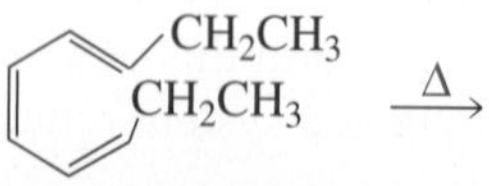 [triene with CH_2CH_3, CH_2CH_3] $\xrightarrow{\Delta}$

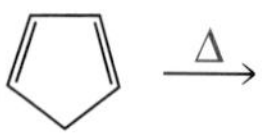

c. [triene with CH_2CH_3, CH_2CH_3] $\xrightarrow{h\nu}$

d. 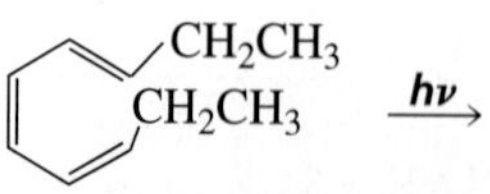 [triene with CH_2CH_3, CH_2CH_3] $\xrightarrow{h\nu}$

23. Chorismate mutase is an enzyme that promotes a pericyclic reaction by forcing the substrate to assume the conformation needed for the reaction. The product of the pericyclic reaction is prephenate, which is subsequently converted into the amino acids phenylalanine and tyrosine. What kind of a pericyclic reaction does chorismate mutase catalyze?

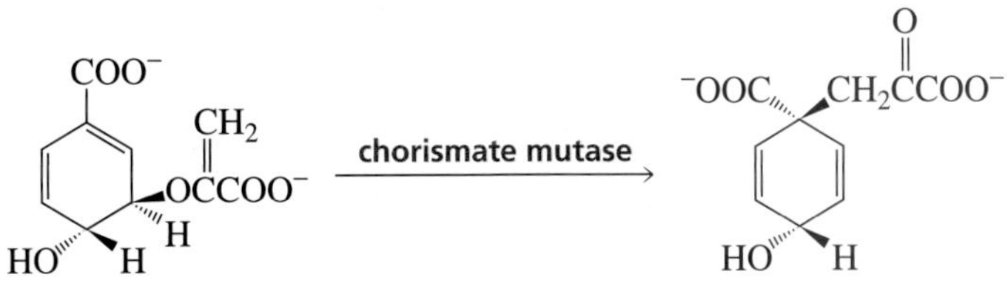

chorismate —chorismate mutase→ prephenate

24. Account for the difference in the products of the following reactions:

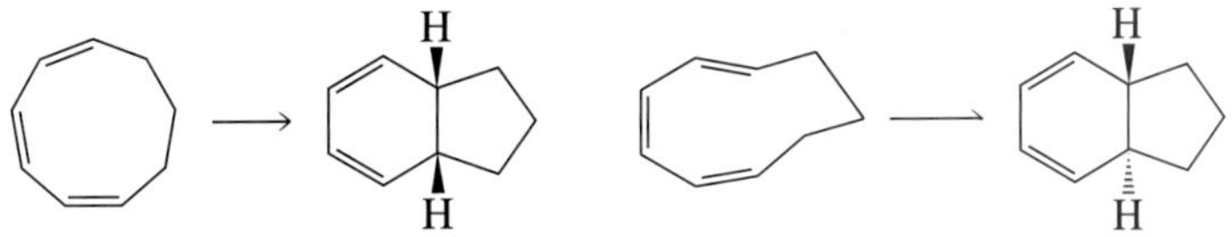

25. Show how norbornane could be prepared from cyclopentadiene.

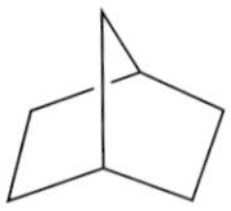

norbornane

26. Give the product formed when each of the following compounds undergoes an electrocyclic reaction
   a. under thermal conditions
   b. under photochemical conditions

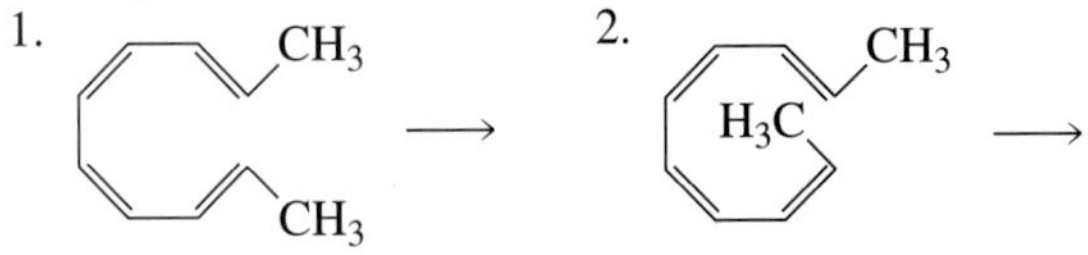

27. Give the product of each of the following reactions:

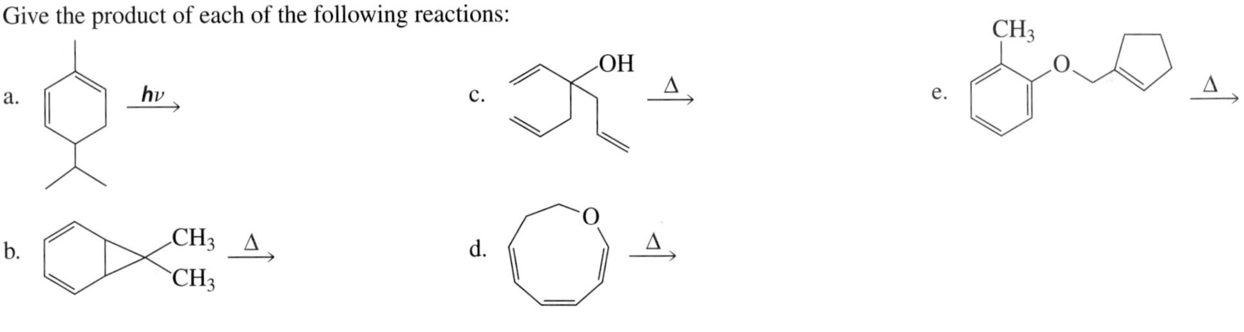

28. Which is the product of the following [1,3] sigmatropic rearrangement, A or B?

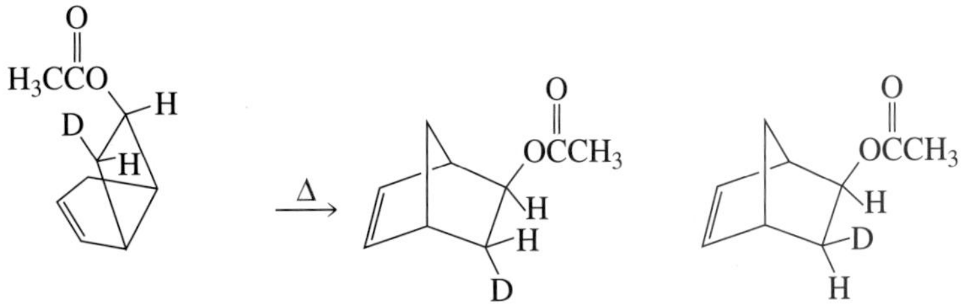

29. Dewar benzene is a highly strained isomer of benzene. In spite of its thermodynamic instability, it is very stable kinetically. It will rearrange to benzene, but only if heated to a very high temperature. Why is it kinetically stable?

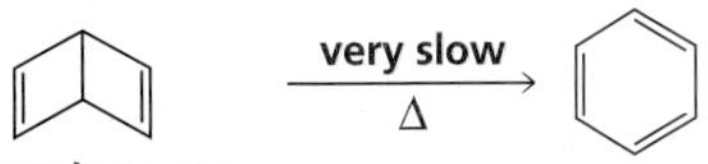

Dewar benzene —very slow, Δ→ benzene

30. If the following compounds are heated, one will form one product from a [1,3] sigmatropic rearrangement and the other will form two products from two different [1,3] sigmatropic rearrangements. Give the products of the reactions.

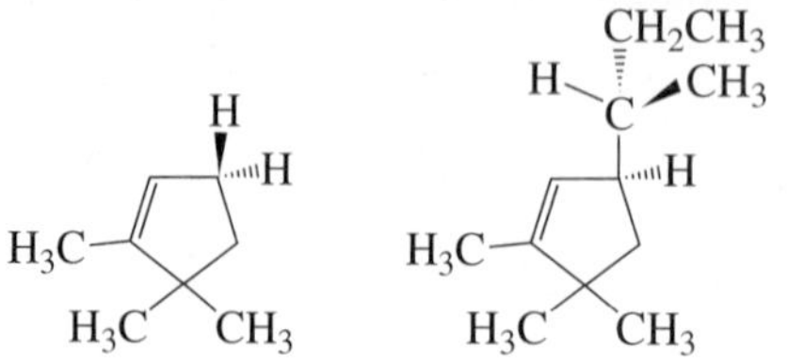

31. When the following compound is heated, a product is formed that shows an infrared absorption band at 1715 cm^{-1}. Draw the structure of the product.

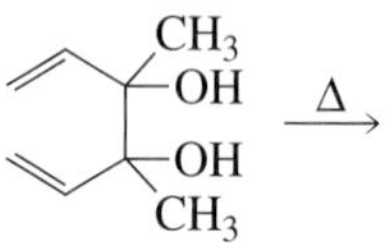

32. Two products are formed in the following [1,7] sigmatropic rearrangement, one due to hydrogen migration and the other to deuterium migration. Show the configuration of the products by replacing A and B with the appropriate substituents (H or D).

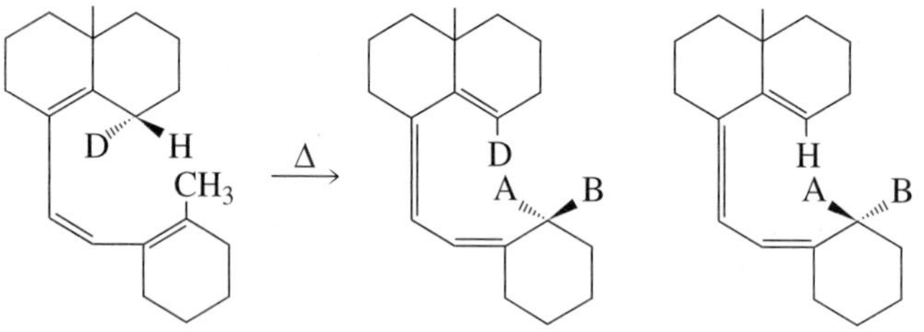

33. a. Propose a mechanism for the following reaction. (*Hint:* An electrocyclic reaction is followed by a Diels–Alder reaction.)
b. What would be the reaction product if *trans*-2-butene were used instead of ethene?

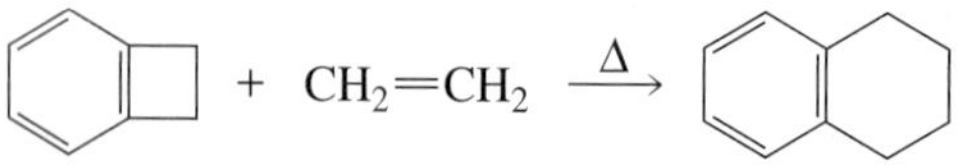

$$+ \ CH_2{=}CH_2 \xrightarrow{\Delta}$$

34. Explain why two different products are formed from disrotatory ring closure of (2*E*,4*Z*,6*Z*)-octatriene, but only one product is formed from disrotatory ring closure of (2*E*,4*Z*,6*E*)-octatriene.

35. Give the product of each of the following sigmatropic rearrangements:

a. $C_6H_5OCH_2CH{=}CHCH_3$ — [3,3] sigmatropic rearrangement, Δ →

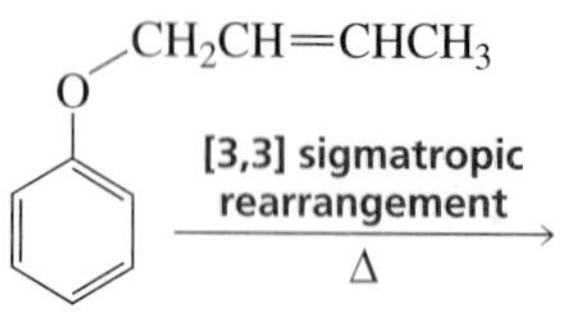

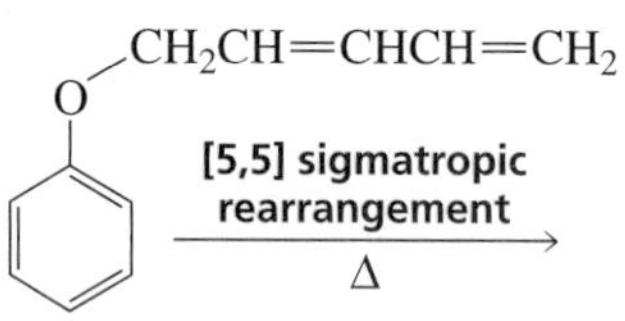

b. $C_6H_5OCH(CH_3)CH{=}CH_2$ — [3,3] sigmatropic rearrangement, Δ →

c. $C_6H_5OCH_2CH{=}CHCH{=}CH_2$ — [5,5] sigmatropic rearrangement, Δ →

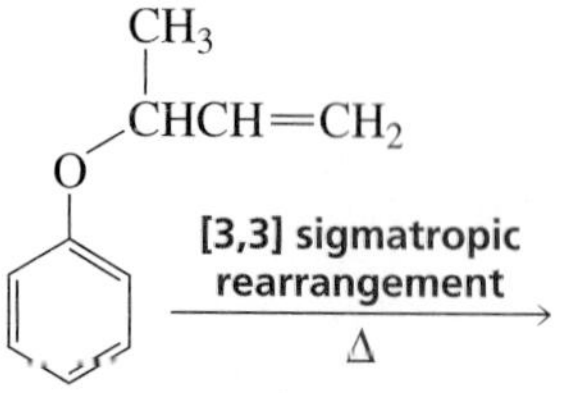

d. $C_6H_5OCH(CH_3)CH{=}CHCH{=}CH_2$ — [5,5] sigmatropic rearrangement, Δ →

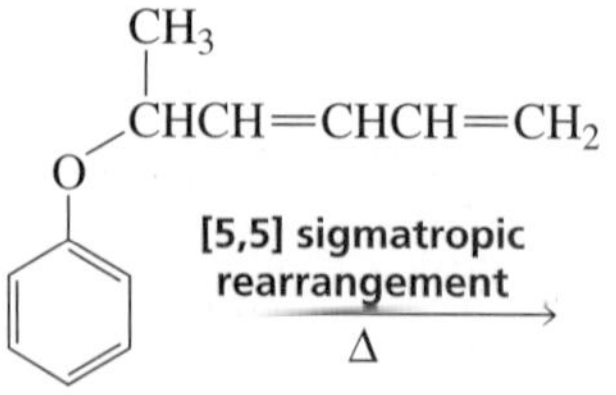

36. *cis*-3,4-Dimethylcyclobutene undergoes thermal ring opening to form the two products shown. One of the products is formed in 99% yield, the other in 1% yield. Which is which?

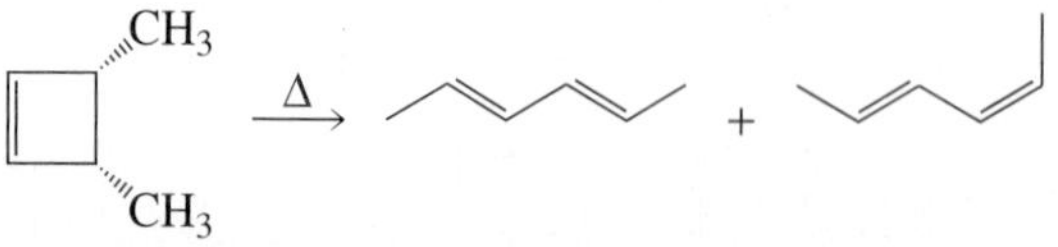

37. If isomer A is heated to about 100 °C, a mixture of isomers A and B is formed. Explain why there is no trace of isomer C or D.

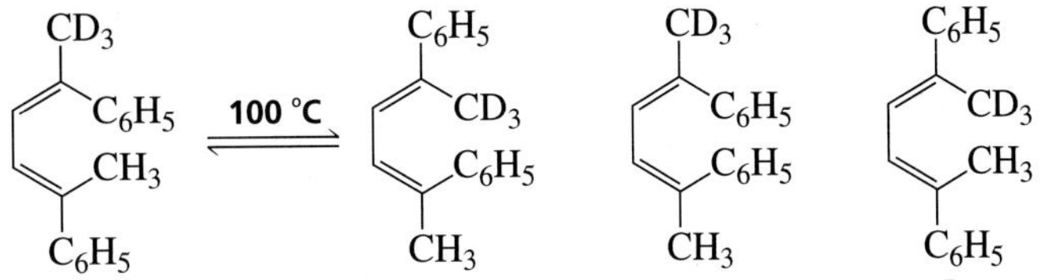

38. Propose a mechanism for the following reaction:

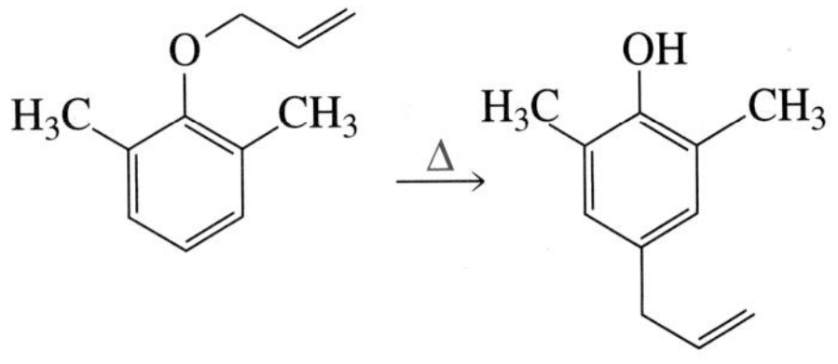

39. Compound A will not undergo a ring-opening reaction under thermal conditions, but compound B will. Explain.

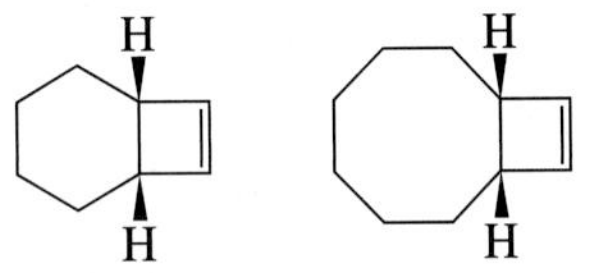

40. Professor Perry C. Click found that heating any one of the following isomers resulted in scrambling of the deuterium to all three positions on the five-membered ring. Propose a mechanism to account for this observation.

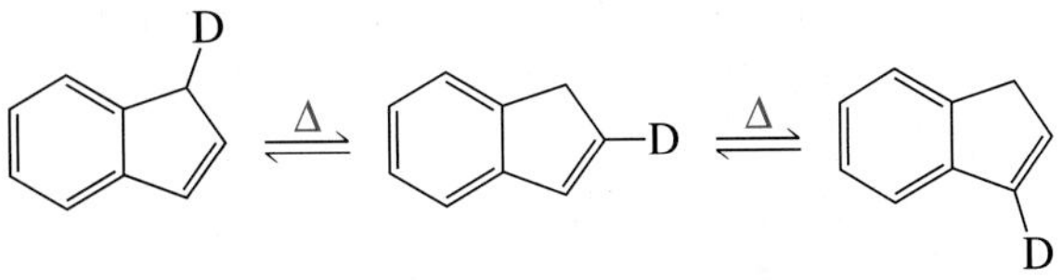

41. How could the following transformation be carried out, using only heat or light?

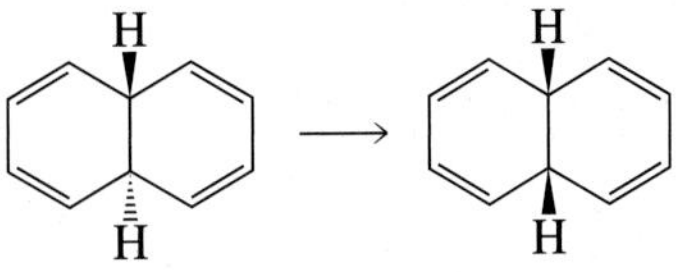

42. Show the steps involved in the following reaction:

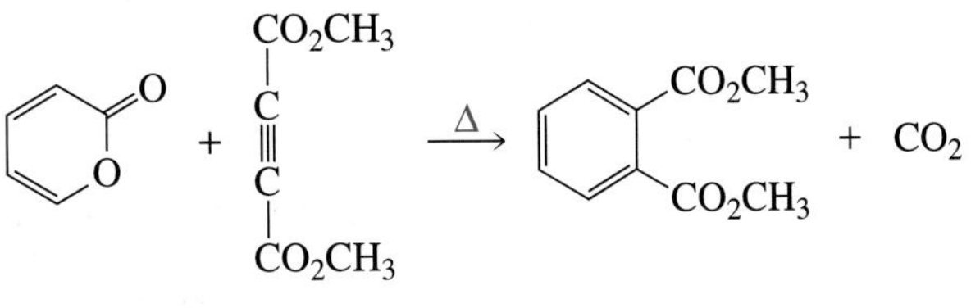

43. Propose a mechanism for the following reaction:

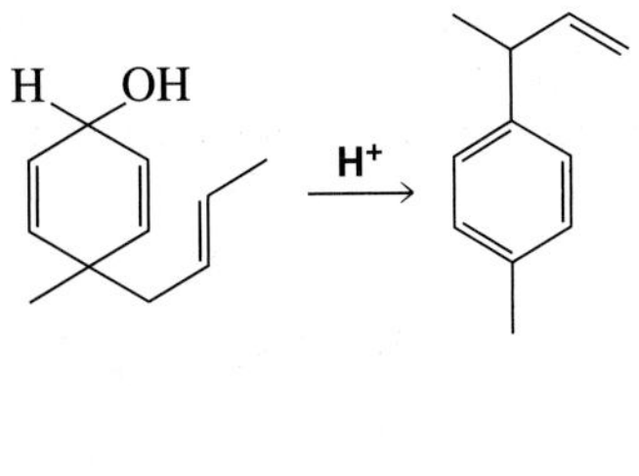

30 The Organic Chemistry of Drugs

Discovery and Design

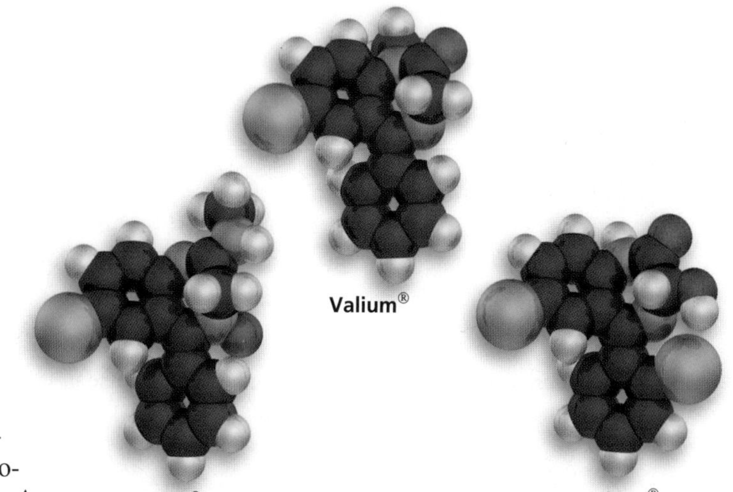

A **drug** is any absorbed substance that changes or enhances a physical or psychological function in the body. A drug can be a gas, a liquid, or a solid and can have a simple structure or a complicated one. Drugs have been used by humans for thousands of years to alleviate pain and illness. By trial and error, people learned which herbs, berries, roots, and bark could be used for medicinal purposes. The knowledge about natural medicines was passed down from generation to generation without any understanding of how the drugs actually worked. Those who dispensed the drugs—medicine men, shamans, and witch doctors—were important members of every civilization. However, the drugs available to them were just a small fraction of the drugs available to us today.

Even at the beginning of the twentieth century, there were no drugs for the dozens of functional, degenerative, neurological, and psychiatric disorders; no hormone therapies; no vitamins; and—most significantly—no effective drug for the cure of any infectious disease. Local anesthetics had just been discovered, and there were only two analgesics to relieve major pain. One reason that families had many children was because some of the children were bound to succumb to childhood diseases. Life spans were generally short. In 1900, for example, the average life expectancy in the United States was 46 years for a man and 48 years for a woman. In 1920, about 80 of every 100,000 children died before their fifteenth birthday, most as a result of infections in their first year of life. Now there is a drug for treating almost every disease, and this is reflected by current life expectancies: 74 years for a man and 79 years for a woman. Now only about four of every 100,000 children die before the age of 15—mainly from cancer, accidents, and inherited diseases.

The shelves of a typical modern pharmacy are stocked with almost 2000 preparations, most of which contain a single active ingredient, usually an organic compound. These medicines can be swallowed, injected, inhaled, or absorbed through the skin. In 2001, more than 3.1 billion prescriptions were dispensed in the United States. The most widely *prescribed drugs* are listed in Table 30.1 in order of the number of prescriptions

Table 30.1 The Most Widely Prescribed Drugs in the United States

Proprietary name	Generic name	Structure	Use
Hydrocodone with APAP	Hydrocodone with APAP	NCH_3; CH_3O; O; O; O; $CH_3\overset{O}{\overset{\|\|}{C}}NH$–C₆H₄–OH	Analgesic
Lipitor®	Atorvastatin	HOC(=O); OH OH; F; N; $(CH_3)_2CH$; NH–C(=O)	Statin (cholesterol-reducing drug)
Premarin®	Conjugated estrogens	H_3C; O; HO	Hormone replacement therapy
Tenormin®	Atenolol	$H_2N\overset{O}{\overset{\|\|}{C}}CH_2$–C₆H₄–$OCH_2\underset{OH}{CH}CH_2NHCH(CH_3)_2$	β-adrenergic blocking agent (antiarrhythmic, antihypertensive)
Synthroid®	Levothyroxine	I; I; HO–; –O–; I; I; $–CH_2\underset{NH_2}{CH}\overset{O}{\overset{\|\|}{C}}O^-Na^+$	Treatment of hypothyroidism
Zithromax®	Azithromycin	CH_3; H_3C; N; CH_3; OH; CH_3; HO; H_3C; OH; H_3C; CH_3; N; HO; H_3C; O; O; O; CH_3; CH_3CH_2; O; O; OCH_3; CH_3; CH_3; O; OH; CH_3	Antibiotic
Lasix®	Furosemide	COOH; $NHCH_2$; O; H_2NSO_2; Cl	Diuretic

Table 30.1 (continued)

Proprietary name	Generic name	Structure	Use
Trimox®	Amoxicillin	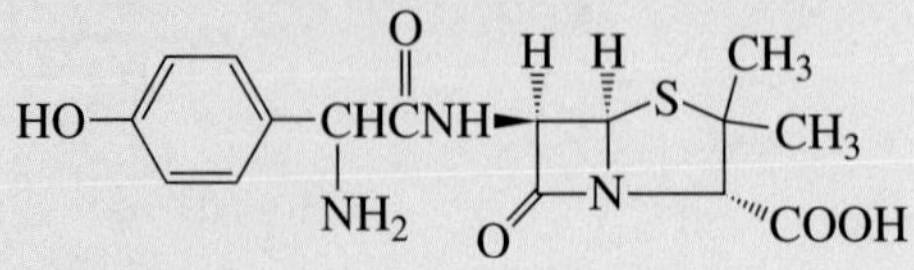	Antibiotic
Norvasc®	Amlodipine	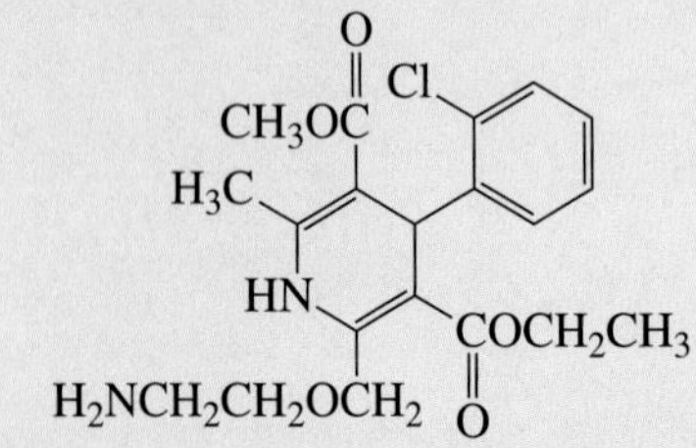	Calcium channel blocker (antihypertensive)
Xanax®	Alprazolam	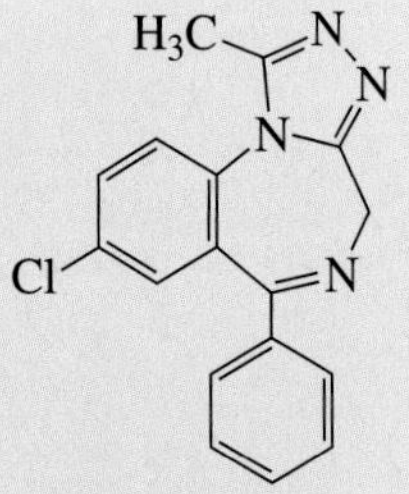	Tranquilizer
Albuterol®	Albuterol	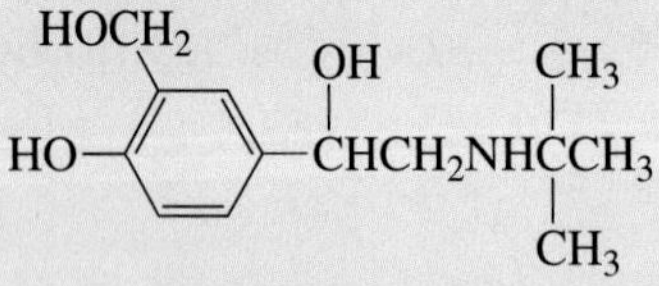	Bronchodilator
Claritin®	Loratadine	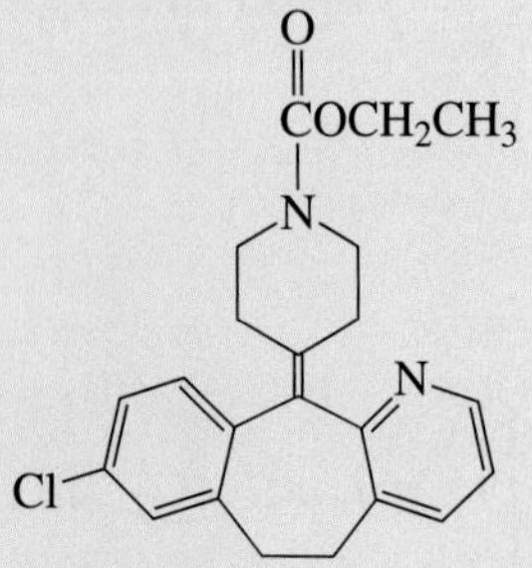	Antihistamine
Hydrodiuril®	Hydrochlorothiazide (HCTZ)	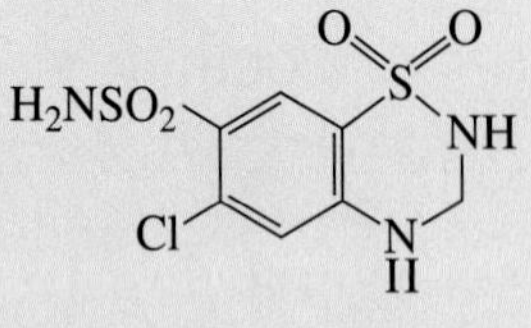	Diuretic
Prilosec®	Omeprazole	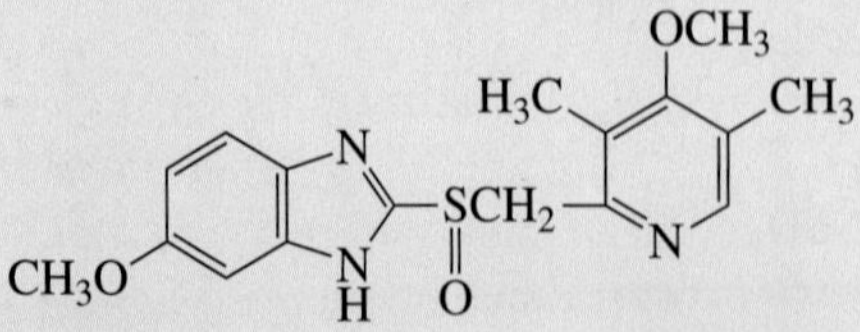	Treatment of stomach acid reflux

written. Antibiotics are the most widely prescribed class of drugs in the world. In the developed world, heart drugs are the most widely prescribed class, partly because they are generally taken for the remainder of the patient's life. In recent years, prescriptions for psychotropic drugs have decreased as doctors have become more aware of problems associated with addiction, and prescriptions for asthma have increased, reflecting a greater incidence (or awareness) of the disease. The U.S. market accounts for 50% of pharmaceutical sales. The aging U.S. population (by 2005, nearly 30% will be over 50 years old) will lead to increased demand for treating conditions such as high cholesterol levels, hypertension, diabetes, osteoarthritis, and menopause symptoms.

Throughout this book we have encountered many drugs, vitamins, and hormones, and in many cases we have discussed the mechanism by which each compound produces its physiological effect. Table 30.2 lists some of these compounds, their uses, and where in the text they are discussed. Now we will take a look at how drugs are discovered and how they are named, and we will examine some of the techniques currently used by scientists in their search for new drugs.

Table 30.2 Drugs, Hormones, and Vitamins Discussed in Earlier Chapters

Drug, hormone, or vitamin	Remark	Section or chapter
Sulfa drugs	The first antibiotics	Section 25.8
Tetracycline	Broad-spectrum antibiotic	Section 5.8
Puromycin	Broad-spectrum antibiotic	Section 27.13
Nonactin	Ionophorous antibiotic	Section 12.9
Penicillin	Antibiotic	Sections 17.4 and 17.16
Gentamicin	Antibiotic	Section 22.19
Gramicidin S	Antibiotic	Section 23.8
Aspirin	Analgesic, anti-inflammatory agent	Sections 19.9 and 26.5
Enkephalins	Painkillers	Section 23.8
Diethyl ether	Anesthetic	Section 12.6
Sodium pentothal	Sedative hypnotic	Section 12.6
5-fluorouracil	Anticancer agent	Section 25.8
Methotrexate	Anticancer agent	Section 25.8
Taxol®	Anticancer agent	Section 18.12
Thalidomide	Sedative with teratogenic side effects	Section 5.15
AZT*	Anti-AIDS agent	Section 27.17
Warfarin	Anticoagulant	Section 25.9
Epinephrine	Vasoconstrictor, bronchodilator	Section 10.11
Vitamins		Chapter 25, Sections 9.8, 18.11, 21.11, 22.19, 26.4, and 26.7
Hormones		Sections 23.8 and 26.9–26.11

* 3′-Azido-2′-deoxythymidine

30.1 Naming Drugs

The most accurate names of drugs are the chemical names that define their structures. However, these names are too long and complicated to appeal to physicians or the general public. The pharmaceutical company is allowed to choose the proprietary name for a drug it develops. A **proprietary name (trade name** or **brand name)** identifies a commercial product and distinguishes it from other products. A proprietary name can be used only by the owner of the registered **trademark**, which can be a name, a symbol, or a picture. It is in the best interest of the company to choose a name that is easy to remember and pronounce so that when the patent expires, the public will continue to request the drug by its proprietary name.

Each drug is also given a **generic name** that any pharmaceutical company can use to identify the product. The pharmaceutical company that develops a drug is allowed to choose the generic name from a list of 10 names provided by an independent group. It is in the best interest of the company to choose the generic name that is hardest to pronounce and the least likely to be remembered, so that physicians and consumers will continue to use the familiar proprietary name. Proprietary names must always be capitalized, whereas generic names are not capitalized.

Drug manufacturers are permitted to patent and retain exclusive rights to the drugs they develop. A patent is valid for 20 years. Once the patent expires, other drug companies can market the drug under the generic name or under their own brand name, which is called a branded generic name. For example, the antibiotic ampicillin is sold as Penbritin® by the company that held the original patent. Now that the patent has expired, the drug is sold by other companies as Ampicin, Ampilar, Amplital, Binotal, Nuvapen, Pentrex, Ultrabion, Viccillin, and 30 other branded generic names.

The over-the-counter drugs that line the shelves of drugstores are available without prescription. They are often mixtures containing one or more active ingredients (generic or proprietary drugs), plus sweeteners and inert fillers. For example, the preparations called Advil® (Whitehall Laboratories), Motrin® (Upjohn), and Nuprin® (Bristol-Meyers Squibb) all contain ibuprofen, a mild analgesic and anti-inflammatory drug. Ibuprofen was patented in Britain in 1964 by Boots, Inc., and the U.S. Food and Drug Administration (FDA) approved its use as a nonprescription drug in 1984.

30.2 Lead Compounds

▲ Foxglove

The goal of the medicinal chemist is to find compounds that have potent effects on given diseases, with minimum side effects. In other words, a drug must react selectively with its target and have minimal negative effects. A drug must get to the right place in the body, at the right concentration, and at the right time. Therefore, a drug must have the appropriate solubility to allow it to be transported to the target cell. If it is taken orally, the drug must be insensitive to the acid conditions of the stomach, and it also must resist enzymatic degradation by the liver before it reaches its target. Finally, it must eventually be either excreted as is or degraded to harmless compounds that can be excreted.

Medicinal agents used by humans since ancient times provided the starting point for the development of our current arsenal of drugs. The active ingredients were isolated from the herbs, berries, roots, and bark used in traditional medicine. Foxglove, for instance, furnished digitoxin, a cardiac stimulant. The bark of the cinchona tree yielded quinine for relief from malaria. Willow bark contains salicylates used to control fever and pain. The sticky juice of the oriental opium poppy provided morphine for severe pain and codeine for the control of cough. By 1882, more than 50 different herbs were commonly used to make medicines. Many of these herbs were grown in the gardens of religious establishments used to treat the sick.

Scientists still search the world for plants and berries and the oceans for flora and fauna that might yield new medicinal compounds. Taxol®, a compound isolated from the bark of the Pacific yew tree, is a relatively recently recognized anticancer agent (Section 18.12).

Once a naturally occurring drug is isolated and its structure determined, it can serve as a prototype in a search for other biologically active compounds. The prototype is called a **lead compound** (i.e., a compound that plays a leading role in the search). Analogs of the lead compound are synthesized in order to find one that might have improved therapeutic properties or fewer side effects. The analog may have a different substituent than the lead compound, a branched chain instead of a straight chain, or a different ring system. Changing the structure of the lead compound is called **molecular modification**.

30.3 Molecular Modification

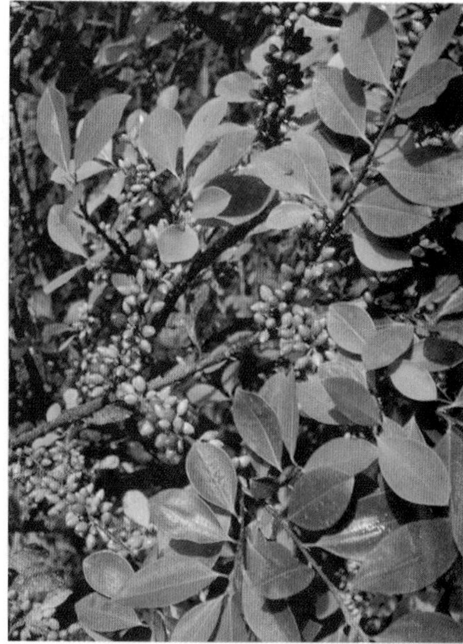
▲ Coca leaves

A classic example of molecular modification is the development of synthetic local anesthetics from cocaine. Cocaine comes from the leaves of *Erythroxylon coca*, a bush native to the highlands of the South American Andes. Cocaine is a highly effective local anesthetic, but it produces disturbing effects on the central nervous system (CNS), ranging from initial euphoria to severe depression. By dissecting the cocaine molecule step by step—removing the methoxycarbonyl group and cleaving the seven-membered-ring system—scientists identified the portion of the molecule that carries the local anesthetic activity without the damaging CNS effects. This knowledge gave an improved lead compound—an ester of benzoic acid, with the alcohol component of the ester having a terminal tertiary amino group.

cocaine
lead compound

improved lead compound

Hundreds of esters were then synthesized, resulting in esters with substituents on the aromatic ring, esters with the alkyl groups bonded to the nitrogen, and esters with the length of the connecting alkyl chain modified. Successful anesthetics obtained through molecular modification were benzocaine, a topical anesthetic, and procaine, commonly known by the trade name Novocain®.

Benzocaine®

procaine
Novocain®

lidocaine
Xylocaine®

Because the ester group of procaine is hydrolyzed relatively rapidly by serum esterases (enzymes that catalyze ester hydrolysis), procaine has a short half-life. Therefore, compounds with less easily hydrolyzed amide groups were synthesized (Section 17.6). In this way, lidocaine, one of the most widely used injectable anesthetics, was discovered. The rate at which lidocaine is hydrolyzed is further decreased by its two *ortho*-methyl substituents, which provide steric hindrance to the vulnerable carbonyl group.

Later, physicians recognized that the action of an anesthetic administered in vivo (in a living organism) could be lengthened considerably if it were administered along with epinephrine. Because epinephrine is a vasoconstrictor, it reduces the blood supply, allowing the drug to remain at its targeted site for a longer period.

In screening the structurally modified compounds for biological activity, scientists were surprised to find that replacing the ester linkage of procaine with an amide linkage led to a compound—procainamide hydrochloride—that had activity as a cardiac depressant as well as activity as a local anesthetic. Procainamide hydrochloride is currently used clinically as an antiarrhythmic.

procainamide hydrochloride

Morphine, the most widely used analgesic for severe pain, is the standard by which other painkilling medications are measured. Although scientists have learned how to synthesize morphine, all commercial morphine is obtained from opium, the juice from a species of poppy. Morphine occurs in opium to the extent of 10%. Methylating one of the OH groups of morphine produces codeine, which has one-tenth the analgesic activity of morphine. Codeine profoundly inhibits the cough reflex. Although 3% of opium is codeine, most commercial codeine is obtained by methylating morphine. Esterifying one of the OH groups of morphine produces a compound with a similar reduced potency. Acetylating both OH groups forms heroin, which is much more potent than morphine. Heroin is less polar than morphine, so it crosses the blood–brain barrier more rapidly, resulting in a more rapid "high." Heroin has been banned in most countries because it is widely abused. It is synthesized by using acetic anhydride to acetylate morphine. Therefore, both heroin and acetic acid are formed as products. Drug enforcement agencies use dogs trained to recognize the pungent odor of acetic acid.

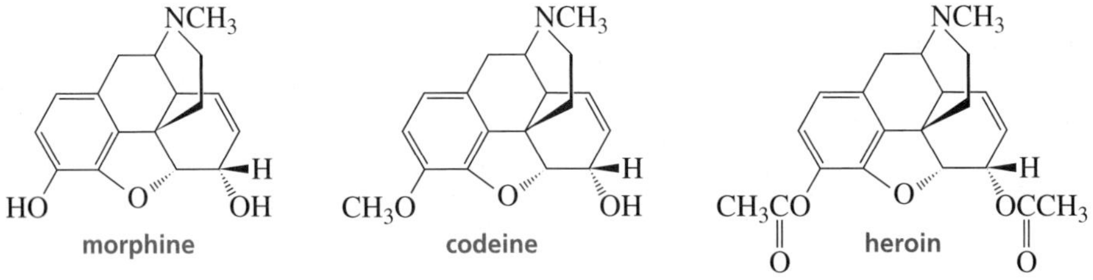

morphine codeine heroin

Molecular modification of codeine led to dextromethorphan, the active ingredient in most cough medicines. Etorphine was synthesized when scientists realized that analgesic potency was related to the drug's ability to bind hydrophobically to the opiate receptor (Section 30.6). Etorphine is about 2000 times more potent than morphine, but it is not safe for use by humans. It has been used to tranquilize elephants and other large animals. Pentazocine is useful in obstetrics because it dulls the pain of labor, but does not depress the respiration of the infant as morphine does.

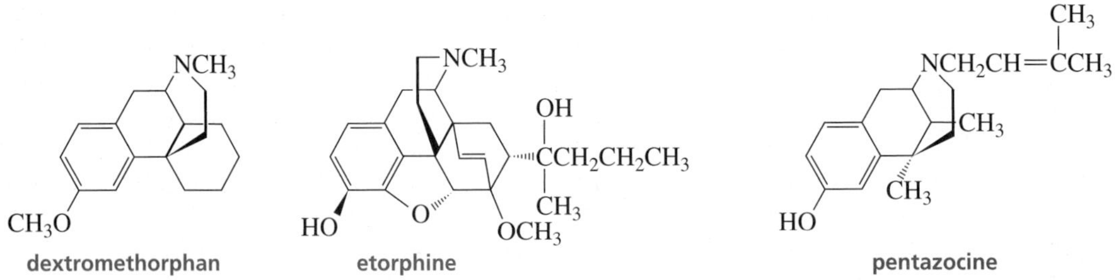

dextromethorphan etorphine pentazocine

Methadone was synthesized by German scientists in 1944 in an attempt to find a drug to treat muscle spasms. (It was originally called "Adolphile" in honor of Adolph Hitler.) It was not recognized until 10 years later—after building molecular models—that methadone and morphine have similar shapes. In contrast to morphine, methadone can be administered orally. Methadone has a considerably longer half-life (24–26 hours) than morphine (2–4 hours). Cumulative effects are seen with repeated doses of methadone, so lower doses and longer intervals between doses are possible. Consequently, methadone is used to treat chronic pain and the withdrawal symptoms of heroin addicts. Reducing the carbonyl group of methadone and acetylating it forms α-acetylmethadol. The levo $(-)$ isomer of this compound can suppress withdrawal symptoms for 72 hours (Section 5.13). When Darvon® (isomethadone) was introduced, it was initially thought to be the long-sought-after nonaddicting painkiller. However, it was later found to have no therapeutic advantage over less toxic and more effective analgesics.

methadone

α-acetylmethadol

isomethadone
Darvon®

Notice that morphine and all the compounds prepared by molecular modification of morphine have a structural feature in common—an aromatic ring attached to a quaternary carbon that is attached to a tertiary amine two carbons away.

30.4 Random Screening

The lead compound for the development of most drugs is found by screening thousands of compounds randomly. A **random screen**, also known as a **blind screen**, is a search for a pharmacologically active compound without having any information about which chemical structures might show activity. The first blind screen was carried out by Paul Ehrlich, who was searching for a "magic bullet" against trypanosomes—the microorganisms that cause African sleeping sickness. After testing more than 900 compounds against trypanosomes, Ehrlich tested some of his compounds against other bacteria. Compound 606 (salvarsan) was found to be dramatically effective against the microorganisms (spirochetes) that cause syphilis.

An important part of random screening is recognizing an effective compound. This requires the development of an assay for the desired biological activity. Some assays can be done in vitro (in glass)—for example, searching for a compound that will inhibit a particular enzyme. Others are done in vivo—for instance, searching for a compound that will save a mouse from a lethal dose of a virus. One problem with in vivo assays is that drugs can be metabolized differently by different animals. Thus, an effective drug in a mouse may be less effective or even useless in a human. Another problem involves regulating the dosages of both the virus and the drug. If the dosage of the virus is too high, it might kill the mouse in spite of the presence of a biologically active compound that could save the animal. If the dosage of the potential drug is too high, the drug might kill the mouse whereas a lower dosage would have saved it.

The observation that azo dyes effectively dyed wool fibers (animal protein) gave scientists the idea that such dyes might selectively bind to bacterial proteins, too. Well

Paul Ehrlich (1854–1915) *was a German bacteriologist. He received a medical degree from the University of Leipzig and was a professor at the University of Berlin. In 1892, he developed an effective diphtheria antitoxin. For his work on immunity, he received the 1908 Nobel Prize in physiology or medicine, together with Ilya Ilich Mechnikov.*

Ilya Ilich Mechnikov (1845–1916), *later known as Elié Metchnikoff, was born in the Ukraine. He was the first scientist to recognize that white blood cells were important in resistance to infection. Metchnikoff succeeded Pasteur as director of the Pasteur Institute.*

Gram-negative bacteria have an outer membrane that covers their cell walls; gram-positive bacteria do not have an outer membrane, but tend to have thicker and more rigid cell walls. The two can be distinguished by a stain, invented by Hans Christian Gram, that turns gram-negative bacteria pink and gram-positive bacteria purple.

over 10,000 dyes were screened in vitro in antibacterial tests, but none showed any antibiotic activity. Some scientists argued that the dyes should be screened in vivo because what physicians really needed were antibacterial agents that would cure infections in humans and animals, not in test tubes.

In vivo studies were done in mice that had been infected with a bacterial culture. Now the luck of the investigators improved. Several dyes turned out to counteract gram-positive infections. The least toxic of these, Prontosil (a bright red dye), became the first drug to treat bacterial infections.

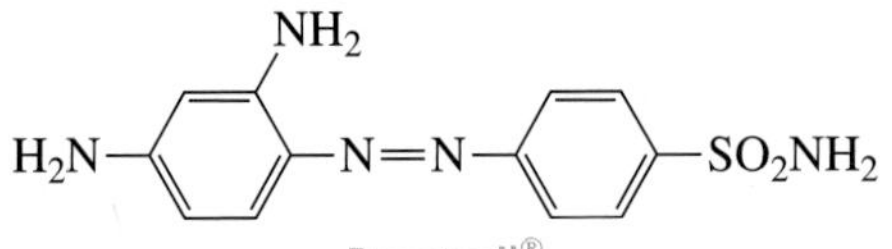

NH_2

H_2N—⟨benzene⟩—N=N—⟨benzene⟩—SO_2NH_2

Prontosil®

Gerhard Domagk (1895–1964) *was a research scientist at I. G. Farbenindustrie, a German manufacturer of dyes and other chemicals. He carried out studies that showed Prontosil to be an effective antibacterial agent. His daughter, who was dying of a streptococcal infection as a result of cutting her finger, was the first patient to receive the drug and be cured by it (1935). Prontosil received wider fame when it was used to save the life of Franklin D. Roosevelt, Jr., son of the U.S. president. Domagk received the Nobel Prize in physiology or medicine in 1939, but Hitler did not allow Germans to accept Nobel Prizes because Carl von Ossietsky, a German who was in a concentration camp, had been awarded the Nobel Prize for peace in 1935. Domagk was eventually able to accept the prize in 1947.*

The fact that Prontosil was inactive in vitro but active in vivo should have suggested that the dye was converted to an active compound by the mammalian organism, but this did not occur to the bacteriologists, who were content to have found a useful antibiotic. When scientists at the Pasteur Institute later investigated Prontosil, they noted that mice given the drug did not excrete a red compound. Urine analysis showed that the mice excreted *para*-acetamidobenzenesulfonamide, a colorless compound. Chemists knew that anilines are acetylated in vivo, so they prepared the nonacetylated compound (sulfanilamide). When sulfanilamide was tested in mice infected with streptococcus, all the mice were cured, whereas untreated control mice died. Sulfanilamide was the first of the sulfa drugs.

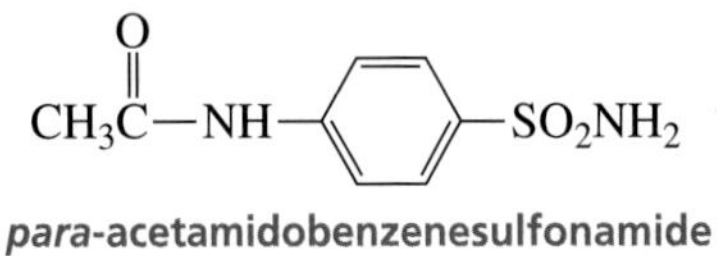

$CH_3C(=O)$—NH—⟨benzene⟩—SO_2NH_2

para-acetamidobenzenesulfonamide

H_2N—⟨benzene⟩—SO_2NH_2

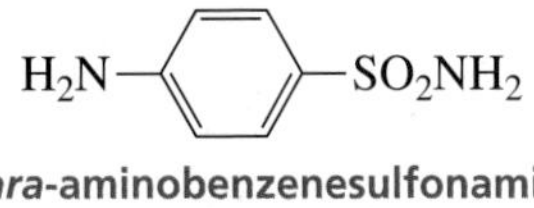

para-aminobenzenesulfonamide
sulfanilamide

We have seen that sulfanilamide acts by inhibiting the bacterial enzyme that incorporates *para*-aminobenzoic acid into folic acid (Section 25.8). Thus, sulfanilamide is a *bacteriostatic* drug, not a *bactericidal* drug. A **bacteriostatic drug** inhibits the further growth of bacteria, whereas a **bactericidal drug** kills the bacteria. Sulfanilamide inhibits the enzyme because the sulfonamide and carboxylic acid groups have similar sizes. Many successful drugs have been designed by using similar isosteric (like-size) replacements (Sections 25.8 and 25.9).

30.5 Serendipity in Drug Development

Many drugs have been discovered accidentally. Nitroglycerin, the drug used to relieve the symptoms of angina pectoris (heart pain), was discovered when workers handling nitroglycerin in the explosives industry experienced severe headaches. Investigation revealed that the headaches were caused by nitroglycerin's ability to produce a marked dilation of blood vessels. The pain associated with an angina attack results from the inability of the blood vessels to supply the heart adequately with blood. Nitroglycerin relieves the pain by dilating cardiac blood vessels.

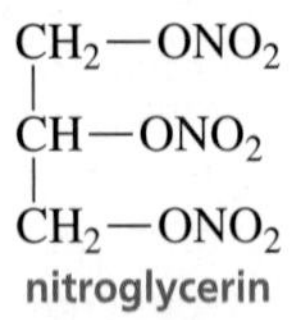

$$\begin{array}{l} CH_2-ONO_2 \\ | \\ CH-ONO_2 \\ | \\ CH_2-ONO_2 \end{array}$$

nitroglycerin

The tranquilizer Librium® is another drug that was discovered accidentally. Leo Sternbach synthesized a series of quinazoline 3-oxides, but none of them showed any pharmacological activity. One of the compounds was not submitted for testing because it was not the quinazoline 3-oxide he had set out to synthesize. Two years after the project was abandoned, a laboratory worker came across this compound while cleaning up the lab, and Sternbach decided that he might as well submit it for testing before it was thrown away. The compound was shown to have tranquilizing properties and, when its structure was investigated, was found to be a benzodiazepine 4-oxide. Methylamine, instead of displacing the chloro substituent to form a different quinazoline 3-oxide, had added to the imine group of the six-membered ring, causing the ring to open and reclose to a seven-membered ring. The compound was given the trade name Librium® when it was put into clinical use in 1960.

an addition reaction occurred

CH_3NH_2

a quinazoline 3-oxide

CH_3NH_2 ✗

a substitution reaction did not occur

a benzodiazepine 4-oxide
chlordiazepoxide
Librium® (1960)

Leo H. Sternbach *was born in Austria in 1908. In 1918, after World War I and the dissolution of the Austro-Hungarian Empire, Sternbach's father moved to Krakow in the re-created Poland and obtained a concession to open a pharmacy. As a pharmacist's son, Sternbach was accepted at the Jagiellonian University School of Pharmacy, where he received both a master's degree in pharmacy and a Ph.D. in chemistry. With discrimination against Jewish scientists growing in Eastern Europe in 1937, Sternbach moved to Switzerland to work with Ružička (p. 1088) at the ETH (the Swiss Federal Institute of Technology). In 1941, Hoffmann–LaRoche brought Sternbach and several other scientists out of Europe. Sternbach became a research chemist at LaRoche's U.S. headquarters in Nutley, New Jersey, where he later became director of medicinal chemistry.*

Librium® was structurally modified in an attempt to find other tranquilizers. One successful modification produced Valium®, a tranquilizer almost 10 times more potent than Librium®. Currently, there are eight benzodiazepines in clinical use as tranquilizers in the United States and some 15 others abroad. Xanax® is one of the most widely prescribed medications; Rohypnol® is one of the so-called date-rape drugs.

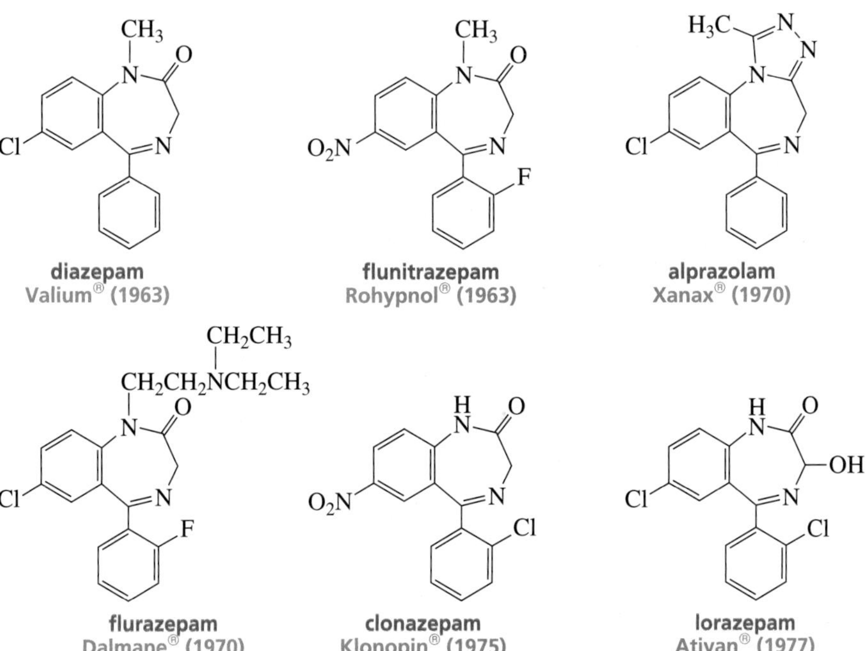

diazepam Valium® (1963)
flunitrazepam Rohypnol® (1963)
alprazolam Xanax® (1970)
flurazepam Dalmane® (1970)
clonazepam Klonopin® (1975)
lorazepam Ativan® (1977)

30.6 Receptors

Many drugs exert their physiological effects by binding to a specific cellular binding site called a **receptor**. That is why a small amount of a drug can bring about a measurable effect. Drug receptors are often glycoproteins or lipoproteins, which explains why different enantiomers of a drug have different effects (Section 5.15). Some receptors are part of cell membranes, while others are found in the cytoplasm—the material outside the nucleus. Because not all cells have the same receptors, drugs have considerable specificity. For example, epinephrine has intense effects on cardiac muscle, but almost no effect on muscle in other parts of the body. Nucleic acids—particularly DNA—also act as receptors for certain kinds of drugs.

A drug interacts with its receptor by means of the same kinds of bonding interactions—hydrogen bonding, electrostatic attractions, and van der Waals interactions—that we encountered in other examples of molecular recognition (Section 24.8). The most important factor in bringing together a drug and a receptor is a snug fit: The greater the affinity of a drug for its binding site, the higher is the drug's potential biological activity. Two drugs for which DNA is a receptor are chloroquine (an antimalarial) and 3,6-diaminoacridine (an antibacterial). These flat cyclic compounds can slide into the DNA double helix between base pairs—like a card being inserted into a deck of playing cards—and interfere with the normal replication of DNA.

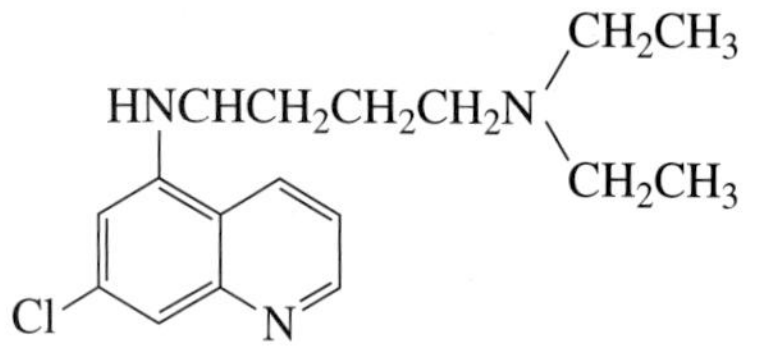

$HNCHCH_2CH_2CH_2N(CH_2CH_3)_2$ (Cl, N)

chloroquine

H_2N (N) NH_2

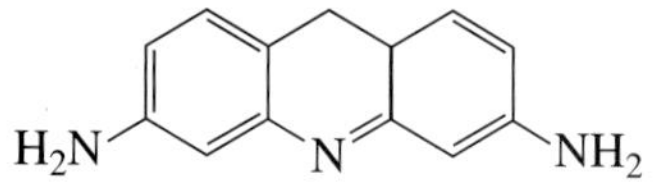

3,6-diaminoacridine

Knowing something about the molecular basis of drug action—such as how a drug interacts with a receptor—allows scientists to design and synthesize compounds that might have a desired biological activity. For example, when excess histamine is produced by the body, it causes the symptoms associated with the common cold and allergic responses. This is thought to be the result of the protonated ethylamino group anchoring the histamine molecule to a negatively charged portion of the histamine receptor.

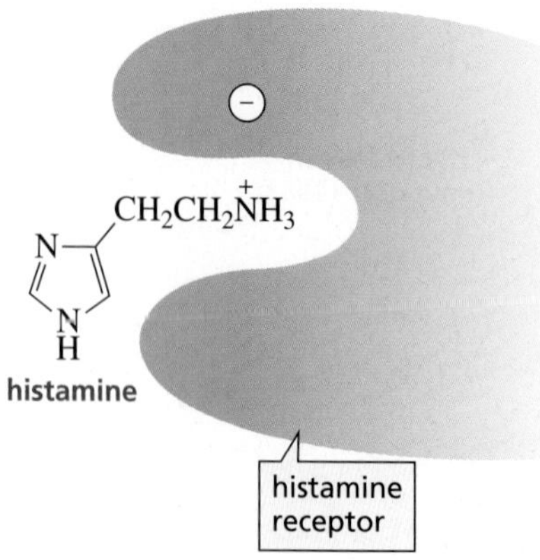

Drugs that interfere with the natural action of histamine—called antihistamines—bind to the histamine receptor but do not trigger the same response as histamine. Like histamine, these drugs have a protonated amino group that binds to the receptor. The drugs also have bulky groups that keep the histamine molecule from approaching the receptor.

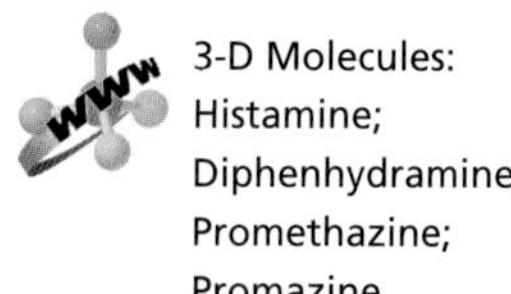

3-D Molecules: Histamine; Diphenhydramine; Promethazine; Promazine

antihistamines

$(C_6H_5)_2CHOCH_2CH_2\overset{+}{N}H(CH_3)_2$

diphenhydramine
Benadryl®

$NCH_2CH(CH_3)\overset{+}{N}H(CH_3)_2$ (phenothiazine ring, S)

promethazine
Promine®

$NCH_2CH_2CH_2\overset{+}{N}H(CH_3)_2$ (phenothiazine ring, S)

promazine
Talofen®

Acetylcholine is a neurohormone that enhances peristalsis, wakefulness, and memory and is essential for nerve transmission. A deficiency of brain cell receptors that bind acetylcholine—cholinergic receptors—contributes to the characteristic loss of memory in Alzheimer's disease. Cholinergic receptors are structurally similar to those that bind histamine. Therefore, antihistamines and cholinergic agents show overlapping activities. As a result, the antihistamine diphenhydramine has been used to treat insomnia and to combat motion sickness.

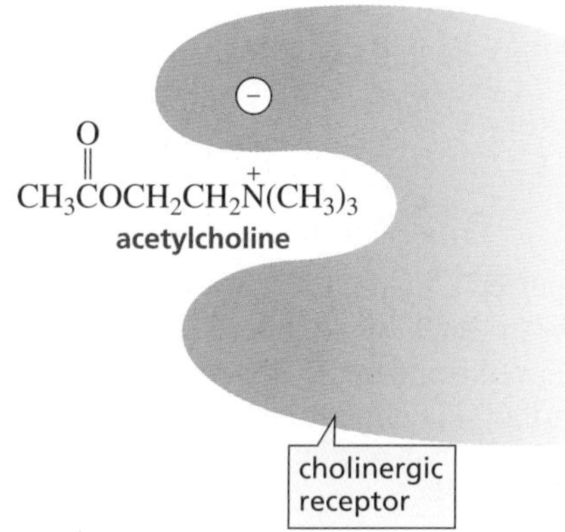

Excess histamine production by the body also causes the hypersecretion of stomach acid by the cells of the stomach lining, leading to the development of ulcers. The antihistamines that block the histamine receptors—thereby preventing the allergic responses associated with excess histamine production—have no effect on HCl production. This fact led scientists to conclude that a second kind of histamine receptor triggers the release of acid into the stomach.

Because 4-methylhistamine was found to cause weak inhibition of HCl secretion, it was used as a lead compound. About 500 molecular modifications were performed over a 10-year period before four clinically useful antiulcer agents were found. Two of these are Tagamet® and Zantac®. Notice that steric blocking of the receptor site is not a factor in these compounds. Compared with the antihistamines, the effective antiulcer drugs have more polar rings and longer side chains.

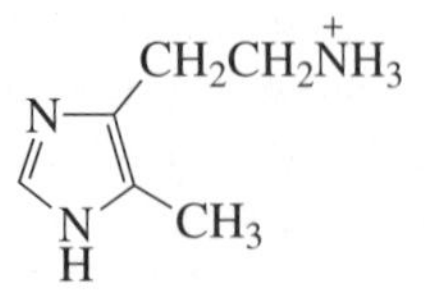

4-methylhistamine

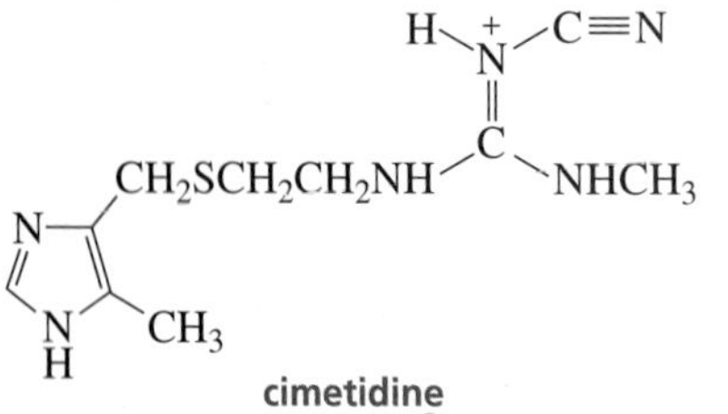

cimetidine
Tagamet®

ranitidine
Zantac®

Tagamet® has the same imidazole ring as 4-methylhistidine, but it has a sulfur atom and a functional group based on guanidine (Section 23.1). Zantac® has a different heterocyclic ring, and although its side chain is similar to that of Tagamet®, it does not contain a guanidino group.

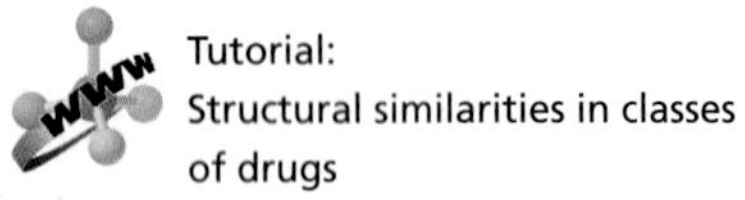

Tutorial:
Structural similarities in classes of drugs

Observations that serotonin was implicated in the generation of migraine attacks led to the development of drugs that bind to serotonin receptors. Sumatriptan, introduced in 1991, relieved not only the pain associated with migraines, but also many of migraine's other symptoms, including sensitivity to light and sound and nausea.

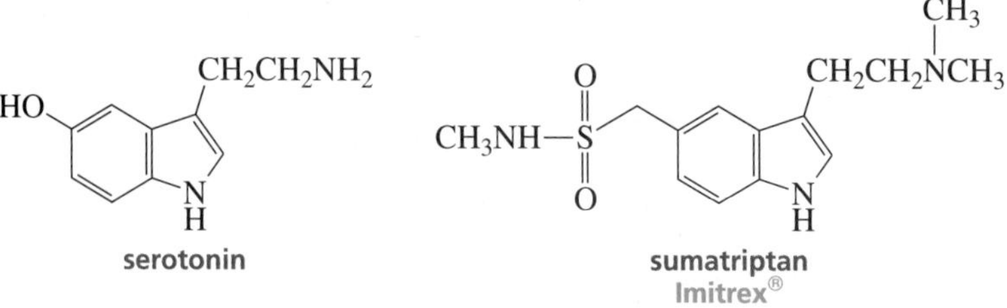

serotonin

sumatriptan
Imitrex®

The success of sumatriptan spurred a search for other antimigraine agents by molecular modification, and three new triptans were introduced in 1997 and 1998. These second-generation triptans showed some improvements over sumatriptan—specifically, a longer half-life, reduced cardiac side effects, and improved CNS penetration.

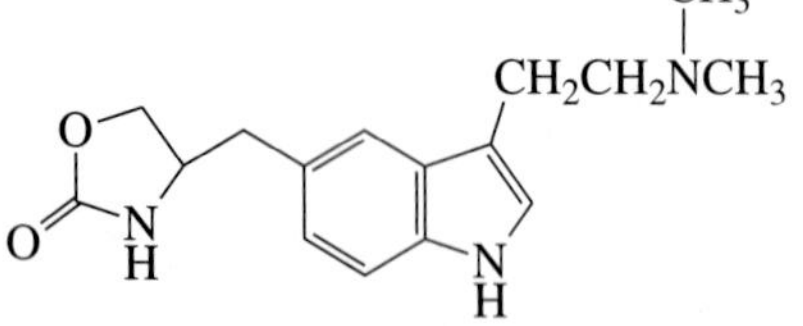

zolmitriptan
Zomig®

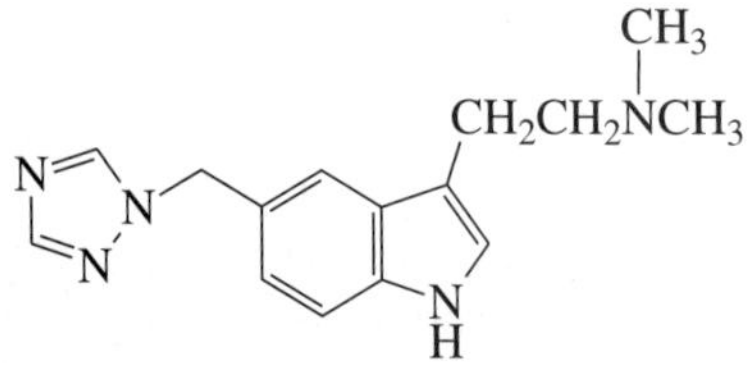

rizatriptan
Maxalt®

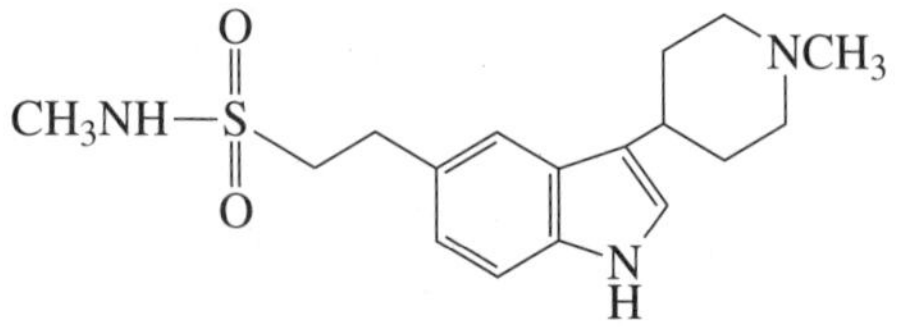

naratriptan
Amerge®

In screening modified compounds, it is not unusual to find a compound with a completely different pharmacological activity than the lead compound. For example, a molecular modification of a sulfonamide, an antibiotic, led to tolbutamide, a drug with hypoglycemic activity (Section 25.8).

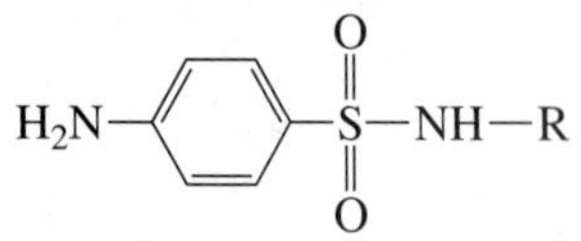

a sulfonamide

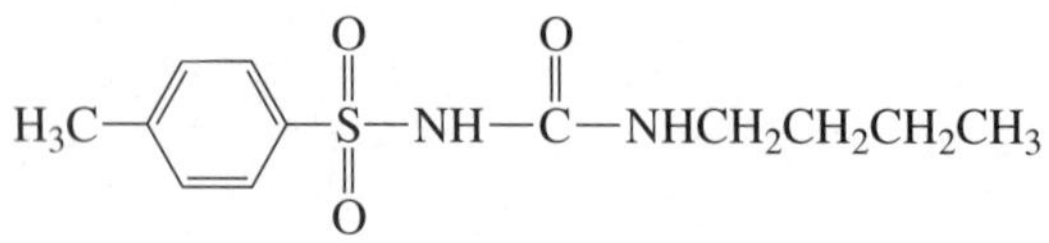

tolbutamide

administered directly to the patient, it would be nonspecific, reacting with whatever nucleophile it first encountered.

When two drugs are given simultaneously to a patient, their combined effect can be additive, antagonistic, or synergistic. That is, the effect of two drugs used in combination can be equal to, less than, or greater than the sum of the effects obtained when the drugs are administered individually. Administering penicillin and the sulfone in combination results in **drug synergism**: The sulfone inhibits the penicillinase, so penicillin won't be destroyed and will be able to inhibit the enzyme that synthesizes bacterial cell walls.

Another reason to administer two drugs in combination is that if some bacteria are resistant to one of them, the second drug will minimize the chance that the resistant strain will proliferate. For example, two antimicrobial agents, isoniazid and rifampin, are given in combination to treat tuberculosis.

isoniazid
Nydrazid®

rifampin
Rifadin®

3-D Molecules: Rifampin; Isoniazid

Typically, it takes a bacterial strain 15 to 20 years to evolve resistance to antibiotics. The fluoroquinolones, the last class of antibiotics to be discovered until very recently, were discovered more than 30 years ago, so **drug resistance** has become an increasingly important problem in medicinal chemistry. More and more bacteria have become resistant to all antibiotics—even vancomycin, until recently the antibiotic of last resort.

The antibiotic activity of the fluoroquinolones results from their ability to inhibit DNA gyrase, an enzyme required for transcription (Section 27.10). Fortunately, the bacterial and mammalian forms of the enzyme are sufficiently different that the fluoroquinolones inhibit only the bacterial enzyme.

There are many different fluoroquinolones. All have fluorine substituents, which increase the lipophilicity of the drug, enabling it to penetrate into tissues and cells. If either the carboxyl group or the double bond in the 4-pyridinone ring is removed, all activity is lost. By changing the substituents on the piperazine ring, excretion of the drug can be shifted from the liver to the kidney, which is useful to patients with impaired liver function. The substituents on the piperazine ring also affect the half-life of the drug.

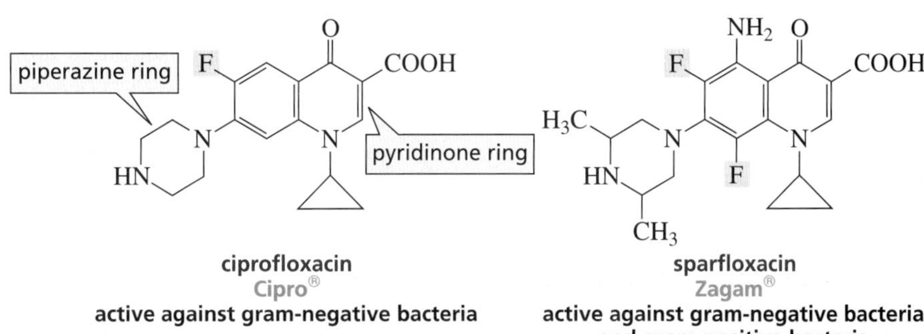

ciprofloxacin
Cipro®
active against gram-negative bacteria

sparfloxacin
Zagam®
active against gram-negative bacteria and gram-positive bacteria

The approval of Zyvox® by the FDA in April 2000 was met with great relief by the medical community. Zyvox® is the first in a new family of antibiotics: the oxazolidinones. In clinical trials, Zyvox® was found to cure 75% of the patients infected with bacteria that had become resistant to all other antibiotics.

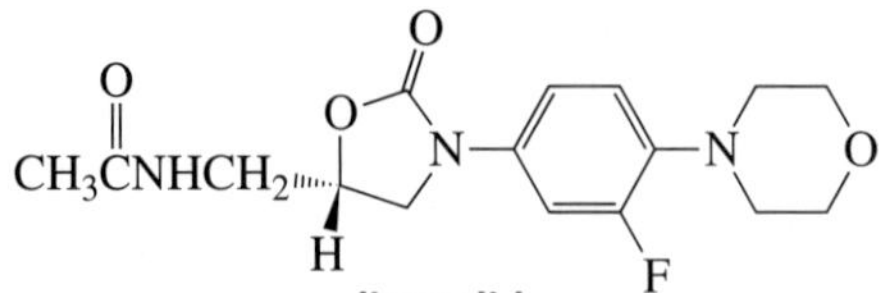

CH_3CNHCH_2 (linezolid structure)

linezolid
Zyvox®

Zyvox® is a synthetic compound designed by scientists to inhibit bacterial growth at a point different from that at which any other antibiotic exerts its effect. Zyvox® inhibits the initiation of protein synthesis by preventing the formation of the complex between the first amino-acid-bearing tRNA, mRNA, and the 30S ribosome (Section 27.13). Because of the drug's new mode of activity, resistance is expected to be rare at first and, hopefully, slow to emerge.

30.8 Designing a Suicide Substrate

It is important for a drug to have a minimum of undesirable side effects. A drug must be administered in sufficient quantity to achieve a therapeutic effect; however, too much of a drug can be lethal. The therapeutic index of a drug is the ratio of the lethal dose to the therapeutic dose. The higher the **therapeutic index**, the greater is the margin of safety of the drug.

Penicillin is an effective antibiotic that has a high therapeutic index because it interferes with cell wall synthesis—and bacterial cells have walls, but human cells do not. What else is characteristic about cell walls that could lead to the design of an antibiotic? We know that enzymes and other proteins are polymers of L-amino acids. Cell walls, however, contain both L-amino acids and D-amino acids. Therefore, if the racemization of naturally occurring L-amino acids to mixtures of L- and D-amino acids could be prevented, D-amino acids would not be available for incorporation into cell walls, and bacterial cell wall synthesis could be stopped.

We have seen that amino acid racemization is catalyzed by an enzyme that requires pyridoxal phosphate as a coenzyme (Section 25.6). What we need, then, is a compound that will inhibit this enzyme. Because the natural substrate for the enzyme is an amino acid, an amino acid analog should be a good inhibitor.

The first step in racemization is removal of the α-hydrogen of the amino acid. If the inhibitor has a leaving group on the β-carbon, the electrons left behind when the proton is removed can displace the leaving group instead of being delocalized into the pyridine ring. (Compare the mechanism shown here with that shown for racemization in Section 25.6.) Transimination with the enzyme forms an α,β-unsaturated amino acid that reacts irreversibly with the imine formed by the enzyme and the coenzyme. Because the enzyme is now bound to the coenzyme in an amine linkage rather than in an imine linkage, the enzyme can no longer undergo a transimination reaction with its amino acid substrate. The enzyme has thus been irreversibly inactivated. This is another example of an inhibitor that does not become chemically active until it is at the active site of the targeted enzyme.

an elimination reaction

an α, β-unsaturated amino acid

transimination with $E{-}(CH_2)_4NH_2$

an amine linkage

inactivated enzyme

30.9 Quantitative Structure–Activity Relationships (QSAR)

The enormous cost involved in synthesizing and testing thousands of modified compounds in an attempt to find an active drug led scientists to develop a more rational approach—called **rational drug design**—to the design of biologically active molecules. They realized that if a physical or chemical property of a series of compounds could be correlated with biological activity, they would know what property of the drug was related to that particular activity. Armed with this knowledge, scientists could design compounds that would have a good chance of exhibiting the desired activity. This strategy would be a great improvement over the random approach to molecular modification that traditionally had been employed.

The first hint that a physical property of a drug could be related to biological activity appeared almost 100 years ago when scientists recognized that chloroform ($CHCl_3$), diethyl ether, cyclopropane, and nitrous oxide (N_2O) were all useful general anesthetics. Clearly, the chemical structures of these diverse compounds could not account for their similar pharmacological effects. Instead, some physical property must explain the similarity of their biological activities.

In the early 1960s, Corwin Hansch postulated that the **biological activity** of a drug depended on two processes. The first is *distribution*: A drug must be able to get from the point where it enters the body to the receptor where it exerts its effect. For example, an anesthetic must be able to cross the aqueous milieu (blood) and penetrate the lipid barrier of nerve cell membranes. The second process is *binding*: When a drug reaches its receptor, it must interact properly with it.

Corwin H. Hansch *was born in North Dakota in 1918. He received a B.S. from the University of Illinois and a Ph.D. from New York University. He has been a professor of chemistry at Pomona College since 1946.*

Chloroform, diethyl ether, cyclopropane, and nitrous oxide were each put into a mixture of 1-octanol and water. 1-Octanol was chosen as the nonpolar solvent because, with its long chain and polar head group, it is a good model of a biological membrane. When the amount of drug dissolving in each of the layers was measured, it turned out that they all had a similar **distribution coefficient** (the ratio of the amount dissolving in 1-octanol to the amount dissolving in water). In other words, the distribution coefficient could be related to biological activity. Compounds with lower distribution coefficients—more polar compounds—could not penetrate the nonpolar cell membrane; compounds with greater distribution coefficients—more nonpolar compounds—could not cross the aqueous phase. This meant that the distribution coefficient of a compound could be used to determine whether a compound should be tested in vivo. The technique of relating a property of a series of compounds to biological activity is known as a **quantitative structure–activity relationship (QSAR)**.

Determining the physical property of a drug cannot take the place of in vivo testing because how the drug will behave once it reaches a suitable receptor cannot be predicted by the distribution coefficient alone. Nevertheless, QSAR analysis provides a way to lead chemists to structures with the greatest probability of having the desired biological activity. In this way, molecular modification that would lead to compounds without the desired activity can be avoided.

In the following example, QSAR was useful not only in determining the structure of a potentially active drug but also in determining something about the structure of the receptor site. A series of substituted 2,4-diaminopyrimidines, used as inhibitors of dihydrofolate reductase (Section 25.8), was investigated.

NH_2, R, N, H_2N, N, R′

a 2,4-diaminopyrimidine

The potency of the inhibitors could be described by the equation

$$\text{potency} = 0.80\pi - 7.34\sigma - 8.14$$

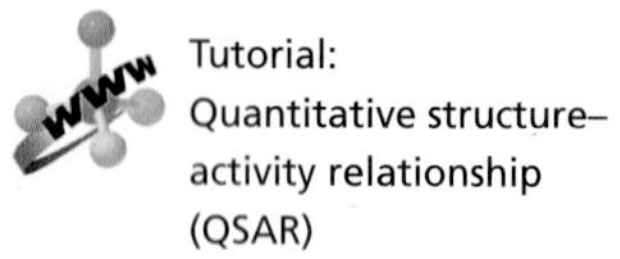
Tutorial: Quantitative structure–activity relationship (QSAR)

where σ and π are substituent parameters.

The σ parameter is a measure of the electron-donating or electron-withdrawing ability of the substituents R and R′. The negative coefficient of σ indicates that potency is increased by electron donation (Chapter 17, Problem 75). The fact that increasing the basicity of the drug increases its potency suggests that the protonated drug is more active than the nonprotonated drug.

The π parameter is a measure of the hydrophobicity of the substituents. Potency was found to be better related to π when the π value for the more hydrophobic of the two substituents was used, rather than the sum of the π values for both substituents. This finding suggests that the receptor has a hydrophobic pocket that can accommodate one, but not both, substituents.

In a search for a new analgesic, the potency of the drug was found to be described by the following equation, where *HA* indicates whether R is a hydrogen bond acceptor and where *B* is a steric factor:

R, COOH, N, O

$$\textbf{potency} = -4.45 - 0.73HA + 6.5B - 1.55(B)^2$$

Analysis indicated that the vinyl substituted compound should be prepared.

In a search for a drug to be used to treat leukemia, a QSAR analysis showed that the antileukemic activity of a series of substituted triazines was related to the electron-donating ability of the substituent. When, however, another QSAR analysis showed that the toxicity of these compounds was also related to the electron-donating ability of the substituent, it was decided that it would be fruitless to continue synthesizing and testing this class of compounds.

In addition to solubility and substituent parameters, some of the properties that have been correlated with biological activity are oxidation–reduction potential, molecular size, interatomic distance between functional groups, degree of ionization, and configuration.

30.10 Molecular Modeling

Because the shape of a molecule determines whether it will be recognized by a receptor and, therefore, whether it will exhibit biological activity, compounds with similar biological activity often have similar structures. Because computers can draw molecular models of compounds on a video display and move them around to assume different conformations, computer **molecular modeling** allows more rational drug design. There are computer programs that allow chemists to scan existing collections of thousands of compounds to find those with appropriate structural and conformational properties.

Any compound that shows promise can be drawn on a computer display, along with the three-dimensional image of a receptor site. For example, the binding of netropsin, an antibiotic with a wide range of antimicrobial activity, to the minor groove of DNA is shown in Figure 30.1.

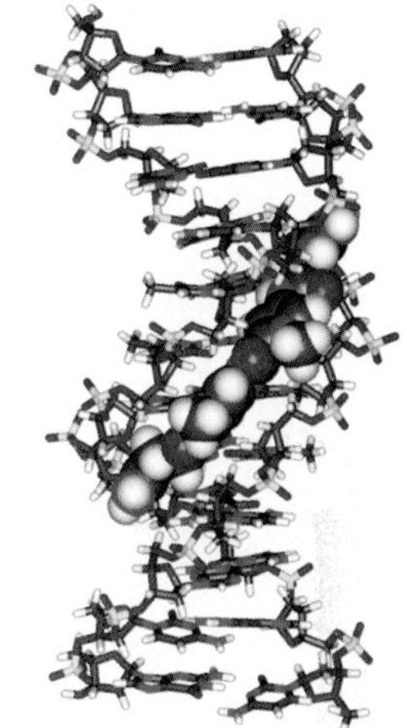

▲ **Figure 30.1**
The antibiotic netropsin bound in the minor groove of DNA.

The fit between the compound and the receptor may suggest modifications that can be made to the compound that result in more favorable binding. In this way, the selection of compounds to be synthesized for the purpose of screening for biological activity can be more rational and will allow pharmacologically active compounds to be discovered more rapidly. The technique will become more valuable as scientists learn more about receptor sites.

30.11 Combinatorial Organic Synthesis

The need for large collections of compounds that can be screened for biological activity in the constant search for new drugs has led organic chemists to a synthetic strategy that employs the concept of mass production. This strategy, called **combinatorial organic synthesis**, involves the synthesis of a large group of related compounds—known as a library—by covalently connecting sets of building blocks of various structures. For example, if a compound can be synthesized by connecting three different building blocks, and if each set of building blocks contains 10 interchangeable compounds, then 1000 $(10 \times 10 \times 10)$ different compounds can be prepared. This approach clearly mimics nature, which uses amino acids and nucleic acids as building blocks to synthesize an enormous number of different proteins and nucleic acids.

The first requirement in combinatorial synthesis is the availability of an assortment of reactive small molecules to be used as building blocks. Because of the ready availability of amino acids, combinatorial synthesis made its first appearance in the creation of peptide libraries. Peptides, however, have limited use as therapeutic agents because they generally cannot be taken orally and are rapidly metabolized. Currently, organic chemists create libraries of small organic molecules for use in a combinatorial approach for modifying *lead compounds* or as a complement to *rational drug design*.

An example of the approach used in combinatorial synthesis is the creation of a library of benzodiazepines. These compounds can be thought of as originating from three different sets of building blocks: a substituted 2-aminobenzophenone, an amino acid, and an alkylating agent.

R^4, R^1, R^3, R^2 — a benzodiazepine ⟹ R^1, NH_2, R^2 — a 2-aminobenzophenone; ^-O, R^3, H_3N^+ — an amino acid; R^4—X — an alkylating agent

The 2-aminobenzophenone is attached to a solid support in a manner that allows it to be readily removed by acid hydrolysis (Figure 30.2). The amino acid—*N*-protected and activated by being converted into an acyl fluoride—is then added. After the amide is formed, the protecting group is removed and the seven-membered ring is created as a result of imine formation. A base is added to remove the amide hydrogen, forming a nucleophile that reacts with the added alkylating agent. The final product is then removed from the solid support.

In order to synthesize a library of these compounds, the solid support containing the 2-aminobenzophenone can be divided into several portions, and a different amino acid can then be added to each portion. Each ring-closed product can also be divided into several portions, and a different alkylating agent can be added to each portion. In this way, many different benzodiazepines can be prepared. Note that while a combinatorial synthesis does not have to have one of the reactants anchored to a solid support, such a support improves yields because none of the product is lost during the purification step.

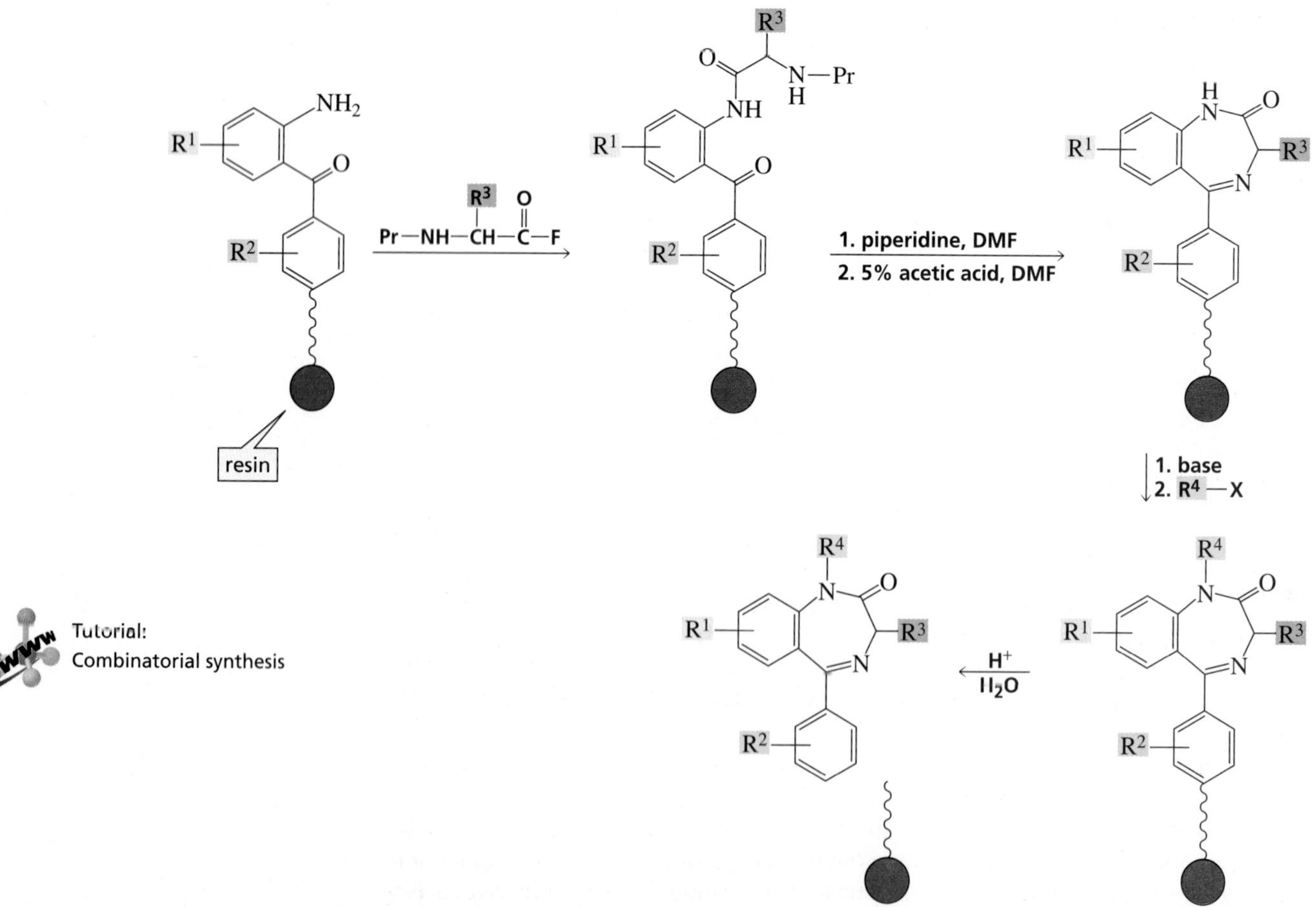

▲ **Figure 30.2**
Combinatorial organic synthesis of benzodiazepines.

30.12 Antiviral Drugs

Relatively few clinically useful drugs have been developed for viral infections. This slow progress is due to the nature of viruses and the way they replicate. Viruses are smaller than bacteria. A virus consists of nucleic acid—either DNA or RNA—surrounded by a coat of protein. A virus penetrates a host cell or merely injects its nucleic acid into the cell. In either case, the nucleic acid is transcribed and is integrated into the host genome.

Most **antiviral drugs** are analogs of nucleosides, interfering with DNA or RNA synthesis. In this way, they prevent the virus from replicating. For example, acyclovir, the drug used against herpes viruses, has a three-dimensional shape similar to guanine. Acyclovir can, therefore, fool the virus into incorporating it instead of guanine into the virus's DNA. Once this happens, the DNA strand can no longer grow, because acyclovir lacks a 3′-OH group. The terminated DNA binds to DNA polymerase and inactivates it irreversibly (Section 27.7).

Cytarabine, used for acute myelocytic leukemia, competes with cytosine for incorporation into viral DNA. Cytarabine contains an arabinose rather than a ribose (Table 22.1). The 2′-OH group in the β-position prevents the bases in DNA from stacking properly (Section 27.7).

Ribavirin is a broad-spectrum antiviral agent that inhibits viral mRNA synthesis (Section 27.10). A step in the metabolic pathway responsible for the synthesis of guanosine triphosphate (GTP) converts inosine monophosphate (IMP) into xanthosine monophosphate (XMP). Ribarvirin is a competitive inhibitor of the enzyme that catalyzes that step. Thus, ribarvirin interferes with the synthesis of GTP and, therefore, with all nucleic acid synthesis.

Idoxuridine is approved in the United States only for the topical treatment of ocular infections, although it is used for herpes infections in other countries. Idoxuridine has an iodo group in place of the methyl group of thymine. The drug is incorporated into DNA in place of thymine. Chain elongation can continue because idoxuridine has a 3′-OH group. The DNA that has incorporated idoxuridine, however, is more easily broken and is also not transcribed properly.

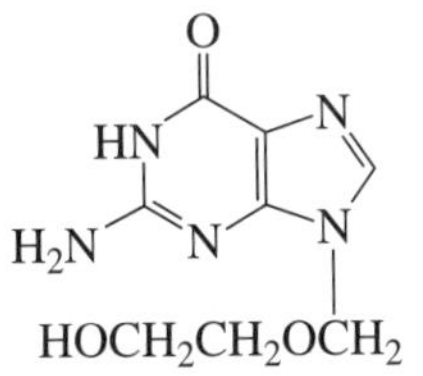

$HOCH_2CH_2OCH_2$
acyclovir
Aclovir®
used against herpes simplex infections

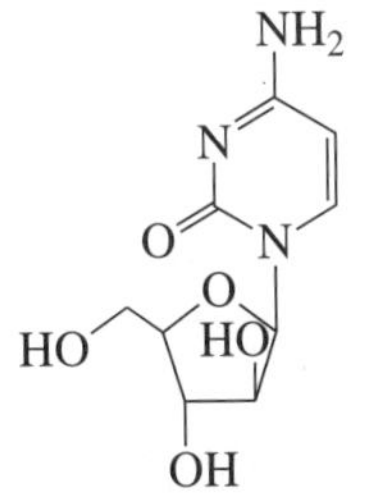

cytarabine
Cytosar®
used against acute myelocytic leukemia

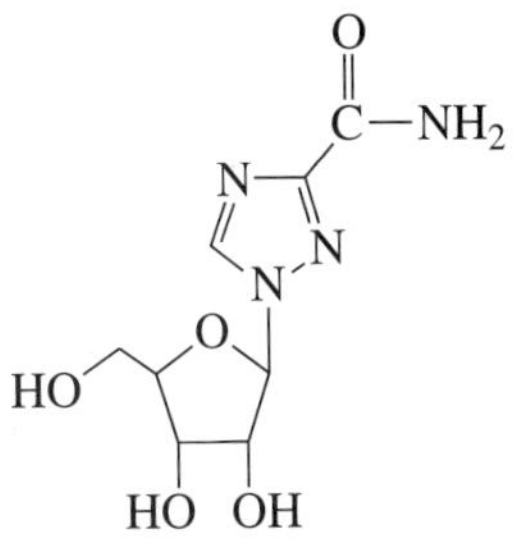

ribavirin
Viramid®
a broad-spectrum antiviral agent

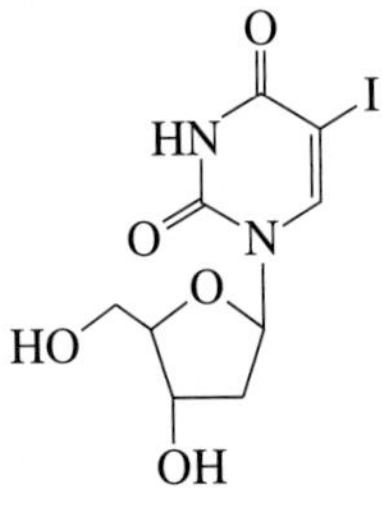

idoxuridine
Herplex®
approved for topical ophthalmic use

30.13 Economics of Drugs • Governmental Regulations

The average cost of launching a new drug is \$100–\$500 million. This cost has to be amortized quickly by the manufacturer because the starting date of a patent is the date the drug is first discovered. A patent is good for 20 years from the date it is applied for, but because it takes an average of 12 years to market a drug after its initial discovery, the patent protects the discoverer of the drug for an average of only 8 years. It is only during the 8 years of patent protection that marketing the drug can provide enough profit so that costs can be recovered and income can be generated to carry out research on new drugs. In addition, the average lifetime of a drug is only 15 to 20 years. After that time, it is generally replaced by a newer and improved drug. Only about 1 in 3 drugs actually makes a profit for the company.

Why does it cost so much to develop a new drug? First of all, the Food and Drug Administration (FDA) has very high standards that must be met before it approves a drug for a particular use. Before the U.S. government became involved with the regulation of drugs, it was not uncommon for charlatans to dispense useless and even harmful medical preparations. Starting in 1906, Congress passed laws governing the manufacture, distribution, and use of drugs. These laws are amended from time to time to reflect changing situations. The present law requires that all new drugs be thoroughly tested for effectiveness and safety before they are used by physicians.

An important factor leading to the high price of many drugs is the low success rate in progressing from the initial concept to an approved product: Only 1 or 2 of every 100 compounds tested become lead compounds; out of a 100 structural modifications of a lead compound, only 1 is worthy of further study; and only 10% of these compounds actually become marketable drugs.

ORPHAN DRUGS

Because of the high cost associated with developing a drug, pharmaceutical companies are reluctant to carry out research on drugs for rare diseases. Even if a company were to find a drug, there would be no way to recoup its expenditure, because of the limited demand. In 1983, the U.S. Congress passed the Orphan Drugs Act. The act creates public subsidies to fund research and provides tax credits for up to 50% of the costs of developing and marketing drugs—called **orphan drugs**—for diseases or conditions that affect fewer than 200,000 people. In addition, the company that develops the drug has four years of exclusive marketing rights if the drug is nonpatentable. In the 10 years prior to the passage of this act, fewer than 10 orphan drugs were developed. Today, there are more than 100, and some 600 others are in development.

Drugs originally developed as orphan drugs include AZT (to treat AIDS), Taxol® (to treat ovarian cancer), Exosurf Neonatal® (to treat respiratory distress syndrome in infants), and Opticrom® (to treat corneal swelling).

Summary

A **drug** is a compound that interacts with a biological molecule, triggering a physiological response. Each drug has a **proprietary name** that can be used only by the owner of the patent, which is valid for 20 years. Once a patent expires, other drug companies can market the drug under a **generic name** that can be used by any pharmaceutical company. Drugs that no one wants to develop because they would be used for diseases or conditions that affect fewer than 200,000 people are called **orphan drugs**. The Food and Drug Administration (FDA) has very high standards that must be met before it approves a drug for a particular use. The average cost of launching a new drug is about $230 million.

The prototype for a new drug is called a **lead compound**. Changing the structure of a lead compound is called **molecular modification**. A **random screen** (or **blind screen**) is a search for a pharmacologically active lead compound without having any information about what structures might show activity. The technique of relating a property of a series of compounds to biological activity is known as a **quantitative structure–activity relationship (QSAR)**.

Many drugs exert their physiological effects by binding to a specific cellular binding site called a **receptor**. Some drugs act by **inhibiting enzymes** or by binding to nucleic acids. Most **antiviral drugs** are analogs of nucleosides, interfering with DNA or RNA synthesis and thereby preventing the virus from replicating.

A **bacteriostatic drug** inhibits the further growth of bacteria; a **bactericidal drug** kills the bacteria. In recent years, many bacteria have become resistant to all antibiotics, so **drug resistance** has become an increasingly important problem in medicinal chemistry.

The therapeutic index of a drug is the ratio of the lethal dose to the therapeutic dose. The higher the **therapeutic index**, the greater is the margin of safety of the drug. The effect of two drugs used in combination is called **drug synergism**.

Large collections of compounds that can be screened for biological activity are prepared by **combinatorial organic synthesis**—the synthesis of a group of related compounds by covalently connecting sets of building blocks.

Key Terms

antiviral drug (p. 1225)
bactericidal drug (p. 1212)
bacteriostatic drug (p. 1212)
biological activity (p. 1221)
blind screen (p. 1211)
brand name (p. 1207)
combinatorial organic synthesis (p. 1223)
distribution coefficient (p. 1222)
drug (p. 1204)
drug resistance (p. 1219)
drug synergism (p. 1219)
generic name (p. 1208)
lead compound (p. 1208)
molecular modeling (p. 1223)
molecular modification (p. 1208)
orphan drug (p. 1226)
proprietary name (p. 1207)
quantitative structure–activity relationship (QSAR) (p. 1222)
random screen (p. 1211)
rational drug design (p. 1221)
receptor (p. 1214)
suicide inhibitor (p. 1218)
therapeutic index (p. 1220)
trademark (p. 1207)
trade name (p. 1207)

Problems

1. What is the chemical name of each of the following drugs?
   a. benzocaine
   b. procaine

2. Based on the lead compound for the development of procaine and lidocaine, propose structures for other compounds that you would like to see tested for use as anesthetics.

3. Which of the following compounds is more likely to exhibit activity as a tranquilizer?

   CH_3, N, O, CH_3O, N — or — CH_3, N, O, CH_3C, O, N

4. Which compound is more likely to be a general anesthetic?

   $CH_3CH_2CH_2OH$ or $CH_3OCH_2CH_3$

5. What accounts for the ease of imine formation between penicillinase and the sulfone antibiotic that counteracts penicillin resistance?

6. For each of the following pairs of compounds, indicate the compound that you would expect to be a more potent inhibitor of dihydrofolate reductase:

   a. NH_2, Cl, N, H_2N, N, CH_2CHCH_3, CH_3 — or — NH_2, NO_2, N, H_2N, N, CH_2CHCH_3, CH_3

   b. NH_2, CH_2CH_3, N, H_2N, N, CH_2CHCH_3, CH_3 — or — NH_2, CH_3, N, H_2N, N, $CH_2CHCH_2CH_3$, CH_3

7. The lethal dose of tetrahydrocannabinol in mice is 2.0 g/kg, and the therapeutic dose is 20 mg/kg. The lethal dose of sodium pentothal in mice is 100 mg/kg, and the therapeutic dose is 30 mg/kg. Which is the safer drug?

8. The following compound is a suicide inhibitor of the enzyme that catalyzes amino acid racemization:

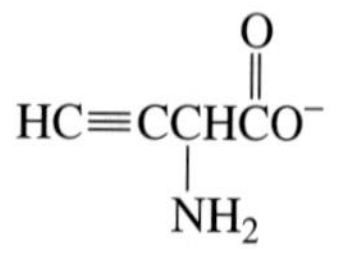

$$HC{\equiv}C\underset{NH_2}{\underset{|}{C}}H\overset{O}{\overset{\|}{C}}O^-$$

Propose a mechanism that explains how this compound irreversibly inactivates the enzyme.

9. Explain how each of the antiviral drugs shown in Section 30.12 differs from the naturally occurring nucleoside that it most closely resembles.

10. Show a mechanism for the formation of a benzodiazepine 4-oxide from the reaction of a quinazoline 3-oxide with methylamine.

11. Show how Valium® could be synthesized from benzoyl chloride, *para*-chloroaniline, methyl iodide, and the ethyl ester of glycine.

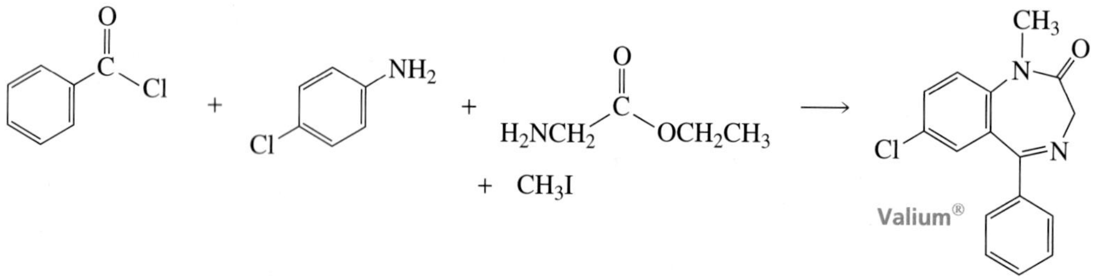

C_6H_5COCl + $p\text{-}ClC_6H_4NH_2$ + $H_2NCH_2\overset{O}{\overset{\|}{C}}OCH_2CH_3$ + CH_3I ⟶ Valium®

12. Show how Tagamet® could be synthesized from the indicated starting materials.

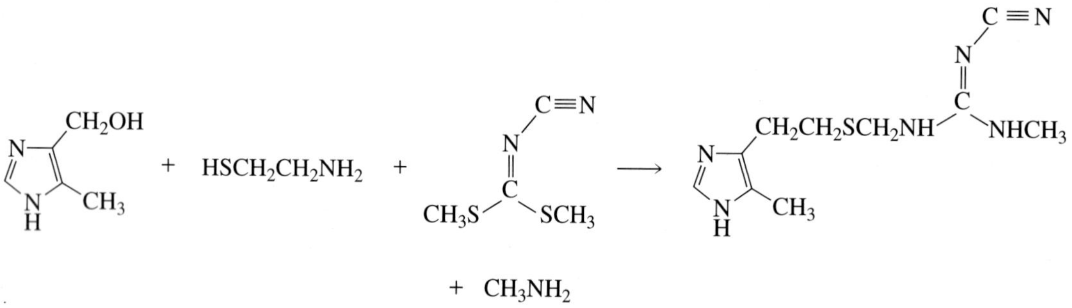

4-(hydroxymethyl)-5-methylimidazole (CH_2OH, CH_3) + $HSCH_2CH_2NH_2$ + $(CH_3S)_2C{=}N{-}C{\equiv}N$ + CH_3NH_2 ⟶ imidazole with CH_3 and $CH_2CH_2SCH_2NH{-}C({=}N{-}C{\equiv}N){-}NHCH_3$

Appendix I

Physical Properties of Organic Compounds

Physical Properties of Alkenes

Name	Structure	mp (°C)	bp (°C)	Density (g/mL)
Ethene	$CH_2{=}CH_2$	−169	−104	
Propene	$CH_2{=}CHCH_3$	−185	−47	
1-Butene	$CH_2{=}CHCH_2CH_3$	−185	−6.3	
1-Pentene	$CH_2{=}CH(CH_2)_2CH_3$		30	0.641
1-Hexene	$CH_2{=}CH(CH_2)_3CH_3$	−138	64	0.673
1-Heptene	$CH_2{=}CH(CH_2)_4CH_3$	−119	94	0.697
1-Octene	$CH_2{=}CH(CH_2)_5CH_3$	−101	122	0.715
1-Nonene	$CH_2{=}CH(CH_2)_6CH_3$	−81	146	0.730
1-Decene	$CH_2{=}CH(CH_2)_7CH_3$	−66	171	0.741
cis-2-Butene	*cis*-$CH_3CH{=}CHCH_3$	−180	37	0.650
trans-2-Butene	*trans*-$CH_3CH{=}CHCH_3$	−140	37	0.649
Methylpropene	$CH_2{=}C(CH_3)_2$	−140	−6.9	0.594
cis-2-Pentene	*cis*-$CH_3CH{=}CHCH_2CH_3$	−180	37	0.650
trans-2-Pentene	*trans*-$CH_3CH{=}CHCH_2CH_3$	−140	37	0.649
Cyclohexene		−104	83	0.811

Physical Properties of Alkynes

Name	Structure	mp (°C)	bp (°C)	Density (g/mL)
Ethyne	$HC{\equiv}CH$	−82	−84.0	
Propyne	$HC{\equiv}CCH_3$	−101.5	−23.2	
1-Butyne	$HC{\equiv}CCH_2CH_3$	−122	8.1	
2-Butyne	$CH_3C{\equiv}CCH_3$	−24	27	0.694
1-Pentyne	$HC{\equiv}C(CH_2)_2CH_3$	−98	39.3	0.695
2-Pentyne	$CH_3C{\equiv}CCH_2CH_3$	−101	55.5	0.714
3-Methyl-1-butyne	$HC{\equiv}CCH(CH_3)_2$		29	0.665
1-Hexyne	$HC{\equiv}C(CH_2)_3CH_3$	−132	71	0.715
2-Hexyne	$CH_3C{\equiv}C(CH_2)_2CH_3$	−92	84	0.731
3-Hexyne	$CH_3CH_2C{\equiv}CCH_2CH_3$	−101	81	0.725
1-Heptyne	$HC{\equiv}C(CH_2)_4CH_3$	−81	100	0.733
1-Octyne	$HC{\equiv}C(CH_2)_5CH_3$	−80	127	0.747
1-Nonyne	$HC{\equiv}C(CH_2)_6CH_3$	−50	151	0.757
1-Decyne	$HC{\equiv}C(CH_2)_7CH_3$	−44	174	0.766

Physical Properties of Cyclic Saturated Alkanes

Name	mp (°C)	bp (°C)	Density (g/mL)
Cyclopropane	−128	−33	
Cyclobutane	−80	−12	
Cyclopentane	−94	50	0.751
Cyclohexane	6.5	81	0.779
Cycloheptane	−12	118	0.811
Cyclooctane	14	149	0.834
Methylcyclopentane	−142	72	0.749
Methylcyclohexane	−126	100	0.769
cis-1,2-Dimethylcyclopentane	−62	99	0.772
trans-1,2-Dimethylcyclopentane	−120	92	0.750

Physical Properties of Ethers

Name	Structure	mp (°C)	bp (°C)	Density (g/mL)
Dimethyl ether	CH_3OCH_3	−141	−24.8	
Diethyl ether	$CH_3CH_2OCH_2CH_3$	−116	34.6	0.706
Dipropyl ether	$CH_3(CH_2)_2O(CH_2)_2CH_3$	−123	88	0.736
Diisopropyl ether	$(CH_3)_2CHOCH(CH_3)_2$	−86	69	0.725
Dibutyl ether	$CH_3(CH_2)_3O(CH_2)_3CH_3$	−98	142	0.764
Divinyl ether	$CH_2{=}CHOCH{=}CH_2$		35	
Diallyl ether	$CH_2{=}CHCH_2OCH_2CH{=}CH_2$		94	0.830
Tetrahydrofuran	O	−108	66	0.889
Dioxane	O O	12	101	1.034

Physical Properties of Alcohols

Name	Structure	mp (°C)	bp (°C)	Solubility (g/100 g H_2O at 25 °C)
Methanol	CH_3OH	−97.8	64	∞
Ethanol	CH_3CH_2OH	−114.7	78	∞
1-Propanol	$CH_3(CH_2)_2OH$	−127	97.4	∞
1-Butanol	$CH_3(CH_2)_3OH$	−90	118	7.9
1-Pentanol	$CH_3(CH_2)_4OH$	−78	138	2.3
1-Hexanol	$CH_3(CH_2)_5OH$	−52	157	0.6
1-Heptanol	$CH_3(CH_2)_6OH$	−36	176	0.2
1-Octanol	$CH_3(CH_2)_7OH$	−15	196	0.05
2-Propanol	$CH_3CHOHCH_3$	−89.5	82	∞
2-Butanol	$CH_3CHOHCH_2CH_3$	−115	99.5	12.5
2-Methyl-1-propanol	$(CH_3)_2CHCH_2OH$	−108	108	10.0
2-Methyl-2-propanol	$(CH_3)_3COH$	25.5	83	∞
3-Methyl-1-butanol	$(CH_3)_2CH(CH_2)_2OH$	−117	130	2
2-Methyl-2-butanol	$(CH_3)_2COHCH_2CH_3$	−12	102	12.5
2,2-Dimethyl-1-propanol	$(CH_3)_3CCH_2OH$	55	114	∞
Allyl alcohol	$CH_2{=}CHCH_2OH$	−129	97	∞
Cyclopentanol	C_5H_9OH	−19	140	s. sol.
Cyclohexanol	$C_6H_{11}OH$	24	161	s. sol.
Benzyl alcohol	$C_6H_5CH_2OH$	−15	205	4

Physical Properties of Alkyl Halides

Name	bp (°C)			
	Fluoride	*Chloride*	*Bromide*	*Iodide*
Methyl	−78.4	−24.2	3.6	42.4
Ethyl	−37.7	12.3	38.4	72.3
Propyl	−2.5	46.6	71.0	102.5
Isopropyl	−9.4	34.8	59.4	89.5
Butyl	32.5	78.4	100	130.5
Isobutyl		68.8	90	120
sec-Butyl		68.3	91.2	120.0
tert-Butyl		50.2	73.1	dec.
Pentyl	62.8	108	130	157.0
Hexyl	92	133	154	179

Physical Properties of Amines

Name	Structure	mp (°C)	bp (°C)	Solubility (g/100 g H_2O at 25 °C)
Primary Amines				
Methylamine	CH_3NH_2	−93	−6.3	v. sol.
Ethylamine	$CH_3CH_2NH_2$	−81	17	∞
Propylamine	$CH_3(CH_2)_2NH_2$	−83	48	∞
Isopropylamine	$(CH_3)_2CHNH_2$	−95	33	∞
Butylamine	$CH_3(CH_2)_3NH_2$	−49	78	v. sol.
Isobutylamine	$(CH_3)_2CHCH_2NH_2$	−85	68	∞
sec-Butylamine	$CH_3CH_2CH(CH_3)NH_2$	−72	63	∞
tert-Butylamine	$(CH_3)_3CNH_2$	−67	46	∞
Cyclohexylamine	$C_6H_{11}NH_2$	−18	134	s. sol.
Secondary Amines				
Dimethylamine	$(CH_3)_2NH$	−93	7.4	v. sol.
Diethylamine	$(CH_3CH_2)_2NH$	−50	55	10.0
Dipropylamine	$(CH_3CH_2CH_2)_2NH$	−63	110	10.0
Dibutylamine	$(CH_3CH_2CH_2CH_2)_2NH$	−62	159	s. sol.
Tertiary Amines				
Trimethylamine	$(CH_3)_3N$	−115	2.9	91
Triethylamine	$(CH_3CH_2)_3N$	−114	89	14
Tripropylamine	$(CH_3CH_2CH_2)_3N$	−93	157	s. sol.

Physical Properties of Benzene and Substituted Benzenes

Name	Structure	mp (°C)	bp (°C)	Solubility (g/100 g H_2O at 25 °C)
Aniline	$C_6H_5NH_2$	−6	184	3.7
Benzene	C_6H_6	5.5	80.1	s. sol.
Benzaldehyde	C_6H_5CHO	−26	178	s. sol.
Benzamide	$C_6H_5CONH_2$	132	290	s. sol.
Benzoic acid	C_6H_5COOH	122	249	0.34
Bromobenzene	C_6H_5Br	−30.8	156	insol.
Chlorobenzene	C_6H_5Cl	−45.6	132	insol.
Nitrobenzene	$C_6H_5NO_2$	5.7	210.8	s. sol.
Phenol	C_6H_5OH	43	182	s. sol.
Styrene	$C_6H_5CH{=}CH_2$	−30.6	145.2	insol.
Toluene	$C_6H_5CH_3$	−95	110.6	insol.

Physical Properties of Carboxylic Acids

Name	Structure	mp (°C)	bp (°C)	Solubility (g/100 g H_2O at 25 °C)
Formic acid	$HCOOH$	8.4	101	∞
Acetic acid	CH_3COOH	16.6	118	∞
Propionic acid	CH_3CH_2COOH	−21	141	∞
Butanoic acid	$CH_3(CH_2)_2COOH$	−5	162	∞
Pentanoic acid	$CH_3(CH_2)_3COOH$	−34	186	4.97
Hexanoic acid	$CH_3(CH_2)_4COOH$	−4	202	0.97
Heptanoic acid	$CH_3(CH_2)_5COOH$	−8	223	0.24
Octanoic acid	$CH_3(CH_2)_6COOH$	17	237	0.068
Nonanoic acid	$CH_3(CH_2)_7COOH$	15	255	0.026
Decanoic acid	$CH_3(CH_2)_8COOH$	32	270	0.015

Physical Properties of Dicarboxylic Acids

Name	Structure	mp (°C)	Solubility (g/100 g H_2O at 25 °C)
Oxalic acid	$HOOCCOOH$	189	s
Malonic acid	$HOOCCH_2COOH$	136	v. sol.
Succinic acid	$HOOC(CH_2)_2COOH$	185	s. sol.
Glutaric acid	$HOOC(CH_2)_3COOH$	98	v. sol.
Adipic acid	$HOOC(CH_2)_4COOH$	151	s. sol.
Pimelic acid	$HOOC(CH_2)_5COOH$	106	s. sol.
Phthalic acid	1,2-$C_6H_4(COOH)_2$	231	s. sol.
Maleic acid	*cis*-$HOOCCH{=}CHCOOH$	130.5	v. sol.
Fumaric acid	*trans*-$HOOCCH{=}CHCOOH$	302	s. sol.

Physical Properties of Acyl Chlorides and Acid Anhydrides

Name	Structure	mp (°C)	bp (°C)
Acetyl chloride	CH_3COCl	−112	51
Propionyl chloride	CH_3CH_2COCl	−94	80
Butyryl chloride	$CH_3(CH_2)_2COCl$	−89	102
Valeryl chloride	$CH_3(CH_2)_3COCl$	−110	128
Acetic anhydride	$CH_3(CO)O(CO)CH_3$	−73	140
Succinic anhydride	(cyclic structure: O=C–O–C=O five-membered ring)		120

Physical Properties of Esters

Name	Structure	mp (°C)	bp (°C)
Methyl formate	$HCOOCH_3$	−100	32
Ethyl formate	$HCOOCH_2CH_3$	−80	54
Methyl acetate	CH_3COOCH_3	−98	57.5
Ethyl acetate	$CH_3COOCH_2CH_3$	−84	77
Propyl acetate	$CH_3COO(CH_2)_2CH_3$	−92	102
Methyl propionate	$CH_3CH_2COOCH_3$	−87.5	80
Ethyl propionate	$CH_3CH_2COOCH_2CH_3$	−74	99
Methyl butyrate	$CH_3CH_2CH_2COOCH_3$	−84.8	102.3
Ethyl butyrate	$CH_3CH_2CH_2COOCH_2CH_3$	−93	121

Physical Properties of Amides

Name	Structure	mp (°C)	bp (°C)
Formamide	$HCONH_2$	3	200 d*
Acetamide	CH_3CONH_2	82	221
Propanamide	$CH_3CH_2CONH_2$	80	213
Butanamide	$CH_3(CH_2)_2CONH_2$	116	216
Pentanamide	$CH_3(CH_2)_3CONH_2$	106	232

*d means the substance decomposes.

Physical Properties of Aldehydes

Name	Structure	mp (°C)	bp (°C)	Solubility (g/100 g H_2O at 25 °C)
Formaldehyde	$HCHO$	−92	−21	v. sol.
Acetaldehyde	CH_3CHO	−121	21	∞
Propionaldehyde	CH_3CH_2CHO	−81	49	16
Butyraldehyde	$CH_3(CH_2)_2CHO$	−96	75	7
Pentanal	$CH_3(CH_2)_3CHO$	−92	103	s. sol.
Hexanal	$CH_3(CH_2)_4CHO$	−56	131	s. sol.
Heptanal	$CH_3(CH_2)_5CHO$	−43	153	0.1
Octanal	$CH_3(CH_2)_6CHO$		171	insol.
Nonanal	$CH_3(CH_2)_7CHO$		192	insol.
Decanal	$CH_3(CH_2)_8CHO$	−5	209	insol.
Benzaldehyde	C_6H_5CHO	−26	178	0.3

Physical Properties of Ketones

Name	Structure	mp (°C)	bp (°C)	Solubility (g/100 g H_2O at 25 °C)
Acetone	CH_3COCH_3	−95	56	∞
2-Butanone	$CH_3COCH_2CH_3$	−86	80	25.6
2-Pentanone	$CH_3CO(CH_2)_2CH_3$	−78	102	5.5
2-Hexanone	$CH_3CO(CH_2)_3CH_3$	−57	127	1.6
2-Heptanone	$CH_3CO(CH_2)_4CH_3$	−36	151	0.4
2-Octanone	$CH_3CO(CH_2)_5CH_3$	−16	173	insol.
2-Nonanone	$CH_3CO(CH_2)_6CH_3$	−7	195	insol.
2-Decanone	$CH_3CO(CH_2)_7CH_3$	14	210	insol.
3-Pentanone	$CH_3CH_2COCH_2CH_3$	−40	102	4.8
3-Hexanone	$CH_3CH_2CO(CH_2)_2CH_3$		123	1.5
3-Heptanone	$CH_3CH_2CO(CH_2)_3CH_3$	−39	149	0.3
Acetophenone	$CH_3COC_6H_5$	19	202	insol.
Propiophenone	$CH_3CH_2COC_6H_5$	18	218	insol.

Appendix II

pK_a Values

Compound	pK_a	Compound	pK_a	Compound	pK_a
$CH_3C{\equiv}\overset{+}{N}H$	−10.1	$O_2N{-}C_6H_4{-}\overset{+}{N}H_3$	1.0	$CH_3{-}C_6H_4{-}COH$ (C=O)	4.3
HI	−10	pyrimidinium ion (ring N–H, +)	1.0	$CH_3O{-}C_6H_4{-}COH$ (C=O)	4.5
HBr	−9	Cl_2CHCOH (C=O)	1.3	$C_6H_5{-}\overset{+}{N}H_3$	4.6
$CH_3CH{=}\overset{+}{O}H$	−8	HSO_4^-	2.0	CH_3COH (C=O)	4.8
$CH_3C({=}\overset{+}{O}H)CH_3$	−7.3	H_3PO_4	2.1	quinolinium ion (ring N–H, +)	4.9
HCl	−7	purinium ion ($H\overset{+}{N}$)	2.5	$CH_3{-}C_6H_4{-}\overset{+}{N}H_3$	5.1
$CH_3C({=}\overset{+}{O}H)OCH_3$	−6.5	FCH_2COH (C=O)	2.7	pyridinium ion (ring N–H, +)	5.2
$CH_3C({=}\overset{+}{O}H)OH$	−6.1	$ClCH_2COH$ (C=O)	2.8	$CH_3O{-}C_6H_4{-}\overset{+}{N}H_3$	5.3
H_2SO_4	−5	$BrCH_2COH$ (C=O)	2.9	$CH_3C(CH_3){=}\overset{+}{N}HCH_3$	5.5
protonated pyrrole (ring $\overset{+}{N}H$, CH_2)	−3.8	ICH_2COH (C=O)	3.2	CH_3CCH_2CH (both C=O)	5.9
$CH_3CH_2\overset{+}{O}(H)CH_2CH_3$	−3.6	HF	3.2	$HO\overset{+}{N}H_3$	6.0
$CH_3CH_2\overset{+}{O}H_2$	−2.4	HNO_2	3.4	H_2CO_3	6.4
$CH_3\overset{+}{O}H_2$	−2.5	$O_2N{-}C_6H_4{-}COH$ (C=O)	3.4	imidazolium ion (HN, NH, +)	6.8
H_3O^+	−1.7	$HCOH$ (C=O)	3.8	H_2S	7.0
HNO_3	−1.3	$Br{-}C_6H_4{-}\overset{+}{N}H_3$	3.9	$O_2N{-}C_6H_4{-}OH$	7.1
CH_3SO_3H	−1.2	$Br{-}C_6H_4{-}COH$ (C=O)	4.0	$H_2PO_4^-$	7.2
$C_6H_5{-}SO_3H$	−0.60	$C_6H_5{-}COH$ (C=O)	4.2	$C_6H_5{-}SH$	7.8
$CH_3C({=}\overset{+}{O}H)NH_2$	0.0				
F_3CCOH (C=O)	0.2				
Cl_3CCOH (C=O)	0.64				
pyridinium $\overset{+}{N}{-}OH$	0.79				

[a] pK_a values are for the red H in each structure

pK_a Values (continued)

Compound	pK_a	Compound	pK_a	Compound	pK_a
aziridinium, $\overset{+}{N}H_2$ in three-membered ring	8.0	$C_6H_{11}\overset{+}{N}H_3$ (cyclohexyl)	10.7	CH_3CHO	17
$H_2N\overset{+}{N}H_3$	8.1	$(CH_3)_2\overset{+}{N}H_2$	10.7	$(CH_3)_3COH$	18
$CH_3C(=O)OOH$	8.2	piperidinium, $\overset{+}{N}H_2$ in six-membered ring	11.1	$CH_3C(=O)CH_3$	20
$CH_3CH_2NO_2$	8.6	$CH_3CH_2\overset{+}{N}H_3$	11.0	$CH_3C(=O)OCH_2CH_3$	24.5
$CH_3C(=O)CH_2C(=O)CH_3$	8.9	pyrrolidinium, $\overset{+}{N}H_2$ in five-membered ring	11.3	$HC\equiv CH$	25
$HC\equiv N$	9.1	HPO_4^{2-}	12.3	$CH_3C\equiv N$	25
morpholinium, O and $\overset{+}{N}H_2$ in six-membered ring	9.3	CF_3CH_2OH	12.4	$CH_3C(=O)N(CH_3)_2$	30
$Cl-C_6H_4-OH$ (para)	9.4	$CH_3CH_2OC(=O)CH_2C(=O)OCH_2CH_3$	13.3	NH_3	36
$\overset{+}{N}H_4$	9.4	$HC\equiv CCH_2OH$	13.5	pyrrolidine, NH in five-membered ring	36
$HOCH_2CH_2\overset{+}{N}H_3$	9.5	$H_2NC(=O)NH_2$	13.7	CH_3NH_2	40
$H_3\overset{+}{N}CH_2C(=O)O^-$	9.8	$(CH_3)_3\overset{+}{N}CH_2CH_2OH$	13.9	$C_6H_5-CH_3$	41
C_6H_5-OH	10.0	imidazole (N, NH ring)	14.4	C_6H_6	43
$CH_3-C_6H_4-OH$ (para)	10.2	CH_3OH	15.5	$CH_2{=}CHCH_3$	43
HCO_3^-	10.2	H_2O	15.7	$CH_2{=}CH_2$	44
CH_3NO_2	10.2	CH_3CH_2OH	16.0	cyclopropane	46
$H_2N-C_6H_4-OH$ (para)	10.3	$CH_3C(=O)NH_2$	16	CH_4	50
CH_3CH_2SH	10.5	$C_6H_5C(=O)CH_3$	16.0	CH_3CH_3	50
$(CH_3)_3\overset{+}{N}H$	10.6	pyrrole (NH in five-membered ring)	~17		
$CH_3C(=O)CH_2C(=O)OCH_2CH_3$	10.7				
$CH_3\overset{+}{N}H_3$	10.7				

Appendix III
Derivations of Rate Laws

How to Determine Rate Constants

A **reaction mechanism** is a detailed analysis of how the chemical bonds (or the electrons) in the reactants rearrange to form the products. The mechanism for a given reaction must obey the observed rate law for the reaction. A **rate law** tells how the rate of a reaction depends on the concentration of the species involved in the reaction.

First-Order Reaction

The rate is proportional to the concentration of one reactant:

$$\mathrm{A} \xrightarrow{k_1} \text{products}$$

Rate law: $\quad \text{rate} = k_1[\mathrm{A}]$

To determine the first-order rate constant (k_1),

Change in the concentration of A with respect to time:

$$\frac{-d[\mathrm{A}]}{dt} = k_1[\mathrm{A}]$$

Let a = the initial concentration of A;
let x = the concentration of A that has reacted up to time t.
Therefore, the concentration of A left at time $t = (a - x)$.
Substituting into the previous equation gives

$$\frac{-d(a-x)}{dt} = k_1(a-x)$$

$$\frac{-da}{dt} + \frac{dx}{dt} = k_1(a-x)$$

$$0 + \frac{dx}{dt} = k_1(a-x)$$

$$\frac{dx}{(a-x)} = k_1\,dt$$

Integrating the previous equation yields

$$-\ln(a-x) = k_1 t + \text{constant}$$

At $t = 0, x = 0$; therefore,

$$\text{constant} = -\ln a$$

$$-\ln(a-x) = k_1 t - \ln a$$

$$\ln\frac{a}{a-x} = k_1 t$$

$$\ln\frac{a-x}{a} = -k_1 t$$

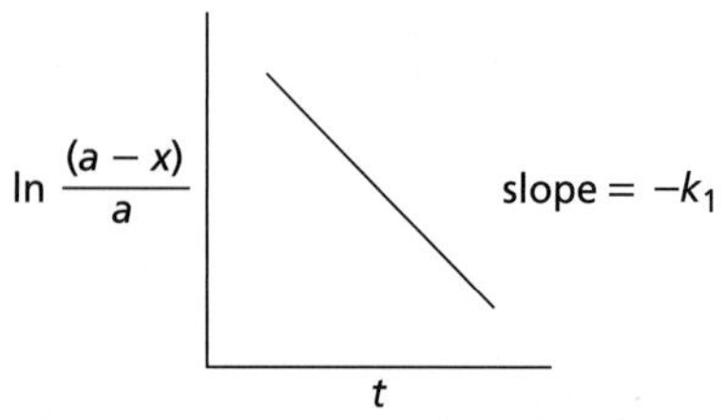

Half-Life of a First-Order Reaction

The **half-life** $(t_{1/2})$ of a reaction is the time it takes for half the reactant to react (or for half the product to form). To derive the half-life of a reactant in a first-order reaction, we begin with the equation

$$\ln \frac{a}{(a - x)} = k_1 t$$

At $t_{1/2}$, $x = \frac{a}{2}$; therefore,

$$\ln \frac{a}{\left(a - \frac{a}{2}\right)} = k_1 t_{1/2}$$

$$\ln \frac{a}{\frac{a}{2}} = k_1 t_{1/2}$$

$$\ln 2 = k_1 t_{1/2}$$

$$0.693 = k_1 t_{1/2}$$

$$t_{1/2} = \frac{0.693}{k_1}$$

Notice that the half-life of a first-order reaction is independent of the concentration of the reactant.

Second-Order Reaction

The rate is proportional to the concentration of two reactants:

$$A + B \xrightarrow{k_2} \text{products}$$

Rate law: $\text{rate} = k_2[A][B]$ (k_2 is the rate constant)

To determine the second-order rate constant (k_2),

Change in the concentration of A with respect to time:

$$\frac{-d[A]}{dt} = k_2[A][B]$$

Let a = the initial concentration of A;
let b = the initial concentration of B;
let x = the concentration of A that has reacted at time t.

Therefore, the concentration of A left at time $t = (a - x)$, and the concentration of B left at time $t = (b - x)$.

Substitution gives

$$\frac{dx}{dt} = k_2(a - x)(b - x)$$

For the case where $a = b$ (this condition can be arranged experimentally),

$$\frac{dx}{dt} = k_2(a - x)^2$$

$$\frac{dx}{(a - x)^2} = k_2\, dt$$

Integrating the equation gives

$$\frac{1}{(a-x)} = k_2t + \text{constant}$$

At $t = 0$, $x = 0$; therefore,

$$\text{constant} = \frac{1}{a}$$

$$\frac{1}{(a-x)} - \frac{1}{a} = k_2t$$

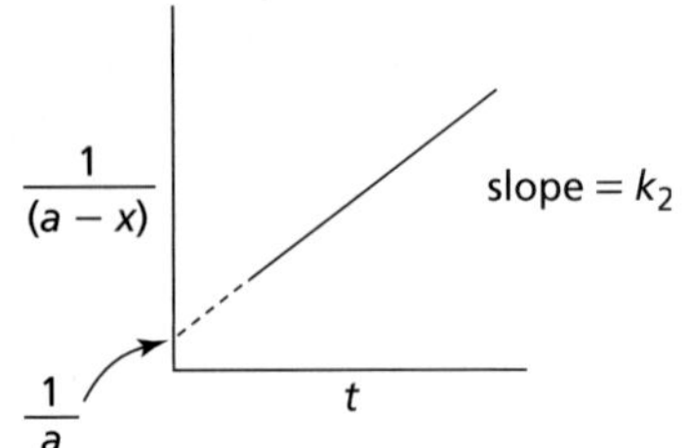

Half-Life of a Second-Order Reaction

$$\frac{1}{(a-x)} - \frac{1}{a} = k_2t$$

At $t_{1/2}$, $x = \frac{a}{2}$; therefore,

$$\frac{1}{a} = k_2t_{1/2}$$

$$t_{1/2} = \frac{1}{k_2a}$$

Pseudo-First-Order Reaction

It is easier to determine a first-order rate constant than a second-order rate constant because the kinetic behavior of a first-order reaction is independent of the initial concentration of the reactant. Therefore, a first-order rate constant can be determined without knowing the initial concentration of the reactant. The determination of a second-order rate constant requires not only that the initial concentration of the reactants be known but also that the initial concentrations of the two reactants be identical in order to simplify the kinetic equation.

However, if the concentration of one of the reactants in a second-order reaction is much greater than the concentration of the other, the reaction can be treated as a first-order reaction. Such a reaction is known as a **pseudo-first-order reaction** and is given by

$$\frac{-d[\text{A}]}{dt} = k_2[\text{A}][\text{B}]$$

If $[\text{B}] \gg [\text{A}]$, then

$$\frac{-d[\text{A}]}{dt} = k_2'[\text{A}]$$

The rate constant obtained for a pseudo-first-order reaction (k_2') includes the concentration of B, but k_2 can be determined by carrying out the reaction at several different concentrations of B and determining the slope of a plot of the observed rate versus [B].

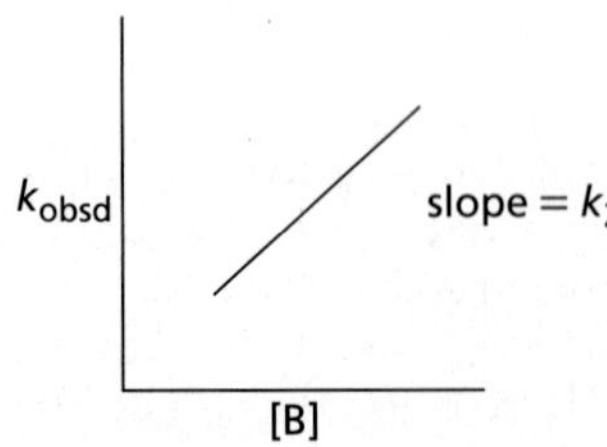

Appendix IV

Summary of Methods Used to Synthesize a Particular Functional Group

SYNTHESIS OF ACETALS

1. Acid-catalyzed reaction of an aldehyde with two equivalents of an alcohol (18.7, 18.8).

SYNTHESIS OF ACID ANHYDRIDES

1. Reaction of an acyl halide with a carboxylate ion (17.8).
2. Preparation of a cyclic anhydride by heating a dicarboxylic acid (17.21).

SYNTHESIS OF ACYL CHLORIDES OR ACYL BROMIDES

1. Reaction of a carboxylic acid with $SOCl_2$, PCl_3, or PBr_3 (17.20).

SYNTHESIS OF ALCOHOLS

1. Acid-catalyzed hydration of an alkene (4.5).
2. Oxymercuration–demercuration of an alkene (4.8).
3. Hydroboration–oxidation of an alkene (4.9).
4. Reaction of an alkyl halide with HO^- (10.4).
5. Reaction of a Grignard reagent with an epoxide (12.11).
6. Reduction of an aldehyde, a ketone, an acyl chloride, an anhydride, an ester, or a carboxylic acid (18.5, 20.1).
7. Reaction of a Grignard reagent with an aldehyde, a ketone, an acyl chloride, or an ester (18.4).
8. Reduction of a ketone with $NaBH_4$ in cold aqueous ethanol in the presence of cerium trichloride (20.1).
9. Cleavage of an ether with HI or HBr (12.6).
10. Reaction of an organozinc reagent with an aldehyde or a ketone (p. 834).

SYNTHESIS OF ALDEHYDES

1. Hydroboration–oxidation of a terminal alkyne with disiamylborane followed by $H_2O_2 + HO^- + H_2O$ (6.7).
2. Oxidation of a primary alcohol with pyridinium chlorochromate (20.2).
3. Swern oxidation of a primary alcohol with dimethyl sulfoxide, oxalyl chloride, and triethylamine (20.2).
4. Rosenmund reduction: catalytic hydrogenation of an acyl chloride (20.1).
5. Reaction of an acyl chloride with lithium tri(*tert*-butoxy)aluminum hydride (20.1).
6. Reaction of an ester with diisobutylaluminum hydride (DIBALH) (20.1).
7. Cleavage of a 1,2-diol with periodic acid (20.7).
8. Ozonolysis of an alkene, followed by workup under reducing conditions (20.8).

SYNTHESIS OF ALKANES

1. Catalytic hydrogenation of an alkene or an alkyne (4.11, 6.8, 20.1).
2. Reaction of a Grignard reagent with a source of protons (12.11).
3. Wolff–Kishner or Clemmensen reduction of an aldehyde or a ketone (15.15, 18.6).
4. Reduction of a thioacetal or thioketal with H_2 and Raney nickel (18.9).
5. Reaction of a Gilman reagent with an alkyl halide (12.12).
6. Preparation of a cyclopropane by the reaction of an alkene with a carbene (4.9).

SYNTHESIS OF ALKENES

1. Elimination of hydrogen halide from an alkyl halide (11.1, 11.2, 11.3).
2. Acid-catalyzed dehydration of an alcohol (12.5).
3. Hofmann elimination reaction: elimination of a proton and a tertiary amine from a quaternary ammonium hydroxide (21.5).
4. Exhaustive methylation of an amine, followed by a Hofmann elimination reaction (21.5).
5. Hydrogenation of an alkyne with Lindlar catalyst to form a cis alkene (6.8, 20.1).
6. Reduction of an alkyne with Na (or Li) and liquid ammonia to form a trans alkene (6.8, 20.1).
7. Formation of a cyclic alkene using a Diels–Alder reaction (8.8, 29.4).
8. Wittig reaction: reaction of an aldehyde or a ketone with a phosphonium ylide (18.10).
9. Reaction of a Gilman reagent with a halogenated alkene (12.12).
10. Heck reaction couples a vinyl halide with an alkene in a basic solution in the presence of $Pd(PPh_3)_4$ (12.12).
11. Stille reaction couples a vinyl halide with a stannane in the presence of $Pd(PPh_3)_4$ (12.12).
12. Suzuki reaction couples a vinyl halide with an organoborane in the presence of $Pd(PPh_3)_4$ (12.12).

SYNTHESIS OF ALKYL HALIDES

1. Addition of hydrogen halide (HX) to an alkene (4.1).
2. Addition of HBr + peroxide (4.10).
3. Addition of halogen to an alkene (4.7).
4. Addition of hydrogen halide or a halogen to an alkyne (6.5).
5. Radical halogenation of an alkane, an alkene, or an alkyl benzene (9.2, 9.5).
6. Reaction of an alcohol with hydrogen halide, $SOCl_2$, PCl_3, or PBr_3 (12.1, 12.3).
7. Reaction of a sulfonate ester with halide ion (12.4).
8. Cleavage of an ether with HI or HBr (12.6).
9. Halogenation of an α-carbon of an aldehyde, a ketone, or a carboxylic acid (19.4, 19.5).
10. Hunsdiecker reaction: reaction of a carboxylic acid with Br_2 and Ag_2O (19.17).

SYNTHESIS OF ALKYNES

1. Elimination of hydrogen halide from a vinyl halide (11.10).
2. Two successive eliminations of hydrogen halide from a vicinal dihalide or a geminal dihalide (11.10).
3. Reaction of an acetylide ion (formed by removing a proton from a terminal alkyne) with an alkyl halide (6.10).

SYNTHESIS OF AMIDES

1. Reaction of an acyl chloride, an acid anhydride, or an ester with ammonia or with an amine (17.8, 17.9, 17.10).
2. Reaction of a carboxylic acid and an amine with dicyclohexylcarbodiimide (23.9, 23.10).
3. Reaction of a nitrile with a secondary or tertiary alcohol (p. 728).

SYNTHESIS OF AMINES

1. Reaction of an alkyl halide with NH_3, RNH_2, or R_2NH (10.4).
2. Reaction of an alkyl halide with azide ion, followed by reduction of the alkyl azide (10.4).
3. Reduction of an imine, a nitrile, or an amide (20.1).
4. Reductive amination of an aldehyde or a ketone (21.8).
5. Gabriel synthesis of primary amines: reaction of a primary alkyl halide with potassium phthalimide (17.17, 21.8).
6. Reduction of a nitro compound (16.2).
7. Condensation of a secondary amine and formaldehyde with a carbon acid (p. 837).

SYNTHESIS OF AMINO ACIDS

1. Hell–Volhard–Zelinski reaction: halogenation of a carboxylic acid, followed by treatment with excess NH_3 (23.6).
2. Reductive amination of an α-keto acid (23.6).

SYNTHESIS OF CARBOXYLIC ACIDS

1. Oxidation of a primary alcohol (20.2).
2. Oxidation of an aldehyde (20.3).
3. Ozonolysis of a monosubstituted alkene or a 1,2-disubstituted alkene, followed by workup under oxidizing conditions (20.8).

4. Ozonolysis of an alkyne (20.9).
5. Oxidation of an alkyl benzene (16.2).
6. Hydrolysis of an acyl halide, an acid anhydride, an ester, an amide, or a nitrile (17.8, 17.9, 17.10, 17.15, 17.18).
7. Haloform reaction: reaction of a methyl ketone with excess Br_2 (or Cl_2 or I_2) + HO^- (19.4).
8. Reaction of a Grignard reagent with CO_2 (18.4).
9. Malonic ester synthesis (19.18).
10. Favorskii reaction: reaction of an α-haloketone with hydroxide ion (p. 838).

SYNTHESIS OF CYANOHYDRINS

1. Reaction of an aldehyde or a ketone with sodium cyanide and HCl (18.4).

SYNTHESIS OF 1,2-DIOLS

1. Reaction of an epoxide with water (12.7).
2. Reaction of an alkene with osmium tetroxide or potassium permanganate (20.6).

SYNTHESIS OF DISULFIDES

1. Mild oxidation of a thiol (23.7).

SYNTHESIS OF ENAMINES

1. Reaction of an aldehyde or a ketone with a secondary amine (18.6).

SYNTHESIS OF EPOXIDES

1. Reaction of an alkene with a peroxyacid (20.4).
2. Reaction of a halohydrin with hydroxide ion (p. 480).
3. Reaction of an aldehyde or a ketone with a sulfonium ylide (p. 783).

SYNTHESIS OF ESTERS

1. Reaction of an acyl halide or an acid anhydride with an alcohol (17.8, 17.9).
2. Acid-catalyzed reaction of an ester or a carboxylic acid with an alcohol (17.10, 17.14).
3. Reaction of an alkyl halide or a sulfonate ester with a carboxylate ion (10.4, 12.4).
4. Oxidation of a ketone (30.3).
5. Preparation of a methyl ester by the reaction of a carboxylate ion with diazomethane (16.12).

SYNTHESIS OF ETHERS

1. Acid-catalyzed addition of an alcohol to an alkene (4.5).
2. Alkoxymercuration–demercuration of an alkene (4.8).
3. Williamson ether synthesis: reaction of an alkoxide ion with an alkyl halide (11.9).
4. Formation of symmetrical ethers by heating an acidic solution of a primary alcohol (12.5).

SYNTHESIS OF HALOHYDRINS

1. Reaction of an alkene with Br_2 (or Cl_2) and H_2O (4.7).
2. Reaction of an epoxide with a hydrogen halide (12.7).

SYNTHESIS OF IMINES

1. Reaction of an aldehyde or a ketone with a primary amine (18.6).

SYNTHESIS OF KETALS

1. Acid-catalyzed reaction of a ketone with two equivalents of an alcohol (18.7, 18.8).

SYNTHESIS OF KETONES

1. Mercuric acid-catalyzed hydration of an alkyne (6.6).
2. Hydroboration–oxidation of an alkyne (6.7).
3. Oxidation of a secondary alcohol (20.2).
4. Cleavage of a 1,2-diol with periodic acid (20.7).
5. Ozonolysis of an alkene (20.8).
6. Friedel–Crafts acylation of an aromatic ring (15.13).
7. Preparation of a methyl ketone by the acetoacetic ester synthesis (19.19).
8. Reaction of a Gilman reagent with an acyl chloride (12.12).
9. Preparation of a cyclic ketone by the reaction of the next-size-smaller cyclic ketone with diazomethane (p. 786).

SYNTHESIS OF α,β-UNSATURATED KETONES

1. Elimination from an α-haloketone (19.6).
2. Selenenylation of a ketone, followed by oxidative elimination (p. 838).

SYNTHESIS OF NITRILES

1. Reaction of an alkyl halide with cyanide ion (10.4).

SYNTHESIS OF SUBSTITUTED BENZENES

1. Halogenation with Br_2 or Cl_2 and a Lewis acid (15.10).
2. Nitration with $HNO_3 + H_2SO_4$ (15.11).
3. Sulfonation: reaction with H_2SO_4 (15.12).
4. Friedel–Crafts acylation (15.13).
5. Friedel–Crafts alkylation (15.14, 15.15).
6. Sandmeyer reaction: reaction of an arenediazonium salt with CuBr, CuCl, or CuCN (16.10).
7. Formation of a phenol by reaction of an arenediazonium salt with water (16.10).
8. Formation of an aniline by reaction of a benzyne intermediate with $^-NH_2$ (16.14).
9. Reaction of a Gilman reagent with an aryl halide (12.12).
10. Heck reaction: couples a benzyl halide or an aryl halide or a triflate with an alkene in a basic solution in the presence of $Pd(PPh_3)_4$ (12.12).
11. Stille reaction: couples a benzyl halide or an aryl halide or a triflate with a stannane in the presence of $Pd(PPh_3)_4$ (12.12).
12. Suzuki reaction: couples a benzyl or aryl halide with an organoborane in the presence of $Pd(PPh_3)_4$ (12.12).

SYNTHESIS OF SULFIDES

1. Reaction of a thiol with an alkyl halide (10.4, 12.10).
2. Catalytic hydrogenation of a disulfide (23.7).

SYNTHESIS OF THIOLS

1. Reaction of an alkyl halide with hydrogen sulfide (10.4).

Appendix V

Summary of Methods Employed to Form Carbon–Carbon Bonds

1. Reaction of an acetylide ion with an alkyl halide or a sulfonate ester (6.10, 10.4, 12.4).
2. Diels–Alder and other cycloaddition reactions (8.8, 29.4).
3. Reaction of a Grignard reagent with an epoxide (12.7).
4. Friedel–Crafts alkylation and acylation (15.13, 15.14, 15.15).
5. Reaction of a cyanide ion with an alkyl halide or a sulfonate ester (10.4, 12.4).
6. Reaction of a cyanide ion with an aldehyde or a ketone (18.4).
7. Reaction of a Grignard reagent with an aldehyde, a ketone, an ester, an amide, an epoxide, or CO_2 (12.7, 18.4).
8. Reaction of an organozinc reagent with an aldehyde or a ketone (p. 834).
9. Reaction of an alkene with a carbene (4.9).
10. Reaction of a lithium dialkylcuprate with an α,β-unsaturated ketone or an α,β-unsaturated aldehyde (18.13).
11. Aldol addition (19.11, 19.12, 19.13, 19.16).
12. Claisen condensation (19.14, 19.15, 19.16).
13. Perkin condensation (p. 834)
14. Knoevenagel condensation (p. 834)
15. Malonic ester synthesis and acetoacetic ester synthesis (19.18, 19.19).
16. Michael addition reaction (19.10).
17. Alkylation of an enamine (19.9).
18. Alkylation of the α-carbon of a carbonyl compound (19.8).
19. Reaction of a Gilman reagent with an aryl halide or a halogenated alkene (12.12).
20. Heck reaction: couples a vinyl, a benzyl, or an aryl halide, or a triflate with an alkene in a basic solution in the presence of $Pd(PPh_3)_4$ (12.12).
21. Stille reaction: couples a vinyl, a benzyl, or an aryl halide, or a triflate with a stannane in the presence of $Pd(PPh_3)_4$ (12.12).
22. Suzuki reaction: couples a vinyl, a benzyl, or an aryl halide with an organoborane in the presence of $Pd(PPh_3)_4$ (12.12).

Appendix VI
Spectroscopy Tables

Mass Spectrometry

Common fragment ions*

m/z	Ion	*m/z*	Ion
14	CH_2	46	NO_2
15	CH_3	47	CH_2SH, CH_3S
16	O	48	$CH_3S + H$
17	OH	49	CH_2Cl
18	H_2O, NH_4	51	CHF_2
19	F, H_3O	53	C_4H_5
26	$C{\equiv}N$	54	$CH_2CH_2C{\equiv}N$
27	C_2H_3	55	C_4H_7, $CH_2{=}CHC{=}O$
28	C_2H_4, CO, N_2, $CH{=}NH$	56	C_4H_8
29	C_2H_5, CHO	57	C_4H_9, $C_2H_5C{=}O$
30	CH_2NH_2, NO	58	$CH_3\overset{O}{\overset{\|}{C}}CH_2$ + H, $C_2H_5CHNH_2$, $(CH_3)_2NCH_2$, $C_2H_5NHCH_2$, C_2H_2S
31	CH_2OH, OCH_3		
32	O_2 (air)		
33	SH, CH_2F		
34	H_2S	59	$(CH_3)_2COH$, $CH_2OC_2H_5$, $\overset{O}{\overset{\|}{C}}OCH_3$, $\underset{NH_2}{CH_2C{=}O}$ + H, CH_3OCHCH_3,
35	Cl		
36	HCl		
39	C_3H_3		
40	$CH_2C{=}N$		
41	C_3H_5, $CH_2C{=}N$ + H, C_2H_2NH	60	CH_3CHCH_2OH CH_2COOH + H, CH_2ONO
42	C_3H_6		
43	C_3H_7, $CH_3C{=}O$, C_2H_5N		
44	$CH_2CH{=}O$ + H, CH_3CHNH_2, CO_2, $NH_2C{=}O$, $(CH_3)_2N$		
45	CH_3CHOH, CH_2CH_2OH, CH_2OCH_3, COOH, CH_3CHO + H		

*All of these ions have a single positive charge.

Mass Spectrometry

Common fragment lost

Molecular ion minus	Fragment lost	Molecular ion minus	Fragment lost
1	H	43	C_3H_7, $CH_3C(=O)$, $CH_2{=}CHO$, HCNO, $CH_3 + CH_2{=}CH_2$
15	CH_3	44	$CH_2{=}CHOH$, CO_2, N_2O, $CONH_2$, $NHCH_2CH_3$
17	HO	45	CH_3CHOH, CH_3CH_2O, CO_2H, $CH_3CH_2NH_2$
18	H_2O	46	$H_2O + CH_2{=}CH_2$, CH_3CH_2OH, NO_2
19	F	47	CH_3S
20	HF	48	CH_3SH, SO, O_3
26	$CH{\equiv}CH$, $C{\equiv}N$	49	CH_2Cl
27	$CH_2{=}CH$, $HC{\equiv}N$	51	CHF_2
28	$CH_2{=}CH_2$, CO, (HCN + H)	52	C_4H_4, C_2N_2
29	CH_3CH_2, CHO	53	C_4H_5
30	NH_2CH_2, CH_2O, NO	54	$CH_2{=}CHCH{=}CH_2$
31	OCH_3, CH_2OH, CH_3NH_2	55	$CH_2{=}CHCHCH_3$
32	CH_3OH, S	56	$CH_2{=}CHCH_2CH_3$, $CH_3CH{=}CHCH_3$
33	HS, (CH_3 and H_2O)	57	C_4H_9
34	H_2S	58	NCS, NO + CO, CH_3COCH_3
35	Cl	59	$CH_3OC(=O)$, $CH_3C(=O)NH_2$
36	HCl, 2 H_2O	60	C_3H_7OH
37	HCl + H		
38	C_3H_2, C_2N, F_2		
39	C_3H_3, HC_2N		
40	$CH_3C{\equiv}CH$		
41	$CH_2{=}CHCH_2$		
42	$CH_2{=}CHCH_3$, $CH_2{=}C{=}O$, cyclopropane (CH_2–CH_2 bridged by CH_2), NCO		

^{1}H NMR Chemical Shifts

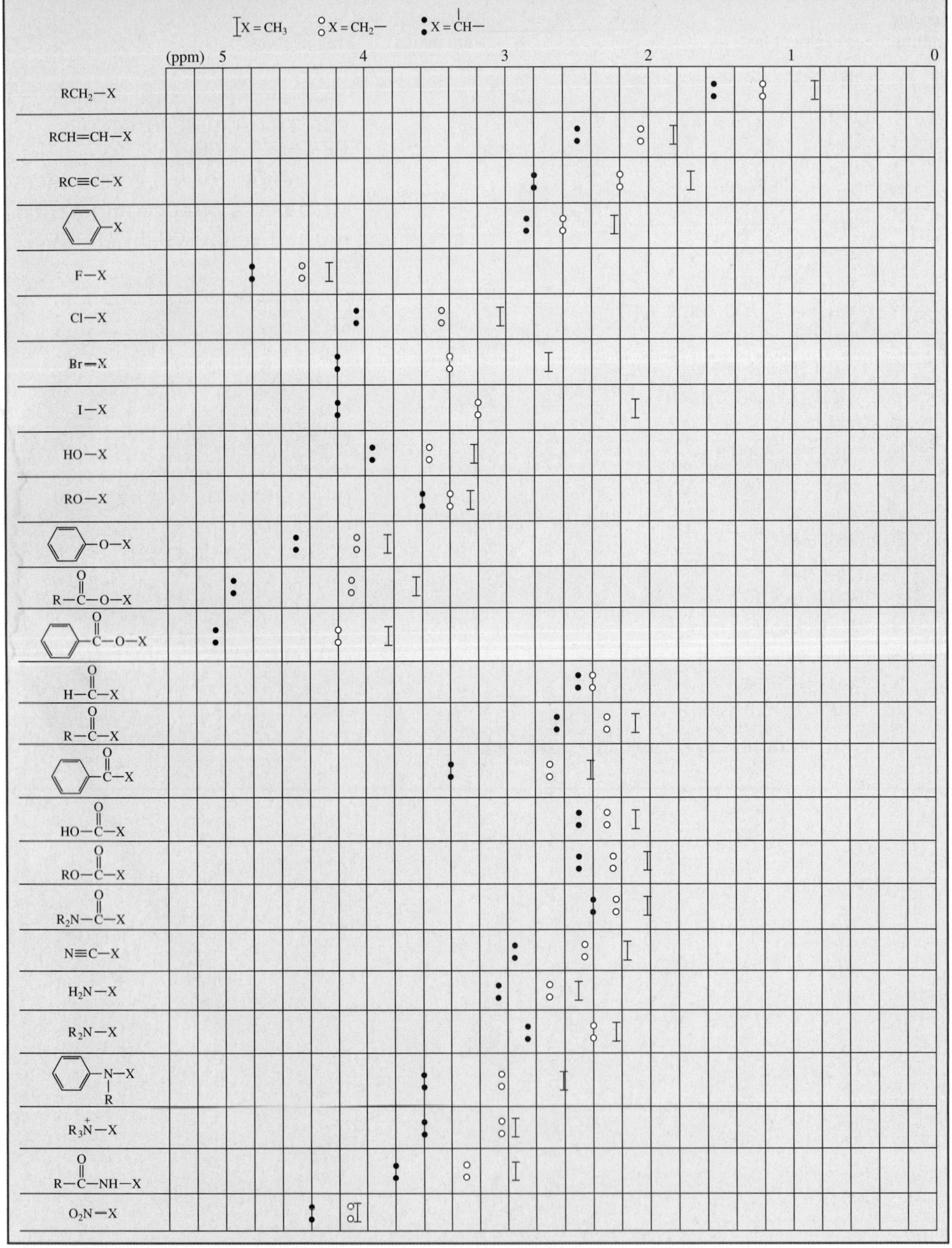

I $X = CH_3$ ○ $X = CH_2-$ ● $X = \overset{|}{C}H-$

(ppm) 5 4 3 2 1 0

RCH_2-X

$RCH{=}CH-X$

$RC{\equiv}C-X$

C_6H_5-X

$F-X$

$Cl-X$

$Br-X$

$I-X$

$HO-X$

$RO-X$

C_6H_5-O-X

$R-C(=O)-O-X$

$C_6H_5-C(=O)-O-X$

$H-C(=O)-X$

$R-C(=O)-X$

$C_6H_5-C(=O)-X$

$HO-C(=O)-X$

$RO-C(=O)-X$

$R_2N-C(=O)-X$

$N{\equiv}C-X$

H_2N-X

R_2N-X

$C_6H_5-N(R)-X$

$R_3\overset{+}{N}-X$

$R-C(=O)-NH-X$

O_2N-X

Characteristic Infrared Group Frequencies (S = strong, M = medium, W = weak). (Courtesy of N.B. Colthup, Stamford Research Laboratories, American Cyanamid Company, and the editor of the *Journal of the Optical Society*.) Overtone bands are marked 2ν.

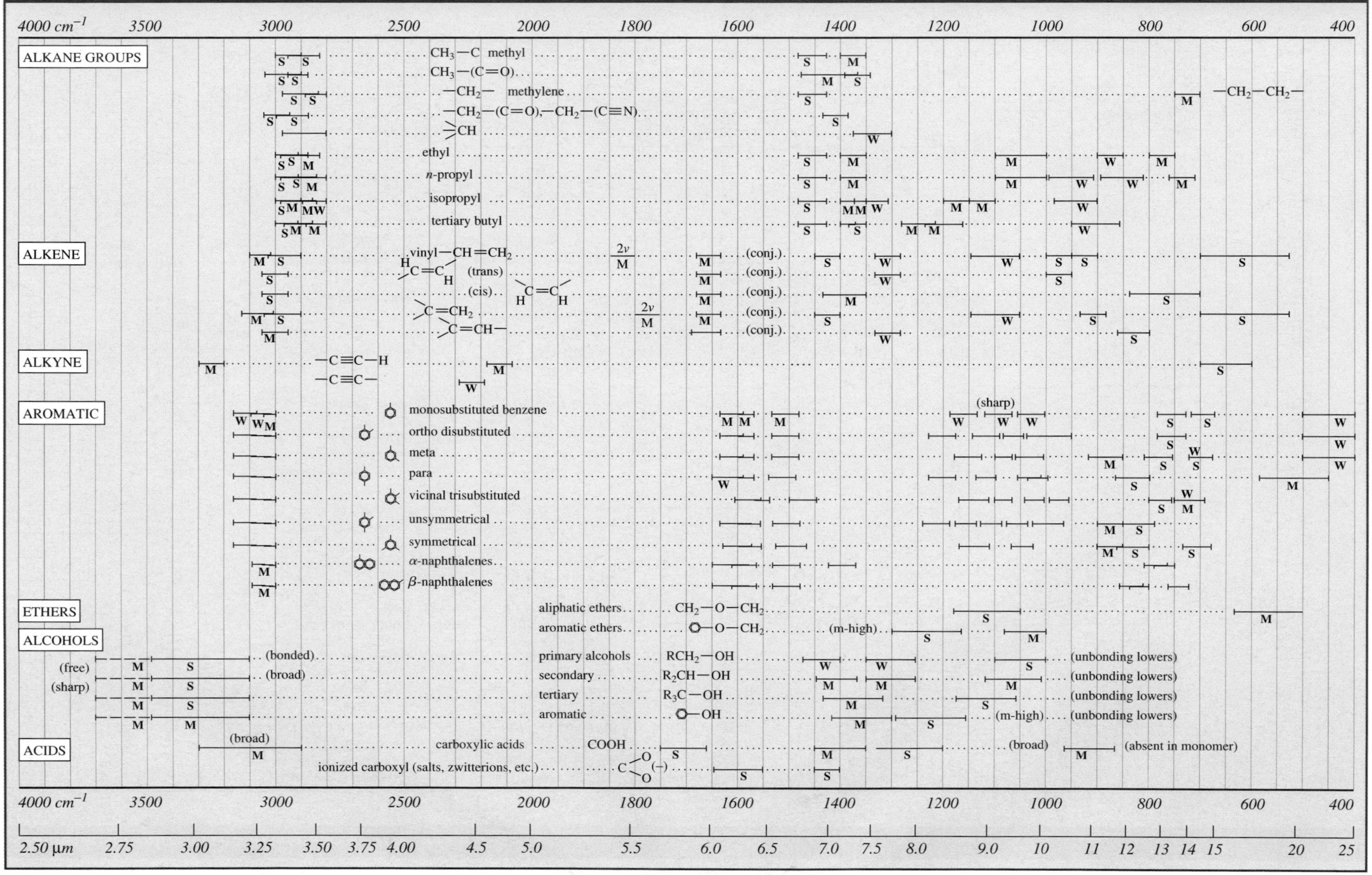

4000 cm^{-1} 3500 3000 2500 2000 1800 1600 1400 1200 1000 800 600 400

ALKANE GROUPS

- CH_3—C methyl
- CH_3—(C=O)
- —CH_2— methylene (—CH_2—CH_2—)
- —CH_2—(C=O), —CH_2—(C≡N)
- ≥CH
- ethyl
- *n*-propyl
- isopropyl
- tertiary butyl

ALKENE

- vinyl —CH=CH_2 (2ν) (conj.)
- HC=CH (trans) (conj.)
- HC=CH (cis) (conj.)
- >C=CH_2 (2ν) (conj.)
- >C=CH— (conj.)

ALKYNE

- —C≡C—H
- —C≡C—

AROMATIC

- monosubstituted benzene (sharp)
- ortho disubstituted
- meta
- para
- vicinal trisubstituted
- unsymmetrical
- symmetrical
- α-naphthalenes
- β-naphthalenes

ETHERS

- aliphatic ethers CH_2—O—CH_2
- aromatic ethers Ar—O—CH_2 (m-high)

ALCOHOLS

- (free) (sharp) (bonded) primary alcohols RCH_2—OH (unbonding lowers)
- (broad) secondary R_2CH—OH (unbonding lowers)
- tertiary R_3C—OH (unbonding lowers)
- aromatic Ar—OH (m-high) (unbonding lowers)

ACIDS

- (broad) carboxylic acids COOH (broad) (absent in monomer)
- ionized carboxyl (salts, zwitterions, etc.) $C(O)O^{(-)}$

4000 cm^{-1} 3500 3000 2500 2000 1800 1600 1400 1200 1000 800 600 400

2.50 μm 2.75 3.00 3.25 3.50 3.75 4.00 4.5 5.0 5.5 6.0 6.5 7.0 7.5 8.0 9.0 10 11 12 13 14 15 20 25

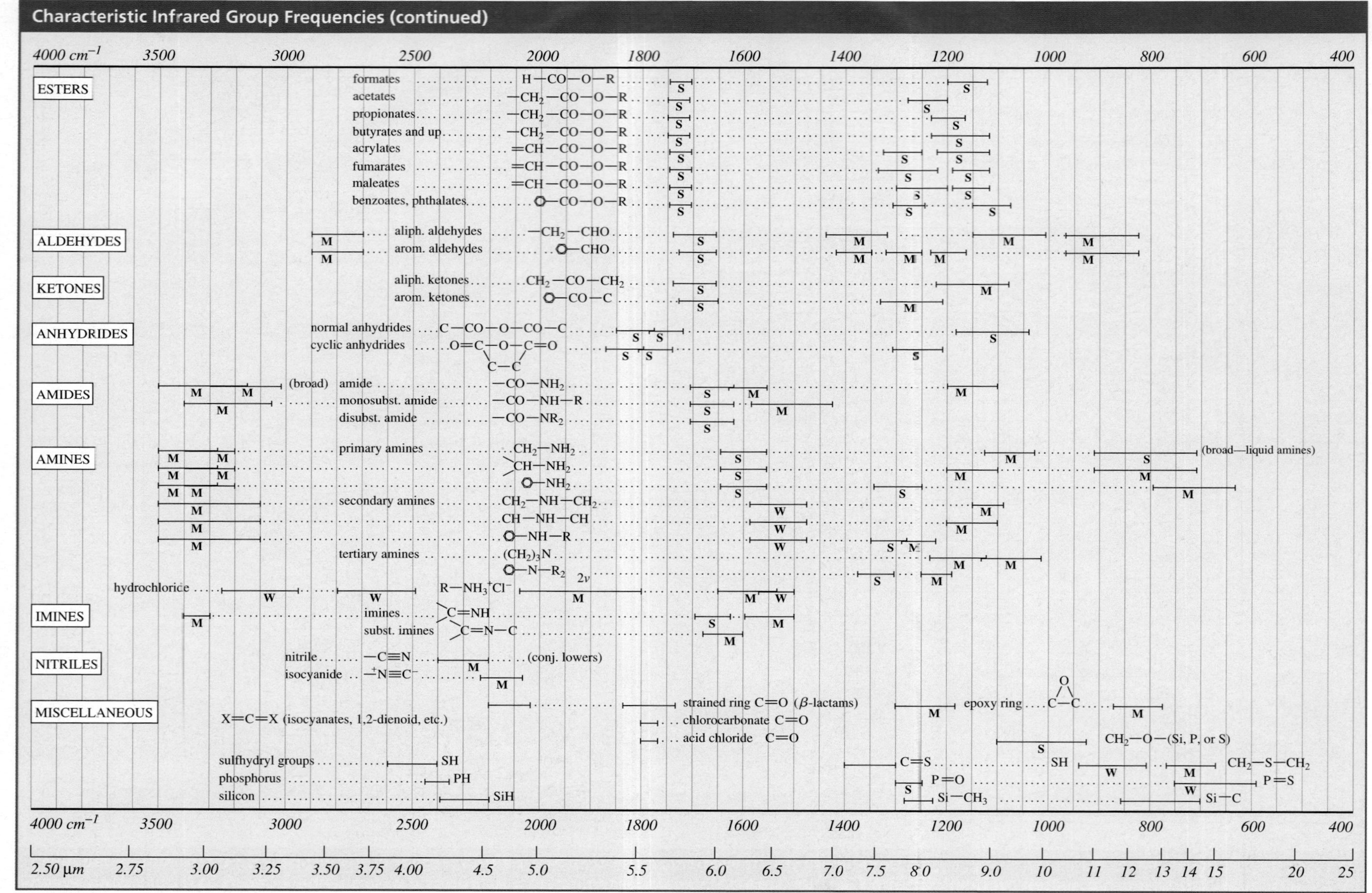

Characteristic Infrared Group Frequencies (continued)
4000 cm^{-1}
3500
3000
2500
2000
1800
1600
1400
1200
1000
800
600
400
ESTERS
formates
H—CO—O—R
acetates
—CH_2—CO—O—R
propionates
—CH_2—CO—O—R
butyrates and up
—CH_2—CO—O—R
acrylates
=CH—CO—O—R
fumarates
=CH—CO—O—R
maleates
=CH—CO—O—R
benzoates, phthalates
—CO—O—R
ALDEHYDES
aliph. aldehydes
—CH_2—CHO
arom. aldehydes
—CHO
KETONES
aliph. ketones
CH_2—CO—CH_2
arom. ketones
—CO—C
ANHYDRIDES
normal anhydrides
C—CO—O—CO—C
cyclic anhydrides
O=C—O—C=O
C—C
AMIDES
(broad)
amide
—CO—NH_2
monosubst. amide
—CO—NH—R
disubst. amide
—CO—NR_2
AMINES
primary amines
CH_2—NH_2
CH—NH_2
—NH_2
secondary amines
CH_2—NH—CH_2
CH—NH—CH
—NH—R
tertiary amines
$(CH_2)_3N$
—N—R_2
2ν
(broad—liquid amines)
hydrochloride
R—$NH_3^+Cl^-$
IMINES
imines
C=NH
subst. imines
C=N—C
NITRILES
nitrile
—C≡N
isocyanide
$^+N≡C^-$
(conj. lowers)
MISCELLANEOUS
X=C=X (isocyanates, 1,2-dienoid, etc.)
strained ring C=O (β-lactams)
chlorocarbonate C=O
acid chloride C=O
epoxy ring
O
C—C
sulfhydryl groups
SH
phosphorus
PH
P=O
P=S
silicon
SiH
Si—CH_3
Si—C
C=S
CH_2—O—(Si, P, or S)
CH_2—S—CH_2
S
M
W
2.50 μm
2.75
3.00
3.25
3.50
3.75
4.00
4.5
5.0
5.5
6.0
6.5
7.0
7.5
8.0
9.0
10
11
12
13
14
15
20
25

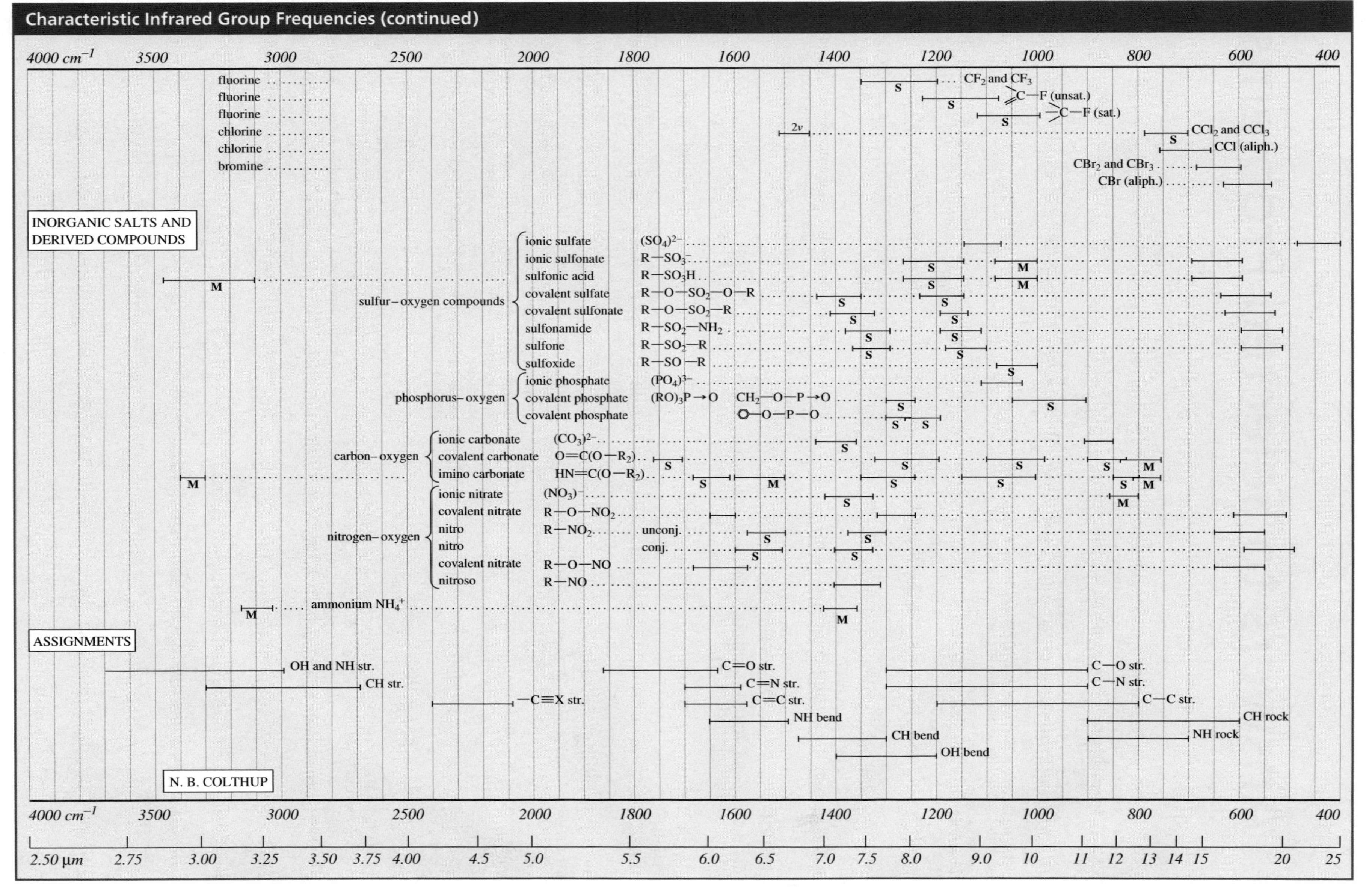
Characteristic Infrared Group Frequencies (continued)
4000 cm^{-1} 3500 3000 2500 2000 1800 1600 1400 1200 1000 800 600 400
fluorine
fluorine
fluorine
chlorine
chlorine
bromine
CF$_2$ and CF$_3$
C—F (unsat.)
C—F (sat.)
2ν
CCl$_2$ and CCl$_3$
CCl (aliph.)
CBr$_2$ and CBr$_3$
CBr (aliph.)
INORGANIC SALTS AND DERIVED COMPOUNDS
sulfur–oxygen compounds
ionic sulfate (SO$_4$)$^{2-}$
ionic sulfonate R—SO$_3^-$
sulfonic acid R—SO$_3$H
covalent sulfate R—O—SO$_2$—O—R
covalent sulfonate R—O—SO$_2$—R
sulfonamide R—SO$_2$—NH$_2$
sulfone R—SO$_2$—R
sulfoxide R—SO—R
phosphorus–oxygen
ionic phosphate (PO$_4$)$^{3-}$
covalent phosphate (RO)$_3$P→O CH$_2$—O—P→O
covalent phosphate ⌬—O—P—O
carbon–oxygen
ionic carbonate (CO$_3$)$^{2-}$
covalent carbonate O=C(O—R$_2$)
imino carbonate HN=C(O—R$_2$)
nitrogen–oxygen
ionic nitrate (NO$_3$)$^-$
covalent nitrate R—O—NO$_2$
nitro R—NO$_2$ unconj.
nitro conj.
covalent nitrate R—O—NO
nitroso R—NO
ammonium NH$_4^+$
S
M
ASSIGNMENTS
OH and NH str.
CH str.
—C≡X str.
C=O str.
C=N str.
C=C str.
NH bend
CH bend
OH bend
C—O str.
C—N str.
C—C str.
CH rock
NH rock
N. B. COLTHUP
4000 cm^{-1} 3500 3000 2500 2000 1800 1600 1400 1200 1000 800 600 400
2.50 μm 2.75 3.00 3.25 3.50 3.75 4.00 4.5 5.0 5.5 6.0 6.5 7.0 7.5 8.0 9.0 10 11 12 13 14 15 20 25

Answers to Selected Problems

CHAPTER 1

1-1. 8 + 8, 8 + 9, 8 + 10 **1-2.** 4*s* **1-3.** Cl $1s^22s^22p^63s^23p^5$; Br $1s^22s^22p^63s^23p^64s^23d^{10}4p^5$; I $1s^22s^22p^63s^23p^64s^23d^{10}4p^65s^24d^{10}5p^5$ **1-5. a.** KCl **b.** Cl_2 **1-6. a.** LiH and HF **b.** Its hydrogen has the greatest electron density. **c.** HF **1-9. a.** oxygen **b.** oxygen **c.** oxygen **d.** hydrogen **1-13.** yes **1-14. a.** π^* **b.** σ^* **c.** σ^* **d.** σ **1-15.** The C—C bonds are formed by sp^3–sp^3 overlap; the C—H bonds are formed by sp^3–*s* overlap. **1-16.** >104°5 <109°5 **1-17.** hydrogens **1-18.** most = water, least = methane **1-19. a.** relative lengths: $Br_2 > Cl_2$; relative strengths: $Cl_2 > Br_2$ **b.** relative lengths; HBr > HCl > HF; relative strengths: HF > HCl > HBr **1-20.** σ **1-21.** sp^2–sp^2, because sp^2 = 33.3% character, sp^3 = 25% *s* character—the more *s* character, the stronger is the bond **1-25. a, e, g, h** **1-26. a. 1.** $^+NH_4$ **2.** HCl **3.** H_2O **4.** H_3O^+ **b. 1.** $^-NH_2$ **2.** Br^- **3.** NO_3^- **4.** HO^- **1-28. a.** 5.2 **b.** 3.4×10^{-3} **1-29.** 8.16×10^{-8} **1-31. a.** CH_3COO^- **b.** $^-NH_2$ **c.** H_2O

1-32. $CH_3NH^- > CH_3O^- > CH_3NH_2 > CH_3C(=O)O^- > CH_3OH$

1-33. a. 2.0×10^5 **b.** 3.2×10^{-7} **c.** 1.0×10^{-5} **d.** 4.0×10^{-13}

1-34. a. $CH_3OCH_2CH_2OH$ **b.** $CH_3CH_2CH_2\overset{+}{O}H_2$ **c.** $CH_3CH_2OCH_2CH_2OH$ **d.** $CH_3CH_2C(=O)OH$

1-35. $CH_3CH(F)CH_2OH > CH_3CH(Cl)CH_2OH > CH_2(Cl)CH_2CH_2OH > CH_3CH_2CH_2OH$

1-36. a. $CH_3CH(Br)C(=O)O^-$ **b.** $CH_3CH(Cl)CH_2C(=O)O^-$ **c.** $CH_3CH_2C(=O)O^-$ **d.** $CH_3C(=O)CH_2CH_2O^-$

1-38. a. F^- **b.** I^- **1-39. a.** oxygen **b.** H_2S **c.** CH_3SH **1-40. a.** $CH_3C{\equiv}\overset{+}{N}H$ **b.** CH_4 or CH_3CH_3 **c.** $F_3CC(=O)OH$ **d.** sp^2 **e.** $sp > sp^2 > sp^3$ **f.** $sp > sp^2 > sp^3$ **g.** HNO_3

1-42. a. $^-O-C(=O)-O^- \longleftrightarrow O{=}C(-O^-)-O^- \longleftrightarrow {}^-O-C(-O^-){=}O$

b. $^-O-\overset{+}{N}(=O)-O^- \longleftrightarrow O{=}\overset{+}{N}(-O^-)-O^- \longleftrightarrow {}^-O-\overset{+}{N}(-O^-){=}O$

1-43. 10.4 **1-44. a.** 10.4 **b.** 2.7 **c.** 4.9 **d.** 7.3 **e.** 9.3 **1-45. a. 1.** neutral **2.** neutral **3.** 1/2 neutral and 1/2 charged **4.** charged **5.** charged **6.** charged **7.** charged **b. 1.** charged **2.** charged **3.** charged **4.** charged **5.** 1/2 charged and 1/2 neutral **6.** neutral **7.** neutral **1-46. a. 1.** 4.9 **2.** 10.7 **b. 1.** >6.9 **2.** <8.7

CHAPTER 2

2-1. 1. $CH_3CH_2CH_2CH_2Br$ butyl bromide or *n*-butyl bromide; $CH_3CH(CH_3)CH_2Br$ isobutyl bromide; $CH_3CH_2CH(Br)CH_3$ *sec*-butyl bromide; $CH_3C(CH_3)_2Br$ *tert*-butyl bromide

2-2. c

2-3. a. $CH_3CH(CH_3)OH$ **b.** $CH_3CH(CH_3)CH_2CH_2F$ **c.** $CH_3CH_2CH(CH_3)I$ **d.** $CH_3C(CH_3)_2CH_2Cl$ **e.** $CH_3C(CH_3)_2NH_2$ **f.** $CH_3CH_2CH_2CH_2CH_2CH_2CH_2CH_2Br$

2-4. a. $CH_3CH(CH_3)CH(CH_3)CH_2CH_2CH_3$ **d.** $CH_3C(CH_3)_2CH_2CH(CH_2CH_2CH_3)CH_2CH_2CH_2CH_3$

b. $CH_3CH(CH_3)CH_2C(CH_3)(CH(CH_3)_2)—CH(CH_3)CH_2CH_3$ **e.** $CH_3CH(CH_3)CH_2CH(CH_2CH(CH_3)_2)CH(CH_3)CH_2CH_2CH_3$

c. $CH_3CH_2CH_2C(CH_2CH_3)_2CH_2CH_2CH_2CH_2CH_2CH_3$ **f.** $CH_3CH_2CH_2CH(C(CH_3)_3)CH_2CH_2CH_2CH_3$ [substituent printed as $(CH_3)_2CCH_3$]

2-6. a. 2,2,4-trimethylhexane **b.** 2,2-dimethylbutane **c.** 3,3-diethylhexane **d.** 2,5-dimethylheptane **e.** 3,3-diethyl-4-methyl-5-propyloctane **f.** 3-methyl-4-propylheptane **g.** 5-ethyl-4,4-dimethyloctane **h.** 4-isopropyloctane

2-8. a. [skeletal structure ending in OH] **b.** [skeletal structure] **c.** [skeletal structure with Br] **d.** [skeletal structure ending in OCH_3]

2-9. a. 1-ethyl-2-methylcyclopentane **b.** ethylcyclobutane **c.** 4-ethyl-1,2-dimethylcyclohexane **d.** 3,6-dimethyldecane **e.** 2-cyclopropylpentane **f.** 1-ethyl-3-isobutylcyclohexane **g.** 5-isopropylnonane **h.** 1-*sec*-butyl-4-isopropylcyclohexane **2-10. a.** *sec*-butyl chloride, 2-chlorobutane, **secondary** **b.** isohexyl chloride, 1-chloro-4-methylpentane, **primary** **c.** cyclohexyl bromide, bromocyclohexane, **secondary** **d.** isopropyl fluoride, 2-fluoropropane, **secondary** **2-12. a. 1.** methoxyethane **2.** ethoxyethane **3.** 4-methoxyoctane **4.** 1-propoxybutane **5.** 2-isopropoxypentane **6.** 1-isopropoxy-3-methylbutane **b.** no **c. 1.** ethyl methyl ether **2.** diethyl ether **4.** butyl propyl ether **5.** isopentyl isopropyl ether **2-14. a.** 1-pentanol, **primary** **b.** 4-methylcyclohexanol, **secondary** **c.** 5-chloro-2-methyl-2-pentanol, **tertiary** **d.** 5-methyl-3-hexanol, **secondary** **e.** 2,6-dimethyl-4-octanol, **secondary** **f.** 4-chloro-3-ethylcyclohexanol, **secondary**

2-15. a. $CH_3C(CH_3)(OH)CH_2CH_2CH_3$ **2-methyl-2-pentanol** **b.** $CH_3CH_2C(CH_3)(OH)CH_2CH_3$ **3-methyl-3-pentanol** **c.** $CH_3C(CH_3)(OH)—CH(CH_3)CH_3$ **2,3-dimethyl-2-butanol**

2-16. a. hexylamine, 1-hexanamine, **b.** butylpropylamine, *N*-propyl-1-butanamine **c.** *sec*-butylisobutylamine, *N*-isobutyl-2-butanamine **d.** ethylmethylpropylamine, *N,N*-diethyl-1-propanamine **e.** cyclohexylamine. cyclohexanamine

2-17. a. $CH_3CH(CH_3)CH_2NHCH_2CH_2CH_3$ **b.** $CH_3CH_2NHCH_2CH_3$ **c.** $CH_3CH(CH_3)CH_2CH_2CH_2CH_2NH_2$ **d.** $CH_3CH_2CH_2N(CH_3)CH_2CH_2CH_3$ **e.** $CH_3CH_2CH(CH_2CH_3)N(CH_3)CH_3$ [printed: $CH_3CH_2\overset{CH_2CH_3}{CH}NCH_3$ with CH_3 on N] **f.** cyclohexyl–$N(CH_3)CH_2CH_3$

2-18. **a.** 6-methyl-1-heptanamine, isooctylamine, primary **b.** 3-methyl-*N*-propyl-1-butanamine, isopentylpropylamine, secondary **c.** *N*-ethyl-*N*-methylethanamine, diethylmethylamine, tertiary **d.** 2,5-dimethylcyclohexanamine, no common name, primary **2-19.** **a.** 104.5° **b.** 107.3° **c.** 104.5° **d.** 109.5° **2-20.** **a.** 1, 4, 5 **b.** 1, 2, 4, 5, 6

2-22. HO–(OH)–OH > (OH)–OH > (OH) > (NH_2) > >

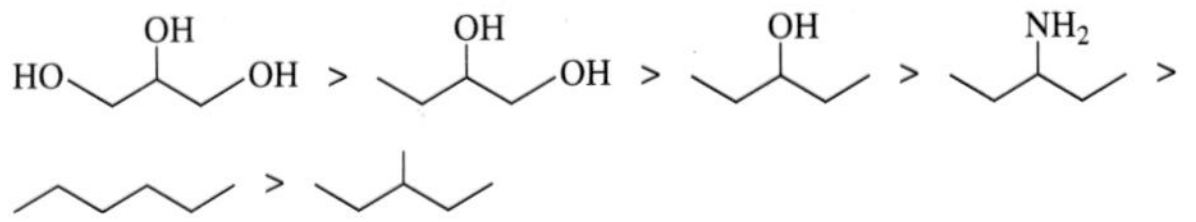

2-24. **a.** $HOCH_2CH_2CH_2OH > CH_3CH_2CH_2OH > CH_3CH_2CH_2CH_2OH > CH_3CH_2CH_2CH_2Cl$

b. 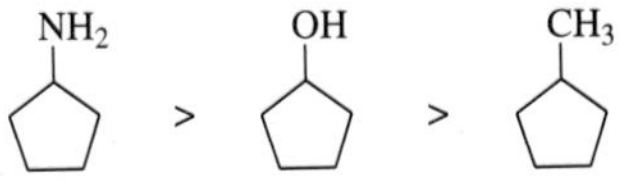 (NH_2)cyclopentane > (OH)cyclopentane > (CH_3)cyclopentane

2-25. ethanol

2-27. **a.** Newman projection: front CH_3, H, H; back CH_3, H, CH_2CH_3 **b.** 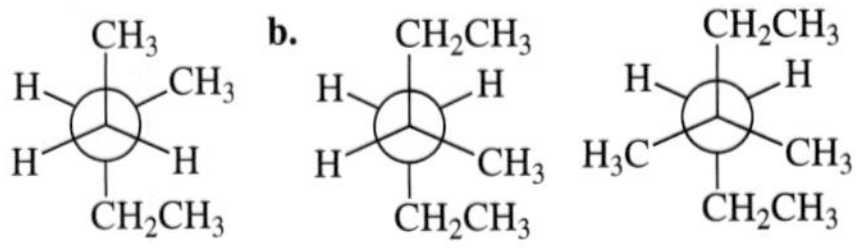 front CH_2CH_3, H, H; back H, CH_3, CH_2CH_3; CH_2CH_3, H, H; H_3C, CH_3, CH_2CH_3

2-28. **a.** 135° **b.** 140° **2-29.** hexethal **2-31.** 6.2 kcal/mol **2-32.** 0.13 kcal/mol **2-33.** 84% **2-35.** *cis*-1-*tert*-butyl-3-methylcyclohexane **2-36.** **a.** cis **b.** cis **c.** cis **d.** trans **e.** trans **f.** trans **2-38.** **a.** one equatorial and one axial **b.** both equatorial and both axial **c.** both equatorial and both axial **d.** one equatorial and one axial **e.** one equatorial and one axial **f.** both equatorial and both axial **2-39.** **a.** 3.6 kcal/mol **b.** 0

CHAPTER 3

3-1. **a.** C_5H_8 **b.** C_4H_6 **c.** $C_{10}H_{16}$ **d.** C_8H_{10} **3-2.** **a.** 3 **b.** 4 **c.** 1 **d.** 3 **e.** 13

3-4. **a.** 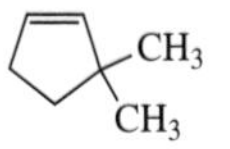 3,3-dimethylcyclopentene (CH_3, CH_3) **c.** $CH_3CH_2OCH{=}CH_2$

b. $BrCH_2CH_2CH_2C(CH_3){=}C(CH_3)CH_3$ **d.** $CH_2{=}CHCH_2OH$

3-5. **a.** 4-methyl-2-pentene **b.** 2-chloro-3,4-dimethyl-3-hexene **c.** 1-bromocyclopentene **d.** 1-bromo-4-methyl-3-hexene **e.** 1,5-dimethylcyclohexene **f.** 1-butoxy-1-propene **3-6.** **a.** 5 **b.** 4 **c.** 4 **d.** 6

3-7. **a.** 1 and 3 **b.** (1.) $H_3C(H)C{=}C(H)CH_2CH_3$ cis; $H_3C(H)C{=}C(H)CH_2CH_3$ trans

(3.) $H_3C(H)C{=}C(H)CH_3$ cis; $H_3C(H)C{=}C(H)CH_3$ trans

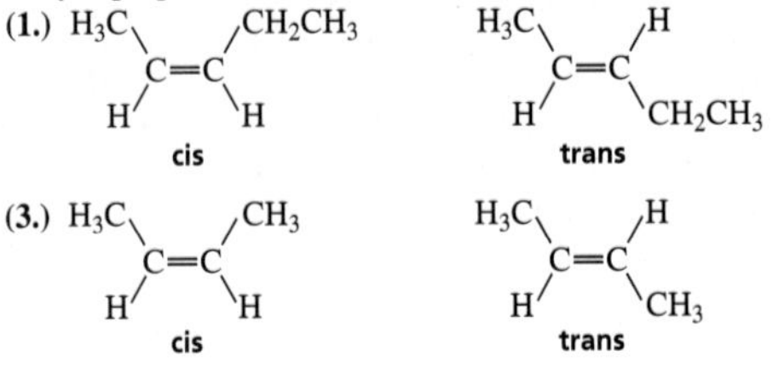

3-8. C **3-11.** nucleophiles: H^- CH_3O^- $CH_3C{\equiv}CH$ NH_3; electrophiles: $AlCl_3$ $CH_3\overset{+}{C}HCH_3$ **3-13.** **a.** all **b.** *tert*-butyl **c.** *tert*-butyl **d.** −1.7 kcal/mol or −7.1 kJ/mol **3-15.** **a.** 1. A + B ⇌ C **b.** None 2. A + B ⇌ C

3-16. **a.** 1. $\Delta G° = -15$; $K_{eq} = 6.5 \times 10^{10}$ 2. $\Delta G° = -16$; $K_{eq} = 1.8 \times 10^8$ **b.** the greater the temperature, the more negative is the $\Delta G°$ **c.** the greater the temperature, the smaller is the K_{eq} **3-17.** **a.** −20 kcal/mol **b.** −35 kcal/mol **c.** exothermic **d.** exergonic **3-18.** **a.** a and b **b.** b **c.** c **3-21.** decreasing; increasing **3-22.** **a.** it will decrease it **b.** it will increase it **3-23.** **a.** the first reaction **b.** the first reaction **3-25.** **a.** first step **b.** revert to reactants **c.** second step **3-26.** **a.** 1 **b.** 2 **c.** k_2 **d.** k_{-1} **e.** k_{-1} **f.** B → C **g.** C → B

CHAPTER 4

4-1. **a.** $CH_3CH_2\overset{+}{C}(CH_3)CH_3$ > $CH_3CH_2\overset{+}{C}HCH_3$ > $CH_3CH_2CH_2\overset{+}{C}H_2$

b. $CH_3CH(CH_3)CH_2\overset{+}{C}H_2$ > $CH_3CH(Cl)CH_2\overset{+}{C}H_2$ > $CH_3CH(F)CH_2\overset{+}{C}H_2$

4-2. **a.** none **b.** ethyl cation **4-3.** **a.** products **b.** reactants **c.** reactants **d.** products

4-4. **a.** $CH_3CH_2CH(Br)CH_3$ **c.** 1-bromo-1-methylcyclopentane (CH_3, Br) **e.** 1-bromo-1-methylcyclohexane (CH_3, Br)

b. $CH_3CH_2C(CH_3)(Br)CH_3$ **d.** $CH_3C(CH_3)(Br)CH_2CH_2CH_3$ **f.** $CH_3CH_2CH(Br)CH_3$

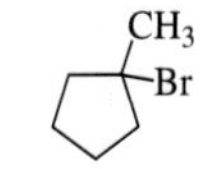

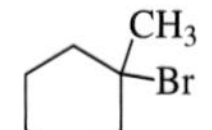

4-5. **a.** $CH_2{=}C(CH_3)CH_3$ **c.** 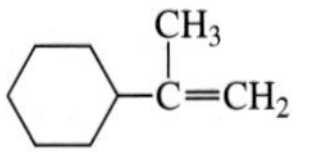 cyclohexyl–$C(CH_3){=}CH_2$

b. 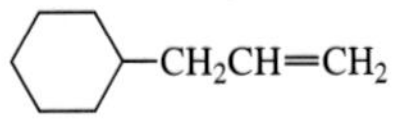 cyclohexyl–$CH_2CH{=}CH_2$ **d.** 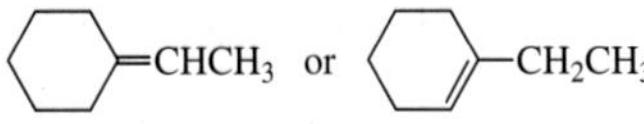 cyclohexylidene$=CHCH_3$ or cyclohexenyl–CH_2CH_3

4-6. **a.** $CH_3CH_2C(CH_3){=}CH_2$ **b.** 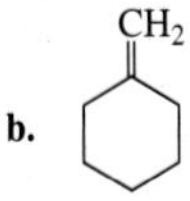 methylenecyclohexane (CH_2)

4-7. greater than −2.5 and less than 15 **4-8.** **a.** 3 **b.** 2 **c.** neutral alcohol **d.** second and third steps in the forward direction

4-15. $CH_3C(CH_3)(Br)CH_2CH_3$ **4-19.** $CH_3CH_2CH(Cl)CH_2I$

4-20. **a.** $CH_2(Br)CH(Br)CH_2CH_3$ **c.** $CH_2(Br)CH(OCH_2CH_3)CH_2CH_3$ **4-23.** 2/3 mole

b. $CH_2(Br)CH(OH)CH_2CH_3$ **d.** $CH_2(Br)CH(OCH_3)CH_2CH_3$

4-24. **a.** $CH_3CH(CH_3)CH(OH)CH_3$ **b.** 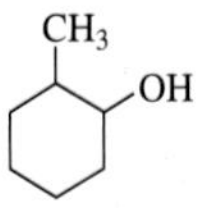 2-methylcyclohexanol (CH_3, OH)

4-27. A

4-28. **a.** 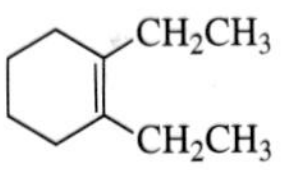 1,2-diethylcyclohexene (CH_2CH_3, CH_2CH_3) **b.** 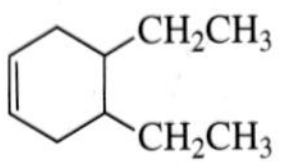 diethylcyclohexene (CH_2CH_3, CH_2CH_3) **c.** 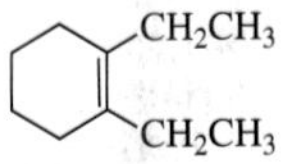 diethylcyclohexene (CH_2CH_3, CH_2CH_3)

4-29. *trans*-3-hexene > *cis*-3-hexene > *cis*-2,5-dimethyl-3-hexene > 1-hexene

CHAPTER 5

5-1. **a.** $CH_3CH_2CH_2OH$ $CH_3CH(OH)CH_3$ $CH_3CH_2OCH_3$ **b.** 7

5-3. **a.** P, F, J, L, G, R, Q, N, Z **b.** T, M, O, A, U, V, C, D, E, H, I, X, Y **5-4.** a, c, and f **5-6.** a, c, and f **5-8.** **a.** *R* **b.** *R* **c.** *R* **d.** *R* **5-9.** **a.** enantiomers **b.** enantiomers **c.** enantiomers **d.** enantiomers

5-10. **a.** $-CH_2OH$ (1) $-CH_3$ (3) $-CH_2CH_2OH$ (2) $-H$ (4)

b. $-CH{=}O$ (2) $-OH$ (1) $-CH_3$ (4) $-CH_2OH$ (3)

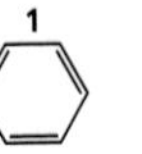

c. $-CH(CH_3)_2$ (2) $-CH_2CH_2Br$ (3) $-Cl$ (1) $-CH_2CH_2CH_2Br$ (4)

d. $-CH{=}CH_2$ (2) $-CH_2CH_3$ (3) phenyl (1) $-CH_3$ (4)

5-11. **a.** *S* **b.** *R* **c.** *S* **d.** *S* **5-12.** +67° **5-13.** **a.** levorotatory **b.** dextrorotatory **5-14.** **a.** −24° **b.** 0° **5-15.** **a.** +79° **b.** 0° **c.** −79° **5-16.** 98.5% dextrorotatory; 1.5% levorotatory **5-19.** **a.** enantiomers **b.** identical **c.** diastereomers **5-20.** **a.** 8 **b.** $2^8 = 256$ **c.** 1 **5-24.** 1-chloro-1-methylcyclooctane, *cis*-1-chloro-5-methylcyclooctane, *trans*-1-chloro-5-methylcyclooctane **5-26.** b, d, and f **5-30.** left = *R*; right = *R*

5-32. **a.** 4 **b.** 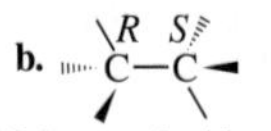 C—C (R, S) **5-33.** **a.** (2*R*,3*R*)-2,3-dichloropentane **b.** (2*R*,3*R*)-2-bromo-3-chloropentane **c.** (1*R*, 3*S*)-1,3-cyclopentanediol **d.** (3*R*,4*S*)-3-chloro-4-methylhexane **5-35.** *R* **5-36.** **a.** *R* **b.** *R* **c.** *S* **d.** *S* **5-37.** b **5-38.** >99% **5-39.** **a.** enantiotopic **b.** diastereotopic **c.** neither **d.** diastereotopic **5-41.** **a.** no **b.** no **c.** no **d.** yes **e.** no **f.** no

5-44. **a.** cyclohexanol (OH) **b.** HO, CH_2CH_3 cyclohexanes (two stereoisomers) **c.** HO, H_3C, CH_3 cyclopentanes (two stereoisomers) **d.** CH_3—C(H)(OH)—CH_2CH_3 and CH_3—C(H)(HO)—CH_2CH_3

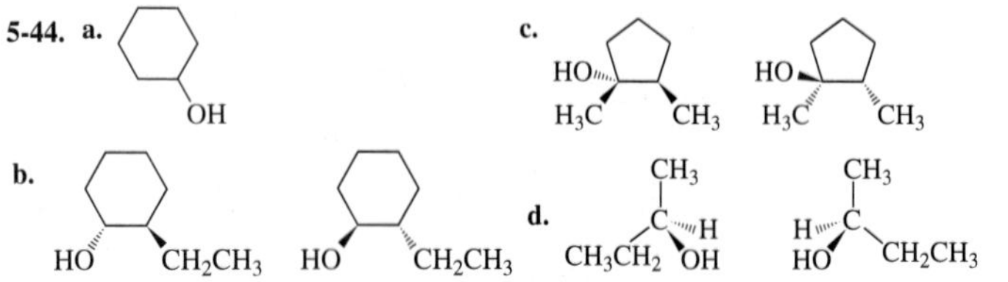

5-51. **a.** 1-bromo-2-chloropropane **b.** equal amounts of *R* and *S*
5-52. **a.** (*R*)-malate and (*S*)-malate **b.** (*R*)-malate and (*S*)-malate

CHAPTER 6

6-1. $C_{14}H_{20}$ **6-2.** **a.** $ClCH_2CH_2C{\equiv}CCH_2CH_3$ **b.** 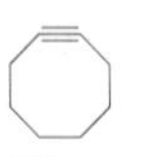 cyclooctyne

c. $CH_3CH(CH_3)C{\equiv}CH$ **d.** $HC{\equiv}CCH_2Cl$ **e.** $HC{\equiv}CCH_2C(CH_3)_2CH_3$ **f.** $CH_3C{\equiv}CCH_3$

6-3.

$HC{\equiv}CCH_2CH_2CH_2CH_3$ — 1-hexyne, butylacetylene

$CH_3C{\equiv}CCH_2CH_2CH_3$ — 2-hexyne, methylpropylacetylene

$CH_3CH_2C{\equiv}CCH_2CH_3$ — 3-hexyne, diethylacetylene

$HC{\equiv}CCH(CH_3)CH_2CH_3$ — 3-methyl-1-pentyne, *sec*-butylacetylene

$HC{\equiv}CCH_2CH(CH_3)CH_3$ — 4-methyl-1-pentyne, isobutylacetylene

$CH_3CH(CH_3)C{\equiv}CCH_3$ — 4-methyl-2-pentyne, isopropylmethylacetylene

$CH_3C(CH_3)_2C{\equiv}CH$ — 3,3-dimethyl-1-butyne, *tert*-butylacetylene

6-4. **a.** 5-bromo-2-pentyne **b.** 6-bromo-2-chloro-4-octyne **c.** 1-methoxy-2-pentyne **d.** 3-ethyl-1-hexyne

6-6. **a.** sp^2–sp^2 **b.** sp^2–sp^3 **c.** sp–sp^2 **d.** sp–sp^3 **e.** sp–sp **f.** sp^2–sp^2 **g.** sp^2–sp^3 **h.** sp–sp^3 **i.** sp^2–sp

6-7. If the less stable reactant has the more stable transition state or if the less stable reactant has the less stable transition state and the difference is stabilities of the reactants is less than the difference in stabilities of the transition states, or if the difference in the stabilities of the reactants is greater than the difference in the stabilities of the transition states.

6-8. **a.** $BrCH{=}CHCH_3$ **b.** $CH_2{=}C(Br)CH_3$ **c.** $CH_3CBr_2CH_3$ **d.** $CH_3C(Br){=}C(Br)CH_3$ **e.** $CH_3CH_2CBr_2CH_3$ **f.** $CH_3CBr_2CH_2CH_2CH_3$ + $CH_3CH_2CBr_2CH_2CH_3$

6-9. **a.** $H_3C(Br)C{=}C(Br)CH_3$ (trans) **b.** $H_3C(H)C{=}C(CH_3)(Br)$ + $H_3C(H)C{=}C(Br)(CH_3)$

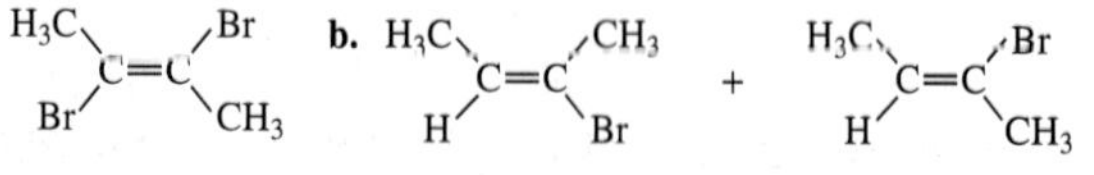

6-10. $CH_3CH_2C(=O)CH_2CH_2CH_2CH_3$ and $CH_3CH_2CH_2C(=O)CH_2CH_2CH_3$

6-11. **a.** $CH_3C{\equiv}CH$ **b.** $CH_3CH_2C{\equiv}CCH_2CH_3$ **c.** 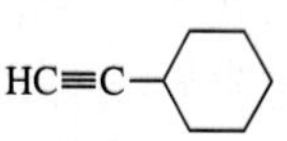 HC≡C–cyclohexyl

6-12. **a.** $CH_2{=}C(OH)CH_3$ **b.** $CH_3CH{=}C(OH)CH_2CH_2CH_3$ and $CH_3CH_2C(OH){=}CHCH_2CH_3$ **c.** 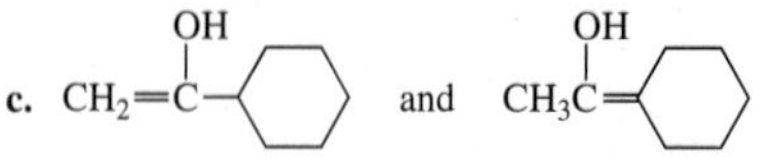 $CH_2{=}C(OH)$–cyclohexyl and $CH_3C(OH){=}$cyclohexylidene

6-13. **a.** **1.** $CH_3CH_2C(=O)CH_3$ **2.** $CH_3CH_2CH_2CH(=O)$ **b.** **1.** $CH_3CH_2C(=O)CH_3$ **2.** $CH_3CH_2C(=O)CH_3$ **c.** **1.** $CH_3CH_2CH_2C(=O)CH_3$ and $CH_3CH_2C(=O)CH_2CH_3$ **2.** $CH_3CH_2CH_2C(=O)CH_3$ and $CH_3CH_2C(=O)CH_2CH_3$

6-14. ethyne (acetylene)

6-15. **a.** $CH_3CH_2CH_2C{\equiv}CH$ or $CH_3CH_2C{\equiv}CCH_3$ $\xrightarrow[\text{Pt}]{H_2}$ **b.** $CH_3C{\equiv}CCH_3$ $\xrightarrow[\text{Lindlar catalyst}]{H_2}$ **c.** $CH_3CH_2C{\equiv}CCH_3$ $\xrightarrow[NH_3]{\text{Na}}$ **d.** $CH_3CH_2CH_2CH_2C{\equiv}CH$ $\xrightarrow[\text{Lindlar catalyst}]{H_2}$ or $\xrightarrow[NH_3]{\text{Na}}$

6-16. 25 **6-17.** **a.** $CH_3\overset{+}{C}H_2$ **b.** $H_2C{=}\overset{+}{C}H$
6-18. The carbanion that would be formed is a stronger base than the amide ion.
6-19. **a.** $CH_3CH_2CH_2\overset{-}{C}H_2 > CH_3CH_2CH{=}\overset{-}{C}H > CH_3CH_2C{\equiv}\overset{-}{C}$ **b.** $CH_3C{\equiv}\overset{-}{C} > CH_3CH_2O^- > {}^-NH_2 > F^-$

CHAPTER 7

7-1. **a.** **1.** 3 **2.** 2 **b.** **1.** 5 **2.** 5
7-3. **a.** all are the same length **b.** 2/3 of a negative charge

7-7. **a.** $CH_3CH_2CH{=}CH\dot{C}H_2$ **b.** $CH_3C(=O)CH{=}CHCH_3$ **c.** $CH_3C(O^-){=}CHCH_3$ **d.** $CH_3{-}C({=}\overset{+}{N}H_2){-}NH_2$

7-8. **a.** $CH_3\overset{+}{N}H\overset{}{C}H_2$ **b.** 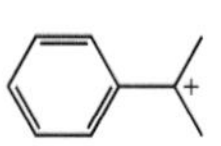 phenyl–$\overset{+}{C}(CH_3)_2$ **c.** $CH_3\overset{+}{O}CH_2$ **d.** 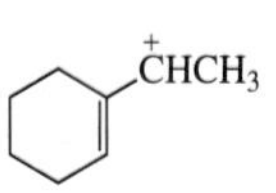 cyclohexenyl–$\overset{+}{C}HCH_3$ **e.** $CH_3OCH{=}CH\overset{+}{C}H_2$

7-11. **a.** $CH_3CH{=}CHOH$ **b.** $CH_3C(=O)OH$ **c.** $CH_3CH{=}CHOH$ **d.** $CH_3CH{=}CH\overset{+}{N}H_3$

7-12. **a.** ethylamine **b.** ethoxide ion **c.** ethoxide ion

7-13. C_6H_5–COOH > C_6H_5–OH > C_6H_5–CH_2OH

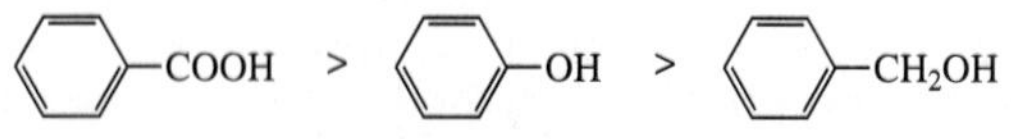

7-14. ψ_3 = 3; ψ_4 = 4 **7-15.** **a.** bonding = ψ_1 and ψ_2; antibonding = ψ_3 and ψ_4 **b.** symmetric = ψ_1 and ψ_3; asymmetric = ψ_2 and ψ_4 **c.** HOMO = ψ_2; LUMO = ψ_3 **d.** HOMO = ψ_3; LUMO = ψ_4 **e.** If the HOMO is symmetric, the LUMO is asymmetric and vice versa. **7-16.** **a.** bonding = ψ_1, ψ_2, and ψ_3; antibonding = ψ_4, ψ_5, and ψ_6 **b.** symmetric = ψ_1, ψ_3, and ψ_5; asymmetric = ψ_2, ψ_4, and ψ_6 **c.** HOMO = ψ_3; LUMO = ψ_4 **d.** HOMO = ψ_4; LUMO = ψ_5 **e.** If the HOMO is symmetric, the LUMO is asymmetric and vice versa. **7-17.** **a.** ψ_1 = 3; ψ_2 = 2 **b.** ψ_1 = 7; ψ_2 = 6

CHAPTER 8

8-1. **a.** 1,5-cyclooctadiene **b.** 1-hepten-4-yne **c.** 4-methyl-1,4-hexadiene **d.** 5-vinyl-5-octen-1-yne **e.** 1,6-dimethyl-1,3-cyclohexadiene **f.** 3-butyn-1-ol **g.** 1,3,5-heptatriene **h.** 2,4-dimethyl-4-hexen-1-ol

8-3. $CH_3C(CH_3){=}CHCH{=}C(CH_3)CH_3$ > $CH_3CH{=}CHCH{=}CHCH_3$ >
2,5-dimethyl-2,4-hexadiene > 2,4-hexadiene >

$CH_3CH{=}CHCH{=}CH_2$ > $CH_2{=}CHCH_2CH{=}CH_2$
1,3-pentadiene > 1,4-pentadiene

8-4.

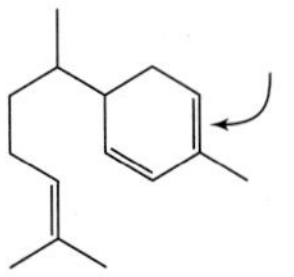

8-8. a. $CH_3CHClCHClCH{=}CHCH_3$ + $CH_3CHClCH{=}CHCHClCH_3$

b. $CH_3CHBrCH{=}C(CH_3)CH_2CH_3$ + $CH_3CH{=}CHCBr(CH_3)CH_2CH_3$

c. $CH_3CH_2CBr(CH_2CH_3){-}C{=}CHCH_3$ + $CH_3CH_2C(CH_2CH_3){=}C(CH_3)CHBrCH_3$

d. 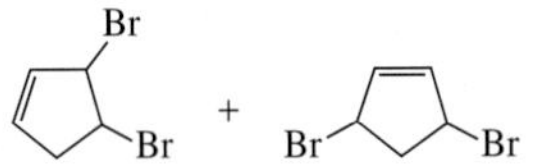 Br ... Br + Br ... Br

8-9. a. Addition at C-1 forms the more stable carbocation. **b.** To cause the 1,2- and, 1,4-products to be different. **8-10. a.** formation of the carbocation **b.** reaction of the carbocation with the nucleophile

8-13. a. 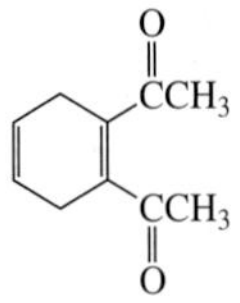 (cyclohexadiene with two CCH_3(=O) groups) **c.** 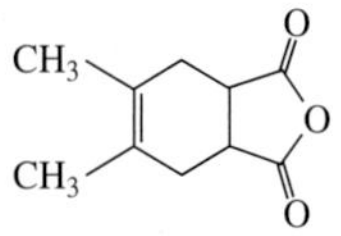 CH_3, CH_3 (anhydride)

b. 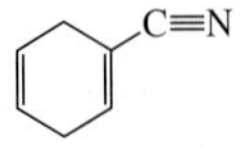 C≡N **d.** 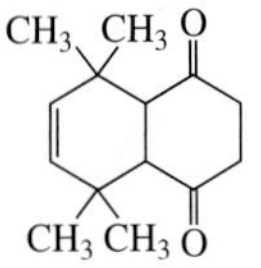 CH_3 CH_3 O ... CH_3 CH_3 O

8-14. a. The product is a meso compound. **b.** The product is a racemic mixture.

8-15. CH_3O ... CH(=O)

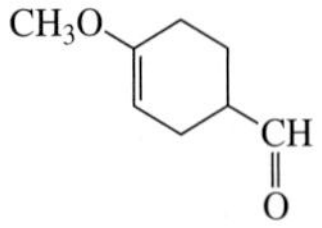

8-16. c. 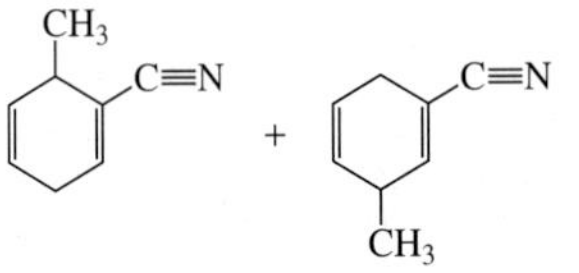 CH_3, C≡N + C≡N, CH_3

d. 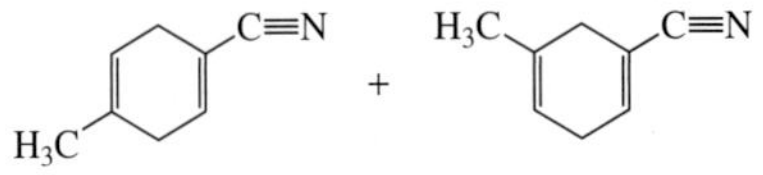 C≡N, H_3C + H_3C, C≡N

8-17. a and **d**

8-19. a. 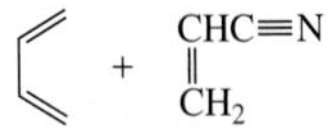 + $CHC{\equiv}N$ ‖ CH_2

b. 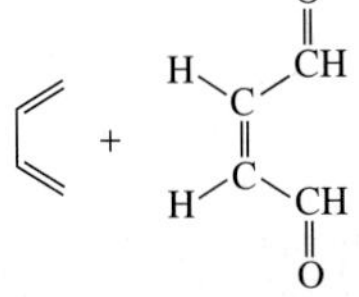 + H, CH(=O), C=C, H, CH(=O)

c. + $CH_3OC(=O)C{\equiv}CC(=O)OCH_3$

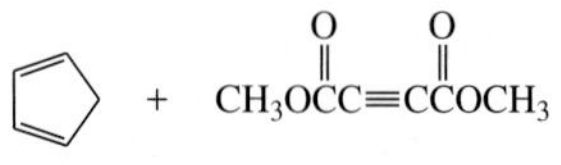

d. 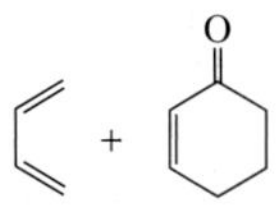 + (cyclohexenone)

e. 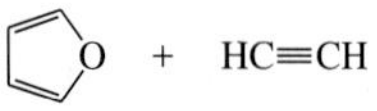 (furan) + $HC{\equiv}CH$

f. 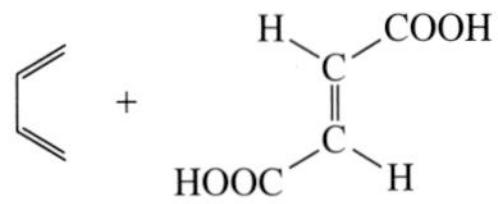 + H, COOH, C=C, HOOC, H

8-21. 4.1×10^{-5} M

8-22.

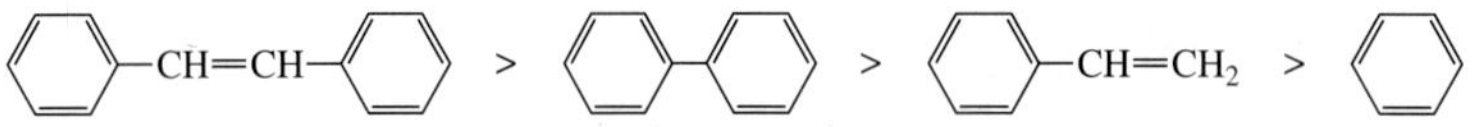

Ph–CH=CH–Ph > biphenyl > Ph–$CH{=}CH_2$ > benzene

CHAPTER 9

9-5.

a. 3	**d.** 1	**g.** 2
b. 3	**e.** 5	**h.** 1
c. 5	**f.** 5	**i.** 4

9-6.

a. $CH_3CH_2CH_2CH_2CH_2Cl$ 21% $CH_3CH_2CH_2CHClCH_3$ 53% $CH_3CH_2CHClCH_2CH_3$ 26%

b. $ClCH_2CH(CH_3)CH_2CH_2CH(CH_3)CH_3$ 32% $CH_3CCl(CH_3)CH_2CH_2CH(CH_3)CH_3$ 27% $CH_3CH(CH_3)CHClCH_2CH(CH_3)CH_3$ 41%

c. $ClCH_2CH(CH_3)CH_2CH_2CH_3$ 21% $CH_3CCl(CH_3)CH_2CH_2CH_3$ 17% $CH_3CH(CH_3)CHClCH_2CH_3$ 26%

$CH_3CH(CH_3)CH_2CHClCH_3$ 26% $CH_3CH(CH_3)CH_2CH_2CH_2Cl$ 10%

9-7. a. $CH_3CH_2CHBrCH_3$ $CH_3CH_2CH_2CH_2Br$
4 × 82 = 328 6 × 1 = 6

$$\frac{328}{328 + 6} = \frac{328}{334} = 0.98 = 98\%$$

b. $CH_3CBr(CH_3)CH_2CH_2C(CH_3)_2CH_3$
1 × 1600 = 1600

other products: 9 × 1 = 9; 2 × 82 = 164; 2 × 82 = 164; 6 × 1 = 6; total 1943 (1600)

$$\frac{1600}{1943} = 0.82 = 82\%$$

9-8. 99.4% (vs. 36%)

9-9.

a. $CH_3CH_2CH_2CH_2CH_2Br$ 1% $CH_3CH_2CHBrCH_2CH_3$ 33% $CH_3CH_2CH_2CHBrCH_3$ 66%

b. $BrCH_2CH(CH_3)CH_2CH_2CH(CH_3)CH_3$ 0.3% $CH_3CBr(CH_3)CH_2CH_2CH(CH_3)CH_3$ 90.4% $CH_3CH(CH_3)CHBrCH_2CH(CH_3)CH_3$ 9.3%

g. $BrCH_2CHCH_2CH_2CH_3$ (CH_3 above; 0.3%) $CH_3CCH_2CH_2CH_3$ (CH_3 above, Br below; 82.6%) $CH_3CHCHCH_2CH_3$ (CH_3 above, Br below; 8.5%)

$CH_3CHCH_2CHCH_3$ (CH_3 above, Br below; 8.5%) $CH_3CH_2CH_2CH_2CH_2CH_2Br$ (0.2%)

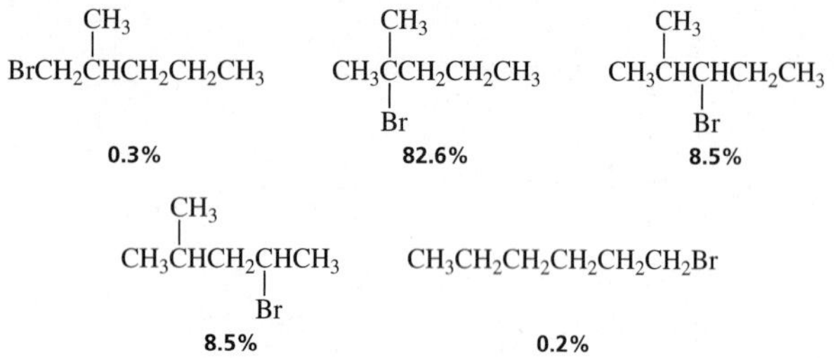

9-10. a. chlorination **b.** bromination **c.** no preference

9-13. a. 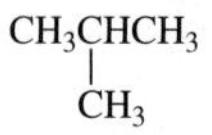 CH_3CHCH_3 (with CH_3) **b.** $CH_3CH_2CH_2CH_3$

9-14. a. (1) [3-bromo-1-methylcyclohexene, CH_3, Br] + [CH_3, Br on cyclohexene] (2) [CH_3, Br, Br on cyclohexane] (3) [CH_3, Br on cyclohexane] (4) [CH_3, Br on cyclohexane]

b. (1) [CH_3, H, Br stereoisomers of bromomethylcyclohexene; Br, CH_3 / H_3C, Br]

(2) [H, Br / Br, CH_3 and Br, H / CH_3, Br chair structures] (3) [H_3C, Br on cyclohexane]

(4) [H, H / Br, CH_3; H, H / CH_3, Br; Br, H / H, CH_3; H, Br / CH_3, H]

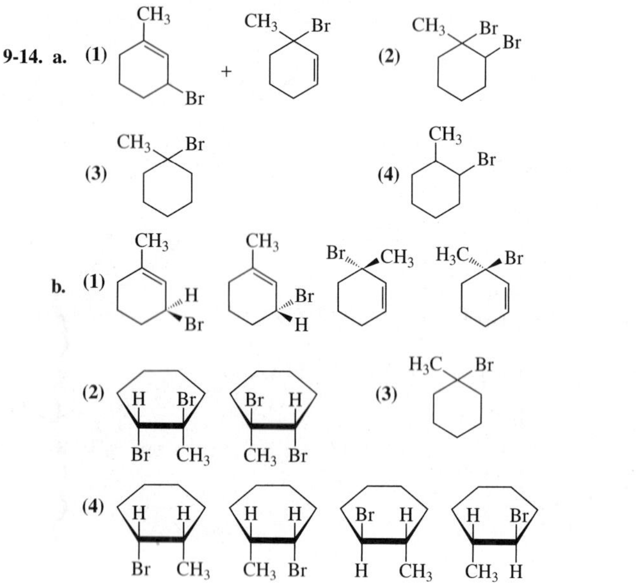

CHAPTER 10

10-1. decrease **10-2.**

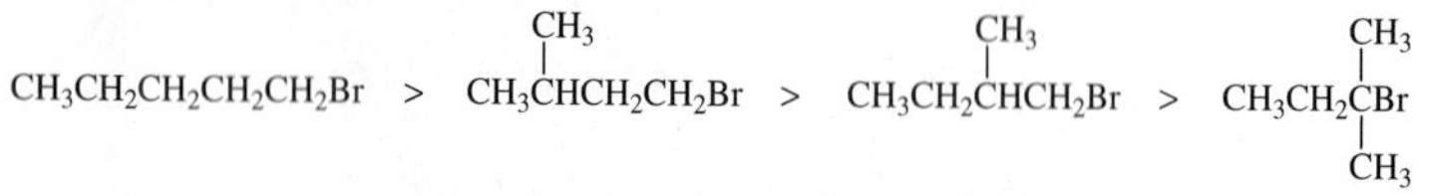

$CH_3CH_2CH_2CH_2CH_2Br$ > $CH_3CHCH_2CH_2Br$ (CH_3) > $CH_3CH_2CHCH_2Br$ (CH_3) > CH_3CH_2CBr (CH_3, CH_3)

10-3. b. (*S*)-2-butanol **c.** *R*)-2-hexanol **d.** 3-pentanol **10-4. a.** RO^- **b.** RS^-

10-5. a. aprotic **b.** aprotic **c.** protic **d.** aprotic

10-7. a. $CH_3CH_2Br + HO^-$ **c.** $CH_3CH_2Cl + CH_3S^-$

d. $CH_3CH_2Br + I^-$

b. 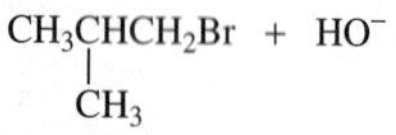 $CH_3CHCH_2Br + HO^-$ (CH_3)

10-11. a. $CH_3CH_2OCH_3$ **c.** $CH_3CH_2\overset{+}{N}(CH_3)_3\ Br^-$

b. $CH_3CH_2N_3$ **d.** $CH_3CH_2SCH_2CH_3$

10-13. CH_3CBr (CH_3, CH_3) > CH_3CHBr (CH_3) > $CH_3CH_2CH_2Br$ > CH_3Br

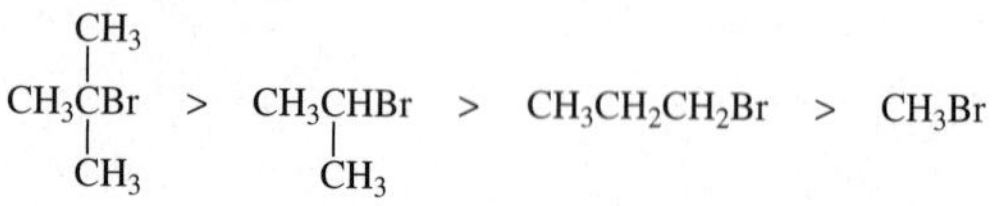

10-15.

$CH_3CH_2CCH_2CH_3$ (CH_3, Br) > $CH_3CHCH_2CH_2CH_3$ (Br) > $CH_3CHCH_2CH_2CH_3$ (Cl) > $ClCH_2CH_2CH_2CH_2CH_3$

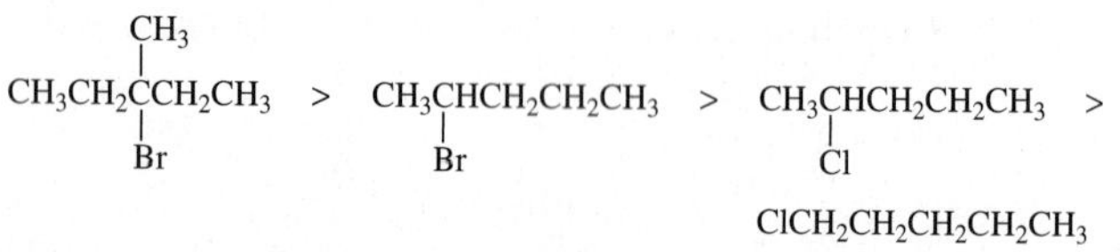

10-16. a, b, c, e **10-18.** *trans*-4-methylcyclohexanol **10-20.** a and b

10-21. *trans*-4-bromo-2-hexene

10-22. a. $CH_3CH_2CH_2I$ **b.** CH_3OCH_2Cl

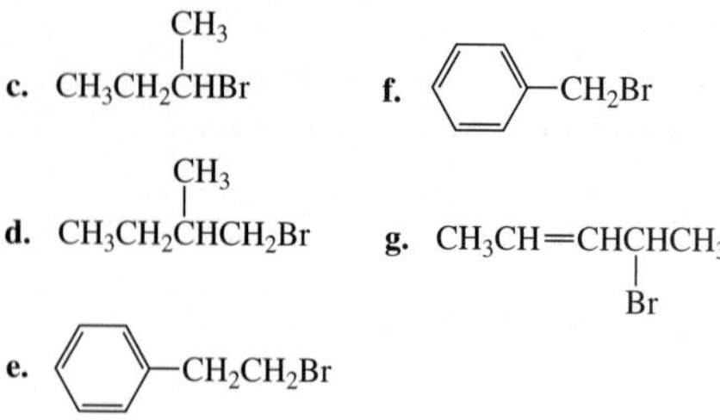

c. CH_3CH_2CHBr (CH_3) **f.** [benzene]$-CH_2Br$

d. $CH_3CH_2CHCH_2Br$ (CH_3) **g.** $CH_3CH{=}CHCHCH_3$ (Br)

e. [benzene]$-CH_2CH_2Br$

10-23. a. $CH_3CH_2CH_2I$ **e.** [benzene]$-CH_2CHCH_3$ (Br)

b. CH_3OCH_2Cl **f.** [benzene]$-CH_2Br$

c. equally reactive **g.** $CH_3CH{=}CHCHCH_3$ (Br)

d. $CH_3CH_2CH_2CHBr$ (CH_3)

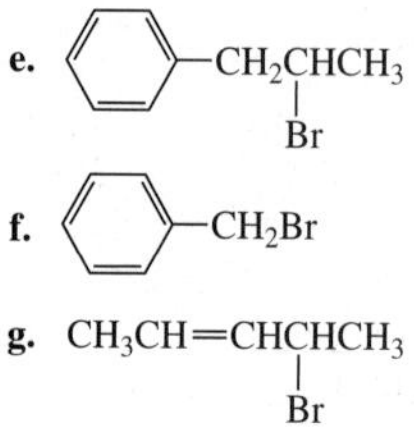

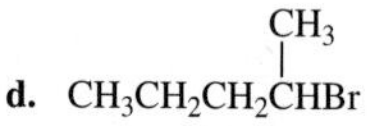

10-24. a. 1-methyoxy-2-butene **b.** 1-methoxy-2-butene and 3-methoxy-1-butene

10-27. a. decrease **b.** decrease **c.** increase

10-28. a. $CH_3Br + HO^- \longrightarrow CH_3OH + Br^-$

b. $CH_3I + HO^- \longrightarrow CH_3OH + I^-$

c. $CH_3Br + NH_3 \longrightarrow CH_3\overset{+}{N}H_3 + Br^-$

d. $CH_3Br + HO^- \xrightarrow{\text{DMSO}} CH_3OH + Br^-$

e. $CH_3Br + NH_3 \xrightarrow{\text{EtOH}} CH_3\overset{+}{N}H_3 + Br^-$

10-30. dimethyl sulfoxide **10-31.** HO^- in 50% water/50% ethanol

CHAPTER 11

11-1. a. $CH_3CH_2CHCH_3$ (Br) **c.** $CH_3CH_2CH_2CCH_3$ (CH_3, Br)

b. [bromocyclohexane] **d.** $CH_3CCH_2CH_2Cl$ (CH_3, CH_3)

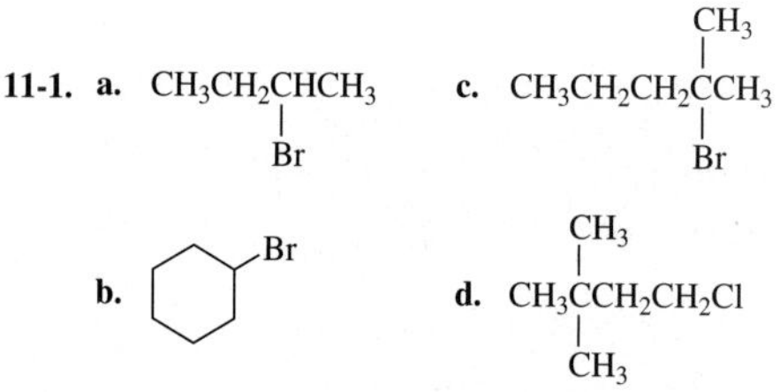

11-3. a. $CH_3CH{=}CHCH_3$ **d.** $CH_3CH{=}CHCH{=}CH_2$

b. $CH_2{=}CHCH_2CH_3$ **e.**  [cyclohexene]

c. $CH_3C{=}CHCH_2CH_3$ (CH_3) **f.** $CH_3CHCH{=}CHCH_3$ (CH_3)

11-4. a. $CH_3CHCHCH_2CH_3$ (CH_3, Br) **c.** $CH_3CH_2CHCH_2CH_3$ (Br)

b. [bromocycloheptene] **d.** [benzene]$-CH_2CHCH_2CH_3$ (Br)

11-5. $CH_3C{=}CCH_2CH_3$ (CH_3, CH_3) > $CH_3CHC{=}CHCH_3$ (CH_3, CH_3) > $CH_3CHCCH_2CH_3$ ($=CH_2$, CH_3)

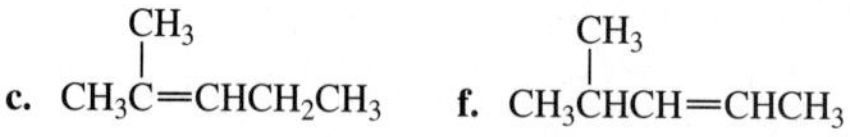

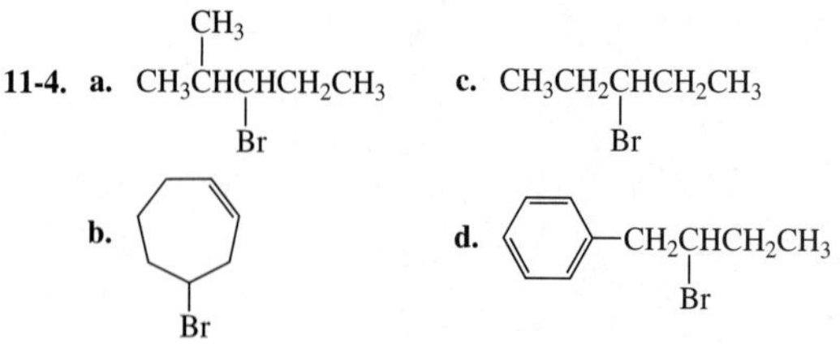

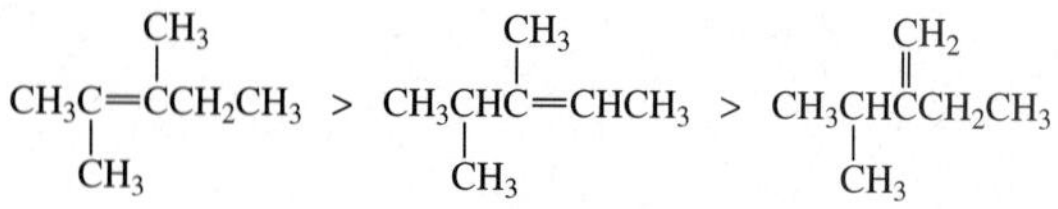

11-8. a. E2 $CH_3CH{=}CHCH_3$ **c.** E1 $CH_3C{=}CH_2$ (CH_3)

b. E1 $CH_3CH{=}CHCH_3$ **d.** E2 $CH_3C{=}CH_2$ (CH_3)

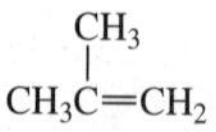

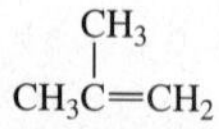

e. E1 $CH_3\overset{CH_3}{C}=\underset{CH_3}{C}CH_3$ **f.** E2 $CH_3\overset{CH_3}{\underset{CH_3}{C}}CH=CH_2$

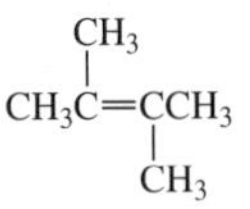

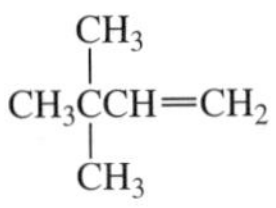

11-9. a. 96% **b.** 1.2%

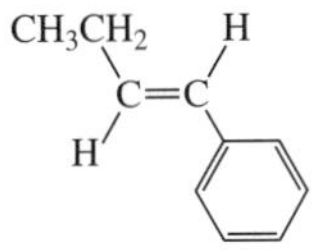

11-10. a. 1. 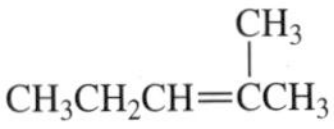 $CH_3CH_2CH=\overset{CH_3}{C}CH_3$ **2.** $CH_3CH_2(H)C=C(H)C_6H_5$

3. $CH_3CH_2(H)C=C(H)CH=CH_2$

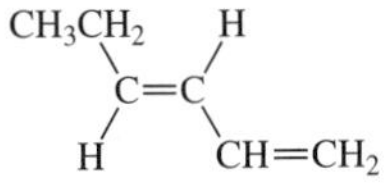

b. no

11-11. b. 2-methyl-2-pentene **c.** (*E*)-3-methyl-3-heptene **d.** 1-methylcyclohexene
11-15. left to right 2, 5, 3, 1, 4 (1 is the most reactive, 5 the least reactive)
11-16. ~1 **11-17.** it will increase **11-19. a. 1.** no reaction **2.** no reaction **3.** substitution and elimination **4.** substitution and elimination **b. 1.** primarily substitution **2.** substitution and elimination **3.** substitution and elimination **4.** elimination **11-20. a.** S_N2 difficult because of steric hindrance; no S_N1 because of primary carbocation **b.** No E2 because there are no hydrogens on a β-carbon; no E1 because of primary carbocation.
11-21. a. $CH_3CH=CH_2$ **b.** $CH_3CH_2CH=CH_2$

11-23. $CH_3\overset{CH_3}{\underset{CH_3}{C}}-OH$ + $CH_3\overset{CH_3}{\underset{CH_3}{C}}-OCH_2CH_3$ + $CH_3\overset{CH_3}{C}=CH_2$

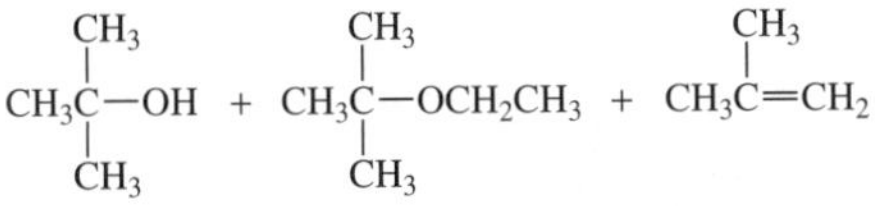

11-26. a. 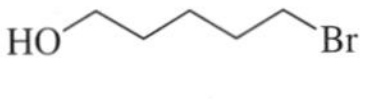 HO~~~~~Br **c.** HO~~~~~~Br

b. 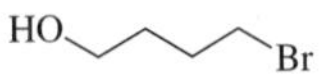 HO~~~~Br

CHAPTER 12

12-1. 3° > 1° > 2°

12-2. a. 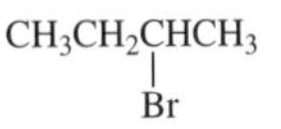 $CH_3CH_2\underset{Br}{C}HCH_3$ **c.** $CH_3\overset{CH_3}{\underset{Br}{C}}-\underset{CH_3}{C}HCH_3$

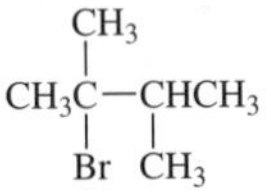

b. 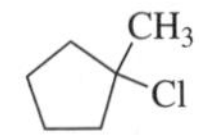 1-chloro-1-methylcyclopentane (CH₃, Cl on ring carbon) **d.** 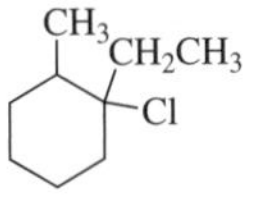 cyclohexane with CH₃, and CH₂CH₃ and Cl on adjacent ring carbon

12-7. 1-bromo-1-methylcyclopentane (Br, CH₃ on ring carbon)

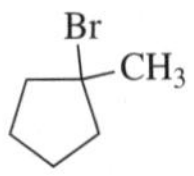

12-10. a. 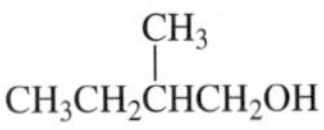 $CH_3CH_2\overset{CH_3}{C}HCH_2OH$ **c.** $CH_3CH_2\underset{OH}{C}HCH_2CH_3$

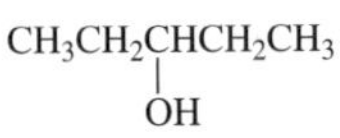

b. 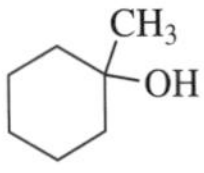 cyclohexane with CH₃ and OH on same carbon **d.** 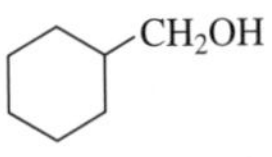 cyclohexane–CH₂OH

12-12. a. 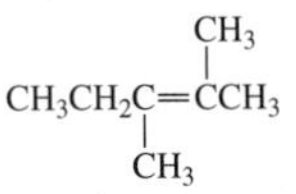 $CH_3CH_2\overset{CH_3}{C}=\underset{CH_3}{C}CH_3$ **b.** cyclopentene with CH₂CH₃ **c.** 1,3-cyclohexadiene

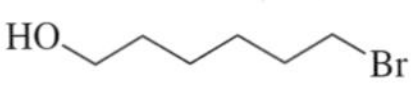

d. 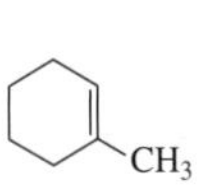 cyclohexene with CH₃ **e.** 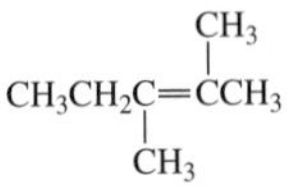 $CH_3CH_2\overset{CH_3}{C}=\underset{CH_3}{C}CH_3$ **f.** 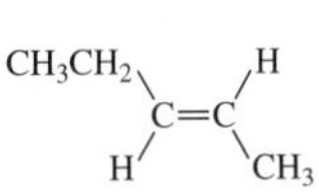 $CH_3CH_2(H)C=C(H)CH_3$

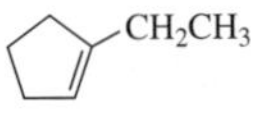

12-14. No, F^- is too poor a nucleophile.

12-16. a. 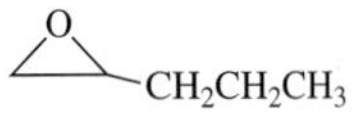 epoxide–$CH_2CH_2CH_3$ **c.** 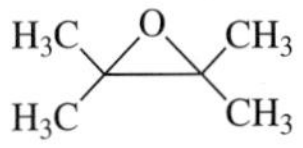 epoxide with H_3C, H_3C on one carbon and CH_3, CH_3 on the other

b. 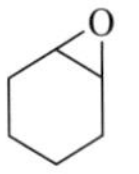 cyclohexene oxide **d.** 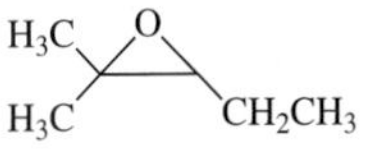 epoxide with H_3C, H_3C on one carbon and CH_2CH_3 on the other

12-18. a. 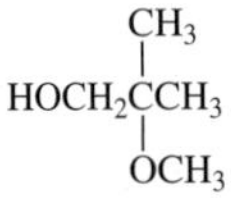 $HOCH_2\overset{CH_3}{\underset{OCH_3}{C}}CH_3$ **c.** $HO\underset{CH_3}{C}H-\overset{OCH_3}{\underset{CH_3}{C}}CH_3$

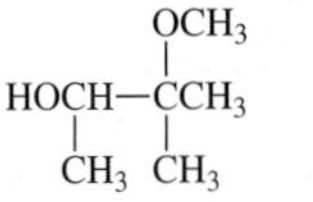

b. 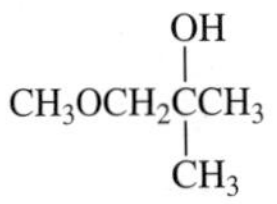 $CH_3OCH_2\overset{OH}{\underset{CH_3}{C}}CH_3$ **d.** 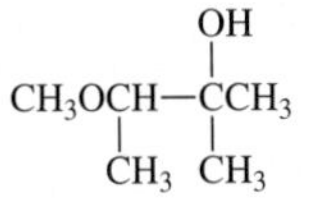 $CH_3O\underset{CH_3}{C}H-\overset{OH}{\underset{CH_3}{C}}CH_3$

12-19. noncyclic ether

12-21. 4-methylphenol (OH and CH₃ para on benzene)

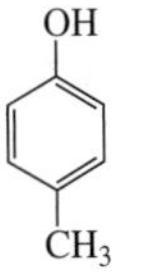

12-22. The products are the same.

12-28. a. $CH_3CH_2CH_2CH_2CH_2OH$ **c.** 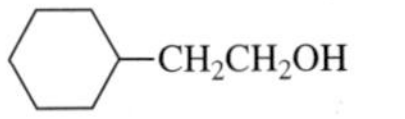 cyclohexane–CH_2CH_2OH

b. 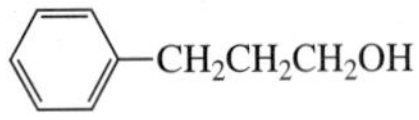 benzene–$CH_2CH_2CH_2OH$

12-31. They all occur. **12-32.** $(CH_3)_4Si$

12-35. a. 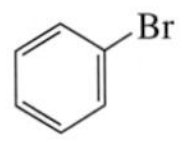 benzyl bromide (benzene–CH₂Br) **b.** 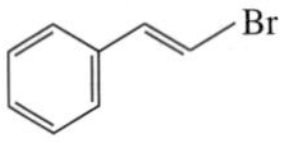 benzene–CH=CH–CH₂Br **c.** 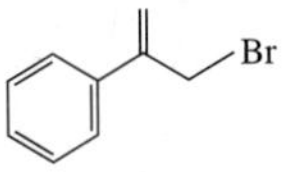 benzene–C(=CH₂)–CH₂Br

12-36. 1-pentyne

12-37. $CH_3\overset{O}{\overset{\|}{C}}$–C₆H₄–Br and $CH_2=CH$–C₆H₄–OCH_3

$CH_3\overset{O}{\overset{\|}{C}}$–C₆H₄–$CH=CH_2$ and Br–C₆H₄–OCH_3

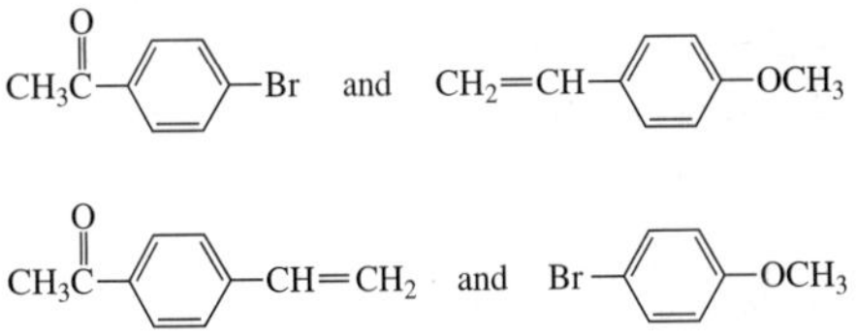

CHAPTER 13

13-2. $m/z = 57$ **13-5.** 8 **13-6.** C_6H_{14} **13-8. a.** 2-methoxy-2-methylpropane **b.** 2-methoxybutane **c.** 1-methoxybutane

13-9. $CH_2=\overset{+}{\ddot{O}}H$

13-11. a. 2-pentanone **b.** 3-pentanone **13-13. a.** 3-pentanol **b.** 2-pentanol
13-14. a. 2000 cm^{-1} **b.** 8 μm **c.** 2 μm **13-15. a.** 2500 cm^{-1} **b.** 50 μm
13-16. C=O
13-17. a. 1. C≡C **2.** C—H stretch **3.** C≡N
b. 1. C—O **2.** C—C
13-18. a. carbon–nitrogen stretch of an amide **b.** carbon–oxygen stretch of a phenol **c.** carbon–oxygen double-bond stretch of a ketone **d.** carbon–oxygen stretch **13-19.** sp^3

13-20. a. δ-valerolactone > cyclohexanone > δ-valerolactam (NH)

b. γ-butyrolactone > unsaturated lactone > unsaturated lactone

c. 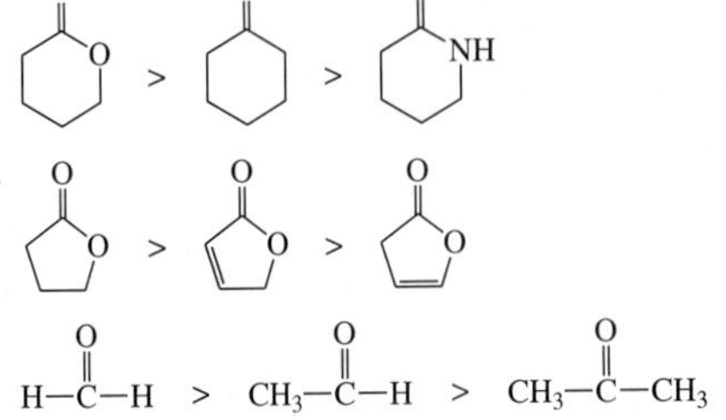 $H-\overset{O}{\overset{\|}{C}}-H$ > $CH_3-\overset{O}{\overset{\|}{C}}-H$ > $CH_3-\overset{O}{\overset{\|}{C}}-CH_3$

13-21. ethanol dissolved in carbon disulfide **13-24. a.** ketone **b.** tertiary amine
13-27. 2-butyne, H_2, Cl_2, ethene **13-28.** *trans*-2-hexene **13-29.** methyl vinyl ketone

CHAPTER 14

14-1. 43 MHz **14-2. a.** 8.46 T **b.** 11.75 T **14-3. a.** 2 **b.** 1 **c.** 3 **d.** 4 **e.** 3 **f.** 4 **g.** 3 **h.** 3 **i.** 5 **j.** 4 **k.** 2 **l.** 3 **m.** 3 **n.** 1 **o.** 3

14-5. 1 2 3

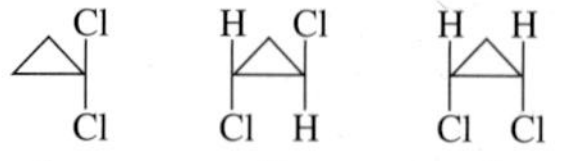

14-6. a. 2.0 ppm **b.** 2.0 ppm **c.** 200 Hz **14-7. a.** 1.5 ppm **b.** 30 hertz **14-8.** to the right of the TMS signal **14-9. a.** in each structure it is the protons on the methyl group on the left-hand side of the structure **b.** in each structure it is the proton(s) on the carbon on the right-hand side of the structure **14-10.** first spectrum = 1-iodopropane

14-11.

a. $CH_3CHCHBr$ (Br Br) **c.** $CH_3CH_2CHCH_3$ (Cl) **e.** $CH_3CH_2CH{=}CH_2$

b. CH_3CHOCH_3 (CH_3) **d.** $CH_3CHCCH_2CH_3$ (O; CH_3)

14-12. a. $CH_3CH_2CH_2Cl$ **b.** $CH_3CH_2CHCH_3$ (Cl) **c.** CH_3CHCH (O)

14-14. The compounds have different integration ratios: 2 : 9, 1 : 3, and 2 : 1

14-16. CH_3–C_6H_4–CH_3

14-17. 9.25 ppm = hydrogens that protrude out; −2.88 ppm = hydrogens that point in **14-19. a.** 2-chloropropanoic acid **b.** 3-chloropropanoic acid

14-24. a. propyl benzene **b.** 3-pentanone **c.** ethyl benzoate

14-26. a. $CH_3OCH_2CH_2OCH_3$ **c.** HC(O)–C_6H_4–CH(O)

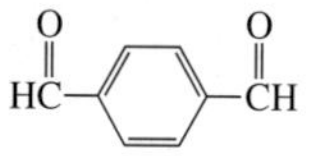

b. $CH_3CCH_2CH_2CCH_3$ (O, O)

or

$CH_3OCH_2C{\equiv}CCH_2OCH_3$

or

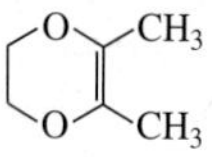

(dioxane ring with two CH_3)

14-29. CH_3O–C_6H_4–CH_3 **14-33.** pure ethanol

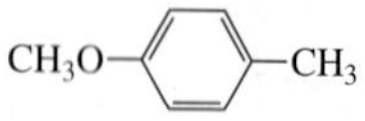

14-35. propanamide

CHAPTER 15

15-1. a. 4 **b.** It will be aromatic if it is cyclic, if it is planar, and if every carbon in the ring has a *p* orbital. **15-2.** b, c, e, g **15-5.** quinoline = sp^2; indole = *p*; imidazole = 1 is sp^2 and 1 is *p*; purine = 3 are sp^2 and 1 is *p*; pyrimidine = both are sp^2 **15-7.** Cyclopentadiene has the lower pK_a. **15-10. a.** Cyclopropane has the lower pK_a. **b.** 3-bromocyclopropene **15-11.** a **15-12.** one bonding, two nonbonding, one antibonding; two π electrons in the bonding MO and one in each of the nonbonding MOs **15-13.** no

15-15. a. $CH_3CHCH_2CH_2CH_2CH_3$ (phenyl) **c.** $CH_3CH_2CHCH_2CH_3$ (CH_2–phenyl)

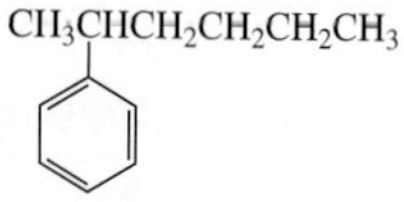

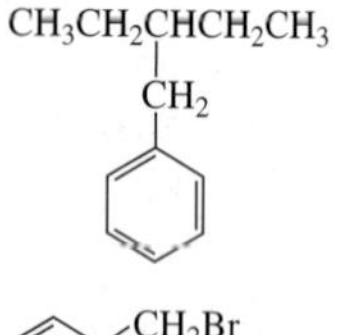

b. C_6H_5–CH_2OH **d.** C_6H_5–CH_2Br

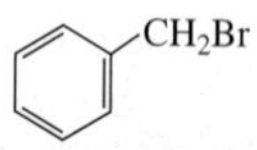

15-17. both will react at about the same rate

15-23. a. ethylbenzene **b.** isopropylbenzene **c.** *sec*-butylbenzene **d.** *tert*-pentylbenzene **e.** *tert*-butylbenzene **f.** 3-phenylpropene

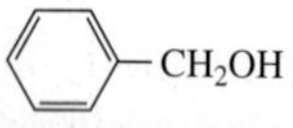

CHAPTER 16

16-1. a. *ortho*-ethylphenol **b.** *meta*-bromochlorobenzene **c.** *meta*-bromobenzaldehyde **d.** *ortho*-ethyltoluene

16-2. a. (CH_3, NH_2) **b.** (CH_3, OH) **c.** (CH_3, CH_3) **d.** (SO_3H, CH_3)

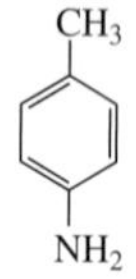

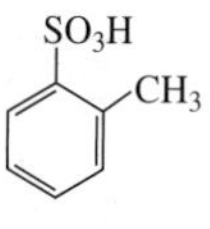

16-3.

a. (CH_3, Cl) **b.** (OH, Br) **c.** (NH_2, NO_2) **d.** (C≡N, Cl)

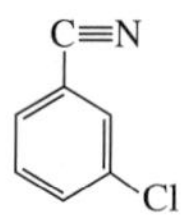

e. (OH, Br, I) **f.** (Cl, Cl) **g.** (HC=O, NO_2, NO_2) **h.** (CH_3, CH_3)

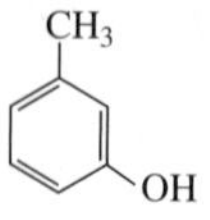

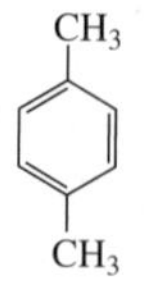

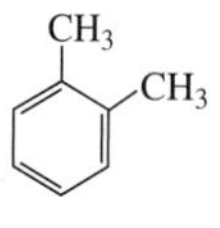

16-4. a. 1,3,5-tribromobenzene **b.** 3-nitrophenol **c.** *para*-bromotoluene **d.** 1,2-dichlorobenzene

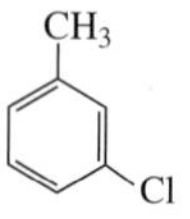

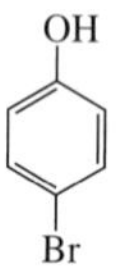

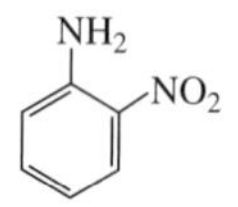

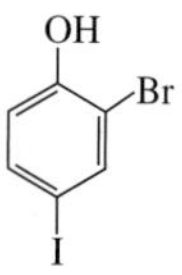

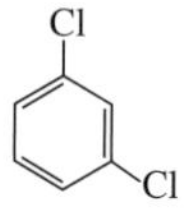

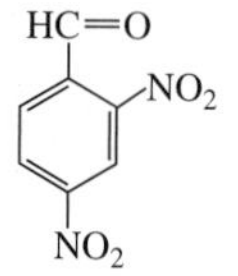

16-5. a. (COOH) **b.** (CH_2OCH_3)

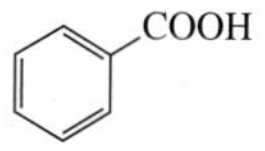

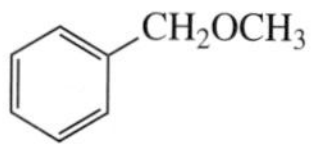

c. (COOH, COOH) **d.** ($CH_2CH_2NH_2$)

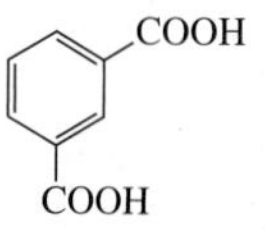

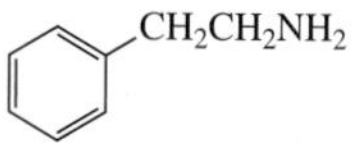

16-7. a. donates electrons by resonance and withdraws electrons inductively **b.** donates electrons inductively **c.** withdraws electrons by resonance and withdraws electrons inductively **d.** donates electrons by resonance and withdraws electrons inductively **e.** donates electrons by resonance and withdraws electrons inductively **f.** withdraws electrons inductively **16-8. a.** phenol > toluene > benzene > bromobenzene > nitrobenzene **b.** toluene > chloromethylbenzene > dichloromethylbenzene > difluoromethylbenzene

16-11. a. ($CH_2CH_2CH_3$, NO_2) + ($CH_2CH_2CH_3$, NO_2) **d.** (SO_3H, NO_2)

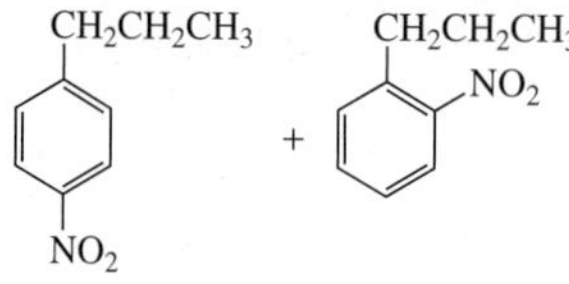

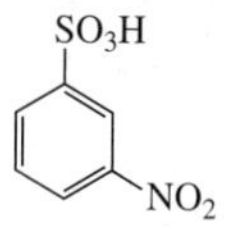

b. (Br, NO_2) + (Br, NO_2) **e.** (NO_2) + (NO_2)

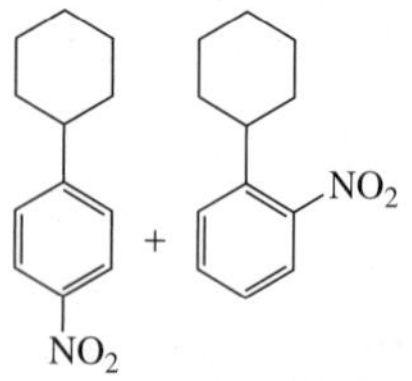

c. (CH=O, NO_2) **f.** (C≡N, NO_2)

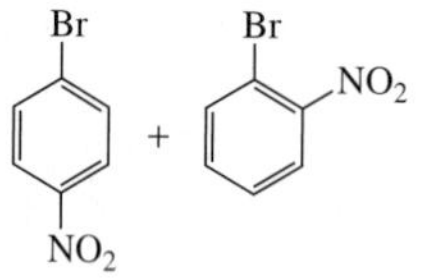

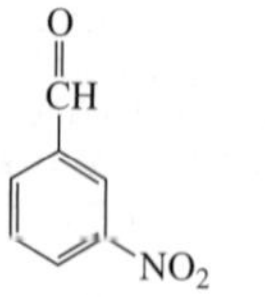

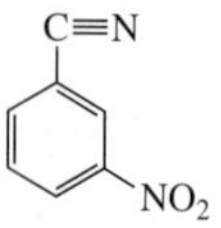

16-12. They are all meta directors.

16-13. a. $ClCH_2COH$ (O) **c.** $H_3\overset{+}{N}CH_2COH$ (O)

b. O_2NCH_2COH (O) **d.** (COOH)

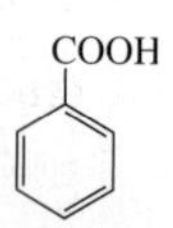

e. $HOCCH_2COH$ (with C=O on both carbons)

f. $HCOH$ (with C=O)

g. FCH_2COH (with C=O)

h. COOH on a benzene ring, Cl *para*

16-15. **a.** no reaction **b.** NH_2 on a benzene ring, with Br at both *ortho* positions and at *para* **c.** no reaction **d.** toluene with CH_3 *para* (*p*-xylene) + toluene with CH_3 *ortho* (*o*-xylene)

16-18. **a.** OCH_3 on a benzene ring, NO_2 *ortho*, F *para* **b.** COOH on a benzene ring, a second COOH *ortho*, Cl *para* **c.** COOH on a benzene ring, Br *meta*, Cl *para*

16-26. CH_3CHCH_3 (Cl on C-2) + CH_3CHCH_3 (OH on C-2) + $CH_3CH{=}CH_2$ + N_2

16-29. **a.** 1-chloro-2,4-dinitrobenzene > *p*-chloronitrobenzene > chlorobenzene **b.** chlorobenzene > *p*-chloronitrobenzene > 1-chloro-2,4-dinitrochlorobenzene

CHAPTER 17

17-2. **a.** butanenitrile, propyl cyanide **b.** ethanoic propanoic anhydride, acetic propionic anhydride **c.** potassium butanoate; potassium butyrate **d.** pentanoyl chloride, valeryl chloride **e.** isobutyl butanoate, isobutyl butyrate **f.** *N*,*N*-dimethylhexanamide **g.** γ-butyrolactam **h.** cyclopentanecarboxylic acid **i.** β-methyl-δ-valerolactone **17-4.** C—O bond in an alcohol: no double-bond character **17-5.** The shortest bond has the highest frequency.

a. $CH_3{-}C({=}O){-}O{-}CH_3$, bonds labeled: C=O 3, C—O (carbonyl carbon–O) 2, O—CH_3 1
1 = longest
3 = shortest

b. $CH_3{-}C({=}O){-}O{-}CH_3$, bonds labeled: C=O 1, C—O (carbonyl carbon–O) 2, O—CH_3 3
1 = highest frequency
3 = lowest frequency

17-7. **a.** no reaction **b.** CH_3COH (with C=O) **c.** $CH_3{-}C({=}O){-}O{-}C({=}O){-}CH_3$ **d.** no reaction

17-8. true

17-9. **a.** new **b.** no reaction **c.** mixture of two

17-13. **a.** $CH_3CH_2CH_2OH$ **b.** $CH_3CH_2NH_2$ **c.** $(CH_3)_2NH$ **d.** H_2O **e.** CH_3CO^- (with C=O) **f.** HO— benzene ring —NO_2 (*para*)

17-17. **a.** carbonyl group of the ester is realtively unreactive, nucleophile is relatively unreactive, leaving group is a strong base **b.** aminolysis

17-19. **a.** protonated carboxylic acid, tetrahedral intermediate I, tetrahedral intermediate III, H_3O^+, CH_3OH^+ **b.** Cl^-, carboxylic acid, tetrahedral intermediate II, H_2O, CH_3OH **c.** H_3O^+ **d.** H_3O^+ if excess water was used; $CH_3OH_2^+$ if not. **17-22.** **a.** $CH_3CH_2CH_2O^-$ **b.** H^+ would make the amine unreactive by protonating it. HO^- would form the hydrolysis product. RO^- would form an ester. **17-23.** methyl butanoate and methanol

17-24. **a.** the alcohol **b.** the carboxylic acid

17-31.

CH_3CNH— benzene ring —NO_2 (*para*) > CH_3CNH— benzene ring with NO_2 *meta* > CH_3CNH— benzene ring > CH_3CNH— cyclohexane ring (each with C=O)

17-32. **a.** pentyl bromide **b.** isohexyl bromide **c.** benzyl bromide **d.** cyclohexyl bromide

17-34. **a.** $CH_3CH_2CH_2Br$ **b.** CH_3CHCH_2Br (CH_3 on C-2) **c.** cyclohexane ring—Br

17-37. **a.** $(CH_3CH_2)_2NCN(CH_2CH_3)_2$ (with C=O) **b.** $CH_3COCCH_2COCCH_3$ (with four C=O) **c.** $H_2NCNHCH_3$ (with C=O) **d.** no reaction **e.** $HOCOH$ (with C=O) + $^+NH_4$ ⇌ CO_2 + H_2O **f.** CH_3CH_2— six-membered cyclic anhydride (ring O, two C=O)

CHAPTER 18

18-2. **a.** 3-methylpentanal, β-methylvaleraldehyde **b.** 4-heptanone, dipropyl ketone **c.** 2-methyl-4-heptanone, isobutyl propyl ketone **d.** 4-phenylbutanal, γ-phenylbutyraldehyde **e.** 4-ethylhexanal, γ-ethylcaproaldehyde **f.** 1-hepten-3-one, butyl vinyl ketone **18-3.** **a.** 6-hydroxy-3-heptanone **b.** 2-oxocyclohexylmethanenitrile **c.** 3-formylpentanamide

18-4. **a.** 2-heptanone **b.** *para*-nitroacetophenone **18-5.** **a.** two; (*R*)-3-methyl-3-hexanol and (*S*)-3-methyl-3-hexanol **b.** one; 2-methyl-2-pentanol

18-6.

$CH_3CCH_2CH_3$ (with C=O) + $CH_3CH_2CH_2MgBr$

$CH_3CH_2CCH_2CH_2CH_3$ (with C=O) + CH_3MgBr

18-7. carboxylate ion + alkane

18-9. 2-propanol and 3-pentanol

18-14. **a.** CH_3CHCH_2OH (CH_3 on C-2) **b.** cyclohexane ring—OH **c.** benzene ring—CH_2OH **d.** benzene ring—$CHCH_3$ (OH on CH)

18-15. **a.** benzene ring—$CNHCH_3$ (with C=O) **b.** CH_3CNH_2 (with C=O) **c.** $CH_3CNHCH_2CH_3$ (with C=O) **d.** $CH_3CN(CH_2CH_3)_2$ (with C=O)

18-17. **a.** 1×10^{-12} **b.** 1×10^{-9} **c.** 3.1×10^{-3}

18-20.

a. cyclopentane ring$=NCH_2CH_3$ + H_2O **b.** cyclopentene ring—$N(CH_2CH_3)_2$ + H_2O **c.** benzene ring—$C({=}N(CH_2)_5CH_3)CH_3$ + H_2O **d.** benzene ring—$C(CH_3){=}N$—cyclohexane ring + H_2O

18-25. the ketone with the nitro substituents

18-28. **a.** cyclohexane ring with OH and, on the adjacent carbon, CH_2OH **b.** $NaBH_4$

18-31. **a.** 26% **b.** 17%

18-34. **a.** benzene ring—$C(OH)(CH_3)CH_2CH_3$, *S* **b.** CH_3—$C(OH)(H)$—benzene ring, *R* **c.** $CH_3CCH_2CH_2CH_3$ (OH and CH_3 on C-2)
compound does not have an asymmetric carbon
d. $CH_3CH_2CH_2$—$C(OH)(CH_3)$—CH_2CH_3, *S*

18-37. nonsterically hindered ketone

CHAPTER 19

19-2. **a.** $CH_3\overset{O}{C}CH_2C{\equiv}N$ a β-keto nitrile **b.** $CH_3O\overset{O}{C}CH_2\overset{O}{C}OCH_3$ a β-diester

19-3. **a.** $CH_3\overset{O}{C}H > HC{\equiv}CH > CH_2{=}CH_2 > CH_3CH_3$

b. $CH_3\overset{O}{C}CH_2\overset{O}{C}CH_3 > CH_3\overset{O}{C}CH_2\overset{O}{C}OCH_3 > CH_3O\overset{O}{C}CH_2\overset{O}{C}OCH_3 > CH_3\overset{O}{C}CH_3$

c. 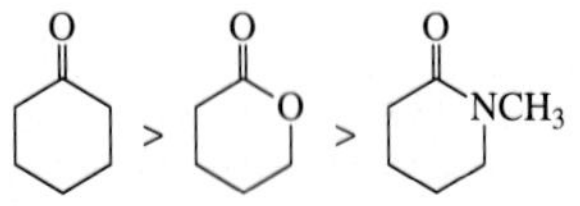 O > O, O > O, NCH₃

19-5. **a.** $CH_3CH{=}\overset{OH}{C}CH_2CH_3$ **b.** 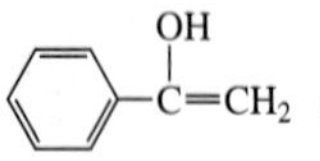 $C_6H_5\text{-}\overset{OH}{C}{=}CH_2$ **c.** OH

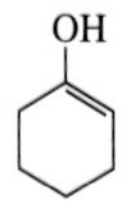

d. 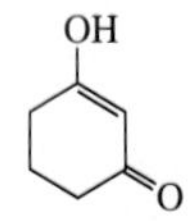 OH, O and OH, O — more stable

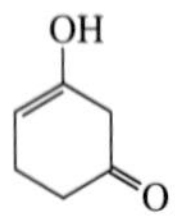

e. $CH_3CH_2\overset{OH}{C}{=}CH\overset{O}{C}CH_2CH_3$ and $CH_3CH{=}\overset{OH}{C}CH_2\overset{O}{C}CH_2CH_3$ — more stable

f. 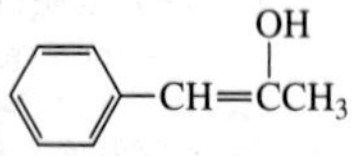 $C_6H_5\text{-}CH{=}\overset{OH}{C}CH_3$ and $C_6H_5\text{-}CH_2\overset{OH}{C}{=}CH_2$ — more stable

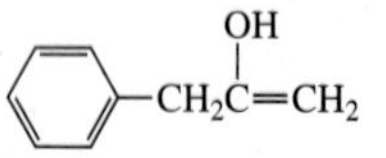

19-8. The rate-determining step must be removal of the proton from the α-carbon of the ketone.

19-12. D, O, D, D, D

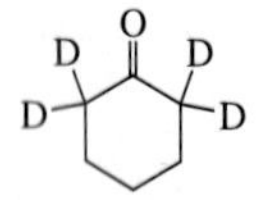

19-16. **a.** 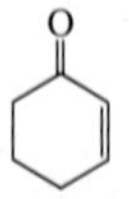 O $CH_3\overset{O}{C}CH_2\overset{O}{C}CH_3$ HO^-

b. $CH_3\overset{O}{C}CH{=}CH_2$ $CH_3CH_2O\overset{O}{C}CH_2\overset{O}{C}OCH_2CH_3$ $CH_3CH_2O^-$

19-18. **a.** $CH_3CH_2CH_2\overset{O}{C}H$ **c.** 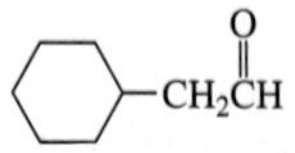 $C_6H_{11}\text{-}CH_2\overset{O}{C}H$

b. $CH_3\overset{O}{C}CH_3$ **d.** $CH_3CH_2\overset{O}{C}CH_2CH_3$

19-24. **a.** $CH_3CH_2CH_2CH_2\overset{O}{C}\underset{CH_2CH_3}{CH}\overset{O}{C}OCH_3$ **b.** $CH_3\underset{CH_3}{CH}CH_2\overset{O}{C}\underset{CH(CH_3)_2}{CH}\overset{O}{C}OCH_2CH_3$

19-29. **a.** 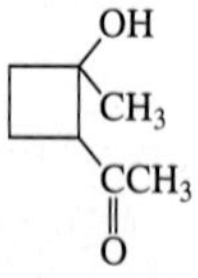 OH, CH₃, $\overset{O}{C}CH_3$ **b.** 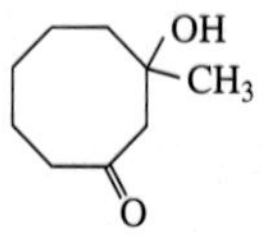 OH, CH₃, O

19-32. **a.** 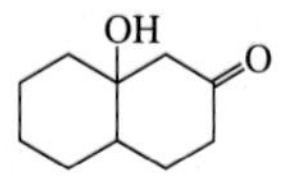 OH, O **c.** 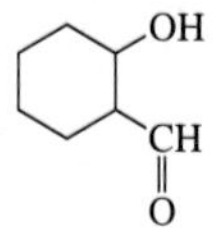 OH, CH, O

b. 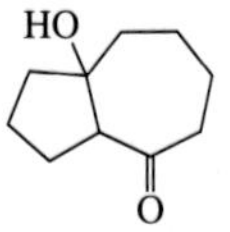 HO, O **d.** 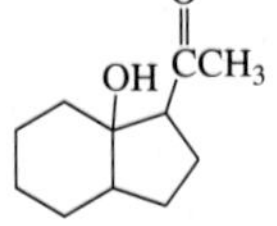 OH, $\overset{O}{C}CH_3$

19-34. **a** and **d**; **b** doesn't have a carboxyl group. **19-35.** **a.** methyl bromide **b.** methyl bronude (twice) **c.** benzyl bromide **d.** isobutyl bromide **19-38.** **a.** ethyl bromide **b.** pentyl bromide **c.** benzyl bromide **19-41.** 7 **19-42.** **a.** 3 **b.** 7

CHAPTER 20

20-1. **a.** reduction **b.** neither **c.** oxidation **d.** oxidation **e.** reduction **f.** neither

20-2. **a.** $CH_3CH_2CH_2CH_2CH_2OH$ **e.** no reaction
b. $CH_3CH_2CH_2CH_2NH_2$ **f.** CH_3CH_2OH
c. 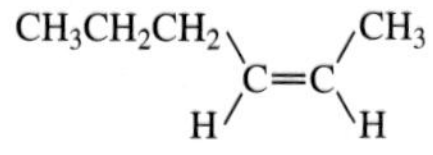 $CH_3CH_2CH_2(H)C{=}C(H)CH_3$ **g.** $CH_3\overset{O}{C}H$
d. 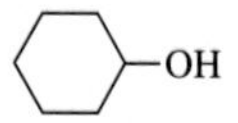 $C_6H_{11}\text{-}OH$ **h.** 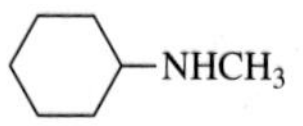 $C_6H_{11}\text{-}NHCH_3$

20-4. **a.** 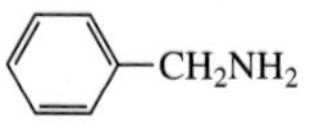 $C_6H_5\text{-}CH_2NH_2$ **d.** 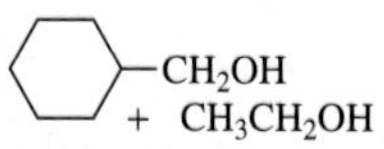 $C_6H_{11}\text{-}CH_2OH + CH_3CH_2OH$
b. 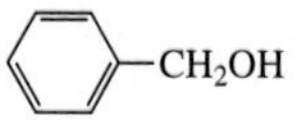 $C_6H_5\text{-}CH_2OH$ **e.** $CH_2CH_2CH_2NHCH_2CH_3$
c. $CH_3CH_2\overset{OH}{C}HCH_2CH_3$ **f.** $CH_3CH_2CH_2CH_2OH$

20-8. **a.**
1. $CH_3CH_2\overset{O}{C}CH_2CH_3$ 4. $CH_3\overset{O}{C}CH_2\overset{O}{C}CH_2CH_3$
2. $CH_3CH_2CH_2CH_2\overset{O}{C}OH$ 5. 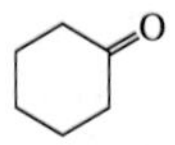 O
3. no reaction 6. $HO\overset{O}{C}CH_2CH_2\overset{O}{C}OH$

b.
1. $CH_3CH_2\overset{O}{C}CH_2CH_3$ 4. $CH_3\overset{O}{C}CH_2\overset{O}{C}CH_2CH_3$
2. $CH_3CH_2CH_2CH_2\overset{O}{C}H$ 5. 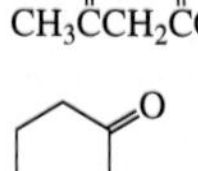 O
3. no reaction 6. $H\overset{O}{C}CH_2CH_2\overset{O}{C}H$

20-12. **a.** cyclohexene **b.** 1-butene **c.** *trans*-2-pentene **d.** *cis*-2-pentene
20-17. It is stereoselective and stereospecific, but not enantioselective.

20-18. **a.** $CH_3\overset{CH_3}{\underset{OH}{C}}\text{—}\underset{OH}{CH}CH_2CH_3$ **b.** 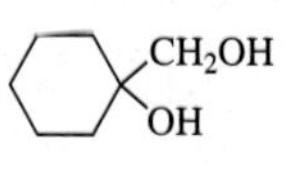 CH₂OH, OH

20-20.

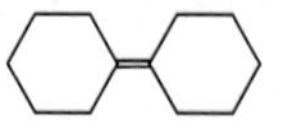

20-21. **a.** 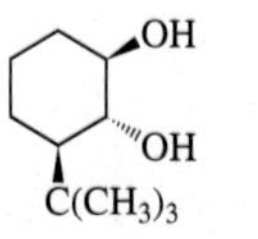 OH, OH, C(CH₃)₃ **b.** 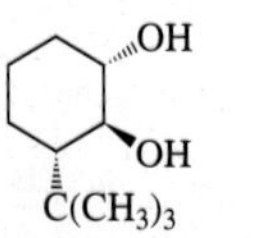 OH, OH, C(CH₃)₃

20-24. **a.** 2,3-dimethyl-2-butene **b.** *cis*-4-octene and *trans*-4-octene
20-26. whether the alkene has the cis or the trans configuration

20-29. **a.** cyclohexyl–C≡CH **b.** $CH_3CH_2C{\equiv}CCH_2CH_2CH_2C{\equiv}CCH_2CH_3$

CHAPTER 21

21-1. **a.** 2,2-dimethylaziridine **b.** 4-ethylpiperidine **c.** 3-methylazacyclobutane **d.** 2-methylthiacyclopropane **e.** 2,3-dimethyltetrahydrofuran **f.** 2-ethyloxacyclobutane **21-2.** The electron-withdrawing oxygen stabilizes the conjugate base.

21-3. **a.** (quinuclidin-3-one structure) **b.** pK_a ~ 8 **c.** 3-chloroquinuclidine

21-7. $CH_3C(CH_3){=}CH_2$ + $(CH_3)_2NCH_2CH_2CH_3$

21-8. **a.** $CH_3CH{=}CH_2$ + $N(CH_3)_3$ **c.** methylenecyclohexane + $N(CH_3)_3$

b. $CH_2{=}CHCH(CH_3)CH_2CH_2N(CH_3)CH_3$ **d.** $CH_3N(CH_3)CH_2CH(CH_3)CH_2CH{=}CH_2$

21-11. carbanion-like

21-12. **a.** $CH_3N(CH_3)OH$ + $CH_2{=}CHCH_3$ **c.** $CH_2{=}CH_2$ + $CH_3N(OH)CH_2CH(CH_3)CH_3$

b. $CH_3N(OH)C_6H_5$ + $CH_2{=}CHCH_3$ **d.** $CH_2{=}CHCH_2CH_2CH_2CH_2N(OH)CH_3$

21-13. a secondary amine and a tertiary amine

21-21. 4-methyl-N-ethylpyridinium iodide (CH_2CH_3, I^-) > 4-methylpyridine > 3-methylpyridine

21-22. 4-bromo-1-methylimidazole (Br, NCH_3)

21-23. pyrrole (N–H) > imidazole (NH) > benzene

21-24. Imidazole forms intramolecular hydrogen bonds that *N*-methylimidazole cannot form. **21-25.** 24% **21-26.** yes

CHAPTER 22

22-1. D-Ribose is an aldopentose. D-Sedoheptulose is a ketoheptose. D-Mannose is an aldohexose. **22-3.** **a.** L-glyceraldehyde **b.** L-glyceraldehyde **c.** D-glyceraldehyde **22-4.** **a.** enantiomers **b.** diastereomers **22-5.** **a.** D-ribose **b.** L-talose **c.** L-allose **22-6.** **a.** (2*R*,3*S*,4*R*,5*R*)-2,3,4,5,6-pentahydroxyhexanal **b.** (2*S*,3*R*,4*S*,5*S*)-2,3,4,5,6-pentahydroxyhexanal **22-7.** D-psicose **22-8.** **a.** A ketoheptose has four asymmetric carbons ($2^4 = 16$ stereoisomers). **b.** An aldoheptose has five asymmetric carbons ($2^5 = 32$ stereoisomers). **c.** A ketotriose has no asymmetric carbons; therefore, it has no stereoisomers. **22-9.** **a.** D-iditol **b.** D-iditol and D-gulitol **22-10.** **a.** **1.** D-altrose **2.** L-galactose **b.** **1.** D-tagatose **2.** D-fructose **22-12.** D-tagatose, D-galactose, and D-talose **22-13.** **a.** L-gulose **b.** L-gularic acid **c.** D-allose and L-allose, D-altrose and D-talose, L-altrose and L-talose, D-galactose and L-galactose **22-14.** **a.** D-arabinose and D-ribulose **b.** D-allose and D-psicose **c.** L-gulose and L-sorbose **d.** D-talose and D-tagatose **22-15.** D-gulose and D-idose **22-16.** **a.** D-gulose and D-idose **b.** L-xylose and L-lyxose **22-17.** **a.** D-glucose and D-mannose **b.** D-erythrose and D-threose **c.** L-allose and L-altrose **22-19.** **A** = D-glucose **B** = D-mannose **C** = D-arabinose **D** = D-erythrose **22-23.** **a.** the OH group at C-2 **b.** the OH group at C-2, C-3, and C-4 **c.** the OH group at C-3 and C-1 **22-24.** A protonated amine is not a nucleophile. **22-26.** an L-aldohexose **22-29.** **a.** They cannot receive type A, B, or AB blood because these have sugar components that type O blood does not have. **b.** They cannot give blood to those with type A, B, or O blood because AB blood has sugar components that these other blood types do not have.

CHAPTER 23

23-2. **a.** (*R*)-alanine **b.** (*R*)-aspartate **c.** The α-carbon of all the D-amino acids except cysteine has the *R*-configuration. **23-3.** Ile, Thr
23-4. because of the electron-withdrawing ammonium group

23-6.
a. $HOOCCH_2CH_2CH(\overset{+}{N}H_3)COOH$ **c.** $^-OOCCH_2CH_2CH(\overset{+}{N}H_3)COO^-$
b. $HOOCCH_2CH_2CH(\overset{+}{N}H_3)COO^-$ **d.** $^-OOCCH_2CH_2CH(NH_2)COO^-$

23-9. **a.** 5.43 **b.** 10.76 **c.** 5.68 **d.** 2.98 **23-10.** **a.** Asp **b.** Arg **c.** Asp **d.** Met **23-12.** Leucine and isoleucine have similar polarities and pI values, so they show up as one spot. **23-15.** His > Val > Ser > Asp **23-18.** **a.** L-Ala, L-Asp, L-Glu **b.** L-Ala and D-Ala, L-Asp and D-Asp, L-Glu and D-Glu **23-20.** **a.** $20^8 = 25{,}600{,}000{,}000$ **b.** 20^{100} **23-21.** the bonds on either side of the α-carbon **23-26.** **a.** 5.8% **b.** 4.4% **23-29.** Gly-Arg-Trp-Ala-Glu-Leu-Met-Pro-Val-Asp **23-31.** Edman's reagent would release two amino acids in approximately equal amounts. **23-33.** Leu-Tyr-Lys-Arg-Met-Phe-Arg-Ser **23-34.** 110 Å in an α-helix and 260 Å in a straight chain **23-35.** nonpolar groups on the outside and polar groups on the inside **23-36.** **a.** cigar-shaped protein **b.** subunit of a hexamer

CHAPTER 24

24-1. $\Delta H^\ddagger, E_a, \Delta S^\ddagger, \Delta G^\ddagger, k_{rate}$ **24-5.** Hydroxide ion acts as a nucleophile.
24-7. 2.5×10^6 **24-10.** close to one

24-11. cyclohexanes with OH/SC_6H_5, SC_6H_5/OH, OCH_2CH_3/SC_6H_5, SC_6H_5/OCH_2CH_3

24-13. The nitro groups cause the phenoxide ion to be a better leaving group than the carboxylate ion. **24-15.** Ser-Ala-Phe **24-17.** Arginine forms a direct hydrogen bond, lysine forms an indirect hydrogen bond. **24-19.** NAM **24-20.** The acid denatures the enzyme. **24-22.** lysine **24-25.** Putting a substituent on cysteine could interfere with binding or catalysis of the substrate.

CHAPTER 25

25-1. It increases the susceptibility of the carbonyl carbon to nucleophilic attack, increases the nucleophilicity of water, and stabilizes the negative charge on the transition state. **25-2.** pyrophosphate **25-3.** **a.** 7 **b.** 3 isolated from 2 others
25-12. **a.** pyridoxal phosphate and α-ketoglutarate

b. (ring with OH, HO, and CH_2COO^-) **c.** *S*-adenosylmethionine

25-13. alanine and aspartate
25-19.

$$CH_3CH_2\overset{O}{\overset{\|}{C}}SCoA \xrightarrow[\text{biotin}]{\text{E}} CH_3CH(COO^-)\overset{O}{\overset{\|}{C}}SCoA \xrightarrow[\text{coenzyme } B_{12}]{\text{E}} CH_2(COO^-)CH_2\overset{O}{\overset{\|}{C}}SCoA$$

25-20. the methylene group of N^5,N^{10}-methylene-THF

CHAPTER 26

26-2. hexanoic acid, 3 malonic acids, and glutaric acid **26-3.** no **26-4.** glyceryl tripalmitate **26-6.** Integral proteins will have a higher percentage of nonpolar amino acids. **26-7.** The bacteria could synthesize phosphoacylglycerols with more saturated fatty acids. **26-12.** The two halves are synthesized in a head-to-tail fashion and then joined together in a tail-to-tail linkage. **26-20.** cis fused; trans fused **26-21.** a β-substituent **26-24.** Two are axial substituents and one is an equatorial substituent.

CHAPTER 27

27-4. It is a stable base **27-5.** phosphocreatine **27-12. a.** 3′—C—C—T—G—T—T—A—G—A—C—G—5′ **b.** guanine **27-13. a.** 25 Å **b.** 34 Å **c.** 46 Å **27-15.** 10^{10} Å **27-18.** Met-Asp-Pro-Val-Ile-Lys-His **27-19.** Met-Asp-Pro-Leu-Leu-Asn **27-21.** 5′—G-C-A-T-G-G-A-C-C-C-C-G-T-T-A-T-T-A-A-A-C-A-C—3′
27-23.

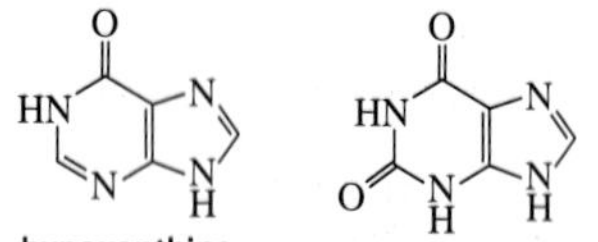

hypoxanthine xanthine

27-25. a

CHAPTER 28

28-1. a. $CH_2{=}CHCl$ **b.** $CH_2{=}C(CH_3)COCH_3$ (with C=O) **c.** $CF_2{=}CF_2$

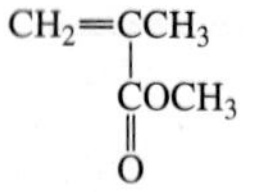

28-2. poly(vinyl chloride) **28-5.** beach balls

28-7. a. $CH_2{=}CH$–C$_6$H$_4$–OCH$_3$ > $CH_2{=}CH$–C$_6$H$_4$–CH$_3$ > $CH_2{=}CH$–C$_6$H$_4$–NO$_2$

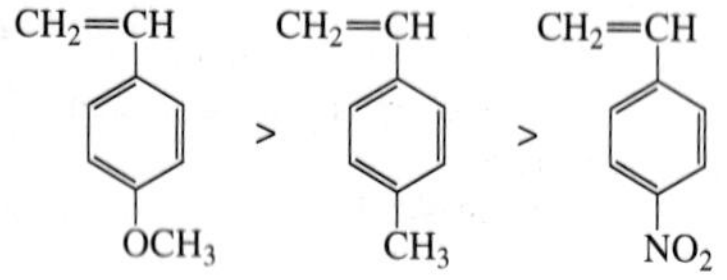

b. $CH_2{=}CHOCCH_3$ (C=O) > $CH_2{=}CHCH_3$ > $CH_2{=}CHCOCH_3$ (C=O)

c. $CH_2{=}C(CH_3)$–C$_6$H$_5$ > $CH_2{=}CH$–C$_6$H$_5$

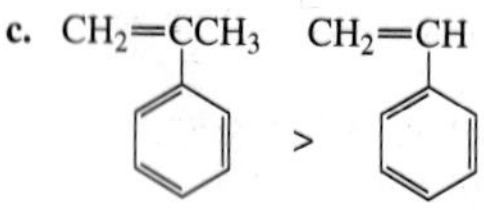

28-8. a. $CH_2{=}CH$–C$_6$H$_4$–NO$_2$ > $CH_2{=}CH$–C$_6$H$_4$–CH$_3$ > $CH_2{=}CH$–C$_6$H$_4$–OCH$_3$

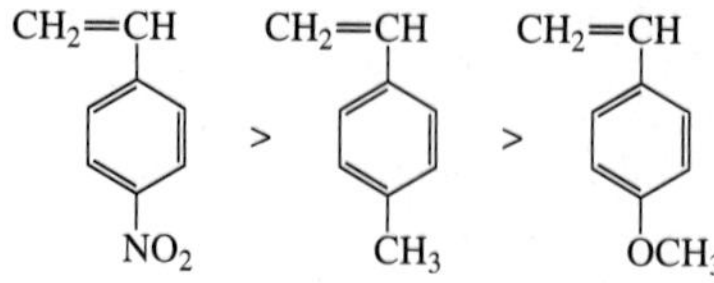

b. $CH_2{=}CHC{\equiv}N > CH_2{=}CHCl > CH_2{=}CHCH_3$

28-11. a. $CH_2{=}C(CH_3)CH_3 + BF_3$ **c.** (ethylene oxide) + CH_3O^-

b. $CH_2{=}CH$–N(pyrrolidine) + BF_3 **d.** $CH_2{=}CHCOCH_3$ (C=O) + BuLi

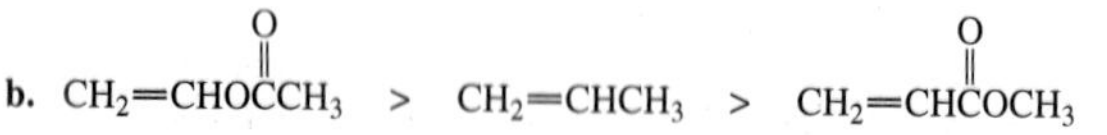

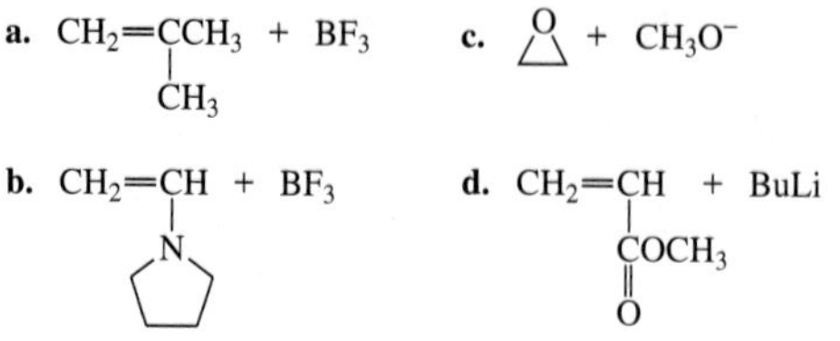

28-12.

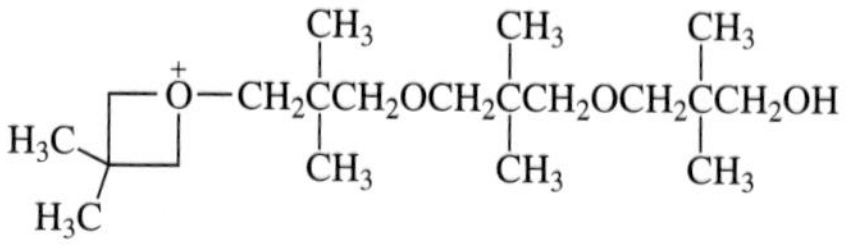

H_3C, H_3C (oxetane ring) $\overset{+}{O}$—$CH_2C(CH_3)_2CH_2OCH_2C(CH_3)_2CH_2OCH_2C(CH_3)_2CH_2OH$

28-15. a. $-NHCH_2CH_2CH_2\overset{O}{\overset{\|}{C}}NHCH_2CH_2CH_2\overset{O}{\overset{\|}{C}}-$

b. (2-pyrrolidinone: cyclic C=O, NH)

c. $-NH(CH_2)_4NH\overset{O}{\overset{\|}{C}}CH_2CH_2\overset{O}{\overset{\|}{C}}NH(CH_2)_4NH\overset{O}{\overset{\|}{C}}CH_2CH_2\overset{O}{\overset{\|}{C}}-$

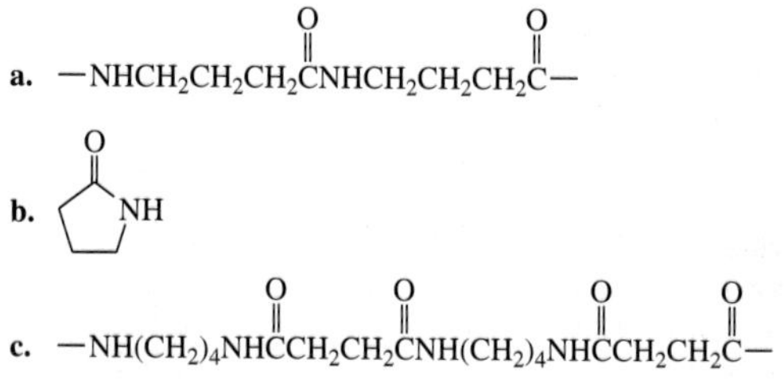

CHAPTER 29

29-1. a. electrocyclic reaction **b.** sigmatropic rearrangement **c.** cycloaddition reaction **d.** cycloaddition reaction **29-2. a.** bonding orbitals = ψ_1, ψ_2, ψ_3; antibonding orbitals = ψ_4, ψ_5, ψ_6; **b.** ground-state HOMO = ψ_3; ground-state LUMO = ψ_4 **c.** excited-state HOMO = ψ_4; excited-state LUMO = ψ_5 **d.** symmetric orbitals = ψ_1, ψ_3, ψ_5; asymmetric orbitals = ψ_2, ψ_4, ψ_6 **e.** The HOMO and LUMO have opposite symmetries. **29-3. a.** 8 **b.** ψ_4 **c.** 8 **29-6. a.** controtatory **b.** trans **c.** disrotatory **d.** cis **29-7. a.** correct **b.** correct **c.** correct **29-8. 1. a.** conrotatory **b.** trans **2. a.** disrotatory **b.** cis **29-11.** yes
29-13.

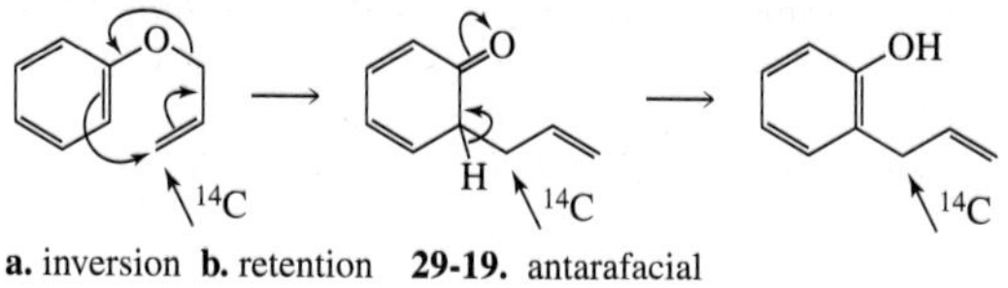

O → O, H → OH; ^{14}C ^{14}C ^{14}C

29-18. a. inversion **b.** retention **29-19.** antarafacial

Glossary

absolute configuration the three-dimensional structure of a chiral compound. The configuration is designated by *R* or *S*.

absorption band a peak in a spectrum that occurs as a result of the absorption of energy.

acetal $\mathrm{R\overset{\displaystyle OR}{\underset{\displaystyle OR}{C}}H}$

acetoacetic ester synthesis synthesis of a methyl ketone, using ethyl acetoacetate as the starting material.

achiral (optically inactive) an achiral molecule has a conformation with a superimposable mirror image.

acid a substance that donates a proton.

acid anhydride $\mathrm{R{-}\overset{\overset{\displaystyle O}{\|}}{C}{-}O{-}\overset{\overset{\displaystyle O}{\|}}{C}{-}R}$

acid–base reaction a reaction in which an acid donates a proton to a base or accepts a share in a base's electrons.

acid catalyst a catalyst that increases the rate of a reaction by donating a proton.

acid-catalyzed reaction a reaction catalyzed by an acid.

acid dissociation constant a measure of the degree to which an acid dissociates in solution.

activating substituent a substituent that increases the reactivity of an aromatic ring. Electron-donating substituents activate aromatic rings toward electrophilic attack, and electron-withdrawing substituents activate aromatic rings toward nucleophilic attack.

active site a pocket or cleft in an enzyme where the substrate is bound.

acyclic noncyclic.

acyl adenylate a carboxylic acid derivative with AMP as the leaving group.

acyl–enzyme intermediate an intermediate formed when an amino acid residue of an enzyme is acetylated.

acyl group a carbonyl group bonded to an alkyl group or to an aryl group.

acyl halide $\mathrm{R{-}\overset{\overset{\displaystyle O}{\|}}{C}{-}Cl}$

acyl phosphate a carboxylic acid derivative with a phosphate leaving group.

acyl pyrophosphate a carboxylic acid derivative with a pyrophosphate leaving group.

1,2-addition (direct addition) addition to the 1- and 2-positions of a conjugated system.

1,4-addition (conjugate addition) addition to the 1- and 4-positions of a conjugated system.

addition polymer (chain-growth polymer) a polymer made by adding monomers to the growing end of a chain.

addition reaction a reaction in which atoms or groups are added to the reactant.

adrenal cortical steroids glucocorticoids and mineralocorticoids.

alcohol a compound with an OH group in place of one of the hydrogens of an alkane; (ROH).

alcoholysis reaction with an alcohol.

aldaric acid a dicarboxylic acid with an OH group bonded to each carbon. Obtained by oxidizing the aldehyde and primary alcohol groups of an aldose.

aldehyde 

alditol a compound with an OH group bonded to each carbon. Obtained by reducing an aldose or a ketose.

aldol addition a reaction between two molecules of an aldehyde (or two molecules of a ketone) that connects the α-carbon of one with the carbonyl carbon of the other.

aldol condensation an aldol addition followed by the elimination of water.

aldonic acid a carboxylic acid with an OH group bonded to each carbon. Obtained by oxidizing the aldehyde group of an aldose.

aldose a polyhydroxyaldehyde.

aliphatic a nonaromatic organic compound.

alkaloid a natural product, with one or more nitrogen heteroatoms, found in the leaves, bark, or seeds of plants.

alkane a hydrocarbon that contains only single bonds.

alkene a hydrocarbon that contains a double bond.

alkoxymercuration addition of alcohol to an alkene, using a mercuric salt of a carboxylic acid as a catalyst.

alkylation reaction a reaction that adds an alkyl group to a reactant.

alkyl halide a compound with a halogen in place of one of the hydrogens of an alkane.

alkyl substituent (alkyl group) formed by removing a hydrogen from an alkane.

alkyl tosylate an ester of *para*-toluenesulfonic acid.

alkyne a hydrocarbon that contains a triple bond.

allene a compound with two adjacent double bonds.

allyl group $CH_2{=}CHCH_2{-}$

allylic carbon an sp^3 carbon adjacent to a vinylic carbon.

allylic cation a species with a positive charge on an allylic carbon.

alpha olefin a monosubstituted olefin.

alternating copolymer a copolymer in which two monomers alternate.

ambident nucleophile a nucleophile with two nucleophilic sites.

amide $\mathrm{R{-}\overset{\overset{\displaystyle O}{\|}}{C}{-}NH_2,\ R{-}\overset{\overset{\displaystyle O}{\|}}{C}{-}NHR,\ R{-}\overset{\overset{\displaystyle O}{\|}}{C}{-}NR_2}$

amine a compound with a nitrogen in place of one of the hydrogens of an alkane; (RNH_2, R_2NH, R_3N).

amine inversion the configuration of an sp^3 hybridized nitrogen with a nonbonding pair of electrons rapidly turns inside out.

amino acid an α-aminocarboxylic acid. Naturally occurring amino acids have the L configuration.

amino acid analyzer an instrument that automates the ion-exchange separation of amino acids.

amino acid residue a monomeric unit of a peptide or protein.

aminolysis reaction with an amine.

amino sugar a sugar in which one of the OH groups is replaced by an NH_2 group.

amphoteric compound a compound that can behave either as an acid or as a base.

anabolic steroids steroids that aid in the development of muscle.

anabolism reactions living organisms carry out in order to synthesize complex molecules from simple precursor molecules.

anchimeric assistance (intramolecular catalysis) catalysis in which the catalyst that facilitates the reaction is part of the molecule undergoing reaction.

androgens male sex hormones.

angle strain the strain introduced into a molecule as a result of its bond angles being distorted from their ideal values.

angstrom unit of length; 100 picometers $= 10^{-8}$ cm $=$ 1 angstrom

angular methyl group a methyl substituent at the 10- or 13-position of a steroid ring system.

anion-exchange resin a positively charged resin used in ion-exchange chromatography.

anionic polymerization chain-growth polymerization in which the initiator is a nucleophile; the propagation site therefore is an anion.

annulation reaction a ring-forming reaction.
annulene a monocyclic hydrocarbon with alternating double and single bonds.
anomeric carbon the carbon in a cyclic sugar that is the carbonyl carbon in the open-chain form.
anomeric effect the preference for the axial position shown by certain substituents bonded to the anomeric carbon of a six-membered-ring sugar.
anomers two cyclic sugars that differ in configuration only at the carbon that is the carbonyl carbon in the open-chain form.
antarafacial bond formation formation of two σ bonds on opposites sides of the π system.
antarafacial rearrangement rearrangement in which the migrating group moves to the opposite face of the π system.
anti addition an addition reaction in which two substituents are added at opposite sides of the molecule.
antiaromatic a cyclic and planar compound with an uninterrupted ring of *p* orbital-bearing atoms containing an even number of pairs of π electrons.
antibiotic a compound that interferes with the growth of a microorganism.
antibodies compounds that recognize foreign particles in the body.
antibonding molecular orbital a molecular orbital that results when two atomic orbitals with opposite signs interact. Electrons in an antibonding orbital decrease bond strength.
anticodon the three bases at the bottom of the middle loop in tRNA.
anti conformer the most stable of the staggered conformers.
anti elimination an elimination reaction in which the two substituents that are eliminated are removed from opposite sides of the molecule.
antigene agent a polymer designed to bind to DNA at a particular site.
antigens compounds that can generate a response from the immune system.
anti-periplanar parallel substituents on opposite sides of a molecule.
antisense agent a polymer designed to bind to mRNA at a particular site.
antisense strand (template strand) the strand in DNA that is read during transcription.
antiviral drug a drug that interferes with DNA or RNA synthesis in order to prevent a virus from replicating.
apoenzyme an enzyme without its cofactor.
applied magnetic field the externally applied magnetic field.
aprotic solvent a solvent that does not have a hydrogen bonded to an oxygen or to a nitrogen.
aramide an aromatic polyamide.
arene oxide an aromatic compound that has had one of its double bonds converted to an epoxide.
aromatic a cyclic and planar compound with an uninterrupted ring of *p* orbital-bearing atoms containing an odd number of pairs of π electrons.
Arrhenius equation relates the rate constant of a reaction to the energy of activation and to the temperature at which the reaction is carried out $(k = Ae^{-E_a/RT})$.
aryl group a benzene or a substituted-benzene group.
asymmetric carbon a carbon bonded to four different atoms or groups.
asymmetric molecular orbital a molecular orbital in which the left (or top) half is not a mirror of the right (or bottom) half.
atactic polymer a polymer in which the substituents are randomly oriented on the extended carbon chain.
atomic number the number of protons (or electrons) that the neutral atom has.
atomic orbital an orbital associated with an atom.
atomic weight the average mass of the atoms in the naturally occurring element.
aufbau principle states that an electron will always go into that orbital with the lowest available energy.
automated solid-phase peptide synthesis an automated technique that synthesizes a peptide while its C-terminal amino acid is attached to a solid support.
autoradiograph the exposed photographic plate obtained in autoradiography.
autoradiography a technique used to determine the base sequence of DNA.
auxochrome a substituent that when attached to a chromophore, alters the λ_{max} and intensity of absorption of UV/Vis radiation.
axial bond a bond of the chair conformation of cyclohexane that is perpendicular to the plane in which the chair is drawn (an up–down bond).
aziridine a three-membered-ring compound in which one of the ring atoms is a nitrogen.
azo linkage an —N=N— bond.
back-side attack nucleophilic attack on the side of the carbon opposite the side bonded to the leaving group.
bactericidal drug a drug that kills bacteria.
bacteriostatic drug a drug that inhibits the further growth of bacteria.
Baeyer–Villiger oxidation oxidation of aldehydes or ketones with H_2O_2 to form carboxylic acids or esters, respectively.
banana bond the σ bonds in small rings that are weaker as a result of overlapping at an angle rather than overlapping head-on.
base[1] a substance that accepts a proton.
base[2] a heterocyclic compound (a purine or a pyrimidine) in DNA and RNA.
base catalyst a catalyst that increases the rate of a reaction by removing a proton.
base peak the peak with the greatest abundance in a mass spectrum.
basicity the tendency of a compound to share its electrons with a proton.
Beer–Lambert law relationship among the absorbance of UV/Vis light, the concentration of the sample, the length of the light path, and the molar absorptivity $(A = cl\varepsilon)$.
bending vibration a vibration that does not occur along the line of the bond. It results in changing bond angles.
benzoyl group ring bonded to a carbonyl group.
benzyl group C_6H_5—CH_2—
benzylic carbon an sp^3 hybridized carbon bonded to a benzene ring.
benzylic cation a compound with a positive charge on a benzylic carbon.
benzyne intermediate a compound with a triple bond in place of one of the double bonds of benzene.
bicyclic compound a compound containing two rings that share at least one carbon.
bifunctional molecule a molecule with two functional groups.
bile acids steroids that act as emulsifying agents so that water-insoluble compounds can be digested.
bimolecular reaction (second-order reaction) a reaction whose rate depends on the concentration of two reactants.
biochemistry (biological chemistry) the chemistry of biological systems.
biodegradable polymer a polymer that can be broken into small segments by an enzyme-catalyzed reaction.
bioorganic compound an organic compound found in biological systems.
biopolymer a polymer that is synthesized in nature.
biosynthesis synthesis in a biological system.
biotin the coenzyme required by enzymes that catalyze carboxylation of a carbon adjacent to an ester or a keto group.
Birch reduction the partial reduction of benzene to 1,4-cyclohexadiene.
blind screen (random screen) the search for a pharmacologically active compound without any information about which chemical structures might show activity.
block copolymer a copolymer in which there are regions (blocks) of each kind of monomer.
blue shift a shift to a shorter wavelength.
boat conformation the conformation of cyclohexane that roughly resembles a boat.
boiling point the temperature at which vapor pressure equals atmospheric pressure.
bonding molecular orbital a molecular orbital that results when two in-phase atomic orbitals interact. Electrons in a bonding orbital increase bond strength.
bond length the internuclear distance between two atoms at minimum energy (maximum stability).
brand name (proprietary name, trade name) identifies a commercial product and distinguishes it from other products. It can be used only by the owner of the registered trademark.
bridged bicyclic compound a bicyclic compound in which rings share two nonadjacent carbons.
Brønsted acid a substance that donates a proton.
Brønsted base a substance that accepts a proton.
buffer an acid and its conjugate base.
carbanion a compound containing a negatively charged carbon.
carbene a species with a carbon that has a nonbonded pair of electrons and an empty orbital $(H_2C:)$
carbocation a species containing a positively charged carbon.
carbocation rearrangement the rearrangement of a carbocation to a more stable carbocation.

carbohydrate a sugar or a saccharide. Naturally occurring carbohydrates have the D configuration.
α-carbon a carbon adjacent to a carbonyl carbon.
carbon acid a compound containing a carbon that is bonded to a relatively acidic hydrogen.
carbonyl addition (direct addition) nucleophilic addition to the carbonyl carbon.
carbonyl carbon the carbon of a carbonyl group.
carbonyl compound a compound that contains a carbonyl group.
carbonyl group a carbon doubly bonded to an oxygen.
carbonyl oxygen the oxygen of a carbonyl group.
carboxyl group COOH
carboxylic acid $R-\overset{\overset{\displaystyle O}{\|}}{C}-OH$
carboxylic acid derivative a compound that is hydrolyzed to a carboxylic acid.
carboxyl oxygen the single-bonded oxygen of a carboxlic acid or an ester.
carotenoid a class of compounds (a tetraterpene) responsible for the red and orange colors of fruits, vegetables, and fall leaves.
catabolism reactions living organisms carry out in order to break down complex molecules into simple molecules and energy.
catalyst a species that increases the rate at which a reaction occurs without being consumed in the reaction. Because it does not change the equilibrium constant of the reaction, it does not change the amount of product that is formed.
catalytic antibody a compound that facilitates a reaction by forcing the conformation of the substrate in the direction of the transition state.
catalytic hydrogenation the addition of hydrogen to a double or a triple bond with the aid of a metal catalyst.
cation-exchange resin a negatively charged resin used in ion-exchange chromatography.
cationic polymerization chain-growth polymerization in which the initiator is an electrophile; the propagation site therefore is a cation.
cephalin a phosphoacylglycerol in which the second OH group of phosphate has formed an ester with ethanolamine.
cerebroside a sphingolipid in which the terminal OH group of sphingosine is bonded to a sugar residue.
chain-growth polymer (addition polymer) a polymer made by adding monomers to the growing end of a chain.
chain transfer a growing polymer chain reacts with a molecule XY in a manner that allows X to terminate the chain, leaving behind Y to initiate a new chain.
chair conformation the conformation of cyclohexane that roughly resembles a chair. It is the most stable conformation of cyclohexane.
chemical exchange the transfer of a proton from one molecule to another.
chemically equivalent protons protons with the same connectivity relationship to the rest of the molecule.
chemical shift the location of a signal in an NMR spectrum. It is measured downfield from a reference compound (most often, TMS).
chiral (optically active) a chiral molecule has a nonsuperimposable mirror image.
chiral auxiliary an enantiomerically pure compound that, when attached to a reactant, causes a product with a particular configuration to be formed.
chirality center a tetrahedral atom bonded to four different groups.
cholesterol a steroid that is the precursor of all other animal steroids.
chromatography a separation technique in which the mixture to be separated is dissolved in a solvent and the solvent is passed through a column packed with an absorbent stationary phase.
chromophore the part of a molecule responsible for a UV or visible spectrum.
cine substitution substitution at the carbon adjacent to the carbon that was bonded to the leaving group.
***s*-cis conformation** the conformation in which two double bonds are on the same side of a single bond.
cis fused two cyclohexane rings fused together such that if the second ring were considered to be two substituents of the first ring, one substituent would be in an axial position and the other would be in an equatorial position.
cis isomer the isomer with identical substituents on the same side of the double bond.
cis trans isomers geometric isomers.
citric acid cycle (Krebs cycle) a series of reactions that converts the acetyl group of acetyl-CoA into two molecules of CO_2.
Claisen condensation a reaction between two molecules of an ester that connects the α-carbon of one with the carbonyl carbon of the other and eliminates an alkoxide ion.
Claisen rearrangement a [3,3] sigmatropic rearrangement of an allyl vinyl ether.
α-cleavage homolytic cleavage of an alpha substituent.
Clemmensen reduction a reaction that reduces the carbonyl group of a ketone to a methylene group using Zn(Hg)/HCl.
codon a sequence of three bases in mRNA that specifies the amino acid to be incorporated into a protein.
coenzyme a cofactor that is an organic molecule.
coenzyme A a thiol used by biological organisms to form thioesters.
coenzyme B_{12} the coenzyme required by enzymes that catalyze certain rearrangement reactions.
cofactor an organic molecule or a metal ion that certain enzymes need to catalyze a reaction.
coil conformation (loop conformation) that part of a protein that is highly ordered, but not in an α-helix or a β-pleated sheet.
combination band occurs at the sum of two fundamental absorption frequencies ($v_1 + v_2$).
combinatorial library a group of structurally related compounds.
combinatorial organic synthesis the synthesis of a library of compounds by covalently connecting sets of building blocks of varying structure.
common intermediate an intermediate that two compounds have in common.
common name nonsystematic nomenclature.
competitive inhibitor a compound that inhibits an enzyme by competing with the substrate for binding at the active site.
complete racemization the formation of a pair of enantiomers in equal amounts.
complex carbohydrate a carbohydrate containing two or more sugar molecules linked together.
concerted reaction a reaction in which all the bond-making and bond-breaking processes occur in one step.
condensation polymer (step-growth polymer) a polymer made by combining two molecules while removing a small molecule (usually water or an alcohol).
condensation reaction a reaction combining two molecules while removing a small molecule (usually water or an alcohol).
conducting polymer a polymer that can conduct electricity.
configuration the three-dimensional structure of a particular atom in a compound. The configuration is designated by *R* or *S*.
configurational isomers stereoisomers that cannot interconvert unless a covalent bond is broken. Cis–trans isomers and optical isomers are configurational isomers.
conformation the three-dimensional shape of a molecule at a given instant that can change as a result of rotations about σ bonds.
conformational analysis the investigation of the various conformations of a compound and their relative stabilities.
conformers different conformations of a molecule.
conjugate acid a species accepts a proton to form its conjugate acid.
conjugate addition 1,4-addition.
conjugate base a species loses a proton to form its conjugate base.
conjugated double bonds double bonds separated by one single bond.
conrotatory ring closure achieves head-to-head overlap of *p* orbitals by rotating the orbitals in the same direction.
conservation of orbital symmetry theory a theory that explains the relationship between the structure and configuration of the reactant, the conditions under which a pericyclic reaction takes place, and the configuration of the product.
constitutional isomers (structural isomers) molecules that have the same molecular formula but differ in the way their atoms are connected.
contributing resonance structure (resonance contributor, resonance structure) a structure with localized electrons that approximates the structure of a compound with delocalized electrons.
convergent synthesis a synthesis in which pieces of the target compound are individually prepared and then assembled.
Cope elimination reaction elimination of a proton and a hydroxyl amine from an amine oxide.
Cope rearrangement a [3,3] sigmatropic rearrangement of a 1,5-diene.
copolymer a polymer formed from two or more different monomers.

corrin ring system a porphyrin ring system without one of the methine bridges.
COSY spectrum a 2-D NMR spectrum that shows coupling between sets of protons.
coupled protons protons that split each other. Coupled protons have the same coupling constant.
coupling constant the distance (in hertz) between two adjacent peaks of a split NMR signal.
coupling reaction a reaction that joins two alkyl groups.
covalent bond a bond created as a result of sharing electrons.
covalent catalysis (nucleophilic catalysis) catalysis that occurs as a result of a nucleophile forming a covalent bond with one of the reactants.
Cram's rule the rule used to determine the major product of a carbonyl addition reaction in a compound with a chirality center adjacent to the carbonyl group.
cross-conjugation nonlinear conjugation.
crossed (mixed) aldol addition an aldol addition in which two different carbonyl compounds are used.
cross-linking connecting polymer chains by intermolecular bond formation.
crown ether a cyclic molecule that contains several ether linkages.
crown–guest complex the complex formed when a crown ether binds a substrate.
cryptand a three-dimensional polycyclic compound that binds a substrate by encompassing it.
cryptate the complex formed when a cryptand binds a substrate.
crystallites regions of a polymer in which the chains are highly ordered.
C-terminal amino acid the terminal amino acid of a peptide (or protein) that has a free carboxyl group.
cumulated double bonds double bonds that are adjacent to one other.
Curtius rearrangement conversion of an acyl chloride into a primary amine with the use of azide ion ($^-N_3$).
cyanohydrin $\mathrm{R\overset{\overset{OH}{|}}{\underset{\underset{C\equiv N}{|}}{C}}R(H)}$
cycloaddition reaction a reaction in which two π-bond-containing molecules react to form a cyclic compound.
[4 + 2] cycloaddition reaction a cycloaddition reaction in which four π electrons come from one reactant and two π electrons come from the other reactant.
cycloalkane an alkane with its carbon chain arranged in a closed ring.
deactivating substituent a substituent that decreases the reactivity of an aromatic ring. Electron-withdrawing substituents deactivate aromatic rings toward electrophilic attack, and electron-donating substituents deactivate aromatic rings toward nucleophilic attack.
deamination loss of ammonia.
decarboxylation loss of carbon dioxide.
degenerate orbitals orbitals that have the same energy.
dehydration loss of water.
dehydrogenase an enzyme that carries out an oxidation reaction by removing hydrogen from the substrate.
dehydrohalogenation elimination of a proton and a halide ion.
delocalization energy (resonance energy) the extra stability a compound achieves as a result of having delocalized electrons.
delocalized electrons electrons that are shared by more than two atoms.
denaturation destruction of the highly organized tertiary structure of a protein.
deoxygenation removal of an oxygen from a reactant.
deoxyribonucleic acid (DNA) a polymer of deoxyribonucleotides.
deoxyribonucleotide a nucleotide in which the sugar component is D-2′-deoxyribose.
deoxy sugar a sugar in which one of the OH groups has been replaced by an H.
DEPT ^{13}C NMR spectrum a series of four spectra that distinguishes among $-CH_3$, $-CH_2$, and $-CH$ groups.
depurination elimination of a purine ring.
detergent a salt of a sulfonic acid.
deuterium kinetic isotope effect ratio of the rate constant obtained for a compound containing hydrogen and the rate constant obtained for an identical compound in which one or more of the hydrogens have been replaced by deuterium.
dextrorotatory the enantiomer that rotates polarized light in a clockwise direction.
diastereomer a configurational stereoisomer that is not an enantiomer.
diastereotopic hydrogens two hydrogens bonded to a carbon that when replaced in turn with a deuterium, result in a pair of diastereomers.
1,3-diaxial interaction the interaction between an axial substituent and the other two axial substituents on the same side of the cyclohexane ring.
diazonium ion $Ar\overset{+}{N}\equiv N$ or $R\overset{+}{N}\equiv N$.
diazonium salt a diazonium ion and an anion ($Ar\overset{+}{N}\equiv N\ X^-$).
Dieckmann condensation an intramolecular Claisen condensation.
dielectric constant a measure of how well a solvent can insulate opposite charges from one another.
Diels–Alder reaction a [4 + 2] cycloaddition reaction.
diene a hydrocarbon with two double bonds.
dienophile an alkene that reacts with a diene in a Diels–Alder reaction.
β-diketone a ketone with a second carbonyl group at the β-position.
dimer a molecule formed by the joining together of two identical molecules.
dinucleotide two nucleotides linked by phosphodiester bonds.
dipeptide two amino acids linked by an amide bond.
dipole–dipole interaction an interaction between the dipole of one molecule and the dipole of another.
dipole moment (μ) a measure of the separation of charge in a bond or in a molecule.
direct addition 1,2-addition.
direct displacement mechanism a reaction in which the nucleophile displaces the leaving group in a single step.
direct substitution substitution at the carbon that was bonded to the leaving group.
disaccharide a compound containing two sugar molecules linked together.
disconnection breaking a bond to carbon to give a simpler species.
disproportionation transfer of a hydrogen atom by a radical to another radical, forming an alkane and an alkene.
disrotatory ring closure achieves head-to-head overlap of *p* orbitals by rotating the orbitals in opposite directions.
dissociation energy the amount of energy required to break a bond, or the amount of energy released when a bond is formed.
dissolving-metal reduction a reduction brought about by the use of sodium or lithium metal dissolved in liquid ammonia.
distribution coefficient the ratio of the amounts of a compound dissolving in each of two solvents in contact with each other.
disulfide bridge a disulfide ($-S-S-$) bond in a peptide or protein.
DNA (deoxyribonucleic acid) a polymer of deoxyribonucleotides.
double bond a σ bond and a π bond between two atoms.
doublet an NMR signal split into two peaks.
doublet of doublets an NMR signal split into four peaks of approximately equal height. Caused by splitting a signal into a doublet by one hydrogen and into another doublet by another (nonequivalent) hydrogen.
drug a compound that reacts with a biological molecule, triggering a physiological effect.
drug resistance biological resistance to a particular drug.
drug synergism when the effect of two drugs used in combination is greater than the sum of the effects obtained when the drugs are administered individually.
eclipsed conformation a conformation in which the bonds on adjacent carbons are aligned as viewed looking down the carbon–carbon bond.
***E* conformation** the conformation of a carboxylic acid or carboxylic acid derivative in which the carbonyl oxygen and the substituent bonded to the carboxyl oxygen or nitrogen are on opposite sides of the single bond.
Edman's reagent phenyl isothiocyanate. A reagent used to determine the N-terminal amino acid of a polypeptide.
effective magnetic field the magnetic field that a proton "senses" through the surrounding cloud of electrons.
effective molarity the concentration of the reagent that would be required in an intermolecular reaction for it to have the same rate as an intramolecular reaction.
***E* isomer** the isomer with the high-priority groups on opposite sides of the double bond.
elastomer a polymer that can stretch and then revert to its original shape.
electrocyclic reaction a reaction in which a π bond in the reactant is lost so that a cyclic compound with a new σ bond can be formed.
electromagnetic radiation radiant energy that displays wave properties.

electron affinity the energy given off when an atom acquires an electron.
electronegative element an element that readily acquires an electron.
electronegativity tendency of an atom to pull electrons toward itself.
electronic transition promotion of an electron from its HOMO to its LUMO.
electron sink site to which electrons can be delocalized.
electrophile an electron-deficient atom or molecule.
electrophilic addition reaction an addition reaction in which the first species that adds to the reactant is an electrophile.
electrophilic aromatic substitution a reaction in which an electrophile substitutes for a hydrogen of an aromatic ring.
electrophilic catalysis catalysis in which the species that facilitates the reaction is an electrophile.
electrophoresis a technique that separates amino acids on the basis of their pI values.
electropositive element an element that readily loses an electron.
electrostatic attraction attractive force between opposite charges.
electrostatic catalysis stabilization of a charge by an opposite charge.
elemental analysis a determination of the relative proportions of the elements present in a compound.
α-elimination removal of two atoms or groups from the same carbon.
β-elimination removal of two atoms or groups from adjacent carbons.
elimination reaction a reaction that involves the elimination of atoms (or molecules) from the reactant.
empirical formula formula giving the relative numbers of the different kinds of atoms in a molecule.
enamine an α,β-unsaturated tertiary amine.
enantiomerically pure containing only one enantiomer.
enantiomeric excess (optical purity) how much excess of one enantiomer is present in a mixture of a pair of enantiomers.
enantiomers nonsuperimposable mirror-image molecules.
enantioselective reaction a reaction that forms an excess of one enantiomer.
enantiotopic hydrogens two hydrogens bonded to a carbon that is bonded to two other groups that are nonidentical.
endergonic reaction a reaction with a positive $\Delta G°$.
endo a substituent is endo if it is closer to the longer or more unsaturated bridge.
endopeptidase an enzyme that hydrolyzes a peptide bond that is not at the end of a peptide chain.
endothermic reaction a reaction with a positive $\Delta H°$.
enkephalins pentapeptides synthesized by the body to control pain.
enolization keto–enol interconversion.
enthalpy the heat given off ($-\Delta H°$) or the heat absorbed ($+\Delta H°$) during the course of a reaction.
entropy a measure of the freedom of motion in a system.
enzyme a protein that is a catalyst.
epimerization changing the configuration of a chirality center by removing a proton from it and then reprotonating the molecule at the same site.
epimers monosaccharides that differ in configuration at only one carbon.
epoxidation formation of an epoxide.
epoxide (oxirane) an ether in which the oxygen is incorporated into a three-membered ring.
epoxy resin substance formed by mixing a low-molecular-weight prepolymer with a compound that forms a cross-linked polymer.
equatorial bond a bond of the chair conformer of cyclohexane that juts out from the ring in approximately the same plane that contains the chair.
equilibrium constant the ratio of products to reactants at equilibrium or the ratio of the rate constants for the forward and reverse reactions.
equilibrium control thermodynamic control.
E1 reaction a first-order elimination reaction.
E2 reaction a second-order elimination reaction.
erythro enantiomers the pair of enantiomers with similar groups on the same side as drawn in a Fischer projection.
essential amino acid an amino acid that humans must obtain from their diet because they cannot synthesize it at all or cannot synthesize it in adequate amounts.
essential oils fragrances and flavorings isolated from plants that do not leave residues when they evaporate. Most are terpenes.
ester $\mathrm{R{-}\overset{\overset{\displaystyle O}{\|}}{C}{-}OR}$
estrogens female sex hormones.
ether a compound containing an oxygen bonded to two carbons (ROR).
eukaryote a unicellular or multicellular body whose cell or cells contain a nucleus.
excited-state electronic configuration the electronic configuration that results when an electron in the ground-state electronic configuration has been moved to a higher energy orbital.
exergonic reaction a reaction with a negative $\Delta G°$.
exhaustive methylation reaction of an amine with excess methyl iodide to form a quaternary ammonium iodide.
exo a substituent is exo if it closer to the shorter or more saturated bridge.
exon a stretch of bases in DNA that are a portion of a gene.
exopeptidase an enzyme that hydrolyzes a peptide bond at the end of a peptide chain.
exothermic reaction a reaction with a negative $\Delta H°$.
experimental energy of activation ($E_a = \Delta H^{\ddagger} - RT$) a measure of the approximate energy barrier to a reaction. (It is approximate because it does not contain an entropy component.)
extrusion reaction a reaction in which a neutral molecule (e.g., CO_2, CO, or N_2) is eliminated from a molecule.
fat a triester of glycerol that exists as a solid at room temperature.
fatty acid a long-chain carboxylic acid.
Favorskii reaction reaction of an α-haloketone with hydroxide ion.
fibrous protein a water-insoluble protein in which the polypeptide chains are arranged in bundles.
fingerprint region the right-hand third of an IR spectrum where the absorption bands are characteristic of the compound as a whole.
first-order rate constant the rate constant of a first-order reaction.
first-order reaction (unimolecular reaction) a reaction whose rate depends on the concentration of one reactant.
Fischer esterification reaction the reaction of a carboxylic acid with alcohol in the presence of an acid catalyst to form an ester.
Fischer projection a method of representing the spatial arrangement of groups bonded to a chirality center. The chirality center is the point of intersection of two perpendicular lines; the horizontal lines represent bonds that project out of the plane of the paper toward the viewer, and the vertical lines represent bonds that point back from the plane of the paper away from the viewer.
flagpole hydrogens (transannular hydrogens) the two hydrogens in the boat conformation of cyclohexane that are closest to each other.
flavin adenine dinucleotide (FAD) a coenzyme required in certain oxidation reactions. It is reduced to $FADH_2$, which can act as a reducing agent in another reaction.
flavin mononucleotide (FMN) a coenzyme required in certain oxidation reactions. It is reduced to $FMNH_2$, which can act as a reducing agent in another reaction.
formal charge the number of valence electrons − (the number of nonbonding electrons $+1/2$ the number of bonding electrons).
Fourier transform NMR a technique in which all the nuclei are excited simultaneously by an rf pulse, their relaxation is monitored, and the data are mathematically converted to a spectrum.
free energy of activation ($\Delta G^{\ddagger}$) the true energy barrier to a reaction.
free-induction decay relaxation of excited nuclei.
frequency the velocity of a wave divided by its wavelength (in units of cycles/s).
Friedel–Crafts acylation an electrophilic substitution reaction that puts an acyl group on a benzene ring.
Friedel–Crafts alkylation an electrophilic substitution reaction that puts an alkyl group on a benzene ring.
frontier orbital analysis determining the outcome of a pericyclic reaction with the use of frontier orbitals.
frontier orbitals the HOMO and the LUMO.
frontier orbital theory a theory that, like the conservation of orbital symmetry theory, explains the relationships among reactant, product, and reaction conditions in a pericyclic reaction.
functional group the center of reactivity in a molecule.
functional group interconversion the conversion of one functional group into another functional group.
functional group region the left-hand two-thirds of an IR spectrum where most functional groups show absorption bands.
furanose a five-membered-ring sugar.
furanoside a five-membered-ring glycoside.

fused bicyclic compound a bicyclic compound in which the rings share two adjacent carbons.

Gabriel synthesis conversion of an alkyl halide into a primary amine, using phthalimide as a starting material.

gauche X and Y are gauche to each other in this Newman projection:

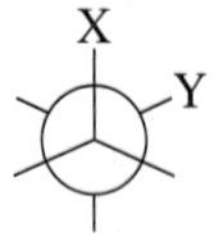

X

Y

gauche conformer a staggered conformer in which the largest substituents are gauche to each other.

gauche interaction the interaction between two atoms or groups that are gauche to each other.

***gem*-dialkyl effect** two alkyl groups on a carbon, the effect of which is to increase the probability that the molecule will be in the proper conformation for ring closure.

***gem*-diol (hydrate)** a compound with two OH groups on the same carbon.

geminal coupling the mutual splitting of two nonidentical protons bonded to the same carbon.

geminal dihalide a compound with two halogen atoms bonded to the same carbon.

gene a segment of DNA.

general-acid catalysis catalysis in which a proton is transferred to the reactant during the slow step of the reaction.

general-base catalysis catalysis in which a proton is removed from the reactant during the slow step of the reaction.

generic name a commercially nonrestricted name for a drug.

gene therapy a technique that inserts a synthetic gene into the DNA of an organism that is defective in that gene.

genetic code the amino acid specified by each three-base sequence of mRNA.

geometric isomers cis–trans (or *E,Z*) isomers.

Gibbs standard free-energy change ($\Delta G°$) the difference between the free-energy content of the products and the free-energy content of the reactants at equilibrium under standard conditions (1 M, 25 °C, 1 atm).

Gilman reagent an organocuprate, prepared from the reaction of an organolithium reagent with cuprous iodide, used to replace a halogen with an alkyl group.

globular protein a water-soluble protein that tends to have a roughly spherical shape.

gluconeogenesis the synthesis of D-glucose from pyruvate.

glycol a compound containing two or more OH groups.

glycolysis (glycolytic cycle) the sequence of reactions that converts D-glucose into two molecules of pyruvate.

glycoprotein a protein that is covalently bonded to a polysaccharide.

glycoside the acetal of a sugar.

***N*-glycoside** a glycoside with a nitrogen instead of an oxygen at the glycosidic linkage.

glycosidic bond the bond between the anomeric carbon and the alcohol in a glycoside.

α-1,4′-glycosidic linkage a glycosidic linkage between the C-1 oxygen of one sugar and the C-4 of a second sugar with the oxygen atom of the glycosidic linkage in the axial position.

β-1,4′-glycosidic linkage a glycosidic linkage between the C-1 oxygen of one sugar and the C-4 of a second sugar with the oxygen atom of the glycosidic linkage in the equatorial position.

graft copolymer a copolymer that contains branches of a polymer of one monomer grafted onto the backbone of a polymer made from another monomer.

Grignard reagent the compound that results when magnesium is inserted between the carbon and halogen of an alkyl halide. (RMgBr, RMgCl).

ground-state electronic configuration a description of which orbitals the electrons of an atom or molecule occupy when all of the electrons of atoms are in their lowest-energy orbitals.

Hagemann's-ester a compound prepared by treating a mixture of formaldehyde and ethylacetoacetate with base and then with acid and heat.

half-chair conformation the least stable conformation of cyclohexane.

haloform reaction the reaction of a halogen and HO^- with a methyl ketone.

halogenation reaction with halogen (Br_2, Cl_2, I_2).

halohydrin an organic molecule that contains a halogen and an OH group on adjacent carbons.

Hammond postulate states that the transition state will be more similar in structure to the species (reactants or products) that it is closer to energetically.

Haworth projection a way to show the structure of a sugar; the five- and six-membered rings are represented as being flat.

head-to-tail addition the head of one molecule is added to the tail of another.

heat of combustion the amount of heat given off when a carbon-containing compound reacts completely with O_2 to form CO_2 and H_2O

heat of formation the heat given off when a compound is formed from its elements under standard conditions.

heat of hydrogenation the heat ($-\Delta H°$) released in a hydrogenation reaction.

Heck reaction couples an aryl, benzyl, or vinyl halide or triflate with an alkene in a basic solution in the presence of $Pd(PPh_3)_4$.

Heisenberg uncertainty principle states that both the precise location and the momentum of an atomic particle cannot be simultaneously determined.

α-helix the backbone of a polypeptide coiled in a right-handed spiral with hydrogen bonding occurring within the helix.

Hell–Volhard–Zelinski (HVZ) reaction heating a carboxylic acid with Br_2 + P in order to convert it into an α-bromocarboxylic acid.

hemiacetal

OH
|
RCH
|
OR

hemiketal

OH
|
RCR
|
OR

Henderson–Hasselbalch equation $pK_a = pH + \log[HA]/[A^-]$

heptose a monosaccharide with seven carbons.

HETCOR spectrum a 2-D NMR spectrum that shows coupling between protons and the carbons to which they are attached.

heteroatom an atom other than carbon or hydrogen.

heterocyclic compound (heterocycle) a cyclic compound in which one or more of the atoms of the ring are heteroatoms.

heterogeneous catalyst a catalyst that is insoluble in the reaction mixture.

heterolytic bond cleavage (heterolysis) breaking a bond with the result that both bonding electrons stay with one of the atoms.

hexose a monosaccharide with six carbons.

high-energy bond a bond that releases a great deal of energy when it is broken.

highest occupied molecular orbital (HOMO) the molecular orbital of highest energy that contains an electron.

high-resolution NMR spectroscopy NMR spectroscopy that uses a spectrometer with a high operating frequency.

Hofmann degradation exhaustive methylation of an amine, followed by reaction with Ag_2O, followed by heating to achieve a Hofmann elimination reaction.

Hofmann elimination (anti-Zaitsev elimination) a hydrogen is removed from the β-carbon bonded to the most hydrogens.

Hofmann elimination reaction elimination of a proton and a tertiary amine from a quaternary ammonium hydroxide.

Hofmann rearrangement conversion of an amide into an amine by using Br_2/HO.

holoenzyme an enzyme plus its cofactor.

homogeneous catalyst a catalyst that is soluble in the reaction mixture.

homolog a member of a homologous series.

homologous series a family of compounds in which each member differs from the next by one methylene group.

homolytic bond cleavage (homolysis) breaking a bond with the result that each of the atoms gets one of the bonding electrons.

homopolymer a polymer that contains only one kind of monomer.

homotopic hydrogens two hydrogens bonded to a carbon bonded to two other groups that are identical.

Hoogsteen base pairing the pairing between a base in a synthetic strand of DNA and a base pair in double-stranded DNA.

Hooke's law an equation that describes the motion of a vibrating spring.

hormone an organic compound synthesized in a gland and delivered by the bloodstream to its target tissue.

Hückel's rule states that, for a compound to be aromatic, its cloud of electrons must contain $(4n + 2)\pi$ electrons, where n is an integer. This is the same as saying that the electron cloud must contain an odd number of pairs of π electrons.

human genome the total DNA of a human cell.

Hund's rule states that when there are degenerate orbitals, an electron will occupy an empty orbital before it will pair up with another electron.
Hunsdiecker reaction conversion of a carboxylic acid into an alkyl halide by heating a heavy metal salt of the carboxylic acid with bromine or iodine.
hybrid orbital an orbital formed by mixing (hybridizing) orbitals.

hydrate (*gem*-diol) $\overset{\displaystyle OH}{\underset{\displaystyle OH}{R\overset{|}{\underset{|}{C}}R(H)}}$

hydrated water has been added to a compound.
hydration addition of water to a compound.
hydrazone $R_2C{=}NNH_2$
hydride ion a negatively charged hydrogen.
1,2-hydride shift the movement of a hydride ion from one carbon to an adjacent carbon.
hydroboration–oxidation the addition of borane to an alkene or an alkyne, followed by reaction with hydrogen peroxide and hydroxide ion.
hydrocarbon a compound that contains only carbon and hydrogen.
α-hydrogen usually, a hydrogen bonded to the carbon adjacent to a carbonyl carbon.
hydrogenation addition of hydrogen.
hydrogen bond an unusually strong dipole–dipole attraction (5 kcal/mol) between a hydrogen bonded to O, N, or F and the nonbonding electrons of an O, N, or F of another molecule.
hydrogen ion (proton) a positively charged hydrogen.
hydrolysis reaction with water.
hydrophobic interactions interactions between nonpolar groups. These interactions increase stability by decreasing the amount of structured water (increasing entropy).
hyperconjugation delocalization of electrons by overlap of carbon–hydrogen or carbon–carbon σ bonds with an empty p orbital.
imine $R_2C{=}NR$
inclusion compound a compound that specifically binds a metal ion or an organic molecule.
induced-dipole–induced-dipole interaction an interaction between a temporary dipole in one molecule and the dipole the temporary dipole induces in another molecule.
induced-fit model a model that describes the specificity of an enzyme for its substrate: The shape of the active site does not become completely complementary to the shape of the substrate until after the enzyme binds the substrate.
inductive electron donation donation of electrons through σ bonds.
inductive electron withdrawal withdrawal of electrons through a σ bond.
inflection point the midpoint of the flattened-out region of a titration curve.
informational strand (sense strand) the strand in DNA that is not read during transcription; it has the same sequence of bases as the synthesized mRNA strand (with a U, T difference).
infrared radiation electromagnetic radiation familiar to us as heat.
infrared spectroscopy uses infrared energy to provide a knowledge of the functional groups in a compound.
infrared (IR) spectrum a plot of percent transmission versus wave number (or wavelength) of infrared radiation.
initiation step the step in which radicals are created, or the step in which the radical needed for the first propagation step is created.
in-line displacement mechanism nucleophilic attack on a phosphorus concerted with breaking a phosphoanhydride bond.
interchain disulfide bridge a disulfide bridge between two cysteine residues in different peptide chains.
intermediate a species formed during a reaction and that is not the final product of the reaction.
intermolecular reaction a reaction that takes place between two molecules.
internal alkyne an alkyne with the triple bond not at the end of the carbon chain.
intimate ion pair pair such that the covalent bond that joined the cation and the anion has broken, but the cation and anion are still next to each other.
intrachain disulfide bridge a disulfide bridge between two cysteine residues in the same peptide chain.
intramolecular catalysis (anchimeric assistance) catalysis in which the catalyst that facilitates the reaction is part of the molecule undergoing reaction.
intramolecular reaction a reaction that takes place within a molecule.
intron a stretch of bases in DNA that contain no genetic information.
inversion of configuration turning the configuration of a carbon inside out like an umbrella in a windstorm, so that the resulting product has a configuration opposite that of the reactant.
iodoform test addition of I_2/HO^- to a methyl ketone forms a yellow precipitate of triiodomethane.
ion–dipole interaction the interaction between an ion and the dipole of a molecule.
ion-exchange chromatography a technique that uses a column packed with an insoluble resin to separate compounds on the basis of their charges and polarities.
ionic bond a bond formed through the attraction of two ions of opposite charges.
ionization energy the energy required to remove an electron from an atom.
ionophore a compound that binds metal ions tightly.
iron protoporphyrin IX the porphyrin ring system of heme plus an iron atom.
isoelectric point (pI) the pH at which there is no net charge on an amino acid.
isolated double bonds double bonds separated by more than one single bond.
isomers nonidentical compounds with the same molecular formula.
isoprene rule rule expressing the head-to-tail linkage of isoprene units.
isopropyl split a split in the IR absorption band attributable to a methyl group. It is characteristic of an isopropyl group.
isotactic polymer a polymer in which all the substituents are on the same side of the fully extended carbon chain.
isotopes atoms with the same number of protons, but different numbers of neutrons.
iterative synthesis a synthesis in which a reaction sequence is carried out more than once.
IUPAC nomenclature systematic nomenclature of chemical compounds.
Kekulé structure a model that represents the bonds between atoms as lines.

ketal $\overset{\displaystyle OR}{\underset{\displaystyle OR}{R\overset{|}{\underset{|}{C}}R}}$

keto–enol tautomerism (keto–enol interconversion) interconversion of keto and enol tautomers.
keto–enol tautomers a ketone and its isomeric α,β-unsaturated alcohol.
β-keto ester an ester with a second carbonyl group at the β-position.

ketone 

ketose a polyhydroxyketone.
Kiliani–Fischer synthesis a method used to increase the number of carbons in an aldose by one, resulting in the formation of a pair of C-2 epimers.
kinetic control when a reaction is under kinetic control, the relative amounts of the products depend on the rates at which they are formed.
kinetic isotope effect a comparison of the rate of reaction of a compound with the rate of reaction of an identical compound in which one of the atoms has been replaced by an isotope.
kinetic product the product that is formed the fastest.
kinetic resolution separation of enantiomers on the basis of the difference in their rate of reaction with an enzyme.
kinetics the field of chemistry that deals with the rates of chemical reactions.
kinetic stability chemical reactivity, indicated by $\Delta G^{\ddagger}$. If $\Delta G^{\ddagger}$ is large, the compound is kinetically stable (not very reactive). If $\Delta G^{\ddagger}$ is small, the compound is kinetically unstable (highly reactive).
Knoevenagel condensation a condensation of an aldehyde or ketone with no α hydrogens and a compound with an α-carbon flanked by two electron-withdrawing groups.
Kolbe–Schmitt carboxylation reaction a reaction that uses CO_2 to carboxylate phenol.
Krebs cycle (citric acid cycle, tricarboxylic acid cycle, TCA cycle) a series of reactions that convert the acetyl group of acetyl-CoA into two molecules of CO_2.
lactam a cyclic amide.
lactone a cyclic ester.
λ_{max} the wavelength at which there is maximum UV/Vis absorbance.
lead compound the prototype in a search for other biologically active compounds.

leaning when a line drawn over the outside peaks of an NMR signal points in the direction of the signal given by the protons that cause the splitting.
leaving group the group that is displaced in a nucleophilic substitution reaction.
Le Châtelier's principle states that if an equilibrium is disturbed, the components of the equilibrium will adjust in a way that will offset the disturbance.
lecithin a phosphoacylglycerol in which the second OH group of phosphate has formed an ester with choline.
levorotatory the enantiomer that rotates polarized light in a counterclockwise direction.
Lewis acid a substance that accepts an electron pair.
Lewis base a substance that donates an electron pair.
Lewis structure a model that represents the bonds between atoms as lines or dots and the valence electrons as dots.
ligation sharing of nonbonding electrons with a metal ion.
linear combination of atomic orbitals (LCAO) the combination of atomic orbitals to produce a molecular orbital.
linear conjugation the atoms in the conjugated system are in a linear arrangement.
linear synthesis a synthesis that builds a molecule step by step from starting materials.
lipid a water-insoluble compound found in a living system.
lipid bilayer two layers of phosphoacylglycerols arranged so that their polar heads are on the outside and their nonpolar fatty acid chains are on the inside.
lipoate a coenzyme required in certain oxidation reactions.
living polymer a nonterminated chain-growth polymer that remains active. This means that the polymerization reaction can continue upon the addition of more monomer.
localized electrons electrons that are restricted to a particular locality.
lock-and-key model a model that describes the specificity of an enzyme for its substrate: The substrate fits the enzyme as a key fits a lock.
London forces Induced-dipole–induced-dipole interactions.
lone-pair electrons (nonbonding electrons) valence electrons not used in bonding.
long-range coupling splitting of a proton by a proton more than three σ bonds away.
loop conformation (coil conformation) that part of a protein that is highly ordered, but not in an α-helix or β-pleated sheet.
lowest unoccupied molecular orbital (LUMO) the molecular orbital of lowest energy that does not contain an electron.
Lucas test a test that determines whether an alcohol is primary, secondary, or tertiary.
magnetic anisotropy the term used to describe the greater freedom of a π electron cloud to move in response to a magnetic field as a consequence of the greater polarizability of π electrons compared with σ electrons.
magnetic resonance imaging (MRI) NMR used in medicine. The difference in the way water is bound in different tissues produces a variation in signal between organs as well as between healthy and diseased tissue.
magnetogyric ratio a property (measured in rad $T^{-1}s^{-1}$) that depends on the magnetic properties of a particular kind of nucleus.
major groove the wider and deeper of the two alternating grooves in DNA.
malonic ester synthesis the synthesis of a carboxylic acid, using diethyl malonate as the starting material.
Mannich reaction condensation of a secondary amine and formaldehyde with a carbon acid.
Markovnikov's rule the actual rule is "When a hydrogen halide adds to an asymmetrical alkene, the addition occurs such that the halogen attaches itself to the sp^2 carbon of the alkene bearing the lowest number of hydrogen atoms." A more universal rule is "The electrophile adds to the sp^2 carbon that is bonded to the greater number of hydrogens."
mass number the number of protons plus the number of neutrons in an atom.
mass spectrometry provides a knowledge of the molecular weight, molecular formula, and certain structural features of a compound.
mass spectrum a plot of the relative abundance of the positively charged fragments produced in a mass spectrometer versus their m/z values.
materials science the science of creating new materials to be used in place of known materials such as metal, glass, wood, cardboard, and paper.
Maxam–Gilbert sequencing a technique used to sequence restriction fragments.
McLafferty rearrangement rearrangement of the molecular ion of a ketone. The bond between the α- and β-carbons breaks, and a γ-hydrogen migrates to the oxygen.
mechanism-based inhibitor (suicide inhibitor) a compound that inactivates an enzyme by undergoing part of its normal catalytic mechanism.
mechanism of a reaction a description of the step-by-step process by which reactants are changed into products.
melting point the temperature at which a solid becomes a liquid.
membrane the material that surrounds a cell in order to isolate its contents.
mercaptan (thiol) the sulfur analog of an alcohol (RSH).
meso compound a compound that contains chirality centers and a plane of symmetry.
metabolism reactions living organisms carry out in order to obtain the energy and to synthesize the compounds they require.
meta-directing substituent a substituent that directs an incoming substituent meta to an existing substituent.
metal-activated enzyme an enzyme that has a loosely bound metal ion.
metal-ion catalysis catalysis in which the species that facilitates the reaction is a metal ion.
metalloenzyme an enzyme that has a tightly bound metal ion.
methine hydrogen a tertiary hydrogen.
methylene group a CH_2 group.
1,2-methyl shift the movement of a methyl group with its bonding electrons from one carbon to an adjacent carbon.
micelle a spherical aggregation of molecules, each with a long hydrophobic tail and a polar head, arranged so that the polar head points to the outside of the sphere.
Michael reaction the addition of an α-carbanion to the β-carbon of an α,β-unsaturated carbonyl compound.
minor groove the narrower and more shallow of the two alternating grooves in DNA.
mixed (crossed) aldol addition an aldol addition in which two different carbonyl compounds are used.
mixed anhydride an acid anhydride with two different R groups.

$$R-\overset{\overset{\displaystyle O}{\|}}{C}-O-\overset{\overset{\displaystyle O}{\|}}{C}-R'$$

mixed Claisen condensation a Claisen condensation in which two different esters are used.
mixed triacylglycerol a triacylglycerol in which the fatty-acid components are different.
molar absorptivity the absorbance obtained from a 1.00 M solution in a cell with a 1.00-cm light path.
molecular ion (parent ion) peak in the mass spectrum with the greatest m/z.
molecular modeling computer-assisted design of a compound with particular structural characteristics.
molecular modification changing the structure of a lead compound.
molecular orbital an orbital associated with a molecule.
molecular orbital theory describes a model in which the electrons occupy orbitals as they do in atoms, but with the orbitals extending over the entire molecule.
molecular recognition the recognition of one molecule by another as a result of specific interactions; for example, the specificity of an enzyme for its substrate.
molozonide an unstable intermediate containing a five-membered ring with three oxygens in a row that is formed from the reaction of an alkene with ozone.
monomer a repeating unit in a polymer.
monosaccharide (simple carbohydrate) a single sugar molecule.
monoterpene a terpene that contains 10 carbons.
MRI scanner an NMR spectrometer used in medicine for whole-body NMR.
multiplet an NMR signal split into more than seven peaks.
multiplicity the number of peaks in an NMR signal.
multistep synthesis preparation of a compound by a route that requires several steps.
mutarotation a slow change in optical rotation to an equilibrium value.
$N + 1$ rule an ^1H NMR signal for a hydrogen with N equivalent hydrogens bonded to an adjacent carbon is split into $N + 1$ peaks. A ^{13}C NMR signal for a carbon bonded to N hydrogens is split into $N + 1$ peaks.
natural-abundance atomic weight the average mass of the atoms in the naturally occurring element.
natural product a product synthesized in nature.
neurotransmitter a compound that transmits nerve impulses.

nicotinamide adenine dinucleotide (NAD^+) a coenzyme required in certain oxidation reactions. It is reduced to NADH, which can act as a reducing agent in another reaction.
nicotinamide adenine dinucleotide phosphate ($NADP^+$) a coenzyme required in certain oxidation reactions. It is reduced to NADPH, which can act as a reducing agent in another reaction.
NIH shift the 1,2-hydride shift of a carbocation (obtained from an arene oxide) that leads to an enone.
nitration substitution of a nitro group (NO_2) for a hydrogen of a benzene ring.
nitrile a compound that contains a carbon–nitrogen triple bond ($RC{\equiv}N$).
nitrosamine (N-nitroso compound) $R_2NN{=}O$
NMR spectroscopy the absorption of electromagnetic radiation to determine the structural features of an organic compound. In the case of NMR spectroscopy, it determines the carbon–hydrogen framework.
node that part of an orbital in which there is zero probability of finding an electron.
nominal mass mass rounded to the nearest whole number.
nonbonding electrons (lone-pair electrons) valence electrons not used in bonding.
nonbonding molecular orbital the *p* orbitals are too far apart to overlap significantly, so the molecular orbital that results neither favors nor disfavors bonding.
nonpolar covalent bond a bond formed between two atoms that share the bonding electrons equally.
nonreducing sugar a sugar that cannot be oxidized by reagents such as Ag^+ and Cu^+. Nonreducing sugars are not in equilibrium with the open-chain aldose or ketose.
normal alkane (straight-chain alkane) an alkane in which the carbons form a contiguous chain with no branches.
***N*-terminal amino acid** the terminal amino acid of a peptide (or protein) that has a free amino group.
nucleic acid the two kinds of nucleic acid are DNA and RNA.
nucleophile an electron-rich atom or molecule.
nucleophilic acyl substitution reaction a reaction in which a group bonded to an acyl or aryl group is substituted by another group.
nucleophilic addition–elimination–nucleophilic addition reaction a nucleophilic addition reaction that is followed by an elimination reaction that is followed by a nucleophilic addition reaction. Acetal formation is an example: An alcohol adds to the carbonyl carbon, water is eliminated, and a second molecule of alcohol adds to the dehydrated product.
nucleophilic addition–elimination reaction a nucleophilic addition reaction that is followed by an elimination reaction. Imine formation is an example: An amine adds to the carbonyl carbon, and water is eliminated.
nucleophilic addition reaction a reaction that involves the addition of a nucleophile to a reagent.
nucleophilic aromatic substitution a reaction in which a nucleophile substitutes for a substituent of an aromatic ring.
nucleophilic catalysis (covalent catalysis) catalysis that occurs as a result of a nucleophile forming a covalent bond with one of the reactants.
nucleophilic catalyst a catalyst that increases the rate of a reaction by acting as a nucleophile.
nucleophilicity a measure of how readily an atom or a molecule with a pair of nonbonding electrons attacks an atom.
nucleophilic substitution reaction a reaction in which a nucleophile substitutes for an atom or a group.
nucleoside a heterocyclic base (a purine or a pyrimidine) bonded to the anomeric carbon of a sugar (D-ribose or D-2′-deoxyribose).
nucleotide a heterocycle attached in the β-position to a phosphorylated ribose.
observed rotation the amount of rotation observed in a polarimeter.
octet rule states that an atom will give up, accept, or share electrons in order to achieve a filled shell. Because a filled second shell contains eight electrons, this is known as the octet rule.
off-resonance decoupling the mode in ^{13}C NMR spectroscopy in which spin–spin splitting occurs between carbons and the hydrogens attached to them.
oil a triester of glycerol that exists as a liquid at room temperature.
olefin an alkene.
oligomer a protein with more than one peptide chain.
oligonucleotide 3 to 10 nucleotides linked by phosphodiester bonds.
oligopeptide 3 to 10 amino acids linked by amide bonds.
oligosaccharide 3 to 10 sugar molecules linked by glycosidic bonds.
open-chain compound an acyclic compound.
operating frequency the frequency at which an NMR spectrometer operates.
optical isomers stereoisomers that contain chirality centers.
optically active rotates the plane of polarized light.
optically inactive does not rotate the plane of polarized light.
optical purity (enantiomeric excess) how much excess of one enantiomer is present in a mixture of a pair of enantiomers.
orbital the volume of space around the nucleus in which an electron is most likely to be found.
orbital hybridization mixing of orbitals.
organic compound a compound that contains carbon.
organic synthesis preparation of organic compounds from other organic compounds.
organometallic compound a compound containing a carbon–metal bond.
oriented polymer a polymer obtained by stretching out polymer chains and putting them back together in a parallel fashion.
orphan drugs drugs for diseases or conditions that affect fewer than 200,000 people.
ortho-para-directing substituent a substituent that directs an incoming substituent ortho and para to an existing substituent.
osazone the product obtained by treating an aldose or a ketose with excess phenylhydrazine. An osazone contains two imine bonds.
overtone band an absorption that occurs at a multiple of the fundamental absorption frequency ($2v_l$, $3v_l$).
oxidation loss of electrons by an atom or a molecule.
oxidation–reduction reaction (redox reaction) a reaction that involves the transfer of electrons from one species to another.
oxidative cleavage an oxidation reaction that cuts the reactant into two or more pieces.
oxime $R_2C{=}NOH$
oxirane (epoxide) an ether in which the oxygen is incorporated into a three-membered ring.
oxonium ion a compound with a positively charged oxygen.
oxyanion a compound with a negatively charged oxygen.
oxymercuration addition of water using a mercuric salt of a carboxylic acid as a catalyst.
ozonide the five-membered-ring compound formed as a result of the rearrangement of a molozonide.
ozonolysis reaction of a carbon–carbon double or triple bond with ozone.
packing the fitting of individual molecules into a frozen crystal lattice.
paraffin an alkane.
parent hydrocarbon the longest continuous carbon chain in a molecule.
parent ion (molecular ion) peak in the mass spectrum with the greatest m/z.
partial hydrolysis a technique that hydrolyzes only some of the peptide bonds in a polypeptide.
partial racemization formation of a pair of enantiomers in unequal amounts.
Pauli exclusion principle states that no more than two electrons can occupy an orbital and that the two electrons must have opposite spin.
pentose a monosaccharide with five carbons.
peptide polymer of amino acids linked together by amide bonds. A peptide contains fewer amino acid residues than a protein does.
peptide bond the amide bond that links the amino acids in a peptide or protein.
peptide nucleic acid (PNA) a polymer containing both an amino acid and a base designed to bind to specific residues on DNA or mRNA.
Perkin condensation a condensation of an aromatic aldehyde and acetic acid.
pericyclic reaction a concerted reaction that takes place as the result of a cyclic rearrangement of electrons.
peroxyacid a carboxylic acid with an OOH group instead of an OH group.
perspective formula a method of representing the spatial arrangement of groups bonded to a chirality center. Two bonds are drawn in the plane of the paper; a solid wedge is used to depict a bond that projects out of the plane of the paper toward the viewer, and a hatched wedge is used to represent a bond that projects back from the plane of the paper away from the viewer.
pH the pH scale is used to describe the acidity of a solution ($pH = -\log[H^+]$).
pH-activity profile a plot of the activity of an enzyme as a function of the pH of the reaction mixture.
phase-transfer catalysis catalysis of a reaction by providing a way to bring a polar reagent into a nonpolar phase so that the reaction between a polar and a nonpolar compound can occur.

phase-transfer catalyst a compound that carries a polar reagent into a nonpolar phase.

phenone $C_6H_5\overset{O}{\overset{\|}{C}}R$

phenyl group $C_6H_5—$

phenylhydrazone $R_2C{=}NNHC_6H_5$

pheromone a compound secreted by an animal that stimulates a physiological or behavioral response from a member of the same species.

phosphatidic acid a phosphoacylglycerol in which only one of the OH groups of phosphate is in an ester linkage.

phosphoacylglycerol (phosphoglyceride) a compound formed when two OH groups of glycerol form esters with fatty acids and the terminal OH group forms a phosphate ester.

phosphoanhydride bond the bond holding two phosphoric acid molecules together.

phospholipid a lipid that contains a phosphate group.

phosphoryl transfer reaction the transfer of a phosphate group from one compound to another.

photochemical reaction a reaction that takes place when a reactant absorbs light.

photosynthesis the synthesis of glucose and O_2 from CO_2 and H_2O.

pH-rate profile a plot of the observed rate of a reaction as a function of the pH of the reaction mixture.

pi (π) bond a bond formed as a result of side-to-side overlap of *p* orbitals.

pi-complex a complex formed between an electrophile and a triple bond.

pinacol rearrangement rearrangement of a vicinal diol.

pK_a describes the tendency of a compound to lose a proton ($pK_a = -\log K_a$, where K_a is the acid dissociation constant).

plane of symmetry an imaginary plane that bisects a molecule into mirror images.

plasticizer an organic molecule that dissolves in a polymer and allows the polymer chains to slide by each other.

β-pleated sheet the backbone of a polypeptide that is extended in a zigzag structure with hydrogen bonding between neighboring chains.

polar covalent bond a bond formed by the unequal sharing of electrons.

polarimeter an instrument that measures the rotation of polarized light.

polarizability an indication of the ease with which the electron cloud of an atom can be distorted.

polarized light light that oscillates only in one plane.

polar reaction the reaction between a nucleophile and an electrophile.

polyamide a polymer in which the monomers are amides.

polycarbonate a step-growth polymer in which the dicarboxylic acid is carbonic acid.

polyene a compound that has several double bonds.

polyester a polymer in which the monomers are esters.

polymer a large molecule made by linking monomers together.

polymer chemistry the field of chemistry that deals with synthetic polymers; part of the larger discipline known as materials science.

polymerization the process of linking up monomers to form a polymer.

polynucleotide many nucleotides linked by phosphodiester bonds.

polypeptide many amino acids linked by amide bonds.

polysaccharide a compound containing more than 10 sugar molecules linked together.

polyunsaturated fatty acid a fatty acid with more than one double bond.

polyurethane a polymer in which the monomers are urethanes.

porphyrin ring system consists of four pyrrole rings joined by one-carbon bridges.

primary alcohol an alcohol in which the OH group is bonded to a primary carbon.

primary alkyl halide an alkyl halide in which the halogen is bonded to a primary carbon.

primary alkyl radical a radical with the unpaired electron on a primary carbon.

primary amine an amine with one alkyl group bonded to the nitrogen.

primary carbocation a carbocation with the positive charge on a primary carbon.

primary carbon a carbon bonded to only one other carbon.

primary hydrogen a hydrogen bonded to a primary carbon.

primary structure (of a nucleic acid) the sequence of bases in a nucleic acid.

primary structure (of a protein) the sequence of amino acids in a protein.

principle of microscopic reversibility states that the mechanism for a reaction in the forward direction has the same intermediates and the same rate-determining step as the mechanism for the reaction in the reverse direction.

prochiral carbonyl carbon a carbonyl carbon that will become a chirality center if it is attacked by a group unlike any of the groups already bonded to it.

prochirality center a carbon bonded to two hydrogens that will become a chirality center if one of the hydrogens is replaced by deuterium.

prokaryote a unicellular body without a nucleus.

promoter site a short sequence of bases at the beginning of a gene.

propagating site the reactive end of a chain-growth polymer.

propagation step in the first of a pair of propagation steps, a radical (or an electrophile or a nucleophile) reacts to produce another radical (or an electrophile or a nucleophile) that reacts in the second to produce the radical (or the electrophile or the nucleophile) that was the reactant in the first propagation step.

proprietary name (trade name, brand name) identifies a commercial product and distinguishes it from other products.

pro-*R* hydrogen replacing this hydrogen with deuterium creates a chirality center with the *R* configuration.

pro-*S* hydrogen replacing this hydrogen with deuterium creates a chirality center with the *S* configuration.

prostacyclin a lipid, derived from arachidonic acid, that dilates blood vessels and inhibits platelet aggregation.

prosthetic group a tightly bound coenzyme.

protecting group a reagent that protects a functional group from a synthetic operation that it would otherwise not survive.

protein a polymer containing 40 to 4000 amino acids linked by amide bonds.

protic solvent a solvent that has a hydrogen bonded to an oxygen or a nitrogen.

proton a positively charged hydrogen (hydrogen ion); a positively charged particle in an atomic nucleus.

proton-decoupled ^{13}C NMR spectrum a ^{13}C NMR spectrum in which all the signals appear as singlets because there is no coupling between the nucleus and its bonded hydrogens.

proton transfer reaction a reaction in which a proton is transferred from an acid to a base.

protoporphyrin IX the porphyrin ring system of heme.

proximity effect an effect caused by one species being close to another.

pseudo-first-order reaction a second-order reaction in which the concentration of one of the reactants is much greater than the other, allowing the reaction to be treated as a first-order reaction.

pyranose a six-membered-ring sugar.

pyranoside a six-membered-ring glycoside.

pyridoxal phosphate the coenzyme required by enzymes that catalyze certain transformations of amino acids.

quantitative structure–activity relationship (QSAR) the relation between a particular property of a series of compounds and their biological activity.

quantum numbers numbers arising from the quantum mechanical treatment of an atom that describe the properties of the electrons in the atom.

quartet an NMR signal split into four peaks.

quaternary ammonium ion an ion containing a nitrogen bonded to four alkyl groups (R_4N^+).

quaternary ammonium salt a quaternary ammonium ion and an anion ($R_4N^+X^-$).

quaternary structure a description of the way the individual polypeptide chains of a protein are arranged with respect to each other.

racemic mixture (racemate, racemic modification) a mixture of equal amounts of a pair of enantiomers.

radical an atom or a molecule with an unpaired electron.

radical addition reaction an addition reaction in which the first species that adds is a radical.

radical anion a species with a negative charge and an unpaired electron.

radical cation a species with a positive charge and an unpaired electron.

radical chain reaction a reaction in which radicals are formed and react in repeating propagating steps.

radical inhibitor a compound that traps radicals.

radical initiator a compound that creates radicals.

radical polymerization chain-growth polymerization in which the initiator is a radical; the propagation site is therefore a radical.

radical reaction a reaction in which a new bond is formed by using one electron from one reagent and one electron from another reagent.
radical substitution reaction a substitution reaction that has a radical intermediate.
random coil the conformation of a totally denatured protein.
random copolymer a copolymer with a random distribution of monomers.
random screen (blind screen) the search for a pharmacologically active compound without any information about what chemical structures might show activity.
rate constant a measure of how easy or difficult it is to reach the transition state of a reaction (to get over the energy barrier to the reaction).
rate-determining step (rate-limiting step) the step in a reaction that has the transition state with the highest energy.
rational drug design designing drugs with a particular structure to achieve a specific purpose.
***R* configuration** after assigning relative priorities to the four groups bonded to a chirality center, if the lowest priority group is on a vertical axis in a Fischer projection (or pointing away from the viewer in a perspective formula), an arrow drawn from the highest priority group to the next-highest-priority group goes in a clockwise direction.
reaction coordinate diagram describes the energy changes that take place during the course of a reaction.
reactivity–selectivity principle states that the greater the reactivity of a species, the less selective it will be.
receptor site the site at which a drug binds in order to exert its physiological effect.
redox reaction (oxidation–reduction reaction) a reaction that involves the transfer of electrons from one species to another.
red shift a shift to a longer wavelength.
reducing sugar a sugar that can be oxidized by reagents such as Ag^4 or Br_2. Reducing sugars are in equilibrium with the open-chain aldose or ketose.
reduction gain of electrons by an atom or a molecule.
reductive amination the reaction of an aldehyde or a ketone with ammonia or with a primary amine in the presence of a reducing agent (H_2/Raney Ni).
reference compound a compound added to a sample whose NMR spectrum is to be taken. The positions of the signals in the NMR spectrum are measured from the position of the signal given by the reference compound.
Reformatsky reaction reaction of an organozinc reagent with an aldehyde or a ketone.
regioselective reaction a reaction that leads to the preferential formation of one constitutional isomer over another.
relative configuration the configuration of a compound relative to the configuration of another compound.
relative rate obtained by dividing the actual rate constant by the rate constant of the slowest reaction in the group being compared.
replication the synthesis of identical copies of DNA.
replication fork the position on DNA at which replication begins.
resolution of a racemic mixture separation of a racemic mixture into the individual enantiomers.
resonance a compound with delocalized electrons is said to have resonance.
resonance contributor (resonance structure, contributing resonance structure) a structure with localized electrons that approximates the true structure of a compound with delocalized electrons.
resonance electron donation donation of electrons through *p* orbital overlap with neighboring π bonds.
resonance electron withdrawal withdrawal of electrons through *p* orbital overlap with neighboring π bonds.
resonance energy (delocalization energy) the extra stability associated with a compound as a result of its having delocalized electrons.
resonance hybrid the actual structure of a compound with delocalized electrons; it is represented by two or more structures with localized electrons.
resonances NMR absorption signals.
restriction endonuclease an enzyme that cleaves DNA at a specific base sequence.
restriction fragment a fragment that is formed when DNA is cleaved by a restriction endonuclease.
retrosynthesis (retrosynthetic analysis) working backwards (on paper) from the target molecule to available starting materials.
retrovirus a virus whose genetic information is stored in its RNA.
rf radiation radiation in the radiofrequency region of the electromagnetic spectrum.
ribonucleic acid (RNA) a polymer of ribonucleotides.
ribonucleotide a nucleotide in which the sugar component is D-ribose.
ribosome a particle composed of about 40% protein and 60% RNA on which protein biosynthesis takes place.
ribozyme an RNA molecule that acts as a catalyst.
ring current the movement of π electrons around an aromatic benzene ring.
ring-expansion rearrangement rearrangement of a carbocation in which the positively charged carbon is bonded to a cyclic compound and, as a result of rearrangement, the size of the ring increases by one carbon.
ring-flip (chair–chair interconversion) the conversion of the chair conformer of cyclohexane into the other chair conformer. Bonds that are axial in one chair conformer are equatorial in the other.
ring opening polymerization a chain-growth polymerization that involves opening the ring of the monomer.
Ritter reaction reaction of a nitrile with a secondary or tertiary alcohol to form a secondary amide.
RNA (ribonucleic acid) a polymer of ribonucleotides.
RNA splicing the step in RNA processing that cuts out nonsense bases and splices informational pieces together.
Robinson annulation a Michael reaction followed by an intramolecular aldol condensation.
Rosenmund reduction reduction of an acyl chloride to an aldehyde by using H_2 and a deactivated palladium catalyst.
Ruff degradation a method used to shorten an aldose by one carbon.
Sandmeyer reaction the reaction of an aryl diazonium salt with a cuprous salt.
saponification hydrolysis of an ester (such as a fat) under basic conditions.
saturated hydrocarbon a hydrocarbon that is completely saturated (i.e., contains no double or triple bonds) with hydrogen.
Schiemann reaction the reaction of an arenediazonium salt with HBF_4.
Schiff base $R_2C{=}NR$
***S* configuration** after assigning relative priorities to the four groups bonded to a chirality center, if the lowest priority group is on a vertical axis in a Fischer projection (or pointing away from the viewer in a perspective formula), an arrow drawn from the highest priority group to the next-highest priority group goes in a counterclockwise direction.
secondary alcohol an alcohol in which the OH group is bonded to a secondary carbon.
secondary alkyl halide an alkyl halide in which the halogen is bonded to a secondary carbon.
secondary alkyl radical a radical with the unpaired electron on a secondary carbon.
secondary amine an amine with two alkyl groups bonded to the nitrogen.
secondary carbocation a carbocation with the positive charge on a secondary carbon.
secondary carbon a carbon bonded to two other carbons.
secondary hydrogen a hydrogen bonded to a secondary carbon.
secondary structure a description of the conformation of the backbone of a protein.
second-order rate constant the rate constant of a second-order reaction.
second-order reaction (bimolecular reaction) a reaction whose rate depends on the concentration of two reactants.
sedimentation constant designates where a species sediments in an ultracentrifuge.
selection rules the rules that determine the outcome of a pericyclic reaction.
selenenylation reaction conversion of an α-bromoketone into an α,β-unsatured ketone via the formation of a selenoxide.
semicarbazone $R_2C{=}NNH\overset{\overset{\displaystyle O}{\|}}{C}NH_2$
semiconservative replication the mode of replication that results in a daughter molecule of DNA having one of the original DNA strands plus a newly synthesized strand.
sense strand (informational strand) the strand in DNA that is not read during transcription; it has the same sequence of bases as the synthesized mRNA strand (with a U, T difference).
separated charges a positive and a negative charge that can be neutralized by the movement of electrons.
sesquiterpene a terpene that contains 15 carbons.

shielding phenomenon caused by electron donation to the environment of a proton. The electrons shield the proton from the full effect of the applied magnetic field. The more a proton is shielded, the farther to the right its signal appears in an NMR spectrum.
sigma (σ) bond a bond with a cylindrically symmetrical distribution of electrons.
sigmatropic rearrangement a reaction in which a σ bond is broken in the reactant, a new σ bond is formed in the product, and the π bonds rearrange.
Simmons–Smith reaction formation of a cyclopropane using $CH_2I_2 +$ Zn(Cu).
simple carbohydrate (monosaccharide) a single sugar molecule.
simple triacylglycerol a triacylglycerol in which the fatty acid components are the same.
single bond a σ bond.
singlet an unsplit NMR signal.
site-specific mutagenesis a technique that substitutes one amino acid of a protein for another.
site-specific recognition recognition of a particular site on DNA.
skeletal structure shows the carbon–carbon bonds as lines and does not show the carbon–hydrogen bonds.
S_NAr reaction a nucleophilic aromatic substitution reaction.
S_N1 reaction a unimolecular nucleophilic substitution reaction.
S_N2 reaction a bimolecular nucleophilic substitution reaction.
soap a sodium or potassium salt of a fatty acid.
solid-phase synthesis a technique in which one end of the compound being synthesized is covalently attached to a solid support.
solvation the interaction between a solvent and another molecule (or ion).
solvent-separated ion pair the cation and anion are separated by a solvent molecule.
solvolysis reaction with the solvent.
specific-acid catalysis catalysis in which the proton is fully transferred to the reactant before the slow step of the reaction takes place.
specific-base catalysis catalysis in which the proton is completely removed from the reactant before the slow step of the reaction takes place.
specific rotation the amount of rotation that will be caused by a compound with a concentration of 1.0 g/mL in a sample tube 1.0 dm long.
spectroscopy study of the interaction of matter and electromagnetic radiation.
sphingolipid a lipid that contains sphingosine.
sphingomyelin a sphingolipid in which the terminal OH group of sphingosine is bonded to phosphocholine or phosphoethanolamine.
spin-coupled ^{13}C NMR spectrum a ^{13}C NMR spectrum in which each signal of a carbon is split by the hydrogens bonded to that carbon.
spin coupling the atom that gives rise to an NMR signal is coupled to the rest of the molecule.
spin decoupling the atom that gives rise to an NMR signal is decoupled from the rest of the molecule.
spin–spin coupling the splitting of a signal in an NMR spectrum described by the $N + 1$ rule.
α-spin state nuclei in this spin state have their magnetic moments oriented in the same direction as the applied magnetic field.
β-spin state nuclei in this spin state have their magnetic moments oriented opposite the direction of the applied magnetic field.
spirocyclic compound a bicyclic compound in which the rings share one carbon.
splitting diagram a diagram that describes the splitting of a set of protons.
squalene a triterpene that is a precursor of steroid molecules.
stacking interactions van der Waals interactions between the mutually induced dipoles of adjacent pairs of bases in DNA.
staggered conformation a conformation in which the bonds on one carbon bisect the bond angle on the adjacent carbon when viewed looking down the carbon–carbon bond.
step-growth polymer (condensation polymer) a polymer made by combining two molecules while removing a small molecule (usually of water or an alcohol).
stereochemistry the field of chemistry that deals with the structures of molecules in three dimensions.
stereoelectronic effects the combination of steric effects and electronic effects.
stereogenic center (stereocenter) an atom at which the interchange of two substituents produces a stereoisomer.
stereoisomers isomers that differ in the way their atoms are arranged in space.
stereoselective reaction a reaction that leads to the preferential formation of one stereoisomer over another.
stereospecific reaction a reaction in which the reactant can exist as stereoisomers and each stereoisomeric reactant leads to a different stereoisomeric product or set of products.
steric effects effects due to the fact that groups occupy a certain volume of space.
steric hindrance refers to bulky groups at the site of a reaction that make it difficult for the reactants to approach each other.
steric strain (van der Waals strain, van der Waals repulsion) the repulsion between the electron cloud of an atom or a group of atoms and the electron cloud of another atom or group of atoms.
steroid a class of compounds that contains a steroid ring system.
Stille reaction couples an aryl, a benzyl, or a vinyl halide or triflate with a stannane in the presence of $Pd(PPh_3)_4$.
stop codon a codon at which protein synthesis is stopped.
Stork enamine reaction uses an enamine as a nucleophile in a Michael reaction.
straight-chain alkane (normal alkane) an alkane in which the carbons form a contiguous chain with no branches.
Strecker synthesis a method used to synthesize an amino acid: An aldehyde reacts with NH_3, forming an imine that is attacked by cyanide ion. Hydrolysis of the product gives an amino acid.
stretching frequency the frequency at which a stretching vibration occurs.
stretching vibration a vibration occurring along the line of a bond.
structural isomers (constitutional isomers) molecules that have the same molecular formula but differ in the way their atoms are connected.
structural protein a protein that gives strength to a biological structure.
α-substituent a substituent on the side of a steroid ring system opposite that of the angular methyl groups.
β-substituent a substituent on the same side of a steroid ring system as that of the angular methyl groups.
α-substitution reaction a reaction that puts a substituent on an α-carbon in place of an α-hydrogen.
substrate the reactant of an enzyme-catalyzed reaction.
subunit an individual chain of an oligomer.
suicide inhibitor (mechanism-based inhibitor) a compound that inactivates an enzyme by undergoing part of its normal catalytic mechanism.
sulfide (thioether) the sulfur analog of an ether (RSR).
sulfonate ester the ester of a sulfonic acid (RSO_2OR).
sulfonation substitution of a hydrogen of a benzene ring by a sulfonic acid group ($-SO_3H$).
suprafacial bond formation the formation of two σ bonds on the same side of the π system.
suprafacial rearrangement rearrangement in which the migrating group remains on the same face of the π system.
Suzuki reaction couples an aryl, a benzyl, or a vinyl halide with an organoborane in the presence of $Pd(PPh_3)_4$.
symmetrical anhydride an acid anhydride with identical R groups:

$$R-\overset{\overset{\displaystyle O}{\|}}{C}-O-\overset{\overset{\displaystyle O}{\|}}{C}-R$$

symmetrical ether an ether with two identical substituents bonded to the oxygen.
symmetric molecular orbital a molecular orbital in which the left half is a mirror image of the right half.
symmetry-allowed pathway a pathway that leads to overlap of in-phase orbitals.
symmetry-forbidden pathway a pathway that leads to overlap of out-of-phase orbitals.
syn addition an addition reaction in which two substituents are added to the same side of the molecule.
syndiotactic polymer a polymer in which the substituents regularly alternate on both sides of the fully extended carbon chain.
syn elimination an elimination reaction in which the two substituents that are eliminated are removed from the same side of the molecule.
syn-periplanar parallel substituents on the same side of a molecule.

synthetic equivalent the reagent actually used as the source of a synthon.
synthetic polymer a polymer that is not synthesized in nature.
synthetic tree an outline of the available routes to get to a desired product from available starting materials.
synthon a fragment of a disconnection.
systematic nomenclature nomenclature based on structure.
target molecule desired end product of a synthesis.
tautomerism interconversion of tautomers.
tautomers rapidly equilibrating isomers that differ in the location of their bonding electrons.
template strand (antisense strand) the strand in DNA that is read during transcription.
terminal alkyne an alkyne with the triple bond at the end of the carbon chain.
termination step when two radicals combine to produce a molecule in which all the electrons are paired.
terpene a lipid, isolated from a plant, that contains carbon atoms in multiples of five.
terpenoid a terpene that contains oxygen.
tertiary alcohol an alcohol in which the OH group is bonded to a tertiary carbon.
tertiary alkyl halide an alkyl halide in which the halogen is bonded to a tertiary carbon.
tertiary alkyl radical a radical with the unpaired electron on a tertiary carbon.
tertiary amine an amine with three alkyl groups bonded to the nitrogen.
tertiary carbocation a carbocation with the positive charge on a tertiary carbon.
tertiary carbon a carbon bonded to three other carbons.
tertiary hydrogen a hydrogen bonded to a tertiary carbon.
tertiary structure a description of the three-dimensional arrangement of all the atoms in a protein.
tetraene a hydrocarbon with four double bonds.
tetrahedral bond angle the bond angle (109.5°) formed by adjacent bonds of an sp^3 hybridized carbon.
tetrahedral carbon an sp^3 hybridized carbon; a carbon that forms covalent bonds by using four sp^3 hybridized orbitals.
tetrahedral intermediate the intermediate formed in a nucleophilic acyl substitution reaction.
tetrahydrofolate (THF) the coenzyme required by enzymes that catalyze reactions that donate a group containing a single carbon to their substrates.
tetraterpene a terpene that contains 40 carbons.
tetrose a monosaccharide with four carbons.
therapeutic index the ratio of the lethal dose of a drug to the therapeutic dose.
thermal cracking using heat to break a molecule apart.
thermal reaction a reaction that takes place without the reactant having to absorb light.
thermodynamic control when a reaction is under thermodynamic control, the relative amounts of the products depend on their stabilities.
thermodynamic product the most stable product.
thermodynamics the field of chemistry that describes the properties of a system at equilibrium.
thermodynamic stability is indicated by $\Delta G°$. If $\Delta G°$ is negative, the products are more stable than the reactants. If $\Delta G°$ is positive, the reactants are more stable than the products.
thermoplastic polymer a polymer that has both ordered crystalline regions and amorphous noncrystalline regions.
thermosetting polymers cross-linked polymers that, after they are hardened, cannot be remelted by heating.
thiamine pyrophosphate (TPP) the coenzyme required by enzymes, which catalyze a reaction that transfers a two-carbon fragment to a substrate.
thiirane a three-membered-ring compound in which one of the ring atoms is a sulfur.
thin-layer chromatography a technique that separates compounds on the basis of their polarity.
thioester the sulfur analog of an ester:

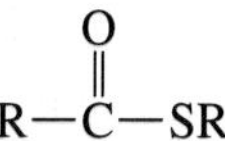

thioether (sulfide) the sulfur analog of an ether (RSR).
thiol (mercaptan) the sulfur analog of an alcohol (RSH).
threo enantiomers the pair of enantiomers with similar groups on opposite sides when drawn in a Fischer projection.
titration curve a plot of pH versus added equivalents of hydroxide ion.
Tollens test an aldehyde can be identified by observing the formation of a silver mirror in the presence of Tollens' reagent (Ag_2O/NH_3).
torsional strain the repulsion felt by the bonding electrons of one substituent as they pass close to the bonding electrons of another substituent.
trademark a registered name, symbol, or picture.
trade name (proprietary name, brand name) identifies a commercial product and distinguishes it from other products.
transamination a reaction in which an amino group is transferred from one compound to another.
transannular hydrogens (flagpole hydrogens) the two hydrogens in the boat conformation of cyclohexane that are closest to each other.
***s*-trans conformation** a conformation in which two double bonds are on opposite sides of a single bond.
transcription the synthesis of mRNA from a DNA blueprint.
transesterification reaction the reaction of an ester with an alcohol to form a different ester.
trans fused two cyclohexane rings fused together such that if the second ring were considered to be two substituents of the first ring, both substituents would be in equatorial positions.
transimination the reaction of a primary amine with an imine to form a new imine and a primary amine derived from the original imine.
transition metal catalyst a catalyst containing a transition metal, such as $Pd(PPh_3)_4$, that is used in coupling reactions.
trans isomer the isomer with identical substituents on opposite sides of the double bond.
transition state the highest point on a hill in a reaction coordinate diagram. In the transition state, bonds in the reactant that will break are partially broken and bonds in the product that will form are partially formed.
transition state analog a compound that is structurally similar to the transition state of an enzyme-catalyzed reaction.
translation the synthesis of a protein from an mRNA blueprint.
transmetallation metal exchange.
triacylglycerol the compound formed when the three OH groups of glycerol are esterified with fatty acids.
triene a hydrocarbon with three double bonds.
trigonal planar carbon an sp^2 hybridized carbon.
triose a monosaccharide with three carbons.
tripeptide three amino acids linked by amide bonds.
triple bond a σ bond plus two π bonds.
triplet an NMR signal split into three peaks.
triterpene a terpene that contains 30 carbons.
twist-boat conformation (skew-boat conformation) a conformation of cyclohexane.
ultraviolet light electromagnetic radiation with wavelengths ranging from 180 to 400 nm.
umpolung reversing the normal polarity of a functional group.
unimolecular reaction (first-order reaction) a reaction whose rate depends on the concentration of one reactant.
unsaturated hydrocarbon a hydrocarbon that contains one or more double or triple bonds.
unsymmetrical ether an ether with two different substituents bonded to the oxygen.
urethane a compound with a carbonyl group that is both an amide and an ester.
UV/Vis spectroscopy the absorption of electromagnetic radiation in the ultraviolet and visible regions of the spectrum; used to determine information about conjugated systems.
valence electron an electron in an unfilled shell.
valence shell electron-pair repulsion (VSEPR) model combines the concept of atomic orbitals with the concept of shared electron pairs and the minimization of electron pair repulsion.
van der Waals forces (London forces) induced-dipole–induced-dipole interactions.
van der Waals radius a measure of the effective size of an atom or a group. A repulsive force occurs (van der Waals repulsion) if two atoms approach each other at a distance less than the sum of their van der Waals radii.

vector sum takes into account both the magnitudes and the directions of the bond dipoles.
vicinal dihalide a compound with halogens bonded to adjacent carbons.
vicinal diol (vicinal glycol) a compound with OH groups bonded to adjacent carbons.
vinyl group $CH_2{=}CH{-}$
vinylic carbon a carbon in a carbon–carbon double bond.
vinylic cation a compound with a positive charge on a vinylic carbon.
vinylic radical a compound with an unpaired electron on a vinylic carbon.
vinylogy transmission of reactivity through double bonds.
vinyl polymer a polymer in which the monomers are ethylene or a substituted ethylene.
visible light electromagnetic radiation with wavelengths ranging from 400 to 780 nm.
vitamin a substance needed in small amounts for normal body function that the body cannot synthesize at all or cannot synthesize in adequate amounts.
vitamin KH_2 the coenzyme required by the enzyme that catalyzes the carboxylation of glutamate side chains.
vulcanization increasing the flexibility of rubber by heating it with sulfur.
wave equation an equation that describes the behavior of each electron in an atom or a molecule.
wave functions a series of solutions of a wave equation.
wavelength distance from any point on one wave to the corresponding point on the next wave (usually in units of μm or nm).
wavenumber the number of waves in 1 cm.
wax an ester formed from a long-chain carboxylic acid and a long-chain alcohol.
wedge-and-dash structure a method of representing the spatial arrangement of groups. Wedges are used to represent bonds that point out of the plane of the paper toward the viewer, and dashed lines are used to represent bonds that point back from the plane of the paper away from the viewer.
Williamson ether synthesis formation of an ether from the reaction of an alkoxide ion with an alkyl halide.
Wittig reaction the reaction of an aldehyde or a ketone with a phosphonium ylide, resulting in the formation of an alkene.
Wolff–Kishner reduction a reaction that reduces the carbonyl group of a ketone to a methylene group with the use of NH_2NH_2/HO^-.
Woodward–Fieser rules allow the calculation of the λ_{max} of the $\pi \longrightarrow \pi^*$ transition for compounds with four or fewer conjugated double bonds.
Woodward–Hoffmann rules a series of selection rules for pericyclic reactions.
ylide a compound with opposite charges on adjacent covalently bonded atoms with complete octets.
Zaitsev's rule the more substituted alkene product is obtained by removing a proton from the β-carbon that is bonded to the fewest hydrogens.
Z conformation the conformation of a carboxylic acid or a carboxylic acid derivative in which the carbonyl oxygen and the substituent bonded to the carboxyl oxygen or nitrogen are on the same side of the single bond.
Ziegler–Natta catalyst an aluminum–titanium initiator that controls the stereochemistry of a polymer.
Z isomer the isomer with the high-priority groups on the same side of the double bond.
zwitterion a compound with a negative charge and a positive charge on nonadjacent atoms.

Photo Credits

CHAPTER 1 p. 4 (bot.) UPI/Corbis p. 5 (top) AP/Wide World Photos p. 5 (bot.) Getty Images Inc./Hulton Archive Photos p. 6 (bot.) Photographer Unknown/Courtesy of the Archives of the Institute for Advanced/Study. p. 8 (bot.) Paul Silverman/ Fundamental Photographs p. 9 Photograph of the Albert Einstein Memorial Sculpture (Copyright Robert Berks, 1978) at the National Academy of Sciences. Credit: National Academy of Sciences. p. 12 Cornell University/Courtesy AIP Emilio Segre Visual Archives. p. 13 Courtesy of the Bancroft Library/University of California, Berkeley. p. 26 (bot.) Joe Mc Nally/Corbis/Sygma p. 52 (ctr. left) Richard Megna/Fundamental Photographs

CHAPTER 2 p. 82 (ctr.) Photo by Gen. Stab. Lit. Anst./AIP Emilio Segre Visual Archives p. 87 (top) Simon Fraser/Science Photo Library/Photo Researchers, Inc.

CHAPTER 3 p. 111 (bot. rt.) Hans Reinhard/Okapia/Photo Researchers, Inc. p. 127 (rt.) American Institute of Physics/Emilio Segre Visual Archives

CHAPTER 4 p. 143 Judith Olah p. 149 (top) Edgar Fahs Smith Collection/University of Pennsylvania/Van Pelt Library. p. 171 (bot.) AP/Wide World Photos

CHAPTER 5 p. 187 (bot.) Brown Brothers p. 192 (ctr.) Diane Schiumo/Fundamental Photographs p. 194 Nobelstiftelsen/© The Nobel Foundation p. 212 Getty Images Inc./Hulton Archive Photos

CHAPTER 6 p. 255 (top rt.) Reuters/Getty Images Inc. - Hulton Archive Photos p. 258 Getty Images Inc./Hulton Archive Photos

CHAPTER 7 p. 266 Corbis

CHAPTER 8 p. 300 Jim Zipp/Photo Researchers, Inc. p. 313 (ctr.) AP/Wide World Photos p. 314 (top) AP/Wide World Photos p. 324 (ctr.) Perkin Elmer, Inc. p. 327 (top rt.) Jerry Alexander/Getty Images Inc. - Stone Allstock

CHAPTER 9 p. 354 (bot. rt.) NASA/Goddard Space Flight Center p. 354 (bot. left) NASA/Goddard Space Flight Center p. 354 (ctr.) Philippe Plailly/Science Photo Library/Photo Researchers, Inc.

CHAPTER 10 p. 361 (top rt.) Mike Severns/Tom Stack & Associates, Inc. p. 366 (ctr.) Edgar Fahs Smith Collection/University of Pennsylvania Van Pelt Library p. 374 (top left) Science Photo Library/Photo Researchers, Inc. p. 381 (bot. rt.) AP/Wide World Photos

CHAPTER 11 p. 404 (left) Butlerov Institute of Chemistry p. 419 (top rt.) Dr. Paula Bruice p. 419 (top rt.) Dr. Paula Bruice p. 419 (top rt.) Dr. Paula Bruice p. 419 (top rt.) Dr. Paula Bruice p. 425 Edgar Fahs Smith Collection/University of Pennsylvania Van Pelt Library

CHAPTER 12 p. 453 (bot.) Corbis p. 461 (ctr.) Getty Images Inc./Hulton Archive Photos p. 463 (top rt.) Professor Donald J. Cram p. 468 (ctr. left) Science Photo Library/Photo Researchers, Inc.

CHAPTER 13 p. 503 Dr. Jeremy Burgess/Science Photo Library/Photo Researchers, Inc.

CHAPTER 14 Fig. 14.27 Reproduced with permission from F. A. Bovey, Nuclear Magnetic-Resonance Spectroscopy. New York: Academic Press, 1968. Fig. 14.41(a) Scott Camazine/Photo Researchers, Inc. Fig. 14.41(b) Simon Fraser/Science Photo Library/Photo Researchers, Inc. Fig. 14.31 Reproduced with permission of J. Houser. p. 529 (top ctr.) CORBIS

CHAPTER 15 p. 597 (top left) Gunter Marx/Corbis p. 597 (top rt.) Ken Eward/Science Source/Photo Researchers, Inc. p. 612 (top) Edgar Fahs Smith Collection/University of Pennsylvania Van Pelt Library p. 612 (bot.) Edgar Fahs Smith Collection/University of Pennsylvania Van Pelt Library

CHAPTER 17 p. 680 Kathleen Campbell/Getty Images, Inc-Liaison p. 701 (bot.) Corbis

CHAPTER 18 p. 761 (bot. rt.) Fotoatelier Tita Binz p. 764 (top left) Grant Heilman Photography, Inc.

CHAPTER 19 p. 805 Gilbert Stork

CHAPTER 20 p. 853 German Information Center

CHAPTER 21 p. 889 Edgar Fahs Smith Collection/University of Pennsylvania Van Pelt Library p. 911 John Gerlach/DRK Photo

CHAPTER 22 p. 945 (left) Dr. Paula Bruice p. 945 (rt.) Dr. Paula Bruice p. 948 (bot.) Stanley Breeden/DRK Photo p. 948 (ctr.) Visuals Unlimited p. 949 (top rt.) Dr. Jeremy Burgess/Science Photo Library/Photo Researchers, Inc. p. 953 Science Photo Library/Photo Researchers, Inc.

CHAPTER 23 p. 980 (bot. left) D. Goldberg/Corbis/Sygma p. 983 (bot. rt.) AP/Wide World Photos

CHAPTER 24 p. 1001 Corbis Fig. 24.8 Copyright by Irving Geis

CHAPTER 25 p. 1035 (ctr. left) Corbis p. 1035 (ctr. rt.) Edgar Fahs Smith Collection, University of Pennsylvania Van Pelt Library p. 1062 (top left) UPI/Corbis-Bettman/Corbis p. 1070 (bot.) Eric L. Heyer/Grant Heilman Photography, Inc.

CHAPTER 26 p. 1077 (bot.) Dr. Paula Bruice p. 1078 (ctr. left) David Cavagnaro p. 1078 (top left) Scott Camazine/Photo Researchers, Inc. p. 1079 Tom Tietz/Getty Images Inc./Stone Allstock p. 1081 (bot.) Kim Heacox/Getty Images Inc./Stone Allstock p. 1083 (bot.) Mary Teresa Giancoli p. 1085 (top) Getty Images Inc./Hulton Archive Photos p. 1088 Corbis p. 1099 The University of Texas Southwestern Medical Center at Dallas

CHAPTER 27 p. 1107 (top) Photo Researchers, Inc. p. 1118 (bot.) Science Photo Library/Photo Researchers, Inc. p. 1128 (bot.) AP/Wide World Photos p. 1131 (bot. left) Dr. Gopal Murti/Science Photo Library/Custom Medical Stock Photo, Inc. p. 1131 (top) Dr. Paula Bruice p. 1131 (bot. rt.) Photo Researchers, Inc.

CHAPTER 28 p. 1149 (bot.) Richard Megna, Fundamental Photographs p. 1160 (bot.) Billy Hustace/Getty Images Inc./Stone Allstock p. 1164 (ctr.) Stephen Frisch, Stock Boston

CHAPTER 29 p. 1177 Roald Hoffmann/Photo by Vivian Torrence p. 1178 (top) Courtesy of the Harvard University Archives p. 1197 (bot.) Gregory K. Scott/Photo Researchers, Inc.

CHAPTER 30 p. 1208 Bill Ivy, Getty Images Inc./Stone Allstock p. 1209 Gregory G. Dimijian/Photo Researchers, Inc. p. 1211 AP/Wide World Photos p. 1213 Hoffmann-La Roche, Inc.

Index

P

Periodic Table of the Elements

	Main groups		Transition metals										Main groups					
	1A[a] 1	2A 2	3B 3	4B 4	5B 5	6B 6	7B 7	8B 8	8B 9	8B 10	1B 11	2B 12	3A 13	4A 14	5A 15	6A 16	7A 17	8A 18
1	1 **H** 1.00794																	2 **He** 4.002602
2	3 **Li** 6.941	4 **Be** 9.012182											5 **B** 10.811	6 **C** 12.0107	7 **N** 14.0067	8 **O** 15.9994	9 **F** 18.998403	10 **Ne** 20.1797
3	11 **Na** 22.989770	12 **Mg** 24.3050											13 **Al** 26.981538	14 **Si** 28.0855	15 **P** 30.973761	16 **S** 32.065	17 **Cl** 35.453	18 **Ar** 39.948
4	19 **K** 39.0983	20 **Ca** 40.078	21 **Sc** 44.955910	22 **Ti** 47.867	23 **V** 50.9415	24 **Cr** 51.9961	25 **Mn** 54.938049	26 **Fe** 55.845	27 **Co** 58.933200	28 **Ni** 58.6934	29 **Cu** 63.546	30 **Zn** 65.39	31 **Ga** 69.723	32 **Ge** 72.64	33 **As** 74.92160	34 **Se** 78.96	35 **Br** 79.904	36 **Kr** 83.80
5	37 **Rb** 85.4678	38 **Sr** 87.62	39 **Y** 88.90585	40 **Zr** 91.224	41 **Nb** 92.90638	42 **Mo** 95.94	43 **Tc** [98]	44 **Ru** 101.07	45 **Rh** 102.90550	46 **Pd** 106.42	47 **Ag** 107.8682	48 **Cd** 112.411	49 **In** 114.818	50 **Sn** 118.710	51 **Sb** 121.760	52 **Te** 127.60	53 **I** 126.90447	54 **Xe** 131.293
6	55 **Cs** 132.90545	56 **Ba** 137.327	71 **Lu** 174.967	72 **Hf** 178.49	73 **Ta** 180.9479	74 **W** 183.84	75 **Re** 186.207	76 **Os** 190.23	77 **Ir** 192.217	78 **Pt** 195.078	79 **Au** 196.96655	80 **Hg** 200.59	81 **Tl** 204.3833	82 **Pb** 207.2	83 **Bi** 208.98038	84 **Po** [208.98]	85 **At** [209.99]	86 **Rn** [222.02]
7	87 **Fr** [223.02]	88 **Ra** [226.03]	103 **Lr** [262.11]	104 **Rf** [261.11]	105 **Db** [262.11]	106 **Sg** [266.12]	107 **Bh** [264.12]	108 **Hs** [269.13]	109 **Mt** [268.14]	110 [271.15]	111 [272.15]	112 [277]		114 [285]		116 [289]		

*Lanthanide series	57 ***La** 138.9055	58 **Ce** 140.116	59 **Pr** 140.90765	60 **Nd** 144.24	61 **Pm** [145]	62 **Sm** 150.36	63 **Eu** 151.964	64 **Gd** 157.25	65 **Tb** 158.92534	66 **Dy** 162.50	67 **Ho** 164.93032	68 **Er** 167.259	69 **Tm** 168.93421	70 **Yb** 173.04
†Actinide series	89 **†Ac** [227.03]	90 **Th** 232.0381	91 **Pa** 231.03588	92 **U** 238.02891	93 **Np** [237.05]	94 **Pu** [244.06]	95 **Am** [243.06]	96 **Cm** [247.07]	97 **Bk** [247.07]	98 **Cf** [251.08]	99 **Es** [252.08]	100 **Fm** [257.10]	101 **Md** [258.10]	102 **No** [259.10]

[a]The labels on top (1A, 2A, etc.) are common American usage. The labels below these (1, 2, etc.) are those recommended by the International Union of Pure and Applied Chemistry.

The names and symbols for elements 110 and above have not yet been decided.

Atomic weights in brackets are the masses of the longest-lived or most important isotope of radioactive elements.

Further information is available at http://www.shef.ac.uk/chemistry/web-elements/

The production of element 116 was reported in May 1999 by scientists at Lawrence Berkeley National Laboratory.

Common Functional Groups

Alkane	RCH_3
Alkene	$R_2C{=}CR_2$ (internal), $R_2C{=}CH_2$ (terminal)
Alkyne	$RC{\equiv}CR$ (internal), $RC{\equiv}CH$ (terminal)
Nitrile	$RC{\equiv}N$
Ether	$R{-}O{-}R$
Thiol	$RCH_2{-}SH$
Sulfide	$R{-}S{-}R$
Disulfide	$R{-}S{-}S{-}R$
Epoxide	three-membered ring containing O
Aniline	$C_6H_5{-}NH_2$
Phenol	$C_6H_5{-}OH$
Carboxylic acid	$R{-}C({=}O){-}OH$
Acyl chloride	$R{-}C({=}O){-}Cl$
Acid anhydride	$R{-}C({=}O){-}O{-}C({=}O){-}R$
Ester	$R{-}C({=}O){-}OR$
Amide	$R{-}C({=}O){-}NH_2$, $-NHR$, $-NR_2$
Aldehyde	$R{-}C({=}O){-}H$
Ketone	$R{-}C({=}O){-}R$

	primary	secondary	tertiary
Alkyl halide	$R{-}CH_2{-}X$ X = F, Cl, Br, or I	$R_2CH{-}X$	$R_3C{-}X$
Alcohol	$R{-}CH_2{-}OH$	$R_2CH{-}OH$	$R_3C{-}OH$
Amine	$R{-}NH_2$	R_2NH	R_3N